Standard
Handbook
of
Engineering
Calculations

OTHER McGRAW-HILL HANDBOOKS OF INTEREST

Standard
Handbook
of Engineering
Calculations

Tyler G. Hicks, P.E., Editor

International Engineering Associates
Member, American Society of Mechanical Engineers,
Institute of Electrical and Electronics Engineers,
United States Naval Institute,
International Oceanographic Foundation

McGraw-Hill Book Company

New York   St. Louis   San Francisco   Auckland   Düsseldorf   Johannesburg
Kuala Lumpur   London   Mexico   Montreal   New Delhi   Panama
Paris   São Paulo   Singapore   Sydney   Tokyo   Toronto

Library of Congress Cataloging in Publication Data

Hicks, Tyler Gregory
  Standard handbook of engineering calculations.

  (McGraw-Hill handbook series)
  Includes bibliographical references.
  1. Engineering—Tables, calculations, etc.
I. Title. II. Title: Handbook of engineering
calculations.
TA151.H52   620'.0021'2   73-130674
ISBN 0-07-028734-1

13 14 15 16   KPKP   7832109

*The editors for this book were Harold B. Crawford and
Stanley E. Redka, the designer was Naomi Auerbach, and its
production was supervised by George E. Oechsner. It was set
in Linofilm Caledonia by The European Printing Corporation
Limited.*

It was printed and bound by The Kingsport Press.

Contents

Contributors and Advisers

In preparing the various sections of this Handbook the following individuals either contributed sections, or portions of sections, or advised the editor or contributors, or both, on the optimum content of specific sections. The affiliations shown are those prevailing at the time of the preparation of the contributed material or the recommendations as to section content.

OTHMAR H. AMMANN   *Senior Partner, Ammann & Whitney*
FREDERICK S. BARTON   *Director, Hewlett-Packard Limited*
HAROLD BECHER   *President, Strato Missiles, Inc.*
EDMUND B. BESSELIEVRE, P.E.   *Consultant, Forrest & Cotton, Inc.*
ALEXANDER G. CHRISTIE   *Consulting Engineer*
ROBERT L. DAVIDSON   CHEMICAL ENGINEERING
ANDREW W. EDWARDS   *Power Engineer, Westinghouse Electric Corporation*
TYLER G. HICKS, P.E.   *Consulting Engineer, International Engineering Associates*
EDGAR J. KATES, P.E.   *Consulting Engineer*
MAX KURTZ, P.E.   *Consulting Engineer*
SAMUEL C. LIND   *Consultant, United States Atomic Energy Commission*
JOSEPH MITTLEMAN   *Senior Associate Editor, Design Theory,* ELECTRONICS *magazine*
GEORGE M. MUSCHAMP   *Consulting Engineer, Honeywell, Inc.*
RUFUS OLDENBURGER   *Professor, Purdue University*
ZVI PRIHAR   *Scientific Advisor, Operations Research, Inc.*
RAYMOND J. ROARK   *Professor, University of Wisconsin*
JOHN P. ROEDEL   *Research Engineer, The Boeing Company*
HAROLD L. RORDEN   *Consulting Engineer, American Electric Power Serviceo Corporation*
LYMAN F. SCHEEL   *Consulting Engineer*
B. G. A. SKROTZKI, P.E.   POWER *magazine*
S. W. SPIELVOGEL   *Piping Engineering Consultant*
FREDERICK W. SUHR   *Consulting Engineer, General Electric Company*
BERNARD TICHAZ   *Director, George G. Sharp Inc.*
MAXWELL M. UPSON   *Consulting Engineer, Raymond International, Inc.*

Preface

This is a handbook of specific engineering calculation procedures that presents to its users more than two thousand step-by-step calculation procedures for solving almost all routine, and many nonroutine, problems met in everyday engineering practice. Having this Handbook on his desk, the engineer or scientist will be able to solve most of the applied problems he meets in his daily activities of design, operation, analysis, or economic evaluation.

The step-by-step calculation procedures are arranged so they can be followed by anyone with an engineering or scientific background. Each worked-out problem presents *fully explained and illustrated steps* for solving similar problems in industrial, research, government, academic, or license-examination situations. For any applied problem, all the Handbook user need do is place his calculation sheets alongside this Handbook and follow the step-by-step procedure line for line to obtain the desired solution to his actual problem. By following the calculation procedures in this Handbook, the engineer or scientist will obtain accurate results in minimum time with the least effort.

The Handbook opens with a discussion of how to use the book most profitably. In this discussion, the user is given many useful hints on general calculation procedures applicable to all engineering disciplines. Also included in this discussion is a description of how to use modern computing equipment in solving applied engineering problems. Both digital and analog computers are discussed, as well as more familiar computing equipment. All engineers using computers in solving applied problems will find this discussion a helpful guide in their work. The discussion will be particularly useful to those engineers and scientists who did not have any formal training in using computers for solving applied problems.

This Handbook contains twelve sections, which follow the discussion of how to use the book. Each of the twelve sections is devoted to a particular branch of engineering—aeronautical and astronautical, architectural, chemical, civil, control, electrical, electronic, economic, marine, mechanical, nuclear, and sanitary. The procedures and worked-out problems presented in each section were selected and reviewed by a panel of specialists familiar with the discipline covered by the particular section.

In choosing the procedures and worked-out problems, these specialists used a number of guidelines, including: (1) What are the most common applied problems that must be solved in this discipline? (2) What are the most accurate methods for solving these problems? (3) What other problems might be met in this discipline? When the answers to these and other related questions were obtained, the procedures and worked-out problems were chosen. Thus, the Handbook represents a cross section of the thinking of a large number of experienced practicing engineers, project directors, and educators.

To those who might claim that the use of step-by-step solution procedures and worked-out examples makes engineering "too easy," the editor points out that for many years engineering educators have recognized the importance and value of problem solving in the development of engineering judgment and experience. Problems courses have been popular in numerous engineering schools for many years and are still given in many schools. However, with the greater emphasis on engineering science in most engineering schools, there is less time for the problems courses. The result is that many of today's graduates can benefit from a more extensive study of specific problem-solving procedures.

Experienced engineers in all fields are often called on to solve problems outside the area of their specialty. When such a request is made, the engineer can hardly refuse. Thus, a mechanical engineer might be asked to determine the size of a beam or beams to carry a given load. Although the engineer might refer to the text he used in school (if the text is still available), he will in general want a more direct procedure for solving such a problem. He will find the correct procedure in the present Handbook. This Handbook is thus useful to all engineers who must work outside the areas of their direct competence. It will, of course, also help an engineer solve problems with which he is less familiar in his own specialty. And with the increased crossing of responsibilities between scientists and engineers, the scientist will find many useful procedures herein.

Covering a wide range of applied problems, the Handbook will also be a major asset to all professional-engineering license-examination candidates, civil service applicants, and a variety of other job seekers who must pass an examination before being employed. Technicians studying to advance themselves will find many valuable procedures in this Handbook.

Throughout the Handbook, emphasis is placed on the solution of *applied* engineering problems. There is hardly a problem of common occurrence in the fields covered that is not solved in the Handbook. Each of the specialists who contributed to the Handbook has reviewed the problems presented in his area and approved their inclusion. The Handbook will therefore be a useful tool to practicing engineers and scientists everywhere, in all fields of modern engineering practice. It will also be of great value to designers, draftsmen, and students who require a working knowledge of the best solution procedures for applied engineering problems in a variety of fields.

Even though modern computers and computing equipment are replacing manual calculation methods, this equipment cannot be used unless the correct solution procedure is known. Thus, this Handbook will provide the engineer or computer programmer with the procedure to follow for the solution of a given problem. Knowing the best procedure, he can readily program the computer to solve the problem.

In preparing a handbook of this magnitude, the editor, advisers, and contributors drew heavily on numerous sources of information. Most of these are cited as references or credited throughout the Handbook, but space limitations preclude mentioning them all. The editor, advisers, and contributors wish to acknowledge their indebtedness to these sources and to express their gratitude.

Lastly, every effort has been made to ensure the accuracy of each calculation procedure in this Handbook. Should any readers discover errors of any kind, the editor will be grateful if they are called to his attention so that corrections can be made in future printings of the Handbook.

Tyler G. Hicks

Acknowledgments

The contributors and advisers consulted hundreds of sources when preparing the material for inclusion in this Handbook. Besides using the books and other publications listed as references at the beginning of each major section of the Handbook, the contributors and advisers consulted and drew material from technical magazines and journals, trade-association standards, engineering and scientific papers, industrial and engineering catalogs, and a variety of similar publications. Most of these are noted in appropriate places throughout the Handbook. Additional acknowledgments, listed in the order received, are given below.

Data and charts credited to the Hydraulic Institute are reprinted from the *Hydraulic Institute Standards,* Twelfth Edition, copyright 1969 by the Hydraulic Institute, and from the *Pipe Friction Manual,* Third Edition, copyright 1961 by the Hydraulic Institute, 122 East 42nd Street, New York, N.Y. 10017. Data on diesel engine cooling systems are reprinted from *Marine Diesel Standard Practices,* Third Edition, copyright 1971 by the Diesel Engine Manufacturers Association, 122 East 42nd Street, New York, N.Y. 10017. Data on minimum requirements for plumbing are drawn from the *American Standard National Plumbing Code* with permission of the publisher, The American Society of Mechanical Engineers.

Specific firms, trade associations, and publications which were extremely helpful in supplying data for various sections of the Handbook include Martin Marietta Corporation; *Electronic Design* magazine; Dresser Industries Inc. — Dresser Industrial Valve and Instrument Division; Ingersoll-Rand Company; Anaconda American Brass Company; Waterloo Register Division — Dynamics Corporation of America; ITT Hammel-Dahl; *Mechanical Engineering,* a monthly publication of The American Society of Mechanical Engineers; McQuay, Inc.; The G. C. Breidert Co.; Modine

Manufacturing Company; Rubber Manufacturers Association; Condenser Service & Engineering Co., Inc.; Armstrong Machine Works; American Air Filter Company; Crane Company; *Machine Design* magazine; The RAND Corporation; Texas Instruments Incorporated; McGraw-Hill Publications Company, McGraw-Hill, Inc.; Morse Chain Company; Grinnell Corporation; General Electric Company; The B. F. Goodrich Company; American Standard Inc.; the American Society of Heating, Refrigerating and Air-Conditioning Engineers; International Engineering Associates; Taylor Instrument Process Control Division of Sybron Corporation; Clark-Reliance Corporation; American Society for Testing and Materials; Acoustical and Insulating Materials Association; W. S. Dickey Clay Manufacturing Co.; Flexonics Division, Universal Oil Products Co.; Dunham-Bush, Inc.; Carrier Air Conditioning Company; National Industrial Leather Association; Worthington Corporation; Goulds Pumps, Inc. Illustrations and problems credited to Carrier Air Conditioning Company are copyrighted by Carrier Air Conditioning Company.

Individuals who were helpful to the editor of this Handbook at one or more times during its preparation include Lyman F. Scheel, Consulting Engineer; Jack Jaklitsch, Editor, *Mechanical Engineering;* Spencer A. Tucker, Martin & Tucker; Paul V. DeLuca, Porta Systems Corp.; Professor Steven Edelglass, Cooper Union; Professor William Vopat, Cooper Union; Professor Theodore Baumeister, Columbia University; Frederick S. Merritt, Consulting Engineer; James J. O'Connor, Editor, *Power* magazine; Nathan R. Grossner, Consulting Engineer; Nicholas P. Chironis, *Product Engineering;* Franklin D. Yeaple, *Product Engineering;* John D. Constance, Consulting Engineer; John R. Miller, Texas Instruments Incorporated; Rupert Le Grand, *American Machinist;* Ronald G. Kogan, United Computing Systems, Inc.; Al Brons, Flexonics Div., Universal Oil Products Company; Carl W. MacPhee, ASHRAE *Guide and Data Book;* Frank P. Anderson, Secretary, Hydraulic Institute; Joseph Mittleman, *Electronics;* and numerous working engineers and scientists in firms and universities in the United States and abroad.

How to Use This Handbook

There are two ways to enter this Handbook to obtain the maximum benefit from the time invested. The first entry is through the Index; the second is through the table of contents of the section covering the discipline, or related discipline, concerned. Each method is discussed in detail below.

Index. Great care and considerable time were expended on preparation of the Index of this Handbook so that it would be of maximum use to every reader. As a general guide, enter the Index using the generic term for the type of calculation procedure being considered. Thus, for the design of a beam, enter at *beam design.* From here, progress to the specific type of beam being considered—such as a *steel I beam.* Once the page number or numbers of the appropriate Calculation Procedure are determined, turn to them to find the step-by-step instructions and worked-out example that can be followed to solve the problem quickly and accurately.

Contents. The Contents of each section lists the titles of the Calculation Procedures contained in that section. Where extensive use of any section is contemplated, the editor suggests that the reader might benefit from an occasional glance at the table of contents of that section. Such a glance will give the user of this Handbook an understanding of the breadth and coverage of a given section, or a series of sections. Then, when he turns to this Handbook for assistance, the reader will be able more rapidly to find the Calculation Procedure he seeks.

Calculation Procedures. Each Calculation Procedure is a unit in itself. However, any given Calculation Procedure will contain subprocedures which might be useful to the reader. Thus, a Calculation Procedure on pump selection will contain subprocedures on pipe friction loss, pump static and dynamic heads, etc. Should the reader of this Handbook wish to make a computation using any of such subprocedures, he will find the worked-out steps which are presented both useful and precise. Hence, the Handbook contains numerous valuable procedures that are useful in solving a variety of applied engineering problems.

One other important point that should be noted about the Calculation Procedures presented in this Handbook is that many of the Calculation Procedures are equally applicable in a variety of disciplines. Thus, a beam-selection procedure can be used for civil-, chemical-, mechanical-, electrical-, and nuclear-engineering activities, as well as some others. Hence, the reader might consider a temporary neutrality for his particular specialty when using the Handbook because the Calculation Procedures are designed for universal use.

Any of the Calculation Procedures presented can be programmed on a computer. Such programming permits rapid solution of a variety of design problems. With the growing use of low-cost time sharing, more engineering design problems are being solved using a remote terminal in the engineering office. The editor hopes that engineers throughout the world will make greater use of desk calculators and digital and analog computers in solving applied engineering problems. This modern equipment promises greater speed and accuracy for nearly all the complex design problems that must be solved in today's world of engineering.

Section 1

Civil Engineering

MAX KURTZ, P.E.
Consulting Engineer

Section Advisers:

OTHMAR H. AMMANN *Senior Partner, Ammann & Whitney*
MAXWELL M. UPSON *Consulting Engineer, Raymond International, Inc.*

REFERENCES: American Concrete Institute—*Building Code Requirements for Reinforced Concrete*; American Institute of Steel Construction—*Manual of Steel Construction*; National Forest Products Association—*National Design Specification for Stress-Grade Lumber and its Fastenings*; Abbett—*American Civil Engineering Practice*, Wiley; Gaylord and Gaylord—*Structural Engineering Handbook*, McGraw-Hill; LaLonde and Janes—*Concrete Engineering Handbook*, McGraw-Hill; Lincoln Electric Co.—*Procedure Handbook of Arc Welding Design and Practice*; Merritt—*Standard Handbook for Civil Engineers*, McGraw-Hill; Timber Engineering Company—*Timber Design and Construction Handbook*, McGraw-Hill; U.S. Department of Agriculture, Forest Products Laboratory—*Wood Handbook (Agriculture Handbook 72)*, GPO; Urquhart—*Civil Engineering Handbook*, McGraw-Hill; Borg and Gennaro—*Advanced Structural Analysis*, Van Nostrand; Gerstle—*Basic Structural Design*, McGraw-Hill; Jensen—*Applied Strength of Materials*, McGraw-Hill; Kurtz—*Comprehensive Structural Design Guide*, McGraw-Hill; Roark—*Formulas for Stress and Strain*, McGraw-Hill; Seely—*Resistance of Materials*, Wiley; Shanley—*Mechanics of Materials*, McGraw-Hill; Timoshenko and Young—*Theory of Structures*, McGraw-Hill; Beedle, et al.—*Structural Steel Design*, Ronald; Grinter—*Design of Modern Steel Structures*, Macmillan; Lothers—*Advanced Design in Structural Steel*, Prentice-Hall; Beedle—*Plastic Design of Steel Frames*, Wiley; Canadian Institute of Timber Construction—*Timber Construction*; Scofield and O'Brien—*Modern Timber Engineering*, Southern Pine Association; Dunham—*Theory and Practice of Reinforced Concrete*, McGraw-Hill; Winter, et al.—*Design of Concrete Structures*, McGraw-Hill; Viest, Fountain, and Singleton—*Composite Construction in Steel and Concrete*, McGraw-Hill; Chi and Biberstein—*Theory of Prestressed Concrete*, Prentice-Hall; Connolly—*Design of Prestressed Concrete Beams*, McGraw-Hill; Evans and Bennett—*Pre-stressed Concrete*, Wiley; Libby—*Prestressed Concrete*, Ronald; Magnel—*Prestressed Concrete*, McGraw-Hill; Gennaro—*Computer Methods in Solid Mechanics*, Macmillan; Laursen—*Matrix Analysis of Structures*, McGraw-Hill; Weaver—*Computer Programs for Structural Analysis*, Van Nostrand; Brenkert—*Elementary Theoretical Fluid Mechanics*, Wiley; Daugherty and Franzini—*Fluid Mechanics with Engineering Applications*, McGraw-Hill; King and Brater—*Handbook of Hydraulics*, McGraw-Hill; Li and Lam—*Principles of Fluid Mechanics*, Addison-Wesley; Sabersky and Acosta—*Fluid Flow*, Macmillan; Streeter—*Fluid Mechanics*, McGraw-Hill; Allen—*Railroad*

Curves and Earthwork, McGraw-Hill; Davis, Foote, and Kelly—*Surveying: Theory and Practice,* McGraw-Hill; Hickerson—*Route Surveys and Design,* McGraw-Hill; Hosmer and Robbins—*Practical Astronomy,* Wiley; Jones—*Geometric Design of Modern Highways,* Wiley; Meyer—*Route Surveying,* International Textbook; American Association of State Highway Officials—*A Policy on Geometric Design of Rural Highways;* Chellis—*Pile Foundations,* McGraw-Hill; Goodman and Karol—*Theory and Practice of Foundation Engineering,* Macmillan; Huntington—*Earth Pressures and Retaining Walls,* Wiley; Ritter and Paquette—*Highway Engineering,* Ronald; Scott and Schoustra—*Soil: Mechanics and Engineering,* McGraw-Hill; Spangler—*Soil Engineering,* International Textbook; Teng—*Foundation Design,* Prentice-Hall; Terzaghi and Peck—*Soil Mechanics in Engineering Practice,* Wiley; U.S. Department of the Interior, Bureau of Reclamation—*Earth Manual,* GPO.

Principles of Statics; Geometrical Properties of Areas

If a body remains in equilibrium under a system of forces, the following conditions obtain:

1. The algebraic sum of the components of the forces in any given direction is zero.
2. The algebraic sum of the moments of the forces with respect to any given axis is zero.

The above statements are verbal expressions of the *equations of equilibrium.* In the absence of any notes to the contrary, a clockwise moment is considered positive; a counterclockwise moment, negative.

GRAPHICAL ANALYSIS OF A FORCE SYSTEM

The body in Fig. 1a is acted on by forces *A*, *B*, and *C*, as shown. Draw the vector representing the equilibrant of this system.

Calculation Procedure:

1. Construct the system force line

In Fig. 1b, draw the vector chain *A-B-C*, which is termed the *force line.* The vector extending from the initial point to the terminal point of the force line represents the

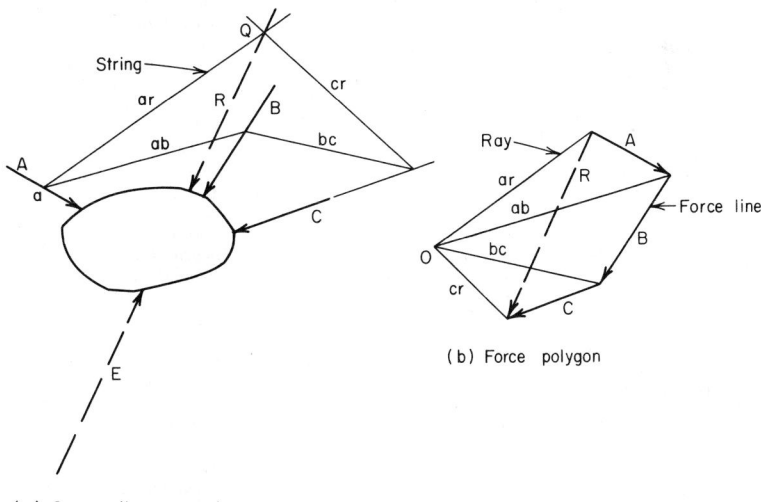

(a) Space diagram and string polygon

(b) Force polygon

Fig. 1 Equilibrant of force system.

resultant R. In any force system, the resultant R is equal to and collinear with the equilibrant E but acts in the opposite direction. The equilibrant of a force system is a single force that will balance the system.

2. Construct the system rays

Selecting an arbitrary point O as the pole, draw the rays from O to the ends of the vectors and label them as shown in Fig. 1b.

3. Construct the string polygon

In Fig. 1a, construct the string polygon as follows: At an arbitrary point a on the action line of force A, draw strings parallel to the rays ar and ab. At the point where the string ab intersects the action line of force B, draw a string parallel to ray bc. At the point where string bc intersects the action line of force C, draw a string parallel to cr. The intersection point Q of ar and cr lies on the action line of R.

4. Draw the vector for the resultant and equilibrant

In Fig. 1a, draw the vector representing R. Establish the magnitude and direction of this vector from the force polygon. The action line of R passes through Q.

Lastly, draw a vector equal to and collinear with that representing R but opposite in direction. This vector represents the equilibrant E.

Related Calculations: Use this general method for any force system acting in a single plane. With a large number of forces, the resultant of a smaller number of forces can be combined with the remaining forces to simplify the construction.

ANALYSIS OF STATIC FRICTION

The bar in Fig. 2a weighs 100 lb and is acted on by a force P that makes an angle of 55° with the horizontal. The coefficient of friction between the bar and the inclined plane is 0.20. Compute the minimum value of P required (a) to prevent the bar from sliding down the plane; (b) to cause the bar to move upward along the plane.

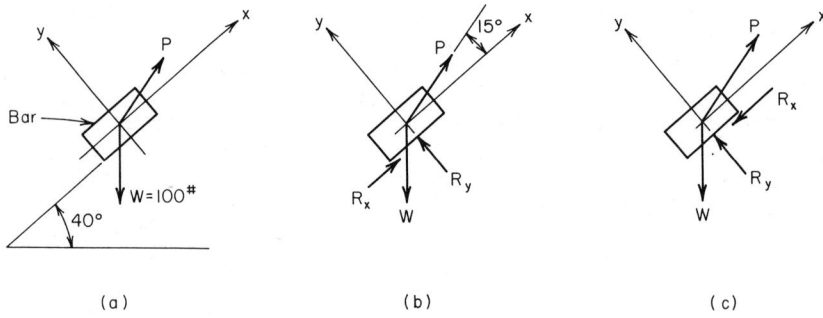

(a)   (b)   (c)

Fig. 2 Body on inclined plane.

Calculation Procedure:

1. Select coordinate axes

Establish coordinate axes x and y through the center of the bar, parallel and perpendicular to the plane, respectively.

2. Draw a free-body diagram of the system

In Fig. 2b, draw a free-body diagram of the bar. The bar is acted on by its weight W, the force P, and the reaction R of the plane on the bar. Show R resolved into its x and y components, the former being directed upward.

3. Resolve the forces into their components

The forces W and P are the important ones in this step, and they must be resolved into their x and y components. Thus;

$$W_x = -100 \sin 40° = -64.3 \text{ lb} \qquad W_y = -100 \cos 40° = -76.6 \text{ lb}$$

$$P_x = P \cos 15° = 0.966P \qquad P_y = P \sin 15° = 0.259P$$

4. Apply the equations of equilibrium

Consider that the bar remains at rest and apply the equations of equilibrium. Thus:

$$\Sigma F_x = R_x + 0.966P - 64.3 = 0 \qquad R_x = 64.3 - 0.966P$$

$$\Sigma F_y = R_y + 0.259P - 76.6 = 0 \qquad R_y = 76.6 - 0.259P$$

5. Assume maximum friction exists and solve for the applied force

Assume that R_x, which represents the frictional resistance to motion, has its maximum potential value. Apply $R_x = \mu R_y$, where μ = coefficient of friction. Then, $R_x = 0.20R_y = 0.20(76.6 - 0.259P) = 15.32 - 0.052P$. Substituting for R_x from Step 4, $64.3 - 0.966P = 15.32 - 0.052P$; $P = 53.6$ lb.

6. Draw a second free-body diagram

In Fig. 2c, draw a free-body diagram of the bar, R_x being directed downward.

7. Solve as in Steps 1 through 5

As before, $R_y = 76.6 - 0.259P$. Also, the absolute value of $R_x = 0.966P - 64.3$. But $R_x = 0.20R_y = 15.32 \times 0.052P$. Then, $0.966P - 64.3 = 15.32 - 0.052P$; $P = 78.2$ lb.

ANALYSIS OF A STRUCTURAL FRAME

The frame in Fig. 3a consists of two inclined members and a tie rod. What is the tension in the rod when a load of 1,000 lb is applied at the hinged apex? Neglect the weight of the frame and consider the supports to be smooth.

Calculation Procedure:

1. Draw a free-body diagram of the frame

Since friction is absent in this frame, the reactions at the supports are vertical. Draw a free-body diagram as in Fig. 3b.

With the free-body diagram shown, compute the distances x_1 and x_2. Since the frame forms a 3-4-5 right triangle, $x_1 = 16(4/5) = 12.8$ ft; $x_2 = 12(3/5) = 7.2$ ft.

2. Determine the reactions on the frame

Take moments with respect to A and B to obtain the reactions. Or:

$$\Sigma M_B = 20R_L - 1,000(7.2) = 0; R_L = 360 \text{ lb};$$

$$\Sigma M_A = 1,000(12.8) - 20R_R = 0; R_R = 640 \text{ lb}.$$

3. Determine the distance y, Fig. 3c

Draw a free-body diagram of member AC, Fig. 3c. Compute $y = 13(3/5) = 7.8$ ft.

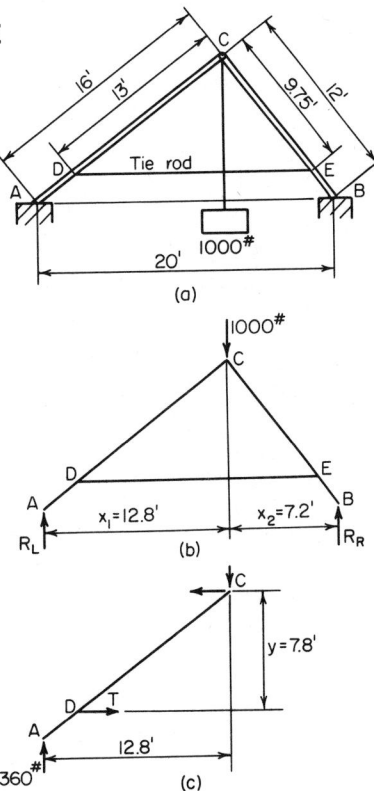

Fig. 3

4.   *Compute the tension in the tie rod*

Take moments with respect to C to find the tension T in the tie rod. Or:

$$\Sigma M_C = 360(12.8) - 7.8T = 0; T = 591 \text{ lb}.$$

5.   *Verify the computed result*

Draw a free-body diagram of member BC and take moments with respect to C. The result verifies that computed above.

GRAPHICAL ANALYSIS OF A PLANE TRUSS

Apply a graphical analysis to the cantilever truss in Fig. $4a$ to evaluate the forces induced in the truss members.

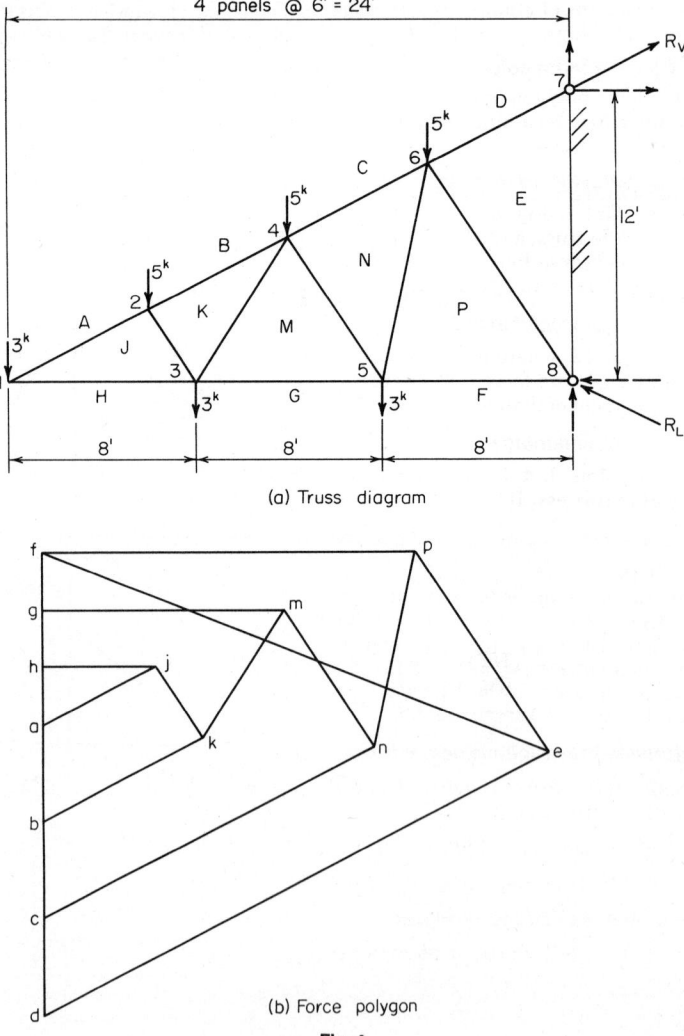

(a) Truss diagram

(b) Force polygon

Fig. 4

Calculation Procedure:

1. Label the truss for analysis

Divide the space around the truss into regions bounded by the action lines of the external and internal forces. Assign an uppercase letter to each region, Fig. 4.

2. Determine the reaction force

Take moments with respect to joint 8, Fig. 4, to determine the horizontal component of the reaction force R_U. Then compute R_U. Thus: $\Sigma M_8 = 12R_{UH} - 3(8 + 16 + 24) - 5(6 + 12 + 18) = 0$; $R_{UH} = 27$ kips to the right.

Since R_U is collinear with the force DE, $R_{UV}/R_{UH} = 12/24$; $R_{UV} = 13.5$ kips up; $R_U = 30.2$ kips.

3. Apply the equations of equilibrium

Use the equations of equilibrium to find R_L. Or: $R_{LH} = 27$ kips to the left; $R_{LV} = 10.5$ kips up; $R_L = 29.0$ kips.

4. Construct the force polygon

Draw the force polygon in Fig. 4b by using a suitable scale and drawing vector fg to represent force FG. Next, draw vector gh to represent force GH, and so forth. Omit the arrowheads on the vectors.

5. Determine the forces in the truss members

Starting at joint 1, Fig. 4b, draw a line through a in the force polygon parallel to member AJ in the truss, and one through h parallel to member HJ. Designate the point of intersection of these lines as j. Now, vector aj represents the force in AJ, and vector hj represents the force in HJ.

6. Analyze the next joint in the truss

Proceed to joint 2, where there are now only two unknown forces — BK and JK. Draw a line through b in the force polygon parallel to BK, and one through j parallel to JK. Designate the point of intersection as k. The forces BK and JK are thus determined.

7. Analyze the remaining joints

Proceed to joints 3, 4, 5, and 6, in that order, and complete the force polygon by continuing the process. If the construction is accurately performed, the vector pe will parallel the member PE in the truss.

8. Determine the magnitude of the internal forces

Scale the vector lengths to obtain the magnitude of the internal forces. Tabulate the results as in Table 1.

TABLE 1   Forces in Truss Members (Fig. 4)

Member	Force, kips
AJ	+6.7
BK	+9.5
CN	+19.8
DE	+30.2
HJ	−6.0
GM	−13.0
FP	−20.0
JK	−4.5
KM	+8.1
MN	−8.6
NP	+10.4
PE	−12.6

9. *Establish the character of the internal forces*

To determine whether an internal force is one of tension or compression, proceed this way: Select a particular joint and proceed around the joint in a clockwise direction, listing the letters in the order in which they appear. Then refer to the force polygon pertaining to that joint, and proceed along the polygon in the same order. This procedure shows the direction in which the force is acting at that joint.

For instance, proceeding around joint 4, CNMKB is obtained. Tracing a path along the force polygon in the order in which the letters appear, force CN is found to act upward to the right; NM acts upward to the left; MK and KB act downward to the left. Therefore, CN, MK, and KB are directed away from the joint, Fig. 4; this condition discloses that they are tensile forces. NM is directed toward the joint; it is therefore compressive.

The validity of this procedure lies in the drawing of the vectors representing external forces while proceeding around the truss in a clockwise direction. Tensile forces are shown with a positive sign in Table 1; compressive forces are shown with a negative sign.

Related Calculations: Use this general method for any type of truss.

TRUSS ANALYSIS BY THE METHOD OF JOINTS

Applying the method of joints, determine the forces in the truss in Fig. 5a. The inclined load at joint 4 has a horizontal component of 4 kips and a vertical component of 3 kips.

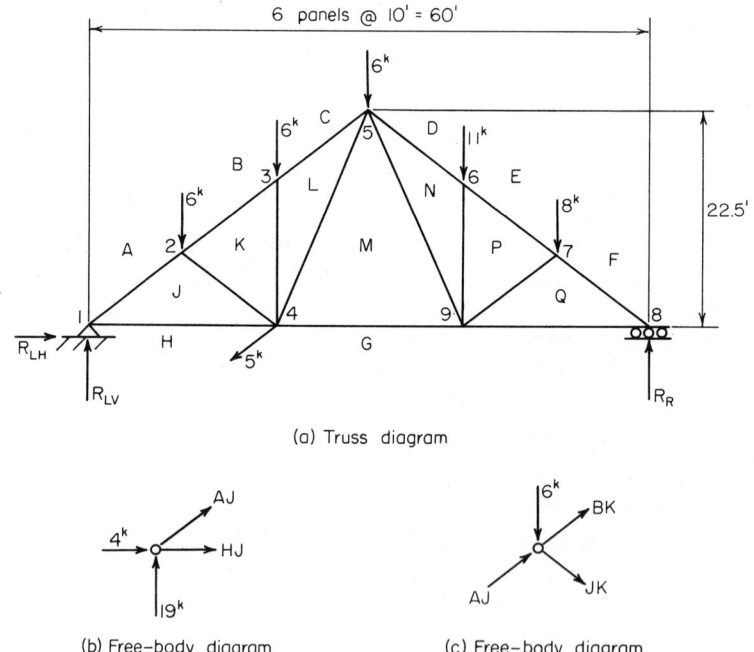

(a) Truss diagram

(b) Free-body diagram of joint 1

(c) Free-body diagram of joint 2

Fig. 5

Calculation Procedure:

1. *Compute the reactions at the supports*

Using the usual analysis techniques, $R_{LV} = 19$ kips; $R_{LH} = 4$ kips; $R_R = 21$ kips.

2. List each truss member and its slope

Table 2 shows each truss member and its slope.

3. Determine the forces at a principal joint

Draw a free-body diagram, Fig. 5b, of the pin at joint 1. For the free-body diagram, assume that the unknown internal forces AJ and HJ are tensile. Apply the equations of equilibrium to evaluate these forces, using the subscripts H and V, respectively, to identify the horizontal and vertical components. Thus: $\Sigma F_H = 4.0 + AJ_H + HJ = 0$; $\Sigma F_V = 19.0 + AJ_V = 0$; $\therefore AJ_V = -19.0$ kips; $AJ_H = -19.0/0.75 = -25.3$ kips. Substituting in the first equation, $HJ = 21.3$ kips.

The algebraic signs disclose that AJ is compressive and HJ is tensile. Record these results in Table 2, showing the tensile forces as positive and compressive forces as negative.

TABLE 2 Forces in Truss Members (Fig. 5)

Member	Slope	Horizontal component	Vertical component	Force, kips
AJ	0.75	25.3	19.0	-31.7
BK	0.75	21.3	16.0	-26.7
CL	0.75	21.3	16.0	-26.7
DN	0.75	22.7	17.0	-28.3
EP	0.75	22.7	17.0	-28.3
FQ	0.75	28.0	21.0	-35.0
HJ	0.0	21.3	0.0	$+21.3$
GM	0.0	16.0	0.0	$+16.0$
GQ	0.0	28.0	0.0	$+28.0$
JK	0.75	4.0	3.0	-5.0
KL	∞	0.0	6.0	-6.0
LM	2.25	5.3	12.0	$+13.1$
MN	2.25	6.7	15.0	$+16.4$
NP	∞	0.0	11.0	-11.0
PQ	0.75	5.3	4.0	-6.7

4. Determine the forces at another joint

Draw a free-body diagram of the pin at joint 2, Fig. 5c. Show the known force AJ as compressive, and assume that the unknown forces BK and JK are tensile. Apply the equations of equilibrium, expressing the vertical components of BK and JK in terms of their horizontal components. Thus: $\Sigma F_H = 25.3 + BK_H + JK_H = 0$; $\Sigma F_V = -6.0 + 19.0 + 0.75BK_H - 0.75JK_H = 0$.

Solve these simultaneous equations to obtain: $BK_H = -21.3$ kips; $JK_H = -4.0$ kips; $BK_V = -16.0$ kips; $JK_V = -3.0$ kips. Record these results in Table 2.

5. Continue the analysis at the next joint

Proceed to joint 3. Since there are no external horizontal forces at this joint, $CL_H = BK_H = 21.3$-kips compression. Also, $KL = 6$-kips compression.

6. Proceed to the remaining joints in their numbered order

Thus, *Joint 4:* $\Sigma F_H = -4.0 - 21.3 + 4.0 + LM_H + GM = 0$; $\Sigma F_V = -3.0 - 3.0 - 6.0 + LM_V = 0$; $LM_V = 12.0$ kips; $LM_H = 12.0/2.25 = 5.3$ kips. Substituting in the first equation, $GM = 16.0$ kips.

Joint 5: $\Sigma F_H = 21.3 - 5.3 + DN_H + MN_H = 0$; $\Sigma F_V = -6.0 + 16.0 - 12.0 - 0.75DN_H - 2.25MN_H = 0$; $DN_H = -22.7$ kips; $MN_H = 6.7$ kips; $DN_V = -17.0$ kips; $MN_V = 15.0$ kips.

Joint 6: $EP_H = DN_H = 22.7$-kips compression; $NP = 11.0$-kips compression.

Joint 7: $\Sigma F_H = 22.7 - PQ_H + FQ_H = 0$; $\Sigma F_V = -8.0 - 17.0 - 0.75PQ_H - 0.75FQ_H = 0$; $PQ_H = -5.3$ kips; $FQ_H = -28.0$ kips; $PQ_V = -4.0$ kips; $FQ_V = -21.0$ kips.

Joint 8: $\Sigma F_H = 28.0 - GQ = 0$; $GQ = 28.0$ kips; $\Sigma F_V = 21.0 - 21.0 = 0$.
Joint 9: $\Sigma F_H = -16.0 - 6.7 - 5.3 + 28.0 = 0$; $\Sigma F_V = 15.0 - 11.0 - 4.0 = 0$.

7. Complete the computation

Compute the values in the last column of Table 2 and enter them as shown.

TRUSS ANALYSIS BY THE METHOD OF SECTIONS

Using the method of sections, determine the forces in members BK and LM in Fig. 5a.

Calculation Procedure:

1. Draw a free-body diagram of one portion of the truss

Cut the truss at the plane aa, Fig. 6a, and draw a free-body diagram of the left part of the truss. Assume that BK is tensile.

2. Determine the magnitude and character of the first force

Take moments with respect to joint 4. Since each half of the truss forms a 3-4-5 right triangle, $d = 20(3/5) = 12$ ft; $\Sigma M_4 = 19(20) - 6(10) + 12BK = 0$; $BK = -26.7$ kips.

The negative result signifies that the assumed direction of BK is incorrect; the force is therefore compressive.

3. Use an alternative solution

Alternatively, resolve BK (again assumed tensile) into its horizontal and vertical components at joint 1. Take moments with respect to joint 4. (A force may be resolved into its components at any point on its action line.) Then: $\Sigma M_4 = 19(20) + 20BK_V - 6(10) = 0$; $BK_V = -16.0$ kips; $BK = -16.0(5/3) = -26.7$ kips.

4. Draw a second free-body diagram of the truss

Cut the truss at plane bb, Fig. 6b, and draw a free-body diagram of the left part. Assume LM is tensile.

5. Determine the magnitude and character of the second force

Resolve LM into its horizontal and vertical components at joint 4. Take moments with respect to joint 1: $\Sigma M_1 = 6(10 + 20) + 3(20) - 20LM_V = 0$; $LM_V = 12.0$ kips; $LM_H = 12.0/2.25 = 5.3$ kips; $LM = 13.1$ kips.

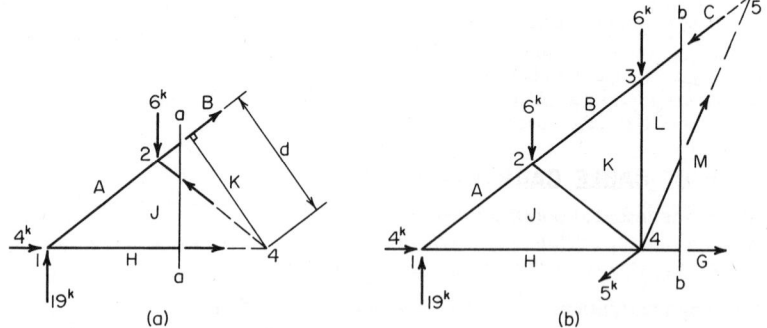

(a)   (b)

Fig. 6

REACTIONS OF A THREE-HINGED ARCH

The parabolic arch in Fig. 7 is hinged at A, B, and C. Determine the magnitude and direction of the reactions at the supports.

Calculation Procedure:

1. Consider the entire arch as a free body and take moments

Since a moment cannot be transmitted across a hinge, the bending moments at A, B, and C are zero. Resolve the reactions R_A and R_C, Fig. 7, into their horizontal and vertical components.

Considering the entire arch ABC as a free body, take moments with respect to A and C. Thus: $\Sigma M_A = 8(10) + 10(25) + 12(40) + 8(56) - 5(25.2) - 72R_{CV} - 10.8R_{CH} = 0$; or $72R_{CV} + 10.8R_{CH} = 1{,}132$, Eq. (a). Also, $\Sigma M_C = 72R_{AV} - 10.8R_{AH} - 8(62) - 10(47) - 12(32) - 8(16) - 5(14.4) = 0$; or $72R_{AV} - 10.8R_{AH} = 1{,}550$, Eq. (b).

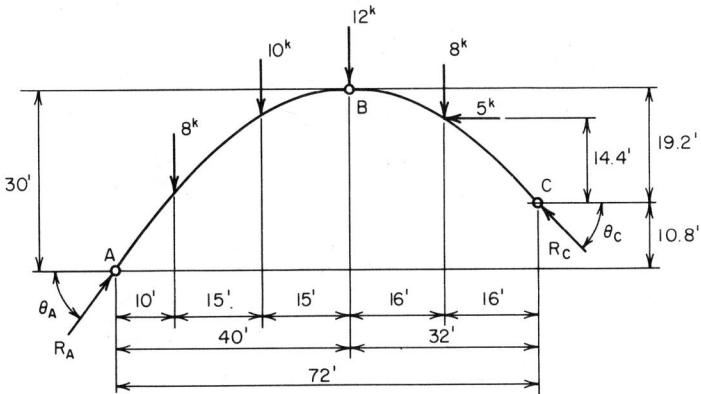

Fig. 7   Parabolic arch.

2. Consider a segment of the arch and take moments

Considering the segment BC as a free body, take moments with respect to B. Or: $\Sigma M_B = 8(16) + 5(4.8) - 32R_{CV} + 19.2R_{CH} = 0$; or $32R_{CV} - 19.2R_{CH} = 152$, Eq. (c).

3. Consider another segment and take moments

Considering segment AB as a free body, take moments with respect to B: $\Sigma M_B = 40R_{AV} - 30R_{AH} - 8(30) - 10(15) = 0$; or $40R_{AV} - 30R_{AH} = 390$, Eq. (d).

4. Solve the simultaneous moment equations

Solve Eqs. (b) and (d) to determine R_A; solve Eqs. (a) and (c) to determine R_C. Thus: $R_{AV} = 24.4$ kips; $R_{AH} = 19.6$ kips; $R_{CV} = 13.6$ kips; $R_{CH} = 14.6$ kips. Then, $R_A = [(24.4)^2 + (19.6)^2]^{0.5} = 31.3$ kips. Also, $R_C = [(13.6)^2 + (14.6)^2]^{0.5} = 20.0$ kips. And, $\theta_A = \arctan (24.4/19.6) = 51°14'$; $\theta_C = \arctan (13.6/14.6) = 42°58'$.

LENGTH OF CABLE CARRYING KNOWN LOADS

A cable is supported at points P and Q, Fig. 8a, and carries two vertical loads as shown. If the tension in the cable is restricted to 1,800 lb, determine the minimum length of cable required to carry the loads.

Calculation Procedure:

1. Sketch the loaded cable

Assume a position of the cable, such as $PRSQ$, Fig. 8a. In Fig. 8b, locate points P' and Q', corresponding to P and Q, respectively, in Fig. 8a.

2. Take moments with respect to an assumed point

Assume that the maximum tension of 1,800 lb occurs in segment PR, Fig. 8. The reaction at P, which is collinear with PR, is therefore 1,800 lb. Compute the true perpendicular distance m from Q to PR by taking moments with respect to Q. Or: $\Sigma M_Q = 1{,}800m - 500(35) - 750(17) = 0$; $m = 16.8$ ft. This dimension establishes the true position of PR.

3. Start the graphical solution of the problem

In Fig. 8b, draw a circular arc having Q' as center and a radius of 16.8 ft. Draw a line through P' tangent to this arc. Locate R' on this tangent at a horizontal distance of 15 ft from P'.

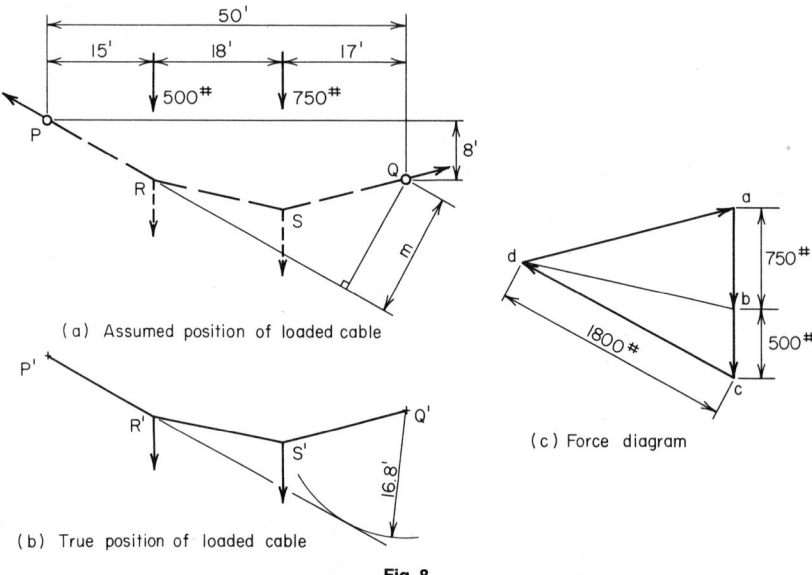

(a) Assumed position of loaded cable

(b) True position of loaded cable

(c) Force diagram

Fig. 8

4. Draw the force vectors

In Fig. 8c, draw vectors ab, bc, and cd to represent the 750-lb load, the 500-lb load, and the 1,800-lb reaction at P, respectively. Complete the triangle by drawing vector da, which represents the reaction at Q.

5. Check the tension assumption

Scale da to ascertain if it is less than 1,800 lb. This is found to be so, and the assumption that the maximum tension exists in PR is thus validated.

6. Continue the construction

Draw a line through Q' in Fig. 8b parallel to da in Fig. 8c. Locate S' on this line at a horizontal distance of 17 ft from Q.

7. Complete the construction

Draw $R'S'$ and db. Test the accuracy of the construction by determining whether these lines are parallel.

8. Determine the required length of the cable

Obtain the required length of the cable by scaling the lengths of the segments in Fig. 8b. Thus: $P'R' = 17.1$ ft; $R'S' = 18.4$ ft; $S'Q' = 17.6$ ft; length of cable $= 53.1$ ft.

PARABOLIC CABLE TENSION AND LENGTH

A suspension bridge has a span of 960 ft and a sag of 50 ft. Each cable carries a load of 1.2 klf (kips per linear foot) uniformly distributed along the horizontal. Compute the tension in the cable at mid-span and at the supports, and determine the length of the cable.

Calculation Procedure:

1. Compute the tension at mid-span

A cable carrying a load uniformly distributed along the horizontal assumes the form of a parabolic arc. Referring to Fig. 9 which shows such a cable having supports at the

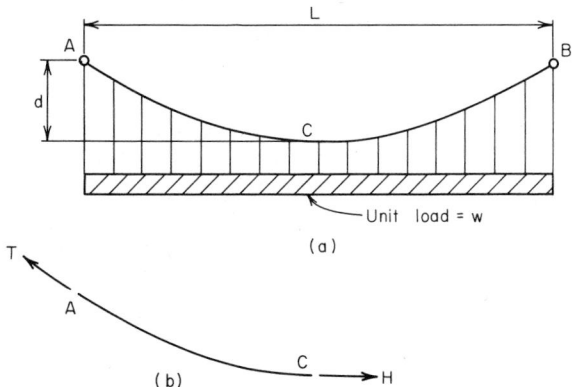

Fig. 9   Cable supporting load uniformly distributed along horizontal.

same level, the tension at mid-span is $H = wL^2/8d$, where H = mid-span tension, kips; w = load on a unit horizontal distance, klf; L = span, ft; d = sag, ft. Substituting, $H = 1.2(960)^2/8(50) = 2,765$ kips.

2. Compute the tension at the supports

Use the relation $T = [H^2 + (wL/2)^2]^{0.5}$, where T = tension at supports, kips; other symbols as before. Thus, $T = [(2,765)^2 + (1.2 \times 480)^2]^{0.5} = 2,824$ kips.

3. Compute the length of the cable

When d/L is 1/20, or less, the cable length can be approximated from $S = L + 8d^2/3L$, where S = cable length, ft. Thus, $S = 960 + 8(50)^2/3(960) = 966.94$ ft.

CATENARY CABLE SAG AND DISTANCE BETWEEN SUPPORTS

A 500-ft-long cable weighing 3 plf (pounds per linear foot) is supported at two points lying in the same horizontal plane. If the tension at the supports is 1,800 lb, find the sag of the cable and the distance between the supports.

Calculation Procedure:

1. Compute the catenary parameter

A cable of uniform cross section carrying only its own weight assumes the form of a catenary. Using the notation of the previous Procedure, the catenary parameter c is found from $d + c = T/w = 1,800/3 = 600$ ft. Then, $c = [(d+c)^2 - (S/2)^2]^{0.5} = [(600)^2 - (250)^2]^{0.5} = 545.4$ ft.

2. Compute the cable sag

Since $d + c = 600$ ft, and $c = 545.4$ ft, $d = 600 - 545.4 = 54.6$ ft.

3. Compute the span length

Use the relation $L = 2c \ln (d+c+0.5S)/c$, or $L = 2(545.4) \ln (600+250)/545.4 = 484.3$ ft.

STABILITY OF A RETAINING WALL

Determine the factor of safety (FS) against sliding and overturning of the concrete retaining wall in Fig. 10. The concrete weighs 150 pcf (pounds per cubic foot), the earth 100 pcf, the coefficient of friction is 0.6, and the coefficient of active earth pressure is 0.333.

Calculation Procedure:

1. Compute the vertical loads on the wall

Select a 1-ft length of wall as typical of the entire structure. The horizontal pressure of the confined soil varies linearly with the depth and is represented by the triangle BGF, Fig. 10.

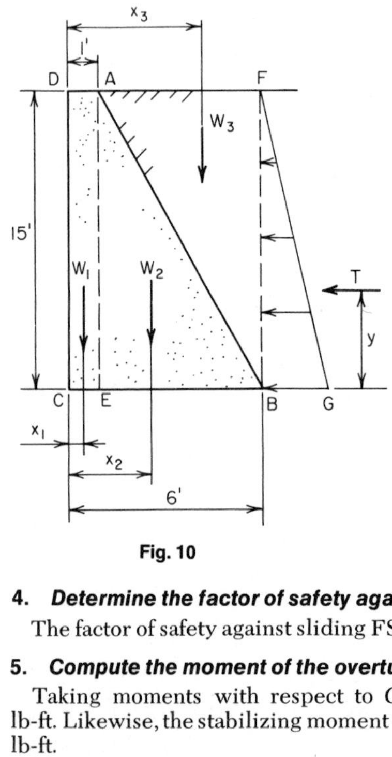

Resolve the wall into the elements $AECD$ and AEB; pass the vertical plane BF through the soil. Calculate the vertical loads and locate their resultants with respect to the toe C. Thus: $W_1 = 15(1)(150) = 2,250$ lb; $W_2 = 0.5(15)(5)(150) = 5,625$; $W_3 = 0.5(15)(5)(100) = 3,750$. Then, $\Sigma W = 11,625$ lb. Also, $x_1 = 0.5$ ft; $x_2 = 1 + 0.333(5) = 2.67$ ft; $x_3 = 1 + 0.667(5) = 4.33$ ft.

Fig. 10

2. Compute the resultant horizontal soil thrust

Compute the resultant horizontal thrust T lb of the soil by applying the coefficient of active earth pressure. Determine the location of T. Thus: $BG = 0.333(15)(100) = 500$ plf; $T = 0.5(15)(500) = 3,750$ lb; $y = 0.333(15) = 5$ ft.

3. Compute the maximum frictional force preventing sliding

The maximum frictional force F_m lb $= \mu(\Sigma W)$, where μ = coefficient of friction. Or, $F_m = 0.6 (11,625) = 6,975$ lb.

4. Determine the factor of safety against sliding

The factor of safety against sliding FSS $= F_m/T = 6,975/3,750 = 1.86$.

5. Compute the moment of the overturning and stabilizing forces

Taking moments with respect to C, the overturning moment $= 3,750(5) = 18,750$ lb-ft. Likewise, the stabilizing moment $= 2,250(0.5) + 5,625(2.67) + 3,750(4.33) = 32,375$ lb-ft.

6. Compute the factor of safety against overturning

The factor of safety against overturning FSO = stabilizing moment, lb-ft/overturning moment, lb-ft $= 32,375/18,750 = 1.73$.

ANALYSIS OF A SIMPLE SPACE TRUSS

In the space truss shown in Fig. 11a, A lies in the xy plane, B and C lie on the z axis, and D lies on the x axis. A horizontal load of 4,000 lb lying in the xy plane is applied at A. Determine the force induced in each member by applying the method of joints, and verify the results by taking moments with respect to convenient axes.

Calculation Procedure:

1. Determine the projected length of members

Let d_x, d_y, and d_z denote the length of a member as projected on the x, y, and z axis, respectively. Record in Table 3 the projected lengths of each member. Record the remaining values as they are obtained.

2. Compute the true length of each member

Use the equation $d = [d_x^2 + d_y^2 + d_z^2]^{0.5}$, where $d =$ the true length of a member.

3. Compute the ratio of the projected length to the true length

For each member, compute the ratios of the three projected lengths to the true length. For example, for member AC: $d_z/d = 6/12.04 = 0.498$.

These ratios are termed *direction cosines* because each represents the cosine of the angle between the member and the designated axis, or an axis parallel thereto.

Since the axial force in each member has the same direction as the member itself, a direction cosine also represents the ratio of the component of a force along the designated axis to the total force in the member. For instance, let AC denote the force in member AC, and let AC_x denote its component along the x axis. Then: $AC_x/AC = d_x/d = 0.249$.

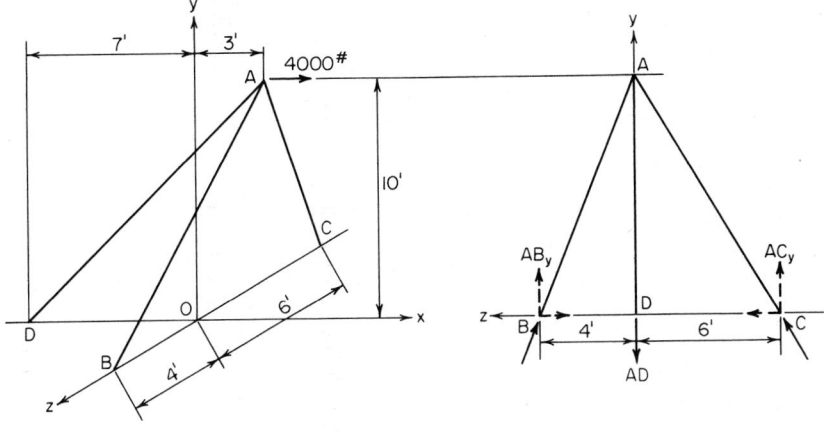

| (a) Isometric view of space truss | (b) View normal to yz plane |

Fig. 11

4. Determine the component forces

Consider joint A as a free body, and assume that the forces in the three truss members are tensile. Equate the sum of the forces along each axis to zero. For instance, if the truss members are in tension, the x components of these forces are directed to the left, and $\Sigma F_x = 4{,}000 - AB_x - AC_x - AD_x = 0$.

Express each component in terms of the total force to obtain: $\Sigma F_x = 4{,}000 - 0.268AB - 0.249AC - 0.707AD = 0$; $\Sigma F_y = -0.894AB - 0.831AC - 0.707AD = 0$; $\Sigma F_z = 0.358AB - 0.498AC = 0$.

TABLE 3   Data for Space Truss (Fig. 11)

Member	AB	AC	AD
d_x, ft	3	3	10
d_y	10	10	10
d_z	4	6	0
d	11.18	12.04	14.14
d_x/d	0.268	0.249	0.707
d_y/d	0.894	0.831	0.707
d_z/d	0.358	0.498	0
Force, lb	−3,830	−2,750	+8,080

5. Solve the simultaneous equations in Step 4 to evaluate the forces in the truss members

A positive result in the solution signifies tension; a negative result, compression. Thus: $AB = 3{,}830$-lb compression; $AC = 2{,}750$-lb compression; $AD = 8{,}080$-lb tension. To verify these results, it is necessary to select moment axes yielding equations independent of those previously developed.

6. Resolve the reactions into their components

In Fig. 11*b*, show the reactions at the supports B, C, and D, each reaction being numerically equal to and collinear with the force in the member at that support. Resolve these reactions into their components.

7. Take moments about a selected axis

Take moments with respect to the axis through C parallel to the x axis. (Since the x components of the forces are parallel to this axis, their moments are zero.) Then: $\Sigma M_{Cx} = 10AB_y - 6AD_y = 10(0.894)(3{,}830) - 6(0.707)(8{,}080) = 0$.

8. Take moments about another axis

Take moments with respect to the axis through D parallel to the x axis. Or: $\Sigma M_{Dx} = 4AB_y - 6AC_y = 4(0.894)(3{,}830) - 6(0.831)(2{,}750) = 0$.

The computed results are thus substantiated.

ANALYSIS OF A COMPOUND SPACE TRUSS

The compound space truss in Fig. 12*a* has the dimensions shown in the orthographic projections, Fig. 12*b* and 12*c*. A load of 5,000 lb, which lies in the xy plane and makes an angle of 30° with the vertical, is applied at A. Determine the force induced in each member, and verify the results.

Calculation Procedure:

1. Compute the true length of each truss member

Since the truss and load system are symmetrical with respect to the xy plane, the internal forces are also symmetrical. As one component of an internal force becomes known, it will be convenient to calculate the other components at once, as well as the total force.

Record in Table 4 the length of each member as projected on the coordinate axes. Calculate the true length of each member using geometric relations.

2. Resolve the applied load into its x and y components

Use only the absolute values of the forces. Thus: $P_x = 5{,}000 \sin 30° = 2{,}500$ lb; $P_y = 5{,}000 \cos 30° = 4{,}330$ lb.

3. Compute the horizontal reactions

Compute the horizontal reactions at D and at line CC', Fig. 12*b*. Thus: $\Sigma M_{CC'} = 4{,}330 (12) - 2{,}500(7) - 10H_1 = 0$; $H_1 = 3{,}446$ lb; $H_2 = 3{,}446 - 2{,}500 = 946$ lb.

4. Compute the vertical reactions

Consider the equilibrium of joint D and the entire truss when computing the vertical reactions. In all instances, assume that an unknown internal force is tensile. Thus, at joint D: $\Sigma F_x = -H_1 + 2BD_x = 0$; $BD_x = 1{,}723$-lb tension; $BD_y = 1{,}723(10/12) = 1{,}436$ lb; likewise, $\Sigma F_y = V_1 - 2BD_y = V_1 - 2(1{,}436) = 0$; $V_1 = 2{,}872$ lb.

For the entire truss, $\Sigma F_y = V_1 + V_2 - 4{,}330 = 0$; $V_2 = 1{,}458$ lb.

The z components of the reactions are not required in this solution. Thus, the remaining calculations for BD are: $BD_z = 1{,}723(4/12) = 574$ lb; $BD = 1{,}723(16.12/12) = 2{,}315$ lb.

5. Compute the unknown forces using the equilibrium of a joint

Calculate the forces AC and BC by considering the equilibrium of joint C. Thus: $\Sigma F_x = 0.5H_2 + AC_x + BC = 0$, Eq. (*a*); $\Sigma F_y = 0.5V_2 - AC_y = 0$, Eq. (*b*). From Eq. (*b*),

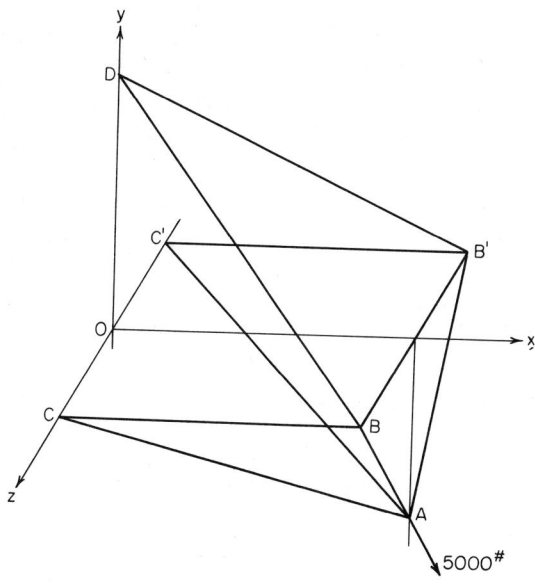

(a) Isometric view of space truss

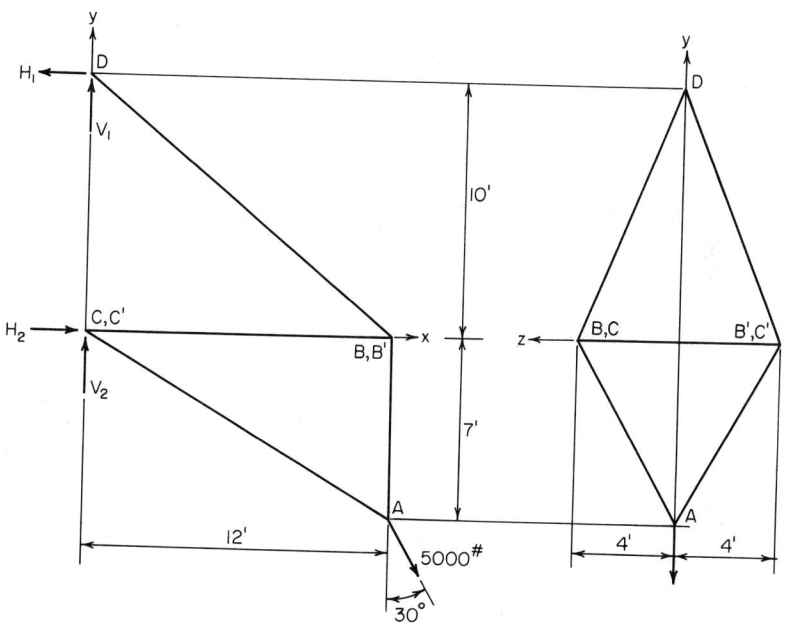

(b) View normal to xy plane

(c) View normal to yz plane

Fig. 12

1-21

$AC_y = 729$-lb tension. Then, $AC_x = 729(12/7) = 1,250$ lb. From Eq. (a), $BC = 1,723$-lb compression. Then, $AC_z = 729(4/7) = 417$ lb; $AC = 729(14.46/7) = 1,506$ lb.

6. Compute another set of forces by considering joint equilibrium

Calculate the forces AB and BB' by considering the equilibrium of joint B. Thus: $\Sigma F_y = BD_y - AB_y = 0$; $AB_y = 1,436$-lb tension; $AB_z = 1,436(4/7) = 821$ lb; $AB = 1,436$ $(8.06/7) = 1,653$ lb; $\Sigma F_z = -AB_z - BD_z - BB' = 0$; $BB' = 1,395$-lb compression.

All the internal forces are now determined. Show in Table 4 the tensile forces as positive, and the compressive forces as negative.

7. Check the equilibrium of the first joint considered

The first joint considered was A. Thus: $\Sigma F_x = -2AC_x + 2,500 = -2(1,250) + 2,500 = 0$; $\Sigma F_y = 2AB_y + 2AC_y - 4,330 = 2(1,436) + 2(729) - 4,330 = 0$. Since the summation of forces for both axes is zero, the joint is in equilibrium.

8. Check the equilibrium of the second joint

Check the equilibrium of joint B by taking moments of the forces acting on this joint with respect to the axis through A parallel to the x axis, Fig. 12c. Thus: $\Sigma M_{Ax} = -7BB' + 7BD_z + 4BD_y = -7(1,395) + 7(574) + 4(1,436) = 0$.

TABLE 4　Data for Space Truss (Fig. 12)

Member	AB	AC	BC	BD	BB'
d_x, ft	0	12	12	12	0
d_y	7	7	0	10	0
d_z	4	4	0	4	8
d	8.06	14.46	12.00	16.12	8
F_x, lb	0	1,250	1,723	1,723	0
F_y	1,436	729	0	1,436	0
F_z	821	417	0	574	1,395
F	+1,653	+1,506	−1,723	+2,315	−1,395

9. Check the equilibrium of the right-hand part of the structure

Cut the truss along a plane parallel to the yz plane. Check the equilibrium of the right-hand part of the structure. $\Sigma F_x = -2BD_x + 2BC - 2AC_x + 2,500 = -2(1,723) + 2(1,723) - 2(1,250) + 2,500 = 0$; $\Sigma F_y = 2BD_y + 2AC_y - 4,330 = 2(1,436) + 2(729) - 4,330 = 0$. The calculated results are thus substantiated in these equations.

GEOMETRICAL PROPERTIES OF AN AREA

Calculate the polar moment of inertia of the area in Fig. 13 (a) with respect to its centroid, and (b) with respect to point A.

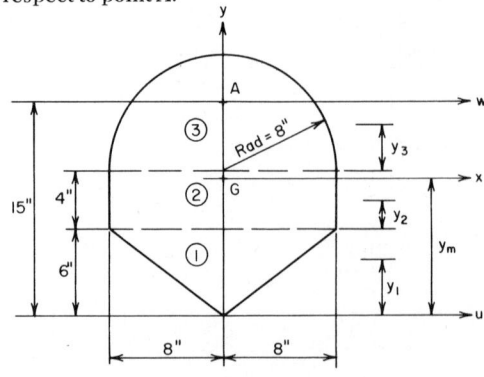

Fig. 13

Calculation Procedure:

1. Establish the area axes

Set up the horizontal and vertical coordinate axes u and y, respectively.

2. Divide the area into suitable elements

Using the American Institute of Steel Construction (AISC) *Manual*, obtain the properties of the elements 1, 2, and 3, Fig. 13, after locating the horizontal centroidal axis of each element. Thus: $y_1 = \frac{2}{3}(6) = 4$ in.; $y_2 = 2$ in.; $y_3 = 0.424(8) = 3.4$ in.

3. Locate the horizontal centroidal axis of the entire area

Let x denote the horizontal centroidal axis of the entire area. Locate this axis by computing the statical moment of the area with respect to the u axis. Thus:

Element	Area, sq in.	×	Arm, in.	=	Moment, in.^{3}
1	0.5(6)(16) = 48		4	=	192
2	4(16) = 64		8	=	512
3	1.57(8)^2 = 100.5		13.4	=	1,347
Total	212.5				2,051

Then, $y_m = 2{,}051/212.5 = 9.7$ in. Since the area is symmetrical with respect to the y axis, this is also a centroidal axis. The intersection point G of the x and y axes is therefore the centroid of the area.

4. Compute the distance between the centroidal axis and the reference axis

Compute k, the distance between the horizontal centroidal axis of each element and the x axis. Only absolute values are required. Thus: $k_1 = 9.7 - 4.0 = 5.7$ in.; $k_2 = 9.7 - 8.0 = 1.7$ in.; $k_3 = 13.4 - 9.7 = 3.7$ in.

5. Compute the moment of inertia of the entire area — x axis

Let I_0 denote the moment of inertia of an element with respect to its horizontal centroidal axis, and A, its area. Compute the moment of inertia I_x of the entire area with respect to the x axis by applying the transfer equation $I_x = \Sigma I_0 + \Sigma A k^2$. Thus:

Element	I_0, in.^{4}	Ak^2, in.^{4}
1	1/36(16)(6)^3 = 96	48(5.7)^2 = 1,560
2	1/12(16)(4)^3 = 85	64(1.7)^2 = 185
3	0.110(8)^4 = 451	100.5(3.7)^2 = 1,376
Total	632	3,121

Then, $I_x = 632 + 3{,}121 = 3{,}753$ in.^{4}.

6. Determine the moment of inertia of the entire area — y axis

For this computation, subdivide element 1 into two triangles having the y axis as a base. Thus:

Element	I about y axis, in.^{4}
1	$2(\frac{1}{12})(6)(8)^3 =$ 512
2	$\frac{1}{12}(4)(16)^3 =$ 1,365
3	$\frac{1}{2}(0.785)(8)^4 =$ 1,607
	$I_y = 3{,}484$

7. Compute the polar moment of inertia of the area

Apply the equation for the polar moment of inertia J_G with respect to G: $J_G = I_x + I_y = 3,753 + 3,484 = 7,237$ in.^{4}.

8. Determine the moment of inertia of the entire area — w axis

Apply the equation in Step 5 to determine the moment of inertia I_w of the entire area with respect to the horizontal axis w through A. Thus: $k = 15.0 - 9.7 = 5.3$ in; $I_w = I_x + Ak^2 = 3,753 + 212.5(5.3)^2 = 9,722$ in.^{4}.

9. Compute the polar moment of inertia

Compute the polar moment of inertia of the entire area with respect to A. Or: $J_A = I_w + I_y = 9,722 + 3,484 = 13,206$ in.^{4}.

PRODUCT OF INERTIA OF AN AREA

Calculate the product of inertia of the isosceles trapezoid in Fig. 14 with respect to the rectangular axes u and v.

Calculation Procedure:

1. Locate the centroid of the trapezoid

Using the AISC *Manual* or another suitable reference: h = centroid distance from the axis, Fig. 14, $= (9/3)([2 \times 5 + 10]/[5 + 10]) = 4$ in.

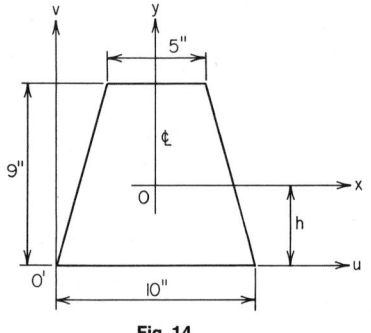

Fig. 14

2. Compute the area and product of inertia P_{xy}

The area of the trapezoid is $A = \frac{1}{2}(9)(5 + 10) = 67.5$ sq in. Since the area is symmetrically disposed with respect to the y axis, the product of inertia with respect to the x and y axes is $P_{xy} = 0$.

3. Compute the product of inertia by applying the transfer equation

The transfer equation for the product of inertia is $P_{uv} = P_{xy} + Ax_m y_m$, where x_m and y_m are the coordinates of $0'$ with respect to the centroidal x and y axes, respectively. Thus: $P_{uv} = 0 + 67.5(-5)(-4) = 1,350$ in.^{4}.

PROPERTIES OF AN AREA WITH RESPECT TO ROTATED AXES

In Fig. 15, x and y are rectangular axes through the centroid of the isosceles triangle; x' and y' are axes parallel to x and y, respectively; x'' and y'' are axes making an angle of 30° with x' and y', respectively. Compute the moments of inertia and the product of inertia of the triangle with respect to the x'' and y'' axes.

Calculation Procedure:

1. Compute the area of the figure

The area of this triangle $= 0.5(\text{base})(\text{altitude}) = 0.5 \ (8)(9) = 36$ sq in.

2. Compute the properties of the area with respect to the x and y axes

Using conventional moment-of-inertia relations, $I_x = bd^3/36 = 8(9)^3/36 = 162$ in.^{4}; $I_y = b^3 d/48 = (8)^3(9)/48 = 96$ in.^{4}. By symmetry, the product of inertia with respect to the x and y axes is $P_{xy} = 0$.

3. Compute the properties of the area with respect to the x' and y' axes

Using the usual moment-of-inertia relations, $I_{x'} = I_x + Ay_m^2 = 162 + 36(6)^2 = 1,458$ in.^{4}; $I_{y'} = I_y + Ax_m^2 = 96 + 36(7)^2 = 1,860$ in.^{4}; $P_{x'y'} = P_{xy} + Ax_m y_m = 0 + 36(7)(6) = 1,512$ in.^{4}.

4. Compute the properties of the area with respect to the x″ and y″ axes

For the x'' axis, $I_{x''} = I_{x'} \cos^2\theta + I_{y'} \sin^2\theta - P_{x'y'} \sin 2\theta = 1,458(0.75) + 1,860(0.25) - 1,512(0.866) = 249$ in.[4]

For the y'' axis, $I_{y''} = I_{x'} \sin^2\theta + I_{y'} \cos^2\theta + P_{x'y'} \sin 2\theta = 1,458(0.25) + 1,860(0.75) + 1,512(0.866) = 3,069$ in.[4]

The product of inertia, $P_{x''y''} = P_{x'y'} \cos 2\theta + [(I_{x'} - I_{y'})/2] \sin 2\theta = 1,512(0.5) + [(1,458 - 1,860)/2] \, 0.866 = 582$ in.[4]

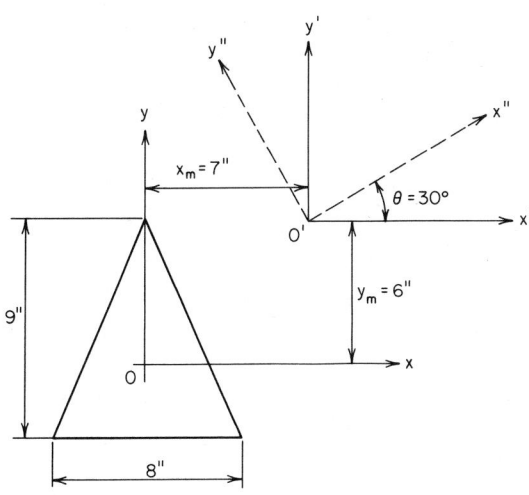

Fig. 15

Analysis of Stress and Strain

The notational system for axial stress and strain used in this section is: $A = $ cross-sectional area of a member; $L = $ original length of the member; $\Delta l = $ increase in length; $P = $ axial force; $s = $ axial stress; $\epsilon = $ axial strain $= \Delta l/L$; $E = $ modulus of elasticity of material $= s/\epsilon$. The units used for each of these factors are given in the Calculation Procedure. In all instances, it is assumed that the induced stress is below the proportional limit. The basic stress and elongation equations used are: $s = P/A$; $\Delta l = sL/E = PL/AE$. For steel, $E = 30 \times 10^6$ psi.

STRESS CAUSED BY AN AXIAL LOAD

A concentric load of 20,000 lb is applied to a hanger having a cross-sectional area of 1.6 sq in. What is the axial stress in the hanger?

Calculation Procedure:

1. Compute the axial stress

Use the general stress relation $s = P/A = 20,000/1.6 = 12,500$ psi.

Related Calculations: Use this general stress relation for a member of any cross-sectional shape, provided the area of the member can be computed and the member is made of only one material.

DEFORMATION CAUSED BY AN AXIAL LOAD

A member having a length of 16 ft and a cross-sectional area of 2.4 sq in. is subjected to a tensile force of 30,000 lb. If $E = 15 \times 10^6$ psi, how much does this member elongate?

Calculation Procedure:

1. Apply the general deformation equation

The general deformation equation is $\Delta l = PL/AE = 30{,}000(16)(12)/2.4(15 \times 10^6) = 0.16$ in.

Related Calculations: Use this general deformation equation for any material whose modulus of elasticity is known. For composite materials, this equation must be altered before it can be used.

DEFORMATION OF A BUILT-UP MEMBER

A member is built up of three bars placed end to end, the bars having the lengths and cross-sectional areas shown in Fig. 16. The member is placed between two rigid surfaces and axial loads of 30 kips and 10 kips are applied at A and B, respectively. If $E = 2{,}000$ ksi, determine the horizontal displacement of A and B.

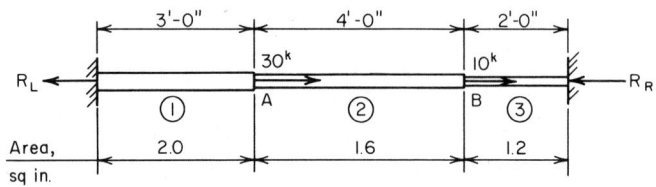

Fig. 16

Calculation Procedure:

1. Express the axial force in terms of one reaction

Let R_L and R_R denote the reactions at the left and right ends, respectively. Assume that both reactions are directed to the left. Consider a tensile force as positive and a compressive force as negative. Consider a deformation positive if the body elongates and negative if the body contracts.

Express the axial force P in each bar in terms of R_L because both reactions are assumed to be directed toward the left. Use subscripts corresponding to the bar numbers, Fig. 16. Thus, $P_1 = R_L$; $P_2 = R_L - 30$; $P_3 = R_L - 40$.

2. Express the deformation of each bar in terms of the reaction and modulus of elasticity

Thus, $\Delta l_1 = R_L(36)/2.0E = 18R_L/E$; $\Delta l_2 = (R_L - 30)(48)/1.6E = (30R_L - 900)/E$; $\Delta l_3 = (R_L - 40)24/1.2E = (20R_L - 800)/E$.

3. Solve for the reaction

Since the ends of the member are stationary, equate the total deformation to zero, and solve for R_L. Thus: $\Delta l_t = (68R_L - 1{,}700)/E = 0$; $R_L = 25$ kips. The positive result confirms the assumption that R_L is directed to the left.

4. Compute the displacement of the points

Substitute the computed value of R_L in the first two equations of Step 2 and solve for the displacement of the points A and B. Thus: $\Delta l_1 = 18(25)/2{,}000 = 0.225$ in.; $\Delta l_2 = [30(25) - 900]/2{,}000 = -0.075$ in.

Combining these results, the displacement of $A = 0.225$ in. to the right; the displacement of $B = 0.225 - 0.075 = 0.150$ in. to the right.

5. Verify the computed results

To verify this result, compute R_R and determine the deformation of bar 3. Thus: $\Sigma F_H = -R_L + 30 + 10 - R_R = 0$; $R_R = 15$ kips. Since bar 3 is in compression, $\Delta l_3 = -15(24)/1.2(2{,}000) = -0.150$ in. Therefore, B is displaced 0.150 in. to the right. This verifies the result obtained in Step 4.

REACTIONS AT ELASTIC SUPPORTS

The rigid bar in Fig. 17a is subjected to a load of 20,000 lb applied at D. It is supported by three steel rods, 1, 2, and 3, Fig. 17a. These rods have the following relative cross-sectional areas: $A_1 = 1.25$; $A_2 = 1.20$; $A_3 = 1.00$. Determine the tension in each rod caused by this load, and locate the center of rotation of the bar.

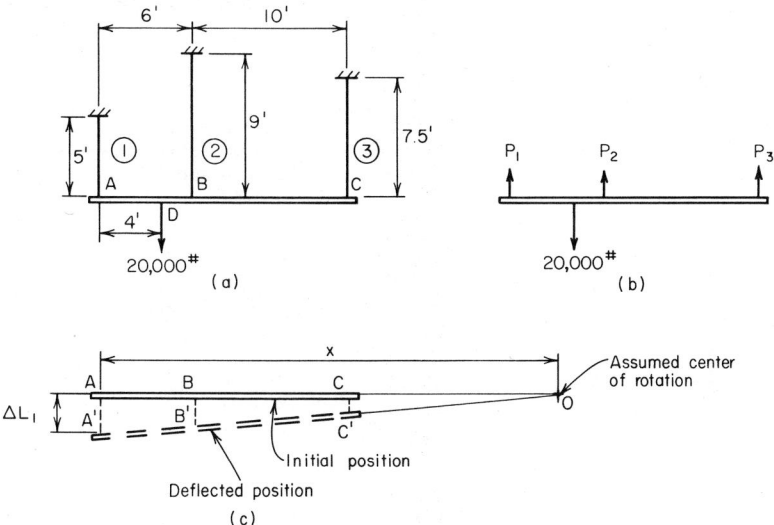

Fig. 17   Bar supported by three hangers.

Calculation Procedure:

1.   Draw a free-body diagram; apply the equations of equilibrium

Draw the free-body diagram, Fig. 17b, of the bar. Apply the equations of equilibrium: $\Sigma F_V = P_1 + P_2 + P_3 - 20{,}000 = 0$, or $P_1 + P_2 + P_3 = 20{,}000$, Eq. (a); also, $\Sigma M_C = 16P_1 + 10P_2 - 20{,}000(12) = 0$, or $16P_1 + 10P_2 = 240{,}000$, Eq. (b).

2.   Establish the relations between the deformations

Selecting an arbitrary center of rotation O, show the bar in its deflected position, Fig. 17c. Establish the relationships among the three deformations. Thus, by similar triangles, $(\Delta l_1 - \Delta l_2)/(\Delta l_2 - \Delta l_3) = 6/10$, or $10\Delta l_1 - 16\Delta l_2 + 6\Delta l_3 = 0$, Eq. (c).

3.   Transform the deformation equation to an axial-force equation

By substituting axial-force relations in Eq. (c), the following equation is obtained: $10P_1(5)/1.25E - 16P_2(9)/1.20E + 6P_3(7.5)/E = 0$, or $40P_1 - 120P_2 + 45P_3 = 0$, Eq. (c′).

4.   Solve the simultaneous equations developed

Solve the simultaneous equations (a), (b), and (c′) to obtain: $P_1 = 11{,}810$ lb; $P_2 = 5{,}100$ lb; $P_3 = 3{,}090$ lb.

5.   Locate the center of rotation

To locate the center of rotation, compute the relative deformation of rods 1 and 2. Thus: $\Delta l_1 = 11{,}810(5)/1.25E = 47{,}240/E$; $\Delta l_2 = 5{,}100(9)/1.20E = 38{,}250/E$.
In Fig. 17c, by similar triangles, $x/(x-6) = \Delta l_1/\Delta l_2 = 1.235$; $x = 31.5$ ft.

6.   Verify the computed values of the tensile forces

To verify the computed values of the tensile forces, calculate the moment with respect to A of the applied and resisting forces. Thus: $M_{Aa} = 20{,}000(4) = 80{,}000$ lb-ft;

$M_{Ar} = 5,100(6) + 3,090(16) = 80,000$ lb-ft. Since the moments are equal, the results are verified.

ANALYSIS OF CABLE SUPPORTING A CONCENTRATED LOAD

A cold-drawn steel wire $\frac{1}{4}$ in. in diameter is stretched tightly between two points lying on the same horizontal plane 80 ft apart. The stress in the wire is 50,000 psi. A load of 200 lb is suspended at the center of the cable. Determine the sag of the cable and the final stress in the cable. Verify that the results obtained are compatible.

Calculation Procedure:

1. Derive the stress and strain relations for the cable

Referring to Fig. 18, L = distance between supports, ft; P = load applied at center of cable span, lb; d = deflection of cable center, ft; ϵ = strain of cable caused by P; s_1 and

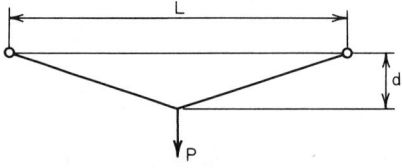

s_2 = initial and final tensile stress in cable, respectively, psi.

Refer to the geometry of the deflection diagram. Taking into account that d/L is extremely small, derive the following approximations: $s_2 = PL/4Ad$, Eq. (a); $\epsilon = 2(d/L)^2$, Eq. (b).

Fig. 18   Deflection of cable under concentrated load.

2. Relate stress and strain

Express the increase in stress caused by P in terms of ϵ, and apply the above two equations to derive $2E(d/L)^3 + s_1(d/L) = P/4A$, Eq. (c).

3. Compute the deflection at the center of the cable

Using Eq. (c), $2(30)(10)^6(d/L)^3 + 50,000d/L = 200/4(0.049)$; $d/L = 0.0157$; $\therefore d = 0.0157(80) = 1.256$ ft.

4. Compute the final tensile stress

Write Eq. (a) as $s_2 = (P/4A)/(d/L) = 1,020/0.0157 = 65,000$ psi.

5. Verify the results computed

To demonstrate that the results are compatible, accept the computed value of d/L as correct. Then apply Eq. (b) to find the strain, and compute the corresponding stress. Thus: $\epsilon = 2(0.0157)^2 = 4.93 \times 10^{-4}$; $s_2 = s_1 + E\epsilon = 50,000 + 30 \times 10^6 \times 4.93 \times 10^{-4} = 64,800$ psi. This agrees closely with the previously calculated stress of 65,000 psi.

DISPLACEMENT OF TRUSS JOINT

In Fig. 19a, the steel members AC and BC both have a cross-sectional area of 1.2 sq in. If a load of 20 kips is suspended at C, how much is joint C displaced?

Calculation Procedure:

1. Compute the length of each member and the tensile forces

Consider joint C as a free body to find the tensile force in each member. Thus: $L_{AC} = 192$ in.; $L_{BC} = 169.7$ in; $P_{AC} = 14,640$ lb; $P_{BC} = 17,930$ lb.

2. Determine the elongation of each member

Use the relation $\Delta l = PL/AE$. Thus: $\Delta l_{AC} = 14,640(192)/1.2(30 \times 10^6) = 0.0781$ in.; $\Delta l_{BC} = 17,930(169.7)/1.2(30 \times 10^6) = 0.0845$ in.

3. Construct the Williott displacement diagram

Selecting a suitable scale, construct the Williott displacement diagram as follows: Draw (Fig. 19b) line Ca parallel to member AC, with $Ca = 0.0781$ in. Similarly, draw Cb parallel to member BC, with $Cb = 0.0845$ in.

4. Determine the displacement

Erect perpendiculars to Ca and Cb at a and b, respectively. Designate the intersection point of these perpendiculars as C'.

Line CC' represents, both in magnitude and direction, the approximate displacement of joint C under the applied load. Scaling distance CC' to obtain the displacement shows that the displacement of $C = 0.134$ in.

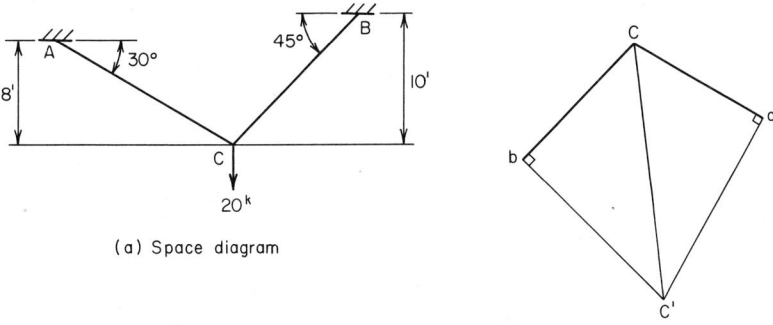

(a) Space diagram

(b) Displacement diagram

Fig. 19

AXIAL STRESS CAUSED BY IMPACT LOAD

A body weighing 18 lb falls 3 ft before contacting the end of a vertical steel rod. The rod is 5 ft long and has a cross-sectional area of 1.2 sq in. If the entire kinetic energy of the falling body is absorbed by the rod, determine the stress induced in the rod.

Calculation Procedure:

1. State the equation for the induced stress

Equate the energy imparted to the rod to the potential energy lost by the falling body. Or, $s = (P/A)\{1 + [1 + 2Eh/(LP/A)]^{0.5}\}$, where $h =$ vertical displacement of body, ft.

2. Substitute the numerical values

Thus, $P/A = 18/1.2 = 15$ psi; $h = 3$ ft; $L = 5$ ft; $2Eh/(LP/A) = 2(30 \times 10^6)(3)/5(15) = 2{,}400{,}000$. Then, $s = 23{,}250$ psi.

Related Calculations: Where the deformation of the supporting member is negligible in relation to the distance h, as it is in the present instance, the following approximation is used: $s = (2PEh/AL)^{0.5}$.

STRESSES ON AN OBLIQUE PLANE

The prism $ABCD$ in Fig. 20a has the principal stresses of 6,300 and 2,400 psi tension. Applying both the analytical and graphical methods, determine the normal and shearing stress on plane AE.

Calculation Procedure:

1. Compute the stresses using the analytical method

A *principal stress* is a normal stress not accompanied by a shearing stress. The plane on which the principal stress exists is termed a *principal plane*. For a condition of plane stress, it can be shown that there are two principal planes through every point in a stressed body and that these planes are mutually perpendicular. Moreover, one principal stress is the maximum normal stress existing at that point; the other is the minimum normal stress.

Let s_x and $s_y =$ the principal stress in the x and y direction, respectively; $s_n =$ normal stress on the plane making an angle θ with the y axis; $s_s =$ shearing stress on this plane. All stresses are expressed in psi, and all angles are in degrees. Tensile stresses are positive; compressive stresses are negative.

Applying the usual stress equations: $s_n = s_y + (s_x - s_y) \cos^2 \theta$; $s_s = \frac{1}{2}(s_x - s_y) \sin 2\theta$. Substituting: $s_n = 2,400 + (6,300 - 2,400)0.766^2 = 4,690$ psi tension; and $s_s = \frac{1}{2}(6,300 - 2,400)0.985 = 1,920$ psi.

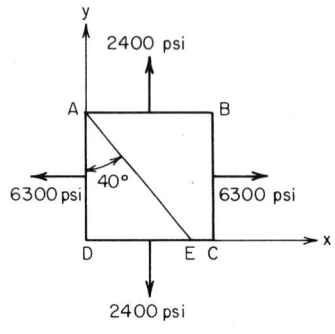 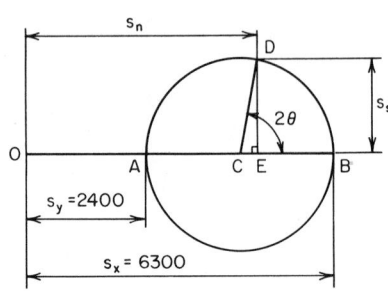

(a) Stresses on prism   (b) Mohr's circle of stress

Fig. 20

2. Apply the graphical method of solution

Construct, in Fig. 20b, Mohr's circle of stress thus: (a) Using a suitable scale, draw $OA = s_y$, and $OB = s_x$. (b) Draw a circle having AB as its diameter. (c) Draw the radius CD making an angle of $2\theta = 80°$ with AB. (d) Through D, drop a perpendicular DE to AB. Then, $OE = s_n$ and $ED = s_s$. (e) Scale the two last lengths to obtain the normal and shearing stresses on plane AE.

Related Calculations: The normal stress may also be computed from $s_n = (s_x + s_y)0.5 + (s_x - s_y)0.5 \cos 2\theta$.

EVALUATION OF PRINCIPAL STRESSES

The prism $ABCD$ in Fig. 21a is subjected to the normal and shearing stresses shown. Construct the Mohr's circle to determine the principal stresses at A, and locate the principal planes.

Calculation Procedure:

1. Draw the lines representing the normal stresses, Fig. 21b

Through the origin O, draw a horizontal base line. Locate points E and F such that $OE = 8,400$ psi and $OF = 2,000$ psi. Since both normal stresses are tensile, E and F lie to the right of O. Note that the construction required here is the converse of that required in the previous Calculation Procedure.

2. Draw the lines representing the shearing stresses

Construct the vertical lines EG and FH such that $EG = 3,600$ psi, and $FH = -3,600$ psi.

3. Continue the construction

Draw line GH to intersect the base line at C.

4. Construct Mohr's circle

Draw a circle having GH as diameter, intersecting the base line at A and B. Then lines OA and OB represent the principal stresses.

5. Scale the diagram

Scale OA and OB to obtain: $f_{max} = 10,020$ psi; $f_{min} = 380$ psi. Both stresses are tension.

6. Determine the stress angle

Scale angle BCG and measure it as 48°22'. The angle between the x axis, on which the maximum stress exists, and the side AD of the prism, is one-half of BCG.

7. Construct the x and y axes

In Fig. 21a, draw the x axis making a counterclockwise angle of 24°11' with AD. Draw the y axis perpendicular thereto.

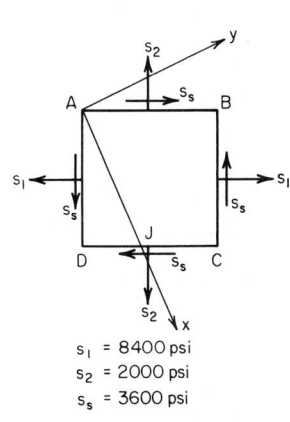

s_1 = 8400 psi
s_2 = 2000 psi
s_s = 3600 psi

(a) Stresses on prism

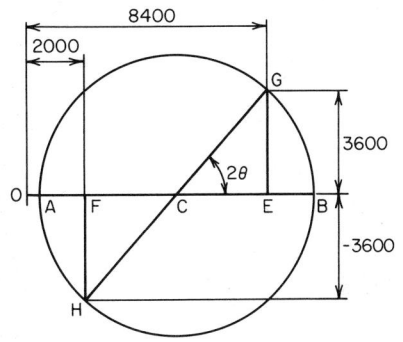

(b) Mohr's circle of stress

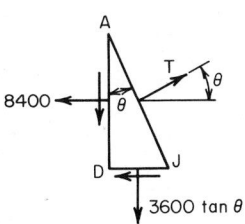

(c) Free-body diagram of ADJ

Fig. 21

8. Verify the locations of the principal planes

Consider ADJ as a free body. Set the length AD equal to unity. In Fig. 21c, since there is no shearing stress on AJ, $\Sigma F_H = T \cos \theta - 8,400 - 3,600 \tan \theta = 0$; $T \cos \theta = 8,400 + 3,600(0.45) = 10,020$ psi. The stress on $AJ = T/AJ = T \cos \theta = 10,020$ psi.

HOOP STRESS IN THIN-WALLED CYLINDER UNDER PRESSURE

A steel pipe 5 ft in diameter and 3/8 in. thick sustains a fluid pressure of 180 psi. Determine the hoop stress, the longitudinal stress, and the increase in diameter of this pipe. Use 0.25 for Poisson's ratio.

Calculation Procedure:

1. Compute the hoop stress

Use the relation $s = pD/2t$, where s = hoop or tangential stress, psi; p = radial pressure, psi; D = internal diameter of cylinder, in.; t = cylinder wall thickness, in. Thus, for this cylinder, $s = 180(60)/2(3/8) = 14,400$ psi.

2. Compute the longitudinal stress

Use the relation $s' = pD/4t$, where s' = longitudinal stress, i.e., the stress parallel to the longitudinal axis of the cylinder, psi; other symbols as before. Substituting, $s' = 7,200$ psi.

3. Compute the increase in the cylinder diameter

Use the relation, $\Delta D = (D/E)(s - vs')$, where v = Poisson's ratio. Thus: $\Delta D = 60(14,400 - 0.25 \times 7,200)/(30 \times 10^6) = 0.0252$ in.

STRESSES IN PRESTRESSED CYLINDER

A steel ring having an internal diameter of 8.99 in. and a thickness of $\frac{1}{4}$ in. is heated and allowed to shrink over an aluminum cylinder having an external diameter of 9.00 in., and a thickness of $\frac{1}{2}$ in. After the steel cools, the cylinder is subjected to an internal pressure of 800 psi. Find the stresses in the two materials. For aluminum, $E = 10 \times 10^6$ psi.

Calculation Procedure:

1. Compute the radial pressure caused by prestressing

Use the relation $p = 2\Delta D/[D^2(1/t_a E_a + 1/t_s E_s)]$, where p = radial pressure resulting from prestressing, psi; other symbols the same as in the previous Calculation Procedure, with the subscripts a and s referring to aluminum and steel, respectively. Thus, $p = 2(0.01)/\{9^2[1/(0.5 \times 10 \times 10^6) + 1/(0.25 \times 30 \times 10^6)]\} = 741$ psi.

2. Compute the corresponding prestresses

Using the subscripts 1 and 2 to denote the stresses caused by prestressing and internal pressure, respectively, $s_{a1} = pD/2t_a$, where the symbols are the same as in the previous Calculation Procedure. Thus, $s_{a1} = 741(9)/2(0.5) = 6,670$ psi compression. Likewise, $s_{s1} = 741(9)/2(0.25) = 13,340$ psi tension.

3. Compute the stresses caused by internal pressure

Use the relation $s_{s2}/s_{a2} = E_s/E_a$ or, for this cylinder, $s_{s2}/s_{a2} = (30 \times 10^6)/(10 \times 10^6) = 3$. Next, compute s_{a2} from $t_a s_{a2} + t_s s_{s2} = pD/2$, or $s_{a2} = 800(9)/[2(0.5 + 0.25 \times 3)] = 2,880$-psi tension. Also, $s_{s2} = 3(2,880) = 8,640$-psi tension.

4. Compute the final stresses

Sum the results in Steps 2 and 3 to obtain the final stresses. Or: $s_{a3} = 6,670 - 2,880 = 3,790$-psi compression; $s_{s3} = 13,340 + 8,640 = 21,980$-psi tension.

5. Check the accuracy of the results

Ascertain whether the final diameters of the steel ring and aluminum cylinder are equal. Thus, setting $s' = 0$ in $\Delta D = (D/E)(s - vs')$, then $\Delta D_a = -3,790(9)/(10 \times 10^6) = -0.0034$ in., $D_a = 9.0000 - 0.0034 = 8.9966$ in. Likewise, $\Delta D_s = 21,980(9)/(30 \times 10^6) = 0.0066$ in.; $D_s = 8.99 + 0.0066 = 8.9966$ in. Since the computed diameters are equal, the results are valid.

HOOP STRESS IN THICK-WALLED CYLINDER

A cylinder having an internal diameter of 20 in. and an external diameter of 36 in. is subjected to an internal pressure of 10,000 psi and an external pressure of 2,000 psi, as shown in Fig. 22. Determine the hoop stress at the inner and outer surfaces of the cylinder.

Calculation Procedure:

1. Compute the hoop stress at the inner surface of the cylinder

Use the relation $s_i = [p_1(r_1^2 + r_2^2) - 2p_2 r_2^2]/(r_2^2 - r_1^2)$, where s_i = hoop stress at inner surface, psi; p_1 = internal pressure, psi; r_1 = internal radius, in.; r_2 = external radius, in.; p_2 = external pressure, psi. Substituting, $s_i = [10,000(100+324)-2(2,000)(324)]/(324-100) = 13,100$-psi tension.

2. Compute the hoop stress at the outer cylinder surface

Use the relation $s_0 = [2p_1r_1{}^2 - p_2(r_1{}^2+r_2{}^2)]/(r_2{}^2-r_1{}^2)$, where the symbols are as before. Substituting, $s_0 = [2(10,000)(100)-2,000(100+324)]/(324-100) = 5,100$ psi tension.

3. Check the accuracy of the results

Use the relation $s_ir_1 - s_0r_2 = [(r_2-r_1)/(r_2+r_1)](p_1r_1+p_2r_2)$. Substituting the known values verifies the earlier calculations.

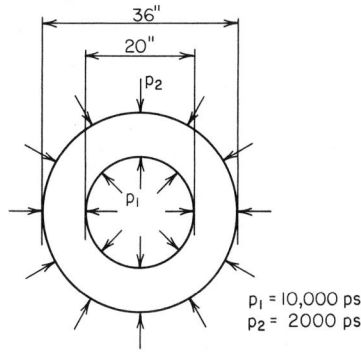

Fig. 22   Thick-walled cylinder under internal and external pressure.

THERMAL STRESS RESULTING FROM HEATING A MEMBER

A steel member 18 ft long is set snugly between two walls and heated 80°F. If each wall yields 0.015 in., what is the compressive stress in the member? Use a coefficient of thermal expansion of $6.5 \times 10^{-6}/°F$ for steel.

Calculation Procedure:

1. Compute the thermal expansion of the member without restraint

Replace the true condition of partial restraint with the following equivalent conditions: The member is first allowed to expand freely under the temperature rise and is then compressed to its true final length.

To compute the thermal expansion without restraint, use the relation $\Delta L = cL\Delta T$, where c = coefficient of thermal expansion,/°F; ΔT = increase in temperature, °F; L = original length of member, in.; ΔL = increase in length of the member, in. Substituting, $\Delta L = 6.5(10^{-6})(18)(12)(80) = 0.1123$ in.

2. Compute the linear restraint exerted by the walls

The walls yield $2(0.015) = 0.030$ in. Thus, the restraint exerted by the walls is $\Delta L_w = 0.1123 - 0.030 = 0.0823$ in.

3. Determine the compressive stress

Use the relation $s = E\Delta L/L$, where the symbols are as given earlier. Thus, $s = 30(10^6)(0.0823)/[18(12)] = 11,430$ psi.

THERMAL EFFECTS IN COMPOSITE MEMBER HAVING ELEMENTS IN PARALLEL

A $\frac{1}{2}$-in.-diameter Copperweld bar consists of a steel core $\frac{3}{8}$ in. in diameter and a copper skin $\frac{1}{16}$ in. thick. What is the elongation of a 1-ft length of this bar and what is the internal force between the steel and copper arising from a temperature rise of 80°F? Use the following values for thermal expansion coefficients: $c_s = 6.5 \times 10^{-6}$; $c_c = 9.0 \times 10^{-6}$, where the subscripts s and c refer to steel and copper, respectively. Also, $E_c = 15 \times 10^6$ psi.

Calculation Procedure:

1. Determine the cross-sectional areas of the metals

The total area $A = 0.1963$ sq in. The area of the steel $A_s = 0.1105$ sq in. By difference, the area of the copper $A_c = 0.0858$ sq in.

2.   *Determine the coefficient of expansion of the composite member*

Weight the coefficients of expansion of the two members according to their respective AE values. Thus:

$A_s E_s$ (relative) $= 0.1105 \times 30 \times 10^6 = 3315$	
$A_c E_c$ (relative) $= 0.0858 \times 15 \times 10^6 = 1287$	
Total	4602

Then, the coefficient of thermal expansion of the composite member is $c = (3315c_s + 1287c_c)/4{,}602 = 7.2 \times 10^{-6}/°F$.

3.   *Determine the thermal expansion of the 1-ft section*

Using the relation $\Delta L = cL\Delta T$, $\Delta L = 7.2(10^{-6})(12)(80) = 0.00691$ in.

4.   *Determine the expansion of the first material without restraint*

Using the same relation as in Step 3 for copper *without* restraint, $\Delta L_c = 9.0(10^{-6}) \times (12)(80) = 0.00864$ in.

5.   *Compute the restraint of the first material*

The copper is restrained to the amount computed in Step 3. Thus, the restraint exerted by the steel is $\Delta L_{cs} = 0.00864 - 0.00691 = 0.00173$ in.

6.   *Compute the restraining force exerted by the second material*

Use the relation $P = (A_c E_c \Delta L_{cs})/L$, where the symbols are as given before. Or, $P = [1{,}287{,}000(0.00173)]/12 = 185$ lb.

7.   *Verify the results obtained*

Repeat Steps 4, 5, and 6 with the two materials interchanged. Or: $\Delta L_s = 6.5(10^{-6}) \times (12)(80) = 0.00624$ in.; $\Delta L_{sc} = 0.00691 - 0.00624 = 0.00067$ in. Then, $P = 3{,}315{,}000(0.00067)/12 = 185$ lb, as before.

THERMAL EFFECTS IN COMPOSITE MEMBER HAVING ELEMENTS IN SERIES

The aluminum and steel bars in Fig. 23 have cross-sectional areas of 1.2 and 1.0 sq in., respectively. The member is restrained against lateral deflection. A temperature rise of 100°F causes the length of the member to increase to 42.016 in. Determine the stress and deformation of each bar. For aluminum, $E = 10 \times 10^6$, $c = 13.0 \times 10^{-6}$; for steel, $c = 6.5 \times 10^{-6}$.

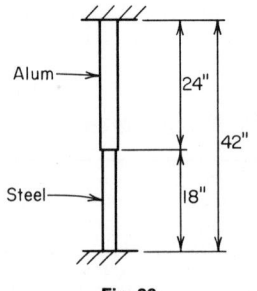

Fig. 23

Calculation Procedure:

1.   *Express the deformation of each bar resulting from the temperature change and the compressive force*

The temperature rise causes the bar to expand, whereas the compressive force resists this expansion. Thus, the net expansion is the difference between these two changes, or $\Delta L_a = cL\Delta T - PL/AE$, where the subscript a refers to the aluminum bar; other symbols the same as given earlier. Substituting: $\Delta L_a = 13.0 \times 10^{-6}(24)(100) - P(24)/[1.2(10 \times 10^6)] = (31{,}200 - 2P)10^{-6}$, Eq. (*a*). Likewise, for steel: $\Delta L_s = 6.5 \times 10^{-6}(18)(100) - P(18)/[1.0(30 \times 10^6)] = (11{,}700 - 0.6P)10^{-6}$, Eq. (*b*).

2.   *Sum the results in Step 1 to obtain the total deformation of the member*

Set the result equal to 0.016 in.; solve for P. Or: $\Delta L = (42{,}900 - 2.6P)10^{-6} = 0.016$ in.; $P = (42{,}900 - 16{,}000)/2.6 = 10{,}350$ lb.

3. Determine the stresses and deformation

Substitute the computed value of P in the stress equation $s = P/A$. For aluminum: $s_a = 10{,}350/1.2 = 8{,}630$ psi. Then $\Delta L_a = (31{,}200 - 2 \times 10{,}350)10^{-6} = 0.0105$ in. Likewise, for steel: $s_s = 10{,}350/1.0 = 10{,}350$ psi; and $\Delta L_s = (11{,}700 - 0.6 \times 10{,}350)10^{-6} = 0.0055$ in.

SHRINK-FIT STRESS AND RADIAL PRESSURE

An open steel cylinder having an internal diameter of 4 ft and a wall thickness of $\frac{5}{16}$ in. is to be heated to fit over an iron casting. The internal diameter of the cylinder before heating is $\frac{1}{32}$ in. less than that of the casting. How much must the temperature of the cylinder be increased to provide a clearance of $\frac{1}{32}$ in. all around between the cylinder and casting? If the casting is considered rigid, what stress will exist in the cylinder after it cools, and what radial pressure will it then exert on the casting?

Calculation Procedure:

1. Compute the temperature rise required

Use the relation $\Delta T = \Delta D/cD$, where $\Delta T =$ temperature rise required, °F; $\Delta D =$ change in cylinder diameter, in.; $c =$ coefficient of expansion of the cylinder $= 6.5 \times 10^{-6}/°F$; $D =$ cylinder internal diameter before heating, in. Thus: $\Delta T = (3/32)/[6.5 \times 10^{-6}(48)] = 300°F$.

2. Compute the hoop stress in the cylinder

Upon cooling, the cylinder has a diameter $\frac{1}{32}$ in. larger than originally. Compute the hoop stress from $s = E\Delta D/D = 30 \times 10^6(1/32)/48 = 19{,}500$ psi.

3. Compute the associated radial pressure

Use the relation $p = 2ts/D$, where $p =$ radial pressure, psi; other symbols as given earlier. Thus: $p = 2(5/16)(19{,}500)/48 = 254$ psi.

TORSION OF A CYLINDRICAL SHAFT

A torque of 8,000 lb-ft is applied at the ends of a 14-ft-long cylindrical shaft having an external diameter of 5 in. and an internal diameter of 3 in. What is the maximum shearing stress and the angle of twist of the shaft if the modulus of rigidity of the shaft is 6×10^6 psi?

Calculation Procedure:

1. Compute the polar moment of inertia of the shaft

For a hollow circular shaft, $J = (\pi/32)(D^4 - d^4)$, where $J =$ polar moment of inertia of a transverse section of the shaft with respect to the longitudinal axis, in.^{4}; $D =$ external diameter of shaft, in.; $d =$ internal diameter of shaft, in. Substituting: $J = (\pi/32)(5^4 - 3^4) = 53.4$ in.^{4}.

2. Compute the shearing stress in the shaft

Use the relation $s_s = TR/J$, where $s_s =$ shearing stress, psi; $T =$ applied torque, lb-in; $R =$ radius of shaft, in. Thus: $s_s = [(8{,}000)(12)(2.5)]/53.4 = 4{,}500$ psi.

3. Compute the angle of twist of the shaft

Use the relation $\theta = TL/JG$, where $\theta =$ angle of twist, radian; $L =$ shaft length, in.; $G =$ modulus of rigidity, psi. Thus: $\theta = (8{,}000)(12)(14)(12)/53.4(6{,}000{,}000) = 0.050$ radians, or 2.9°.

ANALYSIS OF A COMPOUND SHAFT

The compound shaft in Fig. 24 was formed by rigidly joining two solid segments. What torque may be applied at B, Fig. 24, if the shearing stress is not to exceed 15,000 psi in the steel and 10,000 psi in the bronze? $G_s = 12 \times 10^6$ psi; $G_b = 6 \times 10^6$ psi.

Calculation Procedure:

1. Determine the relationship between the torque in the shaft segments

Since the segments AB and BC, Fig. 24, are twisted through the same angle, the torque applied at the junction of these segments is distributed in proportion to their relative rigidities. Using the subscripts s and b to denote steel and bronze, respectively, $\theta = T_s L_s / J_s G_s = T_b L_b / J_b G_b$, where the symbols are as given in the previous Calculation Procedure. Solving for $T_s = (5/4.5)(3^4/4^4)(12/6)T_b = 0.703T_b$.

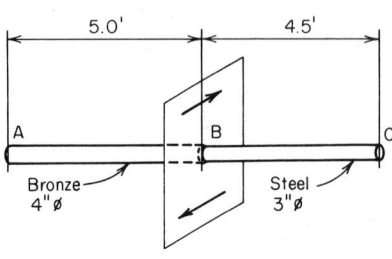

Fig. 24   Compound shaft.

2. Establish the relationship between the shearing stresses

For steel, $s_{ss} = 16T/\pi D^3$, where the symbols are as given earlier. Thus: $s_{ss} = 16(0.703T_b)/\pi 3^3$. Likewise, for bronze, $s_{sb} = 16T_b/\pi 4^3$; $\therefore s_{ss} = 0.703(4^3/3^3)s_{sb} = 1.67s_{sb}$.

3. Compute the allowable torque

Ascertain which material limits the capacity of the member, and compute the allowable torque by solving the shearing-stress equation for T.

If the bronze were stressed to 10,000 psi, inspection of the above relations shows that the steel would be stressed to 16,700 psi, which exceeds the allowed 15,000 psi. Hence, the steel limits the capacity. Substituting the allowed shearing stress of 15,000 psi: $T_s = 15,000\pi(3^3)/16(12) = 6,630$ lb-ft; also, $T_b = 6,630/0.703 = 9,430$ lb-ft. Then, $T = 6,630 + 9,430 = 16,060$ lb-ft.

Stresses in Flexural Members

In the analysis of beam action, the general assumption is that the beam is in a horizontal position and carries vertical loads lying in an axis of symmetry of the transverse section of the beam.

The vertical shear V at a given section of the beam is the algebraic sum of all vertical forces to the left of the section, an upward force being considered positive.

The bending moment M at a given section of the beam is the algebraic sum of the moments of all forces to the left of the section with respect to that section, a clockwise moment being considered positive.

If the proportional limit of the beam material is not exceeded, the bending stress (also called the flexural, or fiber, stress) at a section varies linearly across the depth of the section, being zero at the neutral axis. A positive bending moment induces compressive stresses in the fibers above the neutral axis and tensile stresses in the fibers below. Consequently, the elastic curve of the beam is concave upward where the bending moment is positive.

SHEAR AND BENDING MOMENT IN A BEAM

Construct the shear and bending-moment diagrams for the beam in Fig. 25. Indicate the value of the shear and bending moment at all significant sections.

Calculation Procedure:

1. Replace the distributed load on each interval with its equivalent concentrated load

Where the load is uniformly distributed, this equivalent load acts at the center of the interval of the beam. Thus: $W_{AB} = 2(4) = 8$ kips; $W_{BC} = 2(6) = 12$ kips; $W_{AC} = 8 + 12 = 20$ kips; $W_{CD} = 3(15) = 45$ kips; $W_{DE} = 1.4(5) = 7$ kips.

2. Determine the reaction at each support

Take moments with respect to the other support. Thus: $\Sigma M_D = 25R_A - 6(21) - 20(20) - 45(7.5) + 7(2.5) + 4.2(5) = 0$; $\Sigma M_A = 6(4) + 20(5) + 45(17.5) + 7(27.5) + 4.2(30) - 25R_D = 0$. Solving, $R_A = 33$ kips; $R_D = 49.2$ kips.

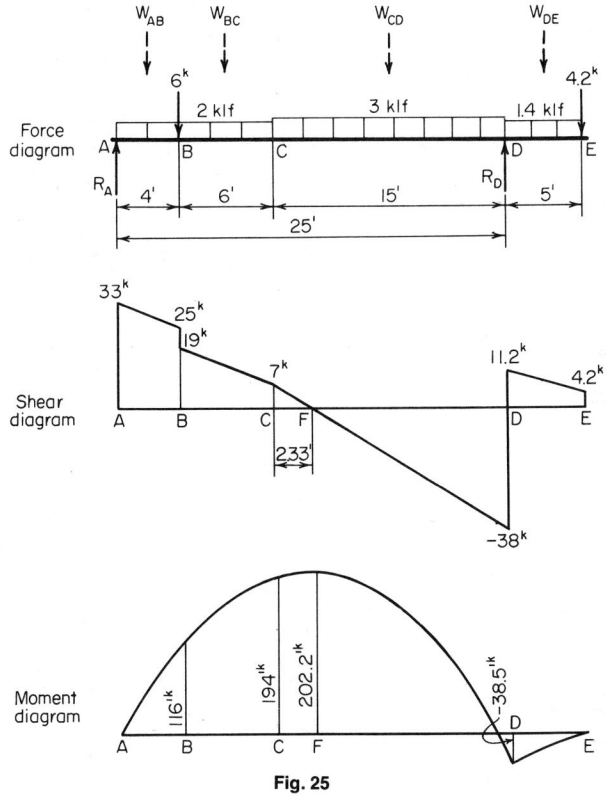

Fig. 25

3. Verify the computed results and determine the shears

Ascertain that the algebraic sum of the vertical forces is zero. If this is so, the computed results are correct.

Starting at A, determine the shear at every significant section, or directly to the left or right of that section, if a concentrated load is present. Thus: V_A at right = 33 kips; V_B at left = $33 - 8 = 25$ kips; V_B at right = $25 - 6 = 19$ kips; $V_C = 19 - 12 = 7$ kips; V_D at left = $7 - 45 = -38$ kips; V_D at right = $-38 + 49.2 = 11.2$ kips; V_E at left = $11.2 - 7 = 4.2$ kips; V_E at right = $4.2 - 4.2 = 0$.

4. Plot the shear diagram

Plot the points representing the forces in the previous step in the shear diagram. Since the loading between the significant sections is uniform, connect these points with straight lines. In general, the slope of the shear diagram is given by $dV/dx = -w$, where w = unit load at the given section; x = distance from left end to the given section.

5. Determine the bending moment at every significant section

Starting at A, determine the bending moment at every significant section. Thus: $M_A = 0$; $M_B = 33(4) - 8(2) = 116$ ft-kips; $M_C = 33(10) - 8(8) - 6(6) - 12(3) = 194$ ft-kips. Similarly, $M_D = -38.5$ ft-kips; $M_E = 0$.

6. *Plot the bending-moment diagram*

Plot the points representing the values in Step 5 in the bending-moment diagram, Fig. 25. Complete the diagram by applying the slope equation $dM/dx = V$, where V denotes the shear at the given section. Since this shear varies linearly between significant sections, the bending-moment diagram comprises a series of parabolic arcs.

7. *Alternatively, apply a moment theorem*

Use the theorem: If there are no externally applied moments in an interval 1-2 of the span, the difference between the bending moments is $M_2 - M_1 = \int_1^2 V\,dx$ = the area under the shear diagram across the interval.

Calculate the areas under the shear diagram to obtain the following results: $M_A = 0$; $M_B = M_A + \frac{1}{2}(4)(33+25) = 116$ ft-kips; $M_C = 116 + \frac{1}{2}(6)(19+7) = 194$ ft-kips; $M_D = 194 + \frac{1}{2}(15)(7-38) = -38.5$ ft-kips; $M_E = -38.5 + \frac{1}{2}(5)(11.2+4.2) = 0$.

8. *Locate the section at which the bending moment is maximum*

As a corollary of the equation in Step 6, the maximum moment occurs where the shear is zero or passes through zero under a concentrated load. Therefore, $CF = 7/3 = 2.33$ ft.

9. *Compute the maximum moment*

Using the computed value for CF, $M_F = 194 + \frac{1}{2}(2.33)(7) = 202.2$ ft-kips.

BEAM BENDING STRESSES

A beam having the trapezoidal cross section shown in Fig. 26a carries the loads indicated in Fig. 26b. What is the maximum bending stress at the top and at the bottom of this beam?

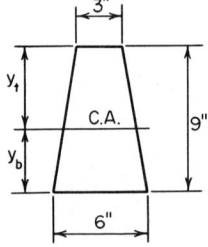

(a) Transverse section

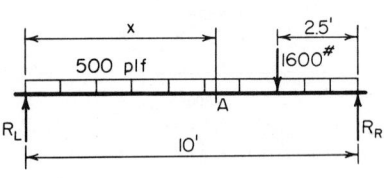

(b) Force diagram

Fig. 26

Calculation Procedure:

1. *Compute the left reaction and the section at which the shear is zero*

The left reaction $R_L = \frac{1}{2}(10)(500) + 1600(2.5/10) = 2{,}900$ lb. The section A at which the shear is zero is $x = 2{,}900/500 = 5.8$ ft.

2. *Compute the maximum moment*

Use the relation $M_A = \frac{1}{2}(2{,}900)(5.8) = 8{,}410$ lb-ft $= 100{,}900$ lb-in.

3. *Locate the centroidal axis of the section*

Use the AISC *Manual* for properties of the trapezoid. Or $y_t = (9/3)[(2\times6+3)/(6+3)] = 5$ in.; $y_b = 4$ in.

4. *Compute the moment of inertia of the section*

Using the AISC *Manual*, $I = (9^3/36)[(6^2 + 4\times6\times3 + 3^2)/(6+3)] = 263.3$ in.⁴.

5. Compute the stresses in the beam

Use the relation $f = My/I$, where f = bending stress in a given fiber, psi; y = distance from neutral axis to given fiber, in. Thus: $f_{top} = 100{,}900(5)/263.3 = 1{,}916$-psi compression, $f_{bottom} = 100{,}900(4)/263.3 = 1{,}533$-psi tension.

In general, the maximum bending stress at a section where the moment is M is given by $f = Mc/I$, where c = distance from the neutral axis to the outermost fiber, in. For a section that is symmetrical about its centroidal axis, it is convenient to use the section modulus S of the section, this being defined as $S = I/c$. Then $f = M/S$.

ANALYSIS OF A BEAM ON MOVABLE SUPPORTS

The beam in Fig. 27a rests on two movable supports. It carries a uniform live load of w plf and a uniform dead load of $0.2w$ plf. If the allowable bending stresses in tension and compression are identical, determine the optimal location of the supports.

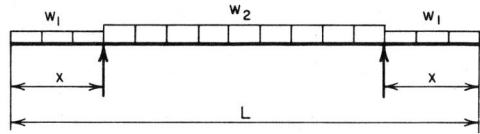

(a) Loads carried by overhanging beam

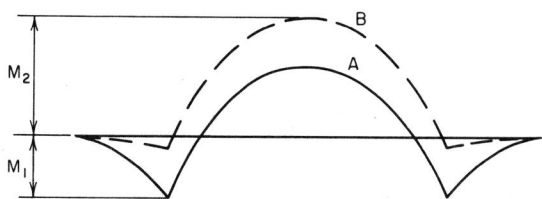

Diagram A: Full load on entire span
Diagram B: Dead load on overhangs; full load between supports

(b) Bending-moment diagrams

Fig. 27

Calculation Procedure:

1. Place full load on the overhangs and compute the negative moment

Refer to the moment diagrams. For every position of the supports, there is a corresponding maximum bending stress. The position for which this stress has the smallest value must be identified.

As the supports are moved toward the interior of the beam, the bending moments between the supports diminish in algebraic value. The optimal position of the supports is that for which the maximum potential negative moment M_1 is numerically equal to the maximum potential positive moment M_2. Thus, $M_1 = -1.2w(x^2/2) = -0.6wx^2$.

2. Place only the dead load on the overhangs and the full load between the supports. Compute the positive moment.

Sum the areas under the shear diagram to compute M_2. Thus, $M_2 = \frac{1}{2}[1.2w(L/2 - x)^2 - 0.2wx^2] = w(0.15L^2 - 0.6Lx + 0.5x^2)$.

3. Equate the absolute values of M_1 and M_2 and solve for x

Substituting: $0.6x^2 = 0.15L^2 - 0.6Lx + 0.5x^2$; $x = L(\overline{10.5}^{0.5} - 3) = 0.240L$.

FLEXURAL CAPACITY OF A COMPOUND BEAM

A 16WF45 steel beam in an existing structure was reinforced by welding an ST6WF20 to the bottom flange, as in Fig. 28. If the allowable bending stress is 20,000 psi, determine the flexural capacity of the built-up member.

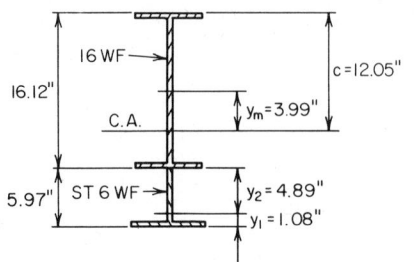

Fig. 28 Compound beam.

Calculation Procedure:

1. Obtain the properties of the elements

Using the AISC *Manual*, determine the following properties. For the 16WF45, $d = 16.12$ in.; $A = 13.24$ sq in.; $I = 583$ in.^{4}. For the ST6WF20, $d = 5.97$ in.; $A = 5.89$ sq in.; $I = 14$ in.^{4}; $y_1 = 1.08$ in.; $y_2 = 5.97 - 1.08 = 4.89$ in.

2. Locate the centroidal axis of the section

Locate the centroidal axis of the section with respect to the centerline of the 16WF45, and compute the distance c from the centroidal axis to the outermost fiber. Thus, $y_m = 5.89[(8.06 + 4.89)]/(5.89 + 13.24) = 3.99$ in. Then: $c = 8.06 + 3.99 = 12.05$ in.

3. Find the moment of inertia of the section with respect to its centroidal axis

Use the relation $I_0 + Ak^2$ for each member and take the sum for the two members to find I for the built-up beam. Thus, for the 16WF45: $k = 3.99$ in.; $I_0 + Ak^2 = 583 + 13.24 (3.99)^2 = 793$ in.^{4}. For the ST6WF20: $k = 8.06 - 3.99 + 4.89 = 8.96$ in.; $I_0 + Ak^2 = 14 + 5.89(8.96)^2 = 487$ in.^{4}. Then: $I = 793 + 487 = 1,280$ in.^{4}.

4. Apply the moment equation to find the flexural capacity

Use the relation $M = fI/c = 20,000(1,280)/12.05(12) = 177,000$ lb-ft.

ANALYSIS OF A COMPOSITE BEAM

An 8×12 in. timber beam (exact size) is reinforced by the addition of a $7 \times \frac{1}{2}$ in. steel plate at the top and a 7-in. 9.8-lb steel channel at the bottom, as shown in Fig. 29a. The

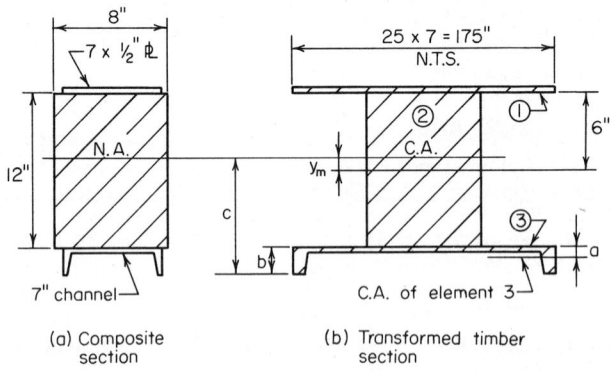

(a) Composite section

(b) Transformed timber section

Fig. 29

allowable bending stresses are 22,000 psi for steel and 1,200 psi for timber. The modulus of elasticity of the timber is 1.2×10^6 psi. How does the flexural strength of the reinforced beam compare with that of the original timber beam?

Calculation Procedure:

1. Compute the rigidity of the steel compared with that of the timber

Let n = the relative rigidity of the steel and timber. Then, $n = E_s/E_t = (30 \times 10^6)/(1.2 \times 10^6) = 25$.

2. Transform the composite beam to an equivalent homogeneous beam

To accomplish this transformation, replace the steel with timber. Sketch the cross section of the transformed beam as in Fig. 29b. Determine the sizes of the hypothetical elements by retaining the dimensions normal to the axis of bending but multiplying the dimensions parallel to this axis by n.

3. Record the properties of each element of the transformed section

Element 1: $A = 25(7)(\frac{1}{2}) = 87.5$ sq in.; I_0 is negligible.
Element 2: $A = 8(12) = 96$ sq in.; $I_0 = \frac{1}{12}(8)12^3 = 1,152$ in.^{4}.
Element 3: Refer to the AISC *Manual* for the data; $A = 25(2.85) = 71.25$ sq in.; $I_0 = 25(0.98) = 25$ in.^{4}; $a = 0.55$ in.; $b = 2.09$ in.

4. Locate the centroidal axis of the transformed section

Take statical moments of the areas with respect to the centerline of the 8×12 in. rectangle. Then, $y_m = [87.5(6.25) - 71.25(6.55)]/(87.5 + 96 + 71.25) = 0.31$ in. The neutral axis of the composite section is at the same location as the centroidal axis of the transformed section.

5. Compute the moment of inertia of the transformed section

Apply the relation in Step 3 of the previous Calculation Procedure. Then compute the distance c to the outermost fiber. Thus, $I = 1,152 + 25 + 87.5(6.25 - 0.31)^2 + 96(0.31)^2 + 71.25(6.55 + 0.31)^2 = 7,626$ in.^{4}. Also, $c = 0.31 + 6 + 2.09 = 8.40$ in.

6. Determine which material limits the beam capacity

Assume that the steel is stressed to capacity, and compute the corresponding stress in the transformed beam. Thus, $f = 22,000/25 = 880$ psi $< 1,200$ psi.

In the actual beam, the maximum timber stress, which occurs at the back of the channel, is even less than 880 psi. Therefore, the strength of the member is controlled by the allowable stress in the steel.

7. Compare the capacity of the original and reinforced beams

Let the subscripts 1 and 2 denote the original and reinforced beams, respectively. Compute the capacity of these members, and compare the results. Thus: $M_1 = fI/c = [1,200(1,152)]/6 = 230,000$ lb-in.; $M_2 = [880(7,626)]/8.40 = 799,000$ lb-in.; $M_2/M_1 = 799,000/230,000 = 3.47$. Thus, the reinforced beam is nearly $3\frac{1}{2}$ times as strong as the original beam, before reinforcing.

BEAM SHEAR FLOW AND SHEARING STRESS

A timber beam is formed by securely bolting a 3×6 in. member to a 6×8 in. member (exact size), as shown in Fig. 30. If the beam carries a uniform load of 600 plf on a simple span of 13 ft, determine the longitudinal shear flow and the shearing stress at the juncture of the two elements at a section 3 ft from the support.

Calculation Procedure:

1. Compute the vertical shear at the given section

Shear flow is the shearing force acting on a unit distance. In the present instance, the shearing force on an area having the same width as the beam and a length of 1 in. measured along the beam span is required.

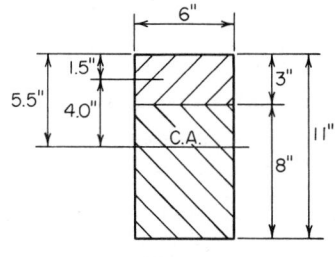

Fig. 30

Using dimensions and data from Fig. 30, $R = \frac{1}{4}(600)(13) = 3,900$ lb; $V = 3,900 - 3(600) = 2,100$ lb.

2. Compute the moment of inertia of the cross section
$I = (\frac{1}{12})(bd^3) = (\frac{1}{12})(6)(11)^3 = 666$ in.^{4}.

3. Determine the statical moment of the cross-sectional area
Calculate the statical moment Q of the cross-sectional area above the plane under consideration with respect to the centroidal axis of the section. Thus, $Q = Ay = 3(6)(4) = 72$ in.^{3}.

4. Compute the shear flow
Compute the shear flow q using $q = VQ/I = [2,100(72)]/666 = 227$ pli.

5. Compute the shearing stress
Use the relation $v = q/t = VQ/It$, where $t =$ width of the cross section at the given plane. Then, $v = 227/6 = 38$ psi.

Note that v represents both the longitudinal and the transverse shearing stress at a particular point. This is based on the principle that the shearing stresses at a given point in two mutually perpendicular directions are equal.

LOCATING THE SHEAR CENTER OF A SECTION

A cantilever beam carries the load shown in Fig. 31a, and has the transverse section shown in Fig. 31b. Locate the shear center of the section.

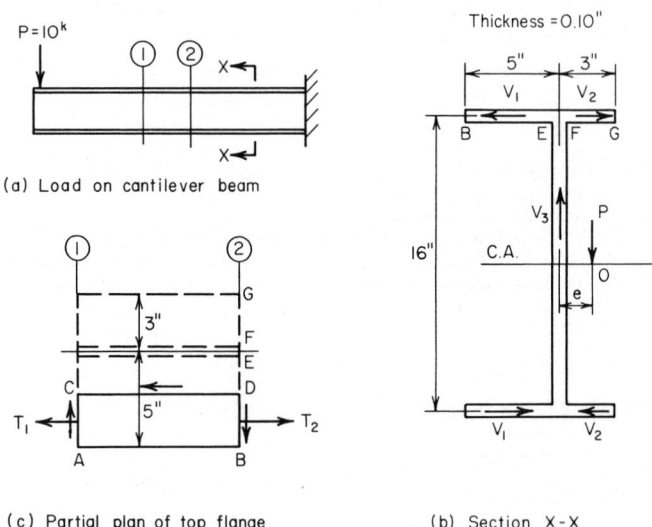

(a) Load on cantilever beam

(c) Partial plan of top flange

(b) Section X-X

Fig. 31

Calculation Procedure:

1. Construct a free-body diagram of a portion of the beam
Consider that the transverse section of a beam is symmetrical solely about its horizontal centroidal axis. If bending of the beam is not to be accompanied by torsion, the vertical shearing force at any section must pass through a particular point on the centroidal axis designated as the *shear, or flexural, center*.

Cut the beam at section 2, and consider the left portion of the beam as a free body.

In Fig. 31b, indicate the resisting shearing forces V_1, V_2, and V_3, that the right-hand portion of the beam exerts on the left-hand portion at section 2. Obtain the directions of V_1 and V_2 this way: Isolate the segment of the beam contained between sections 1 and 2; then isolate a segment $ABDC$ of the top flange, as shown in Fig. 31c. Since the bending stresses at section 2 exceed those at section 1, the resultant tensile force T_2 exceeds T_1. The resisting force on CD is therefore directed to the left. From the equation of equilibrium $\Sigma M = 0$, it follows that the resisting shears on AC and BD have the indicated directions to constitute a clockwise couple.

This analysis also reveals that the shearing stress varies linearly from zero at the edge of the flange to a maximum value at the juncture with the web.

2.   Compute the shear flow

Determine the shear flow at E and F, Fig. 31, by setting Q in $q = VQ/I$ equal to the statical moment of the overhanging portion of the flange. (For convenience, use the dimensions to the centerline of the web and flange.) Thus: $I = \frac{1}{12}(0.10)(16)^3 + 2(8) \times (0.10)(8)^2 = 137$ in.^{4}; $Q_{BE} = 5(0.10)(8) = 4.0$ in.^{3}; $Q_{FG} = 3(0.10)(8) = 2.4$ in.^{3}; $q_E = VQ_{BE}/I = 10,000(4.0)/137 = 292$ pli; $q_F = 10,000(2.4)/137 = 175$ pli.

3.   Compute the shearing forces on the transverse section

Since the shearing stress varies linearly across the flange, $V_1 = \frac{1}{2}(292)(5) = 730$ lb; $V_2 = \frac{1}{2}(175)(3) = 263$ lb; $V_3 = P = 10,000$ lb.

4.   Locate the shear center

Take moments of all forces acting on the left-hand portion of the beam with respect to a longitudinal axis through the shear center O. Thus: $V_3e + 16(V_2 - V_1) = 0$, or $10,000e + 16(263 - 730) = 0$; $e = 0.747$ in.

5.   Verify the computed values

Check the computed values of q_E and q_F by considering the bending stresses directly. Apply the equation $\Delta f = Vy/I$, where Δf = increase in bending stress per unit distance along the span at distance y from the neutral axis. Then, $\Delta f = 10,000$ (8)/137 = 584 psi/in.

In Fig. 31c, set $AB = 1$ in. Then: $q_E = 584(5)(0.10) = 292$ pli; $q_F = 584(3)(0.10) = 175$ pli.

Although a particular type of beam (cantilever) was selected here for illustrative purposes and a numerical value was assigned to the vertical shear, note that the value of e is independent of the type of beam, form of loading, or magnitude of the vertical shear. The location of the shear center is a geometrical characteristic of the transverse section.

BENDING OF A CIRCULAR FLAT PLATE

A circular steel plate 2 ft in diameter and $\frac{1}{2}$-in. thick, simply supported along its periphery, carries a uniform load of 20 psi distributed over the entire area. Determine the maximum bending stress and deflection of this plate, using 0.25 for Poisson's ratio.

Calculation Procedure:

1.   Compute the maximum stress in the plate

If the maximum deflection of the plate is less than about one-half the thickness, the effects of diaphragm behavior may be disregarded.

Compute the maximum stress using the relation $f = (\frac{3}{8})(3 + \nu)w(R/t)^2$, where R = plate radius, in.; t = plate thickness, in.; ν = Poisson's ratio. Thus, $f = (\frac{3}{8})(3.25)(20)(12/0.5)^2 = 14,000$ psi.

2.   Compute the maximum deflection of the plate

Use the relation $y = (1 - \nu)(5 + \nu)fR^2/2(3 + \nu)Et = 0.75(5.25)(14,000)(12)^2/2(3.25)(30 \times 10^6)(0.5) = 0.081$ in. Since the deflection is less than one-half the thickness, the foregoing equations are valid in this case.

BENDING OF A RECTANGULAR FLAT PLATE

A 2×3 ft rectangular plate, simply supported along its periphery, is to carry a uniform load of 8 psi distributed over the entire area. If the allowable bending stress is 15,000 psi, what thickness plate is required?

Calculation Procedure:

1. Select an equation for the stress in the plate

Use the approximation $f = a^2b^2w/2(a^2+b^2)t^2$, where a and b denote the length of the plate sides, in.

2. Compute the required plate thickness

Solve the equation in Step 1 for t. Thus: $t^2 = a^2b^2w/2(a^2+b^2)f = 2^2(3)^2(144)(8)/2(2^2+3^2)(15,000) = 0.106$; $t = 0.33$ in.

COMBINED BENDING AND AXIAL LOAD ANALYSIS

A post having the cross section shown in Fig. 32 carries a concentrated load of 100 kips applied at R. Determine the stress induced at each corner.

Calculation Procedure:

1. Replace the eccentric load with an equivalent system

Use a concentric load of 100 kips and two couples producing the following moments with respect to the coordinate axes:
$M_x = 100,000(2) = 200,000$ lb-in. $M_y = 100,000(1) = 100,000$ lb-in.

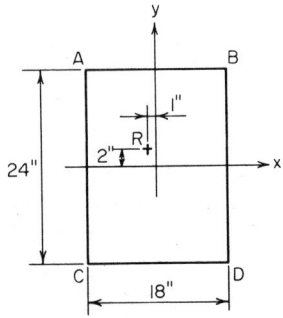

Fig. 32 Transverse section of a post.

2. Compute the section modulus

Determine the section modulus of the rectangular cross section with respect to each axis. Thus: $S_x = (\frac{1}{6})bd^2 = (\frac{1}{6})(18)(24)^2 = 1,728$ in.^{3}; $S_y = (\frac{1}{6})(24)(18)^2 = 1,296$ in.^{3}.

3. Compute the stresses produced

Compute the uniform stress caused by the concentric load and the stresses at the edges caused by the bending moments. Thus: $f_1 = P/A = 100,000/[18(24)] = 231$ psi; $f_x = M_x/S_x = 200,000/1,728 = 116$ psi; $f_y = M_y/S_y = 100,000/1,296 = 77$ psi.

4. Determine the stress at each corner

Combine the results obtained in Step 3 to obtain the stress at each corner. Thus: $f_A = 231+116+77 = 424$ psi; $f_B = 231+116-77 = 270$ psi; $f_C = 231-116+77 = 192$ psi; $f_D = 231-116 -77 = 38$ psi. These stresses are all compressive because a positive stress is considered compressive, whereas a tensile stress is negative.

5. Check the computed corner stresses

Use the following equation that applies to the special case of a rectangular cross section; $f = (P/A)[1 \pm 6e_x/d_x \pm 6e_y/d_y]$, where e_x and e_y = eccentricity of load with respect to the x and y axes, respectively; d_x and d_y = side of rectangle, in., normal to x and y axes, respectively. Solving for the quantities within the brackets, $6e_x/d_x = 6(2)/24 = 0.5$; $6e_y/d_y = [6(1)]/18 = 0.33$. Then, $f_A = 231(1+0.5+0.33) = 424$ psi; $f_B = 231(1+0.5-0.33) = 270$ psi; $f_C = 231(1-0.5+0.33) = 192$ psi; $f_D = 231(1-0.5-0.33) = 38$ psi. These results verify those computed in Step 4.

FLEXURAL STRESS IN CURVED MEMBER

The ring in Fig. 33 has an internal diameter of 12 in. and a circular cross section of 4-in. diameter. Determine the normal stress at A and at B, Fig. 33.

Calculation Procedure:

1. Determine the geometrical properties of the cross section

The area of the cross section is $A = 0.7854$ $(4)^2 = 12.56$ sq in.; the section modulus is $S = 0.7854(2)^3 = 6.28$ in.^{3}. With $c = 2$ in., the radius of curvature to the centroidal axis of this section is $R = 6 + 2 = 8$ in.

2. Compute the R/c ratio and determine the correction factors

Refer to a table of correction factors for curved flexural members, such as Roark–*Formulas for Stress and Strain,* and extract the correction factors at the inner and outer surface associated with the R/c ratio. Thus: $R/c = 8/2 = 4$; $k_i = 1.23$; $k_0 = 0.84$.

3. Determine the normal stress

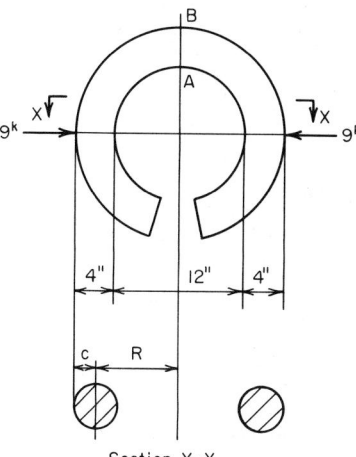

Section X-X

Fig. 33 Curved member in bending.

Find the normal stress at A and B caused by an equivalent axial load and moment. Thus: $f_A = P/A + k_i(M/S) = 9{,}000/12.56 + 1.23(9{,}000 \times 8)/6.28 = 14{,}820$-psi compression; $f_B = 9{,}000/12.56 - 0.84(9{,}000 \times 8)/6.28 = 8{,}930$-psi tension.

SOIL PRESSURE UNDER DAM

A concrete gravity dam has the profile shown in Fig. 34. Determine the soil pressure at the toe and heel of the dam when the water surface is level with the top.

Calculation Procedure:

1. Resolve the dam into suitable elements

The soil prism underlying the dam may be regarded as a structural member subjected to simultaneous axial load and bending, the cross section of the member being identical with the bearing surface of the dam. Select a 1-ft length of dam as representing the entire structure. The weight of the concrete is 150 pcf.

Resolve the dam into the elements AED and $EBCD$. Compute the weight of each element and locate the resultant of the weight with respect to the toe. Thus: $W_1 = \frac{1}{2}(12)$

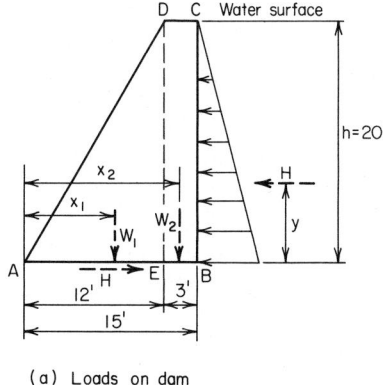

(a) Loads on dam   (b) Soil pressure under dam

Fig. 34

$(20)(150) = 18,000 \text{ lb}; \quad W_2 = 3(20)(150) = 9,000 \text{ lb}; \quad \Sigma W = 18,000 + 9,000 = 27,000 \text{ lb}.$ Then: $x_1 = (2/3)(12) = 8.0 \text{ ft}; x_2 = 12 + 1.5 = 13.5 \text{ ft}.$

2. Find the magnitude and location of the resultant of the hydrostatic pressure

Calling the resultant $H = \frac{1}{2}wh^2 = \frac{1}{2}(62.4)(20)^2 = 12,480 \text{ lb}$, where w = weight of water, pcf, and h = water height, ft, then $y = (\frac{1}{3})(20) = 6.67 \text{ ft}.$

3. Compute the moment of the loads with respect to the base centerline

Thus, $M = 18,000(8 - 7.5) + 9,000(13.5 - 7.5) - 12,480(6.67) = 20,200 \text{ lb-ft}$ counterclockwise.

4. Compute the section modulus of the base

Use the relation $S = (\frac{1}{6})bd^2 = (\frac{1}{6})(1)(15)^2 = 37.5 \text{ ft}^3.$

5. Determine the soil pressure at the dam toe and heel

Compute the soil pressure caused by the combined axial load and bending. Thus: $f_1 = \Sigma W/A + M/S = 27,000/15 + 20,200/37.5 = 2,339 \text{ psf}; f_2 = 1,800 - 539 = 1,261 \text{ psf}.$

6. Verify the computed results

Locate the resultant R of the trapezoidal pressure prism and take its moment with respect to the centerline of the base. Thus: $R = 27,000 \text{ lb}; m = (15/3)[(2 \times 1,261 + 2,339) /(1,261 + 2,339)] = 6.75 \text{ ft}., M_R = 27,000(7.50 - 6.75) = 20,200 \text{ lb-ft}.$ Since the applied and resisting moments are numerically equal, the computed results are correct.

LOAD DISTRIBUTION IN PILE GROUP

A continuous wall is founded on three rows of piles spaced 3 ft apart. The longitudinal pile spacing is 4 ft in the front and center rows and 6 ft in the rear row. The resultant of vertical loads on the wall is 20,000 plf and lies 3 ft 3 in. from the front row. Determine the pile load in each row.

Calculation Procedure:

1. Identify the "repeating group" of piles

The concrete footing, Fig. 35a, binds the piles, causing the surface along the top of the piles to remain a plane as bending occurs. Therefore, the pile group may be regarded as a structural member subjected to axial load and bending, the cross section of the member being the aggregate of the cross sections of the piles.

Indicate the "repeating group" as shown in Fig. 35b.

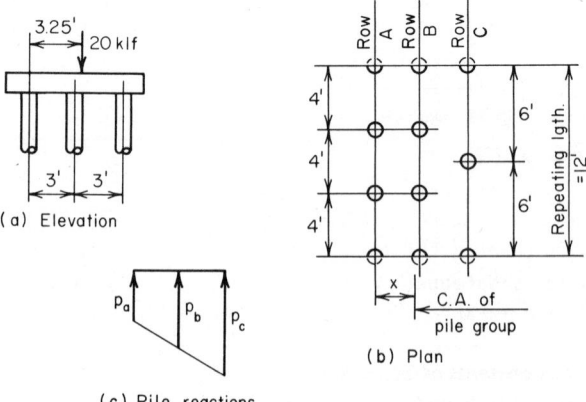

(a) Elevation

(b) Plan

(c) Pile reactions

Fig. 35

2. Determine the area of the pile group and the moment of inertia

Calculate the area of the pile group; locate its centroidal axis; find the moment of inertia. Since all the piles have the same area, set the area of a single pile equal to unity. Then, $A = 3 + 3 + 2 = 8$.

Take moments with respect to row A. Thus: $8x = 3(0) + 3(3) + 2(6)$; $x = 2.625$ ft. Then, $I = 3(2.625)^2 + 3(0.375)^2 + 2(3.375)^2 = 43.9$.

3. Compute the axial load and bending moment on the pile group

The axial load $P = 20,000(12) = 240,000$ lb; then, $M = 240,000(3.25 - 2.625) = 150,000$ lb-ft.

4. Determine the pile load in each row

Find the pile load in each row resulting from the combined axial load and moment. Thus, $P/A = 240,000/8 = 30,000$ lb per pile; then, $M/I = 150,000/43.9 = 3,420$. Also, $p_a = 30,000 - 3,420(2.625) = 21,020$ lb per pile; $p_b = 30,000 + 3,420(0.375) = 31,280$ lb per pile; $p_c = 30,000 + 3,420(3.375) = 41,540$ lb per pile.

5. Verify the above results

Compute the total pile reaction, the moment of the applied load, and the pile reaction with respect to row A. Thus, $R = 3(21,020) + 3(31,280) + 2(41,540) = 239,980$ lb; then, $M_a = 240,000(3.25) = 780,000$ lb-ft and $M_r = 3(31,280)(3) + 2(41,540)(6) = 780,000$ lb-ft. Since $M_a = M_r$, the computed results are verified.

Deflection of Beams

In this Handbook the slope of the elastic curve at a given section of a beam is denoted by θ, and the deflection, in in., by y. The slope is considered positive if the section rotates in a clockwise direction under the bending loads. A downward deflection is considered positive. In all instances, the beam is understood to be prismatic, if nothing is stated to the contrary.

DOUBLE-INTEGRATION METHOD OF DETERMINING BEAM DEFLECTION

The simply supported beam in Fig. 36 is subjected to a counterclockwise moment N applied at the right-hand support. Determine the slope of the elastic curve at each support and the maximum deflection of the beam.

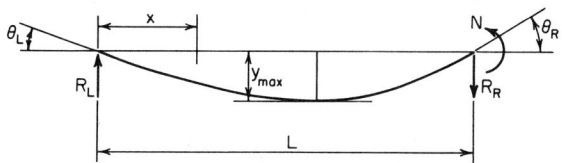

Fig. 36   Deflection of simple beam under end moment.

Calculation Procedure:

1. Evaluate the bending moment at a given section

Make this evaluation in terms of the distance x from the left-hand support to this section. Thus: $R_L = N/L$; $M = Nx/L$.

2. Write the differential equation of the elastic curve; integrate twice

Thus: $EI\, d^2y/dx^2 = -M = -Nx/L$; $EI\, dy/dx = EI\theta = -Nx^2/2L + c_1$; $EIy = -Nx^3/6L + c_1x + c_2$.

3. Evaluate the constants of integration

Apply the following boundary conditions: When $x = 0$, $y = 0$; $\therefore c_2 = 0$; when $x = L$, $y = 0$; $\therefore c_1 = NL/6$.

4. Write the slope and deflection equations

Substitute the constant values found in Step 3 in the equations developed in Step 2. Thus: $\theta = (N/6EIL)(L^2 - 3x^2)$; $y = (Nx/6EIL)(L^2 - x^2)$.

5. Find the slope at the supports

Substitute the values $x = 0$, $x = L$ in the slope equation to determine the slope at the supports. Thus: $\theta_L = NL/6EI$; $\theta_R = -NL/3EI$.

6. Solve for the section of maximum deflection

Set $\theta = 0$ and solve for x to locate the section of maximum deflection. Thus: $L^2 - 3x^2 = 0$; $x = L/3^{0.5}$. Substituting in the deflection equation, $y_{max} = NL^2/9EI3^{0.5}$.

MOMENT-AREA METHOD OF DETERMINING BEAM DEFLECTION

Use the moment-area method to determine the slope of the elastic curve at each support and the maximum deflection of the beam shown in Fig. 36.

Calculation Procedure:

1. Sketch the elastic curve of the member and draw the M/EI diagram

Let A and B denote two points on the elastic curve of a beam. The moment-area method is based on the following theorems:

The difference between the slope at A and that at B is numerically equal to the area of the M/EI diagram within the interval AB.

The deviation of A from a tangent to the elastic curve through B is numerically equal to the statical moment of the area of the M/EI diagram within the interval AB with respect to A. This tangential deviation is measured normal to the unstrained position of the beam.

Draw the elastic curve and the M/EI diagram as shown in Fig. 37.

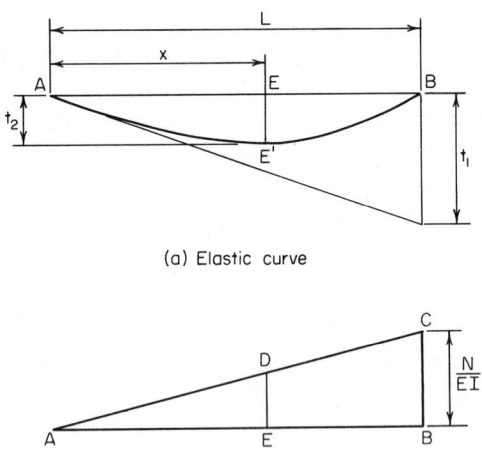

(a) Elastic curve

(b) M/EI diagram

Fig. 37

2. Calculate the deviation t_1 of B from the tangent through A

Thus, $t_1 = $ moment of $\triangle ABC$ about $BC = (NL/2EI)(L/3) = NL^2/6EI$. Also, $\theta_L = t_1/L = NL/6EI$.

3. Determine the right-hand slope in an analogous manner

4. Compute the distance to the section where the slope is zero

Area $\triangle AED$ = area $\triangle ABC(x/L)^2 = Nx^2/2EIL$; $\theta_E = \theta_L$ − area $\triangle AED = NL/6EI -$ $Nx^2/2EIL = 0$; $x = L/3^{0.5}$.

5. Evaluate the maximum deflection

Evaluate y_{max} by calculating the deviation t_2 of A from the tangent through E', Fig. 37. Thus: Area $\triangle AED = \theta_L = NL/6EI$; $y_{max} = t_2 = (NL/6EI)(2x/3) = (NL/6EI)(2L/3 \times 3^{0.5}) = NL^2/9EI3^{0.5}$, as before.

CONJUGATE-BEAM METHOD OF DETERMINING BEAM DEFLECTION

The overhanging beam in Fig. 38 is loaded in the manner shown. Compute the deflection at C.

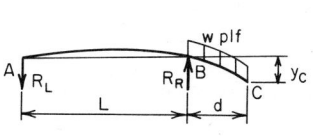
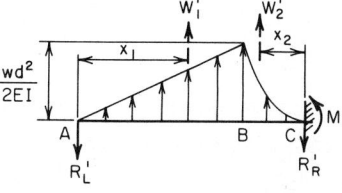

(a) Force diagram of given beam   (b) Force diagram of conjugate beam

Fig. 38   Deflection of overhanging beam.

Calculation Procedure:

1. Assign supports to the conjugate beam

If a conjugate beam of identical span as the given beam is loaded with the M/EI diagram of the latter, the shear V' and bending moment M' of the conjugate beam are equal, respectively, to the slope θ and deflection y at the corresponding section of the given beam.

Assign supports to the conjugate beam that are compatible with the end conditions of the given beam. At A, the given beam has a specific slope but zero deflection. Correspondingly, the conjugate beam has a specific shear but zero moment; i.e., it is simply supported at A.

At C, the given beam has a specific slope and a specific deflection. Correspondingly, the conjugate beam has both a shear and a bending moment; i.e., it has a fixed support at C.

2. Construct the M/EI diagram of the given beam

Load the conjugate beam with this area. The moment at B is $-wd^2/2$; the moment varies linearly from A to B and parabolically from C to B.

3. Compute the resultant of the load in selected intervals

Compute the resultant W_1' of the load in interval AB and the resultant W_2' of the load in the interval BC. Locate these resultants. (Refer to the AISC *Manual*, for properties of the complement of a half parabola.) Then: $W_1' = (L/2)(wd^2/2EI) = wd^2L/4EI$; $x_1 = \frac{2}{3}L$; $W_2' = (d/3)(wd^2/2EI) = wd^3/6EI$; $x_2 = \frac{3}{4}d$.

4. Evaluate the conjugate-beam reaction

Since the given beam has zero deflection at B, the conjugate beam has zero moment at this section. Evaluate the reaction R_L' accordingly. Thus: $M_B' = -R_L'L + W_1'L/3 = 0$; $R_L' = W_1'/3 = wd^2L/12EI$.

5. Determine the deflection

Determine the deflection at C by computing M_c'. Thus: $y_c = M_c' = -R_L'(L+d) + W_1'(d+L/3) + W_2'(3d/4) = wd^3(4L+3d)/24EI$.

UNIT-LOAD METHOD OF COMPUTING BEAM DEFLECTION

The cantilever beam in Fig. 39a carries a load that varies uniformly from w plf at the free end to zero at the fixed end. Determine the slope and deflection of the elastic curve at the free end.

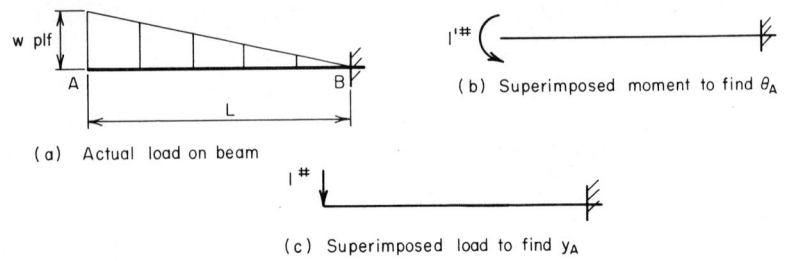

(a) Actual load on beam

(b) Superimposed moment to find θ_A

(c) Superimposed load to find y_A

Fig. 39

Calculation Procedure:

1. Apply a unit moment to the beam

Apply a counterclockwise unit moment at A, Fig. 39b. (This direction is selected because it is known that the end section rotates in this manner.) Let x = distance from A to given section; w_x = load intensity at the given section; M and m = bending moment at the given section induced by the actual load and by the unit moment, respectively.

2. Evaluate the moments in Step 1

Evaluate M and m. By proportion, $w_x = [w(L-x)]/L$; $M = -(x^2/6)(2w+w_x) = -(wx^2/6)[2+(L-x)/L] = -wx^2(3L-x)/6L$; $m = -1$.

3. Apply a suitable slope equation

Use the equation $\theta_A = \int_0^L (Mm/EI)\,dx$. Then, $EI\theta_A = \int_0^L [wx^2(3L-x)/6L]\,dx = (w/6L)\int_0^L(3Lx^2-x^3)\,dx = (w/6L)(3Lx^3/3 - x^4/4)]_0^L = (w/6L)(L^4 - L^4/4)$; thus, $\theta_A = \frac{1}{8}wL^3/EI$ counterclockwise. This is the slope at A.

4. Apply a unit load to the beam

Apply a unit downward load at A as shown in Fig. 39c. Let m' denote the bending moment at a given section induced by the unit load.

5. Evaluate the bending moment induced by the unit load; find the deflection

Apply $y_A = \int_0^L (Mm'/EI)\,dx$. Then: $m' = -x$; $EIy_A = \int_0^L [wx^3(3L-x)/6L]\,dx = (w/6L)\int_0^L x^3(3L-x)\,dx$; $y_A = (11/120)wL^4/EI$.

The first equation in Step 3 is a statement of the work performed by the unit moment at A as the beam deflects under the applied load. The left-hand side of this equation expresses the external work, and the right-hand side expresses the internal work. These work equations consitute a simple proof of Maxwell's theorem of reciprocal deflections, which is presented in a later Calculation Procedure.

DEFLECTION OF A CANTILEVER FRAME

The prismatic rigid frame $ABCD$, Fig. 40a, carries a vertical load P at the free end. Determine the horizontal displacement of A by means of both the unit-load method and the moment-area method.

Calculation Procedure:

1. Apply a unit horizontal load

Apply the unit horizontal load at A, directed to the right.

2. Evaluate the bending moments in each member

Let M and m denote the bending moment at a given section caused by the load P and by the unit load, respectively. Evaluate these moments in each member, considering a moment positive if it induces tension in the outer fibers of the frame. Thus:

Member AB: Let x denote the vertical distance from A to a given section. Then: $M = 0;\ m = x.$

Member BC: Let x denote the horizontal distance from B to a given section. Then: $M = Px;\ m = a.$

Member CD: Let x denote the vertical distance from C to a given section. Then: $M = Pb;\ m = a - x.$

3. Evaluate the required deflection

Calling the required deflection Δ, apply: $\Delta = \int (Mm/EI)\,dx$; $EI\Delta = \int_0^b Pax\,dx + \int_0^c Pb(a-x)\,dx = Pax^2/2]_0^b + Pb(ax - x^2/2)]_0^c = Pab^2/2 + Pabc - Pbc^2/2$; $\Delta = (Pb/2EI)(ab + 2ac - c^2).$

If this value is positive, A is displaced in the direction of the unit load—i.e., to the right. Draw the elastic curve in hyperbolic fashion, Fig. 40b. The above three steps constitute the unit-load method of solving this problem.

4. Construct the bending-moment diagram

Draw the diagram as shown in Fig. 40c.

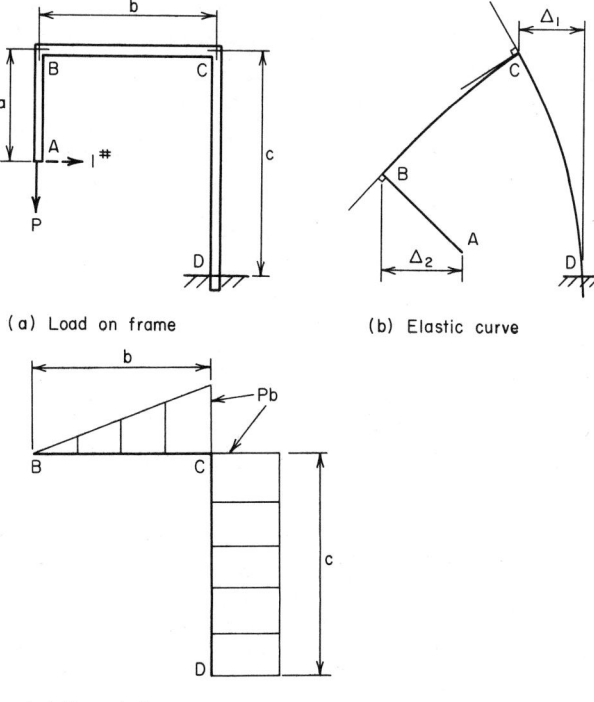

(a) Load on frame   (b) Elastic curve

(c) Moment diagram

Fig. 40

5. Compute the rotation and horizontal displacement by the moment-area method

Determine the rotation and horizontal displacement of C. (Consider only absolute values.) Since there is no rotation at D, $EI\theta_C = Pbc$; $EI\Delta_1 = Pbc^2/2.$

6. *Compute the rotation of one point relative to another and the total rotation*

Thus: $EI\theta_{BC} = Pb^2/2$; $EI\theta_B = Pbc + Pb^2/2 = Pb(c + b/2)$.

The horizontal displacement of B relative to C is infinitesimal.

7. *Compute the horizontal displacement of one point relative to another*

Thus, $EI\Delta_2 = EI\theta_B a = Pb(ac + ab/2)$.

8. *Combine the computed displacements to obtain the absolute displacement*

Thus: $EI\Delta = EI(\Delta_2 - \Delta_1) = Pb(ac + ab/2 - c^2/2)$; $\Delta = (Pb/2EI)(2ac + ab - c^2)$.

Statically Indeterminate Structures

A structure is said to be *statically determinate* if its reactions and internal forces may be evaluated by applying solely the equations of equilibrium, and *statically indeterminate* if such is not the case. The analysis of an indeterminate structure is performed by combining the equations of equilibrium with the known characteristics of the deformation of the structure.

SHEAR AND BENDING MOMENT OF A BEAM ON A YIELDING SUPPORT

The beam in Fig. 41a has an *EI* value of 35×10^9 lb-in.^{2} and bears on a spring at B that has a constant of 100 kips/in., i.e., a force of 100 kips will compress the spring 1 in.

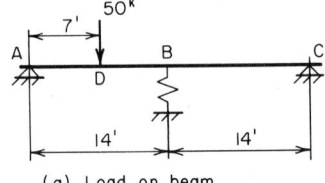

(a) Load on beam

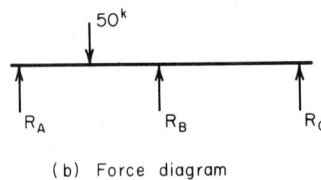

(b) Force diagram

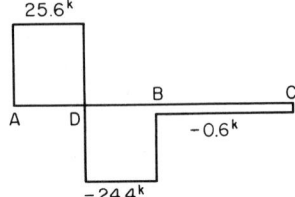

(c) Shear diagram

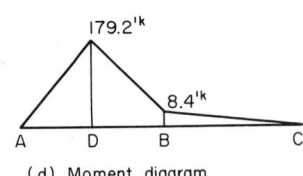

(d) Moment diagram

Fig. 41

Neglecting the weight of the member, construct the shear and bending-moment diagrams.

Calculation Procedure:

1. *Draw the free-body diagram of the beam*

Draw this diagram, Fig. 41b. Consider this as a simply supported member carrying a 50-kip load at D and an upward load R_B at its center.

2. *Evaluate the deflection*

Evaluate the deflection at B by applying the equations presented for Cases 7 and 8 in the AISC *Manual*. With respect to the 50-kip load, $b = 7$ ft and $x = 14$ ft. If y is in in. and

R_B is in lb, $y = 50,000(7)(14)(28^2 - 7^2 - 14^2)1,728/[6(35)(10)^9 28] - R_B(28)^3 1,728/[48(35) \times (10)^9] = 0.776 - (2.26/10^5)R_B$.

3. Express the deflection in terms of the spring constant
The deflection at B is, by proportion: $y/1 = R_B/100,000$; $y = R_B/100,000$.

4. Equate the two deflection expressions and solve for the upward load
Thus: $R_B/10^5 = 0.776 - (2.26/10^5)R_B$; $R_B = 0.776(10)^5/3.26 = 23,800$ lb.

5. Calculate the reactions R_A and R_C by taking moments
$\Sigma M_C = 28R_A - 50,000(21) + 23,800(14) = 0$; $R_A = 25,600$ lb; $\Sigma M_A = 50,000(7) - 23,800(14) - 28R_C = 0$; $R_C = 600$ lb.

6. Construct the shear and moment diagrams
Construct these diagrams as shown in Fig. 41. Then: $M_D = 7(25,600) = 179,200$ lb-ft; $M_B = 179,200 - 7(24,400) = 8,400$ lb-ft.

MAXIMUM BENDING STRESS IN BEAMS JOINTLY SUPPORTING A LOAD

In Fig. 42a, a 16WF40 beam and a 12WF31 beam cross each other at the vertical line V, the bottom of the 16-in. beam being $\frac{3}{8}$ in. above the top of the 12-in. beam before the load

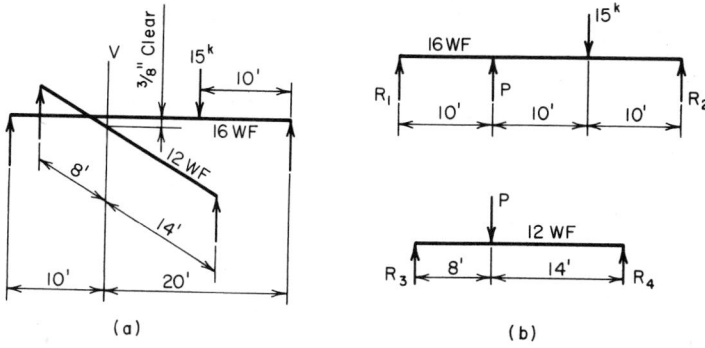

(a)   (b)

Fig. 42   Load carried by two beams.

is applied. Both members are simply supported. A column bearing on the 16-in. beam transmits a load of 15 kips at the indicated location. Compute the maximum bending stress in the 12-in beam.

Calculation Procedure:

1. Determine if the upper beam engages the lower beam
To ascertain whether the upper beam engages the lower one as it deflects under the 15-kip load, compute the deflection of the 16-in. beam at V if the 12-in. beam were absent. This distance is 0.74 in. Consequently, the gap between the members is closed and the two beams share the load.

2. Draw a free-body diagram of each member
Let P denote the load transmitted to the 12-in. beam by the 16-in. beam (or the reaction of the 12-in. beam on the 16-in. beam). Draw, in Fig. 42b, a free-body diagram of each member.

3. Evaluate the deflection of the beams
Evaluate, in terms of P, the deflections y_{12} and y_{16} of the 12-in. and 16-in. beam, respectively, at line V.

4. **Express the relationship between the two deflections**

Thus, $y_{12} = y_{16} - 0.375$.

5. **Replace the deflections in Step 4 with their values as obtained in Step 3**

After substituting these deflections, solve for P.

6. **Compute the reactions of the lower beam**

Once the reactions of the lower beam are computed, obtain the maximum bending moment. Then compute the corresponding flexural stress.

THEOREM OF THREE MOMENTS

For the two-span beam in Fig. 43a, compute the reactions at the supports. Apply the theorem of three moments to arrive at the results.

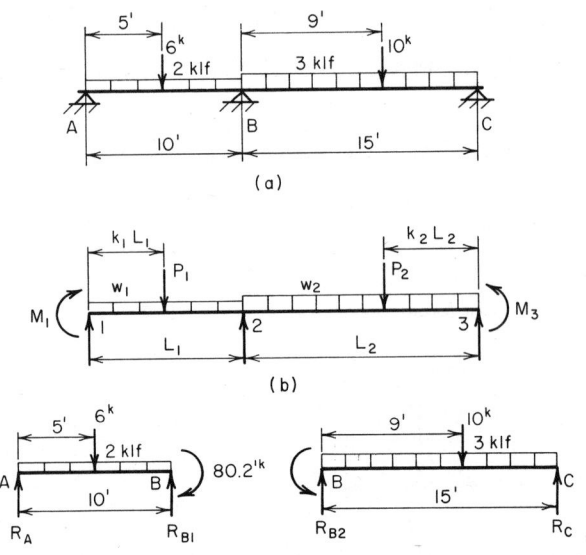

Fig. 43

Calculation Procedure:

1. **Using the bending-moment equation, determine M_B**

Figure 43b represents a general case. For a prismatic beam, the bending moments at the three successive supports are related by $M_1 L_1 + 2M_2(L_1 + L_2) + M_3 L_2 = -\frac{1}{4}w_1 L_1{}^3 - \frac{1}{4}w_2 L_2{}^3 - P_1 L_1{}^2(k_1 - k_1{}^3) - P_2 L_2{}^2(k_2 - k_2{}^3)$. Substituting in this equation: $M_1 = M_3 = 0$; $L_1 = 10$ ft; $L_2 = 15$ ft; $w_1 = 2$ klf; $w_2 = 3$ klf; $P_1 = 6$ kips; $P_2 = 10$ kips; $k_1 = 0.5$; $k_2 = 0.4$; $2M_B(10 + 15) = -\frac{1}{4}(2)(10)^3 - \frac{1}{4}(3)(15)^3 - 6(10)^2(0.5 - 0.125) - 10(15)^2(0.4 - 0.064)$; $M_B = -80.2$ ft-kips.

2. **Draw a free-body diagram of each span**

Figure 43c shows the free-body diagrams.

3. **Take moments with respect to each support to find the reactions**

Span AB: $\Sigma M_A = 6(5) + 2(10)(5) + 80.2 - 10R_{B1} = 0$; $R_{B1} = 21.02$ kips. $\Sigma M_B = 10R_A - 6(5) - 2(10)(5) + 80.2 = 0$; $R_A = 4.98$ kips.

Span BC: $\Sigma M_B = -80.2 + 10(9) + 3(15)(7.5) - 15R_C = 0$; $R_C = 23.15$ kips; $\Sigma M_C = 15R_{B2} - 80.2 - 10(6) - 3(15)(7.5) = 0$; $R_{B2} = 31.85$ kips; $R_B = 21.02 + 31.85 = 52.87$ kips.

THEOREM OF THREE MOMENTS: BEAM WITH OVERHANG AND FIXED END

Determine the reactions at the supports of the continuous beam in Fig. 44a. Use the theorem of three moments.

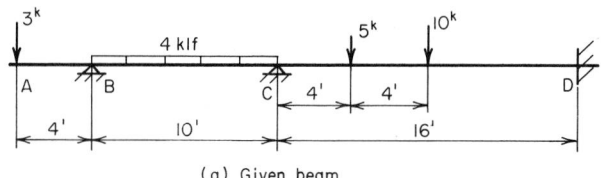

(a) Given beam

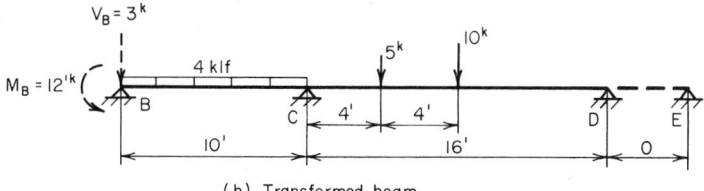

(b) Transformed beam

Fig. 44

Calculation Procedure:

1. Transform the given beam to one amenable to analysis by the theorem of three moments

Perform the following operations to transform the beam:

a. Remove the span AB and introduce the shear V_B and moment M_B that the load on AB induces at B, as shown in Fig. 44b.

b. Remove the fixed support at D and add the span DE of zero length, with a hinged support at E.

For the interval BD, the transformed beam is then identical in every respect with the actual beam.

2. Apply the equation for the theorem of three moments

Consider span BC as span 1 and CD as span 2. For the 5-kip load, $k_2 = 12/16 = 0.75$; for the 10-kip load, $k_2 = 8/16 = 0.5$. Then, $-12(10) + 2M_C(10 + 16) + 16M_D = -\frac{1}{4}(4)(10)^3 - 5(16)^2(0.75 - 0.422) - 10(16)^2(0.5 - 0.125)$. Simplifying, $13M_C + 4M_D = -565.0$, Eq. (a).

3. Apply the moment equation again

Considering CD as span 1 and DE as span 2, apply the moment equation again. Or, for the 5-kip load, $k_1 = 0.25$; for the 10-kip load, $k_1 = 0.5$. Then, $16M_C + 2M_D(16 + 0) = -5(16)^2(0.25 - 0.016) - 10(16)^2(0.50 - 0.125)$. Simplifying, $M_C + 2M_D = -78.7$, Eq. (b).

4. Solve the moment equations

Solving Eqs. (a) and (b): $M_C = -37.1$ ft-kips; $M_D = -20.8$ ft-kips.

5. Determine the reactions using a free-body diagram

Find the reactions by drawing a free-body diagram of each span and taking moments with respect to each support. Thus: $R_B = 20.5$ kips; $R_C = 32.3$ kips; $R_D = 5.2$ kips.

BENDING-MOMENT DETERMINATION BY MOMENT DISTRIBUTION

Using moment distribution, determine the bending moments at the supports of the member in Fig. 45. The beams are rigidly joined at the supports and are composed of the same material.

Calculation Procedure:

1. Calculate the flexural stiffness of each span

Using K to denote the flexural stiffness, $K = I/L$ if the far end remains fixed during moment distribution; $K = 0.75I/L$ if the far end remains hinged during moment distribution. Then: $K_{AB} = 270/18 = 15$; $K_{BC} = 192/12 = 16$; $K_{CD} = 0.75(240/20) = 9$. Record all the values on the drawing as they are obtained.

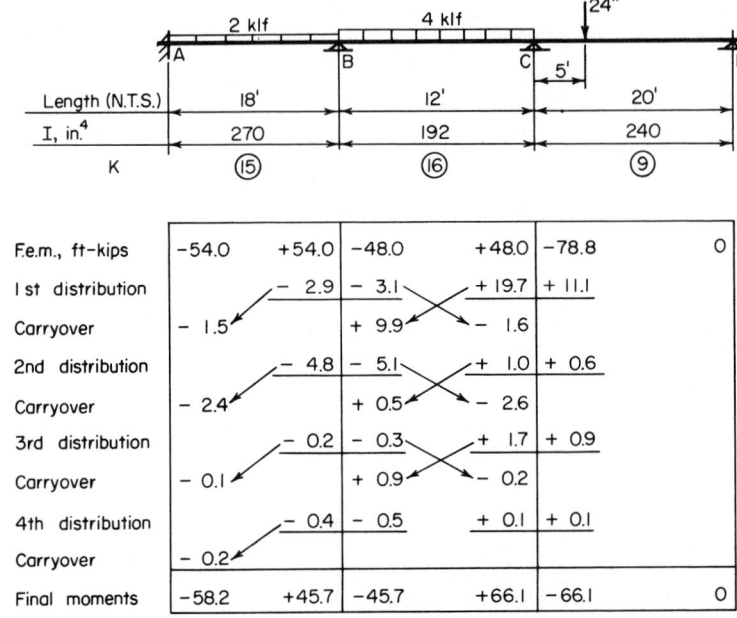

Fig. 45   Moment distribution.

2. For each span, calculate the required fixed-end moments at those supports that will be considered fixed

These are the *external* moments with respect to the span; a clockwise moment is considered positive. (For additional data, refer to Cases 14 and 15 in the AISC *Manual*.) Then: $M_{AB} = -wL^2/12 = -2(18)^2/12 = -54.0$ ft-kips; $M_{BA} = +54.0$ ft-kips. Similarly, $M_{BC} = -48.0$ ft-kips; $M_{CB} = +48.0$ ft-kips; $M_{CD} = -24(15)(5)(15+20)/2(20)^2 = -78.8$ ft-kips.

3. Calculate the unbalanced moments

Computing the unbalanced moments at B and C: At B, $+54.0 - 48.0 = +6.0$ ft-kips; at C, $+48.0 - 78.8 = -30.8$ ft-kips.

4. Apply balancing moments; distribute them in proportion to the stiffness of the adjoining spans

Apply the balancing moments at B and C and distribute them to the two adjoining spans in proportion to their stiffness. Thus: $M_{BA} = -6.0(15/31) = -2.9$ ft-kips; $M_{BC} = -6.0(16/31) = -3.1$ ft-kips; $M_{CB} = +30.8(16/25) = +19.7$ ft-kips; $M_{CD} = +30.8(9/25) = +11.1$ ft-kips.

5. Perform the "carry-over" operation for each span

To do this: Take one-half the distributed moment applied at one end of the span, and add this to the moment at the far end if that end is considered to be fixed during moment distribution.

6. Perform the second cycle of moment balancing and distribution

Thus: $M_{BA} = -9.9(15/31) = -4.8$; $M_{BC} = -9.9(16/31) = -5.1$; $M_{CB} = +1.6(16/25) = +1.0$; $M_{CD} = +1.6(9/25) = +0.6$.

7. Continue the foregoing procedure until the carry-over moments become negligible

Total the results to obtain the following bending moments: $M_A = -58.2$ ft-kips; $M_B = -45.7$ ft-kips; $M_C = -66.1$ ft-kips.

ANALYSIS OF A STATICALLY INDETERMINATE TRUSS

Determine the internal forces of the truss in Fig. 46a. The cross-sectional areas of the members are given in Table 5.

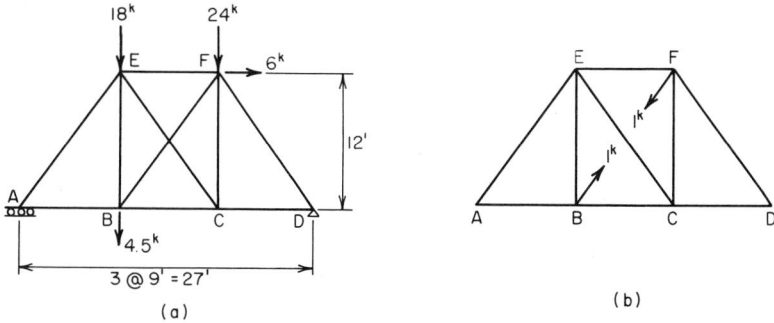

Fig. 46 Statically indeterminate truss.

Calculation Procedure:

1. Test the structure for statical determinateness

Apply the following criterion. Let j = number of joints; m = number of members; r = number of reactions. Then, if $2j = m + r$, the truss is statically determinate; if $2j < m + r$, the truss is statically indeterminate and the deficiency represents the degree of indeterminateness.

In this truss, $j = 6$, $m = 10$, $r = 3$, consisting of a vertical reaction at A and D and a horizontal reaction at D. Thus: $2j = 12$; $m + r = 13$. The truss is therefore statically indeterminate to the first degree; i.e., there is *one* redundant member.

The method of analysis comprises the following steps: Assume a value for the internal force in a particular member, and calculate the relative displacement Δ_i of the two ends of that member caused solely by this force. Now remove this member to secure a determinate truss and calculate the relative displacement Δ_a caused solely by the applied loads. The true internal force is of such magnitude that $\Delta_i = -\Delta_a$.

2. Assume a unit force for one member

Assume for convenience that the force in BF is 1 kip tension. Remove this member, and replace it with the assumed 1-kip force that it exerts at joints B and F, as shown in Fig. 46b.

3. Calculate the force induced in each member solely by the unit force

Calling the induced force U, produced solely by the unit tension in BF, record the results in Table 5, considering tensile forces as positive and compressive forces as negative.

4. Calculate the force induced in each member solely by the applied loads

With BF eliminated, calculate the force S induced in each member solely by the applied loads.

5.   Evaluate the true force in the selected member

Use the relation $BF = -(\Sigma SUL/AE)/(\Sigma U^2L/AE)$. The numerator represents Δ_a; the denominator represents Δ_i for a 1-kip tensile force in BF. Since E is constant, it cancels out. Substituting the values in Table 5, $BF = -(-266.5/135.5) = 1.97$ kips. The positive result confirms the assumption that BF is tensile.

TABLE 5   Forces in Truss Members (Fig. 46)

Member	A, sq in.	L, in.	U, kips	S, kips	U^2L/A	SUL/A	S', kips
AB	5	108	0	+15.25	0	0	+15.25
BC	5	108	−0.60	+15.25	+7.8	−197.6	+14.07
CD	5	108	0	+13.63	0	0	+13.63
EF	4	108	−0.60	−13.63	+9.7	+220.8	−14.81
BE	4	144	−0.80	+4.50	+23.0	−129.6	+2.92
CF	4	144	−0.80	+2.17	+23.0	−62.5	+0.59
AE	6	180	0	−25.42	0	0	−25.42
BF	5	180	+1.00	0	+36.0	0	+1.97
CE	5	180	+1.00	−2.71	+36.0	−97.6	−0.74
DF	6	180	0	−32.71	0	0	−32.71
Total	....	....	....	....	+135.5	−266.5	

6.   Evaluate the true force in each member

Use the relation $S' = S + 1.97U$, where $S' =$ true force. The results are shown in Table 5.

Moving Loads and Influence Lines

ANALYSIS OF BEAM CARRYING MOVING CONCENTRATED LOADS

The loads shown in Fig. 47a traverse a beam of 40-ft simple span while their spacing remains constant. Determine the maximum bending moment and maximum shear induced in the beam during transit of these loads. Disregard the weight of the beam.

Calculation Procedure:

1.   Determine the magnitude of the resultant and its location

Since the member carries only concentrated loads, the maximum moment at any instant occurs under one of these loads. Thus, the problem is to determine the position of the load system that causes the *absolute* maximum moment.

The magnitude of the resultant R is $R = 10 + 4 + 15 = 29$ kips. To determine the location of R, take moments with respect to A, Fig. 47. Thus: $\Sigma M_A = 29AD = 4(5) + 15(17)$. Solving, $AD = 9.48$ ft.

2.   Assume several trial load positions

Assume that the maximum moment occurs under the 10-kip load. Place the system in the position shown in Fig. 47b, with the 10-kip load as far from the adjacent support as the resultant is from the other support. Repeat this procedure for the two remaining loads.

3.   Determine the support reactions for the trial load positions

For these three trial positions, calculate the reaction at the support adjacent to the load under consideration. Determine whether the vertical shear is zero or changes sign at this load. Thus, position 1: $R_L = 29(15.26)/40 = 11.06$ kips. Since the shear does not change sign at the 10-kip load, this position lacks significance.

Position 2: $R_L = 29(17.76)/40 = 12.88$ kips. The shear changes sign at the 4-kip load.

Position 3: $R_R = 29(16.24)/40 = 11.77$ kips. The shear changes sign at the 15-kip load.

4. *Compute the maximum bending moment associated with positions having a change in the shear sign*

This applies to positions 2 and 3. The absolute maximum moment is the larger of these values. Thus, position 2: $M = 12.88(17.76) - 10(5) = 178.7$ ft-kips. Position 3: $M = 11.77(16.24) = 191.1$ ft-kips. Thus, $M_{max} = 191.1$ ft-kips.

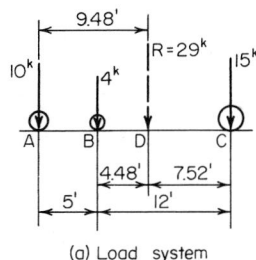

(a) Load system

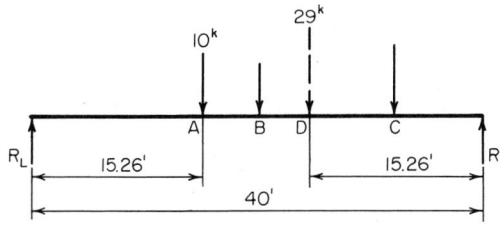

(b) Position 1, for 10-kip load

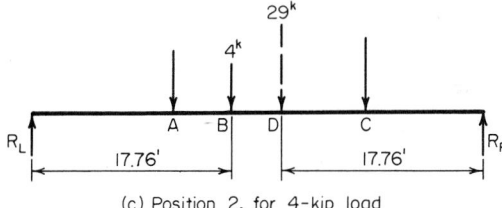

(c) Position 2, for 4-kip load

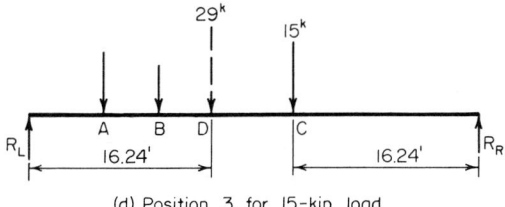

(d) Position 3, for 15-kip load

Fig. 47

5. *Determine the absolute maximum shear*

For absolute maximum shear, place the 15-kip load an infinitesimal distance to the left of the right-hand support. Then, $V_{max} = 29(40 - 7.52)/40 = 23.5$ kips.

When the load spacing is large in relation to the beam span, the absolute maximum moment may occur when only part of the load system is on the span. This possibility requires careful investigation.

INFLUENCE LINE FOR SHEAR IN A BRIDGE TRUSS

The Pratt truss in Fig. 48a supports a bridge at its bottom chord. Draw the influence line for shear in panel cd caused by a moving load traversing the bridge floor.

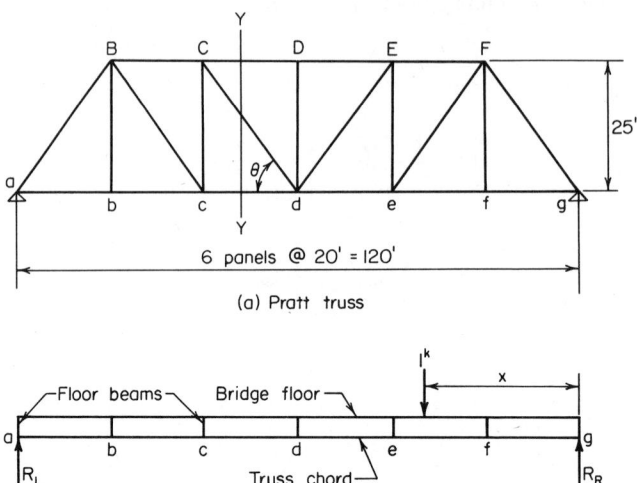

(a) Pratt truss

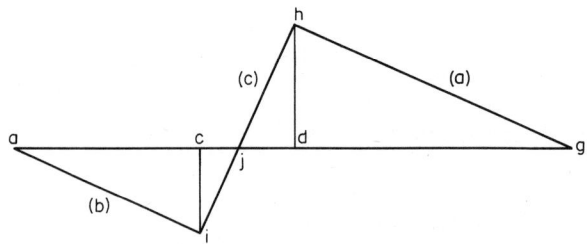

(b) Transmission of load through floor beams

(c) Influence line for shear in panel cd

Fig. 48

Calculation Procedure:

1. Compute the shear in the panel being considered with a unit load to the right of the panel

Cut the truss at section YY. The algebraic sum of vertical forces acting on the truss at panel points to the left of YY is termed the *shear* in panel cd.

Consider that a moving load traverses the bridge floor from right to left and that the portion of the load carried by the given truss is 1 kip. This unit load is transmitted to the truss as concentrated loads at two adjacent bottom-chord panel points, the latter being components of the unit load. Let x denote the instantaneous distance from the right-hand support to the moving load.

Place the unit load to the right of d, as shown in Fig. 48b, and compute the shear V_{cd} in panel cd. The truss reactions may be obtained by considering the unit load itself rather than its panel-point components. Thus: $R_L = x/120$; $V_{cd} = R_L = x/120$, Eq. (a).

2. *Compute the panel shear with the unit load to the left of the panel considered*

Placing the unit load to the left of c, $V_{cd} = R_L - 1 = x/120 - 1$, Eq. (b).

3. *Determine the panel shear with the unit load within the panel*

Place the unit load within panel cd. Determine the panel-point load P_c at c and compute V_{cd}. Thus: $P_c = (x - 60)/20 = (x/20) - 3$. $V_{cd} = R_L - P_c = x/120 - (x/20 - 3) = -x/24 + 3$, Eq. ($c$).

4. *Construct a diagram representing the shear associated with every position of the unit load*

Apply the foregoing equations to represent the value of V_{cd} associated with every position of the unit load. This diagram, Fig. 48c, is termed an *influence line*. The point j at which this line intersects the base is referred to as the *neutral point*.

5. *Compute the slope of each segment of the influence line*

Line a, $dV_{cd}/dx = 1/120$; line b, $dV_{cd}/dx = 1/120$; line c, $dV_{cd}/dx = -1/24$. Lines a and b are therefore parallel because they have the same slope.

FORCE IN TRUSS DIAGONAL CAUSED BY A MOVING UNIFORM LOAD

The bridge floor in Fig. 48a carries a moving uniformly distributed load. The portion of the load transmitted to the given truss is 2.3 klf. Determine the limiting values of the force induced in member Cd by this load.

Calculation Procedure:

1. *Locate the neutral point and compute dh*

The force in Cd is a function of V_{cd}. Locate the neutral point j in Fig. 48c and compute dh. From Eq. (c) of the previous Calculation Procedure, $V_{cd} = -jg/24 + 3 = 0$; $jg = 72$ ft. From Eq. (a) of the previous Procedure, $dh = 60/120 = 0.5$.

2. *Determine the maximum shear*

To secure the maximum value of V_{cd}, apply uniform load continuously in the interval jg. Compute V_{cd} by multiplying the area under the influence line by the intensity of the applied load. Thus, $V_{cd} = \frac{1}{2}(72)(0.5)(2.3) = 41.4$ kips.

3. *Determine the maximum force in the member*

Use the relation $Cd_{max} = V_{cd}(\csc \theta)$, where $\csc \theta = [(20^2 + 25^2)/25^2]^{0.5} = 1.28$. Then, $Cd_{max} = 41.4(1.28) = 53.0$ kips tension.

4. *Determine the minimum force in the member*

To secure the minimum value of V_{cd}, apply uniform load continuously in the interval aj. Perform the final calculation by proportion. Thus, $Cd_{min}/Cd_{max} = $ area $aij/$area $jhg = -(2/3)^2 = -4/9$. Then, $Cd_{min} = -(4/9)(53.0) = 23.6$ kips compression.

FORCE IN TRUSS DIAGONAL CAUSED BY MOVING CONCENTRATED LOADS

The truss in Fig. 49a supports a bridge that transmits the moving-load system shown in Fig. 49b to its bottom chord. Determine the maximum tensile force in De, Fig. 49.

Calculation Procedure:

1. *Locate the resultant of the load system*

The force in De, Fig. 49, is a function of the shear in panel de. This shear will be calculated without recourse to a set rule in order to show the principles involved in designing for moving loads.

To locate the resultant of the load system, take moments with respect to load 1. Thus, $R = 50$ kips. Then: $\Sigma M_1 = 12(6) + 18(16) + 15(22) = 50x$; $x = 13.8$ ft.

2. Construct the influence line for V_{de}

In Fig. 49c, draw the influence line for V_{de}. Assume right-to-left locomotion and express the slope of each segment of the influence line. Thus: Slope of ik = slope of $ma = 1/200$; slope of $km = -7/200$.

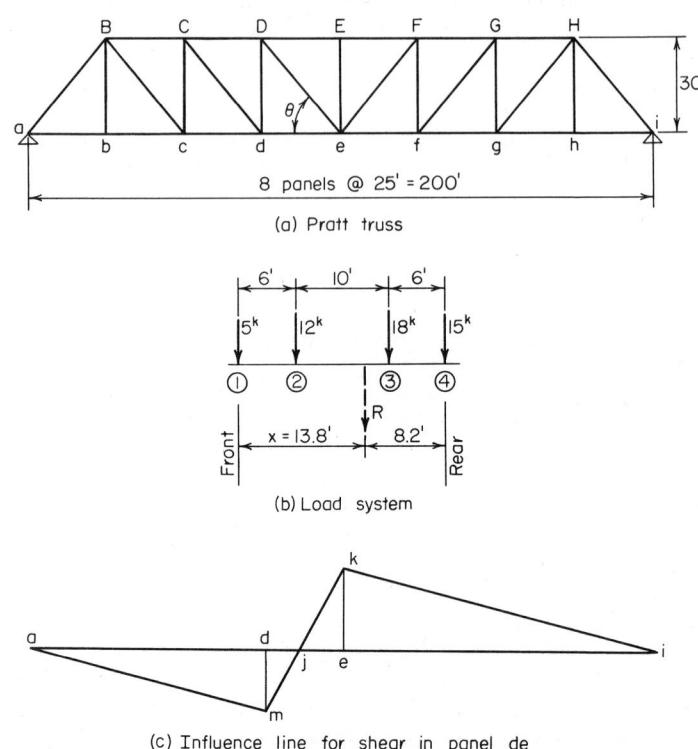

(a) Pratt truss

(b) Load system

(c) Influence line for shear in panel de

Fig. 49

3. Assume a load position and determine if V_{de} increases or decreases

Consider that load 1 lies within panel de and the remaining loads lie to the right of this panel. From the slope of the influence line, ascertain whether V_{de} increases or decreases as the system is displaced to the left. Thus: $dV_{de}/dx = 5(-7/200) + 45(1/200) > 0$; $\therefore V_{de}$ increases.

4. Repeat the foregoing calculation with other assumed load positions

Consider that loads 1 and 2 lie within the panel de and the remaining loads lie to the right of this panel. Repeat the foregoing calculation. Thus: $dV_{de}/dx = 17(-7/200) + 33(1/200) < 0$; $\therefore V_{de}$ decreases.

From these results it is concluded that as the system moves from right to left, V_{de} is maximum at the instant that load 2 is at e.

5. Place the system in the position thus established and compute V_{de}

Thus, $R_L = 50(100 + 6 - 13.8)/200 = 23.1$ kips. The load at panel point d is $P_d = 5(6)/25 = 1.2$ kips. $V_{de} = 23.1 - 1.2 = 21.9$ kips.

6. Assume left-to-right locomotion; proceed as in Step 3

Consider that load 4 is within panel de and the remaining loads are to the right of this panel. Proceeding as in Step 3, $dV_{de}/dx = 15(7/200) + 35(-1/200) > 0$.

It is therefore concluded that as the system moves from left to right, V_{de} is maximum at the instant that load 4 is at e.

7. Place the system in the position thus established and compute V_{de}

Thus: $V_{de} = R_L = [50(100 - 8.2)]/200 = 23.0$ kips; $\therefore V_{de,\text{max}} = 23.0$ kips.

8. Compute the maximum tensile force in De

Using the same relation as in Step 3 of the previous Calculation Procedure, csc $\theta = [(25^2 + 30^2)/30^2]^{0.5} = 1.30$; then, $De = 23.0(1.30) = 29.9$ kips tension.

INFLUENCE LINE FOR BENDING MOMENT IN BRIDGE TRUSS

The Warren truss in Fig. 50a supports a bridge at its top chord. Draw the influence line for the bending moment at b caused by a moving load traversing the bridge floor.

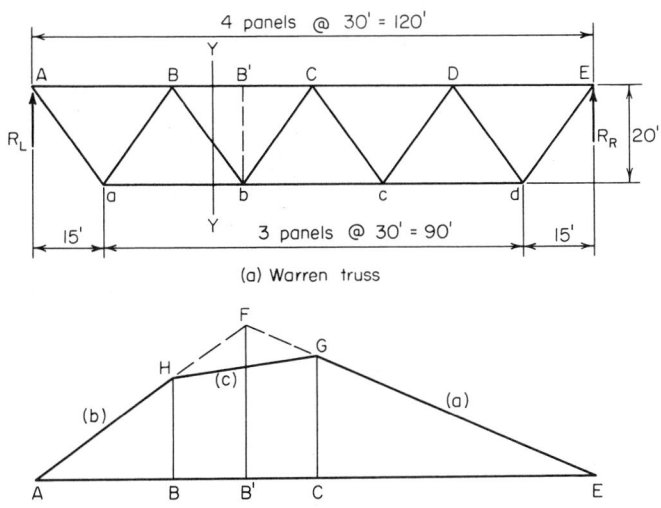

(a) Warren truss

(b) Influence line for bending moment at b

Fig. 50

Calculation Procedure:

1. Place the unit load in position and compute the bending moment

The moment of all forces acting on the truss at panel points to the left of b with respect to b is termed the *bending moment* at that point. Assume that the load transmitted to the given truss is 1 kip and let x denote the instantaneous distance from the right-hand support to the moving load.

Place the unit load to the right of C and compute the bending moment M_b. Thus: $R_L = x/120$; $M_b = 45R_L = 3x/8$, Eq. (a).

2. Place the unit load on the other side and compute the bending moment

Placing the unit load to the left of B and computing M_b, $M_b = 45R_L - (x - 75) = -5x/8 + 75$, Eq. (b).

3. Place the unit load within the panel; compute the panel-point load and bending moment

Place the unit load within panel BC. Determine the panel-point load P_B and compute M_b. Thus: $P_B = (x - 60)/30 = x/30 - 2$; $M_b = 45R_L - 15P_B = 3x/8 - 15(x/30 - 2) = -x/8 + 30$, Eq. (c).

4.   Applying the foregoing equations, draw the influence line

Figure 50b shows the influence line for M_b. Computing the significant values, $CG = (3/8)(60) = 22.50$ ft-kips; $BH = -(5/8)(90)+75 = 18.75$ ft-kips.

5.   Compute the slope of each segment of the influence line

This computation is made for subsequent reference. Thus: Line a, $dM_b/dx = 3/8$; line b, $dM_b/dx = -5/8$; line c, $dM_b/dx = -1/8$.

FORCE IN TRUSS CHORD CAUSED BY MOVING CONCENTRATED LOADS

The truss in Fig. 50a carries the moving-load system shown in Fig. 51. Determine the maximum force induced in member BC during transit of the loads.

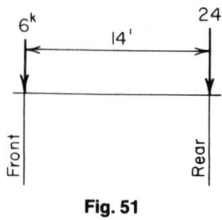

Fig. 51

Calculation Procedure:

1.   Assume that locomotion proceeds from right to left and compute the bending moment

The force in BC is a function of the bending moment M_b at b. Refer to the previous Calculation Procedure for the slope of each segment of the influence line. Study of these slopes shows that M_b increases as the load system moves until the rear load is at C, the front load being 14 ft to the left of C.

Calculate the value of M_b corresponding to this load disposition by applying the computed properties of the influence line. Thus, $M_b = 22.50(24) + (22.50 - 1/8 \times 14)(6) = 664.5$ ft-kips.

2.   Assume that locomotion proceeds from left to right and compute the bending moment

Study shows that M_b increases as the system moves until the rear load is at C, the front load being 14 ft to the right of C. Calculate the corresponding value of M_b. Thus, $M_b = 22.50(24) + (22.50 - 3/8 \times 14)(6) = 643.5$ ft-kips. $\therefore M_{b,\text{max}} = 664.5$ ft-kips.

3.   Determine the maximum force in the member

Cut the truss at plane YY. Determine the maximum force in BC by considering the equilibrium of the left part of the structure. Thus, $\Sigma M_b = M_b - 20BC = 0$; $BC = 664.5/20 = 33.2$-kips compression.

INFLUENCE LINE FOR BENDING MOMENT IN THREE-HINGED ARCH

The arch in Fig. 52a is hinged at A, B, and C. Draw the influence line for bending moment at D and locate the neutral point.

Calculation Procedure:

1.   Start the graphical construction

Draw a line through A and C intersecting the vertical line through B at E. Draw a line through B and C intersecting the vertical line through A at F. Draw the vertical line GH through D.

Let θ denote the angle between AE and the horizontal. Lines through B and D perpendicular to AE (omitted for clarity) make an angle θ with the vertical.

2.   Resolve the reaction into components

Resolve the reaction at A into the components R_1 and R_2 acting along AE and AB, respectively, Fig. 52.

3.   Determine the value of the first reaction

Let x denote the horizontal distance from the right-hand support to the unit load, where x has any value between 0 and L. Evaluate R_1 by equating the bending moment at B to zero. Thus: $M_B = R_1 b \cos \theta - x = 0$; solving, $R_1 = x/b \cos \theta$.

4. Evaluate the second reaction

Place the unit load within the interval CB. Evaluate R_2 by equating the bending moment at C to zero. Thus: $M_c = R_2 d = 0$; $\therefore R_2 = 0$.

5. Calculate the bending moment at D when the unit load lies within the interval CB

Thus, $M_D = -R_1 v \cos\theta = -(v\cos\theta/b\cos\theta)x$, or $M_D = -vx/b$, Eq. (a). When $x = m$, $M_D = -vm/b$.

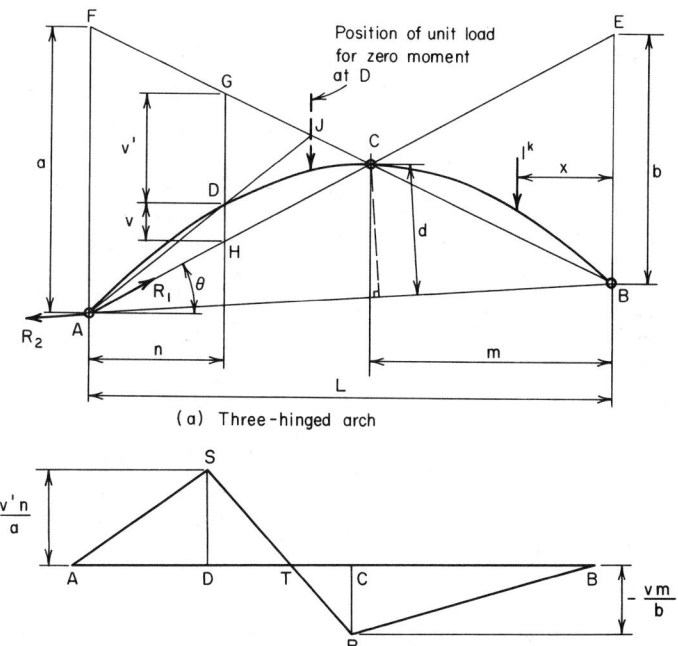

(a) Three-hinged arch

(b) Influence line for bending moment at D

Fig. 52

6. Place the unit load in a new position and determine the bending moment

Place the unit load within the interval AD. Working from the right-hand support, proceed in an analogous manner to arrive at the following result: $M_D = v'(L-x)/a$, Eq. (b). When $x = L - n$, $M_D = v'n/a$.

7. Place the unit load within another interval and evaluate the second reaction

Place the unit load within the interval DC and evaluate R_2. Thus: $M_c = R_2 d - (x - m) = 0$; solving, $R_2 = (x - m)/d$.

Since both R_1 and R_2 vary linearly with respect to x, it follows that M_D is also a linear function of x.

8. Complete the influence line

In Fig. 52b, draw lines BR and AS to represent Eqs. (a) and (b), respectively. Draw the straight line SR, thus completing the influence line. The point T at which this line intersects the base is termed the *neutral point*.

9. Locate the neutral point

To locate T, draw a line through A and D in Fig. 52a intersecting BF at J. The neutral point in the influence line lies vertically below J; i.e., M_D is zero when the action line of the unit load passes through J.

The proof is as follows: Since $M_D = 0$ and there are no applied loads in the interval AD, it follows that the total reaction at A is directed along AD. Similarly, since $M_C = 0$, and there are no applied loads in the interval CB, it follows that the total reaction at B is directed along BC. Because the unit load and the two reactions constitute a balanced system of forces, they are collinear. Therefore, J lies on the action line of the unit load.

Alternatively, the location of the neutral point may be established by applying the geometrical properties of the influence line.

DEFLECTION OF A BEAM UNDER MOVING LOADS

The moving-load system in Fig. 53a traverses a beam on a simple span of 40 ft. What disposition of the system will cause the maximum deflection at mid-span?

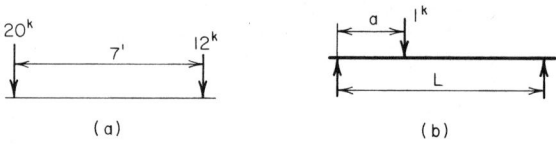

(a)   (b)

Fig. 53

Calculation Procedure:

1. Develop the equations for the mid-span deflection under a unit load

The maximum deflection will manifestly occur when the two loads lie on opposite sides of the centerline of the span. In calculating the deflection at mid-span caused by a load applied at any point on the span, it is advantageous to apply Maxwell's theorem of reciprocal deflections, which states the following: *The deflection at A caused by a load at B equals the deflection at B caused by this load at A.*

In Fig. 53b, consider the beam on a simple span L to carry a unit load applied at a distance a from the left-hand support. By referring to Case 7 of the AISC *Manual* and applying the principle of reciprocal deflections, derive the following equations for the mid-span deflection under the unit load. When $a < L/2$, $y = (3L^2a - 4a^3)/48EI$. When $a > L/2$, $y = [3L^2(L - a) - 4(L - a)^3]/48EI$.

2. Position the system for purposes of analysis

Position the system in such a manner that the 20-kip load lies to the left of center and the 12-kip load to the right of center. For the 20-kip load, set $a = x$. For the 12-kip load, $a = x + 7$; $L - a = 40 - (x + 7) = 33 - x$.

3. Express the total mid-span deflection in terms of x

Substitute in the preceding equations. Combining all constants into a single term k, $ky = 20(3 \times 40^2x - 4x^3) + 12[3 \times 40^2(33 - x) - 4(33 - x)^3]$.

4. Solve for the unknown distance

Set $dy/dx = 0$ and solve for x. Thus, $x = 17.46$ ft.

For maximum deflection, position the load system with the 20-kip load 17.46 ft from the left-hand support.

Riveted and Welded Connections

In the design of riveted and welded connections in this Handbook, the American Institute of Steel Construction *Specification for the Design, Fabrication and Erection of Structural Steel for Buildings* will be applied. This is presented in Part 5 of the *Manual of Steel Construction*.

The structural members considered here are made of ASTM A36 steel having a yield-point stress of 36,000 psi. (The yield-point stress is denoted by F_y in the *Specification*.) All connections considered here are made with A141 hot-driven rivets or fillet welds of A233 Class E60 series electrodes.

From the *Specification*, the allowable stresses are: Tensile stress in connected member, 22,000 psi; shearing stress in rivet, 15,000 psi; bearing stress on projected area of rivet, 48,500 psi; stress on throat of fillet weld, 13,600 psi.

Let n denote the number of sixteenths included in the size of a fillet weld. For example, for a $\frac{3}{8}$ in. weld, $n = 6$. Then, weld size $= n/16$. And, throat area per lin. in. of weld $= 0.707n/16 = 0.0442n$ sq in. Also, capacity of weld $= 13,600(0.0442n) = 600n$ pli.

As shown in Fig. 54, a rivet is said to be in *single shear* if the opposing forces tend to shear the shank along one plane and in *double shear* if they tend to shear it along two

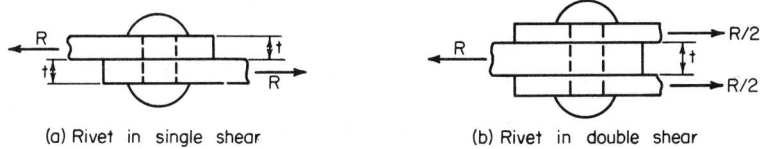

(a) Rivet in single shear   (b) Rivet in double shear

Fig. 54

planes. The symbols R_{ss}, R_{ds}, and R_b are used here to designate the shearing capacity of a rivet in single shear, the shearing capacity of a rivet in double shear, and the bearing capacity of a rivet, respectively, expressed in lb.

CAPACITY OF A RIVET

Determine the values of R_{ss}, R_{ds}, and R_b for a $\frac{3}{4}$ in. and $\frac{7}{8}$ in. rivet.

Calculation Procedure:

1. Compute the cross-sectional area of the rivet

For the $\frac{3}{4}$ in. rivet: Area $= A = 0.785(0.75)^2 = 0.4418$ sq in. Likewise, for the $\frac{7}{8}$ in. rivet: $A = 0.785(0.875)^2 = 0.6013$ sq in.

2. Compute the single and double shearing capacity of the rivet

Let t denote the thickness, in in., of the connected member, as shown in Fig. 54. Multiply the stressed area by the allowable stress to determine the shearing capacity of the rivet. Thus, for the $\frac{3}{4}$ in. rivet: $R_{ss} = 0.4418(15,000) = 6,630$ lb; $R_{ds} = 2(0.4418)(15,000) = 13,250$ lb. Note that the factor of 2 is used for a rivet in double shear.

Likewise, for the $\frac{7}{8}$ in. rivet: $R_{ss} = 0.6013(15,000) = 9,020$ lb; $R_{ds} = 2(0.6013)(15,000) = 18,040$ lb.

3. Compute the rivet bearing capacity

The effective bearing area of a rivet of diameter d in. $= dt$. Thus, for the $\frac{3}{4}$ in. rivet: $R_b = 0.75t(48,500) = 36,380t$ lb. For the $\frac{7}{8}$ in. rivet: $R_b = 0.875t(48,500) = 42,440t$ lb. By substituting the value of t in either relation, the numerical value of the bearing capacity could be obtained.

INVESTIGATION OF A LAP SPLICE

The hanger in Fig. 55a is spliced with nine $\frac{3}{4}$ in. rivets in the manner shown. Compute the load P that may be transmitted across the joint.

Calculation Procedure:

1. Compute the capacity of the joint in shear and bearing

There are three criteria to be considered: (a) the shearing strength of the connection, (b) the bearing strength of the connection, and (c) the tensile strength of the net section of the plate at each row of rivets.

Since the load is concentric, assume that the load transmitted through each rivet is $\frac{1}{9}P$. As plate A, Fig. 55, deflects, it bears against the upper half of each rivet. Consequently, the reaction of the rivet on plate A is exerted *above* the horizontal diametral plane of the rivet.

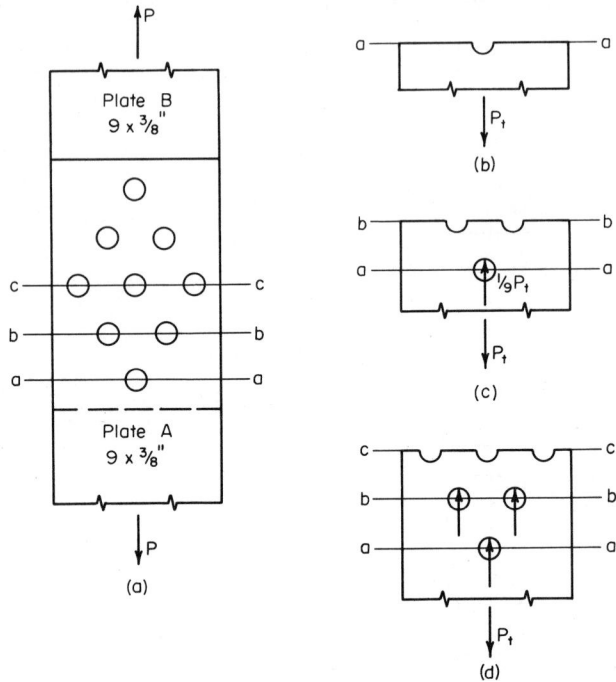

Fig. 55

Computing the capacity of the joint in shear and in bearing: $P_{ss} = 9(6{,}630) = 59{,}700$ lb; $P_b = 9(36{,}380)(0.375) = 122{,}800$ lb.

2. Compute the tensile capacity of the plate

The tensile capacity P_t lb of plate A, Fig. 55, is required. In structural fabrication, rivet holes are usually punched $\frac{1}{16}$ in. larger than the rivet diameter. However, to allow for damage to the adjacent metal caused by punching, the *effective* diameter of the hole is considered to be $\frac{1}{8}$ in. larger than the rivet diameter.

Refer to Fig. 55b, c, and d. Equate the tensile stress at each row of rivets to 22,000 psi to obtain P_t. Thus, at aa, residual tension $= P_t$; net area $= (9-0.875)(0.375) = 3.05$ sq in. The stress $s = P_t/3.05 = 22{,}000$ psi; $P_t = 67{,}100$ lb.

At bb, residual tension $= \frac{8}{9}P_t$; net area $= (9-1.75)(0.375) = 2.72$ sq in.; $s = \frac{8}{9}P_t/2.72 = 22{,}000$; $P_t = 67{,}300$ lb.

At cc, residual tension $= \frac{2}{3}P_t$; net area $= (9-2.625)(0.375) = 2.39$ sq in.; $s = \frac{2}{3}P_t/2.39 = 22{,}000$; $P_t = 78{,}900$ lb.

3. Select the lowest of the five computed values as the allowable load

Thus, $P = 59{,}700$ lb.

DESIGN OF A BUTT SPLICE

A tension member in the form of a $10 \times \frac{1}{2}$ in. steel plate is to be spliced with $\frac{7}{8}$ in. rivets. Design a butt splice for the maximum load the member may carry.

Calculation Procedure:

1. Establish the design load

In a butt splice, the load is transmitted from one member to another through two auxiliary plates called *cover*, *strap*, or *splice* plates. The rivets are therefore in double shear.

Establish the design load, P lb, by computing the allowable load at a cross section having one rivet hole. Thus: net area $= (10 - 1)(0.5) = 4.5$ sq in. Then, $P = 4.5(22,000) = 99,000$ lb.

2. Determine the number of rivets required

Applying the values of rivet capacity found in an earlier Calculation Procedure in this section of the Handbook, determine the number of rivets required. Thus, since the rivets are in double shear: $R_{ds} = 18,040$ lb; $R_b = 42,440\,(0.5) = 21,220$ lb; then, $99,000/18,040 = 5.5$ rivets; use the next largest whole number, or 6 rivets.

3. Select a trial pattern for the rivets; investigate the tensile stress

Conduct this investigation of the tensile stress in the main plate at each row of rivets.

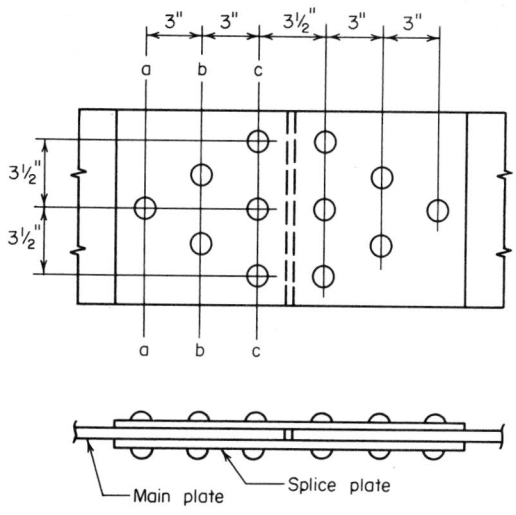

Fig. 56

The trial pattern is shown in Fig. 56. The rivet spacing satisfies the requirements of the AISC *Specification*. Record the calculations as shown below.

Section	Residual tension in main plate, lb	÷	Net area sq in.	=	Stress, psi
aa	99,000		4.5		22,000
bb	82,500		4.0		20,600
cc	49,500		3.5		14,100

Study of the above computations shows that the rivet pattern is satisfactory.

4. Design the splice plates

To the left of the centerline, each splice plate bears against the *left* half of the rivet. Therefore, the entire load has been transmitted to the splice plates at *cc*, which is the critical section. Thus: tension in splice plate $= \frac{1}{2}(99{,}000) = 49{,}500$ lb; plate thickness required $= 49{,}500/22{,}000(7) = 0.321$ in. Make the splice plates $10 \times \frac{3}{8}$ in.

DESIGN OF A PIPE JOINT

A steel pipe 5 ft 6 in. in diameter must withstand a fluid pressure of 225 psi. Design the pipe and the longitudinal lap splice using $\frac{3}{4}$ in. rivets.

Calculation Procedure:

1. Evaluate the hoop tension in the pipe

Let L denote the length, Fig. 57, of the *repeating group* of rivets. In the present instance, this equals the rivet pitch. In Fig. 57, let T denote the hoop tension, lb, in

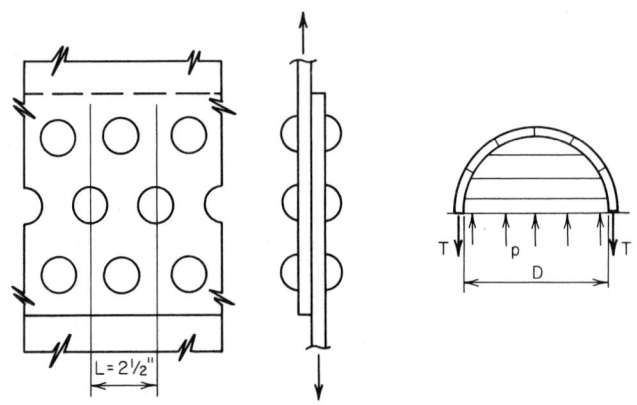

(a) Longitudinal pipe joint

(b) Free-body diagram of upper half of pipe and contents

Fig. 57

the distance L. Evaluate the tension using $T = pDL/2$, where $p =$ internal pressure, psi; $D =$ inside diameter of pipe, in.; $L =$ length considered, in. Thus, $T = 225(66)L/2 = 7{,}425L$.

2. Determine the required number of rows of rivets

Adopt, tentatively, the minimum allowable pitch, which is 2 in. for $\frac{3}{4}$ in. rivets. Then establish a feasible rivet pitch. From an earlier Calculation Procedure in this Section, $R_{ss} = 6{,}630$ lb. Then, $T = 7{,}425(2) = 6{,}630n$; $n = 2.24$. Use the next largest whole number of rows, or three rows of rivets. Also, $L_{max} = 3(6{,}630)/7{,}425 = 2.68$ in. Use a $2\frac{1}{2}$ in. pitch, as shown in Fig. 57a.

3. Determine the plate thickness

Establish the thickness t in. of the steel plates by equating the stress on the net section to its allowable value. Since the holes will be drilled, take $\frac{13}{16}$ in. as their diameter. Then: $T = 22{,}000t(2.5 - 0.81) = 7{,}425(2.5)$; $t = 0.50$ in.; use $\frac{1}{2}$ in. plates. Also, $R_b = 36{,}380(0.5) > 6{,}630$ lb. The rivet capacity is therefore limited by shear, as assumed.

MOMENT ON RIVETED CONNECTION

The channel in Fig. 58a is connected to its supporting column with $\frac{3}{4}$ in. rivets and resists the couple indicated. Compute the shearing stress in each rivet.

Calculation Procedure:

1. Compute the polar moment of inertia of the rivet group

The moment causes the channel, Fig. 58, to rotate about the centroid of the rivet. group and thereby exert a tangential thrust on each rivet. This thrust is directly proportional to the radial distance to the center of the rivet.

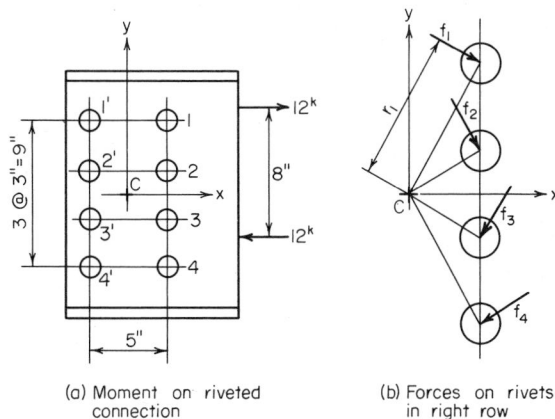

(a) Moment on riveted connection

(b) Forces on rivets in right row

Fig. 58

Establish coordinate axes through the centroid of the rivet group. Compute the polar moment of inertia of the group with respect to an axis through its centroid, taking the cross-sectional area of a rivet as unity. Thus, $J = \Sigma(x^2 + y^2) = 8(2.5)^2 + 4(1.5)^2 + 4(4.5)^2 = 140$ in.^{2}

2. Compute the radial distance to each rivet

Using the right-angle relationship, $r_1 = r_4 = (2.5^2 + 4.5^2)^{0.5} = 5.15$ in.; $r_2 = r_3 = (2.5^2 + 1.5^2)^{0.5} = 2.92$ in.

3. Compute the tangential thrust on each rivet

Use the relation $f = Mr/J$. Since $M = 12,000(8) = 96,000$ lb-in., $f_1 = f_4 = 96,000(5.15)/140 = 3,530$ lb; and $f_2 = f_3 = 96,000(2.92)/140 = 2,000$ lb. The directions are shown in Fig. 58b.

4. Compute the shearing stress

Using $s = P/A$, $s_1 = s_4 = 3,530/0.442 = 7,990$ psi; also, $s_2 = s_3 = 2,000/0.442 = 4,520$ psi.

5. Check the rivet forces

Check the rivet forces by summing their moments with respect to an axis through the centroid. Thus: $M_1 = M_4 = 3,530(5.15) = 18,180$ in.-lb; $M_2 = M_3 = 2,000(2.92) = 5,840$ in.-lb. Then, $\Sigma M = 4(18,180) + 4(5,840) = 96,080$ in.-lb.

ECCENTRIC LOAD ON RIVETED CONNECTION

Calculate the maximum force exerted on a rivet in the connection shown in Fig. 59a.

Calculation Procedure:

1. Compute the effective eccentricity

To account implicitly for secondary effects associated with an eccentrically loaded connection, the AISC *Manual* recommends replacing the true eccentricity with an *effective* eccentricity.

To compute the effective eccentricity, use $e_e = e_a - (1+n)/2$, where e_e = effective eccentricity, in.; e_a = actual eccentricity of the load, in.; n = number of rivets in a vertical row. Substituting, $e_e = 8 - (1+3)/2 = 6$ in.

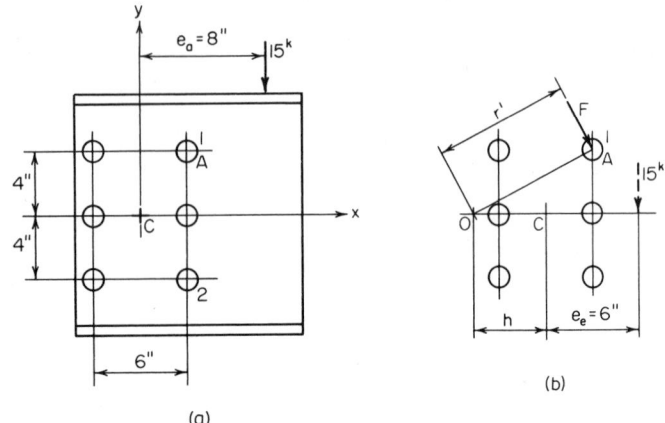

(a)

(b)

Fig. 59

2. Replace the eccentric load with an equivalent system

The equivalent system is comprised of a concentric load P lb and a clockwise moment M in.-lb. Thus, $P = 15,000$ lb, $M = 15,000(6) = 90,000$ in.-lb.

3. Compute the polar moment of inertia of the rivet group

Compute the polar moment of inertia of the rivet group with respect to an axis through its centroid. Thus, $J = \Sigma(x^2 + y^2) = 6(3)^2 + 4(4)^2 = 118$ in.^{2}.

4. Resolve the tangential thrust on each rivet into its horizontal and vertical components

Resolve the tangential thrust f lb on each rivet caused by the moment into its horizontal and vertical components f_x and f_y, respectively. These forces are as follows: $f_x = My/J$ and $f_y = Mx/J$. Computing these forces for rivets 1 and 2, Fig. 59: $f_x = [90,000(4)]/118 = 3,050$ lb; $f_y = [90,000(3)]/118 = 2,290$ lb.

5. Compute the thrust on each rivet caused by the concentric load

This thrust is $f_y' = 15,000/6 = 2,500$ lb.

6. Combine the foregoing results to obtain the total force on the rivets being considered

The total force F lb on rivets 1 and 2 is desired. Thus, $F_x = f_x = 3,050$ lb; $F_y = f_y + f_y' = 2,290 + 2,500 = 4,790$ lb. Then, $F = [(3,050)^2 + (4,790)^2]^{0.5} = 5,680$ lb.

The above six steps comprise Method 1. A second way of solving this problem, Method 2, is presented below.

The total force on each rivet may also be found by locating the *instantaneous center of rotation* associated with this eccentric load and treating the connection as if it were subjected solely to a moment, Fig. 59b.

7. Locate the instantaneous center of rotation

To locate this center, apply the relation $h = J/e_e N$, where N = total number of rivets; other relations as given earlier. Then, $h = 118/6(6) = 3.28$ in.

8. Compute the force on the rivets

Considering rivets 1 and 2, use the equation $F = Mr'/J$, where r' = distance from the instantaneous center of rotation O to the center of the given rivet, in. For rivets 1 and 2,

$r' = 7.45$ in. Then, $F = 90,000(7.45)/118 = 5,680$ lb. The force on rivet 1 has an action line normal to the radius OA.

DESIGN OF A WELDED LAP JOINT

The 5 in. leg of a $5 \times 3 \times \frac{3}{8}$ in. angle is to be welded to a gusset plate, as shown in Fig. 60. The member will be subjected to repeated variation in stress. Design a suitable joint.

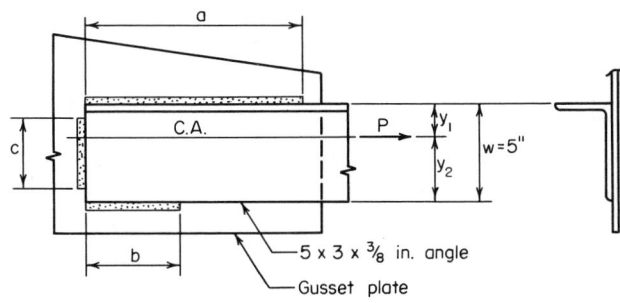

Fig. 60

Calculation Procedure:

1. Determine the properties of the angle

In accordance with the AISC *Specification*, arrange the weld to have its centroidal axis coincide with that of the member. Refer to the AISC *Manual* to obtain the properties of the angle. Thus: $A = 2.86$ sq in.; $y_1 = 1.70$ in.; $y_2 = 5.00 - 1.70 = 3.30$ in.

2. Compute the design load and required weld length

The design load P lb. $= As = 2.86(22,000) = 62,920$ lb. The AISC *Specification* restricts the weld size to $\frac{5}{16}$ in. Hence, the weld capacity, pli $= 5(600) = 3,000$ pli; $L =$ weld length, in. $= P/$capacity, pli $= 62,920/3,000 = 20.97$ in.

3. Compute the joint dimensions

In Fig. 60, set $c = 5$ in. and compute a and b by applying the following equations: $a = (Ly_2/w) - c/2$; $b = (Ly_1/w) - c/2$. Thus, $a = (20.97 \times 3.30)/5 - \frac{5}{2} = 11.34$ in.; $b = (20.97 \times 1.70)/5 - \frac{5}{2} = 4.63$ in. Make $a = 11.5$ in. and $b = 5$ in.

ECCENTRIC LOAD ON A WELDED CONNECTION

The bracket in Fig. 61 is connected to its support with a $\frac{1}{4}$-in. fillet weld. Determine the maximum stress in the weld.

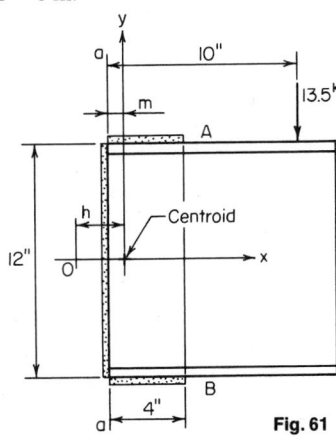

Fig. 61

Calculation Procedure:

1. Locate the centroid of the weld group

Refer to the previous eccentric-load Calculation Procedure. This situation is analogous to that. Determine the stress by locating the instantaneous center of rotation. The maximum stress occurs at A and B, Fig. 61.

Considering the weld as concentrated along the edge of the supported member, locate the centroid of the weld group by taking moments with respect to line aa, Fig. 61. Thus: $m = 2(4)(2)/(12+2\times4) = 0.8$ in.

2. *Replace the eccentric load with an equivalent concentric load and moment*

Thus: $P = 13,500$ lb; $M = 124,200$ in.-lb.

3. *Compute the polar moment of inertia of the weld group*

This moment should be computed with respect to an axis through the centroid of the weld group. Thus: $I_x = (1/12)(12)^3 + 2(4)(6)^2 = 432$ in.^{3}; $I_y = 12(0.8)^2 + 2(1/12)(4)^3 + 2(4)(2-0.8)^2 = 29.9$ in.^{3}. Then, $J = I_x + I_y = 461.9$ in.^{3}.

4. *Locate the instantaneous center of rotation O*

This center is associated with this eccentric load by applying the equation $h = J/eL$, where e = eccentricity of load, in. and L = total length of weld, in. Thus, $e = 10 - 0.8 = 9.2$ in.; $L = 12 + 2(4) = 20$ in.; then, $h = 461.9/[9.2(20)] = 2.51$ in.

5. *Compute the force on the weld*

Use the equation $F = Mr'/J$, pli, where r' = distance from the instantaneous center of rotation to the given point, in. At A and B, $r' = 8.28$ in.; then, $F = [124,200(8.28)]/461.9 = 2,230$ pli.

6. *Calculate the corresponding stress on the throat*

Thus, $s = P/A = 2,230/[0.707(0.25)] = 12,600$ psi, where the value 0.707 is the sine of 45°, the throat angle.

Steel Beams and Plate Girders

In the following Calculation Procedures the design of steel members will be executed in accordance with the *Specification for the Design, Fabrication and Erection of Structural Steel for Buildings* of the American Institute of Steel Construction. This specification is presented in the AISC *Manual of Steel Construction*.

Most allowable stresses are functions of the yield-point stress, denoted as F_y in the *Manual*. The Appendix of the *Specification* presents the allowable stresses associated with each grade of structural steel together with tables intended to expedite the design. The *Commentary* in the *Specification* explains the structural theory underlying the *Specification*.

Unless otherwise noted, the structural members considered here are understood to be made of ASTM A36 steel, having a yield-point stress of 36,000 psi.

The notational system used conforms with that adopted earlier but it is augmented to include the following: A_f = area of flange, sq in.; A_w = area of web, sq in.; b_f = width of flange, in.; d = depth of section, in.; d_w = depth of web, in.; t_f = thickness of flange, in.; t_w = thickness of web, in.; L' = unbraced length of compression flange, in.; f_y = yield-point stress, psi.

MOST ECONOMIC SECTION FOR A BEAM WITH A CONTINUOUS LATERAL SUPPORT UNDER A UNIFORM LOAD

A beam on a simple span of 30 ft carries a uniform superimposed load of 1,650 plf. The compression flange is laterally supported along its entire length. Select the most economic section.

Calculation Procedure:

1. *Compute the maximum bending moment and the required section modulus*

Assume that the beam weighs 50 plf and satisfies the requirements of a compact section as set forth in the *Specification*.

The maximum bending moment is $M = (1/8)wL^2 = (1/8)(1,700)(30)^2(12) = 2,295,000$ in.-lb.

Referring to the *Specification* shows that the allowable bending stress is 24,000 psi. Then, $S = M/f = 2,295,000/24,000 = 95.6$ in.[3].

2.   *Select the most economic section*

Refer to the AISC *Manual* and select the most economic section. Use 18WF55; $S = 98.2$ in.[3]; section compact. The disparity between the assumed and actual beam weight is negligible.

A second method for making this selection is shown below.

3.   *Calculate the total load on the member*

Thus, the total load $= W = 30(1,700) = 51,000$ lb.

4.   *Select the most economic section*

Refer to the tables of allowable uniform loads in the *Manual* and select the most economic section. Thus: use 18WF55; $W_{allow} = 52,000$ lb. The capacity of the beam is therefore slightly greater than required.

MOST ECONOMIC SECTION FOR A BEAM WITH INTERMITTENT LATERAL SUPPORT UNDER UNIFORM LOAD

A beam on a simple span of 25 ft carries a uniformly distributed load, including the estimated weight of the beam, of 45 kips. The member is laterally supported at 5-ft intervals. Select the most economic member (*a*) using A36 steel; (*b*) using A242 steel, having a yield-point stress of 50,000 psi when the thickness of the metal is $\frac{3}{4}$ in. or less.

Calculation Procedure:

1.   *Using the AISC allowable-load tables, select the most economic member made of A36 steel*

After a trial section has been selected, it is necessary to compare the unbraced length L' of the compression flange with the properties L_c and L_u of that section in order to establish the allowable bending stress. The variables are defined thus: $L_c =$ maximum unbraced length of the compression flange if the allowable bending stress $= 0.66f_y$, measured in ft; $L_u =$ maximum unbraced length of the compression flange, ft, if the allowable bending stress is to equal $0.60f_y$.

The values of L_c and L_u associated with each rolled section made of the indicated grade of steel are recorded in the allowable-uniform-load tables of the AISC *Manual*. The L_c value is established by applying the definition of a *laterally supported* member as presented in the *Specification*. The value of L_u is established by applying a formula given in the *Specification*.

There are four conditions relating to the allowable stress, as follows.

Condition	Allowable stress, psi
Compact section; $L' \leq L_c$	$0.66f_y$
Compact section; $L_c < L' \leq L_u$	$0.60f_y$
Noncompact section; $L' \leq L_u$	$0.60f_y$
$L' > L_u$	Apply the *Specification* formula —use the larger value obtained when the two formulas given are applied

The values of allowable uniform load given in the AISC *Manual* apply to beams of A36 steel satisfying the first or third condition above, depending on whether the section is compact or noncompact.

Referring to the table in the *Manual*, we see that the most economic section made of

A36 steel is 16WF45; $W_{allow} = 46$ kips, where W_{allow} = allowable load on the beam, kips. Also, $L_c = 7.6 > 5$. Hence, the beam is acceptable.

2. Compute the equivalent load for a member of A242 steel

To apply the AISC *Manual* tables to choose a member of A242 steel, assume that the shape selected will be compact. Transform the actual load to an equivalent load by applying the conversion factor 1.38 — i.e., the ratio of the allowable stresses. The conversion factors are recorded in the *Manual* tables. Thus, equivalent load = 45/1.38 = 32.6 kips.

3. Determine the lightest satisfactory section

Enter the *Manual* allowable-load table with the load value computed in Step 2 and select the lightest section that appears to be satisfactory. Try 16WF36; $W_{allow} = 36$ kips. However, this section is noncompact in A242 steel, and the equivalent load of 32.6 kips is not valid for this section.

4. Revise the equivalent load

To determine whether the 16WF36 will suffice, revise the equivalent load. Check the L_u value of this section in A242 steel. Then: equivalent load = 45/1.25 = 36 kips, $L_u = 6.3 > 5$ ft; use 16WF36.

5. Verify the second part of the design

To verify the second part of the design, calculate the bending stress in the 16WF36, using $S = 56.3$ in.^{3} from the *Manual*. Thus: $M = (1/8)WL = (1/8)(45,000)(25)(12) = 1,688,000$ in.-lb.; $f = M/S = 1,688,000/56.3 = 30,000$ psi. This stress is acceptable.

DESIGN OF A BEAM WITH REDUCED ALLOWABLE STRESS

The compression flange of the beam in Fig. 62a will be braced only at points A, B, C, D, and E. Using AISC data, a designer has selected a 21WF55 section for the beam. Verify the design.

Calculation Procedure:

1. Calculate the reactions; construct the shear and bending-moment diagrams

The results of this step are shown in Fig. 62.

2. Record the properties of the selected section

Using the AISC *Manual*, record the following properties of the 21WF55 section: $S = 109.7$ in.^{3}; $I_y = 44.0$ in.^{4}; $b_f = 8.215$ in.; $t_f = 0.522$ in.; $d = 20.80$ in.; $t_w = 0.375$ in.; $d/A_f = 4.85$/in.; $L_c = 8.9$ ft; $L_u = 9.4$ ft.

Since $L' > L_u$, the allowable stress must be reduced in the manner prescribed in the *Manual*.

3. Calculate the radius of gyration

Calculate the radius of gyration with respect to the y axis of a T section comprising the compression flange and one-sixth the web, neglecting the area of the fillets. Referring to Fig. 63, $A_f = 8.215(0.522) = 4.29$ sq in.; $(1/6)A_w = (1/6)(19.76)(0.375) = 1.24$; $A_T = 5.53$ sq in.; $I_T = 0.5I_y$ of the section = 22.0 in.^{4}; $r = [22.0/5.53]^{0.5} = 1.99$ in.

4. Calculate the allowable stress in each interval between lateral supports

By applying the provisions of the *Manual*, calculate the allowable stress in each interval between lateral supports, and compare this with the actual stress. For A36 steel, the *Manual* Formula (4) reduces to $f_1 = 22,000 - 0.679(L'/r)^2/C_b$ psi. By *Manual* Formula (5), $f_2 = 12,000,000/(L'd/A_f)$ psi. Set the allowable stress equal to the greater of these values.

For interval AB: $L' = 8$ ft $< L_c$; $\therefore f_{allow} = 24,000$ psi; $f_{max} = 148,000(12)/109.7 = 16,200$ psi — this is acceptable.

For interval BC: $L'/r = 15(12)/1.99 = 90.5$; $M_1/M_2 = 95/-148 = -0.642$; $C_b = 1.75 - 1.05(-0.642) + 0.3(-0.642)^2 = 2.55$; $\therefore$ set $C_b = 2.3$; $f_1 = 22,000 - 0.679(90.5)^2/2.3 =$

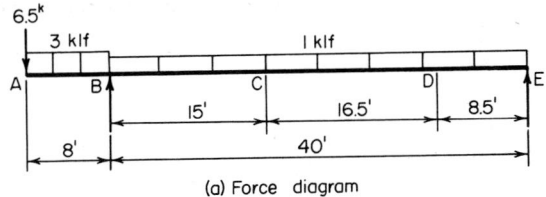

(a) Force diagram

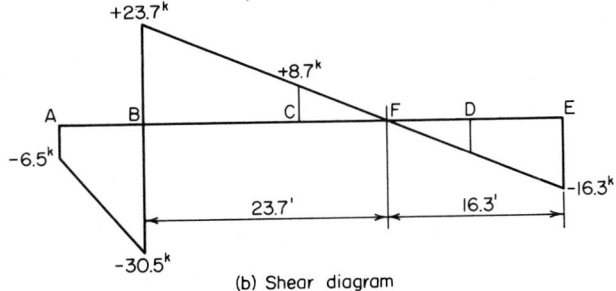

(b) Shear diagram

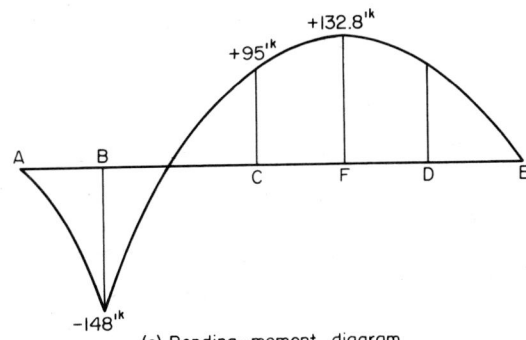

(c) Bending-moment diagram

Fig. 62

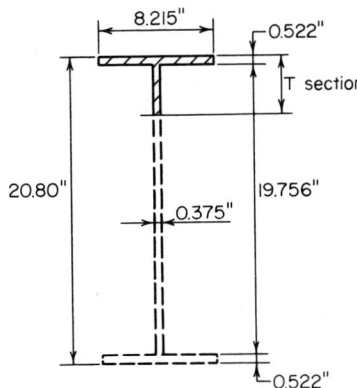

Fig. 63 Dimensions of 21WF55.

19,600 psi; $f_2 = 12,000,000/15(12)(4.85) = 13,700$ psi; $f_{max} = 16,200 < 19,600$ psi. This is acceptable.

Interval CD: Since the maximum moment occurs within the interval rather than at a boundary section, $C_b = 1$; $L'/r = 16.5(12)/1.99 = 99.5$; $f_1 = 22,000 - 0.679(99.5)^2 = 15,300$ psi; $f_2 = 12,000,000/16.5(12)(4.85) = 12,500$ psi; $f_{max} = 132,800(12)/109.7 = 14,500 < 15,300$ psi. This stress is acceptable.

Interval DE: The allowable stress is 24,000 psi, and the actual stress is considerably below this value. The 21WF55 is therefore satisfactory. Where deflection is the criterion, the member should be checked using the *Specification*.

DESIGN OF A COVER-PLATED BEAM

Following the fabrication of an 18WF60 beam, a revision was made in the architectural plans, and the member must now be designed to support the loads shown in Fig. 64a. Cover plates are to be welded to both flanges to develop the required strength. Design

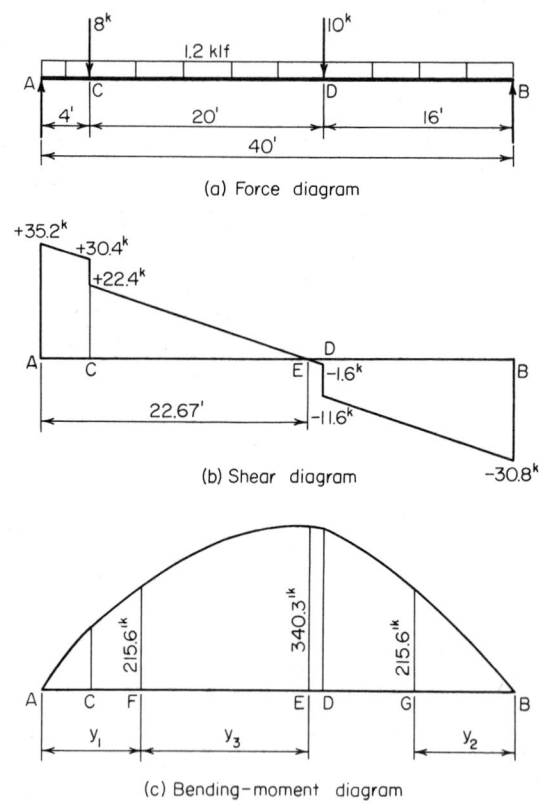

(a) Force diagram

(b) Shear diagram

(c) Bending-moment diagram

Fig. 64

these plates and their connection to the WF shape, using fillet welds of A233 Class E60 series electrodes. The member has continuous lateral support.

Calculation Procedure:

1. Construct the shear and bending-moment diagrams

These are shown in Fig. 64. Also, $M_E = 340.3$ ft-kips.

2. Calculate the required section modulus, assuming the built-up section will be compact

The section modulus $S = M/f = 340.3(12)/24 = 170.2$ in.^{3}.

3. Record the properties of the beam section

Refer to the AISC *Manual* and record the following properties for the 18WF60: $d = 18.25$ in.; $b_f = 7.56$ in.; $t_f = 0.695$ in.; $I = 984$ in.^{4}; $S = 107.8$ in.^{3}.

4. Select a trial section

Apply the approximation $A = 1.05(S - S_{WF})/d_{WF}$, where $A =$ area of one cover plate, sq in.; $S =$ section modulus required, in.^{3}; $S_{WF} =$ section modulus of wide-flange shape, in.^{3}; $d_{WF} =$ depth of wide-flange shape, in. Then, $A = [1.05(170.2 - 107.8)]/18.25 = 3.59$ sq in.

Try $10 \times \frac{3}{8}$ in. plates with $A = 3.75$ sq in. Since the beam flange is 7.5 in. wide, ample space is available to accommodate the welds.

5. Ascertain whether the assumed size of the cover plates satisfies the AISC Specification

Using the appropriate AISC *Manual* section, $7.56/0.375 = 20.2 < 32$, which is acceptable; $\frac{1}{2}(10 - 7.56)/0.375 = 3.25 < 16$, which is acceptable.

6. Test the adequacy of the trial section

Calculate the section modulus of the trial section. Referring to Fig. 65a, $I = 984 + 2(3.75)(9.31)^2 = 1,634$ in.^{4}; $S = I/c = 1,634/9.5 = 172.0$ in.^{3}. The reinforced section is therefore satisfactory.

7. Locate the points at which the cover plates are not needed

To locate the points at which the cover plates may theoretically be dispensed with, calculate the moment capacity of the wide-flange shape alone. Thus, $M = fS = [24(107.8)]/12 = 215.6$ ft-kips.

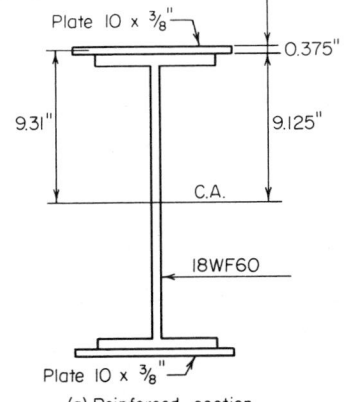

Plate 10 x $\frac{3}{8}$"

0.375"

9.31"

9.125"

C.A.

18WF60

Plate 10 x $\frac{3}{8}$"

(a) Reinforced section

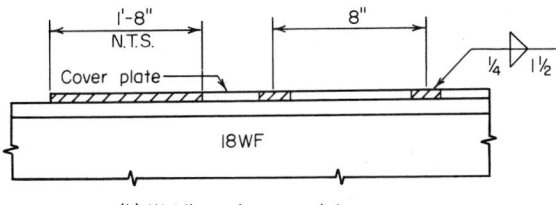

1'-8"

N.T.S.

8"

Cover plate

$\frac{1}{4}$ ⟍ $1\frac{1}{2}$

18WF

(b) Welding of cover plates

Fig. 65

8.   Locate the points at which the computed moment occurs

These points are F and G, Fig. 64. Thus, $M_F = 35.2y_1 - 8(y_1 - 4) - \frac{1}{2}(1.2y_1^2) = 215.6$; $y_1 = 8.25$ ft; $M_G = 30.8y_2 - \frac{1}{2}(1.2y_2^2) = 215.6$; $y_2 = 8.36$ ft.

Alternatively, locate F by considering the area under the shear diagram between E and F. Thus: $M_F = 340.3 - \frac{1}{2}(1.2y_3^2) = 215.6$; $y_3 = 14.42$ ft; $y_1 = 22.67 - 14.42 = 8.25$ ft.

For symmetry, center the cover plates about mid-span, placing the theoretical cutoff points at 8 ft 3 in. from each support.

9.   Calculate the axial force in the cover plate

Calculate the axial force P lb in the cover plate at its end by computing the mean bending stress. Determine the length of fillet weld required to transmit this force to the wide-flange shape. Thus: $f_{mean} = My/I = [215,600(12)(9.31)]/1,634 = 14,740$ psi. Then, $P = Af_{mean} = 3.75(14,740) = 55,280$ lb. Use a $\frac{1}{4}$ in. fillet weld, which satisfies the requirements of the *Specification*. The capacity of the weld = 4(600) = 2,400 pli. Then the length L required for this weld is $L = 55,280/2,400 = 23.0$ in.

10.   Extend the cover plates

In accordance with the *Specification*, extend the cover plates 20 in. beyond the theoretical cutoff point at each end and supply a continuous $\frac{1}{4}$ in. fillet weld along both edges in this extension. This requirement yields 40 in. of weld as compared with the 23 in. needed to develop the plate.

11.   Calculate the horizontal shear flow at the inner surface of the cover plate

Choose F or G, whichever is larger. Design the intermittent fillet weld to resist this shear flow. Thus: $V_F = 35.2 - 8 - 1.2(8.25) = 17.3$ kips; $V_G = -30.8 + 1.2(8.36) = -20.8$ kips. Then, $q = VQ/I = [20,800(3.75)(9.31)]/1,634 = 444$ pli.

The *Specification* calls for a minimum weld length of 1.5 in. Let s denote the center-to-center spacing as governed by shear. Then, $s = [2(1.5)(2,400)]/444 = 16.2$ in. However, the *Specification* imposes additional restrictions on the weld spacing. To preclude the possibility of error in fabrication, provide an identical spacing at the top and bottom. Thus, $s_{max} = 21(0.375) = 7.9$ in. Therefore, use a $\frac{1}{4}$-in. fillet weld, 1.5 in. long, 8 in. on centers, as shown in Fig. 65b.

DESIGN OF A CONTINUOUS BEAM

The beam in Fig. 66a is continuous from A to D and is laterally supported at 5-ft intervals. Design the member.

Calculation Procedure:

1.   Find the bending moments at the interior supports; calculate the reactions and construct shear and bending-moment diagrams

The maximum moments are $+101.7$ ft-kips and -130.2 ft-kips.

2.   Calculate the modified maximum moments

Calculate these moments in the manner prescribed in the AISC *Specification*. The clause covering this calculation is based on the post-elastic behavior of a continuous beam. (Refer to a later Calculation Procedure for an analysis of this behavior.)

Modified maximum moments: $+101.7 + 0.1(0.5)(115.9 + 130.2) = +114.0$ ft-kips; $0.9(-130.2) = -117.2$ ft-kips; design moment = 117.2 ft-kips.

3.   Select the beam size

Thus, $S = M/f = [117.2(12)]/24 = 58.6$ in.^{3}. Use 16WF40 with $S = 64.4$ in.^{3}; $L_c = 7.6$ ft.

SHEARING STRESS IN A BEAM—EXACT METHOD

Calculate the maximum shearing stress in an 18WF55 beam at a section where the vertical shear is 70 kips.

Calculation Procedure:

1. Record the relevant properties of the member

The shearing stress is a maximum at the centroidal axis and is given by $v = VQ/It$. The statical moment of the area above this axis is found by applying the properties of

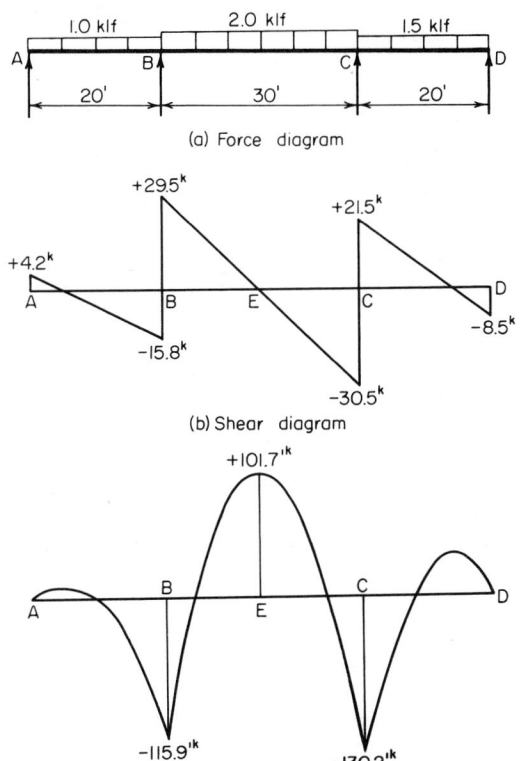

(a) Force diagram

(b) Shear diagram

(c) Bending–moment diagram

Fig. 66

the ST9WF27.5, which are presented in the AISC *Manual*. Note that the T section considered is one-half the wide-flange section being used. See Fig. 67.

The properties of these sections are: $I_{WF} = 890$ in.^{4}; $A_T = 8.10$ sq in.; $t_w = 0.39$ in.; $y_m = 9.06 - 2.16 = 6.90$ in.

2. Calculate the shearing stress at the centroidal axis

Substituting, $Q = 8.10(6.90) = 55.9$ in.^{3}; then $v = 70,000(55.9)/[890(0.39)] = 11,270$ psi.

SHEARING STRESS IN A BEAM — APPROXIMATE METHOD

Solve the previous Calculation Procedure using the approximate method of determining the shearing stress in a beam.

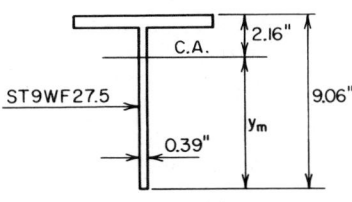

Fig. 67

Calculation Procedure:

1. Assume that the vertical shear is resisted solely by the web

Consider the web as extending the full depth of the section and the shearing stress as uniform across the web. Compare the results obtained by the exact and the approximate methods.

2. Compute the shear stress

Taking the depth of the web as 18.12 in., $v = 70,000/[18.12(0.39)] = 9,910$ psi. Thus, the ratio of the computed stresses is $11,270/9,910 = 1.14$.

Since the error inherent in the approximate method is not unduly large, this method is applied in assessing the shear capacity of a beam. The allowable shear V for each rolled section is recorded in the allowable-uniform-load tables of the AISC *Manual*.

The design of a rolled section is governed by the shearing stress only in those instances where the ratio of maximum shear to maximum moment is extraordinarily large. This condition exists in a heavily loaded short-span beam and a beam that carries a large concentrated load near its support.

MOMENT CAPACITY OF A WELDED PLATE GIRDER

A welded plate girder is composed of a $66 \times \frac{3}{8}$ in. web plate and two $20 \times \frac{3}{4}$ in. flange plates. The unbraced length of the compression flange is 18 ft. If $C_b = 1$, what bending moment can this member resist?

Calculation Procedure:

1. Compute the properties of the section

The tables in the AISC *Manual* are helpful in calculating the moment of inertia. Thus: $A_f = 15$ sq in.; $A_w = 24.75$ sq in.; $I = 42,400$ in.^{4}; $S = 1,256$ in.^{3}.

For the T section comprising the flange and one-sixth the web, $A = 15 + 4.13 = 19.13$ sq in.; then, $I = (1/12)(0.75)(20)^3 = 500$ in.^{4}; $r = (500/19.13)^{0.5} = 5.11$ in.; $L'/r = 18(12)/5.11 = 42.3$.

2. Ascertain if the member satisfies the AISC Specification

Let h denote the clear distance between flanges, in. Then: flange, $\frac{1}{2}(20)/0.75 = 13.3 < 16$ — this is acceptable; web, $h/t_w = 66/0.375 = 176 < 320$ — this is acceptable.

3. Compute the allowable bending stress

Use $f_1 = 22,000 - 0.679(L'/r)^2/C_b$, or $f_1 = 22,000 - 0.679(42.3)^2 = 20,800$ psi; $f_2 = 12,000,000/(L'd/A_f) = 12,000,000(15)/[18(12)(67.5)] = 12,300$ psi. Therefore, use 20,800 psi because it is the larger of the two stresses.

4. Reduce the allowable bending stress in accordance with the AISC Specification

Using the equation given in the *Manual*, $f_3 = 20,800\{1 - 0.005(24.75/15)[176 - 24,000/(20,800)^{0.5}]\} = 20,600$ psi.

5. Determine the allowable bending moment

Use $M = f_3 S = 20.6(1,256)/12 = 2,156$ ft-kips.

ANALYSIS OF A RIVETED PLATE GIRDER

A plate girder is composed of: one web plate $48 \times \frac{3}{8}$ in.; four flange angles $6 \times 4 \times \frac{3}{4}$ in.; two cover plates $14 \times \frac{1}{2}$ in. The flange angles are set 48.5 in. back to back with their 6-in. legs outstanding; they are connected to the web plate by $\frac{7}{8}$ in. rivets. If the member has continuous lateral support, what bending moment may be applied? What spacing of flange-to-web rivets is required in a panel where the vertical shear is 180 kips?

Calculation Procedure:

1. Obtain the properties of the angles from the AISC Manual

Record the angle dimensions as shown in Fig. 68.

2. Check the cover plates for compliance with the AISC Specification

The cover plates are found to comply with the pertinent sections of the *Specification*.

3. Compute the gross flange area and rivet-hole area

Ascertain whether the *Specification* requires a reduction in the flange area. Thus: gross flange area $= 2(6.94)+7.0 = 20.88$ sq in.; area of rivet holes $= 2(\frac{1}{2})(1)+4(3/4)(1) =$

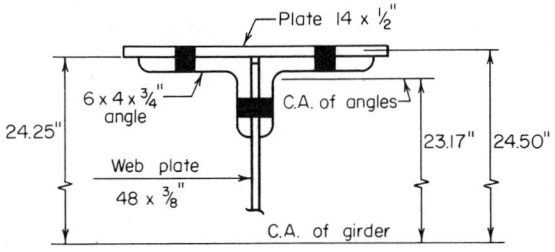

Fig. 68

4.00 sq in.; allowable area of holes $= 0.15(20.88) = 3.13$. The excess area $=$ hole area $-$ allowable area $= 4.00-3.13 = 0.87$ sq in. Consider that this excess area is removed from the outstanding legs of the angles, at both the top and the bottom.

4. Compute the moment of inertia of the net section

One web plate, I_0	3,456 in.^{4}
Four flange angles, I_0	35
$Ay^2 = 4(6.94)(23.17)^2$	14,900
Two cover plates:	
$Ay^2 = 2(7.0)(24.50)^2$	8,400
I of gross section	26,791
Deduct $2(0.87)(23.88)^2$ for excess area	991
I of net section	25,800 in.^{4}

5. Establish the allowable bending stress

Use the *Specification*. Thus: $h/t_w = (48.5-8)/0.375 < 24,000/(22,000)^{0.5}$; $\therefore$ use 22,000 psi. Also, $M = fI/c = [22(25,800)]/[24.75(12)] = 1,911$ ft-kips.

6. Calculate the horizontal shear flow to be resisted

Q of flange $= 13.88(23.17)+7.0(24.50)-0.87(23.88) = 472$ in.^{3}; $q = VQ/I = [180,000 (472)]/25,800 = 3,290$ pli.

From a previous Calculation Procedure, $R_{ds} = 18,040$ lb; $R_b = 42,440(0.375) = 15,900$ lb; $s = 15,900/3,290 = 4.8$ in., where $s =$ allowable rivet spacing, in. Therefore, use a $4\frac{3}{4}$-in. rivet pitch. This satisfies the requirements of the *Specification*.

NOTE: To determine the allowable rivet spacing, divide the horizontal shear flow into the rivet capacity.

DESIGN OF A WELDED PLATE GIRDER

A plate girder of welded construction is to support the loads shown in Fig. 69*a*. The distributed load will be applied to the top flange, thereby offering continuous lateral support. At its ends, the girder will bear on masonry buttresses. The total depth of the girder is restricted to approximately 70 in. Select the cross section, establish the spacing of the transverse stiffeners, and design both the intermediate stiffeners and the bearing stiffeners at the supports.

Calculation Procedure:

1. Construct the shear and bending-moment diagrams

These diagrams are shown in Fig. 69.

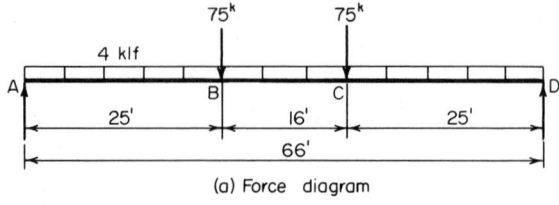

(a) Force diagram

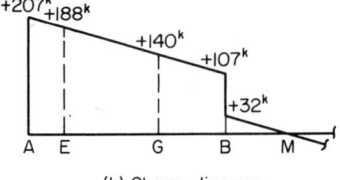

(b) Shear diagram

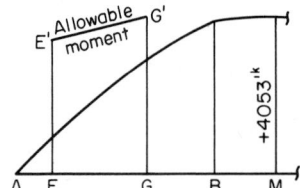

(c) Bending-moment diagram

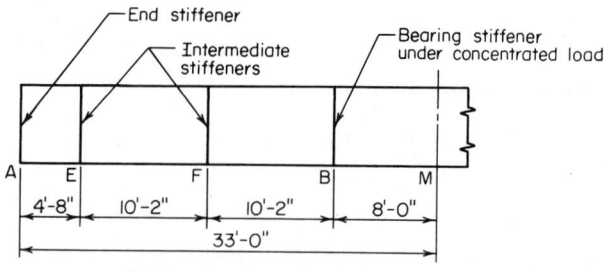

(d) Spacing of stiffeners

Fig. 69

2. Choose the web-plate dimensions

Since the total depth is limited to about 70 in., use a 68-in. deep web plate. Determine the plate thickness using the *Specification* limits, which are a slenderness ratio h/t_w of 320. However, if an allowable bending stress of 22,000 psi is to be maintained, the *Specification* imposes an upper limit of $24,000/(22,000)^{0.5} = 162$. Then, $t_w = h/162 = 68/162 = 0.42$ in.; use a $\frac{7}{16}$ in. plate. Hence, the area of the web $A_w = 29.75$ sq in.

3. Select the flange plates

Apply the approximation $A_f = Mc/2fy^2 - A_w/6$, where y = distance from the neutral axis to the centroidal axis of the flange, in.

Assume 1-in. flange plates. Then, $A_f = [4,053(12)(35)/[2(22)(34.5)^2] - 29.75/6 = 27.54$ sq in. Try $22 \times 1\frac{1}{4}$ in. plates with $A_f = 27.5$ sq in. The width-thickness ratio of projection $= 11/1.25 = 8.8 < 16$. This is acceptable.

Thus, the trial section will be: one web plate $68 \times \frac{7}{16}$ in.; two flange plates $22 \times 1\frac{1}{4}$ in.

4. Test the adequacy of the trial section

For this test, compute the maximum flexural and shearing stresses. Thus, $I = (1/12)(0.438)(68)^3 + 2(27.5)(34.63)^2 = 77,440$ in.^{3}; $f = Mc/I = 4,053(12)(35.25)/77,440 = 22.1$ ksi. This is acceptable. Also, $v = 207/29.75 = 6.96 < 14.5$ ksi. This is acceptable. Hence, the trial section is satisfactory.

5. Determine the distance of the stiffeners from the girder ends

Refer to Fig. 69d for the spacing of the intermediate stiffeners. Establish the length of the end panel AE. The *Specification* stipulates that the smaller dimension of the end panel shall not exceed $11,000(0.438)/(6,960)^{0.5} = 57.8 < 68$ in. Therefore, provide stiffeners at 56 in. from the ends.

6. Ascertain whether additional intermediate stiffeners are required

See if stiffeners are required in the interval EB by applying the *Specification* criteria.

Stiffeners are not required when $h/t_w < 260$ and the shearing stress within the panel is below the value given by either of two equations in the *Specification*, whichever equation applies. Thus: $EB = 396 - (56 + 96) = 244$ in.; $h/t_w = 68/0.438 = 155 < 260$; this is acceptable. Also, $a/h = 244/68 = 3.59$.

In lieu of solving either of the equations given in the *Specification*, enter the table of a/h, h/t_w values given in the AISC *Manual* to obtain the allowable shear stress. Thus, with $a/h > 3$ and $h/t_w = 155$, $v_{\text{allow}} = 3.45$ ksi from the table.

At E, $V = 207 - 4.67(4) = 188$ kips; $v = 188/29.75 = 6.32 > 3.45$ ksi; therefore, intermediate stiffeners are required in EB.

7. Provide stiffeners and investigate the suitability of their tentative spacing

Provide stiffeners at F, the center of EB. See if this spacing satisfies the *Specification*. Thus: $[260/(h/t_w)]^2 = (260/155)^2 = 2.81$; $a/h = 122/68 = 1.79 < 2.81$. This is acceptable. Entering the table referred to in Step 6 with $a/h = 1.79$ and $h/t_w = 155$, $v_{\text{allow}} = 7.85 > 6.32$. This is acceptable.

Before concluding that the stiffener spacing is satisfactory, it is necessary to investigate the combined shearing and bending stress and the bearing stress in interval EB.

8. Analyze the combination of shearing and bending stress

This analysis should be made throughout EB in the light of the *Specification* requirements. The net effect is to reduce the allowable bending moment whenever $V > 0.6V_{\text{allow}}$. Thus, $V_{\text{allow}} = 7.85(29.75) = 234$ kips; and $0.6(234) = 140$ kips.

In Fig. 69b, locate the boundary section G where $V = 140$ kips. The allowable moment must be reduced to the left of G. Thus, $AG = (207 - 140)/4 = 16.75$ ft; $M_G = 2,906$ ft-kips; $M_E = 922$ ft-kips. At G, $M_{\text{allow}} = 4,053$ ft-kips. At E, $f_{\text{allow}} = [0.825 - 0.375(188/234)](36) = 18.9$ ksi; $M_{\text{allow}} = 18.9(77,440)/[35.25(12)] = 3,460$ ft-kips.

In Fig. 69c, plot points E' and G' to represent the allowable moments and connect these points with a straight line. In all instances, $M < M_{\text{allow}}$.

9. Use an alternative procedure, if desired

As an alternative procedure in Step 8, establish the interval within which $M > 0.75M_{\text{allow}}$ and reduce the allowable shear in accordance with the equation given in the *Specification*.

10. Compare the bearing stress under the uniform load with the allowable stress

The allowable stress given in the *Specification* is: $f_{b,\text{allow}} = [5.5 + 4/(a/h)^2]10,000/(h/t_w)^2$ ksi or, for this girder, $f_{b,\text{allow}} = [5.5 + 4/1.79^2]10,000/155^2 = 2.81$ ksi. Then, $f_b = 4/[12(0.438)] = 0.76$ ksi. This is acceptable. The stiffener spacing in interval EB is therefore satisfactory in all respects.

11. *Investigate the need for transverse stiffeners in the center interval*

Considering the interval BC, $V = 32$ kips; $v = 1.08$ ksi; $a/h = 192/68 = 2.82 \approx [260/(h/t_w)]^2$.

The *Manual* table used in Step 6 shows that $v_{\text{allow}} > 1.08$ ksi; $f_{b,\text{allow}} = [5.5 + 4/2.82^2] \times 10{,}000/155^2 = 2.49 > 0.76$ ksi. This is acceptable. Since all requirements are satisfied, stiffeners are not needed in interval BC.

12. *Design the intermediate stiffeners in accordance with the* Specification

For the interval EB, the preceding calculations yield these values: $v = 6.32$ ksi; $v_{\text{allow}} = 7.85$ ksi. Enter the table mentioned in Step 6 with $a/h = 1.79$ and $h/t_w = 155$ to obtain the percentage of web area, shown in italics in the table. Thus, A_{st} required $= 0.0745(29.75)(6.32/7.85) = 1.78$ sq in. Try two $4 \times \frac{1}{4}$ in. plates; $A_{st} = 2.0$ sq in.; width-thickness ratio $= 4/0.25 = 16$. This is acceptable. Also, $(h/50)^4 = (68/50)^4 = 3.42$ in.^{4}; $I = (1/12)(0.25)(8.44)^3 = 12.52 > 3.42$ in.^{4}. This is acceptable.

The stiffeners must be in intimate contact with the compression flange, but they may terminate $1\frac{3}{4}$ in. from the tension flange. The connection of the stiffeners to the web must transmit the vertical shear specified in the *Specification*.

13. *Design the bearing stiffeners at the supports*

Use the directions given in the *Specification*. The stiffeners are considered to act in conjunction with the tributary portion of the web to form a column section, as shown in Fig. 70. Thus, area of web $= 5.25(0.438) = 2.30$ sq in. Assume an allowable stress of 20 ksi. Then, plate area required $= (207/20) - 2.30 = 8.05$ sq in.

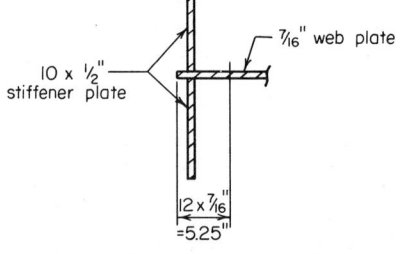

Try two plates $10 \times \frac{1}{2}$ in. and compute the column capacity of the section. Thus, $A = 2(10)(0.5) + 2.30 = 12.30$ sq in.; $I = (1/12)(0.5)(20.44)^3 = 356$ in.^{4}; $r = (356/12.30)^{0.5} = 5.38$ in.; $L/r = 0.75(68)/5.38 = 9.5$.

Enter the table of slenderness ratio and allowable stress in the *Manual* with the slenderness ratio of 9.5 and obtain an allowable stress of 21.2 ksi. Then, $f = 207/12.30 = 16.8 < 21.2$ ksi. This is acceptable.

Fig. 70   Effective column section.

Compute the bearing stress in the stiffeners. In computing the bearing area, assume that each stiffener will be clipped 1 in. to clear the flange-to-web welding. Then, $f = 207/[2(9)(0.5)] = 23$ ksi. The *Specification* provides an allowable stress of 33 ksi.

The $10 \times \frac{1}{2}$ in. stiffeners at the supports are therefore satisfactory with respect to both column action and bearing.

Steel Columns and Tension Members

The general remarks appearing at the opening of the previous part apply to this part as well.

A column is a compression member having a length that is very large in relation to its lateral dimensions. The *effective* length of a column is the distance between adjacent points of contraflexure in the buckled column or in the imaginary extension of the buckled column, as shown in Fig. 71. The column length is denoted by L, and the effective length by KL. Recommended design values of K are given in the AISC *Manual*.

The capacity of a column is a function of its effective length and the properties of its cross section. It therefore becomes necessary to formulate certain principles pertaining to the properties of an area.

Consider that the moment of inertia I of an area is evaluated with respect to a group of concurrent axes. There is a distinct value of I associated with each axis, as given by earlier equations in this section. The *major* axis is the one for which I is maximum; the *minor* axis is the one for which I is minimum. The major and minor axes are referred to collectively as the *principal* axes.

With reference to the equation given earlier, namely, $I_{x''} = I_{x'} \cos^2 \theta + I_{y'} \sin^2 \theta - P_{x'y'} \sin 2\theta$, the orientation of the principal axes relative to the given x' and y' axes is found by differentiating $I_{x''}$ with respect to θ, equating this derivative to zero, and solving for θ to obtain $\tan 2\theta = 2P_{x'y'}/(I_{y'} - I_{x'})$, Fig. 15.

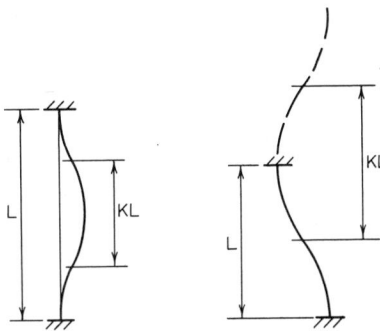

The following statements are corollaries of this equation:

1. The principal axes through a given point are mutually perpendicular, since the two values of θ that satisfy this equation differ by 90°.

2. The product of inertia of an area with respect to its principal axes is zero.

3. Conversely, if the product of inertia of an area with respect to two mutually perpendicular axes is zero, these are principal axes.

4. An axis of symmetry is a principal axis, for the product of inertia of the area with respect to this axis and one perpendicular thereto is zero.

Fig. 71   Effective column lengths.

Let A_1 and A_2 denote two areas, both of which have a radius of gyration r with respect to a given axis. The radius of gyration of their composite area is found in this manner: $I_c = I_1 + I_2 = A_1 r^2 + A_2 r^2 = (A_1 + A_2) r^2$. But $A_1 + A_2 = A_c$. Substituting, $I_c = A_c r^2$; therefore $r_c = r$.

This result illustrates the following principle: If the radii of gyration of several areas with respect to a given axis are all equal, the radius of gyration of their composite area equals that of the individual areas.

The equation $I_x = \Sigma I_0 + \Sigma A k^2$, when applied to a single area, becomes $I_x = I_0 + A k^2$. Then, $A r_x^2 = A r_0^2 + A k^2$, or $r_x = (r_0^2 + k^2)^{0.5}$. If the radius of gyration with respect to a centroidal axis is known, the radius of gyration with respect to an axis parallel thereto may be readily evaluated by applying this relationship.

The Euler equation for the strength of a slender column reveals that the member tends to buckle about the minor centroidal axis of its cross section. Consequently, all column-design equations, both those for slender members and those for intermediate-length members, relate the capacity of the column to its minimum radius of gyration. The first step in the investigation of a column therefore consists in identifying the minor centroidal axis and evaluating the corresponding radius of gyration.

CAPACITY OF A BUILT-UP COLUMN

A compression member consists of two 15-in. 40-lb channels laced together and spaced 10 in. back to back with flanges outstanding, as shown in Fig. 72. What axial load may this member carry if its effective length is 22 ft?

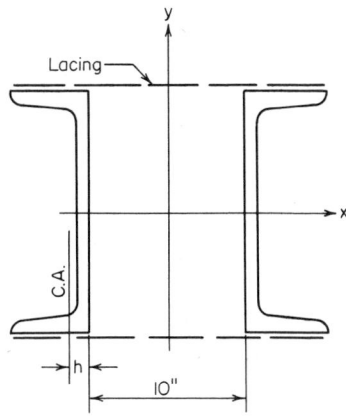

Fig. 72   Built-up column.

Calculation Procedure:

1.   *Record the properties of the individual channel*

Since x and y are axes of symmetry, they are the principal centroidal axes. However, it is not readily apparent which of these is the minor axis, and it is therefore necessary to calculate both r_x and r_y. The symbol r, without a subscript, will be used to denote the *minimum* radius of gyration, in in.

Using the AISC *Manual*, the channel properties are: $A = 11.70$ sq in.; $h = 0.78$ in.; $r_1 = 5.44$ in.; $r_2 = 0.89$ in.

2. Evaluate the minimum radius of gyration of the built-up section; determine the slenderness ratio

Thus, $r_x = 5.44$ in.; $r_y = (r_2^2 + 5.78^2)^{0.5} > 5.78$ in.; therefore, $r = 5.44$ in.; $KL/r = 22(12)/5.44 = 48.5$.

3. Determine the allowable stress in the column

Enter the *Manual* slenderness-ratio allowable-stress table with a slenderness ratio of 48.5 to obtain the allowable stress $f = 18.48$ ksi. Then, the column capacity $= P = Af = 2(11.70)(18.48) = 432$ kips.

CAPACITY OF A DOUBLE-ANGLE STAR STRUT

A star strut is composed of two $5 \times 5 \times 3/8$ in. angles intermittently connected by $\frac{3}{8}$ in. batten plates in both directions. Determine the capacity of the member for an effective length of 12 ft.

Calculation Procedure:

1. Identify the minor axis

Refer to Fig. 73a. Since p and q are axes of symmetry, they are the principal axes; p is manifestly the minor axis because the area lies closer to p than q.

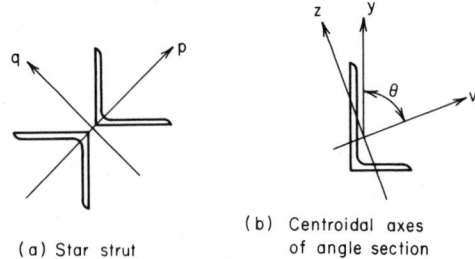

(a) Star strut

(b) Centroidal axes of angle section

Fig. 73

2. Determine r_v^2

Refer to Fig. 73b, where v is the major and z the minor axis of the angle section. Apply $I_{x'} = I_{x'} \cos^2 \theta + I_{y'} \sin^2 \theta - P_{x'y'} \sin 2\theta$, and set $P_{vz} = 0$ to obtain $r_y^2 = r_v^2 \cos^2 \theta + r_z^2 \sin^2 \theta$; therefore $r_v^2 = r_y^2 \sec^2 \theta - r_z^2 \tan^2 \theta$. For an equal-leg angle, $\theta = 45°$, and this equation reduces to $r_v^2 = 2r_y^2 - r_z^2$.

3. Record the member area and compute r_v

From the *Manual*, $A = 3.61$ sq in.; $r_y = 1.56$ in.; $r_z = 0.99$ in.; $r_v = (2 \times 1.56^2 - 0.99^2)^{0.5} = 1.97$ in.

4. Determine the minimum radius of gyration of the built-up section; compute the strut capacity

Thus, $r = r_p = 1.97$ in.; $KL/r = 12(12)/1.97 = 73$. From the *Manual*, $f = 16.12$ ksi. Then, $P = Af = 2(3.61)(16.12) = 116$ kips.

SECTION SELECTION FOR A COLUMN WITH TWO EFFECTIVE LENGTHS

A 30-ft-long column is to carry a 200-kip load. The column will be braced about both principal axes at top and bottom and braced about its minor axis at mid-height. Architectural details restrict the member to a nominal depth of 8 in. Select a section of A242 steel by consulting the allowable-load tables in the AISC *Manual* and then verify the design.

Calculation Procedure:

1. Select a column section

Refer to Fig. 74. The effective length with respect to the minor axis may be taken as 15 ft. Then, $K_xL = 30$ ft and $K_yL = 15$ ft.

The allowable column loads recorded in the *Manual* tables are calculated on the premise that the column tends to buckle about the minor axis. In the present instance, however, this premise is not necessarily valid. It is expedient for design purposes to conceive of a uniform-strength column; i.e., one for which K_x and K_y bear the same ratio as r_x and r_y, thereby endowing the column with an identical slenderness ratio with respect to the two principal axes.

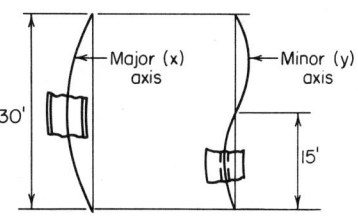

Fig. 74 Effective column lengths.

Select a column section on the basis of the K_yL value; record the value of r_x/r_y of this section. Using linear interpolation in the *Manual* table shows that an 8WF40 column has a capacity of 200 kips when $K_yL = 15.3$ ft; at the bottom of the table it is found that $r_x/r_y = 1.73$.

2. Compute the value of K_xL associated with a uniform-strength column and compare this with the actual value

Thus, $K_xL = 1.73(15.3) = 26.5 < 30$ ft. The section is therefore inadequate.

3. Try a specific column section of larger size

Trying 8WF48, the capacity = 200 kips when $K_yL = 17.7$ ft. For uniform strength, $K_xL = 1.74(17.7) = 30.8 > 30$ ft. The 8WF48 therefore appears to be satisfactory.

4. Verify the design

To verify the design, record the properties of this section and compute the slenderness ratios. For this grade of steel and thickness of member, the yield-point stress is 50 ksi, as given in the *Manual*. Thus, $A = 14.11$ sq in.; $r_x = 3.61$ in.; $r_y = 2.08$ in. Then, $K_xL/r_x = 30(12)/3.61 = 100$; $K_yL/r_y = 15(12)/2.08 = 87$.

5. Determine the allowable stress and member capacity

From the *Manual*, $f = 14.71$ ksi with a slenderness ratio of 100. Then, $P = 14.11 (14.71) = 208$ kips. Therefore, use 8WF48 because the capacity of the column exceeds the intended load.

STRESS IN COLUMN WITH PARTIAL RESTRAINT AGAINST ROTATION

The beams shown in Fig. 75a are rigidly connected to a 14WF95 column of 28-ft height that is pinned at its foundation. The column is held at its upper end by cross bracing

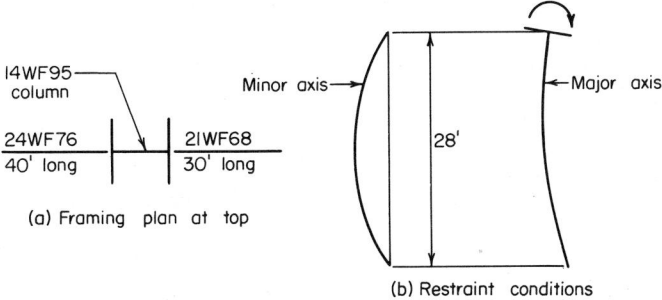

(a) Framing plan at top

(b) Restraint conditions

Fig. 75

lying in a plane normal to the web. Compute the allowable axial stress in the column in the absence of bending stress.

Calculation Procedure:

1. Draw schematic diagrams to indicate the restraint conditions
Show these conditions in Fig. 75b. The cross bracing prevents sidesway at the top solely with respect to the minor axis, and the rigid beam-to-column connections afford partial fixity with respect to the major axis.

2. Record the I_x values of the column and beams

Section	I_x, in.^{4}
14WF95	1,064
24WF76	2,096
21WF68	1,478

3. Calculate the rigidity of the column relative to that of the restraining members at top and bottom
Thus, $I_c/L_c = 1,064/28 = 38$. At the top, $\Sigma(I_g/L_g) = 2,096/40 + 1,478/30 = 101.7$. At the top, the rigidity $G_t = 38/101.7 = 0.37$.

In accordance with the instructions in the *Manual*, set the rigidity at the bottom $G_b = 10$.

4. Determine the value of K_x
Using the *Manual* alignment chart, determine that $K_x = 1.77$.

5. Compute the slenderness ratio with respect to both principal axes and find the allowable stress
Thus, $K_x L/r_x = 1.77(28)(12)/6.17 = 96.4$; $K_y L/r_y = 28(12)/3.71 = 90.6$.

Using the larger value of the slenderness ratio, find from the *Manual* the allowable axial stress in the absence of bending $= f = 13.43$ ksi.

LACING OF BUILT-UP COLUMN

Design the lacing bars and end tie plates of the member in Fig. 72. The lacing bars will be connected to the channel flanges with $\frac{1}{2}$ in. rivets.

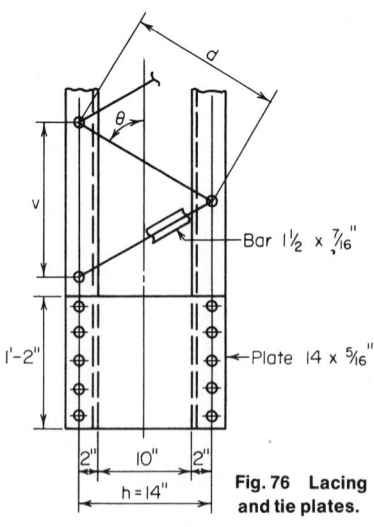

Fig. 76 Lacing and tie plates.

Bar $1\frac{1}{2} \times \frac{7}{16}$"

Plate 14 x $\frac{5}{16}$"

1'-2"

2"   10"   2"

h = 14"

Calculation Procedure:

1. Establish the dimensions of the lacing system to conform to the AISC Specification
The function of the lacing bars and tie plates is to preserve the integrity of the column and to prevent local failure.

Refer to Fig. 76. The standard gage in 15-in. channel = 2 in., from the AISC *Manual*. Then, $h = 14 < 15$ in.; therefore, use single lacing.

Try $\theta = 60°$; then, $v = 2(14) \cot 60° = 16.16$ in. Set $v = 16$ in.; therefore, $d = 16.1$ in. For the built-up section, $KL/r = 48.5$; for the single channel, $KL/r = 16/0.89 < 48.5$. This is acceptable. The spacing of the bars is therefore satisfactory.

2. Design the lacing bars
The lacing system must be capable of transmitting an assumed transverse shear

equal to 2 percent of the axial load; this shear is carried by two bars, one on each side. A lacing bar is classified as a secondary member. To compute the transverse shear, assume that the column will be loaded to its capacity of 432 kips.

Then, force per bar = $\frac{1}{2}(0.02)(432)(16.1/14) = 5.0$ kips. Also, $L/r \leq 140$; therefore $r = 16.1/140 = 0.115$ in.

For a rectangular section of thickness t, $r = 0.289t$. Then, $t = 0.115/0.289 = 0.40$ in. Set $t = \frac{7}{16}$ in.; $r = 0.127$ in.; $L/r = 16.1/0.127 = 127$; $f = 9.59$ ksi; $A = 5.0/9.59 = 0.52$ sq in. From the *Manual*, the minimum width required for $\frac{1}{2}$-in. rivets = $1\frac{1}{2}$ in. Therefore, use a flat bar $1\frac{1}{2} \times \frac{7}{16}$ in.; $A = 0.66$ sq in.

3. Design the end tie plates in accordance with the Specification

The minimum length = 14 in.; $t = 14/50 = 0.28$. Therefore, use plates $14 \times \frac{5}{16}$ in. The rivet pitch is limited to six diameters, or 3 in.

SELECTION OF A COLUMN WITH A LOAD AT AN INTERMEDIATE LEVEL

A column of 30-ft length carries a load of 130 kips applied at the top and a load of 56 kips applied to the web at mid-height. Select an 8-in. column of A242 steel, using $K_xL = 30$ ft and $K_yL = 15$ ft.

Calculation Procedure:

1. Compute the effective length of the column with respect to the major axis

The following procedure affords a rational method of designing a column subjected to a load applied at the top and another load applied approximately at the center. Let m = load at intermediate level, kips per total load, kips. Replace the factor K with a factor K' defined by $K' = K(1 - m/2)^{0.5}$. Thus, for this column, $m = 56/186 = 0.30$. And, $K'_xL = 30(1 - 0.15)^{0.5} = 27.6$ ft.

2. Select a trial section on the basis of the K_yL value

From the AISC *Manual* for an 8WF40, capacity = 186 kips when $K_yL = 16.2$ ft, and $r_x/r_y = 1.73$.

3. Determine if the selected section is acceptable

Compute the value of K_xL associated with a uniform-strength column and compare this with the actual effective length. Thus, $K_xL = 1.73(16.2) = 28.0 > 27.6$ ft. Therefore, the 8WF40 is acceptable.

DESIGN OF AN AXIAL MEMBER FOR FATIGUE

A web member in a welded truss will sustain precipitous fluctuations of stress caused by moving loads. The structure will carry three load systems having the following characteristics.

| System | Force induced in member, kips | | No. of times applied |
	Maximum compression	Maximum tension	
A	46	18	60,000
B	40	9	1,000,000
C	32	8	2,500,000

The effective length of the member is 11 ft. Design a double-angle member.

Calculation Procedure:

1. Calculate for each system the design load and indicate the yield-point stress on which the allowable stress is based

The design of members subjected to a repeated variation of stress is regulated by the AISC *Specification*. For each system, calculate the design load and indicate the yield-point stress on which the allowable stress is based. Where the allowable stress is less than that normally permitted, increase the design load proportionately to compensate for this reduction. Let $+$ denote tension and $-$ denote compression. Then:

System	Design load, kips	Yield-point stress, ksi
A	$-46 - \frac{2}{3}(18) = -58$	36
B	$-40 - \frac{2}{3}(9) = -46$	33
C	$1.5(-32 - \frac{3}{4} \times 8) = -57$	33

2. Select a member for system A and determine if it is adequate for system C

From the AISC *Manual*, try two angles $4 \times 3\frac{1}{2} \times \frac{3}{8}$ in., long legs back to back; the capacity is 65 kips. Then, $A = 5.34$ sq in.; $r = r_x = 1.25$ in.; $KL/r = 11(12)/1.25 = 105.6$.

From the *Manual*, for a yield-point stress of 33 ksi, $f = 11.76$ ksi. Then, the capacity $P = 5.34(11.76) = 62.8 > 57$ kips. This is acceptable. Therefore, use two angles $4 \times 3\frac{1}{2} \times \frac{3}{8}$ in., long legs back to back.

INVESTIGATION OF A BEAM COLUMN

A 12WF53 column with an effective length of 20 ft is to carry an axial load of 160 kips and the end moments indicated in Fig. 77. The member will be secured against sidesway in both directions. Is the section adequate?

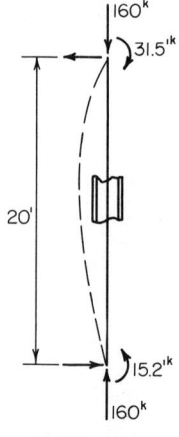

Fig. 77 Beam column.

Calculation Procedure:

1. Record the properties of the section

The simultaneous set of values of axial stress and bending stress must satisfy the inequalities set forth in the AISC *Specification*.

The properties of the section are: $A = 15.59$ sq in.; $S_x = 70.7$ in.³; $r_x = 5.23$ in.; $r_y = 2.48$ in. Also, from the *Manual*: $L_c = 10.8$ ft; $L_u = 21.7$ ft.

2. Determine the stresses listed below

The stresses that must be determined are the axial stress f_a; the bending stress f_b; the axial stress F_a, which would be permitted in the absence of bending, and the bending stress F_b, which would be permitted in the absence of axial load. Thus, $f_a = 160/15.59 = 10.26$ ksi; $f_b = 31.5(12)/70.7 = 5.35$ ksi; $KL/r = 240/2.48 = 96.8$; therefore, $F_a = 13.38$ ksi. $L_u < KL < L_c$; therefore, $F_b = 22$ ksi. (Although this consideration is irrelevant in the present instance, note that the *Specification* establishes two maximum d/t ratios for a compact section. One applies to a beam, the other to a beam column.)

3. Calculate the moment coefficient C_m

Since the algebraic sign of the bending moment remains unchanged, M_1/M_2 is positive. Thus, $C_m = 0.6 + 0.4(15.2/31.5) = 0.793$.

4. Apply the appropriate criteria to test the adequacy of the section

Thus, $f_a/F_a = 10.26/13.38 = 0.767 > 0.15$. The following requirements therefore apply: $f_a/F_a + [C_m/(1 - f_a/F'_e)](f_b/F_b) \leq 1$; $f_a/0.6f_y + f_b/F_b \leq 1$, where $F'_e = 149,000/(KL/r)^2$ ksi and KL and r are evaluated with respect to the plane of bending.

Evaluating, $F'_e = 149,000(5.23)^2/240^2 = 70.76$ ksi; $f_a/F'_e = 10.26/70.76 = 0.145$. Substituting in the first requirements equation, $0.767 + (0.793/0.855)(5.35/22) = 0.993$. This is acceptable. Substituting in the second requirements equation, $(10.26/22) + (5.35/22) = 0.709$. This section is therefore satisfactory.

APPLICATION OF BEAM-COLUMN FACTORS

For the previous Calculation Procedure, investigate the adequacy of the 12WF53 section by applying the values of the beam-column factors B and a given in the AISC *Manual*.

Calculation Procedure:

1. Record the basic values of the previous Calculation Procedure

The beam-column factors were devised in an effort to reduce the labor entailed in analyzing a given member as a beam column when $f_a/F_a > 0.15$. They are defined by: $B = A/S$ per in.; $a = 0.149 \times 10^6 I$ in.^{4}.

Let P denote the applied axial load and P_{allow} the axial load that would be permitted in the absence of bending. The equations given in the previous Procedure may be transformed to: $P + BMC_m(F_a/F_b)a/[a - P(KL)^2] \leqslant P_{allow}$, and $(PF_a/0.6f_y) + BMF_a/F_b \leqslant P_{allow}$, where KL, B, and a are evaluated with respect to the plane of bending.

The basic values of the previous Procedure are: $P = 160$ kips; $M = 31.5$ ft-kips; $F_b = 22$ ksi; $C_m = 0.793$.

2. Obtain the properties of the section

From the *Manual* for a 12WF53, $A = 15.59$ sq in.; $B_x = 0.221$ per in.; $a_x = 63.5 \times 10^6$ in.^{4}. Then, when $KL = 20$ ft, $P_{allow} = 209$ kips.

3. Substitute in the first transformed equation

Thus, $F_a = P_{allow}/A = 209/15.59 = 13.41$ ksi, $P(KL)^2 = 160(240)^2 = 9.22 \times 10^6$ kip-in.^{2}, and $a_x/[a_x - P(KL)^2] = 63.5/(63.5 - 9.22) = 1.17$; then $160 + 0.221(31.5)(12)(0.793)$ $(13.41/22)(1.17) = 207 < 209$ kips. This is acceptable.

4. Substitute in the second transformed equation

Thus, $160(13.41/22) + 0.221(31.5)(12)(13.41/22) = 148 < 209$ kips. This is acceptable. The 12WF53 section is therefore satisfactory.

NET SECTION OF A TENSION MEMBER

The $7 \times \frac{1}{4}$ in. plate in Fig. 78 carries a tensile force of 18,000 lb and is connected to its support with three $\frac{3}{4}$ in. rivets in the manner shown. Compute the maximum tensile stress in the member.

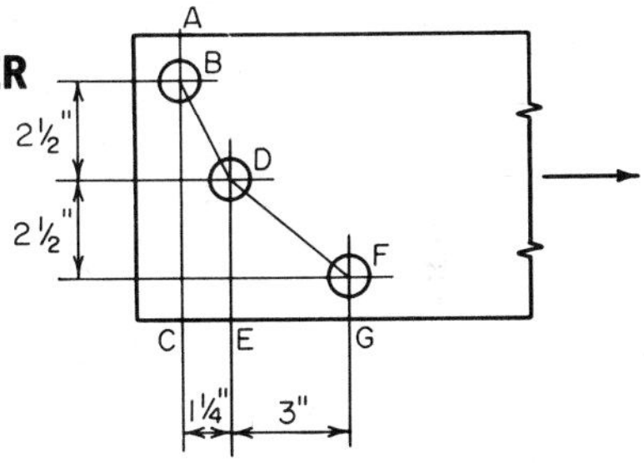

Calculation Procedure:

1. Compute the net width of the member at each section of potential rupture

The AISC *Specification* prescribes the manner of calculating the net section of a tension member. The effective diameter of the holes is considered to be $\frac{1}{8}$ in. greater than that of the rivets.

After computing the net width of each section, select the minimum value as the effective width. The *Specification* imposes an upper limit of 85 percent of the gross width.

Referring to Fig. 78: from B to D, $s = 1.25$ in., $g = 2.5$ in.; from D to F, $s = 3$ in., $g = 2.5$ in.; $w_{AC} = 7 - 0.875 = 6.12$ in.; $w_{ABDE} = 7 - 2(0.875) + 1.25^2/4(2.5) = 5.41$ in.; $w_{ABDFG} = 7 - 3(0.875) + (1.25^2 \times 4 \times 2.5) + (3^2/4 \times 2.5) = 5.43$ in.; $w_{max} = 0.85(7) = 5.95$ in. Selecting the lowest value, $w_{eff} = 5.41$ in.

2. Compute the tensile stress on the effective net section

Thus, $f = 18,000/5.41(0.25) = 13,300$ psi.

Fig. 78

DESIGN OF A DOUBLE-ANGLE TENSION MEMBER

The bottom chord of a roof truss sustains a tensile force of 141 kips. The member will be spliced with $\frac{3}{4}$-in. rivets as shown in Fig. 79a. Design a double-angle member and specify the minimum rivet pitch.

Calculation Procedure:

1. Show one angle in its developed form

Cut the outstanding leg and position it to be coplanar with the other one, as shown in Fig. 79b. The gross width of the angle w_g is the width of the equivalent plate thus formed; it equals the sum of the legs of the angle less the thickness.

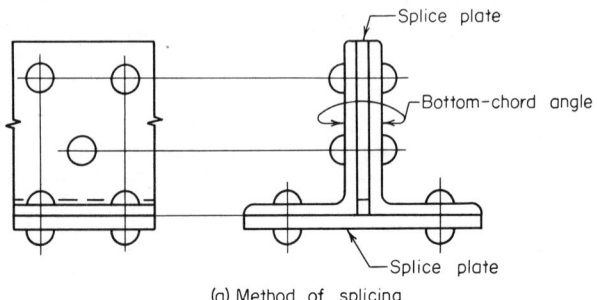

(a) Method of splicing

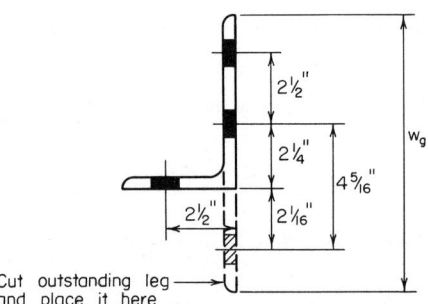

(b) Development of angle for net section

Fig. 79

2. Determine the gross width in terms of the thickness

Assume tentatively that 2.5 rivet holes will be deducted to arrive at the net width. Express w_g in terms of the thickness t of each angle. Then, net area required $= 141/22 = 6.40$ sq in.; also, $2t(w_g - 2.5 \times 0.875) = 6.40$; $w_g = (3.20/t) + 2.19$.

3. Assign trial thickness values and determine the gross width

Construct a tabulation of the computed values. Then select the most economical size of member. Thus:

t, in.	w_g, in.	$w_g + t$, in.	Available size, in.	Area, sq in.
$\frac{1}{2}$	8.59	9.09	$6 \times 3\frac{1}{2} \times \frac{1}{2}$	4.50
$\frac{7}{16}$	9.50	9.94	$6 \times 4 \times \frac{7}{16}$	4.18
$\frac{3}{8}$	10.72	11.10	None	

The most economical member is the one with the least area. Therefore, use two angles $6 \times 4 \times \frac{7}{16}$ in.

4. Record the standard gages

Refer to the *Manual* for the standard gages and record the values shown in Fig. 79*b*.

5. Establish the rivet pitch

Find the minimum value of s to establish the rivet pitch. Thus, net width required $= \frac{1}{2}[6.40/(7/16)] = 7.31$ in.; gross width $= 6 + 4 - 0.44 = 9.56$ in. Then, $9.56 - 3(0.875) + [s^2/(4 \times 2.5)] + [s^2/(4 \times 4.31)] = 7.31$; $s = 1.55$ in.

For convenience, use the standard pitch of 3 in. This results in a net width of 7.29 in.; the deficiency is negligible.

Plastic Design of Steel Structures

Consider that a structure is subjected to a gradually increasing load until it collapses. When the yield-point stress first appears, the structure is said to be in a state of *initial yielding*. The load that exists when failure impends is termed the *ultimate load*.

Elastic design considers that a structure has been loaded to capacity when it attains initial yielding, on the theory that plastic deformation would annul the utility of the structure. Plastic design, on the other hand, recognizes that a structure may be loaded beyond initial yielding if:

1. The tendency of the fiber at the yield-point stress toward plastic deformation is resisted by the adjacent fibers.

2. Those parts of the structure that remain in the elastic-stress range are capable of supporting this incremental load.

The ultimate load is reached when these conditions cease to exist and the structure therefore collapses.

Thus, elastic design is concerned with an allowable *stress*, which equals the yield-point stress divided by an appropriate factor of safety. In contrast, plastic design is concerned with an allowable *load*, which equals the ultimate load divided by an appropriate factor called the *load factor*. In reality, however, the distinction between elastic and plastic design has become rather blurred because specifications that ostensibly pertain to elastic design make covert concessions to plastic behavior. Several of these will be underscored in the Calculation Procedures that follow.

In the plastic analysis of flexural members, the following simplifying assumptions are made:

1. As the applied load is gradually increased, a state is eventually reached at which all fibers at the section of maximum moment are stressed to the yield-point stress, either in tension or compression. The section is then said to be in a state of *plastification*.

2. While plastification is proceeding at one section, the adjacent sections retain their linear-stress distribution.

Although the foregoing assumptions are fallacious, they introduce no appreciable error.

When plastification is achieved at a given section, no additional bending stress may be induced in any of its fibers, and the section is thus rendered impotent to resist any incremental bending moment. As loading continues, the beam behaves as if it had been constructed with a hinge at the given section. Consequently, the beam is said to have developed a *plastic hinge* (in contradistinction to a true hinge) at the plastified section.

The *yield moment* M_y of a beam section is the bending moment associated with initial yielding. The plastic moment M_p is the bending moment associated with plastification.

The *plastic modulus* Z of a beam section, which is analogous to the section modulus used in elastic design, is defined by $Z = M_p/f_y$, where f_y denotes the yield-point stress. The *shape factor* SF is the ratio of M_p to M_y, being so named because its value depends on the shape of the section. Then, SF $= M_p/M_y = f_y Z/f_y S = Z/S$.

In the following Calculation Procedures, it is understood that the members are made of A36 steel.

ALLOWABLE LOAD ON BAR SUPPORTED BY RODS

A load is applied to a rigid bar that is symmetrically supported by three steel rods as shown in Fig. 80. The cross-sectional areas of the rods are: rods A and C, 1.2 sq in.; rod B, 1.0 sq in. Determine the maximum load that may be applied (a) using elastic design with an allowable stress of 22,000 psi; (b) using plastic design with a load factor of 1.85.

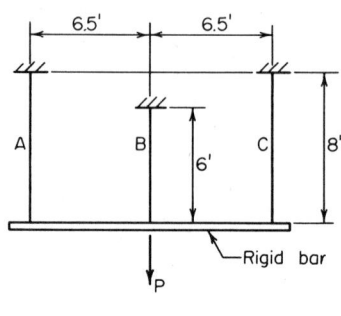

Fig. 80

Calculation Procedure:

1. Express the relationships among the tensile stresses in the rods

The symmetrical disposition causes the bar to deflect vertically without rotating, thereby elongating the three rods by the same amount. As the first method of solving this problem, assume that the load is gradually increased from zero to its allowable value.

Expressing the relationships among the tensile stresses, $\Delta L = s_A L_A/E = s_B L_B/E = s_C L_C/E$; therefore, $s_A = s_C$, and $s_A = s_B L_B/L_A = 0.75 s_B$ for this arrangement of rods. Since s_B is the maximum stress, the allowable stress first appears in rod B.

2. Evaluate the stresses at the instant the load attains its allowable value

Calculate the load carried by each rod and sum these loads to find P_{allow}. Thus: $s_B = 22,000$ psi; $s_A = 0.75(22,000) = 16,500$ psi; $P_A = P_C = 16,500(1.2) = 19,800$ lb; $P_B = 22,000(1.0) = 22,000$ lb; $P_{\text{allow}} = 2(19,800) + 22,000 = 61,600$ lb.

Next, consider that the load is gradually increased from zero to its ultimate value. When rod B attains its yield-point stress, its tendency to deform plastically is inhibited by rods A and C because the rigidity of the bar constrains the three rods to elongate uniformly. The structure therefore remains stable as the load is increased beyond the elastic range until rods A and C also attain their yield-point stress.

3. Find the ultimate load

To find the ultimate load P_u, equate the stress in each rod to f_y, calculate the load carried by each rod, and sum these loads to find the ultimate load P_u. Thus, $P_A = P_C = 36,000(1.2) = 43,200$ lb; $P_B = 36,000(1.0) = 36,000$ lb; $P_u = 2(43,200) + 36,000 = 122,400$ lb.

4. Apply the load factor to establish the allowable load

Thus, $P_{\text{allow}} = P_u/\text{LF} = 122,400/1.85 = 66,200$ lb.

DETERMINATION OF SECTION SHAPE FACTORS

Without applying the equations and numerical values of the plastic modulus given in the AISC *Manual*, determine the shape factor associated with the following sections: a rectangle, a circle, and a 16WF40. Explain why the circle has the highest and the wide-flange section the lowest factor of the three.

Calculation Procedure:

1. Calculate M_y for each section

Use the equation $M_y = S f_y$ for each section. Thus, for a rectangle, $M_y = b d^2 f_y/6$. For a circle, using the properties of a circle as given in the *Manual*, $M_y = \pi d^3 f_y/32$. For a 16WF40, $A = 11.77$ sq in., $S = 64.4$ in.^{3}, and $M_y = 64.4 f_y$.

2. Compute the resultant forces associated with plastification

Referring to Fig. 81, the resultant forces are C and T. Once these forces are known, their action lines and M_p should be computed.

Thus, for a rectangle, $C = bdf_y/2$, $a = d/2$, and $M_p = aC = bd^2f_y/4$. For a circle, $C = \pi d^2 f_y/8$, $a = 4d/3\pi$, and $M_p = aC = d^3 f_y/6$. For a 16WF40, $C = \frac{1}{2}(11.77f_y) = 5.885f_y$.

To locate the action lines, refer to the *Manual* and note the position of the centroidal axis of the ST8WF20 section, i.e., a section half the size of that being considered. Thus, $a = 2(8.00 - 1.82) = 12.36$ in.; $M_p = aC = 12.36(5.885f_y) = 72.7f_y$.

3. Divide M_p by M_y to obtain the shape factor

For a rectangle, $SF = (bd^2/4)/(bd^2/6) = 1.50$. For a circle, $SF = (d^3/6)/(\pi d^3/32) = 1.70$. For a 16WF40, $SF = 72.7/64.4 = 1.13$.

4. Explain the relative values of the shape factor

To explain the relative values of the shape factor, express the resisting moment contributed by a given fiber at plastification and at initial yielding, and compare the results. Let dA denote the area of the given fiber and y its distance from the neutral axis. At plastification, $dM_p = f_y y\, dA$. At initial yielding, $f = f_y y/c$; $dM_y = f_y y^2\, dA/c$; $dM_p/dM_y = c/y$.

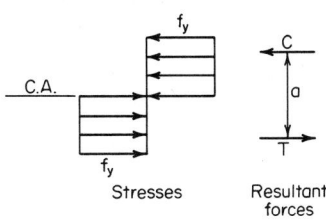

Fig. 81 Conditions at section of plastification.

Comparing a circle and a hypothetical wide-flange section having the same area and depth, the circle is found to have a larger shape factor because of its relatively low values of y.

As this analysis demonstrates, the process of plastification mitigates the detriment that accrues from placing any area near the neutral axis, since the stress at plastification is independent of the position of the fiber. Consequently, a section that is relatively inefficient with respect to flexure has a relatively high shape factor. The AISC *Specification* for elastic design implicitly recognizes the value of the shape factor by assigning an allowable bending stress of $0.75f_y$ to rectangular bearing plates and $0.90f_y$ to pins.

DETERMINATION OF ULTIMATE LOAD BY THE STATICAL METHOD

The 18WF45 beam in Fig. 82a is simply supported at A and fixed at C. Disregarding the beam weight, calculate the ultimate load that may be applied at B (*a*) by analyzing the behavior of the beam during its two phases; (*b*) by analyzing the bending moments that exist at impending collapse. (The first part of the solution illustrates the post-elastic behavior of the member.)

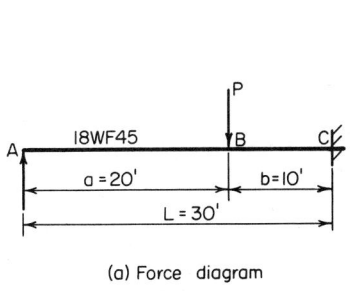

(a) Force diagram

(b) Bending–moment diagram

Fig. 82

Calculation Procedure:

1. Calculate the ultimate-moment capacity of the member

Part a: As the load is gradually increased from zero to its ultimate value, the beam passes through two phases. During phase 1, *the elastic phase*, the member is restrained against rotation at C. This phase terminates when a plastic hinge forms at that end. During phase 2—*the post-elastic, or plastic, phase*—the member functions as a simply supported beam. This phase terminates when a plastic hinge forms at B, since the member then becomes unstable.

Using data from the AISC *Manual*, $Z = 89.6$ in.^{3}. Then, $M_p = f_y Z = 36(89.6)/12 = 268.8$ ft-kips.

2. Calculate the moment BD

Let P_1 denote the applied load at completion of phase 1. In Fig. 82b, construct the bending-moment diagram $ADEC$ corresponding to this load. Evaluate P_1 by applying the equations for Case 14 in the AISC *Manual*. Calculate the moment BD. Thus, $CE = -ab(a + L)P_1/2L^2 = -20(10)(50)P_1/[2(900)] = -268.8$; $P_1 = 48.38$ kips; $BD = ab^2(a+2L)P_1/2L^3 = 20(100)(80)(48.38)/[2(27,000)] = 143.3$ ft-kips.

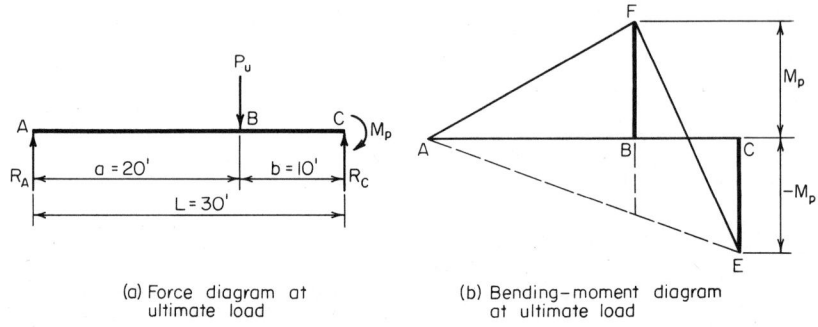

(a) Force diagram at ultimate load

(b) Bending–moment diagram at ultimate load

Fig. 83

3. Determine the incremental load at completion of phase 2

Let P_2 denote the incremental applied load at completion of phase 2, i.e., the actual load on the beam minus P_1. In Fig. 82b, construct the bending-moment diagram $AFEC$ that exists when phase 2 terminates. Evaluate P_2 by considering the beam as simply supported. Thus, $BF = 268.8$ ft-kips; $DF = 268.8 - 143.3 = 125.5$ ft-kips; but $DF = abP_2/L = 20(10)P_2/30 = 125.5$; $P_2 = 18.82$ kips.

4. Sum the results to obtain the ultimate load

Thus, $P_u = 48.38 + 18.82 = 67.20$ kips.

5. Construct the force and bending-moment diagrams for the ultimate load

Part b: The following considerations are crucial: The bending-moment diagram always has vertices at B and C, and formation of two plastic hinges will cause failure of the beam. Therefore, the plastic moment occurs at B and C at impending failure. *The sequence in which the plastic hinges are formed at these sections is immaterial.*

These diagrams are shown in Fig. 83. Express M_p in terms of P_u, and evaluate P_u. Thus, $BF = 20R_A = 268.8$; therefore $R_A = 13.44$ kips. Also, $CE = 30R_A - 10P_u = 30 \times (13.44) - 10P_u = -268.8$; $P_u = 67.20$ kips.

Here is an alternative method. $BF = (abP_u/L) - aM_p/L = M_p$, or $20(10)P_u/30 = 50M_p/30$; $P_u = 67.20$ kips.

The solution method used in part b is termed the *statical* or *equilibrium* method. As this solution demonstrates, it is unnecessary to trace the stress history of the member as it passes through its successive phases, as was done in part a; the analysis can be confined to the conditions that exist at impending failure. This procedure also illustrates the following important characteristics of plastic design:

1. Plastic design is far simpler than elastic design.

2. Plastic design yields results that are much more reliable than those secured through elastic design. For example, assume that the support at C does not completely inhibit rotation at that end. This departure from design conditions will invalidate the elastic analysis but will in no way affect the plastic analysis.

DETERMINING THE ULTIMATE LOAD BY THE MECHANISM METHOD

Use the mechanism method to solve the problem given in the previous Calculation Procedure.

Calculation Procedure:

1. Indicate, in hyperbolic manner, the virtual displacement of the member from its initial to a subsequent position

To the two phases of beam behavior previously considered, it is possible to add a third. Consider that when the ultimate load is reached, the member is subjected to an incremental deflection. This will result in collapse, but the behavior of the member can be analyzed during an infinitesimally small deflection from its stable position. This is termed a *virtual* deflection or displacement.

Since the member is incapable of supporting any load beyond that existing at completion of phase 2, this virtual deflection is not characterized by any change in bending stress. Rotation therefore occurs solely at the real and plastic hinges. Thus, during phase 3, the member behaves as a mechanism (i.e., a constrained chain of pin-connected rigid bodies, or links).

In Fig. 84, indicate, in hyperbolic manner, the virtual displacement of the member from its initial position ABC to a subsequent position $AB'C$. Use dots to represent plastic hinges. (The initial position may be represented by a straight line for simplicity because the analysis is concerned solely with the deformation that occurs *during* phase 3.)

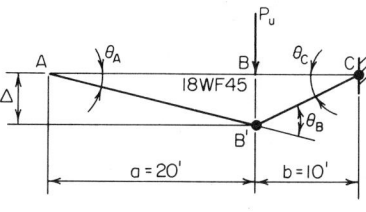

Fig. 84

2. Express the linear displacement under the load and the angular displacement at every plastic hinge

Use a convenient unit to express these displacements. Thus, $\Delta = a\theta_A = b\theta_C$; therefore, $\theta_C = a\theta_A/b = 2\theta_A$; $\theta_B = \theta_A + \theta_C = 3\theta_A$.

3. Evaluate the external and internal work associated with the virtual displacement

The work performed by a constant force equals the product of the force and its displacement parallel to its action line. Also, the work performed by a constant moment equals the product of the moment and its angular displacement. Work is a positive quantity when the displacement occurs in the direction of the force or moment. Thus, the external work $W_E = P_u\Delta = P_u a\theta_A = 20P_u\theta_A$. And the internal work $W_I = M_p(\theta_B + \theta_C) = 5M_p\theta_A$.

4. Equate the external and internal work to evaluate the ultimate load

Thus, $20P_u\theta_A = 5M_p\theta_A$; $P_u = (5/20)(268.8) = 67.20$ kips.

The solution method used here is also termed the *virtual-work*, or *kinematic*, method.

ANALYSIS OF A FIXED-ENDED BEAM UNDER CONCENTRATED LOAD

If the beam in the two previous Calculation Procedures is fixed at A as well as at C, what is the ultimate load that may be applied at B?

Calculation Procedure:

1. Determine when failure impends

When hinges form at A, B, and C, failure impends. Repeat Steps 3 and 4 of the previous Calculation Procedure, modifying the calculations to reflect the revised conditions. Thus: $W_E = 20P_u\theta_A$; $W_I = M_p(\theta_A + \theta_B + \theta_C) = 6M_p\theta_A$; $20P_u\theta_A = 6M_p\theta_A$; $P_u = (6/20)(268.8) = 80.64$ kips.

2. *Analyze the phases through which the member passes*

This member passes through three phases until the ultimate load is reached. Initially, it behaves as a beam fixed at both ends, then as a beam fixed at the left end only, and finally as a simply supported beam. However, as already discussed, these considerations are extraneous in plastic design.

ANALYSIS OF A TWO-SPAN BEAM WITH CONCENTRATED LOADS

The continuous 18WF45 beam in Fig. 85 carries two equal concentrated loads having the locations indicated. Disregarding the weight of the beam, compute the ultimate value of these loads, using both the statical and the mechanism method.

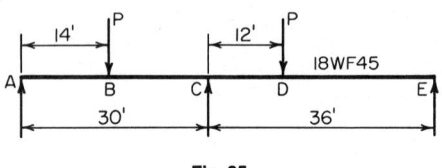

Fig. 85

Calculation Procedure:

1. *Construct the force and bending-moment diagrams*

The continuous beam becomes unstable when a plastic hinge forms at C and at another section. The bending-moment diagram has vertices at B and D, but it is not readily apparent at which of these sections the second hinge will form. The answer is found by assuming a plastic hinge at B and at D, in turn, computing the corresponding

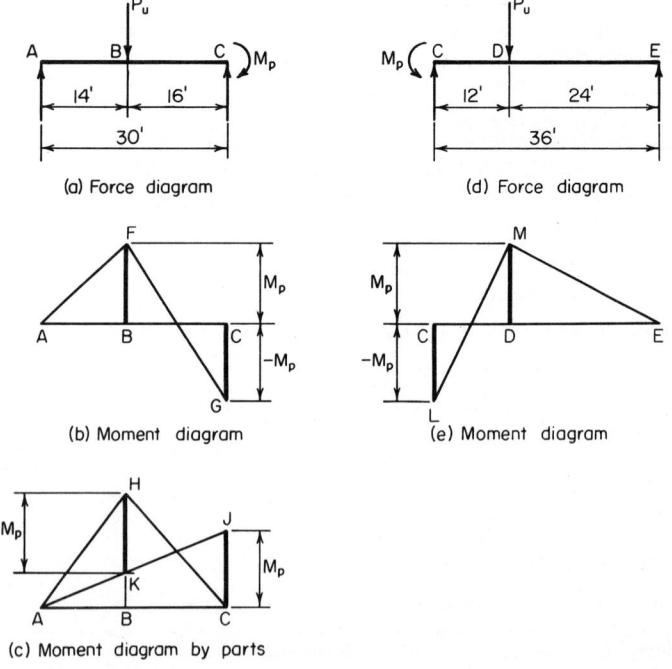

Fig. 86

value of P_u and selecting the lesser value as the correct result. Part a will use the statical method; part b the mechanism method.

Assume, for part a, a plastic hinge at B and C. In Fig. 86, construct the force diagram and bending-moment diagram for span AC. The moment diagram may be drawn in the manner shown in Fig. 86b or c, whichever is preferred. In Fig. 86c, ACH represents the moments that would exist in the absence of restraint at C, and ACJ represents, in absolute value, the moments induced by this restraint. Compute the load P_u associated with the assumed hinge location. From previous Calculation Procedures, $M_p = 268.8$ ft-kips; then $M_B = (14 \times 16P_u/30) - 14M_p/30 = M_p$; $P_u = 44(268.8)/224 = 52.8$ kips.

2. Assume another hinge location and compute the ultimate load associated with this location

Now assume a plastic hinge at C and D. In Fig. 86, construct the force diagram and bending-moment diagram for CE. Computing the load P_u associated with this assumed location, $M_D = (12 \times 24P_u/36) - 24M_p/36 = M_p$; $P_u = 60(268.8)/288 = 56.0$ kips.

3. Select the lesser value of the ultimate load

The correct result is the lesser of these alternative values. Or, $P_u = 52.8$ kips. At this load, plastic hinges exist at B and C but not at D.

4. For the mechanism method, assume a plastic-hinge location

It will be assumed that plastic hinges are located at B and C, Fig. 87. Evaluate P_u. Thus, $\theta_C = 14\theta_A/16$; $\theta_B = 30\theta_A/16$; $\Delta = 14\theta_A$; $W_E = P_u\Delta = 14P_u\theta_A$; $W_I = M_p(\theta_B + \theta_C) = 2.75M_p\theta_A$; $14P_u\theta_A = 2.75M_p\theta_A$; $P_u = 52.8$ kips.

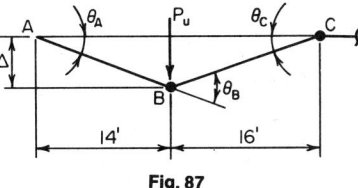

Fig. 87

5. Assume a plastic hinge at another location

Select C and D for the new location. Repeat the above procedure. The result will be identical with that in Step 2.

SELECTION OF SIZES FOR A CONTINUOUS BEAM

Using a load factor of 1.70, design the member to carry the working loads (with beam weight included) shown in Fig. 88a. The maximum length that can be transported is 60 ft.

Calculation Procedure:

1. Determine the ultimate loads to be supported

Since the member must be spliced, it will be economical to adopt the following design:

a. Use the particular beam size required for each portion, considering that the two portions will fail simultaneously at ultimate load. Therefore, three plastic hinges will exist at failure—one at the interior support and one in the interior of each span.

b. Extend one beam beyond the interior support, splicing the member at the point of contraflexure in the adjacent span. Since the maximum simple-span moment is greater for AB than for BC, it is logical to assume that for economy the left beam rather than the right one should overhang the support.

Multiply the working loads by the load factor to obtain the ultimate loads to be supported. Thus, $w = 1.2$ klf; $w_u = 1.70(1.2) = 2.04$ klf; $P = 10$ kips; $P_u = 1.70(10) = 17$ kips.

2. Construct the ultimate-load and corresponding bending-moment diagram for each span

Set the maximum positive moment M_D in span AB and the negative moment at B equal to one another in absolute value.

3. Evaluate the maximum positive moment in the left span

Thus, $R_A = 45.9 - M_B/40$; $x = R_A/2.04$; $M_D = \frac{1}{2}R_A x = R_A^2/4.08 = M_B$. Substitute the value of R_A and solve. Thus, $M_D = 342$ ft-kips.

An indirect but less cumbersome method consists of assigning a series of trial values to M_B and calculating the corresponding value of M_D, continuing the process until the required equality is obtained.

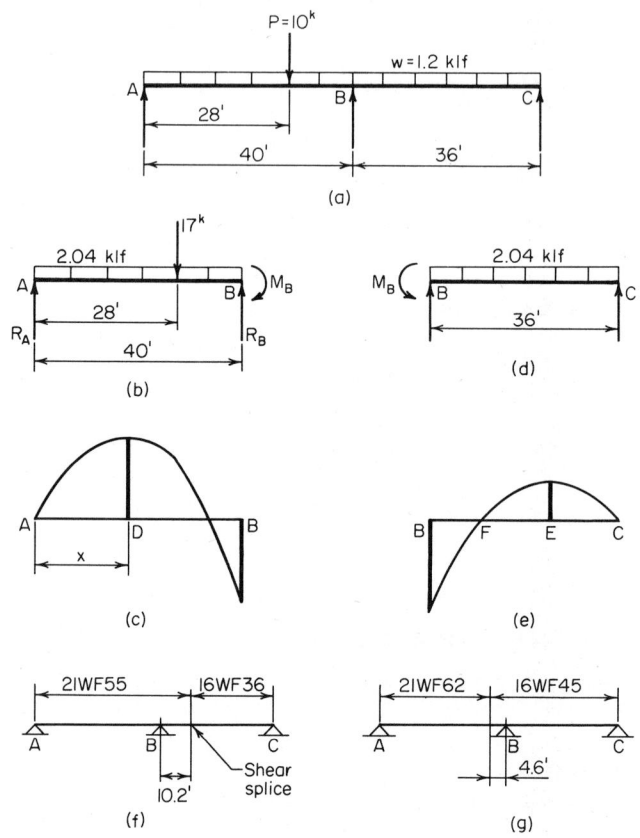

Fig. 88

4. Select a section to resist the plastic moment

Thus, $Z = M_p/f_y = 342(12)/36 = 114$ in.³. Referring to the AISC *Manual*, use a 21WF55 with $Z = 125.4$ in.³.

5. Evaluate the maximum positive moment in the right span

Equate M_B to the true plastic-moment capacity of the 21WF55. Evaluate the maximum positive moment M_E in span BC and locate the point of contraflexure. Thus, $M_B = -36(125.4)/12 = -376.2$ ft-kips; $M_E = 169.1$ ft-kips; $BF = 10.2$ ft.

6. Select a section to resist the plastic moment

The moment to be resisted is M_E. Thus, $Z = 169.1(12)/36 = 56.4$ in.³. Use 16WF36 with $Z = 63.9$ in.³.

The design is summarized in Fig. 88f. By inserting a hinge at F, the continuity of the member is destroyed and its behavior is thereby modified under gradually increasing load. However, the ultimate-load conditions, which constitute the only valid design criteria, are not affected.

7. *Alternatively, design the member with the right-hand beam overhanging the support*

Compare the two designs for economy. The latter design is summarized in Fig. 88g. The total beam weight associated with each scheme is as shown in the following table.

Design 1	Design 2
55(50.2) = 2,761 lb	62(35.4) = 2,195 lb
36(25.8) = 929	45(40.6) = 1,827
Total 3,690 lb	4,022 lb

For completeness, the column sizes associated with the two schemes should also be compared.

MECHANISM-METHOD ANALYSIS OF A RECTANGULAR PORTAL FRAME

Calculate the plastic moment and the reactions at the supports at ultimate load of the prismatic frame in Fig. 89a. Use a load factor of 1.85, and apply the mechanism method.

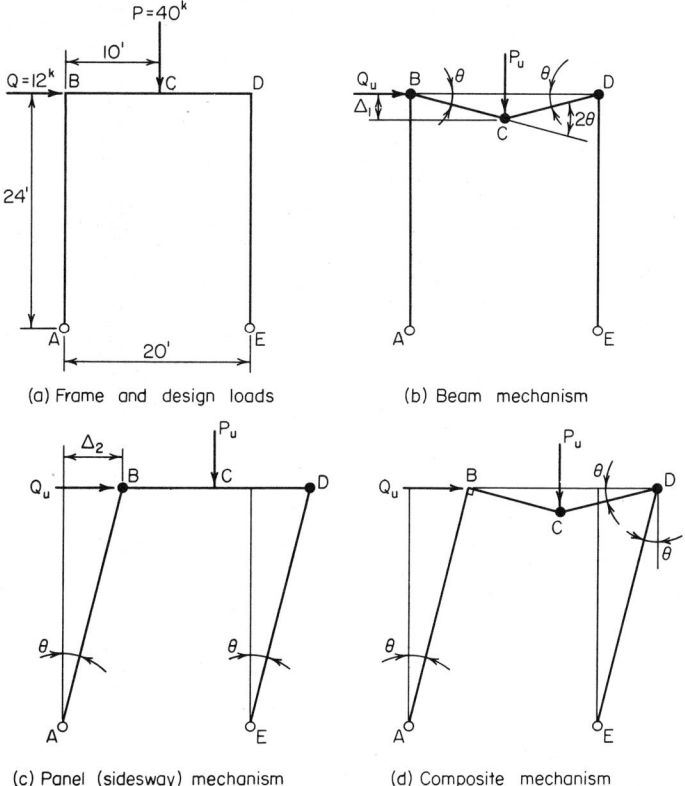

(a) Frame and design loads   (b) Beam mechanism

(c) Panel (sidesway) mechanism   (d) Composite mechanism

Fig. 89

Calculation Procedure:

1. Compute the ultimate loads to be resisted

There are three potential modes of failure to consider:

a. Failure of the beam BD through the formation of plastic hinges at B, C, and D, Fig. 89b.

b. Failure by sidesway through the formation of plastic hinges at B and D, Fig. 89c.

c. A composite of the foregoing modes of failure, characterized by the formation of plastic hinges at C and D.

Since the true mode of failure is not readily discernible, it is necessary to analyze each of the foregoing. The true mode of failure is the one that yields the highest value of M_p.

Although the work quantities are positive, it is advantageous to supply each angular displacement with an algebraic sign. A rotation is considered positive if the angle on the interior side of the frame increases. The algebraic sum of the angular displacements must equal zero.

Computing the ultimate loads to be resisted, $P_u = 1.85(40) = 74$ kips; $Q_u = 1.85(12) = 22.2$ kips.

2. Assume the mode of failure in Fig. 89b and compute M_p

Thus, $\Delta_1 = 10\theta$; $W_E = 74(10\theta) = 740\theta$. Then indicate in a tabulation, such as that below, where the plastic moment occurs. Include all significant sections for completeness.

Section	Angular displacement	Moment	W_I
A			
B	$-\theta$	M_p	$M_p\theta$
C	$+2\theta$	M_p	$2M_p\theta$
D	$-\theta$	M_p	$M_p\theta$
E	....	...	.....
Total..	....	...	$4M_p\theta$

Then, $4M_p\theta = 740\theta$; $M_p = 185$ ft-kips.

3. Repeat the foregoing procedure for failure by sidesway

Thus, $\Delta_2 = 24\theta$; $W_E = 22.2(24\theta) = 532.8\theta$.

Section	Angular displacement	Moment	W_I
A	$-\theta$		
B	$+\theta$	M_p	$M_p\theta$
C			
D	$-\theta$	M_p	$M_p\theta$
E	$+\theta$		
Total...	...	...	$2M_p\theta$

Then, $2M_p\theta = 532.8\theta$; $M_p = 266.4$ ft-kips.

4. Assume the composite mode of failure and compute M_p

Since this results from superposition of the two preceding modes, the angular displacements and the external work may be obtained by adding the algebraic values previously found. Thus, $W_E = 740\theta + 532.8\theta = 1{,}272.8\theta$. Then the tabulation is as shown below.

Section	Angular displacement	Moment	W_I
A	$-\theta$		
B			
C	$+2\theta$	M_p	$2M_p\theta$
D	-2θ	M_p	$2M_p\theta$
E	$+\theta$		
Total . . .	. . .	. . .	$4M_p\theta$

Then, $4M_p\theta = 1{,}272.8\theta$; $M_p = 318.2$ ft-kips.

5. Select the highest value of M_p as the correct result

Thus, $M_p = 318.2$ ft-kips. The structure fails through the formation of plastic hinges at C and D. That a hinge should appear at D rather than at B is plausible when it is considered that the bending moments induced by the two loads are of like sign at D but of opposite sign at B.

6. Compute the reactions at the supports

Draw a free-body diagram of the frame at ultimate load, Fig. 90. Compute the reactions at the supports by applying the computed values of M_C and M_D. Thus, $\Sigma M_E = 20V_A + 22.2(24) - 74(10) = 0$; $V_A = 10.36$ kips; $V_E = 74 - 10.36 = 63.64$ kips; $M_C = 10V_A + 24H_A = 103.6 + 24H_A = 318.2$; $H_A = 8.94$ kips; $H_E = 22.2 - 8.94 = 13.26$ kips; $M_D = -24H_E = -24(13.26) = -318.2$ ft-kips. Thus, the results are verified.

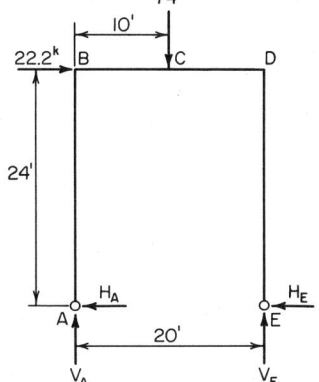

ANALYSIS OF A RECTANGULAR PORTAL FRAME BY THE STATICAL METHOD

Compute the plastic moment of the frame in Fig. 89a by using the statical method.

Calculation Procedure:

1. Determine the relative values of the bending moments

Consider a bending moment as positive if the fibers on the interior side of the neutral plane are in tension. Consequently, as the mechanisms in Fig. 89 reveal, the algebraic sign of the plastic moment at a given section agrees with that of its angular displacement during collapse.

Fig. 90

Determine the relative values of the bending moments at B, C, and D. Refer to Fig. 90. As previously found by statics, $V_A = 10.36$ kips, $M_B = 24H_A$, $M_C = 24H_A + 10V_A$; therefore, $M_C = M_B + 103.6$, Eq. (a). Also, $M_D = 24H_A + 20V_A - 74(10)$; $M_D = M_B - 532.8$, Eq. (b); or $M_D = M_C - 636.4$, Eq. (c).

2. Assume the mode of failure in Fig. 89b

This requires that $M_B = M_D = -M_p$. This relationship is incompatible with Eq. (b), and the assumed mode of failure is therefore incorrect.

3. Assume the mode of failure in Fig. 89c

This requires that $M_B = M_p$, and $M_C < M_p$; therefore, $M_C < M_B$. This relationship is incompatible with Eq. (a), and the assumed mode of failure is therefore incorrect.

By a process of elimination, it has been ascertained that the frame will fail in the manner shown in Fig. 89d.

4. Compute the value of M_p for the composite mode of failure

Thus, $M_C = M_p$, and $M_D = -M_p$. Substitute these values in Eq. (c). Or, $-M_p = M_p - 636.4$; $M_p = 318.2$ ft-kips.

THEOREM OF COMPOSITE MECHANISMS

By analyzing the calculations in the Calculation Procedure before the last one, establish a criterion to determine when a composite mechanism is significant (i.e., under what conditions it may yield an M_p value greater than that associated with the basic mechanisms).

Calculation Procedure:

1. Express the external and internal work associated with a given mechanism

Thus, $W_E = e\theta$, and $W_I = iM_p\theta$, where the coefficients e and i are obtained by applying the mechanism method. Then, $M_p = e/i$.

2. Determine the significance of mechanism sign

Let the subscripts 1 and 2 refer to the basic mechanisms and the subscript 3 to their composite mechanism. Then, $M_{p1} = e_1/i_1$; $M_{p2} = e_2/i_2$.

When the basic mechanisms are superposed, the values of W_E are additive. If the two mechanisms do not produce rotations of opposite sign at any section, the values of W_I are also additive, and $M_{p3} = e_3/i_3 = (e_1 + e_2)/(i_1 + i_2)$. This value is intermediate between M_{p1} and M_{p2}, and the composite mechanism therefore lacks significance. But if the basic mechanisms produce rotations of opposite sign at any section whatsoever, M_{p3} *may* exceed both M_{p1} and M_{p2}. In summary:

A composite mechanism is significant only if the two basic mechanisms of which it is composed produce rotations of opposite sign at any section.

This theorem, which establishes a necessary but not sufficient condition, simplifies the analysis of a complex frame by enabling the engineer to discard the nonsignificant composite mechanisms at the outset.

ANALYSIS OF AN UNSYMMETRICAL RECTANGULAR PORTAL FRAME

The frame in Fig. 91a sustains the ultimate loads shown. Compute the plastic moment and ultimate-load reactions.

Calculation Procedure:

1. Determine the solution method to use

Apply the mechanism method. In Fig. 91b, indicate the basic mechanisms.

2. Identify the significant composite mechanisms

Apply the theorem of the previous Calculation Procedure. Using this theorem, identify the significant composite mechanisms. Thus, for mechanisms 1 and 2: The rotations at B are of opposite sign; their composite therefore warrants investigation.

For mechanisms 1 and 3: There are no rotations of opposite sign; their composite therefore fails the test. For mechanisms 2 and 3: The rotations at B are of opposite sign; their composite therefore warrants investigation.

3. Evaluate the external work associated with each mechanism

Mechanism	W_E
1	$80\Delta_1 = 80(10\theta) = 800\theta$
2	$20\Delta_2 = 20(15\theta) = 300\theta$
3	300θ
4	$1{,}100\theta$
5	600θ

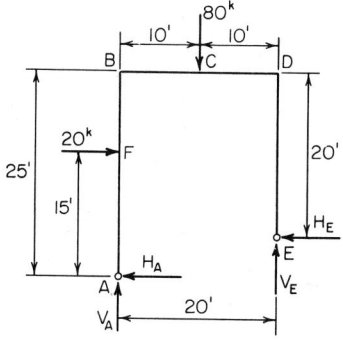

(a) Frame and ultimate loads

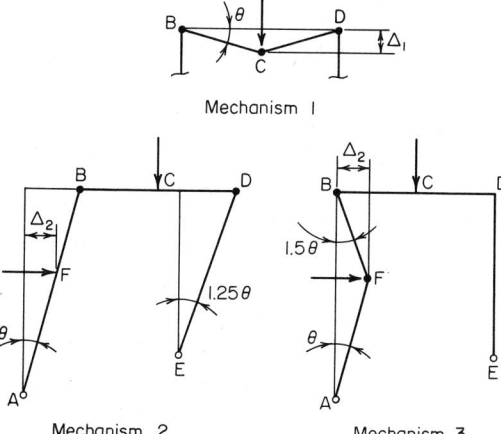

Mechanism 1

Mechanism 2   Mechanism 3
(b) Basic mechanisms

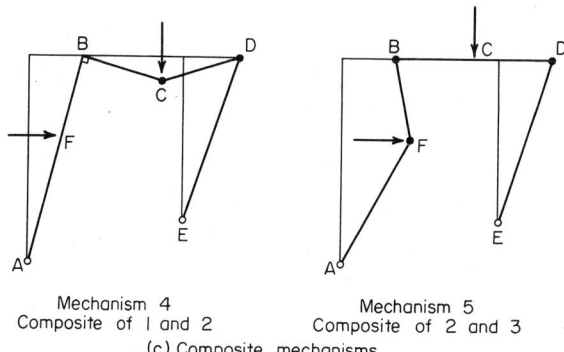

Mechanism 4   Mechanism 5
Composite of 1 and 2   Composite of 2 and 3
(c) Composite mechanisms

Fig. 91

4. List the sections at which plastic hinges form; record the angular displacement associated with each mechanism

Use a list such as the one in the following table.

Mechanism	Section			
	B	C	D	F
1	$-\theta$	$+2\theta$	$-\theta$	
2	$+\theta$	.....	-1.25θ	
3	-1.5θ	.....	.....	$+2.5\theta$
4	.....	$+2\theta$	-2.25θ	
5	-0.5θ	.....	-1.25θ	$+2.5\theta$

5. Evaluate the internal work associated with each mechanism

Equate the external and internal work to find M_p. Thus, $M_{p1} = 800/4 = 200$; $M_{p2} = 300/2.25 = 133.3$; $M_{p3} = 300/4 = 75$; $M_{p4} = 1,100/4.25 = 258.8$; $M_{p5} = 600/4.25 = 141.2$. Equate the external and internal work to find M_p.

6. Select the highest value as the correct result

Thus, $M_p = 258.8$ ft-kips. The frame fails through the formation of plastic hinges at C and D.

7. Determine the reactions at ultimate load

To verify the foregoing solution, ascertain that the bending moment does not exceed M_p in absolute value anywhere in the frame. Refer to Fig. 91a.

Thus, $M_D = -20H_E = -258.8$; therefore $H_E = 12.94$ kips; $M_C = M_D + 10V_E = 258.8$; therefore, $V_E = 51.76$ kips; then, $H_A = 7.06$ kips; $V_A = 28.24$ kips.

Check the moments. Thus: $\Sigma M_E = 20V_A + 5H_A + 20(10) - 80(10) = 0$; this is correct. Also, $M_F = 15H_A = 105.9$ ft-kips $< M_p$. This is correct. Lastly, $M_B = 25H_A - 20(10) = -23.5$ ft-kips $> -M_p$. This is correct.

ANALYSIS OF GABLE FRAME BY STATICAL METHOD

The prismatic frame in Fig. 92a carries the ultimate loads shown. Determine the plastic moment by applying the statical method.

Calculation Procedure:

1. Compute the vertical shear V_A and the bending moment at every significant section, assuming $H_A = 0$

Thus, $V_A = 41$ kips. Then: $M_B = 0$; $M_C = 386$; $M_D = 432$; $M_E = 276$; $M_F = -100$.

Note that failure of the frame will result from the formation of two plastic hinges. It is helpful, therefore, to construct a "projected" bending-moment diagram as an aid in locating these hinges. The computed bending moments are used in plotting the projected bending-moment diagram.

2. Construct a projected bending-moment diagram

To construct this diagram, consider the rafter BD to be projected onto the plane of column AB, and the rafter FD to be projected onto the plane of column GF. Juxtapose the two halves, as shown in Fig. 92b. Plot the values calculated in Step 1 to obtain the bending-moment diagram corresponding to the assumed condition of $H_A = 0$.

The bending moments caused solely by a specific value of H_A are represented by an isosceles triangle with its vertex at D'. The true bending moments are obtained by superposition. It is evident by inspection of the diagram that plastic hinges form at D and F and that H_A is directed to the right.

3. *Evaluate the plastic moment*

Apply the true moments at D and F. Thus, $M_D = M_p$ and $M_F = -M_p$; therefore, $432 - 37H_A = -(-100 - 25H_A)$; $H_A = 5.35$ kips and $M_p = 234$ ft-kips.

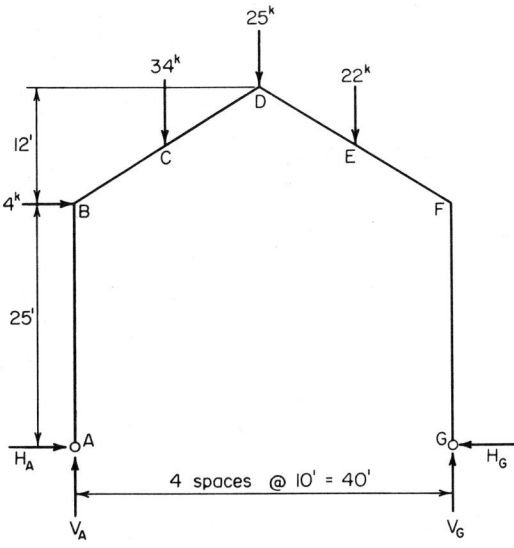

(a) Frame and ultimate loads

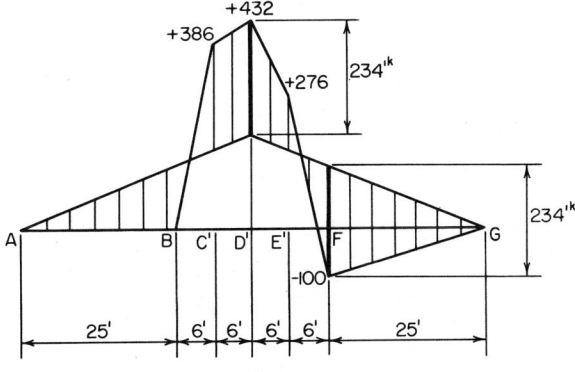

(b) Projected bending-moment diagram

Fig. 92

THEOREM OF VIRTUAL DISPLACEMENTS

In Fig. 93a, point P is displaced along a virtual (infinitesimally small) circular arc PP' centered at O and having a central angle θ. Derive expressions for the horizontal and vertical displacement of P in terms of the given data. (These expressions will later be applied in analyzing a gable frame by the mechanism method.)

Calculation Procedure:

1. Construct the displacement diagram

In Fig. 93b, let r_h = length of horizontal projection of OP; r_v = length of vertical projection of OP; Δ_h = horizontal displacement of P; Δ_v = vertical displacement of P.

In Fig. 93c, construct the displacement diagram. Since PP' is infinitesimally small, replace this circular arc with the straight line PP'' that is tangent to the arc at P and therefore normal to radius OP.

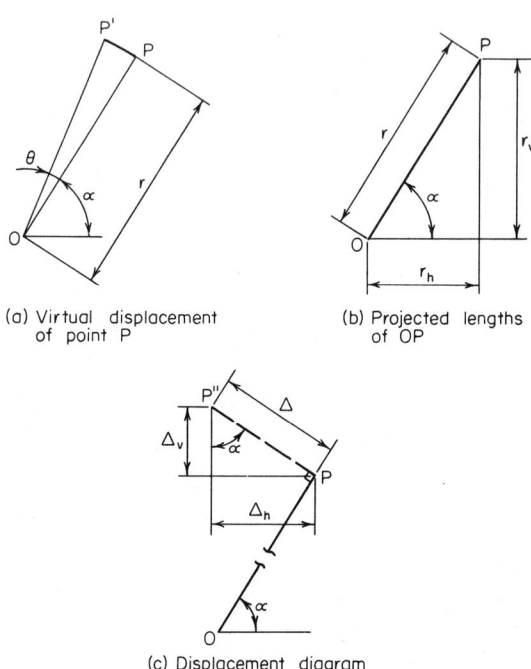

(a) Virtual displacement　　　　　(b) Projected lengths
　　of point P　　　　　　　　　　　　of OP

(c) Displacement diagram

Fig. 93

2. Evaluate Δ_h and Δ_v, considering only absolute values

Since θ is infinitesimally small, set $PP'' = r\theta$; $\Delta_h = PP'' \sin \alpha = r\theta \sin \alpha$; $\Delta_v = PP'' \cos \alpha = r\theta \cos \alpha$. But $r \sin \alpha = r_v$ and $r \cos \alpha = r_h$; therefore, $\Delta_h = r_v\theta$ and $\Delta_v = r_h\theta$.

These results may be combined and expressed verbally as: If a point is displaced along a virtual circular arc, its displacement as projected on the u axis equals the displacement angle times the length of the radius as projected on an axis normal to u.

GABLE-FRAME ANALYSIS USING THE MECHANISM METHOD

For the frame in Fig. 92a, assume that plastic hinges form at D and F. Calculate the plastic moment associated with this assumed mode of failure by applying the mechanism method.

Calculation Procedure:

1. Indicate the frame configuration following a virtual displacement

During collapse, the frame consists of three rigid bodies: ABD, DF, and GF. To evaluate the external and internal work performed during a virtual displacement, it is necessary to locate the instantaneous center of rotation of each body.

In Fig. 94, indicate by dash lines the configuration of the frame following a virtual displacement. In Fig. 94, D is displaced to D', and F to F'. Draw a straight line through A and D intersecting the prolongation of GF at H.

Since A is the center of rotation of ABD, DD' is normal to AD and HD; since G is the center of rotation of GF, FF' is normal to GF and HF. Therefore, H is the instantaneous center of rotation of DF.

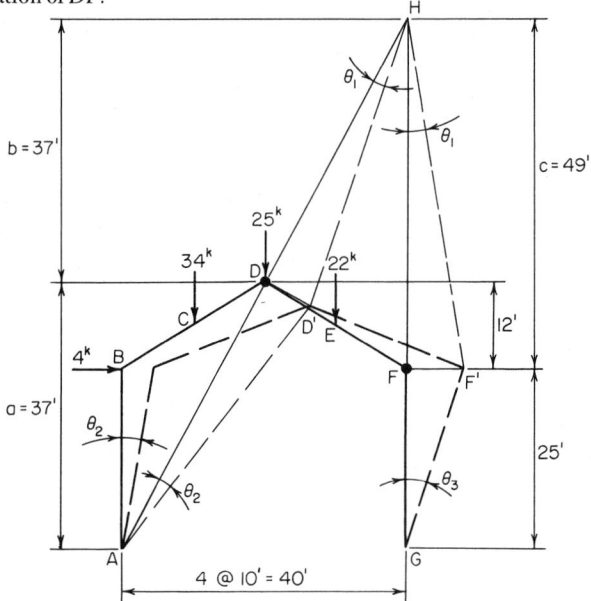

Fig. 94   Virtual displacement of frame.

2.   *Record the pertinent dimensions and rotations*

Record the dimensions a, b, and c in Fig. 94 and express θ_2 and θ_3 in terms of θ_1. Thus, $\theta_2/\theta_1 = HD/AD$; $\therefore \theta_2 = \theta_1$. Also, $\theta_3/\theta_1 = HF/GF = 49/25$; $\therefore \theta_3 = 1.96\,\theta_1$.

3.   *Determine the angular displacement and evaluate the internal work*

Determine the angular displacement (in absolute value) at D and F and evaluate the internal work in terms of θ_1. Thus, $\theta_D = \theta_1 + \theta_2 = 2\theta_1$; $\theta_F = \theta_1 + \theta_3 = 2.96\theta_1$. Then, $W_I = M_p(\theta_D + \theta_F) = 4.96 M_p \theta_1$.

4.   *Apply the theorem of virtual displacements to determine the displacement of each applied load*

Determine the displacement of each applied load in the direction of the load. Multiply the displacement by the load to obtain the external work. Record the results as shown in the following table:

Section	Load, kips	Displacement in direction of load, ft	External work, ft-kips
B	4	$\Delta_h = 25\,\theta_2 = 25\,\theta_1$	$100\,\theta_1$
C	34	$\Delta_v = 10\,\theta_2 = 10\,\theta_1$	$340\,\theta_1$
D	25	$\Delta_v = 20\,\theta_1$	$500\,\theta_1$
E	22	$\Delta_v = 10\,\theta_1$	$220\,\theta_1$
Total ...	...	....	$1{,}160\,\theta_1$

5. *Equate the external and internal work to find M_p*

Thus, $4.96M_p\theta_1 = 1{,}160\,\theta_1$; $M_p = 234$ ft-kips.

Other modes of failure may be assumed and the corresponding value of M_p computed in the same manner. It will be found that the failure mechanism analyzed in this Procedure (plastic hinges at D and F) yields the highest value of M_p and is therefore the true mechanism.

REDUCTION IN PLASTIC-MOMENT CAPACITY CAUSED BY AXIAL FORCE

A 10WF45 beam-column is subjected to an axial force of 84 kips at ultimate load. (*a*) Applying the exact method, calculate the plastic moment this section can develop with respect to the major axis. (*b*) Construct the interaction diagram for this section and then calculate the plastic moment by assuming a linear-interaction relationship that approximates the true relationship.

Calculation Procedure:

1. *Record the relevant properties of the member*

Let P = applied axial force, kips; P_y = axial force that would induce plastification if acting alone, kips = Af_y; M_p' = plastic-moment capacity of the section in combination with P, ft-kips.

A typical stress diagram for a beam-column at plastification is shown in Fig. 95*a*. To simplify the calculations, resolve this diagram into the two parts shown at the right. This procedure is tantamount to assuming that the axial load is resisted by a central core and the moment by the outer segments of the section, although in reality they are jointly resisted by the integral action of the entire section.

From the AISC *Manual*, for a 10WF45: $A = 13.24$ sq in.; $d = 10.12$ in.; $t_f = 0.618$ in.; $t_w = 0.350$ in.; $d_w = 10.12 - 2(0.618) = 8.884$ in.; $Z = 55.0$ in.³.

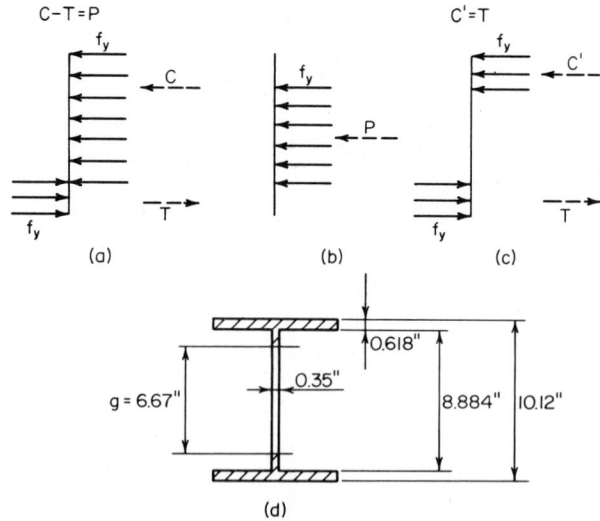

Fig. 95

2. *Assume that the central core that resists the 84-kip load is encompassed within the web; determine the core depth*

Calling the depth of the core g, refer to Fig. 95*d*. Then, $g = 84/[0.35(36)] = 6.67 <$ 8.884 in.

3. Compute the plastic modulus of the core, the plastic modulus of the remaining section, and the value of M'_p

Using data from the *Manual* for the plastic modulus of a rectangle, $Z_c = \frac{1}{4}t_w g^2 = \frac{1}{4}(0.35)$ $(6.67)^2 = 3.9$ in.^{3}; $Z_r = 55.0 - 3.9 = 51.1$ in.^{3}; $M'_p = 51.1(36)/12 = 153.3$ ft-kips. This constitutes the solution of part a. The solution of part b is given in Steps 4 through 6.

4. Assign a series of values to the parameter g and compute the corresponding sets of values of P and M'_p

Apply the results to plot the interaction diagram in Fig. 96. This comprises the parabolic curves CB and BA, where the points A, B, and C correspond to the conditions $g = 0$, $g = d_w$, and $g = d$, respectively.

The interaction diagram is readily analyzed by applying the following relationships: $dP/dg = f_y t$; $dM'_p/dg = -\frac{1}{2}f_y tg$; $\therefore dP/dM'_p = -2/g$. This result discloses that the change in slope along CB is very small, and the curvature of this arc is therefore negligible.

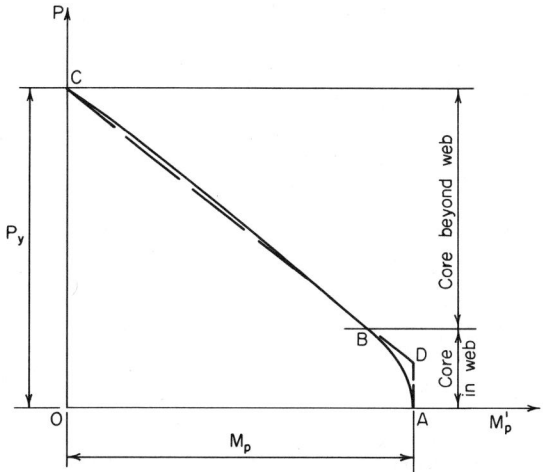

Fig. 96   Interaction diagram for axial force and moment.

5. Replace the true interaction diagram with a linear one

Draw a vertical line $AD = 0.15P_y$, and then draw the straight line CD, Fig. 96. Establish the equation of CD. Thus, slope of $CD = -0.85P_y/M_p$; $\therefore P = P_y - 0.85P_y M'_p/M_p$, or $M'_p = 1.18(1 - P/P_y)M_p$.

The provisions of one section of the AISC *Specification* are based on the linear-interaction diagram.

6. Ascertain whether the data are represented by a point on AD or CD; calculate M'_p accordingly

Thus, $P_y = Af_y = 13.24(36) = 476.6$ kips; $P/P_y = 84/476.6 = 0.176$; therefore apply the last equation given in Step 5. Thus, $M_p = 55.0(36)/12 = 165$ ft-kips; $M'_p = 1.18(1 - 0.176)$ $(165) = 160.4$ ft-kips. This result differs from that in part a by 4.6 percent.

Timber Engineering

In designing timber members, the following references are often used: *Wood Handbook*, Forest Products Laboratory, U.S. Department of Agriculture, and *National Design Specification for Stress-Grade Lumber and Its Fastenings*, National Forest Products Association. The members are assumed to be continuously dry and subject to normal loading conditions.

For most species of lumber, the true or *dressed* dimensions are less than the nominal dimensions by the following amounts: $\frac{3}{8}$ in. for dimensions less than 6 in.; $\frac{1}{2}$ in. for dimensions of 6 in. or more. The average weight of timber is 40 pcf. The width and depth of the transverse section will be denoted by b and d, respectively.

BENDING STRESS AND DEFLECTION OF WOOD JOISTS

A floor is supported by 3×8 in. wood joists spaced 16 in. on centers with an effective span of 10 ft. The total floor load transmitted to the joists is 107 psf. Compute the maximum bending stress and initial deflection, using $E = 1,760,000$ psi.

Calculation Procedure:

1. Calculate the beam properties or extract them from a table

Thus, $A = 2\text{-}\frac{5}{8}(7\frac{1}{2}) = 19.7$ sq in.; beam weight $= (A/144)$ (lumber density, pcf) $= (19.7/144)(40) = 5$ plf; $I = (1/12)(2\text{-}\frac{5}{8})(7\frac{1}{2})^3 = 92.3$ in.^{4}; $S = 92.3/3.75 = 24.6$ in.^{3}.

2. Compute the unit load carried by the joists

Thus, the unit load $w = 107(1.33) + 5 = 148$ plf, where the factor 1.33 is the width, ft, of the floor load carried by each joist, and the value of $5 = $ the beam weight, plf.

3. Compute the maximum bending stress in the joist

Thus, the bending moment in the joist is $M = (1/8)wL^2 12$, where $M = $ bending moment, in.-lb; $L = $ joist length, ft. Substituting, $M = (1/8)(148)(10)^2(12) = 22,200$ in.-lb. Then for the stress in the beam, $f = M/S$, where $f = $ stress, psi and $S = $ beam section modulus, in.^{3}; or $f = 22,200/24.6 = 902$ psi.

4. Compute the initial deflection at mid-span

Using the AISC *Manual* deflection equation, the deflection Δ in. $= (5/384)wL^4/EI$, where $I = $ section moment of inertia, in.^{4}; other symbols as before. Substituting, $\Delta = 5(148)(10)^4(1,728)/[384(1,760,000)(92.3)] = 0.205$ in. In this relation, the factor 1,728 converts cu ft to cu in.

SHEARING STRESS CAUSED BY STATIONARY CONCENTRATED LOAD

A 3×10 in. beam on a span of 12 ft carries a concentrated load of 2,730 lb located 2 ft from the support. If the allowable shearing stress is 120 psi, determine whether this load is excessive. Neglect the beam weight.

Calculation Procedure:

1. Calculate the reaction at the adjacent support

In a rectangular section the shearing stress varies parabolically with the depth and has the maximum value of $v = 1.5V/A$, where $V = $ shear, lb.

The *Wood Handbook* notes that checks are sometimes present near the neutral axis of timber beams. The vitiating effect of these checks is recognized in establishing the allowable shearing stresses. However, these checks also have a beneficial effect, for they modify the shear distribution and thereby reduce the maximum stress. The amount of this reduction depends on the position of the load. The maximum shearing stress to be applied in design is given by $v = 10(a/d)^2v'/\{9[2 + (a/d)^2]\}$, where $v = $ true maximum shearing stress, psi; $v' = $ nominal maximum stress computed from $1.5V/A$; $a = $ distance from load to adjacent support, in.

Computing the reaction R at the adjacent support, $R = V_{max} = 2,730(12 - 2)/12 = 2,275$ lb. Then, $v' = 1.5V/A = 1.5(2,275)/24.9 = 137$ psi.

2. Find the design stress

Using the equation given in Step 1, $(a/d)^2 = (24/9.5)^2 = 6.38$; $v = 10(6.38)(137)/[9(8.38)] = 116 < 120$ psi. The load is therefore not excessive.

SHEARING STRESS CAUSED BY MOVING CONCENTRATED LOAD

A 4×12 in. beam on a span of 10 ft carries a total uniform load of 150 plf and a moving concentrated load. If the allowable shearing stress is 130 psi, what is the allowable value of the moving load as governed by shear?

Calculation Procedure:

1. Calculate the reaction at the support

The transient load induces the absolute maximum shearing stress when it lies at a certain critical distance from the support rather than directly above it. This condition results from the fact that as the load recedes from the support, the reaction decreases but the shear-redistribution effect becomes less pronounced. The approximate method of analysis recommended in the *Wood Handbook* affords an expedient means of finding the moving-load capacity.

Place the moving load P at a distance of $3d$ or $\frac{1}{4}L$ from the support, whichever is less. Calculate the reaction at the support, disregarding the load within a distance of d therefrom.

Thus, $3d = 2.9$ ft and $\frac{1}{4}L = 2.5$ ft; then, $R = V_{max} = 150(5-0.96) + \frac{3}{4}P = 610 + \frac{3}{4}P$.

2. Calculate the allowable shear

Thus, $V_{allow} = \frac{2}{3}vA = \frac{2}{3}(130)(41.7) = 3,610$ lb. Then $610 + \frac{3}{4}P = 3,610$; $P = 4,000$ lb.

STRENGTH OF DEEP WOODEN BEAMS

If the allowable bending stress in a shallow beam is 1,500 psi, what is the allowable bending moment in a 12×20 in. beam?

Calculation Procedure:

1. Calculate the depth factor F

An increase in depth of a rectangular beam is accompanied by a decrease in the modulus of rupture. For beams more than 16 in. deep, it is necessary to allow for this reduction in strength by introducing a *depth factor F*.

Thus, $F = 0.81(d^2 + 143)/(d^2 + 88)$, where $d =$ dressed depth of beam, in. Substituting, $F = 0.81(19.5^2 + 143)/(19.5^2 + 88) = 0.905$.

2. Apply the result of Step 1 to obtain the moment capacity

Use the relation $M = FfS$, where the symbols are as given earlier. Thus, $M = 0.905 \times (1.5)(728.8)/12 = 82.4$ ft-kips.

DESIGN OF A WOOD-PLYWOOD BEAM

A girder having a 36-ft span is to carry a uniform load of 550 plf, which includes its estimated weight. Design a box-type member of glued construction, using the allowable stresses given in the table below. The modulus of elasticity of both materials is 1,760,000 psi, and the ratio of deflection to span cannot exceed 1/360. Architectural details limit the member depth to 40 in.

	Lumber	Plywood
Tension, psi	1,500	2,000
Compression parallel to grain, psi	1,350	1,460
Compression normal to grain, psi..........	390	405
Shear parallel to plane of plies, psi	...	72*
Shear normal to plane of plies, psi	...	192

*Use 36 psi at contact surface of flange and web to allow for stress concentration.

Calculation Procedure:

1. Compute the maximum shear and bending moment

Thus, $V = \frac{1}{2}(550)(36) = 9,900$ lb; $M = \frac{1}{8}(wL^2)12 = \frac{1}{8}(550)(36)^2 12 = 1,070,000$ in.-lb. To preclude the possibility of field error, make the tension and compression flanges alike.

2. Calculate the beam depth for a balanced condition

Assume that the member precisely satisfies the requirements for flexure and deflection, and calculate the depth associated with this balanced condition. To allow for the deflection caused by shear, which is substantial when a thin web is used, increase the deflection as computed in the conventional manner by one-half. Thus, $M = fI/c = 2fI/d = 2,700I/d$, Eq. (a). $\Delta = (7.5/48)L^2 M/EI = L/360$, Eq. (b). Substitute in Eq. (b) the value of M given by Eq. (a); solve for d to obtain $d = 37.3$ in. Use the permissible depth of 40 in. As a result of this increase in depth, a section that satisfies the requirement for flexure will satisfy the requirement for deflection as well.

3. Design the flanges

Approximate the required area of the compression flange; design the flanges. For this purpose, assume that the flanges will be $5\frac{1}{2}$ in. deep. The lever arm of the resultant forces in the flanges will be 34.8 in., and the average fiber stress will be 1,165 psi. Then, $A = 1,070,000/[1,165(34.8)] = 26.4$ sq in. Use three 2×6 in. sections with glued vertical laminations for both the tension and compression flange. Then, $A = 3(8.93) = 26.79$ sq in.; $I_0 = 3(22.5) = 67.5$ in.^{4}.

4. Design the webs

Use the approximation, $t_w = 1.25V/dv_n = 1.25(9,900)/40(192) = 1.61$ in. Try two $\frac{7}{8}$-in.-thick plywood webs. A catalog of plywood properties reveals that the $\frac{7}{8}$-in. member consists of seven plies and that the parallel plies have an aggregate thickness of 0.5 in. Draw the trial section as shown in Fig. 97.

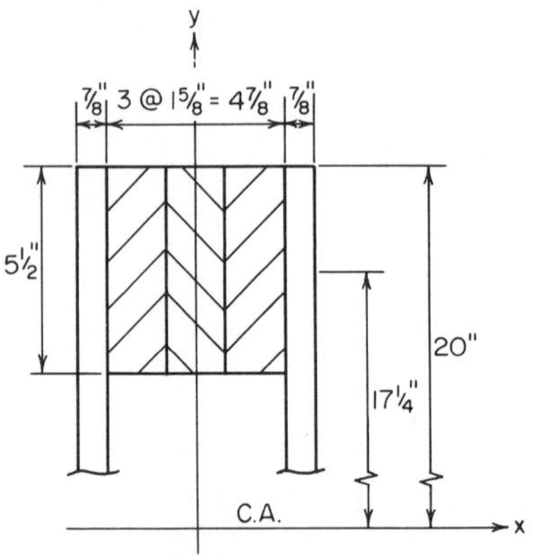

Fig. 97

5. Check the bending stress in the member

For simplicity, disregard the webs in evaluating the moment of inertia. Thus, the moment of inertia of the flanges $I_f = 2(67.5 + 26.79 \times 17.25^2) = 16,080$ in.^{4}; then the stress $f = Mc/I = 1,070,000(20)/16,080 = 1,330 < 1,350$ psi. This is acceptable.

6. Check the shearing stress at the contact surface of the flange and web

Use the relation $Q_f = Ad = 26.79(17.25) = 462$ in.^{3}. The q per surface $= VQ_f/2I_f = 9,900(462)/[2(16,080)] = 142$ pli. Assume that the shearing stress is uniform across the surface, and apply 36 psi, as noted earlier, as the allowable stress. Then, $v = 142/5.5 = 26 < 36$ psi. This is acceptable.

7. Check the shearing stress in the webs

For this purpose, include the webs in evaluating the moment of inertia but apply solely the area of the parallel plies. At the neutral axis: $Q = Q_f + Q_w = 462 + 2(0.5)(20)(10) = 662$ in.^{3}; $I = I_f + I_w = 16,080 + 2(1/12)(0.5)(40)^3 = 21,410$ in.^{4}. Then, $v = VQ/It = 9,900(662)/[21,410(2)(0.875)] = 175 < 192$ psi. This is acceptable.

8. Check the deflection, applying the moment of inertia of only the flanges

Thus, $\Delta = (7.5/384)wL^4/EI_f = 7.5(550)(36)^4(1,728)/[384(1,760,000)(16,080)] = 1.10$ in.; $\Delta/L = 1.10/[36(12)] < 1/360$. This is acceptable and the trial section is therefore satisfactory in all respects.

9. *Establish the allowable spacing of the bridging*

To do this, compare the moments of inertia with respect to the principal axes. Thus, $I_y = 2(1/12)(5.5)(4.875)^3 + 2(0.5)(40)(2.875)^2 = 433$ in.^{4}; then $I_x/I_y = 16,080/433 = 37.1$.

For this ratio, the *Wood Handbook* specifies that "the beam should be restrained by bridging or other bracing at intervals of not more than 8 ft.".

DETERMINING THE CAPACITY OF A SOLID COLUMN

An 8×10 in. column has an unbraced length of 10 ft 6 in. The allowable compressive stress is 1,500 psi and $E = 1,760,000$ psi. Calculate the allowable load on this column (a) by applying the recommendations of the *Wood Handbook*; (b) by applying the provisions of the *National Design Specification*.

Calculation Procedure:

1. *Record the properties of the member; evaluate K; classify the column*

Let $L =$ unbraced length of column, in.; $d =$ smaller side of rectangular section, in.; $f_c =$ allowable compressive stress parallel to the grain in short column of the same species, psi; $f =$ allowable compressive stress parallel to grain in column under investigation, psi.

The *Wood Handbook* divides columns into three categories: short, intermediate, and long. Let K denote a parameter defined by the equation $K = 0.64(E/f_c)^{0.5}$.

The range of the slenderness ratio and the allowable stress for each category of column is: *short column, $L/d \leq 11$ and $f = f_c$; intermediate column, $11 < L/d \leq K$ and $f = f_c[1 - \frac{1}{3}(L/d/K)^4]$; long column, $L/d > K$ and $f = 0.274E/(L/d)^2$.*

For this column, the area $A = 71.3$ sq in., using the dressed dimensions. Then, $L/d = 126/7.5 = 16.8$. Also, $K = 0.64(1,760,000/1,500)^{0.5} = 21.9$. Therefore, this is an intermediate column because L/d lies between K and 11.

2. *Compute the capacity of the member*

Use the relation: capacity, lb, $= P = Af = 71.3(1,500)[1 - \frac{1}{3}(16.8/21.9)^4] = 94,600$ lb. This constitutes the solution to part a, using data from the *Wood Handbook*. For part b, data from the *National Design Specification* will be used.

3. *Compute the capacity of the column*

Determine the stress from $f = 0.30E/(L/d)^2 = 0.30(1,760,000)/16.8^2 = 1,870$ psi. Setting $f = 1,500$ psi, $P = Af = 71.3(1,500) = 107,000$ lb. Note that the smaller stress value is used when computing the column capacity.

DESIGN OF A SOLID WOODEN COLUMN

A 12-ft-long wooden column supports a load of 98 kips. Design a solid section in the manner recommended in the *Wood Handbook* using $f_c = 1,400$ psi and $E = 1,760,000$ psi.

Calculation Procedure:

1. *Assume that d = 7.5 in., and classify the column*

Thus, $L/d = 144/7.5 = 19.2$ and $K = 0.64(1,760,000/1,400)^{0.5} = 22.7$. This is an intermediate column if the assumed dimension is correct.

2. *Compute the required area and select a section*

For an intermediate column, the stress $f = 1,400[1 - \frac{1}{3}(19.2/22.7)^4] = 1,160$ psi. Then, $A = P/f = 98,000/1,160 = 84.5$ sq in.

Study of the required area shows that an 8×12 in. column having an area of 86.3 sq in. should be used.

INVESTIGATION OF A SPACED COLUMN

The wooden column in Fig. 98 is composed of three 3×8 in. sections. Determine the capacity of the member if $f_c = 1,400$ psi and $E = 1,760,000$ psi.

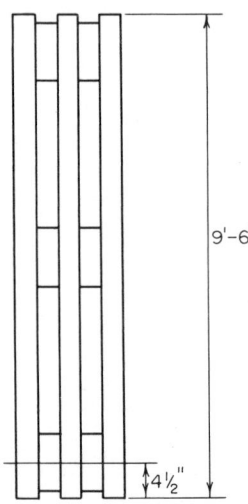

9'-6"

4½"

Fig. 98   Spaced column.

Calculation Procedure:

1.   Record the properties of the elemental section

In analyzing a spaced column, it is necessary to assess both the aggregate strength of the elements and the strength of the built-up section. The end spacer blocks exert a restraining effect on the elements and thereby enhance their capacity. This effect is taken into account by multiplying the modulus of elasticity by a *fixity factor F*.

The area of the column $A = 19.7$ sq in., when the dressed sizes are used. Also, $L/d = 114/2.625 = 43.4$; $F = 2.5$; $K = 0.64(2.5 \times 1,760,000/1,400)^{0.5} = 35.9$. Therefore, this is a long column.

2.   Calculate the aggregate strength of the elements

Thus, $f = 0.274E/(L/d)^2$ for a long column, or $f = 0.274(2.5)(1,760,000)/(43.4)^2 = 640$ psi. $P = 3(19.7)(640) = 37,800$ lb.

3.   Repeat the foregoing steps for the built-up member

Thus, $L/d = 114/7.5 = 15.2$; $K = 22.7$; therefore, this is an intermediate column. Then, $f = 1,400[1 - \frac{1}{3}(15.2/22.7)^4] = 1,306 > 640$ psi.

The column capacity is therefore limited by the elements and $P = 37,800$ lb.

COMPRESSION ON AN OBLIQUE PLANE

Determine if the joint in Fig. 99 is satisfactory with respect to bearing if the allowable compressive stresses are 1,400 and 400 psi parallel and normal to the grain, respectively.

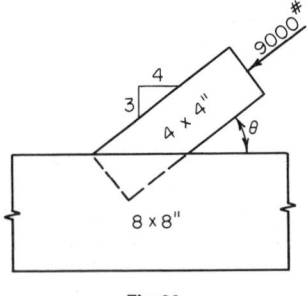

9000#

4
3

4 × 4"

θ

8 × 8"

Fig. 99

Calculation Procedure:

1.   Compute the compressive stress

Thus, $f = P/A = 9,000/3.625^2 = 685$ psi.

2.   Compute the allowable compression stress in the main member

Apply Hankinson's equation: $N = PQ/(P \sin^2 \theta + Q \cos^2 \theta)$, where P = allowable compressive stress parallel to grain, psi; Q = allowable compressive stress normal to grain, psi; N = allowable compressive stress inclined to the grain, psi; θ = angle between action line of N and direction of grain. Thus, $\sin^2 \theta = 0.36$, $\cos^2 \theta = (4/5)^2 = 0.64$; then $N = 1,400(400)/(1,400 \times 0.36 + 400 \times 0.64) = 737 > 685$ psi. Therefore, the joint is satisfactory.

3.   Alternatively, solve Hankinson's equation by using the nomogram in the Wood Handbook

DESIGN OF A NOTCHED JOINT

In Fig. 100, $M1$ is a 4×4, $F = 5,500$ lb, and $\phi = 30°$. The allowable compressive stresses are $P = 1,200$ psi and $Q = 390$ psi. The projection of $M1$ into $M2$ is restricted to a vertical distance of 2.5 in. Design a suitable notch.

Calculation Procedure:

1. Record the values of the trigonometric functions of ϕ and $\phi/2$

The most feasible type of notch is the one shown in Fig. 100, in which AC and BC bisect the angles between the intersecting edges. The allowable bearing pressures on these faces are therefore identical for the two members.

With $\phi = 30°$, $\sin 30° = 0.500$; $\sin 15° = 0.259$; $\cos 15° = 0.966$; $\tan 15° = 0.268$.

2. Find the lengths AC and BC

Express these two lengths as functions of AB. Or, $AB = b/\sin \phi$; $AC = (b \sin \phi/2)/\sin \phi$; $BC = (b \cos \phi/2)/\sin \phi$; $AC = 3.625(0.259/0.500) = 1.9$ in.; $BC = 3.625(0.966/0.500) = 7.0$ in. The projection into $M2$ is therefore not excessive.

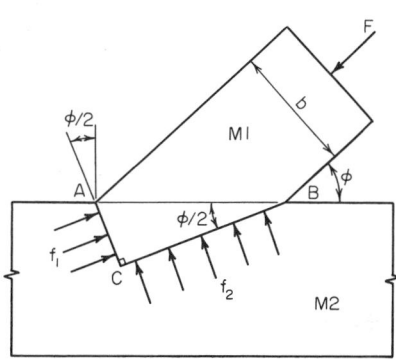

3. Evaluate the stresses f_1 and f_2

Resolve F into components parallel to AC and BC. Thus, $f_1 = F \sin \phi/A(\tan \phi/2)$; $f_2 = F \sin \phi(\tan \phi/2)/A$, where $A =$ cross-sectional area of $M1$. Substituting, $f_1 = 783$ psi; $f_2 = 56$ psi.

Fig. 100

4. Calculate the allowable stresses

Compute the allowable stresses N_1 and N_2 on AC and BC, respectively, and compare these with the actual stresses. Thus, using Hankinson's equation from the previous Calculation Procedure, $N_1 = 1,200(390)/(1,200 \times 0.259^2 + 390 \times 0.966^2) = 1,053$ psi. This is acceptable because it is greater than the actual stress. Also, $N_2 = 1,200(390)/(1,200 \times 0.966^2 + 390 \times 0.259^2) = 408$ psi. This is also acceptable and the joint is therefore satisfactory.

ALLOWABLE LATERAL LOAD ON NAILS

In Fig. 101, the Western-hemlock members are connected with six $50d$ common nails. Calculate the lateral load P that may be applied to this connection.

Calculation Procedure:

1. Determine the member group

The capacity of this connection is calculated in conformity with Part VIII of the *National Design Specification*. Refer to the *Specification* to ascertain the classification of this species. Western hemlock is in Group III.

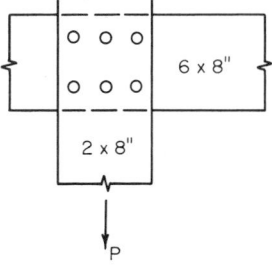

2. Determine the properties of the nail

Refer to the *Specification* to determine the properties of the nail. Calculate the penetration-diameter ratio, and compare this value with that stipulated in the *Specification*. Thus, length = 5.5 in.; diameter = 0.244 in.; penetration/diameter ratio = $(5.5 - 1.63)/0.244 = 15.9 > 13$. This is acceptable.

Fig. 101

3. Find the capacity of the connection

Using the *Specification*, find the capacity of the nail. Then, the capacity of the connection = $P = 6(165) = 990$ lb.

CAPACITY OF LAG SCREWS

In Fig. 102, the cottonwood members are connected with three $\frac{5}{8}$-in. lag screws, 8 in. long. Determine the load P that may be applied to this connection.

Calculation Procedure:

1. Determine the member Group

The *National Design Specification* shows that cottonwood is classified in Group IV.

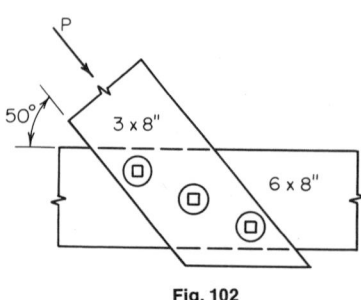

Fig. 102

2. Find the allowable screw loads

The *National Design Specification* gives the following values for each screw: allowable load parallel to grain = 550 lb; allowable load normal to grain = 330 lb.

3. Compute the allowable load on the connection

Use the Scholten nomogram, or $N = PQ/(P \sin^2 \theta + Q \cos^2 \theta)$, with $\theta = 50°$, and solve as given earlier. Either solution gives $P = 3(395) = 1,185$ lb.

DESIGN OF A BOLTED SPLICE

A 6×12 in. southern-pine member carrying a tensile force of 56 kips parallel to the grain is to be spliced with steel side plates. Design the splice.

Calculation Procedure:

1. Determine the number of bolts, and bolt size, required

Find the bolt capacity from the *National Design Specification*. The *Specification* allows a 25 percent increase in capacity of the parallel-to-grain loading when steel plates are used as side members.

Determine the number of bolts from $n = P/\text{capacity per bolt}$, lb, where P = load, lb. Assuming $\frac{7}{8}$ in.-diameter bolts, $n = 56,000/[3,940(1.25)] = 11.4$; use 12 bolts. The value 1.25 in the denominator is the increase in bolt load mentioned above.

As a trial, use three rows of four bolts each, as shown in Fig. 103.

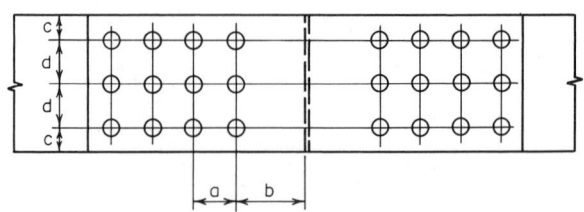

Fig. 103

2. Determine if the joint complies with the Specification

Assume $\frac{15}{16}$ in.-diameter bolt holes. The gross area of the dressed lumber is 63.25 sq in. The net area = gross area − area of the bolt holes = $63.25 - 3(0.94)(5.5) = 47.74$ sq in. The bearing area under the bolts = number of bolts (bolt diameter, in.)(width, in.) = $12(0.875)(5.5) = 57.75$ sq in. The ratio of the net to bearing area is $47.74/57.75 = 0.83 > 0.80$. This is acceptable, according to the *Specification*. The joint is therefore satisfactory and the assumptions are usable in the design.

3. Establish the longitudinal bolt spacing

Using the *Specification*, $a = 4(\frac{7}{8}) = 3.5$ in.; $b_{\min} = 7(\frac{7}{8}) = 6\frac{1}{8}$ in.

4. Establish the transverse bolt spacing

Using the *Specification*, $L/D = 5.5/(\frac{7}{8}) = 6.3 > 6$. Make $c = 2$ in. and $d = 3\frac{3}{4}$ in.

INVESTIGATION OF A TIMBER-CONNECTOR JOINT

The members in Fig. 104a have the following sizes: A, 4×8 in.; B, 3×8 in. They are connected by six 4-in. split-ring connectors, in the manner shown. The lumber is dense structural redwood. Investigate the adequacy of this joint and establish the spacing of the connectors.

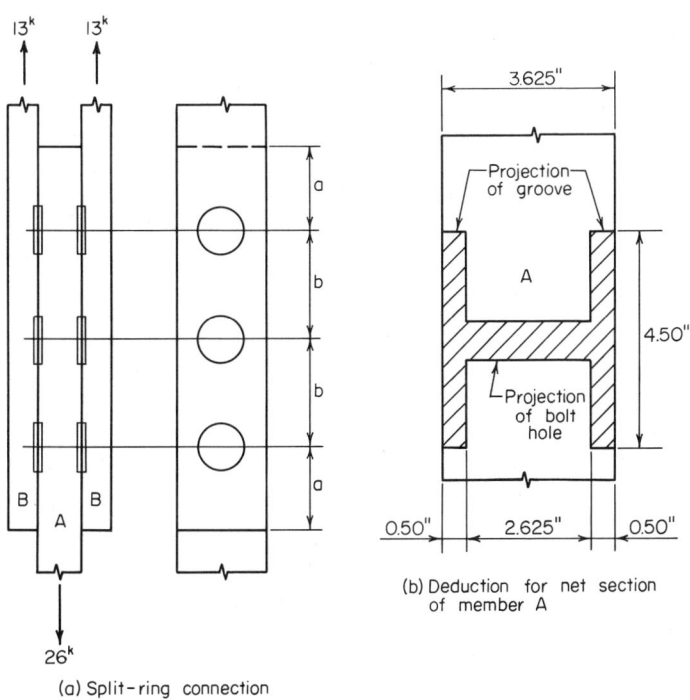

(a) Split-ring connection

(b) Deduction for net section of member A

Fig. 104

Calculation Procedure:

1. Determine the allowable stress

The *National Design Specification* shows that the allowable stress is 1,700 psi.

2. Find the lumber Group

The *Specification* shows this species is classified in Group C.

3. Compute the capacity of the connectors

The *Specification* shows that the capacity of a connector in parallel-to-grain loading for Group C lumber is 4,380 lb. With six connectors, the total capacity is 6(4,380) = 26,280 lb. This is acceptable.

The *Specification* requires a minimum edge distance of $2\frac{3}{4}$ in. The edge distance in the present instance is $3\frac{3}{4}$ in.

4. Calculate the net area of member A

Apply the dimensions of the groove, which are recorded in the *Specification*. Referring to Fig. 104b, gross area = 27.19 sq in. The projected area of the groove and bolt hole = 4.5(1.00) + 0.813(2.625) = 6.63 sq in. The net area = 27.19 − 6.63 = 20.56 sq in.

5. Calculate the stress at the net section; compare with the allowable stress

The stress f = load, lb/net area, sq in. = 26,000/20.56 = 1,260 psi. From the *Specification*, the allowable stress is $f_{\text{allow}} = (\frac{7}{8})(1,700) = 1,488$ psi. Also from the *Specification*, $f_{\text{allow}} = 1,650$ psi. The joint is therefore satisfactory in all respects.

6. Establish the connector spacing

Using the *Specification*, apply the recorded values without reduction because the connectors are stressed almost to capacity. Thus, $a = 7$ in. and $b = 9$ in.

Reinforced Concrete

The design of reinforced-concrete members in this Handbook will be executed in accordance with the specification titled *Building Code Requirements for Reinforced Concrete* of the American Concrete Institute (ACI). The ACI *Reinforced Concrete Design Handbook* contains many useful tables that expedite design work. The designer should thoroughly familiarize himself with this handbook and use the tables it contains whenever possible.

The spacing of steel reinforcing bars in a concrete member is subject to the restrictions imposed by the ACI *Code*. With reference to the beam and slab shown in Fig. 105, the reinforcing steel is assumed for simplicity to be concentrated at its centroidal axis, and the effective depth of the flexural member is taken as the distance from the extreme compression fiber to this axis. (The term *depth* will hereafter refer to the *effective* rather than the overall depth of the beam.) For design purposes, it is usually assumed that the distance from the exterior surface to the center of the first row of steel bars is $2\frac{1}{2}$ in. in a beam with web stirrups, 2 in. in a beam without stirrups, and 1 in. in a slab. Where two rows of steel bars are provided, it is usually assumed that the distance from the exterior surface to the centroidal axis of the reinforcement is $3\frac{1}{2}$ in. The ACI *Handbook* gives the minimum beam widths needed to accommodate various combinations of bars in one row.

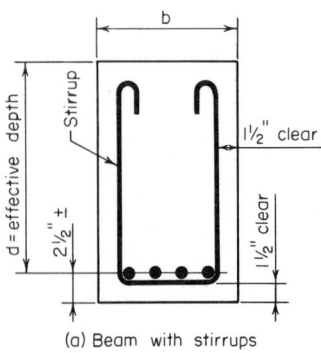

(a) Beam with stirrups

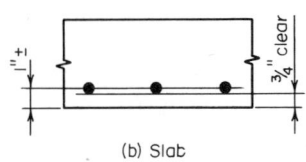

(b) Slab

Fig. 105   Spacing of reinforcing bars.

In a well-proportioned beam, the width-depth ratio lies between 0.5 and 0.75. The width and overall depth are usually an even number of inches.

The basic notational system pertaining to reinforced concrete beams is: f_c' = ultimate compressive strength of concrete, psi; f_c = maximum compressive stress in concrete, psi; f_s = tensile stress in steel, psi; f_y = yield-point stress in steel, psi; ϵ_c = strain of extreme compression fiber; ϵ_s = strain of steel; b = beam width, in.; d = beam depth, in.; A_s = area of tension reinforcement, sq in.; p = tension-reinforcement ratio, A_s/bd; q = tension-reinforcement index, pf_y/f'_c; n = ratio of modulus of elasticity of steel to that of concrete, E_s/E_c; C = resultant compressive force on transverse section, lb; T = resultant tensile force on transverse section, lb.

Where the subscript b is appended to a symbol, it signifies that the given quantity is evaluated at balanced-design conditions.

Design of Flexural Members by Ultimate-strength Method

In the ultimate-strength design of a reinforced-concrete structure, as in the plastic design of a steel structure, the capacity of the structure is found by determining the load that will cause failure and dividing this result by the prescribed load factor. The load at impending failure is termed the *ultimate load*, and the maximum bending moment associated with this load is called the *ultimate moment*.

Since the tensile strength of concrete is relatively small, it is generally disregarded entirely in analyzing a beam. Consequently, the effective beam section is considered to comprise the reinforcing steel and the concrete on the compression side of the neutral axis, the concrete between these component areas serving merely as the ligature of the member.

The following notational system is applied in ultimate strength design: a = depth of compression block, in.; c = distance from extreme compression fiber to neutral axis, in.; ϕ = capacity-reduction factor.

Where the subscript u is appended to a symbol, it signifies that the given quantity is evaluated at ultimate load.

For simplicity, Fig. 106, designers assume that when the ultimate moment is attained at a given section, there is a uniform stress in the concrete extending across a depth a,

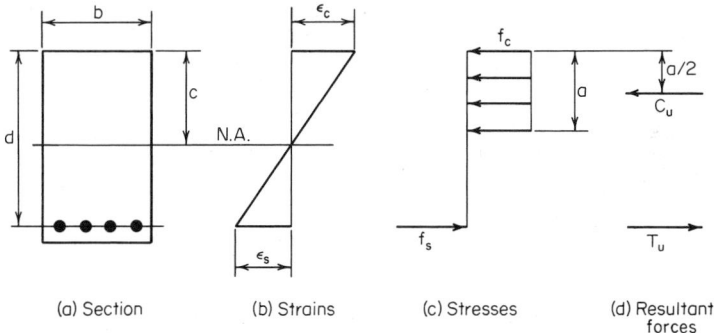

(a) Section   (b) Strains   (c) Stresses   (d) Resultant forces

Fig. 106   Conditions at ultimate moment.

and that $f_c = 0.85f'_c$, and $a = k_1c$, where k_1 has the value stipulated in the ACI *Code*.

A reinforced-concrete beam has three potential modes of failure: crushing of the concrete, which is assumed to occur when ϵ_c reaches the value of 0.003; yielding of the steel, which begins when f_s reaches the value f_y; and the simultaneous crushing of the concrete and yielding of the steel. A beam that tends to fail by the third mode is said to be in *balanced design*. If the value of p exceeds that corresponding to balanced design (i.e., if there is an excess of reinforcement), the beam tends to fail by crushing of the concrete. But if the value of p is less than that corresponding to balanced design, the beam tends to fail by yielding of the steel.

Failure of the beam by the first mode would occur precipitously and without warning, whereas failure by the second mode would occur gradually, offering visible evidence of progressive failure. Therefore, to ensure that yielding of the steel would occur prior to failure of the concrete, the ACI *Code* imposes an upper limit of $0.75p_b$ on p.

To allow for material imperfections, defects in workmanship, etc., the *Code* introduces the capacity-reduction factor ϕ. A section of the *Code* sets $\phi = 0.90$ with respect to flexure and $\phi = 0.85$ with respect to diagonal tension, bond, and anchorage.

The basic equations for the ultimate-strength design of a rectangular beam reinforced solely in tension are:

$$C_u = 0.85abf'_c \qquad T_u = A_sf_y \qquad (1)$$

$$q = \frac{(A_s/\,bd)\,f_y}{f'_c} \qquad (2)$$

$$a = 1.18qd \qquad c = 1.18qd/k_1 \qquad (3)$$

$$M_u = \phi A_s f_y \left(d - \frac{a}{2} \right) \tag{4}$$

$$M_u = \phi A_s f_y d (1 - 0.59q) \tag{5}$$

$$M_u = \phi b d^2 f'_c q (1 - 0.59q) \tag{6}$$

$$A_s = \frac{bdf_c - [(bdf_c)^2 - 2bf_c M_u/\phi]^{0.5}}{f_y} \tag{7}$$

$$p_b = \frac{0.85 k_1 f'_c}{f_y} \frac{87,000}{87,000 + f_y} \tag{8}$$

$$q_b = 0.85 k_1 \left(\frac{87,000}{87,000 + f_y} \right) \tag{9}$$

In accordance with the *Code,*

$$q_{max} = 0.75 \, q_b = 0.6375 k_1 \left(\frac{87,000}{87,000 + f_y} \right) \tag{10}$$

Figure 107 shows the relationship between M_u and A_s for a beam of given size. As A_s increases, the internal forces C_u and T_u increase proportionately, but M_u increases by a smaller proportion because the action line of C_u is depressed. The M_u-A_s diagram

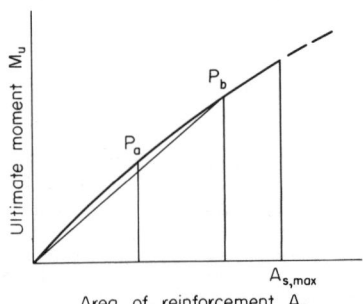

Fig. 107

is parabolic, but its curvature is small. Comparing the coordinates of two points P_a and P_b, the following result is obtained, in which the subscripts correspond to that of the given point:

$$M_{ua}/A_{sa} > M_{ub}/A_{sb} \qquad \text{where } A_{sa} < A_{sb} \tag{11}$$

CAPACITY OF A RECTANGULAR BEAM

A rectangular beam having a width of 12 in. and an effective depth of 19.5 in. is reinforced with steel bars having an area of 5.37 sq in. The beam is made of 2,500-psi concrete and the steel has a yield-point stress of 40,000 psi. Compute the ultimate moment this beam may resist (a) without referring to any design tables and without applying the basic equations of ultimate-strength design except those that are readily apparent; (b) by applying the basic equations.

Calculation Procedure:

1. *Compute the area of reinforcement for balanced design*

Use the relation $\epsilon_s = f_y/E_s = 40,000/29,000,000 = 0.00138$. For balanced design, $c/d = \epsilon_c/(\epsilon_c + \epsilon_s) = 0.003/(0.003 + 0.00138) = 0.685$. Solving for c using the relation for c/d, $c = 13.36$ in. Also, $a = k_1c = 0.85(13.36) = 11.36$ in. Then: $T_u = C_u = ab(0.85)$ $f'_c = 11.36(12)(0.85)(2,500) = 290,000$ lb; $A_s = T_u/f_y = 290,000/40,000 = 7.25$ sq in.; and $0.75A_s = 5.44$ sq in. In the present instance, $A_s = 5.37$ sq in. This is acceptable.

2. Compute the ultimate-moment capacity of this member
   Thus: $T_u = A_sf_y = 5.37(40,000) = 215,000$ lb; $C_u = ab(0.85)f'_c = 25,500a = 215,000$ lb; $a = 8.43$ in.; $M_u = \phi T_u(d - a/2) = 0.90(215,000)(19.5 - 8.43/2) = 2,960,000$ in.-lb. These two steps comprise the solution to part a. The next two steps comprise the solution of part b.

3. Apply Eq. (10); ascertain if the member satisfies the Code
   Thus, $q_{max} = 0.6375k_1(87,000)/(87,000 + f_y) = 0.6375(0.85)(87/127) = 0.371$; $q = (A_s/bd)f_y/f'_c = (5.37/12 \times 19.5)40/2.5 = 0.367$. This is acceptable.

4. Compute the ultimate-moment capacity
   Applying Eq. (5), $M_u = \phi A_sf_yd(1 - 0.59q) = 0.90(5.37)(40,000)(19.5)(1 - 0.59 \times 0.367)$ $= 2,960,000$ in.-lb. This agress exactly with the result computed in Step 2.

DESIGN OF A RECTANGULAR BEAM

A beam on a simple span of 20 ft is to carry a uniformly distributed live load of 1,670 plf and a dead load of 470 plf, which includes the estimated weight of the beam. Architectural details restrict the beam width to 12 in. and require that the depth be made as small as possible. Design the section, using $f'_c = 3,000$ psi and $f_y = 40,000$ psi.

Calculation Procedure:

1. Compute the ultimate load for which the member is to be designed
   The beam depth is minimized by providing the maximum amount of reinforcement permitted by the *Code*. From the previous Calculation Procedure, $q_{max} = 0.371$.
   Use the load factors given in the *Code*: $w_{DL} = 470$ plf; $w_{LL} = 1,670$ plf; $L = 20$ ft. Then, $w_u = 1.5(470) + 1.8(1,670) = 3,710$ plf; $M_u = \frac{1}{8}(3,710)(20)^2 12 = 2,230,000$ in.-lb.

2. Establish the beam size
   Solve Eq. (6) for d. Thus, $d^2 = M_u/\phi bf'_c q(1 - 0.59q) = 2,230,000/[0.90(12)(3,000) \times (0.371)(0.781)]$; $d = 15.4$ in.
   Set $d = 15.5$ in. Then the corresponding reduction in the value of q is negligible.

3. Select the reinforcing bars
   Using Eq. (2), $A_s = qbdf'_c/f_y = 0.371(12)(15.5)(3/40) = 5.18$ sq in. Use four No. 9 and two No. 7 bars, for which $A_s = 5.20$ sq in. This group of bars cannot be accommodated in the 12-in. width and must therefore be placed in two rows. The overall beam depth will therefore be 19 in.

4. Summarize the design
   Thus: beam size, 12×19 in.; reinforcement, four No. 9 and two No. 7 bars.

DESIGN OF THE REINFORCEMENT IN A RECTANGULAR BEAM OF GIVEN SIZE

A rectangular beam 9 in. wide and 13.5 in. effective depth is to sustain an ultimate moment of 95 ft-kips. Compute the area of reinforcement, using $f'_c = 3,000$ psi and $f_y = 40,000$ psi.

Calculation Procedure:

1. Investigate the adequacy of the beam size
   From previous Calculation Procedures, $q_{max} = 0.371$. By Eq. (6), $M_{u,max} = 0.90 \times (9)(13.5)^2(3)(0.371)(0.781) = 1,280$ in.-kips. $M_u = 95(12) = 1,140$ in.-kips. This is acceptable.

2. Apply Eq. (7) to evaluate A_s

Thus, $f_c = 0.85(3) = 2.55$ ksi; $bdf_c = 9(13.5)(2.55) = 309.8$ kips; $A_s = [309.8 - (309.8^2 - 58,140)^{0.5}]/40 = 2.88$ sq in.

CAPACITY OF A T BEAM

Determine the ultimate moment that may be resisted by the T beam in Fig. 108a, if $f'_c = 3,000$ psi and $f_y = 40,000$ psi.

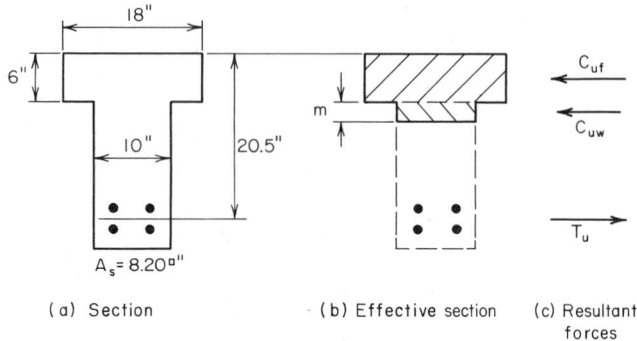

(a) Section   (b) Effective section   (c) Resultant forces

Fig. 108

Calculation Procedure:

1. Compute T_u and the resultant force that may be developed in the flange

Thus, $T_u = 8.20(40,000) = 328,000$ lb; $f_c = 0.85(3,000) = 2,550$ psi; $C_{uf} = 18(6)(2,550) = 275,400$ lb. Since $C_{uf} < T_u$, the deficiency must be supplied by the web.

2. Compute the resultant force developed in the web, and the depth of the stress block in the web

Thus, $C_{uw} = 328,000 - 275,400 = 52,600$ lb; $m =$ depth of the stress block $= 52,600/[2,550(10)] = 2.06$ in.

3. Evaluate the ultimate-moment capacity

Thus, $M_u = 0.90[275,400(20.5 - 3) + 52,600(20.5 - 6 - 1.03)] = 4,975,000$ in.-lb.

4. Determine if the reinforcement complies with the Code

Let $b' =$ width of web, in.; $A_{s1} =$ area of reinforcement needed to resist the compressive force in the overhanging portion of the flange, sq in.; $A_{s2} =$ area of reinforcement needed to resist the compressive force in the remainder of the section, sq in. Then, $p_2 = A_{s2}/b'd$; $A_{s1} = 2,550(6)(18 - 10)/40,000 = 3.06$ sq in.; $A_{s2} = 8.20 - 3.06 = 5.14$ sq in. Then, $p_2 = 5.14/[10(20.5)] = 0.025$.

A section of the ACI *Code* subjects the reinforcement ratio p_2 to the same restriction as that in a rectangular beam. By Eq. (8), $p_{2.max} = 0.75p_b = 0.75(0.85)(0.85)(3/40)(87/127) = 0.0278 > 0.025$. This is acceptable.

CAPACITY OF A T BEAM OF GIVEN SIZE

The T beam in Fig. 109 is made of 3,000-psi concrete and $f_y = 40,000$ psi. Determine the ultimate-moment capacity of this member if reinforced in tension only.

Calculation Procedure:

1. Compute C_{u1}, $C_{u2,max}$, and s_{max}

Let the subscript 1 refer to the overhanging portion of the flange and the subscript 2 refer to the remainder of the compression zone. Then, $f_c = 0.85(3,000) = 2,550$ psi;

$C_{u1} = 2,550(5)(16-10) = 76,500$ lb. From the previous Calculation Procedure, $p_{2,\max} = 0.0278$. Then: $A_{s2,\max} = 0.0278(10)(19.5) = 5.42$ sq in.; $C_{u2,\max} = 5.42(40,000) = 216,800$ lb; $s_{\max} = 216,800/[10(2,550)] = 8.50$ in.

2. Compute the ultimate-moment capacity
Thus, $M_{u,\max} = 0.90[76,500(19.5-5/2) + 216,800(19.5-8.50/2)] = 4,145,000$ in.-lb.

DESIGN OF REINFORCEMENT IN A T BEAM OF GIVEN SIZE

The T beam in Fig. 109 is to resist an ultimate moment of 3,960,000 in.-lb. Determine the required area of reinforcement, using $f'_c = 3,000$ psi and $f_y = 40,000$ psi.

Calculation Procedure:

1. Obtain a moment not subject to reduction
From the previous Calculation Procedure, the ultimate-moment capacity of this member is 4,145,000 in.-lb. To facilitate the design, divide the given ultimate moment M_u by the capacity-reduction factor to obtain a moment M'_u that is not subject to reduction. Thus, $M'_u = 3,960,000/0.9 = 4,400,000$ in.-lb.

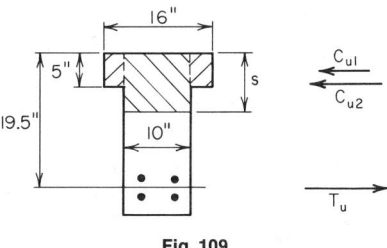

Fig. 109

2. Compute the value of s associated with the given moment
From Step 2 in the previous Calculation Procedure, $M'_{u1} = 1,300,000$ in.-lb. Then, $M'_{u2} = 4,400,000 - 1,300,000 = 3,100,000$ in.-lb. But $M'_{u2} = 2,550(10s)(19.5-s/2)$; solving, $s = 7.79$ in.

3. Compute the area of reinforcement
Thus, $C_{u2} = M'_{u2}/(d-\tfrac{1}{2}s) = 3,100,000/(19.5-3.90) = 198,700$ lb. From Step 1 of the previous Calculation Procedure, $C_{u1} = 76,500$ lb; $T_u = 76,500 + 198,700 = 275,200$ lb; $A_s = 275,200/40,000 = 6.88$ sq in.

4. Verify the solution
To verify the solution, compute the ultimate-moment capacity of the member. Use the notational system given in earlier Calculation Procedures. Thus, $C_{uf} = 16(5)(2,550) = 204,000$ lb; $C_{uw} = 275,200 - 204,000 = 71,200$ lb; $m = 71,200/[2,550(10)] = 2.79$ in.; $M_u = 0.90[204,000(19.5-2.5) + 71,200(19.5-5-1.40)] = 3,960,000$ in.-lb. Thus, the result is verified because the computed moment equals the given moment.

REINFORCEMENT AREA FOR A DOUBLY REINFORCED RECTANGULAR BEAM

A beam that is to resist an ultimate moment of 690 ft-kips is restricted to a 14-in. width and 24-in. total depth. Using $f'_c = 5,000$ psi and $f_y = 50,000$ psi, determine the area of reinforcement.

Calculation Procedure:

1. Compute the values of q_b, q_{max}, and p_{max} for a singly reinforced beam
As the following calculations will show, it is necessary to reinforce the beam both in tension and in compression. In Fig. 110, let A_s = area of tension reinforcement, sq in.; A'_s = area of compression reinforcement, sq in.; d' = distance from compression face of concrete to centroid of compression reinforcement, in.; f_s = stress in tension steel, psi; f'_s = stress in compression steel, psi; ϵ'_s = strain in compression steel; $p = A_s/bd$; $p' = A'_s/bd$; $q = pf_y/f'_c$; M_u = ultimate moment to be resisted by member, in.-lb; M_{u1} = ultimate-moment capacity of member if reinforced solely in tension; M_{u2} = increase in

ultimate-moment capacity resulting from use of compression reinforcement; $C_{u1} =$ resultant force in concrete, lb; $C_{u2} =$ resultant force in compression steel, lb.

If $f'_s = f_y$, the tension reinforcement may be resolved into two parts having areas of $A_s - A'_s$ and A'_s. The first part, acting in combination with the concrete, develops the moment M_{u1}. The second part, acting in combination with the compression reinforcement, develops the moment M_{u2}.

To ensure that failure will result from yielding of the tension steel rather than crushing of the concrete, the ACI *Code* limits $p - p'$ to a maximum value of $0.75p_b$, where p_b

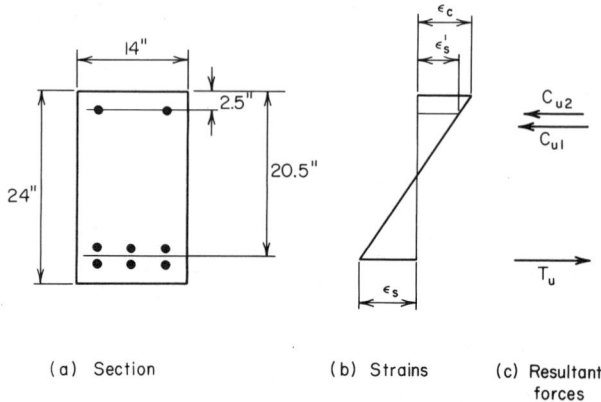

(a) Section   (b) Strains   (c) Resultant
                                                              forces

Fig. 110   Doubly reinforced rectangular beam.

has the same significance as for a singly reinforced beam. Thus the *Code* in effect permits setting $f'_s = f_y$ if inception of yielding in the compression steel will precede or coincide with failure of the concrete at balanced-design ultimate moment. This, however, introduces an inconsistency, for the limit imposed on $p - p'$ precludes balanced design.

Using Eq. (9), $q_b = 0.85(0.80)(87/137) = 0.432$; $q_{max} = 0.75(0.432) = 0.324$; $p_{max} = 0.324(5/50) = 0.0324$.

2.   Compute M_{u1}, M_{u2}, and C_{u2}

Thus, $M_u = 690,000(12) = 8,280,000$ in.-lb. Since two rows of tension bars are probably required, $d = 24 - 3.5 = 20.5$ in. By Eq. (6), $M_{u1} = 0.90(14)(20.5)^2(5,000) \times (0.324)(0.809) = 6,940,000$ in.-lb; $M_{u2} = 8,280,000 - 6,940,000 = 1,340,000$ in.-lb; $C_{u2} = M_{u2}/(d - d') = 1,340,000/(20.5 - 2.5) = 74,400$ lb.

3.   Compute the value of ϵ'_s under the balanced-design ultimate moment

Compare this value with the strain at incipient yielding. By Eq. (3), $c_b = 1.18q_b d/k_1 = 1.18(0.432)(20.5)/0.80 = 13.1$ in.; $\epsilon'_s/\epsilon_c = (13.1 - 2.5)/13.1 = 0.809$; $\epsilon'_s = 0.809(0.003) = 0.00243$; $\epsilon_y = 50/29,000 = 0.0017 < \epsilon'_s$. The compression reinforcement will therefore yield before the concrete fails, and $f'_s = f_y$ may be used.

4.   Alternatively, test the compression steel for yielding

Apply $p - p' \geq 0.85k_1 f'_c d'(87,000)/[f_y d(87,000 - f_y)]$, Eq. (12). If this relation obtains, the compression steel will yield. The value of the right-hand member is $0.85(0.80)(5/50)(2.5/20.5)(87/37) = 0.0195$. From the preceeding calculations, $p - p' = 0.0324 > 0.0195$. This is acceptable.

5.   Determine the areas of reinforcement

Using Eq. (2), $A_s - A'_s = q_{max}bdf'_c/f_y = 0.324(14)(20.5)(5/50) = 9.30$ sq in.; $A'_s = C_{u2}/\phi f_y = 74,400/[0.90(50,000)] = 1.65$ sq in.; $A_s = 9.30 + 1.65 = 10.95$ sq in.

6. Verify the solution

Apply the following equations for the ultimate-moment capacity: $a = (A_s - A'_s)f_y/[0.85f'_cb]$, Eq. (13); $a = 9.30(50,000)/[0.85(5,000)(14)] = 7.82$ in. Also, $M_u = \phi f_y[(A_s - A'_s)(d - a/2) + A'_s(d - d')]$, Eq. (14); $M_u = 0.90(50,000)(9.30 \times 16.59 + 1.65 \times 18) = 8,280,000$ in.-lb, as before. Therefore, the solution has been verified.

DESIGN OF WEB REINFORCEMENT

A 15-in.-wide 22.5-in.-effective-depth beam carries a uniform ultimate load of 10.2 klf. The beam is simply supported and the clear distance between supports is 18 ft. Using $f'_c = 3,000$ psi and $f_y = 40,000$ psi, design web reinforcement in the form of vertical U stirrups for this beam.

Calculation Procedure:

1. Construct the shearing-stress diagram for half-span

The ACI *Code* provides two alternative methods for computing the allowable shearing stress on an unreinforced web. The more precise method recognizes the contribution of both the shearing stress and flexural stress on a cross section in producing diagonal tension. The less precise and more conservative method restricts the shearing stress to a stipulated value that is independent of the flexural stress.

For simplicity, the latter method is adopted here. A section of the *Code* sets $\phi = 0.85$ with respect to the design of web reinforcement. Let $v_u =$ nominal ultimate shearing stress, psi; $v_c =$ shearing stress resisted by concrete, psi; $v'_u =$ shearing stress resisted by the web reinforcement, psi; $A_v =$ total cross-sectional area of stirrup, sq in.; $V_u =$ ultimate vertical shear at section, lb; $s =$ center-to-center spacing of stirrups, in.

The shearing-stress diagram for half-span is shown in Fig. 111. Establish the region *AF* within which web reinforcement is required. The *Code* sets the allowable shearing

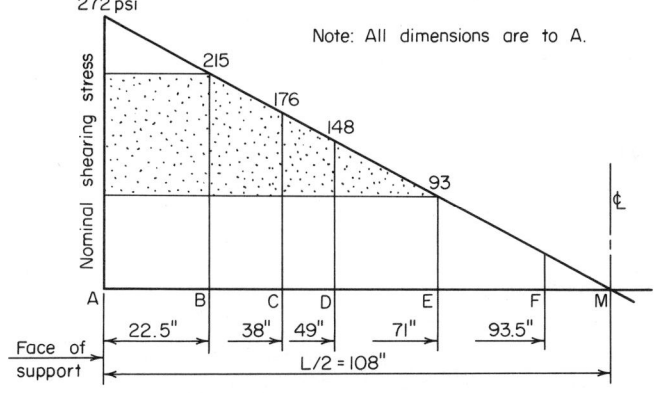

Fig. 111   Shearing-stress diagram.

stress in the concrete at $v_c = 2\phi(f'_c)^{0.5}$, Eq. (15). The equation for nominal ultimate shearing stress is $v_u = V_u/bd$, Eq. (16). Then, $v_c = 2(0.85)(3,000)^{0.5} = 93$ psi.

At the face of the support, $V_u = 9(10,200) = 91,800$ lb; $v_u = 91,800/15(22.5) = 272$ psi. The slope of the shearing-stress diagram $= -272/108 = -2.52$ psi/in. At distance d from the face of the support, $v_u = 272 - 22.5(2.52) = 215$ psi; $v'_u = 215 - 93 = 122$ psi.

Let E denote the section at which $v_u = v_c$. Then, $AE = (272 - 93)/2.52 = 71$ in. A section of the *Code* requires that web reinforcement be continued for a distance d beyond the section where $v_u = v_c$; $AF = 71 + 22.5 = 93.5$ in.

2. Check the beam size for Code compliance

Thus, $v_{u,max} = 10\phi(f'_c)^{0.5} = 466 > 215$ psi. This is acceptable.

3. Select the stirrup size

Equate the spacing near the support to the minimum practical value, which is generally considered to be 4 in. The equation for stirrup spacing is $s = \phi A_v f_y / v'_u b$, Eq. (17). Then, $A_v = s v'_u b / \phi f_y = 4(122)(15)/[0.85(40,000)] = 0.215$ sq in. Since each stirrup is bent into the form of a U, the total cross-sectional area is twice that of a straight bar. Use No. 3 stirrups for which $A_v = 2(0.11) = 0.22$ sq in.

4. Establish the maximum allowable stirrup spacing

Apply the criteria of the *Code*, or $s_{max} = d/4$ if $v_u > 6\phi(f'_c)^{0.5}$. The right-hand member of this inequality has the value 279 psi, and this limit therefore does not apply. Then, $s_{max} = d/2 = 11.25$ in., or $s_{max} = A_v/0.0015b = 0.22/[0.0015(15)] = 9.8$ in. The latter limit applies, and the stirrup spacing will therefore be restricted to 9 in.

5. Locate the beam sections at which the required stirrup spacing is 6 in. and 9 in.

Use Eq. (17). Then, $\phi A_v f_y / b = 0.85(0.22)(40,000)/15 = 499$ lb/in. At C: $v'_u = 499/6 = 83$ psi; $v_u = 83 + 93 = 176$ psi; $AC = (272 - 176)/2.52 = 38$ in. At D: $v'_u = 499/9 = 55$ psi; $v_u = 55 + 93 = 148$ psi; $AD = (272 - 148)/2.52 = 49$ in.

6. Devise a stirrup spacing conforming to the computed results

The following spacing, which requires 17 stirrups for each half of the span, is satisfactory and conforms with the foregoing results.

Quantity	Spacing, in.	Total, in.	Distance from last stirrup to face of support, in.
1	2	2	2
9	4	36	38
2	6	12	50
5	9	45	95

DETERMINATION OF BOND STRESS

A beam of 4,000-psi concrete has an effective depth of 15 in. and is reinforced with four No. 7 bars. Determine the ultimate bond stress at a section where the ultimate shear is 72 kips. Compare this with the allowable stress.

Calculation Procedure:

1. Determine the ultimate shear flow h_u

The adhesion of the concrete and steel must be sufficiently strong to resist the horizontal shear flow. Let u_u = ultimate bond stress, psi; V_u = ultimate vertical shear, lb; Σo = sum of perimeters of reinforcing bars, in. Then, the ultimate shear flow at any plane between the neutral axis and the reinforcing steel is $h_u = V_u/(d - a/2)$.

In conformity with the notational system of the working-stress method, the distance $d - a/2$ is designated as jd. Dividing the shear flow by the area of contact in a unit length and introducing the capacity-reduction factor, $u_u = V_u/\phi \Sigma ojd$, Eq. (18). A section of the ACI *Code* sets $\phi = 0.85$ with respect to bond, and j is usually assigned the approximate value of 0.875 when applying this equation.

2. Calculate the bond stress

Thus, $\Sigma o = 11.0$ in., from the ACI *Handbook*. Then, $u_u = 72,000/[0.85(11.0)(0.875) \times (15)] = 587$ psi.

The allowable stress is given in the *Code* as $u_{u,allow} = 9.5(f'_c)^{0.5}/D$, but not above 800 psi, Eq. (19). Thus, $u_{u,allow} = 9.5(4,000)^{0.5}/0.875 = 687$ psi.

DESIGN OF INTERIOR SPAN OF A ONE-WAY SLAB

A floor slab that is continuous over several spans carries a live load of 120 psf and a dead load of 40 psf, exclusive of its own weight. The clear spans are 16 ft. Design the interior span, using $f'_c = 3,000$ psi and $f_y = 50,000$ psi.

Calculation Procedure:

1. Find the minimum thickness of the slab as governed by the Code

Referring to Fig. 112, the maximum potential positive or negative moment may be found by applying the type of loading that will induce the critical moment and then evaluating this moment. However, such an analysis is time-consuming. Hence, it is wise to apply the moment equations recommended in the ACI *Code* whenever the span

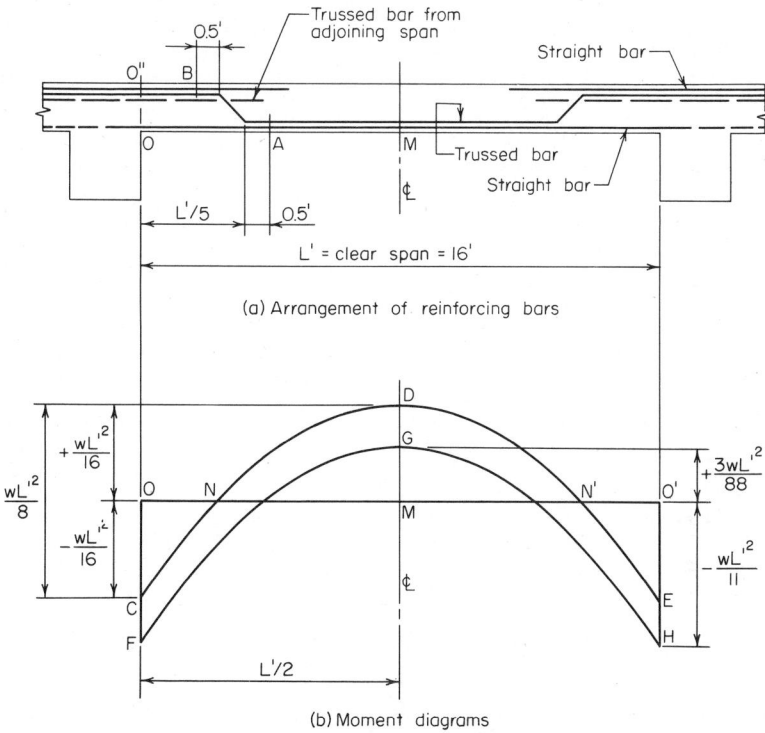

(a) Arrangement of reinforcing bars

(b) Moment diagrams

Fig. 112

and loading conditions satisfy the requirements given therein. The slab is designed by considering a 12-in. strip as an individual beam, making $b = 12$ in.

Assuming that $L = 17$ ft, the minimum thickness of the slab is $t_{\min} = L/35 = 17(12)/35 = 5.8$ in.

2. Assuming a slab thickness, compute the ultimate load on the member

Tentatively assume $t = 6$ in. Then, beam weight $= (6/12)(150 \text{ pcf}) = 75$ plf. Also, $w_u = 1.5(40 + 75) + 1.8(120) = 390$ plf.

3. Compute the shearing stress associated with the assumed beam size

From the *Code* for an interior span, $V_u = \frac{1}{2}w_u L' = \frac{1}{2}(390)(16) = 3,120$ lb; $d = 6 - 1 = 5$ in.; $v_u = 3,120/[12(5)] = 52$ psi; $v_c = 93$ psi. This is acceptable.

4. Compute the two critical moments

Apply the appropriate moment equations. Compare the computed moments with the moment capacity of the assumed beam size to ascertain if the size is adequate. Thus, $M_{u,neg} = (\frac{1}{11})w_u L'^2 = (\frac{1}{11})(390)(16)^2(12) = 108{,}900$ in.-lb, where the value 12 converts the dimension to inches. Then, $M_{u,pos} = \frac{1}{16}w_u L'^2 = 74{,}900$ in.-lb. By Eq. (10), $q_{max} = 0.6375(0.85)(87/137) = 0.344$. By Eq. (6), $M_{u,allow} = 0.90(12)(5)^2(3{,}000)(0.344)(0.797) = 222{,}000$ in.-lb. This is acceptable. The slab thickness will therefore be made 6 in.

5. Compute the area of reinforcement associated with each critical moment

Using Eq. (7), $bdf_c = 12(5)(2.55) = 153.0$ kips; then, $2bf_c M_{u,neg}/\phi = 2(12)(2.55)(108.9)/0.90 = 7{,}405$ kips^2; $A_{s,neg} = [153.0 - (153.0^2 - 7{,}405)^{0.5}]/50 = 0.530$ sq in. Similarly, $A_{s,pos} = 0.353$ sq in.

6. Select the reinforcing bars and locate the bend points

For positive reinforcement, use No. 4 trussed bars 13 in. on centers, alternating with No. 4 straight bars 13 in. on centers, thus obtaining $A_s = 0.362$ sq in.

For negative reinforcement, supplement the trussed bars over the support with No. 4 straight bars 13 in. on centers, thus obtaining $A_s = 0.543$ sq in.

The trussed bars are usually bent upward at the fifth points, as shown in Fig. 112a. The reinforcement satisfies a section of the ACI *Code* which requires that "at least . . . one-fourth the positive moment reinforcement in continuous beams shall extend along the same face of the beam into the support at least 6 in."

7. Investigate the adequacy of the reinforcement beyond the bend points

In accordance with the *Code*, $A_{min} = A_t = 0.0020bt = 0.0020(12)(6) = 0.144$ sq in.

A section of the *Code* requires that reinforcing bars be extended beyond the point at which they become superfluous with respect to flexure a distance equal to the effective depth or 12 bar diameters, whichever is greater. In the present instance, extension $= 12(0.5) = 6$ in. Therefore, the trussed bars in effect terminate as positive reinforcement at section A, Fig. 112. Then, $L'/5 = 3.2$ ft; $AM = 8 - 3.2 - 0.5 = 4.3$ ft.

The conditions immediately to the left of A are: $M_u = M_{u,pos} - \frac{1}{2}w_u(AM)^2 = 74{,}900 - \frac{1}{2}(390)(4.3)^2(12) = 31{,}630$ in.-lb; $A_{s,pos} = 0.181$ sq in.; $q = 0.181(50)/[12(5)(3)] = 0.0503$. By Eq. (5), $M_{u,allow} = 0.90(0.181)(50{,}000)(5)(0.970) = 39{,}500$ in.-lb. This is acceptable.

Alternatively, Eq. (11) may be applied to obtain the following conservative approximation: $M_{u,allow} = 74{,}900(0.181)/0.353 = 38{,}400$ in.-lb.

The trussed bars in effect terminate as negative reinforcement at B, where $O''B = 3.2 - 0.33 - 0.5 = 2.37$ ft. The conditions immediately to the right of B are $|M_u| = M_{u,neg} - 12(3{,}120 \times 2.37 - \frac{1}{2} \times 390 \times 2.37^2) = 33{,}300$ in.-lb. Then, $A_{s,neg} = 0.362$ sq in. As a conservative approximation, $M_{u,allow} = 108{,}900(0.362)/0.530 = 74{,}400$ in.-lb. This is acceptable.

8. Locate the point at which the straight bars at the top may be discontinued

9. Investigate the bond stresses

In accordance with Eq. (19), $u_{u,allow} = 800$ psi.

If *CDE* in Fig. 112b represents the true moment diagram, the bottom bars are subjected to bending stress in the interval NN'. Manifestly, the maximum bond stress along the bottom occurs at these boundary points (points of contraflexure), where the shear is relatively high and the straight bars alone are present. Thus: $MN = 0.354L'$; V_u at N/V_u at support $= 0.354L'/0.5L' = 0.71$; V_u at $N = 0.71(3{,}120) = 2{,}215$ lb. By Eq. (18), $u_u = V_u/\phi\Sigma ojd = 2{,}215/[0.85(1.45)(0.875)(5)] = 411$ psi. This is acceptable. It is apparent that the maximum bond stress in the top bars has a smaller value.

ANALYSIS OF A TWO-WAY SLAB BY THE YIELD-LINE THEORY

The slab in Fig. 113a is simply supported along all four edges and is isotropically reinforced. It supports a uniformly distributed ultimate load of w_u psf. Calculate the ultimate unit moment m_u for which the slab must be designed.

Calculation Procedure:

1. Draw line GH perpendicular to AE at E; express distances b and c in terms of a

Consider a slab to be reinforced in orthogonal directions. If the reinforcement in one direction is identical with that in the other direction, the slab is said to be *isotropically reinforced*; if the reinforcements differ, the slab is described as *orthogonally anisotropic*. In the former case, the capacity of the slab is identical in all directions; in the latter case, the capacity has a unique value in every direction. In this instance, assume that the slab size is excessive with respect to balanced design, the result being that the failure of the slab will be characterized by yielding of the steel.

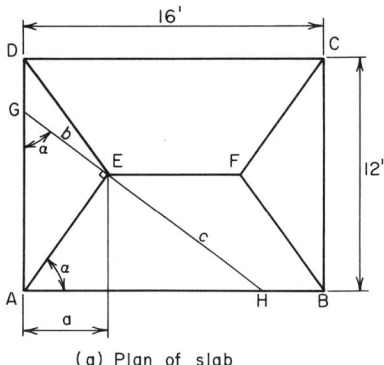

(a) Plan of slab

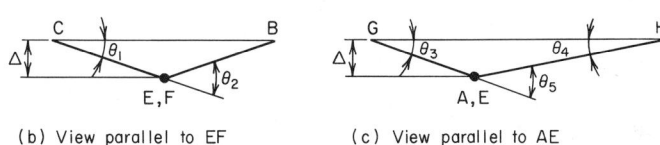

(b) View parallel to EF   (c) View parallel to AE

Fig. 113   Analysis of two-way slab by mechanism method.

In a steel beam, a plastic hinge forms at a *section*; in a slab, a plastic hinge is assumed to form along a *straight line*, termed a *yield line*. It is plausible to assume that by virtue of symmetry of loading and support conditions the slab in Fig. 113a will fail by the formation of a central yield line EF and diagonal yield lines such as AE, the ultimate moment at these lines being positive. The ultimate *unit* moment m_u is the moment acting on a unit length.

Although it is possible to derive equations that give the location of the yield lines, this procedure is not feasible because the resulting equations would be unduly cumbersome. The procedure followed in practice is to assign a group of values to the distance a and to determine the corresponding values of m_u. The true value of m_u is the highest one obtained. Either the statical or mechanism method of analysis may be applied; the latter will be applied here.

Expressing the distances b and c in terms of a: $\tan \alpha = 6/a = AE/b = c/AE$; $b = aAE/6$; $c = 6AE/a$.

2. Find the rotation of the plastic hinges

Allow line EF to undergo a virtual displacement Δ after the collapse load is reached. During the virtual displacement, the portions of the slab bounded by the yield lines and the supports rotate as planes. Referring to Fig. 113b and c: $\theta_1 = \Delta/6$; $\theta_2 = 2\theta_1 = \Delta/3 = 0.333\Delta$; $\theta_3 = \Delta/b$; $\theta_4 = \Delta/c$; $\theta_5 = \Delta(1/b + 1/c) = (\Delta/AE)(6/a + a/6)$.

3. Select a trial value of a and evaluate the distances and angles

Using $a = 4.5$ ft as the trial value, $AE = (a^2 + 6^2)^{0.5} = 7.5$ ft; $b = 5.63$ ft; $c = 10$ ft; $\theta_5 = (\Delta/7.5)(6/4.5 + 4.5/6) = 0.278\Delta$.

4. Develop an equation for the external work W_E performed by the uniform load on a surface that rotates about a horizontal axis

In Fig. 114, consider that the surface ABC rotates about axis AB through an angle θ while carrying a uniform load of w psf. For the elemental area dA, the deflection, total

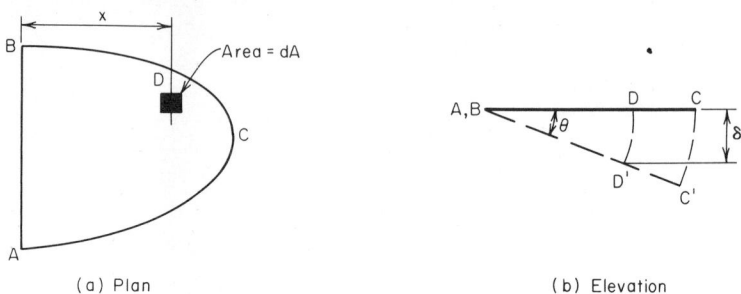

           (a) Plan   (b) Elevation

Fig. 114

load, and external work are: $\delta = x\theta$; $dW = w\,dA$; $dW_E = \delta\,dW = x\theta w\,dA$. The total work for the surface is: $W_E = w\theta \int x\,dA$, or $W_E = w\theta Q$, Eq. (20), where $Q =$ statical moment of total area, with respect to the axis of rotation.

5. Evaluate the external and internal work for the slab

Using the assumed value, $a = 4.5$ ft, $EF = 16 - 9 = 7$ ft. The external work for the two triangles is $2w_u(\Delta/4.5)(\frac{1}{3})(12)(4.5)^2 = 18w_u\Delta$. The external work for the two trapezoids is $2w_u(\Delta/6)(\frac{1}{6})(16 + 2 \times 7)(6)^2 = 60w_u\Delta$. Then, $W_E = w_u\Delta(18 + 60) = 78w_u\Delta$; $W_I = m_u(7\theta_2 + 4 \times 7.5\theta_5) = 10.67m_u\Delta$.

6. Find the value of m_u corresponding to the assumed value of a

Equate the external and internal work to find this value of m_u. Thus, $10.67m_u\Delta = 78w_u\Delta$; $m_u = 7.31w_u$.

7. Determine the highest value of m_u

Assign other trial values to a and find the corresponding values of m_u. Continue this procedure until the highest value of m_u is obtained. This is the true value of the ultimate unit moment.

Design of Flexural Members by the Working-Stress Method

As demonstrated earlier, the analysis or design of a composite beam by the working-stress method is most readily performed by transforming the given beam to an equiv-

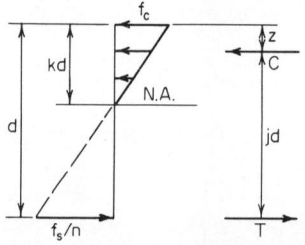

Fig. 115 Stresses and resultant forces.

alent homogeneous beam. In the case of a reinforced-concrete member, the transformation is made by replacing the reinforcing steel with a strip of concrete having an area nA_s and located at the same distance from the neutral axis as the steel. This substitute concrete is assumed capable of sustaining tensile stresses.

The following symbols, shown in Fig. 115, are to be added to the notational system given earlier: $kd =$ distance from extreme compression fiber to neutral axis, in.; $jd =$ distance between action lines of C and T, in.; $z =$ distance from extreme compression fiber to action line of C, in.

The basic equations for the working-stress design of a rectangular beam reinforced solely in tension are:

$$k = f_c/(f_c + f_s/n) \tag{21}$$

$$j = 1 - k/3 \tag{22}$$

$$M = Cjd = \tfrac{1}{2} f_c kjbd^2 \tag{23}$$

$$M = (\tfrac{1}{6}) f_c k(3 - k)bd^2 \tag{24}$$

$$M = Tjd = f_s A_s jd \tag{25}$$

$$M = f_s pjbd^2 \tag{26}$$

$$M = \frac{f_s k^2(3 - k)bd^2}{6n(1 - k)} \tag{27}$$

$$p = \frac{f_c k}{2f_s} \tag{28}$$

$$p = \frac{k^2}{2n(1 - k)} \tag{29}$$

$$k = [2pn + (pn)^2]^{0.5} - pn \tag{30}$$

For a given set of values of f_c, f_s, and n, M is directly proportional to the beam property bd^2. Let K denote the constant of proportionality. Then, $M = Kbd^2$, Eq. (31), where $K = \tfrac{1}{2} f_c kj = f_s pj$, Eq. (32).

The allowable flexural stress in the concrete and the value of n, which are functions of the ultimate strength f_c', are given in the ACI *Code*, as is the allowable flexural stress in the steel. In all instances in the following procedures, the assumption is that the reinforcement is intermediate-grade steel having an allowable stress of 20,000 psi.

Consider that the load on a beam is gradually increased until a limiting stress is induced. A beam that is so proportioned that the steel and concrete simultaneously attain their limiting stress is said to be in *balanced design*. For each set of values of f_c' and f_s, there is a corresponding set of values of K, k, j, and p associated with balanced design. These values are recorded in Table 6.

TABLE 6   Values of Design Parameters at Balanced Design

f_c' and n	f_c	f_s	K	k	j	p
2,500 10	1,125	20,000	178	0.360	0.880	0.0101
3,000 9	1,350	20,000	223	0.378	0.874	0.0128
4,000 8	1,800	20,000	324	0.419	0.860	0.0188
5,000 7	2,250	20,000	423	0.441	0.853	0.0248

In Fig. 116, AB represents the stress line of the transformed section for a beam in balanced design. If the area of reinforcement is increased while the width and depth remain constant, the neutral axis is depressed to O', and $A'O'B$ represents the stress line under the allowable load. But if the width is increased while the depth and area of reinforcement remain constant, the neutral axis is elevated to O'', and $AO''B'$ represents the stress line under the allowable load. This analysis leads to these conclusions: If the reinforcement is in excess of that needed for balanced design, the concrete is the first material to reach its limiting stress under a gradually increasing load. If the beam size is in excess of that needed for balanced design, the steel is the first material to reach its limiting stress.

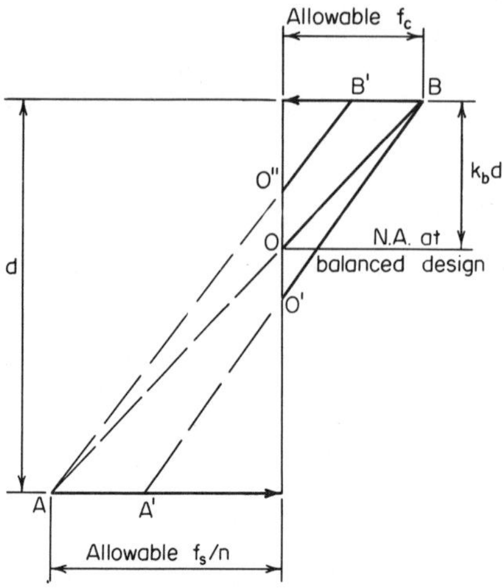

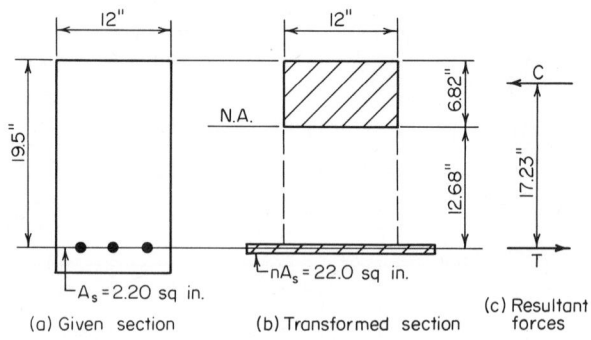

Fig. 116   Stress diagrams.

STRESSES IN A RECTANGULAR BEAM

A beam of 2,500-psi concrete has a width of 12 in. and an effective depth of 19.5 in. It is reinforced with one No. 9 and two No. 7 bars. Determine the flexural stresses caused by a bending moment of 62 ft-kips (a) without applying the basic equations of reinforced-concrete beam design; (b) by applying the basic equations.

Calculation Procedure:

1. Record the pertinent beam data

Thus: $f_c' = 2,500$ psi; $\therefore n = 10$; $A_s = 2.20$ sq in.; $nA_s = 22.0$ sq in. Then, $M = 62,000$ $(12) = 744,000$ in.-lb.

2. Transform the given section to an equivalent homogeneous section, as in Fig. 117b

3. Locate the neutral axis of the member

The neutral axis coincides with the centroidal axis of the transformed section. To locate the neutral axis, set the statical moment of the transformed area with respect to

Fig. 117

its centroidal axis equal to zero. Or: $12(kd)^2/2 - 22.0(19.5 - kd) = 0$; $kd = 6.82$; $d - kd = 12.68$ in.

4. Calculate the moment of inertia of the transformed section

Then evaluate the flexural stresses by applying the stress equation. Or: $I = (\frac{1}{3})(12)$ $(6.82)^3 + 22.0(12.68)^2 = 4,806$ in.^{4} $f_c = Mkd/I = 744,000(6.82)/4,806 = 1,060$ psi; $f_s = 10(744,000)(12.68)/4,806 = 19,600$ psi.

5. *Alternatively, evaluate the stresses by computing the resultant forces C and T*

Thus: $jd = 19.5 - 6.82/3 = 17.23$ in.; $C = T = M/jd = 744,000/17.23 = 43,200$ lb. But $C = \frac{1}{2}f_c(6.82)12$; $\therefore f_c = 1,060$ psi, and $T = 2.20f_s$; $\therefore f_s = 19,600$ psi. This concludes part a of the solution. The next step constitutes the solution to part b.

6. *Compute pn and then apply the basic equations in the proper sequence*

Thus: $p = A_s/bd = 2.20/[12(19.5)] = 0.00940$; $pn = 0.0940$. Then using Eq. (30), $k = [0.188 + (0.094)^2]^{0.5} - 0.094 = 0.350$. Using Eq. (22), $j = 1 - 0.350/3 = 0.883$. By Eq. (23), $f_c = 2M/kjbd^2 = 2(744,000)/[0.350(0.883)(12)(19.5)^2] = 1,060$ psi. By Eq. (25), $f_s = M/A_s jd = 744,000/[2.20(0.883)(19.5)] = 19,600$ psi.

CAPACITY OF A RECTANGULAR BEAM

The beam in Fig. 118a is made of 2,500-psi concrete. Determine the flexural capacity of the member (a) without applying the basic equations of reinforced-concrete beam design; (b) by applying the basic equations.

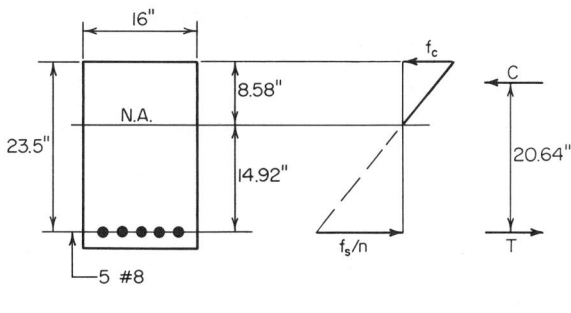

(a) Section   (b) Stresses and resultant forces

Fig. 118

Calculation Procedure:

1. *Record the pertinent beam data*

Thus, $f'_c = 2,500$ psi; $\therefore f_{c,\text{allow}} = 1,125$ psi; $n = 10$; $A_s = 3.95$ sq in.; $nA_s = 39.5$ sq in.

2. *Locate the centroidal axis of the transformed section*

Thus, $16(kd)^2/2 - 39.5(23.5 - kd) = 0$; $kd = 8.58$ in.; $d - kd = 14.92$ in.

3. *Ascertain which of the two allowable stresses governs the capacity of the member*

For this purpose, assume that $f_c = 1,125$ psi. By proportion, $f_s = 10(1,125)(14.92/8.58) = 19,560 < 20,000$ psi. Therefore, concrete stress governs.

4. *Calculate the allowable bending moment*

Thus, $jd = 23.5 - 8.58/3 = 20.64$ in.; $M = Cjd = \frac{1}{2}(1,125)(16)(8.58)(20.64) = 1,594,000$ in.-lb; or $M = Tjd = 3.95(19,560)(20.64) = 1,594,000$ in.-lb. This concludes part a of the solution. The next step comprises part b.

5. *Compute p and compare with p_b to identify the controlling stress*

Thus, from Table 6, $p_b = 0.0101$; then $p = A_s/bd = 3.95/16(23.5) = 0.0105 > p_b$. Therefore, concrete stress governs.

Applying the basic equations in the proper sequence, $pn = 0.1050$; by Eq. (30), $k = [0.210 + 0.105^2]^{0.5} - 0.105 = 0.365$; by Eq. (24), $M = (\frac{1}{6})(1,125)(0.365)(2.635)(16)(23.5)^2 = 1,593,000$ in.-lb. This agrees closely with the previously computed value of M.

DESIGN OF REINFORCEMENT IN A RECTANGULAR BEAM OF GIVEN SIZE

A rectangular beam of 4,000-psi concrete has a width of 14 in. and an effective depth of 23.5 in. Determine the area of reinforcement if the beam is to resist a bending moment of (a) 220 ft-kips; (b) 200 ft-kips.

Calculation Procedure:

1. Calculate the moment capacity of this member at balanced design

Record the following values: $f_{c, \text{allow}} = 1,800$ psi; $n = 8$. From Table 6, $j_b = 0.860$; $K_b = 324$ psi; $M_b = K_b b d^2 = 324(14)(23.5)^2 = 2,505,000$ in.-lb.

2. Determine which material will be stressed to capacity under the stipulated moment

For part a, $M = 220,000(12) = 2,640,000$ in.-lb $> M_b$. This result signifies that the beam size is deficient with respect to balanced design, and the concrete will therefore be stressed to capacity.

3. Apply the basic equations in proper sequence to obtain A_s

By Eq. (24), $k(3-k) = 6M/f_c b d^2 = 6(2,640,000)/[1800(14)(23.5)^2] = 1.138$; $k = 0.446$. By Eq. (29), $p = k^2/[2n(1-k)] = 0.446^2/[16(0.554)] = 0.0224$; $A_s = pbd = 0.0224(14)(23.5) = 7.37$ sq in.

4. Verify the result by evaluating the flexural capacity of the member

For part b, compute A_s by the exact method and then describe the approximate method used in practice.

5. Determine which material will be stressed to capacity under the stipulated moment

$M = 200,000(12) = 2,400,000$ in.-lb $< M_b$. This result signifies that the beam size is excessive with respect to balanced design, and the steel will therefore be stressed to capacity.

6. Apply the basic equations in proper sequence to obtain A_s

Using Eq. (27), $k^2(3-k)/(1-k) = 6nM/f_s b d^2 = 6(8)(2,400,000)/[20,000(14)(23.5)^2] = 0.7448$; $k = 0.411$. By Eq. (22), $j = 1 - 0.411/3 = 0.863$. By Eq. (25), $A_s = M/f_s j d = 2,400,000/[20,000(0.863)(23.5)] = 5.92$ sq in.

7. Verify the result by evaluating the flexural capacity of this member

The value of j obtained in Step 6 differs negligibly from the value $j_b = 0.860$. Consequently, in those instances where the beam size is only moderately excessive with respect to balanced design, the practice is to consider that $j = j_b$ and to solve Eq. (25) directly on this basis. This practice is conservative and it obviates the need for solving a cubic equation, thus saving time.

DESIGN OF A RECTANGULAR BEAM

A beam on a simple span of 13 ft is to carry a uniformly distributed load, exclusive of its own weight, of 3,600 plf and a concentrated load of 17,000 lb applied at mid-span. Design the section, using $f'_c = 3,000$ psi.

Calculation Procedure:

1. Record the basic values associated with balanced design

There are two methods of allowing for the beam weight: (a) to determine the bending moment with an estimated beam weight included; (b) to determine the beam size required to resist the external loads alone and then increase the size slightly. The latter method will be used here.

From Table 6, $K_b = 223$ psi; $p_b = 0.0128$; $j_b = 0.874$.

2. Calculate the maximum moment caused by the external loads

Thus, the maximum moment $M_e = \frac{1}{4}PL + \frac{1}{8}wL^2 = \frac{1}{4}(17,000)(13)(12) + (\frac{1}{8})(3,600)(13)^2 (12) = 1,576,000$ in.-lb.

3. Establish a trial beam size

Thus, $bd^2 = M/K_b = 1,576,000/223 = 7,067$ in.^{3}. Setting $b = (\frac{2}{3})d$, the results obtained are: $b = 14.7$ in.; $d = 22.0$ in. Try $b = 15$ in. and $d = 22.5$ in., producing an overall depth of 25 in. if the reinforcing bars may be placed in one row.

4. Calculate the maximum bending moment with the beam weight included; determine if the trial section is adequate

Thus, beam weight $= 15(25)(150)/144 = 391$ plf; $M_w = (\frac{1}{8})(391)(13)^2(12) = 99,000$ in.-lb; $M = 1,576,000 + 99,000 = 1,675,000$ in.-lb; $M_b = K_b bd^2 = 223(15)(22.5)^2 = 1,693,000$ in.-lb. The trial section is therefore satisfactory because it has adequate capacity.

5. Design the reinforcement

Since the beam size is slightly excessive with respect to balanced design, the steel will be stressed to capacity under the design load. Equation (25) is therefore suitable for this calculation. Thus, $A_s = M/f_s jd = 1,675,000/[20,000(0.874)(22.5)] = 4.26$ sq in.

An alternative method of calculating A_s is to apply the value of p_b while setting the beam width equal to the dimension actually required to produce balanced design. Thus, $A_s = 0.0128(15)(1,675)(22.5)/1,693 = 4.27$ sq in.

Use one No. 10 and three No. 9 bars, for which $A_s = 4.27$ sq in. and $b_{min} = 12.0$ in.

6. Summarize the design

Thus: beam size, 15×25 in.; reinforcement, one No. 10 and three No. 9 bars.

DESIGN OF WEB REINFORCEMENT

A beam 14 in. wide and 18.5 in. effective depth carries a uniform load of 3.8 klf and a concentrated mid-span load of 2 kips. The beam is simply supported and the clear distance between supports is 13 ft. Using $f'_c = 3,000$ psi, and an allowable stress f_v in the stirrups of 20,000 psi, design web reinforcement in the form of vertical U stirrups.

Calculation Procedure:

1. Construct the shearing-stress diagram for half-span

The design of web reinforcement by the working-stress method parallels the design by the ultimate-strength method, which was given earlier. Let $v =$ nominal shearing stress, psi; $v_c =$ shearing stress resisted by concrete; $v' =$ shearing stress resisted by web reinforcement.

The ACI *Code* provides two alternative methods of computing the shearing stress

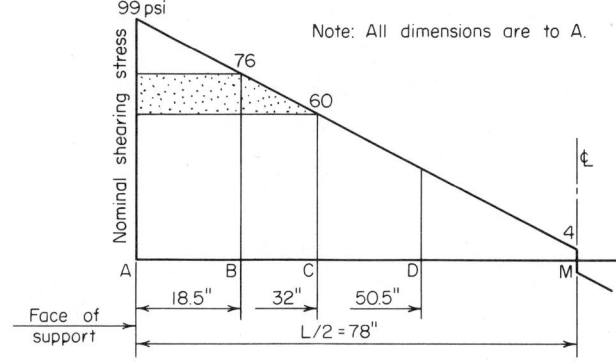

Fig. 119 Shearing-stress diagram.

that may be resisted by the concrete. The simpler method will be used here. This sets $v_c = 1.1(f_c')^{0.5}$, Eq. (33). The equation for nominal shearing stress is $v = V/bd$, Eq. (34).

The shearing-stress diagram for a half-span is shown in Fig. 119. Establish the region AD within which web reinforcement is required. Thus, $v_c = 1.1(3,000)^{0.5} = 60$ psi. At the face of the support, $V = 6.5(3,800) + 1,000 = 25,700$ lb; $v = 25,700/[14(18.5)] = 99$ psi.

At mid-span, $V = 1,000$ lb; $v = 4$ psi; slope of diagram $= -(99-4)/78 = -1.22$ psi/in. At distance d from the face of the support, $v = 99 - 18.5(1.22) = 76$ psi; $v' = 76 - 60 = 16$ psi; $AC = (99-60)/1.22 = 32$ in.; $AD = AC + d = 32 + 18.5 = 50.5$ in.

2. **Check the beam size for compliance with the Code**

Thus, $v_{max} = 5(f_c')^{0.5} = 274 > 76$ psi. This is acceptable.

3. **Select the stirrup size**

Use the method given earlier in the ultimate-strength Calculation Procedure to select the stirrup size, establish the maximum allowable spacing, and devise a satisfactory spacing.

CAPACITY OF A T BEAM

Determine the flexural capacity of the T beam in Fig. 120a, using $f_c' = 3,000$ psi.

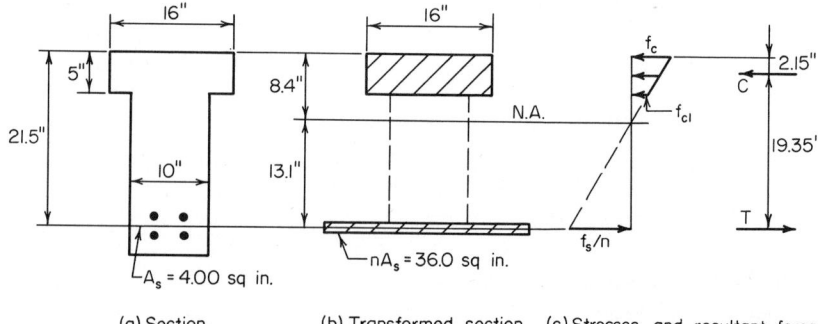

(a) Section   (b) Transformed section   (c) Stresses and resultant forces

Fig. 120

Calculation Procedure:

1. **Record the pertinent beam values**

The neutral axis of a T beam often falls within the web. However, to simplify the analysis, the resisting moment developed by the concrete lying between the neutral axis and the flange is usually disregarded. Let A_f denote the flange area. The pertinent beam values are: $f_{c,\,allow} = 1,350$ psi; $n = 9$; $k_b = 0.378$; $nA_s = 9(4.00) = 36.0$ sq in.

2. **Tentatively assume that the neutral axis lies in the web**

Locate this axis by taking statical moments with respect to the top line. Thus: $A_f = 5(16) = 80$ sq in.; $kd = [80(2.5) + 36.0(21.5)]/(80 + 36.0) = 8.40$ in.

3. **Identify the controlling stress**

Thus: $k = 8.40/21.5 = 0.391 > k_b$; therefore, concrete stress governs.

4. **Calculate the allowable bending moment**

Using Fig. 120c, $f_{c1} = 1,350(3.40)/8.40 = 546$ psi; $C = \frac{1}{2}(80)(1,350 + 546) = 75,800$ lb. The action line of this resultant force lies at the centroidal axis of the stress trapezoid. Thus, $z = (\frac{5}{3})(1,350 + 2 \times 546)/(1,350 + 546) = 2.15$ in.; or $z = (\frac{5}{3})(8.40 + 2 \times 3.40)/(8.40 + 3.40) = 2.15$ in. $M = Cjd = 75,800(19.35) = 1,467,000$ in.-lb.

5. *Alternatively, calculate the allowable bending moment by assuming that the flange extends to the neutral axis*

Then apply the necessary correction. Let C_1 = resultant compressive force if the flange extended to the neutral axis, lb; C_2 = resultant compressive force in the imaginary extension of the flange, lb. Then, $C_1 = \frac{1}{2}(1,350)(16)(8.40) = 90,720$ lb; $C_2 = 90,720(3.40/8.40)^2 = 14,860$ lb; $M = 90,720(21.5 - 8.40/3) - 14,860(21.5 - 5 - 3.40/3) = 1,468,000$ in.-lb.

DESIGN OF A T BEAM HAVING CONCRETE STRESSED TO CAPACITY

A concrete girder of 2,500-psi concrete has a simple span of 22 ft and is built integrally with a 5-in. slab. The girders are spaced 8 ft on centers; the overall depth is restricted to 20 in. by headroom requirements. The member carries a load of 4,200 plf exclusive of the weight of its web. Design the section, using tension reinforcement only.

Calculation Procedure:

1. *Establish a tentative width of web*

Since the girder is built integrally with the slab that it supports, the girder and slab constitute a structural entity in the form of a T beam. The effective flange width is established by applying the criteria given in the ACI *Code*, and the bending stress in the flange is assumed to be uniform across a line parallel to the neutral axis. Let A_f = area of flange, sq in.; b = width of flange, in.; b' = width of web, in.; t = thickness of flange, in.; s = center-to-center spacing of girders.

To establish a tenative width of web, try $b' = 14$ in. Then, weight of web = $14(15)(150)/144 = 219$, say 220 plf; $w = 4,200 + 220 = 4,420$ plf.

Since two rows of bars are probably required, $d = 20 - 3.5 = 16.5$ in. The critical shear value is $V = w(0.5L - d) = 4,420(11 - 1.4) = 42,430$ lb; $v = V/b'd = 42,430/[14(16.5)] = 184$ psi. From the *Code*, $v_{max} = 5(f'_c)^{0.5} = 250$ psi. This is acceptable.

Upon designing the reinforcement, consider whether it is possible to reduce the width of the web.

2. *Establish the effective width of the flange according to the* Code

Thus, $\frac{1}{4}L = \frac{1}{4}(22)(12) = 66$ in.; $16t + b' = 16(5) + 14 = 94$ in.; $s = 8(12) = 96$ in.; therefore $b = 66$ in.

3. *Compute the moment capacity of the member at balanced design*

Compare the result with the moment in the present instance to identify the controlling stress. Using Fig. 120 as a guide, $k_b d = 0.360(16.5) = 5.94$ in.; $A_f = 5(66) = 330$ sq in.; $f_{c1} = 1,125(0.94)/5.94 = 178$ psi; $C_b = T_b = \frac{1}{2}(330)(1,125 + 178) = 215,000$ lb; $z_b = (\frac{5}{3})(5.94 + 2 \times 0.94)/(5.94 + 0.94) = 1.89$ in.; $jd = 14.61$ in.; $M_b = 215,000(14.61) = 3,141,000$ in.-lb; $M = (\frac{1}{8})(4,420)(22)^2(12) = 3,209,000$ in.-lb.

The beam size is slightly deficient with respect to balanced design, and the concrete will therefore be stressed to capacity under the stipulated load. In Fig. 121, let AOB represent the stress line associated with balanced design and $A'O'B$ represent the stress line in the present instance. (The magnitude of AA' is exaggerated for clarity.)

4. *Develop suitable equations for the beam*

Referring to Fig. 121, $T = T_b + bt^2x/2d$, Eq. (35), where T and T_b = tensile force in present instance and at balanced design, respectively. And, $M = M_b + bt^2(3d - 2t)x/6d$, Eq. (36).

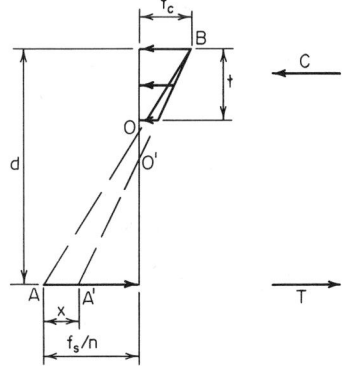

Fig. 121 Stress diagram for T beam.

5. Apply the equations from Step 4

Thus, $M - M_b = 68,000$ in.-lb. By Eq. (36), $x = 68,000(6)(16.5)/[66(25)(49.5 - 10)] = 103$ psi; $f_s = 20,000 - 10(103) = 18,970$ psi; by Eq. (35), $T = 215,000 + 66(25)(103)/33 = 220,200$ lb.

6. Design the reinforcement; establish the web width

Thus: $A_s = 220,200/18,970 = 11.61$ sq in. Use five No. 11 and three No. 10 bars, placed in two rows. Then, $A_s = 11.61$ sq in.; $b'_{min} = 14.0$ in. It is therefore necessary to maintain the 14-in. width.

7. Summarize the design

Width of web: 14 in.; reinforcement: five No. 11 and three No. 10 bars.

8. Verify the design by computing the capacity of the member

Thus: $nA_s = 116.1$ sq in.; $kd = [330(2.5) + 116.1(16.5)]/(330 + 116.1) = 6.14$ in.; $k = 6.14/16.5 = 0.372 > k_b$; therefore, concrete is stressed to capacity. Then, $f_s = 10(1,125)(10.36)/6.14 = 18,980$ psi; $z = (\frac{5}{3})(6.14 + 2 \times 1.14)/(6.14 + 1.14) = 1.93$ in.; $jd = 14.57$ in.; $M_{allow} = 11.61(18,980)(14.57) = 3,210,000$ in.-lb. This is acceptable.

DESIGN OF A T BEAM HAVING STEEL STRESSED TO CAPACITY

Assume that the girder in the previous Calculation Procedure carries a total load, including the weight of the web, of 4,100 plf. Compute the area of reinforcement.

Calculation Procedure:

1. Identify the controlling stress

Thus, $M = (\frac{1}{8})(4,100)(22)^2(12) = 2,977,000$ in.-lb. From the previous Calculation Procedure, $M_b = 3,141,000$ in.-lb. Since $M_b > M$, the beam size is slightly excessive with respect to balanced design, and the steel will therefore be stressed to capacity under the stipulated load.

2. Compute the area of reinforcement

As an approximation, this area may be found by applying the value of jd associated with balanced design, although it is actually slightly larger. From the previous Calculation Procedure, $jd = 14.61$ in. Then, $A_s = 2,977,000/[20,000(14.61)] = 10.19$ sq in.

3. Verify the design by computing the member capacity

Thus, $nA_s = 101.9$ sq in.; $kd = (330 \times 2.5 + 101.9 \times 16.5)/(330 + 101.9) = 5.80$ in.; $z = (\frac{5}{3})(5.80 + 2 \times 0.80)/(5.80 + 0.80) = 1.87$ in.; $jd = 14.63$ in.; $M_{allow} = 10.19(20,000)(14.63) = 2,982,000$ in.-lb. This is acceptable.

REINFORCEMENT FOR DOUBLY REINFORCED RECTANGULAR BEAM

A beam of 4,000-psi concrete that will carry a bending moment of 230 ft-kips is restricted to a 15-in. width and a 24-in. total depth. Design the reinforcement.

Calculation Procedure:

1. Record the pertinent beam data

In Fig. 122, where the imposed moment is substantially in excess of that corresponding to balanced design, it is necessary to reinforce the member in compression as well as tension. The loss in concrete area caused by the presence of the compression reinforcement may be disregarded.

Since plastic flow generates a transfer of compressive stress from the concrete to the steel, the ACI *Code* provides that "in doubly reinforced beams and slabs, an effective modular ratio of $2n$ shall be used to transform the compression reinforcement and compute its stress, which shall not be taken as greater than the allowable tensile stress." This procedure is tantamount to considering that the true stress in the compression reinforcement is twice the value obtained by assuming a linear stress distribution.

Let A_s = area of tension reinforcement, sq in.; A_s' = area of compression reinforcement, sq in.; f_s = stress in tension reinforcement, psi; f_s' = stress in compression reinforcement, psi; C' = resultant force in compression reinforcement, lb; M_1 = moment

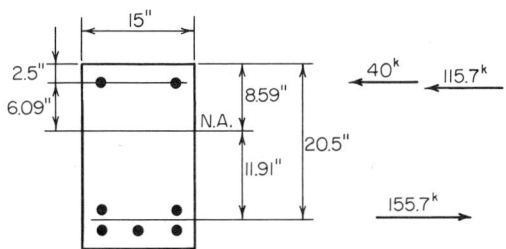

Fig. 122   Doubly reinforced beam.

capacity of member if reinforced solely in tension to produce balanced design; M_2 = incremental moment capacity resulting from use of compression reinforcement.

The data recorded for the beam are: f_c = 1,800 psi; n = 8; K_b = 324 psi; k_b = 0.419; j_b = 0.860; M = 230,000(12) = 2,760,000 in.-lb.

2.   Ascertain if one row of tension bars will suffice

Assume tentatively that the presence of the compression reinforcement does not appreciably alter the value of j. Then, jd = 0.860(21.5) = 18.49 in.; A_s = $M/f_s jd$ = 2,760,000/[20,000(18.49)] = 7.46 sq in. This area of steel cannot be accommodated in the 15-in. beam width, and two rows of bars are therefore required.

3.   Evaluate the moments M_1 and M_2

Thus, d = 24 − 3.5 = 20.5 in.; M_1 = $K_b bd^2$ = 324(15)(20.5)^2 = 2,040,000 in.-lb; M_2 = 2,760,000 − 2,040,000 = 720,000 in.-lb.

4.   Compute the forces in the reinforcing steel

For convenience, assume that the neutral axis occupies the same position as it would in the absence of compression reinforcement. For M_1, arm = $j_b d$ = 0.860(20.5) = 17.63 in; for M_2, arm = 20.5 − 2.5 = 18.0 in.; T = (2,040,000/17.63) + 720,000/18.0 = 155,700 lb; C' = 40,000 lb.

5.   Compute the areas of reinforcement and select the bars

Thus: A_s = T/f_s = 155,700/20,000 = 7.79 sq in.; kd = 0.419(20.5) = 8.59 in.; $d − kd$ = 11.91 in. By proportion, f_s' = 2(20,000)(6.09)/11.91 = 20,500 psi; therefore, set f_s' = 20,000 psi. Then, A_s' = C'/f_s' = 40,000/20,000 = 2.00 sq in. Thus: tension steel − five No. 11 bars, A_s = 7.80 sq in.; compression steel − two No. 9 bars, A_s = 2.00 sq in.

DEFLECTION OF A CONTINUOUS BEAM

The continuous beam in Fig. 123a and b carries a total load of 3.3 klf. When considered as a T beam, the member has an effective flange width of 68 in. Determine the deflection of the beam upon application of full live load, using f_c' = 2,500 psi and f_y = 40,000 psi.

Calculation Procedure:

1.   Record the areas of reinforcement

At support: A_s = 4.43 sq in. (top); A_s' = 1.58 sq in. (bottom). At center: A_s = 3.16 sq in. (bottom).

2. Construct the bending-moment diagram

Apply the ACI equation for maximum mid-span moment. Referring to Fig. 123c:
$M_1 = (\frac{1}{8})wL'^2 = (\frac{1}{8})3.3(22)^2 = 200$ ft-kips; $M_2 = (\frac{1}{16})wL'^2 = 100$ ft-kips; $M_3 = 100$ ft-kips.

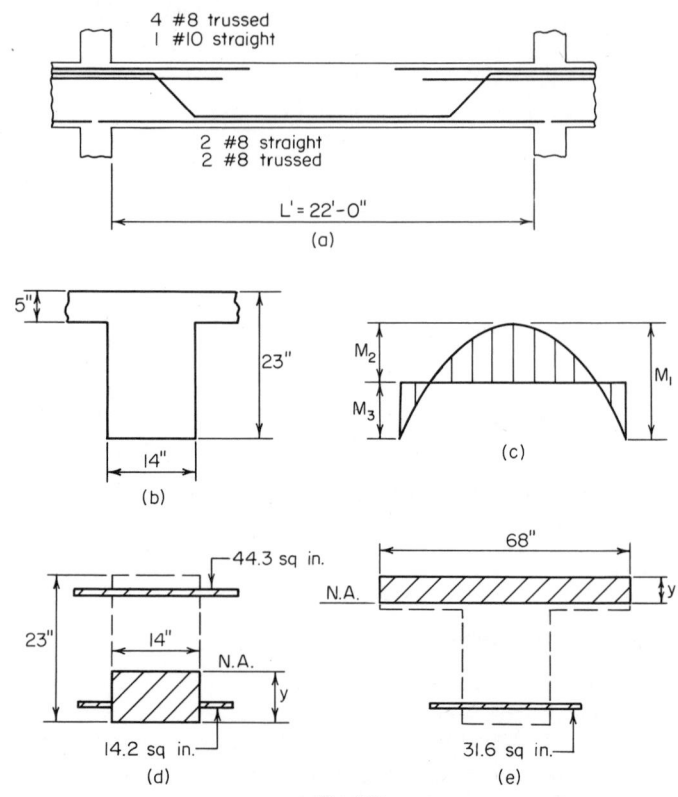

Fig. 123

3. Determine upon what area the moment of inertia should be based

Apply the criterion set forth in the ACI *Code* to determine whether the moment of inertia is to be based upon the transformed gross section or the transformed cracked section. At the support: $pf_y = 4.43(40,000)/[14(20.5)] = 617 > 500$. Therefore, use the cracked section.

4. Determine the moment of inertia of the transformed cracked section at the support

Referring to Fig. 123d, $nA_s = 10(4.43) = 44.3$ sq in.; $(n-1)A'_s = 9(1.58) = 14.2$ sq in. The statical moment with respect to the neutral axis is: $Q = -\frac{1}{2}(14y^2) + 44.3(20.5-y) - 14.2(y-2.5) = 0$; $y = 8.16$ in. The moment of inertia with respect to the neutral axis is $I_1 = (\frac{1}{3})14(8.16)^3 + 14.2(8.16 - 2.5)^2 + 44.3(20.5 - 8.16)^2 = 9{,}737$ in.^{4}

5. Calculate the moment of inertia of the transformed cracked section at the center

Referring to Fig. 123e, and assuming tentatively that the neutral axis falls within the flange, $nA_s = 10(3.16) = 31.6$ sq in. The statical moment with respect to the neutral axis is $Q = \frac{1}{2}(68y^2) - 31.6(20.5-y) = 0$; $y = 3.92$ in. The neutral axis therefore falls within the flange, as assumed. The moment of inertia with respect to the neutral axis is $I_2 = (\frac{1}{3})68(3.92)^3 + 31.6(20.5 - 3.92)^2 = 10{,}052$ in.^{4}

6. *Calculate the deflection at mid-span*

Use the equation $\Delta = (L'^2/EI)(5M_1/48 - M_3/8)$, Eq. (37), where I = average moment of inertia, in.^{4}. Thus, $I = \frac{1}{2}(9,737 + 10,052) = 9,895$ in.^{4}; $E = 145^{1.5} \times 33(f'_c)^{0.5} = 57,600 \times (2,500)^{0.5} = 2,880,000$ psi. Then, $\Delta = [22^2 \times 1,728/(2,880 \times 9,895)](5 \times 200/48 - 100/8) = 0.244$ in.

Where the deflection under sustained loading is to be evaluated, it is necessary to apply the factors recorded in the ACI *Code*.

Design of Compression Members by Ultimate-strength Method

The notational system is: P_u = ultimate axial compressive load on member, lb; P_b = ultimate axial compressive load at balanced design, lb; P_0 = allowable ultimate axial compressive load in absence of bending moment, lb; M_u = ultimate bending moment in member, lb-in.; M_b = ultimate bending moment at balanced design; d' = distance from exterior surface to centroidal axis of adjacent row of steel bars, in.; t = overall depth of rectangular section or diameter of circular section, in.

A compression member is said to be *spirally reinforced* if the longitudinal reinforcement is held in position by spiral hooping, and *tied* if this reinforcement is held by means of intermittent lateral ties.

The presence of a bending moment in a compression member serves to reduce the ultimate axial load that the member may carry. In compliance with the ACI *Code*, it is necessary to design for a minimum bending moment equal to that caused by an eccentricity of $0.05t$ for spirally reinforced members and $0.10t$ for tied members. Thus, every compression member that is designed by the ultimate-strength method must be treated as a beam column. This type of member is considered to be in balanced design if failure would be characterized by the simultaneous crushing of the concrete, which is assumed to occur when $\epsilon_c = 0.003$, and incipient yielding of the tension steel, which occurs when $f_s = f_y$. The ACI *Code* sets $\phi = 0.75$ for spirally reinforced members and $\phi = 0.70$ for tied members.

ANALYSIS OF A RECTANGULAR MEMBER BY INTERACTION DIAGRAM

A short tied member having the cross section shown in Fig. 124a is to resist an axial load and a bending moment that induces compression at A and tension at B. The

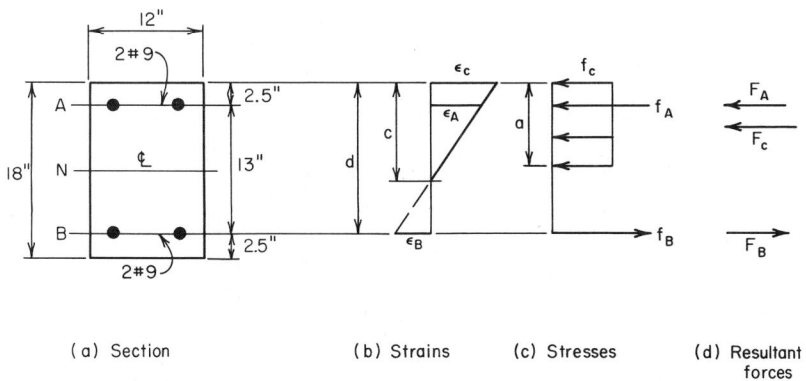

(a) Section   (b) Strains   (c) Stresses   (d) Resultant forces

Fig. 124

member is made of 3,000-psi concrete and the steel has a yield point of 40,000 psi. By starting with $c = 8$ in. and assigning progressively higher values to c, construct the interaction diagram for this member.

Calculation Procedure:

1. Compute the value of c associated with balanced design

An interaction diagram, as the term is used here, is one in which every point on the curve represents a set of simultaneous values of the ultimate moment and allowable ultimate axial load. Let ϵ_A and ϵ_B = strain of reinforcement at A and B, respectively; ϵ_c = strain of extreme fiber of concrete; f_A and f_B = stress in reinforcement at A and B, respectively, psi; F_A and F_B = resultant force in reinforcement at A and B, respectively; F_c = resultant force in concrete, lb.

Compression will be considered positive and tension negative. For simplicity, disregard the slight reduction in concrete area caused by the steel at A.

Referring to Fig. 124b, compute the value of c associated with balanced design. Computing P_b and M_b, $c_b/d = 0.003/(0.003 + f_y/E_s) = 87,000/(87,000 + f_y)$; $c_b = 10.62$ in. Then, $\epsilon_A/\epsilon_B = (10.62 - 2.5)/(15.5 - 10.62) > 1$; therefore, $f_A = f_y$; $a_b = 0.85(10.62) = 9.03$ in.; $F_c = 0.85(3,000)(12a_b) = 276,300$ lb; $F_A = 40,000(2.00) = 80,000$ lb; $F_B = -80,000$ lb; $P_b = 0.70(276,300) = 193,400$ lb. Also, $M_b = 0.70[F_c(t-a)/2 + (F_A - F_B)(t-2d')/2]$, Eq. (38). Thus, $M_b = 0.70[276,300(18 - 9.03)/2 + 160,000(6.5)] = 1,596,000$ in.-lb.

When $c > c_b$, the member fails by crushing of the concrete; when $c < c_b$, it fails by yielding of the reinforcement at line B.

2. Compute the value of c associated with incipient yielding of the compression steel

Then compute the corresponding values of P_u and M_u. Since ϵ_A and ϵ_B are numerically equal, the neutral axis lies at N. Thus, $c = 9$ in.; $a = 0.85(9) = 7.65$ in.; $F_c = 30,600(7.65) = 234,100$ lb; $F_A = 80,000$ lb; $F_B = -80,000$ lb; $P_u = 0.70 (234,100) = 163,900$ lb; $M_u = 0.70(234,100 \times 5.18 + 160,000 \times 6.5) = 1,577,000$ in.-lb.

3. Compute the minimum value of c at which the entire concrete area is stressed to 0.85f'_c

Compute the corresponding values of P_u and M_u. Thus, $a = t = 18$ in.; $c = 18/0.85 = 21.18$ in.; $f_B = \epsilon_c E_s(c - d)/c = 87,000(21.18 - 15.5)/21.18 = 23,300$ psi; $F_c = 30,600(18) = 550,800$ lb; $F_A = 80,000$ lb; $F_B = 46,600$ lb; $P_u = 0.70(550,800 + 80,000 + 46,600) = 474,200$ lb; $M_u = 0.70(80,000 - 46,600)6.5 = 152,000$ in.-lb.

4. Compute the value of c at which $M_u = 0$; compute P_0

The bending moment vanishes when F_B reaches 80,000 lb. From the calculation in Step 3: $f_B = 87,000(c - d)/c = 40,000$ psi; therefore, $c = 28.7$ in.; $P_0 = 0.70(550,800 + 160,000) = 497,600$ lb.

5. Assign other values to c and compute P_u and M_u

Assigning values to c ranging from 8 to 28.7 in., typical calculations are: when $c = 8$ in.; $f_B = -40,000$ psi; $f_A = 40,000(5.5/7.5) = 29,300$ psi; $a = 6.8$ in.; $F_c = 30,600(6.8) = 208,100$ lb; $P_u = 0.70(208,100 + 58,600 - 80,000) = 130,700$ lb; $M_u = 0.70 \times (208,100 \times 5.6 + 138,600 \times 6.5) = 1,446,000$ in.-lb.

When $c = 10$ in., $f_A = 40,000$ psi; $f_B = -40,000$ psi; $a = 8.5$ in.; $F_c = 30,600(8.5) = 260,100$ lb; $P_u = 0.70(260,100) = 182,100$ lb; $M_u = 0.70(260,100 \times 4.75 + 160,000 \times 6.5) = 1,593,000$ in.-lb.

When $c = 14$ in., $f_B = 87,000(14 - 15.5)/14 = -9,320$ psi; $a = 11.9$ in.; $F_c = 30,600 (11.9) = 364,100$ lb; $P_u = 0.70(364,100 + 80,000 - 18,600) = 297,900$ lb; $M_u = 0.70 (364,100 \times 3.05 + 98,600 \times 6.5) = 1,226,000$ in.-lb.

6. Plot the points representing computed values of P_u and M_u in the interaction diagram

Figure 125 shows these points. Pass a smooth curve through these points. It is important to note that when $P_u < P_b$, a reduction in M_u is accompanied by a reduction in the allowable load P_u.

AXIAL-LOAD CAPACITY OF RECTANGULAR MEMBER

The member analyzed in the previous Calculation Procedure is to carry an eccentric longitudinal load. Determine the allowable ultimate load if the eccentricity as measured from N is (a) 9.2 in.; (b) 6 in.

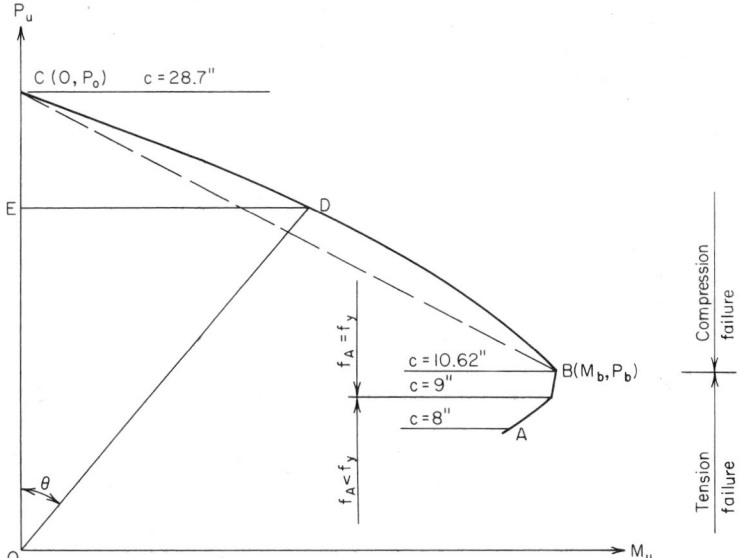

Fig. 125 Interaction diagram.

Calculation Procedure:

1. Evaluate the eccentricity associated with balanced design

Let e denote the eccentricity of the load and e_b the eccentricity associated with balanced design. Then, $M_u = P_u e$. In Fig. 125, draw an arbitrary radius vector OD; then $\tan \theta = ED/OE$ = eccentricity corresponding to point D.

Proceeding along the interaction diagram from A to C, the value of c increases and the value of e decreases. Thus, c and e vary in the reverse manner. To evaluate the allowable loads, it is necessary to identify the portion of the interaction diagram to which each eccentricity applies.

From the computations of the previous Calculation Procedure, $e_b = M_b/P_b = 1,596,000/193,400 = 8.25$ in. This result discloses that an eccentricity of 9.2 in. corresponds to a point on AB, and an eccentricity of 6 in. corresponds to a point on BC.

2. Evaluate P_u when $e = 9.2$ in.

It was found that $c = 9$ in. is a significant value. The corresponding value of e is $1,577,000/163,900 = 9.62$ in. This result discloses that in the present instance $c > 9$ in. and consequently $f_A = f_y$; $F_A = 80,000$ lb; $F_B = -80,000$ lb; $F_c = 30,600a$; $P_u/0.70 = 30,600a$; $M_u/0.70 = 30,600a(18 - a)/2 + 160,000(6.5)$; $e = M_u/P_u = 9.2$ in. Solving, $a = 8.05$ in., $P_u = 172,400$ lb.

3. Evaluate P_u when $e = 6$ in.

To simplify this calculation, the ACI *Code* permits replacement of curve BC in the interaction diagram with a straight line through B and C. The equation of this line is $P_u = P_o - (P_o - P_b)M_u/M_b$, Eq. (39). By replacing M_u with $P_u e$, the following relation is obtained: $P_u = P_o/[1 + (P_o - P_b)e/M_b]$, Eq. (39a). In the present instance, $P_o = 497,600$ lb; $P_b = 193,400$ lb; $M_b = 1,596,000$ in.-lb; solving, $P_u = 232,100$ lb.

ALLOWABLE ECCENTRICITY OF A MEMBER

The member analyzed in the previous two Calculation Procedures is to carry an ultimate longitudinal load of 150 kips that is eccentric with respect to axis N. Determine the maximum eccentricity with which the load may be applied.

Calculation Procedure:

1. Express P_u in terms of c, and solve for c

From the preceding Calculation Procedures it is seen that the value of c corresponding to the maximum eccentricity lies between 8 and 9 in., and therefore $f_A < f_y$. Thus: $f_B = -40,000$ psi; $f_A = 40,000(c-2.5)/(15.5-c)$; $F_c = 30,600(0.85c) = 26,000c$; $150,000 = 0.70\{26,000c + 80,000[(c-2.5)/(15.5-c)-1]\}$; $c = 8.60$ in.

2. Compute M_u and evaluate the eccentricity

Thus, $a = 7.31$ in.; $F_c = 223,700$ lb; $f_A = 35,360$ psi; $M_u = 0.70(223,700 \times 5.35 + 150,700 \times 6.5) = 1,523,000$ in.-lb; $e = M_u/P_u = 10.15$ in.

Design of Compression Members by Working-stress Method

The notational system is: A_g = gross area of section, sq in.; A_s = area of tension reinforcement, sq in.; A_{st} = total area of longitudinal reinforcement, sq in.; D = diameter of circular section, in.; $p_g = A_{st}/A_g$; P = axial load on member, lb; f_s = allowable stress in longitudinal reinforcement, psi; $m = f_y/0.85f_c'$.

The working-stress method of designing a compression member is essentially an adaptation of the ultimate-strength method. The allowable ultimate loads and bending moments are reduced by applying an appropriate factor of safety, and certain simplifications in computing the ultimate values are introduced.

The allowable concentric load on a short spirally reinforced column is $P = A_g(0.25f_c' + f_s p_g)$, or $P = 0.25f_c'A_g + f_s A_{st}$, Eq. (40), where $f_s = 0.40f_y$, but not to exceed 30,000 psi.

The allowable concentric load on a short tied column is $P = 0.85A_g(0.25f_c' + f_s p_g)$, or $P = 0.2125f_c'A_g + 0.85f_s A_{st}$, Eq. (41).

A section of the ACI *Code* provides that p_g may range from 0.01 to 0.08. However, in the case of a circular column in which the bars are to be placed in a single circular row, the upper limit of p_g is often governed by clearance. This section of the *Code* also stipulates that the minimum bar size to be used is No. 5 and requires a minimum of six bars for a spirally reinforced column and four bars for a tied column.

DESIGN OF A SPIRALLY REINFORCED COLUMN

A short circular column, spirally reinforced, is to support a concentric load of 420 kips. Design the member, using $f_c' = 4,000$ psi and $f_y = 50,000$ psi.

Calculation Procedure:

1. Assume $p_g = 0.025$ and compute the diameter of the section

Thus, $0.25f_c' = 1,000$ psi; $f_s = 20,000$ psi. By Eq. (40), $A_g = 420/(1 + 20 \times 0.025) = 280$ sq in. Then, $D = (A_g/0.785)^{0.5} = 18.9$ in. Set $D = 19$ in., making $A_g = 283$ sq in.

2. Select the reinforcing bars

The load carried by the concrete = 283 kips. The load carried by the steel = $420 - 283 = 137$ kips. Then, the area of the steel is $A_{st} = 137/20 = 6.85$ sq in. Use seven No. 9 bars, each having an area of 1 sq in. Then, $A_{st} = 7.00$ sq in. The *Reinforced Concrete Handbook* shows that a 19-in. column can accommodate 11 No. 9 bars in a single row.

3. Design the spiral reinforcement

The portion of the column section bounded by the outer circumference of the spiral is termed the *core* of the section. Let A_c = core area, sq in.; D_c = core diameter, in.;

a_s = cross-sectional area of spiral wire, sq in.; g = pitch of spiral, in.; p_s = ratio of volume of spiral reinforcement to volume of core.

The ACI *Code* requires 1.5-in. insulation for the spiral, with g restricted to a maximum of $D_c/6$. Then, $D_c = 19 - 3 = 16$ in.; $A_c = 201$ sq in.; $D_c/6 = 2.67$ in. Use a 2.5-in. spiral pitch. Taking a 1-in. length of column, p_s = volume of spiral/volume of core = $(a_s \pi D_c/g)/(\pi D_c^2/4)$; $a_s = g D_c p_s/4$, Eq. (42). The required value of p_s as given by the ACI *Code* is $p_s = 0.45(A_g/A_c - 1)f'_c/f_y$, Eq. (43), or $p_s = 0.45(283/201 - 1)4/50 = 0.0147$; $a_s = 2.5(16)(0.0147)/4 = 0.147$ sq in. Use $\frac{1}{2}$-in.-diameter wire with $a_s = 0.196$ sq in.

4. Summarize the design

Thus: column size, 19-in. diameter; longitudinal reinforcement, seven No. 9 bars; spiral reinforcement, $\frac{1}{2}$-in.-diameter wire, 2.5-in. pitch.

ANALYSIS OF A RECTANGULAR MEMBER BY INTERACTION DIAGRAM

A short tied member having the cross section shown in Fig. 126 is to resist an axial load and a bending moment that induces rotation about axis N. The member is made of 4,000-psi concrete and the steel has a yield point of 50,000 psi. Construct the interaction diagram for this member.

Calculation Procedure:

1. Compute P_a and M_f

Consider a composite member of two materials having equal strength in tension and compression, the member being subjected to an axial load P and bending moment M that induce the allowable stress in one or both materials. Let: P_a = allowable axial load in absence of bending moment, as computed by dividing the allowable ultimate load by a factor of safety; M_f = allowable bending moment in absence of axial load, as computed by dividing the allowable ultimate moment by a factor of safety.

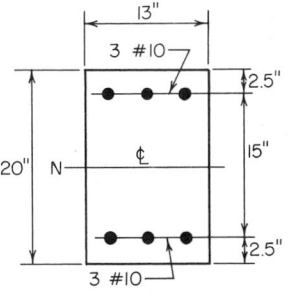

Fig. 126

Find the simultaneous allowable values of P and M by applying the interaction equation $P/P_a + M/M_f = 1$, Eq. (44). Alternate forms of this equation are $M = M_f(1 - P/P_a)$, $P = P_a(1 - M/M_f)$, Eq. (44a); $P = P_a M_f/(M_f + P_a M/P)$, Eq. (44b).

Equation (44) is represented by line AB in Fig. 127; it is also valid with respect to a reinforced-concrete member for a certain range of values of P and M. This equation is not applicable in the following instances: (a) If M is relatively small, Eq. (44) yields a value of P in excess of that given by Eq. (41). Therefore, the interaction diagram must contain line CD, which represents the maximum value of P.

(b) If M is relatively large, the section will crack, and the equal-strength assumption underlying Eq. (44) becomes untenable.

Let point E represent the set of values of P and M that will cause cracking in the extreme concrete fiber, and let P_b = axial load represented by point E, and M_b = bending moment represented by point E; M_o = allowable bending moment in reinforced-concrete member in absence of axial load, as computed by dividing the allowable ultimate moment by a factor of safety. (M_o differs from M_f in that the former is based on a cracked section and the latter on an uncracked section. The subscript b as used by the ACI *Code* in the present instance does *not* refer to balanced design. However, its use serves to illustrate the analogy with ultimate-strength analysis.) Let F denote the point representing M_o.

For simplicity, the interaction diagram is assumed to be linear between E and F. The interaction equation for a cracked section may therefore be expressed in any of the following forms: $M = M_o + (P/P_b)(M_b - M_o)$; $P = P_b(M - M_o)/(M_b - M_o)$, Eq. (45a); $P = P_b M_o/(M_o - M_b + P_b M/P)$, Eq. (45b).

The ACI *Code* gives the following approximations: For spiral columns: $M_o = 0.12A_{st}f_y D_s$, Eq. (46a), where D_s = diameter of circle through center of longitudinal reinforcement. For symmetrical tied columns: $M_o = 0.40A_s f_y(d - d')$, Eq. (46b). For unsymmetrical tied columns: $M_o = 0.40A_s f_y jd$, Eq. (46c). For symmetrical spiral

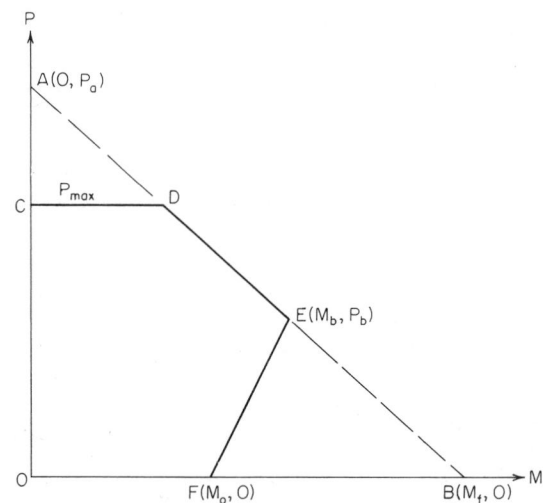

Fig. 127   Interaction diagram.

columns: $M_b/P_b = 0.43p_g m D_s + 0.14t$, Eq. (47a). For symmetrical tied columns: $M_b/P_b = d(0.67p_g m + 0.17)$, Eq. (47b). For unsymmetrical tied columns: $M_b/P_b = [p'm(d - d') + 0.1d]/[(p' - p)m + 0.6]$, Eq. (47c), where p' = ratio of area of compression reinforcement to effective area of concrete. The value of P_a is taken as $P_a = 0.34f'_c A_g(1 + p_g m)$, Eq. (48).

The value of M_f is found by applying the section modulus of the transformed uncracked section, using a modular ratio of $2n$ to account for stress transfer between steel and concrete engendered by plastic flow. (If the steel area is multiplied by $2n - 1$, allowance is made for the reduction of the concrete area.)

Computing P_a and M_f, $A_g = 260$ sq in.; $A_{st} = 7.62$ sq in.; $p_g = 7.62/260 = 0.0293$; $m = 50/[0.85(4)] = 14.7$; $p_g m = 0.431$; $n = 8$; $P_a = 0.34(4)(260)(1.431) = 506$ kips.

The section modulus to be applied in evaluating M_f is found thus: $I = (\frac{1}{12})(13)(20)^3 + 7.62(15)(7.5)^2 = 15,100$ in.^{4}; $S = I/c = 15,100/10 = 1,510$ in.^{3}; $M_f = Sf_c = 1,510(1.8) = 2,720$ in.-kips.

2. Compute P_b and M_b

By Eq. (47b), $M_b/P_b = 17.5(0.67 \times 0.431 + 0.17) = 8.03$ in. By Eq. (44b), $P_b = P_a M_f/(M_f + 8.03P_a) = 506 \times 2,720/(2,720 + 8.03 \times 506) = 203$ kips; $M_b = 8.03(203) = 1,630$ in.-kips.

3. Compute M_o

By Eq. (46b), $M_o = 0.40(3.81)(50)(15) = 1,140$ in.-kips.

4. Compute the limiting value of P

As established by Eq. (41), $P_{max} = 0.2125(4)(260) + 0.85(20)(7.62) = 351$ kips.

5. Construct the interaction diagram

The complete diagram is shown in Fig. 127.

AXIAL-LOAD CAPACITY OF A RECTANGULAR MEMBER

The member analyzed in the previous Calculation Procedure is to carry an eccentric longitudinal load. Determine the allowable load if the eccentricity as measured from N is (a) 10 in.; (b) 6 in.

Calculation Procedure:

1. Evaluate P when e = 10 in.

As the preceding calculations show, the eccentricity corresponding to point E in the interaction diagram is 8.03 in. Consequently, an eccentricity of 10 in. corresponds to a point on EF, and an eccentricity of 6 in. corresponds to a point on ED.

By Eq. (45b), $P = 203(1,140)/(1,140 - 1,630 + 203 \times 10) = 150$ kips.

2. Evaluate P when e = 6 in.

By Eq. (44b), $P = 506(2,720)/(2,720 + 506 \times 6) = 239$ kips.

Design of Column Footings

A reinforced-concrete footing supporting a single column differs from the usual type of flexural member in the following respects: It is subjected to bending in all directions, the ratio of maximum vertical shear to maximum bending moment is very high, and it carries a heavy load concentrated within a small area. The consequences are as follows: The footing requires two-way reinforcement, its depth is determined by shearing rather than by bending stress, the punching-shear stress below the column is usually more critical than the shearing stress that results from ordinary beam action, and the design of the reinforcement is controlled by the bond stress as well as the bending stress.

Since the footing weight and soil pressure are collinear, the former does not contribute to the vertical shear or bending moment. It is convenient to visualize the footing as being subjected to an upward load transmitted by the underlying soil and a downward reaction supplied by the column, this being of course an inversion of the true form of loading. The footing thus functions as an overhanging beam. The effective depth of footing is taken as the distance from the top surface to the center of the upper row of bars, the two rows being made identical to avoid confusion.

Refer to Fig. 128, which shows a square footing supporting a square, symmetrically located concrete column. Let P = column load, kips; p = net soil pressure (that caused by the column load alone), psf; A = area of footing, sq ft; L = side of footing, ft; h = side of column, in.; d = effective depth of footing, ft; t = thickness of footing, ft; f_b = bearing stress at interface of column, psi; v_1 = nominal shearing stress under column, psi; v_2 =

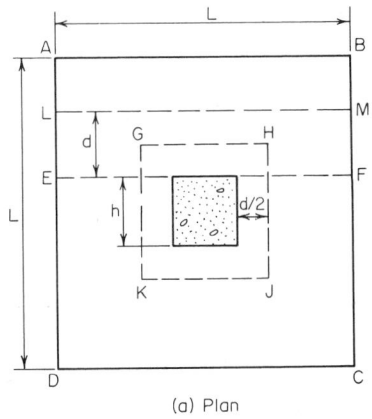

(a) Plan

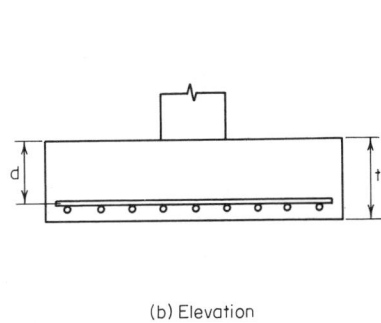

(b) Elevation

Fig. 128

nominal shearing stress caused by beam action, psi; b_o = width of critical section for v_1, ft; V_1 and V_2 = vertical shear at critical section for stresses v_1 and v_2, respectively.

In accordance with the ACI *Code*, the critical section for v_1 is the surface $GHJK$, the sides of which lie at a distance $d/2$ from the column faces. The critical section for v_2 is plane LM, located at a distance d from the face of the column. The critical section for bending stress and bond stress is plane EF through the face of the column. In calculating v_2, f, and u, no allowance is made for the effects of the orthogonal reinforcement.

DESIGN OF AN ISOLATED SQUARE FOOTING

A 20-in. square tied column reinforced with eight No. 9 bars carries a concentric load of 380 kips. Design a square footing by the working-stress method using these values: the allowable soil pressure is 7,000 psf, $f'_c = 3,000$ psi, and $f_s = 20,000$ psi.

Calculation Procedure:

1. Record the allowable shear, bond, and bearing stresses

From the ACI *Code* table: $v_1 = 110$ psi; $v_2 = 60$ psi; $f_b = 1,125$ psi; $u = 4.8(f'_c)^{0.5}$/bar diameter = 264/bar diameter.

2. Check the bearing pressure on the footing

Thus, $f_b = 380/[20(20)] = 0.95 < 1.125$ ksi. This is acceptable.

3. Establish the length of footing

For this purpose, assume the footing weight is 6 percent of the column load. Then, $A = 1.06(380)/7 = 57.5$ sq ft. Make $L = 7$ ft 8 in. = 7.67 ft; $A = 58.8$ sq ft.

4. Determine the effective depth as controlled by v_1

Apply $(4v_1 + p)d^2 + h(4v_1 + 2p)d = p(A - h^2)$, Eq. (49). Verify the result after applying this equation. Thus: $p = 380/58.8 = 6.46$ ksf; $v_1 = 0.11(144) = 15.84$ ksf; $69.8d^2 + 127.1d = 361.8$; $d = 1.54$ ft. Checking in Fig. 128, $GH = 1.67 + 1.54 = 3.21$ ft; $V_1 = 6.46(58.8 - 3.21^2) = 313$ kips; $v_1 = V_1/b_o d = 313/[4(3.21)(1.54)] = 15.83$ ksf. This is acceptable.

5. Establish the thickness and true depth of footing

Compare the weight of the footing with the assumed weight. Allowing 3 in. for insulation and assuming the use of No. 8 bars, $t = d + 4.5$ in. Then, $t = 1.54(12) + 4.5 = 23.0$ in. Make $t = 24$ in.; $d = 19.5$ in. = 1.63 ft. The footing weight = 58.8(2)(0.150) = 17.64 kips. The assumed weight = 0.06(380) = 22.8 kips. This is acceptable.

6. Check v_2

In Fig. 128, $AL = [(7.67 - 1.67)/2] - 1.63 = 1.37$ ft; $V_2 = 380(1.37/7.67) = 67.9$ kips; $v_2 = V_2/Ld = 67,900/[92(19.5)] = 38 < 60$ psi. This is acceptable.

7. Design the reinforcement

In Fig. 128, $EA = 3.00$ ft; $V_{EF} = 380(3.00/7.67) = 148.6$ kips; $M_{EF} = 148.6(\frac{1}{2})(3.00)(12) = 2,675$ in.-kips; $A_s = 2,675/[20(0.874)(19.5)] = 7.85$ sq in. Try 10 No. 8 bars each way. Then, $A_s = 7.90$ sq in.; $\Sigma o = 31.4$ in.; $u = V_{EF}/\Sigma ojd = 148,600/[31.4(0.874)(19.5)] = 278$ psi; $u_{allow} = 264/1 = 264$ psi.

The bond stress at EF is slightly excessive. However, the ACI *Code*, in sections based on ultimate-strength considerations, permits disregarding the local bond stress if the average bond stress across the length of embedment is less than 80 percent of the allowable stress. Let L_e denote this length. Then, $L_e = EA - 3 = 33$ in.; $0.80u_{allow} = 211$ psi; $u_{av} = A_s f_s/L_e \Sigma o = 0.79(20,000)/[33(3.1)] = 154$ psi. This is acceptable.

8. Design the dowels to comply with the Code

The function of the dowels is to transfer the compressive force in the column reinforcing bars to the footing. Since this is a tied column, assume the stress in the bars is 0.85(20,000) = 17,000 psi. Try eight No. 9 dowels with $f_y = 40,000$ psi. Then, $u = 264/(9/8) = 235$ psi; $L_e = 1.00(17,000)/[235(3.5)] = 20.7$ in. Since the footing can provide a

21-in. embedment length, the dowel selection is satisfactory. Also, the length of lap = 20(9/8) = 22.5 in.; length of dowels = 20.7 + 22.5 = 43.2, say 44 in. The footing is shown in Fig. 129.

COMBINED FOOTING DESIGN

An 18-in.-square exterior column and a 20-in.-square interior column carry loads of 250 kips and 370 kips, respectively. The column centers are 16 ft apart, and the footing cannot project beyond the face of the exterior column. Design a combined rectangular footing by the working-stress method, using $f_c' = 3,000$ psi, $f_s = 20,000$ psi, and an allowable soil pressure of 5,000 psf.

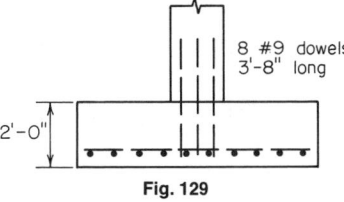

Fig. 129

Calculation Procedure:

1. Establish the length of footing, applying the criterion of uniform soil pressure under total live and dead loads

In many instances, the exterior column of a building cannot be individually supported because the required footing would project beyond the property limits. It then becomes necessary to use a combined footing that supports the exterior column and the adjacent interior column, the footing being so proportioned that the soil pressure is approximately uniform.

The footing dimensions are shown in Fig. 130a, and the reinforcement in Fig. 131. It is convenient to visualize the combined footing as being subjected to an upward load transmitted by the underlying soil and reactions supplied by the columns. The member thus functions as a beam that overhangs one support. However, since the footing is considerably wider than the columns, there is a transverse bending as well as longitudinal bending in the vicinity of the columns. For simplicity, assume that the transverse bending is confined to the regions bounded by planes AB and EF and by planes GH and NP, the distance m being $h/2$ or $d/2$, whichever is smaller.

In Fig. 130a, let Z denote the location of the resultant of the column loads. Then, $x = 370(16)/(250 + 370) = 9.55$ ft. Since Z is to be the centroid of the footing, $L = 2(0.75 + 9.55) = 20.60$ ft. Set $L = 20$ ft 8 in., but use the value 20.60 ft in the stress calculations.

2. Construct the shear and bending-moment diagrams

The net soil pressure per foot of length = 620/20.60 = 30.1 klf. Construct the diagrams as shown in Fig. 130.

3. Establish the footing thickness

Use $(Pv_2 + 0.17VL + Pp')d - 0.17Pd^2 = VLp'$, Eq. (50), where P = aggregate column load, kips; V = maximum vertical shear at a column face, kips; p' = gross soil pressure, ksf.

Assume that the longitudinal steel is centered $3\frac{1}{2}$ in. from the face of the footing. Then: $P = 620$ kips; $V = 229.2$ kips; $v_2 = 0.06(144) = 8.64$ ksf; $9,260d - 105.4d^2 = 23,608$; $d = 2.63$ ft; $t = 2.63 + 0.29 = 2.92$ ft. Set $t = 2$ ft 11 in.; $d = 2$ ft $7\frac{1}{2}$ in.

4. Compute the vertical shear at distance d from the column face

Establish the width of the footing. Thus: $V = 229.2 - 2.63(30.1) = 150.0$ kips; $v = V/Wd$, or $W = V/vd = 150/[8.64(2.63)] = 6.60$ ft. Set $W = 6$ ft 8 in.

5. Check the soil pressure

The footing weight = 20.67(6.67)(2.92)(0.150) = 60.4 kips; $p' = (620 + 60.4)/[(20.67)(6.67)] = 4.94 < 5$ ksf. This is acceptable.

6. Check the punching shear

Thus, $p = 4.94 - 2.92(0.150) = 4.50$ ksf. At $C1$: $b_o = (18 + 31.5) + 2(18 + 15.8) = 117$ in., $V = 250 - 4.50(49.5)(33.8)/144 = 198$ kips; $v_1 = 198{,}000/[117(31.5)] = 54 < 110$ psi; this is acceptable.

At $C2$: $b_o = 4(20 + 31.5) = 206$ in.; $V = 370 - 4.50(51.5)^2/144 = 287$ kips; $v_1 = 287{,}000/[206(31.5)] = 44$ psi. This is acceptable.

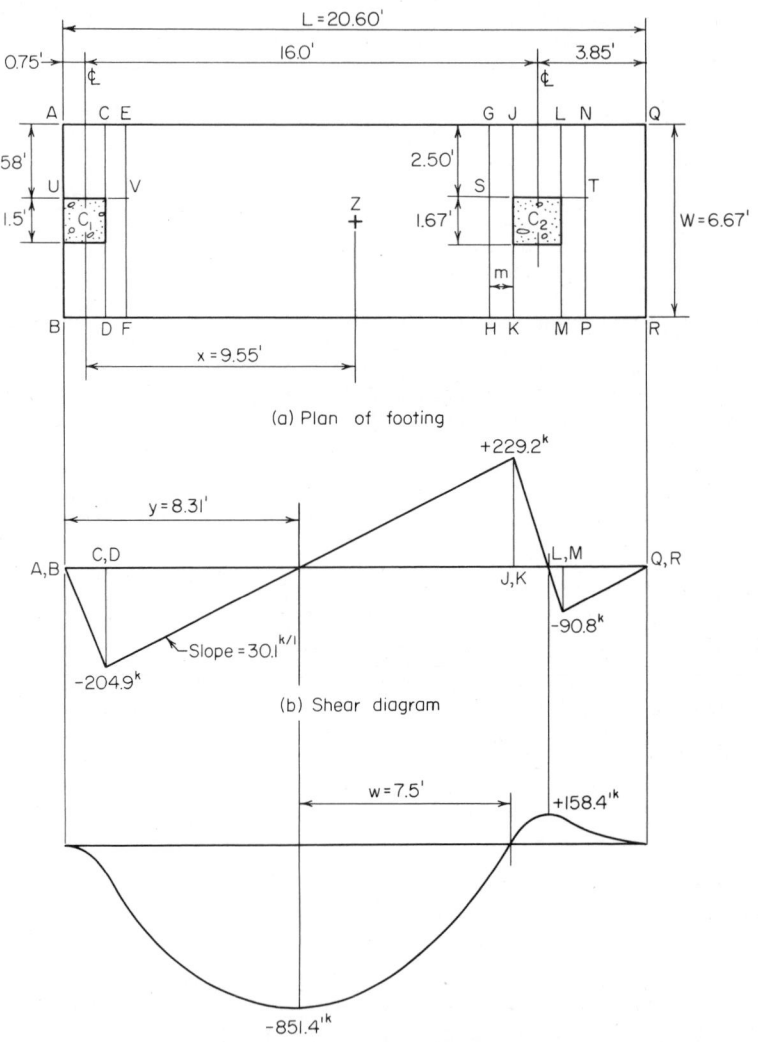

(a) Plan of footing

(b) Shear diagram

(c) Bending-moment diagram

Fig. 130

7. Design the longitudinal reinforcement for negative moment

Thus, $M = 851{,}400$ ft-lb $= 10{,}217{,}000$ in.-lb; $M_b = 223(80)(31.5)^2 = 17{,}700{,}000$ in.-lb. Therefore, the steel is stressed to capacity, and $A_s = 10{,}217{,}000/[20{,}000(0.874)(31.5)] = 18.6$ sq in. Try 15 No. 10 bars with $A_s = 19.1$ sq in.; $\Sigma o = 59.9$ in.

The bond stress is maximum at the point of contraflexure, where $V = 15.81(30.1) - 250 = 225.9$ kips; $u = 225,900/[59.9(0.874)(31.5)] = 137$ psi; $u_{allow} = 3.4(3,000)^{0.5}/1.25 = 149$ psi. This is acceptable.

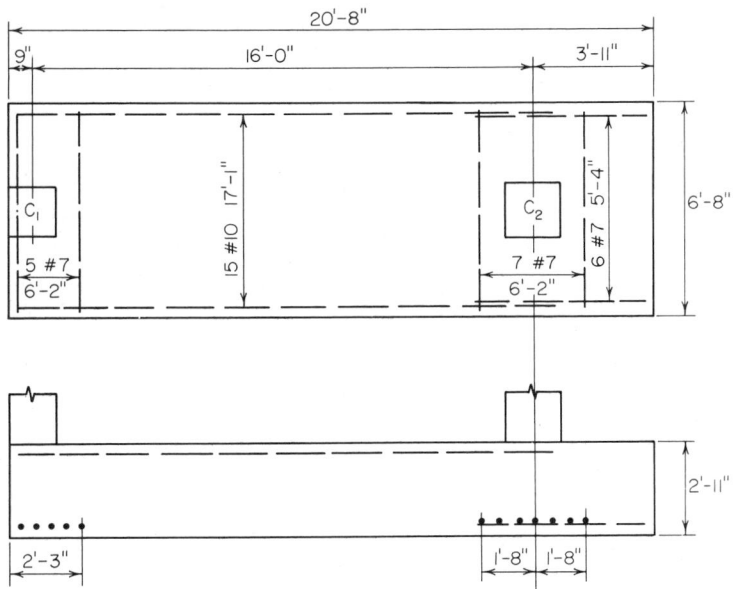

Fig. 131

8. Design the longitudinal reinforcement for positive moment

For simplicity, design for the maximum moment rather than the moment at the face of the column. Then, $A_s = 158,400(12)/[20,000(0.874)(31.5)] = 3.45$ sq in. Try six No. 7 bars with $A_s = 3.60$ sq in., $\Sigma o = 16.5$ in. Take LM as the critical section for bond, and $u = 90,800/[16.5(0.874)(31.5)] = 200$ psi; $u_{allow} = 4.8(3,000)^{0.5}/0.875 = 302$ psi. This is acceptable.

9. Design the transverse reinforcement under the interior column

For this purpose, consider member $GNPH$ as an independent isolated footing. Then, $V_{ST} = 370(2.50/6.67) = 138.8$ kips; $M_{ST} = \frac{1}{2}(138.8)(2.50)(12) = 2,082$ in.-kips. Assume $d = 35 - 4.5 = 30.5$ in.; $A_s = 2,082,000/[20,000(0.874)(30.5)] = 3.91$ sq in. Try seven No. 7 bars; $A_s = 4.20$ sq in.; $\Sigma o = 19.2$ in.; $u = 138,800/[19.2(0.874)(30.5)] = 271$ psi; $u_{allow} = 302$ psi. This is acceptable.

Since the critical section for shear falls outside the footing, shearing stress is not a criterion in this design.

10. Design the transverse reinforcement under the exterior column; disregard eccentricity

Thus, $V_{UV} = 250(2.58/6.67) = 96.8$ kips; $M_{UV} = \frac{1}{2}(96.8)(2.58)(12) = 1,498$ in.-kips; $A_s = 2.72$ sq in. Try five No. 7 bars; $A_s = 3.00$ sq in.; $\Sigma o = 13.7$ in.; $u = 96,800/[13.7 \times (0.874)(31.5)] = 257$ psi. This is acceptable.

Cantilever Retaining Walls

Retaining walls having a height ranging from 10 to 20 ft are generally built as reinforced-concrete cantilever members. As shown in Fig. 132, a cantilever wall comprises a vertical stem to retain the soil, a horizontal base to support the stem, and in many instances a key that projects into the underlying soil to augment the resistance to sliding. Adequate drainage is an essential requirement, because the accumulation of water or ice behind the wall would greatly increase the horizontal thrust.

The calculation of earth thrust in this section is based on Rankine's theory, which is developed in a later Calculation Procedure. When a live load—termed a *surcharge*—is applied to the retained soil, it is convenient to replace this load with a hypothetical equivalent prism of earth. Referring to Fig. 132, consider a portion QR of the wall, R being at distance y below the top. Take the length of wall normal to the plane of the drawing as 1 ft. Let: $T =$ resultant earth thrust on QR; $M =$ moment of this thrust with respect to R; $h =$ height of equivalent earth prism that replaces surcharge; $w =$ unit weight of earth; $C_a =$ coefficient of active earth pressure; $C_p =$ coefficient of passive earth pressure. Then, $T = \frac{1}{2}C_a wy(y + 2h)$, Eq. (51); $M = (\frac{1}{6})C_a wy^2(y + 3h)$, Eq. (52).

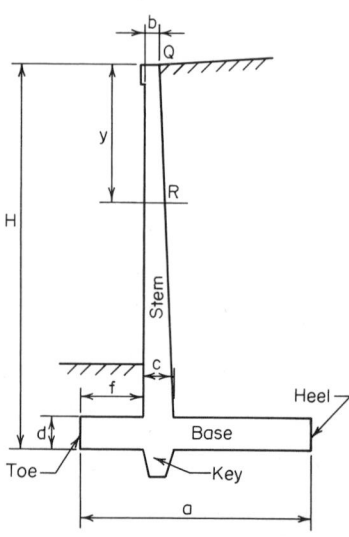

Fig. 132   Cantilever retaining wall.

DESIGN OF A CANTILEVER RETAINING WALL

Applying the working-stress method, design a reinforced-concrete wall to retain an earth bank 14 ft high. The top surface is horizontal and supports a surcharge of 500 psf. The soil weighs 130 pcf and its angle of internal friction is 35°; the coefficient of friction of soil and concrete is 0.5. The allowable soil pressure is 4,000 psf, $f_c' = 3,000$ psi and $f_y = 40,000$ psi. The base of the structure must be set 4 ft below ground level to clear the frost line.

Calculation Procedure:

1.   Secure a trial section of the wall

Apply these relations: $a = 0.60H$; $b \geqslant 8$ in.; $c = d = b + 0.045h$; $f = a/3 - c/2$.

The trial section is shown in Fig. 133a, and the reinforcement is shown in Fig. 134. As the calculation will show, it is necessary to provide a key to develop the required resistance to sliding. The sides of the key are sloped to ensure that the surrounding soil will remain undisturbed during excavation.

2.   Analyze the trial section for stability

The requirements are that there be a factor of safety (FS) against sliding and overturning of at least 1.5 and that the soil pressure have a value lying between 0 and 4,000 psf. Using the equation developed later in this Handbook, $h =$ surcharge/soil weight = $500/130 = 3.85$ ft; $\sin 35° = 0.574$; $\tan 35° = 0.700$; $C_a = 0.271$; $C_p = 3.69$; $C_a w = 35.2$ pcf; $C_p w = 480$ pcf; $T_{AB} = \frac{1}{2}(35.2)18(18 + 2 \times 3.85) = 8,140$ lb; $M_{AB} = (\frac{1}{6})35.2(18)^2(18 + 3 \times 3.85) = 56,200$ ft-lb.

The critical condition with respect to stability is that in which the surcharge extends to G. The moments of the stabilizing forces with respect to the toe are computed in Table 7. In Fig. 133c, $x = 81,030/21,180 = 3.83$ ft; $e = 5.50 - 3.83 = 1.67$ ft. The fact that the resultant strikes the base within the middle third attests to the absence of uplift. By $f = P/A(1 \pm 6e_x/d_x \pm 6e_y/d_y)$, $p_a = (21,180/11)(1 + 6 \times 1.67/11) = 3,680$ psf; $p_b = (21,180/11)(1 - 6 \times 1.67/11) = 171$ psf. Check: $x = (11/3)(3,680 + 2 \times 171)/(3,680 + 171) = 3.83$ ft, as before. Also, $p_c = 2,723$ psf; $p_d = 2,244$ psf; FS against overturning = $137,230/56,200 = 2.44$. This is acceptable.

Lateral displacement of the wall produces sliding of earth on earth to the left of C and

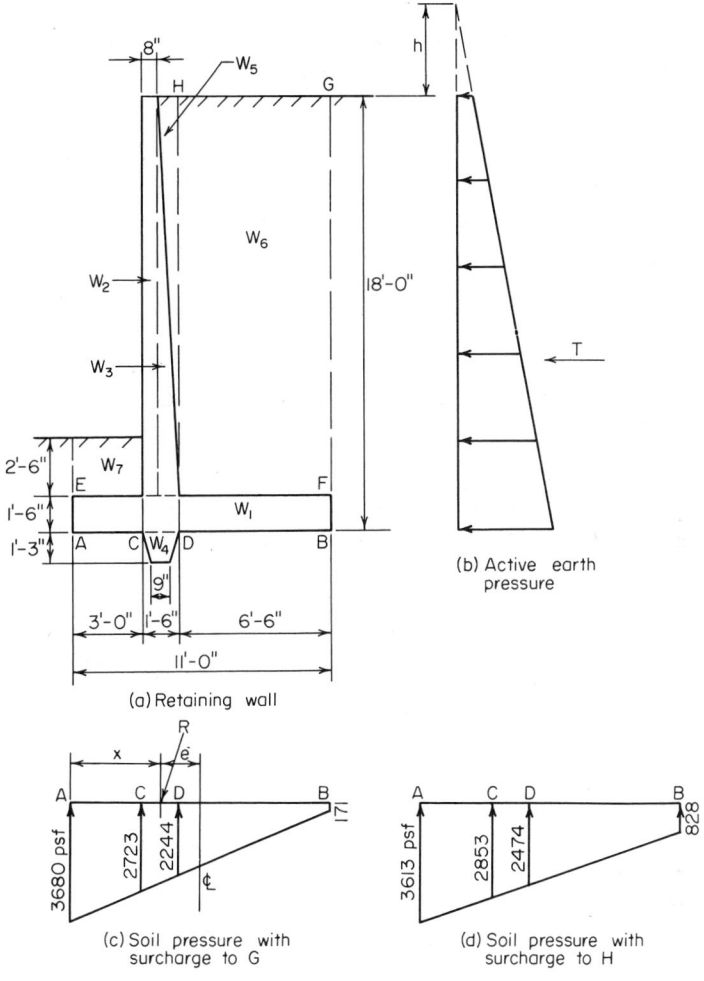

(a) Retaining wall

(b) Active earth pressure

(c) Soil pressure with surcharge to G

(d) Soil pressure with surcharge to H

Fig. 133

TABLE 7   Stability of Retaining Wall

Force, lb		Arm, ft	Moment, ft-lb
W_1 1.5(11)(150)	= 2,480	5.50	13,640
W_2 0.67(16.5)(150)	= 1,650	3.33	5,500
W_3 0.5(0.83)(16.5)(150)	= 1,030	3.95	4,070
W_4 1.25(1.13)(150)	= 210	3.75	790
W_5 0.5(0.83)(16.5)(130)	= 890	4.23	3,760
W_6 6.5(16.5)(130)	= 13,940	7.75	108,000
W_7 2.5(3)(130)	= 980	1.50	1,470
Total................	21,180	...	137,230
Overturning moment...........		...	56,200
Net moment about A		...	81,030

of concrete on earth to the right of C. In calculating the passive pressure, the layer of earth lying above the base will be disregarded, since its effectiveness is unknown. The resistance to sliding is: friction, A to C, Fig. 133, $\frac{1}{2}(3,680+2,723)(3)(0.700) = 6,720$ lb; friction, C to B, $\frac{1}{2}(2,723+171)(8)(0.5) = 5,790$ lb; passive earth pressure, $\frac{1}{2}(480)(2.75)^2 = 1,820$ lb. The total resistance to sliding is the sum of these three items, or 14,330 lb. Thus, the FS against sliding is $14,330/8,140 = 1.76$. This is acceptable because it exceeds 1.5. Hence the trial section is adequate with respect to stability.

3. Calculate the soil pressures when the surcharge extends to H

Thus: $W_s = 500(6.5) = 3,250$ lb; $\Sigma W = 21,180+3,250 = 24,430$ lb; $M_a = 81,030+3,250(7.75) = 106,220$ ft-lb; $x = 106,220/24,430 = 4.35$ ft; $e = 1.15$ ft; $p_a = 3,613$ psf; $p_b = 828$ psf; $p_c = 2,853$ psf; $p_d = 2,474$ psf.

4. Design the stem

At the base of the stem, $y = 16.5$ ft and $d = 18-3.5 = 14.5$ in.; $T_{EF} = 7,030$ lb; $M_{EF} = 538,000$ in.-lb. The allowable shear at a distance d above the base is $V_{\text{allow}} = vbd = 60(12)(14.5) = 10,440$ lb. This is acceptable. Also, $M_b = 223(12)(14.5)^2 = 563,000$ in.-lb; therefore, the steel is stressed to capacity, and $A_s = 538,000[20,000(0.874)(14.5)] = 2.12$ sq in. Use No. 9 bars $5\frac{1}{2}$ in. on centers. Thus, $A_s = 2.18$ sq in.; $\Sigma o = 7.7$ in.; $u = 7,030/[7.7(0.874)(14.5)] = 72$ psi; $u_{\text{allow}} = 235$ psi. This is acceptable.

Alternate bars will be discontinued at the point where they become superfluous. As the following calculations demonstrate, the theoretical cutoff point lies at $y = 11$ ft 7 in.,

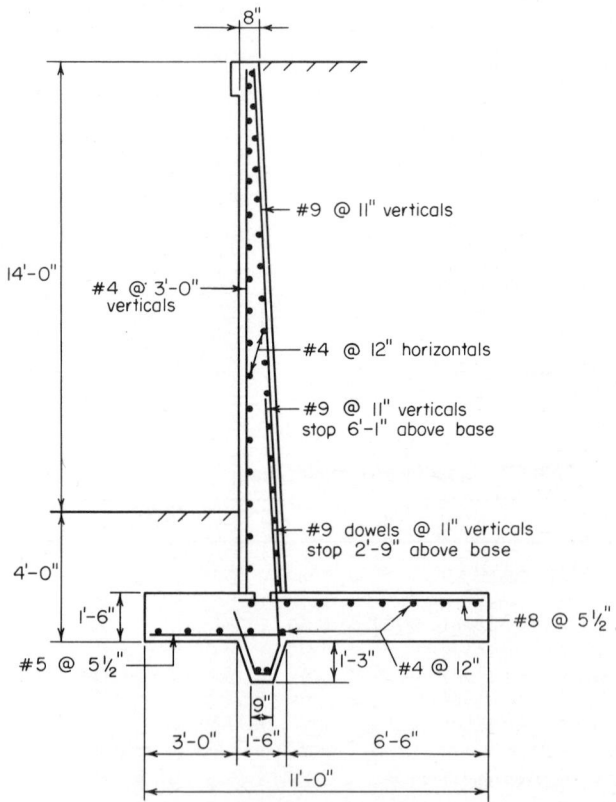

Fig. 134

where $M = 218,400$ in.-lb; $d = 4.5 + 10(11.58/16.5) = 11.52$ in.; $A_s = 218,400/[20,000 \times (0.874)(11.52)] = 1.08$ sq in. This is acceptable. Also, $T = 3,930$ lb; $u = 101$ psi. This is acceptable. From the ACI *Code*, anchorage $= 12(9/8) = 13.5$ in.

The alternate bars will therefore be terminated at 6 ft 1 in. above the top of the base. The *Code* requires that special precautions be taken where more than half the bars are spliced at a point of maximum stress. To circumvent this requirement, the short bars can be extended into the footing; only the long bars therefore require splicing. For the dowels, $u_{\text{allow}} = 0.75(235) = 176$ psi; length of lap $= 1.00(20,000)/[176(3.5)] = 33$ in.

5. Design the heel

Let V and M denote the shear and bending moment, respectively, at section D. Case 1: surcharge extending to G—downward pressure $p = 16.5(130) + 1.5(150) = 2,370$ psf; $V = 6.5[2,370 - \frac{1}{2}(2,244 + 171)] = 7,560$ lb; $M = 12(6.5)^2[\frac{1}{2} \times 2,370 - \frac{1}{6}(2,244 + 2 \times 171)] = 383,000$ in.-lb.

Case 2: surcharge extending to $H - p = 2,370 + 500 = 2,870$ psf; $V = 6.5[2,870 - \frac{1}{2}(2,474 + 828)] = 7,920$ lb $< V_{\text{allow}}$; $M = 12(6.5)^2[\frac{1}{2} \times 2,870 - \frac{1}{6}(2,474 + 2 \times 828)] = 379,000$ in.-lb.; $A_s = 2.12(383/538) = 1.51$ sq in.

To maintain uniform bar spacing throughout the member, use No. 8 bars $5\frac{1}{2}$ in. on centers. In the heel, tension occurs at the *top* of the slab, and $A_s = 1.72$ sq in.; $\Sigma o = 6.9$ in.; $u = 91$ psi; $u_{\text{allow}} = 186$ psi. This is acceptable.

6. Design the toe

For this purpose, assume the absence of backfill on the toe, but disregard the minor modification in the soil pressure that results. Let V and M denote the shear and bending moment, respectively, at section C, Fig. 133. The downward pressure $p = 1.5(150) = 225$ psf.

Case 1: surcharge extending to G, Fig. 133 — $V = 3[\frac{1}{2}(3,680 + 2,723) - 225] = 8,930$ lb; $M = 12(3)^2[(\frac{1}{6})(2,723 + 2 \times 3,680) - \frac{1}{2}(225)] = 169,300$ in.-lb.

Case 2: surcharge extending to H, Fig. 133 — $V = 9,020$ lb $< V_{\text{allow}}$; $M = 169,300$ in.-lb; $A_s = 2.12(169,300/538,000) = 0.67$ sq in. Use No. 5 bars $5\frac{1}{2}$ in. on centers. Then: $A_s = 0.68$ sq in.; $\Sigma o = 4.3$ in.; $u = 166$ psi; $u_{\text{allow}} = 422$ psi. This is acceptable.

The stresses in the key are not amenable to precise evaluation. Reinforcement is achieved by extending the dowels and short bars into the key and bending them.

In addition to the foregoing reinforcement, No. 4 bars are supplied to act as temperature reinforcement and spacers for the main bars, as shown in Fig. 134.

Prestressed Concrete

Prestressed-concrete construction is designed to enhance the suitability of concrete as a structural material by inducing prestresses opposite in character to the stresses resulting from gravity loads. These prestresses are created by the use of steel wires or strands, called *tendons*, that are incorporated in the member and subjected to externally applied tensile forces. This prestressing of the steel may be performed either before or after pouring of the concrete. Thus, two methods of prestressing a concrete beam are available: *pretensioning* and *post-tensioning*.

In pretensioning, the tendons are prestressed to the required amount by means of hydraulic jacks, their ends are tied to fixed abutments, and the concrete is then poured around the tendons. When hardening of the concrete has advanced to the required state, the tendons are released. The tendons now tend to contract longitudinally to their original length and to expand laterally to their original diameter, both these tendencies being opposed by the surrounding concrete. As a result of the longitudinal restraint, the concrete exerts a tensile force on the steel and the steel exerts a compressive force on the concrete. As a result of the lateral restraint, the tendons are deformed to a wedge shape across a relatively short distance at each end of the member. It is within this distance, termed the *transmission length*, that the steel becomes bonded to the concrete and the two materials exert their prestressing forces upon one another. However, unless greater precision is warranted, it is assumed for simplicity that the prestressing forces act at the end sections.

The tendons may be placed either in a straight line or in a series of straight-line segments, being deflected at designated points by means of holding devices. In the latter case, prestressing forces between steel and concrete occur both at the ends and at these deflection points.

In post-tensioning, the procedure usually consists of encasing the tendons in metal or rubber hoses, placing these in the forms, and then pouring the concrete. When the concrete has hardened, the tendons are tensioned and anchored to the ends of the concrete beam by means of devices called end anchorages. If the hoses are to remain in the member, the void within the hose is filled with grout. Post-tensioning has two important advantages when compared with pretensioning: it may be performed at the job site, and it permits the use of parabolic tendons.

The term *at transfer* refers to the instant at which the prestressing forces between steel and concrete are developed. (In post-tensioning, where the tendons are anchored to the concrete one at a time, these forces in reality are developed in steps.) Assume for simplicity that the tendons are straight and that the resultant prestressing force in these tendons lies below the centroidal axis of the concrete section. At transfer, the member cambers (deflects upward), remaining in contact with the casting bed only at the ends. Thus, the concrete beam is compelled to resist the prestressing force and to support its own weight simultaneously.

At transfer, the prestressing force in the steel diminishes because the concrete contracts under the imposed load. The prestressing force continues to diminish as time elapses as a result of the relaxation of the steel and the shrinkage and plastic flow of the concrete subsequent to transfer. To be effective, prestressed-concrete construction therefore requires the use of high-tensile steel in order that the reduction in prestressing force may be small in relation to the initial force. In all instances, it shall be assumed that the ratio of final to initial prestressing force is 0.85. Moreover, to simplify the stress calculations, it shall also be assumed that the full initial prestressing force exists at transfer and that the entire reduction in this force occurs during some finite interval following transfer.

There are therefore two loading states that must be considered in the design: the initial state, in which the concrete sustains the initial prestressing force and the beam weight; the final state, in which the concrete sustains the final prestressing force, the beam weight, and all superimposed loads. Consequently, the design of a prestressed-concrete beam differs from that of a conventional type of beam in the respect that designers must consider two stresses at each point, the initial stress and the final stress, and these must fall between the allowable compressive and tensile stresses. A beam is said to be in *balanced design* if the critical initial and final stresses coincide precisely with the allowable stresses.

The term *prestress* designates the stress induced by the *initial* prestressing force. The terms *prestress shear* and *prestress moment* refer to the vertical shear and bending moment, respectively, that the initial prestressing force induces in the concrete at a given section.

The *eccentricity* of the prestressing force is the distance from the action line of this resultant force to the centroidal axis of the section. Assume that the tendons are subjected to a uniform prestress. The locus of the centroid of the steel area is termed the *trajectory* of the steel or of the prestressing force.

The sign convention is as follows: The eccentricity is positive if the action line of the prestressing force lies below the centroidal axis. The trajectory has a positive slope if it inclines downward to the right. A load is positive if it acts downward. The vertical shear at a given section is positive if the portion of the beam to the left of this section exerts an upward force on the concrete. A bending moment is positive if it induces compression above the centroidal axis and tension below it. A compressive stress is positive; a tensile stress, negative.

The notational system is as follows. *Cross-sectional properties:* A = gross area of section, sq in.; A_s = area of prestressing steel, sq in.; d = effective depth of section at ultimate strength, in.; h = total depth of section, in.; I = moment of inertia of gross area, in.^{4}; y_b = distance from centroidal axis to bottom fiber, in.; S_b = section modulus with respect to bottom fiber = I/y_b, in.^{3}; k_b = distance from centroidal axis to lower kern point, in.; k_t = distance from centroidal axis to upper kern point, in. *Forces and moments:* F_i = initial prestressing force, lb; F_f = final prestressing force, lb; $\eta = F_f/F_i$; e = eccentricity of prestressing force, in.; e_{con} = eccentricity of prestressing force having concordant trajectory; θ = angle between trajectory (or tangent to trajectory) and horizontal line; m = slope of trajectory; w = vertical load exerted by curved tendons on concrete in unit distance; w_w = unit beam weight; w_s = unit superimposed load; w_{DL} = unit dead load; w_{LL} = unit live load; w_u = unit ultimate load; V_p = prestress shear; M_p = prestress moment; M_w = bending moment due to beam weight; M_s = bending moment due to superimposed load; C_u = resultant compressive force at ultimate load; T_u = resultant tensile force at ultimate load. *Stresses:* f'_c = ultimate compressive strength of concrete, psi; f'_{ci} = compressive strength of concrete at transfer; f'_s = ultimate strength of prestressing steel; f_{su} = stress in prestressing steel at ultimate load; f_{bp} = stress in bottom fiber due to initial prestressing force; f_{bw} = bending stress in bottom fiber due to beam weight; f_{bs} = bending stress in bottom fiber due to superimposed loads; f_{bi} = stress in bottom fiber at initial state = $f_{bp} + f_{bw}$; f_{bf} = stress in bottom fiber at final state = $\eta f_{bp} + f_{bw} + f_{bs}$; f_{cai} = initial stress at centroidal axis. *Camber:* Δ_p = camber due to initial prestressing force, in.; Δ_w = camber due to beam weight; Δ_i = camber at initial state; Δ_f = camber at final state.

The symbols that refer to the bottom fiber are transformed to their counterparts for the top fiber by replacing the subscript b with t. For example, f_{ti} denotes the stress in the top fiber at the initial state.

DETERMINATION OF PRESTRESS SHEAR AND MOMENT

The beam in Fig. 135a is simply supported at its ends and prestressed with an initial force of 300 kips. At section C, the eccentricity of this force is 8 in., and the slope of the trajectory is 0.014. (In the drawing, vertical distances are exaggerated in relation to horizontal distances.) Find the prestress shear and prestress moment at C.

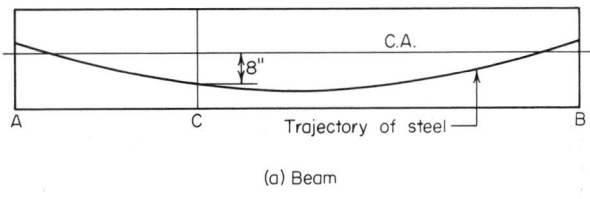

(a) Beam

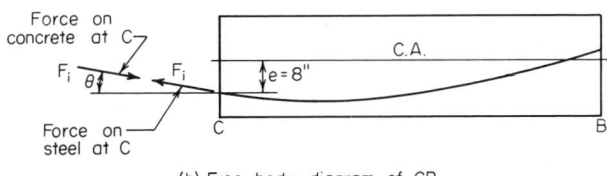

(b) Free–body diagram of CB

Fig. 135

Calculation Procedure:

1. Analyze the prestressing forces

If the composite concrete-and-steel member is regarded as a unit, the prestressing forces that the steel exerts on the concrete are purely internal. Therefore, if a beam is simply supported, the prestressing force alone does not induce any reactions at the supports.

Refer to Fig. 135b, and consider the forces acting on the beam segment CB solely as a result of F_i. The left portion of the beam exerts a tensile force F_i on the tendons.

Since CB is in equilibrium, it follows that the left portion also induces compressive stresses on the concrete at C, these stresses having a resultant that is numerically equal to and collinear with F_i.

2. Express the prestress shear and moment in terms of F_i

Using the sign convention described, express the prestress shear and moment in terms of F_i and θ. (The latter is positive if the slope of the trajectory is positive.) Thus: $V_p = -F_i \sin\theta$; $M_p = -F_i e \cos\theta$.

3. Compute the prestress shear and moment

Since θ is minuscule, apply these approximations: $\sin\theta = \tan\theta$, and $\cos\theta = 1$. Then, $V_p = -F_i \tan\theta$, Eq. (53). Or, $V_p = -300,000(0.014) = -4,200$ lb.

Also, $M_p = -F_i e$, Eq. (54). Or, $M_p = -300,000(8) = -2,400,000$ in.-lb.

STRESSES IN A BEAM WITH STRAIGHT TENDONS

A 12×18 in. rectangular beam is subjected to an initial prestressing force of 230 kips applied 3.3 in. below the center. The beam is on a simple span of 30 ft and carries a superimposed load of 840 plf. Determine the initial and final stresses at the supports and at mid-span. Construct diagrams to represent the initial and final stresses along the span.

Calculation Procedure:

1. Compute the beam properties

Thus, $A = 12(18) = 216$ sq in.; $S_b = S_t = (\frac{1}{6})(12)(18)^2 = 648$ in.³; $w_w = (216/144)(150) = 225$ plf.

2. Calculate the prestress in the top and bottom fibers

Since the section is rectangular, apply $f_{bp} = (F_i/A)(1 + 6e/h) = (230,000/216)(1 + 6 \times 3.3/18) = +2,236$ psi; $f_{tp} = (F_i/A)(1 - 6e/h) = -106$ psi.

For convenience, record the stresses in Table 8 as they are obtained.

3. Determine the stresses at mid-span due to gravity loads

Thus: $M_s = (\frac{1}{8})(840)(30)^2(12) = 1,134,000$ in.-lb; $f_{bs} = -1,134,000/648 = -1,750$ psi; $f_{ts} = +1,750$ psi. By proportion, $f_{bw} = -1,750(225/840) = -469$; $f_{tw} = +469$ psi.

4. Compute the initial and final stresses at the supports

Thus, $f_{bi} = +2,236$ psi; $f_{ti} = -106$ psi; $f_{bf} = 0.85(2,236) = +1,901$ psi; $f_{tf} = 0.85(-106) = -90$ psi.

TABLE 8 Stresses in Prestressed-Concrete Beam

	At support		At mid-span	
	Bottom fiber	Top fiber	Bottom fiber	Top fiber
(a) Initial prestress, psi	+2,236	−106	+2,236	−106
(b) Final prestress, psi	+1,901	−90	+1,901	−90
(c) Stress due to beam weight, psi	....	....	−469	+469
(d) Stress due to superimposed load, psi	....	....	−1,750	+1,750
Initial stress: (a)+(c)	+2,236	−106	+1,767	+363
Final stress: (b)+(c)+(d)	+1,901	−90	−318	+2,129

5. Determine the initial and final stresses at mid-span

Thus: $f_{bi} = +2,236 - 469 = +1,767$ psi; $f_{ti} = -106 + 469 = +363$ psi; $f_{bf} = +1,901$ $-469 - 1,750 = -318$ psi; $f_{tf} = -90 + 469 + 1,750 = +2,129$ psi.

6. Construct the initial-stress diagram

In Fig. 136a, construct the initial-stress diagram A_tA_bBC at the support and the initial-stress diagram M_tM_bDE at mid-span. Draw the parabolic arcs BD and CE. The stress diagram at an intermediate section Q is obtained by passing a plane normal to the longitudinal axis. The offset from a reference line through B to the arc BD represents the value of f_{bw} at that section.

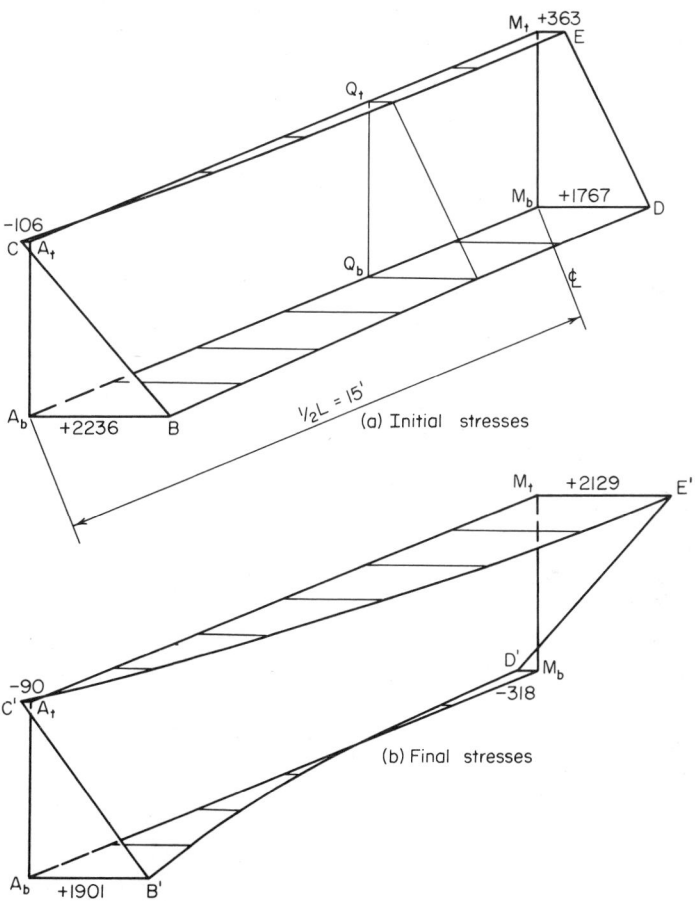

Fig. 136   Isometric stress diagrams for half-span.

7. Construct the final-stress diagram

Construct Fig. 136b in an analogous manner. The offset from a reference line through B' to the arc $B'D'$ represents the value of $f_{bw} + f_{bs}$ at the given section.

8. Alternatively, construct composite stress diagrams for the top and bottom fibers

The diagram pertaining to the bottom fiber is shown in Fig. 137. The difference between the ordinates to DE and AB repsesents f_{bi}, and the difference between the ordinates to FG amd AC represents f_{bf}.

This Procedure illustrates the following principles relevant to a beam with straight tendons carrying a uniform load: At transfer, the critical stresses occur at the supports; under full design load, the critical stresses occur at mid-span if the allowable final stresses exceed η times the allowable initial stresses in absolute value.

The primary objective in prestressed-concrete design is to maximize the capacity of a given beam by maximizing the absolute values of the prestresses at the section having the greatest superimposed-load stresses. The three Procedures that follow, when taken as a unit, illustrate the manner in which the allowable prestresses may be increased numerically by taking advantage of the beam-weight stresses, which are opposite in character to the prestresses. The next Procedure will also demonstrate that when a beam is not in balanced design, there is a range of values of F_i that will enable the member to carry this maximum allowable load. In summary, it may be stated that the objective is to maximize the capacity of a given beam and to provide the minimum prestressing force associated with this capacity.

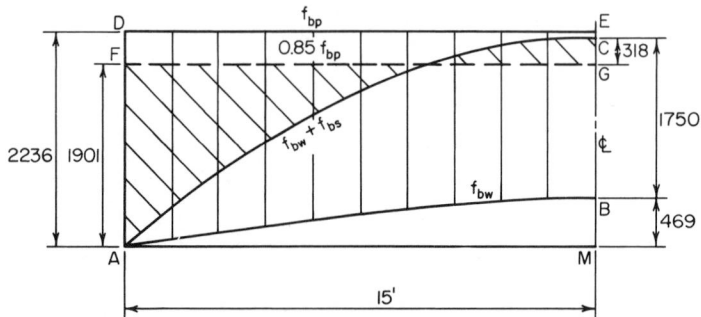

Fig. 137   Stresses in bottom fiber along half-span.

DETERMINATION OF CAPACITY AND PRESTRESSING FORCE FOR A BEAM WITH STRAIGHT TENDONS

An 8×10 in. rectangular beam, simply supported on a 20-ft span, is to be prestressed by means of straight tendons. The allowable stresses are: *initial*, $+2,400$ and -190 psi; *final*, $+2,250$ and -425 psi. Evaluate the allowable unit superimposed load, the maximum and minimum prestressing force associated with this load, and the corresponding eccentricities.

Calculation Procedure:

1. Compute the beam properties
$A = 80$ sq in.; $S = 133$ in.^{2}; $w_w = 83$ plf.

2. Compute the stresses at mid-span due to the beam weight
Thus, $M_w = (\frac{1}{8})(83)(20)^2(12) = 49,800$ in.-lb; $f_{bw} = -49,800/133 = -374$ psi; $f_{tw} = +374$ psi.

3. Set the critical stresses equal to their allowable values to secure the allowable unit superimposed load
Use Fig. 136 or 137 as a guide. At support: $f_{bi} = +2,400$ psi; $f_{ti} = -190$ psi; at mid-span, $f_{bf} = 0.85(2,400) - 374 + f_{bs} = -425$ psi; $f_{tf} = 0.85(-190) + 374 + f_{ts} = +2,250$ psi. Also, $f_{bs} = -2,091$ psi; $f_{ts} = +2,038$ psi.

Since the superimposed-load stresses at top and bottom will be numerically equal, the latter value governs the beam capacity. Or $w_s = w_w f_{ts}/f_{tw} = 83(2,038/374) = 452$ plf.

4. Find $F_{i,\max}$ and its eccentricity
The value of w_s was found by setting the critical value of f_{ti} and of f_{tf} equal to their respective allowable values. However, since S_b is excessive for the load w_s, there is

flexibility with respect to the stresses at the bottom. The designer may set the critical value of either f_{bi} or f_{bf} equal to its allowable value, or produce some intermediate condition. As shown by the calculations in Step 3, f_{bf} may vary within a range of $2,091 - 2,038 = 53$ psi. Refer to Fig. 138, where the lines represent the stresses indicated.

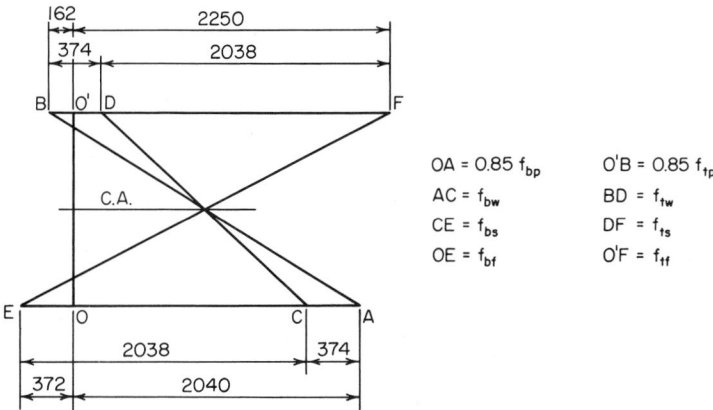

OA = 0.85 f_{bp} O'B = 0.85 f_{tp}
AC = f_{bw} BD = f_{tw}
CE = f_{bs} DF = f_{ts}
OE = f_{bf} O'F = f_{tf}

Fig. 138   Stresses at mid-span under maximum prestressing force.

Points B and F are fixed, but points A and E may be placed anywhere within the 53-psi range. To maximize F_i, place A at its limiting position to the right; i.e., set the critical value of f_{bi} rather than that of f_{bf} equal to the allowable value. Then, $f_{cai} = F_{i,max}/A = \frac{1}{2}(2,400 - 190) = +1,105$ psi; $F_{i,max} = 1,105(80) = 88,400$ lb; $f_{bp} = 1,105 + 88,400e/133 = +2,400$; $e = 1.95$ in.

5. Find $F_{i,min}$ and its eccentricity

For this purpose, place A at its limiting position to the left. Then, $f_{bp} = 2,400 - (53/0.85) = +2,338$ psi; $f_{cai} = +1,074$ psi; $F_{i,min} = 85,920$ lb; $e = 1.96$ in.

6. Verify the value of $F_{i,max}$ by checking the critical stresses

At support: $f_{bi} = +2,400$ psi; $f_{ti} = -190$ psi. At mid-span: $f_{bf} = +2,040 - 374 - 2,038 = -372$ psi; $f_{tf} = -162 + 374 + 2,038 = +2,250$ psi.

7. Verify the value of $F_{i,min}$ by checking the critical stresses

At support: $f_{bi} = +2,338$ psi; $f_{ti} = -190$ psi. At mid-span: $f_{bf} = 0.85(2,338) - 374 - 2,038 = -425$ psi; $f_{tf} = +2,250$ psi.

BEAM WITH DEFLECTED TENDONS

The beam in the previous Calculation Procedure is to be prestressed by means of tendons that are deflected at the quarter points of the span, as shown in Fig. 139a. Evaluate the allowable unit superimposed load, the magnitude of the prestressing force, the eccentricity e_1 in the center interval, and the maximum and minimum allowable values of the eccentricity e_2 at the supports. What increase in capacity has been obtained by deflecting the tendons?

Calculation Procedure:

1. Compute the beam-weight stresses at B

In the composite stress diagram, Fig. 139b, the difference between an ordinate to EFG and the corresponding ordinate to AHJ represents the value of f_{ti} at the given section. It is apparent that if AE does not exceed HF, then f_{ti} does not exceed HF in absolute value anywhere along the span. Therefore, for the center interval BC, the critical stresses at transfer occur at the boundary sections B and C. Analogous observations apply to Fig. 139c.

Computing the beam-weight stresses at B, $f_{bw} = (\frac{3}{4})(-374) = -281$ psi; $f_{tw} = +281$ psi.

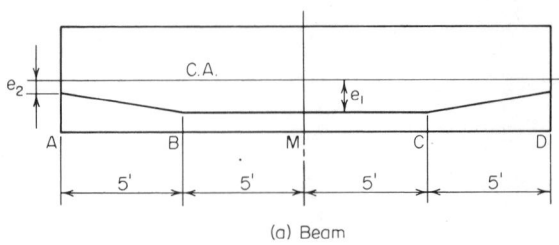

(a) Beam

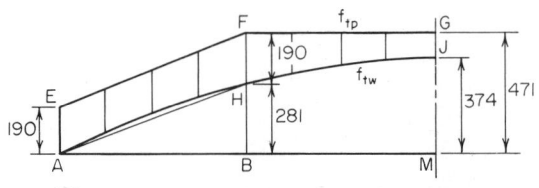

(b) Absolute values of f_{ti} along half-span

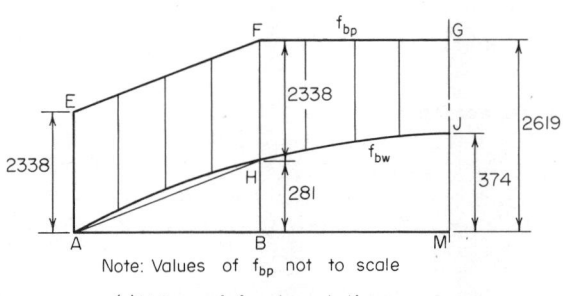

Note: Values of f_{bp} not to scale

(c) Values of f_{bi} along half-span

Fig. 139

2. Tentatively set the critical stresses equal to their allowable values to secure the allowable unit superimposed load

Thus, at B: $f_{bi} = f_{bp} - 281 = +2,400$; $f_{ti} = f_{tp} + 281 = -190$; $f_{bp} = +2,681$ psi; $f_{tp} = -471$ psi.

At M: $f_{bf} = 0.85(2,681) - 374 + f_{bs} = -425$; $f_{tf} = 0.85(-471) + 374 + f_{ts} = +2,250$; $f_{bs} = -2,330$ psi; $f_{ts} = +2,277$ psi. The latter value controls.

Also, $w_s = 83(2,277/374) = 505$ plf; $505/452 = 1.12$. The capacity is increased 12 percent.

When the foregoing calculations are compared with those in the previous Calculation Procedure, it is seen that the effect of deflecting the tendons is to permit an increase of 281 psi in the absolute value of the prestress at top and bottom. The accompanying increase in f_{ts} is $0.85(281) = 239$ psi.

3. Find the minimum prestressing force and the eccentricity e_1

Examination of Fig. 138 shows that f_{cai} is not affected by the form of trajectory used. Therefore, as in the previous Calculation Procedure, $F_i = 85,920$ lb; $f_{tp} = 1,074 - (85,920e_1)/133 = -471$; $e_1 = 2.39$ in.

Although not required, the value of $f_{bp} = 1,074 + [1,074 - (-471)] = +2,619$ psi, or $f_{bp} = 2,681 - 53/0.85 = +2,619$ psi.

4. Establish the allowable range of values of e_2

At the supports, the tendons may be placed an equal distance above or below the center. Then, $e_{2.\max} = 1.96$ in.; $e_{2.\min} = -1.96$ in.

BEAM WITH CURVED TENDONS

The beam in the second previous Calculation Procedure is to be prestressed by tendons lying in a parabolic arc. Evaluate the allowable unit superimposed load, the magnitude of the prestressing force, the eccentricity of this force at mid-span, and the increase in capacity accruing from the use of curved tendons.

Calculation Procedure:

1. Tentatively set the initial and final stresses at mid-span equal to their allowable values to secure the allowable unit superimposed load

Since the prestressing force has a parabolic trajectory, lines EFG in Fig. 139b and c will be parabolic in the present case. Therefore, it is possible to achieve the full allowable initial stresses at mid-span. Thus, $f_{bi} = f_{bp} - 374 = +2,400$; $f_{ti} = f_{tp} + 374 = -190$; $f_{bp} = +2,774$ psi; $f_{tp} = -564$ psi; $f_{bf} = 0.85(2,774) - 374 + f_{bs} = -425$; $f_{tf} = 0.85 (-564) + 374 + f_{ts} = +2,250$; $f_{bs} = -2,409$ psi; $f_{ts} = +2,356$ psi. The latter value controls.

Also, $w_s = 83(2,356/374) = 523$ plf; $523/452 = 1.16$. Thus the capacity is increased 16 percent.

When the foregoing calculations are compared with those in the earlier Calculation Procedure, it is seen that the effect of using parabolic tendons is to permit an increase of 374 psi in the absolute value of the prestress at top and bottom. The accompanying increase in f_{ts} is $0.85(374) = 318$ psi.

2. Find the minimum prestressing force and its eccentricity at mid-span

As before, $F_i = 85,920$ lb; $f_{tp} = 1,074 - 85,920e/133 = -564$; $e = 2.54$ in.

DETERMINATION OF SECTION MODULI

A beam having a cross-sectional area of 500 sq in. sustains a beam-weight moment equal to 3,500 in.-kips at mid-span and a superimposed moment that varies parabolically from 9,000 in.-kips at mid-span to 0 at the supports. The allowable stresses are: *initial*, $+2,400$ and -190 psi; *final*, $+2,250$ and -200 psi. The member will be prestressed by tendons deflected at the quarter points. Determine the section moduli corresponding to balanced design, the magnitude of the prestressing force, and its eccentricity in the center interval. Assume that the calculated eccentricity is attainable (i.e., that the centroid of the tendons will fall within the confines of the section while satisfying insulation requirements).

Calculation Procedure:

1. Equate the critical initial stresses, and the critical final stresses, to their allowable values

Let M_w and M_s denote the indicated moments at mid-span; the corresponding moments at the quarter point are three-fourths as large. The critical initial stresses occur at the quarter point, while the critical final stresses occur at mid-span. After equating the stresses to their allowable values, solve the resulting simultaneous equations to find the section moduli and prestresses. Thus: *stresses in bottom fiber*, $f_{bi} = f_{bp} - 0.75M_w/S_b = +2,400$; $f_{bf} = 0.85f_{bp} - M_w/S_b - M_s/S_b = -200$. Solving, $S_b = (M_S + 0.3625M_w)/2,240 = 4,584$ in.³, and $f_{bp} = +2,973$ psi; *stresses in top fiber*, $f_{ti} = f_{tp} + 0.75 M_w/S_t = -190$; $f_{tf} = 0.85f_{tp} + M_w/S_t + M_s/S_t = +2,250$. Solving, $S_t = (M_s + 0.3625M_w)/2,412 = 4,257$ in.³, and $f_{tp} = -807$ psi.

2. Evaluate F_i and e

In this instance, e denotes the eccentricity in the center interval. Thus: $f_{bp} = F_i/A + F_i e/S_b = +2{,}973$; $f_{tp} = F_i/A - F_i e/S_t = -807$; $F_i = (2{,}973S_b - 807S_t)A/(S_b + S_t) = 576{,}500$ lb; $e = 2{,}973S_b/F_i - S_b/A = 14.47$ in.

3. Alternatively, evaluate F_i by assigning an arbitrary depth to the member

Thus, set $h = 10$ in.; $y_b = S_t h/(S_b + S_t) = 4.815$ in.; $f_{cai} = f_{bp} - (f_{bp} - f_{tp})y_b/h = 2{,}973 - (2{,}973 + 807)0.4815 = +1{,}153$ psi; $F_i = 1{,}153(500) = 576{,}500$ lb.

EFFECT OF INCREASE IN BEAM SPAN

Consider that the span of the beam in the previous Calculation Procedure increases by 10 percent, thereby causing the mid-span moment due to superimposed load to increase by 21 percent. Show that the member will be adequate with respect to flexure if all cross-sectional dimensions are increased by 7.2 percent. Compute the new eccentricity in the center interval and compare this with the original value.

Calculation Procedure:

1. Calculate the new section properties and bending moments

Thus: $A = 500(1.072)^2 = 575$ sq in.; $S_b = 4{,}584(1.072)^3 = 5{,}647$ in.^{3}; $S_t = 4{,}257(1.072)^3 = 5{,}244$ in.^{3}; $M_s = 9{,}000(1.21) = 10{,}890$ in.-kips; $M_w = 3{,}500(1.072)^2(1.21) = 4{,}867$ in.-kips.

2. Compute the required section moduli, prestresses, prestressing force, and its eccentricity in the central interval, using the same sequence as in the previous Calculation Procedure

Thus: $S_b = 5{,}649$ in.^{3}; $S_t = 5{,}246$ in.^{3} Both these values are acceptable. Then: $f_{bp} = +3{,}046$ psi; $f_{tp} = -886$ psi; $F_i = 662{,}800$ lb; $e = 16.13$ in. The eccentricity has increased by 11.5 percent.

In practice, it would be more efficient to increase the vertical dimensions more than the horizontal dimensions. Nevertheless, it is clear that as the span increases, the eccentricity increases more rapidly than the depth.

EFFECT OF BEAM OVERLOAD

The beam in the second previous Calculation Procedure is subjected to a 10 percent overload. How does the final stress in the bottom fiber compare with that corresponding to the design load?

Calculation Procedure:

1. Compute the value of f_{bs} under design load

Thus, $f_{bs} = -M_s/S_b = -9{,}000{,}000/4{,}584 = -1{,}963$ psi.

2. Compute the increment of f_{bs} caused by overload and the revised value of f_{bf}

Thus, $\Delta f_{bs} = 0.10(-1{,}963) = -196$ psi; $f_{bf} = -200 - 196 = -396$ psi. Therefore, a 10 percent overload virtually doubles the tensile stress in the member.

PRESTRESSED-CONCRETE-BEAM-DESIGN GUIDES

On the basis of the previous Calculation Procedures, what conclusions may be drawn that will serve as guides in the design of prestressed-concrete beams?

Calculation Procedure:

1. Evaluate the results obtained with different forms of tendons

The capacity of a given member is increased by using deflected rather than straight tendons, and the capacity is maximimzed by using parabolic tendons. (However, in the case of a pretensioned beam, an economy analysis must also take into account the expense incurred in deflecting the tendons.)

2. Evaluate the prestressing force

For a given ratio of y_b/y_t, the prestressing force that is required to maximize the capacity of a member is a function of the cross-sectional area and the allowable stresses. It is independent of the form of the trajectory.

3. Determine the effect of section moduli

If the section moduli are in excess of the minimum required, the prestressing force is minimized by setting the critical values of f_{bf} and f_{ti} equal to their respective allowable values. In this manner, points A and B in Fig. 138 are placed at their limiting positions to the left.

4. Determine the most economical short-span section

For a short-span member, an I section is most economical because it yields the required section moduli with the minimum area. Moreover, since the required values of S_b and S_t differ, the area should be disposed unsymmetrically about mid-depth to secure these values.

5. Consider the calculated value of e

Since an increase in span causes a greater increase in the theoretical eccentricity than in the depth, the calculated value of e is not attainable in a long-span member because the centroid of the tendons would fall beyond the confines of the section. For this reason, long-span members are generally constructed as T sections. The extensive flange area elevates the centroidal axis, thus making it possible to secure a reasonably large eccentricity.

6. Evaluate the effect of overload

A relatively small overload induces a disproportionately large increase in the tensile stress in the beam and thus introduces the danger of cracking. Moreover, owing to the presence of many variable quantities, there is not a set relationship between the beam capacity at allowable final stress and the capacity at incipient cracking. It is therefore imperative that every prestressed-concrete beam be subjected to an ultimate-strength analysis to ensure that the beam provides an adequate factor of safety.

KERN DISTANCES

The beam in Fig. 140 has the following properties: $A = 850$ sq in.; $S_b = 11,400$ in.³; $S_t = 14,400$ in.³ A prestressing force of 630 kips is applied with an eccentricity of 24 in. at the section under investigation. Calculate f_{bp} and f_{tp} by expressing these stresses as functions of the kern distances of the section.

Calculation Procedure:

1. Consider the prestressing force to be applied at each kern point and evaluate the kern distances

Let Q_b and Q_t denote the points at which a compressive force must be applied to induce a zero stress in the top and bottom fiber, respectively. These are referred to as the *kern points* of the section, and the distances k_b and k_t from the centroidal axis to these points are called the *kern distances*.

Consider the prestressing force to be applied at each kern point in turn. Set the stresses f_{tp} and f_{bp} equal to zero to evaluate the kern distances k_b and k_t, respectively. Thus: $f_{tp} = F_i/A - F_i k_b/S_t = 0$, Eq. (a); $f_{bp} = F_i/A - F_i k_t/S_b = 0$, Eq. (b). Then, $k_b = S_t/a$ and $k_t = S_b/A$, Eq. (55). And, $k_b = 14,400/850 = 16.9$ in.; $k_t = 11,400/850 = 13.4$ in.

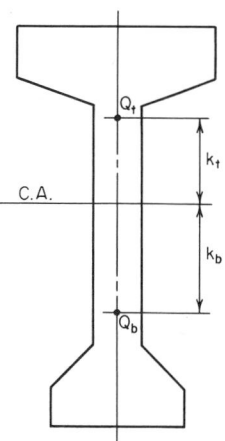

Fig. 140   Kern points.

2. Express the stresses f_{bp} and f_{tp} associated with the actual eccentricity as functions of the kern distances

By combining the stress equations with Eqs. (a) and (b), the following equations are obtained: $f_{bp} = F_i(k_t + e)/S_b$ and $f_{tp} = F_i(k_b - e)/S_t$, Eq. (56). Substituting numerical values, $f_{bp} = 630,000(13.4 + 24)/11,400 = +2,067$ psi; $f_{tp} = 630,000(16.9 - 24)/14,400 = -311$ psi.

3. Alternatively, derive Eq. (56) by considering the increase in prestress caused by an increase in eccentricity

Thus, $\Delta f_{bp} = F_i \Delta e/S_b$; therefore, $f_{bp} = F_i(k_t + e)/S_b$.

MAGNEL DIAGRAM CONSTRUCTION

The data pertaining to a girder having curved tendons are: $A = 500$ sq in.; $S_b = 5,000$ in.^{3}; $S_t = 5,340$ in.^{3}; $M_w = 3,600$ in.-kips; $M_s = 9,500$ in.-kips. The allowable stresses are: *initial*, $+2,400$ and -190 psi; *final*, $+2,250$ and -425 psi. (a) Construct the Magnel diagram for this member. (b) Determine the minimum prestressing force and its eccentricity by referring to the diagram. (c) Determine the prestressing force if the eccentricity is restricted to 18 in.

Calculation Procedure:

1. Set the initial stress in the bottom fiber at mid-span equal to or less than its allowable value, and solve for the reciprocal of F_i

In this situation, the superimposed load is given and the sole objective is to minimize the prestressing force. The Magnel diagram is extremely useful for this purpose because it brings into sharp focus the relationship between F_i and e. In this Procedure, let f_{bi}, f_{bf}, and so forth represent the *allowable* stresses.

Thus, $1/F_i \geq (k_t + e)/(M_w + f_{bi}S_b)$, Eq. (57a).

2. Set the final stress in the bottom fiber at mid-span equal to or algebraically greater than its allowable value, and solve for the reciprocal of F_i

Thus, $1/F_i \leq \eta(k_t + e)/(M_w + M_s + f_{bf}S_b)$, Eq. (57b).

3. Repeat the foregoing procedure with respect to the top fiber

Thus, $1/F_i \geq (e - k_b)/(M_w - f_{ti}S_t)$, Eq. (57c); and $1/F_i \leq \eta(e - k_b)/(M_w + M_s - f_{tf}S_t)$, Eq. (57d).

4. Substitute numerical values, expressing F_i in thousands of kips

Thus, $1/F_i \geq (10 + e)/15.60$, Eq. (a); $1/F_i \leq (10 + e)/12.91$, Eq. (b); $1/F_i \geq (e - 10.68)/4.61$, Eq. (c); $1/F_i \leq (e - 10.68)/1.28$, Eq. (d).

5. Construct the Magnel diagram

In Fig. 141, consider the foregoing relationships as equalities and plot the straight lines that represent them. Each point on these lines represents a set of values of $1/F_i$ and e at which the designated stress equals its allowable value.

When the section moduli are in excess of those corresponding to balanced design, as they are in the present instance, line b makes a greater angle with the e axis than does a, and line d makes a greater angle than does c. From the sense of each inequality, it follows that $1/F_i$ and e may have any set of values represented by a point within the quadrilateral $CDEF$ or on its circumference.

6. To minimize F_i, determine the coordinates of point E at the intersection of lines b and c

Thus, $1/F_i = (10 + e)/12.91 = (e - 10.68)/4.61$; solving, $e = 22.2$ in.; $F_i = 401$ kips.

The Magnel diagram confirms the third design guide presented earlier in this section.

7. For the case where e is restricted to 18 in., minimize F_i by determining the ordinate of point G on line b

Thus, in Fig. 141, $1/F_i = (10 + 18)/12.91$; $F_i = 461$ kips.

The Magnel diagram may be applied to a beam having deflected tendons by substituting for M_w in Eqs. (57a) and (57c) the beam-weight moment at the deflection point.

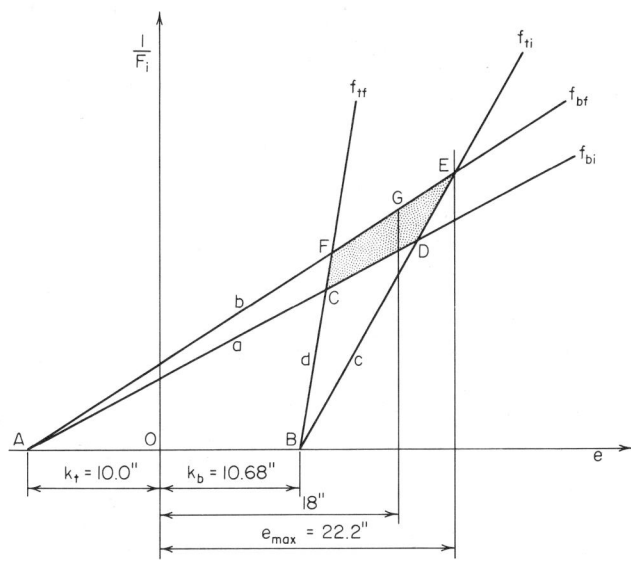

Fig. 141   Magnel diagram.

CAMBER OF A BEAM AT TRANSFER

The following pertain to a simply supported prismatic beam: $L = 36$ ft; $I = 40,000$ in.^{4}; $f'_{ci} = 4,000$ psi; $w_w = 340$ plf; $F_i = 430$ kips; $e = 8.8$ in. at mid-span. Calculate the camber of the member at transfer under each of these conditions: (a) the tendons are straight across the entire span; (b) the tendons are deflected at the third points and the eccentricity at the supports is zero; (c) the tendons are curved parabolically and the eccentricity at the supports is zero.

Calculation Procedure:

1. Evaluate E_c at transfer using the ACI Code

Review the moment-area method of calculating beam deflections, which is summarized earlier in this Handbook. Consider an upward displacement (camber) as positive, and let the symbols Δ_p, Δ_w, and Δ_i, which are defined earlier, refer to the camber at mid-span.

Thus, using the ACI Code, $E_c = (145)^{1.5}(33)(4,000)^{0.5} = 3,644,000$ psi.

2. Construct the prestress-moment diagrams associated with the three cases described

See Fig. 142. By symmetry, the elastic curve corresponding to F_i is horizontal at mid-span. Consequently, Δ_p equals the deviation of the elastic curve at the support from the tangent to this curve at mid-span.

3. Using the literal values shown in Fig. 142, develop an equation for Δ_p by evaluating the tangential deviation; substitute numerical values

Thus, case a: $\Delta_p = ML^2/8E_cI$, Eq. (58), or $\Delta_p = 430{,}000(8.8)(36)^2(144)/[8(3{,}644{,}000)(40{,}000)] = 0.61$ in. For case b: $\Delta_p = M(2L^2 + 2La - a^2)/24E_cI$, Eq. (59), or $\Delta_p = 0.52$ in. For case c: $\Delta_p = 5ML^2/48E_cI$, Eq. (60), or $\Delta_p = 0.51$ in.

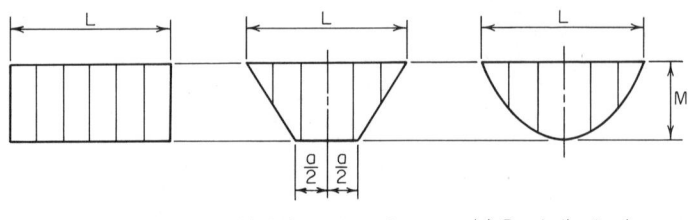

(a) Straight tendons   (b) Deflected tendons   (c) Parabolic tendons

Fig. 142   Prestress-moment diagrams.

4. Compute Δ_w

Thus, $\Delta_w = -5w_wL^4/384E_cI = -0.09$ in.

5. Combine the foregoing results to obtain Δ_i

Thus: case a, $\Delta_i = 0.61 - 0.09 = 0.52$ in.; case b, $\Delta_i = 0.52 - 0.09 = 0.43$ in.; case c, $\Delta_i = 0.51 - 0.09 = 0.42$ in.

DESIGN OF A DOUBLE-T ROOF BEAM

The beam in Fig. 143 was selected for use on a simple span of 40 ft to carry the following loads: roofing 12 psf; snow 40 psf; total 52 psf. The member will be pretensioned with straight 7-wire strands, $\frac{7}{16}$ in. diameter, having an area of 0.1089 sq in. each and an ultimate strength of 248,000 psi. The concrete strengths are $f'_c = 5{,}000$ psi and $f'_{ci} = 4{,}000$ psi. The allowable stresses are: *initial*, $+2{,}400$ and -190 psi; *final*, $+2{,}250$ and -425 psi. Investigate the adequacy of this section and design the tendons. Compute the camber of the beam after the concrete has hardened and all dead loads are present. For this calculation, assume that the final value of E_c is one-third of that at transfer.

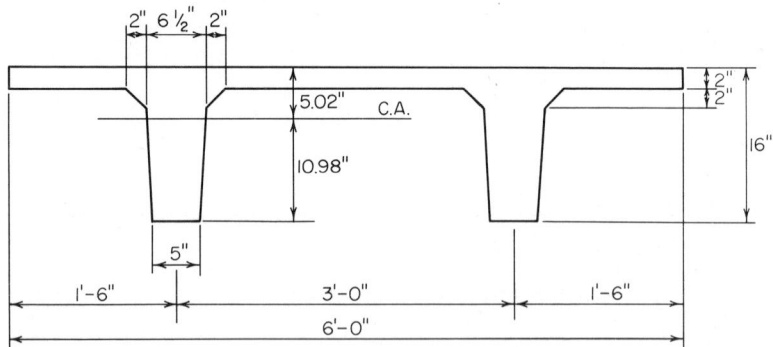

Fig. 143   Double-T roof beam.

Calculation Procedure:

1. Compute the properties of the cross section

Let f_{bf} and f_{tf} denote the respective stresses at *mid-span* and f_{bi} and f_{ti} denote the respective stresses *at the support*. Previous Calculation Procedures demonstrated that where the section moduli are excessive, the minimum prestressing force is obtained by setting f_{bf} and f_{ti} equal to their allowable values.

Thus: $A = 316$ sq in.; $I = 7,240$ in.^{4}; $y_b = 10.98$ in.; $y_t = 5.02$ in.; $S_b = 659$ in.^{3}; $S_t = 1,442$ in.^{3}; $w_w = (316/144)150 = 329$ plf.

2. Calculate the total mid-span moment due to gravity loads and the corresponding stresses

Thus: $w_s = 52(6) = 312$ plf; $w_w = 329$ plf; and $M_w + M_s = (\frac{1}{8})(641)(40)^2(12) = 1,538,000$ in.-lb; $f_{bw} + f_{bs} = -1,538,000/659 = -2,334$ psi; $f_{tw} + f_{ts} = +1,538,000/1,442 = +1,067$ psi.

3. Determine whether the section moduli are excessive

Do this by setting f_{bf} and f_{ti} equal to their allowable values and computing the corresponding values of f_{bi} and f_{tf}. Thus, $f_{bf} = 0.85f_{bp} - 2,334 = -425$; therefore, $f_{bp} = +2,246$ psi; $f_{ti} = f_{tp} = -190$ psi; $f_{bi} = f_{bp} = +2,246 < 2,400$ psi. This is acceptable. Also, $f_{tf} = 0.85(-190) + 1,067 = +905 < 2,250$ psi; this is acceptable. The section moduli are therefore excessive.

4. Find the minimum prestressing force and its eccentricity

Refer to Fig. 144. Thus, $f_{bp} = +2,246$; $f_{tp} = -190$ psi; slope of $AB = 2,246 - (-190)/16 = 152.3$ psi/in.; $F_i/A = CD = 2,246 - 10.98(152.3) = 574$ psi; $F_i = 574(316) = 181,400$ lb; slope of $AB = F_i e/I = 152.3$; $e = 152.3(7,240)/181,400 = 6.07$ in.

5. Determine the number of strands required and establish their disposition.

In accordance with the ACI *Code*, allowable initial force per strand = $0.1089(0.70)(248,000) = 18,900$ lb; number required = $181,400/18,900 = 9.6$. Therefore, use 10 strands (5 in each web) stressed to 18,140 lb each.

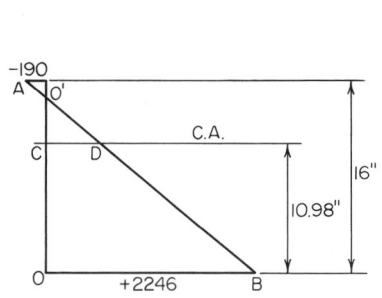

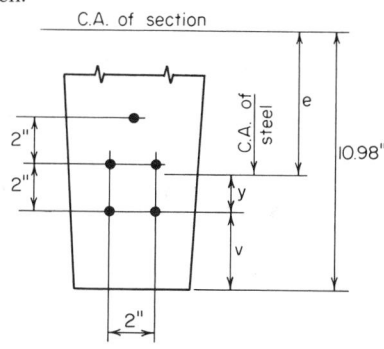

Fig. 144 Prestress diagram.   **Fig. 145 Location of tendons.**

Referring to the ACI *Code* for the minimum clear distance between the strands, the allowable center-to-center spacing = $4(\frac{7}{16}) = 1\frac{3}{4}$ in. Use a 2-in. spacing. In Fig. 145, locate the centroid of the steel, or $y = (2 \times 2 + 1 \times 4)/5 = 1.60$ in.; $v = 10.98 - 6.07 - 1.60 = 3.31$ in.; set $v = 3\frac{5}{16}$ in.

6. Calculate the allowable ultimate moment of the member in accordance with the ACI Code

Thus, $A_s = 10(0.1089) = 1.089$ sq in.; $d = y_t + e = 5.02 + 6.07 = 11.09$ in.; $p = A_s/bd = 1.089/[72(11.09)] = 0.00137$.

Compute the steel stress and resultant tensile force at ultimate load. Or, $f_{su} = f'_s(1 - 0.5pf'_s/f'_c)$, Eq. (61). Or, $f_{su} = 248,000(1 - 0.5 \times 0.00137 \times 248,000/5,000) = 240,000$ psi; $T_u = A_s f_{su} = 1.089(240,000) = 261,400$ lb.

Compute the depth of the compression block. This depth, a, is found from $C_u = 0.85(5,000)(72a) = 261,400$ lb; $a = 0.854$ in.; $jd = d - a/2 = 10.66$ in.; $M_u = \phi T_u jd = 0.90(261,400)(10.66) = 2,500,000$ in.-lb.

Calculate the steel index to ascertain that it is below the limit imposed by the ACI *Code*, or $q = pf_{su}/f'_c = 0.00137\,(240,000)/5,000 = 0.0658 < 0.30$. This is acceptable.

7. Calculate the required ultimate-moment capacity as given by the ACI Code

Thus, $w_{DL} = 329 + 12(6) = 401$ plf; $w_{LL} = 40(6) = 240$ plf; $w_u = 1.5w_{DL} + 1.8w_{LL} = 1,034$ plf; M_u required $= (\frac{1}{8})(1,034)(40)^2(12) = 2,480,000 < 2,500,000$ in.-lb. The member is therefore adequate with respect to its ultimate-moment capacity.

8. Calculate the maximum and minimum area of web reinforcement in the manner prescribed in the ACI Code

Since the maximum shearing stress does not vary linearly with the applied load, the shear analysis is performed at ultimate-load conditions. Let $A_v =$ area of web reinforcement placed perpendicular to the longitudinal axis; $V'_c =$ ultimate-shear capacity of concrete; $V'_p =$ vertical component of F_f at the given section; $V'_u =$ ultimate shear at given section; $s =$ center-to-center spacing of stirrups; $f'_{pc} =$ stress due to F_f, evaluated at the centroidal axis, or at the junction of the web and flange when the centroidal axis lies in the flange.

Calculate the ultimate shear at the critical section, which lies at a distance $d/2$ from the face of the support. Then, distance from mid-span to the critical section $= \frac{1}{2}(L - d) = 19.54$ ft; $V'_u = 1,034(19.54) = 20,200$ lb.

Evaluate V'_c by solving the following equations and selecting the smaller value. $V'_{ci} = 1.7b'd(f'_c)^{0.5}$, Eq. (62), where $d =$ effective depth, in.; $b' =$ width of web at centroidal axis, in.; $b' = 2(5 + 1.5 \times 10.98/12) = 12.74$ in.; $V'_{ci} = 1.7(12.74)(11.09)(5,000)^{0.5} = 17,000$ lb. Also, $V'_{cw} = b'd(3.5f'^{0.5}_c + 0.3f'_{pc}) + V'_p$, Eq. (63), where $d =$ effective depth or 80 percent of the overall depth, whichever is greater, in. Thus, $d = 0.80(16) = 12.8$ in.; $V'_p = 0$. From Step 4, $f'_{pc} = 0.85(574) = +488$ psi; $V'_{cw} = 12.74(12.8)(3.5 \times 5,000^{0.5} + 0.3 \times 488) = 64,300$ lb; therefore, $V'_c = 17,000$ lb.

Calculate the maximum web-reinforcement area by applying the following equation: $A_v = s(V'_u - \phi V'_c)/\phi d f_y$, Eq. (64), where $d =$ effective depth at section of maximum moment, in. Use $f_y = 40,000$ psi and set $s = 12$ in. Then, $A_v = 12(20,200 - 0.85 \times 17,000)/[0.85(11.09)(40,000)] = 0.184$ sq in./ft. This is the area required at the ends.

Calculate the minimum web-reinforcement area by applying $A_v = (A_s/80)(f'_s/f_y)(s/b'd^{0.5})$, Eq. (65), or $A_v = (1.089/80)(248,000/40,000)\,12/(12.74 \times 11.09)^{0.5} = 0.085$ sq in./ft.

9. Calculate the camber under full dead load

From the previous Procedure, $E_c = (\frac{1}{3})(3.644)(10)^6 = 1.215 \times 10^6$ psi; $E_cI = 1.215(10)^6(7,240) = 8.8 \times 10^9$ lb-in.²; $\Delta_{DL} = -5(401)(40)^4(1,728)/[384(8.8)(10)^9] = -2.62$ in. By Eq. (58), $\Delta_p = 0.85(181,400)(6.07)(40)^2(144)/[8(8.8)(10)^9] = 3.06$ in. $\Delta = 3.06 - 2.62 = 0.44$ in.

DESIGN OF A POST-TENSIONED GIRDER

The girder in Fig. 146 has been selected for use on a 90-ft simple span to carry the following superimposed loads: dead load, 1,160 plf, live load, 1,000 plf. The girder will be post-tensioned with Freyssinet cables. The concrete strengths are $f'_c = 5,000$ psi and $f'_{ci} = 4,000$ psi. The allowable stresses are: *initial*, $+2,400$ and -190 psi; *final*, $+2,250$ and -425 psi. Complete the design of this member, and calculate the camber at transfer.

Calculation Procedure:

1. Compute the properties of the cross section

Since the tendons will be curved, the initial stresses at mid-span may be equated to the allowable values. The properties of the cross section are: $A = 856$ sq in.; $I = 394,800$ in.⁴; $y_b = 34.6$ in.; $y_t = 27.4$ in.; $S_b = 11,410$ in.³; $S_t = 14,410$ in.³; $w_w = 892$ plf.

2. Calculate the stresses at mid-span caused by gravity loads

Thus: $f_{bw} = -950$ psi; $f_{bs} = -2,300$ psi; $f_{tw} = +752$ psi; $f_{ts} = +1,820$ psi.

3. Test the section adequacy

To do this, equate f_{bf} and f_{ti} to their allowable values and compute the corresponding values of f_{bi} and f_{tf}. Thus: $f_{bf} = 0.85f_{bp} - 950 - 2,300 = -425$; $f_{ti} = f_{tp} + 752 = -190$; therefore, $f_{bp} = +3,324$ psi and $f_{tp} = -942$ psi; $f_{bi} = +3,324 - 950 = +2,374 < 2,400$ psi. This is acceptable. And, $f_{tf} = 0.85(-942) + 752 + 1,820 = +1,771 < 2,250$ psi. This is acceptable. The section is therefore adequate.

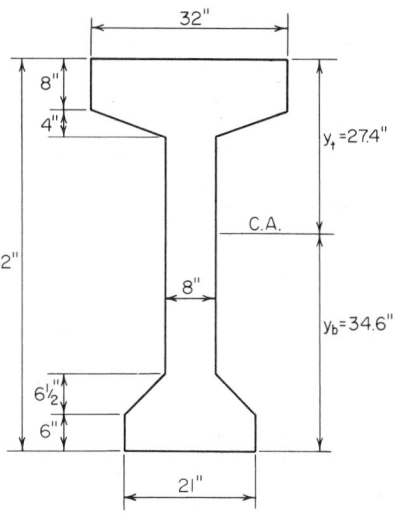

4. Find the minimum prestressing force and its eccentricity at mid-span

Do this by applying the prestresses found in Step 3. Refer to Fig. 147. Slope of $AB = [3,324 - (-942)]/62 = 68.8$ psi/in.; $F_i/A = CD = 3,324 - 34.6(68.8) = 944$ psi; $F_i = 944(856) = 808,100$ lb; slope of $AB = F_ie/I = 68.8$; $e = 68.8(394,800)/808,100 = 33.6$ in. Since $y_b = 34.6$ in., this eccentricity is excessive.

Fig. 146

5. Select the maximum feasible eccentricity; determine the minimum prestressing force associated with this value

Try $e = 34.6 - 3.0 = 31.6$ in. To obtain the minimum value of F_i, equate f_{bf} to its allowable value. Check the remaining stresses. As before, $f_{bp} = +3,324$ psi. But $f_{bp} = F_i/856 + 31.6F_i/11,410 = +3,324$; therefore $F_i = 844,000$ lb. Also, $f_{tp} = -865$ psi; $f_{bi} = +2,374$ psi; $f_{ti} = -113$ psi; $f_{tf} = +1,837$ psi.

6. Design the tendons and establish their pattern at mid-span

Refer to a table of the properties of Freyssinet cables, and select 12/0.276 cables. The designation indicates that each cable consists of 12 wires of 0.276-in. diameter. The ultimate strength is 236,000 psi. Then, $A_s = 0.723$ sq in. per cable. Outside diameter of cable = $1\frac{5}{8}$ in. Recommended final prestress = 93,000 lb per cable; initial prestress = 93,000/0.85 = 109,400 lb per cable. Therefore, use eight cables at an initial prestress of 105,500 lb each.

A section of the ACI *Code* requires a minimum cover of $1\frac{1}{2}$ in., and another section permits the ducts to be bundled at the center. Try the tendon pattern shown in Fig. 148. Thus, $y = [6(2.5) + 2(4.5)]/8 = 3.0$ in. This is acceptable.

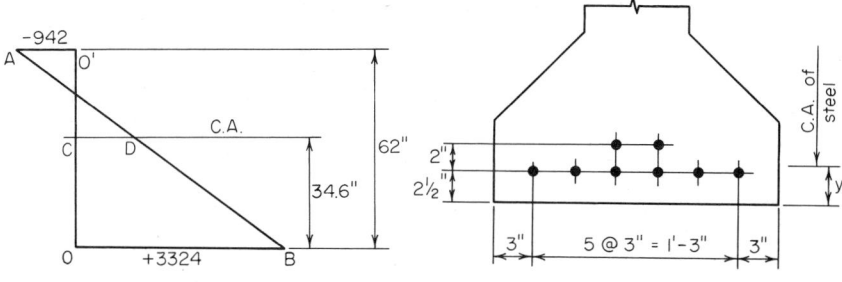

Fig. 147 Prestress diagram.   **Fig. 148 Location of tendons at mid-span.**

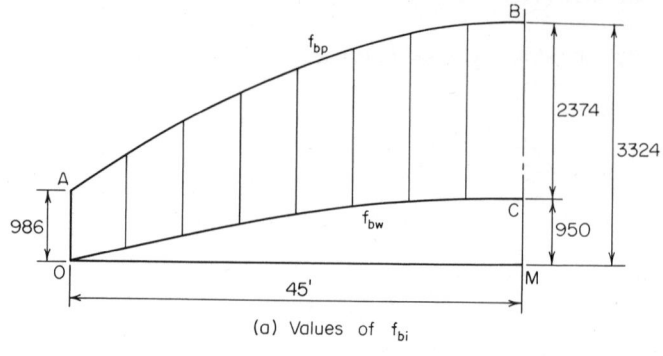

(a) Values of f_{bi}

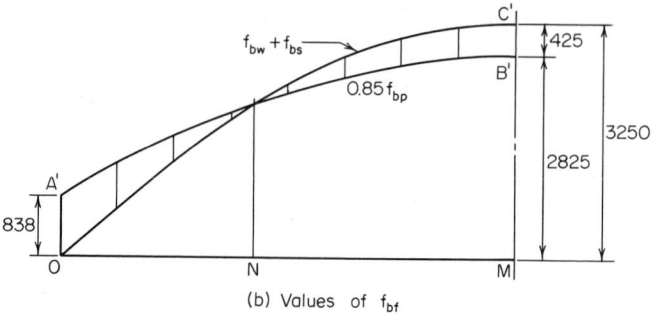

(b) Values of f_{bf}

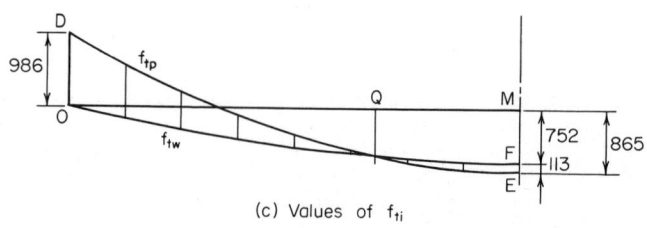

(c) Values of f_{ti}

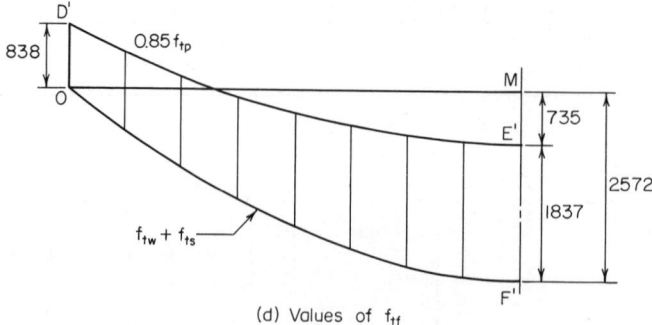

(d) Values of f_{tf}

Fig. 149

7. Establish the trajectory of the prestressing force

Construct stress diagrams to represent the initial and final stresses in the bottom and top fibers along the entire span.

For convenience, set $e = 0$ at the supports. The prestress at the ends is therefore $f_{bp} = f_{tp} = 844,000/856 = +986$ psi. Since e varies parabolically from maximum at mid-span to zero at the supports, it follows that the prestresses also vary parabolically.

In Fig. 149a, draw the parabolic arc AB with summit at B to represent the absolute value of f_{bp}. Draw the parabolic arc OC in the position shown to represent f_{bw}. The vertical distance between the arcs at a given section represents the value of f_{bi}; this value is maximum at mid-span.

In Fig. 149b, draw $A'B'$ to represent the absolute value of the final prestress; draw OC' to represent the absolute value of $f_{bw} + f_{bs}$. The vertical distance between the arcs represents the value of f_{bf}. This stress is compressive in the interval ON and tensile in the interval NM.

Construct Fig. 149c and d in an analogous manner. The stress f_{ti} is compressive in the interval OQ.

8. Calculate the allowable ultimate moment of the member

The mid-span section is critical in this respect. Thus, $d = 62 - 3 = 59.0$ in.; $A_s = 8(0.723) = 5.784$ sq in.; $p = A_s/bd = 5.784/[32(59.0)] = 0.00306$.

Apply Eq. (61), or $f_{su} = 236,000(1 - 0.5 \times 0.00306 \times 236,000/5,000) = 219,000$ psi. Also, $T_u = A_s f_{su} = 5.784(219,000) = 1,267,000$ lb. The concrete area under stress = $1,267,000/[0.85(5,000)] = 298$ sq in. This is the shaded area in Fig. 150, as the following calculation proves: $32(9.53) - 4.59(1.53) = 305 - 7 = 298$ sq in.

Locate the centroidal axis of the stressed area, or $m = [305(4.77) - 7(9.53 - 0.51)]/298 = 4.67$ in.; $M_u = \phi T_u jd = 0.90(1,267,000)(59.0 - 4.67) = 61,950,000$ in.-lb.

Calculate the steel index to ascertain that it is below the limit imposed by the ACI Code. Refer to Fig. 150. Or, area of $ABCD = 8(9.53) = 76.24$ sq in. The steel area A_{sr} that is required to balance the force on this web strip is $A_{sr} = 5.784(76.24)/298 = 1.48$ sq in.; $q = A_{sr} f_{su}/b'df'_c = 1.48(219,000)/[8(59.0)(5,000)] = 0.137 < 0.30$. This is acceptable.

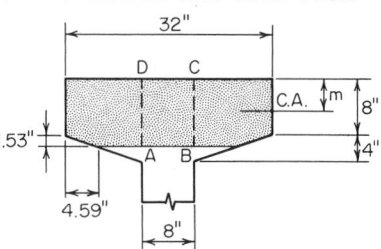

Fig. 150 Concrete area under stress at ultimate load.

9. Calculate the required ultimate-moment capacity as given by the ACI Code

Thus, $w_u = 1.5(892 + 1,160) + 1.8(1,000) = 4,878$ plf; M_u required $= (\frac{1}{8})(4,878)(90)^2(12) = 59,270,000$ in.-lb. This is acceptable. The member is therefore adequate with respect to its ultimate-moment capacity.

10. Design the web reinforcement

Follow the procedure given in Step 8 of the previous Calculation Procedure.

11. Design the end block

This is usually done by applying isobar charts to evaluate the tensile stresses caused by the concentrated prestressing forces. Refer to Winter, et al.—*Design of Concrete Structures*, McGraw-Hill.

12. Compute the camber at transfer

Referring to earlier Procedures in this section, $E_c I = 3.644(10)^6(394,800) = 1.44 \times 10^{12}$ lb-in.^{2}. Also, $\Delta_w = -5(892)(90)^4(1,728)/[384(1.44)(10)^{12}] = -0.91$ in. Apply Eq. (60). Or, $\Delta_p = 5(844,000)(31.6)(90)^2(144)/[48(1.44)(10)^{12}] = 2.25$ in.; $\Delta_i = 2.25 - 0.91 = 1.34$ in.

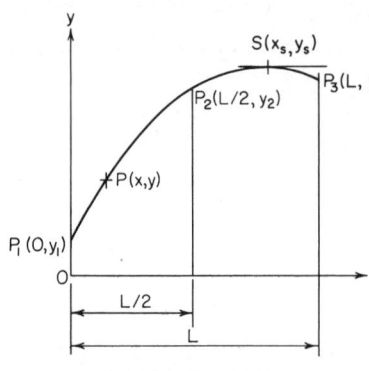

Fig. 151   Parabolic arc.

PROPERTIES OF A PARABOLIC ARC

Figure 151 shows the literal values of the coordinates at the ends and at the center of the parabolic arc $P_1P_2P_3$. Develop equations for y, dy/dx, and d^2y/dx^2 at an arbitrary point P. Find the slope of the arc at P_1 and P_3 and the coordinates of the summit S. (This information is required for the analysis of beams having parabolic trajectories.)

Calculation Procedure:

1. Select a slope for the arc
Let m denote the slope of the arc.

2. Present the results

The equations are:

$$y = 2(y_1 - 2y_2 + y_3)\left(\frac{x}{L}\right)^2 - (3y_1 - 4y_2 + y_3)\frac{x}{L} + y_1 \tag{66}$$

$$m = \frac{dy}{dx} = 4(y_1 - 2y_2 + y_3)\left(\frac{x}{L^2}\right) - \frac{3y_1 - 4y_2 + y_3}{L} \tag{67}$$

$$\frac{dm}{dx} = \frac{d^2y}{dx^2} = \frac{4}{L^2}(y_1 - 2y_2 + y_3) \tag{68}$$

$$m_1 = \frac{-(3y_1 - 4y_2 + y_3)}{L} \tag{69a}$$

$$m_3 = \frac{y_1 - 4y_2 + 3y_3}{L} \tag{69b}$$

$$x_s = \frac{(L/4)(3y_1 - 4y_2 + y_3)}{y_1 - 2y_2 + y_3} \tag{70a}$$

$$y_s = \frac{-(1/8)(3y_1 - 4y_2 + y_3)^2}{y_1 - 2y_2 + y_3} + y_1 \tag{70b}$$

ALTERNATIVE METHODS OF ANALYZING A BEAM WITH PARABOLIC TRAJECTORY

The beam in Fig. 152 is subjected to an initial prestressing force of 860 kips on a parabolic trajectory. The eccentricities at the left end, mid-span, and right end, respectively,

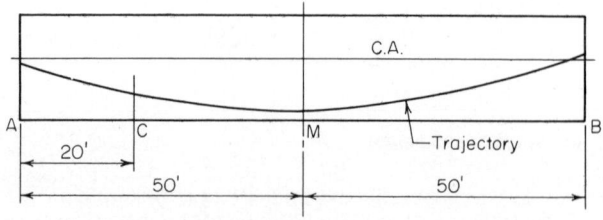

Fig. 152

are: $e_a = 1$ in.; $e_m = 30$ in.; $e_b = -3$ in. Evaluate the prestress shear and prestress moment at section C by each of the following methods: (a) by applying the properties of the trajectory at C; (b) by considering the prestressing action of the steel on the concrete in the interval AC.

Calculation Procedure:

1. Compute the eccentricity and slope of the trajectory at C

Use Eq. (66) and (67). Let m denote the slope of the trajectory. This is positive if the trajectory slopes downward to the right. Thus: $e_a - 2e_m + e_b = 1 - 60 - 3 = -62$ in.; $3e_a - 4e_m + e_b = 3 - 120 - 3 = -120$ in.; $e_c = 2(-62)(20/100)^2 + 120(20/100) + 1 = 20.04$ in.; $m_c = 4(-62/12)(20/100^2) - (-120/12 \times 100) = 0.0587$.

2. Compute the prestress shear and moment at C

Thus: $V_{pc} = -m_c F_i = -0.0587(860,000) = -50,480$ lb; $M_{pc} = -F_i e = -860,000(20.04) = -17,230,000$ in.-lb. This concludes the solution to part a.

3. Evaluate the vertical component w of the radial force on the concrete in a unit longitudinal distance

An alternative approach to this problem is to analyze the forces that the tendons exert on the concrete in the interval AC, namely, the prestressing force transmitted at the end and the radial forces resulting from curvature of the tendons.

Consider the component w to be positive if directed downward. In Fig. 153, $V_{pr} - V_{pq} = -F_i(m_r - m_q)$; therefore, $\Delta V_p/\Delta x = -F_i \Delta m/\Delta x$. Apply Eq. (68). Or, $dV_p/dx = -F_i dm/dx = -(4F_i/L^2)(e_a - 2e_m + e_b)$; but $dV_p/dx = -w$; therefore, $w = F_i dm/dx = (4F_i/L^2)(e_a - 2e_m + e_b)$, Eq. (71).

This result discloses that when the trajectory is parabolic, w is uniform across the span. The radial forces are always directed toward the center of curvature, since the tensile forces applied at their ends tend to straighten the tendons. In the present instance, $w = (4F_i/100^2)(-62/12) = -0.002067F_i$ plf.

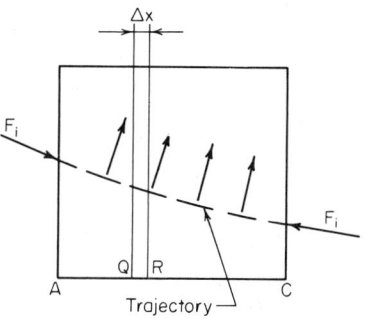

Fig. 153 Free-body diagram of concrete.

4. Find the prestress shear at C

By Eq. (69a), $m_a = -[-120/(100 \times 12)] = 0.1$; $V_{pa} = -0.1F_i$; $V_{pc} = V_{pa} - 20w = F_i(-0.1 + 20 \times 0.002067) = -0.0587F_i = -50,480$ lb.

5. Find the prestress moment at C

Thus, $M_{pc} = M_{pa} + V_{pa}(240) - 20w(120) = F_i(-1 - 0.1 \times 240 + 20 \times 0.002067 \times 120) = -20.04F_i = -17,230,000$ in.-lb.

PRESTRESS MOMENTS IN A CONTINUOUS BEAM

The continuous prismatic beam in Fig. 154 has a prestressing force of 96 kips on a parabolic trajectory. The eccentricities are: $e_a = -0.40$ in.; $e_d = +0.60$ in.; $e_b = -1.20$

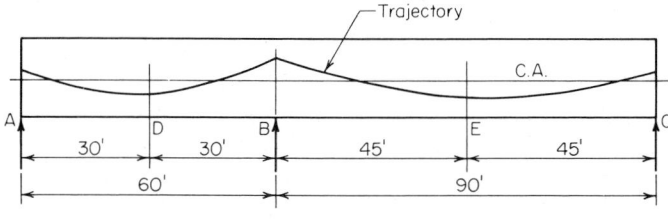

Fig. 154

in.; $e_e = +0.64$ in.; $e_c = -0.60$ in. Construct the prestress-moment diagram for this member, indicating all significant values.

Calculation Procedure:

1. Find the value of $wL^2/4$ for each span by applying Eq. (71)

Refer to Fig. 155. Since the members AB and BC are constrained to undergo an identical rotation at B, there exists at this section a bending moment M_{kb} in addition to that resulting from the eccentricity of F_i. The moment M_{kb} induces reactions at the supports. Thus, at every section of the beam there is a moment caused by continuity of the member as well as the moment $-F_i e$. The moment M_{kb} is termed the *continuity*

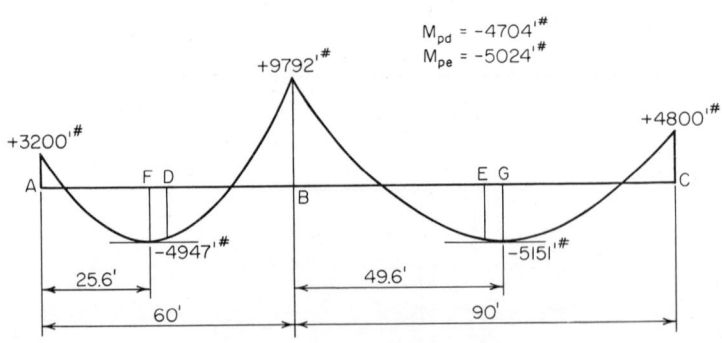

moment; its numerical value is directly proportional to the distance from the given section to the end support.

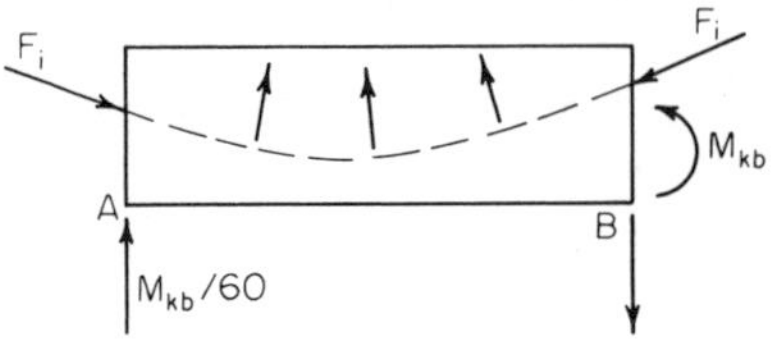

The continuity moment may be evaluated by adopting the second method of solution in the previous Calculation Procedure, since this renders the continuous member amenable to analysis by the theorem of three moments or moment distribution.

Fig. 155   Free-body diagram of concrete.

Determining $wL^2/4$ for each span: span AB, $w_1 L_1^2/4 = F_i - 0.40 - 1.20 - 1.20) = -2.80F_i$ in.-lb; span BC, $w_2 L_2^2/4 = F_i(-1.20 - 1.28 - 0.60) = -3.08F_i$ in.-lb.

2. Determine the true prestress moment at B in terms of F_i

Apply the theorem of three moments; by subtraction, find M_{kb}. Thus, $M_{pa}L_1 + 2M_{pb}(L_1 + L_2) + M_{pc}L_2 = -w_1 L_1^3/4 - w_2 L_2^3/4$. Substitute the value of L_1 and L_2, in ft, and divide each term by F_i, or $0.40(60) + (2M_{pb} \times 150)/F_i + 0.60(90) = 2.80(60) + 3.08(90)$. Solving $M_{pb} = 1.224F_i$ in.-lb. Also, $M_{kb} = M_{pb} - (-F_i e_b) = F_i(1.224 - 1.20) = 0.024F_i$. Thus, the continuity moment at B is positive.

3. Evaluate the prestress moment at the supports and at mid-span

Using ft-lb in the moment evaluation, $M_{pa} = 0.40(96,000)/12 = 3,200$ ft-lb; $M_{pb} = 1.224(96,000)/12 = 9,792$ ft-lb; $M_{pc} = 0.60(96,000)/12 = 4,800$ ft-lb; $M_{pd} = -F_i e_d + M_{kd} = F_i(-0.60 + \frac{1}{2} \times 0.024)/12 = -4,704$ ft-lb; $M_{pe} = F_i(-0.64 + \frac{1}{2} \times 0.024)/12 = -5,024$ ft-lb.

4. Construct the prestress-moment diagram

Figure 156 shows this diagram. Apply Eq. (70) to locate and evaluate the maximum negative moments. Thus, $AF = 25.6$ ft; $BG = 49.6$ ft; $M_{pf} = -4,947$ ft-lb; $M_{pg} = -5,151$ ft-lb.

Fig. 156   Prestress-moment diagram.

PRINCIPLE OF LINEAR TRANSFORMATION

For the beam in Fig. 154, consider that the parabolic trajectory of the prestressing force is displaced thus: e_a and e_c are held constant as e_b is changed to -2.0 in., the eccentricity at any intermediate section being decreased algebraically by an amount directly proportional to the distance from that section to A or C. Construct the prestress-moment diagram.

Calculation Procedure:

1. Compute the revised eccentricities

The modification described is termed a *linear transformation* of the trajectory. Two methods will be presented. Steps 1 through 4 comprise method 1; the remaining steps comprise method 2.

The revised eccentricities are: $e_a = -0.40$ in.; $e_d = +0.20$ in.; $e_b = -2.00$ in.; $e_e = +0.24$ in.; $e_c = -0.60$ in.

2. Find the value of $wL^2/4$ for each span

Applying Eq. (71): span AB, $w_1L_1^2/4 = F_i(-0.40-0.40-2.00) = -2.80F_i$; span BC, $w_2L_2^2/4 = F_i(-2.00-0.48-0.60) = -3.08F_i$.

These results are identical with those obtained in the previous Calculation Procedure. The change in e_b is balanced by an equal change in $2e_d$ and $2e_e$.

3. Determine the true prestress moment at B by applying the theorem of three moments; then find M_{kb}

Refer to Step 2 in the previous Calculation Procedure. Since the linear transformation of the trajectory has not affected the value of w_1 and w_2, the value of M_{pb} remains constant. Thus, $M_{kb} = M_{pb} - (-F_ie_b) = F_i(1.224-2.0) = -0.776F_i$.

4. Evaluate the prestress moment at mid-span

Thus, $M_{pd} = -F_ie_d + M_{kd} = F_i(-0.20-\frac{1}{2} \times 0.776)/12 = -4,704$ ft-lb; $M_{pe} = F_i(-0.24 - \frac{1}{2} \times 0.776)/12 = -5,024$ ft-lb.

These results are identical with those in the previous Calculation Procedure. The change in the eccentricity moment is balanced by an accompanying change in the continuity moment. Since three points determine a parabolic arc, the prestress moment diagram coincides with that in Fig. 156. This constitutes the solution by method 1.

5. Evaluate the prestress moments

Do this by replacing the prestressing system with two hypothetical systems that jointly induce eccentricity moments identical with those of the true system.

Let e denote the original eccentricity of the prestressing force at a given section and Δe the change in eccentricity that results from the linear transformation. The final eccentricity moment is $-F_i(e + \Delta e) = -(F_ie + F_i\Delta e)$.

Consider the beam as subjected to two prestressing forces of 96 kips each. One has the parabolic trajectory described in the previous Calculation Procedure; the other has the linear trajectory shown in Fig. 157, where $e_a = 0$, $e_b = -0.80$ in., and $e_c = 0$. Under the latter prestressing system, the tendons exert three forces on the concrete—one at each end and one at the deflection point above the interior support caused by the change in direction of the prestressing force.

The horizontal component of the prestressing force is considered equal to the force itself; it therefore follows that the force acting at the deflection point has no horizontal component.

Since the three forces that the tendons exert on the concrete are applied directly at the supports, their vertical components do not induce bending. Similarly, since the forces at A and C are applied at the centroidal axis, their horizontal components do not induce bending. Consequently, the prestressing system having the trajectory shown in Fig. 157 does not cause any prestress moments whatsoever. The prestress moments for the beam in the present instance are therefore identical with those for the beam in the previous Calculation Procedure.

The second method of analysis is preferable to the first because it is general. The first method demonstrates the equality of prestress moments before and after the linear transformation where the trajectory is parabolic; the second method demonstrates this equality without regard to the form of trajectory.

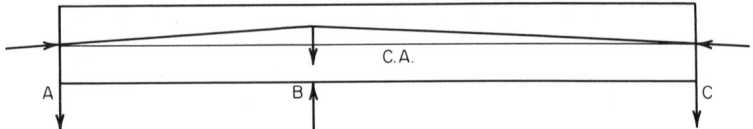

Fig. 157 Hypothetical prestressing system and forces exerted on concrete.

In the present Calculation Procedure, the extremely important *principle of linear transformation* for a two-span continuous beam was developed. This principle states: The prestress moments remain constant when the trajectory of the prestressing force is transformed linearly. The principle is frequently applied in plotting a trial trajectory for a continuous beam.

Two points warrant emphasis. First, in a linear transformation, the eccentricities at the end supports remain constant. Second, the hypothetical prestressing systems introduced in Step 5 of this Calculation Procedure are equivalent to the true system solely with respect to bending stresses; the axial stress F_i/A under the hypothetical systems is double that under the true system.

CONCORDANT TRAJECTORY OF A BEAM

Referring to the beam in the second previous Calculation Procedure, transform the trajectory linearly to obtain a concordant trajectory.

Calculation Procedure:

1. *Calculate the eccentricities of the concordant trajectory*

Two principles apply here. (*a*) In a continuous beam, the prestress moment M_p consists of two elements: a moment $-F_i e$ due to eccentricity and a moment M_k due to continuity. The continuity moment varies linearly from zero at the ends to its maximum numerical value at the interior support.

(*b*) In a linear transformation, the change in $-F_i e$ is offset by a compensatory change in M_k, with the result that M_p remains constant.

It is possible to transform a given trajectory linearly to obtain a new trajectory having the characteristic that $M_k = 0$ along the entire span, and therefore $M_p = -F_i e$. The latter is termed a *concordant trajectory*. Since M_p retains its original value, the concordant trajectory corresponding to a given trajectory is found simply by equating the final eccentricity to $-M_p/F_i$.

Refer to Fig. 154, and calculate the eccentricities of the concordant trajectory. As before, $e_a = -0.40$ in. and $e_c = -0.60$ in. Then, $e_d = 4,704(12)/96,000 = +0.588$ in.; $e_b = -9,792(12)/96,000 = -1.224$ in.; $e_e = 5,024(12)/96,000 = +0.628$ in.

2. *Analyze the eccentricities*

All eccentricities have thus been altered by an amount directly proportional to the distance from the adjacent end support to the given section, and the trajectory has undergone a linear transformation. The advantage accruing from plotting a concordant trajectory is shown in the next Calculation Procedure.

DESIGN OF TRAJECTORY TO OBTAIN ASSIGNED PRESTRESS MOMENTS

The prestress moments shown in Fig. 156 are to be obtained by applying an initial prestressing force of 72 kips with an eccentricity of -2 in. at B. Design the trajectory.

Calculation Procedure:

1. Plot a concordant trajectory

Set $e = -M_p/F_i$, or $e_a = -3,200(12)/72,000 = -0.533$ in.; $e_d = +0.784$ in.; $e_b = -1.632$ in.; $e_e = +0.837$ in.; $e_c = -0.800$ in.

2. Set e_b = desired eccentricity and transform the trajectory linearly

Thus, $e_a = -0.533$ in.; $e_c = -0.800$ in.; $e_d = +0.784 - \frac{1}{2}(2.000 - 1.632) = +0.600$ in.; $e_e = +0.837 - 0.184 = +0.653$ in.

EFFECT OF VARYING ECCENTRICITY AT END SUPPORT

For the beam in Fig. 154, consider that the parabolic trajectory in span AB is displaced thus: e_b is held constant as e_a is changed to -0.72 in., the eccentricity at every inter-mediate section being decreased algebraically by an amount directly proportional to the distance from that section to B. Compute the prestress moment at the supports and at mid-span caused by a prestressing force of 96 kips.

Calculation Procedure:

1. Apply the revised value of e_a ; repeat the calculations of the earlier procedure

Thus, $M_{pa} = 5,760$ ft-lb; $M_{pd} = -3,680$ ft-lb; $M_{pb} = 9,280$ ft-lb; $M_{pe} = -5,280$ ft-lb; $M_{pc} = 4,800$ ft-lb.

The change in prestress moment caused by the displacement of the trajectory varies linearly across each span. Figure 158 compares the original and revised moments along span AB. This constitutes method 1.

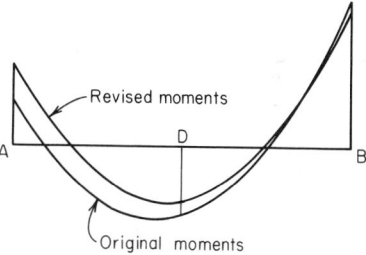

Fig. 158   Prestress-moment diagrams.

2. Replace the prestressing system with two hypothetical systems that jointly in-duce eccentricity moments identical with those of the true system

This constitutes method 2. For this purpose, consider the beam to be subjected to two prestressing forces of 96 kips each. One has the parabolic trajectory described in the earlier Procedure; the other has a trajectory that is linear in each span, the eccentricities being $e_a = -0.72 - (-0.40) = -0.32$ in., $e_b = 0$, and $e_c = 0$.

3. Evaluate the prestress moments induced by the hypothetical system having the linear trajectory

The tendons exert a force on the concrete at A, B, and C, but only the force at A causes bending moment.

Thus, $M_{pa} = -F_i e_a = -96,000(-0.32)/12 = 2,560$ ft-lb. Also, $M_{pa}L_1 + 2M_{pb}(L_1 + L_2) + M_{pc}L_2 = 0$. But $M_{pc} = 0$; therefore, $M_{pb} = -512$ ft-lb, $M_{pd} = \frac{1}{2}(2,560 - 512) = 1,024$ ft-lb, $M_{pe} = \frac{1}{2}(-512) = -256$ ft-lb.

4. Find the true prestress moments by superposing the two hypothetical systems

Thus: $M_{pa} = 3,200 + 2,560 = 5,760$ ft-lb; $M_{pd} = -4,704 + 1,024 = -3,680$ ft-lb; $M_{pb} = 9,792 - 512 = 9,280$ ft-lb; $M_{pe} = -5,024 - 256 = -5,280$ ft-lb; $M_{pc} = 4,800$ ft-lb.

DESIGN OF TRAJECTORY FOR A TWO-SPAN CONTINUOUS BEAM

A T beam that is continuous across two spans of 120 ft each is to carry a uniformly distributed live load of 880 plf. The cross section has these properties: $A = 1,440$ sq in.; $I = 752,000$ in.^{4}; $y_b = 50.6$ in.; $y_t = 23.4$ in. The allowable stresses are: *initial*, $+2,400$ and -60 psi; *final*, $+2,250$ and -60 psi. Assume that the minimum possible distance from the extremity of the section to the centroidal axis of the prestressing steel is 9 in. Determine the magnitude of the prestressing force and design the parabolic trajectory (a) using solely prestressed reinforcement; (b) using a combination of prestressed and nonprestressed reinforcement.

Calculation Procedure:

1. Compute the section moduli, kern distances, and beam weight

For part a, an exact design method consists of these steps: First, write equations for the prestress moment, beam-weight moment, maximum and minimum potential superimposed-load moment, expressing each moment in terms of the distance from a given section to the adjacent exterior support. Second, apply these equations to identify the sections at which the initial and final stresses are critical. Third, design the prestressing system to restrict the critical stresses to their allowable range. Whereas the exact method is not laborious when applied to a prismatic beam carrying uniform loads, this Procedure adopts the conventional, simplified method for illustrative purposes. This consists of dividing each span into a suitable number of intervals and analyzing the stresses at each boundary section.

For simplicity, set the eccentricity at the ends equal to zero. The trajectory will be symmetrical about the interior support, and the vertical component w of the force exerted by the tendons on the concrete in a unit longitudinal distance will be uniform across the entire length of member. Therefore, the prestress-moment diagram has the same form as the bending-moment diagram of a nonprestressed prismatic beam continuous over two equal spans and subjected to a uniform load across its entire length. It follows as a corollary that the prestress moments at the boundary sections previously referred to have specific *relative* values, although their absolute values are functions of the prestressing force and its trajectory.

The following steps constitute a methodical procedure: Evaluate the relative prestress moments and select a trajectory having ordinates directly proportional to these moments. The trajectory thus fashioned is concordant. Compute the prestressing force that is required to restrict the stresses to the allowable range. Then transform the concordant trajectory linearly to secure one that lies entirely within the confines of the section. Although the number of satisfactory concordant trajectories is infinite, the one to be selected is that which requires the minimum prestressing force. Therefore, the selection of the trajectory and the calculation of F_i are blended into one operation.

Divide the left span into five intervals, as shown in Fig. 159. (The greater the number of intervals chosen, the more reliable are the results.)

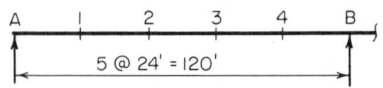

Fig. 159   Division of span into intervals.

Computing the moduli, kern distances, and beam weight: $S_b = 14{,}860$ in.^{3}; $S_t = 32{,}140$ in.^{3}; $k_b = 22.32$ in.; $k_t = 10.32$ in.; $w_w = 1{,}500$ plf.

2. Record the bending-moment coefficients C_1, C_2, and C_3

Use Table 9 to record these coefficients at the boundary sections. The subscripts refer to these conditions of loading: 1, load on entire left span and none on right span; 2, load on entire right span and none on left span; 3, load on entire length of beam.

To obtain these coefficients, refer to the AISC *Manual*, Case 29, which respresents condition 1. Thus, $R_1 = (\tfrac{7}{16})wL$; $R_3 = -(\tfrac{1}{16})wL$. At section 3, for example, $M_1 = (\tfrac{7}{16})wL$ $(0.6L) - \tfrac{1}{2}w(0.6L)^2 = [7(0.6) - 8(0.36)]wL^2/16 = 0.0825wL^2$; $C_1 = M_1/wL^2 = +0.0825$.

To obtain condition 2, interchange R_1 and R_3. At section 3, $M_2 = -(\tfrac{1}{16})wL(0.6L) = -0.0375wL^2$; $C_2 = -0.0375$; $C_3 = C_1 + C_2 = +0.0825 - 0.0375 = +0.0450$.

These moment coefficients may be applied without appreciable error to find the maximum and minimum potential live-load bending moments at the respective sections. The values of C_3 also represent the relative eccentricities of a concordant trajectory.

Since the gravity loads induce the maximum positive moment at section 2 and the maximum negative moment at section B, the prestressing force and its trajectory will be designed to satisfy the stress requirements at these two sections. (However, the stresses at all boundary sections will be checked.) The Magnel diagram for section 2 is similar to that in Fig. 141, but that for section B is much different.

TABLE 9   Calculations for Two-span Continuous Beam: Part a

	Section	1	2	3	4	B
1	C_1	$+0.0675$	$+0.0950$	$+0.0825$	$+0.0300$	-0.0625
2	C_2	-0.0125	-0.0250	-0.0375	-0.0500	-0.0625
3	C_3	$+0.0550$	$+0.0700$	$+0.0450$	-0.0200	-0.1250
4	f_{bw}	-959	$-1,221$	-785	$+349$	$+2,180$
5	f_{bs1}	-691	-972	-844	-307	$+640$
6	f_{bs2}	$+128$	$+256$	$+384$	$+512$	$+640$
7	f_{tw}	$+444$	$+565$	$+363$	-161	$-1,008$
8	f_{ts1}	$+319$	$+450$	$+390$	$+142$	-296
9	f_{ts2}	-59	-118	-177	-237	-296
10	e_{con}	$+17.19$	$+21.87$	$+14.06$	-6.25	-39.05
11	f_{bp}	$+2,148$	$+2,513$	$+1,903$	$+318$	$-2,243$
12	f_{tp}	$+185$	$+16$	$+298$	$+1,031$	$+2,215$
13	$0.85f_{bp}$	$+1,826$	$+2,136$	$+1,618$	$+270$	$-1,906$
14	$0.85f_{tp}$	$+157$	$+14$	$+253$	$+876$	$+1,883$

At mid-span: $C_3 = +0.0625$ and $e_{con} = +19.53$

3. Compute the value of C_3 at mid-span
   Thus, $C_3 = +0.0625$.

4. Apply the moment coefficients to find the gravity-load stresses
   Record the results in Table 9. Thus: $M_w = C_3(1,500)(120)^2(12) = 259,200,000C_3$ in.-lb; $f_{bw} = -259,200,000C_3/14,860 = -17,440C_3$; $f_{bs1} = -10,230C_1$; $f_{bs2} = -10,230C_2$; $f_{tw} = 8,065C_3$; $f_{ts1} = 4,731C_1$; $f_{ts2} = 4,731C_2$.
   Since S_t far exceeds S_b, it is manifest that the prestressing force must be designed to confine the bottom-fiber stresses to the allowable range.

5. Consider that a concordant trajectory has been plotted. Express the eccentricity at section B relative to that at section 2
   Thus, $e_b/e_2 = -0.1250/+0.0700 = -1.786$; therefore, $e_b = -1.786e_2$.

6. Determine the allowable range of values of f_{bp} at sections 2 and B
   Referring to Fig. 160: at section 2, $f_{bp} \leq +3,621$ psi, Eq. (a); $0.85f_{bp} \geq 1,221 + 972 - 60$; therefore, $f_{bp} \geq +2,509$ psi, Eq. (b). At section B, $f_{bp} \geq -2,240$ psi, Eq. (c); $0.85f_{bp} \leq -(2,180 + 1,280) + 2,250$; $f_{bp} \leq -1,424$ psi, Eq. (d).

7. Substitute numerical values in Eq. (56), expressing e_b in terms of e_2
   The values obtained are: $1/F_i \geq (k_t + e_2)/3,621S_b$, Eq. (a'); $1/F_i \leq (k_t + e_2)/2,509S_b$, Eq. (b'); $1/F_i \geq (1.786e_2 - k_t)/2,240S_b$, Eq. (c'); $1/F_i \leq (1.786e_2 - k_t)/1,424S_b$, Eq. (d').

8. Obtain the composite Magnel diagram
   Considering the relations in Step 7 as equalities, plot the straight lines representing them to obtain the composite Magnel diagram in Fig. 161. The slopes of the lines have these relative values: $m_a = 1/3,621$; $m_b = 1/2,509$; $m_c = 1.786/2,240 = 1/1,254$; $m_d = 1.786/1,424 = 1/797$. The shaded area bounded by these lines represents the region of permissible sets of values of e_2 and $1/F_i$.

9. Calculate the minimum allowable value of F_i and the corresponding value of e_2
   In the composite Magnel diagram, this set of values is represented by point A. Therefore, consider Eqs. (b') and (c') as equalities and solve for the unknowns. Or, $(10.32 + e_2)/2,509 = (1.786e_2 - 10.32)/2,240$; solving, $e_2 = 21.87$ in. and $F_i = 1,160,000$ lb.

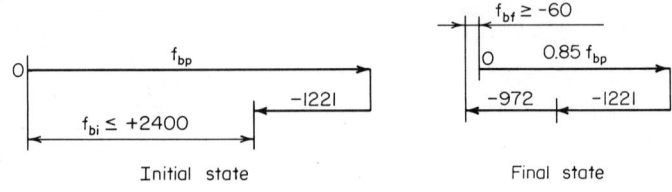

(a) Limiting values of f_{bp} at section 2

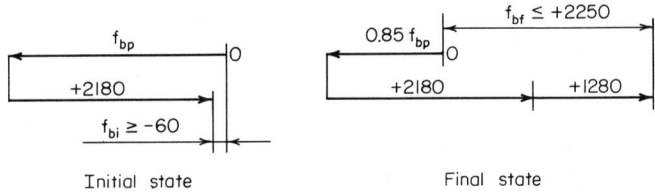

(b) Limiting values of f_{bp} at section B

Fig. 160

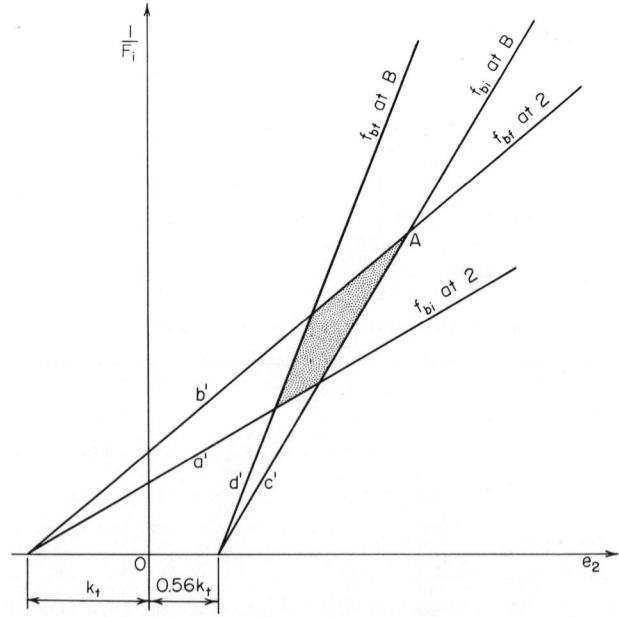

Fig. 161   Composite Magnel diagram.

10. *Plot the concordant trajectory*

Do this by applying the values of C_3 appearing in Table 9; for example, $e_1 = +21.87 \times (0.0550)/0.0700 = +17.19$ in. At mid-span, $e_m = +21.87(0.0625)/0.0700 = +19.53$ in.

Record the eccentricities on line 10 of the table. It is apparent that this concordant trajectory is satisfactory in the respect that it may be linearly transformed to one falling within the confines of the section; this is proved in Step 14.

11. *Apply Eq. (56) to find f_{bp} and f_{tp}*

Record the results in Table 9. For example, at section 1, $f_{bp} = 1,160,000(10.32 + 17.19)/14,860 = +2,148$ psi; $f_{tp} = 1,160,000(22.32 - 17.19)/32,140 = +185$ psi.

12. *Multiply the values of f_{bp} and f_{tp} by 0.85 and record the results*

These results appear in Table 9.

13. *Investigate the stresses at every boundary section*

In calculating the final stresses, apply the live-load stress that produces a more critical condition. Thus, at section 1, $f_{bi} = -959 + 2,148 = +1,189$ psi; $f_{bf} = -959 - 691 + 1,826 = +176$ psi; $f_{ti} = +444 + 185 = +629$ psi; $f_{tf} = +444 + 319 + 157 = +920$ psi. At section 2: $f_{bi} = -1,221 + 2,513 = +1,292$ psi; $f_{bf} = -1,221 - 972 + 2,136 = -57$ psi; $f_{ti} = +565 + 16 = +581$ psi; $f_{tf} = +565 + 450 + 14 = +1,029$ psi. At section 3: $f_{bi} = -785 + 1,903 = +1,118$ psi; $f_{bf} = -785 - 844 + 1,618 = -11$ psi; $f_{ti} = +363 + 298 = +661$ psi; $f_{tf} = +363 + 390 + 253 = +1,006$ psi. At section 4: $f_{bi} = +349 + 318 = +667$ psi; $f_{bf} = +349 - 307 + 270 = +312$ psi, or $f_{bf} = +349 + 512 + 270 = +1,131$ psi; $f_{ti} = -161 + 1,031 = +870$ psi; $f_{tf} = -161 - 237 + 876 = +478$ psi, or $f_{tf} = -161 + 142 + 876 = +857$ psi. At section B: $f_{bi} = +2,180 - 2,243 = -63$ psi; $f_{bf} = +2,180 + 1,280 - 1,906 = +1,554$ psi; $f_{ti} = -1,008 + 2,215 = +1,207$ psi; $f_{tf} = -1,008 - 592 + 1,883 = +283$ psi.

In all instances, the stresses lie within the allowable range.

14. *Establish the true trajectory by means of a linear transformation*

The imposed limits are: $e_{max} = y_b - 9 = 41.6$ in.; $e_{min} = -(y_t - 9) = -14.4$ in.

Any trajectory that falls between these limits and that is obtained by linearly transforming the concordant trajectory is satisfactory. Set $e_b = -14$ in. and compute the eccentricity at mid-span and the maximum eccentricity.

Thus, $e_m = +19.53 + \frac{1}{2}(39.05 - 14) = +32.06$ in. By Eq. (70b), $e_s = -(\frac{1}{8})(-4 \times 32.06 - 14)^2/(-2 \times 32.06 - 14) = +32.4 < 41.6$ in. This is acceptable. This constitutes the solution to part a of the Procedure. Steps 15 through 20 constitute the solution to part b.

15. *Assign eccentricities to the true trajectory and check the maximum eccentricity*

The preceding calculation shows that the maximum eccentricity is considerably below the upper limit set by the beam dimensions. Refer to Fig. 161. If the restrictions imposed by line c' are removed, e_2 may be increased to the value corresponding to a maximum eccentricity of 41.6 in. and the value of F_i is thereby reduced. This revised set of values will cause an excessive initial tensile stress at B, but the condition can be remedied by supplying nonprestressed reinforcement over the interior support. Since the excess tension induced by F_i extends across a comparatively short distance, the saving accruing from the reduction in prestressing force will more than offset the cost of the added reinforcement.

Assigning the following eccentricities to the true trajectory and checking the maximum eccentricity by applying Eq. (70b), $e_a = 0$; $e_m = +41$ in.; $e_b = -14$ in.; $e_s = -(\frac{1}{8}) \times (-4 \times 41 - 14)^2/(-2 \times 41 - 14) = +41.3$ in. This is acceptable.

16. *To analyze the stresses, obtain a hypothetical concordant trajectory by linearly transforming the true trajectory*

Let y denote the upward displacement at B. Apply the coefficients C_3 to find the eccentricities of the hypothetical trajectory. Thus, $e_m/e_b = (41 - \frac{1}{2}y)/(-14 - y) = +0.0625/-0.1250$; $y = 34$ in.; $e_a = 0$; $e_m = +24$ in.; $e_b = -48$ in.; $e_1 = -48(+0.0550)/-0.1250 = +21.12$ in.; $e_2 = +26.88$ in.; $e_3 = +17.28$ in.; $e_4 = -7.68$ in..

17. Evaluate F_i by substituting in relation (b') of Step 7

Thus, $F_i = 2,509(14,860)/(10.32 + 26.88) = 1,000,000$ lb. Hence, the introduction of nonprestressed reinforcement served to reduce the prestressing force by 14 percent.

18. Calculate the prestresses at every boundary section; then find the stresses at transfer and under design load

Record the results in Table 10. (At sections 1 through 4, the final stresses were determined by applying the values on lines 5 and 8 in Table 9. The slight discrepancy between the final stress at 2 and the allowable value of -60 psi arises from the degree of precision in the calculations.)

With the exception of f_{bi} at B, all stresses at the boundary sections lie within the allowable range.

TABLE 10 Calculations for Two-span Continuous Beam: Part b

Section	1	2	3	4	B
e_{con}	$+21.12$	$+26.88$	$+17.28$	-7.68	-48.00
f_{bp}	$+2,116$	$+2,503$	$+1,857$	$+178$	$-2,535$
f_{tp}	$+37$	-142	$+157$	$+933$	$+2,188$
$0.85f_{bp}$	$+1,799$	$+2,128$	$+1,578$	$+151$	$-2,155$
$0.85f_{tp}$	$+31$	-121	$+133$	$+793$	$+1,860$
f_{bi}	$+1,157$	$+1,282$	$+1,072$	$+527$	-355
f_{bf}	$+149$	-65	-51	$+193$	$+1,305$
f_{ti}	$+481$	$+423$	$+520$	$+772$	$+1,180$
f_{tf}	$+794$	$+894$	$+886$	$+774$	$+260$

19. Locate the section at which $f_{bi} = -60$ psi

Since f_{bp} and f_{bw} vary parabolically across the span, their sum f_{bi} also varies in this manner. Let x denote the distance from the interior support to a given section. Apply Eq. (66) to find the equation for f_{bi}, using the initial-stress values at sections B, 3, and 1. Or, $-355 - 2 \times 1,072 + 1,157 = -1,342;$ $3(-355) - 4(1,072) + 1,157 = -4,196;$ $f_{bi} = -2,684(x/96)^2 + 4,196x/96 - 355$. When $f_{bi} = -60$, $x = 7.08$ ft. The tensile stress at transfer is therefore excessive in an interval of only 14.16 ft.

20. Design the nonprestressed reinforcement over the interior support

As in the preceding Procedures, the member must be investigated for ultimate-strength capacity. The calculation pertaining to any quantity that varies parabolically across the span may be readily checked by verifying that the values at uniformly spaced sections have equal "second differences." For example, with respect to the values of f_{bi} recorded in Table 10, the verification is:

$$+1,157 \qquad +1,282 \qquad +1,072 \qquad +527 \qquad -355$$
$$-125 \qquad +210 \qquad +545 \qquad +882$$
$$+335 \qquad +335 \qquad +337$$

The values on the second and third lines represent the differences between successive values on the preceding line.

REACTIONS FOR A CONTINUOUS BEAM

With reference to the beam in the previous Calculation Procedure, compute the reactions at the supports caused by the initial prestressing force designed in part a.

Calculation Procedure:

1. Determine what causes the reactions at the supports

As shown in Fig. 155, the reactions at the supports result from the continuity at B, and $R_a = M_{kb}/L$.

2. Compute the continuity moment at B; then find the reactions

Thus, $M_p = -F_i e + M_k = -F_i e_{con}$; $M_k = F_i(e - e_{con}) = 1,160(-14 + 39.05) = 29,060$ in.-kips. $R_a = 29,060/[120(12)] = 20.2$ kips; $R_B = -40.4$ kips.

Design of Highway Bridges

Where a bridge is supported by steel trusses, the stresses in the truss members are determined by applying the rules formulated in the truss Calculation Procedures given earlier in this Handbook.

The following Procedures show the design of a highway bridge supported by concrete or steel girders. Except for the deviations indicated, the *Standard Specifications for Highway Bridges*, published by the American Association of State Highway Officials (AASHO), are applied.

The AASHO *Specification* recognizes two forms of truck loading: the H loading, and the HS loading. Both are illustrated in the *Specification*. For a bridge of relatively long span, it is necessary to consider the possibility that several trucks will be present simultaneously. To approximate this condition, the AASHO *Specification* offers various lane loadings, and it requires that the bridge be designed for the lane loading if this yields greater bending moments and shears than does the corresponding truck loading.

In designing the bridge members, it is necessary to modify the wheel loads to allow for the effects of dynamic loading and the lateral distribution of loads resulting from the rigidity of the floor slab.

The basic notational system is: DF = factor for lateral distribution of wheel loads; IF = impact factor; P = resultant of group of concentrated loads.

The term *live load* as used in the following material refers to the wheel load after correction for distribution but before correction for impact.

DESIGN OF A T-BEAM BRIDGE

A highway bridge consisting of a concrete slab and concrete girders is to be designed for these conditions: loading, HS20-44; clear width, 28 ft; effective span, 54 ft; concrete strength, 3,000 psi; reinforcement, intermediate grade. The slab and girders will be poured monolithically, and the slab will include a $\frac{3}{4}$ in. wearing surface. In addition, the design is to make an allowance of 15 psf for future paving. Design the slab and the cross section of the interior girders.

Calculation Procedure:

1. Record the allowable stresses and modular ratio given in the AASHO Specification

Refer to Fig. 162, which shows the spacing of the girders and the dimensions of the members. The sizes were obtained by a trial-and-error method. Values from the

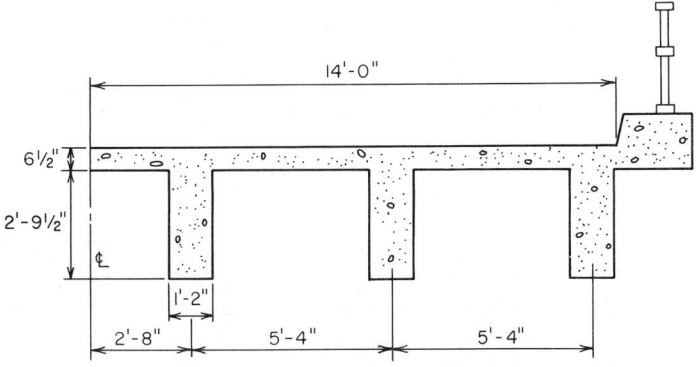

Fig. 162 Transverse section of T-beam bridge.

Specification are: $n = 10$ in stress calculations; $f_c = 0.4f'_c = 1,200$ psi; for beams with web reinforcement, $v_{max} = 0.075f'_c = 225$ psi; $f_s = 20,000$ psi; $u = 0.10f'_c = 300$ psi.

2. Compute the design coefficients associated with balanced design

Thus, $k = 1,200/(1,200 + 2,000) = 0.375$, using Eq. (21). Using Eq. (22), $j = 1 - 0.125 = 0.875$. By Eq. (32), $K = \frac{1}{2}(1,200)(0.375)(0.875) = 197$ psi.

3. Establish the wheel loads and critical spacing associated with the designated vehicular loading

As shown in the AASHO *Specification*, the wheel-load system comprises two loads of 16 kips each and one load of 4 kips. Since the girders are simply supported, an axle spacing of 14 ft will induce the maximum shear and bending moment in these members.

4. Verify that the slab size is adequate and design the reinforcement

The AASHO *Specification* does not present moment coefficients for the design of continuous members. The positive and negative reinforcement will be made identical, using straight bars for both. Apply a coefficient of $\frac{1}{10}$ in computing the dead-load moment. The *Specification* provides that the span length S of a slab continuous over more than two supports be taken as the clear distance between supports.

In computing the effective depth, disregard the wearing surface, assume the use of No 6 bars, and allow 1 in. for insulation, as required by AASHO. Then, $d = 6.5 - 0.75 - 1.0 - 0.38 = 4.37$ in.; $w_{DL} = (6.5/12)(150) + 15 = 96$ plf; $M_{DL} = (\frac{1}{10})w_{DL}S^2 = (\frac{1}{10})(96) \times (4.17)^2 = 167$ ft-lb; $M_{LL} = 0.8(S + 2)P_{20}/32$, by AASHO, or $M_{LL} = 0.8(6.17)(16,000)/32 = 2,467$ ft-lb. Also by AASHO, IF $= 0.30$; $M_{total} = 12(167 + 1.30 \times 2,467) = 40,500$ in.-lb. The moment corresponding to balanced design is $M_b = K_bbd^2 = 197(12)(4.37)^2 = 45,100$ in.-lb. The concrete section is therefore excessive, but a 6-in. slab would be inadequate. The steel is stressed to capacity at design load. Or, $A_s = 40,500/(20,000 \times 0.875 \times 4.37) = 0.53$ sq in. Use No. 6 bars 10 in. on centers, top and bottom.

The transverse reinforcement resists the tension caused by thermal effects and by load distribution. By AASHO, $A_t = 0.67(0.53) = 0.36$ sq in. Use five No. 5 bars in each panel, for which $A_t = 1.55/4.17 = 0.37$ sq in.

5. Calculate the maximum live-load bending moment in the interior girder caused by the moving-load group

The method of positioning the loads to evaluate this moment is described in an earlier Calculation Procedure in this Handbook. The resultant, Fig. 163, has this loca-

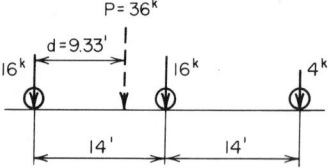

Fig. 163　Load group and its resultant.

tion: $d = [16(14) + 4(28)]/(16 + 16 + 4) = 9.33$ ft. Place the loads in the position shown in Fig. 164a. The maximum live-load bending moment occurs under the center load.

The AASHO prescribes a distribution factor of $S/6$ in the present instance, where S denotes the spacing of girders. However, a factor of $S/5$ will be applied here. Then DF $= 5.33/5 = 1.066$; $16 \times 1.066 = 17.06$ kips; $4 \times 1.066 = 4.26$ kips; $P = 2(17.06) + 4.26 = 38.38$ kips; $R_L = 38.38(29.33)/54 = 20.85$ kips. The maximum live-load moment is $M_{LL} = 20.85(29.34) - 17.06(14) = 372.8$ ft-kips.

6. Calculate the maximum live-load shear in the interior girder caused by the moving-load group

Place the loads in the position shown in Fig. 164b. Do not apply lateral distribution to the load at the support. Then, $V_{LL} = 16 + 17.06(40/54) + 4.26(26/54) = 30.69$ kips.

7. Verify that the size of the girder is adequate and design the reinforcement

Thus, $w_{DL} = 5.33(96) + 14(33.5/144)(150) = 1,000$ plf; $V_{DL} = 27$ kips; $M_{DL} = (\frac{1}{8})(1)(54)^2 = 364.5$ ft-kips. By AASHO, IF $= 50/(54 + 125) = 0.28$; $V_{total} = 27 + 1.28 \times (30.69) = 66.28$ kips; $M_{total} = 12(364.5 + 1.28 \times 372.8) = 10,100$ in.-kips.

In establishing the effective depth of the girder, assume that No. 4 stirrups will be supplied and that the main reinforcement will consist of three rows of No. 11 bars. AASHO requires $1\frac{1}{2}$-in. insulation for the stirrups and a clear distance of 1 in. between rows of bars. However, 2 in. of insulation will be provided in this instance, and the center-to-center spacing of rows will be taken as 2.5 times the bar diameter. Then, $d = 5.75 + 33.5 - 2 - 0.5 - 1.375(0.5 + 2.5) = 32.62$ in.; $v = V/b'jd = 66,280/(14 \times 0.875 \times 32.62) = 166 < 225$ psi. This is acceptable.

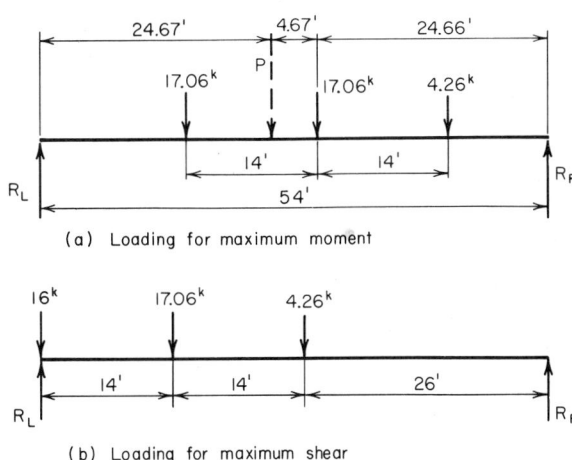

(a) Loading for maximum moment

(b) Loading for maximum shear

Fig. 164

Compute the moment capacity of the girder at balanced design. Since the concrete is poured monolithically, the girder and slab function as a T beam. Refer to Fig. 120 and its Calculation Procedure.

Thus, $k_b d = 0.375(32.62) = 12.23$ in.; $12.23 - 5.75 = 6.48$ in. At balanced design, $f_{c1} = 1,200(6.48/12.23) = 636$ psi. The effective flange width of the T beam as governed by AASHO is 64 in., and $C_b = 5.75(64)(\frac{1}{2})(1.200 + 0.636) = 338$ kips; $jd = 32.62 - (5.75/3) \times (1,200 + 2 \times 636)/(1,200 + 636) = 30.04$ in.; $M_b = 338(30.04) = 10,150$ in.-kips. The concrete section is therefore slightly excessive, and the steel is stressed to capacity, or $A_s = 10,100/20(30.04) = 16.8$ sq in. Use 11 No. 11 bars, arranged in three rows.

AASHO requires that the girders be tied together by diaphragms to obtain lateral rigidity of the structure.

COMPOSITE STEEL-AND-CONCRETE BRIDGE

The bridge shown in cross section in Fig. 165 is to carry an HS20-44 loading on an effective span of 74 ft 6 in. The structure will be unshored during construction. The concrete strength is 3,000 psi, and the entire slab is considered structurally effective; the allowable bending stress in the steel is 18,000 psi. The dead load carried by the composite section is 250 plf. Preliminary design calculations indicate that the interior girder is to consist of a 36WF150 and a cover plate $10 \times 1\frac{1}{2}$ in. welded to the bottom flange. Determine whether the trial section is adequate and complete the design.

Calculation Procedure:

1. *Record the relevant properties of the 36WF150*

The design of a composite bridge consisting of a concrete slab and steel girders is governed by specific articles in the AASHO *Specification*.

Composite behavior of the steel and concrete is achieved by adequately bonding the materials to function as a flexural unit. Loads that are present before the concrete has hardened are supported by the steel member alone; loads that are applied after hardening are supported by the composite member. Thus, the steel alone supports the concrete slab, and the steel and concrete jointly support the wearing surface.

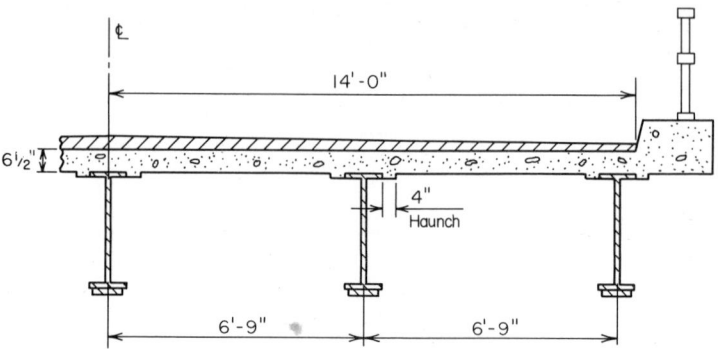

Fig. 165 Transverse section of composite bridge.

Plastic flow of the concrete under sustained load generates a transfer of compressive stress from the concrete to the steel. Consequently, the stresses in the composite member caused by dead load are analyzed by using a modular ratio three times the value that applies for transient loads.

If a wide-flange shape is used without a cover plate, the neutral axis of the composite section is substantially above the center of the steel, and the stress in the top steel fiber is therefore far below that in the bottom fiber. Use of a cover plate depresses the neutral axis, reduces the disparity between these stresses, and thereby results in a more economical section. Let y' = distance from neutral axis of member to given point, in absolute value; $\bar{y}$ = distance from centroidal axis of WF shape to neutral axis of member. The subscripts b, ts, and tc refer to the bottom of member, top of steel, and top of concrete, respectively. The superscripts c and n refer to the composite and noncomposite member, respectively.

The relevant properties of the 36WF150 are: $A = 44.16$ sq in.; $I = 9{,}012$ in.^{4}; $d = 35.84$ in.; $S = 503$ in.^{3}; flange thickness $= 1$ in., approximately.

2. Compute the section moduli of the noncomposite section where the cover plate is present

To do this, compute the statical moment and moment of inertia of the section with respect to the center of the WF shape; record the results in Table 11. Referring to Fig.

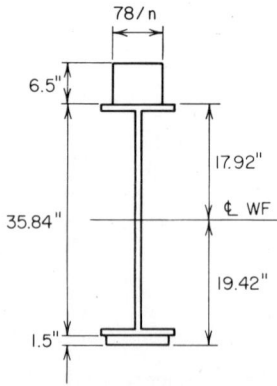

Fig. 166 Transformed section.

166: $\bar{y} = -280/59.16 = -4.73$ in.; $y'_b = 19.42 - 4.73 = 14.69$ in.; $y'_{ts} = 17.92 + 4.73 = 22.65$ in. By the moment-of-inertia equation: $I = 5{,}228 + 9{,}012 - 59.16(4.73)^2 = 12{,}916$ in.^{4}; $S_b = 879$ in.^{3}; $S_{ts} = 570$ in.^{3}.

3. Transform the composite section, with cover plate included, to an equivalent homogeneous section of steel, and compute the section moduli

In accordance with AASHO, the effective flange width is $12(6.5) = 78$ in. Using the method of an earlier Calculation Procedure, when $n = 30$: $\bar{y} = 78/76.06 = 1.03$ in.; $y'_b = 19.42 + 1.03 = 20.45$ in.; $y'_{ts} = 17.92 - 1.03 = 16.89$ in.; $y'_{tc} = 16.89 + 6.50 = 23.39$ in.; $I = 12{,}802 + 9{,}072 - 76.06(1.03)^2 = 21{,}793$ in.^{4}; $S_b = 1{,}066$ in.^{3}; $S_{ts} = 1{,}290$ in.^{3}; $S_{tc} = 932$ in.^{3}.

When $n = 10$: $\bar{y} = 7.22$ in.; $y'_b = 26.64$ in.; $y'_{ts} = 10.70$ in.; $y'_{tc} = 17.20$ in.; $I = 27{,}950 + 9{,}191 - 109.86(7.22)^2 = 31{,}414$ in.^{4}; $S_b = 1{,}179$ in.^{3}; $S_{ts} = 2{,}936$ in.^{3}; $S_{tc} = 1{,}826$ in.^{3}.

4. Transform the composite section, exclusive of the cover plate, to an equivalent homogeneous section of steel, and compute the values shown below

Thus, when $n = 30$: $y_b' = 23.78$ in.; $y_{ts}' = 12.06$ in.; $I = 14,549$ in.^{4}; $S_b = 612$ in.^{3}. When $n = 10$: $y_b' = 29.23$ in.; $y_{ts}' = 6.61$ in.; $I = 19,779$ in.^{4}; $S_b = 677$ in.^{3}.

TABLE 11   Calculations for Girder with Coverplate

	A	y	Ay	Ay^2	I_o
Noncomposite:					
36WF150	44.16	0	0	0	9,012
Coverplate	15.00	−18.67	−280	5,228	0
Total...........	59.16	....	−280	5,228	9,012
Composite, $n = 30$:					
Steel (total).......	59.16	....	−280	5,228	9,012
Slab	16.90	21.17	358	7,574	60
Total...........	76.06	....	78	12,802	9,072
Composite, $n = 10$:					
Steel (total).......	59.16	....	−280	5,228	9,012
Slab	50.70	21.17	1,073	22,722	179
Total...........	109.86	....	793	27,950	9,191

5. Compute the dead load carried by the noncomposite member

Thus,

Beam..........................	150 plf
Cover plate	51
Slab: 0.54(6.75)(150)	547
Haunch: 0.67(0.083)(150).........	8
Diaphragms (approximate).......	12
Shear connectors (approximate) ..	6
Total	774 plf; say 780 plf

6. Compute the maximum dead-load moments

Thus, $M_{DL}^c = (\frac{1}{8})(0.250)(74.5)^2(12) = 2,080$ in.-kips; $M_{DL}^n = (\frac{1}{8})(0.780)(74.5)^2(12) = 6,490$ in.-kips.

7. Compute the maximum live-load moment, with impact included

In accordance with the AASHO, the distribution factor is: $DF = 6.75/5.5 = 1.23$; $IF = 50/(74.5 + 125) = 0.251$; and $16(1.23)(1.251) = 24.62$ kips; $4(1.23)(1.251) = 6.15$ kips; $P_{LL+I} = 2(24.62) + 6.15 = 55.39$ kips. Refer to Fig. 164a as a guide. Then, $M_{LL+I} = 12[(55.39 \times 39.58 \times 39.58/74.5) - 24.62(14)] = 9,840$ in.-kips.

For convenience, the foregoing results are summarized in the tabulation below.

	M, in.-kips	S_b, in.^{3}	S_{ts}, in.^{3}	S_{tc}, in.^{3}
Noncomposite	6,490	879	570	
Composite, dead loads	2,080	1,066	1,290	932
Composite, moving loads ..	9,840	1,179	2,936	1,826

8. Compute the critical stresses in the member

To simplify the calculations, consider the sections of maximum live-load and dead-load stresses to be coincident. Then, $f_b = (6,490/879) + (2,080/1,066) + (9,840/1,179) = 17.68$ ksi; $f_{ts} = (6,490/570) + (2,080/1,290) + (9,840/2,936) = 16.35$ ksi; $f_{tc} = [2,080/(30 \times 932)] + [9,840/(10 \times 1,826)] = 0.61$ ksi. The section is therefore satisfactory.

9. Determine the theoretical length of cover plate

Let K denote the theoretical cutoff point at the left end. Let $L_c =$ length of cover plate exclusive of the development length; $b =$ distance from left support to K; $m = L_c/L$; $d =$ distance from heavier exterior load to action line of resultant, as shown in Fig. 163; $r = 2d/L$.

From these definitions: $b = (L - L_c)/2 = L(1 - m)/2$; $m = 1 - b/0.5L$. The maximum moment at K due to live load and impact is: $M_{LL+I} = (P_{LL+I}L)(1 - r + rm - m^2)/4$, Eq. (72). The diagram of dead-load moment is a parabola having its summit at mid-span.

To locate K, equate the bottom-fiber stress immediately to the left of K, where the cover plate is inoperative, to its allowable value. Or, $(P_{LL+I}L)/4 = 55.39(74.5)(12)/4 = 12,380$ in.-kips; $d = 9.33$ ft; $r = 18.67/74.5 = 0.251$; $6,490(1 - m^2)/503 + 2,080(1 - m^2)/612 + 12,380(0.749 + 0.251m - m^2)/677 = 18$ ksi; $m = 0.659$; $L_c = 0.659(74.5) = 49.10$ ft.

The plate must be extended toward each support and welded to the WF shape to develop its strength.

10. Verify the result obtained in Step 9

Thus, $b = \frac{1}{2}(74.5 - 49.10) = 12.70$ ft. At K: $M_{DL}^n = 12(\frac{1}{2} \times 74.5 \times 0.780 \times 12.70 - \frac{1}{2} \times 0.780 \times 12.70^2) = 3,672$ in.-kips; $M_{DL}^c = 3,672(250/780) = 1,177$ in.-kips. The maximum moment at K due to the moving-load system occurs when the heavier exterior load lies directly at this section. Also, $M_{LL+I} = 55.39(74.5 - 12.70 - 9.33)(12.70)(12)/74.5 = 5,945$ in.-kips; $f_b = (3,672/503) + (1,177/612) + (5,945/677) = 18.0$ ksi. This is acceptable.

11. Compute V_{DL} and V_{LL+I} at the support and at K

At the support: $V_{DL}^c = \frac{1}{2}(0.250 \times 74.5) = 9.31$ kips; IF $= 0.251$.

Consider that the load at the support is not subject to distribution. Applying the necessary correction, the following is obtained: $V_{LL+I} = 55.39(74.5 - 9.33)/74.5 - 16(1.251)(0.23) = 43.8$ kips. At K: $V_{DL}^c = 9.31 - 12.70(0.250) = 6.13$ kips; IF $= 50/(61.8 + 125) = 0.268$; $P_{LL+I} = 36(1.268)(1.23) = 56.15$ kips; $V_{LL+I} = 56.15(74.5 - 12.70 - 9.33)/74.5 = 39.55$ kips.

12. Select the shear connectors and determine the allowable pitch p at the support and immediately to the right of K

Assume use of $\frac{3}{4}$ in. studs, 4 in. high, with four studs in each transverse row, as shown in Fig. 167. The capacity of a connector as established by AASHO is $110d^2(f_c')^{0.5} = 110 \times 0.75^2(3,000)^{0.5} = 3,390$ lb. The capacity of a row of connectors $= 4(3,390) = 13,560$ lb.

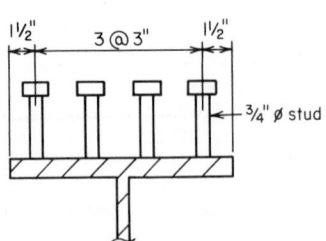

The shear flow at the bottom of the slab is found by applying $q = VQ/I$, or $q_{DL}^c = 9,310(16.90)(12.06 + 3.25)/14,549 = 166$ pli; $q_{LL+I} = 43,850(50.70)(6.61 + 3.25)/19,779 = 1,108$ psi; $p = 13,560/(166 + 1,108) = 10.6$ in.

Directly to the right of K: $q_{DL}^c = 6,130(16.90)(16.89 + 3.25)/21,793 = 96$ pli; $q_{LL+I} = 39,550(50.70)(10.70 + 3.25)/31,414 = 890$ pli; $p = 13,560/(96 + 890) = 13.8$ in.

It is necessary to determine the allowable pitch at other sections and to devise a suitable spacing of connectors for the entire span.

Fig. 167   Shear connectors.

13. Design the weld connecting the cover plate to the WF shape

The calculations for shear flow are similar to those in Step 12. The live-load deflection of an unshored girder is generally far below the limit imposed by AASHO. However, where an investigation is warranted, the deflection at mid-span may be calculated by assuming for simplicity that the position of loads for maximum deflection coincides with the position for maximum moment. The theorem of reciprocal deflections, presented in an earlier Calculation Procedure, may conveniently be applied in calculating this deflection. The girders are usually tied together by diaphragms at the ends and at third points to obtain lateral rigidity of the structure.

Fluid Mechanics

Hydrostatics

The notational system used in hydrostatics is: W = weight of floating body, lb; V = volume of displaced liquid, cu ft; w = specific weight of liquid, pcf; for water w = 62.4 pcf, unless another value is specified.

BUOYANCY AND FLOTATION

A timber member 12 ft long with a cross-sectional area of 90 sq in. will be used as a buoy in salt water. What volume of concrete must be fastened to one end so that 2 ft of the member will be above the surface? Use these specific weights: timber = 38 pcf; salt water = 64 pcf; concrete = 145 pcf.

Calculation Procedure:

1. Express the weight of the body and the volume of the displaced liquid in terms of the volume of concrete required

Archimedes' principle states that a body immersed in a liquid is subjected to a vertical buoyant force equal to the weight of the displaced liquid. In accordance with the equations of equilibrium, the buoyant force on a floating body equals the weight of the body. Therefore, $W = Vw$, Eq. (73). Let x denote the volume of concrete.

Then, $W = (90/144)(12)(38) + 145x = 285 + 145x$; $V = (90/144)(12 - 2) + x = 6.25 + x$.

2. Substitute in Eq. (73) and solve for x

Thus, $285 + 145x = (6.25 + x)64$; $x = 1.42$ cu ft.

HYDROSTATIC FORCE ON A PLANE SURFACE

In Fig. 168, AB is the side of a vessel containing water, and CDE is a gate located in this plane. Find the magnitude and location of the resultant thrust of the water on the gate when the liquid surface is 2 ft above the apex.

Calculation Procedure:

1. State the equations for the resultant magnitude and position

In Fig. 168, FH denotes the centroidal axis of area CDE that is parallel to the liquid surface, and G denotes the point of application of the resultant force. Point G is termed the *pressure center*.

Let A = area of given surface, sq ft; P = hydrostatic force on given surface, lb; p_m = mean pressure on surface, psf; y_{CA} and y_{PC} = vertical distance from centroidal axis and pressure center, respectively, to liquid surface, ft; z_{CA} and z_{PC} = distance along plane of given surface from the centroidal axis and pressure center, respectively, to line of intersection of this plane and the liquid surface, ft; I_{CA} = moment of inertia of area with respect to its centroidal axis, ft⁴.

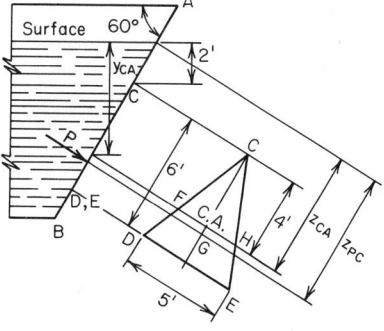

Fig. 168 Hydrostatic thrust on plane surface.

Consider an elemental surface of area dA at a vertical distance y below the liquid surface. The hydrostatic force dP on this element is normal to the surface and has the magnitude $dP = wy\,dA$, Eq. (74).

By applying Eq. (74), develop the following equations for the magnitude and position of the resultant force on the entire surface: $P = wy_{CA}A$, Eq. (75); $z_{PC} = I_{CA}/Az_{CA} + z_{CA}$, Eq. (76).

2. Compute the required values and solve the equations in Step 1

Thus: $A = \frac{1}{2}(5)(6) = 15$ sq ft; $y_{CA} = 2 + 4\sin 60° = 5.464$ ft; $z_{CA} = 2\csc 60° + 4 = 6.309$ ft; $I_{CA}/A = (bd^3/36)/(bd/2) = d^2/18 = 2$ sq ft; $P = 62.4(5.464)(15) = 5{,}114$ lb; $z_{PC} = 2/6.309 + 6.309 = 6.626$ ft; $y_{PC} = 6.626\sin 60° = 5.738$ ft. By symmetry, the pressure center lies on the centroidal axis through C.

An alternative equation for P is $P = p_m A$, Eq. (77). Equation (75) shows that the mean pressure occurs at the centroid of the area. The above two steps constitute Method 1 for solving this problem. The next three steps consitute Method 2.

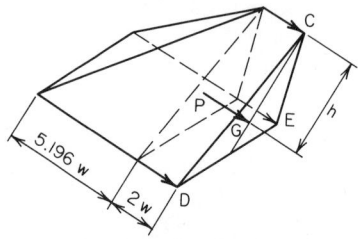

Fig. 169 Pressure prism.

3. Now construct the pressure "prism" associated with the area

In Fig. 169, construct the pressure prism associated with area CDE. The pressures are: at apex, $p = 2w$; at base, $p = (2 + 6\sin 60°)w = 7.196w$.

The force P equals the volume of this prism, and its action line lies on the centroidal plane parallel to the base. For convenience, resolve this prism into a triangular prism and rectangular pyramid, as shown.

4. Determine P by computing the volume of the pressure prism

Thus, $P = Aw[2 + \frac{2}{3}(5.196)] = Aw(2 + 3.464) = 15(62.4)(5.464) = 5{,}114$ lb.

5. Find the location of the resultant thrust

Compute the distance h from the top line to the centroidal plane. Then find y_{PC}. Or, $h = 2(\frac{2}{3})(6) + 3.464(\frac{3}{4})(6)/5.464 = 4.317$ ft; $y_{PC} = 2 + 4.317\sin 60° = 5.738$ ft.

HYDROSTATIC FORCE ON A CURVED SURFACE

The cylinder in Fig. 170a rests on an inclined plane and is immersed in liquid up to its top, as shown. Find the hydrostatic force on a 1-ft length of cylinder in terms of w and the radius R; locate the pressure center.

Calculation Procedure:

1. Evaluate the horizontal and vertical component of the force dP on an elemental surface having a central angle dθ

Refer to Fig. 170b. Adopt this sign convention: A horizontal force is positive if directed to the right; a vertical force is positive if directed upward. The first three steps constitute Method 1.

Evaluating dP, $dP_H = wR^2(\sin\theta - \sin\theta\cos\theta)\,d\theta$; $dP_V = wR^2(-\cos\theta + \cos^2\theta)\,d\theta$.

2. Integrate these equations to obtain the resultant forces PH and PV ; then find P

$P_H = wR^2(-\cos\theta + \frac{1}{2}\cos^2\theta)\,]_0^{7\pi/6} = wR^2[-(-0.866-1) + \frac{1}{2}(0.75-1)] = 1.741wR^2$, to right; $P_V = wR^2(-\sin\theta + \frac{1}{2}\theta + \frac{1}{4}\sin 2\theta)\,]_0^{7\pi/6} = wR^2(0.5 + 1.833 + 0.217) = 2.550wR^2$, upward; $P = wR^2(1.741^2 + 2.550^2)^{0.5} = 3.087wR^2$.

3. Determine the value of θ at the pressure center

Since each elemental force dP passes through the center of the cylinder, the resultant force P also passes through the center. Thus, $\tan(180° - \theta_{PC}) = P_H/P_V = 1.741/2.550$; $\theta_{PC} = 145°41'$.

4. Evaluate P_H and P_V

Apply these principles: P_H = force on an imaginary surface obtained by projecting the wetted surface on a vertical plane; $P_V = \pm$ weight of real or imaginary liquid lying between the wetted surface and the liquid surface. Use the plus sign if the *real* liquid lies below the wetted surface and the minus sign if it lies above this surface.

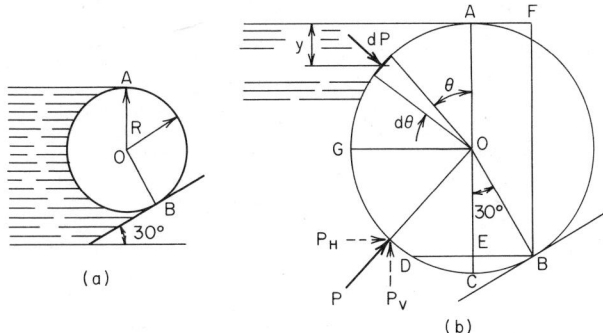

Fig. 170

Then P_H = force, to right, on AC + force, to left, on EC = force, to right, on AE; $AE = 1.866R$; $p_m = 0.933\,wR$; $P_H = 0.933\,wR(1.866R) = 1.741\,wR^2$; P_V = weight of imaginary liquid above GCB − weight of real liquid above GA = weight of imaginary liquid in cylindrical sector $AOBG$ and in prismoid $AOBF$. Volume of sector $AOBG =$ $[(7\pi/6)/2\pi](\pi R^2) = 1.833R^2$; volume of prismoid $AOBF = \frac{1}{2}(0.5R)(R + 1.866R) = 0.717R^2$; $P_V = wR^2(1.833 + 0.717) = 2.550wR^2$.

STABILITY OF A VESSEL

The boat in Fig. 171 is initially floating upright in fresh water. The total weight of the boat and cargo is 182 long tons; the center of gravity lies on the longitudinal (i.e., the fore-and-aft) axis of the boat and 8.6 ft above the bottom. A wind causes the boat to

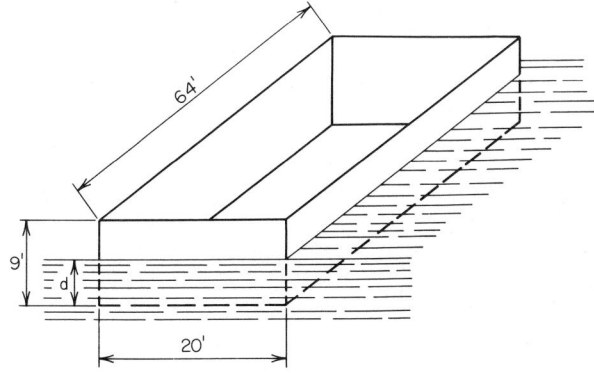

Fig. 171

list through an angle of 6° while the cargo remains stationary relative to the boat. Compute the righting or upsetting moment (*a*) without applying any set equation; (*b*) by applying the equation for metacentric height.

Calculation Procedure:

1. Compute the displacement volume and draft when the boat is upright

The buoyant force passes through the center of gravity of the displaced liquid; this point is termed the *center of buoyancy*. Figure 172 shows the cross section of a boat ro-

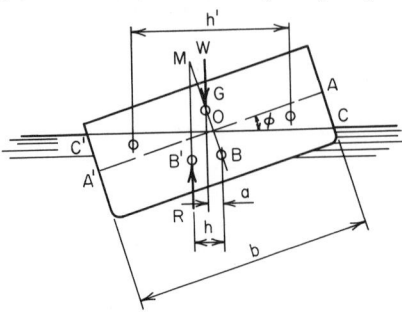

tated through an angle ϕ. The center of buoyancy for the upright position is B; B' is the center of buoyancy for the position shown, and G is the center of gravity of the boat and cargo.

In the position indicated in Fig. 172, the weight W and buoyant force R constitute a couple that tends to restore the boat to its upright position when the disturbing force is removed; their moment is therefore termed *righting*. When these forces constitute a couple that increases the rotation, their moment is said to be *upsetting*. The wedges OAC and $OA'C'$ are termed the *wedge of emersion* and *wedge of immersion*, respectively. Let $h =$

Fig. 172   Location of resultant forces on inclined vessel.

horizontal displacement of center of buoyancy caused by rotation; $h' =$ horizontal distance between centroids of wedge of emersion and wedge of immersion; $V' =$ volume of wedge of emersion (or immersion). Then, $h = V'h'/V$, Eq. (78).

The displacement volume and the draft when the boat is upright are: $W = 182(2{,}240) = 407{,}700$ lb; $V = W/w = 407{,}700/62.4 = 6{,}530$ cu ft; $d = 6{,}530/[64(20)] = 5.10$ ft.

2. Find h using Eq. (78)

Since ϕ is relatively small, apply this approximation: $h' = 2b/3 = 2(20)/3 = 13.33$ ft, $h = \frac{1}{2}(10)(10 \tan 6°)(13.33)/[5.10(20)] = 0.687$ ft.

3. Compute the horizontal distance a, Fig. 172

Thus, $BG = 8.6 - \frac{1}{2}(5.10) = 6.05$ ft; $a = 6.05 \sin 6° = 0.632$ ft.

4. Compute the moment of the vertical forces

Thus, $M = W(h - a) = 407{,}700(0.055) = 22{,}400$ ft-lb. Since $h > a$, this moment is righting. This constitutes the solution to part a. The remainder of this Procedure is concerned with part b.

In Fig. 172, let M denote the point of intersection of the vertical line through B' and the line BG prolonged. M is termed the *metacenter* associated with this position, and the distance GM is called the *metacentric height*. BG is positive if G is above B, and GM is positive if M is above G. Thus, the moment of vertical forces is righting or upsetting depending on whether the metacentric height is positive or negative, respectively.

5. Find the lever arm of the vertical forces

Use the relation for metacentric height, $GM = (I_{WL}/V \cos \phi) - BG$, Eq. (79), where $I_{WL} =$ moment of inertia of original waterline section about axis through O. Or, $I_{WL} = (1/12)(64)(20)^3 = 42{,}670$ ft^4; $GM = 42{,}670/6{,}530 \cos 6° - 6.05 = 0.52$ ft; $h - a = 0.52 \sin 6° = 0.054$ ft, which agrees closely with the previous result.

Mechanics of Incompressible Fluids

The notational system is: $a =$ acceleration; $A =$ area of stream cross section; $C =$ discharge coefficient; $D =$ diameter of pipe or depth of liquid in open channel; $F =$ force; $g =$ gravitational acceleration; $H =$ total head, or total specific energy; $h_F =$ loss of head between two sections caused by friction; $h_L =$ total loss of head between two sections; $h_V =$ difference in velocity heads at two sections if no losses occur; $L =$ length of stream between two sections; $M =$ mass of body; $N_R =$ Reynolds number;

p = pressure; Q = volumetric rate of flow, or discharge; s = hydraulic gradient = $-dH/dL$; T = torque; V = velocity; w = specific weight; z = elevation above datum plane; ρ = density (mass per unit volume); μ = dynamic (or absolute) viscosity; ν = kinematic viscosity = μ/ρ; τ = shearing stress. The units used for each symbol are given in the Calculation Procedure where the symbol is used.

If the discharge of a flowing stream of liquid remains constant, the flow is termed *steady*. Let the subscripts 1 and 2 refer to cross sections of the stream, 1 being the upstream section. From the definition of steady flow, it follows that $Q = A_1V_1 = A_2V_2 =$ constant, Eq. (80). This is termed the *equation of continuity*. Where no statement is made to the contrary, it is understood that the flow is steady.

Conditions at two sections may be compared by applying the following equation, which is a mathematical statement of Bernoulli's theorem: $V_1^2/2g + p_1/w + z_1 = V_2^2/2g + p_2/w + z_2 + h_L$, Eq. (81). The terms on each side of this equation represent, in their order of appearance, the *velocity head*, *pressure head*, and *potential head* of the liquid. Alternatively, they may be considered to represent forms of specific energy, namely, kinetic, pressure, and potential energy.

The force causing a change in velocity is evaluated by applying the basic equation $F = Ma$, Eq. (82).

Consider that liquid flows from section 1 to section 2 in a time interval t. At any instant, the volume of liquid bounded by these sections is Qt. The force required to change the velocity of this body of liquid from V_1 to V_2 is found from: $M = Qwt/g$; $a = (V_2 - V_1)/t$. Substituting in Eq. (82), $F = Qw(V_2 - V_1)/g$; or $F = A_1V_1w(V_2 - V_1)/g = A_2V_2w(V_2 - V_1)/g$, Eq. (83).

VISCOSITY OF FLUID

Two horizontal circular plates 9 in. in diameter are separated by an oil film 0.08 in. thick. A torque of 0.25 ft-lb applied to the upper plate causes that plate to rotate at a constant angular velocity of 4 rps relative to the lower plate. Compute the dynamic viscosity of the oil.

Calculation Procedure:

1. Develop equations for the force and torque

Consider that the fluid film in Fig. 173a is in motion and that a fluid particle at boundary A has a velocity dV relative to a particle at B. The shearing stress in the fluid is: $\tau = \mu \, dV/dx$, Eq. (84).

Figure 173b shows a cross section of the oil film, the shaded portion being an elemental surface. Let m = thickness of film; R = radius of plates; ω = angular velocity of one plate relative to the other; dA = area of elemental surface; dF = shearing force

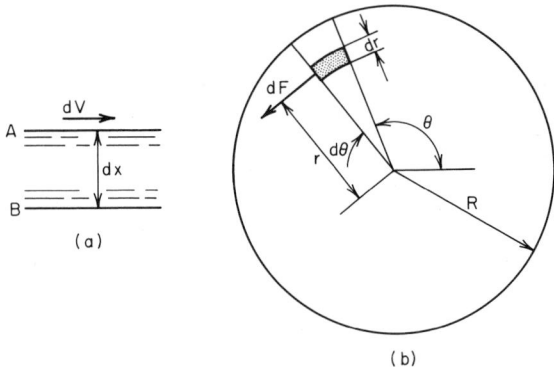

(a)

(b)

Fig. 173

on elemental surface; dT = torque of dF with respect to the axis through the center of the plate.

Applying Eq. (84), develop these equations: $dF = 2\pi\omega\mu r^2 \, dr \, d\theta/m$; $dT = r \, dF = 2\pi\omega\mu r^2 \, dr \, d\theta/m$.

2. Integrate the foregoing equation to obtain the resulting torque; solve for μ

Thus, $\mu = Tm/\pi^2\omega R^4$, Eq. (85). $T = 0.25$ ft-lb; $m = 0.08$ in.; $\omega = 4$ rps; $R = 4.5$ in.; $\mu = 0.25(0.08)(12)^3/\pi^2(4)(4.5)^4 = 0.00214$ lb-sec/ft².

APPLICATION OF BERNOULLI'S THEOREM

A steel pipe is discharging 10 cfs of water. At section 1, the pipe diameter is 12 in., the pressure is 18 psi, and the elevation is 140 ft. At section 2, farther downstream, the pipe diameter is 8 in., and the elevation is 106 ft. If there is a head loss of 9 ft between these sections due to pipe friction, what is the pressure at section 2?

Calculation Procedure:

1. Tabulate the given data

Thus: $D_1 = 12$ in.; $D_2 = 8$ in.; $p_1 = 18$ psi; $p_2 = ?$; $z_1 = 140$ ft; $z_2 = 106$ ft.

2. Compute the velocity at each section

Applying Eq. (80), $V_1 = 10/0.785 = 12.7$ fps; $V_2 = 10/0.349 = 28.7$ fps.

3. Compute p_2 by applying Eq. (81)

Thus, $(p_2 - p_1)/w = (V_1^2 - V_2^2)/2g + z_1 - z_2 - h_F = (12.7^2 - 28.7^2)/64.4 + 140 - 106 - 9 = 14.7$ ft; $p_2 = 14.7(62.4)/144 + 18 = 24.4$ psi.

FLOW THROUGH A VENTURI METER

A venturi meter of 3-in. throat diameter is inserted in a 6-in.-diameter pipe conveying fuel oil having a specific gravity of 0.94. The pressure at the throat is 10 psi and that at an upstream section 6 in. higher than the throat is 14.2 psi. If the discharge coefficient of the meter is 0.97, compute the flow rate in gpm.

Calculation Procedure:

1. Record the given data, assigning the subscript 1 to the upstream section and 2 to the throat

The loss of head between two sections can be taken into account by introducing a *discharge coefficient* C. This coefficient represents the ratio between the actual discharge Q and the discharge Q_i that would occur in the absence of any losses. Then, $Q = CQ_i$, or $(V_2^2 - V_1^2)/2g = C^2 h_V$.

Recording the given data: $D_1 = 6$ in.; $p_1 = 14.2$ psi; $z_1 = 6$ in.; $D_2 = 3$ in.; $p_2 = 10$ psi; $z_2 = 0$; $C = 0.97$.

2. Express V_1 in terms of V_2 and develop velocity and flow relations

Thus, $V_2 = C\{2gh_V/[1 - (A_2/A_1)^2]\}^{0.5}$, Eq. (86a); also, $Q = CA_2\{2gh_V/[1 - (A_2/A_1)^2]\}^{0.5}$, Eq. (86b). If V_1 is negligible, these relations reduce to $V_2 = C(2gh_V)^{0.5}$, Eq. (87a) and $Q = CA_2(2gh_V)^{0.5}$, Eq. (87b).

3. Compute h_V by applying Eq. (81)

Thus, $h_V = (p_1 - p_2)/w + z_1 - z_2 = 4.2(144)/[0.94(62.4)] + 0.5 = 10.8$ ft.

4. Compute Q by applying Eq. (86b)

Thus, $(A_2/A_1)^2 = (D_2/D_1)^4 = \frac{1}{16}$; $A_2 = 0.0491$ sq ft; and $Q = 0.97(0.0491)[64.4 \times 10.8/(1 - \frac{1}{16})]^{0.5} = 1.30$ cfs or, converting to gpm using the conversion factor of 1 cfs = 449 gpm, the flow rate is $1.30(449) = 584$ gpm.

FLOW THROUGH AN ORIFICE

Compute the discharge through a 3-in.-diameter square-edged orifice if the water on the upstream side stands 4 ft 8 in. above the center of the orifice.

Calculation Procedure:

1. Determine the discharge coefficient

For simplicity, the flow through a square-edged orifice discharging to the atmosphere is generally computed by equating the area of the stream to the area of the opening and then setting the discharge coefficient $C = 0.60$ to allow for contraction of the issuing stream. (The area of the issuing stream is about 0.62 times that of the opening.)

2. Compute the flow rate

Since the velocity of approach is negligible, use Eq. (87b). Or, $Q = 0.60(0.0491)$ $(64.4 \times 4.67)^{0.5} = 0.511$ cfs.

FLOW THROUGH THE SUCTION PIPE OF A DRAINAGE PUMP

Water is being evacuated from a sump through the suction pipe shown in Fig. 174. The entrance-end diameter of the pipe is 3 ft; the exit-end diameter, 1.75 ft. The exit pressure is 12.9 in. mercury vacuum. The head loss at the entry is one-fifteenth of the velocity head at that point, and the head loss in the pipe due to friction is one-tenth of the velocity head at the exit. Compute the discharge flow rate.

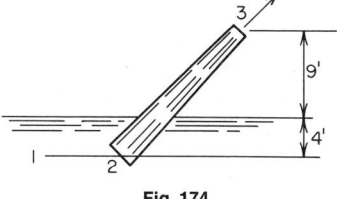

Fig. 174

Calculation Procedure:

1. Convert the pressure head to feet of water

The discharge may be found by comparing the conditions at an upstream point 1, where the velocity is negligible, with the conditions at point 3, Fig. 174. Select the elevation of point 1 as the datum.

Converting the pressure head at point 3 to feet of water and using the specific gravity of mercury as 13.6, $p_3/w = -(12.9/12)13.6 = -14.6$ ft.

2. Express the velocity head at 2 in terms of that at 3

By the equation of continuity, $V_2 = A_3 V_3 / A_2 = (1.75/3)^2 V_3 = 0.34 V_3$.

3. Evaluate V_3 by applying Eq. (81); then determine Q

Thus, $V_1^2/2g + p_1/w + z_1 = V_3^2/2g + p_3/w + z_3 + (\frac{1}{15})V_2^2/2g + (\frac{1}{10})V_3^2/2g$, or $0 + 4 + 0 = (V_3^2/2g) - 14.6 + 13 + (V_3^2/2g)(\frac{1}{15} \times 0.34^2 + \frac{1}{10})$; $V_3 = 18.0$ fps; then, $Q_3 = A_3 V_3 = 0.785 \times (1.75)^2(18.0) = 43.3$ cfs.

POWER OF A FLOWING LIQUID

A pump is discharging 8 cfs of water. Gages attached immediately upstream and downstream of the pump indicate a pressure differential of 36 psi. If the pump efficiency is 85 percent, what is the horsepower output and input?

Calculation Procedure:

1. Evaluate the increase in head of the liquid

Power is the rate of performing work, or the amount of work performed in a unit time. If the fluid flows with a specific energy H, the total energy of the fluid discharged in a unit time is QwH. This expression thus represents the work that the flowing fluid can perform in a unit time and therefore the power associated with this discharge. Since one horsepower = 550 ft-lb/sec, hp $= QwH/550$, Eq. (88).

In this situation, the power developed by the pump is desired. Therefore, H must be equated to the specific energy added by the pump.

To evaluate the increase in head, consider the differences of the two sections being considered. Since both sections have the same velocity and elevation, only their pressure heads differ. Thus, $p_2/w - p_1/w = 36(144)/62.4 = 83.1$ ft.

2. **Compute the horsepower output and input**

Thus, $hp_{out} = 8(62.4)(83.1)/550 = 75.4$ hp. And, $hp_{in} = 75.4/0.85 = 88.7$ hp.

DISCHARGE OVER A SHARP-EDGED WEIR

Compute the discharge over a sharp-edged rectangular weir 4 ft high and 10 ft long, with two end contractions, if the water in the canal behind the weir is 4 ft 9 in. high. Disregard the velocity of approach.

Calculation Procedure:

1. Adopt a standard relation for this weir

The discharge over a sharp-edged rectangular weir without end contractions in which the velocity of approach is negligible is given by the Francis formula as $Q = 3.33bh^{1.5}$, Eq. (89a), where b = length of crest and h = head on weir, i.e., the difference between the elevation of the crest and that of the water surface upstream of the weir.

2. Modify the Francis equation for end contractions

With two end contractions, the discharge of the weir is $Q = 3.33(b - 0.2h)h^{1.5}$, Eq. (89b). Substituting the given values, $Q = 3.33(10 - 0.2 \times 0.75)0.75^{1.5} = 21.3$ cfs.

LAMINAR FLOW IN A PIPE

A tank containing crude oil discharges 340 gpm through a steel pipe 220 ft long and 8 in. in diameter. The kinematic viscosity of the oil is 0.002 ft²/sec. Compute the difference in elevation between the liquid surface in the tank and the pipe outlet.

Calculation Procedure:

1. Identify the type of flow in the pipe

To investigate the discharge in a pipe, it is necessary to distinguish between two types of fluid flow—*laminar* and *turbulent*. Laminar (or *viscous*) flow is characterized by the telescopic sliding of one circular layer of fluid past the adjacent layer, each fluid particle traversing a straight line. The velocity of the fluid flow varies parabolically from zero at the pipe wall to its maximum value at the pipe center, where it equals twice the mean velocity.

Turbulent flow is characterized by the formation of eddy currents, each fluid particle traversing a sinuous path.

In any pipe the type of flow is assertained by applying a dimensionless index termed the *Reynolds number*, defined as $N_R = DV/\nu$, Eq. (90). Flow is considered laminar if $N_R < 2,100$, and turbulent if $N_R > 3,000$.

In laminar flow the head loss due to friction is $h_F = 32L\nu V/gD^2$, Eq. (91a), or $h_F = (64/N_R)(L/D)(V^2/2g)$, Eq. (91b). Let 1 denote a point on the liquid surface and 2 a point at the pipe outlet. The elevation of 2 will be taken as datum.

To identify the type of flow, compute N_R. Thus, $D = 8$ in.; $L = 220$ ft; $\nu = 0.002$ ft²/sec; $Q = 340/449 = 0.757$, converting from gpm to cfs. Then, $V = Q/A = 0.757/0.349 = 2.17$ fps. And $N_R = 0.667(2.17)/0.002 = 724$. Therefore, the flow is laminar because N_R is less than 2,100.

2. Express all losses in terms of the velocity head

By Eq. (91b), $h_F = (64/724)(220/0.667)V^2/2g = 29.2V^2/2g$. Where $L/D > 500$, the following may be regarded as negligible in comparison with the loss due to friction: loss at pipe entrance, losses at elbows, velocity head at the discharge, etc. In the present instance, include the secondary items. The loss at the pipe entrance is $h_E = 0.5V^2/2g$. The total loss is $h_L = 29.7V^2/2g$.

3. Find the elevation of 1 by applying Eq. (81)

Thus, $z_1 = (V_2^2/2g) + h_L = 30.7V_2^2/2g = 30.7(2.17)^2/64.4 = 2.24$ ft.

TURBULENT FLOW IN PIPE—APPLICATION OF DARCY-WEISBACH FORMULA

Water is pumped at the rate of 3 cfs through an 8-in. fairly smooth pipe 2,600 ft long to a reservoir where the water surface is 180 ft higher than the pump. Determine the gage pressure at the pump discharge.

Calculation Procedure:

1. Compute h_F

Turbulent flow in a pipe flowing full may be investigated by applying the Darcy-Weisbach formula for friction head: $h_F = fLV^2/2gD$, Eq. (92), where f is a friction factor. However, since the friction head does not vary precisely in the manner implied by this equation, f is dependent on D and V, as well as the degree of roughness of the pipe. Values of f associated with a given set of values of the independent quantities may be obtained from Fig. 175.

Accurate equations for h_F are the following: *Extremely smooth pipes*: $h_F = 0.30LV^{1.75}/1,000D^{1.25}$, Eq. (93a). *Fairly smooth pipes*: $h_F = 0.38LV^{1.86}/1,000D^{1.25}$, Eq. (93b). *Rough*

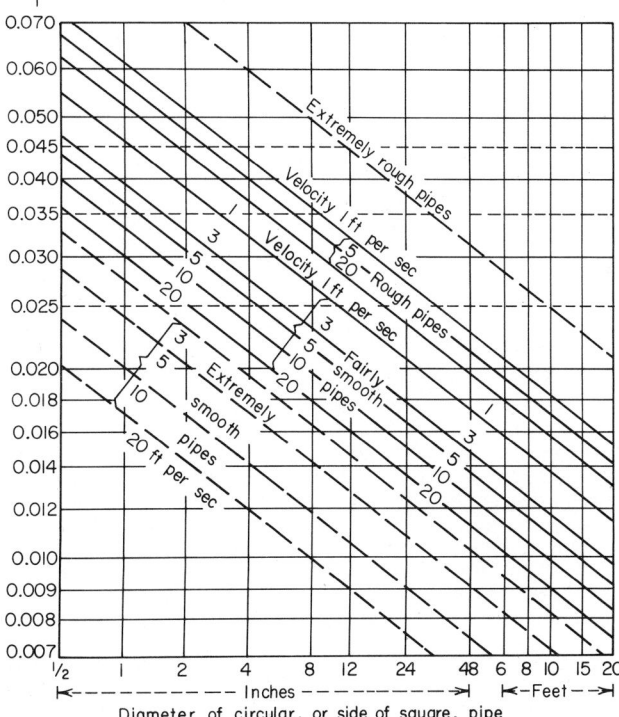

Fig. 175   Flow of water in pipes. (*From E. W. Schoder and F. M. Dawson, "Hydraulics," McGraw-Hill Book Company, New York, 1934. By permission of the publishers.*)

pipes: $h_F = 0.50LV^{1.95}/1,000D^{1.25}$, Eq. (93c). *Extremely rough pipes*: $h_F = 0.69LV^2/1,000D^{1.25}$, Eq. (93d).

Using Eq. (93b), $V = Q/A = 3/0.349 = 8.60$ fps; $h_F = 0.38(2.6)(8.60)^{1.86}/0.667^{1.25} = 89.7$ ft.

2. Alternatively, determine h_F using Eq. (92)

First obtain the appropriate f value from Fig. 175, or $f = 0.020$ for this pipe. Then, $h_F = 0.020(2,600/0.667)(8.60^2/64.4) = 89.6$ ft.

3. Compute the pressure at the pump discharge

Use Eq. (81). Since $L/D > 500$, ignore the secondary items. Then, $p_1/w = z_2 + h_F = 180 + 89.6 = 269.6$ ft; $p_1 = 269.6(62.4)/144 = 117$ psi.

DETERMINATION OF THE FLOW IN A PIPE

Two reservoirs are connected by a 7,000-ft fairly smooth cast-iron pipe 10 in. in diameter. The difference in elevation of the water surfaces is 90 ft. Compute the discharge to the lower reservoir.

Calculation Procedure:

1. Determine the fluid velocity and flow rate

Since the secondary items are negligible, the entire head loss of 90 ft results from friction. Using Eq. (93b) and solving for V, $90 = 0.38(7)V^{1.86}/0.833^{1.25}$; $V = 5.87$ fps. Then, $Q = VA = 5.87(0.545) = 3.20$ cfs.

2. Alternatively, assume a value of f and compute V

Referring to Fig. 175, select a value for f. Then compute V by applying Eq. (92). Then compare the value of f corresponding to this result with the assumed value of f. If the two values differ appreciably, assume a new value of f and repeat the computation. Continue this process until the assumed and actual values of f agree closely.

PIPE-SIZE SELECTION USING THE MANNING FORMULA

A cast-iron pipe is to convey water at 3.3 cfs on a grade of 0.001. Applying the Manning formula with $n = 0.013$, determine the required size of pipe.

Calculation Procedure:

1. Compute the pipe diameter

The Manning formula, which is suitable for both open and closed conduits, is: $V = 1.486R^{2/3}s^{1/2}/n$, Eq. (94), where n = roughness coefficient; R = hydraulic radius = ratio of cross-sectional area of pipe to the wetted perimeter of the pipe; s = hydraulic gradient = dH/dL. If the flow is uniform, i.e., the area and therefore the velocity are constant along the stream, the loss of head equals the drop in elevation, and the grade of the conduit is s.

For a circular pipe flowing full, Eq. (94) becomes $D = (2.159Qn/s^{1/2})^{3/8}$, Eq. (94a). Substituting numerical values, $D = (2.159 \times 3.3 \times 0.013/0.001^{1/2})^{3/8} = 1.50$ ft. Therefore, use an 18-in.-diameter pipe.

LOSS OF HEAD CAUSED BY SUDDEN ENLARGEMENT OF PIPE

Water flows through a pipe at 4 cfs. Compute the loss of head resulting from a change in pipe size if (a) the pipe diameter increases abruptly from 6 to 10 in.; (b) the pipe diameter increases abruptly from 6 to 8 in. at one section and then from 8 to 10 in. at a section farther downstream.

Calculation Procedure:

1. Evaluate the pressure-head differential required to decelerate the liquid

Where there is an abrupt increase in pipe size, the liquid must be decelerated upon entering the larger pipe, since the fluid velocity varies inversely with area. Let the subscript 1 refer to a section immediately downstream of the enlargement, where the higher velocity prevails, and let the subscript 2 refer to a section farther downstream, where deceleration has been completed. Disregard the frictional loss.

Using Eq. (83), $p_2/w = p_1/w + (V_1V_2 - V_2^2)/g$.

2. Combine the result of Step 1 with Eq. (81)

The result is Borda's formula for the head loss h_E caused by sudden enlargement of the pipe cross section: $h_E = (V_1 - V_2)^2/2g$, Eq. (95).

As this investigation shows, only part of the drop in velocity head is accounted for by a gain in pressure head. The remaining head h_E is dissipated through the formation of eddy currents at the entrance to the larger pipe.

3. Compute the velocity in each pipe

Thus:

Pipe diam, in.	Pipe area, sq ft	Fluid velocity, fps
6	0.196	20.4
8	0.349	11.5
10	0.545	7.3

4. Find the head loss for part a

Thus, $h_E = (20.4 - 7.3)^2/64.4 = 2.66$ ft.

5. Find the head loss for part b

Thus, $h_E = [(20.4 - 11.5)^2 + (11.5 - 7.3)^2]/64.4 = 1.50$ ft. Comparison of these results indicates that the eddy-current loss is attenuated if the increase in pipe size occurs in steps.

DISCHARGE OF LOOPING PIPES

A pipe carrying 12.5 cfs of water branches into three pipes of the following diameters and lengths: $D_1 = 6$ in.; $L_1 = 1,000$ ft; $D_2 = 8$ in.; $L_2 = 1,300$ ft; $D_3 = 10$ in.; $L_3 = 1,200$ ft. These pipes rejoin at their downstream ends. Compute the discharge in the three pipes, considering each as fairly smooth.

Calculation Procedure:

1. Express Q as a function of D and L

Since all fluid particles have the same energy at the juncture point, irrespective of the loops they traversed, it follows that the head losses in the three loops are equal. The flow thus divides itself in a manner that produces equal values of h_F in the loops.

Transforming Eq. (93b), $Q = kD^{2.67}/L^{0.538}$, Eq. (96), where k is a constant.

2. Establish the relative values of the discharges; then determine the actual values

Thus, $Q_2/Q_1 = (8/6)^{2.67}/1.3^{0.538} = 1.87$; $Q_3/Q_1 = (10/6)^{2.67}/1.2^{0.538} = 3.55$. Then, $Q_1 + Q_2 + Q_3 = Q_1(1 + 1.87 + 3.55) = 12.5$ cfs. Solving, $Q_1 = 1.95$ cfs; $Q_2 = 3.64$ cfs; $Q_3 = 6.91$ cfs.

FLUID FLOW IN BRANCHING PIPES

The pipes AM, MB, and MC in Fig. 176 have the diameters and lengths indicated. Compute the water flow in each pipe if the pipes are considered rough.

Calculation Procedure:

1. Write the basic equations governing the discharges

Let the subscripts 1, 2 and 3 refer to AM, MB, and MC, respectively. Then, $h_{F1} + h_{F2} = 110$; $h_{F1} + h_{F3} = 150$, Eq. (a); $Q_1 = Q_2 + Q_3$, Eq. (b).

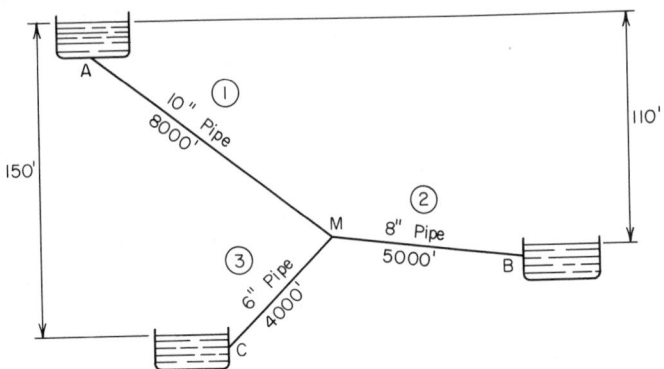

Fig. 176   Branching pipes.

2. Transform Eq. (93c)

The transformed equation is $Q = 38.7D^{2.64}(h_F/L)^{0.513}$, Eq. (97).

3. Assume a trial value for h_{F1} and find the discharge. Test the result

Use Eqs. (a) and (97) to find the discharges. Test the results for compliance with Eq. (b). Assuming $h_{F1} = 70$ ft, then $h_{F2} = 40$ ft and $h_{F3} = 80$ ft; $Q_1 = 38.7(0.833)^{2.64}(70/8,000)^{0.513} = 2.10$ cfs. Similarly, $Q_2 = 1.12$ cfs and $Q_3 = 0.83$ cfs; $Q_2 + Q_3 = 1.95 < Q_1$. The assumed value of h_{F1} is excessive.

4. Make another assumption for h_{F1} and the corresponding revisions

Assume $h_{F1} = 66$ ft. Then, $Q_1 = 2.10(66/70)^{0.513} = 2.04$ cfs. Similarly, $Q_2 = 1.18$ cfs; $Q_3 = 0.85$ cfs; $Q_2 + Q_3 = 2.03$ cfs. These results may be accepted as sufficiently precise.

UNIFORM FLOW IN OPEN CHANNEL—DETERMINATION OF SLOPE

It is necessary to convey 1,200 cfs of water from a dam to a power plant in a canal of rectangular cross section, 24 ft wide and 10 ft deep, having a roughness coefficient of 0.016. The canal is to flow full. Compute the required slope of the canal in ft/mile.

Calculation Procedure:

1. Apply Eq. (94)

Thus, $A = 24(10) = 240$ sq ft; wetted perimeter $= WP = 24 + 2(10) = 44$ ft; $R = 240/44 = 5.45$ ft; $V = 1,200/240 = 5$ fps; $s = (nV/1.486R^{2/3})^2 = [0.016 \times 5/(1.486 \times 5.45^{2/3})]^2 = 0.000302$; slope $= 0.000302(5,280$ ft/mile$) = 1.59$ ft/mile.

REQUIRED DEPTH OF CANAL FOR SPECIFIED FLUID FLOW RATE

A trapezoidal canal is to carry water at 800 cfs. The grade of the canal is 0.0004; the bottom width is 25 ft; the slope of the sides is $1\frac{1}{2}$ horizontal to 1 vertical; the roughness coefficient is 0.014. Compute the required depth of the canal, to the nearest tenth of a foot.

Calculation Procedure:

1. Transform Eq. (94) and compute $AR^{2/3}$

Thus, $AR^{2/3} = nQ/1.486s^{1/2}$, Eq. (94b). Or, $AR^{2/3} = 0.014(800)/[1.486(0.0004)^{1/2}] = 377$.

2. Express the area and wetted perimeter in terms of D, Fig. 177

Side of canal = $D(1^2 + 1.5^2)^{0.5} = 1.80D$. $A = D(25 + 1.5D)$; WP = $25 + 3.60D$.

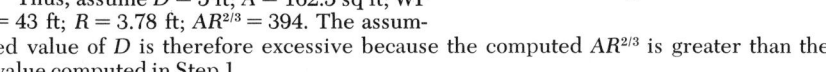

Fig. 177

3. Assume trial values of D until Eq. (94b) is satisfied

Thus, assume $D = 5$ ft; $A = 162.5$ sq ft; WP = 43 ft; $R = 3.78$ ft; $AR^{2/3} = 394$. The assumed value of D is therefore excessive because the computed $AR^{2/3}$ is greater than the value computed in Step 1.

Next, assume a lower value for D, or $D = 4.9$ ft; $A = 158.5$ sq ft; WP = 42.64 ft; $R = 3.72$ ft; $AR^{2/3} = 381$. This is acceptable. Therefore, $D = 4.9$ ft.

ALTERNATE STAGES OF FLOW; CRITICAL DEPTH

A rectangular channel 20 ft wide is to discharge 500 cfs of water having a specific energy of 4.5 ft-lb/lb. (a) Using $n = 0.013$, compute the required slope of the channel. (b) Compute the maximum potential discharge associated with the specific energy of 4.5 ft-lb/lb. (c) Compute the minimum specific energy required to maintain a flow of 500 cfs.

Calculation Procedure:

1. Evaluate the specific energy of an elemental mass of liquid at a distance z above the channel bottom

To analyze the discharge conditions at a given section in a channel, it is advantageous to evaluate the specific energy (or head) by taking the elevation of the bottom of the channel *at the given section* as datum. Assume a uniform velocity across the section, and let D = depth of flow, ft; H_e = specific energy as computed in the prescribed manner; Q_u = discharge through a unit width of channel, cfs/ft.

Evaluating the specific energy of an elemental mass of liquid at a given distance z above the channel bottom, $H_e = Q_u^2/2gD^2 + D$, Eq. (98). Thus, H_e is constant across the entire section. Moreover, if the flow is uniform, as it is in the present instance, H_e is constant along the entire stream.

2. Apply the given values and solve for D

Thus, $H_e = 4.5$ ft-lb/lb; $Q_u = 500/20 = 25$ cfs/ft. Rearrange Eq. (98) to obtain $D^2(H_e - D) = Q_u^2/2g$, Eq. (98a). Or, $D^2(4.5 - D) = 25^2/64.4 = 9.705$. This cubic equation has two positive roots, $D = 1.95$ ft and $D = 3.84$ ft. There are therefore two stages of flow that accommodate the required discharge with the given energy. (The third root of the equation is $D = -1.29$ ft, an impossible condition.)

3. Compute the slope associated with the computed depths

Using Eq. (94), at the lower stage: $D = 1.95$ ft; $A = 20(1.95) = 39.0$ sq ft; WP = $20 + 2(1.95) = 23.9$ ft; $R = 39.0/23.9 = 1.63$ ft; $V = 25/1.95 = 12.8$ fps; $s = (nV/1.486R^{2/3})^2 = (0.013 \times 12.8/1.486 \times 1.63^{2/3})^2 = 0.00654$.

At the upper stage: $D = 3.84$ ft; $A = 20(3.84) = 76.8$ sq ft; WP = $20 + 2(3.84) = 27.68$ ft; $R = 76.8/27.68 = 2.77$ ft; $V = 25/3.84 = 6.51$ fps; $s = [0.013 \times 6.51/(1.486 \times 2.77^{2/3})]^2 = 0.000834$. This constitutes the solution to part a.

4. Plot the D-Q_u curve

For part b, consider H_e as remaining constant at 4.5 ft-lb/lb while Q_u varies. Plot the D-Q_u curve as shown in Fig. 178a. The depth that provides the maximum potential discharge is called the *critical depth* with respect to the given specific energy.

5. Differentiate Eq. (98) to find the critical depth; then evaluate $Q_{u,max}$

Differentiating Eq. (98) and setting $dQ_u/dD = 0$, the critical depth $D_c = \frac{2}{3}(H_e)$, Eq. (99). Or $D_c = \frac{2}{3}(4.5) = 3.0$ ft; $Q_{u,max} = [64.4(4.5 \times 3.0^2 - 3.0^3)]^{0.5} = 29.5$ cfs/ft; $Q_{max} = 29.5(20) = 590$ cfs. This constitutes the solution to part b.

6. Plot the D-H_e curve

For part c, consider Q_u as remaining constant at 25 cfs/ft while H_c varies. Plot the D-H_e curve as shown in Fig. 178b. (This curve is asymptotic with the straight lines $D = H_e$ and $D = 0$.) The depth at which the specific energy is minimum is called the *critical depth* with respect to the given unit discharge.

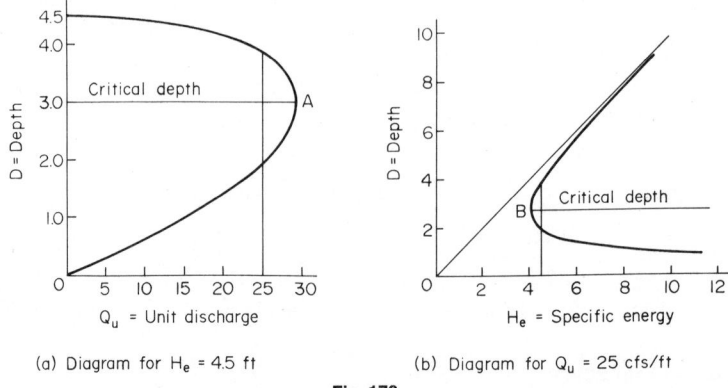

(a) Diagram for H_e = 4.5 ft   (b) Diagram for Q_u = 25 cfs/ft

Fig. 178

7. Differentiate Eq. (98) to find the critical depth; then evaluate H_{e,min}

Differentiating, $D_c = (Q_u^2/g)^{1/3}$. Or $D_c = (25^2/32.2)^{1/3} = 2.69$ ft, Eq. (100); then $H_{e,min} = 25^2/[64.4(2.69)^2] + 2.69 = 4.03$ ft-lb/lb.

The values of D as computed in part a coincide with those obtained by referring to the two graphs in Fig. 178. The equations derived in this Procedure are valid solely for rectangular channels, but analogous equations pertaining to other channel profiles may be derived in a similar manner.

DETERMINATION OF HYDRAULIC JUMP

Water flows over a 100-ft-long dam at 7,500 cfs. The depth of tailwater on the level apron is 9 ft. Determine the depth of flow immediately upstream of the hydraulic jump.

Calculation Procedure:

1. Find the difference in hydrostatic forces per unit width of channel required to decelerate the liquid

Referring to Fig. 179, *hydraulic jump* designates an abrupt transition from lower-stage to upper-stage flow caused by a sharp decrease in slope, sudden increase in roughness, encroachment of backwater, or some other factor. The deceleration of liquid requires an increase in hydrostatic pressure, but only part of the drop in velocity head is accounted for by a gain in pressure head. The excess head is dissipated in the formation of a turbulent standing wave. Thus, the phenomenon of hydraulic jump resembles

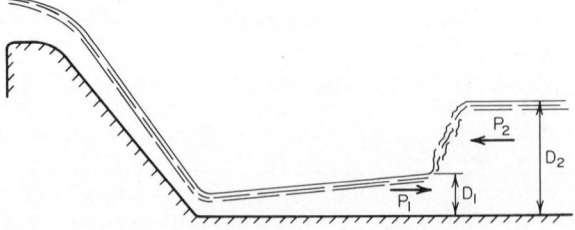

Fig. 179   Hydraulic jump on apron of dam.

the behavior of liquid in a pipe at a sudden enlargement, as analyzed in an earlier Calculation Procedure.

Let D_1 and D_2 denote the depth of flow immediately upstream and downstream of the jump, respectively. Then $D_1 < D_c < D_2$. Refer to Fig. 178b. Since the hydraulic jump is accompanied by a considerable drop in energy, the point on the D-H_e diagram that represents D_2 lies both above and to the left of that representing D_1. Therefore, the upstream depth is less than the depth that would exist in the absence of any loss.

Using literal values, apply Eq. (83) to find the difference in hydrostatic forces per unit width of channel that is required to decelerate the liquid. Solve the resulting equation for D_1. Or, $D_1 = -D_2/2 + (2V_2{}^2D_2/g + D_2{}^2/4)^{0.5}$, Eq. (101).

2. Substitute numerical values in Eq. (101)

Thus, $Q_u = 7,500/100 = 75$ cfs/ft; $V_2 = 75/9 = 8.33$ fps; $D_1 = -\tfrac{9}{2} + (2 \times 8.33^2 \times 9/32.2 + 9^2/4)^{0.5} = 3.18$ ft.

RATE OF CHANGE OF DEPTH IN NONUNIFORM FLOW

The unit discharge in a rectangular channel is 28 cfs/ft. The energy gradient is 0.0004, and the grade of the channel bed is 0.0010. Determine the rate at which the depth of flow is changing in the downstream direction (i.e., the grade of the liquid surface with respect to the channel bed) at a section where the depth is 3.2 ft.

Calculation Procedure:

1. Express H as a function of D

Let H = total specific energy at a given section as evaluated by selecting a fixed horizontal reference plane; L = distance measured in downstream direction; z = elevation of given section with respect to datum plane; s_b = grade of channel bed = $-dz/dL$; s_e = energy gradient = $-dH/dL$.

Express H as a function of D by annexing the potential-energy term to Eq. (98). Thus, $H = Q_u{}^2/2gD^2 + D + z$, Eq. (102).

2. Differentiate this equation with respect to L to obtain the rate of change of D; substitute numerical values

Differentiating, $dD/dL = (s_b - s_e)/(1 - Q_u{}^2/gD^3)$, Eq. (103a); or in accordance with Eq. (100), $dD/dL = (s_b - s_e)/(1 - D_c{}^3/D^3)$, Eq. (103b). Substituting, $Q_u{}^2/gD^3 = 28^2/(32.2 \times 3.2^3) = 0.743$; $dD/dL = (0.0010 - 0.0004)/(1 - 0.743) = 0.00233$ ft/ft. The depth is increasing in the downstream direction, and the water is therefore being decelerated.

As Eq. (103b) reveals, the relationship between the actual depth at a given section and the critical depth serves as a criterion in ascertaining whether the depth is increasing or decreasing.

DISCHARGE BETWEEN COMMUNICATING VESSELS

In Fig. 180, liquid is flowing from tank A to tank B through an orifice near the bottom. The area of the liquid surface is 200 sq ft in A and 150 sq ft in B. Initially, the difference in water levels is 14 ft and the discharge is 2 cfs. Assuming that the discharge coefficient remains constant, compute the length of time required for the water level in tank A to drop 1.8 ft.

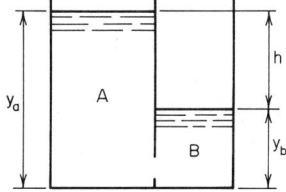

Fig. 180

Calculation Procedure:

1. By expressing the change in h during an elemental time interval, develop the time-interval equation

Let A_a and A_b denote the area of the liquid surface in tanks A and B, respectively; let the subscripts 1 and 2 refer to the beginning and end, respectively, of a time interval t. Then, $t = 2A_aA_b(h_1 - [h_1h_2]^{0.5})/[Q_1(A_a + A_b)]$, Eq. (104).

2. Find the value of h when y_a diminishes by 1.8 ft

Thus, $\Delta y_b = (-A_a/A_b)(\Delta y_a) = -(200/150)(-1.8) = 2.4$ ft; $\Delta h = \Delta y_a - \Delta y_b = -1.8 - 2.4$ $= -4.2$ ft; $h_1 = 14$ ft; $h_2 = 14 - 4.2 = 9.8$ ft.

3. Substitute numerical values in Eq. (104) and solve for t

Thus, $t = 2(200)(150)(14 - [14 \times 9.8]^{0.5})/[2(200 + 150)] = 196$ sec $= 3.27$ min.

VARIATION IN HEAD ON A WEIR WITHOUT INFLOW TO THE RESERVOIR

Water flows over a weir of 60-ft length from a reservoir having a surface area of 50 acres. If the inflow to the reservoir ceases when the head on the weir is 2 ft, what will the head be at the expiration of 1 hr? Consider that the instantaneous discharge is given by Eq. (89a).

Calculation Procedure:

1. Develop the time-interval equation

Let A = surface area of reservoir; C = numerical constant in discharge equation; subscripts 1 and 2 refer to the beginning and end, respectively, of a time interval t. By expressing the change in head during an elemental time interval, $t = 2A/Cb(1/h_2^{0.5} - 1/h_1^{0.5})$, Eq. (105).

2. Substitute numerical values in Eq. (105); solve for h_2

Thus, $A = 50(43,560) = 2,178,000$ sq ft; $t = 3,600$ sec; solving, $h_2 = 1.32$ ft.

VARIATION IN HEAD ON A WEIR WITH INFLOW TO THE RESERVOIR

Water flows over an 80-ft-long weir from a reservoir having a surface area of 6,000,000 sq ft while the rate of inflow to the reservoir remains constant at 2,175 cfs. How long will it take for the head on the weir to increase from zero to 95 percent of its maximum value? Consider that the instantaneous rate of flow over the weir is $3.4bh^{1.5}$.

Calculation Procedure:

1. Compute the maximum head on the weir by equating outflow to inflow

The water in the reservoir reaches its maximum height when equilibrium is achieved; i.e., when the rate of outflow equals the rate of inflow. Let Q_i = rate of inflow; Q_o = rate of outflow at a given instant; t = time elapsed since the start of the outflow.

Equating outflow to inflow, $3.4(80h_{max}^{1.5}) = 2,175$; $h_{max} = 4.0$ ft; $0.95h_{max} = 3.8$ ft.

2. Using literal values, determine the time interval dt during which the water level rises a distance dh

Thus, with C = numerical constant in the discharge equation, $dt = [A/(Q_i - Cbh^{1.5})]$ $\times dh$, Eq. (106). The right side of this equation is not amenable to direct integration. Consequently, the only feasible way of computing the time is to perform an approximate integration.

3. Obtain the approximate value of the required time

Select suitable increments of h, calculate the corresponding increments of t, and total the latter to obtain an approximate value of the required time. In calculating Q_o, apply the mean value of h associated with each increment.

The precision inherent in the result thus obtained depends on the judgement used in selecting the increments of h, and a clear visualization of the relationship between h and t is therefore essential. Let $m = dt/dh = A/(Q_i - Cbh^{1.5})$. The m-h curve is shown in Fig. 181a. Then, $t = \int m \, dh$ = area between the m-h curve and h axis.

This area is approximated by summing the areas of the rectangles as indicated in Fig. 181, the length of each rectangle being equal to the value of m at the center of the interval. Note that as h increases, the increments Δh should be made progressively smaller to minimize the error introduced in the procedure.

Select the increments shown in Table 12, and perform the indicated calculations. The symbols h_b and h_m denote the values of h at the beginning and center, respectively, of an interval. The following calculations for the third interval serve as an illustration of

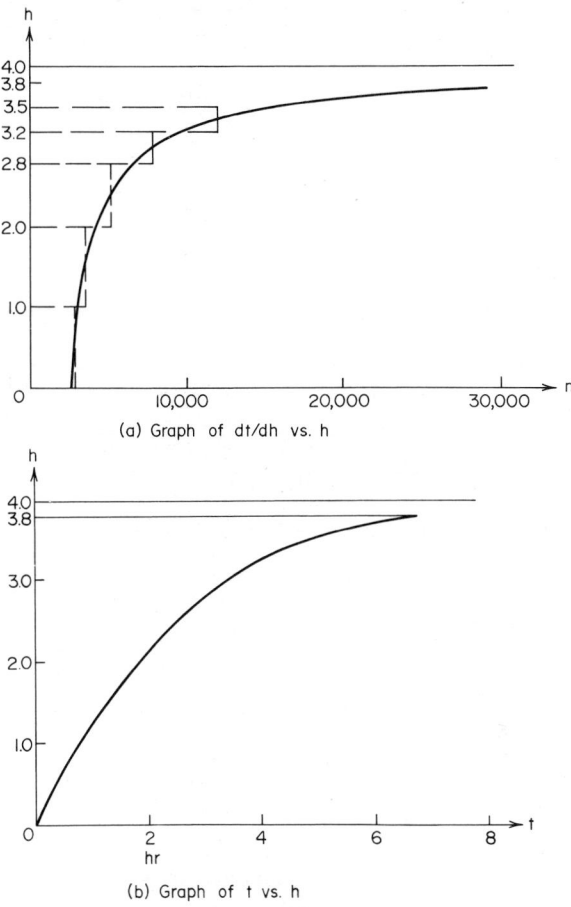

(a) Graph of dt/dh vs. h

(b) Graph of t vs. h

Fig. 181

TABLE 12 Approximate Integration of Eq. (106)

Δh, ft	h_b, ft	h_m, ft	m, sec/ft	Δt, sec
1.0	0	0.5	2,890	2,890
1.0	1.0	1.5	3,580	3,580
0.8	2.0	2.4	5,160	4,130
0.4	2.8	3.0	7,870	3,150
0.3	3.2	3.35	11,830	3,550
0.2	3.5	3.6	18,930	3,790
0.1	3.7	3.75	30,000	3,000
Total	...	...	...	24,090

the method: $h_m = \frac{1}{2}(2.0 + 2.8) = 2.4$ ft; $m = 6,000,000/(2,175 - 3.4 \times 80 \times 2.4^{1.5}) = 5,160$ sec/ft; $\Delta t = m \, \Delta h = 5,160(0.8) = 4,130$ sec. From Table 12, the required time is $t = 24,090$ sec $= 6$ hr 41.5 min.

The t-h curve is shown in Fig. 181b. The time required for the water to reach its maximum height is difficult to evaluate with precision because m becomes infinitely large as h approaches h_{max}; that is, the water level rises at an imperceptible rate as it nears its limiting position.

DIMENSIONAL-ANALYSIS METHODS

The velocity of a raindrop in still air is known or assumed to be a function of these quantities: gravitational acceleration, drop diameter, dynamic viscosity of the air, and the density of both the water and the air. Develop the dimensionless parameters associated with this phenomenon.

Calculation Procedure:

1. Using a generalized notational system, record the units in which the six quantities of this situation are expressed

Dimensional analysis is an important tool both in theoretical investigations and in experimental work because it clarifies the relationships intrinsic in a given situation.

A quantity that appears in every dimensionless parameter is termed *repeating*; a quantity that appears in only one parameter is termed *nonrepeating*. Since the engineer is usually more accustomed to dealing with units of force rather than of mass, the force-length-time system of units will be applied here. Let F, L, and T denote a unit of force, length, and time, respectively.

Using this generalized notational system, it is convenient to write the appropriate British units and then replace these with the general units. For example, with respect to acceleration: British units, ft/sec^2; general units, L/T^2 or LT^{-2}. Similarly, with respect to density (w/g): British units, (lb/ft^3)/(ft/sec^2); general units, FL^{-3}/LT^{-2} or $FL^{-4}T^2$.

The results are shown in the following table.

Quantity	Units
V = velocity of raindrop	LT^{-1}
g = gravitational acceleration	LT^{-2}
D = diameter of drop	L
μ_a = air viscosity	$FL^{-2}T$
ρ_w = water density	$FL^{-4}T^2$
ρ_a = air density	$FL^{-4}T^2$

2. Compute the number of dimensionless parameters present

This phenomenon contains six physical quantities and three units. Therefore, as a consequence of Buckingham's pi theorem, the number of dimensionless parameters is $6 - 3 = 3$.

3. Select the repeating quantities

The number of repeating quantities must equal the number of units (three here). These quantities should be independent and they should collectively contain all the units present. The quantities g, D, and μ_a satisfy both requirements and are therefore selected as the repeating quantities.

4. Select the dependent variable V as the first nonrepeating quantity

Then write $\pi_1 = g^x D^y \mu_a^z V$, Eq. ($a$), where π_1 is a dimensionless parameter and x, y, and z are unknown exponents that may be evaluated by experiment.

5. Transform Eq. (a) to a dimensional equation

Do this by replacing each quantity with the units in which it is expressed. Then perform the necessary expansions and multiplications. Or, $F^0 L^0 T^0 = (LT^{-2})^x L^y L^z$ $(FL^{-2}T)^z LT^{-1}, F^0 L^0 T^0 = F^z L^{x+y-2z+1} T^{-2x+z-1}$, Eq. ($b$).

Every equation must be dimensionally homogeneous; i.e., the units on one side of the equation must be consistent with those on the other side. Therefore, the exponent of a unit on one side of Eq. (b) must equal the exponent of that unit on the other side.

6. Evaluate the exponents x, y, and z

Do this by applying the principle of dimensional homogeneity to Eq. (b). Thus, $0 = z$; $0 = x + y - 2z + 1$; $0 = -2x + z - 1$. Solving these simultaneous equations yields: $x = -\frac{1}{2}$; $y = -\frac{1}{2}$; $z = 0$.

7. Substitute these values in Eq. (a)

Thus, $\pi_1 = g^{-1/2}D^{-1/2}V$, or $\pi_1 = V/(gD)^{1/2}$

8. Follow the same procedure for the remaining nonrepeating quantities

Select ρ_w and ρ_a in turn as the nonrepeating quantities. Follow the same procedure as before to obtain the following dimensionless parameters: $\pi_2 = \rho_w(gD^3)^{1/2}/\mu_a$, and $\pi_3 = \rho_a(gD^3)^{1/2}/\mu_a = (gD^3)^{1/2}/\nu_a$, where $\nu_a = $ kinematic viscosity of air.

HYDRAULIC SIMILARITY AND CONSTRUCTION OF MODELS

A dam discharges 36,000 cfs of water, and a hydraulic jump occurs on the apron. The power loss resulting from this jump is to be determined by constructing a geometrically similar model having a scale of 1 : 12. (a) Determine the required discharge in the model. (b) Determine the power loss on the dam if the power loss on the model is found to be 0.18 hp.

Calculation Procedure:

1. Determine the value of Q_m

Two systems are termed similar if their corresponding variables have a constant ratio. A hydraulic model and its prototype must possess three forms of similarity: geometric, or similarity of shape; kinematic, or similarity of motion; and dynamic, or similarity of forces.

In the present instance, the ratio associated with the geometric similarity is given; i.e., the ratio of a linear dimension in the model to the corresponding linear dimension in the prototype. Let r_g denote this ratio, and let the subscripts m and p refer to the model and prototype, respectively.

Apply Eq. (89a) to evaluate Q_m. Or $Q = C_1 bh^{1.5}$, where C_1 is a constant. Then, $Q_m/Q_p = (b_m/b_p)(h_m/h_p)^{1.5}$. But, $b_m/b_p = h_m/h_p = r_g$; therefore, $Q_m/Q_p = r_g^{2.5} = (\frac{1}{12})^{2.5} = 1/499$; $Q_m = 36{,}000/499 = 72$ cfs.

2. Evaluate the power loss on the dam

Apply Eq. (88) to evaluate the power loss on the dam. Thus, hp $= C_2 Qh$, where C_2 is a constant. Then, hp$_p$/hp$_m = (Q_p/Q_m)(h_p/h_m)$. But $Q_p/Q_m = (1/r_g)^{2.5}$, and $h_p/h_m = 1/r_g$; therefore, hp$_p$/hp$_m = (1/r_g)^{3.5} = 12^{3.5} = 5{,}990$. Hence, hp$_p = 5{,}990(0.18) = 1{,}078$ hp.

Surveying and Route Design

PLOTTING A CLOSED TRAVERSE

Complete the following table for a closed traverse.

Course	Bearing	Length, ft
a	N32°27'E	110.8
b		83.6
c	S8°51'W	126.9
d	S73°31'W	
e	N18°44'W	90.2

Calculation Procedure:

1. Draw the known courses; then form a closed traverse

Refer to Fig. 182a. A line PQ is described by expressing its length L and its bearing α with respect to a reference meridian NS. For a closed traverse, such as $abcde$ in Fig. 182b, the algebraic sum of the latitudes and the algebraic sum of the departures must

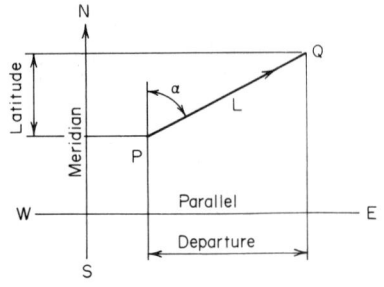

(a) Latitude and departure

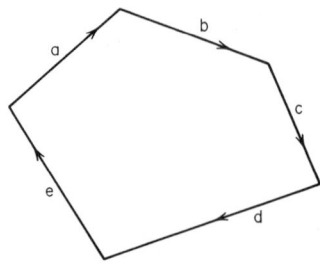

(b) Closure of traverse

Fig. 182

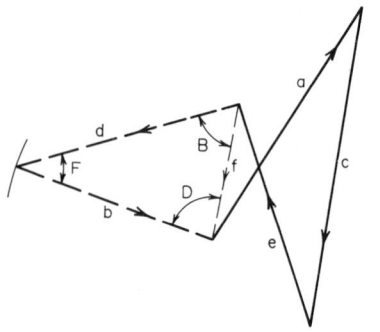

Fig. 183

equal zero. A positive latitude corresponds to a northerly bearing, and a positive departure corresponds to an easterly bearing.

In Fig. 183, draw the known courses a, c, and e. Then introduce the hypothetical course f to form a closed traverse.

2. Calculate the latitude and departure of the courses

Use these relations: latitude = $L \cos \alpha$, Eq. (107); departure = $L \sin \alpha$, Eq. (108). Computing the results for courses a, c, e, and f, we have the values shown in the following table.

Course	Latitude, ft	Departure, ft
a	+93.5	+59.5
c	−125.4	−19.5
e	+85.4	−29.0
Total . . .	+53.5	+11.0
f	−53.5	−11.0

3. Find the length and bearing of f

Thus, $\tan \alpha_f = 11.0/53.5$; therefore, the bearing of $f = S11°37'W$; length of $f = 53.5/\cos \alpha_f = 54.6$ ft.

4. Complete the layout

Complete Fig. 183 by drawing line d through the upper end of f with the specified bearing and by drawing a circular arc centered at the lower end of f having a radius equal to the length of b.

5. Find the length of d and the bearing of b

Solve the triangle fdb to find the length of d and the bearing of b. Thus, $B = 73°31' - 11°37' = 61°54'$. By the law of sines, $\sin F = f \sin B/b = 54.6 \sin 61°54'/83.6$; $F = 35°11'$; $D = 180° - (61°54' + 35°11') = 82°55'$; $d = b \sin D/\sin B = 83.6 \sin 82°55'/\sin 61°54' = 94.0$ ft; $\alpha_b = 180° - (73°31' + 35°11') = 71°18'$. The bearing of $b = S71°18'E$.

AREA OF TRACT WITH RECTILINEAR BOUNDARIES

The balanced latitudes and departures of a clased transit-and-tape traverse are recorded in the table below. Compute the area of the tract by the *DMD* method.

Course	Latitude, ft	Departure, ft
AB	− 132.3	− 135.6
BC	+ 9.6	− 77.5
CD	+ 97.9	− 198.5
DE	+ 161.9	+ 143.6
EF	− 35.3	+ 246.7
FA	− 101.8	+ 21.3

Calculation Procedure:

1. Plot the tract

Refer to Fig. 184. The sum of m_1 and m_2 is termed the *double meridian distance* (DMD) of course AB. Let D denote the departure of a course. Then, $\text{DMD}_n = \text{DMD}_{n-1} + D_{n-1} + D_n$, Eq. (109), where the subscripts refer to two successive courses.

The area of trapezoid $ABba$, which will be termed the *projection area* of AB, equals half the product of the DMD and latitude of the course. A projection area may be either positive or negative.

Plot the tract in Fig. 185. Since D is the most westerly point, pass the reference meridian through D, thus causing all DMDs to be positive.

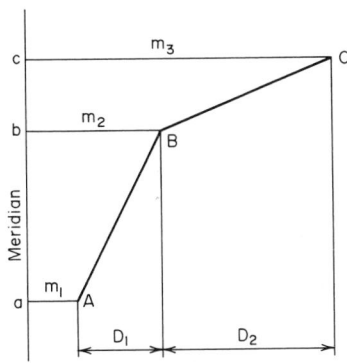

Fig. 184   Double meridian distance.

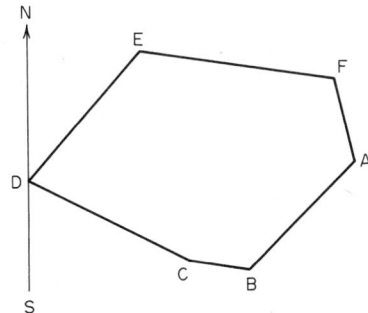

Fig. 185

2. Establish the DMD of each course by successive applications of Eq. (109)

Thus, $\text{DMD}_{DE} = 143.6$ ft; $\text{DMD}_{EF} = 143.6 + 143.6 + 246.7 = 533.9$ ft; $\text{DMD}_{FA} = 533.9 + 246.7 + 21.3 = 801.9$ ft; $\text{DMD}_{AB} = 801.9 + 21.3 - 135.6 = 687.6$ ft; $\text{DMD}_{BC} = 687.6 - 135.6 - 77.5 = 474.5$ ft; $\text{DMD}_{CD} = 474.5 - 77.5 - 198.5 = 198.5$ ft. This is acceptable.

3. Calculate the area of the tract

Use the following theorem: The area of a tract is numerically equal to the aggregate of the projection areas of its courses. The results of this calculation are shown in the following table.

Course	Latitude × DMD = 2 × Projection area		
AB	− 132.3	687.6	− 90,970
BC	+ 9.6	474.5	+ 4,555
CD	+ 97.9	198.5	+ 19.433
DE	+ 161.9	143.6	+ 23,249
EF	− 35.3	533.9	− 18,847
FA	− 101.8	801.9	− 81,634
Total			− 144,214

$$\text{Area} = \tfrac{1}{2}(144{,}214) = 72{,}107 \text{ sq ft.}$$

PARTITION OF A TRACT

The tract in the previous Calculation Procedure is to be divided into two parts by a line through E, the part to the west of this line having an area of 30,700 sq ft. Locate the dividing line.

Calculation Procedure:

1. Ascertain the location of the dividing line EG

This Procedure requires the solution of an oblique triangle. Refer to Fig. 186. It will be necessary to apply the following equations, which may be readily developed by drawing the altitude BD: area $= \tfrac{1}{2}bc \sin A$, Eq. (110); $\tan C = c \sin A / (b - c \cos A)$, Eq. (111). In Fig. 187, let EG represent the dividing line of this tract. By scaling the dimensions and making preliminary calculations or by using a planimeter, ascertain that G lies between B and C.

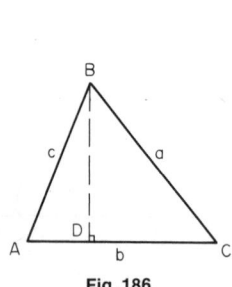

Fig. 186

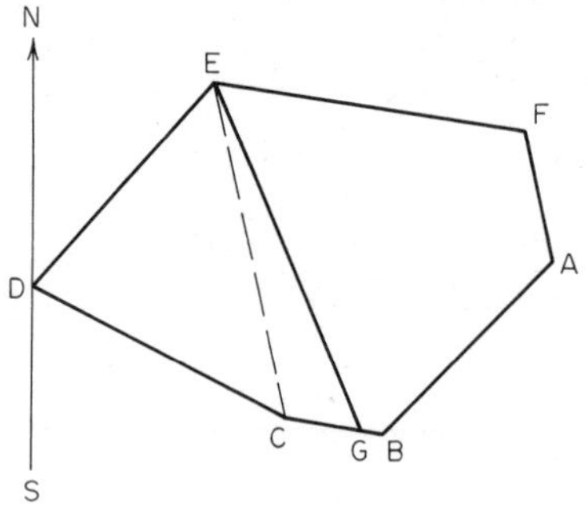

Fig. 187   Partition of tract.

2. Establish the properties of the hypothetical course EC

By balancing the latitudes and departures of DEC, latitude of $EC = -(+161.9 + 97.9) = -259.8$ ft; departure of $EC = -(+143.6 - 198.5) = +54.9$ ft; length of $EC = (259.8^2 + 54.9^2)^{0.5} = 265.5$ ft. Then, $\text{DMD}_{DE} = 143.6$ ft; $\text{DMD}_{EC} = 143.6 + 143.6 + 54.9 = 342.1$ ft; $\text{DMD}_{CD} = 342.1 + 54.9 - 198.5 = 198.5$ ft. This is acceptable.

3. Determine angle GCE by finding the bearings of courses EC and BC

Thus: $\tan \alpha_{EC} = 54.9/259.8$; bearing of $EC = S11°55.9'E$; $\tan \alpha_{BC} = 77.5/9.6$; bearing of $BC = N82°56.3'W$; angle $GCE = 180° - (82°56.3' - 11°55.9') = 108°59.6'$.

4. Determine the area of triangle GCE

Calculate the area of triangle DEC; then find the area of triangle GCE by subtraction. Thus:

Course	Latitude	× DMD	= 2 × Projection area
CD	+97.9	198.5	+19,433
DE	+161.9	143.6	+23,249
EC	−259.8	342.1	−88,878
Total			−46,196

Area of $DEC = \frac{1}{2}(46,196) = 23,098$ sq ft; area of $GCE = 30,700 - 23,098 = 7,602$ sq ft.

5. Solve triangle GCE completely

Apply Eqs. (110) and (111). To ensure correct substitution, identify the corresponding elements, making A the known angle GCE and c the known side EC. Thus:

Fig. 186	Fig. 187	Known values	Calculated values
A	GCE	108°59.6′	
B	CEG	....	11°21.6′
C	EGC	....	59°38.8′
a	EG	....	291.0 ft
b	GC	....	60.6 ft
c	EC	265.5 ft	

By Eq. (110), $7,602 = \frac{1}{2}GC(265.5 \sin 108°59.6′)$; solving, $GC = 60.6$ ft. By Eq. (111), tan $EGC = 265.5 \sin 108°59.6′/(60.6 - 265.5 \cos 108°59.6′)$; $EGC = 59°38.8′$. By the law of sines, $EG/\sin GCE = EC/\sin EGC$; $EG = 291.0$ ft; $CEG = 180° - (108°59.6′ + 59°38.8′) = 11°21.6′$.

6. Find the bearing of course EG

Thus, $\alpha_{EG} = \alpha_{EC} + CEG = 11°55.9′ + 11°21.6′ = 23°17.5′$; bearing of $EG = $ S23°17.5′E.

The surveyor requires the length and bearing of EG to establish this line of demarcation. He is able to check the accuracy of both the fieldwork and the office calculations by ascertaining that the point G established in the field falls on BC and that the measured length of GC agrees with the computed value.

AREA OF TRACT WITH MEANDERING BOUNDARY: OFFSETS AT IRREGULAR INTERVALS

The offsets below were taken from stations on a traverse line to a meandering stream, all data being in feet. What is the encompassed area?

Station..	0+00	0+25	0+60	0+75	1+10
Offset....	29.8	64.6	93.2	58.1	28.5

Calculation Procedure:

1. Assume a rectilinear boundary between successive offsets; develop area equations

Refer to Fig. 188. When a tract has a meandering boundary, this boundary is approximated by measuring the perpendicular offsets of the boundary from a straight

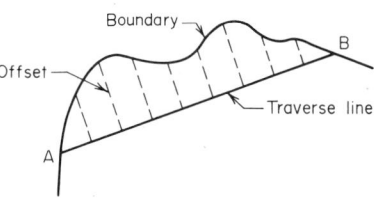

Fig. 188   Tract with irregular boundary.

line AB. Let d_r denote the distance along the traverse line between the first and the rth offset, and let $h_1, h_2, \ldots, h_n$ denote the offsets.

Developing the area equations, area $= \frac{1}{2}[d_2(h_1-h_3)+d_3(h_2-h_4)+\cdots+d_{n-1}(h_{n-2}-h_n)+d_n(h_{n-1}+h_n)]$, Eq. (112). Or, area $= \frac{1}{2}[h_1d_2+h_2d_3+h_3(d_4-d_2)+h_4(d_5-d_3)+\cdots+h_n(d_n-d_{n-1})]$, Eq. (113).

2. Determine the area using Eq. (112)

Thus,   area $= \frac{1}{2}[25(29.8-93.2)+60(64.6-58.1)+75(93.2-28.5)+110(58.1+28.5)]$
$= 6,590$ sq ft.

3. Determine the area using Eq. (113)

Thus,   area $= \frac{1}{2}[29.8 \times 25 + 64.6 \times 60 + 93.2(75-25) + 58.1(110-60) + 28.5(110-75)] = 6,590$ sq ft. Hence, both equations yield the same result. The second equation has a distinct advantage over the first because it has only positive terms.

DIFFERENTIAL LEVELING PROCEDURE

Complete the following level notes, and show an arithmetic check. All data are in ft.

Point	BS	HI	FS	Elevation
BM42	2.076	...	....	180.482
TP1	3.408	...	8.723	
TP2	1.987	...	9.826	
TP3	2.538	...	10.466	
TP4	2.754	...	8.270	
BM43	....	...	11.070	

Calculation Procedure:

1. Obtain the elevation for each point

Differential leveling is used to ascertain the difference in elevation between two successive benchmarks by finding the elevations of several convenient intermediate points, called *turning points* (TP). In Fig. 189, consider that the instrument is set up at $L1$ and C is selected as a turning point. The rod reading AB represents the backsight (BS) of BM1, and rod reading CD represents the foresight (FS) of TP1. The elevation of BD represents the height of instrument (HI). The instrument is then set up at $L2$ and rod readings CE and FG are taken. Let a and b designate two successive turning points. Then, elevation$_a$ + BS$_a$ = HI, Eq. (114); HI − FS$_b$ = elevation$_b$, Eq. (115); therefore, elevation BM2 − elevation BM1 = ΣBS − ΣFS, Eq. (116).

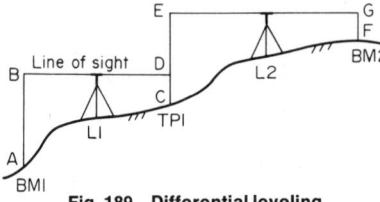

Fig. 189   Differential leveling.

Apply Eqs. (114) and (115) successively to obtain the elevations recorded in the accompanying table.

Point	BS	HI	FS	Elevation
BM42	2.076	182.558	....	180.482
TP1	3.408	177.243	8.723	173.835
TP2	1.987	169.404	9.826	167.417
TP3	2.538	161.476	10.466	158.938
TP4	2.754	155.960	8.270	153.206
BM43	....	....	11.070	144.890
Total ..	12.763	....	48.355	

2. *Verify the result by summing the backsights and foresights*

Substitute the results in Eq. (116). Or, $144.890 - 180.482 = 12.763 - 48.355 = -35.592$.

STADIA SURVEYING

The following stadia readings were taken with the instrument at a station of elevation 483.2 ft, the height of instrument being 5 ft. The stadia interval factor is 100, and the value of C is negligible. Compute the horizontal distances and elevations.

Point	Rod intercept, ft	Vertical angle
1	5.46	$+2°40'$ on 8 ft
2	6.24	$+3°12'$ on 3 ft
3	4.83	$-1°52'$ on 4 ft

Calculation Procedure:

1. State the equations used in stadia surveying

Refer to Fig. 190 for the notational system pertaining to stadia surveying. The transit is set up over a reference point O, the rod is held at a control point N, and the telescope is sighted at a point Q on the rod. P and R represent the apparent locations of the stadia hairs on the rod.

The first column in these notes presents the rod intercept s, and the second column presents the vertical angle α and the distance NQ. Then, $H = Ks \cos^2 \alpha + C \cos \alpha$, Eq. (117); $V = \frac{1}{2}Ks \sin 2\alpha + C \sin \alpha$, Eq. (118). Elevation of $N =$ elevation of $O + OM + V - NQ$, Eq. (119), where $K =$ stadia interval factor; $C =$ distance from center of instrument to principal focus.

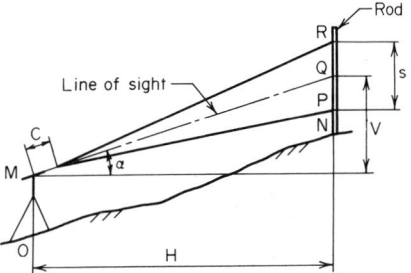

Fig. 190 Stadia surveying.

2. Substitute numerical values in the above equations

Substituting, the results obtained are as shown in the table.

Point	H, ft	V, ft	Elevation, ft
1	544.8	25.4	505.6
2	622.0	34.8	520.0
3	482.5	-15.7	468.5

VOLUME OF EARTHWORK

Figure 191a and b represents two highway cross sections 100 ft apart. Compute the volume of earthwork to be excavated, in cu yd. Apply both the average-end-area method and the prismoidal method.

Calculation Procedure:

1. Resolve each section into an isosceles trapezoid and a triangle; record the relevant dimensions

Let A_1 and A_2 denote the areas of the end sections, L the intervening distance, and V the volume of earthwork to be excavated or filled.

Method 1: The average-end-area method equates the average area to the mean of the two end areas. Then, $V = L(A_1 + A_2)/2$, Eq. (120).

Figure 191c shows the first section resolved into an isosceles trapezoid and a triangle, along with the relevant dimensions.

2. Compute the end areas, and apply Eq. (120)

Thus: $A_1 = [24(40+64) + (32-24)64]/2 = 1,504$ sq ft; $A_2 = [36(40+76) + (40-36)76]/2 = 2,240$ sq ft; $V = 100(1,504+2,240)/[2(27)] = 6,933$ cu yd.

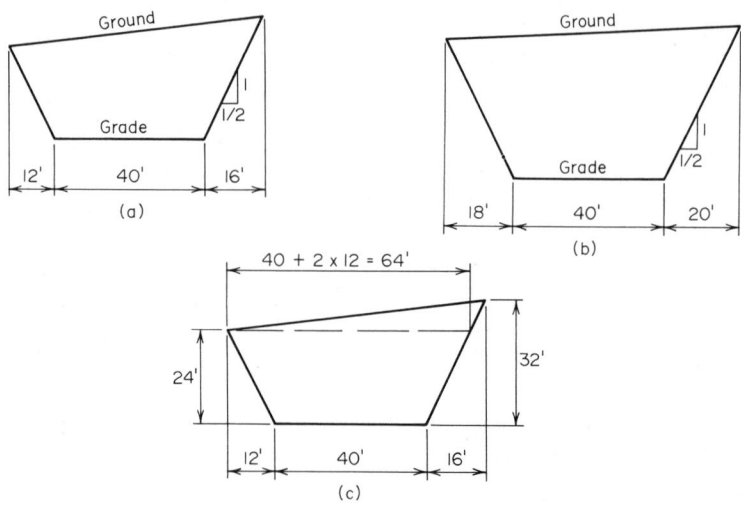

Fig. 191

3. Apply the prismoidal method

Method 2: The prismoidal method postulates that the earthwork between the stations is a prismoid (a polyhedron having its vertices in two parallel planes). The volume of a prismoid is $V = L(A_1 + 4A_m + A_2)/6$, Eq. (121), where A_m = area of center section.

Compute A_m. Note that each coordinate of the center section of a prismoid is the arithmetical mean of the corresponding coordinates of the end sections. Thus, $A_m = [30(40+70) + (36-30)70]/2 = 1,860$ sq ft.

4. Compute the volume of earthwork

Using Eq. (121), $V = 100(1,504 + 4 \times 1,860 + 2,240)/[6(27)] = 6,904$ cu yd.

DETERMINATION OF AZIMUTH OF A STAR BY FIELD ASTRONOMY

An observation of the Sun was made at a latitude of 41°20'N. The altitude of the center of the Sun, after correction for refraction and parallax, was 46°48'. By consulting a solar ephemeris, it was found that the declination of the Sun at the instant of observation was 7°58'N. What was the azimuth of the Sun?

Calculation Procedures:

1. Calculate the azimuth of the body

Refer to Fig. 192. The *celestial sphere* is an imaginary sphere on the surface of which the celestial bodies are assumed to be located; this sphere is of infinite radius and has the earth as its center. The *celestial equator*, or *equinoctial*, is the great circle along which the Earth's equatorial plane intersects the celestial sphere. The *celestial axis* is the prolongation of the Earth's axis of rotation. The *celestial poles* are the points at which the celestial axis pierces the celestial sphere. An *hour circle*, or *meridian*, is a great circle that passes through the celestial poles.

The *zenith* and *nadir* of an observer are the points at which the vertical (plumb) line at the observer's site pierces the celestial sphere, the former being visible and the latter invisible to the observer. A *vertical circle* is a great circle that passes through the observer's zenith and nadir. The *observer's meridian* is the meridian that passes through the observer's zenith and nadir; it is both a meridian and a vertical circle.

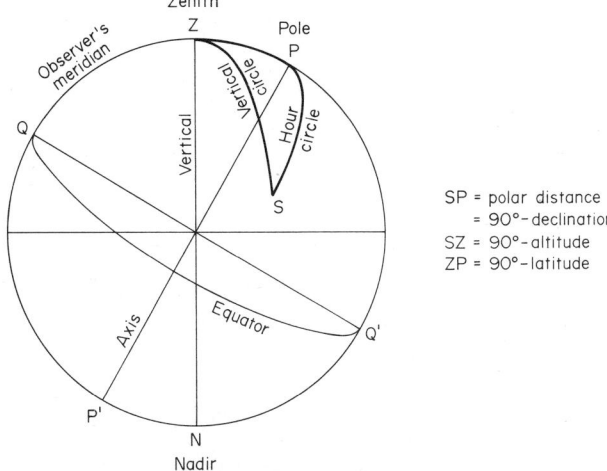

SP = polar distance
   = 90°- declination
SZ = 90°- altitude
ZP = 90°- latitude

Fig. 192   The celestial sphere.

In Fig. 192, P is the celestial pole, S is the apparent position of a star on the celestial sphere, and Z is the observer's zenith.

The coordinates of a celestial body *relative to the observer* are: the *azimuth*, which is the angular distance from the observer's meridian to the vertical circle through the body as measured along the observer's horizon in a clockwise direction; and the *altitude*, which is the angular distance of the body from the observer's horizon as measured along a vertical circle.

The *absolute* coordinates of a celestial body are: the *right ascension*, which is the angular distance between the vernal equinox and the hour circle through the body as measured along the celestial equator; and the *declination*, which is the angular distance of the body from the celestial equator as measured along an hour circle.

The relative coordinates of a body at a given instant are obtained by observation; the absolute coordinates are obtained by consulting an almanac of astronomical data. The latitude of the observer's site equals the angular distance of the observer's zenith from the celestial equator as measured along the meridian. In the astronomical triangle PZS in Fig. 192, the arcs represent the indicated coordinates and angle Z represents the azimuth of the body as measured from the north.

Calculating the azimuth of the body, $\tan^2 \frac{1}{2}Z = \sin (S-L) \sin (S-h)/\cos S \cos (S-p)$, Eq. (122), where L = latitude of site; h = altitude of star; p = polar distance = $90° -$ declination; $S = \frac{1}{2}(L+h+p)$; $L = 41°20'$; $h = 46°48'$; $p = 90° - 7°58' = 82°02'$; $S = \frac{1}{2}(L+h+p) = 85°05'$; $S-L = 43°45'$; $S-h = 38°17'$; $S-p = 3°03'$.
Then:

$\log \sin 43°45' =$		9.839800
$\log \sin 38°17' =$		9.792077
		9.631877
$\log \cos 85°05' =$	8.933015	
$\log \cos \ \ 3°03' =$	9.999384	8.932399
$2 \log \tan \frac{1}{2}Z =$		0.699478
$\log \tan \frac{1}{2}Z =$		0.349739
$\frac{1}{2}Z = 65°55'03.5''$	$Z = 131°50'07''$	

2. Verify the solution by calculating Z in an alternative manner

To do this, introduce an auxiliary angle M, defined by $\cos^2 M = \cos p / \sin h \sin L$, Eq. (123). Then, $\cos (180° - Z) = \tan h \tan L \sin^2 M$, Eq. (124). Then

$\log \cos 82°02' =$		9.141754
$\log \sin 46°48' =$	9.862709	
$\log \sin 41°20' =$	9.819832	9.682541
$2 \log \cos M =$		9.459213
$\log \cos M =$		9.729607
$\log \sin M =$		9.926276
$2 \log \sin M =$		9.852552
$\log \tan 46°48' =$		0.027305
$\log \tan 41°20' =$		9.944262
$\log \cos (180° - Z) =$		9.824119

$Z = 131°50'07''$, as before

TIME OF CULMINATION OF A STAR

Determine the Eastern Standard Time (75th meridian time) of the upper culmination of Polaris at a site having a longitude 81°W of Greenwich. Reference to an almanac shows that the Greenwich Civil Time (GCT) of upper culmination for the date of observation is $3^h 20^m 05^s$.

Calculation Procedure:

1. Convert the longitudes to the hr-min-sec system

The rotation of the Earth causes a star to appear to describe a circle on the celestial sphere centered at the celestial axis. The star is said to be *at culmination* or *transit* when it appears to cross the observer's meridian.

In Fig. 193, P and M represent the position of Polaris and the mean Sun, respectively, when Polaris is at the Greenwich meridian, and P' and M' represent the position of these bodies when Polaris is at the observer's meridian. The distances h and h' represent, respectively, the time of culmination of Polaris at Greenwich and at the observer's site, measured from local noon. Since the apparent velocity of the mean Sun is less than that of the stars, h' is less than h, the difference being approximately 10 sec/hr of longitude.

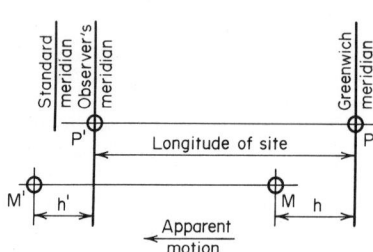

Fig. 193   Culmination of Polaris.

Converting the longitudes, 360° corresponds to 24 hr; therefore, 15° corresponds to 1 hr. Longitude of site = 81° = 5.4^h = $5^h 24^m 00^s$; standard longitude = 75° = 5^h.

2. Calculate the time of upper culmination at the site

Then correct this result to Eastern Standard Time. Since the standard meridian is east of the observer's meridian, the standard time is greater. Thus:

GCT of upper culmination at Greenwich...	$3^h 20^m 05^s$
Correction for longitude, 5.4×10 sec	54^s
Local civil time of upper culmination at site	$3^h 19^m 11^s$
Correction to standard meridian..........	$24^m 00^s$
EST of upper culmination at site	$3^h 43^m 11^s$ A.M.

PLOTTING A CIRCULAR CURVE

A horizontal circular curve having an intersection angle of 28° is to have a radius of 1,200 ft. The PC is at station 82 + 30. (*a*) Determine the tangent distance, long chord, middle ordinate, and external distance. (*b*) Determine all the data necessary to stake the curve if the *chord* distance between successive stations is to be 100 ft. (*c*) Calculate all the data necessary to stake the curve if the *arc* distance between successive stations is to be 100 ft.

Calculation Procedure:

1. Determine the geometric properties of the curve

Refer to Fig. 194. A is termed the *point of curve* (PC), B is the *point of tangent* (PT), and V the *point of intersection* (PI) or vertex. The notational system is: Δ = intersection angle = angle between back and forward tangents = central angle AOB; R = radius of curve; T = tangent distance = $AV = VB$; C = long chord = AB; M = middle ordinate = DC; E = external distance = CV.

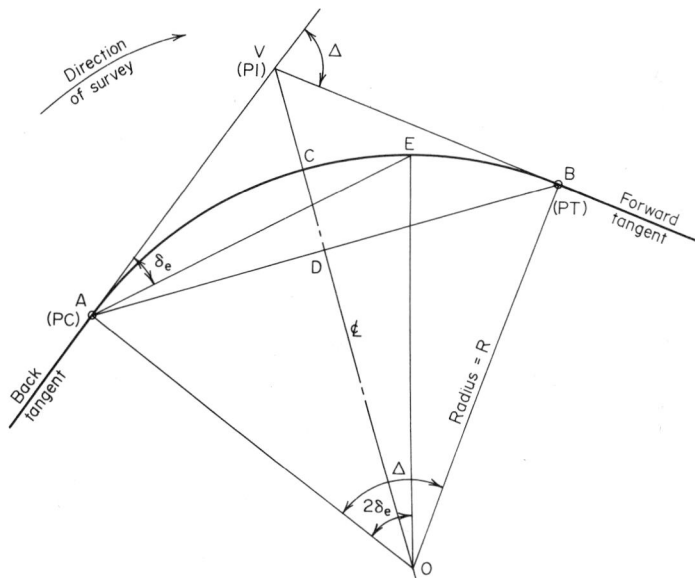

Fig. 194   Circular curve.

From the geometrical relationships, $T = R \tan \frac{1}{2}\Delta$, Eq. (125); $T = 1{,}200(0.2493) = 299.2$ ft; also, $C = 2R \sin \frac{1}{2}\Delta$, Eq. (126); $C = 2(1{,}200)(0.2419) = 580.6$ ft. And, $M = R(1 - \cos\frac{1}{2}\Delta)$, Eq. (127), $M = 1{,}200(1 - 0.9703) = 35.6$ ft. Lastly, $E = R \tan \frac{1}{2}\Delta \tan \frac{1}{4}\Delta$, Eq. (128). $E = 1{,}200(0.2493)(0.1228) = 36.7$ ft.

2. Verify the results in Step 1

Use the Pythagorean theorem on triangle ADV. Or, $AD = \frac{1}{2}(580.6) = 290.3$ ft; $DV = 35.6 + 36.7 = 72.3$ ft; then, $290.3^2 + 72.3^2 = 89{,}500$ ft² to the nearest hundred; $299.2^2 = 89{,}500$ ft²; this is acceptable.

3. Calculate the degree of curve D

Part b. In Fig. 194, let E represent a station along the curve. Angle VAE is termed the *deflection angle* δ_e of this station; it is equal to one-half the central angle AOE. In the field, the curve is staked by setting up the transit at the PC and then locating each station by means of its deflection angle and its chord distance from the preceding station.

Calculate the *degree of curve D*. This is the central angle formed by the radii to two successive stations or what is the same in this instance, the central angle subtended by a *chord* of 100 ft. Then, sin $\frac{1}{2}D = 50/R$, Eq. (129); $\frac{1}{2}D = $ arcsin $50/1,200 = $ arcsin 0.04167; $\frac{1}{2}D = 2°23.3'$; $D = 4°46.6'$.

4. Determine the station at the PT

Number of stations on the curve $= 28°/4°46.6' = 5.862$; station of PT $= (82+30) + (5+86.2) = 88+16.2$.

5. Calculate the deflection angle of station 83 and the difference between the deflection angles of station 88 and the PT

For simplicity, assume that central angles are directly proportional to their corresponding chord lengths; the resulting error is negligible. Then, $\delta_{83} = 0.70(2°23.3') = 1°40.3'$; $\delta_{PT} - \delta_{88} = 0.162(2°23.3') = 0°23.2'$.

6. Calculate the deflection angle of each station

Do this by adding $\frac{1}{2}D$ to that of the preceding station. Record the results thus:

Station	Deflection angle
82+30	0
83	1°40.3'
84	4°03.6'
85	6°26.9'
86	8°50.2'
87	11°13.5'
88	13°36.8'
88+16.2	14°00'

7. Calculate the degree of curve in the present instance

Part c. Since the subtended central angle is directly proportional to its arc length, $D/100 = 360/2\pi R$; therefore, $D = 18,000/\pi R = 5,729.58/R$ degrees, Eq. (130). Then, $D = 5,729.58/1,200 = 4.7747° = 4°46.5'$.

8. Repeat the calculations in Steps 4, 5 and 6

INTERSECTION OF CIRCULAR CURVE AND STRAIGHT LINE

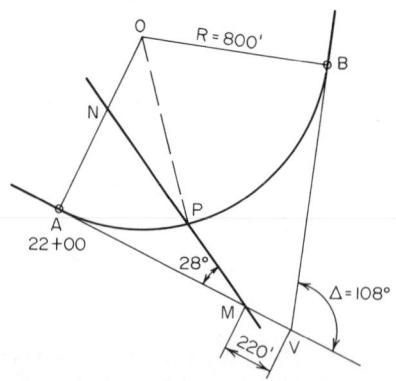

In Fig. 195, *MN* represents a straight railroad spur that intersects the curved highway route *AB*. Distances on the route are measured along the arc. Applying the recorded data, determine the station of the intersection point *P*.

Calculation Procedure:

1. Apply trigonometric relationships to determine three elements in triangle ONP

Draw line *OP*. The problem resolves itself into the calculation of the central angle *AOP*, and this may be readily found by solving the oblique triangle *ONP*. Applying trigonometric relationships, $AV = T = 800 \tan 54° = 1,101.1$ ft; $AM = $

Fig. 195 Intersection of curve and straight line.

$1,101.1 - 220 = 881.1$ ft; $AN = AM$ tan $28° = 468.5$ ft; $ON = 800 - 468.5 = 331.5$ ft; $OP = 800$ ft; $ONP = 90° + 28° = 118°$.

2. Establish the station of P

Solve triangle ONP to find the central angle; then calculate arc AP and establish the station of P. By the law of sines: sin $OPN =$ sin $ONP(ON)/OP$; therefore, $OPN = 21°$-$27.7'$; $AOP = 180° - (118° + 21°27.7') = 40°32.3'$; arc $AP = 2\pi(800)(40°32.3')/360° = 566.0$ ft; station of $P = (22 + 00) + (5 + 66) = 27 + 66$.

REALIGNMENT OF CIRCULAR CURVE BY DISPLACEMENT OF FORWARD TANGENT

In Fig. 196, the horizontal circular curve AB has a radius of 720 ft and an intersection angle of 126°. The curve is to be realigned by rotating the forward tangent through an angle of 22° to the new position $V'B$ while maintaining the PT at B. Compute the radius and locate the PC of the new curve.

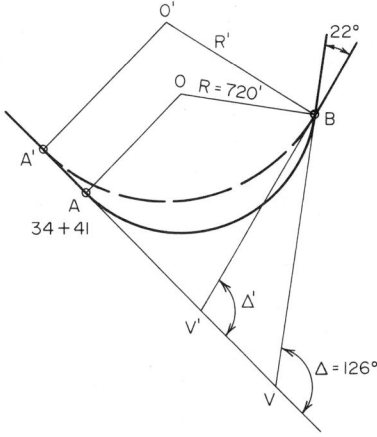

Calculation Procedure:

1. Find the tangent distance of the new curve

Solve triangle $BV'V$ to find the tangent distance of the new curve and the location of V'. Thus, $\Delta' = 126° - 22° = 104°$; $VB = 720$ tan $63° = 1,413.1$ ft. By the law of sines, $V'B = 1,413.1$ sin $126°/$sin $104° = 1,178.2$ ft; $V'V = 1,413.1$ sin $22°/$sin $104° = 545.6$ ft.

2. Compute the radius R'

By Eq. (125), $R' = 1,178.2$ cot $52° = 920.5$ ft.

Fig. 196   Displacement of forward tangent.

3. Determine the station of A'

Thus, $AV = VB = 1,413.1$ ft; $A'V' = V'B = 1,178.2$ ft; $A'A = A'V' + V'V - AV = 310.7$ ft; station of new PC $= (34 + 41) - (3 + 10.7) = 31 + 30.3$.

4. Verify the foregoing results

Draw the long chords AB and $A'B$. Then apply the computed value of R' to solve triangle $BA'A$ and thereby find $A'A$. By Eq. (126), $A'B = 2R'$ sin $\frac{1}{2}\Delta' = 1,450.7$ ft; $AA'B = \frac{1}{2}\Delta' = 52°$; $A'AB = 180° - \frac{1}{2}\Delta = 117°$; $ABA' = 180° - (52° + 117°) = 11°$. By the law of sines, $A'A = 1,450.7$ sin $11°/$sin $117° = 310.7$ ft. This is acceptable.

CHARACTERISTICS OF A COMPOUND CURVE

The tangents to a horizontal curve intersect at an angle of 68°22'. To fit the curve to the terrain, it is necessary to use a compound curve having tangent lengths of 955 ft and 800 ft, as shown in Fig. 197. The minimum allowable radius is 1,000 ft. Compute the larger radius and the two central angles.

Calculation Procedure:

1. Calculate the latitudes and departures of the known sides

A *compound curve* is a curve that comprises two successive circular arcs of unequal radii that are tangent at their point of intersection, the centers of the arcs lying on the same side of their common tangent. (Where the centers lie on opposite sides of this tangent, the curve is termed a *reversed curve*.) In Fig. 197, C is the point of intersection of the arcs, and DE is the common tangent.

This situation will be analyzed without applying any set equation to illustrate the general method of solution for compound and reversed curves. There are two unknown quantities: the radius R_1 and a central angle. (Since $\Delta_1 + \Delta_2 = \Delta$, either central angle may be considered the unknown.)

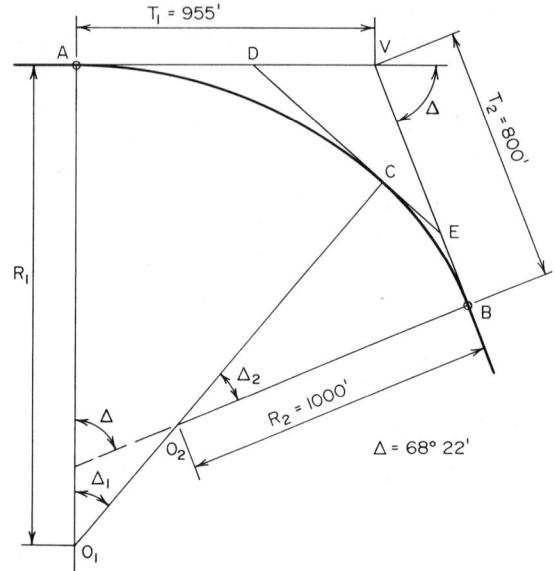

Fig. 197 Compound curve.

If the polygon $AVBO_2O_1$ is visualized as a closed traverse, the latitudes and departures of its sides calculated, and the sum of the latitudes and sum of the departures equated to zero, two simultaneous equations containing these two unknowns are obtained. For convenience, select O_1A as the reference meridian. Then:

Side	Length	Bearing	Latitude	Departure
AV	955	90°	0	+955.00
VB	800	21°38′	−743.65	+294.93
BO_2	1,000	68°22′	−368.67	−929.56
Total	. . .	. . .	−1,112.32	+320.37

2. Express the latitudes and departures of the unknown sides in terms of R_1 and Δ_1

Thus, side O_2O_1: length $= R_1 - 1,000$; bearing $= \Delta_1$; latitude $= -(R_1 - 1,000) \cos \Delta_1$; departure $= -(R_1 - 1,000)\sin \Delta_1$.

Also, side O_1A: length $= R_1$; bearing $= 0$; latitude $= R_1$; departure $= 0$.

3. Equate the sum of the latitudes and sum of the departures to zero; express Δ_1 as a function of R_1

Thus, $\Sigma\text{lat} = R_1 - (R_1 - 1,000)\cos \Delta_1 - 1,112.32 = 0$; $\cos \Delta_1 = (R_1 - 1,112.32)/(R_1 - 1,000)$, or $1 - \cos \Delta_1 = 112.32/(R_1 - 1,000)$, Eq. ($a$). Also, $\Sigma\text{dep} = -(R_1 - 1,000) \sin \Delta_1 + 320.37 = 0$; $\sin \Delta_1 = 320.37/(R_1 - 1,000)$, Eq. ($b$).

4. Divide Eq. (a) by Eq. (b) and determine the central angles

Thus, $(1 - \cos \Delta_1)/\sin \Delta_1 = \tan \tfrac{1}{2}\Delta_1 = 112.32/320.37$; $\tfrac{1}{2}\Delta_1 = 19°19′13″$; $\Delta_1 = 38°38′26″$; $\Delta_2 = 68°22′ - \Delta_1 = 29°43′34″$.

5. **Substitute the value of Δ_1 in Eq. (b) to find R_1**
   Thus, $R_1 = 1,513.06$ ft.

6. **Verify the foregoing results by analyzing triangle DEV**
   Thus, $AD = R_1 \tan \frac{1}{2}\Delta_1 = 530.46$ ft; $DV = 955 - 530.46 = 424.54$ ft; $EB = R_2 \tan \frac{1}{2}\Delta_2 = 265.40$ ft; $VE = 800 - 265.40 = 534.60$ ft; $DE = 530.46 + 265.40 = 795.86$ ft. By the law of cosines, $\cos \Delta = -(DV^2 + VE^2 - DE^2)/[2(DV)(VE)]$; $\Delta = 68°22'$. This is correct.

ANALYSIS OF A HIGHWAY TRANSITION SPIRAL

A horizontal circular curve for a highway is to be designed with transition spirals. The PI is at station $34 + 93.81$ and the intersection angle is $52°48'$. In accordance with the governing design criteria, it has been decided that the spirals are to be 350 ft long and the degree of curve of the circular curve is to be 6° (arc definition). The approach spiral will be staked by setting the transit at the TS and locating 10 stations on the spiral by means of their deflection angles from the main tangent. Compute all data needed for staking the approach spiral. Also, compute the long tangent, short tangent, and external distance.

Calculation Procedure:

1. Calculate the basic values

In the design of a road, a spiral is interposed between a straight-line segment and a circular curve to effect a gradual transition from rectilinear to circular motion, and vice versa. The type of spiral most frequently used is the clothoid, which has the property that the curvature at a given point is directly proportional to the distance from the start of the curve to the given point, measured along the curve.

Refer to Fig. 198. The key points are identified by the following notational system: PI = point of intersection of main tangents; TS = point of intersection of main tangent and approach spiral (tangent-to-spiral point); SC = point of intersection of approach spiral and circular curve (spiral-to-curve point); CS = point of intersection of circular curve and departure spiral (curve-to-spiral point); ST = point of intersection of departure spiral and main tangent (spiral-to-tangent point); PC and PT = points at which tangents to the circular curve prolonged are parallel to the main tangents (also referred to as the *offsets*). Distances are designated in the following manner: L_s = length of spiral from TS to SC; L = length of spiral from TS to given point on spiral; R_c = radius of circular curve; R = radius of curvature at given point on spiral; T_s = length of main tangent from TS to PI; E_s = external distance; i.e., distance from PI to midpoint of circular curve.

In addition to the foregoing, there is a long tangent (LT), short tangent (ST), and long chord (LC), as indicated with respect to the departure spiral.

Place the origin of coordinates at the TS and the x axis on the main tangent. Then: x_c and y_c = coordinates of SC; k and p = abscissa and ordinate, respectively, of PC. The coordinates of the SC and PC are useful as parameters in the calculation of required distances. The distance p is termed the *throw* or *shift* of the curve; it represents the displacement of the circular curve from the main tangent resulting from interposition of the spiral.

The basic angles are as follows: Δ = angle between main tangents, or intersection angle; Δ_c = angle between radii at SC and CS, or central angle of circular curve; θ_s = angle between radii of spiral at TS and SC, or central angle of entire spiral; D_c = degree of curve of circular curve; D = degree of curve at given point on spiral; δ_s = deflection angle of SC from main tangent, with transit at TS; δ = deflection angle of given point on spiral from main tangent, with transit at TS.

Although extensive tables of spiral values have been compiled, this example will be solved without recourse to these tables in order to illuminate the relatively simple mathematical relationships that inhere in the clothoid. Consider that a vehicle starts at the TS and traverses the approach spiral at constant speed. The degree of curve, which is zero at the TS, increases at a uniform rate to become D_c at the SC. The basic equations are as follows: $\theta_s = L_s D_c/200 = L_s/2R_c$, Eq. (131); $x_c = L_s(1 - \theta_s^2/10)$, Eq. (132); $y_c = L_s(\theta_s/3 - \theta_s^3/42)$, Eq. (133); $k = x_c - R_c \sin \theta_s$, Eq. (134); $p = y_c - R_c(1 - \cos \theta_s)$, Eq. (135); $\delta_s = L_s/6R_c = \theta_s/3$, Eq. (136); $y = (L/L_s)^3 y_c$, Eq. (137); $\delta = (L/L_s)^2 \delta_s$, Eq. (138); $T_s = (R_c + p) \tan \frac{1}{2}\Delta + k$, Eq. (139); $E_s = (R_c + p)(\sec \frac{1}{2}\Delta - 1) + p$, Eq. (140); $\text{LT} = x_c - y_c \cot \theta_s$, Eq. (141); $\text{ST} = y_c \csc \theta_s$, Eq. (142). Even though several of the foregoing equations are actually approximations, their use is valid when the value of D_c is relatively small.

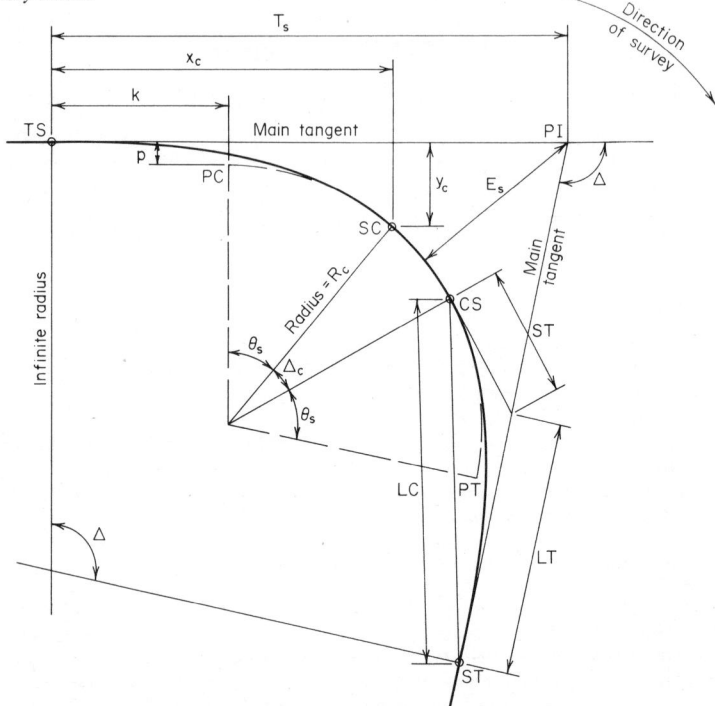

Fig. 198   Notational system for transition spirals.

Calculating the basic values: $\Delta = 52°48'$; $L_s = 350$ ft; $D_c = 6°$; $\theta_s = L_s D_c/200 = 350(6)/200 = 10.5° = 10°30'$, or $\theta_s = 10.5(0.017453) = 0.18326$ rad; $\Delta_c = 52°48' - 2(10°30') = 31°48'$; $D_c = 6(0.017453) = 0.10472$ rad; $R_c = 100/D_c = 954.93$ ft; $x_c = 350(1 - 0.18326^2/10) = 348.83$ ft; $y_c = 350(0.18326/3 - 0.18326^2/42) = 21.33$ ft; $k = 348.83 - 954.93 \sin 10°30' = 174.80$ ft; $p = 21.33 - 954.93(1 - \cos 10°30') = 5.34$ ft.

2. Locate the TS and SC

Thus, $T_s = (954.93 + 5.34)\tan 26°24' + 174.80 = 651.47$; station of TS = $(34 + 93.81) - (6 + 51.47) = 28 + 42.34$; station of SC = $(28 + 42.34) + (3 + 50.00) = 31 + 92.34$.

3. Calculate the deflection angles

Thus, $\delta_s = 10°30'/3 = 3°30' = 3.5°$. Apply Eq. (138) to find the deflection angles at the intermediate stations. For example, for point 7, $\delta = 0.7^2(3.5°) = 1.715° = 1°42.9'$.

Record the results in Table 13. The chord lengths between successive stations differ from the corresponding arc lengths by negligible amounts, and therefore each chord length may be taken as 35.00 ft.

4. Compute the LT, ST, and E_s

Thus, $\text{LT} = 348.83 - 21.33 \cot 10°30' = 233.75$ ft; $\text{ST} = 21.33 \csc 10°30' = 117.04$ ft; $E_s = (954.93 + 5.34)(\sec 26°24' - 1) + 5.34 = 117.14$ ft.

5. Verify the last three calculations by substituting in the following test equation

Thus, $(ST + R_c \tan \frac{1}{2}\Delta_c)/\cos \frac{1}{2}\Delta = (T_s - LT)/\cos \frac{1}{2}\Delta_c = [E_s - R_c(\sec \frac{1}{2}\Delta_c - 1)]/\sin \theta_s$, Eq. (143).

TABLE 13 Deflection Angles on Approach Spiral

Point	Station	Deflection angle
TS	28 + 42.34	0
1	77.34	0°02.1′
2	29 + 12.34	0°08.4′
3	47.34	0°18.9′
4	82.34	0°33.6′
5	30 + 17.34	0°52.5′
6	52.34	1°15.6′
7	87.34	1°42.9′
8	31 + 22.34	2°14.4′
9	57.34	2°50.1′
SC	92.34	3°30.0′

TRANSITION SPIRAL: TRANSIT AT INTERMEDIATE STATION

Referring to the transition spiral in the previous Calculation Procedure, assume that lack of visibility from the TS makes these setups necessary: Points 4, 5, 6, and 7 will be located with the transit at point 3; points 8 and 9 and the SC will be located with the transit at point 7. Compute the orientation and deflection angles.

Calculation Procedure:

1. Consider that the transit is set up at point 3 and a backsight is taken to the TS; find the orientation angle

In Fig. 199, assume that the spiral has been staked up to P with the transit set up at the TS and that the remainder of the spiral is to be staked with the transit set up at P. Deflection angles are measured from the tangent through P (the *local* tangent). The

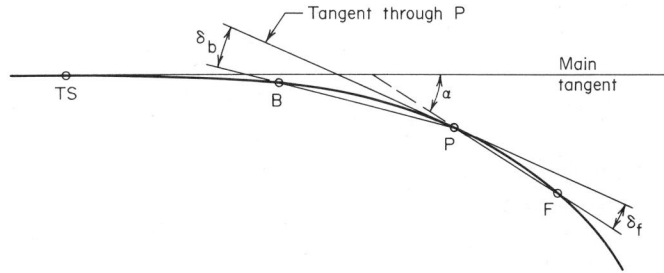

Fig. 199 Deflection angles from local tangent to spiral.

instrument is oriented by backsighting to a preceding point B and then turning the angle δ_b. The orientation angle to B and deflection angle to a subsequent point F are:
$\delta_b = (2L_p + L_b)(L_p - L_b)\theta_s/3L_s^2$, Eq. (144); $\delta_f = (2L_p + L_f)(L_f - L_p)\theta_s/3L_s^2$, Eq. (145).

If B, P, and F are points obtained by dividing the spiral into an integral number of arcs, these equations may be converted to these more suitable forms: $\delta_b = (2n_p + n_b)(n_p - n_b)\theta_s/3n_s^2$, Eq. (144a); $\delta_f = (2n_p + n_f)(n_f - n_p)\theta_s/3n_s^2$, Eq. (145a), where n denotes the number of arcs to the designated point.

Applying Eq. (144a) to find the orientation angle, and using data from the previous Calculation Procedure, $\theta_s = 10.5°$, $\theta_s/3n_s^2 = 10.5°/3(10^2) = 0.035° = 2.1'$; $n_b = 0$; $n_p = 3$; $\delta_b = 6(3)(2.1') = 0°37.8'$.

2. Find the deflection angles from the tangent through point 3

Thus, using Eq. (145a): point 4, $\delta = (6+4)(2.1') = 0°21'$; point 5, $\delta = (6+5)(2)(2.1') = 0°46.2'$; point 6, $\delta = (6+6)(3)(2.1') = 1°15.6'$; point 7, $\delta = (6+7)(4)(2.1') = 1°49.2'$.

3. Consider that the transit is set up at point 7 and a backsight is taken to point 3; compute the orientation angle

Thus: $n_b = 3$; $n_p = 7$; $\delta_b = (14+3)(4)(2.1') = 2°22.8'$.

4. Compute the deflection angles from the tangent through point 7

Thus: point 8, $\delta = (14+8)(2.1') = 0°46.2'$; point 9, $\delta = (14+9)(2)(2.1') = 1°36.6'$; SC, $\delta = (14+10)(3)(2.1') = 2°31.2'$.

5. Test the results obtained

In Fig. 199, extend chord PF to its intersection with the main tangent, and let α denote the angle between these lines. Then, $\alpha = (n_f^2 + n_f n_p + n_p^2)\theta_s/3n_s^2$, Eq. (146). This result should equal the sum of the angles applied in staking the curve from the TS to F. This Procedure will be shown with respect to point 9.

For point 9, let P and F refer to points 7 and 9, respectively. Then, $\alpha = (9^2 + 9 \times 7 + 7^2)(2.1') = 6°45.3'$. Summing the angles leading from the TS to point 9, we get the results shown in the following table.

Deflection angle from main tangent to point 3	0°18.9'
Orientation angle at point 3	0°37.8'
Deflection angle from local tangent to point 7	1°49.2'
Orientation angle at point 7	2°22.8'
Deflection angle from local tangent to point 9	1°36.6'
Total.	6°45.3'

This test may be applied to each deflection angle beyond point 3.

PLOTTING A PARABOLIC ARC

A grade of -4.6 percent is followed by a grade of $+1.8$ percent, the grades intersecting at station $54 + 20$ of elevation 296.30 ft. The change in grade is restricted to 2 percent in 100 ft. Compute the elevation of every 50-ft station on the parabolic curve, and locate the sag (lowest point of the curve). Apply both the average-grade method and the tangent-offset method.

Calculation Procedure:

1. Compute the required length of curve

Using the notation in Figs. 200 and 201, $G_a = -4.6$ percent; $G_b = +1.8$ percent; $r = $ rate of change in grade $= 0.02$ percent/ft; $L = (G_b - G_a)/r = [1.8 - (-4.6)]/0.02 = 320$ ft.

2. Locate the PC and PT

The station of the PC = station of the $PI - L/2 = (54+20) - (1+60) = 52 + 60$; station of $PT = (54+20) + (1+60) = 55 + 80$; elevation of PC = elevation of $PI - G_a L/2 = 296.30 + 0.046(160) = 303.66$ ft; elevation of PT $= 296.30 + 0.018(160) = 299.18$ ft.

3. Use the average-grade method to find the elevation of each station

Calculate the grade at the given station; calculate the average grade between the PI and that station, and multiply the average grade by the horizontal distance to find the ordinate. Equations used in analyzing a parabolic arc are: $y = rx^2/2 + G_ax$, Eq. (147); $G = rx + G_a$, Eq. (148); $y = (G_a + G)x/2$, Eq. (149); $DT = -rx^2/2$, Eq. (150a); $DT = (G_c - G_d)x/2$, Eq. (150b), where DT = distance in Fig. 201.

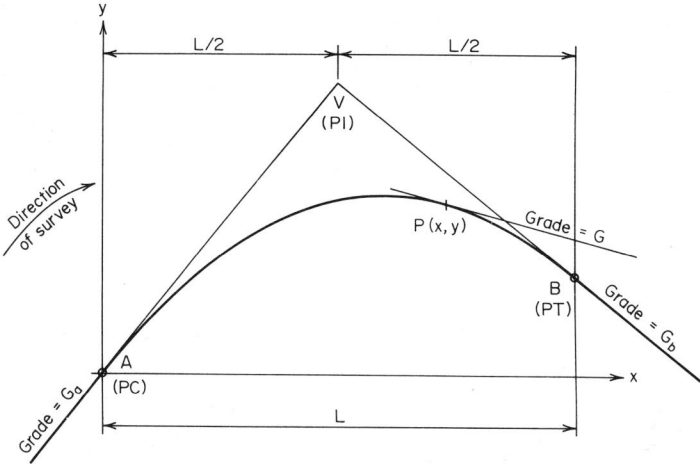

Fig. 200   Parabolic arc.

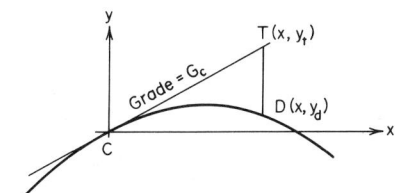

Fig. 201   Tangent offset.

Applying Eq. (148) with respect to station $53 + 00$, $x = 40$ ft, $G = 0.0002(40) - 0.046 = -0.038$; $G_{av} = (-0.046 - 0.038)/2 = -0.042$; $y = -0.042(40) = -1.68$ ft; elevation = $303.66 - 1.68 = 301.98$ ft. Perform these calculations for each station and record the results in tabular form as shown below.

Station	x, ft	G	G_{av}	y, ft	Elevation, ft
$52 + 60$	0	-0.046	-0.046	0	303.66
$53 + 00$	40	-0.038	-0.042	-1.68	301.98
$53 + 50$	90	-0.028	-0.037	-3.33	300.33
$54 + 00$	140	-0.018	-0.032	-4.48	299.18
$54 + 50$	190	-0.008	-0.027	-5.13	298.53
$55 + 00$	240	$+0.002$	-0.022	-5.28	298.38
$55 + 50$	290	$+0.012$	-0.017	-4.93	298.73
$55 + 80$	320	$+0.018$	-0.014	-4.48	299.18

4.   *Verify the foregoing results*

Apply the principle that for a uniform horizontal spacing the "second differences" between the ordinates are equal. The results are shown in the table below.

Calculation of Differences

Elevations, ft	First differences, ft	Second differences, ft
301.98		
	1.65	
300.33		0.50
	1.15	
299.18		0.50
	0.65	
298.53		0.50
	0.15	
298.38		0.50
	−0.35	
298.73		

5. Apply the tangent-offset method to find the elevation of each station

Since this method is based on Eq. (147), substitute directly in that equation. For the present case, the equation becomes $y = rx^2/2 + G_a x = 0.0001x^2 - 0.046x$. Record the calculations for y in tabular form. The results, as shown below, agree with those obtained by the average-grade method.

Tangent-Offset Method

Station	x, ft	$0.0001x^2$, ft	$0.046x$, ft	y, ft
52+60	0	0	0	0
53+00	40	0.16	1.84	−1.68
53+50	90	0.81	4.14	−3.33
54+00	140	1.96	6.44	−4.48
54+50	190	3.61	8.74	−5.13
55+00	240	5.76	11.04	−5.28
55+50	290	8.41	13.34	−4.93
55+80	320	10.24	14.72	−4.48

6. Locate the sag S

Since the grade is zero at this point, Eq. (148) yields $G_s = rx_s + G_a = 0$; therefore $x_s = -G_a/r = -(-0.046/0.0002) = 230$ ft; station of sag $= (52+60) + (2+30) = 54+90$; $G_{av} = \frac{1}{2}G_a = -0.023$; $y_s = -0.023(230) = -5.29$ ft; elevation of sag $= 303.66 - 5.29 = 298.37$ ft.

7. Verify the location of the sag

Do this by ascertaining that the offsets of the PC and PT from the tangent through S, which is horizontal, satisfy the tangent-offset principle. From the preceding results, tangent offset of PC $= 5.29$ ft; tangent offset of PT $= 5.29 - 4.48 = 0.81$ ft; distance to PC $= 230$ ft; distance to PT $= 320 - 230 = 90$ ft; $5.29/0.81 = 6.53$; $230^2/90^2 = 6.53$. Therefore, the results are verified.

LOCATION OF A SINGLE STATION ON A PARABOLIC ARC

The PC of a vertical parabolic curve is at station 22+00 of elevation 165.30, and the grade at the PC is +3.2 percent. The elevation of the station 24+00 is 168.90 ft. What is the elevation of station 25+50?

Calculation Procedure:

1. Compute the offset of P_1 from the tangent through the PC

Refer to Fig. 202. The tangent-offset principle offers the simplest method of solution. Thus: $x_1 = 200$ ft; $y_1 = 168.90 - 165.30 = 3.60$ ft; $Q_1 T_1 = 200(0.032) = 6.40$ ft; $P_1 T_1 = 6.40 - 3.60 = 2.80$ ft.

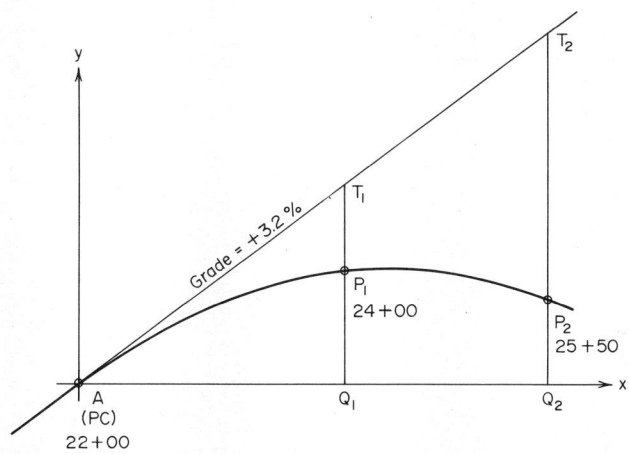

Fig. 202

2. Compute the offset of P_2 from the tangent through the PC; find the elevation of P_2

Thus: $x_2 = 350$ ft; $P_2 T_2 / P_1 T_1 = x_2^2 / x_1^2$; $P_2 T_2 = 2.80(350/200)^2 = 8.575$ ft; $Q_2 T_2 = 350(0.032) = 11.2$ ft; $Q_2 P_2 = 11.2 - 8.575 = 2.625$ ft; elevation of $P_2 = 165.30 + 2.625 = 167.925$ ft.

LOCATION OF A SUMMIT

An approach grade of $+1.5$ percent intersects a grade of -2.5 percent at station $29 + 00$ of elevation 226.30 ft. The connecting parabolic curve is to be 800 ft long. Locate the summit.

Calculation Procedure:

1. Locate the PC

Draw a freehand sketch of the curve, and record all values in the sketch as they are obtained. Thus, station of PC = station of PI $- L/2 = 25 + 00$; elevation of PC = 226.30 $- 400(0.015) = 220.30$ ft.

2. Calculate the rate of change in grade; locate the summit

Apply the average-grade method to locate the summit. Thus, $r = (-2.5 - 1.5)/800 = -0.005$ percent/ft.

Place the origin of coordinates at the PC. By Eq. (148), $x_s = -G_a/r = 1.5/0.005 = 300$ ft. From the PC to the summit, $G_{av} = \frac{1}{2}G_a = 0.75$ percent. Then $y_s = 300(0.0075) = 2.25$ ft; station of summit $= (25 + 00) + (3 + 00) = 28 + 00$; elevation of summit $= 220.30 + 2.25 = 222.55$ ft. The summit can also be located by the tangent-offset method.

PARABOLIC CURVE TO CONTAIN A GIVEN POINT

A grade of -1.6 percent is followed by a grade of $+3.8$ percent, the grades intersecting at station $42 + 00$ of elevation 210.00 ft. The parabolic curve connecting these grades is to pass through station $42 + 60$ of elevation 213.70 ft. Compute the required length of curve.

Calculation Procedure:

1. Compute the tangent offsets; establish the horizontal location of P in terms of L

Refer to Fig. 203, where P denotes the specified point. The given data enable computation of the tangent offsets CP and DP, thus giving a relationship between the horizontal distances from A to P and from P to B. Since the distance from the centerline of curve to P is known, the length of curve may readily be found.

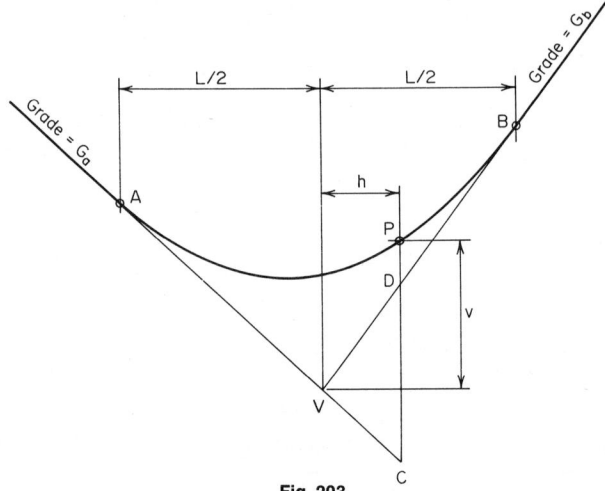

Fig. 203

Computing the tangent offsets, $CP = v - G_a h$ and $DP = v - G_b h$; but $CP/DP = (L/2 + h)^2/(L/2 - h)^2 = (L + 2h)^2/(L - 2h)^2$; therefore, $(L + 2h)/(L - 2h) = [(v - G_a h)/(v - G_b h)]^{1/2}$, Eq. (151).

2. Substitute numerical values and solve for L

Thus, $G_a = -1.6$ percent; $G_b = +3.8$ percent; $h = 60$ ft; $v = 3.70$ ft; then $(L + 120)/(L - 120) = [(3.7 \times 0.016 \times 60)/(3.7 - 0.038 \times 60)]^{1/2} = 1.81$; solving, $L = 416$ ft.

3. Verify the solution

There are many ways of verifying the solution. The simplest way is to compare the offsets of P and B from a tangent through A. By Eq. (150b), offset of B from tangent through $A = 208(0.016 + 0.038) = 11.232$ ft. From the preceding calculations, offset of P from tangent through $A = 4.66$ ft; $4.66/11.232 = 0.415$; $(208 + 60)^2/(416)^2 = 0.415$. This is acceptable.

SIGHT DISTANCE ON A VERTICAL CURVE

A vertical summit curve has tangent grades of $+ 2.6$ and $- 1.5$ percent. Determine the minimum length of curve that is needed to provide a sight distance of 450 ft to an object 4 in. in height. Assume that the eye of the motorist is 4.5 ft above the roadway.

Calculation Procedure:

1. State the equation for minimum length when S < L

The vertical curvature of a road must be limited to ensure adequate visibility across the summit. Consequently, the distance across which a given change in grade may be effected is subject to a lower limit imposed by the criterion of sight distance.

Let S denote the required sight distance and L the minimum length of curve. In Fig. 204, let E denote the position of the motorist's eye and P the top of an object. Assume that the curve has the maximum allowable curvature, so that the distance from E to P equals S.

Applying Eq. (150a), $L = AS^2/\{100[(2h_1)^{1/2} + (2h_2)^{1/2}]^2\}$, Eq. (152), where $A = G_a - G_b$, in percent.

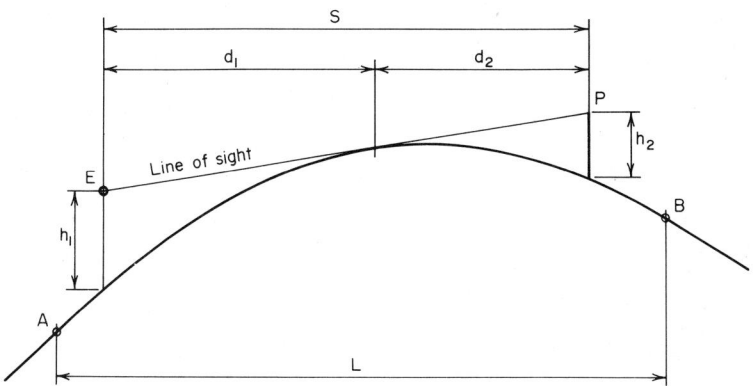

Fig. 204   Visibility on vertical summit curve.

2. State the equation for L when S > L

Thus, $L = 2S - 200(h_1^{1/2} + h_2^{1/2})^2/A$, Eq. (153).

3. Assuming, tentatively, that S < L, compute L

Thus: $h_1 = 4.5$ ft; $h_2 = 4$ in. $= 0.33$ ft; $A = 2.6 + 1.5 = 4.1$ percent; $L = 4.1(450)^2/[100(9^{1/2} + 0.67^{1/2})^2] = 570$ ft. Therefore, the assumption that $S < L$ is valid because $450 < 570$.

MINE SURVEYING: GRADE OF DRIFT

A vein of ore has a strike of S38°20′E and a northeasterly dip of 33°14′. What is the grade of a drift having a bearing of S43°10′E?

Calculation Procedure:

1. Express β as a function of α and θ

A vein of ore is generally assumed to have plane faces. The *strike*, or *trend*, of the vein is the bearing of any horizontal line in a face, and the *dip* is the angle of inclination of its face. A *drift* is a slightly sloping passage that follows the vein. Any line in a plane perpendicular to the horizontal is a *dip line*. The dip line is the steepest line in a plane, and the dip of the plane equals the angle of inclination of this line. With reference to the inclined plane $ABCD$ in Fig. 205a, let $\alpha =$ dip of plane; $\beta =$ angle of inclination of arbitrary line AG; $\theta =$ angle between horizontal projections of AG and dip line.

Expressing β as a function of α and θ: $\tan \beta = AF/GF$; $\tan \alpha = AF/DF$; $\tan \beta/\tan \alpha = DF/GF = \cos \theta$; $\tan \beta = \tan \alpha \cos \theta$, Eq. (154).

2. Find the grade of the drift

Apply Eq. (154). In Fig. 205b, OA is a horizontal line in the vein, OB is the horizontal projection of the drift, and the arrow indicates the direction of dip. Then, angle $COD = 43°10′ - 38°20′ = 4°50′$; $\theta =$ angle $CDO = 90° - 4°50′ = 85°10′$; $\tan \beta = \tan 33°14′ \cos 85°10′ = 0.0552$; grade of drift $= 5.52$ percent.

3. Alternatively, solve without the use of trigonometric functions

In Fig. 205*b*, set *OD* = 100 ft; let *D'* denote the point on the face of the vein vertically below *D*. Then, *CD* = 100 sin 4°50' = 8.426 ft; drop in elevation from *O* to *D'* = drop in elevation from *C* to *D'* = 8.426 tan 33°14' = 5.52 ft. Therefore, grade = 5.52 percent.

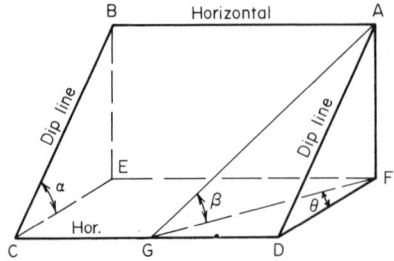

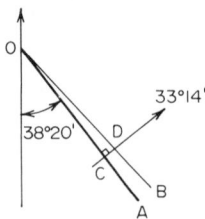

(a) Isometric view of inclined plane   (b) Strike-and-dip diagram (plan)

Fig. 205

DETERMINING STRIKE AND DIP FROM TWO APPARENT DIPS

Three points on the hanging wall (upper face) of a vein of ore have been located by vertical boreholes. These points, designated *P*, *Q*, and *R*, have these relative positions: *P* is 142 ft above *Q* and 130 ft above *R*; horizontal projection of *PQ*, length = 180 ft and bearing = S55°32'W; horizontal projection of *PR*, length = 220 ft and bearing = N19°26'W. Determine the strike and dip of the vein by both graphical construction and trigonometric calculations.

Calculation Procedure:

1. Plot the given data for the graphical procedure

In Fig. 206*a*, draw lines *PQ* and *PR* in plan in accordance with the given data for their horizontal projections. The angle of inclination of any line other than a dip line is an *apparent dip* of the vein.

2. Draw the elevations

In Fig. 206*b* and *c*, draw elevations normal to *PQ* and *PR*, respectively, locating the points in accordance with the given differences in elevation. Find the points *S* and *T* lying on *PQ* and *PR*, respectively, at an arbitrary distance *v* below *P*.

3. Draw the representation of the strike of the vein

Locate points *S* and *T* in Fig. 206*a*, and connect them with a straight line. This line is horizontal, and its bearing φ therefore represents the strike of the vein.

4. Draw an edge view of the vein

In Fig. 206*d*, draw an elevation parallel to *ST*. Since this is an edge view of one line in the face, it is an edge view of the vein itself; it therefore represents the dip α of the vein in its true magnitude.

5. Determine the strike and dip

Scale angles φ and α, respectively. In Fig. 206*a*, the direction of dip is represented by the arrow perpendicular to *ST*.

6. Draw the dip line for the trigonometric solution

In Fig. 207, draw an isometric view of triangle *PST*, and draw the dip line *PW*. Its angle of inclination α equals the dip of the vein. Let *O* denote the point on a vertical line through *P* at the same elevation as *S* and *T*. Let β_1 and β_2 denote the angle of inclination of PS and PT, respectively; and let $\theta = $ angle SOT, $\theta_1 = $ angle SOW, $\theta_2 = $ angle TOW, $m = \tan \beta_2 / \tan \beta_1$.

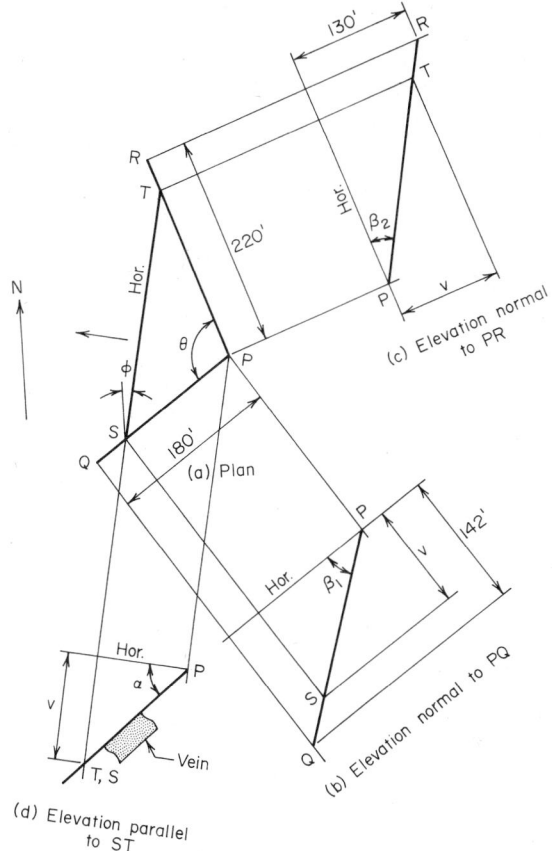

Fig. 206   Determination of strike and dip by orthographic projections.

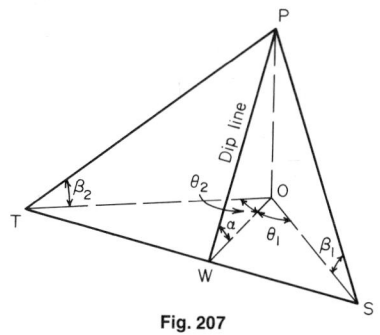

Fig. 207

7.   Express θ_1 in terms of the known angles β_1, β_2, and θ

Then substitute numerical values to find the strike ϕ of the vein. Thus, $\tan \theta_1 = (m - \cos \theta)/\sin \theta$, Eq. (155). For this vein, $m = \tan \beta_2 / \tan \beta_1 = 130(180)/[220(142)] = 0.749040$;

$\theta = 180° - (55°32' + 19°26') = 105°02'$. Substituting, $\tan \theta_1 = (0.749040 + 0.259381)/ 0.965775$; $\theta_1 = 46°14'15''$; $\phi = 55°32' + 46°14'15'' - 90° = 11°46'15''$; strike of vein = N11°46'15''E.

8. Compute the dip of the vein

Use Eq. (154), considering PS as the line of known inclination. Thus, $\tan \alpha = \tan \beta_1/\cos \theta_1$; $\alpha = 48°45'25''$.

9. Verify these results

Apply Eq. (154), considering PT as the line of known inclination. Thus: $\theta_2 = \theta - \theta_1 = 105°02' - 46°14'15'' = 58°47'45''$; $\tan \alpha = \tan \beta_2/\cos \theta_2$; $\alpha = 48°45'25''$. This value agrees with the earlier computed value.

DETERMINATION OF STRIKE, DIP, AND THICKNESS FROM TWO SKEW BOREHOLES

In Fig. 208a, A and B represent points on the earth's surface through which skew boreholes were sunk to penetrate a vein of ore. Point B is 110 ft due south of A. The data for these boreholes are as follows. *Borehole through A*: surface elevation = 870 ft; inclination = 49°; bearing of horizontal projection = N58°30'E. The hanging wall and footwall (lower face of vein) were struck at distances of 55 ft and 205 ft, respectively, measured along the borehole. *Borehole through B*: surface elevation = 842 ft; inclination = 73°; bearing of horizontal projection = S44°50'E. The hanging wall and footwall were struck at distances of 98 ft and 182 ft, respectively, measured along the borehole. Determine the strike, dip, and thickness of the vein by both graphical construction and trigonometric calculations.

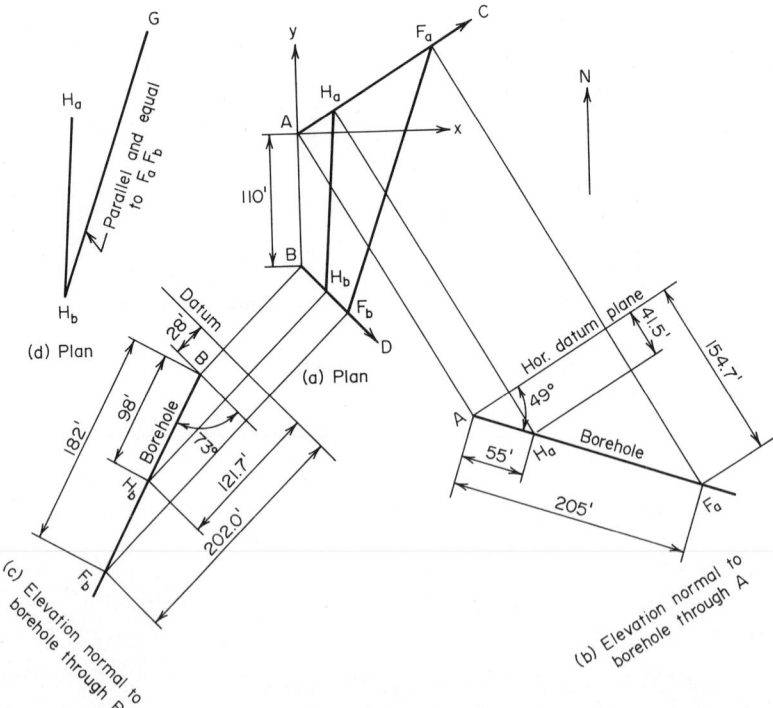

Fig. 208

Calculation Procedure:

1. Draw horizontal projections of the boreholes

Since the vein is assumed to have uniform thickness, the hanging wall and footwall are parallel. Two straight lines determine a plane. In the present instance, two points on the hanging wall and two points on the footwall are given, enabling one line to be drawn in each of two parallel planes. These planes may be located using these principles:

  a. Consider a plane P and line L parallel to each other. If through any point on P a line is drawn parallel to L, this line lies in plane P.

  b. Lines that are parallel and equal in length appear to be parallel and equal in length in all orthographic views.

These principles afford a means of locating a second line in the hanging wall or footwall.

Applying the specified bearings, draw the horizontal projections AC and BD of the boreholes in Fig. 208a.

2. Locate the points of intersection with the hanging wall and footwall in elevation

In Fig. 208b, draw an elevation normal to the borehole through A; locate the points of intersection H_a and F_a with the hanging wall and footwall, respectively. Select the horizontal plane through A as datum.

3. Repeat the foregoing construction with respect to borehole through B

This construction is shown in Fig. 208c.

4. Locate the points of intersection in plan

In Fig. 208a, locate the points of intersection. Draw lines H_aH_b and F_aF_b. The former lies in the hanging wall and the latter in the footwall. To avoid crowding, reproduce the plan of line H_aH_b in Fig. 208d.

5. Draw the plan of a line H_bG that is parallel and equal in length to F_aF_b

Do this by applying the second principle given above. In accordance with principle a, H_bG lies in the hanging wall, and this plane is therefore determined. The ensuing construction parallels that in the previous Calculation Procedure.

6. Establish a system of rectangular coordinate axes

Use A as the origin, Fig. 208a. Make x the east-west axis, y the north-south axis, and z the vertical axis.

7. Apply the given data to obtain the coordinates of the intersection points and point G

For example, with respect to F_a, $y = 205 \cos 49° \cos 58°30'$. The coordinates of G are obtained by adding to the coordinates of H_b the differences between the coordinates of F_a and F_b. The results are shown in the following table.

Point	x	y	z
H_a	30.8	18.9	−41.5
H_b	20.2	−130.3	−121.7
F_a	114.7	70.3	−154.7
F_b	37.5	−147.7	−202.0
G	97.4	87.7	−74.4

8. For convenience, reproduce the plan of the intersection points, and G

This is shown in Fig. 209a.

9. Locate the point S at the same elevation as G

In Fig. 209b, draw an elevation normal to H_aH_b, and locate the point S on this line at the same elevation as G.

10. Establish the strike of the plane

Locate S in Fig. 209a, and draw the horizontal line SG. Since both S and G lie on the hanging wall, the strike of this plane is now established.

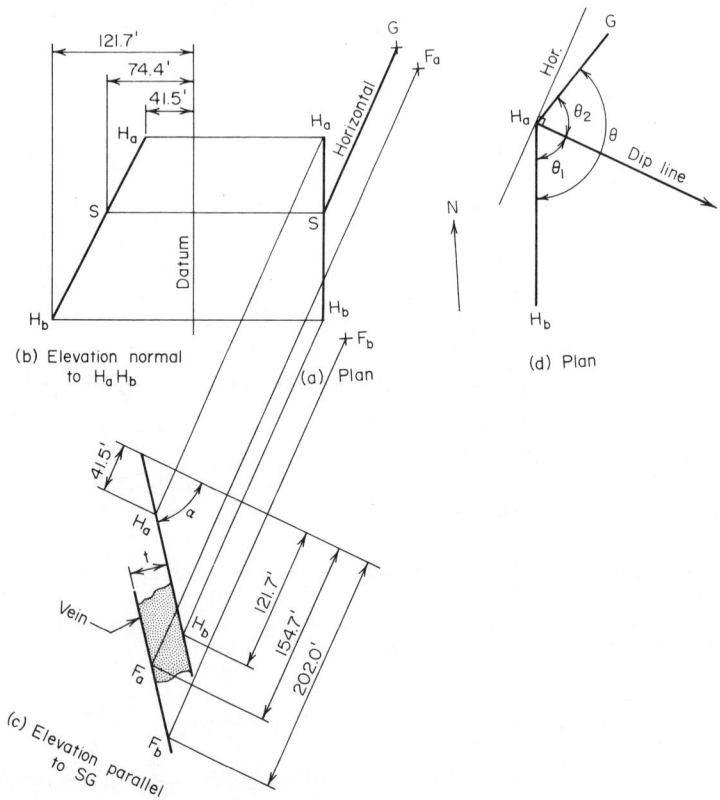

Fig. 209

11. Complete the graphical solution

In Fig. 209c, draw an elevation parallel to SG. The line through H_a and H_b and that through F_a and F_b should be parallel to each other. This drawing is an edge view of the vein, and it presents the dip α and thickness t in their true magnitude. The graphical solution is now completed.

12. Reproduce the plan view

For convenience, reproduce the plan of H_a, H_b, and G in Fig. 209d. Draw the horizontal projection of the dip line, and label the angles as indicated.

13. Compute the lengths of lines $H_a H_b$ and $H_a G$

Compute these lengths as projected on each coordinate axis and as projected on a horizontal plane. Use absolute values. Thus, line $H_a H_b$: $L_x = 30.8 - 20.2 = 10.6$ ft; $L_y = 18.9 - (-130.3) = 149.2$ ft; $L_z = -41.5 - (-121.7) = 80.2$ ft; $L_{\text{hor}} = (10.6^2 + 149.2^2)^{0.5} = 149.6$ ft. Line $H_a G$: $L_x = 97.4 - 30.8 = 66.6$ ft; $L_y = 87.7 - 18.9 = 68.8$ ft; $L_z = -41.5 - (-74.4) = 32.9$ ft; $L_{\text{hor}} = (66.6^2 + 68.8^2)^{0.5} = 95.8$ ft.

14. **Compute the bearing and inclination of lines H_aH_b and H_aG**

Let $\phi_1 =$ bearing of H_aH_b; $\phi_2 =$ bearing of H_aG; $\beta_1 =$ angle of inclination of H_aH_b; $\beta_2 =$ angle of inclination of H_aG. Then, $\tan \phi_1 = 10.6/149.2$; $\phi_1 = $ S4°04'W; $\tan \phi_2 = 66.6/68.8$; $\phi_2 = $ N44°04'E; $\tan \beta_1 = 80.2/149.6$; $\tan \beta_2 = 32.9/95.8$.

15. **Compute angle θ shown in Fig. 209d; determine the strike of the vein, using Eq. (155)**

Thus, $\theta = 180° + \phi_1 - \phi_2 = 140°00'$; $m = \tan \beta_2/\tan \beta_1 = 0.6406$; $\tan \theta_1 = (m - \cos 140°00')/\sin 140°00'$; $\theta_1 = 65°26'$; $\theta_2 = 74°34'$. The bearing of the horizontal projection of the dip line $= \theta_1 - \phi_1 = $ S61°22'E; therefore, the strike of the vein $=$ N28°38'E.

16. **Compute the dip α of the vein**

Using Eq. (154), $\tan \alpha = \tan \beta_1/\cos \theta_1$; $\alpha = 52°12'$; or $\tan \alpha = \tan \beta_2/\cos \theta_2$; $\alpha = 52°14'$. This slight discrepancy between the two computed values falls within the tolerance of these calculations. Use the average value $\alpha = 52°13'$.

17. **Establish the relationship between the true thickness t of a vein and its apparent thickness t' as measured along a skew borehole**

Refer to Figs. 208 and 209. Let $\delta =$ angle of inclination of borehole; $\gamma =$ angle in plan between downward-sloping segments of borehole and dip line of vein. Then, $t = t'$ $(\cos \alpha \sin \delta - \sin \alpha \cos \delta \cos \gamma)$, Eq. (156).

18. **Find the true thickness using Eq. (156)**

Thus, borehole through A: $\delta = 49°$; $\gamma = 180° - (58°30' + 61°22') = 60°08'$; $t' = 205 - 55 = 150$ ft; $t = 150(\cos 52°13' \sin 49° - \sin 52°13' \cos 49° \cos 60°08') = 30.6$ ft. For the borehole through B: $\delta = 73°$; $\gamma = 61°22' - 44°50' = 16°32'$; $t' = 182 - 98 = 84$ ft; $t = 84(\cos 52°13' \sin 73° - \sin 52°13' \cos 73° \cos 16°32') = 30.6$ ft. This agrees with the value previously computed

Soil Mechanics

The basic notational system used is: $c =$ unit cohesion; $s =$ specific gravity; $V =$ volume; $W =$ total weight; $w =$ specific weight; $\phi =$ angle of internal friction; $\tau =$ shearing stress; $\sigma =$ normal stress.

COMPOSITION OF SOIL

A specimen of moist soil weighing 122 g has an apparent specific gravity of 1.82. The specific gravity of the solids is 2.53. After the specimen is oven-dried, the weight is 104 g. Compute the void ratio, porosity, moisture content, and degree of saturation of the original mass.

Calculation Procedure:

1. **Compute the weight of moisture, volume of mass, and volume of each ingredient**

In a three-phase soil mass, the voids, or pores, between the solid particles are occupied by moisture and air. A mass that contains moisture but not air is termed *fully saturated*; this constitutes a two-phase system. The term *apparent specific gravity* denotes the specific gravity of the mass.

Let the subscripts s, w, and a refer to the solids, moisture, and air, respectively. Where a subscript is omitted, the reference is to the entire mass. Also, let $e =$ void ratio $= (V_w + V_a)/V_s$; $n =$ porosity $= (V_w + V_a)/V$; MC $=$ moisture content $= W_w/W_s$; $S =$ degree of saturation $= V_w/(V_w + V_a)$.

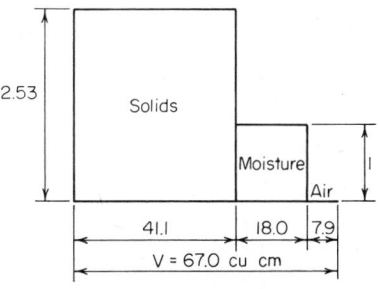

Fig. 210   Soil ingredients.

Refer to Fig. 210. A horizontal line represents volume, a vertical line represents specific gravity, and the area of a rectangle represents the weight of the respective ingredient in grams.

Computing weight and volume, $W = 122$ g; $W_s = 104$ g; $W_w = 122 - 104 = 18$ g; $V = 122/1.82 = 67.0$ cu cm; $V_s = 104/2.53 = 41.1$ cu cm; $V_w = 18.0$ cu cm; $V_a = 67.0 - (41.1 + 18.0) = 7.9$ cu cm.

2. Compute the properties of the original mass

Thus, $e = 100(18.0 + 7.9)/41.1 = 63.0$ percent; $n = 100(18.0 + 7.9)/67.0 = 38.7$ percent; MC $= 100(18)/104 = 17.3$ percent; $S = 100(18.0)/(18.0 + 7.9) = 69.5$ percent. The factor of 100 is used to convert to percent.

SPECIFIC WEIGHT OF SOIL MASS

A specimen of sand has a porosity of 35 percent, and the specific gravity of the solids is 2.70. Compute the specific weight of this soil in pcf in the saturated and in the submerged state.

Calculation Procedure:

1. Compute the weight of the mass in each state

Set $V = 1$ cu cm. The (apparent) weight of the mass when submerged equals the true weight less the buoyant force of the water. Thus, $V_w + V_a = nV = 0.35$ cu cm; $V_s = 0.65$ cu cm. In the saturated state: $W = 2.70(0.65) + 0.35 = 2.105$ g. In the submerged state: $W = 2.105 - 1 = 1.105$ g; or $W = (2.70 - 1)0.65 = 1.105$ g.

2. Find the weight of the soil in pcf

Multiply the foregoing values by 62.4 to find the specific weight of the soil in pcf. Thus: saturated, $w = 131.4$ pcf; submerged, $w = 69.0$ pcf.

ANALYSIS OF QUICKSAND CONDITIONS

Soil having a void ratio of 1.05 contains particles having a specific gravity of 2.72. Compute the hydraulic gradient that will produce a quicksand condition.

Calculation Procedure:

1. Compute the minimum gradient causing quicksand

As water percolates through soil, the head that induces flow diminishes in the direction of flow as a result of friction and viscous drag. The drop in head in a unit distance is termed the *hydraulic gradient*. A quicksand condition exists when water that is flowing upward has a sufficient momentum to float the soil particles.

Let i denote the hydraulic gradient in the vertical direction and i_c the minimum gradient that causes quicksand. Equate the buoyant force on a soil mass to the submerged weight of the mass to find i_c. Or $i_c = (s_s - 1)/(1 + e)$, Eq. (157). For this situation, $i_c = (2.72 - 1)/(1 + 1.05) = 0.84$.

MEASUREMENT OF PERMEABILITY BY FALLING-HEAD PERMEAMETER

A specimen of soil is placed in a falling-head permeameter. The specimen has a cross-sectional area of 66 sq cm and a height of 8 cm; the standpipe has a cross-sectional area of 0.48 sq cm. The head on the specimen drops from 62 to 40 cm in 1 hr 18 min. Determine the coefficient of permeability of the soil, in cm/min.

Calculation Procedure:

1. Using literal values, equate the instantaneous discharge in the specimen to that in the standpipe

The velocity at which water flows through a soil is a function of the *coefficient of permeability*, or *hydraulic conductivity*, of the soil. By Darcy's law of laminar flow,

$v = ki$, Eq. (158), where i = hydraulic gradient; k = coefficient of permeability; v = velocity.

In a falling-head permeameter, water is allowed to flow vertically from a standpipe through a soil specimen. Since the water is not replenished, the water level in the standpipe drops as flow continues, and the velocity is therefore variable. Let A = cross-sectional area of soil specimen; a = cross-sectional area of standpipe; h = head on specimen at given instant; h_1 and h_2 = head at beginning and end, respectively, of time interval T; L = height of soil specimen; Q = discharge at a given instant.

Using literal values, $Q = Aki = -a\,dh/dt$.

2. Evaluate k

Since the head h is dissipated in flow through the soil, $i = h/L$. Substituting and re-arranging, $(Ak/L)\,dT = -a\,dh/h$; integrating, $AkT/L = a \ln (h_1/h_2)$, where ln denotes the natural logarithm. Then, $k = (aL/AT) \ln (h_1/h_2)$, Eq. (159). Substituting, $k = (0.48 \times 8/66 \times 78) \ln (62/40) = 0.000326$ cm/min.

CONSTRUCTION OF FLOW NET

State the Laplace equation as applied to two-dimensional flow of moisture through a soil mass, and list three methods of constructing a flow net that are based on this equation.

Calculation Procedure:

1. Plot flow lines and equipotential lines

The path traversed by a water particle flowing through a soil mass is termed a *flow line*, *stream line*, or *path of percolation*. A line that connects points in the soil mass at which the head on the water has some assigned value is termed an *equipotential line*. A diagram consisting of flow lines and equipotential lines is called a *flow net*.

In Fig. 211a, where water flows under a dam under a head H, lines AB and CD are flow lines and EF and GH are equipotential lines.

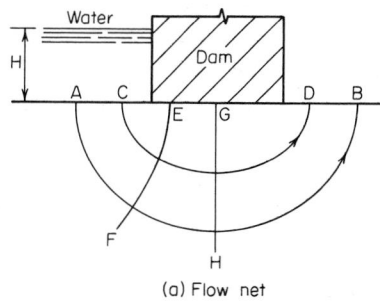

(a) Flow net

(b) Relaxation grid

Fig. 211

2. Discuss the relationship of flow and equipotential lines

Since a water particle flowing from one equipotential line to another of smaller head will traverse the shortest path, it follows that flow lines and equipotential lines intersect at right angles, thus forming a system of orthogonal curves. In a flow net, the equipotential lines should be so spaced that the difference in head between successive lines is a constant, and the flow lines should be so spaced that the discharge through the space between successive lines is a constant. A flow net constructed in compliance with these rules illustrates the basic characteristics of the flow. For example, a close spacing of equipotential lines signifies a rapid loss of head in that region.

3. Write the velocity equation

Let h denote the head on the water at a given point. Equation (158) can be written as $v = -k\,dh/dL$, Eq. (158a), where dL denotes an elemental distance along the flow line.

4. State the particular form of the general Laplace equation

Let x and z denote a horizontal and vertical coordinate axis, respectively. By investigating the two-dimensional flow through an elemental rectangular prism of homogeneous, isentropic soil, and combining Eq. (158a) with the equation of continuity, the particular form of the general Laplace equation, $\partial^2 h/\partial x^2 + \partial^2 h/\partial z^2 = 0$, Eq. (160), is obtained.

This equation is analogous to the equation for the flow of an electric current through a conducting sheet of uniform thickness and the equation of the trajectory of principal stress. (This is a curve that is tangent to the direction of a principal stress at each point along the curve. Refer to earlier Calculation Procedures for a discussion of principal stresses.)

The seepage of moisture through soil may therefore be investigated by analogy with either the flow of an electric current or the stresses in a body. In the latter method, it is merely necessary to load a body in a manner that produces identical boundary conditions and then to ascertain the directions of the principal stresses.

5. Apply the principal-stress analogy

Refer to Fig. 211a. Consider the surface directly below the dam to be subjected to a uniform pressure. Principal-stress trajectories may be readily constructed by applying the principles of elasticity. In the flow net, flow lines correspond to the minor-stress trajectories and equipotential lines correspond to the major-stress trajectories. In this case, the flow lines are ellipses having their foci at the edges of the base of the dam and the equipotential lines are hyperbolas.

A flow net may also be constructed by an approximate, trial-and-error procedure based on the method of relaxation. Consider that the area through which discharge occurs is covered with a grid of squares, a part of which is shown in Fig. 211b. If it is assumed that the hydraulic gradient is constant within each square, Eq. (160) leads to $h_1 + h_2 + h_3 + h_4 - 4h_0 = 0$, Eq. (161).

Trial values are assigned to each node in the grid, and the values are adjusted until a consistent set of values is obtained. With the approximate head at each node thus established, it becomes a simple matter to draw equipotential lines. The flow lines are then drawn normal thereto.

SOIL PRESSURE CAUSED BY POINT LOAD

A concentrated vertical load of 6 kips is applied at the ground surface. Compute the vertical pressure caused by this load at a point 3.5 ft below the surface and 4 ft from the action line of the force.

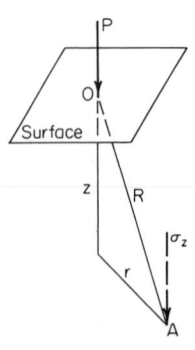

Fig. 212

Calculation Procedure:

1. Sketch the load conditions

Figure 212 shows the load conditions. In Fig. 212, O denotes the point at which the load is applied and A denotes the point under consideration. Let R denote the length of OA and r and z denote the length of OA as projected on a horizontal and vertical plane, respectively.

2. Determine the vertical stress σ_z at A

Apply the Boussinesq equation $\sigma_z = 3Pz^3/2\pi R^5$, Eq. (162). Thus, with $P = 6,000$ lb, $r = 4$ ft, $z = 3.5$ ft, $R = (4^2 + 3.5^2)^{0.5} = 5.32$ ft; then $\sigma_z = 3(6,000)(3.5)^3/[2\pi(5.32)^5] = 28.8$ psf.

Although the Boussinesq equation is derived by assuming an idealized homogeneous mass, its results agree reasonably well with those obtained experimentally.

VERTICAL FORCE ON RECTANGULAR AREA CAUSED BY POINT LOAD

A concentrated vertical load of 20 kips is applied at the ground surface. Determine the resultant vertical force caused by this load on a rectangular area 3×5 ft that lies 2 ft below the surface and has one vertex on the action line of the applied force.

Calculation Procedure:

1. State the equation for the total force

Refer to Fig. 213a where A and B denote the dimensions of the rectangle, H its distance from the surface, and F the resultant vertical force. Establish rectangular coordinate axes along the sides of the rectangle, as shown. Let $C = A^2 + H^2$, $D = B^2 + H^2$, $E = A^2 + B^2 + H^2$, $\theta = \sin^{-1} H(E/CD)^{0.5}$ deg.

The force dF on an elemental area dA is given by the Boussinesq equation as $dF = (3Pz^3/2\pi R^5)\, dA$, where $z = H$ and $R = (H^2 + x^2 + y^2)^{0.5}$. Integrate this equation to obtain an equation for the total force F. Set $dA = dx\, dy$; then, $F/P = 0.25 - \theta/360° + (ABH/2\pi E^{0.5})(1/C + 1/D)$, Eq. (163).

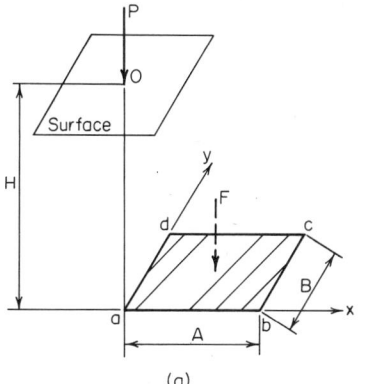

Fig. 213

2. Substitute numerical values and solve for F

Thus, $A = 3$ ft; $B = 5$ ft; $H = 2$ ft; $C = 13$; $D = 29$; $E = 38$; $\theta = \sin^{-1} 0.6350 = 39.4°$; $F/P = 0.25 - 0.109 + 0.086 = 0.227$; $F = 20(0.227) = 4.54$ kips.

The resultant force on an area such as $abcd$, Fig. 213b, may be found by expressing the area in this manner: $abcd = ebhf - eagf + fhcj - fgdj$. The forces on the areas on the right side of this equation are superposed to find the force on $abcd$. Various diagrams and charts have been devised to expedite the calculation of vertical soil pressure.

VERTICAL PRESSURE CAUSED BY RECTANGULAR LOADING

A rectangular concrete footing 6×8 ft carries a total load of 180 kips, which may be considered to be uniformly distributed. Determine the vertical pressure caused by this load at a point 7 ft below the center of the footing.

Calculation Procedure:

1. State the equation for σ_z

Referring to Fig. 214, let p denote the uniform pressure on the rectangle $abcd$ and σ_z the resulting vertical pressure at a point A directly below a vertex of the rectangle. Then stating the equation, $\sigma_z/p = 0.25 - \theta/360 + (ABH/2\pi E^{0.5})(1/C + 1/D)$, Eq. (164).

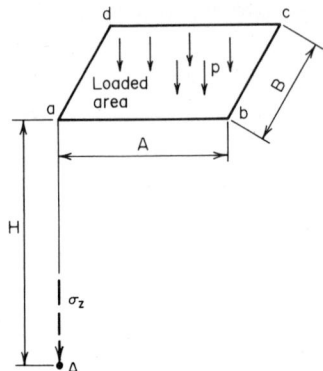

Fig. 214

2. *Substitute given values and solve for* σ_z

Resolve the given rectangle into four rectangles having a vertex above the given point. Then $p = 180,000/[6(8)] = 3,750$ psf. With $A = 3$ ft; $B = 4$ ft; $H = 7$ ft; $C = 58$; $D = 65$; $E = 74$; $\theta = \sin^{-1} 0.9807 = 78.7°$; $\sigma_z/p = 4(0.25 - 0.218 + 0.051) = 0.332$; $\sigma_z = 3,750(0.332) = 1,245$ psf.

APPRAISAL OF SHEARING CAPACITY OF SOIL BY UNCONFINED COMPRESSION TEST

In an unconfined compression test on a soil sample, it was found that when the axial stress reached 2,040 psf, the soil ruptured along a plane making an angle of 56° with the horizontal. Find the cohesion and angle of internal friction of this soil by constructing Mohr's circle.

Calculation Procedure:

1. *Construct Mohr's circle in Fig. 215b*

Failure of a soil mass is characterized by the sliding of one part past the other; the failure is therefore one of shear. Resistance to sliding occurs from two sources: cohesion of the soil and friction.

Consider that the shearing stress at a given point exceeds the cohesive strength. It is usually assumed that the soil has mobilized its maximum potential cohesive resistance plus whatever frictional resistance is needed to prevent failure. The mass therefore remains in equilibrium if the ratio of the computed frictional stress to the normal stress is below the coefficient of internal friction of the soil.

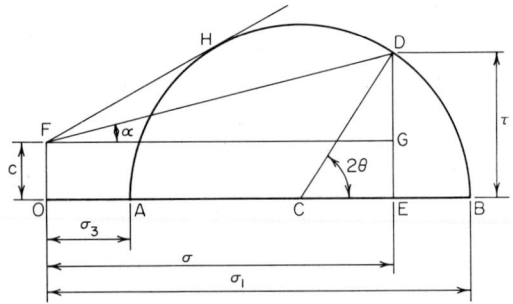

(a) Mohr's diagram for triaxial-stress condition

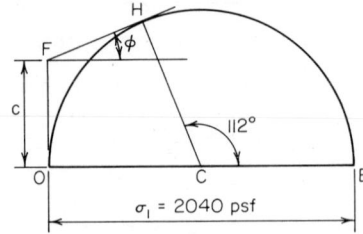

(b) Mohr's diagram for unconfined compression test

Fig. 215

Consider a soil prism in a state of triaxial stress. Let Q denote a point in this prism and P a plane through Q. Let c = unit cohesive strength of soil; σ = normal stress at Q on plane P; σ_1 and σ_3 = maximum and minimum normal stress at Q, respectively; τ = shearing stress at Q, on plane P; θ = angle between P and the plane on which σ_1 occurs; ϕ = angle of internal friction of the soil.

For an explanation of Mohr's circle of stress, refer to an earlier Calculation Procedure; then refer to Fig. 215a. The shearing stress ED on plane P may be resolved into the cohesive stress EG and the frictional stress GD. Therefore, $\tau = c + \sigma \tan \alpha$. The maximum value of α associated with point Q is found by drawing the tangent FH.

Assume that failure impends at Q. Two conclusions may be drawn: the angle between FH and the base line OAB equals ϕ, and the angle between the plane of impending rupture and the plane on which σ_1 occurs equals one-half angle BCH. (A soil mass that is on the verge of failure is said to be in *limit equilibrium*.)

In an unconfined compression test, the specimen is subjected to a vertical load without being restrained horizontally. Therefore, σ_1 occurs on a horizontal plane.

Constructing Mohr's circle in Fig. 215b, apply these values: $\sigma_1 = 2,040$ psf; $\sigma_3 = 0$; angle $BCH = 2(56°) = 112°$.

2. Construct a tangent to the circle

Draw a line through H tangent to the circle. Let F denote the point of intersection of the tangent and the vertical line through O.

3. Measure OF and the angle of inclination of the tangent

The results obtained are: $OF = c = 688$ psf; $\phi = 22°$.

In general, in an unconfined compression test, $c = \frac{1}{2}\sigma_1 \cot \theta'$; $\phi = 2\theta' - 90°$, Eq. (165), where θ' denotes the angle between the plane of failure and the plane on which σ_1 occurs. In the special case where frictional resistance is negligible, $\phi = 0$; $c = \frac{1}{2}\sigma_1$.

APPRAISAL OF SHEARING CAPACITY OF SOIL BY TRIAXIAL COMPRESSION TEST

Two samples of a soil were subjected to triaxial compression tests, and it was found that failure occurred under the following principal stresses: sample 1, $\sigma_1 = 6,960$ psf and $\sigma_3 = 2,000$ psf; sample 2, $\sigma_1 = 9,320$ psf and $\sigma_3 = 3,000$ psf. Find the cohesion and angle of internal friction of this soil, both trigonometrically and graphically.

Calculation Procedure:

1. State the equation for the angle ϕ

Trigonometric method: Let S and D denote the sum and difference, respectively, of the stresses σ_1 and σ_3. By referring to Fig. 215a, develop this equation, $D - S \sin \phi = 2c \cos \phi$, Eq. (166). Since the right-hand member represents a constant that is characteristic of the soil, $D_1 - S_1 \sin \phi = D_2 - S_2 \sin \phi$, or $\sin \phi = (D_2 - D_1)/(S_2 - S_1)$, Eq. (167), where the subscripts correspond to the sample numbers.

2. Evaluate ϕ and c

Using Eq. (167), $S_1 = 8,960$ psf; $D_1 = 4,960$ psf; $S_2 = 12,320$ psf; $D_2 = 6,320$ psf; $\sin \phi = (6,320 - 4,960)/(12,320 - 8,960)$; $\phi = 23°53'$. Evaluating c using Eq. (166), $c = \frac{1}{2}(D \sec \phi - S \tan \phi) = 729$ psf.

3. For the graphical solution, use the Mohr's circle

Draw the Mohr's circle associated with each set of principal stresses, as shown in Fig. 216.

4. Draw the envelope; measure its angle of inclination

Draw the envelope (common tangent) FHH', and measure OF and the angle of inclination of the envelope. In practice, three or four samples should be tested and the average value of ϕ and c determined.

EARTH THRUST ON RETAINING WALL CALCULATED BY RANKINE'S THEORY

A retaining wall supports sand weighing 100 pcf and having an angle of internal friction of 34°. The back of the wall is vertical, and the surface of the backfill is inclined at an angle of 15° with the horizontal. Applying Rankine's theory, calculate the active earth pressure on the wall at a point 12 ft below the top.

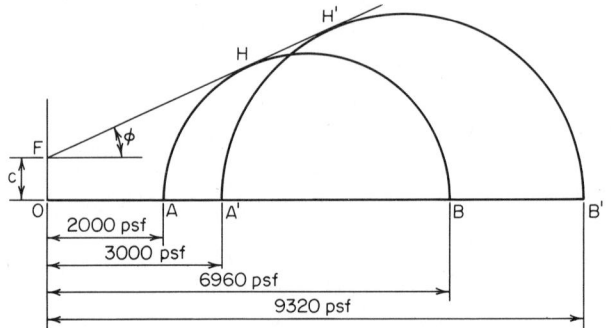

Fig. 216   Composite Mohr's diagram for triaxial compression tests.

Calculation Procedure:

1. Construct the Mohr's circle associated with the soil prism

Rankine's theory of earth pressure applies to a uniform mass of dry cohesionless soil. This theory considers the state of stress at the instant of impending failure caused by a slight yielding of the wall. Let h = vertical distance from soil surface to a given point, ft; p = resultant pressure on a vertical plane at the given point, psf; ϕ = ratio of shearing stress to normal stress on given plane; θ = angle of inclination of earth surface. The quantity o may also be defined as the tangent of the angle between the resultant stress on a plane and a line normal to this plane; it is accordingly termed the *obliquity* of the resultant stress.

Consider the elemental soil prism $abcd$ in Fig. 217a, where faces ab and dc are parallel to the surface of the backfill and faces bc and ad are vertical. The resultant pressure p_v on ab is vertical and p is parallel to the surface. Thus, the resultant stresses on ab and bc have the same obliquity, namely, tan θ. (Stresses having equal obliquities are called *conjugate* stresses.) Since failure impends, there is a particular plane for which the obliquity is tan ϕ.

In Fig. 217b, construct Mohr's circle associated with this soil prism. The procedure is: Using a suitable scale, draw line OD making an angle θ with the base line, where

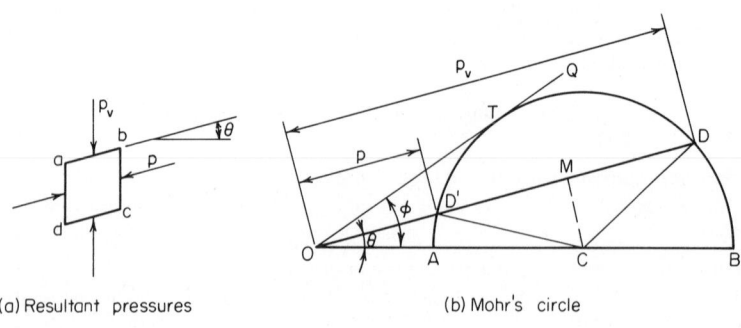

(a) Resultant pressures   (b) Mohr's circle

Fig. 217

OD represents p_v. Draw line OQ making an angle ϕ with the base line. Draw a circle that has its center C on the base line, passes through D, and is tangent to OQ. Line OD' represents p. Draw CM perpendicular to OD.

2. Using the Mohr's circle, state the equation for p

Thus, $p = \{[\cos \theta - (\cos^2 \theta - \cos^2 \phi)^{0.5}]wh\}/[\cos \theta + (\cos^2 \theta - \cos^2 \phi)^{0.5}]$, Eq. (168). Substituting, $w = 100$ pcf; $h = 12$ ft; $\theta = 15°$; $\phi = 34°$; $p = 0.321(100)(12) = 385$ psf.

The lateral pressure that accompanies a slight displacement of the wall *away from* the retained soil is termed *active pressure*; that which accompanies a slight displacement of the wall *toward* the retained soil is termed *passive pressure*. By an analogous procedure, the passive pressure is $p = \{[\cos \theta + (\cos^2 \theta - \cos^2 \phi)^{0.5}]wh\}/[\cos \theta - (\cos^2 \theta - \cos^2 \phi)^{0.5}]$, Eq. (169).

The equations of active and passive pressure are often written as $p_a = C_awh$; $p_p = C_pwh$, Eq. (170), where the subscripts identify the type of pressure and C_a and C_p are the coefficients appearing in Eqs. (168) and (169), respectively.

In the special case where $\theta = 0$, these coefficients reduce to: $C_a = (1 - \sin \phi)/(1 + \sin \phi) = \tan^2 (45° - \tfrac{1}{2}\phi)$, Eq. (171); $C_p = (1 + \sin \phi)/(1 - \sin \phi) = \tan^2 (45° + \tfrac{1}{2}\phi)$, Eq. (172). The planes of failure make an angle of $45° + \tfrac{1}{2}\phi$ with the principal planes.

EARTH THRUST ON RETAINING WALL CALCULATED BY COULOMB'S THEORY

A retaining wall 20 ft high supports sand weighing 100 pcf and having an angle of internal friction of 34°. The back of the wall makes an angle of 8° with the vertical; the surface of the backfill makes an angle of 9° with the horizontal. The angle of friction between the sand and wall is 20°. Applying Coulomb's theory, calculate the total thrust of the earth on a 1-ft length of the wall.

Calculation Procedure:

1. Determine the resultant pressure P of the wall

Refer to Fig. 218a. Coulomb's theory postulates that as the wall yields slightly, the soil tends to rupture along some plane BC through the heel.

Let δ denote the angle of friction between the soil and wall. As shown in Fig. 218b, the wedge ABC is held in equilibrium by three forces: the weight W of the wedge, the

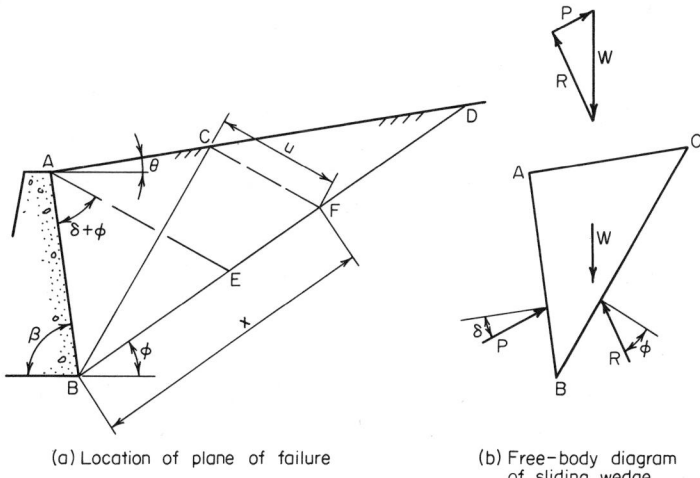

(a) Location of plane of failure

(b) Free-body diagram of sliding wedge

Fig. 218

resultant pressure R of the soil beyond the plane of failure, and the resultant pressure P of the wall, which is equal and opposite to the thrust exerted by the earth on the wall. The forces R and P have the directions indicated in Fig. 218b. By selecting a trial wedge and computing its weight, the value of P may be found by drawing the force polygon. The problem is to identify the wedge that yields the maximum value of P.

In Fig. 218a, perform this construction: Draw a line through B at an angle ϕ with the horizontal, intersecting the surface at D. Draw line AE making an angle $\delta + \phi$ with the back of the wall; this line makes an angle $\beta - \delta$ with BD. Through an arbitrary point C on the surface, draw CF parallel to AE. Triangle BCF is similar to the triangle of forces in Fig. 218b. Then, $P = Wu/x$, where $W = w$(area ABC).

2. Set $dP/dx = 0$ and state Rebhann's theorem

This theorem states: The wedge that exerts the maximum thrust on the wall is that for which triangles ABC and BCF have equal areas.

3. Considering BC as the true plane of failure, develop equations for x^2, u, and P

Thus, $x^2 = BE(BD)$, Eq. (173); $u = AE(BD)/(x + BD)$, Eq. (174); $P = \frac{1}{2}wu^2 \sin (\beta - \delta)$, Eq. (175).

4. Evaluate P using the foregoing equations

Thus: $\phi = 34°$; $\delta = 20°$; $\theta = 9°$; $\beta = 82°$; $\angle ABD = 64°$; $\angle BAE = 54°$; $\angle AEB = 62°$; $\angle BAD = 91°$; $\angle ADB = 25°$; $AB = 20 \csc 82° = 20.2$ ft. In triangle ABD: $BD = AB \sin 91°/\sin 25° = 47.8$ ft. In triangle ABE: $BE = AB \sin 54°/\sin 62° = 18.5$ ft; $AE = AB \sin 64°/\sin 62° = 20.6$ ft; $x^2 = 18.5(47.8)$; $x = 29.7$ ft; $u = 20.6(47.8)/(29.7 + 47.8) = 12.7$ ft; $P = \frac{1}{2}(100)(12.7)^2 \sin 62°$; $P = 7{,}120$ lb per ft of wall.

5. Alternatively, determine u graphically

Do this by drawing Fig. 218a to a suitable scale.

There are many situations that do not lend themselves to analysis by Rebhann's theorem. For instance, the backfill may be nonhomogeneous, the earth surface may not be a plane, a surcharge may be applied over part of the surface, etc. In these situations, graphical analysis gives the simplest solution. Select a trial wedge, compute its weight and the surcharge it carries, and find P by constructing the force polygon as shown in Fig. 218b. After several trial wedges have been investigated, the maximum value of P will become apparent.

If the backfill is cohesive, the active pressure on the retaining wall is reduced. However, in view of the difficulty of appraising the cohesive capacity of a disturbed soil, most designers prefer to disregard cohesion.

EARTH THRUST ON TIMBERED TRENCH CALCULATED BY GENERAL WEDGE THEORY

A timbered trench of 12-ft depth retains a cohesionless soil having a horizontal surface. The soil weighs 100 pcf, its angle of internal friction is $26°30'$, and the angle of friction between the soil and timber is $12°$. Applying Terzaghi's general wedge theory, compute the total thrust of the soil on a 1-ft length of trench. Assume that the resultant acts at mid-depth.

Calculation Procedure:

1. Start the graphical construction

Refer to Fig. 219. The soil behind a timbered trench and that behind a cantilever retaining wall tend to fail by dissimilar modes, for in the former case the soil is restrained against horizontal movement at the surface by bracing across the trench. Consequently, the soil behind a trench tends to fail along a curved surface that passes through the base and is vertical at its intersection with the ground surface. At impending failure, the resultant force dR acting on any elemental area on the failure surface makes an angle ϕ with the normal to this surface.

The general wedge theory formulated by Terzaghi postulates that the arc of failure is a logarithmic spiral. Let v_o denote a reference radius vector and v denote the radius vector to a given point on the spiral. The equation of the curve is $r = r_o e^{\alpha \tan \phi}$, Eq. (176), where r_o = length of v_o; r = length of v; α = angle between v_o and v; e = base of natural logarithms = 2.718 . . . .

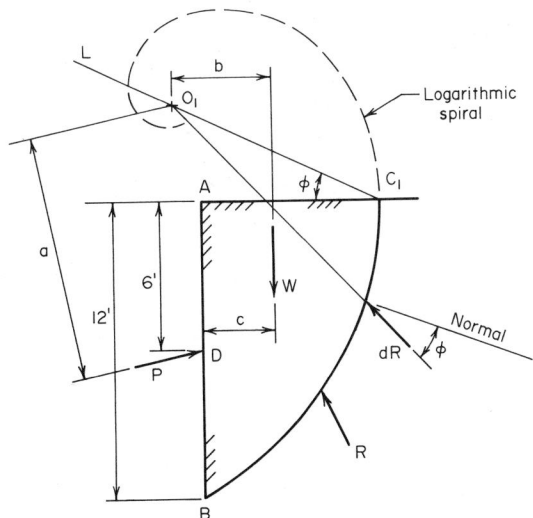

Fig. 219   General wedge theory applied to timbered trench.

The property of this curve that commends it for use in this analysis is that at every point the radius vector and the normal to the curve make an angle ϕ with one another. Therefore, if it is assumed that the failure line is defined by Eq. (176), it follows that the action line of the resultant force dR at any point is a radius vector or, in other words, the action line passes through the center of the spiral. Consequently, the action line of the total resultant force R also passes through the center.

The pressure distribution on the wall departs radically from a hydrostatic one, and the resultant thrust P is applied at a point considerably above the lower third point of the wall. Terzaghi recommends setting the ratio BD/AB at between 0.5 and 0.6.

Perform the following construction: Using a suitable scale, draw line AB to represent the side of the trench, and draw a line to represent the ground surface. At mid-depth, draw the action line of P at an angle of 12° with the horizontal.

On a sheet of transparent paper, draw the logarithmic spiral representing Eq. (176), setting $\phi = 26°30'$ and assigning any convenient value to r_o. Designate the center of the spiral as O.

Select a point C_1 on the ground surface and draw a line L through C_1 at an angle ϕ with the horizontal. Superimpose the drawing containing the spiral on the main drawing, orienting it in such a manner that O lies on L and the spiral passes through B and C_1. On the main drawing, indicate the position of the center of the spiral, and designate this point as O_1. Line AC_1 is normal to the spiral at C_1 because it makes an angle ϕ with the radius vector, and the spiral is therefore vertical at C_1.

2. *Compute the total weight W of the soil above the failure line*

Draw the action line of W by applying these approximations: area of wedge = $\frac{2}{3}(AB)$ AC_1, and $c = 0.4AC_1$, Eq. (177). Scale the lever arms a and b.

3. *Evaluate P by taking moments with respect to O_1*

Since R passes through this point, $P = bW/a$, Eq. (178).

4. *Select a second point C_2 on the ground surface; repeat the foregoing procedure*

5. Continue this process until the maximum value of P is obtained

After investigating this problem intensively, Peckworth concluded that the distance AC to the true failure line varies between $0.4h$ and $0.5h$, where h is the depth of the trench. It is therefore advisable to select some point that lies within this range as the first trial position of C.

THRUST ON A BULKHEAD

The retaining structure in Fig. 220a supports earth that weighs 114 pcf in the dry state, is 42 percent porous, and has an angle of internal friction of 34° in both the dry and submerged state. The backfill carries a surcharge of 320 psf. Applying Rankine's theory, compute the total pressure on this structure between A and C.

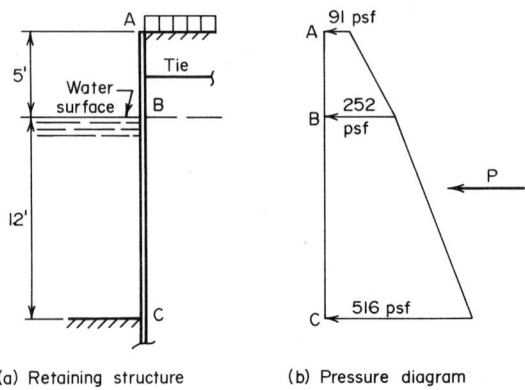

(a) Retaining structure   (b) Pressure diagram

Fig. 220

Calculation Procedure:

1. Compute the specific weight of the soil in the submerged state

The lateral pressure of the soil below the water level consists of two elements: the pressure exerted by the solids and that exerted by the water. The first element is evaluated by applying the appropriate equation with w equal to the weight of the soil in the submerged state; the second element is assumed to be the full hydrostatic pressure, as though the solids were not present. Since there is water on both sides of the structure, the hydrostatic pressures balance one another and may therefore be disregarded.

In calculating the forces on a bulkhead, it is assumed that the pressure distribution is hydrostatic (i.e., that the pressure varies linearly with the depth), although this is not strictly true with regard to a flexible wall.

Computing the specific weight of the soil in the submerged state, $w = 114 - (1 - 0.42)62.4 = 77.8$ pcf.

2. Compute the vertical pressure at A, B, and C caused by the surcharge and weight of solids

Thus, $p_A = 320$ psf; $p_B = 320 + 5(114) = 890$ psf; $p_C = 890 + 12(77.8) = 1,824$ psf.

3. Compute the Rankine coefficient of active earth pressure

Determine the lateral pressure at A, B, and C. Since the surface is horizontal, Eq. (171) applies, with $\phi = 34°$. Refer to Fig. 220b. Then, $C_a = \tan^2 (45° - 17°) = 0.283$; $p_A = 0.283(320) = 91$ psf; $p_B = 252$ psf; $p_C = 516$ psf.

4. Compute the total thrust between A and C

Thus, $P = \frac{1}{2}(5)(91 + 252) + \frac{1}{2}(12)(252 + 516) = 5,466$ lb.

CANTILEVER BULKHEAD ANALYSIS

Sheet piling is to function as a cantilever retaining wall 5 ft high. The soil weighs 110 pcf-and its angle of internal friction is 32°; the backfill has a horizontal surface. Determine the required depth of penetration of the bulkhead.

Calculation Procedure:

1. Take moments with respect to C to obtain an equation for the minimum value of d

Refer to Fig. 221a and consider a 1-ft length of wall. Assume that the pressure distribution is hydrostatic and apply Rankine's theory.

The wall pivots about some point Z near the bottom. Consequently, passive earth pressure is mobilized to the left of the wall between B and Z and to the right of the wall between Z and C. Let P = resultant active pressure on wall; R_1 and R_2 = resultant passive pressure above and below center of rotation, respectively.

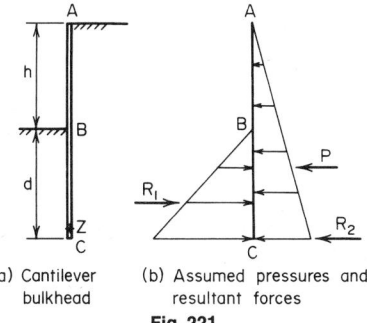

(a) Cantilever bulkhead

(b) Assumed pressures and resultant forces

Fig. 221

The position of Z may be found by applying statics. But to simplify the calculations, these assumptions are made: The active pressure extends from A to C; the passive pressure to the left of the wall extends from B to C; and R_2 acts at C. Figure 221b illustrates these assumptions.

Taking moments with respect to C and substituting values for R_1 and R_2, $d = h/[(C_p/C_a)^{1/3} - 1]$, Eq. (179).

2. Substitute numerical values and solve for d

Thus, $45° + \frac{1}{2}\phi = 61°$; $45° - \frac{1}{2}\phi = 29°$. By Eqs. (171) and (172), $C_p/C_a = (\tan 61°/\tan 29°)^2 = 10.6$; $d = 5/[(10.6)^{1/3} - 1] = 4.2$ ft. Add 20 percent of the computed value to provide a factor of safety and to allow for the development of R_2. Thus, penetration = 4.2(1.2) = 5.0 ft.

ANCHORED BULKHEAD ANALYSIS

Sheet piling is to function as a retaining wall 20 ft high, anchored by tie rods placed 3 ft from the top at an 8-ft spacing. The soil weighs 110 pcf and its angle of internal friction is 32°. The backfill has a horizontal surface and carries a surcharge of 200 psf. Applying the equivalent-beam method, determine the depth of penetration to secure a fixed earth support, the tension in the tie rod, and the maximum bending moment in the piling.

Calculation Procedure:

1. Locate C and construct the net-pressure diagram for AC

Refer to Fig. 222a. The depth of penetration is readily calculated if stability is the sole criterion. However, when the depth is increased beyond this minimum value, the tension in the rod and the bending moment in the piling are reduced; the net result is a saving in material despite the increased length.

Investigation of this problem discloses that the most economical depth of penetration is that for which the tangent to the elastic curve at the lower end passes through the anchorage point. If this point is considered as remaining stationary, this condition can be described as one in which the elastic curve is vertical at D, the surrounding soil acting as a fixed support. Whereas an equation can be derived for the depth associated with this condition, such an equation is too cumbersome for rapid solution.

It has been found that when the elastic curve is vertical at D, the lower point of contraflexure lies close to the point where the net pressure (the difference between active pressure to the right and passive pressure to the left of the wall) is zero. By assuming that the point of contraflexure and the point of zero pressure are in fact coincident, this problem is transformed into one that is statically determinate. The method of analysis based on this assumption is termed the *equivalent-beam* method.

When the piling is driven to a depth greater than the minimum needed for stability, it deflects in such a manner as to mobilize passive pressure to the right of the wall at its lower end. However, the same simplifying assumption concerning the pressure distribution as made in the previous Calculation Procedure will be made here.

Let C denote the point of zero pressure. Consider a 1-ft length of wall and let $T =$ reaction at anchorage point and $V =$ shear at C.

Locate C and construct the net-pressure diagram for AC as shown in Fig. 222*b*.

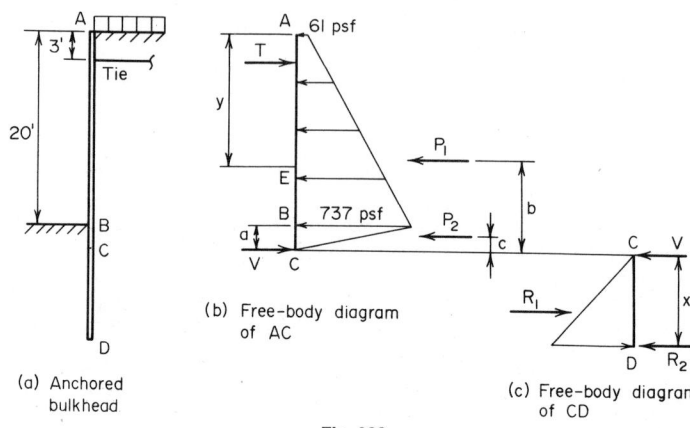

(a) Anchored bulkhead

(b) Free-body diagram of AC

(c) Free-body diagram of CD

Fig. 222

Thus, $w = 110$ pcf and $\phi = 32°$. Then: $C_a = \tan^2(45° - 16°) = 0.307$; $C_p = \tan^2(45° + 16°) = 3.26$; $C_p - C_a = 2.953$; $p_A = 0.307(200) = 61$ psf; $p_B = 61 + 0.307(20)(110) = 737$ psf; $a = 737/[2.953(110)] = 2.27$ ft.

2. Calculate the resultant forces P_1 and P_2

Thus, $P_1 = \frac{1}{2}(20)(61 + 737) = 7,980$ lb; $P_2 = \frac{1}{2}(2.27)(737) = 836$ lb; $P_1 + P_2 = 8,816$ lb.

3. Equate the bending moment at C to zero to find T, V, and the tension in the tie rod

Thus: $b = 2.27 + (20/3)(737 + 2 \times 61)/(737 + 61) = 9.45$ ft; $c = \frac{2}{3}(2.27) = 1.51$ ft; $\Sigma M_C = 19.27T - 9.45(7,980) - 1.51(836) = 0$; $T = 3,980$ lb; $V = 8,816 - 3,980 = 4,836$ lb. The tension in the rod $= 3,980(8) = 31,840$ lb.

4. Construct the net-pressure diagram for CD

Refer to Fig. 222*c* and calculate the distance x. (For convenience, Fig. 222*c* is drawn to a different scale from that of Fig. 222*b*.) Thus: $p_D = 2,953(110x) = 324.8x$; $R_1 = \frac{1}{2}(324.8x^2) = 162.4x^2$; $\Sigma M_D = R_1 x/3 - Vx = 0$; $R_1 = 3V$; $162.4x^2 = 3(4,836)$; $x = 9.45$ ft.

5. Establish the depth of penetration

To provide a factor of safety and to compensate for the slight inaccuracies inherent in this method of analysis, increase the computed depth by about 20 percent. Thus, penetration $= 1.20(a + x) = 14$ ft.

6. Locate the point of zero shear; calculate the piling maximum bending moment

Refer to Fig. 222*b*. Locate the point E of zero shear. Thus: $p_E = 61 + 0.307(110y) = 61 + 33.77y$; $\frac{1}{2}y(p_A + p_E) = T$; or $\frac{1}{2}y(122 + 33.77y) = 3,980$; $y = 13.6$ ft, and $p_E = 520$ psf; $M_{max} = M_E = 3,980[10.6 - (13.6/3)(520 + 2 \times 61)/(520 + 61)] = 22,300$ ft-lb per ft of piling. Since the tie rods provide intermittent rather than continuous support, the piling sustains biaxial bending stresses.

STABILITY OF SLOPE BY METHOD OF SLICES

Investigate the stability of the slope in Fig. 223 by the method of slices (also known as the Swedish method). The properties of the upper and lower soil strata, designated as A and B, respectively, are: $A - w = 110$ pcf, $c = 0$, $\phi = 28°$; $B - w = 122$ pcf, $c = 650$ psf, $\phi = 10°$. Stratum A is 36 ft deep. A surcharge of 8,000 plf is applied 20 ft from the edge.

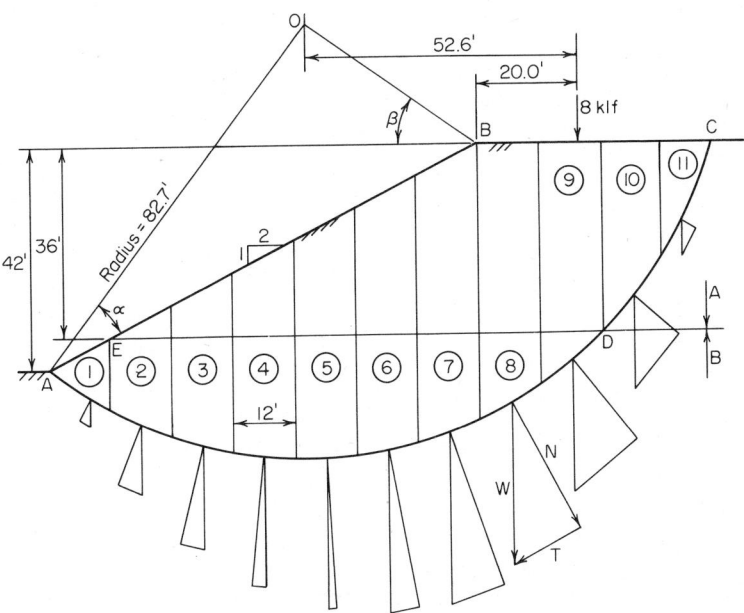

Fig. 223   Analysis of stability of slope by slices.

Calculation Procedure:

1. Locate the center of the trial arc of failure passing through the toe

It is assumed that failure of an embankment occurs along a circular arc, the prism of soil above the failure line tending to rotate about an axis through the center of the arc. However, there is no direct method of identifying the arc along which failure is most likely to occur, and it is therefore necessary to resort to a cut-and-try procedure.

Consider a soil mass having a thickness of 1 ft normal to the plane of the drawing; let O denote the center of a trial arc of failure that passes through the toe. For a given inclination of embankment, Fellenius recommends certain values of α and β in locating the first trial arc.

Locate O by setting $\alpha = 25°$ and $\beta = 35°$.

2. Draw the arc AC and the boundary line ED of the two strata

3. Compute the length of arc AD

Scale the radius of the arc and the central angle AOD, and compute the length of the arc AD. Thus, radius = 82.7 ft; arc $AD = 120$ ft.

4. Determine the distance horizontally from O to the applied load

Scale the horizontal distance from O to the applied load. This distance is 52.6 ft.

5. Divide the soil mass into vertical strips

Starting at the toe, divide the soil mass above AC into vertical strips of 12-ft width, and number the strips. For simplicity, consider that D lies on the boundary line between strips 9 and 10, although this is not strictly true.

6. Determine the volume and weight of soil in each strip

By scaling the dimensions or using a planimeter, determine the volume of soil in each strip; then compute the weight of soil. For instance, for strip 5: volume of soil $A = 252$ cu ft; volume of soil $B = 278$ cu ft; weight of soil $= 252(110) + 278(122) = 61,600$ lb. Record the results in Table 14.

TABLE 14   Stability Analysis of Slope

Strip	Weight, kips	Normal component, kips	Tangential component, kips
1	10.3	8.9	−5.2
2	28.1	26.0	−10.7
3	41.9	40.6	−10.4
4	53.0	52.7	−5.5
5	61.6	61.5	2.6
6	67.7	66.5	12.8
7	71.0	67.0	23.4
8	67.1	58.8	32.4
9	54.8	43.0	34.0
10	38.3	24.9	29.1
11	14.3	7.0	12.5
Total, 1 to 9		425.0	
Total, 10 and 11		31.9	
Grand total		456.9	115.0

7. Draw a vector below each strip

This vector represents the weight of the soil in the strip. (Theoretically, this vector should lie on the vertical line through the center of gravity of the soil, but such refinement is not warranted in this analysis. For the interior strips, place each vector on the vertical centerline.)

8. Resolve the soil weights vectorially into components normal and tangential to the circular arc

9. Scale the normal and tangential vectors; record the results in Table 14

10. Total the normal forces acting on soils A and B; total the tangential forces

Failure of the embankment along arc AC would be characterized by the clockwise rotation of the soil prism above this arc about an axis through O, this rotation being induced by the unbalanced tangential force along the arc and by the external load. Therefore, consider a tangential force as positive if its moment with respect to an axis through O is clockwise and negative if this moment is counterclockwise. In the method of slices, it is assumed that the lateral forces on each soil strip approximately balance each other.

11. Evaluate the moment tending to cause rotation about O

In the absence of external loads, $DM = r\Sigma T$, Eq. (180), where DM = disturbing moment; r = radius of arc; ΣT = algebraic sum of tangential forces.
In the present instance, $DM = 82.7(115) + 52.6(8) = 9,930$ ft-kips.

12. *Sum the frictional and cohesive forces to find the maximum potential resistance to rotation; determine the stabilizing moment*

In general, $F = \Sigma N \tan \phi$; $C = cL$, Eq. (181); $SM = r(F + C)$, Eq. (182), where $F =$ frictional force; $C =$ cohesive force; $\Sigma N =$ sum of normal forces; $L =$ length of arc along which cohesion exists; $SM =$ stabilizing moment.

In the present instance, $F = 425 \tan 10° + 31.9 \tan 28° = 91.9$ kips; $C = 0.65(120) = 78.0$; total of $F + C = 169.9$ kips; $SM = 82.7(169.9) = 14,050$ ft-kips.

13. *Compute the factor of safety against failure*

The factor of safety, $FS = SM/DM = 14,050/9,930 = 1.41$.

14. *Select another trial arc of failure; repeat the foregoing procedure*

15. *Continue this process until the minimum value of FS is obtained*

The minimum allowable factor of safety is generally regarded as 1.5.

STABILITY OF SLOPE BY ϕ-CIRCLE METHOD

Investigate the stability of the slope in Fig. 224 by the ϕ-circle method. The properties of the soil are: $w = 120$ pcf; $c = 550$ psf; $\phi = 4°$.

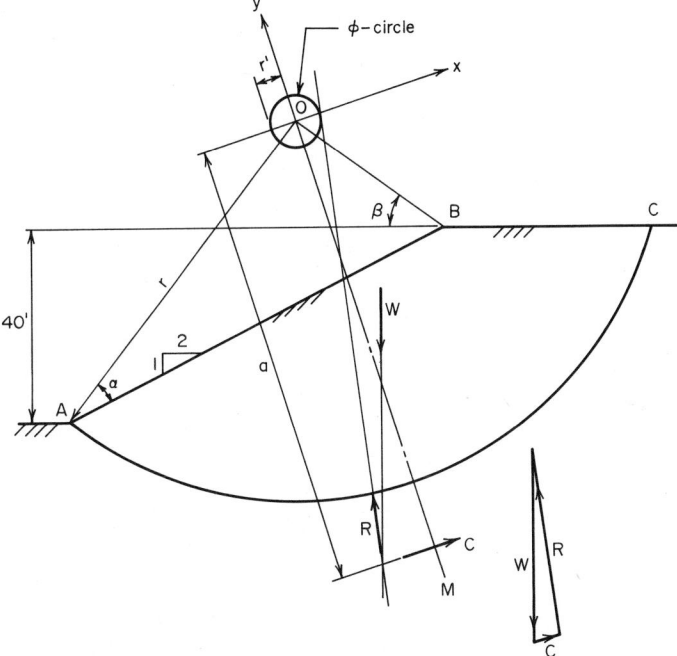

Fig. 224 Analysis of stability of slope by ϕ-circle method.

Calculation Procedure:

1. Locate the first trial position

The ϕ-circle method of analysis formulated by Krey is useful where standard conditions are encountered. In contrast to the assumption concerning the stabilizing forces stated earlier, the ϕ-circle method assumes that the soil has mobilized its maximum potential *frictional* resistance plus whatever cohesive resistance is needed to prevent failure. A comparison of the maximum available cohesion with the required cohesion serves as an index of the stability of the embankment.

In Fig. 224, O is the center of an assumed arc of failure AC. Let W = weight of soil mass above arc AC; R = resultant of all normal and frictional forces existing along arc AC; C = resultant cohesive force developed; L_a = length of arc AC; L_c = length of chord AC. The soil above the arc is in equilibrium under the forces W, R, and C. Since W is known in magnitude and direction, the magnitude of C may be readily found if the directions of R and C are determined.

Locate the first trial position of O by setting $\alpha = 25°$ and $\beta = 35°$.

2. Draw the arc AC and the radius OM bisecting this arc

3. Establish rectangular coordinate axes at O, making OM the y axis

4. Obtain the needed basic data

Scale the drawing or make the necessary calculations. Thus, $r = 78.8$ ft; $L_a = 154.6$ ft; $L_c = 131.0$ ft; area above arc = 4,050 sq ft; $W = 4,050(120) = 486,000$ lb; horizontal distance from A to centroid of area = 66.7 ft.

5. Draw the vector representing W

Since the soil is homogeneous, this vector passes through the centroid of the area.

6. State the equation for C; locate its action line

Thus, $C = C_x = cL_c$, Eq. (183). The action line of C is parallel to the x axis. Determine the distance a by taking moments about O. Thus: $M = aC = acL_c$; $a = (L_a/L_c)r$, Eq. (184); $a = (154.6/131.0)78.8 = 93.0$ ft. Draw the action line of C.

7. Locate the action line of R

For this purpose, consider the resultant force dR acting on an elemental area. Its action line is inclined at an angle ϕ with the radius at that point, and therefore the perpendicular distance r' from O to this action line is $r' = r \sin \phi$, Eq. (185). Thus, r' is a constant for the arc AC. It follows that regardless of the position of dR along this arc, its action line is tangent to a circle centered at O and having a radius r'; this is called the ϕ circle or friction circle. It is plausible to conclude that the action line of the total resultant is also tangent to this circle.

Draw a line tangent to the ϕ circle and passing through the point of intersection of the action lines of W and C. This is the action line of R. (The moment of R about O is counterclockwise, since its frictional component opposes clockwise rotation of the soil mass.)

8. Using a suitable scale, determine the magnitude of C

Draw the triangle of forces; obtain the magnitude of C by scaling. Thus, $C = 67,000$ lb.

9. Calculate the maximum potential cohesion

Apply Eq. (183), equating c to the unit cohesive capacity of the soil. Thus, $C_{max} = 550(131) = 72,000$ lb. This result indicates a relatively low factor of safety. Other arcs of failure should be investigated in the same manner.

ANALYSIS OF FOOTING STABILITY BY TERZAGHI'S FORMULA

A wall footing carrying a load of 58 klf rests on the surface of a soil having these properties: $w = 105$ pcf; $c = 1,200$ psf; $\phi = 15°$. Applying Terzaghi's formula, determine the minimum width of footing required to ensure stability, and compute the soil pressure associated with this width.

Calculation Procedure:

1. Equate the total active and passive pressures and state the equation defining conditions at impending failure

While several methods of analyzing the soil conditions under a footing have been formulated, the one proposed by Terzaghi is gaining wide acceptance.

The soil underlying a footing tends to rupture along a curved surface, but the Terzaghi method postulates that this surface may be approximated by straight-line segments without introducing any significant error. Thus, in Fig. 225, the soil prism OAB tends to heave by sliding downward along OA under active pressure and sliding upward along AB against passive pressure. As stated in an earlier Calculation Procedure, these planes of failure make an angle of $\alpha = 45° + \frac{1}{2}\phi$ with the principal planes.

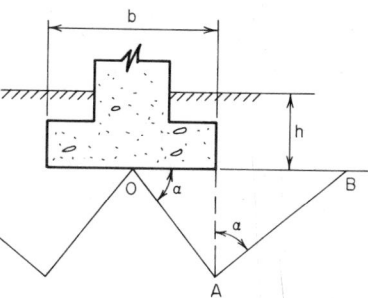

Fig. 225   Failure of soil under footing in accordance with Terzaghi's assumption.

Let b = width of footing; h = distance from ground surface to bottom of footing; p = soil pressure directly below footing. By equating the total active and passive pressures, state the following equation defining the conditions at impending failure: $p = wh\tan^4\alpha + wb(\tan^5\alpha - \tan\alpha)/4 + 2c(\tan\alpha + \tan^3\alpha)$, Eq. (186).

2.   Substitute numerical values; solve for b; evaluate p

Thus, $h = 0$; $p = 58/b$; $\phi = 15°$; $\alpha = 45° + 7°30' = 52°30'$; $58/b = 0.105b(3.759 - 1.303)/4 + 2(1.2)(1.303 + 2.213)$; $b = 6.55$ ft; $p = 58,000/6.55 = 8,850$ psf.

SOIL CONSOLIDATION AND CHANGE IN VOID RATIO

In a laboratory test, a load was applied to a soil specimen having a height of 30 in. and a void ratio of 96.0 percent. What was the void ratio when the load settled $\frac{1}{2}$ in.?

Calculation Procedure:

1.   Construct a diagram representing the volumetric composition of the soil in the original and final states

According to the Terzaghi theory of consolidation, the compression of a soil mass under an increase in pressure results primarily from the expulsion of water from the pores. At the instant the load is applied, it is supported entirely by the water, and the hydraulic gradient thus established induces flow. However, the flow in turn causes a continuous transfer of load from the water to the solids.

Equilibrium is ultimately attained when the load is carried entirely by the solids, and the expulsion of the water then ceases. The time rate of expulsion, and therefore of consolidation, is a function of the permeability of the soil, the number of drainage faces, etc. Let H = original height of soil stratum; s = settlement; e_1 = original void ratio; e_2 = final void ratio. Using the given data, construct the diagram in Fig. 226 representing the volumetric composition of the soil in the original and final states.

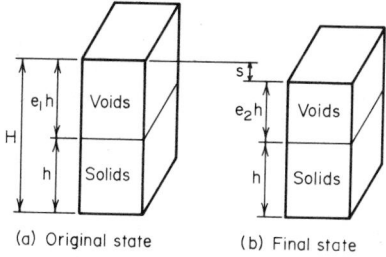

(a) Original state   (b) Final state

Fig. 226

2. State the equation relating the four defined quantities

Thus, $s = H(e_1 - e_2)/(1 + e_1)$, Eq. (187).

3. Solve for e_2

Thus: $H = 30$ in.; $s = 0.50$ in.; $e_1 = 0.960$; $e_2 = 92.7$ percent.

COMPRESSION INDEX AND VOID RATIO OF A SOIL

A soil specimen under a pressure of 1,200 psf is found to have a void ratio of 103 percent. If the compression index is 0.178, what will be the void ratio when the pressure is increased to 5,000 psf?

Calculation Procedure:

1. Define the compression index

By testing a soil specimen in a consolidometer, it is possible to determine the void ratio associated with a given compressive stress. When the sets of values thus obtained are plotted on semilogarithmic scales (void ratio vs. logarithm of stress), the resulting diagram is curved initially but becomes virtually a straight line beyond a specific point. The slope of this line is termed the *compression index*.

2. Compute the soil void ratio

Let C_c = compression index; e_1 and e_2 = original and final void ratio, respectively; σ_1 and σ_2 = original and final normal stress, respectively.

Write the equation of the straight-line portion of the diagram, or $e_1 - e_2 = C_c \log (\sigma_2/\sigma_1)$, Eq. (188). Substituting and solving, $1.03 - e_2 = 0.178 \log (5,000/1,200)$; $e_2 = 92.0$ percent. Note that the logarithm is taken to the base 10.

SETTLEMENT OF FOOTING

An 8-ft-square footing carries a load of 150 kips that may be considered uniformly distributed, and it is supported by the soil strata shown in Fig. 227. The silty clay has a compression index of 0.274; its void ratio prior to application of the load is 84 percent. Applying the unit weights indicated in Fig. 227, calculate the settlement of the footing caused by consolidation of the silty clay.

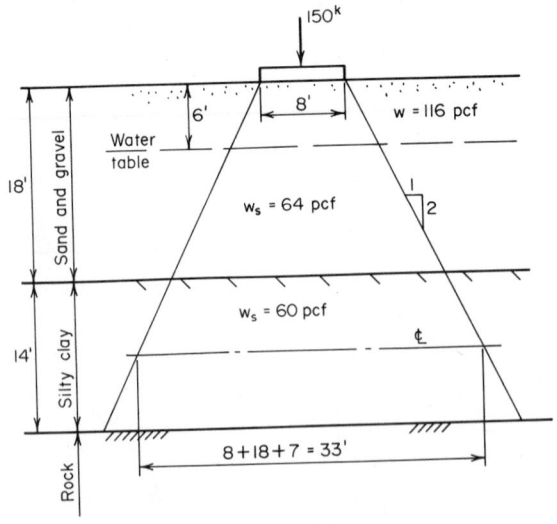

Fig. 227 Settlement of footing.

Calculation Procedure:

1. Compute the vertical stress at mid-depth before and after application of the load

To simplify the calculations, assume that the load is transmitted through a truncated pyramid having side slopes of 2 to 1 and that the stress is uniform across a horizontal plane. Take the stress at mid-depth of the silty-clay stratum as the average for that stratum.

Compute the vertical stress σ_1 and σ_2 at mid-depth before and after application of the load, respectively. Thus: $\sigma_1 = 6(116) + 12(64) + 7(60) = 1{,}884$ psf; $\sigma_2 = 1{,}884 + 150{,}000/33^2 = 2{,}022$ psf.

2. Compute the footing settlement

Combine Eqs. (187) and (188) to obtain $s = HC_c \log (\sigma_2/\sigma_1)/(1 + e_1)$, Eq. (189). Solving, $s = 14(0.274) \log (2{,}022/1{,}884)/(1 + 0.84) = 0.064$ ft $= 0.77$ in.

DETERMINATION OF FOOTING SIZE BY HOUSEL'S METHOD

A square footing is to transmit a load of 80 kips to a cohesive soil, the settlement being restricted to $\frac{3}{8}$ in. Two test footings were loaded at the site until the settlement reached this value. The results were as shown in the following table.

Footing size	Load, lb
1 ft 6 in. × 2 ft	14,200
3 ft × 3 ft	34,500

Applying Housel's method, determine the size of the footing in plan.

Calculation Procedure:

1. Determine the values of p and s corresponding to the allowable settlement

Housel considers that the ability of a cohesive soil to support a footing stems from two sources: bearing strength and shearing strength. This concept is embodied in $W = Ap + Ps$, Eq. (190), where W = total load; A = area of contact surface; P = perimeter of contact surface; p = bearing stress directly below footing; s = shearing stress along perimeter.

Applying the given data for the test footings: footing 1, $A = 3$ sq ft, $P = 7$ ft; footing 2, $A = 9$ sq ft, $P = 12$ ft. Then: $3p + 7s = 14{,}200$; $9p + 12s = 34{,}500$; $p = 2{,}630$ psf; $s = 900$ plf.

2. Compute the size of the footing to carry the specified load

Let x denote the side of the footing. Then, $2{,}630x^2 + 900(4x) = 80{,}000$; $x = 4.9$ ft. Make the footing 5 ft sq.

APPLICATION OF PILE-DRIVING FORMULA

A 16×16 in. pile of 3,000-psi concrete, 45 ft long, is reinforced with eight No. 7 bars. The pile is driven by a double-acting steam hammer. The weight of the ram is 4,600 lb and the energy delivered is 17,000 ft-lb per blow. It is found that the average penetration caused by the final blows is 0.42 in. Compute the bearing capacity of the pile by applying Redtenbacker's formula and using a factor of safety of 3.

Calculation Procedure:

1. Find the weight of the pile and the area of the transformed section

The work performed in driving a pile into the soil is a function of the reaction of the soil on the pile and the properties of the pile. Therefore, the soil reaction may be evaluated if the work performed by the hammer is known. Let A = cross-sectional area of pile; E = modulus of elasticity; h = height of fall of ram; L = length of pile; P = allowable load on pile; R = reaction of soil on pile; s = penetration per blow; W = weight of falling ram; w = weight of pile.

Redtenbacker developed the following equation by taking these quantities into consideration: the work performed by the soil in bringing the pile to rest; the work performed in compressing the pile; and the energy delivered to the pile $-Rs + R^2L/2AE = W^2h/(W + w)$, Eq. (191).

Finding the weight of the pile and the area of the transformed section, w = 16(16)(0.150)(45)/144 = 12 kips. The area of a No. 7 bar = 0.60 sq in.; n = 9; A = 16(16) + 8(9 − 1)0.60 = 294 sq in.

2. Apply Eq. (191) to find R; evaluate P

Thus, s = 0.42 in.; L = 540 in.; E_c = 3,160 ksi; W = 4.6 kips; Wh = 17 ft-kips = 204 in.-kips. Substituting, $0.42R + 540R^2/[2(294)(3,160)] = 4.6(204)/(4.6 + 12)$; R = 84.8 kips; $P = R/3 = 28.3$ kips.

CAPACITY OF A GROUP OF FRICTION PILES

A structure is to be supported by 12 friction piles of 10-in. diameter. These will be arranged in four rows of three piles each at a spacing of 3 ft in both directions. A test pile was found to have an allowable load of 32 kips. Determine the load that may be carried by this pile group.

Calculation Procedure:

1. State a suitable equation for the load

When friction piles are compactly spaced, the area of soil that is needed to support an individual pile overlaps that needed to support the adjacent ones. Consequently, the capacity of the group is less than the capacity obtained by aggregating the capacities of the individual piles. Let P = capacity of group; P_i = capacity of single pile; m = number of rows; n = number of piles per row; d = pile diameter; s = center-to-center spacing of piles; $\theta = \tan^{-1} d/s$ deg. A suitable equation using these variables is the Converse-Labarre equation $P/P_i = mn - (\theta/90°)[m(n − 1) + n(m − 1)]$, Eq. (192).

2. Compute the load

Thus: P_i = 32 kips; m = 4; n = 3; d = 10 in.; s = 36 in.; $\theta = \tan^{-1} 10/36 = 15.5°$. Then: $P/32 = 12 - (15.5/90)(4 \times 2 + 3 \times 3)$; P = 290 kips.

LOAD DISTRIBUTION AMONG HINGED BATTER PILES

Figure 228a shows the relative positions of four steel bearing piles that carry the indicated load. The piles, which may be considered as hinged at top and bottom, have identical cross sections and the following relative effective lengths: A, 1.0; B, 0.95; C, 0.93; D, 1.05. Outline a graphical procedure for determining the load transmitted to each pile.

Calculation Procedure:

1. Subject the structure to a load for purposes of analysis

Steel and timber piles may be considered to be connected to the concrete pier by frictionless hinges, and bearing piles that extend a relatively short distance into compact soil may be considered to be hinge-supported by the soil.

Since four unknown quantities are present, the structure is statically indeterminate. A solution to this problem therefore requires an analysis of the deformation of the structure.

As the load is applied, the pier, assumed to be infinitely rigid, rotates to some new position. This displacement causes each pile to rotate about its base and to undergo an axial strain. The contraction or elongation of each pile is directly proportional to the perpendicular distance p from the axis of rotation to the longitudinal axis of that pile.

Let P denote the load induced in the pile. Then, $P = \Delta LAE/L$. Since ΔL is proportional to p and AE is constant for the group, this equation may be transformed to $P = kp/L$, Eq. (193), where k is a constant of proportionality.

If the center of rotation is established, the pile loads may therefore be found by scaling the p distances. Westergaard devised a simple graphical method of locating the center of rotation. This method entails the construction of string polygons, described in the first Calculation Procedure of this Handbook.

In Fig. 228a, select any convenient point a on the action line of the load. Consider the structure to be subjected to a load H_a that causes the pier to rotate about a as a center. The object is to locate the action line of this hypothetical load.

It is often desirable to visualize that a load is applied to a body at a point that in reality lies outside the body. This condition becomes possible if the designer annexes to the body an infinitely rigid arm containing the given point. Since this arm does not deform, the stresses and strains in the body proper are not modified.

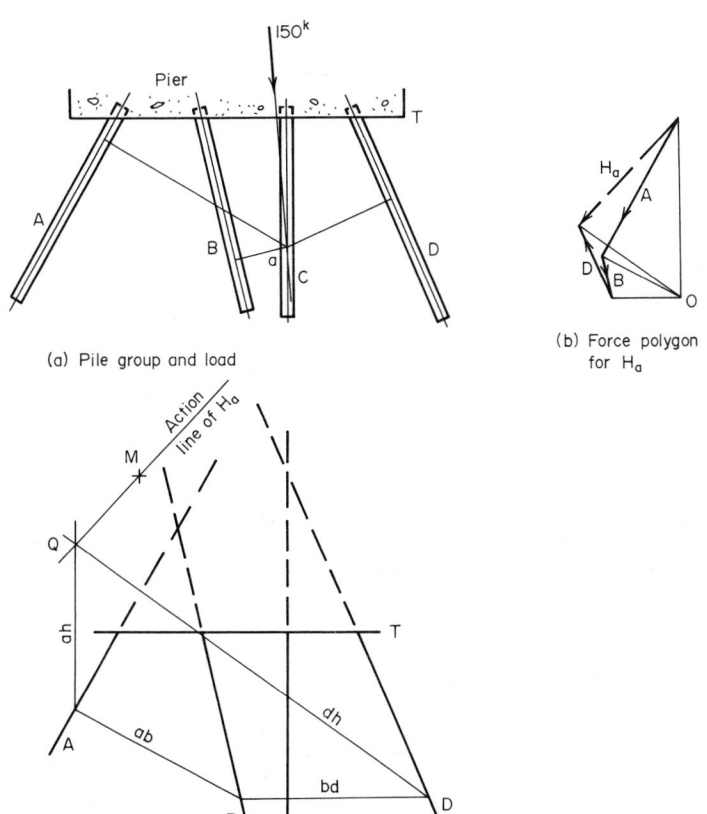

(a) Pile group and load

(b) Force polygon for H_a

(c) Construction to locate action line of H_a

Fig. 228

2. Scale the perpendicular distance from a to the longitudinal axis of each pile; divide this distance by the relative length of the pile

In accordance with Eq. (193), the quotient represents the relative magnitude of the load induced in the pile by the load H_a. If rotation is assumed to be counterclockwise, piles A and B are in compression and D is in tension.

3. Using a suitable scale, construct the force polygon

This polygon is shown in Fig. 228*b*. Construct this polygon by applying the results obtained in Step 2. This force polygon yields the direction of the action line of H_a.

4. In Fig. 228b, select a convenient pole O and draw rays to the force polygon

5. Construct the string polygon shown in Fig. 228c

The action line of H_a passes through the intersection point Q of rays ah and dh, and its direction appears in Fig. 228*b*. Draw this line.

6. Select a second point on the action line of the load

Choose point b on the action line of the 150-kip load and consider the structure to be subjected to a load H_b that causes the pier to rotate about b as center.

7. Locate the action line of H_b

Repeat the foregoing procedure to locate the action line of H_b in Fig. 228*c*. (The construction has been omitted for clarity.) Study of the diagram shows that the action lines of H_a and H_b intersect at M.

8. Test the accuracy of the construction

Select a third point c on the action line of the 150-kip load and locate the action line of the hypothetical load H_c causing rotation about c. It is found that H_c also passes through M. In summary, these hypothetical loads causing rotation about specific points on the action line of the true load are all concurrent.

Thus, M is the center of rotation of the pier under the 150-kip load. This conclusion stems from the following analysis: Load H_a applied at M causes zero deflection at a. Therefore, in accordance with Maxwell's theorem of reciprocal deflections, if the true load is applied at a it will cause zero deflection at M in the direction of H_a. Similarly, if the true load is applied at b, it will cause zero deflection at M in the direction of H_b. Thus, M remains stationary under the 150-kip load; i.e., M is the center of rotation of the pier.

9. Scale the perpendicular distance from M to the longitudinal axis of each pile

Divide this distance by the relative length of the pile. The quotient represents the relative magnitude of the load induced in the pile by the 150-kip load.

10. Using a suitable scale, construct the force polygon by applying the results from Step 9

If the work is accurate, it will be found that the resultant of these relative loads is parallel to the true load.

11. Scale the resultant; compute the factor needed to correct this value to 150 kips

12. Multiply each relative pile load by this correction factor to obtain the true load induced in the pile

LOAD DISTRIBUTION AMONG PILES WITH FIXED BASES

Assume that the piles in Fig. 228*a* penetrate a considerable distance into a compact soil and may therefore be regarded as restrained against rotation at a certain level. Outline a procedure for determining the axial load and bending moment induced in each pile.

Calculation Procedure:

1. State the equation for the length of a dummy pile

Since the Westergaard construction presented in the previous Calculation Procedure applies solely to hinged piles, the group of piles now being considered is not directly amenable to analysis by this method.

As shown in Fig. 229*a*, the pile *AB* functions in the dual capacity of a column and cantilever beam. In Fig. 229b, let A' denote the position of A following application of the load. If secondary effects are disregarded, the axial force P transmitted to this pile is a function of Δ_y, and the transverse force S is a function of Δ_x.

Consider that the fixed support at B is replaced with a hinged support and a pile AC of identical cross section is added perpendicular to AB, as shown in Fig. 229b. If pile

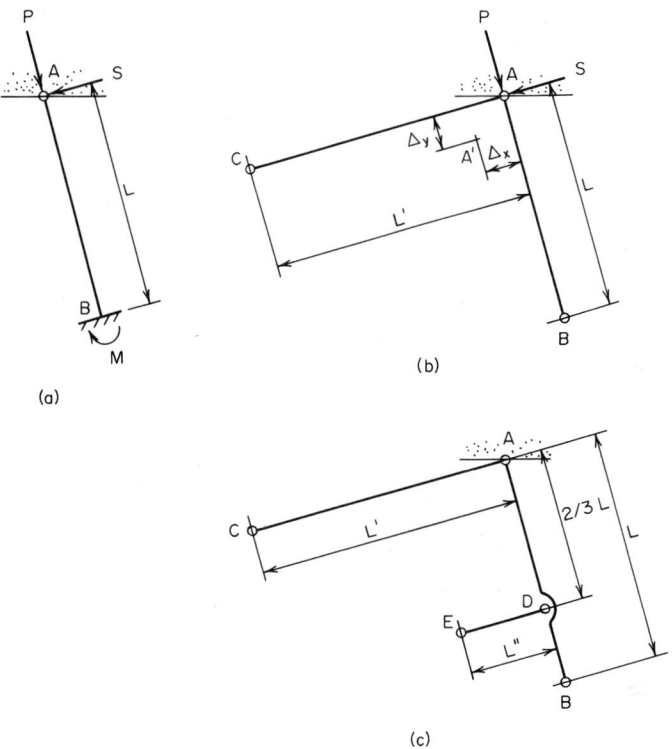

Fig. 229   Real and dummy piles.

AC deforms an amount Δ_x under an axial force S, the forces transmitted by the pier at each point of support are not affected by this modification of supports. The added pile is called a *dummy* pile. Thus, the given pile group may be replaced with an equivalent group consisting solely of hinged piles.

Stating the equation for the length L' of the dummy pile, equate the displacement Δ_x in the equivalent pile group to that in the actual group. Or, $\Delta_x = SL'/AE = SL^3/3EI$; $L' = AL^3/3I$, Eq. (194).

2.   *Replace all fixed supports in the given pile group with hinged supports*

Add the dummy piles. Compute the lengths of these piles by applying Eq. (194).

3.   *Determine the axial forces induced in the equivalent pile group*

Using the given load, apply Westergaard's construction, as described in the previous Calculation Procedure.

4.   *Remove the dummy piles; restore the fixed supports*

Then compute the bending moments at these supports by applying the equation $M = SL$.

LOAD DISTRIBUTION AMONG PILES FIXED AT TOP AND BOTTOM

Assume that the piles in Fig. 228*a* may be regarded as having fixed supports both at the pier and at their bases. Outline a procedure for determining the axial load and bending moment induced in each pile.

Calculation Procedure:

1. Describe how dummy piles may be used

A pile made of reinforced concrete and built integrally with the pier is restrained against rotation relative to the pier. As shown in Fig. 229*c*, the fixed supports of pile *AB* may be replaced with hinges provided that dummy piles *AC* and *DE* are added, the latter being connected to the pier by means of a rigid arm through *D*.

2. Compute the lengths of the dummy piles

If *D* is placed at the lower third point as indicated, the lengths to be assigned to the dummy piles are $L' = AL^3/3I$ and $L'' = AL^3/9I$, Eq. (195). Replace the given group of piles with its equivalent group, and follow the method of solution in the previous Calculation Procedure.

Section 2

Architectural Engineering

MAX KURTZ, P.E.
TYLER G. HICKS, P.E.

ARCHITECTURAL ENGINEERING

REFERENCES: Aluminum Association—*Aluminum Construction Manual*; American Concrete Institute—*Building Code Requirements for Reinforced Concrete*; American Institute of Steel Construction—*Manual of Steel Construction*; American Iron and Steel Institute—*Light Gage Cold-Formed Steel Design Manual*; National Forest Products Association—*National Design Specification for Stress-Grade Lumber and its Fastenings*; Beedle, et al.—*Structural Steel Design*, Ronald; Ferguson—*Reinforced Concrete Fundamentals*, Wiley; Gaylord and Gaylord—*Design of Steel Structures*, McGraw-Hill; Grinter—*Design of Modern Steel Structures*, Macmillan; Hoadley—*Essentials of Structural Design*, Wiley; Leontovich—*Frames and Arches*, McGraw-Hill; Lothers—*Advanced Design in Structural Steel*, Prentice-Hall; Norris et al.—*Structural Design for Dynamic Loads*, McGraw-Hill; Salvadori and Levy—*Structural Design in Architecture*, Prentice-Hall; Sanks—*Statically Indeterminate Structural Analysis*, Ronald; Sutherland and Bowman—*Structural Theory*, Wiley; Timoshenko and Young—*Vibration Problems in Engineering*, Van Nostrand; Viest, Fountain, and Singleton—*Composite Construction in Steel and Concrete*, McGraw-Hill; Weidlinger—*Aluminum in Modern Architecture*, Vol. II, Reinhold; Williams and Harris—*Structural Design in Metals*, Ronald; Winter, et al.—*Design of Concrete Structures*, McGraw-Hill.

The basic principles of structural design are developed and applied in Sec. 1, Civil Engineering.

In the following Calculation Procedures, structural steel members are designed in accordance with the *Specification for the Design, Fabrication and Erection of Structural Steel for Buildings* of the American Institute of Steel Construction. In the absence of any statement to the contrary, it is to be understood that the structural-steel members are made of ASTM A36 steel, which has a yield-point stress of 36,000 psi.

Reinforced-concrete members are designed in accordance with the specification *Building Code Requirements for Reinforced Concrete* of the American Concrete Institute.

DESIGN OF AN EYEBAR

A hanger is to carry a load of 175 kips. Design an eyebar of A440 steel.

Calculation Procedure:

1. Record the yield-point stresses of the steel

Refer to Fig. 1 for the notational system. Let the subscripts 1 and 2 refer to cross sections through the body of the bar and through the center of the pin hole, respectively.

Eyebars are generally flame-cut from plates of high-strength steel. The design provisions of the AISC *Specification* reflect the results of extensive testing of such members. A section of the *Specification* permits a tensile stress of $0.60f_y$ at 1 and $0.45f_y$ at 2, where f_y denotes the yield-point stress.

From the AISC *Manual* for A440 steel:
If $t \leqslant 0.75$ in., $f_y = 50$ ksi.
If $0.75 < t \leqslant 1.5$ in., $f_y = 46$ ksi.
If $1.5 < t \leqslant 4$ in., $f_y = 42$ ksi.

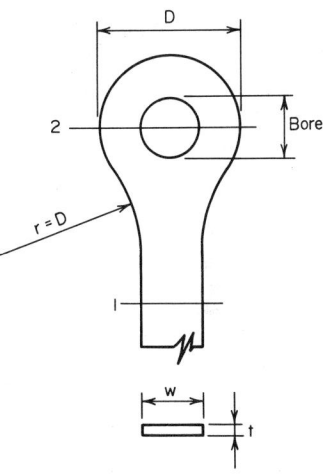

Fig. 1   Eyebar hanger.

2. Design the body of the member, using a trial thickness

The *Specification* restricts the ratio w/t to a value of 8. Compute the capacity P of a $\frac{3}{4}$ in. eyebar of maximum width. Thus: $w = 8(\frac{3}{4}) = 6$ in.; $f = 0.6(50) = 30$ ksi; $P = 6(0.75) \times (30) = 135$ kips. This is not acceptable because the desired capacity is 175 kips. Hence, the required thickness therefore exceeds the trial value of $\frac{3}{4}$ in. With t greater than $\frac{3}{4}$ in., the allowable stress at 1 is $0.60f_y$, or $0.60(46\,\text{ksi}) = 27.6$ ksi; say 27.5 for design use. At 2 the allowable stress is $0.45(46) = 20.7$ ksi; say 20.5 ksi for design purposes.

To determine the required area at 1, use the relation $A_1 = P/f$, where $f =$ allowable stress as computed above. Thus, $A_1 = 175/27.5 = 6.36$ sq in. Use a plate $6\frac{1}{2} \times 1$ in. in which $A_1 = 6.5$ sq in.

3. Design the section through the pin hole

The AISC *Specification* limits the pin diameter to a minimum value of $7\,w/8$. Select a pin diameter of 6 in. The bore will then be $6\frac{3}{32}$ in. diameter. The net width required will be $P/ft = 175/[20.5(1.0)] = 8.54$ in.; $D_{min} = 6.03 + 8.54 = 14.57$ in. Set $D = 14\frac{3}{4}$ in.; $A_2 = 1.0(14.75 - 6.03) = 8.72$ sq in.; $A_2/A_1 = 1.34$. This result is satisfactory, because the ratio of A_2/A_1 must lie between 1.33 and 1.50.

4. Determine the transition radius r

In accordance with the *Specification*, set $r = D = 14\frac{3}{4}$ in.

ANALYSIS OF A STEEL HANGER

A $12 \times \frac{1}{2}$ in. steel plate is to support a tensile load applied 2.2 in. from its center. Determine the ultimate load.

Calculation Procedure:

1. Determine the distance x

The plastic analysis of steel structures is developed in Sec. 1 of this Handbook. Figure 2a is the load diagram and Fig. 2b is the stress diagram at plastification. The latter may be replaced for convenience with the stress diagram in Fig. 2c, where $T_1 = C$; $P_u =$ ultimate load; $e =$ eccentricity; $M_u =$ ultimate moment $= P_u e$; $f_y =$ yield-point stress; $d =$ depth of section; $t =$ thickness of section.

Using Fig. 2c, $P_u = T_2 = f_y t(d-2x)$, Eq. (1). Also, $T_1 = f_y tx$; $M_u = P_u e = T_1(d-x)$; $x = d/2 + e - [(d/2+e)^2 - ed]^{0.5}$ Eq. (2). Or, $x = 6 + 2.2 - [(6+2.2)^2 - 2.2 \times 12]^{0.5} = 1.81$ in.

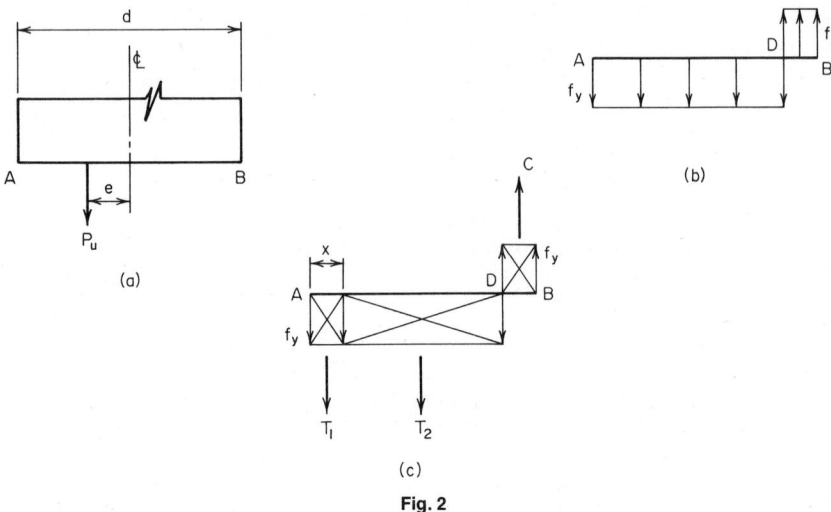

Fig. 2

2. Find P_u

By Eq. (1), $P_u = 36,000(0.50)(12-3.62) = 151,000$ lb.

ANALYSIS OF A GUSSET PLATE

The gusset plate in Fig. 3 is $\frac{1}{2}$ in. thick and connects three web members to the bottom chord of a truss. The plate is subjected to the indicated ultimate forces, and transfer of these forces from the web members to the plate is completed at section *a-a*. Investigate the adequacy of this plate. Use 18,000 psi as the yield-point stress in shear, and disregard interaction of direct stress and shearing stress in computing the ultimate-load and ultimate-moment capacity.

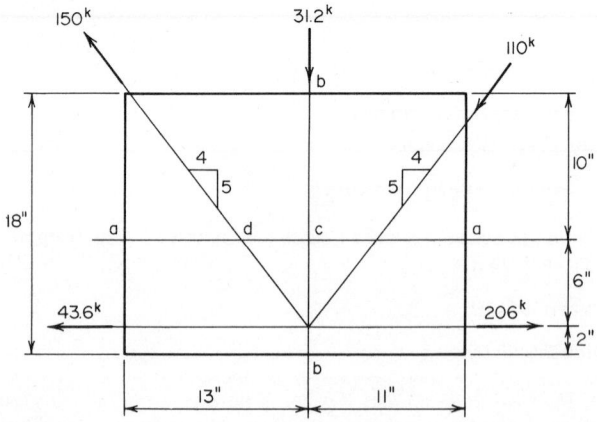

Fig. 3   Gusset plate.

Calculation Procedure:

1. Resolve the diagonal forces into their horizontal and vertical components

Let H_u and V_u denote the ultimate shearing force on a horizontal and vertical plane, respectively. Resolving the diagonal forces into their horizontal and vertical components, $(4^2 + 5^2)^{0.5} = 6.40$. Horizontal components: $150(4/6.40) = 93.7$ kips; $110(4/6.40) = 68.7$ kips. Vertical components: $150(5/6.40) = 117.1$ kips; $110(5/6.40) = 85.9$ kips.

2. Check the force system for equilibrium

Thus, $\Sigma F_H = 206.0 - 43.6 - 93.7 - 68.7 = 0$; this is satisfactory, as is $\Sigma F_V = 117.1 - 85.9 - 31.2 = 0$.

3. Compare the ultimate shear at section a-a with the allowable value

Thus, $H_u = 206.0 - 43.6 = 162.4$ kips. To compute $H_{u,\text{allow}}$, assume that the shearing stress is equal to the yield-point stress across the entire section. Then $H_{u,\text{allow}} = 24(0.5) \times (18) = 216$ kips. This is satisfactory.

4. Compare the ultimate shear at section b-b with the allowable value

Thus, $V_u = 117.1$ kips; $V_{u,\text{allow}} = 18(0.5)(18) = 162$ kips. This is satisfactory.

5. Compare the ultimate moment at section a-a with the plastic moment

Thus, $cd = 4(6)/5 = 4.8$ in.; $M_u = 4.8(117.1 + 85.9) = 974$ in.-kips. Or, $M_u = 6(206 - 43.6) = 974$ in.-kips. To find the plastic moment M_p, use the relation $M_p = f_y bd^2/4$, or $M_p = 36(0.5)(24)^2/4 = 2,592$ in.-kips. This is satisfactory.

6. Compare the ultimate direct force at section b-b with the allowable value

Thus, $P_u = 93.7 + 43.6 = 137.3$ kips; or $P_u = 206.0 - 68.7 = 137.3$ kips; $e = 9 - 2 = 7$ in. By Eq. (2), $x = 9 + 7 - [(9 + 7)^2 - 7 \times 18]^{0.5} = 4.6$ in. By Eq. (1), $P_{u,\text{allow}} = 36,000 \times (0.5)(18 - 9.2) = 158.4$ kips. This is satisfactory.

On horizontal sections above a-a, the forces in the web members have not been completely transferred to the gusset plate, but the eccentricities are greater than those at a-a. Therefore, the calculations in Step 5 should be repeated with reference to one or two sections above a-a before any conclusion concerning the adequacy of the plate is drawn.

DESIGN OF A SEMIRIGID CONNECTION

A 14WF38 beam is to be connected to the flange of a column by a semirigid connection that transmits a shear of 25 kips and a moment of 315 in.-kips. Design the connection for the moment, using A141 shop rivets and A325 field bolts of $\frac{7}{8}$-in. diameter.

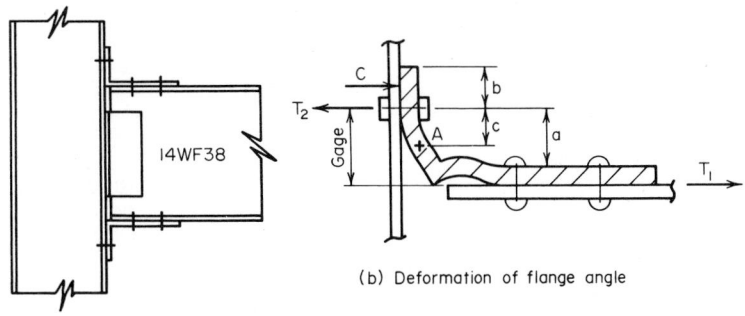

(b) Deformation of flange angle

(a) Semirigid connection

Fig. 4 (a) **Semirigid connection;** (b) **deformation of flange angle.**

Calculation Procedure:

1. Record the relevant properties of the 14WF38

A semirigid connection is one that offers only partial restraint against rotation. For a relatively small moment, a connection of the type shown in Fig. 4a will be adequate. In designing this type of connection, it is assumed for simplicity that the moment is resisted entirely by the flanges; and the force in each flange is found by dividing the moment by the beam depth.

Figure 4b indicates the assumed deformation of the upper angle, A being the point of contraflexure in the vertical leg. Since the true stress distribution cannot be readily ascertained, it is necessary to make simplifying assumptions. The following equations evolve from a conservative analysis of the member: $c = 0.6a$; $T_2 = T_1(1 + 3a/4b)$.

Study shows that use of an angle having two rows of bolts in the vertical leg would be unsatisfactory because the bolts in the outer row would remain inactive until those in the inner row yielded. If the two rows of bolts are required, the flange should be connected by means of a tee rather than an angle.

The following notational system will be used with reference to the beam dimensions: b = flange width; d = beam depth; t_f = flange thickness; t_w = web thickness.

Recording the relevant properties of the 14WF38: $d = 14.12$ in.; $t_f = 0.513$ in. (Obtain these properties from a table of structural-shape data.)

2. Establish the capacity of the shop rivets and field bolts used in transmitting the moment

From the AISC Specification: rivet capacity in single shear = 0.6013(15) = 9.02 kips; rivet capacity in bearing = 0.875(0.513)(48.5) = 21.77 kips; bolt capacity in tension = 0.6013(40) = 24.05 kips.

3. Determine the number of rivets required in each beam flange

Thus, T_1 = moment/d = 315/14.12 = 22.31 kips; no. of rivets = T_1/rivet capacity in single shear = 22.31/9.02 = 2.5; use four rivets, the next highest even number.

4. Assuming tentatively that one row of field bolts will suffice, design the flange angle

Try an angle $8 \times 4 \times \frac{3}{4}$ in., 8 in. long, having a standard gage of $2\frac{1}{2}$ in. in the vertical leg. Compute the maximum bending moment M in this leg. Thus, $c = 0.6(2.5 - 0.75) = 1.05$ in.; $M = T_1c = 23.43$ in.-kips. Then apply the relation $f = M/S$ to find the flexural stress. Or, $f = 23.43/[(\frac{1}{6})(8)(0.75)^2] = 31.24$ ksi.

Since the cross section is rectangular, the allowable stress is 27 ksi, as given by the AISC Specification. (The justification for allowing a higher flexural stress in a member of rectangular cross section as compared with a wide-flange member is presented in the previous section of this Handbook.)

Try a $\frac{7}{8}$ in. angle, with $c = 0.975$ in.; $M = 21.75$ in.-kips; $f = 21.75/(\frac{1}{6})(8)(0.875)^2 = 21.3$ ksi. This is an acceptable stress.

5. Check the adequacy of the two field bolts in each angle

Thus, $T_2 = 22.31[1 + 3 \times 1.625/(4 \times 1.5)] = 40.44$ kips; the capacity of two bolts = 2(24.05) = 48.10 kips. Hence the bolts are acceptable because their capacity exceeds the load.

6. Summarize the design

Use angles $8 \times 4 \times \frac{7}{8}$ in., 8 in. long. In each angle, use four rivets for the beam connection and two bolts for the column connection. For transmitting the shear, the standard web connection for a 14-in. beam shown in the AISC Manual is satisfactory.

RIVETED MOMENT CONNECTION

An 18WF60 beam frames to the flange of a column and transmits a shear of 40 kips and a moment of 2,500 in.-kips. Design the connection, using $\frac{7}{8}$ in.-diameter rivets of A141 steel for both the shop and field connections.

Calculation Procedure:

1. *Record the relevant properties of the 18WF60*

The connection is shown in Fig. 5a. Referring to the row of rivets in Fig. 5b, consider that there are n rivets having a uniform spacing p. The moment of inertia and section modulus of this rivet group with respect to its horizontal centroidal axis are: $I = p^2 n \times (n^2 - 1)/12$; $S = pn(n+1)/6$, Eq. (3).

Recording the properties of the 18WF60, $d = 18.25$ in.; $b = 7.558$ in.; $k = 1.18$ in.; $t_f = 0.695$ in.; $t_w = 0.416$ in.

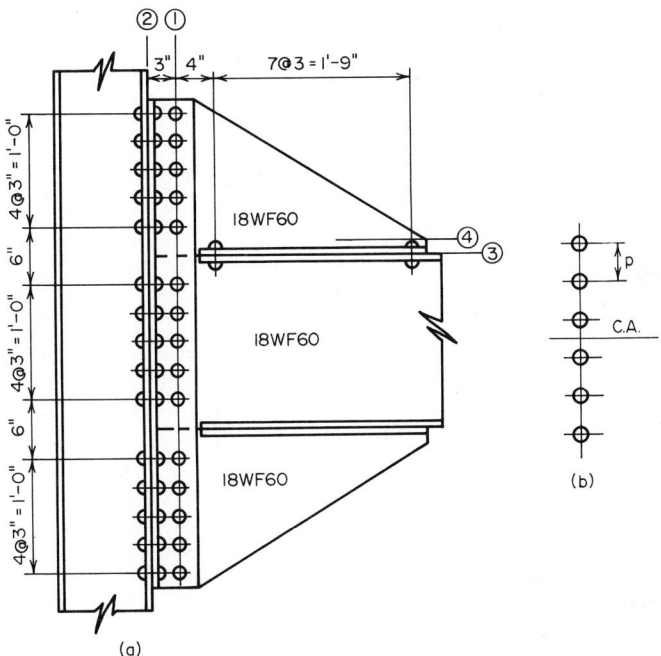

Fig. 5   Riveted moment connection.

2. *Establish the capacity of a rivet*

Thus: single shear, 9.02 kips; double shear, 18.04 kips; bearing on beam web, $0.875 \times (0.416)(48.5) = 17.65$ kips.

3. *Determine the number of rivets required on line 1 as governed by the rivet capacity*

Try 15 rivets having the indicated disposition. Apply Eq. (3) with $n = 17$; then make the necessary correction. Thus, $I = 9(17)(17^2 - 1)/12 - 2(9)^2 = 3,510$ in.^{2}; $S = 3,510/24 = 146.3$ in.

Let F denote the force on a rivet, and let the subscripts x and y denote the horizontal and vertical component, respectively. Thus, $F_x = M/S = 2,500/146.3 = 17.09$ kips; $F_y = 40/15 = 2.67$ kips; $F = (17.09^2 + 2.67^2)^{0.5} = 17.30 < 17.65$. Therefore, this is acceptable.

4. *Compute the stresses in the web plate at line 1*

The plate is considered continuous; the rivet holes are assumed to be 1 in. in diameter for the reasons explained in the previous section of this Handbook.

The total depth of the plate is 51 in.; the area and moment of inertia of the net section are: $A_n = 0.416(51 - 15 \times 1) = 14.98$ sq in.; $I_n = (1/12)(0.416)(51)^3 - 1.0(0.416)(3,510) = 3,138$ in.^{4}.

Apply the general shear equation. Since the section is rectangular, the maximum shearing stress is $v = 1.5V/A_n = 1.5(40)/14.98 = 4.0$ ksi. The AISC *Specification* gives an allowable stress of 14.5 ksi.

The maximum flexural stress is $f = Mc/I_n = 2,500(25.5)/3,138 = 20.3 < 27$ ksi. This is acceptable. The use of 15 rivets is therefore satisfactory.

5. Compute the stresses in the rivets on line 2

The center of rotation of the angles cannot be readily located because it depends on the amount of initial tension to which the rivets are subjected. For a conservative approximation, assume that the center of rotation of the angles coincides with the horizontal centroidal axis of the rivet group. The forces are: $F_x = 2,500/[2(146.3)] = 8.54$ kips; $F_y = 40/30 = 1.33$ kips. The corresponding stresses in tension and shear are: $s_t = F_x/A = 8.54/0.6013 = 14.20$ ksi; $s_s = F_y/A = 1.33/0.6013 = 2.21$ ksi. The *Specification* gives $s_{t,\text{allow}} = 28 - 1.6(2.21) > 20$ ksi. This is acceptable.

6. Select the size of the connection angles

The angles are designed by assuming a uniform bending stress across a distance equal to the spacing p of the rivets; the maximum stress is found by applying the tensile force on the extreme rivet.

Try $4 \times 4 \times \frac{3}{4}$ in. angles, with a standard gage of $2\frac{1}{2}$ in. in the outstanding legs. Assuming the point of contraflexure to have the location specified in the previous Calculation Procedure, $c = 0.6(2.5 - 0.75) = 1.05$ in.; $M = 8.54(1.05) = 8.97$ in.-kips; $f = 8.97/[(\frac{1}{6})(3)(0.75)^2] = 31.9 > 27$ ksi. Use $5 \times 5 \times \frac{7}{8}$ in. angles, with a $2\frac{1}{2}$-in. gage in the outstanding legs.

7. Determine the number of rivets required on line 3

The forces on the rivets above this line are shown in Fig. 6a. The resultant forces are: $H = 64.11$ kips; $V = 13.35$ kips. Let M_3 denote the moment of H with respect to line 3. Then, $a = \frac{1}{2}(24 - 18.25) = 2.88$ in.; $M_3 = 633.3$ in.-kips.

With reference to Fig. 6b, the tensile force F_y in the rivet is usually limited by the bending capacity of the beam flange. As shown in the AISC *Manual*, the standard gage in the 18WF60 is $3\frac{1}{2}$ in. Assume that the point of contraflexure in the beam flange lies midway between the center of the rivet and the face of the web. Referring to Fig. 4b, $c = \frac{1}{2}(1.75 - 0.416/2) = 0.771$ in.; $M_{\text{allow}} = fS = 27(\frac{1}{6})(3)(0.695)^2 = 6.52$ in.-kips. If the compressive force C is disregarded, $F_{y,\text{allow}} = 6.52/0.771 = 8.46$ kips.

Try 16 rivets. The moment on the rivet group is $M = 633.3 - 13.35(14.5) = 440$ in.-kips. By Eq. (3), $S = 2(3)(8)(9)/6 = 72$ in. Also, $F_y = 440/72 + 13.35/16 = 6.94 < 8.46$ kips. This is acceptable. (The value of F_y corresponding to 14 rivets is excessive.)

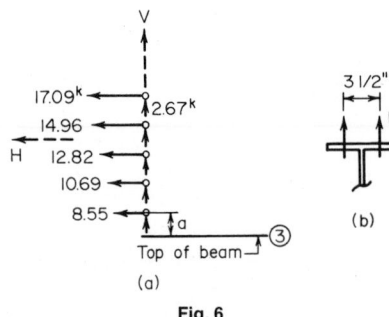

Fig. 6

The rivet stresses are: $s_t = 6.94/0.6013 = 11.54$ ksi; $s_s = 64.11/[16(0.6013)] = 6.67$ ksi. From the *Specification*, $s_{t,\text{allow}} = 28 - 1.6(6.67) = 17.33$ ksi. This is acceptable. The use of 16 rivets is therefore satisfactory.

8. Compute the stresses in the bracket at the toe of the fillet (line 4)

Since these stresses are seldom critical, take the length of the bracket as 24 in. and disregard the eccentricity of V. Then, $M = 633.3 - 64.11(1.18) = 558$ in.-kips; $f = 558/[(\frac{1}{6})(0.416)(24)^2] + 13.35/[0.416(24)] = 15.31$ ksi. This is acceptable. Also, $v = 1.5(64.11)/[0.416(24)] = 9.63$ ksi. This is also acceptable.

DESIGN OF A WELDED FLEXIBLE BEAM CONNECTION

An 18WF64 beam is to be connected to the flange of its supporting column by means of a welded framed connection, using E60 electrodes. Design a connection to transmit a reaction of 40 kips. The AISC table of welded connections may be applied in selecting the connection, but the design must be verified by computing the stresses.

Calculation Procedure:

1. Record the pertinent properties of the beam

It is necessary to investigate both the stresses in the weld and the shearing stress in the beam induced by the connection. The framing angles must fit between the fillets of the beam. Recording the properties, $T = 15\frac{3}{8}$ in.; $t_w = 0.403$ in.

2. Select the most economical connection from the AISC Manual

The most economical connection is: angles $3 \times 3 \times \frac{5}{16}$ in., 12 in. long; weld size $\frac{3}{16}$ in. for connection to beam web; $\frac{1}{4}$ in. for connection to the supporting member.

According to the AISC table, weld A has a capacity of 40.3 kips and weld B has a capacity of 42.8 kips. The minimum web thickness required is 0.25 in. The connection is shown in Fig. 7a.

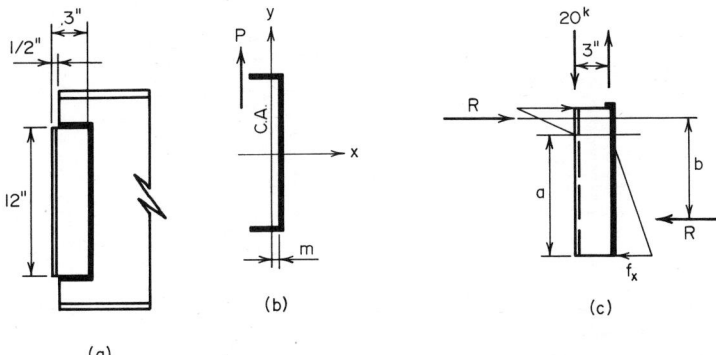

(a)

(b)

(c)

Fig. 7 Welded flexible beam connection.

3. Compute the unit force in the shop weld

The shop weld connects the angles to the beam web. Refer to Sec. 1 of this Handbook for two Calculation Procedures for analyzing welded connections.

The weld for one angle is shown in Fig. 7b. The allowable force, as given in Sec. 1, is: $m = 2(2.5)(1.25)/[2(2.5) + 12] = 0.37$ in.; $P = 20,000$ lb; $M = 20,000(3 - 0.37) = 52,600$ in.-lb; $I_x = (\frac{1}{12})(12)^3 + 2(2.5)(6)^2 = 324$ in.^{3}; $I_y = 12(0.37)^2 + 2(\frac{1}{12})(2.5)^3 + 2(2.5) \times (0.88)^2 = 8$ in.^{3}; $J = 324 + 8 = 332$ in.^{3}; $f_x = My/J = 52,600(6)/332 = 951$ pli; $f_y = Mx/J = 52,600(2.5)(0.37)/332 = 337$ pli; $f'_y = 20,000/(2 \times 2.5 + 12) = 1,176$ pli; $F_x = 951$ pli; $F_y = 337 + 1,176 = 1,513$ pli; $F = (951^2 + 1,513^2)^{0.5} = 1,787 < 1,800$, which is acceptable.

4. Compute the shearing stress in the web

The allowable stress given in the AISC *Manual* is 14,500 psi. The two angles transmit a unit shearing force of 3,574 pli to the web. The shearing stress is $v = 3,574/0.403 = 8,870$ psi, which is acceptable.

5. Compute the unit force in the field weld

The field weld connects the angles to the supporting member. As a result of the 3-in. eccentricity on the outstanding legs, the angles tend to rotate about a neutral axis located near the top, bearing against the beam web above this axis and pulling away from the web below this axis. Assume that the distance from the top of the angle to the neutral axis is one-sixth of the length of the angle. The resultant forces are shown in

Fig. 7c. Then, $a = (\frac{5}{6})12 = 10$ in.; $b = (\frac{2}{3})12 = 8$ in.; $R = 20{,}000(3)/8 = 7{,}500$ lb; $f_x = 2R/a = 1{,}500$ pli; $f_y = 20{,}000/12 = 1{,}667$ pli; $F = (1{,}500^2 + 1{,}667^2)^{0.5} = 2{,}240 < 2{,}400$ pli, which is acceptable. The weld is returned a distance of $\frac{1}{2}$ in. across the top of the angle, as shown in the AISC *Manual*.

DESIGN OF A WELDED SEATED BEAM CONNECTION

A 27WF94 beam with a reaction of 77 kips is to be supported on a seat. Design a welded connection, using E60 electrodes.

Calculation Procedure:

1. Record the relevant properties of the beam

Refer to the AISC *Manual*. The connection will consist of a horizontal seat plate and a stiffener plate below the seat, as shown in Fig. 8a. Recording the relevant properties of the 27WF94, $k = 1.44$ in.; $b = 9.99$ in.; $t_f = 0.747$ in.; $t_w = 0.490$ in.

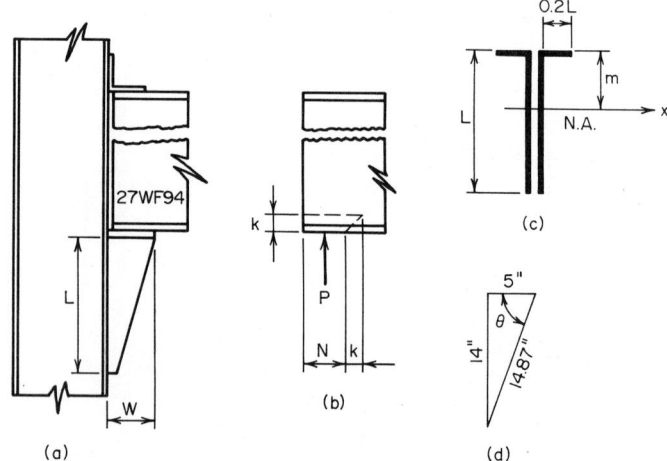

Fig. 8   Welded seated beam connection.

2. Compute the effective length of bearing

Equate the compressive stress at the toe of the fillet to its allowable value of 27 ksi as given in the AISC *Manual*. Assume that the reaction distributes itself through the web at an angle of 45°. Refer to Fig. 8b. Then, $N = P/27t_w - k$, or $N = 77/27(0.490) - 1.44 = 4.38$ in.

3. Design the seat plate

As shown in the AISC *Manual*, the beam is set back about $\frac{1}{2}$ in. from the face of the support. Make $W = 5$ in. The minimum allowable distance from the edge of the seat plate to the edge of the flange equals the weld size plus $\frac{5}{16}$ in. Make the seat plate 12 in. long; its thickness will be made the same as that of the stiffener.

4. Design the weld connecting the stiffener plate to the support

The stresses in this weld are not amenable to precise analysis. The stiffener rotates about a neutral axis, bearing against the support below this axis and pulling away from the support above this axis. Assume for simplicity that the neutral axis coincides with the centroidal axis of the weld group; the maximum weld stress occurs at the top. A weld length of 0.2L is supplied under the seat plate on each side of the stiffener. Refer to Fig. 8c.

Compute the distance e from the face of the support to the center of the bearing, measuring N from the edge of the seat. Thus, $e = W - N/2 = 5 - 4.38/2 = 2.81$ in.; $P = 77$ kips; $M = 77(2.81) = 216.4$ in.-kips; $m = 0.417L$; $I_x = 0.25L^3$; $f_1 = Mc/I_x = 216.4$ $(0.417L)/0.25L^3 = 361.0/L^2$ kli; $f_2 = P/A = 77/2.4L = 32.08/L$ kli. Use a $\frac{5}{16}$-in. weld, which has a capacity of 3 kli. Then, $F^2 = f_1^2 + f_2^2 = 130{,}300/L^4 + 1{,}029/L^2 \leq 3^2$. This equation is satisfied by $L = 14$ in.

5. Determine the thickness of the stiffener plate

Assume this plate is triangular, Fig. 8d. The critical section for bending is assumed to coincide with the throat of the plate, and the maximum bending stress may be obtained by applying $f = (P/tW \sin^2 \theta)(1 + 6e'/W)$, where $e' =$ distance from center of seat to center of bearing.

Using an allowable stress of 22,000 psi, $e' = e - 2.5 = 0.31$ in.; $t = \{77/[22 \times 5(14/14.87)^2]\}(1 + 6 \times 0.31/5) = 1.08$ in.

Use a $1\frac{1}{8}$-in. stiffener plate. The shearing stress in the plate caused by the weld is $v = 2(3{,}000)/1.125 = 5{,}330 < 14{,}500$ psi, which is acceptable.

DESIGN OF A WELDED MOMENT CONNECTION

A 16WF40 beam frames to the flange of a 12WF72 column and transmits a shear of 42 kips and a moment of 1,520 in.-kips. Design a welded connection, using E60 electrodes.

Calculation Procedure:

1. Record the relevant properties of the two sections

In designing a welded moment connection, it is assumed for simplicity that the beam flanges alone resist the bending moment. Consequently, the beam transmits three forces to the column: the tensile force in the top flange, the compressive force in the bottom flange, and the vertical load. Although the connection is designed ostensibly on an elastic design basis, it is necessary to consider its behavior at ultimate load, since a plastic hinge would form at this joint. The connection is shown in Fig. 9.

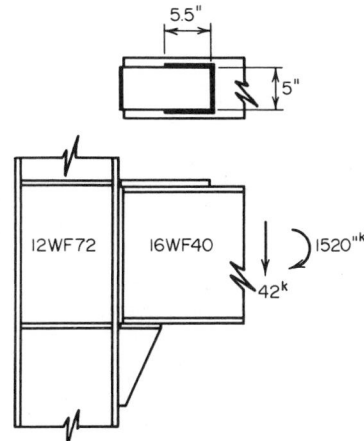

Recording the relevant properties of the sections: for the 16WF40, $d = 16.00$ in.; $b = 7.00$ in.; $t_f = 0.503$ in.; $t_w = 0.307$ in.; $A_f = 7.00(0.503) = 3.52$ sq in. For the 12WF72, $k = 1.25$ in.; $t_f = 0.671$ in.; $t_w = 0.403$ in.

2. Investigate the need for column stiffeners, design the stiffeners if they are needed

The forces in the beam flanges introduce two potential modes of failure: crippling of the column web caused by the compressive force, and fracture of the weld transmitting the tensile force as a result of the bending of the column flange. The AISC *Specification*

Fig. 9   Welded moment connection.

establishes the criteria for ascertaining whether column stiffeners are required. The first criterion is obtained by equating the compressive stress in the column web at the toe of the fillet to the yield-point stress f_y; the second criterion was obtained empirically. At the ultimate load, the capacity of the unreinforced web $= (0.503 + 5 \times 1.25)0.430f_y = 2.904f_y$; capacity of beam flange $= 3.52f_y$; $0.4(A_f)^{0.5} = 0.4(3.52)^{0.5} = 0.750 > 0.671$ in.

Stiffeners are therefore required opposite both flanges of the beam. The required area is $A_{st} = 3.52 - 2.904 = 0.616$ sq in. Make the stiffener plates $3\frac{1}{2}$ in. wide to match the beam flange. From the AISC, $t_{min} = 3.5/8.5 = 0.41$ in. Use two $3\frac{1}{2} \times \frac{1}{2}$ in. stiffener plates opposite both beam flanges.

3. Design the connection plate for the top flange

Compute the flange force by applying the total depth of the beam. Thus, $F = 1,520/16.00 = 95$ kips; $A = 95/22 = 4.32$ in.

Since the beam flange is 7 in. wide, use a plate 5 in. wide and $\frac{7}{8}$ in. thick, for which $A = 4.38$ sq in. This plate is butt-welded to the column flange and fillet-welded to the beam flange. In accordance with the AISC *Specification*, the minimum weld size is $\frac{5}{16}$ in. and the maximum size is $\frac{13}{16}$ in. Use a $\frac{5}{8}$ in. weld, which has a capacity of 6,000 pli. Then, length of weld $= 95/6 = 15.8$ in., say 16 in. To ensure that yielding of the joint at ultimate load will occur in the plate rather than in the weld, the top plate is left unwelded for a distance approximately equal to its width, as shown in Fig. 9.

4. Design the seat

The connection plate for the bottom flange requires the same area and length of weld as does the plate for the top flange. The stiffener plate and its connecting weld are designed in the same manner as presented in the previous Calculation Procedure.

RECTANGULAR KNEE OF RIGID BENT

Figure 10a is the elevation of the knee of a rigid bent. Design the knee to transmit an ultimate moment of 8,100 in.-kips.

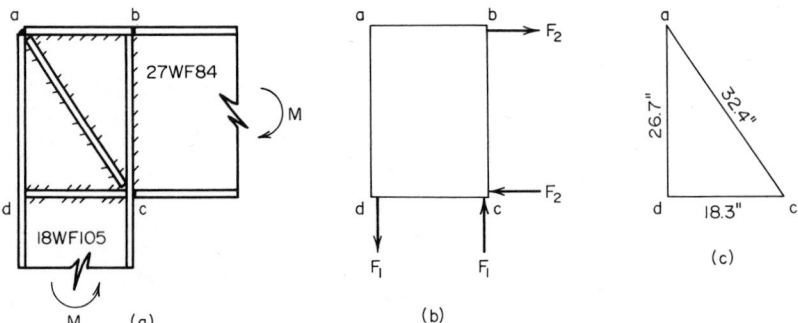

Fig. 10   Rectangular knee.

Calculation Procedure:

1. Record the relevant properties of the two sections

Refer to the AISC *Specification* and *Manual*. It is assumed that the moment in each member is resisted entirely by the flanges and that the distance between the resultant flange forces is 0.95 times the depth of the member.

Recording the properties of the members: for the 18WF105, $d = 18.32$ in.; $b_f = 11.79$ in.; $t_f = 0.911$ in.; $t_w = 0.554$ in.; $k = 1.625$ in. For the 27WF84, $d = 26.69$ in.; $b_f = 9.96$ in.; $t_f = 0.636$ in.; $t_w = 0.463$ in.

2. Compute F_1

Thus, $F_1 = M_u/0.95d = 8,100/[0.95(18.32)] = 465$ kips.

3. Determine if web stiffeners are needed to transmit F_1

The shearing stress is assumed to vary linearly from zero at a to its maximum value at d. The allowable average shearing stress is taken as $f_y/(3)^{0.5}$, where f_y denotes the yield-point stress. The capacity of the web $= 0.554(26.69)(36/3^{0.5}) = 307$ kips. Therefore use diagonal web stiffeners.

4. Design the web stiffeners

Referring to Fig. 10c, $ac = (18.3^2 + 26.7^2)^{0.5} = 32.4$ in. The force in the stiffeners $= (465 - 307)32.4/26.7 = 192$ kips. (The same result is obtained by computing F_2 and considering the capacity of the web across ab.) Then, $A_{st} = 192/36 = 5.33$ sq in. Use two plates $4 \times \frac{3}{4}$ in.

5. Design the welds, using E60 electrodes

The AISC *Specification* stipulates that the weld capacity at ultimate load is 1.67 times the capacity at the working load. Consequently, the ultimate-load capacity is 1,000 pli times the number of sixteenths in the weld size. The welds are generally designed to develop the full moment capacity of each member. Refer to the AISC *Specification*.

Weld at ab. This weld transmits the force in the flange of the 27-in. member to the web of the 18-in. member. Then, $F = 9.96(0.636)(36) = 228$ kips, weld force $= 228/[2(d - 2t_f)] = 228/[2(18.32 - 1.82)] = 6.91$ kli. Use a $\frac{7}{16}$ in. weld.

Weld at bc. Use a full-penetration butt weld.

Weld at ac. Use the minimum size of $\frac{1}{4}$ in. The required total length of weld is $L = 192/4 = 48$ in.

Weld at dc. Let F_3 denote that part of F_2 that is transmitted to the web of the 18-in. member through bearing, and let F_4 denote the remainder of F_2. Force F_3 distributes itself through the 18-in. member at 45° angles, and the maximum compressive stress occurs at the toe of the fillet. Find F_3 by equating this stress to 36 ksi; or $F_3 = 36(0.554)(0.636 + 2 \times 1.625) = 78$ kips. To evaluate F_4, apply the moment capacity of the 27-in. member. Or $F_4 = 228 - 78 = 150$ kips.

The minimum weld size of $\frac{1}{4}$ in. is inadequate. Use a $\frac{5}{16}$ in. weld. The required total length is $L = 150/5 = 30$ in.

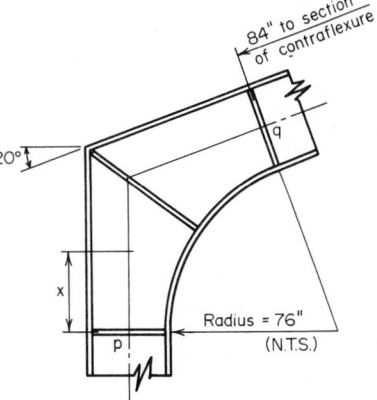

CURVED KNEE OF RIGID BENT

In Fig. 11 the rafter and column are both 21WF82, and the ultimate moment at the two sections of tangency $-p$ and $q-$ is 6,600 in.-kips. The section of contraflexure in each member lies at a distance of 84 in. from the section of tangency. Design the knee.

Calculation Procedure:

Fig. 11 Curved knee.

1. Record the relevant properties of the members

Refer to the Commentary in the AISC *Manual*. The notational system is the same as that used in the *Manual*, plus a = distance from section of contraflexure to section of tangency; b = member flange width; x = distance from section of tangency to given section; M = ultimate moment at given section; M_p = plastic-moment capacity of knee at given section.

Assume that the moment gradient dM/dx remains constant across the knee. The web thickness of the knee is made equal to that of the main material. The flange thickness of the knee, however, must exceed that of the main material, for this reason: As x increases, both M and M_p increase, but the former increases at a faster rate when x is small. The critical section occurs where $dM/dx = dM_p/dx$.

An exact solution to this problem is possible, but the resulting equation is rather cumbersome. An approximate solution is given in the AISC *Manual*.

Recording the relevant properties of the 21WF82: $d = 20.86$ in.; $b = 8.96$ in.; $t_f = 0.795$ in.; $t_w = 0.499$ in.

2. Design the cross section of the knee, assuming tentatively that flexure is the sole criterion

Use a trial thickness of $\frac{1}{2}$ in. for the web plate and a 9-in. width for the flange plate. Then, $a = 84$ in.; $n = a/d = 84/20.86 = 4.03$. From the AISC *Manual*, $m = 0.14 \pm; t' = t(1 + m) = 0.795(1.14) = 0.906$ in. Make the flange plate 1 in. thick.

3. Design the stiffeners; investigate the knee for compliance with the AISC Commentary

From the Commentary, *Item 5:* Provide stiffener plates at the sections of tangency and at the center of the knee. Make the stiffener plates $4 \times \frac{7}{8}$ in., one on each side of the web.

Item 3: Thus, $\phi = \frac{1}{2}(90° - 20°) = 35°$; $\phi = 35/57.3 = 0.611$ rad; $L = R\phi = 76(0.611) = 46.4$ in.; or $L = \pi R(70°/360°) = 46.4$ in.; $L_{cr} = 6b = 6(9) = 54$ in., which is acceptable.

Item 4: Thus, $b/t' = 9$; $2R/b = 152/9 = 16.9$, which is acceptable.

BASE PLATE FOR STEEL COLUMN CARRYING AXIAL LOAD

A 14WF53 column carries a load of 240 kips and is supported by a footing made of 3,000-psi concrete. Design the column base plate.

Calculation Procedure:

1. Compute the required area of the base plate; establish the plate dimensions

Refer to the base-plate diagram in the AISC *Manual.* The column load is assumed to be uniformly distributed within the indicated rectangle, and the footing reaction is assumed to be uniformly distributed across the base plate. The required thickness of the plate is established by computing the bending moment at the circumference of the indicated rectangle. Let $f =$ maximum bending stress in plate; $p =$ bearing stress; $t =$ thickness of plate.

The ACI *Code* permits a bearing stress of 750 psi if the entire concrete area is loaded and 1,125 psi if one-third of this area is loaded. Applying the 750-psi value, plate area = load, lb/750 = 240,000/750 = 320 sq in.

The dimensions of the 14WF53 are: $d = 13.94$ in.; $b = 8.06$ in.; $0.95d = 13.24$ in.; $0.80b = 6.45$ in. For economy, the projections m and n should be approximately equal. Set $B = 15$ in. and $C = 22$ in.; then, area = 15(22) = 330 sq in.; $p = 240,000/330 = 727$ psi.

2. Compute the required thickness of the base plate

Thus, $m = \frac{1}{2}(22 - 13.24) = 4.38$ in., which governs. Also, $n = \frac{1}{2}(15 - 6.45) = 4.28$ in.

The AISC *Specification* permits a bending stress of 27,000 psi in a rectangular plate. The maximum bending stress is: $f = M/S = 3pm^2/t^2$; $t = m(3p/f)^{0.5} = 4.38(3 \times 727/27,000)^{0.5} = 1.24$ in.

3. Summarize the design

Thus, $B = 15$ in.; $C = 22$ in.; $t = 1\frac{1}{4}$ in.

BASE FOR STEEL COLUMN WITH END MOMENT

A steel column of 14-in. depth transmits to its footing an axial load of 30 kips and a moment of 1,100 in.-kips in the plane of its web. Design the base, using A307 anchor bolts and 3,000-psi concrete.

Calculation Procedure:

1. Record the allowable stresses and modular ratio

Refer to Fig. 12. If the moment is sufficiently large, it causes uplift at one end of the plate and thereby induces tension in the anchor bolt at that end. A rigorous analysis of the stresses in a column base transmitting a moment is not possible. For simplicity, compute the stresses across a horizontal plane through the base plate by treating this as the cross section of a reinforced-concrete beam, the anchor bolt on the tension side acting as the reinforcing steel. The effects of initial tension in the bolts will be disregarded.

The anchor bolts are usually placed $2\frac{1}{2}$ or 3 in. from the column flange. Using a plate of 26-in. depth as shown in Fig. 12a, let $A_s =$ anchor-bolt cross-sectional area; $B =$ base-plate width; $C =$ resultant compressive force on base plate; $T =$ tensile force in anchor bolt; f_s = stress in anchor bolt; p = maximum bearing stress; p' = bearing stress at column face; t = base-plate thickness.

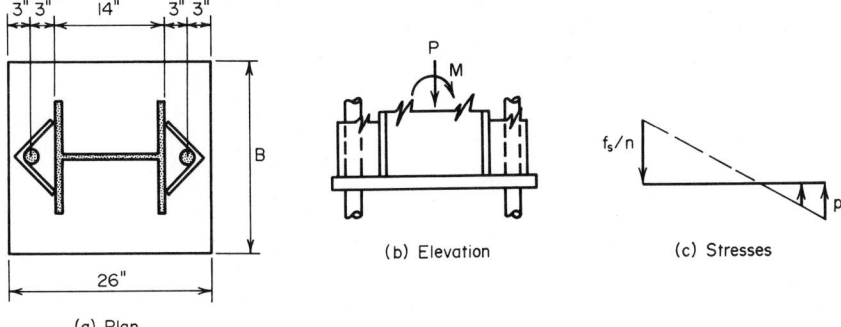

(a) Plan

(b) Elevation

(c) Stresses

Fig. 12   Anchor-bolt details. (a) Plan; (b) elevation; (c) stresses.

Recording the allowable stresses and modular ratio using the ACI *Code*, $p = 750$ psi and $n = 9$. From the AISC *Specification*, $f_s = 14{,}000$ psi; the allowable bending stress in the plate is 27,000 psi.

2.   Construct the stress and force diagrams

These are shown in Fig. 13. Then $f_s/n = 14/9 = 1.555$ ksi; $kd = 23(0.750/2.305) = 7.48$ in.; $jd = 23 - 7.48/3 = 20.51$ in.

3.   Design the base plate

Thus, $C = \frac{1}{2}(7.48)(0.750B) = 2.805B$. Take moments with respect to the anchor bolt, or $\Sigma M = 30(10) + 1{,}100 - 2.805B(20.51) = 0$; $B = 24.3$ in.

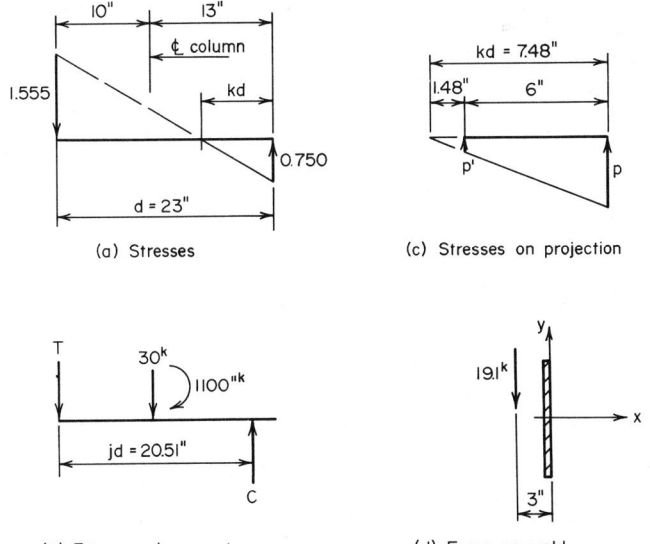

(a) Stresses

(c) Stresses on projection

(b) Forces and moment

(d) Force on weld

Fig. 13   (a) Stresses; (b) forces and moment; (c) stresses on projection; (d) force on weld.

Assume that the critical bending stress in the base plate occurs at the face of the column. Compute the bending moment at the face for a 1-in. width of plate. Referring to Fig. 13c, $p' = 0.750(1.48/7.48) = 0.148$ ksi; $M = (6^2/6)(0.148 + 2 \times 0.750) = 9.89$ in.-kips; $t^2 = 6M/27 = 2.20$ in.^{2}; $t = 1.48$ in. Make the base plate 25 in. wide and $1\frac{1}{2}$ in. thick.

4. Design the anchor bolts

From the calculation in Step 3, $C = 2.805B = 2.805(24.3) = 68.2$ kips; $T = 68.2 - 30 = 38.2$ kips; $A_s = 38.2/14 = 2.73$ sq in. Refer to the AISC *Manual*. Use $2\frac{1}{4}$-in. anchor bolts, one on each side of the flange. Then, $A_s = 3.02$ sq in.

5. Design the anchorage for the bolts

The bolts are held by angles welded to the column flange, as shown in Fig. 12 and in the AISC *Manual*. Use $\frac{1}{2}$-in. angles 12 in. long. Each line of weld resists a force of $\frac{1}{2}T$. Refer to Fig. 13d and compute the unit force F at the extremity of the weld. Thus: $M = 19.1(3) = 57.3$ in.-kips; $S_x = (\frac{1}{6})(12)^2 = 24$ sq in.; $F_x = 57.3/24 = 2.39$ kli; $F_y = 19.1/12 = 1.59$ kli; $F = (2.39^2 + 1.59^2)^{0.5} = 2.87$ kli. Use a $\frac{5}{16}$-in. fillet weld of E60 electrodes, which has a capacity of 3 kli.

GRILLAGE SUPPORT FOR COLUMN

A steel column in the form of a 14WF320 reinforced with two $20 \times 1\frac{1}{2}$ in. cover plates carries a load of 2,790 kips. Design the grillage under this column, using an allowable bearing stress of 750 psi on the concrete. The space between the beams will be filled with concrete.

Calculation Procedure:

1. Establish the dimensions of the grillage

Refer to Fig. 14. A load of this magnitude cannot be transmitted from the column to its footing through the medium of a base plate alone. It is therefore necessary to interpose steel beams between the base plate and the footing; these may be arranged in one tier or in two orthogonal tiers. Integrity of each tier is achieved by tying the beams together by pipe separators. This type of column support is termed a *grillage*. In designing the grillage, it is assumed that bearing pressures are uniform across each surface under consideration.

The area of grillage required = load, kips/allowable stress, ksi = 2,790/0.750 = 3,720 sq in. Set $A = 60$ in., and $B = 62$ in., giving an area of 3,720 sq in., as required.

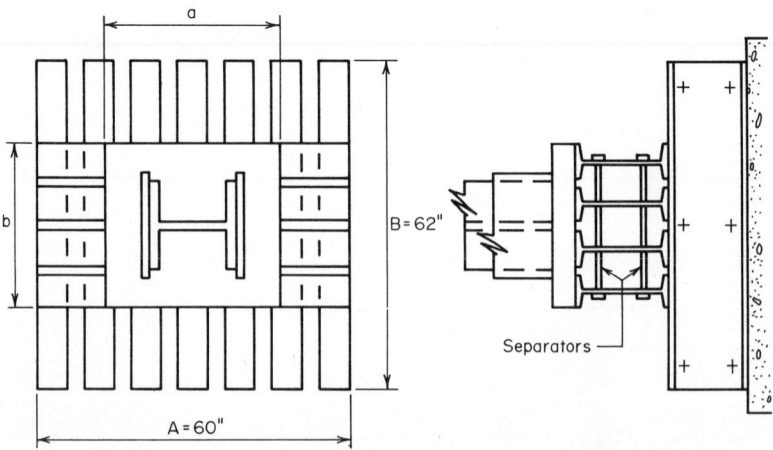

Fig. 14 Grillage under column.

2. Design the upper-tier beams

There are three criteria: bending stress, shearing stress, and compressive stress in the web at the toe of the fillet. The concrete between the beams supplies lateral restraint, and the allowable bending stress is therefore 24 ksi.

Since the web stresses are important criteria, a grillage is generally constructed of I beams rather than wide-flange beams to take advantage of the thick webs of I sections. The design of the beams requires the concurrent determination of the length a of the base plate. Let f = bending stress; f_b = compressive stress in web at fillet toe; v = shearing stress; P = load carried by single beam; S = section modulus of single beam; k = distance from outer surface of beam to toe of fillet; t_w = web thickness of beam; a_1 = length of plate as governed by flexure; a_2 = length of plate as governed by compressive stress in web.

Select a beam size on the basis of stresses f and f_b, and then investigate v. The maximum bending moment occurs at the center of the span; its value is $M = P(A-a)/8 = fS$; therefore, $a_1 = A - 8fS/P$.

At the toe of the fillet, the load P is distributed across a distance $a+2k$. Then, $f_b = P/(a+2k)t_w$; therefore, $a_2 = P/f_b t_w - 2k$. Try four beams; then $P = 2,790/4 = 697.5$ kips; $f = 24$ ksi; $f_b = 27$ ksi. Upon substitution, the foregoing equations reduce to: $a_1 = 60 - 0.275S$; $a_2 = 25.8/t_w - 2k$.

Select the trial beam sizes shown in the accompanying table and calculate the corresponding values of a_1 and a_2. Thus:

Size	S, in.^{3}	t_w, in.	k, in.	a_1, in.	a_2, in.
18I54.7	88.4	0.460	1.375	35.7	53.3
18I70.0	101.9	0.711	1.375	32.0	33.6
20I65.4	116.9	0.500	1.563	27.9	48.5
20I75.0	126.3	0.641	1.563	25.3	37.1

Try 18I70.0, with $a = 34$ in. The flange width is 6.25 in. The maximum vertical shear occurs at the edge of the plate; its magnitude is: $V = P(A-a)/2A = 697.5(60-34)/[2(60)] = 151.1$ kips; $v = 151.1/[18(0.711)] = 11.8 < 14.5$ ksi, which is acceptable.

3. Design the base plate

Refer to the second previous Calculation Procedure. To permit the deposition of concrete, allow a minimum space of 2 in. between the beam flanges. The minimum value of b is therefore $b = 4(6.25) + 3(2) = 31$ in.

The dimensions of the effective bearing area under the column are: $0.95(16.81 + 2 \times 1.5) = 18.82$ in.; $0.80(20) = 16$ in. The projections of the plate are: $(34-18.82)/2 = 7.59$ in.; $(31-16)/2 = 7.5$ in.

Therefore, keep $b = 31$ in., because this results in a well-proportioned plate. The pressure under the plate $= 2,790/[34(31)] = 2.65$ ksi. For a 1-in.-width plate, $M = \frac{1}{2}(2.65)(7.59)^2 = 76.33$ in.-kips; $S = M/f = 76.33/27 = 2.827$ in.^{3}; $t = (6S)^{0.5} = 4.12$ in.

Plate thicknesses within this range vary by $\frac{1}{8}$-in. increments, as stated in the AISC *Manual*. However, a section of the AISC *Specification* requires that plates over 4 in. thick be planed at all bearing surfaces. Set $t = 4\frac{1}{2}$ in. to allow for the planing.

4. Design the beams at the lower tier

Try seven beams. Thus, $P = 2,790/7 = 398.6$ kips; $M = 398.6(62-31)/8 = 1,545$ in.-kips; $S_3 = 1,545/24 = 64.4$ in.^{3}.

Try 15I50.0. Then, $S = 64.2$ in.^{3}; $t_w = 0.550$ in.; $k = 1.25$ in.; $b = 5.64$ in. The space between flanges is $[60 - 7 \times 5.64]/6 = 3.42$ in. This result is satisfactory. Then, $f_b = 398.6/[0.550(31 + 2 \times 1.25)] = 21.6 < 27$ ksi, which is satisfactory; $V = 398.6(62-31)/[2(62)] = 99.7$ kips; $v = 99.7/[15(0.550)] = 12.1 < 14.5$, which is satisfactory.

5. *Summarize the design*

Thus: $A = 60$ in.; $B = 62$ in.; base plate is $31 \times 34 \times 4\frac{1}{2}$ in.; upper-tier steel, four beams 18I70.0; lower-tier steel, seven beams 15I50.0.

WIND-STRESS ANALYSIS BY PORTAL METHOD

The bent in Fig. 15 resists the indicated wind loads. Applying the portal method of analysis, calculate all shears, end moments, and axial forces.

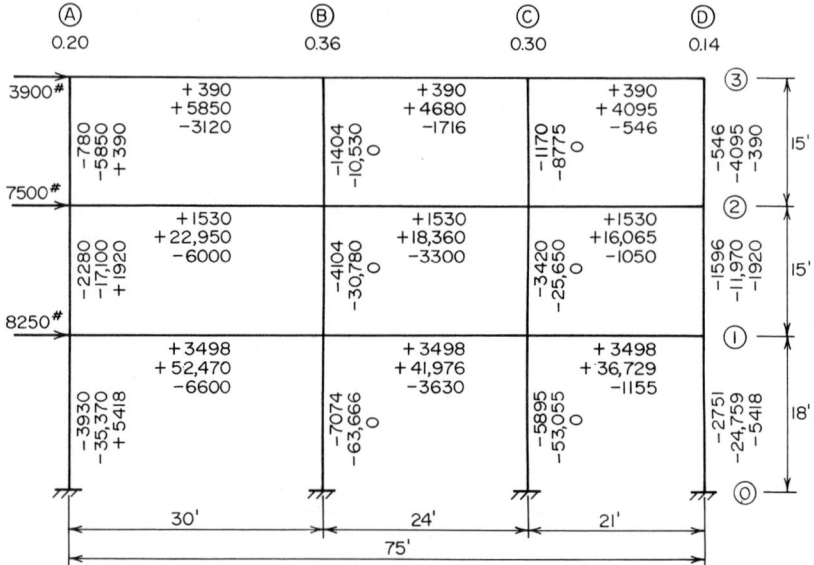

Note. Data recorded in following order: shear, end moments, axial force.

Fig. 15   Wind-stress analysis by portal method.

Calculation Procedure:

1. *Compute the shear factor for each column*

The portal method is an approximate and relatively simple method of wind-stress analysis that is frequently applied to regular bents of moderate height. It considers the bent to be composed of a group of individual portals and makes the following assumptions. (1) The wind load is distributed among the aisles of the bent in direct proportion to their relative widths. (2) The point of contraflexure in each member lies at its center.

Because of the first assumption, the shear in a given column is directly proportional to the average width of the adjacent aisles. (An alternative form of the portal method assumes that the wind load is distributed uniformly among the aisles, irrespective of their relative widths.)

In this analysis, we consider the *end moments* of a member, i.e., the moments exerted at the ends of the member by the joints. The sign conventions used are as follows. An end moment is positive if it is clockwise. The shear is positive if the lateral forces exerted on the member by the joints constitute a couple having a counterclockwise moment. An axial force is positive if it is tensile.

Figure 16a and b represents a beam and column, respectively, having positive end moments and positive shear. By applying the second assumption, $M_a = M_b = M$, Eq. (*a*); $V = 2M/L$, or $M = VL/2$, Eq. (*b*); $H = 2M/L$, or $M = HL/2$, Eq. (*c*). In Fig. 15, the calculated data for each member are recorded in the order indicated.

The shear factor equals the ratio of the average width of the adjacent aisles to the total width. Or: line A, 15/75 = 0.20; line B, (15 + 12)/75 = 0.36; line C, (12 + 10.5)/75 = 0.30; line D, 10.5/75 = 0.14. For convenience, record these values in Fig. 15.

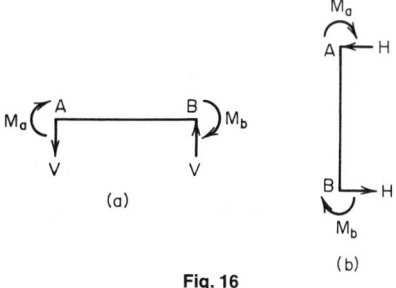

Fig. 16

2. **Compute the shear in each column**

For instance: column A-2-3, $H = -3,900(0.20) = -780$ lb; column C-1-2, $H = -(3,900 + 7.500)0.30 = -3,420$ lb.

3. **Compute the end moments of each column**

Apply Eq. (*c*). For instance: column A-2-3, $M = \frac{1}{2}(-780)15 = -5,850$ ft-lb; column D-0-1, $M = \frac{1}{2}(-2,751)18 = -24,759$ ft-lb.

4. **Compute the end moments of each beam**

Do this by equating the algebraic sum of end moments at each joint to zero. For instance, at line 3: $M_{AB} = 5,850$ ft-lb; $M_{BC} = -5,850 + 10,530 = 4,680$ ft-lb; $M_{CD} = -4,680 + 8,775 = 4,095$ ft-lb. At line 2: $M_{AB} = 5,850 + 17,100 = 22,950$ ft-lb; $M_{BC} = -22,950 + 30,780 + 10,530 = 18,360$ ft-lb.

5. **Compute the shear in each beam**

Do this by applying Eq. (*b*). For instance, beam B-2-C, $V = 2(18,360)/24 = 1,530$ lb.

6. **Compute the axial force in each member**

Do this by drawing free-body diagrams of the joints and applying the equations of equilibrium. It is found that the axial forces in the interior columns are zero. This condition stems from the first assumption underlying the portal method and the fact that each interior column functions both as the leeward column of one portal and the windward column of the adjacent portal.

The absence of axial forces in the interior columns in turn results in the equality of the shear in the beams at each tier. Thus, the calculations associated with the portal method of analysis are completely self-checking.

WIND-STRESS ANALYSIS BY CANTILEVER METHOD

For the bent in Fig. 17, calculate all shears, end moments, and axial forces induced by the wind loads by applying the cantilever method of wind-stress analysis. For this purpose, assume that the columns have equal cross-sectional areas.

Calculation Procedure:

1. **Compute the shear and moment on the bent at mid-height of each horizontal row of columns**

The cantilever method, which is somewhat more rational than is the portal method, considers that the bent behaves as a vertical cantilever. Consequently, the direct stress in a column is directly proportional to the distance from the column to the centroid of the combined column area. As in the portal method, the assumption is made that the point of contraflexure in each member lies at its center. Refer to the previous Calculation Procedure for the sign convention.

Computing the shear and moment on the bent at mid-height, we have the following. Upper row: $H = 3,900$ lb; $M = 3,900(7.5) = 29,250$ ft-lb, Center row: $H = 3,900 + 7,500$

= 11,400 lb; $M = 3,900(22.5) + 7,500(7.5) = 144,000$ ft-lb, Lower row: $H = 11,400 + 8,250 = 19,650$ lb; $M = 3,900(39) + 7,500(24) + 8,250(9) = 406,400$ ft-lb, or $M = 144,000 + 11,400(16.5) + 8,250(9) = 406,400$ ft-lb, as before.

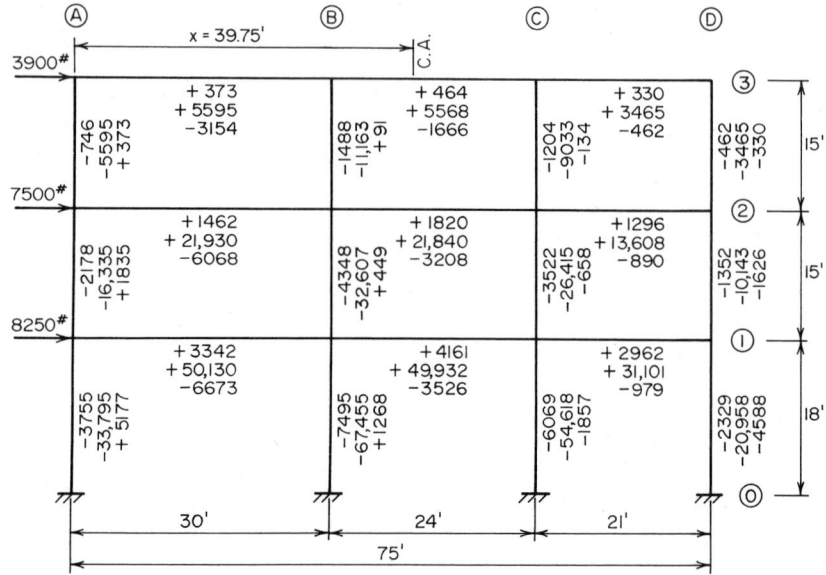

Note. Data recorded in following order: shear, end moments, axial force.

Fig. 17   Wind-stress analysis by cantilever method.

2. Locate the centroidal axis of the combined column area and compute the moment of inertia of the area with respect to this axis

Take the area of one column as a unit. Then: $x = (30 + 54 + 75)/4 = 39.75$ ft; $I = 39.75^2 + 9.75^2 + 14.25^2 + 35.25^2 = 3,121$ ft².

3. Compute the axial force in each column

Use the equation, $f = My/I$. The y/I values are:

	A	B	C	D
y	39.75	9.75	−14.25	−35.25
y/I	0.01274	0.00312	− 0.00457	−0.01129

Then: column A-2-3, $P = 29,250(0.01274) = 373$ kips; column B-0-1, $P = 406,400 (0.00312) = 1,268$ kips.

4. Compute the shear in each beam by analyzing each joint as a free body

Thus: beam A-3-B, $V = 373$ lb; beam B-3-C, $V = 373 + 91 = 464$ lb; beam C-3-D, $V = 464 - 134 = 330$ lb; beam A-2-B, $V = 1,835 - 373 = 1,462$ lb; beam B-2-C, $V = 1,462 + 449 - 91 = 1,820$ lb.

5. Compute the end moments of each beam

Apply Eq. (b) of the previous Calculation Procedure. Or for beam A-3-B, $M = \frac{1}{2}(373)(30) = 5,595$ ft-lb.

6. Compute the end moments of each column

Do this by equating the algebraic sum of the end moments at each joint to zero.

7. Compute the shear in each column

Apply Eq. (*c*) of the previous Calculation Procedure. The sum of the shears in each horizontal row of columns should equal the wind load above that plane. For instance, for the center row, $\Sigma H = -(2,178 + 4,348 + 3,522 + 1,352) = -11,400$ lb, which is correct.

8. Compute the axial force in each beam by analyzing each joint as a free body

Thus: beam A-3-B, $P = -3,900 + 746 = -3,154$ lb; beam B-3-C, $P = -3,154 + 1,488 = -1,666$ lb.

WIND-STRESS ANALYSIS BY SLOPE-DEFLECTION METHOD

Analyze the bent in Fig. 18*a* by the slope-deflection method. The moment of inertia of each member is shown in the drawing.

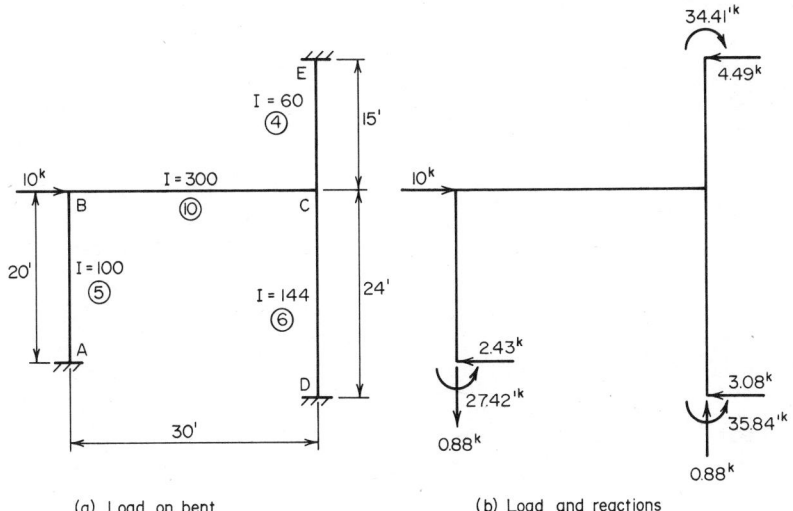

(a) Load on bent   (b) Load and reactions

Fig. 18   (a) Load on bent; (b) load and reactions.

Calculation Procedure:

1. Compute the end rotations caused by the applied moments and forces; superpose the rotation caused by the transverse displacement

This method of analysis has not been extensively applied in the past because the arithmetical calculations involved become voluminous where the bent contains many joints. However, the increasing use of computers in structural design is overcoming this obstacle and stimulating a renewed interest in the method.

Figure 19 is the elastic curve of a member subjected to moments and transverse

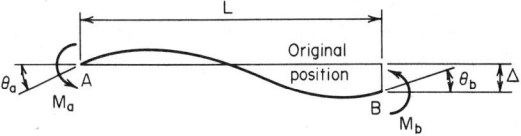

Fig. 19   Elastic curve of beam.

forces applied solely at its ends. The sign convention is as follows: an end moment is positive if it is clockwise; an angular displacement is positive if the rotation is clockwise; the transverse displacement Δ is positive if it rotates the member in a clockwise direction.

Computing the end rotations, $\theta_a = (L/6EI)(2M_a - M_b) + \Delta/L$; $\theta_b = (L/6EI)(-M_a + 2M_b) + \Delta/L$. These results may be obtained by applying the moment-area method or unit-load method given in Sec. 1 of this Handbook.

2. Solve the foregoing equations for the end moments

Thus, $M_a = (2EI/L)(2\theta_a + \theta_b - 3\Delta/L)$, and $M_b = (2EI/L)(\theta_a + 2\theta_b - 3\Delta/L)$, Eq. (4). These are the basic slope-deflection equations.

3. Compute the value of I/L for each member of the bent

Let K denote this value, which represents the relative stiffness of the member. Thus: $K_{ab} = 100/20 = 5$; $K_{cd} = 144/24 = 6$; $K_{bc} = 300/30 = 10$; $K_{ce} = 60/15 = 4$. These values are recorded in circles in Fig. 18.

4. Apply Eq. (4) to each joint in turn

When the wind load is applied, the bent will deform until the horizontal reactions at the supports total 10 kips. It is evident, therefore, that the end moments of a member are functions of the *relative* rather than the absolute stiffness of that member. Therefore, in writing the moment equations, the coefficient $2EI/L$ may be replaced with I/L; to view this in another manner, $E = \frac{1}{2}$.

Disregard the deformation associated with axial forces in the members and assume that joints B and C remain in a horizontal line. The symbol M_{ab} denotes the moment exerted on member AB at joint A. Thus: $M_{ab} = 5(\theta_b - 3\Delta/20) = 5\theta_b - 0.75\Delta$; $M_{dc} = 6(\theta_c - 3\Delta/24) = 6\theta_c - 0.75\Delta$; $M_{ec} = 4(\theta_c + 3\Delta/15) = 4\theta_c + 0.80\Delta$; $M_{ba} = 5(2\theta_b - 3\Delta/20) = 10\theta_b - 0.75\Delta$; $M_{cd} = 6(2\theta_c - 3\Delta/24) = 12\theta_c - 0.75\Delta$; $M_{ce} = 4(2\theta_c + 3\Delta/15) = 8\theta_c + 0.80\Delta$; $M_{bc} = 10(2\theta_b + \theta_c) = 20\theta_b + 10\theta_c$; $M_{cb} = 10(\theta_b + 2\theta_c) = 10\theta_b + 20\theta_c$.

5. Write the equations of equilibrium for the joints and for the bent

Thus: joint B, $M_{ba} + M_{bc} = 0$, Eq. (a); joint C, $M_{cb} + M_{cd} + M_{ce} = 0$, Eq. (b). Let H denote the horizontal reaction at a given support. Consider a horizontal force positive if directed toward the right. Then, $H_a + H_d + H_e + 10 = 0$, Eq. (c).

6. Express the horizontal reactions in terms of the end moments

Rewrite Eq. (c). Or, $(M_{ab} + M_{ba})/20 + (M_{dc} + M_{cd})/24 - (M_{ec} + M_{ce})/15 + 10 = 0$, or $6M_{ab} + 6M_{ba} + 5M_{dc} + 5M_{cd} - 8M_{ec} - 8M_{ce} = -1{,}200$, Eq. (c').

7. Rewrite Eqs. (a), (b), and (c') by replacing the end moments with the expressions obtained in Step 4

Thus, $30\theta_b + 10\theta_c - 0.75\Delta = 0$, Eq. (A); $10\theta_b + 40\theta_c + 0.05\Delta = 0$, Eq. (B); $90\theta_b - 6\theta_c - 29.30\Delta = -1{,}200$, Eq. (C).

8. Solve the simultaneous equations in Step 7 to obtain the relative values of θ_b, θ_c, and Δ

Thus: $\theta_b = 1.244$; $\theta_c = -0.367$; $\Delta = 44.85$.

9. Apply the results in Step 8 to evaluate the end moments

The values, in ft-kips, are: $M_{ab} = -27.42$; $M_{dc} = -35.84$; $M_{ec} = 34.41$; $M_{ba} = -21.20$; $M_{cd} = -38.04$; $M_{ce} = 32.94$; $M_{bc} = 21.21$; $M_{cb} = 5.10$.

10. Compute the shear in each member by analyzing the member as a free body

The shear is positive if the transverse forces exert a counterclockwise moment. Thus: $H_{ab} = (M_{ab} + M_{ba})/20 = -2.43$ kips; $H_{cd} = -3.08$ kips; $H_{ce} = 4.49$ kips; $V_{bc} = 0.88$ kips.

11. Compute the axial force in AB and BC

Thus: $P_{ab} = 0.88$ kips; $P_{bc} = -7.57$ kips. The axial forces in EC and CD are found by equating the elongation of one to the contraction of the other.

12. *Check the bent for equilibrium*

The forces and moments acting on the structure are shown in Fig. 18*b*. It is found that the three equations of equilibrium are satisfied.

WIND DRIFT OF A BUILDING

Figure 20*a* is the partial elevation of the steel framing of a skyscraper. The wind shear directly above line 11 is 40 kips, and the wind force applied at lines 11 and 12 is 4 kips each. The members represented by solid lines have the moments of inertia shown in Table 1, and the structure is to be analyzed for wind stress by the portal method. Compute the wind drift for the bent bounded by lines 11 and 12; that is, find the horizontal displacement of the joints on line 11 relative to those on line 12 as a result of wind.

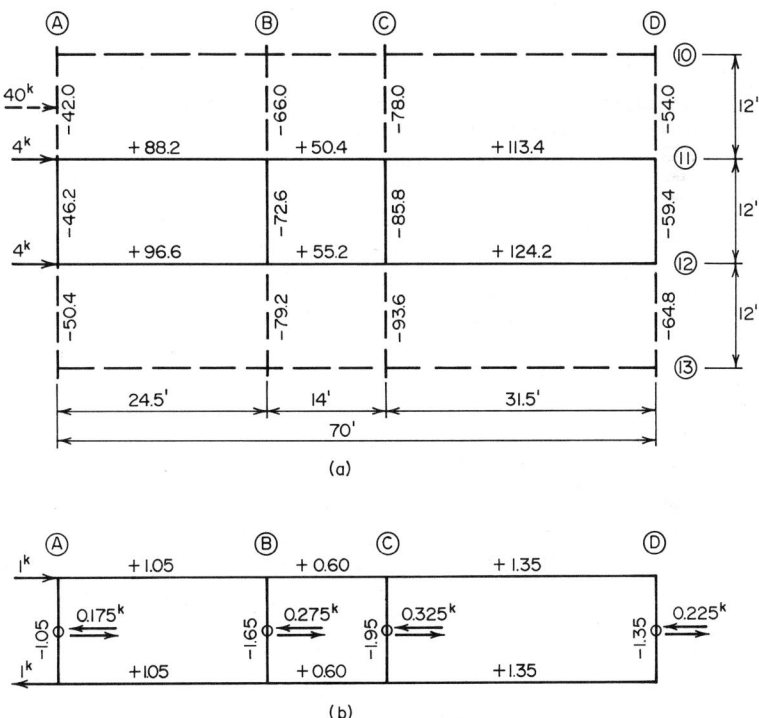

Fig. 20

Calculation Procedure:

1. *Using the portal method of wind-stress analysis, compute the shear in each column caused by the unit loads*

Apply the unit-load method, presented in Sec. 1 of this Handbook. For this purpose, consider that unit horizontal loads are applied to the structure in the manner shown in Fig. 20*b*.

The results obtained in Steps 1, 2, and 3 below, are recorded in Fig. 20*b*. To apply the portal method of wind-stress analysis, see the fourteenth Calculation Procedure in this Section.

2. *Compute the end moments of each column caused by the unit loads*

TABLE 1   Calculation of Wind Drift

Member	I, in.^{4}	L, ft	M_e, ft-kips	m_e, ft-kips	$M_e m_e L/I$
A-11-12	1,500	12	46.2	1.05	0.39
B-11-12	1,460	12	72.6	1.65	0.98
C-11-12	1,800	12	85.8	1.95	1.12
D-11-12	2,000	12	59.4	1.35	0.48
A-11-B	660	24.5	88.2	1.05	3.44
B-11-C	300	14	50.4	0.60	1.41
C-11-D	1,400	31.5	113.4	1.35	3.44
A-12-B	750	24.5	96.6	1.05	3.31
B-12-C	400	14	55.2	0.60	1.16
C-12-D	1,500	31.5	124.2	1.35	3.52
Total ...	...	...	...	...	19.25

3.   Equate the algebraic sum of end moments at each joint to zero; from this find the end moments of the beams caused by the unit loads

4.   Find the end moments of each column

Multiply the results obtained in Step 2 by the wind shear in each panel to find the end moments of each column in Fig. 20a. For instance, the end moments of column C-11-12 are $-1.95(44) = -85.8$ ft-kips. Record the result in Fig. 20a.

5.   Find the end moments of the beams caused by the true loads

Equate the algebraic sum of end moments at each joint to zero to find the end moments of the beams caused by the true loads.

6.   State the equation for wind drift

In Fig. 21, M_e and m_e denote the end moments caused by the true loads and unit loads, respectively. Then, the wind drift $\Delta = \Sigma M_e m_e L / 3EI$, Eq. (5).

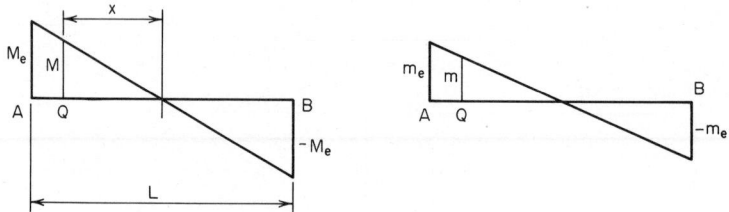

Fig. 21   Bending-moment diagrams.

7.   Compute the wind drift by completing Table 1

In recording end moments, algebraic signs may be disregarded because the product $M_e m_e$ is always positive. Taking the total of the last column in Table 1, $\Delta = 19.25(12)^3/[3(29)(10)^3] = 0.382$ in. For dimensional homogenity, the left side of Eq. (5) must be multiplied by 1 kip. The product represents the external work performed by the unit loads.

REDUCTION IN WIND DRIFT BY USING DIAGONAL BRACING

With reference to the previous Calculation Procedure, assume that the wind drift of the bent is to be restricted to 0.20 in. by introducing diagonal bracing between lines B and C. Design the bracing, using the gross area of the member.

Calculation Procedure:

1. State the change in length of the brace

The bent will be reinforced against lateral deflection by a pair of diagonal cross braces, each brace being assumed to act solely as a tension member. Select the lightest single-angle member that will satisfy the stiffness requirements; then compute the wind drift of the reinforced bent.

Assume that the bent in Fig. 22 is deformed in such a manner that B is displaced a horizontal distance Δ relative to D. Let $A =$ cross-sectional area of member CB; $P =$ axial force in CB; $P_h =$ horizontal component of P; $\delta L =$ change in length of CB. From the geometry of the diagram in Fig. 22, $\delta L = \Delta \cos \theta = a\Delta/L$, approximately.

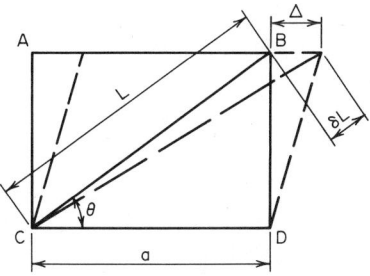

Fig. 22

2. Express P_h in terms of Δ

Thus, $P = aAE\Delta/L^2$; $P_h = P \cos \theta = Pa/L$; then, $P_h = a^2AE\Delta/L^3$, Eq. (6).

3. Select a trial size for the diagonal bracing; compute the tensile capacity

A section of the AISC *Specification* limits the slenderness ratio for bracing members in tension to 300, and another section provides an allowable stress of 22 ksi. Thus, $L^2 = 14^2 + 12^2 = 340$ ft²; $L = 18.4$ ft; $r_{min} = (18.4 \times 12)/300 = 0.74$ in.

Try a $4 \times 4 \times \frac{1}{4}$ in. angle; $r = 0.79$ in.; $A = 1.94$ sq in.; $P_{max} = 1.94(22) = 42.7$ kips.

4. Compute the wind drift if the assumed size of bracing is used

By Eq. (6), $P_h = \{196/[(340)(18.4)(12)]\}1.94(29)(10)^3\Delta = 147\Delta$ kips. The wind shear resisted by the columns of the bent is reduced by P_h, and the wind drift is reduced proportionately.

From the previous Calculation Procedure, the following values are obtained: without diagonal bracing, $\Delta = 0.382$ in.; with diagonal bracing, $\Delta = 0.382(44 - P_h)/44 = 0.382 - 1.28\Delta$. Solving, $\Delta = 0.168 < 0.20$ in., which is acceptable.

5. Check the axial force in the brace

Thus, $P_h = 147(0.168) = 24.7$ kips; $P = P_hL/a = 24.7(18.4)/14 = 32.5 < 42.7$ kips, which is satisfactory. Therefore, the assumed size of the member is satisfactory.

LIGHT-GAGE STEEL BEAM WITH UNSTIFFENED FLANGE

A beam of light-gage cold-formed steel consists of two $7 \times 1\frac{1}{2}$ in. by No. 12 gage channels connected back to back to form an I section. The beam is simply supported on a 16-ft span, has continuous lateral support, and carries a total dead load of 50 plf. The live-load deflection is restricted to 1/360 of the span. If the yield-point stress f_y is 33,000 psi, compute the allowable unit live load for this member.

Calculation Procedure:

1. Record the relevant properties of the section

Apply the AISI *Specification for the Design of Light Gage Cold-Formed Steel Structural Members*. This is given in the AISI publication *Light Gage Cold-Formed Steel Design Manual*. Use the same notational system except denote the flat width of an element by g rather than w.

The publication mentioned above provides a basic design stress of 20,000 psi for this grade of steel. However, since the compression flange of the given member is unstiffened in accordance with the definition in one section of the publication, it may be necessary to reduce the allowable compressive stress. A table in the *Manual* gives the dimensions, design properties, and allowable stress of each section, but the allowable stress will be computed independently in this Calculation Procedure.

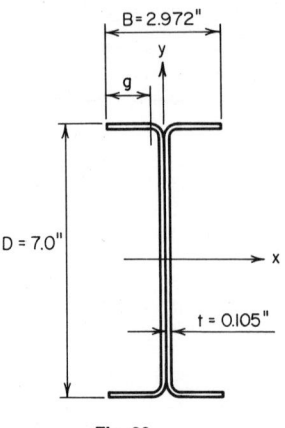

Let V = maximum vertical shear; M = maximum bending moment; w = unit load; f_b = basic design stress; f_c = allowable bending stress in compression; v = shearing stress; Δ = maximum deflection.

Recording the relevant properties of the section as shown in Fig. 23, $I_x = 12.4$ in.^{4}; $S_x = 3.54$ in.^{3}; $R = \frac{3}{16}$ in.

2. Compute f_c

Thus, $g = B/2 - t - R = 1.1935$ in.; $g/t = 1.1935/0.105 = 11.4$. From the *Manual*, the allowable stress corresponding to this ratio is $f_c = 1.667 f_b - 8{,}640 - [(f_b - 12{,}950)g/t]/15 = 1.667(20{,}000) - 8{,}640 - (20{,}000 - 12{,}950)11.4/15 = 19{,}340$ psi.

3. Compute the allowable unit live load if flexure is the sole criterion

Thus: $M = f_c S_x = 19{,}340(3.54)/12 = 5{,}700$ ft-lb; $w = 8M/L^2 = 8(5{,}700)/16^2 = 178$ plf; $w_{LL} = 178 - 50 = 128$ plf.

Fig. 23

4. Investigate the deflection under the computed live load

Using $E = 29{,}500{,}000$ psi, as given in the AISI *Manual*, $\Delta_{LL} = 5 w_{LL} L^4/384 E I_x = 5(128)(16)^4(12)^3/[384(29.5)(10)^6 12.4] = 0.516$ in.; $\Delta_{LL,allow} = 16(12)/360 = 0.533$ in., which is satisfactory.

5. Investigate the shearing stress under the computed total load

Refer to the AISI *Specification*. For the individual channel: $h = D - 2t = 6.79$ in.; $h/t = 64.7$; $64{,}000{,}000/64.7^2 > \frac{2}{3} f_b$; therefore, $v_{allow} = 13{,}330$ psi; the web area = $0.105(6.79) = 0.713$ sq in.; $V = \frac{1}{4}(178)16 = 712$ lb; $v = 712/0.713 < v_{allow}$, which is satisfactory. The allowable unit live load is therefore 128 plf.

LIGHT-GAGE STEEL BEAM WITH STIFFENED COMPRESSION FLANGE

A beam of light-gage cold-formed steel has a hat cross section, 8×12 in. by No. 12 gage, as shown in Fig. 24. The beam is simply supported on a span of 13 ft. If the yield-point stress is 33,000 psi, compute the allowable unit load for this member and the corresponding deflection.

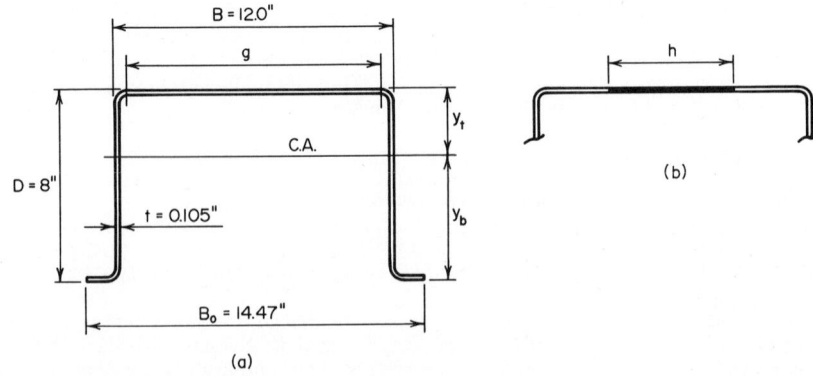

(a)

(b)

Fig. 24

Calculation Procedure:

1. Record the relevant properties of the entire cross-sectional area

Refer to the AISI *Specification* and *Manual*. The allowable load is considered to be the ultimate load that the member will carry divided by a load factor at 1.65. At ultimate load, the bending stress varies considerably across the compression flange. To surmount the difficulty that this condition introduces, the AISI *Specification* permits the designer to assume that the stress is uniform across an *effective flange width* to be established in the prescribed manner. The investigation is complicated by the fact that the effective flange width and the bending stress in compression are interdependent quantities, for the following reason. The effective width depends on the compressive stress; the compressive stress, which is less than the basic design stress, depends on the location of the neutral axis; the location of the neutral axis in turn depends on the effective width.

The beam deflection is also calculated by establishing an effective flange width. However, since the beam capacity is governed by stresses at the ultimate load and the beam deflection is governed by stresses at working load, the effective widths associated with these two quantities are unequal.

A table in the AISI *Manual* contains two design values that afford a direct solution to this problem. However, the values will be computed independently here to demonstrate how they are obtained. The notational system presented in the previous Calculation Procedure will be used, as well as A' = area of cross section exclusive of compression flange; H = statical moment of cross-sectional area with respect to top of section; y_b and y_t = distance from centroidal axis of cross section to bottom and top of section, respectively.

Using the AISI *Manual* to determine the relevant properties of the entire cross-sectional area, as shown in Fig. 24: $A = 3.13$ sq in.; $y_b = 5.23$ in.; $I_x = 26.8$ in.^{4}; $R = \frac{3}{16}$ in.

2. Establish the value of f_c for load determination

Use the relation $(8,040t^2/f_c^{0.5})\{1 - 2,010/[(f_c^{0.5}g)/t]\} = (H/D)(f_c+f_b)/f_c - A'$. Substituting, $g = B - 2(t+R) = 12.0 - 2(0.105 + 0.1875) = 11.415$ in.; $g/t = 108.7$; $gt = 1.20$ sq in.; $A' = 3.13 - 1.20 = 1.93$ sq in.; $y_t = 8.0 - 5.23 = 2.77$ sq in.; $H = 3.13(2.77) = 8.670$ in.^{3}. The foregoing equation then reduces to $(88.64/f_c^{0.5})(1 - 18.49/f_c^{0.5}) = 1.084(f_c + 20,000)/f_c - 1.93$. By successive approximations, $f_c = 14,800$ psi.

3. Compute the corresponding effective flange width for load determination in accordance with the AISI Manual

Thus, $b = (8,040t/f_c^{0.5})\{1 - 2,010/[(f_c^{0.5}g)/t]\} = (8,040 \times 0.105/14,800^{0.5})[1 - 2,010/(14,800^{0.5} \times 108.7)] = 5.885$ in.

4. Locate the centroidal axis of the cross section having this effective width; check the value of f_c

Refer to Fig. 24b. Thus: $h = g - b = 11.415 - 5.885 = 5.530$ in.; $ht = 0.581$ sq in.; $A = 3.13 - 0.581 = 2.549$ sq in.; $H = 8.670$ in.^{3}; $y_t = 8.670/2.549 = 3.40$ in.; $y_b = 4.60$ in.; $f_c = y_t f_b/y_b = 3.40(20,000)/4.60 = 14,800$ psi, which is satisfactory.

5. Compute the allowable load

The moment of inertia of the net section may be found by applying the value of the gross section and making the necessary corrections. Applying $S_x = I_x/y_b$, $I_x = 26.8 + 3.13(3.40 - 2.77)^2 - 0.581(3.40 - 0.053)^2 = 21.53$ in.^{4}. Then, $S_x = 21.53/4.60 = 4.68$ in.^{3}. This value agrees with that recorded in the AISI *Manual*.

Then, $M = f_b S_x = 20,000(4.68)/12 = 7,800$ ft-lb; $w = 8M/L^2 = 8(7,800)/13^2 = 369$ plf.

6. Establish the value of f_c for deflection determination

Apply: $(10,320t^2/f_c^{0.5})[1 - 2,580/(f_c^{0.5}g/t)] = (H/D)(f_c+f_b)/f_c - A'$, or $(113.8/f_c^{0.5}) \times (1 - 23.74/f_c^{0.5}) = 1.084(f_c + 20,000)/f_c - 1.93$. By successive approximation, $f_c = 13,300$ psi.

7. Compute the corresponding effective flange width for deflection determination

Thus, $b = (10,320t/f_c^{0.5})[1 - 2,580/(f_c^{0.5}g/t)] = (10,320 \times 0.105/13,300^{0.5})[1 - 2,580/(13,300^{0.5} \times 108.7)] = 7.462$ in.

8. Locate the centroidal axis of the cross section having this effective width; check the value of f_c

Thus: $h = 11.415 - 7.462 = 3.953$ in.; $ht = 0.415$ sq in.; $A = 3.13 - 0.415 = 2.715$ sq in.; $H = 8.670$ in.^{3}: $y_t = 8.670/2.715 = 3.19$ in.; $y_b = 4.81$ in.; $f_c = (3.19/4.81)20,000 = 13,300$ psi, which is satisfactory.

9. Compute the deflection

For the net section, $I_x = 26.8 + 3.13(3.19 - 2.77)^2 - 0.415(3.19 - 0.053)^2 = 23.3$ in.^{4}. This value agrees with that tabulated in the AISI *Manual*. The deflection is $\Delta = 5wL^4/384EI_x = 5(369)(13)^4(12)^3/[384(29.5)(10)^6 23.3] = 0.345$ in.

STEEL BEAM ENCASED IN CONCRETE

A concrete floor slab is to be supported by steel beams spaced 10 ft on centers and having a span of 28 ft 6 in. The beams will be encased in concrete with a minimum cover of 2 in. all around; they will remain unshored during construction. The slab has been designed as $4\frac{1}{2}$ in. thick, with $f_c' = 3,000$ psi. The loading includes the following: live load, 120 psf; finished floor and ceiling, 25 psf. The steel beams have been tentatively designed as 16WF40. Review the design.

Calculation Procedure:

1. Record the relative properties of the section and the allowable flexural stresses

In accordance with the AISC *Specification*, the member may be designed as a composite steel-and-concrete beam, reliance being placed upon natural bond of the two materials to obtain composite action. Refer to Sec. 1 of this Handbook for the design of a composite bridge member. In the design of a composite building member, the effects of plastic flow are usually disregarded. Since the slab is poured monolithically, the composite member is considered to be continuous. Apply the following equations in computing bending moments in the composite beams: at mid-span, $M = (\frac{1}{20}) wL^2$; at support, $M = -(\frac{1}{12}) wL^2$.

The subscripts c, ts, and bs refer to the extreme fiber of concrete, top of steel, and bottom of steel, respectively. The superscripts c and n refer to the composite and noncomposite sections, respectively.

Recording the properties of the 16WF40, $A = 11.77$ sq in., $d = 16.00$ in., $I = 515.5$ in.^{4}, $S = 64.4$ in.^{3}, flange width = 7 in. By the AISC *Specification*, $f_s = 24,000$ psi. By the ACI *Code*, $n = 9$ and $f_c = 1,350$ psi.

2. Transform the composite section in the region of positive moment to an equivalent section of steel; compute the section moduli

Refer to Fig. 25a and the AISC *Specification*. Use the gross concrete area. Then: effective flange width, $\frac{1}{4}L = \frac{1}{4}(28.5)12 = 85.5$ in.; spacing of beams = 120 in.; $16t + 11 = 16(4.5) + 11 = 83$ in.; this governs. Transformed width = $83/9 = 9.22$ in.

Assume that the neutral axis lies within the flange, and take statical moments with respect to this axis; or $\frac{1}{2}(9.22y^2) - 11.77(10 - y) = 0$; $y = 3.93$ in.

Compute the moment of inertia. Slab: $(\frac{1}{3})9.22(3.93)^3 = 187$ in.^{4}. Beam: $515.5 + 11.77 \times (10 - 3.93)^2 = 949$ in.^{4}; $I = 187 + 949 = 1,136$ in.^{4}; $S_c = 1,136/3.93 = 289.1$ in.^{3}; $S_{bs} = 1,136/14.07 = 80.7$ in.^{3}.

3. Transform the composite section in the region of negative moment to an equivalent section of steel; compute the section moduli

Referring to Fig. 25b, transformed width = $11/9 = 1.22$ in. Take statical moments with respect to the neutral axis. Or $11.77(10 - y) - \frac{1}{2}(1.22y^2) = 0$; $y = 7.26$ in. Compute the moment of inertia. Thus, slab: $(\frac{1}{3})1.22(7.26)^3 = 155.6$ in.^{4}. Beam: $515.5 + 11.77(10 - $

$7.26)^2 = 603.9$ in.^{4}; $I = 155.6 + 603.9 = 759.5$ in.^{4}. Then: $S_c = 759.5/7.26 = 104.6$ in.^{3}; $S_{ts} = 759.5/10.74 = 70.7$ in.^{3}.

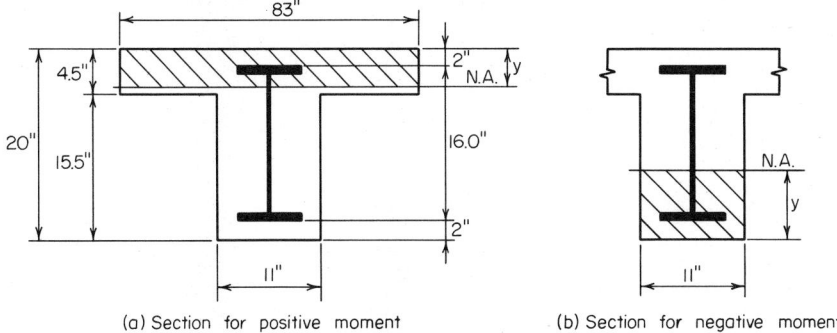

(a) Section for positive moment   (b) Section for negative moment

Fig. 25   Steel beam encased in concrete.

4.   Compute the bending stresses at mid-span

The loads carried by the noncomposite member are: slab, $(4.5)150(10)/12 = 563$ plf; stem, $11(15.5)150/144 = 178$ plf; steel, 40 plf; total $= 563 + 178 + 40 = 781$ plf. The load carried by the composite member $= 145(10) = 1,450$ plf. Then: $M^n = (\frac{1}{8})781(28.5)^2 12 = 951,500$ in.-lb; $M^c = (\frac{1}{20}) 1,450(28.5)^2 \, 12 = 706,600$ in.-lb; $f_c = 706,600/[289.1(9)] = 272$ psi, which is acceptable. Also, $f_{bs} = (951,500/64.4) + (706,600/80.7) = 23,530$ psi, which is acceptable.

5.   Compute the bending stresses at the support

Thus, $M^c = 706,600(20/12) = 1,177,700$ in.-lb; $f_c = 1,177,700/[104.6(9)] = 1,251$ psi, which is satisfactory. Also, $f_{ts} = 1,177,700/70.7 = 16,660$ psi, which is acceptable. The design is therefore satisfactory with respect to flexure.

6.   Investigate the composite member with respect to horizontal shear in the concrete at the section of contraflexure

Assume that this section lies at a distance of $0.2L$ from the support. The shear at this section is $V^c = 1,450(0.3)(28.5) = 12,400$ lb.

Refer to Sec. 1 of this Handbook. Where the bending moment is positive, the critical plane for horizontal shear is considered to be the surface $abcd$ in Fig. 26a. For simplicity, however, compute the shear flow at the neutral axis. Apply the relation $q = VQ/I$, where $Q = \frac{1}{2}(9.22)(3.93)^2 = 71.20$ in.^{3}, and $q = 12,400(71.20)/1,136 = 777$ pli.

Resistance to shear flow is provided by the bond between the steel and concrete along bc and by the pure-shear strength of the concrete along ab and cd. (The term *pure shear* is used to distinguish this from shear that is used as a measure of diagonal tension.) The allowable stresses in bond and pure shear are usually taken as $0.03f_c'$ and

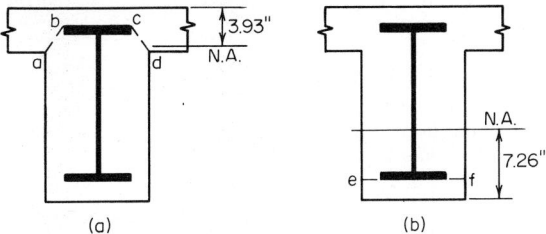

(a)   (b)

Fig. 26   Critical planes for horizontal shear.

$0.12f_c'$, respectively. Thus: $bc = 7$ in.; $ab = (2.5^2 + 2^2)^{0.5} = 3.2$ in.; $q_{allow} = 7(90) + 2(3.2)\times$ $360 = 2,934$ pli, which is satisfactory.

7. Investigate the composite member with respect to horizontal shear in the concrete at the support

The critical plane for horizontal shear is ef in Fig. 26b. Thus: $V^c = 1,450(0.5)28.5 = 20,660$ lb; $Q = 1.22(2)(7.26 - 1) = 15.27$ in.^{3}; $q \doteq 20,660(15.27)/759.5 = 415$ pli; $q_{allow} = 7(90) + 2(2)360 = 2,070$ pli, which is satisfactory.

Mechanical shear connectors are not required to obtain composite action, but the beam is wrapped with wire mesh.

COMPOSITE STEEL-AND-CONCRETE BEAM

A concrete floor slab is to be supported by steel beams spaced 11 ft on centers and having a span of 36 ft. The beams will be supplied with shear connectors to obtain composite action of the steel and concrete. The slab will be 5 in. thick and made of 3,000-psi concrete. Loading includes the following: live load, 200 psf; finished floor, ceiling, and partition, 30 psf. In addition, each girder will carry a dead load of 10 kips applied as a concentrated load at mid-span prior to hardening of the concrete. Conditions at the job site preclude the use of temporary shoring. Design the interior girders, limiting the overall depth of steel to 20 in., if possible.

Calculation Procedure:

1. Compute the unit loads w_1, w_2, and w_3

Refer to the AISC *Specification* and *Manual*. Although ostensibly applying the elastic-stress method, the design of a composite steel-and-concrete beam in reality is based upon the ultimate-strength behavior of the member. Loads that are present before the concrete has hardened are supported by the steel member alone; loads that are present after the concrete has hardened are considered to be supported by the composite member, regardless of whether these loads originated before or after hardening. The effects of plastic flow are disregarded.

The subscripts 1, 2, and 3 refer, respectively, to the following loads: dead loads applied before hardening of the concrete; dead loads applied after hardening of the concrete; and live loads. The subscripts b, ts, and tc refer to the bottom of the member, top of the steel, and top of the concrete, respectively. The superscripts c and n refer to the composite and noncomposite member, respectively.

Computing the unit loads for a slab weight of 63 plf and an assumed steel weight of 80 plf, $w_1 = 63(11) + 80 = 773$ plf; $w_2 = 30(11) = 330$ plf; $w_3 = 200(11) = 2,200$ plf.

2. Compute all bending moments required in the design

Thus, $M_1 = 12[(\frac{1}{8})0.773(36)^2 + \frac{1}{4}(10)36] = 2,583$ in.-kips; $M_2 = (\frac{1}{8})0.330(36)^2 12 = 642$ in.-kips; $M_3 = (\frac{1}{8})2.200(36)^2 12 = 4,277$ in.-kips; $M^c = 2,583 + 642 + 4,277 = 7,502$ in.-kips; $M^n = 2,583$ in.-kips; $M_{DL} = 2,583 + 642 = 3,225$ in.-kips; $M_{LL} = 4,277$ in.-kips.

3. Compute the required section moduli with respect to the steel, using an allowable bending stress of 24 ksi

In the composite member, the maximum steel stress occurs at the bottom; in the noncomposite member, it occurs at the top of the steel if a bottom-flange cover plate is used.

Thus: composite section, $S_b = 7,502/24 = 312.6$ in.^{3}; noncomposite section, $S_{ts} = 2,583/24 = 107.6$ in.^{3}.

4. Select a trial section by tentatively assuming that the composite-design tables in the AISC Manual are applicable

The *Manual* shows that a composite section consisting of a 5-in. concrete slab, an 18WF55 steel beam, and a cover plate having an area of 9 sq in. provides $S_b = 317.5$ in.^{3}. The noncomposite section provides $S_{ts} = 113.7$ in.^{3}.

Since unshored construction is to be used, the section must conform with the *Manual* equation $1.35 + 0.35 M_{LL}/M_{DL} = 1.35 + 0.35(4,277/3,225) = 1.81$. And, $S_b{}^c/S_b{}^n = 317.5/213.6 = 1.49$, which is satisfactory.

The flange width of the 18WF55 is 7.53 in. The minimum allowable distance between the edge of the cover plate and the edge of the beam flange equals the size of the fillet weld plus $\tfrac{5}{16}$ in. Use a 9×1 in. plate. The steel section therefore coincides with that presented in the AISC *Manual*, which has a cover plate thickness t_p of 1 in. The trial section is therefore 18WF55; cover plate is 9×1 in.

5. Check the trial section

The AISC composite-design tables are constructed by assuming that the effective flange width of the member equals 16 times the slab thickness plus the flange width of the steel. In the present instance, the effective flange width, as governed by the AISC, is: $\tfrac{1}{4}L = \tfrac{1}{4}(36)12 = 108$ in.; spacing of beams = 132 in.; $16t + 7.53 = 16(5) + 7.53 = 87.53$ in., which governs.

The cross-section properties in the AISC table may therefore be applied. The moment of inertia refers to an equivalent section obtained by transforming the concrete to steel. Refer to Sec. 1 of this Handbook. Thus: $y_{tc} = 5 + 18.12 + 1 - 16.50 = 7.62$ in.; $S_{tc} = I/y_{tc} = 5,242/7.62 = 687.9$ in.³. From the ACI *Code*, $f_c = 1,350$ psi and $n = 9$. Then, $f_c = M^c/n S_{tc} = 7,502,000/[9(687.9)] = 1,210$ psi, which is satisfactory.

6. Record the relevant properties of the 18WF55

Thus, $A = 16.19$ sq in; $d = 18.12$ in.; $I = 890$ in.⁴; $S = 98.2$ in.³; flange thickness $= 0.630$ in.

7. Compute the section moduli of the composite section where the cover plate is absent

To locate the neutral axis, take statical moments with respect to the center of the steel. Thus, transformed flange width $= 87.53/9 = 9.726$ in. Further:

Element	A, sq in.	y, in.	Ay, in.³	Ay², in.⁴	I_o, in.⁴
18WF55	16.19	0	0	0	890
Slab	48.63	11.56	562.2	6,499	101
Total	64.82	...	562.2	6,499	991

Then, $\bar{y} = 562.2/64.82 = 8.67$ in.; $I = 6,499 + 991 - 64.82(8.67)^2 = 2,618$ in.⁴; $y_b = 9.06 + 8.67 = 17.73$ in.; $y_{tc} = 9.06 + 5 - 8.67 = 5.39$ in.; $S_b = 2,618/17.73 = 147.7$ in.³; $S_{tc} = 2,618/5.39 = 485.7$ in.³.

8. Verify the value of S_b

Apply the value of the K factor in the AISC table. This factor is defined by $K^2 = 1 - S_b$ without plate/S_b with plate. The S_b value without the plate $= 317.5(1 - 0.73^2) = 148$ in.³, which is satisfactory.

9. Establish the theoretical length of the cover plate

In Fig. 27, let C denote the section at which the cover plate becomes superfluous with respect to flexure. Then, for the composite section: $w = 0.773 + 0.330 + 2.200 =$

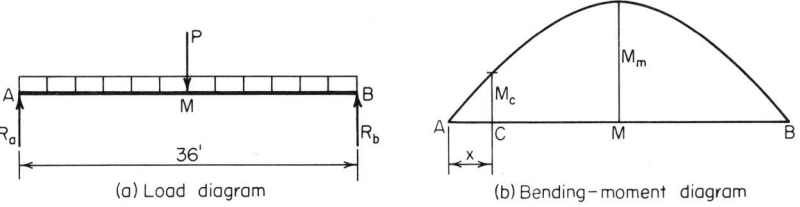

(a) Load diagram   (b) Bending-moment diagram

Fig. 27 (a) Load diagram; (b) bending-moment diagram.

3.303 klf; $P = 10$ kips; $M_m = 7,502$ in.-kips; $R_a = 64.45$ kips. The allowable values of M_c are, for concrete, $M_c = 485.7(9)1.35/12 = 491.8$ ft-kips, and, for steel, $M_c = 147.7(24)/12 = 295.4$ ft-kips, which governs. Then: $R_a x - \frac{1}{2} w x^2 = 295.4$; $x = 5.30$ ft. The theoretical length $= 36 - 2(5.30) = 25.40$ ft.

For the noncomposite section, investigate the stresses at the section C previously located. Thus: $w = 0.773$ klf; $P = 10$ kips; $R_a = 18.91$ kips; $M_c = 18.91(5.30) - \frac{1}{2}(0.773) \times 5.30^2 = 89.4$ ft-kips; $f_b = f_{ts} = 89.4(12)/98.2 = 10.9$ ksi, which is satisfactory.

10. Determine the axial force F in the cover plate at its end by computing the mean bending stress

Thus: $f_{mean} = M y_{mean}/I = 295.4(12)(16.50 - 0.50)/5,242 = 10.82$ ksi; $F = A f_{mean} = 9(10.82) = 97.4$ kips. Alternatively, calculate F by applying the factor $12Q/I$ recorded in the AISC table. Thus, $F = 12QM/I = 0.33(295.4) = 97.5$ kips.

11. Design the weld required to develop the cover plate at each end

Use fillet welds of E60 electrodes, placed along the sides but not along the end of the plate. The AISC *Specification* requires a minimum weld of $\frac{5}{16}$ in. for a 1-in. plate; the capacity of this weld is 3,000 pli. Then, length $= 97,400/3,000 = 32.5$ in. However, the AISC requires that the plate be extended a distance of 18 in. beyond the theoretical cutoff point, thus providing 36 in. of weld at each end.

12. Design the intermittent weld

The vertical shear at C is: $V_c = R_a - 5.30w = 64.45 - 5.30(3.303) = 46.94$ kips; $q = VQ/I = 46,940(0.33)/12 = 1,290$ pli. The AISC calls for a minimum weld length of $1\frac{1}{2}$ in. Let s denote the center-to-center spacing. Then, $s = 2(1.5)3,000/1,290 = 7.0$ in. The AISC imposes an upper limit of 24 times the thickness of the thinner part joined, or 12 in. Thus, $s_{max} = 24(0.63) > 12$ in. Use a 7-in. spacing at the ends and increase the spacing as the shear diminishes.

13. Design the shear connectors

Use $\frac{3}{4}$-in. studs, 3 in. high. The design of the connectors is governed by the AISC *Specification*. The capacity of the stud $= 11.5$ kips. From the AISC table, $V_h = 453.4$ kips. Total number of studs required $= 2(453.4)/11.5 = 80$. These are to be equally spaced.

DESIGN OF A CONCRETE JOIST IN A RIBBED FLOOR

The concrete floor of a building will be constructed by using removable steel pans to form a one-way ribbed slab. The loads are: live load, 80 psf; allowance for movable partitions, 20 psf; plastered ceiling, 10 psf; wood floor with sleepers in cinder-concrete fill, 15 psf. The joists will have a clear span of 17 ft and be continuous over several spans. Design the interior joist by the ultimate-strength method, using $f'_c = 3,000$ psi and $f_y = 40,000$ psi.

Calculation Procedure:

1. Compute the ultimate load carried by the joist

A one-way ribbed floor consists of a concrete slab supported by closely spaced members called *ribs*, or *joists*. The joists in turn are supported by steel or concrete girders that frame to columns. Manufacturers' engineering data present the dimensions of steel-pan forms that are available and the average weight of floor corresponding to each form.

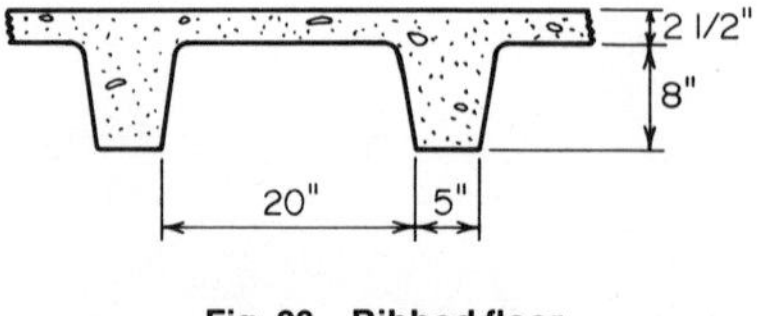

Fig. 28   Ribbed floor.

Try the cross section shown in Fig. 28, which has an average weight of 54 psf. Although the forms are tapered to facilitate removal, assume for design purposes that the joist has a constant width of 5 in. The design of a ribbed floor is governed by the ACI

Code. The ultimate-strength design of reinforced-concrete members is covered in Sec. 1 of this Handbook.

Referring to the ACI *Code,* compute the ultimate load carried by the joist. Or, $w_u = 2.08[1.5(54+20+10+15)+1.8(80)] = 608$ plf.

2. Determine if the joist is adequate with respect to shear

Since the joist is too narrow to permit the use of stirrups, the shearing stress must be limited to the value given in the ACI *Code.* Or, $v_c = 1.1(2\phi)(f'_c)^{0.5} = 1.1(2)(0.85) \times (3,000)^{0.5} = 102$ psi.

Assume that the reinforcement will consist of No. 4 bars. With $\frac{3}{4}$ in. for fireproofing, as required by the ACI *Code,* $d = 8+2.5-1.0 = 9.5$ in. The vertical shear at a distance d from the face of the support is $V_u = (8.5-0.79)608 = 4,690$ lb.

The critical shearing stress computed as required by the ACI *Code* is $v_u = V_u/bd = 4,690/[5(9.5)] = 99$ psi $< v_c$, which is satisfactory.

3. Compute the ultimate moments to be resisted by the joist

Do this by applying the moment equations given in the ACI *Code.* Or, $M_{u,\text{pos}} = (\frac{1}{16})608(17)^2 12 = 132,000$ in.-lb; $M_{u,\text{neg}} = (\frac{1}{11})608(17)^2 12 = 192,000$ in.-lb.

Where the bending moment is positive, the fibers above the neutral axis are in compression, and the joist and tributary slab function in combination to form a T beam. Where the bending moment is negative, the joist functions alone.

4. Determine if the joist is capable of resisting the negative moment

Use the equation $q_{\max} = 0.6375k_1 87,000/(87,000+f_y)$, or $q_{\max} = 0.6375(0.85)87,000/127,000 = 0.371$. By Eq. (6) of Sec. 1, $M_u = \phi bd^2 f'_c q(1-0.59q)$, or $M_u = 0.90(5)9.5^2 \times (3,000)0.371(0.781) = 353,000$ in.-lb, which is satisfactory.

5. Compute the area of negative reinforcement

Use Eq. (7) of Sec. 1 of this Handbook. Or, $f_c = 0.85(3) = 2.55$ ksi; $bdf_c = 5(9.5)2.55 = 121.1$; $2bf_c M_u/\phi = 2(5)2.55(192)/0.90 = 5,440$; $A_s = [121.1-(121.1^2-5,440)^{0.5}]/40 = 0.63$ sq in.

6. Compute the area of positive reinforcement

Since the stress block lies wholly within the flange, apply Eq. (7) of Sec. 1, with $b = 25$ in. Or, $bdf_c = 605.6$; $2bf_c M_u/\phi = 18,700$; $A_s = [605.6-(605.6^2-18,700)^{0.5}]/40 = 0.39$ sq in.

7. Select the reinforcing bars and locate the bend points

For positive reinforcement, use two No. 4 bars, one straight and one trussed, to obtain $A_s = 0.40$ sq in. For negative reinforcement, supplement the two trussed bars over the support with one straight No. 5 bar to obtain $A_s = 0.71$ sq in.

To locate the bend points of the trussed bars and to investigate the bond stress, follow the method given in Sec. 1 of this Handbook.

DESIGN OF A STAIR SLAB

The concrete stair shown in elevation in Fig. 29a, which has been proportioned in conformity with the requirements of the local building code, is to carry a live load of 100 psf. The slab will be poured independently of the supporting members. Design the slab by the working-stress method, using $f'_c = 3,000$ psi and $f_s = 20,000$ psi.

Calculation Procedure:

1. Compute the unit loads

The working-stress method of designing reinforced-concrete members is presented in Sec. 1 of this Handbook. The slab is designed as a simply supported beam having a span equal to the horizontal distance between the center of supports. For convenience, consider a strip of slab having a width of 1 ft.

Assume that the slab will be 5.5 in. thick, the thickness of the stairway slab being measured normal to the soffit. Compute the average vertical depth in Fig. 29b. Thus: sec $\theta = 1.25$; $h = 5.5(1.25) + 3.75 = 10.63$ in. For the stairway, $w = 100 + 10.63(150)/12 = 233$ plf; for the landing, $w = 100 + 5.5(150)/12 = 169$ plf.

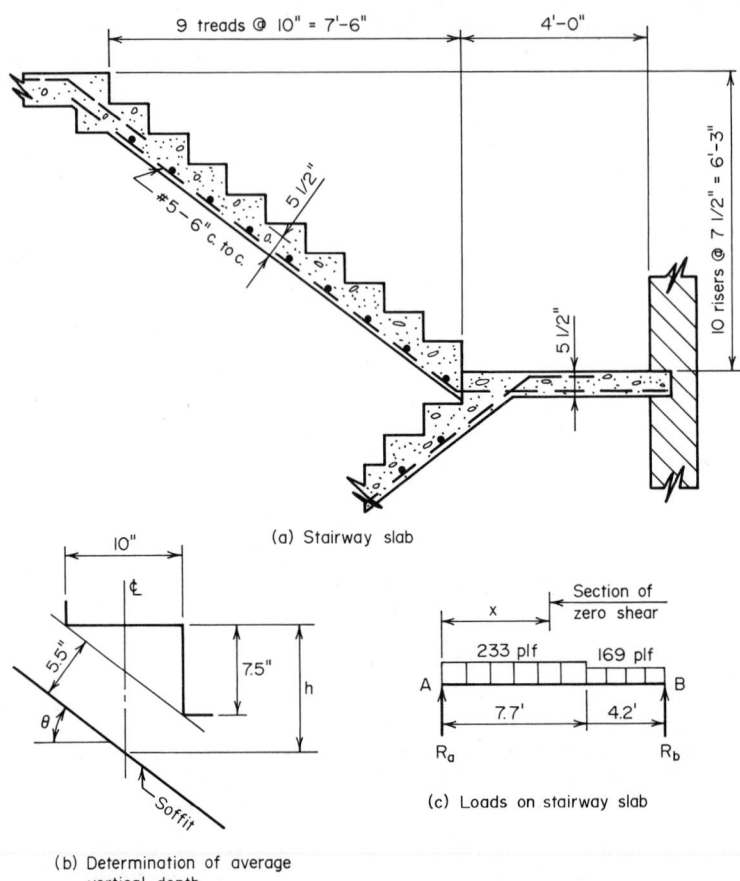

(b) Determination of average vertical depth

Fig. 29 *(a) Stairway slab; (b) determination of average vertical depth; (c) loads on stairway slab.*

2. Compute the maximum bending moment in the slab

Construct the load diagram shown in Fig. 29c, adding about 5 in. to the clear span to obtain the effective span. Thus: $R_a = [169(4.2)2.1 + 233(7.7)8.05]/11.9 = 1,339$ lb; $x = 1,339/233 = 5.75$ ft; $M_{max} = \frac{1}{2}(1,339)5.75(12) = 46,200$ in.-lb.

3. Design the reinforcement

Refer to Table 6 in Sec. 1 to obtain the following values: $K_b = 223$ psi; $j = 0.874$. Assume an effective depth of 4.5 in. By Eq. (31), the moment capacity of the member at balanced design is $M_b = K_b bd^2 = 223(12)4.5^2 = 54,190$ in.-lb. The steel is therefore stressed to capacity. (Upon investigation, a 5-in. slab is found to be inadequate.) By Eq. (25), $A_s = M/f_s jd = 46,200/[20,000(0.874)4.5] = 0.587$ sq in.

Use No. 5 bars, 6 in. on centers, to obtain $A_s = 0.62$ sq in. In addition, place one No. 5 bar transversely under each tread to assist in distributing the load and to serve as temperature reinforcement. Since the slab is poured independently of the supporting members, it is necessary to furnish dowels at the construction joints.

FREE VIBRATORY MOTION OF A RIGID BENT

The bent in Fig. 30 is subjected to a horizontal load P applied suddenly at the top. Using literal values, determine the frequency of vibration of the bent. Make these simplifying assumptions: The girder is infinitely rigid; the columns have negligible mass; damping forces are absent.

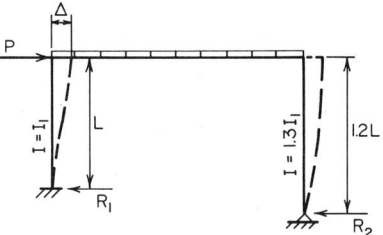

Fig. 30   Vibrating bent.

Calculation Procedure:

1. Compute the spring constant

The amplitude (maximum horizontal displacement of the bent from its position of static equilibrium) is a function of the energy imparted to the bent by the applied load. The frequency of vibration is independent of this energy. To determine the frequency, it is necessary to find the *spring constant* of the vibrating system. This is the static force that is required at the top to cause a horizontal displacement of one unit. Since the girder is considered to be infinitely rigid, the elastic curves of the columns are vertical at the top. Let f = frequency; k = spring constant; M = total mass of girder and bodies supported by girder.

Using Cases 22 and 23 in the AISC *Manual*, when $\Delta = 1$, the horizontal reactions are: $R_1 = 12EI_1/L^3$; $R_2 = 3E(1.3I_1)/(1.2L)^3 = 2.26EI_1/L^3$; $k = R_1 + R_2 = 14.26EI_1/L^3$.

2. Compute the frequency of vibration

Use the equation $f = (1/2\pi)(k/M)^{0.5} = (1/2\pi)(14.26\,EI_1/ML^3)^{0.5} = 0.601(EI_1/ML^3)^{0.5}$ cps.

PLUMBING AND DRAINAGE

REFERENCES: Manas — *National Plumbing Code Handbook*, McGraw-Hill; Babbitt — *Plumbing*, McGraw-Hill; American Standards Institute — *National Plumbing Code*; National Bureau of Standards — *Water Distributing Systems for Buildings*; National Association of Plumbing Contractors — *Water Supply Piping for the Plumbing System*; Copper and Brass Research Association — *Brass Pipe Handbook for Plumbing Installations* and *Copper Tube Handbook on Plumbing and Heating*; Nielsen — *Standard Plumbing Engineering Design*, McGraw-Hill.

DETERMINATION OF PLUMBING-SYSTEM PIPE SIZES

A two-story industrial plant has the following plumbing fixtures: first floor — six wall-lip urinals, three valve-operated water closets, three large-size lavatories, and six showers, each with a separate head; second floor — three wall-lip urinals, three valve-operated water closets, three large-size lavatories, and three showers, each with a separate head. Size the waste and vent stacks and the building house drain for this system. Use the *National Plumbing Code* (NPC) as the governing code for the plant locality. The branch piping and house drain will be pitched ¼ in. per ft of length.

Calculation Procedure:

1. Select the upper-floor branch layout

Sketch the layout of the proposed plumbing system, beginning with the upper, or second, floor. Figure 1 shows a typical plumbing-system sketch. Assume in this plant that the second-floor urinals, water closets, and lavatories are served by one branch

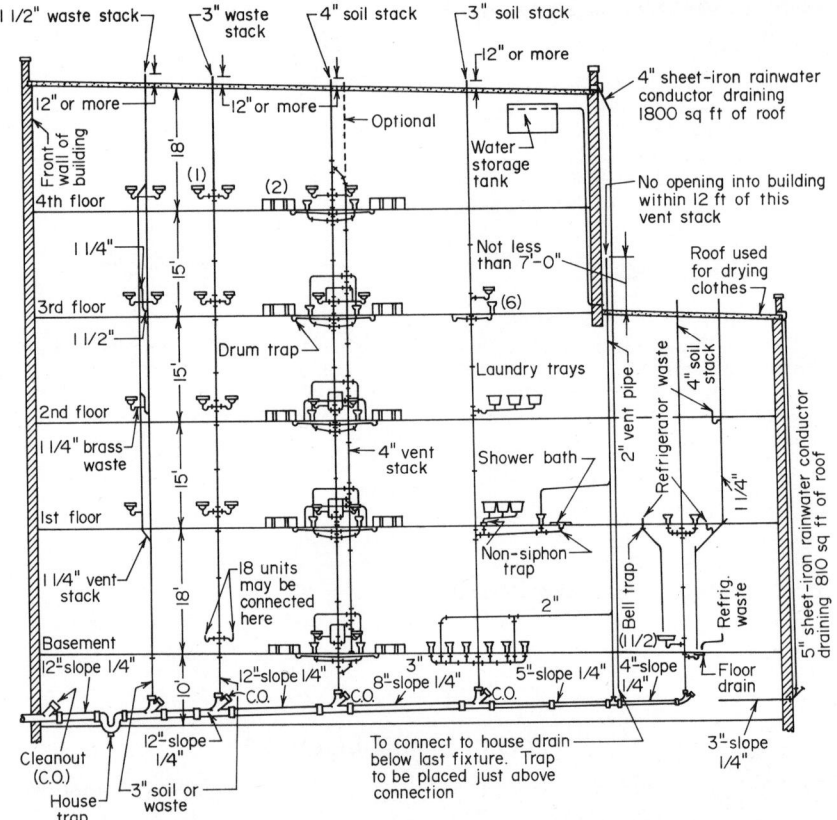

Fig. 1 Typical plumbing layout diagram for a multistory building.

drain and the showers by another branch. Both branch drains will discharge into a vertical soil stack.

2. Compute the upper-floor branch fixture units

List each plumbing fixture as in Table 1.

Obtain the data for each numbered column of Table 1 in the following manner. (1) List the number of the floor being studied, and number of each branch drain from the system sketch. Since it was earlier decided to use two branch drains, number them accordingly. (2) List the name of each fixture that will be used. (3) List the number of each type of fixture that will be used. (4) Obtain from the *National Plumbing Code*, or Table 2, the number of *fixture units per fixture*, i.e., the average discharge, during use, of an arbitrarily selected fixture, such as a lavatory or toilet. Once this value is established in a plumbing code, the discharge rates of other types of fixtures are stated in terms of the basic unit. Plumbing codes adopted by various localities usually list the fixture units they recommend in a tabulation similar to Table 2. (5) Multiply the number of fixtures, column 3, by the fixture units, column 4, to obtain the result in column 5. Thus, for the urinals, (three urinals)(four fixture units per urinal fixture) = twelve fixture units. Find the sum of the fixture units for each branch.

3. Size the upper-floor branch pipes

Refer to the *National Plumbing Code*, or Table 3, for the number of fixture units each branch can have connected to it. Thus, Table 3 shows that a 4-in. branch pipe

TABLE 1   Floor-Fixture Analysis

(1) Floor	(2) Fixture name	(3) No. of fixtures	(4) Fixture units per fixture	(5) Total no. of fixture units
Floor (2)	Urinals wall-lip	3	4	12
Branch drain 1	Water closets, valve-operated	3	8	24
	Lavatories, large-size	3	2	6
Total	. . .	9	. . .	42
Branch drain 2	Showers	3	3	9
Total	. . .	3	. . .	9
Floor 1	Urinals, wall-lip	6	4	24
	Water closets, valve-operated	3	8	24
Branch drain 3	Lavatories, large-size	3	2	6
Total	. . .	12	. . .	54
Branch drain 4	Showers	6	3	18
Total	. . .	6	. . .	18

TABLE 2   Fixture Units per Fixture or Group*

Fixture type	Fixture-unit value as load factors	Minimum size of trap, in.
One bathroom group consisting of water closet, lavatory, and bathtub or shower stall	Tank water closet, 6; Flush-valve water closet, 8	
Bathtub† (with or without overhead shower). . .	2	$1\frac{1}{2}$
Bathtub†.	3	2
Bidet.	3	Nominal, $1\frac{1}{2}$
Combination sink and tray.	3	$1\frac{1}{2}$
Combination sink and tray with food-disposal		Separate traps, $1\frac{1}{2}$
unit	4	
Dental unit or cuspidor	1	$1\frac{1}{4}$
Dental lavatory.	1	$1\frac{1}{4}$
Drinking fountain	$\frac{1}{2}$	1

TABLE 2   Fixture Units per Fixture or Group (Continued) *

Fixture type	Fixture-unit value as load factors	Minimum size of trap, in.
Dishwasher, domestic	2	$1\frac{1}{2}$
Floor drains‡	1	2
Kitchen sink, domestic......................	2	$1\frac{1}{2}$
Kitchen sink, domestic, with food-waste grinder.................................	3	$1\frac{1}{2}$
Lavatory§.................................	1	Small PO, $1\frac{1}{4}$
Lavatory§.................................	2	Large PO, $1\frac{1}{2}$
Lavatory, barber, beauty parlor..............	2	$1\frac{1}{2}$
Lavatory, surgeon's........................	2	$1\frac{1}{2}$
Laundry tray (one or two compartments)	2	$1\frac{1}{2}$
Shower stall, domestic......................	2	2
Showers (group) per head...................	3	
Sinks:		
Surgeon's...............................	3	$1\frac{1}{2}$
Flushing rim (with valve)	8	3
Service (trap standard)....................	3	3
Service (P trap)..........................	2	2
Pot, scullery, etc.........................	4	$1\frac{1}{2}$
Urinal, pedestal, siphon jet, blowout	8	Nominal, 3
Urinal, wall lip............................	4	$1\frac{1}{2}$
Urinal stall, washout.......................	4	2
Urinal trough (each 2-ft section)	2	$1\frac{1}{2}$
Wash sink (circular or multiple) each set of faucets.................................	2	Nominal, $1\frac{1}{2}$
Water closet, tank-operated.................	2	Nominal, 3
Water closet, valve-operated................	8	3

*From *National Plumbing Code.*
†A shower head over a bathtub does not increase the fixture value.
‡Size of floor drain shall be determined by the area of surface water to be drained.
§Lavatories with $1\frac{1}{4}$- or $1\frac{1}{2}$-in. trap have the same load value; larger PO (plumbing orifice) plugs have greater flow rate.

must be used for branch drain 1 because no more than twenty fixture units can be connected to the next smaller, or 3-in. pipe. Hence, branch drain 1 will use a 4-in. pipe because it serves forty-two fixture units, Step. 2.

Branch drain 2 serves 9 fixture units, Step 2. Hence, a $2\frac{1}{2}$-in. branch pipe will be suitable because it can serve 12 fixture units, or less, Table 3.

4. Size the upper-floor stack

The two horizontal branch drains are sloped towards a vertical *stack* pipe that conducts the waste and water from the upper floors to the sewer. Use Table 3 to size the stack, which is three stories high, including the basement. The total number of second-floor fixture units the stack must serve is $42 + 9 = 51$. Hence, for a 4-in. stack, Table 3 must be used.

5. Size the upper-story vent pipe

Each branch drain on the upper floor must be vented. However, the stack can be extended upwards and each branch vent connected to it, if desired. Use the *NPC,* or Table 4, to determine the vent size.

As a guide, the diameter of a branch vent or vent stack is one-half or more of the branch or stack it serves, but not less than $1\frac{1}{4}$ in. Thus branch drain 1 would have a $4/2 = 2$-in. vent, whereas branch drain 2 would have a $2\frac{1}{2}/2 = 1\frac{1}{4}$-in. vent.

TABLE 3   Horizontal Fixture Branches and Stacks*

Diameter of pipe, in.	Any horizontal† fixture branch	One stack of three stories in height or three intervals	More than three stories in height	
			Total for stack	Total at one story or branch interval
1¼	1	2	2	1
1½	3	4	8	2
2	6	10	24	6
2½	12	20	42	9
3	20‡	30§	60§	16‡
4	160	240	500	90
5	360	540	1,100	200
6	620	960	1,900	350
8	1,400	2,200	3,600	600
10	2,500	3,800	5,600	1,000
12	3,900	6,000	8,400	1,500
15	7,000			

Maximum number of fixture units that may be connected to

*From *National Plumbing Code.*
†Does not include branches of the building drain.
‡Not over two water closets.
§Not over six water closets.

TABLE 4   Sizes of Building Drains and Sewers*

Maximum number of fixture units that may be connected to any portion† of the building drain or the building sewer

Diameter of pipe, in.	Fall per ft			
	1/16 in.	1/8 in.	1/4 in.	1/2 in.
2	...	...	21	26
2½	...	...	24	31
3	...	20‡	27‡	36‡
4	...	180	216	250
5	...	390	480	575
6	...	700	840	1,000
8	1,400	1,600	1,920	2,300
10	2,500	2,900	3,500	4,200
12	3,900	4,600	5,600	6,700
15	7,000	8,300	10,000	12,000

*From *National Plumbing Code.*
†Includes branches of the building drain.
‡Not over two water closets.

6. Select the lower-floor branch layout

Assume that the six urinals, three water closets, and three lavatories are served by one branch drain and the six showers by another. Indicate these on the system sketch. Further, arrange both branch drains so that they discharge into the vertical stack serving the second floor.

7. Compute the lower-floor branch fixture units

Use the same procedure as in Step 2, listing the fixtures and their respective fixture units in the lower part of Table 1.

8. Size the lower-floor branch pipes

Using Table 3, branch drain 3 must be 4 in. because it serves a total of 54 fixture units. Branch 4 must be 3 in. because it serves a total of 18 fixture units.

9. Size the lower-floor stack

The lower-floor stack serves both the upper- and lower-floor branch drains, or a total of $42 + 9 + 54 + 18 = 123$ fixture units. Table 3 shows that a 4-in. stack will be satisfactory.

10. Size the lower-floor vents

Using the one-half rule of Step 5, the vent for branch drain 3 must be 2 in., whereas that for branch drain 4 must be $1\frac{1}{2}$ in.

11. Size the building drain

The building drain serves all the fixtures installed in the building and slopes downward toward the city sewer. Hence, the total number of fixture units it serves is $42 + 9 + 54 + 18 = 123$. This is the same as the vertical stack. Table 4 shows that a 4-in. drain that is sloped $\frac{1}{4}$-in./ft will serve 216 fixture units. Thus, a 4-in. drain will be satisfactory. The house trap that is installed in the building drain should also be a 4-in. unit.

Related Calculations: Where a local plumbing code exists, use it instead of the *NPC*. If no local code exists, follow the *NPC* for all classes of buildings. Use the general method given here to size the various pipes in the system. Select piping materials (cast iron, copper, clay, steel, brass, wrought iron, lead, etc.) in accordance with the local or *NPC* recommendations. Where the house drain is below the level of the public sewer line, it is often arranged to discharge into a suitably sized *sump pit*. Sewage is discharged from the sump pit to the public sewer by a pneumatic ejector or motor-driven pump.

DESIGN OF ROOF AND YARD RAINWATER DRAINAGE SYSTEMS

An industrial plant is 300 ft long and 100 ft wide. The roof of the building is flat except for a 50-ft-long, 100-ft-wide, 80-ft-high machinery room at one end of the roof. Size the leaders and horizontal drains for this roof for a maximum rainfall of 4 in./hr. What size storm drain is needed if the drain is sloped $\frac{1}{4}$ in. per ft of length?

Calculation Procedure:

1. Sketch the building roof

Figure 2 shows the building roof and machinery-room roof. Indicate on the sketch the major dimensions of the roof and machinery room.

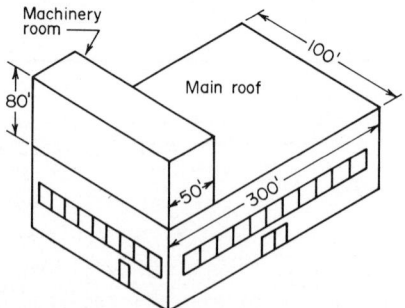

Fig. 2   Building roof areas.

2. Compute the roof area to be drained

Two roof areas must be drained—the machinery-room roof and the main roof. The respective areas are: Machinery-room-roof area $= 50 \times 100 = 5,000$ sq ft; main-roof area $= 250 \times 100 = 25,000$ sq ft.

The wall of the machinery room facing the main roof will also collect rain to some extent. This must be taken into consideration when sizing the roof leaders. Do this by computing the area of the wall facing the main roof and adding one-half this area to the main-roof area. Thus, wall area =

$80 \times 100 = 8,000$ sq ft. Adding half this area to the main-roof area gives $25,000 + 8,000/2 = 29,000$ sq ft.

3. Select the leader size for each roof

Decide whether the small-roof area, i.e., the machinery-room roof, will be drained by separate leaders to the ground or to the main-roof area. If the small-roof area is drained separately, treat it as a building unto itself. Where the small roof drains onto the main roof, add the two roof areas together to determine the leader size.

Treating the two roofs as separate units, Table 5 shows that a 5-in. leader is needed for the 5,000 sq ft machinery-room roof. This same table shows that an 8-in. leader is needed for the 29,000 sq ft main roof, including the machinery-room wall.

TABLE 5 Sizes of Vertical Leaders and Horizontal Storm Drains*

Vertical leaders	
Size of leader or conductor,† in.	Maximum projected roof area, sq ft
2	720
$2\frac{1}{2}$	1,300
3	2,200
4	4,600
5	8,650
6	13,500
8	29,000

Horizontal storm drains			
Diameter of drain, in.	Maximum projected roof area for drains of various slopes, sq ft		
	$\frac{1}{8}$-in. slope	$\frac{1}{4}$-in. slope	$\frac{1}{2}$-in. slope
3	822	1,160	1,644
4	1,880	2,650	3,760
5	3,340	4,720	6,680
6	5,350	7,550	10,700
8	11,500	16,300	23,000
10	20,700	29,200	41,400
12	33,300	47,000	66,600
15	59,500	84,000	119,000

*From *National Plumbing Code*.
†The equivalent diameter of square or rectangular leader may be taken as the diameter of that circle that may be inscribed within the cross-sectional area of the leader.

4. Size the storm drain for each roof

The lower portion of Table 5 shows that a 6-in. storm drain is needed for the 5,000 sq ft roof. A 10-in. storm drain, Table 5, is needed for the 29,000 sq ft main roof.

When any storm drain is connected to a building sanitary drain or storm sewer a trap should be used at the inlet to the sanitary drain or storm sewer. The trap prevents sewer gases entering the storm leaders.

Related Calculations: Size roof leaders in strict accordance with the *National Plumbing Code* (*NPC*) or the local applicable code. Undersized roof leaders are dangerous because they can cause water buildup on a roof, leading to excessive roof loads. Where gutters are used on a building, size them in accordance with Table 6.

TABLE 6   Size of Gutters*

Diameter of gutter,† in.	Maximum projected roof area for gutters of various slopes, sq ft			
	$\frac{1}{16}$-in. slope	$\frac{1}{8}$-in. slope	$\frac{1}{4}$-in. slope	$\frac{1}{2}$-in. slope
3	170	240	340	480
4	360	510	720	1,020
5	625	880	1,250	1,770
6	960	1,360	1,920	2,770
7	1,380	1,950	2,760	3,900
8	1,990	2,800	3,980	5,600
10	3,600	5,100	7,200	10,000

*From *National Plumbing Code*.
†Gutters other than semicircular may be used provided they have an equivalent cross-sectional area.

Where a roof leader discharges into a sanitary drain, convert the roof area to equivalent fixture units to determine the load on the sanitary drain. To convert roof area to fixture units, take the first 1,000 sq ft of roof area as equivalent to 256 fixture units when designing for a maximum rainfall of 4 in./hr. Where the total roof area exceeds 1,000 sq ft, divide the remaining roof area by 3.9 sq ft per fixture unit to determine the fixture load for the remaining area.

Thus, the machinery-room roof in the above plant is equivalent to $256 + 4,000/3.9 = 1,281$ fixture units. The main roof and machinery-room wall is equivalent to $256 + 28,000/3.9 = 7,436$ fixture units. These roofs, if taken together, would place a total load of $1,281 + 7,436 = 8,717$ fixture units on a sanitary drain.

Where the rainfall differs from 4 in./hr, compute the load on the drain in the same way as described above. Choose the drain size from the appropriate table. Then multiply the drain size by the ratio: actual maximum rainfall, in./4. If the drain size obtained is nonstandard, as will often be the case, use the next *larger* standard drain size. Thus, with a 6-in. rainfall and a 5-in. leader based on the 4-in. rainfall tables, leader size = $(5)(6/4) = 7.5$ in. Since this is not a standard size, use the next larger size, or 8 in. Roof areas should be drained as quickly as possible to prevent excessive structural stresses caused by water accumulations on the roof.

To compute the required size of drains for paved areas, yards, courts, and courtyards, use the same procedure and tables as for roofs. Where the rainfall differs from 4 in., apply the conversion ratio discussed in the previous paragraph. Note that the flow capacity of floor and roof drains must equal, or exceed, the flow capacity of the leader to which either unit is connected.

SIZING COLD- AND HOT-WATER-SUPPLY PIPING

An industrial building has the following plumbing fixtures: 2 showers, 200 private lavatories, 200 service sinks, 20 public lavatories, 1 dishwasher, 25 flush-valve water closets, and 20 stall urinals. Size the cold- and hot-water piping for these fixtures using an upfeed system. The highest fixture is 50 ft above the water main. The minimum water pressure available in the water main is 60 psi; the pressure loss in the water meter is 8.3 psi.

Calculation Procedure:

1. *Sketch the proposed piping system*

Draw a single-line diagram of the proposed cold- and hot-water piping. Thus, Fig. 3*a* shows the proposed basement layout of the water piping and Fig. 3*b* shows two of the

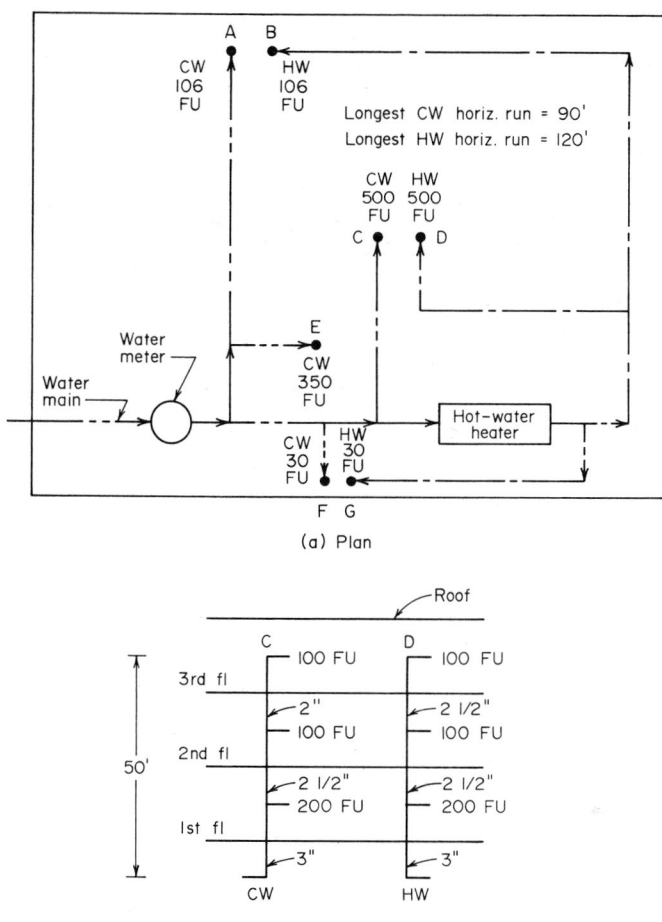

Fig. 3 (a) Plan of industrial-plant water piping; (b) elevation of building water-supply risers.

risers used in this industrial plant. Indicate on each branch line the "weight" in *fixture units* of fixtures served and the required water flow. Table 7 shows the rate of flow and required pressure during flow to different types of fixtures.

2. Compute the demand weight of the fixtures

List the fixtures as in Table 8. Alongside the name and number of each fixture, list the demand weight for cold or hot water, or both, from Table 9. Note that when a fixture has both a cold-water and hot-water supply, only three-fourths of the fixture weight listed in Table 9 is used for each cold-water and each hot-water outlet. Thus, with a total demand weight of 1 for a private lavatory, the cold-water demand weight is 0.75 (1) = 0.75 fixture units, and the hot-water demand weight is 0.75(1) = 0.75 fixture units.

Find the product of the number of each type of fixture and the demand weight per fixture for cold and hot water and enter the result in the last two columns of Table 8. The sum of the cold- and hot-water-fixture demand weights, 986 and 636 fixture units, respectively, gives the total demand weight for the building, in fixture units, except for the dishwasher.

TABLE 7 Rate of Flow and Required Pressure during Flow for Different Fixtures*

Fixture	Flow pressure,† psi	Flow rate, gpm
Ordinary basin faucet..............	8	3.0
Self-closing basin faucet	12	2.5
Sink faucet, ⅜ in.	10	4.5
Sink faucet, ½ in.	5	4.5
Bathtub faucet	5	6.0
Laundry-tub cock, ½ in.	5	5.0
Shower	12	5.0
Ball cock for closet	15	3.0
Flush valve for closet.............	10–20	15–40‡
Flush valve for urinal.............	15	15.0
Garden hose, 50 ft and sill cock	30	5.0

*From *National Plumbing Code*.
†Flow pressure is the pressure in the pipe at the entrance to the particular fixture considered.
‡Wide range due to variation in design and type of flush-valve closets.

TABLE 8 Fixture Demand Weight

Fixture name	No. of fixtures	Demand weight per fixture in fixture units		Total fixture demand weight in fixture units	
		Cold water	Hot water	Cold water	Hot water
Shower	2	3	3	6	6
Lavatory, private	200	0.75	0.75	150	150
Lavatory, public	20	1.5	1.5	30	30
Sink, service	200	2.25	2.25	450	450
Dishwasher	1	...	*	...	...
Water closet, flush-valve........	25	10	...	250	...
Urinal, stall	20	5	...	100	...
Total.............	...	...	...	986	636

*Not given in *National Plumbing Code* tabulation.

3. Compute the building water demand

Using Fig. 4a, enter at the bottom with the number of fixture units and project vertically upwards to the curve. At the left read the demand − 210 gpm of cold water and 160 gpm of hot water, excluding the dishwasher.

Table 9 shows that a dishwasher serving 500 people in an industrial plant requires 250 gph with a demand factor of 0.40. This is equivalent to a demand of (demand, gph) (demand factor), or $(250)(0.40) = 100$ gph, or 100 gph/(60 min/hr) = 1.66 gpm; say 2.0 gpm. Hence, the total hot-water demand is $160 + 2 = 162$ gpm. The total building water demand is therefore $210 + 162 = 372$ gpm.

4. Compute the allowable piping pressure drop

The minimum inlet water pressure generally recommended for a plumbing fixture is 8 psi, although some authorities use a lower limit of 5 psi. Flushometers normally

TABLE 9 Demand Weight of Fixtures in Fixture Units*

Fixture or group†	Occupancy	Type of supply control	Weight in fixture units‡
Water closet	Public	Flush valve	10
Water closet	Public	Flush tank	5
Pedestal urinal........	Public	Flush valve	10
Stall or wall urinal	Public	Flush valve	5
Stall or wall urinal	Public	Flush tank	3
Lavatory..............	Public	Faucet	2
Bathtub	Public	Faucet	4
Shower head..........	Public	Mixing valve	4
Service sink	Office, etc.	Faucet	3
Kitchen sink	Hotel or restaurant	Faucet	4
Water closet	Private	Flush valve	6
Water closet	Private	Flush tank	3
Lavatory..............	Private	Faucet	1
Bathtub	Private	Faucet	2
Shower head..........	Private	Mixing valve	2
Bathroom group	Private	Flush valve for closet	8
Bathroom group	Private	Flush tank for closet	6
Separate shower	Private	Mixing valve	2
Kitchen sink	Private	Faucet	2
Laundry trays (one to three)	Private	Faucet	3
Combination fixture ..	Private	Faucet	3

*From *National Plumbing Code.* For supply outlets likely to impose continuous demands, estimate continuous supply separately and add to total demand for fixtures.

†For fixtures not listed, weights may be assumed by comparing the fixture to a listed one using water in similar quantities and at similar rates.

‡The given weights are for total demand. For fixtures with both hot and cold water supplies, the weights for maximum separate demands may be taken as three-fourths the listed demand for supply.

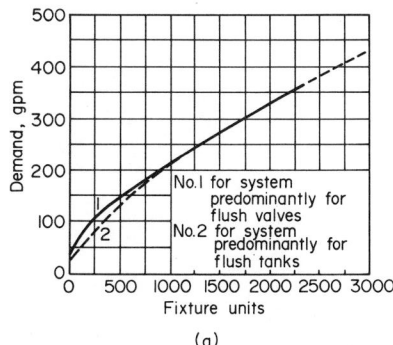

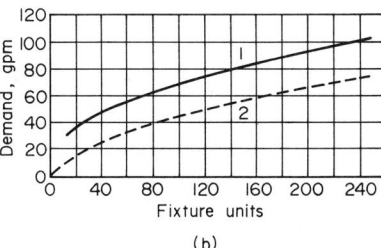

Fig. 4 *(a)* **Domestic water demand for various fixtures;** *(b)* **enlargement of low-demand portion of a.**

require an inlet pressure of 15 psi. Table 7 lists the usual inlet pressure and flow rates required for various plumbing fixtures.

Assume a 15-psi inlet pressure at the highest fixture. This fixture is 50 ft above the water main, Fig. 3. To convert elevation in feet to pressure in psi, multiply by 0.434, or

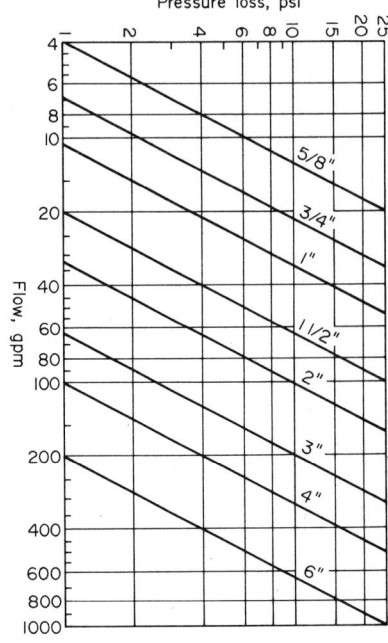

Pressure loss, psi

Flow, gpm

Fig. 5   Pressure loss in disk-type water meters.

(50 ft)(0.434) = 21.7 psi. Lastly, the pressure loss in the water meter is 8.3 psi, as given in the problem statement. Thus, the pressure loss in this or any other water-supply system, not considering piping friction loss, = fixture inlet pressure, psi, + vertical elevation loss, psi, + water-meter pressure loss, psi = 15 + 21.7 + 8.3 = 45 psi. Hence, the pressure available to overcome the piping frictional resistance = 60 − 45 = 15 psi.

NOTE: The pressure loss in water meters of various sizes can be obtained from manufacturers' engineering data, or Fig. 5, for disk-type meters.

5.   Compute the allowable friction loss in the piping

Figure 3a shows that the longest horizontal run of pipe is 90 + 50 = 140 ft. Allowing 50 percent of the straight run for the equivalent length of valves and fittings in the longest run and riser, the total equivalent length of cold-water piping is 140 + 0.50 = 210 ft.

Compute the allowable friction loss per 100 ft of cold-water pipe from F = 100 (pressure available to overcome piping frictional resistance, psi)/equivalent length of cold-water piping, ft. Or F = 100 (15)/210 = 7.14 psi per 100 ft; use 7.0 psi per 100 ft for design purposes.

Using the same procedure for the hot-water pipe, $F = 100(15)/255 = 5.88$ psi per 100 ft; use 5.75 psi per 100 ft. Reducing the design pressure loss for the cold- and hot-water piping design pressure loss to the next lower convenient pressure is done only to save time. If desired, the actual computed value can be used. *Never* round off to the next higher convenient pressure loss because this can lead to undersized pipes and reduced flow from the fixture.

6.   Size the water main

Step 3 shows that the total building water demand is 372 gpm. Using the cold-water friction loss of 7.0 psi per 100 ft, enter Fig. 6 at the bottom at 7.0 and project vertically upwards to 372 gpm. Read the main size as 4 in. This size would be run to the water heater, Fig. 3, unless the run was extremely long. With a long run, the main size would be reduced after each branch takeoff to the risers to reduce the cost of the piping.

7.   Compute the water flow in each riser

List the risers in Fig. 3 as shown in Table 10. Alongside the letter identifying a riser, list the water it handles (hot or cold), the number of fixture units served by the riser, and the flow, gpm. Find the flow in gpm by entering Fig. 4 with the number of fixture units served by the riser and projecting up to the flush-valve curve. Read the gpm at the left of Fig. 4.

8.   Choose the riser size

Enter the pressure loss, psi per 100 ft, found in Step 5 alongside each riser, Table 10. Using Fig. 4 and the appropriate pressure loss, size each riser and enter the chosen size in Table 10. Thus, riser A conveys 70 gpm with a pressure loss of 7.0 psi per 100 ft. Figure 4 shows that a 2-in. riser is suitable. When Fig. 4 indicates a pipe size that is between two standard pipe sizes, use the next *larger* pipe size.

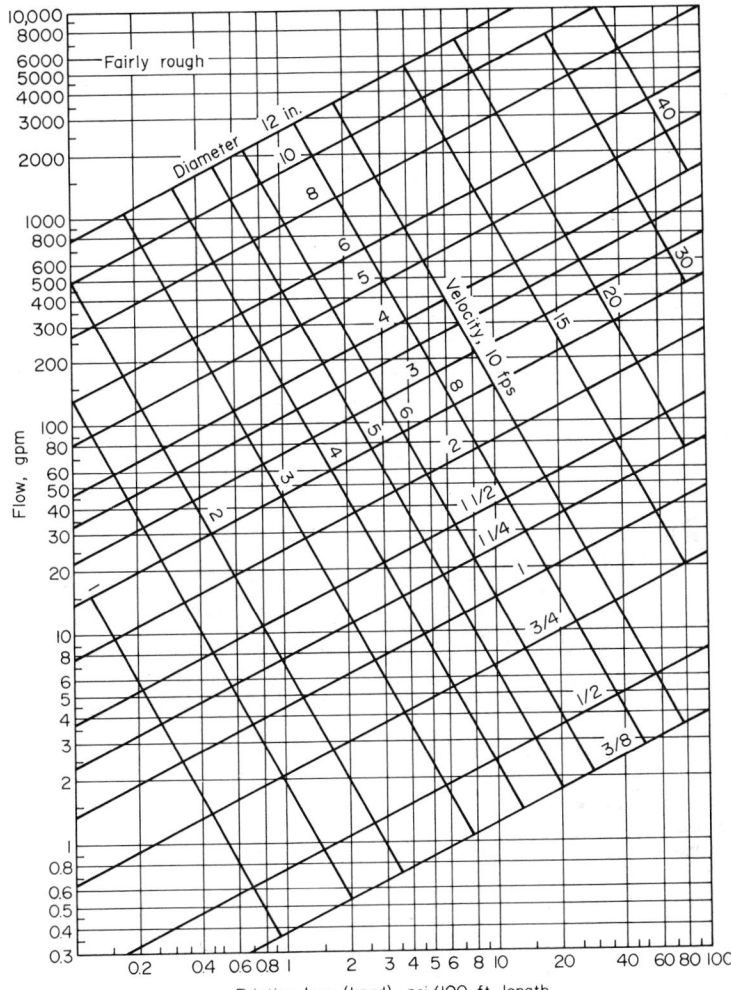

Fig. 6   Chart for selecting water-pipe size for various flow rates.

TABLE 10   Riser Sizing Calculations

Riser	Type*	Fixture units	Gpm	Pressure loss, psi	Riser size, in.
A	CW	106	70	7.0	2
B	HW	106	70	5.75	2½
C	CW	500	150	7.0	3
D	HW	500	150	5.75	3
E	CW	350	130	7.0	2½
F	CW	30	40	7.0	2
G	HW	30	40	5.75	2

*CW stands for cold water; HW stands for hot water.

9. Choose the fixture supply-pipe size

Use Table 11 as a guide for choosing the fixture supply-pipe size. Note that these tabulated sizes are the minimum recommended. Where the supply-pipe run is more than 3 ft, or where more than one fixture is served, use a larger size.

TABLE 11   Minimum Sizes for Fixture-Supply Pipes*

Type of Fixture or Device	Pipe Size, in.	Type of Fixture or Device	Pipe Size, in.
Bath tubs	$\frac{1}{2}$	Shower (single head)	$\frac{1}{2}$
Combination sink and tray....	$\frac{1}{2}$	Sinks (service, slop).........	$\frac{1}{2}$
Drinking fountain	$\frac{3}{8}$	Sinks, flushing rim..........	$\frac{3}{4}$
Dishwasher (domestic).......	$\frac{1}{2}$	Urinal (flush tank)	$\frac{1}{2}$
Kitchen sink, residential	$\frac{1}{2}$	Urinal (direct flush valve)	$\frac{3}{4}$
Kitchen sink, commercial	$\frac{3}{4}$	Water closet (tank type)	$\frac{3}{8}$
Lavatory...................	$\frac{3}{8}$	Water closet (flush valve type)	1
Laundry tray, one, two or		Hose bibs..................	$\frac{1}{2}$
three compartments	$\frac{1}{2}$	Wall hydrant...............	$\frac{1}{2}$

*From *National Plumbing Code.*

10. Select the hot-water-heater capacity

Table 12 shows that the demand factor for a hot-water heater in an industrial plant is 0.40 times the hourly hot-water demand. Step 3 shows that the total hot-water demand is 162 gpm, or 162(60) = 9,720 gph. Therefore, this hot-water heater must have a heating coil capable of heating at least 0.4(9,720) = 3,888 gph; say 3,900 gph.

TABLE 12   Hot-water Demand per Fixture for Various Building Types
(Based on average conditions for types of buildings listed, gallons of water per hour per fixture at 140°F)

Type of fixture	Apartment house	Hospital	Hotel	Industrial plant	Office building
Basins, private lavatories.......	2	2	2	2	2
Basins, public lavatories	4	6	8	12	6
Showers.....................	75	75	75	225	
Slop sinks...................	20	20	30	20	15
Dishwashers (per 500 people) ..	250	250	250	250	250
Pantry sinks.................	5	10	10		
Demand factor	0.30	0.35	0.25	0.40	0.30
Storage factor	1.25	0.60	0.80	1.00	2.00

The storage capacity should equal the product of (hourly water demand)(storage factor from Table 12). Thus, storage capacity = 9,720(1.0) = 9,720 gal. Table 13 shows the usual hot-water temperature used for various services in different types of structures.

TABLE 13   Hot-Water Temperatures for Various Services, F

Cafeterias (serving areas)	130
Lavatories and showers.............	130
Slop sinks (floor cleaning)..........	150
Slop sinks (other cleaning)	130
Cafeteria kitchens.................	130 + steam

Related Calculations: Size the risers serving each floor using the same procedure as in Steps 6 and 7. Thus, risers C and D are each 3 in. up to the first-floor branch. Between this and the second-floor branch, a $2\frac{1}{2}$-in. riser is needed. Between the second and third floors, a 2-in. cold-water riser and a $2\frac{1}{2}$-in. hot-water riser is needed.

In a *downfeed* water-supply system, an elevated roof tank generally supplies cold water to the fixtures. To provide a 15-psi inlet pressure to the highest fixtures, the bottom of the tank must be (15 psi)(2.31 ft of water per psi) = 34.6 ft above the fixture inlet Where this height cannot be obtained because the building design prohibits it, tank-type fixtures requiring only a 3-psi or (3 psi)(2.31) = 6.93-ft elevation at the fixture inlet may be used on the upper floors. Valve-type fixtures are used on the lower floors where the tank elevation provides the required 15-psi inlet pressure.

To design a downfeed system: (*a*) Compute the pressure available at the highest fixture resulting from the tank elevation from psi = 0.434 (tank elevation above inlet to highest fixture, ft). (*b*) Subtract the required inlet pressure to the highest fixture from the pressure obtained in *a*. (*c*) Compute the pressure available to overcome the friction in 100 ft of piping using the method of Step 5 of the upfeed design procedure, substituting the value found in item *b*. (*d*) Size the main from the tank so it is large enough to provide the needed flow to all the upper- and lower-floor fixtures. (*e*) Note that the pressure in each supply main increases as the distance from the tank bottom becomes greater. Thus, the hydraulic pressure increases 0.43 psi per ft of distance from the tank bottom. Usual design practice allows a 15-psi drop through the fittings and valves in the main. The remaining pressure produced by the tank elevation is then available for overcoming pipe friction.

Note that both cold and hot water can be supplied from separate overhead tanks. However, hot water is usually supplied from the building basement by means of a pump. In exceptionally high buildings, water tanks may be located on several intermediate floors as well as the roof. Hot-water heaters may also be located on intermediate floors, although the usual location is in the building basement.

In a *zoned* system, one water tank and one set of hot-water heaters serves several floors or one or more wings of a building. The piping in each zone is designed as described above, using the appropriate method for an upfeed or downfeed system.

To provide hot water as soon as possible after a fixture is opened, the water may be continuously recirculated to the fixtures, Fig. 7. Recirculation is used with both upfeed

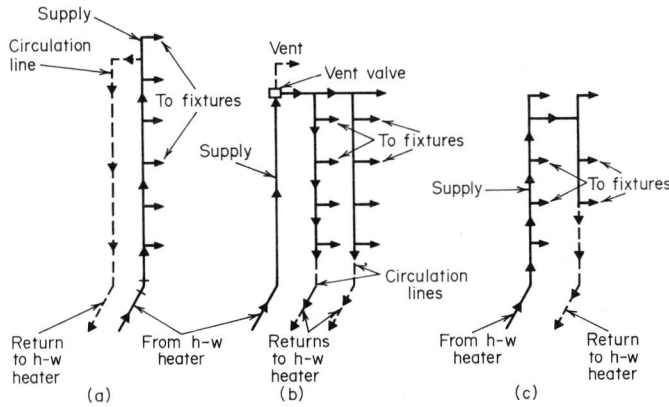

Fig. 7   Hot-water piping systems.

and downfeed systems. To determine the required hot-water temperature in a system, use Table 13, which shows the usual hot-water temperatures used for various services in buildings of different types. Hot-water piping is generally insulated to reduce heat losses.

SPRINKLER-SYSTEM SELECTION AND DESIGN

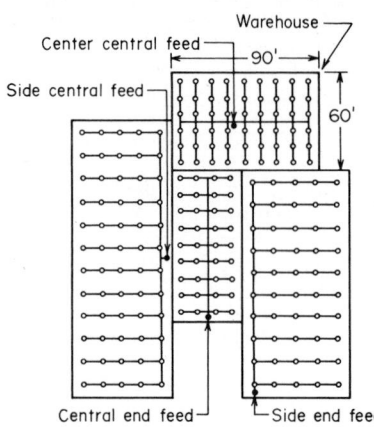

Fig. 8 Typical arrangement of sprinkler piping.

Select and design a sprinkler system for the warehouse building shown in Fig. 8. The materials stored in this warehouse are not flammable. The warehouse is built of fire-resistive materials.

Calculation Procedure:

1. Determine the type of occupancy of the building

The classifications of occupancy used by the National Board of Fire Underwriters (NBFU) are: (1) light hazard, such as apartment houses, asylums, clubhouses, colleges, churches, dormitories, hospitals, hotels, libraries, museums, office buildings, and schools; (2) ordinary hazard, such as mercantile buildings, warehouses, manufacturing plants, and occupancies not classed as light or extra hazardous; (3) extra-hazard occupancies are those buildings or portions of buildings where the inspection agency having jurisdiction determines that the hazard is severe.

Since this is a warehouse used to store nonflammable materials, it can be tentatively classed as an ordinary-hazard occupancy.

2. Compute the number of sprinkler heads required

Consult the local fire-prevention code and the fire underwriters regarding the type, size, and materials required for sprinkler systems. Typical codes recommend that each sprinkler head in an ordinary-hazard fire-resistive building protect not more than 100 sq ft and that the sprinkler branch pipes, and the sprinklers themselves, be not more than 12 ft apart, center-to-center.

The area of the warehouse floor is 90(60) = 5,400 sq ft. With each sprinkler protecting 100 sq ft of area, the number of sprinkler heads required is 5,400 sq ft/100 sq ft per sprinkler = 54 heads.

3. Sketch the sprinkler layout

If the warehouse has a centrally located support column or piping cluster, a center central feed pipe, Fig. 8, can be used. Assuming the sprinkler branch pipes are spaced on 10-ft centers, sketch the branches and heads as shown in Fig. 8. Use a small circle to indicate each sprinkler head.

Space the end sprinkler heads and branch pipes away from the walls by an amount equal to one-half the center-to-center distance between branch pipes. Thus, the end sprinkler heads and branch pipes will be 10/2 = 5 ft from the walls.

4. Size the branch and main sprinkler pipes

Use the local code, Fig. 9, or Table 14. Thus, Table 14 shows that a 1¼-in. branch line will be suitable for three sprinklers in an ordinary-hazard occupancy such as this warehouse. Hence, each branch line having three sprinklers will be this size.

The horizontal overhead main supplying the branches will progressively decrease in diameter as it runs further from the vertical center central feed and serves fewer sprinklers. To the right of the vertical feed, the horizontal main serves 30 sprinklers. Table 14 shows that a 3-in. pipe can serve up to 40 sprinklers. Hence, this size will be used because the next smaller size, 2½-in., can serve only 20 sprinklers.

Since the first branch has six sprinklers, a 3-in. pipe is still needed for the main because Table 14 shows that a 2½-in. pipe can serve only 20, or fewer, sprinklers. However, beyond the second branch, the diameter of the horizontal main can be reduced to

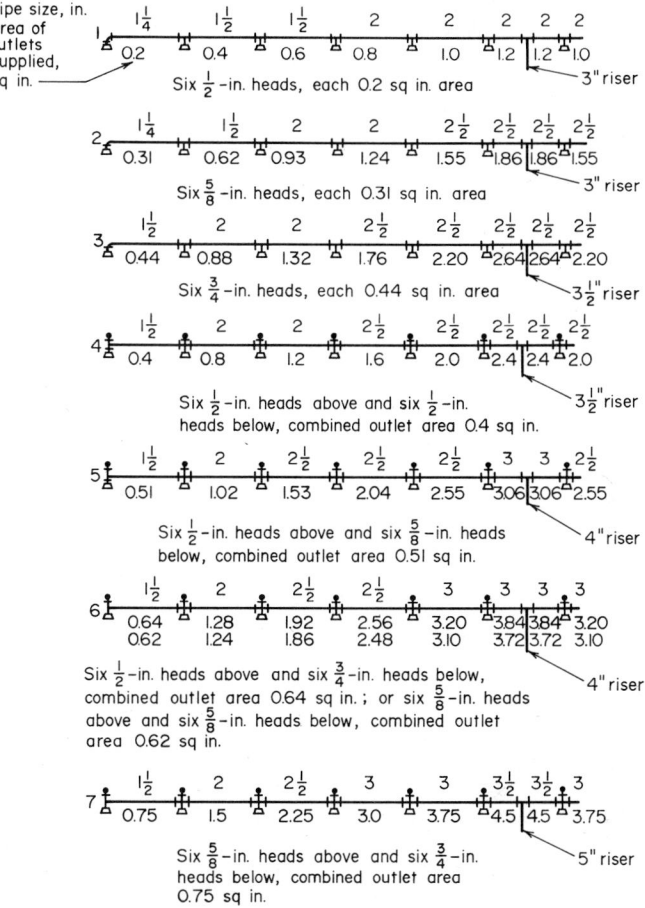

Fig. 9   Sprinkler pipe sizes.

$2\frac{1}{2}$ in. because the number of sprinklers served is $30 - 12 = 18$. Beyond the fourth branch, the main size can be reduced to 2 in. because only six sprinklers are served. Size the left-hand horizontal main in the same way.

The vertical center central feed pipe serves 54 sprinklers. Hence, a $3\frac{1}{2}$-in. pipe must be used, according to Table 14.

5.   Choose the primary and secondary water supply

Usual codes require that each sprinkler system have two water supplies. The primary supply should be automatic and must have sufficient capacity and pressure to serve the system. Local codes usually specify the minimum pressure and capacity acceptable for sprinklers serving various occupancies.

The secondary supply is often a motor-driven automatically controlled fire pump supplied from a water main or taking its suction under pressure from a storage system having sufficient capacity to meet the water requirements of the structure protected.

For light-hazard occupancy, the pump should have a capacity of at least 250 gpm; when supplying both sprinklers and hydrants, the capacity of the pump should be at least 500 gpm. Where the occupancy is classed as an ordinary hazard, as this warehouse

TABLE 14   Pipe-Size Schedule for Typical Sprinkler Installations

Occupancy and Pipe Size, in.	No. of Sprinklers
Light Hazard	
1	2
$1\frac{1}{4}$	3
$1\frac{1}{2}$	5
2	10
$2\frac{1}{2}$	40
3	No limit
Ordinary Hazard	
1	2
$1\frac{1}{4}$	3
$1\frac{1}{2}$	5
2	10
$2\frac{1}{2}$	20
3	40
$3\frac{1}{2}$	65
4	100
5	160
6	250
Extra Hazard	
1	1
$1\frac{1}{4}$	2
$1\frac{1}{2}$	5
2	8
$2\frac{1}{2}$	15
3	27
$3\frac{1}{2}$	40
4	55
5	90
6	150

is, the capacity of the pump should be at least 500 gpm, or 750 gpm, depending on whether or not hydrants are supplied in addition to sprinklers. For extra-hazard occupancy, consult the underwriter and local fire-protection authorities.

Related Calculations: For fire-resistive construction and light-hazard occupancy, the area protected by each sprinkler should not exceed 196 sq ft, and the center-to-center distance of the sprinkler pipes and sprinklers themselves should not exceed 14 ft. For extra-hazard occupancy, the area protected by each sprinkler should not exceed 90 sq ft; the distance between pipes and between sprinklers should not be more than 10 ft. Local fire-protection codes and underwriters' requirements cover other types of construction, including mill, semimill, open-joist, and joist-type with a sheathed or plastered ceiling.

For protection of structures against exposure to fires, outside sprinklers may be used. They can be arranged to protect cornices, windows, side walls, ridge poles, mansard roofs, etc. They are also governed by underwriters' requirements. Figure 9 shows the pipe sizes used for sprinklers protecting outside areas of buildings, including cornices, windows, side walls, etc.

Four common types of automatic sprinkler systems are in use today: (1) wet-pipe, (2) dry-pipe, (3) preaction, and (4) deluge. The type of system used depends on a number of factors, including occupancy classification, local-code requirements, and the requirements of the building fire underwriters. Since the requirements vary from one area to another, no attempt will be made here to list those of each locality or underwriter. The Standards of the National Board of Fire Underwriters, as recommended by the National Fire Protection Association, are excerpted instead because they are so widely used that they are applicable for the majority of buildings. In general, the type of sprinkler chosen does not change the design procedure given above. Figure 10

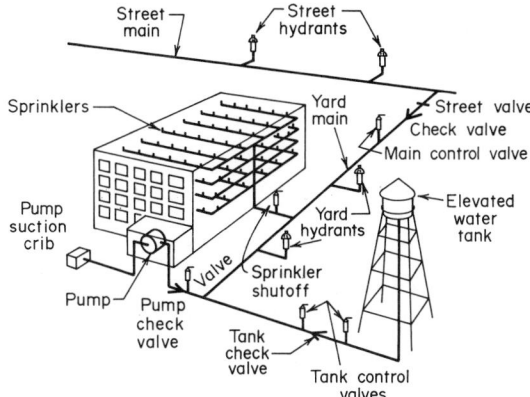

Fig. 10   Water-supply piping for sprinklers.

shows a typical layout of the water-supply piping for an industrial-plant sprinkler system. Figure 11 shows how sprinklers are positioned with respect to a building ceiling.

Use the same general design procedure presented here for sprinklers in other types of buildings — hotels, office buildings, schools, churches, dormitories, colleges, museums, libraries, clubhouses, hospitals, and asylums.

NOTE: *Do not finalize a sprinkler-system design until after it is approved by local fire authorities and the fire underwriters insuring the building.*

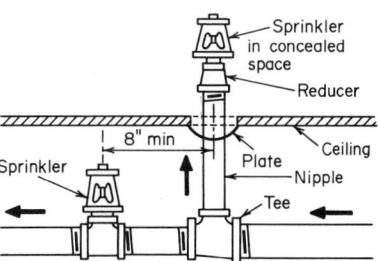

Fig. 11   Sprinkler positioning with respect to a building ceiling.

SIZING GAS PIPING FOR HEATING AND COOKING

An industrial building has two 8-gpm water heaters and ten ranges, each of which has four top burners and one oven burner. What maximum gas consumption must be provided for if carbureted water gas is used as the fuel? Determine the pressure in the longest run of gas pipe in this building if the total equivalent length of pipe in the longest run is 150 ft and the specific gravity of the gas is 0.60 relative to air. What would the pressure loss of a 0.35-gravity gas be?

Calculation Procedure:

1. Compute the heat input to the appliances

Table 15 lists the typical heat input to various gas-burning appliances. Using the tabulated data for the 8-gpm water heaters and the 4-burner stove, the maximum heat input = 2(300,000) + 10(62,500) = 1,225,000 Btu/hr. The gas-supply pipe must handle sufficient gas to supply this heat input because all burners might be operated simultaneously.

TABLE 15   Heat Input to Common Appliances

Unit	Approx. Input, Btu/hr
Water heater, side-arm or circulating type	25,000
Water heater, automatic instantaneous:	
4 gpm	150,000
6 gpm	225,000
8 gpm	300,000
Refrigerator	2,500
Ranges, domestic:	
four top burners, 1 over burner...............	62,500
four top burners, two oven burners...........	82,500

2. Compute the required gas-flow rate

Table 16 shows that the heating value of carbureted water gas is 508 Btu/cu ft. Using a value of 500 Btu/cu ft to provide a modest safety factor, gas flow required, cu ft = maximum heat input required, Btu per hr/(fuel heating value, Btu/cu ft) = 1,225,000/ 500 = 2,450 cu ft/hr.

TABLE 16   Typical Heating Values of Commercial Gases

Gas	Net Heating Value, Btu/cu ft
Natural gas (Los Angeles)	971
Natural gas (Pittsburgh).........................	1,021
Coke-oven gas	514
Carbureted water gas............................	508
Commerical propane	2,371
Commercial butane..............................	2,977

3. Compute the pressure loss in the gas pipe

The longest equivalent run is 150 ft. Gas flows through the pipe at the rate of 2,450 cu ft/hr. Enter Table 17 at this flow rate, or at the next *larger* tabulated flow rate, and project horizontally to the first pressure drop listed, or 3.5 in./100 ft in a 2-in. pipe.

The pressure loss listed in Table 17 is for 100 ft of pipe if the gas has a specific gravity of 0.6 in relation to air. To find the pressure loss in 150 ft of pipe, use the relation actual pressure loss, in. of water = (tabulated pressure loss, in. per 100 ft)(actual pipe length, ft/tabulated pipe length, ft) = 3.5(150/100) = 5.25 in. of water. Since the actual flow rate is less than 3,000 cu ft/hr, the actual pressure drop will be less than computed.

4. Compute the pressure loss of the lighter gas

To correct for a gas of a different specific gravity, multiply the actual gas flow by the appropriate factor from Table 18. Thus, equivalent flow rate for this plant when a gas of 0.5 gravity is flowing is 2,450(0.77) = 1,882 cu ft per hr; say 2,000 cu ft/hr.

Entering Table 17 shows that a flow of 2,000 cu ft/hr will have a pressure loss of 6.3 in. of water in 100 ft of 1½-in. pipe. In 150 ft of 1½-in. pipe, the pressure loss will be 6.3(150/100) = 9.45 in. of water. Increasing the pipe size to 2 in. would reduce the pressure loss to 2.25 in. of water.

Related Calculations: When gas flows upward in a vertical pipe to serve upper floors, there is a *gain* in the gas pressure, if the gas is lighter than air. Table 19 shows the gain in gas pressure per 100 ft of rise in a vertical pipe for gases of various specific gravities. This pressure gain must be recognized when designing piping systems.

As with piping and fixtures for plumbing systems, gas piping and fixtures are subject to code regulations in most cities and towns. Natural and manufactured gases are

TABLE 17   Capacities of Gas Pipes (Losses of pressure are shown in inches of water per 100 ft of pipe, due to the flow of gas with a specific gravity of 0.6 with respect to air)*

Rate of flow of cu ft hr	Size of pipe, in.											
	$\frac{3}{8}$	$\frac{1}{2}$	$\frac{3}{4}$	1	$1\frac{1}{4}$	$1\frac{1}{2}$	2	$2\frac{1}{2}$	3	$3\frac{1}{2}$	4	5
10	0.14	0.09										
20	0.59	0.19										
50	...	1.0	0.14	0.08								
100	...	...	0.59	0.16	0.04							
150	...	...	1.02	0.36	0.08							
200	...	...	2.20	0.65	0.15	0.06						
259	...	...	3.5	1.00	0.23	0.10						
500	...	...	...	4.1	0.90	0.40	0.09					
750	...	...	...	9.9	2.1	0.89	0.21	0.09				
1,000	...	...	...	...	...	1.6	0.80	0.15	0.05			
1,500	...	...	...	...	8.8	3.6	0.86	0.33	0.11	0.05		
2,000	...	...	...	...	...	6.3	1.50	0.60	0.18	0.08	0.05	
3,000	...	...	...	...	...	...	3.5	0.94	0.44	0.19	0.10	
5,000	...	...	...	...	...	...	8.8	3.6	1.2	0.54	0.34	0.08
6,000	...	...	...	...	...	...	...	5.4	1.7	0.76	0.40	0.13
8,000	...	...	...	...	...	...	...	...	3.0	1.3	0.72	0.23
10,000	...	...	...	...	...	...	...	...	5.2	2.1	1.2	0.36
20,000	...	...	...	...	...	...	...	...	...	8.2	4.0	1.4
30,000	...	...	...	...	...	...	...	...	...	...	10.0	3.2
40,000	...	...	...	...	...	...	...	...	...	...	...	5.7
50,000	...	...	...	...	...	...	...	...	...	...	...	9.0

*To determine head losses for other lengths of pipe, multiply the head losses in this table by the length of the pipe and divide by 100. For head losses due to flow of gases with specific gravity other than 0.6, use the figures given in Table 18.

TABLE 18   Factors by Which Flows in Table 17 Must Be Multiplied for Gases of Other Specific Gravity*

Specific gravity of gas	0.35	0.40	0.45	0.50	0.55	0.60	0.65	0.70
Factor	0.77	0.82	0.87	0.91	0.96	1.00	1.04	1.08

*From "Standards for Piping, Appliances, and Fittings for City Gas," Pamphlet 54, National Board Fire Underwriters, Sept. 1, 1943.

TABLE 19   Changes in Pressure in Gas Pipes

Gain in pressure per 100 ft of rise in vertical pipe, in. of water	0.96	0.89	0.81	0.74	0.66	0.59	0.52	0.44
Specific gravity of gas compared to air	0.35	0.40	0.45	0.50	0.55	0.60	0.65	0.70

widely used in stoves, water heaters, and space heaters of many designs. Since gas can form explosive mixtures when mixed with air, gas piping must be absolutely tight and free of leaks at all times. Usual codes cover every phase of gas-piping size, installation, and testing. The local code governing a particular building should be carefully followed during design and installation.

For gas supply, the usual practice is for the public-service gas company to run its pipes into the building cellar, terminating with a brass shutoff valve and gas meter inside the cellar wall. From this point, the plumbing contractor or gas-pipe fitter runs lines through the building to the various fixture outlets. When the pressure of the gas supplied by the public-service company is too high for the devices in the building, a pressure-reducing valve can be installed near the point where the line enters the building. The valve is usually supplied by the gas company.

Besides municipal codes governing the design and installation of gas piping and devices, the gas company serving the area will usually have a number of regulations that must be followed. In general, gas piping should be run in such a manner that it is unnecessary to locate the meter near a boiler, under a window or steps, or in any other area where it may be easily damaged. Where multiple-meter installations are used, the piping must be plainly marked by means of a metal tag showing which part of the building is served by the particular pipe. When two or more meters are used in a building to supply separate consumers, there should be no interconnection on the outlet side of the meters.

Materials used for gas piping include black iron, steel, and wrought iron. Copper tubing is also finding some use, and the values listed in Table 17 apply to it as well as Schedule 40 (standard weight) pipe made of the materials listed above. Use the procedure given here to size gas pipes for industrial, commercial, and residential installations.

SWIMMING-POOL SELECTION, SIZING, AND SERVICING

Choose a swimming pool to serve 140 bathers with facilities for diving and swimming contests. Size the pumps for the pool. Select a suitable water-treatment system and the inlet and outlet pipe sizes. Determine the size of heater required for the pool, should heating of the water be required.

Calculation Procedure:

1. Compute the swimming-pool area required

Usual swimming pools are sized in accordance with the recommendations of the Joint Committee on Swimming Pools, which uses 25 sq ft per bather as a desirable pool area. With 140 bathers, the recommended area = 25(140) = 3,500 sq ft.

2. Choose the pool dimensions

Use Table 20 as a guide to usual pool dimensions. This tabulation shows that a 105-ft-long by 35-ft-wide pool is suitable for 147 bathers. Since the next smaller pool will handle only 108 bathers, the larger pool must be used.

To provide for swimming contests, lanes at least 7 ft wide are required. Thus, this pool could have 35 ft/7 ft = 5 lanes for swimming contests. If more lanes are desired, the pool width must be increased, if there is sufficient space. Also, consideration of the pool length is required if swimming meets covering a specified distance are required. Assume that the 105 × 35 ft pool chosen earlier is suitable with respect to contests and space.

To provide for diving contests, a depth of more than 9 ft is recommended at the deep end of the pool. Table 20 shows that this pool has an actual maximum depth of 10 ft, which makes it better suited for diving contests. Some swimming specialists recomment a depth of at least 10 ft for diving contests. Assume, therefore, that the 10 ft is acceptable for this pool.

The pool will have a capacity of 155,600 gal of water, Table 20. If installed indoors, the pool would probably be faced with tile or glazed brick. An outdoor pool of this size is usually constructed of concrete and the walls are a smooth finish.

3. Determine the pump capacity required

To keep the pool water as pure as possible, three *turnovers* — i.e., the number of times the water in the pool is changed each 24-hr day — are generally used. This means

TABLE 20 Dimensions of official swimming pools

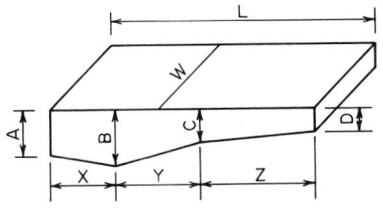

Pool capacity, gal	Bathing load, persons	Bathing capacity per day*	A	B	C	D	X	Y	Z	L	W
							Dimensions, ft				
55,000	48	418	8	9	5	3.25	15	20	25	60	20
80,800	75	607	8	9.5	5	3.25	15	20	40	75	25
120,000	108	900	8	9.5	5	3.25	18	25	47	90	30
155,600	147	1,170	8	10	5	3.25	18	25	62	105	35
207,600	192	1,555	8	10	5	3.25	20	30	70	120	40
254,000	243	1,905	8	10	5	3.25	20	30	85	135	45
306,000	300	2,300	8	10	5	3.25	20	30	100	150	50
422,400	432	3,170	8	10	5	3.25	20	30	130	180	60
558,000	590	4,180	8	10	5	3.25	20	30	160	210	70

*Based on 8-hr turnover.

that the water will be changed once each 24 hr/3 changes = 8 hr. The water is changed by recirculating it through filters, a chlorinator, strainer, and heater. Thus, the pump must handle water at the rate of pool capacity, gal/8 hr = 155,600/8 = 19,450 gph, or 19,450/60 min/hr = 324.1 gpm; say 325 gpm.

4. Choose the pump discharge head

Motor-driven centrifugal pumps find almost universal application for swimming pools. Reciprocating pumps are seldom suitable because they produce pulsations in the delivery pipe and pool filters. Either single- or double-suction single-stage centrifugal pumps can be used. The double-suction design is usually preferred because the balanced impeller causes less wear.

The discharge head that a swimming-pool circulating pump must develop is a function of the resistance of the piping, fittings, heater, and filters. Of these four, the heater and filters produce the largest head loss.

The usual swimming-pool heater causes a head loss of up to 10 ft of water. Sand filters cause a head loss of about 50 ft of water, whereas diatomaceous earth filters cause a head loss of about 90 ft of water. To choose the pump discharge head, find the sum of the pump suction lift, piping and fitting head loss, and heater and filter head loss. Add a 10 per cent allowance for overload. The result is the required pump discharge head in feet of water.

Most pools are equipped with two identical circulating pumps. The spare pump ensures constant operation of the pool should one pump fail. Also, the spare pump permits regular maintenance of the other pump.

5. Compute the quantity of make-up water required

Swimmers splash water over the gutter line of the pool. This water is drained away to the sewer in some pools; in others the water is treated and returned to the pool for reuse. Gutter drains are usually spaced at 15-ft intervals.

Since the pool waterline is level with the gutter, every swimmer who enters the pool displaces some water, which enters the gutter and is drained away. This drainage must be made up by the pool recirculating system.

The water displaced by a swimmer is approximately equal to his weight. Assuming each swimmer weighs 160 lb, he will displace this weight of water into the gutter. Since 1 gal of water weighs 8.33 lb, each swimmer will displace 160 lb/(8.33 lb/gal) = 19.2 gal. With a maximum of 140 swimmers in the pool, the total quantity of water displaced is (140)(19.2) = 2,695 gph, or 2,695/(60 min/hr) = 44.9; say 45 gpm.

Thus, to keep this pool operating, the water-supply system must be capable of delivering at least 45 gpm. This quantity of water can come from a city water system. a well, or recirculation of the gutter water after purification.

6. Compute the required filter-bed area

Two types of filters are used in swimming pools — (1) sand, and (2) diatomaceous earth. In either type, a flow rate of 2 to 4 gpm/sq ft is generally used. The lower flow rates — 2 to 2.5 gpm/sq ft are usually preferred. Assuming that a flow rate of 2.5 gpm/sq ft is used and that 325 gpm flows through the filters, as computed in Step 3, the filter-bed area required is 325 gpm/(2.5 gpm/sq ft) = 130 sq ft.

Two filters are generally used in swimming pools to ensure continuity of service and backwashing of one filter while the other is in use. The required area of 130 sq ft could then be divided between the two filters. Some pools use three, or more, filters. Regardless of how many filters are used, the required area can be evenly divided amongst them.

7. Choose the number of water inlets and outlets for the pool

The pool inlets supply the recirculation water required. Usual practice rates each inlet for a flow of 10 to 20 gpm. Assuming a 10-gpm flow for this pool, the number of inlets required is 325 gpm/10 gpm per inlet = 32.5; say 32.

Locate the inlets around the periphery of the pool and on each end. Space the inlets so that they provide an even distribution of the water. In general, inlets should not be located more than 30 ft apart.

Size the pool drain to release the water in the pool within the desired time interval — usually 4 to 12 hr. Since there is no harm in emptying a pool quickly — if the sewer into which the pool discharges has sufficient capacity — size the discharge line liberally. Thus, a 12-in. discharge line can handle about 2,000 gpm when draining a swimming pool.

8. Compute the quantity of disinfectant required

Chlorine, bromine, and ozone are some of the disinfectants used in swimming pools. Chlorine is probably the most popular. It is used in quantities sufficient to maintain 0.5 ppm chlorine in the water. Since this pool contains 155,600 gal of water that is recirculated three times per 24-hr day, the quantity of chlorine disinfectant that must be added each day is (155,600 gal) (3 changes per day)(8.33 lb/gal)(0.5 lb of chlorine per 10^6 lb of water) = 1.95 lb of chlorine per day. The required chlorine can be pumped into the pool inlet water or fed from cylinders.

9. Size the water heater for the pool

The usual swimming-pool heater has a heating capacity, in gph, which is 10 times the gpm rating of the circulating pump. Since the circulating pump for this pool is rated at 325 gpm, the heater should have a capacity of 10(325) = 3,250 gph.

Since the entering water temperature may be as low as 40 F in the winter, the heater should be chosen for this entering temperature. The outlet temperature of the water should be at least 80 F. To heat the entire contents of the pool from 40 F to about 70 F, at least 48 hr is generally allowed. Instantaneous hot-water heaters are usually chosen for swimming-pool service.

10. Select the backwash sump pump

When the filter backwash flow cannot be discharged directly to a sewer, the usual practice is to pipe the backwash to a sump in the pool machinery room. The accumulated backwash is then pumped to the sewer by a sump pump mounted in the sump. The sump should be large enough to store sufficient backwash to prevent overflowing. Assuming that either of the 130 sq ft filter beds is backwashed with a flow of 12.5 gpm per sq ft of filter-bed area, the quantity of water entering the sump will be (12.5) (130) = 1,725 gpm. If there is room for a 5-ft-deep, 8-ft-wide, and 5-ft-long sump, its capacity will be $5 \times 8 \times 5 \times 7.5$ gal/cu ft = 1,500 gal. The difference, of $1,625 - 1,500 = 125$ gpm, must be discharged by the pump to prevent overflow of the sump. A 150-gpm sump pump should probably be chosen to provide a margin of safety. Further, it is usual practice to install duplicate sump pumps to ensure pool operation in the event one pump fails. Where water is collected from other drains and discharged to the sump, the pump capacity may have to be increased accordingly.

Related Calculations: Use the general procedure given here to choose swimming pools and their related equipment for schools, recreation centers, hotels, motels, cities, towns, etc. Wherever possible, follow the recommendations of local codes and of the Joint Committee on Swimming Pools.

HEATING, VENTILATING, AND AIR CONDITIONING

REFERENCES: Carrier Air Conditioning Company—*Handbook of Air Conditioning System Design*, McGraw-Hill; American Society of Heating, Refrigerating, and Air Conditioning Engineers —*Guide and Data Book, Fundamentals and Equipment* and *Guide and Data Book, Applications*; Buffalo Forge Company—*Fan Engineering*; Strock and Koral—*Handbook of Heating, Air Conditioning, and Ventilation*, Industrial Press; Carrier, Cherne, Grant, and Roberts—*Modern Air Conditioning, Heating and Ventilating*, Pitman; Emerick—*Heating Design and Practice*, McGraw-Hill; Holmes—*Air Conditioning in Summer and Winter*, McGraw-Hill; The Trane Company—*Air-Conditioning Manual*.

BUILDING OR STRUCTURE HEAT-LOSS DETERMINATION

An industrial building has 8-in.-thick uninsulated brick walls, a 2-in.-thick-concrete uninsulated roof, and a concrete floor. The building is 150 ft long, 75 ft wide, and 15 ft high. Each long wall contains eight 5×10 ft double-glass windows, and each short wall contains two 4×8 ft double-glass doors. What is the heat loss of this building per hour if the required indoor temperature is 70 F and the design outside temperature is 0 F? How much will the heat loss increase if infiltration causes two air changes per hour?

Calculation Procedure:

1. Compute the heat loss through the glass

The usual heat-loss computation begins with the glass areas of a building. Hence, this procedure will be followed here. However, a heat-loss computation can be started with any part of the building, provided each part of the structure is eventually considered.

To compute the heat loss through a building surface, use the general relation $H_L = UA \, \Delta t$, where H_L = heat loss, Btu/hr, through the surface; U = overall coefficient of heat transmission for the material, Btu/(hr)(F)(sq ft); A = area of heat-transmission surface, sq ft; Δt = temperature difference, $F = t_i - t_o$, where t_i = inside temperature, F; t_o = outside temperature, F. Find U from Table 1 for the material in question.

This building has sixteen 5×10 ft double-glass windows, and four 4×8 ft double-glass doors. Hence, the total glass area = $16 \times 5 \times 10 + 4 \times 4 \times 8 = 928$ sq ft. The value of U for double-glass is, from Table 1, 0.45. Thus, $H_L = UA \, \Delta t = 0.45(928)(70 - 0) = 29,200$ Btu/hr.

TABLE 1   Typical Overall Coefficients of Heat Transmission [Btu/(sq ft)(hr)(F)]

Building surface	Type	Type of insulation
Walls	8-in. brick	0.50
Roof	2-in. concrete	0.82
Windows	Single glass	1.13
	Double glass	0.45

2. Compute the heat loss through the building walls

Use the same relation as in Step 1, substituting the wall heat-transfer coefficient and wall area. Thus, $U = 0.50$, and $A = 2 \times 150 \times 15 + 2 \times 75 \times 15 - 928 = 5{,}822$ sq ft. Then, $H_L = UA \, \Delta t = 0.50(5{,}822)(70 - 0) = 204{,}000$ Btu/hr.

3. Compute the heat loss through the building roof

Use the same relation as in Step 1, substituting the roof heat-transfer coefficient and roof area. Thus, $U = 0.82$ and $A = 150 \times 75 = 11{,}250$ sq ft. Then, $H_L = UA \, \Delta t = 0.82 \times (11{,}250)(70 - 0) = 646{,}000$ Btu/hr.

4. Compute the total heat loss of the building

The total heat loss of a building is the sum of the individual heat losses of the walls, glass areas, roof, and floor. In large buildings the heat loss through concrete floors is usually negligible and can be ignored. Hence, the total heat loss of this building caused by transmission through the building surfaces is $H_T = 29{,}200 + 204{,}000 + 646{,}000 = 879{,}200$ Btu/hr.

5. Compute the infiltration heat loss

The building volume is $150 \times 75 \times 15 = 168{,}500$ cu ft. With two air changes per hour, the volume of infiltration air that must be heated is $2 \times 168{,}500 = 337{,}000$ cfh. The heat that must be supplied to raise the temperature of this air is $H_i = (\text{cfh})(\Delta t)/55 = (337{,}000)(70 - 0)/55 = 429{,}000$ Btu/hr. Thus, the total heat loss of this building, including infiltration, is $H_T = 879{,}200 + 429{,}000 = 1{,}308{,}200$ Btu/hr.

Related Calculations: Determine the design outdoor temperature for a given locality from Baumeister – *Standard Handbook for Mechanical Engineers* or the ASHRAE *Guide and Data Book*, published by the American Society of Heating, Refrigerating and Air-Conditioning Engineers. Both these works are also suitable sources of comprehensive listings of U values for various materials and types of building construction. Since the winter-design outdoor temperature is usually for nighttime conditions, no credit is taken for heat given off by machinery, lights, people, etc., unless the structure will always operate on a 24-hr basis. The safest design ignores these heat sources because the machinery in the building may be removed, the operating cycle changed, or the heat sources eliminated in some other way. However, where an internal heat source of any kind will be a permanent part of a building, simply subtract the hourly heat release, Btu/hr, from the total building heat loss, Btu/hr. The result is the net heat loss of the building and is used in choosing the heating equipment for the building.

Most heat-loss calculations for large structures are made on a form available from heat-equipment manufacturers. Such a form helps organize the calculations. The steps followed, however, are identical to those given above. Another advantage of the calculation form is that it helps the designer remember the various items – walls, roof, glass, infiltration, etc. – that he must consider.

When some areas in a structure will be kept at a lower temperature than others, compute the heat loss from one area to another in the manner shown above. Substitute the lower indoor temperature for the outdoor design temperature. For areas exposed to prevailing winds, some manufacturers recommend increasing the computed heat loss by 10 percent. Thus, if the north wall of a building is exposed to the prevailing winds and its heat loss is 50,000 Btu/hr, this heat loss would be increased to 1.1(50,000) = 55,000 Btu/hr.

HEATING-SYSTEM SELECTION AND ANALYSIS

Choose a heating system suitable for an industrial plant consisting of a production area 150 ft long and 75 ft wide and an office area 75 ft wide and 60 ft long. The heat loss from the production area is 1,396,000 Btu/hr; the heat loss from the office area is 560,000 Btu/hr. Indoor design temperature for both areas is 70 F; outdoor design temperature is 0 F. What will the fuel consumption of the chosen heating system be if the annual degree-days for the area in which the plant is located is 6,000? Compare the annual fuel consumption of gas, oil, and coal.

Calculation Procedure:

1. Choose the type of heating system to use

Table 2 lists the various types of heating systems used today and typical applications for them. Study of this tabulation shows that steam unit heaters would probably be best for the production area because it is relatively large and open. Either a forced warm-air or a two-pipe steam heating system could be used for the office area. Since the production area will use steam unit heaters, a two-pipe steam heating system would probably be best for the office area. As steam unit heaters are almost universally two-pipe, the same method of supply and return is best chosen for the office system.

TABLE 2 Typical Applications of Heating Systems

System type	Fuel*	Typical applications
Gravity warm air....	G, O, C	Small residences, wooden or masonry.
Forced warm air	G, O, C	Small and large residences, wooden or masonry; small and medium-sized industrial plants, offices
Steam heating:		
a. One-pipe	G, O, C	Small residences, wooden or masonry
b. Two-pipe†	G, O, C	Small and large residences, wooden or masonry; small and large industrial plants, offices. High-pressure systems (30 to 150 psig) may be used in large industrial buildings having unit heaters or fan units. Unit heaters are used for large, open areas
Hot water:		
a. Gravity	G, O, C	Small residences, wooden or masonry
b. Forced‡........	G, O, C	Small and large residences; small and large industrial plants
c. Radiant	G, O, C	Small and large residences and plants
Electric	Electricity	Small residences and plants

*G − gas; O − oil; C − coal
†May be low-pressure, two-pipe vapor; two-pipe vacuum; two-pipe subatmospheric; two-pipe orifice; high-pressure.
‡May be one-pipe; two-pipe.

Note that Table 2 lists six different types of two-pipe steam heating systems. Choice of a particular type of two-pipe system depends on a number of factors, including economics, steam pressure required for non-heating—i.e., process—services in the building, type and pressure rating of boiler used, etc. Where high-pressure steam—30 to 150 psig, or higher—is used for process, the two-pipe system fitted with pressure-reducing valves between the process and heating mains is often an economical choice.

Hot-water heating is unsuitable for this building because unit heaters are required for the production area. An inlet air temperature less than 30 F is generally not recommended for hot-water unit heaters. Since the inlet-air temperature can be as low as 0 F in this plant, hot-water unit heaters could not be used.

2. Compute the annual fuel consumption of the system

Use the degree-day method to compute the annual fuel consumption. To apply the degree-day method, substitute the appropriate values in $F_g = DU_gRC$, for gas heating; $F_o = DU_oH_TC/1{,}000$, for oil; $F_c = DU_cH_TC/1{,}000$, for coal, where F = fuel consumption, the type of fuel being identified by the subscript, g for gas, o for oil, c for coal, with the unit of consumption being the therm for gas, gal for oil, and lb for coal; D = degree-days during the heating season; U = unit fuel consumption, the unit again being identified by the subscript; R = sq ft EDR (equivalent direct radiation) in the heating system; C = a correction factor for the outdoor design temperatures. Values of U and C are given in Table 3. Note the fuel heating values on which this table is based. For other heating values, see Related Calculations, below.

TABLE 3 Factors for Estimating Fuel Consumption*

Fuel	Consumption per degree-day	Heating conditions	
		Steam	Hot water
Gas	Therms (100,000 Btu)	0.00127 (300 sq ft EDR) 0.00121 (300–700 sq ft EDR) 0.00116 (700 sq ft EDR)	0.000743 (500 sq ft EDR) 0.000709 (500–1,000 sq ft EDR) 0.000675 (1,200 sq ft EDR)
		Heating-plant efficiency	
Oil (heating value = 141,000 Btu/gal)	Gal per 1,000 Btu/hr heat loss	70% 0.00437	80% 0.00383
Coal (heating value = 12,000 Btu/lb)	Lb per 1,000 Btu/hr heat loss	70% 0.0507	80% 0.0444

Outside-design-temperature correction factor					
Outside design temperature, F	-20	-10	0	$+10$	$+20$
Correction factor	0.778	0.875	1.000	1.167	1.400

*ASHRAE *Guide and Data Book* with permission.

For gas heating, $F_g = DU_gRC$, therms, where 1 therm = 100,000 Btu. To select the correct value for U_g, compute the sq ft EDR (see the next Calculation Procedure, p. 2-67) from $H_T/240$, or sq ft EDR = 1,956,000/240 = 8,150. Hence, $U_g = 0.00116$ from Table 3. From the same table, $C = 1.00$ for an outdoor design temperature of 0 F. Hence, $F_g = (6{,}000)(0.00116)(8{,}150)(1.00) = 56{,}800$ therms. Assuming gas having a heating value h_v of 1,000 Btu/cu ft is burned in this heating system, the annual gas consumption in cu ft is $F_g \times 10^5/h_v$. Or, $56{,}800 \times 10^5/1{,}000 = 5{,}680{,}000$ cu ft.

Using Table 3 for oil heating, and assuming a heating plant efficiency of 70 percent, $U_o = 0.00437$ gal per 1,000 Btu per hr heat loss, and $C = 1.00$ for the design conditions. Then, $F_o = DU_oH_TC/1{,}000 = (6{,}000)(0.00437)(1{,}956{,}000)(1.00)/1{,}000 = 51{,}400$ gal.

Using Table 3 for coal heating and assuming a heating-plant efficiency of 70 percent, $U_c = 0.507$ lb coal per 1,000 Btu/hr heat loss, and $C = 1.00$ for the design conditions. Then, $F_c = DU_cH_TC/1{,}000 = (6{,}000)(0.0507)(1{,}956{,}000)(1.00)/1{,}000 = 595{,}000$ lb, or 297 tons of 2,000 lb each.

Related Calculations: When the outdoor design temperature is above or below 0 F, a correction factor must be applied as shown in the equations given in Step 2, above. The appropriate correction factor is given in Table 3. Fuel-consumption values listed in Table 3 are based on an indoor design temperature of 70 F and an outdoor design temperature of 0 F.

The oil-consumption values are based on oil having a heating value of 141,000 Btu/gal. Where oil having a different heating value is burned, multiply the fuel-consumption value selected by 141,000/(heating value, Btu/gal), of oil burned.

The coal-consumption values are based on coal having a heating value of 12,000 Btu/lb. Where coal having a different heating value is burned, multiply the fuel-consumption value selected by 12,000/(heating value, Btu/lb), of coal burned.

For example, if the oil burned in the heating system in Step 2 has a heating value of 138,000 Btu/gal, $U_o = 0.00437(141,000/138,000) = 0.00446$. And if the coal has a heating value of 13,500 Btu/lb, $U_c = 0.0507(12,000/13,500) = 0.0451$.

Steam consumption of a heating system can also be computed using the degree-day method. Thus, the weight of steam, W lb, required for the degree-day period—from a day to an entire heating season—is $W = 24H_rD/1,000$, where all the symbols are as given earlier. Thus, for the industrial building analyzed above, $W = 24(1,956,000) \times (6,000)/1,000 = 281,500,000$ lb of steam per heating season. The denominator in this equation is based on low-pressure steam having an enthalpy of vaporization of approximately 1,000 Btu/lb. Where high-pressure steam is used and the enthalpy of vaporization is lower, substitute the actual enthalpy in the denominator to obtain more accurate results.

Steam consumption for building heating purposes can also be given in lb of steam per 1,000 cu ft of building space per degree-day, and lb of steam per 1,000 sq ft of EDR per degree-day. On the building-volume basis, steam consumption in the United States can range from a low of 0.130 to a high of 2.07 lb of steam per 1,000 cu ft/degree-day, depending on the building type (apartment house, bank, church, department store, garage, hotel, or office building), and the building location (southwest or far north). On a sq ft of EDR basis, steam consumption can range from a low of 21 lb per 1,000 sq ft EDR per degree-day to a high of 120 lb per 1,000 sq ft EDR per degree-day, depending on the building type and location. The ASHRAE *Guide and Data Book* lists typical steam-consumption values for buildings of various types in different locations.

REQUIRED CAPACITY OF A UNIT HEATER

An industrial building is 150 ft long, 75 ft wide, and 30 ft high. The heat loss from the building is 350,000 Btu/hr. Choose suitable unit heaters for this building if 5-psig steam and 200-F hot water are available to supply heat. Air enters the unit heater at 0 F; an indoor temperature of 70 F is desired in the building. What capacity unit heaters are needed if 20,000 cfm is exhausted from the building?

Calculation Procedure:

1. Compute the total heat loss of the building

The heat loss through the building walls and roof is given as 350,000 Btu/hr. However, there is an additional heat loss caused by infiltration of outside air into the building. Compute this loss as follows.

Find the cubic content of the building from volume, cu ft = LWH, where L = building length, ft; W = building width, ft; H = building height, ft. Thus, volume = (150)(75)(30) = 337,500 cu ft.

Determine the heat loss caused by infiltration by estimating the number of air changes per hour caused by leakage of air into and out of the building. For the usual industrial building, one to two air changes per hour are produced by infiltration. Assuming one air change per hour, the quantity of infiltration air that must be heated from the outside to the inside temperature = building volume = 337,500 cfh. Had two air changes per hour been assumed, the quantity of infiltration air that must be heated = 2 × building volume.

Compute the heat required to raise the temperature of the infiltration air from the outside to the inside temperature from $H_i = (\text{cfh})(\Delta t)/55$, where H_i = heat required to raise the temperature of the air. Btu/hr, through Δt, where $\Delta t = t_i - t_o$, and t_i = inside temperature, F; t_o = outside temperature, F. For this building, $H_i = 337,500(70-0)/55$

$= 429,000$ Btu/hr. Hence, the total heat loss of this building $H_t = 350,000 + 429,000$ Btu/hr, without exhaust ventilation.

2. Determine the extra heat load caused by exhausting air

When air is exhausted from a building, an equivalent amount of air must be supplied by infiltration or ventilation. In either case, an amount of air equal to that exhausted must be heated from the outside temperature to the room temperature.

With an exhaust rate of 10,000 cfm = 60(10,000) = 600,000 cfh, the heat required is $H_e = (\text{cfh})(\Delta t)/55$, where $H_e =$ heat required to raise the temperature of the air that replaces the exhaust air, Btu/hr. Or, $H_e = 600,000(70 - 0)/55 = 764,000$ Btu/hr.

To determine the total heat loss from a building when both infiltration and exhaust occur, add the larger of the two heat requirements—infiltration or exhaust—to the heat loss caused by transmission through the building walls and roof. Since, in this building, $H_e > H_i$, $H_t = 350,000 + 764,000 = 1,114,000$ Btu/hr.

3. Choose the location and number of unit heaters

This building is narrow—i.e., it is half as wide as it is long. In such a building, three vertical-discharge unit heaters, Fig. 1, will provide good distribution of the heated air. With three heaters, the capacity of each should be 764,000/3 = 254,667 Btu/hr without ventilation, and 1,114,000/3 = 371,333 Btu/hr with ventilation. Once the capacity of each unit heater is chosen, the spread diameter of the heated air discharged by the heater can be checked to determine if it is sufficient to provide the desired comfort.

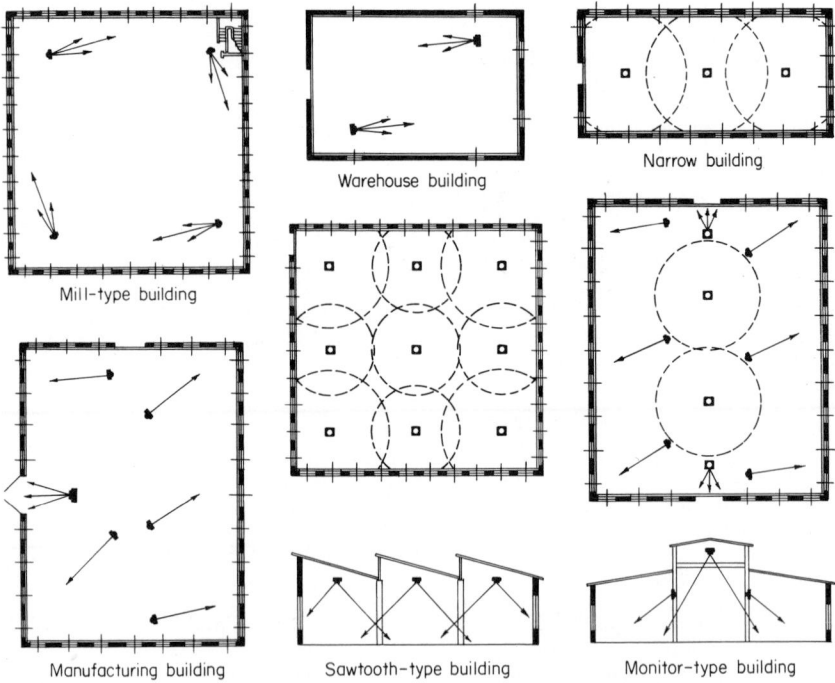

Fig. 1   Recommended arrangements of unit heaters in various types of buildings. (*Modine Manufacturing Company.*)

4. Select the capacity of the unit heaters

Use the engineering data published by a unit-heater manufacturer, such as Table 4, to determine the final air temperature, Btu delivered per hour, cubic feet of air handled, and the quantity of condensate formed. Thus, Table 4 shows that vertical-discharge Model D unit heater delivers 277,900 Btu/hr when the entering air is at 0 F and the heating steam is at a pressure of 5 psig. This heater discharges 3,400 cfm of heated air at 76 F and forms 290 lb/hr of condensate. The capacity table for a horizontal-discharge unit heater is similar to Table 4. When the building is ventilated, a Model E unit heater delivering 388,400 Btu/hr could be used with entering air at 0 F. This heater, as Table 4 shows, delivers 4,920 cfm of air at 73 F and forms 404 lb of condensate.

5. Check the spread diameter produced by the heater

Table 4 shows the different spread diameters—i.e., diameter of the heated-air blast at the floor level for different mounting heights of the unit heater. Thus, at a 14-ft mounting level above the floor, Model D will produce a spread diameter of 48 ft or 54 ft, depending on the type of outlet cone used. These spread diameters are based on 2-psig steam and 60-F room temperature. For 5-psig steam and 70-F room temperature, multiply the tabulated spread diameter and mounting height by the correction factor shown at the bottom of Table 4. Thus, for Model D, spread diameter = 1.07(48) = 51.4 ft and 1.07(54) = 57.8 ft, whereas the mounting height could be 1.07(14) = 14.98 ft.

Find the spread diameter for Model E in the same way. Or, 56 ft and 62 ft at a 15-ft height with 2-psig steam. With 5-psig steam, the spread diameters are 1.07(56) = 59.9 ft and 1.07(62) = 66.4 ft, whereas the mounting height could be 1.07(15) = 16 ft.

TABLE 4   Typical Vertical-Delivery Unit-Heater Capacities
(5-psig steam supply)

Model	Mtg. ht., ft	Spread diam., ft	Motor speed, rpm	Cfm*	0-F entering air			50-F entering air		
					Btu/hr	Final temp., F	Cond., lb/hr	Btu/hr	Final temp., F	Condensate, lb/hr
A	9	22, 24	1,620	690	55,600	75	59	41,800	110	44
B	10	31, 33	1,135	1,420	114,900	75	120	86,400	110	90
C	11	38, 40	1,135	2,350	186,800	74	194	140,400	109	146
D	14	48, 54	1,135	3,400	277,900	76	290	208,900	111	218
E	15	56, 62	1,135	4,920	388,400	73	404	292,000	109	304

Mounting height correction factors

Steam press., psig	Water temp., F	Normal room temp., F		
		60	70	80
. . .	210	1.05	1.10	1.20
0–5	220	1.00	1.07	1.14
6–15	. . .	0.88	0.94	1.00
16–30	. . .	0.77	0.81	0.86
31–50	. . .	0.70	0.73	0.77

*Cfm capacity at *final* air temperature. For horizontal discharge unit heaters the cfm capacity is usually stated at the *entering* air temperature.

6. Compute the hot-water-heater capacity required

Study several manufacturers' engineering data to determine the capacity of a suitable hot-water unit heater. This study will show that hot-water unit heaters are generally not available for inlet-air temperatures less than 30 F. Since the inlet-air temperature in this building is 0 F, a hot-water unit heater would be unsuitable; hence, it cannot be used. The *minimum* outlet-air temperature often recommended for unit heaters is 95 F.

Related Calculations: Use the same general method to choose horizontal-delivery unit heaters. The heated air delivered by these units travels horizontally or can be deflected downward toward the floor. Tables in manufacturers' engineering data list the heat-throw distance for horizontal-delivery unit heaters.

Standard ratings of steam unit heaters are given for 2-psig steam and 60-F entering air; hot-water unit heaters for 180-F entering water and 60-F entering air. For other steam pressures, water temperatures, or entering-air temperatures, *divide* the Btu per hr at stated conditions by the appropriate correction factor from Table 5 to obtain the required rating at *standard* conditions. Thus, a steam unit heater rated at 18,700 Btu/hr at 20-psig and 70-F entering air has a standard rating of 18,700/1.178 = 15,900 Btu/hr, closely, at 2-psig and 60-F entering air temperature, using the correction factor from Table 5. Conversely, a steam unit heater rated at 228,000 Btu/hr at 2 psig and 60 F will deliver 228,000(1.421) = 324,000 Btu/hr at 50-psig and 70-F entering air. Electric and gas-fired unit heaters are also rated on the basis of 60-F entering air.

TABLE 5　Unit-Heater Conversion Factors

Steam pressure, psig	Entering-air temperature, F			
	40	50	60	70
	Horizontal-delivery steam heaters			
2	1.153	1.076	1.000	0.927
5	1.209	1.131	1.055	0.981
10	1.288	1.209	1.132	1.057
20	1.413	1.333	1.254	1.178
40	1.593	1.510	1.430	1.351
50	1.664	1.582	1.500	1.421
Entering-water temperature, F	Horizontal-delivery hot-water heaters			
150	0.911	0.790	0.676	0.568
160	1.027	0.900	0.783	0.670
170	1.142	1.012	0.890	0.773
180	1.262	1.127	1.000	0.880
190	1.384	1.245	1.115	0.990
200	1.503	1.365	1.231	1.102

When a unit heater is supplied both outside air and recirculated air from within the building, use the temperature of the combined air streams as the inlet-air temperature. Thus, with 1,000 cfm of 10-F outside air and 4,000 cfm of 70-F recirculated air, the temperature of the air entering the heater $t_e = (t_{oa}cfm_{oa} + t_r cfm_r)/cfm_t$, where t_e = temperature of air entering heater, F; t_{oa} = temperature of outside air, F; cfm_{oa} = quantity of outside air entering the heater, cfm; t_r = temperature of room air entering heater, F; cfm_r = quantity of room air entering heater, cfm; cfm_t = total cfm entering heater = $cfm_{oa} + cfm_r$. Thus, $t_e = (10 \times 1,000 + 70 \times 4,000)/5,000 = 58$ F. This relation is valid for both steam and hot-water unit heaters.

To find the approximate outlet-air temperature of a unit heater when the capacity, cfm, and entering-air temperature are known, use the relation $t_o = t_i + (460 + t_i)/[575 cfm/(Btu/hr)] - 1$, where t_o = unit-heater outlet-air temperature, F; t_i = temperature of air entering the heater, F; cfm = quantity of air passing through the heater, Btu/hr = rated capacity of the unit heater. Thus, for a heater rated at 73,500 Btu/hr, 1,530 cfm, with 50-F entering air temperature, $t_o = 50 + (460 + 50)/(575 \times 1,530/73,500) - 1 = 94.1$ F. The unit-heater capacity used in this equation should be the capacity at standard conditions — 2-psig steam or 180-F water and 60-F entering-air temperature. Results obtained with this equation are only approximate. In any event, the outlet-air temperature should never be less than the room-air temperature. For the actual outlet temperature of a specific unit heater, refer to the manufacturers' engineering data.

The air discharged by a unit heater should be at a temperature greater than the room temperature because air in motion tends to chill the occupants of a room. Choose the outlet temperature by referring to the heated-air velocity. Thus, with a velocity of 20 fpm, the air temperature should be at least 76 F at the heater outlet. As the air velocity increases, higher air temperatures are required.

The outlet-air velocity and distance of blow of typical unit heaters are shown in Table 6.

TABLE 6   Unit-Heater-Outlet Velocity and Blow Distance

Unit-heater type	Outlet velocity, fpm	Blow distance, ft
Centrifugal fan	1,500–2,500	20–200
Horizontal propeller fan ..	400–1,000	30–100
Vertical propeller fan	1,200–2,200	70

When a unit heater of any type discharges against an external resistance, such as a duct or grille, its heating capacity and air capacity in cfm, as compared to standard conditions, are reduced. However, the final air temperature usually increases a few degrees over that at standard conditions.

To convert the rated output of any steam unit heater to sq ft of equivalent direct radiation (abbreviated sq ft EDR), divide the unit-heater rated capacity in Btu/hr by 240 Btu/(hr)(sq ft EDR). Thus, a unit heater rated at 240,000 Btu/hr has a heat output of 240,000/240 = 1,000 sq ft EDR. For hot-water unit heaters, use the conversion factor of 150 Btu/(hr)(sq ft EDR).

To determine the rate of condensate formation in a steam unit heater, divide the rated output in Btu/hr by an enthalpy of 930 Btu per lb of steam. Most unit-heater rating tables list the rate of condensate formation for each heater. Table 7 shows typical pipe sizes recommended for various condensate loads of steam unit heaters. Thus, with 1,000 lb/hr of condensate and 30-psig steam supply to the unit heater, the supply main should be $2\frac{1}{2}$-in. nominal diameter and the return main should be $1\frac{1}{4}$-in. nominal diameter.

TABLE 7   Typical Steam Unit-Heater Pipe Diameters*

Condensate, lb/hr	Steam-supply pressure, psig					
	5		30		80	
	Supply	Return†	Supply	Return	Supply	Return
100	2	1	$1\frac{1}{4}$	$\frac{3}{4}$	1	$\frac{3}{4}$
200	$2\frac{1}{2}$	$1\frac{1}{4}$	$1\frac{1}{4}$	$\frac{3}{4}$	$1\frac{1}{4}$	$\frac{3}{4}$
300	3	$1\frac{1}{2}$	$1\frac{1}{2}$	1	$1\frac{1}{4}$	$\frac{3}{4}$
400	3	2	2	1	$1\frac{1}{2}$	1
600	$3\frac{1}{2}$	2	2	$1\frac{1}{4}$	2	1
800	4	$2\frac{1}{2}$	$2\frac{1}{2}$	$1\frac{1}{4}$	2	$1\frac{1}{4}$
1,000	5	$2\frac{1}{2}$	$2\frac{1}{2}$	$1\frac{1}{4}$	2	$1\frac{1}{4}$

*Modine Manufacturing Company.
†Gravity return.

Figure 1 shows how unit heaters of any type should be located in buildings of various types. The diagrams are also useful in determining the approximate number of heaters needed once the heat loss is known. Locate unit heaters so that the following general conditions prevail, if possible.

Unit Heater Type	Desirable Conditions
Horizontal delivery	Discharge should wipe exposed walls at an angle of about 30°. With multiple units, the air streams should support each other.
Vertical delivery	With only vertical units, the airstream should blanket exposed walls with warm air.

The unit-heater arrangements shown in Fig. 1 illustrate a number of important principles.[1] The basic principle of unit-heater location is shown in the *mill-type* building. Here the heated-air flow from each unit heater supports the air flow from the other unit heaters and tends to set up a general circulation of air in the space heated. In the *warehouse*-building arrangement, maximum area coverage is obtained with a minimum number of units. The *narrow* building uses vertical-discharge unit heaters that blanket the building walls with warmed air.

In the *manufacturing* building, circular air movement is sacrificed to offset a large roof-heat loss and to permit short runouts from a single steam main. Note how a long-throw unit heater blankets a frequently used doorway.

Vertical-discharge unit heaters are used in the medium-height *sawtooth-type* building shown in Fig. 1. The *monitor-type*-building installation combines both horizontal and vertical unit heaters. Horizontal-discharge unit heaters are located in the low-ceiling areas and vertical-discharge units in the high-ceiling areas above the craneway. Much of the data in this Procedure were supplied by Modine Manufacturing Company.

STEAM CONSUMPTION OF HEATING APPARATUS

Determine the probable steam consumption of the non-space-heating equipment in a building equipped with the following: three bain-maries, each 100 sq ft, two 50-gal coffee urns, one jet-type dishwasher, one plate warmer having a 60 cu ft volume, two steam tables, each having an area of 50 sq ft, and one water still having a capacity of 75 gph. The available steam pressure is 40 psig; the kitchen equipment will operate at 20 psig and the water still at 40 psig.

Calculation Procedure:

1. Determine steam consumption of the equipment

The general procedure in determining heating-equipment steam consumption is to obtain engineering data from the manufacturer of the unit. When these data are unavailable, Table 8 will provide enough information for a reasonably accurate first ap-

[1]Modine Manufacturing Company.

TABLE 8   Typical Steam Consumption of Heating Equipment

Equipment	Steam, lb/hr at 20–40 psig
Bain-marie, per sq ft of surface.	3.0
Coffee urns, per gal.	2.5–3.0
Dishwashers, jet-type.	60
Plate warmer, per 20 cu ft	30
Soup or stock kettle, 60 gal; 40 gal	60; 45
Steam table, per sq ft of surface	1.5
Vegetable steamer, per compartment	
5-lb press	30
Water still, per gal capacity per hr	9

proximation. Hence, this tabulation will be used here to show how the data are applied.

Since the supply steam pressure is 40-psig, a pressure-reducing valve will have to be used between the steam main and the kitchen supply main. The capacity of this valve depends on the steam consumption of the equipment. Hence, the valve capacity cannot be determined until the equipment steam consumption is known.

During equipment operation the steam consumption is different from the consumption during startup. Since the operating consumption must be known before the starting consumption can be computed, the former will be determined first.

Using data from Table 8, the three 100 sq ft bain-maries will require (3 lb/hr)(3 units) (100 sq ft each) = 900 lb/hr of steam. The two 50-gal coffee urns require (2.75 lb/hr) (2 units)(50 gal per unit) = 275 lb/hr, using the average steam consumption. One jet-type dishwasher will require 60 lb/hr of steam. A 60 cu ft plate warmer will require (60 cu ft)(1 unit) [(30 lb/hr)/20 cu ft unit] = 90 lb/hr. Two 50 sq ft steam tables require (1.5 lb/hr)(2 units)(50 sq ft per unit) = 150 lb/hr. The 75 gph water still will consume (9 lb/hr)(1 unit)(75 gph) = 675 lb/hr. Hence, the total operating consumption of 20-psig steam is 900 + 275 + 60 + 90 + 150 = 1,475 lb/hr. The water still will consume 675 lb/hr of 40-psig steam.

Since the 20-psig steam must pass through the pressure-reducing valve before entering the 20-psig main, the required *operating* capacity of this valve is 1,475 lb/hr. However, the total steam consumption during operation, without an allowance for condensation in the pipelines, is 1,475 + 675 = 2,150 lb/hr.

2. Compute the system condensation losses

Condensation losses can range from 25 to 50 percent of the steam supplied, depending on the type of insulation used on the piping, the ambient temperature in the locality of the pipe, and the degree of superheat, if any, in the steam. Since the majority of the steam used in this building is 20-psig steam reduced in pressure from 40 psig and there will be a small amount of superheating during pressure reduction, a 25 percent allowance for pipe condensation is probably adequate. Hence, the total operating steam consumption = (1.25)(2,150) = 2,688 lb/hr.

3. Compute the startup steam consumption

During equipment startup there is additional condensation caused by the cold metal and, possibly, some cold products in the equipment. Therefore, the startup steam consumption is different from the operating consumption.

One rule of thumb estimates the startup steam consumption as two times the operating consumption. Thus, using this rule of thumb, startup steam consumption = 2(2,688) = 5,376 lb/hr. Note that this consumption rate is of relatively short duration because the metal parts are warmed rapidly. However, the pressure-reducing valve must be sized for this flow rate unless slower warming is acceptable.

The actual rate of condensate formation can be computed if the weight of the equipment, the specific heat of the materials of construction, and initial and final temperature of the equipment are known. Use the relation steam condensation, lb/hr = $60 Ws(\Delta t)/h_{fg}T$, where W = weight of equipment and piping being heated, lb; s = specific heat of the equipment and piping, Btu/(lb)(F); Δt = temperature rise of the equipment from the cold to the hot state, F; h_{fg} = latent heat of vaporization of the heating steam, Btu/lb; T = heating period, min.

This relation assumes that the final temperature of the equipment approximately equals the temperature of the heating steam. Where the specific heat of the equipment is different from that of the piping, solve for the steam-condensation rate of each unit and sum the results. Where products in the equipment must be heated, use the same relation but substitute the product weight and specific heat.

Related Calculations: Use the general method given here to compute the steam consumption of any type of industrial equipment for which the unit steam consumption is known or can be determined.

SELECTION OF AIR HEATING COILS

Select a steam heating coil to heat 80,000 cfm of outside air from 10 to 150 F when using steam at 15 psia. The heated air will be used for factory space heating. Illustrate how a steam coil is piped. Show the steps for choosing a hot-water heating coil.

Calculation Procedure:

1. Compute the required face area of the coil

If the coil-face air velocity is not given, a suitable air velocity must be chosen. In usual air-conditioning and heating practice, the air velocity across the face of the coil can range from 300 to 1,000 fpm with 500, 800, and 1,000 fpm being common choices. The higher velocities—up to 1,000 fpm—are used for industrial installations where noise is not a critical factor. Assume a coil face velocity of 800 fpm for this installation.

Compute the required face area from $A_c = cfm/V_a$, where A_c = required coil face area, sq ft; cfm = quantity of air to be heated by the coil, cfm, at 70 F and 29.92 in. Hg; V_a = air velocity through the coil, fpm. To correct the air quantity to standard conditions, when the air is being delivered at nonstandard conditions, multiply the flow in cfm at the other temperature by the appropriate factor from Table 9. Thus, with the incoming air at 10 F, cfm at 70 F and 29.92 in. Hg = $(1.128)(80,000) = 90,400$ cfm. Hence, $A_c = 90,400/800 = 112.8$ sq ft.

TABLE 9 Air-Volume Conversion Factors*

(The air volume in heating and air-conditioning applications is usually measured in cfm at 70 F and 29.92 in. Hg. When the actual air volume is at another temperature, multiply the cfm or velocity by the factor given for the temperature of the air. The result is the cfm or velocity at standard conditions and is the value used when entering heater-capacity tables and the friction chart, Fig. 2)

Air temp, F	Factor	Air temp, F	Factor	Air temp, F	Factor	Air temp, F	Factor
0	1.152	90	0.964	180	0.828	270	0.726
10	1.128	100	0.946	190	0.815	280	0.716
20	1.104	110	0.930	200	0.803	290	0.706
30	1.082	120	0.914	210	0.791	300	0.697
40	1.060	130	0.898	220	0.779	310	0.688
50	1.039	140	0.883	230	0.768	320	0.679
60	1.019	150	0.869	240	0.757	330	0.671
70	1.000	160	0.855	250	0.746	340	0.662
80	0.981	170	0.841	260	0.736	350	0.654

*McQuay, Inc.

2. Compute the coil outlet temperature

The capacity, final temperature, and condensate formation rate for steam heating coils for air-conditioning and heating systems are usually based on steam supplied at 5 psia and inlet air at 0 F. At other steam pressures a correction factor must be applied to the tabulated outlet temperature for 5-psia coils with 0-F inlet air.

Table 10 shows an excerpt from a typical coil-rating table and excerpts from coil correction-factor tables. To use such a tabulation for a coil supplied steam at 5 psia, enter at the air-inlet temperature and coil face velocity. Find the final air temperature equal to, or higher than, the required final air temperature. Opposite this read the number of rows of tubes required. Thus, in a 5-psia coil with 0-F inlet air, 800-fpm face velocity, and a 165-F final air temperature, Table 10 shows that five rows of tubes would be required. This table shows that the coil forms condensate at the rate of 149.8 lb per hr per sq ft of net fin area when the final air temperature is 166 F. The coil thus chosen is the first-trial coil, which must be checked against the actual steam conditions as described below.

When a coil is supplied steam at a pressure different from 5 psia, multiply the final air temperature given in the 5-psia table for 0 F, at the face velocity being used, by the correction factor given in Table 10 for the actual steam pressure and actual inlet-air temperature. Thus, for this coil, which is supplied steam at 15 psia and has an inlet-air temperature of 10 F, the temperature correction factor from Table 10 is 1.056. Add the product of the correction factor and the tabulated final air temperature to the inlet-air temperature to obtain the actual final air temperature. Several trials may be necessary before the desired outlet temperature is obtained.

The desired final air temperature for this coil is 150 F. Using the 124-F final air temperature from Table 10 as the first-trial valve, the actual final air temperature = (1.056)(124) + 10 = 141 F. This is too low. Trying the next higher final air temperature, the actual final air temperature = (1.056)(148) + 10 = 166.5 F. This is higher than required, but the steam supply can be reduced to produce the desired final air temperature. Thus, the coil will be four rows of tubes deep, as Table 10 shows. Hence, the five rows of coils originally indicated will not be needed. Instead, four rows will suffice.

TABLE 10 Steam-Heating-Coil Final Temperatures and Condensate-Formation Rate

Inlet-air temp., F	Rows of tubes	Face velocity, fpm 800	
		Final air temp., F	Condensate, lb/hr
0	1	51	46.2
0	2	51	83.7
0	3	124	111.7
0	4	148	132.9
0	5	166	149.8
0	6	180	162.4

Temperature-rise correction factor

Actual inlet-air temp., F	Steam pressure, psia		
	10	15	20
0	1.054	1.100	1.139
10	1.010	1.056	1.095
20	0.966	1.011	1.051

Condensate correction factors

	10	15	20
0	1.063	1.117	1.165
10	1.019	1.072	1.120
20	0.974	1.027	1.075

3. Compute the quantity of condensate produced by the coil

Use the same general procedure as in Step 2, or actual condensate formed, lb/(hr)/(sq ft) = [lb/(hr)(sq ft)] (condensate from 5-psia, 0-F table)(correction factor from Table 10); or for this coil, (132.9)(1.072) = 142.47 lb/(hr)/(sq ft). Since the coil has a net fin face area of 112.8 sq ft, the total actual condensate formed = (112.8)(142.47) = 16,100 lb/hr.

4. Determine the coil friction loss

Most manufacturers publish a chart or table of coil friction losses in coils having various face velocities and tube rows. Thus, Fig. 2 shows that a coil having a face velocity of 800 fpm and four rows of tube has a friction loss of 0.45 in. of water. Figure 2 is a typical friction-loss chart and can be safely used for all routine preliminary coil selections. However, when the final choice of a heating coil is made, use the friction chart or table prepared by the manufacturer of the coil chosen.

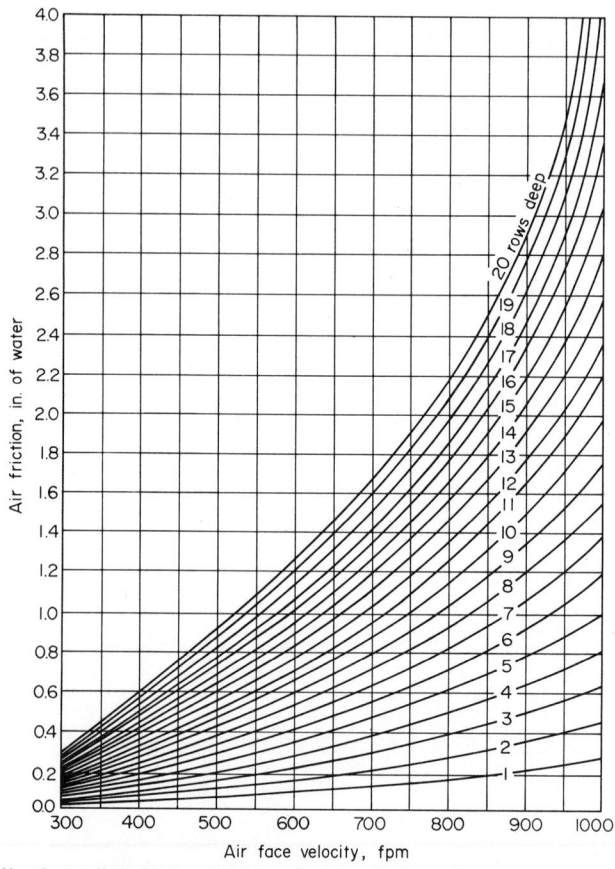

Fig. 2   Heating-coil air-friction chart for air at standard conditions of 70 F and 29.92 in. Hg. (*McQuay, Inc.*)

5. Determine the coil dimensions

Refer to the manufacturer's engineering data for the dimensions of the coil chosen. Each manufacturer has certain special construction features. Hence, there will be some variation in dimensions from one manufacturer to another.

6. Indicate how the coil will be piped

The ASHRAE *Guide and Data Book* shows piping arrangements for low- and high-pressure steam heating coils as recommended by various coil manufacturers. Follow the recommendations of the manufacturer whose coil is actually used when the final selection is made.

Related Calculations: Typical variables met in heating-coil selection are: the *face velocity*, which varies from 300 to 1,000 fpm, with the higher velocities being used for industrial applications, the lower velocities for non-industrial applications; the *final air temperature*, which ranges between 50 and 300 F, the lower temperatures being used for ventilation, the higher ones for heating; *steam pressures*, which vary from 2 to 150 psig, with the lower pressures—2 to 15 psig—being the most popular for heating.

Hot-water heating coils are also used for air heating. The general selection procedure is: (a) Compute the heating capacity, Btu/hr, required from (1.08)(temperature rise of air, F)(cfm heated). (b) Compute the coil face area required, sq ft, from cfm/face velocity, fpm. Assume a suitable face velocity using the guide given above. (c) Compute the logarithmic mean-effective-temperature difference across the coil using the method given elsewhere in this Handbook. (d) Compute the required hot-water flow rate, gpm, from Btu per hr heating capacity/(500)(temperature drop of water, F). The usual temperature drop of the hot water during passage through the heating coil is 20 F, with water supplied at 150 to 225 F. (e) Determine the tube water velocity, fps, from (8.33)(gpm)/(384)(number of tubes in heating coil). The number of tubes in the coil is obtained by making a preliminary selection of the coil using heating-capacity tables similar to Table 10. The usual hot-water heating coil has a water velocity between 2 and 6 fps. (f) Compute the number of tube rows required from Btu per hr heating capacity/(face area, sq ft)(logarithmic mean-effective-temperature difference from Step 3)(K factor from the manufacturer's engineering data). (g) Compute the coil air resistance or friction loss using the manufacturer's chart or table. The usual friction loss ranges between 0.375 to 0.675 in. of water for commercial applications to about 1.0 in. of water for industrial installations.

RADIANT-HEATING-PANEL CHOICE AND SIZING

One room of a building has a heat loss of 13,900 Btu/hr. Choose and size a radiant heating panel suitable for this room. Illustrate the trial method of panel choice. The floor of the room is made of wooden blocks.

Calculation Procedure:

1. Choose the type and location of the heating coil

Compute the heat loss for a given room or building using the method given earlier in this section. Once the heat loss is known, choose the type of heating panel to use — ceiling or floor. In some rooms or buildings a combination of floor and ceiling panels may prove more effective than either type used alone. Wall panels are also used but not so extensively as floor and ceiling panels.

In general, ceiling panels are imbedded in concrete or plaster, as are wall panels. Floor panels are almost always imbedded in concrete. Hence, use of another type of floor — block, tile, wood, or metal — may rule out the use of floor panels. Since this room has a wooden-block floor, a ceiling panel will be chosen.

2. Size the heating panel

Table 11 shows the maximum Btu per hr heat output of $\frac{3}{8}$-in. copper-tube ceiling panels. The $\frac{3}{8}$-in. size is popular; however other sizes — $\frac{1}{2}$-, $\frac{5}{8}$-, and $\frac{7}{8}$-in. diameter — are also used, depending on the heat load served. Using $\frac{3}{8}$-in. tubing on 6-in. centers imbedded in a plaster ceiling will provide a heat output of 60 Btu per hr per sq ft of tubing, as shown in Table 11.

To obtain the area of the heated panel, A sq ft, use the relation $A = H_L/P$, where H_L = room or building heat loss, Btu/hr; P = panel maximum heat output, Btu/(hr) (sq ft). For this room, $A = 13,900/60 = 232$ sq ft.

3. Determine the total length of tubing required

Use the appropriate tube-length factor from Table 11, or $(2.0)(232) = 464$ lin ft of $\frac{3}{8}$-in. copper tubing.

4. Find the maximum panel tube length

To stay within the commercial limits of smaller-size hot-water circulating pumps, the maximum tube lengths per panel circuit given in Table 11 are generally used. This tabulation shows that the maximum panel unit tube length for $\frac{3}{8}$-in. tubing on 6-in. centers is 165 ft. Such a length will not require a pump head of more than 4 ft, excluding the head loss in the mains.

TABLE 11   Heating Panel Characteristics

(Maximum Btu/hr per sq ft of $\frac{3}{8}$-in. copper tube embedded in ceiling plaster)		(Maximum Btu/hr per sq ft of copper tube embedded in concrete floor	
	Approx.		*Approx.*
$4\frac{1}{2}$ in. center-to-center	75	$\frac{1}{2}$ in. 9 in. center-to-center...........	50
6 in. center-to-center	60	$\frac{3}{4}$ in. 9 in. center-to-center...........	50
9 in. center-to-center	45	$\frac{3}{4}$ in. 12 in. center-to-center...........	50
		1 in. 12 in. center-to-center...........	50

Total length of tube required, ft.†

2.7	Where $4\frac{1}{2}$-in. centers are required
2	Where 6-in. centers are required
1.3	Where 9-in. centers are required
1	Where 12-in. centers are required

Maximum panel unit tube length‡

Ceilings				Floors			
Nominal size, in.	Centers, in.	Btu/hr per ft of tube	Ft	Nominal size, in.	Centers, in.	Btu/hr per ft of tube	Ft
$\frac{3}{8}$	$4\frac{1}{2}$	27	175	$\frac{1}{2}$	9	38	220
$\frac{3}{8}$	6	30	165	$\frac{3}{4}$	9	38	400
$\frac{3}{8}$	9	34	150	$\frac{3}{4}$	12	50	350
				1	12	50	550

Number of panel circuits of maximum length§

Mains	Ceiling	Floor		
Diam. in.	$\frac{3}{8}$ in. − $4\frac{1}{2}$, 6, 9 center-to-center	$\frac{1}{2}$ in. − 9 center-to-center	$\frac{3}{4}$ in. − 9, 12 center-to-center	1 in. − 12 center-to-center
2	27	16	8	5
$1\frac{1}{2}$	12	7	4	2
$1\frac{1}{4}$	8	5	2	1
1	4	3	1	1
$\frac{3}{4}$	2	1		

†To arrive at the required lin ft of tube per panel, multiply the sq ft of heated panel by the factors given.

‡To keep within the commercial limits of the smaller size pumps, as an example the above maximum tube lengths per panel circuit are suggested. These lengths alone will require not more than about a 4-ft head. This does not include loss in mains.

§Use the information given in the section on maximum panel unit tube length, as given above, which can be supplied, allowing about 0.5 ft head required per 100 ft of main (supply and return).

5.   Determine the number of panels required

Find the number of panels required by dividing the linear tubing length needed by the maximum unit tube length, or 464/165 = 2.81. Use three panels, the next larger whole number, because partial panels cannot be used. To conserve tubing and reduce the first cost of the installation, three panels, each having 155 linear ft of piping, would be used. Note that the tubing length chosen for the actual panel must be *less than* the maximum length listed in Table 11.

6. Find the required piping main size required

Determine the number of panels required in the remainder of the building. Use Table 11 to select the proper main size for a pressure loss of about 0.5 ft per 100 ft of the main, including the supply and return lines. Thus, if 12 ceiling panels of maximum length were used in the building, a 1½-in. main would be used.

7. Use the trial method to choose the main size

If the size of the main for the panels required cannot be found in Table 11, compute the total Btu required for the panels from Table 11. Then, by trial, find the size of the main from Table 11 that will deliver approximately, and preferably somewhat more, than this total Btu/hr requirement.

For instance, suppose an industrial building requires the following panel circuits:

No. of panel circuits	Tube size and location	Btu required from Table 11 (number of circuits × Btu per ft of tube × maximum panel tube length)
4	¾-in. tubes on 4½-in. centers, ceiling	$4 \times 27 \times 175 =$ 18,900
1	½-in. tube on 9-in. centers, floor	$1 \times 38 \times 220 =$ 8,360
1	1-in. tube on 12-in. centers, floor	$1 \times 50 \times 550 =$ 27,500
3	¾-in. tubes on 9-in. centers, floor	$3 \times 38 \times 400 =$ 45,600
9		Total Btu/hr required = 100,360

Trial 1: Assume 100 ft of 1½-in. main is used for seven floor circuits that are each 220 ft long and made of ½-in. tubing on 9-in. centers. From Table 11 the output of these 7 floor circuits is 7 circuits × 38 Btu per hr per ft of tube × 220 ft of tubing = 58,520 Btu/hr. Since this output is considerably less than the required output of 100,360 Btu/hr, a 1½-in. main is not large enough.

Trial 2: Assume 100 ft of 2-in. main for 16 circuits that are each 220 ft long and are made of ½-in. tubing on 9-in. centers. Then, as in trial 1, 16 × 38 × 220 = 133,760 Btu/hr delivered. Hence, a 2-in. main is suitable for the nine panels listed above.

Related Calculations: Use the same general method given here for heating panels imbedded in the concrete floor of a building. The liquid used in most panel heating systems is water at about 130 F. This warm water produces a panel temperature of about 85 F in floors and about 115 F in ceilings. The maximum water temperature at the boiler is seldom allowed to exceed 150 F.

When the first-floor ceiling of a multistory building is not insulated, the floor above a ceiling panel develops about 17 Btu/(sq ft)(hr) from the heated panel below. If this type of construction is used, the radiation into the room above can be deducted from the heat loss computed for that room. It is essential, however, to calculate only the heat output in the floor area directly above the heated panel.

Standard references, such as the ASHRAE *Guide and Data Book* present heat-release data for heating panels in both graphical and tabular form. Data obtained from charts are used in the same way as described above for the tabular data. Tubing data for radiant heating are available from Anaconda American Brass Company.

SNOW-MELTING HEATING-PANEL CHOICE AND SIZING

Choose and size a snow-melting panel to melt a maximum snowfall of 3 in./hr in a parking lot that has an area of 1,000 sq ft. Heat losses downward, at the edges, and back of the slab are about 25 percent of the heat supplied; also, there is an atmospheric evaporation loss of 15 percent of the heat supplied. The usual temperature during snowfalls in the locality of the parking lot is 32 F.

Calculation Procedure:

1. Compute the hourly snowfall weight rate

The density of snow varies from about 3 lb/cu ft at 5 F to about 7.8 lb/cu ft at 34 F. Assuming a density of 7.3 lb/cu ft for this installation, the hourly snowfall weight rate per sq ft is (area, sq ft) (depth, ft)(density) = (1.0)(3/12)(7.3) = 1.83 lb/(hr)(sq ft). In this computation, the rate of fall of 3 in./hr is converted to a depth in ft by dividing by 12 in./ft.

2. Compute the heat required for snow melting

The heat of fusion of melting snow is 144 Btu/lb. Since the snow accumulates at the rate of 1.83 lb/(hr)(sq ft), the amount of heat that must be supplied to melt the snow is (1.83)(144) = 264 Btu/(sq ft)(hr).

Of the heat supplied, the percent lost is 25 + 15 = 40 percent as given. Of this total loss, 25 percent is lost downward and 15 percent is lost to the atmosphere. Hence, the total heat that must be supplied is (1.0 + 0.40)(264) = 370 Btu/(hr)(sq ft). With an area of 1,000 sq ft to be heated, the panel system must supply (1,000)(370) = 370,000 Btu/hr.

3. Determine the length of pipe or tubing required

Consult the ASHRAE *Guide and Data Book* or manufacturer's engineering data to find the heat output per ft of tubing or pipe length. Suppose the heat output is 50 Btu per hr per ft of tube. Then the length of tubing required is 370,000/50 = 7,400 ft.

Some manufacturers rate their pipe or tubing on the basis of the rainfall equivalent of the snowfall and the wind velocity across the heated surface. Where this method is used, compute the heat required to melt the snow as the equivalent amount of heat to vaporize the water. This is $Q_e = 1,074(0.002V + 0.055)(0.185 - v_a)$, where Q_e = heat required to vaporize the water, Btu/(hr)(sq ft); V = wind velocity over the heated surface, mph; v_a = vapor pressure of the atmospheric air, in. Hg.

4. Determine the quantity of heating liquid required

Use the relation gpm = $0.125 H_t/dc \Delta t$, for ethylene glycol, the most commonly used heating liquid. In this relation, H_t = total heat required for snow melting, Btu/hr; d = density of the heating liquid, lb/cu ft; c = specific heat of the heating liquid, Btu/(lb)(F); Δt = temperature loss of the heating liquid during passage through the heating coil — usually taken as 15 to 20 F.

Assuming that a 60 percent ethylene glycol solution is used for heating, d = 68.6 lb/cu ft; c = 0.75 Btu/(lb)(F). Since the piping must supply 370,000 Btu/hr, gpm = (0.125)(370,000)/(68.6)(0.75)(20) = 45 gpm when the temperature loss of the heating liquid is 20 F.

5. Size the heater for the system

The heater must provide at least 370,000 Btu/hr to the ethylene glycol. If the heater has an overall efficiency of 60 percent, then the required heat input to deliver 370,000 Btu/hr is 370,000/0.60 = 617,000 Btu/hr.

To avoid a long warm-up time at the start of a snowfall, the usual practice is to operate the system for several hours prior to an expected snowfall. The heating liquid temperature during warmup is kept at about 100 F and the pump is operated at half-speed.

Related Calculations: Use this general method to size snow-melting systems for sidewalks, driveways, loading docks, parking lots, storage yards, roads, and similar areas. To prevent an excessive warm-up load on the system, provide for prestorm operation. Without prestorm operation, the load on the heater can be twice the normal hourly load.

Copper tubing and steel pipe are the most commonly used heating elements. For properties of tubing and piping important in snow-melting calculations, see Baumeister — *Standard Handbook for Mechanical Engineers.*

HEAT RECOVERY FROM LIGHTING SYSTEMS FOR SPACE HEATING

Determine the quantity of heat obtainable from 30 water-cooled fluorescent luminaries rated at 200 watts each if the entering-water temperature is 70 F and the water flow rate is 1.0 gpm. How much heat can be recovered from 10 air-cooled fluorescent luminaries rated at 100 watts each?

Calculation Procedure:

1. Compute the water-cooled luminaire heat recovery

Luminaire manufacturers publish heat-recovery data in chart form, Fig. 3. This chart shows that with a flow rate of 0.5 gpm and a 70-F entering-water temperature, the heat recovery from a luminaire is 74 percent of the total input to the fixture.

For a group of lighting fixtures, total input, watts = (number of fixtures)(rating per fixture, watts). Or, for this installation, input, watts = (30)(200) = 6,000 watts. Since 74 percent of this input is recoverable by the cooling water, recovered input = 0.74 (6,000) = 4,400 watts.

To convert incandescent lighting watts to Btu/hr, multiply by 3.4. Where fluorescent lights are used, apply a factor of 1.25 to include the heat gain in the lamp ballast. Thus, the heat available for recovery from these fluorescent lamps is (3.4)(1.25)(4,440) = 18,900 Btu/hr.

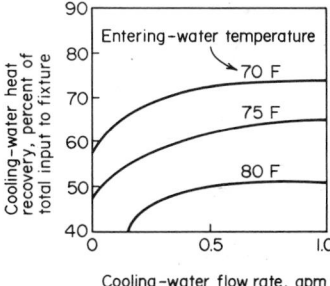

Fig. 3 Heat recovery in water-cooled lighting fixtures.

2. Compute the temperature rise of the water

Find the temperature rise of the water from $\Delta t = $ (Btu/hr)/500 gpm, where $\Delta t = $ temperature rise of the water, F; Btu/hr = heat available; gpm = water flow rate through the luminaire. Or, $\Delta t = 18,900/500(1.0) = 37.8$ F. A temperature rise of this magnitude is seldom used in practice. However, this calculation shows the large amount of heat recoverable with water-cooled lighting fixtures.

3. Compute the heat recoverable with air cooling

In the usual air-cooled luminaire, 50 to 70 per cent of the input energy is recoverable. Assuming a 60 percent recovery, with a total input of (10)(100) = 1,000 watts, the energy recoverable is 0.60(1,000) = 600 watts. Converting to the heat recoverable, (3.4)(1.25)(600) = 2,545 Btu/hr.

Related Calculations: Heat recovery from lighting fixtures is receiving increasing attention for many different structures because substantial fuel savings are possible. The water or air heated by the lighting is used to heat the supply or return air supplied to the conditioned space. Where the heat recovered must be rejected, as in the summer, either a cooling tower (for water) or an air-cooled condenser (for air) may be used.

Other popular sources of heat are refrigeration condensers and electric motors. In many installations the heat is absorbed from the condenser or motors, or both, by air.

AIR-CONDITIONING-SYSTEM HEAT-LOAD DETERMINATION—GENERAL METHOD

Show how to compute the total heat load for an air-conditioned industrial building fitted with windows having shades, internal heat loads from people and machines, and heat transmission gains through the walls, roof, and floor. Use the ASHRAE *Guide and Data Book* as a data source.

Calculation Procedure:

1. Determine the design outdoor and indoor conditions

Refer to the ASHRAE *Guide and Data Book* (called *Guide* hereinafter) for the state and city in which the building is located. Read from the *Guide* table for the appropriate city the design outdoor dry-bulb and wet-bulb temperatures. At the same time, determine from the *Guide* the indoor design conditions—temperature and relative humidity for the type of application being considered. The *Guide* lists a variety of typical applications such as apartment houses, motels, hotels, industrial plants, etc. It also lists the average summer wind velocity for a variety of locations. Where the exact location of a plant is not tabulated in the *Guide*, consult the nearest local branch of the weather bureau for information on the usual summer outdoor high and low dry- and wet-bulb temperatures, relative humidity, and velocity.

2. Compute the sunlight heat gain

The sunlight heat gain results from the solar radiation through the glass in the building's windows and the materials of construction in certain of the building's walls. If the glass or wall of a building is shaded by an adjacent solid structure, the sunlight heat gain for that glass and wall is usually neglected. The same is true for the glass and wall of the building facing the north.

Compute the glass sunlight heat gain, Btu/hr, from (glass area, sq ft)(equivalent temperature difference from the appropriate *Guide* table)(factor for shades, if any are used). The equivalent temperature difference is based on the time of day and orientation of the glass with respect to the points of the compass. A latitude correction factor may have to be applied if the building is located in a tropical area. Use the equivalent temperature difference for the time of day on which the heat-load estimate is based. Several times may be chosen to determine at which time the greatest heat gain occurs. Where shades are used in the building, choose a suitable shade factor from the appropriate *Guide* table and insert it in the equation above.

Compute the sunlight heat gain, Btu/hr, for the appropriate walls and the roof from (wall area, sq ft)(equivalent temperature difference from *Guide* for walls) [coefficient of heat transmission for the wall, Btu/(hr)(sq ft)(F)]. For the roof, find the heat gain, Btu/hr, from (roof area, sq ft)(equivalent temperature difference from the appropriate *Guide* table for roofs)[coefficient of heat transmission for the roof, Btu/(hr)(sq ft)(F)].

3. Compute the transmission heat gain

All the glass in the building windows is subject to transmission of heat from the outside to the inside as a result of the temperature difference between the outdoor and indoor dry-bulb temperatures. This transmission gain is commonly called the *all-glass gain*. Find the all-glass transmission heat gain, Btu/hr, from (total window-glass area, sq ft)(outdoor design dry-bulb temperature, F — indoor design dry-bulb temperature, F)[coefficient of heat transmission of the glass, Btu/(hr)(sq ft)(F) from the appropriate *Guide* table].

Compute the heat transmission, Btu/hr, through the shaded walls, if any, from (total shaded wall area, sq ft)(equivalent temperature difference for shaded wall from the appropriate *Guide* table, F)[coefficient of heat transmission of the wall material, Btu/(hr)(sq ft)(F) from the appropriate *Guide* table)].

Where the building has a machinery room or utility room that is not air conditioned and is next to a conditioned space, and the temperature in the utility room is higher than in the conditioned space, find the heat gain, Btu/hr, from (area of utility- or machine-room partition, sq ft)(utility- or machine-room dry-bulb temperature, F — conditioned-space dry-bulb temperature, F)[coefficient of heat transmission of the utility- or machine-room partition, Btu/(hr)(sq ft)(F) from the appropriate *Guide* table].

For buildings having a floor contacting the earth, or over unventilated and unheated basements, there is generally *no* heat gain through the floor because the ground is usually at a lower temperature than the floor. Where the floor is above the ground and in contact with the outside air, find the heat gain, Btu/hr, through the floor from (floor area, sq ft)(design outside dry-bulb temperature, F − design inside dry-bulb temperature, F) [coefficient of heat transmission of the floor material, Btu/(hr)(sq ft)(F) from the appropriate *Guide* table]. When a machine room or utility room is below the floor, use the same relation but substitute the machine-room or utility-room dry-bulb temperature for the design outside dry-bulb temperature.

For floors above the ground, some designers reduce the difference between the design outdoor and indoor dry-bulb temperatures by 5 F; other designers use the shaded-wall equivalent temperature difference from the *Guide*. Either method, or that given above, will provide safe results.

4. Compute the infiltration heat gain

Use the relation infiltration heat gain, Btu/hr = (window crack length, ft)(window infiltration, cu ft per ft of crack per min from the appropriate *Guide* table)(design outside dry-bulb temperature, F − design inside dry-bulb temperature, F)(1.08). Three aspects of this computation require explanation.

The window crack length used is usually one-half the total crack length in all the windows. Infiltration through cracks is caused by the wind acting on the building. Since the wind cannot act on all sides of a building at once, one-half the total crack length is generally used (but never less than one-half) in computing the infiltration heat gain. Note that the crack length varies with different types of windows. Thus the *Guide* gives, for metal sash, crack length = total perimeter of the movable section. For double-hung windows, the crack length = three times the width plus twice the height.

The window infiltration rate, cfm per ft of crack, is given in the *Guide* for various wind velocities. Some designers use the infiltration rate for a wind velocity of 10 mph; others use 5 mph. The factor 1.08 converts the computed infiltration to Btu/hr.

5. Compute the outside-air bypass heat load

Some outside air may be needed in the conditioned space to ventilate fumes, odors, and other undesirables in the conditioned space. This ventilation air imposes a cooling or dehumidifying load on the air conditioner because the heat or moisture, or both, must be removed from the ventilation air. Most air conditioners are arranged to permit some outside air to bypass the cooling coils. The bypassed outdoor air becomes a load within the conditioned space similar to infiltration air.

Determine heat load, Btu/hr, of the outside air bypassing the air conditioner from (cfm of ventilation air)(design outdoor dry-bulb-temperature, F − design indoor dry-bulb temperature, F)(air-conditioner bypass factor)(1.08).

Find the ventilation-air quantity by multiplying the number of people in the conditioned space by the cfm per person recommended by the *Guide*. The cfm per person can range from a minimum of 5 to a high of 50 where heavy smoking is anticipated. If industrial processes within the conditioned space require ventilation, the air may be supplied by increasing the outside air flow, by a local exhaust system at the process, or a combination of both. Regardless of the method used, outside air must be introduced to make up for the air exhausted from the conditioned space. The sum of the air required for people and processes is the total ventilation-air quantity.

Until the air conditioner is chosen, its bypass factor is unknown. However, to solve for the outside-air bypass heat load, a bypass factor must be applied. Table 12 shows typical bypass factors for various applications.

6. Compute the heat load from internal heat sources

Within an air-conditioned space, heat is given off by people, lights, appliances, machines, pipes, etc. Find the sensible heat, Btu/hr, given off by people by taking the product (number of people in the air-conditioned space)(sensible heat release per person, Btu/hr, from the appropriate *Guide* table). The sensible heat release per person varies with the activity of each person (seated, at rest, doing heavy work) and the room dry-bulb temperature. Thus, at 80 F, a person doing heavy work in a factory will give off 465 Btu/hr sensible heat; seated at rest in a theater at 80 F, a person will give off 195 Btu/hr sensible heat.

Find the heat, Btu/hr, given off by electric lights from (wattage rating of all installed

TABLE 12   Typical Bypass Factors*

For various applications

Coil bypass factor†	Type of application	Example
0.30–0.50	A *small* total load or a load that is somewhat larger with a low sensible heat factor (high latent load)	Residence
0.20–0.30	Typical comfort application with a *relatively small* total load or a low sensible heat factor with a somewhat larger load	Residence; small retail shop; factory
0.10–0.20	Typical comfort application	Department store; bank; factory
0.05–0.10	Applications with high internal sensible loads or requiring a large amount of outdoor air for ventilation	Department store; Restaurant; factory
0–0.10	All outdoor air applications	Hospital; operating room; factory

For finned coils

Depth of coils, rows	Without sprays		With sprays†	
	8 fins/in.	14 fins/in.	8 fins/in.	14 fins/in.
	Velocity, fpm			
	300–700	300–700	300–700	300–700
2	0.42–0.55	0.22–0.38		
3	0.27–0.40	0.10–0.23		
4	0.19–0.30	0.50–0.14	0.12–0.22	0.03–0.10
5	0.12–0.23	0.02–0.09	0.08–0.14	0.01–0.08
6	0.08–0.18	0.01–0.06	0.06–0.11	0.01–0.05
8	0.03–0.08	· · ·	0.02–0.05	

*Carrier Air Conditioning Company.
†The bypass factor with spray coils is decreased because spray provides more surface for contacting the air.

lights) (3.4). Where the installed lighting capacity is expressed in kw, use the factor 3,413 instead of 3.4.

For electric motors, find the heat, Btu/hr, given off from (total installed motor hp) (2,546)/motor efficiency expressed as a decimal. The usual efficiency assumed for electric motors is 85 percent.

Many other sensible-heat-generating devices may be used in an air-conditioned space. These devices include restaurant, beauty-shop, hospital, gas-burning, and kitchen appliances. The *Guide* lists the heat given off by a variety of devices, as well as pipes, tanks, pumps, etc.

7. *Compute the room sensible heat*

Find the sum of the sensible heat gains computed in Step 2 (sunlight heat gain); 3, (transmission heat gain); 4, (infiltration heat gain); 5, (outside-air heat gain; 6, (internal heat sources). This sum is the room sensible-heat subtotal.

A further sensible heat gain may result from: (*a*) supply-duct heat gain, (*b*) supply-duct-leakage loss, and (*c*) air-conditioning-fan horsepower. To the sum of these losses, a safety factor is usually added in the form of a percentage, since all the losses are also generally expressed as a percentage. The *Guide* provides means to estimate each loss and the safety factor. Assuming the sum of the losses and safety factor is x %, the room sensible heat load, Btu/hr, is $(1 + 0.01 \times x)$(room sensible heat subtotal, Btu/hr).

8. *Compute the room latent heat load*

The room latent heat load results from the moisture entering the room with the infiltration air and bypass ventilation air, the moisture given off by room occupants, and any other moisture source such as open steam kettles, sterilizers, etc.

Find the infiltration-air latent heat load, Btu/hr, from (cfm infiltration)(moisture content of outside air at design outdoor conditions, g/lb − moisture content of the conditioned air at the design indoor conditions, g/lb)(0.68). Use a similar relation for the bypass ventilation air, or Btu/hr latent heat = (cfm ventilation)(moisture content of the outside air at design outdoor conditions, g/lb − moisture content of conditioned air at design indoor conditions, g/lb)(bypass factor)(0.68).

Find the latent heat gain from the room occupants, Btu/hr, from (number of occupants in the conditioned space)(latent heat gain, Btu/hr per person, from the appropriate *Guide* table). Be sure to choose the latent heat gain that applies to the activity *and* conditioned-room dry-bulb temperature.

Nonhooded restaurant, hospital, laboratory, and similar equipment produces both a sensible and latent heat load in the conditioned space. Consult the *Guide* for the latent heat load for each type of unit in the space. Find the latent heat load of these units, Btu/hr, from (number of units of each type)(latent heat load, Btu/hr, per unit).

Take the sum of the latent heat loads for infiltration, ventilation bypass air, people, and devices. This sum is the room latent heat subtotal, if water-vapor transmission through the building surfaces is neglected.

Water vapor flows through building structures, resulting in a latent heat load whenever a vapor-pressure difference exists across a structure. The latent heat load from this source is usually insignificant in comfort applications and need be considered only in low or high dew-point applications. Compute the latent heat gain from this source using the appropriate *Guide* table and add it to the room latent heat subtotal.

Factors for the supply-duct leakage loss and for a safety margin are usually applied to the above sum. When all the latent heat subtotals are summed, the result is the room latent heat total.

9. *Compute the outside-air heat*

Air brought in for space ventilation imposes a sensible and latent heat load on the air-conditioning apparatus. Compute the sensible heat load, Btu/hr, from (cfm outside air)(design outdoor dry-bulb temperature − design indoor dry-bulb temperature)(1 − bypass factor)(1.08). Compute the latent heat load, Btu/hr, from (cfm outside air) (design outdoor moisture content, g/lb − design indoor moisture content, g/lb)(1 − bypass factor)(0.68). Apply percentage factors for return duct heat and leakage gain, pump horsepower, dehumidifier, and piping loss; see the *Guide* for typical values.

10. *Compute the grand-total heat and refrigeration tonnage*

Take the sum of the room total heat and outside-air heat. The result is the grand-total heat load of the space, Btu/hr.

Compute the refrigeration load, tons, from (grand-total heat, Btu/hr)/(12,000 Btu per hr per ton of refrigeration). A refrigeration system having the next higher standard rating is generally chosen.

11. *Compute the sensible heat factor and the apparatus dew point*

For any air-conditioning system, sensible heat factor = (room sensible heat, Btu/hr)/(room total heat, Btu/hr). Using a psychrometric chart and the known room conditions, find the apparatus dew point. An alternative, and quicker, way to find the apparatus dew point is to use the Carrier *Handbook of Air Conditioning System Design* tables.

12. *Compute the quantity of dehumidified air required*

Determine the dehumidified air temperature rise, F, from (1 − bypass factor)(design indoor temperature, F − apparatus dew point, F). Compute the dehumidified air quantity, cfm, from (room sensible heat, Btu/hr)/1.08 (dehumidified air temperature rise, F).

Related Calculations: The general procedure given above is valid for all types of air-conditioned spaces — offices, industrial plants, residences, hotels, apartment houses, motels, etc. Use the ASHRAE *Guide and Data Book* or the Carrier *Handbook of Air Conditioning System Design* as a source of data for the various calculations. Application of this method to an actual building is shown in the next Calculation Procedure.

In actual design work, a calculations form incorporating the calculations shown above is generally used. Such forms are obtainable from equipment manufacturers. Since the usual form does not provide any explanation of the calculations, the present Calculation Procedure is a useful guide to using the form. Refer to the Carrier *Handbook of Air Conditioning System Design* for one such calculation form.

AIR-CONDITIONING-SYSTEM HEAT-LOAD DETERMINATION — NUMERICAL COMPUTATION

Determine the required capacity of an air-conditioning system to serve the industrial building shown in Fig. 4. The outside walls are 8-in. brick with an interior finish of $\frac{3}{8}$-in. gypsum lath plastered and furred. A 6-in. plain poured-concrete partition separates the machinery room from the conditioned space. The roof is 6-in. concrete covered with $\frac{1}{2}$-in.-thick insulating board and the floor is 2-in. concrete. The windows are double-hung, metal-frame locked units with light-colored shades three-quarters drawn. Internal heat loads are: 100 people doing light assembly work; 25 1-hp motors running continuously at full load; 20,000 watts of light kept on at all times. The building is located in Port Arthur, Texas, at about 30° north latitude. The desired indoor design conditions are 80 F dry bulb, 67 F wet bulb, and 51 percent relative humidity. Air-conditioning equipment will be located in the machinery room. Use the general method given in the previous Calculation Procedure.

Calculation Procedure:

1. *Determine the design outdoor and indoor conditions*

The ASHRAE *Guide and Data Book* lists the design dry-bulb temperature in common use for Port Arthur, Texas, as 95 F and the design wet-bulb temperature as 79 F. Design indoor temperature and humidity conditions are given; if they were not given, the recommended conditions given in the *Guide* for an industrial building housing light assembly work would be used.

2. *Compute the sunlight heat gain*

The east, south, and west windows and walls of the building are subject to sunlight heat gains. North-facing walls are neglected because the sunlight heat gain is usually less than the transmission heat gain. Reference to the *Guide* table for sunlight radiation through glass shows that the largest amount of heat radiation occurs through the east and west walls. The maximum radiation is 181 Btu per hr per sq ft of glass area at 8 A.M. for the east wall and the same for the west wall at 4 P.M. Radiation through the glass in the south wall never reaches this magnitude. Hence, only the east or west wall need be considered. Since the west wall has 22 windows compared with 20 in the east wall, the west-wall sunlight heat gain will be used because it has a *larger* heat gain. (If both walls had an equal number of windows, either wall could be used.) When the window shades are normally three-quarters drawn, a value 0.6 times the tabulated sunlight radiation can be used. Hence, the west-glass sunlight heat gain = (22 windows) (5 × 8 ft each)(181)(0.6) = 95,600 Btu/hr.

For the same time of day, 4 P.M., the *Guide* table shows that the east-glass radiation is 0 Btu/(hr)(sq ft) and the south glass is 2 Btu/(hr)(sq ft). Hence, the south-glass sunlight heat gain = (10 windows, (5 × 8 ft each)(2)(0.6) = 480 Btu/hr.

The same three walls, and the roof, are also subject to sunlight heat gains. Reference to the *Guide* shows that with 8-in. walls the temperature difference resulting from sunlight heat gains is 15 F for south walls and 20 F for east and west walls. At 4 P.M. the roof temperature difference is given as 40 F. Hence, sunlight gain, south wall = (wall area, sq ft)(temperature difference, F)[wall coefficient of heat transfer, Btu/(hr) (sq ft)(F)], or (1,100)(15)(0.30) = 4,950 Btu/hr. Likewise, the east-wall sunlight heat gain = (1,300)(20)(0.30) = 7,800 Btu/hr; the west-wall sunlight heat gain = (1,220) (20)(0.30) = 7,320 Btu/hr; the roof sunlight heat gain = (21,875)(40)(0.33) = 289,000 Btu/hr. Note that the wall and roof coefficients of heat transfer are obtained from the appropraite *Guide* table.

The sum of the sunlight heat gains gives the total sunlight gain, or 405,150 Btu/hr.

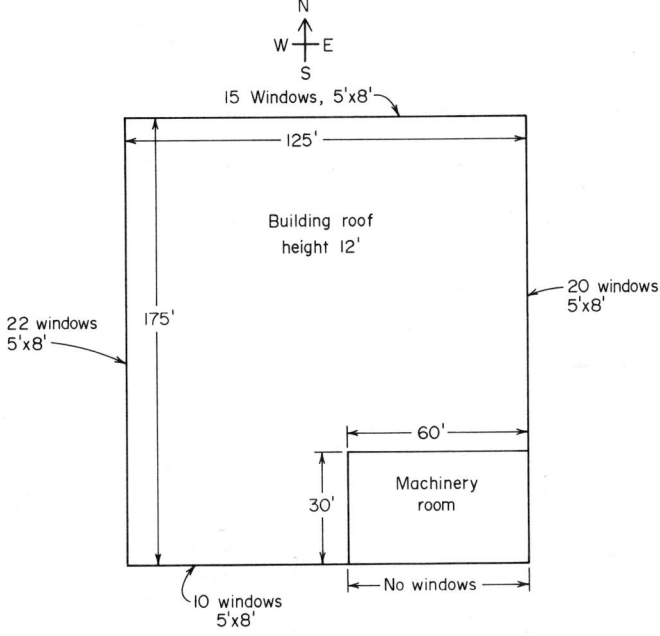

N wall area = 125 (12) − 15 (40) = 900 sq ft

S wall area = 125 (12) − 5 x 8 x 10 = 1100 sq ft

E wall area = (175)(12) − 5 x 8 x 20 = 1300 sq ft

W wall area = (175)(12) − 5 x 8 x 22 = 1220 sq ft

Roof area = (175)(125) = 21,875 sq ft

Glass area = (20 x 40) + (10 x 40) + 22 (40) + 15 (40) = 2680 sq ft

Partition area = 90 x 12 = 1080 sq ft

Fig. 4   Industrial building layout.

3. Compute the glass transmission heat gain

All the glass in the building is subject to a transmission heat gain. Find the all-glass transmission heat gain, Btu/hr, from (total glass area, sq ft)(outdoor design dry-bulb temperature, F − indoor design dry-bulb temperature, F)[coefficient of heat transmission of glass, Btu/(hr)(sq ft)(F), from *Guide*], or $(2,680)(95 − 80)(1.13) = 45,400$ Btu/hr.

The transmission heat gain of the south, east, and west walls can be neglected because the sunlight heat gain is greater. Hence, only the north-wall transmission heat gain need be computed. For unshaded walls, the transmission heat gain, Btu/hr is (wall area, sq ft)(design outdoor dry-bulb temperature, F − design indoor dry-bulb temperature, F)[coefficient of heat transmission, Btu/(hr)(sq ft)(F)] $= (900)(95 − 80)(0.30) = 4,050$ Btu/hr.

The heat gain from the ground can be neglected because the ground is usually at a lower temperature than the floor. Thus, the total transmission heat gain is the sum of the individual gains, or 66,530 Btu/hr.

4. Compute the infiltration heat gain

The total crack length for double-hung windows is $3 \times$ width $+ 2 \times$ height, or (67 windows)$[(3 \times 5) + (2 \times 8)] = 2,077$ ft. Using one-half the total length, or $2,077/2 = 1,039$ ft, and a wind velocity of 10 mph, the leakage, cfm, is (crack length, ft)(leakage, per ft of crack) $= (1,039)(0.75) = 770$ cfm, or $60(779) = 46,740$ cfh.

The heat gain due to infiltration through the window cracks is (leakage, cfm)(design outdoor dry-bulb temperature, F − design indoor dry-bulb temperature, F)(1.08), or $(779)(95 − 80)(1.08) = 12,610$ Btu/hr.

5. Compute the outside-air bypass heat load

For factories, the *Guide* recommends a ventilation air quantity of 10 cfm per person. Local codes may require a larger quantity; hence the codes should be checked before making a final choice of the ventilation-air quantity used per person. Since there are 100 people in this factory, the required ventilation quantity is $100(10) = 1,000$ cfm. Next, the bypass factor for the air-conditioning equipment must be chosen.

Table 12 shows that the usual factory air-conditioning equipment has a bypass factor ranging between 0.10 and 0.20. Assume a value of 0.10 for this installation.

The heat load, Btu/hr, of the outside air bypassing the air conditioner is (cfm of ventilation air)(design outdoor dry-bulb temperature, F − design indoor dry-bulb temperature, F)(air-conditioner bypass factor)(1.08). Hence, $(1,000)(95 − 80)(0.10)(1.08) = 1,620$ Btu/hr.

6. Compute the heat load from internal heat sources

The internal heat sources in this building are people, lights, and motors. Compute the sensible heat load of the people from Btu/hr = (number of people in the air-conditioned space)(sensible heat release per person, Btu/hr, from the appropriate *Guide* table). Thus, for this building with an 80-F indoor dry-bulb temperature and 100 occupants doing light assembly work, the heat load produced by people $= (100)(210) = 21,000$ Btu/hr.

The motor heat load, Btu/hr, is (motor hp)(2,546)/motor efficiency. Assuming an 85 percent motor efficiency, the motor heat load $= (25)(2,546)/0.85 = 75,000$ Btu/hr. Thus, the total internal heat load $= 21,000 + 68,000 + 75,000 = 164,000$ Btu/hr.

7. Compute the room sensible heat

Find the sum of the sensible heat gains computed in Steps 2, 3, 4, 5, and 6. Thus, sensible heat load $= 405,150 + 66,530 + 12,610 + 1,620 + 164,000 = 649,910$ Btu/hr; say 650,000 Btu/hr. Using an assumed safety factor of 5 percent to cover the various losses that may be encountered in the system, room sensible heat $= (1.05)(650,000) = 682,500$ Btu/hr.

8. Compute the room latent heat load

The room latent load results from the moisture entering the air-conditioned space with the infiltration and bypass air, moisture given off by room occupants, and any other moisture sources.

Find the infiltration heat load, Btu/hr, from (cfm infiltration)(moisture content of outside air at design outdoor conditions, g/lb — moisture content of the conditioned air at the design indoor conditions, g/lb)(0.68). Using a psychrometric chart or the *Guide* thermodynamic tables, infiltration latent heat load = $(779)(124 - 78)(0.68) = 24,400$ Btu/hr.

Using a similar relation for the ventilation air, or Btu/hr latent heat = (cfm ventilation air)(moisture content of outside air at design outdoor conditions, g/lb — moisture content of the conditioned air at the design indoor conditions, g/lb)(bypass factor)(0.68). Or, $(1,000)(124 - 78)(0.10)(0.68) = 3,130$ Btu/hr.

The latent heat gain from room occupants, Btu/hr = (number of occupants in the conditioned space)(latent heat gain, Btu per hr per person, from the appropriate *Guide* table). Or, $(100)(450) = 45,000$ Btu/hr.

Find the latent heat gain subtotal by taking the sum of the above heat gains, or $24,400 + 3,130 + 45,000 = 72,530$ Btu/hr. Using an allowance of 5 percent for supply-duct leakage loss and a safety margin, latent heat gain = $(1.05)(72,530) = 76,157$ Btu/hr.

9. Compute the outside heat

Compute the sensible heat load of the outside ventilation air from Btu/hr = (cfm outside ventilation air)(design outdoor dry-bulb temperature — design indoor dry-bulb temperature)(1 — bypass factor)(1.08). For this system, with 1,000 cfm outside air ventilation, sensible heat = $(1.000)(95 - 80)(1 - 0.10)(1.08) = 14,600$ Btu/hr.

Compute the latent heat load of the outside ventilation air from (cfm outside air) × design outdoor moisture content, g/lb — design indoor moisture content, g/lb)(1 — bypass factor)(0.68). Using the moisture content from Step 8, $(1,000)(124 - 78)(1 - 0.1) \times (0.68) = 28,200$ Btu/hr.

10. Compute the grand-total heat and refrigeration tonnage

Take the sum of the room total heat and the outside-air sensible and latent heat. The result is the grand-total heat load of the space, Btu/hr.

The room total heat = room-sensible-heat total + room-latent-heat total = $682,500 + 76,157 = 768,657$ Btu/hr. Then, the grand-total heat = $768,657 + 14,600 + 28,200 = 811,457$ Btu/hr; say 811,500 Btu/hr.

Compute the refrigeration load, tons, from (grand-total heat, Btu/hr)/(12,000 Btu per hr per ton of refrigeration), or $811,500/12,000 = 67.6$ tons; say 70 tons.

The quantity of cooling water required for the refrigeration-system condenser, Q gpm = 30 × tons of refrigeration/condenser water-temperature rise, F. Assuming a 75-F entering-water temperature and 95-F leaving-water temperature, which are typical values for air-conditioning practice, $Q = 30(70)/95 - 75 = 105$ gpm.

11. Compute the sensible heat factor and apparatus dew point

For any air-conditioning system, sensible heat factor = (room sensible heat, Btu/hr)/(room total heat, Btu/hr) = $682,500/768,657 = 0.888$.

The *Guide* or the Carrier *Handbook of Air Conditioning Design* gives an apparatus dew point of 58 F, closely.

12. Compute the quantity of dehumidified air required

Determine the dehumidified air temperature rise first from F = (1 — bypass factor) (design indoor temperature, F — apparatus dew point, F), or F = $(1 - 0.1)(80 - 58) = 19.8$ F.

Next, compute the dehumidified air quantity, cfm from (room sensible heat, Btu/hr)/ 1.08(dehumidified air temperature rise, F), or $(682,500)/1.08(10.8) = 34,400$ cfm.

Related Calculations: Use this general procedure for any type of air-conditioned building or space — industrial, office, hotel, motel, apartment house, residence, laboratories, school, etc. Use the ASHRAE *Guide and Data Book* or the Carrier *Handbook of Air Conditioning System Design* as a source of data for the various calculations. When comparing the various values from the *Guide* used in this procedure, note that there may be slight changes in certain tabulated values from one edition of the *Guide* to the next. Hence, the values shown may differ slightly from those in the current edition of the *Guide*. This should not cause concern, because the procedure is the same regardless of the values used.

AIR-CONDITIONING-SYSTEM COOLING-COIL SELECTION

Select an air-conditioning cooling coil to cool 15,000 cfm of air from 85-F dry bulb, 67-F wet bulb, 57-F dew point, 38 percent relative humidity, to a dry-bulb temperature of 65 F with cooling water at 50 F. Suppose the air were cooled below the dew point of the entering air. How would the calculation procedure differ?

Calculation Procedure:

1. Compute the weight of air to be cooled

From a psychrometric chart, Fig. 5, find the specific volume of the entering air as 13.75 cu ft/lb. To convert the air flow in cfm to lb/hr, use the relation lb/hr = 60 cfm/v_s, where cfm = air flow in cfm; v_s = specific volume of the entering air, cu ft/lb. Hence for this cooling coil, lb/hr = 60(15,000)/13.75 = 65,500 lb/hr.

Where a cooling coil is rated by the manufacturer for air at 70-F dry bulb and 50 percent relative humidity, as is often done, use the relation lb/hr = 4.45(cfm). Thus, if this coil were rated for air at 70-F dry bulb, the weight of air to be cooled would be lb/hr = 4.45(15,000) = 66,800.

Since this air quantity is somewhat greater than when the entering-air specific volume is used, and since cooling coils are often rated on the basis of 70-F dry-bulb air, this quantity, 66,800 lb/hr, will be used. The procedure is the same in either case. (NOTE: 4.45 = 60/13.5 cu ft/lb, the specific volume of air at 70 F and 50 percent relative humidity.)

2. Compute the quantity of heat to be removed

Use the relation $H_r = ws\Delta t$, where H_r = heat to be removed from the air, Btu/hr; w = weight of air cooled, lb/hr; s = specific heat of air = 0.24 Btu/(lb)(F); Δt = temperature drop of the air = entering dry-bulb temperature, F − leaving dry-bulb temperature, F. For this coil, H_r = (66,800)(0.24)(85 − 65) = 321,000 Btu/hr.

3. Compute the quantity of cooling water required

The quantity of cooling water required in gpm = $H_r/500\ \Delta t_w$, where Δt_w = leaving-water temperature, F − entering-water temperature, F. Since the leaving-water temperature is not known, a value must be assumed. The usual temperature rise of water during passage through an air-conditioning cooling coil is 4 to 12 F. Assuming a 10-F rise, which is a typical value, gpm required = 321,000/500(10) = 64.2 gpm.

4. Determine the logarithmic mean temperature difference

Use Fig. 4 in the Heat Transfer section of this Handbook to determine the logarithmic mean temperature difference for the cooling coil. In this chart, greatest terminal difference = entering-air temperature, F − leaving-water temperature, F = 85 − 60 = 25 F, and least terminal temperature difference = leaving-air temperature, F − entering water temperature, F = 65 − 50 = 15 F. Entering Fig. 4 at these two temperature values gives a logarithmic mean temperature difference (LMTD) of 19.5 F.

5. Compute the coil core face area

The coil core face area is the area exposed to the air flow; it does not include the area of the mounting flanges. Compute the coil core face area, sq ft, from $A_c = cfm/V_a$, where V_a = air velocity through the coil, fpm. The usual air velocity through the coil, often termed face velocity, ranges from 300 to 800 fpm, although special designs may use velocities down to 200 fpm or up to 1,200 fpm. Assuming a face velocity of 500 fpm, A_c = 15,000/500 = 30 sq ft.

6. Select the cooling coil for the load

Using the engineering data provided by the manufacturer whose coil is to be used, choose the coil. Table 13 summarizes typical engineering data provided by a coil manufacturer. This table shows that two 15.4 sq ft coils placed side by side will provide

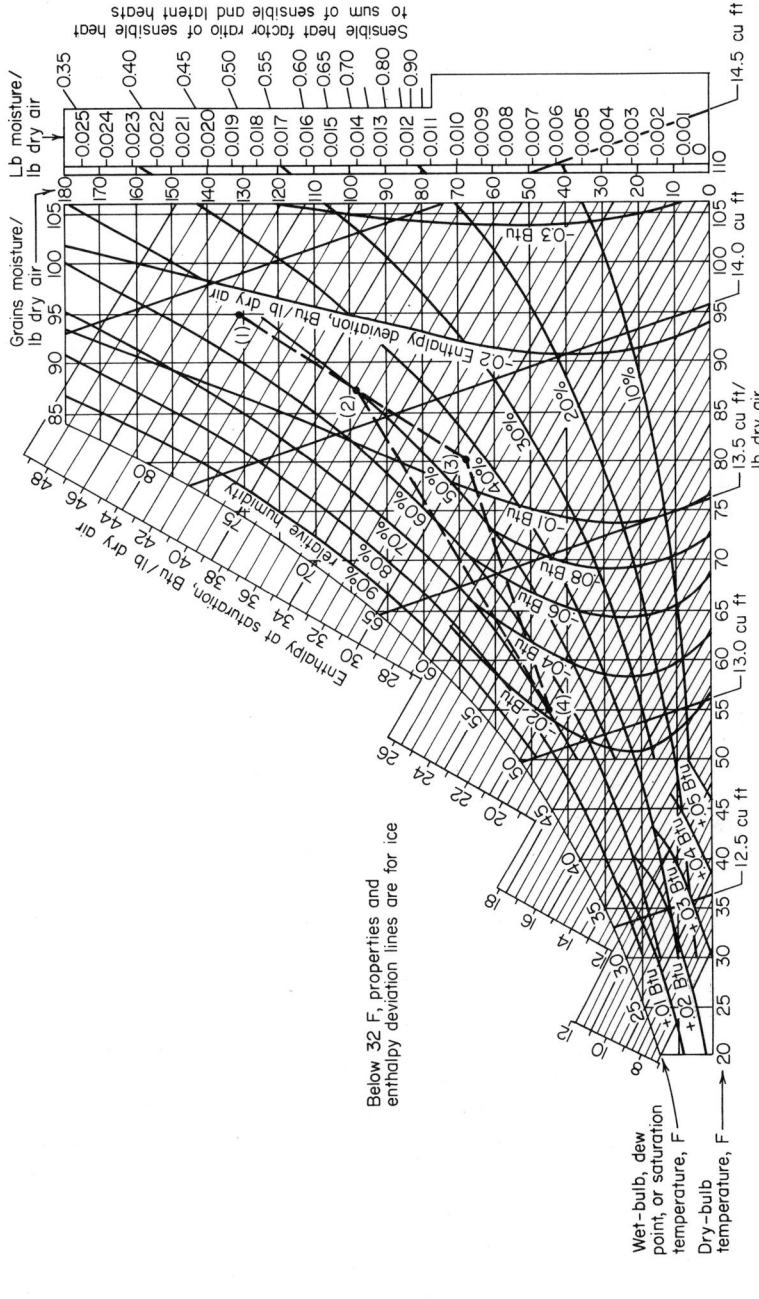

Fig. 5   Psychrometric chart for normal temperatures. (*Carrier Air Conditioning Company.*)

TABLE 13 Typical Cooling-Coil Characteristics

Face area, sq ft	14.0	15.4	17.9
Tube length, ft-in.	5–6	6–0	7–0
Water velocity, fpm	3.59 (gpm)	3.59 (gpm)	3.59 (gpm)

Coil heat-transfer factors, k = Btu per hr per sq ft face area deg
LMTD row

Water velocity, fpm	Air velocity, fpm		
	400	500	550
113	154	162	170
115	158	167	175
117	161	172	176

Coil water pressure drop, ft of water per row

Tube length, ft-in.	Water velocity, fpm		
	90	120	150
5–6	0.16	0.26	0.38
6–0	0.18	0.29	0.42
7–0	0.21	0.32	0.46

Header water pressure drop, ft of water

Coil type	Water velocity, fpm		
	90	120	150
A	0.26	0.48	0.72
B	0.34	0.62	0.92

a total face area of $2 \times 15.4 = 30.8$ sq ft. Hence, the actual air velocity through the coil is $V_a = cfm/A_c = 15,000/30.8 = 487$ fpm.

7. *Compute the water velocity in the coil*

Table 13 shows that the coil water velocity, fpm = 3.59 (*gpm*). Since the required water flow is, from Step 3, 64.2 gpm, the water flow for each unit will be half this, or $64.2/2 = 32.1$. Hence, the water velocity = 3.59(32.1) = 115.2 fpm.

8. *Determine the coil heat-transfer factors*

Table 13 lists typical heat-transfer factors for various water and air velocities. Interpolating between 400 and 500 fpm air velocities at a water velocity of 115 fpm gives a heat transfer factor of $k = 165$ Btu per hr per sq ft face area deg LMTD row. The increase in velocity from 15 to 15.2 fpm is so small that it can be ignored. If the actual velocity is midway, or more, between the two tabulated velocities, interpolate vertically also.

9. *Compute the number of tube rows required*

Use the relation number of tube rows = $H_r/(\text{LMTD})(A_c)(k)$. Thus, number of rows = $321,000/(19.5)(30.8)(165) = 3.24$, or four rows, the next larger *even* number.

Water cooling coils for air-conditioning service are usually built in units having two, four, six, or eight rows of coils. If the above calculation indicates that an odd number of coils should be used—i.e., the result was 2.24 rows, use the next smaller or larger *even* number of rows after increasing or decreasing the air and water velocity. Thus, to decrease the air velocity, use a coil having a larger face area. Recompute the air velocity and water velocity; find the new heat-transfer factor and the required number of rows. Continue doing this until a suitable number of rows is obtained. Usually, only one recalculation is necessary.

10. Determine the coil water-pressure drop

Table 13 shows the water-pressure drop, ft of water, for various tube lengths and water velocities. Interpolating between 90 and 120 fpm for a 6-ft-long tube gives a pressure drop of 0.27 ft of water per row at a water velocity of 115.2 fpm. Since the coil has four rows, total tube pressure drop = 4(0.27) = 1.08 ft of water.

There is also a water pressure drop in the coil headers. Table 13 lists typical values. Interpolating between 90 and 120 fpm for a B-type coil gives a header pressure loss of 0.57 ft of water at a water velocity of 115.2 fpm. Hence, the total pressure loss in the coil is 1.08 + 0.57 = 1.65 ft of water = coil loss + header loss.

11. Determine the coil resistance to air flow

Table 14 lists the resistance of coils having two to six rows of tubes and various air velocities. Interpolating for four tube rows gives a resistance of 0.225 in. H_2O for an air velocity of 487 fpm. The increase in resistance with a wet tube surface is, from Table 14, 28 percent at a 500-fpm air velocity. This occurs when the air is cooled below the entering-air dew point and is discussed in Step 13.

TABLE 14   Typical Cooling-Coil Resistance Characteristics

Air-flow resistance, in. H_2O for 70-F air

No. of tube rows	Air-face velocity, fpm		
	400	500	600
2	0.081	0.122	0.164
4	0.162	0.234	0.318
6	0.234	0.344	0.472
8	0.312	0.454	0.622

Resistance increase due to wet tube surface, percent

32	28	24

Coil cooling capacity, Btu per hr per sq ft face area and final air temperature, F

Air velocity, fpm	Number of tube rows	Entering-air temperature, F	Entering-water temperature, F	
			45	50
500	4	85 {	11,900 63	10,300 65

12. Check the coil selection in a coil-rating table

Many manufacturers publish precomputed coil-rating tables as part of their engineering data. Table 14 shows a portion of one such table. This tabulation shows that with an air velocity of 500 fpm, 4 tube rows, an entering-air temperature of 85 F, and an entering-water temperature of 50 F, the cooling coil has a cooling capacity of 10,300 Btu/(hr) (sq ft) and a final air temperature of 65 F. Since the actual air velocity of 487 fpm is close to 500 fpm, the tabulated cooling capacity closely approximates the actual cooling capacity. Hence, the required heat-transfer area is $A_c = H_r/10,300 = 321,000/10,300 = 31.1$ sq ft. This agrees closely with the area of 30.8 sq ft found in Step 6.

In actual practice, designers use a coil-cooling-capacity table whenever it is available. However, the procedure given in Steps 1 through 11 is also used when an exact analysis of a coil is desired or when a capacity table is not available.

13. *Compute the heat removal when cooling below the dew point*

When the temperature of the air leaving the cooling coil is lower than the dew point of the entering air, $H_r =$ (weight of air cooled, lb)(total heat of entering air at its wet-bulb temperature, Btu/lb − total heat of the leaving air at its wet-bulb temperature, Btu/lb). Once H_r is known, follow all the steps given above except that (a) a correction must be applied in Step 11 for a wet tube surface. Obtain the appropriate correction factor from the manufacturer's engineering data and apply it to the air-flow-resistance data, for the coil selected. (b) Also, the usual coil-rating table presents only the sensible-heat capacity of the coil. Where the ratio of sensible heat removed to latent heat removed is more than 2:1, the usual coil-rating table can be used. If the ratio is less than 2:1, use the procedure in Steps 1 through 13.

Related Calculations: Use the method given here in Steps 1 through 11 for any finned-type cooling coil mounted perpendicular to the air flow and having water as the cooling medium where the final air temperature leaving the cooling coil is *higher than* the dew point of the entering air. Follow Step 13 for cooling below the dew point of the entering air.

Cooling and dehumidifying coils used in air-conditioning systems generally serve the following ranges of variables: (1) dry-bulb temperature of the entering air is 60 to 100 F; wet-bulb temperature of entering air is 50 to 80 F; (2) coil core face velocity can range from 200 to 1,200 fpm, with 500 to 800 fpm being the most common velocity for comfort cooling applications; (3) entering-water temperature ranges from 40 to 65 F; (4) the water-temperature rise ranges from 4 to 12 F during passage through the coil; (5) the water velocity ranges from 2 to 6 fps.

To choose an air-cooling coil using a direct-expansion refrigerant, follow the manufacturer's engineering data. Since most of the procedures are empirical, it is difficult to generalize about which procedure to use. However, the usual range of the volatile refrigerant temperature at the coil suction outlet is 25 to 55 F. Where chilled water is circulated through the coil, the usual quantity range is 2 to 6 gpm/ton.

MIXING OF TWO AIR STREAMS

An air-conditioning system is designed to deliver 100,000 cfm of air to a conditioned space. Of this total, 90,000 cfm is recirculated indoor air at 72 F and 40 percent relative humidity; 10,000 cfm is outdoor air at 0 F. What is the enthalpy, temperature, moisture content, and relative humidity of the resulting air mixture? If air enters the room from the outlet grille at 60 F after leaving the apparatus at a 50-F dew point and the return air is at 75 F, what proportion of conditioned air and bypassed return air must be used to produce the desired outlet temperature at the grille?

Calculation Procedure:

1. *Determine the proportions of each air stream*

Use the relations $p_r = r/t$ and $p_0 = o/t$, where $p_r =$ percent recirculated room air, expressed as a decimal; $r =$ recirculated air quantity, cfm; $t =$ total air quantity, cfm; $p_0 =$ percent outside air, expressed as a decimal; $o =$ outside air quantity, cfm. For this system, $p_r = 90,000/100,000 = 0.90$, or 90 percent; $p_0 = 10,000/100,000 = 0.10$, or 10 percent.

2. *Determine the enthalpy of each air stream*

Use a psychrometric chart or table to find the enthalpy of the recirculated indoor air as 24.6 Btu/lb.

The enthalpy of the outdoor is 0.0 Btu/lb because in considering heating or humidifying processes in winter it is always safest to assume that the outdoor air is completely dry. This condition represents the greatest heating and humidifying load because the enthalpy and the water-vapor content of the air are at a minimum when the air is considered dry at the outdoor temperature.

3. Determine the moisture content of each air stream

The moisture content of the indoor air at a 72-F dry-bulb temperature and 40 percent relative humidity is, from the psychrometric chart, Fig. 5, 47.2 grains/lb. From a psychrometric table the moisture content of the 0-F outdoor air, which is assumed to be completely dry, is 0.0 grains/lb.

4. Compute the enthalpy of the air mixture

Use the relation $h_m = (oh_0 + rh_r)/t$, where h_m = enthalpy of mixture, Btu/lb; h_0 = enthalpy of the outside air, Btu/lb; h_r = enthalpy of the recirculated room air, Btu/lb; other symbols as before. Hence, $h_m = (10,000 \times 0 + 90,000 \times 24.6)/100,000 = 22.15$ Btu/lb.

5. Compute the temperature of the air mixture

Use a similar relation to that in Step 4, substituting the air temperature for the enthalpy. Or, $t_m = (ot_0 + rt_r)/t$, where t_m = mixture temperature, F; t_0 = temperature of outdoor air, F; t_r = temperature of recirculated room air, F; other symbols as before. Hence, $t_m = (10,000 \times 0 + 90,000 \times 72)/100,000 = 64.9$ F.

6. Compute the moisture content of the air mixture

Use a similar relation to that in Step 4, substituting the moisture content for the enthalpy. Or, $g_m = (og_0 + rg_r)/t$, where g_m = grains of moisture per lb of mixture; g_0 = grains of moisture per lb of outdoor air; g_r = grains of moisture per lb of recirculated room air; other symbols as before. Thus, $g_m = (10,000 \times 0 + 90,000 \times 47.2)/100,000 = 42.5$ grains/lb.

7. Determine the relative humidity of the mixture

Enter the psychrometric chart at the temperature of the mixture, 64.9 F, and the moisture content, 42.5 grains/lb. At the intersection of the two lines, find the relative humidity of the mixture as 47 percent relative humidity.

8. Determine the required air proportions

Set up an equation in which x = the proportion of conditioned air required to produce the desired outlet temperature at the grille and y the proportion of bypassed air required. The air quantities will also be proportional to the dry-bulb temperatures of each air stream. Since the dew point of the air leaving an air-conditioning apparatus = dry-bulb temperature of the air, $50x + 75y = 60(x + y)$, or $15y = 10x$. Also, the sum of the two air streams $x + y = 1$. Substituting and solving for x and y, $x = 60$ percent; $y = 40$ percent. Multiplying the actual air quantity supplied to the room by the percentage representing the proportion of each air stream will give the actual cfm required for supply and bypass air.

Related Calculations: Use this general procedure to determine the properties of any air mixture in which two air streams are mixed without compression, expansion, or other processes involving a marked changed in the pressure or volume of either or both air streams.

SELECTION OF AN AIR-CONDITIONING SYSTEM FOR A KNOWN LOAD

Choose the type of air-conditioning system to use for comfort conditioning of a factory having a heat gain that varies from 500,000 to 750,000 Btu/hr depending on the outdoor temperature and the conditions inside the building. Indicate why the chosen system is preferred.

TABLE 15 Systems and Applications*

	Individual room or zone unit systems				Central station apparatus systems							
	DX self-contained		All-water		All-air					Air-water	
			Room fan-coil		Single air stream					Primary air systems	
	Room	Zone					Reheat				
Applications	⅓ to 2 tons	2 tons and over	Recirculating air	With outdoor air	Variable volume	Bypass	At terminal	Zone in duct	Multizone single duct	Secondary water H-V H-P induction	Room fan-coil with outside air
Page	(9-8)	(9-8)	(9-8)	(9-8)	(9-9)	(9-9)	(9-10)	(9-10)	(9-10)	(9-11)	(9-8)
Single-purpose occupancies											
Residential:											
Medium .... (9-13)	x										
Large .... (9-13)		x							x		
Restaurants:											
Medium .... (9-13)		x	x								
Large .... (9-13)		x				x	x	x	x		
Variety & Speciality shops .... (9-13)		x				x					
Bowling alleys .... (9-14)		x				x					
Radio and TV studios:											
Small .... (9-14)		x				x			x		
Large .... (9-14)		x				x		x	x		
Country clubs .... (9-14)		x				x		x	x		
Funeral homes .... (9-14)		x							x		
Beauty salons .... (9-14)	x	x									
Barber shops .... (9-15)	x	x									

Churches	(9–15)							x
Theaters	(9–15)							
Auditoriums	(9–15)							
Dance and roller skating pavilions	(9–15)			x			x	
Factories (comfort)	(9–15)			x			x	

Multi-purpose occupancies

Office buildings	(9–16)		x				x	x
Hotels, dormitories	(9–18)	x						x
Motels	(9–18)	x						
Apartment buildings	(9–18)		x					x
Hospitals	(9–18)	x	x		x			x
Schools and colleges	(9–19)	x	x	x	x			
Museums	(9–20)			x	x	x		
Libraries:								
Standard	(9–20)	x		x		x		x
Rare books	(9–20)		x		x			
Department stores	(9–19)	x	x					
Shopping centers	(9–19)	x	x				x	
Laboratories:								
Small	(9–20)	x		x	x		x	x
Large building	(9–20)		x				x	x
Marine	(9–21)		x	x				

NOTES:

1. Systems checked for a particular application are the systems most commonly *used.* Economics and design objectives dictate the choice and deviations of systems listed above, other systems as listed in note 2, and some entirely new systems.

2. There are several systems used on many of these applications when higher quality air conditioning is desired (often at higher expense). They are dual-duct (9–11), dual conduit (9–9), three-pipe induction and fan-coil (9–11), four-pipe induction and fan-coil (9–11), and panel-air (9–12).

3. Numbers in parentheses are page numbers of the text describing the particular system or application.

*Carrier Air Conditioning Co. – *Handbook of Air Conditioning System Design,* McGraw-Hill. The page numbers refer to that book.

Calculation Procedure:

1. Review the types of air-conditioning systems available

Table 15 summarizes the various types of air-conditioning systems *commonly used* for different applications. Economics and special design objectives dictate the final choice and modifications of the systems listed. Where higher quality air conditioning is desired (often at a higher cost), certain other systems may be considered. These are dual-duct, dual-conduit, three-pipe induction and fan-coil, four-pipe induction and fan-coil, and panel-air systems.

Study of Table 15 shows that four main types of air-conditioning systems are popular: direct-expansion (termed DX), all-water, all-air, and air-water. These classifications indicate the methods used to obtain the final within-the-space cooling and heating. The air surrounding the occupant is the end medium that is conditioned.

2. Select the type of air-conditioning system to use

Table 15 indicates that direct-expansion and all-air air-conditioning systems are *commonly used* for factory comfort conditioning. The load in the factory being considered may range from 500,000 to 750,000 Btu/hr. This is the equivalent of a maximum cooling load of 750,000/12,000 Btu/hr/ton of refrigeration = 62.5 tons of refrigeration.

Where a building has a varying heat load, bypass control wherein neutral air is recirculated from the conditioned space while the amount of cooling air is reduced is often used. With this arrangement, the full quantity of supply air is introduced to the cooled area at all times during system operation.

Self-contained direct-expansion systems can serve large factory spaces. Their choice over an all-air bypass system is largely a matter of economics and design objectives.

Where reheat is required, this may be provided by a reheater in a zone duct. Reheat control maintains the desired dry-bulb temperature within a space by replacing any decrease in sensible loads by an artificial heat load. Bypass control maintains the desired dry-bulb temperature within the space by modulating the amount of air to be cooled. Since the bypass all-air system is probably less costly for this building, it will be the first choice. A complete economic analysis would be necessary before this conclusion could be accepted as fully valid.

Related Calculations: Use this general method to make a preliminary choice of the 11 different types of air-conditioning systems for the 31 applications listed in Table 15. Where additional analytical data for comparison of systems are required, consult Carrier Air Conditioning Company's *Handbook of Air Conditioning System Design* or the ASHRAE *Guide and Data Book*.

SIZING LOW-VELOCITY AIR-CONDITIONING-SYSTEM DUCTS— EQUAL-FRICTION METHOD

An industrial air-conditioning system requires 36,000 cfm of air. This low-velocity system will be fitted with enough air outlets to distribute the air uniformly throughout the conditioned space. The required operating pressure for each duct outlet is 0.20 in. wg. Determine the duct sizes required for this system by using the equal-friction method of design. What is the required fan static discharge pressure?

Calculation Procedure:

1. Sketch the duct system

The required air quantity, 36,000 cfm, must be distributed in approximately equal quantities to the various areas in the building. Sketch the proposed duct layout as shown in Fig. 6. Locate air outlets as shown to provide air to each area in the building.

Determine the required capacity of each air outlet from air quantity required, cfm/ number of outlets, or outlet capacity = 36,000/18 = 2,000 cfm per outlet. This is within the usual range of many commercially available air outlets. Where the required capacity per outlet is extremely large—say 10,000 cfm—or extremely small—say 5 cfm— change the number of outlets shown on the duct sketch to obtain an air quantity within the usual capacity range of commercially available outlets. Relocate each outlet so it serves approximately the same amount of floor area as each of the other outlets in the system. Thus, the duct sketch serves as a trial-and-error analysis of the outlet location and capacity.

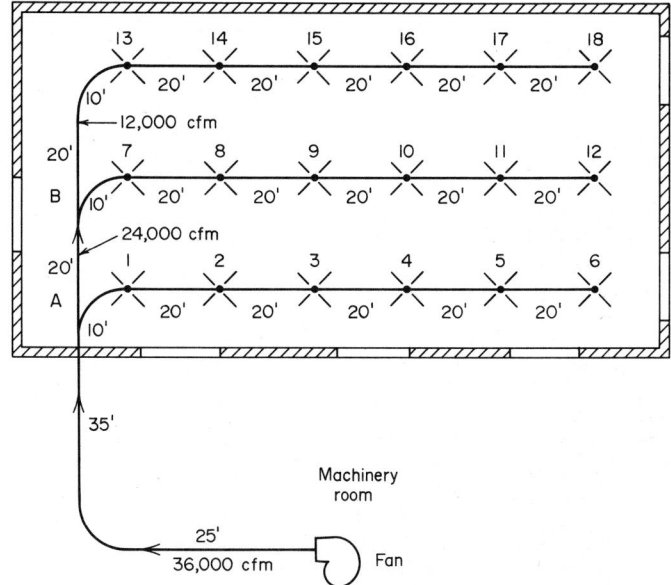

Fig. 6 Duct-system layout.

Where a building area requires a specific amount of air, select one or more outlets to supply this air. Size the remaining outlets by the method described above, after subtracting the quantity of air supplied through the outlets already chosen.

2. Determine the required outlet operating pressure

Consult the manufacturer's engineering data for the required operating pressure of each outlet. Where possible, try to use the same type of outlets throughout the system. This will reduce the initial investment. Assume that the required outlet operating pressure is 0.20-in. wg for each outlet in this system.

3. Choose the air velocity for the main duct

Use Table 16 to determine a suitable air velocity for the main duct of this system. Table 16 shows that an air velocity up to 2,500 fpm can be used for main ducts where noise is the controlling factor; 3,000 fpm where duct friction is the controlling factor. A velocity of 2,500 fpm will be used for the main duct in this installation.

4. Determine the dimensions of the main duct

The required duct area A sq ft = cfm/fpm = 36,000/2,500 = 14.4 sq ft. A nearly square duct, i.e., a duct 46 × 45 in., has an area of 14.38 sq ft and is a good first choice for this system because it closely approximates the outlet size of a standard centrifugal fan. Where possible, use a square main duct to simplify fan connections. Thus a 46 × 46 in. duct might be the final choice for this system.

5. Determine the main-duct friction loss

Convert the duct area to the equivalent diameter in in., d, using $d = 2(144A/\pi)^{0.5} = 2(144 \times 14.4/\pi)^{0.5} = 51.5$ in.

Enter Fig. 7 at 36,000 cfm and project horizontally to a round-duct diameter of 51.5 in. At the top of Fig. 7 read the friction loss as 0.13 in. wg per 100 ft of equivalent duct length.

TABLE 16   Recommended Maximum Duct Velocities for Low-Velocity Systems, fpm*

Application	Controlling factor noise generation— main ducts	Controlling factor — duct friction			
		Main ducts		Branch ducts	
		Supply	Return	Supply	Return
Residences	600	1,000	800	600	600
Apartments, hotel bedrooms, hospital bedrooms	1,000	1,500	1,300	1,200	1,000
Private offices, directors rooms, libraries ..	1,200	2,000	1,500	1,600	1,200
Theatres, auditoriums	800	1,300	1,100	1,000	800
General offices, high-class restaurants, high-class stores, banks.............	1,500	2,000	1,500	1,600	1,200
Average stores, cafeterias	1,800	2,000	1,500	1,600	1,200
Industrial	2,500	3,000	1,800	2,200	1,500

*Carrier Air Conditioning Company

6. Size the branch ducts

For many common air-conditioning systems the equal-friction method is used to size the ducts. In this method the supply, exhaust, and return-air ducts are sized so they have the same friction loss per ft of length for the entire system. The equal-friction method is superior to the velocity-reduction method of duct sizing because the former requires less balancing for symmetrical layouts.

The usual procedure in the equal-friction method is to select an initial air velocity in the main duct near the fan, using the sound level as the limiting factor. With this initial velocity and the design air flow rate, the required duct diameter is found, as in Steps 4 and 5, above. Once the duct diameter is known, the friction loss is found from Fig. 7, as in Step 5. This same friction loss is then maintained throughout the system and the equivalent round-duct diameter is chosen from Fig. 7.

To expedite equal-friction calculations, Table 17 is often used instead of the friction chart. This provides the same duct sizes. Duct areas are determined from Table 17, and

TABLE 17   Percent Section Area in Branches for Maintaining Equal Friction*

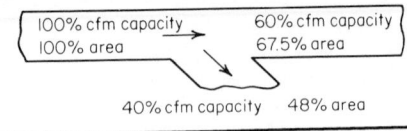

Cfm Capacity, %	Duct Area, %	Cfm Capacity, %	Duct Area, %	Cfm Capacity, %	Duct Area, %	Cfm Capacity, %	Duct Area, %
1	2.0	26	33.5	51	59.0	76	81.0
2	3.5	27	34.5	52	60.0	77	82.0
3	5.5	28	35.5	53	61.0	78	83.0
4	7.0	29	36.5	54	62.0	79	84.0
5	9.0	30	37.5	55	63.0	80	84.5
6	10.5	31	39.0	56	64.0	81	85.5
7	11.5	32	40.0	57	65.0	82	86.0
8	13.0	33	41.0	58	65.5	83	87.0
9	14.5	34	42.0	59	66.5	84	87.5
10	16.5	35	43.0	60	67.5	85	88.5

TABLE 17   Percent Section Area in Branches for Maintaining Equal Friction* (Continued)

Cfm Capacity, %	Duct Area, %	Cfm Capacity, %	Duct Area, %	Cfm Capacity, %	Duct Area, %	Cfm Capacity, %	Duct Area, %
11	17.5	36	44.0	61	68.0	86	89.5
12	18.5	37	45.0	62	69.0	87	90.0
13	19.5	38	46.0	63	70.0	88	90.5
14	20.5	39	47.0	64	71.0	89	91.5
15	21.5	40	48.0	65	71.5	90	92.0
16	23.0	41	49.0	66	72.5	91	93.0
17	24.0	42	50.0	67	73.5	92	94.0
18	25.0	43	51.0	68	74.5	93	94.5
19	26.0	44	52.0	69	75.5	94	95.0
20	27.0	45	53.0	70	76.5	95	96.0
21	28.0	46	54.0	71	77.0	96	96.5
22	29.5	47	55.0	72	78.0	97	97.5
23	30.5	48	56.0	73	79.0	98	98.0
24	31.5	59	57.0	74	80.0	99	99.0
25	32.5	50	58.0	75	80.5	100	100.0

*Carrier Air Conditioning Company.

the area found is converted to a round-, rectangular-, or square-duct size suitable for the installation. This procedure of duct sizing automatically reduces the air velocity in the direction of air flow. Hence, the equal-friction method will be used for this system.

Compute the duct areas using Table 17. Tabulate the results using the duct run having the highest resistance. The friction loss through all elbows and fittings in the section must be included. The total friction loss in the duct having the highest resistance is the loss the fan must overcome.

Inspection of the duct layout, Fig. 6, shows that the duct run from the fan to outlet 18 probably has the highest resistance because it is the longest run. Tabulate the results as shown.

(1) Duct section	(2) Air quantity, cfm	(3) Cfm capacity, percent	(4) Duct, percent	(5) Area, sq ft	(6) Duct size, in.
Fan to A	36,000	100	100	14.4	46×45
A–B	24,000	67	73.5	10.6	39×39
B–13	12,000	33	41.0	5.9	30×29
13–14	10,000	28	35.5	5.1	27×27
14–15	8,000	22	29.5	4.3	25×25
15–16	6,000	17	24.0	3.5	23×22
16–17	4,000	11	17.5	2.5	20×18
17–18	2,000	6	10.5	1.5	15×15

The values in this tabulation are found as follows. Column 1 lists the longest duct run in the system. In column 2, the air leaving the outlets in branch A, or (6 outlets) (2,000 cfm per outlet) = 12,000 cfm, is subtracted from the quantity of air, 36,000 cfm, discharged by the fan to give the air quantity flowing from A–B. A similar procedure is followed for each successive duct and air quantity.

Column 3 is found by dividing the air quantity in each branch listed in columns 1 and 2 by 36,000, the total air flow, and multiplying the result by 100. Thus, for run B–13, column 3 = 12,000 (100)/36,000 = 33 percent.

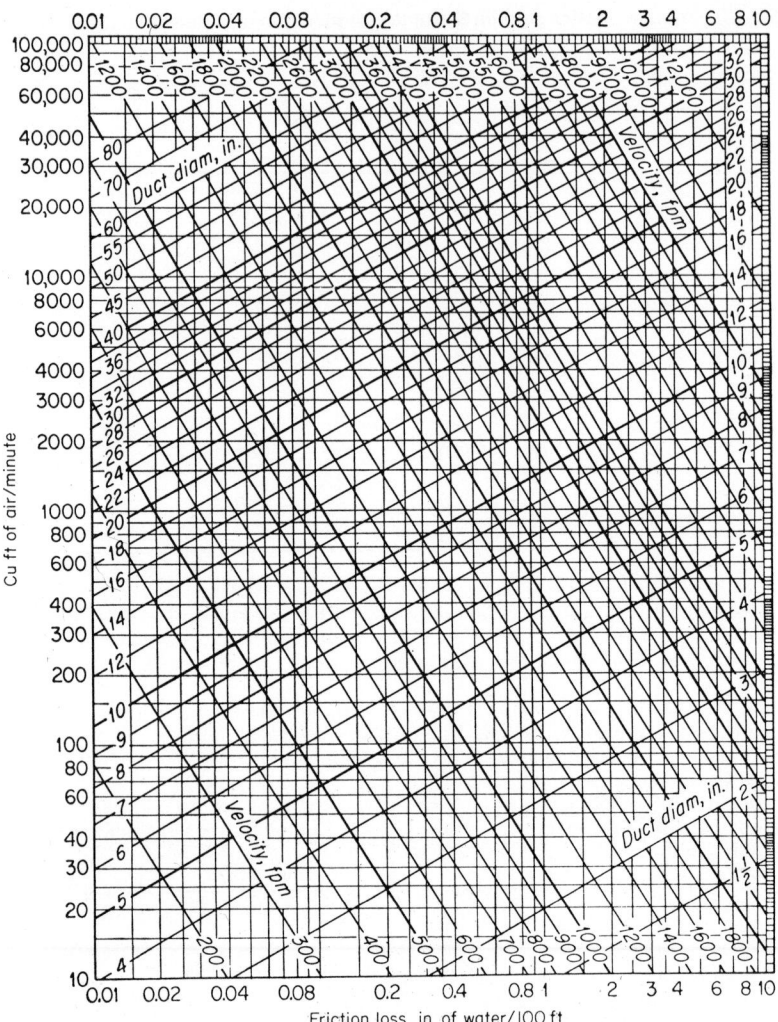

Friction loss for usual air conditions. This chart applies to smooth round galvanized iron ducts. See table below for corrections to apply when using other pipe.

Type of pipe	Degree of roughness	Velocity, fpm	Roughness factor (Use as multiplier)
Concrete.........	Medium rough	1,000–2,000	1.4
Riveted steel	Very rough	1,000–2,000	1.9
Tubing	Very smooth	1,000–2,000	0.9

Fig. 7   Friction loss in round ducts.

Column 4 values are found from Table 17. Enter that table with the cfm capacity from column 3 and read the duct area, percent. Thus, for branch 13–14 with 28 percent cfm capacity, the duct area from Table 17 is 35.5 percent. Determine the duct area, sq ft, column 6, by taking the product, line by line, of column 4 and the main duct area, sq ft. Thus, for branch 13–14, duct area = $(0.355)(14.4) = 5.1$ sq ft. Convert the duct area, sq ft, to a nearly square, or a square, duct by finding two dimensions that will produce the desired area.

Duct sections A through 6 and B through 12 have the same dimensions as the corresponding duct sections B through 18.

7. Find the total duct friction loss

Examination of the duct sketch, Fig. 6, indicates that the duct run from the fan to outlet 18 has the highest resistance. Compute the total duct run length and the equivalent length of the two elbows in the run thus as shown.

(1)	(2)	(3)	(4)
Duct section	System part	Length, ft	Elbow equivalent length, ft
Fan to A	Duct	60	
	Elbow	...	30
A–B	Duct	20	
B–13	Duct	30	
	Elbow	...	15
13–14	Duct	20	
14–15	Duct	20	
15–16	Duct	20	
16–17	Duct	20	
17–18	Duct	20	
Total	...	210	45

Note several facts about this calculation. The duct lengths, column 3, are determined from the system sketch, Fig. 6. The equivalent length of the duct elbows, column 4, is determined from the *Guide* or Carrier *Handbook of Air Conditioning Design*. The total equivalent duct length = column 3 + column 4 = $210 + 45 = 255$ ft.

8. Compute the duct friction loss

Use the general relation $h_T = Lf$, where h_T = total friction loss in duct, in. wg; L = total equivalent duct length, ft; f = friction loss for the system, in. wg per 100 ft. With the friction loss of 0.13 in. wg per 100 ft, as determined in Step 5, $h_T = (229/100)(0.13) = 0.2977$ in. wg; say 0.30 in. wg.

9. Determine the required fan static discharge pressure

The total static pressure required at the fan discharge = outlet operating pressure + duct loss − velocity regain between first and last sections of the duct, all expressed in in. wg. The first two variables in this relation are already known. Hence, only the velocity regain need be computed.

The velocity, v fpm, of air in any duct is v = cfm/duct area, sq ft. For duct section A, $v = 36,000/14.4 = 2,500$ fpm; for the last duct section, 17–18, $v = 2,000/1.5 = 1,333$ fpm.

When the fan discharge velocity is higher than the duct velocity in an air-conditioning system, use this relation to compute the static pressure regain; $R = 0.75 [(v_f/4,000)^2 - (v_d/4,000)^2]$, where R = regain, in. wg; v_f = fan outlet velocity, fpm; v_d = duct velocity, fpm. Thus, for this system, $R = 0.75 [(2,500/4,000)^2 - (1,333/4,000)^2] = 0.21$ in. wg.

With the regain known, compute the total static pressure required as $0.20 + 0.30 - 0.21 = 0.29$ in. wg. A fan having a static discharge pressure of at least 0.30 in. wg would probably be chosen for this system.

If the fan outlet velocity exceeded the air velocity in duct section A, the air velocity in this section would be used instead of the air velocity in the last duct section. Thus, in this circumstance, the last section becomes the duct connected to the fan outlet.

Related Calculations: Where the velocity in the fan outlet duct is *higher* than the fan outlet velocity, use the relation $l = 1.1 [(v_d/4,000)^2 - (v_f/4,000)^2]$, where l = loss, in. wg. This loss is the additional static pressure required of the fan. Hence, this loss must be *added* to the outlet operating pressure and the duct loss to determine the total static pressure required at the fan discharge.

The equal-friction method does not satisfy the design criteria of uniform static pressure at all branches and air terminals. To obtain the proper air quantity at the beginning of each branch, it is necessary to include a splitter damper to regulate the flow to the branch. It may also be necessary to have a control device (vanes, volume damper, or adjustable-terminal volume control) to regulate the flow at each terminal for proper air distribution.

The *velocity-reduction method* of duct design is not too popular because it requires a broad background of duct-design experience and knowledge to be within reasonable accuracy. It should be used for only the simplest layouts. Splitters and dampers should be included for balancing purposes.

To apply the velocity-reduction method: (1) Select a starting velocity at the fan discharge. (2) Make arbitrary reductions in velocity down the duct run. The starting velocity should not exceed the values in Table 16. Obtain the equivalent round-duct diameter from Fig. 7. Compute the required duct area from the round-duct diameter, and from this the duct dimensions, as shown in Steps 4 and 5 above. (3) Determine the required fan static discharge pressure for the supply by using the longest run of duct, including all elbows and fittings. Note, however, that the longest run is not necessarily the run with the greatest friction loss, as shorter runs may have more elbows, fittings, and restrictions.

The equal-friction and velocity-reduction methods of air-conditioning-system-duct design are applicable only to low-velocity systems—i.e., systems in which the maximum air velocity is 3,000 fpm, or less. The methods presented in this Calculation Procedure are those used by the Carrier Air Conditioning Company at the time of this writing.

SIZING LOW-VELOCITY AIR-CONDITIONING DUCTS— STATIC-REGAIN METHOD

Using the same data as in the previous Calculation Procedure, an air velocity of 2,500 fpm in the main duct section, an unvaned elbow radius of $R/D = 1.25$, and an operating pressure of 0.20 in. wg for each outlet, size the system ducts using the static-regain method of design for low-velocity systems.

Calculation Procedure:

1. Compute the fan outlet duct size

The fan outlet duct, also called the main duct section, will have an air velocity of 2,500 fpm. Hence, the required duct area is A = cfm/fpm = 36,000/2,500 = 14.4 sq ft. This corresponds to a round-duct diameter of $d = 2(144A/\pi)^{0.5} = 2(144 \times 14.4/\pi)^{0.5} = 51.5$ in. A nearly square duct, i.e., a duct 46 × 45 in. has an area of 14.38 sq ft and is a good first choice for this system because it closely approximates the outlet size of a standard centrifugal fan.

Where possible, use a square main duct to simplify fan connections. Thus, a 46 × 46-in. duct might be the final choice of this system.

2. Compute the main-duct friction loss

Using Fig. 7, find the main-duct friction loss as 0.13 in. wg per 100 ft of equivalent duct length for a flow of 36,000 cfm and a diameter of 51.5 in.

3. Determine the friction loss up to the first branch duct

The length of the main duct between the fan and the first branch is 25 + 35 = 60 ft. The equivalent length of the elbow is, from the *Guide* or Carrier *Handbook of Air*

Conditioning Design, 26 ft. Hence, the total equivalent length = 60 + 30 = 90 ft. The friction loss is then $h_T = Lf = (90/100)(0.13) = 0.117$ in. wg.

4. Size the longest duct run

The longest duct run is from A to outlet 18, Fig. 6. Size this duct using the following tabulation, preparing it as described below.

(1) Section number	(2) Air flow, cfm	(3) Equivalent length, ft	(4) L/Q ratio	(5) Velocity, fpm	(6) Duct area, sq ft (2)/(5)	(7) Duct size, in.
Fan to A	36,000	86	. . .	2,500	14.4	46 × 45
A–B	24,000	20	0.034	2,410	9.95	38 × 38
B–13	12,000	26*	0.088	2,200	5.45	28 × 28
13–14	10,000	20	0.072	2,040	4.90	27 × 27
14–15	8,000	20	0.083	1,850	4.33	25 × 25
15–16	6,000	20	0.098	1,700	3.53	24 × 23
16–17	4,000	20	0.130	1,520	2.63	20 × 19
17–18	2,000	20	0.195	1,300	1.54	15 × 15
B–7	12,000	25*	. . .	. . .	. . .	28 × 28
7–8	10,000	20	. . .	. . .	. . .	27 × 27
8–9	8,000	20	. . .	. . .	. . .	25 × 25
9–10	6,000	20	. . .	. . .	. . .	24 × 23
10–11	4,000	20	. . .	. . .	. . .	20 × 19
11–12	2,000	20	. . .	. . .	. . .	15 × 15
A–1	12,000	25*	. . .	. . .	. . .	28 × 28
1–2	10,000	20	. . .	. . .	. . .	27 × 27
2–3	8,000	20	. . .	. . .	. . .	25 × 25
3–4	6,000	20	. . .	. . .	. . .	24 × 23
4–5	4,000	20	. . .	. . .	. . .	20 × 19
5–6	2,000	20	. . .	. . .	. . .	15 × 15

*See text

List in column 1 the various duct sections in the longest duct run, as shown in Fig. 6. In column 2 list the air quantity flowing through each duct section. Tabulate in column 3 the equivalent length of each duct. Where a fitting is in the duct section, as in B–13, assume a duct size and compute the equivalent length using the *Guide* or Carrier fitting table. When the duct section does not have a fitting, as with section 13–14, the equivalent length equals the distance between the centerlines of two adjacent outlets.

Next, determine the L/Q ratio for each duct section using Fig. 8. Enter Fig. 8 at the air quantity in the duct and project vertically upward to the curve representing the equivalent length of the duct. At the left read the L/Q ratio for this section of the duct. Thus, for duct section 13–14, $Q = 10,000$ cfm and $L = 20$ ft. Entering the chart as detailed above shows that $L/Q = 0.72$. Proceed in this manner, determining the L/Q ratio for each section of the duct in the longest duct run.

Determine the velocity of the air in the duct by using Fig. 9. Enter Fig. 9 at the L/Q ratio for the duct section—say 0.072 for section 13–14. Find the intersection of the L/Q curve with the velocity curve for the preceeding duct section—2,200 fpm for section 13–14. At the bottom of Fig. 9 read the velocity in the duct section—i.e., after the previous outlet and in the duct section under consideration. Enter this velocity in column 5. Proceed in this manner, determining the velocity in each section of the duct in the longest duct run.

Determine the required duct area, sq ft, from cfm/fpm, or column 2/column 5, and insert the result in column 6. Find the duct size, column 7, by converting the required

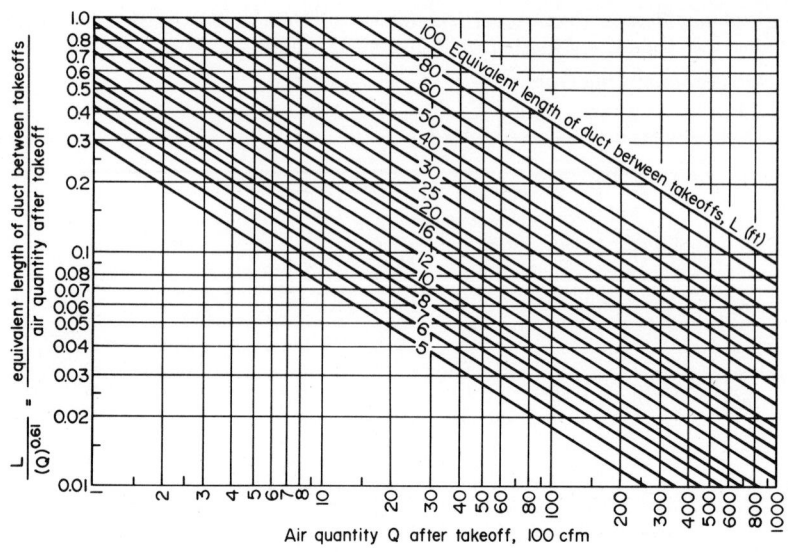

Fig. 8 L/Q **ratio for air ducts.** (*Carrier Air Conditioning Company.*)

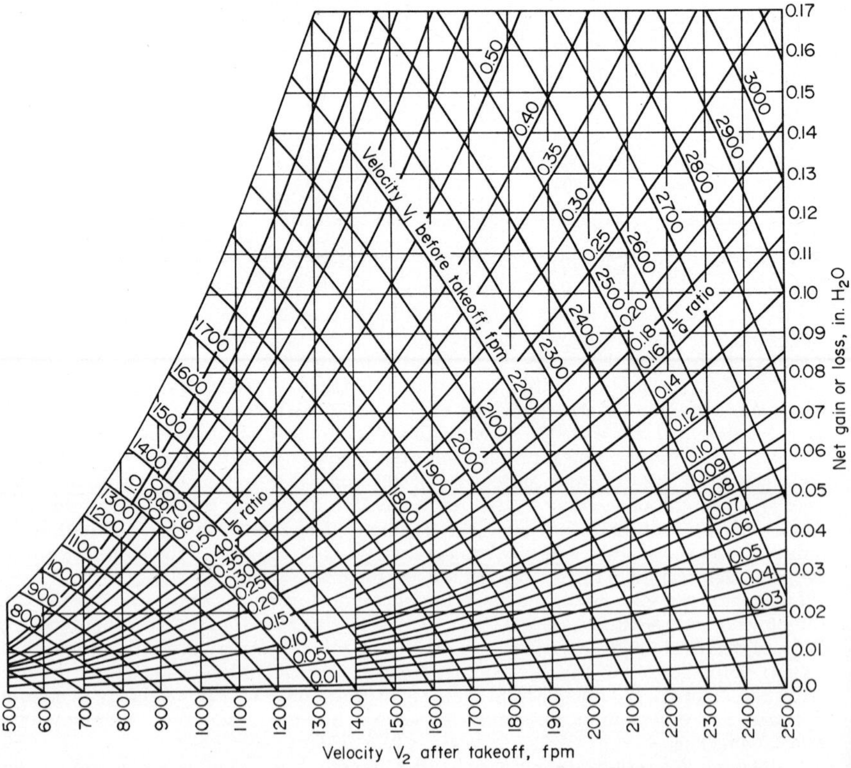

Fig. 9   **Low-velocity static regain in air ducts.** (*Carrier Air Conditioning Company.*)

duct area to a square- or rectangular-duct dimension. Thus a 27 × 17-in. square duct has a cross-sectional area slightly greater than 4.90 sq ft.

4. Determine the sizes of the other ducts in the system

Since the ducts in runs *A* and *B* are symmetrical with the duct containing the outlets in the longest run, they can be given the same size when the same quantity of air flows through them. Thus, duct section 7–8 is sized the same as section 13–14 because the same quantity of air — 10,000 cfm — is flowing through both sections.

Where the duct section contains a fitting — as *B*–7 and *A*–1 — assume a duct size and find the equivalent length using the *Guide* or Carrier fitting table. These sections are marked with an asterisk.

5. Determine the required fan discharge pressure

The total pressure required at the fan discharge equals the sum of the friction loss in the main duct plus the terminal operating pressure. Hence the required fan static discharge pressure = 0.117 + 0.20 = 0.317 in. wg.

Related Calculations: The basic principle of the static-regain method is to size a duct run so that the increase in static pressure (regain due to the reduction in velocity) at each branch or air terminal just offsets the friction loss in the succeeding section of duct. The static pressure is then the same before each terminal and at each branch.

As a *general* guide to the results obtained with the static-regain and equal-friction duct-design methods, the following should be helpful.

	Static regain	Equal friction
Main-duct sizes	Same	Same
Branch-duct sizes	Larger	Smaller
Sheet-metal weight . .	Greater	Less
Fan horsepower	Less	Greater
Balancing time	Less	Greater
Operating costs	Less	Greater

Note that these tabulated results are *general* and may not necessarily apply to every system. The method presented in this Calculation Procedure is that used by the Carrier Air Conditioning Company at the time of this writing.

HUMIDIFIER SELECTION FOR DESIRED ATMOSPHERIC CONDITIONS

A paper mill has a storeroom with a volume of 500,000 cu ft. The lowest recorded outdoor temperature in the mill locality is 0 F. What capacity humidifier is required for this storeroom if a 70-F dry-bulb temperature and a 65 percent relative humidity are required in it? Moisture absorption by the paper products in the room is estimated to be 450 lb/hr. The storeroom ventilating system produces three air changes per hour. What capacity humidifier is required if the room temperature is maintained at 60 F and 65 percent relative humidity? The products release 400 lb/hr of moisture. Steam at 25 psig is available for humidification. The outdoor air has a relative humidity of 50 percent and a minimum temperature of 5 F.

Calculation Procedure:

1. Determine the outdoor design temperature

When choosing a humidifier, the usual procedure is to add 10 F to the minimum outdoor recorded temperature because this temperature level seldom lasts more than a few hours. The result is the design outdoor temperature. Thus, for this mill, design outdoor temperature = 0 + 10 = 10 F.

TABLE 18 Steam required for humidification at 70 F*

(Pounds of steam per hour required per 1,000 cu ft of space to secure desired indoor relative humidity at 70 F, with various outdoor temperatures. Assuming two air changes per hour and outdoor relative humidity of 75 percent.)

Outdoor temp, F	Relative humidity desired indoors, percent														
	25	30	35	40	45	50	55	60	65	70	75	80	85	90	95
50	...	...	...	0.045	0.155	0.271	0.386	0.501	0.616	0.731	0.846	0.961	1.076	1.191	1.306
40	...	0.077	0.192	0.307	0.422	0.537	0.652	0.767	0.882	1.000	1.112	1.228	1.343	1.458	1.573
30	0.158	0.273	0.388	0.503	0.619	0.734	0.849	0.964	1.079	1.194	1.309	1.424	1.539	1.654	1.769
20	0.309	0.424	0.539	0.654	0.770	0.883	0.999	1.115	1.230	1.345	1.460	1.575	1.690	1.805	1.920
10	0.409	0.524	0.639	0.754	0.869	0.985	1.099	1.215	1.330	1.445	1.560	1.675	1.790	1.905	2.020
0	0.474	0.589	0.704	0.819	0.934	1.049	1.164	1.279	1.394	1.509	1.625	1.740	1.854	1.970	2.085
−10	0.514	0.629	0.744	0.860	0.975	1.090	1.205	1.320	1.435	1.550	1.665	1.780	1.895	2.010	2.125
−20	0.540	0.655	0.770	0.885	1.000	1.115	1.230	1.345	1.460	1.575	1.689	1.805	1.921	2.036	2.150

*Armstrong Machine Works

2. *Compute the weight of moisture required for humidification*

Enter Table 18 at an outdoor temperature of 10 F and project across to the desired relative humidity, 65 percent. Read the quantity of steam required as 1.330 lb per hr per 1,000 cu ft of room volume for two air changes per hour. Since this room has three air changes per hour, the quantity of moisture required is $(3/2)(1.330) = 1.995$ lb per hr per 1,000 cu ft of volume.

The amount of moisture in the form of steam required for this storeroom = (room volume, cu ft/1,000)(lb of steam per hr per 1,000 cu ft) = $(500,000/1,000)(1.995) =$ 997.5 lb for humidification of the air. However, the products in the storeroom absorb 450 lb of moisture per hr. Hence, the total moisture quantity required = moisture for air humidification + moisture absorbed by products = $997.5 + 450.0 = 1,447.5$ lb/hr; say 1,450 lb/hr for humidifier sizing purposes.

3. *Select a suitable humidifier*

Table 19 lists typical capacities for humidifiers having orifices of various sizes and different steam pressures. Study of Table 19 shows that one $1\frac{1}{4}$-in. orifice humidifier and two $\frac{3}{8}$-in. orifice humidifiers will discharge $1,130 + (2)(174) = 1,478$ lb/hr of steam when the steam supply pressure is 25 psig. Since the required capacity is 1,450 lb/hr, these humidifiers may be acceptable.

TABLE 19   Humidifier Capacities*

Steam pressure, psig	Orifice size, in.					
	$\frac{17}{64}$	$\frac{7}{32}$	$\frac{7}{16}$	$\frac{3}{8}$	$1\frac{1}{4}$	$1\frac{7}{64}$
2	25	· · ·	60	· · ·	145	25
3	32	· · ·	76	· · ·	210	32
4	37	· · ·	88	· · ·	280	37
5	42	· · ·	100	· · ·	340	42
6	46	· · ·	110	· · ·	410	46
7	50	· · ·	118	· · ·	460	50
8	54	· · ·	126	· · ·	510	54
9	57	· · ·	133	· · ·	560	57
10	60	· · ·	140	· · ·	610	60
11	64	· · ·	147	· · ·	650	64
12	67	· · ·	153	· · ·	700	67
13	70	· · ·	160	· · ·	740	70
14	72	· · ·	165	· · ·	780	72
15	74	56	170	138	810	74
20	· · ·	65	· · ·	158	980	80
25	· · ·	71	· · ·	174	1,130	90
30	· · ·	76	'· · ·	190	1,280	100

*Armstrong Machine Works.

†Continuous discharge capacity with steam pressures as indicated. No allowance for pressure drop after solenoid valve opens.

Large-capacity steam humidifiers usually must depend on existing ducts or large floor-type unit heaters for distribution of the moisture. When such means of distribution are not available, choose a larger number of smaller-capacity humidifiers and arrange them as shown in Fig. 10c. Thus, if $\frac{3}{8}$-in.-orifice humidifiers were selected, the number required would be (moisture needed, lb/hr)/(humidifier capacity, lb/hr) = $1,450/174 = 8.33$, or 9 humidifiers.

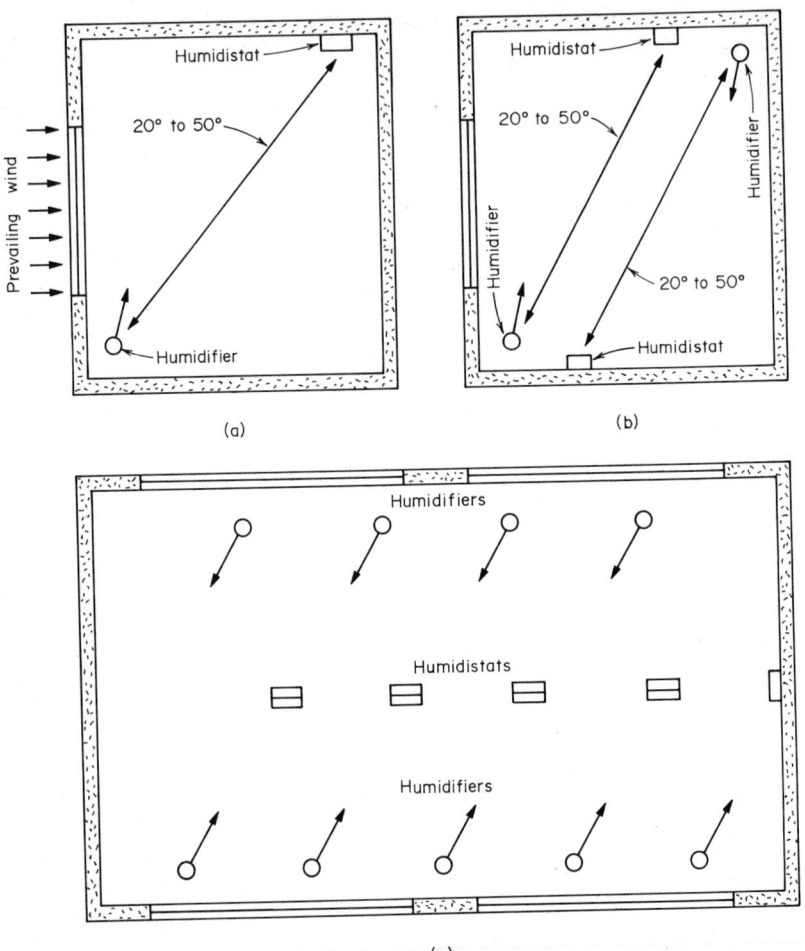

(a)   (b)

(c)

Fig. 10 *(a)* Location of a single humidifier; *(b)* location of two humidifiers; *(c)* location of multiple humidifiers.

4. Choose a humidifier for the other operating conditions

Where the desired room temperature is different from 70 F, use Table 20 instead of Table 18. Enter Table 20 at the desired room temperature, 60 F, and read the moisture content of saturated air at this temperature, as 5.795 grains/cu ft. The outdoor air at $5 + 10 = 15$ F contains, as Table 20 shows, 0.984 grains of moisture per cu ft when fully saturated.

Find the moisture content of the air at the room and the outdoor conditions from moisture content, grains/cu ft = (relative humidity of the air, expressed as a decimal) (moisture content of saturated air, grains/cu ft). For the 60-F, 65 percent relative humidity room air, moisture content = (0.65)(5.795) = 3.77 grains/cu ft. For the 15-F 50 percent relative humidity outdoor air, moisture content = (0.50)(0.984) = 0.492 grains/cu ft. Thus, the humidifier must add the difference, or 3.77 − 0.492 = 3.278 grains/cu ft.

This storeroom has a volume of 500,000 cu ft and three air changes per hour. Thus, the weight of moisture that must be added per hour is (number of air changes per hr) (volume, cu ft)(grains per cu ft of air)/7,000 grains/lb or, for this storeroom, (3)(500,000)

TABLE 20   Moisture content of saturated air
(Grains of water per cu ft of air)

F	Grains	F	Grains	F	Grains
−50	0.0283	25	1.558	100	19.95
−45	0.0386	30	1.946	105	22.95
−40	0.0525	35	2.376	110	26.34
−35	0.0708	40	2.863	115	30.13
−30	0.0945	45	3.436	120	34.38
−25	0.1261	50	4.106	125	39.13
−20	0.1662	55	4.889	130	44.41
−15	0.2182	60	5.795	135	50.30
−10	0.2847	65	6.845	140	56.81
− 5	0.3692	70	8.055	145	64.04
0	0.475	75	9.448	150	71.99
5	0.609	80	11.04	155	80.77
10	0.776	85	12.87	160	90.43
15	0.984	90	14.94	165	101.0
20	1.242	95	17.28	170	112.6

$(3.278)/7,000 = 701$ lb/hr, excluding the product load. Since the product load is 400 lb/hr, the total humidification load is $701 + 400 = 1,101$ lb/hr. Choose the humidifiers for these conditions in the same way as described in Step 4.

Related Calculations: Use the method given here to choose a humidifier for any normal industrial or comfort application. Table 21 summarizes typical recommended humidities and temperatures for a variety of industrial operations. The relative humidity maintained in industrial plants is extremely important because it can control the moisture content of hygroscopic materials.

Where the number of hourly air changes is not specified, assume two air changes, except in cotton mills where three or four may be necessary. If the plant ventilating system provides more than two air changes per hour, use the actual number of changes when computing the required humidifier capacity.

Many types of manufactured goods and raw materials absorb or release moisture during processing and storage. Since product quality usually depends directly on the moisture content, carefully controlled humidity will often reduce the number of rejects. The room humidifier must supply sufficient moisture for humidification of the air, plus any moisture absorbed by the products or materials in the room. Where these products or materials continuously release moisture to the atmosphere in the room, the quantity released can be subtracted from the moisture required for humidification. However, this condition can seldom be relied on. The usual procedure then is to select the humidifier on the basis of the moisture required for humidification of the air. The humidistat controls the operation of the humidifier, shutting it off when the products release enough moisture to supply the room requirements.

Correct locations for one or more humidifiers are shown in Fig. 10. Proper location of humidifiers is necessary if the design is to take advantage of the prevailing wind in the plant locality. Also, correct location provides a uniform, continuous circulation of air throughout the humidified area.

When only one humidifier is used, it is placed near the prevailing wind wall and arranged to discharge parallel to the wall exposed to the prevailing wind, Fig. 10a. Two humidifiers, Fig. 10b, are generally located in opposite corners of the manufacturing space and their discharges are used to produce a rotary air motion. Installations using more than two humidifiers generally have a slightly greater number of humidifiers on the windward wall to take advantage of the natural air drift from one side of the room to the other.

TABLE 21   Recommended Industrial Humidities and Temperatures

Industry	Degrees, F	Relative humidity, %
Ceramics:		
Drying refractory shapes...............	110–150	50–60
Molding room.......................	80	60
Confectionery:		
Chocolate covering...................	62–65	50–55
Hard-candy making..................	70–80	30–50
Storage............................	60–68	50–65
Electrical:		
Manufacture of cotton-covered wire.....	60–80	60–70
Storage, general.....................	60–80	35–50
Food storage:		
Apple	31–34	75–85
Citrus fruit..........................	32	80
Egg	30	80
Grain..............................	60	30–45
Meat ripening.......................	40	80
Sugar..............................	80	35
Paper products:		
Binding............................	70	45
Folding............................	77	65
Printing	75	60–78
Storage............................	75–80	40–60
Textile:		
Cotton carding	75–80	50–55
Cotton spinning.....................	60–80	50–70
Cotton weaving	68–75	85
Rayon spinning......................	70	85
Rayon throwing	70	60
Silk processing......................	75–80	60–70
Wool carding........................	75–80	65–70
Wool spinning.......................	75–80	55–60
Wool weaving.......................	75–80	50–55
Miscellaneous:		
Laboratory, analytical................	60–70	60–70
Munitions, fuse loading...............	70	55
Cigar and cigarette making...........	70–75	55–65

Pipe spray humidifiers as shown in Fig. 11 unless the manufacturer advises otherwise. Size the return lines as shown in Table 22.

Humidistats to start and stop the flow of moisture into the room may be either electrically or air (hygrostat) operated, according to the type of activities in the space. Where electric switches and circuits might cause a fire hazard, use an air-operated hygrostat instead of a humidistat. Locate either type of control to one side of the humidifying moisture stream, 20 to 50 ft away.

USE OF THE PSYCHROMETRIC CHART IN AIR-CONDITIONING CALCULATIONS

Determine the properties of air at 80-F dry-bulb (db) temperature and 65-F wet-bulb (wb) temperature using the psychrometric chart. Determine the same properties of air

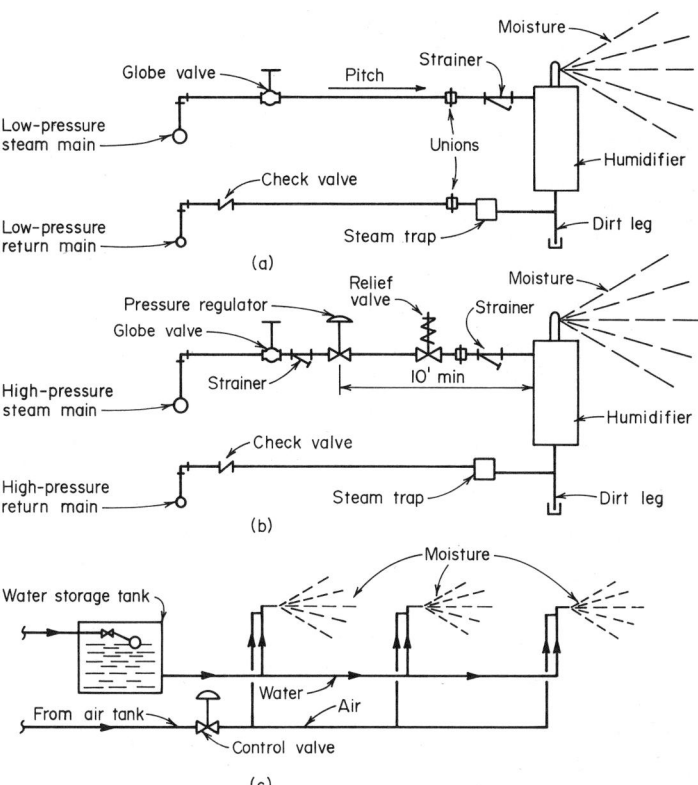

Fig. 11 Piping for spray-type humidifiers. (a) low-pressure steam; (b) high-pressure steam; (c) water spray.

TABLE 22   Steam- and Return-Pipe Sizes

Steam or condensate flow, lb/hr	Steam pressure, psig					Length of return pipe, ft	
	5	10	15	50	100	100	200
100	$1\frac{1}{2}$	$1\frac{1}{4}$	$1\frac{1}{4}$	1	1	1	1
150	2	$1\frac{1}{2}$	$1\frac{1}{2}$	$1\frac{1}{4}$	1	$1\frac{1}{4}$	$1\frac{1}{4}$
200	2	2	2	$1\frac{1}{4}$	$1\frac{1}{4}$	$1\frac{1}{4}$	$1\frac{1}{4}$
300	$2\frac{1}{2}$	2	2	$1\frac{1}{2}$	$1\frac{1}{4}$	$1\frac{1}{2}$	$1\frac{1}{2}$
400	3	$2\frac{1}{2}$	$2\frac{1}{2}$	2	$1\frac{1}{2}$	$1\frac{1}{2}$	2
500	3	$2\frac{1}{2}$	$2\frac{1}{2}$	2	$1\frac{1}{2}$	2	2
750	$3\frac{1}{2}$	3	3	$2\frac{1}{2}$	2	2	2
1,000	$3\frac{1}{2}$	3	3	$2\frac{1}{2}$	2	$2\frac{1}{2}$	$2\frac{1}{2}$
1,250	4	$3\frac{1}{2}$	$3\frac{1}{2}$	3	$2\frac{1}{2}$	$2\frac{1}{2}$	$2\frac{1}{2}$
1,500	5	$3\frac{1}{2}$	$3\frac{1}{2}$	3	$2\frac{1}{2}$	3	3
2,000	5	4	4	3	3	3	3
3,000	6	$4\frac{1}{2}$	$4\frac{1}{2}$	4	3	$3\frac{1}{2}$	$3\frac{1}{2}$
4,000	6	5	5	4	$3\frac{1}{2}$	4	4
5,000	8	6	6	5	4	4	5

if the wet-bulb temperature is 75 F and the dew-point temperature is 67 F. Show on the psychrometric chart an air-conditioning process in which outside air at 95-F db and 80-F wb is mixed with return air from the room at 80-F db and 65-F wb. Air leaves the conditioning apparatus at 55-F db and 50-F wb.

Calculation Procedure:

1. Determine the relative humidity of the air

Using Fig. 5, enter the bottom of the chart at the first dry-bulb temperature, 80 F, and project vertically upward until the slanting 65-F wet-bulb temperature line is intersected. At the intersection, or *state point*, read the relative humidity as 45 percent on the sloping curve. Note that the number representing the wet-bulb temperature appears on the saturation, or 100 percent relative humidity, curve and that the wet-bulb temperature line is a straight line sloping downward from left to right. The relative humidity curves slope upward from left to right and have the percent relative humidity marked on them.

When the wet-bulb and dew-point temperatures are given, enter the psychrometric chart at the wet-bulb temperature, 75 F, on the saturated curve. From here project downward along the wet-bulb temperature line until the horizontal line representing the dew-point temperature, 67 F, is intersected. At the intersection, or state point, read the dry-bulb temperature as 94.7 F on the bottom scale of the chart. Read the relative humidity at the intersection as 40.05 percent because the intersection is very close to the 40 percent relative humidity curve.

2. Determine the moisture content of the air

Read the moisture content of the air in grains on the right-hand scale by projecting horizontally from the intersection, or state point. Thus, for the first condition of 80-F dry bulb and 65-F wet bulb, projection to the right-hand scale gives a moisture content of 68.5 grains per lb of dry air.

For the second condition, 75-F wet bulb and 67-F dew point, projection to the right-hand scale gives a moisture content of 99.2 grains/lb.

3. Determine the dew point of the air

This applies to the first condition only because the dew point is known for the second condition. From the intersection of the dry-bulb temperature, 80 F, and the wet-bulb temperature, 65 F, that is, the state point, project horizontally to the left to read the dew point on the horizontal intersection with the saturation curve as 56.8 F. Note that the temperatures plotted along the saturation curve correspond to both the wet-bulb and dew point temperatures.

4. Determine the enthalpy of the air

Find the enthalpy (also called *total heat*) by reading the value on the sloping line on the central scale above the saturation curve at the state point for the air. Thus, for the first condition, 80-F dry bulb and 65-F wet bulb, the enthalpy is 30 Btu/lb. The enthalpy value on the psychrometric chart includes the heat of 1 lb of dry air and the heat of the moisture in the air — in this case, 68.5 grains of water vapor.

For the second condition, 75-F wet bulb and 67-F dew point, read the enthalpy as 38.5 Btu/lb at the state point.

5. Determine the specific volume of the air

The specific volume lines slope downward from left to right from the saturation curve to the horizontal dry-bulb temperature. Values of specific volume increase by 0.5 cu ft/lb between each line.

For the first condition, 80-F dry-bulb and 65-F wet-bulb temperature, the stage point lies just to the right of the 13.8 line, giving a specific volume of 13.81 cu ft/lb. For the second condition, 75-F wet-bulb and 67-F dew-point temperatures, the specific volume, read in the same way, is 14.28 cu ft/lb.

The weight of the air-vapor mixture can be found from $1.000 + 68.5$ grains per lb of air/(7,000 grains/lb) = 1.0098 lb for the first condition and $1.000 + 99.2/7,000 = 1.0142$

lb. In both these calculations the 1.000 lb represents the weight of the *dry* air and 68.5 and 99.2 grains represent the weight of the moisture for each condition.

6. *Determine the vapor pressure of the moisture in the air*

Read the vapor pressure by projecting horizontally from the state point to the extreme left-hand scale. Thus, for the first condition the pressure of the water vapor is 0.228 psi. For the second condition the pressure of the water vapor is 0.328 psi.

7. *Plot the air-conditioning process on the psychrometric chart*

Air-conditioning processes are conveniently represented on the psychrometric chart. To represent any process, locate the various state points on the chart and convert the points by means of lines representing the process.

Thus, for the air-conditioning process being considered here, start with the outside air at 95-F db and 80-F wb and plot point 1, Fig. 5, at the intersection of the two temperature lines. Next, plot point 3, the return air from the room at 80-F db and 65-F wb. Point 2 is obtained by computing the final temperature of two air streams that are mixed, using the method of the Calculation Procedure given earlier in this Section. Plot point 4 using the given leaving temperatures for the apparatus, 55-F db and 50-F wb.

The process in this system is as follows: Air is supplied to the conditioned space along line 4-3. During passage along this line on the chart, the air absorbs heat and moisture from the room. While passing from point 3 to 2, the air absorbs additional heat and moisture while mixing with the warmer outside air. From point 1 to 2, the outside air is cooled while it is mixed with the indoor air. At point 2, the air enters the conditioning apparatus, is cooled, and has its moisture content reduced.

Related Calculations: Use the psychrometric chart for all applied air-conditioning problems where graphic representation of the state of the air or a process will save time. At any given state point of air, the relative humidity in percent can be computed from (partial pressure of the water vapor at the dew-point temperature, psia + partial pressure of the water vapor at saturation corresponding to the dry-bulb temperature of the air, psia)(100). Determine the partial pressures from a table of air properties or from the steam tables.

In an *air washer* the temperature of the entering air is reduced. Well-designed air washers produce a leaving-air dry-bulb temperature that equals the wet-bulb and dew-point temperatures of the leaving air. The humidifier portion of an air-conditioning apparatus adds moisture to the air while the dehumidifier removes moisture from the air. In an ideal air washer, adiabatic cooling is assumed to occur.

Using the methods of Step 7, any basic air-conditioning process can be plotted on the psychrometric chart. Once a process is plotted, the state points for the air are easily determined from the psychrometric chart.

When making air-conditioning computations keep these facts in mind: (1) The total enthalpy, sometimes termed total heat, varies with the wet-bulb temperature of the air. (2) The sensible heat of air depends on the dry-bulb temperature of the air; the enthalpy of vaporization, also called the latent heat, depends on the dew-point temperature of the air; the dry-bulb, wet-bulb, and dew-point temperatures of air are the same for a saturated mixture. (3) The dew-point temperature of air is fixed by the amount of moisture present in the air.

DESIGNING HIGH-VELOCITY AIR-CONDITIONING DUCTS

Design a high-velocity air-distribution system for the duct arrangement shown in Fig. 12, if the required total air flow is 5,000 cfm.

Calculation Procedure:

1. *Determine the main-duct friction loss*

Many high-velocity air-conditioning systems are designed for a main-header velocity of 4,000 fpm and a friction loss of 1.0 in. H_2O per 100 ft of equivalent duct length.

The fan usualy discharges into a combined air-diffuser noise-attenuator in which the static pressure of the air increases. This pressure increase must be considered in the choice of the fan-outlet static pressure but the duct friction loss must be calculated first, as shown below.

Determine the main-duct friction loss by assuming a 1-in. per 100 ft static pressure loss for the main duct and a fan-outlet and main-duct velocity of 4,000 fpm. Size the duct by using the equal-friction method. Thus, for the 300-ft equivalent-length main duct in Fig. 12, the friction pressure loss will be (300 ft)(1.0 in. per 100 ft) = 3.0 in. H_2O.

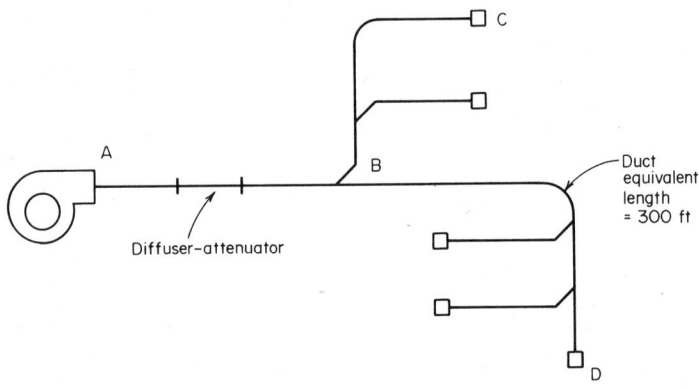

Fig. 12   High-velocity air-duct-system layout.

2.   *Compute the required fan-outlet pressure*

The total friction loss in the duct = duct friction, in. H_2O + diffuser-attenuator static pressure, in. H_2O. In typical installations the diffuser-attenuator static pressure varies from 0.3 to 0.5 in. H_2O. This is the inlet pressure required to force air through the diffuser-attenuator with all outlets open. Using a value of 0.5, the total friction loss in the duct = 3.0 + 0.5 = 3.5 in.

At the fan outlet the required static pressure is less than the total friction loss in the main duct because there is static regain at each branch takeoff to the outlets. This static regain is produced by the reduction in velocity that occurs at each takeoff from the main duct. There is a recovery of static pressure (velocity regain) at the takeoff that offsets the friction loss in the succeeding duct section.

Assume that the velocity in branch C, Fig. 12, is 2,000 fpm. This is the usual maximum velocity in takeoffs to terminals. Then, the maximum static regain that could occur $R = (v_i/4,005)^2 - (v_f/4,000)^2$, where R = static regain, in. of H_2O; v_i = initial velocity of the air, fpm; v_f = final velocity of the air, fpm. For this system with an initial velocity of 4,000 fpm and a final velocity of 2,000 fpm, $R = (4,000/4,005)^2 - (2,000/4,005)^2 = 0.75$.

The maximum static regain is seldom achieved. Actual static regains range from 0.5 to 0.8 of the maximum. Using a value of 0.8, the actual static regain = 0.8(0.75) = 0.60 in. H_2O. This static regain occurs at point B, the takeoff, and reduces the required fan discharge pressure to total friction loss in the duct − static regain at first takeoff = 3.5 − 0.60 = 2.9 in. H_2O. Thus, a fan developing a static discharge pressure of 3.0 in. H_2O would probably be chosen for this system.

3.   *Find the branch-duct pressure loss*

To find the branch-duct pressure loss, find the pressure in the main duct at the takeoff point. Use the standard duct-friction chart, Fig. 7, to determine the pressure loss from the fan to the takeoff point. Subtract the sum of this loss and the diffuser-attenuator static pressure from the fan static discharge pressure. The result is the pressure available to force air through the branch duct. Size the branch duct by using the equal-friction method.

Related Calculations: Note that the design of a high-velocity duct system (i.e., a system design in which the air velocities and static pressures are higher than in conventional systems) is basically the same as a low-velocity duct system designed for static regain. The air velocity is reduced at each takeoff to the riser and air terminals. Design of any high-velocity duct system involves a compromise between the reduced duct sizes (with a saving in materials, labor, and space costs) and higher fan horsepower.

Class II centrifugal fans, Table 23, are generally required for the higher static pressures used in high-velocity air-conditioning systems. Extra care must be taken in duct layout and construction. The high-velocity ducts are usually sealed to prevent air leakage that may cause objectionable noise. Round ducts are preferred to rectangular because of the greater rigidity of the round duct.

TABLE 23 Classes of Construction for Centrifugal Fans

Class	Maximum Total Pressure, in. H_2O
I	$3\frac{3}{4}$ – standard
II	$6\frac{3}{4}$ – standard
III	$12\frac{3}{4}$ – standard
IV	More than $12\frac{3}{4}$ – recommended

Use as many symmetrical duct runs as possible when designing high-velocity duct systems. The greater the system symmetry, the less time required for duct design, layout, balancing, construction, and installation.

The initial starting velocity used in the supply header depends on the number of hours of operation. To achieve an economic balance between first cost and operating cost, lower air velocities in the header are recommended for 24-hr operation, where space permits. Table 24 shows typical air velocities used in high-velocity air-conditioning systems. Use this tabulation to select suitable velocities for the main and branch ducts in high-velocity systems.

TABLE 24 Typical High-Velocity-System Air Velocities*

	Velocity, fpm
Header or main duct:	
12-hr operation	3,000–4,000
24-hr operation	2,000–3,500
Branch ducts:†	
90° conical tee	4,000–5,000
90° tee	3,500–4,000

*Carrier Air Conditioning Company.
†Branches are defined as a branch header or riser having four to five, or more, takeoffs to terminals.

Carrier Air Conditioning Company recommends that the following factors be considered when laying out header ductwork for high-velocity air-conditioning systems.

1. The design friction losses from the fan discharge to a point immediately upstream of the first riser takeoff from each branch header should be as nearly equal as possible.

2. To satisfy principle 1 above when applied to multiple headers leaving the fan, and to take maximum advantage of the allowable high velocity, adhere to the following basic rule whenever possible: Make as nearly equal as possible the ratio of the total

equivalent length of each header run (fan discharge to the first riser takeoff) to the initial header diameter (L/D ratio). Thus, the longest header run should preferably have the highest air quantity so that the highest velocities can be used throughout.

3. Unless space conditions dictate otherwise, use a 90° tee or 90° conical tee for the takeoff from the header rather than a 45° tee. Fittings of 90° provide more uniform pressure drops to the branches throughout the system. Also, the first cost is lower.

AIR-CONDITIONING-SYSTEM OUTLET- AND RETURN-GRILLE SELECTION

Choose an air grille to deliver 425 cfm of air to a broadcast studio having a 12-ft ceiling height. The room is 10 ft long and 10 ft wide. Specify the temperature difference to use, the air velocity, grille static resistance, size, and face area.

Calculation Procedure:

1. Choose the outlet-grille velocity

The air velocity specified for an outlet grille is a function of the type of room in which the grille is used. Table 25 lists typical maximum outlet-air velocities used in grilles serving various types of rooms. Assuming a velocity of 350 fpm for the outlet grille in this broadcast studio, compute the grille area required from $A = cfm/v$, where $A =$ grille area, sq ft; $cfm =$ air flow through the grille, cm; $v =$ air velocity, fpm. Hence, $A = 425/350 = 1.214$ sq ft.

Table 25   Typical Air-Outlet Velocities

Type of Room	Maximum Velocity, fpm
Broadcasting studio	300–500
Apartments	500–750
Private residences	500–750
Churches	500–750
Hotel bedrooms	500–750
Legitimate theatres	500–750
Private offices	500–750
Movie theaters	1,000
General offices	1,200–1,500
Stores – upper floors	1,500
Stores – main floors	2,000

2. Select the outlet-grille size

Use the selected manufacturer's engineering data, such as that in Table 26. Examination of this table shows that there is no grille rated at 425 cfm. Hence, the next larger

Table 26   Air-Grille-Selection Table*

Air flow, cmf	Wall area per outlet, sq ft		Throw, ft†	Min. ceiling height, ft, for temp. difference of			Air velocity, fpm	Grille static resist., in., H$_2$O	Outlet size, in.	Grille face area	
	Max.	Min.		15 F	10 F	25 F				Sq ft	Sq in.
306	25	8	S – 11	13	14	15	250	0.005	24 × 8	1.224	176
			F – 6	10	10	10					
459	57	17	S – 20	16	17	18	375	0.01	24 × 8	1.224	176
			F – 10	10	10	11					

*Waterloo Register
†S – straight; F – fan spread.

capacity, 459 cfm, must be used. This grille, as the third column from the right of Table 26 shows, is 24 in. wide and 8 in. high.

3. Choose the grille throw distance

Throw is the horizontal distance the air will travel after leaving the grille. With a *fan spread*, the throw of this grille is 10 ft, Table 26. This throw is sufficient if the duct containing the grille is located at any point in the room—i.e., along one wall, in the center, etc. If desired, the grille can be adjusted to reduce the throw, but the throw cannot be increased beyond the distance tabulated. Hence, a fan-spread grille will be used.

4. Select the grille-mounting height

The grille-mounting height is a function of several factors—the difference between the temperature of the entering air and the room air, the room-ceiling height, and the air *drop*, i.e., the distance the air falls from the time it passes through the outlet until it reaches the end of the throw.

Assuming a temperature difference of 20 F between the entering air and the room air, Table 26 shows that the minimum ceiling height for this grille is 10 ft. Since the room is 12 ft high, the grille can be mounted at any distance above the floor of 10 ft or higher.

5. Determine the actual air velocity in the grille

Table 26 shows that the actual air velocity in the grille is 375 fpm. Table 25 shows that an air velocity of 300 to 500 fpm is suitable for broadcast studios. Hence, this grille is acceptable. If the actual velocity at the grille outlet is higher than that recommended in Table 25, a larger grille giving a velocity within the recommended range would have to be chosen.

6. Determine the grille static resistance

Table 26 shows that the grille static resistance is 0.01 in. H_2O. This is within the usual static resistance range of outlet grilles.

7. Determine the outlet-grille area

Table 26 shows that the outlet grille has an area of 1.224 sq ft, or 176 sq in. This agrees well with the area computed in Step 1, or 1.214 sq ft.

8. Select the air-return grille

Table 27 shows typical air velocities used for return grilles in various locations. Assuming that the air is returned through a wall louvre, a velocity of 500 fpm might be used. Hence, using the equation of Step 1, grille area $A = cfm/v = 425/500 = 0.85$ sq ft.

If a lattice-type return intake having a free area of 60 percent is used, Table 28 shows that the pressure drop during passage of the air through the grille is 0.04 in. H_2O. Locate the return grille away from the supply grille to prevent short circuiting of the air and excessive noise. The pressure losses in Table 27 are typical for return grilles. Choice of the pressure drop to use is generally left with the system designer.

Related Calculations: Use this general method to choose outlet and return grilles for industrial, commercial, and domestic applications. Be certain not to exceed the tabulated velocities where noise is a factor in an installation. Excessive noise can lead to complaints from the room occupants.

The outlet-table excerpt presented here is typical of the table arrangement used by many manufacturers. Hence, the general procedure given for selecting an outlet is similar to that for any other manufacturer's outlet.

Many modern-design ceiling outlets are built so that the leaving air entrains some of the room air. The air being discharged by the outlet is termed *primary air* and the room air is termed *secondary air*. The induction ratio R_i = total air, cfm/primary air, cfm. Typical induction ratios run in the range of 30 percent.

For a given room, the total air in circulation, cfm = (outlet cfm)(induction ratio). Also, average room air velocity, fpm = 1.4 (total cfm in circulation)/area of wall, sq ft, opposite the outlet or outlets. The wall area in the last equation is the *clear* wall area. Any obstructions must be deducted. The multiplier 1.4 allows for blocking caused by the air stream. Where the room circulation factor K must be computed, use the relation K = average room air velocity, fpm/1.4(induction ratio). The ideal room-air velocity for most applications is 25 fpm. However, velocities up to 300 fpm are used in some factory air-conditioning applications.

TABLE 27 Lattice-Type Return-Grille Pressure Drop, in. H_2O

Free area of grille, percent	Face velocity, fpm		
	400	500	600
50	0.04	0.06	0.09
60	0.03	0.04	0.06
70	0.02	0.03	0.05
80	0.01	0.02	0.03

Return-intake-air velocities*

Intake location	Velocity over gross area, fpm
Above occupied zone	800 and up
In occupied zone:	
Not near seats	600–800
Near seats...........................	400–600
Door or wall louvers	500–700
Undercut door (through undercut area)...	600

*ASHRAE *Guide*

The types of outlets commonly used today are grille (perforated, fixed-bar, adjustable-bar), slotted, ejector, internal induction, pan, diffuser, and perforated ceiling. Choice of a given type depends on the room ceiling height, desired air-temperature difference, blow, drop, and spread, as well as other factors that are a function of the room, air quantity, and the activities in the room.

As a general guide to outlet selection, use the following pointers: (1) Choose the number of outlets for each room after considering the quantity of air required, throw or diffusion distance available, ceiling height, obstructions, etc. (2) Try to arrange the outlets symmetrically in the space available as shown by the room floor plan.

TABLE 28   Approximate Pressure Drop for Lattice Return Intakes*

Percent free area	Face velocity, fpm						
	400	500	600	700	800	900	1,000
50	0.06	0.09	0.13	0.17	0.22	0.28	0.35
60	0.04	0.06	0.09	0.12	0.16	0.20	0.24
70	0.03	0.05	0.07	0.09	0.12	0.15	0.18
80	0.02	0.03	0.05	0.07	0.09	0.11	0.14

*ASHRAE *Guide and Data Book*

SELECTING ROOF VENTILATORS FOR BUILDINGS

A 10-bay building is 200 ft long, 100 ft wide, 50 ft high to the top of the pitched roof, and 35 ft high to the eaves. The building houses 15 turbine-driven generators and is classed as an engine room. Choose enough roof ventilators to produce a suitable number of air changes in the building. During reduced-load operating periods between 12 midnight and 7 A.M. on weekdays, and on weekends, only half the full-load air changes are required. The prevailing summer-wind velocity against the long side of the building is 10 mph. The total available open-window area on each long side is 300 sq ft. The minimum difference between the outdoor and indoor temperatures will be 40 F.

Calculation Procedure:

1. Determine the cubic volume of the building

When computing the cubic volume of a pitched-roof building, the usual procedure is to assume an average height from the eaves to the ridge. Since this building has a 15-ft-high ridge from the eaves, the average height = 15/2 = 7.5 ft. As the height from the ground to the eaves is 35 ft, the building height to be used in the volume computation is $35 + 7.5 = 42.5$ ft. Hence, the volume of the building, cu ft = V = length × width × average height, all measured in ft = $200 \times 100 \times 42.5 = 850,000$ cu ft.

2. Determine the number of air changes required

Table 29 shows that four to six air changes per hr are normally recommended for engine rooms. Using five air changes per hr will probably be satisfactory, and the roof ventilators will be chosen on this basis. During the early morning, and on weekends, 2.5 air changes will be satisfactory, since only half the normal number of air changes are needed during these periods.

3. Compute the required hourly air flow

The required hourly air flow, cfh = Q = (number of air changes per hr)(building volume, cu ft) = $(5)(850,000) = 4,250,000$ cfh. During the early morning hours and on weekends when 2.5 air changes are used, $Q = 2.5(850,000) = 2,125,000$ cfh.

4. Compute the air flow produced by natural ventilation

The ASHRAE *Guide and Data Book* lists the prevailing winter- and summer-wind velocities for a variety of locations. Usual practice, when designing natural-ventilation systems, is to use one-half the tabulated wind velocity for the season being considered. Since summer ventilation is usually of greater importance than winter ventilation, one-half the prevailing summer-wind velocity is generally used in natural-ventilation calculations. As the prevailing summer-wind velocity in this locality is 8 mph, a velocity of $8/2 = 4$ mph will be used when computing the air flow produced by the wind.

TABLE 29 Number of Air Changes Required per Hour*

Auditoriums and assembly rooms ...	10–15	Libraries................	3
Boiler rooms	10–15	Machine shops...........	6
Churches	10–15	Paint shops..............	10–15
Engine rooms	4–6	Paper mills..............	15–20
Factory buildings (general)........	4	Pump rooms.............	8–10
Factory buildings (where excessive	15–20	Railroad shops...........	4
conditions of fumes, moisture, etc.,		Schools	10–12
are present)		Textile mills (general).....	4
Foundries.......................	12	Textile mill dye houses....	15–20
Garages.........................	10–15	Theaters	5–8
General offices	3	Waiting rooms...........	4
Hotel dining rooms	4	Warehouses.............	4
Hotel kitchens...................	10–20	Wood-working shops......	8
Laundries.......................	15–25		

*DeBothezat Fans Division, AMETEK Inc.

Use the relation $Q = VAE$ to find the air flow produced by the wind. In this relation, Q = air flow produced by the wind, cfm; V = design wind velocity, fpm = 88 × mph; A = free area of the air-inlet openings, sq ft; E = effectiveness of the air-inlet openings, − use 0.50 to 0.60 for openings perpendicular to the wind and 0.25 to 0.35 for diagonal winds.

Assuming $E = 0.50$, $Q = VAE = (4 \times 88)(300)(0.50) = 52{,}800$ cfm, or $60(52{,}800) = 3{,}168{,}000$ cfh. Step 3 shows that the required air flow is 4,250,000 cfh when all turbines are operating. Hence, the air flow produced by natural ventilation is inadequate for full-load operation. However, since the required flow of 2,125,000 cfh for the early-morning hours and weekends is less than the natural-ventilation flow of 3,168,000 cfh, natural ventilation may be acceptable during these periods.

5. *Determine the number of stationary-type ventilators needed*

A stationary-type roof ventilator − i.e., one that depends on the wind and air-temperature difference to produce the desired air movement − may be suitable for this application. If the stationary-type is not suitable, a powered-fan-type roof ventilator will be investigated and must be used. The Breidert-type ventilator will be investigated here because the procedure is similar to that used for other stationary-type roof ventilators.

Stationary ventilators produce air flow out of a building by two means: (*a*) suction caused by wind action across the ventilators, and (*b*) the stack effect caused by the temperature difference between the inside and outside air.

Figure 13 shows the air velocity produced in a stationary Breidert ventilator by winds of various velocities. Thus, using the average 5-mph wind assumed earlier for this building, Fig. 13 shows that the air velocity through the ventilator produced by this wind velocity is 220 fpm, closely.

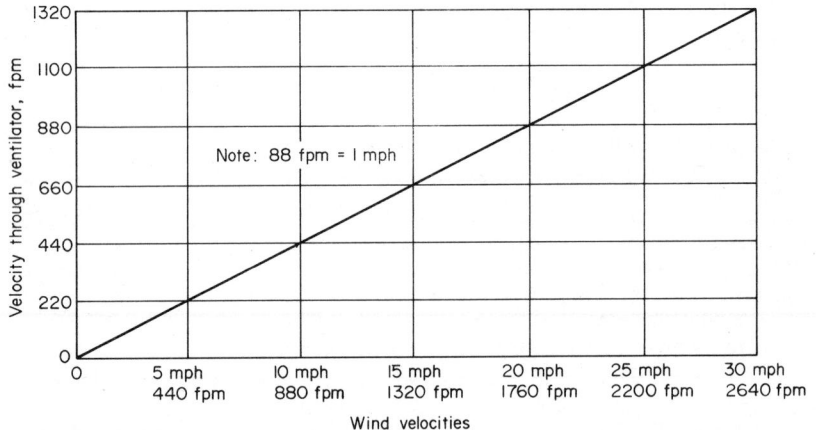

Wind velocities

Fig. 13   Roof-ventilator air-exhaust capacity for various wind velocities. Add the extra velocity for temperature difference given in Table 30. (*G. C. Breidert Co.*)

Table 30 shows that a 1.0 sq ft ventilator installed on a 50-ft-high building having an air-temperature difference of 40 F will produce, due to the stack effect, an air-flow velocity of 482 fpm. Hence, the total velocity through this ventilator resulting from the wind and stack action is $220 + 482 = 702$ fpm.

Since air flow, cfm = (air velocity, fpm)(area of ventilator opening, sq ft), an air flow of $(702)(1.0) = 702$ cfm will be produced by each sq ft of ventilator-neck or inlet-duct area. Thus, to produce a flow of 4,250,000 cfh/(60 min/hr) = 70,700 cfm will require a ventilator area of $70{,}700/702 = 101$ sq ft. A 48-in. Breidert ventilator has a neck area of 12.55 sq ft. Hence, a $101/12.55 = 8.05$ or, say eight ventilators will be required. Alternatively, the Breidert capacity table in the engineering data prepared by the manufacturer shows that a 48-in. ventilator has a ventilating capacity of 8,835 cfm when used for a

5-mph, 50-ft-high, 40-F temperature-difference application. Using this capacity, the number of ventilators required = 70,700/8,835 = 8.02; say eight ventilators. These ventilators will be suitable for both full- and part-load operation.

TABLE 30   Flow of Air in Natural-Draft Flues, cfm/sq ft*

Difference in temperature, F	Height of flue, ft, same as height of room or building								
	10	15	20	30	40	50	60	80	100
10	108	133	153	188	217	242	264	306	342
15	133	162	188	230	265	297	325	375	420
20	153	188	217	265	306	342	373	435	485
25	171	210	242	297	342	383	420	485	530
30	188	230	265	325	375	419	461	530	594
40	216	265	305	374	431	482	529	608	680
50	242	297	342	419	484	541	594	680	768
60	266	327	376	460	532	595	650	747	842

*G. C. Breidert Co.

6. Determine the number of powered ventilators needed

Powered ventilators are equipped with single- or two-speed fans to produce a positive air flow independent of wind velocity and stack effect. For this reason, some engineers prefer powered ventilators where it is essential that air movement out of the building be maintained at all times.

Two-speed powered ventilators are usually designed so that the reduced-speed rpm is approximately one-half the full-speed rpm. The air flow at half-speed is about one-half that at full speed.

Checking the capacity table of a typical powered-ventilator manufacturer shows that ventilator capacities range from about 2,100 cfm for a 21-in.-diameter unit at a ⅛-in. static pressure difference to about 24,000 cfm for a 36-in.-diameter ventilator at the same pressure difference. Using a 27-in.-diameter powered ventilator which has a capacity of 14,900 cfm, the number required is 70,700 cfm/14,900 cfm per unit = 4.76, or five.

7. Choose the type of ventilator to use

Either a stationary or powered ventilator might be chosen for this application. Since a large amount of heat is generated in an engine room, the powered ventilator would probably be a better choice because there would be less chance of overheating during periods of little or no wind.

Related Calculations: Use the general method given here when choosing stationary or powered ventilators for any of the 25 applications listed in Table 29. The usual practice is to locate one ventilator in each bay or sawtooth of a building.

VIBRATION-ISOLATOR SELECTION FOR AN AIR CONDITIONER

Choose a vibration isolator for a packaged air conditioner operating at 1,800 rpm. What minimum mounting deflection is required if the air conditioner is mounted on a basement floor? On an upper-story floor made of light concrete?

Calculation Procedure:

1. Determine the suggested isolation efficiency

Table 31 lists the suggested isolation efficiency for various components used in air-conditioning and refrigeration systems. This tabulation shows that the suggested isolation efficiency for a packaged air conditioner is 90 percent. This means that the vibration isolator or mounting should absorb 90 percent, or more, of the vibration caused by the machine. At this efficiency only 10 percent of the machine vibration would be transmitted to the supporting structure.

TABLE 31   Suggested Isolation Efficiencies

Equipment	*Installed Efficiency, %*
Absorption units.....................................	95
Steam generators....................................	95
Centrifugal compressors	98
Reciprocating compressors:	
Up to 15 hp..	85
20–60 hp..	90
75–150 hp...	95
Packaged air conditioners	90
Centrifugal fans:	
800 rpm and above; all diameters	90–95
350–800 rpm; all diameters	70–90
200–350 rpm; 48-in.diameter or smaller	*
200–350 rpm; 54-in.-diameter or larger..............	70–80
Centrifugal pumps..................................	95
Cooling towers	85
Condensers..	80
Fan coil units.......................................	80
Piping...	95

*Installed for noise isolation only.

2. *Determine the static deflection caused by the vibration*

Use Fig. 14 to find the static deflection caused by the vibration. Enter at the bottom of Fig. 14 at 1,800 rpm the disturbing frequency, and project vertically upward to the 90 percent efficiency curve. At the left read the static deflection as 0.11 in.

3. *Select the type of vibration isolator to use*

Project to the right from the intersection with the efficiency curve, Fig. 14, to read the type of isolator to use. Thus, neoprene pads or neoprene-in-shear mounts will safely absorb up to 0.25-in. static deflection. Hence, either type of isolator mounting could be used.

4. *Check the isolator selection*

Use Table 32 to check the theoretical isolation efficiency of the mounting chosen. Enter at the top at the rpm of the machine and project vertically downward until an efficiency equal to, or greater than, that desired is intersected.

For this machine operating at 1,800 rpm, single-deflection rubber mountings have an efficiency of 94 percent. Since neoprene is also called synthetic rubber, the isolator choice is acceptable because it yields a higher efficiency than required.

5. *Determine the minimum mounting deflection required*

Table 33 lists the minimum mounting deflection required at various operating speeds for machines installed on various types of floors. Thus, at 1,800 rpm, machines mounted on a basement floor must have isolator mountings that will absorb deflections up to 0.10 in. Since the neoprene mountings chosen in Step 3 will absorb up to 0.25 in. deflection, they will be acceptable for use on a basement-mounted machine.

For mounting on a light-concrete upper-story floor, Table 33 shows that the mounting must be able to absorb a deflection of 0.80 in. for machines operating at 1,800 rpm.

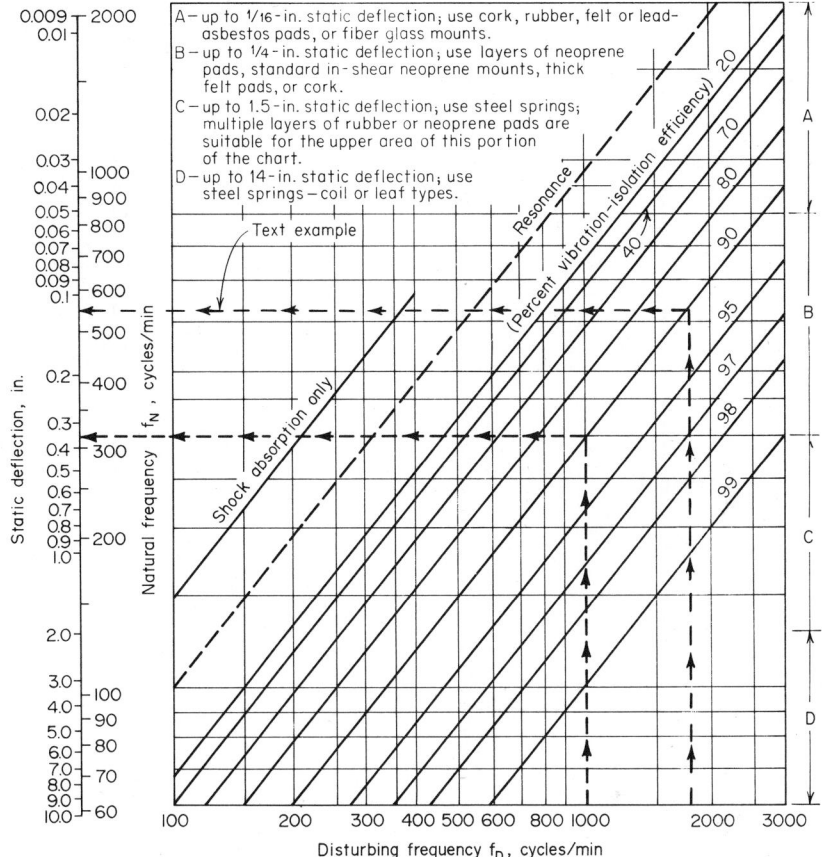

Fig. 14   Vibration-isolator deflection for various disturbing frequencies. (*Power.*)

Since the neoprene isolators can absorb only 0.25 in., another type of mounting is needed if the machine is installed on an upper floor. Figure 14 shows that steel springs will absorb up to 1.5 in. static deflection. Hence, this type of mounting would be used for machines installed on upper floors of the building.

Related Calculations: Use this general procedure for engines, compressors, turbines, pumps, fans and similar rotating and reciprocating equipment. Note that the suggested isolation efficiencies in Table 31 are for air-conditioning equipment located in critical areas of buildings—such as offices, hospitals, etc. In noncritical areas, such as basements or warehouses, an isolation efficiency of 70 percent may be acceptable. Note that the efficiencies given in Table 31 are useful as general guides for all types of rotating machinery.

SELECTION OF NOISE-REDUCTION MATERIALS

A concrete-walled test laboratory is 25 ft long, 20 ft wide, and 10 ft high. The laboratory is used for testing chipping hammers. What noise reduction can be achieved in this laboratory by lining it with acoustical materials?

TABLE 32   Theoretical Vibration-Isolation Efficiencies*

Isolation material	Average static deflection	Average natural frequency	Efficiencies, percent									
			350 rpm	500 rpm	600 rpm	800 rpm	1,000 rpm	1,200 rpm	1,500 rpm	1,800 rpm	3,000 rpm	3,600 rpm
2-in.-thick standard-density cork	0.08	By test 1,420	...	...	...	...	...	...	...	...	72	82
Type W waffle pad	Curvature corrected, 0.035	1,000	...	...	...	...	...	...	20	55	87	92
Two layers of W waffle pad	Curvature corrected, 0.070	710	...	...	...	...	...	46	71	82	93	96
Single-deflection rubber mountings	0.20	420	...	...	...	62	79	86	91	94	98	99
Double-deflection rubber mountings	0.40	300	...	44	67	84	90	93	96	97	99	Almost perfect
Standard spring mountings	1.00	188	70	85	89	94	96	97	98	99	Almost perfect	Almost perfect
Double-deflection rubber and spring mountings	1.40	160	75	89	93	96	97	98	99	Almost perfect	Almost perfect	Almost perfect

*Power magazine.

TABLE 33 Minimum Mounting Deflections

Operating speed,, rpm	Basement— negligible floor deflection, in.	Rigid concrete floor, in.	Upper story— light- concrete floor, in.	Wood floor, in.
300	1.50	3.00	3.50	4.00
500	0.63	1.25	1.65	1.95
800	0.25	0.60	1.00	1.25
1,200	0.20	0.45	0.80	1.00
1,800	0.10	0.35	0.80	1.00
3,600	0.03	0.20	0.80	1.00
7,200	0.03	0.20	0.80	1.00

TABLE 34 Power and Intensity of Noise Sources

Sound source	Power range, watts	Decibel range (10^{-13} watts)
Ram jet..........................	100,000.0	180
Turbojet with 7,000-lb thrust	10,000.0	170
	1000.0	160
4-propeller airliner................	100.0	150
75-piece orchestra, pipe organ; small aircraft engine............	10.0	140
Chipping hammer..............	1.0	130
Piano, blaring radio...............	0.1	120
Centrifugal ventilating fan at 13,000 cfm...................	0.01	110
Automobile on roadway; vane-axial ventilating fan.........	0.001	100
Subway car, air drill	0.0001	90
Conversational voice; traffic on street corner............	0.000,01	80
Street noise, average radio	0.000,001	70
Typical office.....................	0.000,000,1	60
	0.000,000,01	50
Very soft whisper	0.000,000,001	40

Reference values relate decibel scales

Db scale	Definition	Reference quantity
Sound-power level	$PWL = 10 \log \dfrac{W}{W \, re}$	$W \, re = 10^{-13}$ watt
Sound-intensity level	$IL = 10 \log \dfrac{I}{I \, re}$	$I \, re = 10^{-12}$ watt/m² = 10^{-16} watt/cm²
Sound-pressure level	$SPL = 10 \log \dfrac{P^2}{P^2 \, re} = 20 \log \dfrac{P}{P \, re}$	$P \, re = 0.000,02$ newton/m² = 0.0002 microbar = 0.0002 dyne/cm²

Power magazine.

Calculation Procedure:

1. Determine the noise level of devices in the room

Table 34 shows that a chipping hammer produces noise in the 130-db range. Hence, the noise level of this room can be assumed to be 130 db. This is rated as deafening by various authorities. Therefore, some kind of sound-absorption material is needed in this room if the uninsulated walls do not absorb enough sound.

2. Compute the total sound absorption of the room

The *sound-absorption coefficient* of bare concrete is 0.1. This means that 1 percent of the sound produced in the room is absorbed by the bare concrete walls.

To find the total sound absorption by the walls and ceiling, find the product of the total area exposed to the sound and the sound-absorption coefficient of the material. Thus, concrete area, excluding the floor but including the ceiling = two walls (25 ft long × 10 ft high) + two walls (20 ft wide × 10 ft high) + one ceiling (25 ft long × 20 ft wide) = 1,400 sq ft. Then, the total sound absorption = (1,400)(0.1) = 140.

3. Compute the total sound absorption with acoustical materials

Table 35 lists the sound- or noise-reduction coefficients for various acoustical materials. Assume that the four walls and ceiling are insulated with membrane-faced mineral-fiber tile having a sound-absorption coefficient of 0.90, from Table 35.

Then, using the procedure of Step 2, total noise reduction = (1,400 sq ft)(0.90) = 1,260.

4. Compute the noise reduction resulting from insulation use

Use the relation noise reduction, db = 10 log (total absorption *after* treatment/total absorption *before* treatment) = 10 log (1,260/140) = 9.54 db. Thus, the sound level in the room would be reduced to 130.00 − 9.54 = 120.46 db. This is a reduction of 9.54/130 = 0.0733, 7.33 percent. To obtain a further reduction of the noise in this room, the floor could be insulated or, preferably, the noise-producing device could be redesigned so it gives off less noise.

Related Calculations: Use this general procedure to determine the effectiveness of acoustical materials used in any room in a building, on a ship, in an airplane, etc.

Table 35   Noise-Reduction Coefficients*

Type	Material	Noise-reduction-coefficient range ($\frac{3}{4}$-in. thick)
1	Regularly perforated cellulose-fiber tile	0.65–0.85
2	Randomly perforated cellulose-fiber tile	0.60–0.75
3	Textured, perforated, fissured, cellulose tile	0.50–0.70
4	Cellulose-fiber lay-in panels	0.50–0.60†
5	Perforated mineral-fiber tile	0.65–0.85
6	Fissured mineral-fiber tile	0.65–0.80
7	Textured, perforated or smooth mineral-fiber tile	0.65–0.85
8	Membrane-faced mineral-fiber tile	0.30–0.90
9	Mineral-fiber lay-in panels	0.20–0.90
10	Perforated-metal pans with mineral-fiber pads	0.60–0.80†
11	Perforated-metal lay-in panels with mineral-fiber pads	0.75–0.85
12	Mineral-fiber tile — fire-resistive assemblies	0.55–0.90
13	Mineral-fiber lay-in panels — fire-resistive units	0.65–0.75‡
14	Perforated-asbestos panels with mineral-fiber pads	0.65–0.75§
15	Sound-absorbent duct lining	0.65–0.75†
16	Special acoustical panels and materials	

*Acoustical and Insulating Materials Association.
†Noise-reduction coefficient 1-in. thick.
‡Noise-reduction coefficient $\frac{5}{8}$-in. thick.
§Noise-reduction coefficient $\frac{13}{16}$-in. thick.

Section 3

Mechanical Engineering

TYLER G. HICKS, P.E.
Consulting Engineer

EDGAR J. KATES, P.E.
Consulting Engineer

B. G. A. SKROTZKI, P.E.
Power Magazine

RAYMOND J. ROARK
Professor, University of Wisconsin

S. W. SPIELVOGEL
Piping Engineering Consultant

RUFUS OLDENBURGER
Professor, Purdue University

ALEXANDER G. CHRISTIE
Consulting Engineer

ROBERT L. DAVIDSON
Chemical Engineering

LYMAN F. SCHEEL
Consulting Engineer

MACHINE DESIGN AND ANALYSIS

REFERENCES: Roark—*Formulas for Stress and Strain*, McGraw-Hill; Church—*Mechanical Vibrations*, Wiley; *Machinery's Handbook*, Industrial Press; Johnson—*Optimum Design of Mechanical Elements*, Wiley; Slaymaker—*Mechanical Design and Analysis*, Wiley; Chironis—*Spring Design and Application*, McGraw-Hill; Spotts—*Design of Machine Elements*, Prentice-Hall; AGMA *Standards Books*, American Gear Manufacturers Association; Doughtie and Vallance —*Design of Machine Members*, McGraw-Hill; Buckingham—*Manual of Gear Design*, Industrial Press; Fuller—*Theory and Practice of Lubrication for Engineers*, Wiley; Dudley—*Gear Handbook*, McGraw-Hill; Churchman—*Prediction and Optimal Decision*, Wiley; Crandall—*Engineering Analysis*, McGraw-Hill; Ver Planck and Teare—*Engineering Analysis*, Wiley; Wahl—*Mechanical Springs*, McGraw-Hill; Haberman—*Engineering Systems Analysis*, Merrill; Shigley—*Mechanical Engineering Design*, McGraw-Hill; Ryder—*Creative Engineering Analysis*, Prentice-Hall; Baumeister and Marks—*Standard Handbook for Mechanical Engineers*, McGraw-Hill; Church—*Kinematics of Machines*, Wiley; Carmichael—*Kent's Mechanical Engineers' Handbook*, Wiley; Faires—*Design of Machine Elements*, Macmillan; Black—*Machine Design*, McGraw-Hill; Maleev —*Machine Design*, International; Bradford and Eaton—*Machine Design*, Wiley; Dudley—*Practical Gear Design*, McGraw-Hill; Shigley—*Simulation of Mechanical Systems*, McGraw-Hill.

ENERGY STORED IN A ROTATING FLYWHEEL

A 48-in.-diameter spoked steel flywheel having a 12-in.-wide × 10-in.-deep rim rotates at 200 rpm. How long a cut can be stamped in a 1-in.-thick aluminum plate if the stamping energy is obtained from this flywheel? The ultimate shearing strength of the aluminum is 40,000 psi.

Calculation Procedure:

1. Determine the kinetic energy of the flywheel

In routine design calculations, the weight of a spoked or disk flywheel is assumed to be concentrated in the rim of the flywheel. The weight of the spokes or disk is neglected. When computing the kinetic energy of the flywheel, the weight of a rectangular, square, or circular rim is assumed to be concentrated at the horizontal centerline. Thus, for this rectangular rim, the weight is concentrated at a radius of $(48/2 - 10/2) =$ 19 in. from the centerline of the shaft to which the flywheel is attached.

Then, the kinetic energy $K = Wv^2/2g$, where $K =$ kinetic energy of the rotating shaft, ft-lb; $W =$ flywheel weight of flywheel rim, lb; $v =$ velocity of flywheel at the horizontal centerline of the rim, fps. The velocity of a rotating rim is $v = 2\pi RD/60$, where $\pi =$ 3.1416; $R =$ rotational speed, rpm; $D =$ distance of the rim horizontal centerline from the center of rotation, ft. For this flywheel, $v = 2\pi(200)(19/12)/60 = 33.2$ fps.

The rim of the flywheel has a volume of (rim height, in.)(rim width, in.)(rim circumference measured at the horizontal centerline, in.), or $(10)(12)(2\pi)(19) = 14,350$ in.^{3}. Since machine steel weighs 0.28 lb/in.^{3}, the weight of the flywheel rim is $(14,350)(0.28)$ $= 4,010$ lb. Then, $K = (4,010)(33.2)^2/[2(32.2)] = 68,700$ ft-lb.

2. Compute the dimensions of the hole that can be stamped

A stamping operation is a shearing process. The area sheared is the product of the plate thickness and the length of the cut. Each sq in. of the sheared area offers a resistance equal to the ultimate shearing strength of the material punched.

During stamping, the force exerted by the stamp varies from a maximum F lb at the point of contact to 0 lb when the stamp emerges from the metal. Thus, the average force during stamping is $(F + 0)/2 = F/2$. The work done is the product of $F/2$ and the distance through which this force moves, or the plate thickness t in. Therefore, the maximum length that can be stamped is that which occurs when the full kinetic energy of the flywheel is converted to stamping work.

With a 1-in.-thick aluminum plate, the work done W ft-lb = (force, lb)(distance, ft). The work done when all the flywheel kinetic energy is used is $W = K$. Substituting the kinetic energy from Step 1, $W = K = 68,700 = (F/2)(1/12)$; and solving for the force, $F = 1,650,000$ lb.

The force F also equals the product of the plate area sheared and the ultimate shearing strength of the material stamped. Thus, $F = lts_u$, where l = length of cut, in.; t = plate thickness, in.; s_u = ultimate shearing strength of the material. Substituting the known values and solving for l, $l = 1,650,000/(1)(40,000)] = 41.25$ in.

Related Calculations: The length of cut computed above can be distributed in any form—square, rectangular, circular, or irregular. This method is suitable for computing the energy stored in a flywheel used for any purpose. Use the general procedure in Step 2 for computing the principal dimension in blanking, punching, piercing, trimming, bending, forming, drawing, or coining.

SHAFT TORQUE, HORESPOWER, AND DRIVER EFFICIENCY

A 4-in.-diameter shaft is driven at 3,600 rpm by a 400-hp motor. The shaft drives a 48-in.-diameter chain sprocket having an output efficiency of 85 percent. Determine the torque in the shaft, the output force on the sprocket, and the power delivered by the sprocket.

Calculation Procedure:

1. Compute the torque developed in the shaft

For any shaft driven by any driver, the torque developed T lb-in. = 63,000 hp/R, where hp = horsepower delivered to, or by, the shaft; R = shaft rotative speed, rpm. Thus, the torque developed by this shaft is $T = (63,000)(400)/3,600 = 7,000$ lb-in.

2. Compute the sprocket output force

The force developed at the output surface, tooth, or other part of a rotating member is given by $F = T/r$, where F = force developed, lb; r = radius arm of the force, in. In this drive the radius is $48/2 = 24$ in. Hence, $F = 7,000/24 = 291$ lb.

3. Compute the power delivered by the sprocket

The work input to this shaft is 400 hp. But the work output is less than the input because the efficiency is less than 100 percent. Since efficiency = work output, hp/work input, hp, the work output, hp = (work input, hp) (efficiency), or output hp = (400)(0.85) = 340 hp.

Related Calculations: Use this procedure for any shaft driven by any driver—electric motor, steam turbine, internal-combustion engine, gas turbine, belt, chain, sprocket, etc. When computing the radius of toothed or geared members, use the pitch-circle or pitch-line radius.

PULLEY AND GEAR LOADS ON SHAFTS

A 500-rpm shaft is fitted with a 30-in.-diameter pulley weighing 250 lb. This pulley delivers 35 hp to a load. The shaft is also fitted with a 24-in. pitch-diameter gear weighing 200 lb. This gear delivers 25 hp to a load. Determine the concentrated loads produced on the shaft by the pulley and the gear.

Calculation Procedure:

1. Determine the pulley concentrated load

The largest concentrated load caused by the pulley occurs when the belt load acts vertically downward. Then the total pulley concentrated load is the sum of the belt load and pulley weight.

For a pulley in which the tension of the tight side of the belt is twice the tension in the slack side of the belt, the maximum belt load is $F_p = 3T/r$, where F_p = tension force, lb, produced by the belt load; T = torque acting on the pulley, lb-in.; r = (pulley radius, in.). The torque acting on a pulley is found from $T = 63,000$ hp/R, where hp = horsepower delivered by pulley; R = shaft rpm.

For this pulley, $T = 63,000(35)/500 = 4,410$ lb-in. Then, $F_p = 3(4,410)/15 = 882$ lb. Hence, the total pulley concentrated load $= 882 + 250 = 1,132$ lb.

2. Determine the gear concentrated load

With a gear, the turning force acts only on the teeth engaged with the meshing gear. Hence, there is no slack force as in a belt. Therefore, $F_g = T/r$, where $F_g =$ gear tooth-thrust force, lb; $r =$ gear pitch radius, in.; other symbols as before. The torque acting on the gear is found in the same way as for the pulley.

Thus, $T = 63,000(25)/500 = 3,145$ lb-in. Then, $F_g = 3,145/12 = 263$ lb. Hence, the total gear concentrated load is $263 + 200 = 463$ lb.

Related Calculations: Use this procedure to determine the concentrated load produced by any type of gear (spur, herringbone, worm, etc.), pulley (flat, V-, or chain belt), sprocket, or their driving member. When the power-transmission belt or chain leaves the belt or sprocket at an angle other than the vertical, take the vertical component of the pulley force and add it to the pulley weight to determine the concentrated load.

SHAFT REACTIONS AND BENDING MOMENTS

A 30-ft-long steel shaft weighting 150 lb per ft of length has a 500-lb concentrated gear load 10 ft from the left end of the shaft, and a 2,000-lb concentrated pulley load 15 ft from the right end of the shaft. Determine the end reactions and maximum bending moment in this shaft.

Calculation Procedure:

1. Draw a sketch of the shaft

Figure 1a shows a sketch of the shaft. Label the left- and right-hand reactions L_R and R_R, respectively.

2. Compute the shaft end reactions

Take moments about R_R to determine the magnitude of L_R. Since the shaft has a uniform weight per ft of length, assume that the total weight of the shaft is concentrated at its midpoint. Then $30L_R - 500(20) - 50(30)(15) - 2,000$ $(15) = 0$; $L_R = 3,583.33$ lb. Take moments about L_R to determine R_R. Or, $30R_R - 500(10) - 150(30)$ $(15) = 2,000(15) = 0$; $R_R = 3,416.67$ lb. Alternatively, the first reaction found could be subtracted from the sum of the vertical loads, or $(500 + 30 \times 150 + 2,000) - 3,583.33 = 3,416.67$ lb. However, taking moments about each support permits checking the result, because the sum of the reactions should equal the sum of the vertical loads, including the weight of the shaft.

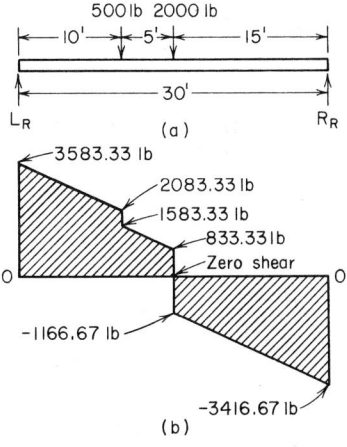

3. Compute the maximum bending moment

The maximum bending moment in a shaft occurs where the shear is zero. Find the vertical shear at each point of applied load or reaction by taking the algebraic sum of the vertical forces to the left and right of the load. Use a plus sign for upward forces and a minus sign for downward forces. Designate each shear force by V with a subscript number showing its location, in ft, along the shaft from the left end. Use L and R to indicate if the shear is to the left or right of a load. The shear at the left-hand reaction is $V_{LR} = +3,583.33$ lb; $V_{10L} = 3,583.33 - 10 \times 150 = 2,083.33$ lb, where the product $10 \times 150 =$ the weight of the shaft from the point V_{LR} to the 500-lb load. At this load,

Fig.1 Shaft bending-moment diagram.

$V_{10R} = 2,083.33 - 500 = 1,583.33$ lb. To the right of the 500-lb load, at the 2,000-lb load, $V_{20L} = 1,583.33 - 5 \times 150 = 833.33$ lb. To the right of the 2,000-lb load, $V_{20R} = 833.33 - 2,000 = -1,166.67$ lb. At the left of V_{R_R}, $V_{30L} = -1,166.67 - 15 \times 150 = -3,416.67$ lb. At the right-hand end of the shaft $V_{R30R} = -3,416.67 + 3,416.67 = 0$.

Draw the shear diagram, Fig. 1b. This diagram shows that zero shear occurs at a point 15 ft from the left-hand reaction. Hence, the maximum bending moment M_m on this shaft is $M_m = 3,583.33(15) - 500(5) - 150(15)(7.5) = 34,430$ lb-ft.

Related Calculations: Use this procedure for shafts made of any metal—steel, bronze, aluminum, plastic, etc.—if the shaft is of uniform cross section. For nonuniform shafts, use the procedures discussed later in this section of the handbook.

SOLID AND HOLLOW SHAFTS IN TORSION

A solid steel shaft will transmit 500 hp at 3,600 rpm. What diameter shaft is required if the allowable stress in the shaft is 12,500 psi? What diameter hollow shaft is needed to transmit the same horsepower if the inside diameter of the shaft is 1.0 in.?

Calculation Procedure:

1. Compute the torque in the solid shaft

For any solid shaft, the torque T lb-in. $= 63,000\, hp/R$, where $R =$ shaft rpm. Thus, $T = 63,000(500)/3,600 = 8.750$ lb-in.

2. Compute the required shaft diameter

For any solid shaft, the required diameter d in. $= 1.72(T/s)^{1/3}$, where $s =$ allowable stress in shaft, psi. Thus, for this shaft, $d = 1.72(8,750/12,500)^{1/3} = 1.526$ in.

3. Analyze the hollow shaft

The usual practice is to size hollow shafts such that the ratio q of the inside diameter d_i in. to the outside diameter d_o in. is $1:2$ to $1:3$, or some intermediate value. With a q in this range the shaft will have sufficient thickness to prevent failure in service.

Assume $q = d_i/d_o = \frac{1}{2}$. Then, with $d_i = 1.0$ in., $d_o = d_i/q$, or $d_o = 1.0/0.5 = 2.0$ in. With $q = 1/3$, $d_o = 1.0/0.33 = 3.0$ in.

4. Compute the stress in each hollow shaft

For the hollow shaft $s = 5.1T/d_o^3(1 - q^4)$, where the symbols are defined above. Thus, for the 2-in. outside-diameter shaft, $s = 5.1(8,750)/[8(1 - 0.0625)] = 5,950$ psi.

By inspection, it can be seen that the stress in the 3-in. outside-diameter shaft will be lower because the torque is constant. Thus, $s = 5.1(8,750)/[27(1 - 0.0123)] = 1,672$ psi.

5. Choose the hollow shaft outside diameter

Use a trial-and-error procedure to choose the hollow shaft outside diameter. Since it is already known that the stress in the 2-in. outside-diameter shaft, 5,950 psi, is less than half the allowable stress of 12,500 psi, select a smaller outside diameter and compute the stress while holding the inside diameter constant.

Thus, with a 1.5-in. shaft and the same inside diameter, $s = 5.1(8,750)/[3.38(1 - 0.197)] = 16.430$ psi. This exceeds the allowable stress.

Try a larger outside diameter, 1.75 in., to find the effect on the stress, or $s = 5.1(8,750)/[5.35(1 - 0.107)] = 9,350$. This is lower than the allowable stress.

Since a 1.5-in. shaft has a 16,430-psi stress and a 1.75-in. shaft has a 9,350-psi stress, a shaft of intermediate size will have a stress approaching 12,500 psi. Trying 1.625 in., $s = 5.1(8,750)/[4.4(1 - 0.143)] = 11,820$ psi. This is within 680 psi of the allowable stress and is close enough for usual design calculations.

Related Calculations: Use this procedure to find the diameter of any solid or hollow shaft acted on *only* by torsional stress. Where bending and torsion occur, use the next Calculation Procedure. Find the allowable torsional stress for various materials in Baumeister—*Standard Handbook for Mechanical Engineers.*

SOLID SHAFTS IN BENDING AND TORSION

A 30-ft-long solid shaft weighing 150 lb/ft is fitted with a pulley and gear as shown in Fig. 2. The gear delivers 100 hp to the shaft while driving the shaft at 500 rpm. Determine the required diameter of the shaft if the allowable stress is 10,000 psi.

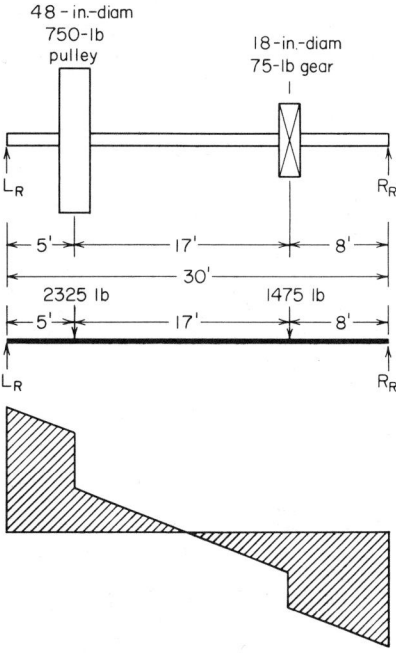

Calculation Procedure:

1. Compute the pulley and gear concentrated loads

Using the method of the previous Calculation Procedures, $T = 63,000\ hp/R = 63,000$ $(100)/500 = 12,600$ lb-in. Assuming that the maximum tension of the tight side of the belt is twice the tension of the slack side, the maximum belt load is $R_y = 3T/r = 3$ $(12,600)/24 = 1,575$ lb. Hence, the total pulley concentrated load = belt load + pulley weight = $1,575 + 750 = 2,325$ lb.

The gear concentrated load is found from $F_g = T/r$, where the torque is the same as computed for the pulley, or $F_y = 12,600/9 = 1,400$ lb. Hence, the total gear concentrated load is $1,400 + 75 = 1,475$ lb.

Draw a sketch of the shaft showing the two concentrated loads in position, Fig. 2.

Fig. 2   Solid-shaft bending moments.

2. Compute the end reactions of the shaft

Take moments about R_R to determine L_R, using the method of the previous Calculation Procedures. Thus, $L_R(30) - 2,325(25) - 1,475(8) - 150(30)(15) = 0$; $L_R = 4,580$ lb. Taking moments about L_R to determine R_R, $R_R(30) - 1,475(22) - 2,325(5) - 150(30)(15) = 0$; $R_R = 3,720$ lb. Checking by taking the sum of the upward forces, $4,580 + 3,720 = 8,300$ lb = the sum of the downward forces, or $2,325 + 1,475 + 4,500 = 8,300$ lb.

3. Compute the vertical shear acting on the shaft

Using the method of the previous Calculation Procedure, $V_{l_R} = 4,580$ lb; $V_{5L} = 4,580 - 5(150) = 3,830$ lb; $V_{5R} = 3,830 - 2,325 = 1,505$ lb; $V_{22L} = 1,505 - 17(150) = -1,045$ lb. $V_{22R} = -1,045 - 1,475 = -2,520$ lb; $V_{30L} = -2,520 - 8(150) = -3,720$ lb; $V_{30}R = -3,720 + 3,720$ lb $= 0$.

4. Find the maximum bending moment on the shaft

Draw the shear diagram shown in Fig. 2. Determine the point of zero shear by scaling it from the shear diagram or setting up an equation thus: positive shear $- x(150$ lb/ft$) = 0$, where the positive shear is the last recorded plus value, V_{5R} in this shaft, and $x =$ distance from V_{5R} where the shear is zero. Substituting values, $1,505 - 150x = 0$; $x = 10.03$ ft. Then, $M_m = 4,580(15.03) - 2,325(10.03) - (150)(5 + 10.03)[(5 + 10.03)/2] = 28,50$ lb-ft.

5. Determine the required shaft diameter

Use the method of *maximum shear theory* to size the shaft. Determine the equivalent torque T_e from $T_e = (M_m^2 + T^2)^{0.5}$, where M_m is the maximum bending moment, lb-ft, acting on the shaft, and T is the maximum torque acting on the shaft. For this shaft, $T_e = [28,500^2 + (12,600/12)^2]^{0.5} = 28,500$ lb-ft, where the torque in lb-in. is divided by 12 to convert it to lb-ft. To convert T_e to $T_{e'}$ lb-in., multiply by 12.

Once the equivalent torque is known, the shaft diameter d in. is computed from $d = 1.72(T_{e'}/s)^{1/3}$, where $s =$ allowable stress in the shaft. For this shaft, $d = 1.72(28,500 (12)/10,000)^{1/3} = 5.59$ in. Use a 6.0-in.-diameter shaft.

Related Calculations: Use this procedure for any solid shaft of uniform cross section made of metal—steel, aluminum, bronze, brass, etc. The equation used in Step 4 to determine the location of zero shear is based on a strength-of-materials principle: When zero shear occurs between two concentrated loads, find its location by dividing the last *positive* shear by the uniform load. If desired, the maximum principal stress theory can be used to combine the bending and torsional stresses in a shaft. The results obtained approximate those of the maximum shear theory.

EQUIVALENT BENDING MOMENT AND IDEAL TORQUE FOR A SHAFT

A 2-in.-diameter solid steel shaft has a maximum bending moment of 6,000 lb-in. and an applied torque of 3,000 lb-in. Is this shaft safe if the maximum allowable bending stress is 10,000 psi? What is the ideal torque for this shaft?

Calculation Procedure:

1. Compute the equivalent bending moment

The equivalent bending moment M_e lb-in. for a solid shaft is $M_e = 0.5[M + (M^2 + T^2)^{0.5}]$, where $M =$ maximum bending moment acting on the shaft, lb-in; $T =$ maximum torque acting on the shaft, lb-in. For this shaft, $M_e = 0.5[6,000 + (6,000^2 + 3,000^2)^{0.5}] = 6,355$ lb-in.

2. Compute the stress in the shaft

Use the flexure relation $s = Mc/I$, where $s =$ stress developed in the shaft, psi; $M = M_e$ for a shaft; $I =$ section moment of inertia of the shaft about the neutral axis; in.^{4}; $c =$ distance from shaft neutral axis to outside fibers, in. For a circular shaft, $I = \pi d^4/64 = \pi(2)^4/64 = 0.785$ in.^{4}; $c = d/2 = 2/2 = 1.0$. Then, $s = Mc/I = (6,355)(1.0)/0.785 = 8,100$ psi. Thus, the actual bending stress is 1,900 psi less than the maximum allowable bending stress. Therefore the shaft is safe. Alternatively, compute the maximum equivalent bending moment from $M_e = sI/c = (10,000)(0.785)/1.0 = 7,850$ lb-in. This is $7,850 - 6,355 = 1,495$ lb-in. greater than the actual equivalent bending moment. Hence, the shaft is safe.

3. Compute the ideal torque for the shaft

The ideal torque T_i lb-in. for a shaft is $T_i = M + (M^2 + T^2)^{0.5}$, where M and T are the bending and torsional moments, respectively, acting on the shaft, lb-in. For this shaft, $T_i = 6,000 + (6,000^2 + 3,000^2)^{0.5} = 12,710$ lb-in.

Related Calculations: Use this procedure for any shaft of uniform cross section made of metal—steel, aluminum, bronze, brass, etc.

TORSIONAL DEFLECTION OF SOLID AND HOLLOW SHAFTS

What diameter solid steel shaft should be used for a 500-hp 250-rpm application if the allowable torsional deflection is 1°, the maximum allowable stress is 10,000 psi, and the modulus of rigidity is 13×10^6 psi? What diameter hollow steel shaft should be used if the ratio of the inside diameter to the outside diameter is 1:3, the allowable deflection is 1°, the allowable stress is 10,000 psi, and the modulus of rigidity is 13×10^6 psi? What shaft has the greatest weight?

Calculation Procedure:

1. Determine the torque acting on the shaft

For any shaft, $T = 63,000 \, hp/R$; or for this shaft, $T = 63,000(500)/250 = 126,000$ lb-in.

2. Compute the required diameter of the solid shaft

For a solid metal shaft, $d = (584Tl/G\alpha)^{1/3}$, where $l =$ shaft length expressed as a number of shaft diameters, in.; $G =$ modulus of rigidity, psi; $\alpha =$ angle of torsional deflection,°.

Usual specifications for noncritical applications of shafts require that the torsional deflection not exceed 1° in a shaft having a length equal to 20 diameters. Using this length, $d = [584 \times 126,000 \times 20/(13 \times 10^6 \times 1.0)]^{1/3} = 4.84$ in. Use a 5-in.-diameter shaft.

3. Compute the outside diameter of the hollow shaft

Assume that the shaft has a length equal to 20 diameters. Then, for a hollow shaft, $d = [584Tl/G\alpha(1 - q^4)]^{1/3}$, where $q = d_i/d_o$, and $d_i =$ inside diameter of shaft, in.; $d_o =$ outside diameter of shaft, in. For this shaft, $d = \{584 \times 126,000 \times 20/(13 \times 10^6 \times 1.0) [1 - (\frac{1}{3})^4]\}^{1/3} = 4.86$ in. Use a 5-in. outside-diameter shaft. The inside diameter would be $5.0/3 = 1.667$ in.

4. Compare the weight of the shafts

Steel weighs approximately 480 lb/cu ft. To find the weight of each shaft, compute its volume in cu ft and multiply it by 480. Thus, for the 5-in.-diameter solid shaft, weight $= (\pi 5^2/4)(5 \times 20)(480)/1,728 = 540$ lb. The 5-in. outside-diameter hollow shaft weighs $(\pi 5^2/4 - \pi 1.667^2/4)(5 \times 20)(480)/1,728 = 242$ lb. Thus, the hollow shaft weighs less than half the solid shaft. It would, however, probably be more expensive to manufacture because drilling the central hole could be costly.

Related Calculations: Use this procedure to determine the steady-load torsional deflection of any shaft of uniform cross section made of any metal—steel, bronze, brass, aluminum, Monel, etc. The assumed torsional deflection of 1° for a shaft that is 20 times as long as the shaft diameter is typical for routine applications. Special shafts may be designed for considerably less torsional deflection.

DEFLECTION OF A SHAFT CARRYING CONCENTRATED AND UNIFORM LOADS

A 2-in.-diameter steel shaft is 6 ft long between bearing centers and turns at 500 rpm. The shaft carries a 600-lb concentrated gear load 3 ft from the left-hand center. Determine the deflection of the shaft if the modulus of elasticity E of the steel is 30×10^6 psi. What would the shaft deflection be if the load were 2 ft from the left-hand bearing? The shaft weighs 10 lb per ft of length.

Calculation Procedure:

1. Compute the deflection caused by the concentrated load

When a beam carries both a concentrated and a uniformly distributed load, compute the deflection for each load separately and find the sum. This sum is the total deflection caused by the two loads.

For a beam carrying a concentrated load, the deflection Δ in. $= Wl^3/48EI$, where $W =$ concentrated load, lb; $l =$ length of beam, in.; $E =$ modulus of elasticity, psi; $I =$ moment of inertia of shaft cross section, in.⁴. For a circular shaft, $I = \pi d^4/64 = \pi(2)^4/64 = 0.7854$ in.⁴. Then, $\Delta = 600(72)^3/[48(30)(10^6)(0.7854)] = 0.198$ in. The deflection per ft of shaft length is $\Delta_f = 0.198/6$ ft $= 0.033$ in. for the concentrated load.

2. Compute the deflection due to shaft weight

For a shaft of uniform weight, $\Delta = 5wl^3/384EI$, where $w =$ total distributed load $=$ weight of shaft, lb. Thus, $\Delta = 5(60)(72)^3/[384(30 \times 10^6)(0.7854)] = 0.0129$ in. The deflection per ft of shaft length is $\Delta_f = 0.0129/6 = 0.00214$ in.

3. Determine the total deflection of the shaft

The total deflection of the shaft is the sum of the deflections caused by the concentrated and uniform loads, or $\Delta_t = 0.198 + 0.0129 = 0.2109$ in. The total deflection per ft of length is $0.033 + 0.00214 = 0.03514$ in.

Usual design practice limits the transverse deflection of a shaft of any diameter to 0.01 in. per ft of shaft length. The deflection of this shaft is $3\frac{1}{2}$ times this limit. Therefore, the shaft diameter must be increased if this limit is not to be exceeded.

Using a 3-in.-diameter shaft weighing 25 lb/ft, and computing the deflection in the same way, the total transverse deflection is 0.0453 in., and the total deflection per ft of shaft length is 0.00755 in. This is within the desired limits. By reducing the assumed shaft diameter in 1/8-in. decrements and computing the deflection per ft of length, a deflection closer to the limit can be obtained.

4. Compute the total deflection for the noncentral load

For a noncentral load, $\Delta = (Wc'/3EIl)[(cl/3 + cc'/3)^3]^{0.5}$ where $c =$ distance of concentrated load from left-hand bearing, in.; $c' =$ distance of concentrated load from right-hand bearing, in. Thus $c + c' = l$, and for this shaft $c = 24$ in. and $c' = 48$ in. Then, $\Delta = [600 \times 48/(3 \times 30 \times 10^6 \times 0.7854 \times 72)]\{[24 \times (72/3) + 24 \times (48/3)]^3\}^{0.5} = 0.169$ in.

The deflection caused by the weight of the shaft is the same as computed in Step 2, or 0.0129 in. Hence, the total shaft deflection is $0.169 + 0.0129 = 0.1819$ in. The deflection per ft of shaft length is $0.1819/6 = 0.0303$ in. Again, this exceeds 0.01 in./ft.

Using a 3-in.-diameter shaft as in Step 3 shows that the deflections can be reduced to within the desired limits.

Related Calculations: Use this procedure for any metal shaft—aluminum, brass, bronze, etc.—that is uniformly loaded or carries a concentrated load.

SELECTION OF KEYS FOR MACHINE SHAFTS

Select a key for a 4-in.-diameter shaft transmitting 1,000 hp at 1,000 rpm. The allowable shear stress in the key is 15,000 psi and the allowable compressive stress is 30,000 psi. What type of key should be used if the allowable shear stress is 5,000 psi and the allowable compressive stress is 20,000 psi?

Calculation Procedure:

1. Compute the torque acting on the shaft

The torque acting on the shaft is $T = 63,000 \ hp/R$, or $T = 63,000(1,000/1,000) = 63,000$ lb-in.

2. Determine the shear force acting on the key

The shear force $F_s$lb acting on a key is $F_s = T/r$, where $T =$ torque acting on shaft, lb-in.; $r =$ shaft radius, in. Thus, $T = 63,000/2 = 31,500$ lb.

3. Select the type of key to use

When a key is designed so that its allowable shear stress is approximately one-half its allowable compressive stress, a square key—i.e., a key having its height equal to its width—is generally chosen. For other values of the stress ratio, a flat key is generally chosen.

Since the stress ratio for this key is $15,000/30,000 = 0.5$, a square key will be used.

Determine the dimensions of the key from Baumeister and Marks—*Standard Handbook for Mechanical Engineers.* This handbook shows that a 4-in.-diameter shaft should have a square key 1 in. wide × 1 in. high.

4. Determine the required length of the key

The length of a key l in. based on the allowable shear stress is $l = F_s/w_k S_s$, where $w_k =$ width of key, in. Thus, $l = 31,500/[(1)(15,000)] = 2.1$ in.; say $2\frac{1}{8}$ in.

5. Check the key length for the compressive load

The length of a key l in. based on the allowable compressive stress is $l = 2F_s/ts_c$, where $t =$ key thickness, in.; $s_c =$ allowable compressive stress, psi. Thus, $l = 2(31,500)/[(1)(30,000)] = 2.1$ in. This agrees with the key length based on the allowable shear stress. The key length found in Steps 4 and 5 should agree if the key is square in cross section.

6. *Determine key size for other stress values*

When the allowable shear stress does not equal one-half the allowable compressive stress for a shaft key, a flat key is generally used. A flat key has a width greater than its height.

Find the recommended dimensions for a flat key from Baumeister and Marks — *Standard Handbook for Mechanical Engineers*. This handbook shows that a 4-in.-diameter shaft will use a 1-in.-wide × $\frac{3}{4}$-in.-thick flat key.

The length of the key based on the allowable shear stress is $l = F_s/w_k s_s = 31,500/(1)$ $(5,000) = 6.31$ in. Use a $6\frac{5}{16}$-in.-long key.

Checking the key length based on the allowable compressive stress, $l = 2F_s/ts_c = 2(31,500)/[(0.75)(20,000)] = 4.2$ in. Use the longer length, $6\frac{5}{16}$ in., because the shorter key would be overloaded in compression.

Related Calculations: Use this procedure for shafts and keys made of any metal — steel, bronze, brass, stainless steel, etc. The dimensions of shaft keys can also be found in *ANSI Standard B17f, Woodruff Keys, Keyslots and Cutters*. Woodruff keys are used only for light-torque applications.

SELECTING A LEATHER BELT FOR POWER TRANSMISSION

Choose a leather belt to transmit 50 hp from a 1,750-rpm squirrel-cage compensator-starting motor through a 12-in.-diameter pulley in an oily atmosphere. What belt width is needed with a 50-hp internal-combustion engine fitted with a 1,750-rpm 12-in.-diameter pulley operating in an oily atmosphere?

Calculation Procedure:

1. *Determine the belt speed*

The speed of a belt S expressed in ft/min, is found from $S = \pi RD$, where $R = $ rpm of driving or driven pulley; $D = $ diameter, ft, of driving or driven pulley. Thus, for this belt, $S = \pi(1,750)(12/12) = 5,500$ fpm.

2. *Determine the belt thickness needed*

Use the National Industrial Leather Association recommendations. Enter Table 1 at the bottom at a belt speed of 5,500 fpm — i.e., between 4,000 and 6,000 fpm — and project horizontally to the next *smaller* pulley diameter than that actually used. Thus, entering at the line marked 4,000 − 6,000 fpm and projecting to the 10-in.-minimum-diameter pulley, since a 12-in. pulley is used, shows that a 23/64-in.-thick double-ply heavy belt should be used. Read the belt thickness and type at the top of the column in which the next smaller pulley diameter appears.

3. *Determine the belt capacity factors*

Enter the body of Table 1 at a belt speed of 5,500 fpm — i.e., between 5,000 and 6,000 fpm — and project to the double-ply heavy column. Interpolating by eye gives a belt capacity factor of $K_c = 14.8$.

4. *Determine the belt correction factors*

Table 2 lists motor, pulley-diameter, and operating-condition correction factors, respectively. Thus, from Table 2, the motor correction factor $M = 1.5$ for a squirrel-cage compensator-starting motor. Also from Table 2, the smaller pulley diameter correction factor $P = 0.7$; and $F = 1.35$ for an oily atmosphere.

5. *Compute the required belt width*

The required belt width, in., is $W = hp\,MF/K_c P$, where $hp = $ horsepower transmitted by the belt; the other factors are as given above. For this belt, then, $W = (50)(1.5)(1.35)/[(14.8)(0.7)] = 9.7$ in. Thus, a 10-in.-wide belt would be used because belts are commercially available in 1-in.-width increments.

TABLE 1   Leather-Belt Capacity Factors

Belt speed, fpm	Single ply		Double ply		Triple ply	
	11/64 in. medium	13/64 in. heavy	20/64 in. medium	23/64 in. heavy	30/64 in. medium	34/64 in. heavy
1,000	1.8	2.1	3.1	3.6	4.1	4.5
2,000	3.5	4.1	6.0	6.9	8.1	8.9
3,000	5.2	5.9	8.7	10.0	11.6	12.8
4,000	6.4	7.4	10.9	12.6	14.5	16.0
5,000	7.4	8.4	12.5	14.3	16.5	18.2
6,000	7.8	8.9	13.2	15.2	17.6	19.3

Minimum pulley diameters, in.

Up to 2,500	2.5	3	5*	8*	16†	20†
2,500–4,000	3	3.5	6*	9*	18†	22‡
4,000–6,000	3.5	4	7*	10*	20†	24‡

*For belts 8 in. and over, add 2 in. to pulley diameter.
†For belts 8 in. and over, add 4 in. to pulley diameter.

TABLE 2   Leather-Belt Correction Factors

Correction Factor

Characteristics or condition of motor and starter:
Squirrel-cage, compensator-starting motor . . . . $M = 1.5$
Squirrel-cage, line-starting . . . . $M = 2.0$
Slip-ring, high starting torque . . . . $M = 2.5$
Diameter of small pulley, in.:
4 and under . . . . $P = 0.5$
4.5 to 8 . . . . $P = 0.6$
9 to 12 . . . . $P = 0.7$
13 to 16 . . . . $P = 0.8$
17 to 30 . . . . $P = 0.9$
Over 30 . . . . $P = 1.0$
Operating conditions:
Oily, wet, or dusty atmosphere . . . . $F = 1.35$
Vertical drives . . . . $F = 1.2$
Jerky loads . . . . $F = 1.2$
Shock and reversing loads . . . . $F = 1.4$

6.   *Determine the belt width for the engine drive*

For a double-ply belt driven by a driver other than an electric motor, $W = 2,750$ hp/dR, where $d =$ driving pulley diameter, in.; $R =$ driving pulley, rpm. Thus, $W = 2,750(50)/[(12)(1,750)] = 6.54$ in. Hence, a 7-in.-wide belt would be used.

For a single-ply belt the above equation becomes $W = 1,925 \, hp/dR$.

Related Calculations: Note that the relations in Steps 1, 5, and 6 can be solved for any unknown variable when the other factors in the equations are known. Where the hp rating of a belt material is available from the manufacturer's catalog or other published data find the required width from $W = hp_b F/K_c P$, where $hp_b =$ hp rating of the belt material, as stated by the manufacturer; other symbols as before. To find the tension

T_b lb in a belt, solve $T_b = 33,000\ hp/S$ where S = belt speed, fpm. The tension per in. of belt width is $T_{bi} = T_b/W$. Where the belt speed exceeds 6,000 fpm, consult the manufacturer.

SELECTING A RUBBER BELT FOR POWER TRANSMISSION

Choose a rubber belt to transmit 15 hp from a 7-in.-diameter pulley driven by a shunt-wound dc motor. The pulley speed is 1,300 rpm and the belt drives an electric generator. The arrangement of the drive is such that the arc of contact of the belt on the pulley is 220°.

Calculation Procedure:

1. Determine the belt service factor

The belt *service factor* allows for the typical conditions met in the use of a belt with a given driver and driven machine or device. Table 3 lists typical service factors S_f used by the B. F. Goodrich Company. Entering Table 3 at the type of driver, a shunt-wound dc motor, and projecting downward to the driven machine, an electric generator, shows that $S_f = 1.2$.

TABLE 3   Service Factors S

| Application | Squirrel-cage ac motor | | Wound rotor ac motor (slip ring) | Single-phase capacitor motor | Dc shunt-wound motor | Diesel engine, four or more cylinders, above 700 rpm |
	Normal torque, line start	High torque				
Agitators.............	1.0–1.2	1.2–1.4	1.2			
Compressors.........	1.2–1.4	...	1.4	1.2	1.2	1.2
Belt conveyors (ore, coal, sand)	...	1.4	...	...	1.2	
Screw conveyors	...	1.8	...	...	1.6	
Crushing machinery..	...	1.6	1.6	...	...	1.4–1.6
Fans, centrifugal	1.2	...	1.4	...	1.4	1.4
Fans, propeller.......	1.4	2.0	1.6	...	1.6	1.6
Generators and exciters...........	1.2	...	...	...	1.2	2.0
Line shafts..........	1.4	...	1.4	1.4	1.4	1.6
Machine tools........	1.0–1.2	...	1.2–1.4	1.0	1.0–1.2	
Pumps, centrifugal ...	1.2	1.4	1.4	1.2	1.2	
Pumps, reciprocating.	1.2–1.4	...	1.4–1.6	...	...	1.8–2.0

2. Determine the arc-of-contact factor

A rubber belt can contact a pulley in a range from about 140 to 220°. Since the hp-capacity ratings for belts are based on an arc of contact of 180°, a correction factor must be applied for other arcs of contact.

Table 4 lists the arc-of-contact correction factor C_c. Thus, for an arc of contact of 220°, $C_c = 1.12$.

TABLE 4   Arc of Contact Factor K — Rubber Belts

Arc of contact, °......	140	160	180	200	220
Factor K............	0.82	0.93	1.00	1.06	1.12

3. Compute the belt speed

The belt speed is $S = \pi RD$, where S = belt speed, fpm; R = pulley rpm; D = pulley diameter, ft. For this pulley, $S = \pi(1,300)(7/12) = 2,380$ fpm.

4. Choose the minimum pulley diameter and belt ply

Table 5 lists minimum recommended pulley diameters, belt material, and number of plies for various belt speeds. Choose the pulley diameter and number of plies for the

TABLE 5 Minimum Pulley Diameters — Rubber Belts

	Ply	Belt speed, fpm										
		500	1,000	1,500	2,000	2,500	3,000	4,000	5,000	6,000	7,000	8,000
32-oz fabric	3	4	4	4	4	5	5	5	6	6		
	4	4	5	6	6	7	7	8	9	10		
	5	6	7	9	10	10	11	12	13	14		
	6	9	10	11	13	14	14	16	18	19		
	7	13	14	16	17	18	19	21	22	24		
	8	18	19	21	22	23	24	25	27	29		
32-oz hard fabric	3	3	3	3	4	4	4	4	5	5	6	7
	4	4	4	5	5	6	6	7	7	8	9	12
	5	5	6	7	8	8	9	10	11	12	13	16
	6	6	8	10	11	11	12	13	15	16	18	21
	7	10	12	14	15	15	16	17	19	20	22	26
	8	14	16	17	18	19	20	21	23	24	27	31
	9	18	20	21	22	23	24	25	27	28	31	36
	10	22	24	25	26	27	28	29	31	33	35	41
No. 70 rayon cord.	3	5	6	7	7	8	8	9	10	11	12	13
	4	7	8	9	9	10	11	12	12	14	15	17
	5	9	10	11	12	13	13	15	16	17	19	21
	6	13	14	15	16	16	17	18	19	21	23	25
	7	16	17	18	19	20	21	22	23	24	26	29
	8	19	20	22	23	23	24	25	26	28	30	33

next *higher* belt speed when the computed belt speed falls between two tabulated values. Thus, for a belt speed of 2,380 fpm, use a 7-in.-diameter pulley as listed under 2,500 fpm. The corresponding material specifications are found in the left-hand column and are 4 plies, 32-oz fabric.

5. Determine the belt hp rating

Enter Table 6 at 32-oz 4-ply materials specifications and project horizontally to the belt speed. This occurs between the tabulated speeds of 2,000 and 2,500 fpm. Interpolating, $[(2,500 - 2,380)/(2,500 - 2,000)](4.4 - 3.6) = 0.192$. Hence, the hp rating of the belt hp_{bi} is $4.400 - 0.192 = 4.208$ hp per in. of width.

6. Determine the required belt width

The required belt width W in. $= hp \, S_f/hp_{bi}C_c$, or $W = (15)(1.2)/[(4.208)(1.12)] = 3.82$ in. Use a 4-in.-wide belt.

Related Calculations: Use this procedure for rubber-belt drives of all types. For additional service factors, consult the engineering data published by B. F. Goodrich Company, The Goodyear Tire and Rubber Company, United States Rubber Company, etc.

SELECTING A V-BELT FOR POWER TRANSMISSION

Choose a V-belt to drive a 0.75-hp stoker at about 900 rpm from a 1,750-rpm motor. The stoker is fitted with a 3-in.-diameter sheave and the motor with a 6-in.-diameter sheave.

TABLE 6 Horsepower Ratings of Rubber Belts
(Hp = horsepower per inch of belt width for 180° wrap)

	Ply	Belt speed, fpm										
		500	1,000	1,500	2,000	2,500	3,000	4,000	5,000	6,000	7,000	8,000
32-oz fabric	3	0.7	1.4	2.1	2.7	3.3	3.9	4.9	5.6	6.0		
	4	0.9	1.9	2.8	3.6	4.4	5.2	6.5	7.4	7.9		
	5	1.2	2.3	3.4	4.5	5.5	6.5	8.1	9.2	9.8		
	6	1.4	2.8	4.1	5.4	6.6	7.8	9.6	11.0	11.7		
	7	1.6	3.2	4.7	6.2	7.7	9.0	11.2	12.8	13.6		
	8	1.8	3.6	5.3	7.0	8.7	10.2	12.7	14.6	15.5		
32-oz hard fabric	3	0.7	1.5	2.2	2.9	3.5	4.1	5.1	5.8	6.2	6.1	5.5
	4	1.0	2.0	3.0	3.9	4.7	5.5	6.8	7.8	8.3	8.1	7.3
	5	1.3	2.5	3.7	4.9	5.9	6.9	8.5	9.8	10.3	9.1	9.0
	6	1.5	3.0	4.5	5.9	7.1	8.3	10.2	11.7	12.3	12.1	10.7
	7	1.7	3.5	5.2	6.9	8.3	9.7	11.9	13.6	14.3	14.1	12.4
	8	1.9	4.0	5.9	7.9	9.5	11.1	13.6	15.5	16.3	16.0	14.1
	9	2.1	4.5	6.6	8.9	10.6	12.4	15.3	17.4	18.3	17.9	15.8
	10	2.3	5.0	7.3	9.8	11.7	13.7	17.0	19.3	20.3	19.8	17.5
No. 70 rayon cord....	3	1.6	3.1	4.6	6.0	7.3	8.6	10.6	12.0	12.7	12.3	10.7
	4	2.1	4.1	6.1	8.0	9.8	11.5	14.5	16.6	17.8	17.8	16.4
	5	2.6	5.1	7.6	10.1	12.3	14.5	18.3	21.1	23.0	23.5	22.2
	6	3.1	6.2	9.2	12.1	14.8	17.5	22.1	25.7	28.1	28.9	27.9
	7	3.6	7.2	10.7	14.1	17.4	20.4	26.0	30.3	33.2	34.5	33.7
	8	4.1	8.2	12.2	16.2	19.9	23.4	29.8	34.8	38.4	40.0	39.4

The distance between the sheave shaft centerlines is 18 in. The stoker handles soft coal free of hard lumps.

Calculation Procedure:

1. Determine the design horsepower for the belt

V-belt manufacturers publish service factors for belts used in various applications. Table 7 shows that a stoker is classed as heavy service and has a service factor of 1.4 to 1.6. Using the lower value, because the stoker handles soft coal free of hard lumps, the design horsepower for the belt is found by taking the product of the rated horsepower of the device driven by the belt and the service factor, or (0.75 hp)(1.4 service factor) = 1.05 hp. The belt must be capable of transmitting this, or a greater, horsepower.

TABLE 7 Service Factors for V-Belt Drives

Typical machines	Type of service	Service factors
Domestic washing machines, domestic ironers, advertising display fixtures, small fans and blowers...................	Light	1.0–1.2
Fans and blowers (heavy rotors), centrifugal pumps, oil burners, home workshop machines.....................	Medium	1.2–1.4
Stokers, reciprocating pumps and compressors, refrigerators, drill presses, grinders, lathes, meat slicers, machines for industrial use...............	Heavy	1.4–1.6

2. Determine the belt speed and arc of contact

The belt speed $S = \pi RD$, where R = sheave rpm; D = sheave pitch diameter, ft = (sheave outside diameter, in. $-2X)/12$, where $2X$ = sheave dimension from Table 8. Before solving this equation, an assumption about the cross-sectional width of the belt must be made because $2X$ varies from 0.10 to 0.30 in. and the exact cross section of the belt that will be used is not yet known. A value of $X = 0.15$ in. is usually a safe assumption. It corresponds to a $3L$ belt cross section. Using $X = 0.15$ and the diameter and speed of the larger sheave, $S = \pi(1,750)(6.0 - 0.15)/12 = 2,675$ fpm.

TABLE 8   Sheave Dimensions—Light-Duty V-Belts

Belt cross section	Sheave effective OD, in.	Groove angle,°	W, in.	D, in	$2X$, in.
2L	Under 1.5	32	0.240	0.250	0.10
	1.5–1.99	34	0.243		
	2.0–2.5	36	0.246		
	Over 2.5	38	0.250		
3L	Under 2.2	32	0.360	0.406	0.15
	2.2–3.19	34	0.364		
	3.2–4.2	36	0.368		
	Over 4.20	38	0.372		
4L	Under 2.65	30	0.485	0.490	0.20
	2.65–3.24	32	0.490		
	3.25–5.65	34	0.494		
	Over 5.65	38	0.504		
5L	Under 3.95	30	0.624	0.580	0.30
	3.95–4.94	32	0.630		
	4.95–7.35	34	0.637		
	Over 7.35	38	0.650		

Compute the belt arc of contact from arc of contact,$° = 180 - [60(d_1 - d_s)/l]$, where d_1 = large sheave nominal diameter, in.; d_s = small sheave nominal diameter, in.; l = distance between shaft centers, in. For this drive, arc = $180 - [60(6-3)/18] = 170°$. An arc-of-contact correction factor must be applied when computing the belt power capacity. Read this correction factor from Table 9 as $C_c = 0.98$ for a V-sheave to V-sheave drive and a 170° arc of contact. NOTE: If desired, the pitch diameters can be used in the above relation in place of the nominal diameters.

TABLE 9   Correction Factors for Arc of Contact—V-Belt Drives

Arc of contact,°	Correction factor		Arc of contact,°	Correction factor	
	V to V	V to flat*		V to V	V to flat*
180	1.00	0.75	130	0.86	0.86
170	0.98	0.77	120	0.82	0.82
160	0.95	0.80	110	0.78	0.78
150	0.92	0.82	100	0.74	0.74
140	0.89	0.84	90	0.69	0.69

*A V-to-flat drive has a small sheave and a larger-diameter flat pulley.

3. *Select the belt to be used*

The 2X value used in Step 3 corresponds to a 3L cross-section belt. Check the power capacity of this belt by entering Table 10 at a belt speed of 2,800 fpm, the next larger tabulated speed, and projecting across to the appropriate small-sheave diameter—3 in. or larger. Read the belt horsepower rating as 0.87 hp. This is considerably less than the required capacity of 1.05 hp computed in Step 1. Therefore, the 3L belt is unsatisfactory.

Try a 4L belt, Table 11, following the same procedure. A 4L belt with a 3-in.-diameter small sheave has a rating of 1.16 hp. Correct this for the actual arc of contact by multiplying by C_c, or $(1.16)(0.98) = 1.137$. Thus, the belt is suitable for the design hp value of 1.05 hp.

As a final check, compute the actual belt speed using the actual 2X value from Table 8. Thus, for a 4L belt on a 6-in. sheave, $2X = 0.20$, and $S = \pi(1,750)(6-0.20)/12 = 2,660$ fpm. Hence, use of 2,800 fpm in selecting the belt was a safe assumption. Note that the

TABLE 10   Hp Ratings of 3L Cross-Section V-Belts
(Based on 180° arc of contact on small sheave)

Belt speed, fpm	Effective OD of small sheave, in.			
	$1\frac{1}{2}$	2	$2\frac{1}{2}$	3 and larger
200	0.05	0.08	0.10	0.11
400	0.08	0.14	0.18	0.20
600	0.11	0.20	0.25	0.29
800	0.12	0.24	0.31	0.36
1,000	0.13	0.28	0.37	0.43
1,200	0.14	0.32	0.43	0.50
1,400	0.15	0.35	0.48	0.56
1,600	0.15	0.38	0.52	0.62
1,800	0.14	0.41	0.57	0.67
2,000	0.13	0.43	0.60	0.72
2,200	0.12	0.44	0.64	0.77
2,400	0.10	0.45	0.66	0.81
2,600	0.07	0.46	0.69	0.84
2,800	0.04	0.46	0.70	0.87
3,000	0.01	0.45	0.72	0.89
3,200	⋯	0.44	0.72	0.91
3,400	⋯	0.42	0.72	0.92
3,600	⋯	0.39	0.71	0.92
3,800	⋯	0.36	0.69	0.92
4,000	⋯	0.31	0.67	0.91
4,200	⋯	0.26	0.64	0.89
4,400	⋯	0.21	0.60	0.86
4,600	⋯	0.14	0.55	0.82
4,800	⋯	0.06	0.49	0.77
5,000	⋯	⋯	0.42	0.72
5,200	⋯	⋯	0.34	0.65
5,400	⋯	⋯	0.26	0.58
5,600	⋯	⋯	0.16	0.49
5,800	⋯	⋯	0.05	0.39
6,000	⋯	⋯	⋯	0.29

TABLE 11   Hp Ratings of 4L Cross-Section V-Belts
(Based on 180° arc of contact on small sheave)

Belt speed, fpm	Effective OD of small sheave, in.				
	2	$2\frac{1}{2}$	3	$3\frac{1}{2}$	4 and larger
200	0.07	0.13	0.16	0.18	0.21
400	0.12	0.23	0.31	0.36	0.40
600	0.15	0.32	0.43	0.51	0.57
800	0.17	0.40	0.54	0.64	0.73
1,000	0.18	0.46	0.65	0.78	0.88
1,200	0.17	0.51	0.74	0.89	1.01
1,400	0.16	0.56	0.82	1.01	1.14
1,600	0.14	0.60	0.90	1.11	1.26
1,800	0.11	0.62	0.96	1.19	1.37
2,000	0.08	0.64	1.02	1.28	1.47
2,200	0.04	0.67	1.08	1.37	1.58
2,400	...	0.68	1.12	1.43	1.66
2,600	...	0.66	1.16	1.50	1.75
2,800	...	0.65	1.18	1.54	1.81
3,000	...	0.63	1.19	1.58	1.87
3,200	...	0.60	1.20	1.61	1.92
3,400	...	0.55	1.19	1.63	1.96
3,600	...	0.50	1.16	1.64	1.98
3,800	...	0.43	1.13	1.63	2.00
4,000	...	0.35	1.09	1.61	2.00
4,200	...	0.24	1.03	1.58	1.98
4,400	...		0.96	1.53	1.95
4,600	...	0.01	0.87	1.46	1.91
4,800	...	...	0.76	1.39	1.85
5,000	...	...	0.65	1.30	1.78
5,200	...	...	0.51	1.19	1.70
5,400	...	...	0.36	1.07	1.59
5,600	...	...	0.18	0.91	1.46
5,800	...	...	...	0.72	1.29
6,000	...	...	...	0.54	1.11

difference between the belt speed based on the assumed value of 2X, 2,675 fpm, and the actual belt speed, 2,660 fpm, is about 0.5 percent. This is negligible.

Related Calculations: Use this procedure when choosing a single V-belt for a drive. Where multiple belts are used, follow the steps given in the next Calculation Procedure. The data presented for single V-belts is abstracted from *Standards for Light-duty or Fractional-horsepower V-Belts*, published by the Rubber Manufacturers Association.

SELECTING MULTIPLE V-BELTS FOR POWER TRANSMISSION

Choose the type and number of V-belts needed to drive an air compressor from a 5-hp wound-rotor ac motor when the motor speed is 1,800 rpm and the compressor speed is

600 rpm. The pitch diameter of the large sheave is 20 in.; and the distance between shaft centers is 36.0 in.

Calculation Procedure:

1. Choose the V-belt section

Determine the design horsepower of the drive by finding the product of the service factor and the rated horsepower. Use Table 3 to find the service factor. The value of this factor is 1.4 for a compressor driven by a wound-rotor ac motor. Thus, the design horsepower = (5.0 hp)(1.4 service factor) = 7.0 hp.

Enter Fig. 3 at 7.0 hp and project up to the small sheave speed, 600 rpm. Read the belt cross section as type *B*.

2. Determine the small-sheave pitch diameter

Use the speed ratio of the shafts to determine the diameter of the small sheave. The speed ratio of the shafts is the ratio of the speed of the high-speed shaft to the low-speed shaft, or 1,800/600 = 3.0. The sheave pitch diameters have the same ratio, or $20/PD_s = 3$; $PD_s = 20/3$ = 6.67 in.

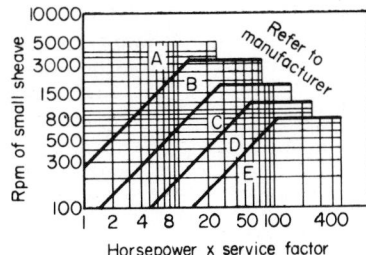

Fig. 3 V-belt cross-section for required hp rating.

3. Compute the belt speed

The belt speed is $S = \pi RD$, where R = small-sheave rpm; D = small-sheave pitch diameter, ft. Thus, $S = \pi(1,800)(6.67/12) = 3,140$ fpm.

4. Determine the belt horsepower rating

A tabulation of allowable belt horsepower ratings is used to determine the rating of a specific belt. To enter this table, the belt speed and the small-sheave equivalent diameter must be known.

Find the equivalent diameter d_e of the small sheave by taking the product of the small-sheave pitch diameter and the diameter factor, Table 2. Thus, for a speed range of 3.0 the small-diameter factor = 1.14, from Table 2. Hence, $d_e = (6.67)(1.14) = 7.6$ in.

Enter Table 12A at a belt speed of 3,200 fpm and $d_e = 7.6$ in. In the last column read

TABLE 12 Small-Diameter Factors—Multiple V-Belts

Speed Ratio Range	Small-Diameter Factor
1.000–1.019	1.00
1.020–1.032	1.01
1.033–1.055	1.02
1.056–1.081	1.03
1.082–1.109	1.04
1.110–1.142	1.05
1.143–1.178	1.06
1.179–1.222	1.07
1.223–1.274	1.08
1.275–1.340	1.09
1.341–1.429	1.10
1.430–1.562	1.11
1.563–1.814	1.12
1.815–2.948	1.13
2.949 and over	1.14

TABLE 12A Horsepower Ratings for Premium-quality B-section V-belts

Belt speed, fpm	Equivalent diameter d_e		
	6.2	6.6	7.0 and over
2,800	5.27	5.65	5.99
3,000	5.49	5.90	6.26
3,200	5.68	6.12	6.50
3,400	5.85	6.31	6.73

the belt horsepower rating as 6.5 hp. This rating must be corrected for the arc of contact and the belt length.

The arc of contact $= 180 - [60(d_l - d_s)/l)]$, where d_l and $d_s =$ large- and small-sheave pitch diameters, respectively, in.; $l =$ distance between sheave shaft centers, in. Thus, arc of contact $= 180 - [60(20 - 6.67)/36)] = 157.8°$. Using Table 4 and interpolating, the arc-of-contact correction factor $C_c = 0.94$.

Compute the belt pitch lengh, in., from $L = 2l + 1.57(d_l + d_s) + (d_l - d_s)^2/4l$, where all the symbols are as given earlier. Thus, $L = 2(36) + 1.57(20 + 6.67) + (20 - 6.67)^2/ [4(36)] = 115.1$ in.

Enter Table 13 by interpolating between the standard belt lengths of 105 and 120 in. and find the length correction factor for a B cross-section belt as 1.06.

Find the product of the rated horsepower of the belt and the two correction factors — arc of contact and belt length, or $(6.5)(0.94)(1.06) = 6.47$ hp.

TABLE 13 Length Correction Factors — Multiple V-Belts

Standard length designation	Belt cross section				
	A	B	C	D	E
26	0.81				
33	0.86				
38	0.88	0.83			
46	0.92	0.87			
51	0.94	0.89	0.80		
55	0.96	0.90			
62	0.99	0.93			
66	1.00	0.94			
71	1.01	0.95			
78	1.03	0.98			
81	...	0.98	0.89		
85	1.05	0.99	0.90		
96	1.08	...	0.92		
105	1.10	1.04	0.94		
120	1.13	1.07	0.97	0.86	
136	...	1.09	0.99		
158	...	1.13	1.02	0.92	
173	...	1.15	1.04	0.93	
195	...	1.18	1.07	0.96	0.92
240	...	1.22	1.11	1.00	0.96
300	...	1.27	1.16	1.05	1.01
360	...	...	1.21	1.09	1.05
420	...	...	1.24	1.12	1.09
540	...	...	...	1.18	1.14
660	...	...	...	1.23	1.19

5. *Choose the number of belts required*

The design horsepower, Step 1, is 7.0 hp. Thus, 7.0 design hp/6.47 rated belt hp = 1.08 belts; use two belts. Choose the next *larger* number of belts whenever a fractional number is indicated.

Related Calculations: The tables and data used here are based on engineering information which is available from and updated by the Mechanical Power Transmission

Association, and the Rubber Manufacturers Association. Similar engineering data are published by the various V-belt manufacturers. Data presented here may be used when manufacturer's engineering data are not available.

SELECTION OF A WIRE-ROPE DRIVE

Choose a wire-rope drive for a 3,000-lb traction-type freight elevator designed to lift freight or passengers totaling 4,000 lb. The vertical lift of the elevator is 500 ft and the rope velocity is 750 fpm. The traction-type elevator sheaves are designed to accelerate the car to full speed in a distance of 60 ft, when starting from a stopped position. A 48-in.-diameter sheave is used for the elevator.

Calculation Procedure:

1. Select the number of hoisting ropes to use

The number of ropes required for an elevator is usually fixed by state or city laws. Check the local ordinances before choosing the number of ropes. Usual laws require at least four ropes for a freight elevator. Assume four ropes are used for this elevator.

2. Select the rope size and strength

Standard "blue-center" steel hoisting rope is a popular choice, as is "plow-steel" and "mild plow-steel" rope. Assume that four $\frac{9}{16}$-in. 6-strand 19-wires-per-strand blue-center steel ropes will be suitable for this car. The 6×19 rope is commonly used for freight and passenger elevators. Once the rope size is chosen, its strength can be checked against the actual load. The breaking strength of $\frac{9}{16}$-in. 6×19 blue-center steel rope is 13.5 tons and its weight is 0.51 lb/ft. These values are tabulated in Baumeister and Marks — *Standard Handbook for Mechanical Engineers* and in rope manufacturers' engineering data.

3. Compute the total load on each rope

The weight of the car and its contents is $3,000 + 4,000 = 7,000$ lb. With four ropes, the load per rope is $7,000/[4(2,000 \text{ lb/ton})] = 0.875$ tons.

With a 500-ft lift the length of each rope would be about equal to the lift height. Hence, with a rope weight of 0.51 lb/ft, total weight of the rope $= (0.51)(500)/2,000 = 0.127$ ton.

Acceleration of the car from the stopped condition places an extra load on the rope. The rate of acceleration of the car is found from $a = v^2/2d$, where $a =$ car acceleration, ft/sec^2; $v =$ final velocity of the car, fps; $d =$ distance through which the acceleration occurs, ft. For this car, $a = (750/60)^2/[2(60)] = 1.3$ ft/sec^2. The value 60 in the numerator of the above relation converts fpm to fps.

The rope load caused by acceleration of the car is $L_r = Wa/(\text{number of ropes})(2,000 \text{ lb/ton})(g = 32.2 \text{ ft/sec}^2)$, where $L_r =$ rope load, tons; $W =$ weight of car and load, lb. Thus, $L = (7,000)(1.3)/[(4)(2,000)(32.2)] = 0.03351$ ton per rope.

The rope load caused by acceleration of the rope is $L_r = Wa/32.2$, where $W =$ weight of rope, tons. Or, $L_r = (0.127)(1.3)/32.2 = 0.0512$ ton.

When the rope bends around the sheave, another load is produced. This bending load is, in lb, $F_b = AE_r d_w/d_s$, where $A =$ rope area, sq in.; $E_r =$ modulus of elasticity of the whole rope $= 12 \times 10^6$ psi for steel rope; $d_w =$ rope diameter, in.; $d_s =$ sheave diameter, in. Thus, for this rope, $F_b = (0.0338)(12 \times 10^6)(0.120)/48 = 1,014$ lb, or 0.507 ton.

The total load on the rope is the sum of the individual loads, or $0.875 + 0.127 + 0.0351 + 0.051 + 0.507 = 1.545$ tons. Since the rope has a breaking strength of 13.5 tons, the factor of safety FS $=$ breaking strength, tons/rope load, tons $= 13.5/1.545 = 8.74$. The usual minimum acceptable factor of safety for elevator ropes is 8.0. Hence, this rope is satisfactory.

Related Calculations: Use this general procedure when choosing wire-rope drives for mine hoists, inclined-shaft hoists, cranes, derricks, car pullers, dredges, well drilling, etc. When choosing *standard hoisting rope*, which is the type most commonly used, the sheave diameter should not be less than $30d_w$; the recommended diameter is $45d_w$.

For *haulage rope* use $42d_w$ and $72d_w$, respectively; for special flexible *hoisting rope*, use $18d_w$ and $27d_w$ sheaves.

SPEEDS OF GEARS AND GEAR TRAINS

A gear having 60 teeth is driven by a 12-tooth gear turning at 800 rpm. What is the speed of the driven gear? What would be the speed of the driven gear if a 24-tooth idler gear were placed between the driving and driven gear? What would be the speed of the driven gear if two 24-tooth idlers were used? What is the direction of rotation of the driven gear when one and two idlers are used? A 24-tooth driving gear turning at 600 rpm meshes with a 48-tooth compound gear. The second gear of the compound gear has 72 teeth and drives a 96-tooth gear. What is the speed and direction of rotation of the 96-tooth gear?

Calculation Procedure:

1. *Compute the speed of the driven gear*

For any two meshing gears, the speed ratio $R_D/R_d = N_d/N_D$, where $R_D =$ rpm of driving gear; $R_d =$ rpm of driven gear; $N_d =$ number of teeth in driven gear; $N_D =$ number of teeth in driving gear. Substituting the given values, $R_D/R_d = N_d/N_D$, or $800/R_d = 60/12$; $R_d = 160$ rpm.

2. *Determine the effect of one idler gear*

An idler gear has *no* effect on the speed of the driving or driven gear. Thus, the speed of each gear would remain the same, regardless of the number of teeth in the idler gear. An idler gear is generally used to reduce the required diameter of the driving and driven gears on two widely separated shafts.

3. *Determine the effect of two idler gears*

The effect of more than one idler is the same as that of a single idler – i.e., the speed of the driving and driven gears remains the same, regardless of the number of idlers used.

4. *Determine the direction of rotation of the gears*

Where an odd number of gears is used in a gear train, the first and last gears turn in the *same* direction. Thus, with one idler, one driver, and one driven gear, the driver and driven gear turn in the *same* direction because there are three gears – i.e., an odd number – in the gear train.

Where an even number of gears is used in a gear train, the first and last gears turn in the *opposite* direction. Thus, with two idlers, one driver, and one driven gear, the driver and driven gear turn in the *opposite* direction because there are four gears – i.e., an even number – in the gear train.

5. *Determine the compound-gear output speed*

A compound gear has two gears keyed to the same shaft. One of the gears is driven by another gear; the second gear of the compound set drives another gear. In a compound gear train, the product of the number of teeth of the driving gears and the rpm of the first driver equals the product of the number of teeth of the driven gears and the rpm of the last driven gear.

In this gearset, the first driver has 24 teeth and the second driver has 72 teeth. The rpm of the first driver is 600. The driven gears have 48 and 96 teeth, respectively. Speed of the final gear is unknown. Applying the above rule, $(24)(72)(600) = (48)(96)(R_d)$; $R_d = 215$ rpm.

Apply the rule in Step 4 to determine the direction of rotation of the final gear. Since the gearset has an even number of gears, four, the final gear revolves in the opposite direction from the first driving gear.

Related Calculations: Use the general procedure given here for gears and gear trains having spur, bevel, helical, spiral, worm, or hypoid gears. Be certain to determine the correct number of teeth and the gear rpm before substituting values in the given equations.

SELECTION OF GEAR SIZE AND TYPE

Select the type and size of gears to use for a 100-cfm reciprocating air compressor driven by a 50-hp electric motor. The compressor and motor shafts are on parallel axes 21 in. apart. The motor shaft turns at 1,800 rpm while the compressor shaft turns at 300 rpm. Is the distance between the shafts sufficient for the gears chosen?

Calculation Procedure:

1. Choose the type of gears to use

Table 14 lists the kinds of gears in common use for shafts having parallel, intersecting, and nonintersecting axes. Thus, Table 14 shows that for shafts having parallel axes, spur or helical, external or internal, gears are commonly chosen. Since external gears are simpler to apply than internal gears, the external type is chosen wherever

TABLE 14   Types of Gears in Common Use*

Parallel axes	Intersecting axes	Nonintersecting parallel axes
Spur, external	Straight bevel	Crossed-helical
Spur, internal	Zerol† bevel	Single-enveloping worm
Helical, external	Spiral bevel	Double-enveloping worm
Helical, internal	Face gear	Hypoid

*From Darle W. Dudley, *Practical Gear Design*, McGraw-Hill, 1954.
†Registered trademark of the Gleason Works.

possible. Internal gears are the planetary type and are popular for applications where limited space is available. Space is not a consideration in this application; hence, an external spur gearset will be used.

Table 15 lists factors to consider when selecting gears by the characteristics of the application. As with Table 14, the data in Table 15 indicate that spur gears are suitable for this drive. Table 16, based on the convenience of the user, also indicates that spur gears are suitable.

2. Compute the pitch diameter of each gear

The distance between the driving and driven shafts is 21 in. This distance is approximately equal to the sum of the driving gear pitch radius r_D in. and the driven gear pitch radius r_d in. Or $d_D + r_d = 21$ in.

In this installation the driving gear is mounted on the motor shaft and turns at 1,800 rpm. The driven gear is mounted on the compressor shaft and turns at 300 rpm. Thus, the speed ratio of the gears (R_D, driver rpm/R_d, driven rpm) = (1,800/300) = 6. For a spur gear, $R_D/R_d = r_d/r_D$, or $6 = r_d/r_D$, and $r_d = 6r_D$. Hence, substituting in $r_D + r_d = 21$, $r_D + 6r_D = 21$; $r_D = 3$ in. Then, $3 + r_d = 21$, $r_d = 18$ in. The respective pitch diameters of the gears are: $d_D = 2 \times 3 = 6.0$ in.; $d_d = 2 \times 18 = 36.0$ in.

3. Determine the number of teeth in each gear

The number of teeth in a spur gearset, N_D and N_d, can be approximated from the ratio $R_D/R_d = N_d/N_D$, or $1,800/300 = N_d/N_D$; $N_d = 6N_D$. Hence, the driven gear will have approximately six times as many teeth as the driving gear.

As a trial, assume that $N_d = 72$ teeth; then $N_D = N_d/6 = 72/6 = 12$ teeth. This assumption must now be checked to determine if the gears will give the desired output speed. Since $R_D/R_d = N_d/N_D$, or $1,800/300 = 72/12$; $6 = 6$. Thus, the gears will provide the desired speed change.

The distance between the shafts is 21 in. $= r_D + r_d$. This means that there is no clearance when the gears are meshed. Since all gears require some clearance, the

TABLE 15   Gear Drive Selection by Application Characteristics*

Characteristic	Type of gearbox	Kind of teeth	Range of use
High power	Simple branched or epicyclic	Helical	Up to 40,000 hp per single mesh; over 60,000 hp in MDT designs; up to 40,000 hp in epicyclic units
	Simple, branched or epicyclic	Spur	Up to 4,000 hp per single mesh; up to 10,000 hp in an epicyclic
	Simple	Spiral bevel	Up to 15,000 hp per single mesh
		Zerol bevel	Up to 1,000 hp per single mesh
High efficiency	Simple	Spur, helical or bevel	Over 99 percent efficiency in the most favorable cases — 98 percent efficiency is typical
Light weight	Epicyclic	Spur or helical	Outstanding in airplane and helicopter drives
	Branched-MDT	Helical	Very good in marine main reductions
	Differential	Spur or helical	Outstanding in high torque actuating devices
		Bevel	Automobiles, trucks, and instruments
Compact	Epicyclic	Spur or helical	Good in aircraft nacelles
	Simple	Worm-gear	Good in high-ratio industrial speed reducers
	Simple	Spiroid	Good in tools and other applications
	Simple	Hypoid	Good in auto and truck rear ends plus other applications

Precision	Simple	Worm-gear	Widely used in machine-tool index drives
	Simple	Hypoid	Used in certain index drives for machine tools
	Simple or branched	Helical	A favorite for high-speed, high-accuracy power gears
	Simple	Spur	Widely used in radar pedestal gearing, gun control drives, navigation instruments, and many other applications
	Simple	Spiroid	Used where precision and adjustable backlash are needed

*Mechanical Engineering, November, 1965.

TABLE 16 Gear Drive Selection for the Convenience of the User*

Consideration	Kind of teeth	Typical applications	Comments
Cost	Spur	Toys, clocks, instruments, industrial drives, machine tools, transmissions, military equipment, household applications, rocket boosters	Very widely used in all manner of applications where power and speed requirements are not too great – parts are often mass-produced at very low cost per part
Ease of use	Spur or helical	Change gears in machine tools, vehicle transmissions where gear shifting occurs	Ease of changing gears to change ratio is important

TABLE 16 Gear Drive Selection for the Convenience of the User* (Continued)

Consideration	Kind of teeth	Typical applications	Comments
Simplicity	Worm-gear	Speed reducers	High ratio drive obtained with only two gear parts
	Crossed-helical	Light power drives	No critical positioning required in a right-angle drive
	Face gear	Small power drives	Simple and easy to position for a right-angle drive
	Helical	Marine main drive units for ships, generator drives in power plants	Helical teeth with good accuracy and a design that provides good axial overlap mesh very smoothly
Noise	Spiral bevel	Main drive units for aircraft, ships, and many other applications	Helical type of tooth action in a right-angle power drive
	Hypoid	Automotive rear axle	Helical type of tooth provides high overlap
	Worm-gear	Small power drives in marine, industrial, and household appliance applications	Overlapping, multiple tooth contacts
	Spiroid-gear	Portable tools, home appliances	Overlapping, multiple tooth contacts

Mechanical Engineering, November, 1965.

shafts will have to be moved apart slightly to provide this clearance. If the shafts cannot be moved apart, the gear diameter must be reduced. In this installation, however, the electric-motor driver can probably be moved a fraction of an in. to provide the desired clearance.

4. Choose the final gear size

Refer to a catalog of stock gears. From this catalog choose a driving and a driven gear having the required number of teeth and the required pitch diameter. If gears of the exact size required are not available, pick the nearest suitable stock sizes.

Check the speed ratio using the procedure in Step 3. As a general rule, stock gears having a slightly different number of teeth or a somewhat smaller or larger pitch diameter will provide nearly the desired speed ratio. When suitable stock gears are not available in one catalog, refer to one or more other catalogs. If suitable stock gears are still not available, and if the speed ratio is a critical factor in the selection of the gear, custom-sized gears may have to be manufactured.

Related Calculations: Use this general procedure to choose gear drives employing any of the twelve types of gears listed in Table 14. Table 17 lists typical gear selections based on the arrangement of the driving and driven equipment. These tables are the work of Darle W. Dudley.

GEAR SELECTION FOR LIGHT LOADS

Detail a generalized gear-selection procedure useful for spur, rack, spiral miter, miter, bevel, helical, and worm gears. Assume that the drive horsepower and speed ratio are known.

Calculation Procedure:

1. Choose the type of gear to use

Use Table 14 of the previous Calculation Procedure as a general guide to the type of gear go use. Make a tentative choice of the gear type.

2. Select the pitch diameter of the pinion and gear

Compute the pitch diameter of the pinion from $d_p = 2c/(R+2)$, where d_p = pitch diameter, in., of the pinion, which is the *smaller* of the two gears in mesh; c = center distance between the gear shafts, in.; R = gear ratio = larger rpm, number of teeth, or pitch diameter + smaller rpm, smaller number of teeth, or smaller pitch diameter.

Compute the pitch diameter of the gear, which is the *larger* of the two gears in mesh, from $d_g = d_p R$.

3. Determine the diametral pitch of the drive

Tables 18 to 21 show typical diametral pitches for various horsepower ratings and gear materials. Enter the appropriate table at the horsepower that will be transmitted and select the diametral pitch of the pinion.

4. Choose the gears to use

Enter a manufacturer's engineering tabulation of gear properties and select the pinion and gear for the horsepower and rpm of the drive. Note that the rated horsepower of the pinion and the gear must equal, or exceed, the rated horsepower of the drive at this specified input and output rpm.

5. Compute the actual center distance

Find half the sum of the pitch diameter of the pinion and the pitch diameter of the gear. This is the actual center-to-center distance of the drive. Compare this value with the available space. If the actual center distance exceeds the allowable distance, try to rearrange the drive or select another type of gear and pinion.

6. Check the drive speed ratio

Find the actual speed ratio by dividing the number of teeth in the gear by the number of teeth in the pinion. Compare the actual ratio with the desired ratio. If there

TABLE 17 Gear Drive Selection by Arrangement of Driving and Driven Equipment*

Kind of teeth	Axes	Gearbox type	Type of tooth contact	Generic family
Spur†	Parallel	Simple (pinion and gear), epicyclic (planetary, star, solar), branched systems, idler for reverse	Line	Coplanar
Helical† (single or double helical, herringbone)	Parallel	Simple, epicyclic, branched	Overlapping line	Coplanar
Bevel..........	Right-angle or angular, but intersecting	Simple, epicyclic, branched	(Straight) line, (zerol)‡ line, (spiral) overlapping	Coplanar
Worm	Right-angle, nonintersecting	Simple	(Cylindrical) overlapping line (double-enveloping),§ overlapping line	Nonplanar
Crossed-helical	Right-angle or skew, nonintersecting	Simple	Point	Nonplanar
Face gear	Right-angle, intersecting	Simple	Line (overlapping if helical)	Coplanar
Hypoid.........	Right-angle or angular, non-intersecting	Simple	Overlapping line	Nonplanar
Spiroid, helicon, ... planoid	Right-angle, nonintersecting	Simple	Overlapping line	Nonplanar

*Mechanical Engineering, November, 1965.
†These kinds of teeth are often used to change rotary motion to linear motion by use of a pinion and rack.
‡Zerol is a registered trademark of the Gleason Works, Rochester, New York.
§The most widely used double-enveloping worm gear is the Cone-Drive type.

TABLE 18 Spur-Gear Pitch Selection Guide*
(20° pressure angle)

Gear diametral pitch	Pinion hp	Gear hp
20	0.04–1.69	0.13–0.96
16	0.09–2.46	0.22–1.61
12	0.24–5.04	0.43–3.16
10	0.46–6.92	0.70–5.12
8	0.88–10.69	1.11–7.87
6	1.84–16.63	2.28–12.39
5	3.04–24.15	3.72–17.19
4	5.29–34.83	6.36–25.17
3	13.57–70.46	15.86–51.91

*Morse Chain Company.

TABLE 19 Miter and Bevel-Gear Pitch Selection Guide*
(20° pressure angle)

	Steel spiral miter	
Gear diametral pitch	Hp, hardened gear	Hp, unhardened gear
18	0.07–0.70	0.04–0.42
12	0.15–1.96	0.09–1.17
10	0.50–4.53	0.30–2.70
8	1.56–7.15	0.93–4.26
7	1.93–9.30	1.15–5.54
6	3.18–12.91	1.90–7.70
5	5.83–16.98	3.47–10.12
	Steel miter	
20	. . .	0.01–0.12
16	0.07–0.73	0.02–0.72
14	. . .	0.04–0.37
12	0.14–2.96	0.07–1.77
10	0.39–3.47	0.23–2.07
8	1.06–7.59	0.63–4.52
6	2.44–9.89	1.45–8.77
5	4.77–13.00	2.66–8.42
4	7.79–21.28	4.64–20.71

Steel and cast-iron bevel gears	
Ratio	Hp
1.5 : 1	0.04–2.34
2 : 1	0.01–12.09
3 : 1	0.04–8.32
4 : 1	0.05–10.60
6 : 1	0.07–2.16

*Morse Chain Company.

TABLE 20 Helical-Gear Pitch Selection Guide*

Gear Diametral Pitch	Hp, Hardened-Steel Gear
20	0.04–1.80
16	0.08–2.97
12	0.22–5.87
10	0.37–8.29
8	0.66–11.71 (1-in. face)
	0.49–9.07 (¾-in. face)
6	1.44–19.15 (1¼-in. face)
	1.15–15.91 (1-in. face)

*Morse Chain Company.

TABLE 21 Worm-Gear Pitch Selection Guide*

Gear diametral pitch	Hp, bronze gears		
	Single	Double	Quadruple
16	0.01–0.29	0.02–0.75	0.03–1.42
12	0.04–0.64	0.05–1.21	0.05–3.11
10	0.06–0.97	0.08–2.49	0.13–4.73
8	0.11–1.51	0.15–3.95	0.08–7.69
6	0.23–3.44	0.33–8.01	0.40–14.42
		Triple	
5	0.51–4.61	1.10–10.53	
4	0.66–6.74		
3	1.31–11.98		

*Morse Chain Company.

is a major difference, change the number of teeth in the pinion or gear or both.

Related Calculations: Use this general procedure to select gear drives for loads up to the ratings shown in the accompanying tables. For larger loads, use the procedures given elsewhere in this Section.

SELECTION OF GEAR DIMENSIONS

A mild-steel 20-tooth 20° full-depth-type spur-gear pinion turning at 900 rpm must transmit 50 hp to a 300-rpm mild-steel gear. Select the number of gear teeth, diametral pitch of the gear, width of the gear face, the distance between the shaft centers, and the dimensions of the gear teeth. The allowable stress in the gear teeth is 8,000 psi.

Calculation Procedure:

1. Compute the number of teeth on the gear

For any gearset, $R_D/R_d = N_d/N_d$, where R_D = rpm of driver; R_d = rpm of driven gear; N_d = number of teeth on the driven gear; N_D = number of teeth on driving gear. Thus, $900/300 = N_d/20$; $N_d = 60$ teeth.

2. Compute the diametral pitch of the gear

The diametral pitch of the gear must be the same as the diametral pitch of the pinion if the gears are to run together. If the diametral pitch of the pinion is known, assume that the diametral pitch of the gear equals that of the pinion.

When the diametral pitch of the pinion is not known, use a modification of the Lewis formula, as shown in the next Calculation Procedure, to compute the diametral pitch. Thus, $P = (\pi \, SaYv/33,000 \, hp)^{0.5}$, where all the symbols are as in the next Calculation Procedure, except that $a = 4$ for machined gears. Obtain $Y = 0.421$ for 60 teeth in a 20° full-depth gear from Baumeister and Marks — *Standard Handbook for Mechanical Engineers*. Assume that $v =$ pitch-line velocity $= 1,200$ fpm. This is a typical reasonable value for v. Then, $P = [\pi \times 8,000 \times 4 \times 0.421 \times 1,200/(33,000 \times 50)]^{0.5} = 5.56$; say 6, because diametral pitch is expressed as a whole number whenever possible.

3. Compute the gear face width

Spur gears often have a face width equal to about four times the circular pitch of the gear. Circular pitch $p_c = \pi/P = \pi/6 = 0.524$. Hence, the face width of the gear $= 4 \times 0.524 = 2.095$ in.; say $2\frac{1}{8}$ in., to the nearest eight of an in.

4. Determine the distance between the shaft centers

Find the exact shaft centerline distance from $d_c = (N_p + N_g)/2P$, where $N_p =$ number of teeth on pinion gear; $N_g =$ number of teeth on gear. Thus, $d_c = (20 + 60)/[2(6)] = 6.66$ in.

5. Compute the dimensions of the gear teeth

Use AGMA *Standards*, Dudley — *Gear Handbook*, or the engineering tables published by gear manufacturers. Each of these sources provides a list of factors by which either the circular or diametral pitch can be multiplied to obtain the various dimensions of the teeth in a gear or pinion. Thus, for a 20° full-depth spur gear, using the circular pitch of 0.524, computed in Step 3 we have the following:

	Factor		Circular pitch		Dimension, in.
Addendum	=	0.3183	×	0.524	=	0.1668
Dedendum	=	0.3683	×	0.524	=	0.1930
Working depth	=	0.6366	×	0.524	=	0.3336
Whole depth	=	0.6866	×	0.524	=	0.3598
Clearance	=	0.05	×	0.524	=	0.0262
Tooth thickness	=	0.50	×	0.524	=	0.262
Width of space	=	0.52	×	0.524	=	0.2725
Backlash = width of space − tooth thickness	=	0.2725 − 0.262			=	0.0105

The dimensions of the pinion teeth are the same as those of the gear teeth.

Related Calculations: Use this general procedure to select the dimensions of helical, herringbone, spiral, and worm gears. Refer to the AGMA *Standards* for suitable factors and typical allowable working stresses for each type of gear and gear material.

HORSEPOWER RATING OF GEARS

What is the strength horsepower rating, durability horsepower rating, and service horsepower rating of a 600-rpm 36-tooth 1.75-in. face-width 14.5° full-depth 6-in. pitch-diameter pinion driving a 150-tooth 1.75-in. face width 14.5° full-depth 25-in. pitch-diameter gear if the pinion is made of SAE 1040 steel 245 BHN and the gear is made of cast steel 0.35/0.45 carbon 210 BHN when the gearset operates under intermittent heavy shock loads for 3 hr/day under fair lubrication conditions? The pinion is driven by an electric motor.

Calculation Procedure:

1. Compute the strength horsepower using the Lewis formula

The widely used Lewis formula gives the strength horsepower, $hp_s = SYFK_v v/33,000$ P, where S = allowable working stress of gear material, psi; Y = tooth form factor (also called the Lewis factor); F = face width, in.; K_v = dynamic load factor = $600/(600 + v)$ for metal gears, $0.25 + 150/(200 + v)$ for nonmetallic gears; v = pitchline velocity, fpm = (pinion pitch diameter, in.)(pinion rpm)(0.262); P = diametral pitch, in., = number of teeth/pitch diameter, in. Obtain values of S and Y from tables in Baumeister and Marks—*Standard Handbook for Mechanical Engineers*, or AGMA *Standards Books*, or gear manufacturers' engineering data. Compute the strength horsepower for the pinion and gear separately.

Using one of the above references for the pinion, $S = 25,000$ psi and $Y = 0.298$. The pitchline velocity for the metal pinion is $v = (6.0)(600)(0.262) = 944$ fpm. Then, $K_v = 600/(600 + 944) = 0.388$. The diametral pitch of the pinion is $P = N_p/d_p$, where N_p = number of teeth on pinion; d_p = diametral pitch of pinion, in. Or $P = 36/6 = 6$.

Substituting the above values in the Lewis formula, $hp_s = (25,000)(0.298)(1.75)$ $(0.388)(944)/[(33,000)(6)] = 24.117$ hp for the pinion.

Using the Lewis formula and the same procedure for the 150-tooth gear, $hp_s = (20,000)(0.374)(1.75)(0.388)(944)/[(33,000)(6)] = 24.2$ hp. Thus, the strength horsepower of the gear is greater than that of the pinion.

2. Compute the durability horsepower

The durability horsepower of spur gears is found from $hp_d = F_i K_r D_o C_r$ for $20°$ pressure-angle full-depth or stub teeth. For $14.5°$ full-depth teeth, multiply hp_d by 0.75. In this relation, F_i = face-width and inbuilt factor from AGMA *Standards*; K_r = factor for tooth form, materials, and ratio of gear to pinion from AGMA *Standards*; $D_o = (d^2_p R_p/158,000)(1 - v^{0.5}/84)$, where d_p = pinion pitch diameter, in.; R_p = pinion rpm; v = pinion pitchline velocity, fpm, as computed in Step 1; C_r = factor to correct for increased stress at the start of single-tooth contact as given by AGMA *Standards*.

Using appropriate values from these standards for low-speed gears of double speed reductions, $hp_d = (0.75)(1.46)(387)(0.0865)(1.0) = 36.6$ hp.

3. Compute the gearset service rating

Determine, by inspection, which is the lowest computed value for the gearset—the strength or durability horsepower. Thus, Step 1 shows that the strength horsepower $hp_s = 13.78$ hp of the pinion is the lowest computed value. Use this lowest value when computing the gear-train service rating.

Using the AGMA *Standards*, determine the service factors for this installation. The load service factor for heavy shock loads and 3 hr/day intermittent operation with an electric-motor drive is 1.5 from the *Standards*. The lubrication factor for a drive operating under fair conditions is, from the *Standards*, 1.25. To find the service rating, divide the lowest computed horsepower by the product of the load and lubrication factors; or service rating = $13.78/[(1.5)(1.25)] = 7.35$ hp.

Were this gearset operated only occasionally (0.5 hr or less per day), the service rating could be determined by using the lower of the two computed strength horse-powers; in this case 13.78 hp. Apply only the load service factor, or 1.25 for occasional heavy shock loads. Thus, the service rating for these conditions = $13.78/1.25 = 11$ hp.

Related Calculations: Similar AGMA gear construction-material, tooth-form, face-width, tooth-stress, service, and lubrication tables are available for rating helical, double-helical, herringbone, worm, straight-bevel, spiral bevel, and Zerol gears. Follow the general procedure given here. Be certain, however, to use the applicable values from the appropriate AGMA tables. In general, choose suitable stock gears first; then check the horsepower rating as detailed above.

MOMENT OF INERTIA OF A GEAR DRIVE

A 12-in. outside-diameter 36-tooth steel pinion gear having a 3-in. face width is mounted on a 2-in.-diameter 36-in.-long steel shaft turning at 600 rpm. The pinion drives a

200-rpm 36-in. outside-diameter 108-tooth steel gear mounted on a 12-in.-long 2-in.-diameter steel shaft that is solidly connected to a 24-in.-long 4-in.-diameter shaft. What is the moment of inertia of the high-speed and low-speed assemblies of this gearset?

Calculation Procedure:

1. Compute the moment of inertia of each gear

The moment of inertia of a cylindrical body about its longitudinal axis is $I_i = WR^2$, where I_i = moment of inertia of a cylindrical body, in.^{4} per in. of length; W = weight of cylinder material, lb/in.^{3}; R = radius of cylinder to its outside surface, in. For a steel shaft or gear, this relation can be simplified to $I_i = D^4/35.997$, where D = shaft or gear diameter, in. When computing I for a gear, treat it as a solid blank of material. This is a safe assumption.

Thus, for the 12-in.-diameter pinion, $I = (12)^4/35.997 = 576.05$ in.^{4} per in. of length. Since the gear has a 3-in. face width, the moment of inertia for the total length $I_t = (3.0\text{ in.})(576.05) = 1,728.15$ in.^{4}.

For the 36-in. gear, $I_i = (36)^4/35.997 = 46,659.7$ in.^{4} per in. of length. With a 3-in. face width, $I_t = (3.0)(46,659.7) = 139,979.1$ in.^{4}.

2. Compute the moment of inertia of each shaft

Follow the same procedure as in Step 1. Thus, for the 36-in.-long 2-in.-diameter pinion shaft, $I_t = (2^4/35.997)(36) = 16.0$ in.^{4}.

For the 12-in.-long 2-in.-diameter portion of the gear shaft, $I_t = (2^4/35.997)(12) = 5.33$ in.^{4}. For the 24-in.-long 4-in.-diameter portion of the gear shaft, $I_t = (4^4/35.997)(24) = 170.69$ in.^{4}. The total moment of inertia of the gear shaft equals the sum of the individual moments, or $I_t = 5.33 + 170.69 = 176.02$ in.^{4}.

3. Compute the high-speed assembly moment of inertia

The effective moment of inertia at the high-speed assembly input = $I_{thi} = I_{th} + I_{tl}/(R_h/R_l)^2$, where I_{th} = moment of inertia of high-speed assembly, in.^{4}; I_{tl} = moment of inertia of low-speed assembly, in.^{4}; R_h = high speed, rpm; R_l = low speed, rpm. To find I_{th} and I_{tl} take the sum of the shaft and gear moments of inertia for the high-speed and low-speed assemblies, respectively. Or $I_{th} = 16.0 + 1,728.15 = 1,744.15$ in.^{4}; $I_{tl} = 176.02 + 139,979.1 = 140,155.1$ in.^{4}.

Then, $I_{thi} = 1,744.15 + 140,155.1/(600/200)^2 = 17,324.2$ in.^{4}.

4. Compute the low-speed assembly moment of inertia

The effective moment of inertia at the low-speed assembly output, $I_{tlo} = I_{tl} + I_{th}(R_h/R_l)^2 = 140,155.1 + (1,744.15)(600/200)^2 = 155,852.5$ in.^{4}.

Note that $I_{thi} \neq I_{tlo}$. One value is approximately nine times that of the other. Thus, when stating the moment of inertia of a gear drive, be certain to specify whether the given value applies to the high- or low-speed assembly.

Related Calculations: Use this procedure for shafts and gears made of any metal — aluminum, brass, bronze, chromium, copper, cast iron, magnesium, nickel, tungsten, etc. Compute WR^2 for steel and multiply the result by [weight of shaft material, lb/in.^{3}/0.283].

BEARING LOADS IN GEARED DRIVES

A geared drive transmits a torque of 48,000 lb-in. Determine the resulting bearing load in the drive shaft if a 12-in. pitch-radius spur gear having a 20° pressure angle is used. A helical gear having a 20° pressure angle and 14.5° spiral angle transmits a torque of 48,000 lb-in. Determine the bearing load it produces if the pitch radius is 12 in. Determine the bearing load in a straight bevel gear having the same proportions as the helical gear above, except that the pitch cone angle is 14.5°. A worm having an efficiency of 70 percent and a 30° helix angle drives a gear having a 20° normal pressure angle. Determine the bearing load when the torque is 48,000 lb-in. and the worm pitch radius is 12 in.

Calculation Procedure:

1. Compute the spur-gear bearing load

The tangential force acting on a spur-gear tooth is $F_t = T/r$, where $F_t =$ tangential force, lb; $T =$ torque, lb-in; $r =$ pitch radius, in. For this gear, $F_t = T/r = 48,000/12 = 4,000$ lb. This force is tangent to the pitch-diameter circle of the gear.

The separating force acting on a spur-gear tooth perpendicular to the tangential force is $F_s = F_t \tan \alpha$, where $\alpha =$ pressure angle, °. For this gear, $F_s = (4,000)(0.364) = 1,456$ lb.

Find the resultant force R_f lb from $R_f = (F_t^2 + F_s^2)^{0.5} = (4,000^2 + 1,456^2)^{0.5} = 4,260$ lb. This is the bearing load produced by the gear.

2. Compute the helical-gear bearing load

The tangential force acting on a helical gear is $F_t = T/r = 48,000/12 = 4,000$ lb. The separating force, acting perpendicular to the tangential force, is $F_s = F_t \tan \alpha / \cos \beta$, where $\beta =$ the spiral angle,°. For this gear, $F_s = (4,000)(0.364)/0.968 = 1,503$ lb. The resultant bearing load, which is a side thrust, is $R_f = (4,000^2 + 1,503^2)^{0.5} = 4,380$ lb.

Helical gears produce an end thrust as well as the side thrust just computed. This end thrust is given by $F_e = F_t \tan \beta$, or $F_e = (4,000)(0.259) = 1,036$ lb. The end thrust of the driving helical gear is equal and opposite to the end thrust of the driven helical gear when the teeth are of the opposite hand in each gear.

3. Compute the bevel-gear bearing load

The tangential force acting on a bevel gear is $F_t = T/r = 48,000/12 = 4,000$ lb. The separating force is $F_s = F_t \tan \alpha \cos \theta$, where $\theta =$ pitch cone angle,°. For this gear, $F_s = (4,000)(0.364)(0.968) = 1,410$ lb.

Bevel gears produce an end thrust similar to helical gears. This end thrust is $F_e = F_t \tan \alpha \sin \theta$, or $F_e = (4,000)(0.364)(0.25) = 364$ lb. The side thrust in a bevel gear is $F_t = 4,000$ lb and acts tangent to the pitch-diameter circle. The resultant is an end thrust produced by F_s and F_e, or $R_f = (F_s^2 + F_e^2)^{0.5} = (1,410^2 + 364^2)^{0.5} = 1,458$ lb. In a bevel-gear drive, F_t is common to both gears, F_s becomes F_e on the mating gear, and F_e becomes F_s on the mating gear.

4. Compute the worm-gear bearing load

The worm tangential force is $F_t = T/r$, or $F_t = 48,000/12 = 4,000$ lb. The separating force is $F_s = F_t E \tan \alpha / \sin \phi$, where $E =$ worm efficiency expressed as a decimal; $\phi =$ worm helix or lead angle,°. Thus, $F_s = (4,000)(0.70)(0.364)/0.50 = 2,040$ lb.

The worm end thrust force is $F_e = F_t E \cotan \phi = (4,000)(0.70)(1.732) = 4,850$ lb. This end thrust acts perpendicular to the separating force. Thus the resultant bearing load $R_f = (F_s^2 + F_e^2)^{0.5} = (2,040^2 + 4,850^2)^{0.5} = 5,260$ lb.

Forces developed by the gear are equal and opposite to those developed by the worm tangential force is cancelled by the gear tangential force.

Related Calculations: Use these procedures to compute the bearing loads in any type of geared drive – open, closed, or semi-enclosed – serving any type of load. Computation of the bearing load is a necessary step in bearing selection.

FORCE RATIO OF GEARED DRIVES

A geared hoist will lift a maximum load of 1,000 lb. The hoist is estimated to have friction and mechanical losses of 5 percent of the maximum load. How much force is required to lift the maximum load if the drum on which the lifting cable reels is 10 in. in diameter and the driving gear is 50 in. in diameter? If the load is raised at a velocity of 100 fpm, what is the hp input? What is the driving-gear tooth load if this gear turns at 191 rpm? A 15-in. triple-reduction hoist has three driving gears of 48, 42, and 36 in. diameters, respectively, and two pinions of 12 and 10 in. diameter. What force is required to lift a 1,000-lb load if friction and mechanical losses are 10 percent?

Calculation Procedure:

1. Compute the total load on the hoist

The friction and mechanical losses *increase* the maximum load on the drum. Thus, the total load on the drum = maximum lifting load, lb + friction and mechanical losses, lb = $1{,}000 + 1{,}000(0.05) = 1{,}050$ lb.

2. Compute the required lifting force

Find the lifting force from $L/D_g = F/d_d$, where L = total load on hoist, lb; D_g = diameter of driving gear, in.; F = lifting force required, lb; d_d = diameter of lifting drum, in. For this hoist, $1{,}050/50 = F/10$; $F = 210$ lb.

3. Compute the horsepower input

Find the horsepower input from $hp = Lv/33{,}000$, where v = load velocity, fpm. Thus, $hp = (1{,}050)(100)/33{,}000 = 3.19$ hp.

Where the mechanical losses are not added to the load before computing the horse-power, use the equation $hp = Lv/(1.00 - \text{losses})(33{,}000)$. Thus, $hp = (1{,}000)(100)/(1 - 0.05)(33{,}000) = 3.19$ hp, as before.

4. Compute the driving-gear tooth load

Assume that the entire load is carried by one tooth. Then, the tooth load L_t lb = $33{,}000\, hp/v_g$, where v_g = peripheral velocity of the driving gear, fpm. With a diameter of 50 in. and a speed of 191 rpm, $v_g = \pi D_g R/12$, where R = gear rpm. Or, $v_g = \pi(50)(191)/12 = 2{,}500$ fpm. Then, $L_t = (33{,}000)(3.19)/2{,}500 = 42.1$ lb. This is a nominal tooth-load value.

5. Compute the triple-reduction hoisting force

Use the equation from Step 2, but substitute the product of the three driving-gear diameters for D_g and the three driven-gear diameters for d_d. The total load = $1{,}000 + 0.10(1{,}000) = 1{,}100$ lb. Then, $L/D_g = F/d_d$, or $1{,}100/(48 \times 42 \times 36) = F/(15 \times 12 \times 10)$; $F = 27.2$ lb. Thus, the triple-reduction hoist reduces the required lifting force to about one-tenth that required by a double-reduction hoist (Step 2).

Related Calculations: Use this procedure for geared hoists of all types. Where desired, the number of gear teeth can be substituted for the driving- and driven-gear diameters in the force equation in Step 2.

DETERMINATION OF GEAR BORE DIAMETER

Two helical gears transmit 500 hp at 3,600 rpm. What should the bore diameter of each gear be if the allowable stress in the gear shafts is 12,500 psi? How should the gears be fastened to the shafts? The shafts are solid in cross section.

Calculation Procedure:

1. Compute the required hub bore diameter

The hub bore diameter must at least equal the outside diameter of the shaft, unless the gear is a press- or shrink-fit on the shaft. Regardless of how the gear is attached to the shaft, the shaft must be large enough to transmit the rated torque at the allowable stress.

Use the method of Step 2 of "Solid and Hollow Shafts in Torsion" in this Section to compute the required shaft diameter, after finding the torque using the method of Step 1 in the same Procedure. Thus, $T = 63{,}000\, hp/R = (63{,}000)(500)/3{,}600 = 8{,}750$ lb-in. Then, $d = 1.72(T/s)^{1/3} = 1.72(8{,}750/12{,}500)^{1/3} = 1.526$ in.

2. Determine how the gear should be fastened to the shaft

First decide if the gears are to be permanently fastened or removable. This decision is usually based on the need for gear removal for maintenance or replacement. Re-movable gears can be fastened by a key, setscrew, spline, pin, clamp, or a taper and screw. Large gears transmitting 100 hp, or more, are usually fitted with a key for easy removal. See "Selection of Keys for Machine Shafts" in this Section for the steps in choosing a key.

Permanently fastened gears can be shrunk, pressed, cemented, or riveted to the shaft. Shrink fits generally transmit more torque before slippage occurs than do press fits. With either type of fastening, interference is necessary, i.e., the gear bore is made smaller than the shaft outside diameter.

Baumeister and Marks—*Standard Handbook for Mechanical Engineers* shows that press or shrink fits on shafts of 1.19 to 1.58 in. diameter should have an interference ranging from 0.3 to 4.0 thousandths of an in. on the diameter, depending on the class of fit desired.

Related Calculations: Use this general procedure for any type of gear—spur, helical, herringbone, worm, etc. Never reduce the shaft diameter below that required by the stress equation, Step 1. Thus, if interference is provided by the shaft diameter, *increase* the diameter; do not reduce it.

TRANSMISSION GEAR RATIO FOR A GEARED DRIVE

A four-wheel vehicle must develop a drawbar pull of 17,500 lb. The engine, which develops 500 hp and drives through a gear transmission a 34-tooth spiral bevel pinion gear which meshes with a spiral bevel gear having 51 teeth. This gear is keyed to the drive shaft of the 48-in.-diameter rear wheels of the vehicle. What transmission gear ratio should be used if the engine develops maximum torque at 1,500 rpm? Select the axle diameter for an allowable torsional stress of 12,500 psi. The efficiency of the bevel-gear differential is 80 percent.

Calculation Procedure:

1. Compute the torque developed at the wheel

The wheel torque = (drawbar pull, lb)(moment arm, ft), where the moment arm = wheel radius, ft. For this vehicle having a wheel radius of 24 in., or 24/12 = 2 ft, the wheel torque = (17,500)(2) = 35,000 lb-ft.

2. Compute the torque developed by the engine

The engine torque $T = 63,000 \ hp/R$, or $T = (63,000)(500)/1,500 = 21,000$ lb-ft, where R = rpm.

3. Compute the differential speed ratio

The differential speed ratio = $N_g/N_p = 51/34 = 1.5$, where N_g = number of gear teeth; N_p = number of pinion teeth.

4. Compute the transmission gear ratio

For any transmission gear, its ratio = (output torque, lb-ft)/(input torque, lb-ft) (differential speed ratio)(differential efficiency), or transmission gear ratio = 35,000/ [(21,000)(1.5)(0.80)] = 1.388. Thus, a transmission with a 1.388 ratio will give the desired output torque at the rated engine speed.

5. Determine the required shaft diameter

Use the relation $d = 1.72(T/s)^{1/3}$ from the previous Calculation Procedure to determine the axle diameter. Since the axle is transmitting a total torque of 35,000 lb-ft, each of the two rear wheels develops a torque of 35,000/2 = 17,500 lb-ft, and $d = 1.72(17,500/12,500)^{1/3} = 1.34$ in.

Related Calculations: Use this general procedure for any type of differential— worm-gear, herringbone-gear, helical-gear, or spiral-gear—connected to any type of differential. The output torque can be developed through a wheel, propeller, impeller, or any other device. Note that although this vehicle has two rear wheels, the total drawbar pull is developed by *both* wheels. Either wheel delivers *half* the drawbar pull. If the total output torque were developed by only one wheel, its shaft diameter would be $d = 1.72(35,000/12,500)^{1/3} = 1.69$ in.

EPICYCLIC GEAR TRAIN SPEEDS

Figure 4 shows several typical arrangements of epicyclic gear trains. The number of teeth and the rpm of the driving arm are indicated in each diagram. Determine the driven-member rpm for each set of gears.

Calculation Procedure:

1. Compute the spur-gear speed

For a gear arranged as in Fig. 4a, $R_d = R_D(1 + N_s/N_d)$, where R_d = driven member rpm; R_D = driving-member rpm; N_s = number of teeth on the stationary gear; N_d = number of teeth on the driven gear. Using the values given for this gear and since the arm is the driving member, $R_d = 40(1 + 84/21)$; $R_d = 200$ rpm.

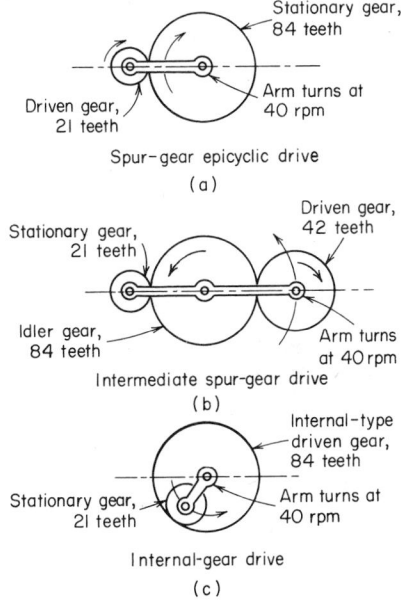

Fig. 4   Epicyclic gear trains.

Note how the driven-gear speed is attained. During one planetary rotation around the stationary gear, the driven gear will rotate axially on its shaft. The number of times the driven gear rotates on its shaft $= N_s/N_d = 84/21 = 4$ times per planetary rotation about the stationary gear. While rotating on its shaft, the driven gear makes a planetary rotation around the fixed gear. So while rotating axially on its shaft four times, the driven gear makes one additional planetary rotation about the stationary gear. Its total axial and planetary rotation is $4 + 1 = 5$ rpm per rpm of the arm. Thus, the gear ratio $G_r = R_D/R_d = 40/200 = 1:5$.

2. Compute the idler gear train speed

The idler gear, Fig. 4b, turns on its shaft while the arm rotates. Movement of the idler gear causes rotation of the driven gear. For an epicyclic gear train of this type, $R_d = R_D(1 - N_s/N_d)$, where the symbols are the same as defined in Step 1. Thus, $R_d = 40(1 - 21/42) = 20$ rpm.

3. Compute the internal gear drive speed

The arm of the internal gear drive, Fig. 4c, turns and carries the stationary gear with it. For a gear train of this type, $R_d = R_D(1 - N_s/N_d)$, or $R_d = 40(1 - 21/84) = 30$ rpm.

Where the internal gear is the driving gear that turns the arm, making the arm the driven member, the velocity equation becomes $R_d = R_D N_D/(N_D + N_s)$, where R_D = driving-member rpm; N_D = number of teeth on the driving member.

Related Calculations: The arm was the driving member for each of the gear trains considered here. However, any gear can be made the driving member if desired. Use the same relations as given above, but substitute the gear rpm for R_D. Thus, a variety of epicyclic gear problems can be solved using these relations. Where unusual epicyclic gear configurations are encountered, refer to Dudley—*Gear Handbook* for a tabular procedure for determining the gear ratio.

PLANETARY-GEAR-SYSTEM SPEED RATIO

Figure 5 shows several arrangements of important planetary-gear systems using internal ring gears, planet gears, sun gears, and one or more carrier arms. Determine the output rpm for each set of gears.

Calculation Procedure:

1. Determine the planetary-gear output speed

For the planetary-gear drive, Fig. 5a, the gear ratio $G_r = (1 + N_4N_=/N_3N_1)/(1 - N_4N_2/N_5N_1)$, where $N_1, N_2, \ldots, N_5 =$ number of teeth, respectively, on each of the gears $1, 2, \ldots, 5$. Also, for any gearset, the gear ratio $G_r =$ input rpm/output rpm, or $G_r =$ driver rpm/driven rpm.

With ring gear 2 fixed and ring gear 5 the output gear, Fig. 5a, and the number of teeth shown, $G_r = \{1 + (33)(74)/[(9)(32)]\}/\{1 - (33)(74)/[(175)(32)]\} = -541.667$. The minus sign indicates that the output shaft revolves in a direction *opposite* to the input shaft. Thus, with an input speed of 5,000 rpm, $G_r =$ input rpm/output rpm; output rpm = input rpm/G_r, or output rpm = 5,000/541.667 = 9.24 rpm.

2. Determine the coupled planetary drive output speed

The drive, Fig. 5b, has the coupled ring gear 2, the sun gear 3, the coupled planet carriers C and C',, and the fixed ring gear 4. The gear ratio is $G_r = (1 - N_2N_4/N_1N_3)$, where the symbols are the same as before. Find the output speed for any given number of teeth by first solving for G_r and then solving $G_r =$ input rpm/output rpm.

With the number of teeth shown, $G_r = \{1 - (75)(75)/[(32)(12)]\} = -13.65$. Then, output rpm = input rpm/G_r = 1,200/13.65 = 87.9 rpm.

Two other arrangements of coupled planetary drives are shown, Fig. 5c and d. Compute the output speed in the same manner as described above.

3. Determine the fixed-differential output speed

Figure 5e and f shows two typical fixed-differential planetary drives. Compute the output speed in the same manner as Step 2.

4. Determine the triple planetary output speed

Figure 5g shows three typical triple planetary drives. Compute the output speed in the same manner as Step 2.

5. Determine the output speed of other drives

Figure 5h, i, j, k, and l shows the gear ratio and arrangement for the following drives: compound spur-bevel gear, plancentric, wobble gear, double eccentric, and Humpage's bevel gears. Compute the output speed for each in the same manner as Step 2.

Related Calculations: Planetary and sun-gear calculations are simple once the gear ratio is determined. The gears illustrated here[1] comprise an important group in the planetary and sun-gear field. For other gear arrangements, consult Dudley's *Gear Handbook*.

SELECTION OF A RIGID FLANGE-TYPE SHAFT COUPLING

Choose a steel flange-type coupling to transmit a torque of 15,000 lb-in. between two $2\frac{1}{2}$-in.-diameter steel shafts. The load is uniform and free of shocks. Determine how many bolts are needed in the coupling if the allowable bolt shear stress is 3,000 psi. How thick must the coupling flange be and how long should the coupling hub be if the allowable stress in bearing for the hub is 20,000 psi, and in shear 6,000 psi? The allowable shear stress in the key is 12,000 psi. There is no thrust force acting on the coupling.

Calculation Procedure:

1. Choose the diameter of the coupling bolt circle

Assume a bolt-circle diameter for the coupling. As a first choice, assume the bolt-circle diameter is three times the shaft diameter, or $3 \times 2.5 = 7.5$ in. This is a reasonable first assumption for most commercially available couplings.

[1]John H. Glover, "Planetary Gear Systems," *Product Engineering*, Jan. 6, 1964.

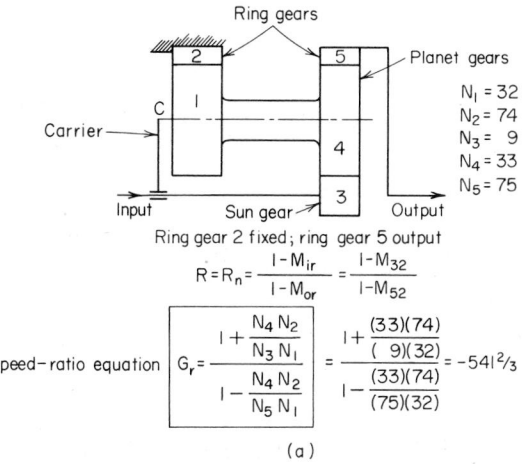

Ring gears

Planet gears

$N_1 = 32$
$N_2 = 74$
$N_3 = 9$
$N_4 = 33$
$N_5 = 75$

Carrier

Input   Sun gear   Output

Ring gear 2 fixed; ring gear 5 output

$$R = R_n = \frac{1 - M_{ir}}{1 - M_{or}} = \frac{1 - M_{32}}{1 - M_{52}}$$

Speed-ratio equation
$$G_r = \frac{1 + \dfrac{N_4 N_2}{N_3 N_1}}{1 - \dfrac{N_4 N_2}{N_5 N_1}} = \frac{1 + \dfrac{(33)(74)}{(9)(32)}}{1 - \dfrac{(33)(74)}{(75)(32)}} = -541\,{}^2/_3$$

(a)

Coupled planetary drives

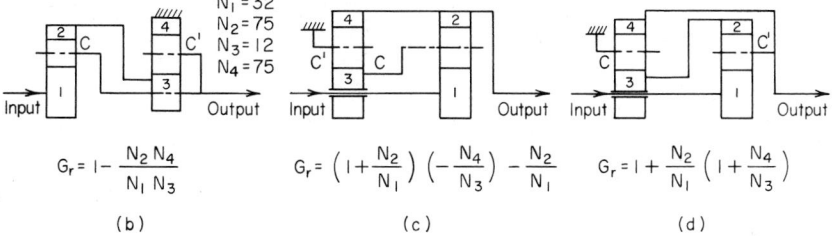

$N_1 = 32$
$N_2 = 75$
$N_3 = 12$
$N_4 = 75$

Input   Output

$$G_r = 1 - \frac{N_2 N_4}{N_1 N_3}$$

(b)

Input   Output

$$G_r = \left(1 + \frac{N_2}{N_1}\right)\left(-\frac{N_4}{N_3}\right) - \frac{N_2}{N_1}$$

(c)

Input   Output

$$G_r = 1 + \frac{N_2}{N_1}\left(1 + \frac{N_4}{N_3}\right)$$

(d)

Fixed-differential drives

Output is difference between speeds of two parts leading to high reduction ratios

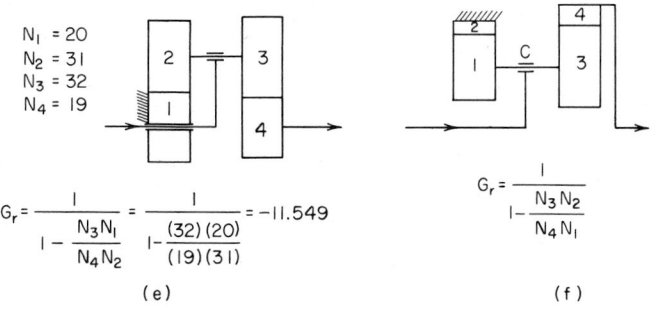

$N_1 = 20$
$N_2 = 31$
$N_3 = 32$
$N_4 = 19$

$$G_r = \frac{1}{1 - \dfrac{N_3 N_1}{N_4 N_2}} = \frac{1}{1 - \dfrac{(32)(20)}{(19)(31)}} = -11.549$$

(e)

$$G_r = \frac{1}{1 - \dfrac{N_3 N_2}{N_4 N_1}}$$

(f)

Fig. 5   Planetary gear systems. *(Product Engineering.)*

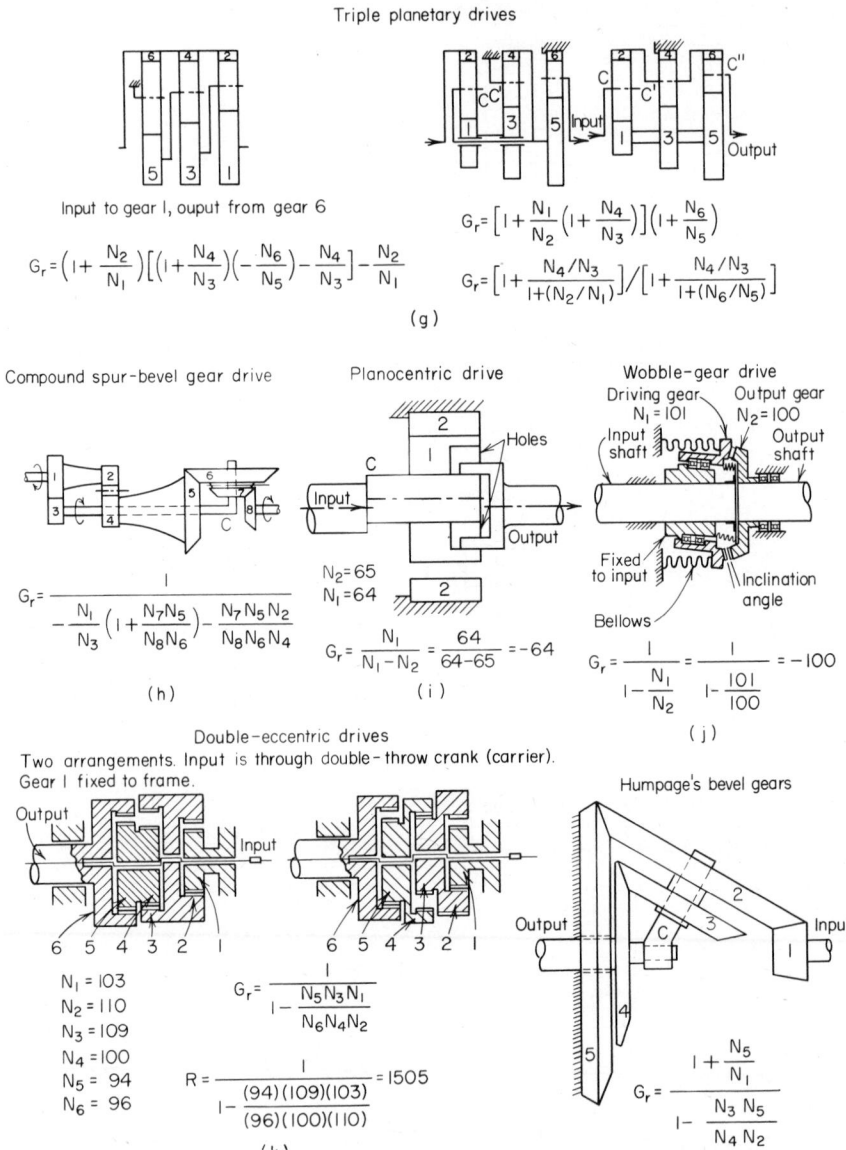

Triple planetary drives

Input to gear 1, ouput from gear 6

$$G_r = \left(1 + \frac{N_2}{N_1}\right)\left[\left(1 + \frac{N_4}{N_3}\right)\left(-\frac{N_6}{N_5}\right) - \frac{N_4}{N_3}\right] - \frac{N_2}{N_1}$$

$$G_r = \left[1 + \frac{N_1}{N_2}\left(1 + \frac{N_4}{N_3}\right)\right]\left(1 + \frac{N_6}{N_5}\right)$$

$$G_r = \left[1 + \frac{N_4/N_3}{1 + (N_2/N_1)}\right] \Big/ \left[1 + \frac{N_4/N_3}{1 + (N_6/N_5)}\right]$$

(g)

Compound spur-bevel gear drive

$$G_r = \cfrac{1}{-\cfrac{N_1}{N_3}\left(1 + \cfrac{N_7 N_5}{N_8 N_6}\right) - \cfrac{N_7 N_5 N_2}{N_8 N_6 N_4}}$$

(h)

Planocentric drive

Holes

$N_2 = 65$
$N_1 = 64$

$$G_r = \frac{N_1}{N_1 - N_2} = \frac{64}{64 - 65} = -64$$

(i)

Wobble-gear drive

Driving gear Output gear
$N_1 = 101$ $N_2 = 100$

Input shaft   Output shaft

Fixed to input   Inclination angle

Bellows

$$G_r = \cfrac{1}{1 - \cfrac{N_1}{N_2}} = \cfrac{1}{1 - \cfrac{101}{100}} = -100$$

(j)

Double-eccentric drives

Two arrangements. Input is through double-throw crank (carrier). Gear 1 fixed to frame.

Output   Input

6 5 4 3 2 1   6 5 4 3 2 1

$N_1 = 103$
$N_2 = 110$
$N_3 = 109$
$N_4 = 100$
$N_5 = 94$
$N_6 = 96$

$$G_r = \cfrac{1}{1 - \cfrac{N_5 N_3 N_1}{N_6 N_4 N_2}}$$

$$R = \cfrac{1}{1 - \cfrac{(94)(109)(103)}{(96)(100)(110)}} = 1505$$

(k)

Humpage's bevel gears

Output   Input

$$G_r = \cfrac{1 + \cfrac{N_5}{N_1}}{1 - \cfrac{N_3 \, N_5}{N_4 \, N_2}}$$

(l)

Fig. 5 Planetary gear systems. *(Product Engineering.)*

2. Compute the shear force acting at the bolt circle

The shear force F_s lb acting at the bolt-circle radius r_b in. is $F_s = T/r_b$, where $T =$ torque on shaft, lb-in. Or, $F_s = 15,000/(7.5/2) = 4,000$ lb.

3. Determine the number of coupling bolts needed

When the allowable shear stress in the bolts is known, compute the number of bolts N required from $N = 8F_s/\pi d^2 s_s$, where $d =$ diameter of each coupling bolt, in.; $s_s =$ allowable shear stress in coupling bolts, psi.

The usual bolt diameter in flanged, rigid couplings ranges from $\frac{1}{4}$ to 2 in., depending on the torque transmitted. Assuming that $\frac{1}{2}$-in.-diameter bolts are used in this coupling, $N = 8(4,000)/[\pi(0.5)^2(3,000)] = 13.58$; say 14 bolts.

Most flanged, rigid couplings have two to eight bolts, depending on the torque transmitted. A coupling having 14 bolts would be a poor design. To reduce the number of bolts, assume a larger diameter — say 0.75 in. Then, $N = 8(4,000)/[\pi(0.75)^2(3,000)] = 6.03$; say eight bolts, because an odd number of bolts is seldom used in flanged couplings.

Determine the shear stress in the bolts by solving the above equation for $s_s = 8F_s/\pi d^2 N = 8(4,000)/[\pi(0.75)^2(8)] = 2,265$ psi. Thus, the bolts are not overstressed, because the allowable stress is 3,000 psi.

4. Compute the coupling flange thickness required

The flange thickness t in. for an allowable bearing stress s_b psi is $t = 2F_s/Nds_b = 2(4,000)/[(8)(0.75)(20,000)] = 0.0666$ in. This thickness is much less than the usual thickness used for flanged couplings manufactured for off-the-shelf use.

5. Determine the hub length required

The hub length is a function of the key length required. Assuming a $\frac{3}{4}$-in.-square key, compute the hub length l in. from $l = 2F_{ss}/t_k s_b$, where $F_{ss} =$ force acting at shaft outer surface, lb; $t_k =$ key thickness, in. The force $F_{ss} = T/r_h$, where $r_h =$ inside radius of hub, in., = shaft radius = $2.5/2 = 1.25$ in. for this shaft. Then, $F_{ss} = 15,000/1.25 = 12,000$ lb-in. Then, $l = 2(12,000)/[(0.75)(20,000)] = 1.6$ in.

When the allowable design stress for bearing, 20,000 psi here, is less than half the allowable design stress for shear, 12,000 psi here, the longest key length is obtained when the bearing stress is used. Thus, it is not necessary to compute the thickness needed to resist the shear stress for this coupling. If it is necessary to compute this thickness, find the force acting at the surface of the coupling hub from $F_h = T/r_h$, where $r_h =$ hub radius, in. Then, $t_s = F_h/\pi d_h s_s$, where $d_h =$ hub diameter, in.; $s_s =$ allowable hub shear stress, psi.

Related Calculations: Couplings offered as standard parts by manufacturers are usually of sufficient thickness to prevent fatigue failure.

Since each half of the coupling transmits the total torque acting, the length of the key must be the same in each coupling half. The hub diameter of the coupling is usually 2 to 2.5 times the shaft diameter, and the coupling lip is generally made the same thickness as the coupling flange. The Procedure given here can be used for couplings made of any metallic material.

SELECTION OF A FLEXIBLE COUPLING FOR A SHAFT

Choose a stock flexible coupling to transmit 15 hp from a 1,000-rpm four-cylinder gasoline engine to a dewatering pump turning at the same rpm. The pump runs 8 hr/day and is an uneven load because debris may enter the pump. The pump and motor shafts are each 1.0 in. in diameter. Maximum misalignment of the shafts will not exceed 0.5°. There is no thrust force acting on the coupling, but the end float or play may reach $\frac{1}{16}$ in.

Calculation Procedure:

1. Choose the type of coupling to use

Consult Table 22 or the engineering data published by several coupling manufacturers. Make a tentative choice from Table 22 of the type of coupling to use, based on the maximum misalignment expected and the tabulated end-float capacity of the coupling. Thus, a roller-chain-type coupling (one in which the two flanges are connected by a double roller chain) will be chosen from Table 22 for this drive because it can accommodate 0.5° of misalignment and an end float of up to $\frac{1}{16}$ in.

TABLE 22   Allowable Flexible Coupling Misalignment

Coupling type	Angular misalignment,°	Parallel misalignment	End float, in.
Plastic chain	Up to 1.0	0.005 in.	$\frac{1}{16}$
Roller chain.......	Up to 0.5	2% of chain pitch	$\frac{1}{16}$
Silent chain.......	Up to 0.5	2% of chain pitch	$\frac{1}{4}$ to $\frac{3}{4}$
Neoprene biscuit..	Up to 5.0	0.01 to 0.05 in.	Up to $\frac{1}{4}$
Radial............	Up to 0.5	0.01 to 0.02 in.	Up to $\frac{1}{16}$

2. Choose a suitable service factor

Table 23 lists typical service factors for roller-chain-type flexible couplings. Thus, for a four-cylinder gasoline engine driving an uneven load, the service factor SF = 2.5.

TABLE 23   Flexible Coupling Service Factors*

Type of drive			Type of load
Engine†, less than than six cylinders	Engine, six cylinders or more	Electric motor; steam turbine	
2.0	1.5	1.0	Even load, 8 hr/day; nonreversing, low starting torque
2.5	2.0	1.5	Uneven load, 8 hr/day; moderate shock or torque, nonreversing
3.0	2.5	2.0	Heavy shock load, 8 hr/day; reversing under full load, high starting torque

*Morse Chain Company.
†Gasoline or diesel.

3. Apply the service factor chosen

Multiply the horsepower or torque to be transmitted by the service factor to obtain the coupling design horsepower or torque. Or, coupling design $hp = (15)(2.5) = 37.5$ hp.

4. Select the coupling to use

Refer to the coupling design horsepower rating table in the manufacturer's engineering data. Enter the table at the shaft rpm and project to a design horsepower slightly greater than the value computed in Step 3. Thus, in Table 24 a typical rating tabulation shows that a coupling design horsepower rating of 38.3 hp is the next higher value above 37.5 hp.

5. *Determine if the coupling bore is suitable*

Table 24 shows that a coupling suitable for 38.3 hp will have a maximum bore diameter up to 1.75 in., and a minimum bore diameter of 0.625 in. Since the engine and pump shafts are each 1.0 in. in diameter, the coupling is suitable.

The usual engineering data available from manufacturers also includes the stock keyway sizes, coupling weight, and the principal dimensions of the coupling. Check the overall dimensions of the coupling to determine if the coupling will fit the available space. Where the coupling bore diameter is too small to fit the shaft, choose the next larger coupling. If the dimensions of the coupling make it unsuitable for the available space, choose a different type or make a coupling.

TABLE 24   Flexible Coupling Hp Ratings*

	Rpm		Bore diameter, in.	
800	1,000	1,200	Maximum	Minimum
16.7	19.9	23.2	1.25	0.5
32.0	38.3	44.5	1.75	0.625
75.9	90.7	105.0	2.25	0.75

*Morse Chain Company.

Related Calculations: Use the general procedure given here to select any type of flexible coupling using flanges, springs, roller chain, preloaded biscuits, etc., to transmit torque. Be certain to apply the service factor recommended by the manufacturer. Note that biscuit-type couplings are rated in hp/100 rpm. Thus, a biscuit-type coupling rated at 1.60 hp/100 rpm and a maximum allowable speed of 4,800 rpm could transmit a maximum of $(1.60 \text{ hp})(4,800/100) = 76.8$ hp.

SELECTION OF A SHAFT COUPLING FOR TORQUE AND THRUST LOADS

Select a shaft coupling to transmit 500 hp and a thrust of 12,500 lb at 100 rpm from a six-cylinder diesel engine. The load is an even one, free of shock.

Calculation Procedure:

1. *Compute the torque acting on the coupling*

Use the relation $T = 5252\, hp/R$ to determine the torque, where T = torque acting on coupling, lb-ft; hp = horsepower transmitted by the coupling; R = shaft rotative speed, rpm. For this coupling, $T = (5,252)(500)/100 = 26,260$ lb-ft.

2. *Find the service torque*

Multiply the torque T by the appropriate service factor from Table 23. This table shows that a service factor of 1.5 is suitable for an even load, free of shock. Thus, the service torque = $(26,260 \text{ lb-ft})(1.5) = 39,390$ lb-ft; say 39,500 lb-ft.

3. *Choose a suitable coupling*

Enter Fig. 6 at the torque on the left and project horizontally to the right. Using the known thrust, 12,500 lb, enter Fig. 6 at the

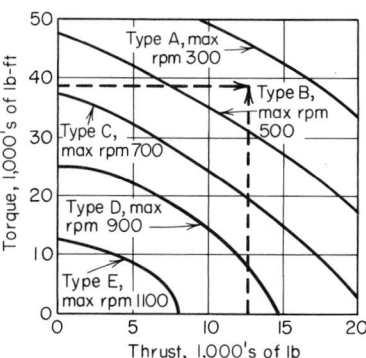

Fig. 6   Shaft-coupling characteristics.

bottom and project vertically upward until the torque line is intersected. Choose the coupling model represented by the next higher curve. This shows that a type A coupling having a maximum allowable speed of 300 rpm will be suitable. If the plotted maximum rpm is lower than the actual rpm of the coupling, use the next plotted coupling type rated for the actual, or a higher, rpm.

When choosing a specific coupling, use the manufacturer's engineering data. This will resemble Fig. 6, or will be a tabulation of the ranges plotted.

Related Calculations: Use this procedure to select couplings for industrial and marine drives where both torque and thrust must be accommodated. See the *Marine Engineering* section of this Handbook for an accurate way to compute the thrust produced on a coupling by a marine propeller. Always check to see that the coupling bore is large enough to accommodate the connected shafts. Where the bore is too small, use the next larger coupling.

HIGH-SPEED POWER-COUPLING CHARACTERISTICS

Select the type of power coupling to transmit 50 hp at 200 rpm if the angular misalignment varies from a minimum of 0° to a maximum of 45°. Determine the effect of angular misalignment on the shaft position, speed, and acceleration at angular misalignments of 30 and 45°.

Calculation Procedure:

1. *Determine the type of coupling to use*

Table 25, developed by N. B. Rothfuss, lists the operating characteristics of eight types of high-speed couplings. Study of this table shows that a universal joint is the only type of coupling among those listed that can handle an angular misalignment of 45°. Further study shows that a universal coupling has a suitable speed and hp range for the load being considered. The other items tabulated are not factors in this application. Therefore, a universal coupling will be suitable. Table 26 compares the functional characteristics of the couplings. Data shown supports the choice of the universal joint.

2. *Determine the shaft position error*

Table 27, developed by David A. Lee, shows the output variations caused by misalignment between the shafts. Thus, at 30° angular misalignment, the position error is 4°06′42″. This means that the output shaft position shifts from $-4°06′42″$ to $+4°06′42″$ twice each revolution. At a 45° misalignment the position error, Table 27, is 9°52′26″. The shift in position is similar to that occurring at 30° angular misalignment.

3. *Compute the output-shaft speed variation*

Table 27 shows that at 30° angular misalignment the output-shaft speed variation is ± 15.47 percent. Thus, the output-shaft speed varies between $200(1.00 \mp 0.1547) = 169.06$ rpm and 230.94 rpm. This speed variation occurs *twice* per revolution.

For a 45° angular misalignment the speed variation, determined in the same way, is 117.16 rpm to 282.84 rpm. This speed variation also occurs twice per revolution.

4. *Determine output-shaft acceleration*

Table 27 lists the ratio of maximum output-shaft acceleration A to the square of the input speed ω^2 expressed in radians. To convert rpm to radians/sec, use $rps = 0.1047$ rpm $= 0.1047(200) = 20.94$ radians/sec.

For 30° angular misalignment, A/ω^2 from Table $27 = 0.294571$. Thus, $A = \omega^2 (0.294571) = (20.94)^2(0.294571) = 129.6$ radians/sec². This means that a constant input speed of 200 rpm produces an output acceleration ranging from -129.6 to $+129.6$ rad/sec², and back, at a frequency of 2(200 rpm) = 400 cycles/min.

At a 45° angular misalignment, the acceleration range of the output shaft, determined in the same way, is -346 to $+346$ radians/sec². Thus, the acceleration range at the larger shaft angle misalignment is 2.67 times that at the smaller, 30°, misalignment.

TABLE 25 Operating Characteristics of Couplings*

	Contoured diaphragm	Axial spring	Laminated disk	Universal joint	Ball-race	Gear	Chain	Elastomeric
Speed range, rpm....	0-60,000	0-8,000	0-20,000	0-8,000	0-8000	0-25,000	0-6,300	0-6,000
Horsepower range, hp/100 rpm	1-500	1-9,000	1-100	1-100	1-100	1-2,000	1-200	0-400
Angular misalignment,° ..	0-8.0	0-2.0	0-1.5	0-45	0-40	0-3	0-2	0-4
Parallel misalignment, in.	0-0.10	0-0.10	0-0.10	None	None	0-0.10	0-0.10	0-0.10
Axial movement, in	0-0.20	0-1.0	0-0.20	None	Some	0-2.0	0-1.0	0-0.30
Ambient temperature, F........	900	Varies	900	Varies	Varies	Varies	Varies	Varies
Ambient pressure, psia. ..	Sea level to zero	Varies	Sea level to zero	Varies	Varies	Varies	Varies	Varies

*Product Engineering.

TABLE 26 Functional Characteristics of Couplings *

	Contoured diaphragm	Axial spring	Laminated disk	Universal joint	Ball-race	Gear	Chain	Elastomeric
No lubrication	√	…	√	…	…	…	…	√
No backlash	√	√	√	†	†	…	…	√
Constant velocity ratio	√	√	…	‡	√	‡	‡	√
Containment	√	…	†	…	†	†	†	…
Angular only	√	…	√	√	√	√	√	√
Axial and angular	√	…	√	…	√	√	√	√
Axial and parallel	√	√	√	…	√	√	√	√
Axial, angular, and parallel	√	√	√	…	…	√	√	√
High temperature	√	…	√	…	…	…	…	…
High altitude	√	…	√	…	…	…	…	…
High torsional spring rate	√	…	√	√	√	√	√	…
Low bending moment	√	√	√	√	√	√	√	√
No relative movement	√	…	…	…	…	…	…	√

*Product Engineering.
†Zero backlash and containment can be obtained by special design.
‡Constant velocity ratio at small angles can be closely approximated.

Related Calculations: Table 25 is useful for choosing any of seven other types of high-speed couplings. The eight couplings listed in this table are popular for high-horse-power applications. All are classed as rigid types, as distinguished from entirely flexible connectors such as flexible cables.

Values listed in Table 25 are nominal ones that may be exceeded by special designs. These values are guideposts rather than fixed; in borderline cases, consult the manufacturer's engineering data. Table 26 compares the functional characteristics of the couplings and is useful to the designer who is seeking a unit with specific operating characteristics. Note that the values in Table 25 are maximum and not additive. In other words, a coupling *cannot* be operated at the maximum angular and parallel misalignment and at the maximum horsepower and speed simultaneously—although in some cases the combination of maximum angular misalignment, maximum horse-power, and maximum speed would be acceptable. Where shock loads are anticipated, apply a suitable correction factor, as given in earlier Calculation Procedures, to the horsepower to be transmitted before entering Table 25.

TABLE 27   Universal Joint Output Variations*

Misalignment angle,°	Maximum position error	Maximum speed error, percent	Ratio A/W^2
5	0°06′34″	0.382	0.011747
10	0°26′18″	1.543	0.030626
15	0°59′36″	3.526	0.069409
20	1°46′54″	6.418	0.124966
25	2°48′42″	10.338	0.198965
30	4°06′42″	15.470	0.294571
35	5°42′20″	22.077	0.417232
40	7°36′43″	30.541	0.576215
45	9°52′26″	41.421	0.787200

*Caused by misalignment of the shaft. Table from *Machine Design.*

SELECTION OF ROLLER AND INVERTED-TOOTH (SILENT) CHAIN DRIVES

Choose a roller chain and the sprockets to transmit 6 hp from an electric motor to a propeller fan. The speed of the motor shaft is 1,800 rpm and of the driven shaft, 900 rpm. How long will the chain be if the centerline distance between the shafts is 30 in?

Calculation Procedure:

1. *Determine, and apply, the load service factor*

Consult the manufacturer's engineering data for the appropriate load service factor. Table 28 shows several typical load ratings (smooth, moderate shock, heavy shock) for various types of driven devices. Use the load rating and the type of drive to determine the service factor. Thus, a propeller fan is rated as a heavy shock load. For this type of load and an electric-motor drive, the load service factor is 1.5, from Table 28.

Apply the load service factor by taking the product of it and the horsepower transmitted, or $(1.5)(6\text{ hp}) = 9.0\text{ hp}$. The roller chain and sprockets must have enough strength to transmit this horsepower.

2. *Choose the chain and number of teeth in the small sprocket*

Using the manufacturer's engineering data, enter the horsepower rating table at the small-sprocket rpm and project to a horsepower value equal to, or slightly greater than, the required rating. At this horsepower rating, read the number of teeth in the small sprocket, which is also listed in the table. Thus, in Table 29, which is an excerpt from a typical horsepower rating tabulation, 9.0 hp is not listed at a speed of 1,800 rpm. However, the next higher horsepower rating, 9.79 hp, will be satisfactory. The table shows that at this power rating, 16 teeth are used in the small sprocket.

This sprocket is a good choice because most manufacturers recommend that at least 16 teeth be used in the smaller sprocket, except at low speeds (100 to 500 rpm).

3. *Determine the chain pitch and number of strands*

Each horsepower rating table is prepared for a given chain pitch, number of chain strands, and various types of lubrication. Thus, Table 29 is for standard single-strand

TABLE 28   Roller Chain Loads and Service Factors*

Load rating	
Driven device	Type of load
Agitators (paddle or propeller)	Smooth
Brick and clay machinery	Heavy shock
Compressors (centrifugal and rotary) ...	Moderate shock
Conveyors (belt)	Smooth
Crushing machinery..................	Heavy shock
Fans (centrifugal)	Moderate shock
Fans (propeller).....................	Heavy shock
Generators and exciters...............	Moderate shock
Laundry machinery	Moderate shock
Mills...............................	Heavy shock
Pumps (centrifugal, rotary)............	Moderate shock
Textile machinery....................	Smooth

Service factor			
Type of load	Internal-combustion engine		Electric motor or turbine
	Hydraulic drive	Mechanical drive	
Smooth.........	1.0	1.2	1.0
Moderate shock.	1.2	1.4	1.3
Heavy shock....	1.4	1.7	1.5

*Excerpted from Morse Chain Company data.

TABLE 29   Roller Chain Hp Rating*
(Single-strand, $\frac{3}{8}$-in.-pitch roller chain)

No. of teeth in small sprocket	Small sprocket rpm		
	1,500	1,800	2,100
14	10.7	8.01	6.34
15	11.9	8.89	7.03
16	13.1	9.79	7.74
17	14.3	10.7	

*Excerpted from Morse Chain Company data.

$\frac{5}{8}$-in. pitch roller chain. The 9.79-hp rating at 1,800 rpm for this chain is with type III lubrication – oil bath or oil slinger – with the oil level maintained in the chain casing at a predetermined height. See the manufacturer's engineering data for the other types of lubrication (manual, drip, and oil stream) requirements.

4. Compute the drive speed ratio

For a roller-chain drive, the speed ratio $S_r = R_h/R_l$, where R_h = rpm of high-speed shaft; R_l = rpm of low-speed shaft. For this drive, $S_r = 1,800/900 = 2$.

5. Determine the number of teeth in the large sprocket

To find the number of teeth in the large sprocket, multiply the number of teeth in the small sprocket, found in Step 2, by the speed ratio, found in Step 4. Thus, the number of teeth in the large sprocket = $(16)(2) = 32$.

6. Select the sprockets

Refer to the manufacturer's engineering data for the dimensions of the available sprockets. Thus, one manufacturer supplies the following sprockets for $\frac{5}{8}$-in. pitch single-strand roller chain: 16 teeth, OD = 3.517 in., bore = $\frac{5}{8}$ in.; 32 teeth, OD = 6.721 in., bore = $\frac{5}{8}$ or $\frac{3}{4}$ in. When choosing a sprocket, be certain to refer to data for the size and type of chain selected in Step 3, because each sprocket is made for a specific type of chain. Choose the type of hub – setscrew, keyed, or taper-lock bushing – based on the torque that must be transmitted by the drive. See earlier Calculation Procedures in this section for data on key selection.

7. Determine the length of the chain

Compute the chain length in pitches L_p from $L_p = 2C + (S/2) + K/C$, where C = shaft center distance, in./chain pitch, in.; S = sum of the number of teeth in the small and large sprocket; K = a constant from Table 30, obtained by entering this table with the

TABLE 30 Roller Chain Length Factors*

D	K	D	K
1	0.03	11	3.06
2	0.10	12	3.65
3	0.23	13	4.28
4	0.41	14	4.96
5	0.63	15	5.70
6	0.91	16	6.48
7	1.24	17	7.32
8	1.62	18	8.21
9	2.05	19	9.14
10	2.53	20	10.13

*Excerpted from Morse Chain Company data.

value D = number of teeth in large sprocket – number of teeth in small sprocket. For this drive, $C = 30/0.625 = 48$; $S = 16 + 32 = 48$; $D = 32 - 16 = 16$; $K = 6.48$ from Table 30. Then, $L_p = 2(48) + (48/2) + 6.48/48 = 120.135$ pitches. However, a chain cannot contain a fractional pitch; therefore, use the next higher number of pitches, or $L_p = 121$ pitches.

Convert the length in pitches to length in in., L_i, by taking the product of the chain pitch p in. and L_p. Or $L_i = L_p p = (121)(0.625) = 75.625$ in.

Related Calculations: At low-speed ratios, large-diameter sprockets can be used to reduce the roller-chain pull and bearing loads. At high-speed ratios, the number of teeth in the high-speed sprocket may have to be kept as small as possible to reduce the chain pull and bearing loads. The Morse Chain Company states: Ratios over 7:1 are generally not recommended for single-width roller-chain drives. Very slow speed drives (10 to 100 rpm) are often practical with as few as 9 or 10 teeth in the small sprocket, allowing ratios up to 12:1. In all cases where ratios exceed 5:1, the designer should consider the possibility of using compound drives to obtain maximum service life.

When selecting standard inverted-tooth (silent) chain and high-velocity inverted-tooth silent-chain drives, follow the same general procedures as given above, except for the following changes.

Standard inverted-tooth silent chain: (a) Use a minimum of 17 teeth, and an odd number of teeth on one sprocket, where possible. This increases the chain life. (b) To achieve minimum noise, select sprockets having 23, or more, teeth. (c) Use the proper service factor for the load, as given in the manufacturer's engineering data. (d) Where a long or fixed-center drive is necessary, use a sprocket or shoe idler where the largest amount of slack occurs. (e) Do not use an idler to reduce the chain wrap on small-diameter sprockets. (f) Check to see that the small-diameter sprocket bore will fit the high-speed shaft. Where the high-speed shaft diameter exceeds the maximum bore available for the chosen smaller sprocket, increase the number of teeth in the sprocket or choose the next larger chain pitch. (This general procedure also applies to roller-chain sprockets.) (g) Compute the chain design horsepower from (drive hp)(chain service factor). (h) Select the chain pitch, number of teeth in the small sprocket, and the chain *width* from the manufacturer's rating table. Thus, if the chain design horsepower = 36 hp and the chain is rated at 4 hp per in. of width, the required chain width = 36 hp/4 hp/in. = 9 in.

High-velocity inverted-tooth silent chain: (a) Use a minimum of 25 teeth and an odd number of teeth on one sprocket, where possible. This increases the chain life. (b) To achieve minimum noise, select sprockets with 27, or more, teeth. (c) Use a larger service factor than the manufacturer's engineering data recommends, if trouble-free drives are desired. (d) Use a wider chain than needed, if an increased chain life is wanted. Note that the chain width is computed in the same way as described in item *h* above. (e) If a longer center distance between the drive shafts is desired, select a larger chain pitch (usual pitches are ¾, 1, 1½, or 2 in.). (f) Provide a means to adjust the center-line distance between the shafts. Such an adjustment *must* be provided in vertical drives. (g) Try to use an even number of pitches in the chain to avoid an offset link.

CAM CLUTCH SELECTION AND ANALYSIS

Choose a cam-type clutch to drive a centrifugal pump. The clutch must transmit 125 hp at 1,800 rpm to the pump, which starts and stops 40 times per hr throughout its 12 hr/day, 360 day/year operating period. The life of the pump will be 10 years.

Calculation Procedure:

1. Compute the maximum torque acting on the clutch

Compute the torque acting on the clutch from $T = 5,252 \, hp/R$, where the symbols are the same as in the previous Calculation Procedure. Thus, for this clutch, $T = (5,252) \times (125)/1,800 = 365$ lb-ft.

2. Analyze the torque acting on the clutch

For installations free of shock loads during starting and stopping, the running torque is the maximum torque that acts on the clutch. But if there is a shock load during starting or stopping, or at other times, the shock torque must be added to the running torque to determine the total torque acting. Compute the shock torque using the relation in Step 1 and the actual hp and speed developed by the shock load.

3. Compute the total number of load applications

With 40 starts and stops (cycles) per hr, a 12-hr day, and 360 operating days per year, the number of cycles per year is (40 cycles/hr)(12 hr/day)(360 days/year) = 172,800. In 10 years, the clutch will undergo (172,800 cycles/year)(10 years) = 1,728,000 cycles.

4. Choose the clutch size

Enter Fig. 7 at the maximum torque, 365 lb-ft, on the left, and the number of load cycles, 1,728,000, on the bottom. Project horizontally and vertically until the point of intersection is reached. Select the clutch represented by the next higher curve. Thus a type A clutch would be used for this load. (Note that the clutch capacity could be tabulated instead of plotted but the results would be the same.)

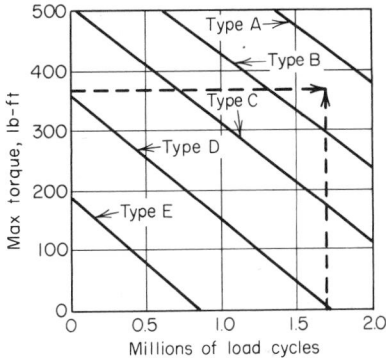

Fig. 7   Cam-type-clutch selection chart.

5. Check the clutch dimensions

Determine if the clutch bore will accommodate the shafts. If the clutch bore is too small, choose the next larger clutch size. Also check to see if the clutch will fit into the available space.

Related Calculations: Use this general procedure to select cam-type clutches for business machines, compressors, conveyors, cranes, food processing, helicopters, fans, aircraft, printing machinery, pumps, punch presses, speed reducers, looms, grinders, etc. When choosing a specific clutch, use the manufacturer's engineering data to determine the clutch size to select.

TIMING-BELT DRIVE SELECTION AND ANALYSIS

Choose a toothed timing belt to transmit 20 hp from an electric motor to a rotary mixer for liquids. The motor shaft turns at 1,750 rpm and the mixer shaft is to turn at 600 rpm ± 20 rpm. This drive will operate 12 hr/day, 7 days/week. Determine the type of timing belt to use and the driving and driven pulley diameters if the shaft centerline distance is about 27 in.

Calculation Procedure:

1. Choose the service factor for the drive

Timing-belt manufacturers publish service factors in their engineering data based on: (*a*) the type of prime mover, (*b*) the type of driven machine (compressor, mixer, pump, etc.), (*c*) type of drive (speedup), and (*d*) drive conditions (continuous operation, use of an idler, etc.).

Usual service factors for any type of driver range from 1.3 to 2.5 for various types of driven machines. Correction factors for speedup drives range from 0 to 0.40; the specific value chosen is *added* to the machine-drive correction factor. Drive conditions, such as 24-hr continuous operation, or the use of an idler pulley on the drive, cause an additional 0.2 to be added to the correction factor. Seasonal or intermittent operation *reduces* the machine-drive factor by 0.2.

Look up the service factor in Table 31, if the manufacturer's engineering data are not readily available. Table 31 gives safe data for usual timing-belt applications and is suitable for preliminary selection of belts. Where a final choice is being made, use the manufacturer's engineering data.

For a liquid mixer shock-free load, use a service factor of 2.0 from Table 31, since there are no other features which would require a larger value.

2. Compute the design horsepower for the belt

The design horsepower $hp_d = hp_l \times SF$, where hp_l = load horsepower; SF = service factor. Thus, for this drive, $hp_d = (20)(2) = 40$ hp.

3. Compute the drive speed ratio

The drive speed ratio $S_r = R_h/R_l$, where R_h = rpm of high-speed shaft; R_l = rpm of low-speed shaft. For this drive $S_r = 1,750/600 = 2.92:1$ the rated rpm. If the driven-pulley speed falls 10 rpm, $S_r = 1,750/580 = 3.02:1$. Thus, the speed ratio may vary between 2.92 and 3.02.

TABLE 31 Typical Timing-Belt Service Factors*

Type of drive	Type of load	Service factor
Electric motors, hydraulic motors, internal-combustion engines, line shafts	Shock-free	2.0
	Shocks	2.5
	Continuous operation or idler use	2.7
	Speed-up	3.0

*Use only for preliminary selection of belt. From Morse Chain Company data.

4. Choose the timing-belt pitch

Enter Table 32, or the manufacturer's engineering data, at the design horsepower and project to the driver rpm. Where the exact value of the design horsepower is not tabulated, use the next higher tabulated value. Thus, for this 1,750-rpm drive having a design horsepower of 40, Table 32 shows that a $\frac{7}{8}$-in. pitch belt is required. This value is found by entering Table 32 at the next higher design horsepower, 50, and projecting to the 1,750-rpm column. If 40 hp were tabulated, the table would be entered at this value.

TABLE 32 Typical Timing-Belt Pitch*

Design hp	Speed of high-speed shaft, rpm				
	3,500	1,750	1,160	870	690
0.5	$\frac{3}{8}$	$\frac{3}{8}$	$\frac{3}{8}$	$\frac{3}{8}, \frac{1}{2}$	$\frac{3}{8}, \frac{1}{2}$
1	$\frac{3}{8}$	$\frac{3}{8}, \frac{1}{2}$	$\frac{3}{8}, \frac{1}{2}$	$\frac{3}{8}, \frac{1}{2}$	$\frac{1}{2}$
5	$\frac{1}{2}$	$\frac{1}{2}$	$\frac{1}{2}$	$\frac{1}{2}$	$\frac{1}{2}, \frac{7}{8}$
10	$\frac{1}{2}$	$\frac{1}{2}$	$\frac{1}{2}$	$\frac{1}{2}, \frac{7}{8}$	$\frac{1}{2}, \frac{7}{8}$
20	$\frac{1}{2}$	$\frac{1}{2}, \frac{7}{8}$	$\frac{1}{2}, \frac{7}{8}$	$\frac{7}{8}$	$\frac{7}{8}$
25	$\frac{1}{2}, \frac{7}{8}$	$\frac{1}{2}, \frac{7}{8}$	$\frac{7}{8}$	$\frac{7}{8}$	$\frac{7}{8}$
50	$\frac{1}{2}, \frac{7}{8}$	$\frac{7}{8}$	$\frac{7}{8}, 1\frac{1}{4}$	$\frac{7}{8}, 1\frac{1}{4}$	$\frac{7}{8}, 1\frac{1}{4}$
60 and up	$\frac{7}{8}$	$\frac{7}{8}$	$\frac{7}{8}, 1\frac{1}{4}$	$\frac{7}{8}, 1\frac{1}{4}$	$1\frac{1}{4}$

*Morse Chain Company.

5. Choose the number of teeth for the high-speed sprocket

Enter Table 33, or the manufacturer's engineering data, at the timing-belt pitch and project across to the rpm of the high-speed shaft. Opposite this value read the minimum number of sprocket teeth. Thus, for a 1,750-rpm $\frac{7}{8}$-in. pitch timing belt, Table 32 shows that the high-speed sprocket should not have less than 24 teeth, nor a pitch diameter less than 6.685 in. (If a smaller diameter sprocket were used, the belt service life would be reduced.)

6. Select a suitable timing belt

Enter Table 34, or the manufacturer's engineering data, at either the exact speed ratio, if tabulated, or the nearest value to the speed-ratio range. For this drive, having a

TABLE 33 Minimum Number of Sprocket Teeth*

Belt pitch, in.	High-speed shaft rpm	Minimum sprocket pitch distance, in.	No. of teeth
$\frac{3}{8}$	3,500	1.910	16
	1,750	1.671	14
	1,160	1.432	12
$\frac{1}{2}$	3,500	3.501	20
	1,750	3.183	18
	1,160	2.865	16
$\frac{7}{8}$	3,500	7.241	26
	1,750	6.685	24
	1,160	6.127	22
$1\frac{1}{4}$	3,500	10.345	26
	1,750	9.549	24
	1,160	8.753	22

*Morse Chain Company.

ratio of 2.92 to 3.02, the nearest value in Table 34 is 3.00. This table shows that with a 24-tooth driver and a 72-tooth driven sprocket, a center distance of 27.17 in. is obtainable. Since a center distance of about 27 in. is desired, this belt is acceptable.

Where an exact center distance is specified, several different sprocket combinations may have to be tried before a belt having a suitable center distance is obtained.

TABLE 34 Timing-Belt Center Distances*

Speed ratio	No. of sprocket teeth		Center distance, in.		
	Driver	Driven	XH	XH	XH
2.80	30	84	22.81	30.11	37.30
3.00	24	72	27.17	34.34	41.46
3.20	30	96	19.19	26.84	34.19

*Morse Chain Company.

7. Determine the required belt width

Each center distance listed in Table 34 corresponds to a specific pitch and type of belt construction. The belt construction is often termed XL, L, H, XH, and XXH. Thus, the belt chosen in Step 6 is an XH construction.

Refer now to Table 35 or the manufacturer's engineering data. Table 35 shows that a 2-in.-wide belt will transmit 38 hp at 1,750 rpm. This is too low, because the design horsepower rating of the belt is 40 hp. A 3-in.-wide belt will transmit 60 hp. Therefore, a 3-in. belt should be used because it can safely transmit the required horsepower.

If five, or less, teeth are in mesh when a timing belt is installed, the width of the belt must be increased to ensure sufficient load-carrying ability. To determine the required belt width to carry the load, divide the belt width by the appropriate factor given below.

Teeth in mesh	5	4	3	2
Factor.........	0.80	0.60	0.40	0.20

Thus, a 3-in. belt with four teeth in mesh would have to be widened to $3/0.60 = 5.0$ in. to carry the desired load.

Related Calculations: Use this procedure to select timing belts for any of these drives: agitators, mixers, centrifuges, compressors, conveyors, fans, blowers, generators (electric), exciters, hammer mills, hoists, elevators, laundry machinery, line shafts, machine tools, paper-manufacturing machinery, printing machinery, pumps, saw mills, textile machinery, woodworking tools, etc. For exact selection of a specific make of belt, consult the manufacturer's tabulated or plotted engineering data.

TABLE 35   Belt Horsepower Ratings*
($\frac{7}{8}$-in. pitch XH)

No. of teeth in high-speed sprocket	Belt width, in.	Sprocket rpm			
		1,500	1,700	1,750	2,000
24	2	34	37	38	43
	3	53	59	60	67
	4	75	83	85	95

*Morse Chain Company.

GEARED SPEED REDUCER SELECTION AND APPLICATION

Select a speed reducer to lift a sluice gate weighing 200 lb through a distance of 6 ft in 5 sec or less. The door must be opened and closed 12 times per hr. The drive for the door lifter is a 1,150-rpm electric motor that operates 10 hr/day.

Calculation Procedure:

1. Choose the type of speed reducer to use

There are many types of speed reducers available for industrial drives. Thus, a roller-chain with different size sprockets, a V-belt drive, or a timing-belt drive might be considered for a speed-reduction application because all will reduce the speed of a driven shaft. Where a load is to be raised, geared speed reducers are often selected because they provide a positive drive without slippage. Also, modern geared drives are compact, efficient units that are easily connected to an electric motor. For these reasons, a right-angle worm-gear speed reducer will be tentatively chosen for this drive. If upon investigation this type of drive proves unsuitable, another type will be chosen.

2. Determine the torque that the speed reducer must develop

A convenient way to lift a sluice door is by means of a roller chain attached to a bracket on the door and driven by a sprocket keyed to the speed reducer output shaft. As a trial, assume that a 12-in.-diameter sprocket is used.

The torque T lb-in. developed by sprocket $= T = Wr$, where $W =$ weight lifted, lb; $r =$ sprocket radius, in. For this sprocket, assuming that the starting friction in the sluice-door guides produces an additional load of 50 lb, $T = (200 + 50)(6) = 1,500$ lb-in.

3. Compute the required rpm of the output shaft

The door must be lifted 6 ft in 5 sec. This is a speed of (6 ft × 60 sec/min)/5 sec = 72 fpm. The circumference of the sprocket is $\pi d = \pi(1.0) = 3.142$ ft. To lift the door at a speed of 72 fpm, the output shaft must turn at a speed of fpm/(ft/revolution) = 72/3.142 = 22.9 rpm. Since a slight increase in the speed of the door is not objectionable, assume that the output shaft turns at 23 rpm.

4. Apply the drive service factor

The AGMA *Standard Practice for Single and Double Reduction Cylindrical Worm and Helical Worm Speed Reducers* lists service factors for geared speed reducers driven by electric motors and internal-combustion engines. These factors range between a low of 0.80 for an electric motor driving a machine producing a uniform load for occasional 0.5-hr service to a high of 2.25 for a single-cylinder internal-combustion engine driving a heavy shock load 24 hr/day. The service factor for this drive, assuming a heavy shock load during opening and closing of the sluice gate, would be 1.50 for 10 hr/day operation. Thus, the drive must develop a torque of at least (load torque, lb-in.)(service factor) = (1,500)(1.5) = 2,250 lb-in.

5. Choose the speed reducer

Refer to Table 36 or the manufacturer's engineering data. Table 36 shows that a single-reduction worm-gear speed reducer having an input of 1.24 hp will develop 2,300 lb-in. of torque at 23 rpm. This is an acceptable speed reducer because the required output torque is 2,250 lb-in. at 23 rpm. Also, the allowable overhung load, 1,367 lb, is adequate for the sluice-gate weight. A 1.5 hp motor would be chosen for this drive.

Related Calculations: Use this general procedure to select geared speed reducers (single- or double-reduction worm gears, single-reduction helical gears, gearmotors, and miter boxes) for machinery drives of all types, including pumps, loaders, stokers,

TABLE 36   Speed Reducer Torque Ratings*
(Single-reduction worm gear)

Input hp at 1,150 rpm	Drive output		
	Rpm	Torque, lb-in.	OHL,[†] lb
1.54	28.7	2,416	1,367
1.24	23.0	2,300	1,367
0.93	19.2	1,970	1,367

*Extracted from Morse Chain Company data.
†Allowable overhung load on drive, lb.

welding positioners, fans, blowers, machine tools, etc. The starting friction load, applied to the drive considered in this procedure, is typical for applications where a heavy friction load is likely to occur. In rotating machinery of many types, the starting friction load is usually nil, except where the drive is connected to a loaded member, such as a conveyor belt. Where a clutch disconnects the driver from the load, there is negligible starting friction.

Well-designed geared speed reducers generally will not run at temperatures higher than 100 F *above* the prevailing ambient temperature, measured in the lubricant sump. At higher operating temperatures the lubricant may break down, leading to excessive wear. Fan-cooled speed reducers can carry heavier loads than noncooled reducers without overheating.

POWER TRANSMISSION FOR A VARIABLE-SPEED DRIVE

Choose the power-transmission system for a three-wheeled contractor's vehicle designed to carry a load of 1,000 lb at a speed of 8 mph over rough terrain. The vehicle tires will be 16 in. in diameter, and the engine driving the vehicle will operate continuously. The empty vehicle weighs 600 lb, and the engine being considered has a maximum speed of 4,200 rpm.

Calculation Procedure:

1. Compute the horsepower required to drive the vehicle

Compute the required driving horsepower from $hp = 1.25 \, Wmph/1,750$, where $W =$ total weight of *loaded* vehicle, lb; $mph =$ maximum loaded vehicle speed, mph. Thus, for this vehicle, $hp = 1.25(1,000 + 600)(8)/1,750 = 9.15$ hp.

2. Determine the maximum vehicle wheel speed

Compute the maximum wheel rpm from, $rpm_w =$ (maximum vehicle speed, mph) $\times$ (5,280 ft/mile)/15.72 (tire rolling diameter, in.). Or, $rpm_w = (8)(5,280)/[(15.72)(16)] =$ 167.8 rpm.

3. Select the power transmission for the vehicle

Refer to engineering data published by drive manufacturers. Choose a drive suitable for the anticipated load. The load on a typical contractor's vehicle is one of sudden starts and stops. Also, the drive must be capable of transmitting the required horsepower. A 10-hp drive would be chosen for this vehicle.

Small vehicles are often belt-driven by means of an infinitely variable transmission. Such a drive, having an overdrive or speed-increase ratio of 1:1.5 or 1:1, would be suitable for this vehicle. From the manufacturer's engineering data, a drive having an input rating of 10 hp will be suitable for momentary overloads of up to 25 percent. The operating temperature of any part of the drive should never exceed 250 F. For best results, the drive should be operated at temperatures well below this limit.

4. Compute the required output-shaft speed reduction

To obtain the maximum power output from the engine, the engine should operate at its maximum rpm when the vehicle is traveling at its highest speed. This prevents lugging of the engine at lower speeds.

The transmission transmits power from the engine to the driving axle. Usually, however, the transmission cannot provide the needed speed reduction between the engine and the axle. Therefore, a speed-reduction gear is needed between the transmission and the axle. The transmission chosen for this drive could provide a 1:1 speed ratio, or a 1:1.5 speed ratio. Assume that the 1:1.5 speed ratio is chosen to provide higher speeds at the maximum vehicle load. Then, speed reduction required = (maximum engine speed, rpm)(transmission ratio)/(maximum wheel rpm) = (4,200)(1.5)/167.8 = 37.6.

Check the manufacturer's engineering data for the ratios of available geared speed reducers. Thus, a study of one manufacturer's data shows that a speed-reduction ratio of 38 is available using a single-reduction worm-gear drive. This drive would be suitable if it were rated at 10 hp or higher. Check to see that the gear has a suitable horsepower rating before making the final selection.

Related Calculations: Use the general procedure given here to choose power transmissions for small vehicles compressors, hoists, lawn mowers, machine tools, conveyors, pumps, snow sleds, and similar equipment. For non-vehicle drives, substitute the maximum rpm of the driven machine for the maximum wheel velocity in Steps 2, 3, and 4.

BEARING-TYPE SELECTION FOR A KNOWN LOAD

Choose a suitable bearing for a 3-in.-diameter 100-rpm shaft carrying a total radial load of 12,000 lb. A reasonable degree of shaft misalignment must be allowed by the bearing. Quiet operation of the shaft is desired. Lubrication will be intermittent.

Calculation Procedure:

1. Analyze the desired characteristics of the bearing

There are two major types of bearings available to the designer—*rolling* and *sliding*. Rolling bearings are of two types—*ball* and *roller*. Sliding bearings are also of two types—*journal* for radial loads and *thrust* for axial loads only, or for combined axial and radial loads.   Table 37 shows the principal characteristics of rolling and sliding bearings. Based on the data in Table 37, a sliding bearing would be suitable for this application because it has a *fair* misalignment tolerance and a *quiet* noise level. Both these factors are key considerations in the bearing choice.

TABLE 37   Key Characteristics of Rolling and Sliding Bearings*

	Rolling	Sliding
Life......................	Limited by fatigue properties of bearing metal	Unlimited, except for cyclic loading
Load:		
Unidirectional...........	Excellent	Good
Cyclic...................	Good	Good
Starting.................	Excellent	Poor
Unbalance...............	Excellent	Good
Shock...................	Good	Fair
Emergency..............	Fair	Fair
Speed limited by...........	Centrifugal loading and material surface speeds	Turbulence and temperature rise
Starting friction............	Good	Poor
Cost......................	Intermediate, but standardized, varying little with quantity	Very low in simple types or in mass production
Space requirements (radial bearing):		
Radial dimension.........	Large	Small
Axial dimension..........	$\frac{1}{8}$ to $\frac{1}{2}$ shaft diameter	$\frac{1}{4}$ to 2 times shaft diameter
Misalignment tolerance.....	Poor in ball bearings except where designed for at sacrifice of load capacity; good in spherical roller bearings; poor in cylindrical roller bearings	Fair
Noise.....................	May be noisy, depending upon quality and resonance of mounting	Quiet
Damping..................	Poor	Good
Low-temperature starting...	Good	Poor
High-temperature operation	Limited by lubricant	Limited by lubricant
Type of lubricant...........	Oil or grease	Oil, water, other liquids, grease, dry lubricants, air, or gas
Lubrication, quantity required.................	Very small, except where large amounts of heat must be removed	Large, except in low-speed boundary-lubrication types
Type of failure.............	Limited operation may continue after fatigue failure but not after lubricant failure	Often permits limited emergency operation after failure
Ease of replacement........	Function of type of installation; usually shaft need not be replaced	Function of design and installation; split bearings used in large machines

Product Engineering.

2. Choose the bearing materials

Table 38 shows that a porous-bronze bearing, suitable for intermittent lubrication, can carry a maximum pressure load of 4,000 psi at a maximum shaft speed of 1,500 fpm. Using the relation $l = L/Pd$, where l = bearing length, in., L = load, lb; d = shaft diameter, in., the required length of this sleeve bearing is $l = L/Pd = 12,000/[(4,000)(3)] = 1$ in.

Compute the shaft surface speed V fpm from $V = \pi dR/12$, where d = shaft diameter, in.; R = shaft rpm. Thus, $V = \pi(3)(100)/12 = 78.4$ fpm.

With the shaft speed known, the PV, or pressure-velocity, value of the bearing can be computed. For this bearing, with an operating pressure of 4,000 psi, PV = (4,000) ×

TABLE 38   Materials for Sleeve Bearings*

(Cost figures are for a 1-in. sleeve bearing ordered in quantity)

	Maximum load, psi	Maximum speed, fpm	PV limit, psi fpm	Maximum operating temperature, F	Cost, $
Porous bronze	4,000	1,500	50,000	150	0.11
Porous iron	8,000	800	50,000	150	0.09
Teflon fabric	60,000	50	25,000	500	0.04
Phenolic.............	6,000	2,500	15,000	200	0.05
Wood...............	6,000	2,000	15,000	150	0.40
Carbon-graphite	600	2,500	10,000	750	0.39
Reinforced Teflon	2,500	2,500	10,000	500	0.45
Nylon	1,000	1,000	3,000	200	0.04
Delrin..............	1,000	1,000	3,000	180	0.03
Lexan	1,000	1,000	3,000	220	0.05
Teflon..............	500	100	1,000	500	1.00

Product Engineering.

(78.4) = 313,600 psi fpm. This is considerably in excess of the PV limit of 50,000 psi fpm listed in Table 38. To come within the recommended PV limit, the operating pressure of the bearing must be reduced.

Assume an operating pressure of 600 psi. Then, $l = L/Pd = 12,000/[(600)(3)] = 6.67$ in.; say 7 in. The PV value of the bearing then is $(600)(78.4) = 47,000$ psi fpm. This is a satisfactory value for a porous-bronze bearing because the recommended limit is 50,000 psi fpm.

3.   *Check the selected bearing size*

The sliding bearing chosen will have a diameter somewhat in excess of 3 in. and a length of 7 in. If this length is too great to fit in the allowable space, another bearing material will have to be studied, using the same procedure. Figure 8 shows the space occupied by rolling and sliding bearings of various types.

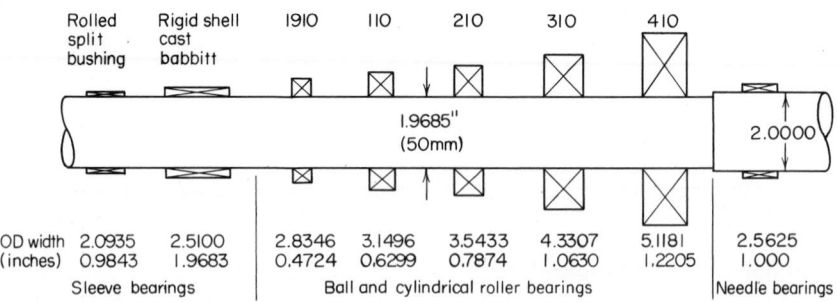

Fig. 8   Relative space requirements of sleeve and rolling-element bearings to carry the same diameter shaft. *(Product Engineering.)*

Table 39 shows the load-carrying capacity and maximum operating temperatures for oil-film journal sliding bearings that are regularly lubricated. These bearings are termed *full film* because they receive a supply of lubricant at regular intervals. Surface speeds of 20,000 to 25,000 fpm are common for industrial machines fitted with these bearings. This corresponds closely to the surface speed for ball and roller bearings.

TABLE 39   Oil-Film Journal Bearing Characteristics*

Bearing material	Load-carrying capacity, psi	Maximum operating temperature, F
Tin-base babbitt.......	800–1,500	300
Lead-base babbitt	800–1,200	300
Alkali-hardened steel..	1,200–1,500	500
Cadmium base	1,500–2,000	500
Copper-lead	1,500–2,500	350
Tin bronze............	4,000	500 +
Lead bronze	3,000–4,000	450
Aluminum alloy	4,000	250
Silver (overplated)	4,000	500
Three-component bearings babbitt-surfaced............	2,000–4,000	225–300

Product Engineering.

4. *Evaluate oil-film bearings*

Oil-film sliding bearings are chosen using the method of the next Calculation Procedure. The bearing size is made large enough so that the maximum operating temperature listed in Table 39 is not exceeded. Table 40 lists typical design load limits for oil-film bearings in various services. Figure 9 shows the typical temperature limits for rolling and sliding bearings made of various materials.

5. *Evaluate rolling bearings*

Rolling bearings have lower starting friction (coefficient of friction $f = 0.002$ to 0.005) than sliding bearings ($f = 0.15$ to 0.30). Thus, the rolling bearing is preferred for applications requiring low starting torque (integral-horsepower electric motors up to 500 hp, jet engines, etc.). By pumping oil into a sliding bearing, its starting coefficient of friction can be reduced to nearly zero. This arrangement is used in large electric generators and certain mill machines.

TABLE 40   Typical Design Load Limits for Oil Film Bearings*

Bearing	Maximum Load on Projected Area, $L \times D$, psi
Electric motors...........................	200
Steam turbines...........................	300
Automotive engines:	
Main bearings.........................	3,500
Connecting rods.......................	5,000
Diesel engines:	
Main bearings.........................	3,000
Connecting rods.......................	4,500
Railroad car axles........................	350
Steel mill roll necks:	
Steady................................	2,000
Peak..................................	5,000

Product Engineering.

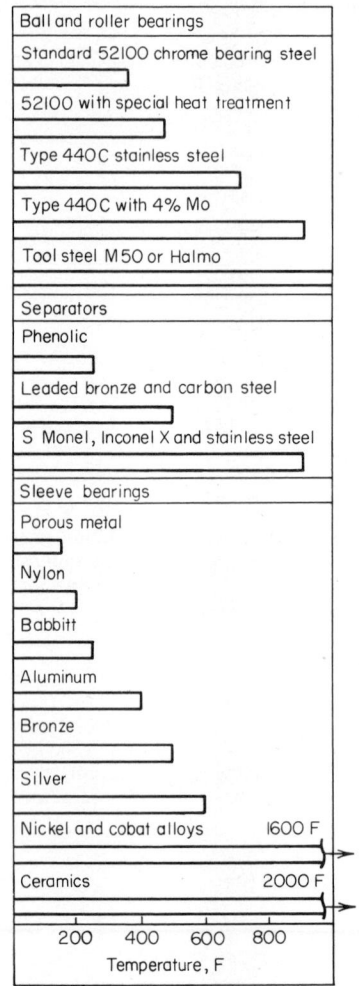

Fig. 9 Bearing temperature limits. *(Product Engineering.)*

The running friction of rolling bearings is in the range of $f = 0.001$ to 0.002. For oil-film sliding bearings, $f = 0.002$ to 0.005.

Rolling bearings are more susceptible to dirt than are sliding bearings. Also, rolling bearings are inherently noisy. Oil-film bearings are relatively quiet, but they may allow higher amplitudes of shaft vibration.

Table 41 compares the size, load capacity, and cost of rolling bearings of various types. Briefly, ball bearings vs. roller bearings may be compared thus: ball bearings (a) run at higher speeds without undue heating, (b) cost less per lb of load-carrying capacity for light loads, (c) have friction torque at light loads, (d) are available in a wider variety of sizes, (e) can be made in smaller sizes, and (f) have seals and shields for easy lubrication. Roller bearings (a) can carry heavier loads, (b) are less expensive for larger sizes and heavier loads, (c) are more satisfactory under shock and impact loading, and (d) may have lower friction at heavy loads. Table 42 shows the speed limit, termed the dR limit (equals bearing shaft bore d in mm multiplied by the shaft rpm R) for ball and roller bearings. Speeds higher than those shown in Table 42 may lead to early bearing failures. Since the dR limit is proportional to the shaft surface speed, the dR value gives an approximate measure of the bearing power loss and temperature rise.

Related Calculations: Use this general procedure to select shaft bearings for any type of regular service conditions. For unusual service — i.e., excessively high or low operating temperatures, large loads, etc. — consult the specific selection procedures given elsewhere in this section.

Note that the PV value of a sliding bearing can also be expressed as $\text{PV} = (L/dl) \times (\pi dR/12) = (\pi LR/12l)$. The bearing load and shaft speed are usually fixed by other requirements of a design. Where the PV equation is solved for the bearing length l and the bearing is too long to fit the available space, select a bearing material having a higher allowable PV value.

SHAFT BEARING LENGTH AND HEAT GENERATION

How long should a sleeve-type bearing be if the combined weight of the shaft and gear tooth load acting on the bearing is 200 lb? The shaft is 1 in. in diameter and is oil lubricated. What is the rate of heat generation in the bearing when the shaft turns at 600 rpm? How much above an ambient room temperature of 70 F will the temperature of the bearing rise during operation in still air? In moving air?

Calculation Procedure:

1. *Compute the required length of the bearing*

The required length l of a sleeve bearing carrying a load of L lb is $l = L/Pd$, where l = bearing length, in.; L = bearing load, lb = bearing reaction force, lb; P = allowable

TABLE 41   Relative Load Capacity, Cost, and Size of Rolling Bearings*

Bearing type (for 50 mm bore)	Radial capacity	Axial capacity	Cost	Outer diameter	Width
Ball bearings:					
Deep groove (Conrad) .	1.0	1.0	1.0	1.0	1.0
Filling notch..........	1.2	Low	1.2	1.0	1.0
Double row...........	1.5	1.1	2.2	1.0	1.6
Angular contact	1.1	1.9	1.6	1.0	1.0
Duplex..............	1.8	1.9	2.0	1.0	2.0
Self-aligning..........	0.7	0.2	1.3	1.0	1.0
Ball thrust	0	0.9	0.8	0.7	0.8
Roller bearings:					
Cylindrical	1.6	· · ·	1.9	1.0	1.0
Tapered..............	1.3	0.8	0.9	1.0	1.0
Spherical............	3.0	1.1	5.0	1.0	1.5
Needle	1.0	0	0.3	0.5	1.6
Flat thrust	0	4.0	3.8	0.8	0.9

Product Engineering.

TABLE 42   Speed Limits for Ball and Roller Bearings*

Lubrication	DN Limit, mm × rpm
Oil:	
Conventional bearing designs ..	300,000–350,000
Special finishes and separators..	1,000,000–1,500,000
Grease:	
Conventional bearing designs ..	250,000–300,000
Silicone grease................	150,000–200,000
Special finishes and separators high-speed greases	500,000–600,000

Product Engineering.

mean bearing pressure, psi (ranges from 25 to 2,500 psi for normal service and up to 8,000 psi for severe service) on the projected bearing area, sq in. $= ld$; $d =$ shaft diameter, in. Thus, for this bearing, assuming an allowable mean bearing pressure of 400 psi, $l = L/Pd = 2,000/[(400)(1)] = 5$ in.

2. *Compute the rate of bearing heat generation*

The rate of heat generation in a plain sleeve bearing is given by $h = fLdR/3,000$, where $h =$ rate of heat generation in the bearing, Btu/min; $f =$ bearing coefficient of friction for the lubricant used; $R =$ shaft rpm; other symbols as in Step 1.

The coefficient of friction for oil-lubricated bearings ranges from 0.005 to 0.030, depending on the lubricant viscosity, shaft rpm, and mean bearing pressure. Using a value of $f = 0.020$, $h = (0.020)(200)(1)(600)/3,000 = 0.8$ Btu/min, or $H = 0.8(60$ min/hr$) = 48.0$ Btu/hr.

3. Compute the bearing wall area

The wall area A of a small sleeve-type bearing, such as a pillow block fitted with a bushing, is $A = (10$ to $15)dl/144$, where $A =$ bearing wall area, sq ft; other symbols as before. For larger bearing pedestals fitted with a cast-iron or steel bearing shell, the factor in this equation varies from 18 to 25.

Since this is a small bearing having a 1-in.-diameter shaft, the first equation with a factor of 15 to give a larger wall area, can be used. The value of 15 was chosen to ensure adequate radiating surface. Where space or weight is a factor, the value of 10 might be chosen. Intermediate values might be chosen for other conditions. Substituting, $A = (15)(1)(5)/144 = 0.521$ sq ft.

4. Determine the bearing temperature rise

In *still air*, a bearing will dissipate $H = 2.2A\,(t_w - t_a)$ Btu/hr, where $t_w =$ bearing wall temperature, F; t_a ambient air temperature, F; other symbols as before. Since H and A are known, the temperature rise can be found by solving for $t_w - t_a = H/2.2A = 48.0/[(2.2)(0.521)] = 41.9F$, and $t_w = 41.9 + 70 = 111.9$ F. This is a low enough temperature for safe operation of the bearing. The maximum allowable bearing operating temperature for sleeve bearings using normal lubricants is usually assumed to be 200 F. To reduce the operating temperature of a sleeve bearing, the bearing wall area must be increased, the shaft speed decreased, or the bearing load reduced.

In *moving air*, the heat dissipation from a sleeve-type bearing is $H = 6.5A(t_w - t_a)$. Solving for the temperature rise as before, $t_w - t_a = H/6.5A = 48.0/[(6.5)(0.521)] = 14.2$ F, and $t_w = 14.2 + 70 = 84.2$ F. This is a moderate operating temperature that could be safely tolerated by any of the popular bearing materials.

Related Calculations: Use this procedure to analyze sleeve-type bearings used for industrial line shafts, marine propeller shafts, conveyor shafts, etc. Where the ambient temperature varies during bearing operation, use the highest ambient temperature expected, when computing the bearing operating temperature.

ROLLER-BEARING OPERATING-LIFE ANALYSIS

A machine must have a shaft of about 5.5 in. in diameter. Choose a roller bearing for this 5.5-in.-diameter shaft that turns at 1,000 rpm while carrying a radial load of 20,000 lb. What is the expected life of this bearing?

Calculation Procedure:

1. Determine the bearing life in revolutions

The operating life of rolling-type bearings is often stated in millions of revolutions. Find this life from $R_L = (C/L)^{10/3}$, where $R_L =$ bearing operating life, millions of revolutions; $C =$ dynamic capacity of the bearing, lb; $L =$ applied radial load on bearing, lb.

Obtain the dynamic capacity of the bearing being considered by consulting the manufacturer's engineering data. Usual values of dynamic capacity range between 2,500 lb and 750,000 lb, depending on the bearing design, type, and bore. For a typical 5.5118-in.-bore roller bearing, $C = 92,400$ lb.

With C known, compute $R_L = (C/L)^{10/3} = (92,400/20,000)^{10/3} = 162 \times 10^6$ revolutions.

2. Determine the bearing life in hr

The minimum life of a bearing in millions of revolutions, R_L, is related to its life in hr, h, by the expression $R_L = 60Rh/10^6$, where $R =$ shaft speed, rpm. Solving, $h = 10^6 R_L / 60R = (10^6)(162)/[(60)(1,000)] = 2,700$ hr, minimum life.

Related Calculations: This procedure is useful for those situations where a bearing must fit a previously determined shaft diameter or fit in a restricted space. In these circumstances, the bearing size cannot be varied appreciably and the machine designer is interested in knowing the minimum probable life that a given size of bearing will have. Use this procedure whenever the bearing size is approximately predetermined by the installation conditions in motors, pumps, engines, portable tools, etc. Obtain the dynamic capacity of any bearing under consideration from the manufacturer's engineering data.

ROLLER-BEARING CAPACITY REQUIREMENTS

A machine must be fitted with a roller bearing that will operate at least 30,000 hr without failure. Select a suitable bearing for this machine in which the shaft operates at 3,600 rpm and carries a radial load of 5,000 lb.

Calculation Procedure:

1. Determine the bearing life in revolutions

Use the relation $R_L = 60\,Rh/10^6$, where the symbols are the same as in the previous Calculation Procedure. Thus, $R_L = 60(3,600)(30,000)/10^6$; $R_L = 6,480$ million revolutions.

2. Determine the required dynamic capacity of the bearing

Use the relation $R_L = (C/L)^{10/3}$, where the symbols are the same as in the previous Calculation Procedure. Solving, $C = L(R_L)^{3/10} = (5,000)(6,480)^{3/10} = 69,200$ lb.

Choose a bearing of suitable bore having a dynamic capacity of 69,200 lb, or more. Thus, a typical 5.9055-in.-bore roller bearing has a dynamic capacity of 72,400 lb. It is common practice to undercut the shaft to suit the bearing bore, if such a reduction in the shaft does not weaken the shaft. Use the manufacturer's engineering data when choosing the actual bearing that will be used.

Related Calculations: This procedure shows a situation where the life of the bearing is of greater importance than its size. Such a situation is common where the reliability of a machine is a key factor in its design. A dynamic rating of a given amount — say 72,400 lb — means that if in a large group of bearings of this size each bearing has a 72,400-lb load applied to it, 90 percent of the bearings in the group will complete, or exceed, one million revolutions before the first evidence of fatigue occurs. This average life of the bearing is the number of revolutions that 50 percent of the bearings will complete, or exceed, before the first evidence of fatigue develops. The average life is about 3.5 times the minimum life.

Use this procedure to choose bearings for motors, engines, turbines, portable tools, etc. Where extreme reliability is required, some designers choose a bearing having a much larger dynamic capacity than calculations show is required.

RADIAL LOAD RATING FOR ROLLING BEARINGS

A mounted rolling bearing is fitted to a shaft driven by a 4-in.-wide double-ply leather belt. The shaft is subjected to moderate shock loads about one-third of the time while operating at 300 rpm. An operating life of 40,000 hr is required of the bearing. What is the required radial capacity of the bearing? The bearing has a normal rated life of 15,000 hr at 500 rpm. The weight of the pulley and shaft is 145 lb.

Calculation Procedure:

1. Determine the bearing operating factors

To determine the required radial capacity of a rolling bearing, a series of operating factors must be applied to the radial load acting on the shaft. These factors are: life factor f_L; operating factor f_O; belt tension factor f_B; speed factor f_S. Obtain each of the four factors from the manufacturer's engineering data because there may be a slight variation in the factor value between different bearing makers. Where a given factor does not apply to the bearing being considered, omit the factor from the calculation.

2. Determine the bearing life factor

Rolling bearings are normally rated for a certain life, expressed in hours. If a different life for the bearing is required, a life factor must be applied. The bearing being considered here has a normal rated life of 15,000 hr. The manufacturer's engineering data shows that for a mounted bearing which must have a life of 40,000 hr, a life factor $f_L = 1.340$ should be used. For this particular make of bearing, f_L varies from 0.360 at 500 hr to 1.700 at a 100,000 hr life for mounted units. At 15,000 hr, $f_L = 1.000$.

3. Determine the bearing operating factor

A rolling-bearing operating factor is used to show the effect of peak and shock loads on the bearing. Usual operating factors vary from 1.00 for steady loads with any amount of overload to 2.00 for bearings with heavy shock loads throughout their operating period. For this bearing with moderate shock loads about one-third of the time, $f_O = 1.32$.

A combined operating factor, obtained by taking the product of two applicable factors, is used in some circumstances. Thus, when the load is an oscillating type, an additional factor of 1.25 must be applied. This type of load occurs in certain linkages and pumps. When the outer race of the bearing revolves, as in sheaves, truck wheels, or gyrating loads, an additional factor of 1.2 is used. To find the combined operating factor, first find the normal operating factor, as described earlier. Then take the product of the normal and the additional operating factors. The result is the combined operating factor.

4. Determine the bearing belt-tension factor

When the bearing is used on a belt-driven shaft, a belt-tension factor must be applied. Usual values of this factor range from 1.0 for a chain drive to 2.30 for a single-ply leather belt. For a double-ply leather belt, $f_B = 2.0$.

5. Determine the bearing speed factor

Rolling bearings are rated at various speeds. When the shaft operates at a speed different from the rated speed, a speed factor f_S must be applied. Since this shaft operates at 300 rpm, while the rated speed of the bearing is 500 rpm, $f_S = 0.860$, from the manufacturer's engineering data. For a 500-rpm bearing, f_S varies from 0.245 at 5 rpm to 1.87 at 4,000 rpm.

6. Determine the radial load on the bearing

The radial load produced by a leather belt can vary from 130 lb/in. of width for normal-tension belts to 450 lb per in. of width for very tight belts. Assuming normal tension, the radial load for a double-ply leather belt is, from engineering data, 180 lb per in. of width. Since this belt is 4 in. wide, the radial belt load = 4(180) = 720 lb. The total radial load R_T is the sum of the belt, shaft, and pulley loads, or $R_T = 720 + 145 = 865$ lb.

7. Compute the required radial capacity of the bearing

The required radial capacity of a bearing $R_C = R_T f_L f_O f_B f_S = (865)(1.340)(1.32)(2.0)(0.86) = 2,630$ lb.

8. Select a suitable bearing

Enter the manufacturer's engineering data at the shaft rpm (300 rpm for this shaft) and project to a bearing radial capacity equal to, or slightly greater than, the computed required radial capacity. Thus, one make of bearing, suitable for 2- and $2\frac{3}{16}$-in.-diameter shafts, has a radial capacity of 2,710 at 300 rpm. This is close enough for general selection purposes.

Related Calculations: Use this general procedure for any type of rolling bearing. When comparing different makes of rolling bearings, be sure to convert them to the same life expectancies before making the comparison. Use the life-factor table presented in engineering handbooks or a manufacturer's engineering data for each bearing to convert the bearings being considered to equal lives.

ROLLING-BEARING CAPACITY AND RELIABILITY

What is the required basic load rating of a ball bearing having an equivalent radial load of 3,000 lb if the bearing must have a life of 400×10^6 revolutions at a reliability of 0.92? The ratio of the average life to the rating life of the bearing is 5.0. Show how the required basic load rating is determined for a roller bearing.

Calculation Procedure:

1. Compute the required basic load rating

Use the Weibull two-parameter equation, which for a life ratio of 5 is $L_e/L_B = (1.898/R_L^{0.333})(\ln 1/R_e)^{0.285}$, where L_e = equivalent radial load on the bearing, lb; L_B = the required basic load rating of the bearing, lb, to give the desired reliability at the stated life; R_L = bearing operating life, millions of revolutions, $\ln$ = natural or Naperian logarithm to the base e; R_e = required reliability, expressed as a decimal.

Substituting, $(3,000/L_B) = (1.898/400^{0.333})(\ln 1/0.92)^{0.285}$; $L_B = 23,425$ lb. Thus, a bearing having a basic load rating of at least 23,425 lb would provide the desired reliability. Select a bearing having a load rating equal to, or slightly in excess of, this value. Use the manufacturer's engineering data as the source of load-rating data.

2. Compute the roller-bearing basic load rating

Use the following form of the Weibull equation for roller bearings: $L_e/L_B = (1.780/R_L^{0.30})(\ln 1/R_e)^{0.257}$. Substitute values as in Step 1 and solve for L_B as in Step 1.

Related Calculations: Use the Weibull equation as given here when computing bearing life in the range of 0.9, and higher. The ratio of average life/rating life = 5 is usual for commercially available bearings.[1]

POROUS-METAL BEARING CAPACITY AND FRICTION

Determine the load capacity, ψ and coefficient of friction of a porous-metal bearing for a 1-in.-diameter shaft, 1 in. bearing length, 0.2-in.-thick bearing, 0.001 in. radial clearance, metal permeability $\phi = 5 \times 10^{-10}$ in., shaft speed = 1,500 rpm, eccentricity ratio $\epsilon = 0.8$, and an SAE-30 mineral-oil lubricant with a viscosity of 6×10^{-6} lb-sec/sq in.

Calculation Procedure:

1. Sketch the bearing and shaft

Figure 10 shows the bearing and shaft with the various known dimensions indicated by the identifying symbols given above.

2. Compute the load capacity factor

The load capacity factor $\psi = \phi H/C^3$, where ϕ = metal permeability, in.; H = bearing thickness, in.; C = radial clearance of bearing, $R_b - r = 0.001$ in. for this bearing. Hence, $\psi = (5 \times 10^{-10})(0.2)/(0.001)^3 = 0.10$.

3. Compute the bearing thickness-length ratio

The thickness-length ratio = $H/b = 0.2/1.0 = 0.2$.

4. Determine the $S(d/b)^2$ value for the bearing

In the $S(d/b)^2$ value for the bearing, S = the Summerfeld number for the bearing; the other values are as shown in Fig. 10.

Using the ψ, ϵ, and H/b values, enter Fig. 11a, and read $S(b/d)^2 = 1.4$ for an eccentricity ratio of $\epsilon = e/C = 0.8$. Substitute in this equation the known values for b and d and solve for S, or $S(1.0/1.0)^2 = 1.4$; $S = 1.4$.

5. Compute the bearing load capacity

Find the bearing load capacity from $S = (L/R_i \eta b)(C/r)^2$, where η = lubricant viscosity, lb-sec/sq in.; r = shaft radius, in.; R_i = shaft velocity, in. sec. Solving for $L = (SR_i \eta b)/$

[1]C. Mischke, "Bearing Reliability and Capacity," *Machine Design*, September 30, 1965.

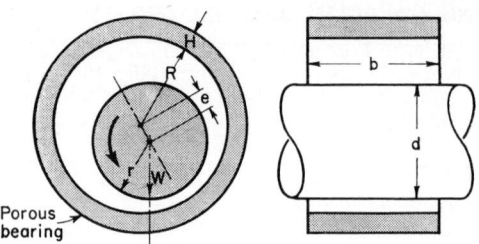

Fig. 10   Typical porous-metal bearing. *(Product Engineering.)*

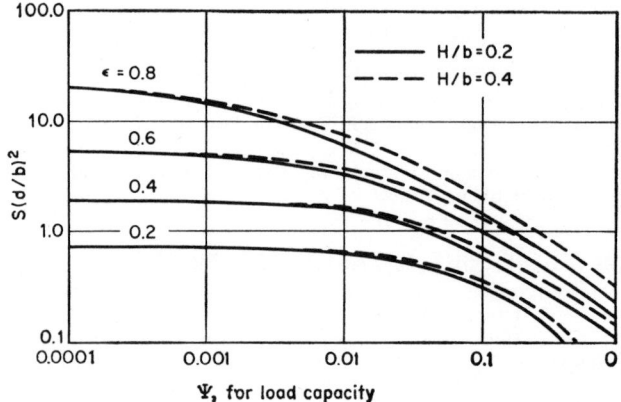

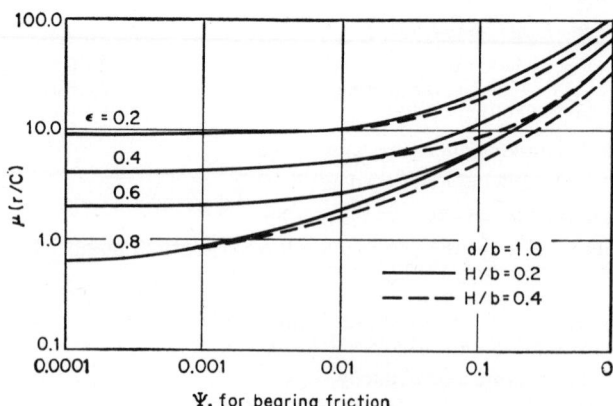

Fig. 11   (a) Bearing load capacity factors; (b) bearing friction factors. *(Product Engineering.)*

$(C/r)^2 = (1.4)(78.5)(6 \times 10^{-6})(1.0)/(0.001/0.5)^2 = 164.7$ lb. (The shaft rotative velocity must be expressed in in./sec in this equation because the lubricant viscosity is given in lb-sec/sq in.)

6. Determine the bearing coefficient of friction

Enter Fig. 11b with the known values of ψ, ϵ, and H/b and read $u(r/C) = 7.4$. Substitute in this equation the known values for r and C and solve for the bearing coefficient of friction μ, or $\mu = 7.4C/r = (7.4)(0.001)/0.5 = 0.0148$.

Related Calculations: Porous-metal bearings are similar to conventional sliding-journal bearings except that the pores contain an additional supply of lubricant to replace that which may be lost during operation. The porous-metal bearing is useful in assemblies where there is not enough room for a conventional lubrication system, or where there is a need for improved lubrication during the starting and stopping of a machine. The permeability of the finished porous metal greatly influences the ability of a lubricant to work its way through the pores. Porous-metal bearings are used in railroad axle supports, water pumps, generators, machine tools, and other equipment. Use the procedure given here when choosing porous-metal bearings for any of these applications. The method given here was developed by Professor W. T. Rouleau of Carnegie Institute of Technology, and C. A. Rhodes, Senior Research Engineer, Jet Propulsion Laboratory, California Institute of Technology.

HYDROSTATIC THRUST BEARING ANALYSIS

An oil-lubricated hydrostatic thrust bearing must support a load of 107,700 lb. This vertical bearing has an outside diameter of 16 in. and a recess diameter of 10 in. What oil pressure and flow rate are required to maintain a 0.006-in. lubricant film thickness with an SAE-20 oil having an absolute viscosity of $\eta = 42.4 \times 10^{-7}$ lb/sec/sq in. if the shaft turns at 750 rpm? What is the pumping loss and the viscous friction loss? What is the optimum lubricant-film thickness?

Calculation Procedure:

1. Determine the required lubricant-supply pressure

The design equations and methods developed at Franklin Institute by Dudley Fuller, Professor of Mechanical Engineering, Columbia University, are applicable to vertical hydrostatic bearings, Fig. 12, using oil, grease or gas lubrication. By substituting the appropriate value for the lubricant viscosity, the same set of design equations can be used for any of the lubricants listed above. These equations are accurate, simple, and reliable; they are therefore used here.

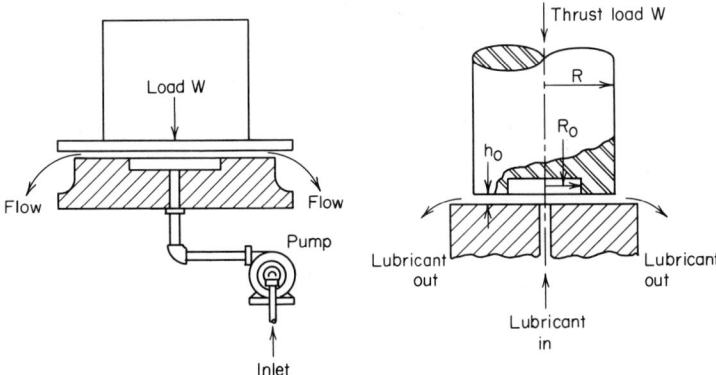

Fig. 12 Hydrostatic thrust bearings. (*Product Engineering.*)

Solve Fuller's applied load equation, $L = (p_i\pi/2)\{r^2 - r_i^2/[\ln(r/r_i)]\}$, for the lubricant-supply inlet pressure. In this equation, L = applied load on the bearing, lb; p_i = lubricant-supply inlet pressure, psi; r = shaft radius, in.; r_i = recess or step radius, in.; $\ln$ = natural or Naperian logarithm to the base e. Solving for $p_i = 2L/\pi\{r^2 - r_i^2/[\ln(r/r_i)]\}$ = $2(107,700)/\pi[8^2 - 5^2/(\ln 8/5)]$, or $p_i = 825$ psi.

2. Compute the required lubricant flow rate

Using Fuller's flow-rate equation, $Q = p_i\pi h^3/6\eta \ln r/r_i$, where Q = lubricant flow rate, in.^{3}/sec; h = lubricant-film thickness, in.; η = lubricant absolute viscosity, lb-sec/sq in.; other symbols as in Step 1. Thus, with $h = 0.006$ in., $Q = (825\,\pi)(0.006)^3/[6(42.4 \times 10^{-6})(0.470)] = 46.85$ in.^{3}/sec.

3. Compute the pumping loss

The pumping loss results from the work necessary to force the lubricant radially outward through the film space, or $H_p = Q(p_i - p_o)$, where H_p = power required to pump the lubricant = pumping loss, in.-lb/sec; p_o = lubricant outlet pressure, psi; other symbols as in Step 1. For circular thrust bearings it can be assumed that the lubricant outlet pressure p_0 is negligible, or $p_0 = 0$. Then, $H_p = 46.85(825 - 0) = 38,680$ in.-lb/sec = 38,680/[550 ft-lb/(min)(hp)](12 in./ft) = 5.86 hp.

4. Compute the viscous friction loss

The viscous-friction-loss equation developed by Fuller is $H_f = [(R^2\eta/(58.05h)](r^4 - r_o^4)$, where H_f = viscous friction loss, in.-lb/sec; R = shaft rpm; other symbols as in Step 1. Thus, $H_f = \{(750)^2(42.4 \times 10^{-7})/[(58.05)(0.006)]\}(8^4 - 5^4)$, or $H_f = 23,770$ in.-lb/sec = 23,770/[(550)(12)] = 3,60 hp.

5. Compute the optimum lubricant film thickness

The film thickness that will produce a minimum combination of pumping loss and friction loss can be evaluated by determining the minimum point of the curve representing the sum of the respective energy losses (pumping and viscous friction) when plotted against film thickness.

With the shaft speed, lubricant viscosity, and bearing dimensions constant at the values given in the problem statement, the viscous-friction loss becomes $H_f = 0.0216/h$ for this bearing. Substitute various values for h ranging between 0.001 and 0.010 in. (the usual film-thickness range) and solve for H_f. Plot the results as shown in Fig. 13.

Combine the lubricant-flow and pumping-loss equations to express H_p in terms of the lubricant film thickness, or $H_p = (1,000h)^3/36.85$, for this bearing with a pump having an efficiency of 100 percent. For a pump with a 50 percent efficiency, this equation becomes $H_p = (1,000h)^3/18.42$. Substitute various values of h ranging between 0.001 and 0.010 and plot the results as in Fig. 13 for pumps with 100 and 50 percent efficiencies, respectively. Figure 13 shows that for a 100 percent efficient pump, the minimum total energy loss occurs at a film thickness of 0.004 in. For 50 percent efficiency, the minimum total energy loss occurs at 0.0035-in. film thickness, Fig. 13.

Related Calculations: Similar equations developed by Fuller can be used to analyze hydrostatic thrust bearings of other configurations. Figures 14 and 15 show the equations for modified square bearings and circular-sector bearings. To apply these equations, use the same general procedures shown above. Note, however, that each equation uses a factor K obtained from the respective design chart.

Also note that a hydrostatic bearing uses an externally fed pressurized fluid to keep two bearing surfaces *completely* separated. Compared with hydrodynamic bearings, in which the pressure is self-induced by the rotation of the shaft, hydrostatic bearings have: (1) lower friction, (2) higher load-carrying capacity, (3) a lubricant-film thickness insensitive to shaft speed, (4) a higher spring constant, which leads to a self-centering effect, and (5) a relatively thick lubricant film permitting cooler operation at high shaft speeds. Hydrostatic bearings are used in rolling mills, instruments, machine tools, radar, telescopes, and other applications.

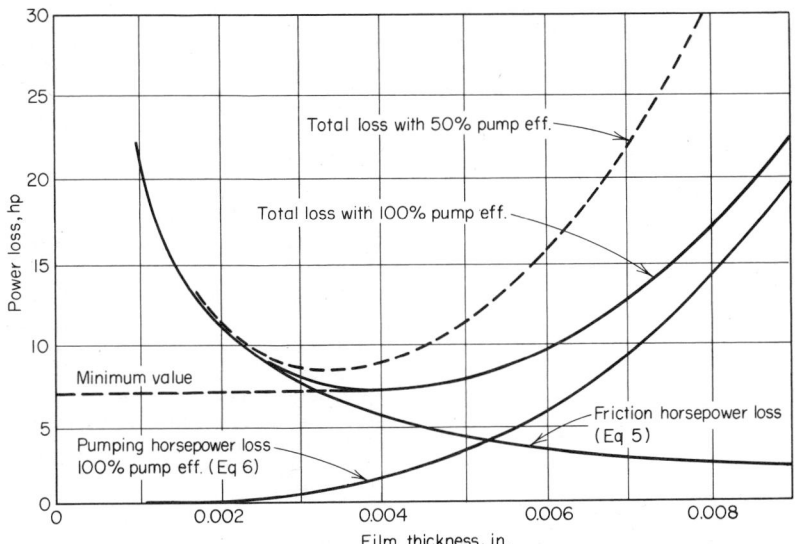

Fig. 13 Oil-film thickness for minimum power loss in a hydrostatic thrust bearing. (*Product Engineering.*)

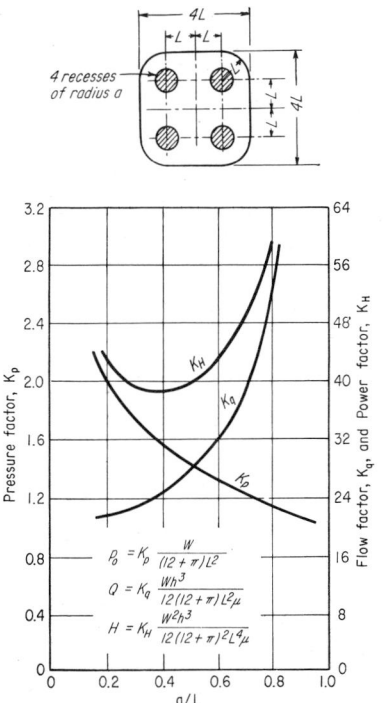

Fig. 14 Constants and equations for modified square hydrostatic bearings. (*Product Engineering.*)

$$P_o = K_p \frac{W}{(12 + \pi)L^2}$$

$$Q = K_q \frac{Wh^3}{12(12 + \pi)L^2\mu}$$

$$H = K_H \frac{W^2h^3}{12(12 + \pi)^2L^4\mu}$$

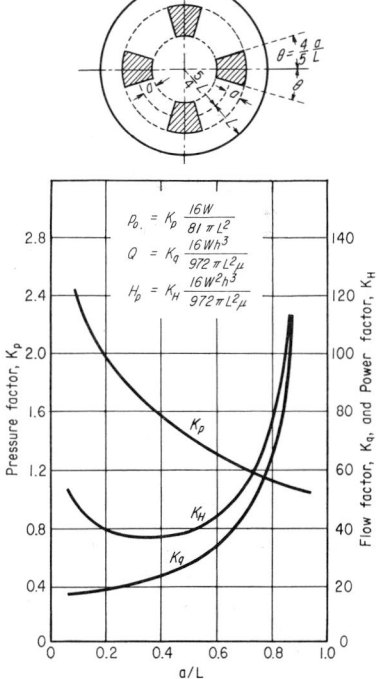

Fig. 15 Constants and equations for circular-sector hydrostatic bearings. (*Product Engineering.*)

$$P_o = K_p \frac{16W}{81\pi L^2}$$

$$Q = K_q \frac{16Wh^3}{972\pi L^2\mu}$$

$$H_p = K_H \frac{16W^2h^3}{972\pi L^2\mu}$$

HYDROSTATIC JOURNAL BEARING ANALYSIS

A 4.000 in. metal shaft rests in a journal bearing having an internal diameter of 4.012 in. The lubricant is SAE-30 oil at 100 F having a viscosity of 152×10^{-7} reyns. This lubricant is supplied under pressure through a groove at the lowest point in the bearing. The length of the bearing is 6 in.; the length of the groove is 3 in.; and the load on the bearing is 3,600 lb. What lubricant-inlet pressure and flow rate is required to raise the shaft 0.002 in. and 0.004 in?

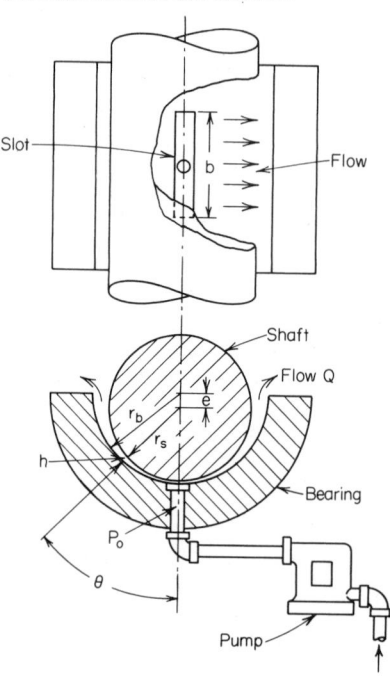

Fig. 16 Hydrostatic journal bearing.
(Product Engineering.)

Calculation Procedure:

1. Determine the radial clearance and clearance modulus

The design equations and methods developed at Franklin Institute by Dudley Fuller, Professor of Mechanical Engineering, Columbia University, are applicable to hydrostatic journal bearings using oil, grease, or gas lubrication. These equations are accurate, simple, and reliable; therefore they are used here.

Using Fuller's method, the radial clearance c in. $= r_b - r_s$, Fig. 16, where $r_b =$ bearing internal radius, in.; $r_s =$ shaft radius, in. Or, $c = (4.012/2) - (4.000/2) = 0.006$ in.

Next, compute the clearance modulus m from $m = c/r_s = 0.006/2 = 0.003$ in./in. Typical values of m range from 0.005 to 0.003 in./in. for hydrostatic journal bearings.

2. Compute the shaft eccentricity in the clearance space

The numerical parameter used to describe the eccentricity of the shaft in the bearing clearance space is the ratio $\epsilon = 1 - (h/mr)$, where $h =$ shaft clearance, in., during operation. With a clearance of $h = 0.002$ in., $\epsilon = 1 - [0.002/(0.003 \times 2)] = 0.667$. With a clearance of $h = 0.004$ in., $\epsilon = 1 - [0.004/(0.003 \times 2)] = 0.333$.

3. Compute the eccentricity constants

The eccentricity constant $A_k = 12[2 - \epsilon/(1 - \epsilon)^2] = 12[2 - 0.667/(1 - 0.667)^2] = 144.6$. A second eccentricity constant B is given by $B_k = 12 \{\epsilon(4 - \epsilon^2)/[2(1 - \epsilon^2)^2] + 2 + \epsilon^2/(1 - \epsilon^2)^{2.5} \times \arctan [1 + \epsilon/(1 - \epsilon^2)^{0.5}]\}$. Since this relation is awkward to handle, Fig. 17 was developed by Fuller. From Fig. 17, $B_k = 183$.

4. Compute the required lubricant flow rate

The lubricant flow rate is found from Q in.^{3}/sec $= 2Lm^3r_s/\eta\, A_k$ where $L =$ load acting on shaft, lb; $\eta =$ lubricant viscosity, reyns. For the bearing with $h = 0.002$ in., $Q = 2(3,600)(0.003)^3(2)/[(152 \times 10^{-7})(144.6)] = 0.177$ in.^{3}/sec, or 0.0465 gpm.

5. Compute the required lubricant inlet pressure

The lubricant-inlet pressure is found from $p_i = QB/2bm^3r_s^2$, where $p_i =$ lubricant inlet pressure, psi; other symbols as before. Thus, $p_i = (152 \times 10^{-7})(0.177)(183)/[(2)(3)(0.003)^3(2)^2] = 759$ psi.

6. Analyze the larger-clearance bearing

Use the same procedure as in Steps 1 through 5. Then, $\epsilon = 0.333$; $A = 45.0$; $B = 42.0$; $Q = 0.560$ in.^{3}/sec $= 0.1472$ gpm; $p_i = 551$ psi.

Related Calculations: Note that the closer the shaft is to the center of the bearing, the smaller the lubricant pressure required and the larger the oil flow. If the larger flow requirements can be met, the design with the thicker oil film is usually preferred, because it has a greater ability to absorb shock loads and tolerate thermal change.

Use the general design procedure given here for any applications where a hydrostatic journal bearing is applicable and there is no thrust load.

HYDROSTATIC MULTIDIRECTION BEARING ANALYSIS

Determine the lubricant pressure and flow requirements for the multidirection hydrostatic bearing shown in Fig. 18 if the vertical coplanar forces acting on the plate are 164,000 lb upward and downward, respectively. The lubricant viscosity $\eta = 393 = 10^{-7}$ reyns, film thickness = 0.005 in., $L = 7$ in., $a = 3.5$ in. What would be the effect of decreasing the film thickness h on one side of the plate by 0.001-in. increments from 0.005 to 0.002 in? What is the bearing stiffness?

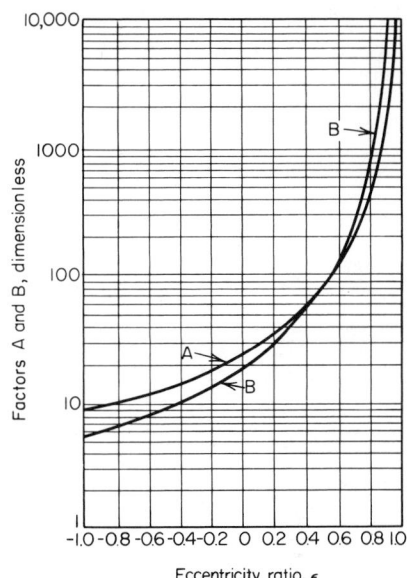

Fig. 17 Constants for hydrostatic journal bearing oil flow and load capacity. (*Product Engineering.*)

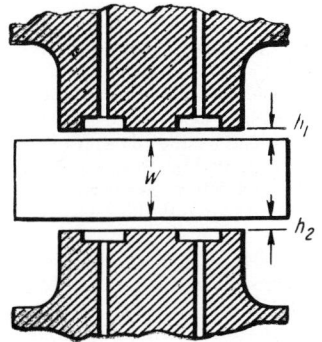
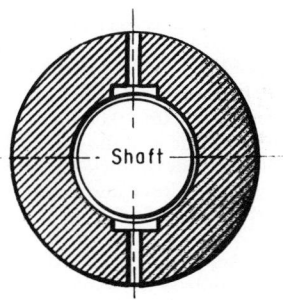

Fig. 18 Double-acting hydrostatic thrust bearing. (*Product Engineering.*)

Calculation Procedure:

1. *Compute the required lubricant inlet pressure*

Using Fuller's method, Fig. 19 shows that the bearing has four pressure pads to support the plate loads. Figure 19 also shows the required pressure, flow, and power equations, and the appropriate constants for these equations. The inlet-pressure equation is $p_i = K_p L_s/16L^2$, where p_i = required lubricant inlet pressure, psi; K_p = pressure constant from Fig. 19; L_s = plate load, lb; L = bearing length, in.

Find K_p from Fig. 19 after setting up the ratio $a/L = 3.5/7.0 = 0.5$, where a = one-half the pad length, in. Then, $K_p = 1.4$. Hence $p_i = (1.4)(164,000)/[(16)(7)^2] = 293$ psi.

2. Compute the required lubricant flow rate

From Fig. 19, the required lubricant flow rate $Q = K_q L_s h^3 / 192 L^2 \eta$, where K_q = flow constant; h = lubricant-film thickness, in.; other symbols as before. For $a/L = 0.5$, $K_q = 36$. Then, $Q = (36)(164,000)(0.005)^3 / [(192)(7)^2(393 \times 10^{-7})] = 1.99$ in.^{3}sec. This can be rounded off to 2.0 in.^{3}/sec for usual design calculations.

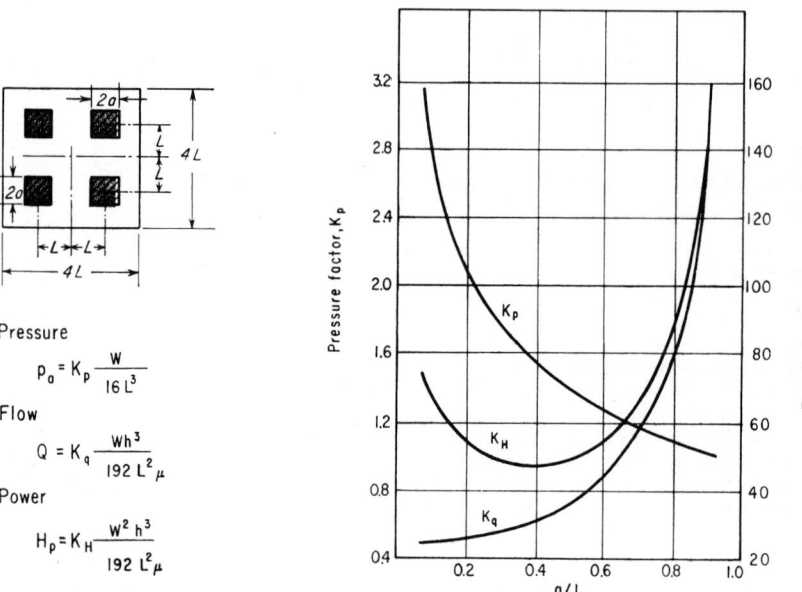

Pressure

$$p_a = K_p \dfrac{W}{16 L^3}$$

Flow

$$Q = K_q \dfrac{W h^3}{192 L^2 \mu}$$

Power

$$H_p = K_H \dfrac{W^2 h^3}{192 L^2 \mu}$$

Fig. 19 Dimensions and equations for thrust-bearing design. *(Product Engineering.)*

3. Compute the pressure and load for other plate clearances

The sum of the plate lubricant film thicknesses, $h_1 + h_2$, Fig. 18, is a constant. For this bearing, $h_1 + h_2 = 0.005 + 0.005 = 0.010$ in. With no load on the plate, if oil is pumped into both bearing faces at the rate of 2.0 in.^{3}/sec, the maximum recess pressure will be 293 psi. The force developed on each face will be 164,000 lb. Since the lower face is pushed up with this force and the top face is pushed down with the same force, the net result is zero.

With a downward external load imposed on the plate such that the lower film thickness h_2 is reduced to 0.004 in., the upper film thickness will become 0.006 in., since $h_1 + h_2 = 0.010$ in. = a constant for this bearing. If the lubricant flow rate is held constant at 2.0 in.^{3}/sec, $K_q = 36$ from Fig. 19, since $a/L = 0.5$. With these constants, the load equation becomes $L_s = 0.0205/h^3$ and the inlet-pressure equation becomes $p_i = L_s/560$.

Using these equations, compute the upper and lower loads and inlet pressures for h_2 and h_1 ranging from 0.004 to 0.002 in. and 0.006 to 0.008 in., respectively. Tabulate the results as shown in Table 43. Note that the allowable load is computed using the respective film thickness for the lower and upper part of the plate. The same is true of the lubricant inlet pressure, except that the corresponding load is used instead.

Thus, for $h_2 = 0.004$ in., $L_{s2} = 0.0205/(0.004)^3 = 320,000$ lb. Then, $p_{i2} = 320,000/560 = 571$ psi. For $h_1 = 0.006$ in., $L_{s1} = 0.0205/(0.006)^3 = 95,000$ lb. Then, $p_{i1} = 95,000/560 = 169.6$, say 170 psi.

The load difference $L_{s2} - L_{s1}$ = the bearing load capacity. For the film thicknesses considered above, $L_{s2} - L_{s1} = 320,000 - 95,000 = 225,000$ lb. Load capacities for various other film thicknesses are also shown in Table 43.

TABLE 43 Load Capacity of Dual-Direction Bearing

Film thickness, in.		Inlet pressure, psi		Load, lb		Load capacity, lb
h_2	h_1	p_{i2}	p_{i1}	L_{s2}	L_{s1}	$L_{s2} - L_{s1}$
0.005	0.005	293	293	164,000	164,000	0
0.004	0.006	571	170	320,000	95,000	225,000
0.003	0.007	1,360	106	760,000	59,700	700,300
0.002	0.008	4,570	71	2,560,000	40,000	2,520,000

4. *Determine the bearing stiffness*

Plot the net load capacity of this bearing vs. the lower film thickness, Fig. 20. A tangent to the curve at any point indicates the stiffness of this bearing. Draw a tangent through the origin where $h_2 = h_1 = 0.005$ in. The slope of this tangent = vertical value/horizontal value = $(725,000 - 0)/(0.005 - 0.002) = 241,000,000$ lb/in. = the bearing stiffness. This means that a load of 241,000,000 lb would be required to displace the plate 1.0 in. Since the plate cannot move this far, a load of 241,000 lb. would move the plate 0.001 in.

If the lubricant flow rate to each face of the bearing were doubled, to $Q = 4.0$ in.^{3}/sec, the stiffness of the bearing would increase to 333,000,000 lb/in., as shown in Fig. 20.

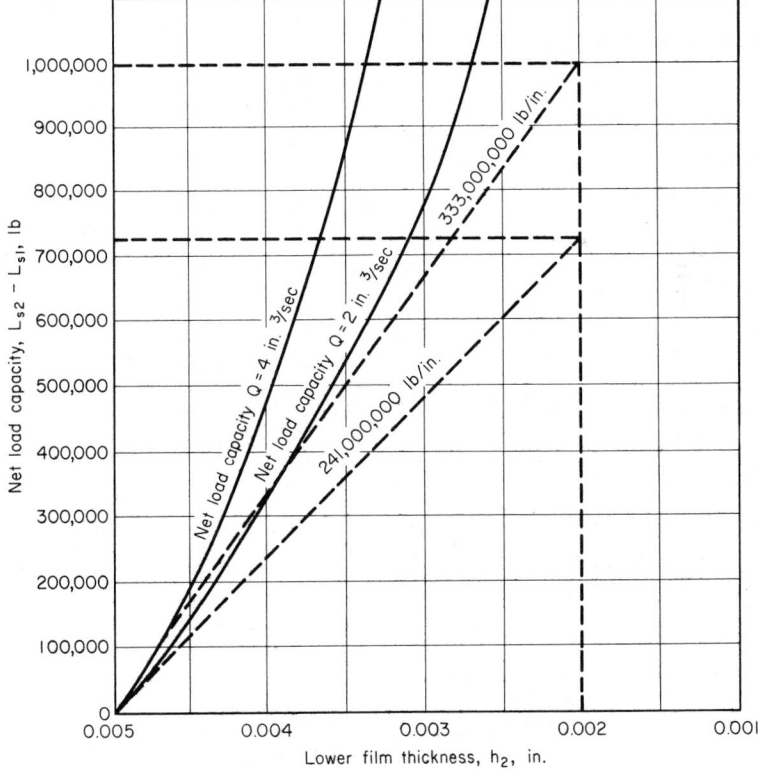

Fig. 20 Net load capacity for a double-direction thrust bearing. (*Product Engineering.*)

This means that an additional load of 333,000 lb would displace the plate 0.001 in. The stiffness of a hydrostatic bearing can be controlled by suitable design and a wide range of stiffness values can be designed into the bearing system.

Related Calculations: Hydrostatic bearings of the design shown here are useful for a variety of applications. Journal bearings for multidirectional loads are analyzed in a manner similar to that described here.

LOAD CAPACITY OF GAS BEARINGS

Determine the load capacity and bearing stiffness of a hydrostatic air bearing using 70-F air if the bearing orifice radius = 0.0087 in., the radial clearance $h = 0.0015$ in., the bearing diameter $d = 3.00$ in., the bearing length $L = 3$ in., the total number of air orifices $N = 8$, the ambient air pressure $p_a = 14.7$ psia, the air supply pressure $p_s = 15$ psig $= 29.7$ psia, the (gas constant)(total temperature) $= RT = 1.322 \times 10^8$ sq in./sec^2, $\epsilon =$ the eccentricity ratio $= 0.30$, air viscosity $\eta = 2.82 \times 10^{-9}$ lb-sec/sq in., the orifice coefficient $\alpha = 0.63$, and the shaft speed $\omega = 2,100$ radians/sec $= 20,000$ rpm.

Calculation Procedure:

1. Compute the bearing factors Λ and Λ_T

For a hydrostatic gas bearing, $\Lambda = (6\eta\omega/p_a)(d/zh)^2 = [(6)(2.82 \times 10^{-9})(2,100)/14.7]$ $[3/(2 \times 0.0015)]^2 = 2.41$.

Also, $\Lambda_T = 6\eta Na^2\alpha(RT)^{0.5}/p_a h^3 (6)(2.82 \times 10^{-9})(8)(0.0087)^2(0.63)(1.322 \times 10^8)^2/[(14.7)(0.0015)^3] = 1.5$.

2. Determine the dimensionless load

Since $L/d = 3/3 = 1$, and $p_s/p_a = 29.7/14.7 \approx 2$, use the first chart[1] in Fig. 21. Before entering the chart, compute $1/\Lambda = 1/2.41 = 0.415$. Then, from the chart, the dimensionless load $= 0.92 = L_d$.

3. Compute the bearing load capacity

The bearing load capacity $L_s = L_d p_a LD\epsilon = (0.92)(14.7)(3.00)(3.00)(0.3) = 36.5$ lb. If there were no shaft rotation, $\Lambda = 0$, $\Lambda_T = 1.5$, and $L_d = 0.65$, from the same chart. Then, $L_s = L_d p_a LD\epsilon = (0.65)(14.7)(3.00)(3.00)(0.3) = 25.8$ lb. Thus, rotation of the shaft increases the load-carrying ability of the bearing by $[(36.5 - 25.8)/25.8](100) = 41.5$ percent.

4. Compute the bearing stiffness

For a hydrostatic gas bearing, the bearing stiffness $B_s = L_s/h\epsilon = 36.5/[(0.0015)(0.3)] = 81,200$ lb/in.

Related Calculations: Use this procedure for the selection of gas bearings where the four charts presented here are applicable. The data summarized in these charts results from computer solutions of the complex equations for "hybrid" gas bearings. the work was done by Mechanical Technology Inc., headed by Beno Sternlicht.

SPRING SELECTION FOR A KNOWN LOAD AND DEFLECTION

Give the steps in choosing a spring for a known load and an allowable deflection. Show how the type and size of spring is determined.

Calculation Procedure:

1. Determine the load that must be handled

A spring may be required to absorb the force produced by a falling load or the recoil of a mass, to mitigate a mechanical shock load, to apply a force or torque, to isolate vibration, to support moving masses, or to indicate or control a load or torque. Analyze

[1]"Gas Bearings," *Product Engineering*, July 8, 1963.

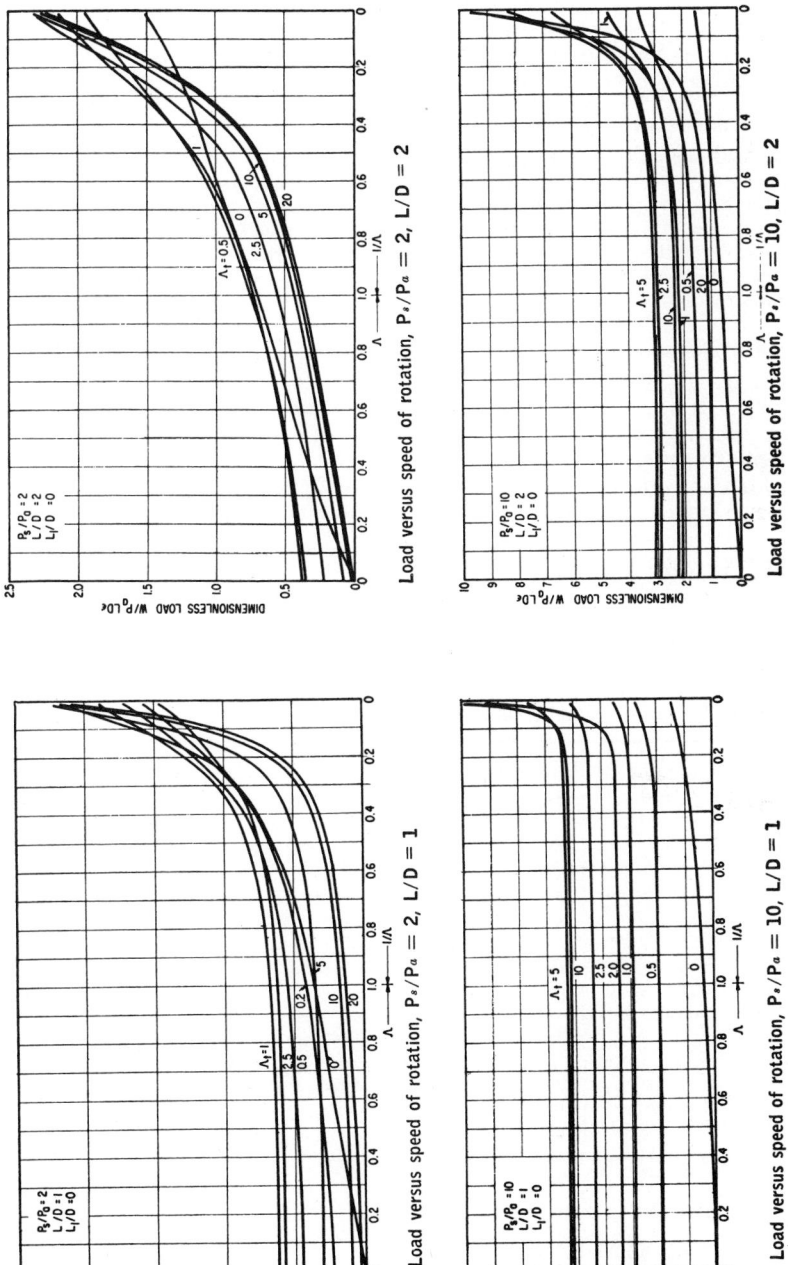

Fig. 21 Gas-bearing constants. (*Product Engineering.*)

the load to determine the magnitude of the force that is acting and the distance through which it acts.

Once the magnitude of the force is known, determine how it might be absorbed—by compression or extension (tension) of a spring. In some applications, either compression or extension of the spring is acceptable.

2. Determine the distance through which the load acts

The load member usually moves when it applies a force to the spring. This movement can be in a vertical, horizontal, or angular direction, or it may be a rotation. With the first type of movement, a *compression*, or *tension*, spring is generally chosen. With a torsional movement, a *torsion-type* spring is usually selected. Note that the movement in either case may be negligible—i.e., the spring applies a large restraining force—or the movement may be large, with the spring exerting only a nominal force compared with the load.

3. Make a tentative choice of spring type

Refer to Table 44, entering at the type of load. Based on the information known about the load, make a tentative choice of the type of spring to use.

4. Compute the spring size and stress

Use the methods given in the following Calculation Procedures to determine the spring dimensions, stress, and deflection.

5. Check the suitability of the spring

Determine (*a*) if the spring will fit in the allowable space, (*b*) the probable spring life, (*c*) the spring cost, and (*d*) the spring reliability. Based on these findings, use the spring chosen, if it is satisfactory. If the spring is unsatisfactory, choose another type of spring from Table 44 and repeat the study.

TABLE 44   Metal Spring Selection Guide

Type of load	Suitable spring type	Relative magnitude of load on spring	Deflection absorbed
Compression	Helical	Small to large	Small to large
	Leaf	Large	Moderate
	Flat	Small to large	Small to large
	Belleville	Small to large	Moderate
	Ring	Large	Small
Tension	Helical	Small to large	Small to large
	Leaf	Large	Moderate
	Flat	Small to large	Small to large
Torsion	Helical torsion	Small to large	Small to large
	Spiral	Moderate	Moderate
	Torsion bar	Large	Small

SPRING WIRE LENGTH AND WEIGHT

How long a wire is needed to make a helical spring having a mean coil diameter of 0.820 in. if there are 5 coils in the spring? What will this spring weigh if it is made of oil-tempered spring steel 0.055 in. in diameter?

Calculation Procedure:

1. Compute the spring wire length

Find the spring wire length from $l = \pi n d_m$, where l = wire length, in.; n = number of coils in the spring; d_m = mean coil diameter of the spring, in. Thus, for this spring, $l = \pi(5)(0.820) = 12.9$ in.

2. Compute the weight of the spring

Find the spring weight from $w = 0.224\ ld^2$, where w = spring weight, lb; d = spring wire diameter, in. For this spring, $w = 0.224(12.9)(0.055)^2 = 0.00874$ lb.

Related Calculations: The weight equation in Step 2 is valid for springs made of oil-tempered spring steel, chrome vanadium steel, silica-manganese steel, and silicon-chromium steel. For stainless steels, use a constant of 0.228 in place of 0.224 in the equation. The relations given in this procedure are valid for any spring having a continuous coil — helical, spiral, etc. Where a number of springs x are to be made, simply multiply the length and weight of each by the number to be made to determine the total wire length required, and the weight of the wire.

HELICAL COMPRESSION AND TENSION SPRING ANALYSIS

Determine the dimensions of a helical compression spring to carry a 5,000-lb load if it is made of hard-drawn steel wire having an allowable shear stress of 65,000 psi. The spring must fit in a 2-in.-diameter hole. What is the deflection of the spring? The spring operates at atmospheric temperature and the shear modulus of elasticity is 5×10^6 psi.

Calculation Procedure:

1. Choose the tentative dimensions of the spring

Since the spring must fit inside a 2-in.-diameter hole, the mean diameter of the coil d_m should not exceed about 1.75 in. Use this as a trial mean diameter and compute the wire diameter from $d = (8Ld_m/\pi s_s)^{1/3}$, where d = spring-wire diameter, in.; L = load on spring, lb; s_s = allowable shear stress in the spring material, psi. Thus, $d = [8 \times 5,000 \times 1.75/(\pi \times 65,000)]^{1/3} = 0.7$ in. The outside diameter d_o of the spring will therefore be $d_o = d_m + 2(d/2) = d_m + d = 1.75 + 0.70 = 2.45$ in. But the spring must fit a 2-in.-diameter hole. Hence, a smaller value of d_m must be tried.

Using $d_m = 1.5$ in. and following the same procedure, $d = [8 \times 5,000 \times 1.50/(\pi \times 65,000)]^{1/3} = 0.665$ in. Then, $d_o = 1.5 + 0.665 = 2.165$ in., which is still too large.

Using $d_m = 1.25$ in., $d = [8 \times 5,000 \times 1.25/(\pi \times 65,000)]^{1/3} = 0.625$ in. Then, $d_o = 1.25 + 0.625 = 1.875$ in. Since this is nearly 2 in., the spring will probably be acceptable. If desired, several other d_m values could be tried until a spring with a desired d_o more nearly equal to 2.0 in. was obtained. Use this procedure where a specific outside diameter is required for a spring.

2. Compute the spring deflection

The deflection of a helical compression spring is given by $f = 64\ n r_m^3 L/d^4 Gk$, where f = spring deflection, in.; n = number of coils in the spring; r_m = mean radius of spring coil, in.; G = shear modulus of elasticity of spring material, psi; k = spring curvature correction factor = $[(4c - 1)/(4c - 4)] + 0.615/c$ for heavily coiled springs, where $c = 2r_m/d$. For lightly coiled springs, $k = 1.0$.

Assuming $k = 1.0$ and $n = 10$ coils, $f = (64 \times 10 \times 0.625^3 \times 5,000)/(0.625^4 \times 5 \times 10^6 \times 10^6 \times 1.0) = 0.196$ in. This is a modest deflection for a load of 5,000 lb and indicates that the spring is heavily coiled. Therefore, as a check, the value of k should be determined and the deflection computed again.

Thus, $c = 2r_m/d = 2(1.25/2)/0.625 = 2.0$. Note that $r_m = d_m/2 = 1.25/2$ in this calculation for the value of c. Then, $k = [(4c - 1)/(4c - 4)] + 0.615/c = [(4 \times 2 - 1)/(4 \times 2 - 4)] + 0.615/2 = 2.0575$. Hence, the assumed value of $k = 1.0$ was inaccurate for this spring. Using the computed value of k, $f = (64 \times 10 \times 0.625^3 \times 5,000)/(0.625^4 \times 4 \times 10^6 \times 2.0575) = 0.0954$ in.

The number of coils n assumed for this spring is based on past experience with similar springs. However, where past experience does not exist, several trial values of n can be used until a spring of suitable deflection and length is obtained.

Related Calculations: Use this general procedure to analyze helical coil compression or tension springs. As a general guide, the outside diameter of a spring of this type is taken as (0.96)(hole diameter). The active solid height of a compression-type spring, i.e., the height of the spring when fully closed by the load, usually is nd, or (0.9) (final height when compressed by the design load).

SELECTION OF HELICAL COMPRESSION AND TENSION SPRINGS

Choose a helical compression spring to carry a 90-lb load with a stress of 50,000 psi and a deflection of about 2.0 in. The spring should fit in a 3.375-in.-diameter hole. The spring operates at about 70 F. How many coils will the spring have? What will the free length of the spring be?

Calculation Procedure:

1. Determine the spring outside diameter

Using the usual relation between spring outside diameter and hole diameter, $d_o = 0.96d_h$, where d_h = hole diameter, in. Thus, $d_o = 0.96(3.375) = 3.24$ in.; say 3.25 in.

2. Determine the required wire diameter

The equations in the previous Calculation Procedure can be used to determine the required wire diameter, if desired. However, the usual practice is to select the wire diameter using precomputed tabulations of spring properties, charts of spring properties, or a special slide rule available from some spring manufacturers. The tabular solution will be used here because it is one of the most popular methods.

Table 45 shows typical loads and spring rates for springs of various outside diameters and wire diameters based on a corrected shear stress of 100,000 psi and a shear modulus of $G = 11.5 \times 10^6$ psi.

TABLE 45   Load and Spring Rates for Helical Compression and Tension Springs*

Spring wire diameter, in.	Outside diameter of spring coil, in.						
	1	1.5	2	2.5	3	3.25	3.5
0.095	32.4†	16.5					
	158	16.9					
0.125	72.5	49.4	37.6	30.2			
	525	135	53.3	26.2			
0.156	138	94.8	72.4	58.5	48.7		
	1,412	351	137	66.3	37.2		
0.207	307	217	166	134	113	104	97.2
	5,266	1,218	457	218	121	93.6	74.1
0.250	...	375	290	235	198	183	170
		2,874	1,051	492	270	208	163
0.283	...	534	415	338	285	263	247
		5,154	1,820	849	460	352	276
0.3125	...	706	555	453	381	355	330
		8,143	2,842	1,311	705	541	424

*After H. F. Ross, "Application of Tables for Helical Compression and Extension Spring Design," *Transactions ASME*, vol. 69, p. 727.
†First figure given is loads in lb. at 100,000-psi stress. Second figure is spring rate in lb per in. per coil. $G = 11.5 \times 10^6$ psi.

Before using Table 45, the actual load must be corrected for the tabulated stress. Do this by taking the product of (actual load, lb)(table stress, psi/allowable spring stress, psi). For this spring, tabular load, lb = (90)(100,000/50,000) = 180 lb. This means that a 90-lb load at a 50,000-psi stress corresponds to a 180-lb load at 100,000 psi stress.

Enter Table 45 at the spring outside diameter, 3.25 in., and project vertically downward in this column until a load of approximately 180 lb is intersected. At the left read the wire diameter. Thus, with a 3.25-in. outside diameter and 183-lb load, the required wire diameter is 0.250 in.

3. Determine the number of coils required

The allowable spring deflection is 2.0 in.; and the spring rate per single coil, Table 45, is 208 lb/in. at a tabular stress of 100,000 psi. Using the relation, deflection f in. = load, lb/(desired spring rate, lb/in.), S_R; or $2.0 = 90/S_R$; $S_R = 90/2.0 = 45$ lb/in.

4. Compute the number of coils in the spring

The number of active coils in a spring is n = (tabular spring rate, lb/in.)/(desired spring rate, lb/in.). For this spring, $n = 208/45 = 4.62$, say 5 active coils.

5. Determine the spring free length

Find the approximate length of the spring in its free, expanded condition from l in. = $(n+i)d+f$, where l = approximate free length of spring, in.; i = number of inactive coils in the spring; other symbols as before. Assuming two inactive coils for this spring, $l = (5+2)(0.25)+2 = 3.75$ in.

Related Calculations: Similar design tables are available for torsion springs, spiral springs, coned-disk (Belleville) springs, ring springs, and rubber springs. These design tables can be found in engineering handbooks and in spring manufacturers' engineering data. Likewise, spring design charts are available from many of these same sources. Spring design slide rules are generally available free of charge to design engineers from spring manufacturers.

SIZING HELICAL SPRINGS FOR OPTIMUM DIMENSIONS AND WEIGHT

Determine the dimensions of a helical spring having the minimum material volume if the initial load on the spring is 15 lb, the mean coil diameter is 1.02 in., the spring stroke is 1.16 in., the final spring stress is 100,000 psi, and the spring modulus of torsion is 11.5×10^6 psi.

Calculation Procedure:

1. Compute the minimum spring volume

Use the relation $v_m = 8fLG/s_f^2$, where v_m = minimum volume of spring, in.³; f = spring stroke, in.³; L = initial load on spring, lb; G = modulus of torsion of spring material, psi; s_f = final stress in spring, psi. For this spring, $v_m = 8(1.16)(15)(11.5 \times 10^6)/(100,000)^2 = 0.16$ in.³.

2. Compute the required spring wire diameter

Find the wire diameter from $d = (16Ld_m/\pi s_f)^{1/3}$, where d = wire diameter, in.; d_m = mean diameter of spring, in.; other symbols as before. For this spring, $d = [16 \times 15 \times 1.02/(\pi \times 100,000)]^{1/3} = 0.092$ in.

3. Find the number of active coils in the spring

Use the relation $n = 4v_m/\pi^2 d^2 d_m$, where n = number of active coils; other symbols as before. Thus, $n = 4(0.16)/[\pi^2(0.092)^2(1.02)] = 7.5$ coils.

4. Determine the active solid height of the spring

The solid height $H_s = (n+1)d$, in., or $H_s = (7.5+1)(0.092) = 0.782$ in. For a practical design, allow a 10 percent clearance between the solid height and the minimum-compressed height H_c. Thus, $H_c = 1.1H_s = 1.1(0.782) = 0.860$ in. The assembled height $H_a = H_c + f = 0.860 + 1.16 = 2.020$ in.

5. Compute the spring load-deflection rate

The load-deflection rate $R = Gd^4/8d_m^3n$, where R = load-deflection rate, lb/in.; other symbols as before. Thus, $R = (11.5 \times 10^6)(0.092)^4/[8(1.02)^3(7.5)] = 12.9$ lb/in.

The initial deflection of the spring is $f_i = L/R$ in., or $f_i = 15/12.9 = 1.163$ in. Since the free height of a spring $H_f = H_a + f_i$, the free height of this spring is $H_f = 2.020 + 1.163 = 3.183$ in.

Related Calculations: The above procedure for determining the minimum spring volume can be used to find the minimum spring weight by relating the spring weight W lb to the density of the spring material ρ lb/in.^{3} in the following manner: For the required initial load, L_l lb, $W_{min} = \rho(8fL_lG/s_f^2)$. For the required energy capacity E in.-lb, $W_{min} = \rho(4EG/S_f^2)$. For the required final load L_2 lb, $W_{min} = \rho(2f_2L_2G/s_f^2)$.

The above procedure assumes the spring ends are open and not ground. For other types of end conditions, the minimum spring volume will be greater by the following amount: For squared (closed) ends, $v_m = 0.5\pi^2d^2d_m$. For ground ends, $v_m = 0.25\pi^2d^2d_m$. The methods presented here were developed by Henry Swieskowski and reported in *Product Engineering*.

SELECTION OF SQUARE AND RECTANGULAR WIRE HELICAL SPRINGS

Choose a square wire spring to support a load of 500 lb with a deflection of not more than 1.0 in. The spring must fit in a 4.25-in.-diameter hole. The modulus of rigidity for the spring material is $G = 11.5 \times 10^6$ psi. What is the shear stress in the spring? Determine the corrected shear stress for this spring.

Calculation Procedure:

1. Determine the spring dimensions

Assume that a 4-in.-diameter square-bar spring is used. Such a spring will fit in the 4.25-in.-diameter hole with a small amount of room to spare.

As a trial, assume that the width of the spring wire = 0.5 in. = a. Since the spring is square, the height of the spring wire = 0.5 in. = b.

With a 4-in. outside diameter and spring wire width of 0.5 in., the mean radius of the spring coil $r_m = 1.75$ in. This is the radius from the center of the spring to the center of the spring wire coil.

2. Compute the spring deflection

The deflection of a square-wire tension spring is $f = 45Lr_m^3n/Ga^4$, where f = spring deflection, in., L = load on spring, lb; n = number of coils in spring; other symbols as before. To solve this equation, the number of coils must be known. Assume, as a trial value, 5 coils. Then, $f = 45(500)(1.75)^3(5)/[(11.5 \times 10^6)(0.5)^4] = 0.838$ in. Since a deflection of not more than 1.0 in. is permitted, this spring is probably acceptable.

3. Compute the shear stress in the spring

Find the shear stress in a square-bar spring from $S_s = 4.8Lr_m/a^3$, where S_s = spring shear stress, psi; other symbols as before. For this spring, $S_s = (4.8)(500)(1.75)/(0.5)^3 = 33,600$ psi. This is within the allowable limits for usual spring steel.

4. Determine the corrected shear stress

Find the shear stress in a square-bar spring from $S_s = 4.8Lr_m/a^3$, where S_s = spring correction factor $k = 1 + (1.2/c) + (0.56/c^2) + (0.5/c^3)$, where $c = 2r_m/a$. For this spring, $c = (2 \times 1.75)/0.5 = 7.0$. Then, $k = 1 + (1.2/7) + (0.56/7^2) + (0.5/7^3) = 1.184$. Hence, the corrected shear stress is $S_s' = ks_s$, or $S_s = (1.184)(33,600) = 39,800$ psi. This is still within the limits for usual spring steel.

Related Calculations: Use a similar procedure to select rectangular wire springs.

Once the dimensions are selected, compute the spring deflection from $f = 19.6\,Lr_m^3 n/Gb^3(a-0.566)$, where all the symbols are as given earlier in this Calculation Procedure. Compute the uncorrected shear stress from $S_s = Lr_m(3a+1.8b)/a^2b^2$. To correct the stress, use the Liesecke correction factor given in Wahl—*Mechanical Springs*. For most selection purposes, the uncorrected stress is satisfactory.

CURVED SPRING DESIGN ANALYSIS

Find the maximum load P, maximum deflection F, and spring constant C for the curved rectangular-wire spring shown in Fig. 22 if the spring variables expressed in metric units are: $E = 14{,}500\ \text{kg/mm}^2$, $S_b = 55\ \text{kg/mm}^2$, $b = 1.20\ \text{mm}$, $h = 0.30\ \text{mm}$, $r_1 = 0.65\ \text{mm}$, $r_2 = 1.75\ \text{mm}$, $L = 9.7\ \text{mm}$, $u_1 = 1.7\ \text{mm}$, and $u_2 = 5.6\ \text{mm}$.

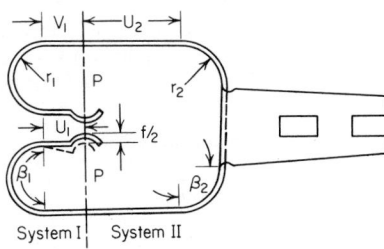

Fig. 22 Typical curved spring. (*Product Engineering.*)

Calculation Procedure:

1. Divide the spring into analyzable components

Using Fig. 23, developed by J. Palm and K. Thomas of West Germany, as a guide, divide the spring to be analyzed into two or more analyzable components, Fig. 22.

Spring type	Spring deflection	Spring force and bending stresses		
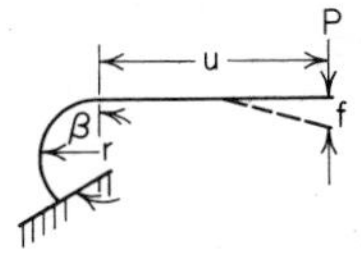 A	$F_1 = \dfrac{KPr^3}{3EI}(m+\beta)^3$ where $\alpha = \beta$ for finding K	When $\alpha = 0°$ to $90°$ $P = \dfrac{S\sigma}{u+\sin\beta}$ $\sigma = \dfrac{Pr(m+\sin\beta)}{S}$		When $\alpha = 90°$ to $180°$ $P = \dfrac{S\sigma}{u+r}$ $\sigma = \dfrac{Pr(m+1)}{S}$
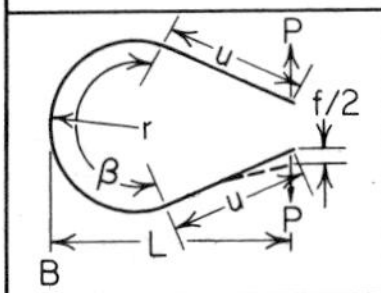 B	$F_2 = \dfrac{2KPr^3}{3EI}(m+\dfrac{\beta}{2})^3$ where $\alpha = \dfrac{\beta}{2}$ for finding K	$P = \dfrac{S\sigma}{L}$		
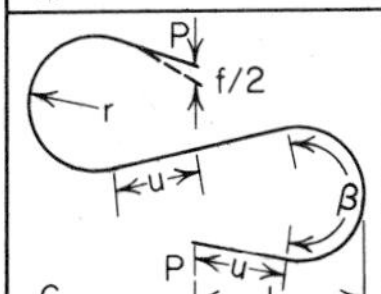 C	$F_3 = 2F_2 = \dfrac{4KPr^3}{3EI}(m+\dfrac{\beta}{2})^3$ where $\alpha = \dfrac{\beta}{2}$ for finding K	$\sigma = \dfrac{PL}{S}$		
D / E	$F_4 = F_5 = \dfrac{P}{3EI}\left[2Kr^3(m+\dfrac{\beta}{2})^3+(v-u)^3\right]$ where $\alpha = \dfrac{\beta}{2}$ for finding K	$P = \dfrac{S\sigma}{\lambda} = \dfrac{P\lambda}{S}$		

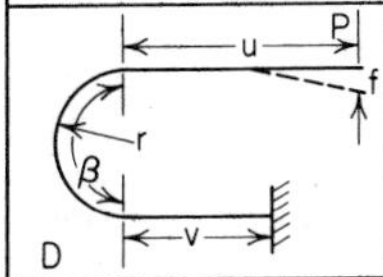

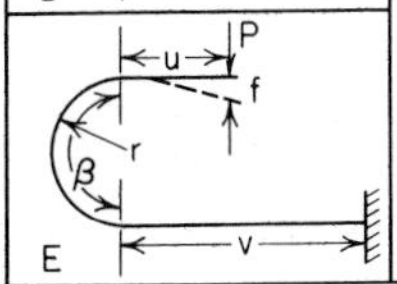

	First condition	Second condition	λ
	$u \geq v$	- - -	$u+r$
	$u < v$	$(u-v) < (u+r)$	$u+r$
	$u < v$	$(v-u) > (u+r)$	$v-u$
	$u = 0$	$v \leq r$	r
	$u = 0$	$v > r$	v

Fig. 23 Deflection, force, and stress relations for curved springs. (*Product Engineering.*)

Thus, the given spring can be divided into two springs—a type D (Fig. 23), called system I, and a type A (Fig. 23), called system II.

2. Compute the spring force

The spring force $P = P_I = P_{II}$. Since $(u_2 + r_2) > (u_1 + r_1)$, the spring in system II exerts a larger force. From Fig. 23 for $\beta = 90°$, $P = S\sigma_{max}/(u_2 + r_2)$, where S = section modulus, mm³, of the spring wire. Since $S = bh^2/6$ for a rectangle, $P = bh^2\sigma_{max}/6(u_2 + r_2)$ where b = spring wire width, mm; h = spring wire height, mm; σ_{max} = maximum bending stress in the spring, kg/mm³; other symbols as given in Fig. 22. Then, $P = (1.20)(0.30)^2 \times (55)/[6(5.6 + 1.75)] = 0.135$ kg.

3. Compute the spring deflection

The total deflection of the springs is $F = (2F_I + F_{II})$, where F = spring deflection, mm, and the subscripts refer to each spring system. Taking the sum of the deflections as given in Fig. 23, $F = (2P/3EI)[2K_1r_1^3(m_1 + \beta_1/2)^2 + (v_1 - u_1)^3 + K_2r_2^3(m_2 + \beta_2)^3]$, where E = Young's modulus, kg/mm²; I = spring wire moment of inertia mm⁴; K = correction factor for the spring from Fig. 24, where the subscripts refer to the radius being considered in the relation u/r; $m = u/r$; β = angle of spring curvature, radians. Where the subscripts 1 and 2 are used in this equation, they refer to the respective radius identified by this subscript. Since $I = bh^3/12$ for a rectangle, or $I = (1.20)(0.30)^3/12 = 0.0027$ mm⁴, $F = \{[2(0.135)/[(3)(14,500)(0.0027)]\}[2(0.92)(0.65)(2.62 + 1.57)^3 + 0 + 0.94(1.75)^3(3.2 + 1.57)^3] = 1.34$ mm.

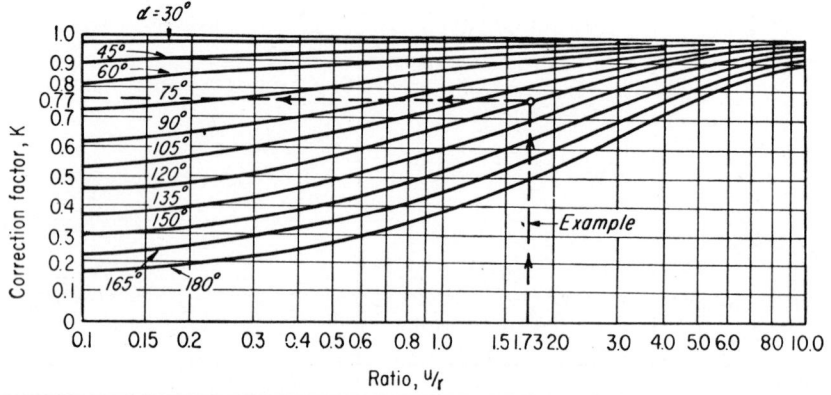

Fig. 24   Correction factors for curved springs. (*Product Engineering.*)

4. Compute the spring constant

The spring constant $C = P/F = 0.135 = 0.135/1.34 = 0.101$ kg/mm.

Related Calculations: The relations given here can also be used for round-wire springs. For accurate results, h/r for flat springs and d_o/r for round-wire springs should be less than 0.6. The various symbols used in this Calculation Procedure are defined in the text and illustrations. Since the equations given here analyze the springs and do not contain any empirical constants, the equations can be used, as presented, for both metric and English units. Where a round spring is analyzed, $h = b = d_o$, where d_o = spring outside diameter, mm or in.

ROUND- AND SQUARE-WIRE HELICAL TORSION-SPRING SELECTION

Choose a round-music-wire torsion spring to handle a moment load of 15.0 lb·in. through a deflection angle of 250°. The mean diameter of the spring should be about 1.0 in. to satisfy the space requirements of the design. Determine the required diameter of the spring wire, the stress in the wire, and the number of turns required in the spring. What is the maximum moment and angular deflection the spring can handle? What is the maximum moment and deflection without permanent set?

Calculation Procedure:

1. Select a suitable wire diameter

To reduce the manufacturing cost of a spring, a wire of standard diameter should be used, whenever possible, for the spring, Fig. 25. Usual torsion-spring wire diameters and the side of square-wire springs range from 0.02 to 0.60 in., depending on the moment the spring must carry and the angular deflection.

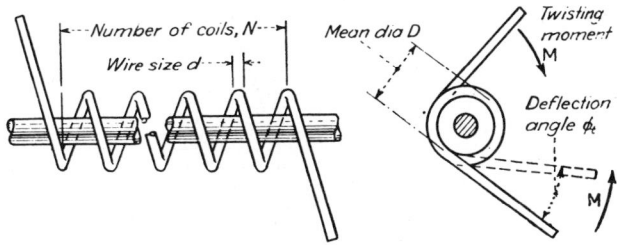

Fig. 25   Typical torsion spring. *(Product Engineering.)*

Assume a wire diameter of 0.10 in. and a bending stress of 150,000 psi as trial values for this spring. (Typical round-wire and square-wire torsion-spring bending stresses range from 100,000 to 200,000 psi, depending on the material used in the spring.)

Compute the twisting moment corresponding to the assumed stress from $M_i = \pi d^3 S_b/32$, where M_i = twisting moment load, lb-in.; d = spring wire diameter, in.; S_b = bending stress in spring, psi. Thus, $M_i = \pi(0.10)^3(150,000)/32 = 14.7$ lb-in. This is very close to the actual moment load of 15.0 lb-in. Therefore, the assumed spring diameter and bending stress are acceptable, thus far.

2. Compute the actual spring stress

Use the following relation to find the actual bending stress S_b psi in the spring: S_b = (actual spring moment lb-in./computed spring moment, lb-in.)(assumed stress, psi); $S_b = (15.0/14.7)(150,000) = 153,000$ psi.

3. Check the actual vs. recommended spring stress

Enter Fig. 26 at the wire diameter of 0.10 in. and project vertically upward to the music-wire curve to read the recommended bending stress for music wire as 159,000 psi. Since the actual stress, 153,000 psi, is less than but reasonably close to the recommended stress, the selected wire diameter is acceptable for the planned load on the spring. This chart and Calculation Procedure were developed by H. F. Ross and reported in *Product Engineering*.

4. Determine the angular deflection per spring coil

Compute the angular deflection per coil from $\phi = 360\, S_b d_m/Ed$, where ϕ = angular deflection per spring coil, °; d_m = mean diameter of spring, in.; E = Young's modulus for spring material = 30×10^6 psi for spring steel; other symbols as before. Thus, using the *assumed* bending stress in the spring, $\phi = 360(150,000)(1.0)/(30 \times 10^6)(0.1) = 18°$. This value is the maximum safe deflection per coil for the spring.

5. Compute the number of coils required

The number of coils n required in a helical torsion spring is $n = \phi_t$(assumed stress, psi)/ϕ(actual stress, psi), where ϕ_t = total angular deflection of spring, °; ϕ = maximum safe deflection per coil, °. Thus, $n = 250(150,000)/[(18)(153,000)] = 13.6$ coils; use 14 coils.

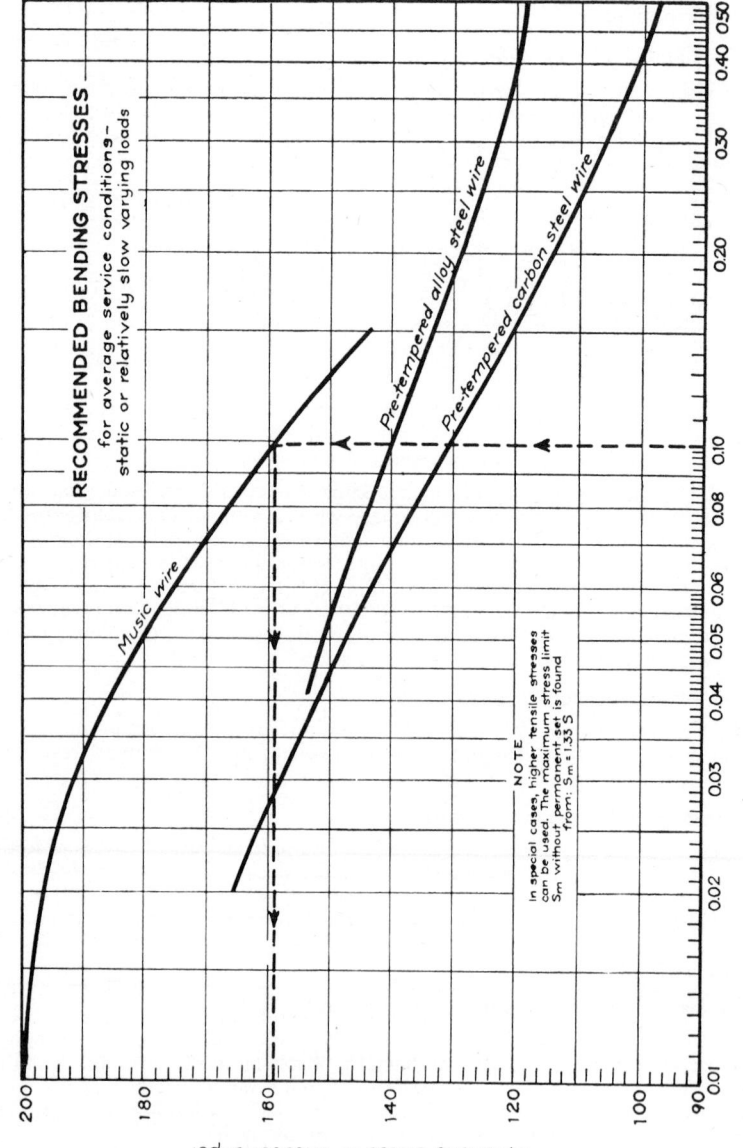

Fig. 26   Recommended bending stresses for torsion springs. (*Product Engineering.*)

6. Determine the maximum moment the spring can handle

On the basis of the maximum recommended stress, the moment can be increased to M_i = (maximum recommended stress, psi/assumed stress, psi)(actual moment, lb-in.). Read the maximum recommended stress from Fig. 26 as 159,000 psi for 0.1-in.-diameter music wire, as in Step 3. Thus, $M_i = (159,000/150,000)(14.7) = 15.6$ lb-in.

7. Compute the maximum angular deflection

The maximum angular deflection per coil is ϕ = (maximum recommended stress, psi/assumed stress, psi)(computed angular deflection per coil, °) = $(159,000/150,000) \times (18) = 19.1°$ per coil.

8. Determine the special-case moment and deflection

The maximum moment M_{max} and deflection ϕ_{max}, without permanent set, can be one-third greater than in Steps 6 and 7, or $M_{max} = 15.6(1.33) = 20.8$ lb-in., and $\phi_{max} = 19.1(1.33) = 25.5°$ per coil. These maximum values allow for overloads on the spring.

Related Calculations: Use the same procedure for square-wire helical torsion springs, but substitute the length of the side of the square for d in each equation where d appears.

TORSION-BAR SPRING ANALYSIS

What must the diameter of a torsion bar be if it is to have a spring rate of 2,400 lb-in./radian and a total angle of twist of 0.20 radians? The bar is made of 302 stainless steel, which has a proportional limit in tension of 35,000 psi, and G = torsional modulus of elasticity = 10^7 psi. The length of the torsion bar is 26.0 in. and it is solid throughout. What size square torsion bar would be required? What size equilateral triangular section would be required? What is the energy storage of each bar form?

Calculation Procedure:

1. Determine the proportional limit in shear

For stainless steel, the proportional limit S_s psi in shear is 0.55 times that in tension, or $S_s = 0.55(35,000) = 19,250$ psi.

2. Compute the required diameter of the bar

Use the relation $d = 2S_s l/G\theta$, where d = torsion-bar diameter, in.; l = torsion-bar length, in., θ = total angle of twist of torsion bar, radians; other symbols as before. Thus, $d = 2(19,250)(26.0)/[10^7(0.20)] = 0.50$ in.

3. Compute the square-bar size

Use the relation $d = 1.482S_s l/G\theta$, where d = side of the square bar, in. Thus, $d = 1.482(19,250)(26.0)/[10^7(0.2)] = 0.371$ in.

4. Compute the triangular bar size

Use the relation $d = 2.31S_s l/G\theta$, where d = side of the triangular bar, in. Thus, $d = 2.31(19,250)(26.0)/[10^7(0.2)] = 0.578$ in.

5. Compute the energy storage of each bar

For a solid circular torsion spring, the energy storage $e = S_s^2/4G$, where e = energy storage in the bar, in.-lb/in.³. Thus, $e = (19,250)^2/[4(10^7)] = 9.25$ in.-lb/in.³.

For a square bar, $e = S_s^2/6.48G$, where the symbols are the same as before, or, $e = (19,250)^2/[6.48(10^7)] = 5.71$ in.-lb/in.³.

For a triangular bar, $e = S_s^2/7.5G$, where the symbols are the same as before. Or, $e = (19,250)^2/[7.5(10^7)] = 4.94$ in.-lb/in.³.

Related Calculations: Use this procedure for torsion-bar springs made of any metal. The energy-storage capacity of various springs in terms of the spring weight is:

Type of Spring	Energy Storage, in.-lb per lb of Spring
Leaf......................	300–450
Helical round-wire coil	700–1,100
Torsion-bar...............	1,000–1,500
Volute....................	500–1,000
Rubber in shear...........	2,000–4,000

The analyses in this Calculation Procedure are based on the work of Donald Bastow and D. A. Derse reported in *Product Engineering.*

MULTIRATE HELICAL SPRING ANALYSIS

Determine the required spring rates, number of coils, coil clearances, and free length of two helical coil springs if spring 1 has a preload of 1.2 lb and spring 2 a preload of 19.1 lb in a double preload mechanism. The rod is to deflect 0.46 in. before building up to the preload of 19.1 lb. Total deflection is to be 3.0 in. with a load of 78 lb. The mean spring diameter $d_m = 1.29$ in. for both springs; the wire diameter $d = 0.148$ in. for spring 1; $d = 0.156$ in. for spring 2; $G = 11.5 \times 10^6$ for both springs.

Calculation Procedure:

1. Determine the spring rate for each spring

The spring rate, lb/in., $R_s =$ (preload spring 2, lb − preload spring 1, lb)/deflection, in., before full preload. Thus, for spring 1; $R_{s1} = (19.1 - 1.2)/0.46 = 38.9$ lb/in.

For the combination of the two springs R_{st}, using a similar relation, $R_{st} = (78 - 19.1)/(3.0 - 0.46) = 23.1$ lb/in.

For spring 2, $R_{s2} = R_{s1}R_{st}/(R_{s1} - R_{st})$, where the symbols are the same as before. Or, $R_{s2} = (38.9)(23.1)/(38.0 - 23.1) = 56.9$ lb/in.

2. Check the spring rate against the spring deflection

The deflection, in., is $f = L/R$, where $L =$ load on the spring lb; $R =$ spring rate, lb/in. Thus, for spring 1, $f_1 = (78 - 1.2)/38.9 = 1.97$ in. For spring 2, $f_2 = L_2/R_{s2} = (78 - 19.1)/56.9 = 1.03$ in. For the two springs, $f_t = f_1 + f_2 = 1.97 + 1.03 = 3.00$ in. This agrees with the allowable deflection of 3 in. at the full load of 78 lb. Therefore, the computed spring rates and preloads are acceptable.

3. Compute the number of coils for each spring

The number of coils $n = Gd^4/8d_m^3R$, where the symbols are as defined before. Thus, $n_1 = (11.5 \times 10^6)(0.148)^4/[8(1.29)^3(38.9)] = 8.25$ coils. And, $n_2 = (11.5 \times 10^6)(0.156)^4/[8(1.29)^3(56.9)] = 7$ coils.

4. Compute the solid height of each spring

Allowing one inactive coil for each end of each spring, so that the ends may be squared and ground, the solid height $h_s = d$(number of coils + 2). Or $h_{s1} = (0.148) \times (8.25 + 2) = 1.517$ in. And, $h_{s2} = (0.156)(7 + 2) = 1.404$ in.

5. Determine the coil clearances

Assume a coil clearance of three times the spring wire diameter. Then the coil clearance c in. for each spring is $c_1 = (3)(0.148) = 0.444$ in., and $c_2 = (3)(0.156) = 0.468$ in.

6. **Compute the free length of each spring**

The free length of a helical spring = l_f = solid height + coil clearance + deflection + [preload, lb/(spring rate, lb/in.)]. For spring 1, l_{f1} = 1.517 + 0.444 + 1.970 + (1.2/38.9) = 3.962 in. For spring 2, l_{f2} = 1.404 + 0.468 + 1.030 + (19.1/56.9) = 3.235 in.

Related Calculations: Use this procedure for springs made of any metal. This analysis is based on the work of K. A. Flesher, as reported in *Product Engineering*.

BELLEVILLE SPRING ANALYSIS FOR SMALLEST DIAMETER

What is the minimum outside radius r_o and thickness t for a steel Belleville spring that carries a load of 1,000 lb at a maximum compressive stress of 200,000 psi when compressed flat?

Calculation Procedure:

1. **Determine the spring radius ratio and the height-thickness ratio**

The radius ratio $r_r = r_o/r_i$ for a Belleville spring, when r_o = outside radius of spring, in.; r_i = inside radius of spring, in. = radius of hole in spring, in. Table 46 summarizes recommended values for the radius ratio for various values of the height-thickness ratio to produce the smallest diameter spring. In general, an r_r value of 1.75 usually produces a spring of suitably small size. When r_r = 1.75, Table 46 shows that the height-thickness ratio h/t with both values expressed in in. is 1.5. Assume that these two values are valid and proceed with the calculation.

TABLE 46   Design Constants for Belleville Springs*

h/t	r_o/r_i
1.00	1.25
1.25	1.50
1.50	1.75
1.75	2.00
2.00	2.50

Product Engineering.

2. **Determine the spring outside radius**

Table 47 shows the stress constant, $r_o s_c/L^{0.5}$, where s_c = maximum compressive stress on the top surface at the inner edge, psi, Fig. 27; L = total axial load on spring, lb. For h/t = 1.5 and r_r = 1.75, the stress constant $r_o s_c/L^{0.5}$ = −1,905. Solving, r_o = −1,905$L^{0.5}$/s_c. Substituting the given values, r_o = −1,905(1,000)$^{0.5}$/−200,000 = 3.01 in. The negative sign is used for the spring stress because it is a compressive stress.

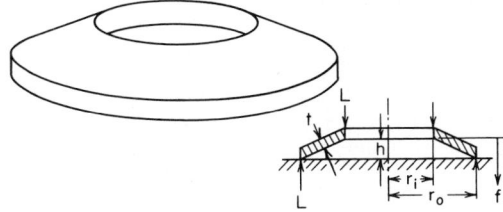

Fig. 27   Belleville spring appearance and dimensions. *(Product Engineering.)*

3. **Determine the radius of the hole in the spring**

For this Belleville spring, $r_i = r_o/r_r$ = 3.01/1.75 = 1.72 in.

4. Compute the spring thickness

The thickness of a Belleville spring is given by $t = (s_c r_o^2/KE)^{0.5}$, where K = a stress constant from Table 47; E = modulus of elasticity of the spring material, psi; other symbols as before. Thus, with $E = 30 \times 10^6$, $t = [200,000 \times 3.01^2/(5.6279 \times 30 \times 10^6)]^{0.5} = 0.1037$ in.

TABLE 47   Stress Constants for Belleville Springs*

h/t	$r_0/r_1 - 1.75$	
	K	$r_o s_c/L^{0.5}$
1.00	−3.2455	−1,346
1.25	−4.3734	−1,622
1.50	−5.6279	−1,905
1.75	−7.0090	−2,197

Product Engineering.

5. Compute the spring height

Since $h/t = 1.5$ for this spring, $h = 1.5(0.1037) = 0.156$ in.

Related Calculations: Professor M. F. Spotts developed the analytical procedure and data presented here. His studies show that space is usually the limiting factor in spring selection and the designer generally must determine the minimum permissible outside diameter of the spring to carry a given load at a specified stress. Further, the ratio of the outside to the inside diameter for the smallest spring is about 1.75, assuming that the load spring is compressed nearly flat, which is the usual design assumption. A value of h/t of 1.5 is recommended for most spring applications. Belleville springs are used in disk brakes, the preloading of bolted assemblies, ball bearings, etc. The analysis presented here is useful for all usual applications of Belleville springs.

RING-SPRING DESIGN ANALYSIS

Determine the major dimensions of a ring spring made of material having an allowable stress of 175,000 psi, $E = 20 \times 10^6$, a coefficient of friction of 0.12, an inside diameter of 7.0 in., an outside diameter of 9.0 in. or less, a taper angle of 14°, an axial load of 56 tons, and a deflection of not more than 8.0 in.

Calculation Procedure:

1. Determine the inner-ring dimensions

For the usual ring spring, the ring height h in. is 15 percent of the allowable outside diameter, or $(0.15)(9.0) = 1.35$ in., Fig. 28. The axial gap between the rings g in. is usually 25 percent of the ring height.

Compute the area of the internal ring from $A_i = L/\pi K_c s_i$, where A_i = area of internal ring, sq in.; L = axial load on spring, lb; K_c = spring constant from Fig. 29; s_i = allowable stress in the inner ring of the spring, psi. With a coefficient of friction $\mu = 0.12$ and a taper angle of 14°, $K_c = 0.38$. Then, $A_i = (56 \times 2,000)/[\pi(0.38)(175,000)] = 0.537$ sq in.

The width w_i in. of the inner ring is $w_i = [A_i - (h_i^2 \tan \theta)/4]/h_i$, where $\tan \theta$ = tangent of taper angle; h_i = height of inner ring, in. Thus, $w_i = [0.537 - (1.35^2 \tan 14°)/4]/1.35 = 0.314$ in.

Use a trial-and-error process to determine the dimensions of the outer ring. Do this by assuming a cross-sectional area for the outer ring; then compute whether the outside diameter and stress meet the specifications for the spring.

Assume that $A_o = 0.609$ sq in. Then, $s_o = L/\pi A_o K_c$, where s_o = stress in outer ring, psi; other symbols as before. Solving, $s_o = (56 \times 2,000)/[\pi(0.609)(0.38)] = 154,200$ psi. This stress is within the allowable limits.

In the usual ring spring, $h_o = h_i = 1.35$ in. for this spring. Then, using a relation similar to that for the inner ring, $w_o = [A_o - (h_o^2 \tan \theta)/4]/h_o = [0.609 - (1.35^2 \tan 14°)/4]/1.35 = 0.366$ in.

Find the outside diameter of the ring from $d_o = d_i + 2w_i + 2w_o + (h - g) \tan \theta$, where d_o = outside diameter of outer ring, in.; d_i = inside diameter of inner ring, in.; g = axial gap of rings, in. = 25 percent of ring height for this spring, or $0.25(1.35) = 0.3375$ in.

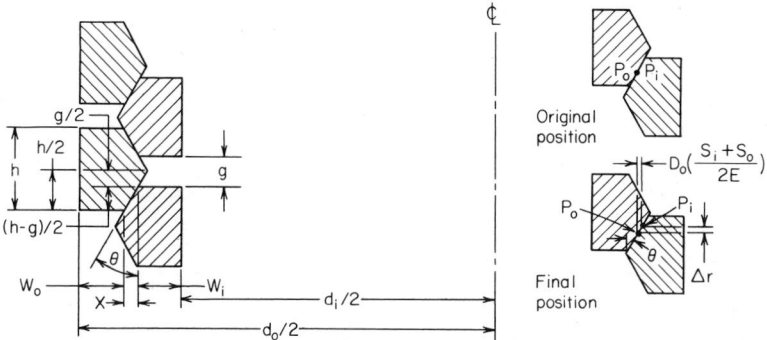

Fig. 28   Ring-spring positions and dimensions. (*Product Engineering.*)

Hence, $d_o = 7.0 + 2(0.314) + 2(0.366) + (1.35 - 0.3375) \tan 14° = 8.613$ in. This is close enough to the maximum allowable outside diameter of 9 in. to be acceptable. Were the value of d_o unacceptable, another value of A_o would be assumed and the calculation repeated until the stress and d_o values were acceptable.

3.   Compute the number of rings required

Find the axial deflection per ring f in. from $f = d_a[(s_i + s_o)/2E] \cot \theta$, where d_a = mean diameter of the spring, in.; E = modulus of elasticity of the spring material, psi. Compute $d_a = [(d_o - 2w_o) + (d_i + 2w_i)]/2 = [(8.613 - 2 \times 0.366) + (7.0 + 2 \times 0.314)]/2 = 7.755$ in. Then, $f = 7.755[(175.000 + 154,200)/(2 \times 29 \times 10^6)] \cot 14° = 0.176$ in. Since the axial deflection must not exceed 8 in., the number of rings required = axial deflection, in./deflection per ring, in. = $8.0/0.176 = 45.5$, or 46 rings. Figure 30 shows the spring dimensions.

Related Calculations: Ring springs are suitable for pipe-vibration isolation, shock absorbers, plows, trench diggers, railroad couplers, etc. The recommended approximate proportions of ring springs are: (1) Compressed height should be at least four

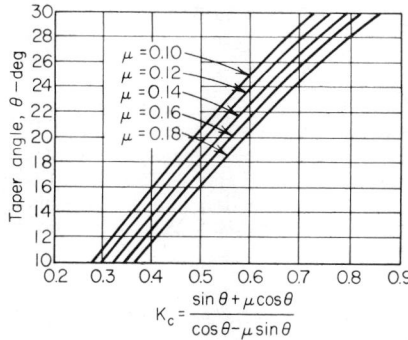

Fig. 29   Ring-spring compression constant in terms of the taper angle for various values of the coefficient of friction. (*Product Engineering.*)

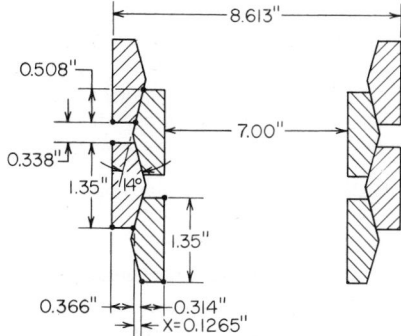

Fig. 30   Dimensions of a typical ring spring. (*Product Engineering.*)

times the deflection of the spring. (2) Ring height should be 15 to 20 percent of the ring outside diameter. (3) Spring outside diameter and height are usually as large as space permits. (4) Thin ring sections are preferred to thick ones. (5) Ring taper should be 1 to 4. (6) Coefficient of friction for ring springs varies from 0.10 to 0.18. (7) Allowable spring stresses are 160,000 psi for nonmachined steel; 200,000 psi for machined steel. For vibratory loads, the allowable stress is about one-half these values. (8) Load capacities of ring springs vary between 2 and 150 tons. (9) Spring deflections vary between 1 in. and 1 ft. (10) The equations given above can be used for spring design or for analysis of an existing spring.

The design method given here was developed by Tyler G. Hicks and reported in *Product Engineering*.

LIQUID-SPRING SELECTION

Select a liquid spring to absorb a 50,000-lb load with a 5-in. stroke. The rod diameter is 1 in. What is the probable temperature rise per stroke? Compare this spring with metal-coil, Belleville, and ring springs.

Calculation Procedure:

1. Compute the liquid volume required

Assume that the final pressure of the compressed liquid is 50,000 psi and that the liquid is compressed 18 percent on application of full load on the spring. This means that 82 percent $(100 - 18)$ of the original volume remains after application of the load.

Compute the liquid volume required from $v = \pi S d^2/4c$, where v = liquid volume required, in.^{3}; S = stroke length, in.; d = rod diameter, in.; c = liquid compressibility, expressed as a decimal. Thus, $v = \pi(5)(1)^2/[4(0.18)] = 21.8$ in.^{3}.

2. Determine the cylinder length

In a liquid spring, the cylinder inside diameter d_i is usually greater than that of the rod. Assuming an inside diameter of 1.8 in. for the cylinder, length = $4v/\pi d_i^2$, where d_i = cylinder inside diameter, in.; other symbols as before. For this cylinder, length = $4(21.8)/[\pi(1.8)^2] = 8.56$ in.; say 8.6 in.

3. Determine the cylinder dimensions

With a 1.8-in. inside diameter, a 3-in. outside diameter will be required, based on the usual cylinder proportions. Allowing 3.4 in. for the cylinder ends and seals and 5 in. for the stroke, the total length of the cylinder will be $8.6 + 3.4 + 5.0 = 17.0$ in.

4. Compute the cylinder temperature rise

Assume that the average friction load is 10 percent of the load on the spring, or $0.1 \times (50,000) = 5,000$ lb. A friction load of 10 percent is typical for liquid springs.

The energy absorbed per stroke of the spring is $e = Fl$, where e = energy absorbed, ft-lb; F = friction force, lb; l = stroke length, ft. For this spring, $e = 5,000(5/12) = 2,085$ ft-lb. Since 778.2 ft-lb = 1 Btu, $e = 2,085/778.2 = 2.68$ Btu.

An assembly of the dimensions computed in Step 3 will weigh about 35 lb and will have an average overall specific heat of 0.15 Btu/(lb)(F). Hence, the temperature rise per stroke will be: Btu of heat generated per stroke/[(specific heat)(cylinder weight, lb)] = $2.68/[(0.15)(35)] = 0.51°$/stroke. A temperature rise of this magnitude is easily dissipated by the external surfaces of the cylinder. But a smaller liquid spring under rapidly fluctuating loads may have an excessive temperature rise. Each spring must be analyzed separately.

5. Compare the various types of springs

Using previously presented Calculation Procedures, Table 48 can be constructed. This table and the spring analysis given above is based on the work of Lloyd M. Polentz, Consulting Engineer. The tabulation shows that the liquid spring is the shortest and has the smallest diameter for the load in question. Figure 31 shows typical liquid springs; Fig. 32 shows the compressibility of liquids used in various liquid springs.

Related Calculations: Use the method given here to select liquid springs for applications in any of a variety of services where a large load must be absorbed. The seals at the cylinder ends must be absolutely tight. Liquid springs are best applied in atmospheres where the temperature variation is minimal.

TABLE 48 Performance of Four Typical Spring Types*

| | Coil | Nested | | Liquid |
		Belleville washers	Tapered rings	
Useful range:				
Low load	1 oz	20 lb	2 ton	100 lb
High load	10 ton	100 ton	150 ton	200 ton
Force vs.				
deflection	Low to high	High	High	Medium to high
Stroke..........	Short to long	Short	Short to medium	Short to long
Damping ability ..	Low	Low	Low	Low to high
Relative cost......	Low	Low	Medium	High

Product Engineering.

NOTE: An example: For 50,000-lb load, 5-in. stroke:

Size, in.	Length	68	37	24	17
	Diameter	11.5	8	5	3

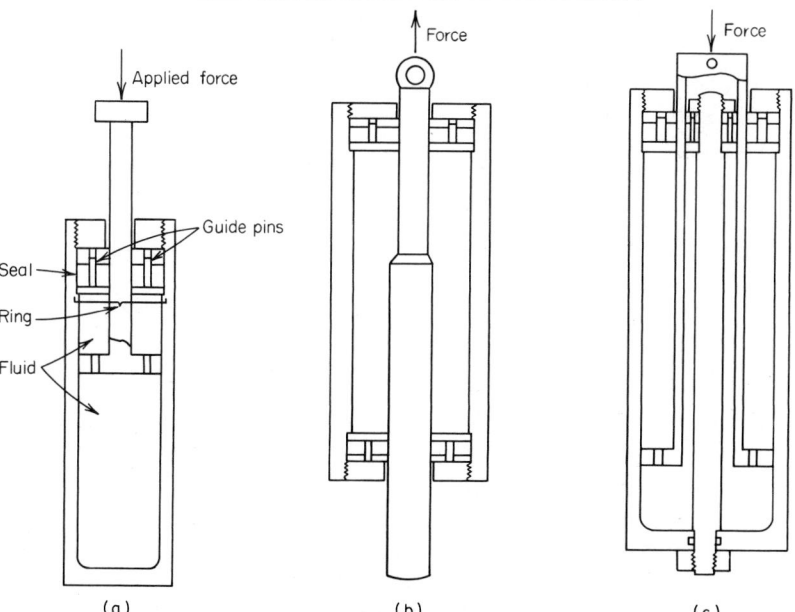

Fig. 31 Typical liquid springs. (a) General design; (b) tension-type; (c) long-stroke type.
(Product Engineering.)

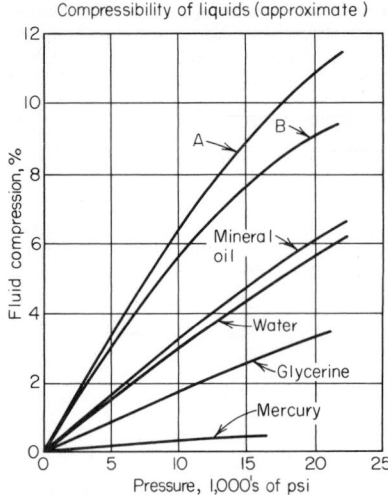

Compressibility of liquids (approximate)

Fig. 32 Common fluids for liquid springs are Dow-Corning type F-4029, curve A, and type 200, curve B. *(Product Engineering.)*

SELECTION OF AIR-SNUBBER DASHPOT DIMENSIONS

Determine the required orifice area, peak actuator pressure, peak negative acceleration, and the time required for the stroke of a 3-in.^{3} capacity air snubber if the total load mass $M = 0.1$ lb-sec^2/in.; the snubber pressure $P_i = 100$ psi; piston area $A_p = 3$ sq in.; initial snubber active length $S = 1.0$ in.; initial piston velocity $v_i = 100$ in./sec; piston velocity at the end of travel $v = 29$ in./sec; constant external force on snubber $F = 150$ lb; initial gas temperature $T_i = 530R$; gas constant $R = [639.6$ in. lb/lb°(R)](air); $C_D =$ orifice discharge coefficient = 0.9 dimensionless.

Calculation Procedure:

1. Compute the snubber dimensionless parameters

The first dimensionless parameter $K_E =$ stored energy/kinetic energy $= P_iV_i/Mv_i^2 = (100)(3)/[(0.1)(100)^2] = 0.3$. The next parameter $K_F =$ constant external force/initial pressure force $= F/P_iA_p = 150/[(100)(3)] = 0.5$. The third parameter $K_v =$ piston velocity at end of stroke/initial piston velocity $= v/v_i = 20/100 = 0.20$.

2. Determine the actual value of the orifice parameter

The parameter $K_w =$ initial orifice flow/initial displacement flow $= w_i/\rho A_p v_i$, where $\rho =$ gas density, lb/in.^{3}. Figure 33 gives values of K_w for $K_F = 0$ and $K_F = 1.0$. However, K_F for this snubber $= 0.5$. Therefore, it is necessary to interpolate between the charts for $K_F = 0$ and $K_F = 1.0$.

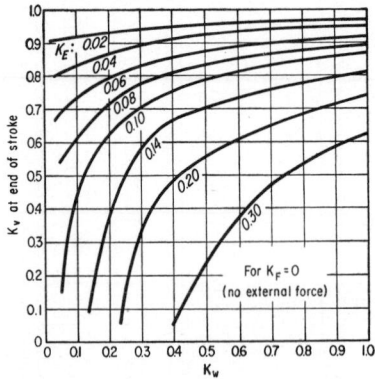

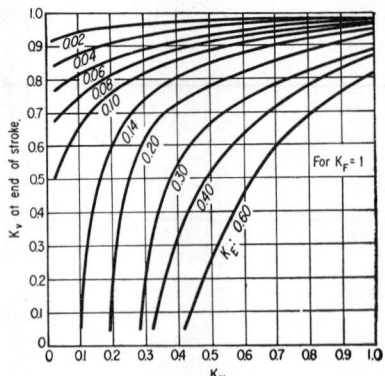

Fig. 33   Impact velocity vs. orifice flow (dimensionless) for air snubber. *(Product Engineering.)*

Interpolate by constructing a chart, Fig. 34, using values of K_w read from each chart in Fig. 33. Thus, when $K_F = 1$, $K_v = 0.2$, $K_E = 0.3$, $K_w = 0.295$. After the curve is constructed, read $K_w = 0.375$ for $K_f = 0.5$.

3. Compute the true flow through the orifice

The true initial flow rate w_i lb/sec $= K_w(P_i/RT_i)A_p v_i$, where all the symbols are as defined earlier. Thus, $w_i = (0.375)[100/(639.6)(530)](3)(100) = 0.0332$ lb/sec.

4. Compute the required orifice area

Use the equation $A_o = w_i/P_i C_D\{(kg/RT_i)[2/(k+1)](k+1)/(k-1)\}^{0.5}$, where $k = 1.4$; $g = 32.2$ ft/sec^2; other symbols as before. Thus, $A_o = 0.0332/[100(0.9)]\{(1.4 \times 32.2/639.6 \times 530)[(2/(1.4+1)](1.4+1)/(1.4-1)\}^{0.5} = 0.016$ sq in.

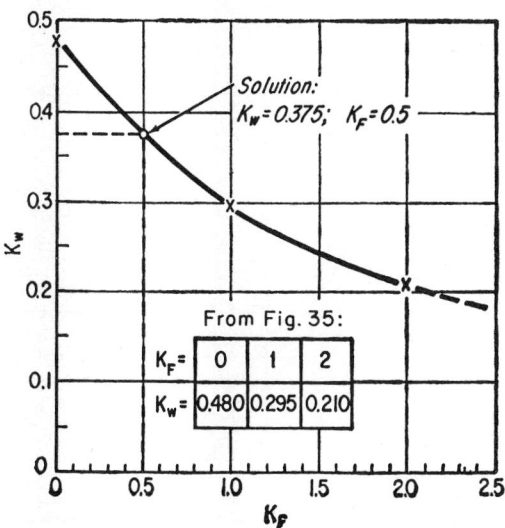

Fig. 34 Cross plot for an air snubber. (*Product Engineering.*)

5. Determine the maximum pressure at the end of the stroke

Read, from Fig. 35, $K_{P,\max} = 10.2$ for $K_F = 0.5$, $K_w = 0.375$. Then, the true $P_{\max}$ at end of stroke $= K_P P_i = 10.2(100) = 1,020$ psi.

6. Determine the maximum acceleration of the piston

For an air snubber, $K_{a,\max} = K_F - K_P = 0.5 - 10.2 = -9.7$. Also, the maximum acceleration $a_{\max} = K_a P_i A_p/M = (-9.7)(100)(3.0)/0.1 = -29,100$ in./sec^2.

7. Determine the approximate travel time of the piston

The travel time for the piston is $t = K_t S/v_i$, where $t =$ travel time, sec. Or, $t = 0.95 \times (1.0)/100 = 0.0095$ sec, assuming $K_t = 0.95$.

Related Calculations: The equations in this Procedure were developed by Tom Carey and T. T. Hadeler, and are based on these assumptions: (1) they apply only to a piston-orifice-type dashpot; (2) the piston is firmly stopped at the end of the stroke and does not rebound, oscillate, or bounce; (3) friction is zero; (4) the external force is constant; (5) $k = 1.4$, which means that the equations are valid for air, hydrogen, nitrogen, oxygen, and any other gas having a specific-heat ratio of about 1.4; (6) the contained air or gas is ideal; (7) compression is adiabatic; (8) flow through the bleed orifice is critical (a valid assumption except when the dashpot initial pressure is atmospheric, as in screen-door snubbers). When actual friction exists, there is a slight increase in the value of K_t. Figure 36 shows a simplified design of a typical air snubber.

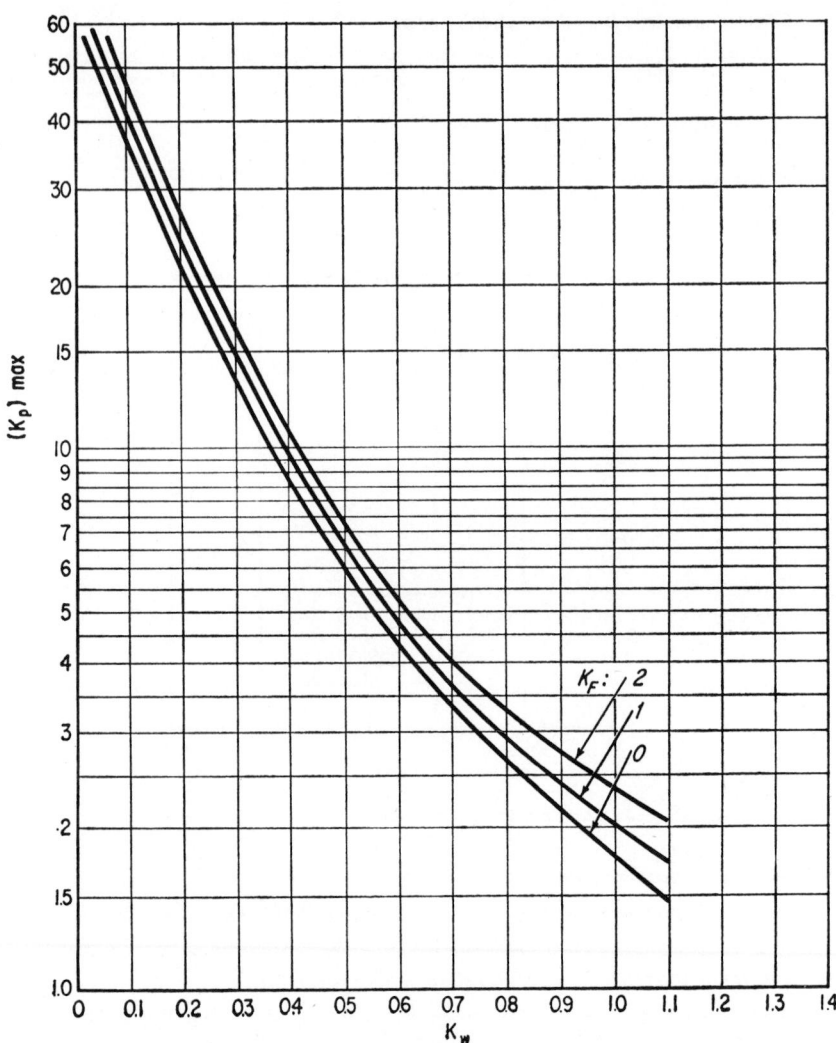

Fig. 35   Maximum pressure at impact (dimensionless). *(Product Engineering.)*

DESIGN ANALYSIS OF FLAT REINFORCED-PLASTIC SPRINGS

A large shaker unit in a vibrating screen system is supported on a series of six steel leaf springs, each a cantilever 6 in. wide by 0.125 in. thick. How thick should a single epoxy-glass leaf spring of the same width be if it is to replace the composite steel spring? The cantilever is 30 in. long with a 24-in. free length; maximum deflection = 0.375 in.; axial load per spring = 2,500 lb; safety factor = 8; $E = 4.5 \times 10^6$ psi for the plastic spring; ultimate flexure strength = 100,000 psi.

Calculation Procedure:

1. Compute the spring thickness for minimum bending stress

The equation for the thickness giving the minimum bending stress is $t = (4Ll^2/wE)^{1/3}$, where t = spring thickness, in.; L = axial load on spring, lb; l = spring free length, in.; w = spring width, in.; E = modulus of elasticity of spring material, psi. For this spring, $t = [4 \times 2,500 \times 24^2/(6 \times 4.5 \times 10^6)]^{1/3} = 0.598$; say 0.6 in.

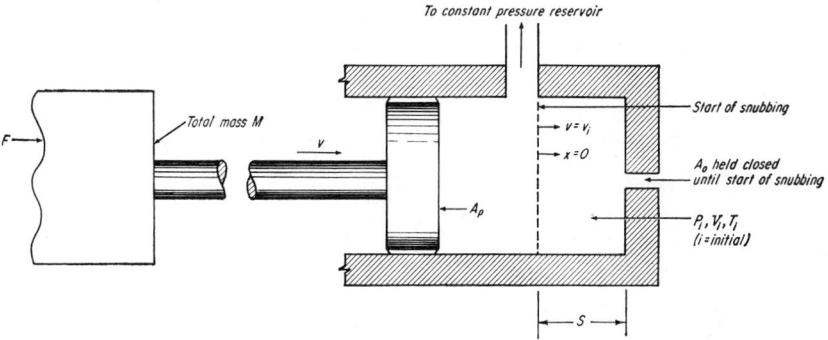

Fig. 36 Simplified air-snubber design. *(Product Engineering.)*

2. Determine the total combined stress in the beam

The maximum combined stress $s_B = (3Et/l^2 + 6L/wt^2)f$, where f = spring deflection, in.; other symbols as before. Thus, $s_B = \{e \times 4.5 \times 10^6 \times 0.6/[24^2 + 6 \times 2,500/(6 \times 0.6^2)]\} \times (0.375) = 7,875$ psi.

The total stress in the spring $s_T = s_B + L/A$, where A = spring cross-sectional area. Thus, $s_T = 7,875 + 2,500/(6 \times 0.6) = 8,570$ psi. The allowable stress = ultimate flexure strength, psi/factor of safety = 100,000/8 = 12,500 psi. Since $s_T < 12,500$ psi, the dimensions of the spring are satisfactory.

3. Check the critical buckling stress of the spring

To prevent buckling of the spring, the following must hold, $L/A \leq \pi^2 Et^2/36^2 = s_{CR}$, or $2,500/(6 \times 0.6) \leq \pi^2 \times 4.5 \times 10^6 \times 0.6^2/(3 \times 24^2) = 694$ psi $< 9,240$ psi. Hence, the spring dimensions are satisfactory.

4. Determine if the computed thickness gives adequate stiffness

For a plastic spring to have a stiffness equal to a steel spring having n leaves, the plastic spring should have a thickness of $t = t_s(nE_s/E)^{1/3}$, where t_s and E_s refer to the thickness, in., and modulus of elasticity, psi, respectively, of the steel spring; other symbols as before. Thus, $t = 0.125[6 \times 30 \times 10^6/(4.5 \times 10^6)]^{1/3} = 0.43$ in. Since $t = 0.60$ in., as computed in Step 1, the plastic spring is slightly too stiff.

5. Check the thickness required for equivalent thickness

Using the equations for s_B and s_T from Step 2, and the equation for s_{CR} from Step 3, compute the respective stresses for values of t less than, and greater than, 0.43. Thus:

t	s_B	Q/A	s_T	s_{CR}
0.250	17,158	1,670	18,828	1,665
0.375	9,960	1,110	11,070	3,600
0.500	8,143	833	8,976	6,400
0.600	7,875	695	8,570	9,240
0.625	7,900	666	8,566	10,000
1.000	9.727	416	10,143	25,650

Plot the results as in Fig. 37. This plot clearly shows that $t = 0.43$ in. gives $s_T < 12,500$ psi, and $Q/A < s_{CR}$. Hence, this thickness is satisfactory.

6 Determine if a thinner spring can be used

A thinner spring will save money. From Fig. 37, $t = 0.375$ in. gives an s_T value well below the maximum design stress and the actual spring stress is one-third the critical buckling stress. If tests on a 0.375-in.-thick spring show no serious disruption of harmonic operation, then specify the thinner material to lower the cost. Otherwise, use the thicker 0.43-in. spring.

Related Calculations: Use this procedure for unidirectional, cross-plied, or isotropic ply plastic springs. Obtain the allowable stress for the spring from the plastic manufacturer. The method given here is the work of L.A. Heggernes, reported in *Product Engineering*.

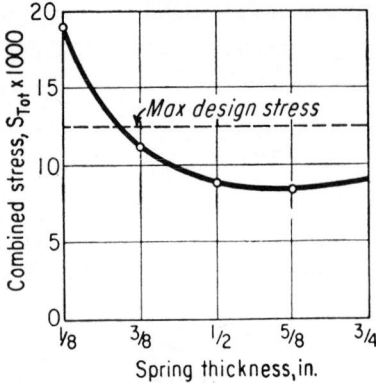

Fig. 37 Combined stress in a plastic spring. (*Product Engineering.*)

LIFE OF CYCLICALLY LOADED MECHANICAL SPRINGS

What is the probable life in cycles of a Belleville spring under a bending load if it is made of carbon steel having a Rockwell hardness of C48?

Calculation Procedure:

1. Determine the spring material tensile strength

Enter Fig. 38 at the Rockwell hardness C48 and project vertically upward to read the tensile strength of the carbon-steel spring material as 235,000 psi.

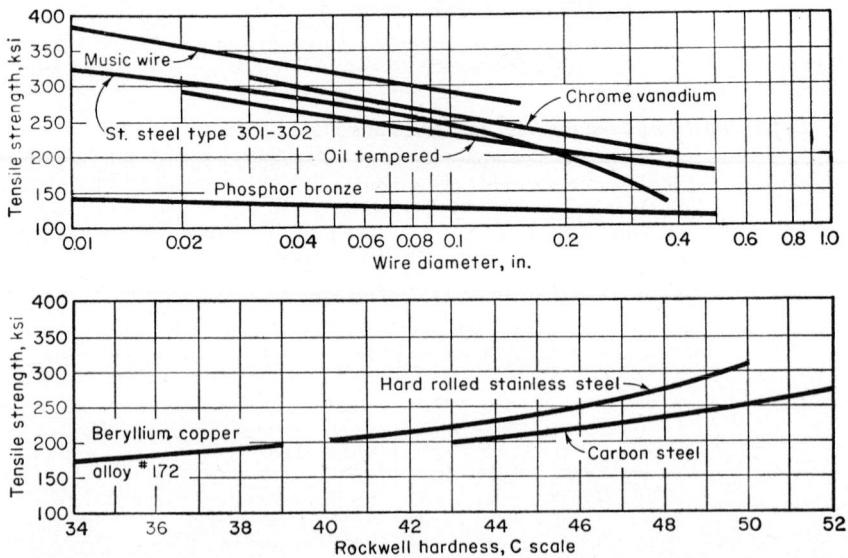

Fig. 38 (a) Tensile strength of spring wire; (b) tensile strength of spring strip. (*Product Engineering.*)

2. Compute the actual stress in the spring

Using the spring dimensions and the equations presented in the Belleville spring Calculation Procedure, compute the actual stress in the spring. For the spring in question, the actual stress is found to be 150,000 psi. This is $150,000(100)/235,000 = 63.8$, say 64 percent of the spring material tensile strength.

3. Estimate the spring cycle life

Enter the upper part of Table 49 for springs in bending. This tabulation shows that at a stress of 65 percent of the tensile strength, the spring will have a life between 10,000 and 100,000 stress cycles. Actual test of the spring caused failure at about 100,000 cycles.

Related Calculations: Use this procedure for helical torsion springs, cantilever springs, wave washers, flat springs, motor springs, helical compression and extension springs, torsion bars, and Belleville springs. Be sure to enter the proper portion of Table 49 when finding the approximate number of repetitive stress cycles. The method presented here is the work of George W. Kuasz and William R. Johnson and is reported in *Product Engineering*.

TABLE 49   Design Stresses for Springs*

No. of Repetitive Stress Cycles	*Maximum Design Stress (Percent of the Tensile Strength Shown in Charts)*
Design Stress for Springs in Bending†	
10,000	80
	65‡
100,000	53
1,000,000	50
10,000,000	48
Design Stress for Springs in Torsion§	
10,000	45
	35‡
100,000	35
1,000,000	33
10,000,000	30

*Product Engineering.
†For example, helical torsion springs, Bellevilles, cantilever springs, wave washers, flat springs, and motor springs.
‡For stainless-steel and phosphor-bronze materials. Tests show that such materials have low yield points.
§For example, helical compression springs, helical extension springs, and torsion bars.

SHOCK-MOUNT DEFLECTION AND SPRING RATE

Determine the maximum probable acceleration, the shock isolator deflection, and the isolator spring rate for a 25-lb piece of electronic equipment which drops from a 24-in.-high tailgate of a truck onto a concrete road. The product lands on one corner point, and should be considered as rigid steel for analysis purposes. In its carton the load will be supported by 16 shock isolators.

Calculation Procedure

1. Compute the acceleration of the load

Use the relation $g = (72/t)(h)^{0.5}$, where g = load acceleration in g ($1g = 32.2$ ft/sec^2 at sea level); t = shock-rise time, msec, from Table 50; h = drop height, in. From Table

TABLE 50   Typical Value for Shock-Time Rise

Condition	Shock-time rise, msec	
	Flat face	Point
Rigid steel against concrete......................	1	
Rigid steel against wood or mastic	2–3	5–6
Steel or aluminum against compact earth.................	2–4	6–8
Steel or aluminum against sand ...	5–6	15
Product case against mud	15	20
Product case against 1-in. felt.....	20	30

Note: Mass of struck surface is assumed to be at least 10 times the striking mass. Point contact with spherical radius of 1 in.

50, $t = 2$ msec for rigid steel making point contact with concrete. Then, $g = (72/2) \times (24)^{0.5} = 176.5\,g$.

2. Compute the isolator deflection

Use the relation $d = 2h/(g-1)$, where $d =$ isolator deflection, in.; other symbols as before. For this load, $d = [2 \times 24/(176.5-1)] = 0.273$ in.

3. Compute the required specific spring rate for the isolator

Use the relation $K = g/d$, where $D =$ isolator specific spring rate, lb/(in.)(lb). Thus, $K = 176.5/0.273 = 646$ lb/(in.)(lb).

4. Determine the required spring rate per isolator

With n shock isolators, the required spring rate, lb/in., per isolator is $k = KW/n$, where $W =$ weight of part, lb; $n =$ number of isolators used in the carton. Thus, $k = (646)(25)/16 = 1,020$ lb/in.

Related Calculations: Some of the largest stock loads encountered by equipment occur during transportation. Thus, vertical accelerations on the body of a 2-ton truck traveling at 30 mph on good pavement range from 1 to 2 g, with a rise time of 10 to 15 msec. Higher speeds, rougher roads, stiffer truck springs, and careless driving all decrease the rise time and thus double or triple the acceleration loads.

The highest acceleration forces in railroad freight cars occur during humping, when the impact loads on a product container may range from 4.5 to 28 g.

For most components that are sensitive to shock, suppliers include maximum safe acceleration loads in engineering data. Maximum allowable loads on vacuum tubes are 2 to 5 g; relays may withstand higher accelerations, depending on the type and direction of the acceleration. Transistors have low mass and good rigidity, and are highly resistant to shock when properly supported. Ball-bearing races may be indented by the balls; sleeve bearings are usually much more resistant to shock.

The function of a shock mount is to provide enough protection to avoid damage under expected conditions. But overdesign can be costly, both in the design of the product and in the shock-mount components. Underdesign can lead to failures of the shock mount in service and possible damage to the product. Therefore, careful design of shock mounts is important. The method presented here is the work of Raymond T. Magner, reported in *Product Engineering*.

CLUTCH SELECTION FOR SHAFT DRIVE

Choose a clutch to connect a 50-hp internal-combustion engine to a 300 rpm single-acting reciprocating pump. Determine the general dimensions of the clutch.

Calculation Procedure:

1. Choose the type of clutch for the load

Table 51 shows typical applications for the major types of clutches. Where economy is the prime consideration, a positive-engagement or a cone-type friction clutch would be chosen. Since a reciprocating pump runs at a slightly varying speed, a centrifugal clutch is not suitable. For greater dependability a disk or plate friction clutch is more desirable than a cone clutch. Assume that dependability is more important than economy and choose a disk-type friction clutch.

TABLE 51   Clutch Characteristics

Type of clutch	*Typical Applications**
Friction:	
Cone	Varying loads; 0 to 200 hp; losing popularity for many applications, particularly in the higher hp ranges
Disk or plate	Varying loads; 0 to 500 hp; widely used; more popular than the cone clutch
Rim:	
Band	Varying loads; 0 to 100 hp; not too widely used
Overrunning.	Constant or moderately varying loads; 0 to 200 hp; engages in one direction; freewheels in the opposite direction
Centrifugal	Constant loads; 0 to 50 hp
Inflatable	Varying loads; 0 to 5,000 hp; compressed air inflates clutch; have 360-deg friction surface
Magnetic	Varying loads; 0 to 10,000 hp; high speeds; also used where disk clutch would be overloaded
Positive-engagement	Nonslip operation; low-speed (10 to 150 rpm) engagement; has sudden starting action
Fluid.	Large, varying loads; 0 to 10,000 hp; variable-speed output; can produce a desired slip
Electromagnetic	Large, varying loads; 0 to 10,000 hp; variable-speed output; characteristics similar to fluid clutches

*Clutch capacity depends on the design, materials of constructions, type of load, shaft speed, and operating conditions. The applications and capacity ranges given here are typical but should not be taken as the only uses for which the listed clutches are suitable.

2. Determine the required clutch torque starting capacity

A clutch must start its load from a stopped condition. Under these circumstances the instantaneous torque may be two, three, or four times the running torque. Therefore, the usual clutch is chosen so it has a torque capacity of at least twice the running torque. For internal-combustion engine drives, a starting torque of three to four times the running torque is generally used. Assume 3.5 times is used for this engine and pump combination. This is termed the *clutch starting factor*.

Since $T = 63,000 \, hp/R$, where T = torque, lb = in.; hp = horsepower transmitted; R = shaft rpm, $T = 63,000(50)/300 = 10,500$ lb-in. This is the required starting torque capacity of the clutch.

3. Determine the total required clutch torque capacity

In addition to the clutch starting factor, a service factor is also usually applied. Table 52 lists typical clutch service factors. This tabulation shows that the service factor for a single-reciprocating pump is 2.0. Hence, the total required clutch torque capacity = required starting torque capacity × service factor = 10,500 × 2.0 = 21,000 lb-in. torque capacity.

4. Choose a suitable clutch for the load

Consult a manufacturer's engineering data sheet listing clutch torque capacities for clutches of the type chosen in Step 1 of this Procedure. Choose a clutch having a

TABLE 52   Clutch Service Factors

Type of Service	Service Factor
Driver:	
Electric motor:	
Steady load	1.0
Fluctuating load.	1.5
Gas engine:	
Single cylinder.	1.5
Multiple cylinder	1.0
Diesel engine:	
High-speed	1.5
Large, slow-speed	2.0
Driven machine:	
Generator:	
Steady load	1.0
Fluctuating load.	1.5
Blower	1.0
Compressor, depending on number of	
cylinders	2.0–2.5
Pumps:	
Centrifugal	1.0
Reciprocating, single-acting	2.0
Reciprocating, double-acting	1.5
Lineshaft	1.5
Wood-working machinery.	1.75
Hoists, elevators, cranes, and shovels . . .	2.0
Hammer mills, ball mills, and crushers . .	2.0
Brick machinery	3.0
Rock crushers.	3.0

rated torque equal to or greater than that computed in Step 3. Table 53 shows a portion of a typical engineering data sheet. A size 6 clutch would be chosen for this drive.

Related Calculations: Use the general method given here to select clutches for industrial, commercial, marine, automotive, tractor, and similar applications. Note that engineering data sheets often list the clutch rating in terms of torque, lb-in., and hp/100 rpm.

Friction clutches depend, for their load-carrying ability, on the friction and pressure between two mating surfaces. Usual coefficients of friction for friction clutches range between 0.15 and 0.50 for dry surfaces, 0.05 and 0.30 for greasy surfaces, and 0.05 and 0.25 for lubricated surfaces. The allowable pressure between the surfaces ranges from a low of 8 psi to a high of 300 psi.

TABLE 53   Clutch Ratings

Clutch number	Torque rating, lb-in.	Hp/100 rpm
1	2,040	3
2	4,290	6
3	8,150	12
4	13,300	21
5	19,700	31
6	35,200	55
7	44,000	69

BRAKE SELECTION FOR A KNOWN LOAD

Choose a suitable brake to stop a 50-hp motor automatically when power is cut off. The motor must be brought to rest within 40 sec after power is shut off. The load inertia, including the brake rotating member, will be about 200 lb-ft²; the shaft being braked turns at 1,800 rpm. How many revolutions will the shaft turn before stopping? How much heat must the brake dissipate? The brake operates once per min.

Calculation Procedure:

1. Choose the type of brake to use

Table 54 shows that a shoe-type electric brake is probably the best choice for stopping a load when the braking force must be applied automatically. The only other possible choice—the eddy-current brake—is generally used for larger loads than this brake will handle.

TABLE 54 Mechanical and Electrical Brake Characteristics

Type of Brake	Typical Characteristics
Block...............	Wooden or cast-iron shoe bearing on iron or steel wheel; double blocks prevent bending of shaft; used where economy is prime consideration; leverage 5 : 1
Band	Asbestos fabric bearing on metal wheels; fabric may be reinforced with copper wire and impregnated with asphalt; bands are faced with wooden blocks; used where economy is a major consideration; leverage 10 : 1
Cone	Friction surface attached to metal cone; popular for cranes; coefficient of friction = 0.08 to 0.10; useful for intermittent braking applications
Disk................	Have one or more flat braking surfaces; effective for large loads; continuous application
Internal-shoe	Popular for vehicles where shaft rotation occurs in both directions; self-energizing, i.e., friction makes shoe follow rotating brake drum; capable of large braking power
Eddy-current	Used for flywheels requiring quick braking and where large kinetic energy of rotating masses precludes use of block brakes because of excessive heating
Electric, shoe-type ..	Used where automatic application of brake is required as soon as power is turned off; spring-activated brake shoes apply the braking action
Electric, friction-disk type..........	Best for duty cycles requiring a number of stops and starts per min; may have one, or multiple, disks

2. Compute the average brake torque required to stop the load

Use the relation $T_a = Wk^2n/308t$, where T_a = average torque required to stop the load, lb-ft; Wk^2 = load inertia, including brake rotating member, lb-ft², n = shaft speed prior to braking, rpm; t = required or desired stopping time, sec. For this brake, $T_a = (200)(1,800)/[308(40)] = 29.2$ lb-ft, or 351 lb-in.

3. Apply a service factor to the average torque

A service factor varying from 1.0 to 4.0 is usually applied to the average torque to ensure that the brake is of sufficient size for the load. Applying a service factor of 1.5 for this brake, the required capacity = 1.5(351) = 526 in.-lb.

4. Choose the brake size

Use an engineering data sheet from the selected manufacturer to choose the brake size. Thus, one manufacturer's data shows that a 16-in.-diameter brake will adequately handle the load.

5. *Compute the revolutions prior to stopping*

Use the relation, $R_s = tn/120$, where R_s = number of revolutions prior to stopping; other symbols as before. Thus, $R_s = (40)(1800)/120 = 600$ revolutions.

6. *Compute the heat the brake must dissipate*

Use the relation $H = 1.7\ FWk^2(n/100)^2$, where H = heat generated at friction surfaces, ft-lb/min; F = number of duty cycles per min; other symbols as before. Thus, $H = 1.7(1)(200)(1,800/100)^2 = 110,200$ ft-lb/min.

7. *Determine if the brake temperature will rise*

From the manufacturer's data sheet, find the heat dissipation capacity of the brake while operating and while at rest. For a 16-in. shoe-type brake, one manufacturer gives an operating heat dissipation $H_o = 150,000$ ft-lb/min, and an at-rest heat dissipation of $H_v = 35,000$ ft-lb/min.

Apply the cycle time for the event; i.e., the brake operates for 40 sec, or 40/60 of the time, and is at rest for 20 sec, or 20/60 of the time. Hence, the heat dissipation of the brake is $(150,000)(40/60) + (35,000)(20/60) = 111,680$ ft-lb/min. Since the heat dissipation, 111,680 ft-lb/min, exceeds the heat generated, 110,200 ft-lb/min, the temperature of the brake will remain constant. If the heat generated exceeded the heat dissipated, the brake temperature would rise constantly during operation.

Brake temperatures higher than 250 F can reduce brake life. In the 250- to 300-F range, periodic replacement of the brake friction surfaces may be necessary. Above 300 F, forced-air cooling of the brake is usually necessary.

Related Calculations: Because electric brakes are finding wider industrial use, Tables 55 and 56, summarizing their performance characteristic and ratings, are presented here for easy reference.

The coefficient of friction for brakes must be carefully chosen; otherwise the brake may "grab," i.e., attempt to stop the load instantly instead of slowly. Usual values for the coefficient of friction range between 0.08 and 0.50.

The methods given above can be used to analyze brakes applied to hoists, elevators, vehicles, etc. Where Wk^2 is not given, estimate it, using the moving parts of the brake and load as a guide to the relative magnitude of load inertia. The method presented here is the work of Joseph F. Peck, reported in *Product Engineering*.

MECHANICAL BRAKE SURFACE AREA AND COOLING TIME

How much radiating surface must a brake drum have if it absorbs 20 hp, operates for half the use cycle, and cannot have a temperature rise greater than 300 F? How long will it take this brake to cool to a room temperature of 75 F if the brake drum is made of cast iron and weighs 100 lb?

Calculation Procedure:

1. *Compute the required radiating area of the brake*

Use the relation $A = 42.4\ hpF/K$, where A = required brake radiating area, sq in.; hp = horsepower absorbed by the brake; F = brake load factor = operating portion of use cycle; K = constant = Ct_r, where C = radiating factor from Table 57; t_r = brake temperature rise, F. For this brake, assuming a full 300 F-temperature rise, and using data from Table 57, $A = 42.4(20)(0.5)/[(0.00083)(300)] = 1,702$ sq in.

TABLE 57 Brake Radiating Factors

Temperature Rise of Brake, F	Radiating Factor, C
100	0.00060
200	0.00075
300	0.00083
400	0.00090

Brake type	Operational mode On-off	Operational mode Continuous	Torque adjustment	Torque-control range	Wear adjustment	Residual drag	Heat dissipation	Instant stop	Cushioned stop	Retard (drag)	Hold	Fail-safe brake	On-off duty-cycling capability
Magnetic particle...	Yes	Yes	Electrical	Wide	Nonwearing	High	Limited	No	Yes	Yes	Yes	No	Limited by heat-dissipation capability to low inertia loads
Eddy current, air cooled.......	Yes	Yes	Electrical	Wide	Nonwearing	Moderate to low	Good	No	Yes	Yes	No	No	Limited to long time cycles
Eddy current, water cooled.....	Yes	Yes	Electrical	Wide	Nonwearing	High to moderate	Excellent	No	Yes	Yes	No	No	Limited to long time cycles
Single-disk friction, electrically actuated..........	Yes	Yes	Electrical	Wide	Self-compensating	None	Excellent	Yes	Yes	Yes	Yes	No	Excellent—up to several hundred stops per min
Multidisk friction, electrically actuated, direct acting..........	Yes	No	Electrical	Moderate		Low	Limited	Yes	Yes	No	Yes	No	Same as comparable size electric motor: 12 stops per min (maximum)
Multidisk friction, electrically actuated, indirect acting..........	Yes	No	Mechanical	Limited	Mechanical			Yes	Semisoft	No	Yes	No	Same as comparable size electric motor: 12 stops min (maximum)
Multidisk friction, spring actuated...	Yes	No	Mechanical	Limited	Mechanical	Low	Limited	Yes	Semisoft	No	Yes	Yes	Same as comparable size electric motor: 12 stops per min (maximum)
Shoe brake, spring actuated........	Yes	No	Mechanical	Limited	Mechanical	None	Good	Yes	Semisoft	No	Yes	Yes	Generally not over three stops per min without derating

TABLE 56 Representative Range of Ratings and Dimensions for Electric Brakes

Brake type	Hp	Torque, maximum, lb-ft	Shaft speed, maximum, rpm	Diameter, in.	Length, in.	Inertia of rotating member, lb-ft^2
Magnetic particle brakes	1/50–25	0.6–150	1,000–2,000	2–10	2–6	1.5×10^{-4}–0.27
Eddy current brakes:						
Air cooled	3/4–75	5–1,740	2,000–900	$6\frac{1}{2}$–$24\frac{3}{4}$	$9\frac{1}{2}$–$43\frac{1}{2}$	0.12–100
Water cooled	40–800	130–4,600	1,800–1,200	$14\frac{3}{4}$–$36\frac{1}{2}$	$18\frac{1}{2}$–43	8.5–725
Friction disk brakes:						
Single disk, electrically actuated	1/50–200	0.17–700	10,000–1,800	$1\frac{1}{2}$–$15\frac{1}{4}$	$1\frac{1}{2}$–$4\frac{1}{2}$	0.000125–3
Multiple disk, electrically actuated	1/4–2,000	3–15,000	5,000–750	$2\frac{1}{4}$–21	2–8	Up to 90
Multiple disk, spring actuated	1/4–2,000	4–7,500	5,000–1,200	4–29	$2\frac{1}{2}$–$16\frac{1}{2}$	
Shoe brakes, spring actuated	1–2,500	3–10,000	10,000–1,200	2–28	$4\frac{1}{2}$–12	0.023–485 lb-ft^2

2. *Compute the brake cooling time*

Use the relation $t = (cW \ln t_r)/K_c A$, where t = brake cooling time, min; c = specific heat of brake-drum material, Btu/(lb)(F); W = weight of brake drum, lb; t_r = drum temperature rise, F; $\ln$ = log to base e = 2.71828; K_c = a constant varying from 0.4 to 0.8; other symbols as before. Using $K_c = 0.4$, $c = 0.13$, $t = (0.13 \times 100 \ln 300)/[(0.4)(1,702)] = 0.1088$ min.

Related Calculations: Use this procedure for friction brakes used to stop loads that are lifted or lowered, as in cranes, moving vehicles, rotating cylinders, and similar loads.

INVOLUTE SPLINE SIZE FOR KNOWN LOAD

Choose the type and size of involute spline to transmit a torque of 10,000 lb-in. from an electric motor to a centrifugal pump. What is the required face width and number of teeth for this spline?

Calculation Procedure:

1. *Select the type of spline to use*

Involute splines are usually chosen for industrial drives because this type transmits more torque for its size than a parallel-side spline does. The involute spline has almost no speed limitation, being used at speeds of 10,000 rpm and higher. Further, an involute spline can be cut and measured by the same machines that cut and measure gear teeth. A spline, however, differs from a gear in that the spline has no rolling action and all teeth are in contact at once.

Involute splines may be either *flexible* or *fixed*. Flexible splines allow some rocking motion; and under torque, the teeth slip axially to accommodate axial expansion or runout. Fixed splines allow no relative or rocking motion between the internal and external teeth. The fixed-type spline can be either shrink-fitted or loosely fitted together. For a centrifugal-pump drive, the flexible-type spline is generally preferred. Therefore, a flexible involute spline will be chosen for this drive. A standard commercial grade will be acceptable.

2. *Determine the pitch diameter of the spline*

Enter Fig. 39 at a torque of 10,000 lb-in. and project vertically upward to the curve marked *Commercial flexible*. From the intersection with this curve, project horizon-

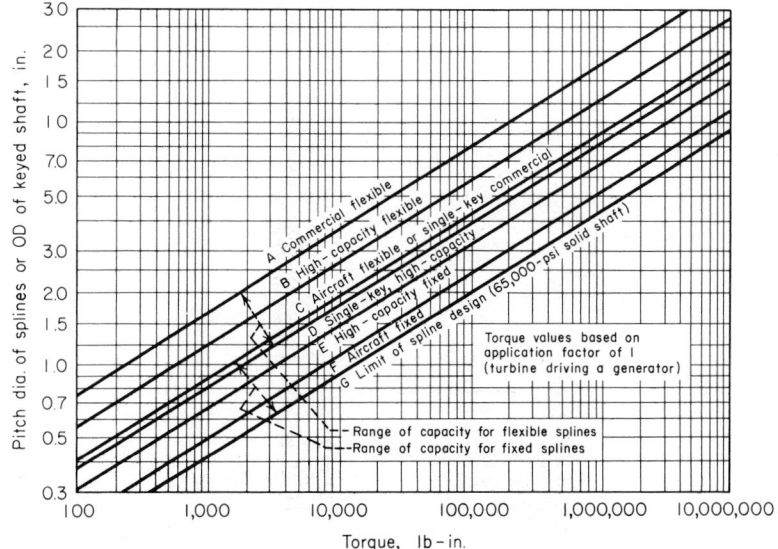

Fig. 39 Spline size based on diameter-torque relationships. (*Product Engineering.*)

tally to the left to find the required spline pitch diameter as 3.75 in. This is also the required outside diameter of a keyed shaft to transmit the same torque.

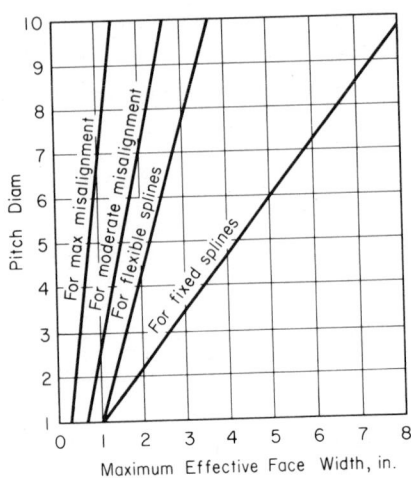

Maximum Effective Face Width, in.

Fig. 40 Face width of splines for various applications. *(Product Engineering.)*

3. Determine the maximum effective face width

Enter Fig. 40 at the pitch diameter of 3.75 in. and project horizontally to the curve marked *for flexible splines*. From the intersection, project vertically downward to read the maximum effective face width as 1.75 in.

4. Choose the number of teeth for the spline

Table 58 lists the recommended minimum number of teeth for an involute spline. Cost and manufacturing considerations determine the number of teeth to use, because the number of teeth chosen has no effect on tooth stress. An even number of teeth should be used whenever possible. When a large number of teeth are used on a spline, the root diameter of the external member is greater, tool design is easier, and lubrication is improved. Generally, however, the cost of the spline increases with a larger number of teeth.

For industrial drives, where the spline cost is usually more important than the weight of the spline, or the space it occupies, a tooth with a 20° pressure angle is generally chosen. The nominal tooth depth, compared with gear teeth, is 75 percent. Using these data and a pitch diameter of 3.75 in., as determined in Step 2, shows that 32 teeth should be used.

Related Calculations: Involute splines for use in aircraft applications generally have a 30° pressure angle and 50 percent depth. In automotive service, shaved splines having

TABLE 58   Recommended Minimum Numbers of Spline Teeth*

Pitch diameter	Broaching			Shaping				Shaving or grinding	
Angle	30°	20° or 14½°		30°	25°	25°	14½°	25°	20°
Depth	50%	50%	30%	50%	70%	75%	30%	70%	75%
0.5	6	10	10	12	18				
0.8	8	12	10	14	20	22	16		
1.0	8	12	10	16	20	24	16		
2.0	8	12	12	20	20	24	24		
4.0	10	16	16	24	24	32	24	48	48
8.0	20	20	24	32	32	40	32	56	56
12.0	30	30	36	36	36	48	36	60	60

Product Engineering.

the same proportions as the industrial splines mentioned above are often used. Rolled splines having 30 or 40° pressure angles and 50 and 40 percent depth, respectively, are also used. ANSA standards covering involute and straight-sided splines are available.

The method presented here is the work of Darle W. Dudley, reported in *Product Engineering.*

FRICTION DAMPING FOR SHAFT VIBRATION

Design a dry friction (also termed coulomb friction) sleeve for a shaft transmitting power to an air compressor. The shaft has an outside diameter of 7.5 in. and a length of 8 ft, and it drives the compressor as shown in Fig. 41. The angular value of torsional vibration should be limited to 10 percent of the steady displacement caused by the mean torque in the shaft. The compressor torque is 800,000 lb-in.

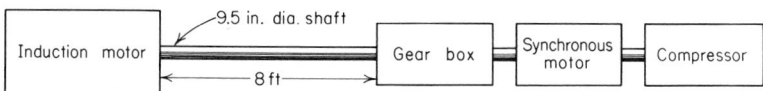

Fig. 41   Transmission system designed for friction-damping. (*Product Engineering.*)

Calculation Procedure:

1. Compute the required damping ratio

To apply the friction-damping technique to a shaft, a sleeve, Fig. 42, is added which is attached to the shaft at one end, A. The sleeve is extended along the shaft and makes contact with some point on the shaft through the disk, Fig. 42. This disk may be welded to or tightly pressed on the shaft and snugly fits into the sleeve.

In most dry-friction damping, about 3 percent of the damping takes place per cycle. If the forcing torque were reduced for one cycle, the strain energy would drop to 97 percent of its maximum value and the angular displacement, θ, of the shaft would drop to 0.97θ. Hence, the forcing torque must be such as to increase the angular displacement by an amount, or (in the absence of damping): $\Delta\theta = 0.03\theta$ per cycle $= 0.015\theta$ for a half cycle.

Compute the damping ratio for the system from $R = 1 - \Delta\theta/\theta$, where $R =$ damping ratio. Thus, $R = 1 - 0.015\theta/0.1\theta = 0.85$. The value 0.1θ is used in the denominator because the design requires that $\Delta\theta$ be limited to 10 percent of the steady displacement θ, which results from the mean torque in the shaft.

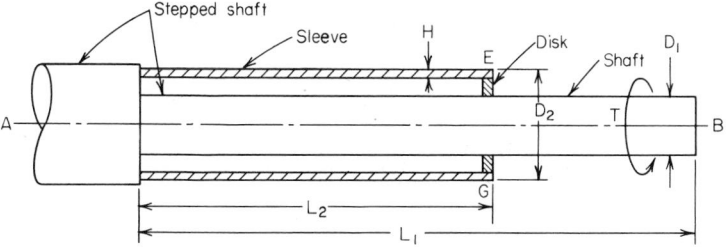

Fig. 42   Thin sleeve added to rotating shaft reduces torsional vibrations. (*Product Engineering.*)

2. Determine the shaft damping/critical damping value

With $R = 0.85$, enter Fig. 43, and find $m = 5.2$, where $m = D_1/8HC^3 =$ ratio of the torsional stiffness of the shaft to that of the sleeve, a dimensional constant; $D_1 =$ shaft diameter, in.; $H =$ thickness of the sleeve wall, in.; $C = D_2/D_1$, where $D_2 =$ outside diameter of sleeve, in. Thus, damping/critical damping value $= 0.026$, or 2.6 percent, from Fig. 43, assuming $L_1/L_2 = 1.0$.

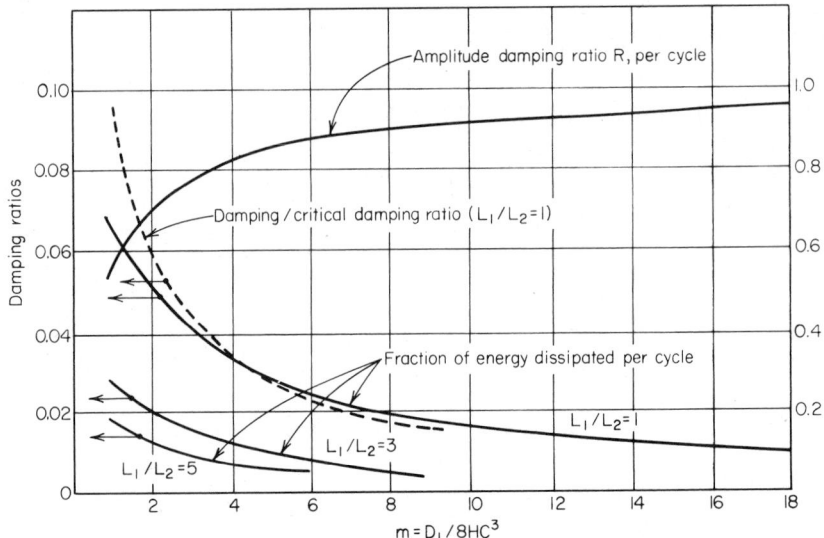

Fig. 43   Design chart for friction damping. *(Product Engineering.)*

3. Select the sleeve outside diameter

Since $m = D_1/8HC^3 = 5.2$, and $D_1 = 7.5$ in., $HC^3 = 0.1802$, by substitution of the value of D_1 and m in this relation.

Choose how HC^3 is to be made up. Assuming $D_2 = 2D_1$, $C = D_2/D_1 = 2.0$. Since $HC^3 = 0.1802$, $H = -0.1802/2^3 = 0.0225$ in. This provides a sleeve thickness of about 24 gage. The sleeve will weigh only about 2.7 percent of the shaft weight. Thus, a 10:1 reduction in the vibration is obtained with very little extra weight.

4. Compute the resisting torque of the system

The ratio of resisting frictional torque applied by the sleeve T_r lb-in. to the applied torque on the shaft T lb-in. is $T_r/T = 1[1-(1-1/ar)^{0.5}]$, where $a = L_1/L_2$; $r = 1+m$. Or, $T_r/T = 1\{1-[1/(1 \times 6.2)]^{0.5}\} = 0.09$. Since the compressor torque $T = 800,000$ lb-in., $T_r = 0.09(800,000)72,000$ lb-in.

5. Compute the friction force on the sleeve

In Step 4 the sleeve diameter D_2 was chosen as $2D_1$. Since $D_1 = 7.5$ in., $2D_1 = 15$ in. The frictional torque acts, through the disk, over the circumference of the inner surface of the sleeve. The diameter of the inner surface of the sleeve is $15.0 - 2(0.0225) = 14.955$ in., using the sleeve thickness obtained in Step 3. The circumference of the inner surface of the sleeve is $14.955\pi = 47.0$ in. Hence, the friction force acting on the sleeve $F_f = T_r/\text{circumference, in.} = 72,000/47.0 = 1,532$ lb.

6. Determine the disk normal force

Assume that the disk has a coefficient of friction of 0.6. Then, the normal force acting on the sleeve is $F_n = F_f/f$, where $f = $ coefficient of friction, or $F_n = 1,532/0.6 = 2,550$ lb.

Related Calculations: Dry-friction damping can be applied to industrial machines of many types, military equipment (submarines, missiles, aircraft), internal-combustion engines, and similar machinery. Vibration amplitudes in a shaft become a problem when the shaft length-to-thickness ratio L_1/D_1 becomes large. Although shaft diameter can be increased to reduce the ratio, this adds to the weight and cost of the machine. Here are several useful design pointers:

(1) If weight is a primary objective, make the damping-sleeve diameter as large

as possible. (2) If weight is not important, use a sleeve diameter only slightly larger than the shaft diameter. (3) Sleeve length can vary from 0.1 to 1.0 shaft length. With short sleeves, be sure the sleeve has sufficient rigidity and stiffness. (4) Reduce the sleeve wall thickness at the end of the sleeve in contact with the disk so that the contact pressure will not induce large stresses in the shaft. The method presented here was developed by Burt Zimmerman, and reported in *Product Engineering*.

DESIGNING PARTS FOR EXPECTED LIFE

A machined and ground rod has an ultimate strength of $s_u = 90{,}000$ psi, and a yield strength $s_y = 60{,}000$ psi. It is grooved by grinding and has a stress concentration factor of $K_f = 1.5$. The expected loading in bending is 10,000 to 60,000 psi for 0.5 percent of the time; 20,000 to 50,000 psi for 9.5 percent of the time; 20,000 to 45,000 psi for 20 percent of the time; 30,000 to 40,000 psi for 30 percent of the time; 30,000 to 35,000 psi for 40 percent of the time. What is the expected fatigue life of this part in cycles?

Calculation Procedure:

1. Determine the material endurance limit

For $s_u = 90{,}000$ psi, the endurance limit of the material $s_e = 40{,}000$ psi, closely, from Fig. 44.

2. Compute the equivalent completely reversed stress

The largest equivalent completely reversed stress for each load in bending is $s_F = (s_e/s_y)s_a + K_f s_a$ psi, where $s_a =$ average or steady stress, psi; other symbols as before. Since $s_e/s_y = 40{,}000/60{,}000 = \frac{2}{3}$, and $K_f = 1.5$, $s_v = \frac{2}{3}s_a + 1.5 s_a$. Then,

$$s_{v1} = (\tfrac{2}{3})(35{,}000) + (1.5)(25{,}000) = 60{,}830 \text{ psi}$$

$$s_{v2} = (\tfrac{2}{3})(35{,}000) + (1.5)(15{,}000) = 45{,}830 \text{ psi}$$

$$s_{v3} = (\tfrac{2}{3})(32{,}500) + (1.5)(12{,}500) = 40{,}420 \text{ psi}$$

$$s_{v4} = (\tfrac{2}{3})(35{,}000) + (1.5)(5{,}000) = 30{,}830 \text{ psi}$$

$$s_{v5} = (\tfrac{2}{3})(32{,}500) + (1.5)(2{,}500) = 25{,}420 \text{ psi}$$

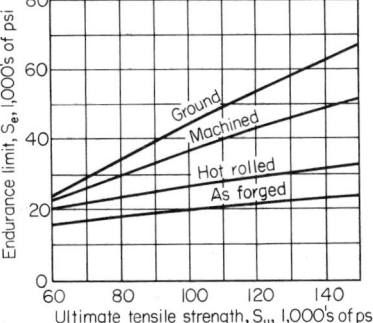

Fig. 44 Relationship between endurance limit and ultimate tensile strength. (*Product Engineering.*)

3. Compute the fatigue life for the initial stress

The initial stress is s_{v1}; the fatigue life at this stress is, in cycles, $N_1 = 1{,}000$ $(s_u/s_{vi})^{3/(\log(s_u/s_e))}$, where $s_{vi} =$ equivalent completely reversed stress, psi; other symbols as before. Taking the first value of $s_{vi} = s_{v1} = 60{,}830$ psi, $N_1 = 1{,}000(90{,}000/60{,}830)^{3/\log 2.25} = 28{,}100$ cycles.

4. Compute the exponent for the fatigue-life equation

The exponent for the fatigue-life equation is $2.55/\log s_u/s_e = 2.55/(\log 90{,}000/40{,}000) = 7.2406$.

5. Compute the factors for the fatigue-life equation

The factors needed for the fatigue-life equation are s_{v1}/s_{vi}, $(s_{vi}/s_{v1})^{2.55/\log s_u/s_e}$, and $\alpha_i(s_{vi}/s_{v1})^{2.55/\log(s_u/s_e)}$. In these factors, the value of $s_{vi} = s_{v2}$, s_{v3}, and so forth, as summarized in Table 59. The value $\alpha_i =$ percent-time duration of a stress, expressed as a decimal. The numerical values computed are summarized in Table 59.

6. Compute the part fatigue life in cycles

The part fatigue life, in cycles, $N = N_1/\alpha_1 + \alpha_2[1/(s_{v1}/s_{v2})]^{2.55 \log (s_u/s_e)} + 3[1/(s_{v1}/s_{v3})]^{2.55 \log(s_u/s_e)} + \cdots$ for each bending load. In this equation, $\alpha_1, \alpha_2, \ldots =$ percent-time duration of a stress, expressed as a decimal, the subscript referring to the stress mentioned in Step 2, above.

Since Table 59 summarizes the denominator of the fatigue-life equation, $N = N_1/0.03050 = 28,100/0.03050 = 922,000$ cycles.

TABLE 59   Values for Cycles-to-Failure Analysis

s_{vi}	s_{v1}/s_{vi}	$(s_{vi}/s_1)^{7.2406}$	i	$i(s_{vi}/s_{v1})^{7.2406}$
60,830	1.00	1.00	0.005	0.00500
45,830	1.3273	0.12873	0.095	0.01223
40,420	1.5051	0.05179	0.200	0.01036
30,830	1.9730	0.007296	0.300	0.00219
25,420	2.3934	0.001802	0.400	0.00072
Total	$\cdots$	$\cdots$	$\cdots$	0.03050

TABLE 60   Load-Stress K Factors for Various Materials*

Roller 1		Roller 2	K of roller 1 at number of cycles			
Material	Hardness	Material	10^6	10^7	4×10^7	10^8
Gray cast iron	130–180 Bhn		4,000	2,000	1,300	
GM Meehanite	190–240 Bhn	Same as roller 1	4,000	2,500	1,950	
Nodular cast iron	207–241 Bhn		10,000	5,600	$\cdots$	3,400
Gray cast iron . . . .·. . .	270–290 Bhn		7,500	5,300	4,200	
Gray cast iron phosphate-coated. . . .	140–160 Bhn		2,600	1,400	1,000	
	160–190 Bhn		3,200	1,900	1,300	
	270–290 Bhn		5,500	4,000	3,100	
SAE 1020 steel phosphate-coated. . . .	130–150 Bhn		4,500	2,700	1,700	
SAE 4150 steel chromium-plated. . . .	270–300 Bhn	Carbon tool steel 60–62 Re	13,500	11,000	$\cdots$	9,000
SAE 6150 steel	270–300 Bhn		2,600	1,300		
SAE 1020 steel induction-hardened .	45–55 RC		21,000	14,500	$\cdots$	10,000
SAE 1340 steel case-hardened	50–58 RC		26,000	20,000	$\cdots$	15,000
Phosphor bronze	67–77 Bhn		3,600	1,600	1,000	
Yellow brass	Drawn		5,600	3,000	2,000	
	Extruded		4,500	2,400	1,700	
Zinc diecasting.	$\cdots$		1,100	500	320	
Laminated-graphitized phenolic.	$\cdots$		1,700	1,300	1,000	
Cast aluminum SAE 39	60–65 Bhn	Gray cast iron 340–360 Bhn	1,200	500	300	

Product Engineering. Based on data presented by W. C. Cram at a University of Michigan symposium on surface damage.

Related Calculations: Data on the endurance limit, yield point, and ultimate strength of ferrous materials is tabulated in Baumeister and Marks—*Standard Handbook for Mechanical Engineers*. The equations presented in this Calculation Procedure hold true for both simple and complex loading. These equations can be used for analysis of an existing part, or for design of a part to fail after a selected number of cycles. The later procedure is sometimes used for components in an assembly in which the principal part has an accurately known life.

The method presented here is the work of Professor M. F. Spotts, reported in *Product Engineering*.

WEAR LIFE OF ROLLING SURFACES

Determine the maximum allowable bearing load in various bearing materials to avoid pronounced wear before 40×10^6 stress cycles if the roller bearing made of these materials has these dimensions: outside diameter = 4.3307 in., bore = 2.3622 in., width = 0.866 in., inner-race radius $r_2 = 1.439$ in., roller diameter = 0.468 in., roller width $f = 0.468$ in., number of rollers $n = 16$. Materials being considered are gray cast iron, Meehanite; and hardened-steel rollers on cast-iron races, on heat-treated cast-iron races, on heat-treated medium-steel races, and on carburized low-carbon steel races.

Calculation Procedure:

1. Determine the load-stress factor for each material

Table 60 lists the load-stress factor K for various materials at varying load cycles. Thus for gray cast iron, $K = 1,300$ at 40×10^6 cycles.

2. Compute the maximum allowable bearing load

Use the relation $F = nfK/5(1/r_1 + 1/r_2)$, where all symbols are defined in the problem statement above. Thus, $F = (16)(0.468)(1,300)/[5(1/0.234 + 1/1.439)] = 391$ lb for gray cast iron.

For the other materials, using the appropriate K value from Table 60 and the above Procedure, the allowable load is as follows:

Bearing Materials	Allowable Load, lb
Meehanite rollers and races	587
Hardened-steel rollers, cast-iron races	300
Hardened-steel rollers, heat-treated cast-iron races	933
Hardened-steel rollers, heat-treated medium-steel races	2,700
Hardened-steel rollers, carburized low-carbon steel races	3,912

Related Calculations: The same, or similar, procedures can be used for computing the wear life of gears, cams, bearings, clutches, chains, and other devices having rolling surfaces. Thus, in a joint composed of a pin having radius r_1 in a hole of radius r_2, $F = fK/(1/r_1 - 1/r_2)$. This relation also applies to a roller chain. For a cam, $F = fK \cos \infty/(1/\rho - 1/r)$, where ∞ = cam pressure angle,°; ρ = cam radius of contact, in.; r = radius of contact in.

The method presented here is the work of Professor Donald J Myatt, reported in *Product Engineering*.

FACTOR OF SAFETY AND ALLOWABLE STRESS IN DESIGN

Determine the cross-section dimension b in. of a uniform square bar of structural steel having a tensile yield strength $s_y = 33,000$ psi, if the bar carries a center load of

1,000 lb as a beam with a span of 24 in. with simply supported ends. Use both the allowable-stress and ultimate-strength methods to determine the required dimension.

Calculation Procedure:

1. Design first on the basis of allowable stress

From Table 61, in the section on buildings, the working or allowable stress $s_w = 0.6s_y = 0.6(33,000) = 19,800$ psi.

2. Compute the maximum bending moment

For a central load and simple supports, the maximum bending moment (in.-lb) is, from earlier sections of this Handbook, $M_m = (W/2)(L/2)$, where W = load on beam, lb; L = length of beam, in. Thus, $M_m = (1,000/2)(24/2) = 6,000$ in.-lb.

3. Compute the required cross-section dimension

For a beam of square cross section, $I/c = [b(b)^3/12]/b/2 = b^3/6$. Also, $s_w = M_m c/I$. Substituting and solving for b gives $b^3 = (6,000)(6)/19,800$; $b = 1.22$ in.

4. Design the beam on the basis of ultimate strength and a factor of safety.

The safety factor from Table 61 is 1.70 for beams. This safety factor is applied by designing the beam to fail just under a load of $1.7(1,000 \text{ lb}) = 1,700$ lb. Thus, to generalize, the design failure load = (factor of safety)(load on part, lb) = W_u.

For structural steel and other materials capable of fully plastic behavior, a simple beam or other member will collapse when the maximum moment equals the *plastic moment* M_p, which is developed when the stress throughout the section becomes equal to s_y. Hence, $M_p = s_y z$, where Z = plastic modulus = arithmetical sum of the statical moments of the upper and lower parts of the cross section about the horizontal axis that divides the area in half. For a square with its edges horizontal and vertical, as assumed here, $Z = (\frac{1}{2}b^2)(\frac{1}{4}b) + (\frac{1}{2}b^2)(\frac{1}{4}b) = \frac{1}{4}b^3$. Hence, $M_p = 33,000(\frac{1}{4}b^3)$.

Set M_p = the ultimate bending moment, or $(W_u/2)(L/2)$, where W_u = design failure load = (factor of safety)(load on part, lb) = $(1.70)(1,000) = 1,700$ lb for this beam. Thus, $M_p = 33,000(\frac{1}{4}b^3) = (1,700/2)(24/2)$; $b = 1.07$ in.

Using the allowable stress, $b = 1.22$ in., as compared with 1.07 in. The difference in results for the two methods (about 12 percent) can be traced to the ratio of Z to I/c, called the *shape factor*. For the case under consideration, this ratio is $(\frac{1}{4}b^3)/(\frac{1}{6}b^3) = 1.5$.

5. Determine the beam size for a vertical diagonal

Here $Z = 0.2357b^3$, and $I/c = 0.1178b^3$, and the shape factor = $Z/I/c = 0.2357/0.1178 = 2.0$.

Designing by allowable stress, using Steps 1, 2, and 3, gives $(0.1178b^3)(19,800) = 6,000$; $b = 1.37$ in. Designing by ultimate strength gives $(0.2357b^3)(33,000) = 10,200$; $b = 1.095$ in.

This computation shows that a more economical design is generally obtained by ultimate-strength design, with the advantage becoming greater as the shape factor becomes larger.

Related Calculations: For conventional structural sections, such as I beams, the shape factor is not much greater than 1, and the two methods yield about the same result for statically determinate problems, such as this one. If the problem involves a statically indeterminate beam or a rigid frame, the advantage of the ultimate-strength method becomes more apparent because it takes account of the fact that collapse cannot occur, as a rule, until the plastic moment is developed at each of two or more sections.

The method given here is the work of Professor Raymond J. Roark, reported in *Product Engineering.*

RUPTURE FACTOR AND ALLOWABLE STRESS IN DESIGN

Determine the proper thickness t in. of a circular plate 40 in. in diameter if the edge of the plate is simply supported and the plate carries a uniformly distributed pressure of 200 psi. The plate is made of cast iron having an ultimate strength of 50,000 psi.

TABLE 61 Illustrative Allowable Stresses and Factors of Safety*

(s_w = allowable or working stress; s_u = ultimate tensile strength; s_y = tensile yield strength; s' = modulus of rupture in cross-bending of rectangular bar; s'_s = ultimate shear strength; s'_c = ultimate compressive strength; s_c = endurance limit or endurance strength for specified life; n = dividing factor applied to s_u, s_y, or s_c to obtain s_w.)

Application	Materials	Allowable stress s_w	Approximate factor of safety
Buildings and other structures	Structural steel	Direct tension $s_w = 0.6s_y$; $0.45s_y$ on net section at pin holes; bending $s_w = 0.6s_y$; shear $s_w = 0.45s_y$ on rivets, $0.40s_y$ on girder webs; bearing $s_w = 1.35s_y$ on rivets in double shear; $1s_y$ in single shear	1.70 for beams; 1.85 for continuous frames
	Structural aluminum 6061-T6, 6062-T6 $s_u = 38,000$, $s_y = 35,000$	Direct tension $s_w = 17,000$; bending structural shapes, $s_w = 17,000$; rectangular sections $s_w = 23,000$; shear $s_w = 10,000$ on cold-driven rivets, 11,000 on girder webs; bearing $s_w = 30,000$ on rivets	1.8 for beams; 2 for columns
	Reinforced concrete	Bending, compression in concrete, $s_w = 0.45s'_c$; bending, tension in steel, $s_w = 0.40s_y$; bending, tension in plain concrete footings, $s_w = 0.03s'_c$; shear, on concrete in unreinforced web, $s_w = 0.03s'_c$; compression in concrete column, $s_w = 0.225s'_c$	
	Wood	Bending, $s_w = \frac{1}{6}s'$; long compression, $s_w = \frac{1}{6}s'_c$; transverse-compression, $s_w = \frac{1}{2}$ elastic limit, (400 for Douglas fir); shear parallel to grain $s_w = 120$ for Douglas fir	6 in general
Bridges	Structural metals and reinforced concrete	s_w about $0.9s_w$ for buildings	2.05
	Wood	s_w same as for buildings	6

3-117

TABLE 61   Illustrative Allowable Stresses and Factors of Safety* (Continued)

Application	Materials	Allowable stress s_w	Approximate factor of safety
Machinery	Steel (shafts, etc.)	Steady tension, compression or bending, $s_w = s_y/n$; pure shear, $s_w = s_y/2n$; tension s, plus shear s_s; $s_t{}^2 + 4s_s{}^2 \leqq s_y/n$; alternating stress s_a plus mean stress s_m: point representing an alternating stress $nk_f s_a$ and a mean stress ns_m must lie below the Goodman diagram, Fig. 1	n usually between 1.5 and 2
	Steel (SAE 1095), leaf springs, thickness $= t$ in., $t > 0 < 0.10$	Static loading, $s_w = 230{,}000 - 1{,}000{,}000t$; variable loading, 10^7 cycles, $s_w = 200{,}000 - 800{,}000t$; dynamic loading, 10^7 cycles, $s_w = 155{,}000 - 600{,}000t$	
	Steel (wire, ASTM-A228, helical springs)	$s_w = 100{,}000$ (for $d = 0.2$ in.; 10^7 cycles repeated stress)	
Pressure vessels (unfired)	Carbon steel	Membrane stress, $s_w = 0.211 s_u$; membrane plus discontinuity stresses, $s_w = 0.9 s_y$ or $0.6 s_u$	5
	Alloy steels	Membrane stress, $s_w = 0.25 s_u$; membrane plus discontinuity stresses, $s_w = 0.95 s_y$ or $0.6 s_u$	4
	Cast iron	Membrane stress, $s_w = 0.1 s_u$; bending stress $s_w = 0.15 s_u$	10, 6.67
	Nonferrous metals	Same rule as alloy steels	4
Airplanes	Aluminum alloy and steel	Ultimate strength design	1.5 against ultimate 1 against yield
	Wood	Ultimate strength design,	

*From Roark – *Formulas for Stress and Strain*, 4th ed., McGraw-Hill.

TABLE 62 Values of the Rupture Factor for Brittle Materials*

Form of member and manner of loading	Rupture factor; ratio of computed maximum stress at rupture to ultimate tensile strength		Ratio of computed maximum stress at rupture to modulus of rupture in bending or torsion	
	Cast iron	Plaster	Cast iron	Plaster
1. Rectangular beam, end support, center loading $l/d = 8$ or more............	1.70	1.60	1	1
2. Solid circular plate, edge support, uniform loading, $a/t = 10$ or more.........	1.9	1.71	...	1.07
3. Solid circular plate edge support, uniform loading on concentric circular area............	$2.4 - 0.5(r_0/a)^{1/6}$	$2.2 - 0.5(r_0/a)^{1/6}$	$1.40 - 0.3(r_0/a)^{1/6}$	$1.4 - 0.3(r_0/a)^{1/6}$

*From Roark—*Formulas for Stress and Strain*, 4th ed., McGraw-Hill, 1964.

Calculation Procedure:

1. Design on the basis of allowable stress

From Table 61 of the previous Calculation Procedure, note in the section for pressure vessels that the value of the working or allowable stress s_w psi for tension due to bending is $0.15 s_u$, where $s_u =$ ultimate tensile strength of the material, psi. Thus, $s_w = 0.15(50,000) = 7,500$ psi.

2. Compute the required plate thickness

The maximum stress for a simply supported plate is, from Roark—*Formulas for Stress and Strain*, $s_{max} = (3W/8\pi t^2)(3+v)$, where $W =$ total load on plate, lb; $t =$ plate thickness, in.; $v =$ Poisson's ratio. For this plate, $W = (200 \text{ psi})(\pi)(20)^2 = 251,330$ lb. Assuming $v = 0.3$, and solving for t, $7,500 = [(3)(251,330)/8\pi t^2](3+0.3)$; $t = 3.63$ in.

3. Design on the basis of ultimate strength

For ultimate-strength design, from Table 61 of the previous Calculation Procedure, use the value of 6.67 for the factor of safety. Design the plate to break, theoretically, under a load. Hence, the breaking load would be (factor of safety) W, or $6.67(251,330) = 1,667,000$ lb.

4. Apply the rupture factor to the design

With a brittle material like cast iron, the concept of the plastic moment does not apply. Use instead the *rupture factor*, which is the ratio of the calculated maximum tensile stress at rupture to the ultimate tensile strength, both expressed in psi. Whereas the rupture factor must be determined experimentally, a number of typical values are given in Table 62 for a variety of cases.

For case 2, which corresponds to this problem, the rupture factor for cast iron is $R_i = 1.9$. Using this in the same equation for s_w as in Step 1, $s_w = 1.9(50,000) = 95,000$ psi. Then, using the procedure and equation in Step 2, but substituting the breaking load for the plate, $95,000 = [(3)(1,677,000)/8\pi t^2](3+0.3)$; $t = 2.63$ in.

Related Calculations: The reliability of the solution using the ultimate-strength design technique depends on the accuracy of the rupture factor. When experimental or tabulated values of R_i are not available, the modulus of rupture may be used in place of $R_i s_u$. Typical values of Poisson's ratio v for various materials are given in Table 63.

The method presented here is the work of Professor Raymond J. Roark, reported in *Product Engineering*.

TABLE 63 Poisson's Ratio for Various Materials

Material	Poisson's Ratio
Aluminum:	
Cast.......................................	0.330
Wrought.............................	0.330
Brass, cast, 66% Cu, 34% Zn	0.350
Bronze, cast, 85% Cu, 7.2% Zn, 6.4% Sn	0.358
Cast iron.....................................	0.260
Copper, pure	0.337
Phosphor bronze, cast, 92.5% Cu, 7.0% Sn, 0.5% Ph	0.380
Steel:	
Soft	0.300
1% C.......................................	0.287
Cast.......................................	0.280
Tin, cast, pure......	0.330
Wrought iron	0.280
Nickel......................................	0.310
Zinc..	0.210

FORCE AND SHRINK FIT STRESS, INTERFERENCE, AND TORQUE

A 0.5-in-thick steel band having a modulus of elasticity of $E = 30 \times 10^6$ psi is to be forced on a 4-in.-diameter steel shaft. The maximum allowable stress in the band is 24,000 psi. What interference should be used between the band and the shaft? How much torque can the fit develop if the band is 3 in. long and the coefficient of friction is 0.20?

Calculation Procedure:

1. Compute the required interference

Use the relation $i = sd/E$, where i = the required interference to produce the maximum allowable stress in the band, in.; s = stress in band or hub, psi; d = shaft diameter, in.; E = modulus of elasticity of band or hub, psi. For this fit, $i = (24,000)(4.0)/(30 \times 10^6) = 0.0032$ in.

2. Compute the torque the fit will develop

Use the relation $T = Eitl\pi f$, where T = fit torque, lb-in.; t = band or hub thickness, in.; l = band or hub length, in.; f = coefficient of friction between the materials. For this joint, $T = 30 \times 10^6 \times 0.0032 \times 0.5 \times 3.0 \times \pi \times 0.20 = 90,432$ lb-in.

Related Calculations: Use this general procedure for either shrink or press fits. The axial force required for a press fit of two members made of the same material is F_a = axial force for the press fit, lb; p_c = radial pressure between the two members, psi = $iE(d_c^2 - d_i^2)(d_o^2 - d_c^2)/zd_c^3(d_o^2 - d_i^2)$, where d_o = outside diameter of the external member, in.; d_c = nominal diameter of the contact surfaces, in.; d_i = inside diameter of the inner member, in.

HYDRAULIC SYSTEM PUMP AND DRIVER SELECTION

Choose the pump and the driver horsepower for a rubber-tired tractor bulldozer having four-wheel drive. The hydraulic system must propel the vehicle, operate the dozer, and drive the winch. Each main wheel will be driven by a hydraulic motor at a maximum wheel speed of 59.2 rpm and a maximum torque of 30,000 lb-in. The wheel speed at maximum torque will be 29.6 rpm; maximum torque at low speed will be 74,500 lb-in. The tractor speed must be adjustable in two ways: for overall forward and reverse motion; and for turning, where the outside wheels turn at a faster rate than do the inside ones. Other operating details are given in the appropriate design steps below.

Calculation Procedure:

1. Determine the propulsion requirements of the system

Usual output requirements include speed, torque, force, and power for each function of the system, through the full capacity range.

First analyze the *propel* power requirements. For any propel condition, $hp = Tn/63,000$, where hp = horsepower required; T = torque, lb-in. at n rpm. Thus, at maximum speed, $hp = (30,000)(59.2)/63,000 = 28.2$ hp. At maximum torque, $hp = (74,500) \times (29.6)/63,000 = 35.0$; at maximum speed and maximum torque, $hp = (74,500)(59.2)/63,000 = 70.0$.

The drive arrangement for a bulldozer generally uses hydraulic motors geared down to wheel speed. Choose a 3,000-rpm step-variable-type motor for each wheel of the vehicle. Then each motor will operate at either of two displacements. At maximum vehicle loads, the higher displacement is used to provide maximum torque at low speed; at light loads, where a higher speed is desired, the lower displacement, producing reduced torque, is used.

Determine from a manufacturer's engineering data the motor specifications. For each of these motors the specifications might be: maximum displacement, 2.1 in.^{3}/

revolution; rated pressure, 6,000 psi; rated speed, 3,000 rpm; power output at rated speed and pressure, 90.5 hp; torque at rated pressure, 1,900 lb-in.

The gear reduction ratio GR between each motor and wheel = (output torque required, lb-in.)/(input torque, lb-in. × gear reduction efficiency). Assuming a 92 percent gear reduction efficiency, a typical value, GR = 74,500/(1,900 × 0.92) = 42.6 : 1. Hence, the maximum motor speed = wheel speed × GR = 59.2 × 42.6 = 2,520 rpm. At full torque the motor speed is, using the same relation, 29.6 × 42.6 = 1,260 rpm.

The required oil flow for the four motors is, at 1,260 rpm, in.³/revolution × 4 motors × rpm/(231 in.³/gal) = 2.1 × 4 × 1,260/231 = 45.8 gpm. With a 10 percent leakage allowance, the required flow = 50 gpm, closely, or 50/4 = 12.5 gpm per motor.

As computed above, the power output per motor is 35 hp. Thus, the four motors will have a total output of 4(35) = 140 hp.

2. Determine the linear auxiliary power requirements

The dozer uses a linear power output. Two hydraulic cylinders each furnish a maximum force of 10,000 lb to the dozer at a maximum speed of 10 in./sec. Assuming that the maximum operating pressure of the system is 3,500 psi, the piston area required per cylinder is: force developed, lb/operating pressure, psi = 10,000/3,500 = 2.86 sq in., or about a 2-in. cylinder bore. With a 2-in. bore, the operating pressure could be reduced in the inverse ratio of the piston areas. Or, $2.86/2^2\pi/4 = p/3,500$, where p = cylinder operating pressure, psi. Hence, p = 3,180 psi; say 3,200 psi.

Using a 2-in. bore cylinder, the required oil flow, gal, to each cylinder = (cylinder volume, in.³)(stroke length, in.)/231 in.³/gal = $(2^2\pi/4)(10)/231$ = 0.1355 gps, or 0.1355 × (60 sec/min) = 8.15 gpm per cylinder, or 16.3 gpm for two cylinders. The power input to the two cylinders is hp = gpm × psi/1,714 = 16.3(3,200)/1,714 = 30.4 hp.

3. Determine rotary auxiliary power requirements

The winch will be turned by one hydraulic motor. This winch must exert a maximum line pull of 20,000 lb at a maximum linear speed of 280 fpm with a maximum drum torque of 200,000 lb-in. at a drum speed of 53.5 rpm.

Compute the drum hp from $hp = Tn/63,000$, where the symbols are the same as in Step 1. Or, hp = (200,000)(53.5)/63,000 = 170 hp.

Choose a hydraulic motor having these specifications: displacement = 6 in.³/revolution; rated pressure = 6,000 psi; rated speed = 2,500 rpm; output torque at rated pressure = 5,500 lb-in.; power output at rated speed and pressure = 218 hp. This power output rating is somewhat greater than the computed rating but it allows some overloading.

The gear reduction ratio GR between the hydraulic motor and winch drum, based on the maximum motor torque, is GR = (output torque required, lb-in.)/(torque at rated pressure, lb-in., × reduction gear efficiency) = 200,000/(5,500 × 0.92) = 39.5 : 1. Hence, using this ratio, the maximum motor speed = 53.5 × 39.5 = 2,110 rpm. Oil flow rate to the motor = in.³/revolution × rpm/231 = 6 × 2,110/231 = 54.8 gpm, without leakage. With 5 percent leakage, flow rate = 1.05(54.8) = 57.2 gpm.

4. Categorize the required power outputs

List the required outputs and the type of motion required—rotary or linear. Thus: *propel* = rotary; *dozer* = linear; *winch* = rotary.

5. Determine the total number of simultaneous functions

There are two simultaneous functions: (*a*) propel motors and dozer cylinders; (*b*) propel motors at slow speed and drive the winch.

For function *a*, maximum oil flow = 50 + 16.3 = 66.3 gpm; maximum propel motor pressure = 6,000 psi; maximum dozer cylinder pressure = 3,200 psi. Data for function *a* came from previous steps in this Calculation Procedure.

For function *b*, the maximum oil flow need not be computed because it will be less than for function *a*.

6. Determine the number of series nonsimultaneous functions

These are the dozer, propel, and winch functions.

7. Determine the number of parallel simultaneous functions

These are the propel and dozer functions.

8. Establish function priority

The propel and dozer functions have priority over the winch function.

9. Size the piping and values

Table 64 lists the normal functions required in this machine and the type of valve that would be chosen for each function. Each of the valves incorporates additional functions: The step variable selector valve has a built-in check valve; the propel directional valve and winch directional valve have built-in relief valves and motor overload valves; the dozer directional valve has a built-in relief valve and a fourth position called "float." In the float position, all ports are interconnected, allowing the dozer blade to move up or down as the ground contour varies.

TABLE 64   Hydraulic-System Valving and Piping

Valving	
Function	Type of Valve
Step variable selector....	three-way, two-position
Propel directional	four-way, three-position, tandem-center
Winch directional	four-way, three-position, tandem-center
Dozer directional	four-way, four-position

Piping			
Branch of circuit	Propel motor	Dozer cylinder	Winch motor
Maximum flow, gpm	12.5	16.3	57.2
Maximum pressure, psi	6,000	3,200	6,000
Tube size, in	$\frac{3}{4}$	$\frac{3}{4}$	$1\frac{1}{2}$
Tube material, ASTM......	4130	4130	4130
Tube wall, in.	0.120	0.109	0.250

10. Determine the simultaneous power requirements

These are: Horsepower for propel and dozer = (gpm)(pressure, psi)/1,714 for the propel and dozer functions, or (50)(6,000)/1,714 + (16.3)(3,200)/1,714 = 205.4 hp. Winch horsepower, using the same relation, is (57.2)(6,000)/1,714 = 200 hp. Since the propel-dozer functions do not operate at the same time as the winch, the prime mover power need be only 205.4 hp.

11. Plan the specific circuit layouts

To provide independent simultaneous flow to each of the four propel motors, plus the dozer cylinders, choose two split-flow piston-type pumps having independent outlet ports. Split the discharge of each pump into three independent flows. Two pumps rated at 66.3/2 = 33.15 gpm each at 6,000 psi will provide the needed oil.

When steering the vehicle, additional flow is required by the outside wheels. Design the circuit so oil will flow from three pump pistons to each wheel motor. Four pistons of one split-flow pump are connected through check valves to all four motors. With this arrangement, oil will flow to the motors with the least resistance.

To make use of all or part of the oil from the propel-dozer circuits for the winch circuit, the outlet series ports of the propel and dozer valves are connected into the winch circuit, since the winch circuit is inoperative only when both the propel *and* the dozer are operating. When only the propel function is in operation, the winch is able to operate slowly but at full torque.

12.   *Investigate adjustment of the winch gear ratio*

As computed in Step 3, the winch gear ratio is based on torque. Now, because a known gpm is available for the winch motor from the propel and dozer circuits when these are not in use, the gear ratio can be based on the motor speed resulting from the available gpm.

Flow from the propel and dozer circuit = 66.3 gpm; winch motor speed = 2,450 rpm; required winch drum speed = 53.5 rpm. Thus, GR = 2,450/53.5 = 45.8 : 1.

With the proposed circuit, the winch gear reduction should be increased from 39.5 : 1 to 45.8 : 1. The winch circuit pressure can be reduced to (39.5/45.8)(6,000) = 5,180 psi. The required size of the winch oil tubing can be reduced to 0.219 in.

13.   *Select the prime mover hp*

Using a mechanical efficiency of 89 percent, the prime mover for the pumps should be rated at 205.4/0.89 = 230 hp. The prime mover chosen for vehicles of this type is usually a gasoline or diesel engine.

Related Calculations: The method presented here is also valid for fixed equipment using a hydraulic system — such as presses, punches, and balers. Other applications for which the method can be used include aircraft, marine, and on-highway vehicles. Use the method presented in an earlier section of this Handbook to determine the required size of the connecting tubing.

The procedure presented above is the work of Wes Master, reported in *Product Engineering*.

METALWORKING

REFERENCES—Le Grand—*American Machinist's Handbook*, McGraw-Hill; Boston—*Metal Processing*, Wiley; Nordhoff—*Machine-Shop Estimating*, McGraw-Hill; *Machinery's Handbook*, Industrial Press; *Welding Handbook*, American Welding Society; ASTME—*Tool Engineer's Handbook*, McGraw-Hill; *Procedure Handbook of Arc Welding Design and Practice*, The Lincoln Electric Company; Black—*Theory of Metal Cutting*, McGraw-Hill; Doyle—*Manufacturing Processes and Materials for Engineers*, Prentice-Hall; Brierly and Siekmann—*Machining Principles and Cost Control*, McGraw-Hill; Reason—*The Measurement of Surface Texture*, Cleaver-Hume; Bolz—*Production Processes: Their Influence on Design*, Penton; Harris—*A Handbook of Woodcutting*, HMSO, London; *Application Data, Cemented Carbides, Cemented Oxides*, Metallurgical Products Department, General Electric Company; Wood—*Final Report on Advanced Theoretical Formability Manufacturing Technology*, LTV, Inc., and USAF; Maynard—*Handbook of Business Administration*, McGraw-Hill; Niedzwiedzki—*Manual of Machinability and Tool Evaluation*, Huebner Publications, Cleveland; Hendriksen—*Chipbreakers*, The National Machine Tool Builders Association; ASTME—*Fundamentals of Tool Design*, Prentice-Hall; Crane—*Plastic Working of Metals and Power Press Operations*, Wiley; Jones—*Die Design and Die Making Practice*, Industrial Press; DeGarmo—*Materials and Processes in Manufacturing*, Macmillan; Jevons—*The Metallurgy of Deep Drawing and Pressing*, Wiley; Stanley—*Punches and Dies*, McGraw-Hill.

TOTAL ELEMENT TIME AND TOTAL OPERATION TIME

The observed times for a turret-lathe operation are: (1) material to bar stop, 0.0012 hr; (2) index turret, 0.0010 hr; (3) point material, 0.0005 hr; (4) index turret, 0.0012 hr;

(5) turn 0.300-in.-diameter part, 0.0075 hr; (6) clear hexagonal turret, 0.0009 hr; (7) advance cross-slide tool, 0.0008 hr; (8) cutoff part, 0.0030 hr; (9) aside with part, 0.0005 hr. What is the total element time? What is the total operation time if 450 parts are processed? Pointing of the material was later found unnecessary. What effect does this have on the element and operation total time?

Calculation Procedure:

1. Compute the total element time

Compute the total element time by finding the sum of each of the observed times in the operation, or summing steps 1 through 9: $0.0012 + 0.0010 + 0.0005 + 0.0012 + 0.0075 + 0.0009 + 0.0008 + 0.0030 + 0.0005 = 0.0166 \, \text{hr} = 0.0166 \, (60 \, \text{min/hr}) = 0.996 \, \text{min}$ per element.

2. Compute the total operation time

The total operation time = (element time, hr)(number of parts processed). Or, $(0.0166)(450) = 7.47 \, \text{hr}$.

3. Compute the time savings on deletion of one step

When one step is deleted, two or more times are usually saved. These times are the machine preparation and machine working times. In this process, they are Steps 2 and 3. Subtract the sum of these times from the total element time, or $0.0166 - (0.0010 + 0.0005) = 0.0151 \, \text{hr}$. Thus, the total element time decreases by $0.0015 \, \text{hr}$. The total operation time will now be $(0.0151)(450) = 6.795 \, \text{hr}$, or a reduction of $(0.0015)(450) = 0.6750 \, \text{hr}$. Checking, $7.470 - 6.795 = 0.675 \, \text{hr}$.

Related Calculations: Use this procedure for any multiple-step metalworking operation in which one or more parts are processed. These processes may be turning, boring, facing, threading, tapping, drilling, milling, profiling, shaping, grinding, broaching, hobbing, cutting, etc. The time elements used may be from observed or historical data.

CUTTING SPEEDS FOR VARIOUS MATERIALS

What spindle rpm is needed to produce a cutting speed of 150 fpm on a 2-in.-diameter bar? What is the cutting speed of a tool passing through 2.5-in.-diameter material at 200 rpm? Compare the required rpm of a turret-lathe cutter with the available spindle speeds.

Calculation Procedure:

1. Compute the required spindle rpm

In a rotating tool, the spindle rpm $R = 12C/\pi d$, where C = cutting speed, fpm; d = work diameter, in. For this machine, $R = 12(150)/\pi(2) = 286 \, \text{rpm}$.

2. Compute the tool cutting speed

For a rotating tool, $C = R\pi d/12$. Thus, for this tool, $C = (200)(\pi)(2.5)/12 = 131 \, \text{fpm}$.

The cutting-speed equation is sometimes simplified to $C = Rd/4$. Using this equation for the above machine, $C = 200(2.5)/4 = 125 \, \text{fpm}$. In general, it is wiser to use the exact equation.

3. Compare the required rpm with the available rpm

Consult the machine nameplate, American Machinist's Handbook, or a manufacturer's catalog to determine the available spindle rpm for a given machine. Thus, one Warner and Swasey turret lathe has a spindle speed of 282 rpm compared with the 286 rpm required in Step 1. The part could be cut at this lower spindle speed but the time required would be slightly greater because the available spindle speed is $286 - 282 = 4$ rpm less than the computed spindle speed.

When preparing job-time estimates, be certain to use the available spindle speed, because this is frequently less than the computed spindle speed. As a result, the actual cutting time will be longer when the available spindle speed is lower.

Related Calculations: Use this procedure for a cutting tool having a rotating cutter, such as a lathe, boring mill, automatic screw machine, etc. Tables of cutting speeds for various materials—metals, plastics, etc.—are available in the *American Machinist's Handbook*, as are tables of spindle rpm and cutting speed.

DEPTH OF CUT AND CUTTING TIME FOR A KEYWAY

What depth of cut is needed for a $\frac{3}{4}$-in.-wide keyway in a 3-in.-diameter shaft? The keyway length is 2 in. How long will it take to mill this keyway with a 24-tooth cutter turning at 130 rpm if the feed is 0.005 per tooth?

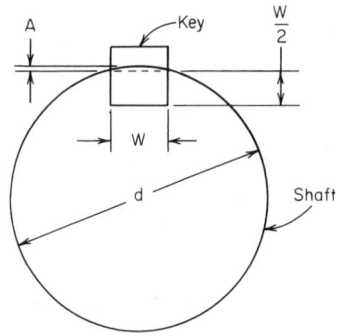

Fig. 1   Keyway dimensions.

Calculation Procedure:

1.   Sketch the shaft and keyway

Figure 1 shows the shaft and keyway. Note that the depth of cut D in. $=(W/2)+A$, where $W =$ keyway width, in.; $A =$ distance from the key horizontal centerline to the top of the shaft, in.

2.   Compute the distance from the centerline to the shaft top

For a machined keyway, $A = [d-(d^2-W^2)^{0.5}]/2$, where $d =$ shaft diameter, in. With the given dimensions, $A = [3-(3^2-0.75^2)^{0.5}]/2 = 0.045$ in.

3.   Compute the depth of cut for the keyway

The depth of cut $D = (W/2)+A = 0.75/2+0.045 = 0.420$ in.

4.   Compute the keyway cutting time

For a single milling cutter, cutting time, min $=$ length of cut, in./(feed per tooth) $\times$ (number of teeth on cutter)(cutter rpm). Thus, for this keyway, cutting time $= 2.0/[(0.005)(24)(130)] = 0.128$ min.

Related Calculations: Use this procedure for square or rectangular keyways. For Woodruff key-seat milling, use the same cutting-time equation as in Step 4. A Woodruff key seat is almost a semicircle, being one-half the width of the key *less than* a semicircle. Thus, a $\frac{9}{16}$-in.-deep Woodruff key seat containing a $\frac{3}{4}$-in.-wide key will be $3/8/2 = \frac{3}{16}$ in. less than a semicircle. The key seat would be cut with a cutter having a radius of $\frac{9}{16}+\frac{3}{16}=\frac{12}{16}$, or $\frac{3}{4}$ in.

MILLING-MACHINE TABLE FEED AND CUTTER APPROACH

A 12-tooth milling cutter turns at 400 rpm and has a feed of 0.006 per tooth per revolution. What table feed is needed? If this cutter is 8 in. in diameter and is facing a 2-in.-wide part, determine the cutter approach.

Calculation Procedure:

1.   Compute the required table feed

For a milling machine, the table feed F_T in./min $= f_t nR$, where $f_t =$ feed per tooth per revolution; $n =$ number of teeth in cutter; $R =$ cutter rpm. For this cutter, $F_T = (0.006) \times (12)(400) = 28.8$ in./min.

2. Compute the cutter approach

The approach of a milling cutter A_c in. $= 0.5D_c - 0.5(D_c{}^2 - w^2)^{0.5}$, where $D_c =$ cutter diameter, in.; $w =$ width or face of cut, in. For this cutter, $A_c = 0.5(8) - 0.5(8^2 - 2^2)^{0.5} = 0.53$ in.

Related Calculations: Use this procedure for any milling cutter whose dimensions and speed are known. These cutters can be used for metals, plastics, and other non-metallic materials.

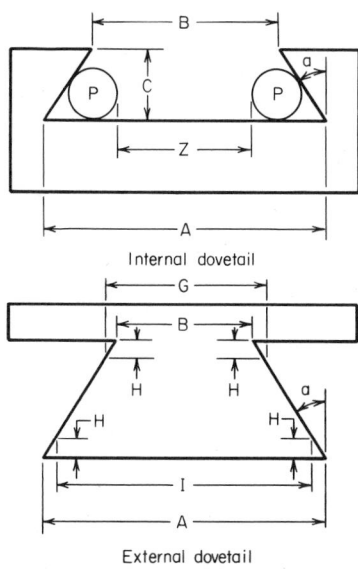

Internal dovetail

External dovetail

Fig. 2 Dovetail dimensions.

DIMENSIONS OF TAPERS AND DOVETAILS

What is the taper per ft TPF and taper per in. TPI of an 18-in.-long part having a large diameter d_l of 3 in. and a small diameter d_s of 1.5 in? What is the length of a part with the same large and small diameters as the above part if the TPF is 3 in./ft? Determine the dimensions of the dovetail in Fig. 2 if $B = 2.15$ in., $C = 0.60$ in., and $a = 30°$. A $\frac{3}{8}$-in.-diameter plug is used to measure the dovetail.

Calculation Procedure:

1. Compute the taper of the part

For a round part TPF in./ft $= (d_l - d_s)/L$, where $L =$ length of part, in.; other symbols as defined above. Thus, for this part, TPF $= 12(3.0 - 1.5)/18 = 1$ in./ft.

The taper per in. TPI in./in. $= (d_l - d_s)/L_2$ or $(3.0 - 1.5)/18 = 0.0833$ in./in.

The taper of round parts may also be expressed as the angle measured from the shaft centerline, that is, one-half the included angle between the tapered surfaces of the shaft.

2. Compute the length of the tapered part

Converting the first equation of Step 1, $L = 12(d_l - d_s)/$TPF. Or, $L = 12(3.0 - 1.5)/3.0 = 6$ in.

3. Compute the dimensions of the dovetail

For external and internal dovetails, Fig. 2, with all dimensions except the angles in in., $A = B + CF = I + HF$; $B = A - CF = G - HF$; $E = P[\cot(90 + a/2)] + P$; $D = P \times [\cot(90 - a/2)] + P$; $F = 2 \tan a$; $Z = A - D$. Note that $P =$ diameter of plug used to measure the dovetail, in.

With the given dimensions, $A = B + CF$, or $A = 2.15 + (0.60)(2 \times 0.577) = 2.84$ in. Since the plug P is $\frac{3}{8}$ in. in diameter, $D = P[\cot(90 - a/2)] + P = 0.375[\cot(90 - 30/2)] + 0.375 = 1.025$ in. Then, $Z = A - D = 2.840 - 1.025 = 1.815$ in. Also, $E = P[\cot(90 + a/2)] + P = 0.375[\cot(90 + 30/2)] + 0.375 = 0.591$ in.

With flat-cornered dovetails, as at I and G, and $H = \frac{1}{8}$ in., $A = I + HF$. Solving for I, $I = A - HF = 2.84 - (0.125)(2 \times 0.577) = 2.696$ in. Then, $G = B + HF = 2.15 + (0.125) \times (2 \times 0.577) = 2.294$.

Related Calculations: Use this procedure for tapers and dovetails in any metallic and nonmetallic material. When a large number of tapers or dovetails must be computed, use the appropriate tables in the *American Machinist's Handbook*.

ANGLE AND LENGTH OF CUT FROM GIVEN DIMENSIONS

What angle must a cutting tool be set at to cut the part in Fig. 3? How long is the cut in this part?

Calculation Procedure:

1. Compute the angle of the cut

Use trigonometry to compute the angle of the cut. Thus, $\tan a =$ opposite side/ adjacent side $= (8-5)/6 = 0.5$. From a table of trigonometric functions, $a =$ cutting angle $= 26° 34'$, closely.

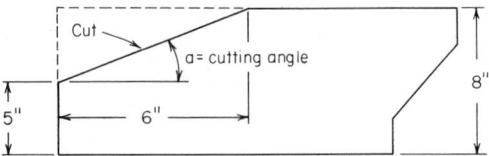

Fig. 3 Length of cut of a part.

2. Compute the length of the cut

Use trigonometry to compute the length of cut. Thus, $\sin a =$ opposite side/hypotenuse, or $0.4472 = (8-5)/$hypotenuse; length of cut $=$ length of hypotenuse $= 3/0.4472 = 6.7$ in.

Related Calculations: Use this general procedure to compute the angle and length of cut for any metallic or nonmetallic part.

TOOL FEED RATE AND CUTTING TIME

A part 3.0 in. long is turned at 100 rpm. What is the feed rate if the cutting time is 1.5 min? How long will it take to cut a 7.0 in.-long part turning at 350 rpm if the feed is 0.020 in./revolution? How long will it take to drill a 5-in.-deep hole with a drill speed of 1,000 rpm and a feed of 0.0025 in./revolution?

Calculation Procedure:

1. Compute the tool feed rate

For a tool cutting a rotating part, $f = L/Rt$, where $t =$ cutting time, min. For this part, $f = 3.0/[(100)(1.5)] = 0.02$ in./revolution.

2. Compute the cutting time for the part

Transpose the equation in Step 1 to yield $t = L/Rf$, or $t = 7.0/[(350)(0.020)] = 1.0$ min.

3. Compute the drilling time for the part

Drilling time is computed using the equation of Step 2, or $t = 5.0/[(1,000)(0.0025)] = 2.0$ min.

Related Calculations: Use this procedure to compute the tool feed, cutting time, and drilling time in any metallic or nonmetallic material. Where many computations must be made, use the feed-rate and cutting-time tables in the *American Machinist's Handbook*.

TRUE UNIT TIME, MINIMUM LOT SIZE, AND TOOL-CHANGE TIME

What is the machine unit time to work 25 parts if the setup time is 75 min and the unit standard time is 5.0 min? If one machine tool has a setup standard time of 9 min and a unit standard time of 5.0 min, how many pieces must be handled if a machine with a setup standard of 60 min and a unit standard time of 2.0 min is to be more economical? Determine the minimum lot size for an operation requiring 3 hr to set up if the unit standard time is 2.0 min and the maximum increase in the unit standard may not exceed 15 percent. Find the unit time to change a lathe cutting tool if the operator takes 5 min to change the tool and the tool cuts 1.0 min/cycle and has a life of 3 hr.

Calculation Procedure:

1. Compute the true unit time

The true unit time for a machine $T_u = (S_u/N) + U_s$, where $S_u =$ setup time, min; $N =$ number of pieces in lot; $U_s =$ unit standard time, min. For this machine, $T_u = (75/75) + 5.0 = 6.0$ min.

2. Determine the most economical machine

Call one machine X, the other Y. Then, (unit standard time of X, min)(number of pieces) + (setup time of X, min) = (unit standard time of Y, min)(number of pieces) + (setup time of Y, min). For these two machines, since the number of pieces Z is unknown, $5.0Z + 9 = 2.0Z + 60$. Solving, $Z = 17$ pieces. Thus, machine Y will be more economical when 17 or more pieces are made.

3. Compute the minimum lot size

The minimum lot size $M = S_u/U_sK$, where $K =$ allowable increase in unit-standard time, percent. For this run, $M = (3 \times 60)/[(2.0)(0.15)] = 600$ pieces.

4. Compute the unit tool-changing time

The unit tool-changing time U_t to change from dull to sharp tools is $U_t = T_cC_t/l$, where $T_c =$ total time to change tool, min; $C_t =$ time tool is in use during cutting cycle, min; $l =$ life of tool, min. For this lathe, $U_t = (5)(1)/[(3)(60)] = 0.0278$ min.

Related Calculations: Use these general procedures to find true unit time, the most economical machine, minimum lot size, and unit tool-changing time for any type of machine tool – drill, lathe, milling machine, hobs, shapers, thread chasers, etc.

TIME REQUIRED FOR TURNING OPERATIONS

Determine the time to turn a 3-in.-diameter brass bar down to a $2\frac{1}{2}$-in. diameter with a spindle speed of 200 rpm and a feed of 0.020 per revolution if the length of cut is 4 in. Show how the turning-time relation can be used for relief turning, pointing of bars, internal and external chamfering, hollow mill work, knurling, and forming operations.

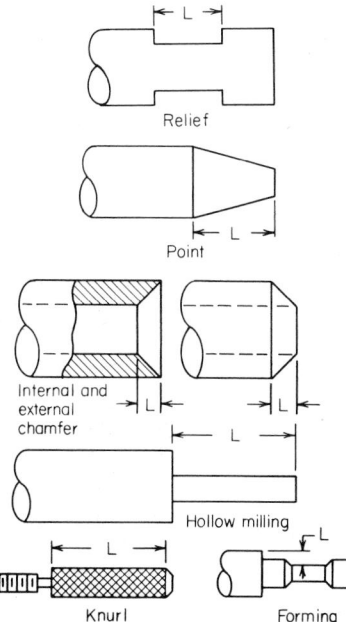

Fig. 4 **Turning operations.**

Calculation Procedure:

1. Compute the turning time

For a turning operation, the time to turn T_t min $= L/fR$, where $L =$ length of cut, in.; $f =$ feed, in./revolution; $R =$ work rpm. For this part, $T_t = 4/[0.02)(200)] = 1.00$ min.

2. Develop the turning relation for other operations

For *relief turning* use the same relation as in Step 1. Length of cut is the length of the relief, Fig. 4. A small amount of time is also required to hand-feed the tool to the minor diameter of the relief. This time is best obtained by observation of the operation.

The time required to *point a bar*, called *pointing*, is computed using the relation in Step 1. The length of cut is the distance from the end of the bar to the end of the tapered point, measured parallel to the axis of the bar, Fig. 4.

Use the relation in Step 1 to compute the time to cut an internal or external chamfer. The length of cut of a chamfer is the horizontal distance L, Fig. 4.

A hollow mill reduces the external diameter of a part. The cutting time is computed using the relation in Step 1. The length of cut is shown in Fig. 4.

Compute the time to knurl using the relation in Step 1. The length of cut is shown in Fig. 4.

Compute the time for forming using the relation in Step 1. The length of cut is shown in Fig. 4.

TIME AND POWER TO DRILL, BORE, COUNTERSINK, AND REAM

Determine the time and power required to drill a 3-in.-deep hole in an aluminum casting if a $\frac{3}{4}$-in.-diameter drill turning at 1,000 rpm in used and the feed is 0.030 in. per revolution. Show how the drilling relations can be used for boring, countersinking, and reaming. How long will it take to drill a hole through a 6-in.-thick piece of steel if the cone height of the drill is 0.5 in., the feed is 0.002 in./revolution, and the drill speed is 100 rpm?

Calculation Procedure:

1. Compute the time required for drilling

The time required to drill T_d min $= L/fR$, where $L =$ depth of hole $=$ length of cut, in. In most drilling calculations, the height of the drill cone (point) is ignored. (Where the cone height is used, follow the procedure in Step 4.) For this hole, $T_d = 3/[(0.030) \times (1,000)] = 0.10$ min.

2. Compute the power required to drill the hole

The power required to drill, in hp, is $hp = 1.3LfCK$, where $C =$ cutting speed, fpm, sometimes termed surface f/min, $sfpm = \pi DR/12$; $K =$ power constant from Table 1. For an aluminum casting, $K = 3$. Then, $hp = (1.3)(3)(0.030)(\pi \times 0.75 \times 1,000/12)(3) = 66.0$ hp. The factor 1.3 is used to account for dull tools and for overcoming friction in the machine.

TABLE 1   Power Constants for Machining

Material	Power Constant
Carbon steel C1010 to C1025	6
Manganese steel T1330 to T1350	9
Nickel steel 2015 to 2320	7
Molybdenum........................	9
Chromium	10
Stainless steels	11
Cast iron:	
Soft................................	3
Medium	3
Hard...............................	4
Aluminum alloys:	
Castings	3
Bar................................	4
Copper	4
Brass (except manganese).............	4
Monel metal........................	10
Magnesium alloys....................	3
Malleable iron:	
Soft................................	3
Medium	4
Hard...............................	5

3. Adapt the drill relations to other operations

The time and power required for boring are found using the two relations given above. The length of the cut = length of the bore. Also use these relations for undercutting, sometimes called internal relieving, and for counterboring. These same relations are also valid for countersinking, center drilling, start or spot drilling, and reaming. In reaming, the length of cut is the total depth of the hole reamed.

4. Compute the time for drilling a deep hole

With parts having a depth of 6 in. or more, compute the drilling time from $T_d = (L+h)/fR$, where h = cone height, in. For this hole, $T_d = (6+0.5)/[(0.002)(100)] = 32.25$ min. This compares with $T_d = L/fR = 6/[(0.002)(100)] = 30$ min when the height of the drill cone is ignored.

TIME REQUIRED FOR FACING OPERATIONS

How long will it take to face a part on a lathe if the length of cut is 4 in., the feed is 0.020 in./revolution, and the spindle speed is 50 rpm? Determine the facing time if the same part is faced by an eight-tooth milling cutter turning at 1,000 rpm and having a feed of 0.005 in. per tooth per revolution. What table feed is required if the cutter is turning at 50 rpm? What is the feed per tooth with a table feed of 4.0 in./min? What added table travel is needed when a 4-in.-diameter cutter is cutting a 4-in.-wide piece of work?

Calculation Procedure:

1. Compute the lathe facing time

For lathe facing, the time to face T_f min = L/fR, where the symbols are the same as given for previous Calculation Procedures in this Section. For this part, $T_f = 4/[(0.02)(50)] = 4.0$ min.

2. Compute the facing time using a milling cutter

With a milling cutter, $T_f = L/f_t nR$, where f_t = feed per tooth, in./revolution; n = number of teeth on cutter; other symbols as before. For this part, $T_f = 4/[(0.005)(8) \times (1,000)] = 0.10$ min.

3. Compute the required table feed

In a milling machine, the table feed F_t in./min = $f_t nR$. For this machine, $F_t = (0.005) \times (8)(50) = 2.0$ in./min.

4. Compute the feed per tooth

For a milling machine, the feed per tooth, in./revolution, $f_t = F_t/Rn$. In this machine, $f_t = 4.0/[(50)(8)] = 0.01$ in./revolution.

5. Compute the added table travel

In face milling, the added table travel A_t in. = $0.5[(D_c - (D_c^2 - W^2)^{0.5}]$, where the symbols are the same as given earlier. For this cutter and work, $A_t = 0.5[4 - (4^2 - 4^2)^{0.5}] = 2.0$ in.

THREADING AND TAPPING TIME

How long will it take to cut a 4-in.-long thread at 100 rpm if the rod will have 12 threads per inch and a button die is used? The die is backed off at 200 rpm. What would the threading time be if a self-opening die were used instead of a button die? What would the threading time be for a single-pointed threading tool if the part being threaded is aluminum and the back-off speed is twice the threading speed? The rod is 1 in. in diameter. How long will it take to tap a 2-in.-deep hole with a 1-14 solid tap turning at 100 rpm? How long will it take to millthread a 1-in.-diameter bolt having 15 threads per in. 3 in. long if a 4-in.-diameter 20-flute thread-milling hob turning at 80 rpm with a 0.003-in. feed is used?

Calculation Procedure:

1. Compute the button-die threading time

For a multiple-pointed tool, the time to thread $T_t = Ln_t/R$, where L = length of cut = length of thread measured parallel to thread longitudinal axis, in.; n_t = number of threads per in. For this button die, $T_t = (4)(12)/100 = 0.48$ min. This is the time required to cut the thread.

Compute the back-off time B min from $B = Ln_t/R_B$, where R_B = back-off rpm, or $B = (4)(12)/200 = 0.24$ min. Hence, the total time to cut and back off $= T_t + B = 0.48 + 0.24 = 0.72$ min.

2. Compute the self-opening die threading time

With a self-opening die, the die opens automatically when it reaches the end of the cut thread and is withdrawn instantly. Therefore, the back-off time is negligible. Hence the time to thread $= T_t = Ln_t/R = (4)(12)/100 = 0.48$ min. One cut is usually sufficient to make a suitable thread.

3. Compute the single-pointed tool cutting time

With a single-pointed tool, more than one cut is usually necessary. Table 2 lists the number of cuts needed with a single-pointed tool working on various materials. The maximum cutting speed for threading and tapping is also listed.

Table 2 shows that four cuts are needed for an aluminum rod when a single-pointed tool is used. Before computing the cutting time, compute the cutting speed to determine if it is within the recommended range given in Table 2. From a previous Calculations Procedure, $C = R\pi d/12$, or $C = (100)(\pi)(1.0)/12 = 26.2$ fpm. Since this is less than maximum recommended speed of 30 rpm, the work speed is acceptable.

Compute the time to thread from $T_t = Ln_t c/R$, where c = number of cuts to thread, from Table 2. For this part, $T_t = (4)(12)(4)/100 = 1.92$ min.

TABLE 2 Number of Cuts and Cutting Speed for Dies and Taps

	No. of cuts*	Cutting speed, fpm†
Aluminum	4	30
Brass (commercial) . .	3	30
Brass (naval)	4	30
Bronze (ordinary) . . .	5	30
Bronze (hard)	7	20
Copper	5	20
Drill rod	8	10
Magnesium	4	30
Monel (bar)	8	10
Steel (mild)	5	20
Steel (medium)	7	10
Steel (hard)	8	10
Steel (stainless)	8	10

*Single-pointed threading tool; maximum spindle speed, 250 rpm.
†Maximum recommended speed for single- and multiple-pointed tools; maximum spindle speed for multiple-pointed tools = 150 rpm for dies and taps.

If the tool is backed off at twice the threading speed, and the back-off time $B = Ln_t c/R_B$, $B = (4)(12)(4)/200 = 0.96$ min. Hence, the total time to thread and back off $= T_t + B = 1.92 + 0.96 = 2.88$ min. In some shapes, a single-pointed tool may not be backed off; the tool may instead be repositioned. The time required for this approximates the back-off time.

4. Compute the tapping time

The time to tap $T_t \min = Ln_t/R$. With a solid tap, the tool is backed out at twice the tapping speed. With a collapsing tap, the tap is withdrawn almost instantly without reversing the machine or tap.

For this hole, $T_t = (2)(14)/100 = 0.28$ min. The back-off time $B = Ln_t/R_B = (2)(14)/200 = 0.14$ min. Hence, the total time to tap and back off $= T_t + B = 0.28 + 0.14 = 0.42$ min.

The maximum spindle speed for tapping should not exceed 250 rpm. Use the cutting-speed values given in Table 2 when computing the desirable speed for various materials.

5. Compute the thread-milling time

The time for thread milling is $T_t = L/fnR$, where $L =$ length of cut, in. $=$ circumference of work, in.; $f =$ feed per flute, in.; $n =$ number of flutes on hob; $R =$ hob rpm. For this bolt, $T_t = 3.1416/[(0.003)(20)(80)] = 0.655$ min.

Note that neither the length of the threaded portion nor the number of threads per in. enters into the calculation. The thread hob covers the entire length of the threaded portion and completes the threading in one revolution of the work head.

TURRET-LATHE POWER INPUT

How much power is required to drive a turret lathe making a $\frac{1}{2}$-in.-deep cut in cast iron if the feed is 0.015 in./revolution, the part is 2.0 in. in diameter, and its speed is 382 rpm? How many 1.5-in.-long parts can be cut from a 10-ft-long bar if a $\frac{1}{4}$-in.-cutoff tool is used? Allow for end squaring.

Calculation Procedure:

1. Compute the surface speed of the part

The cutting, or surface, speed, as given in a previous Calculation Procedure, is $C = R\pi d/12$, or $C = (382)(\pi)(2.0)/12 = 200$ fpm.

2. Compute the power input required

For a turret lathe, the hp input $hp = 1.33\, DfCK$, where $D =$ cut depth, in.; $f =$ feed, in./revolution; $K =$ material constant from Table 3. For cast iron, $K = 3.0$. Then, $hp = (1.33)(0.5)(0.015)(200)(3.0) = 5.98\, hp$; say 6.0 hp.

TABLE 3   Turret-Lathe Power Constant

Material	Constant, K
Bronze	3
Cast iron.....................	3
SAE steels:	
1020.........................	6
1045.........................	8
3250.........................	9
4150.........................	9
4615.........................	6
X1315	6
Straight tubing	6
Steel castings and forgings	9
Heat-treated steels:	
4150.........................	10
52100	10

3. *Compute the number of parts that can be cut*

Allow 2 in. on the bar end for checking and $\frac{1}{2}$ in. on the opposite end for squaring. With an original length of 10 ft = 120 in., this leaves $120 - 2.5 = 117.5$ in. for cutting.

Each part cut will be 1.5 in. long $+ 0.25$ in. for the cutoff, or 1.75 in. of stock. Hence, the number of pieces which can be cut = $117.5/1.75 = 67.1$, or 67 pieces.

Related Calculations: Use the procedure given here to find the turret-lathe power input for any of the materials and similar materials, listed in Table 3. The parts cutoff computation can be used for any material — metallic or nonmetallic. Be sure to allow for the width of the cutoff tool.

TIME TO CUT A THREAD ON AN ENGINE LATHE

How long will it take an engine lathe to cut an acme thread having a length of 5 in., a major diameter of 2 in., four threads per in., a depth of 0.1350 in., a cutting speed of 70 fpm, and a depth of cut of 0.005 in. per pass if the material cut is medium steel? How many passes of the tool are required?

Calculation Procedure:

1. *Compute the cutting time*

For an acme, square, or worm thread cut on an engine lathe, the total cutting time T_t min excluding the tool positioning time, is found from $T_t = Ld_tDn_t/4Cd_c$, where L = thread length, in.. measured parallel to the thread longitudinal axis; d_t = thread major diameter, in.; D = depth of thread, in.; n_t = number of threads per in.; C = cutting speed, fpm; d_c = depth of cut per pass, in.

For this acme thread, $T_t = (5)(2)(0.1350)(4)/[4(70)(0.005)] = 3.85$ min. To this must be added the time required to position the tool for each pass.

2. *Compute the number of tool passes required*

The depth of cut per pass is 0.005 in. A total depth of 0.1350 in. must be cut. Therefore, the number of passes required = total depth cut, in./depth of cut per pass, in. = $0.1350/0.005 = 27$ passes.

Related Calculations: Use this procedure for threads cut in ferrous and nonferrous metals. Table 4 shows typical cutting speeds.

TABLE 4   Thread Cutting Speeds

Material°	Cutting Speed, fpm
Soft nonferrous metals...........	250
Mild steel......................	100
Medium steel...................	75
Hard steel.....................	50

°Average depth of cut = 0.005 in. for ferrous metals; 0.010 in. for nonferrous metals.

TIME TO TAP WITH A DRILLING MACHINE

How long will it take to tap a 4-in.-deep hole with a $1\frac{1}{2}$-in.-diameter tap having six threads per in. if the tap turns at 75 rpm?

Calculation Procedure:

1. *Compute the tap surface speed*

Using the method of a previous Calculation Procedure, $C = R\pi d/12 = (75)(\pi)(1.5)/12 = 29.5$ fpm.

2. *Compute the time to tap the hole and withdraw the tool*

For tapping with a drilling machine, $T_t = Dn_tD_c\pi/8C$, where D = depth of cut = depth of hole tapped, in.; n_t = number of threads per in.; D_c = cutter diameter, in. = tap diameter, in. For this hole, $T_t = (4)(6)(1.5)\pi/[8(29.5)] = 0.48$ min, which is the time required to tap and withdraw the tool.

Related Calculations: Use this procedure for tapping ferrous and nonferrous metals on a drill press. The recommended tap surface speed for various metals is: aluminum, soft brass, ordinary bronze, soft cast iron, and magnesium: 30 fpm; naval brass, hard bronze, medium cast iron, copper, and mild steel: 20 fpm; hard cast iron, medium steels, and hard stainless steel: 10 fpm.

MILLING CUTTING SPEED, TIME, FEED, TEETH NUMBER, AND HORSEPOWER

What is the cutting speed of a 12-in.-diameter milling cutter turning at 190 rpm? How many teeth are needed in the cutter at this speed if the feed is 0.010 in. per tooth, the depth of cut is 0.075 in., the length of cut is 5 in., the power available at the cutter is 12 hp, and the mill is cutting hard malleable iron? How long will it take the mill to make this cut? What is the maximum feed rate that can be used? What is the power input to the cutter if a 20-hp machine is used?

Calculation Procedure:

1. *Compute the cutter cutting speed*

For a milling cutter, use the simplified relation $C = Rd/4$, where the symbols are as given earlier in this Section. Or, $C = (190)(12)/4 = 570$ fpm.

2. *Compute the number of cutter teeth required*

For a carbide cutter, $n = K_mhp_c/Df_tLR$, where n = number of teeth on cutter; K_m = machinability constant or K factor from Table 5; hp_c = horsepower available at the milling cutter; D = depth of cut, in.; f_t = cutter feed, in. per tooth; L = length of cut, in.; R = cutter rpm.

Table 5 shows that $K_m = 0.90$ for malleable iron. Then, $n = (0.90)(12)/[(0.075)(0.01) \times (5)(190)] = 15.15$; say 16 teeth. For general-purpose use, the Metal Cutting Institute recommends that $n = 1.5$(cutter diameter, in.) for cutters having a diameter of more than 3 in. For this cutter, $n = 1.5(12) = 16$ teeth. This agrees with the number of teeth computed with the cutter equation.

3. *Compute the milling time*

For a milling machine, the time to cut T_t min = L/f_tnR, where L = length of cut, in.; f_t = feed per tooth, in. per tooth per revolution; n = number of teeth on the cutter; R = cutter rpm. Thus, the time to cut is $T_t = 5/[(0.01)(16)(190)] = 0.164$ min.

4. *Compute the maximum feed rate*

For a milling machine, the maximum feed rate f_m in./min = K_mhp_c/DL, where L = length of cut; other symbols are the same as in Step 2. Thus, $f_m = (0.90)(12)/[(0.075)(5)] = 28.8$ in./min.

5. *Compute the power input to the machine*

The power available at the cutter is 12 hp. The power required $hp_c = DLnRf_t/K_m$, where all symbols are as given above. Thus, $hp_c = (0.075)(5)(16)(190)(0.01)/0.90 = 12.68$ hp. This is slightly more than the available horsepower.

Milling machines have overall efficiencies ranging from a low of 40 percent to a high of 80 percent, Table 6. Assume a machine efficiency of 65 percent. Then, the required power input is 12.68/0.65 = 19.5. Therefore, a 20- or 25-hp machine will be satisfactory, depending on its actual operating efficiency.

TABLE 5   Machinability Constant K_m

Aluminum.............	2.28
Brass, soft	2.00
Bronze, hard..........	1.40
Bronze, very hard......	0.65
Cast iron, soft.........	1.35
Cast iron, hard.........	0.85
Cast iron, chilled.......	0.65
Cast magnesium	2.50
Malleable iron.........	0.90
Steel, soft.............	0.85
Steel, medium	0.65
Steel, hard.............	0.48
Steel:	
100 Brinell	0.80
150 Brinell	0.70
200 Brinell	0.65
250 Brinell	0.60
300 Brinell	0.55
400 Brinell	0.50

TABLE 6   Typical Milling-Machine Efficiencies

Rated Hp of Machine	Overall Efficiency, percent
3	40
5	48
7.5	52
10	52
15	52
20	60
25	65
30	70
40	75
50	80

Related Calculations: After selecting a feed rate, check it against the suggested feed per tooth for milling various materials given in the *American Machinist's Handbook*. Use the method of a previous Calculation Procedure in this Section to determine the cutter approach. With the approach known, the maximum chip thickness, in. = (cutter approach, in.)(table advance per tooth, in.)/(cutter radius, in.). Also, the feed per tooth in. = (feed rate, in./min)/[(cutter rpm)(number of teeth on cutter)].

GANG-, MULTIPLE-, AND FORM-MILLING CUTTING TIME

How long will it take to gang mill a part if three cutters are used with a spindle speed of 70 rpm, and there are 12 teeth on the smallest cutter, a feed of 0.015 in./revolution, and a length of cut of 8 in.? What will be the unit time to multiple mill four keyways if each of the four cutters has 20 teeth, the feed is 0.008 in. per tooth, spindle speed is 150 rpm, and the keyway length is 3 in.? Show how the cutting time for form milling is computed, and how the cutter diameter for straddle milling is computed.

Calculation Procedure:

1. Compute the gang-milling cutting time

For any gang-milling operation, using the dimensions of the smallest cutter, the time to cut $T_t = L/f_t nR$, where L = length of cut, in.; f_t = feed per tooth, in./revolution; n = number of teeth on cutter; R = spindle rpm. For this part, $T_t = 8/[(0.015)(12)(70)] = 0.635$ min.

Note that in all gang-milling cutting-time calculations, the number of teeth and feed of the *smallest* cutter are used.

2. Compute the multiple-milling cutting time

In multiple milling, the cutting time $T_t = L/f_t nR_m$, where n = number of milling cutter used. In multiple milling, the cutting time is termed the unit time. For this machine, $T_t = 3/[(0.008)(20)(150)(4)] = 0.0303$ min.

3. Show how form milling time is computed

Form-milling cutters are used on surfaces that are neither flat nor square. The cutters used for form milling resemble other milling cutters. The cutting time is therefore computed from $T_t = L/f_t nR$, where all symbols are the same as in Step 1.

4. Show how the cutter diameter is computed for straddle milling

In straddle milling, the cutter diameter must be large enough to permit the work to pass under the cutter arbor. The minimum-diameter cutter to straddle mill a part is in. = (diameter of arbor, in.) + 2 (face of cut, in. + 0.25). The 0.25 in. is the allowance for clearance of the arbor.

Related Calculations: Use the equation of Step 1 to compute the cutting time for metal slitting, screw slotting, angle milling, T-slot milling, Woodruff key-seat milling, and profiling and routing of parts. In T-slot milling, two steps are required—milling of the vertical member and milling of the horizontal member. Compute the milling time of each; the sum of the two is the total milling time.

SHAPER AND PLANER CUTTING SPEED, STROKES, CYCLE TIME, POWER

What is the cutting speed of a shaper making 54 strokes/min if the stroke length is 6 in? How many strokes/min should the ram of a shaper make if it is shaping a 12-in.-long aluminum bar at a cutting speed of 200 fpm? How long will it take to make a cut across a 12-in. face of a cast-iron plate if the feed is 0.050 in./stroke and the ram makes 50 strokes/min? What is the cycle time of a planer if its return speed is 200 fpm, the acceleration-deceleration constant is 0.05, and the cutting speed is 100 fpm? What is the planer power input if the depth of cut is $\frac{1}{8}$ in. and the feed is $\frac{1}{16}$ in./stroke?

Calculation Procedure:

1. Compute the shaper cutting speed

For a shaper, the cutting speed, fpm, is $C = SL/6$, where S = strokes/min; L = length of stroke, in., where the cutting-stroke time = return-stroke time. Thus, for this shaper, $C = (54)(6)/6 = 54$ fpm.

2. Compute the shaper stroke rate

Transpose the equation of Step 1 to $S = 6C/L$. Then, $S = 6(200)/12 = 100$ strokes/min.

3. Compute the shaper cutting time

For a shaper the cutting time, min, is $T_t = L/fS$, where L = length of cut, in.; f = feed, in./stroke; S = strokes/min. Thus, for this shaper, $T_t = 12/[(0.05)(50)] = 4.8$ min. Multiply T_t by the number of strokes needed; the result is the total cutting time, min.

4. Compute the planer cycle time

The cycle time for a planer, min, $= (L/C) + (L/R_c) + k$, where $R_c =$ cutter return speed, fpm; $k =$ acceleration-deceleration constant. Since the cutting speed is 100 fpm and the return speed is 200 fpm, then the cycle time $= (12/100) + (12/200) + 0.05 = 0.23$ min,

5. Compute the power input to the planer

Table 7 lists typical power factors for planers planing cast iron and steel. To find the horsepower required, multiply the power factor by the cutter speed, fpm. For the planar in Step 3 with a cutting speed of 100 fpm and a power factor of 0.0235 for a $\frac{1}{8}$-in.-deep cut and a $\frac{1}{16}$-in. feed, $hp_{input} = (0.0235)(100) = 2.35$ hp,

For steel up to 40 points carbon, multiply the above result by 2; for steel above 40 points carbon, multiply by 2.25.

TABLE 7   Horsepower Factors for Planers*

Depth of cut, in.	Feed, in./stroke		
	$\frac{1}{32}$	$\frac{1}{16}$	$\frac{1}{8}$
$\frac{1}{8}$	0.0115	0.0235	0.047
$\frac{1}{4}$	0.023	0.047	0.094
$\frac{3}{8}$	0.035	0.070	0.141
$\frac{1}{2}$	0.047	0.094	0.189
$\frac{5}{8}$	0.063	0.118	0.236
$\frac{3}{4}$	0.080	0.142	0.284
$\frac{7}{8}$	0.087	0.165	0.331
1	0.094	0.189	0.378

*Excerpted from the Cincinnati Planer Company and *American Machinist's Handbook*.

Related Calculations: Where a shaper has a cutting stroke time that does not equal the return-stroke time, compute its cutting speed from $C = SL/(12)$(cutting-stroke time, min/sum of cutting- and return-stroke time, min). Thus, if the shaper in Step 1 has a cutting-stroke time of 0.8 min and a return-stroke time of 0.4 min, $C = (54)(6)/[(12) \times (0.8/1.2)] = 40.5$ fpm

GRINDING FEED AND WORK TIME

What is the feed of a centerless grinding operation if the regulating wheel is 8 in. in diameter and turns at 100 revolutions/min at an angle of inclination of 5°? How long will it take to rough grind on an external cylindrical grinder a brass shaft that is 3.0 in. diameter and 12 in. long, if the feed is 0.003 in., the spindle speed is 20 rpm, the grinding-wheel width is 3 in. and the diameter is 8 in., and the total stock on the part is 0.015 in? How long would it take to make a finishing cut on this grinder with a feed of 0.001 in., stock of 0.010 in., and a cutting speed of 100 fpm?

Calculation Procedure:

1. Compute the feed rate for centerless grinding

In centerless grinding, the feed in./min $f = \pi dR \sin \infty$, where $\pi = 3.1416$; $d =$ diameter of the regulating wheel, in.; $R =$ regulating wheel rpm; $\infty =$ angle of inclination of the regulating wheel, °. For this grinder, $f = \pi(8)(100)(\sin 5°) = 219$ in./min. Centerless grinders will grind as many as 50,000 1-in. parts per hr.

2. **Compute the rough-grinding time**

The rough-grinding time T_t min $= Lt_s d/2WfC$, where $L =$ length of ground part, in.; $t_s =$ total stock on part, in.; $d =$ diameter of part, in.; $W =$ width of grinding-wheel face, in.; $f =$ feed, in.; $C =$ cutting speed, fpm.

Compute the cutting speed first because it is not known. Using the method of previous Calculation Procedures, $C = \pi dR/12 = \pi(8)(20)/12 = 42$ fpm. Then, $T_t = (12)(0.015)(3)/[2(3)(0.003)(42)] = 0.714$ min.

3. **Compute the finish-grinding time**

For finish grinding, use the same equation as in Step 2, except that the factor 2 is omitted from the denominator. Thus, $T_t = Lt_s d/WfC$, or $T_t = (12)(0.010)(3)/[(3)(0.001)(100)] = 1.2$ min.

Related Calculations: Use the same equations as in Steps 1 and 4 for internal cylindrical grinding. In surface grinding, about 250 sq in./min can be ground 0.001 in. deep if the material is hard. For soft materials, about 1,000 sq in. and 0.001 in. deep can be ground per min.

In honing cast iron, the average stock removal is 0.006 to 0.008 in./(ft)(min). With hard steel or chrome plate, the rate of honing averages 0.003 to 0.004 in./(ft)(min).

BROACHING TIME AND PRODUCTION RATE

How long will it take to broach a medium-steel part if the cutting speed is 20 fpm, the return speed is 100 fpm, and the stroke length is 36 in? What will the production rate be if starting and stopping occupy 2 sec and loading 5 sec with an efficiency of 85 percent?

Calculation Procedure:

1. **Compute the broaching time**

The broaching time T_t min $= (L/C) + (L/R_c)$, where $L_f =$ length of stroke, ft; $C =$ cutting speed, fpm; $R_c =$ return speed, fpm; for this work, $T_t = (3/20) + (3/100) = 0.18$ min.

2. **Compute the production rate**

In a complete cycle of the broaching machine there are three steps: (a) broaching; (b) starting and stopping; (c) loading. The cycle time, at 100 percent efficiency, is the sum of these three steps, or $(0.18 \times 60) + 2 + 5 = 17.8$ sec, where the factor 60 converts 0.18 min to sec. At 85 percent efficiency, the cycle time is greater, or $17.8/0.85 = 20.9$ sec. Since there are 3,600 sec in an hr, production rate $= 3,600/20.9 = 172$ pieces per hr.

HOBBING, SPLINING, AND SERRATING TIME

How long will it take to hob a 36-tooth 12-pitch brass spur gear having a tooth length of 1.5 in. using a 2.75-in. hob? The whole depth of the gear tooth is 0.1798 in. How many teeth should the hob have? Hob feed is 0.084 in./revolution. What would the cutting time be for a 47° helical gear? How long will it take to spline hob a brass shaft which is 2.0 in. diameter, has 12 splines, each 10 in. long, if the hob diameter is 3.0 in., cutter feed is 0.050 in., cutter speed is 120 rpm, and spline depth is 0.15 in? How long will it take to hob 48 serrations on a 2-in.-diameter brass shaft if each serration is 2 in. long, the 18-flute hob is 2.5 in. in diameter, approach is 0.30 in., feed per flute is 0.008 in., and hob speed is 250 rpm?

Calculation Procedure:

1. **Compute the hob approach**

The hob approach A_c in. $= \sqrt{d_g(D_c - d_g)}$, where $d_g =$ whole depth of gear tooth, in.; $D_c =$ hob diameter, in. For this hob, $A_c = \sqrt{0.1798(2,7500 - 0.1798)} = 0.68$ in.

2. Determine the cutting speed of the hob

Table 8 shows that a cutting speed of $C = 150$ fpm is generally used for brass gears. With a 2.75-in.-diameter hob, this corresponds to a hob rpm of $R = 12C/\pi D_c = (12)(150)/[\pi(2.75)] = 208$ rpm.

TABLE 8 Gear-Hobbing Cutting Speeds

Spur gears		Helical gears[*]	
Gear material	Cutting speed, fpm	Angle, °	Percentage of feed to use
Brass.............	150	0–36	100
Fiber.............	150	36–48	80
Cast iron (soft)	100	48–60	67
Steel (mild)	100	60–70	50
Steel (medium)....	75	70–90	33
Steel (hard)	50		

[*]Reduce feed by percentage shown when cutting helical gears.

3. Compute the hobbing time

The time to hob a spur gear $T_t \min = N(L + A_c)/fR$, where $N =$ number of teeth in gear to be cut; $L =$ length of a tooth in the gear, in.; $A_c =$ hob approach, in.; $f =$ hob feed, in./revolution; $R =$ hob revolutions/min. For this spur gear, $T_t = (36)(1.5 + 0.68)/[(0.084)(208)] = 4.49$ min.

4. Compute the cutting time for a helical gear

Table 8 shows that the feed for a 47° helical gear should be 80 percent of that for a spur gear. Using the relation in Step 3, $T_t = (36)(1.5 + 0.68)/[(0.80)(0.084)(208)] = 5.61$ min.

5. Compute the time to spline hob

Use the same procedure as for hobbing. Thus, $A_c = \sqrt{d_g(D_c - d_g)} = \sqrt{0.15(3.0 - 0.15)} = 0.654$ in. Then, $T_t = N(L + A)/fR$, where $N =$ number of splines; $L =$ length of spline, in.; other symbols as before. For this shaft, $T_t = (12)(10 + 0.654)/[(0.05)(120)] = 21.3$ min.

6. Compute the time to serrate

The time to hob serrations $T_t \min = N(L + A)/fnR$, where $N =$ number of serrations; $L =$ length of serration, in.; $n =$ number of flutes on hob; other symbols as before. For this shaft, $T_t = (48)(2 + 0.30)/[(0.008)(18)(250)] = 3.07$ min.

TIME TO SAW METAL WITH POWER AND BAND SAWS

How long will it take to saw a rectangular piece of alloy-plate aluminum 6 in. wide and 2 in. thick if the length of cut is 6 in., the power hacksaw makes 120 strokes/min, and the average feed per stroke is 0.0040 in? What would the sawing time be if a band saw with a 200-fpm cutting speed, 16 teeth per in., and a 0.0003-in. feed per tooth is used?

Calculation Procedure:

1. Compute the sawing time for a power saw

For a power saw with positive feed, the time to saw $T_t \min = L/Sf$, where $L =$ length of cut, in.; $S =$ strokes/min of saw blade; $f =$ feed per stroke, in. In this saw, $T_t = (6)/[(120)(0.0040)] = 12.5$ min.

2. Compute the band-saw cutting time

For a band saw, the sawing time T_t min $= L/12Cnf$, where $L =$ length of cut, in.; $C =$ cutting speed fpm; $n =$ number of saw teeth per in.; $f =$ feed, in. per tooth. With this band saw, $T_t = (6)/[(12)(200)(16)(0.0003)] = 0.521$ min.

Related Calculations: When nested round, square, or rectangular bars are to be cut, use the greatest *width* of the nested bars as the length of cut in either of the above equations.

OXYACETYLENE CUTTING TIME AND GAS CONSUMPTION

How long will it take to make a 96-in.-long cut in a 1-in.-thick steel plate by hand and by machine? What will the oxygen and acetylene consumption be for each cutting method?

Calculation Procedure:

1. Compute the cutting time

For any flame cutting, the cutting time T_t min $= L/C$, where $L =$ length of cut, in.; $C =$ cutting speed, in./min, from Table 9. With manual cutting, $T_t = 96/8 = 12$ min, using the lower manual cutting speed given in Table 9. At the higher manual cutting speed, $T_t = 96/12 = 8$ min. With machine cutting, $T_t = 96/14 = 6.86$ min, using the lower machine cutting speed in Table 9. At the higher machine cutting speed, $T_t = 96/18 = 5.34$ min.

2. Compute the gas consumption

From Table 9 the oxygen consumption is 130 to 200 cfh. Thus, actual consumption, cu ft $=$ (cutting time, min/60) (consumption, cfh) $= (12/60)(130) = 26$ cu ft at the minimum cutting speed and minimum oxygen consumption. For this same speed with maximum oxygen consumption, actual cu ft used $= (12/60)(200) = 40$ cu ft.

Compute the acetylene consumption in the same manner, or $(12/60)(13) = 2.6$ cu ft, and $(12/60)(16) = 3.2$ cu ft. Use the same procedure to compute the acetylene and oxygen consumption at the higher cutting speeds.

TABLE 9   Oxyacetylene Cutting Speed and Gas Consumption

Metal thickness, in.	Speed, in./min		Gas consumption, cfh	
	Manual	Machine	Oxygen	Acetylene
0.25	16–18	20–26	50–90	8–11
0.50	12–15	17–22	90–125	10–13
1	8–12	14–18	130–200	13–16
2	5–7	10–13	200–300	16–20
4	4–5	7–9	300–400	21–26
6	3–4	5–7	400–500	26–32
8	3–6	4–6	500–650	28–35
10	2–3	3–4	700–1,000	30–38
12	2.5–3.5	3–4	720–880	42–52

Related Calculations: Use the procedure given here for computing the cutting time and gas consumption when cutting steel, wrought iron, or cast iron. Thicknesses ranging up to 5 ft are economically cut using an oxyacetylene torch. Alloying elements in steel may require preheating of the metal to permit cutting. To compute the gas required per lineal foot, divide the actual consumption for the length cut, in in., by 12.

COMPARISON OF OXYACETYLENE AND ELECTRIC-ARC WELDING

Determine the time required to weld a 4-ft-long seam in a $\frac{3}{8}$-in.-thick plate using the oxyacetylene and electric-arc methods. How much oxygen and acetylene are required? What weight of electrode will be used? What is the electric-power consumption? Assume that one weld bead is run in the joint.

Calculation Procedure:

1. Compute the welding time

For any welding operation, the time required to weld T_t min $= L/C$, where $L =$ length of weld, in.; $C =$ welding speed, in./min. When oxyacetylene welding is used, $T_t = 48/1.0 = 48$ min, when a welding speed of 1.0 in./min is used. With electric-arc welding, $T_t = 48/18 = 2.66$ min when the welding speed is 18 in./min per bead. For plate thicknesses under 1 in., typical welding speeds are in the 1 to 2 in./min range for oxyacetylene and 18 in./min for electric arc. For thicker plates, consult *The Welding Handbook*, American Welding Society.

2. Compute the gas consumption

Gas consumption for oxyacetylene welding is given in cu ft per ft of weld. Using values from *The Welding Handbook*, or a similar reference, oxygen consumption = (cu ft O_2 per ft of weld)(length of weld, ft); acetylene consumption = (cu ft acetylene per ft of weld)(length of weld, ft). For this weld, with only one bead, oxygen consumption = (10.0)(4) = 40 cu ft; acetylene consumption = (9.0)(4) = 36 cu ft.

3. Compute the weight of electrode required

The Welding Handbook tabulates the weight of electrode for various types of welds — square grooves, 90° V-grooves, etc., per ft of weld. Then, electrode weight required, lb = (rod consumption, lb per ft of weld)(weld length, ft).

For oxyacetylene welding, the electrode weight required, using data from *The Welding Handbook*, is (0.597)(4) = 2.388 lb. For electric-arc welding, weight = (0.18)(4) = 0.72 lb.

4. Compute the electric-power consumption

In electric-arc welding the power consumption, kw = (v)(amp)/(1,000)(efficiency). *The Welding Handbook* shows that for a $\frac{3}{8}$-in.-thick plate, v = 40, amp = 450, efficiency = 60 percent. Then, power consumption = (40)(450)/[(1,000)(0.60)] = 30 kw.

Related Calculations: Where more than one pass or bead is required, multiply the time for one bead by the number of beads deposited. If only 50 percent penetration is required for the bead, the welding speed will be twice that where full penetration is required.

PRESSWORK FORCE FOR SHEARING AND BENDING

What is the press force required to shear an 8-in.-long 0.5-in.-thick piece of annealed bronze having a shear strength of 16.0 tons/sq in? What is the stripping load? Determine the force required to produce a U bend in this piece of bronze if the unsupported length is 4 in., the bend length is 6 in., and the ultimate tensile strength is 32.0 tons/sq in.

Calculation Procedure:

1. Compute the required shearing force

For any metal in which a straight cut is made, the required shearing force, tons $= F = Lts$, where $L =$ length of cut, in.; $t =$ metal thickness, in.; $s =$ shear strength of metal being cut, tons/sq in. Where round, elliptical, or other shaped holes are being cut, substitute the sum of the circumferences of all the holes for L in this equation. For this press, $F = (8)(0.5)(16.0) = 64$ tons.

2. Compute the stripping load

For the typical press, the stripping load is 3.5 percent of the required shearing force, or $(0.035)(64) = 2.24$ tons.

3. Compute the required bending force

When U bends or channels are pressed in a metal, $F = 2Lt^2s_t/W$, where $s_t =$ ultimate tensile strength of the metal, tons/sq in.; $W =$ width of unsupported metal, in. = distance between the vertical members of a channel or U bend, measured to the *outside* surfaces, in. For this U bend, $F = 2(6)(0.5)^2(32)/4 = 24$ tons.

Related Calculations: Right-angle edge bends require a bending force of $F = Lt^2s_t/2W$, while free V bends with a centrally located load require a bending force of $F = Lt^2s_t/W$. All symbols are the same as given in Steps 1 and 2.

MECHANICAL-PRESS MIDSTROKE CAPACITY

Determine the maximum permissible midstroke capacity of single- and twin-driven 2-in.-diameter crankshaft presses if the stroke of the slide is 12 in. for each.

Calculation Procedure:

1. Compute the single-driven press capacity

For a single-driven crankshaft press with a heat-treated 0.35 to 0.45 percent carbon-steel crankshaft having a shear strength of 6 tons/sq in., the maximum permissible midstroke capacity F tons $= 2.4d^3/S$, where $d =$ shaft diameter at main bearing, in.; $S =$ stroke length, in., or $F = (2.4)(2)^3/12 = 1.6$ tons.

2. Compute the twin-driven press capacity

Twin-driven presses with main (bull) gears on each end of the crankshaft have a maximum permissible midstroke capacity of $F = 3.6d^3/S$, when the shaft shearing strength is 9 tons/sq in. For this press, $F = 3.6(2)^3/12 = 2.4$ tons.

Related Calculations: Use the equation in Step 2 to compute the maximum permissible midstroke capacity of all wide (right-to-left) double-crank presses. Since gear eccentric presses are built in competition with crankshaft presses, their midstroke pressure capacity is within the same limits as in crankshaft presses. The diameters of the fixed pins on which the gear eccentrics revolve are usually made the same as the crankshaft in crankshaft presses of the same rated capacity.

STRIPPING SPRINGS FOR PRESSWORKING METALS

Determine the force required to strip the work from a punch if the length of cut is 5.85 in. and the stock is 0.25 in. thick. How many springs are needed for the punch if the force per inch deflection of the spring is 100 lb?

Calculation Procedure:

1. Compute the required stripping force

The required stripping force F_p lb needed to strip the work from a punch is $F_p = Lt/0.00117$, where $L =$ length of cut, in.; $t =$ thickness of stock cut, in. For this punch, $F_p = (5.85)(0.25)/0.00117 = 1,250$ lb.

2. Compute the number of springs required

Only the first $\frac{1}{8}$-in. deflection of the spring can be used in the computation of the stripping force produced by the spring. Thus, for this punch, number of springs required = stripping force, lb/force, lb, to produce $\frac{1}{8}$-in. deflection of the spring, or $1,250/100 = 12.5$ springs. Since a fractional number of springs cannot be used, 13 springs would be selected.

Related Calculations: In high-speed presses, the springs should not be deflected more than 25 percent of their free length. For heavy, slow-speed presses, the total deflection should not exceed 37.5 percent of the free length of the spring. The stripping force for aluminum alloys is generally taken as one-eighth the maximum blanking pressure.

BLANKING, DRAWING, AND NECKING METALS

What is the maximum blanking force for an aluminum part if the length of the cut is 30 in., the metal is 0.125 in. thick, and the yield strength is 2.5 tons/sq in? How much force is required to draw a 12-in.-diameter 0.25-in.-thick stainless steel shell if the yield strength is 15 tons/sq in? What force is required to neck a 0.125-in.-thick aluminum shell from a 3- to a 2-in. diameter if the necking angle is 30° and the ultimate compressive strength of the material is 14 tons/sq in?

Calculation Procedure:

1. Compute the maximum blanking force

The maximum blanking force for any metal is given by $F = Lts$, where F = blanking force, tons; L = length of cut, in. (= circumference of part, in.); t = metal thickness, in.; s = yield strength of metal, tons/sq in. For this part, $F = (30)(0.125)(2.5) = 0.375$ tons.

2. Compute the maximum drawing force

Use the same equation as in Step 1, substituting the drawing-edge length or perimeter (circumference of part) for L. Thus, $F = (12\pi)(0.25)(15) = 141.5$ tons.

3. Compute the required necking force

The force required to neck a shell is $F = ts_c(d_1 - d_s)/\cos$ necking angle, where F = necking force, tons; t = shell thickness, in.; s_c = ultimate compressive strength of the material, tons/sq in.; d_1 = large diameter of shell, i.e., the diameter *before* necking, in.; d_s = small diameter of shell, i.e., the diameter *after* necking, in. For this shell, $F = (0.125)(14)(3.0 - 2.0)/\cos 30° = 2.02$ tons.

Related Calculations: Table 10 presents typical yield strengths of various metals which are blanked or drawn in metalworking operations. Use the given strength as shown above.

TABLE 10 Metal Yield Strength

Metal	Yield Strength, tons/sq in.
Aluminum, 2S annealed	2.5
Aluminum, 24S heat-treated	23.0
Low brass, $\frac{1}{4}$ hard	24.5
Yellow brass, annealed	10.0
Cold-rolled steel, $\frac{1}{4}$ hard	16.0
Stainless steel, 18-8	15.0

NOTE: As a general rule, the necking angle should not exceed 35°.

METAL PLATING TIME AND WEIGHT

How long will it take to electroplate a 0.004-in.-thick zinc coating on a metal plate if a current density of 25 amp/sq ft is used at an 80 percent plating efficiency? How much zinc is required to produce a 0.001-in.-thick coating on an area of 60 sq ft?

Calculation Procedure:

1. Compute the metal plating time

The plating time T_p min $= 60\,An/A_a e$, where $A =$ amp/sq ft required to deposit 0.001 in. of metal at 100 percent cathode efficiency; $n =$ number of thousandths of in. actually deposited; $A_a =$ current actually supplied, amp/sq ft; $e =$ plating efficiency, expressed as a decimal. Table 11 gives typical values of A for various metals used in electroplating.

TABLE 11   Electroplating Current and Metal Weight

Metal	Amp-hr to deposit 0.001 in./sq ft at 100 percent efficiency	Metal density, lb/in.^{3}
Antimony, Sb	10.40	0.241
Cadmium, Cd	9.73	0.312
Chromium, Cr	51.80	0.256
Cobalt(ous), Co	19.00	0.322
Copper(ous), Cu	8.89	0.322
Copper(ic), Cu	17.80	0.322
Gold(ous), Au	6.20	0.697
Gold(ic), Au	18.60	0.697
Nickel, Ni	19.00	0.322
Platinum, Pt	27.80	0.775
Silver, Ag	6.20	0.380
Tin(ous), Sn	7.80	0.264
Tin(ic), Sn	15.60	0.264
Zinc, Zn	14.30	0.258

For plating zinc, using the value in Table 11, $T_p = 60(14.3)(4)/[(25)(0.80)] = 171.5$ min, or $171.5/60 = 2.86$ hr.

2. Compute the weight of metal required

The plating metal weight $=$ (area plated, sq in.) (plating thickness, in.)(plating metal density, lb/in.^{3}). For this plating job, using the density of zinc from Table 11, the plating metal weight $= (60 \times 144)(0.004)(0.258) = 8.91$ lb of zinc. In this calculation the value 144 is used to convert 60 sq ft to sq in.

Related Calculations: The efficiency of finishing cathodes is high, ranging from 80 percent to nearly 100 percent. Where the actual efficiency is unknown, assume a value of 80 percent and the results obtained will be safe for most situations.

SHRINK- AND EXPANSION-FIT ANALYSES

To what temperature must an SAE 1010 steel ring 24 in. in inside diameter be raised above a 68-F room temperature to expand it 0.004 in. if the linear coefficient of expansion of the steel is 0.0000068 in./(in.)(F)? To what temperature must a 2-in.-diameter SAE steel shaft be reduced to fit it into a 1.997-in.-diameter hole for an expansion fit? What cooling medium should be used?

Calculation Procedure:

1. Compute the required shrink-fit temperature rise

The temperature needed to expand a metal ring a given amount before making a shrink fit is given by $T = E/Kd$, where $T =$ temperature rise above room temperature, F; $K =$ linear coefficient of expansion of the metal ring, in./(in.)(F); $d =$ ring internal diameter, in. For this ring, $T = 0.004/[(0.0000068)(24)] = 21.5$ F. With a room temperature of 68 F, the final temperature of the ring must be $68 + 21.5 = 89.5$ F, or higher.

2. Compute the temperature required for an expansion fit

Nitrogen, air, and oxygen in liquid form have a low boiling point, as does dry ice (solid carbon dioxide). Nitrogen and dry ice are considered the safest cooling media for expansion fits because both are relatively inert. Liquid nitrogen boils at -320.4 F and dry ice at -109.3 F. At -320 F, liquid nitrogen will reduce the diameter of metal parts by the amount shown in Table 12. Dry ice will reduce the diameter by about one-third the values listed in Table 12.

Using liquid nitrogen, the diameter of a 2-in. round shaft will be reduced by (2.0) $(0.0022) = 0.0044$ in., using the value for SAE steels from Table 12. Thus, the diameter of the shaft at -320.4 F will be $2.0000 - 0.0044 = 1.9956$ in. Since the hole is 1.997 in. in diameter, the liquid nitrogen will reduce the shaft size sufficiently.

If dry ice were used, the shaft diameter would be reduced $0.0044/3 = 0.00146$ in., giving a final shaft diameter of $2.00000 - 0.00146 = 1.99854$ in. This is too large to fit into a 1.997-in.-diameter hole. Thus, dry ice is unsuitable as a cooling medium.

TABLE 12   Metal Shrinkage with Nitrogen Cooling

Metal	Shrinkage, in. per in. of Shaft Diameter
Magnesium alloys	0.0046
Aluminum alloys	0.0042
Copper alloys	0.0033
Cr=Ni alloys (18-8 to 18-12)	0.0029
Monel metals	0.0023
SAE steels	0.0022
Cr steels (5 to 27% Cr)	0.0019
Cast iron (not alloyed)	0.0017

PRESS-FIT FORCE, STRESS, and SLIPPAGE TORQUE

What force is required to press a 4-in. outside-diameter-cast-iron hub on a 2-in. outside-diameter steel shaft if the allowance is 0.001 in. interference per in. of shaft diameter, the length of fit is 6 in., and the coefficient of friction is 0.15? What is the maximum tensile stress at the hub bore? What torque is required to produce complete slippage of the hub on the shaft?

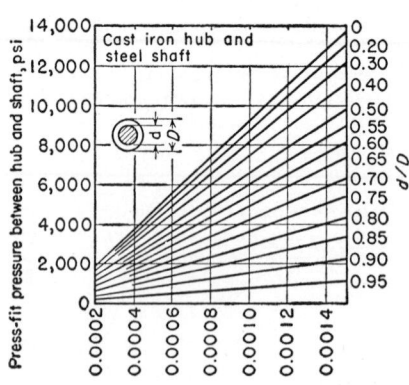

Fig. 5 **Press-fit pressures between steel hub and shaft.**

Calculation Procedure:

1. Determine the unit press-fit pressure

Figure 5 shows that with an allowance of 0.001 in. interference per in. of shaft diameter and a shaft-to-hub diameter ratio of $2/4 = 0.5$, the unit press-fit pressure between the hub and the shaft is $p = 6,800$ psi.

2. Compute the press-fit force

The press-fit force F tons $= \pi f p d L/2,000$, where $f =$ coefficient of friction between hub and shaft; p = unit press-fit pressure, psi; d = shaft diameter, in.; L = length of fit, in. For this press fit, $F = (\pi)(0.15)(6,800)(2.0)(6)/2,000 = 19.25$ tons.

3. Determine the hub bore stress

Use Fig. 6 to determine the hub bore stress. Enter the bottom of Fig. 6 at 0.0010 in. interference allowance per in. of shaft diameter and project vertically to $d/D = 0.5$. At the left read the hub stress as 11,600 psi.

4. Compute the slippage torque

The torque, in.-lb, required to produce complete slippage of a press fit is $T = 0.5$ $\pi f p L d^2$, or $T = 0.5(3.1416)(0.15)(6,800)(6)(2)^2 = 38,450$ in.-lb.

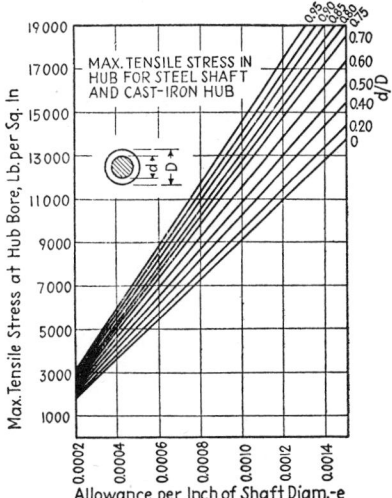

Fig. 6 Variation in tensile stress in cast-iron hub in press-fit allowance.

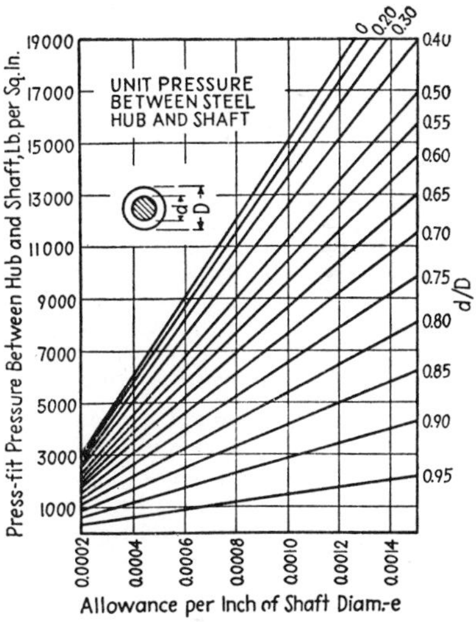

Fig. 7 Press-fit pressures between cast-iron hub and shaft.

Related Calculations: Figure 7 shows the press-fit pressures existing with a steel hub on a steel shaft. The three charts presented in this Calculation Procedure are useful for many different press fits, including those using a hollow shaft having an internal diameter less than 25 percent of the external diameter and for all solid steel shafts.

LEARNING-CURVE ANALYSIS AND CONSTRUCTION

A short-run metalworking job requires five operators. The longest individual learning time for the new task is 3 days; 2 days are allowed for group familiarization with the task. If the normal output is 1,000 units per 8-hr day, determine the daily allowance per operator when the standard for 100 percent performance is 0.8 man-hour per 100 units produced.

Calculation Procedure:

1. Plot the learning curve

A learning curve shows the improvement that occurs with repetition of a task. Figure 8 is a typical learning curve with the learning period, days, plotted against the percent of methods-time-measurement (MTM) determined normal task. The shape of the curve, once determined for a given operation, does not change. The horizontal scale

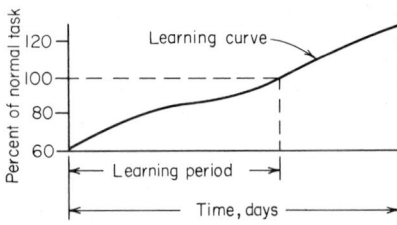

Fig. 8 Typical learning curve for a metal-working task.

division is, however, changed to suit the minimum learning period for 100 percent performance. Thus, for a 3-day learning period the horizontal scale becomes 3 days. The coordinate at each of these three points (i.e., days) becomes the minimum expected task for each day. Performance above these tasks rates a bonus. The base of 60 percent of normal performance for the first day of learning for all jobs is attainable and meets management's minimum requirements.

2. Determine the learning period to allow

(a) Find the learning time, by test, for each work station in the group. (b) Select the longest individual learning time—in this instance, 3 days. (c) Add a group familiarization allowance when the group exceeds three operators—2 days here. (d) Find the sum of $b + \bar{c}$, or $3 + 2 = 5$ days. This is the learning period to allow.

3. Find the task for each day

Divide the horizontal learning-period axis into five parts, one part for each of the 5 learning-period days allowed. Draw an ordinate for each day and read the percent task for that day at the intersection with the learning curve, or: 60.0; 70.5; 75.5; 80.0; 87 percent for days 1, 2, 3, 4, and 5, respectively.

4. Compute the daily task and daily allowance

With a normal (100 percent) task of 1,000 units for an 8-hr day, set up a table (like Table 13) of daily tasks and time allowance during the learning period. Begin with a column listing the number of learning days. In the next column, list the percent learning performance read from Fig. 8. Find the daily task in units, column 3 of Table 13, by taking the product of 1,000 units and the percent learning performance, column 2, expressed as a decimal. Lastly, compute the daily allowance per operator by finding the product of 8 hr and the difference between 1.00 and the percent learning performance; i.e., for day 1: $(1.00 - 0.60)(8) = 3.2$ hr. Tabulate the results in the fourth column of Table 13.

TABLE 13   Learning-Curve Analysis

Learning days	Percent learning performance	Daily task, units	Daily allowance per operator, in units
1	60.0	600	$40\% \times 8 = 3.20$
2	70.5	705	$29.5\% \times 8 = 2.36$
3	75.5	755	$24.5\% \times 8 = 1.96$
4	80.0	800	$20.0\% \times 8 = 1.60$
5	87.0	870	$13.0\% \times 8 = 1.04$
6	100.0	1,000	$0\% \times 8 = 0$

5. Compute the incentive pay for the group

In this plant the incentive pay is found by taking the product of the production in units and the standard set for 100 percent performance, or 0.80 man-hour per 100 units produced. Thus, production of 600 units on day 1 will earn $(600/100)(0.80) = 4.8$-hr pay for each group. Add to this the learning allowance of 3.2 hr for day 1 and each group has earned $4.8 + 3.2 = 8$-hr pay for 8-hr work.

If the group produced 700 units during day 1, it would earn $(700/100)(0.80) = 5.6$-hr pay at this standard. With the learning allowance of 3.2 hr, the daily earnings would be

5.6+3.2 = 8.8-hr pay for 8-hr work. This is exactly what is desired. The operator is rewarded for learning quickly.

Related Calculations: Select the length of the learning period for any new short-run task by conferring with representatives of the manufacturing, industrial engineering, and industrial relations departments. A simple operation that will be performed 1,000 to 2,000 times in an 8-hr period would require a 3-day learning period. This is considered the minimum time for bringing such an operation up to normal speed. This is also true if a small group (three or less operators) perform equally simple operations. With larger groups (four or more operators), both simple and complex operations require an additional allowance for operators to adjust themselves to each other. Two days is a justified allowance for up to 15 operators learning to cooperate with one another under incentive conditions.

To prepare a plant-wide learning curve, keep records of the learning rates for a number of short-run tasks. Combine these data to prepare a typical learning curve for a particular plant. The method developed here was first described in *Factory*, now *Modern Manufacturing*, magazine.

LEARNING-CURVE EVALUATION OF MANUFACTURING TIME

A metalworking process requires 1.00 hr for manufacture of the first unit of a production run. If the operator has an improvement or learning rate of 90 percent, determine the time required to manufacture the 2nd, 4th, 8th, and 16th units. What is the cummulative average unit time for the 16th unit? If 100 units are manufactured, what is the cummulative average time for the 100th unit? What is the unit manufacturing time for the 100th item?

Calculation Procedure:

1. Compute the unit time for the production cycle

The learning curve relates the production time to the number of units produced. When the number of units produced doubles, the time required to produce the unit representing the doubled quantity is: (Learning rate, percent)(time, hr or min, to produce the unit representing one-half the doubled quantity). Or, for the production line being considered here:

Unit Number	Production Time, hr
1	1.00
2	0.90(1.00) = 0.900
4	0.90(0.90) = 0.810
8	0.90(0.81) = 0.729
16	0.90(0.729) = 0.656

2. Compute the cumulative average unit time

The cumulative average unit time for any unit in a production run = (Σ unit time for each item in the run)/(number of items in the run). Thus, computing the time for items 1 through 16 as shown in Step 1, and taking the sum, the cumulative average unit time = 12.044 hr/16 units = 0.752 hr.

3. Compute the cumulative average time for the 100th unit

Set up a ratio of the learning factor for the 100th unit/learning factor for the 16th unit, and multiply the ratio by the cumulative average 16th unit time. Or, using the factors from Table 14, (0.497/0.656)(0.752 hr) = 0.570 hr.

4. Compute the unit time for the 100th unit

Using the factor for the 90 percent learning curve in Table 14, the unit time for the 100th unit made = (1.00 hr)(0.497) = 0.497 hr.

Related Calculations: When using learning curves, be extremely careful to distinguish between *unit time* and *cumulative average unit time*. The unit time is the time required to make a particular unit in a production run — say the 10th, 16th, etc. Thus, a unit time of 0.5 hr for the 16th unit in a production run means that the time required to make the 16th unit is 0.5 hr. The 15th unit will require *more* time to make it; the 17th unit will require *less* time.

TABLE 14   Learning-Curve Factors

No. of units	Learning rate, percent		
	85	90	95
	Time or cost, percent of unit 1	Time or cost, percent of unit 1	Time or cost, percent of unit 1
1	1.000	1.000	1.000
2	0.850	0.900	0.950
4	0.723	0.810	0.903
8	0.614	0.729	0.857
16	0.522	0.656	0.815
32	0.444	0.591	0.774
64	0.377	0.531	0.735
100	0.340	0.497	0.711

Learning-curve slopes

Learning rate, percent	Curve slope
70	−0.514
75	−0.415
80	−0.322
85	−0.234
90	−0.152
95	−0.074

The cumulative average unit time is the *average* time to manufacture a given number of identical items. To obtain the cumulative average unit time for any given number of items, take the sum of the time required for each item up to and including that item and divide the sum by the number of items.

Either the *unit time or cumulative average unit time* can be used in manufacturing time or cost estimates, so long as the estimator knows which time value he is using. Failure to recognize the respective time values can result in serious errors.

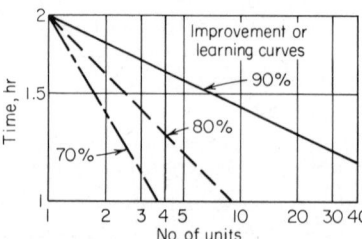

Fig. 9   Learning curves plotted on log–log scale.

A learning curve plotted on log-log coordinates is a straight line, Fig. 9. The slope of typical learning curves is listed in Table 14. Since a learning curve slopes downward — i.e., the unit manufacturing time decreases as more units are produced — the slope is expressed as a negative value.

Typical improvement or learning rates are: machining, drilling, etc., 90 to 95 percent; short-cycle bench assembly, 85 to 90 percent; equipment maintenance, 75 to 80 percent; electronics assembly and welding, 80 to 90 percent; general assembly, 70 to 80 percent. When an operation consists of several tasks having different learning rates, compute the overall learning rate for the task by taking the sum of the product of each learning rate (LR) and the percentage of the total task it represents. Thus, with $LR_1 = 0.90$ for 60 percent of the total task; $LR_2 = 0.80$ for 20 percent of the total task; $LR_3 = 0.70$ for 10 percent of the total task, the overall learning rate $LR = 0.90(0.60) + 0.80(0.20) + 0.70(0.10) = 0.77$.

Note that in machine-paced operations—i.e., those in which the speed of the machine controls the operator's activities—there is less chance for the operator to learn. Hence, the learning rate will be higher—90 to 95 percent—than in worker-paced operations that have learning rates of 70 to 80 percent. When learning or improvement ceases, the operator has reached the level-off point and the task cannot be performed any more rapidly. The ratio set up in Step 3 can use any two items in a production run, provided that the cumulative average time for the smaller item is multiplied by the ratio.

DETERMINING BRINELL HARDNESS

A 3,000-kg load is put on a 10-mm diameter ball to determine the Brinell hardness of a steel. The ball produces a 4-mm-diameter indentation in 30 sec. What is the Brinell hardness of the steel?

Calculation Procedure:

1. Determine the Brinell hardness using an exact equation

The standard equation for determining the Brinell hardness is $BHN = F/(\pi d_1/2)(d_1 - \sqrt{d_1^2 - d_s^2})$, where $F =$ force on ball, kg; $d_1 =$ ball diameter, mm; $d_s =$ indentation diameter, mm. For this test, $BHN = 3,000/(\pi \times 10/2)(10 - \sqrt{10^2 - 4^2}) = 229$.

2. Compute the Brinell hardness using an approximate equation

One useful approximate equation for Brinell hardness is $BHN = (4F/\pi d_s^2) - 10$. For this test, $BHN = (4 \times 3,000/\pi \times 4^2) - 10 = 228.5$. This compares favorably with the exact formula. For Brinell hardness exceeding 200, the approximate equation gives results that are less than 0.1 percent in error.

Related Calculations: Use this procedure for iron, steel, brass, bronze, and other hard or soft metals. A 500-kg test load is used for soft metals (brass, bronze, etc.). For Brinell hardness above 500, use a tungsten-carbide ball. The metal tested should be at least 10 times as thick as the indentation depth and wide enough so that no metal flows towards the edges of the specimen. The metal surface must be clean and free of defects.

ECONOMICAL CUTTING SPEEDS AND PRODUCTION RATES

A cutting tool used to cut beryllium costs $6 with its shank and can be reground for reuse five times. The average tool-changing time is 5 min. What is the most economical cutting speed if the machine labor rate is $3/hr and the overhead is 200 percent? What is the cutting speed for the maximum production rate? The cost of regrinding the tool is 35 cents per edge.

Calculation Procedure:

1. Determine the tool cost factor

The cost factor $(T_c + Y/X)$ for a tool is composed of $T_c =$ time to change tool, min; $Y =$ tool cost per cutting edge, including prorated initial cost plus reconditioning costs, cents; $X =$ machining rate, including labor and overhead, cents/min.

For this tool, $T_c = 5$ min. The tool can be reground five times after its original use, giving a total of $5 + 1 = 6$ cutting edges (five regrindings + the original edge) during its life. Since the tool costs $6 new, the prorated cost per edge = $6/6 edges = $1, or 100 cents. The regrinding cost = 35 cents per edge; thus $Y = 100 + 35 = 135$ cents per edge.

With a machine labor rate of \$3/hr and an overhead factor of 200 percent, or 2.00(\$3) = \$6, the value of X = machining rate = \$3 + \$6 = \$9, or 900 cents/hr, or 900/60 = 15 cents/min. Then, the cost factor $(T_c + Y/X) = (5 + 135/15) = 14$.

2.   Determine the cutting speed for minimum cost

Enter Fig. 10 at a cost factor of 14 and project vertically upward until the cutting-speed curve is intersected. At the left, read the cutting speed for minimum tool cost as 320 surface fpm.

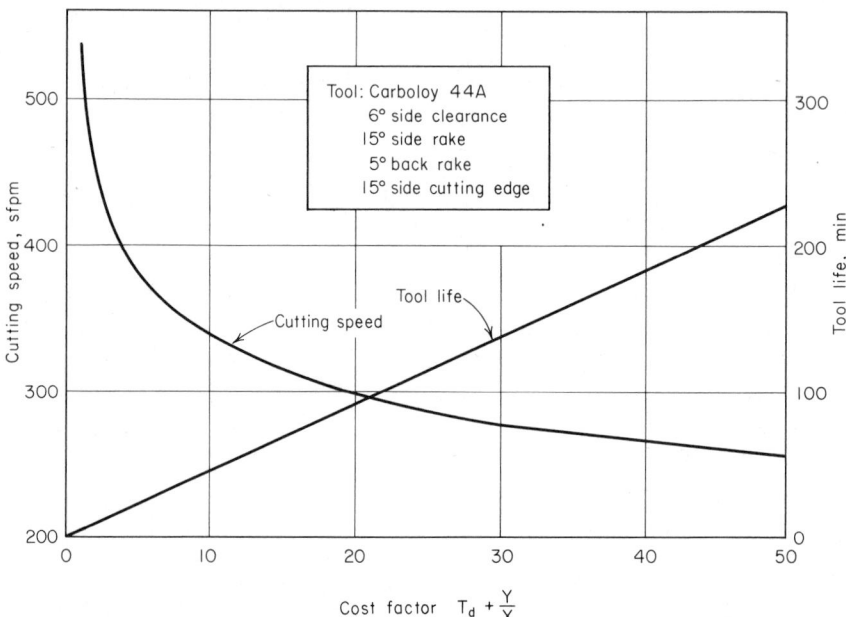

Fig. 10   Optimum cutting-speed chart. (*American Machinist.*)

3.   Determine the probable tool life

Project upward from the cost factor of 14 in Fig. 10 to the tool-life curve. At the right, read the tool life as 66 min.

4.   Determine the speed and life for the maximum production rate

Substitute the value of T_c for the cost factor $(T_c + Y/X)$ on the horizontal scale of Fig. 10. As before, read the cutting speed and tool life at the intersection with the respective curves. The plotted values apply when the chip-removal suction devices will operate efficiently at the cutting speeds indicated by the curves. Thus, with $T_c = 5$, the cutting speed is 370 surface fpm, and the tool life is 30 min.

Related Calculations: Figure 10, and similar optimum cutting-speed charts, are plotted for a specific land wear—in this case 0.010 in. For a land wear of 0.015 in., multiply the cutting speeds obtained from Fig. 10 by 1.13. However, a land wear of 0.015 in. is not recommended because the wear rates are accelerated. If Carboloy 883 tools are used in place of the 44A grade plotted in Fig. 10, multiply the cutting speeds obtained from this chart by 1.12 for a 0.010-in. wear land, or 1.26 for a 0.015-in. wear land.

Charts similar to Fig. 10 for other tool materials can be obtained from tool manufacturers. Do not use Fig. 10 for any tool material other than Carboloy 44A. The method presented here is the work of D. R. Walker and J. Gubas, as reported in *American Machinist.*

OPTIMUM LOT SIZE IN MANUFACTURING

A manufacturing plant has a demand for 900 of its products per month on which the setup cost is $10. The cost of each unit is $5; the annual inventory charge is 12 percent/year of the average $ value held in stock; the period for which the demand has occurred is 1/12 year. What is optimum manufacturing lot size? Plot a cost chart for this plant.

Calculation Procedure:

1. Determine the optimum manufacturing lot size

Optimum manufacturing lot size can be found from: {2(demand, units, per period) (cost per setup, $)/[(demand period, fraction of a yr)($ cost per unit)(annual inventory charge, percent of average $ value held in stock)]}$^{0.5}$. For this run, optimum lot size = {2(900)($10)/(1/12)($5)(0.12)]}$^{0.5}$ = 600 units.

2. Plot a cost chart for this plant

Figure 11 shows a typical cost chart. Plot each curve using production runs of 100, 200, 300, 400, ..., 1,600 units. The values for each curve are determined from: inven-

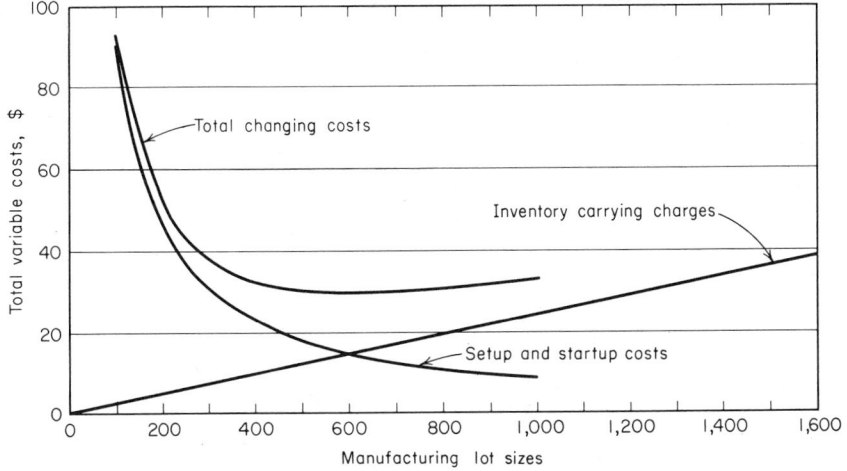

Fig. 11   Changing costs associated with manufacturing lot sizes. (*American Machinist.*)

tory carrying charges = (number of units in run)($ cost per unit)(annual inventory charge, percent)(demand period)/2; setup and startup costs = (demand during period, units) ($ cost of setup)/(number of units in run); total changing costs = inventory carrying charges + setup and startup costs.

The units for these equations are the same as given in Step 1. Note that the total-changing-costs curve is a minimum at the point where the inventory-carrying-charges curve and setup-and-startup-costs curve intersect. Also, the two latter curves intersect at the optimum manufacturing lot size — 600 units, as computed in Step 1.

Related Calculations: Economical lot-size relations are readily adaptable to machine-shop computations. With only slight changes, the same principles can be applied to determination of optimum-quantity purchases. The procedure described here is the work of I. Heitner, as reported in *American Machinist.*

PRECISION DIMENSIONS AT VARIOUS TEMPERATURES

A magnesium workpiece with a dimension of 12.5000 to 12.4996 in. is at a temperature of 85 F after machining. The steel gage with which the dimensions of the workpiece will be checked is at 75 F. The workpiece must be gaged immediately to determine if further grinding is necessary. Tolerance on the work is ± 0.0002 in. What should the dimensions of the workpiece be if there is not enough time available to allow the gage and workpiece temperatures to equalize? The standard reference temperature is 68 F.

Calculation Procedure:

1. Compute the actual work dimension

The temperature of the workpiece is 85 F, or 85 F $-$ 68 F $= 17$ F above the standard reference temperature of 68 F. Since the actual temperature of the workpiece is greater than the standard temperature, the dimensions of the workpiece will be larger than at the standard temperature because the part expands as its temperature increases.

To find the amount by which the workpiece will be oversize at the actual temperature, multiply the nominal dimension of the workpiece, 12.5 in., by the coefficient of linear expansion of the material and by the difference between the actual and standard temperatures. For this magnesium workpiece, the average oversize amount at the actual temperature, 85 F, is $(12.5)(14.4 \times 10^{-6})(85 - 68) = 0.003060$ in.

2. Compute the actual gage dimension

Compute the actual gage dimension in a similar way, using the same dimension, 12.5 in., but the coefficient of linear expansion of the gage material, steel, and the gage temperature, 75 F. Or, $(12.5)(6.4 \times 10^{-6})(75 - 68) = 0.000560$ in.

3. Compute the workpiece dimension to check

The workpiece dimension, corrected for tolerance, plus the difference between the oversize amounts computed in Steps 1 and 2 is the dimension to which the part should be checked at the existing shop and gage temperature.

Applying the tolerance, ± 0.0002 in., to the drawing dimension, $12.5000 - 12.4996$, gives a drawing dimension of 12.4998 ± 0.0002 in. Adding the difference between the oversize dimensions, or $0.003060 - 0.000560 = 0.002500$ to 12.4998 ± 0.0002, gives a checking dimension of 12.5023 ± 0.0002 in. If shop personnel check the workpiece at this dimension they will have full confidence that it will be the right size.

Related Calculations: This procedure can be used for any metal — bronze, aluminum, cast iron, etc. — for which the coefficient of linear expansion is known. Obtain the coefficient from Baumeister and Marks — *Standard Handbook for Mechanical Engineers* or a similar reference. When a workpiece is at a temperature *less than* the National Bureau of Standards standard of 68 F, the part contracts instead of expanding. The dimension change computed in Step 1 is then negative. This is also true of the gage, if it is at a temperature of less than 68 F. Note that the tolerance is constant regardless of the actual temperature of the part.

The procedure given here is the work of H. K. Eitelman, as reported in *American Machinist*.

HORSEPOWER REQUIRED FOR METALWORKING

What is the input horsepower required for machining, on a geared-head lathe, a 4-in.-diameter piece of AISI 4140 steel having a hardness of 260 BHN if the depth of cut is 0.25 in., the cutting speed is 300 fpm, and the feed per revolution is 0.025 in?

Calculation Procedure:

1. Determine the metal removal rate

Compute the metal removal rate MRR in.^{3}/min from MRR $= 12fDC$, where $f =$ tool feed rate, in./revolution; $D =$ depth of cut, in.; $C =$ cutting speed, fpm. For this workpiece, MRR $= 12(0.025)(0.25)(300) = 22.5$ in.^{3}/min.

2. Determine the unit horsepower required

Table 15 lists the average unit horsepower required for cutting various metals. The unit horsepower hp_u is the power required to remove 1 in.^{3} of metal per min at 100 per-

TABLE 15   Average Unit Hp Factors*

	Ferrous metals and alloys					
Material classification	Brinell hardness number					
	150–175	176–200	201–250	251–300	301–350	351–400
AISI 1010–1025	0.58	0.67				
AISI 1030–1055	0.58	0.67	0.80	0.96		
AISI 1060–1095	...	...	0.75	0.88	1.0	
AISI 1112–1120	0.50					
AISI 1314–1340	0.42	0.46	0.50			
AISI 1330–1350	...	0.67	0.75	0.92	1.1	
AISI 2015–2115	0.67					
AISI 2315–2335	0.54	0.58	0.62	0.75	0.92	
AISI 2340–2350	...	0.50	0.58	0.70	0.83	1.0
AISI 2512–2515	0.50	0.58	0.67	0.80	0.92	
AISI 3115–3130	0.50	0.58	0.70	0.83	1.0	
AISI 3160–3450	...	0.50	0.62	0.75	0.87	1.0
AISI 4130–4345	...	0.46	0.58	0.70	0.83	1.0
AISI 4615–4820	0.46	0.50	0.58	0.70	0.83	0.87
AISI 5120–5150	0.46	0.50	0.62	0.75	0.87	1.0
AISI 52100	...	0.58	0.67	0.83	1.0	
AISI 6115–6140.....	0.46	0.54	0.67	0.83	1.0	
AISI 6145–6195.....	...	0.70	0.83	1.0	1.2	1.3
Plain cast iron	0.30	0.33	0.42	0.50		
Alloy cast iron	0.30	0.42	0.54			
Malleable iron......	0.42					
Cast steel...........	0.62	0.67	0.80			

High temperature alloys			Nonferrous metals and alloys	
Material classification	BHN	Unit hp		Unit hp
A 286	165	0.82	Brass:	
A 286	285	0.93	Hard................	0.83
Chromoloy ...	200	0.78	Medium.............	0.50
Chromoloy ...	310	1.18	Soft................	0.33
Hastelloy-B...	230	1.10	Free machining.......	0.25
Inco 700......	330	1.12	Bronze:	
Inco 702......	230	1.10	Hard................	0.83
M-252........	230	1.10	Medium.............	0.50
M-252........	310	1.20	Soft................	0.33
Ti-150A	340	0.65	Copper (pure)	0.90
U-500	375	1.10	Aluminum:	
4340	200	0.78	Cast.................	0.25
4340	340	0.93	Hard (rolled)	0.33
			Monel (rolled)	1.0
			Zinc alloy (die cast)......	0.25

*General Electric Company.

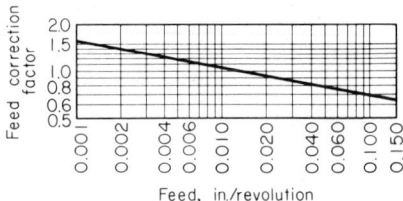

Fig. 12 Feed correction factors based on normal tool geometries. *(General Electric Co.)*

cent efficiency of the machine. Table 15 shows that AISI 4130 to 4345 of 250 to 300 BHN, the range into which AISI 4140 260 BHN falls, has a unit hp of 0.70.

The unit horsepower must be corrected for feed. Using Fig. 12 and a feed of 0.025 in./revolution, the correction factor is found to be 0.90. Thus, the true unit horsepower = $(0.70)(0.90) = 0.63$ hp/(in.^{3})(min).

3. Compute the horsepower required at the cutter

The horsepower required at the cutter $hp_c = (hp_u)(\text{MRR})$, or $hp_c = (0.63)(22.5) = 14.18$ hp.

4. Compute the motor horsepower required

The power required at the cutter is the input necessary after allowing for losses in gears, bearings, and other parts of the drive. Table 16 lists typical overall machine-tool efficiencies. A gear-head lathe has an efficiency of 70 percent. Thus, $hp_m = hp_c/e$, where e = machine-tool efficiency, expressed as a decimal. Or, $hp_m = 14.18/0.70 = 20.25$ hp. A 20-hp motor would be satisfactory for this machine.

Related Calculations: Use this procedure for single or multiple tools. When more than one tool is working at the same time, compute hp_c for each tool and add the individual values to find the total hp_c. Divide the total hp_c by the machine efficiency to determine the required motor horsepower hp_m. This procedure makes ample allowance for dulling of the tools.

Compute metal removal rates for other operations as follows. *Face milling:* MRR = WDF_T, where W = width of cut, in., D = depth of cut, in., F_T = table feed, in./min.

TABLE 16   Efficiencies of Metalworking Machines*

Typical overall machine-tool efficiency values (except milling machines), percent	Typical overall efficiencies for milling machines	
	Rated hp of machine	Overall efficiency, percent
Direct spindle drive.............. 90	3	40
	5	48
One-belt drive................... 85	7.5	52
Two-belt drive.................. 70	10	52
	15	52
Geared head..................... 70	20	60
	25	65
	30	70
	40	75
	50	80

*General Electric Company.

Slot milling: MRR = WDF_T, where all symbols are as before. *Planing or shaping:* MRR = $DfLS$, where D = depth of cut, in., f = feed, in. per stroke or revolution, L = length of workpiece, in., S = strokes/min. *Multiple tools:* MRR = $(d_1{}^2 - d_s{}^2)\pi fR/4$, where d_1 = original diameter of workpiece, in., *before* cutting, d_s = workpiece diameter *after* cutting, in., R = rpm of workpiece, other symbols as before.

The procedure given here is the work of Robert G. Brierley and H. J. Siekmann as reported in *Machining Principles and Cost Control.*

CUTTING SPEED FOR LOWEST-COST MACHINING

What is the optimal cutting speed for a part if the maximum feed for which an accept-able finish is obtained at 169 rpm of the workpiece is 0.011 in./revolution when the cost of labor and overhead is $0.24/hr, the number of pieces produced per tool change is 15, the cost per tool change is $0.62, and the length of cut is 8 in?

Calculation Procedure:

1. Compute the optimization factor

When determining the lowest-cost machining speed for an operation, one popular procedure is to choose any speed and feed at which the operation meets the finish requirements. If desired, the speed and feed at which the operation is now running might be chosen. Keeping the speed constant, the feed is increased to the maximum value for which the finish is acceptable. This is the optimal value for the feed and is called the *optimal feed*. The number of pieces produced under these conditions is measured between tool changes. Then the optimal cutting speed is computed from: optimal cutting speed, rpm = (chosen speed, rpm)(optimization factor).

For any operation, the optimization factor = {(labor and overhead cost, $/min)(num-ber of pieces per tool change)(length of cut, in.)/[(3)(cost per tool change, $)(chosen speed, rpm)(optimal feed, in./revolution)]}$^{-4}$. Substitute the given values. Thus, the optimization factor = $\{(0.24)(15)(8)/[(3)(0.62)(169)(0.011)]\}^{-4} = 1.7$.

2. Compute the optimal cutting speed

Using the relation given in Step 1, optimal cutting speed = (chosen speed, rpm)(opti-mization factor) = $(169)(1.7) = 287$ rpm.

Related Calculations: The relation given in Step 1 for the optimization factor is valid for carbide tools. It can be modified to apply to high-speed tools by changing the fourth root to an eight root and changing the 3 in the denominator to 7.

REORDER QUANTITY FOR OUT-OF-STOCK PARTS

A metalworking process uses 10 parts during the lead time. How many parts should be reordered if an out-of-stock situation can be accepted for 10 percent of the time? For 35 percent of the time?

TABLE 17   Out-of-Stock Factors*

Acceptable Percent of Stock-Outs	Out-of-Stock Factor
50	0.00
45	0.13
40	0.26
35	0.39
25	0.68
15	1.04
10	1.29
5	1.65
4	1.76
3.5	1.82
3	1.89
2	2.06
1	2.33
0	4.0

*Nyles V. Reinfeld in *American Machinist*.

Calculation Procedure:

1. *Determine the out-of-stock factor*

Table 17 lists out-of-stock factors for various times during which a part might be out of stock. Thus, the acceptable out-of-stock factor for 10 percent is 1.29, and for 35 percent it is 0.39.

2. *Compute the reorder quantity*

For any manufacturing process, reorder quantity = (out-of-stock factor)(usage during lead time)$^{0.5}$ + (usage during lead time). Thus, the reorder point for this process with an acceptable out-of-stock factor for 10 percent is $(1.29)(10)^{0.5} + (10) = 14.08$ parts; say 15 parts. With 35 percent, reorder point = $(0.29)(10)^{0.5} + 10 = 11.23$, or 12 parts.

Related Calculations: Use this procedure for any types of parts ordered from either an internal or external source. In general, reducing the allowable stock-out time will increase the time during which a process using the parts can operate.

SAVINGS WITH MORE MACHINABLE MATERIALS

What are the gross and net savings made with a more machinable material that reduces the production time by 36 sec per part when 800 lb of steel are required for 1,000 parts and the total machine operating cost is $6/hr? The more machinable material costs 4 cents/lb more than the less machinable material and 5,000 parts are produced per day.

Calculation Procedure:

1. *Compute the gross savings possible*

The gross saving possible in a machining operation when a more machinable material is used is: gross saving, cents/lb = (machining time saved with new material, sec per piece)(total cost of operating machine, cents/hr)/[(3.6)(weight of material to make 1,000 pieces, lb)]. For this operation, the gross saving = $(36)(600)/[(3.6)(800)] = 7.5$ cents/lb. With a production rate of 5,000 parts per day, the gross saving is (5,000 parts) (800 lb/1,000 parts)(7.5 cents/lb) = 30,000 cents, or $300.

2. *Compute the net savings possible*

The more machinable materials cost 4 cents more per lb than the less machinable material. Hence, the net saving is $(7.5 - 4.0) = 3.5$ cents/lb, or (5,000 parts)(800 lb/1,000 parts)(3.5 cents/lb) = 14,000 cents or $140.

Related Calculations: Use this general procedure for parts made of any material — steel, brass, bronze, aluminum, plastic, etc.

TIME REQUIRED FOR THREAD MILLING

How long will it take to thread-mill a $2\frac{7}{8}$-in.-diameter hard steel bolt with a $2\frac{1}{2}$-in.-diameter 18-flute hob?

Calculation Procedure:

1. *Determine the cutting speed and feed of the hob*

Table 18 lists typical cutting speeds and feeds for various materials. For hard steel, the usual cutting speed in thread milling is 50 fpm and the feed per flute is 0.002 in.

2. *Compute the time required for thread milling*

The time required for thread milling T_t min = $\pi d/fnR$, where d = work diameter, in.; f = feed, in. per flute; n = number of flutes on hob; R = hob rpm. From a previous Calculation Procedure, $R = 12C/\pi d$, where C = hob cutting speed, fpm; d = hob diameter, in. For this hob, $R = (12)(50)/[\pi(2.5)] = 76.4$ rpm. Then, $T_t = \pi(2\frac{7}{8})/[(0.002)(18)(76.4)] = 3.29$ min.

Related Calculations: Use this procedure for any metallic or nonmetallic material – aluminum, brass, mild steel, medium steel, hard steel, plastics, etc.

TABLE 18   Thread-Milling Speeds and Feeds

Material threaded	Speed, fpm	Feed, in. per flute
Aluminum..........	500	0.0015
Brass..............	250	0.0015
Mild steel	100	0.0020
Medium steel.......	75	0.0020
Hard steel..........	50	0.0020

DRILL PENETRATION RATE AND CENTERLESS GRINDER FEED RATE

What is the drill penetration rate when a drill turns at 1,000 rpm and has a feed of 0.006 in./revolution? What is the feed rate of a centerless grinder having a 12-in.-diameter regulating wheel running at 60 rpm if the angle of inclination between the regulating and grinding wheel is 5°?

Calculation Procedure:

1. Compute the rate of drill penetration

The rate of drill penetration P in./min $= fR$, where f = drill feed, in./revolution; R = drill rpm. For this drill, $P = (0.006)(1,000) = 6.0$ per min.

2. Compute the grinder feed rate

The work feed f in./min in a centerless grinder is $f = \pi dR \sin a$, where d = regulating-wheel diameter, in.; R = regulating-wheel rpm; a = angle of inclination between the regulating and grinding wheel,°. For this grinder, $f = \pi(12)(60)(\sin 5°) = 197.6$ in./min.

BENDING, DIMPLING, AND DRAWING METAL PARTS

What is the minimum bend radius R in. for 0.02-gage Vascojet 1000 metal if it is bent transversely to an angle of 130°? What is the minimum radius R in. of a bend in 0.040-gage Rene 41 metal bent longitudinally at an angle of 52° at room temperature? Determine the maximum length of dimple flange H in. for AM-350 metal at 500 F when the bend angle is 42° and the edge radius R is 0.250 in. Determine the maximum blank diameter and maximum cup depth for drawing Rene 41 metal at 400 F when using a die diameter of 10 in. and 0.063-gage material. Figure 13a, b, and c shows the anticipated manufacturing conditions.

Calculation Procedure:

1. Compute the minimum bend radius

Table 19 shows that the critical bend angle (i.e., maximum bend angle α without breakage) for Vascojet 1000 metal is 118°. Hence, the required bend angle is greater than the critical bend angle. Therefore, the required bend limit equals the critical bend limit, and $R/T = 1.30$, from Table 19. Hence, the minimum radius $R_m = (R/T)(T) = (1.30)(0.02) = 0.026$ in.

With Rene 41 metal, bent longitudinally at room temperature, the critical bend angle is 122°, from Table 19. Since the required bend angle of 52° is less than critical, find the

R/T value in the right-hand portion of Table 19. When the actual bend angle is between two tabulated angles, interpolate thus:

Angle, °	R/T Value
60	0.37
52	
45	0.22

$[(52-45)/(60-45)](0.37-0.22) = 0.07$. Then, R/T for $52° = 0.22+0.07 = 0.29$. With R/T known for $52°$, compute the minimum radius $R_m = (R/T)(T) = (0.29)(0.040) = 0.0116$ in.

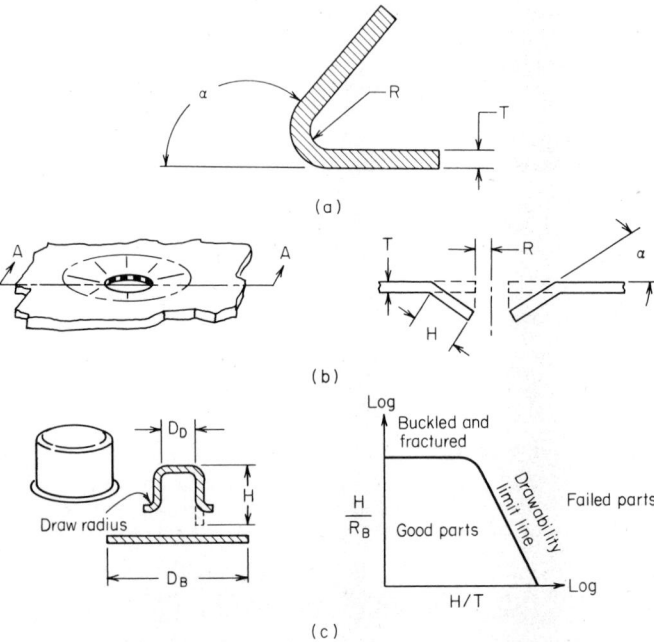

(a)

(b)

(c)

Fig. 13 (a) Brake-bent part shape and parameters; (b) ram-coin dimpling setup; (c) drawing set-up. (*American Machinist.*)

2. Determine the dimple-flange length

Table 20 shows typical dimpling limits to avoid radial splitting at the edge of the hole of various modern materials. With a bend angle between the tabulated angles, interpolate thus:

Angle, °	H/R Value
45	1.10
42	
40	1.43

$[(42-40)/(45-40)](1.10-1.43) = -0.132$, and $H/R = 1.43+(-0.132) = 1.298$ at $42°$. Then, the maximum dimple-flange length $H_m = (H/R)(R) = (1.298)(0.250) = 0.325$ in.

3. Determine the maximum blank diameter

Table 21 lists the drawing limits for flat-bottom cups made of various modern materials. For this cup, $D_D/T = 10/0.063 = 158.6$; say 159. The corresponding D_B/T and H/D_D

TABLE 19 Brake-Bent Parts Parameters*

Material	L/T	F	∞	R/T	R/T for angles ∞ below critical						
					30	45	60	75	90	105	120
2024–O Al.......	L/T	RT	130	0.30	0.04	0.08	0.12	0.17	0.22	0.26	0.28
HM21XA–T8	L/T	RT	72	5.50	1.55	3.40	5.16	5.50			
Titanium (6Al–4V)	L/T	RT	68	5.70	2.05	4.20	5.50	5.70			
Titanium (13V–11Cr–3Al).....	L/T	RT	105	2.40	0.34	0.68	1.16	1.80	2.25	2.40	
Vascojet 1000	L/T	RT	118	1.30	0.18	0.38	0.64	0.92	1.13	1.26	1.30
USS 12 MoV.....	L/T	RT	119	1.20	0.16	0.34	0.60	0.84	1.04	1.16	1.20
17–7 PH.........	L/T	RT	122	0.80	0.10	0.22	0.37	0.54	0.66	0.75	0.79
AM–350.........	L/T	RT	122	0.80	0.10	0.22	0.37	0.54	0.66	0.75	0.79
PH 15–7 Mo	L/T	RT	121	0.86	0.11	0.23	0.42	0.60	0.72	0.80	0.84
A–286...........	L/T	RT	124	0.66	0.07	0.15	0.29	0.43	0.54	0.62	0.65
Hastelloy X......	L/T	RT	120	1.00	0.12	0.26	0.47	0.67	0.84	0.95	1.00
Inconel X........	L/T	RT	124	0.64	0.06	0.14	0.28	0.41	0.52	0.60	0.63
Rene 41	L	RT	122	0.80	0.10	0.22	0.37	0.54	0.66	0.75	0.79
Rene 41	T	RT	113	1.64	0.28	0.53	0.84	1.16	1.44	1.58	1.64
J–1570	L	RT	124	0.68	0.08	0.16	0.30	0.45	0.56	0.64	0.67
J–1570	T	RT	122	0.80	0.10	0.22	0.37	0.54	0.66	0.75	0.79
L–605	L/T	RT	120	1.00	0.12	0.26	0.47	0.67	0.84	0.95	1.00
Molybdenum (0.5 Ti)........	L	RT	118	1.30	0.18	0.38	0.64	0.92	1.13	1.26	1.30
Molybdenum (0.5 Ti)........	T	RT	107	2.20	0.32	0.64	1.05	1.60	2.02	2.18	2.20

American Machinist, LTV, Inc.; USAF.
NOTE: L/T = grain direction, where L = longitudinal and T = transverse; F = bending temperature; ∞ = critical bend; R/T = critical bend limits.

TABLE 20 Dimpling Limits to Avoid Radial Splitting at Hole Edge*

Material	Temperature, F	Dimpling limit H/R				
		Standard, for various bend angles a; above and below standard bend angle				
		30°	35°	40°	45°	50°
2024-T3	70	2.15	1.60	1.20	0.93	0.80
Ti-8-1-1	70	1.88	1.42	1.08	0.82	0.70
TZM Moly	70	1.98	1.50	1.12	0.87	0.73
Cb-752	70	2.28	1.70	1.30	0.98	0.83
PH 15-7 Mo	500	2.43	1.84	1.40	1.07	0.90
AM-350	500	2.46	1.87	1.43	1.10	0.93
Ti-8-1-1	1200	2.30	1.72	1.30	1.00	0.85
Ti-13-11-3	1200	2.58	1.95	1.48	1.15	0.95

American Machinist, LTV, Inc.; USAF.

TABLE 21   Drawing Limits for Flat-Bottom Cups*

Material	Temperature ratio, F		Die to blank diameter ratios D_B/D_D; cup depth ratios H/D_D For various D_D/T ratios							
			25	50	100	150	200	250	300	400
2024-T3	70	D_B/D_D	1.91	1.87	1.82	1.67	1.50	1.41	1.41	1.09
		H/D_D	0.65	0.63	0.56	0.40	0.30	0.25	0.22	0.13
	600	D_B/D_D	2.44	2.58	2.22	1.89	1.66	1.55	1.47	1.40
		H/D_D	1.24	1.43	0.95	0.61	0.45	0.37	0.31	0.24
T1-8-1-1	70	D_B/D_D	1.72	1.63	1.63	1.50	1.37	1.31	1.32	1.29
		H/D_D	0.46	0.44	0.41	0.30	0.23	0.19	0.17	0.14
	1200	D_B/D_D	1.92	1.96	1.79	1.57	1.50	1.39	1.38	1.27
		H/D_D	0.67	0.71	0.59	0.40	0.32	0.30	0.23	0.17
TZM MOLY	70	D_B/D_D	2.00	2.08	2.00	1.73	1.57	1.52	1.42	1.49
		H/D_D	0.75	0.82	0.70	0.48	0.35	0.30	0.25	0.19
Cb-752	70	D_B/D_D	1.91	1.96	1.78	1.61	1.43	1.43	1.37	1.30
		H/D_D	0.64	0.67	0.57	0.41	0.31	0.26	0.22	0.18
	400	D_B/D_D	2.19	2.30	2.04	1.75	1.59	1.46	1.40	1.36
		H/D_D	0.92	1.02	0.76	0.51	0.38	0.37	0.32	0.21
Ph 15-7 Mo	500	D_B/D_D	2.39	2.47	2.16	1.86	1.64	1.54	1.46	1.38
		H/D_D	1.15	1.26	0.87	0.57	0.42	0.34	0.29	0.23
Am-350	500	D_B/D_D	2.22	2.18	2.00	1.71	1.54	1.42	1.40	1.30
		H/D_D	0.97	0.95	0.75	0.50	0.37	0.30	0.26	0.20
A-286	1000	D_B/D_D	2.22	2.46	2.16	1.85	1.64	1.49	1.41	1.36
		H/D_D	1.00	1.21	0.87	0.57	0.42	0.34	0.28	0.22
Rene 41	400	D_B/D_D	2.22	2.22	1.92	1.73	1.52	1.48	1.42	1.33
		H/D_D	0.97	0.97	0.73	0.51	0.37	0.31	0.27	0.21
L-605		D_B/D_D	2.22	2.29	2.00	1.68	1.54	1.45	1.44	1.38
		H/D_D	0.97	1.05	0.74	0.47	0.35	0.30	0.26	0.21
T1-13-11-3	1200	$D_B D/$	2.38	2.53	2.28	1.92	1.67	1.58	1.45	1.44
		H/D_D	1.15	1.34	0.94	0.60	0.44	0.35	0.30	0.24
Tungsten	600	D_B/D_D	2.08	2.11	1.98	1.66	1.53	1.46	1.38	1.34
		H/D_D	0.83	0.87	0.69	0.45	0.35	0.29	0.24	0.20

American Machinist, LTV, Inc., U.S. Air Force.

ratios are not tabulated. Therefore, interpolate between D_D/T values of 150 and 200. Thus, for Rene 41:

D_D/T	D_B/D_D
200	1.52
159	
150	1.73

$[(159-150)/(200-150)](1.52-1.73) = -0.0378$, and $D_B/D_D = 1.73 + (-0.0378) = 1.692$, when $D_D/T = 159$. Then, the *maximum* value of $D_{Bm} = (D_B/D_D)(D_D) = (1.692)(10) = 16.92$ in.

Interpolating as above, $H/D_D = 0.48$ when $D_D/T = 159$. Then the maximum height $H_m = (H/D_D)(D) = (0.48)(10) = 4.8$ in.

Related Calculations: The procedures given here are typical of those used for the newer "exotic" metals developed for use in aerospace, cryogenic, and similar advanced technologies. The three tables presented here were developed by LTV, Inc., for the U.S. Air Force, and reported in *American Machinist.*

BLANK DIAMETERS FOR ROUND SHELLS

What blank diameter D in. is required for the round shells in Fig. 14a and b if $d = 12$ in., $d_1 = 12$ in., $d_2 = 14$ in., and $h = 14$ in?

Calculation Procedure:

1. Compute the plain-cup blank diameter

Figure 14a shows the plain cup. Compute the required blank diameter from $D = (d^2 + 4dh)^{0.5} = (12^2 + 4 \times 12 \times 4)^{0.5} = 18.33$ in.

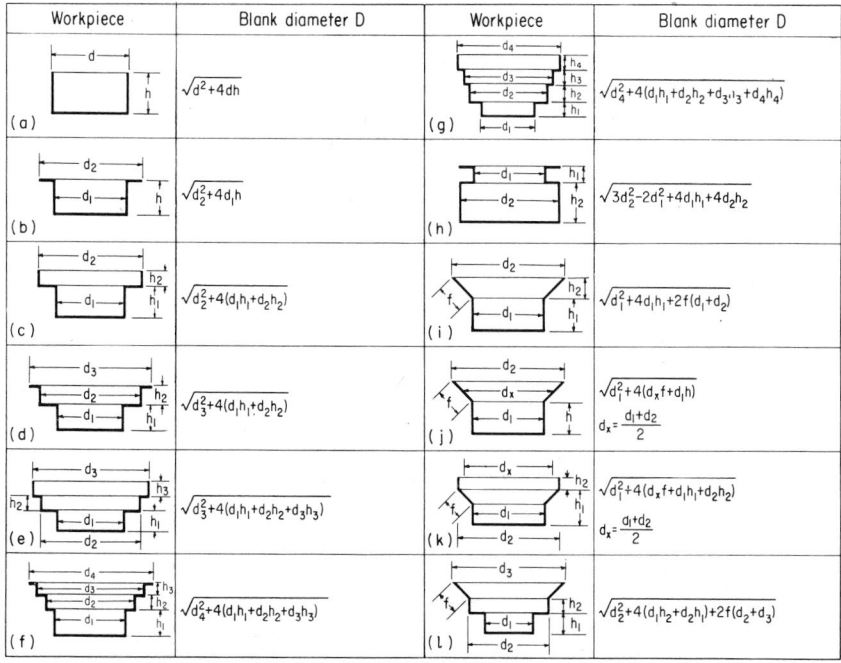

Fig. 14   Blank diameters for round shells. (*American Machinist.*)

2. Compute the flanged-cup blank diameter

Figure 14b shows the flanged cup. Compute the required blank diameter from $D = (d_2{}^2 + fd_1h)^{0.5} = (14^2 + 4 \times 12 \times 4)^{0.5} = 19.7$ in.

Related Calculations: Figure 14 gives the equations for computing 12 different round-shell blank diameters. Use the same general procedures as in Steps 1 and 2 above. These equations were derived by Ferene Kuchta, Mechanical Engineer, J. Wiss & Sons Co., and reported in *American Machinist*.

BREAK-EVEN CONSIDERATIONS IN MANUFACTURING OPERATIONS

A manufacturing plant has the net sales and fixed and variable expenses shown in Table 22. What is the break-even point for this plant in units and sales? Plot a conventional and an alternative break-even chart for this plant.

Calculation Procedure:

1. Compute the break-even units

Use the relation BE_u = fixed expenses, \$/[(sales income per unit, \$ − variable costs per unit, \$)], where BE_u = break even in units. Substituting, $BE_u = \$400{,}000/(\$20 - \$12) = 50{,}000$ units.

2. Compute the break-even sales

Two methods can be used to compute the break-even sales. With the break-even units known from Step 1, $BE_s = BE_u$(unit sales price, \$), where BE_s = break-even sales, \$. Substituting, $BE_s = (50{,}000)(\$20) = \$1{,}000{,}000$.

Alternatively, compute the profit-volume, or PV, ratio: PV = (sales, \$ − variable costs, \$)/sales, \$ = (\$1{,}200{,}000 − \$720{,}000)/\$1{,}200{,}000 = 0.40. Then BE_s = fixed costs,

TABLE 22 Manufacturing Business Income and Expenses

Condensed Income Statement
For year ending Dec. 31, 19−

	Variable	Fixed	
Net sales (60,000 units @ \$20 per unit)			\$1,200,000
Less costs and expenses:	*Variable*	*Fixed*	
Direct material	\$195,000	...	
Direct labor	215,000	...	
Manufacturing expenses.............	100,000	\$200,000	
Selling expenses....................	50,000	150,000	
General and administrative expenses .	160,000	50,000	
Total.............................	720,000	400,000	1,120,000
Net profit before Federal income taxes......................			80,000

\$/PV = \$400,000/0.40 = \$1,000,000. This is identical to the break-even sales computed in the previous paragraph.

3. Draw the conventional and alternative break-even charts

Figure 15a shows the conventional break-even chart for this plant. Construct this chart by drawing the horizontal line F for the fixed expenses, the solid sloping line I for the income or sales, and the dotted sloping line C for the total costs. Note that the vertical axis is for the expenses and income and the horizontal axis is for the sales, all measured in monetary units.

The break-even point is at the intersection of the income and total-cost curves, point B, Fig. 15a. Projecting vertically downward shows that point B corresponds to a sales volume of \$1,000,000, as computed in Step 2.

Alternative break-even charts are shown in Fig. 15b and c. Both these charts are constructed in a manner similar to Fig. 15a.

Related Calculations: Break-even computations are valuable tools for analyzing any manufacturing operation. The concepts are also applicable to other business activities. Thus, typical PV values for various types of businesses are:

Business	Typical activity	Typical PV
Consumer appliances........	Fully automated; high volume output	0.15–0.25
Standard centrifugal pumps	Batch output in large volume	0.20–0.30
Acid-handling centrifugal pumps	Batch output in small volume	0.25–0.35
Standard prototype, one-of-a-kind	Ships, machine tools	0.30–0.40
Special-design one-of-a-kind prototype........	Buildings, factories	0.35–0.50

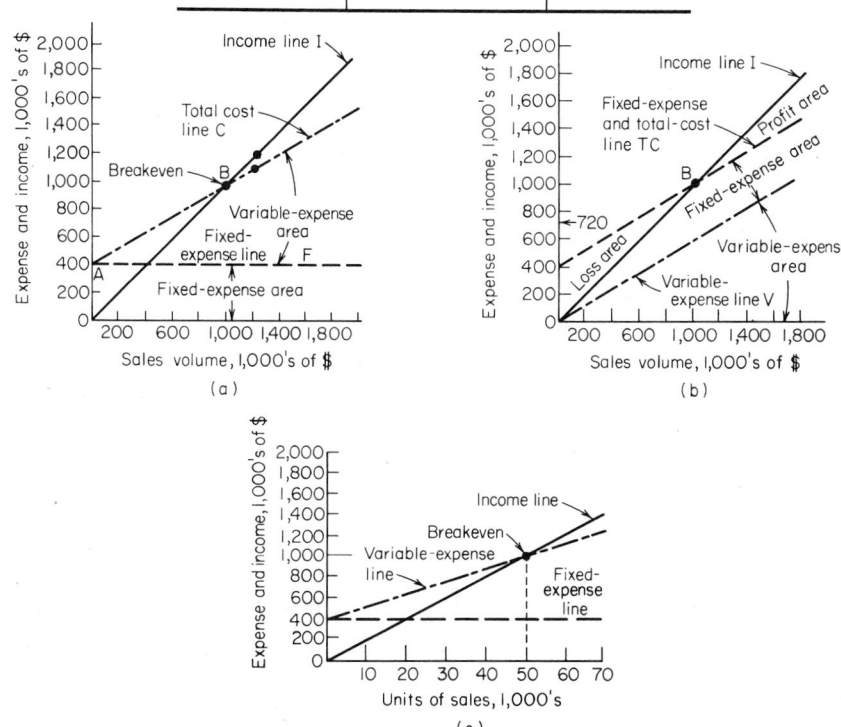

Fig. 15   Three forms of the break-even chart as used in metalworking activities.

COMBUSTION

REFERENCES: Johnson and Auth—*Fuels and Combustion Handbook*, McGraw-Hill; Babcock & Wilcox Company—*Steam: Its Generation and Use*; Combustion Engineering Corporation—*Combustion Engineering*; Gaffert—*Steam Power Stations*, McGraw-Hill; Skrotzki and Vopat—*Applied Energy Conversion*, McGraw-Hill; Popovich-Hering—*Fuels and Lubricants*, Wiley; ASME—*Power Test Code for Steam Boilers*; Moore—*Coal*, Wiley; Moore—*Liquid Fuels*, The Technical Press, Ltd., London; American Gas Association—*Combustion*; Dunstan—*Science of Petroleum*, Oxford, London; Trinks—*Industrial Furnaces*, Wiley; Perry—*Chemical Engineers' Handbook*, McGraw-Hill.

COMBUSTION OF COAL FUEL IN A FURNACE

A coal has the following ultimate analysis (or percent by weight): C = 0.8339; H_2 = 0.0456; O_2 = 0.0505; N_2 = 0.0103; S = 0.0064; ash = 0.0533; total = 1.000 lb. This coal is burned in a steam-boiler furnace. Determine the weight of air required for theoretically perfect combustion, the weight of gas formed per pound of coal burned, and the volume of flue gas, at the boiler exit temperature of 600 F per lb of coal burned; air required with 20 percent excess air, and the volume of gas formed with this excess; the CO_2 percentage in the flue gas on a dry and wet basis.

Calculation Procedure:

1. Compute the weight of oxygen required per lb of coal

To find the weight of oxygen required for theoretically perfect combustion of coal, set up the following tabulation, based on the ultimate analysis of the coal:

Element	× Molecular-weight ratio	= Lb O_2 required
C; 0.8339	× 32/12	= 2.2237
H_2; 0.0456	× 16/2	= 0.3648
O_2; 0.0505; decreases external O_2 required		=
N_2; 0.0103 is inert in combustion and is ignored		
S; 0.0064	× 32/32	= 0.0064
Ash 0.0533 is inert in combustion and is ignored		
Total 1.0000		
lb external O_2 per lb fuel		= 2.5444

Note that of the total oxygen needed for combustion, 0.0505 lb is furnished by the fuel itself and is assumed to reduce the total external oxygen required by the amount of oxygen present in the fuel. The molecular-weight ratio is obtained from the equation for the chemical reaction of the element with oxygen in combustion. Thus, for carbon $C + O_2 \rightarrow CO_2$, or $12 + 32 = 44$, where 12 and 32 are the molecular weights of C and O_2, respectively.

2. Compute the weight of air required for perfect combustion

Air at sea level is a mechanical mixture of various gases, principally 23.2 percent oxygen and 76.8 percent nitrogen by weight. The nitrogen associated with the 2.5444 lb of oxygen required per lb of coal burned in this furnace is the product of the ratio of the nitrogen and oxygen weights in the air and 2.5444, or (2.5444) (0.768/0.232) = 8.4219 lb. Then the weight of air required for perfect combustion of one lb of coal = sum of nitrogen and oxygen required = 8.4219 + 2.5444 = 10.9663 lb of air per lb of coal burned.

3. Compute the weight of the products of combustion

Find the products of combustion by addition:

Fuel constituents	+ Oxygen	→	Products of combustion
C; 0.8339	+ 2.2237	→	CO_2 = 2.0576 lb
H; 0.0456	+ 0.3648	→	H_2O = 0.4104
O_2; 0.0505; this is *not* a product of combustion			
N_2; 0.0103; inert but passes through furnace			= 0.0103
S; 0.0064	+ 0.0064	→	SO_2 = 0.0128
Outside nitrogen from Step 2			= N_2 = 8.4219
Lb of flue gas per lb of coal burned			= 11.9130

4. Convert the flue-gas weight to volume

Use Avogadro's law, which states that under the same conditions of pressure and temperature, 1 mole (the molecular weight of a gas expressed in lb) of any gas will occupy the same volume.

At 14.7 psia and 32 F, 1 mole of any gas occupies 359 cu ft. The volume per lb of any gas at these conditions can be found by dividing 359 by the molecular weight of the gas and correcting for the gas temperature by multiplying the volume by the ratio of the absolute flue-gas temperature and the atmospheric temperature. To change the weight analysis (Step 3) of the products of combustion to volumetric analysis, set up the calculation thus:

Products	Weight, lb	Molecular weight	Temperature correction		Volume at 600 F, cu ft
CO_2	3.0576	44	(359/44)(3.0576)(2.15)	=	53.8
H_2O	0.4104	18	(359/18)(0.4104)(2.15)	=	17.6
Total N_2	8.4322	28	(359/28)(8.4322)(2.15)	=	233.0
SO_2	0.0128	64	(359/64)(0.0128)(2.15)	=	0.17
Cu ft of flue gas per lb of coal burned				=	304.57

In this calculation, the temperature correction factor 2.15 = absolute flue-gas temperature, R/absolute atmospheric temperature, R = (600 + 460)/(32 + 460). The total weight of N_2 in the flue gas is the sum of the N_2 in the combustion air and the fuel, or 8.4219 + 0.0103 = 8.4322 lb. This value is used in computing the flue-gas volume.

5. Compute the CO content of the flue gas

The volume of CO_2 in the products of combustion at 600 F is 53.8 cu ft, as computed in Step 4; and the total volume of the combustion products is 304.57 cu ft. Therefore, the percent CO_2 on a wet basis (i.e., including the moisture in the combustion products) = cu ft CO_2/total cu ft = 53.8/304.57 = 0.1765, or 17.65 percent.

The percent CO_2 on a dry, or Orsat, basis is found in the same manner except that the weight of H_2O in the products of combustion, 17.6 lb from Step 4, is subtracted from the total gas weight. Or, percent CO_2, dry, or Orsat basis = (53.8)/(304.57 − 17.6) = 0.1875, or 18.75 percent.

6. Compute the air required with the stated excess flow

With 20 percent excess air, the air flow required = (0.20 + 1.00)(air flow with no excess) = 1.20(10.9663) = 13.1596 lb of air per lb of coal burned. The air flow with no excess is obtained from Step 2.

7. Compute the weight of the products of combustion

The excess air passes through the furnace without taking part in the combustion and increases the weight of the products of combustion per lb of coal burned. Therefore, the weight of the products of combustion is the sum of the weight of the combustion products without the excess air and the product of (percent excess air)(air for perfect combustion, lb); or using the weights from Steps 3 and 2, respectively, $= 11.9130 + (0.20)(10.9663) = 14.1063$ lb of gas per lb of coal burned with 20 percent excess air.

8. Compute the volume of the combustion products and the percent CO_2

The volume of the excess air in the products of combustion is obtained by converting from the weight analysis to the volumetric analysis and correcting for temperature as in Step 4, using the air weight from Step 2 for perfect combustion and the excess-air percentage, or $(10.9663)(0.20)(359/28.95)(2.15) = 58.5$ cu ft. In this calculation the value 28.95 is the molecular weight of air. The total volume of the products of combustion is the sum of the column for perfect combustion, Step 4, and the excess-air volume, above, or $304.57 + 58.5 = 363.07$ cu ft.

Using the procedure in Step 5, the percent CO_2, wet basis $= 53.8/363.07 = 14.8$ percent. The percent CO_2, dry basis $= 53.8/(363.07 - 17.6) = 15.6$ percent.

Related Calculations: Use the method given here when making combustion calculations for any type of coal—bituminous, semibituminous, lignite, anthracite, cannel, or coking—from any coal field in the world used in any type of furnace—boiler, heater, process, or waste-heat. When the air used for combustion contains moisture, as is usually true, this moisture is added to the combustion-formed moisture appearing in the products of combustion. Thus, for 80-F air of 60 percent relative humidity, the moisture content is 0.013 lb per lb of dry air. This amount appears in the products of combustion for each lb of air used and is a commonly assumed standard in combustion calculations.

COMBUSTION OF FUEL OIL IN A FURNACE

A fuel oil has the following ultimate analysis: $C = 0.8543$; $H_2 = 0.1131$; $O_2 = 0.0270$; $N_2 = 0.0022$; $S = 0.0034$; total $= 1.0000$. This fuel oil is burned in a steam-boiler furnace. Determine the weight of air required for theoretically perfect combustion, the weight of gas formed per lb of oil burned, and the volume of flue gas, at the boiler exit temperature of 600 F, per lb of oil burned; the air required with 20 percent excess air, and the volume of gas formed with this excess; the CO_2 percentage in the flue gas on a dry and wet basis.

Calculation Procedure:

1. Compute the weight of oxygen required per lb of oil

The same general steps as given in the previous Calculation Procedure will be followed. Consult that Procedure for a complete explanation of each step.

Using the molecular weight of each element,

Element	×	Molecular-weight ratio	=	Lb O_2 required
C; 0.8543	×	32/12	=	2.2781
H_2; 0.1131	×	16/2	=	0.9048
O_2; 0.0270; decreases external O_2 required			=	−0.0270
N_2; 0.0022 is inert in combustion and is ignored				
S; 0.0034	×	32/32	=	0.0034
Total 1.0000				
Lb external O_2 per lb fuel			=	3.1593

2. Compute the weight of air required for perfect combustion

The weight of nitrogen associated with the required oxygen = (3.1593)(0.768/0.232) = 10.458 lb. The weight of air required = 10.4583 + 3.1593 = 13.6176 lb per lb of oil burned.

3. Compute the weight of the products of combustion

As before:

Fuel constituents	+	Oxygen	=	Products of combustion
C; 0.8543 + 2.2781	=	3.1324	=	CO_2
H_2; 0.1131 + 0.9148	=	1.0179	=	H_2O
O_2; 0.270; *not* a product of combustion				
N_2; 0.0022; inert but passes through furnace	=	0.0022	=	N_2
S; 0.0034 + 0.0034	=	0.0068	=	SO_2
Outside N_2 from Step 2	=	10.458	=	N_2
Lb of flue gas per lb of oil burned	=	14.6173		

4. Convert the flue-gas weight to volume

As before:

Products	Weight, lb	Molecular weight	Temperature correction		Volume at 600 F, cu ft
CO_2	3.1324	44	(359/44)(3.1324)(2.15)	=	55.0
H_2O	1.0179	18	(359/18)(1.0179)(2.15)	=	43.5
N_2 (total)	10.460	28	(359/28)(10.460)(2.15)	=	288.5
SO_2	0.0068	64	(359/64)(0.0068)(2.15)	=	0.82
Cu ft of flue gas per lb of oil burned				=	387.82

In this calculation, the temperature correction factor 2.15 = absolute flue-gas temperature, R/absolute atmospheric temperature, R = (600 + 460)/(32 + 460). The total weight of N_2 in the flue gas is the sum of the N_2 in the combustion air and the fuel, or 10.4580 + 0.0022 = 10.4602 lb.

5. Compute the CO_2 content of the flue gas

The CO_2, wet basis, = 55.0/387.82 = 0.142, or 14.2 percent.
The CO_2, dry basis, = 55.0/(387.2 − 43.5) = 0.160, or 16.0 percent.

6. Compute the air required with stated excess flow

The lb of air per lb of oil with 20 percent excess air = (1.20)(13.6176) = 16.3411 lb of air per lb of oil burned.

7. Compute the weight of the products of combustion

The weight of the products of combustion = product weight for perfect combustion, lb + (percent excess air)(air for perfect combustion, lb) = 14.6173 + (0.20)(13.6176) = 17.3408 lb of flue gas per lb of oil burned with 20 percent excess air.

8. Compute the volume of the combustion products and the percent CO_2

The volume of excess air in the products of combustion is found by converting from the weight to the volumetric analysis and correcting for temperature as in Step 4, using the air weight from Step 2 for perfect combustion and the excess-air percentage, or $(13.6176)(0.20)(359/28.95)(2.15) = 72.7$ cu ft. Add this to the volume of the products of combustion found in Step 4, or $387.82 + 72.70 = 460.52$ cu ft.

Using the procedure in Step 5, the percent CO_2, wet basis, $= 55.0/460.52 = 0.1192$, 11.92 percent. The percent CO_2, dry basis, $= 55.0/(460.52 - 43.5) = 0.1318$, or 13.18 percent.

Related Calculations: Use the method given here when making combustion calculations for any type of fuel oil—paraffin-base, asphalt-base, Bunker C, No. 2, 3, 4, or 5—from any source, domestic or foreign, in any type of furnace—boiler, heater, process, or waste-heat. When the air used for combustion contains moisture, as is usually true, this moisture is added to the combustion-formed moisture appearing in the products of combustion. Thus, for 80-F air of 60 percent relative humidity, the moisture content is 0.013 lb per lb of dry air. This amount appears in the products of combustion for each lb of air used and is a commonly assumed standard in combustion calculations.

COMBUSTION OF NATURAL GAS IN A FURNACE

A natural gas has the following volumetric analysis at 60 F: $CO_2 = 0.004$; $CH_4 = 0.921$; $C_2H_6 = 0.041$; $N_2 = 0.034$; total $= 1.000$. This natural gas is burned in a steam-boiler furnace. Determine the weight of air required for theoretically perfect combustion, the weight of gas formed per lb of natural gas burned, and the volume of the flue gas, at the boiler exit temperature of 650 F, per lb of natural gas burned; air required with 20 percent excess air, and the volume of gas formed with this excess; CO_2 percentage in the flue gas on a dry and wet basis.

Calculation Procedure:

1. Compute the weight of oxygen required per lb of gas

The same general steps as given in the previous Calculation Procedures will be followed, except that they will be altered to make allowances for the differences between natural gas and coal.

The composition of the gas is given on a volumetric basis, which is the usual way of expressing a fuel-gas analysis. To use the volumetric-analysis data in combustion calculations, they must be converted to a weight basis. This is done by dividing the weight of each component by the total weight of the gas. A volume of 1 cu ft of the gas is used for this computation. Find the weight of each component and the total weight of 1 cu ft as follows, using the properties of the combustion elements and compounds given in Table 1:

Component	Percent by volume	Density, lb/cu ft	Component weight, lb = column 1 × column 2
CO_2	0.004	0.1161	0.0004644
CH_4	0.921	0.0423	0.0389583
C_2H_6	0.041	0.0792	0.0032472
N_2	0.034	0.0739	0.0025026
Total	1.000		0.0451725 lb/cu ft

$$\text{Percent } CO_2 = 0.0004644/0.0451725 = 0.01026, \text{ or } 1.03 \text{ percent}$$
$$\text{Percent } CH_4 \text{ by weight} = 0.0389583/0.0451725 = 0.8625 \text{ or } 86.25 \text{ percent}$$
$$\text{Percent } C_2H_6 \text{ by weight} = 0.0032472/0.0451725 = 0.0718, \text{ or } 7.18 \text{ percent}$$
$$\text{Percent } N_2 \text{ by weight} = 0.0025026/0.0451725 = 0.0554, \text{ or } 5.54 \text{ percent}$$

The sum of the weight percentages $= 1.03 + 86.25 + 7.18 + 5.54 = 100.00$. This sum checks the accuracy of the weight calculation, because the sum of the weights of the component parts should equal 100 percent.

Next, find the oxygen required for combustion. Since both the CO_2 and N_2 are inert, they do not take part in the combustion; they pass through the furnace unchanged. Using the molecular weights of the remaining compounds in the gas and the weight percentages, we have:

Compound	×	Molecular-weight ratio	=	Lb O_2 required
CH_4; 0.8625	×	64/16	=	3.4500
C_2H_6; 0.0718	×	112/30	=	0.2920
Lb external O_2 required per lb fuel			=	3.7420

In this calculation, the molecular-weight ratio is obtained from the equation for the combustion chemical reaction, or $CH_4 + 2O_2 = CO_2 + 2H_2O$, that is, $16 + 64 = 44 + 36$, and $C_2H_6 + \frac{7}{2}O_2 = 2CO_2 + 3H_2O$, that is, $30 + 112 = 88 + 54$. See Table 2 from these and other useful chemical reactions in combustion.

2. Compute the weight of air required for perfect combustion

The weight of nitrogen associated with the required oxygen $= (3.742)(0.768/0.232) = 12.39$ lb. The weight of air required $= 12.39 + 3.742 = 16.132$ lb per lb of gas burned.

3. Compute the weight of the products of combustion

Use the relation:

Fuel constituents	+	Oxygen	=	Products of combustion
CO_2; 0.0103; inert but passes through the furnace			=	0.010300
CH_4; 0.8625	+	3.45	=	4.312500
C_2H_6; 0.003247	+	0.2920	=	0.032447
N_2; 0.0554; inert but passes through the furnace			=	0.055400
Outside N_2 from Step 2			=	12.390000
Lb of flue gas per lb of natural gas burned			=	16.800347

4. Convert the flue-gas weight to volume

The products of complete combustion of any fuel that does not contain sulfur are CO_2, H_2O, and N_2. Using the combustion equation in Step 1, compute the products of combustion thus: $CH_4 + 2O_2 = CO_2 + H_2O$; $16 + 64 = 44 + 36$; or the CH_4 burns to CO_2 in the ratio of one part CH_4 to 44/16 parts CO_2. Since, from Step 1, there is 0.03896 lb CH_4 per cu ft of natural gas, this forms $(0.03896)(44/16) = 0.1069$ lb of CO_2. Likewise, for C_2H_6, $(0.003247)(88/30) = 0.00952$ lb. The total CO_2 in the combustion products $= 0.00464 + 0.1069 + 0.00952 = 0.11688$ lb, where the first quantity is the CO_2 in the fuel.

Using a similar procedure for the H_2O formed in the products of combustion by CH_4, $(0.03896)(36/16) = 0.0875$ lb. For C_2H_6, $(0.003247)(54/30) = 0.005816$ lb. The total H_2O in the combustion products $= 0.0875 + 0.005816 = 0.093316$ lb.

Step 2 shows that 12.39 lb of N_2 is required per lb of fuel. Since 1 cu ft of the fuel weighs 0.04517 lb, the volume of gas which weighs 1 lb is $1/0.04517 = 22.1$ cu ft. Therefore, the weight of N_2 per cu ft of fuel burned $= 12.39/22.1 = 0.560$ lb. This, plus the weight of N_2 in the fuel, Step 1, is $0.560 + 0.0025 = 0.5625$ lb of N_2 in the products of combustion.

Next, find the total weight of the products of combustion by taking the sum of the

TABLE 1　Properties of Combustion Elements*

Element or compound	For-mula	Mole-cular weight	At 14.7 psia, 60°F Weight, lb/cu ft	At 14.7 psia, 60°F Volume, cu ft/lb	Nature Gas or solid	Nature Com-busti-ble	Heat value, Btu Per lb	Heat value, Btu Per cu ft at 14.7 psia, 60 F	Heat value, Btu Per mole cule
Carbon...........	C	12	...	...	S	Yes	14,540	...	174,500
Hydrogen	H₂	2.02†	0.0053	188	G	Yes	61,000	325	123,100
Sulphur..........	S	32	...	...	S	Yes	4,050	...	129,600
Carbon monoxide ..	CO	28	0.0739	13.54	G	Yes	4,380	323	122,400
Methane	CH₄	16	0.0423	23.69	G	Yes	24,000	1,012	384,000
Acetylene	C₂H₂	26	0.0686	14.58	G	Yes	21,500	1,483	562,000
Ethylene..........	C₂H₄	28	0.0739	13.54	G	Yes	22,200	1,641	622,400
Ethane...........	C₂H₆	30	0.0792	12.63	G	Yes	22,300	1,762	668,300
Oxygen	O₂	32	0.0844	11.84	G				
Nitrogen	N₂	28	0.0739	13.52	G				
Air‡..............	...	29	0.0765	13.07	G				
Carbon dioxide	CO₂	44	0.1161	8.61	G				
Water............	H₂O	18	0.0475	21.06	G				

*P. W. Swain and L. N. Rowley, "Library of Practical Power Engineering" (collection of articles published in *Power*).

†For most practical purposes, the value of 2 is sufficient.

‡The molecular weight of 29 is merely the weighted average of the molecular weight of the constituents.

CO_2, H_2O, and N_2 weights, or $0.11688 + 0.09332 + 0.5625 = 0.7727$ lb. Now convert each weight to cu ft at 650 F, the temperature of the combustion products, or:

Products	Weight, lb	Molecular weight	Temperature correction		Volume at 650 F, cu ft
CO_2;	0.11688	44	(379/44)(0.11688)(2.255)	=	2.265
H_2O;	0.09332	18	(379/18)(0.09332)(2.255)	=	4.425
N_2 (total)	0.5625	28	(379/28)(0.5625)(2.255)	=	17.190
Cu ft of flue gas per cu ft of natural-gas fuel				=	23.880

In this calculation, the value of 379 is used in the molecular-weight ratio because at 60 F and 14.7 psia the volume of 1 lb of any gas $= 379/$gas molecular weight. The fuel gas used is initially at 60 F and 14.7 psia. The ratio $2.255 = (650 + 460)/(32 + 460)$.

5.　Compute the CO_2 content of the flue gas

The CO_2, wet basis, $= 2.265/23.88 = 0.947$, or 9.47 percent. The CO_2, dry basis, $= 2.265/(23.88 - 4.425) = 0.1164$, or 11.64 percent.

6.　Compute the air required with the stated excess flow

The lb of air per lb of natural gas with 20 percent excess air $= (1.20)(16.132) = 19.3584$ lb of air per lb of natural gas, or $19.3584/22.1 = 0.875$ lb of air per cu ft of natural gas. See Step 4 for an explanation of the value 22.1

7.　Compute the weight of the products of combustion

Weight of the products of combustion = product weight for perfect combustion, lb + (percent excess air) (air for perfect combustion, lb) $= 16.80 + (0.20)(16.132) = 20.03$ lb.

8. Compute the volume of the combustion products and the percent CO_2

The volume of excess air in the products of combustion is found by converting from the weight to the volumetric analysis and correcting for temperature as in Step 4 using the air weight from Step 2 for perfect combustion and the excess-air percentage, or $(16.132/22.1)(0.20)(379/28.95)(2.255) = 4.31$ cu ft. Add this to the volume of the products of combustion found in Step 4, or $23.88 + 4.31 = 28.19$ cu ft.

Using the Procedure in Step 5, the percent CO_2, wet basis, $= 2.265/28.19 = 0.0804$, or 8.04 percent. The percent CO_2, dry basis, $= 2.265/(28.19 - 4.425) = 0.0953$, or 9.53 percent.

Related Calculations: Use the method given here when making combustion calculations for any type of gas used as a fuel—natural gas, blast-furnace gas, coke-oven gas, producer gas, water gas, sewer gas—from any source, domestic or foreign, in any type of furnace—boiler, heater, process, or waste-heat. When the air used for combustion contains moisture, as is usually true, this moisture is added to the combustion-formed moisture appearing in the products of combustion. Thus, for 80-F air of 60 percent relative humidity, the moisture content is 0.013 lb per lb of dry air. This amount appears in the products of combustion for each lb of air used and is a commonly assumed standard in combustion calculations.

COMBUSTION OF WOOD FUEL IN A FURNACE

The weight analysis of a yellow-pine wood fuel is: $C = 0.490$; $H_2 = 0.074$; $O_2 = 0.406$; $N_2 = 0.030$. Determine the weight of oxygen and air required with perfect combustion and with 20 percent excess air. Find the weight and volume of the products of combustion under the same conditions, and the wet and dry CO_2. The flue-gas temperature is 600 F. The air supplied for combustion has a moisture content of 0.013 lb per lb of dry air.

Calculation Procedure:

1. Compute the weight of oxygen required per lb of wood

The same general steps as given in earlier Calculation Procedures will be followed; consult them for a complete explanation of each step. Using the molecular weight of each element, we have:

Element	×	Molecular-weight ratio	= Lb O_2 required
C; 0.490	×	32/12	= 1.307
H_2; 0.074	×	16/2	= 0.592
O_2; 0.406; decreases external O_2 required			= −0.406
N_2; 0.030 inert in combustion			
Total 1.000			
Lb external O_2 per lb fuel			= 1.493

2. Compute the weight of air required for complete combustion

The weight of nitrogen associated with the required oxygen $= (1.493)(0.768/0.232) = 4.95$ lb. The weight of air required $= 4.95 + 1.493 = 6.443$ lb per lb of wood burned, if the air is dry. But the air contains 0.013 lb of moisture per lb of air. Hence, the total weight of the air $= 6.443 + (0.013)(6.443) = 6.527$ lb.

3. Compute the weight of the products of combustion

Use the following relation:

Fuel constituents	+Oxygen	= Products of combustion
C; 0.490	+1.307	= 1.797 = CO_2
H_2; 0.074	+0.592	= 0.666 = H_2O
O_2; not a product of combustion		
N_2; inert but passes through the furnace		= 0.030 = N_2
Outside N_2 from Step 2		= 4.950 = N_2
Outside moisture from Step 2		= 0.237
Lb of flue gas per lb of wood burned		= 8.280

4. Convert the flue-gas weight to volume

Use, as before, the following tabulation:

Products	Weight, lb	Molecular weight	Temperature correction	Volume at 600 F, cu ft
CO_2	1.797	44	(359/44)(1.797)2.15 =	31.5
H_2O(fuel)	0.666	18	(359/18)(0.666)(2.15) =	28.6
N_2(total)	4.980	28	(359/28)(4.980)(2.15) =	137.2
H_2O(outside air)	0.837	18	(359/18)(0.837)(2.15) =	35.9
Cu ft of flue gas per lb of oil				233.2

In this calculation the temperature correction factor 2.15 = (absolute flue-gas temperature, R)/(absolute atmospheric temperature, R) = (600 + 460)/(32 + 460). The total weight of N_2 is the sum of the N_2 in the combustion air and the fuel.

5. Compute the CO_2 content of the flue gas

The CO_2, wet basis, = 31.5/233.2 = 0.135, or 13.5 percent. The CO_2, dry basis, = 31.5/(233.2 − 28.6 − 35.9) = 0.187, or 18.7 percent.

6. Compute the air required with the stated excess flow

The lb of air per lb of wood with 20 percent excess air = (1.20)(6.527) = 7.832 lb of air per lb of wood burned.

7. Compute the weight of the products of combustion

The weight of the products of combustion = product weight for perfect combustion, lb + (percent excess air)(air for perfect combustion, lb) = 8.280 + (0.20)(6.527) = 9.585 lb of flue gas per lb of wood burned with 20 percent excess air.

8. Compute the volume of the combustion products and the percent CO_2

The volume of the excess air in the products of combustion is found by converting from the weight to the volumetric analysis and correcting for temperature as in Step 4, using the air weight from Step 2 for perfect combustion and the excess-air percentage, or (6.527)(0.20)(359/28.95)(2.15) = 34.8 cu ft. Add this to the volume of the products of combustion found in Step 4, or 233.2 + 34.8 = 268.0 cu ft.

Using the Procedure in Step 5, the percent CO_2, wet basis, = 31.5/268 = 0.1174, or 11.74 percent. The percent CO_2, dry basis, = 31.5/(268 − 28.6 − 35.9 − 0.20 × 0.837) = 0.155, or 15.5 percent. In the dry-basis calculation, the factor (0.20)(0.837) is the outside moisture in the excess air.

Related Calculations: Use the method given here when making combustion calculations for any type of wood or wood-like fuel − spruce, cypress, maple, oak, sawdust, wood shavings, tanbark, bagesse, peat, charcoal, redwood, hemlock, fir, ash, birch, cottonwood, elm, hickory, walnut, chopped trimmings, hogged fuel, straw, corn, cotton-seed hulls, city refuse—in any type of furnace—boiler, heating, process, or waste-heat; most of these fuels contain a small amount of ash—usually less than 1 percent. This was ignored in this Calculation Procedure because it does not take part in the combustion.

MOLAL METHOD OF COMBUSTION ANALYSIS

A coal fuel has this ultimate analysis: C = 0.8339 H_2 = 0.0456; O_2 = 0.0505; N_2 = 0.0103; S = 0.0064; ash = 0.0533; total = 1.000. This coal is completely burned in a boiler furnace. Using the molal method, determine the weight of air required per lb of coal with complete combustion. How much air is needed with 25 percent excess air? What is the weight of the combustion products with 25 percent excess air? The combustion air contains 0.013 lb of moisture per lb of air.

Calculation Procedure:

1. *Convert the ultimate analysis to mols*

A mol of any substance is an amount of the substance having a weight equal to the molecular weight of the substance. Thus, a mol of carbon is 12 lb of carbon, because the molecular weight of carbon is 12. To convert an ultimate analysis of a fuel to mols, assume that 100 lb of the fuel is being considered. Set up a tabulation thus:

Ultimate analysis, %	Weight, lb	Molecular weight	Mols per 100-lb fuel
C = 0.8339 ..	83.39	12	6.940
H_2 = 0.0456 ..	4.56	2	2.280
O_2 = 0.0505 ..	5.05	32	0.158
N_2 = 0.0103 ..	1.03	28	0.037
S = 0.0064 ..	0.64	32	
Ash = 0.0533 ..	5.33	Inert	
Totals.......	100.00	...	9.435

2. *Compute the mols of oxygen for complete combustion*

From Table 2, the burning of carbon to carbon dioxide requires 1 mol of carbon and 1 mol of oxygen, yielding 1 mol of CO_2. Using the molal equations in Table 2 for the other elements in the fuel, set up a tabulation thus, entering the product of columns 2 and 3 in column 4:

(1) Element	(2) Mols per 100-lb fuel	(3) Mols O_2 per 100-lb fuel	(4) Total mols O_2
C	6.940	1.00	6.940
H_2	2.280	0.5	1.140
O_2	0.158	Reduces O_2 required	−0.158
N_2	0.037	Inert in combustion	
S	0.020	1.00	0.020
Total mols of O_2 required	...	...	7.942

TABLE 2   Chemical Reactions

Combustible substance	Reaction	Mols	Lb*
Carbon to carbon monoxide...	$C + \frac{1}{2}O_2 = CO$	$1 + \frac{1}{2} = 1$	$12 + 16 = 28$
Carbon to carbon dioxide	$C + O_2 = CO_2$	$1 + 1 = 1$	$12 + 16 = 28$
Carbon monoxide to carbon dioxide..................	$CO + \frac{1}{2}O_2 = CO_2$	$1 + \frac{1}{2} = 1$	$28 + 16 = 44$
Hydrogen	$H_2 + \frac{1}{2}O_2 = H_2O$	$1 + \frac{1}{2} = 1$	$2 + 16 = 18$
Sulphur to sulphur dioxide....	$S + O_2 = SO_2$	$1 + 1 = 1$	$32 + 32 = 64$
Sulphur to sulphur trioxide ...	$S + \frac{3}{2}O_2 = SO_3$	$1 + \frac{3}{2} = 1$	$32 + 48 = 80$
Methane	$CH_4 + 2O_2 = CO_2 + 2H_2O$	$1 + 2 = 1 + 2$	$16 + 64 = 44 + 36$
Ethane......................	$C_2H_6 + \frac{7}{2}O_2 = 2CO_2 + 3H_2O$	$1 + \frac{7}{2} = 2 + 3$	$30 + 112 = 88 + 54$
Propane.....................	$C_3H_8 + 5O_2 = 3CO_2 + 4H_2O$	$1 + 5 = 3 + 4$	$44 + 160 = 132 + 72$
Butane	$C_4H_{10} + \frac{13}{2}O_2 = 4CO_2 + 5H_2O$	$1 + \frac{13}{2} = 4 + 5$	$58 + 208 = 176 + 90$
Acetylene	$C_2H_2 + \frac{5}{2}O_2 = 2CO_2 + H_2O$	$1 + \frac{5}{2} = 2 + 2$	$26 + 80 = 88 + 18$
Ethylene	$C_2H_4 + 3O_2 = 2CO_2 + 2H_2O$	$1 + 3 = 2 + 2$	$28 + 96 = 88 + 36$

*Substitute the molecular weights in the reaction equation to secure lb. The lb on each side of the equation must balance.

3.   Compute the mols of air for complete combustion

Set up a similar tabulation for air, thus:

(1) Element	(2) Mols per 100-lb fuel	(3) Mols air per 100-lb fuel	(4) Total mols air
C	6.940	4.76	33.050
H₂	2.280	2.38	5.430
O₂	0.158	Reduces O_2 required	−0.752
N₂	0.037	Inert in combustion	
S	0.020	4.76	0.095
Total mols of air required		...	37.823

In this tabulation, the factors in column 3 are constants used for computing the total mols of air required for complete combustion of each of the fuel elements listed. These factors are given in the Babcock & Wilcox Company — *Steam: Its Generation and Use* and similar treatises on fuels and their combustion. A tabulation of these factors is given in Table 3, herewith.

An alternative, and simpler, way of computing the mols of air required is to convert the required O_2 to the corresponding N_2 and find the sum of the O_2 and N_2. Or, 3.76 $O_2 = N_2$; $N_2 + O_2 =$ mols of air required. The factor 3.76 converts the required O_2 to the corresponding N_2. These two relations were used to convert the 0.158 mols of O_2 in the above tabulation to mols of air.

Using the same relations and the mols of O_2 required from Step 2, (3.76)(7.942) = 29.861 mols of N_2. Then $29.861 + 7.942 = 37.803$ mols of air, which agrees closely with the 37.823 mols computed in the tabulation. The difference of 0.02 mol is traceable to slide-rule readings.

4.   Compute the air required with the stated excess air

With 25 percent excess, the air mols of air required for combustion = (125/100)(37.823) = 47.24 mols.

TABLE 3 Molal Conversion Factors

Element or compound	Mols per mol of combustible for complete combustion; no excess air					
	For combustion			Combustion products		
	O_2	N_2	Air	CO_2	H_2O	N_2
Carbon,* C............	1.0	3.76	4.76	1.0	...	3.76
Hydrogen, H_2.........	0.5	.188	2.38·	...	1.0	1.88
Oxygen, O_2...........						
Nitrogen, N_2						
Carbon monoxide, CO..	0.5	1.88	2.38	1.0	...	1.88
Carbon dioxide, CO_2 ...						
Sulfur,* S	1.0	3.76	4.76	1.0	...	3.76
Methane, CH_4	2.0	7.53	0.53	1.0	2.0	7.53
Ethane, C_2H_6	3 5	13.18	16.68	2.0	3.0	13.18

*In molal calculations, carbon and sulfur are considered as gases.

5. Compute the mols of combustion products

Using data from Table 3, and recalling that the products of combustion of a sulfur-containing fuel are CO_2, H_2O, and SO_2 and that N_2 and excess O_2 pass through the furnace, set up a tabulation thus:

(1) Mols per 100 lb fuel	(2) Mols per mol of combustible	(3) Mols of combustion products per 100 lb of fuel
CO_2; 6.940.....................	1	6.940
H_2O; 2.280 + (47.24) (0.021 + 0.158)...............	...	3.430
SO_2; 0.020.....................	1	0.202
N_2; (47.24)(0.79)...............	...	37.320
Excess O_2; (1.25)(7.942) − 7.942...	...	1.986

Total mols, wet combustion products = 49.878
Total mols, dry combustion products = 49.878 − 3.232
= 46.646

In this calculation, the total mols of CO_2 is obtained from Step 2. The mols of H_2 in 100 lb of the fuel, 2.280, is assumed to form H_2O. In addition, the air from Step 4, 47.24 mols, contains 0.013 lb of moisture per lb of air. This moisture is converted to mols by dividing the molecular weight of air, 28.95, by the molecular weight of water, 18, and multiplying the result by the moisture content of the air, or (28.95/18)(0.013) = 0.0209; say 0.021 mols of water per mol of air. The product of this and the mols of air gives the total mols of moisture (water) in the combustion products per 100 lb of fuel fired. To this is added the mols of O_2, 0.158, per 100 lb of fuel, because this oxygen is assumed to unite with hydrogen in the air to form water. The nitrogen in the products of combustion is that portion of the mols of air required, 47.24 mols from Step 4, times the proportion of N_2 in the air, or 0.79. The excess O_2 passes through the furnace and adds to the combustion products and is computed as shown in the tabulation. Subtracting the total moisture, 3,430 mols, from the total (or wet) combustion products gives the mols of dry combustion products.

Related Calculations: Use this method for molal combustion calculations for all types of fuels—solid, liquid, and gaseous—burned in any type of furnace—boiler, heater, process, or waste-heat. Select the correct factors from Table 3.

STEAM BOILER HEAT BALANCE DETERMINATION

A steam generator having a maximum rated capacity of 60,000 lb/hr is operating at 45,340 lb/hr, delivering 125-psig 400-F steam with a feedwater temperature of 181 F. At this generating rate, the boiler requires 4,370 lb per hr of West Virginia bituminous coal having a heating value of 13,850 Btu/lb on a dry basis. The ultimate fuel analysis is: $C = 0.7757$; $H_2 = 0.0507$; $O_2 = 0.0519$; $N_2 = 0.0120$; $S = 0.0270$; ash $= 0.0827$; total $= 1.0000$. The coal contains 1.61 percent moisture. The boiler-room intake air and the fuel temperature $= 79$ F dry bulb, 71 F wet bulb. The flue-gas temperature is 500 F and the analysis of the flue gas shows these percentages: $CO_2 = 12.8$; $CO = 0.4$; $O_2 = 6.1$; $N_2 = 80.7$; total $= 100.0$. Measured ash and refuse $= 9.42$ percent of dry coal; combustible in ash and refuse $= 32.3$ percent. Compute a heat balance for this boiler based on these test data. The boiler has four water-cooled furnace walls.

Calculation Procedure:

1. Determine the heat input to the boiler

In a boiler heat balance the input is usually stated in Btu per lb of fuel as fired. Therefore, input = heating value of fuel = 13,850 Btu/lb.

2. Compute the output of the boiler

The output of any boiler = Btu per lb of fuel + the losses. In this Step the first portion of the output, Btu per lb of fuel, will be computed. The losses will be computed in Step 3.

First find W_s, lb of steam produced per lb of fuel fired. Since 45,340 lb per hr of steam are produced when 4,370 lb per hr of fuel are fired, $W_s = 45,340/4,370 = 10.34$ lb of steam per lb of fuel.

Once W_s is known, the output h_1 Btu per lb of fuel can be found from $h_1 = W_s(h_s - h_w)$, where $h_s =$ enthalpy of steam leaving the superheater, or boiler if a super-heater is not used; $h_w =$ enthalpy of feedwater, Btu/lb. For this boiler with steam at 125 psig ($= 139.7$ psia) and 400 F, $h_s = 1221.2$ Btu/lb, and $h_w = 180.92$ Btu/lb, from the steam tables. Then, $h_1 = 10.34(1221.2 - 180.92) = 10,766.5$ Btu per lb of coal.

3. Compute the dry flue-gas loss

For any boiler, the dry flue-gas loss h_2 Btu per lb of fuel is given by $h_2 = 0.24W_g \times (T_g - T_a)$, where $W_g =$ lb of dry flue gas per lb of fuel; $T_g =$ flue-gas exit temperature, F; $T_a =$ intake-air temperature, F.

Before W_g can be found, however, it must be determined if any excess air is passing through the boiler. Compute the excess air, if any, from excess air, percent, $= 100 (O_2 - \frac{1}{2}CO)/(0.264N_2 - [O_2 - \frac{1}{2}CO])$, where the symbols refer to the elements in the flue-gas analysis. Substituting, values from the flue-gas analysis, excess air $= 100(6.1 - 0.2)/[0.264 \times 80.7 - [6.1 - 0.2]]) = 38.4$ percent.

Using the method given in earlier Calculation Procedures, find the air required for complete combustion as 10.557 lb per lb of coal. With 38.4 percent excess air, the additional air required $= (10.557)(0.384) = 4.053$ lb per lb of fuel.

From the same computation in which the air required for complete combustion was determined, the lb of *dry* flue gas per lb of fuel $= 11.018$. Then, the total flue gas at 38.4 percent excess air $= 11.018 + 4.053 = 15.071$ lb per lb of fuel.

With a flue-gas temperature of 500 F and an intake-air temperature of 79 F, $h_2 = 0.24(15.071)(500 - 70) = 1,524$ Btu per lb of fuel.

4. Compute the loss due to evaporation of hydrogen-formed water

Hydrogen in the fuel is burned forming H_2O. This water is evaporated by heat in the fuel, and less heat is available for producing steam. This loss, h_3 Btu per lb of fuel $= 9H(1,089 - T_f + 0.46T_g)$, where $H =$ percent H_2 in the fuel $\div 100$; $T_f =$ temperature of fuel *before* combustion, F; other symbols as before. For this fuel with 5.07 percent H_2, $h_3 = 9(5.07/100)(1,089 - 79 + 0.46 \times 500) = 565.8$ Btu per lb of fuel.

5. Compute the loss from evaporation of fuel moisture

This loss, h_4 Btu per lb of fuel $= W_{mf}(1,089 - T_f + 0.46T_g)$, where $W_{mf} =$ lb of moisture per lb of fuel; other symbols as before. Since the fuel contains 1.61 percent moisture, in terms of *dry* coal this is $(1.61)/(100 - 1.61) = 0.0164$, or 1.64 percent. Then, $h_4 = (1.64/100)(1,089 - 79 + 0.46 \times 500) = 20.34$ Btu per lb of fuel.

6. Compute the loss from moisture in the air

This loss, h_5 Btu per lb of fuel $= 0.46W_{ma}(T_g - T_a)$, $W_{ma} =$ (lb of water per lb of dry air)(lb air supplied per lb fuel). From a psychrometric chart, the weight of moisture per lb of air at a 79-F dry-bulb and 71-F wet-bulb temperature is 0.014. The combustion calculation, Step 3, shows that the total air required with 38.4 percent excess air $= 10.557 + 4.053 = 14.61$ lb of air per lb of fuel. Then, $W_{ma} = (0.014)(14.61) = 0.2045$ lb of moisture per lb of air. And, $h_5 = (0.46)(0.2045)(500 - 79) = 39.6$ Btu per lb of fuel.

7. Compute the loss from incomplete combustion of C to CO_2 in the stack

This loss, h_6 Btu per lb of fuel $= [CO/(CO + CO_2)](C) (10,190)$, where CO and CO_2 are the percent by volume of these compounds in the flue gas by Orsat analysis; $C =$ lb carbon per lb of coal. With the given flue-gas analysis and the coal ultimate analysis, $h_6 = 0.4/(0.4 + 12.8)[(77.57)/(100)](10,190) = 239.5$ Btu per lb of fuel.

8. Compute the loss due to unconsumed carbon in the refuse

This loss, h_7 Btu per lb of fuel $= W_c(14,150)$, where $W_c =$ lb of unconsumed carbon in refuse per lb of fuel fired. With an ash and refuse of 9.42 percent of the dry coal and combustible in the ash and refuse of 32.3 percent, $h_7 = (9.42/100)(32.3/100)(14,150) = 430.2$ Btu per lb of fuel.

9. Find the radiation loss in the boiler furnace

Use the American Boiler and Affiliated Industries (ABAI) chart, or the manufacturer's engineering data to approximate the radiation loss in the boiler. Either source will show that the radiation loss is 1.09 percent of the gross heat input. Since the gross heat input is 13,850 Btu per lb of fuel, the radiation loss $= (13,850)(1.09/100) = 151.0$ Btu per lb of fuel.

10. Summarize the losses; find the unaccounted-for loss

Set up a tabulation thus, entering the various losses computed earlier:

Item	Btu per lb fuel	Percent
1. Input	13,850.0	100.0
2. Output	10,770.0	77.75
Losses:		
3. Flue gas......	1,524.0	11.00
4. Hydrogen	565.8	4.09
5. Water-fuel....	20.3	0.15
6. Water-air.....	39.6	0.29
7. CO	239.5	1.73
8. Carbon-ash...	430.2	3.11
9. Radiation.....	151.0	1.09
10. Unaccounted .	109.6	0.79
Total.........	13,850.0	100.00

The unaccounted-for loss is found by summing all the other losses, 3 through 9, and subtracting from 100.00.

Related Calculation: Use this method to compute the heat balance for any type of boiler—watertube or firetube—in any kind of service—power, process, or heating—using any kind of fuel—coal, oil, gas, wood, or refuse. Note that Step 3 shows how to compute excess air from an Orsat flue-gas analysis.

POWER GENERATION

REFERENCES: Babcock & Wilcox Company—*Steam: Its Generation and Use*; Combustion Engineering Corporation—*Combustion Engineering*; Skrotzki and Vopat—*Power Station Engineering and Economy*, McGraw-Hill; Heat Exchange Institute—*Steam Surface Condenser Standards*; Gaffert—*Steam Power Stations*, McGraw-Hill; ASME—*Test Code for Steam Generating Units*; Potter—*Steam Power Plants*, Ronald; Smith and Stinson—*Fuels and Combustion*, McGraw-Hill; Zerban and Nye—*Steam Power Plants*, International Textbook; Sorenson—*Gas Turbines*, Ronald; Salisbury—*Steam Turbines and Their Cycles*, Wiley; Dusinberre—*Gas Turbine Power*, International Textbook; Zemansky—*Heat and Thermodynamics*, McGraw-Hill; Jakob—*Heat Transfer*, Wiley; McAdams—*Heat Transmission*, McGraw-Hill; Buffalo Forge Company—*Fan Engineering*; Church—*Steam Turbines*, McGraw-Hill; Bleeder Heater Manufacturers Association, Inc.—*Standards*; Tubular Exchanger Manufacturers Association—*Standards*.

STEAM MOLLIER DIAGRAM AND STEAM TABLE USE

(1) Determine from the Mollier diagram for steam: (*a*) the enthalpy of 100 psi a saturated steam; (*b*) the enthalpy of 10 psia steam containing 40 percent moisture; (*c*) the enthalpy of 100 psia steam at 600 F. (2) Determine from the steam tables: (*a*) the enthalpy, specific volume, and entropy of steam at 145.3 psig; (*b*) the enthalpy and specific volume of superheated steam at 1,100 psia and 600 F; (*c*) the enthalpy and specific volume of high-pressure steam at 7,500 psia and 1200 F; (*d*) the enthalpy, specific volume, and entropy of 10-psia steam containing 40 percent moisture.

Calculation Procedure:

1. Use the pressure and saturation (or moisture) lines to find enthalpy

(*a*). Enter the Mollier diagram by finding the 100-psia pressure line, Fig. 1. In the Mollier diagram for steam, the pressure lines slope upward to the right from the lower left-hand corner. For saturated steam, the enthalpy is read at the intersection of the pressure line with the saturation curve *cef*, Fig. 1.

Thus, project along the 100-psia pressure curve, Fig. 1, until it intersects the saturation curve, point *g*. From here project horizontally to the left-hand scale of Fig. 1 and read the enthalpy of 100-psia saturated steam as 1,187 Btu/lb. (The Mollier diagram in Fig. 1 has fewer grid divisions than large-scale diagrams to permit easier location of the major elements of the diagram.)

(*b*). On a Mollier diagram, the enthalpy of wet steam is found at the intersection of the saturation pressure line with the percent moisture curve corresponding to the amount of moisture in the steam. In a Mollier diagram for steam, the moisture curves slope downward to the right from the saturated liquid line *cd*, Fig. 1.

To find the enthalpy of 10-psia steam containing 40 percent moisture, project along the 10-psia saturation pressure line until the 40 percent moisture curve is intersected, Fig. 1. From here project horizontally to the left-hand scale and read the enthalpy of 10-psia wet steam containing 40 percent moisture as 750 Btu/lb.

2. Find the steam properties from the steam tables

(*a*). Steam tables normally list absolute pressures or temperature in F as one of their arguments. Therefore, when the steam pressure is given in terms of a gage reading, it must be converted to an absolute pressure value before the table can be entered. To convert gage pressure to absolute pressure, add 14.7 to the gage pressure, or $p_a = p_g + 14.7$. In this instance, $p_a = 145.3 + 14.7 = 160.0$ psia. Once the absolute pressure is known, enter the saturation-pressure table of the steam table at this value and project horizontally to the desired values. For 160-psia steam, using the ASME or Keenan and Keyes—*Thermodynamic Properties of Steam*, the enthalpy of evaporation $h_{fg} = 859.2$ Btu/lb; enthalpy of saturated vapor $h_g = 1195.1$ Btu/lb, read from the respective

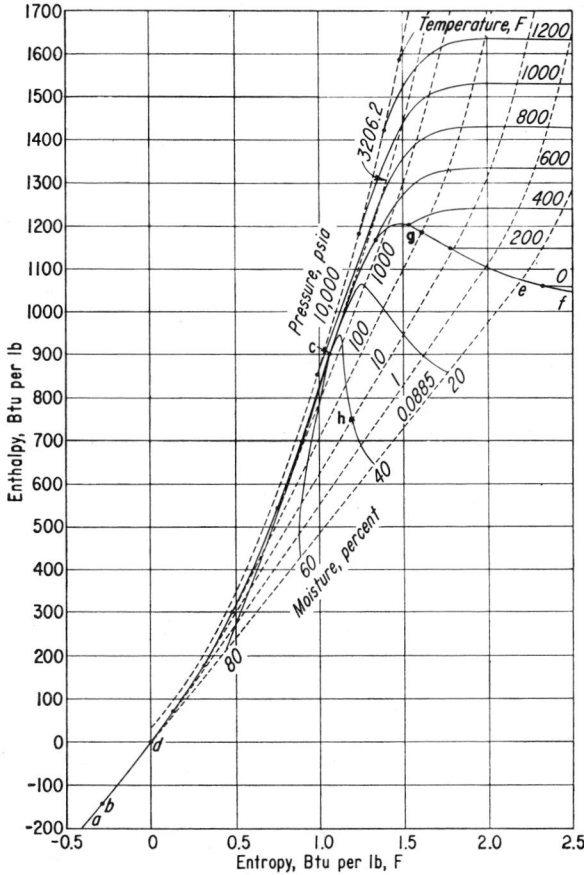

Fig. 1   Simplified Mollier diagram for steam.

columns of the steam tables. The specific volume v_g of the saturated vapor of 160-psia steam is, from the tables, 2.834 cu ft/lb, and the entropy s_g is 1.5640 Btu/(lb)(F).

(*b*). Every steam table contains a separate tabulation of the properties of superheated steam. To enter the superheated steam table, two arguments are needed—the absolute pressure and the temperature of the steam in F. To determine the properties of 1,100-psia 600-F steam, enter the superheated steam table at the given absolute pressure and project horizontally from this absolute pressure (1,100 psia) to the column corresponding to the superheated temperature (600 F) to read the enthalpy of the superheated vapor as $h = 1,236.7$ Btu/lb, and the specific volume of the superheated vapor $v = 0.4532$ cu ft/lb.

(*c*). For high-pressure steam use the ASME—*Steam Table*, entering it in the same manner as the superheated steam table. Thus, for 7,500-psia 1200-F steam, the enthalpy of the superheated vapor is 1,474.9 Btu/lb and the specific volume of the superheated vapor is 0.1060 cu ft/lb.

(d). To determine the enthalpy, specific volume, and entropy of wet steam having y percent moisture by using the steam tables instead of the Mollier diagram, apply these relations: $h = h_g - yh_{fg}/100$; $v = v_g - yv_{fg}/100$; $s = s_g - ys_{fg}/100$, where y = percent moisture expressed as a whole number. For 10-psia steam containing 40 percent moisture, obtain the needed values—h_g, h_{fg}, v_g, v_{fg}, s_g, and s_{fg} from the *saturation-pressure steam table* and substitute in the above relations. Thus,

$$h = 1,143.3 - \frac{40(982.1)}{100} = 750.5 \text{ Btu/lb}$$

$$v = 38.42 - \frac{40(38.40)}{100} = 23.06 \text{ cu ft/lb}$$

$$s = 1.7876 - \frac{40(1.5041)}{100} = 1.1860 \text{ Btu/(lb)(F)}$$

Note that the Keenan and Keyes *Thermodynamic Properties of Steam* do not tabulate v_{fg}. Therefore this value must be obtained by subtraction of the tabulated values, or $v_{fg} = v_g - v_f$. The v_{fg} value thus obtained is used in the relation for the volume of the wet steam. For 10-psia steam containing 40 percent moisture, $v_g = 38.42$ cu ft/lb, $v_f = 0.017$ cu ft/lb. Then, $v_{fg} = 38.42 - 0.017 = 28.403$ cu ft/lb.

In some instances, the *quality* of steam may be given instead of its moisture content in percent. The quality of steam is the percent vapor in the mixture. In the above calculation, the quality of the steam is 60 per cent because 40 percent is moisture. Thus, quality $= 1 - m$, where m = percent moisture, expressed as a decimal.

INTERPOLATION OF STEAM TABLE VALUES

(a) Determine the enthalpy, specific volume, entropy, and temperature of saturated steam at 151 psia. (b) Determine the enthalpy, specific volume, entropy, and pressure of saturated steam at 261 F. (c) Determine the pressure of steam at 1000 F if its specific volume is 2.6150 cu ft/lb. (d) Determine the enthalpy, specific volume, and entropy of 300-psia steam at 567.22 F.

Calculation Procedure:

1. Use the saturation-pressure steam table

(a) Study of the saturation-pressure table shows that there is no pressure value for 151 psia listed. Therefore, it will be necessary to interpolate between the *next higher* and *next lower* tabulated pressure values. In this instance these values are 152 and 150 psia, respectively. The pressure for which properties are being found (151 psia) is called the *intermediate pressure*. At 152 psia: $h_g = 1,194.3$; $v_g = 2.977$; $s_g = 1.5683$; $t = 359.46$ F. At 150 psia: $h_g = 1,194.1$; $v_g = 3.015$; $s_g = 1.5694$; $t = 358.42$ F.

For the enthalpy, note that as the pressure increases, so does h_g. Therefore, the enthalpy at 151 psia (the intermediate pressure) will equal the enthalpy at 150 psia (the lower pressure used in the interpolation) *plus* the proportional change (difference between the intermediate pressure and the lower interpolating pressure) for a 1-psia pressure increase. Or, at any higher pressure: $h_{gi} = h_{g1} + [(p_i - p_1)/(p_h - p_1)](h_h - h_1)$, where h_{gi} = enthalpy at the intermediate pressure; h_{g1} = enthalpy at the *lower* pressure used in the interpolation; h_h = enthalpy at the *higher* pressure used in the interpolation; p_i = intermediate pressure; p_h and p_1 = higher and lower pressures used in the interpolation. Thus, using the enthalpy values obtained from the steam table for 150 and 152 psia, $h_{gi} = 1,194.1 + [(151 - 150)/(152 - 150)](1,194.3 - 1,194.1) = 1,194.2$ Btu/lb at 151 psia saturated.

Next study the steam table to determine the direction of change of specific volume between the lower and higher pressures. This study shows that the specific volume *decreases* as the pressure increases. Therefore, the specific volume at 151 psia (the intermediate pressure) will equal the specific volume at 150 psia (the lower pressure used in the interpolation) *minus* the proportional change (difference between the intermediate pressure and the lower interpolating pressure) for a 1-psia pressure increase. Or, at any higher pressure, $v_{gi} = v_{gi} - [(p_i - p_1)/(p_h - p_1)](v_l - v_h)$, where the subscripts are the same as above and v = specific volume at the respective pressure. With the volume values obtained from the steam tables for 150 and 152 psia, $v_{gi} = 3.015 - [(151 - 150)/(152 - 150)](3.015 - 2.977) = 2.996$ cu ft/lb at 151 psia saturated.

Study of the steam table for the direction of entropy change shows that entropy, like specific volume, *decreases* as the pressure increases. Therefore, the entropy at 151 psia (the intermediate pressure) will equal the entropy at 150 psia (the lower pressure used in the interpolation) *minus* the proportional change (difference between the intermediate pressure and the lower interpolating pressure) for a 1-psia pressure increase. Or, at any higher pressure, $s_{gi} = s_{g1} - [(p_i - p_1)/(p_h - p_1)](s_1 - s_h) = 1.5164 - [(151 - 150)/(152 - 150)](1.5694 - 1.5683) = 1.56885$ Btu/(lb)(F) at 151 psia saturated.

Study of the steam table for the direction of temperature change shows that the saturation temperature, like entha'py, *increases* as the pressure increases. Therefore, the temperature at 151 psia (the intermediate pressure) will equal the temperature at 150 psia (the lower pressure used in the interpolation) *plus* the proportional change (difference between the intermediate pressure and the lower interpolating pressure) for a 1-psia increase. Or, at any higher pressure, $t_{gi} = t_{g1} + [(p_i - p_1)/(p_h - p_1)] \times (t_h - t_1) = 358.42 + [(151 - 150)/(152 - 150)](359.46 - 358.42) = 358.94$ F at 151 psia saturated.

2. Use the saturation-temperature steam table

(b) Study of the saturation-temperature table shows that there is no temperature value of 261 F listed. Therefore, it will be necessary to interpolate between the *next higher* and *next lower* tabulated temperature values. In this instance these values are 262 and 260 F, respectively. The temperature for which properties are being found (261 F) is called the *intermediate temperature*.

At 262 F: $h_g = 1,168.0$ $v_g = 11.396$ $s_g = 1.6833$ $p_g = 36.646$

At 260 F: $h_g = 1,167.3$ $v_g = 11.763$ $s_g = 1.6860$ $p_g = 35.429$

For enthalpy, note that as the temperature increases, so does h_g. Therefore, the enthalpy at 261 F (the intermediate temperature) will equal the enthalpy at 260 F (the lower temperature used in the interpolation) *plus* the proportional change (difference between the intermediate temperature and the lower interpolating temperature) for a 1-F temperature increase. Or, at any higher temperature, $h_{gi} = h_{g1} + [(t_i - t_1)/(t_h - t_1)] \times (h_h - h_1)$, where h_{g1} = enthalpy at the *lower* temperature used in the interpolation; h_h = enthalpy at the *higher* temperature used in the interpolation; t_i = intermediate temperature; t_h and t_1 = higher and lower temperatures used in the interpolation. Thus, using the enthalpy values obtained from the steam table for 260 and 262 F, $h_{gi} = 1,167.3 + [(261 - 260)/(262 - 260)](1,168.0 - 1,167.3) = 1,167.65$ Btu/lb at 261 F saturated.

Next study the steam table to determine the direction of change of specific volume between the lower and higher temperatures. This study shows that the specific volume *decreases* as the pressure increases. Therefore, the specific volume at 261 F (the intermediate temperature) will equal the specific volume at 260 F (the lower temperature used in the interpolation) *minus* the proportional change (difference between the intermediate temperature and the lower interpolating temperature) for a 1-F temperature increase. Or, at any higher temperature, $v_{gi} = v_{gl} - [(t_i - t_l)/(t_h - t_l)](v_l - v_h) = 11.763 - [(261 - 260)/(262 - 260)](11.763 - 11.396) = 11.5795$ cu ft/lb at 261 F saturated.

Study of the steam table for the direction of entropy change shows that entropy, like specific volume, *decreases* as the temperature increases. Therefore, the entropy at

261 F (the intermediate temperature) will equal the entropy at 260 F (the lower temperature used in the interpolation) *minus* the proportional change (difference between the intermediate temperature and the lower interpolating temperature) for a 1-F temperature increase. Or, at any higher temperature, $s_{gi} = s_{gl} - [(t_i - t_l)/(t_h - t_l)]$ $(s_l - s_h) = 1.6860 - [(261 - 260)/(262 - 260)](1.6860 - 1.6833) = 1.68465$ Btu/(lb)(F) at 261 F.

Study of the steam table for the direction of pressure change shows that the saturation pressure, like enthalpy, *increases* as the temperature increases. Therefore, the pressure at 261 F (the intermediate temperature) will equal the pressure at 260 F (the lower temperature used in the interpolation) *plus* the proportional change (difference between the intermediate temperature and the lower interpolating temperature) for a 1-F temperature increase. Or, at any higher temperature, $p_{gi} = p_{gl} + [(t_i - t_l)/(t_h - t_l)](p_h - p_l) = 35.429 + [(261 - 260)/(262 - 260)](36.646 - 35.429) = 36.0375$ psia at 261 F saturated.

3. Use the superheated-steam table

(c) Choose the superheated-steam table for steam at 1000 F and 2.6150 cu ft/lb because the highest temperature at which saturated steam can exist is 705.4 F. This is also the highest temperature tabulated in some saturated-temperature tables. Therefore, the steam is superheated when at a temperature of 1000 F.

Look down the 1000-F columns in the superheat table until a specific volume value of 2.6150 is found. This occurs between 325 psia ($v = 2.636$) and 330 psia ($v = 2.596$). Since there is no volume value exactly equal to 2.6150 tabulated, it will be necessary to interpolate. List the values from the steam tables thus:

$$p = 325 \text{ psia} \qquad t = 1000 \text{ F} \qquad v = 2.636$$
$$p = 330 \text{ psia} \qquad t = 1000 \text{ F} \qquad v = 2.596$$

Note that as the pressure rises, at constant temperature, the volume *decreases*. Therefore, the intermediate (or unknown) pressure is found by subtracting from the higher interpolating pressure (330 psia in this instance) the product of the proportional change in specific volume and the difference in the pressures used for the interpolation. Or, $p_{gi} = p_h - [(v_i - v_h)/(v_l - v_h)](p_h - p_l)$, where the subscripts h, l, and i refer to the high, low, and intermediate (or unknown) pressures, respectively. In this instance, $p_{gi} = 330 - [(2.615 - 2.596)/(2.636 - 2.596)](330 - 325) = 327.62$ psia at 1000 F and a specific volume of 2.6150 cu ft/lb.

4. Use the superheated-steam table

(d) When given a steam pressure and temperature, determine, before performing any interpolation, the state of the steam. Do this by entering the saturation-pressure table at the given pressure and noting the saturation temperature. If the given temperature exceeds the saturation temperature, the steam is superheated. In this instance, the saturation-pressure table shows that at 300-psia the saturation temperature is 417.33 F. Since the given temperature of the steam is 567.22 F, the steam is superheated because its actual temperature is greater than the saturation temperature.

Enter the superheated steam table at 300 psia and find the next temperature *lower* than 567.22 F; this is 560 F. Also find the next *higher* temperature; this is 580 F. Tabulate the enthalpy, specific volume, and entropy for each of these temperatures thus:

$$t = 560 \text{ F} \qquad h = 1,292.5 \qquad v = 1.9128 \qquad s = 1.6054$$
$$t = 580 \text{ F} \qquad h = 1,303.7 \qquad v = 1.9594 \qquad s = 1.6163$$

Use the same procedures for each property—enthalpy, specific volume, and entropy—as given in Step 2 above but change the sign between the lower volume and entropy and the proportional factor (temperature in this instance), because for superheated steam the volume and entropy *increase* as the steam temperature increases. Thus:

$$h_{gi} = 1,292.5 + \frac{567.22 - 560}{580 - 560}(1,303.7 - 1,292.5) = 1,296.6 \text{ Btu/lb}$$

$$v_{gi} = 1.9128 + \frac{567.22 - 560}{580 - 560}(1.9594 - 1.9128) = 1.9296 \text{ cu ft/lb}$$

$$s_{gi} = 1.6054 + \frac{567.22 - 560}{580 - 560}(1.6163 - 1.6054) = 1.6093 \text{ Btu/(lb)(F)}$$

NOTE: Also observe the direction of change of a property *before* interpolating. Use a *plus* or *minus* sign between the higher interpolating value and the proportional change depending on whether the tabulated value increases (+) or decreases (−).

CONSTANT-PRESSURE STEAM PROCESS

Three pounds of wet steam, containing 15 percent moisture and initially at a pressure of 400 psia, expand at constant pressure ($P = C$) to 600 F. Determine the initial temperature T_1; enthalpy H_1; internal energy E_1; volume V_1; entropy S_1; final entropy H_2; internal energy E_2; volume V_2; entropy S_2; heat added to the steam Q_1; work output W_2; change in internal energy ΔE; change in specific volume ΔV; change in entropy ΔS.

Calculation Procedure:

1. Determine the initial steam temperature from the steam tables

Enter the saturation-pressure table at 400 psia and read the saturation temperature as 444.59 F.

2. Correct the saturation values for the moisture of the steam in the initial state

Sketch the process on a pressure-volume (P-V), Mollier (H-S), or temperature-entropy (T-S) diagram, Fig. 2. In state 1, y = moisture content = 15 percent. Using the appropriate values from the saturation-pressure steam table for 400 psia, correct them for a moisture content of 15 percent:

$$H_1 = h_g - yh_{fg} = 1{,}204.5 - 0.15(780.5) = 1{,}087.4 \text{ Btu/lb}$$
$$E_1 = u_g - yu_{fg} = 1{,}118.5 - 0.15(695.9) = 1{,}015.1 \text{ Btu/lb}$$
$$V_1 = v_g - yv_{fg} = 1.1613 - 0.15(1.1420) = 0.990 \text{ cu ft/lb}$$
$$S_1 = s_g - ys_{fg} = 1.4844 - 0.15(0.8630) = 1.2945 \text{ Btu/(lb)(F)}$$

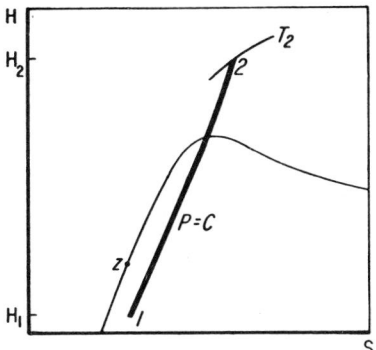
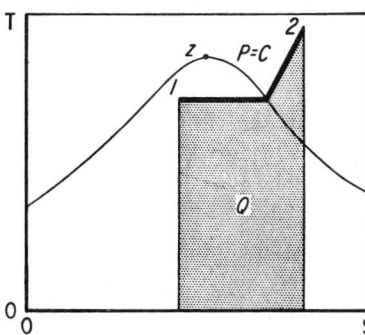

Fig. 2   Constant-pressure process.

3. Determine the steam properties in the final state

Since this is a constant-pressure process, the pressure in state 2 is 400 psia, the same as state 1. The final temperature is given as 600 F. This is greater than the saturation temperature of 444.59 F. Hence, the steam is superheated when in state 2. Use the superheated steam tables, entering at 400 psia and 600 F. At this condition, $H_2 = 1,306.9$ Btu/lb; $V_2 = 1.477$ cu ft/lb. Then, $E_2 = h_{2g} - P_2V_2/J = 1,306.9 - 400(144)(1.477)/778 = 1,197.5$ Btu/lb. In this equation, the constant 144 converts psia to psf, and J = mechanical equivalent of heat = 778 ft-lb/Btu. From the steam tables, $S_2 = 1.5894$ Btu/(lb)(F).

4. Compute the process inputs, outputs, and changes

$W_2 = (P_1/J)(V_2 - V_1)m = [400(144)/778](1.4770 - 0.9900)(3) = 108.1$ Btu. In this equation, m = weight of steam used in the process = 3 lb. Then:

$$Q_1 = (H_2 - H_1)m = (1,306.9 - 1,087.4)(3) = 658.5 \text{ Btu}$$
$$\Delta E = (E_2 - E_1)m = (1,197.5 - 1,014.1)(3) = 550.2 \text{ Btu}$$
$$\Delta V = (V_2 - V_1)m = (1.4770 - 0.9900)(3) = 1.461 \text{ cu ft}$$
$$\Delta S = (S_2 - S_1)m = (1.5894 - 1.2945)(3) = 0.8847 \text{ Btu/F}$$

5. Check the computations

The work output W_2 should equal the change in internal energy plus the heat input, or $W_2 = E_1 - E_2 + Q_1 = -550.2 + 658.5 = 108.3$ Btu. This value very nearly equals the computed value of $W_2 = 108.1$ Btu and is close enough for all normal engineering computations. The difference can be traced to slide-rule reading errors. In computing the work output, the internal-energy change has a negative sign because there is a decrease in E during the process.

Related Calculations: Use this Procedure for all constant-pressure steam processes.

CONSTANT-VOLUME STEAM PROCESS

Five pounds of wet steam initially at 120 psia with 30 percent moisture are heated at constant volume ($V = C$) to a final temperature of 1000 F. Determine the initial temperature T_1; enthalpy H_1; internal energy E_1; volume V_1; final pressure P_2; enthalpy H_2; internal energy E_2; volume V_2; heat added Q_1; work output W; change in internal energy ΔE, volume ΔV, and entropy ΔS.

Calculation Procedure:

1. Determine the initial steam temperature from the steam tables

Enter the saturation-pressure table at 120 psia, the initial pressure, and read the saturation temperature, $T_1 = 341.25$ F.

2. Correct the saturation values for the moisture in the steam in the initial state

Sketch the process on P-V, H-S, or T-S diagrams, Fig. 3. Using the appropriate values from the saturation pressure table for 120 psia, correct them for a moisture content of 30 percent:

$$H_1 = h_g - yh_{fg} = 1,190.4 - 0.3(877.9) = 927.0 \text{ Btu/lb}$$
$$E_1 = u_g - yu_{fg} = 1,107.6 - 0.3(795.6) = 868.9 \text{ Btu/lb}$$
$$V_1 = v_g - yv_{fg} = 3.7280 - 0.3(3.7101) = 2.6150 \text{ cu ft/lb}$$
$$S_1 = s_g - ys_{fg} = 1.5878 - 0.3(1.0962) = 1.2589 \text{ Btu/(lb)(F)}$$

3. Determine the steam volume in the final state

$T_2 = 1000$ F, given. Since this is a constant-volume process, $V_2 = V_1 = 2.6150$ cu ft/lb. The total volume of the vapor equals the product of the specific volume and the number of pounds of vapor used in the process, or total volume = 2.6150(5) = 13.075 cu ft.

4. Determine the final steam pressure

The final steam temperature (1000 F) and the final steam volume (2.6150 cu ft) are known. To determine the final steam pressure, find in the steam tables the state corresponding to the above temperature and specific volume. Since a temperature of 1000 F is higher than any saturation temperature (705.4 F is the highest saturation

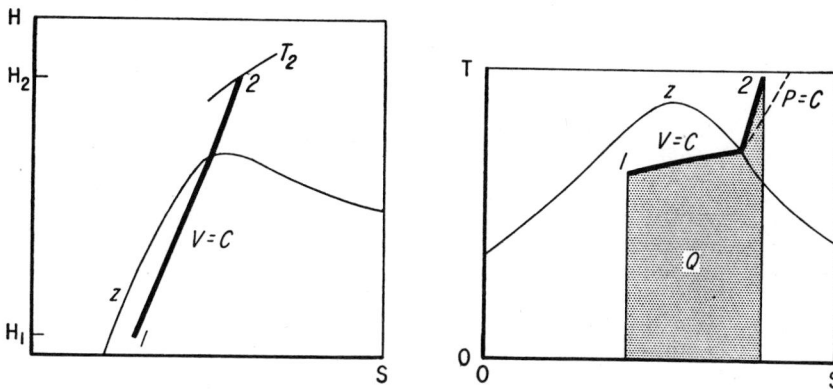

Fig. 3   Constant-volume process.

temperature for saturated steam), the steam in state 2 must be superheated. Therefore, the superheated steam tables must be used to determine P_2.

Enter the 1000-F column in the steam table and look for a superheated-vapor specific volume of 2.6150 cu ft/lb. At a pressure of 325 psia,

$$v = 2.636 \qquad h = 1,524.5 \qquad s = 1.7863$$

and at a pressure of 330 psia

$$v = 2.596 \qquad h = 1,524.4 \qquad s = 1.7845$$

Thus, 2.6150 lies between 325 and 330 psia. To determine the pressure corresponding to the final volume, it is necessary to interpolate between the specific-volume values of $P_2 = 330 - [(2.615 - 2.596)/(2.636 - 2.596)](330 - 325) = 327.62$ psia. In this equation, the volume values correspond to the upper (330 psia), lower (325 psia), and unknown pressures.

5.   Determine the final enthalpy, entropy, and internal energy

The final enthalpy can be interpolated in the same manner, using the enthalpy at each volume instead of the pressure. Thus $H_2 = 1,524.5 - [(2.615 - 2.596)/(2.636 - 2.596)](1,524.5 - 1,524.4) = 1,524.45$ Btu/lb. Since the difference in enthalpy between the two pressures is only 0.1 Btu/lb (= 1,524.5 - 1,524.4), the enthalpy at 327.62 psia could have been assumed = the enthalpy at the lower pressure (325 psia), or 1,524.4 Btu/lb, and the error would have been only 0.05 Btu/lb, which is negligible. However, where the enthalpy values vary by more than 1.0 Btu/lb, interpolate as shown, if accurate results are desired.

Find S_2 by interpolating between pressures, or

$$S_2 = 1.7863 - \frac{327.62 - 325}{330 - 325}(1.7863 - 1.7845) = 1.7854 \text{ Btu/(lb)(F)}$$

$$E_2 = H_2 - P_2 V_2/J = 1524.4 - \frac{327.62(144)(2.615)}{778} = 1,365.9 \text{ Btu/lb}$$

6.   Compute the changes resulting from the process

$Q_1 = (E_2 - E_1)m = (1,365.9 - 868.9)(5) = 2,485$ Btu; $\Delta S = (S_2 - S_1)m = (1.7854 - 1.2589)(5) = 2.6325$ Btu/F.

By definition, $W = 0$; $\Delta V = 0$; $\Delta E = Q_1$. Note that the curvatures of the constant-volume line on the T-S chart, Fig. 3, are different from the constant-pressure line,

Fig. 2. Adding heat Q_1 to a constant-volume process affects only the internal energy. The total entropy change must take into account the total steam mass $m = 5$ lb.

Related Calculations: Use this general procedure for all constant-volume steam processes.

CONSTANT-TEMPERATURE STEAM PROCESS

Six lb of wet steam initially at 1,200-psia and 50 percent moisture expand at constant temperature $(T = C)$ to 300 psia. Determine the initial temperature T_1; enthalpy H_1; internal energy E_1; specific volume V_1; entropy S_1; final temperature T_2; enthalpy H_2; internal energy E_2; volume V_2; entropy S_2; heat added Q_1; work output W_2; change in internal energy ΔE, volume ΔV, and entropy ΔS.

Calculation Procedure:

1. Determine the initial steam temperature from the steam tables

Enter the saturation-pressure table at 1,200 psia, the initial pressure, and read the saturation temperature $T_1 = 567.22$ F.

2. Correct the saturation values for the moisture in the steam in the initial state

Sketch the process on P-V, H-S, or T-S diagrams, Fig. 4. Using the appropriate values from the saturation-pressure table for 1,200 psia, correct them for the moisture content of 50 percent.

$$H_1 = h_g - y_1 h_{fg} = 1,183.4 - 0.5(611.7) = 877.5 \text{ Btu/lb}$$
$$E_1 = u_g - y_1 u_{fg} = 1,103.0 - 0.5(536.3) = 834.8 \text{ Btu/lb}$$
$$V_1 = u_g - y_1 u_{fg} = 0.3619 - 0.5(0.3396) = 0.1921 \text{ cu ft/lb}$$
$$S_1 = s_g - y_1 s_{fg} = 1.3667 - 0.5(0.5956) = 1.0689 \text{ Btu/(lb)(F)}$$

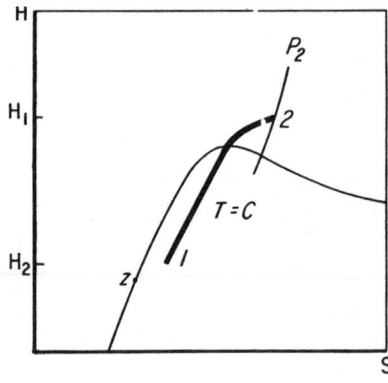
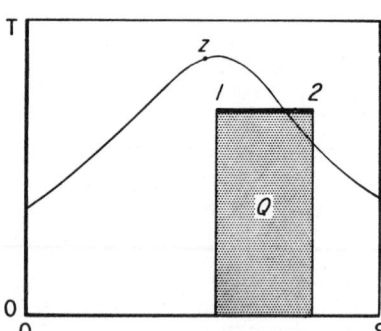

Fig. 4   Constant-temperature process.

3. Determine the steam properties in the final state

Since this is a constant-temperature process, $T_2 = T_1 = 567.22$ F; $P_2 = 300$ psia, given. The saturation temperature of 300-psia steam is 417.33 F. Therefore, the steam is superheated in the final state because 567.22 F > 417.33 F, the saturation temperature.

To determine the final enthalpy, entropy, and specific volume it is necessary to interpolate between the known final temperature and the nearest tabulated temperatures greater and less than the final temperature. Or,

At $T = 560$ F:
$v = 1.9128$ $\quad$ $h = 1,292.5$ $\quad$ $s = 1.6054$
At $T = 580$ F:
$v = 1.9594$ $\quad$ $h = 1,303.7$ $\quad$ $s = 1.6163$

Then,

$$H_2 = 1{,}292.5 + \frac{567.22 - 560}{580 - 560}(1{,}303.7 - 1{,}292.5) = 1{,}296.5 \text{ Btu/lb}$$

$$S_2 = 1.6054 + \frac{567.22 - 560}{580 - 560}(1.6163 - 1.6054) = 1.6093 \text{ Btu/(lb)(F)}$$

$$V_2 = 1.9128 + \frac{567.22 - 560}{580 - 560}(1.9594 - 1.9128) = 1.9296 \text{ cu ft/lb}$$

$$E_2 = H_2 - P_2 V_2/J = 1{,}296.5 - 300(144)(1.9296)/778 = 1{,}109.3 \text{ Btu/lb}$$

4. Compute the process changes

$Q_1 = T(S_2 - S_1)_m$, where $T_1 =$ absolute initial temperature, R. $Q_1 = (567.22 + 460)$ $(1.6093 - 1.0689)(6) = 3{,}330$ Btu. Then,

$$\Delta E = E_2 - E_1 = 1{,}109.3 - 834.8 = 274.5 \text{ Btu/lb}$$
$$\Delta H = H_2 - H_1 = 1{,}296.5 - 877.5 = 419.0 \text{ Btu/lb}$$
$$W_2 = (Q_1 - \Delta E)m = (555 - 274.5)(6) = 1.683 \text{ Btu}$$
$$\Delta S = S_2 - S_1 = 1.6093 - 1.0689 = 0.5404 \text{ Btu/(lb)(F)}$$
$$\Delta V = V_2 - V_1 = 1.9296 - 0.1921 = 1.7375 \text{ cu ft/lb}$$

Related Calculations: Use this Procedure for any constant-temperature steam process.

CONSTANT-ENTROPY STEAM PROCESS

Ten pounds of steam expand under two conditions — nonflow and steady flow — at constant entropy. $(S = C)$ from an initial pressure of 2,000 psia and a temperature of 800 F to a final pressure of 2 psia. In the steady-flow process, assume that the initial kinetic energy $E_{k1} =$ the final kinetic energy E_{k2}. Determine the initial enthalpy H_1; internal energy E_1; volume V_1; entropy S_1; final temperature T_2; percent moisture y; enthalpy H_2; internal energy E_2; volume V_2; entropy S_2; change in internal energy ΔE, enthalpy ΔH, entropy ΔS, volume ΔV; heat added Q_1; and work output W_2.

Calculation Procedure:

1. Determine the initial enthalpy, volume, and entropy from the steam tables

Enter the superheated-vapor table at 2,000 psia and 800 F and read $H_1 = 1{,}335.5$ Btu/lb; $V_1 = 0.3074$ cu ft/lb; $S_1 = 1.4576$ Btu/(lb)(F).

2. Compute the initial internal energy

$$E_1 = H_1 - \frac{P_1 V_1}{J} = 1{,}335.5 - \frac{2{,}000(144)(0.3074)}{778} = 1{,}221.6 \text{ Btu/lb}$$

3. Determine the vapor properties in the final state

Sketch the process on P-V, H-S, or T-S diagrams, Fig. 5. Note that the expanded steam is wet in the final state because the 2-psia pressure line is under the saturation curve on the H-S and T-S diagrams. Therefore, the vapor properties in the final state must be corrected for the moisture content. Read, from the saturation-pressure steam table, the liquid and vapor properties at 2 psia. Tabulate these properties thus:

$s_f = 0.1749$	$s_{fg} = 1.7451$	$s_g = 1.9200$
$h_f = 93.99$	$h_{fg} = 1022.2$	$h_g = 1{,}116.2$
$u_f = 93.98$	$u_{fg} = 957.9$	$u_g = 1{,}051.9$
$v_f = 0.016$	$v_{fg} = 173.71$	$v_g = 173.73$

Since this is a constant-entropy process, $S_2 = S_1 = s_g - y_2 s_{fg}$. Solve for y_2, the percent moisture in the final state. Or, $y_2 = (s_g - S_1)/s_{fg} = (1.9200 - 1.4576)/1.7451 = 0.265$, or 26.5 percent.

Then,

$$H_2 = h_g - y_2 h_{fg} = 1{,}116.2 - 0.265(1{,}022.2) = 845.3 \text{ Btu/lb}$$
$$E_2 = u_g - y_2 u_{fg} = 1{,}051.9 - 0.265(957.9) = 798.0 \text{ Btu/lb}$$
$$V_2 = v_g - y_2 v_{fg} = 173.73 - 0.265(173.71) = 127.7 \text{ cu ft/lb}$$

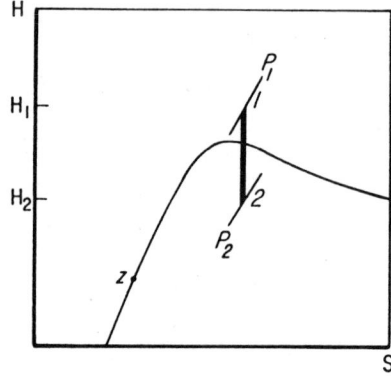 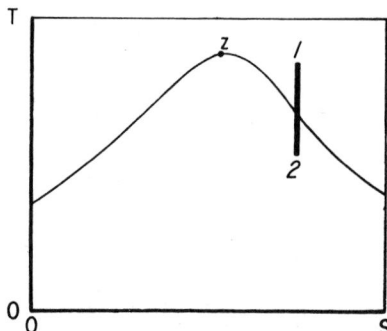

Fig. 5   Constant-entropy process.

4.   *Compute the changes resulting from the process*

The *total* change in properties is for 10 lb of steam, the quantity used in this process. Thus,

$$\Delta E = (E_1 - E_2)m = (1{,}221.6 - 798.0)(10) = 4{,}236 \text{ Btu}$$
$$\Delta H = (H_1 - H_2)m = (1{,}335.5 - 845.3)(10) = 4{,}902 \text{ Btu}$$
$$\Delta S = (S_1 - S_2)m = (1.4576 - 1.4576)(10) = 0 \text{ Btu/F}$$
$$\Delta V = (V_1 - V_2)m = (0.3074 - 127.7)(10) = -1{,}274 \text{ cu ft}$$

$Q_1 = 0$ Btu. (By definition, there is no transfer of heat in a constant-entropy process.) Nonflow $W_2 = \Delta E = 4{,}236$ Btu. Steady flow $W_2 = \Delta H = 4{,}902$ Btu.

NOTE: In a constant-entropy process, the nonflow work depends on the change in internal energy. The steady-flow work depends on the change in enthalpy, and is larger than the nonflow work by the amount of the change in the flow work.

IRREVERSIBLE ADIABATIC EXPANSION OF STEAM

Ten pounds of steam undergo a steady-flow expansion from an initial pressure of 2,000 psia and a temperature 800 F to a final pressure of 2 psia at an expansion efficiency of 75 percent. In this steady flow, assume $E_{k1} = E_{k2}$. Determine ΔE, ΔH, ΔS, ΔV, Q, and W_2.

Calculation Procedure:

1.   *Determine the initial vapor properties from the steam tables*

Enter the superheated-vapor tables at 2,000 psia and 800 F and read $H_1 = 1{,}335.5$ Btu/lb; $V_1 = 0.3074$ cu ft/lb; $E_1 = 1{,}221.6$ Btu/lb; $S_1 = 1.4576$ Btu/(lb)(F).

2.   *Determine the vapor properties in the final state*

Sketch the process on P-V, H-S, or T-S diagrams, Fig. 6. Note that the expanded steam is wet in the final state because the 2-psia pressure line is under the saturation curve on the H-S and T-S diagrams. Therefore, the vapor properties in the final state must be corrected for the moisture content. However, the actual final enthalpy cannot be determined until after the expansion efficiency $[H_1 - H_2(H_1 - H_{2s})]$ is evaluated.

To determine the final enthalpy H_2, another enthalpy H_{2s} must first be computed by assuming a constant-entropy expansion to 2 psia and a temperature of 126.08 F. Enthalpy H_{2s} will then be that corresponding to a constant-entropy expansion into the wet region and the percent moisture will be that corresponding to the final state. This percentage

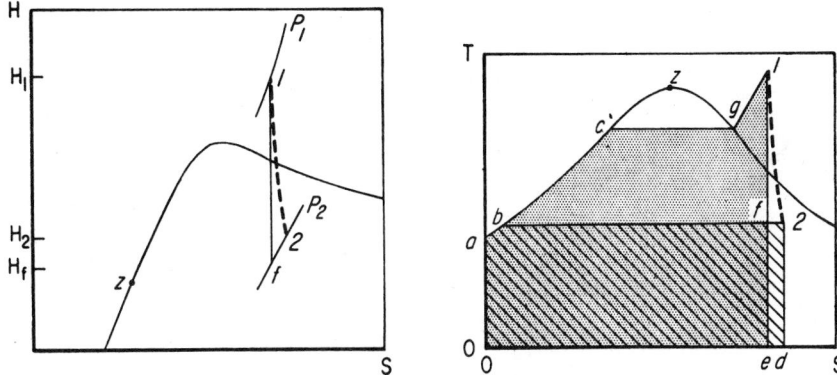

Fig. 6 Irreversible adiabatic process.

is determined by finding the ratio of $s_g - S_1$ to s_{fg}, or $y_{2s} = s_g - S_1/s_{fg} = 1.9200 - 1.4576/1.7451 = 0.265$, where s_g and s_{fg} are entropies at 2 psia. Then, $H_{2s} = h_g - y_{2s}h_{fg} = 1,116.2 - 0.265(1,022.2) = 845.3$ Btu/lb. In this relation, h_g and h_{fg} are enthalpies at 2 psia.

The expansion efficiency, given as 0.75, then is $H_1 - H_2/(H_1 - H_{2s}) = $ actual work/ideal work $= 0.75 = 1,335.5 - H_2/(1,335.5 - 845.3)$. Solve for $H_2 = 967.9$ Btu/lb.

Next, read from the saturation-pressure steam table the liquid and vapor properties at 2 psia. Tabulate these properties thus:

$$h_f = 93.99 \qquad h_{fg} = 1,022.2 \qquad h_g = 1,116.2$$
$$s_f = 0.1749 \qquad s_{fg} = 1.7451 \qquad s_g = 1.9200$$
$$u_f = 93.98 \qquad u_{fg} = 957.9 \qquad u_g = 1,051.9$$
$$v_f = 0.016 \qquad v_{fg} = 173.71 \qquad v_g = 173.73$$

Since the actual final enthalpy H_2 is different from H_{2s}, the final actual moisture y_2 must be computed using H_2. Or $y_2 = h_g - H_2/h_{fg} = 1,116.1 - 967.9/1,022.2 = 0.1451$. Then,

$$E_2 = u_g - y_2u_{fg} = 1,051.9 - 0.1451(957.9) = 912.9 \text{ Btu/lb}$$
$$V_2 = v_g - y_2v_{fg} = 173.73 - 0.1451(173.71) = 148.5 \text{ cu ft/lb}$$
$$S_2 = s_g - y_2s_{fg} = 1.9200 - 0.1451(1.7451) = 1.6668 \text{ Btu/(lb)(F)}$$

3. *Compute the changes resulting from the process*

The total change in properties is for 10 lb of steam, the quantity used in this process. Thus:

$$\Delta E = (E_1 - E_2)m = (1,221.6 - 912.9)(10) = 3,087 \text{ Btu}$$
$$\Delta H = (H_1 - H_2)m = (1,335.5 - 967.9)(10) = 3,676 \text{ Btu}$$
$$\Delta S = (S_2 - S_1)m = (1.6668 - 1.4576)(10) = 2.092 \text{ Btu/F}$$
$$\Delta V = (V_2 - V_1)m = (148.5 - 0.3074)(10) = 1,482 \text{ cu ft}$$
$$Q = 0; \text{ by definition } W_2 = \Delta H = 3,676 \text{ Btu for the steady-flow process.}$$

IRREVERSIBLE ADIABATIC STEAM COMPRESSION

Two pounds of saturated steam at 120 psia with 80 percent quality undergo nonflow adiabatic compression to a final pressure of 1,700 psia at 75 percent compression efficiency. Determine the final steam temperature T_2; change in internal energy ΔE; change in entropy ΔS; work input W; and heat input Q.

Calculation Procedure:

1. Determine the vapor properties in the initial state

From the saturation-pressure steam tables, $T_1 = 341.25$ F at a pressure of 120 psia saturated. With $x_1 = 0.8$, $E_1 = u_f + x_1 u_{fg} = 312.05 + 0.8(795.6) = 948.5$ Btu/lb, using inter-nal-energy values from the steam tables. The initial entropy is $S_1 = s_f + x_1 s_{fg} = 0.4916 + 0.8(1.0962) = 1.3686$ Btu/(lb)(F).

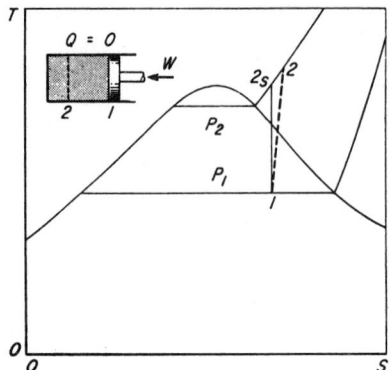

2. Determine the vapor properties in the final state

Sketch a T-S diagram of the process, Fig. 7. Assume a constant-entropy compression from the initial to the final state. Then, $S_{2s} = S_1 = 1.3686$ Btu/(lb)(F).

The final pressure, 1,700 psia, is known, as is the final entropy, 1.3686 Btu/(lb)(F), with constant-entropy expansion. The T-S dia-gram, Fig. 7, shows that the steam is super-heated in the final state. Enter the super-heated steam table at 1,700 psia and project across to an entropy of 1.3686 and read the final steam temperature as 650 F. (In most cases, the final entropy would not exactly equal a tabulated value and it would be nec-essary to interpolate between tabulated entropy values to determine the intermediate pressure value.)

Fig. 7 Irreversible adiabatic compression process.

From the same table, at 1,700 psia and 650 F, $H_{2s} = 1,214.4$ Btu/lb; $V_{2s} = 0.2755$ cu ft/lb. Then, $E_{2s} = H_{2s} - P_2 V_{2s}/H = 1,214.4 - 1,700(144)(0.2755)/778 = 1,127.8$ Btu/lb. Since E_1 and E_{2s} are known, the ideal work W can be computed. Or, $W = E_{2s} - E_1 = 1,127.8 - 948.5 = 179.3$ Btu/lb.

3. Compute the vapor properties of the actual compression

Since the compression efficiency is known, the actual final internal energy can be found from compression efficiency = ideal W/actual $W = E_{2s} - E_1/(E_2 - E_1)$, or $0.75 = 1,127.8 - 948.5/(E_2 - 948.5)$; $E_2 = 1,187.6$ Btu/lb. Then, $E = (E_2 - E_1)m = (1,187.6 - 948.5)(2) = 478.2$ Btu, for 2 lb of steam. The actual work input $W = \Delta E = 478.2$ Btu. By definition, $Q = 0$.

Lastly, the actual final temperature and entropy must be computed. The final actual internal energy $E_2 = 1,187.6$ Btu/lb is known. Also, the $T - S$ diagram, Fig. 7, shows that the steam is superheated. However, the superheated steam tables do not list the internal energy of the steam. Therefore, it is necessary to assume a final temperature for the steam and then compute its internal energy. The computed value is compared with the known internal energy and the next assumption is adjusted as necessary. Thus, assume a final temperature of 720 F. This assumption is higher than the ideal final temperature of 650 F because the $T - S$ diagram, Fig. 7, shows that the actual final temperature is higher than the ideal final temperature. Using values from the superheated steam table for 1,700 psia and 720 F,

$$E = H - \frac{PV}{J} = 1,288.4 - \frac{1,700(144)(0.3283)}{778} = 1,185.1 \text{ Btu/lb}$$

This value is *less than* the actual internal energy of 1,187.6 Btu/lb. Therefore, the actual temperature must be higher than 720 F, since the internal energy increases with temperature. To obtain a higher value for the internal energy to permit interpolation between the lower, actual, and higher values, assume a higher final temperature – in this case the next temperature listed in the steam table, or 740 F. Then, for 1,700 psia and 740 F,

$$E = 1305.8 - \frac{1,700(144)(0.3410)}{778} = 1,198.5 \text{ Btu/lb}$$

This value is *greater than* the actual internal energy of 1,187.6 Btu/lb. Therefore, the actual final temperature of the steam lies somewhere between 720 and 740 F. Interpolate between the known internal energies to determine the final steam temperature and final entropy. Or,

$$T_2 = 720 + \frac{1,187.6 - 1,185.1}{1,198.5 - 1,185.1}(740 - 720) = 723.7 \text{ F}$$

$$S_2 = 1.4333 + \frac{1,187.6 - 1,185.1}{1,198.5 - 1,185.1}(1.4480 - 1.4333) = 1.4360, \text{ Btu/(lb)(F)}$$

$$\Delta S = (S_2 - S_1)m = (1.4360 - 1.3686)(2) = 0.1348 \text{ Btu/F}$$

Note that the final actual steam temperature is 73.7 F higher than that (650 F) for the ideal compression.

Related Calculations: Use this procedure for any irreversible adiabatic steam process.

THROTTLING PROCESSES FOR STEAM AND WATER

A throttling process begins at 500 psia and ends at 14.7 psia with (*a*) steam at 500 psia and 500 F, (*b*) steam at 500 psia and 4 percent moisture, (*c*) steam at 500 psia with 50 percent moisture, and (*d*) saturated water at 500 psia. Determine the final enthalpy H_2, temperature T_2, and moisture content y_2, for each process.

Calculation Procedure:

1. *Compute the final-state conditions of the superheated steam*

(*a*) From the superheated steam table for 500 psia and 500 F, $H_1 = 1,231.3$ Btu/lb. By definition of a throttling process, $H_1 = H_2 = 1,231.3$ Btu/lb. Sketch the *T-S* diagram for a throttling process (*a*), Fig. 8.

To determine the final temperature, enter the superheated steam table at 14.7 psia, the final pressure, and project across to an enthalpy value equal to or less than the known enthalpy, 1,231.3 Btu/lb. (The superheated steam table is used because the *T-S* diagram, Fig. 8, shows that the steam is superheated in the final state.) At 14.7 psia there is no tabulated enthalpy value that exactly equals 1,230 Btu/lb. The next lower value is 1,230 Btu/lb at $T = 380$ F. The next higher value at 14.7 psia is 1,239.9 Btu/lb at $T = 400$ F. Interpolate between these enthalpy values to find the final steam temperature. Or,

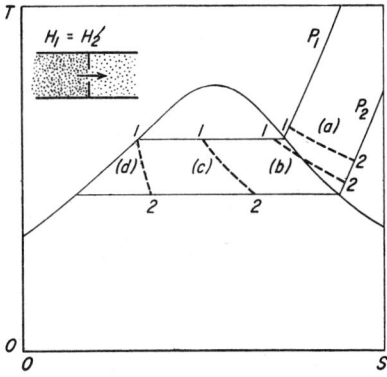

Fig. 8 Throttling process for steam.

$$T_2 = 380 + \frac{1,231.3 - 1,230.5}{1,239.9 - 1,230.5}(400 - 380)$$
$$= 381.7 \text{ F}$$

The steam does not contain any moisture in the final state because it is superheated.

2. Compute the final-state conditions of the slightly wet steam

(b) Determine the enthalpy of 500-psia saturated steam from the saturation-pressure steam table:

$$h_g = 1,204.4 \text{ Btu/lb} \qquad h_{fg} = 755.0 \text{ Btu/lb}$$

Correct the enthalpy for moisture:

$$H_1 = h_g - y_1 h_{fg} = 1,204.4 - 0.04(755.0) = 1,174.2 \text{ Btu/lb}$$

Then, by definition, $H_2 = H_1 = 1,174.2$ Btu/lb.

Determine the final condition of the throttled steam (wet, saturated, or superheated) by studying the T-S diagram. If a diagram was not drawn, enter the saturation-pressure steam table at 14.7 psia, the final pressure, and check the tabulated h_g. If the tabulated h_g is less than H_1, the throttled steam is superheated. If the tabulated h_g is less than H_1, the throttled steam is saturated. Examination of the saturation-pressure steam table shows that the throttled steam is superheated because $H_1 > h_g$.

Next, enter the superheated steam table to find an enthalpy value of H_1 at 14.7 psia. There is no value equal to 1,174.2 Btu/lb. The next lower value is 1,173.8 Btu/lb at $T = 260$ F. The next higher value at 14.7 psia is 1,183.3 Btu/lb at $T = 280$ F. Interpolate between these enthalpy values to find the final steam temperature. Or,

$$T_2 = 260 + \frac{1,174.2 - 1,173.8}{1,183.3 - 1,173.8}(280 - 260) = 260.8 \text{ F.}$$

This is higher than the temperature of saturated steam at 14.7 psia (212 F), giving further proof that the throttled steam is superheated. The throttled steam therefore does not contain any moisture.

3. Compute the final-state conditions of the very wet steam

(c) Determine the enthalpy of 500-psia saturated steam from the saturation-pressure steam table. Or $h_g = 1,204.4$ Btu/lb; $h_{fg} = 755.0$ Btu/lb. Correct the enthalpy for moisture:

$$H_1 = H_2 = h_g - y_1 h_{fg} = 1,204.4 - 0.5(755.0) = 826.9 \text{ Btu/lb}$$

Then, by definition, $H_2 = H_1 = 826.9$ Btu/lb.

Compare the final enthalpy $H_2 = 826.9$ Btu/lb, with the enthalpy of saturated steam at 14.7 psia, or 1,150.4 Btu/lb. Since the final enthalpy is less than the enthalpy of saturated steam at the same pressure, the throttled steam is wet. Since $H_1 = h_g - y_2 h_{fg}$, $y_2 = (h_g - H_1)/h_{fg}$. With a final pressure of 14.7 psia, use h_g and h_{fg} values at this pressure. Or,

$$y_2 = \frac{1,150.4 - 826.9}{970.3} = 0.3335$$

The final temperature of the steam T_2 is the same as the saturation temperature at the final pressure of 14.7 psia, or $T_2 = 212$ F.

4. Compute the final-state conditions of saturated water

(d) Determine the enthalpy of 500-psia saturated water from the saturation-pressure steam table at 500 psia: $H_1 = h_f = 449.4$ Btu/lb $= H_2$, by definition. The T-S diagram, Fig. 8, shows that the throttled water contains some steam vapor. Or, comparing the final enthalpy of 449.4 Btu/lb with the enthalpy of saturated liquid at the final pressure, 14.7 psia, 180.07 Btu/lb, shows that the liquid contains some vapor in the final state because its enthalpy is greater.

Since $H_1 = H_2 = h_g - y_2 h_{fg}$, $y_2 = (h_g - H_1)/h_{fg}$. Using enthalpies at 14.7 psia of $h_g = 1,150.4$ Btu/lb, and $h_{fg} = 970.3$ Btu/lb from the saturation-pressure steam table, $y_2 = 1,150.4 - 449.4/970.3 = 0.723$.

The final temperature of the steam is the same as the saturation temperature at the final pressure of 14.7 psia, or $T_2 = 212$ F.

NOTE: Calculation b shows that when starting with slightly wet steam, it can be throttled (expanded) through a large enough pressure range to produce superheated steam. This procedure is often used in a throttling calorimeter to determine the initial quality of steam in a pipe. When very wet steam is throttled, calculation c, the net effect may be to produce drier steam at a lower pressure. Throttling saturated water, calculation c, can produce partial or complete flashing of the water to steam. All these processes find many applications in power generation and process-steam plants.

REVERSIBLE HEATING PROCESS FOR STEAM

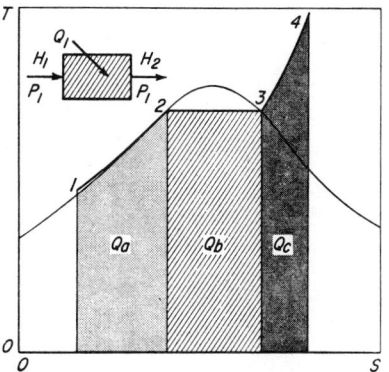

Subcooled water at 1,500 psia and 140 F, state 1, Fig. 9 is heated at constant pressure to state 4, superheated steam at 1,500 psia and 1000 F. Find the heat added (a) to raise the compressed liquid to saturation temperature, (b) to vaporize the saturated liquid to saturated steam, (c) to superheat the steam to 1000 F, (d) Q_1, ΔV, and ΔS from state 1 to state 4.

Fig. 9   Reversible heating process.

Calculation Procedure:

1.   Sketch the T-S diagram for this process

Figure 9 is typical of a steam boiler and superheater. Feedwater fed to a boiler is usually a subcooled liquid. If the feedwater pressure is relatively high, subcooling must be taken into account if accurate results are desired. Some authorities recommend that at pressures below 400 psia, subcooling be ignored and values from the saturated-steam table used. This means that the enthalpies and other properties listed in the steam table corresponding to the actual water temperature are sufficiently accurate. But above 400 psia, the compressed-liquid table should be used.

2.   Determine the initial properties of the liquid

(a) In the saturation-temperature steam table read, at 140 F, $h_f = 107.89$ Btu/lb; $p_f = 2.889$ psia; $v_f = 0.01629$ cu ft/lb; $s_f = 0.1984$ Btu/(lb)(F).

Next, the enthalpy, volume, and entropy of the water at 1,500 psia and 140 F must be found. Since the water is at a much higher pressure than that corresponding to its temperature (1,500 versus 2.889 psia), the compressed liquid portion of the steam table must be used. Study of this table shows that the three desired properties are plotted for 32, 100, and 200 F, and higher temperatures. However, 140 F is not included. Therefore, it is necessary to interpolate between 100 and 200 F. Thus, at 1,500 psia in the compressed-liquid table:

Property	Temperature		Interpolation
	100 F	200 F	
$(h - h_f)$	+3.99	+3.36	+3.74
$(v - v_f)10^5$	−7.5	−8.1	−7.7
$(s - s_f)10^3$	−0.86	−1.79	−1.23

Each property is interpolated in the following way:

$$(h - h_f) \qquad 3.99 - \frac{3.99 - 3.36}{200 - 100}(140 - 100) = 3.99 - 0.25 \qquad = 3.74$$

$$(v - v_f)10^5 \quad -7.5 - \frac{8.1 - 7.5}{200 - 100}(140 - 100) = 7.5 - 0.24 \qquad = -7.74$$

$$(s - s_f)10^3 \quad -0.86 - \frac{1.79 - 0.86}{200 - 100}(140 - 100) = -0.86 - 0.37 = -1.23$$

These interpolated values must now be used to correct the saturation data at 140 F to the actual subcooled state-1 properties. Thus, at 1,500 psia and 140 F,

$$H_1 = h_f + \text{interpolated } h, \text{ or } H_1 = 107.89 + 3.74 = 111.63 \text{ Btu/lb}$$

$$V_1 = v_f - \frac{\text{interpolated } v}{10^5} = 0.01629 - \frac{7.74}{10^5} = 0.01621 \text{ cu ft/lb}$$

$$S_1 = s_f - \frac{\text{interpolated } s}{10^3} = 0.1984 - \frac{1.23}{10^3} = 0.1972 \text{ Btu/(lb)(F)}$$

3. Compute the heat added to raise the compressed liquid to the saturation temperature

From the saturation-pressure steam table for 1,500 psia, the enthalpy of the saturated liquid $H_2 = 611.6$ Btu/lb. The heat added Q_a to raise the compressed liquid to the saturation temperature is $Q_a = H_2 - H_1 = 611.6 - 111.6 = 500$ Btu/lb.

4. Compute the heat added to vaporize the saturated liquid

(b) Read from the saturation-pressure steam table the enthalpy of saturated vapor at 1,500 psia; $H_3 = 1,167.9$ Btu/lb. Then, the heat added to vaporize the saturated water $Q_b = H_3 - H_2 = 1,167.9 - 611.6 = 556.3$ Btu/lb.

5. Compute the heat added to superheat the steam

(c) Find in the superheated steam table for 1,500 psia and 1000 F the properties of the superheated steam: $H_4 = 1,490.1$ Btu/lb; $V_4 = 0.5390$ cu ft/lb; $S_4 = 1.6001$ Btu/(lb)(F). Then the heat added to superheat the saturated steam $Q_4 = H_4 - H_3 = 1,490.1 - 1,167.9 = 322.2$ Btu/lb.

6. Determine the property changes during the process

(d) $Q_1 = Q_a + Q_b + Q_c = H_4 - H_1 = 1490.1 - 111.6 = 1,378.5$ Btu/lb
$\Delta V = V_4 - V_1 = 0.5390 - 0.01621 = 0.5228$ cu ft/lb
$\Delta S = S_4 - S_1 = 1.6001 - 0.1972 = 1.4029$ Btu/(lb)(F)

BLEED STEAM REGENERATIVE CYCLE LAYOUT AND *T-S* PLOT

Sketch the cycle layout, *T-S* diagram, and energy flow chart for a regenerative bleed steam turbine plant having three feedwater heaters and four feed pumps. Write the equations for the work output, available energy, and the energy rejected to the condenser.

Calculation Procedure:

1. Sketch the cycle layout

Figure 10 shows a typical practical regenerative cycle having three feedwater heaters and four feedwater pumps. Number each point where steam enters and leaves the turbine and where steam enters or leaves the condenser and boiler. Also number the points in the feedwater cycle where feedwater enters and leaves a heater. Indicate the heater steam flow by m with a subscript corresponding to the heater number. Use W_p

and a suitable subscript to indicate the pump work for each feed pump, except the last, which is labeled W_{pF}. The heat input to the steam generator is Q_a; the work output of the steam turbine is W_e; the heat rejected by the condenser is Q_r.

2. Sketch the T-S diagram for the cycle

To analyze any steam cycle, trace the flow of 1 lb of steam through the system. Thus, in this cycle, 1 lb of steam leaves the steam generator at point 2 and flows to the turbine. From state 2 to 3, 1 lb of steam expands at constant entropy (assumed) through the turbine, producing work output $W_1 = H_2 - H_3$, represented by area 1-a-2-3 on the T-S

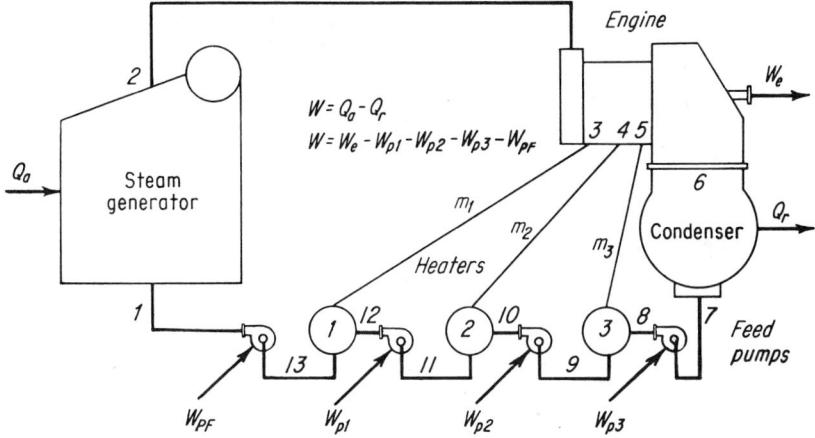

Fig. 10   Regenerative steam cycle uses bleed steam.

diagram, Fig. 11a. At point 3, some steam is bled from the turbine to heat the feedwater passing through heater 1. The quantity of steam bled, m_1 lb is less than the 1 lb flowing between points 2 and 3. Plot states 2 and 3 on the T-S diagram, Fig. 11a.

From point 3 to 4, the quantity of steam flowing through the turbine is $1 - m_1$ lb. This steam produces work output $W_2 = H_3 - H_4$. Plot point 4 on the T-S diagram. Then, area 1-3-4-12 represents the work output W_2, Fig. 11a.

At point 4, steam is bled to heater 2. The weight of this steam is m_2 lb. From point 4, the steam continues to flow through the turbine to point 5, Fig. 11a. The weight of steam flowing between points 4 and 5 is $1 - m_1 - m_2$ lb. Plot point 5 on the T-S diagram, Fig. 11a. The work output between points 4 and 5, $W_3 = H_4 - H_5$, is represented by area 4-5-10-11 on the T-S diagram.

At point 5, steam is bled to heater 3. The weight of this bleed steam is m_3 lb. From point 5, steam continues to flow through the turbine to exhaust at point 6, Fig. 11a. The weight of steam flowing between points 5 and 6 is $1 - m_1 - m_2 - m_3$ lb. Plot point 6 on the T-S diagram, Fig. 11a.

The work output between points 5 and 6 is $W_4 = H_5 - H_6$, represented by area 5-6-7-9 on the T-S diagram, Fig. 11a. Area Q_r represents the heat given up by 1 lb of exhaust steam. Similarly, the area marked Q_a represents the heat absorbed by 1 lb of water in the steam generator.

3. Alter the T-S diagram to show actual cycle conditions

Q_a as plotted in Fig 11a is true for this cycle since 1 lb of water flows through the steam generator and the first section of the turbine. But Q_r is much too large; only $1 - m_1 - m_2 - m_3$ lb of steam flows through the condenser. Likewise, the net areas for W_2, W_3, and W_4, Fig. 11a, are all too large, because less than 1 lb of steam flows through the respective turbine sections. The area for W_1, however, is true.

A true *proportionate-area* diagram can be plotted by applying the factors for actual

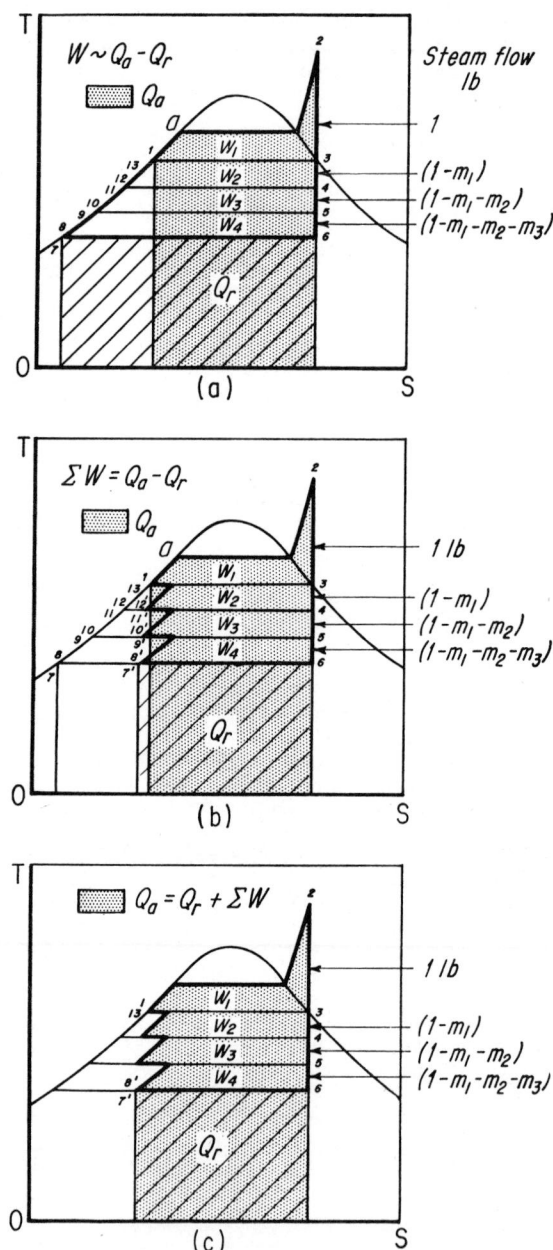

Fig. 11   (a) T-S chart for the bleed-steam regenerative cycle in Fig. 10; (b) actual fluid flow in the cycle; (c) alternative plot of (b).

flow, as in Fig. 11b. W_2, outlined by the heavy lines, equals the similarly labeled area in Fig. 11a, multiplied by $1 - m_1$. The states marked 11' and 12', Fig. 11b, are not true state points because of the ratioing factor applied to the area for W_2. The true state points 11 and 12 of the liquid before and after heater pump 3 stay as shown in Fig. 11a.

Apply $1 - m_1 - m_2$ to W_3 of Fig. 11a to obtain the proportionate area of Fig. 11b; to obtain W_4, multiply by $1 - m_1 - m_2 - m_3$. Multiplying by this factor also gives Q_r. Then all the areas in Fig. 11b will be in proper proportion for 1 lb of steam entering the turbine throttle but less in other parts of the cycle.

In Fig. 11b, the work can be measured by the difference of the area Q_a and the area Q_r. There is no simple net area left, because the areas coincide on only two sides. But the area enclosed by the heavy lines *is* the total net work W for the cycle, equal to the sum of the work produced in the various sections of the turbine, Fig. 11b. Then Q_a is the alternate area $Q_r + W_1 + W_2 + W_3 + W_4$, as shaded in Fig. 11c.

The saw-tooth appearance of the liquid-heating line shows that as the number of heaters in the cycle increases, the heating line approaches a line of constant entropy. The best number of heaters for a given cycle depends on the steam state of the turbine inlet. Many medium-pressure and medium-temperature cycles use five to six heaters. High-pressure and high-temperature cycles use as many as nine heaters.

4. Draw the energy-flow chart

Choose a suitable scale for the heat content of 1 lb of steam leaving the steam generator. A typical scale is 0.375 in. per 1,000 Btu/lb. Plot the heat content of 1 lb of steam vertically on line 2-2, Fig. 12. Using the same scale, plot the heat content in energy streams m_1, m_2, m_3, W_e, W, W_p, W_{pF}, and so forth. In some cases, as W_{p1}, W_{p2}, and so forth, the energy stream may be so small that it is impossible to plot it to scale. In these instances a single thin line is used. The completed diagram, Fig. 12, provides a useful concept of the distribution of the energy in the cycle.

Related Calculations: The Procedure given here can be used for all regenerative cycles, provided that the equations are altered to allow for more, or fewer, heaters and pumps. The following Calculation Procedure shows the application of this method to an actual regenerative cycle.

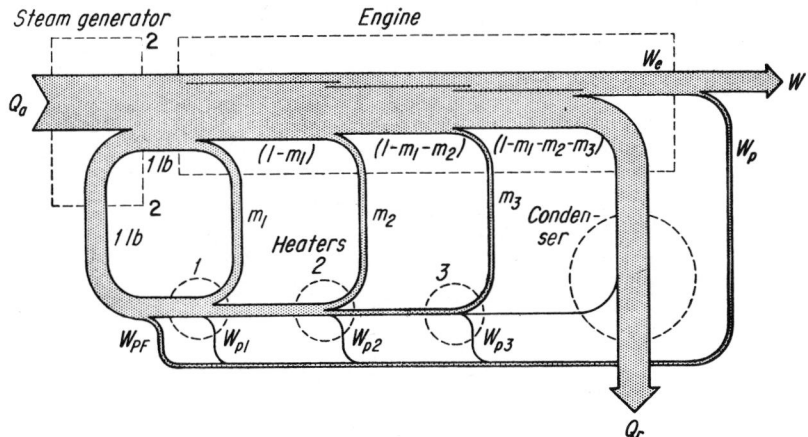

Fig. 12   Energy-flow chart of cycle in Fig. 10.

BLEED REGENERATIVE STEAM CYCLE ANALYSIS

Analyze the bleed regenerative cycle shown in Fig. 13, determining the heat balance for each heater, plant thermal efficiency, turbine or engine thermal efficiency, plant heat rate, turbine or engine heat rate, and turbine or engine steam rate. Throttle steam pressure is 2,000 psia at 1000 F; steam-generator efficiency = 0.88; station auxiliary steam consumption (excluding pump work) = 6 percent of the turbine or engine output; engine efficiency of each turbine or engine section = 0.80; turbine or engine cycle has three feedwater heaters and bleed-steam pressures as shown in Fig. 13; exhaust pressure to condenser is 1 in. Hg absolute.

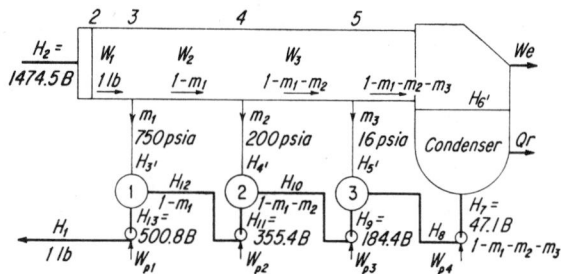

Fig. 13   Bleed regenerative steam cycle.

Calculation Procedure:

1. *Determine the enthalpy of the steam at the inlet of each heater and the condenser*

From a superheated-steam table, find the throttle enthalpy $H_2 = 1,474.5$ Btu/lb at 2,000 psia and 1000 F. Next find the throttle entropy $S_2 = 1.5603$ Btu/(lb)(F), at the same conditions in the superheated-steam table.

Plot the throttle steam conditions on a Mollier chart, Fig. 14. Assume that the steam expands from the throttle conditions at constant entropy = constant S to the inlet of the first feedwater heater, 1, Fig. 13. Plot this constant S expansion by drawing the straight

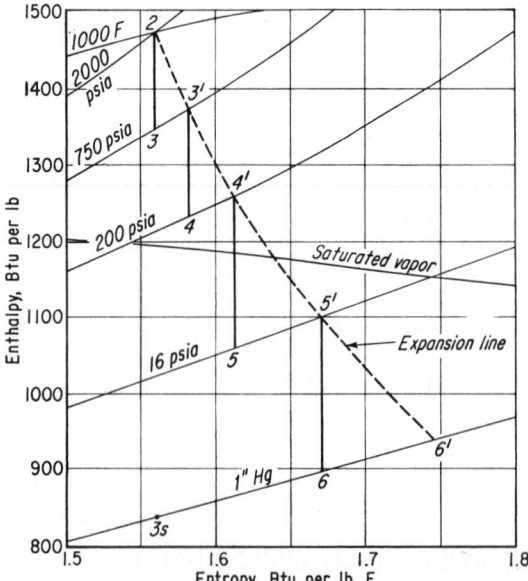

Fig. 14   Mollier-chart plot of the cycle in Fig. 13.

vertical line 2-3 on the Mollier chart, Fig. 14, between the throttle condition and the heater inlet pressure of 750 psia.

Read on the Mollier chart $H_3 = 1,346.7$ Btu/lb. Since the engine or turbine efficiency $e_e = H_2 - H_3/(H_2 - H_3) = 0.8 = 1,474.5 - H_3/(1,474.5 - 1,346.7)$; H_3 = actual enthalpy of the steam at the inlet to heater $1 = 1,474.5 - 0.8(1,474.5 - 1,346.7) = 1,372.2$ Btu/lb. Plot this enthalpy point on the 750-psia pressure line of the Mollier chart, Fig. 14. Read the entropy at the heater inlet from the Mollier chart as $S_{3'} = 1.5819$ Btu/(lb)(F) at 750 psia and 1,372.2 Btu/lb.

Assume constant-S expansion from $H_{3'}$ to H_4 at 200 psia, the inlet pressure for feed-water heater 2. Draw the vertical straight line $3' - 4$ on the Mollier chart, Fig. 14. Using a procedure similar to that for heater 1, $H_{4'} = H_{3'} = e_e(H_{3'} - H_4) = 1,372.2 - 0.8$ $(1,372.2 - 1,230.0) = 1,258.4$ Btu/lb. This is the actual enthalpy of the steam at the inlet to heater 2. Plot this enthalpy on the 200-psia pressure line of the Mollier chart and find $S_{4'} = 1.613$ Btu/(lb)/(F), Fig. 14.

Using the same procedure with constant-S expansion from $H_{4'}$, find $H_5 = 1,059.5$ Btu/lb at 16 psia, the inlet pressure to heater 3. Next find $H_{5'} = H_{4'} - e_e(H_{4'} - H_5)$ $= 1,258.4 - 0.8(1,258.4 - 1,059.5) = 1,099.2$ Btu/lb. From the Mollier chart find $S_{5'} = 1.671$ Btu/(lb)(F), Fig. 14.

Using the same procedure with constant-S expansion from $H_{5'}$ to H_6, find $H_6 = 898.2$ Btu/lb at 1-in. Hg absolute, the condenser inlet pressure. Then, $H_{6'} = H_{5'} - e_e(H_{5'} - H_6) = 1,099.2 - 0.8(1,099.2 - 898.2) = 938.4$ Btu/lb, actual enthalpy of the steam at the condenser inlet. Find, on the Mollier chart, the moisture in the turbine exhaust $= 15.1$ percent.

2. Determine the overall engine efficiency

Overall engine efficiency e_e is higher than the engine-section efficiency because there is partial available-energy recovery between sections. Constant-S expansion from the throttle to the 1-in. Hg absolute exhaust gives H_{3S}, Fig. 14, as 838.3 Btu/lb, assuming that all the steam flows to the condenser. Then, overall $e_e = H_2 - H_{6'}/(H_2 - H_{3S}) = 1,474.5 - 938.4/1,474.5 - 838.3 = 0.8425$, or 84.25 percent, compared with 0.8, or 80 percent, for individual engine sections.

3. Compute the bleed-steam flow to each feedwater heater

For each heater, energy in = energy out. Also, the heated condensate leaving each heater is a saturated liquid at the heater bleed-steam pressure. To simplify this calculation, assume negligible steam pressure drop between the turbine bleed point and the heater inlet. This assumption is permissible when the distance between the heater and bleed point is small. Determine the pump work by using the chart accompanying the Compressed Liquid Table in Keenan and Keyes – *Thermodynamic Properties of Steam*, or the ASME *Steam Tables*.

For heater 1, energy in = energy out, or $H_{3'}m_1 + H_{12}(1 - m_1) = H_{13}$, where m = bleed-steam flow to the feedwater heater, lb per lb of throttle steam flow. (The subscript refers to the heater under consideration.) Then, $H_{3'}m_1 + (H_{11} + W_{p2})(1 - m_1) = H_{13}$, where W_{p2} = work done by pump 2, Fig. 13, in Btu/lb per lb throttle flow. Then, 1,372.2 $m_1 + (355.4 + 1.7)(1 - m_1) = 500.8$; $m_1 = 0.1416$ lb per lb throttle flow; $H_1 = H_{13} + W_{p1}$ $= 500.8 + 4.7 = 505.5$ Btu/lb, where W_{p1} = work done by pump 1, Fig. 13. For each pump, find the pump work from the chart accompanying the Compressed Liquid Table in Keenan and Keyes – *Steam Tables* by entering the chart at the heater inlet pressure and projecting vertically at constant entropy to the heater outlet pressure, which equals the next heater inlet pressure. Read the enthalpy values at the respective pressures and subtract the smaller from the larger to obtain the pump work during passage of the feedwater through the pump from the lower to the higher pressure. Thus, $W_{p2} = 1.7 - 0.0 = 1.7$ Btu/lb, using enthalpy values for 200 psia and 750 psia, the heater inlet and discharge pressures, respectively.

For heater 2, energy in = energy out, or $H_{4'}m_2 + H_{10}(1 - m_1 - m_2) = H_{11}(1 - m_1)$ $H_{4'}m_2 + (H_9 + W_{p3})(1 - m_1 - m_2) = H_{11}(1 - m_1)$ 1,258.4$m_2 + (184.4 + 0.5)(0.8584 - m_2) =$ 355.4(0.8584) $m_2 = 0.1365$ lb per lb throttle flow.

For heater 3, energy in = energy out, or $H_{5'}m_3 + H_8(1 - m_1 - m_2 - m_3) = H_9(1 - m_1$

$-m_2)H_{5'}m_3 + (H_7 + W_{p4})(1 - m_1 - m_2 - m_3) = H_9(1 - m_1 - m_2) \ 1,099.2\,m_3 + (47.1 + 0.1)$
$(0.7210 - m_3) = 184.4(0.7219)m_3 = 0.0942$ lb per lb throttle flow.

4. Compute the turbine work output

The work output per section W Btu is $W_1 = H_2 - H_{3'} = 1,474.5 - 1,372.1 = 102.3$ Btu, using the previously computed enthalpy values. Also, $W_2 = (H_{3'} - H_{4'})(1 - m_1) = (1,372.2 - 1,258.4)(1 - 0.1416) = 97.7$ Btu; $W_3 = (H_{4'} - H_{5'})(1 - m_1 - m_2) = (1,258.4 - 1,099.2)(1 - 0.1416 - 0.1365) = 115.0$ Btu; $W_4 = (H_{5'} - H_{6'})(1 - m_1 - m_2 - m_3) = (1,099.2 - 938.4)(1 - 0.1416 - 0.1365 - 0.0942) = 100.9$ Btu. The total work output of the turbine $= W_e = \Sigma W = 102.3 + 97.7 + 115.0 + 100.9 = 415.9$ Btu. The total $W_p = \Sigma W_p = W_{p1} + W_{p2} + W_{p3} + W_{p4} = 4.7 + 1.7 + 0.5 + 0.1 = 7.0$ Btu.

Since the station auxiliaries consume 6 percent of W_e, the auxiliary consumption $= 0.06(415.9) = 25.0$ Btu. Then, net station work $w = 415.9 - 7.0 - 25.0 = 383.9$ Btu.

5. Check the turbine work output

The heat added to the cycle Q_a Btu/lb $= H_2 - H_1 = 1,474.5 - 505.5 = 969.0$ Btu. The heat rejected from the cycle Q_r Btu/lb $= (H_{6'} - H_7)(1 - m_1 - m_2 - m_3) = (938.4 - 47.1)(0.6277) = 559.5$ Btu. Then $W_e - W_p = Q_a - Q_r = 969.0 - 559.5 = 409.5$ Btu.

Compare this with $W_c - W_p$ computed earlier, or $415.9 - 7.0 = 408.9$ Btu, or a difference of $409.5 - 408.9 = 0.6$ Btu. This is an accurate check; the difference of 0.6 Btu comes from errors in Mollier-chart and slide-rule readings. Assume 408.9 Btu as correct because it is the lower of the two values.

6. Compute the plant and turbine efficiencies

Plant energy input $= Q_a/e_b$, where $e_b =$ boiler efficiency. Then, plant energy input $= 969.0/0.88 = 1,101.0$ Btu. Plant thermal efficiency $= W/(Q_a/e_b) = 383.9/1,101.0 = 0.3486$. Turbine thermal efficiency $= W_e/Q_a = 415.9/969.0 = 0.4292$. Plant heat rate $= 3413/0.3486 = 9,790$ Btu/kw-hr, where $3,413 =$ Btu/kw-hr. Turbine heat rate $= 3,413/0.4292 = 7,950$ Btu/kw-hr. Turbine throttle steam rate $=$ (turbine heat rate)/$(H_2 - H_1) = 7,950/(1,474.5 - 505.5) = 8.21$ lb/kw-hr.

Related Calculations: Using the procedures given, the following values can be computed for any actual steam cycle: engine or turbine efficiency e_e; steam enthalpy at the main-condenser inlet; bleed-steam flow to a feedwater heater; turbine or engine work output per section; total turbine or engine work output; station auxiliary power consumption; net station work output; plant energy input; plant thermal efficiency; turbine or engine thermal efficiency; plant heat rate; turbine or engine heat rate; turbine throttle heat rate. To compute any of these values, use the equations given and insert the applicable variables.

REHEAT-STEAM CYCLE PERFORMANCE

A reheat-steam cycle has a 2,000-psia throttle pressure at the turbine inlet and a 400-psia reheat pressure. Throttle and reheat temperature of the steam is 1000 F; condenser pressure is 1 in. Hg absolute; engine efficiency of the high-pressure and low-pressure turbines is 80 percent. Find the cycle thermal efficiency.

Calculation Procedure:

1. Sketch the cycle layout and cycle T-S diagram

Figures 15 and 16 show the cycle layout and T-S diagram with each important point numbered. Use a cycle layout and T-S diagram for every calculation of this type, because it reduces the possibility of errors.

2. Determine the throttle-steam properties from the steam tables

Use the superheated-steam tables, entering at 2,000 psia and 1000 F to find throttle-steam properties. Applying the symbols of the T-S diagram in Fig. 16, $H_2 = 1,474.5$ Btu/lb; $S_2 = 1.5603$ Btu/(lb)(F).

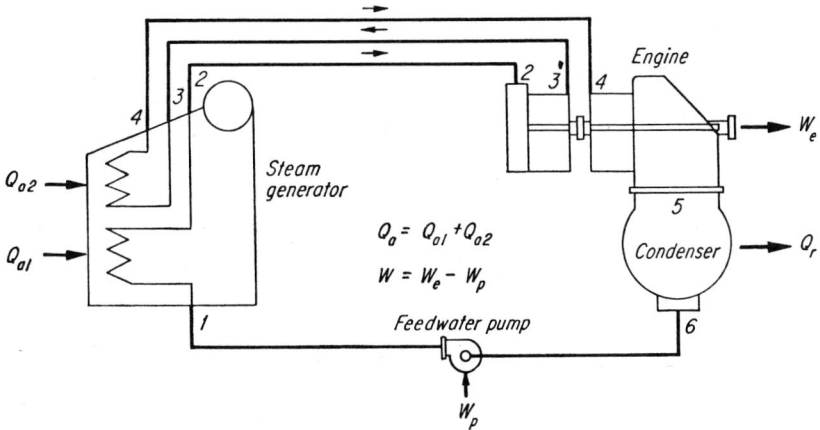

Fig. 15   Typical steam reheat cycle.

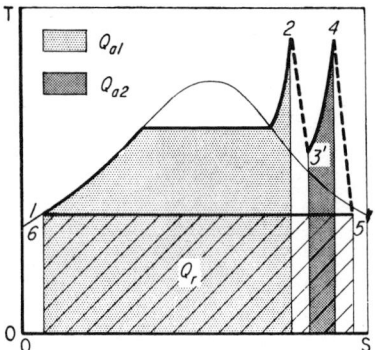

Fig. 16   Irreversible expansion in reheat cycle.

3. *Find the reheat-steam enthalpy*

Assume a constant-entropy expansion of the steam from 2,000 psia to 400 psia. Trace this expansion on a Mollier (*H-S*) chart, Fig. 1, where a constant-entropy process is a vertical line between the initial (2,000 psia) and reheat (400 psia) pressures. Read on the Mollier Chart $H_3 = 1,276.8$ Btu/lb at 400 psia.

4. *Compute the actual reheat properties*

The ideal enthalpy drop, throttle to reheat, $= H_2 - H_3 = 1,474.5 - 1,276.8 = 197.7$ Btu/lb. The actual enthalpy drop = (ideal drop)(turbine efficiency) $= H_2 - H_{3'} = 197.5$ (0.8) $= 158.2$ Btu/lb $= W_{e1} =$ work output in the high-pressure section of the turbine.

Once W_{e1} is known, $H_{3'}$ can be computed from $H_{3'} = H_2 - W_{e1} = 1,474.5 - 158.2 = 1,316.3$ Btu/lb.

The steam now returns to the boiler and leaves at condition 4, where $P_4 = 400$ psia; $T_4 = 1000$ F; $S_4 = 1.7623$; $H_4 = 1,522.4$ Btu/lb, from the superheated-steam table.

5. *Compute the exhaust-steam properties*

Use the Mollier chart and an assumed constant-entropy expansion to 1 in. Hg absolute to determine the ideal exhaust enthalpy, or $H_5 = 947.4$ Btu/lb. The ideal work of the low-pressure section of the turbine is then $H_4 - H_5 = 1,522.4 - 947.4 = 575.0$ Btu/lb.

The actual work output of the low-pressure section of the turbine is $W_{e2} = H_4 - H_{5'} = 575.0(0.8) = 460.8$ Btu/lb.

Once W_{e2} is known, $H_{5'}$ can be computed from $H_{5'} = H_4 - W_{e2} = 1,522.4 - 460.0 = 1,062.4$ Btu/lb.

The enthalpy of the saturated liquid at the condenser pressure is found in the saturation-pressure steam table at 1 in. Hg absolute $= H_6 = 47.1$ Btu/lb.

The pump work W_p from the compressed-liquid table diagram in the stream tables is $W_p = 5.5$ Btu/lb. Then the enthalpy of the water entering the boiler $H_1 = H_6 + W_p = 47.1 + 5.5 = 52.6$ Btu/lb.

6. *Compute the cycle thermal efficiency*

For any reheat cycle,

$$e = \text{cycle thermal efficiency} =$$
$$\frac{(H_2 - H_{3'}) + (H_4 - H_{5'}) - W_p}{(H_2 - H_1) + (H_4 - H_{3'})} = \frac{(1,474.5 - 1,316.3) + (1,522.4 - 1,062.4) - 5.5}{(1,474.5 - 52.6) + (1,522.4 - 1,316.3)}$$
$$= 0.3766, \text{ or } 37.66 \text{ percent}$$

Figure 17 is an energy-flow diagram for the reheat cycle analyzed here. This diagram shows that the fuel burned in the steam generator to produce energy-flow Q_{a1} is the largest part of the total energy input. The cold-reheat line carries the major share of energy leaving the high-pressure turbine.

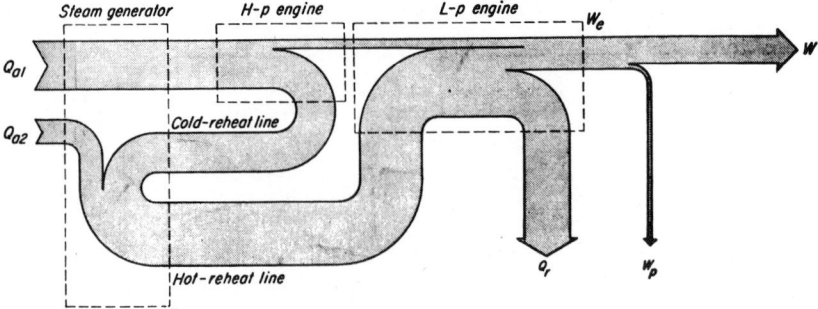

Fig. 17   Energy-flow diagram for reheat cycle in Fig. 15.

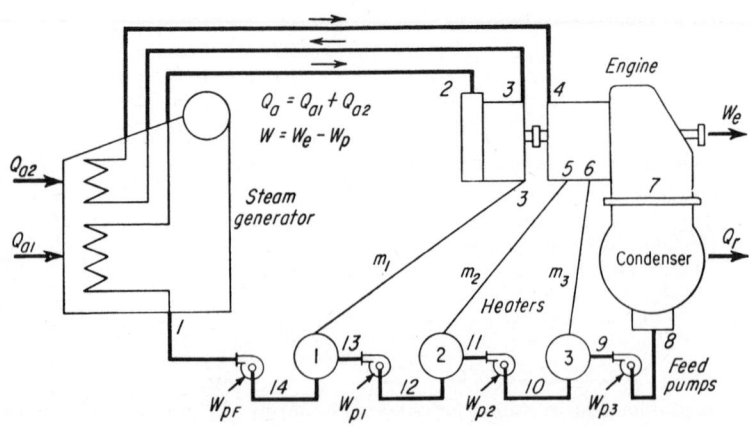

Fig. 18   Combined reheat and bleed-regenerative cycle.

Related Calculations: Reheat-regenerative cycles are used in some large power plants. Figure 18 shows a typical layout for such a cycle having three stages of feedwater heating and one stage of reheating. The heat balance for this cycle is computed as shown above, with the bleed-flow terms, m, computed by setting up an energy balance around each heater, as in earlier Calculation Procedures.

Using a T-S diagram, Fig. 19, the cycle thermal efficiency is:

$$e = \frac{W}{Q_a} = \frac{Q_a - Q_r}{Q_a} = 1 - \frac{Q_r}{Q_{a1} + Q_{a2}}$$

Based on 1 lb of working fluid entering the steam generator and turbine throttle,

$$Q_r = (1 - m_1 - m_2 - m_3)(H_7 - H_8)$$
$$Q_{a1} = (H_2 - H_1)$$
$$Q_{a2} = (1 - m_1)(H_4 - H_3)$$

Figure 20 shows the energy-flow chart for this cycle.

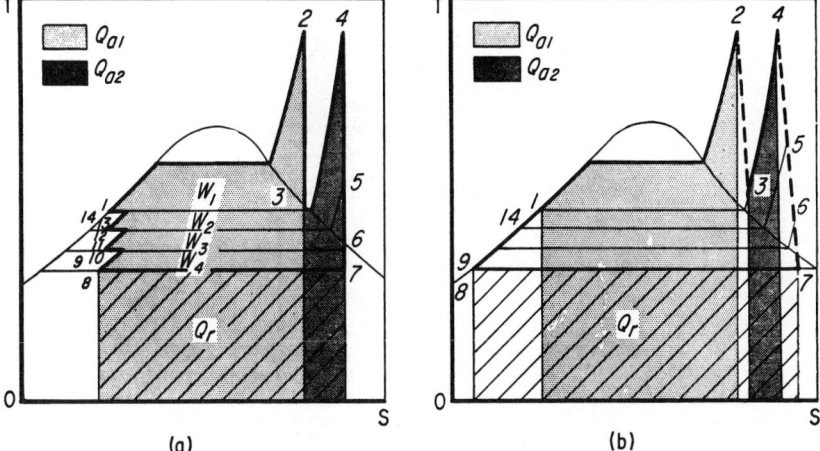

(a)   (b)

Fig. 19 (a) T-S diagram for ideal reheat-regenerative-bleed cycle; (b) T-S diagram for actual cycle.

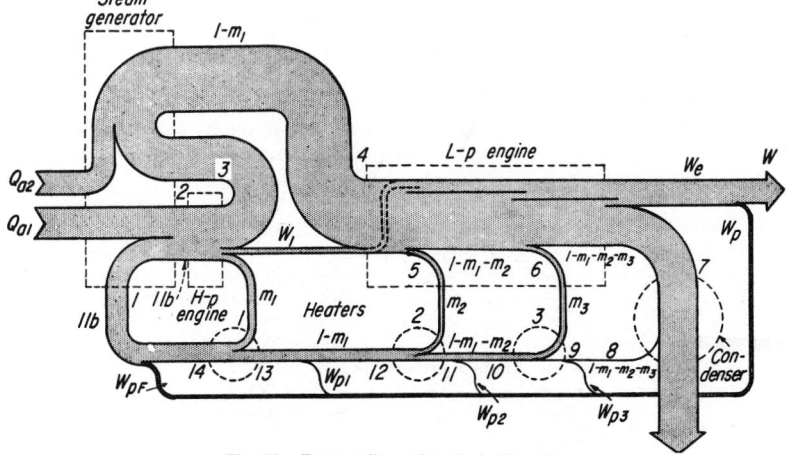

Fig. 20 Energy flow of cycle in Fig. 18.

Some high-pressure plants use two stages of reheating, Fig. 21, to raise the cycle efficiency. With two stages of reheating, the maximum number generally used, and taking values from Fig. 21,

$$e = \frac{(H_2 - H_3) + (H_4 - H_5) + (H_6 - H_7) - W_p}{(H_2 - H_1) + (H_4 - H_3) + (H_6 - H_5)}$$

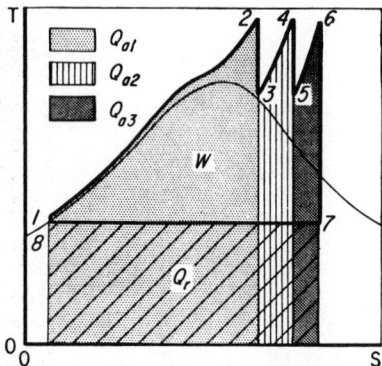

Fig. 21 T-S diagram for multiple reheat stages.

MECHANICAL-DRIVE STEAM-TURBINE POWER-OUTPUT ANALYSIS

Show the effect of turbine engine efficiency on the condition lines of a turbine having engine efficiencies of 100 (isentropic expansion), 75, 50, 25, and 0 percent. How much of the available energy is converted to useful work for each engine efficiency? Sketch the effect of different steam inlet pressures on the condition line of a single-nozzle turbine at various loads. What is the available energy, Btu per lb of steam, in a non-condensing steam turbine having an inlet pressure of 1,000 psia and an exhaust pressure of 100 psig? How much work will this turbine perform if the steam flow rate to it is 1,000 lb/sec and the engine efficiency is 40 percent?

Calculation Procedure:

1. Sketch the condition lines on the Mollier chart

Draw on the Mollier chart for steam initial- and exhaust-pressure lines, Fig. 22, and the initial-temperature line. For an isentropic expansion, the entropy is constant during the expansion and the engine efficiency = 100 percent. The expansion or condition line is a vertical trace from h_1 on the initial-pressure line to h_{2s} on the exhaust-pressure line. Draw this line as shown in Fig. 22.

For zero percent engine efficiency, the other extreme in the efficiency range, $h_1 = h_2$ and the condition line is a horizontal line. Draw this line as shown in Fig. 22.

Between zero and 100 percent engine efficiency, the condition lines become more nearly vertical as the engine efficiency approaches 100 percent, or an isentropic expansion. Draw the condition lines for 25, 50, and 75 percent efficiency, as shown in Fig. 22.

For the isentropic expansion, the available energy = $h_1 - h_{2s}$, Btu per lb of steam. This is the energy that an ideal turbine would make available.

For actual turbines, the enthalpy at the exhaust pressure $h_2 = h_1 -$ (available energy) (engine efficiency)/100, where available energy = $h_1 - h_{2s}$ for an ideal turbine working between the same initial and exhaust pressures. Thus, the available energy converted to useful work for any engine efficiency = (ideal available energy, Btu/lb)(engine efficiency, percent)/100. Using this relation, the available energy at each of the given engine efficiencies is found by substituting the ideal available energy and the actual engine efficiency.

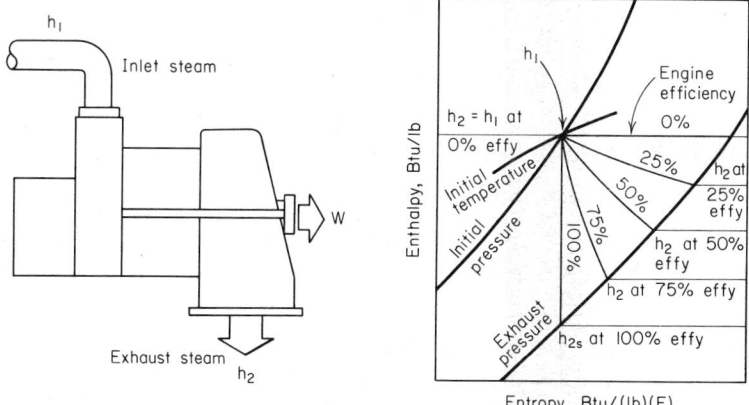

Fig. 22   Mollier chart of turbine condition lines.

2.   Sketch the condition lines for various throttle pressures

Draw the throttle- and exhaust-pressure lines on the Mollier chart, Fig. 23. Since the inlet control valve throttles the steam flow as the load on the turbine decreases, the pressure of the steam entering the turbine nozzle is lower at reduced loads. Show this throttling effect by indicating the lower inlet pressure lines, Fig. 23, for the reduced loads. Note that the lowest inlet pressure occurs at the minimum plotted load—25 percent of full load—and the maximum inlet pressure at 125 percent of full load. As the turbine inlet steam pressure decreases, so does the available energy, because the exhaust enthalpy rises with decreasing load.

3.   Compute the turbine available energy and power output

Use a noncondensing-turbine performance chart, Fig. 24, to determine the available energy. Enter the bottom of the chart at 1,000 psia and project vertically upward until

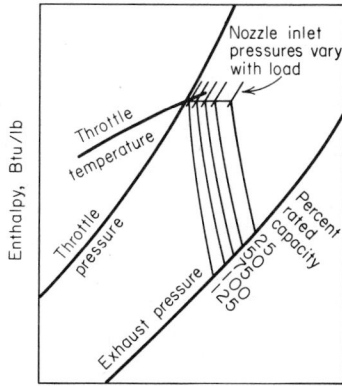

Fig. 23   Turbine condition line shifts as the inlet steam pressure varies.

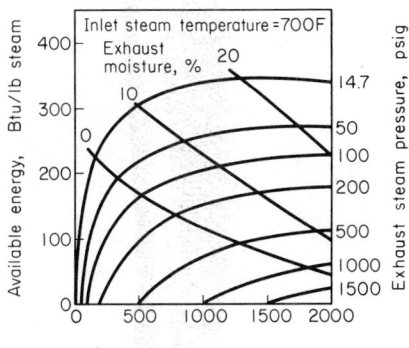

Fig. 24   Available energy in turbine depends on the initial steam state and the exhaust pressure.

the 100-psig exhaust-pressure curve is intersected. At the left, read the available energy as 205 Btu per lb of steam.

With the available energy, flow rate, and engine efficiency known, the work output = (available energy, Btu/lb)(flow rate, lb/sec)(engine efficiency/100)/550 ft-lb/(sec)(hp). For this turbine, work output = (205 Btu/lb)(1,000 lb/sec)(40/100)/550 = 149 hp.

Related Calculations: Use the steps given here to analyze single-stage noncondensing mechanical-drive turbines for stationary, portable, or marine applications. Performance curves like Fig. 24 are available from turbine manufacturers. Single-stage noncondensing turbines are used for feed-pump, draft-fan, and auxiliary-generator drive.

CONDENSING STEAM-TURBINE POWER-OUTPUT ANALYSIS

What is the available energy in steam supplied to a 5,000-kw turbine if the inlet steam conditions are 1,000 psia and 800 F and the turbine exhausts at 1 in. Hg absolute? Determine the theoretical and actual heat rate of this turbine if its engine efficiency is 74 percent. What is the full-load output and steam rate of the turbine?

Calculation Procedure:

1. Determine the available energy in the steam

Enter Fig. 25 at the bottom at 1,000-psia inlet pressure and project vertically upward to the 800-F 1-in. exhaust-pressure curve. At the left, read the available energy as 545 Btu per lb of steam.

2. Determine the heat rate of the turbine

Enter Fig. 26 at an initial steam temperature of 800 F and project vertically upward to the 1,000-psia 1-in. curve. At the left, read the theoretical heat rate as 8,400 Btu/kw-hr.

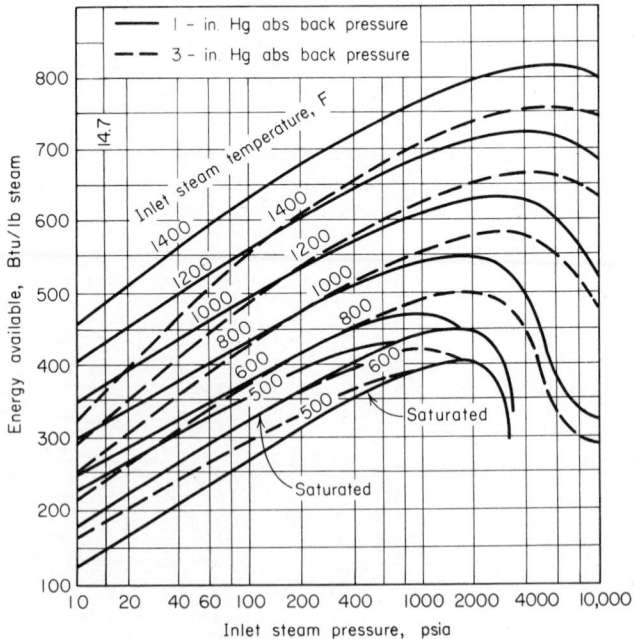

Fig. 25   Available energy for typical condensing turbines.

When the theoretical heat rate is known, the actual heat rate is found from: actual heat rate HR Btu/kw-hr = (theoretical heat rate, Btu/kw-hr)/(engine efficiency). Or, actual HR = 8,400/0.74 = 11,350 Btu/kw-hr.

3. Compute the full-load output and steam rate

The energy convertèd to work, Btu per lb of steam, = (available energy, Btu per lb of steam)(engine efficiency) = (545)(0.74) = 403 Btu per lb of steam.

For any prime mover driving a generator, the full-load output, Btu, = (generator kw rating)(3,413 Btu/kw-hr) = (5,000)(3,413) = 17,060,000 Btu/hr.

The steam flow = (full-load output, Btu/hr)/(work output, Btu/lb) = 17,060,000/403 = 42,300 lb per hr of steam. Then, the full-load steam rate of the turbine, lb/kw-hr = (steam flow, lb/hr)/(kw output at full load) = 42,300/5,000 = 8.46 lb/kw-hr.

Related Calculations: Use this general procedure to determine the available energy, theoretical and actual heat rates, and full-load output and steam rate for any stationary, marine, or portable condensing steam turbine operating within the ranges of Figs. 25 and 26. If the actual performance curves are available, use them instead of Figs. 25 and 26. The curves given here are suitable for all preliminary estimates for condensing turbines operating with exhaust pressures of 1 or 3 in. Hg absolute. Many modern turbines operate under these conditions.

STEAM-TURBINE REGENERATIVE-CYCLE PERFORMANCE

When throttle steam is at 1,000 psia and 800 F and the exhaust pressure is 1 in. Hg absolute, a 5,000-kw condensing turbine has an actual heat rate of 11,350 Btu/kw-hr. Three feedwater heaters are added to the cycle, Fig. 27, to heat the feedwater to 70

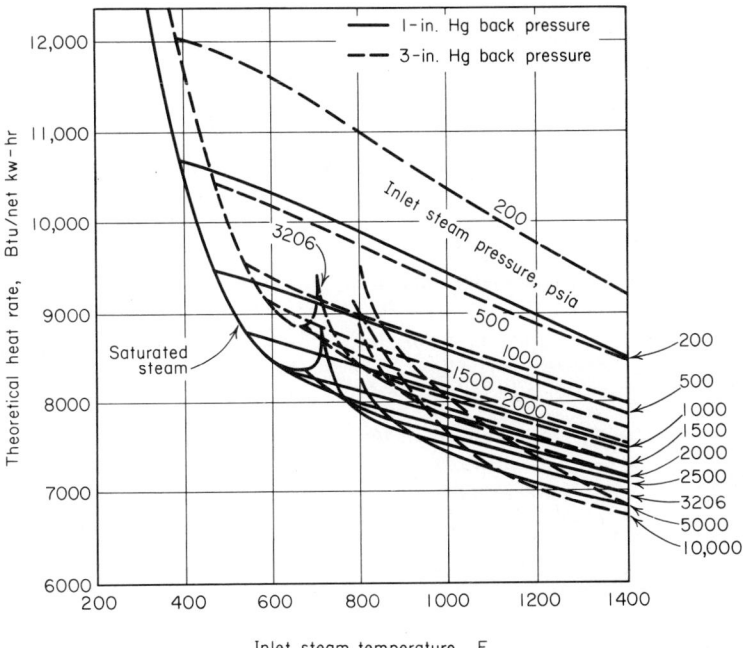

Fig. 26   Theoretical heat rate for condensing turbines.

percent of the maximum possible enthalpy rise. What is the actual heat rate of the turbine? If 10 heaters instead of 3 were used and the water enthalpy was raised to 90 percent of the maximum possible rise in these 10 heaters, would the reduction in the actual heat rate be appreciable?

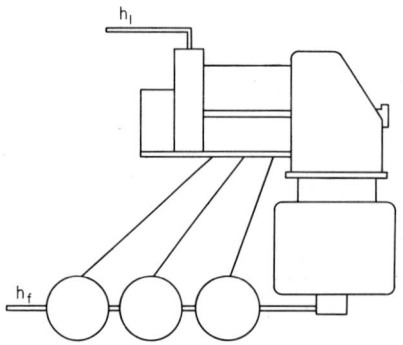

Fig. 27 Regenerative feedwater heating cycle.

Calculation Procedure:

1. Determine the actual enthalpy rise of the feedwater

Enter Fig. 28 at the throttle pressure of 1,000 psia and project vertically upward to the 1 in. Hg absolute back-pressure curve. At the left, read the maximum possible feedwater enthalpy rise as 495 Btu/lb. Since the actual rise is limited to 70 percent of the maximum possible rise by the conditions of the design, the actual enthalpy rise = (495) (0.70) = 346.5 Btu/lb.

2. Determine the heat-rate and heater-number correction factors

Find the theoretical reduction in straight-condensing (no regenerative heaters) heat rates from Fig. 29. Enter the bottom of Fig. 29 at the inlet steam temperature, 800 F, and project vertically upward to the 1,000-psia, 1-in. Hg back-pressure curve. At the left, read the reduction in straight-condensing heat rate as 14.8 percent.

Next, enter Fig. 30 at the bottom at 70 percent of maximum possible rise in feedwater enthalpy and project vertically to the three-heater curve. At the left, read the reduction in straight-condensing heat rate for the number of heaters and actual enthalpy rise as 0.71.

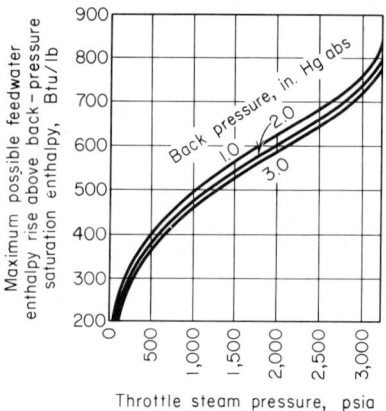

Throttle steam pressure, psia

Fig. 28 Feedwater enthalpy rise.

3. Apply the heat-rate and heater-number correction factors

Full-load regenerative-cycle heat rate, Btu/kw-hr, = (straight-condensing heat rate, Btu/kw-hr) [1 − (heat-rate correction factor)(heater-number correction factor)] = (13,350)[1 − (0.148)(0.71)] = 10,160 Btu/kw-hr.

4. Find and apply the correction factors for the larger number of heaters

Enter Fig. 30 at 90 percent of the maximum possible enthalpy rise and project vertically to the 10-heater curve. At the left, read the heat-rate reduction for the number of heaters and actual enthalpy rise as 0.89.

Using the heat-rate correction factor from Step 2 and 0.89, found above, the full-load 10-heater regenerative-cycle heat rate = (11,350)[1 − (0.148)(0.89)] = 9,850 Btu/kw-hr, using the same procedure as in Step 3. Thus, adding 10 − 3 = 7 heaters reduces the heat rate by 10,160 − 9,850 = 310 Btu/kw-hr. This is a reduction of 3.05 percent.

To determine if this reduction in heat rate is appreciable, the carrying charges on the extra heaters, piping, and pumps must be compared with the reduction in annual fuel costs resulting from the lower heat rate. If the fuel saving is greater than the

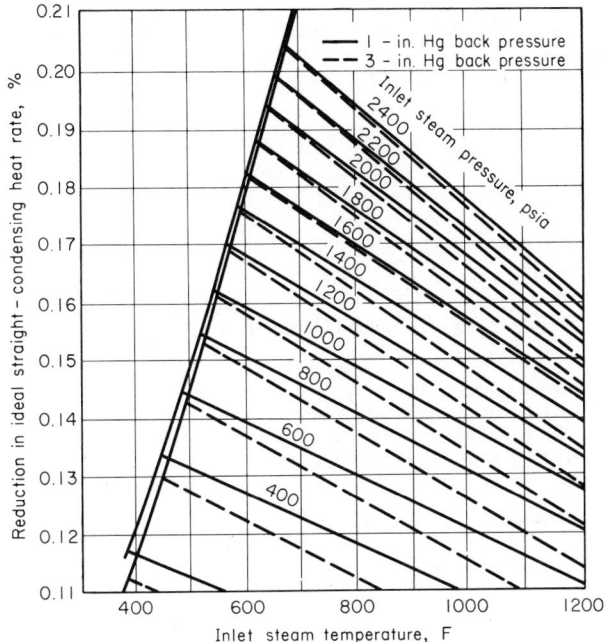

Fig. 29   Reduction in straight-condensing heat rate obtained by regenerative heating.

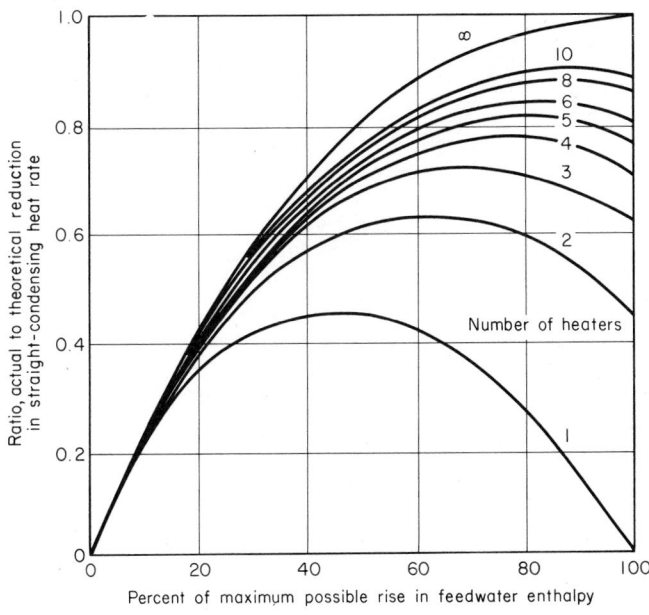

Fig. 30   Maximum possible rise in feedwater enthalpy varies with the number of heaters used.

carrying charges, the larger number of heaters can usually be justified. In this case, tripling the number of heaters would probably increase the carrying charges to a level exceeding the fuel savings. Therefore, the reduction in heat rate is probably not appreciable.

Related Calculations: Use the procedure given here to compute the actual heat rate of steam-turbine regnerative cycles for stationary, marine, and portable installations. Where necessary, use the steps of the previous Procedure to compute the actual heat rate of a straight-condensing cycle before applying the present Procedure. The performance curves given here are suitable for first approximations in situations where actual performance curves are unavailable.

REHEAT-REGENERATIVE STEAM-TURBINE HEAT RATES

What are the net and gross heat rates of a 300-kw reheat turbine having an initial steam pressure of 3,500 psig with initial and reheat steam temperatures of 1,000 F with 1.5 in. Hg absolute back pressure and six stages of regenerative feedwater heating? Compare this heat rate with that of 3,500 psig 600-mw cross-compound four-flow turbine with 3,600/1,800 revolutions/min shafts at a 300-mw load.

Calculation Procedure:

1. Determine the reheat-regenerative heat rate

Enter Fig. 31 at 3,500-psig initial steam pressure and project vertically to the 300-mw capacity net-heat-rate curve. At the left, read the net heat rate as 7,680 Btu/kw-hr. On the same vertical line, read the gross heat rate as 7,350 Btu/kw-hr. The gross heat

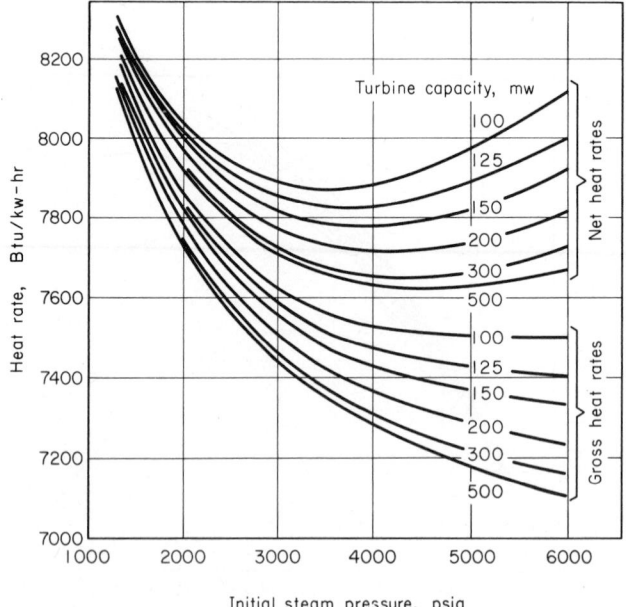

Fig. 31   Full-load heat rates for steam turbines with six feedwater heaters, 1000/1000-F steam, 1.5-in. Hg abs exhaust pressure.

rate is computed using the generator-terminal output; the net heat rate is computed after deducting the feedwater-pump energy input from the generator output.

2. Determine the cross-compound turbine heat rate

Enter Fig. 32 at 350 mw at the bottom and project vertically upward to 1.5 in. Hg exhaust pressure midway between the 1- and 2-in. Hg curves. At the left, read the net heat rate as 7,880 Btu/kw-hr. Thus, the reheat-regenerative unit has a lower net heat rate. Even at full rated load of the cross-compound turbine, its heat rate is higher than the reheat unit.

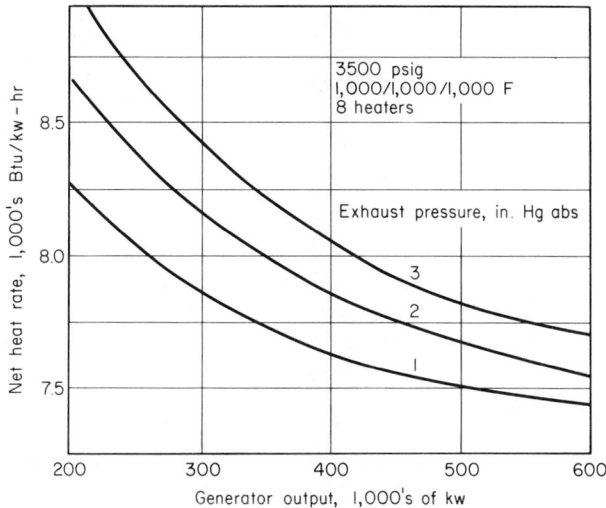

Fig. 32   Heat rate of a cross-compound four-flow steam turbine with 3,600/1,800 revolution/min shafts.

Related Calculations: Use this general procedure for comparing stationary and marine high-pressure steam turbines. The curves given here are typical of those supplied by turbine manufacturers for their turbines.

STEAM-TURBINE–GAS-TURBINE CYCLE ANALYSIS

Sketch the cycle layout, T-S diagram, and energy-flow chart for a combined steam-turbine–gas-turbine cycle having one stage of regenerative feedwater heating and one stage of economizer feedwater heating. Compute the thermal efficiency and heat rate of the combined cycle.

Calculation Procedure:

1. Sketch the cycle layout

Figure 33 shows the cycle. Since the gas-turbine exhaust-gas temperature is usually higher than the bleed-steam temperature, the economizer is placed after the regenerative feedwater heater. The feedwater will be progressively heated to a higher temperature during passage through the regenerative heater and the gas-turbine economizer. The cycle shown here is only one of many possible combinations of a steam plant and a gas turbine.

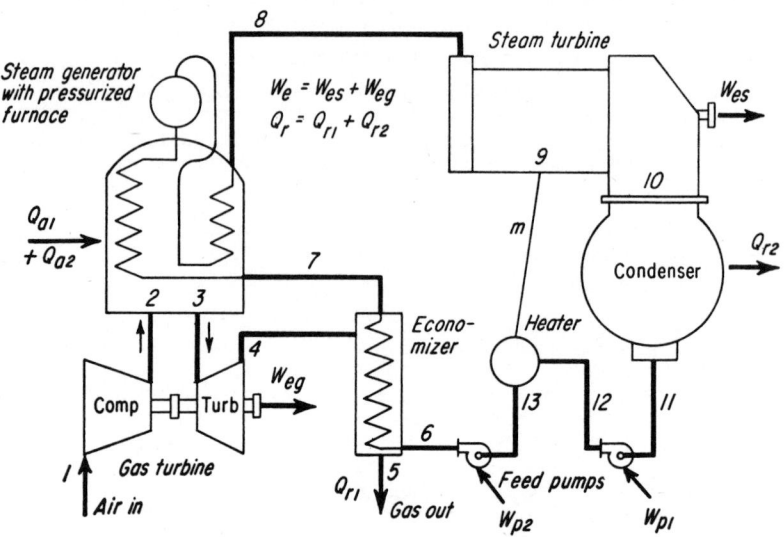

Fig. 33   Combined gas-turbine–steam-turbine cycle.

2.   Sketch the T-S diagram

Figure 34 shows the T-S diagram for the combined gas-turbine steam-turbine cycle. There is irreversible heat transfer Q_T from the gas-turbine exhaust to the feedwater in the economizer, which helps reduce the required energy input Q_{a2}.

3.   Sketch the energy-flow chart

Choose a suitable scale for the energy input and proportion the energy flow to each of the other portions of the cycle. Use a single line when the flow is too small to plot to scale. Figure 35 shows the energy-flow chart.

4.   Determine the thermal efficiency of the cycle

Since $e = W/Q_a$, $e = Q_a - Q_r/Q_a = 1 - [Q_{r1} + Q_{r2}/(Q_{a1} + Q_{a2})]$, using the notation in Figs. 33, 34, and 35.

The relative weight of the gas w_g to 1 lb of water must be computed by taking an energy balance about the economizer. Or, $H_7 - H_6 = w_g(H_4 - H_5)$. Using the actual values for the enthalpies, solve this equation for w_g.

With w_g known, the other factors in the efficiency computation are

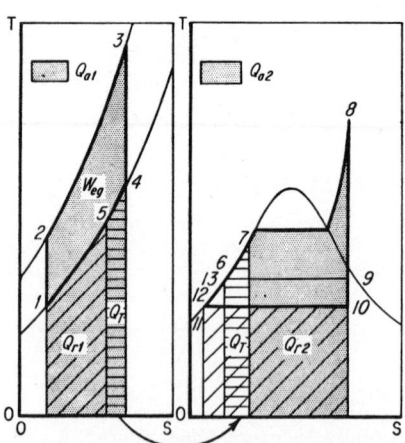

Fig. 34   T-S charts for combined gas-turbine-steam-turbine cycle have irreversible heat transfer Q from gas-turbine exhaust to the feedwater.

$$Q_{r1} = w_g(H_5 - H_1)$$
$$Q_{r2} = (1 - m)(H_{10} - H_{11})$$
$$Q_{a1} = w_g(H_3 - H_2)$$
$$Q_{a2} = H_8 - H_7$$

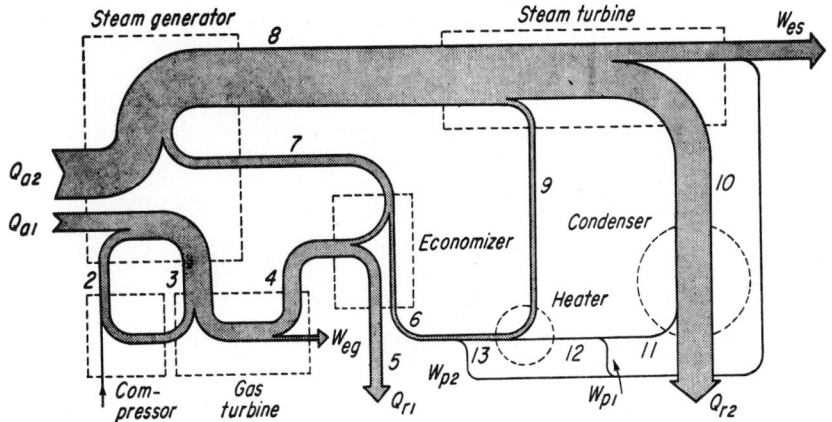

Fig. 35   Energy-flow chart of the gas-turbine–steam-turbine cycle in Fig. 33.

The bleed-steam flow m is calculated from an energy balance about the feedwater heater. Note that the units for the above equations can be any of those normally used in steam- and gas-turbine analyses.

STEAM-CONDENSER PERFORMANCE ANALYSIS

(a) Find the required tube surface area for a shell-and-tube-type condenser serving a steam turbine when the quantity of steam condensed S is 25,000 lb/hr; condenser back pressure = 2 in. Hg absolute; steam temperature $t_s = 101.1$ F; inlet water temperature $t_1 = 80$ F; tube length per pass $L = 14$ ft; water velocity $V = 6.5$ fps; number of passes = 2; tube size and gage: $\frac{3}{4}$ in., No. 18 BWG; cleanliness factor = 0.80. (b) Compute the required area and cooling-water flow rate for the same conditions as (a) except that cooling water enters at 85 F. (c) If the steam flow through the condenser in (a) decreases to 15,000 lb/hr, what will be the absolute steam pressure in the condenser shell?

Calculation Procedure:

1. Sketch the condenser, showing flow conditions

(a) Figure 36 shows the condenser and the flow conditions prevailing.

2. Determine the condenser heat-transfer coefficient

Use standard condenser-tube engineering data available from the manufacturer or Heat Exchange Institute. Table 1 and Fig. 37 show typical condenser-tube data used in condenser selection. These data are based on: (a) a *minimum* water velocity of 3 fps through the condenser tubes, (b) a *minimum* absolute pressure of 0.7 in. Hg in the condenser shell, and (c) a *minimum* Δt terminal temperature difference $(t_s - t_2)$ of 5 F. These conditions are typical for power-plant surface condensers.

Enter Fig. 37 at the bottom at the given water velocity, 6.5 fps, and project vertically upward until the $\frac{3}{4}$-in.-OD tube curve is intersected. From this point, project horizontally to the

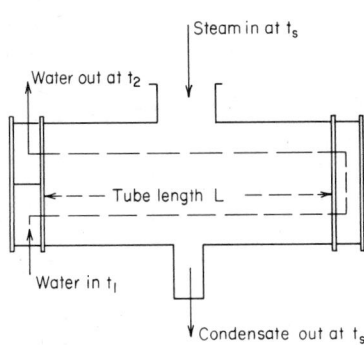

Fig. 36   Temperatures governing condenser performance.

TABLE 1   Standard Condenser Tube Data

Tube OD, in.	Tube gage, BWG	Tube ID, in.	Surface area, sq ft per ft length		Velocity, fps for 1 gpm	Value of k for number of tube passes		
			Outside	Inside		One	Two	Three
$\frac{3}{4}$	18	0.652	0.1963	0.1706	0.9611	0.188	0.377	0.565
	16	0.620	0.1963	0.1613	1.063	0.208	0.417	0.625
	14	0.584	0.1963	0.1528	1.198	0.235	0.470	0.705
$\frac{7}{8}$	18	0.777	0.2291	0.2034	0.6670	0.155	0.310	0.465
	16	0.745	0.2291	0.1951	0.7360	0.168	0.337	0.505
	14	0.709	0.2291	0.1856	0.8126	0.186	0.372	0.558
1	18	0.902	0.2618	0.2361	0.5022	0.131	0.263	0.394
	16	0.870	0.2618	0.2277	0.5398	0.141	0.283	0.424
	14	0.834	0.2618	0.2183	0.5874	0.154	0.308	0.461

left to read the heat-transfer coefficient $U = 690$ Btu/(sq ft)(F)LMTD (log mean temperature difference). Also read from Fig. 37 the temperature correction factor for an inlet-water temperature of 80 F by entering at the bottom at 80 F and projecting vertically upward to the temperature-correction curve. From the intersection with this curve, project to the right to read the correction as 1.04. Correct U for temperature and cleanliness by multiplying the value obtained from the chart by the correction factors, or $U = 690(1.04)(0.80) = 574$ Btu/(sq ft)(hr)(F)LMTD.

3.   *Compute the tube constant*

Read from Table 1, for two passes through $\frac{3}{4}$-in.-OD 18 BWG tubes, $k = $ a constant $= 0.377$. Then, $kL/V = 0.377(14)/6.5 = 0.812$.

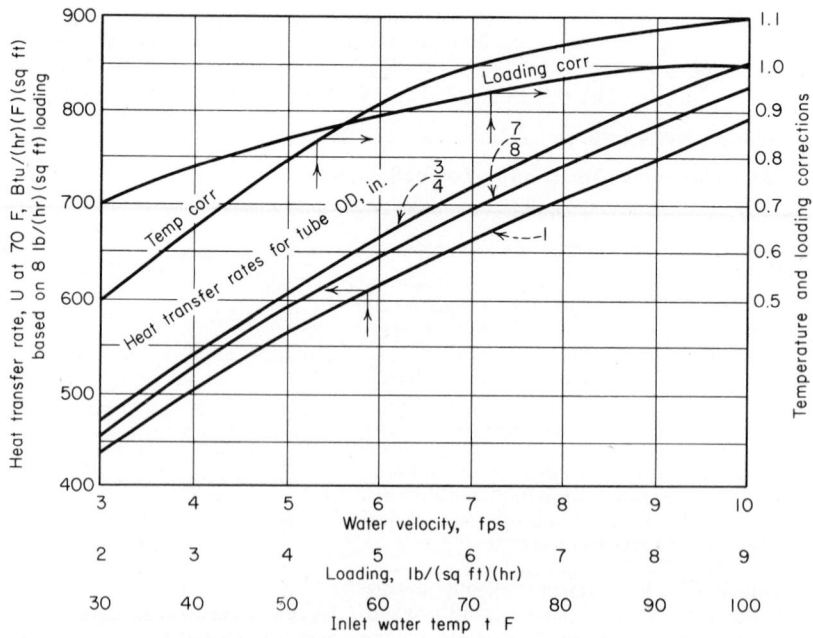

Fig. 37   Heat-transfer and correction curves for calculating surface-condenser performance.

4. Compute the outlet-water temperature

The equation for outlet-water temperature is $t_2 = t_s - [t_s - t_1/e^x]$, where $x = (kL/V)$ $(U/500)$, or $x = 0.812(574/500) = 0.932$. Then, $e^x = 2.7183^{0.932} = 2.54$. With this value known, $t_2 = 101.1 - (101.1 - 80/2.54) = 92.8$ F. Check to see if Δt $(t_s - t_2)$ is less than the minimum 5-F terminal temperature difference. Or $101.1 - 92.8 = 8.3$ F, which is > 5 F.

5. Compute the required tube surface area

The required cooling-water flow, gpm, $= 950S/[500(t_2 - t_1)] = 950(25,000)/[500(92.8 - 80)] = 3,700$ gpm. This equation assumes that 950 Btu is to be removed from each lb of steam condensed. When a different quantity of heat must be removed, use the actual quantity in place of the 950 in this equation.

With the tube constant (kL/V) and cooling-water flow rate known, the required area is computed from $A = (kL/V)(\text{gpm}) = (0.812)(3,700) = 3,000$ sq ft.

Since the value of U was not corrected for condenser loading, it is necessary to check to see if such a correction is needed. Condenser loading $= S/A = 25,000/3,000 = 8.33$ lb/sq ft. Figure 37 shows that no correction (correction factor $= 1.0$) is necessary for loadings greater than 8.0 lb/sq ft. Therefore, the loading for this condenser is satisfactory *without* correction.

This step concludes the general Calculation Procedure for a surface condenser serving any steam turbine. The next Procedure shows the method to follow when a higher cooling-water inlet temperature prevails.

6. Compute the cooling-water outlet temperature

(b) Higher cooling-water inlet temperature.

From Fig. 37 for 85-F cooling-water inlet temperature and a 0.80 cleanliness factor, $U = 690(1.06)(0.80) = 585$ Btu/(sq ft)(hr)(F)LMTD.

Using data from Table 1, the tube constant $kL/V = 0.377(14)/6.5 = 0.812$. Then, $x = (kL/V)(U/500) = 0.812(585/500) = 0.950$. Using this exponent, $e^x = 2.8183^{0.950} = 2.586$. The cooling-water outlet temperature is then $t_2 = t_s - (t_s - t_1/e^x) = 101.1 - (101.1 - 85)/2.586 = 94.9$ F. Check to see if Δt $(t_s - t_2)$ is greater than the minimum 5-F terminal temperature difference. Or, $101.1 - 94.9 = 6.5$, which is > 5 F.

7. Compute the water flow rate, required area, and loading

The required cooling-water flow, gpm, $= 950S/[500(t_2 - t_1)] = 950(25,000)/[500(94.9 - 85)] = 4,800$ gpm.

With the tube constant (kL/V) and cooling-water flow rate known, the required area is computed from $A = (kL/V)(\text{gpm}) = 0.812(4,800) = 3,900$ sq ft. Then, loading $= S/A = 25,000/3,900 = 6.4$ lb/sq ft.

Since the loading is less than 8 lb/sq ft, refer to Fig. 37 to obtain the loading correction factor. Enter at the bottom at 6.4 lb/sq ft and project vertically to the loading curve. At the right, read the loading correction factor as 0.95. Now the value of U already computed must be corrected and all dependent quantities recalculated.

8. Recalculate the condenser proportions

First, correct U for loading. Or $U = 585(0.95) = 555$. Then, $x = 0.812(555/500) = 0.90$; $e^x = 2.7183^{0.90} = 2.46$; $t_2 = 101.1 - (101.1 - 85/2.46) = 94.6$ F. Check $\Delta t = t_s - t_2 = 101.1 - 94.6 = 6.5$ F, which is > 5 F. The cooling-water flow rate, gpm, $= 950(25,000)/[500(94.6 - 85)] = 4,950$ gpm. Then, $A = 0.812(4,950) = 4,020$ sq ft, and loading $= 25,000/4,020 = 6.23$ lb/sq ft.

Check the correction factor for this loading in Fig. 37. The correction factor is 0.94, compared with 0.95 for the first calculation. Since the value of U would be changed only about 1 percent by using the lower correction, the calculations need not be revised further. Where U would change by a larger amount—say 5 percent or more—it would be necessary to repeat the procedure just detailed, applying the new correction factor.

Note that the 5-F increase in cooling-water temperature (from 80 to 85 F) requires an additional 1,020 sq ft of condenser surface and 125 gpm of cooling-water flow to maintain the same back pressure. These increments will vary, depending upon the temperature level at which the increase occurs. The effect of reduced steam flow on the steam pressure in the condenser shell will not be computed because the recalculation above is the last step in part (b) of this procedure.

(c) Reduced steam flow to condenser.

9. Determine the condenser loading

From procedure (a) above, the cooling-water flow $= 3,700$ gpm; condenser surface $A = 3,000$ sq ft. Then, with a 15,000 lb/hr steam flow, loading $= S/A = 15,000/3,000 = 5$ lb/sq ft.

10. Compute the heat-transfer coefficient

Correct the previous heat-transfer rate $U = 690$ Btu/(sq ft)(hr)(F) LMTD for temperature, cleanliness, and loading. Or, $U = 690(1.04)(0.80)(0.89) = 511$ Btu/(sq ft)(hr)(F) LMTD, using correction factors from Fig. 37.

11. Compute the final steam temperature

As before, $x = (kL/V)(U/500) = (0.377)(14/6.5)(511/500) = 0.830$. Then, $\Delta t = t_2 - t_1 = 950 \, S/[500(\text{gpm})] = 950(15,000)/[500(3,700)] = 7.7$ F. With $t_1 = 80$ F, $t_2 = \Delta t + t_1 = 7.7 + 80 = 87.7$ F. Since $t_2 = t_s - (t_s - t_1)/e^x$, $e^x = t_s - t_1/(t_s - t_2)$, or $2.7183^{0.830} = t_s - 80/(t_s - 87.7)$. Solve for t_s; or $t_s = 201.1 - 80/1.294 = 93.6$ F.

At a saturation temperature of 93.6 F, the steam table (saturation temperature) shows that the steam pressure in the condenser shell is 1.59 in. Hg.

Check the Δt terminal temperature difference. Or $\Delta t = t_s - t_2 = 93.6 - 87.7 = 5.9$ F. Since the terminal temperature difference is > 5 F, the calculated performance can be realized.

Related Calculations: The procedures and data given here can be used to compute the required cooling-water flow, cooling-water temperature rise, quantity of steam condensed by a given cooling-water flow rate and temperature rise, required condenser surface area, tube length per pass, water velocity, steam temperature in condenser, cleanliness factor, and heat-transfer rate. Whereas Fig. 37 is suitable for all usual condenser calculations for the ranges given, check the Heat Exchange Institute for any new curves that might have been made available before making the final selection of very large condensers (more than 100,000 lb/hr steam flow).

NOTE: The *design water temperature* used for condensers is either the average summer water temperature or the average annual water temperature, depending on which is higher. The *design steam load* is the maximum steam flow expected at the full-load rating of the turbine or engine. Usual shell-and-tube condensers have tubes that vary in length from about 8 ft in the smallest sizes to about 40 ft, or more, in the largest sizes. Each sq ft of tube surface will condense 7 to 20 lb of steam per hr with a cooling-water circulating rate of 0.1 to 0.25 gpm/lb of steam condensed. The method presented here is the work of Glenn C. Boyer.

STEAM-CONDENSER SELECTION

Select a condenser for a steam turbine exhausting 150,000 lb of steam per hr at 2 in. Hg absolute with a cooling-water inlet temperature of 75 F. Assume an 0.85 condition factor, $\frac{7}{8}$-in. No. 18 BWG tubes, and an 8-fps water velocity. The water supply is restricted. Obtain condenser constants from the Heat Exchange Institute *Steam Surface Condenser Standards*.

Calculation Procedure:

1. Select the $t_s - t_1$ temperature difference

Table 2 shows customary design conditions for steam condensers. With an inlet-water temperature of 75 F and an exhaust steam pressure of 2.0 in. Hg absolute, the customary temperature difference $t_s - t_1 = 26.1$ F. With sufficient water supply and a siphonic circuitry, $(t_2 - t_1)/(t_s - t_1)$ is usually between 0.5 and 0.55; for a restricted water supply or high frictional resistance and static head, the value of this factor ranges from 0.55 to 0.75.

TABLE 2   Typical Design Conditions for Steam Condensers

Cooling water temperature, F	Steam pressure, in. Hg	Temperature difference, $(t_s - t_1)$, F
50	1.0	29.0
55	1.0–1.25	24.0–30.9
60	1.0–1.5	19.0–31.7
65	1.5–1.75	26.7–31.7
70	1.5–2.0	21.7–31.1
75	2.0–2.5	26.1–33.7
80	2.0–4.0	21.1–45.4
85	2.5–4.0	23.7–40.4
90	3.0–4.0	25.1–35.4

2.   Compute the LMTD across the condenser

With 75-F inlet water, $t_s - t_1 = 101.14 - 75 = 26.14$ F, using the steam temperature given in the saturation-pressure steam table. Once $t_s - t_1$ is known, it is necessary to assume a value for the ratio $(t_2 - t_1)/(t_s - t_1)$. As a trial, assume 0.60, since the water supply is restricted. Then, $(t_2 - t_1)/(t_s - t_1) = 0.60 = (t_2 - t_1)/26.14$. Solving, $t_2 - t_1 = 15.68$ F. The difference between the steam temperature t_s and the outlet water temperature t_2 then is $t_s - t_2 = 26.14 - 15.68 = 10.46$ F.   Checking, $t_2 = t_1 + (t_2 - t_1) = 75 + 15.68 = 90.68$ F; $(t_s - t_2) = 101.14 - 90.68 = 10.46$ F. This value is greater than the required minimum value of 5 F for $t_s - t_2$. The assumed ratio 0.60 is therefore satisfactory.

Were $t_s - t_2$ less than 5 F, another ratio value would be assumed and the difference computed again. Continue doing this until a value of $t_s - t_2$ greater than 5 F is obtained. Then, $\text{LMTD} = (t_2 - t_1)/\ln[(t_s - t_1)/(t_s - t_2)]$; $\text{LMTD} = 15.68/\ln(26.1/10.46) = 17.18$ F.

3.   Determine the heat-transfer coefficient

U from the Heat Exchange Institute or manufacturer's data is 740 Btu/(sq ft)(hr)(F) LMTD for a water velocity of 8 fps. If these data are not available, Fig. 37 can be used with complete safety for all preliminary selections.

U must now be corrected for the inlet-water temperature, 75 F, and the condition factor, 0.85, which is a term used in place of the correction factor by some authorities. From Fig. 37, the correction for 75-F inlet water = 1.04. Then, actual $U = 740(1.04)$ $(0.85) = 655$ Btu/(sq ft)(hr)(F)LMTD.

4.   Compute the steam condensation rate

The heat-transfer rate per sq ft of condenser surface with a 17.18-F LMTD is U (LMTD) $= 655(17.18) = 11,252.9$ Btu/(sq ft)(hr).

Condensers serving steam turbines are assumed, for design purposes, to remove 950 Btu per lb of steam condensed. Therefore, the steam condensation rate for any condenser is [Btu/(sq ft)(hr)]/950, or $11,252.9/950 = 11.25$ lb/(hr)(sq ft).

5.   Compute the required surface area and water flow

The required surface area = lb per hr steam flow/[condensation rate, lb/(hr)(sq ft)], or with a 150,000 lb/hr flow, $150,000/11.25 = 13.320$ sq ft.

The water flow rate, gpm, $= 950 \, S/500(t_2 - t_1) = 950(150,000)/[500(15.68)] = 18,200$ gpm.

Related Calculations: See the previous Calculation Procedure for steps in determining the water-pressure loss through a surface condenser.

To choose a surface condenser for a steam engine, use the same procedures as given above, except that the heat removed from the exhaust steam is 1,000 Btu/lb. Use a condition (cleanliness) factor of 0.65 for steam engines because the oil in the exhaust steam fouls the condenser tubes, reducing the rate of heat transfer. The condition (cleanliness) factor for steam turbines is usually assumed to be 0.8 to 0.9 for relatively clean, oil-free cooling water.

At loads greater than 50 percent of the design load, $t_s - t_1$ follows a straight-line relationship. Thus, in the above condenser, $t_s - t_1 = 26.14$ F at the full load of 150,000 lb/hr. If the load falls to 60 percent (90,000 lb/hr), $t_s - t_1 = 26.14(0.60) = 15.7$ F. At 120 percent load (180,000 lb/hr), $t_s - t_1 = 26.14(1.20) = 31.4$ F. This *straight-line law* is valid with constant inlet-water temperature and cooling-water flow rate. It is useful in analyzing condenser operating conditions at other than full load.

Single- or multiple-pass surface condensers may be used in power services. When a liberal supply of water is available, the single-pass condenser is often chosen. With a limited water supply, a two-pass condenser is often chosen.

AIR-EJECTOR ANALYSIS AND SELECTION

Choose a steam-jet air ejector for a condenser serving a 250,000 lb/hr steam turbine exhausting at 2 in. Hg absolute. Determine the number of stages to use, the approximate steam consumption, and the quantity of air and vapor mixture the ejector will handle.

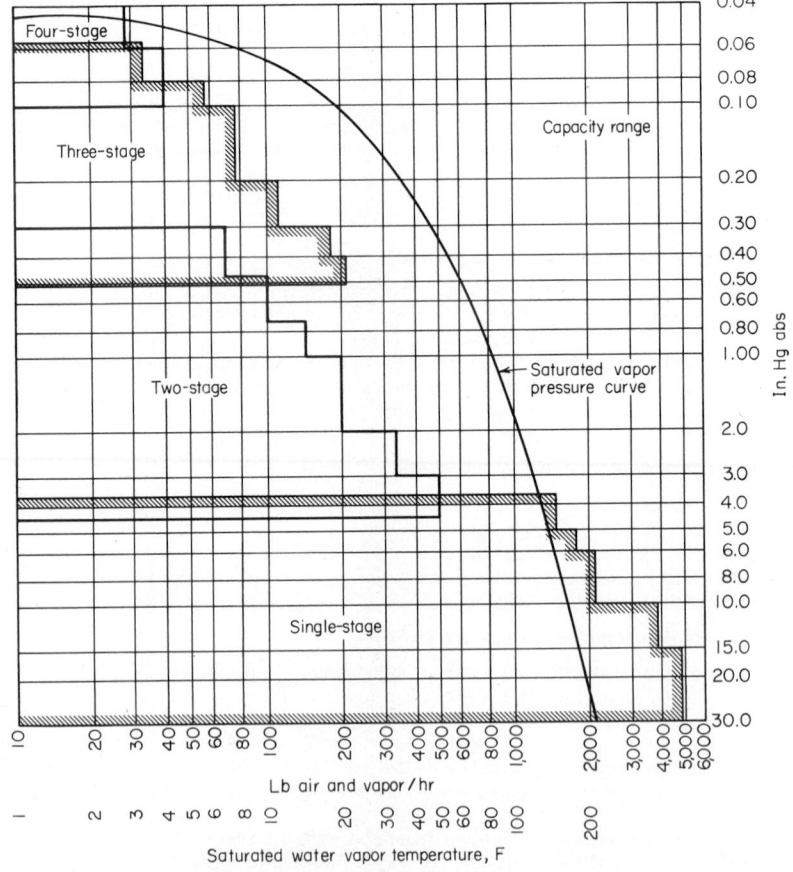

Fig. 38 Steam-ejector capacity-range chart.

Calculation Procedure:

1. Select the number of stages for the ejector

Use Fig. 38 as a preliminary guide to the number of stages required in the ejector. Enter at 2-in. Hg absolute condenser pressure and project horizontally to the stage area. This shows that a two-stage ejector will probably be satisfactory.

Check the number of stages chosen above against the probable overload range of the prime mover by using Fig. 39. Enter at 2-in. Hg absolute condenser pressure and project to the two-stage curve. This curve shows that a two-stage ejector can readily handle a 25 percent overload of the prime mover. Also, the two-stage curve shows that this ejector could handle up to 50 percent overload with an increase in the condenser absolute pressure of only 0.4 in. Hg. This is shown by the pressure, 2.4 in. Hg absolute, at which the two-stage curve crosses the 150 percent overload ordinate, Fig. 39.

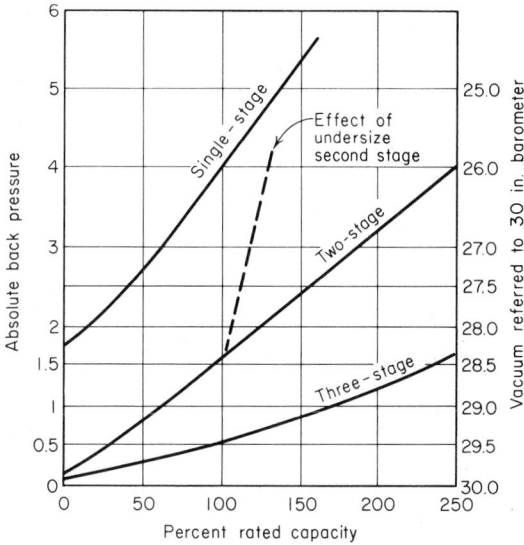

Fig. 39   Steam-jet ejector characteristics.

2. Determine the ejector operating conditions

Use the Heat Exchange Institute or manufacturer's data. Table 3 excerpts data from the Heat Exchange Institute for condensers in the range considered in this Procedure.

Study of Table 3 shows that a two-stage condensing ejector unit serving a 250,000 lb/hr steam turbine will require 450 lb/hr of 300-psig steam. Also, the ejector will handle 7.5 cfm of free, dry air, or 33.75 lb/hr of air. It will remove up to 112.5 lb/hr of an air-vapor mixture.

The actual air leakage into a condenser varies with the absolute pressure in the condenser, the tightness of the joints, and the condition of the tubes. Some authorities cite a maximum leakage of about 250 lb/hr steam flow. At 400,000 lb/hr, the leakage is 160 lb/hr; at 250,000 lb/hr, 130 lb/hr of air-vapor mixture. A condenser in good condition will usually have less leakage.

For an installation in which the manufacturer supplies data on the probable air leakage, use a psychrometric chart to determine the weight of water vapor contained in the air. Thus, at 2-in. Hg absolute and 80 F, each lb of air will carry with it 0.68 lb of water vapor. In a surface condenser into which 20 lb of air leaks, the ejector must handle $20 + 20(0.68) = 33.6$ lb/hr of air-vapor mixture. Table 3 shows that this ejector can readily handle this quantity of air-vapor mixture.

TABLE 3 Air-Ejector Capacities for Surface Condensers for Steam Turbines

(Two-stage condensing ejector unit)

Steam load, lb/hr	Cfm, 70 F free, dry air	Air, lb/hr	Air-vapor mixture at 30 percent dry air, lb/hr	Steam consumption at 300 psig, lb/hr
100,001–250,000	7.5	33.75	112.5	450

Related Calculations: When choosing an air ejector for steam-engine service, double the Heat Exchange Institute steam-consumption estimates. For most low-pressure power-plant service, a two-stage ejector with inter- and after-condensers is satisfactory, although some steam engines operating at higher absolute exhaust pressures require only a single-stage ejector. Twin-element ejectors have two sets of stages; one set serves as a spare and may also be used for capacity regulation in stationary and marine service. The capacity of an ejector is constant for a given steam pressure and suction pressure. Raising the steam pressure will not increase the ejector capacity.

SURFACE-CONDENSER CIRCULATING-WATER PRESSURE LOSS

Determine the circulating-water pressure loss in a two-pass condenser having 12,000 sq fr of condensing surface, a circulating-water flow rate of 10,000 gpm, $\frac{3}{4}$-in. No. 16 BWG tubes, a water flow rate of 7 fps, external friction of 20 ft of water, and a 10-ft-of-water siphonic effect on the circulating-water discharge.

Calculation Procedure:

1. Determine the water flow rate per tube

Use a tabulation of condenser-tube engineering data available from the manufacturer or the Heat Exchange Institute, or compute the water flow rate from the physical dimensions of the tube thus: $\frac{3}{4}$-in. No. 16 BWG tube ID = 0.620 in. from a tabulation of condenser-tube data, such as Table 1. Assume a water velocity of 1 fps. Then a 1-ft length of the tube will contain $(12)(0.620)^2\pi/4 = 3.62$ in.^{3} of water. This quantity of water will flow through the tube for each ft of length per sec of water velocity. The flow per min will be $3.62(60 \text{ sec/min}) = 217.2$ in.^{3}/min. Since 1 U.S. gal = 231 in.^{3}, the gpm flow at a 1 fps velocity = 217.2/231 = 0.94 gpm. With an actual velocity of 7 fps, the water flow rate per tube is $7(0.94) = 6.58$ gpm.

2. Determine the number of tubes and length of water travel

Since the water flow rate through the condenser is 10,000 gpm and each tube conveys 6.58 gpm, the number of tubes = 10,000 gpm/6.58 gpm per tube = 1,520 tubes per pass. Next, the total length of water travel for a condenser having A sq ft of condensing surface is computed from: A (number of tubes)(outside area per lin ft of tube, sq ft). The outside area of each tube can be obtained from a table of tube properties, such as Table 1, or computed from (OD, in.)$(\pi)(12)/144$, or $(0.75)(\pi)(12)/(144) = 0.196$ sq ft/lin ft. Then, total length of water travel = $12,000/[(1,520)(0.196)] = 40.2$ ft. Since the condenser has two passes, the length of tubes per pass = 40.2/2 = 20.1 ft. As each pass has an equal number of tubes, and there are two passes, the total number of tubes in the condenser = 2 passes (1,520 tubes per pass) = 3,040 tubes.

3. Compute the friction loss in the system

Use the Heat Exchange Institute or manufacturer's curves to find the friction loss per ft of condenser tube. At 7 fps, the Heat Exchange Institute curve shows the head loss is 0.4 ft of head per ft of travel for $\frac{3}{4}$-in. No. 16 BWG tubes. With a total length of 40.2 ft, the tube head loss is 0.4(40.2) = 16.1 ft.

Use the Heat Exchange Institute or manufacturer's curves to find the head loss through the condenser water boxes. From the first reference, for a velocity of 7 fps, head loss = 1.4 ft of water for a single-pass condenser. Since this is a two-pass condenser, the total water-box head loss = 2(1.4) = 2.8 ft.

The total condenser friction loss is then the sum of the tube and water-box losses, or 16.1 + 2.8 = 18.9 ft of water. With an external friction loss of 20 ft in the circulating-water piping, the total loss in the system, without siphonic assistance, is 18.9 + 20 = 38.9 ft. Since there is 10 ft of siphonic assistance, the total friction loss in the system *with* siphonic assistance is 38.9 − 10 = 28.9 ft. In choosing a pump to serve this system, the frictional resistance of 28.9 ft would be rounded off to 30 ft and any factor of safety added to this value of head loss.

NOTE: The most economical cooling-water velocity in condenser tubes is 6 to 7 fps; a velocity greater than 8 fps should not be used, unless warranted by special conditions.

SURFACE-CONDENSER WEIGHT ANALYSIS

A turbine exhaust nozzle can support a weight of 100,000 lb. Determine what portion of the total weight of a surface condenser must be supported by the foundation if the weight of the condenser is 275,000 lb, the tubes and water boxes have a capacity of 8,000 gal, and the steam space a capacity of 30,000 gal of water.

Calculation procedure:

1. Compute the maximum weight of the condenser

The maximum weight on a condenser foundation occurs when the shell, tubes, and water boxes are full of water. This condition could prevail during accidental flooding of the steam space or during tests for tube leaks when the steam space is purposefully flooded. In either circumstance, the condenser foundation and spring supports, if used, must be able to carry the load imposed on them. To compute this load, find the sum of the individual weights:

Condenser weight, dry	275,000 lb
Water in tubes and boxes =	
(8.33 lb/gal)(8,000 gal)	66, 640
Water in steam space =	
(8.33)(30,000)	249,900
Maximum weight when full of water .	591,540 lb

2. Compute the foundation load

The turbine nozzle can support 100,000 lb. Therefore, the foundation must support 591,540 − 100,000 = 491,540 lb. For foundation design purposes this would probably be rounded off to 495,000 lb.

Related Calculations: When designing a condenser foundation do the following. (1) Leave enough room at one end to permit withdrawal of faulty tubes and insertion of new tubes. Since some tubes may exceed 40 ft in length, careful planning is needed to provide sufficient installation space. During the design of a power plant, a template representing the tube length is useful for checking the tube clearance on a scale plan and side view of the condenser installation. When there is insufficient room for tube removal with one shape of condenser, try another shape with shorter tubes.

(2) Provide enough headroom under the condenser to produce the required submergence on the condensate-pump impeller. Most condensate pumps require at least 3 ft submergence. If necessary, the condensate pump can be installed in a pit under the condenser, but this should be avoided if possible.

BAROMETRIC-CONDENSER ANALYSIS AND SELECTION

Select a countercurrent barometric condenser to serve a steam turbine exhausting 25,000 lb/hr of steam at 5 in. Hg absolute. Determine the quantity of cooling water required if the water inlet temperature is 50 F. What is the required dry-air capacity of the ejector? What is the required pump head if the static head is 40 ft and the pipe friction is 15 ft of water?

Calculation Procedure:

1. Find the steam properties from the steam tables

At 5 in. Hg absolute, $h_g = 1,119.4$ Btu/lb, from the saturation pressure table. If the condensing water were to condense the steam without subcooling the condensate, the final temperature of the condensate, from the steam tables, would be 133.76 F, corresponding to the saturation temperature. However, subcooling almost always occurs, and the usual practice in selecting a countercurrent barometric condenser is to assume the final condensate temperature t_c will be 5 F below the saturation temperature corresponding to the absolute pressure in the condenser. Using a 5-F difference, $t_c = 133.76 - 5 = 128.76$ F. Interpolating in the saturation temperature steam table, the enthalpy of the condensate h_f at 128.76 F is 96.66 Btu/lb.

2. Compute the quantity of condensing water required

In any countercurrent barometric condenser, the quantity of cooling water Q lb/hr required is $Q = W(h_g - h_f)/(t_c - t_i)$, where $W =$ weight of steam condensed, lb/hr, $t_i =$ cooling-water inlet temperature, F. Then, $Q = 25,000(1,119.4 - 96.66)/(128.76 - 50) = 325,000$ lb/hr. Converting to gpm $= q = 325,000/500 = 650$ gpm.

3. Determine the required ejector dry-air capacity

Use the Heat Exchange Institute or a manufacturer's tabulation of free, dry-air leakage and the allowance for air in the cooling water to determine the required dry-air

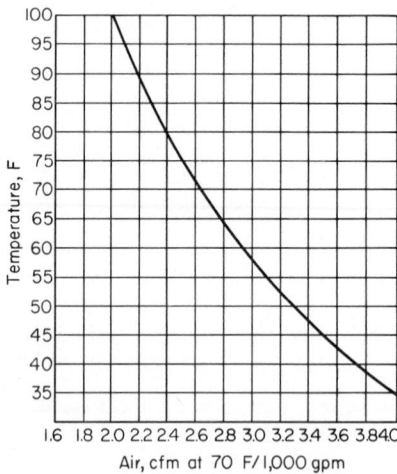

Fig. 40 Allowance for air in condenser injection water.

capacity. Thus, from Table 4, the free, dry-air leakage for a barometric condenser serving a turbine is 3.0 cfm of air and vapor. The allowance for air in the 50-F cooling water is 3.3 cfm of air at 70 F per 1,000 gpm of cooling water, Fig. 40. The total dry-air leakage is the sum, or $3.0 + 3.3 = 6.3$ cfm. Thus, the ejector must be capable of handling at least 6.3 cfm of dry air to serve this barometric condenser at its rated load of 25,000 lb/hr of steam.

Where the condenser will operate at a lower vacuum (i.e., a higher absolute pressure), overloads up to 50 percent may be met. To provide adequate dry-air handling capacity at this overload with the same cooling-water inlet temperature, find the free, dry-air leakage at the higher condensing rate from Table 4 and add this to the previously found allowance for air in the cooling water. Or $4.5 + 3.3 = 7.8$ cfm. An ejector capable of handling up to 10 cfm would be a wise choice for this countercurrent barometric condenser.

4. Determine the pump head required

Since a countercurrent barometric condenser operates at pressures below atmospheric, it assists the cooling-water pump by "sucking" the water into the condenser.

TABLE 4   Free, Dry-Air Leakage

(Cfm at 70 F air and vapor mixture, $7\frac{1}{2}°$ below vacuum temperature)

| Maximum lbs of steam to be condensed per hour | Barometric and Low Level Jet Condensers | | | |
| | Serving turbines | | Serving engines | |
	Cfm	Lb/hr air and vapor	Cfm	Lb/hr air and vapor
5,000 or less	2.2	33.0	4.4	66.0
5,001–25,000	3.0	45.0	6.0	90.0
25,001–75,000	4.5	67.5	9.0	135.0
75,001–150,000	6.5	97.5	13.0	195.0
150,001–250,000	8.5	127.5		
250,001–350,000	10.0	150.0		
350,001–450,000	11.5	172.5		
450,001–600,000	13.5	202.5		
600,001–800,000	16.0	250.0		

The maximum assist that can be assumed is 0.75 V, where V = design vacuum, in. Hg.

In this condenser, with a 26-in. vacuum, the maximum assist is $0.75(26) = 19.5$ in. Hg. Converting to ft of water, using 1.0 in. Hg = 1.134 ft of water, $19.5(1.134) = 22.1$ ft of water. The total head on the pump is then the sum of the static and friction heads *less* 0.75V expressed in feet of water. Or, the total head on the pump = $40 + 15 - 22.1 = 32.9$ ft. A pump with a total head of at least 35 ft of water would be chosen for this condenser. Where corrosion or partial clogging of the piping is expected, a pump with a total head of 50 ft would probably be chosen to ensure sufficient head even though the piping is partially clogged.

Related Calculations: (1) When a condenser serving a steam engine is being chosen, use the appropriate dry-air leakage value from Table 4. (2) For ejector-jet barometric condensers, assume the final condensate temperature t_c as 10 to 20 F *below* the saturation temperature corresponding to the absolute pressure in the condenser. This type of condenser does not use an ejector but it requires 25 to 50 percent more cooling water than the countercurrent barometric condenser for the same vacuum. (3) The total pumping head for an ejector-jet barometric condenser is the sum of the static and friction heads *plus* 10 ft. The additional positive head is required to overcome the pressure loss in the spray nozzles.

COOLING-POND SIZE FOR A KNOWN HEAT LOAD

How many spray nozzles and what surface area is needed to cool 10,000 gpm of water from 120 to 90 F in a spray-type cooling pond if the average wet-bulb temperature is 60 F? What would the approximate dimensions of the cooling pond be? Determine the total pumping head if the static head is 10 ft, the pipe friction is 35 ft of water, and the nozzle pressure is 8 psi.

Calculation Procedure:

1. Compute the number of nozzles required

Assume a water flow of 50 gpm per nozzle; this is a typical flow rate for usual cooling-pond nozzles. Then, the number of nozzles required = 10,000 gpm/50 gpm per nozzle = 200 nozzles. If six nozzles are used in each spray group a series of crossed arms, with each arm containing one or more nozzles, then 200 nozzles/6 nozzles per spray group = 33⅓ spray groups will be needed. Since a partial spray group is seldom used, 34 spray groups would be chosen.

2. Determine the surface area required

Usual design practice is to provide 1 sq ft of pond area per 250 lb of water cooled for water quantities exceeding 1,000 gpm. Thus, in this pond, the weight of water cooled = (10,000 gpm)(8.33 lb/gal)(60 min/hr) = 4,998,000, say 5,000,000 lb/hr. Then, the area required, using 1 sq ft of pond area per 250 lb of water cooled = 5,000,000/250 = 20,000 sq ft.

As a cross-check, use another commonly accepted area value: 125 Btu/(sq ft)(F) difference between the air wet-bulb temperature and the warm entering-water temperature. This is the equivalent of $(120-60)(125) = 7,500$ Btu/sq ft in this spray pond, because the air wet-bulb temperature is 60 F and the warm-water temperature is 120 F. The heat removed from the water is (lb/hr of water)(temperature decrease, F) (specific heat of water) = $(5,000,000)(120-90)(1.0) = 150,000,000$ Btu/hr. Then, area required = [(heat removed, Btu/hr)]/[(heat removal, Btu/sq ft)] = 150,000,000/7,500 = 20,000 sq ft. This checks the previously obtained area value.

3. Determine the spray-pond dimensions

Spray groups on the same header or pipe main are usually arranged on about 12-ft centers with the headers or pipe mains spaced on about 25-ft centers, Fig. 41. Assume that 34 spray groups are used, instead of the required 33⅓, to provide an equal number of groups in two headers and a small extra capacity.

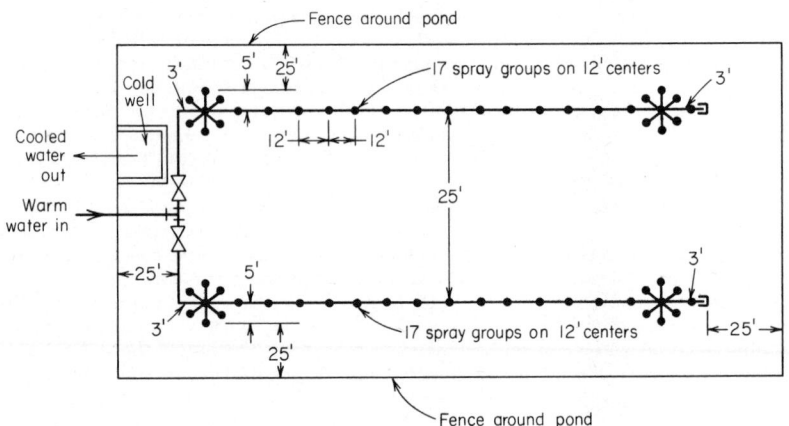

Fig. 41   Spray-pond nozzle and piping layout.

Sketch the spray pond and headers, Fig. 41. This shows that the length of each header will be about 204 ft because there are seventeen 12-ft spaces between spray groups in each header. Allowing 3 ft at each end of a header for fittings and cleanouts gives an overall header length of 210 ft. The distance between headers is 25 ft. Allow 25 ft between the outer sprays and the edge of the pond. This gives an overall width of 85 ft for the pond, assuming the width of each arm in a spray group is 10 ft. The overall length will then be 210 + 25 + 25 = 260 ft. A cold well for the pump suction and suitable valving for control of the incoming water must be provided, as shown in Fig. 41. The water depth in the pond should be 2 to 3 ft.

4. Compute the total pumping head

The total head, all expressed in ft of water, = static head + friction head + required nozzle head = 10 + 35 + 80(0.434) = 48.5 ft of water. A pump having a total head of at least 50 ft of water would be chosen for this spray pond. If future expansion of the pond is anticipated, compute the probable total head required at a future date and choose a pump to deliver that head. Until the pond is expanded, the pump would operate with a throttled discharge. Normal nozzle inlet pressures range from about 6 to 10 psi. Higher pressures should not be used, because there will be excessive spray loss and rapid wear of the nozzles.

Related Calculations: Unsprayed cooling ponds cool 4 to 6 lb of water from 100 to 70 F per sq ft of water surface. An alternative design rule is to assume that the pond will dissipate 3.5 Btu per hr per sq ft water surface per degree difference between the wet-bulb temperature of the air and the entering warm water.

DIRECT-CONTACT FEEDWATER HEATER ANALYSIS

Determine the outlet temperature of water leaving a direct-contact open-type feedwater heater if 250,000 lb/hr of water enter the heater at 100 F. Exhaust steam at 10.3 psig saturated flows to the heater at the rate of 25,000 lb/hr. What saving is obtained by using this heater if the boiler pressure is 250 psia? Determine the approximate volume of the heater if a 2-min storage capacity is provided in it.

Calculation Procedure:

1. Compute the water outlet temperature

Assume the heater is 90 percent efficient. Then $t_o = t_i w_w + 0.9 w_s h_g / (w_w + 0.9 w_s)$, where t_o = outlet water temperature, F; t_i = inlet water temperature, F; w_w = weight of water flowing through heater, lb/hr; 0.9 = heater efficiency, expressed as a decimal; w_s = weight of steam flowing to the heater, lb/hr; h_g = enthalpy of the steam flowing to the heater, Btu/lb.

For saturated steam at 10.3 psig, or $10.3 + 14.7 = 25$ psia, $h_g = 1,160.6$ Btu/lb, from the saturation pressure steam tables. Then,

$$t_o = \frac{100(250,000) + 0.9(25,000)(1160.6)}{250,000 + 0.9(25,000)} = 187.5 \text{ F}$$

2. Compute the savings obtained by feed heating

The percent saving, expressed as a decimal, obtained by heating feedwater is $(h_o - h_i)/(h_b - h_i)$ where h_o and h_i = enthalpy of the water leaving and entering the heater, respectively, Btu/lb; h_b = enthalpy of the steam at the boiler operating pressure, Btu/lb. For this plant, using the steam tables, $h_o - h_i/(h_b - h_i) = 155.44 - 67.97/(1,201.1 - 67.97) = 0.077$, or 7.7 percent.

A popular rule of thumb states that for every 11-F rise in feedwater temperature in a heater, there is approximately a 1 percent saving in the fuel that would otherwise be used to heat the feedwater. Checking the above calculation with this rule of thumb shows reasonably good agreement.

3. Determine the heater volume

With a capacity of W lb/hr of water, the volume of a direct-contact or open-type heater can be approximated from $v = W/10,000$, where v = heater internal volume, cu ft. For this heater, $v = 250,000/10,000 = 25$ cu ft.

Related Calculations: Most direct-contact or open feedwater heaters store a 2-min supply of feedwater when the boiler load is constant and the feedwater supply is all makeup. With little or no makeup, the heater volume is chosen so that there is enough capacity to store 5 to 30 min feedwater for the boiler.

CLOSED FEEDWATER HEATER ANALYSIS AND SELECTION

Analyze and select a closed feedwater heater for the third stage of a regenerative steam-turbine cycle in which the feedwater flow rate is 37,640 lb/hr, the desired temperature rise of the water during flow through the heater is 80 F (from 238 to 318 F), bleed heating steam is at 100 psia and 460 F, drains leave the heater at the saturation temperature corresponding to the heating steam pressure (100 psia), and $\frac{5}{8}$-in. OD admiralty metal tubes with a maximum length of 6 ft are used. Use the *Standards of the Bleeder Heater Manufacturers Association, Inc.*, when analyzing the heater.

Calculation Procedure:

1. Determine the logarithmic mean temperature difference across heater

When computing heat-transfer rates in feedwater heaters, the average film temperature of the feedwater is used. In computing this, the *Standards of the Bleeder Heater Manufacturers Association* specify that the *saturation temperature* of the heating steam be used. At 100 psia, $t_s = 327.81$ F. Then,

$$\text{LMTD} = t_m = \frac{(t_s - t_i) - (t_s - t_o)}{\ln\left[t_s - t_i/(t_s - t_o)\right]}$$

where the symbols are as defined in the previous Calculation Procedure. Thus,

$$t_m = \frac{(327.81 - 238) - (327.81 - 318)}{\ln\left[327.81 - 238/(327.81 - 318)\right]}$$

$$t_m = 36.5 \text{ F}$$

The average film temperature t_f for any closed heater is then

$$t_f = t_s - (0.8t_m)$$
$$t_f = 327.81 - 29.2 = 298.6 \text{ F}$$

2. Determine the overall heat-transfer rate

Assume a feedwater velocity of 8 fps for this heater. This velocity value is typical for smaller heaters handling less than 100,000 lb/hr feedwater flow. Enter Fig. 42

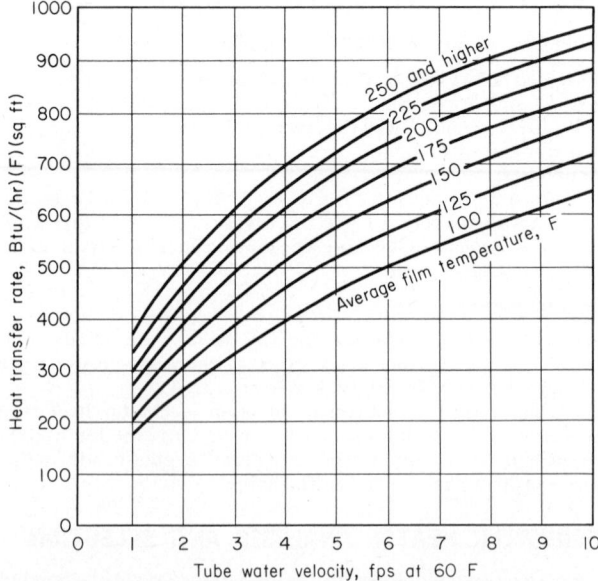

Fig. 42 Heat-transfer rates for closed feedwater heaters. (*Standards of Bleeder Heater Manufacturers Association, Inc.*)

at 8 fps on the lower horizontal scale and project vertically upward to the 250-F average film temperature curve. This curve is used even though $t_f = 298.6$ F, because the *Standards* recommend that heat-transfer rates higher than those for a 250-F film temperature not be used. So, from the 8-fps intersection with the 250-F curve in Fig. 42, project to the left to read U = the overall heat transfer rate = 910 Btu/(hr)(sq ft)(F).

Next, check Table 5 for the correction factor for U. Assume that No. 18 BWG $\frac{5}{8}$-in. OD arsenical copper tubes are used in this exchanger. Then the correction factor from Table 5 is 1.00, and $U_{corr} = 910(1.00) = 910$. If No. 9 BWG tubes are chosen, $U_{corr} = 910(0.85) = 773.5$ Btu/(hr)(sq ft)(F), using the correction factor from Table 5 for arsenical copper tubes.

TABLE 5 Multipliers for Base Heat-Transfer Rates
(For tube OD $\frac{5}{8}$ to 1 in. inclusive.)

BWG	As-Cu	Adm	90/10 Cu-Ni	80/20 Cu-Ni	70/30 Cu-Ni	Monel
18	1.00	1.00	0.97	0.95	0.92	0.89
17	1.00	1.00	0.94	0.91	0.87	0.85
16	1.00	1.00	0.91	0.88	0.84	0.82
15	1.00	0.99	0.89	0.86	0.82	0.79
14	1.00	0.96	0.85	0.82	0.77	0.75
13	0.98	0.93	0.81	0.78	0.73	0.70
12	0.95	0.90	0.77	0.73	0.68	0.65
11	0.92	0.87	0.74	0.70	0.65	0.62
10	0.89	0.83	0.69	0.66	0.60	0.58
9	0.85	0.80	0.65	0.62	0.56	0.54

3. Compute the amount of heat transferred by the heater

The enthalpy of the entering feedwater at 238 F is, from the saturation temperature steam table, $h_{fi} = 206.32$ Btu/lb. The enthalpy of the leaving feedwater at 318 F is, from the same table, $h_{fo} = 288.20$ Btu/lb. Then, the heat transferred H_t Btu/hr is $H_t = w_w (h_{fo} - h_{fi})$, where w_w = feedwater flow rate, lb/hr. Or, $H_t = 37,640(288.20 - 206.32) = 3,080,000$ Btu/hr.

4. Compute the surface area required in the exchanger

The surface area required A sq ft = H_t/Ut_m. Then, $A = 3,080,000/[(910)(36.5)] = 92.7$ sq ft.

5. Determine the number of tubes per pass

Assume that the heater has only one pass and compute the number of tubes required. Once the number of tubes is known, a decision can be made about the number of passes required. In a closed heater, number of tubes = w_w(passes)(cu ft per sec per tube)/[v(sq ft per tube open area)], where w_w = lb/hr of feedwater passing through heater; v = feedwater velocity in tubes, fps.

Since the feedwater enters the heater at 238 F and leaves at 318 F, its specific volume at 278 F, midway between t_i and t_o, can be considered the average specific volume of the feedwater in the heater. From the saturation pressure steam table, $v_f = 0.01691$ cu ft/lb at 278 F. Convert this to cu ft per sec per tube by dividing this specific volume by 3,600 (number of seconds in 1 hr) and multiply by the lb/hr of feedwater per tube. Or, cu ft per sec per tube = $(0.01691/3,600)$(lb per tube per hr).

Since No. 18 BWG 5/8-in. OD tubes are being used, ID = $0.625 - 2$(thickness) = $0.625 - 2(0.049) = 0.527$ in. Then, sq ft open area per tube = $(\pi d^2/4)/144 = 0.7854 \times$

$(0.527)^2/144 = 0.001525$ sq ft per tube. Alternatively, this area could be obtained from a table of tube properties.

With these data, compute the total number of tubes from number of tubes = $[(37,640)(1)(0.01681/3,600)]/[(8)(0.001525)] = 14.49$ tubes.

6. Compute the required tube length

Assume that 14 tubes are used, since the number required is less than 14.5. Then, tube length l ft = A/(number of tubes per pass)(passes)(area per ft of tube). Or, tube length for 1 pass = $92.7/[(14)(1)(0.1636)] = 40.6$ ft. The area per ft of tube length is obtained from a table of tube properties or computed from $12\pi(\text{OD})/144 = 12\pi(0.625)/155 = 0.1636$ sq ft.

7. Compute the actual number of passes and the actual tube length

Since the tubes in this heater cannot exceed 6 ft in length, the number of passes required = (length for one pass, ft)/(maximum allowable tube length, ft) = $40.6/6 = 6.77$ passes. Since a fractional number of passes cannot be used and an even number of passes permits a more convenient layout of the heater, choose eight passes.

Using the same equation for tube length as in Step 6, l = tube length = $92.7/[(14)(8)(0.1636)] = 5.06$ ft.

8. Determine the feedwater pressure drop through heater

In any closed feedwater heater, the pressure loss Δp psi is $\Delta p = F_1 F_2 (L + 5.5D)N/D^{1.24}$, where Δp = pressure drop in the feedwater passing through the heater, psi; F_1 and F_2 = correction fractors from Fig. 43; L = total lin ft of tubing divided by the number of tube holes in one tube sheet; D = tube ID; N = number of passes. When finding F_2, the average water temperature is taken as $t_s - t_m$.

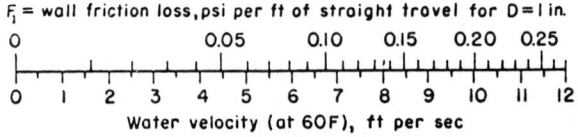

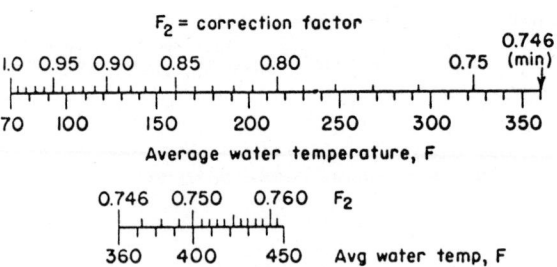

Fig. 43 Correction factors for closed feedwater heaters. (*Standards of Bleeder Heater Manufacturers Association, Inc.*)

For this heater using correction factors from Fig. 43,

$$\Delta p = (0.136)(0.761)\left[\frac{5.06(8)(14)}{(8)(14)} + 5.5(0.527)\right]\frac{(8)}{0.527^{1.24}}$$

$$= 14.6\text{ psi}$$

9. Find the heater shell outside diameter

The total number of tubes in the heater = (number of passes)(tubes per pass) = $8(14) = 112$ tubes. Assume that there is $\frac{3}{8}$-in. clearance between each tube and the tube alongside, above, or below it. Then the pitch or center-to-center distance between tubes = pitch + tube OD = $\frac{3}{8} + \frac{5}{8} = 1$ in.

The number of tubes per sq ft of tube sheet = $166/(\text{pitch})^2$, or $166/(1)^2 = 166$ tubes per sq ft. Since the heater has 112 tubes, the area of the tube sheet = $112/166 = 0.675$ sq ft, or 97 sq in.

The inside diameter of the heater shell = (tube sheet area, sq in./$0.7854)^{0.5} = (97/0.7854)^{0.5} = 11.1$ in. With a 0.25-in.-thick shell, the heater shell OD = $11.1 + 2(0.25) = 11.6$ in.

10. Compute the quantity of heating steam required

Steam enters the heater at 100 psia and 460 F. The enthalpy at this pressure and temperature is, from the superheated-steam table, $h_g = 1{,}258.8$ Btu/lb. The steam condenses in the heater, leaving as condensate at the saturation temperature corresponding to 100 psia, or 327.81 F. The enthalpy of the saturated liquid at this temperature is, from the steam tables, $h_f = 298.4$ Btu/lb.

The heater steam consumption for any closed-type feedwater heater is W lb/hr = $W = w_w(\Delta t)c/(h_g - h_f)$ where Δt = temperature rise of feedwater in heater, F; c = specific heat of feedwater, Btu/(lb)(F). Assume $c = 1.00$ for the temperature range in this heater, and $W = (37{,}640)(318 - 238)(1.00)/(1{,}258.8 - 298.4) = 3{,}140$ lb/hr.

Related Calculations: The Procedure used here can be applied to closed feedwater heaters in stationary and marine service. A similar Procedure is used for selecting hot-water heaters for building, marine, and portable service. Various authorities recommend the following *terminal difference* (heater condensate temperature minus the outlet feedwater temperature) for closed feedwater heaters:

Feedwater Outlet Temperature, F	Terminal Difference, F
86 to 230	5
230 to 300	10
300 to 400	15
400 to 525	20

POWER-PLANT HEATER EXTRACTION-CYCLE ANALYSIS

A steam power plant operates at a boiler-drum pressure of 460 psia, a turbine throttle pressure of 415 psia and 725 F, and a turbine capacity of 10,000 kw (or 13,410 hp). The Rankine-cycle efficiency ratio (including generator losses) is: full load, 75.3 percent; three-quarters load, 74.75 percent; half load, 71.75 percent. The turbine exhaust pressure is 1 in. Hg absolute; steam flow to the steam-jet air ejector is 1,000 lb/hr. Analyze this cycle to determine the possible gains from two stages of extraction for feedwater heating with the first stage a closed heater and the second stage a direct-contact or mixing heater. Use engineering-office methods in analyzing the cycle.

Calculation Procedure:

1. Sketch the power-plant cycle

Figure 44a shows the plant with one closed heater and one direct-contact heater. Values marked on Fig. 44a will be computed as part of this Calculation Procedure. Enter each value on the diagram as soon as it is computed.

2. Compute the throttle flow without feedwater heating extraction

Use the superheated-steam tables to find the throttle enthalpy $h_f = 1{,}375.5$ Btu/lb at 415 psia and 725 F.

Assume an irreversible adiabatic expansion between throttle conditions and the exhaust pressure of 1 in. Hg. Compute the final enthalpy H_{2s} by the same method used in earlier Calculation Procedures by finding y_{2s}, the percent moisture at the exhaust

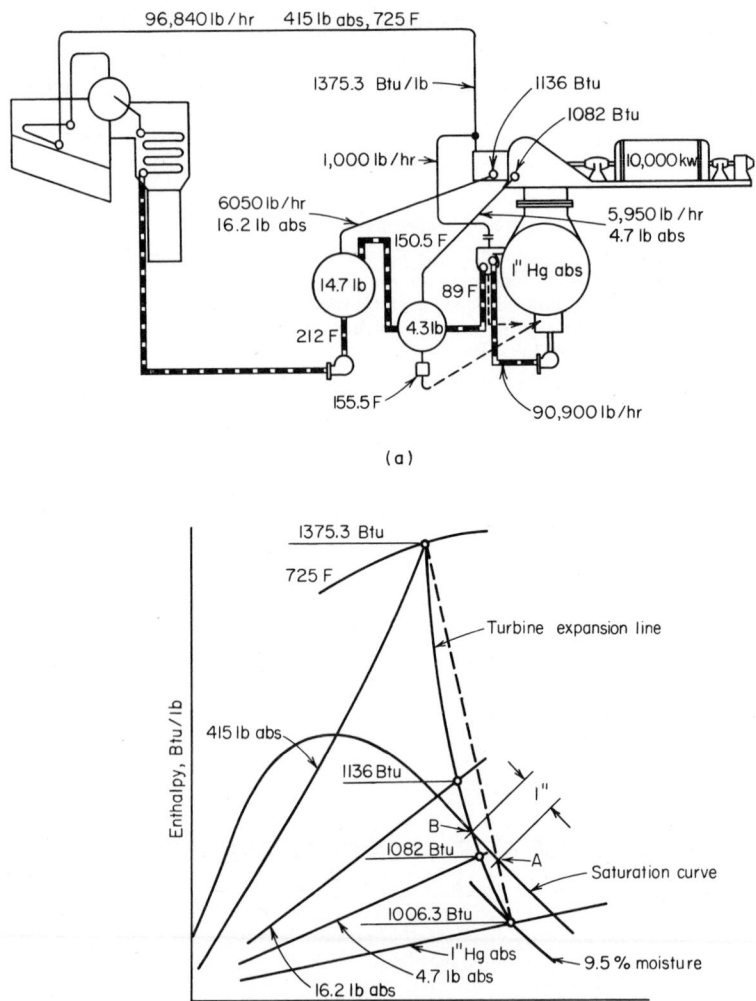

Fig. 44 *(a)* **Two stages of feedwater heating in a steam plant;** *(b)* **Mollier chart of the cycle in** *(a)***.**

conditions with 1 in. Hg absolute exhaust pressure. Do this by setting up the ratio $y_{2s} = (s_g - S_1)/s_{fg}$, where s_g and s_{fg} are entropies at the exhaust pressure; S_1 is entropy at throttle conditions. Using the steam tables, $y_{2s} = 2.0387 - 1.6468/1.9473 = 0.201$. Then, $H_{2s} = h_g - y_{2s}h_{fg}$, where h_g and h_{fg} are enthalpies at 1 in. Hg absolute. Substitute values from the steam table for 1 in. Hg absolute; or $H_{2s} = 1096.3 - 0.201(1,049.2) = 885.3$ Btu/lb.

The available energy in this irreversible adiabatic expansion is the difference between the throttle and exhaust conditions, or $1,375.5 - 885.3 = 490.2$ Btu/lb. The work at full load on the turbine is: (Rankine-cycle efficiency)(adiabatic available energy) $= (0.753)(490.2) = 369.1$ Btu/lb. Enthalpy at the exhaust of the actual turbine = throttle enthalpy minus full-load actual work, or $1,375.5 - 369.1 = 1,006.4$ Btu/lb. Use the Mollier chart to find, at 1.0 in. Hg absolute and 1,006.4 Btu/lb, that the exhaust steam contains 9.5 percent moisture.

Now the turbine steam rate $SR = 3,413$ (actual work output, Btu). Or $SR = 3,413/369.1 = 9.25$ lb/kw-hr. With the steam rate known, the nonextraction throttle flow is $(SR)(kw\ output) = 9.25(10,000) = 92,500$ lb/hr.

3. Determine the heater extraction pressures

With steam extraction from the turbine for feedwater heating, the steam flow to the main condenser will be reduced, even with added throttle flow to compensate for extraction.

Assume that the final feedwater temperature will be 212 F and that the heating range for each heater is equal. Both these assumptions represent typical practice for a moderate-pressure cycle of the type being considered.

Feedwater leaving the condenser hot well at 1 in. Hg absolute is at 79.03 F. This feedwater is pumped through the air-ejector intercondensers and aftercondensers where the condensate temperature will usually rise 5 to 15 F, depending on the turbine load. Assume that there is a 10-F rise in condensate temperature from 79 to 89 F. Then the temperature range for the *two* heaters is $212 - 89 = 123$ F. The temperature rise per heater is $123/2 = 61.5$ F, since there are two heaters and each will have the same temperature rise. Since water enters the first-stage closed heater at 89 F, the exit temperature from this heater is $89 + 61.5 = 150.5$ F.

The second-stage heater is a direct-contact unit operating at 14.7 psia, because this is the saturation pressure at an outlet temperature of 212 F. Assume a 10 percent pressure drop between the turbine and heater steam inlet. This is a typical pressure loss for an extraction heater. Extraction pressure for the second-stage heater is then $1.1(14.7) = 16.2$ psia.

Assume a 5-F terminal difference for the first-stage heater. This is a typical terminal difference, as explained in an earlier Calculation Procedure. The saturated steam temperature in the heater equals the condensate temperature $= 150.5$-F exit temperature $+ 5$-F terminal difference $= 155.5$ F. From the saturation temperature steam table, the pressure at 155.5 F is 4.3 psia. With a 10 percent pressure loss, the extraction pressure $= 1.1(4.3) = 4.73$ psia.

4. Determine the extraction enthalpies

To establish the enthalpy of the extracted steam at each stage, the actual turbine-expansion line must be plotted. Two points, the throttle inlet conditions and the exhaust conditions, are known. Plot these on a Mollier chart, Fig. 44b. Connect these two points by a dashed straight line, Fig. 44b.

Next, measure along the saturation curve 1 in. from the intersection point A back toward the enthalpy coordinate and locate point B. Now draw a gradually sloping line from the throttle conditions to point B; from B increase the slope to the exhaust conditions. The enthalpy of the steam at each extraction point is read where the lines of constant pressure cross the expansion line. Thus, for the second-stage direct-contact heater where $p = 16.2$ psia, $h_g = 1,136$ Btu/lb. For the first-stage closed heater where $p = 4.7$ psia, $h_g = 1,082$ Btu/lb.

When plotting the actual expansion curve, a steeper slope is used between the throttle superheat conditions and the saturation curve of the Mollier chart, because the turbine stages using superheated steam (stages above the saturation curve) are more efficient than stages using wet steam (stages below the saturation curve).

5. Compute the extraction steam flow

To determine the extraction flow rates, two assumptions must be made — condenser steam flow rate and first-stage closed-heater extraction flow rate. The complete cycle will be analyzed and the assumptions checked. If the assumptions are incorrect, new values will be assumed and the cycle analyzed again.

Assume that the condenser steam flow from the turbine is 84,000 lb/hr when operating with extraction. Note that this value is less than the nonextraction flow of 92,500 lb/hr. The reason for this is that extraction of steam will reduce flow to the condenser because the steam is bled from the turbine after passage through the throttle but before the condenser inlet. Then, for the first-stage closed heater, condensate flow, lb/hr:

From condenser	84,000 (assumed)
From steam-jet ejector.....	1,000
From first-stage heater	5,900 (assumed)
Total..................	90,900 lb/hr

The value of 5,900 lb/hr of condensate from the first-stage heater is the second assumption made. Since it will be checked later, an error in the assumption can be detected.

Assume a 2 percent heat radiation loss between the turbine and heater. This is a typical loss. Then:

Steam enthalpy at heater = 1,082(0.98)	= 1,060.4 Btu/lb
Enthalpy of condensate at 155.5 F	= − 123.4
Heat given up per lb of steam condensed	= 937.0 Btu/lb
Enthalpy of feedwater at 150.5 F	= 118.3 Btu/lb
Enthalpy of feedwater to heater at 89 F	= − 57.0
Heat absorbed by feedwater	= 61.3 Btu/lb
Required extraction = (total condensate flow, lb/hr)[(heat absorbed by feedwater, Btu/lb)/ (heat given up per lb of steam condensed, Btu/ lb)], or required extraction = (90,900)(61.3/937)	= 5,950 lb/hr.

Compare the required extraction, 5,950 lb/hr with the assumed extraction, 5,950 lb/ hr, with the assumed extraction, 5,900 lb/hr. The difference is only 50 lb/hr, which is less than 1 percent. Therefore, the assumed flow rate is satisfactory, because estimates within 1 percent are considered sufficiently accurate for all routine analyses.

For the second-stage direct-contact heater, condensate flow, lb/hr:

From first-stage heater	90,900
Steam enthalpy at heater = 1,135(0.98)	= 1,112.3 Btu/lb
Enthalpy of condensate at 212 F	= − 180.0
Heat given up per lb of steam condensed	= 932.3 Btu/lb
Enthalpy of feedwater at 212 F	= 180.0 Btu/lb
Enthalpy of feedwater at 150.5 F	= 118.3
Heat absorbed by feedwater	= 61.7 Btu/lb

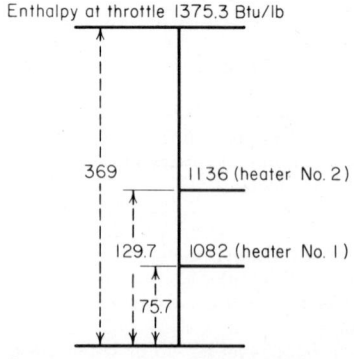

Enthalpy at throttle 1375.3 Btu/lb

369

1136 (heater No. 2)

129.7

1082 (heater No. 1)

75.7

Enthalpy at exhaust 1006.3 Btu/lb

Fig. 45　Diagram of turbine-expansion line.

The required extraction, calculated in the same way as for the first-stage heater, = (90,900) = (90,900)(61.7/932.3) = 6,050 lb/hr.

The computed extraction flow for the second-stage heater is not compared with an assumed value because an assumption was unnecessary.

6.　Compute the actual condenser steam flow

Sketch a vertical line diagram, Fig. 45, showing the enthalpies at the throttle, heaters, and exhaust. From this diagram, the work lost by the extracted steam can be computed. As Fig. 45 shows, the total enthalpy drop from the throttle to the exhaust is 369 Btu. Each lb of extracted steam from the first- and second-stage bleed points causes a work loss of 75.7 Btu/lb and 129.7 Btu/lb, respectively. To carry the same load, 10,000 kw, with extraction, it will be necessary to supply the following additional compensation steam to the turbine throttle: (heater flow, lb/hr)(work loss, Btu/hr)/(total work, Btu/hr). Then,

First-stage closed heater:
  $(5,950)(75.7/369)$ .... = 1,220 lb/hr
Second-stage direct-contact heater:
  $(6,050)(129.7/369)$ .... = 2,120 lb/hr
Total additional throttle flow to compensate for
  extraction .... 3,340 lb/hr

Check the assumed condenser flow using nonextraction throttle flow + additional throttle flow − heater extraction = condenser flow. All quantities are in lb/hr. Set up a tabulation of the flows as follows:

Flow	lb/hr
Throttle; nonextraction	92,500
Added flow (compensation)	3,340
Throttle; extraction	95,840
Extraction $(5,950 + 6,050)$	− 12,000
Condenser flow	83,840

Compare this actual flow, 83,840 lb/hr, with the assumed flow, 84,000 lb/hr. The difference, 160 lb/hr, is less than 1 percent. Since an accuracy within 1 percent is sufficient for all normal power-plant calculations, it is not necessary to recompute the cycle. Had the difference been greater than 1 percent, a new condenser flow would be assumed and the cycle recomputed. Follow this procedure until a difference of less than 1 percent is obtained.

7. Determine the economy of the extraction cycle

For a nonextraction cycle operating in the same pressure range:

Enthalpy of throttle steam	1,375.3 Btu/lb
Enthalpy of condensate at 79 F ...	47.0 Btu/lb
Heat supplied by boiler.	1,328.3 Btu/lb

Heat chargeable to turbine = (throttle flow + air-ejector flow)(heat supplied by boiler)/(kw output of turbine) = $(92,500 + 1,000)(1,328.3)/10,000 = 12,410$ Btu/kw-hr, which is the actual heat rate HR of the nonextraction cycle.

For the extraction cycle using two heaters,

Enthalpy of throttle steam	1,375.3 Btu/lb
Enthalpy of feedwater leaving	
second heater.	180.0 Btu/lb
Heat supplied by boiler.	1,195.3 Btu/lb

As before, heat chargeable to turbine = $(95,840 + 1,000)(1,195.3)/10,000 = 11,580$ Btu/kw-hr. Therefore, the improvement = (nonextraction HR − extraction HR)/nonextraction HR = $(12,410 − 11,580)/12,410 = 0.0662$, or 6.62 percent.

Related Calculations: (1) To determine the percent improvement in a steam cycle resulting from additional feedwater heaters in the cycle, use the same Procedure as given above for three, four, five, six, or more heaters. Plot the percent improvement vs. number of stages of extraction, Fig. 46, to observe the effect of additional heaters. A plot of this type shows the decreasing gains made by additional heaters. Eventually the gains become so small that the added expenditure for an additional heater cannot be justified.

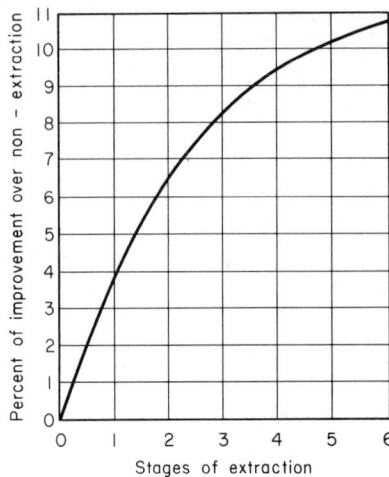

Fig. 46 Percent improvement in turbine heat rate vs. stages of extraction.

(2) Many simple marine steam plants use only two stages of feedwater heating. To analyze such a cycle, use the Procedure given, substituting the hp output for the kw output of the turbine.

(3) Where a marine plant has more than two stages of feedwater heating, follow the Procedure given in 1 above.

STEAM BOILER, ECONOMIZER, AND AIR-HEATER EFFICIENCY

Determine the overall efficiency of a steam boiler generating 56,000 lb/hr of 600-psia 800-F steam. The boiler is continuously blown down at the rate of 2,500 lb/hr. Feedwater enters the economizer at 300 F. The furnace burns 5,958 lb/hr of 13,100 Btu/lb HHV (higher heating value) coal having an ultimate analysis of 68.5% C, 5% H_2, 8.9% O_2, 1.2% N_2, 3.2% S, 8.7% ash, and 4.5% moisture. Air enters the boiler at 63-F dry-bulb and 56-F wet-bulb temperature, with 56 grains of vapor per lb of dry air.

Carbon in the fuel refuse is 7%; refuse is 0.093 lb per lb of fuel. Feedwater leaves the economizer at 370 F. Flue gas enters the economizer at 850 F, has an analysis of 15.8% CO_2, 2.8% O_2, and 81.4% N_2. Air enters the air heater at 63 F with 56 grains of vapor per lb of dry air; air leaves the heater at 480 F. Gas enters the air heater at 570 F, and 14 percent of the air to the furnace comes from the mill fan. Determine the steam generator overall efficiency, economizer efficiency, and air-heater efficiency. Figure 47 shows the steam generator and the flow factors that must be considered.

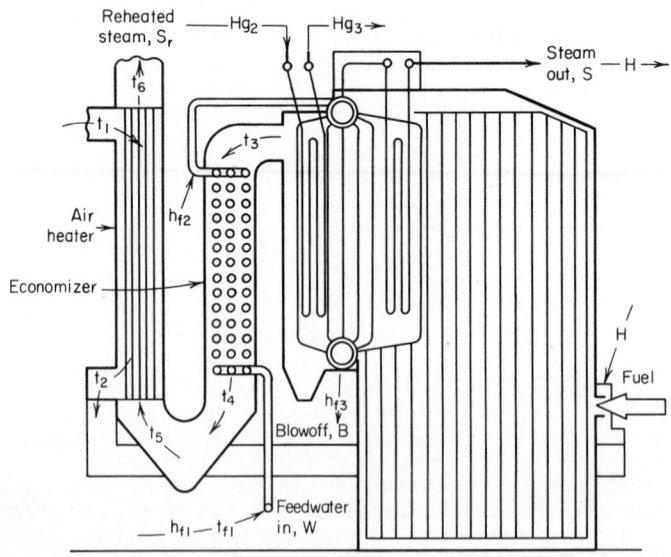

Fig. 47 Points in a steam generator where temperatures and enthalpies are measured in determining the boiler efficiency.

Calculation Procedure:

1. Determine the boiler output, Btu/hr

The boiler output $= S(h_g - h_{f1}) + S_r(h_{g3} - h_{g2}) + B(h_{f3} - h_{f1})$ where $S =$ steam generated, lb/hr; $h_g =$ enthalpy of the generated steam, Btu/lb; $h_{f1} =$ enthalpy of inlet feedwater; $S_r =$ reheated steam flow, lb/hr (if any); $h_{g3} =$ outlet enthalpy of reheated steam; $h_{g2} =$ inlet enthalpy of reheated steam; $B =$ blowoff, lb/hr; $h_{f3} =$ blowoff enthalpy, where all enthalpies are in Btu/lb. Using the appropriate steam table and deleting the reheat factor because there is no reheat, boiler output $= 56,000(1,407.7 - 269.6) + 2,500(471.6 - 269.6) = 64,238,600$ Btu/hr

2. Compute the heat input to the boiler, Btu/hr

The boiler input $= FH$, where $F =$ fuel input lb/hr (as fired); $H =$ higher heating value, Btu/lb (as fired). Or, boiler input $= 5,958(13,100) = 78,049,800$ Btu/hr.

3. Compute the boiler efficiency

The boiler efficiency $=$ (output, Btu/hr)/(input, Btu/hr) $= 64,238,600/78,049,800 = 0.822$, or 82.2 percent.

4. Determine the heat absorbed by the economizer

The heat absorbed by the economizer, Btu/hr, $= w_w(h_{f2} - h_{f1})$, where $w_w =$ feedwater flow, lb/hr; h_{f2} and $h_{f1} =$ enthalpies of feedwater leaving and entering the economizer, respectively, Btu/lb. For this economizer, with the feedwater leaving the economizer at 370 F and entering at 300 F, heat absorbed $= (56,000 + 2,500)(342.79 - 269.59) = 4,283,000$ Btu/hr. Note that the total feedwater flow w_w is the sum of the steam generated and the continuous blowdown rate.

5. Compute the heat available to the economizer

The heat available to the economizer, Btu/hr, $= H_g F$, where $H_g =$ heat available in flue gas, Btu/lb of fuel $=$ heat available in dry gas $+$ heat available in flue-gas vapor, Btu per lb of fuel $= (t_3 - t_{f1})(0.24G) + (t_3 - t_{f1})(0.46)\{M_f + 8.94H_2 + M_a[G - C_b - N_2 - 7.94(H_2 - O_2/8)]\}$, where $G = \{[11CO_2 + 8O_2 + 7(N_2 + CO)]/[3(CO_2 + CO)]\}(C_b + S/2.67) + S/1.60$; $M_f =$ lb of moisture per lb fuel burned; $M_a =$ lb of moisture per lb of dry air to furnace; $C_b =$ lb of carbon burned per lb of fuel burned $= C - RC_r$; $C_r =$ lb of combustible per lb of refuse; $R =$ lb of refuse per lb of fuel; $H_2, N_2, C, O_2, S =$ lb of each element per lb of fuel, as fired; $CO_2, CO, O_2, N_2 =$ percentage parts of volumetric analysis of dry combustion gas entering the economizer. Substituting, $C_b = (0.685 - 0.093)(0.07) = 0.678$ lb per lb fuel $G = [11(0.158) + 8(0.028) + 7(0.814)]/[3(0.158)] \times (0.678 + 0.032/2.67) + 0.032/1.60$; $G = 11.18$ lb per lb fuel. $H_g = (800 - 300)(0.24) \times (11.18) + (800 - 300)(0.46)\{0.045 + (8.9)(0.05) + 56/7,000[11.18 - 0.678 - 0.012 - 7.94 \times (0.05 - 0.089/8)]\}$; $H_g = 1,473$ Btu per lb fuel. Heat available $= H_g F = (1,473)(5,958) = 8,770,000$ Btu/hr.

6. Compute the economizer efficiency

The economizer efficiency $=$ (heat absorbed, Btu/hr)/(heat available, Btu/hr) $= 4,283,000/8,770,000 = 0.488$, or 48.8 percent.

7. Compute the heat absorbed by air heater

The heat absorbed by the air heater, Btu per lb of fuel, $= A_h(t_2 - t_1)(0.24 + 0.46M_a)$, where $A_h =$ air flow through heater, lb per lb fuel $= A - A_m$; $A =$ total air to furnace, lb per lb fuel $= G - C_b - N_2 - 7.94(H_2 - O_2/8)$; $G =$ similar to economizer but based on gas at the furnace exit; $A_m =$ external air supplied by the mill fan or other source, lb per lb of fuel. Substituting, $G = [11(0.16) + 8(0.26) + 7(0.184)]/[3(0.16)](0.678 + 0.032/2.67) + 0.032/1.60$; $G = 11.03$ lb per lb fuel. $A = 11.03 - 0.69 - 0.012 - 7.94(0.05 - 0.089/8)$; $A = 10.02$ lb per lb fuel. Heat absorbed $= (1 - 0.15)(10.02)(480 - 63)(0.24 + (56/7,000)) = 865.5$ Btu per lb fuel.

8. Compute the heat available to the air heater

The heat available to the air heater, Btu/hr, $= (t_5 - t_1)0.24G + (t_5 - t_1)0.46(M_f + 8.94H_2 + M_aA)$. In this relation, all symbols are the same as for the economizer except that G and A are based on the gas entering the heater. Substituting, $G = [11(0.15) + 8(0.036) + 7(0.814)]/[3(0.15)](0.678 + 0.032/2.67) + 0.032/1.60$; $G = 11.72$ lb per lb fuel. $A = 11.72 - 0.69 - 0.012 - 7.94(0.05 - 0.089/8) = 10.71$ lb per lb fuel. Heat available $= (570 - 3)(0.24)(11.72) + (570 - 63)(0.46)[0.045 + 8.94(0.05) + 56/7,000(10.71)] = 1,561$ Btu/lb.

9. Compute the air-heater efficiency

The air-heater efficiency = (heat absorbed, Btu per lb fuel)/(heat available, Btu per lb fuel) $= 865.5/1,561 = 0.554$, or 55.4 percent.

Related Calculations: The above procedure is valid for all types of steam generators, regardless of the kind of fuel used. Where oil or gas is the fuel, alter the combustion calculations to reflect the differences between the fuels. Further, this procedure is also valid for marine and portable boilers.

FIRE-TUBE BOILER ANALYSIS AND SELECTION

Determine the heating surface in an 84-in.-diameter fire-tube boiler 18 ft long having 84 tubes 4 in. ID if 25 percent of the upper shell ends are heat insulated. How much steam is generated if the boiler evaporates 34.5 lb of water per hr per 12 sq ft of heating surface? How much heat is added by the boiler if it operates at 200 psia with 200-F feedwater? What is the factor of evaporation for this boiler? How much hp is developed by the boiler if 7,000,000 Btu/hr is delivered to the water?

Calculation Procedure:

1. Compute the shell area exposed to furnace gas

Shell area $= \pi DL(1 - 0.25)$, where D = boiler diameter, ft; L = shell length, ft; $1 - 0.25$ is the portion of the shell in contact with the furnace gas. Then, shell area $= \pi(84/12)(18)(0.75) = 297$ sq ft.

2. Compute the tube area exposed to furnace gas

Tube area $= \pi dLN$, where d = tube ID, ft; L = tube length, ft; N = number of tubes in boiler. Substituting, tube area $= \pi(4/12)(18)(84) = 1,583$ sq ft.

3. Compute the head area exposed to furnace gas

The area exposed to furnace gas is twice (since there are *two* heads) the exposed head area minus twice the area occupied by the tubes. The exposed head area is the (total area)(1 − portion covered by insulation, expressed as a decimal). Substituting, $2\pi D^2/4 - [(2)(84)\pi d^2/4] = 2\pi/4(84/12)^2(0.75) - [(2)(84)\pi(4/12)^2/4] =$ head area $= 43.1$ sq ft.

4. Find the total heating surface

The total heating surface of any fire-tube boiler is the sum of the shell, tube, and head areas, or $297.0 + 1,583 + 43.1 = 1,923$ sq ft, total heating surface.

5. Compute the quantity of steam generated

Since the boiler evaporates 34.5 lb of water per hr per 12 sq ft of heating surface, the quantity of steam generated = 34.5 (total heating surface, sq ft)/12 $= 34.5(1,923.1)/12 = 5,200$ lb/hr.

NOTE: Evaporation of 34.5 lb/hr from and at 212 F is the definition of the now-discarded term "boiler horsepower." However, this term is still met in some engineering examinations and is used by some manufacturers when comparing the performance of boilers. A term used in lieu of boiler horsepower, with the same definition, is *equivalent evaporation*. Both these terms are falling into disuse, but they are included here because they still find some use today.

6. Determine the heat added by the boiler

Heat added, Btu per lb of steam, $= h_g - h_{f1}$; using steam-table values $= 1{,}198.4 - 167.99 = 1{,}030.41$ Btu/lb. An alternative way of computing heat added $= h_g - $ (feed-water temperature, $F - 32$), where 32 is the freezing temperature of water on the Fahrenheit scale. Using this method, heat added $= 1{,}198.4 - (200 - 32) = 1{,}030.4$ Btu/lb. Thus, both methods give the same results in this case. In general, however, use of steam-table values is preferred.

7. Compute the factor of evaporation

The factor of evaporation is used to convert from the actual to the equivalent evaporation, defined earlier. Or, factor of evaporation = (heat added by boiler, Btu/lb)/970.3, where 970.3 Btu/lb is the heat added to develop 1 boiler hp. Thus, the factor of evaporation for this boiler $= 1030.4/970.3 = 1.066$.

8. Compute the boiler hp output

Boiler hp = (actual evaporation, lb/hr)(factor of evaporation)/34.5. In this relation, the actual evaporation must first be computed. Since the furnace delivers 7,000,000 Btu/hr to the boiler water and the water absorbs 1,030.4 Btu/lb to produce 200-psia steam with 200-F feedwater, the steam generated, lb/hr, = (total heat delivered, Btu/hr)/(heat absorbed, Btu/lb) $= 7{,}000{,}000/1{,}030.4 = 6{,}760$ lb/hr. Then, boiler hp = $(6{,}760)(1.066)/34.5 = 209$ hp.

The rated hp output of horizontal fire-tube boilers with separate supporting walls is based on 12 sq ft of heating surface per boiler hp. Thus, the rated hp of this boiler = $1{,}923.1/12 = 160$ hp. When producing 209 hp, the boiler is operating at 209/160, or 1.305 times its normal rating, or $(100)(1.305) = 130.5$ percent of normal rating.

NOTE: Today most boiler manufacturers rate their boilers in terms of lb/hr of steam generated at a stated pressure. Use this measure of boiler output whenever possible. Inclusion of the term boiler hp in this Handbook does not indicate that the editor favors or recommends use of the term. Instead, the term was included to make the Handbook as helpful as possible to those users who might encounter the term in their work.

SAFETY-VALVE STEAM-FLOW CAPACITY

How much saturated steam at 150 psia can a 2.5-in.-diameter safety valve having a 0.25-in. lift pass if the discharge coefficient of the valve c_d is 0.75? What is the capacity of the same valve if the steam is superheated 100 F above its saturation temperature?

Calculation Procedure:

1. Determine the area of the valve annulus

Annulus area, sq in., $= A = \pi DL$, where $D =$ valve diameter, in.; $L =$ valve lift, in. Annulus area $= \pi(2.5)(0.25) = 1.966$ sq in.

2. Compute the ideal flow for this safety valve

Ideal flow F_i lb/sec for any safety valve handling saturated steam is $F = p_s^{0.97} A/60$, where $p_s =$ saturated-steam pressure, psia. For this valve, $F = (150)^{0.97}(1.966)/60 = 4.24$ lb/sec.

3. Compute the actual flow through the valve

Actual flow $F_a = F_i c_d = (4.24)(0.75) = 3.18$ lb/sec $= (3.18)(3{,}600$ sec/hr$) = 11{,}448$ lb/hr.

4. Determine the superheated-steam flow rate

The ideal superheated-steam flow F_{is} lb/sec is $F_{is} = p_s^{0.97} A/60(1 + 0.00065 t_s)$, where $t_s =$ superheat temperature, above saturation temperature, F. Then, $F_{is} = (150)^{0.97} \times (1.966)/[60(1 + 0.00065 \times 100)] = 3.96$ lb/sec. The actual flow is $F_{as} = F_{is} c_d = (3.96) \times (0.75) = 2.97$ lb/sec $= (2.97)(3{,}600) = 10{,}700$ lb/hr.

Related Calculations: Use this procedure for safety valves serving any type of stationary or marine steam boiler.

SAFETY-VALVE SELECTION FOR A WATERTUBE STEAM BOILER

Select a safety valve for a watertube steam boiler having a maximum rating of 100,000 lb/hr at 800 psia and 900 F. Determine the valve diameter, size of boiler connection for the valve, opening pressure, closing pressure, type of connection, and valve material. The boiler is oil-fired and has a total heating surface of 9,200 sq ft of which 1,000 sq ft is in waterwall surface. Use the ASME *Boiler and Pressure Vessel Code* rules when selecting the valve. Sketch the escape-pipe arrangement for the safety valve.

Calculation Procedure:

1. Determine the minimum valve relieving capacity

Refer to the latest edition of the *Code* for the relieving-capacity rules. Recent editions of the *Code* require that the safety valve have a *minimum* relieving capacity based on the lb of steam generated per hr per sq ft of boiler heating surface and waterwall heating surface. In the edition of the *Code* used in preparing this Handbook, the relieving requirement for oil-fired boilers was 10 lb of steam per hr per sq ft of boiler heating surface, and 16 lb of steam per hr per sq ft of waterwall surface. Thus, the minimum safety-valve relieving capacity for this boiler, based on total heating surface, would be $(8,200)(10) + (1,000)(16) = 92,000$ lb/hr. In this equation, 1,000 sq ft of waterwall surface is deducted from the total heating surface of 9,200 sq ft to obtain the boiler heating surface of 8,200 sq ft.

The minimum relieving capacity based on total heating surface is 92,000 lb/hr; the maximum rated capacity of the boiler is 100,000 lb/hr. Since the *Code* also requires that "the safety valve or valves will discharge all the steam that can be generated by the boiler," the minimum relieving capacity must be 100,000 lb/hr, because this is the maximum capacity of the boiler and it exceeds the valve capacity based on the heating-surface calculation. If the valve capacity based on the heating-surface steam generation were larger than the stated maximum capacity of the boiler, the *Code* heating-surface valve capacity would be used in safety-valve selection.

2. Determine the number of safety valves needed

Study the latest edition of the *Code* to determine the requirements for the number of safety valves. The edition of the *Code* used here requires that "each boiler shall have at least one safety valve and if it [the boiler] has more than 500 sq ft of water heating surface, it shall have two or more safety valves." Thus, at least two safety valves are needed for this boiler. The *Code* further specifies, in the edition used, that "when two or more safety valves are used on a boiler, they may be mounted either separately or as twin valves made by placing individual valves on Y bases, or duplex valves having two valves in the same body casing. Twin valves made by placing individual valves on Y bases, or duplex valves having two valves in the same body, shall be of equal sizes." Also, "when not more than two valves of different sizes are mounted singly, the relieving capacity of the smaller valve shall not be less than 50 percent of that of the larger valve."

Assume that two equal-size valves mounted on a Y base will be used on the steam drum of this boiler. Two or more equal-size valves are usually chosen for the steam drum of a watertube boiler.

Since this boiler handles superheated steam, check the *Code* requirements regarding superheaters. The *Code* states that "every attached superheater shall have one or more safety valves near the outlet." Also, "the discharge capacity of the safety valve, or valves, on an attached superheater may be included in determining the number and size of the safety valves for the boiler, provided there are no intervening valves between the superheater safety valve and the boiler, and provided the discharge capacity of the safety valve, or valves, on the boiler, as distinct from the superheater, is at least 75 percent of the aggregate valve capacity required."

As the safety valves used must handle 100,000 lb/hr, and one or more superheater safety valves are required by the *Code*, assume that the two steam-drum valves will handle, in accordance with the above requirement, 80,000 lb/hr. Assume that one superheater safety valve will be used. Its capacity must then be at least $100,000 - 80,000 = 20,000$ lb/hr. (Use as few superheater safety valves as possible, because this simplifies the installation and reduces cost.) With this arrangement, each steam-drum valve must handle 80,000/2 = 40,000 lb/hr of steam, since there are two safety valves on the steam drum.

3. Determine the valve pressure settings

Consult the *Code*. It requires that "one or more safety valves on the boiler proper shall be set at or below the maximum allowable working pressure." For modern boilers, the maximum allowable working pressure is usually 1.5, or more, times the rated operating pressure in the lower (under 1,000 psia) pressure ranges. To prevent unnecessary operation of the safety valve and to reduce steam losses, the lowest safety-valve setting is usually about 5 percent higher than the boiler operating pressure. For this boiler, the lowest pressure setting would be 800 + 800(0.05) = 840 psia. Round this off to 850 psia (6.25 percent) for ease of selection from the usual safety-valve rating tables. The usual safety-valve pressure setting is between 5 and 10 percent higher than the rated operating pressure of the boiler.

Boilers fitted with superheaters usually have the superheater safety valve set at a lower pressure than the steam-drum safety valve. This arrangement ensures that the superheater safety valve opens first when overpressure occurs. This provides steam flow through the superheater tubes at all times, preventing tube burnout. Therefore, the superheater safety valve in this boiler will be set to open at 850 psia, the lowest opening pressure for the safety valves chosen. The steam-drum safety valves will be set to open at a higher pressure. As decided earlier, the superheater safety valve will have a capacity of 20,000 lb/hr.

Between the steam drum and the superheater safety valve, there is a pressure loss that varies from one boiler to another. The boiler manufacturer supplies a performance chart showing the drum outlet pressure for various percentages of the maximum continuous steaming capacity of the boiler. This chart also shows the superheater outlet pressure for the same capacities. The difference between the drum and superheater outlet pressures for any given load is the superheater pressure loss. Obtain this pressure loss from the performance chart.

Assume, for this boiler, that the superheater pressure loss, plus any pressure losses in the nonreturn valve and dry pipe, at maximum rating, is 60 psia. The steam-drum operating pressure will then be superheater outlet pressure + superheater pressure loss = 800 + 60 = 860 psia. As with the superheater safety valve, the steam-drum safety valve is usually set to open at about 5 percent above the drum operating pressure at maximum steam output. For this boiler then, the drum safety-valve set pressure = 860 + 860(0.05) = 903 psia. Round this off to 900 psia to simplify valve selection.

Some designers add the drum safety-valve blowdown or blowback pressure (difference between the valve opening and closing pressures, psi) to the total obtained above to find the drum operating pressure. However, the 5 percent allowance used above is sufficient to allow for the blowdown in boilers operating at less than 1,000 psia. At pressures of 1,000 psia, and higher, add the drum safety-valve blowdown *and* the 5 percent allowance to the superheater outlet pressure and pressure loss to find the drum pressure.

4. Determine the required valve orifice discharge area

Refer to a safety-valve manufacturer's engineering data listing valve capacities at various working pressures. For the two steam-drum valves, enter the table at 900 psia and project horizontally until a capacity of 40,000 lb/hr, or more, is intersected. Here is an excerpt from a typical manufacturer's capacity table for safety valves handling *saturated steam*:

Set pressure, psia	Orifice area, sq in.		
	0.994	1.431	2.545
890	41,750	60,000	107,200
900	42,200	60,800	108,000
910	42,700	61,600	109,300

Thus, at 900 psia a valve with an orifice area of 0.994 sq in. will have a capacity of 42,200 lb/hr of saturated steam. This is 5.5 percent greater than the required capacity of 40,000 lb/hr for each steam-drum valve. However, the usual selection cannot be made at exactly the desired capacity. Provided that the valve chosen has a *greater* steam-relieving capacity than required, there is no danger of overpressure in the steam drum. Be careful to note that safety valves for *saturated steam* are chosen for the steam drum because superheating of the steam does not occur in the steam drum.

The superheater safety valve must handle 20,000 lb/hr of 850-psia steam at 900 F. Safety valves handling superheated steam have a smaller capacity than when handling saturated steam. To obtain the capacity of a safety valve handling superheated steam, the saturated-steam capacity is multiplied by a correction factor that is less than 1.00. An alternative procedure is to divide the required superheated-steam capacity by the same correction factor to obtain the saturated-steam capacity of the valve. The latter procedure will be used here because it is more direct.

Obtain the correction factor from the safety-valve manufacturer's engineering data by entering at the steam pressure and projecting to the steam temperature, as shown below.

Set pressure, psia	Steam temperature, F				
	860	880	900	920	940
800	0.81	0.80	0.80	0.79	0.78
850	0.82	0.81	0.80	0.79	0.78
900	0.82	0.81	0.80	0.79	0.78

Thus, at 850 psia and 900 F, the correction factor = 0.80. The required saturated steam capacity then = 20,000/0.80 = 25,000 lb/hr.

Refer to the manufacturer's saturated-steam capacity table as before, and at 850 psia, find the closest capacity as 31,500 lb/hr for a 0.785 sq in. orifice. As with the steam-drum valves, the actual capacity of the safety valve is somewhat greater than the required capacity. In general, it is difficult to find a valve with exactly the required steam-relieving capacity.

5. Determine the valve nominal size and construction details

Turn to the data section of the safety-valve engineering manual to find the valve construction features. For the steam-drum valves having 0.994 sq in. orifice areas, the engineering data shows, for 900-psia service, each valve is a 1½-in. unit rated for temperatures up to 1050 F. The inlet is a 900-psi 1½-in. flanged connection and the outlet is a 150-psi 3-in. flanged connection. Materials used in the valve include: body—cast carbon steel; disk seat—stainless steel AISI321. The overall height is 27⅞ in.; dismantling height is 32¾ in.

Similar data for the superheater steam valve show, for a maximum pressure of 900 psia, that it is a 1½-in. unit rated for temperatures up to 1000 F. The inlet is a 900-psi 1½-in. flanged connection and the outlet is a 150-psi 3-in. flanged connection. Materials used in the valve include: body—cast alloy steel, ASTM 217-WC6; spindle—stainless steel; spring—alloy steel; disk seat—stainless steel. Overall height is 21⅜ in.; dismantling height is 25¼ in. Checking the *Code* shows that "every safety valve used on a superheater discharging superheated steam at a temperature over 450 F shall have a casing, including the base, body, bonnet and spindle, of steel, steel alloy, or equivalent heat-resisting material. The valve shall have a flanged inlet connection."

Thus, the superheater valve selected is satisfactory.

6. Compute the steam-drum connection size

The *Code* requires that "when a boiler is fitted with two or more safety valves on one connection, this connection to the boiler shall have a cross-sectional area not less than the combined areas of inlet connections of all the safety valves with which it connects."

The inlet area for each valve $= \pi D^2/4 = \pi(1.5)^2/4 = 1.77$ sq in. For two valves, the total inlet area $= 2(1.77) = 3.54$ sq in. The required minimum diameter of the boiler connection is $d = 2(A/\pi)^{0.5}$, where $A =$ inlet area. Or, $d = 2(3.54/\pi)^{0.5} = 2.12$ in. Select a $2\frac{1}{2} \times 1\frac{1}{2} \times 1\frac{1}{2}$ in. Y for the two steam-drum valves and a $2\frac{1}{2}$-in. steam-drum outlet connection.

7. Compute the safety-valve closing pressure

The *Code* requires safety valves "to close after blowing down not more than 4 percent of the set pressure." For the steam-drum valves the closing pressure will be $900 - (900)(0.04) = 865$ psia. The superheater safety valve will close at $850 - (850)(0.04) = 816$ psia.

8. Sketch the discharge elbow and drip pan

Figure 48 shows a typical discharge elbow and drip-pan connection. Fit all boiler safety valves with escape pipes to carry the steam out of the building and away from personnel. Extend the escape pipe to at least 6 ft above the roof of the building. Use an escape pipe having a diameter equal to the valve outlet size. When the escape pipe is more than 12 ft long, some authorities recommend increasing the escape-pipe diameter by $\frac{1}{2}$ in. for each additional 12-ft length. Excessive escape-pipe length without an increase in diameter can cause a backpressure on the safety valve because of flow friction. The safety valve may then chatter excessively.

Support the escape pipe independently of the safety valve. Fit a drain to the valve body and drip pan as shown in Fig. 48. This prevents freezing of the condensate and

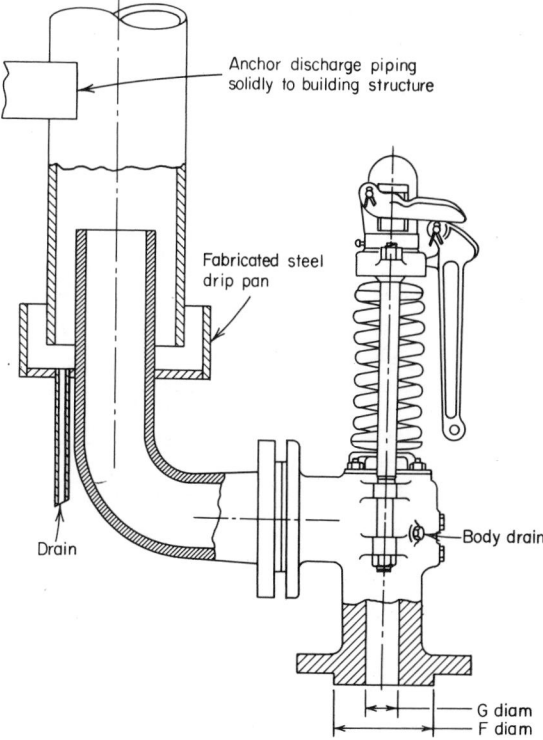

Fig. 48 **Typical boiler safety valve discharge elbow and drip-pan connection.** *(Industrial Valve & Instrument Div. of Dresser Industries Inc.)*

also eliminates the possibility of condensate in the escape pipe raising the valve opening pressure. When a muffler is fitted to the escape pipe, the inlet diameter of the muffler should be the same as, or larger than, the escape-pipe diameter. The outlet area should be greater than the inlet area of the muffler.

Related Calculations: Compute the safety-valve size for fire-tube boilers in the same way as described above, except that the *Code* gives a tabulation of the required area for safety-valve boiler connections based on boiler operating pressure and heating surface. Thus, with an operating pressure of 200 psig and 1,800 sq ft of heating surface, the *Code* table shows that the safety-valve connection should have an area of at least 9.148 sq in. A 3½-in. connection would provide this area; or two smaller connections could be used provided that the sum of their areas exceeded 9.148 sq in.

NOTE: Be sure to select safety valves approved for use under the *Code* or local law governing boilers in the area in which the boiler will be used. Choice of an unapproved valve can lead to its rejection by the bureau or other agency controlling boiler installation and operation.

STEAM-QUALITY DETERMINATION WITH A THROTTLING CALORIMETER

Steam leaves an industrial boiler at 120 psia and 341.25 F. A portion of the steam is passed through a throttling calorimeter and is exhausted to the atmosphere when the barometric pressure is 14.7 psia. How much moisture does the steam leaving the boiler contain if the temperature of the steam at the calorimeter is 240 F?

Calculation Procedure:

1. *Plot the throttling process on the Mollier diagram*

Begin with the end point, 14.7 psia and 240 F. Plot this point on the Mollier diagram as point *A*, Fig. 49. Note that this point is in the superheat region of the Mollier diagram,

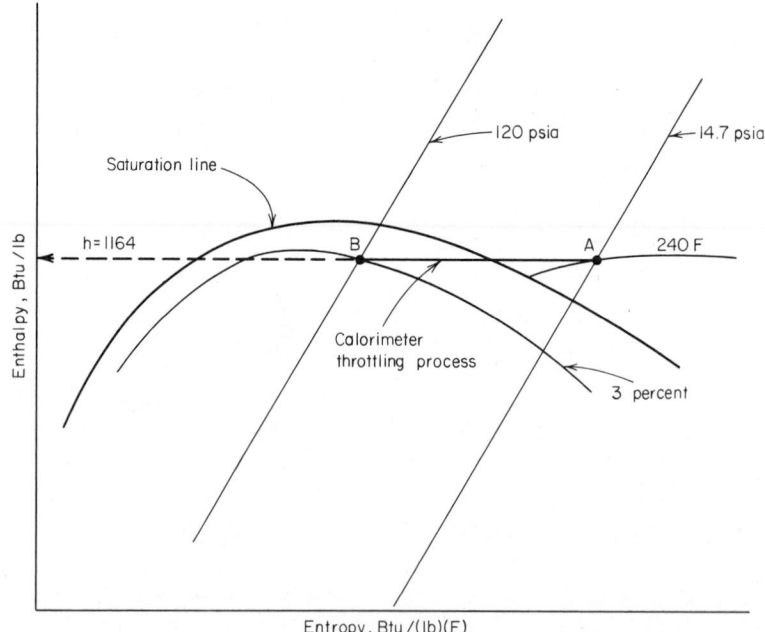

Fig. 49   Mollier-diagram plot of a throttling-calorimeter process.

because steam at 14.7 psia has a temperature of 212 F, whereas the steam in this calorimeter has a temperature of 240 F. The enthalpy of the calorimeter steam is, from the Mollier diagram, 1,164 Btu/lb.

2. Trace the throttling process on the Mollier diagram

In a throttling process, the steam expands at constant enthalpy. Draw a straight, horizontal line from point A to the left on the Mollier diagram until the 120-psia pressure curve is intersected, point B, Fig. 49. Read the moisture content of the steam as 3 percent where the 1,164 Btu/lb horizontal trace A-B, the 120-psia pressure line, and the 3 percent moisture line intersect.

Related Calculations: A throttling calorimeter *must* produce superheated steam at the existing atmospheric pressure if the moisture content of the supply steam is to be found. Where the throttling calorimeter cannot produce superheated steam at atmospheric pressure, connect the calorimeter outlet to an area at a pressure less than atmospheric. Expand the steam from the source and read the temperature at the calorimeter. If the steam temperature is greater than that corresponding to the absolute pressure of the vacuum area—for example, a temperature greater than 133.76 F in an area of 5 in. Hg absolute pressure—follow the same procedure as given above. Point A would then be in the below-atmospheric area of the Mollier diagram. Trace to the left to the origin pressure and read the moisture content as before.

STEAM PRESSURE DROP IN A BOILER SUPERHEATER

What is the pressure loss in a boiler superheater handling $w_s = 200,000$ lb/hr of saturated steam at 500 psia if the desired outlet temperature is 750 F? The steam free-flow area through the superheater tubes A_s sq ft is 0.500, friction factor f is 0.025, tube ID is 2.125 in., developed length l of a tube in one circuit is 150 in., and the tube bend factor B_f is 12.0.

Calculation Procedure:

1. Determine the initial conditions of the steam

To compute the pressure loss in a superheater, the initial specific volume of the steam v_g and the mass-flow ratio w_s/A_s must be known. From the steam table, $v_g = 0.9278$ cu ft/lb at 500 psia saturated. The mass-flow ratio $w_s/A_s = 200,000/0.500 = 400,000$.

2. Compute the superheater entrance and exit pressure loss

Entrance and exit pressure loss p_E psi $= v_f/8(0.00001w_s/A_s) = 0.9278/8[(0.00001) \times (400,000)]^2 = 1.856$ psi.

3. Compute the pressure loss in the straight tubes

Straight-tube pressure loss p_S psi $= v_f lf/\text{ID}(0.00001w_s/A_s)^2 = 0.9278(150)(0.025)/2.125[(0.00001)(400,000)]^2 = 26.2$ psia.

4. Compute the pressure loss in the superheater bends

Bend pressure loss $p_b = 0.0833B_f(0.00001w_s/A_s)^2 = 0.0833(12.0)[(0.00001)(400,000)]^2 = 16.0$ psi.

5. Compute the total pressure loss

The total pressure loss in any superheater is the sum of the entrance, straight-tube, bend, and exit pressure losses. These losses were computed in Steps 2, 3, and 4 above. Therefore, total pressure loss $p_t = 1.856 + 26.2 + 16.0 = 44.056$ psi.

NOTE: Data for superheater pressure-loss calculations are best obtained from the boiler manufacturer. Several manufacturers have useful publications discussing super-heater pressure losses. These are listed in the references at the beginning of this section.

SELECTION OF A STEAM BOILER FOR A GIVEN LOAD

Choose a steam boiler, or boilers, to deliver up to 250,000 lb/hr of superheated steam at 800 psia and 900 F. Determine the type or types of boilers to use, the capacity, type of firing, feedwater-quality requirements, and best fuel if coal, oil, and gas are all available. The normal continuous steam requirement is 200,000 lb/hr.

Calculation Procedure:

1. Select type of steam generator

Use Fig. 50 as a guide to the usual types of steam generators chosen for various capacities and different pressure and temperature conditions. Enter Fig. 50 at the left at 800 psia and project horizontally to the right, along AB, until the 250,000 lb/hr capacity ordinate BC is intersected. At B, the operating point of this boiler, Fig. 50 shows that a watertube boiler should be used.

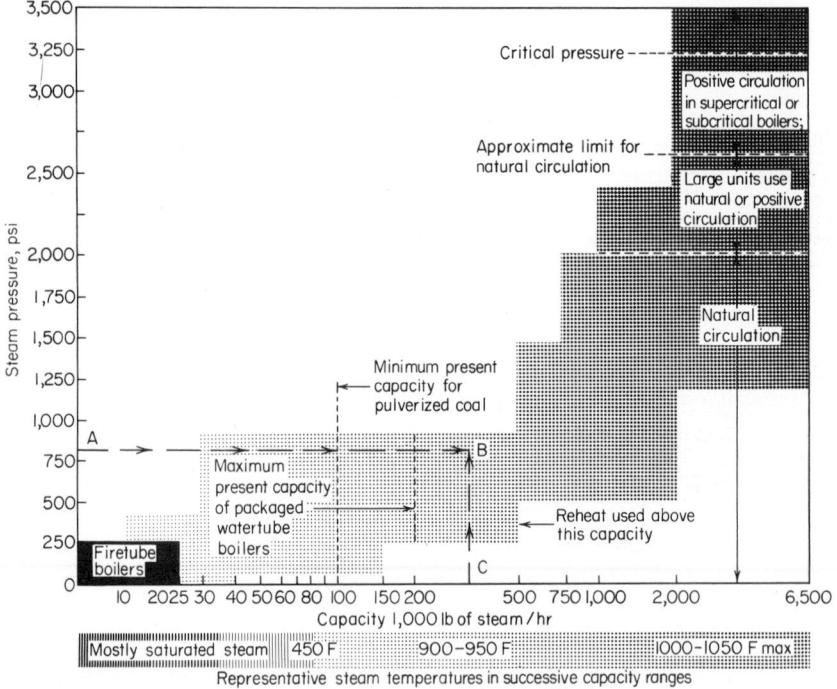

Fig. 50   Typical pressure and capacity relationships for steam generators. *(Power.)*

Boiler units presently available can deliver steam at the desired temperature of 900 F. The required capacity of 250,000 lb/hr is beyond the range of *packaged watertube boilers*—defined by the American Boiler Manufacturer Association as "a boiler equipped and shipped complete with fuel-burning equipment, mechanical-draft equipment, automatic controls, and accessories."

Shop-assembled boilers are larger units, where all assembly is handled in the builder's plant but with some leeway in the selection of controls and auxiliaries. The current maximum capacity of shop-assembled boilers is about 100,000 lb/hr. Thus, a standard-design, larger-capacity boiler is required.

Study manufacturers' engineering data to determine which types of watertube boilers are available for the required capacity, pressure, and temperature. This study reveals that, for this installation, a standard, field-assembled, welded-steel-cased, bent-tube, single-steam-drum boiler with a completely water-cooled furnace would be suitable. This type of boiler is usually fitted with an air heater; and an economizer might also be used. The induced- and forced-draft fans are not integral with the boiler. Capacities of this type of boiler usually available range from 50,000 to 350,000 lb/hr; pressures from 160 to 1,050 psi; steam temperature from saturation to 950 F; fuels — pulverized coal, oil, gas, or a combination; controls — manual to completely automatic; efficiency — to 90 percent.

2. Determine the number of boilers required

The normal continuous steam requirement is 200,000 lb/hr. If a 250,000 lb/hr boiler were chosen to meet the maximum required output, the boiler would normally operate at 200,000/250,000, or 80 percent capacity. Obtain the performance chart, Fig. 51, from the manufacturer and study it. This chart shows that at 80 percent load, the boiler efficiency is about equal to that at 100 percent load. Thus, there will not be any significant efficiency loss when the unit is operated at its normal continuous output. The total losses in the boiler are lower at 80 percent load than at full (100 percent) load.

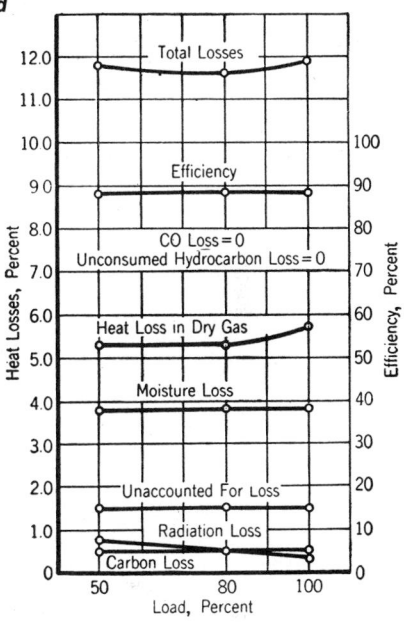

Since there is not a large efficiency decrease at the normal continuous load, and since there are not other factors that require or make more than one boiler desirable, a single boiler unit would be most suitable for this installation. One boiler is more desirable than two or more because installation of a single unit is simpler and maintenance costs are lower. However, where the load fluctuates widely and two or more boilers could best serve the steam demand, the savings in installation and maintenance costs would be insignificant compared with the extra cost of operating a relative large

Fig. 51 Typical watertube steam-generator losses and efficiency.

boiler installed in place of two or more smaller boilers. Therefore, each installation must be carefully analyzed and a decision made on the basis of the existing conditions.

3. Determine the required boiler capacity

The stated steam load is 250,000 lb/hr at maximum demand. Study the installation to determine if the steam demand will increase in the future. Try to determine the rate of increase in the steam demand — for example, installation of several steam-using process units each year during the next few years will increase the steam demand by a predictable amount every year. Using these data, the rate of growth and total steam demand can be estimated for each year. Where the growth will exceed the allowable overload capacity of the boiler — which can vary from 0 to 50 percent, of the full-load rating, depending on the type of unit chosen — consider installing a larger-capacity boiler now to meet future load growth. Where the future load is unpredictable, or where no load growth is anticipated, a unit sized to meet today's load would be satisfactory. If this situation existed in this plant, a 250,000 lb/hr unit would be chosen for the load. Any small temporary overloads could be handled by operating the boiler at a higher output for short periods.

Alternatively, assume that a load of 25,000 lb/hr will be added to the maximum demand on this boiler each year for the next 5 years. This means that in 5 years the maximum demand will be $250,000 + 25,000(5) = 375,000$ lb/hr. This is an overload of $(375,000 - 250,000)/250,000 = 0.50$, or 50 percent. It is unlikely that the boiler could carry a continuous overload of 50 percent. Therefore it might be wise to install a 375,000 lb/hr boiler now to meet present and future demands. Base this decision on the accuracy of the future-demand predictions and the economic advantages or disadvantages of investing more money now for a demand that will not occur until some future date. Refer to the section on *Engineering Economics* for procedures to follow in economic calculations of this type.

Thus, with *no increase* in the future load, a 250,000 lb/hr unit would be chosen. With the *load increase* specified, a 375,000 lb/hr unit would be the choice, if there were no major economic disadvantages.

4. Choose the type of fuel to use

Watertube boilers of the type being considered will economically burn the three fuels available — coal, oil, or gas — either singly or in combination. In the design considered here, the furnace water-cooled surfaces and boiler surface are integral parts of each other. For this reason the boiler is well suited for pulverized coal firing in the 50,000 to 300,000 lb/hr capacity range. Thus, if a 250,000 lb/hr unit were chosen, it could be pulverized coal fired. With a larger unit — 375,000 lb/hr — either pulverized coal, oil, or gas firing might be used. Use an economic comparison to determine which fuel would give the lowest overall operating cost for the life of the boiler.

5. Determine the feedwater-quality requirements

Watertube boilers of all types require careful control of feedwater quality to prevent scale and sludge deposits in tubes and drums. Corrosion of the interior boiler surfaces must be controlled. Where all condensate is returned to the boiler, the makeup water must be treated to prevent the conditions just cited. Therefore, a comprehensive water-treating system must be planned for, particularly if the quality of the raw-water supply is poor.

6. Estimate the boiler space requirements

The space occupied by steam-generating units is an important consideration in plants in municipal areas and where power-plant buildings are presently crowded by existing equipment. The manufacturer's engineering data for this boiler shows that for pulverized coal firing, the hopper-type furnace bottom is best. This data also shows that the smallest boiler with a hopper bottom occupies a space 21 ft wide, 31 ft high, and 14 ft front to rear. The largest boiler occupies a space 21 ft wide, 55 ft high, and 36 ft front to rear. Check these dimensions against the available space to determine if the chosen boiler can be installed without major structural changes. The steel walls permit outdoor or indoor installation with top or bottom support of the boiler optional in either method of installation.

Related Calculations: Use this general procedure to select boilers for industrial, central-station, process, and marine applications.

SELECTING BOILER FORCED- AND INDUCED-DRAFT FANS

Combustion calculations show that an oil-fired watertube boiler requires 200,000 lb/hr of air for combustion at maximum load. Select forced- and induced-draft fans for this boiler if the average temperature of the inlet air is 75 F and the average temperature of the combustion gas leaving the air heater is 350 F with an ambient barometric pressure of 29.9 in. Hg. Pressure losses on the air-inlet side are, in in. H_2O: air heater, 1.5; air-supply ducts, 0.75; boiler windbox, 1.75; burners, 1.25. Draft losses in the boiler and related equipment are, in in. H_2O; furnace pressure, 0.20; boiler, 3.0; superheater, 1.0; economizer, 1.50; air heater, 2.00; uptake ducts and dampers, 1.25. Determine the fan discharge pressure and hp input. The boiler burns 18,000 lb/hr of oil at full load.

Calculation Procedure:

1. Compute the quantity of air required for combustion

The combustion calculations show that 200,000 lb/hr of air are theoretically required for combustion in this boiler. To this theoretical requirement must be added allowances for excess air at the burner and leakage out of the air heater and furnace. Allow 25 percent excess air for this boiler. The exact allowance for a given installation depends on the type of fuel burned. However, a 25 percent excess-air allowance is an average used by power-plant designers for coal, oil, and gas firing. Using this allowance, the required excess air = 200,000(0.25) = 50,000 lb/hr.

Air-heater air leakage varies from about 1 to 2 percent of the theoretically required airflow. Using 2 percent, the air-heater leakage allowance = 200,000(0.02) = 4,000 lb/hr.

Furnace air leakage ranges from 5 to 10 percent of the theoretically required airflow. Using 7.5 percent, the furnace leakage allowance = 200,000(0.075) = 15,000 lb/hr.

The total airflow required is the sum of the theoretical requirement, excess air, and leakage. Or 200,000 + 50,000 + 4,000 + 15,000 = 269,000 lb/hr. The forced-draft fan must supply at least this quantity of air to the boiler. Usual practice is to allow a 10 to 20 percent safety factor for fan capacity to ensure an adequate air supply at all operating conditions. This factor of safety is applied to the total airflow required. Using a 10 percent factor of safety, fan capacity = 269,000 + 269,000(0.1) = 295,900 lb/hr. Round this off to 296,000 lb/hr fan capacity.

2. Express the required airflow in cfm

Convert the required flow in lb/hr to cfm. To do this, apply a factor of safety to the ambient air temperature to ensure an adequate air supply during times of high ambient temperature. At such times, the density of the air is lower and the fan discharges less air to the boiler. The usual practice is to apply a factor of safety of 20 to 25 percent to the known ambient air temperature. Using 20 percent, the ambient temperature for fan selection = 75 + 75(0.20) = 90 F. The density of air at 90 F is 0.0717 lb/cu ft, found in Baumeister — *Standard Handbook for Mechanical Engineers.* Converting, cfm = lb/hr/60 (lb/cu ft) = 296,000/60(0.0717) = 69,400 cfm. This is the minimum capacity the forced-draft fan may have.

3. Determine the forced-draft discharge pressure

The total resistance between the forced-draft fan outlet and furnace is the sum of the losses in the air heater, air-supply ducts, boiler windbox, and burners. For this boiler, the total resistance, in. H_2O = 1.5 + 0.75 + 1.75 + 1.25 = 5.25 in. H_2O. Apply a 15 to 30 percent factor of safety to the required discharge pressure to ensure adequate airflow at all times. Or fan discharge pressure, using a 20 percent factor of safety, = 5.25 + 5.25(0.20) = 6.30 in. H_2O. The fan must therefore deliver at least 69,400 cfm at 6.30 in. H_2O.

4. Compute the power required to drive the forced-draft fan

The air horsepower for any fan = $0.0001753 H_f C$, where H_f = total head developed by fan, in. H_2O; C = airflow, cfm. For this fan, air hp = 0.0001753(6.3)(69,400) = 76.5 hp. Assume or obtain the fan and fan-driver efficiencies at the rated capacity (69,400 cfm) and pressure (6.30 in. H_2O). With a fan efficiency of 75 percent and assuming the fan is driven by an electric motor having an efficiency of 90 percent, the overall efficiency of the fan-motor combination is (0.75)(0.90) = 0.675, or 67.5 percent. Then the motor horsepower required = air horsepower/overall efficiency = 76.5/0.675 = 113.2 hp. A 125-hp motor would be chosen because it is the nearest, next larger, unit readily available. Usual practice is to choose a *larger* driver capacity when the computed capacity is lower than a standard capacity. The next larger standard capacity is generally chosen, except for extremely large fans where a special motor may be ordered.

5. Compute the quantity of flue gas handled

The quantity of gas reaching the induced draft fan is the sum of the actual air required for combustion from Step 1, air leakage in the boiler and furnace, and the weight of

Standard Handbook of Engineering Calculations

fuel burned. With an air leakage of 10 percent in the boiler and furnace (this is a typical leakage factor applied in practice), the gas flow is as follows:

Actual airflow required	296,000 lb/hr
Air leakage in boiler and furnace............	29,600 lb/hr
Weight of oil burned	18,000 lb/hr
Total	343,600 lb/hr

Determine from combustion calculations for the boiler the density of the flue gas. Assume that the combustion calculations for this boiler show that the flue-gas density is 0.045 lb/cu ft at the exit-gas temperature. To determine the exit-gas temperature, apply a 10 percent factor of safety to the given exit temperature, 350 F. Hence, exit-gas temperature = $350 + 350(0.10) = 385$ F. Then, flue-gas flow, cfm, = (flue-gas flow, lb/hr)/(60)(flue-gas density, lb/cu ft) = $343,600/[(60)(0.045)] = 127,000$ cfm. Apply a 10 to 25 percent factor of safety to the flue-gas quantity to allow for increased gas flow. Using a 20 percent factor of safety, the actual flue-gas flow the fan must handle = $127,000 + 127,000(0.20) = 152,400$ cfm; say 152,500 cfm for fan-selection purposes.

6. Compute the induced-draft fan discharge pressure

Find the sum of the draft losses from the burner outlet to the induced-draft fan inlet. These losses are, for this boiler:

Furnace draft loss, in. H_2O	0.20
Boiler draft loss, in. H_2O	3.00
Superheater draft loss, in. H_2O	1.00
Economizer draft loss, in. H_2O	1.50
Air heater draft loss, in. H_2O	2.00
Uptake ducts and damper draft loss, in. H_2O	1.25
Total draft loss, in. H_2O	8.95

Allow a 10 to 25 percent factor of safety to ensure adequate pressure during all boiler loads and furnace conditions. Using a 20 percent factor of safety for this fan, the total actual pressure loss = $8.95 + 8.95(0.20) = 10.74$ in. H_2O. Round this off to 11.0 in. H_2O for fan-selection purposes.

7. Compute the power required to drive the induced-draft fan

As with the forced-draft fan, air horsepower = $0.0001753\ H_fC = 0.0001753(11.0) \times (127,000) = 245$ hp. If the combined efficiency of the fan and its driver, assumed to be an electric motor, is 68 percent, the motor horsepower required = $245/0.68 = 360.5$ hp. A 375-hp motor would be chosen for the fan driver.

8. Choose the fans from a manufacturer's engineering data

Use the next Calculation Procedure to select the fans from the engineering data of an acceptable manufacturer. For larger boiler units, the forced-draft fan is usually a backward-curved blade centrifugal-type unit. Where two fans are chosen to operate in parallel, the pressure curve of each fan should decrease at the same rate near shutoff so that the fans divide the load equally. Be certain that forced-draft fans are heavy-duty units designed for continuous operation with well-balanced rotors. Choose high-efficiency units with self-limiting power characteristics to prevent overloading the driving motor. Airflow is usually controlled by dampers on the fan discharge.

Induced-draft fans handle hot, dusty combustion products. For this reason, extreme care must be used to choose units specifically designed for induced-draft service. The usual choice for large boilers is a centrifugal-type unit with forward- or backward-curved, or flat blades, depending on the type of gas handled. Flat blades are popular when the flue gas contains large quantities of dust. Fan bearings are generally water-cooled.

Related Calculations: Use the Procedure given above for selecting draft fans for all types of boilers—fire-tube, packaged, portable, marine, and stationary. Obtain draft losses from the boiler manufacturer. Compute duct pressure losses using the methods given in later Procedures in this Handbook.

POWER-PLANT FAN SELECTION FROM CAPACITY TABLES

Choose a forced-draft fan to handle 69,400 cfm of 90-F air at 6.30 in. H_2O static pressure and an induced-draft fan to handle 152,500 cfm of 385-F gas at 11.0 in. H_2O static pressure. The boiler that these fans serve is installed at an elevation of 5,000 ft above sea level. Use commercially available capacity tables for making the fan choice. The flue-gas density is 0.045 lb/cu ft at 385 F.

Calculation Procedure:

1. Compute the correction factors for the forced-draft fan

Commercial fan-capacity tables are based on fans handling standard air at 70 F at a barometric pressure of 29.92 in. Hg and having a density of 0.075 lb/cu ft. Where different conditions exist, the fan flow rate must be corrected for temperature and altitude.

Obtain the engineering data for commercially available forced-draft fans and turn to the temperature and altitude correction-factor tables. Pick the appropriate correction factors from these tables for the prevailing temperature and altitude of the installation. Thus, in Table 6, select the correction factors for 90-F air and 5,000-ft altitude. These correction factors are $C_T = 1.018$ for 90-F air and $C_A = 1.095$ for 5,000-ft altitude.

TABLE 6   Fan Correction Factors

Temperature, F	Correction Factor	Altitude, ft	Correction Factor
80	1.009	4,500	1.086
90	1.018	5,000	1.095
100	1.028	5,500	1.106
375	1.255		
400	1.273		
450	1.310		

Find the composite correction factor CCF by taking the product of the temperature and altitude correction factors. Or $CCF = C_T C_A = (1.018)(1.095) = 1.1147$. Now divide the given cfm by the composite correction factor to find the capacity-table cfm. Or, capacity-table cfm = 69,400/1.1147 = 62,250 cfm.

2. Choose the fan size from the capacity table

Turn to the fan-capacity table in the engineering data and look for a fan delivering 62,250 cfm at 6.3 in. H_2O static pressure. Inspection of the table shows that the capacities are tabulated for pressures of 6.0 and 6.5 in. H_2O static pressure. There is no tabulation for 6.3 in. H_2O. The fan must therefore be selected for 6.5 in. H_2O static pressure.

Enter the table at the nearest capacity to that required, 62,250 cfm, as shown in the Table 7. This table, excerpted with permission from the American Standard Inc.

TABLE 7   Typical Fan Capacities

Cfm	Outlet velocity, fpm	Outlet velocity pressure, in. H_2O	6.5-in. H_2O static pressure	
			Rpm	Bhp
61,204	4,400	1.210	1,083	95.45
62,595	4,500	1.266	1,096	99.08
63,975	4,600	1.323	1,109	103.0

engineering data, shows that the nearest capacity of this particular type of fan is 62,595 cfm. The difference, or $62{,}595 - 62{,}250 = 345$ cfm, is only $345/62{,}250 = 0.0055$, or 0.55 percent. This is a negligible difference, and the 62,595-cfm fan is well suited for its intended use. The extra static pressure, $6.5 - 6.3 = 0.2$ in. H_2O, is desirable in a forced-draft fan because furnace or duct resistance may increase during the life of the boiler. Also, the extra static pressure is so small that it will not markedly increase the fan power consumption.

3. Compute the fan speed and power input

Multiply the capacity-table rpm and bhp by the composite correction factor to determine the actual rpm and bhp. Thus, using data from Table 7, the actual rpm = $(1{,}096)(1.1147) = 1{,}221.7$ rpm. Actual bhp = $(99.08)(1.1147) = 110.5$ hp. This is the horsepower input required to drive the fan and is close to the 113.2 hp computed in the previous Calculation Procedure. The actual motor horsepower would be the same in each case because a standard-size motor would be chosen. The difference of $113.2 - 110.5 = 2.7$ hp results from the assumed efficiencies that depart from the actual values. Also, a sea-level altitude was assumed in the previous Calculation Procedure. However, the two methods used show how accurately fan capacity and horsepower input can be estimated by judicious evaluation of variables.

4. Compute the correction factors for the induced-draft fan

The flue-gas density is 0.045 lb/cu ft at 385 F. Interpolate in the temperature correction-factor table because a value of 385 F is not tabulated. Find the correction factor for 385 F thus:
(Actual temperature − lower temperature)/(Higher temperature − lower temperature) × (Higher temperature-correction factor − lower temperature-correction factor) + lower-temperature-correction factor. Or, $[(385 - 375)/(400 - 375)](1.273 - 1.255) + 1.255 = 1.262$.

The altitude-correction factor is 1.095 for an elevation of 5,000 ft, as shown in Table 6.

As for the forced-draft fan, $CCF = C_T C_A = (1.262)\ (1.095) = 1.3819$. Use the CCF to find the capacity-table cfm in the same manner as for the forced-draft fan. Or, capacity-table cfm = (given cfm)/CCF = $152{,}500/1.3819 = 110{,}355$ fcm.

5. Choose the fan size from the capacity table

Check the capacity table to be sure that it lists fans suitable for induced-draft (elevated temperature) service. Turn to the 11-in. static-pressure-capacity table and find a capacity equal to 110,355 cfm. In the engineering data used for this fan, the nearest capacity at 11-in. static pressure is 110,467 cfm, with an outlet velocity of 4,400 cfm, an outlet velocity pressure of 1.210 in. H_2O, a speed of 1,222 rpm, and an input horsepower of 255.5 bhp. The tabulation of these quantities is of the same form as that given for the forced-draft fan, Step 2. The selected capacity of 110,467 cfm is entirely satisfactory because it is only $110{,}467 - 110{,}355/110{,}355 = 0.00101$, or 0.1 percent higher than the desired capacity.

6. Compute the fan speed and power input

Multiply the capacity-table rpm and brake horsepower by the CCF to determine the actual rpm and brake horsepower. Thus, the actual rpm = $(1{,}222)(1.3819) = 1{,}690$ rpm. Actual brake horsepower = $(255.5)(1.3819) = 353.5$ bhp. This is the horsepower input required to drive the fan and is close to the 360.5 hp computed in the previous Calculation Procedure. The actual motor horsepower would be the same in each case because a standard-size motor would be chosen. The difference in horsepower of $360.5 - 353.5 = 7.0$ hp results from the same factors discussed in Step 3.

NOTE: The static pressure is normally used in most fan-selection procedure because this is the pressure value used in computing pressure and draft losses in boilers, economizers, air heaters, and ducts. In any fan system, the total air pressure = static pressure + velocity pressure. However, the velocity pressure at the fan discharge is not considered in draft calculations unless there are factors requiring its evaluation. These requirements are generally related to pressure losses in the fan-control devices.

Related Calculations: Use the fan-capacity table to obtain these additional details of the fan: outlet inside dimensions (length and width), fan-wheel diameter and circumference, fan maximum bhp, inlet area, fan-wheel peripheral velocity, NAFM fan class, and fan arrangement. Use the engineering data containing the fan-capacity table to find the fan dimensions, rotation and discharge designations, shipping weight, and for some manufacturers, prices.

ANALYSIS OF BOILER AIR DUCTS AND GAS UPTAKES

Three oil-fired boilers are supplied air through the breeching shown in Fig. 52a. Each boiler will burn 13,600 lb/hr of fuel oil at full load. The draft loss through each boiler is 8 in. H_2O. Uptakes from the three boilers are connected as shown in Fig. 52b. Determine the draft loss through the entire system if a 50-ft-high metal stack is used and the gas temperature at the stack inlet is 400 F.

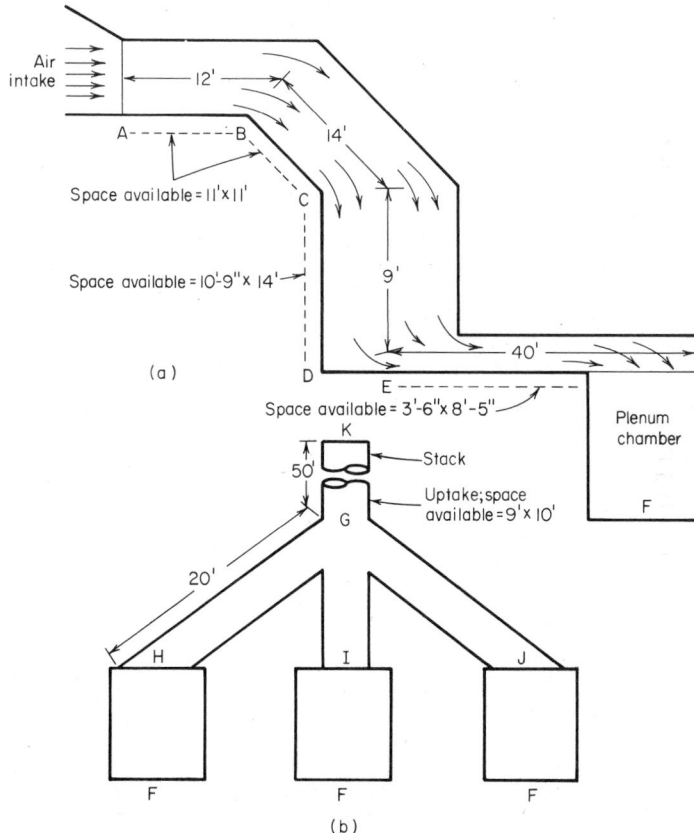

Fig. 52   *(a)* **Boiler intake-air duct;** *(b)* **boiler uptake ducts.**

Calculation Procedure:

1. *Determine the airflow through the breeching*

Compute the airflow required, cu ft per lb of oil burned, using the methods given in earlier Calculation Procedures. For this installation, assume that the combustion calculation shows that 250 cu ft of air per lb of oil burned is required. Then, the total airflow required, cfm, = (number of boilers)(lb/hr oil burned per boiler)(cu ft air per lb oil)/(60 min/hr) = (3)(13,600)(250)/60 = 170,000 cfm.

2. Select the dimensions for each length of breeching duct

With the airflow rate, 170,000 cfm, known, the duct area can be determined by assuming an air velocity and computing the duct area A_d sq ft from A_d = (airflow rate, cfm)/(air velocity, fpm). Once the area is known, the duct can be sized to give this area. Thus, if 9 sq ft is the required duct area, a duct 3×3 ft, or 2×4.5 ft would provide the required area.

In the usual power plant, the room available for ducts limits the maximum allowable duct size. The designer must therefore try to fit a duct of the required area into the available space. This is done by changing the duct height and width until a duct of suitable area fitting the available space is found. If the duct area is reduced below that required, compute the actual air velocity to determine if it exceeds recommended limits.

In this power plant, the space available in the open area between A and C, Fig. 52, is a square 11×11 ft. Allowing a 3-in. clearance around the outside of the duct, and using a square duct, its dimensions would be 10.5×10.5 ft, or a cross-sectional area of $(10.5)(10.5) = 110$ sq ft, closely. With 170,000 cfm flowing through the duct, the air velocity v fpm = (cfm)/A_d = 170,000/110 = 1,545 fpm. This is a satisfactory air velocity because the usual-plant air systems is 1,200 to 3,600 fpm.

Between C and D the open area in this power plant is 10 ft 9 in. by 14 ft. Using the same 3-in. clearance all around the duct, the dimensions of the vertical duct CD are 10.25×13 ft, or a cross-sectional area of $10.25 \times 13 = 133$ sq ft, closely. The air velocity in this section of the duct is v = 170,000/133 = 1,275 fpm. Since it is desirable to maintain, if possible, a constant velocity in all sections of the duct where the space available permits it, the size of this duct might be changed so it equals that of AB, 10.5×10.5 ft. However, the installation costs would probably be high because the limited space available would require alteration of the power-plant structure. Also, the velocity in section CD is above the usual minimum value of 1,200 fpm. For these reasons, the duct will be installed in the 10.25×13 ft size.

Between E and F the vertical distance available for installation of the duct is 3.5 ft and the horizontal distance 8.5 ft. Using the same 3-in. clearance as before gives a 3×8 ft duct size, or a cross-sectional area of $(3)(8) = 24$ sq ft. At E the duct divides into three equal-size branches, one for each boiler, and the same area, 24 sq ft, is available for each branch duct. The flow in any branch duct is then 170,000/3 = 56,700 cfm. The velocity in any of the three equal branches is v = 56,700/24 = 2,360 fpm. When a duct system has two or more equal-size branches, compute the pressure loss in one branch only because the losses in the other branches will be the same. The velocity in branch EF is acceptable because it is within the limits normally used in power-plant practice. At F the air enters a large plenum chamber and its velocity becomes negligible because of the large flow area. The boiler forced-draft fan intakes are connected to the plenum chamber. Each of the three branch ducts feeds into the plenum chamber.

3. Compute the pressure loss in each duct section

Begin the pressure-loss calculations at the system inlet, point A, and work through each section to the stack outlet. This procedure reduces the possibility of error and permits easy review of the calculations for detection of errors. Assign letters to each point of the duct where a change in section dimensions or direction, or both, occur. Use these letters as shown below.

Point A: Assume that 70-F air having a density of 0.075 lb/cu ft enters the system when the ambient barometric pressure is 29.92 in. Hg. Compute the velocity pressure at point A, in in. H_2O, from $p_v = v^2/[3.06(10^4)(460 + t)]$, where t = air temperature, F. Since the velocity of the air at A is 1,545 fpm, $p_v = (1,545)^2/[3.06(10^4)530)] = 0.147$ in. H_2O at 70-F air temperature.

The entrance loss at A, where there is a sharp-edged duct is 0.5 p_v, or 0.5(0.147) = 0.0735 in. H_2O. With a rounded inlet, the loss in velocity pressure would be negligible.

Section AB: There is a pressure loss due to duct friction between A and B, B and C. Also, there is a bend loss at points B and C. Compute the duct friction first.

For any circular duct, the static pressure loss due to friction p_s in. $H_2O = (0.03L/1.24)(v/1,000)^{1.84}$, where $L =$ duct length, ft; $d =$ duct diameter, in. To convert any rectangular or square duct with sides a ft and b ft height and width, respectively, to an equivalent round duct of D ft diameter, use the relation $D = 2ab/(a+b)$. For this duct, $d = 2(10.5)(10.5)/(10.5+10.5) = 10.5$ ft $= 126$ in. $= d$. Since this duct is 12 ft long between A and B, $p_s = (0.03(12)/126^{1.24})(1,545/1,000)^{1.84} = 0.002$ in. H_2O.

Point B: The $45°$ bend at B has, from Baumeister—*Standard Handbook for Mechanical Engineers*, a pressure drop of 60 percent of the velocity head in the duct, or $(0.60)(0.147) = 0.088$ in. H_2O loss.

Section BC: Duct friction in the 14-ft-long downcomer BC is $p_s = (0.03(14)/126^{1.24})(1,545/1,000)^{1.84} = 0.0023$ in. H_2O. Point C: The $45°$ bend at C has a velocity head loss of 60 percent of the velocity pressure. Determine the velocity pressure in this duct in the same manner as for Point A, or $p_v = (1,545)^2/[3.06(10^4)(530)] = 0.147$ in. H_2O, since the velocity at points B and C is the same. Then, the velocity head loss $= (0.60)(0.147) = 0.088$ in. H_2O.

Section CD: The equivalent round-duct diameter is $D = (2)(10.25)(13)/(10.25+13) = 11.45$ ft $= 137.3$ in. Duct friction is then, $p_s = (0.03(9)/137.3^{1.24})(1,275/1,000)^{1.84} = 0.000934$ in. H_2O. Velocity pressure in the duct is $p_v = (1,275)^2/[3.06(10^4)(530)] = 0.100$ in. H_2O. Since there is no room for a transition piece—that is, a duct providing a gradual change in flow area between points C and D—the decrease in velocity pressure from 0.147 in. to 0.100 in., or $0.147 - 0.100 = 0.047$ in. H_2O, is not converted to static pressure and is lost.

Point E: The pressure loss in the right-angle bend at E is, from Baumeister and Marks—*Standard Handbook for Mechanical Engineers*, 1.2 times the velocity head, or $(1.2)(0.1) = 0.12$ in. H_2O. Also, since this is a sharp-edged elbow, there is an additional loss of 50 percent of the velocity head, or $(0.50)(0.10) = 0.05$ in. H_2O. The velocity pressure at point E is $p_v = (2,360)^2/[3.06(10^4)(530)] = 0.343$ in. H_2O.

Section EF: The equivalent round-duct diameter is $D = (2)(3)(8)/(3+8) = 4.36$ ft $= 52.4$ in. Duct friction $p_s = (0.03(40)/52.4^{1.24})(2,360/1,000)^{1.84} = 0.0247$ in. H_2O.

Air entering the large plenum chamber at F loses all its velocity. There is no static-pressure regain; therefore, the velocity-head loss $= 0.348 - 0.0 = 0.348$ in. H_2O.

4. Compute the losses in the uptakes and stack

Convert the airflow of 250 cu ft per lb of fuel oil to lb of air per lb of fuel oil by multiplying by the density, or $250(0.075) = 18.75$ lb of air per lb of fuel oil. The flue gas will contain 18.75 lb of air $+ 1$ lb of oil per lb of fuel burned, or $(18.75 + 1)/18.75 = 1.052$ times as much gas leaves the boiler as air enters; this can be termed the *flue-gas factor*.

Point G: The quantity of flue gas entering the stack from each boiler (corrected to a 400-F outlet temperature) is, using °R, (cfm air to furnace)(stack, R/air, R)(flue-gas factor). Or stack flue-gas flow $= (56,700)[(400+460)/(70+460)](1.052) = 97,000$ cfm per boiler.

The total duct area available for the uptake leading to the stack is 9×10 ft $= 90$ sq ft, based on the clearance above the boilers. The flue-gas velocity for three boilers is $v = (3)(97,000)/90 = 3,235$ fpm. The velocity pressure in the uptake is $p = (3,235)^2/[3.06(10^4)(460+400)] = 0.397$ in. H_2O.

Point H: The flue-gas flow from all the boilers is divided equally between three ducts $-HG, IG, JG$, Fig. 52. It is desirable to maintain the same gas velocity in each duct and have this velocity equal to that in the uptake. The same velocity can be obtained in each duct by making each duct one-third the area of the uptake, or $90/3 = 30$ sq ft. Then, $v = 97,000/30 = 3,235$ fpm in each duct. Since the velocity in each duct equals the velocity in the uptake, the velocity pressure in each duct equals that in the uptake, or 0.397 in. H_2O.

Ducts HG and JG have two $45°$ bends in them, or the equivalent of one $90°$ bend. The velocity-pressure loss in a $90°$ bend is 1.20 times the velocity head in the duct, or for either HG or JG, $(1.20)(0.397) = 0.476$ in. H_2O.

Section HG: The equivalent duct diameter for a 30 sq ft duct is $D = 2(30/\pi)^{0.5} = 6.19$ ft $= 74.2$ in. The duct friction in HG, which equals that in JG, $p_s = (0.03(20)/74.2^{1.24})(3,235/1,000)^{1.84}(530/860) = 0.01536$ in. H_2O, correcting for the flue-gas temperature with the ratio $(70 + 460)/(400 + 460) = 530/860$.

Section GK: The stack joins the uptake at point G. Assume that this installation is designed for a stack gas area of 500 lb of oil per sq ft of stack, or for three boilers, stack area $= (3)(13,600$ lb oil per hr$)/500 = 81.5$ sq ft. The stack diameter will then be $D = 2(8.15/\pi)^{0.5} = 10.18$ ft $= 122$ in.

The gas velocity in the stack is $v = (3)(97,000)/81.5 = 3,570$ fpm. The friction in the stack is $p_s = (0.03(50)/122^{1.24})(3,570/1,000)^{1.84}(530/860) = 0.0194$ in. H_2O.

5. Compute the total losses in the system

Tabulate the individual losses and find the sum as follows:

Point A; entrance loss, in. H_2O	0.0735
Section AB; duct friction	0.0020
Point B; bend loss	0.0880
Section BC; duct friction	0.0023
Point C; bend loss	0.0880
Section CD; duct friction	0.0009
Section CD; velocity-pressure loss	0.470
Point E; bend loss	0.1200
Point E; sharp-edge loss	0.0500
Section EF; duct friction	0.0247
Section EF; plenum velocity-head loss	0.3480
Boiler friction loss	8.0000
Section HG; duct friction	0.0154
Points H and G total bend loss	0.4760
Section GK; stack friction	0.0194
Total loss, in. H_2O	9.3552

The total loss computed here is the minimum static pressure that must be developed by the draft fans or blowers. This total static pressure can be divided between the forced- and induced-draft fans or confined solely to the forced-draft fan in plants not equipped with an induced-draft fan. If only a forced-draft fan is used, its static discharge pressure should be at least 20 percent greater than the losses, or $(1.2)(9.3552) = 11.21$ in. H_2O at a total airflow of 97,000 cfm. If more than one forced-draft fan were used for each boiler, each fan would have a total static pressure of at least 11.21 in. H_2O and a capacity of less than 97,000 cfm. In making the final selection of the fan, the static pressure would be rounded off to 12 in. H_2O.

Where dampers are used for combustion-air control, include the *wide-open* resistance of the dampers when computing the total losses in the system at full load on the boilers. Damper resistance values can be obtained from the damper manufacturer. Note that as the damper is closed to reduce the airflow at lower boiler loads, the resistance through the damper is increased. Check the fan head-capacity curve to determine if the head developed by the fan at lower capacities is sufficient to overcome the greater damper resistance. Since the other losses in the system will decrease with smaller airflow, the fan static pressure is usually adequate.

NOTE: (a) Follow the notation system used here to avoid errors from plus and minus signs applied to atmospheric pressures and draft. Use of the plus and minus signs does not simplify the calculation and can be confusing.

(b) A few designers, reasoning that the pressure developed by a fan varies as the square of the air velocity, *square the percentage safety-factor increase* before multiplying by the static pressure. Thus, in the above forced-draft fan, the static discharge pressure with a 20 percent increase in pressure would be $(1.2)^2(9.3552) = 13.5$ in. H_2O. This procedure provides a wider margin of safety but is not widely used.

(c) Large steam-generating units, some ship propulsion plants, and some packaged boilers use only forced-draft fans. Induced-draft fans are eliminated because there is a saving in the total fan hp required, there is no air infiltration into the boiler setting, and a slightly higher boiler efficiency can be obtained.

(d) The duct system analyzed here is typical of a study-type design where no refinements are used in bends, downcomers, and other parts of the system. This type of system was chosen for the analysis because it shows more clearly the various losses met in a typical duct installation. The system could be improved by using a bellmouthed intake at A, dividing vanes or splitters in the elbows, a transition in the downcomer, and a transition at F. None of these improvements would be expensive and they would all reduce the static pressure required at the fan discharge.

(e) Do not subtract the stack draft from the static pressure the forced- or induced-draft fan must produce. Stack draft can vary considerably, depending on ambient temperature, wind velocity, and wind direction. Therefore, the usual procedure is to ignore any stack draft in fan-selection calculations because this is the safest procedure.

Related Calculations: The procedure given here can be used for all types of boilers fitted with air-supply ducts and uptake breechings — heating, power, process, marine, portable, and packaged.

DETERMINATION OF THE MOST ECONOMICAL FAN CONTROL

Determine the most economical fan control for a forced- or induced-draft fan designed to deliver 140,000 cfm at 14-in. H_2O at full load. Plot the power-consumption curve for each type of control device considered.

Calculation Procedure:

1. Determine the types of controls to consider

There are five types of controls used for forced- and induced-draft fans: (a) a damper in the duct with constant-speed fan drive; (b) two-speed fan driver; (c) inlet vanes or inlet louvres with a constant-speed fan drive; (d) multiple-step variable-speed fan drive; and (e) hydraulic or electric coupling with constant-speed driver giving wide control over fan speed.

2. Evaluate each type of fan control

Tabulate the selection factors influencing the control decision as follows, using the control letters in Step 1:

Control type	Control cost	Required power input	Advantages (A), and disadvantages (D)
a	Low	High	(A) Simplicity; (D) High power input
b	Moderate	Moderate	(A) Lower input power; (D) Higher cost
c	Low	Moderate	(A) Simplicity; (D) ID fan erosion
d	Moderate	Moderate	(D) Complex; also needs dampers
e	High	Low	(A) Simple; no dampers needed

3. Plot the control characteristics for the fans

Draw the fan head-capacity curve for the airflow or gasflow range considered, Fig. 53. This plot shows the maximum capacity of 140,000 cfm and required static head of 14 in. H_2O, point P.

Plot the power-input curve *ABCD* for a constant-speed motor or turbine drive with damper control—type *a*, listed above—after obtaining from the fan manufacturer, or damper builder, the input power required at various static pressures and capacities. Plotting these values gives curve *ABCD*. Fan speed is 1,200 rpm.

Plot the power-input curve *GHK* for a two-speed drive, type *b*. This drive might be a motor with an additional winding, or it might be a second motor for use at reduced boiler capacities. With either arrangement, the fan speed at lower boiler capacities is 900 rpm.

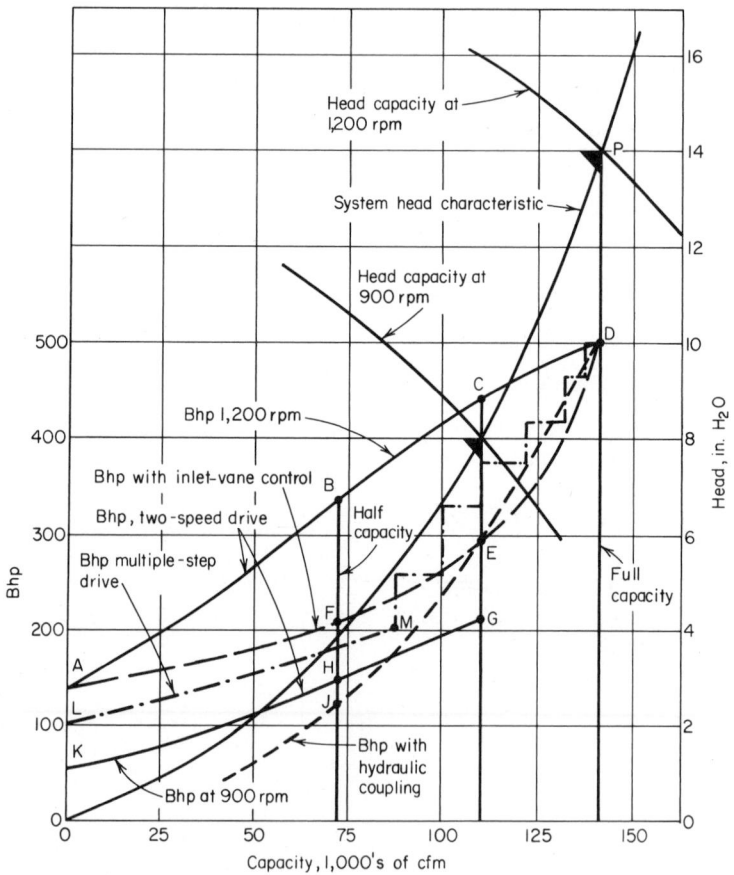

Fig. 53 Power requirements for a fan fitted with different types of controls. (*American Standard Inc.*)

Plot the power-input curve *AFED* for inlet-vane control on the forced-draft fan, or inlet-louvre control on induced-draft fans. The data for plotting this curve can be obtained from the fan manufacturer.

Multiple-step variable-speed fan control, type *d*, is best applied with steam-turbine drives. In a plant with ac auxiliary motor drives, slipring motors with damper integration must be used between steps, making the installation expensive. Although dc motor drives would be less costly, few power plants other than marine propulsion plants have direct current available. And since marine units normally operate at full load 90 percent or more of the time, part-load operating economics are unimportant. If steam-turbine drive will be used for the fans, plot the power-input curve *LMD*, using data from the fan manufacturer.

A hydraulic coupling or electric magnetic coupling, type *e*, with a constant-speed motor drive would have the power-input curve *DEJ*.

Study of the power-input curves shows that the hydraulic and electric couplings have the smallest power input. Their first cost, however, is usually greater than any other types of power-saving devices. To determine the return on any extra investment in power-saving devices, an economic study, including a load-duration analysis of the boiler load, must be made.

4. Compare the return on the extra investment

Compute and tabulate the total cost of each type of control system. Then determine the extra investment for each of the more costly control systems by subtracting the cost of type *a* from the cost of each of the other types. With the extra investment known, compute the lifetime savings in power input for each of the more efficient control methods. With the extra investment and savings resulting from it known, compute the percentage return on the extra investment. Tabulate the findings as in Table 8.

TABLE 8 Fan Control Comparison

	Type of Control Used				
	a	*b*	*c*	*d*	*e*
Total cost, $.............	30,000	50,000	75,000	89,500	98,000
Extra cost, $	...	20,000	25,000	14,500	8,500
Total power saving, $....	...	8,000	6,500	3,000	6,300
Return on extra investment, %	...	40	26	20.7	74.2

In Table 8, considering control type *c*, the extra cost of type *c* over type *b* = $75,000 − 50,000 = $25,000. The total power saving of $6,500 is computed on the basis of the cost of energy in the plant for the life of the control. The return on the extra investment then = $6,500/$25,000 = 0.26, or 26 percent. Type *e* control provides the highest percentage return on the extra investment. It would probably be chosen if the only measure of investment desirability is the return on the extra investment. However, if other criteria are used — such as a minimum rate of return on the extra investment — one of the other control types might be chosen. This is easily determined by studying the tabulation in conjunction with the investment requirement.

Related Calculations: The procedure used here can be applied to heating, power, marine, and portable boilers of all types. Follow the same steps given above, changing the values to suit the existing conditions. Work closely with the fan and drive manufacturer when analyzing drive power input and costs.

SMOKESTACK HEIGHT AND DIAMETER DETERMINATION

Determine the required height and diameter of a smokestack to produce 1.0 in. H_2O draft at sea level if the average air temperature is 60 F; barometric pressure is 29.92 in. Hg; the boiler flue gas enters the stack at 500 F; the flue-gas flow rate is 100 lb/sec; the flue-gas density is 0.045 lb/cu ft and the flue-gas velocity is 30 fps. What diameter and height would be required for this stack if it were located at an elevation 5,000 ft above sea level?

Calculation Procedure:

1. Compute the required stack height

The required stack height S_h ft $= d_s/0.256\, pK$, where $d_s =$ stack draft, in. H_2O; $p =$ barometric pressure, in. Hg; $K = (1/T_a)-(1/T_g)$, where $T_a =$ air temperature, R; T_g

= average temperature of stack gas, R. In applying this equation, the temperature of the gas at the stack outlet must be known to determine the average temperature of the gas in the stack. Since the outlet temperature cannot be measured until after the stack is in use, an assumed outlet temperature must be used for design calculations. The outlet temperature depends on the inlet temperature, ambient air temperature, and materials used in the stack construction. For usual smokestacks, the gas temperature will decrease 100 to 200 F between the stack inlet and outlet. Using a 100-F gas-temperature decrease for this stack, $S_h = (1.0) + 0.256(29.92)(1/520 - 1/910) = 159$ ft. Apply a 10 percent factor of safety. Then, the stack height $= (159)(1.10) = 175$ ft.

2. Compute the required stack diameter

Stack diameter d_s ft is found from $d_s = 0.278(W_g T_g/Vd_g p)^{0.5}$, where $W_g =$ flue-gas flow rate in stack, lb/sec; $V =$ flue-gas velocity in stack, fps; $d_g =$ flue-gas density, lb/cu ft. For this stack, $d_s = 0.278\{(100)(910)/[(30)(0.045)(29.92)]\}^{0.5} = 13.2$ ft, or 13 ft 3 in., rounding off to the nearest in. diameter.

NOTE: (a) Use this Calculation Procedure for any stack material—masonary, steel, brick, or plastic. (b) Most boiler and stack manufacturers use charts based on the equations above to determine the economical height and diameter of a stack. Thus, the Babcock & Wilcox Company, New York, presents four charts for stack sizing, in *Steam: Its Generation and Use*. Combustion Engineering, Inc., also presents four charts for stack sizing, in *Combustion Engineering*. The equations used in the present Calculation Procedure are adequate for a quick, first approximation of stack height and diameter.

3. Compute the required stack height and diameter at 5,000-ft elevation

Fuels require the same amount of oxygen for combustion regardless of the altitude at which they are burned. Therefore, this stack must provide the same draft as at sea level. But as the altitude above sea level increases, more air must be supplied to the fuel to sustain the same combustion rate, because air above sea level contains less oxygen per cu ft than at sea level. To accommodate the larger air and flue-gas flow rate without an increase in the stack friction loss, the stack diameter must be increased.

To determine the required stack height S_e ft at an elevation above sea level, multiply the sea-level height S_h by the ratio of the sea-level and elevated-height barometric pressures, in. Hg. Since the barometric pressure at 5,000 ft is 24.89 in. Hg and the sea-level barometric pressure is 29.92 in. Hg, $S_e = (175)(29.92/24.89) = 210.2$ ft.

The stack diameter d_e ft at an elevation above sea level will vary as the 0.40 power of the ratio of the sea-level and altitude barometric pressures, or $d_e = d_s(p_e/p)^{0.4}$, where $p_e =$ barometric pressure at altitude, in. Hg. For this stack, $d_e = (13.2)(29.92/24.89)^{0.4} = 14.2$ ft, or 14 ft 3 in., rounding off to the nearest in. diameter.

Related Calculations: The Procedure given here can be used for heating, power, marine, industrial, and residential smokestacks or chimneys, regardless of the materials used for construction. When designing smokestacks for use at altitudes above sea level, use Step 3, or substitute the actual barometric pressure at the elevated location in the height and diameter equations of Steps 1 and 2.

POWER-PLANT COAL-DRYER ANALYSIS

A power-plant coal dryer receives 180 tons/hr of wet coal containing 15 percent free moisture. The dryer is arranged to drain 6 percent of the moisture from the coal, and a moisture content of 1 percent is acceptable in the coal delivered to the power plant. Determine the volume and temperature of the drying gas required for the dryer, the total heat, grate area, and combustion-space volume needed. Ambient air temperature during drying is 70 F.

Calculation Procedure:

1. Compute the quantity of moisture to be removed

The total moisture in the coal = 15 percent. Of this, 6 percent is drained and 1 percent can remain in the coal. The amount of moisture to be removed is therefore

$15 - 6 - 1 = 8$ percent. Since 180 tons of coal are received per hr, the quantity of moisture to be removed per min is $[180/(60 \text{ min/hr})](2,000 \text{ lb/ton})(0.08) = 480 \text{ lb/min}$.

2. Compute the airflow required through the dryer

Air enters the dryer at 70 F. Assume that evaporation of the moisture on the coal takes place at 125 F—this is about midway in the usual evaporation temperature range of 110 to 145 F. Determine the moisture content of saturated air at each temperature using the psychrometric chart for air. Thus, for saturated air at 70-F dry-bulb temperature, the weight of the moisture it contains w_m lb of water per lb of dry air $= 0.0159$, whereas at 125 F, $W_m = 0.09537$ lb of water per lb of dry air. The weight of water removed per lb of air passing through the dryer is the difference between the moisture content at the leaving temperature, 125 F, and the entering temperature, 70 F, or $0.09537 - 0.01590 = 0.07947$ lb of water per lb of dry air.

Since air at 70 F has a density of 0.075 lb/cu ft, $1/0.075 = 13.33$ cu ft of air at 70 F must be supplied to absorb 0.07947 lb of water per lb of dry air. With 480 lb/min of water to be evaporated in the dryer, each cu ft of air will absorb $0.07947/13.3 = 0.005945$ lb of moisture, and the total airflow must be $(480 \text{ lb/min})/(0.005945) = 80,800 \text{ cfm}$, assuming a dryer efficiency of 100 percent. However, the usual dryer efficiency is about 75 percent—not 100 percent. Therefore, the total actual airflow through the dryer should be $80,800/0.75 = 107,700 \text{ cfm}$.

NOTE: If desired, a table of moist air properties can be used instead of a psychrometric chart to determine the moisture content of the air at the dryer inlet and outlet conditions. The moisture content is read in the humidity ratio W_s column. See the ASHRAE *Guide and Data Book* for such a tabulation of moist-air properties.

3. Compute the required air temperature

Assume that the heating air enters at a temperature t greater than 125 F. Set up a heat balance such that the heat given up by the air in cooling from t to 125 F = the heat required to evaporate the water on the coal + the heat required to raise the temperature of the coal and water from ambient to the evaporation temperature + radiation losses.

The heat given up by the air, Btu = (cfm)(density of air, lb/cu ft)[specific heat of air, Btu/(lb)(F)](t — evaporation temperature, F). The heat required to evaporate the water, Btu, = (weight of water, lb/min)(h_{fg} at evaporation temperature). The heat required to raise the temperature of the coal and water from ambient to the evaporation temperature, Btu, = (weight of coal, lb/min)(evaporation temperature — ambient temperature) [specific heat of coal, Btu/(lb)(F)] + (weight of water, lb/min)(evaporation temperature — ambient temperature) specific heat of water, Btu/(lb)(F)]. The heat required to make up for radiation losses, Btu, = [(area of dryer insulated surfaces, sq ft)[heat transfer coefficient, Btu/(sq ft)(F)(hr)](t — ambient temperature) + (area of dryer uninsulated surfaces, sq ft)[heat-transfer coefficient, Btu/(sq ft)(F)(hr)](t — ambient temperature)]/ 60.

Compute the heat given up by the air, Btu, as $(107,700)(0.075)(0.24)(t - 70)$, where 0.075 is the air density and 0.24 is the specific heat of the air.

Compute the heat required to evaporate the water, Btu, as $(480)(1,022.9)$, where $1,022.9 = h_{fg}$ at 125 F from the steam tables.

Compute the heat required to raise the temperature of the coal and water from ambient to the evaporation temperature, Btu, as $(6,000)(t - 70)(0.30) + (480)(t - 70)$ (1.0), where 0.30 is the specific heat of the coal and 1.0 is the specific heat of water.

Compute the heat required to make up the radiation losses, assuming 3,000 sq ft of insulated and 1,500 sq ft of uninsulated surface in the dryer, with coefficients of heat transfer of 0.35 and 3.0 for the insulated and uninsulated surfaces, respectively. Then, radiation heat loss, Btu, = $(3,000)(0.35)(t - 70) + (1,500)(3.0)(t - 70)$.

Set up the heat balance thus and solve for t: $(107,700)(0.075)(0.24)(t - 70) = (480)$ $(1,022.9) + (6,000)(125 - 70)(0.30) + (480)(125 - 70)(1.0) + [(3,000)(0.35)(t - 70) + (1,500)(3.0)(t - 70)]/60$; $t = 406$ F. In this heat balance, the factor 60 is divided into the radiation heat loss to convert the flow in Btu/hr to Btu/min because all the other expressions are in Btu/min.

4. Determine the total heat required by the dryer

Using the equation of Step 3 with $t = 406$ F, the total heat $= (107,700)(0.075)(0.24)$ $(406 - 70) = 651,000$ Btu/min, or $60(651,000) = 39,060,000$ Btu/hr.

5. Compute the dryer-furnace grate area

Assume that heat for the dryer is produced from coal having a lower heating value of 13,000 Btu/lb and that 40 lb of coal is burned per hr per sq ft of grate area with a combustion efficiency of 70 percent.

The rate of coal firing $= $ (Btu/min to dryer)/(coal heating value, Btu/lb)(combustion efficiency) $= (651,000)/(13,000)(0.70) = 71.5$ lb/min, or $60(71.5) = 4,990$ lb/hr. Grate area $= 4,990/40 = 124.75$ sq ft; say 125 sq ft.

6. Compute the dryer-furnace volume

The usual heat-release rates for dryer furnaces are about 50,000 Btu per hr per cu ft of furnace volume. For this furnace, which burns 4,900 lb/hr of 13,000 Btu/lb coal, the total heat released is $4,990(13,000) = 64,870,000$ Btu/hr. With an allowable heat release of 50,000 Btu/(hr)(cu ft), the required furnace volume $= 64,870,000/50,000 = 1,297.4$ cu ft; say 1,300 cu ft.

Related Calculations: The general Procedure given here can be used for any air-heated dryer used to dry moist materials. Thus, the Procedure is applicable to chemical, soil, and fertilizer drying, as well as coal drying. In each case, the specific heat of the material dried must be used in place of the specific heat of coal given above.

COAL STORAGE CAPACITY OF PILES AND BUNKERS

Bituminous coal is stored in a 25-ft-high 68.8-ft-diameter circular-base conical pile. How many tons of coal does the pile contain if its base angle is 36°? How much bituminous coal is contained in a 25-ft-high rectangular pile 100 ft long if the pile cross section is a triangle having a 36° base angle?

Calculation Procedure:

1. Sketch the coal pile

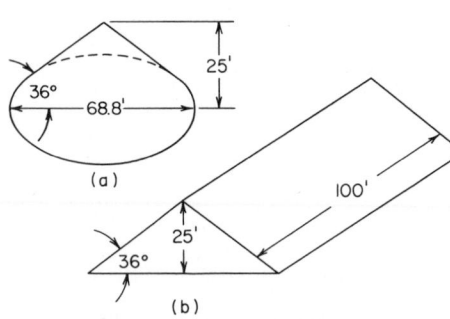

Figures 54a and b show the two coal piles. Indicate the pertinent dimensions—height, the diameter, length, and base angle—on each sketch.

2. Compute the volume of the coal pile

Volume of a right circular cone, cu ft, $= \pi r^2 h/3$, where $r =$ radius, ft; $h =$ cone height, ft. Volume of triangular pile $= bal/2$, where $b =$ base length, ft; $a =$ altitude, ft; $l =$ length of pile, ft.

Fig. 54 (a) Conical coal pile; (b) triangular coal pile.

For this conical pile, volume $= \pi(34.4)^2(25)/3 = 31,000$ cu ft. Since 50 lb of bituminous coal occupy about 1 cu ft of volume, the weight of coal in the conical pile $= (31,000$ cu ft$)(50$ lb/cu ft$) = 1,550,000$ lb, or $(1,550,000$ lb$)/(2,000$ lb/ton$) = 775$ tons.

For the triangular pile, base length $= 2h/\tan 36° = (2)(25)/0.727 = 68.8$ ft. Then, volume $= (68.8)(25)(100)/2 = 86,000$ cu ft. The weight of bituminous coal in the pile is, as for the conical pile, $(86,000)(50) = 4,300,000$ lb, or $(4,300,000$ lb$)/(2,000$ lb/ton$) = 2,150$ tons.

Related Calculations: Use this general Procedure to compute the weight of coal in piles of all shapes, and in bunkers, silos, bins, and similar storage compartments. The Procedure can also be used for other materials—grain, sand, gravel, coke, etc. Be sure to use the correct density when converting the total storage volume to total weight.

Refer to Baumeister—*Standard Handbook for Mechanical Engineers* for a comprehensive tabulation of the densities of various materials.

PROPERTIES OF A MIXTURE OF GASES

A 10 cu ft tank holds 1 lb of hydrogen H_2, 2 lb of nitrogen N_2, and 3 lb of carbon dioxide CO_2, at 70 F. Find the specific volume, pressure, specific enthalpy, internal energy, and specific entropy of the individual gases and of the mixture and the mixture density. Use Avogadro's and Dalton's laws and Keenan and Kaye—*Thermodynamic Properties of Air, Products of Combustion and Component Gases*, Wiley, commonly termed the *Gas Tables*.

Calculation Procedure:

1. Compute the specific volume of each gas

Using H, N, and C as subscripts for the respective gases, the specific volume of any gas v cu ft/lb = total volume of tank, cu ft/weight of gas in tank, lb. Thus, $v_H = 10/1 = 10$ cu ft/lb; $v_N = 10/2 = 5$ cu ft/lb; $v_C = 10/3 = 3.33$ cu ft/lb. Then, the specific volume of the mixture of gases v_t cu ft/lb = total volume of gas in tank, cu ft/sum of weight of individual gases, lb = $10/(1+2+3) = 1.667$ cu ft/lb.

2. Determine the absolute pressure of each gas

Using $P = RTw/v_tM$, where P = absolute pressure of the gas, psfa; R = universal gas constant = 1,545; T = absolute temperature of the gas, R = F + 459.9, usually taken as 460; w = weight of gas in the tank, lb; v_t = total volume of the gas in the tank, cu ft; M = molecular weight of the gas. Thus, $P_H = (1,545)(70+460)(1.0)/[(10)(2.0)] = 40,530$ psfa; $P_N = (1,545)(70+460)(2.0)/[(10)(28)] = 5,850$ psfa; $P_C = (1,545)(70+460)(3.0)/[(10)(44)] = 5,583$ psfa; $P_t = \Sigma P_H, P_N, P_C = 40,530 + 5,850 + 5,583 = 51,963$ psfa.

3. Determine the specific enthalpy of each gas

Refer to the *Gas Tables*, entering the left-hand column of the table at the absolute temperature, 530 R, for the gas being considered. Opposite the temperature, read the specific enthalpy in the h column. Thus, $h_H = 1,796.1$ Btu/lb; $h_N = 131.4$ Btu/lb; $h_C = 90.17$ Btu/lb. The total enthalpy of the mixture of the gases is the sum of the products of the weight of each gas and its specific enthalpy or $(1)(1,796.1) + (2)(131.4) + (3)(90.17) = 2,329.4$ Btu for the 6 lb or 10 cu ft of gas. The specific enthalpy of the mixture is the total enthalpy/gas weight, lb, or $2,329.4/(1+2+3) = 388.2$ Btu per lb of gas mixture.

4. Determine the internal energy of each gas

Using the *Gas Tables* as in Step 3, $E_H = 1,260.0$ Btu/lb; $E_N = 93.8$ Btu/lb; $E_C = 66.3$ Btu/lb. The total energy = $(1)(1,260.0) + (2)(93.8) + (3)(66.3) = 1,646.5$ Btu. The specific enthalpy of the mixture = $1,646.5/(1+2+3) = 274.4$ Btu per lb of gas mixture.

5. Determine the specific entropy of each gas

Using the *Gas Tables* as in Step 3, $S_H = 15.52$ Btu/(lb)(F); $S_N = 1.558$ Btu/(lb)(F); $S_C = 1.114$ Btu/(lb)(F). The entropy of the mixture = $(1)(12.52) + (2)(1.558) + (3)(1.114) = 18.978$. The specific entropy of the mixture = $18.978/(1+2+3) = 3.163$ Btu/(lb)(F) of the gas mixture.

6. Compute the density of the mixture

For any gas, the total density d_t = sum of the densities of the individual gases; and since the density of a gas = 1/specific volume, $d_t = 1/v_t = 1/v_H + 1/v_N + 1/v_C = 1/10 + 1/5 = 1/3.33 = 0.6$ lb per cu ft of mixture. This checks with Step 1, where $v_t = 1.667$ cu ft/lb, and is based on the principle that all gases occupy the same volume.

Related Calculations: Use this method for any number of gases stored in any type of container—steel, plastic, rubber, canvas, etc.—under any pressure from less than atmospheric to greater than atmospheric at any temperature.

REGENERATIVE-CYCLE GAS-TURBINE ANALYSIS

What is the cycle air rate, lb/kw-hr, for a regenerative gas turbine having a pressure
ratio of 5, an air inlet temperature of 60 F, a compressor discharge temperature of
1500 F, and performance in accordance with Fig. 55? Determine the cycle thermal
efficiency and work ratio. What is the power output of a regenerative gas turbine if the
work input to the compressor is 4,400 hp?

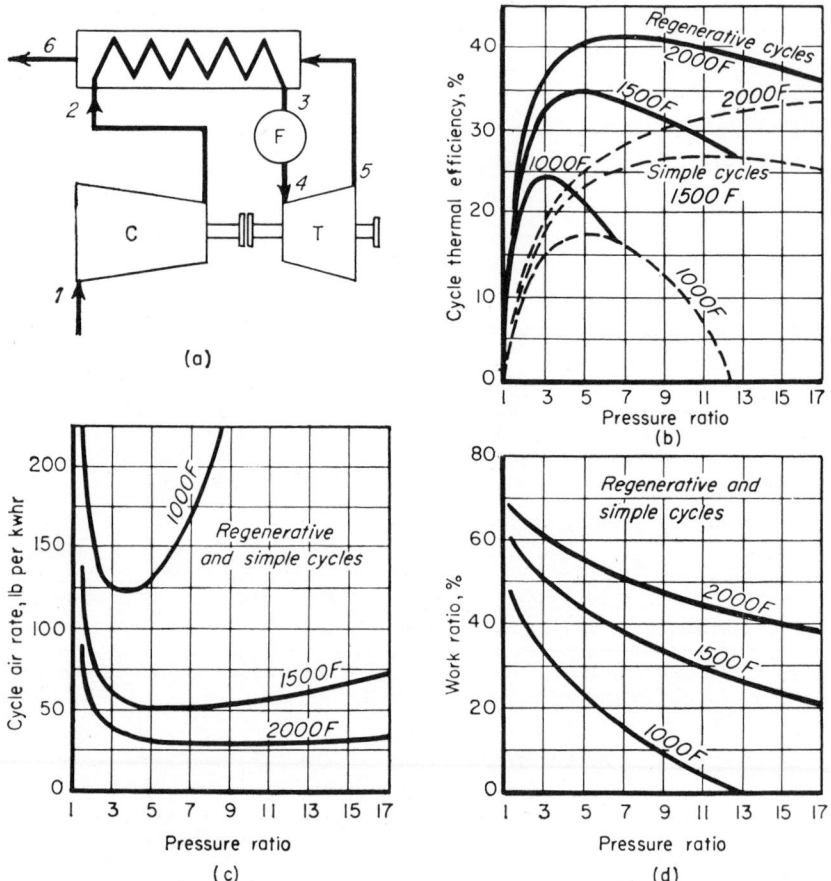

Fig. 55 **(a)** Schematic of regenerative gas turbine; **(b)**, **(c)**, and **(d)** gas-turbine performance
based on a regenerator effectiveness of 70 percent, compressor and turbine efficiency of 85
percent; air inlet = 60 F; no pressure losses.

Calculation Procedure:

1. Determine the cycle rate

Use Fig. 55, entering at the pressure ratio of 5 in Fig. 55c and projecting to the
1500-F curve. At the left, read the cycle air rate as 52 lb/kw-hr.

2. Find the cycle thermal efficiency

Enter Fig. 55b at the pressure ratio of 5 and project vertically to the 1500-F curve.
At left, read the cycle thermal efficiency as 35 percent. Note that this point corresponds
to the maximum efficiency obtainable from this cycle.

3. Find the cycle work ratio

Enter Fig. 55d at the pressure ratio of 5 and project vertically to the 1500-F curve. At the left, read the work ratio as 44 percent.

4. Compute the turbine power output

For any gas turbine, the work ratio, percent, $= 100\,w_c/W_t$, where w_c = work input to the turbine, hp; w_t = work output of the turbine, hp. Substituting, $44 = 100(4,400)/w_t$; $w_t = 100(4,400)/44 = 10,000$ hp.

Related Calculations: Use this general procedure to analyze gas turbines for power-plant, marine, and portable applications. Where the operating conditions are different from those given here, use the manufacturer's engineering data for the turbine under consideration.

Figure 56 shows the effect of turbine-inlet temperature, regenerator effectiveness, and compressor-inlet-air temperature on the performance of a modern gas turbine. Use these curves to analyze the cycles of gas turbines being considered for a particular application if the operating conditions are close to those plotted.

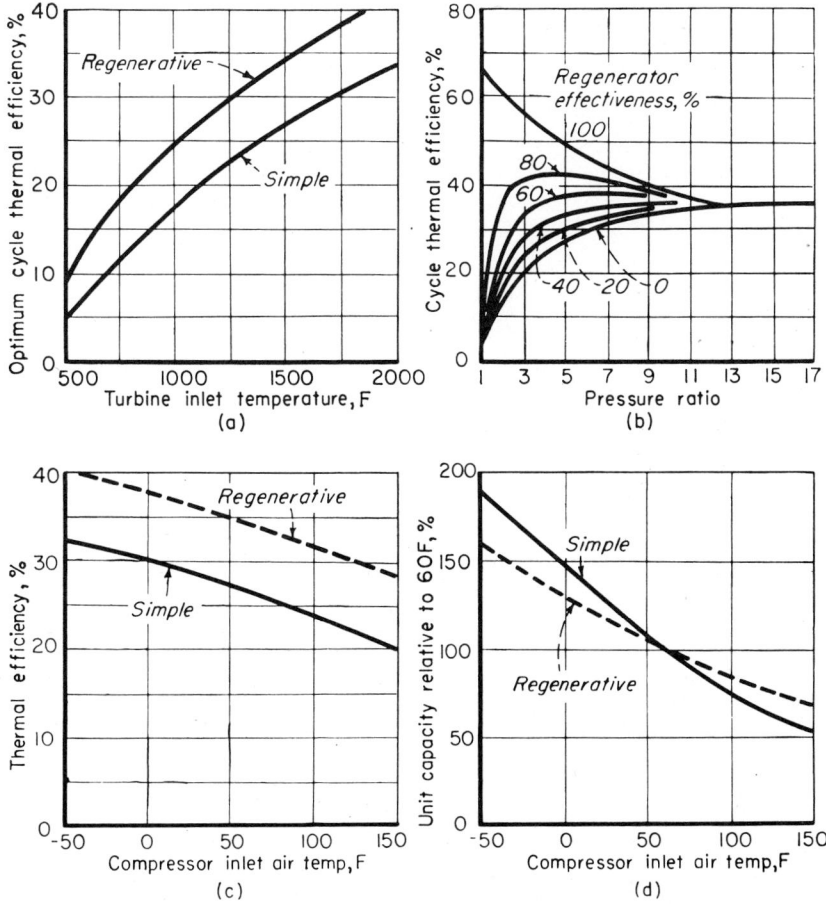

Fig. 56 (a) Effect of turbine-inlet on cycle performance; (b) effect of regenerator effectiveness; (c) effect of compressor-inlet air temperature; (d) effect of inlet-air temperature on turbine-cycle capacity. These curves are based on a turbine and compressor efficiency of 85 percent; a regenerator effectiveness of 70 percent; and a 1500-F inlet gas temperature.

INTERNAL-COMBUSTION ENGINES

REFERENCES—Diesel Engine Manufacturers Association—*Standard Practices for Stationary Diesel Engines*; Lichty—*Internal-Combustion Engines*, McGraw-Hill; Allen—*The Modern Diesel*, Prentice-Hall; Maleev—*Internal-Combustion Engines*, McGraw-Hill; *Diesel Engineering Handbook*, Diesel Publications; Adams—*Elements of Diesel Engineering*, Henley; Severns and Degler —*Steam, Air, and Gas Power*, Wiley; Ricardo—*The High-speed Internal-combustion Engine*, Blackie; Obert—*Internal combustion Engines*, International Textbook; Fors—*Practical Marine Diesel Engineering*, Simmons-Boardman.

DIESEL GENERATING UNIT EFFICIENCY

A 3,000-kw diesel generating unit performs thus: fuel rate, 1.5 bbl of 25° API fuel for a 900-kw-hr output; mechanical efficiency, 82.0 percent; generator efficiency, 92.0 percent. Compute (1) engine fuel rate; (2) engine-generator fuel rate; (3) indicated thermal efficiency; (4) overall thermal efficiency; (5) brake thermal efficiency.

Calculation Procedure:

1. *Compute the engine fuel rate*

The fuel rate of an engine driving a generator is the weight of fuel, lb, used to generate 1 kw-hr at the generator input shaft. Since this engine burns 1.5 bbl of fuel for 900 kw at the generator terminals, the total fuel consumption in gal, is (1.5 bbl)(42 gal/bbl) = 63 gal, at a generator efficiency of 92.0 percent.

To determine the weight of this oil, compute its specific gravity s from $s = 141.5/(131.5 + °API)$, where °API = API gravity of the fuel, °. Hence, $s = 141.5(131.5 + 25) = 0.904$. Since 1 gal of water weighs 8.33 lb at 60 F, 1 gal of this oil weighs (0.904)(8.33) = 7.529 lb. The total weight of fuel used when burning 63 gal is (63 gal)(7.529 lb/gal) = 474.5 lb.

The generator is 92 percent efficient. Hence, the engine actually delivers enough power to generate 900/0.92 = 977 kw-hr at the generator terminals. Thus, the engine fuel rate = 474.5 lb fuel/977 kw-hr = 0.485 lb/kw-hr.

2. *Compute the engine-generator fuel rate*

The engine-generator fuel rate takes these two units into consideration and is the weight of fuel required to generate 1 kw-hr at the generator terminals. Using the fuel-consumption data from Step 1 and the given output of 900 kw, engine-generator fuel rate = 474.5 lb fuel/900 kw-hr output = 0.527 lb/kw-hr.

3. *Compute the indicated thermal efficiency*

Indicated thermal efficiency is the thermal efficiency based on the *indicated* horsepower of the engine. This is the horsepower developed in the engine cylinder. The engine fuel rate, computed in Step 1, is the fuel consumed to produce the brake or shaft horsepower output, after friction losses are deducted. Since the mechanical efficiency of the engine is 82 percent, the fuel required to produce the indicated horsepower is 82 percent of that required for the brake horsepower, or (0.82)(0.485) = 0.398 lb/kw-hr.

The indicated thermal efficiency of an internal-combustion engine driving a generator is $e_i = 3,413/f_i$ HHV, where e_i = indicated thermal efficiency expressed as a decimal; f_i = indicated fuel consumption, lb/kw-hr; HHV = higher heating value of the fuel, Btu/lb.

Compute the HHV for a diesel fuel from HHV = $17,680 + 60 × °API$. For this fuel, HHV = $17,680 + 60(25) = 19,180$ Btu/lb.

With the HHV known, compute the indicated thermal efficiency from $e_i = 3,413/[(0.398)(19,180)] = 0.447$, or 44.7 percent.

4. *Compute the overall thermal efficiency*

The overall thermal efficiency e_o is computed from $e_o = 3{,}413/f_o\text{HHV}$, where $f_o =$ overall fuel consumption, Btu/kw-hr; other symbols as before. Using the engine-generator fuel rate from Step 2, which represents the overall fuel consumption $e_o = 3{,}413/[(0.527)(19{,}180)] = 0.347$, or 34.7 percent.

5. *Compute the brake thermal efficiency*

The engine fuel rate, Step 1, corresponds to the brake fuel rate f_b. Compute the brake thermal efficiency from $e_b = 3{,}413/f_b\text{HHV}$, where $f_b =$ brake fuel rate, Btu/kw-hr; other symbols as before. For this engine-generator set, $e_b = 3{,}413/[(0.485)(19{,}180)] = 0.367$, or 36.7 percent.

Related Calculations: Where the fuel consumption is given or computed in terms of lb/hp-hr, substitute the value of 2,545 Btu/hp-hr in place of the value 3,413 Btu/kw-hr in the numerator of the e_i, e_o, and e_b equations. Compute the indicated, overall, and brake thermal efficiencies as before. Use the same procedure for gas and gasoline engines, except that the higher heating value of the gas or gasoline should be obtained from the supplier or by test.

ENGINE DISPLACEMENT, MEAN EFFECTIVE PRESSURE, AND EFFICIENCY

A 12×18 in. four-cylinder 4-stroke single-acting diesel engine is rated at 200 bhp at 260 rpm. Fuel consumption at rated load is 0.42 lb/bhp-hr. The higher heating value of the fuel is 18,920 Btu/lb. What is the brake mean effective pressure, engine displacement in cfm/bhp, and brake thermal efficiency?

Calculation Procedure:

1. *Compute the brake mean effective pressure*

Compute the brake mean effective pressure for an internal-combustion engine from $bmep = 33{,}000\,bhp_n/LAn$, where $bmep =$ brake mean effective pressure, psi; $bhp_n =$ brake horsepower output delivered per cylinder, hp; $L =$ piston stroke length, ft.; $a =$ piston area, sq in.; $n =$ cycles per min per cylinder $=$ crankshaft rpm for a two-stroke cycle engine, and 0.5 the crankshaft rpm for a 4-stroke cycle engine.

For this engine at its rated bhp, the output per cylinder is 200 bhp/4 cylinders $=$ 50 bhp. Then, $bmep = 33{,}000(50)/[(18/12)(12)^2(\pi/4)(260/2)] = 74.8$ psi. (The factor 12 in the denominator converts the stroke length from in. to ft.)

2. *Compute the engine displacement*

The total engine displacement V_d cu ft is given by $V_d = LAnN$, where $A =$ piston area, sq ft; $N =$ number of cylinders in the engine; other symbols as before. For this engine, $V_d = (18/12)(12/12)^2(\pi/4)(260/2)(4) = 614$ cfm. The displacement is in cfm because the crankshaft speed is in rpm. The factor of 12 in the denominators converts the stroke and area to ft and sq ft, respectively. The displacement per bhp $=$ total displacement, cfm/bhp output of engine $= 614/200 = 3.07$ cfm/bhp.

3. *Compute the brake thermal efficiency*

The brake thermal efficiency e_b of an internal-combustion engine is given by $e_b = 2{,}545/(sfc)(\text{HHV})$, where $sfc =$ specific fuel consumption, lb/bhp-hr; $\text{HHV} =$ higher heating value of fuel, Btu/lb. For this engine, $e_b = 2{,}545/[(0.42)(18{,}920)] = 0.32$, 32.0 percent.

Related Calculations: Use the same procedure for gas and gasoline engines. Obtain the higher heating value of the fuel from the supplier, a tabulation of fuel properties, or by test.

ENGINE MEAN EFFECTIVE PRESSURE AND HORSEPOWER

A 500-hp internal-combustion engine has a brake mean effective pressure of 80 psi at full load. What is the indicated mean effective pressure and friction mean effective pressure if the mechanical efficiency of the engine is 85 percent? What is the indicated horsepower and friction horsepower of the engine?

Calculation Procedure:

1. Determine the indicated mean effective pressure

Indicated mean effective pressure, $imep$ psi, for an internal-combustion engine is found from $imep = bmep/e_m$, where $bmep$ = brake mean effective pressure, psi; e_m = mechanical efficiency, percent, expressed as a decimal. For this engine, $imep = 80/0.85 = 94.1$ psi.

2. Compute the friction mean effective pressure

For an internal-combustion engine, the friction mean effective pressure, $fmep$ psi, is found from: $fmep = imep - bmep$, or $fmep = 94.1 - 80 = 14.1$ psi.

3. Compute the indicated horsepower of the engine

For an internal-combustion engine, the mechanical efficiency $e_m = bhp/ihp$, where ihp = indicated horsepower. Thus, $ihp = bhp/e_m$, or $ihp = 500/0.85 = 588$ ihp.

4. Compute the friction hp of the engine

For an internal-combustion engine, the friction horsepower, $fhp = ihp - bhp$. In this engine, $fhp = 588 - 500 = 88$ fhp.

Related Calculations: Use a similar procedure to determine the *indicated engine efficiency* $e_{ei} = e_i/e$, where e = ideal cycle efficiency; *brake engine efficiency*, $e_{eb} = e_b e$; *combined engine efficiency* or *overall engine thermal efficiency* $e_{eo} = e_o/e$. Note that each of these three efficiencies is an *engine* efficiency and corresponds to an actual thermal efficiency, e_i, e_b, and e_o.

Engine efficiency $e_e = e_t/e$, where e_t = actual *engine* thermal efficiency. Where desired, the respective *actual* indicated brake, or overall, output can be susbstitued for e_i, e_b, and e_o in the numerator of the above equations if the ideal output is substituted in the denominator. The result will be the respective engine efficiency. Output can be expressed in Btu per unit time, or horsepower. Also, e_e = actual *mep*/ideal *mep*, and $e_{ei} = imep$/ideal *mep*; $e_{eb} = bmep$/ideal *mep*; e_{eo} = overall *mep*/ideal *mep*. Further, $e_b = e_m e_i$, and $bmep = e_m imep$. Where the actual heat supplied by the fuel, HHV Btu/lb, is known, compute $e_i e_b$ and e_o by the method given in the previous Calculation Procedure. The above relations apply to any reciprocating internal-combustion engine using any fuel.

SELECTION OF AN INDUSTRIAL INTERNAL-COMBUSTION ENGINE

Select an internal-combustion engine to drive a centrifugal pump handling 2,000 gpm of water at a total head of 350 ft. The pump speed will be 1,750 rpm and it will run continuously. The engine and pump are located at sea level.

Calculation Procedure:

1. Compute the power input to the pump

The power required to pump water is $hp = 8.33GH/33,000e$, where G = water flow, gpm; H = total head on the pump, ft of water; e = pump efficiency, expressed as a decimal. Typical centrifugal pumps have operating efficiencies ranging from 50 to 80 percent, depending on the pump design and condition and liquid handled. Assume that this pump has an efficiency of 70 percent. Then, $hp = 8.33(2,000)(350)/[(33,000)(0.70)] = 252$ hp. Thus, the internal-combustion engine must develop at least 252 hp to drive this pump.

2. Select the internal-combustion engine

Since the engine will run continuously, extreme care must be used in its selection. Refer to a tabulation of engine ratings, such as Table 1. This table shows that a diesel engine that delivers 275 continuous brake horsepower (the nearest tabulated rating equal to or greater than the required input) will be rated at 483 bhp at 1,750 rpm.

The gasoline-engine rating data in Table 1 show that for continuous full load at a given speed, 80 percent of the tabulated power can be used. Thus, at 1,750 rpm, the engine must be rated at 252/0.80 = 315 bhp. A 450-hp unit is the only one shown in Table 1 that would meet the needs. This is too large; refer to another builder's rating table to find an engine rated at 315 to 325 bhp at 1,750 rpm.

TABLE 1   Internal-Combustion Engine Rating Table

Diesel engines

Continuous bhp at given rpm				Rated bhp	No. of cylinders	Cooling*
1,400	1,600	1,750	1,800			
187	214	227	230	300	6	E
230	256	275	280	438	12	R
240	273	295	305	438	12	E

Gasoline engines†

405	430	450	475	595	12	R

*E = heat-exchanger-cooled; R = radiator-cooled.
†Use 80 percent of tabulated power if engine is to run at continuous full load.

The unsuitable capacity range in the gasoline-engine section of Table 1 is a typical situation met when selecting equipment. More time is often spent in finding a suitable unit at an acceptable price than is spent computing the required power output.

Related Calculations: Use this procedure to select any type of reciprocating internal-combustion engine using oil, gasoline, liquefied-petroleum gas, or natural gas for fuel.

ENGINE OUTPUT AT HIGH TEMPERATURES AND HIGH ALTITUDES

An 800-hp diesel engine is operated at an elevation of 10,000 ft above sea level. What is its output at this elevation if the intake air is at 80 F? What will the output at 10,000-ft altitude be if the intake air is at 110 F? What would the output be if this engine were equipped with an exhaust turbine-driven blower?

Calculation Procedure:

1. Compute the engine output at altitude

Diesel engines are rated at sea level at atmospheric temperatures of not more than 90 F. The sealevel rating applies at altitudes up to 1,500 ft. At higher altitudes, a correction factor for elevation must be applied. If the atmospheric temperature is higher than 90 F, a temperature correction must be applied.

Table 2 lists both altitude and temperature correction factors. For an 800-hp engine at 10,000 ft above sea level and 80-F intake air, hp output = (sea-level hp) (altitude correction factor), or output = (800)(0.68) = 544 hp.

2. Compute the engine output at the elevated temperature

When the intake air is at a temperature greater than 90 F, a temperature correction factor must be applied. Then, output = (sea-level hp)(altitude correction factor) (intake-air-temperature correction factor), or output = (800)(0.68)(0.95) = 516 hp, with 110-F intake air.

3. *Compute the output of a supercharged engine*

A different altitude correction factor is used for a supercharged engine, but the same temperature correction factor is applied. Table 2 lists the altitude correction factors for supercharged diesel engines. Thus, for this supercharged engine at 10,000-ft altitude with 80-F intake air, output = (sea-level hp)(altitude correction factor) = (800)(0.74) = 592 hp.

TABLE 2 Correction Factors for Altitude and Temperature

Engine altitude, ft	Engine type	
	Nonsupercharged	Supercharged
2,000	0.980	0.980
3,000	0.935	0.950
4,000	0.895	0.915
5,000	0.855	0.882
6,000	0.820	0.850
7,000	0.780	0.820
8,000	0.745	0.790
9,000	0.712	0.765
10,000	0.680	0.740
12,000	0.612	0.685
14,000	0.550	0.630
16,000	0.494	0.580

Intake temperature, F	Correction factor
90 or less	1.000
95	0.986
100	0.974
105	0.962
110	0.950
115	0.937
120	0.925
125	0.913
130	0.900

At 10,000-ft altitude with 110-F inlet air, output = (sea-level hp)(altitude correction factor)(temperature correction factor) = (800)(0.74)(0.95) = 563 hp.

Related Calculations: Use the same procedure for gasoline, gas, oil, and liquefied-petroleum gas engines. Where altitude correction factors are not available for the type of engine being used, other than a diesel, multiply the engine sea-level brake horsepower by the ratio of the altitude-level atmospheric pressure to the atmospheric pressure at sea level. Table 3 lists the atmospheric pressure at various altitudes.

An engine located below sea level can theoretically develop more power than at sea level because the intake air is denser. However, the greater potential output is generally ignored in engine-selection calculations.

INDICATOR USE ON INTERNAL-COMBUSTION ENGINES

An indicator card taken on an internal-combustion engine cylinder has an area of 5.3 sq in. and a length of 4.95 in. What is the indicated mean effective pressure in this cylinder? What is the indicated horsepower of this four-cycle engine if it has eight

TABLE 3   Atmospheric Pressure at Various Altitudes

Altitude, ft	Pressure, in. Hg
Sea level	29.92
500*	29.38
1,000*	28.86
1,500*	28.33
2,000	27.82
3,000	26.81
4,000	25.84
5,000	24.89
6,000	23.98
8,000	22.22
10,000	20.58
12,000	19.03
14,000	17.57
16,000	16.21

*Considered equivalent to sea level by the Diesel Engine Manufacturers Association if the atmospheric pressure is not less than 28.25 in. Hg.

6-in.-diameter cylinders, an 18-in. stroke, and operates 300 rpm? The indicator spring scale is 100 lb/in.

Calculation Procedure:

1. Compute the indicated mean effective pressure

For any indicator card, $imep$ = (card area, sq in.) (indicator spring scale, lb)/(length of indicator card, in.), where $imep$ = indicated mean effective pressure, psi. Thus, for this engine, $imep = (5.3)(100)/4.95 = 107$ psi.

2. Compute the indicated horsepower

For any reciprocating internal-combustion engine, $ihp = (imep)LAn/33,000$, where ihp = indicated horsepower per cylinder; L = piston stroke length, ft; A = piston area, sq in.; n = number of cycles/min. Thus, for this four-cycle engine where $n = 0.5$ rpm, $ihp = (107)(18/12)(6)^2(\pi/4)(300/2)/33,000 = 20.6$ ihp per cylinder. Since the engine has eight cylinders, total ihp = (8 cylinders) (20.6 ihp per cylinder) = 164.8 ihp.

Related Calculations: Use this procedure for any reciprocating internal-combustion engine using diesel oil, gasoline, kerosene, natural gas, liquefied-petroleum gas, or similar fuel.

ENGINE PISTON SPEED, TORQUE, DISPLACEMENT, AND COMPRESSION RATIO

What is the piston speed of an 18-in.-stroke 300-rpm engine? How much torque will this engine deliver when its output is 800 hp? What is the displacement per cylinder and the total displacement if the engine has eight 12-in.-diameter cylinders? Determine the engine compression ratio if the volume of the combustion chamber is 9 percent of the piston displacement.

Calculation Procedure:

1. Compute the engine piston speed

For any reciprocating internal-combustion engine, piston speed fpm = fpm = $2L(rpm)$, where L = piston stroke length, ft; rpm = crankshaft rotative speed, rpm. Thus, for this engine, piston speed = $2(18/12)(300) = 9,000$ fpm.

2. Determine the engine torque

For any reciprocating internal-combustion engine, $T = 63,000 \ (bhp)/rpm$, where T = torque developed, in.-lb; bhp = engine brake horsepower output; rpm =crankshaft rotative speed, rpm. Or $T = 63,000 \ (800)/300 = 168,000$ in.-lb.

Where a prony brake is used to measure engine torque, apply this relation: $T = (F_b - F_o)r$, where F_b = brake scale force, lb, with engine operating; F_o = brake scale force with engine stopped and brake loose on flywheel; r = brake arm, in., = distance from flywheel center to brake knife edge.

3. Compute the displacement

The displacement per cylinder d_c in.^{3} of any reciprocating internal-combustion engine is $d_c = L_i A_i$ where L_i = piston stroke, in.; A = piston head area, sq in. For this engine, $d_c = (18)(12)^2(\pi/4) = 2,035$ in.^{3} per cylinder.

The total displacement of this eight-cylinder engine is therefore (8 cylinders)(2,035 in.^{3} per cylinder) = 16,280 in.^{3}.

4. Compute the compression ratio

For a reciprocating internal-combustion engine, the compression ratio $r_c = V_b/V_a$, where V_b = cylinder volume at the start of the compression stroke, in.^{3} or cu ft; V_a = combustion-space volume at the end of the compression stroke, in.^{3} or cu ft. When using this relation, both volumes must be expressed in the same units.

In this engine, $V_b = 2,035$ in.^{3}; $V_a = (0.09)(2,035) = 183.15$ in.^{3}. Then, $r_c = 2,035/183.15 = 11.1:1$.

Related Calculations: Use these procedures for any reciprocating internal-combustion engine, regardless of the fuel burned.

INTERNAL-COMBUSTION ENGINE COOLING-WATER REQUIREMENTS

A 1,000-bhp diesel engine has a specific fuel consumption of 0.360 lb of fuel per bhp-hr. Determine the cooling-water flow required if the higher heating value of the fuel is 10,350 Btu/lb. The net heat rejection rates of various parts of the engine are, in percent: jacket water, 11.5; turbocharger, 2.0; lube oil, 3.8; aftercooling, 4.0; exhaust, 34.7; radiation, 7.5; How much 30-psia steam can be generated by the exhaust gas if this is a four-cycle engine? The engine operates at sea level.

Calculation Procedure:

1. Compute the engine heat balance

Determine the amount of heat used to generate 1 bhp-hr from: heat rate, Btu/bhp-hr $= (sfc)(HHV)$, where sfc = specific fuel consumption, lb/bhp-hr; HHV = higher heating value of fuel, Btu/lb. Or, heat rate = $(0.36)(19,350) = 6,967$ Btu/bhp-hr.

Compute the heat balance of the engine by taking the product of the respective heat rejection percentages and the heat rate as follows:

			Btu/bhp-hr
Jacket water	(0.115)(6,967)	=	800
Turbocharger	(0.020)(6,967)	=	139
Lube oil	(0.038)(6,967)	=	264
Aftercooling	(0.040)(6,967)	=	278
Exhaust	(0.347)(6,967)	=	2,420
Radiation	(0.075)(6,967)	=	521
Total heat loss		=	4,422

Then, the power output = $6,967 - 4,422 = 2,545$ Btu/bhp-hr, or $2,545/6,967 = 0.365$, or 36.5 percent. Note that the sum of the heat losses and power generated, expressed in percent, = 100.0.

2. Compute the jacket cooling-water flow rate

The jacket water cools the jackets and the turbocharger. Hence, the heat that must be absorbed by the jacket water, per bhp-hr, is $800 + 139 = 939$ Btu/bhp-hr, using the heat rejection quantities computed in Step 1. When the engine is developing its full rated output of 1,000 bhp, the jacket water must absorb (939 Btu/bhp-hr)(1,000 bhp) = 939,000 Btu/hr.

Apply a safety factor to allow for scaling of the heat-transfer surfaces and other unforeseen difficulties. Most designers use a 10 percent safety factor. Applying this value of the safety factor for this engine, the total jacket-water heat load = $939,000 + (0.10)(939,000) = 1,032,900$ Btu/hr.

Find the required jacket-water flow from $G = H/500\Delta t$, where G = jacket-water flow, gpm, H = heat absorbed by jacket water, Btu/hr; Δt = temperature rise of the water during passage through the jackets, F. The usual temperature rise of the jacket water during passage through a diesel engine is 10 to 20 F. Using 10 F for this engine, $G = 1,032,900/[(500)(10)] = 206.58$ gpm; say 207 gpm.

3. Determine the water quantity for radiator cooling

In the usual radiator cooling system for large engines, a portion of the cooling water is passed through a horizontal or vertical radiator. The remaining water is recirculated, after being tempered by the cooled water. Thus, the radiator must dissipate the jacket, turbocharger, and lube-oil cooler heat.

The lube oil gives off 264 Btu/bhp-hr. With a 10 percent safety factor, the total heat flow is $264 + (0.10)(264) = 290.4$ Btu/bhp-hr. At the rated output of 1,000 bhp, the lube-oil heat load = (290.4 Btu/bhp-hr)(1,000 bhp) = 290,400 Btu/hr. Hence, the total heat load on the radiator = jacket + lube-oil heat load = $1,032,900 + 290,400 = 1,323,300$ Btu/hr.

Radiators (also called fan coolers) serving large internal-combustion engines are usually rated for a 35-F temperature reduction of the water. To remove 1,323,300 Btu/hr with a 35-F temperature decrease will require a flow of $G = H/500\,\Delta t = 1,323,300/[(500)(35)] = 76.1$ gpm.

4. Determine the aftercooler cooling-water quantity

The aftercooler must dissipate 278 Btu/bhp-hr. At an output of 1,000 bhp, the heat load = (278 Btu/bhp-hr)(1,000 bhp) = 278,000 Btu/hr. In general, designers do not use a factor of safety for the aftercooler because there is less chance of fouling or other difficulties.

With a 5-F temperature rise of the cooling water during passage through the aftercooler, the quantity of water required $G = H/500\Delta t = 278,000/[(500)(5)] = 111$ gpm.

5. Compute the quantity of steam generated by the exhaust

Find the heat available in the exhaust using $H_e = Wc\Delta t_e$, where H_e = heat available in the exhaust, Btu/hr; W = exhaust-gas flow, lb/hr; c = specific heat of the exhaust gas = 0.252 Btu/(lb)(F); Δt_e = exhaust-gas temperature at the boiler inlet, F $-$ exhaust-gas temperature at the boiler outlet, F.

The exhaust-gas flow from a four-cycle turbocharged diesel is about 12.5 lb/bhp-hr. At full load this engine will exhaust (12.5 lb/bhp-hr)(1,000 bhp) = 12,500 lb/hr.

The temperature of the exhaust gas will be about 750 F at the boiler inlet, whereas the temperature at the boiler outlet is generally held at 75 F higher than the steam temperature to prevent condensation of the exhaust gas. Steam at 30 psia has a temperature of 250.33 F. Thus, the exhaust-gas outlet temperature from the boiler will be $250.33 + 75 = 325.33$ F; say 325 F. Then, $H_e = (12,500)(0.252)(750 - 325) = 1,375,000$ Btu/hr.

At 30 psia, the enthalpy of vaporization of steam is 945.3 Btu/lb, found in the steam tables. Thus, the exhaust heat can generate $1,375,000/945.3 = 1,415$ lb/hr, if the boiler is 100 percent efficient. With a boiler efficiency of 85 percent, the steam generated = $(1,415 \text{ lb/hr})(0.85) = 1,220$ lb/hr, or $(1,220 \text{ lb/hr})/1,000 \text{ bhp} = 1.22$ lb/bhp-hr.

Related Calculations: Use this procedure for any reciprocating internal-combustion

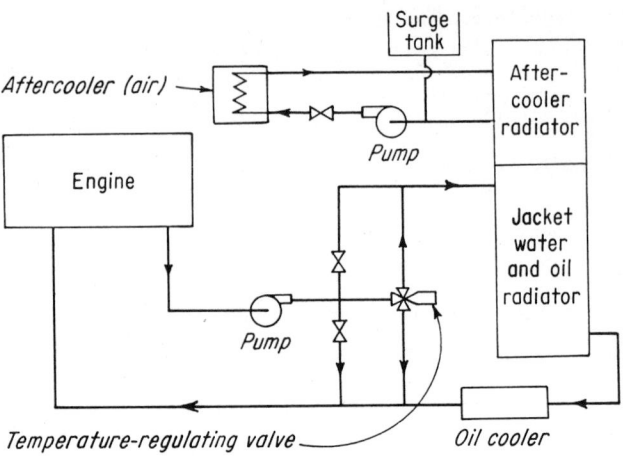

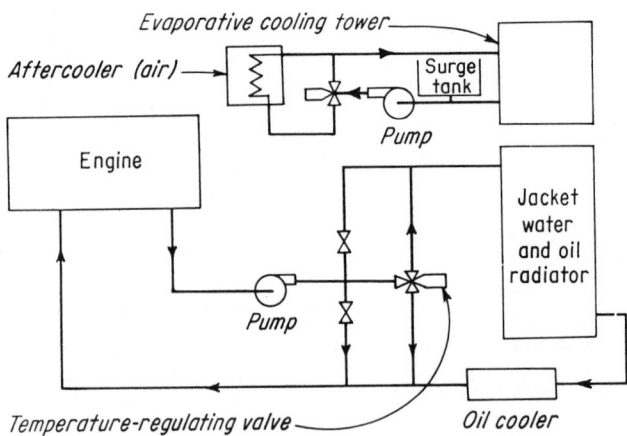

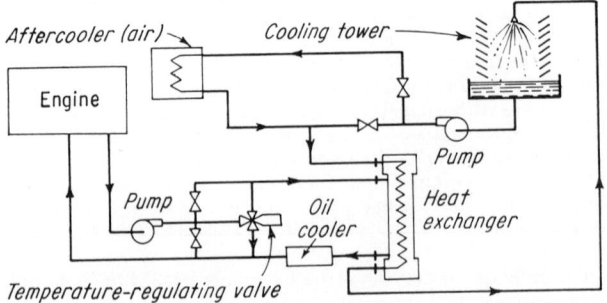

Fig. 1 Internal-combustion engine cooling systems: (a) radiator-type; (b) evaporative cooling tower; (c) cooling tower. (*Power.*)

engine burning gasoline, kerosene, natural gas, liquified-petroleum gas, or similar fuel. Figure 1 shows typical arrangements for a number of internal-combustion engine cooling systems.

When ethylene glycol or another antifreeze solution is used in the cooling system, alter the denominator of the flow equation to reflect the change in specific gravity and specific heat of the antifreeze solution, as compared with water. Thus, with a 50 percent glycol-50 percent water mixture, the flow equation in Step 2 becomes $G = H/436\,\Delta t$. With other solutions, the numerical factor in the denominator will change. This factor = (weight of liquid, lb/gal)(60 min/hr), and the factor converts a flow rate of lb/hr to gpm when divided into the lb/hr flow rate. Slant diagrams, Fig. 2, are often useful for heat-exchanger analysis.

	Engine jackets			Oil cooler			Turbo aftercooler		
Bhp	Btu/hr	Q_{JW} gpm	Q_R gpm	Btu/hr	Q_R gpm	Q_o gpm	Btu/hr	Q_A lb/sec	Q_{AW} gpm
1000	1,032,000	207	–	290,000	75	140	278,000	3.3	110
	1,322,000	–	75						

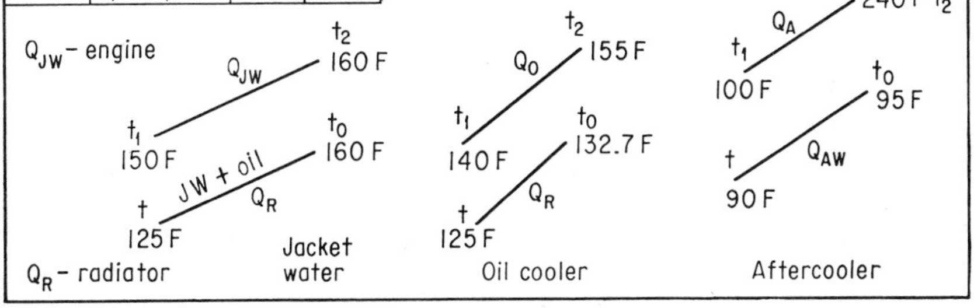

Q_{JW} – engine

Q_{JW} t_2 160 F

t_1 150 F t_o 160 F

JW + oil Q_R

t 125 F   Jacket water

Q_R – radiator

Q_o t_2 155 F

t_1 140 F t_o 132.7 F

Q_R

t 125 F

Oil cooler

Q_A 240 F t_2

t_1 100 F t_o 95 F

t 90 F Q_{AW}

Aftercooler

Fig. 2 Slant diagrams for internal-combustion engine heat exchangers. *(Power.)*

Two-cycle engines may have a larger exhaust-gas flow than four-cycle engines because of the scavenging air. However, the exhaust temperature will usually be 50 to 100 F lower, reducing the quantity of steam generated.

Where a dry exhaust manifold is used on an engine, the heat rejection to the cooling system is reduced by about 7.5 percent. Heat rejected to the aftercooler cooling water is about 3.5 percent of the total heat input to the engine. About 2.5 percent of the total heat input to the engine is rejected by the turbocharger jacket.

The jacket cooling water absorbs 11 to 14 percent of the total heat supplied. From 3 to 6 percent of the total heat supplied to the engine is rejected in the oil cooler.

The total heat supplied to an engine = (engine output, bhp)(heat rate, Btu/bhp-hr). A jacket-water flow rate of 0.25 to 0.60 gpm/bhp is usually recommended. The normal jacket-water temperature rise is 10 F; with a jacket-water outlet temperature of 180 F, or higher, the temperature rise of the jacket water is usually held to 7 F, or less.

To keep the cooling-water system pressure loss within reasonable limits, some designers recommend a pipe velocity, fps, equal to the nominal pipe size used in the system, or 2 fps for 2-in. pipe; 3 fps for 3-in. pipe, etc. The maximum recommended velocity is 10 fps for 10-in., and larger, pipes. Compute the actual pipe diameter d in. from $d = (G/2.5\,v)^{0.5}$, where G = cooling-water flow, gpm; v = water velocity fps.

Air needed for a four-cycle high-output turbocharged diesel engine is about 3.5 cfm/bhp; 4.5 cfm/bhp for two-cycle engines. Exhaust-gas flow is about 8.4 cfm/bhp for a four-cycle diesel engine; 13 cfm/bhp for two-cycle engines. Air velocity in the turbocharger blower piping should not exceed 3,300 fpm; gas velocity in the exhaust system should not exceed 6,000 fpm. The exhaust-gas temperature should not be reduced below 275 F, to prevent condensation.

The method presented here is the work of W. M. Kauffman, reported in *Power*.

DESIGN OF A VENT SYSTEM FOR AN ENGINE ROOM

A radiator-cooled 60-kw internal-combustion engine generating set operates in an area where the maximum summer ambient temperature of the inlet air is 100 F. How much air does this engine need for combustion and for the radiator? What is the maximum permissible temperature rise of the room air? How much heat is radiated by the engine-alternator set if the exhaust pipe is 25 ft long? What capacity exhaust fan is needed for this engine room if the engine room has two windows with an area of 30 sq ft each, and the average height between the air inlet and outlet is 5 ft? Determine the rate of heat dissipation by the windows. The engine is located at sea level.

Calculation Procedure:

1. Determine engine air-volume needs

Table 4 shows typical air-volume needs for internal-combustion engines installed indoors. Thus, a 60-kw set requires 390 cfm for combustion and 6,000 cfm for the radiator. Note that in the smaller ratings, the combustion air needed is 6.5 cfm/kw and the radiator air requirement is 150 cfm/kw.

2. Determine maximum permissible air temperature rise

Table 4 also shows that with an ambient temperature of 95 to 105 F, the maximum permissible room temperature rise is 15 F. When determining this value, be certain to use the highest inlet air temperature expected in the engine locality.

TABLE 4   Total Air Volume Needs*

Set kw	Cfm for combustion	Cfm for radiator
20	130	3,000
30	195	5,000
40	260	5,500
60	390	6,000

Maximum room temperature rise

Maximum ambient temperature of inlet air, F	Room air rise, F
90	20
95–105	15
110–120	10

Power.

3. Determine the heat radiated by the engine

Table 5 shows the heat radiated by typical internal-combustion engine generating sets. Thus, a 60-kw radiator-and fan-cooled set radiates 2,625 Btu/min, when the engine is fitted with a 25-ft-long exhaust pipe and a silencer.

4. Compute the airflow produced by the windows

The two windows can be used to ventilate the engine room. One window will serve as the air inlet; the other as the air outlet. The area of the air outlet must at least equal the air-inlet area. Airflow will be produced by the stack effect resulting from the temperature difference between the inlet and outlet air.

TABLE 5   Heat Radiated from Typical I-C Units*

Alternator, kw	Cooling by radiator and fan				Cooling by radiator, fan, and city water			
	20	30	40	60	20	30	40	60
Engine-alternator set, silencer and 25 ft of exhaust pipe, Btu	1,020	1,471	1,830	2,625	960	1,386	1,701	2,500
Exhaust pipe beyond silencer:								
Length 5 ft	18	17	24	35	16	18	20	22
Length 10 ft	32	33	45	65	29	31	39	40
Length 15 ft	45	48	65	89	41	45	57	55

Power.

The airflow C cfm resulting from the stack effect is $C = 9.4A(h\Delta t_a)^{0.5}$, where $A =$ free area of the air inlet, sq ft; $h =$ height from the middle of the air-inlet opening to the middle of the air-outlet opening, ft; $\Delta t_a =$ difference between the average indoor air temperature at point H and the temperature of the incoming air, F. In this plant, the maximum permissible air temperature rise is 15 F, from Step 2. With a 100-F outdoor temperature, the maximum indoor temperature would be $100 + 15 = 115$ F. Assume that the difference between the temperature of the incoming and outgoing air is 15 F. Then,$C = 9.4(30)(5 \times 15)^{0.5} = 2,445$ cfm.

5. Compute the cooling airflow required

This 60-kw internal-combustion engine generating set radiates 2,625 Btu/hr, Step 3. Compute the cooling airflow required from $C = HK/\Delta t_a$; where $C =$ cooling airflow required, cfm; $H =$ heat radiated by the engine, Btu/min; $K =$ constant from Table 6; other symbols as before. Thus, for this engine with a fan discharge temperature of 111 to 120 F, Table 6, $K = 60$; $\Delta t_a = 15$ F from Step 4. Then, $C = (2,625)(60)/15 = 10,500$ cfm.

The windows provide 2,445 cfm, Step 4, and the engine radiator 6,000 cfm, Step 1, or a total of $2,445 + 6,000 = 8,445$ cfm. Thus, $10,500 - 8,445 = 2,055$ cfm must be removed from the room. The usual method employed to remove the air is an exhaust fan. An exhaust fan with a capacity of 2,100 cfm would be suitable for this engine room.

TABLE 6   Range of Discharge Temperature*

Room Fan Discharge Temperature Range, F	K
80–89	57
90–99	58
100–110	59
111–120	60
121–130	61

Wind to water gage

Wind Velocity, mph	Inlet Pressure Water Gage
60	1.75
30	0.43
60	0.11

* *Power.*

Related Calculations: Use this procedure for engines burning any type of fuel – diesel, gasoline, kerosene, or gas – in any type of enclosed room at sea level or elevations up to 1,000 ft. Where windows or the fan outlet are fitted with louvers, screens, or intake filters, be certain to compute the net free area of the opening. When the radiator fan requires more air than is needed for cooling the room, an exhaust fan is unnecessary.

Be certain to select an exhaust fan with a sufficient discharge pressure to overcome the resistance of exhaust ducts and outlet louvers, if used. A propeller fan is usually chosen for exhaust service. In areas having high wind velocity, an axial-flow fan may be needed to overcome the pressure produced by the wind on the fan outlet.

Table 6 shows the pressure developed by various wind velocities. When the engine is located above sea level, use the multiplying factor in Table 7 to correct the computed air quantities for the lower air density.

TABLE 7   Air Density at Various Elevations*

Elevation above sea level, ft	Multiplying factor, A	Approximate air density percent compared with sea level for same temperature
1,000	1.037	96.5
2,000	1.076	93.0
3,000	1.116	89.6
4,000	1.158	86.4
5,000	1.202	83.2
6,000	1.247	80.2
7,000	1.296	77.2
10,000	1.454	68.8

Power.

An engine radiates 2 to 5 percent of its total heat input. The total heat input = (engine output, bhp) (heat rate, Btu/bhp-hr). Provide 12 to 20 air changes per hr for the engine room. The most effective ventilators are power-driven exhaust fans or roof ventilators. Where the heat load is high, 100 air changes per hr may be provided. Auxiliary-equipment rooms require 10 air changes per hr. Windows, louvers, or power-driven fans are used. A four-cycle engine requires 3 to 3.5 cfm of air per bhp; a two-cycle engine, 4 to 5 cfm per bhp.

The method presented here is the work of John P. Callaghan, reported in *Power.*

DESIGN OF A BYPASS COOLING SYSTEM FOR AN ENGINE

The internal-combustion engine in Fig. 3 is rated at 402 hp at 514 rpm and dissipates 3,500 Btu/bhp-hr at full load to the cooling water from the power cylinders and water-cooled exhaust manifold. Determine the required cooling-water flow rate if there is a 10-F temperature rise during passage of the water through the engine. Size the piping for the cooling system using the head-loss data in Fig. 4, and the pump characteristic curve, Fig. 5. Choose a surge tank of suitable capacity. Determine the net positive suction head requirements for this engine. The total length of straight piping in the cooling system is 45 ft. The engine is located 500 ft above sea level.

Calculation Procedure:

1. Compute the cooling-water quantity required

The cooling-water quantity required is $G = H/500 \, \Delta t$, where G = cooling-water flow, gpm; H = heat absorbed by the jacket water, Btu/hr = (maximum engine hp) (heat

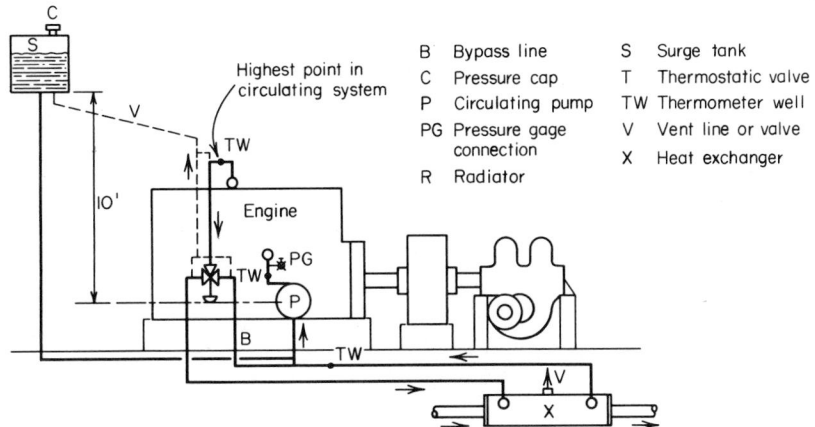

Fig. 3 Engine cooling system hookup. *(Mechanical Engineering.)*

dissipated, Btu/bhp-hr); Δt = temperature rise of the water during passage through the engine, F. Thus, for this engine, $G = (402)(3,500)/[500(10)] = 281$ gpm.

2. Choose the cooling-system valve and pipe size

Obtain the friction head-loss data for the engine, the heat exchanger, and the three-way valve from the manufacturers of the respective items. Most manufacturers have

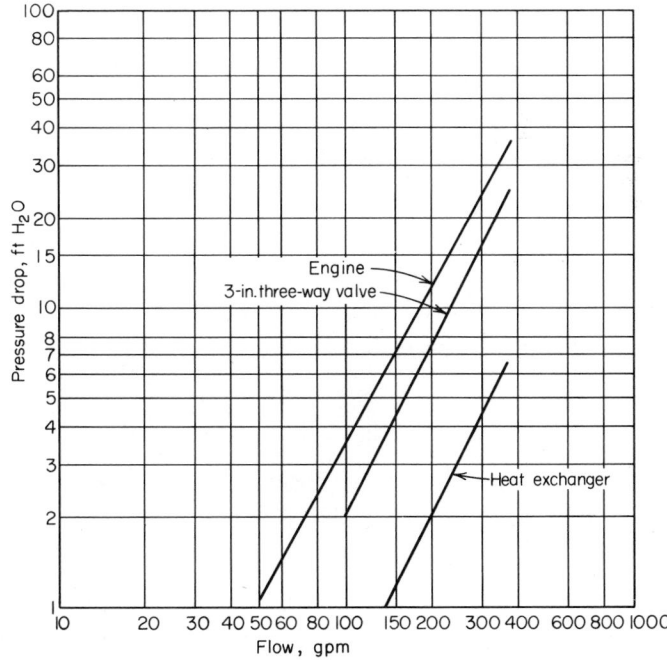

Fig. 4 Head-loss data for engine cooling-system components. *(Mechanical Engineering.)*

curves on tables available for easy use. Plot the head losses, as shown in Fig. 4, for the engine and heat exchanger.

Before the three-way valve head loss can be plotted, a valve size must be chosen. Refer to a three-way valve capacity tabulation to determine a suitable valve size to handle a flow of 281 gpm. One such tabulation recommends a 3-in. valve for a flow of 281 gpm. Obtain the head-loss data for the valve and plot it as shown in Fig. 4.

Next, assume a size for the cooling-water piping. Experience shows that a water velocity of 300 to 600 fpm is satisfactory for internal-combustion engine cooling systems. Using the Hydraulic Institute *Pipe Friction Manual* or Cameron *Hydraulic Data*, enter at 280 gpm, the approximate flow, and choose a pipe size to give a velocity of 400 to 500 fpm — i.e., midway in the recommended range.

Alternatively, compute the approximate pipe diameter, in., from $d = 4.95(\text{gpm}/\text{velocity, fpm})^{0.5}$. With a velocity of 450 fpm, $d = 4.95(281/450)^{0.5} = 3.92$ in.; say 4 in. The *Pipe Friction Manual* shows that the water velocity will be 7.06 fps, or 423.6 fpm in a 4-in. schedule 40 pipe. This is acceptable. Using a 3½-in. pipe would increase the cost because the size is not readily available from pipe suppliers. A 3-in. pipe would give a velocity of 720 fpm, which is too high.

3. Compute the piping-system head loss

Examine Fig. 3, which shows the cooling system piping layout. Three flow conditions are possible: (*a*) all the jacket water passes through the heat exchanger, (*b*) a portion of the jacket water passes through the heat exchanger, and (*c*) none of the jacket water passes through the heat exchanger — instead, all the water passes through the bypass circuit. The greatest head loss usually occurs when the largest amount of water passes through the longest circuit (or flow condition *a*). Compute the head loss for this situation first.

Using the method given in the piping section of this Handbook, compute the equivalent length of the cooling-system fitting and piping, as shown in Table 8. Once the equivalent length of the pipe and fittings is known, compute the head loss in the piping system using the method given in the piping section of this Handbook with a Hazen-Williams constant of $C = 130$, and a rounded-off flow rate of 300 gpm. Summarize the results as shown in Table 8.

The total head loss is produced by the water flow through the piping, fittings, engine, three-way valve, and heat exchanger. Find the head loss for the last components in Fig. 4 for a flow of 300 gpm. List the losses in Table 8 and find the sum of all the losses. Thus, the total circuit head loss is 57.61 ft of water.

Compute the head loss for 0, 0.2, 0.4, 0.6, and 0.8 load on the engine, using the same procedure as in Steps 1, 2, and 3, above. Plot on the pump characteristic curve, Fig. 5, the system head loss for each load. Draw a curve *A* through the points obtained, Fig. 5.

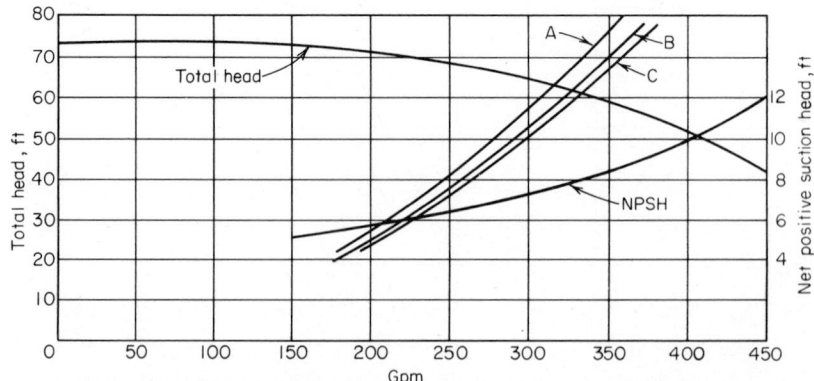

Fig. 5 Pump and system characteristics for engine cooling system. (*Mechanical Engineering.*)

Compute the system head loss for condition b with half the jacket water (150 gpm) passing through the heat exchanger and half (150 gpm) through the bypass circuit. Make the same calculation for 0, 0.2, 0.4, 0.6, and 0.8 load on the engine. Plot the result as curve B, Fig. 5.

Perform a similar calculation for condition c—full flow through the bypass circuit. Plot the results as curve C, Fig. 5.

TABLE 8 Sample Calculation for Full Flow through Cooling Circuit

(Fittings and Piping in Circuit)

Fitting or pipe	Number in circuit	Equivalent length of straight pipe, ft
3-in. elbow	1	5.5
3 × 4 reducer	4	7.2
4-in. elbow	7	50.4
4-in. tee	1	23.0
3-in. pipe.........................	...	0.67
4-in. pipe.........................	...	45.0
Total equivalent length of pipe:		
3-in. pipe, standard weight........	...	13.37
4-in. pipe, standard weight........	...	118.4

Head loss calculation: Calculation for a flow rate of 300 gpm through circuit:

Using the Hazen-Williams friction-loss equation with a C factor of 130 (surface roughness constant), with 300 gpm flowing through the pipe, the head loss in ft per 100 ft of pipe is 21.1 ft and 5.64 ft for the 3-in. and 4-in. pipes, respectively. Thus head loss in piping is:†

$$3 \text{ in. } \frac{21.1}{100} \times 13.37 = 2.83 \text{ ft}$$

$$4 \text{ in. } \frac{5.64}{100} \times 118.4 = 6.68 \text{ ft}$$

From Fig. 5 head loss is:

Through engine..	26.00 ft
Through 3-in. three-way valve...........................	17.50
Through heat exchanger	4.6
Total circuit head loss	57.61 ft

Mechanical Engineering.
†Shaw and Loomis, "Cameron Hydraulic Data Book," Ingersoll-Rand Company, 12th ed., 1951, p. 27.

4. Compute the actual cooling-water flow rate

Find the points of intersection of the pump total-head curve and the three system head-loss curves A, B, and C, Fig. 5. These intersections occur at 314, 325, and 330 gpm, respectively.

The initial design assumed a 10-F temperature rise through the engine with a water flow rate of 281 gpm. Rearranging the equation in Step 1, $\Delta t = H/400G$. Substituting the flow rate for condition a gives an actual temperature rise of $\Delta t = (402)(3,500)/[(500)(314)] = 8.97$ F. If a 180-F-rated thermostatic element is used in the three-way valve, holding the outlet temperature t_o to 180 F, the inlet temperature t_i will be $\Delta t = t_o - t_i = 8.97$; $180 - t_i = 8.97$; $t_i = 171.03$ F.

5. Determine the required surge-tank capacity

The surge tank in a cooling system provides storage space for the increase in volume of the coolant caused by thermal expansion. Compute this expansion from $E = 62.4g\Delta V$, where E = expansion, gal; g = number of gal required to fill the cooling system; ΔV = specific volume, cu ft/lb, of the coolant at the operating temperature − specific volume of the coolant, cu ft/lb, at the filling temperature.

The cooling system for this engine must have a total capacity of 281 gal, Step 1. Round this off to 300 gal for design purposes. The system operating temperature is 180 F and the filling temperature is usually 60 F. Using the steam tables to find the specific volume of the water at these temperatures, $E = 62.4(300)(0.01651 - 0.01604) = 8.8$ gal.

Usual design practice is to provide two to three times the required expansion volume. Thus, a 25-gal tank (nearly three times the required capacity) would be chosen. The extra volume provides for excess cooling water that might be needed to make up water lost through minor leaks in the system.

Locate the surge tank so that it is the highest point in the cooling system. Some engineers recommend that the bottom of the surge tank be at least 10 ft above the pump centerline and connected as close as possible to the pump intake. A 1½- or 2-in. pipe is usually large enough for connecting the surge tank to the system. The line should be sized so that the head loss of the vented fluid flowing back to the pump suction will be negligible.

6. Determine the pump net positive suction head

The pump characteristic curve, Fig. 5, shows the net positive suction head NPSH required by this pump. As the pump discharge rate increases, so does the NPSH. This is typical of a centrifugal pump.

The greatest flow, 330 gpm, occurs in this system when all the coolant is diverted through the bypass circuit, Figs. 4 and 5. At a 330-gpm flow rate through the system, the required NPSH for this pump is 8 ft, Fig. 5. This value is found at the intersection of the 330-gpm ordinate and the NPSH curve.

Compute the existing NPSH, ft, from NPSH $= (H_s - H_f) + 2.31(P_s - P_v)/s$, where H_s = height of minimum surge-tank liquid level above the pump centerline, ft; H_f = friction loss in the suction line from the surge-tank connection to the pump inlet flange, ft of liquid; P_s = pressure in surge tank, or atmospheric pressure at the elevation of the installation, psia; P_v = vapor pressure of the coolant at the pumping temperature, psia; s = specific gravity of the coolant at the pumping temperature.

7. Determine the operating temperature with a closed surge tank

A pressure cap on the surge tank, or a radiator, will permit operation at temperatures above the atmospheric boiling point of the coolant. At a 500-ft elevation, water boils at 210 F. Thus, without a closed surge tank fitted with a pressure cap, the maximum operating temperature of a water-cooled system would be about 200 F.

If a 7-psig pressure cap were used at the 500-ft elevation, then the pressure in the vapor space of the surge tank could rise to $P_s = 14.4 + 7.0 = 21.4$ psia. The steam tables show that water at this pressure boils at 232 F. Checking the NPSH at this pressure shows that NPSH $= (10 - 1.02) + 2.31(21.4 - 21.4)/0.0954 = 8.98$ ft. This is close to the required 8-ft head. However, the engine could be safely operated at a slightly lower temperature − say 225 F.

8. Compute the pressure at the pump suction flange

The pressure at the pump suction flange P psig $= 0.433s(H_s - H_f) = (0.433)(0.974)(10.00 - 1.02) = 3.79$ psig.

A positive pressure at the pump suction is needed to prevent the entry of air along the shaft. To further ensure against air entry, a mechanical seal can be used on the pump shaft in place of packing.

Related Calculations: Use this general procedure when designing the cooling system for any type of reciprocating internal-combustion engine − gasoline, diesel, gas, etc. Where a coolant other than water is used, follow the same procedure but change the value of the constant in the denominator of the equation of Step 1. Thus, for a 50 percent glycol–50 percent water mixture, the constant = 436, instead of 500.

The method presented here is the work of Duane E. Marquis, reported in *Mechanical Engineering.*

HOT-WATER HEAT-RECOVERY SYSTEM ANALYSIS

An internal-combustion engine fitted with a heat-recovery silencer and a jacket-water cooler is rated at 1,000 bhp. It exhausts 13.0 lb of exhaust gas per bhp-hr at 700 F. To what temperature can hot water be heated when 500 gpm of jacket water is circulated through the hookup in Fig. 6 and 100 gpm of 60-F water is heated? The jacket water enters the engine at 170 F and leaves at 180 F.

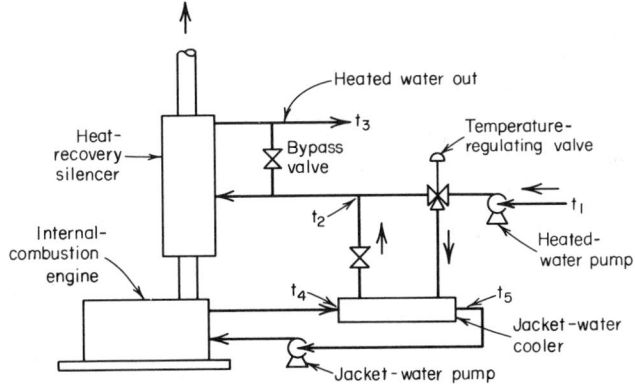

Fig. 6 Internal-combustion engine cooling system.

Calculation Procedure:

1. Compute the exhaust heat recovered

Find the exhaust-heat recovered from $H_e = Wc\Delta t_e$, where the symbols are the same as in the previous Calculation Procedures. Since the final temperature of the exhaust gas is not given, a value must be assumed. Temperatures below 275 F are undesirable because condensation of corrosive vapors in the silencer may occur. Assume that the exhaust-gas outlet temperature from the heat-recovery silencer is 300 F. Then, $H_e = (1,000)(13)(0.252)(700-300) = 1,310,000$ Btu/hr.

2. Compute the heated-water outlet temperature from the cooler

Using the temperature notation in Fig. 6, the heated-water outlet temperature from the jacket-water cooler is $t_z = (w_z/w_1)(t_4 - t_5) + t_1$, where w_1 = heated-water flow, lb/hr; w_z = jacket-water flow, lb/hr; the other symbols are indicated in Fig. 6. To convert gpm of water flow to lb/hr, multiply by 500. Thus, $w_1 = (100 \text{ gpm})(500) = 50,000$ lb/hr, and $w_z = (500 \text{ gpm})(500) = 250,000$ lb/hr. Then, $t_z = (250,000/50,000)(180-170) + 60 = 110$ F.

3. Compute the heated-water outlet temperature from the silencer

The silencer outlet temperature $t_3 = (H_e/w_1) + t_z$, or $t_3 = (1,310,000/50,000) + 110 = 136.2$ F.

Related Calculations: Use this method for any type of engine—diesel, gasoline, or gas—burning any type of fuel. Where desired, a simple heat balance can be set up between the heat-releasing and heat-absorbing sides of the system instead of using the equations given here. However, the equations are faster and more direct.

Standard Handbook of Engineering Calculations

DIESEL FUEL STORAGE CAPACITY AND COST

A diesel power plant will have six 1,000-hp engines and three 600-hp engines. The annual load factor is 85 percent and is nearly uniform throughout the year. What capacity day tanks should be used for these engines? If fuel is delivered every 7 days, what storage capacity is required? Two fuel supplies are available; a 24° API fuel at $0.0825/gal and a 28° API fuel at $0.0910/gal. Which is the better buy?

Calculation Procedure:

1. Compute the engine fuel consumption

Assume, or obtain from the engine manufacturer, the specific fuel consumption of the engine. Typical modern diesel engines have a full-load heat rate of 6,900 to 7,500 Btu/ bhp-hr, or about 0.35 lb of fuel per bhp-hr. Using this value of fuel consumption for the nine engines in this plant, the hourly fuel consumption at 85 percent load factor will be (6 engines)(1,000 hp)(0.35)(0.85) + (3 engines)(600 hp)(0.35)(0.85) = 2,320 lb/hr.

Convert this consumption rate to gph by finding the specific gravity of the diesel oil. The specific gravity $s = 141.5/(131.5 + °API)$. For the 24° API oil, $s = 141.5/(131.5 + 24)$ = 0.910. Since water at 60 F weighs 8.33 lb/gal, the weight of this oil is (0.910)(8.33) = 7.578 lb/gal. For the 28° API oil, $s = 141.5/(131.5 + 28) = 0.887$, and the weight of this oil is (0.887)(8.33) = 7.387 lb/gal. Using the lighter oil, since this will give a larger gph consumption, fuel rate = (2,320 lb/hr)/(7.387 lb/gal) = 315 gph.

The daily fuel consumption is then (24 hr/day)(315 gph) = 7,550 gal/day. In seven days the engines will use (7 days)(7,550 gal/day) = 52,900 gal; say 53,000 gal.

2. Select the tank capacity

The actual fuel consumption is 53,000 gal in 7 days. If fuel is delivered exactly on time every 7 days, a fuel-tank capacity of 53,000 gal would be adequate. However, bad weather, transit failures, strikes, or other unpredictable incidents may delay delivery. Therefore, added capacity must be provided to prevent engine stoppage because of an inadequate fuel supply.

Where sufficient space is available, and local regulations do not restrict the storage capacity installed, use double the required capacity. The reason for this is that the additional storage capacity is relatively cheap compared with the advantages gained. Where space or storage capacity is restricted, use 1½ times the required capacity.

Assuming double capacity is used in this plant, the total storage capacity will be (2)(53,000) = 106,000 gal. At least two tanks should be used, to permit cleaning of one without interrupting engine operation.

Consult the National Board of Fire Underwriters bulletin *Storage Tanks for Flammable Liquids* for rules governing tank materials, location, spacing, and fire-protection devices. Refer to a tank capacity table to determine the required tank diameter and length or height depending on whether the tank is horizontal or vertical. Thus, the Buffalo Tank Corporation *Handbook* shows that a 16.5-ft-diameter 33.5-ft-long horizontal tank will hold 53,600 gal when full. Two tanks of this size would provide the desired capacity. Alternatively, a 35-ft-diameter 7.5-ft-high vertical tank will hold 54,000 gal when full. Two tanks of this size would provide the desired capacity.

Where a tank capacity table is not available, compute the capacity of a cylindrical tank from, capacity, gal, $= 5.87 D^2 L$, where D = tank diameter, ft; L = tank length or height, ft. Consult the NBFU or the tank manufacturer for the required tank wall thickness and vent size.

3. Select the day-tank capacity

Day tanks supply filtered fuel to an engine. The day tank is usually located in the engine room and holds enough fuel for a 4- to 8-hr operation of an engine at full load. Local laws, insurance requirements, or the NBFU may limit the quantity of oil that can be stored in the engine room or a day tank. One day tank is usually used for each engine.

Assume that a 4-hr supply will be suitable for each engine. Then, the day tank capacity for a 1,000-hp engine = (1,000 hp)(0.35 lb fuel per bhp-hr)(4 hr) = 1,400 lb, or 1,400/

7.387 = 189.6 gal, using the lighter weight fuel, Step 1. Thus, one 200-gal day tank would be suitable for each of the 1,000-hp engines.

For the 600-hp engines, the day-tank capacity should be (600 hp)(0.35 lb fuel per bhp-hr)(4 hr) = 840 lb, or 840/7.387 = 113.8 gal. Thus, one 125-gal day tank would be suitable for each of the 600-hp engines.

4. Determine which is the better fuel buy

Compute the higher heating value HHV of each fuel from HHV = 17,645 + 54 (°API), or for 24° fuel, HHV = 17,645 + 54(24) = 18,941 Btu/lb. For the 28° fuel, HHV = 17,645 + 54(28) = 19,157 Btu/lb.

Compare the two oils on the basis of cost per 10,000 Btu, because this is the usual way of stating the cost of a fuel. The weight of each oil was computed in Step 1. Thus the 24° API oil weighs 7.578 lb/gal, while the 28° API oil weighs 7.387 lb/gal.

Then, the cost per 10,000 Btu = (cost, $/gal)/[(HHV, Btu/lb)/10,000](oil weight, lb/gal). For the 24° API oil, cost per 10,000 Btu = $0.0825/[(18.941/10,000)(7.578)] = $0.00574, or 0.574 cents/10,000 Btu. For the 28° API oil, cost per 10,000 Btu = $0.0910/[(19,157/10,000)(7,387)] = $0.00634, or 0.634 cents/10,000 Btu. Thus, the 24° API is the better buy because it costs less per 10,000 Btu.

Related Calculations: Use this method for engines burning any liquid fuel. Be certain to check local laws and the latest NBFU recommendations before ordering fuel storage or day tanks.

POWER INPUT TO COOLING-WATER AND LUBE-OIL PUMPS

What is the required power input to a 200-gpm jacket-water pump if the total head on the pump is 75 ft of water and the pump has an efficiency of 70 percent when it handles (1) freshwater, and saltwater? What capacity lube-oil pump is needed for a four-cycle 500-hp turbocharged diesel engine having oil-cooled pistons? What is the required power input to this pump if the discharge pressure is 80 psi and the efficiency of the pump is 68 percent?

Calculation Procedure:

1. Determine the power input to the jacket-water pump

The power input to jacket-water and raw-water pumps serving internal-combustion engines is often computed from the relation $hp = Gh/Ce$, where hp = hp input; G = water discharged by pump, gpm; h = total head on pump, ft of water; C = constant = 3,960 for fresh water having a density of 62.4 lb/cu ft; 3,855 for saltwater having a density of 64 lb/cu ft.

For this pump handling freshwater, $hp = (200)(75)/(3,960)(0.70) = 5.42$ hp. A 7.5-hp motor would probably be selected to handle the rated capacity plus any overloads.

For this pump handling saltwater, $hp = (200)(75)/[(3,855)(0.70)] = 5.56$ hp. A 7.5-hp motor would probably be selected to handle the rated capacity plus any overloads. Thus, the same motor could drive this pump whether it handles freshwater or saltwater.

2. Compute the lube-oil pump capacity

The lube-oil pump capacity required for a diesel engine is found from $G = H/200 \, \Delta t$, where G = pump capacity, gpm; H = heat rejected to the lube oil, Btu/bhp-hr; Δt = lube-oil temperature rise during passage through the engine, F. Usual practice is to limit the temperature rise of the oil to a range of 20 to 25 F, with a maximum operating temperature of 160 F. The heat rejection to the lube oil can be obtained from the engine heat balance, the engine manufacturer, or *Standard Practices for Stationary Diesel Engines*, published by the Diesel Engine Manufacturers Association. Using a maximum heat rejection rate of 500 Btu/bhp-hr from *Standard Practices* and an oil-temperature rise of 20 F, $G = (500 \, \text{Btu/bhp-hr})(1,000 \, \text{hp})/[(200)(20)] = 125$ gpm.

By using the *lowest* temperature rise and the *highest* heat rejection rate, a safe pump capacity is obtained. Where the pump cost is a critical factor, use a higher temperature rise and a lower heat rejection rate. Thus, with a heat rejection rate of 300 Btu/bhp-hr from *Standard Practices*, the above pump would have a capacity of $G = (300)(1,000)/[(200)(25)] = 60$ gpm, using these values.

3. Compute the lube-oil pump power input

The power input to a separate oil pump serving a diesel engine is given by $hp = Gp/1,720e$, where $G =$ pump discharge rate, gpm; $p =$ pump discharge pressure, psi; $e =$ pump efficiency. For this pump, $hp = (125)(80)/[(1,720)(0.68)] = 8.56$ hp. A 10-hp motor would be chosen to drive this pump.

With a capacity of 60 gpm, the input is $hp = (60)(80)/[(1,720)(0.68)] = 4.1$ hp. A 5-hp motor would be chosen to drive this pump.

Related Calculations: Use this method for any reciprocating diesel engine, two- or four-cycle. Lube-oil pump capacity is generally selected 10 to 15 percent oversize to allow for bearing wear in the engine and wear of the pump moving parts. Always check the selected capacity with the engine builder. Where a bypass-type lube-oil system is used, be sure to have a pump of sufficient capacity to handle *both* the engine and cooler oil flow.

Raw-water pumps are generally duplicates of the jacket-water pump, having the same capacity and head ratings. Then the raw-water pump can serve as a standby jacket-water pump, if necessary.

LUBE-OIL COOLER SELECTION AND OIL CONSUMPTION

A 500-hp internal-combustion engine rejects 300 to 600 Btu/bhp-hr to the lubricating oil. What capacity and type of lube-oil cooler should be used for this engine if 10 percent of the oil is bypassed? If this engine consumes 2 gal of lube oil per 24 hr at full load, determine its lube-oil consumption rate.

Calculation Procedure:

1. Determine the required lube-oil cooler capacity

Base the cooler capacity on the maximum heat rejection rate plus an allowance for overloads. The usual overload allowance is 10 percent of the full-load rating for periods of not more than 2 hr in any 24 hr period.

For this engine, the maximum output with a 10 percent overload is $500 + (0.10)(500) = 550$ hp. Thus, the maximum heat rejection to the lube oil would be $(550 \text{ hp})(600 \text{ Btu/bhp-hr}) = 330,000$ Btu/hr.

2. Choose the type and capacity of the lube-oil cooler

Choose a shell-and-tube type heat exchanger to serve this engine. Long experience with many types of internal-combustion engines shows that the shell-and-tube heat exchanger is well suited for lube-oil cooling.

Select a lube-oil cooler suitable for a heat-transfer load of 330,000 Btu/hr at the prevailing cooling-water temperature difference, which is usually assumed to be 10 F. See previous Calculation Procedures for the steps in selecting a liquid cooler.

3. Determine the lube-oil consumption rate

The lube-oil consumption rate is normally expressed in terms of bhp-hr/gal. Thus, if this engine operates for 24 hr and consumes 2 gal of oil, its lube-oil consumption rate = $(24 \text{ hr})(500 \text{ bhp})/2 \text{ gal} = 6,000$ bhp-hr/gal.

Related Calculations: Use this procedure for any type of internal-combustion engine using any fuel.

QUANTITY OF SOLIDS ENTERING AN INTERNAL-COMBUSTION ENGINE

What weight of solids annually enters the cylinders of a 1,000-hp internal-combustion engine if the engine operates 24 hr/day, 300 days/year in an area having an average dust concentration of 1.6 grains per 1,000 cu ft of air? The engine air rate (displacement) is 3.5 cfm/bhp. What would the dust load be reduced to if an air filter fitted to the engine removed 80 percent of the dust from the air?

Calculation Procedure:

1. Compute the quantity of air entering the engine

Since the engine is rated at 1,000 hp and uses 3.5 cfm/bhp, the quantity of air used by the engine each min is (1,000 hp)(3.5 cfm/hp) = 3,500 cfm.

2. Compute the quantity of dust entering the engine

Each 1,000 cu ft of air entering the engine contains 1.6 grains of dust. Thus, during every min of engine operation, the quantity of dust entering the engine is (3,500/1,000) (1.6) = 5.6 grains. The hourly dust intake = (60 min/hr)(5.6 grains/min) = 336 grains/hr.

During the year the engine operates 24 hr/day for 300 days. Hence, the annual intake of dust is (24 hr/day)(300 days/year)(336 grains/hr) = 2,419,200 grains. Since there are 7,000 grains/lb, the weight of dust entering the engine per year = 2,419,200 grains/ (7,000 grains/lb) = 345.6 lb/year.

3. Compute the filtered dust load

With the air filter removing 80 percent of the dust, the quantity of dust reaching the engine is (1.00 − 0.80) (345.6 lb/year) = 69.12 lb/year. This shows the effectiveness of an air filter in reducing the dust and dirt load on an engine.

Related Calculations: Use this general procedure to compute the dirt load on an engine from any external source.

INTERNAL-COMBUSTION ENGINE PERFORMANCE FACTORS

Discuss and illustrate the important factors in internal-combustion engine selection and performance. In this discussion, consider both large and small engines for a full range of usual applications.

Calculation Procedure:

1. Plot typical engine load characteristics

Figure 7 shows four typical load patterns for internal-combustion engines. A continuous load, Fig. 7a, is generally considered to be heavy-duty, and is often met in engines driving pumps or electric generators.

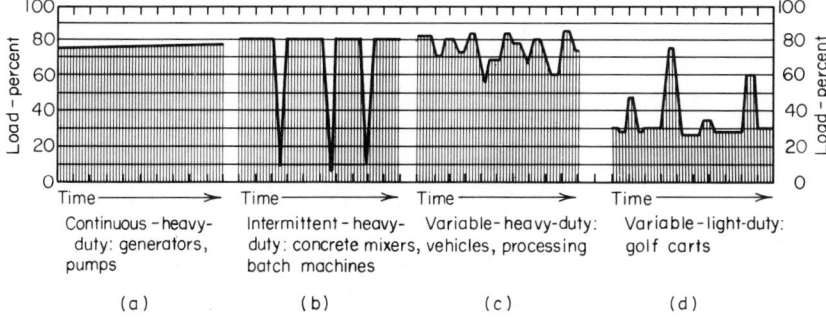

Fig. 7   Typical internal-combustion engine load cycles: (a) continuous, heavy-duty; (b) intermittent, heavy-duty; (c) variable, heavy-duty; (d) variable, light-duty. *(Product Engineering.)*

Intermittent heavy-duty loads, Fig. 7b, are often met in engines driving concrete mixers, batch machines, and similar loads. Variable heavy-duty loads, Fig. 7c, are encountered in large vehicles, process machinery, and similar applications. Variable light-duty loads, Fig. 7d, are met in small vehicles like golf carts, lawn mowers, chain saws, etc.

2. Compute the engine output torque

Use the relation $T = 5{,}250$ bhp/rpm to compute the output torque of an internal-combustion engine. In this relation, bhp = engine bhp being developed at a crankshaft speed having rotating speed of rpm.

3. Compute the hp output required

Knowing the type of load on the engine — generator, pump, mixer, saw blade, etc. — compute the power output required to drive the load at a constant speed. Where a speed variation is expected, as in variable-speed drives, compute the average power needed to accelerate the load between two desired speeds in a given time.

4. Choose the engine output speed

Internal-combustion engines are classified in three speed categories: high (1,500 rpm or more), medium (750 to 1,500 rpm), and low (less than 750 rpm).

Base the speed chosen on the application of the engine. A high-speed engine can be lighter and smaller for the same hp rating, and may cost less than a medium-speed or slow-speed engine serving the same load. But medium-speed and slow-speed engines, although larger, offer a higher torque output for the equivalent hp rating. Other advantages of these two speed ranges include longer service life, and, in some instances, lower maintenance costs.

Usually an application will have its own requirements, such as allowable engine weight, available space, output torque, load speed, and type of service. These requirements will often indicate that a particular speed classification must be used. Where an application has no special speed requirements, the speed selection can be made on the basis of cost (initial, installation, maintenance, and operating cost), type of parts service available, and other local conditions.

5. Analyze the engine output torque required

In some installations, an engine with good lugging power is necessary, especially in tractors, harvesters, and hoists, where the load frequently increases above normal. For good lugging power, the engine should have the inherent characteristic of increasing torque with drooping speed. The engine can then resist the tendency for increased load to reduce the output speed, giving the engine good lugging qualities.

One way to increase the torque delivered to the load is to use a variable-ratio hydraulic transmission. The transmission will amplify the torque so that the engine will not be forced into the lugging range.

Other types of loads, such as generators, centrifugal pumps, air conditioners, and marine drives, may not require this lugging ability. So be certain to consult the engine power curves and torque characteristic curve to determine the speed at which the maximum torque is available.

6. Evaluate the environmental conditions

Internal-combustion engines are required to operate under a variety of environmental conditions. The usual environmental conditions critical in engine selection are: altitude, ambient temperature, dust or dirt, and special or abnormal service. Each of these, except the last, is considered in previous Calculation Procedures.

Special or abnormal service include such applications as fire fighting, emergency flood pumps and generators, and hospital standby service. In these applications, an engine must start and pick up full load without warmup.

7. Compare engine fuels

Table 9 compares four types of fuels and the internal-combustion engines using them. Note that where the cost of the fuel is high, the cost of the engine is low; where the cost of the fuel is low, the cost of the engine is high. This condition prevails for both large and small engines in any service.

TABLE 9 Comparison of Fuels for Internal-Combustion Engines*

| | Storage life (Quantities) | | Consistency, Btu/cu ft | Initial cost of engine, relative | Cost of fuel | Residue | Anti knock rating | Filtering necessary | Weight, lb/gal | Heat content | |
	Small	Large								Btu/vol	Btu/lb
Gasoline ..	Good	Poor (6 months)	Good	Low	High	High	Best is costly	Medium	6.000	123,039 Btu/gal	20,627
Diesel:											
No. 1....	Good	Fair (1 year)	Good	High	Low	Low if properly filtered	...	High	6.850	135,800 Btu/gal	19,750
No. 2....	Good	Fair (1 year)	Good	High	Low	Low if properly filtered	...	High	7.020	139,000 Btu/gal	19,786
Natural gas	Not necessary	Not necessary	Poor	Medium	Medium	Low	High	Very little	...	1,000 Btu/cu ft	
LPG:											
Propane.	Good	Good	Poor	Medium	Medium	Low	Good	Very little	4.235	91,740 Btu/gal	21,308
Butane..	Good	Good	Poor	Medium	Medium	Low	Good	Very little	4.873	103,830 Btu/gal	20,627

Product Engineering.

TABLE 10 Performance Table for Small Internal-Combustion Engines (Less than 7 hp)*

	Variety of models available	Typical weight lb/hp	Operating speeds		Lugging ability	Torque output	Relative life expectancy, hr	Relative cost	Fuel required	Shaft direction	Noise level	Starters	Integral optional Pto's	Ignition	Cost of operation	Variety of options and accessories
			Typical maximum	Typical efficient minimum												
Lightweight: 2-stroke	Narrow	2:1	3,600 (governed) to 7,500	2,000 to 3,000	Poor to fair	Fair	500	Lowest	Gasoline oil mixed	Vertical, horizontal or universal	High	Rope, recoil, impulse	No	Magneto	High	Standard—extremely low custom—wide
4-stroke	Wide	6:1; 10:1	4,000	2,000 to 2,400	Fair to good	Good	500	1 to 2	Gasoline (LPG)	Vertical or horizontal	Moderate	Rope, recoil, impulse, electric	Several	Magneto	Moderate	Standard—wide
Heavyweight: 4-stroke	Wide	11:1; 20:1	4,000	1,600 to 1,800	Good to excellent	Good	7500	2 to 4	Gasoline (LPG)	Vertical or horizontal	Moderate	Rope, recoil, impulse, crank, electric	No	Magneto, distributor	Moderate	Standard—moderately wide
Diesel	Narrow	35:1	2,400	1,500	Excellent	Good	25,000	4	Diesel	Horizontal	Moderate to high	Electric	No	Battery, distributor, glow plugs	Low	Narrow

*Product Engineering.

8. Compare the performance of small engines

Table 10 compares the principal characteristics of small gasoline and diesel engines rated at 7 hp or less. Note that engine life expectancy can vary from 500 to 25,000 hr. With modern, mass-produced small engines it is often just as cheap to use short-life replaceable 2-stroke gasoline engines instead of a single long-life diesel engine. Thus, the choice of a small engine is often based on other considerations, such as ease and convenience of replacement, instead of just hours of life. Chances are, however, that most long-life applications of small engines will still require a long-life engine. But the alternatives must be considered in each case.

Related Calculations: Use the general data presented here when selecting internal-combustion engines having ratings up to 200 hp. For larger engines, other factors such as weight, specific fuel consumption, lube-oil consumption, etc., become important considerations in the choice. The method given here is the work of Paul F. Jacobi, as reported in *Product Engineering*.

AIR AND GAS COMPRESSORS AND VACUUM SYSTEMS

REFERENCES: Van Atta—*Vacuum Science and Engineering*, McGraw-Hill; Dushman—*Scientific Foundations of Vacuum Technique*, Wiley; Guthrie and Wakerling—*Vacuum Equipment and Techniques*, McGraw-Hill; Yarwood—*High Vacuum Techniques*, Wiley; Lewin—*Vacuum Science and Technology*, McGraw-Hill; Pirani and Yarwood—*Principles of Vacuum Engineering*, Reinhold; Reimann—*Vacuum Technique*, Chapman and Hall; Steinherz—*Handbook of High Vacuum Engineering*, Reinhold; Compressed Air and Gas Institute—*Compressed Air and Gas Handbook*; Ingersoll-Rand Company—*Compressed Air Data*.

COMPRESSOR SELECTION FOR COMPRESSED-AIR SYSTEMS

Determine the required capacity, discharge pressure, and type of compressor for an industrial-plant compressed-air system fitted with the tools listed in Table 1. The plant is located at sea level and operates 16 hr/day.

TABLE 1 Typical Computation of Compressed-Air Requirements

Tool	(1) Air consumption, cfm	(2) Number of tools	(3) Air required, cfm (1) × (2)	(4) Load factor	(5) Probable air demand cfm (3) × (4)
Grinding wheel, 6 in.	50	5	250	0.3	75
Rotary sander, 9-in. pad....	55	2	110	0.5	55
Chipping hammers, 13 lb ..	30	8	240	0.4	96
Nut setters, $\frac{5}{16}$ in.	20	10	200	0.6	120
Paint spray	10	1	10	0.1	1
Plug drills	40	3	120	0.2	24
Riveters, 18 lb	35	5	175	0.4	70
Steel drill, $\frac{7}{8}$ in., 25 lb	80	5	400	0.4	160
					601*

*To this value must be added allowance for future needs and expected leakage loss, if any.

Calculation Procedure:

1. Compute the required airflow rate

List all the tools and devices in the compressed-air system that will consume air, Table 1. Then obtain from Table 2 the probable air consumption, cfm, of each tool. Enter this value in column 1, Table 1. Next, list the number of each type of tool that will be used in the system in column 2. Find the maximum probable air consumption of each tool by taking the product, line by line, of columns 1 and 2. Enter the result in column 3, Table 1, for each tool.

TABLE 2   Approximate Air Needs of Pneumatic Tools, Cfm

Grinders:
    6- and 8-in.-diameter wheels 50
    2- and 2½-in.-diameter wheels 14–20
File and burr machines 18
Rotary sanders, 9-in.-diameter pads 55
Sand rammers and tampers:
    1 × 4 in. cylinder 25
    1¼ × 5 in. cylinder 28
    1½ × 6 in. cylinder 39
Chipping hammers:
    10 to 13 lb 28–30
    2 to 4 lb...................... 12
Nut setters:
    To $\frac{5}{16}$ in., 8 lb.................... 20
    ½ to ¾ in., 18 lb 30

Paint spray 2–20
Plug drills.................... 40–50
Riveters:
    $\frac{3}{32}$- to ⅛-in. rivets 12
    Larger, weighing 18 to 22 lb 35
Rivet busters 35–39
Steel drills, rotary motors:
    To ¼ in., weighing 1¼ to 4 lb... 18–20
    ¼ to ⅜ in., weighing 6 to 8 lb ... 20–40
    ½ to ¾ in., weighing 9 to 14 lb..... 70
    ⅞ to 1 in., weighing 25 lb 80
Wood borers to 1 in. diameter,
    weighing 14 lb................ 40

The air consumption values shown in column 3 represent the airflow rate required for continuous operation of each type and number of tools listed. However, few air tools operate continually. To provide for this situation a load factor is generally used when selecting an air compressor.

2. Select the equipment load factor

The equipment load factor = actual air consumption of the tool or device, cfm/full-load continuous air consumption of the tool or device, cfm. Load factors for compressed-air operated devices are usually less than 1.0.

Two variables are involved in the equipment load factor. The first is the *time factor*, or the percent of the total time the tool or device actually uses compressed air. The second is the *work factor*, or percent of maximum possible work output done by the tool. The load factor is the product of these two variables.

Determine the load factor for a given tool or device by consulting the manufacturer's engineering data, or by estimating the factor value by using previous experience as a guide. Enter the load factor in column 4, Table 1. The values shown represent typical load factors encountered in industrial plants.

3. Compute the actual air consumption

Take the product, line by line, of columns 3 and 4, Table 1. Enter the result, i.e., the probable air demand, cfm, in column 5, Table 1. Find the sum of the values in column 5, or 601 cfm. This is the probable air demand of the system.

4. Apply allowances for leakage and future needs

Most compressed-air system designs allow for 10 percent of the required air to be lost through leaks in the piping, tools, hoses, etc. Whereas some designers claim that allowing for leakage is a poor design procedure, observation of many installations indicates that air leakage is a fact of life and must be considered when designing an actual system.

Using a 10 percent leakage factor, the required air capacity = 1.1(601) = 661 cfm.

Future requirements are best estimated by predicting what types of tools and devices will probably be used. Once this is known, prepare a tabulation similar to Table 1, listing the predicted future tools and devices and their air needs. Assume that the future air needs, column 5, are 240 cfm. Then the total required air capacity = 661 + 240 = 901 cfm; say 900 cfm = present requirements + leakage allowance + predicted future needs, all expressed in cfm.

5. Choose the compressor discharge pressure and capacity

In selecting the type of compressor to use, two factors are of key importance: (*a*) discharge pressure required, and (*b*) capacity required.

Most air tools and devices are designed to operate at a pressure of 90 psi at the tool inlet. Hence, usual industrial compressors are rated for a discharge pressure of 100 psi, the extra psi providing for pressure loss in the piping between the compressor and the tools. Since none of the tools used in this plant are specialty items requiring higher than the normal pressure, a 100-psi discharge pressure will be chosen.

Where the future air demands are expected to occur fairly soon — within 2 to 3 years — the general practice is to choose a compressor having the capacity to satisfy present and future needs. Hence, in this case, a 900-cfm compressor would be chosen.

6. Compute the power required to compress the air

Table 3 shows the power required to compress air to various discharge pressures at different altitudes above sea level. Study of this table shows that at sea level a single-stage compressor requires 22.1 bhp/100 cfm when the discharge pressure is 100 psi. A two-stage compressor requires 19.1 bhp under the same conditions. This is a saving of 3.0 bhp/100 cfm. Hence, a two-stage compressor would probably be a better investment because this hp will be saved for the life of the compressor. The usual life of an air compressor is 20 years. Hence, using a two-stage compressor, the approximate required bhp = (900/100)(19.1) = 171.9 bhp; say 175.

TABLE 3 Air Compressor Brake Horsepower Input*

Altitude, ft	Single-stage Discharge pressure, psig			Two-stage Discharge pressure, psig		
	60	80	100	60	80	100
0	16.3	19.5	22.1	14.7	17.1	19.1
2,000	15.9	18.9	21.3	14.3	16.5	18.4
4,000	15.4	18.2	20.6	13.8	15.8	17.7
6,000	15.0	17.6	20.0	13.3	15.2	17.0

Courtesy Ingersoll-Rand. Values shown are the approximate bhp input required per 100 cfm of free air actually delivered. The bhp input can vary considerably with the type and size of compressor.

7. Choose the type of compressor to use

Reciprocating compressors find the widest use for stationary plant air supply. They may be single- or two-stage, air- or water-cooled. Here is a general guide to the types of reciprocating compressors that are satisfactory for various loads and service:

Single-stage air-cooled compressor up to 3 hp, pressures to 150 psi, for light and intermittent running up to 1 hr/day.

Two-stage air-cooled compressor up to 3 hp, pressures to 150 psi, for 4 to 8 hr/day running time.

Single-stage air-cooled compressor up to 15 hp for pressures to 80 psi; above 80 psi, use two-stage air-cooled compressor.

Single-stage horizontal double-acting water-cooled compressor for pressures to 100 psi, horsepowers of 10 to 100, for 24 hr/day or less operating time.

Two-stage, single-acting air-cooled compressor for 10 to 100 hp, 5 to 10 hr/day operation.

Two-stage double-acting water-cooled compressor for 100 hp or more, 24 hr/day, or less operating time.

Using this general guide, choose a two-stage double-acting water-cooled reciprocating compressor, because more than 100 hp input is required and the compressor will operate 16 hr/day.

Rotary compressors are not as widely used for industrial compressed-air systems as reciprocating compressors. The reason for this is that usual rotary compressors discharge at pressures under 100 psi, unless they are multistage units.

Centrifugal compressors are generally used for large airflows—several thousand cfm, or more. Hence, they usually find use for services requiring large air quantities, such as steel-mill blowing, copper conversion, etc. As a general rule, machines discharging at pressures of 35 psi or less are termed *blowers*; machines discharging at pressures greater than 35 psi are termed *compressors*.

Using these facts as a guide enables the designer to choose, as before, a two-stage double-acting water-cooled compressor for this application. Refer to the manufacturer's engineering data for the compressor dimensions and weight.

8. Select the compressor drive

Air compressors can be driven by electric motors, gasoline engines, diesel engines, gas turbines, or steam turbines. The most popular drive for reciprocating air compressors is the electric motor—either direct-connected or belt-connected. Where either a dc or ac power supply is available, the usual choice is an electric-motor drive. However, special circumstances, such as the availability of low-cost fuel, may dictate another choice of drive for economic reasons. Assuming that there are no special economic reasons for choosing another type of drive, an electric motor would be chosen for this installation.

With an ac power supply, the squirrel-cage induction motor is generally chosen for belt-driven compressors. Synchronous motors are also used, particularly when power-factor correction is desired. Motor-driven air compressors generally operate at constant speed and are fitted with cylinder unloaders to vary the quantity of air delivered to the air receiver. A typical power input to a large reciprocating compressor is 22 hp per 100 cfm of free air compressed.

Air compressors are almost always rated in terms of *free air* capacity—i.e., air at the compressor intake location. Since the altitude, barometric pressure, and air temperature may vary at any locality, the term free air does not mean air under standard or uniform conditions. The displacement of an air compressor is the volume of air displaced per unit of time, usually stated in cfm. In a multistage compressor, the displacement is that of the low-pressure cylinder only.

9. Choose the type of air distribution system

Two types of air distribution systems are in use in industrial plants: (a) *central*, and (b) *unit*. In a central system, Fig. 1, one or more large compressors centrally located in the plant supply compressed air to the areas needing it. The supply piping often runs in the form of a loop around the areas needing air.

A unit system, Fig. 2, has smaller compressors located in the areas where air is used. In the usual plant, each compressor serves only the area in which it is located. Emergency connections between the various areas may or may not be installed.

Central systems have been used for many years in large industrial plants and give excellent service. Unit systems are used in both small and large plants but probably find more use in smaller plants today. With the large quantity of air required by this plant, a central system would probably be chosen, unless the air was needed at widely scattered locations in the plant, leading to excessive pressure losses in the distribution piping of a central system. In such a situation, a unit system with the capacity divided between compressors as necessary would be chosen.

Related Calculations: Where possible, choose a larger compressor than the calculations indicate is needed, because air use in industrial plants tends to increase. Avoid choosing a compressor having a free air capacity less than one-third the required free air capacity.

When choosing a water-cooled compressor instead of an air-cooled unit, remember that water cooling is more expensive than air cooling. However, the power input to water-cooled compressors is usually less than to air-cooled compressors of the same

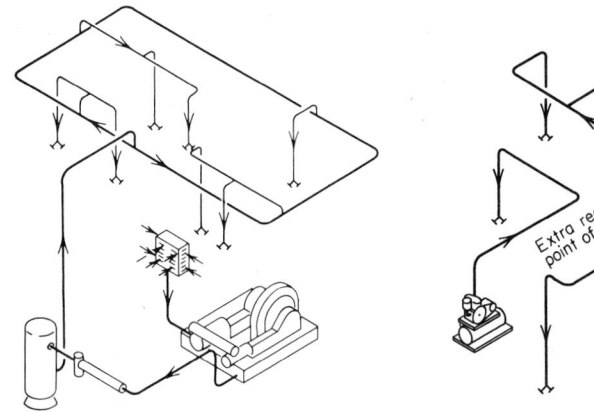

Fig. 1   Central system for compressed-air supply.

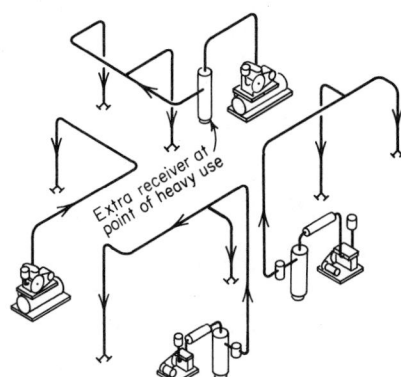

Fig. 2   Unit system for compressed-air supply.

capacity. For either type of cooling, a two-stage compressor, with intercooling, is more economical when the compressor must operate 4 or more hr in a 24-hr period. Table 4 shows the typical cooling-water requirements of various types of water-cooled compressors.

Table 4   Cooling Water Recommended for Intercoolers, Cylinder Jackets, Aftercoolers

Gpm per 100 cfm Actual Free Air	
Intercooler separate.	2.5–2.8
Intercooler and jackets in series	2.5–2.8
Aftercoolers:	
80 to 100 psi, two-stage	1.25
80 to 100 psi, single-stage	1.8
Two-stage jackets alone (both).	0.8
Single-stage jackets:	
40 psi	0.6
60 psi	0.8
80 psi	1.1
100 psi	1.3

When the inlet air temperature is above or below 60 F, the compressor delivery will vary. Table 5 shows the relative delivery of compressors handling air at various inlet temperatures.

TABLE 5 Effect of Initial or Intake Temperature on Delivery of Air Compressors

(Based on a nominal intake temperature of 60 F)

Initial temperature		Relative delivery	Initial temperature		Relative delivery
F	F, absolute		F	F, absolute	
−20	440	1.18	70	530	0.980
−10	450	1.155	80	540	0.961
0	460	1.13	90	550	0.944
10	470	1.104	100	560	0.928
20	480	1.083	110	570	0.912
30	490	1.061	120	580	0.896
32	492	1.058	130	590	0.880
40	500	1.040	140	600	0.866
50	510	1.020	150	610	0.852
60	520	1.00	160	620	0.838

SIZING COMPRESSED-AIR-SYSTEM COMPONENTS

What is the minimum capacity air receiver that should be used in a compressed-air system having a compressor displacing 800 cfm when the intake pressure is 14.7 psia and the discharge pressure is 120 psia? How long will it take for this compressor to pump up a 300 cu ft receiver from 80 to 120 psi if the average volumetric efficiency of the compressor is 68 percent? For how long can an 80-psia tool be operated from a 120-psia, 300 cu ft receiver if the tool uses 10 cfm of free air and the receiver pressure is allowed to fall to 85 psia when the atmospheric pressure is 14.7 psia? What diameter air piston is required to produce a 1,000-lb force if the pressure of the air is 150 psia?

Calculation Procedure:

1. Compute the required volume of the air receiver

Use the relation $V_m = dp_1/p_2$, where V_m = minimum receiver volume needed, cu ft; d = compressor displacement, cfm (use only the first-stage displacement for two-stage compressors); p_1 = compressor intake pressure, psia; p_2 = compressor discharge pressure, psia. Thus, for this compressor, $V_m = 800(14.7/120) = 97$ cu ft. To provide a reserve capacity, a receiver having a volume of 150 or 200 cu ft would probably be chosen.

2. Compute the receiver pump-up time

Use the relation $t = V(p_f - p_i)/14.7 \, de$, where t = receiver pump-up time, min; p_f = final receiver pressure, psia; p_i = initial receiver pressure, psia; d = compressor piston displacement, cfm; e = compressor volumetric efficiency, percent. Thus, $t = 300(120 - 80)/[14.7(800)(0.68)] = 1.5$ min. When the compressor discharge capacity is given in cfm of free air instead of in terms of piston displacement, drop the volumetric efficiency term from the above relation before computing the pump-up time.

3. Compute the air supply time

Use the relation $t_s = V(p_{max} - p_{min})/cp_a$, where t_s = time in min during which the receiver of volume V cu ft will supply air from the receiver maximum pressure p_{max} psia to the minimum pressure p_{min} psia; c = cfm of free air required to operate the tool; p_a = atmospheric pressure, psia. Or, $t_s = 300(120 - 85)/[(10)(14.7)] = 71.5$ min

Note in this relation that p_{min} is the minimum air pressure required to operate the air tool. A higher minimum tank pressure was chosen here because this provides a safer estimate of the time duration for the supply of air. Had the tool operating pressure been chosen instead, the time available, using the same relation, would be $t_s = 81.5$ min.

This calculation shows that it is often wise to install an auxiliary receiver at a distance from the compressor but near the tools drawing large amounts of air. Use of such an auxiliary receiver, particularly near the end of a long distribution line, can often eliminate the need for purchasing another air compressor.

4. Compute the required piston diameter

Use the relation $A_p = F/p_m$, where $A_p =$ required piston area to produce the desired force, sq in.; $F =$ force produced, lb; $p_m =$ maximum air pressure available for the piston, psia. Or, $A_p = 1{,}000/150 = 6.66$ sq in. The piston diameter d is $d = 2(A_p/\pi)^{0.5} = 2.91$ in.

Related Calculations: The air consumption of power tools is normally expressed in cfm of free air at sea level; the actual capacity of any type of air compressor is expressed in the same units. At locations above sea level, the quantity of free air required to operate an air tool increases because the atmospheric pressure is lower. To find the air consumption of an air tool at an altitude above sea level in terms of cfm of free air at the elevation location, multiply the sea-level consumption by the appropriate factor from Table 6. Thus, a tool that consumes 10 cfm of free air at sea level will use $10(1.310) = 13.1$ cfm of 100-psi free air at an 8,000-ft altitude.

TABLE 6   Air-Consumption Altitude Factors

(100-psi air supply)

Altitude, ft	Factor
2,000	1.068
4,000	1.142
6,000	1.224
8,000	1.310
10,000	1.404
12,000	1.506
14,000	1.610

NOTE: For pressure losses in compressed-air piping systems, see the index.

COMPRESSED-AIR RECEIVER SIZE AND PUMP-UP TIME

What is the minimum size receiver that can be used in a compressed-air system having a compressor rated at 800 cfm of free air if the intake pressure is 14.7 psia and the discharge pressure is 120 psia? How long will it take the compressor to pump up the receiver from 60 psia to 120 psia? The compressor is a two-stage water-cooled unit. How much cooling water is required for the intercooler and jacket if they are piped in series and for the aftercooler?

Calculation Procedure:

1. Compute the required minimum receiver volume

For any air compressor, the minimum receiver volume v_m cu ft $= Dp_i/p_d$, where $D =$ compressor displacement, cfm free air (use only the first-stage displacement for multistage compressors); $p_i =$ compressor inlet pressure, psia; $p_d =$ compressor discharge pressure, psia. For this compressor, $v_m = (800)(14.7)/(120) = 98$ cu ft. To provide a reserve supply of air, a receiver having a volume of 150 or 200 cu ft would probably be chosen. Be certain that the receiver chosen is a standard unit; otherwise its cost may be excessive.

2. Compute the pump-up time required

Assume that a 150 cu ft receiver is chosen. Then, for any receiver, the pump-up time t min $= v_r(p_e - p_s)/De$, where v_r = receiver volume, cu ft; p_e = pressure at end of pump up, psia; p_s = pressure at start of pump up, psia; e = compressor volumetric efficiency, expressed as a decimal (0.50 to 0.75 for single-stage and 0.80 to 0.90 for multistage compressors). For this compressor, using a volumetric efficiency of 0.85, $t = (150)(120 - 60)/[(800)(0.85)] = 13.22$ min.

3. Determine the quantity of cooling water required

Use the Compressed Air and Gas Institute (CAGI) cooling-water recommendations given in the *Compressed Air and Gas Handbook*, or Baumeister and Marks — *Standard Handbook for Mechanical Engineers*. For 80- to 125-psig discharge pressure with the intercooler and jacket in series, CAGI recommends a flow of 2.5 to 2.8 gpm per 100 cfm of free air. Using 2.5 gpm, the cooling water required for the intercooler and jackets $= (2.5)(800/100) = 20.0$ gpm. CAGI recommends 1.25 gpm per 100 cfm of free air for an aftercooler serving a two-stage 80- to 125-psig compressor, or $(1.25)(800/100) = 10.0$ gpm for this compressor. Thus, the total quantity of cooling water required for this compressor is $20 + 10 = 30$ gpm.

Related Calculations: Use this Procedure for any type of air compressor serving an industrial, commercial, utility, or residential load of any capacity. Follow CAGI or the manufacturer's recommendations for cooling-water flow rate. When a compressor is located above or below sea level, multiply its rated free air capacity by the appropriate altitude correction factor obtained from the CAGI — *Compressed Air and Gas Handbook* or Baumeister and Marks — *Standard Handbook for Mechanical Engineers*.

VACUUM-SYSTEM PUMP-DOWN TIME

An industrial vacuum system with a 200 cu ft receiver serving cleaning outlets is to operate to within 2.5-in. Hg absolute of the barometer when the barometer is 29.8 in. Hg. How long will it take to evacuate the receiver to this pressure when a single-stage vacuum pump with a displacement of 60 cfm is used? The pump is rated to dead-end at a 29.0-in. Hg vacuum when the barometer is 30.0 in. Hg. The pump volumetric efficiency is shown in Fig. 3.

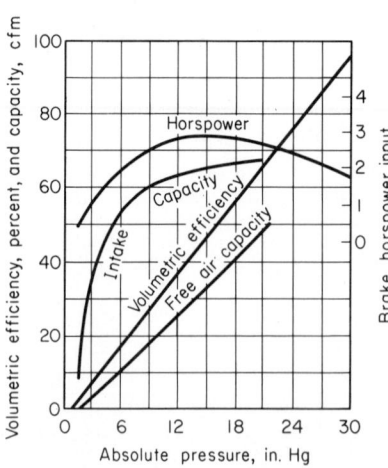

Fig. 3 Capacity, power-input, and efficiency curves for a typical reciprocating vacuum pump.

Calculation Procedure:

1. Compute the pump operating vacuum

The pump must operate to within 2.5 in. of the barometer, or a vacuum of $29.8 - 2.5 = 27.3$ in. Hg.

2. Compute the quantity of free air removed from the receiver

Select a number of absolute pressures between 29.8 in. Hg, the actual barometric pressure, and the final receiver pressure, 2.5 in. Hg, and list them in the first column of a table such as Table 7. Assume equal pressure reductions — say 3 in. Hg — for each step except the last few, where smaller reductions have been assumed to ensure greater accuracy.

Enter in the second column of Table 7 the ratio of the absolute pressure in the receiver to the atmospheric pressure, or P_r/P_a, both expressed in in. Hg. Thus, for the second step, $P_r/P_a = 26.8/29.8 = 0.899$.

The amount of air remaining in the receiver, measured at atmospheric conditions, is then the product of the receiver volume, 200 cu ft, and the ratio of the pressures. Or, for the second pressure reduction, $200(0.899) = 179.8$ cu ft. Enter the result in the third column of Table 7. This computation is a simple application of the gas laws with the receiver temperature assumed constant. Assumption of a constant air temperature is valid because, although the air temperature varies during pumping down, the overall effect is that of a constant temperature.

TABLE 7   Evacuation Time Calculations

Absolute pressure in receiver, in. Hg	P_r/P_a	Cu ft of free air		Average volumetric efficiency, Fig. 2	Free air capacity, cfm	Evacuation time, min
		In receiver	Removed			
29.8	1.000	200.0	0.0			
26.8	0.899	179.8	20.2	0.91	54.6	0.370
23.8	0.798	159.6	20.2	0.81	48.6	0.415
20.8	0.698	139.6	20.0	0.72	43.2	0.464
17.8	0.597	119.4	20.2	0.62	37.2	0.544
14.8	0.496	99.2	20.2	0.52	31.2	0.648
11.8	0.396	79.2	20.0	0.43	25.8	0.776
8.8	0.295	59.0	20.2	0.33	19.8	1.020
5.8	0.195	39.0	20.0	0.23	13.8	1.450
3.8	0.128	25.6	13.4	0.14	8.4	1.596
2.8	0.094	18.8	6.8	0.09	5.4	1.260
2.5	0.084	16.8	2.0	0.07	4.2	0.476
Total time required	...	...	...	...	...	9.019

Find the quantity of air removed from the receiver by successive subtraction of the values in the third column. Thus, for the second pressure step, the air removed from the receiver $= 200.0 - 179.8 = 20.2$ cu ft, and so on for the remaining steps. Enter the result of each subtraction in the fourth column of Table 7.

3. Compute the actual quantity of air handled by the pump

The volumetric efficiency of a vacuum pump varies during each pressure reduction. To simplify the pump-down-time calculation, an average value for the volumetric efficiency can be used for each step in the receiver pressure reduction. Find the average volumetric efficiency for this vacuum pump from Fig. 3. Thus, for the pressure reduction from 29.8 to 26.8 in. Hg, the volumetric efficiency is found at $(29.8 + 26.8)/2 = 28.3$ in. Hg to be 91 percent. Enter this value in the fifth column of Table 7. Follow the same procedure to find the remaining values and enter them as shown.

The actual quantity of free air this vacuum pump can handle is numerically equal to the product of the volumetric efficiency, column 5, Table 7, and the pump piston displacement. Or, for the above pressure reduction, free air capacity $= 0.91(60) = 54.6$ cfm. Enter this result in column 6, Table 7.

4. Compute the pump-down time for each pressure reduction

The second line of Table 7 shows, in column 4, that at an absolute pressure of 26.8 in. Hg, 20.2 cu ft of free air is removed from the receiver. However, the vacuum pump can handle, 54.6 cfm, column 6. Since the time required to remove air from the receiver is, in min, (cu ft removed)/(cylinder capacity, cfm), the time required to remove 20.2 cu ft is $20.2/54.6 = 0.370$ min.

Compute the required time for each pressure step in the same manner. The total pump-down time is then the sum of the individual times, or 9.019 min, column 7,

Table 7. This result is suitable for all usual design purposes because it closely approximates the actual time required, and the errors involved are so slight as to be negligible. Leakage into industrial-plant vacuum systems often equals the volume handled by the vacuum pump.

5. *Use the pump-down time for compressor selection*

To choose an industrial vacuum pump using the pump-down procedure described in Steps 1 to 4, (a) obtain the characteristics curves for several makes and capacities of vacuum pumps; (b) compute the pump-down time for each pump using the procedure in Steps 1 to 4; (c) compute the air inflow to the system, based on the free air capacity of each outlet and the number of outlets in the system; (d) compute how long the pump must run to handle the air inflow; and (e) choose the pump having the shortest running time and smallest required power input.

Thus, with 10 vacuum outlets each having a free air flow of 50 cfh, the total air inflow is 10(50) = 500 cfh. This means that a 200 cu ft receiver would be filled 500/200 = 2.5 times per hr. Since the pump discussed in Steps 1 to 4 requires approximately 9 min to reduce the receiver pressure from atmospheric to 2.5 in. Hg absolute, its running time to serve these outlets would be 9(2.5) = 22.5 min, approximately. The power input to this vacuum pump, Fig. 3, ranges from a minimum of about 1 hp to a maximum of about 3 hp.

If another pump could evacuate this receiver in 6 min and needed only 2.5 hp as the maximum power input, it might be a better choice, provided that its first cost were not several times that of the other pump. Use the methods of engineering economics to compare the economic merits of the two pumps.

Related Calculations: Note carefully that the Procedure given here applies to industrial vacuum systems used for cleaning, maintenance, and similar purposes. The Procedure should not be used for high-vacuum systems applied to production processes, experimental laboratories, etc. Use instead the method given in the next Calculation Procedure in this section.

To be certain that the correct pump-down time is obtained, many engineers include the volume of the system piping in the computation. This is done by computing the volume of all pipes in the system and adding the result to the receiver volume. This, in effect, increases the receiver volume that must be pumped down and gives a more accurate estimate of the probable pump-down time. Some engineers also add a leakage allowance of up to 100 percent of the sum of the receiver and piping volume. Thus, if the piping volume in the above system were 50 cu ft, the total volume to be evacuated would be 2(200 + 50) = 500 cu ft. The factor 2 in this expression was inserted to reflect the 100 percent leakage — i.e., the pump must handle the receiver and piping volume plus the leakage, or twice the sum of the receiver and piping volume.

Some industrial vacuum pumps are standard reciprocating air compressors run in the reverse of their normal direction after slight modification. The vacuum lines are connected to the receiver, from which the compressor takes its suction. After removing air from the receiver, the compressor discharges to the atmosphere.

VACUUM-PUMP SELECTION FOR HIGH-VACUUM SYSTEMS

Choose a mechanical vacuum pump for use in a laboratory fitted with a vacuum system having a total volume, including the piping, of 12,000 cu ft. The operating pressure of the system is 0.10 torr, and the optimum pump-down time is 150 min. (NOTE: 1 torr = 1 mm Hg.)

Calculation Procedure:

1. *Make a tentative choice of pump type*

Mechanical vacuum pumps of the reciprocating type are well suited for system pressures in the 0.0001- to 760-torr range. Hence, this type of pump will be considered first to see if it meets the desired pump-down time.

2. *Obtain the pump characteristic curves*

Many manufacturers publish pump-down factor curves such as those in Fig. 4a and b. These curves are usually published as part of the engineering data for a given line of pumps. Obtain the curves from the manufacturers whose pumps are being considered.

3. *Compute the pump-down time for the pumps being considered*

Three reciprocating pumps can serve this system: (a) a single-stage pump, (b) a compound or two-stage pump, or (c) a combination of a mechanical booster and a single-stage backing or roughing-down pump. Figure 4 gives the pump-down factor for each type of pump.

To use the pump-down factor, apply this relation: $t = VF/d$, where t = pump-down time, min; V = system volume, cu ft; F = pump-down factor for the pump; d = pump displacement, cfm.

Thus, for a single-stage pump, Fig. 4a shows that $F = 10.8$ for a pressure of 0.10 torr. Assuming a pump displacement of 1,000 cfm, $t = 12,000(10.8)/1,000 = 129.6$ min; say 130 min.

For a compound pump, $F = 9.5$ from Fig. 4a. Hence, a compound pump having the same displacement, or 1,000 cfm, will require $t = 12,000(9.5)/1,000 = 114.0$ min.

With a combination arrangement, the backing or roughing pump, a 130-cfm unit, reduces the system pressure from atmospheric, 760 torr, to the economical transition pressure, 15 torr, Fig. 4b. Then the single-stage mechanical booster pump, a 1,200-cfm unit, takes over and in combination with the backing pump reduces the pressure to the desired level, or 0.10 torr. During this part of the cycle, the unit operates as a two-stage pump. Hence, the total pump-down time consists of the sum of the backing-pump and booster-pump times. The pump-down factors are, respectively, 4.2 for the backing pump at 15 torr and 6.9 for the booster pump at 0.10 torr. Hence, the respective pump-down times are: $t_1 = 12,000(4.2)/130 = 388$ min; $t_2 = 12,000(6.9)/1,200 = 69$ min. The total time is thus $388 + 69 = 457$ min.

The pump-down time with the combination arrangement is greater than the optimum 150 min. Where a future lower operating pressure is anticipated, making the combination arrangement desirable, an additional large-capacity single-stage roughing pump can be used to assist the 130-cfm unit. This large-capacity unit is operated until the transition pressure is reached and roughing down is finished. The pump is then shut off and the balance of the pumping down is carried on by the combination unit. This keeps the power consumption at a minimum.

Thus, if a 1,200-cfm single-stage roughing pump were used to reduce the pressure to 15 torr, its pump-down time would be $t = 12,000(4.0)/1,200 = 40$ min. The total pump-down time for the combination would then be $40 + 69 = 109$ min, using the time computed above for the two pumps in combination.

4. *Apply the respective system factors*

Studies and experience show that the calculated pump-down time for a vacuum system must be corrected by an appropriate system factor. This factor makes allowance for the normal outgassing of surfaces exposed to atmospheric air. It also provides a basis for judging whether a system is pumping down normally or whether some problem exists that must be corrected. Table 8 lists typical system factors that have proven reliable in many tests. To use the system factor for any pump, apply it this way: $t_a = tS$, where t_a = actual pump-down time, min; t = computed pump-down time from Step 3, min; S = system factor for the type of pump being considered.

Thus, using the appropriate system factor for each pump, the actual pump-down time for the single-stage mechanical pump is $t_a = 130(1.5) = 195$ min. For the compound mechanical pump, $t_a = 114(1.25) = 142.5$ min. For the combination mechanical booster pump, $t_a = 109(1.35) = 147$ min.

5. *Choose the pump to use*

Based on the actual pump-down time, either the compound mechanical pump or the combination mechanical booster pump can be used. The final choice of the pump

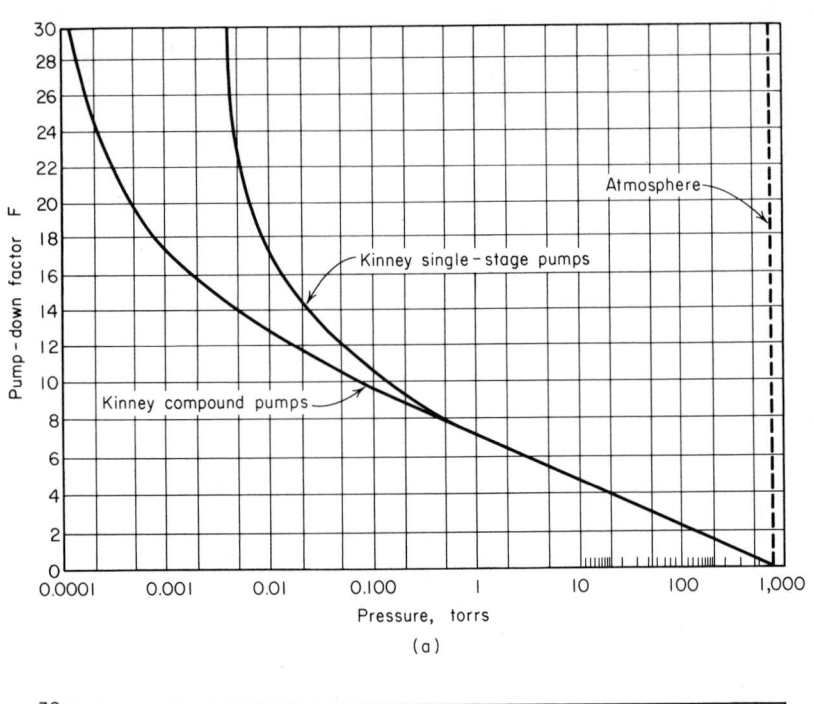

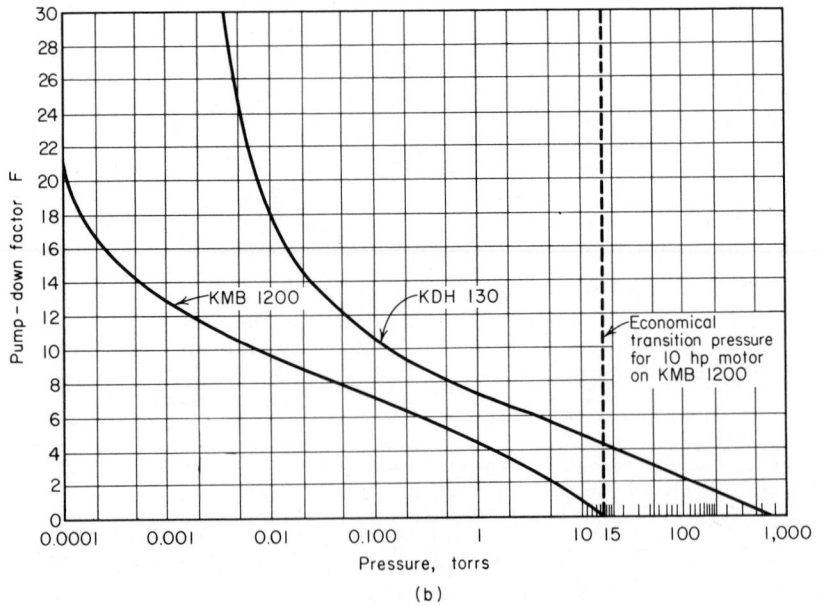

Fig. 4 (a) Pump-down factor for single-stage and compound vacuum pumps; (b) pump-down factor for mechanical booster and backing pump. (*After Kinney Vacuum Div., The New York Air Brake Company, and Van Atta.*)

should take other factors into consideration—first cost, operating cost, maintenance cost, reliability, and probable future pressure requirements in the system. Where future lower pressure requirements are not expected, the compound mechanical pump would be a good choice. However, if lower operating pressures are anticipated in the future, the combination mechanical booster pump would probably be a better choice.

TABLE 8   Recommended System Factors*

Pressure range, torr	System factor		
	Single-stage mechanical pump	Compound mechanical pump	Mechanical booster pump†
760–20	1.0	1.0	
20–1	1.1	1.1	1.15
1–0.5	1.25	1.25	1.15
0.5–0.1	1.5	1.25	1.35
0.1–0.02	...	1.25	1.35
0.02–0.001	...	...	2.0

*From Van Atta—*Vacuum Science and Engineering*, McGraw-Hill.
†Based upon bypass operation until the booster pump is put into operation. Larger system factors apply if rough pumping flow must pass through the idling mechanical booster. Any time needed for operating valves and getting the mechanical booster pump up to speed must also be added.

Van Atta[1] gives the following typical examples of pumps chosen for vacuum systems:

Pressure Range, torr	*Typical Pump Choice*
Down to 50.................	Single-stage oil-sealed rotary; large water or vapor load may require use of refrigerated traps
0.05 to 0.01.................	Single-stage or compound oil-sealed pump plus refrigerated traps, particularly at the lower pressure limit
0.01 to 0.005	Compound oil-sealed plus refrigerated traps, or single-stage pumps backing diffusion pumps if a continuous large evolution of gas is expected
1 to 0.0001...................	Mechanical booster and backing pump combination with interstage refrigerated condenser and cooled vapor trap at the high-vacuum inlet for extreme freedom from vapor contamination
0.0005 and lower.........	Single-stage pumps backing diffusion pumps, with refrigerated traps on the high-vacuum side of the diffusion pumps and possibly between the single-stage and diffusion pumps if evolution of condensable vapor is expected

[1]C. M. Van Atta, —*Vacuum Science and Engineering*, McGraw-Hill Book Company, New York, 1965.

VACUUM-SYSTEM PUMPING SPEED AND PIPE SIZE

A laboratory vacuum system has a volume of 500 cu ft. Leakage into the system is expected at the rate of 0.00035 cfm. What backing pump speed, i.e., displacement, should an oil-sealed vacuum pump serving this system have if the pump blocking pressure is 0.150 mm Hg and the desired operating pressure is 0.0002 mm Hg? What should be the speed of the diffusion pump be? What pipe size is needed for the connecting pipe of the backing pump if it has a displacement or pumping speed of 380 cfm at 0.150 mm Hg and a length of 15 ft?

Calculation Procedure:

1. Compute the required backing pump speed

Use the relation $d_b = G/P_b$, where d_b = backing pump speed or pump displacement, cfm; G = gas leakage or flow rate, mm/cfm. To convert the gas or leakage flow rate to mm/cfm, multiply the cfm by 760 mm, the standard atmospheric pressure, mm Hg. Thus, $d_b = 760(0.00035)/0.150 = 1.775$ cfm.

2. Select the actual backing pump speed

For practical purposes, since gas leakage and outgassing are impossible to calculate accurately, a backing pump speed or displacement of at least twice the computed value, or $2(1.775) = 3.550$ cfm — say 4 cfm — would probably be used.

If this backing pump is to be used for pumping down the system, compute the pump-down time as shown in the previous Calculation Procedure. Should the pump-down time be excessive, increase the pump displacement until a suitable pump-down time is obtained.

3. Compute the diffusion pump speed

The diffusion pump reduces the system pressure from the blocking point, 0.150 mm Hg, to the system operating pressure of 0.0002 mm Hg. (NOTE: 1 torr = 1 mm Hg.) Compute the diffusion pump speed from $d_d = G/P_d$, where d_d = diffusion pump speed, cfm; P_d = diffusion-pump operating pressure, mm Hg. Or $d_d = 760(0.00035)/0.0002 = 1,330$ cfm. To allow for excessive leaks, outgassing, and manifold pressure loss, a 3,000- or 4,000-cfm diffusion pump would be chosen. To ensure reliability of service, two diffusion pumps would be chosen so that one could operate while the other was being overhauled.

4. Compute the size of the connecting pipe

In usual vacuum-pump practice, the pressure drop in pipes serving mechanical pumps is not allowed to exceed 20 percent of the inlet pressure prevailing under steady operating conditions. A correctly designed vacuum system, where this pressure loss is not exceeded, will have a pump-down time which closely approximates that obtained under ideal conditions.

Compute the pressure drop in the high-pressure region of vacuum pumps from $p_d = 1.9 d_b L/d^4$, where p_d = pipe pressure drop, μ; d_b = backing pump displacement or speed, cfm; L = pipe length, ft; d = inside diameter of pipe, in. Since the pressure drop should not exceed 20 percent of the inlet or system operating pressure, the drop for a backing pump is based on its blocking pressure, or 0.150 mm Hg, or 150 μ. Hence $p_d = 0.20(150) = 30\ \mu$. Then, $30 = 1.9(380)(15)/d^4$, and $d = 4.35$ in. Use a 5-in.-diameter pipe.

In the low-pressure region, the diameter of the converting pipe should equal, or be larger than, the pump inlet connection. Whenever the size of a pump is increased, the diameter of the pipe should also be increased to conform with the above guide.

Related Calculations: Use the general Procedures given here for laboratory- and production-type high-vacuum systems.

MATERIALS HANDLING

REFERENCES: Hudson—*Conveyors*, Wiley; Buffalo Forge Company—*Fan Engineering*; Staniar —*Plant Engineering Handbook*, McGraw-Hill; Baumeister and Marks—*Standard Handbook for Mechanical Engineers*, McGraw-Hill.

BULK MATERIAL ELEVATOR AND CONVEYOR SELECTION

Choose a bucket elevator to handle 150 tons/hr of abrasive material weighing 50 lb/cu ft through a vertical distance of 75 ft at a speed of 100 fpm. What hp input is required to drive the elevator? The bucket elevator discharges onto a horizontal conveyor which must transport the material 1,400 ft. Choose the type of conveyor to use and determine the required power input needed to drive it.

Calculation Procedure:

1. Select the type of elevator to use

Table 1 summarizes the various characteristics of bucket elevators used to transport bulk materials vertically. This table shows that a continuous bucket elevator would be a good choice, because it is a recommended type for abrasive materials. The second choice would be a pivoted bucket elevator. However, the continuous bucket type is popular and will be chosen for this application.

TABLE 1   Bucket Elevators

	Centrifugal discharge	Perfect discharge	Continuous bucket	Gravity discharge	Pivoted bucket
Carrying paths	Vertical	Vertical to inclination 15 deg from vertical	Vertical to inclination 15 deg from vertical	Vertical and horizontal	Vertical and horizontal
Capacity range, tons/ hr, material weighing 50 lb/cu ft	78	34	345	191	255
Speed range, fpm	306	120	100	100	80
Location of loading point	Boot	Boot	Boot	On lower horizontal run	On lower horizontal run
Location of discharge point	Over head wheel	Over head wheel	Over head wheel	On horizontal run	On horizontal run
Handling abrasive materials	Not preferred	Not preferred	Recommended	Not recommended	Recommended

SOURCE: Link-Belt Div. of FMC Corp.

2. Compute the elevator height

To allow for satisfactory loading of the bulk material, the elevator length is usually increased by about 5 ft more than the vertical lift. Hence, the elevator height = 75 + 5 = 80 ft.

3. Compute the required power input to the elevator

Use the relation $hp = 2CH/1,000$, where C = elevator capacity, tons/hr; H = elevator height, ft. Thus, for this elevator, $hp = 2(150)(80)/1,000 = 24.0$ hp.

TABLE 2 Conveyor Characteristics

	Belt conveyor	Apron conveyor	Flight conveyor	Drag chain	En masse conveyor	Screw conveyor	Vibratory conveyor
Carrying paths	Horizontal to 18°	Horizontal to 25°	Horizontal to 45°	Horizontal or slight incline, 10°	Horizontal to 90°	Horizontal to 15°; may be used up to 90° but capacity falls off rapidly	Horizontal or slight incline, 5° above or below horizontal
Capacity range, tons/ hr, material weighing 50 lb/cu ft............	2,160	100	360	20	100	150	100
Speed range, fpm.......	600	100	150	20	80	100	40
Location of loading point	Any point	Any point	Any point	Any point	On horizontal runs	Any point	Any point
Location of discharge point	Over end wheel and intermediate points by tripper or plow	Over end wheel	At end of trough and intermediate points by gates	At end of trough	Any point on horizontal runs by gate;	At end of trough and intermediate points by gates	At end of trough
Handling abrasive materials	Recommended	Recommended	Not recommended	Recommended with special steels	Not recommended	Not preferred	Recommended

SOURCE: Link-Belt Div. of FMC Corp.

The power input relation given above is valid for continuous-bucket, centrifugal-discharge, perfect-discharge, and super-capacity elevators. A 25-hp motor would probably be chosen for this elevator.

4. Select the type of conveyor to use

Since the elevator discharges onto the conveyor, the capacity of the conveyor should be the same, per unit time, as the elevator. Table 2 lists the characteristics of various types of conveyors. Study of the tabulation shows that a belt conveyor would probably be best for this application, based on the speed, capacity, and type of material it can handle, hence, it will be chosen for this installation.

5. Compute the required power input to the conveyor

The power input to a conveyor is composed of two portions: (*a*) the power required to move the empty belt conveyor; and (*b*) the power required to move the load horizontally.

Determine from Fig. 1 the power required to move the empty belt conveyor, after choosing the required belt width from Table 3.

Thus, for this conveyor, Table 3 shows that a belt width of 42 in. is required to transport up to 150 tons/hr at a belt speed of 100 fpm. (Note that the next *larger* capacity, 162 tons/hr, is used when the exact capacity required is not tabulated.) Find the horsepower required to drive the empty belt by entering Fig. 1 at the belt distance between centers, 1,400 ft, and projecting vertically upward to the belt width, 42 in. At the left, read the required power input as 7.2 hp.

Compute the power required to move the load horizontally from $hp = (C/100)(0.4 + 0.00345L)$, where $L =$ distance between conveyor centers, ft; other symbols as before. For this conveyor, $hp = (150/100)(0.4 + 0.00325 \times 1,400) = 6.83$ hp. Hence, the total horsepower to drive this horizontal conveyor is $7.2 + 6.83 = 14.03$ hp.

The total horsepower input to this conveyor installation is the sum of the elevator and conveyor belt horsepowers, or $14.03 + 24.0 = 38.03$ hp.

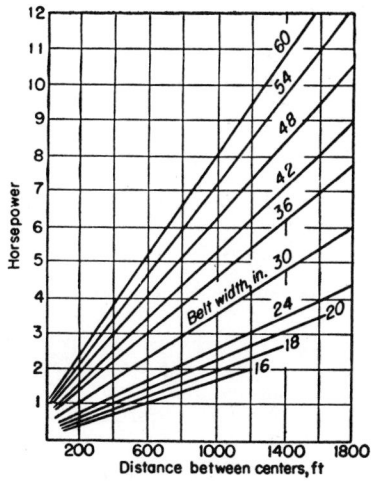

Fig. 1 HP required to move an empty conveyor belt at 100 fpm.

TABLE 3 Capacities of Troughed Rest
(United States Rubber Co., tons/hr with belt speed of 100 fpm)

Belt width, in,	Weight of material, lb/cu ft				Belt width, in.	Weight of material, lb/cu ft			
	30	50	100	150		30	50	100	150
14	10	17	34	51	30	47	79	158	237
16	13	22	44	66	36	69	114	228	342
18	17	28	56	84	42	97	162	324	486
20	20	34	68	102	48	130	215	430	645
24	30	50	100	150	60	207	345	690	1035

Related Calculations: The Procedure given here is valid for conveyors using rubber belts reinforced with cotton duck, open-mesh fabric, cords, or steel wires. It is also valid for stitched-canvas belts, balata belts, and flat-steel belts. The required horsepower input includes any power absorbed by idler pulleys.

Table 4 shows the minimum recommended belt widths for lumpy materials of various sizes. Maximum recommended belt speeds for various materials are shown in Table 5.

When a conveyor belt is equipped with a tripper, the belt must rise about 5 ft above its horizontal plane of travel.

This rise must be included in the vertical lift power input computation. When the tripper is driven by the belt, allow 1 hp for a 16-in. belt, 3 hp for a 36-in. belt, and 7 hp for a 60-in. belt. Where a rotary cleaning brush is driven by the conveyor shaft, allow about the same power input to the brush for belts of various widths.

TABLE 4   Minimum Belt Width for Lumps

Belt width, in	14	16	18	24	30	36	42	48	60
Sized material, in.	2	$2\frac{1}{2}$	3	$4\frac{1}{2}$	7	8	10	12	12
Unsized material, in. ..	3	4	5	8	12	14	20	35	28

TABLE 5   Maximum Belt Speeds, Fpm, for Various Materials

Width of belt, in.	Light or free-flowing materials, grains dry sand, etc.	Moderately free-flowing sand, gravel, fine stone, etc.	Lump coal, coarse stone, crushed ore	Heavy sharp lumpy materials, heavy ores, lump coke
12–14	400	250		
16–18	500	300	250	
20–24	600	400	350	250
30–36	750	500	400	300
42–60	850	550	450	350

SCREW CONVEYOR POWER INPUT AND CAPACITY

What is the required power input for a 100-ft-long screw conveyor handling dry coal ashes having a maximum density of 40 lb/cu ft if the conveyor capacity is 30 tons/hr?

Calculation Procedure:

1. *Select the conveyor diameter and speed*

Refer to a manufacturer's engineering data or Table 6 for a listing of recommended screw conveyor diameters and speeds for various types of materials. Dry coal ashes are commonly rated as group 3 materials, Table 7—i.e., materials with small mixed lumps with fines.

TABLE 6   Screw Conveyor Capacities and Speeds

Material group	Max material density, lb/cu ft	Max rpm for diameters of	
		6 in.	20 in.
1	50	170	110
2	50	120	75
3	75	90	60
4	100	70	50
5	125	30	25

To determine a suitable screw diameter, assume two typical values and obtain the recommended rpm from the sources listed above or Table 6. Thus, the maximum rpm recommended for a 6-in. screw when handling group 3 material is 90, as shown in Table 6; for a 20-in screw, 60 rpm. Assume a 6-in screw as a trial diameter

2. Determine the material factor for the conveyor

A material factor is used in the screw conveyor power input computation to allow for the character of the substance handled. Table 7 lists the material factor for dry ashes as $F = 4.0$. Standard references show that the average weight of dry coal ashes is 35 to 40 lb/cu ft.

TABLE 7 Material Factors for Screw Conveyors

Material group	Material type	Material factor
1	Lightweight: Barley, beans, flour, oats, pulverised coal, etc.	0.5
2	Fines and granular:	
	Coal – slack or fines	0.9
	Sawdust, soda ash	0.7
	Flyash	0.4
3	Small lumps and fines:	
	Ashes, dry alum	4.0
	Salt	1.4
4	Semiabrasives; small lumps	
	Phosphate, cement	1.4
	Clay, limestone;	2.0
	Sugar, white lead	1.0
5	Abrasive lumps:	
	Wet ashes	5.0
	Sewage sludge	6.0
	Flue dust	4.0

3. Determine the conveyor size factor

A size factor that is a function of the conveyor diameter is also used in the power input computation. Table 8 shows that for a 6-in.-diameter conveyor the size factor $A = 54$.

TABLE 8 Screw Conveyor Size Factors

Conveyor diameter, in....	6	9	10	12	16	18	20	24
Size factor	54	96	114	171	336	414	510	690

4. Compute the required power input to the conveyor

Use the relation $hp = 10^{-6}(ALN + CWLF)$, where hp = hp input to the screw conveyor head shaft; A = size factor from Step 3; L = conveyor length, ft; N = conveyor rpm; C = quantity of material handled, cu ft/hr; W = density of material, lb/cu ft; F = material factor from Step 2. For this conveyor, using the data listed above, $hp = 10^{-6}(54 \times 100 \times 60 + 1{,}500 \times 40 \times 100 \times 4.0) = 24.3$ hp. With a 90 percent motor efficiency, the required motor rating would be 24.3/0.90 = 27 hp. A 30-hp motor would be chosen to drive this conveyor. Since this is not an excessive power input, the 6-in. conveyor is suitable for this application.

If the calculation indicates that an excessively large power input—say 50 hp or more

—is required, the larger diameter conveyor should be analyzed. In general, a higher initial investment in conveyor size that reduces the power input will be more than recovered by the savings in power costs.

Related Calculations: Use the Procedure given here for screw or spiral conveyors and feeders handling any material that will flow. The usual screw or spiral conveyor is suitable for conveying materials for distances up to about 200 ft, although special designs can be built for greater distances. Conveyors of this type can be sloped upward to angles of 35° with the horizontal. However, the capacity of the conveyor decreases as the angle of inclination is increased. Thus the reduction in capacity at a 10° inclination is 10 percent over the horizontal capacity; at 35° the reduction is 78 percent.

The capacities of screw and spiral conveyors are generally stated in cu ft/hr of various classes of materials at the maximum recommended shaft rpm. As the size of the lumps in the material conveyed increases, the recommended shaft rpm decreases. The capacity of a screw or spiral conveyor at a lower speed is found from (capacity at given speed, cu ft/hr)(lower speed, rpm/higher speed, rpm). Table 6 shows typical screw conveyor capacities at usual operating speeds.

Various types of screws are used for modern conveyors. These include short-pitch, variable-pitch, cut flights, ribbon, and paddle screws. The Procedure given above also applies to these screws.

DESIGN AND LAYOUT OF PNEUMATIC CONVEYING SYSTEMS

A pneumatic conveying system for handling solids in an industrial exhaust installation contains two grinding-wheel booths and one lead each for a planer, sander, and circular saw. Determine the required duct sizes, resistance, and fan capacity for this pneumatic conveying system.

Calculation Procedure:

1. Sketch the proposed exhaust system

Make a freehand sketch, Fig. 2, of the proposed system. Show the main and branch ducts and the booths and hoods. Indicate all major structural interferences, such as building columns, deep girders, beams, overhead conveyors, piping, etc. Draw the layout approximately to scale.

Mark on the sketch the length of each duct run. Avoid, if possible, vertical drops or rises in the main exhaust duct between the hoods and the fan. Do this by locating the main duct centerline 10 ft or so above the finished floor.

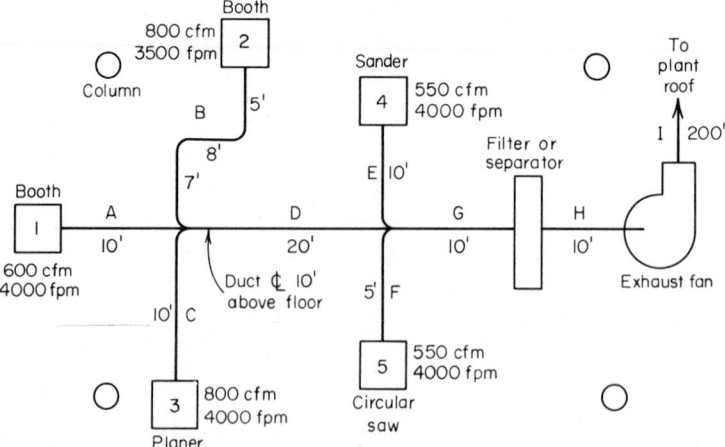

Fig. 2   Exhaust system layout.

TABLE 5 Exhaust System Design Calculations

(1) Booth or hood	(2) Duct run	(3) cfm in duct	(4) Design velocity, fpm	(5) Duct area = Column 3/Column 4 sq ft	(6) Duct diameter, in.	(7) Actual velocity, fpm	(8) Actual velocity pressure in. H₂O	(9) Length of straight duct, ft	(10) Equivalent length of elbows	(11) Total duct length = Column 9+Column 10	(12) Friction per 100 ft of duct, in. H₂O	(13) Actual friction, in. H₂O
1	A	600	4,000	0.150	5	4,300	1.15	10	0	10	5.4	0.54
2	B	800	3,500	0.228	6	4,200	1.0	20	18	38	4.0	1.57
3	C	800	4,000	0.200	6	4,200	1.0	10	6	16	4.0	0.64
	D	2,200	4,000	0.550	10	4,000	1.0	20	0	20	2.1	0.42
4	E	550	4,000	0.137	5	4,000	1.0	10	5	15	4.6	0.69
5	F	550	4,000	0.137	5	4,000	1.0	5	5	10	4.6	0.46
	G	3,300	4,000	0.825	12	4,200	1.0	10	0	10	1.9	0.19
	H	3,300	3,000	1.10	14	3,000	0.55	10	14	24	0.84	0.20
	I	3,300	2,000	1.65	18	2,000	0.25	200	0	200	0.25	0.50

System resistance

		Hood number			
	1	2	3	4	5
Velocity pressure in hood branch, in. H₂O	1.15	1.0	1.0	1.0	1.0
Entrance loss (% of velocity pressure)	50	11	50	60	60
Entrance loss, in. H₂O	0.58	0.11	0.50	0.60	0.60
Branch and main duct resistances — A	0.54				
B		1.57			
C			0.64		
D	0.42	0.42	0.42		
E				0.69	
F					0.46
G	0.19	0.19	0.19	0.19	0.19
H	0.20	0.20	0.20	0.20	0.20
I	0.50	0.50	0.50	0.50	0.50
Collector or filter resistance, in. H₂O	2.00	2.00	2.00	2.00	2.00
Total resistance in each branch, in. H₂O	4.43	4.99	4.45	4.18	3.95

Number each hood or booth and give each duct run an identifying letter. Although it is not absolutely necessary, it is more convenient during the design process to have the hoods in numerical order and the duct runs in alphabetical order.

2. Determine the required air quantities and velocities

Prepare a listing, Columns 1 and 2, Table 9, of the booths, hoods, and duct runs. Enter the required air quantities and velocities for each booth or hood and duct in Table 9, columns 3 and 4. Select the air quantities and velocities from the local code covering industrial exhaust systems, if such a code is available. If a code does not exist, use the ASHRAE *Guide* or Table 10.

Use extreme care in selecting the air quantities and velocities, because insufficient flow may cause dangerous atmospheric conditions. Harmful process wastes in the form of dust, gas, or moisture may injure plant personnel.

TABLE 10   Recommended Exhaust Air Quantities

Operation	Cfm	Branch duct velocity, fpm	Branch duct diameter, in.
Sanding:			
Single drum, (10-in. diameter)	400	4,000	4
Disk	550	4,000	5
Circular saws (16 to 24-in.			
diameter	450	4,000	4.5
Shoe machinery	550	4,000	5
Buffing and polishing wheels			
(16 to 24-in. diameter)........	600	4,500	5
Grinding wheels (16 to 20-in.			
diameter)..................	600	4,500	5
Abrasive blast rooms...........	...	3,500	
Pharmaceuticals..............	...	3,000	

Conveying velocities

Material conveyed	Conveying velocity, fpm
Vapors, gases, fumes, fine dusts........	1,500 to 2,000
Fine dry dusts.......................	3,000
Average industrial dusts	3,500
Coarse particles.....................	3,500 to 4,500
Large particles, heavy loads, moist	
materials, pneumatic conveying	4,500 and higher

3. Size the main and branch ducts

Determine the required duct area by dividing the air quantity, cfm, by the air velocity in the duct, or column 3/column 4, Table 9. Enter the result in column 5, Table 9.

Once the required duct area is known, find from Table 11 the nearest whole-number duct diameter corresponding to the required area. Avoid fractional diameters at this stage of the calculation, because ducts of these sizes are usually more expensive to fabricate. Later, if necessary, two or three duct sizes may be changed to fractional values. By selecting only whole-number diameters in the beginning, the cost of duct fabrication may be reduced somewhat. Enter the duct whole-number diameter in column 6, Table 9.

4. Compute the actual air velocity in the duct

Use Fig. 3 to determine the actual velocity in each duct. Enter the chart at the air quantity corresponding to that in the duct and project vertically to the diameter curve

TABLE 11   Duct Diameters and Areas

Diameter, in.	Area, sq ft	Diameter, in.	Area, sq ft	Diameter, in.	Area, sq ft
1.5	0.0123	17	1.576	52	14.750
2.0	0.0218	18	1.767	54	15.900
2.5	0.0341	19	1.969	56	17.100
3.0	0.0491	20	2.182	58	18.350
3.5	0.0668	21	2.405	60	19.630
4.0	0.0873	22	2.640		
4.5	0.1104	23	2.885		
5.0	0.1364	24	3.142		
5.5	0.1650	25	3.409		
6.0	0.1964	26	3.687		
6.5	0.2304	27	3.976		
7.0	0.2673	28	4.276		
7.5	0.3068	29	4.587		
8.0	0.3491	30	4.909		
8.5	0.3942	32	5.585		
9.0	0.4418	33	5.940		
9.5	0.4923	34	6.305		
10.0	0.5454	36	7.069		
11	0.6600	38	7.876		
12	0.7854	40	8.727		
13	0.9218	42	9.621		
14	1.069	45	11.045		
15	1.227	48	12.566		
16	1.396	50	13.640		

representing the duct size. Read the actual velocity in the duct on the velocity scale and enter the value in column 7 of Table 9.

The actual velocity in the duct should, in all cases, be equal to or greater than the design velocity shown in column 4, Table 9. If the actual velocity is less than the design velocity, decrease the duct diameter until the actual velocity is equal to or greater than the design velocity.

5.   Compute the duct velocity pressure

With the actual velocity known, compute the corresponding velocity pressure in the duct from $h_v = (v/4{,}005)^2$, where h_v = velocity pressure in the duct, in. H_2O; v = air velocity in the duct, fpm. Thus, for duct run A in which the actual air velocity is 4,300 fpm, $h_v = (4{,}300/4{,}005)^2 = 1.15$ in. H_2O. Compute the actual velocity pressure in each duct run and enter the result in column 8, Table 9.

6.   Compute the equivalent length of each duct

Enter the total straight length of each duct, including any vertical drops, in column 9, Table 9. Use accurate lengths, because the system resistance is affected by the duct length.

Next list the equivalent length of each elbow in the duct runs in column 10, Table 9. For convenience, assume that the equivalent length of an elbow is 12 times the duct diameter in ft. Thus, an elbow in a 6-in.-diameter duct has an equivalent resistance of (6 in. diameter/[(12 in./ft)(12)]) = 6 ft of straight duct. When making this calculation, assume that all elbows have a radius equal to twice the diameter of the duct. Consider 45° bends as having the same resistance as 90° elbows. Note that branch ducts are usually arranged to enter the main duct at an angle of 45° or less. These assumptions are valid for all typical industrial exhaust systems and pneumatic conveying systems.

Find the total equivalent length of each duct by taking the sum of columns 9 and 10, Table 9, horizontally, for each duct run. Enter the result in column 11, Table 9.

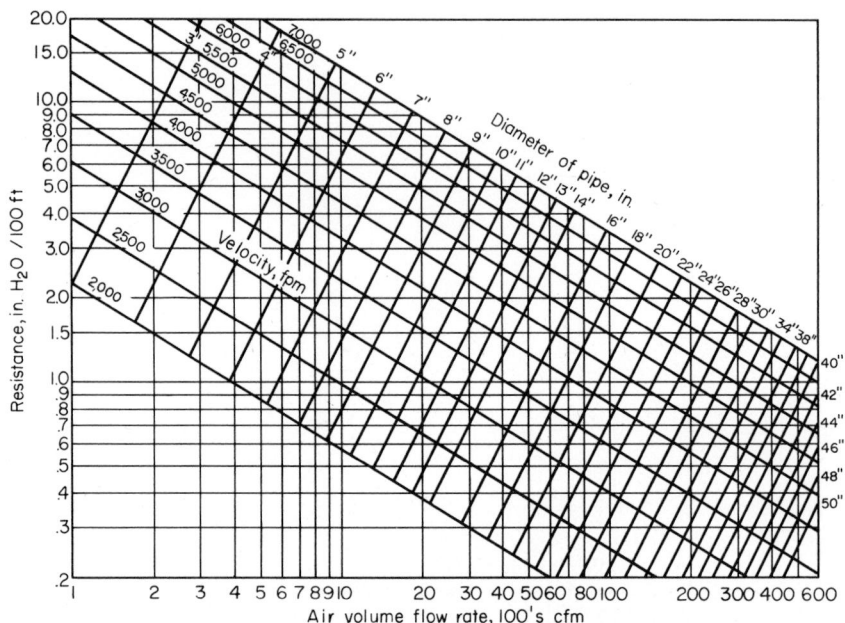

Fig. 3   Duct resistance chart. (*American Air Filter Co.*)

7.   Determine the actual friction in each duct

Using Fig. 3, determine the resistance, in. of H_2O per 100 ft, of each duct by entering with the air quantity and diameter of that duct. Enter the frictional resistance thus found in column 12, Table 9.

Compute actual friction in each duct by multiplying the friction per 100 ft of duct, column 12, Table 9, by the total duct length, column 11, ÷ 100. Thus for duct run A, actual friction = 5.4(10/100) = 0.54 in. H_2O. Compute the actual friction for the other duct runs in the same manner. Tabulate the results in column 13, Table 9.

8.   Compute the hood entrance losses

Hoods are used in industrial exhaust systems to remove vapors, dust, fumes, and other undesirable airborne contaminants from the work area. The hood entrance loss, which depends upon the hood configuration, is usually expressed as a certain percentage of the velocity pressure in the branch duct connected to the hood, Fig. 4. Since the hood entrance loss usually accounts for a large portion of the branch resistance, the entrance loss chosen should always be on the safe side.

List the hood designation number under the "System Resistance" heading, as shown in Table 9. Under each hood designation number, list the velocity pressure in the branch connected to that hood. Obtain this value from column 8, Table 9. List under the velocity pressure, the hood entrance loss from Fig. 4 for the particular type of hood used in that duct run. Take the product of these two values and enter the result under the hood number on the "entrance loss, in. H_2O" line. Thus, for hood 1, entrance loss = 1.15 (0.50) = 0.58 in. H_2O. Follow the same procedure for the other hoods listed.

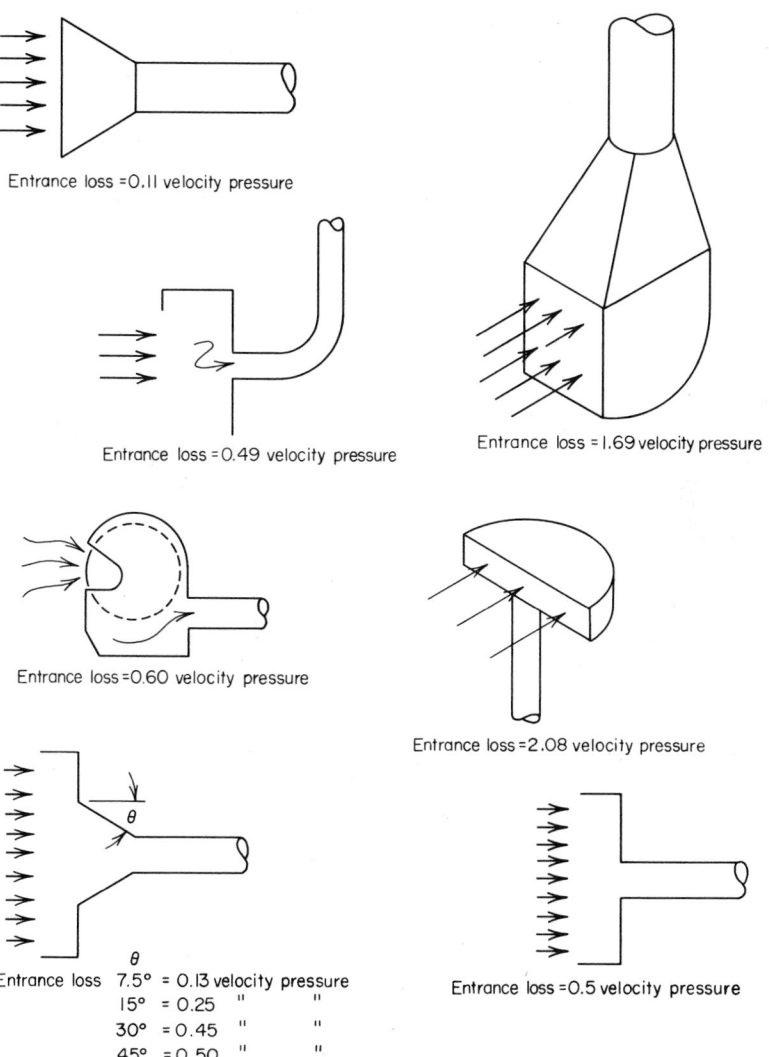

Fig. 4   Entrance losses for various types of exhaust system intakes.

9. Find the resistance of each branch run

List the main and branch runs, *A* through *F*, Table 9. Trace out each main and branch run in Fig. 2 and enter the actual friction listed in column 3 of Table 9. Thus for booth 1, the main and branch runs consist of *A*, *D*, *G*, *H*, and *I*. Insert the actual friction, in. H_2O, as shown in Table 9, or $A = 9.54, D = 0.42, G = 0.19, H = 0.20, I = 0.50$.

Determine the filter friction loss from the manufacturer's engineering data. It is common practice to design industrial exhaust systems on the basis of dirty filters or separators, i.e., the frictional resistance used in the design calculations is the resistance of a filter or separator containing the maximum amount of dust allowable under normal operating conditions. The frictional resistance of dirty filters can vary from 0.5 to 6 in.

H_2O or more. Assume that the frictional resistance of the filter used in this industrial exhaust system is 2.0 in. H_2O.

Add the filter resistance to the main and branch duct resistance as shown in Table 9. Find the sum of each column in the table, as shown. This is the total resistance in each branch, in. H_2O, Table 9.

10. Balance the exhaust system

Inspection of the lower part of Table 9 shows that the computed branch resistances are unequal. This condition is usually encountered during system design. To balance the system, certain duct sizes must be changed to produce equal resistance in all ducts. Or, if possible, certain ducts can be shortened. If duct shortening is not possible, as is often the case, an exhaust fan capable of operating against the largest resistance in a branch can be chosen. If this alternative is selected, special dampers must be fitted to the air inlets of the booths or ducts. For economical system operation, choose the balancing method that permits the exhaust fan to operate against the minimum resistance.

In the system being considered here, a fairly accurate balance can be obtained by decreasing the size of ducts E and F to 4.75 and 4.375 in., respectively. Duct B would be increased to 6.5 in. in diameter.

11. Choose the exhaust fan capacity and static pressure

Find the required exhaust fan capacity in cfm from the sum of the airflows in the ducts, A through H, column 3, Table 9, or 3,300 cfm. Choose a static pressure equal to or greater than the total resistance in the branch duct having the greatest resistance. Since this is slightly less than 4.5 in. H_2O, a fan developing 4.5 in. H_2O static pressure will be chosen. A 10 percent safety factor is usually applied to these values, giving a capacity of 3,600 cfm and a static pressure of 5.0 in. H_2O for this system.

12. Select the duct material and thickness

Galvanized sheet steel is popular for industrial exhaust systems, except where corrosive fumes and gases rule out galvanized material. Under these conditions, plastic, tile, stainless steel, or composition ducts may be substituted for galvanized ducts. Table 12 shows the recommended metal gage for galvanized ducts of various diameters. Do not use galvanized-steel ducts for gas temperatures higher than 400 F.

TABLE 12   Exhaust System Duct Gages

Duct Diameter, in.	Metal Gage
Up to 8	22
9 to 18	20
19 to 30	18
31 in. and larger	16

Hoods should be two gages heavier than the connected branch duct. Use supports not more than 12 ft apart for horizontal ducts up to 8 in. diameter. Supports can be spaced up to 20 ft apart for larger ducts. Fit a duct cleanout opening every 10 ft. Where changes of diameter are made in the main duct, fit an eccentric taper with a length of at least 5 in. for every 1 in. change in diameter. The end of the main duct is usually extended 6 in. beyond the last branch and closed with a removable cap. For additional data on industrial exhaust system design, see the newest issue of the ASHRAE *Guide*.

Related Calculations: Use this Procedure for any type of industrial exhaust system, such as those serving metalworking, woodworking, plating, welding, paint spraying, barrel filling, foundry, crushing, tumbling, and similar operations. Consult the local code or ASHRAE *Guide* for specific airflow requirements for these and other industrial operations.

This design procedure is also valid, in general, for industrial pneumatic conveying systems. For several comprehensive, worked-out designs of pneumatic conveying systems, see Hudson — *Conveyors*, Wiley.

PUMPS AND PUMPING SYSTEMS

REFERENCES: The Hydraulic Institute—*Standards of the Hydraulic Institute*; Allis-Chalmers Manufacturing Company—*Pic-A-Pump*; Hicks and Edwards—*Pump Application Engineering*, McGraw-Hill; Stepanoff—*Centrifugal and Axial Flow Pumps*, Wiley; Karassik and Carter—*Centrifugal Pumps*, McGraw-Hill; Allen—*Using Centrifugal Pumps*, Oxford; Buffalo Pumps—*Centrifugal Pump Application Manual*; Kristal and Annett—*Pumps*, McGraw-Hill; Economy Pumps, Inc.—*Pump Data*; Molloy—*Pumps and Pumping*, Chemical Publishing; Moore, et al.—*The Vertical Pump*, Johnston Pump Company; Karassik—*Engineers' Guide to Centrifugal Pumps*, McGraw-Hill; Kovats and Desmur—*Pompes, Ventilateurs, Compresseurs*, Dunod, Paris; Fuchslocher and Schulz—*Die Pumpen*, Springer-Verlag, Berlin; Pfleiderer—*Die Kreiselpumpen*, Springer-Verlag, Berlin.

SIMILARITY OR AFFINITY LAWS FOR CENTRIFUGAL PUMPS

A centrifugal pump designed for a 1,800-rpm operation and a head of 200 ft has a capacity of 3,000 gpm with a power input of 175 hp. What effect will a speed reduction to 1,200 rpm have on the head, capacity, and power input of the pump? What will be the change in these variables if the impeller diameter is reduced from 12 to 10 in. while the speed is held constant at 1,800 rpm?

Calculation Procedure:

1. Compute the effect of a change in pump speed

For any centrifugal pump in which the effects of fluid viscosity are negligible, or are neglected, the similarity or affinity laws can be used to determine the effect of a speed, power, or head change. For a *constant impeller diameter*, these laws are: $Q_1/Q_2 = N_1/N_2$; $H_1/H_2 = (N_1/N_2)^2$; $P_1/P_2 = (N_1/N_2)^3$. For a *constant speed*, $Q_1/Q_2 = D_1/D_2$; $H_1/H_2 = (D_1/D_2)^2$; $P_1/P_2 = (D_1/D_2)^3$. In both sets of laws, Q = capacity, gpm; N = impeller rpm; D = impeller diameter, in.; H = total head, ft of liquid; P = bhp input. The subscripts 1 and 2 refer to the initial and changed conditions, respectively.

For this pump, with a constant impeller diameter, $Q_1/Q_2 = N_1/N_2$; $3,000/Q_2 = 1,800/1,200$; $Q_2 = 2,000$ gpm. And, $H_1/H_2 = (N_1/N_2)^2 = 200/H_2 = (1,800/1,200)^2$; $H_2 = 88.9$ ft. Also, $P_1/P_2 = (N_1/N_2)^3 = 175/P_2 = (1,800/1,200)^3$; $P_2 = 51.8$ bhp.

2. Compute the effect of a change in impeller diameter

With the speed constant, use the second set of laws. Or for this pump, $Q_1/Q_2 = D_1/D_2$; $3,000/Q_2 = \frac{12}{10}$;　$Q_2 = 2,500$ gpm. And, $H_1/H_2 = (D_1/D_2)^2$; $200/H_2 = (\frac{12}{10})^2$; $H_2 = 138.8$ ft. Also, $P_1/P_2 = (D_1/D_2)^3$; $175/P_2 = (\frac{12}{10})^3$; $P_2 = 101.2$ bhp.

Related Calculations: Use the similarity laws to extend or change the data obtained from centrifugal pump characteristic curves. These laws are also useful in field calculations when the pump head, capacity, speed, or impeller diameter is changed.

The similarity laws are most accurate when the efficiency of the pump remains nearly constant. Results obtained when the laws are applied to a pump having a constant impeller diameter are somewhat more accurate than for a pump at constant speed with a changed impeller diameter. The latter laws are more accurate when applied to pumps having a low specific speed.

If the similarity laws are applied to a pump whose impeller diameter is increased, be certain to consider the effect of the higher velocity in the pump suction line. Use the similarity laws for any liquid whose viscosity remains constant during passage through the pump. However, the accuracy of the similarity laws decreases as the liquid viscosity increases.

SIMILARITY OR AFFINITY LAWS IN CENTRIFUGAL PUMP SELECTION

A test-model pump delivers, at its best efficiency point, 500 gpm at a 350-ft head with a required net positive suction head (NPSH) of 10 ft, a power input of 55 hp at 3,500 rpm, when using a 10.5-in. diameter impeller. Determine the performance of the model at 1,750 rpm. What is the performance of a full-scale prototype pump with a 20-in. impeller operating at 1,170 rpm? What are the specific speeds and the suction specific speeds of the test-model and prototype pumps?

Calculation Procedure:

1. Compute the pump performance at the new speed

The similarity or affinity laws can be stated in general terms, with subscripts p and m for prototype and model, respectively, as $Q_p = K_d^3 K_n Q_m$; $H_p = K_d^2 K_n^2 H_m$; $NPSH_p = K_d^2 K_n^2 NPSH_m$; $P_p = K_d^5 K_n^5 P_m$, where K_d = size factor = prototype dimension/model dimension. The usual dimension used for the size factor is the impeller diameter. Both dimensions should be in the same units of measure. Also, K_n = prototype speed, rpm/model speed, rpm. Other symbols are the same as in the previous Calculation Procedure.

When the model speed is reduced from 3,500 to 1,750 rpm, the pump dimensions remain the same and $K_d = 1.0$; $K_n = 1,750/3,500 = 0.5$. Then $Q = (1.0)(0.5)(500) = 250$ rpm; $H = (1.0)^2(0.5)^2(350) = 87.5$ ft; $NPSH = (1.0)^2(0.5)^2(10) = 2.5$ ft; $P = (1.0)^5(0.5)^3(55) = 6.9$ hp. In this computation, the subscripts were omitted from the equations because the same pump, the test model, was being considered.

2. Compute performance of the prototype pump

First, K_d and K_n must be found. $K_d = 20/10.5 = 1.905$; $K_n = 1,170/3,500 = 0.335$. Then, $Q_p = (1.905)^3(0.335)(500) = 1,158$ gpm; $H_p = (1.905)^2(0.335)^2(350) = 142.5$ ft; $NPSH_p = (1.905)^2(0.335)^2(10) = 4.06$ ft; $P_p = (1.905)^5(0.335)^3(55) = 51.8$ hp.

3. Compute the specific speed and suction specific speed

The specific speed or, as Horwitz[1] says, "more correctly, discharge specific speed," $N_S = N(Q)^{0.5}/(H)^{0.75}$, while the suction specific speed $S = N(Q)^{0.5}/(NPSH)^{0.75}$, where all values are taken at the best efficiency point of the pump.

For the model, $N_S = 3,500(500)^{0.5}/(350)^{0.75} = 965$; $S = 3,500(500)^{0.5}/(10)^{0.75} = 13,900$. For the prototype, $N_S = 1,17.0(1,158)^{0.5}/(142.5)^{0.75} = 965$; $S = 1,170(1,156)^{0.5}/(4.06)^{0.75} = 13,900$. The specific speed and suction specific speed of the model and prototype are equal because these units are geometrically similar or homologous pumps and both speeds are mathematically derived from the similarity laws.

Related Calculations: Use the procedure given here for any type of centrifugal pump where the similarity laws apply. When the term model is used, it can apply to a production test pump or to a standard unit ready for installation. The Procedure presented here is the work of R. P. Horwitz, as reported in *Power* magazine.[1]

SPECIFIC-SPEED CONSIDERATIONS IN CENTRIFUGAL PUMP SELECTION

What is the upper limit of specific speed and capacity of a 1,750-rpm single-stage double-suction centrifugal pump having a shaft that passes through the impeller eye if it handles clear water at 85 F at sea level at a total head of 280 ft with a 10-ft suction lift? What is the efficiency of the pump and its approximate impeller shape?

Calculation Procedure:

1. Determine the upper limit of specific speed

Use the Hydraulic Institute upper specific-speed curve, Fig. 1, for centrifugal pumps or a similar curve, Fig. 2, for mixed- and axial-flow pumps. Enter Fig. 1 at the bottom at 280-ft total head and project vertically upward until the 10-ft suction-lift curve is intersected. From here, project horizontally to the right to read the specific speed $N_S = 2,000$. Figure 2 is used in a similar manner.

[1]R. P. Horwitz, "Affinity Laws and Specific Speed Can Simplify Centrigual Pump Selection," *Power*, November, 1964.

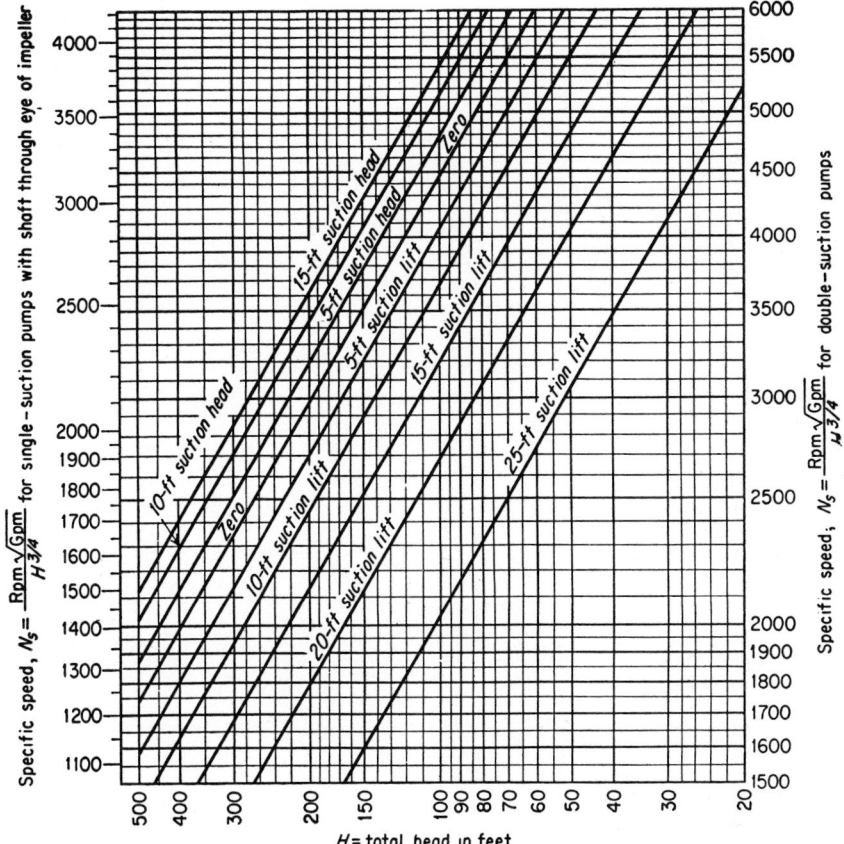

Left axis: Specific speed, $N_S = \dfrac{\text{Rpm}\sqrt{\text{Gpm}}}{H^{3/4}}$ for single-suction pumps with shaft through eye of impeller

Right axis: Specific speed; $N_S = \dfrac{\text{Rpm}\sqrt{\text{Gpm}}}{H^{3/4}}$ for double-suction pumps

H = total head in feet

Fig. 1 Upper limits of specific speeds of single-stage, single- and double-suction centrifugal pumps handling clear water at 85 F at sea level. (*Hydraulic Institute.*)

2. Compute the maximum pump capacity

For any centrifugal, mixed- or axial-flow pump, $N_S = (\text{gpm})^{0.5}(\text{rpm})/H_t^{0.75}$, where H_t = total head on the pump, ft of liquid. Solving for the maximum capacity, gpm = $(N_S H_t^{0.75}/\text{rpm})^2 = (2{,}000 \times 280^{0.75}/1{,}750)^2 = 6{,}040$ gpm.

3. Determine the pump efficiency and impeller shape

Figure 3 shows the general relation between impeller shape, specific speed, pump capacity, efficiency, and characteristic curves. At $N_S = 2{,}000$, efficiency = 87 percent. The impeller, as shown in Fig. 3, is moderately short and has a relatively large discharge area. A cross section of the impeller appears directly under the $N_S = 2{,}000$ ordinate.

Related Calculations: Use the method given here for any type of pump whose variables are included in the Hydraulic Institute curves, Figs. 1 and 2, and in similar curves available from the same source. *Operating specific speed*, computed as above, is sometimes plotted on the performance curve of a centrifugal pump so that the characteristics of the unit can be better understood. *Type specific speed* is the operating specific speed giving maximum efficiency for a given pump and is a number used to identify a pump. Specific speed is important in cavitation and suction-lift studies. The Hydraulic Institute curves, Figs. 1 and 2, give upper limits of speed, head, capacity and suction lift for cavitation-free operation. When making actual pump analyses, be

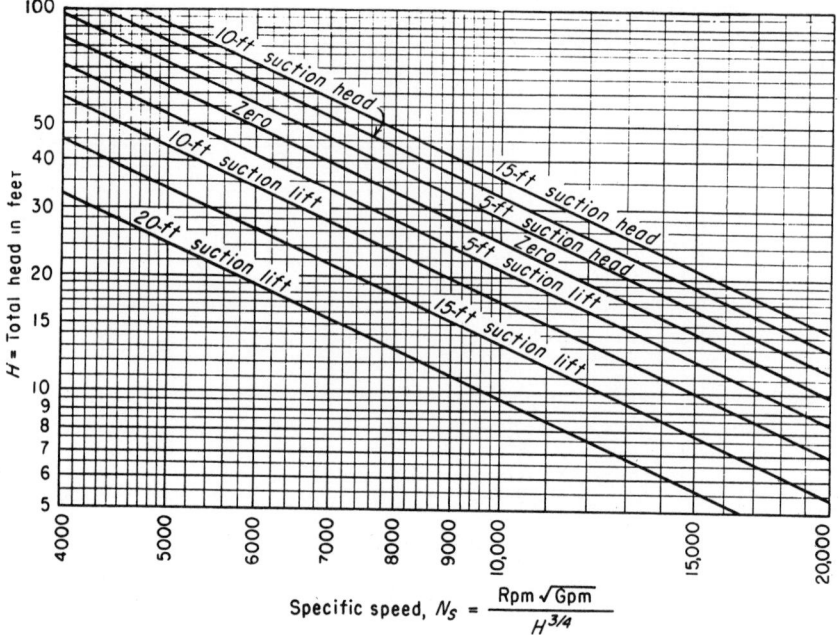

Fig. 2 Upper limits of specific speeds of single-suction mixed-flow and axial-flow pumps. *(Hydraulic Institute.)*

certain to use the curves (Figs. 1 and 2 herewith) in the latest edition of the *Standards of the Hydraulic Institute.*

SELECTING THE BEST OPERATING SPEED FOR A CENTRIFUGAL PUMP

A single-suction centrifugal pump is driven by a 60-cps ac motor. The pump delivers 10,000 gpm of water at a 100-ft head. The available net positive suction head = 32 ft of water. What is the best operating speed for this pump if the pump operates at its best efficiency point?

Calculation Procedure:

1. *Determine the specific speed and suction specific speed*

A-c motors can operate at a variety of speeds, depending on the number of poles. Assume that the motor driving this pump might operate at 870, 1,160, 1,750, or 3,500 rpm. Compute the specific speed $N_S = N(Q)^{0.5}/(H)^{0.75} = N(10,000)^{0.5}/(100)^{0.75} = 3.14N$ and the suction specific speed $S = N(Q)^{0.5}/(NPSH)^{0.75} = N(10,000)^{0.5}/(32)^{0.75} = 7.43N$ for each of the assumed speeds and tabulate the results as follows:

Operating speed, rpm	Required specific speed	Required suction specific speed
870	2,740	6,460
1,160	3,640	8,620
1,750	5,500	13,000
3,500	11,000	26,000

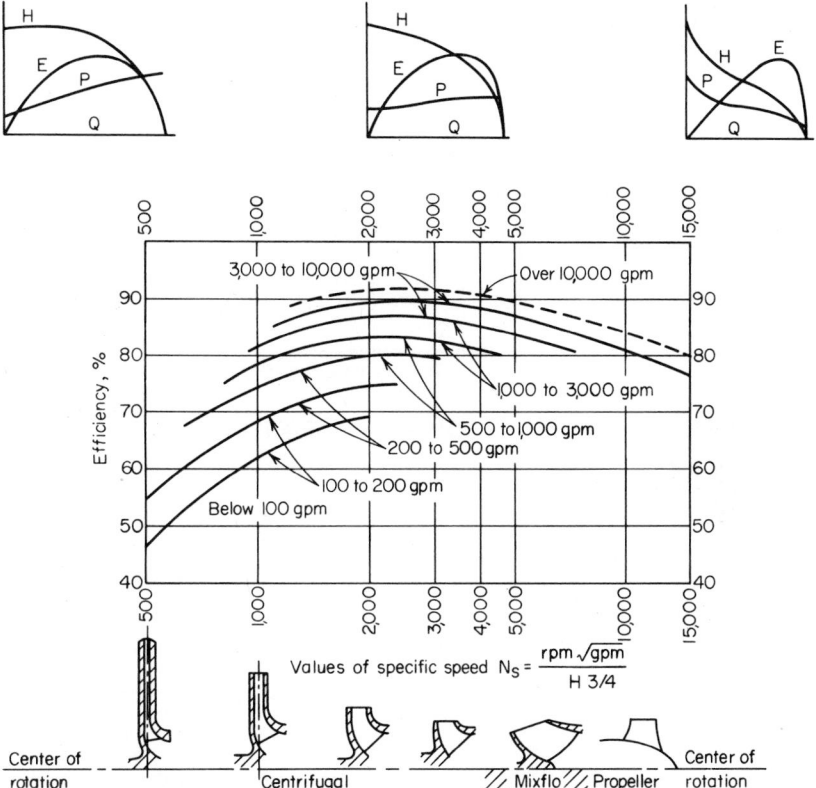

Fig. 3 **Approximate relative impeller shapes and efficiency variations for various specific speeds of centrifugal pumps.** (*Worthington Corporation.*)

2. Choose the best speed for the pump

Analyze the specific speed and suction specific speed at each of the various operating speeds using the data in Tables 1 and 2. These tables show that at 870 and 1,160 rpm, the suction specific-speed rating is poor. At 1,750 rpm, the suction specific-speed

TABLE 1 Pump Types Listed by Specific Speed*

Specific Speed Range	Type of Pump
Below 2,000	Volute, diffuser
2,000–5,000	Turbine
4,000–10,000	Mixed-flow
9,000–15,000	Axial-flow

*Peerless Pump Division, FMC Corporation.

rating is excellent, and a turbine or mixed-flow type pump will be suitable. Operation at 3,500-rpm is unfeasible because a suction specific speed of 26,000 is beyond the range of conventional pumps.

Related Calculations: Use this procedure for any type of centrifugal pump handling water for plant services, cooling, process, fire protection, and similar requirements. This Procedure is the work of R. P. Horwitz, Hydrodynamics Division, Peerless Pump, FMC Corporation, as reported in *Power* magazine.

TABLE 2   Suction Specific-Speed Ratings*

Single-suction pump	Double-suction pump	Rating
Above 11,000	Above 14,000	Excellent
9,000–11,000	11,000–14,000	Good
7,000–9,000	9,000–11,000	Average
5,000–7,000	7,000–9,000	Poor
Below 5,000	Below 7,000	Very poor

*Peerless Pump Division, FMC Corporation.

TOTAL HEAD ON A PUMP HANDLING VAPOR-FREE LIQUID

Sketch three typical pump piping arrangements with static suction lift and submerged, free, and varying discharge head. Prepare similar sketches for the same pump with static suction head. Label the various heads. Compute the total head on each pump if the elevations are as shown in Fig. 4 and the pump discharges a maximum of 2,000 gpm of water through 8-in. schedule 40 pipe. What hp is required to drive the pump? A swing check valve is used on the pump suction line and a gate valve on the discharge line.

Calculation Procedure:

1. Sketch the possible piping arrangements

Figure 4 shows the six possible piping arrangements for the stated conditions of the installation. Label the total static head—i.e., the *vertical* distance from the surface of the source of the liquid supply to the free surface of the liquid in the discharge receiver, or to the point of free discharge from the discharge pipe. When both the suction and discharge surfaces are open to the atmosphere, the total static head equals the vertical difference in elevation. Use the free-surface elevations that cause the maximum suction lift and discharge head—i.e., the *lowest* possible level in the supply tank and the *highest* possible level in the discharge tank or pipe. When the supply source is *below* the pump centerline, the vertical distance is called the *static suction lift*; with the supply *above* the pump centerline, the vertical distance is called *static suction head*. With variable static suction head, use the lowest liquid level in the supply tank when computing total static head. Label the diagrams as shown in Fig. 4.

2. Compute the total static head on the pump

The total static head H_{ts} ft = static suction lift, h_{sl} ft + static discharge head, h_{sd} ft, where the pump has a suction lift, s in Fig. 4a, b, and c. In these installations, H_{ts} = 10 + 100 = 110 ft. Note that the static discharge head is computed between the pump centerline and the water level with an underwater discharge, Fig. 4a; to the pipe outlet with a free discharge, Fig. 4b; and to the maximum water level in the discharge tank, Fig. 4c. When a pump is discharging into a closed compression tank, the total discharge head equals the static discharge head plus the head equivalent, ft of liquid, of the internal pressure in the tank, or 2.31 × tank pressure, psi.

Where the pump has a static suction head, as in Fig. 4d, e, and f, the total static head H_{ts} ft = h_{sd} − static suction head h_{sh} ft. In these installations, H_t = 100 − 15 = 85 ft.

The total static head, as computed above, refers to the head on the pump without liquid flow. To determine the total head on the pump, the friction losses in the piping system during liquid flow must be also determined.

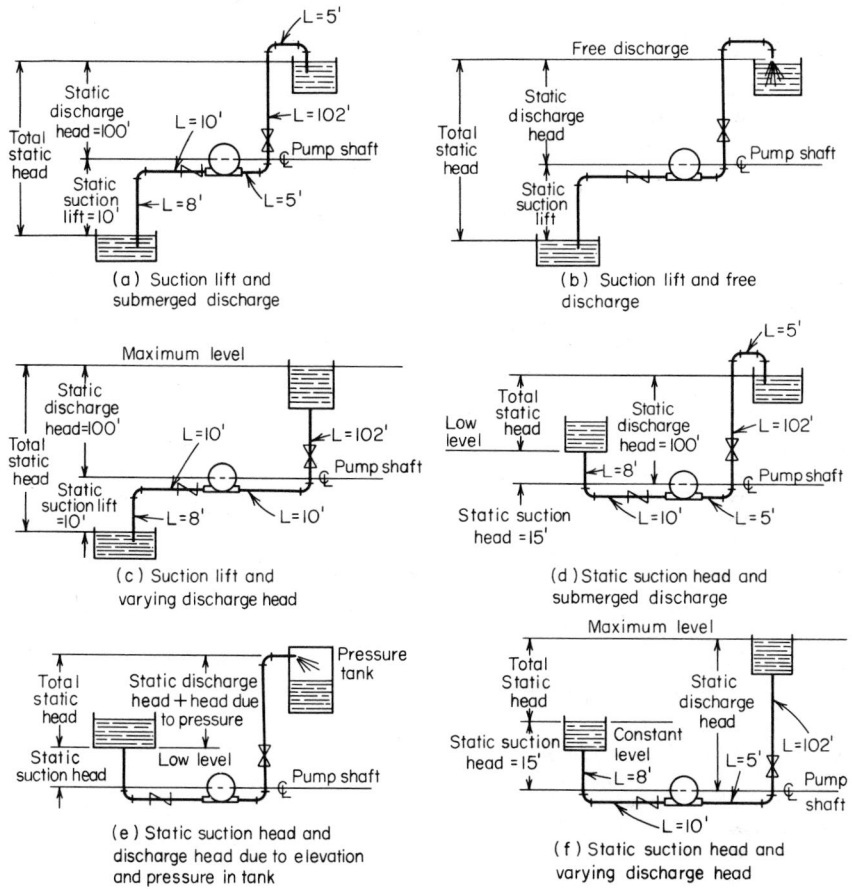

Fig. 4   Typical pump suction and discharge piping arrangements.

3. Compute the piping friction losses

Mark the length of each piece of straight pipe on the piping drawing. Thus, in Fig 4a, the total length of straight pipe L_t ft $= 8+10+5+102+5 = 130$ ft, starting at the suction tank and adding each length until the discharge tank is reached. To the total length of straight pipe must be added the *equivalent* length of the pipe fittings. In Fig. 4a there are four long-radius elbows, one swing check valve, and one globe valve. In addition, there is a minor head loss at the pipe inlet and at the pipe outlet.

The equivalent length of one 8-in.-long-radius elbow is 14 ft of pipe, from Table 3. Since the pipe contains four elbows the total equivalent length $= 4(14) = 96$ ft of straight pipe. The open gate valve has an equivalent resistance of 4.5 ft; and the open swing check valve has an equivalent resistance of 53 ft.

The entrance loss h_e ft, assuming a basket-type strainer is used at the suction-pipe inlet, is h_e ft $= Kv^2/2g$, where $K = $ a constant from Fig. 5; $v = $ liquid velocity, fps; $g = 32.2$ ft/sec². The exit loss occurs when the liquid passes through a sudden enlargement, as from a pipe to a tank. Where the area of the tank is large, causing a final velocity that is zero, $h_{ex} = v^2/2g$.

The velocity v fps in a pipe $=$ gpm/$2.448d^2$. For this pipe, $v = 2,000/[(2.448)(7.98)^2] = 12.82$ fps. Then, $h_e = 0.74(12.82)^2/[2(32.2)] = 1.89$ ft, and $h_{ex} = (12.82)^2/[(2)(32.2)]$

Table 3   Resistance of Fittings and Valves

(Length of straight pipe, ft, giving equivalent resistance)

Pipe size, in.	Standard ell	Medium radius ell	Long-radius ell	45° ell	Tee	Gate valve, open	Globe valve, open	Swing check, open
1	2.7	2.3	1.7	1.3	5.8	0.6	27	6.7
2	5.5	4.6	3.5	2.5	11.0	1.2	57	13
3	8.1	6.8	5.1	3.8	17.0	1.7	85	20
4	11.0	9.1	7.0	5.0	22	2.3	110	27
5	14.0	12.0	8.9	6.1	27	2.9	140	33
6	16.0	14.0	11.0	7.7	33	3.5	160	40
8	21	18.0	14.0	10.0	43	4.5	220	53
10	26	22	17.0	13.0	56	5.7	290	67
12	32	26	20.0	15.0	66	6.7	340	80
14	36	31	23	17.0	76	8.0	390	93
16	42	35	27	19.0	87	9.0	430	107
18	46	40	30	21	100	10.2	500	120
20	52	43	34	23	110	12.0	560	134
24	63	53	40	28	140	14.0	680	160
36	94	79	60	43	200	20.0	1,000	240

= 2.56 ft. Hence, the total length of the piping system in Fig. 4a is $130 + 96 + 4.5 + 53 + 1.89 + 2.56 = 287.95$ ft, say 288 ft.

Use a suitable head-loss equation, or Table 4, to compute the head loss for the pipe and fittings. Enter Table 4 at an 8-in. pipe size and project horizontally across to 2,000 gpm and read the head loss as 5.86 ft of water per 100 ft of pipe.

The total length of pipe and fittings computed above is 288 ft. Then, total friction-head loss with a 2,000-gpm flow is H_f ft = $(5.86)(288/100) = 16.88$ ft.

4. Compute the total head on the pump

The total head on the pump $H_t = H_{ts} + H_f$. For the pump in Fig. 4a, $H_t = 110 + 16.88 = 126.88$ ft., say 127 ft. The total head on the pump in Fig. 4b and c would be the same. Some engineers term the total head on a pump the *total dynamic head* to distinguish between static head (no-flow vertical head) and operating head (rated flow through the pump).

The total head on the pumps in Fig. 4d, c, and f is computed in the same way as described above, except that the total static head is less because the pump has a static suction head—that is, the elevation of the liquid on the suction side reduces the total distance through which the pump must discharge liquid; thus the total static head is less. The static suction head is *subtracted* from the static discharge head to determine the total static head on the pump.

5. Compute the horsepower required to drive the pump

The brake horsepower input to a pump $bhp_i = $ (gpm)$(H_t)(s)/3,960e$, where $s =$ specific gravity of the liquid handled is 1.0. Then, $bhp_i = (2,000)(127)(1.0)/(3,960)(0.70)$ decimal. The usual hydraulic efficiency of a centrifugal pump is 60 to 80 percent; reciprocating pumps, 55 to 90 percent; rotary pumps, 50 to 90 percent. For each class of pump, the hydraulic efficiency decreases as the liquid viscosity increases.

Assume that the hydraulic efficiency of the pump in this system is 70 percent and the

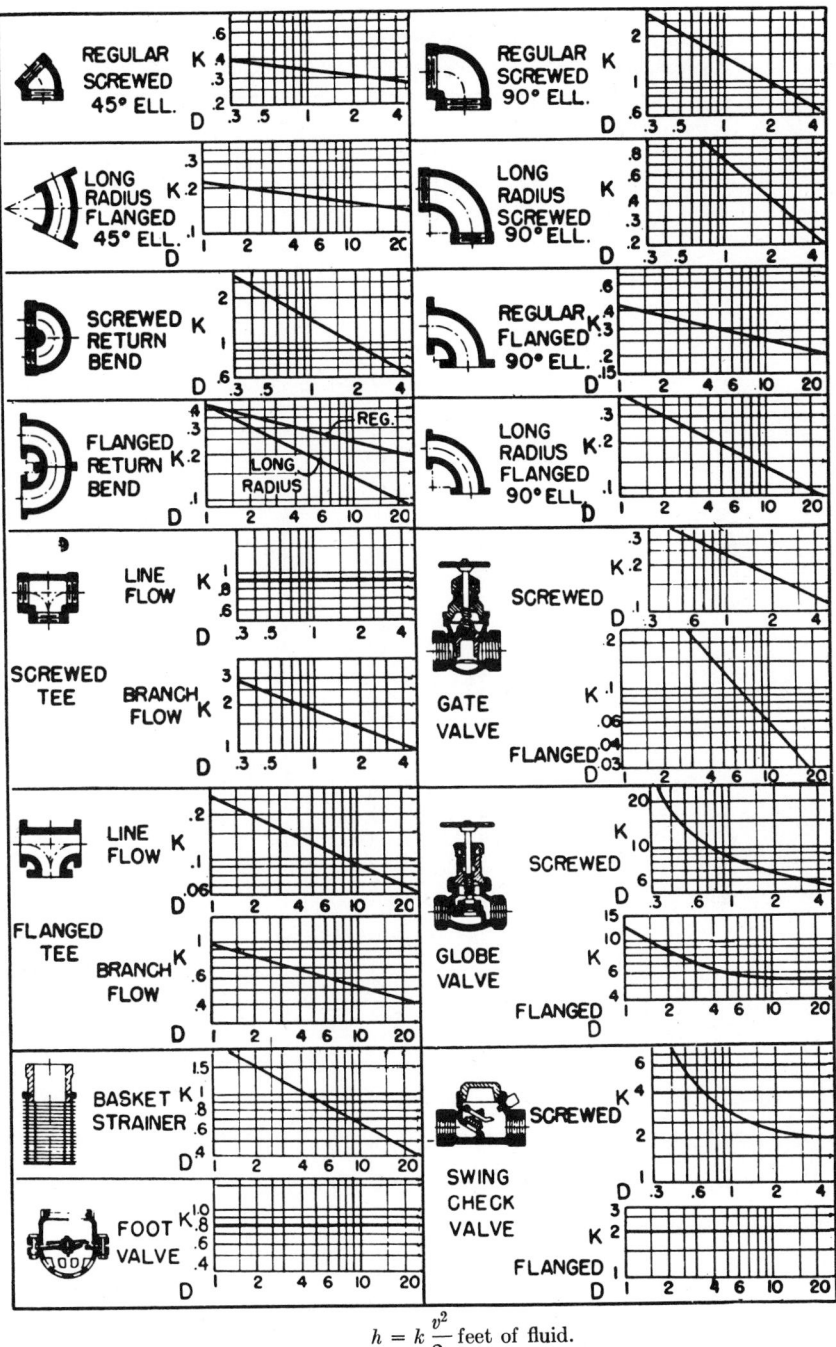

$$h = k \frac{v^2}{2g} \text{ feet of fluid.}$$

Fig. 5 Resistance coefficients of pipe fittings. (*Hydraulic Institute.*)

TABLE 4 Pipe Friction Loss for Water

(Wrought-iron or steel schedule 40 pipe in good condition)

Diameter, in.	Flow, gpm	Velocity ft/sec	Velocity head, ft of water	Friction loss, ft of water per 100 ft pipe
2	50	4.78	0.355	4.67
2	100	9.56	1.42	17.4
2	150	14.3	3.20	38.0
2	200	19.1	5.68	66.3
2	300	28.7	12.8	146
4	200	5.04	0.395	2.27
4	300	7.56	0.888	4.89
4	500	12.6	2.47	13.0
4	1,000	25.2	9.87	50.2
4	2,000	50.4	39.5	196
6	200	2.22	0.0767	0.299
6	500	5.55	0.479	1.66
6	1,000	11.1	1.92	6.17
6	2,000	22.2	7.67	23.8
6	4,000	44.4	30.7	93.1
8	500	3.21	0.160	0.424
8	1,000	6.41	0.639	1.56
8	2,000	12.8	2.56	5.86
8	4,000	25.7	10.2	22.6
8	8,000	51.3	40.9	88.6
10	1,000	3.93	0.240	0.497
10	3,000	11.8	2.16	4.00
10	5,000	19.6	5.99	10.8
10	7,500	29.5	13.5	24.0
10	10,000	39.3	24.0	42.2
12	2,000	5.73	0.511	0.776
12	5,000	14.3	3.19	4.47
12	10,000	28.7	12.8	17.4
12	15,000	43.0	28.7	38.4
12	20,000	57.3	51.1	68.1

specific gravity of the liquid handled is 1.0. Then, $bhp_i = (2,000)(127)(1.0)/(3,960)(0.70)$ $= 91.6$ hp.

The theoretical or *hydraulic horsepower* $hp_h = (gpm)(H_t)(s)/3,960$, or $hp_h = (2,000)$ $\times (127)(1.0)/3,900 = 64.1$ hp.

Related Calculations: Use this Procedure for any liquid—water, oil, chemical, sludge, etc., whose specific gravity is known. When liquids other than water are being pumped, the specific gravity and viscosity of the liquid, as discussed in later Calculation Procedures, must be taken into consideration. The Procedure given here can be used for any class of pump—centrifugal, rotary, or reciprocating.

Note that Fig. 5 can be used to determine the equivalent length of a variety of pipe fittings. To use Fig. 5, simply substitute the appropriate K value in the relation $h = Kv^2/2g$, where $h =$ equivalent length of straight pipe; other symbols as before.

PUMP SELECTION FOR ANY PUMPING SYSTEM

Give a step-by-step procedure for choosing the class, type, capacity, drive, and materials for a pump that will be used in an industrial pumping system.

Calculation Procedure:

1. Sketch the proposed piping layout

Use a single-line diagram, Fig. 6, of the piping system. Base the sketch on the actual job conditions. Show all the piping, fittings, valves, equipment, and other units in the system. Mark the *actual* and *equivalent* pipe length (see the previous Calculation Procedure) on the sketch. Be certain to include all vertical lifts, sharp bends, sudden enlargements, storage tanks, and similar equipment in the proposed system.

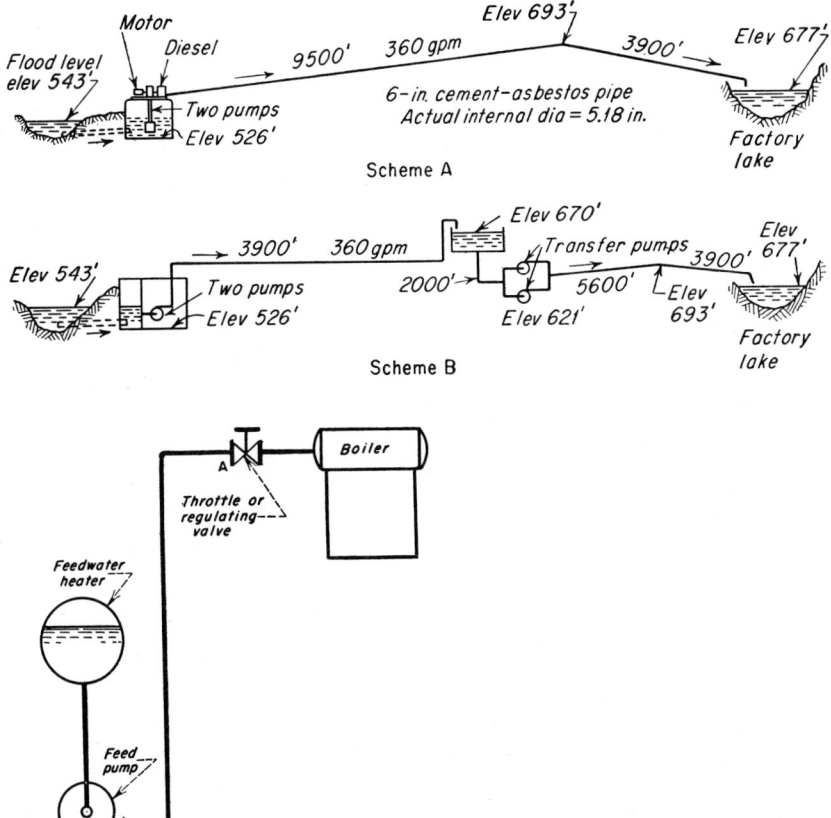

Fig. 6 *(a)* Single-line diagrams for an industrial pipeline; *(b)* single-line diagram of a boiler-feed system. (*Worthington Corporation.*)

2. Determine the required capacity of the pump

The required capacity is the flow rate that must be handled in gpm, million gal/day, cfs, gph, bbl/day, lb/hr, acre-ft/day, mil/hr, or some similar measure. Obtain the required flow rate from the process conditions—for example, boiler feed rate, cooling-water flow rate, chemical feed rate, etc. The required flow rate for any process unit is usually given by the manufacturer or can be computed using the Calculation Procedures given throughout this Handbook.

Once the required flow rate is determined, apply a suitable factor of safety. The value of this factor of safety can vary from a low of 5 percent of the required flow to a high of 50 percent, or more, depending on the application. Typical safety factors are in the 10 percent range. With flow rates up to 1,000 gpm, and in the selection of process pumps, it is common practice to round off a computed required flow rate to the next highest round-number capacity. Thus, with a required flow rate of 450 gpm and a 10 percent safety factor, the flow of $450 + 0.10(450) = 495$ gpm would be rounded off to

---

Summary of Essential Data Required in Selection of Centrifugal Pumps

1. Number of Units Required

2. Nature of the Liquid to Be Pumped
   Is the liquid:
   a. Fresh or salt water, acid or alkali, oil, gasoline, slurry, or paper stock?
   b. Cold or hot and if hot, at what temperature? What is the vapor pressure of the liquid at the pumping temperature?
   c. What is its specific gravity?
   d. Is it viscous or nonviscous?
   e. Clear and free from suspended foreign matter or dirty and gritty? If the latter, what is the size and nature of the solids, and are they abrasive? If the liquid is of a pulpy nature, what is the consistency expressed either in percentage or in lb per cu ft of liquid? What is the suspended material?
   f. What is the chemical analysis, pH value, etc.? What are the expected variations of this analysis? If corrosive, what has been the past experience, both with successful materials and with unsatisfactory materials?

3. Capacity
   What is the required capacity as well as the minimum and maximum amount of liquid the pump will ever be called upon to deliver?

4. Suction Conditions
   Is there:
   a. A suction lift?
   b. Or a suction head?
   c. What are the length and diameter of the suction pipe?

5. Discharge Conditions
   a. What is the static head? Is it constant or variable?
   b. What is the friction head?
   c. What is the maximum discharge pressure against which the pump must deliver the liquid?

6. Total Head
   Variations in items 4 and 5 will cause variations in the total head.

7. Is the service continuous or intermittent?

8. Is the pump to be installed in a horizontal or vertical position? If the latter,
   a. In a wet pit?
   b. In a dry pit?

9. What type of power is available to drive the pump and what are the characteristics of this power?

10. What space, weight, or transportation limitations are involved?

11. Location of installation
    a. Geographical location
    b. Elevation above sea level
    c. Indoor or outdoor installation
    d. Range of ambient temperatures

12. Are there any special requirements or marked preferences with respect to the design, construction, or performance of the pump?

Fig. 7   Typical selection chart for centrifugal pumps. (*Worthington Corporation.*)

500 gpm *before* selecting the pump. A pump of 500-gpm, or larger, capacity would be selected.

3. Compute the total head on the pump

Use the steps given in the previous Calculation Procedure to compute the total head on the pump. Express the result in ft of water—this is the most common way of expressing the head on a pump. Be certain to use the exact specific gravity of the liquid handled when expressing the head in ft of water. A specific gravity less than 1.00 *reduces* the total head when expressed in ft of water; whereas a specific gravity greater than 1.00 *increases* the total head when expressed in ft of water. Note that variations in the suction and discharge conditions can affect the total head on the pump.

4. Analyze the liquid conditions

Obtain complete data on the liquid pumped. These data should include the name and chemical formula of the liquid, maximum and minimum pumping temperature, corresponding vapor pressure at these temperatures, specific gravity, viscosity at the pumping temperature, pH, flash point, ignition temperature, unusual characteristics (such as tendency to foam, curd, crystallize, become gelatinous or tacky), solids content, type of solids and their size, and variation in the chemical analysis of the liquid.

Enter the liquid conditions on a pump selection form like that in Fig. 7. Such forms are available from many pump manufacturers or can be prepared to meet special job conditions.

5. Select the class and type of pump

Three *classes* of pumps are used today—centrifugal, rotary, and reciprocating, Fig. 8. Note that these terms apply only to the mechanics of moving the liquid—not to the service for which the pump was designed. Each class of pump is further subdivided into a number of *types*, Fig. 8.

Use Table 5 as a general guide to the class and type of pump to be used. For example, when a large capacity at moderate pressure is required, Table 5 shows that a centrifugal pump would probably be best. Table 5 also shows the typical characteristics of various classes and types of pumps used in industrial process work.

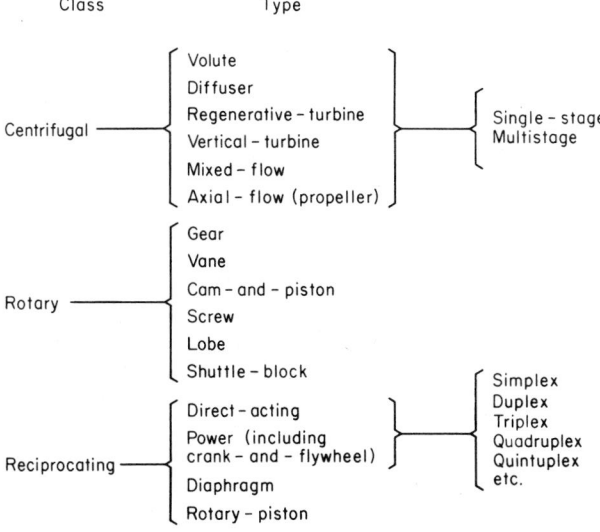

Fig. 8 Modern pump classes and types.

TABLE 5   Characteristics of Modern Pumps

	Centrifugal		Rotary	Reciprocating		
	Volute and diffuser	Axial flow	Screw and gear	Direct-acting steam	Double-acting power	Triplex
Discharge flow.....	Steady	Steady	Steady	Pulsating	Pulsating	Pulsating
Usual maximum suction lift, ft	15	15	22	22	22	22
Liquids handled ...	Clean, clear; dirty, abrasive; liquids with high solids content		Viscous, nonabrasive	Clean and clear		
Discharge pressure range	Low to high		Medium	Low to highest produced		
Usual capacity range	Small to largest available		Small to medium	Relatively small		
How increased head affects:						
Capacity	Decrease		None	Decrease	None	None
Power input ...	Depends on specific speed		Increase	Increase	Increase	Increase
How decreased head affects:						
Capacity	Increase		None	Small increase	None	None
Power input ...	Depends on specific speed		Decrease	Decrease	Decrease	Decrease

Consider the liquid properties when choosing the class and type of pump, because exceptionally severe conditions may rule out one or another class of pump at the start. Thus, screw- and gear-type rotary pumps are suitable for handling viscous, non-abrasive liquid, Table 5. When an abrasive liquid must be handled, either another class of pump or another type of rotary pump must be used.

Also consider all the operating factors related to the particular pump. These factors include the type of service (continuous or intermittent), operating-speed preferences, future load expected and its effect on pump head and capacity, maintenance facilities available, possibility of parallel or series hookup, and other conditions peculiar to a given job.

Once the class and type of pump is selected, consult a rating table (Table 6) or rating chart, Fig. 9, to determine if a suitable pump is available from the manufacturer whose unit will be used. When the hydraulic requirements fall between two standard pump models, it is usual practice to choose the next larger size of pump, unless there is some reason why an exact head and capacity are required for the unit. When one manufacturer does not have the desired unit, refer to the engineering data of other manufacturers. Also keep in mind that some pumps are custom-built for a given job when precise head and capacity requirements must be met.

Other pump data included in manufacturer's engineering information include characteristic curves for various diameter impellers in the same casing, Fig. 10, and variable-speed head-capacity curves for an impeller of given diameter, Fig. 11. Note that the required power input is given in Figs. 9 and 10, and may also be given in Fig. 11. Use of Table 6 is explained in the table.

Performance data for rotary pumps is given in several forms. Figure 12 shows a typical plot of the head and capacity ranges of different types of rotary pumps. Reciprocating-pump capacity data are often tabulated, as in Table 7.

TABLE 6   Typical Centrifugal-Pump Rating Table*

Size, gpm	Total head, ft				
	10	15	20	25	30
2C:					
100	...	...	1,000–0.8	1,060–1.0	1,150–1.2
150	...	...	1,070–1.2	1,150–1.5	1,240–1.7
200	...	...	...	1,290–2.1	1,360–2.4
3CS:					
150	750–0.53	850–0.78	950–1	1,030–1.2	1,100–1.5
200	...	950–1.1	1,010–1.4	1,100–1.7	1,170–2
250	...	...	1,170–1.9	1,190–2.3	1,260–2.6
300	...	...	...	...	1,400–3.5
3CL:					
200	690–0.63	800–0.95	910–1.3	1,010–1.6	1,110–2.05
300	870–1.2	950–1.6	1,000–1.9	1,100–2.4	1,170–2.8
400	...	...	1,200–3.1	1,230–3.7	1,290–4.1
500	...	...	...	...	1,490–5.8
4C:					
400	750–1.3	850–1.8	940–2.4	1,040–3	1,120–3.7
600	...	...	1,080–4	1,170–4.6	1,210–5.5
800	...	...	...	...	1,400–8.4
1¼D:					
25	...	617–0.21	707–0.03	778–0.40	845–0.51
50	...	680–0.37	760–0.49	865–0.63	900–0.76
75	...	...	856–0.78	916–0.94	980–1.1
100					
125					
150					
2DL:					
150	...	820–0.93	850–1.1	930–1.35	990–1.6
200	...	...	970–1.8	1,040–2.1	1,080–2.3
250					
300					

EXAMPLE: 1,080–4 indicates pump speed is 1,080 rpm; actual input required to operate pump is 4 hp.
*Condensed from data of Goulds Pumps, Inc.

6.   *Evaluate the pump chosen for the installation*

Check the specific speed of a centrifugal pump, using the method given in an earlier Calculation Procedure. Once the specific speed is known, the impeller type and approximate operating efficiency can be found from Fig. 3.

Check the piping system, using the method of an earlier Calculation Procedure, to see if the available net positive suction head equals, or is greater than, the required net positive suction head of the pump.

Determine whether a vertical or horizontal pump is more desirable. From the standpoint of floor space occupied, required NPSH, priming, and flexibility in changing the pump use, vertical pumps may be preferable to horizontal designs in some installations. But where headroom, corrosion, abrasion, and ease of maintenance are important factors, horizontal pumps may be preferable.

As a general guide, single-suction centrifugal pumps handle up to 50 gpm at total heads up to 50 ft; either single- or double-suction pumps are used for the flow rates to 1,000 gpm and total heads to 300 ft; beyond these capacities and heads, double-suction or multistage pumps are generally used.

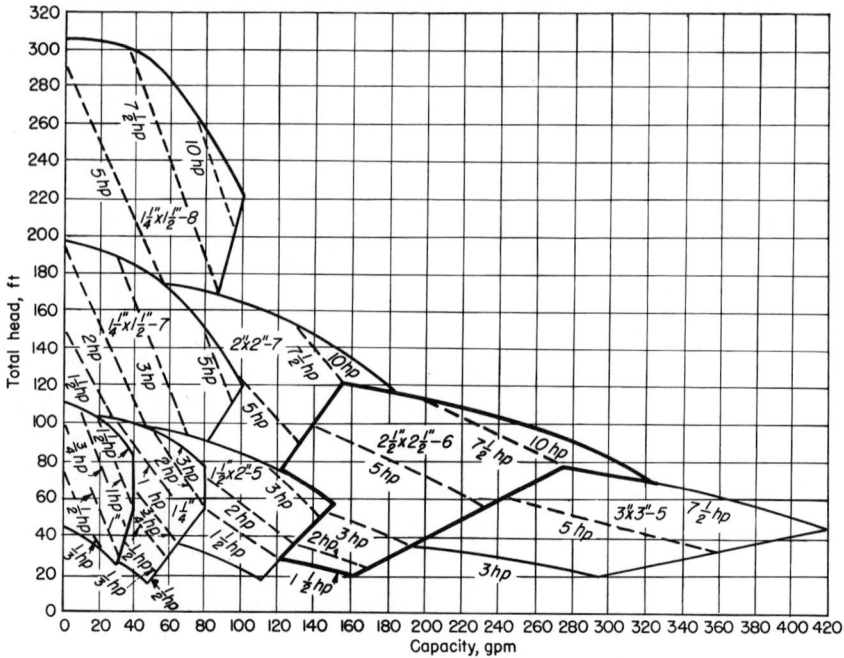

Fig. 9   Composite rating chart for a typical centrifugal pump. (*Goulds Pumps, Inc.*)

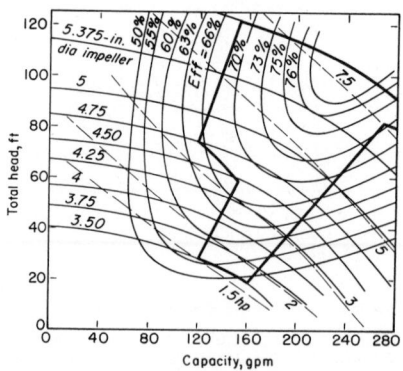

Fig. 10   Pump characteristics when impeller diameter is varied within the same casing.

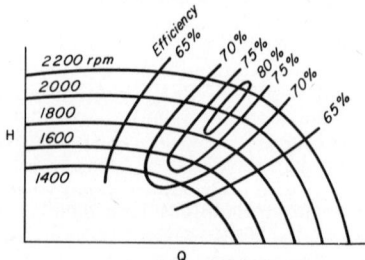

Fig. 11   Variable-speed head-capacity curves for a centrifugal pump.

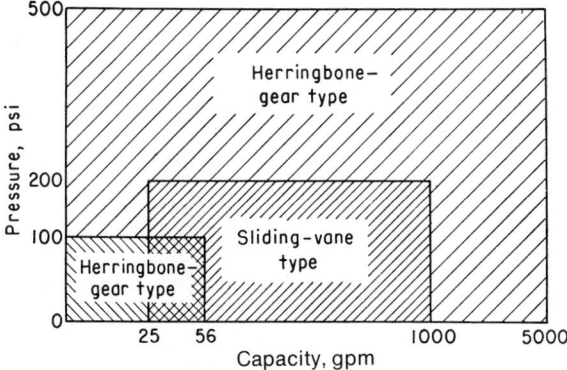

Fig. 12 Capacity ranges of some rotary pumps. (*Worthington Corporation.*)

TABLE 7 Capacities of Typical Horizontal Duplex Plunger Pumps*

Size, in.	Cold-water pressure service		Boiler-feed service		
	Gpm	Piston speed, ft/min	Gpm	Boiler hp	Piston speed, ft/min
6 × 3½× 6	60	60	36	475	36
7½× 4½× 10	124	75	74	975	45
9 × 5 × 10	153	75	92	1,210	45
10 × 6 × 12	235	80	141	1,860	48
12 × 7 × 12	320	80	192	2,530	48
12 × 7½× 15	413	90	248	3,270	54
14 × 8½× 15	530	90	318	4,190	54
17 × 10 × 15	738	90	443	5,830	54
20 × 12 × 15	1,060	90	636	8,390	54

*Courtesy of Worthington Corporation.

Mechanical seals are becoming more popular for all types of centrifugal pumps in a variety of services. Though more costly than packing, the mechanical seal reduces pump maintenance costs.

Related Calculations: Use the Procedure given here to select any class of pump—centrifugal, rotary, or reciprocating—for any type of service—power plant, atomic energy, petroleum processing, chemical manufacture, paper mills, textile mills, rubber factories, food processing, water supply, sewage and sump service, air conditioning and heating, irrigation and flood control, mining and construction, marine services, industrial hydraulics, iron and steel manufacture, etc.

ANALYSIS OF PUMP AND SYSTEM CHARACTERISTIC CURVES

Analyze a set of pump and system characteristic curves for the following conditions: friction losses without static head; friction losses with static head; pump without lift; system with little friction, much static head; system with gravity head; system with different pipe sizes; system with two discharge heads; system with diverted flow; and effect of pump wear on characteristic curve.

Calculation Procedure:

1. Plot the system-friction curve

Without static head, the system-friction curve passes through the origin (0, 0), Fig. 13, because when no head is developed by the pump, flow through the piping is zero. For

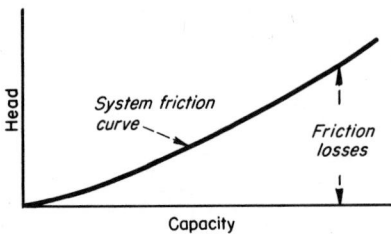

most piping systems the friction-head loss varies as the square of the liquid flow rate in the system. Hence, a system-friction curve, also called a friction-head curve, is parabolic—the friction head increasing as the flow rate or capacity of the system increases. Draw the curve as shown in Fig. 13.

Fig. 13   Typical system-friction curve.

2. Plot the piping system and system-head curve

Figure 14a shows a typical piping system with a pump operating against a static discharge head. Indicate the total static head, Fig. 14b, by a dashed line—in this installation $H_{ts} = 110$ ft. Since static head is a physical dimension, it does not vary with flow rate and is a constant for all flow rates. Draw the dashed line parallel to the abscissa, Fig. 14b.

From the point of no flow—zero capacity—plot the friction-head loss at various flow rates—100, 200, 300 gpm, etc. Determine the friction-head loss by computing it as

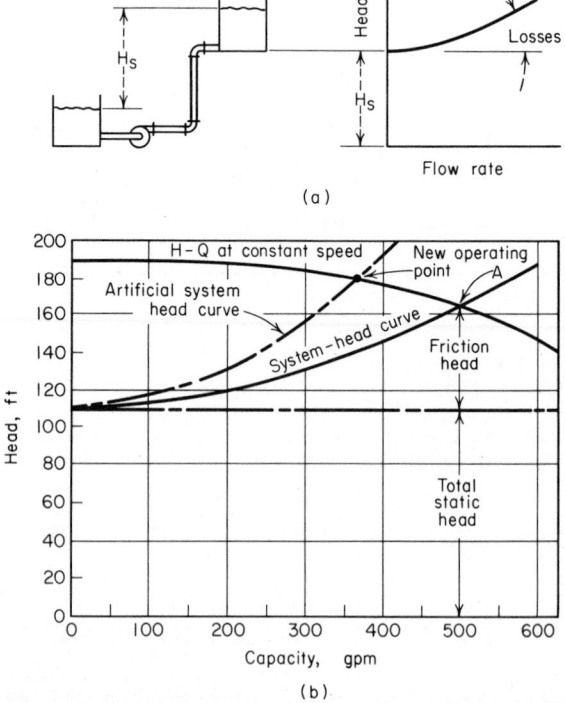

Fig. 14   (a) Significant friction loss and lift; (b) system-head curve superimposed on pump head-capacity curve. (*Peerless Pumps.*)

shown in an earlier Calculation Procedure. Draw a curve through the points obtained. This is called the system-head curve.

Plot the pump head-capacity (H-Q) curve of the pump on Fig. 14b. The H-Q curve can be obtained from the pump manufacturer or from a tabulation of H and Q values for the pump being considered. The point of intersection, A, between the H-Q and system-head curves is the operating point of the pump.

Changing the resistance of a given piping system by partially closing a valve or making some other change in the friction alters the position of the system-head curve and pump operating point. Compute the frictional resistance as before and plot the artificial system-head curve as shown. Where this curve intersects the H-Q curve is the new operating point of the pump. System-head curves are valuable for analyzing the suitability of a given pump for a particular application.

3. Plot the no-lift system-head curve and compute the losses

With no static head or lift, the system-head curve passes through the origin (0,0), Fig. 15. For a flow of 900 gpm in this system, compute the friction loss as follows using the Hydraulic Institute *Pipe Friction Manual* tables or the method of earlier Calculation Procedures:

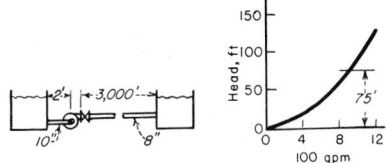

Fig. 15　No lift; all friction head. (*Peerless Pumps.*)

Entrance loss from tank into 10-in. suction pipe, $0.5v^2/2g$	0.10 ft
Friction loss in 2 ft of suction pipe	0.02
Loss in 10-in. 90° elbow at pump	0.20
Friction loss in 3,000 ft of 8-in. discharge pipe	74.50
Loss in fully open 8-in. gate valve	0.12
Exit loss from 8-in. pipe into tank, $v^2/2g$	0.52
Total friction loss	75.46 ft

Compute the friction loss at other flow rates in a similar manner and plot the system-head curve, Fig. 15. Note that if all losses in this system except the friction in the discharge pipe are ignored, the total head would not change appreciably. However, for the purposes of accuracy, all losses should always be computed.

4. Plot the low-friction, high-head system-head curve

The system-head curve for the vertical pump installation in Fig. 16 starts at the total static head, 15 ft, and zero flow. Compute the friction head for 15,000 gpm as follows:

Friction in 20 ft of 24-in. pipe	0.40 ft
Exit loss from 24-in. pipe into tank, $v^2/2g$	1.60
Total friction loss	2.00 ft

Hence, almost 90 percent of the total head of $15 + 2 = 17$ ft at 15,000-gpm flow is static head. But neglect of the pipe friction and exit losses could cause appreciable error during selection of a pump for the job.

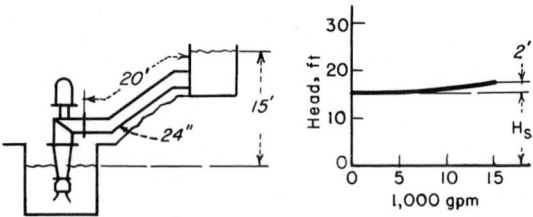

Fig. 16   Mostly lift; little friction head. (*Peerless Pumps.*)

5. Plot the gravity-head system-head curve

In a system with gravity head (also called negative lift), fluid flow will continue until the system friction loss equals the available gravity head. In Fig. 17 the available gravity head is 50 ft. Flows up to 7,200 gpm are obtained by gravity head alone. To obtain larger flow rates, a pump is needed to overcome the friction in the piping between the tanks. Compute the friction loss for several flow rates as follows:

At 5,000 gpm, friction loss in 1,000 ft of
  16-in. pipe... 25 ft
At 7,200 gpm, friction loss = available
  gravity head .... 50 ft
At 13,000 gpm, friction loss............................ 150 ft

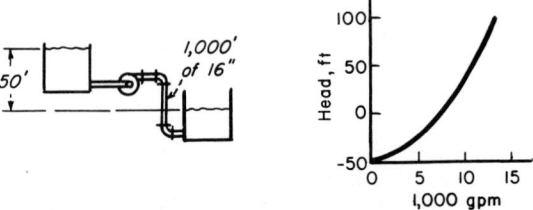

Fig. 17   Negative lift (gravity head). (*Peerless Pumps.*)

Using these three flow rates, plot the system-head curve, Fig. 17.

6. Plot the system-head curves for different pipe sizes

When different diameter pipes are used, the friction-loss-vs.-flow rate is plotted independently for the two pipe sizes. At a given flow rate, the total friction loss for the system is the sum of the loss for the two pipes. Thus, the combined system-head curve represents the sum of the static head and the friction losses for all portions of the pipe.

Figure 18 shows a system with two different pipe sizes. Compute the friction losses as follows:

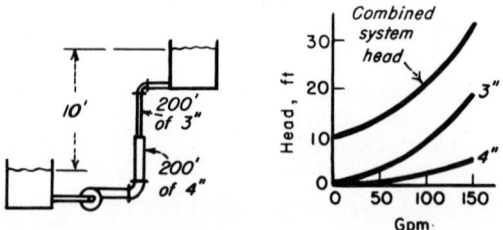

Fig. 18   System with two different pipe sizes. (*Peerless Pumps.*)

At 150 gpm, friction loss in 200 ft of 4-in. pipe	5 ft
At 150 gpm, friction loss in 200 ft of 3-in. pipe	19 ft
Total static head for 3- and 4-in. pipes...........	10 ft
Total head at 150-gpm flow	34 ft

Compute the total head at other flow rates and plot the system-head curve as shown in Fig. 18.

7. Plot the system-head curve for two discharge heads

Figure 19 shows a typical pumping system having two different discharge heads. Plot separate system-head curves when the discharge heads are different. Add the flow rates for the two pipes at the same head to find points on the combined system-head curve, Fig. 19. Thus,

At 550 gpm, friction loss in 1,000 ft of		
8-in. pipe	=	10 ft
At 1,150 gpm, friction	=	38 ft
At 1,150 gpm, friction + lift in pipe 1		
= 38 + 50	=	88 ft
At 550 gpm, friction + lift in pipe 2		
= 10 + 78	=	88 ft

The flow rate for the combined system at a head of 88 ft is $1,150 + 550 = 1,700$ gpm. To produce a flow of 1,700 gpm through this system, a pump capable of developing an 88-ft head is required.

8. Plot the system-head curve for diverted flow

To analyze a system with diverted flow, assume that a constant quantity of liquid is tapped off at the intermediate point. Plot the friction-loss-vs.-flow rate in the normal manner for pipe 1, Fig. 20. Move the curve for pipe 3 to the right at zero head by an amount equal to Q_2, since this represents the quantity passing through pipes 1 and 2 but not through pipe 3. Plot the combined system-head curve by adding, at a given flow rate, the head losses for pipes 1 and 3. With $Q = 300$ gpm, pipe 1 = 500 ft of 10-in. pipe, and pipe 3 = 50 ft of 6-in. pipe.

At 1,500 gpm through pipe 1, friction loss	=	11 ft
Friction loss for pipe 3 (1,500 − 300 = 1,200 gpm)	=	8
Total friction loss at 1,500-gpm delivery	=	19 ft

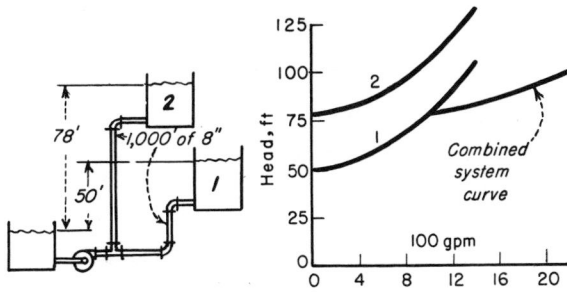

Fig. 19 System with two different discharge heads. (*Peerless Pumps.*)

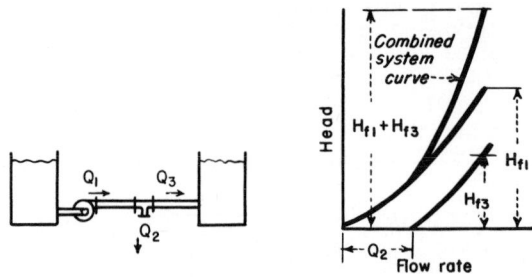

Fig. 20   Part of the fluid flow diverted from the main pipe. (*Peerless Pumps.*)

9.   Plot the effect of pump wear

When a pump wears, there is a loss in capacity and efficiency. The amount of loss depends, however, on the shape of the system-head curve. For a centrifugal pump, Fig. 21, the capacity loss is greater for a given amount of wear if the system-head curve is flat, as compared with a steep system-head curve.

Determine the capacity loss for a worn pump by plotting its H-Q curve. Find this curve by testing the pump at different capacities and plotting the corresponding head.

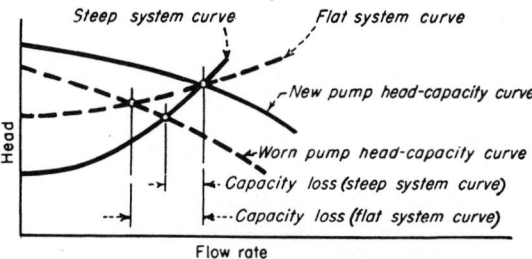

Fig. 21   Effect of pump wear on pump capacity. (*Peerless Pumps.*)

On the same chart, plot the H-Q curve for a new pump of the same size, Fig. 21. Plot the system-head curve and determine the capacity loss as shown in Fig. 21.

Related Calculations: Use the techniques given here for any type of pump—centrifugal, reciprocating, or rotary—handling any type of liquid—oil, water, chemicals, etc. The methods given here are the work of Melvin Mann, as reported in *Chemical Engineering*, and Peerless Pump Div. of FMC Corp.

NET POSITIVE SUCTION HEAD FOR HOT-LIQUID PUMPS

What is the maximum capacity of a double-suction condensate pump operating at 1,750 rpm if it handles 100-F water from a hot well in a condenser having an absolute pressure of 2.0 in. Hg if the pump centerline is 10 ft below the hot-well liquid level and the friction-head loss in the suction piping and fitting is 5 ft of water?

Calculation Procedure:

1.   Compute the net positive suction head on the pump

The net positive suction head h_n on a pump when the liquid supply is *above* the pump inlet = pressure on liquid surface + static suction head − friction-head loss in suction piping and pump inlet − vapor pressure of the liquid, all expressed in ft absolute of liquid handled. When the liquid supply is *below* the pump centerline—i.e., there is a static suction lift—the vertical distance of the lift is *subtracted* from the pressure on the liquid surface instead of added as in the above relation.

The density of 100-F water is 62.0 lb/cu ft, computed as shown in earlier Calculation Procedures in this Handbook. The pressure on the liquid surface, in absolute ft of liquid = (2.0 in. Hg)(1.133)(62.4/62.0. = 2.24 ft. In this calculation, 1.133 = ft of 39.2-F water = 1-in. Hg; 62.4 = lb/cu ft of 39.2-F water. The temperature of 39.2 F is used because at this temperature water has its maximum density. Thus, to convert in. Hg to ft absolute of water, find the product of (in. Hg)(1.133)(water density at 39.2 F)/(water density at operating temperature). Express both density values in the same unit, usually lb/cu ft.

The static suction head is a physical dimension that is measured in ft of liquid at the operating temperature. In this installation, $h_{sh} = 10$ ft absolute.

The friction-head loss is 5 ft of water. When computed using the methods of earlier Calculation Procedures, this head loss is in ft of water at maximum density. To convert to ft absolute, multiply by the ratio of water densities at 39.2 F and the operating temperature, or (5)(62.4/62.0) = 5.03 ft.

The vapor pressure of water at 100 F is 0.949 psia, from the steam tables. Convert any vapor pressure to ft absolute by finding the result of (vapor pressure, psia)(144 sq in./sq ft)/liquid density at operating temperature, or (0.949)(144)/62.0 = 2.204 ft absolute.

With all the heads known, the net positive suction head is $h_n = 2.24 + 10 - 5.03 - 2.204 = 5.01$ ft absolute.

2. Determine the capacity of the condensate pump

Use the Hydraulic Institute curve, Fig. 22, to determine the maximum capacity of the pump. Enter at the left of Fig. 22 at a net positive suction head of 5.01 ft and project

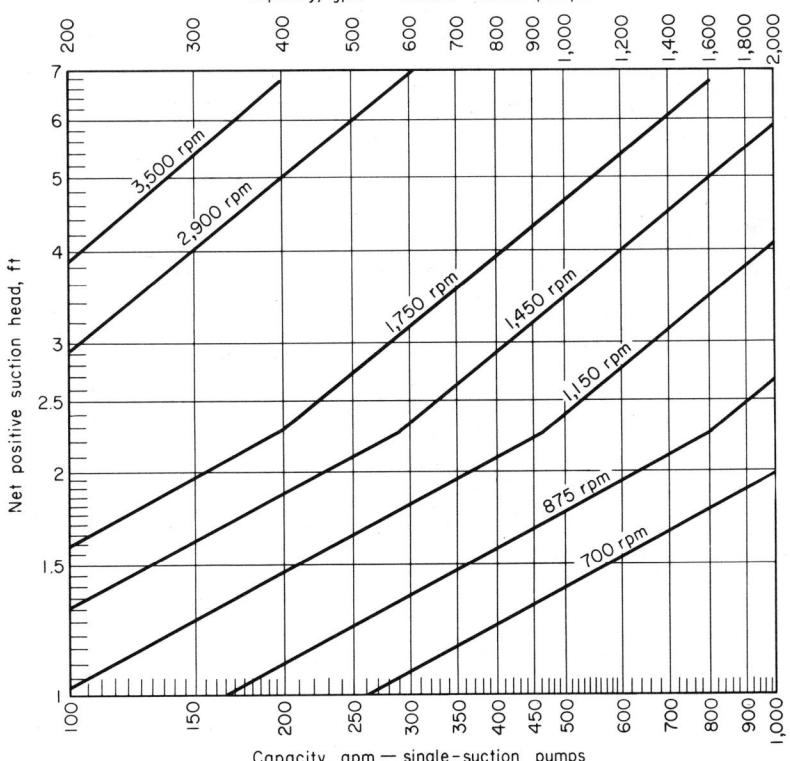

Fig. 22 Capacity and speed limitations of condensate pumps with the shaft through the impeller eye. (*Hydraulic Institute.*)

horizontally to the right until the 3,500-rpm curve is intersected. At the top, read the capacity as 278 gpm.

Related Calculations: Use this procedure for any condensate or boiler-feed pump handling water at an elevated temperature. Consult the *Standards of the Hydraulic Institute* for capacity curves of pumps having different types of construction. In general, pump manufacturers who are members of the Hydraulic Institute rate their pumps in accordance with the *Standards*, and a pump chosen from a catalog capacity table or curve will deliver the stated capacity. A similar procedure is used for computing the capacity of pumps handling volatile petroleum liquids. When using this Procedure, be certain to refer to the latest edition of the *Standards*.

CONDENSATE PUMP SELECTION FOR A STEAM POWER PLANT

Select the capacity for a condensate pump serving a steam power plant having a 140,000 lb/hr exhaust flow to a condenser that operates at an absolute pressure of 1.0 in. Hg. The condensate pump discharges through 4-in. schedule 40 pipe to an air-ejector condenser that has a frictional resistance of 8 ft of water. From here, the condensate flows to and through a low-pressure heater that has a frictional resistance of 12 ft of water and is vented to the atmosphere. The total equivalent length of the discharge piping, including all fittings and bends, is 400 ft, and the suction piping total equivalent length is 50 ft. The inlet of the low pressure heater is 75 ft above the pump centerline, and the condenser hot-well water level is 10 ft above the pump centerline. How much power is required to drive the pump if its efficiency is 70 percent?

Calculation Procedure:

1. Compute the static head on the pump

Sketch the piping system as shown in Fig. 23. Mark the static elevations and equivalent lengths as indicated.

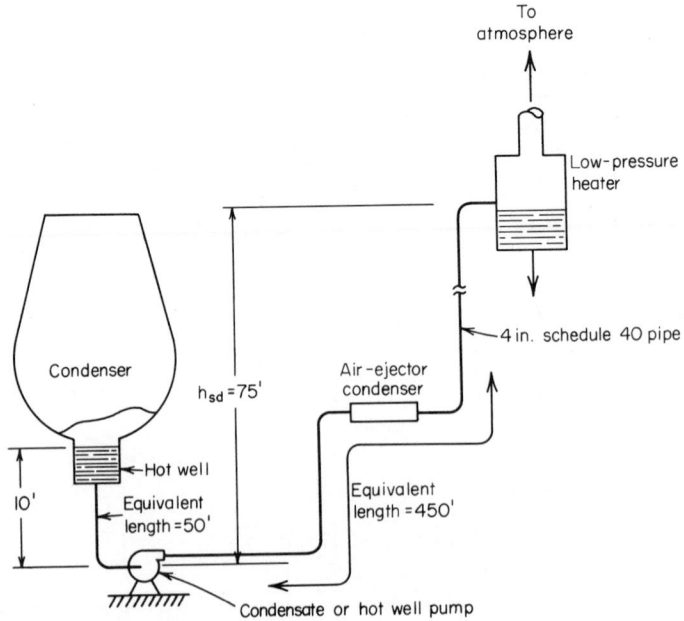

Fig. 23   Condensate pump serving a steam power plant.

The total head on the pump $H_t = H_{ts} + H_f$, where the symbols are the same as in earlier Calculation Procedures. The total static head $H_{ts} = h_{sd} - h_{sh}$. In this installation, $H_{sd} = 75$ ft. To make the calculation simpler, convert all the heads to absolute values. Since the heater is vented to the atmosphere, the pressure acting on the surface of the water in it = 14.7 psia, or 34 ft of water. The pressure acting on the condensate in the hot well is 1 in. Hg = 1.133 ft of water. (An absolute pressure of 1 in. Hg = 1.133 ft of water.) Thus, the absolute discharge static head = $75 + 34 = 109$ ft, whereas the absolute suction head = $10 + 1.13 = 11.13$ ft. Then, $H_{ts} = h_{hd} - h_{sh} = 109.00 - 11.13 = 97.87$ ft, say 98 ft of water.

2. Compute the friction head in the piping system

The total friction head $H_f =$ pipe friction + heater friction. The pipe friction loss is found first, as shown below. The heater friction loss, obtained from the manufacturer or his engineering data, is then added to the pipe-friction loss. Both must be expressed in ft of water.

To determine the pipe friction, use Fig. 24 of this section and Table 22 and Fig. 13 of the *Piping* section of this Handbook in the following manner. Find the product of the liquid velocity, fps, and the pipe internal diameter, in., or vd. With an exhaust flow of 140,000 lb/hr to the condenser, the condensate flow is the same, or 140,000 lb/hr at a temperature of 79.03 F, corresponding to an absolute pressure in the condenser of 1 in. Hg, obtained from the steam tables. The specific volume of the saturated liquid at this temperature and pressure is 0.01608 cu ft/lb. Since 1 gal of liquid occupies 0.13368 cu ft, specific volume, gal/lb, is (0.01608/0.13368) = 0.1202. Therefore, a flow of 140,000 lb/hr = a flow of (140,000)(0.1202) = 16,840 gph, or (16,840/60) = 281 gpm. Then the liquid velocity $v = gpm/2.448d^2 = 281/2.448(4.026)^2 = 7.1$ fps, and the product $vd = (7.1)(4.026) = 28.55$.

Enter Fig. 24 at a temperature of 79 F and project vertically upward to the water curve. From the intersection, project horizontally to the right to $vd = 28.55$, and then

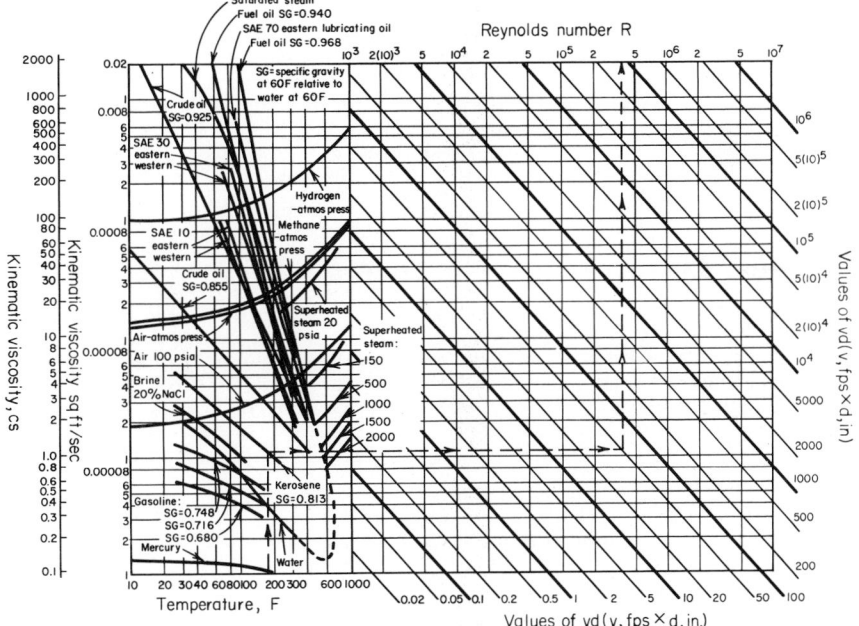

Fig. 24   Kinematic viscosity and Reynolds number chart. (*Hydraulic Institute.*)

vertically upward to read $R = 250{,}000$. Using Table 22 and Fig. 13 of the *Piping* section and $R = 250{,}000$, find the friction factor $f = 0.0185$. Then, the head loss due to pipe friction $H_f = (L/D)(v^2/2g) = 0.0185 \ (450/4.026/12)/[(7.1)^2/2(32.2)] = 19.18$ ft. In this computation, $L = $ total equivalent length of the pipe, pipe fittings, and system valves, or 450 ft.

3. Compute the other head losses in the system

There are two other head losses in this piping system: the entrance loss at the square-edged hot-well pipe leading to the pump and the sudden enlargement in the low-pressure heater. The velocity head $v^2/2g = (7.1)^2/2(32.2) = 0.784$ ft. Using k values from Fig. 5 in this section, $h_e = kv^2/2g = (0.5)(0.784) = 0.392$ ft; $h_{ex} = v^2/2g = 0.784$ ft.

4. Find the total head on the pump

The total head on the pump $H_t = H_{ts} + H_f = 97.87 + 19.18 + 8 + 12 + 0.392 + 0.784 = 138.226$ ft, say 140 ft of water. In this calculation, the 8- and 12-ft head losses are those occurring in the heaters. Using a 25 percent safety factor, total head $= (1.25)(140) = 175$ ft.

5. Compute the horsepower required to drive the pump

The brake horsepower input, $bhp_i = (gpm)(H_t)(s)/3{,}960e$, where the symbols are the same as in earlier Calculation Procedures. At 1 in. Hg, 1 lb of the condensate has a volume of 0.01608 cu ft. Since density $= 1/$specific volume, the density of the condensate $= 1/0.01608 = 62.25$ cu ft/lb. Water having a specific gravity of unity weighs 62.4 cu ft/lb. Hence, the specific gravity of the condensate is $62.25/62.4 = 0.997$. Then, assuming that the pump has an operating efficiency of 70 percent, $bhp_i = (281)(175) \times (0.997)/[3{,}960(0.70)] = 17.7$ bhp.

6. Select the condensate pump

Condensate or hot-well pumps are usually centrifugal units having two or more stages, with the stage inlets opposed to give better axial balance and to subject the sealing glands to positive internal pressure, thereby preventing air leakage into the pump. In the head range developed by this pump—175 ft—two stages are satisfactory. Refer to a pump manufacturer's engineering data for specific stage head ranges. Either a turbine or motor drive can be used.

Related Calculations: Use this procedure to choose condensate pumps for steam plants of any type—utility, industrial, marine, portable, heating, or process—and for combined steam-diesel plants.

MINIMUM SAFE FLOW FOR A CENTRIFUGAL PUMP

A centrifugal pump handles 220-F water and has a shutoff head (with closed discharge valve) of 3,200 ft. At shutoff, the pump efficiency is 17 percent and the input brake horsepower is 210. What is the minimum safe flow through this pump to prevent overheating at shutoff? Determine the minimum safe flow if the NPSH is 18.8 ft of water and the liquid specific gravity is 0.995. If the pump contains 500 lb of water, determine the rate of the temperature rise at shutoff.

Calculation Procedure:

1. Compute the temperature rise in the pump

With the discharge valve closed, the power input to the pump is converted to heat in the casing and causes the liquid temperature to rise. The temperature rise $t = (1 - e) \times H_s/778e$, where $t = $ temperature rise during shutoff, F; $e = $ pump efficiency, expressed as a decimal; $H_s = $ shutoff head, ft. For this pump, $t = (1 - 0.17)(3{,}200)/[778(0.17)] = 20.4$ F.

2. Compute the minimum safe liquid flow

For general-service pumps, the minimum safe flow M gpm = 6.0(bhp input at shut-off)/t. Or, $M = 6.0(210)/20.4 = 62.7$ gpm. This equation includes a 20 percent safety factor.

Centrifugal boiler-feed pumps usually have a maximum allowable temperature rise of 15 F. The minimum allowable flow through the pump to prevent the water temperature from rising more than 15 F is 30 gpm for each 100 bhp input at shutoff.

3. Compute the temperature rise for the operating NPSH

A NPSH of 18.8 ft is equivalent to a pressure of 18.8(0.433)(0.995) = 7.78 psia at 220 F, where the factor 0.433 converts ft of water to psi. At 220 F, the vapor pressure of the water is 17.19 psia, from the steam tables. Thus, the total vapor pressure the water can develop before flashing occurs = NPSH pressure + vapor pressure at operating temperature = 7.78 + 17.19 = 24.97 psia. Enter the steam tables at this pressure and read the corresponding temperature as 240 F. The allowable temperature rise of the water is then 240 − 220 = 20 F. Using the safe-flow relation of step 2, the minimum safe flow is 62.9 gpm.

4. Compute the rate of temperature rise

In any centrifugal pump, the rate of temperature rise t_r F/min = 42.4(bhp input at shutoff)/wc, where w = weight of liquid in the pump, lb; c = specific heat of the liquid in the pump, Btu/(lb)(F) For this pump containing 500 lb of water with a specific heat, $c = 1.0$, $t_r = 42.4(210)/[500(1.0)] = 17.8$ F/min. This is a very rapid temperature rise and could lead to overheating in a few min.

Related Calculations: Use this procedure for any centrifugal pump handling any liquid in any service — power, process, marine, industrial, or commercial. Pump manufacturers can supply a temperature-rise curve for a given model pump if it is requested. This curve is superimposed on the pump characteristic curve and shows the temperature rise accompanying a specific flow through the pump.

SELECTING A CENTRIFUGAL PUMP TO HANDLE A VISCOUS LIQUID

Select a centrifugal pump to deliver 750 gpm of 1,000-SSU oil at a total head of 100 ft. The oil has a specific gravity of 0.90 at the pumping temperature. Show how to plot the characteristic curves when the pump is handling the viscous liquid.

Calculation Procedure:

1. Determine the required correction factors

A centrifugal pump handling a viscous liquid usually must develop a greater capacity and head, and it requires a larger power input than the same pump handling water. With the water performance of the pump known — either from the pump characteristic curves or a tabulation of pump performance parameters — Fig. 25, prepared by the Hydraulic Institute, can be used to find suitable correction factors. Use this chart only within its scale limits; do not extrapolate. Do not use the chart for mixed-flow or axial-flow pumps or for pumps of special design. Use the chart only for pumps handling uniform liquids; slurries, gels, paper stock, etc., may cause incorrect results. In using the chart, the available net positive suction head is assumed adequate for the pump.

To use Fig. 25, enter at the bottom at the required capacity, 750 gpm, and project vertically to intersect the 100-ft head curve, the required head. From here project horizontally to the 1,000-SSU viscosity curve, and then vertically upward to the correction-factor curves. Read $C_E = 0.635$; $C_Q = 0.95$; $C_H = 0.92$ for $1.0Q_{NW}$. The subscripts E, Q, and H refer to correction factors for efficiency, capacity, and head, respectively; and NW refers to the water capacity at a particular efficiency. At maximum efficiency, the water capacity is given as $1.0Q_{NW}$; other efficiencies, expressed by numbers equal to or less than unity, give different capacities.

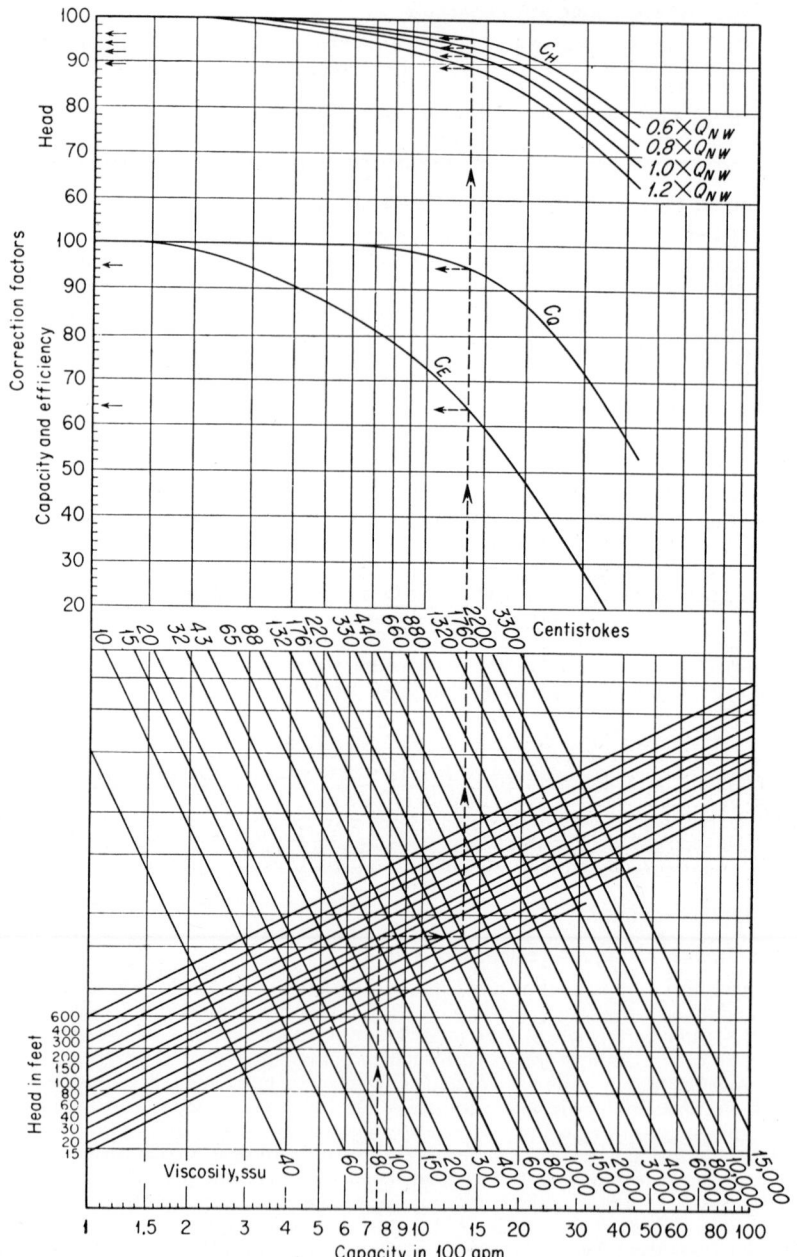

Fig. 25 Correction factors for viscous liquids handled by centrifugal pumps. (*Hydraulic Institute.*)

2. *Compute the water characteristics required*

The water capacity required for the pump $Q_w = Q_v/C_Q$, where $Q_v =$ viscous capacity, gpm. For this pump, $Q_w = 750/0.95 = 790$ gpm. Likewise, water head $H_w = H_v/C_H$, where $H_v =$ viscous head. Or, $H_w = 100/0.92 = 108.8$, say 109 ft of water.

Choose a pump to deliver 790 gpm of water at 109-ft head of water and the required viscous head and capacity will be obtained. Pick the pump so that it is operating at or near its maximum efficiency on water. If the water efficiency $E_w = 81$ percent at 790 gpm for this pump, the efficiency when handling the viscous liquid $E_v = E_w C_E$. Or, $E_v = 0.81(0.635) = 0.515$, or 51.5 percent.

The power input to the pump when handling viscous liquids is given by $P_v = Q_v H_v s/3{,}960 E_v$, where $s =$ specific gravity of the viscous liquid. For this pump, $P_v = (750) \times (100)(0.90)/[3{,}960(0.515)] = 33.1$ hp.

3. *Plot the characteristic curves for viscous-liquid pumping*

Follow these eight steps to plot the complete characteristic curves of a centrifugal pump handling a viscous liquid when the water characteristics are known: (a) Secure a complete set of characteristic curves (H, Q, P, E) for the pump to be used. (b) Locate the point of maximum efficiency for the pump when handling water. (c) Read the pump capacity, Q gpm, at this point. (d) Compute the values of $0.6Q$, $0.8Q$, and $1.2Q$ at the maximum efficiency. (e) Using Fig. 25, determine the correction factors at the capacities in Steps c and d. Where a multistage pump is being considered, use the head per stage (= total pump head, ft/number of stages), when entering Fig. 25. (f) Correct the head, capacity, and efficiency for each of the flow rates in c and d using the correction factors from Fig. 25. (g) Plot the corrected head and efficiency against the corrected capacity, as in Fig. 26. (h) Compute the power input at each flow rate and plot. Draw smooth curves through the points obtained, Fig. 26.

Related Calculations: Use the method given here for any uniform viscous liquid – oil, gasoline, kerosene, mercury, etc – handled by a centrifugal pump. Be careful to use Fig. 25 only within its scale limits; *do not extrapolate.* The method presented here is that developed by the Hydraulic Institute. For new developments in the method, be certain to consult the latest edition of the Hydraulic Institute *Standards.*

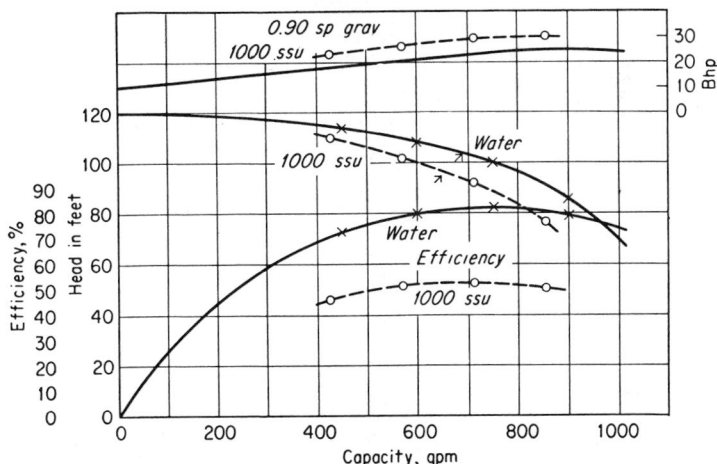

Fig. 26   Characteristic curves for water (solid line) and oil (dashed line). (*Hydraulic Institute.*)

PUMP SHAFT DEFLECTION AND CRITICAL SPEED

What is the shaft deflection and approximate first critical speed of a centrifugal pump if the total combined weight of the pump impellers is 23 lb and the pump manufacturer supplies the engineering data in Fig. 27?

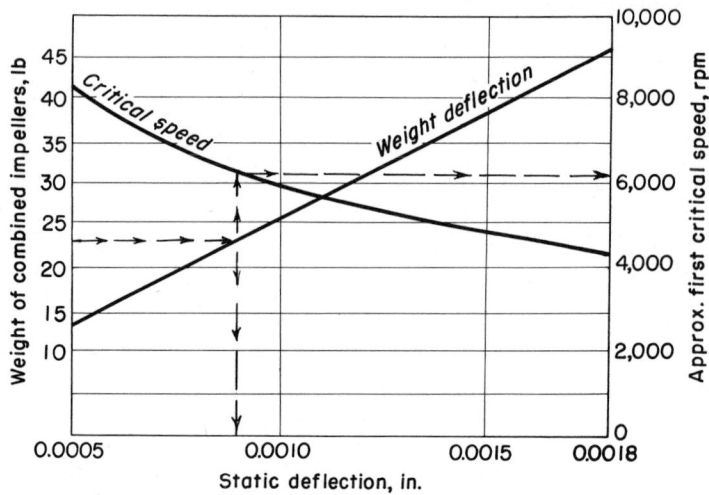

Fig. 27   Pump shaft deflection and critical speed. (*Goulds Pumps, Inc.*)

Calculation Procedure:

1. Determine the deflection of the pump shaft

Use Fig. 27 to determine the shaft deflection. Note that this chart is valid for only one pump or series of pumps and must be obtained from the pump builder. Such a chart is difficult to prepare from test data without extensive test facilities.

Enter Fig. 27 at the left at the total combined weight of the impellers, 23 lb, and project horizontally to the right until the weight-deflection curve is intersected. From the intersection, project vertically downward to read the shaft deflection as 0.0009 in. at full speed.

2. Determine the critical speed of the pump

From the intersection of the weight-deflection curve in Fig. 27 project vertically upward to the critical-speed curve. Project horizontally right from this intersection and read the first critical speed as 6,200 rpm.

Related Calculations: Use this procedure for any class of pump—centrifugal, rotary, or reciprocating—for which the shaft-deflection and critical-speed curves are available. These pumps can be used for any purpose—process, power, marine, industrial, or commercial.

EFFECT OF LIQUID VISCOSITY ON REGENERATIVE PUMP PERFORMANCE

A regenerative (turbine) pump has the water head-capacity and power-input characteristics shown in Fig. 28. Determine the head-capacity and power-input characteristics for four different viscosity oils to be handled by the pump—400, 600, 900, and 1,000 SSU. What effect does increased viscosity have on the performance of the pump?

Calculation Procedure:

1. Plot the water characteristics of the pump

Obtain a tabulation or plot of the water characteristics of the pump from the manufacturer or from his engineering data. With a tabulation of the characteristics, enter the various capacity and power points given and draw a smooth curve through them, Fig. 28.

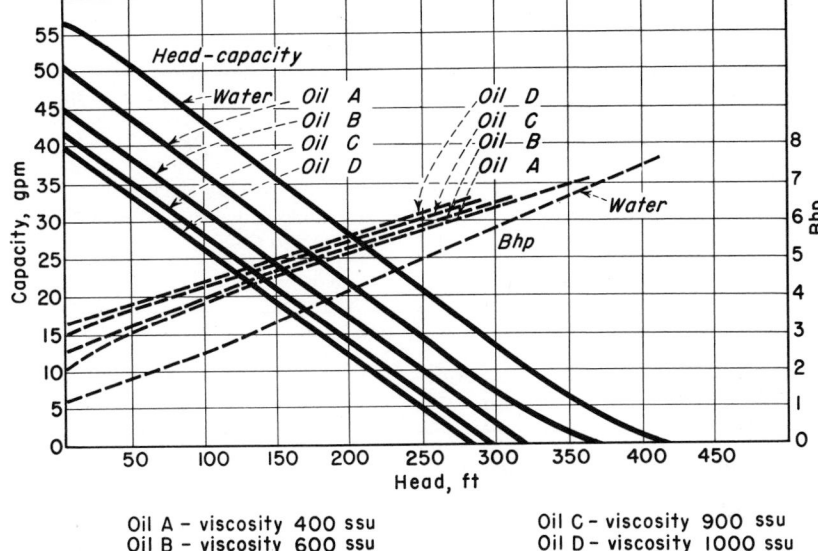

Oil A – viscosity 400 ssu   Oil C – viscosity 900 ssu
Oil B – viscosity 600 ssu   Oil D – viscosity 1000 ssu

Fig. 28   Regenerative pump performance when handling water and oil. (*Aurora Pump Division, The New York Air Brake Company.*)

2.   Plot the viscous-liquid characteristics of the pump

The viscous-liquid characteristics of regenerative-type pumps are obtained by test of the actual unit. Hence, the only source of this information is the pump manufacturer. Obtain these characteristics from the pump manufacturer or his test data and plot them on Fig. 28, as shown, for each oil or other liquid handled.

3.   Evaluate the effect of viscosity on pump performance

Study Fig. 28 to determine the effect of increased liquid viscosity on the performance of the pump. Thus at a given head—say 100 ft—the capacity of the pump decreases as the liquid viscosity increases. At 100-ft head, this pump has a water capacity of 43.5 gpm, Fig. 28. The pump capacity for the various oils at 100-ft head is 36 gpm for 400 SSU; 32 gpm for 600 SSU; 28 gpm for 900 SSU; and 26 gpm for 1,000 SSU, respectively. There is a similar reduction in capacity of the pump at the other heads plotted in Fig. 28. Thus, as a general rule, it can be stated that the capacity of a regenerative pump decreases with an increase in liquid viscosity at constant head. Or conversely, at constant capacity, the head developed decreases as the liquid viscosity increases.

Plots of the power input to this pump show that the input power increases as the liquid viscosity increases.

Related Calculations: Use this procedure for a regenerative-type pump handling any liquid—water, oil, kerosene, gasoline, etc. A decrease in the viscosity of a liquid—as compared with the viscosity of water—will produce the opposite effect from that of increased viscosity.

EFFECT OF LIQUID VISCOSITY ON RECIPROCATING-PUMP PERFORMANCE

A direct-acting steam-driven reciprocating pump delivers 100 gpm of 70-F water when operating at 50 strokes/min. How much 2,000-SSU crude oil will this pump deliver? How much 125-F water will this pump deliver?

Calculation Procedure:

1. Determine the recommended change in pump performance

Reciprocating pumps of any type—direct-acting or power—having any number of liquid-handling cylinders—1 to 5, or more—are usually rated for maximum delivery when handling 250-SSU liquids or 70-F water. At higher liquid viscosities or water temperatures, the speed—strokes or rpm—is reduced. Table 8 shows typical recommended speed-correction factors for reciprocating pumps for various liquid viscosities and water temperatures. This table shows that with a liquid viscosity of 2,000 SSU, the pump speed should be reduced 20 percent. When handling 125-F water, the pump speed should be reduced 25 percent, as shown in Table 8.

TABLE 8   Speed-Correction Factors

Liquid viscosity, SSU......	250	500	1,000	2,000	3,000	4,000	5,000
Speed reduction, %	0	4	11	20	26	30	35
Water temperature, F	70	80	100	125	150	200	250
Speed reduction, %	0	9	18	25	29	34	38

2. Compute the delivery of the pump

The delivery capacity of any reciprocating pump is directly proportional to the number of strokes/min it makes or to its rpm.

When handling 2,000-SSU oil, the pump strokes/min must be reduced 20 percent, or $(50)(0.20) = 10$ strokes/min. Hence, the pump speed will be $50 - 10 = 40$ strokes/min. Since the delivery is directly proportional to speed, the delivery of 2,000-SSU oil $= (40/50)(100) = 80$ gpm.

When handling 125-F water, the pump strokes/min must be reduced 25 percent, or $(50)(0.5) = 12.5$ strokes/min. Hence, the pump speed will be $50.0 - 12.5 = 37.5$ strokes/min. Since the delivery is directly proportional to speed, the delivery of 125-F water $= (37.5/50)(100) = 75$ gpm.

Related Calculations: Use this procedure for any type of reciprocating pump handling liquids falling within the range of Table 8. Such liquids include oil, kerosene, gasoline, brine, water, etc.

EFFECT OF VISCOSITY AND DISSOLVED GAS ON ROTARY PUMPS

A rotary pump handles 8,000-SSU liquid containing 5 percent entrained gas and 10 percent dissolved gas at a 20-in. Hg pump inlet vacuum. The pump is rated at 1,000 gpm when handling gas-free liquids at viscosities less than 600 SSU. What is the output of this pump without slip? With 10 percent slip?

Calculation Procedure:

1. Compute the required speed reduction of the pump

When the liquid viscosity exceeds 600 SSU, many pump manufacturers recommend that the speed of a rotary pump be reduced to permit operation without excessive noise or vibration. The speed reduction usually recommended is shown in Table 9.

With this pump handling 8,000-SSU liquid, a speed reduction of 40 percent is necessary, as shown in Table 9. Since the capacity of a rotary pump varies directly with its speed, the output of this pump when handling 8,000-SSU liquid = (1,000 gpm) $\times (1.0 - 0.40) = 600$ gpm.

2. Compute the effect of gas on the pump output

Entrained or dissolved gas reduces the output of a rotary pump, as shown in Table 10. The gas in the liquid expands when the inlet pressure of the pump is below atmospheric and the gas occupies part of the pump chamber, reducing the liquid capacity.

With a 20-in. Hg inlet vacuum, 5 percent entrained gas, and 10 percent dissolved gas,

TABLE 9 Rotary Pump Speed Reduction for Various Liquid Viscosities

Liquid Viscosity, SSU	*Speed Reduction, Percent of Rated Pump Speed*
600	2
800	6
1,000	10
1,500	12
2,000	14
4,000	20
6,000	30
8,000	40
10,000	50
20,000	55
30,000	57
40,000	60

Table 10 shows that the liquid displacement is 74 percent of the rated displacement. Thus, the output of the pump when handling this viscous, gas-containing liquid will be (600 gpm)(0.74) = 444 gpm without slip.

3. *Compute the effect of slip on the pump output*

Slip reduces rotary-pump output in direct proportion to the slip. Thus, with 10 percent slip, the output of this pump = (444 gpm)(1.0 − 0.10) = 369.6 gpm.

Related Calculations: Use this procedure for any type of rotary pump—gear, lobe, screw, swinging-vane, sliding-vane, or shuttle-block, handling any clear, viscous liquid. Where the liquid is gas-free, apply only the viscosity correction. Where the liquid viscosity is less than 600 SSU but the liquid contains gas or air, apply the entrained or dissolved gas correction, or both corrections.

TABLE 10 Effect of Entrained or Dissolved Gas on the Liquid Displacement of Rotary Pumps*

(Liquid displacement: percent of displacement)

Vacuum at pump inlet, in. Hg	Gas entrainment					Gas solubility					Gas entrainment and gas solubility combined				
	1%	2%	3%	4%	5%	2%	4%	6%	8%	10%	1% 2%	2% 4%	3% 6%	4% 8%	5% 10%
5	99	97½	96½	95	93½	99½	99	98½	97	97½	98½	96½	96	92	91
10	98½	97¼	95½	94	92	99	97½	97	95	95	97½	95	93	90	88¼
15	98	96½	94½	92½	90½	97	96	94	92	90½	96	93	89½	86½	83¼
20	97½	94½	92	89	86½	96	92	89	86	83	94	88	83	78	74
25	94	89	84	79	75½	90	83	76½	71	66	85½	75½	68	61	55

For example: with 5 percent gas entrainment at 15 in. Hg vacuum, the liquid displacement will be 90½ percent of the pump displacement neglecting slip or with 10 percent dissolved gas liquid displacement will be 90½ percent of pump displacement, and with 5 percent entrained gas combined with 10 percent dissolved gas, the liquid displacement will be 83¼ percent of pump displacement.

*Courtesy of Kinney Mfg. Div., The New York Air Brake Co.

SELECTION OF MATERIALS FOR PUMP PARTS

Select suitable materials for the principal parts of a pump handling cold ethylene chloride. Use the Hydraulic Institute recommendations for materials of construction.

Calculation Procedure:

1. Determine which materials are suitable for this pump

Refer to the Data Section of the Hydraulic Institute *Standards*. This section contains a tabulation of hundreds of liquids and the pump construction materials that have been successfully used to handle each liquid.

The table shows that for cold ethylene chloride having a specific gravity of 1.28 an all-bronze pump is satisfactory. In lieu of an all-bronze pump, the principal parts of the pump—casing, impeller, cylinder, and shaft—can be made of one of the following materials: austenitic steels (low-carbon 18–8; 18–8/Mo; highly alloyed stainless); nickel-base alloys containing chromium, molybdenum, and other elements, and usually less than 20 percent iron; or nickel-copper alloy (Monel metal). The order of listing in the *Standards* does not necessarily indicate relative superiority, since certain factors predominating in one instance may be sufficiently overshadowed in others to reverse the arrangement.

2. Choose the most economical pump

Use the methods of earlier Calculation Procedures to select the most economical pump for the installation. Where the corrosion resistance of two or more pumps is equal, the standard pump—in this instance an all-bronze unit—will be the most economical.

Related Calculations: Use this Procedure to select the materials of construction for any class of pump—centrifugal, rotary, or reciprocating—in any type of service—power, process, marine, or commercial. Be certain to use the latest edition of the Hydraulic Institute *Standards*, because the recommended materials may change from one edition to the next.

SIZING A HYDROPNEUMATIC STORAGE TANK

A 200-gpm water pump serves a pumping system. Determine the capacity required for a hydropneumatic tank to serve this system if the allowable high pressure in the tank and system is 60 psig and the allowable low pressure is 30 psig. How many starts per hr will the pump make if the system draws 3,000 gpm from the tank?

Calculation Procedure:

1. Compute the required tank capacity

In the usual hydropneumatic system, a storage-tank capacity in gal of 10 times the pump capacity in gpm is used, if this capacity produces a moderate running time for the pump. Thus, this system would have a tank capacity of $(10)(200) = 2,000$ gal.

2. Compute the quantity of liquid withdrawn per cycle

For any hydropneumatic tank, the withdrawal W expressed as a percentage of the tank total volume is $W = (HU/L) - U$, where H = high pressure in the tank, psia; U = unused liquid, percentage of tank volume; L = low pressure in the tank, psia.

In most hydropneumatic tanks, a reserve of 10 or 20 percent of the total tank volume is kept in the tank to prevent the pump from running dry. Assuming 10 percent for this tank, $W = \{[(H/L) \times 0.9] - 0.9\} + H/L$. Substituting, $W = \{[(74.7/44.7) \times 0.9] - 0.9\} + (74.7/44.7) = 0.357$, or 35.7 percent. Since, the tank capacity is 2,000 gal, $W_g = (0.357)(2,000) = 714$ gal; where W_g = withdrawal in gal of liquid.

3. Compute the pump running time

The pump has a capacity of 200 gpm. Therefore, it will take $714/200 = 3.57$ min to replace the withdrawn liquid. To supply 3,000 gph to the system, the pump must start $3,000/714 = 4.2$, or 5 times per hr. This is acceptable, because a system in which the pump starts six or fewer times per hr is generally thought satisfactory.

Where the pump capacity is insufficient to supply the system demand for short periods, use a 20 percent reserve. Use the same equations as in Step 2 but substitute 0.8 for the 0.9 value.

To compute the unused portion with a 10 percent reserve, solve $U = [(H/L)(0.9 - W)] - (0.9 - W)$. For a 20 percent reserve, substitute 0.8 for the 0.9 value.

Related Calculations: Use this procedure for any liquid system having a hydro-pneumatic tank—well, drinking water, marine, industrial, or process.

PIPING AND FLUID FLOW

REFERENCES: King and Crocker—*Piping Handbook*, McGraw-Hill; USASI—Code for *Pressure Piping* (commonly called the *Piping Code*); ASME—*Fluid Meters—Their Theory and Application*; King and Brater—*Handbook of Hydraulics*, McGraw-Hill; Ingersoll-Rand Company—*Cameron Hydraulic Data*; The Hydraulic Institute—*Standards of the Hydraulic Institute*; Baumeister and Marks—*Standard Handbook for Mechanical Engineers*, McGraw-Hill; Littleton—*Industrial Piping*, McGraw-Hill; The Hydraulic Institute—*Pipe Friction Manual*; Black, Sivalls, and Bryson—*Valve Sizing Book* and *Cv Book*; Fluid Controls Institute—*Recommended Voluntary Standard Formulas for Sizing Control Valves*; Bell—*Petroleum Transportation Handbook*, McGraw-Hill; Perry—*Chemical Engineers' Handbook*, McGraw-Hill; Spielvogel—*Piping Stress Calculations Simplified*, Spielvogel Publishing; Grinnell Company, Inc.—*Piping Design and Engineering*; M. W. Kellogg Co.—*Design of Piping Systems*, Wiley; National Valve and Manufacturing Co.—*Piping Catalog*; McClain—*Fluid Flow in Pipes*, Industrial Press; Tube Turns Division of Chemetron Corp.—*Piping Engineering*.

PIPE-WALL THICKNESS AND SCHEDULE NUMBER

Determine the minimum wall thickness t_m in. and schedule number SN for a branch steam pipe operating at 900 F if the internal steam pressure is 1,000 psia. Use ANSA B31.1 *Code for Pressure Piping* and the ASME *Boiler and Pressure Vessel Code* valves and equations where they apply. Steam flow rate is 72,000 lb/hr.

Calculation Procedure:

1. Determine the required pipe diameter

When the length of pipe is not given, or is as yet unknown, make a first approximation of the pipe diameter using a suitable velocity for the fluid. Once the length of the pipe is known, the pressure loss can be determined. If the pressure loss exceeds a desirable value, the pipe diameter can be increased until the loss is within an acceptable range.

Compute the pipe cross-sectional area a sq in. from $a = 2.4Wv/V$, where W = steam flow rate, lb/hr; v = specific volume of the steam, cu ft/lb; V = steam velocity, fpm. The only unknown in this equation, other than the pipe area, is the steam velocity V. Use Table 1 to find a suitable steam velocity for this branch line.

Table 1 shows that the recommended steam velocities for branch steam pipes range from 6,000 to 15,000 fpm. Assume that a velocity of 12,000 fpm is used in this branch steam line. Then, using the steam table to find the specific volume of steam at 900 F and 1,000 psia, $a = 2.4(72,000)(0.7604)/12,000 = 10.98$ sq in. The inside diameter of the pipe d in. is then $d = 2(a/\pi)^{0.5} = 2(10.98/\pi)^{0.5} = 3.74$ in. Since pipe is not ordinarily made in this fractional internal diameter, round it off to the next larger size, or 4 in. inside diameter.

TABLE 1   Recommended Fluid Velocities in Piping

Service	Velocity of Fluid, fpm
Boiler and turbine leads	6,000–12,000
Steam headers	6,000–8,000
Branch steam lines	6,000–15,000
Feedwater lines	250–850
Exhaust and low-pressure steam lines	6,000–15,000
Pump suction lines	100–300
Bleed steam lines	4,000–6,000
Service water mains	120–300
Vacuum steam lines	20,000–40,000
Steam superheater tubes	2,000–5,000
Compressed-air lines	1,500–2,000
Natural-gas lines (large cross-country)	100–150
Economizer tubes (water)	150–300
Crude-oil lines (6 to 30 in.)	50–350

2.   Determine the pipe schedule number

The ANSA *Code for Pressure Piping*, commonly called the *Piping Code*, defines schedule number as $SN = 1,000 P_i / S$, where $P_i =$ internal pipe pressure, psig; $S =$ allowable stress in the pipe, psi, from *Piping Code*. Table 2 shows typical allowable stress values for pipe in power piping systems. For this pipe, assuming that seamless ferritic alloy steel (1% Cr, 0.55% Mo) pipe is used with the steam at 900 F, $SN =$ $(1,000)(1,014.7)/13,100 = 77.5$. Since pipe is not ordinarily made in this schedule number, use the next *highest* readily available schedule number, or $SN = 80$. (Where large quantities of pipe are required, it is sometimes economically wise to order pipe of the exact SN required. This is not usually done for orders of less than 1,000 ft of pipe.)

3.   Determine the pipe-wall thickness

Enter a tabulation of pipe properties, such as in King and Crocker—*Piping Handbook*, and find the wall thickness for 4-in. SN 80 pipe as 0.337 in.

Related Calculations: Use the method given here for any type of pipe—steam, water, oil, gas, or air—in any service—power, refinery, process, commercial, etc. Refer to the proper section of B31.1 *Code for Pressure Piping* when computing the schedule number, because the allowable stress S varies for different types of service.

The *Piping Code* contains an equation for determining the minimum required pipe-wall thickness based on the pipe internal pressure, outside diameter, allowable stress, a temperature coefficient, and an allowance for threading, mechanical strength, and corrosion. This equation is seldom used in routine piping-system design. Instead, the schedule number as given here is preferred by most designers.

PIPE-WALL THICKNESS DETERMINATION BY PIPING CODE FORMULA

Use the ANSA B31.1 *Code for Pressure Piping* wall-thickness equation to determine the required wall thickness for an 8.625-in. OD ferritic steel plain-end pipe if the pipe is used in 900-F, 900-psig steam service.

Calculation Procedure:

1.   Determine the constants for the thickness equation

Pipe-wall thickness to meet ANSA *Code* requirements for power service is computed from $t_m = \{DP/[2(S + YP)]\} + C$, where $t_m =$ minimum wall thickness, in.; $D =$ outside diameter of pipe, in.; $P =$ internal pressure in pipe, psig; $S =$ allowable stress in pipe material, psi; $Y =$ temperature coefficient; $C =$ end-condition factor, in.

TABLE 2   Allowable Stresses (S Values) Psi for Alloy-Steel Pipe in Power Piping Systems*

(Abstracted from ASME *Power Boiler Code* and *Code for Pressure Piping*, ASA B31.1)

Material	ASTM specification	Grade or symbol	Minimum tensile strength, psi	S values, psi, for metal temperatures not to exceed‡									
				−20 to 650 F	700 F	750 F	800 F	850 F	900 F	950 F	1000 F	1050 F	1100 F
Electric-fusion-welded:‡													
Carbon-molybdenum......	A155	A204A	65,000	14,600	14,600	14,600	14,100	12,950	11,250				
Carbon-molybdenum......	A155	A204B	70,000	15,750	15,750	15,750	15,200	13,500	11,450				
Carbon-molybdenum......	A155	A204C	75,000	16,850	16,850	16,850	16,200	14,300	11,700				
½% Cr, ½% Mo......	A155	A301A	65,000	14,600	14,600	14,600	14,100	12,950	11,250	9,000	5,600		
1% Cr, ½% Mo......	A155	A301B	60,000	13,500	13,500	13,500	13,250	12,750	11,800	9,900	6,750	4,500	2,500
1¼% Cr, ½% Mo......	A155	P11	60,000	13,500	13,500	13,500	13,500	12,950	11,800	9,900	7,000	4,950	3,600
2¼% Cr, 1% Mo......	A155	P22	60,000	13,500	13,500	13,500	13,500	12,950	11,800	9,900	7,000	5,200	3,750
Seamless ferritic steels:													
Carbon-molybdenum......	A335	P1	55,000	13,750	13,750	13,750	13,450	13,150	12,500				
0.65 Cr, 0.55 Mo......	A335	P2	55,000	13,750	13,750	13,750	13,450	13,150	12,500	10,000	6,250	5,000	2,800
1.00 Cr, 0.55 Mo......	A335	P12	60,000	15,000	15,000	15,000	14,750	14,200	13,100	11,000	7,500	5,500	4,000
1.25 Cr, 0.55 Mo......	A335	P11	60,000	15,000	15,000	15,000	15,000	14,400	13,100	11,000	7,800	5,800	4,200
2.25 Cr, 1.00 Mo......	A335	P22	60,000	15,000	15,000	15,000	15,000	14,400	13,100	11,000	7,800	5,500	4,000
3.00 Cr, 0.90 Mo......	A335	P21	60,000	15,000	14,800	14,500	13,900	13,200	12,000	9,000	7,000	5,500	4,000
5% Cr, ½% Mo......	A335	P5a	60,000	See Code	13,400	13,100	12,800	12,400	11,500	10,000	7,300	5,200	3,300
5% Cr, % Mo+Si......	A335	P5b	60,000	See Code	13,400	13,100	12,800	12,400	10,900	9,000	5,500	3,500	2,500
Seamless austenitic steels:													
18% Cr, 10% Ni+Ti......	A158	P8b(321)	75,000	See Code	14,800	14,700	14,550	14,300	14,100	13,850	13,500	13,100	10,300
18% Cr, 10% Ni+Cb......	A158	P8d(347)	75,000	See Code	14,800	14,700	14,550	14,300	14,100	13,850	13,500	13,100	10,300

*King and Crocker—*Piping Handbook.*

†Where welded construction is used, consideration should be given to the possibility of graphite formation in carbon-molybdenum steel above 875 F or in chromium-molybdenum steel containing less than 0.60% chromium above 975 F.

‡The values given are for class 2 A155 pipe. For class 1 pipe that is radiographed, these stresses may be increased by the ratio 0.95/0.90.

Values of S, Y, and C are given in tables in the *Code for Pressure Piping* in the section on Power Piping. Using values from the latest edition of the *Code*, $S = 12,500$ psi for ferritic-steel pipe operating at 900 F; $Y = 0.40$ at the same temperature; $C = 0.065$ in. for plain-end steel pipe.

2. Compute the minimum wall thickness

Substitute the given and *Code* values in the equation in Step 1, or $t_m = [(8.625)(900)]/[2(12,500 + 0.4 \times 900)] + 0.065 = 0.367$ in.

Since pipe mills do not fabricate to precise wall thicknesses, a tolerance above or below the computed wall thickness is required. An allowance must be made in specifying the wall thickness found with this equation by *increasing* the thickness by $12\frac{1}{2}$ percent. Thus, for this pipe, wall thickness $= 0.367 + 0.125(0.367) = 0.413$ in.

Refer to the *Code* to find the schedule number of the pipe. Schedule 60 8-in. pipe has a wall thickness of 0.406 in., and schedule 80 pipe has a wall thickness of 0.500 in. Since the required thickness of 0.413 in. is greater than schedule 60 but less than schedule 80, the higher schedule number, 80, should be used.

3. Check the selected schedule number

From the previous Calculation Procedure, $SN = 1,000P_i/S$. For this pipe, $SN = 1,000(900)/12,500 = 72$. Since piping is normally fabricated for schedule numbers 10, 20, 30, 40, 60, 80, 100, 120, 140, and 160, the next larger schedule number higher than 72, that is 80, will be used. This agrees with the schedule number found in Step 2.

Related Calculations: Use this method in conjunction with the appropriate *Code* equation to determine the wall thickness of pipe conveying air, gas, steam, oil, water, alcohol, or any other similar fluids in any type of service. Be certain to use the correct equation, which in some cases is simpler than that used here. Thus, for lead pipe, $t_m = Pd/2S$, where P = safe working pressure of the pipe, psig; d = inside diameter of pipe, in.; other symbols as before.

When a pipe will operate at a temperature between two tabulated *Code* values, find the allowable stress by interpolating between the tabulated temperature and stress values. Thus, for a pipe operating at 680 F, find the allowable stress at 650 F (= 9,500 psi) and 700 F (= 9,000 psi). Interpolate thus: allowable stress at 680 F = $[(700\,F - 680\,F)/(700\,F - 650\,F)](9,500 - 9,000) + 9,000 = 200 + 9,000 = 9,200$ psi. The same result can be obtained by interpolating downward from 9,500 psi, or allowable stress at 680 F = $9,500 - [(680 - 650)/(700 - 650)](9,500 - 9,000) = 9,200$ psi.

DETERMINING THE PRESSURE LOSS IN STEAM PIPING

Use a suitable pressure-loss chart to determine the pressure loss in 510 ft of 4-in. flanged steel pipe containing two 90° elbows and four 45° bends. The schedule 40 piping conveys 13,000 lb/hr of 40-psig 350-F superheated steam. List other methods of determining the pressure loss in steam piping.

Calculation Procedure:

1. Determine the equivalent length of the piping

The equivalent length of a pipe L_e ft = length of straight pipe, ft + equivalent length of fittings, ft. Using data from the Hydraulic Institute, King and Crocker — *Piping Handbook*, earlier sections of the present Handbook, or Fig. 1, find the equivalent length of a 90° 4-in. elbow as a 10 ft of straight pipe. Likewise, the equivalent length of a 45° bend is 5 ft of straight pipe. Substituting in the above relation and using the straight lengths and the number of fittings of each type, $L_e = 510 + (2)(10) + 4(5) = 550$ ft of straight pipe.

2. Compute the pressure loss using a suitable chart

Figure 2 presents a typical pressure-loss chart for steam piping. Enter the chart at the top left at the superheated steam temperature of 350 F and project vertically downward until the 40-psig superheated steam pressure curve is intersected. From here, project horizontally to the right until the outer border of the chart is intersected. Next,

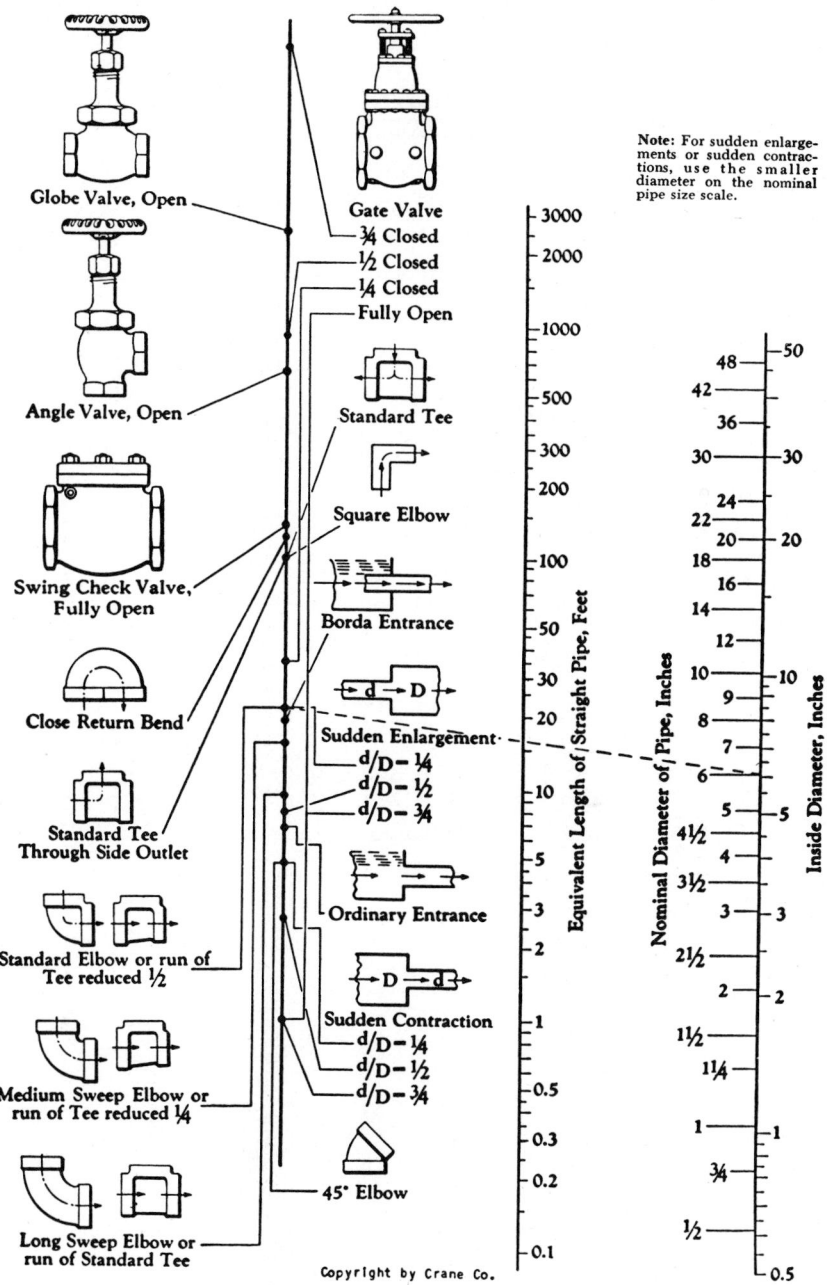

Globe Valve, Open

Angle Valve, Open

Swing Check Valve,
Fully Open

Close Return Bend

Standard Tee
Through Side Outlet

Standard Elbow or run of
Tee reduced ½

Medium Sweep Elbow or
run of Tee reduced ¼

Long Sweep Elbow or
run of Standard Tee

Gate Valve
¾ Closed
½ Closed
¼ Closed
Fully Open

Standard Tee

Square Elbow

Borda Entrance

Sudden Enlargement
d/D - ¼
d/D - ½
d/D - ¾

Ordinary Entrance

Sudden Contraction
d/D - ¼
d/D - ½
d/D - ¾

45° Elbow

Note: For sudden enlargements or sudden contractions, use the smaller diameter on the nominal pipe size scale.

Equivalent Length of Straight Pipe, Feet

3000
2000
1000
500
300
200
100
50
30
20
10
5
3
2
1
0.5
0.3
0.2
0.1

Nominal Diameter of Pipe, Inches

48
42
36
30
24
22
20
18
16
14
12
10
9
8
7
6
5
4½
4
3½
3
2½
2
1½
1¼
1
¾
½

Inside Diameter, Inches

50
30
20
10
5
3
2
1
0.5

Copyright by Crane Co.

Fig. 1 Equivalent length of pipe fittings and valves. (*Crane Company.*)

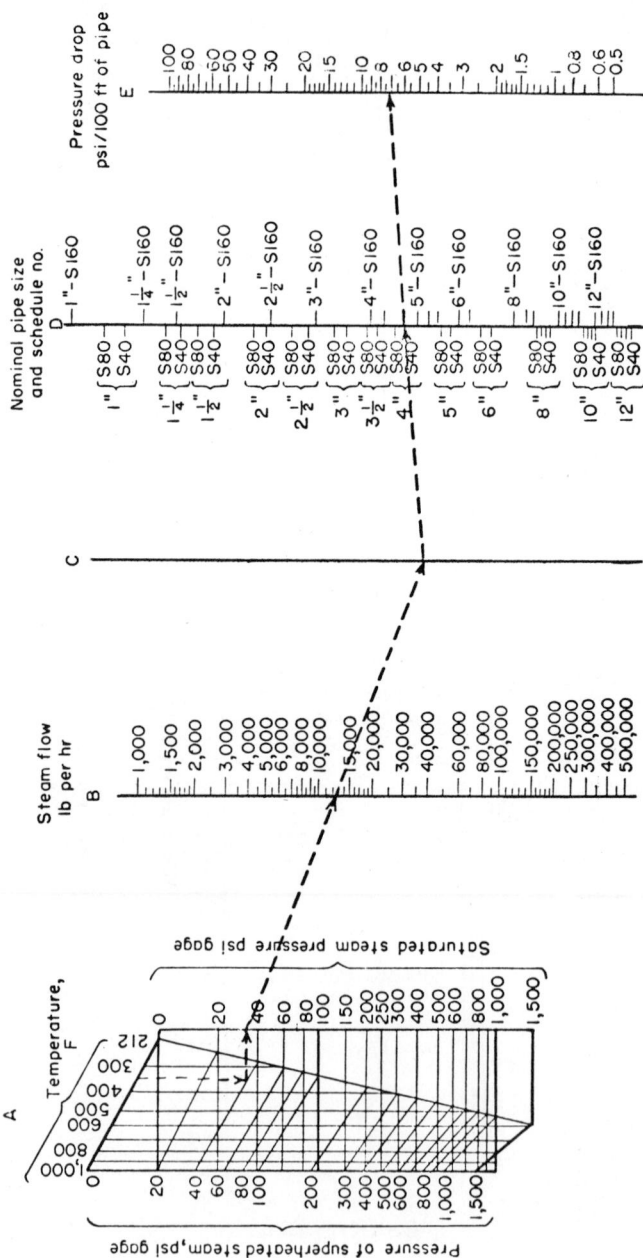

Fig. 2 Pressure loss in steam pipes based on the Fritsche formula. (*Power.*)

project through the steam flow rate, 13,000 lb/hr on scale B, Fig. 2, to the pivot scale C. From this point, project through 4-in. schedule 40 pipe on scale D, Fig. 2. Extend this line to intersect the pressure-drop scale and read the pressure loss as 7.25 psi per 100 ft of pipe.

Since the equivalent length of this pipe is 550 ft, the total pressure loss in the pipe is $(550/100)(7.25) = 39.875$ psi, say 40 psi.

3. List the other methods of computing pressure loss

Numerous pressure-loss equations have been developed to compute the pressure drop in steam piping. Among the better-known equations are those of Unwin, Fritzche, Spitzglass, Babcock, Gutermuth, and others. These equations are discussed in some detail in King and Crocker—*Piping Handbook* and in the engineering data published by valve and piping manufacturers.

Most piping designers use a chart to determine the pressure loss in steam piping because a chart saves time and reduces the effort involved. Further, the accuracy obtained is sufficient for all usual design practice.

Figure 3 is a popular flow chart for determining steam flow rate, pipe size, steam pressure, or steam velocity in a given pipe. Using this chart, the designer can determine any one of the four variables listed above when the other three are known. In solving a problem on the chart in Fig. 3, use the steam-quantity lines to intersect pipe sizes and the steam-pressure lines to intersect steam velocities. Here are two typical applications of this chart.

Example: What size schedule 40 pipe is needed to deliver 8,000 lb/hr of 120-psig steam at a velocity of 5,000 fpm?

Solution: Enter Fig. 3 at the upper left at a velocity of 5,000 fpm and project along this velocity line until the 120-psig pressure line is intersected. From this intersection, project horizontally until the 8,000 lb/hr vertical line is intersected. Read the *nearest* pipe size as 4 in. on the *nearest* pipe-diameter curve.

Example: What is the steam velocity in a 6-in. pipe delivering 20,000 lb/hr of steam at 85 psig?

Solution: Enter the bottom of the chart, Fig. 3, at the flow rate, 20,000 lb/hr, and project vertically upward until the 6-in.-pipe curve is intersected. From this point, project horizontally to the 85-psig curve. At the intersection, read the velocity as 7,350 fpm.

Table 3 shows typical steam velocities for various industrial and commercial applications. Use the given values as guides when sizing steam piping.

TABLE 3 Steam Velocities Used in Pipe Design

Steam condition	Steam pressure, psi	Steam use	Steam velocity, fpm
Saturated.........	0–15	Heating	4,000–6,000
Saturated.........	50–150	Process	6,000–10,000
Superheated......	200 and higher	Boiler leads	10,000–15,000

PIPING WARM-UP CONDENSATE LOAD

How much condensate is formed in 5 min during warm-up of 500 ft of 6-in. schedule 40 steel pipe conveying 215-psia saturated steam if the pipe is insulated with 2 in. of 85 percent magnesia and the minimum external temperature is 35 F?

Calculation Procedure:

1. Compute the amount of condensate formed during pipe warm-up

For any pipe, the condensate formed during warm-up C_h lb/hr $= 60(W_p)(\Delta t)(s)/h_{fg}N$, where W_p = total weight of pipe, lb; Δt = difference between final and initial tempera-

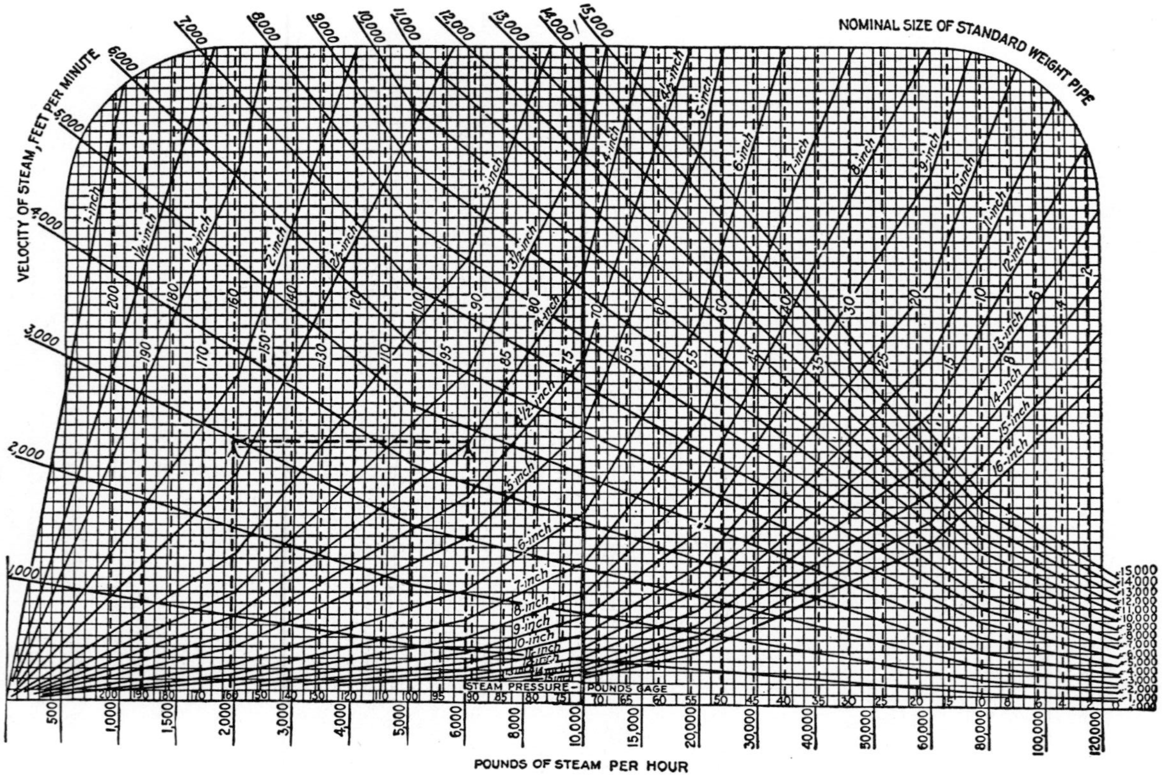

Fig. 3 Spitzglass chart for saturated steam flowing in schedule 40 pipe.

ture of the pipe, F; s = specific heat of pipe material, Btu/(lb)(F); h_{fg} = enthalpy of vaporization of the steam, Btu/lb; N = warm-up time, min.

A table of pipe properties shows that this pipe weighs 18.974 lb/ft. The steam table shows that the temperature of 215-psia saturated steam is 387.89 F, say 388 F; the enthalpy h_{fg} = 837.4 Btu/lb. The specific heat of steel pipe s = 0.144 Btu/(lb)(F). Then, C_h = 60(500 × 18.974)(388 − 35)(0.114)/[(837.4)(5)] = 5,470 lb/hr.

2. Compute the radiation-loss condensate load

Condensate is also formed by radiation of heat from the pipe during warm-up and while the pipe is operating. The warm-up condensate load decreases as the radiation load increases, the peak occurring midway ($2\frac{1}{2}$ min in this case) through the warm-up period. For this reason, one-half the normal radiation load is added to the warm-up load. Where the radiation load is small, it is often disregarded. However, the load must be computed before its magnitude can be determined.

For any pipe, C_r = $(L)(A)(\Delta t)(H)/h_{fg}$, where L = length of pipe, ft; A = external area of pipe, sq ft per ft of length; H = heat loss through bare pipe or pipe insulation, Btu/(sq ft)(hr)(F), from the piping or insulation tables. This 6-in. schedule 40 pipe has an external area A = 1.73 sq ft per ft of length. The heat loss through 2 in. of 85 percent magnesia, from insulation tables, is H = 0.286 Btu/(sq ft)(hr)(F). Then, C_r = (500) × (1.73)(388 − 35)(0.286)/837.4 = 104.2 lb/hr. Adding half the radiation load to the warm-up load gives 5,470 + 52.1 = 5,522.1 lb/hr.

3. Apply a suitable safety factor to the condensate load

Trap manufacturers recommend a safety factor of 2 for traps installed between a boiler and the end of a steam main; traps at the end of a long steam main or ahead of pressure-regulating or shut-off valves usually have a safety factor of 3. Using a safety factor of 3 for this pipe, the steam trap should have a capacity of at least 3(5,522.1) = 16,566.3 lb/hr, say 17,000 lb/hr.

Related Calculations: Use this method to find the warm-up condensate load for any type of steam pipe — main or auxiliary, in power, process, heating, or vacuum service. The same method is applicable to other vapors that form condensate — Dowtherm, refinery vapors, process vapors, and others.

STEAM TRAP SELECTION FOR INDUSTRIAL APPLICATIONS

Select steam traps for the following five types of equipment: (1) where the steam directly heats solid materials as in autoclaves, retorts, and sterilizers; (2) where the steam indirectly heats a liquid through a metallic surface as in heat exchangers and kettles where the quantity of liquid heated is known and unknown; (3) where the steam indirectly heats a solid through a metallic surface as in dryers using cylinders or chambers and platen presses; and (4) where the steam indirectly heats air through metallic surfaces as in unit heaters, pipe coils, and radiators.

Calculation Procedure:

1. Determine the condensate load

The first step in selecting a steam trap for any type of equipment is determination of the condensate load. Use the following general procedure.

a. Solid materials in autoclaves, retorts, and sterilizers. How much condensate is formed when 2,000 lb of solid material with a specific heat of 1.0 is processed in 15 min at 240 F by 25-psig steam from an initial temperature of 60 F in an insulated steel retort?

For this type of equipment, use C = WsP, where C = condensate formed, lb/hr; W = weight of material heated, lb; s = specific heat, Btu/(lb)(F); P = factor from Table 4. Thus, for this application, C = (2,000)(1.0)(0.193) = 386 lb of condensate. Note that P is based on a temperature rise of 240 − 60 = 180 F and a steam pressure of 25 psig. For the retort, using the specific heat of steel from Table 5, C = (4,000)(0.12)(0.193) = 92.6

lb of condensate, say 93 lb. The total weight of condensate formed in 15 min = 386 + 93 = 479 lb. In 1 hr, 479(60/15) = 1,916 lb of condensate is formed.

TABLE 4   Factors, $P = (T - t)/L$, to Find Condensate Load

Pressure, psia	Temperature rise, F								
	40	60	80	100	120	140	160	180	200
5	0.042	0.062	0.083	0.104	0.125	0.146	0.167	0.187	0.208
10	0.042	0.063	0.084	0.105	0.126	0.147	0.168	0.189	0.211
15	0.042	0.064	0.085	0.106	0,127	0.148	0.169	0.191	0.212
20	0.043	0.064	0.085	0.106	0.128	0.149	0.170	0.192	0.213
25	0.043	0.064	0.086	0.107	0.129	0.150	0.172	0.193	0.214
30	0.043	0.065	0.086	0.108	0.129	0.151	0.172	0.194	0.215
35	0.043	0.065	0.087	0.108	0.130	0.152	0.173	0.195	0.216
40	0.044	0.065	0,087	0.109	0.130	0.152	0.174	0.196	0.218
50	0.044	0.066	0.087	0.110	0.132	0.154	0.176	0.198	0.219
60	0.044	0.066	0.088	0.111	0.133	0.155	0.177	0.199	0.221
75	0.045	0.067	0.089	0.112	0.134	0.156	0.179	0.201	0.224
100	0.045	0.068	0.091	0.114	0.136	0.159	0.182	0.204	0.227
125	0.046	0.069	0.092	0.115	0.138	0.162	0.184	0.207	0.230
150	0.047	0.070	0.093	0.117	0.140	0.163	0.187	0.210	0.234
175	0.047	0.071	0.095	0.118	0.142	0.165	0.189	0.213	0.236
200	0.048	0.072	0.096	0.120	0.144	0.167	0.191	0.215	0.239

TABLE 5   Use These Specific Heats When Calculating Condensate Load

Solids		Liquids	
Aluminum	0.23	Alcohol	0.65
Brass	0.10	Carbon tetrachloride	0.20
Copper	0.10	Gasoline	0.53
Glass	0.20	Glycerin	0.58
Iron	0.13	Kerosene	0.47
Steel	0.12	Oils	0.40–0.50

A safety factor must be applied to compensate for radiation and other losses. Typical safety factors used in selecting steam traps are:

Steam mains and headers	2–3
Steam heating pipes	2–6
Purifiers and separators	2–3
Retorts for process	2–4
Unit heaters	3
Submerged pipe coils	2–4
Cylinder dryers	4–10

Using a safety factor of 4 for this process retort, the trap capacity = (4)(1,916) = 7,664 lb/hr, say 7,700 lb/hr.

b(1). Submerged heating surface and a known quantity of liquid. How much condensate forms in the jacket of a kettle when 500 gal of water are heated in 30 min from 72 to 212 F with 50-psig steam?

TABLE 6 Ordinary Ranges of Overall Coefficients of Heat Transfer

Type of heat exchanger	State of controlling resistance		Typical fluid	Typical apparatus
	Free convection, U	Forced convection, U		
Liquid to liquid	25–60	150–300	Water	Liquid-to-liquid heat exchangers
Liquid to liquid	5–10	20–50	Oil	
Liquid to gas*	1–3	2–10	. . . .	Hot-water radiators
Liquid to boiling liquid	20–60	50–150	Water	Brine coolers
Liquid to boiling liquid	5–20	25–60	Oil	
Gas* to liquid	1–3	2–10	. . . .	Air coolers, economizers
Gas* to gas	0.6–2	2–6	. . . .	Steam superheaters
Gas* to boiling liquid	1–3	2–10	. . . .	Steam boilers
Condensing vapor to liquid.	50–200	150–800	Steam to water	Liquid heaters and condensers
Condensing vapor to liquid.	10–30	20–60	Steam to oil	
Condensing vapor to liquid.	40–80	60–150	Organic vapor to water	
Condensing vapor to liquid.	. . . .	15–300	Steam-gas mixture	Steam pipes in air, air heaters
Condensing vapor to gas*	1–2	2–10	. . . .	Scale-forming evaporators
Condensing vapor to boiling liquid .	40–100	. . . .	Steam to water	
Condensing vapor to boiling liquid .	300–800	. . . .	Steam to water	
Condensing vapor to boiling liquid .	50–150	. . . .	Steam to oil	

*At atmospheric pressure.

$U = $ Btu/(hr)(sq. ft.)(F) Under many conditions, either higher or lower values may be realized.

For this type of equipment, $C = GwsP$, where $G =$ gal of liquid heated; $w =$ weight of liquid, lb/gal. Substitute the appropriate values as follows: $C = (500)(8.33)(1.0) \times (0.154) = 641$ lb, or $(641)(60/30) = 1,282$ lb/hr. Using a safety factor of 3, the trap capacity $= (3)(1,282) = 3,846$ lb/hr; say 3,900 lb/hr.

$b(2)$. *Submerged heating surface and an unknown quantity of liquid.* How much condensate is formed in a coil submerged in oil when the oil is heated as quickly as possible from 50 to 250 F by 25-psig steam if the coil has an area of 50 sq ft and the oil is free to circulate around the coil?

For this condition, $C = UAP$, where $U =$ overall coefficient of heat transfer, Btu/(hr)(sq ft)(F), from Table 6; $A =$ area of heating surface, sq ft. With free convection and a condensing-vapor-to-liquid type of heat exchanger, $U = 10$ to 30. Using an average value of $U = 20$, $C = (20)(50)(0.214) = 214$ lb/hr of condensate. Choosing a safety factor 3, the trap capacity $= (3)(214) = 642$ lb/hr; say 650 lb/hr.

$b(3)$. *Submerged surfaces having more area than needed to heat a specified quantity of liquid in a given time with condensate withdrawn as rapidly as formed.* Use Table 7 instead of step $b(1)$ or $b(2)$. Find the condensation rate by multiplying the submerged area by the appropriate factor from Table 7. Use this method for heating water, chemical solutions, oils, and other liquids. Thus, with steam at 100 psig and a temperature of 338 F and heating oil from 50 to 226 F with a submerged surface having an area of 500 sq ft. the mean temperature difference = steam temperature minus the average liquid temperature $= Mtd = 338 - (50 + 226/2) = 200$ F. The factor from Table 7 for 100-psig steam and a 200-F Mtd is 56.75. Thus, the condensation rate $= (56.75) \times (500) = 28,375$ lb/hr. With a safety factor of 2, the trap capacity $= (2)(28,375) = 56,750$ lb/hr.

c. Solids indirectly heated through a metallic surface. How much condensate is formed in a chamber dryer when 1,000 lb of cereal are dried to 750 lb by 10-psig steam? The initial temperature of the cereal is 60 F and the final temperature equals that of the steam.

For this condition, $C = 970(W - D)/h_{fg} + WP$, where $D =$ dry weight of the material, lb; $h_{fg} =$ enthalpy of vaporization of the steam at the trap pressure, Btu/lb. Using the steam tables and Table 4, $C = 970(1,000 - 750)/952 + (1,000)(0.189) = 443.5$ lb/hr of condensate. With a safety factor of 4, the trap capacity $= (4)(443.5) = 1,774$ lb/hr.

d. Indirect heating of air through a metallic surface. How much condensate is formed in a unit heater using 10-psig steam if the entering-air temperature is 30 F and the leaving-air temperature is 130 F? Airflow is 10,000 cfm.

Use Table 8, entering at a temperature difference of 100 F and projecting to a steam pressure of 10 psig. Read the condensate formed as 122 lb/hr per 1,000 cfm. Since 10,000 cfm of air are being heated, the condensation rate $= (10,000/1,000)(122) = 1,220$ lb/hr. With a safety factor of 3, the trap capacity $= (3)(1,220) = 3,660$ lb/hr, say 3,700 lb/hr.

Table 9 shows the condensate formed by radiation from bare iron and steel pipes in still air and with forced air circulation. Thus, with a steam pressure of 100 psig and an initial air temperature of 75 F, 1.05 lb/hr of condensate will be formed per sq ft of heating surface in still air. With forced air circulation, the condensate rate is $(5)(1.05) = 5.25$ lb/hr per sq ft of heating surface.

Unit heaters have a *standard rating* based on 2-psig steam with entering air at 60 F. If the steam pressure or air temperature is different from these standard conditions, multiply the heater Btu/hr capacity rating by the appropriate correction factor form Table 10. Thus, a heater rated at 10,000 Btu/hr with 2-psig steam and 60-F air would have an output of $(1.290)(10,000) = 12,900$ Btu/hr with 40-F inlet air and 10-psig steam. Trap manufacturers usually list heater Btu ratings and recommend trap model numbers and sizes in their trap engineering data. This allows easier selection of the correct trap.

2. Select the trap size based on the load and steam pressure

Obtain a chart or tabulation of trap capacities published by the manufacturer whose trap will be used. Figure 4 is a capacity chart for one type of bucket trap manufactured by Armstrong Machine Works. Table 11 shows typical capacities of impulse traps manufactured by the Yarway Company.

To select a trap from Fig. 4, when the condensation rate is uniform and the pressure

TABLE 7 Condensate Formed in Submerged Steel Heating Elements, lb/(sq ft)(hr)*

Mtd†	Btu/(sq ft)(hr)	Steam pressure, psig																		
		1	2	5	10	15	20	25	50	75	100	150	200	250	300	350	400	450	500	600
10	280	0.28	0.29	0.29	0.29	0.30	0.30	0.30	0.31	0.31	0.32	0.33	0.33	0.34	0.35	0.35	0.36	0.37	0.37	0.38
20	930	0.96	0.97	0.97	0.98	0.99	0.99	1.00	1.02	1.04	1.05	1.08	1.11	1.14	1.16	1.18	1.20	1.22	1.24	1.28
30	1,900	1.96	1.97	1.98	1.99	2.01	2.02	2.03	2.08	2.12	2.16	2.22	2.27	2.32	2.36	2.41	2.45	2.49	2.53	2.61
40	3,100	3.20	3.21	3.23	3.26	3.28	3.30	3.32	3.40	3.46	3.52	3.62	3.72	3.78	3.85	3.92	3.99	3.99	4.13	4.25
50	4,500	4.65	4.66	4.68	4.73	4.76	4.78	4.82	4.93	5.02	5.11	5.25	5.38	5.48	5.58	5.69	5.80	5.89	5.99	6.17
60	6,250	6.45	6.47	6.51	6.57	6.62	6.65	6.69	6.86	6.97	7.10	7.30	7.47	7.62	7.76	7.92	8.06	8.18	8.33	8.58
70	8,000	8.26	8.29	8.33	8.41	8.46	8.52	8.57	8.77	8.93	9.08	9.34	9.56	9.75	9.94	10.15	10.32	10.47	10.65	10.97
80	10,400	10.73	10.77	10.83	10.93	11.00	11.06	11.13	11.40	11.61	11.80	12.14	12.43	12.67	12.92	13.17	13.40	13.62	13.85	14.26
90	12,500	12.90	12.95	13.00	13.13	13.23	13.30	13.37	13.70	13.95	14.20	14.60	14.94	15.24	15.53	15.83	16.13	16.36	16.65	17.15
100	15,000	15.47	15.55	15.62	15.75	15.86	15.95	16.05	16.44	16.75	17.03	17.50	17.90	18.28	18.63	18.97	19.34	19.64	19.97	20.57
125	22,400	23.10	23.20	23.30	23.53	23.70	23.85	23.95	24.55	25.10	24.45	26.15	26.75	27.30	27.80	28.35	28.90	29.30	29.85	20.73
150	30,000	30.95	31.10	31.20	31.50	31.70	31.90	32.10	32.90	33.50	34.10	35.20	35.85	36.60	37.30	38.00	38.70	39.30	39.95	41.15
175	40,000	41.30	41.50	41.65	42.00	42.30	42.60	42.80	43.80	44.30	45.40	46.70	47.80	48.75	49.70	50.65	51.60	52.40	53.30	54.85
290	50,000	51.60	51.80	52.10	52.50	52.85	53.20	53.50	54.80	55.80	56.75	58.30	59.70	60.90	62.10	63.30	64.50	65.50	66.60	68.60
250	82,000	83.70	85.00	85.30	86.20	86.70	87.25	87.75	90.00	91.60	93.10	95.70	98.00	100.00	102.00	103.70	105.70	107.30	109.30	112.50
300	100,000	103.30	103.60	104.00	105.00	105.70	106.30	107.00	109.50	111.60	113.50	116.60	119.40	122.00	124.20	126.50	128.80	131.00	133.30	137.00

*For copper, multiply table data by 2.0. For brass, multiply table data by 1.6.
†Mean temperature difference, F = temperature of steam minus average liquid temperature. Heat transfer data for calculating this table obtained from and used by permission of the American Radiator & Standard Sanitary Corp.

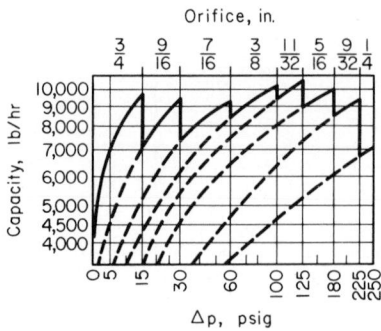

Fig. 4 Capacities of one type of bucket steam trap. (*Armstrong Machine Works.*)

across the trap is constant, enter at the left at the condensation rate—say 8,000 lb/hr (as obtained from Step 1)—and project horizontally to the right to the vertical ordinate representing the pressure across the trap (= Δp = steam-line pressure, psig—return-line pressure with trap valve closed, psig). Assume $\Delta p = 20$ psig for this trap. The intersection of the horizontal 8,000 lb/hr projection and the vertical 20-psig projection is on the sawtooth capacity curve for a trap having a $\frac{9}{16}$-in.-diameter orifice. If these projections intersected beneath this curve, a $\frac{9}{16}$-in. orifice would still be used if the point was between the verticals for this size orifice.

The dashed lines extending downward from the sawtooth curves show the capacity of a trap at reduced Δp. Thus, the capacity of a trap with a $\frac{3}{8}$-in. orifice at $\Delta p = 30$ psig is 6,200 lb/hr, read at the intersection of the 30-psig ordinate and the dashed curve extended from the $\frac{3}{8}$-in. solid curve.

TABLE 8   Steam Condensed by Air, lb/hr/(1,000 cfm)*

Temp difference, F	Pressure, psig				
	5	10	50	100	150
50	61	61	63	66	67
100	120	122	126	132	134
150	180	183	189	198	201
200	240	244	252	264	268
250	300	305	315	330	335
300	360	366	378	396	402

*Based on 0.0192 Btu absorbed per cu ft of saturated air per F at 32 F. For 0 F, multiply by 1.1.

To select an impulse trap from Table 11, enter the table at the trap inlet pressure—say 125 psig—and project to the desired capacity—say 8,000 lb/hr, determined from Step 1. Table 11 shows that a 2-in. trap having an 8,530 lb/hr capacity must be used because the next smallest size has a capacity of 5,165 lb/hr. This capacity is less than that required.

Some trap manufacturers publish capacity tables relating various trap models to specific types of equipment. Such tables simplify trap selection, but the condensation rate must still be computed as given here.

Related Calculations: Use the procedure given here to determine the trap capacity required for any industrial, commercial, or domestic application including acid vats, air dryers, asphalt tanks, autoclaves, baths (dyeing), belt presses, bleach tanks, blenders, bottle washers, brewing kettles, cabinet dryers, calenders, can washers, candy kettles, chamber dryers, chambers (reaction), cheese kettles, coils (cooking, kettle, pipe, tank, tank-car), confectioners' kettles, continuous dryers, conveyor dryers, cookers (non-pressure and pressure), cooking coils, cooking kettles, cooking tanks, cooking vats, cylinder dryers, cylinders (jacketed), double-drum dryers, drum dryers, drums (dyeing), dry cans, dry kilns, dryers (cabinet, chamber, continuous, conveyor, cylinder, drum, festoon, jacketed, linoleum, milk, paper, pulp, rotary, shelf, stretch, sugar, tray, tunnel), drying rolls, drying rooms, drying tables, dye vats, dyeing baths and drums, dryers (package), embossing-press platens, evaporators, feedwater heaters, festoon dryers, fin-type heaters, fourdriniers, fuel-oil pre-heaters, greenhouse coils, heaters

TABLE 9 Condensate Formed by Radiation from Bare Iron and Steel, lb/(sq ft)(hr)*

Air temperature, F	Steam pressure, psig																		
	1	2	5	10	15	20	25	50	75	100	150	200	250	300	350	400	450	500	600
32	0.53	0.54	0.57	0.70	0.74	0.78	0.81	1.02	1.13	1.30	1.46	1.70	1.86	1.97	2.31	2.41	2.52	2.62	3.14
50	0.48	0.49	0.52	0.56	0.68	0.71	0.74	0.87	1.06	1.15	1.38	1.52	1.74	1.85	1.96	2.06	2.41	2.52	2.71
60	0.45	0.46	0.49	0.53	0.56	0.59	0.71	0.84	1.02	1.10	1.34	1.47	1.58	1.80	1.91	2.00	2.35	2.46	2.65
65	0.44	0.45	0.47	0.52	0.55	0.58	0.69	0.82	1.00	1.08	1.32	1.45	1.56	1.77	1.88	1.97	2.07	2.43	2.62
70	0.39	0.40	0.45	0.50	0.53	0.56	0.59	0.80	0.98	1.06	1.21	1.43	1.54	1.75	1.86	1.95	2.04	2.39	2.59
75	0.38	0.39	0.44	0.49	0.52	0.55	0.58	0.77	0.88	1.05	1.19	1.40	1.52	1.62	1.83	1.93	2.02	2.11	2.56
Value of heat loss use (H)	2.6				2.8			3.2		3.5		3.75			4.0			4.5	5.0

*Based on still air; for forced air circulation multiply above values by 5.

TABLE 10   Unit-Heater Correction Factors*

Steam pressure, psia	Temperature of entering air, F					
	0	20	40	60	80	100
2	...	...	1.155	1.000	0.853	0.713
5	1.550	1.370	1.206	1.050	0.901	0.760
10	1.639	1.460	1.290	1.131	0.982	0.838
15	1.708	1.525	1.335	1.194	1.043	0.897
20	1.769	1.584	1.416	1.251	1.097	0.952
30	1.871	1.684	1.509	1.346	1.190	1.042
40	1.959	1.771	1.596	1.430	1.270	1.119
50	2.035	1.845	1.666	1.498	1.338	1.187
60	2.094	1.902	1.725	1.555	1.393	1.239
70	2.157	1.961	1.782	1.610	1.447	1.293
75	2.183	1.990	1.808	1.635	1.472	1.316
80	2.211	2.015	1.836	1.660	1.497	1.342
90	2.258	2.063	1.880	1.705	1.541	1.383
100	2.307	2.108	1.927	1.749	1.581	1.424

*Yarway Corporation.

(steam), heat exchangers, heating coils and kettles, hot-break tanks, hot plates, kettle coils, kettles (brewing, candy, cheese, confectioners', cooking, heating, process), kiers, kilns (dry), liquid heaters, mains (steam), milk-bottle washers, milk-can washers, milk dryers, mixers, molding-press platens, package dryers, paper dryers, percolators, phonograph-record press platens, pipe coils (still- and circulating-air), platens, plating tanks, plywood press platens, preheaters (fuel-oil), preheating tanks, press platens, pressure cookers, process kettles, pulp dryers, purifiers, reaction chambers, retorts, rotary dryers, steam mains (risers, separators), stocking boarders, storage-tank coils, storage water heaters, stretch dryers, sugar dryers, tank-car coils, tire-mold presses, tray dryers, tunnel dryers, unit heaters, vats, veneer press platens, vulcanizers, and water stills. Hospital equipment—such as autocalves and sterilizers—can be analyzed in the same way, as can kitchen equipment—bain marie, compartment cooker, egg boiler, kettles, steam table, and urns; and laundry equipment—blanket dryers, curtain dryers, flatwork ironers, presses (dry-cleaning, laundry), sock forms, starch cookers, tumblers, etc.

When using a trap capacity diagram or table, be sure to determine the basis on which it was prepared. Apply any necessary correction factors. Thus, *cold-water capacity ratings* must be corrected for traps operating at higher condensate temperatures. Correction factors are published in trap engineering data. The capacity of a trap is greater at condensate temperatures less than 212 F, because at or above this temperature condensate forms flash steam when it flows into a pipe or vessel at atmospheric (14.7 psia) pressure. At altitudes above sea level, condensate flashes into steam at a lower temperature, depending upon the altitude.

The method presented here is the work of L. C. Campbell, Yarway Corporation, as reported in *Chemical Engineering*.

SELECTING HEAT INSULATION FOR HIGH-TEMPERATURE PIPING

Select the heat insulation for a 300-ft-long 10-in. turbine lead operating at 570 F for 8,000 hr/year in a 70-F turbine room. How much heat is saved per year by this insulation? The boiler supplying the turbine has an efficiency of 80 percent when burning fuel having a heating value of 14,000 Btu/lb. Fuel costs $6/ton. How much money is saved by the insulation each year? What is the efficiency of the insulation?

TABLE 11 Capacities of Impulse Traps*

(Maximum continuous discharge of condensate in lb/hr. Based on condensate at 30 F below steam temperature.)

Pressure at trap inlet, lb gage	¼ in.	½ in.	¾ in.	1 in.	1¼ in.	1½ in.	2 in.
1	25	50	100	180	270	320	470
2	55	110	190	350	520	660	940
3	85	175	290	525	780	980	1,400
4	115	230	380	690	1,050	1,300	1,820
5	140	280	470	825	1,275	1,580	2,200
6	160	320	560	960	1,480	1,800	2,560
7	190	380	640	1,070	1,640	2,000	2,840
8	210	420	720	1,180	1,780	2,150	3,070
9	225	450	780	1,270	1,870	2,290	3,270
10	235	475	825	1,360	1,910	2,400	3,425
15	285	570	1,010	1,580	2,130	2,760	3,970
20	325	650	1,150	1,725	2,315	3,050	4,340
30	400	800	1,375	1,985	2,650	3,525	5,050
40	450	900	1,575	2,220	2,950	3,950	5,615
50	490	980	1,725	2,425	3,200	4,315	6,100
60	525	1,050	1,850	2,600	3,425	4,630	6,525
70	570	1,140	1,975	2,775	3,660	4,940	6,930
80	600	1,200	2,080	2,925	3,875	5,200	7,260
90	620	1,240	2,175	3,075	4,075	5,440	7,600
100	645	1,290	2,260	3,215	4,280	5,675	7,875
125	700	1,400	2,460	3,550	4,760	6,165	8,530
150	745	1,490	2,640	3,870	5,175	6,630	9,075
200	815	1,630	2,900	4,380	5,960	7,410	9,950
300	915	1,830	3,300	4,920	6,760	8,360	11,220
400	1,020	2,000	3,600	5,360	7,320	9,100	12,220

*Yarway Corporation.

Calculation Procedure:

1. Choose the type of insulation to use

Refer to an insulation manufacturer's engineering data or King and Crocker — *Piping Handbook* for recommendations about a suitable insulation for a pipe operating in the 500- to 600-F range. These references will show that calcium silicate is a popular insulation for this temperature range. Table 12 shows that a thickness of 3 in. is usually recommended for 10-in. pipe operating at 500 to 599 F.

2. Determine heat loss through the insulation

Refer to an insulation manufacturer's engineering data to find the heat loss through 3-in.-thick calcium silicate as 0.200 Btu per hr per sq ft of pipe F. Since 10-in. pipe has an area of 2.817 sq ft per ft of length and since the temperature difference across the pipe is $570 - 70 = 500$ F, the heat loss per hr = $(0.200)(2.817)(500) = 281.7$ Btu per hr per ft of pipe. The heat loss from bare 10-in. pipe with a 500-F temperature difference is, from an insulation manufacturer's engineering data, 4.640 Btu per hr per sq ft of pipe F, or $(4.64)(2.817)(500) = 6,510$ Btu per hr per ft of pipe.

3. *Determine annual heat saving*

The heat saved = bare-pipe loss, Btu/hr − insulated-pipe loss, Btu/hr = 6,510 − 281.7 = 6,228.3 Btu per hr per ft of pipe. Since the pipe is 300 ft long and operates 8,000 hr/year, the annual heat saving = (300)(8,000)(6,228.3) = 14,940,000,000 Btu/year.

TABLE 12 Recommended Thickness of Insulation

Nominal pipe size, in.	Temperature of pipe, F				
	100–199	200–299	300–399	400–499	500–599
	Nominal thickness, in.				
$1\frac{1}{2}$ and under	1	1	1	$1\frac{1}{2}$	$1\frac{1}{2}$
2	1	1	$1\frac{1}{2}$	$1\frac{1}{2}$	2*
$2\frac{1}{2}$ and 3	1	1	$1\frac{1}{2}$	2*	2*
4	1	$1\frac{1}{2}$	$1\frac{1}{2}$	2*	$2\frac{1}{2}$*
5	1	$1\frac{1}{2}$	2	2	$2\frac{1}{2}$*
6	1	$1\frac{1}{2}$	2	$2\frac{1}{2}$*	$2\frac{1}{2}$*
8	$1\frac{1}{2}$	$1\frac{1}{2}$	2	$2\frac{1}{2}$	3
10	$1\frac{1}{2}$	2	$2\frac{1}{2}$	$2\frac{1}{2}$	3
12	$1\frac{1}{2}$	2	$2\frac{1}{2}$	3	3
14 and over	$1\frac{1}{2}$	2	$2\frac{1}{2}$	3	$3\frac{1}{2}$

*Available in single- or double-layer insulation.

4. *Compute the money saved by the heat insulation*

The heat saved in fuel as fired = (annual heat saving, Btu/year)/(boiler efficiency) = 14,940,000,000/0.80 = 18,680,000,000 Btu/year. Weight of fuel saved = (annual heat saving, Btu/year)/(heating value of fuel, Btu/lb)(2,000 lb/ton) = 18,680,000,000/[(14,000)(2,000)] = 667 tons. At \$6/ton, the monetary saving is (\$6)(667) = \$4,002/year.

5. *Determine the insulation efficiency*

Insulation efficiency = (*bare-pipe loss* − *insulated-pipe loss*)/bare-pipe loss, all expressed in Btu/hr, or bare-pipe loss, = (6,510.0 − 281.7)/6,510.0 = 0.957, or 95.7 percent.

Related Calculations: Use this method for any type of insulation—magnesia, fiberglass, asbestos, felt, diatomaceous, mineral wool, etc.—used for piping at elevated temperatures conveying steam, water, oil, gas, or other fluids or vapors. To coordinate and simplify calculations, become familiar with the insulation tables in a reliable engineering handbook or comprehensive insulation catalog. Such familiarity will simplify routine calculations.

ORIFICE METER SELECTION FOR A STEAM PIPE

Steam is metered with an orifice meter in a 10-in. boiler lead having an internal diameter of $d_p = 9.760$ in. Determine the maximum rate of steam flow that can be measured with a steel orifice plate having a diameter of $d_0 = 5.855$ in. at 70 F. The upstream pressure tap is 1D ahead of the orifice, and the downstream tap is 0.5D past the orifice. Steam pressure at the orifice inlet $p_p = 250$ psig; temperature is 640 F. A differential gage fitted across the orifice has a maximum range of 120 in. of water. What is the steam flow rate when the observed differential pressure is 40 in. of water? Use the ASME Research Committee on Fluid Meters method in analyzing the meter. Atmospheric pressure is 14.696 psia.

Calculation Procedure:

1. Determine the diameter ratio and steam density

For any orifice meter, diameter ratio $= \beta =$ meter orifice diameter, in./pipe internal diameter, in. $= 5.855/9.760 = 0.5999$.

Determine the density of the steam by entering the superheated steam table at $250 + 14.696 = 264.696$ psia and 640 F and reading the specific volume as 2.387 cu ft/lb. For steam, the density $= 1/$specific volume $= d_s = 1/2.387 = 0.4193$ lb/cu ft.

2. Determine the steam viscosity and meter flow coefficient

From the ASME publication, *Fluid Meters – Their Theory and Application*, the steam viscosity gu_1 for a steam system operating at 640 F is $gu_1 = 0.0000141$ in.-lb/(F)(sec)(ft^2).

Find the flow coefficient K from the same ASME source by entering the 10-in. nominal pipe diameter table at $\beta = 0.5999$ and projecting to the appropriate Reynolds number column. Assume that the Reynolds number $= 10^7$, approximately, for the flow conditions in this pipe. Then, $K = 0.6486$. Since the Reynolds number for steam pressures above 100 psi ranges from 10^6 to 10^7, this assumption is safe because the value of K does not vary appreciably in this Reynolds number range. Also, the Reynolds number cannot be computed yet because the flow rate is unknown. Therefore, assumption of the Reynolds number is necessary. The assumption will be checked later.

3. Determine the expansion factor and the meter area factor

Since steam is a compressible fluid, the expansion factor Y_1 must be determined. For superheated steam, the ratio of the specific heat at constant pressure c_p to the specific heat at constant volume c_v is $k = c_p/c_v = 1.3$. Also, the ratio of the differential maximum pressure reading h_w, in. of water, to the maximum pressure in the pipe, psia $= 120/246.7 = 0.454$. Using the expansion-factor curve in the ASME *Fluid Meters*, $Y_1 = 0.994$ for $\beta = 0.5999$ and the pressure ratio $= 0.454$. And, from the same reference, the meter area factor $F_a = 1.0084$ for a steel meter operating at 640 F.

4. Compute the rate of steam flow

For square-edged orifices, the flow rate, lb/sec, $= w = 0.0997 F_a K d^2 Y_1 (h_w d_s)^{0.5} = (0.0997)(1.0084)(0.6486)(5.855)^2(0.994)(120 \times 0.4188)^{0.5} = 15.75$ lb/sec.

5. Compute the Reynolds number for the actual flow rate

For any steam pipe, the Reynolds number $R = 48w/d_p gu_1 = 48(15.75)/[(3.1416)(0.760)(0.0000141)] = 1,750,000$.

6. Adjust the flow coefficient for the actual Reynolds number

In Step 2, $R = 10^7$ was assumed and $K = 0.6486$. For $R = 1,750,000$, $K = 0.6489$, from ASME *Fluid Meters*, by interpolation. Then, the actual flow rate $w_h =$ (computed flow rate)(ratio of flow coefficients based on assumed and actual Reynolds numbers) $= (15.75)(0.6489/0.6486)(3,600) = 56,700$ lb/hr, closely, where the value 3,600 is a conversion factor for changing lb/sec to lb/hr.

7. Compute the flow rate for a specific differential gage deflection

For a 40-in. H_2O deflection, F_a is unchanged and $= 1.0084$. The expansion factor changes because $h_w/p_p = 40/264.7 = 0.151$. Using the ASME *Fluid Meters*, $Y_1 = 0.998$. Assuming again that $R = 10^7$, $K = 0.6486$, as before, then $w = (0.0997)(1.0084)(0.6486)(5.855)^2(0.998)(40 \times 0.4188)^{0.5} = 9.132$ lb/sec. Computing the Reynolds number as before, $R = (40)(0.132)/[(3.1416)(0.76)(0.0000141)] = 1,014,000$. The value of K corresponding to this value, as before, is from ASME – *Fluid Meters*, $K = 0.6497$. Therefore, the flow rate for a 40 in. H_2O reading, in lb/hr, $= w_h = (0.132)(0.6497/0.6486)(3,600) = 32,940$ lb/hr.

Related Calculations: Use these steps and the ASME *Fluid Meters* or comprehensive meter engineering tables giving similar data to select or check an orifice meter used in any type of steam pipe – main, auxiliary, process, industrial, marine, heating, or commercial, conveying wet, saturated, or superheated steam.

SELECTION OF A PRESSURE-REGULATING VALVE FOR STEAM SERVICE

Select a single-seat spring-loaded diaphragm-actuated pressure-reducing valve to deliver 350 lb/hr of steam at 50 psig when the initial pressure is 225 psig. Also select an integral pilot-controlled piston-operated single-seat pressure-regulating valve to deliver 30,000 lb/hr of steam at 40 psig with an initial pressure of 225 psig saturated. What size pipe must be used on the downstream side of the valve to produce a velocity of 10,000 fpm? How large should the pressure-regulating valve be if the steam entering the valve is at 225 psig and 600 F?

Calculation Procedure:

1. Compute the maximum flow for the diaphragm-actuated valve

For best results in service, pressure-reducing valves are selected so that they operate 60 to 70 percent open at normal load. To obtain a valve sized for this opening, divide the desired delivery, lb/hr, by 0.7 to obtain the maximum flow expected. For this valve then, the maximum flow = 350/0.7 = 500 lb/hr.

2. Select the diaphragm-actuated valve size

Using a manufacturer's engineering data for an acceptable valve, enter the appropriate valve capacity table at the valve inlet steam pressure, 225 psig, and project to a capacity of 500 lb/hr, as in Table 13. Read the valve size as $\frac{3}{4}$ in. at the top of the capacity column.

TABLE 13 Pressure-Reducing Valve Capacity*

Inlet pressure, psig	Valve capacity, lb/hr				
	Valve size, in.				
	$\frac{1}{2}$	$\frac{3}{4}$	1	$1\frac{1}{4}$	$1\frac{1}{2}$
200	420	460	560	1,200	1,400
225	450	500	600	1,350	1,500
250	485	560	650	1,500	1,625

*Clark-Reliance Corporation.

3. Select the size of the pilot-controlled pressure-regulating valve

Enter the capacity table in the engineering data of an acceptable pilot-controlled pressure-regulating valve, similar to Table 14, at the required capacity, 30,000 lb/hr, and project across until the correct inlet steam pressure column, 225 psig, is intercepted and read the required valve size as 4 in.

Note that it is not necessary to compute the maximum capacity before entering the table, as in Step 1, for the pressure-reducing valve. Also note that a capacity table such

TABLE 14 Pressure-Regulating Valve Capacity*

Steam capacity, lb/hr	Initial steam pressure, psig saturated					
	40	50	100	175	225	300
20,000	6	6	5	4	4	3
30,000	8	8	6	5	4	4
40,000	. . .	8	6	5	5	4

*Clark-Reliance Corporation.

as Table 14 can be used only for valves conveying saturated steam, unless the table notes state that the values listed are valid for other steam conditions.

4. Determine the size of the downstream pipe

Enter Table 14 at the required capacity, 30,000 lb/hr, and project across to the valve *outlet pressure*, 40 psig, and read the required pipe size as 8 in., for a velocity of 10,000 fpm. Thus, the pipe immediately downstream from the valve must be enlarged from the valve size, 4 in., to the required pipe size, 8 in., to obtain the desired steam velocity.

5. Determine the size of the valve handling superheated steam

To determine the correct size of a pilot-controlled pressure-regulating valve handling superheated steam, a correction must be applied. Either a factor may be used or a tabulation of corrected pressures, Table 15. To use Table 15, enter at the valve inlet pressure, 225 psig, and project across to the total temperature, 600 F, to read the corrected pressure, 165 psig. Enter Table 14 at the *next highest* saturated steam pressure, 175 psig, and project down to the required capacity, 30,000 lb/hr, and read the required valve size as 5 in.

TABLE 15 Equivalent Saturated Steam Values for Superheated Steam at Various Pressures and Temperatures*

Steam pressure, psig	Saturated temperature, F	Total temperature, F									
		360	400	440	480	500	600	700	800	900	1000
105	341	101	94	88	83	81	70	62	56	50	45
125	353	124	115	108	101	98	86	76	68	61	56
145	363	...	135	127	119	116	102	90	81	73	67
185	381	...	179	167	157	152	133	118	106	97	88
205	389	...	202	190	175	171	149	133	120	109	100
225	397	...	...	210	196	190	165	147	133	121	110
265	411	...	...	250	233	227	200	177	159	145	133
305	423	...	...	295	275	270	230	205	185	169	155
365	439	...	...	...	335	325	282	250	225	205	190
405	449	...	...	...	385	368	315	282	255	230	210

*Clark-Reliance Corporation.

Related Calculations: To simplify pressure-reducing and pressure-regulating valve selection, become familiar with two or three acceptable valve manufacturer's engineering data. Use the procedures given in the engineering data or those given here to select valves for industrial, marine, utility, heating, process, laundry, kitchen, or hospital service with a saturated or superheated steam supply.

Do not oversize reducing or regulating valves. Oversizing causes chatter and excessive wear.

When an anticipated load on the downstream side will not develop for several months after installation of a valve, fit to the valve a reduced-area disk sized to handle the present load. When the load increases, install a full-size disk. Size the valve for the ultimate load, not the reduced load.

Where there is a wide variation in demand for steam at the reduced pressure, consider installing two regulators piped in parallel. Size the smaller regulator to handle light loads and the larger regulator to handle the difference between 60 percent of the light load and the maximum heavy load. Set the larger regulator to open when the minimum allowable reduced pressure is reached. Then both regulators will be open to handle the heavy load. Be certain to use the actual regulator inlet pressure and not the boiler pressure when sizing the valve if this is different from the inlet pressure. Data in this calculation procedure are based on valves built by the Clark-Reliance Corporation, Cleveland, Ohio.

Some valve manufacturers use the valve flow coefficient C_v for valve sizing. This coefficient is defined as the flow rate, lb/hr, through a valve of given size when the pressure loss across the valve is 1 psi. Tabulations like Tables 13 and 14 incorporate this flow coefficient and are somewhat easier to use. These tables make the necessary allowances for downstream pressures less than the critical pressure (= 0.55 × absolute upstream pressure, psi, for superheated steam and hydrocarbon vapors; and 0.58 × absolute upstream pressure, psi, for saturated steam). The accuracy of these tabulations equals that of valve sizes determined by using the flow coefficient.

HYDRAULIC RADIUS AND LIQUID VELOCITY IN WATER PIPES

What is the velocity of 1,000 gpm of water flowing through a 10-in. inside-diameter cast-iron water main? What is the hydraulic radius of this pipe when it is full of water? When the water depth is 8 in?

Calculation Procedure:

1. Compute the water velocity in the pipe

For any pipe conveying water, the liquid velocity in fps is v fps = gpm/2.448d^2, where d = internal pipe diameter, in. For this pipe, $v = 1,000/[2.448(100)] = 4.08$ fps, or (60)(4.08) = 244.8 fpm.

2. Compute the hydraulic radius for a full pipe

For any pipe, the hydraulic radius is the ratio of the cross-sectional area of the pipe to the wetted perimeter, or $d/4$. For this pipe, when full of water, the hydraulic radius = 10/4 = 2.5.

3. Compute the hydraulic radius for a partially full pipe

Use the hydraulic radius tables in King and Brater—*Handbook of Hydraulics* or compute the wetted perimeter using the geometric properties of the pipe, as in Step 2. Using the King and Brater table, the hydraulic radius = Fd, where F = table factor for the ratio of the depth of water, in./diameter of channel, in. = 8/10 = 0.8. For this ratio, $F = 0.304$. Then, hydraulic radius = (0.304)(10) = 3.04 in.

Related Calculations: Use this method to determine the water velocity and hydraulic radius in any pipe conveying cold water—water supply, plumbing, process, drain, or sewer.

FRICTION-HEAD LOSS IN WATER PIPING OF VARIOUS MATERIALS

Determine the friction-head loss in 2,500 ft of clean 10-in. new tar-dipped cast-iron pipe when 2,000 gpm of cold water is flowing. What is the friction-head loss 20 years later? Use the Hazen-Williams and Manning formulas and compare the results.

Calculation Procedure:

1. Compute the friction-head loss using the Hazen-Williams formula

The Hazen-Williams formula is $h_f = (v/1.318CR_h^{0.63})^{1.85}$, where h_f = friction-head loss per ft of pipe, ft of water; v = water velocity, fps; C = a constant depending on the condition and kind of pipe; R_h = hydraulic radius of pipe, ft.

For a water pipe, v = gpm/2.44d^2; for this pipe, $v = 2,000/[2.448(10)^2] = 8.18$ fps. From Table 16 or King and Crocker—*Piping Handbook*, C for new pipe = 120; for 20-year-old pipe $C = 90$; $R_h = d/4$ for a full-flow pipe = 10/4 = 2.5 in., or 2.5/12 = 0.208 ft. Then, $h_f = (8.18/(1.318 \times 120 \times 0.208^{0.63})^{1.85} = 0.0263$ ft of water per ft of pipe. For 2,500 ft of pipe, the total friction-head loss = 2,500(0.0263) = 65.9 ft of water for the new pipe.

For 20-year-old pipe using the same formula, except with $C = 90$, $h_f = 0.0451$ ft of water per ft of pipe. For 2,500 ft of pipe, the total friction-head loss = 2,500(0.0451) =

TABLE 16 Values of C in Hazen-Williams Formula

Type of Pipe	C°	Type of Pipe	C°
Cement-asbestos	140	Cast iron or wrought iron	100
Asphalt-lined iron or steel	140	Welded or seamless steel	100
Copper or brass	130	Concrete	100
Lead, tin, or glass	130	Corrugated steel	60
Wood stave	110		

°Values of C commonly used for design. The value of C for pipes made of corrosive materials decreases as the age of the pipe increases; the values given are those that apply at an age of 15 to 20 years. For example, the value of C for cast-iron pipes 30 in. in diameter or greater at various ages is approximately as follows: new, 130; 5 years old, 120; 10 years old, 115; 20 years old, 100; 30 years old, 90; 40 years old, 80; and 50 years old, 75. The value of C for smaller-size pipes decreases at a more rapid rate.

112.9 ft of water. Thus, the friction-head loss nearly doubles (from 65.9 to 112.9 ft) in 20 years. This shows that it is wise to design for future friction losses; otherwise pumping equipment may become overloaded.

2. Compute the friction-head loss using the Manning formula

The Manning formula is $h_f = n^2 v^2/2.208 R_h^{4/3}$, where n = a constant depending on the condition and kind of pipe; other symbols as before.

Using $n = 0.011$ for new coated cast-iron pipe from Table 17 or King and Crocker—

TABLE 17 Roughness Coefficients (Manning's n) for Closed Conduits

Type of conduit			Manning's n	
			Good construction°	Fair construction°
Concrete pipe			0.013	0.015
Corrugated metal pipe or pipe arch, 2⅔ × ½ in. corrugation, riveted:				
Plain			0.024	
Paved invert:				
Per cent of circumference paved	25	50		
Depth of flow:				
Full	0.021	0.018		
0.8D	0.021	0.016		
0.6D	0.019	0.013		
Vitrified clay pipe			0.012	0.014
Cast-iron pipe, uncoated			0.013	
Steel pipe			0.011	
Brick			0.014	0.017
Monolithic concrete:				
Wood forms, rough			0.015	0.017
Wood forms, smooth			0.012	0.014
Steel forms			0.012	0.013
Cemented-rubble masonry walls:				
Concrete floor and top			0.017	0.022
Natural floor			0.019	0.025
Laminated treated wood			0.015	0.017
Vitrified-clay liner plates			0.015	

°For poor-quality construction, use larger values of n.

Piping Handbook, $h_f = (0.011)^2(8.18)^2/[2.208(0.208)^{4/3}] = 0.0295$ ft of water per ft of pipe. For 2,500 ft of pipe, the total friction-head loss $= 2,500(0.0295) = 73.8$ ft of water, as compared with 65.9 ft of water computed with the Hazen-Williams formula.

For coated cast-iron pipe in fair condition, $n = 0.013$, and $h_f = 0.0411$ ft of water. For 2,500 ft of pipe, the total friction-head loss $= 2,500(0.0411) = 102.8$ ft of water, as compared with 112.9 ft of water computed with the Hazen-Williams formula. Thus, the Manning formula gives results higher than the Hazen-Williams in one case and lower in another. However, the differences in each case are not excessive; $(73.8 - 65.9)/65.9 = 0.12$, or 12 percent higher, and $(112.9 - 102.8)/102.8 = 0.0983$, or 9.83 percent lower. Both these differences are within the normal range of accuracy expected in pipe friction-head calculations.

Related Calculations: The Hazen-Williams and Manning formulas are popular with many piping designers for computing pressure losses in cold-water piping. To simplify calculations, most designers use the precomputed tabulated solutions available in King and Crocker—*Piping Handbook,* King and Brater—*Handbook of Hydraulics,* and similar publications. In the rush of daily work these precomputed solutions are also preferred over the more complex Darcy-Weisbach equation used in conjunction with the friction factor f, the Reynolds number R, and the roughness-diameter ratio.

Use the method given here for sewer lines, water-supply pipes for commercial, industrial, or process plants, and all similar applications where cold water at temperatures of 33 to 90 F flows through a pipe made of cast iron, riveted steel, welded steel, galvanized iron, brass, glass, wood-stove, concrete, vitrified, common clay, corrugated metal, unlined rock, or enameled steel. Thus, either of these formulas, used in conjunction with a suitable constant, gives the friction-head loss for a variety of piping materials. Suitable constants are given in Tables 16 and 17 and in the above references. For the Hazen-Williams formula, the constant C varies from about 70 to 140, while n in the Manning formula varies from about 0.017 for $C = 70$ to $n = 0.010$ for $C = 140$. Values obtained with these formulas have been used for years with satisfactory results. At present, the Manning formula appears the more popular of the two.

CHART AND TABULAR DETERMINATION OF FRICTION HEAD

Figure 5 shows a process piping system supplying 1,000 gpm of 70-F water. Determine the total friction head using published charts and pipe-friction tables. All the valves and fittings are flanged and the piping is 10-in. steel, schedule 40.

Calculation Procedure:

1. Determine the total length of the piping

Mark the length of each piping run on the drawing after scaling it or measuring it in the field. Determine the total length by adding the individual lengths, starting at the supply source of the liquid. In Fig. 5, beginning at the storage sump, the total length of piping, in ft, $= 10 + 20 + 40 + 50 + 75 + 105 = 300$ ft. Note that the physical length of the fittings is included in the length of each run.

2. Compute the equivalent length of each fitting

The frictional resistance of pipe fittings (elbows, tees, etc.) and valves is greater than the actual length of each fitting. Therefore, the equivalent length of straight piping having a resistance equal to that of the fittings must be determined. This is done by finding the equivalent length of each fitting and taking the sum for all the fittings.

Use the equivalent length table in the pump section of this Handbook or in King and Crocker—*Piping Handbook,* Baumeister—*Standard Handbook for Mechanical Engineers,* or *Standards of the Hydraulic Institute.* Equivalent length values will vary slightly from one reference to another.

Starting at the supply source, as in Step 1, for 10-in. flanged fittings throughout, the equivalent fitting lengths are: bell-mouth inlet, 2.9 ft; 90° ell at pump, 14 ft; gate valve, 3.2 ft; swing check valve, 120 ft; 90° ell, 14 ft; tee, 30 ft; 90° ell, 14 ft; 90° ell, 14 ft; globe valve, 310 ft; swing check valve, 120 ft; sudden enlargement $=$ (liquid velocity, fps)$^2/2g$

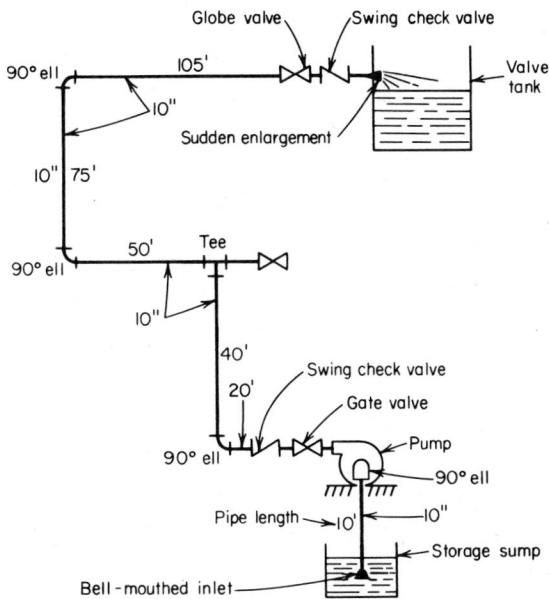

Fig. 5   Typical industrial piping system.

$= (4.07)^2/2(32.2) = 0.257$ ft, where the terminal velocity is zero, as in the tank. Find the liquid velocity as shown in a previous Calculation Procedure in this section. The sum of the fitting equivalent lengths is $2.9 + 14 + 3.2 + 120 + 14 + 30 + 14 + 14 + 310 + 120 + 0.257 = 642.4$ ft. Adding this to the straight length gives a total length of $642.4 + 300 = 942.4$ ft.

3. Compute the friction-head loss using a chart

Figure 6 is a popular friction-loss chart for fairly rough pipe, which is any ordinary pipe after a few years' use. Enter at the left at a flow of 1,000 gpm and project to the right until the 10-in.-diameter curve is intersected. Read the friction-head loss at the top or bottom of the chart as 0.4 psi, closely, per 100 ft of pipe. Therefore, total friction-head loss $= (0.4)(942.4/100) = 3.77$ psi. Converting to ft of water, $(3.77)(2.31) = 8.71$ ft of water.

4. Compute the friction-head loss using tabulated data

Using the *Standards of the Hydraulic Institute* pipe-friction table, the friction head h_f ft of water per 100 ft of pipe $= 0.500$. Hence, the total friction head $= (0.500)(942.4/100) = 4.71$ ft of water. The Institute recommends that 15 percent be added to the tabulated friction head, or $(1.15)(4.71) = 5.42$ ft of water.

Using the friction-head tables in King and Crocker — *Piping Handbook*, the friction head $= 6.27$ ft per 1,000 ft of pipe with $C = 130$ for new, very smooth pipe. For this piping system, the friction-head loss $= (942.4/1,000)(6.27) = 5.91$ ft of water.

5. Use the Reynolds number method to determine the friction head

In this method, the friction factor is determined by using the Reynolds number R and the relative roughness of the pipe ϵ/D, where $\epsilon =$ pipe roughness, ft, and $D =$ pipe diameter, ft.

For any pipe, $R = Dv/\nu$, where $v =$ liquid velocity, fps, and $\nu =$ kinematic viscosity, sq ft/sec. Using King and Brater — *Handbook of Hydraulics*, $v = 4.07$ fps, and $\nu = 0.00001059$ sq ft/sec for water at 70 F. Then, $R = (10/12)(4.07)/0.00001059 = 320,500$.

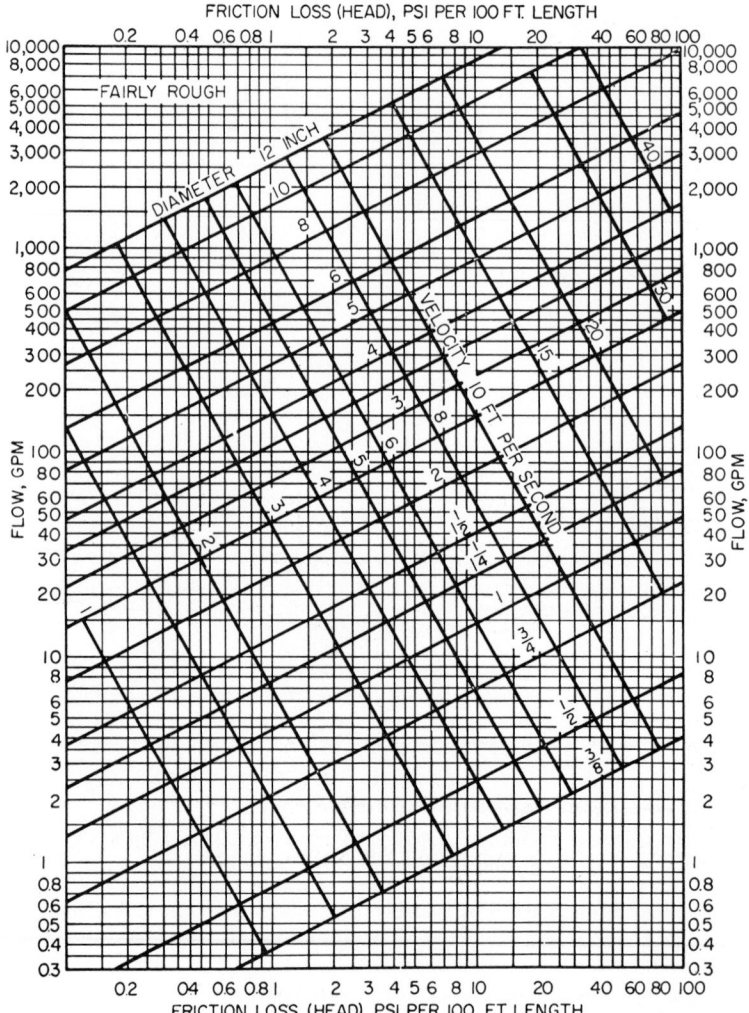

FRICTION LOSS (HEAD), PSI PER 100 FT. LENGTH

FLOW, GPM

FRICTION LOSS (HEAD), PSI PER 100 FT. LENGTH

Fig. 6   Friction loss in water piping.

From Table 18 or the above reference, $\epsilon = 0.00015$, and $\epsilon/D = 0.00015/(10/12) = 0.00018$. Using the Reynolds-number, relative-roughness, friction-factor curve in Fig. 7 or in Baumeister—*Standard Handbook for Mechanical Engineers*, the friction factor $f = 0.016$.

Apply the Darcy-Weisbach equation $h_f = f(l/D)(v^2/2g)$, where l = total pipe length, including the fittings' equivalent length, ft. Then, $h_f = (0.016)(942.4/10/12)(4.07)^2/(2 \times 32.2) = 4.651$ ft of water.

6.   *Compare the results obtained*

Three different friction-head values were obtained: 8.71, 5.91, and 4.651 ft of water. The results show the variations that can be expected with the different methods. Actually, the Reynolds number method is probably the most accurate. As can be seen, the other two methods give safe results—i.e., the computed friction head is higher. The

TABLE 18 Absolute Roughness Classification of Pipe Surfaces for Selection of Friction Factor *f* in Fig. 7

Commercial Pipe Surface (New)	Absolute Roughness ϵ, ft	Commercial Pipe Surface (New)	Absolute Roughness ϵ, ft
Glass, drawn brass, copper, lead......	Smooth	Cast iron	0.00085
Wrought iron, steel.................	0.00015	Wood stave	0.0006–0.003
Asphalted cast iron.................	0.0004	Concrete	0.001–0.01
Galvanized iron....................	0.0005	Riveted steel	0.003–0.03

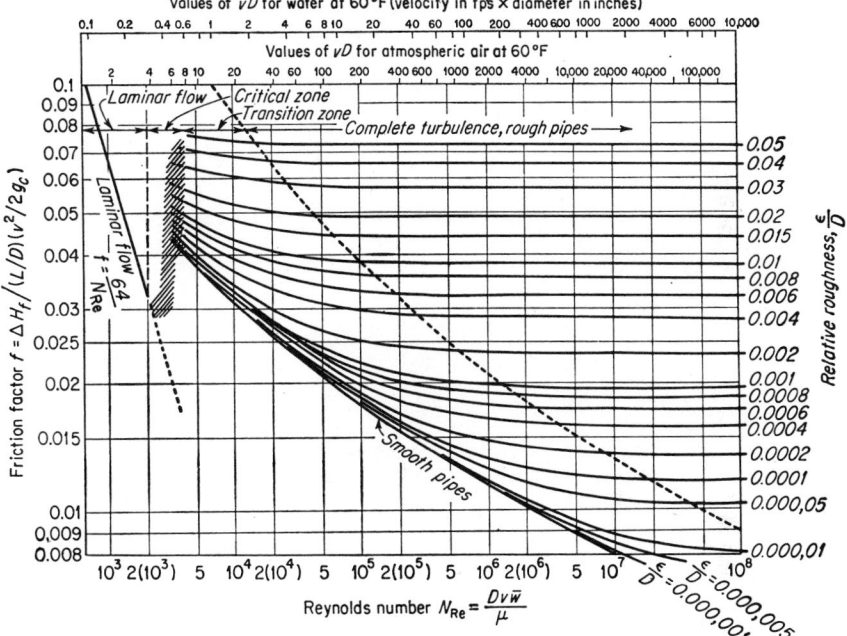

Fig. 7 Friction factors for laminar and turbulent flow.

Pipe Friction Manual, published by the Hydraulic Institute, presents excellent simplified charts for use with the Reynolds number method.

Related Calculations: Use any of these methods to compute the friction-head loss for any type of pipe. The Reynolds number method is useful for a variety of liquids other than water—mercury, gasoline, brine, kerosene, crude oil, fuel oil, and lube oil. It can also be used for saturated and superheated steam, air, methane, and hydrogen.

RELATIVE CARRYING CAPACITY OF PIPES

What is the equivalent steam-carrying capacity of a 24-in.-inside-diameter pipe in terms of a 10-in.-inside-diameter pipe? What is the equivalent water-carrying capacity of a 23-in.-inside-diameter pipe in terms of a 13.25-in.-inside-diameter pipe?

Calculation Procedure:

1. Compute the relative carrying capacity of the steam pipes

For steam, air, or gas pipes, the number N of small pipes of inside diameter d_2 in. equal to one pipe of larger inside diameter d_1 in. is $N = (d_1^3\sqrt{d_2+3.6})/(d_2^3+\sqrt{d_1+3.6})$.

For this piping system, $N = (24^3 + \sqrt{10} + 3.6)/(10^3 + \sqrt{24} + 3.6) = 9.69$, say 9.7. Thus, a 24-in.-inside-diameter steam pipe has a carrying capacity equivalent to 9.7 pipes having a 10-in. inside diameter.

2. Compute the relative carrying capacity of the water pipes

For water, $N = (d_2/d_1)^{2.5} = (23/13.25)^{2.5} = 3.97$. Thus, one 23-in.-inside-diameter pipe can carry as much water as 3.97 pipes of 13.25 in. inside diameter.

Related Calculations: King and Crocker — *Piping Handbook*, and certain piping catalogs (Crane, Walworth, National Valve and Manufacturing Company) contain tabulations of relative carrying capacities of pipes of various sizes. Most piping designers use these tables. However, the equations given here are useful for ranges not covered by the tables and when the tables are unavailable.

PRESSURE-REDUCING VALVE SELECTION FOR WATER PIPING

What size pressure-reducing valve should be used to deliver 1,200 gph of water at 40 psi if the inlet pressure is 140 psi?

Calculation Procedure:

1. Determine the valve capacity required

Pressure-reducing valves in water systems operate best when the nominal load is 60 to 70 percent of the maximum load. Using 60 percent, the maximum load for this valve = $1,200/0.6 = 2,000$ gph.

2. Determine the valve size required

Enter a valve capacity table in suitable valve engineering data at the valve inlet pressure and project to the exact, or next higher, valve capacity. Thus, enter Table 19 at 140 psi and project to the next higher capacity, 2,200 gph, since a capacity of 2,000 gph is not tabulated. Read at the top of the column the required valve size as 1 in.

TABLE 19   Maximum Capacities of Water Pressure-Reducing Valves*

Inlet pressure, psig	Valve capacity, gph				
	Valve size, in.				
	$\frac{1}{2}$	$\frac{3}{4}$	1	$1\frac{1}{4}$	$1\frac{1}{2}$
120	1,500	1,550	2,000	4,500	5,000
140	1,650	1,700	2,200	5,000	5,500
160	1,800	1,850	2,400	5,500	6,000

*Clark-Reliance Corporation.

Some valve manufacturers present the capacity of their valves in graphical instead of tabular form. One popular chart, Fig. 8, is entered at the difference between the inlet and outlet pressures on the abscissa, or $140 - 40 = 100$ psi. Project vertically to the flow rate of $2,000/60 = 33.3$ gpm. Read the valve size on the intersecting valve capacity curve, or on the next curve if there is no intersection with the curve. Figure 8 shows that a 1-in. valve should be used. This agrees with the tabulated capacity.

Related Calculations: Use this method for pressure-reducing valves in any type of water piping — process, domestic, commercial — where the water temperature is 100 F or less. Table 19 is from data prepared by the Clark-Reliance Corporation; Figure 8 is from Foster Engineering Company data.

Some valve manufacturers use the valve flow coefficient C_v for valve sizing. This coefficient is defined as the flow rate, gpm, through a valve of given size when the pressure loss across the valve is 1 psi. Tabulations like Table 19 and flow charts like Fig. 8 incorporate this flow coefficient and are somewhat easier to use. Their accuracy equals that of the flow coefficient method.

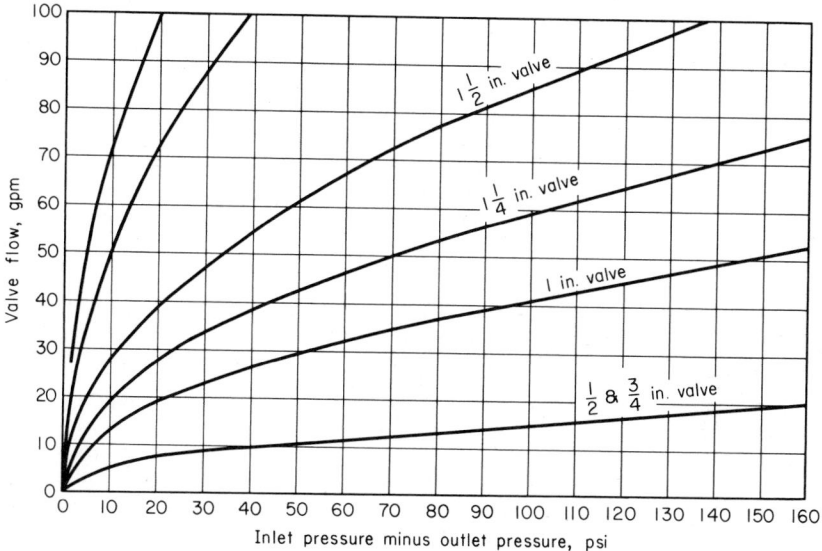

Fig. 8   Pressure-reducing valve flow capacity. (*Foster Engineering Company.*)

SIZING A WATER METER

A 6×4 in. Venturi tube is used to measure water flow rate in a piping system. The dimensions of the meter are: inside pipe diameter $d_p = 6.094$ in.; throat diameter $d = 4.023$ in. The differential pressure is measured with a mercury manometer having water on top of the mercury. The average manometer reading for 1 hr is 10.1 in. of mercury. The temperature of the water in the pipe is 41 F, and that of the room is 77 F. Determine the water flow rate in lb/hr, gph, and gpm. Use the ASME Research Committee on Fluid Meters method in analyzing the meter.

Calculation Procedure:

1.   Convert the pressure reading to standard conditions

The ASME meter equation constant is based on a manometer liquid temperature of 68 F. Therefore, the water and mercury density at room temperature, 77 F, and the water density at 68 F, must be used to convert the manometer reading to standard conditions using the equation $h_w = h_m(m_d - w_d)/w_s$, where $h_w =$ equivalent manometer reading, in. H_2O at 68 F; $h_m =$ manometer reading at room temperature, in. mercury; $m_d =$ mercury density at room temperature, lb/cu ft; $w_d =$ water density at room temperature, lb/cu ft; $w_s =$ water density at standard conditions, 68 F, lb/cu ft. Using density values from the ASME publication—*Fluid Meters: Their Theory and Application,* $h_w = 10.1(844.88 - 62.244)/62.316 = 126.8$ in. of water at 68 F.

2.   Determine the throat-to-pipe diameter ratio

The throat-to-pipe diameter ratio $\beta = 4.023/6.094 = 0.6602$. Then, the ratio $1/(1 - \beta^4)^{0.5} = 1/(1 - 0.6602^4)^{0.5} = 1.1111$.

3. Assume a Reynolds number value and compute the flow rate

The flow equation for a Venturi tube is w lb/hr $= 359.0\ (Cd^2/\sqrt{1-\beta^4})\ (w_{dp}h_w)^{0.5}$, where $C =$ meter discharge coefficient, expressed as a function of the Reynolds number; $w_{dp} =$ density of the water at the pipe temperature, lb/cu ft. With a Reynolds number greater than 250,000, C is a constant. As a first trial, assume $R > 250{,}000$ and $C = 0.984$ from *Fluid Meters*. Then $w = 359.0(0.984)(4.023)^2(1.1111)(62.426 \times 126.8)^{0.5}$ $= 565{,}020$ lb/hr, or $565{,}020/8.33$ lb/gal $= 67{,}800$ gph, or $67{,}800/60$ min/hr $= 1{,}129$ gpm.

4. Check the discharge coefficient by computing the Reynolds numbers

For a water pipe, $R = 48w_s/\pi d_p g u$, where $w_s =$ flow rate, lb/sec $= w/3{,}600$; $u =$ coefficient of absolute viscosity. Using *Fluid Meters* data for water at 41 F, $R = 48(156.95)/[(\pi \times 6.094)(0.001004)] = 391{,}900$. Since C is constant for $R > 250{,}000$, use of $C = 0.984$ is correct, and no adjustment in the computations is necessary. Had the value of C been incorrect, another value would be chosen and the Reynolds number recomputed. Continue this procedure until a satisfactory value for C is obtained.

5. Use an alternative solution to check the results

Fluid Meters gives another equation for Venturi meter flow rate, that is, w lb/sec $= 0.525(Cd^2/\sqrt{1-\beta^4})(w_{dp}[p_1-p_2])^{0.5}$, where (p_1-p_2) is the manometer differential pressure in psi. Using the conversion factor in *Fluid Meters* for converting in. of mercury under water at 77 F to psi, $(p_1-p_2) = (10.1)(0.4528) = 4.573$ psi. Then, $w = (0.525)(0.984)(4.023)^2(1.1111)(62.426 \times 4.573)^{0.5} = 156.9$ lb/sec, or $(156.9)(3{,}600\ \text{sec/hr}) = 564{,}900$ lb/hr, or $564{,}900/8.33$ lb/gal $= 67{,}800$ gph, or $67{,}800/60$ min/hr $= 1{,}129$ gpm. This result agrees with that computed in Step 3 within 1 part in 5,600. This is much less than the probable uncertainties in the values of the discharge coefficient and the differential pressure. The gph and gpm flows agree because it is difficult to read closer values on the usual 10-in. slide rule.

Related Calculations: Use this method for any Venturi tube serving cold-water piping in process, industrial, water-supply, domestic, or commercial service.

EQUIVALENT LENGTH OF A COMPLEX SERIES PIPELINE

Figure 9 shows a complex series pipeline made up of four lengths of different size pipe. Determine the equivalent length of this pipe if each size of pipe has the same friction factor.

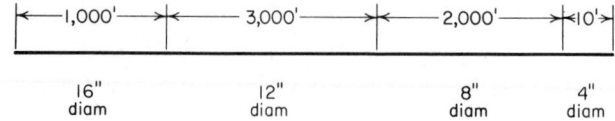

←——1,000'——→	←———— 3,000'———— →	←——— 2,000'——— →	←10'→
16" diam	12" diam	8" diam	4" diam

Fig. 9   Complex series pipeline.

Calculation Procedure:

1. Select the pipe size for expressing the equivalent length

The usual procedure when analyzing complex pipelines is to express the equivalent length in terms of the smallest, or next-to-smallest, diameter pipe. Choose the 8-in. size as being suitable for expressing the equivalent length.

2. Find the equivalent length of each pipe

For any complex series pipeline having equal friction factors in all the pipes, $L_e =$ equivalent length, ft, of a section of constant diameter = (actual length of section, ft) (inside diameter, in., of pipe used to express the equivalent length/inside diameter, in., of section under consideration)[5].

For the 16-in. pipe, $L_e = (1{,}000)(7.981/15.000)^5 = 42.6$ ft. The 12-in. pipe is next; for it $L_e = (3{,}000)(7.981/12.00)^5 = 390$ ft. For the 8-in. pipe, the equivalent length = actual length = 2,000 ft. For the 4-in. pipe, $L_e = (10)(7.981/4.026)^5 = 306$ ft. Then, the total equivalent length of 8-in. pipe = sum of the equivalent lengths = $42.6 + 390 + 2,000 + 306 = 2,738.6$ ft; or rounding off, 2,740 ft of 8-in. pipe will have a frictional resistance equal to the complex series pipeline shown in Fig. 9. To compute the actual frictional resistance, use the methods given in previous Calculation Procedures.

Related Calculations: Use this general procedure for any complex series pipeline conveying water, oil, gas, steam, etc. See King and Crocker—*Piping Handbook* for derivation of the flow equations. Use the tables in King and Crocker to simplify finding the fifth power of the inside diameter of a pipe. The method of the next Calculation Procedure can also be used if a given flow rate is assumed.

Choosing a flow rate of 1,000 gpm and using the tables in the Hydraulic Institute *Pipe Friction Manual* gives an equivalent length of 2,770 ft for the 8-in. pipe. This compares favorably with the 2,740 ft computed above. The difference of 30 ft is negligible and can be accounted for by slide-rule reading variations.

The equivalent length is found by summing the friction-head loss for 1,000 gpm flow for each length of the four pipes — 16, 12, 8, and 4 in. — and dividing this by the friction-head loss for 1,000 gpm flowing through an 8-in. pipe. Be careful to observe the units in which the friction-head loss is stated, because errors are easy to make if the units are ignored.

EQUIVALENT LENGTH OF A PARALLEL PIPING SYSTEM

Figure 10 shows a parallel piping system used to supply water for industrial needs. Determine the equivalent length of a single pipe for this system. All pipes in the system are approximately horizontal.

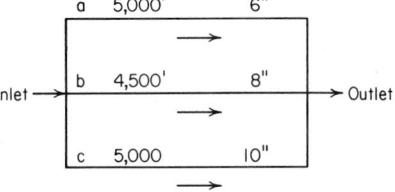

Fig. 10   Parallel piping system.

Calculation Procedure:

1. Assume a total head loss for the system

To determine the equivalent length of a parallel piping system, assume a total head loss for the system. Since this head loss is assumed for computation purposes only, its value need not be exact or even approximate. Assume a total head loss of 50 ft of water for each pipe in this system.

2. Compute the flow rate in each pipe in the system

Assume that the roughness coefficient C in the Hazen-Williams formula is equal for each of the pipes in the system. This is a valid assumption. Using the assumed value of C, compute the flow rate in each pipe. To allow for possible tuberculation of the pipe, assume that $C = 100$.

The Hazen-Williams formula is given in a previous Calculation Procedure and can be used to solve for the flow rate in each pipe. A more rapid way to make the computation is to use the friction-loss tabulations for the Hazen-Williams formula in King and Crocker—*Piping Handbook*, the Hydraulic Institute—*Pipe Friction Manual*, or a similar set of tables.

Using such a set of tables, enter at the friction-head loss equal to 50 ft per 5,000 ft of pipe for the 6-in. line and find the corresponding flow rate Q gpm. Using the Hydraulic Institute tables, $Q_a = 380$ gpm; $Q_b = 850$ gpm; $Q_c = 1,450$ gpm. Hence, the total flow = $\Sigma Q = 380 + 835 + 1,450 = 2,665$ gpm.

3. Find the equivalent size and length of the pipe

Using the Hydraulic Institute tables again, look for a pipe having a 50-ft head loss with a flow of 2,665 gpm. Any pipe having a discharge equal to the sum of the discharge rates for all the pipes, at the assumed friction head, is an equivalent pipe.

Interpolating friction-head values in the 14-in.-outside-diameter (13.126-in.-inside-diameter) table shows that 5,970 ft of this pipe is equivalent to the system in Fig. 10.

This equivalent size can be used in any calculations related to this system — selection of a pump, determination of head loss with longer or shorter mains, etc. If desired, another equivalent-size pipe could be found by entering a different pipe-size table. Thus, 5,570 ft of 12-in. pipe (11.938 in. inside diameter) is also equivalent to this system.

Related Calculations: Use this procedure for any liquid — water, oil, gasoline, brine — flowing through a parallel piping system. The pipes are assumed to be full at all times.

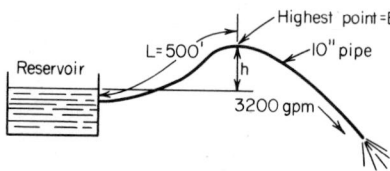

Fig. 11   Liquid siphon piping system.

MAXIMUM ALLOWABLE HEIGHT FOR A LIQUID SIPHON

What is the maximum height h ft, Fig. 11, that can be used for a siphon in a water system if the length of the pipe from the water source to its highest point is 500 ft, the water velocity is 13.0 fps, the pipe diameter is 10 in., and the water temperature is 70 F, if 3,200 gpm is flowing?

Calculation Procedure:

1.   Compute the velocity of the water in the pipe

From an earlier Calculation Procedure, $v = \text{gpm}/2.448d^2$. With an internal diameter of 10.020 in., $v = 3,200/[(2.448)(10.02)^2] = 13.0$ fps.

2.   Determine the vapor pressure of the water

Using a steam table, the vapor pressure of water at 70 F, $p_v = 0.3631$ psia, or (0.3631) (144 sq in/sq ft) = 52.3 psf. The specific volume of water at 70 F is, from a steam table, 0.01606 cu ft/lb. Converting this to density at 70 F, density = $1/0.01606 = 62.2$ lb/cu ft. The vapor pressure in ft of 70-F water is then $f_v = 52.3$ psf/62.2 lb/cu ft = 0.84 ft of water.

3.   Compute or determine the friction-head loss and velocity head

From the reservoir to the highest point of the siphon, B, Fig. 11, the friction head in the pipe must be overcome. Use the Hazen-Williams or a similar formula to determine the friction head, as given in earlier Calculation Procedures, or a pipe-friction table. Using the Hydraulic Institute *Pipe Friction Manual*, $h_f = 4.59$ ft per 100 ft, or (500/100) (4.59) = 22.95 ft. From the same table, velocity head = 2.63 fps.

4.   Determine the maximum height for the siphon

For a siphon handling water, the maximum allowable height, h ft, at sea level with an atmospheric pressure of 14.7 psia = $(14.7 \times 144$ sq in./sq ft/density of water at operating temperature, lb/cu ft) − (vapor pressure of water at operating temperature, ft + 1.5 × velocity head, ft + friction head, ft). For this pipe, $h = (14.7 \times 144/62.2) - (0.84 + 1.5 \times 2.63 + 22.95) = 11.32$ ft. In actual practice, the value of h is taken as 0.75 to 0.8 the computed value. Using 0.75, $h = (0.75)(11.32) = 8.5$ ft.

Related Calculations: Use this procedure for any type of siphon conveying a liquid — water, oil, gasoline, brine, etc. Where the liquid has a specific gravity different from that of water, i.e., less than or greater than 1.0, proceed as above, expressing all heads in ft of liquid handled. Divide the resulting siphon height by the specific gravity of the liquid. At elevations above atmospheric, use the actual atmospheric pressure instead of 14.7 psia.

WATER-HAMMER EFFECTS IN LIQUID PIPELINES

What is the maximum pressure developed in a 200-psi water pipeline if a valve is closed nearly instantly, or pumps discharging into the line are all stopped at the same instant? The pipe is 8-in. schedule 40 steel and the water flow rate is 2,800 gpm. What maximum pressure is developed if the valve closes in 5 sec and the line is 5,000 ft long?

Calculation Procedure:

1. Determine the velocity of the pressure wave

For any pipe, the velocity of the pressure wave during water hammer is found from $v_w = 4,720/(1 + [Kd/Et])^{0.5}$, where v_w = velocity of the pressure wave in the pipeline, fps; K = bulk modulus of the liquid in the pipeline = 300,000 for water; d = internal diameter of pipe, in.; E = modulus of elasticity of pipe material, psi, = 30×10^6 psi for steel; t = pipe-wall thickness, in. For 8-in. schedule 40 steel pipe, using data from a table of pipe properties, $v_w = 4,720/\{1 + [300,000 \times 7.981/(30 \times 10^6 \times 0.322)]\}^{0.5} = 4,225.6$ fps.

2. Compute the pressure increase caused by water hammer

The pressure increase p_1 psi due to water hammer = $v_w v/[32.2(2.31)]$, where v = liquid velocity in the pipeline fps; 32.2 = acceleration due to gravity, fps²; 2.31 ft of water = 1-psi pressure.

For this pipe, $v = 0.4085$ gpm/$d^2 = 0.4085(2,800)/(7.981)^2 = 18.0$ fps. Then, $p_i = (4,225.6)(18)/[32.2(2.31)] = 1,022.56$ psi. The maximum pressure developed in the pipe is then p_i + pipe operating pressure, psi, = $1,022.56 + 200 = 1,222.56$ psi.

3. Compute the hammer pressure rise caused by valve closure

The hammer pressure rise caused by valve closure p_v psi = $2p_iL/v_wT$, where L = pipeline length, ft; T = valve closing time, sec. For this pipeline, $p_v = 2(1,022.56)(5,000)/[(4,225.6)(5)] = 484$ psi. Thus, the maximum pressure in the pipe will be $484 + 200 = 648$ psi.

Related Calculations: Use this procedure for any type of liquid—water, oil, etc.—in a pipeline subject to sudden closure of a valve or stoppage of a pump or pumps. The effects of water hammer can be reduced by relief valves, slow-closing check valves on pump discharge pipes, air chambers, air spill valves, and air injection into the pipeline.

SPECIFIC GRAVITY AND VISCOSITY OF LIQUIDS

An oil has a specific gravity of 0.8000 and a viscosity of 200 SSU (Saybolt Seconds Universal) at 60 F. Determine the API gravity and Bé gravity of this oil at 70 F and its weight in lb/gal. What is the kinematic viscosity in cs? What is the absolute viscosity in cp?

Calculation Procedure:

1. Determine the API gravity of the liquid

For any oil at 60 F, its specific gravity S, in relation to water at 60 F, is $S = 141.5/(131.5 + °API)$; or $°API = (141.5 - 131.5S)/S$. For this oil, $°API = [141.5 - 131.5(0.80)]/0.80 = 45.4 °API$.

2. Determine the Bé gravity of the liquid

For any liquid lighter than water, $S = 140/(130 + Bé)$; or $Bé = (140 - 130S)/S$. For this oil, $Bé = [140 - 130(0.80)]/0.80 = 45$ Bé.

3. Compute the weight per gal of liquid

With a specific gravity of S, the weight of 1 cu ft of oil = (S)(weight of 1 cu ft of fresh water at 60 F) = $(0.80)(62.4) = 49.92$ lb/cu ft. Since 1 gal of liquid occupies 0.13368 cu ft, the weight of this oil per gal is $(49.92(0.13368)) = 6.66$ lb/gal.

4. Compute the kinematic viscosity of the liquid

For any liquid having a viscosity between 32 and 99 SSU, the kinematic viscosity $k = 0.226$ SSU $- 195/$SSU cs. For this oil, $k = 0.226(200) - 195/200 = 44.225$ cs.

5. Convert the kinematic viscosity to absolute viscosity

For any liquid, the absolute viscosity, cp, = (kinematic viscosity, cs)(specific gravity). Thus, for this oil, the absolute viscosity = $(44.225)(0.80) = 35.38$ cp.

Related Calculations: For liquids *heavier* than water, $S = 145/(145 - \text{Bé})$. When the SSU viscosity is greater than 100 sec, $k = 0.220 \text{ SSU} - 135/\text{SSU}$. Use these relations for any liquid—brine, gasoline, crude oil, kerosene, Bunker C, diesel oil, etc. Consult the *Pipe Friction Manual* and King and Crocker—*Piping Handbook* for tabulations of typical viscosities and specific gravities of various liquids.

PRESSURE LOSS IN PIPING HAVING LAMINAR FLOW

Fuel oil at 300 F and having a specific gravity of 0.850 is pumped through a 30,000-ft-long 24-in. pipe at the rate of 500 gpm. What is the pressure loss if the viscosity of the oil is 75 cp?

Calculation Procedure:

1. Determine the type of flow that exists

Flow is laminar (also termed viscous) if the Reynolds number R for the liquid in the pipe is less than 1,200. Tubulent flow exists if the Reynolds number is greater than 2,500. Between these values is a zone in which either condition may exist, depending on the roughness of the pipe wall, entrance conditions, and other factors. Avoid sizing a pipe for flow in this critical zone because excessive pressure drops result without a corresponding increase in the pipe discharge.

Compute the Reynolds number from $R = 3.162G/kd$, where G = flow rate gpm; k = kinematic viscosity of liquid, cs = viscosity z cp/specific gravity of the liquid S; d = inside diameter of pipe, in. From a table of pipe properties, $d = 22.626$ in. Also, $k = z/S$ = $75/0.85 = 88.2$ cs. Then, $R = 3,162(500)/[88.2(22.626)] = 792$. Since $R < 1,200$, laminar flow exists in this pipe.

2. Compute the pressure loss using the Poiseville formula

The Poiseville formula gives the pressure drop p_d psi = $2.73(10^{-4})luG/d^4$, where l = total length of pipe, including equivalent length of fittings, ft; u = absolute viscosity of liquid, cp; G = flow rate gpm; d = inside diameter of pipe, in. For this pipe, $p_d = 2.73$ $(10^{-4})(10,000)(75)(500)/262,078 = 1.17$ psi.

Related Calculations: Use this procedure for any pipe in which there is laminar flow of the liquid. Other liquids for which this method can be used include water, molasses, gasoline, brine, kerosene, and mercury. Table 20 gives a quick summary of various ways in which the Reynolds number can be expressed. The symbols in Table 20, in the order of their appearance, are: D = inside diameter of pipe, ft; v = liquid velocity, fps; ρ = liquid density, lb/cu ft; μ = absolute viscosity of liquid, lb mass/ft-sec; d = inside diameter of pipe, in. From a table of pipe properties, $d = 22.626$ in. Also, $k = z/S$ liquid flow rate, lb/hr; B = liquid flow rate, bbl/hr; k = kinematic viscosity of the liquid, cs; q = liquid flow rate, cfs; Q = liquid flow rate, cfm. Use Table 20 to find the Reynolds number for any liquid flowing through a pipe.

DETERMINING THE PRESSURE LOSS IN OIL PIPES

What is the pressure drop in a 5,000-ft-long 6-in. oil pipe conveying 500 bbl/hr of kerosene having a specific gravity of 0.813 at 65 F, which is the temperature of the liquid in the pipe? The pipe is schedule 40 steel.

Calculation Procedure:

1. Determine the kinematic viscosity of the oil

Use Fig. 12 and Table 21 or the Hydraulic Institute—*Pipe Friction Manual* kinematic viscosity and Reynolds number chart to determine the kinematic viscosity of the liquid. Enter Table 12 at kerosene and find the coordinates as $X = 10.2$, $Y = 16.9$. Using these

TABLE 20   Reynolds Number

Reynolds number R	Numerator				Denominator	
	Coefficient	First symbol	Second symbol	Third symbol	Fourth symbol	Fifth symbol
Dvp/μ	...	Ft	Fps	Lb/cu ft	Lb mass/ ft-sec	
$124dvp/z$	124	In.	Fps	Lb/cu ft	Cp	
$50.7Gp/dz$	50.7	Gpm	Lb/cu ft	...	In.	Cp
$6.32W/dz$	6.32	Lb/hr	...	...	In.	Cp
$35.5Bp/dz$	35.5	Bbl/hr	Lb/cu ft	...	In.	Cp
$7{,}742\,dv/k$	7,742	In.	Fps	...	...	Cp
$3{,}162G/dk$	3,162	Gpm	...	...	In.	Cp
$2{,}214B/dk$	2,214	Bbl/hr	...	...	In.	Cp
$22735qp/dz$	22,735	Cfs	Lb/cu ft	...	In.	Cp
$378.9Qp/dz$	378.9	Cfm	Lb/cu ft	...	In.	Cp

coordinates, enter Fig. 12 and find the absolute viscosity of kerosene at 65 F as 2.4 cp. Using the method of a previous Calculation Procedure, the kinematic viscosity, cs, = absolute viscosity, cp/specific gravity of the liquid = 2.4/0.813 = 2.95 cs. This value agrees closely with that given in the *Pipe Friction Manual*.

2. Determine the Reynolds number of the liquid

The Reynolds number can be found from the *Pipe Friction Manual* chart mentioned in Step 1 or computed from $R = 2{,}214\,B/dk = 2{,}214(500)/[(6.065)(2.95)] = 61{,}900$.

To use the *Pipe Friction Manual* chart, compute the velocity of the liquid in the pipe by converting the flow rate to cfs. Since there are 42 gal/bbl and 1 gal = 0.13368 cu ft, 1 bbl = (42)(0.13368) = 5.6 cu ft. With a flow rate of 500 bbl/hr, the equivalent flow in cu ft = (500)(5.6) = 2,800 cfh, or 2,800/3,600 sec/hr = 0.778 cfs. Since 6-in. schedule 40 pipe has a cross-sectional area of 0.2006 sq ft internally, the liquid velocity, fps, = 0.778/0.2006 = 3.88 fps. Then, the product (velocity, fps)(internal diameter, in.) = (3.88)(6.065) = 23.75. In the *Pipe Friction Manual*, project horizontally from the kerosene specific gravity curve to the vd product of 23.75 and read the Reynolds number as 61,900, as before. In general, the Reynolds number can be found faster by computing it using the appropriate relation given in an earlier Calculation Procedure, unless the flow velocity is already known.

3. Determine the friction factor of this pipe

Enter Fig. 13 at the Reynolds number value of 61,900 and project to the curve 4 as indicated by Table 22. Read the friction factor as 0.0212 at the left. Alternatively, the *Pipe Friction Manual* friction-factor chart could be used, if desired.

4. Compute the pressure loss in the pipe

Use the Fanning formula $p_d = 1.06(10^{-4})f\rho l B^2/d^5$. In this formula, ρ = density of the liquid, lb/cu ft. For kerosene, ρ = (density of water, lb/cu ft)(specific gravity of kerosene) = (62.4)(0.813) = 50.6 lb/cu ft. Then, $p_d = 1.06(10^{-4})(0.0212)(50.6)(5{,}000)$ $(500)^2/8{,}206 = 17.3$ psi.

Related Calculations: The Fanning formula is popular with oil-pipe designers and can be stated in various ways: (1) with velocity v fps, $p_d = 1.29(10^{-3})f\rho v^2 l/d$; (2) with velocity V fpm, $p_d = 3.6(10^{-7})f\rho V^2 l/d$; (3) with flow rate in G gpm, $p_d = 2.15(10^{-4})f\rho lG^2/d^2$; (4) with the flow rate in W lb/hr, $p_d = 3.36(10^{-6})flW^2/d^5\rho$.

Use this procedure for any petroleum product — crude oil, kerosene, benzene, gasoline, naptha, fuel oil, Bunker C, diesel oil toulene, etc. The tables and charts presented here and in the *Pipe Friction Manual* save computation time.

TABLE 21   Viscosities of Liquids
Coordinates for use with Fig. 12.

No.	Liquid	X	Y	No.	Liquid	X	Y
1	Acetaldehyde	15.2	4.8	56	Freon-22	17.2	4.7
	Acetic acid:			57	Freon-13	12.5	11.4
2	100%	12.1	14.2		Glycerol:		
3	70%	9.5	17.0	58	100%	2.0	30.0
4	Acetic anhydride	12.7	12.8	59	50%	6.9	19.6
	Acetone:			60	Heptene	14.1	8.4
5	100%	14.5	7.2	61	Hexane	14.7	7.0
6	35%	7.9	15.0	62	Hydrochloric acid, 31.5%	13.0	16.6
7	Allyl alcohol	10.2	14.3	63	Isobutyl alcohol	7.1	18.0
	Ammonia:			64	Isobutyric acid	12.2	14.4
8	100%	12.6	2.0	65	Isopropyl alcohol	8.2	16.0
9	26%	10.1	13.9	66	Kerosene	10.2	16.9
10	Amyl acetate	11.8	12.5	67	Linseed oil, raw	7.5	27.2
11	Amyl alcohol	7.5	18.4	68	Mercury	18.4	16.4
12	Aniline	8.1	18.7		Methanol:		
13	Anisole	12.3	13.5	69	100%	12.4	10.5
14	Arsenic trichloride	13.9	14.5	70	90%	12.3	11.8
15	Benzene	12.5	10.9	71	40%	7.8	15.5
	Brine:			72	Methyl acetate	14.2	8.2
16	CaCl₂, 25%	6.6	15.9	73	Methyl chloride	15.0	3.8
17	NaCl, 25%	10.2	16.6	74	Methyl ethyl ketone	13.9	8.6
18	Bromine	14.2	13.2	75	Naphthalene	7.9	18.1
19	Bromotoluene	20.0	15.9		Nitric acid:		
20	Butyl acetate	12.3	11.0	76	95%	12.8	13.8
21	Butyl alcohol	8.6	17.2	77	60%	10.8	17.0
22	Butyric acid	12.1	15.3	78	Nitrobenzene	10.6	16.2
23	Carbon dioxide	11.6	0.3	79	Nitrotoluene	11.0	17.0
24	Carbon disulfide	16.1	7.5	80	Octane	13.7	10.0
25	Carbon tetrachloride	12.7	13.1	81	Octyl alcohol	6.6	21.1
26	Chlorobenzene	12.3	12.4	82	Pentachloroethane	10.9	17.3
27	Chloroform	14.4	10.2	83	Pentane	14.9	5.2
28	Chlorosulfonic acid	11.2	18.1	84	Phenol	6.9	20.8
	Chlorotoluene:			85	Phosphorus tribromide	13.8	16.7
29	Ortho	13.0	13.3	86	Phosphorus trichloride	16.2	10.9
30	Meta	13.3	12.5	87	Propionic acid	12.8	13.8
31	Para	13.3	12.5	88	Propyl alcohol	9.1	16.5
32	Cresol, meta	2.5	20.8	89	Propyl bromide	14.5	9.6
33	Cyclohexanol	2.9	24.3	90	Propyl chloride	14.4	7.5
34	Dibromoethane	12.7	15.8	91	Propyl iodide	14.1	11.6
35	Dichloroethane	13.2	12.2	92	Sodium	16.4	13.9
36	Dichloromethane	14.6	8.9	93	Sodium hydroxide, 50%	3.2	25.8
37	Diethyl oxalate	11.0	16.4	94	Stannic chloride	13.5	12.8
38	Dimethyl oxalate	12.3	15.8	95	Sulfur dioxide	15.2	7.1
39	Diphenyl	12.0	18.3		Sulfuric acid:		
40	Dipropyl oxalate	10.3	17.7	96	110%	7.2	27.4
41	Ethyl acetate	13.7	9.1	97	98%	7.0	24.8
	Ethyl alcohol:			98	60%	10.2	21.3
42	100%	10.5	13.8	99	Sulfuryl chloride	15.2	12.4
43	95%	9.8	14.3	100	Tetrachloroethane	11.9	15.7
44	40%	6.5	16.6	101	Tetrachloroethylene	14.2	12.7
45	Ethyl benzene	13.2	11.5	102	Titanium tetrachloride	14.4	12.3
46	Ethyl bromide	14.5	8.1	103	Toluene	13.7	10.4
47	Ethyl chloride	14.8	6.0	104	Trichloroethylene	14.8	10.5
48	Ethyl ether	14.5	5.3	105	Turpentine	11.5	14.9

TABLE 21   Viscosities of Liquids (continued)

No.	Liquid	X	Y	No.	Liquid	X	Y
49	Ethyl formate			106	Vinyl acetate		
50	Ethyl iodide................	14.7	10.3	107	Water......................	10.2	13.0
51	Ethylene glycol.............	6.0	23.6		Xylene:		
52	Formic acid	10.7	15.8	108	Ortho.....................	13.5	12.1
53	Freon-11...................	14.4	9.0	109	Meta.....................	13.9	10.6
54	Freon-12...................	16.8	5.6	110	Para	13.9	10.9
55	Freon-21...................	15.7	7.5				

VISCOSITIES

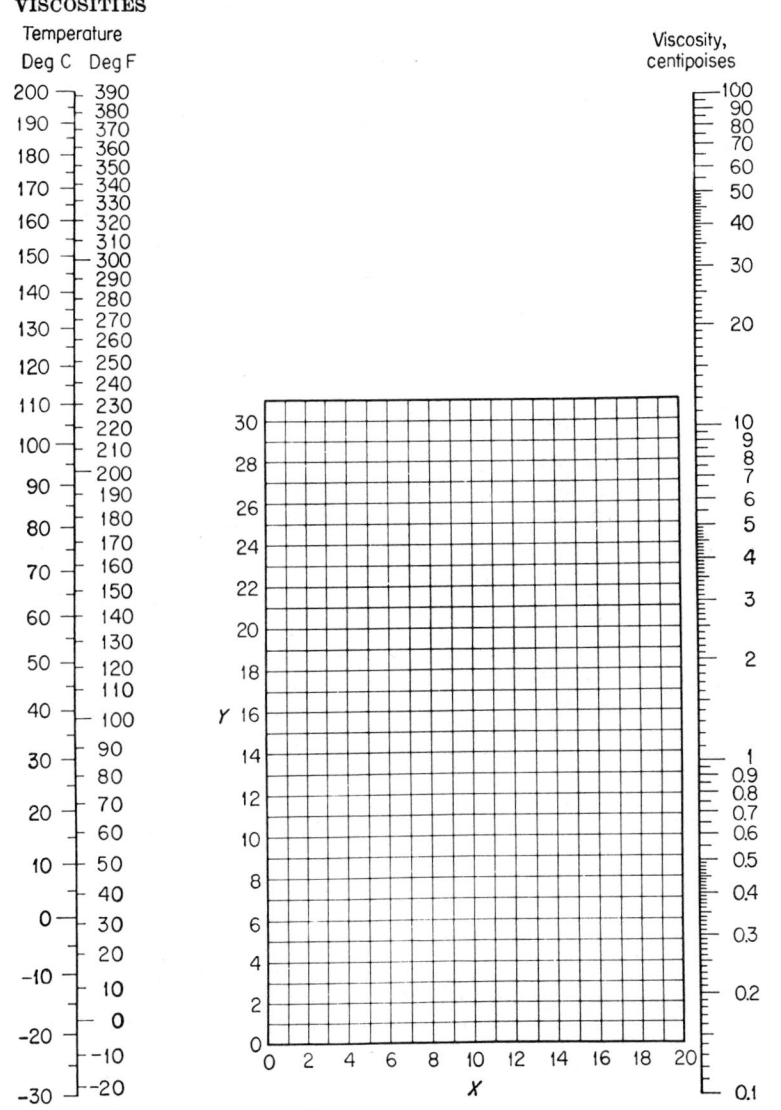

Fig. 12   Viscosities of liquids at 1 atm. For coordinates, see Table 21.

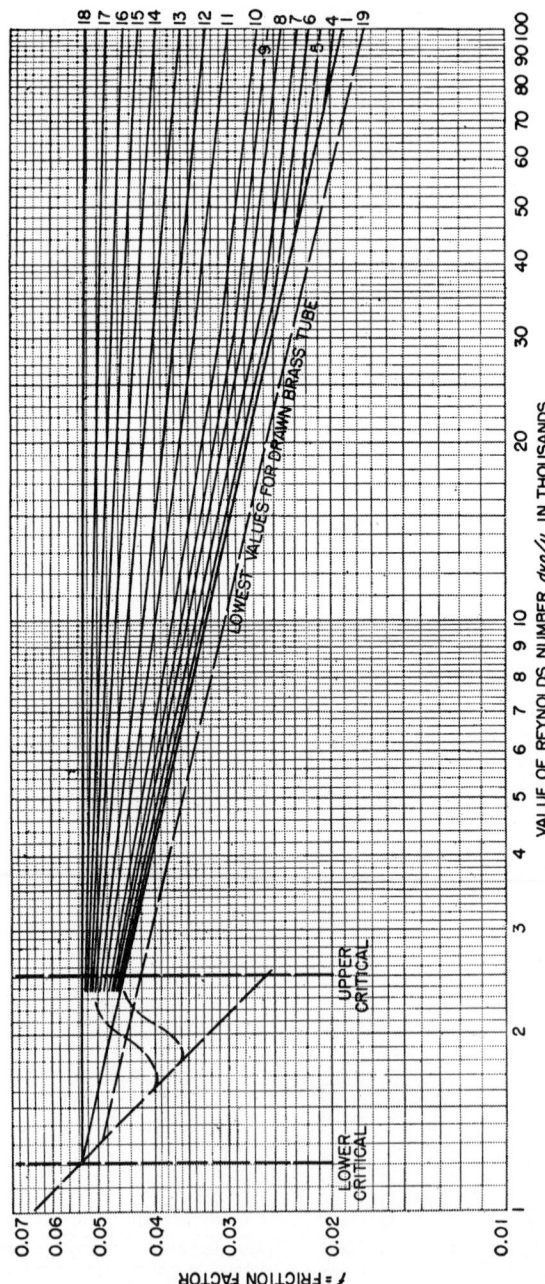

Fig. 13   Friction-factor curves. (*Mechanical Engineering.*)

3-388

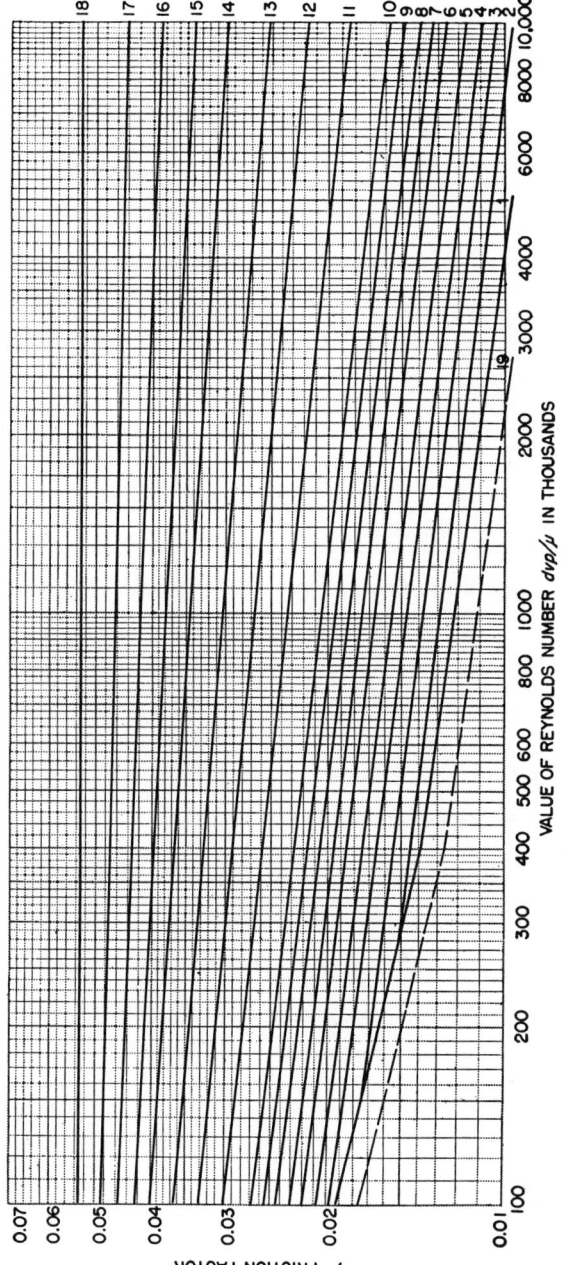

Fig. 13 (continued) Friction factor curves. (*Mechanical Engineering.*)

TABLE 22   Data for Fig. 13

Percent roughness	For value of f see curve	Drawn tubing, brass, tin, lead, glass	Clean steel, wrought iron	Clean, galvanized	Best cast iron	Average cast iron	Heavy riveted, spiral riveted
		Diameter, in. (actual of drawn tubing; nominal of standard weight pipe)					
0.2	1	0.35 up	72				
1.35	4	...	6–12	10–24	20–48	42–96	84–204
2.1	5	...	4–5	6–8	12–16	24–36	48–72
3.0	6	...	2–3	3–5	5–10	10–20	20–42
3.8	7	...	$1\frac{1}{2}$	$2\frac{1}{2}$	3–4	6–8	16–18
4.8	8	...	$1–1\frac{1}{4}$	$1\frac{1}{2}–2$	$2–2\frac{1}{2}$	4–5	10–14
6.0	9	...	$\frac{3}{4}$	$1\frac{1}{4}$	$1\frac{1}{2}$	3	8
7.2	10	...	$\frac{1}{2}$	1	$1\frac{1}{4}$	...	5
10.5	11	...	$\frac{3}{8}$	$\frac{3}{4}$	1	...	4
14.5	12	...	$\frac{1}{4}$	$\frac{1}{2}$	...	...	3
24.0	14	0.125	...	$\frac{3}{8}$			
31.5	16	...	...	$\frac{1}{4}$			
37.5	18	0.0625	...	$\frac{1}{8}$			

FLOW RATE AND PRESSURE LOSS IN COMPRESSED-AIR AND GAS PIPING

Dry air at 80 F and 150 psia flows at the rate of 500 cfm through a 4-in. schedule 40 pipe from the discharge of an air compressor. What is the flow rate in lb/hr and the air velocity in fps? Using the Fanning formula, determine the pressure loss if the total equivalent length of the pipe is 500 ft.

Calculation Procedure:

1. Determine the density of the air or gas in the pipe

For air or a gas, $pV = MRT$, where p = absolute pressure of the gas, psfa; V = volume of M lb of gas, cu ft; M = weight of gas, lb; R = gas constant, ft-lb/(lb)(F); T = absolute temperature of the gas, R. For this installation, using 1 cu ft of air, $M = pV/RT$, $M = (150)(144)/[(53.33)(80+459.7)] = 0.754$ lb/cu ft. The value of R in this equation was obtained from Table 23.

2. Compute the flow rate of the air or gas

For air or a gas, the flow rate W_h lb/hr = (60) (density, lb/cu ft)(flow rate, cfm); or $W_h = (60)(0.754)(500) = 22,620$ lb/hr.

3. Compute the velocity of the air or gas in the pipe

For any air or gas pipe, velocity of the moving fluid v fps = $183.4 W_h/3,600 d^2\rho$, where d = internal diameter of pipe, in.; ρ = density of fluid, lb/cu ft. For this system, $v = (183.4)(22,620)/[(3,600)(4.026)^2(0.754)] = 95.7$ fps.

4. Compute the Reynolds number of the air or gas

The viscosity of air at 80 F is 0.0186 cp, obtained from King and Crocker—*Piping Handbook*, Perry, et al.—*Chemical Engineers' Handbook*, or a similar reference. Then, using the Reynolds number relation given in Table 20, $R = 6.32W/dz = (6.32)(22,620)/[(4.026)(0.0186)] = 3,560,000$.

TABLE 23   Gas Constants

Gas	R, Ft = lb/(lb)(F)	C for critical velocity equation
Air...............	53.33	2,870
Ammonia.........	89.42	2,080
Carbon dioxide ...	34.87	3,330
Carbon monoxide .	55.14	2,820
Ethane...........	50.82	
Ethylene.........	54.70	2,480
Hydrogen........	767.04	750
Hydrogen sulfide .	44.79	
Isobutane	25.79	
Methane	96.18	2,030
Natural gas	...	2,070–2,670
Nitrogen	55.13	2,800
n-butane	25.57	
Oxygen	48.24	2,990
Propane..........	34.13	
Propylene........	36.01	
Sulfur dioxide	23.53	3,870

5.   Compute the pressure loss in the pipe

Using Fig. 13 or the Hydraulic Institute *Pipe Friction Manual*, $f = 0.0142$ to 0.0162 for a 4-in. schedule 40 pipe when the Reynolds number = 3,560,000. Using the Fanning formula from an earlier Calculation Procedure and the higher value of f, $p_d = 3.36$ $(10^{-6}) f l W^2/d^5 \rho$, or $p_d = 3.36(10^{-6})(0.0162)(500)(22,620)^2/[(4.026)^5(0.754)] = 17.52$ psi.

Related Calculations: Use this procedure to compute the pressure loss, velocity, and flow rate in compressed-air and gas lines of any length. Gases for which this procedure can be used include ammonia, carbon dioxide, carbon monoxide, ethane, ethylene, hydrogen, hydrogen sulfide, isobutane, methane, nitrogen, n-butane, oxygen, propane, propylene, and sulfur dioxide.

Alternate relations for computing the velocity of air or gas in a pipe are: $v = 144 W_s/ a\rho$; $v = 183.4 W_s/d^2\rho$; $v = 0.0509 W_s v_g/d^2$, where W_s = flow rate, lb/sec; a = cross-sectional area of pipe, sq in.; v_g = specific volume of the air or gas at the operating pressure and temperature, cu ft/lb.

FLOW RATE AND PRESSURE LOSS IN GAS PIPELINES

Using the Weymouth formula, determine the flow rate in a 10-mile-long 4-in. schedule 40 gas pipeline when the inlet pressure is 200 psig, the outlet pressure is 20 psig, the gas has a specific gravity of 0.80, a temperature of 60 F, and the atmospheric pressure is 14.7 psia.

Calculation Procedure:

1.   Compute the flow rate using the Weymouth formula

The Weymouth formula for flow rate in Q lb/hr is $Q = 28.05[(p_i^2 - p_o^2)d^{5.33}/sL]^{0.5}$, where p_i = inlet pressure, psia; p_o = outlet pressure, psia; d = inside diameter of pipe, in.; s = specific gravity of gas; L = length of pipeline, miles. For this pipe, $Q = 28.05 \times [(214.7^2 - 34.7^2)4.026^{5.33}/0.8 \times 10]^{0.5} = 86,500$ lb/hr.

2.   Determine if the acoustic velocity limits flow

If the outlet pressure of a pipe is less than the critical pressure p_c psia, the flow rate in the pipe cannot exceed that obtained with a velocity equal to the critical or acoustic velocity, i.e., the velocity of sound in the gas. For any gas, $p_c = Q(T_i)^{0.5}/d^2C$, where T_i = inlet temperature, R; C = a constant for the gas being considered.

Using $C = 2,070$ from Table 23, or King and Crocker—*Piping Handbook*, $p_c =$ $(86,500)(60+460)^{0.5}/[(4.026)^2(2,070)] = 58.8$ psia. Since the outlet pressure $p_o = 34.7$ psia, the critical or acoustic velocity limits the flow in this pipe because $p_c > p_o$. When $p_c < p_o$, critical velocity does not limit the flow.

Related Calculations: Where a number of gas pipeline calculations must be made, use the tabulations in King and Crocker—*Piping Handbook* and Bell—*Petroleum Transportation Handbook*. These tabulations will save much time. Other useful formulas for gas flow include the Panhandle, Unwin, Fritzsche, and rational. Results obtained with these formulas agree within satisfactory limits for normal engineering practice.

Where the outlet pressure is unknown, assume a value for it and compute the flow rate that will be obtained. If the computed flow is less than desired, check to see if the outlet pressure is less than the critical. If it is, increase the diameter of the pipe. Use this procedure for natural gas from any gas field, manufactured gas, or any other similar gas.

To find the volume of gas, cu ft, that can be stored per mile of pipe, solve $V_m =$ $1.955p_md^2K$, where p_m = mean pressure in pipe, psia $\approx (p_i + p_o)/2$; $K = (1/Z)^{0.5}$, where Z = supercompressibility factor of the gas, as given in Baumeister—*Standard Handbook for Mechanical Engineers* and Perry's *Chemical Engineer's Handbook*. For exact computation of p_m, use $p_m = (\frac{2}{3})(p_i + p_o - p_ip_o/p_i + p_o)$.

SELECTING HANGERS FOR PIPES AT ELEVATED TEMPERATURES

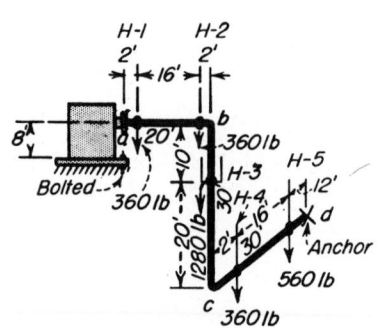

Fig. 14   Typical complex pipe operating at high temperature.

Select the number, capacity, and types of pipe hanger needed to support the 6-in. schedule 80 pipe in Fig. 14 when the installation temperature is 60 F and the operating temperature is 700 F. The pipe is insulated with 85 percent magnesia weighing 11.4 lb/ft. The pipe and unit served by the pipe have a coefficient of thermal expansion of 0.0575 in./ft between the 60-F installation temperature and the 700-F operating temperature.

Calculation Procedure:

1.   Draw a freehand sketch of the pipe expansion

Use Fig. 15 as a guide and sketch the expanded pipe using a dashed line. The sketch need not be exactly to scale; if the proportions are accurate, satisfactory results will be obtained. The shapes shown in Fig. 15 cover the 11 most common situations met in practice.

2.   Tentatively locate the required hangers

Begin by locating hangers H-1 and H-5 close to the supply and using units, Fig. 14. Keeping a hanger close to each unit (boiler, turbine, pump, engine, etc.) prevents overloading the connection on the unit.

Space intermediate hangers H-2, H-3, and H-4 so that the recommended distances in Table 24 or hanger engineering data (e.g., Grinnell Corporation *Pipe Hanger Design and Engineering*) are not exceeded. Indicate the hangers on the piping drawing as shown in Fig. 14.

3.   Adjust the hanger locations to suit structural conditions

Study the building structural steel in the vicinity of the hanger locations and adjust these locations so that each hanger can be attached to a support having adequate strength.

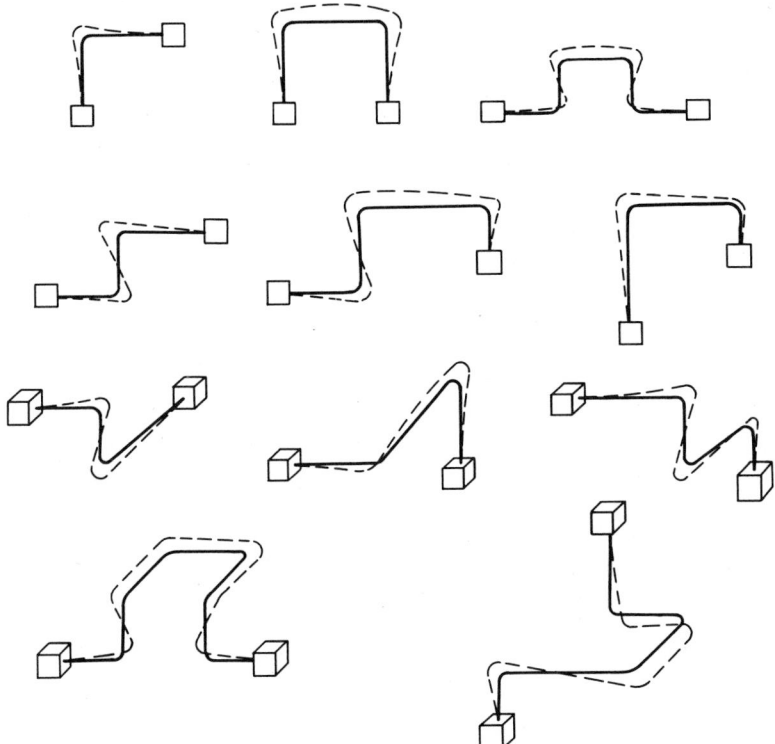

Fig. 15 Pipe shapes commonly used in power and process plants assume the approximate forms shown by the dotted lines when the pipe temperature rises. (*Power.*)

TABLE 24   Maximum Recommended Spacing between Pipe Hangers

Nominal pipe size, in.	1	$1\frac{1}{2}$	2	$2\frac{1}{2}$	3	$3\frac{1}{2}$	4	5	6	8	10	12	14	16	18	20	24
Maximum span, ft	7	9	10	11	12	13	14	16	17	19	22	23	25	27	28	30	32

4. *Compute the load each hanger must support*

From a table of pipe properties, such as in King and Crocker—*Piping Handbook*, find the weight of 6-in. schedule 80 pipe as 28.6 lb/ft. The insulation weighs 11.4 lb/ft, giving a total weight per ft of insulated pipe of $28.6 + 11.4 = 40.0$ lb/ft.

Compute the load on the hangers supporting horizontal pipes by taking half the length of the pipe on each side of the hanger. Thus, for hanger H-1, there is $(2 \text{ ft})(\frac{1}{2}) + (16 \text{ ft}) \times (\frac{1}{2}) = 9$ ft of horizontal pipe, Fig. 14, which it supports. Since this pipe weighs 40 lb/ft, the total load on hanger H-1 is $(9 \text{ ft})(40 \text{ lb/ft}) = 360$ lb. A similar analysis for hanger H-2 shows that it supports $(8 + 1)(40) = 360$ lb.

Hanger H-3, Fig. 14, supports the entire weight of the vertical pipe, 30 ft, plus 1 ft at the top bend and 1 ft at the bottom bend, or a total of $(1 + 30 + 1) = 32$ ft. The total load on hanger H-3 is therefore $(32)(40) = 1,280$ lb.

Hanger H-4 supports $(1+8)(40) = 360$ lb, and hanger H-5 supports $(8+6)(40) = 560$ lb.

As a check, compute the total weight of the pipe and compare it with the sum of the end-point and hanger loads. Thus, there is 100 ft of pipe weighing $(80)(40) = 3,200$ lb. The total load the hangers will support is $360 + 360 + 1,280 + 360 + 560 = 2,920$ lb. The first end point will support $(1)(40) = 40$ lb, and the anchor will support $(6)(40) = 240$ lb. The total hanger and end-point support $= 2,920 + 40 + 240 = 3,200$ lb; therefore, the pipe weight = the hanger load.

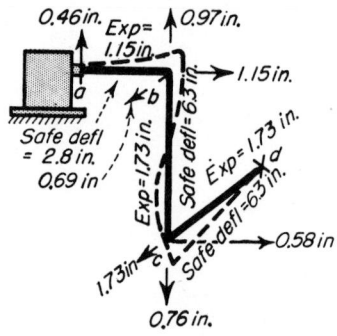

Fig. 16   Expansion of the various parts of the pipe shown in Fig. 14. (*Power.*)

5.   Sketch the shape of the hot pipe

Use Fig. 15 as a guide and draw a dotted outline of the approximate shape the pipe will take when hot. Start with the first corner point nearest the unit on the left, Fig. 16. This point will move away from the unit, as in Fig. 16. Do the same for the first corner point near the other unit served by the pipe and for intermediate corner points. Use arrows to indicate the probable direction of pipe movement at each corner, Fig. 16. When sketching the shape of the hot pipe, remember that a straight pipe expanding against a piece of pipe at right angles to itself will bend the latter. The distance that various lengths of pipe will bend while producing a tensile stress of 14,000 psi is given in Table 25. This stress is a typical allowable value for pipes in industrial systems.

TABLE 25   Deflections that Produce 14,000-psi Tensile Stress in Pipe Legs Acting as Cantilever Beam, Load at Free End

Cantilever length, ft	Nominal pipe size, in.								
	1	2	3	4	6	8	10	12	14
5	0.88	0.49	0.33	0.26	0.17	0.13	0.11	0.09	0.08
10	3.53	1.94	1.33	1.03	0.70	0.54	0.43	0.36	0.33
15	7.94	4.36	2.98	2.32	1.58	1.21	0.97	0.82	0.74
20	14.10	7.76	5.30	4.12	2.80	2.15	1.72	1.45	1.32
25	22.00	12.1	8.28	6.44	4.38	3.35	2.69	2.27	2.07
30	31.75	17.45	11.93	9.26	6.30	4.83	3.87	3.27	2.98

6.   Determine the thermal movement of units served by the pipe

If either or both fixed units (boiler, turbine, etc.) operate at a temperature above or below atmospheric, determine the amount of movement at the flange of the unit to which the piping connects, using the thermal data in Table 26. Do this by applying the thermal expansion coefficient for the metal of which the unit is made. Determine the vertical and horizontal distance of the flange face from the point of no movement of the unit. The point of no movement is the point or surface where the unit is fastened to *cold* structural steel or concrete.

The flange, point a, Fig. 16, is 8 ft above the bolted end of the unit and directly in line with the bolt, Fig. 14. Since the bolt and flange are on a common vertical line, there will not be any *horizontal* movement of the flange because the bolt is the no-movement point of the unit.

Since the flange is 8 ft away from the point of no movement, the amount that the flange will move, in in., = (distance away from point of no movement, ft)(coefficient of thermal expansion, in./ft) = $(8)(0.0575) = 0.46$ in. *away* (up) from the point of no movement. If the unit were operating at a temperature *less than* atmospheric, it would contract and the flange would move *toward* (down) the point of no movement. Mark the flange movement on the piping sketch, Fig. 16.

Anchor d, Fig. 16, does not move because it is attached to either cold structural steel or concrete.

TABLE 26   Thermal Expansion of Pipe (Carbon and Carbon-Moly Steel and WI)

Operating temperature, F	In./ft	
	Installation temperature, F	
	32	60
100	0.005	0.0029
150	0.009	0.0069
200	0.013	0.0108
250	0.017	0.0148
300	0.022	0.0197
350	0.026	0.0237
400	0.030	0.0277
450	0.035	0.0326
500	0.040	0.0376
550	0.045	0.0425
600	0.050	0.0475
650	0.055	0.0525
700	0.060	0.0575
750	0.065	0.0624
800	0.070	0.0674
850	0.075	0.0724
900	0.081	0.0784
950	0.087	0.0843
1000	0.094	0.0913

7.   Compute the amount of expansion in each pipe leg

Expansion of the pipe, in., = (pipe length, ft)(coefficient of linear expansion, in./ft). For length ab, Fig. 14, the expansion = $(20)(0.0575) = 1.15$ in.; for bc, $(30)(0.0575) = 1.73$ in.; for cd, $(30)(0.0575) = 1.73$ in. Mark the amount and direction of expansion on Fig. 16.

8.   Determine the allowable deflection for each pipe leg

Enter Table 25 at the nominal pipe size and find the allowable deflection for a 14,000-psi tensile stress for each pipe leg. Thus, for ab, the allowable deflection = 2.80 in. for a 20-ft-long leg; for bc, 6.30 in. for a 30-ft-long leg; for cd, 6.30 in. for a 30-ft-long leg. Mark these allowable deflections on Fig. 16, using dashed arrows.

9.   Compute the actual vertical and horizontal deflections

Sketch the vertical deflection diagram, Fig. 17a, by drawing a triangle showing the total expansion in each direction in proportion to the length of the parts at right angles to the expansion. Thus, the 0.46-in. upward expansion at the flange, a, is at right angles to leg ab and is drawn as the altitude of the right triangle. Lay off 20 ft, ab, on the base

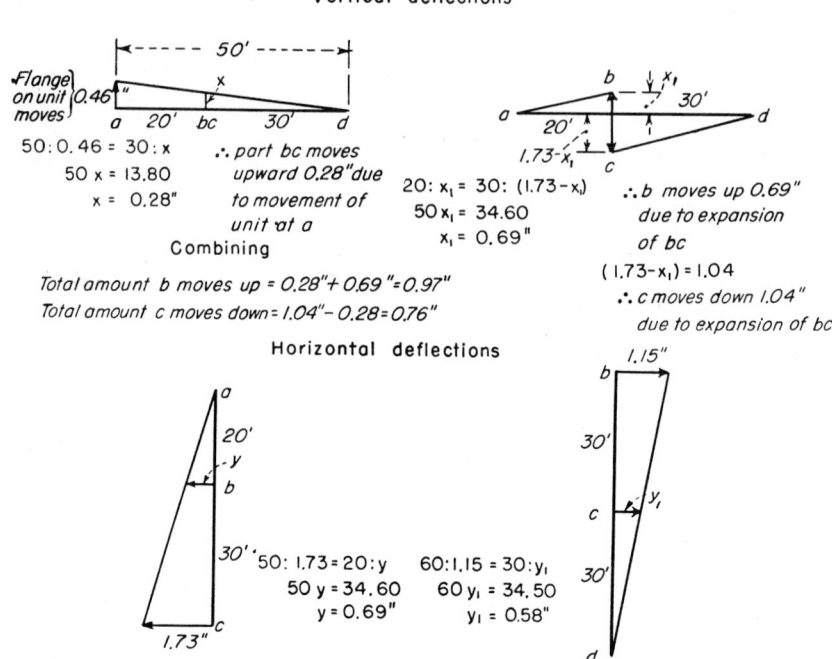

Vertical deflections

50:0.46 = 30:x   ∴ part bc moves

50 x = 13.80   upward 0.28"due

x = 0.28"   to movement of

unit at a

Combining

20: x_1 = 30: (1.73−x_1)

50 x_1 = 34.60

x_1 = 0.69"

∴b moves up 0.69"

due to expansion

of bc

(1.73−x_1) = 1.04

∴ c moves down 1.04"

due to expansion of bc

Total amount b moves up = 0.28"+0.69"=0.97"

Total amount c moves down = 1.04"−0.28"=0.76"

Horizontal deflections

50: 1.73 = 20:y   60:1.15 = 30:y_1

50 y = 34.60   60 y_1 = 34.50

y = 0.69" y_1 = 0.58"

Fig. 17. (a) and (b) Vertical deflection diagrams for the pipe in Fig. 14; **(c) and (d)** horizontal deflection diagrams for the pipe in Fig. 14. (*Power.*)

of the triangle. Since bc is parallel to the direction of the flange movement, it is shown as a point, bc, on the base of this triangle. From point bc, lay off cd on the base of the triangle, Fig. 17a, since it is at right angles to the expansion of point a. Then, by similar triangles, $50:46 = 30:x$; $x = 0.28$ in. Therefore, leg bc moves upward 0.28 in. due to the flange movement at a.

Now draw the deflection diagram, Fig. 17b, showing the upward movement of leg ab and the downward movement of leg cd along the length of each leg, or 20 and 30 ft, respectively. Solve the similar triangles, or $20:x_1 = 30:(1.73−x_1)$; $x_1 = 0.69$ in. Therefore, point b moves *up* 0.69 in. as a result of the expansion of leg bc. Then, $(1.73−x_1) = 1.73−0.69 = 1.04$ in. Thus, point c moves *down* 1.04 in. as a result of the expansion of bc. The total distance b moves up $= 0.28+0.69 = 0.97$ in., whereas the total distance c moves down $= 1.04−0.28 = 0.76$ in. Mark these actual deflections on Fig. 16.

Find the actual horizontal deflections in a similar fashion by constructing the triangle, Fig. 17c, formed by the vertical pipe bc and the horizontal pipe ab. Since point a does not move horizontally but point b does, lay off leg ab at right angles to the direction of movement, as shown. From point b lay off leg bc. Then, since leg bc expands 1.73 in., lay this distance off perpendicular to ac, Fig. 17c. By similar triangles, $(20+30):1.73 = 20:y$; $y = 0.69$ in. Hence, point b deflects 0.69 in. in the direction shown in Fig. 16.

Follow the same procedure for leg cd, constructing the triangle in Fig. 17d. Beginning with point b, lay off legs bc and cd. The altitude of this right triangle is then the distance point c moves when leg ab expands, or 1.15 in. By similar triangles, $(30+30):1.15 = 30:y_1$; $y_1 =$ deflection of point $c = 0.58$ in.

10. *Select the type of pipe hanger to use*

Figure 18 shows several popular types of pipe hangers, together with the movements that they are designed to absorb. For hangers *H*-1 and *H*-2, use type *E*, Fig. 18, because the pipe moves both vertically and horizontally at these points, as Fig. 17 shows. Use type *F*, Fig. 18, for hanger *H*-3, because riser *bc* moves both vertically and horizontally. Hangers *H*-4 and *H*-5 should be type *E*, because they must absorb both horizontal and vertical movements.

Once the hangers are selected from Fig. 18, refer to hanger engineering data for the exact design details of the hangers that will be selected. During the study of the data, look for other hangers that absorb the same movement or movements but may be more adaptable to the existing structural steel conditions.

11. *Select the hanger-rod diameter for each hanger*

Use Table 27 to find the required hanger-rod diameter. Since the pipe operates at 700 F, select the maximum safe load from the 750-F column. Tabulate the loads and diameters as follows:

TABLE 27   Hanger-Rod Load Carrying Capacity (Hot-rolled Steel Rod)

Nominal diameter of rod, in.	Thread root area, sq in.	Maximum safe load on rod, lb, at rod temperature of	
		450 F	750 F
$\frac{3}{8}$	0.068	610	510
$\frac{1}{2}$	0.126	1130	940
$\frac{5}{8}$	0.202	1810	1510
$\frac{3}{4}$	0.302	2710	2260
$\frac{7}{8}$	0.419	3770	3150
1	0.552	4960	4150
$1\frac{1}{8}$	0.693	6230	5200
$1\frac{1}{4}$	0.889	8000	6660
$1\frac{3}{8}$	1.053	9470	7800
$1\frac{1}{2}$	1.293	11,630	9700

Select standard springs for spring-loaded hangers from pipe-hanger engineering data. Springs are listed in the data on the basis of loading per in. of travel. For small movements (less than 1 in.), it is generally desirable to select a lighter spring and precompress it at installation so that it has a light loading. Hanger movement will then load the spring to the desired value. This approach is desirable from another standpoint — any error in estimating hanger movement will not cause as large an unbalanced load on the pipe as would a heavier spring with a greater loading per in. of travel.

Related Calculations: Use this procedure for any type of pipe operating at elevated temperature — steam, oil, water, gas, etc. — serving a load in a power plant, process plant, ship, barge, aircraft, or other type of installation. In piping systems having very little or no increase in temperature during operation, the steps for computing the expansion can be eliminated. In this type of installation, the weight of the piping is the primary consideration in the choice of the hangers.

If desired, hanger loads can also be determined by taking moments about an arbitrarily selected axis on either side of the hanger. This method gives the same results as the procedure used above. The weight of bends is assumed to be concentrated at the center of gravity of each bend, whereas the weight of valves is assumed to be concentrated at the vertical centerline of the valve. Figure 19 shows typical moment arms, *a*

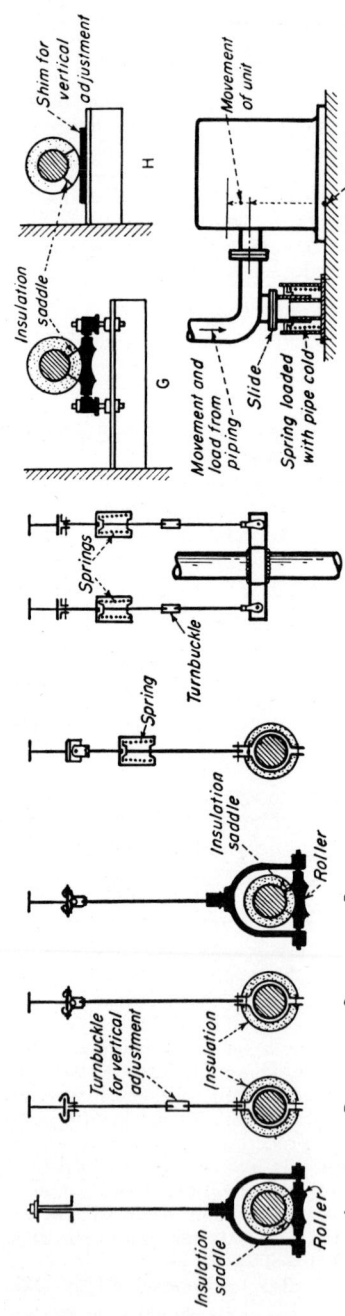

Fig. 18 Pipe hangers chosen depend on the movement expected. Hangers *A* and *B* are suitable for pipe movement in one horizontal direction. Hangers *C* and *D* permit pipe movement in two horizontal directions. Vertical and horizontal movement requires use of hangers such as *E* for horizontal pipes and *F* for vertical pipes. (*G*) Cantilever support; (*H*) sliding movement in two horizontal directions; (*I*) base elbow support. (*Power.*)

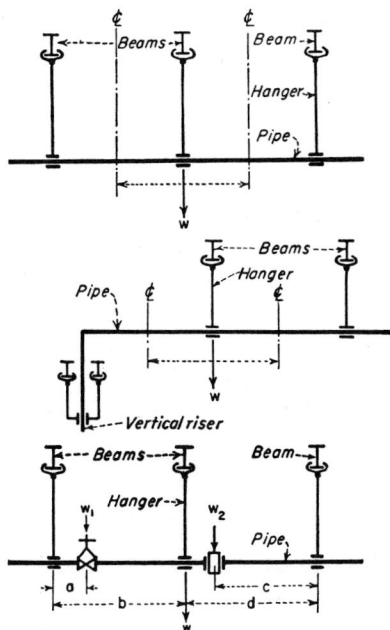

Fig. 19 Compute hanger loads of uniformly loaded pipes as shown. Use beam relations for concentrated loads. (*Power.*)

and c, for valves and other fittings. The moment for W_1 about the hanger to the left of it is W_1a, and the moment for W_2 is W_2c about the hanger to the right of it. The weight of the pipe is assumed to be concentrated at a point midway between the hangers, and the moment is (weight of pipe, lb)(distance between hangers, ft/2). The method given here was developed by Frank Kamarck, Mechanical Engineer, and reported in *Power* magazine.

Hanger	Load, lb	Rod diameter, in.
H-1	360	$\frac{3}{8}$
H-2	360	$\frac{3}{8}$
H-3	1,280	2-$\frac{1}{2}$ each
H-4	360	$\frac{3}{8}$
H-5	560	$\frac{1}{2}$

HANGER SPACING AND PIPE SLOPE FOR AN ALLOWABLE STRESS

An 8-in. schedule 40 water pipe has an allowable bending stress of 10,000 psi. What is the maximum allowable distance between hangers for this pipe? What slope will the allowable hanger span require to prevent pocketing of water in the pipe? Describe the method for computing hanger span and pipe slope for empty pipe. How are hanger distances computed when the pipe contains concentrated loads?

Calculation Procedure:

1. *Compute the allowable span between hangers*

For a pipe filled with water, $S = WL^2/8m$, where S = bending stress in pipe, psi; W = weight of pipe and water lb/lin in.; L = maximum allowable distance between hangers, in.; m = section modulus of pipe, in.^{3}. Using a table of pipe properties, as in King and Crocker—*Piping Handbook*, $L = (8mS/W)^{0.5} = (8 \times 16.81 \times 10,000/4.18)^{0.5} = $ 568 in., or 568/12 = 47.4 ft.

2. Compute the pipe slope required by the span

To prevent pocketing of water or condensate at the low point in the pipe, the pipe must be pitched so that the outlet is lower than the lowest point in the span. When the pipe has no concentrated loads—such as valves, cross connections, or meters—the deflection of the pipe is y in. $= 22.5wl^4/EI$, where w = weight of the pipe and its contents, lb/ft; l = distance between hangers, ft; E = modulus of elasticity of pipe, psi = 30×10^6 for steel; I = moment of inertia of the pipe, in.^{4}. Substituting values, $y = (22.5)(50.24)(47.4)^4/[(30 \times 10^6)(72.5)] = 2.61$ in.

With the deflection y known, the pipe slope, expressed as 1 in. per G ft of pipe length, is 1 in. per G ft = $\frac{1}{4}y$, or $G = (47.4)/[(4)(2.61)] = 4.53$. Thus, a pipe slope of 1 in. in 4.53 ft is necessary to prevent pocketing of the water when the hanger span is 47.4 ft. With this slope, the outlet of the pipe would be 47.4/4.53 = 10.45 in. below the inlet.

3. Compute the empty-pipe hanger span and pipe slope

Use the same procedure as in Steps 1 and 2, except that the empty weight of the pipe is substituted in the equations instead of the weight of the pipe when full of water. For pipes containing steam, gas, or vapor, compute the flowing-fluid weight and add it to the pipe weight. Follow the same procedure for insulated pipes, adding the insulation weight to the pipe weight.

4. Determine the hanger span and slope with concentrated loads

Hanger span and pipe slope can be computed using standard beam relations. However, most piping designers use the deflection chart and deflection factors for concentrated loads in King and Crocker—*Piping Handbook*. The chart and correction factors simplify the calculations considerably. The computation involves only simple multiplication and division.

Related Calculations: Use this procedure for piping in any type of installation—power, process, marine, industrial, or utility, for any type of liquid, vapor, or gas.

EFFECT OF COLD SPRING ON PIPE ANCHOR FORCES AND STRESSES

A carbon molybdenum pipe operates at 800 F and has an anchor force of 5,000 lb and a maximum bending stress s_b of 15,000 psi without cold spring. Compute the anchor force and bending stress in the hot and cold condition when the pipe is cold-sprung an amount equal to the expansion e and $0.5e$. The total expansion of the pipe is 24 in.

Calculation Procedure:

1. Compute the hot-condition force and stress

The allowable cold-spring adjustment is expressed as a ratio $(e - 2S/3)/e$, where e = the total expansion of the pipe, in.; S = cold-spring distance, in. This ratio is multiplied by the original anchor force and bending stress at the maximum operating temperature *without* cold spring to find the anchor force and bending stress *with* cold spring in the hot condition. If the ratio is less than 2/3, the value of 2/3 is used where maximum credit for cold spring is desired.

For this pipe, with maximum cold spring, the ratio = $(24 - 2 \times 24/3)/24 = 1/3$. Since this is less than 2/3, use 2/3. Then, the anchor force $F = (2/3)(5,000) = 3,333$ lb, and the bending stress $s_b = (2/3)(15,000) = 10,000$ psi.

With $S = 0.5e = (0.5)(24) = 12$ in., the ratio = $(24 - 2 \times 12/3)/24 = 2/3$. Hence, $F = (2/3)(5,000) = 3,333$ lb; $s_b = (2/3)(15,000) = 10,000$ psi.

2. Compute the cold-condition force and stress

For the cold condition, the adjustment ratio = $-S/eM_R$, where M_R = modulus ratio for the pipe material = modulus of elasticity, psi, of the pipe material at the operating temperature, F/modulus of elasticity of the pipe material, psi, at 70 F. For this pipe, $M_R = 0.865$, from a table of pipe properties. The minus sign in the ratio indicates that the anchor force and stress are reversed in the cold condition as compared with the hot condition.

For this pipe, with maximum cold spring, the ratio $= -24/[(24)(0.865)] = -1.156$. Then, the anchor force in the cold condition $= (-1.156)(15,000) = -5,790$ lb, and the bending stress $= (-1.156)(15,000) = -17,350$ psi.

With $S = 0.5e = (0.5)(24) = 12$ in., the ratio $= -12/[(24)(0.865)] = -0.578$. Then, the anchor force in the cold condition $(-0.578)(5,000) = -2,895$ lb, and the bending stress $= (-0.578)(15,000) = -8,670$ psi.

These calculations show that cold spring reduces the anchor force and bending stress when the pipe is in the hot condition, Step 1. With a cold spring of one-half the pipe expansion, the anchor force and bending stress are reduced and reversed when in the cold condition, Step 2. When the cold spring equals the expansion, the anchor force and bending stress increase in the cold condition, Step 2.

Related Calculations: Use this procedure for a pipe conveying steam, oil, gas, water, and similar vapors, liquids, and gases.

REACTING FORCES AND BENDING STRESS IN SINGLE-PLANE PIPE BEND

Determine the horizontal and vertical reacting forces in the single-plane pipe bend of Fig. 20 if the pipe is 6-in. schedule 40 carbon steel A106 seamless operating at 500 F. What is the maximum bending stress in the pipe and the resultant reacting or anchor force? Determine the maximum bending stress if a long-radius welded elbow is used at point C, Fig. 20. Use the tabular method of solution.

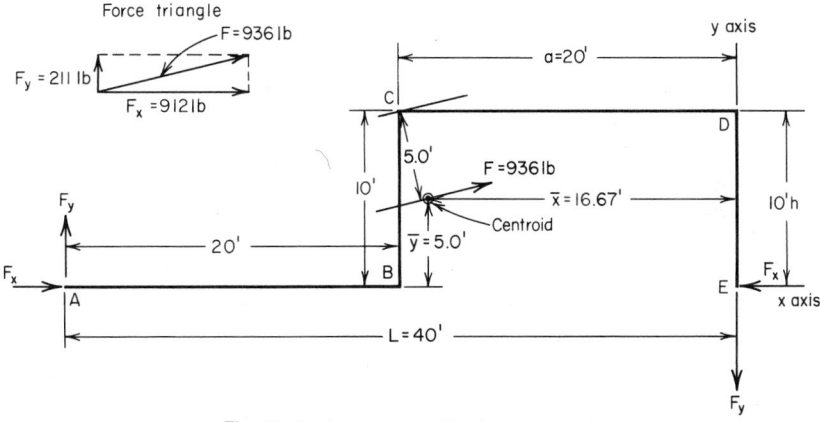

Fig. 20 U-shaped pipe with single tangent.

Calculation Procedure:

1. Compute the horizontal reacting or anchor force

Several methods are available for determining the reacting or anchor forces and maximum bending stress in a single-plane pipe bend. King and Crocker—*Piping Handbook* presents simplified, analytical, and graphical methods for computing forces and stresses in single- and multiplane piping systems. Another useful reference, *Design of Piping Systems*, written by members of the engineering departments of the M. W. Kellogg Company, presents both simplified and analytical methods and an excellent history and discussion of piping flexibility analysis. Probably the simplest method for routine piping flexibility analyses is that developed by the Grinnell

Company, Inc., and S. W. Spielvogel. This method uses tabulated constants for specific pipe shapes in one, two, and three planes. It is satisfactory for the majority of piping problems met in normal engineering practice. To assist the practicing engineer, a number of Grinnell-Spielvogel tabulations for common pipe shapes are included here. For uncommon pipe shapes, refer to Grinnell Company — *Piping Design and Engineering* or to Spielvogel — *Piping Stress Calculations Simplified*. Both these references contain complete tabulations for a variety of pipe shapes.

To apply the Grinnell-Spielvogel solution procedure, compute the horizontal reaction force F_x lb from $F_x = k_x c I_p / L^2$, where K_x = a constant from Table 28 for the bend shape shown in Fig. 20; c = expansion factor = (pipe expansion, in./100 ft)($E M_R$/172,800), where E = modulus of elasticity of the pipe material being used, psi; M_R = modulus ratio = E at the operating temperature, F/E at 70 F; I_p = moment of inertia of pipe cross section, in.[4]; L = length of bend, ft, as shown in Fig. 20.

To enter Table 28 for the shape in Fig. 20, the values of L/a and L/h must be known, or $L/a = 40/20 = 2$; $L/h = 40/10 = 4$. Entering Table 28 at these values, read $k_x = 91$; $k_y = 21$; $K_b = 120$. From the Spielvogel c table, or by computation, $c = 570$ for carbon-steel pipe operating at 500 F. From a table of pipe properties, $I_p = 28.14$ in.[4] for 6-in. schedule 40 pipe. Then, $F_x = (91)(570)(28.14)/(40)^2 = 912$ lb.

2. Compute the vertical reacting or anchor force

Use the same procedure as in Step 1, except that the vertical reacting force F_y lb = $k_y c I_p / L^2$, or $F_y = (21)(570)(28.14)/40)^2 = 211$ lb, using the appropriate value from Table 28.

3. Compute the resultant reacting or anchor force

The resultant reacting or anchor force F lb is found by drawing and solving the force triangle in Fig. 20. Using the pythagorean theorem $F = (912^2 + 211^2)^{0.5} = 936$ lb. Draw the force triangle to scale, as shown in Fig. 20.

4. Compute the maximum bending stress in the pipe

The pipe bending stress s_b psi is found in a similar manner from $s_b = k_b c D / L$, where k_b = bending-stress factor from Table 28; D = outside diameter of pipe, in. For 6-in. schedule 40 pipe, having an outside diameter of 6.625 in., $s_b = (120)(570)(6.625)/40 = 11,330$ psi.

5. Determine the bending stress in the welded elbow

The tables presented here are accurate when all the turns in the piping system analyzed are miters or rigid fittings. When all the turns are welded elbows or bends, the anchor forces derived from Table 28 are accurate for practical systems. The actual forces will be somewhat smaller than the values obtained from Table 28. Stresses in the elbows or bends may, however, exceed the values computed from Table 28 if the stress intensification factor β for these curved sections is > 1. If the proportion of the straight to curved pipe is large, use the following procedure to obtain a close approximation of the stress in the curved section.

Determine the value of β from a table of pipe properties. For a 6-in. schedule 40 long-radius welded elbow, B = 2.22. Therefore, the actual stress may exceed the table-computed stress, because $\beta > 1$.

Lay out the pipe bend to scale and compute the centroid of the bend by taking line moments about the x and y axes, Fig. 20.

	x-axis			y-axis	
AB	(20')(0')	= 0	AB	(20')(30') =	600
BC	(10')(5')	= 50	BC	(10')(20') =	200
CD	(20')(10')	= 200	CD	(20')(10') =	200
DE	(10')(5')	= 50	DE	(10')(0') =	0
	60	300		60	1,000
	$\bar{y} = 300/60 = 5$ ft			$\bar{x} = 1,000/60 = 16.67$ ft	

TABLE 28   U Shape with Single Tangent

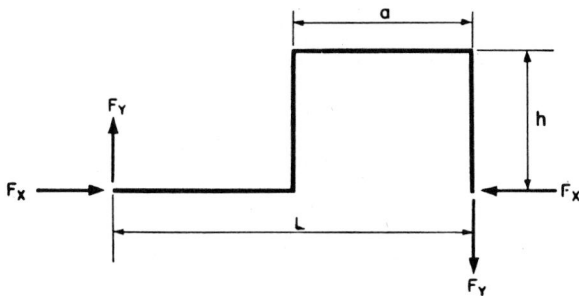

Reacting Force $\qquad F_x = k_x \cdot c \cdot \dfrac{I_P}{L^2}$ lb

Reacting Force $\qquad F_y = k_y \cdot c \cdot \dfrac{I_P}{L^2}$ lb

Maximum Bending Stress $\quad s_B = k_b \cdot c \cdot \dfrac{D}{L}$

I_P in inches^4 $\qquad L$ in feet $\qquad D$ in inches

					L/a				
L/h		1.5			2			3	
	k_x	k_y	k_b	k_x	k_y	k_b	k_x	k_y	k_b
1.0	2.63	0.75	10.5	2.8	1.41	11.3	3.3	2.3	12.5
1.2	4.0	1.27	15.0	4.8	2.28	15	6	3.5	18.3
1.4	6.0	1.79	19.5	7	3.15	20	9	4.8	24.1
1.6	8.0	2.31	24.0	9	4.0	25	12	6.0	30
1.8	11.0	2.83	28.5	12	4.9	30	16	7.2	36
2.0	14.5	3.4	33.6	16	5.8	38	20	8.4	42
2.2	18	4.2	39	20	7.1	44	25	10.0	49
2.4	22	5.0	46	24	8.4	50	30	11.5	56
2.6	26	5.9	53	30	9.8	58	36	13.0	63
2.8	32	6.8	60	37	11.0	66	44	14.5	70
3.0	39	7.7	67	45	12.4	75	53	16	77
3.2	45	8.8	74	52	14.1	82	62	18	87
3.4	51	9.9	81	60	15.8	90	71	20	96
3.6	59	11.0	89	70	17.5	99	81	22	105
3.8	69	12.0	98	80	19	109	94	24	114
4.0	79	13.5	108	91	21	120	108	26	124
4.2	89	15.0	118	102	23	130	121	28	134
4.4	100	16.5	128	114	25	140	135	30	144
4.6	111	18.0	138	128	27	151	150	33	154
4.8	124	19.9	148	142	29	162	167	35	164
5.0	139	21.8	159	156	31	173	185	37	174

In this calculation, the first value is the length of the pipe segment and the second value is the distance of the center of gravity of the segment from the axis. For a straight section of pipe, the center of gravity is taken as the midpoint of the pipe section. The welded elbows are ignored in stress calculations based on table values.

Lay off $\bar{x}$ and $\bar{y}$ to scale, Fig. 20. Scale the distance to the tangent to the centerline of the long-radius elbow at C. This distance $d = 5.0$ ft. The moment at point C, $M_C = Fd$ lb-ft; or $m_C = 12\ Fd$ lb-in., and $m_C = 12(936)(5) = 56{,}160$ lb-in., after force F is transposed from the force triangle to the centroid, Fig. 20.

TABLE 29   90° Turn

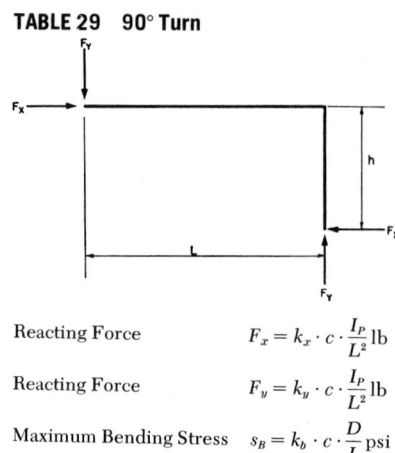

Reacting Force	$F_x = k_x \cdot c \cdot \dfrac{I_p}{L^2}$ lb
Reacting Force	$F_y = k_y \cdot c \cdot \dfrac{I_p}{L^2}$ lb
Maximum Bending Stress	$s_B = k_b \cdot c \cdot \dfrac{D}{L}$ psi

I_p in inches⁴ L in feet D in inches

L/h	k_x	k_y	k_b
1.0	12.0	12.0	36
1.2	17.2	12.5	46
1.4	23.0	13.4	58
1.6	32.0	14.4	71
1.8	42.0	15.4	85
2.0	54.0	16.6	102
2.2	68.3	17.8	120
2.4	84.4	19.2	140
2.6	103	20.6	161
2.8	125	22.0	184
3.0	150	23.5	209
3.2	175	25.0	234
3.4	207	26.5	259
3.6	237	28.0	287
3.8	274	29.5	318
4.0	315	31.5	349
4.2	356	33.0	381
4.4	406	34.6	414
4.6	456	36.2	450
4.8	510	37.8	487
5.0	570	39.5	528

The bending stress at any point in a pipe is $s_b = m\beta/S_m$, where S_m = section modulus of the pipe cross section, in.^{3}. For 6-in. schedule 40 pipe, $S_m = 8.50$ in.^{3}, from a table of pipe properties. Then, $s_b = (56,160)(2.22)/8.50 = 14,700$ psi. This is somewhat greater than the 11,330 psi computed in Step 4 but within the allowable stress of 15,000 psi for seamless carbon steel A106 pipe at 500 F.

By inspection of the scale drawing, Fig. 20, the stress in the long-radius elbows at B and D is less than at C because the moment arm at each of these points is less than at C.

Related Calculations: Tables 29, 30, and 31 present Grinnell-Spielvogel reaction and stress factors for three other single-plane bends — 90° turn, U shape with equal tangents,

TABLE 30   U Shape with Equal Tangents

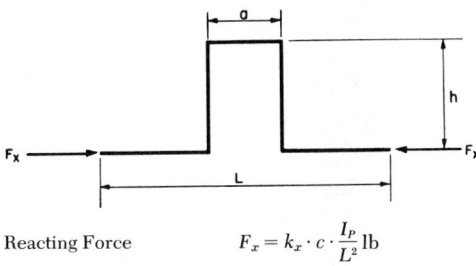

Reacting Force $\qquad\qquad F_x = k_x \cdot c \cdot \dfrac{I_p}{L^2}$ lb

Maximum Bending Stress $\quad s_B = k_b \cdot c \cdot \dfrac{D}{L}$ psi

I_p in inches^4 L in feet D in inches

					L/a					
L/h	2		3		4		5		6	
	k_x	k_b	k_x	k_b	k_x	k_b	k_x	k_b	k_x	k_b
1.0	2.40	7.20	2.46	8.2	2.52	8.82	2.58	9.29	2.64	9.69
1.2	3.70	9.25	4.46	10.9	4.65	12.0	4.78	12.8	4.84	13.3
1.4	5.31	11.37	6.46	13.6	6.79	15.2	6.98	16.3	7.1	17.0
1.6	7.22	13.53	8.46	16.3	8.93	18.4	9.20	19.8	9.5	20.8
1.8	9.45	15.75	10.48	19.0	11.08	21.6	11.42	23.4	11.9	24.7
2.0	12.00	18.00	12.5	21.8	13.24	24.8	13.87	27.1	14.4	28.8
2.2	14.85	20.25	15.8	24.9	16.6	28.5	16.9	31.0	17.5	33.4
2.4	18.00	22.50	19.6	28.0	20.4	32.2	20.8	35.3	21.3	38.0
2.6	21.52	24.83	23.4	31.1	24.4	35.9	25.5	39.7	26.2	42.7
2.8	25.32	27.10	27.3	34.2	28.9	39.7	30.6	44.0	31.7	47.5
3.0	29.45	29.45	31.2	37.4	33.6	43.7	35.8	48.7	37.7	52.7
3.2	33.9	31.8	35.6	40.6	39.0	47.6	41.2	53.3	43.7	58.0
3.4	38.7	34.1	40.0	43.8	44.5	51.6	46.9	58.0	49.5	63.3
3.6	43.7	36.5	46.1	47.0	50.3	55.6	53.0	62.8	57.5	68.7
3.8	49.1	38.8	52.3	50.2	57.0	59.8	60.2	67.6	65.5	74.1
4.0	54.9	41.1	58.5	53.6	64.0	64.0	69.1	72.5	73.6	79.7
4.2	60.8	43.4	64.7	57.0	71.1	68.2	78.1	77.5	82.0	85.2
4.4	67.3	45.9	71.0	60.4	78.9	72.4	87.2	82.5	91.0	90.8
4.6	73.9	48.2	79.1	63.8	87.0	76.6	96.3	87.5	101.7	96.3
4.8	81.0	50.6	87.2	67.3	95.8	80.8	105.4	92.5	112.4	101.9
5.0	88.2	52.9	95.3	70.8	104.6	85.2	114.7	97.8	122.5	107.5

TABLE 31   U Shape Unequal Legs

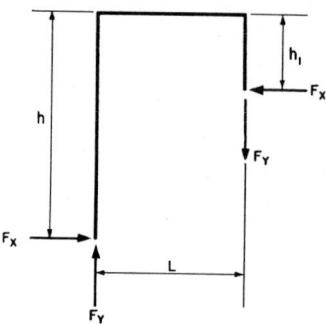

Reacting Force $F_x = k_x \cdot c \cdot \dfrac{I_P}{L^2}$ lb

Reacting Force $F_y = k_y \cdot c \cdot \dfrac{I_P}{L^2}$ lb

Maximum Bending Stress $s_B = k_b \cdot c \cdot \dfrac{D}{L}$ psi

I_P in inches^4 L in feet D in inches

L/h	h/h₁								
	4/3			2			3		
	k_x	k_y	k_b	k_x	k_y	k_b	k_x	k_y	k_b
0.2	0.07	0.6	1.5	0.29	1.8	7	0.53	3.4	11
0.4	0.60	0.7	3.0	0.75	2.0	8	1.4	3.5	13
0.6	1.15	0.8	5.8	1.9	2.2	11	2.7	3.8	15
0.8	2.4	0.9	9.5	3.6	2.5	15	4.8	4.4	20
1.0	4.3	1.2	16	6.2	3.0	21	8	4.9	26
1.2	6	1.4	21	8	3.6	29	10	5.7	34
1.4	9	1.6	28	11	4.2	39	15	6.5	43
1.6	13	1.9	36	18	4.8	50	22	7.3	56
1.8	19	2.1	45	27	5.4	62	32	8.1	72
2.0	27	2.3	58	37	6.0	75	44	8.0	88
2.2	35	2.6	68	48	6.6	90	57	10.0	104
2.4	43	2.9	80	60	7.3	106	71	11.0	121
2.6	52	3.2	93	75	8.0	123	87	12.5	140
2.8	65	3.5	108	91	9.0	142	105	13.5	162
3.0	81	3.8	124	110	10.0	162	128	15	185

and U shape with unequal legs. Use these tables and the factors in them in the same way as described above. Correct for curved elbows in the same manner. The tables can be used for piping conveying steam, water, gas, oil, and similar liquids, vapors, or gases. For bends of different shape, the analytical method must be used.

REACTING FORCES AND BENDING STRESS IN A TWO-PLANE PIPE BEND

Determine the horizontal reacting forces and bending and torsional stresses in the two-plane pipe bend shown in Table 32 if the dimensions of the bend are: $L = 20$ ft; $h = 5$ ft; $a = 5$ ft; $b = 5$ ft. Use the tabular method of solution. The pipe is a 10-in. carbon steel schedule 80 line operating at 750 psig and 750 F. Determine the combined stress in the pipe.

Table 32   Two Plane U with Tangents

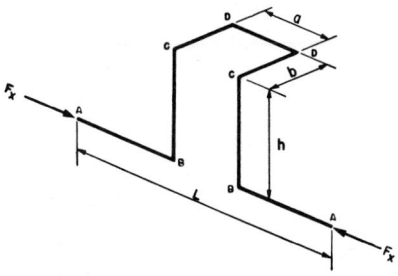

Reacting Force $F_x = k_x \cdot c \cdot \dfrac{I_P}{L^2}$ lb

Bending Stress $s_B = k_B \cdot c \cdot \dfrac{D}{L}$ psi

Torsional Stress $s_T = k_T \cdot c \cdot \dfrac{D}{L}$ psi

I_P in inches^4 L in feet D in inches

L/h	a/b						$L/a = 4$					
	0.25			0.5			1			2		
	k_x	k_b	k_t	k_x	k_b	k_t	k_x	k_b	k_t	k_x	k_b	k_t
1	0.67	$\overset{D}{3.20}$	...	1.22	$\overset{A}{4.35}$	$\overset{A}{0.30}$	1.67	$\overset{A}{5.2}$	$\overset{A}{0.15}$	2.0	$\overset{C}{6.3}$	
2	1.35	$\overset{D}{5.80}$	...	4.30	$\overset{D}{9.96}$	$\overset{D}{2.45}$	6.96	$\overset{A}{11.0}$	...	9.3	$\overset{C}{15.0}$	
3	1.70	$\overset{D}{7.00}$	...	6.23	$\overset{D}{13.8}$	$\overset{D}{2.28}$	14.0	$\overset{D}{16.5}$	6.55	21.2	$\overset{C}{24.0}$	
4	1.88	$\overset{D}{7.44}$	...	7.84	$\overset{D}{16.9}$	$\overset{D}{2.09}$	21.3	$\overset{D}{24.5}$	7.40	36.2	$\overset{C}{30.0}$	
5	2.01	$\overset{D}{7.75}$	...	8.94	$\overset{D}{18.8}$	$\overset{D}{1.89}$	27.8	31.4	$\overset{D}{7.75}$	52.6	$\overset{D}{31.0}$	$\overset{D}{17.3}$

Calculation Procedure:

1. Compute the tabular factors for the pipe bend

To apply the Grinnell-Spielvogel method to two-plane pipe bends, three tabular factors are required: L/a; a/b; and L/h. Using the given values, $L/a = 20/5 = 4$; $a/b = 5/5 = 1$; $L/h = 20/5 = 4$.

2. Determine the force and stress factors for the pipe

From Table 32, for the factors in Step 1, $k_x = 21.3$; $k_b = 24.5$; $k_t = 7.40$.

3. Compute the horizontal reacting force of the bend

The horizontal reacting force $F_x = k_x c I_p / L^2$, where the symbols are the same as in the preceding Calculation Procedure, except for L. Substituting the values for 10-in. carbon-steel schedule 80 pipe operating at 750 F, $F_x = (21.3)(874)(244.9)/(20)^2 = 11,380$ lb.

4. Compute the bending stress in the pipe

The bending stress in the pipe is found from $s_b = k\, cD/L$, where the symbols are the same as in the previous Calculation Procedure. Substituting values $s_b = (24.5)(874)(10.75)/(20) = 11,510$ psi. Table 32 shows that the maximum combined stress in the pipe occurs at the two upper bends, D.

5. Compute the torsional stress in the pipe

The torsional stress in the pipe is found from $s_t = k_t cD/L$, where the symbols are the same as in the previous Calculation Procedure. Substituting values, $s_t = (7.40)(874)(10.75)/20 = 3,475$ psi. Table 32 shows that the maximum combined stress in the pipe occurs at the two upper bends, D.

6. Determine the combined stress in the pipe

For any multiplane piping system, the combined stress s_{co} psi $= 0.5\{s_1 + s_c + [4s_t^2 + (s_1 - s_c)^2]^{0.5}\}$. In this equation, $s_1 = s_b + s_p$, where $s_p =$ pressure due to internal pressure, psi; $s_c =$ circumferential or hoop stress, psi; other symbols are as given earlier. Also, $s_p = pA_i/A_m$, where $p =$ operating pressure, psig; $A_i =$ inside area of pipe cross section, sq in.; $A_m =$ metal area of pipe cross section, sq in. Likewise, $s_c = p(D - t)/2t$, where $D =$ outside diameter of pipe, in.; $t =$ pipe-wall thickness, in.

Computing stress values for this 10-in. schedule 80 carbon-steel pipe operating at 750 psig and 750 F and using values from a table of pipe properties, $s_p = (750)(71.8/18.92) = 2,845$ psi; $s_c = (750)(10.75 - 0.593)/2(0.593) = 6,420$ psi. Then, $s_1 = s_b + s_p = 11,510 + 2,845 = 14,355$ psi, where s_b is from Step 4. Substituting in the combined-stress equation, $s_{co} = 0.5\{14,355 + 6,420 + [4 \times 3,475^2 + (14,355 - 6,420)^2]^{0.5}\} = 16,648$ psi. This is higher than the stress allowed in carbon-steel pipe by the ANSA *Piping Code*, unless the pipe conforms to the special conditions of certain paragraphs of the *Code*.

Related Calculations: Use this procedure for piping conveying steam, water, gas, oil, and similar liquids, vapors, or gases. For bends of different shape, the analytical method is generally used.

REACTING FORCES AND BENDING STRESS IN A THREE-PLANE PIPE BEND

Determine the three reacting forces and moments and bending and torsional stresses in the three-plane pipe bend shown in Table 33 if the dimensions of the bend are $L_1 = 20$ ft, $L_2 = 10$ ft, $L_3 = 5$ ft. The pipe is 10-in. carbon-steel schedule 80 operating at 750 psig and 750 F.

Calculation Procedure:

1. Compute the tabular factors for the pipe bend

Using the Grinnell-Spielvogel method, the two tabular factors required are $m = L_1/L_3$ and $n = L_2/L_3$; or $m = 20/5 = 4$ and $n = 10/5 = 2$.

2. Determine the force and stress factors for the pipe

From Table 33, for the factors in Step 1, $k_b = 8.0$; $k_t = 3.6$; $k_x = 1.48$; $k_y = 0.13$; $k_z = 0.80$; $k_{xy} = 1.3$; $k_{xz} = 1.2$; $k_{yz} = 0.51$.

3. Compute the longitudinal reaction force of the bend

The horizontal reacting force $F_x = k_x C i_p / L_3^2$, where the symbols are the same as in the preceding Calculation Procedure, except for L_3. Substituting values for 10-in. carbon-steel schedule 80 pipe operating at 750 F, $F_x = (1.48)(874)(244.9)/(5)^2 = 12,680$ lb.

TABLE 35   Three-Dimensional 90° Turns

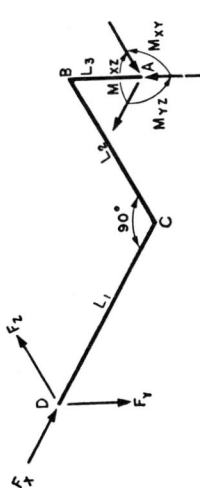

$L_1 \geq L_3$ $L_1 = m$ $\dfrac{L_2}{L_3} = n$

Bending Stress $s_B = k_b \cdot c \cdot \dfrac{D}{L}$ psi

Torsional Stress $s_T = k_t \cdot c \cdot \dfrac{D}{L_3}$ psi

Reacting Force $F_x = k_x \cdot c \cdot \dfrac{I_p}{L_3^2}$ lb

Reacting Force $F_y = k_y \cdot c \cdot \dfrac{I_p}{L_3^2}$ lb

Reacting Force $F_z = k_z \cdot c \cdot \dfrac{I_p}{L_3^2}$ lb

Reacting Moment $M_{xy} = k_{xy} \cdot x \cdot \dfrac{I_p}{L_3}$ ft lb

Reacting Moment $M_{xx} = k_{xx} \cdot c \cdot \dfrac{I_p}{L_3}$ ft lb

Reacting Moment $M_{yz} = k_{yz} \cdot c \cdot \dfrac{I_p}{L_3}$ ft lb

I_p in inches[4] L in feet D in inches

$m = 3$

n	k_b	k_t	k_x	k_y	k_z	k_{xy}	k_{xz}	k_{yz}
0.25	56.7 (A)	3.3 (A)	12.8	1.78	1.04	9.5	1.10	0.59
0.50	40.3 (A)	4.9 (A)	8.7	1.12	1.40	6.7	1.64	0.84
0.75	28.7 (A)	5.0 (A)	6.1	0.77	1.54	4.7	1.68	0.96
1	22.3 (A)	4.86 (A)	4.5	0.54	1.50	3.6	1.62	0.74
2	9.3 (D)	0.15 (D)	1.4	0.22	1.10	1.1	1.00	0.71
3	10.0 (D)	0.24 (D)	0.76	0.13	1.08	0.60	0.74	0.70
4	12.0 (D)	0.11	0.52	0.095	1.14	0.44	0.69	0.82

$m = 4$

k_b	k_t	k_x	k_y	k_z	k_{xy}	k_{xz}	k_{yz}	n
72.3 (A)	5.0 (A)	15.6	1.37	0.84	12.0	1.7	0.49	0.25
50.5 (A)	6.8 (A)	10.5	0.85	1.13	8.4	2.3	0.70	0.50
32.6 (A)	6.7 (A)	6.65	0.52	1.12	5.4	2.2	0.72	0.75
24.0 (A)	6.0 (A)	4.80	0.37	1.10	3.9	2.0	0.61	1
8.0 (A)	3.6 (A)	1.48	0.13	0.80	1.3	1.2	0.51	2
7.26 (D)	0.10 (D)	0.76	0.09	0.65	0.6	0.88	0.42	3
7.98 (D)	0.09 (D)	0.47	0.055	0.64	0.4	0.66	0.45	4

4. Compute the vertical reacting force of the bend

The vertical reacting force $F_y = k_y c I_p L_3{}^2$, where the symbols are the same as in the preceding Calculation Procedure, except for L_3. Substituting values for this pipe, $F_y = (0.13)(874)(244.9)/(5)^2 = 1,115$ lb.

5. Compute the horizontal reacting force of the bend

The horizontal reacting force $F_z = k_z c I_p/L_3{}^2$, where the symbols are the same as in the preceding Calculation Procedure except for K_z and L_3. Substituting values for this pipe, $F_z = (0.80)(874)(244.9)/(5)^2 = 6,850$ lb.

6. Compute the bending and torsional stresses in the pipe

The bending stress $s_b = k_b c D/L_3$, where the symbols are the same as in the preceding Calculation Procedure, except for L_3. Substituting values for this pipe, $s_b = (8.0)(874)(10.75)/5 = 15,020$ psi.

The torsional stress $s_t = k_t c D/L_3$, where the symbols are the same as in the preceding Calculation Procedure, except for L_3. Substituting values for this pipe, $s_t = (3.6)(874)(10.75)/5 = 6,760$ psi.

7. Compute the three reacting moments at the pipe end

For each bending moment M ft-lb $= kc I_p/L_3$, where the symbols are the same as given in the previous steps in this Calculation Procedure, except that k is the appropriate bending-moment factor.

For the xy moment $M_{xy} = (1.3)(874)(244.9)/5 = 55,700$ ft-lb. For the xz moment $M_{xz} = (1.2)(874)(244.9)/5 = 51,400$ ft-lb. For the yz moment $M_{yz} = (0.51)(874)(244.9)/5 = 22,250$ ft-lb.

Related Calculations: Use this procedure for piping conveying steam, water, gas, oil, and similar liquids, vapors, or gases. For bends of different shape, the analytical method must be used. Compute the combined stress in the same way as in Step 6 of the previous Calculation Procedure. Table 33 shows that the maximum combined stress occurs at point A in this piping system.

ANCHOR FORCE, STRESS, AND DEFLECTION OF EXPANSION BENDS

Determine the deflection and anchor force in an 8-in. schedule 40 double-offset expansion U bend having a radius of 64 in. if the bending stress is 10,000 psi. What would the deflection and anchor force be with a bending stress of 15,000 psi if the bend tangents are guided and the pipe is carbon steel operating at 500 F? With a bending stress of 8,000 psi? Tabulate the deflection and anchor-force equations for the popular types of expansion bends when the expanding pipe is guided axially.

Calculation Procedure:

1. Compute the deflection of the pipe bend

For a double-offset expansion U bend, the deflection d in. $= 0.728 R^2 K/D\beta$, where $R =$ bend radius, ft; $K =$ flexibility factor for curved pipe, from a table of pipe properties, or from $K = (12\lambda^2 + 10)/(12\lambda^2 + 1)$, where $\lambda = tR/r^2$, where $t =$ pipe thickness, in., $r = (D - t)/2$; $D =$ outside diameter of pipe, in.; $\beta =$ stress coefficient for curved pipe from a table of pipe properties or from $\beta = (2K/3)[(6\lambda^2 + 5)/18]^{0.5}$ when $\lambda \leqslant 1.47$; $\beta = (12\lambda^2 - 2)/(12\lambda^2 + 1)$ when $\lambda > 1.47$.

For this bend, $R = 64/12 = 5.33$ ft; $K = 1.49$ from a table of pipe properties or by computation; $D = 8.625$ in. from a table of pipe properties; $\beta = 0.86$ from a table of pipe properties, or by computation. Then, $d = (0.728)(5.33)^2(1.49)/[(8.625)(0.86)] = 4.15$ in.

2. Compute the anchor force of the pipe bend

For a double-offset expansion U bend, the anchor force F_x lb $= 976\, I_p/RD\beta$, where $I_p =$ moment of inertia of pipe cross section, in.^{4}. For this pipe, use values from a table of pipe properties; or computing the values, $F_x = (976)(72.5)/[(5.33)(8.625)(0.86)] = 1,790$ lb.

3. Compute the deflection and anchor force for a larger bending stress

With a larger bending stress — 15,000 psi in this instance — and a greater deflection at the higher stress d_h, solve $d_h = (d)$(allowable stress, psi)$/10,000$ M_R, or $d_h = (4.15)(15,000)/[10,000(0.932)] = 6.68$ in.

The anchor force at the larger bending stress is $F_h = F_x d_h M_R/d$, or $F_x = (1,790)(6.68)(0.932)/4.15 = 2,680$ lb.

4. Compute the deflection and anchor force for a smaller bending stress

Use the same equations as in Step 3, except that the lower bending stress is substituted for the higher one. Or, $d_1 = (4.15)(8,000)/[10,000(0.932)] = 3.56$ in., and $F_1 = (1,790)(3.56)(0.932)/4.15 = 1,432$ lb.

5. Tabulate the deflection and anchor-force equations

Related Calculations: Use the Procedures given here for piping conveying steam, water, oil, gas, air, and similar vapors, liquids, and gases. The value of E in Step 5, 29×10^6 psi, is satisfactory for pipes made of carbon steel, carbon moly steel, chromium moly steel, nickel steel, and chromium nickel steel. These materials are commonly used in piping systems requiring expansion bends.

Note that the equations in Step 5 apply to pipe bends having guides to direct the axial expansion of the pipe. This is the usual arrangement used today because unguided bends require too much space. For design of unrestrained bends, multiply d by 1.5 when finding the deflection at the higher stress, as in Step 3, This factor, 1.5, is an approximation, but it is on the safe side in almost every case. The equations given in Step 5 are presented in great detail in Grinnell — *Piping Design and Engineering* and King and Crocker — *Piping Handbook*.

Bend type	Deflection for 10,000 psi s_b; $E = 29 \times 10^6$ psi	Anchor force for 10,000 psi s_b; $E = 29 \times 10^6$ psi
Double-offset U.........	$d = 0.728R^2K/D\beta$	$F_x = 976\,I_p/RD\beta$
Expansion U bend (no tangents)	$d = 0.312R^2K/D\beta$	$F_x = 1,667I_p/RD\beta$
Expansion U bend (tangents $= 2$ ft)	$d = [(0.312R^3 + 0.795R^2 + 0.624R)K + 0.132]/(R+1)D\beta$	$F_x = 1,667I_p/(R+1)D\beta$
Expansion U bend (tangents $= R$)	$d = (0.577 + 0.011)R^2/D\beta$	$F_x = 1,111I_p/RD\beta$
Expansion U bend (tangents $= 2R$)	$d = (0.865K + 0.0662)R^2/D\beta$	$F_x = 833I_p/RD\beta$
Expansion U bend (tangents $= 4R$)	$d = (1.465K + 0.353)R^2/D\beta$	$F_x = 556I_p/RD\beta$
Double-offset U bend	$d = 0.260R^2K/D\beta$	$F_x = 1,209I_p/RD\beta$
Single-offset quarter bend..........	$d = 0.0366R^2K/D\beta$	$F_x = 2,763I_p/RD\beta$
Circle bend.............	$d = 0.312R^2K/D\beta$	$\begin{cases} F_y = 0.066F_x \\ F_x = 1,667I_p/RD\beta \end{cases}$

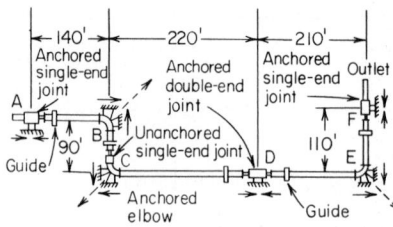

Fig. 21 Slip-type expansion joints in a piping system. (*Yarway Corporation.*)

SLIP-TYPE EXPANSION JOINT SELECTION AND APPLICATION

Select and size slip-type expansion joints for the 20-in. carbon-steel schedule 40 pipeline in Fig. 21 if the pipe conveys 125-psig steam having a temperature of 380 F. The minimum temperature expected in the area where the pipe is installed is 0 F. Determine the anchor loads that can be expected. The steam inlet to the pipe is at A; the outlet is at F.

Calculation Procedure:

1. Determine the expansion of each section of pipe

From Fig. 22, the expansion of steel pipe at 380 F with a 0-F minimum temperature is 3.4 in. per 100 ft of pipe. Expansion of each section of pipe is then e in. = (3.4)(pipe length, ft/100). For AB, e = (3.4)(140/100) = 4.76 in.; for BC, e = (3.4)(90/100) = 3.06 in.; for CD, e = (3.4)(220/100) = 7.48 in.; for DE, e = (3.4)(210/100) = 71.4 in.; for EF, e = (3.4(110/100) = 3.74 in.

2. Select the type and the traverse of each expansion joint

The slip-type expansion joint at A will absorb expansion from only one direction — the right-hand side. This expansion will occur in pipe section AB and is 4.76 in., from Step 1. Therefore, a single-end slip-type expansion joint (one that absorbs expansion on only one side) can be used. The traverse — the amount of expansion a slip joint will absorb — is usually given in multiples of 4 in., that is, 4, 8, and 12 in. Hence, an 8-in. traverse slip-type single-end joint will be suitable at A because the expansion is 4.76 in. A 4-in. traverse joint would be unsatisfactory because it could not absorb at 4.76-in. expansion.

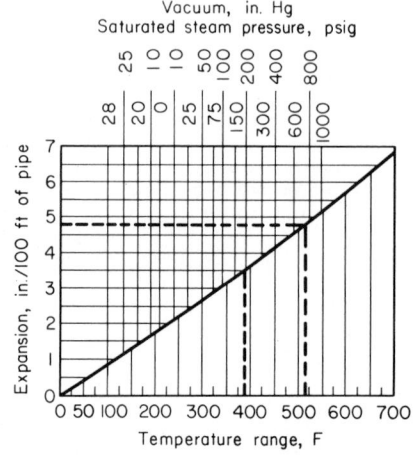

Fig. 22 Expansion of steel pipe. (*Yarway Corporation.*)

The next joint, at C, must absorb the expansion in the vertical pipe BC. Since the elbow beneath the joint is anchored, an unanchored joint can be used. With pipe expansion in only one direction — from B to C — a single-end joint can be used. Since the expansion of section BC is 3.06 in., use a single-end 4-in. traverse slip-type expansion joint, unanchored at C.

The expansion joint at D must absorb expansion from two directions — from C to D and from E to D. Therefore, a double-end joint (one that can absorb expansion on each end) must be used. This double-end joint must be anchored because the pipe expands *away* from the anchored elbow C in section CD and *away* from the anchored elbow E in section DE. In both instances the pipe expands *toward* the expansion joint at D.

The expansion in section CD is, from Step 1, 7.48 in., whereas the expansion in DE is 7.14 in. Therefore, a double-end anchored joint with an 8-in. traverse at *each* end will be suitable.

Since the pipe outlet is at F and there is no anchor in the pipe at F, the expansion joint at this point must be anchored. The pipe section between E and F will expand vertically upward into the joint for a distance of 3.74 in., as computed in Step 1. Therefore, a single-end anchored joint with a 4-in. traverse will be suitable.

3. Compute the anchor loads in the pipeline

Use Fig. 23 to determine the anchor loads on intermediate and end anchors (those where the pipe makes a sharp change in direction). Enter Fig. 23 at the bottom at a pipe size of 20-in. diameter and project vertically upward to the dashed curve labeled intermediate anchor—all pressures. At the left read the anchor load at each intermediate anchor, A, D, and F, as 20,000 lb. Note that the joint expansion load = joint contraction load = 20,000 lb.

The end anchors, B, C, and E, have, from Fig. 23, a possible maximum load of 58,000 lb, found by projecting vertically upward from the 20-in. pipe size to 125-psig steam pressure, which lies midway between the 100- and 150-psig curves. Indicate the possible maximum end-anchor loads by the solid arrows at each elbow, as shown in Fig. 21. The resultant R of the loads at any end anchor is found by the pythagorean theorem to be $R = (58,000^2 + 58,000^2)^{0.5} = 82,200$ lb. Indicate the resultant by a dotted arrow, as shown in Fig. 21.

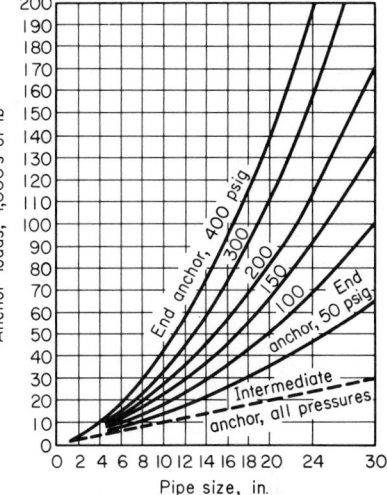

Fig. 23   End- and intermediate-anchor loads in piping systems. (*Yarway Corporation.*)

Contraction loads on the end anchors are in the reverse direction and consist only of friction. This friction load equals the joint expansion load, or 20,000 lb. The resultant of the joint expansion loads is $(20,000^2 + 20,000^2)^{0.5} = 28,350$ lb.

Locate guides within 25 or 12 ft of the expansion joint, depending on the type of packing used, Table 34. These guides should allow free axial movement of the pipe into and out of the joint with minimum friction.

TABLE 34   Guide and Support Spacing

Nominal pipe size, in.	Distance between guide and joint, ft		Distance between guides, ft
	Packing type		
	Gun	Gland	
$1\frac{1}{2}$	8	5	10
2	10	5	13
$2\frac{1}{2}$	11	6	15
3	12	6	19
$3\frac{1}{2}$	13	7	22
	14	7	25
5	15	8	30
6	16	8	35
8	18	9	45
10	20	9	60
12	21	10	70
14	22	10	80
16	23	11	90
18	24	11	100
20	25	12	105
24	26	12	110

Related Calculations: Use this procedure to choose slip-type expansion joints for pipes conveying steam, water, air, oil, gas, and similar vapors, liquids, and gases. In some instances, the gland friction and pressure thrust is used instead of Fig. 23 to determine anchor loads. With either method, the results are about the same.

CORRUGATED EXPANSION JOINT SELECTION AND APPLICATION

Select corrugated expansion joints for the 8-, 6-, and 4-in. carbon-steel pipeline in Fig. 24 if the steam pressure in the pipe is 75 psig, the steam temperature is 340 F, and the installation temperature is 60 F.

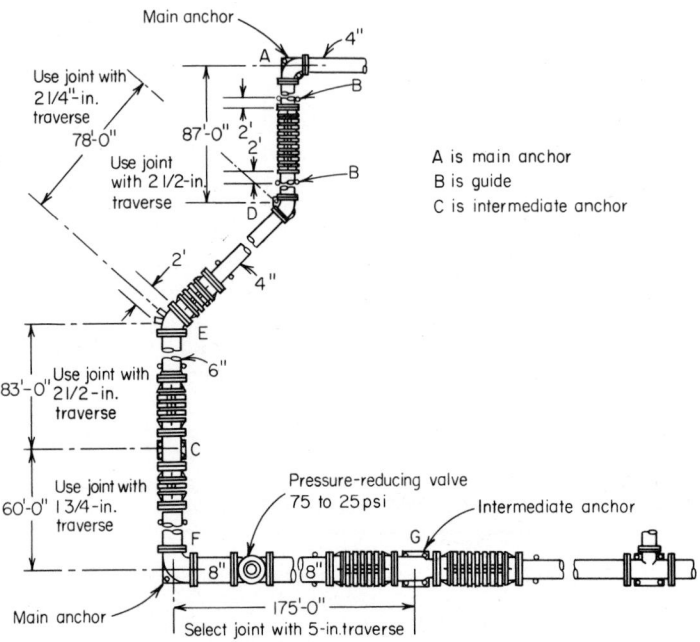

Fig. 24   Piping system fitted with expansion joints. (*Flexonics Division, Universal Oil Products Company.*)

Calculation Procedure:

1. Determine the expansion of each section of pipe

From a table of thermal expansion of pipe, the expansion of carbon-steel pipe at 340 F is 2.717 in./100 ft from 0 to 340 F. Between 0 and 60 F the expansion is 0.448 in./ 100 ft. Hence, the expansion between 60 and 340 F is $2.717 - 0.448 = 2.269$ in./100 ft. This factor can now be applied to each length of pipe by finding the product of (pipe-section length, ft/100)(expansion, in./100 ft) = expansion of section, in. = e.

For section AD, $e = (87/100)(2.269) = 1.97$ in.; for DE, $e = (78/100)(2.269) = 1.77$ in.; for EC, $e = (83/100)(2.269) = 1.88$ in.; for CF, $e = (60/100)(2.269) = 1.36$ in.; for FG, $e = (175/100)(2.269) = 3.97$ in.

In selecting corrugated expansion joints, the usual practice is to increase the computed expansion by a suitable safety factor to allow for any inaccuracies in temperature measurement. Applying a 25 percent safety[1] factor: for AD, $e = (1.97)(1.25) = 2.46$ in.; for DE, $e = (1.77)(1.25) = 2.13$ in.; for EC, $e = (1.88)(1.25) = 2.35$ in.; for CF, $e = (1.36)(1.25) = 1.70$; for FG, $e = (3.97)(1.25) = 4.96$ in.

[1]This value is for illustration purposes only. Contact the expansion-joint manufacturer for the exact value of the safety factor to use.

2. Select the traverse for, and type of, each expansion joint

Obtain corrugated-expansion joint engineering data and select a joint with the next largest traverse for each section of pipe. Thus, traverse $AD \geqslant 2\frac{1}{2}$ in.; traverse $DE \geqslant 2\frac{1}{4}$ in.; traverse $EC \geqslant 2\frac{1}{2}$ in.; traverse $CF \geqslant 1\frac{3}{4}$ in.; traverse $FG \geqslant 5.0$ in.

Two types of expansion joints are commonly used: free-flexing and controlled-flexing. Free-flexing joints are generally used where the pressures in the pipeline are relatively low and the required motion is relatively small. Controlled-flexing expansion joints are generally used for higher pressures and larger motions. Both types of expansion joints are available in stainless steel in both single and dual units. For precise data on a given joint being considered, consult the expansion-joint manufacturer. Corrugated expansion joints are characterized by their freedom from any maintenance needs.

3. Compute the anchor loads in the pipeline

Main anchors are used between expansion joints, as at F and A, Fig. 24, and at turns such as at F and A. The force[1] a main anchor must absorb is given by F_i lb $= F_p + F_e$, where F_p = pressure thrust in the pipe, lb, $= pA$, where p = pressure in pipe, psig; A = effective internal cross-sectional area of expansion joint, sq in.; (see Table 35 for cross-sectional areas of typical corrugated joints); F_e = force required to compress the expansion joint, lb, $= (300$ lb/in.)(joint inside diameter, in.) for stainless-steel self-equalizing joints, and (200 lb/in.)(joint inside diameter, in.) for copper nonequalizing joints. Determining the main anchor force for the 8-in. pipeline, $F_i = (75)(85) + (300) \times (8) = 8{,}775$ lb. In this equation, the area of 85 sq in. in the first term is obtained from Table 35.

TABLE 35   Effective Area of Corrugated Expansion Joints

Joint Inside Diameter, in.	Joint Effective Area, sq in.
3	17.0
4	23.5
5	32.0
6	51.0
8	85.0
10	120.0
12	174.0
14	215.0
16	270.0
18	310.0
20	390.0
24	540.0
30	749.0
36	1187.0
40	2033.0

The total force at a main anchor, as at A and F, Fig. 24, is the vector sum of the forces in each line leading to the anchor. Thus, at F, there is a force of 8,775 lb in the 8-in. line and a force of $F_i = (75)(51) + (300)(6) = 5{,}625$ lb in the 6-in. line connected to the elbow outlet. Since the elbow at F is a right angle, use the pythagorean theorem, or R = resultant anchor force, lb $= (8{,}775^2 + 5{,}625^2)^{0.5} = 10{,}400$ lb.

Where two lines containing corrugated expansion joints are connected by a bend of other than 90°, as at D and E, use a force triangle to determine the anchor force after computing F_i for each pipe. Thus, at E, F_i for the 6-in. pipe $= 5{,}625$ lb, and $F_i = (75) \times$

[1]This is an approximate method for finding the anchor force. For a specific make of expansion joint, consult the joint manufacturer.

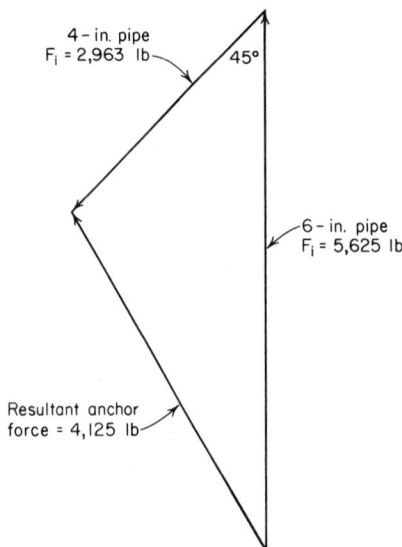

4 - in. pipe
F_i = 2,963 lb

45°

6 - in. pipe
F_i = 5,625 lb

Resultant anchor
force = 4,125 lb

Fig. 25 Force triangle for determining piping anchor forces.

$(23.5) + (300)(4) = 2,963$ lb for the 4-in. pipe. Draw the force triangle in Fig. 25 with the 6-in. pipe F_i and the 4-in. pipe F_i as two sides and the bend angle, 45°, as the included angle. Connect the third side, or resultant, to the ends of the force vectors and scale the resultant as 4,125 lb, or compute the resultant using the law of cosines. Find the resultant force at D in a similar manner as 2,963 lb.

Intermediate anchors, as at C and G, must withstand only one force, the unbalanced (differential) spring force. With approximate force calculations,[1] starting at C, for a 6-in. expansion joint, $F_e = (300)(6) = 1,800$ lb. At G, for an 8-in. expansion joint, $F_e = (300)(8) = 2,400$ lb. Thus, the loads the intermediate anchors must withstand are considerably less than the main anchor loads.

Provide pipe guides at suitable locations in accordance with the joint manufacturer's recommendations and at suitable intervals on the pipeline to prevent any lateral and buckling forces on the joint and adjacent piping. Intermediate anchors between two joints in a straight run of pipe ensure that each joint will absorb its share of the total pipe motion. Slope the pipe in the direction of fluid flow to prevent condensate accumulation. Use enough pipe hangers to prevent sagging of the pipe.

Related Calculations: Use this procedure to choose corrugated-type expansion joints for pipes conveying steam, water, air, oil, gas, and similar vapors, liquids, and gases. When choosing a specific make of corrugated expansion joint, use the manufacturer's engineering data, where available, to determine the maximum allowable traverse. One popular make has a maximum traverse of 7.5 in. or a maximum allowable lateral motion of 1.104 in. in its various joint sizes. The larger the lateral motion, the greater the number of corrugations required in the joint.

In some pipelines there is an appreciable pressure thrust caused by a change in direction of the pipe. This pressure or centrifugal thrust F_c is usually negligible, but the wise designer makes a practice of computing this thrust from $F_c = (2A\rho v^2/32.2) \times (\sin \theta/2)$ lb, where A = inside area of pipe, sq ft; ρ = density of fluid or vapor, lb/cu ft; v = fluid or vapor velocity, fps; θ = change in direction of the pipeline, °.

The number of corrugations required in a joint varies with the expansion and lateral motion to be absorbed. A typical free-flexing joint can absorb 6.25 in. of expansion and a variable amount of lateral motion, depending on joint size and operating condition. Free-flexing joints are commonly built in diameters up to 48 in., while controlled-flexing joints are commonly built in diameters up to 24 in. For a more precise Calculation Procedure, consult the Flexonics Division, Universal Oil Products Company.

DESIGN OF STEAM TRANSMISSION PIPING

Design a steam transmission pipe to supply a load that is 1,700 ft from the power plant. The terrain permits a horizontal run between the power plant and the load. Maximum steam flow required by the load is 300,000 lb/hr, whereas the average steam flow required is estimated as 150,000 lb/hr. The maximum steam pressure at the load must not exceed 150 psia saturated. Superheated steam at 450 psia and 600 F is available at the power plant. Two schemes are proposed for the line: (1) Reducing the steam pres-

[1]Consult the expansion-joint manufacturer for an exact procedure for computing the anchor forces.

sure to 180 psia at the line inlet, thus allowing a $180 - 150 = 30$ psi loss in the 1,700-ft long line. This scheme is called the *nominal pressure-loss line*. (2) Admitting high-pressure steam to the line and thereby allowing the steam pressure to fall to a level slightly greater than 150 psia. Since 600-F steam would probably cause expansion and heat-loss difficulties in the pipe, assume that the inlet temperature of the steam is reduced to 455 F in a desuperheater in the power plant. There is a 10-psi pressure loss between the power plant and the line, reducing the line inlet pressure to 440 psia. Since the pressure can fall about $440 - 150 = 290$ psi, this will be called the *maximum pressure-loss line*. During design, determine which line is the most economical.

Calculation Procedure:

1. Determine the required pipe diameter for each condition

The average steam pressure in the nominal pressure-loss line is (inlet pressure + outlet pressure)/2 = $(180 + 150)/2 = 165$ psia. Use this average pressure to determine the pipe size, because the average pressure is more representative of actual conditions in the pipe. Assume that there will be a 5-psi pressure drop through any expansion bends and other fittings in the pipe. Then, the allowable friction-pressure drop = $30 - 5 = 25$ psi.

Use the Thomas saturated-steam formula to determine the required pipe diameter, or $d = (80,000W/Pv)^{0.5}$, where d = inside pipe diameter, in.; W = weight of steam flowing, lb/min; P = average steam pressure, psia; v = steam velocity, fpm. Assuming a steam velocity of 10,000 fpm, which is typical for a long steam transmission line, $d = [(80,000 \times 300,000/60)/(165 \times 10,000)]^{0.5} = 15.32$ in.

The inside diameter of a schedule 40 16-in.-outside-diameter pipe is, from a table of pipe properties, 15.00 in. Assume that a 16-in. pipe will be used if schedule 40 wall thickness is satisfactory for the nominal pressure-loss line. Note that the larger flow was used in computing the size of this line because a pipe satisfactory for the larger flow will be acceptable for the smaller flow.

The maximum pressure-loss line will have an average pressure that is a function of the inlet pressure at the pressure-reducing valve at the line outlet. Assume that there is a 10-psi drop through this reducing valve. Then steam will enter the valve at $150 + 10 = 160$ psia, and the average line pressure = $(440 + 160)/2 = 300$ psia. Using a higher steam velocity (15,000 fpm) for this maximum pressure-loss line than for the nominal pressure-loss line (10,000 fpm) because there is a larger allowable pressure drop, compute the required inside diameter using the Thomas saturated-steam formula, because the steam has a superheat of only $456.28 - 455.00 = 1.28$ F. Or, $d = [(80,000 \times 300,000/60)/(300 \times 15,000)]^{0.5} = 9.44$ in. Since a 10-in. schedule 40 pipe has an inside diameter of 10.020 in., use this size for the maximum pressure-loss line.

2. Compute the required pipe-wall thickness

As shown in an earlier Calculation Procedure, the schedule number SN = $1,000\,P_i/S$. Assuming that seamless carbon steel ASTM A53 Grade A pipe is used for both lines, the *Piping Code* allows a stress of 12,000 psi for this material at 600 F. Then, SN = $(1.000) \times (435)/12,000 = 36.2$; use schedule 40 pipe, the next largest schedule number for both lines. This computation verifies the assumption in Step 1 of the suitability of schedule 40 for each line.

3. Check the pipeline for critical velocity

In a steam line, $p_c = W'/Cd^2$, where p_c = critical pressure in pipe, psia; W' = steam flow rate, lb/hr; C = constant from King and Crocker — *Piping Handbook*; d = inside diameter of pipe, in.

When the pressure loss in a pipe exceeds 50 to 58 percent of the initial pressure, flow may be limited by the fluid velocity. The limiting velocity that occurs under these conditions is called the *critical velocity*, and the coexisting pipeline pressure, the *critical pressure*.

Critical velocity may limit flow in the 10-in.-maximum pressure-loss line because the terminal pressure of 150 psia is less than 58 percent of 440 psia, the inlet pressure. Use the above equation to find the critical pressure. Or, $p_c = (300,000)/[(75.15)(10.02)^2] =$

39.7 psia, using the constant from the *Piping Handbook* after interpolating for the initial enthalpy of 1,205.4 Btu/lb, which is obtained from steam-table values.

Critical velocity would limit flow if the pipeline terminal pressure were equal to, or less than, 39.7 psia. Since the terminal pressure of 150 psia is greater than 39.7 psia, critical velocity does not limit the steam flow. With smaller flow rates, the critical pressure will be lower because the denominator in the equation remains constant for a given pipe. Hence, the 10-in. line will readily transmit 300,000 lb/hr and smaller flows.

If critical pressure existed in the pipeline, the diameter of the pipe might have to be increased to transmit the desired flow. The 16-in. line does not have to be checked for critical pressure because its final pressure is more than 58 percent of the initial pressure.

4. Compute the heat loss for each line

Assume that 2-in.-thick 85 percent magnesia insulation is used on each line and that the lines will run above the ground in an area having a minimum temperature of 40 F. Set up a computation form as follows:

	16	10
Pipe size, in.	16	10
Steam temperature, F	373	455
Air temperature, F..........................	40	40
Temperature difference, F....................	333	415
Insulation heat loss, Btu/(hr)(ft)(F)*	1.11	0.704
Heat loss, Btu/(hr)(lin ft)....................	370	292
Heat loss, Btu/hr for 1,700 ft	629,000	496,400
Total heat loss, Btu/hr, with a 25 percent safety factor	786,250	620,500
Heat loss, Btu per lb of steam for 300,000 lb/hr flow	2.62	2.07
Heat loss, Btu per lb of steam for the average flow of 150,000 lb/hr	5.24	4.14

*From table of pipe insulation, Ehret Magnesia Manufacturing Company.

In this form, the following computations were made for both pipes: heat loss, Btu/(hr)(lin ft) = [insulation heat loss, Btu/(hr)(ft)(F)](temperature difference, F); heat loss, Btu/hr for 1,700 ft = [heat loss, Btu/(hr)(lin ft)](1,700); total heat loss, Btu/hr, 25 percent safety factor = (heat loss, Btu/hr 1,700 ft)(1.25); heat loss, Btu per lb steam = (total heat loss, Btu/hr, with a 25 percent safety factor)/(300,000 lb steam).

5. Compute the leaving enthalpy of the steam in each line

Acceleration of steam in each line results from an enthalpy decrease of $h_a = (v_2{}^2 - v_1{}^2)/2g(778)$, where h_a = enthalpy decrease, Btu/lb; v_2 and v_1 = final and initial velocity of the steam, respectively, fps; $g = 32.2$ ft/sec^2. The velocity at any point x in the pipe is found from the continuity equation $v_x = (W'v_g)/3,600 A_x$, where v_x = steam velocity, fps, when the steam volume is v_g cu ft/lb, and A_x is the cross-sectional area of pipe, sq ft, at the point being considered.

For the 16-in. nominal pressure-loss line with a flow of 300,000 lb/hr at 180-psia entering and 150-psia leaving pressure, using steam and piping table values, $v_1 = (300,000)(2.53)/[(3,600)(1.23)] = 171.5$ fps, $v_2 = 300,000(3.015)/[(3,600)(1.23)] = 205$ fps. Then, $h_a = [(204.5)^2 - (171.5)^2]/[(64.4)(778)] = 0.2504$ Btu/lb, say 0.25 Btu/lb.

By an identical calculation, $h_a = 3.7$ Btu/lb for the 10-in. maximum pressure-loss line when the leaving steam is assumed to be 150 psia saturated.

Enthalpy of the 180-psia saturated steam entering the 16-in. line is 1,196.9 Btu/lb. Heat loss during 300,000 lb/hr flow is 2.62 Btu/lb, as computed in Step 4. The enthalpy drop of 0.25 Btu/lb accelerates the steam. Hence, the calculated leaving enthalpy is 1,196.9 − (2.62 + 0.25) = 1,194.03 Btu/lb. The enthalpy of the leaving steam at 150-psia saturated is 1,194.1 Btu/lb. To have saturated steam leave the line, (1,194.10 − 1,194.03), or 0.07 Btu/lb must be supplied to the steam. This heat will be obtained from the enthalpy of vaporization given off by condensation of some of the steam in the line.

Make a group of identical calculations for the 10-in. maximum pressure-loss line. The enthalpy of 440-psia 455-F entering steam is 1,205.4 Btu/lb, found by interpolation in the steam tables. Heat loss during 300,000 lb/hr flow is 2.07 Btu/lb. An enthalpy drop of 3.7 Btu/lb accelerates the steam. Hence, the calculated leaving enthalpy = 1,205.4 − (2.07 + 3.7) = 1,199.63 Btu/lb.

The enthalpy of the leaving steam at 150-psia saturated is 1,194.1 Btu/lb. As a result, under maximum flow conditions, the steam will be superheated from the entering point to the leaving point of the line. The enthalpy difference of 5.53 Btu/lb (= 1,199.63 − 1,194.10) produces this superheat. Because the steam is superheated throughout the line length, condensation of the steam will not occur during maximum flow conditions.

For most industrial applications, the steam leaving the line may be considered as saturated at the desired pressure. But for precise temperature regulation, some form of pressure-temperature control must be used at the end of long lines.

During average flow conditions of 150,000 lb/hr, the line heat loss is 4.14 Btu/lb, as computed in Step 4. The enthalpy drop to accelerate the steam is 0.925 Btu/lb. As in the case of maximum flow, the steam is superheated throughout the length of the 10-in. maximum pressure-loss line because the calculated leaving enthalpy is (1,205.40 − 5.07) = 1,200.33 Btu/lb.

6. Compute the quantity of condensate formed in each line

For either line, the quantity of condensate formed, lb/hr, $= C = W'(h_g$ at leaving pressure − calculated leaving h_g)/outlet pressure h_{fg}.

Using computed values from Step 5 and steam-table values, the 16-in. line with 300,000 lb/hr flowing forms $C = (300,000)(0.07)/863.6 = 24.35$, say 24.4 lb/hr of condensate.

Condensation during an average flow of 150,000 lb/hr is found in the same way. The enthalpy drop to accelerate the steam is neglected for average flow in normal pressure-loss lines because the value is generally small. For the 150,000 lb/hr flow, the calculated leaving enthalpy = 1,196.90 − 5.24 = 1,191.66 Btu/lb. Hence, $C = (150,000)(1,194.10 − 1,191.66)/863.6 = 424$, say 425 lb/hr.

The largest amount of condensate is formed during line warm-up. Condensate-removal equipment—traps and related piping—must be sized up on a warm-up basis, not on the average steam flow. Using a warm-up time of 30 min and the method of an earlier Calculation Procedure, the condensate formed in 16-in. schedule 40 pipe weighing 83 lb/ft is, with a 25 percent safety factor to account for radiation, $C = 1.25 \times (60)(83)(1,700)(373 − 40)(0.12)/[(30)(850.8)] = 16,550$ lb/hr. Thus, the trap or traps should have a capacity of about 17,000 lb/hr to remove the condensate during the 30-min warm-up period.

Condensate does not form in the 10-in. maximum pressure-loss line during either maximum or average flow. Warm-up condensate for a 30-min warm-up period and a 25 percent safety factor is $C = 1.25(60)(40.5)(1,700)(455 − 40)(0.12)/[(30)(770.0)] = 11,120$ lb/hr. Thus, the trap or traps, should have a capacity of about 11,500 lb/hr to remove the condensate during the 30-min warm-up period.

In general, traps sized on a warm-up basis have adequate capacity for the condensate formed during the maximum and average flows. However, the condensate formed under all three conditions must be computed to determine the maximum rate of formation for trap and drain-line sizing.

7. Determine the number of plain U bends needed

A 1,700-ft-long steel steam line operating at a temperature in the 400-F range will expand nearly 50 in. during operation. This expansion must be absorbed in some way without damaging the pipe. There are four popular methods for absorbing expansion in long transmission lines—(a) plain U bends, (b) double-offset expansion U bends, (c) slip or corrugated expansion joints, and (d) welded-elbow expansion bends. Each of these will be investigated to determine which is the most economical.

Assume that the governing code for piping design in the locality in which the line will be installed requires that the combined stress resulting from bending and pressure S_{bp} not exceed three-fourths the sum of the allowable stress for the piping material at atmospheric temperature S_a and the allowable stress at the operating temperature S_o of the pipe. This is a common requirement. In equation form, $S_{bp} = 0.75(S_a + S_o)$, where each stress is in psi.

Using allowable stress values from the *Piping Code* or the local code for 16-in. seamless carbon steel ASTM A53 Grade A pipe operating at 373 F, $S_{bp} = 0.75(12,000 + 12,000) = 18,000$ psi.

Determine the longitudinal pressure stress P_L by dividing the end force due to internal pressure F_e lb by the cross-sectional area of the pipe wall a_m sq in., or $P_L = F_e/a_m$. In this equation, $F_e = pa$, where p = pipe operating pressure, psig; a = cross-sectional area of the pipe, sq in. Since the 16-in. line operates at $180 - 14.7 = 165.3$ psig and, from a table of pipe properties, $a = 176.7$ sq in. and $a_m = 24.35$ sq in., $P_L = (165.3)(176.7)/24.35 = 1,197$, say 1,200 psi. The allowable bending stress at 373 F, the pipe operating temperature, is then $S_{np} - P_L = 18,000 - 1,200 = 16,800$ psi.

Assume that the expansion U bend will have a radius of seven times the nominal pipe diameter, or $(7)(16$ in.$) = 112$ in. The allowable bending stress is 16,800 psi. Full *Piping Code* allowable credit will be taken for cold spring—i.e., the pipe will be cut short by 50 percent or more of the computed expansion and sprung into position.

Referring to King and Crocker—*Piping Handbook*, or a similar tabulation of allowable U-bend overall lengths for various operating temperatures, and choosing the length for 400 F, the next higher tabulated temperature greater than the 373-F operating temperature, an allowable length of 157.0 ft is obtained for the bend. Plot a curve of the allowable bend length vs. temperature at 200, 300, 400, and 500 F. From this curve, the allowable bend length at 373 F is found to be 175 ft. This length is based on an allowable pipe stress of 12,000 psi and no cold spring. Since the allowable stress is 16,800 psi and maximum cold spring is used, permitting a length 1.5 times the tabulated length, the total allowable length per bend $= (175.0)(16,800/12,000)(1.5) = 367.5$ ft. With a total length of pipe between the power plant and load of 1,700 ft, the number of bends required $= 1,700/367.5 = 4.64$ bends. Since only a whole number of bends can be used, the next larger whole number, or five bends, would be satisfactory for this 16-in. line. Each bend would have an overall length, Fig. 26, of $1,700/5 = 340$ ft.

Find the actual stress S_a in the pipe when five 340-ft bends are used by setting up a proportion between the tabulated stress and bend length. Thus, the *Piping Handbook* chart is based on a stress of 12,000 psi without cold spring. For this stress, the maximum allowable bend length is 175 ft, found by graphical interpolation of the tabular values, as discussed above. When a 340-ft bend with maximum cold spring is used, the pipe stress is such that the allowable bend length is $340/1.5 = 226.5$ ft. The actual stress in the pipe is therefore $S_a/12,000 = 226.5/175$, or $S_a = 15,520$ psi. This compares favorably with the allowable stress of 16,800 psi. The actual stress is less because the overall bend length was reduced.

Use the *Piping Handbook* or the method of an earlier Calculation Procedure to find the anchor reaction forces for these bends. Using the *Piping Handbook* method with graphical interpolation, the anchor reacting force for a 16-in. schedule 80 bend having a radius of seven times the pipe diameter is 10,550 lb at 373 F, based on a 12,000-psi stress in the pipe. This tabular reaction must be corrected for the actual pipe stress and for schedule 40 pipe instead of schedule 80 pipe. Thus, the actual anchor reaction, lb, = (tabular reaction, lb)(actual stress, psi/tabular stress, psi)(moment of inertia, schedule 40 pipe, in.[4]/moment of inertia of schedule 80 pipe, in.[4]) $= (10,550)(15,520/12,000)(731.9/1,156.6) = 8,650$ lb. With a reaction of this magnitude, each anchor would be designed to withstand a force of 10,000 lb. Good design would locate the bends midway between the anchor points; that is, there would be an anchor at each end of each bend. Adjustment for cold spring is not necessary, because it has negligible effect on anchor forces.

Use the same procedure for the 10-in. maximum pressure-loss line. Using 100-in.-radius bends, seven are required. The bending stress is 14,700 psi and the anchor force is 2,935 lb. Anchors designed to withstand 3,000 lb would be used.

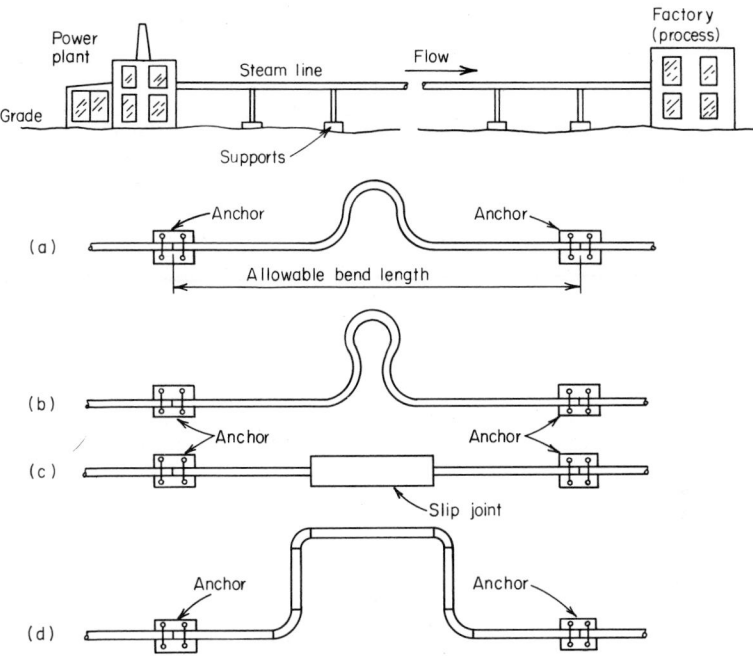

Fig. 26   Process steam line and different schemes for absorbing pipe thermal expansion.

8.   Determine the number of double-offset U bends needed

Using the same procedure and the *Piping Handbook* tabulation similar to that in Step 7, the 16-in. nominal pressure-loss line requires two 850-ft-long 112-in.-radius bends. Stress in the pipe is 15,610 psi and the anchor reaction is 4,780 lb.

The 10-in. maximum pressure-loss line requires five 340-ft-long 70-in.-radius bends. Stress in the pipe is 12,980 psi and the anchor reaction is 2,090 lb.

Note that a smaller number of double-offset U bends are required, two rather than five for the 16-in. pipe, and five rather than seven for the 10-in. pipe. This shows that double-offset U bends can absorb more expansion than plain U bends.

9.   Determine the number of expansion joints needed

For any pipe, the total linear expansion e_t in. at an elevated temperature above 32 F is $e_t = (c_e)(\Delta t)(l)$, where c_e = coefficient of linear expansion, in./(ft)(F); Δt = operating temperature, F − installation temperature, F; l = length of straight pipe, ft. Using King and Crocker − *Piping Handbook* as the source for c_e for both lines, the expansion of the 373-F 16-in. line with a 40-F installation temperature is $e_t = (12)(0.0000069)(373 - 40)(1,700) = 46.8$ in. For the 10-in. 455-F line, $e_t = (12)(0.0000072)(455 - 40)(1,700) = 61$ in. The factor 12 is used in each of these computations because King and Crocker give c_e in in./in.; therefore, the pipe total length must be converted to in. by multiplying by 12.

Double-ended slip-type expansion joints that can absorb up to 24 in. of expansion are available. Hence, the number of joints N needed for each line is: 16-in. line − N = 46.8/24, or 2; 10-in. line − N = 61/24, or 3.

The joints for each line would be installed midway between anchors, Fig. 26. Joints in both lines would be anchored to the ground or a supporting structure. Between the joints, the pipe must be adequately supported and free to move. Roller supports that guide and permit longitudinal movement are usually best for this service. Whereas roller-support friction varies, it is usually assumed to be about 100 lb per support. At least six supports per 100 ft are needed for the 16-in. line and seven per 100 ft for the 10-in. line. Support friction and the number of rollers required is obtained from King and Crocker — *Piping Handbook* or piping engineering data.

The required anchor size and strength depends on the pipe diameter, steam pressure, slip-joint construction, and type of supports used. During expansion of the pipe, friction at the supports and in the joint packing sets up a force that must be absorbed by the anchor. Also, steam pressure in the joint tends to force it apart. The magnitude of these forces is easily computed. With the total force known, a satisfactory anchor can be designed. Slip-joint packing-gland friction varies with different manufacturers, type of joint, and packing used. Gland friction in one popular type of slip joint is about 2,200 lb/in. of pipe diameter. Assuming use of these joints in both lines, compute the anchor forces as follows:

16-in. nominal pressure-loss line	
Support friction = [(1,700 ft)(6 supports per 100 ft)	
(100 lb per support)]/100	= 10,200
Gland friction = (2,200 lb per in. diameter)(16 in.)	= 35,200
Pressure force = (165.3 psig)(176.7 sq in. pipe area)	= 29,200
Total force to be absorbed by anchor	= 74,600 lb
10-in. maximum pressure-loss line	
Support friction = [(1,700)(7)(100)]/100	= 11,900
Gland friction = (2,200)(10)	= 22,000
Pressure force = (425.3)(78.9)	= 33,600
Total force to be absorbed by anchor	= 67,500 lb

Comparing these results shows that the 10-in. line requires smaller anchors than does the 16-in. line. However, the 16-in. line requires only three anchors whereas the 10-in. line needs four anchors. The total cost of anchors for both lines will be about equal because of the difference in size of the anchors for each line.

The advantages of slip joints become apparent when the piping layout is studied. Only a minimum of pipe is needed because the pipe runs in a straight line between the point of supply and point of use. The amount of insulation is likewise a minimum.

Corrugated expansion joints could be used in place of slip-type joints. These would reduce the required anchor size somewhat because there would be no gland friction. The selection procedure resembles that given for slip-type joints.

10. *Select welded-elbow expansion bends*

Use the graphical analysis in King and Crocker — *Piping Handbook* or in any welding fittings engineering data. Using either method shows that three bends of the most economical shape are suitable for the 16-in. line and four for the 10-in. line. The most economical bend is obtained when the bend width, divided by the distance between the anchor points, is 0.50. With these proportions, the longitudinal stress at the top and bottom of the bend is the same. Use of such bends, although desirable, is not always feasible, because existing piping or structures interfere.

When bend dimensions other than the most economical must be used, the maximum longitudinal stress occurs at the top of the bend when the width/anchor distance < 0.5. When this ratio is > 0.5, the maximum stress occurs at the bottom of the bend. Regardless of the bend type — plain U, double-offset U, or welded — the actual stress in the pipe should not exceed 40 percent of the tensile strength of the pipe material.

11. *Determine the materials, quantities and costs*

Set up tabulations showing the materials needed and their cost. Table 36 shows the materials required. Piping length is computed by using standard bend tables available in the cited references.

Table 37 shows the approximate material costs for each pipeline. The costs used in preparing this table were the most accurate available at the time of writing. However,

TABLE 36   Summary of Material Requirements for Various Lines

Means used to absorb expansion	Number of anchors required		Approximate number of supports required		Approximate footage of pipe and insulation required	
	Pipe size, in.					
	10	16	10	16	10	16
Plain U bends.........	9	5	127	120	2,120	1,970
Double-offset U bends............	6	3	119	114	1,985	1,820
Slip joints	4	3	102	102	1,700	1,700
Welding elbows.......	5	4	106	106	1,760	1,760

the actual numerical values given in the table should not be used for similar design work because price changes may cause them to be incorrect. The important findings in such a tabulation are the differences in total cost. These differences will remain substantially constant even though prices change. Hence, if an $8,000 difference exists between two sizes of pipe, this difference will not change appreciably with a moderate rise or fall in unit prices of materials.

Study of Table 37 shows that, in general, lines using double-offset U bends or welding elbows have the lowest material first cost. However, higher first costs do not rule out slip joints or plain U bends. Frequently use of slip joints will eliminate offsets to clear existing buildings or piping because the pipe path is a straight line. Plain U bends have smaller overall heights than double offset U bends. For this reason, the plain bend is often preferable where the pipe is run through congested areas of factories.

In some cases, past piping practice will govern line selection. For instance, in a factory that has made wide use of slip joints, the slightly higher cost of such a line might be overlooked. Preference might also be shown for plain U bends, double-offset U bends, or welded bends.

The values given in Table 37 do not include installation, annual operating costs, or depreciation. These have been omitted because accurate estimates are difficult to make unless actual conditions are known. Thus, installation costs may vary considerably according to who does the work. Annual costs are a function of the allowable depreciation, nature of process served, and location of the line. For a given transmission line of the type considered here, annual costs will usually be less for the smaller line.

The economic analysis, as made by the pipeline designer, should include all costs relative to the installation and operation of the line. The allowable cost of money and recommended depreciation period can be obtained from the accounting department.

12. *Select the most economical pipe size*

Table 37 shows that from the standpoint of first costs, the smaller line is more economical. This lower first cost is not, however, obtained without losing some large-line advantages.

Thus, steam leaves the 16-in. line at 150 psia saturated, the desired outlet condition. Special controls are unnecessary. With the 10-in. line, the desired leaving conditions are not obtained. Slightly superheated steam leaves the line unless special controls are used. Where an exact leaving temperature is needed by the process served, a desuperheater at the end of the 10-in. line will be needed. Neglecting this disadvantage, the 10-in. line is more economical than is the 16-in. line.

Besides lower first cost, the small line loses less heat to the atmosphere, has smaller anchor forces, and does not cause steam condensation during average flows. Lower heat losses and condensation reduce operating costs. Therefore, if special temperature controls are acceptable, the 10-in. maximum pressure-loss line will be a more economical investment.

TABLE 37 Approximate Material Costs for Various Lines

Means used to absorb expansion	Total material cost, $		Condensate removal equipment		Cost of anchors, $		Cost of supports, $		Cost of insulation, $		Cost of pipe and bends or joints, $	
	Pipe size, in.											
	10	16	10	16	10	16	10	16	10	16	10	16
Plain U bends.........	26,500	51,350	2,000	3,000	1,000	1,800	1,800	2,400	3,700	5,650	18,000	38,500
Double-offset U bends...........	23,800	44,000	2,000	3,000	600	800	1,700	2,300	3,500	5,400	16,000	32,500
Slip joints	29,650	51,775	2,000	3,000	400	600	1,500	2,000	3,000	4,675	22,750	41,500
Welding elbows......	23,800	43,975	2,000	3,000	400	800	1,800	2,300	3,600	5,375	16,000	32,500

Such a conclusion neglects the possibility of future plant expansion. Where expansion is anticipated, installation of a small line now and another line later to handle increased steam requirements is uneconomical. Instead, installation of a large nominal pressure-loss line now that can later be operated as a maximum pressure-loss line will be found more economical. Besides the advantage of a single line in crowded spaces, there is a reduction in installation and maintenance costs.

13. Provide for condensate removal

Fit a condensate drip line for every 100 ft of pipe, regardless of size. Attach a trap of suitable capacity (see Step 6) to each drip line. Pitch the steam-transmission pipe toward the trap, if possible. Where the condensate must flow *against* the steam, the steam-transmission pipe *must* be sloped in the direction of condensate flow. Every vertical rise of the main line must also be dripped. Where water is scarce, return the condensate to the boiler.

Related Calculations: Use this method to design long steam, gas, liquid, or vapor lines for factories, refineries, power plants, ships, process plants, steam heating systems, and similar installation. Follow the applicable piping code when designing the pipeline.

STEAM DESUPERHEATER ANALYSIS

A spray- or direct-contact-type desuperheater is to remove the superheat from 100,000 lb/hr of 300-psia 700-F steam. Water at 200 F is available for desuperheating. How much water must be furnished per hr to produce 30-psia saturated steam? How much steam leaves the desuperheater? If a shell-and-tube-type noncontact desuperheater is used, determine the required water flow rate if the overall coefficient of heat transfer $U = 500$ Btu/(hr)(sq ft)(F). How much tube area A is required? How much steam leaves the desuperheater? Assume that the desuperheating water is not allowed to vaporize in the desuperheater.

Calculation Procedure:

1. Compute the heat abosrbed by the water

Water entering the desuperheater must be heated from the entering temperature, 200 F, to the saturation temperature of 300-psia steam, or 417.3 F. Using the steam tables, the sensible heat that must be absorbed by the water $= h_f$ at 417.3 F $- h_f$ at 200 F $= 393.81 - 167.99 = 225.81$ Btu per lb of water used.

Once the desuperheating water is at 417.3 F, the saturation temperature of 300-F steam, the water must be vaporized if additional heat is to be absorbed. From the steam tables, the enthalpy of vaporization at 300 psia, $h_{fg} = 809.0$ Btu/lb. This is the amount of heat the water will absorb when vaporized from 417.3 F.

Superheated steam at 300 psia and 700 F has an enthalpy of $h_g = 1,368.3$ Btu/lb, and the enthalpy of 300-psia saturated steam $h_g = 1,202.8$ Btu/lb. Thus, $1,368.3 - 1,202.8 = 165.5$ Btu/lb must be absorbed by the water to desuperheat the steam from 700 F to saturation at 300 psia.

2. Compute the weight of water required for the spray

The weight of water evaporated by 1 lb of steam while it is being desuperheated = heat absorbed by water, Btu per lb of steam/heat required to evaporate 1 lb of water entering the desuperheater at 200 F, Btu $= 165.5/(225.81 + 809.0) = 0.16$ lb of water. Since 100,000 lb/hr of steam is being desuperheated, the water flow rate required $= (0.16)(100,000) = 16,000$ lb/hr. Water for direct-contact desuperheating can be taken from the feedwater piping or from the boiler.

Note that 16,000 lb/hr of additional steam will leave the desuperheater because the superheated steam is not condensed while being desuperheated. Thus, the total flow from the desuperheater $= 100,000 + 16,000 = 116,000$ lb/hr.

3. Compute the tube area required in the desuperheater

The total heat transferred in the desuperheater, Btu/hr, $= UAt_m$, where $t_m =$ logarithmic mean temperature difference across the heater. Using the method for computing the logarithmic temperature difference given elsewhere in this Handbook, or a graphical solution as in Perry—*Chemical Engineers' Handbook*, $t_m = 134$ F, with desuperheating water entering at 200 F and leaving at 430 F, a temperature about 13 F higher than the leaving temperature of the saturated steam, 417.3 F. Steam enters the desuperheater at 700 F. Assumption of a leaving water temperature 10 to 15 F higher than the steam temperature is usually made to ensure an adequate temperature difference so that the desired heat-transfer rate will be obtained. If the graphical solution is used, the greatest temperature difference then becomes $700 - 200 = 500$ F, and the least temperature difference $= 430 - 417.3 = 12.7$ F.

Then, the heat transferred $= (500)(A)(134)$, whereas the heat given up by the steam is, from Step 1, (100,000 lb/hr)(165.5 Btu/lb). Since the heat transferred = the heat absorbed, $(500)(A)134 = (100,000)(165.5)$; $A = 247$ sq ft, say 250 sq ft.

4. Compute the required water flow

Heat transferred to the water $= (500)(247)(134)$ Btu/hr. The temperature rise of the water during passage through the desuperheater = outlet temperature, F − inlet temperature = outlet temperature, $F = 430 - 200 = 230$ F. Since the specific heat of water $= 1.0$, closely, the heat absorbed by the water = (flow rate, lb/hr)(230)(1.0). Then, the heat transferred = heat absorbed, or $(500)(247)(134)$ = (flow rate, lb/hr)(230)(1.0); flow rate $= 72,000$ lb/hr. Since the water and steam do *not* mix, the steam output of the desuperheater = steam input = 100,000 lb/hr.

Only about 25 percent as much water, 16,000 lb/hr, is required by the direct-contact desuperheater as compared with the indirect desuperheater. The indirect type of superheater requires more cooling water because the enthalpy of vaporization, nearly 1,000 Btu/lb of water, is not used to absorb heat. Some indirect-type desuperheaters are designed to permit the desuperheating water to vaporize. This steam is returned to the boiler. The water consumption determination and the calculation procedure for this type are similar to the spray-type discussed earlier. Where the water does not vaporize, it must be kept at a high enough pressure to prevent vaporization.

Related Calulations: Use this method to analyze steam desuperheaters for any type of steam system—industrial, utility, heating, process, or commercial.

STEAM ACCUMULATOR SELECTION AND SIZING

Select and size a steam accumulator to deliver 10,000 lb/hr of 25-psia steam for peak loads in a steam system. Charging steam is available at 75 psia. Room is available for an accumulator not more than 30 ft long, 20 ft wide, and 20 ft high. How much steam is required for start-up?

Calculation Procedure:

1. Determine the required water capacity of the accumulator

One lb of water stored in this accumulator at 75 psia has a saturated liquid enthalpy $h_f = 277.43$ Btu/lb, from the steam tables; whereas for 1 lb of water at 25 psia, $h_f = 208.42$ Btu/lb. In an accumulator, the stored water flashes to steam when the pressure on the outlet is reduced. For this accumulator, when the pressure on the 75-psia water is reduced to 25 psia by a demand for steam, each lb of stored 75-psia water flashes to steam, releasing $277.43 - 208.42 = 69.01$ Btu/lb.

The enthalpy of vaporization of 25-psia steam $h_{fg} = 952.1$ Atu/lb. Thus, 1 lb of 75-psia water will form $69.01/952.1 = 0.0725$ lb of steam. To supply 10,000 lb/hr of steam, the accumulator must store $10,000/0.0725 = 138,000$ lb/hr of 75-psia water.

Saturated water at 75 psia has a specific volume of 0.01753 cu ft/lb, from the steam tables. Since density $=$ 1/specific volume, the density of 75-psia saturated water $=$ $1/0.01753 = 57$ lb/cu ft. The volume required in the accumulator to store 138,000 lb of 75-psia water = total weight, lb/density of water = $138,000/57 = 2,420$ cu ft.

2. Select the accumulator dimensions

Many steam accumulators are cylindrical because this shape permits convenient manufacture. Other shapes—rectangular, cubic, etc.—may also be used. However, a cylindrical shape will be assumed here because it is the most common.

The usual accumulator that serves as a reserve steam supply between a boiler and a load (often called a Ruths-type accumulator) can safely release steam at the rate of 3 (accumulator storage pressure, psia) lb per sq ft of water surface per hr. Thus, this accumulator can release $(3)(75) = 225$ lb of steam per sq ft of water surface per hr. Since a release rate of 10,000 lb/hr is desired, the surface area required $= 10,000/225 = 445$ sq ft.

Space is available for a 30-ft-long accumulator. A cylindrical accumulator of this length would require a diameter of $445/30 = 14.82$ ft, say 15 ft. When half full of water, the accumulator would have a surface area $(30)(15) = 450$ sq ft.

Once the accumulator dimensions are known, its storage capacity must be checked. The volume of a horizontal cylinder of d-ft diameter and l-ft length $= (\pi d^2/4)(l) = (\pi \times 15^2/4)(30) = 5,300$ cu ft. When half full, this accumulator could store $5,300/2 = 2,650$ cu ft. Since, from Step 1, a capacity of 2,420 cu ft is required, a 15×30 ft accumulator is satisfactory. A water-level controller must be fitted to the accumulator to prevent filling beyond about the midpoint. In this accumulator, the water level could rise to about 60 percent, or $(0.60)(15) = 9$ ft, without seriously reducing the steam capacity. When an accumulator delivers steam from a more-than-half-full condition, its releasing capacity increases as the water level falls to the midpoint, where the release area is a maximum. Since most accumulators function for only short periods—say 5 or 10 min—it is more important that the vessel be capable of delivering the desired rate of flow than that it deliver the last lb of steam in its lb/hr rating.

If the size of the accumulator computed as shown above is unsatisfactory from the standpoint of space, alter the dimensions and recompute the size.

3. Compute the quantity of charging steam required

To start an accumulator, it must first be partially filled with water and then charged with steam at the charging pressure. The usual procedure is to fill the accumulator from the plant feedwater system. Assume that the water used for this accumulator is at 14.7 psia and 212 F and that the accumulator vessel is half-full at the start.

For any accumulator, the weight of charging steam required is found by solving the following heat-balance equation: (weight of starting water, lb)(h_f of starting water, Btu/lb) + (weight of charging steam, lb)(charging steam h_g, Btu/lb) = (weight of charging steam, lb + weight of starting water, lb)(h_f at charging pressure, Btu/lb). For this accumulator with a 75-psia charging pressure and 212-F starting water, the first step is to compute the weight of water in the half-full accumulator. Since, from Step 2, the accumulator must contain 2,420 cu ft of water, this water has a total weight of (volume of water, cu ft)/(specific volume of water, cu ft/lb) = $2,420/0.01672 = 144,600$ lb. However, the accumulator can actually store 2,650 cu ft of water. Hence, the actual weight of water = $2,650/0.1672 = 158,300$ lb. Then, with C = weight of charging steam, lb, $(158,300)(180.07) + (C)(1181.9) = (C + 158,300)(277.43)$; $C = 17,080$ lb of steam.

Once the accumulator is started up, less steam will be required. The exact amount is computed in the same manner, using the steam and water conditions existing in the accumulator.

Related Calculations: Use this method to size an accumulator for any type of steam service—heating, industrial, process, utility. The operating pressure of the accumulator may be greater or less than atmospheric.

SELECTING PLASTIC PIPING FOR INDUSTRIAL USE

Select the material, schedule number, and support spacing for a 1-in.-nominal-diameter plastic pipe conveying ethyl alcohol liquid having a temperature of 75 F and a pressure of 400 psi. What expansion must be anticipated if a 1,000-ft length of the pipe is installed at a temperature of 50 F? How does the cost of this plastic pipe compare with galvanized-steel pipe of the same size and length?

Calculation Procedure:

1. Determine the required schedule number

Refer to Baumeister and Marks — *Standard Handbook for Mechanical Engineers* or a plastic pipe manufacturer's engineering data for the required schedule number. Table 38 shows typical pressure ratings for various sizes and schedule number polyvinyl chloride (plastic) piping.

TABLE 38 Maximum Operating Pressure, PVC Pipe (Normal-impact grade; fluid temperature 75 F, or less).

Pipe size, in.	Schedule 40; plain end	Schedule 80	
		Plain end	Threaded
$\frac{1}{2}$	410 psi	575 psi	330 psi
$\frac{3}{4}$	335 psi	470 psi	285 psi
1	310 psi	435 psi	255 psi
$1\frac{1}{2}$	230 psi	325 psi	205 psi
2	195 psi	280 psi	190 psi
3	185 psi	260 psi	170 psi
4	155 psi	225 psi	160 psi

Table 38 shows that schedule 40 normal-impact grade 1-in. pipe is unsuitable because its maximum operating pressure with fluid at 75 F is 310 psi. Plain-end 1-in. schedule 80 pipe is, however, satisfactory because it can withstand pressures up to 435 psi. Note that threaded schedule 80 pipe can withstand pressures only to 255 psi. Therefore, plain-end normal-impact grade pipe must be used for this installation. High-impact grade pipe, in general, has lower allowable pressure ratings at 75 F because the additive used to increase the impact resistance lowers the tensile strength, temperature, and chemical resistance. Data shown in Table 38 are also presented in graphical form in some engineering data.

2. Select a suitable piping material

Refer to piping engineering data to determine the corrosion resistance of PVC to ethyl alcohol. A Grinnell Company data sheet rates PVC normal-impact and high-impact pipe as having excellent corrosion resistance to ethyl alcohol at 72 and 140 F. Therefore, PVC is a suitable piping material for this liquid at its operating temperature of 75 F.

3. Find the required support spacing

Use a tabulation or chart in the plastic-pipe engineering data to find the required support spacing for the pipe. Be sure to read the spacing under the correct schedule number. Thus, a Grinnell Company plastic-piping tabulation recommends a 5-ft 4-in. spacing for schedule 80 1-in. PVC pipe that weighs 0.382 lb/ft when empty. The pipe hangers should not clamp the pipe tightly; instead, free axial movement should be allowed.

4. Compute the expansion of the pipe

The temperature of the pipe rises from 50 to 75 F when it is put in operation. This is a rise of $75 - 50 = 25$ F. Table 39 shows the thermal expansion of various types of plastic piping.

The thermal expansion of any plastic pipe is found from $E_t = LC\Delta t$, where E_t = total expansion, in.; L = pipe length, ft; C = coefficient of thermal expansion, in.(ft)(F), from Table 39, Δt = temperature change of the pipe, F. For this pipe, $E_t = (1,000)$ $(0.00054)(25) = 13.5$ in. when the temperature rises from 50 to 75 F.

TABLE 39   Thermal Expansion of Plastic Pipe

Piping Material	Expansion, in./(ft)(F)
Butyrate........................	0.00118
Kralastic.......................	0.00067
Polyethylene....................	0.00108
Polyvinyl chloride..............	0.00054
Saran...........................	0.00126

5. Determine the relative cost of the pipe

Check the prices of galvanized-steel and PVC pipe as quoted by various suppliers. These quotations will permit easy comparison. In this case, the two materials will be approximately equal in per-ft cost.

Related Calculations: Use the method given here for selecting plastic pipe for any service — process, domestic, or commercial — conveying any fluid or gas. Note that the maximum operating pressure of plastic piping is normally taken as about 20 percent of the bursting pressure. The allowable operating pressure decreases with an increase in temperature. The maximum allowable operating temperature is usually 150 F. The pressure loss caused by pipe friction in plastic pipe is usually about one-half the pressure loss in galvanized-steel pipe of the same diameter. Pressure loss for plastic piping is computed in the same way as for steel piping.

FRICTION LOSS IN PIPES HANDLING SOLIDS IN SUSPENSION

What is the friction loss in 800 ft of 6-in. schedule 40 pipe when 400 gpm of sulfate paper stock is flowing? The consistency of the sulfate stock is 6 percent.

Calculation Procedure:

1. Determine the friction loss in the pipe

There are few general equations for friction loss in pipes conveying liquids having solids in suspension. Therefore, most practicing engineers use plots of friction loss available in engineering handbooks, *Cameron Hydraulic Data, Standards of the Hydraulic Institute,* and from pump engineering data. Figure 27 shows one set of typical friction-loss curves based on work done at the University of Maine on the data of Brecht and Heller of the Technical College, Darmstadt, Germany, and published by Goulds Pumps, Inc. There is a similar series of curves for commonly used pipe sizes from 2 through 36 in.

Enter Fig. 27 at the pipe flow rate, 400 gpm, and project vertically upward to the 6 percent consistency curve. From the intersection, project horizontally to the left to read the friction loss as 60 ft of liquid per 100 ft of pipe. Since this pipe is 800 ft long, the total friction-head loss in the pipe = (800/100)(60) = 480 ft of liquid flowing.

2. Correct the friction loss for the liquid consistency

Friction-loss factors are usually plotted for one type of liquid, and correction factors are applied to determine the loss for similar, but different, liquids. Thus, with the Goulds charts, a factor of 0.9 is used for soda, sulfate, bleached sulfite, and reclaimed paper stocks. For ground wood, the factor is 1.40.

When the stock consistency is less than 1.5 percent, water-friction values are used. Below a consistency of 3 percent, the velocity of flow should not exceed 10 fps. For suspensions of 3 percent and above, limit the maximum velocity in the pipe to 8 fps.

Since the liquid flowing in this pipe is sulfate stock, use the 0.9 correction factor, or the actual total friction head = (0.9)(480) = 432 ft of sulfate liquid. Note that Fig. 27 shows that the liquid velocity is less than 8 fps.

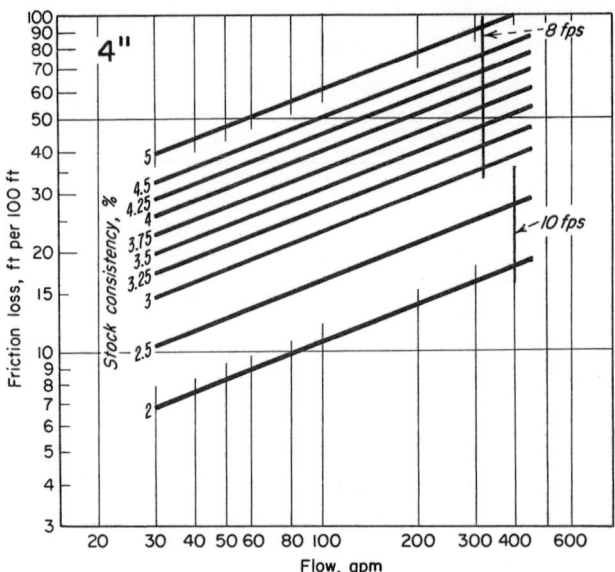

Fig. 27   Friction loss of paper stock in 4-in. steel pipe. *(Goulds Pumps, Inc.)*

Related Calculations: Use this procedure for soda, sulfate, bleached sulfite, and reclaimed and ground-wood paper stock. The values obtained are valid for both suction and discharge piping. The same general procedure can be used for sand mixtures, sewage, slurries, trash, sludge, and foods in suspension in a liquid.

HEAT TRANSFER AND HEAT EXCHANGERS

REFERENCES: McAdams—*Heat Transmission*, McGraw-Hill; Kern—*Process Heat Transfer*, McGraw-Hill; General Electric Company—*Electric Heaters and Heating Devices*; Jakob—*Heat Transfer*, Wiley; Bosworth—*Heat Transfer Phenomena*, Wiley; Kays and London—*Compact Heat Exchangers*, McGraw-Hill; Kraus—*Cooling Electronic Equipment*, Prentice-Hall; Fraas and Ozisik —*Heat Exchanger Design*, Wiley; Heat Transfer Research, Inc.—*Design Manual*: API Standards— *Heat Exchangers for General Refinery Service*; Giedt—*Principles of Engineering Heat Transfer*, Van Nostrand; Eckert and Drake—*Heat and Mass Transfer*, McGraw-Hill; Schneider—*Conduction Heat Transfer*, Addison-Wesley; Kreith—*Principles of Heat Transfer*, International Textbook; Perry—*Chemical Engineers' Handbook*, McGraw-Hill; Carslaw and Jaeger—*Conduction of Heat in Solids*, Oxford; Wilkes—*Heat Insulation*, Wiley.

SELECTING TYPE OF HEAT EXCHANGER FOR A SPECIFIC APPLICATION

Determine the type of heat exchanger to use for each of the following applications: (1) heating oil with steam; (2) cooling internal-combustion engine liquid coolant; (3) evaporating a hot liquid. For each heater chosen, specify the typical pressure range for which the heater is usually built and the typical range of the overall coefficient of heat transfer U.

Calculation Procedure:

1. Determine the heat-transfer process involved

In a heat exchanger, one or more of four processes may occur — heating, cooling, boiling, or condensing. Table 1 lists each of these four processes and shows the usual heat-transfer fluids involved. Thus, the heat exchangers being considered here involve: (a) Oil heater — heating — vapor-liquid; (b) internal-combustion engine coolant — cooling — gas-liquid; (c) hot-liquid evaporation — boiling — liquid-liquid.

2. Specify the heater action and the usual type selected

Using the same indentifying letters for the heaters being selected, Table 1 shows the action and usual type of heater chosen. Thus,

3. Specify the usual pressure range and typical U

Using the same identifying letters for the heaters being selected, Table 1 shows the action and usual type of heater chosen. Thus,

	Action	Type
a.	Steam condensed; oil heated	Shell-and-tube
b.	Air heated; water cooled	Tubes in open air
c.	Waste liquid cooled; water boiled	Shell-and-tube

4. Select the heater for each service

Where the heat-transfer conditions are normal for the type of service met, the type of heater listed in Step 2 can be safely used. When the heat-transfer conditions are unusual, a special type of heater may be needed. To select such a heater, study the data in Table 1 and make a tentative selection. Check the selection using the methods given in the following Calculation Procedures in this section of the Handbook.

Related Calculations: Use Table 1 as a general guide to heat-exchanger selection in any industry — petroleum, chemical, power, marine, textile, lumber, etc. Once the general type of heater and its typical U value are known, compute the required size using the Procedures given later in this section.

	Usual Pressure Range	Typical U Range, $Btu/(hr)(F)(sq\,ft)$
a.	0–500 psia	20–60
b.	0–100 psia	2–10
c.	0–500 psia	40–150

SHELL-AND-TUBE HEAT EXCHANGER SIZE

What is the required heat-transfer area for a parallel-flow shell-and-tube heat exchanger used to heat oil if the entering oil temperature is 60 F, the leaving oil temperature is 120 F, and the heating medium is steam at 200 psia? There is no subcooling of condensate in the heat exchanger. The overall coefficient of heat transfer $U = 25$ Btu/(hr)(F) (sq ft). How much heating steam is required if the oil flow rate through the heater is 100 gpm, the specific gravity of the oil is 0.9, and the specific heat of the oil is 0.5 Btu/(lb)(F)?

Calculation Procedure:

1. Compute the heat-transfer rate of the heater

With a flow rate of 100 gpm or (100 gpm)(60 min/hr) = 6,000 gph, the weight flow rate of the oil, using the weight of water of specific gravity 1.0 as 8.33 lb/gal, is (6,000 gph)(0.9 specific gravity)(8.33 lb/gal) = 45,000 lb/hr, closely.

TABLE 1 Heat-Exchanger Selection Guide*

Heat-transfer fluids	Equipment	Action	Type†	Pressure range‡	Typical range of U§
Liquid-liquid	Boiler-water blowdown exchanger	Blowdown cooled, feedwater heated	S	M, H	50–300
	Laundry-water heat reclaimer	Waste water cooled, feed heated	S	L	30–200
	Service-water heater	Waste liquid cooled, water heated	S	L, H	50–300
Vapor-liquid	Bleeder heater	Steam condensed, feedwater heated	S	L, H	200–800
	Deaerating feed heater	Steam condensed, feedwater heated	M	L, M	DC
	Jet heater	Steam condensed, water heated	M	L	DC
	Process kettle	Steam condensed, liquid heated	S	L, M	100–500
	Oil heater	Steam condensed, oil heated	S	L, M	20–60
	Service-water heater	Steam condensed, water heated	S	L, M	200–800
	Open flow-through heater	Steam condensed, water heated	M	L	DC
	Liquid-sodium steam superheater	Sodium cooled, steam superheated	S	M, H	50–200
Gas-liquid	Waste-heat water heater	Waste gas cooled, water heated	T	L	2–10
	Boiler economizer	Flue gas cooled, feedwater heated	T	M, H	2–10
	Hot-water radiator	Water cooled, air heated	T	L	1–10
Gas-gas	Boiler air heater	Flue gas cooled, combustion air heated	T, R	L	2–10
	Gas-turbine regenerator	Flue gas cooled, combustion air heated	T	L	2–10
Vapor-gas	Boiler superheater	Combustion gas cooled, steam superheated	T	M, H	2–20
	Steam pipe coils	Steam condensed, air heated	T	L, M	2–10
	Steam radiator	Steam condensed, air heated	T	L	2–10
Liquid-liquid	Oil cooler	Water heated, oil cooled	S, D	L, M	20–200
	Water chiller	Refrigerant boiled, water cooled	S	L, M	30–151
	Brine cooler	Refrigerant boiled, brine cooled	S	L, M	30–150
	Transformer-oil cooler	Water heated, oil cooled	S	L, M	20–50
Vapor-liquid	Boiler desuperheater	Boiler water heated, steam desuperheated	S, M	M, H	150–800

Heating (section label spanning Gas-liquid / Gas-gas / Vapor-gas groups)

				Type†	U‡§	
Cooling	Gas-liquid	Compressor intercoolers and aftercoolers	Water heated, compressed air cooled	S	L, H	10–20
		Internal-combustion-engine radiator	Air heated, water cooled	T	L	2–10
		Generator hydrogen, air coolers	Water heated, hydrogen or air cooled	S	L	2–10
		Air-conditioning cooler	Water heated, air cooled	T	L, M	2–10
		Refrigeration heat exchanger	Brine heated, air cooled	T	L, M	2–10
		Refrigeration evaporator	Refrigerant boiled, air cooled	T	L, M	2–10
	Vapor-gas	Boiler desuperheater	Flue gas heated, steam desuperheated	T	M, H	2–8
Boiling	Liquid-liquid	Hot-liquid evaporator	Waste liquid cooled, water boiled	S	L, H	40–150
		Liquid-sodium steam generator	Sodium cooled, water boiled	S	M, H	500–1000
	Vapor-liquid	Evaporator (vacuum)	Steam condensed, water boiled	S	L	400–600
		Evaporator (high pressure)	Steam condensed, water boiled	S	L, M	400–600
		Mercury condenser-boiler	Mercury condensed, water boiled	S	M, H	500–700
	Gas-liquid	Waste-heat steam boiler	Flue gas cooled, water boiled	T	L, H	2–10
		Direct-fired steam boiler	Combustion gas cooled, water boiled	T	L, H	2–10
Condensing	Vapor liquid	Refrigeration condenser	Water heated, refrigerant condensed	S, D	L, M	80–250
		Steam surface condenser	Water heated, steam condensed	S	L	300–800
		Steam mixing condenser	Water heated, steam condensed	M	L	DC
		Intercondenser and aftercondenser	Condensate heated, steam condensed	S	L	15–300
	Vapor-gas	Air-cooled surface condenser	Air heated, steam condensed	T	L	2–16

*Power.

†S – shell-and-tube exchanger; M – direct-contact mixing exchanger; T – tubes in path of moving fluid, or exchanger open to surrounding air; R – regenerative plate-type or simple plate-type exchanger; D – double-tube exchanger.

‡L – highest pressure ranges from 0 to 100 psia; M – highest pressure from 100 to 500 psia; H – 500 psia up.

§Values of U represent range of overall heat-transfer coefficients that might be expected in various exchangers. Coefficients are stated in Btu/(hr)(ft²)(F), per sq ft of heating surface. Total heat transferred in exchanger, in Btu/hr, is obtained by multiplying a specific value of U for that type of exchanger by the surface and the log mean temperature difference. DC indicates direct exchange of heat.

Since the temperature of the oil rises $120 - 60 = 60$ F during passage through the heat exchanger and the oil has a specific heat of 0.50, find the heat-transfer rate of the heater from the general relation $Q = wc\Delta t$, where Q = heat-transfer rate, Btu/hr; w = oil flow rate, lb/hr; c = specific heat of the oil, Btu/(lb)(F); Δt = temperature rise of the oil during passage through the heater. Thus, $Q = (45,000)(0.5)(60) = 1,350,000$ Btu/hr.

2. Compute the heater logarithmic mean temperature difference

The logarithmic mean temperature difference LMTD is found from LMTD = $(G - L)/\ln G/L$, where G = greater terminal temperature difference of the heater, F; L = lower terminal temperature difference of the heater, F; ln = logarithm to the base e. This relation is valid for heat exchangers in which the number of shell passes equals the number of tube passes.

In general, for parallel flow of the fluid streams, $G = T_1 - t_1$ and $L = T_2 - t_2$, where T_1 = heating fluid inlet temperature, F; T_2 = heating fluid outlet temperature, F; t_1 = heated fluid inlet temperature, F; t_2 = heated fluid outlet temperature, F. Figure 1 shows the maximum and minimum terminal temperature differences for various fluid flow paths.

For this parallel-flow exchanger, $G = T_1 - t_1 = 382 - 60 = 322$ F, where 382 F = the temperature of 200-psia saturated steam, from a table of steam properties. Also, $L = T_2 - t_2 = 382 - 120 = 262$ F, where the condensate temperature = the saturated steam temperature because there is no subcooling of the condensate. Then, LMTD = $G - L/\ln G/L = (322 - 262)/\ln 322/262 = 290$ F.

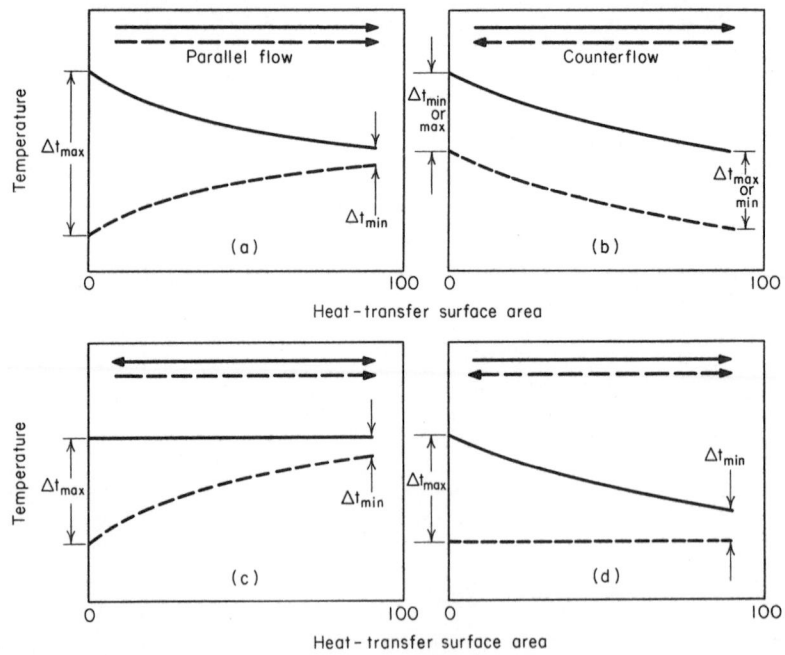

Fig. 1   Temperature relations in typical parallel-flow and counterflow heat exchangers.

3. Compute the required heat-transfer area

Use the relation $A = Q/U \times$ LMTD, where A = required heat-transfer area, sq ft; U = overall coefficient of heat transfer, Btu/(sq ft)(hr)(F). Thus, $A = 1,350,000/[(25)(290)] = 186.4$ sq ft, say 200 sq ft.

4. Compute the required quantity of heating steam

The heat added to the oil = Q = 1,350,000 Btu/hr, from Step 1. The enthalpy of vaporization of 200-psia saturated steam is, from the steam tables, 843.0 Btu/lb. Use the relation $W = Q/h_{fg}$, where W = flow rate of heating steam, lb/hr; h_{fg} = enthalpy of vaporization of the heating steam, Btu/lb. Hence, W = 1,350,000/843.0 = 1,600 lb/hr.

Related Calculations: Use this general Procedure to find the heat-transfer area, fluid outlet temperature, and required heating-fluid flow rate when true parallel flow or counterflow of the fluids occurs in the heat exchanger. When such a true flow does *not* exist, use a suitable correction factor, as shown in the next Calculation Procedure.

The Procedure described here can be used for heat exchangers in power plants, heating systems, marine propulsion, air-conditioning systems, etc. Any heating or cooling fluid—steam, gas, chilled water, etc.—can be used.

To select a heat exchanger using the results of this Calculation Procedure, enter the engineering data tables available from manufacturers at the computed heat-transfer area. Read the heater dimensions directly from the table. Be sure to use the next *larger* heat-transfer area when the exact required area is not available.

When there is little movement of the fluid on either side of the heat-transfer area, such as occurs during heat transmission through a building wall, the arithmetic mean (average) temperature difference can be used instead of the LMTD. Use the LMTD when there is rapid movement of the fluids on either side of the heat-transfer area and a rapid change in temperature in one, or both, fluids. When one of the two fluids is partially, but not totally, evaporated or condensed, the true mean temperature difference is different from the arithmetic mean and the LMTD. Special methods, such as those presented in Perry—*Chemical Engineers' Handbook*, must be used to compute the actual temperature difference under these conditions.

When two liquids or gases with constant specific heats are exchanging heat in a heat exchanger, the area between their temperature curves, Fig. 2, is a measure of the total heat being transferred. Figure 2 shows how the temperature curves vary with the

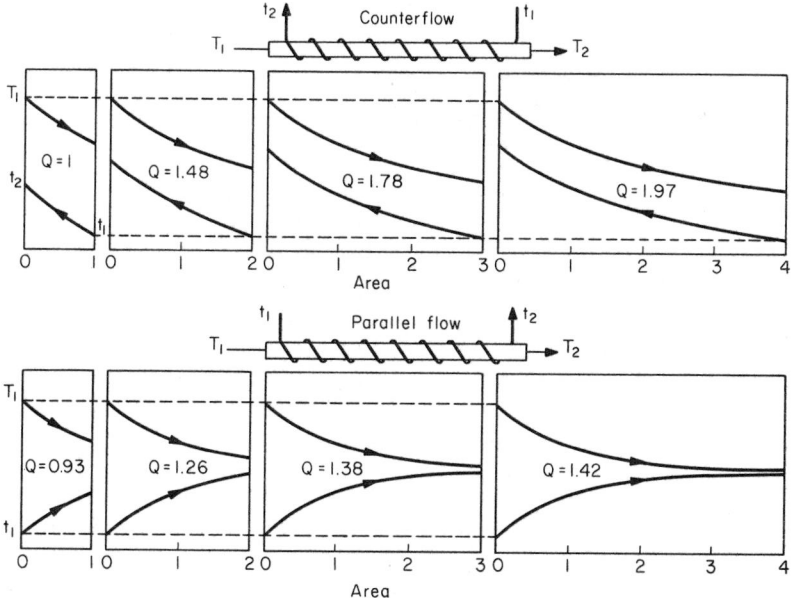

Fig. 2 For certain conditions, the area between the temperature curves measures the amount of heat being transferred.

amount of heat-transfer area for counterflow and parallel-flow exchangers when the fluid inlet temperatures are kept constant. As Fig. 2 shows, the counterflow arrangement is superior.

If enough heating surface is provided, in a counterflow exchanger, the leaving cold-fluid temperature can be raised above the leaving hot-fluid temperature. This cannot be done in a parallel-flow exchanger, where the temperatures can only approach each other regardless of how much surface is used. The counterflow arrangement transfers more heat for given conditions and usually proves more economical to use.

HEAT EXCHANGER ACTUAL TEMPERATURE DIFFERENCE

A counterflow shell-and-tube heat exchanger has one shell pass for the heating fluid and two shell passes for the fluid being heated. What is the actual LMTD for this exchanger if $T_1 = 300$ F, $T_2 = 250$ F, $t_1 = 100$ F, and $t_2 = 230$ F?

Calculation Procedure:

1. Determine how the LTMD should be computed

When the number of shell and tube passes are unequal, true counterflow does not exist in the heat exchanger. To allow for this deviation from true counterflow, a correction factor must be applied to the logarithmic mean temperature difference (LTMD). Figure 3 gives the correction factor to use.

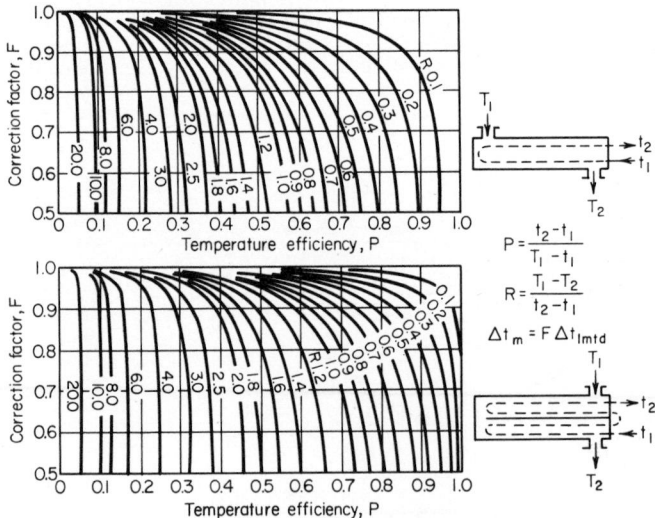

$$P = \frac{t_2 - t_1}{T_1 - t_1}$$

$$R = \frac{T_1 - T_2}{t_2 - t_1}$$

$$\Delta t_m = F \, \Delta t_{lmtd}$$

Fig. 3   Correction factors for LMTD when the heater flow path differs from true counterflow. (*Power.*)

2. Compute the variables for the correction factor

The two variables that determine the correction factor are shown in Fig. 3 as $P = (t_2 - t_1)/(T_1 - t_1)$, and $R = (T_1 - T_2)/(t_2 - t_1)$. Thus, $P = (230 - 100)/(300 - 100) = 0.65$, and $R = (300 - 250)/(230 - 100) = 0.385$. From Fig. 3, the correction factor $F = 0.90$ for these values of P and R.

3. Compute the theoretical LMTD

Use the relation LMTD $= (G - L)/\ln G/L$, where the symbols for counterflow heat exchange are $G = T_2 - t_1$; $L = T_1 - t_2$; $\ln = $ logarithm to the base e. All temperatures in this equation are expressed in F. Thus, $G = 250 - 100 = 150$ F; $L = 300 - 230 = 70$ F. Then, LMTD $= (150 - 70)/(\ln 150/70) = 105$ F.

4. Compute the actual LMTD for this exchanger

The actual LMTD for this or any other heat exchanger is $\text{LMTD}_{actual} = F(\text{LMTD}_{computed})$ $= 0.9(105) = 94.5$ F. Use the actual LMTD to compute the required exchanger heat-transfer area.

Related Calculations: Once the corrected LMTD is known, compute the required heat exchanger size in the manner shown in the previous Calculation Procedure. The method given here is valid for both two- and four-pass shell-and-tube heat exchangers. Figure 4 simplifies the computation of the uncorrected LMTD for temperature differences ranging from 1 to 1000 F. It gives LMTD with sufficient accuracy for all normal industrial and commercial heat exchanger applications. Correction-factor charts for three shell passes, six or more tube passes, four shell passes, and eight or

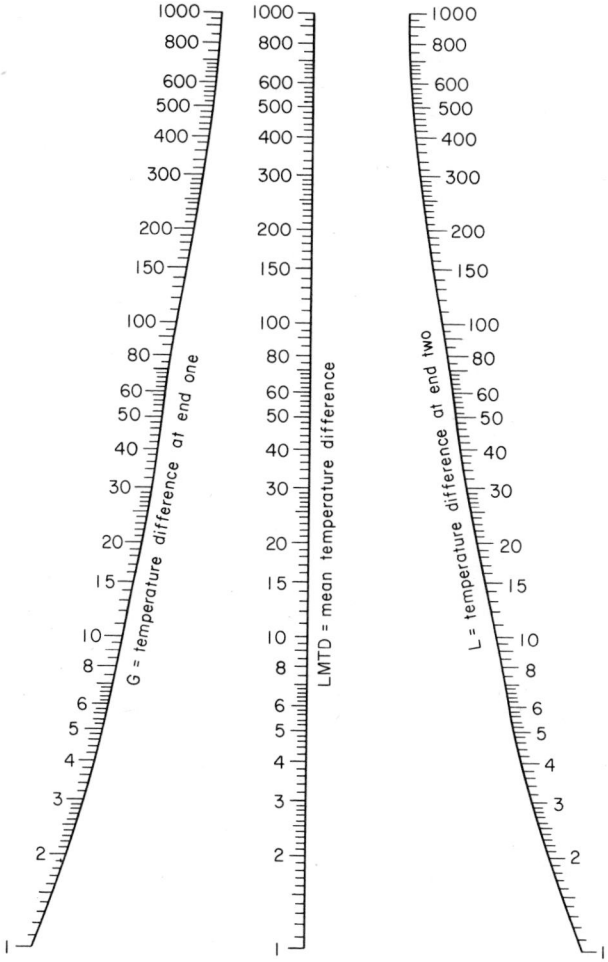

Fig. 4   Logarithmic mean temperature for a variety of heat-transfer applications.

more tube passes are published in the *Standards of the Tubular Exchanger Manufacturers Association.*

FOULING FACTORS IN HEAT EXCHANGER SIZING AND SELECTION

A heat exchanger having an overall coefficient of heat transfer of $U = 100$ Btu/(sq ft)(hr) (F) is used to cool lean oil. What effect will the tube fouling have on the value of U for this exchanger?

Calculation Procedure:

1. Determine the heat exchanger fouling factor

Use Table 2 to determine the fouling factor for this exchanger. Thus, the fouling factor for lean oil = 0.0020.

TABLE 2   Heat Exchanger Fouling Factors*

Fluid Heated or Cooled	Fouling Factor
Fuel oil	0.0055
Lean oil	0.0020
Clean recirculated oil	0.0010
Quench oils	0.0042
Refrigerants (liquid)	0.0011
Gasoline	0.0006
Steam-clean and oil-free	0.0001
Refrigerant vapors	0.0023
Diesel exhaust	0.013
Compressed air	0.0022
Clean air	0.0011
Seawater under 130 F	0.0006
Seawater over 130 F	0.0011
City or well water under 130 F	0.0011
City or well water over 130 F	0.0021
Treated boiler feedwater under 130 F, 3 fps	0.0008
Treated boiler feedwater over 130 F, 3 fps	0.0009
Boiler blowdown	0.0022

*Condenser Service and Engineering Company, Inc.

2. Determine the actual U for the heat exchanger

Enter Fig. 5 at the bottom with the clean heat-transfer coefficient of $U = 100$ and project vertically upward to the 0.002 fouling-factor curve. From the intersection with this curve, project horizontally to the left to read the design or actual heat-transfer coefficient as $U_a = 78$ Btu/(hr)(sq ft)(F). Thus, the fouling of the tubes causes a reduction of the U value of $100 - 78 = 22$ Btu/(hr)(sq ft)(F). This means that the required heat transfer area must be increased by nearly 25 percent to compensate for the reduction in heat transfer caused by fouling.

Related Calculations: Table 2 gives fouling factors for a wide variety of service conditions in applications of many types. Use these factors as described above; or add the fouling factor to the film resistance for the heat exchanger to obtain the total resistance to heat transfer. Then U = the reciprocal of the total resistance. Use the actual value U_a of the heat-transfer coefficient when sizing a heat exchanger. The method given here is that used by Condenser Service and Engineering Company, Inc.

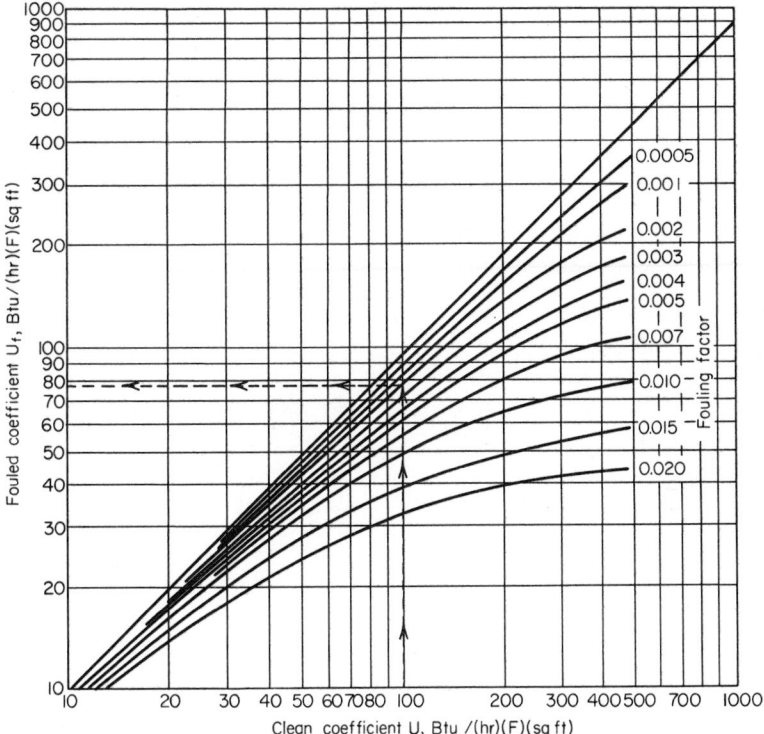

Fig. 5 Effect of heat exchanger fouling on the overall coefficient of heat transfer. (*Condenser Service and Engineering Co., Inc.*)

HEAT TRANSFER IN BAROMETRIC AND JET CONDENSERS

A counterflow barometric condenser must maintain an exhaust pressure of 2 psia for an industrial process. What condensing-water flow rate is required with a cooling-water inlet temperature of 60F; of 80F? How much air must be removed from this barometric condenser if the steam flow rate is 25,000 lb/hr; 250,000 lb/hr?

Calculation Procedure:

1. Compute the required unit cooling water flow rate

Use Fig. 6 as a quick guide to the required cooling-water flow rate for counterflow barometric condensers. Thus, entering the bottom of Fig. 6 at 2-psia exhaust pressure and projecting vertically upward to the 60-F and 80-F cooling-water

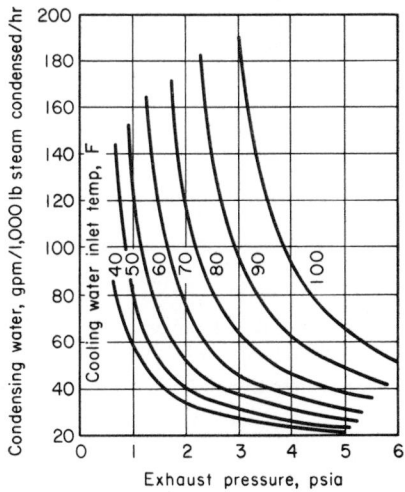

Fig. 6 Barometric condenser condensing-water flow rate.

inlet temperature curves shows that the required flow rate is 52 gpm and 120 gpm, respectively, per 1,000 lb of steam condensed per hr.

2. Compute the total cooling-water flow rate required

Use the relation, total cooling water required, gpm = (unit cooling-water flow rate, gpm per 1,000 lb of steam condensed per hr)(steam flow, lb/hr)/1,000. Or, total gpm = (52)(250,000/1,000) = 13,000 gpm of 60-F cooling water. For 80-F cooling water, total gpm = (120)(250,000/1,000) = 30,000 gpm. Thus, a 20-F rise in the cooling-water temperature raises the flow rate required by 30,000 − 13,000 = 17,000 gpm.

3. Compute the quantity of air that must be handled

With a steam flow of 25,000 lb/hr to a barometric condenser, manufacturers' engineering data show that the quantity of air entering with the steam is 3 cfm; with a steam flow 250,000 lb/hr, air enters at the rate of 10 cfm. Hence, the quantity of air in the steam that must be handled by this condenser is 10 cfm.

Air entering with the cooling water varies from about 2 cfm per 1,000 gpm of 100-F water to 4 cfm per 1,000 gpm at 35 F. Using a value of 3 cfm for this condenser, the quantity of air that must be handled is (cfm per 1,000 gpm)(cooling-water flow rate, gpm)/1,000, or cfm of air = (3)(13,000/1,000) = 39 cfm at 60 F. At 80 F, cfm = (3)(30,000/ 1,000) = 90 cfm.

Hence, the total air quantity that must be handled is 39 + 10 = 49 cfm with 60-F cooling water, and 90 + 10 = 100 cfm with 80-F cooling water. The air is usually removed from the barometric condenser by a two-stage air ejector.

Related Calculations: For help in specifying conditions for parallel-flow and counter-flow barometric condensers, refer to *Standards of Heat Exchange Institute — Barometric and Low-Level Jet Condensers.* Whereas Fig. 6 can be used for a first approximation of the cooling water required for parallel-flow barometric condensers, the results obtained will not be as accurate as for counterflow condensers.

SELECTION OF A FINNED-TUBE HEAT EXCHANGER

Choose a finned-tube heat exchanger for a 1,000-hp four-cycle turbocharged diesel engine having oil-cooled pistons and a cooled exhaust manifold. The heat exchanger will be used only for jacket-water cooling.

Calculation Procedure:

1. Determine the heat exchanger cooling load

The Diesel Engine Manufacturers Association (DEMA) tabulation, Table 3, lists the heat rejection to the cooling system by various types of diesel engines. Table 3 shows that the heat rejection from the jacket water of a four-cycle turbocharged engine having oil-cooled pistons and a cooled manifold is 1,800 to 2,200 Btu/bhp-hr. Using the higher value, the jacket-water heat rejection by this engine is (1,000 bhp)(2,200 Btu/ bhp-hr) = 2,200,000 Btu/hr.

2. Determine the jacket-water temperature rise

DEMA reports that a water temperature rise of 15 to 20 F is common during passage of the cooling water through the engine. The maximum water discharge temperature reported by DEMA ranges from 140 to 180 F. Assume a 20-F water temperature rise and a 160-F water discharge temperature for this engine.

3. Determine the air inlet and outlet temperatures

Refer to weather data for the locality of the engine installation. Assume that the weather data for the locality of this engine shows that the maximum dry-bulb temperature met in summer is 90 F. Use this as the air inlet temperature.

Before the required surface area can be determined, the air outlet temperature from the radiator must be known. This outlet temperature cannot be computed directly. Hence, it must be assumed and a trial calculation made. If the area obtained is too large,

TABLE 3   Approximate Rates of Heat Rejection to Cooling Systems*

	Four-cycle engines			
Engine type	Normally aspirated, dry pistons water-jacketed exhaust manifold, Btu/bhp-hr	Normally aspirated, oil-cooled pistons, water-jacketed manifold, Btu/bhp-hr	Turbo-charged oil-cooled pistons, dry manifold, Btu/bhp-hr	Turbo-charged oil-cooled pistons, cooled manifold, Btu/bhp-hr
Jacket water	2,200–2,600	2,000–2,500	1,450–1,750	1,800–2,200
Lubricating oil	175–350	300–600	300–500	300–500
Raw water	2,375–2,950	2,300–3,100	1,750–2,250	2,100–2,700

	Two-cycle engines		
	Loop scavenging oil-cooled pistons, Btu/bhp-hr	Uniflow scavenging oil-cooled pistons	
Engine type		Opposed piston, Btu/bhp-hr	Valve in head, Btu/bhp-hr
Jacket water	1,300–1,900	1,200–1,600	1,700–2,100
Lubricating oil	500–700	900–1,100	400–750
Raw water........	1,800–2,600	2,100–2,700	2,100–2,850

*Diesel Engine Manufacturers Association.

a higher outlet air temperature must be assumed and the calculation made again. Assume an outlet air temperature of 150 F.

4. Compute the LMTD for the radiator

The largest temperature difference for this exchanger is $160 - 90 = 70$ F, and the smallest temperature difference is $150 - 140 = 10$ F. In the smallest temperature difference expression, 140 F = water discharge temperature from the engine — cooling-water temperature rise during passage through the engine, or $160 - 20 = 140$ F. Then, LMTD $= (70 - 10)/(\ln 70/10) = 30$ F. (Figure 4 could also be used to compute the LMTD.)

5. Compute the required exchanger surface area

Use the relation $A = Q/U \times LMTD$, where A = surface area required, sq ft; Q = rate of heat transfer, Btu/hr; U = overall coefficient of heat transfer, Btu/(hr)(sq ft)(F). To solve this equation, U must be known.

Table 1 in the first Calculation Procedure in this section shows that U ranges from 2 to 10 Btu/(hr)(sq ft)(F) in the usual internal-combustion engine finned-tube radiator. Using a value of 5 for U, $A = 2,200,000/[(5)(30)] = 14,650$ sq ft.

6. Determine the length of finned tubing required

The total area of a finned tube is the sum of the tube and fin area per unit length. The tube area is a function of the tube diameter, whereas the finned area is a function of the number of fins per inch of tube length and the tube diameter.

Assume that 1-in.-diameter tubes having four fins per in. are used in this radiator. A tube manufacturer's engineering data shows that a finned tube of these dimensions has 5.8 sq ft of area per lineal foot of tube.

To compute the lin ft L of finned tubing required, use the relation $L = A/(\text{sq ft/ft})$, or $L = 14,650/5.8 = 2,530$ lin ft of tubing.

6. Compute the number of individual tubes required

Assume a length for the radiator tubes. Typical lengths range between 4 and 20 ft, depending on the size of the radiator. Using a length of 16 ft per tube, the total number of tubes required $= 2,530/16 = 158$ tubes. This number is typical for finned-tube heat exchangers having large (more than 10^6 Btu/hr) heat-transfer rates.

7. Determine the fan horsepower required

The fan horsepower required can be computed by determining the quantity of air that must be moved through the heat exchanger, after assuming a resistance – say 1.0 in. of water – for the exchanger. However, the more common way of determining the fan horsepower is by referring to the manufacturer's engineering data.

Thus, one manufacturer recommends three 5-hp fans for this cooling load and another recommends two 8-hp fans. Hence, about 16 hp is required for the radiator.

Related Calculations: The steps given here are suitable for the initial sizing of finned-tube heat exchangers for a variety of applications. For exact sizing, it may be necessary to apply a correction factor to the LMTD. These correction factors are published in Kern – *Process Heat Transfer*, McGraw-Hill, and McAdams – *Heat Transfer*, McGraw-Hill.

The method presented here can be used for finned-tube heat exchanges used for air heating or cooling, gas heating or cooling, and similar industrial and commercial applications.

SPIRAL-TYPE HEATING COIL SELECTION

How many ft of heating coil are required to heat 1,000 gph of 0.85 specific gravity oil if the specific heat of the oil is 0.50 Btu/(lb)(F), the heating medium is 65-psig steam, and the oil enters at 60 F and leaves at 125 F? There is no subcooling of the condensate.

Calculation Procedure:

1. Compute the LMTD for the heater

Steam at $65 + 14.7 = 79.7$ psia has a temperature of approximately 312 F, as given by the steam tables. Condensate at this pressure has the same approximate temperature. Hence, the entering and leaving temperature of the heating fluid are approximately the same.

Oil enters the heater at 60 F and leaves at 125 F. Therefore, the greater temperature G across the heater is $G = 312 - 60 = 252$ F, and the lesser temperature difference L is $L = 312 - 125 = 187$ F. Hence, the LMTD $= (G - L)/(\ln G/L)$, or $(252 - 187)/(\ln 252/187) = 222$ F. In this relation, $\ln = $ logarithm to the base $e = 2.7183$. (Figure 4 could also be used to determine the LMTD.)

2. Compute the heat required to raise the oil temperature

Water weighs 8.33 lb/gal. Since this oil has a specific gravity of 0.85, it weighs $(8.33) \times (0.85) = 7.08$ lb/gal. With 1,000 gph of oil to be heated, the weight of oil heated is $(1,000$ gph$)(7.08$ lb/gal$) = 7,080$ lb/hr. Since the oil has a specific heat of 0.5 Btu/(lb)(F) and this oil is heated through a temperature range of $125 - 60 = 65$ F, the quantity of heat Q required to raise the temperature of the oil is $Q = (7,080$ lb/hr$)[0.5$ Btu/(lb)(F)$] \times (65$ F$) = 230,000$ Btu/hr.

3. Compute the heat-transfer area required

Use the relation $A = Q/U \times \text{LMTD}$, where $Q = $ heat-transfer rate, Btu/hr; $U = $ overall coefficient of heat transfer, Btu/(hr)(sq ft)(F). For heating oil to 125 F, the U value given in Table 1 is 20 to 60 Btu/(hr)(sq ft)(F). Using a value of $U = 30$ Btu/(hr)(sq ft)(F) to produce a conservatively sized heater, $A = 230,000/[(30)(222)] = 33.4$ sq ft of heating surface.

4. Choose the coil material for the heater

Spiral-type tank heating coils are usually made of steel because this material has a good corrosion resistance in oil. Hence, this coil will be assumed to be made of steel.

5. Compute the heating steam flow required

To determine the steam flow rate required, use the relation $S = Q/h_{fg}$, where $S =$ steam flow, lb/hr; $h_{fg} =$ latent heat of vaporization of the heating steam, Btu/lb, from the steam tables; other symbols as before. Hence, $S = 230,000/901.1 = 256$ lb/hr, closely.

6. Compute the heating coil pipe diameter

Steam heating coils submerged in the liquid being heated are usually chosen for a steam velocity of 4,000 to 5,000 fpm. Compute the heating pipe internal cross-sectional area a sq in. from $a = 2.4Sv_g/V$, where $v_g =$ specific volume of the steam at the coil operating pressure, cu ft/lb., from the steam vables; $V =$ steam velocity in the heating coil, fpm; other symbols as before. Using a steam velocity of 4,000 fpm, $a = 2.4(256) \times (5.47)/4,000 = 0.838$ sq in.

Refer to a tabulation of pipe properties. Such a tabulation shows that the internal transverse area of a schedule 40 1-in. nominal diameter steel pipe is 0.863 sq in. Hence, a 1-in. pipe will be suitable for this heating coil.

7. Determine the length of coil required

A pipe property tabulation shows that 2.9 lin ft of 1-in. schedule 40 pipe has 1.0 sq ft of external area. Hence, the total length of pipe required in this heating coil = (33.1 sq ft)(2.9 ft/sq ft) = 96 ft.

Related Calculations: Use this general Procedure to find the area and length of spiral heating coil required to heat water, industrial solutions, oils, etc. This Procedure can also be used to find the area and length of cooling coils used to cool brine, oils, alcohol, wine, etc. In every case, be certain to substitute the correct specific heat for the liquid being heated or cooled. Consult Perry—*Chemical Engineers' Handbook*, McGraw-Hill; McAdams—*Heat Transmission*, McGraw-Hill; or Kern—*Process Heat Transfer*, McGraw-Hill, for typical values of U.

SIZING ELECTRIC HEATERS FOR INDUSTRIAL USE

Choose the heating capacity of an electric heater to heat a pot containing 600 lb of lead from the charging temperature of 70 F to a temperature of 750 F if 600 lb of the lead is to be melted and heated per hr. The pot is 30 in. in diameter and 18 in. deep.

Calculation Procedure:

1. Compute the heat needed to reach the melting point

When a solid is melted it must first be raised from its ambient or room temperature to the melting temperature. The quantity of heat required is $H = $ (weight of solid, lb)[specific heat of solid, Btu/(lb)(F)]$(t_m - t_i)$, where $H = $ Btu required to raise the temperature of the solid; $t_m = $ melting temperature of the solid, F; $t_i = $ room, charging, or initial temperature of the solid, F.

For this pot with lead having a melting temperature of 620 F and an average specific heat of 0.031 Btu/(lb)(F), $H = (600)(0.031)(620 - 70) = 10,240$ Btu/hr, or 10,240/3,412 Btu/kw-hr = 2.98 kw-hr.

2. Compute the heat required to melt the solid

The heat H_m Btu required to melt a solid is $H_m = $ (weight of solid melted, lb)(heat of fusion of the solid, Btu/lb). Since the heat of fusion of lead is 10 Btu/lb, $H_m = (600)(10) = 6,000$ Btu/hr, or 6,000/3,412 = 1.752 kw-hr.

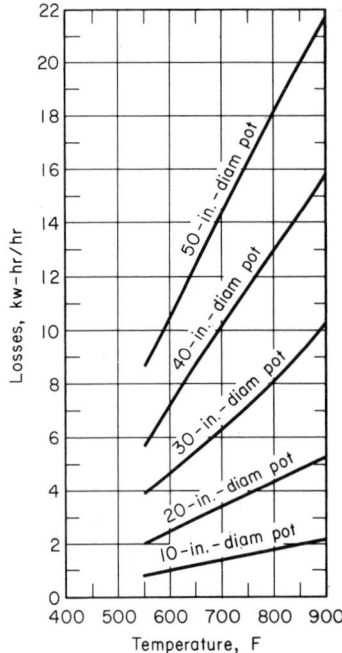

Fig. 7   Heat losses from melting pots. *(General Electric Co.)*

3. Compute the heat required to reach the working temperature

Use the same relation as in Step 1, except that the temperature range is expressed as $t_w - t_m$, where t_w = working temperature of the melted solid. Thus, for this pot, $H = (600)(0.031)(750 - 620) = 2,420$ Btu/hr, or $2,420/3,412 = 0.708$ kw-hr.

4. Determine the heat loss from the pot

Use Fig. 7 to determine the heat loss from the pot. Enter at the bottom of Fig. 7 at 750 F and project vertically upward to the 10-in.-diameter pot curve. At the left, read the heat loss at 7.3 kw-hr/hr.

5. Compute the total heating capacity required

The total heating capacity required is the sum of the individual capacities, or $2.98 + 1.752 + 0.708 + 7.30 = 12.74$ kw-hr. A 15-kw electric heater would be chosen because this is a standard size and it provides a moderate extra capacity for overloads.

Related Calculations: Use this general Procedure to compute the capacity required for an electric heater used to melt a solid of any kind — lead, tin, type metal, solder, etc. When the substance being heated is a liquid — water, dye, paint varnish, oil, etc. — use the relation H = (weight of liquid heated, lb) [specific heat of liquid, Btu/(lb)(F)] (temperature rise desired, F), when the liquid is heated to approximately its boiling temperature, or a lower temperature.

For space heating of commercial and residential buildings, two methods used for computing the approximate wattage required are: (*a*) the watts/cu ft, and (*b*) the "35" method. These are summarized in Table 4. In many cases, the results given by these methods agree closely with more involved calculations. When the desired room tem-

TABLE 4   Two Methods for Determining Wattage for Heating Buildings Electrically*

Watts/cu ft Method	
1. Interior rooms with no or little outside exposure	0.75 to 1.25
2. Average rooms with moderate windows and doors....................................	1.25 to 1.75
3. Rooms with severe exposure and great window and door space..........................	1.0 to 4.0
4. Isolated rooms, cabins, watchhouses, and similar buildings...........................	3.0 to 6.0

The "35" Method	
1. Volume in cu ft for one air change × 0.35 = ...	watts
2. Exposed net wall, roof, or ceiling and floor in sq ft × 3.5 =................................	watts
3. Area of exposed glass and doors in sq ft × 35.0 =	watts

*General Electric Company.

perature is different from 70 F, increase or decrease the required kilowatt capacity proportionately, depending on whether the desired temperature is higher than or less than 70 F.

For heating pipes with electric heaters, use a heater capacity of 0.8 watt/sq ft of uninsulated exterior pipe surface per F temperature difference between the pipe and the surrounding air. If the pipe is insulated with 1 in. of insulation, use 30 percent of this value, or 0.24 watts/(sq ft)(F).

The types of electric heaters used today include immersion (for water, oil, plating liquids, etc.), strip, cartridge, tubular, vane, fin, unit, and edgewound resistor heaters. These heaters are used in a wide variety of applications including liquid heating, gas and air heating, oven warming, deicing, humidifying, plastics heating, pipe heating, etc.

For pipe heating, a tubular heating element can be fastened to the bottom of the pipe and run parallel with it. For large-wattage applications, the heater can be spiraled around the pipe. For temperatures below 165 F, heating cable can be used. Electric heating is often used in place of steam tracing of outdoor pipes.

The Procedure presented above is the work of General Electric Company.

REFRIGERATION

REFERENCES: Emerick—*Heating Design and Practice*, McGraw-Hill; Carrier Air Conditioning Company—*Handbook of Air Conditioning System Design*, McGraw-Hill; Severns and Fellows —*Air Conditioning and Refrigeration*, Wiley; ASHRAE—*Guide and Data Book: Fundamentals and Equipment*; ASHRAE—*Guide and Data Book: Applications*; Strock and Koral—*Handbook of Heating, Air Conditioning, and Ventilation*, Industrial Press; American Blower Corporation—*Air Conditioning & Engineering*; MacIntire-Hutchinson—*Refrigeration Engineering*, Wiley.

REFRIGERATION SYSTEM SELECTION

Choose a refrigeration system for a given load. Show the steps that the designer should follow in choosing a suitable refrigeration system for various types of loads.

Calculation Procedure:

1. Determine the refrigeration load

Use the method given in the next Calculation Procedure. In any refrigeration plant, the total refrigeration load = heat gain from external sources, tons + product load, tons + sensible heat load, tons.

2. Choose the type of refrigeration system to use

Table 1 shows the usual compressor choices for various refrigeration loads. Thus, reciprocating compressors find wide use for refrigeration loads up to 400 tons. Up to loads of about 5 tons, *unit systems* that combine the compressor, drive, evaporator, and condenser in a compact unit are popular. In some instances, larger-capacity unit systems may be available from certain manufacturers. Some large unit systems, called *central-station systems*, are built with capacities of 100 to 150 tons.

From 5 to 400 tons capacity, *built-up central systems* are popular. In these systems, the manufacturer supplies the compressor, evaporator, and condenser as separate units. These are then connected by suitable piping. The refrigeration equipment manufacturer may, or may not, supply the compressor driving unit. This driver may be an electric motor, steam turbine, internal-combustion engine, or some other type of prime mover.

TABLE 1   Typical Refrigeration System Choices*

System load, tons	System type		
	Often used	Occasionally used	Rarely used
0–5	Unit system with reciprocating compressor	Central-station built-up system; reciprocating compressor	Central-station built-up units
5–25	Central-station built-up systems; reciprocating compressor	Central station built-up systems; reciprocating compressor	Absorption or adsorption units
25–50	Central-station built-up systems; reciprocating compressor	Central-station built-up systems; centrifugal compressor	Absorption units
50–400	Central-station built-up systems; reciprocating compressor	Central-station built-up systems; steam-jet and centrifugal compressors	
400 and up	Central system; centrifugal and/or absorption unit	Central station built-up steam-jet unit	

*Adapted from ASHRAE data.

For loads greater than 400 tons, the centrifugal refrigeration compressor is often chosen. Whereas this may be a built-up system, more and more manufacturers today supply completely fabricated systems containing all the needed components, including the controls, driver, etc.

Steam-jet refrigeration units find some application for loads of 50 tons or more. The steam-jet refrigeration system is used for a large number of applications where steam is available. Typical applications include comfort air conditioning, industrial process cooling, and similar service. In recent years, some large office buildings have used steam-jet systems mounted in the building penthouse. These units provide the cooling needed for the building air-conditioning system.

Absorption refrigeration systems were once popular for a variety of cooling tasks in industry, food storage, etc. In recent years, the absorption system has found renewed use in medium- and large-size air-conditioning systems. The usual absorbent used today is lithium bromide; the refrigerant is ordinary tap water. Absorption refrigeration systems are popular in areas where fuel costs are low, electric rates are high, waste steam is available, low-pressure heating boilers are unused during the cooling season, or steam or gas utility companies desire to promote summer loads. Absorption refrigeration systems can be installed in almost any location in a building where the floor is of adequate strength and reasonably level. Absence of heavy moving parts practically eliminates vibration and reduces the noise level to a minimum.

Combination absorption-centrifugal refrigeration systems are well suited for many large-tonnage air-conditioning and industrial loads. These systems are extremely economical where medium- or high-pressure steam is used as the energy source.

Using the expected refrigeration load from Step 1 and the data above, make a preliminary choice of the type of refrigeration systems to use. Remember that the necessity for part-load operation might change the preliminary choice of the system type.

3. Choose the system components

Manufacturer's engineering data generally lists compatible components for a given capacity compressor. These components include the condenser, expansion valve, evaporator, receiver, cooling tower, etc. Later Calculation Procedures in this section give specific instructions for selecting these and the other important components of the system. When a unit system is chosen, the important components are preselected by the manufacturer.

4. Have the system choice verified

Have the manufacturer whose equipment will be used verify the selection for the given load. This ensures a correct choice.

Related Calculations: Use this general Procedure to select the type of refrigeration system serving air-conditioning, product-cooling, liquid-cooling, ice-making, and similar applications in stationary (land) and marine service.

SELECTION OF A REFRIGERATION UNIT FOR PRODUCT COOLING

What capacity and type of refrigeration system is needed for a walk-in cooler having inside dimensions of $8 \times 6 \times 10$ ft if it is insulated with 4-in.-thick cork? The user estimates that a maximum of 400 lb of beef will be placed in the cooler daily, arriving at 70 F. The average hottest summer day in the cooler locality is, according to weather bureau records, 92 F. The meat is to be stored at 36 F. A $\frac{1}{8}$-hp blower circulates air in the cooler. What refrigeration capacity is required for the same cooler, except that the meat is stored at -10 F and the cork insulation is 8 in. thick? Two $\frac{1}{8}$-hp blowers will be used in the cooler.

Calculation Procedure:

1. Compute the outside area of the cooler

The outside dimensions of this cooler are 9 ft high, 7 ft wide, and 11 ft long, including the cork insulation and the supporting structure. Hence, the total outside area of the cooler, including the floor and roof, is $2(9 \times 7) + 2(9 \times 11) + 2(7 \times 11) = 478$ sq ft.

2. Compute the heat gain and service load

There is a heat gain *into* the cooler through the insulated surfaces caused by the difference in the inside and outside temperature. Also, there is a *service load*, that is, a heat gain caused by the opening and shutting of the cooler door. Since meat will be loaded only once per day, it is safe to assume that the service load is a normal one—i.e., the door will be opened less than five times per hr.

For product storage cooling, heat and service load, Btu/hr, = (total outside area of cooler, sq ft)(maximum outside temperature, F − minimum inside temperature, F)(factor from Table 2), or $(478)(92 - 36)(0.110) = 2,944$ Btu/hr.

TABLE 2   Heat leakage factors*

	Btu per hr per deg temperature difference per sq ft of outside surface					
	Insulation thickness, in.					
	1	2	4	6	8	10
Heat leakage only............	0.178	0.127	0.079	0.059	0.046	0.038
Heat leakage plus normal service load	0.216	0.163	0.110	0.090	0.077	0.069

NOTE: Light duty—multiply factor by 0.90; heavy duty—multiply factor by 1.10; single glass—multiply factor by 15; double glass—multiply factor by 6.5; triple glass—multiply factor by 5. If any wall or ceiling of a cooler is exposed to the sun, increase temperature difference by 20° for that wall.

*Brunner Manufacturing Company.

3. *Compute the product heat load*

Use this relation: product heat load, Btu/hr, = (lb of product cooled per hr)(temperature of product entering cooler, F — temperature of product leaving cooler, F)(specific heat of product, Btu/(lb)(F). For this cooler, using the specific heat from Table 3, the product heat load = (400 lb/24 hr)(70 − 36)(0.8) = 453 Btu/hr.

TABLE 3   Typical Specific and Latent Heats* [Btu/(lb)(F)]

Article	Specific heat		Latent heat of freezing	Cold storage temperature
	Above freezing	Below freezing		
Canned goods:				
Fruits	As	Fresh	...	35–40
Meats	As	Fresh	...	35–40
Sardines	0.760	0.410	101.0	35–40
Butter, eggs, etc.:				
Butter	0.302	0.238	18.4	18–20
Cheese	0.480	0.305	50.5	34
Eggs	0.760	0.410	100.0	31
Milk, ice cream	0.900	0.462	124.0	35
Flour, meal (wheat)	0.26–0.38	0.210–0.280	14.4–28.8	36–40
Vegetables:				
Asparagus	0.952	0.482	134.0	34–35
Cabbage	0.928	0.473	131.0	34–35
Carrots	0.864	0.449	119.5	34–35
Celery (edible portion)	0.952	0.482	135.0	34–35
Dried beans	0.300	0.237	18.0	32–45
Dried corn	0.284	0.231	15.1	35–45
Dried peas	0.276	0.224	13.7	35–45
Onions	0.900	0.462	126.0	36
Parsnips	0.864	0.449	119.5	34–35
Potatoes	0.792	0.422	106.5	36–40
Sauerkraut	0.912	0.467	128.0	35
Miscellaneous:				
Cigars, tobacco	...	...	...	35–42
Furs, woolens, etc.	...	...	...	35
Honey	0.344	0.254	25.9	36–40
Hops	...	...	...	32–40
Maple syrup	0.488	0.308	51.8	40–45
Maple sugar	0.240	0.215	7.2	40–45
Poultry dressed iced	0.790	0.421	105.0	28–30
Poultry dry packed	0.720	0.395	93.5	26–28
Poultry scalded	0.800	0.425	108.0	20
Game frozen	0.680	0.380	86.5	15–28
Poultry frozen	0.680	0.380	86.5	15–28
Nuts (dried)	0.21–0.294	0.195–0.244	4.3–14.4	35–40
Water	1.0	0.5	144	...
Meats:				
Fresh (typical only)	0.80	0.404	20–40	
Fruits:				
Fresh (typical only)	0.70	0.387	32–55	

*Brunner Manufacturing Company.

4. Compute the total heat load

The total heat load = sum of heat gain and service load + product heat load + supplementary heat load, Btu/hr, or 2,944 + 453 + 424 = 3,821 Btu/hr.

5. Compute the refrigeration-system capacity required

In cooler operation, it is essential to assure defrosting of the evaporator during the off cycle. To permit this defrosting, select a condensing unit to operate 18 hr per 24-hr day. With an 18-hr operating time, the required condensing-unit capacity to handle the 24-hr load is (24 hr/operating time, hr)(total heat load, Btu/hr) = (24/18)(3,821) = 5,082 Btu/hr.

6. Select the refrigeration unit

Since the required capacity of this refrigeration system is between 0 and 5 tons, the previous Calculation Procedure indicates that a unit system with a reciprocating compressor is the most common type used. Referring to a manufacturer's engineering data shows that a 5,000 Btu/hr 0.5-hp air-cooled unit having a 20-F suction temperature is available. This unit will operate about 18.5 hr/day to carry the actual heat load of 5,082 Btu/hr if the evaporator is chosen on a 16-F (36 F − 20 F) temperature difference between the room and refrigerant.

The exact size of a condensing unit cannot be selected until a choice is made of the evaporating temperature, or suction pressure, at which the compressor is to work. In general, a difference of between 10 and 20 F should be maintained between the product or room temperature and the evaporator temperature. Thus, the 16-F temperature difference used above is within the normal working range.

A better plan for this product cooler would be to select a larger evaporator on a 10-F temperature-difference basis. The running time of the condensing unit would be decreased because of the higher operating suction temperature, that is, 26 instead of 20 F.

A standard refrigeration unit having the same characteristics as the unit described above, except that the evaporating temperature is 25 F, has a capacity of 5,550 Btu/hr with refrigerant 12. The evaporating pressure is 24.6 psi. The compressor is a two-cylinder unit and is belt-driven by an electric motor. A finned-tube air-cooled condenser is used. The receiver is mounted below the compressor, on the same frame. Refrigerant 12 (formerly called Freon-12) is most satisfactory for low-temperature systems.

7. Compute the required capacity at below-freezing temperature

The six steps above are for product storage at temperatures above freezing—i.e., above 32 F. For temperatures below 32 F, the same Procedure is followed except that the product load is computed in three steps—cooling to 32 F; freezing; and cooling to the final temperature. Thus:

Heat and service load = 478[92 − (−10)](0.085) = 4,140 Btu/hr, using the heavy-duty factor from Table 2.

Product load: Cooling to 32 F, (400/24)(70 − 32)(0.8) = 504 Btu/hr. For the freezing process, Btu/hr heat removal = (lb of product cooled per hr)(latent heat or enthalpy of freezing, Btu/lb, from Table 3) or (400 lb/24 hr)(98) = 1,635 Btu/hr. To cool below freezing, Btu/hr = (lb of product cooled per hr)(32 F − temperature of storage room, F) × (specific heat of product at temperature below freezing, Btu/(lb)(F), from Table 3), or (400/24)[32 − (−10)](0.404) = 282 Btu/hr. Then, total product load = 504 + 1,635 + 282 = 2,421 Btu/hr.

The supplementary load with two ⅛-hp blowers is (2)(⅛)(2,545) = 635 Btu/hr.

The total load is thus 4,140 + 2,421 + 635 = 7,196 Btu/hr. Assuming a 16 hr/day operating time for the refrigeration unit, the condensing capacity required is 7,196(24/16) = 10,800 Btu/hr.

Choose an evaporator for a 10-F temperature difference with a capacity of 10,800 Btu/hr at a suction temperature of −20 F. Checking a manufacturer's engineering data shows that a 3-hp air-cooled two-cylinder unit will be suitable.

Related Calculations: Use this general Procedure to choose refrigeration units for stationary, mobile (truck), and marine applications of walk-in coolers, display cases, milk and bottle coolers, ice-cream freezers and hardners, air conditioning, etc. Note that one Procedure is used for applications above 32 F and another is used for applications below 32 F.

In general, choose a unit for a 10- to 20-F difference between the product and evaporator temperature. Thus, where a room is maintained at 40 F, choose a condensing unit capacity corresponding to above a 25-F evaporator temperature. If brine is to be cooled to 5 F, select a condensing unit for about -10 F. Where a high relative humidity is desired in a cold room, select cooling coils with a large surface area so that the minimum operating differential temperature can be maintained between the room and the coil. The Procedure and data given here were published by the Brunner Manufacturing Company based on ASRE data.

ENERGY REQUIRED FOR STEAM-JET REFRIGERATION

A steam-jet refrigeration system operates with an evaporator temperature of 45 F and a chilled-water inlet temperature of 60 F. The condenser operating pressure is 1.135 psia, and the steam-jet ejectors use 3.1 lb of boiler steam per pound of vapor removed from the evaporator. How many pounds of boiler steam are required per ton of refrigeration produced? How much steam is required per hr for a 100-ton capacity steam-jet refrigeration unit?

Calculation Procedure:

1. Determine the system pressures and enthalpies

Using the steam tables, find the following values. At 45 F, $P = 0.1475$ psia; $h_f = 13.06$ Btu/lb; $h_{fg} = 1,068.4$ Btu/lb. At 60 F, $h_f = 28.06$ Btu/lb. At 1.135 psia, $h_f = 73.95$ Btu/lb, where $P =$ absolute pressure, psia; $h_f =$ enthalpy of liquid, Btu/lb; $h_{fg} =$ enthalpy of vaporization, Btu/lb.

2. Compute the chilled-water heat pickup

The chilled-water inlet temperature is 60 F; and the chilled-water outlet temperature is the same as the evaporator temperature, or 45 F, as shown in Fig. 1. Hence, the chilled-water heat pickup = enthalpy at 60 F − enthalpy at 45 F, both expressed in Btu/lb. Or, heat pickup = $28.06 - 13.06 = 15.0$ Btu/lb.

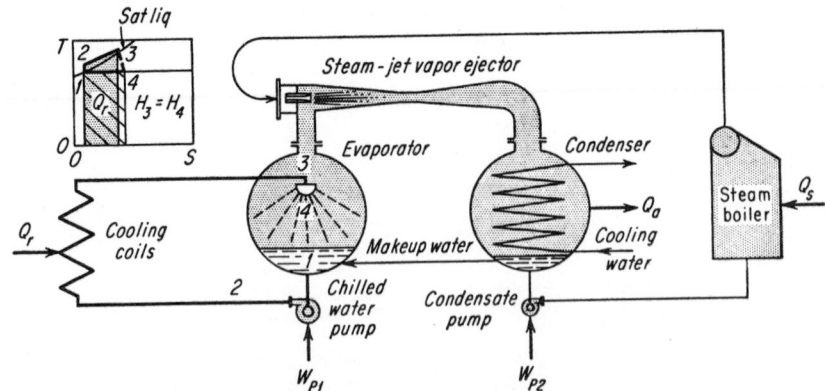

Fig. 1   Steam-jet refrigeration unit and *T-S* diagram of its operating cycle.

3. Compute the required chilled-water flow rate

Since a ton of refrigeration corresponds to a heat removal rate of 12,000 Btu/hr, the chilled-water flow rate, lb per hr per ton of refrigeration = (12,000 Btu/hr)/(chilled-water heat pickup, Btu/lb) = $12,000/15 = 8,000$ lb/(hr)(ton).

4. Compute the quantity of chilled water that vaporizes

Figure 1 shows the three fluid cycles involved: (a) chilled-water flow from the evaporator to the cooling coils and back, (b) chilled-water vapor flow from the evaporator through the ejector to the condenser and back as makeup, and (c) boiler steam flow from the boiler to the ejector to the condenser and back to the boiler as condensate.

Base the calculations on 1 lb of chilled water flowing through the cooling coils. For the throttling process from 3 to 4 in the evaporator, Fig. 1, the enthalpy remains constant but part of the chilled water vaporizes at the lower, or evaporator, pressure; hence, $H_3 = H_4 = h_f + xh_{fg}$, where x = lb of vapor formed per lb of chilled water entering, or $28.06 = 13.06 + x(1,068.4)$; $x = 0.01405$ lb of vapor per lb of chilled water entering. The quantity of chilled water remaining at 1 in the evaporator is $1.0 - 0.01405 = 0.98595$ lb per lb of chilled water recirculating.

5. Compute the quantity of makeup vaporized

Some of the condensate in the condenser returns to the evaporator as makeup, Fig. 1. This makeup throttles into the evaporator and part of it evaporates. Hence, $H_m = h_f + x_m h_{fg}$, where H_m = enthalpy of condensate, Btu/lb; x_m = quantity of makeup vaporized, lb per lb of makeup water. Since the enthalpy of the condensate at the condenser pressure of 1.135 psia is 73.95 Btu/lb, $73.95 = 13.06 + x_m(1068.4)$; $x_m = 0.057$ lb of makeup vaporized per lb of makeup water entering the evaporator.

Makeup vapor simply recirculates between the evaporator and the condenser. So the total makeup water entering the evaporator must replace both the chilled-water vapor and the makeup vapor formed by the two throttling processes.

6. Compute the makeup vapor and water quantities

The lb of makeup vapor per lb of makeup water remaining in the evaporator = $x_m/(1.0 - x_m) = 0.0570/(1.0 - 0.0570) = 0.0604$.

The total makeup water to the evaporator needed to replace the vapor = $x(1 + \text{lb of makeup vapor per lb of makeup water}) = 0.01405(1 + 0.0604) = 0.01491$ lb per lb of chilled water circulating. This is also the vapor removed from the evaporator by the ejector.

7. Compute the total vapor removed from the evaporator

The total vapor removed from the evaporator = (lb of chilled water per hr per ton) × (makeup water per lb of chilled water circulated) = $(8,000)(0.01491) = 119.3$ lb per ton of refrigeration.

8. Compute the boiler steam required

The boiler steam required = (vapor removed from the evaporator, lb per ton of refrigeration)(steam-jet steam, lb per lb of vapor removed from the evaporator) = $(119.3)(3.1) = 370$ lb of boiler steam per ton of refrigeration. For a 100-ton machine, the boiler steam required = $(100)(370) = 37,000$ lb/hr.

Related Calculations: Use this general method for any steam-jet refrigeration system using water and steam to produce a low temperature for air conditioning, product cooling, manufacturing processes, or other applications. Note that any of the eight items computed can be found when the other variables are known.

REFRIGERATION COMPRESSOR CYCLE ANALYSIS

An ammonia refrigeration compressor takes its suction from the evaporator, Fig. 2a, at a temperature of -20 F and a quality of 95 percent. The compressor discharges at a pressure of 100 psia. Liquid ammonia leaves the condenser at 50 F. Find the heat absorbed by the evaporator, the work input to the compressor, the heat rejected to the condenser, the coefficient of performance (COP) of the cycle, horsepower per ton of refrigeration, the quality of the refrigerant at state 2, quantity of refrigerant circulated per ton of refrigeration, required rate of condensing-water flow for a 100-ton load, compressor displacement for a 100-ton capacity; and what cylinder dimensions are required for a 100-ton capacity if the stroke = 1.3(cylinder bore) and the compressor makes 200 rpm.

Calculation Procedure:

1. Compute the enthalpy and entropy at cycle points

Assume a constant-entropy compression process for this cycle. This is the usual procedure when analyzing a refrigeration compressor whose actual performance is not known.

Using Fig. 2b as a guide, $H_3 = h_f + x h_{fg}$, where H_3 = enthalpy at point 3, Btu/lb; h_f = enthalpy of liquid ammonia, Btu/lb from Table 4; h_{fg} = enthalpy of evaporation, Btu/lb, from the same table; x = vapor quality, expressed as a decimal. Since point 3

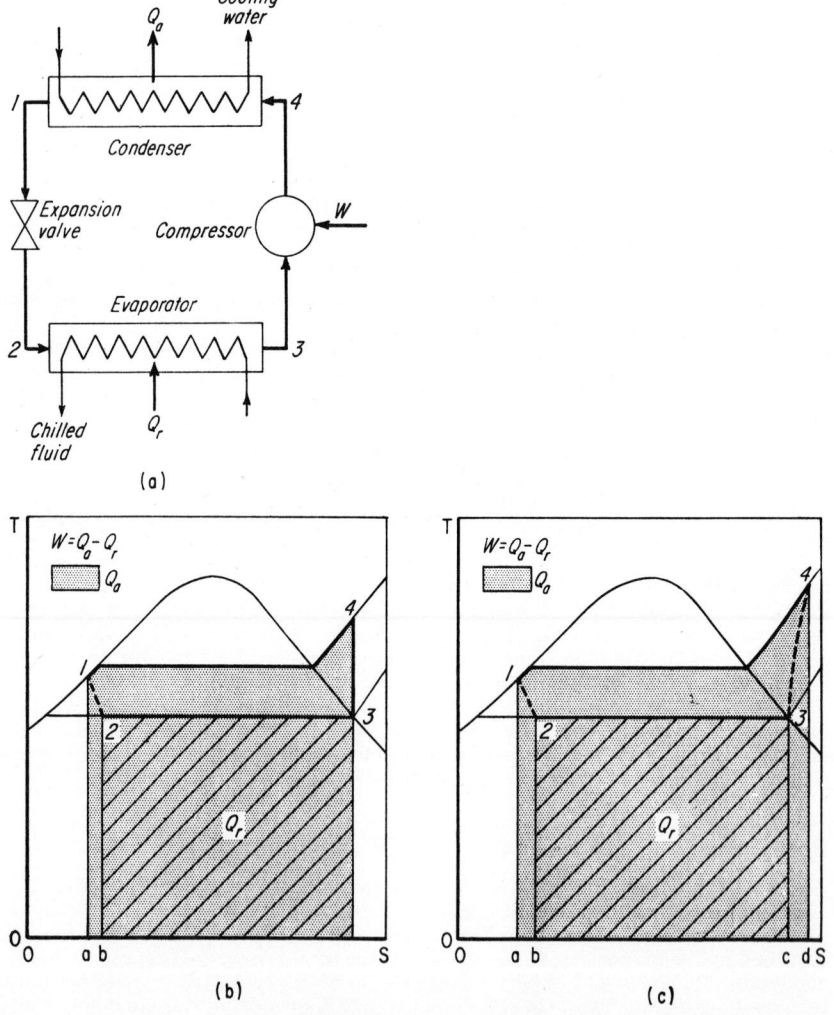

Fig. 2 (a) Components of a vapor refrigeration system; (b) ideal refrigeration cycle *T-S* diagram; (c) actual refrigeration cycle *T-S* diagram.

TABLE 4　Thermodynamic Properties of Ammonia*

Saturated ammonia

Temperature t, F	Pressure p, psia	Volume, cu ft/lb		Enthalpy, Btu/lb			Entropy, Btu/(lb)(F)	
		Liquid v_f	Vapor v_g	Liquid h_f	Evaporation h_{fg}	Vapor h_g	Liquid s_f	Vapor s_g
−60	5.55	0.0227	44.73	−21.2	610.8	589.6	−0.0517	1.4769
−40	10.41	0.0232	24.86	0.0	597.6	597.6	0.0000	1.4242
−20	18.30	0.0237	14.68	21.4	583.6	605.4	0.0497	1.3774
0	30.42	0.0242	9.116	42.9	568.9	611.8	0.0975	1.3352
20	48.21	0.0247	5.910	64.7	553.1	617.8	0.1437	1.2969
40	73.32	0.0253	3.971	86.8	536.2	623.0	0.1885	1.2618
60	107.6	0.0260	2.751	109.2	518.1	627.3	0.2322	1.2294
80	153.0	0.0268	1.955	132.0	498.7	630.7	0.2749	1.1991
100	211.9	0.0272	1.419	155.2	477.8	633.0	0.3166	1.1705
120	286.4	0.0284	1.047	179.0	455.0	634.0	0.3576	1.1427

Superheated ammonia

Temperature, F	15 psia (−27.29-F saturation)			50 psia (21.67-F saturation)			100 psia (56.05-F saturation)		
	v	h	s	v	h	s	v	h	s
−20	18.01	606.4	1.4031						
0	18.92	617.2	1.4272						
20	19.82	627.8	1.4497						
40	20.70	638.2	1.4709	5.988	623.4	1.3046			
60	21.58	648.5	1.4912	6.280	641.2	1.3399	2.985	629.3	1.2409
80	22.44	658.9	1.5108	6.564	652.6	1.3613	3.149	642.6	1.2661
100	23.31	669.2	1.5296	6.843	663.7	1.3816	3.304	655.2	1.2891
120	24.17	679.6	1.5478	7.117	674.7	1.4009	3.454	667.3	1.3104
140	25.03	690.0	1.5655	7.387	685.7	1.4195	3.600	679.2	1.3305
160	25.88	700.5	1.5827	7.655	696.6	1.4374	3.743	690.8	1.3495
180	26.74	711.1	1.5995	7.921	707.5	1.4548	3.883	702.3	1.3678
200	27.59	721.7	1.6158	8.185	718.5	1.4716	4.021	713.7	1.3854
220	28.44	732.4	1.6318	8.448	729.4	1.4880	4.158	725.1	1.4024
240	...	...	...	8.710	740.5	1.5040	4.294	736.5	1.4190
260	...	...	...	8.970	751.6	1.5197	4.428	747.9	1.4350
280	...	...	...	9.230	762.7	1.5350	4.562	759.4	1.4507
300	...	...	...	9.489	774.0	1.5500	4.695	770.8	1.4660

	150 psia (78.81-F saturation)			200 psia (96.34-F saturation)			300 psia (123.21-F saturation)		
80	2.001	631.4	1.2025						
100	2.118	645.9	1.2289	1.520	635.6	1.1809			
120	2.228	659.4	1.2526	1.612	650.9	1.2077			
140	2.334	672.3	1.2745	1.698	665.0	1.2317	1.058	648.7	1.1632
160	2.435	684.8	1.2949	1.780	678.4	1.2537	1.123	664.7	1.1894
180	2.534	696.9	1.3142	1.859	691.3	1.2742	1.183	679.5	1.2129
200	2.631	708.9	1.3327	1.935	703.9	1.2935	1.239	693.5	1.2344
220	2.726	720.7	1.3504	2.009	716.3	1.3120	1.294	706.9	1.2546
240	2.820	732.5	1.3675	2.082	728.4	1.3296	1.346	720.0	1.2736
260	2.912	744.3	1.3840	2.154	740.5	1.3467	1.397	732.9	1.2917
280	3.004	756.0	1.4001	2.225	752.5	1.3631	1.447	745.5	1.3090
300	3.095	767.7	1.4157	2.295	764.5	1.3791	1.496	758.1	1.3257

*Abstracted from National Bureau of Standards *Circular 142.*

represents the suction conditions of the compressor, $H_3 = 21.4 + 0.95(583.6) = 575.8$ Btu/lb.

The entropy at point 3 is $S_3 = s_f + x s_{fg}$, where the subscripts refer to the same fluid states as above, and the S and s values are the entropy. Or, $S_3 = 0.0497 + 0.95(1.3277) = 1.3110$ Btu/(lb)(F).

2. Compute the final cycle temperature and enthalpy

The compressor discharges at 100 psia at an entropy of $S_4 = 1.3110$. Inspection of the saturated ammonia properties, Table 4, shows that at 100 psia the entropy of saturated vapor is less than that computed. Hence, the vapor discharged by the compressor must be superheated.

Enter Table 4 at $S_4 = 1.3110$. Inspection shows that the final cycle temperature T_4 lies between 120 and 130 F because the actual entropy value lies between the entropy values for these two temperatures. Interpolating, $T_4 = 130 - [(S_{130} - S_4)/(S_{130} - S_{120})] \times (130 - 120)$, where the subscripts refer to the respective temperatures. Or, $T_4 = 130 - [(1.3206 - 1.3110)/(1.3206 - 1.3104)](130 - 120) = 120.6$ F.

Interpolating in a similar fashion for the final enthalpy, using the enthalpy at 130 F as the base, $H_4 = 673.3 - [(1.3206 - 1.3110)/(1.3206 - 1.3104)](673.3 - 667.3) = 667.7$ Btu/lb.

3. Compute the heat absorbed by the evaporator

The heat absorbed by the evaporator is $Q_r = H_3 - H_2$, where $Q_r =$ Btu per lb of refrigerant. Or, for this system, $Q_r = 575.8 - 97.9 = 477.9$ Btu/lb.

4. Compute the work input to the compressor

Find the work input to the compressor from $W = H_4 - H_3$, where $W =$ work input, Btu per lb of refrigerant. Or, $W = 667.7 - 575.8 = 91.9$ Btu per lb of refrigerant circulated.

5. Compute heat rejected to the condenser

The heat rejected to the condenser is $Q_a = H_4 - H_b$, where $Q_a =$ heat rejection, Btu per lb of refrigerant. Or, $Q_a = 667.7 - 97.9 = 569.8$ Btu per lb of refrigerant circulated.

6. Compute the coefficient of performance of the machine

For any refrigerating machine, the coefficient of performance $COP = Q_r/W$, where the symbols are as defined earlier. Or $COP = 477.9/91.9 = 5.20$.

7. Compute the horsepower per ton for this system

For any refrigerating system, the horsepower per ton $hp_t = 4.72/COP$. Or, for this system, $hp_t = 4.72/5.20 = 0.908$.

8. Compute the refrigerant quality at the evaporator inlet

At the evaporator inlet, or point 2, the quality of the refrigerant $x = (H_2 - h_f)/h_{fg}$, where the enthalpies are those at -20 F, the evaporator operating temperature. Or, $x = (97.9 - 21.4)/583.6 = 0.1311$, or 13.11 percent quality.

9. Compute the quantity of refrigerant circulated per ton capacity

Find the quantity of refrigerant circulated, lb per min per ton of refrigeration produced from $q_t = 200/Q_r$, or $q_t = 200/477.9 = 0.419$ lb of ammonia per min per ton of refrigeration.

10. Compute the required rate of condensing-water flow

The heat rejected to the condenser Q_a must be absorbed by the condenser cooling water. The quantity of water that must be circulated is $q_w = Q_a/\Delta t$, where $q_w =$ weight of water circulated per lb of refrigerant; $\Delta t =$ temperature rise of the cooling water during passage through the condenser, F. Assuming a 20-F temperature rise of the cooling water, $q_w = 569.8/20 = 28.49$, say 28.5 lb of water per lb of refrigerant circulated.

Since 0.419 lb of ammonia must be circulated per min per ton of refrigeration, Step 9, at a load of 100 tons, the quantity of refrigerant circulated will be $100(0.419) = 41.9$ lb/min. The condenser cooling water required is then $(28.5)(41.9) = 1,191$ lb/min, or $1,191/8.33 = 143.4$ gpm.

11. Compute the compressor displacement

Use the relation $V_d = q_t v_g T$, where V_d = required compressor displacement, cfm; q_t = quantity of refrigerant circulated, lb/(ton)(min); v_g = specific volume of suction gas, cu ft/lb; T = tons of refrigeration capacity. For a 100-ton capacity with the suction gas at -20 F, $V_d = (0.419)(14.68)(100) = 614$ cfm, using the specific volume for -20 F suction gas from Table 4.

12. Compute the compressor cylinder dimensions

For any reciprocating refrigeration compressor, V_d = (shaft rpm)(piston displacement, cu ft/stroke) = v_d or $614 = (200) v_d$; $v_d = 3.07$ cu ft.

Also, $D = (v_d/0.785\, r)^{1/3}$, where D = piston diameter ft; r = ratio of stroke length to cylinder bore. Or, $D = [3.07/(0.785 \times 1.3)]^{1/3} = 1.447$ ft. Then, $L = 1.3D = 1.3(1.447) = 1.88$ ft.

Related Calculations: Use the method given here for any reciprocating compressor using any refrigerant. Note that where the volumetric efficiency E_v of a compressor is given, the actual volume of gas drawn into the cylinder, cu ft, $= E_v \times$ piston displacement, cu ft. When analyzing an actual compressor, be sure to use the enthalpies which actually prevail. Thus, the gas entering the compressor suction may be superheated instead of saturated, as assumed here.

RECIPROCATING REFRIGERATION COMPRESSOR SELECTION

Choose the compressor capacity and hp and determine the heat rejection rate for a 36-ton load, a 30-F evaporator temperature, a 20-F evaporator coil superheat, a suction-line pressure drop of 2 psi, a condensing temperature of 105 F, a compressor speed of 1,750 rpm, a subcooling of the refrigerant of 5 F in the water-cooled condenser, and use of refrigerant 12. Determine the required condensing-water flow rate when the entering water temperature is 70 F. How many gpm of chilled water can be handled if the water temperature is reduced 10 F by the evaporator chiller?

Calculation Procedure:

1. Compute the compressor suction temperature

With refrigerant 12, a pressure change of 1 psi at 0 F is equivalent to a temperature change of 2 F; at 50 F, a 1-psi pressure change is equivalent to a 1-F temperature change, At the evaporator temperature of 30 F, the temperature change is about 1.4 F/psi, obtained by interpolation between the ranges given above. Then, suction temperature F = evaporator temperature, F-(suction-line loss, F/psi)(suction-line pressure drop, psi), or $30 - (1.4 \times 2) = 27.2$ F, say 27 F.

2. Compute the compressor equivalent capacity

To compute the compressor equivalent capacity, two correction factors must be applied. These are (a) the superheat correction factor, and (b) the subcooling correction factor. Both these factors are given in the engineering data available from compressor manufacturers.

To apply correction-factor listings, such as those in Table 5, use the following as guides. (a) Superheating of the suction gas can result from heat pickup by the gas outside the cooled space. Superheating increases the refrigeration compressor capacity 0.3 to 1.0 percent per 10 F when using refrigerants 12 or 500 if the heat absorbed represents useful refrigeration, such as coil superheat, and not superheating from a liquid suction heat exchanger. (b) Subcooling increases the potential refrigeration effect by reducing the percentage of liquid flashed during expansion. For each deg F of subcooling, the compressor capacity is increased about 0.5 percent due to the increased refrigeration effect per lb of refrigerant flow.

Applying guide (a) to a 27-F suction, 20-F superheat, interpolate in Table 5 between the 40- and 50-F actual suction-gas temperature for a 30-F saturated suction temperature, because the actual suction temperature is $27 + 20 = 47$ F and the saturated suction temperature is given as 30 F. Or $(0.987 - 0.979)(47 - 40/50 - 40) = 0.9846$, say 0.985.

TABLE 5   Open Compressor Ratings*

Suction temperature, F	Condensing temperature, 105 F		
	Capacity, tons	Power input, bhp	Heat rejection, tons
10	26.2	41.3	34.9
20	34.0	45.3	43.6
30	43.0	48.6	53.2

Applying guide (b), subcooling = 5 F, given. Then subcooling correction $= 1 - 0.005$ $(15 - 5) = 0.95$, where $0.005 = 0.5$ percent expressed as a decimal; 15 F = the liquid subcooling upon which the compressor capacity is based. This value is given in the compressor rating, Table 6.

With the superheat and subcooling correction factors known, compute the compressor equivalent capacity from (load in tons)/(superheat correction factor)(subcooling correction factor), or $36/[(0.985)(0.95)] = 38.5$ tons.

TABLE 6   Rating Basis and Capacity Multipliers — Refrigerants 12 and 500*

Saturated suction temperature, F	Actual suction gas temperature to compressor, F			
	30	40	50	60
20	0.969	0.978	0.987	0.996
30	0.970	0.979	0.987	0.996
40	...	0.987	0.992	0.997

*Carrier Air Conditioning Company.

3.   Select the compressor unit

Use Table 5. Choose an eight-cylinder compressor. Interpolate for a 27-F suction and 105-F condensing temperature to find: compressor capacity = 40.3 tons; power input = 47.6 bhp; heat rejection = 50.3.

4.   Compute the required condensing-water flow rate

From Step 3, the condensing temperature of the compressor chosen is 105 F. Assume a condenser-water outlet temperature of 95 F, a typical value. Then, the required condenser-water flow rate, gpm, = 24 × condenser load, tons/(condensing-water outlet temperature, F — entering condenser-water temperature, F). Or gpm = 24(50.3)/(95 − 70) = 48.4 gpm. This is within the normal flow for water-cooled condensing units. Thus, city-water quantities range from 1 to 2 gpm/ton; cooling-tower quantities are usually chosen for 3 gpm/ton.

5.   Compute the quantity of chilled water that can be handled

Use the relation chilled-water gpm = 24 × capacity, tons/chilled-water temperature range, or inlet − outlet temperature, F. Since, from Step 3, the compressor capacity is

40.3 tons and the chilled-water temperature range is 10 F, gpm = 24(40.3)/10 = 96.7 gpm.

The temperature of the chilled water leaving the evaporator chiller is selected so it equals the inlet temperature required at the heat-load source. The required inlet temperature is a function of the type of heat exchanger, type of load, and similar factors.

Related Calculations: The standard operating conditions for an air-conditioning refrigeration system, as usually published by the manufacturer, are based on an entering saturated refrigerant vapor temperature of 40 F, an actual entering refrigerant vapor temperature of 55 F, a leaving saturated refrigerant vapor temperature of 105 F, an ambient of 90 F, and no liquid subcooling.

The Air Conditioning and Refrigeration Institute (ARI) Standards for a reciprocating-compressor liquid-chilling package establish a standard rating condition for a water-cooled model of a leaving chilled-water temperature of 44 F, a chilled-water range of 10 F, a 0.0005 fouling factor in the cooler and the condenser, a leaving condenser-water temperature of 95 F, and a condenser-water temperature rise of 10 F. The standard rating conditions for a condenserless model is a leaving chilled-water temperature of 44 F, a chilled-water temperature range of 10 F, a 0.0005 fouling factor in the cooler, and a condensing temperature of 105 or 120 F.

Use these standard rating conditions to make comparisons between compressors. When comparing catalog ratings of compressors of different manufacturers, the rating conditions must be known, particularly the amount of subcooling and superheating needed to produce the capacities shown.

General guides for reciprocating compressors using refrigerants 12, 22, and 500 are as follows.

(1) Lowering the evaporator temperature 10 F from a base of 40 and 105 F reduces the system (evaporator) capacity about 24 percent and at the same time increases the compressor hp/ton about 18 percent.

(2) Increasing the condensing temperature 15 F from a base of 40 and 105 F reduces the capacity about 13 percent and at the same time increases the compressor hp/ton about 27 percent.

(3) In air-conditioning service at normal loads, a piping loss equivalent to approximately 2 F is allowed in the suction piping and 2 F in the hot-gas discharge piping. Thus, when an evaporator requires a refrigerant temperature of 42 F to handle a load, the compressor must be selected for a 40-F suction temperature, or 42 − 2. Correspondingly, if the condenser requires 103 F to reject the proper amount of heat, the compressor must be selected for a 103 + 2 = 105 F condensing temperature.

(4) Compressor manufacturers generally state the operating limits for each compressor in the capacity table describing it. These limits should not be exceeded.

(5) To select a condenser to match a compressor, the heat rejection of the compressor must be known. For an open-type compressor, heat rejection, tons, = 0.212 (compressor power input, bhp) + tons refrigeration capacity of the compressor. For a gas-cooled hermetic-type compressor, heat rejection, tons, = 0.285 (kw input to the compressor) + tons refrigeration capacity. The selection Procedure and other data given here were developed by the Carrier Air Conditioning Company.

CENTRIFUGAL REFRIGERATION MACHINE LOAD ANALYSIS

Select a centrifugal refrigeration machine to cool 720 gpm of chilled water from an entering temperature of 60 F to a leaving temperature of 45 F.

Calculation Procedure:

1. Compute the load on the machine

Use the relation, load, tons, = $gpm \times \Delta t/24$, where gpm = quantity of chilled water cooled, gpm; Δt = temperature reduction of the chilled water during passage through the evaporator chiller, F. For this machine, load = 720(50 − 45)/24 = 450 tons.

2. *Choose the compressor to use*

Table 7 shows typical hermetic centrifugal refrigeration machine ratings. In a hermetic machine, the driver is built into the housing, completely isolating the refrigerant space from the atmosphere. An open machine has a shaft that projects outside the compressor housing. The shaft must be fitted with a suitable seal to prevent refrigerant leakage. Open machines are available in capacities up to approximately 4,500 tons at air-conditioning load temperatures. Hermetic machines are available in capacities up to approximately 2,000 tons capacity.

Study of Table 7 shows that a 450-ton unit is available with a leaving chilled-water temperature of 45 F and a leaving condenser-water temperature of 85 F. If the condenser water were available at temperatures of 75 F or lower, this machine would probably be chosen.

Table 7   Typical Hermetic Centrifugal Machine Ratings*
(Refrigeration capacity, tons)

Leaving chilled-water temperature, F	Leaving condenser water temperature, F		
	85	90	95
44	442†	435	424
45	450	441	430
46	457†	447	435

*Carrier Air Conditioning Company.

†These ratings require less than 330-kw input. All ratings shown are based on a two-pass cooler using 380 to 1,260 gpm and on a two-pass condenser using 430 to 1,430 gpm.

Related Calculations: The factors involved in the selection of a centrifugal machine are load; chilled-water, or brine quantity; temperature of the chilled water or brine; condensing medium (usually water) to be used; quantity of the condensing medium and its temperature; type and quantity of power available; fouling-factor allowance; amount of usable space available; and the nature of the load, whether variable or constant. The final selection is usually based on the least expensive combination of machine and heat rejection device, as well as a reasonable machine operating cost.

Brine cooling normally requires special selection of the machine by the manufacturer. As a general rule, multiple-machine applications are seldom made on normal air-conditioning loads less than about 400 tons.

The optimum machine selection involves matching the correct machine and cooling tower as well as the correct entering chilled-water temperature and temperature reduction. A selection of several machines and cooling towers often results in finding one combination having a minimum first cost. In many instances, it is possible to reduce the condenser-water quantity and increase the leaving condenser-water temperature, resulting in a smaller tower.

Centrifugal refrigeration machines are used for air-conditioning, process, marine, manufacturing, and many other cooling applications throughout industry.

HEAT PUMP CYCLE ANALYSIS AND COMPARISON

Determine the quantity of water required to supply heat to a heat pump that must deliver 70,000 Btu/hr to a building. Refrigerant 12 is used; the temperature of the water in the heat sink is 50 F. Air must be delivered to the heating system at a temperature of 118 F.

Calculation Procedure:

1. Determine the compressor suction temperature to use

To produce sufficient heat transfer between the water and the evaporator, a temperature difference of at least 10 F must exist. With a water temperature of 50 F, this means that a suction temperature 40 F might be satisfactory. A suction temperature of 40 F corresponds to a suction pressure of 51.68 psia, as a table of thermodynamic properties of refrigerant 12 shows.

Since water entering the evaporator heat exchanger cannot be reduced to 40 F, the refrigerant temperature, the actual outlet temperature must be either assumed or computed. Assume that the water leaves the evaporator heat exchanger at 44 F. Then, each lb of water passing through the evaporator yields 50 F − 44 F = 6 Btu. Since 1 gal of water weighs 8.33 lb, the quantity of heat released by the water is (6 Btu/lb) × (8.33 lb/gal) = 49.98 Btu/gal, say 50 Btu/gal.

As an alternative solution, assume a suction temperature of 35 F and an evaporator exit temperature of 39 F. Then, each lb of water will yield 50 − 39 = 11 Btu. This is equal to (11)(8.33) = 91.6 Btu/gal. This comparison indicates that for every deg F the cooling range of the water heat sink is extended, 8.33 additional Btu per gal of water is obtained.

2. Evaluate the effect of suction-temperature decrease

As the compressor suction temperature is reduced, the specific volume of the suction gas increases. This means that the compressor must handle more gas to evaporate the same quantity of refrigerant. However, the displacement of the usual reciprocating compressor used in a heat-pump system cannot be varied easily, if at all, in some designs. Also, at the lower suction temperature, the enthalpy of vaporization of the refrigerant increases only slightly.

Study of a table of thermodynamic properties of refrigerant 12 shows that reducing the suction temperature from 40 to 35 F increases the specific volume from 0.792 to 0.862 cu ft/lb. The enthalpy of vaporization increases from 65.71 to 66.28 Btu/lb, but the total enthalpy decreases from 82.71 to 82.16 Btu/lb. Hence, the advisability of reducing the suction temperature must be carefully investigated before a final decision is made.

3. Determine the required compressor discharge temperature

Air must be delivered to the heating system at 118 F, according to the design requirements. To produce a satisfactory transfer of heat between the condenser and the air, a 10-F temperature difference is necessary. Hence, the compressor discharge temperature must be at least 118 + 10 = 128 F.

Checking a table of thermodynamic properties of refrigerant 12 shows that a temperature of 128 F corresponds to a discharge pressure of 190.1 psia. The table also shows that the enthalpy of the vapor at the 118-F condensing temperature is 90.01 Btu/lb, whereas the enthalpy of the liquid is 35.65 Btu/lb.

With a suction temperature of 40 F, the enthalpy of the vapor is 82.71 Btu/lb. Hence, the heat supplied by the evaporator is enthalpy of vapor at 40 F − enthalpy of liquid at 118 F = 82.71 − 35.65 = 47.06 Btu/lb. This heat is abstracted from the water that is drawn from the heat sink.

The gas leaving the evaporator contains 82.71 Btu/per lb. When this gas enters the condenser it contains 90.64 Btu/lb. The difference, or 90.64 − 82.71 = 7.93 Btu/lb, is added to the gas by the compressor and represents a portion of the work input to the compressor.

4. Compute the evaporator and compressor heat contribution

The total heat delivered to the air = evaporator heat + compressor heat = 47.06 + 7.93 = 54.99 Btu/lb. Then, the evaporator supplies 47.06/54.99 = 0.856, or 85.6 percent of the total heat, and the compressor supplies 7.93/54.99 = 0.144, or 14.4 percent of the total heat.

5. Determine the actual evaporator and compressor heat contribution

Since this heat pump is rated at 70,000 Btu/hr, the evaporator contributes $0.856 \times (70,000) = 59,920$ Btu/hr, and the compressor supplies $0.144(70,000) = 10,080$ Btu/hr. As a check, $59,920 + 10,080 = 70,000$ Btu/hr.

6. Compute the sink water flow rate required

The evaporator obtains its heat, or 59,920 Btu/hr, from the sink water. Since, from Step 1, each gal of water delivers 50 Btu at a 40-F suction temperature, the flow rate required to contribute the evaporator heat is 59,920/50 = 1,198.4 gph, or 1,198.4/60 = 19.9 gpm.

7. Evaluate the lower suction temperature

At 35 F, the evaporator will supply $82.16 - 35.65 = 46.51$ Btu/lb, using the same reasoning as in Step 3. The balance, or $90.64 - 82.16 = 8.48$ Btu/lb, must be supplied by the compressor.

8. Compute the required refrigerant gas flow

At a 40-F suction temperature, a table of thermodynamic properties of refrigerant 12 shows that the specific volume of the gas is 0.792 cu ft/lb. Step 4 shows that the heat pump must deliver 54.99 Btu/lb of refrigerant to the air, or 54.99/0.792 = 69.4 Btu per cu ft of gas. With a total heat requirement of 70,000 Btu/hr, the compressor must handle 70,000/69.4 = 1,010 cu ft/hr.

As noted earlier, the cubic capacity of a reciprocating compressor is a fixed value at a given speed. Hence, a compressor chosen to handle this quantity of gas cannot handle a larger heat load.

At a 35-F suction temperature, using the same Procedure as above, the required heat content of the gas is 54.99/0.862 = 63.7 Btu/cu ft. The compressor capacity must be 70,000/63.7 = 1,099 cu ft/hr.

If the compressor were selected to handle 1,010 cu ft/hr, reducing the suction temperature to 35 F would give a heat capacity of only $(1,010)(63.7) = 64,400$ Btu/hr. This is inadequate because the system requires 70,000 Btu/hr.

9. Compute the water flow rate at the lower suction temperature

Step 6 shows the Procedure for finding the sink water flow rate required at a 40-F suction temperature. Suppose, however, that the heat output of 64,400 Btu/hr at the 35-F suction temperature was acceptable. The evaporator portion of this load is, using the method of Step 4, $(82.16 - 35.65)/54.99 = 0.847$, or 84.7 percent. Hence, the quantity of water required is $(64,400)(0.847)/[(91.6)(60)] = 9.92$ gpm. In this equation, the value of 91.6 Btu/gal is obtained from Step 1. The factor 60 converts from hr to min. Thus, reducing the suction temperature from 40 to 35 F just about halves the water quantity—from 19.9 to 9.92 gpm.

10. Compare the pumping power requirements

The power input to the pump is $hp = 8.33$ (gpm)(head, ft)/33,000(pump efficiency) (motor efficiency). If the total head on the pump is computed as being 40 ft, the efficiency of the pump is 60 percent and the efficiency of the motor is 85 percent, then with a 40-F suction temperature, the pump horsepower is $hp = 8.33(19.9)(40)/[33,000(0.60) \times (0.85)] = 0.394$, say 0.40 hp.

At a 35-F suction temperature with a flow rate of 9.92 gpm and all the other factors being the same, $hp = 8.33(9.92)(40)/[33,000(0.60)(0.85)] = 0.1965$; say 0.20 hp. Thus, the 35-F suction temperature requires only half the pump hp that the 40-F suction temperature requires.

11. Compute the compressor power input and power cost

At the 40-F suction temperature, the compressor delivers 7.93 Btu per lb of refrigerant gas, Step 3. Since the total weight of gas delivered by the compressor per hr is (70,000 Btu/hr)/(54.99 Btu/lb) = 1,272 lb, the compressor's total heat contribution is (1,272 lb) (7.93 Btu/lb) = 10,100 Btu/hr.

With a compressor-driving motor having an efficiency of 85 percent, the hourly motor input is equivalent to 10,100/0.85 = 11,880 Btu/hr, or 11,880/2,545 Btu/hp-hr = 4.66 hp-hr = (4.66)(746 watt-hr/hp-hr) = 3,480 watt-hr, or 3.48 kw-hr. Also, the pump requires 0.4 hp-hr, Step 10, or (0.4)(746) = 299 watt-hr = 0.299 kw-hr. Hence, the total power consumption at a 40-F suction temperature is 3.48 + 0.299 = 3.779 kw-hr. At a power cost of 5 cents/kw-hr, the energy cost is 3.779 (5.0) = 18.9 cents/hr.

With a 35-F suction temperature and using the lower heating capacity obtained with the smaller, fixed-capacity compressor, Step 9, the weight of gas handled by the compressor will be (64,400 Btu/hr)/(54.99 Btu/lb) = 1,172 lb/hr. From Step 7, the compressor must supply 8.48 Btu/lb. Therefore, the compressor's total heat contribution is (1,172 lb)(8.48) = 9,950 Btu/hr. With a motor efficiency of 85 percent, the hourly motor input is equivalent to 9,950/0.85 = 11,700 Btu/hr.

Using the same Procedure as above, the electrical power input to the compressor will be 3.43 kw-hr, while the pump electrical power input is 0.150 kw-hr. The total electrical power input is 3.58 kw-hr at a cost of (3.58)(5.0) = 17.9 cents/hr. Hence, the hourly saving using the 35-F suction temperature is 18.9 − 17.9 = 1.0 cents. Note, however, that the heat output at the 35-F suction temperature, 64,400 Btu/hr, is 5,600 Btu/hr less than at the 40-F suction temperature. If the lower heat output were unacceptable, the higher suction temperature, or a larger compressor, would have to be used. Either alternative would increase the power cost.

Related Calculations: With a water sink as the heat source, the usual water consumption of a heat pump ranges from 1.1 to more than 4 gpm/ton. A consumption range this broad requires that the actual flow rate be computed because a guess could be considerably in error.

Either air or the earth may be used as heat sources instead of water. When the cooling load rather than the heating load establishes the basic equipment size, an ideal situation exists for the use of the heat pump with air as the heat source. This occurs in localities where the minimum outdoor temperature in the winter is 20 F, or higher.

Ground coils can be bulky, costly, and troublesome. One study shows that the temperature difference between the evaporating refrigerant in a ground coil and the surrounding earth is about equal to the number of Btu/hr that may be drawn from each lin ft of coil. Thus, with a temperature difference of 15 F, a 70,000-Btu load of which 85 percent is supplied by the coil, the length of coil needed is (70,000)(0.85)/15 = 3,970 ft.

The coefficient of performance of a heat pump = heat rejected by condenser, Btu/heat equivalent of the net work of compression, Btu. The usual single-stage air-source heat pump has a coefficient of performance ranging between 2.25 and 3.0. The procedure and data presented here were developed by Robert Henderson Emerick, P. E., Consulting Mechanical Engineer.

Section **4**

Electrical Engineering

ANDREW W. EDWARDS
Power Engineer, Westinghouse Electric Corporation

HAROLD L. RORDEN
Consulting Engineer, American Electric Power Service Corporation

FREDERICK W. SUHR
Consulting Engineer, General Electric Company

REFERENCES: Slurzberg and Osterheld—*Essentials of Electricity-Electronics*, McGraw-Hill; Corcoran—*Basic Electrical Engineering*, Wiley; Oppenheimer and Borchers—*Direct and Alternating Currents*, McGraw-Hill; Chang—*Energy Conversion*, Prentice-Hall; Rosenblatt and Friedman—*Direct and Alternating Current Machinery*, McGraw-Hill; Pender—*Electrical Engineers' Handbook*, Wiley; Fink and Carroll—*Standard Handbook for Electrical Engineers*, McGraw-Hill; Timbie-Willson—*Industrial Electricity*, Wiley; Abbott and Steka—*National Electrical Code Handbook*, McGraw-Hill; Richter—*Practical Electrical Wiring*, McGraw-Hill; Steka and Brandon—*NFPA Handbook of the National Electrical Code*, McGraw-Hill; Hubert—*Operational Electricity*, Wiley; Croft and Carr—*American Electrician's Handbook*, McGraw-Hill; Beeman—*Industrial Power Systems Handbook*, McGraw-Hill; Gibbs—*Transformer Principles and Practices*, McGraw-Hill; Libby—*Motor Selection and Application*, McGraw-Hill; U.S. Government Printing Office—*Electric Current Abroad*; IEEE—*Electric Systems for Commercial Buildings*; IEEE—*Electric Power Distribution for Industrial Plants*; *Power* Magazine Editors—*Industrial Electrical Systems*, McGraw-Hill; Dawes—*Electrical Engineering*, McGraw-Hill; NEMA—*Standards for Motors and Generators*; NEMA—*Standards for Industrial Control*; ANSA—*Induction Machines*; ANSA—*Synchronous Motors*; McPartland and Novak—*Electrical Systems Design*, McGraw-Hill; McPartland and Novak—*Electrical Design Details*, McGraw-Hill; McPartland and Novak—*Electrical Equipment Manual*, McGraw-Hill; McPartland and Novak—*Electrical Systems for Power and Light*, McGraw-Hill.

DIRECT-CURRENT-CIRCUIT ANALYSIS

A direct-current circuit contains 15 resistors arranged as shown in Fig. 1. Compute the current flow through, and the voltage drop across, each resistor in this circuit.

Calculation Procedure:

1. Divide the circuit into sections and groups

When analyzing a complex dc combination circuit the simplest procedure is to divide the circuit into sections that can later be combined into single or multiple resistances in series. Once this is done the circuit is much easier to analyze.

Mark on the circuit diagram the sections decided upon. Figure 1 shows that the three

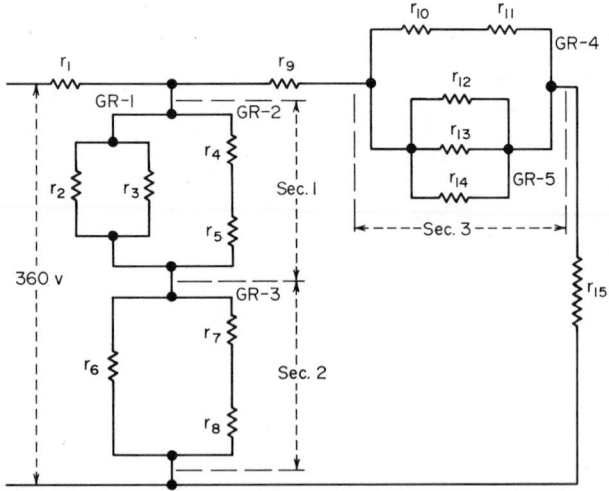

Fig. 1. Direct-current circuit containing numerous resistances (values in ohms): $r_1 = 20$; $r_2 = 1{,}000$; $r_3 = 1{,}500$; $r_4 = 200$; $r_5 = 400$; $r_6 = 600$; $r_7 = 400$; $r_8 = 200$; $r_9 = 20$; $r_{10} = 200$; $r_{11} = 40$; $r_{12} = 100$; $r_{13} = 200$; $r_{14} = 600$; $r_{15} = 52$.

groups of series-parallel resistances can be divided into three sections. Indicate each section as shown.

Further subdivide each section into parallel and series groups. Mark the groups as shown in Fig. 1.

2. Combine the group resistances to obtain a single section resistance

Use the rules for parallel and series circuits to combine the resistance values in each group to obtain one single equivalent resistance for each section. Use R = group resistance, ohms, and r = resistor resistance, ohms.

For group 1, Fig. 1:

$$R_1 = \frac{1}{1/r_2 + 1/r_3} = \frac{1}{1/1,000 + 1/1,500} = 600 \text{ ohms}$$

since r_2 and r_3 are resistances in parallel. For group 2: $R_2 = r_4 + r_5 = 200 + 400 = 600$ ohms, since r_4 and r_5 are resistances in series. For section 1, Fig. 1:

$$R_{S1} = \frac{1}{1/R_1 + 1/R_2} = \frac{1}{1/600 + 1/600} = 300 \text{ ohms}$$

since R_1 and R_2 are two resistances in parallel that replace the former r_2, r_3, r_4, and r_5 resistances. R_{S1} is the equivalent single resistance for groups 1 and 2.

Follow the same procedure for group 3, Fig. 1. Thus, $R_3 = r_7 + r_8 = 400 + 200 = 600$ ohms. Then, since r_6 and group 3 are in parallel,

$$R_{S2} = \frac{1}{1/r_6 + 1/R_3} = \frac{1}{1/600 + 1/600} = 300 \text{ ohms}$$

Follow the same procedure for groups 4 and 5, Fig. 1. Or, $R_4 = r_{10} + r_{11} = 200 + 40 = 240$ ohms.

$$R_5 = \frac{1}{1/r_{12} + 1/r_{13} + 1/r_{14}} = \frac{1}{1/100 + 1/200 + 1/600} = 60 \text{ ohms}$$

Then, for section e,

$$R_{S3} = \frac{1}{1/R_4 + 1/R_5} = \frac{1}{1/240 + 1/60} = 48 \text{ ohms}$$

3. Sketch the equivalent circuit

Sections 1, 2, and 3 have been reduced to equivalent single-series resistances. Sketch them in place as shown in Fig. 2.

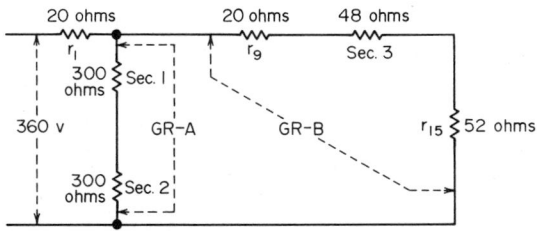

Fig. 2 Direct-current circuit of Fig. 1 reduced to two groups of resistances.

4. Study the equivalent circuit for its makeup

Study of Fig. 2 shows that the circuit now consists of a simple parallel-series circuit— i.e, group A is in parallel with group B while both are in series with r_1.

5. *Analyze the equivalent circuit to find the line resistance*

Using the rules for parallel and series circuits as in Step 2, the resistance of group A is $R_A = R_{S1} + R_{S2} = 300 + 300 = 600$ ohms. Likewise for group B, $R_B = R_9 + R_3 + r_{15} = 20 + 48 + 52 = 120$ ohms.

The equivalent resistance of the parallel circuit consisting of groups A and B is then

$$R_{AB} = \frac{1}{1/R_A + 1/R_B} = \frac{1}{1/600 + 1/120} = 100 \text{ ohms}$$

Lastly, combine this series resistance with the one remaining series resistance r_1 to find the total resistance of the line, or $R_T = r_1 + R_{AB} = 20 + 100 = 120$ ohms. Hence, the total equivalent resistance of this circuit is 120 ohms.

6. *Compute the line current*

Use the relation $I = E/R_T$, where I = line current, amp; E = line voltage; R_T = line resistance, ohms. Or $I = 360/120 = 3$ amp.

7. *Compute the equivalent-circuit current and voltage*

The voltage across any resistance is $e_r = I_r r_r$, where I and r are the current and resistance, respectively, for the resistance in question. Thus, $e_1 = I_T r_1 = 3 \times 20 = 60$ v.

Then, $E_A = E_B = E_T - e_1$ because the voltage acting on groups A and B is that potential remaining after the voltage drop across r_1. Or, $E_A = E_B = 360 - 60 = 300$ v.

The current in each group is $I = E/R$, or $I_A = E_A/R_A = 300/600 = 0.50$ amp; $I_B = E_B/R_B = 300/120 = 2.50$ amp. As a check, $I_T = I_A + I_B = 0.50 + 2.50 = 3.0$ amp, as computed in Step 6.

8. *Compute the section current and voltage*

Using the same reasoning as in Step 7; $e_{S1} = I_A R_{S1} = 0.50 \times 300 = 150$ v; $e_{S2} = I_A R_{S2} = 0.50 \times 300 = 150$ v.

The voltage across group A equals the sum of the voltage drops across the resistances in this group, or $E_A = e_{S1} + e_{S2} = 150 + 150 = 300$ v. This checks with the computation in Step 7.

Using the same procedure for group B: $e_9 = I_B r_9 = 2.50 \times 20 = 50$ v; $e_{S3} = I_B R_{S3} = 2.50 \times 48 = 120$ v; $e_{15} = I_{S3} + e_{15} = 2.50 \times 52 = 130$ v. As a check, $E_B = e_9 + e_{S3} + e_{15} = 50 + 120 + 130 = 300$ v, as previously computed in Step 7.

9. *Compute the original-circuit current and voltage*

The current in each resistor, section, and group of the original circuit is given by the general relation $i = e/r$, where the appropriate voltage and resistance is substituted. Thus, for section 1 of the original circuit; $i_2 = e_{S1}/R_2 = 150/600 = 0.250$ amp; $i_{r2} = e_{S1}/r_2 = 150/1,000 = 0.150$ amp; $i_{r3} = e_{S1}/r_3 = 150/1,500 = 0.100$ amp. Then $I_{S1} = i_2 + i_{r2} + i_{r3} = 0.250 + 0.150 + 0.100 = 0.500$ amp, and $e_{r4} = i_2 r_4 = 0.250 \times 400 = 100$ v; $e_{S1} = e_{r4} + e_{r5} = 50 + 100 = 150$ v.

Likewise, for the voltage across section 1; $e_{r4} = i_2 r_4 = 0.250 \times 200 = 50$ v; $e_{r5} = i_2 r_5 = 0.250 \times 400 = 100$ v; $e_{S1} = e_{r4} + e_{r5} = 50 + 100 = 150$ v.

For section 2 find the current in the same way as for section 1: $i_{r6} = e_{S2}/r_6 = 150/600 = 0.250$ amp; $i_3 = e_{S2}/R_3 = 150/600 = 0.250$ amp; $I_{S2} = i_{r6} + i_3 = 0.250 + 0.250 = 0.500$ amp. The voltage across section 2 is made up of: $e_{r7} = i_3 r_7 = 0.250 \times 400 = 100$ v; $e_{r8} = i_3 r_8 = 0.250 \times 200 = 50$ v; $e_{S2} = e_{r7} + e_{r8} = 100 + 50 = 150$ v.

For section 3; $i_4 = e_{S3}/R_4 = 120/240 = 0.5$ amp; $i_5 = e_{S3}/R_5 = 120/60 = 2.0$ amp; $I_{S3} = i_4 + i_5 = 0.50 + 2.0 = 2.5$ amp. Also; $e_{r10} = i_4 r_{10} = 0.50 \times 200 = 100$ v; $e_{r11} = i_4 r_{11} = 0.50 \times 40 = 20$ v; $e_{S3} = e_{10} + e_{11} = 100 + 20 = 120$ v.

For group 5: $i_{r12} = e_{S3}/r_{12} = 120/100 = 1.20$ amp; $i_{r13} = e_{S3}/r_{13} = 120/200 = 0.60$ amp; $i_{r14} = e_{S3}/r_{14} = 120/600 = 0.20$ amp; $i_5 = i_{r12} + i_{r14} = 1.20 + 0.60 + 0.20 = 2.00$ amp.

Related Calculations: Any reducible dc circuit (i.e., any circuit that can be reduced to one equivalent resistance with a single power source), no matter how complex, can be solved in a similar manner to that described above. When each circuit is solved separately and combined whenever possible, the current through each resistor and the voltage drop across it may be obtained. To analyze such circuits, perform the following steps in the order listed: (1) Combine the resistance values in each group to obtain one single equivalent resistance value for each section. (2) Combine the resistance values of all sections to obtain one single equivalent resistance value for the line. (3) Solve for the line current. (4) Find the current flowing in each resistor. (5) Find the voltage across each resistor. In the analysis of some circuits, Steps 4 and 5 may have to be interchanged.

KIRCHHOFF'S LAWS FOR DC-CIRCUIT ANALYSIS

Analyze the circuit of the previous Calculation Procedure by applying Kirchhoff's laws.

Calculation Procedure:

1. Label all circuit elements as to name and value

Figure 3a shows the circuit of Fig. 2 with its elements identified by symbol and value. The 360-v generator represents the circuit-voltage source.

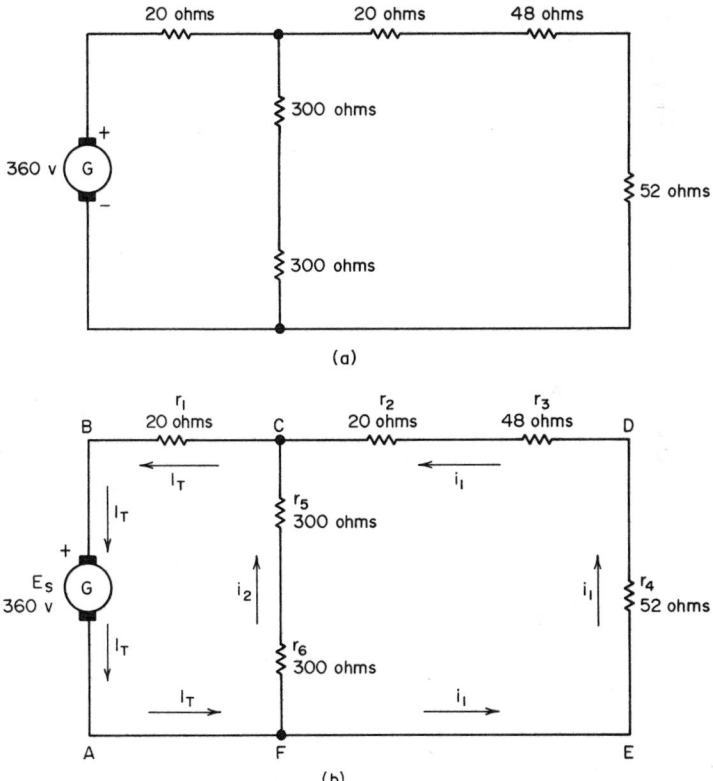

Fig. 3 (a) Typical direct-current circuit, (b) current and loop designations for analysis of a circuit by use of Kirchoff's laws.

2. Label the current direction in each branch of the circuit

Draw an arrow alongside each branch of the circuit to indicate the direction of electron flow in the branch. Use judgement in assuming the probable direction of electron flow in each branch of the circuit, Fig. 3b.

3. Mark all circuit connecting points with a reference letter

Follow a clockwise direction when marking the connecting points with the reference letter, Fig. 3b.

4. Set up current equations at each junction of three or more circuit elements

When setting up the current junction equations, consider the currents entering the junction as algebraically *positive*; consider currents leaving the junction as algebraically *negative*. Thus, at junction C, Fig. 3b, $i_1 + i_2 - I_T = 0$; at junction F, $I_T - i_1 - i_2 = 0$.

In this case, except for the algebraic signs, the two equations are the same. The reason for this condition is that the number of *independent equations* that can be used is always one less than the total number of junctions in a circuit. (Kirchhoff's current law states: *The algebraic sum of all currents at any point in a circuit must be zero.*)

5. Set up voltage equations for each closed path in the circuit

Kirchhoff's voltage law states: *The algebraic sum of all the voltages in any closed path of a circuit must be zero.* So be sure to indicate the polarity of the voltages. When setting up the voltage-loop equations, follow these two rules:

(a) The voltage of a power source is positive when the direction of the path being traced is from the negative to the positive terminal and negative when in the reverse direction.

(b) The polarity of a voltage at a resistor depends on the direction of the electron flow through it. When the indicated direction of electron flow is opposite to the direction in which the voltage loop is being traced, the voltage at a resistor is negative. When the directions are both the same, the voltage is positive.

Using these rules for loop $ABCDEFA$, Fig. 3b, $E_S - I_T r_1 - i_1 r_2 - i_1 r_3 - i_1 r_4 = 0$. Taking the path of this loop in the opposite direction as $AFEDCBA$, then $i_1 r_4 + i_1 r_3 + i_1 r_2 + i_T r_1 - E_S = 0$.

Substitute known circuit-element values in the first equation to obtain $360 - 20 I_T - 120 i_1 = 0$, for loop $ABCDEFA$.

Setting up the voltage equation for loop $ABCFA$ in a similar manner, $E_S - I_T r_1 - i_2 r_5 - i_2 r_6 = 0$. Substituting known circuit-element values, $360 - 20 I_T - 600 i_2 = 0$. Likewise for loop $FCDEF$, $i_2 r_6 + i_2 r_5 - i_1 r_2 - i_1 r_3 - i_1 r_4 = 0$, and $600 i_2 - 120 i_1 = 0$.

6. Solve the independent current and voltage equations simultaneously

When a circuit has n basic unknowns, n equations that contain each unknown at least once are necessary for the solution of the problem. Since there are three basic unknowns in this circuit, namely, I_T, i_1, and i_2, only three equations are needed, if each equation contains each unknown at least once.

From one of the current equations, write one of the unknowns in terms of the other unknown or unknowns. Or, $i_1 + i_2 - I_T = 0$; hence, $I_T = i_1 + i_2$. Substituting $i_1 + i_2$ for I_T in the current equation for loop $AFEDCBA$, $360 - 20(i_1 + i_2) - 120 i_1 = 0$, or $360 - 140 i_1 - 20 i_2 = 0$.

Solve this last equation for one of the unknowns in terms of the other unknown, or $i_1 = 2.57 - 0.143 i_2$. Substitute this quantity in the current equation for loop $FCDEF$, or $600 i_2 - 120(2.57 - 0.143 i_2) = 0$; $i_2 = 0.5$ amp. Substitute this value for i_2 in the first combined equation, or $360 - 140 i_1 - 20 i_2 = 0$, and solve for i_1, or $i_1 = 2.5$ amp. Since i_1 and i_2 are known, the equation $I_T = i_1 + i_2$ can be solved, or $I_T = 2.5 + 0.5 = 3.0$ amp.

7. Compute the unknown voltages using Ohm's law

Thus; $e_{r1} = I_T r_1 = 3 \times 20 = 60$ v; $e_{r2} = i_1 r_2 = 2.5 \times 20 = 50$ v; $e_{r3} = i_1 r_3 = 2.5 \times 48 = 120$ v; $e_{r4} = i_1 r_4 = 2.5 \times 52 = 130$ v; $e_{r5} = i_2 r_5 = 0.5 \times 300 = 150$ v; $e_{r6} = i_2 r_6 = 0.5 \times 300 = 150$ v.

Related Calculations: The above Calculation Procedure is an application of Kirchhoff's laws to a *reducible circuit*. Use the same general method and steps for *irreducible* circuits.

Note that when a circuit has two power sources, such as often occurs in irreducible circuits, the directions of the currents are not known. Hence, current directions must be assumed. If an incorrect direction is chosen, it will show up in the solution as a negative current. However, the magnitude of the current—i.e., the numerical value—will be correct regardless of the direction chosen. Use the negative sign for the current in any remaining calculations. As a final check, substitute the values obtained in any unused current or voltage equations so that all unknown current values are used at least once.

ALTERNATING-CURRENT-CIRCUIT ANALYSIS

A series-parallel alternating-current circuit is connected as in Fig. 4*a*. Find the current, power, apparent power, power factor, and phase angle of the line. Find the voltage and current of each part of the circuit. Draw a vector diagram for the circuit.

Calculation Procedure:

1. Reduce each parallel group to a single value of resistance and reactance

The purpose of this reduction is to produce the same effect, when the circuit elements are connected in series, as the original parallel group. These values are called the *equivalent resistance* and *equivalent reactance*. Do this by (*a*) assigning a convenient assumed voltage to the group for finding the impedance and phase angle of the group, (*b*) finding the equivalent resistance, and (*c*) finding the equivalent reactance.

Assume a 120-v supply. Then, for group I, Fig. 4*a*; $I_R = E/R_1 = 120/30 = 4$ amp; $I_{XL} = E/X_L = 120/40 = 3$ amp; $I_{line} = (I_R^2 + I_{XL}^2)^{0.5} = (4^2 + 3^3)^{0.5} = 5$ amp; $Z_{eq} = E/I_{line} = 120/5 = 24$ ohms, where Z_{eq} = equivalent reactance of group I.

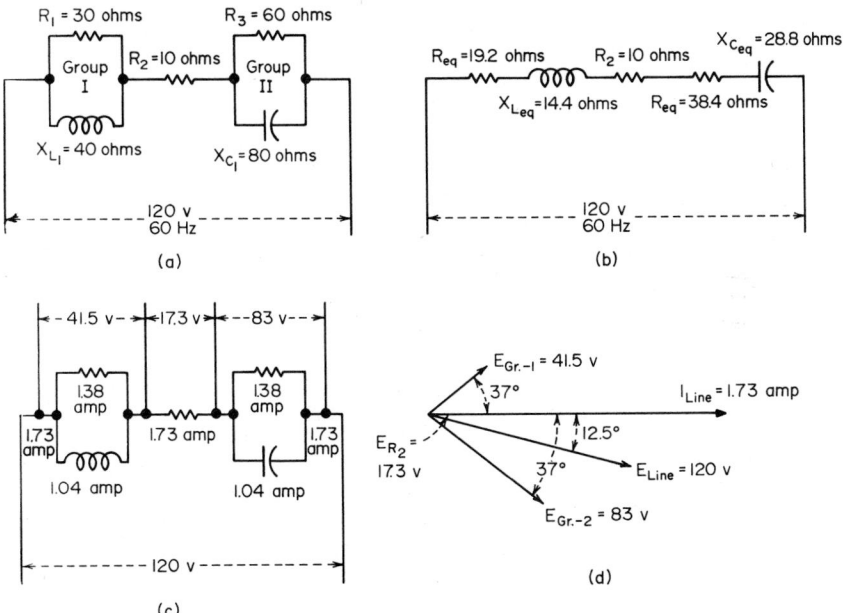

Fig. 4. (*a*) Typical ac circuit; (*b*) circuit converted to series elements; (*c*) voltages across the circuit elements; (*d*) circuit vector diagram.

Find the power factor of group I from $\cos \theta = I_R/I_{line} = \frac{4}{5} = 0.80$; $\sin \theta = I_X/I_{line} = \frac{3}{5} = 0.60$. From a table of trigonometric functions, $\theta = 37°$ lagging.

Use the relation $R_{eq} = Z_{eq} \cos \theta$ to find the equivalent resistance, R_{eq} ohms. Or, $R_{eq} = 24 \times 0.80 = 19.2$ ohms. Likewise, the equivalent reactance X_{eq} is $X_{eq} = Z_{eq} \sin \theta = 24 \times 0.60 = 14.4$ ohms.

Using a similar procedure for group II, Fig. 4a, after assuming a 240-v supply: $I_{R3} = E/R_3 = 240/60 = 4$ amp; $I_{XC} = E/X_C = 240/80 = 3$ amp; $I_{\text{line}} = (I_{R3}^2 + I_{XC}^2)^{0.5} =$ 5 amp; $Z_{\text{eq}} = E/I_{\text{line}} = 240/5 = 48$ ohms; $\cos\theta = I_R/I_{\text{line}} = \frac{4}{5} = 0.80$; $\sin\theta = I_{XC}/I_{\text{line}} = \frac{3}{5} = 0.60$; $\theta = 37°$ leading; $R_{\text{eq}} = Z_{\text{eq}}\cos\theta = 48\times 0.80 = 38.4$ ohms; $X_{\text{eq}} = Z_{\text{eq}}\sin\theta = 48 \times 0.60 = 28.8$ ohms.

2. Draw an equivalent series-circuit diagram

Figure 4b shows the equivalent series-circuit diagram in which the computed equivalent resistance and equivalent reactance replace parallel groups I and II.

3. Compute the desired quantities for the equivalent circuit

Use the relation $Z_{cct} = [(R_{\text{eq}} + R)^2 + (X_{L\text{eq}} - X_{L\text{eq}})^2]^{0.5}$ where $Z_{cct} =$ circuit impedance, ohms. Or, $Z_{cct} = [(19.2 + 38.4 + 10)^2 + (14.4 - 28.8)^2]^{0.5} = 69.1$ ohms. Then, $I_{\text{line}} = E_{\text{line}}/Z_{cct} = 120/69.1 = 1.73$ amp. The line power is $P_{cct} = I^2 R_{cct}$ watts, or $1.73^2 \times (19.2 + 38.4)$ $= 203$ watts. The apparent power, va, of the circuit is $AP_{cct} = E_{\text{line}}I_{\text{line}} = 120 \times 1.73 = 208$ va. The power factor of the circuit is $PF_{cct} = P_{cct}/AP_{cct} = 203/208 = 0.976$, or $\theta_{cct} = 12.5°$ leading.

4. Compute the voltage and current of the original circuit

The voltage at group I $= I_{\text{ine}}Z_{\text{eq}} = 1.73 \times 24 = 41.5$ v; at R_2 the voltage is $I_2 R_2 = 1.73 \times 10 = 17.3$ v. Also, the voltage at group II $= I_{II}Z_{\text{eq}} = 1.73 \times 48 = 83$ v.

The current through R_1 is $I_{R1} = E_{GR1}/R_1 = 41.5/30 = 1.38$ amp; the current through X_{L1} is $I_{XL1} = E_{GR1}/X_{L1} = 41.5/40 = 1.04$ amp. As a check, the line current to group I should be $I_{\text{line}} = (I_{R1}{}^2 + I_{XL1}{}^2)^{0.5} = (1.38^2 + 1.04^2)^{0.5} = 1.73$ amp.

The current through group II $= I_{R2} = I_{\text{line}} = 1.73$ amp. The current through R_3 is $I_{R3} = E_{GRII}/R_3 = 83/60 = 1.38$ amp; also $I_{XC} = E_{GRII}/X_C = 83/80 = 1.04$ amp. Figure 4c shows the voltage and current distribution in the original circuit, and Fig. 4d shows the vector diagram for the circuit. Note that the line voltage, $E_{\text{line}} = 120$ v $=$ the assumed voltage used in Step 1.

Related Calculations: To analyze parallel-series ac circuits, use the following steps: (1) Find the impedance, current, power, $\cos\theta$, and $\sin\theta$ for each parallel branch using the general relations given above. (2) Resolve each current into its in-phase (resistance) component and quadrature (90° angle) or reactance component using the relations $I_R = I\cos\theta$, and $I_X = I\sin\theta$, where $I =$ branch current, $\theta =$ leading or lagging power factor angle. (3) Compute the line current by combining the in-phase (resistance) components and the quadrature (reactance) components of all the branch currents using the general relation $I_{\text{line}} = [(I_1\cos\theta_1 + I_2\cos\theta_2, \ldots)^2 + (\pm I_1\sin\theta_1 \pm I_2\sin\theta_2, \ldots)^2]^{0.5}$ In the expression $\pm I_1\sin\theta_1 \ldots$, use $+$ for leading currents and $-$ for lagging currents. When the sum of the second parentheses in the I_{line} expression is $+$, this indicates that I_{line} is a leading current; a negative sum indicates a lagging line current. (4) Compute the impedance of the circuit. (5) Compute the power taken by the circuit. (6) Compute the va of the circuit. (7) Compute the line power factor. Although this is the recommended solution sequence, it may have to be altered, depending on the characteristics of the circuit analyzed.

VECTOR ALGEBRA IN ALTERNATING-CURRENT-CIRCUIT ANALYSIS

Use vector algebra to analyze the circuit in Fig. 4. Find the impedance, phase angle, current, power factor, and power of the line, and the voltage and current of each part of the circuit.

Calculation Procedure:

1. Compute the impedance of each circuit group

For circuit groups with two components in parallel, use the relation $Z_T = Z_1 Z_2/(Z_1 + Z_2)$, where Z_T is the total impedance, ohms; Z_1 and Z_2 are the impedances, respectively, of the two components in parallel. For group I, using vector notation, $Z_{GI} = [(30 \times j40)/(30 + j40)] \times [(30 - j40)/(30 - j40)] = 19.2 + j14.4 = 24\underline{/37°}$ ohms. For group II, $Z_{GII} = \{(60(-j80))/(60 - j80)\}[(60 + j80)/(60 + j80)] = 38.4 - j28.8 = 48\underline{/-37°}$ ohms.

2. Draw the equivalent series circuit

Using the computed impedances of the parallel groups, draw the equivalent series circuit, Fig. 4b.

3. Compute the line impedance

The line impedance $Z_{\text{line}} = Z_{GI} + R_2 + Z_{GII}$, ohms. Or in words, the line impedance = sum of the parallel-circuit group impedances plus the line resistance. For the series circuit in Fig. 4b, $Z_{\text{line}} = 19.2 + j14.4 + 10 + 38.4 - j28.8 = 67.6 - j14.4 = 69 - \underline{/12°}$ ohms.

4. Compute the line current

Use the relation $I_{\text{line}} = E/Z_{\text{line}}$, ohms. Or, $I_{\text{line}} = 120/69 = 1.74$ amp.

5. Determine the line power factor

Step 3 shows the line phase angle as 12°. Since $PF_{\text{line}} = \cos \theta_{\text{line}} = \cos 12°$, $PF_{\text{line}} = 0.978$.

6. Compute the line power

Use the relation, $P_{\text{line}} = EIPF$, watts, or $120 \times 1.74 \times 0.978 = 204$ watts.

7. Compute the voltage and current in the circuit parts

For R_1, $E_{R1} = E_{XL1} = E_{GI} = I_{\text{line}}Z_{GI} = 1.74 \times 24 = 41.8$ v. Then, $I_{R1} = E_{GI}/R_1 = 41.8/30 = 1.39$ amp. Also, $I_{XL1} = E_{GI}/X_{L1} = 41.8/40 = 1.04$ amp. This completes the voltage and current computations for group I.

For group II: $E_{GII} = I_{\text{line}}R_2 = 1.74 \times 10 = 17.4$ v; $I_{R2} = I_{\text{line}} = 1.74$ amp; $E_{R3} = E_{XC1} = E_{GII} = I_{\text{line}}Z_{GII} = 1.74 \times 48 = 83.6$ v; $I_{R3} = E_{GII}/R_3 = 83.6/60 = 1.39$ amp; $I_{XC1} = E_{GII}/X_{C1} = 83.6/80 = 1.04$ amp.

Related Calculations: Vector algebra saves many steps in complex ac-circuit problems. When polar notation is used, as in this Calculation Procedure, solutions are obtained more rapidly. Note that the circuit analyzed in this Procedure is the circuit analyzed in the previous Procedure.

LIGHTNING-ARRESTER SELECTION AND APPLICATION

Select lightning arresters for a three-phase 13.8-kv industrial plant electrical system fitted with 4.16-kv rotating machines.

Calculation Procedure:

1. Select the distribution-system arrester voltage rating

Table 1 shows the typical voltage ratings of lightning arresters usually chosen for three-phase power systems. Thus, either a 15-kv or a 12-kv arrester may be used for this system, depending on how the system neutral is grounded. Where the type of grounding is not known, or is yet to be chosen, the higher-rated arrester is a safer choice. Also, an effectively grounded neutral may, under fault conditions or other emergencies, leave a portion of the system ungrounded. For these reasons the higher-rated arrester is often preferred. Hence, a 15-kv arrester will be chosen for this distribution system.

2. Choose the type of arrester to use

Table 2 shows the typical arrester types used for various required voltage ratings. Thus, either a distribution- or station-type arrester can be used for a required voltage rating of 3 to 15 kv. Further study of Table 2 shows that protection of industrial-plant electrical systems is usually accomplished by use of a station-type lightning arrester. The reason for this is that the value of the equipment and the importance of uninterrupted service warrants the use of station-type arresters throughout the voltage range of the plant. Hence, this type of arrester will be chosen for this plant.

3. Choose the rotating-machinery arresters

Table 3 shows the voltage rating of lightning arresters used to protect three-phase ac rotating machines. Since it has already been decided, Step 2, to use station-type arresters

TABLE 1   Voltage Ratings of Arresters Usually Selected for Three-Phase Systems

Nominal system voltage, kv	Voltage rating of arrester, kv	
	System neutral ungrounded or resistance grounded	System neutral effectively grounded
0.120/0.208 Y	0.65	0.175
0.240	0.65	0.65
0.480	0.65	0.65
0.600	0.65	0.65
2.4	3	3
2.4/4.16Y	4.5* or 6	3,† 4.5,* or 6
4.16	4.5* or 6	4.5* or 6
4.8	6	4.5* or 6
6.9	7.5* or 9	6
12	15	12
7.2/12.47 Y	15	9† or 12
13.2 (or 13.8)	15	12
23	25	20
34.5	37	30
46	50	40
69	73	60
115	121	97
138	145	121

*The 4.5- and 7.5-kv arresters are available only in the station type.

†The use of these arresters requires an X_0/X_1 ratio less than that necessary to make the system "effectively grounded."

SOURCE: Beeman—*Industrial Power Systems Handbook*, McGraw-Hill.

TABLE 2   Lightning-Arrester Applications

Required Arrester Voltage Rating, kv	Type of Arrester Used
3–15	Distribution or station
20–73	Line or station

Equipment Protected	Typical Arrester Choice
Industrial plant	Station
Liquid-filled transformers and substations rated 1,000 kva and less; short cables between overhead lines and apparatus; small breakers; disconnects	Distribution or line
Rotating machines	Distribution

Table 3 Protective Equipment for Three-Phase AC Rotating Machines

Machine voltage rating (phase-to-phase)	For installation at machine terminals or on machine bus						For installation 1,500 to 2,000 ft out on directly connected exposed overhead lines		
	Protective capacitors			Station-type arresters			Distribution-type arresters		
	Voltage rating	Microfarad per pole	Single-pole units required	Voltage rating		Single-pole units required	Voltage rating		Single-pole units required
				Ungrounded or resistance-grounded system	Effectively grounded system		Ungrounded or resistance-grounded system	Effectively grounded system	
0–650	0–650	1.0	3*	650	650	3*	650	650	3
2,400	2,400	0.5	3*	3,000	3,000	3	3,000	3,000	3
4,160	4,160	0.5	3*	4,500	3,000†	3	6,000	3,000†	3
4,800	4,800	0.5	3	6,000	4,500	3	6,000	6,000	3
6,900	6,900	0.5	3	7,500	6,000	3	9,000	6,000	3
11,500	11,500	0.25	3 or 6‡	12,000	9,000	3	12,000	9,000	3
13,800	13,800	0.25	3 or 6‡	15,000	12,000	3	15,000	12,000	3

*A single three-pole unit is commonly used.

†The use of 3,000-v arresters on a 4,160-v system requires an X_0/X_1 ratio less than that necessary to make the system "effectively grounded."

‡Use six capacitor units (0.5 μf per phase) where both of the following conditions apply: (1) Machine is directly connected to the exposed overhead lines, is connected through an autotransformer, or is connected through a Y-Y transformer with both Y's grounded. (2) Machine is ungrounded, is neutral grounded through a resistance greater than 50 ohms, or is neutral grounded through a reactance greater than 5 ohms (60-cycle basis). In all other cases three capacitor units (0.25 μf per phase) will suffice.

SOURCE: Beeman—*Industrial Power Systems Handbook*, McGraw-Hill.

in this plant, the choice of the rotating-machinery-type arrester is simplified. Thus, all that need be done is to choose the station-type arrester voltage rating.

Table 3 shows that either 4,500-v or 3,000-v station-type arresters are suitable for 4,160-v machines, depending on the type of system grounding. Once again, unless the system is known to be effectively grounded, the higher voltage rating is the safer choice. Thus, either the higher or lower voltage rating would be used, depending on the grounding method employed.

Related Calculations: Other ac equipment that can benefit from lightning-arrester protection includes metal-clad switchgear, generators, transformers, distribution lines, circuit breakers, overhead feeders, etc.

Direct-current motors and generators connected to exposed overhead lines are generally protected by capacitor-type dc arresters. These arresters can be installed at the machine terminals, on the bus, or at the station on each outgoing feeder. Mercury-arc rectifiers and their transformers can be protected by a set of station-type or distribution-type arresters on the supply side of the transformer. If the dc feeders are exposed, suitable dc arresters should also be used at the dc terminals of the rectifier, on the dc bus, or on the exposed dc feeders.

DIRECT-CURRENT-GENERATOR SELECTION

Select an alternator for a new industrial plant having an expected demand for 8,000 kw The load served by the generator requires an input voltage of at least 230 v at all ratings up to full load.

Calculation Procedure:

1. Compute the generator amperage output

Use the relation $I = P/E$, where I = generator amperage output; P = generator power output, watts; E = generator terminal voltage. For this generator, $I = 1,000,000/230 = 4,350$ amp.

2. Select the generator rating

Typical standard ratings for direct-current generators are 1,000, 2,000, 3,000, 4,000, 5,000, 6,000, etc., amp. Since the required amperage output of this generator is between two standard ratings, select the next higher rating, or 5,000 amp. This provides extra capacity for a moderate growth in the generator load.

3. Select the type of generator to use

Two classes of dc generators are used today: (a) separately excited, and (b) self-excited. The self-excited generator is, in general, more popular for power service. Three types of self-excited dc generators are used for power generation—shunt, series, and compound wound. The first two have *drooping* voltage characteristics—i.e., the generator output voltage decreases as the external load increases.

For general-purpose power-supply service, a cumulative-compound-type generator is preferred. This type of generator may be *flat-compounded*, Fig. 5, so that it produces the same voltage at full load as at no load, or it may be *overcompounded* so it produces a higher voltage at full load than at no load, Fig. 5. Other wiring arrangements are *under-compounded* and *differentialcompounded*. Since the load served by this generator requires an input voltage of at least 230 v, either a flat-compounded or an overcompounded generator can be used. To provide for the possibility of load growth, an overcompounded generator would be the best choice because it will provide the desired voltage at all loads within the generator rating.

4. Compute the generator efficiency

The manufacturer's engineering data list the losses in the generator. These losses are usually expressed in watts and are shunt-field loss = shunt-field current × generator voltage rating, armature loss = [shunt-field current + armature current]² × armature resistance, ohms; series-field loss = series-field loss = [series-field current + armature current]²× series field resistance ohms; and the stray power loss.

Assume that the sum of all these losses for this generator is given as 80,000 watts. Then, since generator efficiency = output watts/(output watts + losses, watts), efficiency = 1,000,000/1,000,000 + 80,000 = 0.925, or 92.5 percent.

Related Calculations: Use this general method to choose dc generators for power, emergency marine, and similar applications. Be sure to relate the generator output-

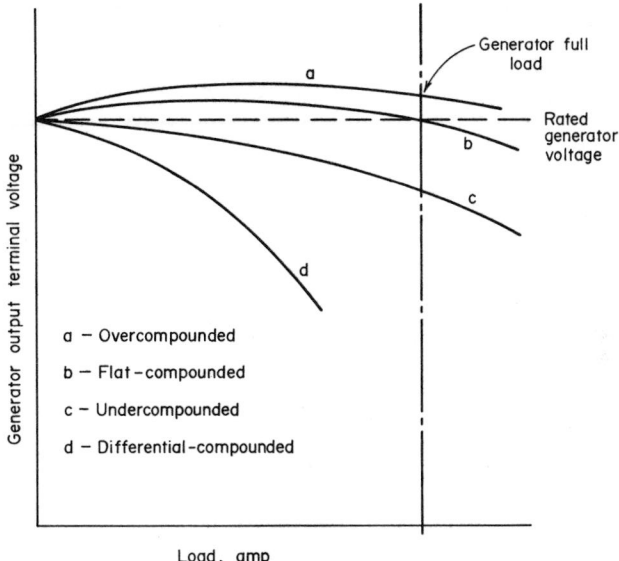

Fig. 5   Voltage characteristics of direct-current generators.

voltage characteristic to the load being served. If this important aspect of dc generator choice is overlooked, the unit chosen may be unsuitable for the load. Also, be sure that the rpm of the generator selected can be obtained using a suitable prime mover.

ALTERNATOR SELECTION FOR A KNOWN LOAD

Select an alternator for a new industrial plant having an expected demand for 8,000 kw at a power factor of 0.8. Continuous operation of the plant is expected and interruptions of generating service must be avoided because they are costly. Hence, reserve capacity must be provided for forced outages and regular maintenance. The plant will have a continuously high load factor.

Calculation Procedure:

1. Select the alternator capacity

In general, the cost per kw of installed generating capacity decreases as the size of the unit increases. Also, the efficiency of steam-driven generating units increases as unit size increases. Operating labor costs are nearly proportional to the number of generating units installed. The efficiency of a generating unit is usually low at light loads and rises to a "best point" somewhere between 75 percent and full load. The exact shape of the efficiency curve of an alternator depends on the type of prime mover and, to a lesser degree, on the size, design, and other features of the alternator and prime mover.

These characteristics of generating units point toward supplying a load with one large alternator. Yet this is rarely possible because of load conditions.

Where continuity of operation is required, or where interruptions are costly, reserve capacity is required, not only for forced outages but for scheduled maintenance outages. It is usually more economical to provide reserve capacity for several smaller generating units than for one large unit. A typical plant might use three units, any two of which can carry the load. This arrangement also provides economically for load growth — adding a unit gives four machines with any three able to carry the full plant load. Where continuity of service is an important consideration, check to see what standby service, if any, the local utility can provide, and what it costs.

Using the above facts as a general guide, make a tentative choice of three units for this 8,000-kw plant. The capacities of the three units chosen will be 6,000, 4,000, and 4,000 kw to fulfill the requirement that any two units be able to carry the plant load. Thus, the capacity of the units, when operated in twos, is $6,000 + 4,000 = 10,000$ kw, or $4,000 + 4,000 = 8,000$ kw.

2. Study the plant load factor

The plant load factor also exerts a strong influence on alternator choice. A low load factor discourages the use of one or two large units because much of the operating time will be on the lower part of the efficiency curve. A special problem in some industries is the weekend load, which may require a special small unit. The load-duration curve provides a valuable insight for relating unit sizes to plant load conditions. Since this plant has a continuously high load factor, the three units will probably be suitable.

3. Select the alternator operating voltage

Figure 6 shows the typical standard voltage ratings for alternators of various capacities. Thus, 4,000-kw alternators are rated between 4.16 and 6.9 kv. In the larger sizes — 5,000 kw and up — 13.8 kv is a standard voltage.

Since any two of these alternators must operate in parallel, the same voltage should be used for each. In this plant the smaller alternators will use 6.8 kv because higher voltages are the trend today. Hence, the large generator should be chosen for the same voltage because it must operate with either smaller unit.

When choosing alternator voltage, keep in mind the trend toward generating at higher voltages (up to 13.8 kv) in industrial plants and distributing at that voltage to secondary substations. Where a later change from delta to Y connections to increase the alternator voltage is planned, be sure the alternator is now connected delta.

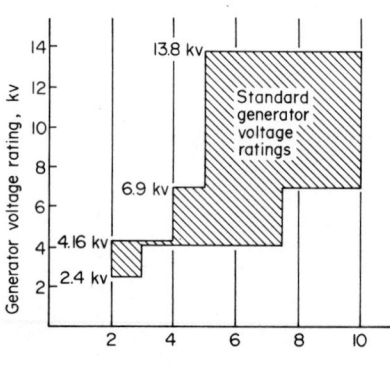

Fig. 6   Standard alternator voltage ratings.

4. Check the generator regulation

The standard alternator regulation of 40 to 45 percent at 0.8 pf will satisfy most industrial-plant needs. Where the plant has one load which is much larger than any other — such as the chipper motor in a paper mill — a different regulation from that cited above might be required. The exact regulation required varies with the motor rating and application.

5. Check the paralleling characteristics of the alternators

To parallel two alternators, the voltage, frequency, waveform, phasing, and phase rotation of both alternators must be the same. During paralleling, slight adjustments can be made in the alternator voltage and frequency. Check the specification sheet of each alternator to see that the requirements listed above are met.

Related Calculations: Use this general method to choose alternators for commercial, marine, portable, and similar applications. Be sure to use the manufacturer's specification sheet when considering a specific alternator.

SELECTING ELECTRIC-MOTOR STARTING AND SPEED CONTROLS

Choose a suitable starter and speed control for a 500-hp wound-rotor ac motor that must have a speed range of 2 to 1 with a capability for low-speed jogging. The motor is to operate at about 1,800 rpm with current supplied at 4,160 v, 60 Hz. An enclosed starter and controller is desirable from the standpoint of protection. What is the actual motor speed if the motor has four poles and a slip of 3 percent?

Calculation Procedure:

1.   *Select the type of starter to use*

Table 4 shows that a magnetic starter is suitable for wound-rotor motors in the 220 to 4,500-v and 5 to 1,000-hp range. Since the motor is in this voltage and horsepower

TABLE 4   Typical Alternating-Current Motor Starters*

Motor type	Starter type	Typical range	
		Voltage	Hp
Squirrel cage ..	Magnetic, full-voltage	110–550	1.5–600
	With fusible or nonfusible disconnect or circuit breaker	208–550	2–200
	Reversible	110–550	1.5–200
	Manual, full-voltage	110–550	1.5–7.5
	Manual, reduced-voltage, autotransformer	220–2,500	5–150
	Magnetic, reduced-voltage, autotransformer	220–5,000	5–1,750
	Magnetic, reduced-voltage, resistor	220–550	5–600
Wound rotor...	Magnetic, primary and secondary control	220–4,500	5–1,000
	Drums and resistors for secondary control	1,000 max	5–750
Synchronous ..	Reduced-voltage, magnetic	220–5,000	25–3,000
	Reduced-voltage, semimagnetic	220–2,500	20–175
	Full-voltage, magnetic	220–5,000	25–3,000
High-capacity induction....	Magnetic, full-voltage	2,300–4,600	Up to 2,250
	Magnetic, reduced-voltage	2,300–4,600	Up to 2,250
High-capacity synchronous	Magnetic, full-voltage	2,300–4,600	Up to 2,500
	Magnetic, reduced-voltage	2,300–4,600	Up to 2,500
High-capacity wound rotor	Magnetic primary and secondary	2,300–4,600	Up to 2,250

*Based on Allis-Chalmers, General Electric, and Westinghouse units.

range, a magnetic starter will probably be suitable. Also, the magnetic starter is available in an enclosed cabinet, making it suitable for this installation.

Table 5 shows that a motor starting torque of approximately 200 per cent of the full-load motor torque and current are obtained on the first point of acceleration.

2. Compute the full-load speed of the motor

Use the relation $S = [(100 - s)/100]120f/n$, where S = motor full-load speed, rpm; s = slip, percent; f = frequency of supply current, Hz; n = number of poles in the motor. For this motor, $S = [(100 - 3)/100]120(60)/4 = 1,750$ rpm.

TABLE 5   Adjustable-Speed Drives

Drive features	Drive types						
	Constant-voltage dc	Adjust.-v dc m-g set	Adjust.-v rectifier	Eddy-current clutch	Wound-rotor ac, standard	Wound-rotor thyratron	Wound-rotor dc-motor set
Power units required	Rectifier, dc motor	Ac motor, dc generator, dc motor	Rectifier, reactor[a], dc motor	Ac motor, eddy-current clutch	Ac motor	Ac motor, thyratrons	Ac motor, dc motor, rectifier
Normal speed range	4-1	8-1 c-t+[b] 4-1 c-hp[c]	8-1 c-t+ 4-1 c-hp[c]	34-1, 2 pole; 17-1, 4 pole	3-1	10-1[c]	3-1
Low speed for jogging......	No[d]	Yes	Yes	Yes	Yes	Yes	Yes
Torque available	c-hp	c-t	c-t	c-t	c-t	c-t	c-t, c-hp
Speed regulation	10–15%	5% with regulator	5% with regulator	2% with regulator	Poor	±3%	5–7½%
Speed control	Field rheostat	Rheostats or pots	Rheostats or pots	Rheostats or pots	Steps, power contactors	Rheostats or pots	Rheostats or pots
Enclosures available	All	All	All	Open[e]	All	All	All
Braking:							
Regen.......	No	Yes	No	No	Yes	Yes	No
Dynamic	Yes	Yes	Yes	No[f]	Yes	Yes	Yes
Multiple operation....	Yes	Yes	Yes	Yes	Yes	Yes	No
Parallel operation....	Yes	Yes	Yes	Yes	No	Yes	Yes
Controlled acceleration, deceleration	Yes	Yes	Yes	Yes	No	Yes	No
Efficiency	80–85%	63–73%	70–80%	80–85%	80–85%	80–85%	80–85%
Top speed at maximum torque	83–87%	60–67%	60–70%	29%	29%	85–90%	73–78%
Rotor inertia[g].....,	100%[h]	100%	100%	75%	90%	90%	175%
Starting torque	200–300%	200–300%	200–300%	200–300%	200%	200–300%	200–300%
Number of comm, rings .	1 comm	2 comm	1 comm	None	1 set rings	1 set rings	1 comm, 1 set rings.

[a]Used only in saturable-reactor designs.
[b]c-t—constant-torque; c-hp—constant horsepower.
[c]Units of 200 to 1 speed range are available.
[d]Low speed can be obtained using armature resistance.
[e]Totally enclosed units must be water- or oil-cooled.
[f]Eddy-current brake may be integral with unit.
[g]Based on standard dc motor.
[h]Normally is a larger dc motor since it has slower base speed.

3. Choose the type of speed control to use

Table 5 summarizes the various types of adjustable-speed drives available today. This listing shows that power-operated contactors used with wound-rotor motors will give a 3 to 1 speed range with low-speed jogging. Since a 2 to 1 speed range is required, the proposed controller is suitable because it gives a wider speed range than needed.

Note from Table 5 that if a wider speed range were required, a thyratron control could produce a range up to 10 to 1 on a wound-rotor motor. Also, a wound-rotor direct-current motor set might be used too. In such an arrangement, an ac and dc motor are combined on the same shaft. The rotor current is converted to dc by external silicon rectifiers and fed back to the dc armature through the commutator.

Related Calculations: Use the two tables presented here to guide the selection of starters and controls for alternating-current motors serving industrial, commercial, marine, portable, and residential applications.

To choose a direct-current motor starter, use Table 6 as a guide.

Speed controls for dc motors can be chosen using Table 7 as a guide. Dc motors are finding increasing use in industry. They are also popular in marine service.

TABLE 6   Direct-Current Motor Starters

Type of Starter	Typical Uses
Across-the-line	Limited to motors of less than than 2 hp
Reduced-voltage, manual-control (face-plate type)	Used for motors up to 50 hp where starting is infrequent
Reduced-voltage, multiple-switch	Motors of more than 50 hp
Reduced-voltage, drum-switch	Large motors; frequent starting and stopping
Reduced-voltage, magnetic-switch	Frequent starting and stopping; large motors

TABLE 7   Direct-Current Motor-Speed Controls

Type of motor	Speed characteristic	Type of control
Series-wound . . .	Varying; wide-speed regulation	Armature shunt and series resistors
Shunt-wound . . .	Constant at selected speed	Armature shunt and series resistors; field weakening; variable armature voltage
Compund-wound	Regulation about 25 per cent	Armature shunt and series resistors; field weakening; variable armature voltage

BASIC SHORT-CIRCUIT CURRENT DETERMINATION

A 50-hp ac motor draws 63 amp at full load. This 40-ohm apparent impedance motor is supplied from an "infinite" bus through a transformer with a rated output of 440 v, 200 amp, and a 0.2-ohm impedance. Determine the short-circuit current flow if a fault occurs between the transformer and the motor. What will be the effect of using a 2,000-amp, 0.02-ohm impedance transformer for the same motor to provide for a load growth?

Calculation Procedure:

1. Sketch the circuit hookup

Figure 7 shows the typical circuit hookup for an installation of this type. The circuit breaker at point X must have a large enough rating to handle the short-circuit current.

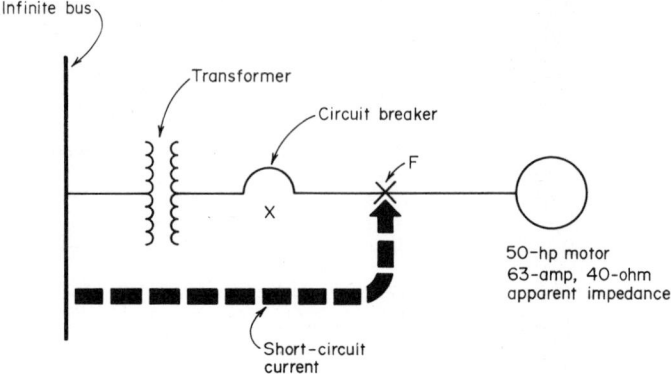

Fig. 7 Typical motor circuit with a step-down transformer and protective circuit breaker.

2. Compute the short-circuit current with the small transformer

With a short circuit at F, the only impedance limiting the short-circuit current flow is the transformer impedance of 0.2 ohm, because the current will take the path of least resistance. The motor apparent impedance of 40 ohm is so much larger than the transformer impedance that the short-circuit current will rush out at F.

Compute the short-circuit current from $I_s = E/Z_t$, where I_s = short-circuit current, amp; E = bus voltage rating, i.e., transformer-output rating, v; Z_t = transformer impedance, ohms. Thus, $I_s = 440/0.2 = 2,200$ amp. Hence, the circuit breaker must handle at least 2,200 amp to protect this circuit.

3. Compute the short-circuit current with the large transformer

Use the same relation as in Step 2. Or, $I_s = 440/0.02 = 22,000$ amp. Thus, the larger transformer, installed to handle the greater load, will require a circuit breaker with a much higher rating. Note that the motor-load current will remain the same, yet the short-circuit current increases tenfold as the system load increases.

Related Calculations: This simple short-circuit computation shows the basic procedure to use. As a circuit and its components become more complex, so do the short-circuit computations. Typical methods are shown in the following Calculation Procedure.

POWER-SYSTEM SHORT-CIRCUIT CURRENT

A three-phase power system has two generating stations that supply one substation. If the generator ratings, line voltages, and line reactances are as listed below, determine the short-circuit current when a fault occurs in the distribution line beyond the substation.

Unit	Rating, kva	Line-to-line Kv	Reactance, ohms
Generator x...	80,000	13.8	2.3
Generator y...	85,000	13.8	1.1
Line xz	80,000	13.8	0.65
Line yz	85,000	13.8	0.40
Line z to short .	150,000	13.8	0.45

Calculation Procedure:

1. Express the system reactances on a per-unit basis

Select a kva base, such as 50,000, 100,000, 200,000 kva, etc. Any easily manipulated value can be used. Once a kva base is chosen, compute the reactance per unit X_{pu} from $X_{pu} = $ (kva base) $X/kv^2x\,1,000$, where $X = $ unit reactance, ohms; $kv = $ line-to-line kilovolts for the unit.

Selecting a kva base of 100,000 kva and using this relation for generator x, $X_{pu} = (100,000)(2.3)/(13.8)^2(1,000) = 1.21$. Using the same kva base, compute the per-unit reactance for the other generator and each line. Draw a single-line diagram of the system. Mark the reactance values on the system single-line diagram, Fig. 8a.

2. Sketch the network representing this power system

Figure 8b shows the network representing this system. It consists of a parallel network of two parts, each part having two X_{pu} reactances in series.

3. Compute the equivalent reactance between the generator and the fault

Use the relation $X_{eq} = (X_x X_y/X_x + X_y) + X_{3A}$, where $X_{eq} = $ equivalent reactance of the network, per unit; $X_x = $ total per unit reactance for leg xz; $X_y = $ total per unit reactance for leg yz; $X_{zA} = $ total per unit reactance for leg zA. Substituting, $X_{eq} = [(1.21+0.341)(0.577+0.21)]/(1.21+0.341+0.577+0.21)+0.236 = 0.759$ per unit.

4. Compute the normal line current at the fault

Use the relation $I_n = $ kva base$/kv\,\sqrt{3}$ for the normal line current at the fault, where $I_n = $ normal line current at the fault, amp; other symbols as before. Thus, $I_f = 100,000/13.8\sqrt{3} = 4,175$ amp.

5. Compute the current in each line at the fault

Use the relation $I_f = I_n/X_{eq}$ for a three-phase short circuit, where $I_f = $ current in each line at the fault, amp; other symbols as before. Thus, $I_f = 4,175/0.759 = 5,500$ amp.

6. Compute the fault current in each line

For generator x and transmission line xz, $I_x = I_f(X_{pux} + X_{puxz})/\Sigma X_{pu}$, where $I_x = $ generator and transmission-line fault current, amp; $X_{pux} = $ generator reactance, per unit; $X_{puxz} = $ transmission-line reactance, per unit; $\Sigma X_{pu} = $ sum of the individual network reactances; other symbols as before. Substituting, $I = 5,500(1.21+0.341)/(1.21+0.341+0.577+0.21) = 3,650$ amp.

Using a similar relation for generator y and transmission-line yz, $I_y = 5,500(0.577+0.21)/(1.21+0.341+0.577+0.21) = 1,850$ amp. As a check, the sum of the two generator and transmission-line currents should equal I_f, or $3,650+1,850 = 5,500$ amp $= I_f$.

Related Calculations: The general method presented here is valid for simple and complex short-circuit calculations. Summarized, this procedure is as follows. (1) Obtain equipment and line reactances from equipment characteristics and tabulated data. (2) Draw the system single-line diagram. (3) Convert reactances to per unit on a base kva. (4) Combine reactances to obtain a reactance diagram without series reactances. (5) Compute the fault current or kva.

To combine reactances, use the equations shown in Fig. 8c.

Lastly, Table 8 shows the short-circuit current theoretically possible at various operating voltages and kva ratings.

TRANSFORMER CHARACTERISTICS AND PERFORMANCE

A 60-HZ 1,000-kva three-winding transformer is rated at 4,800-v primary voltage and 600- and 480-v secondary voltages. The transformer has 800 primary turns, and the rating of each secondary winding is 500 kva. Compute the number of turns in each secondary winding, the rated primary current at unit power factor (pf), rated primary current at 0.8 pf, lagging current, rated current of the 600- and 480-v secondary windings, primary current when the rated current flows in the 480-v winding with pf = 1.0, and the rated current flow in the 600-v winding with a 0.8 lagging pf.

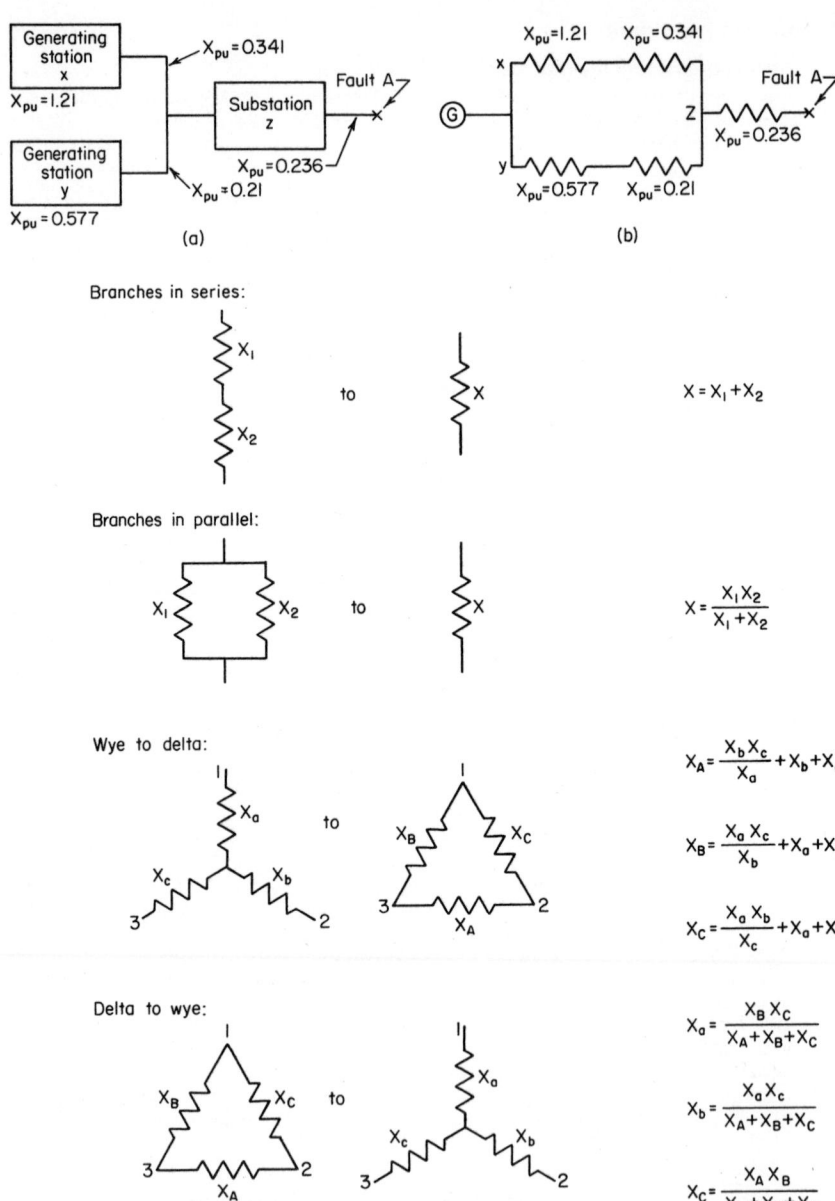

Fig. 8 (a) Electric-power system with two generating stations and one substation; (b) circuit diagram of generating system in a; (c) reactances for various circuit arrangements.

Calculation Procedure:

1. Compute the number of turns in the secondary windings

Use the relation $N_s = v_s N_p / v_p$, where N_s = number of turns in secondary winding; v_s = voltage rating of secondary winding; N_p = number of turns in the primary winding v_p = voltage rating of the primary winding.

TABLE 8 Theoretical Short-Circuit Currents

Short-circuit energy, 3-phase kva	Short-circuit amp per phase for various voltages			
	15,000 v	2,500 v	480 v	208 v
25,000	1,000	5,800	31,000	70,000
50,000	2,000	11,500	63,000	140,000
100,000	4,000	23,000	125,000	280,000

NOTE: These are calculated current values discounting impedance that may be met in typical plant systems on secondaries of unit substations.

For the 600-v secondary winding, $N_s = 600(800)/4,800 = 100$ turns. For the 480-v secondary winding, $N_s = 480(800)/4,800 = 80$ turns.

2. Compute the rated primary current

Use the relation $I_p = va/v_p$, where I_p = primary current, amp; va = transformer volt-ampere rating; v = transformer primary voltage rating. For this transformer at unity power factor, $I_p = 1,000,000/4,800 = 208$ amp.

At 0.8-pf lagging current, the rated primary current is the same as the rated primary current at unity power factor, or 208 amp. The reason for this is that the power factor relates to the transformer or secondary load. The power factor of the load does not affect the rated primary-winding current.

3. Compute the secondary-winding current

Use the relation $I_s = va_s/v_s$, where I_s = secondary current flow, amp; va_s = secondary volt-ampere rating; v_s = secondary voltage rating.

For the 600-v secondary winding, $I_s = 500,000/600 = 833$ amp. For the 480-v secondary winding, $I_s = 500,000/480 = 1,041$ amp.

4. Compute the primary current at the given secondary loads

Use the relation $I_p = (va_{s1}/v_p)(\cos^{-1} pf_1) + (va_{s2}/v_p)(\cos^{-1} pf_2)$, where I_p = primary current, amp; va_{s1} = volt-amperes of the first secondary coil; v_p = transformer primary voltage; $\cos^{-1} pf$ = angle, expressed in degrees, whose cosine = the power factor of the load on the first secondary coil; other symbols are the same, except that they refer to the other secondary coil of the transformer.

Thus, $I_p = (600 \times 833/4,800) \underline{/-36.8°} + (480 \times 1,041/4,800) \underline{/0°} = 104.1 \cos 36.8° - j104.1 \sin 36.8° + 104.1 \cos 0° + 104.1 \sin 0° = 83.4 - j62.4 + 0' + 104.1$, or $I_p = 187.5 - j62.4$. Converting this expression using vector algebra, $I_p = (187.5^2 - 62.4^2)^{0.5} = 197.8$ amp.

Related Calculations: Use this general method to analyze transformers with one or more secondary coil windings used for power, distribution, residential, or commercial service.

TRANSFORMER SELECTION FOR AN INDUSTRIAL LOAD

Select a three-phase transformer for an industrial plant having an expected load of 300 kva located 400 ft from the transformer. The transformer will be located in an area of high humidity. What is the transformer voltage drop, copper loss, and core loss if the primary voltage is 4,160 and the secondary 480 v, and the power factor 0.80? What sound level can be expected with this transformer?

Calculation Procedure:

1. Choose between an outdoor substation and a load center

Outdoor substations were once the popular way of supplying power to industrial plants. Today a load center is often used in place of the outdoor substation. In a load center, the primary power is brought close to the plant load instead of being ended at an outdoor substation.

Load centers are not always the most economical choice, however. Here are general guides for the choice of outdoor substation versus load-center distribution. With a 120/240-v single-phase load, a load center is usually best if the transformer-to-load distance is more than 160 ft for a 25-kva load, more than 90 ft for a 75-kva load, and more than 60 ft for a 200-kva load. For a 480-v three-phase distribution system, load centers are generally best for a 150-kva load more than 400 ft from the transformer; a 300-kva load more than 300 ft; a 750-kva load more than 150 ft from the transformer.

Since this installation serves a 300-kva three-phase load 400 ft from the transformer, a load-center distribution system will probably be best, according to the above guide. Hence, this will be the tenative first choice for this plant.

2. Choose the type of transformer cooling

Basically, two types of transformer cooling are used today: liquid and air. The usual liquid coolants currently used are askarel, a synthetic nonflammable liquid, and mineral oil. The liquid-cooled transformer predominates in sizes over 500 kva. By far the largest number of transformers in terms of kva capacity are oil-cooled.

Air-cooled transformers are termed *dry-type* units. Dry-type transformers can be sealed units, open units, forced-air-cooled, or self- and forced-air-cooled.

Table 9 shows the important factors that should be considered in choosing load-center transformers. Study of this table shows that in areas of high humidity an askarel-cooled transformer is generally chosen.

TABLE 9   Transformer Coolant Characteristics*

Exposure to lightning	Dry	Askarel
Where transformers connect directly to lightning-exposed circuits with only usual lightning protection	No	Yes
If lightning exposure is negligible or if unit is suitably protected, if feeder cable is underground, or if overhead supply having usual lightning protection feeds into primary cable at least 1,500 ft long	Yes	Yes
Location and atmospheric conditions		
Relatively clean atmosphere: assembly plants, etc	Yes	Yes
Bad atmosphere: foundries, cement, flour mills, etc.	No	Yes
Areas of high humidity or of possible flooding	No	Yes
Acid, oil, or corrosive vapors present	No	Yes
Units to be installed in hazardous locations (see National Electrical Code, article 500)	No	Yes
Transformers to be overhead on platforms or roof trusses, all other conditions being satisfactory	Yes	No

*In general, the sealed dry-type transformer can be used under all the given conditions that would rule out ventilated dry-type units.

3. Specify the transformer variables

Table 10 lists many of the variables for load-center transformers. Thus, askarel-filled load-center transformers normally use Class A insulation, have an average temperature rating of 55-C rise and 65-C hottest-spot rise. The weight of an askarel-filled unit is about 1.25 times the weight of an oil-filled transformer; the floor space required by the two is the same.

TABLE 10 Comparison of Load-Center Unit Substation Transformer Sections

Type of transformer	Liquid filled		Dry type		
	Oil	Askarel	Open ventilated	Sealed Class B	Sealed Class H
Impulse strength............	100%	100%	50%*	50%*	50%*
Total loss at 75 C	100%	100%	100%	100%	100%
Insulation..................	Class A	Class A	Class B	Class B	Class H
Temperature ratings:					
Average rise	55 C	55 C	80 C	120 C	120 C*
Hottest-spot rise..........	65 C	65 C	110 C	140 C	140 C*
Audio sound level	X db	X db	(X + 10 db*)	(X + 10 db*)	(X + 10 db*)
Weights	100%	125%	80%	125%	125%
Dimensions:					
Floor space	100%	100%	100%	120%	120%
Height...................	100%	100%	90%	110%	110%
Normally available for application:					
Indoor or outdoor	Outdoor	All	Indoor only	All†	All†
Submersible	Submersible				
Fire and explosion resistant .	No	Yes	Yes	Yes (plus)	Yes (plus)
Maintenance required:					
Liquid..................	Normal	Infrequent	None	None	None
Internal cleaning.........	None	None	Frequent	None	None
External cleaning and painting expense	Normal	Normal	Subnormal	Minimum	Minimum
Special precautions before energizing either initially or after shutdown..........	None	None	Yes	None	None

*Not yet covered by industry standards.
†Applicable for all types of installation assuming no exposure to lightning or assuming adequate protection against impulse voltages can be provided.
SOURCE: Beeman—*Industrial Power Systems Handbook*, McGraw-Hill.

4. Determine the transformer electrical characteristics

Table 11 lists the electrical ratings of typical modern transformers. This table shows that a 300-kva rating is a standard one. Hence, such a unit is readily available from manufacturers.

5. Determine the transformer losses

Table 12 shows typical losses for three-phase distribution transformers with ratings to 400 kva. Thus, at full load on a transformer with a 0.80 power factor, the voltage drop through the transformer is 2.83 percent. The copper loss is 0.92 percent and the core loss is 0.80 percent. Note that Table 12 also shows the impedance for the transformer.

6. Determine the transformer and sound level

Table 13 shows that the average factory has a sound level of 70 to 75 db. Table 14 shows that a 300-kva oil-immersed transformer has a sound level of 55 db. Since the

TABLE 11   Electrical Ratings of Transformers for Power and Distribution Service*

Kva rating		Voltage ratings	
Single-phase	Three-phase	Primary	Secondary
1.5	5.0	120	120
2.5	7.5	240	240
3.0	10	480	480
5.0	15	600	600
7.5	25	2,400	2,400
10	50	2,500	4,160
15	75	4,160	6,900
25	100	4,330	11,500
37.5	150	4,800	13,800
50	200	6,900	23,000
75	300	11,500	34,500
100	450	13,800	46,000
150	600	25,000	69,000
200	750	34,500	
250	1,000	46,000	
300	1,500	69,000	
500	2,000	92,000	
1,000	2,500	115,000	
1,250	3,000	138,000	
1,500	5,000	161,000	

*Partial listing of commercially available units

TABLE 12   Electrical Characteristics of Single-phase and Three-Phase 60-Cycle Distribution Transformers

		Single-phase transformers; voltage rating—2,400 to 120/240 v				
Size, single-phase, kva	Impedance, Percent $Z/\phi°$	Percent voltage drop through transformer with full-load current			Percent Cu loss	Percent core loss
		97 percent pf	80 percent pf	50 percent pf		
3	2.7$/32.90°$	2.56	2.7	2.40	2.27	0.93
5	2.7$/38.40°$	2.46	2.7	2.51	2.12	0.72
7½	2.7$/43.5°$	2.35	2.66	2.58	1.96	0.64
10	2.7$/45.9°$	2.29	2.66	2.62	1.88	0.57
15	2.8$/51.6°$	2.21	2.71	2.76	1.74	0.51
25	2.8$/56°$	2.08	2.64	2.79	1.56	0.46
37½	2.9$/61.8°$	1.95	2.62	2.9	1.37	0.394
50	2.9$/64.7°$	1.84	2.56	2.89	1.24	0.372
75	3.5$/69.4°$	1.99	2.94	3.45	1.24	0.373
100	3.5$/69.9°$	1.96	2.93	3.45	1.20	0.370

TABLE 12   Electrical Characteristics of Single-phase and Three-Phase 60-Cycle Distribution Transformers (Continued)

Size, three-phase, kva	Impedance, percent $Z/\phi°$	Percent voltage drop through transformer with full-load current			Percent Cu loss	Percent core loss
		97 percent pf	80 percent pf	50 percent pf		
5	4.2/41.3°	3.73	4.19	4.00	3.12	1.28
10	3.8/46.80°	3.20	3.74	3.70	2.55	0.88
15	3.75/53.1°	2.91	3.60	3.72	2.2	0.74
25	3.9/57.2°	2.85	3.66	3.89	2.05	0.63
37½	4.02/60°	2.79	3.70	4.02	1.93	0.58
50	4.03/61.8°	2.72	3.66	4.02	1.82	0.51
75	3.54/63.1°	2.32	3.17	3.53	1.60	0.53
100	3.7/65°	2.31	3.26	3.68	1.56	0.57
200	3.92/71.4°	2.11	3.23	3.84	1.25	0.47
300	3.61/75.25°	1.74	2.83	3.46	0.92	0.80
400	3.77/71.9°	1.97	3.09	3.70	1.17	0.75

Three-phase transformers; voltage rating—2,400/4,160 to 240/480 v

TABLE 13   Average Sound Levels (db) for Various Occupancies

Occupancy	Decibel Range
Apartments and hotels	35–45
Average factory	70–75
Classrooms and lecture rooms	35–40
Hospitals, auditoriums, and churches	35–40
Private offices and conference rooms	40–45
Offices:	
Small	53
Medium (3 to 10 desks)	58
Large	64
Factory	61
Stores:	
Average	45–55
Large (5 or more clerks)	61
Residence:	
Without radio	53
With radio, conversation	60
Radio, recording, and television	25–30
Theaters and music rooms	30–35
Street:	
Average	80

NOTE: Manufacturers now sound-rate dry-type transformers to meet or exceed NEMA audible sound level standards. Select a transformer with a decibel rating lower than the ambient sound level of the area in which it is to be installed.

transformer sound level is less than the ambient sound level in the average factory, the transformer chosen is a suitable unit. As a general guide, choose a transformer having a decibel rating *lower* than the ambient sound level of the area in which the transformer will be installed.

TABLE 14   NEMA Audible Sound Levels

For dry-type general-purpose speciality transformers 600 v or less, single or three-phase

Transformer rating, kva	Average sound level, db
0–9	40
10–50	45
51–150	50
151–300	55
301–500	60

Oil-immersed and dry-type self-cooled transformers 15,000 v insulation class and below

Kva	Oil immersed, db	Dry type, db	
		Ventilated	Sealed
0–300	55	58	57
301–500	56	60	59
501–700	57	62	61
701–1,000	58	64	63
1,001–1,500	60	65	64
1,501–2,000	61	66	65
2,001–3,000	63	68	66

Related Calculations: Use this general procedure to choose transformers for industrial, commercial, and residential use. As a general guide, the following definitions are useful. *Power transformers* are used in generating plants to step up voltage and in substations to step down voltage. Power transformers are usually rated at 500 kva and larger.

Distribution transformers step voltage down to 600 or 480 v for industrial use, or 240 and 120 v for residential or commercial use. *Instrument transformers,* classed *potential* or *current,* serve low-voltage meters and relays. *Speciality transformers* include units used to change the voltage for specific applications such as signs, arc lamps, bells, etc.

SYSTEM POWER-FACTOR ANALYSIS

A system has three types of loads — lighting, induction motors, and synchronous motors. Determine the system power factor for the following conditions: lighting load = 400 kva at unity (1.0) power factor; induction-motor load = 700 kva at 0.85 power factor, lagging; synchronous motor load = 300 kva at 0.75 power factor, leading. What is the system power factor?

Calculation Procedure:

1. Determine the lighting-load kvar

The 400-kva lighting load has a unity (1.0) power factor. With a unity power factor, the kva load = the kw load. Also, since there is no reactive current when the power factor is unity, the lighting-load kvar = 0.

2. *Compute the induction-motor kvar*

Find the induction-motor kw load from: $kw = kva \ (pf)$, where $kva =$ induction-motor kva; $pf =$ induction-motor power factor. Or, $kw = 700(0.85) = 595$ kw.

Compute the induction-motor kvar from $kvar = (kva^2 - kw^2)^{0.5}$, or $(700^2 - 595^2)^{0.5} = 370$ kvar.

3. *Compute the synchronous-motor kvar*

As in Step 2, the synchronous-motor $kw = kva \ (pf) = 300(0.75) = 225$ kw. Then, $kvar = (kva^2 - kw^2)^{0.5}$, as in Step 2. Or $kvar = (300^2 - 225^2)^{0.5} = 198.5$ kvar.

4. *Compute the system kw and kvar*

The kw load is the sum of the individual kw loads, or lighting $kw +$ motor $kw = 400 + 595 + 225 = 1,220$ kw. The kvar of the system is found in the same way, except that the lighting-load kvar is zero and the leading (synchronous-motor) kvar offsets the lagging (induction motor) kvar, or $370.0 = 198.5 = 171.5$ kvar.

The reason for taking the difference between the induction- and synchronous-motor kvar is because a synchronous motor can supply kvar to the system.

5. *Compute the system kva and pf*

Use the relation $kva = (kw^2 + kvar^2)^{0.5} = [(1,220)^2 + (171.5)^2]^{0.5} = 1,333$ kva. Compute the system power factor from $pf =$ system $kw/$system $kva = 1,220/1,333 = 0.915$ lagging. The power factor is termed *lagging* because the lagging or induction-motor kvar exceeds the leading or synchronous-motor kvar. Either capacitors, synchronous motors, or synchronous generators could be used to improve the power factor of this system.

Related Calculations: Use this general method to analyze the power factor of any power system – industrial, commercial, or residential.

POWER-FACTOR DETERMINATION AND IMPROVEMENT

Determine the power factor in a 440-v three-phase power system when the load draws 135 amp, 85 kw. If it is desired to improve the power factor of this circuit to 88 percent with capacitors, what capacitor rating is required?

Calculation Procedure:

1. *Compute the circuit kva*

For a three-phase ac circuit, use the relation $kva = \sqrt{3} \ va/1,000$, where kva = kilo-volt-amperes of the circuit; $v =$ circuit voltage; $a =$ circuit amperage. Or, $kva = \sqrt{3} \times (440)(135)/1,000 = 103$ kva.

2. *Compute the circuit power factor*

Use the relation power factor $= pf = kw/kva$, where $kw =$ circuit kilowatt load. Or, $pf = 85/103 = 0.825$, or 82.5 percent.

3. *Compute the circuit kvar*

The total current in an ac circuit is usually made up of two components – (a) power-producing current, and (b) magnetizing current. Other terms used for these currents are *working current* and *reactive current*, respectively. From a power standpoint, the terms used are *true power*, watts or kw, and reactive power, var or kilovars (kvar). The abbreviation var stands for volt-ampere reactive.

In an ac circuit, $kvar = (kva^2 - kw^2)^{0.5}$. Thus, for this circuit, $kvar = (103^2 - 85^2)^{0.5} = 58.5$ kvar.

4. *Compute the kvar at the new power factor*

At the new power factor of 88 percent the circuit kw is the same, or 85. However, the circuit kva will be $kw/$new pf, or $85/0.88 = 96.5$ kva.

The new circuit $kvar = (kva^2 - kw^2)^{0.5} = (96.5^2 - 85^2)^{0.5} = 45.6$ kvar. Thus, the circuit provides 45.6 of the 58.5 kvar required by the load.

5. Compute the required capacitor kvar

The capacitor must provide the difference between the load and circuit kvar, or $58.5 - 45.6 = 12.9$ kvar; say 15 kvar. Thus to improve the power factor of this circuit from 82.5 to 88 percent, a 15 kvar capacitor is required.

Related Calculations: The method given above is useful for determining the capacitor, synchronous-motor, or synchronous-condenser rating required to produce a given power-factor increase. As a general rule, the synchronous condenser is usually too costly a device for power-factor improvement service in industrial plants. Hence, it is seldom used for this purpose in industrial plants.

WIRING-SIZE CHOICE FOR PRIMARY DISTRIBUTION

Determine the proper size of underground cable to use to supply a 1,000-kw load at 80 percent power factor at a distance of 0.4 mile with an allowable voltage drop of 5.0 percent. The receiving-end voltage is 2,200 v, three-phase.

Calculation Procedure:

1. Compute the system kva-miles

Use the relation, kva-miles = (load, kw/pf) (line length, miles), where pf = system power factor. For this system, kva-miles = $(1,000/0.80)(0.4) = 500$ kva-miles.

2. Choose the wire size to use

Table 15 shows that with an 80 percent power factor, 2,400 v, three-phase line, a 4/0 cable will have a 1 percent voltage drop per 174.0 kva-miles. However, the receiving voltage of this system is 2,200 v. To correct for the difference between the actual and tabulated receiving voltage, multiply the tabulated kva-miles by the ratio (actual receiving voltage/tabulated receiving voltage)^2, or $(2,200/2,400)^2(174.0) = 146.2$ kva-miles per 1.0 percent voltage drop.

3. Compute the actual voltage drop

Use the relation actual voltage drop, percent = system kva-miles/kva-miles per 1 percent voltage drop, or $500/146.2 = 3.41$ percent actual voltage drop. Since the allowable voltage drop is 5 percent, this is within the desired range.

4. Check the next smaller size cable

Since there is a difference of $5.00 - 3.41 = 1.59$ percent between the actual and the allowable voltage drop, the next smaller cable might be satisfactory. A smaller cable will cost less and is therefore worth checking.

Using the same procedure as in Step 2, except that a 2/0 10-kv cable is used, kva-mile per 1 percent voltage drop = $(2,200/2,400)^2(118.0) = 99.0$. Then, as in Step 3, actual voltage drop = $500/99.0 = 5.06$ percent. Hence, the smaller cable is unsuitable because it produces a higher voltage drop than allowed.

Related Calculations: Table 15 can be used to determine (1) the voltage drop in an existing circuit when the load is known, (2) the proper size of a conductor to use to limit the voltage drop to a predetermined value, or (3) the proper voltage rating of a new line. The 50 percent power-factor column in Table 15 can be used to calculate the instantaneous drop in voltage caused by the starting of a large motor on the line. The average power factor of a motor is 50 percent when the motor is being started. Note that data are listed in Table 15 for both underground and overhead distribution systems.

WIRE- AND CABLE-SIZE DETERMINATION FOR A KNOWN LOAD

An electrical installation consists of 1 motor and 48 lights. What size wire and fuses are needed for the 50-hp squirrel-cage induction motor that is started at its rated three-phase 440-v 60-Hz current if the motor is 280 ft from the power panel and the voltage drop must not exceed 1 percent of the supply voltage? The lighting load totals 10,000 watts. This load is supplied by two 110-v circuits. The voltage drop in the lighting

TABLE 15 Voltage Regulation of 5- and 10-kv Cables and Overhead-Line-Wire Primary Distribution Voltages, Kva-Miles per 1% Voltage Drop, Balanced Load*

Underground distribution, 5, and 10-kv single-conductor cables, annealed copper		Power factor								
		4,160 v, three-phase			2,400 v, three-phase			2,400 v, single-phase		
Cable size	Voltage rating, kv	97%	80%	50%	97%	80%	50%	97%	80%	50%
		Kva-mile per 1% drop			Kva-mile per 1% drop			Kva-mile per 1% drop		
6	10	79.2	79.2	127.9	26.4	29.9	42.5	13.2	15.0	21.3
6	5	79.3	90.2	128.7	26.4	30.0	42.9	13.2	15.0	21.5
4	10	124.0	135.5	186.0	41.2	45.1	62.0	20.6	22.6	31.0
4	5	124.0	137.0	187.6	41.2	45.6	62.3	20.6	23.0	31.2
2	10	192.0	203.0	260.0	64.0	67.7	86.3	32.0	34.0	43.2
1/0	10	295.0	297.0	355.0	98.2	98.7	118.0	49.0	49.7	59.0
1/0	5	300.0	306.5	377.0	100.0	102.2	125.7	50.0	51.1	62.8
2/0	10	364.0	354.0	407.0	120.7	118.0	135.5	60.6	59.0	68.0
4/0	10	545.0	492.0	525.0	181.7	164.0	175.0	91.0	82.0	87.5
4/0	5	560.0	523.0	572.0	186.0	174.0	191.0	93.2	87.0	95.3
350 Mem	5	852.0	723.0	728.0	282.0	240.0	242.0	142.0	120.5	121.0

Overhead distribution copper-line wire		4,160 v, three-phase			2,400 v, three-phase			2,400 v, single-phase		
Wire size	Equivalent spacing, in.	Kva-mile per 1% drop			Kva-mile per 1% drop			Kva-miles per 1% drop		
6	28	74.0	78.0	99.2	24.7	26.0	33.0	12.3	13.0	16.5
4	28	113.5	112.0	132.2	37.8	37.3	44.0	18.9	18.7	22.1
2	28	169.0	156.0	168.5	56.2	52.0	56.2	28.2	26.0	28.1
1/0	28	248.0	207.0	206.0	82.6	69.0	68.7	41.3	34.5	34.4
2/0	28	298.0	236.0	225.0	99.3	78.7	75.0	49.6	39.4	37.5
4/0	28	415.0	295.0	260.0	138.3	98.3	86.6	69.2	49.2	43.3
350 Mem	28	574.0	363.0	297.0	191.2	121.0	99.0	95.5	60.5	59.4
500 Mem	28	700.0	405.0	318.5	233.3	135.0	106.2	116.6	67.4	53.1

		6,900 v, three-phase			11,950 v, three-phase		
		Kva-miles per 1% drop			Kva-miles per 1% drop		
6	60	205	213	260	615	638	777
4	60	306	302	345	917	905	1032
2	60	447	412	432	1339	1235	1294
1/0	60	627	537	519	1879	1610	1555
2/0	60	739	601	560	2217	1803	1680
3/0	60	859	673	601	2576	2019	1803
4/0	60	988	739	640	2963	2217	1919
250 Mem	60	1097	805	677	3291	2415	2032

*For receiving voltages slightly different from the given values, multiply kva-miles in the tables by the square of the ratio of the new voltage to the voltage in the table. For example, for 4,000 v multiply by $(4,000/4,160)^2 = 0.924$.

circuits must not exceed 1 percent. What size wire is needed if each branch circuit is 170 ft long?

Calculation Procedure:

1. Sketch the circuit layout

Figure 9 shows a single-line diagram of a typical light and power branch-circuit layout. Figure 10 shows the typical equipment used in such a circuit.

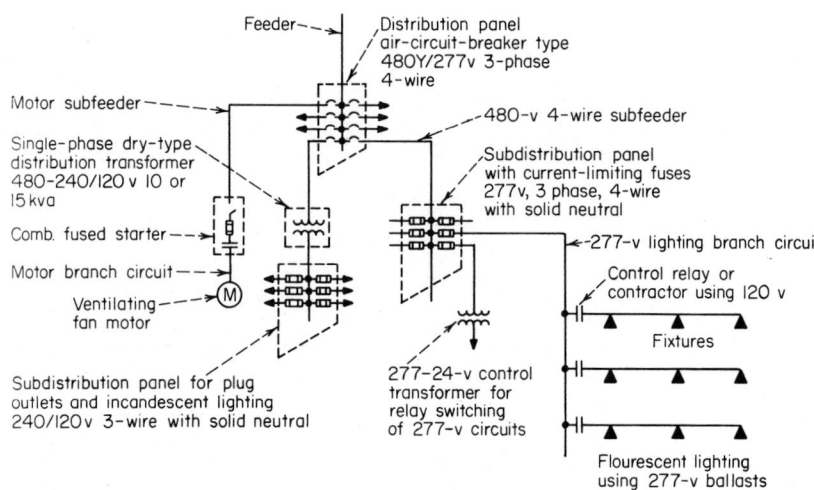

Fig. 9   Typical light and power branch-circuit layout served from a 480 Y/277-v system.

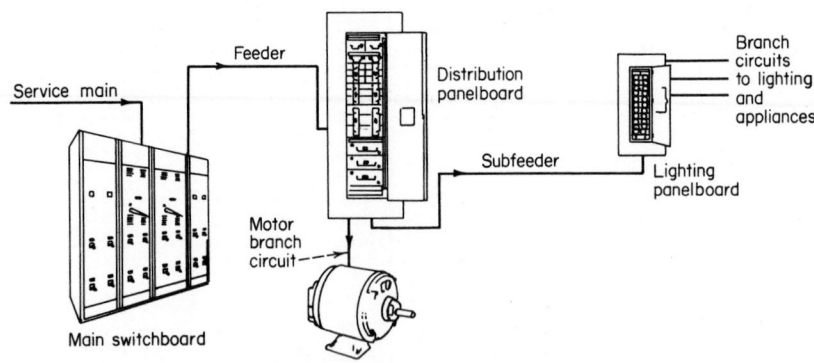

Fig. 10   Typical equipment used in motor and lighting circuits.

2. Determine the motor full-load current

Use the *National Electrical Code* (*NEC*) table of motor full-load current or Table 16 to determine the full-load current of the motor. Thus, Table 16 shows that a 440-v 50-hp induction motor requires 63 amp at full load.

TABLE 16 Electric-Motor Full-Load Current

Full-Load current* direct-current motors

Hp	115 v	230 v	550 v
$\frac{1}{4}$	3	1.5	
$\frac{1}{3}$	3.8	1.9	
$\frac{1}{2}$	5.4	2.7	
$\frac{3}{4}$	7.4	3.7	1.6
1	9.6	4.8	2.0
$1\frac{1}{2}$	13.2	6.6	2.7
2	17	8.5	3.6
3	25	12.5	5.2
5	40	20	8.3
$7\frac{1}{2}$	58	29	12
10	76	38	16
15	112	56	23
20	148	74	31
25	184	92	38
30	220	110	46
40	292	146	61
50	360	180	75
60	430	215	90
75	536	268	111
100	...	355	148
125	...	443	184
150	...	534	220
200	...	712	295

Full-load current† single phase ac motors

Hp	115 v	230 v	440 v
$\frac{1}{6}$	4.4	2.2	
$\frac{1}{4}$	5.8	2.9	
$\frac{1}{3}$	7.2	3.6	
$\frac{1}{2}$	9.8	4.9	
$\frac{3}{4}$	13.8	6.9	
1	16	8	
$1\frac{1}{2}$	20	10	
2	24	12	
3	34	17	
5	56	28	
$7\frac{1}{2}$	80	40	21
10	100	50	26

*These values of full-load current are for motors running at speeds usual for belted motors and motors with normal torque characteristics. Motors built for especially low speeds or high torques may require more running current, in which case the nameplate current rating should be used.

For full-load currents of 208- and 200-volt motors, increase the corresponding 220-volt motor full-load current by 6 and 10 percent, respectively,

†These values are for motors with usual speeds and torque characteristics. Nameplate current values should be used for motors with low speeds or high torques.

To obtain full-load currents of 208- and 200-volt motors, increase 230-volt current values by 10 and 15 percent, respectively.

TABLE 16   Electric-Motor Full-Load Current (Continued)

Three-phase ac motors

Hp	Induction-type squirrel-cage and wound rotor, amp					Synchronous-type unity power factor, amp			
	110 v	220 v	440 v	550 v	2,300 v	220 v	440 v	550 v	2,300 v
½	4	2	1	.8					
¾	5.6	2.8	1.4	1.1					
1	7	3.5	1.8	1.4					
1½	10	5	2.5	2.0					
2	13	6.5	3.3	2.6					
3	...	9	4.5	4					
5	...	15	7.5	6					
7½	...	22	11	9					
10	...	27	14	11					
15	...	40	20	16					
20	...	52	26	21					
25	...	64	32	26	7	54	27	22	5.4
30	...	78	39	31	8.5	65	33	26	6.5
40	...	104	52	41	10.5	86	43	35	8
50	...	125	63	50	13	108	54	44	10
60	...	150	75	60	16	128	64	51	12
75	...	185	93	74	19	161	81	65	15
100	...	246	123	98	25	211	106	85	20
125	...	310	155	124	31	264	132	106	25
150	...	360	180	144	37	...	158	127	30
200	...	480	240	192	49	...	210	168	40

3. Determine the required carrying capacity of the motor circuit

The *NEC* recommends that individual branch circuits to motors have a current-carrying capacity at least 125 percent of the motor full-load running current. Thus, for this 50-hp motor the current-carrying capacity of the branch circuit supplying the motor must be at least 1.25 (63 amp) = 78.7 amp; say 80 amp.

4. Determine the required wire size

Table 17 lists the allowable wire lengths for a 1 percent, or less, voltage drop in 220- to 230-v and 440- to 460-v systems. To use Table 17 for 440- to 460-v three- or four-wire feeders, multiply the tabulated lengths by 2.

Thus, with an 80-amp flow, a 2/0 wire will have a 1 percent voltage drop in a length of 2(151) = 302 ft, using the data from Table 17. Since the motor is 280 ft from the power panel, the voltage drop will be less 1 percent if a 2/0 wire is used.

5. Select the fuse capacity for the motor circuit

For motors of more than 1 hp the *NEC* allows an external overcurrent device actuated by the motor running current and set to open at not more than 125 percent of the motor full-load current for motors with a temperature rise not over 40 C, or 115 percent of the motor full-load current for all other motors.

Where fuses are used to protect a circuit, the fuses should have a rating of at least 300 percent the motor full-load current. Or for this motor 3(63 amp) = 189 amp. A 200-amp fuse would be suitable.

Note that the fuse capacity cannot be arbitrarily increased without a study of the circuit and the devices in it. Thus, a 300-amp fuse could not be used for this circuit because the excessive current might damage the motor before the fuse blew.

Were a circuit breaker used instead of fuses, the breaker would be rated at 1.25(63 amp) = 78.7; say 75 amp. The circuit-breaker rating is reduced to the next lowest standard rating to provide greater protection for the motor. For motors over 1,500 hp, an embedded temperature detector may be used to cause opening of the current supply when the motor temperature becomes excessive.

6. Compute the current flow in the lighting circuits

With two lighting circuits, the load in each circuit = total load, watts/number of circuits, or 11,000/2 = 5,500 watts per circuit.

Use the relation $I = W/v$ to determine the current flow in each lighting circuit. In this relation, I = current flow, amp; w = lighting load of the circuit, watts; v = circuit voltage. Thus, for each of these circuits, $I = 5,500/110 = 50$ amp.

7. Select the wire size for the lighting circuits

Table 17 shows the wire size for 115/230-v single-phase circuits having a 1 percent voltage drop for various lengths. This tabulation is also applicable to 110-v ac circuits.

Inspection of the tabulated values shows that a 1/0 wire is required for a 1 percent voltage drop in a 170-ft-long circuit. Hence, this size wire will be used for the lighting circuit.

Related Calculations: In planning electric circuits, the loading and lengths of feeders and runs between outlets must be related to the voltage drop and the need for spare capacity in the circuit for possible future increases in load. Each lamp, appliance, or other utilization device in the circuit is designed for best performance at a particular operating voltage. Although such devices will operate at voltages on either side of the design value, there generally will be adverse effects when operating at voltages lower than the specified value.

A 1 percent drop in voltage at an incandescent lamp produces about a 3 percent decrease in light output; a 10 percent voltage drop will decrease the output about 30 percent. In resistance-type heating devices, a voltage drop has a similar effect on the heat output. In motor-operated appliances, low voltage to the motor affects the starting and pull-out torque. Also, the current drawn from the line increases with the drop in voltage. Overheating of the windings may result.

The method presented here is valid for lighting, power, heating, and similar circuits. For comprehensive tabulations of wire sizes and allowable loads, consult the latest edition of the *National Electrical Code*. Where a local electrical code governs wiring and system design, be sure to consult it when selecting wire sizes.

PRELIMINARY ELECTRICAL LOAD ESTIMATING

Estimate the lighting and power loads for an electronics factory that is 200 ft long and 100 ft wide. The fluorescent lights in this one-story plant will be located 12 ft above the floor and will provide an illumination level of 50 ft-c. What is the demand amperage if the power is supplied at 13,800 v?

Calculation Procedure:

1. Compute the lighting power requirements

For preliminary electrical load estimates, tabulated average power demands are often sufficient. Thus, Table 18 lists the typical power requirements in terms of the lighting-demand factor for lights mounted at various heights above the floor. To use the values from this table, solve the relation va/sq ft = (lighting-demand factor)(illumination level, ft-c). Thus, for this electronics factory using fluorescent lights mounted 12 ft above the floor, va/sq ft = (0.060)(50) = 3.0. With a plant area of 200 × 100 = 20,000 sq ft, the total *estimated* power requirement for lighting is 3.0(20,000) = 60,000 va.

TABLE 17 Average Circuit Lengths (Feet) for 1 Percent Voltage Drop

Table A. Single-Phase AC Loads
115/230 v, 60 Hz, 100% pf

Amp load	Wire size – Circular mils					Wire size – B & S or A.W.G.									
	500	400	350	300	250	4/0	3/0	2/0	1/0	1	2	3	4	6	8
40	1106	898	788	669	558	475	378	299	239	188	150	119	94	59	38
50	885	719	630	535	447	380	303	240	191	150	120	91	75	47	30
60	737	599	525	446	372	317	252	200	159	125	100	79	62	39	
70	632	513	450	382	319	271	216	171	136	107	86	68	53	34	
80	553	449	394	334	279	238	189	150	119	94	75	59	47		
90	491	399	350	297	248	211	168	133	106	83	67	53	42		
100	442	359	315	267	223	190	151	120	95	75	60	47			
110	402	327	286	243	203	173	138	109	87	68	55				
120	369	299	263	223	186	158	126	100	79	63					
130	340	276	242	206	172	146	116	92	73	58					
140	316	257	225	191	159	136	108	86	68						
150	295	240	210	178	149	127	101	80	64						
160	276	225	197	167	140	119	95	75	60						
170	260	211	185	157	131	112	89	70							
180	246	200	175	148	124	106	84	66							
190	233	189	166	140	117	100	80								
200	221	180	157	134	112	95	76								
210	211	171	150	127	106	90									
220	201	163	143	122	101	86									
230	192	156	137	116	97	83									

240	184	150	131	111	93
250	177	144	126	107	89
260	170	138	121	103	80
270	164	133	117	99	
280	158	128	112	96	
290	152	124	109	92	
300	147	120	105		
310	143	116	102		
320	138	112			
330	134	109			
340	130	106			

Calculations based on copper resistance of 12.5 ohms per CM-ft at 50 C (122 F).
Reactance and impedance losses calculated for each wire.
Conductors closely grouped in metallic conduit.
Balanced three-wire loads: drop is 1.15 v for given length.
Two-wire, 230-v loads: drop is 2.3 v for given length.

TABLE 17 Average Circuit Lengths (Feet) for 1 Percent Voltage Drop (Continued)

Table B. Three-Phase Delta AC Loads
230 v, 60 Hz, 85% pf

Amp load	Wire size — Circular mils					Wire size — B & S or A.W.G.									
	500	400	350	300	250	4/0	3/0	2/0	1/0	1	2	3	4	6	8
40	710	625	584	530	475	429	364	303	253	208	173	139	113	75	49
50	568	500	467	424	380	343	291	242	203	167	139	111	90	60	39
60	473	417	389	353	317	286	243	202	169	139	115	93	75	50	
70	406	357	333	303	271	245	208	173	145	119	99	79	64	43	
80	355	312	292	265	238	214	182	151	127	104	87	69	56		
90	316	278	259	235	211	191	162	134	113	93	77	62	45		
100	284	250	233	212	190	172	146	121	101	83	69	55			
110	258	227	212	193	173	156	132	110	92	76	63				
120	237	208	195	177	158	143	121	101	84	69	58				
130	218	192	180	163	146	132	112	93	78	64					
140	203	179	167	151	136	123	104	86	72						
150	189	168	156	141	127	114	97	81	67						
160	177	156	146	132	119	107	91	76							
170	167	147	137	125	112	101	86	71							
180	158	139	130	118	106	95	81	67							
190	149	132	123	112	100	90	77								
200	142	125	117	106	95	86	73								
210	135	119	111	101	90	82									
220	129	114	106	96	86	78									
230	123	109	101	92	83	75									

240	118	104	97	88	79
250	114	100	93	85	76
260	109	96	90	81	73
270	105	93	86	78	
280	101	89	83	76	
290	98	86	80	73	
300	95	83	78		
310	92	81	75		
320	89	78			
330	86	76			
340	83	73			
350	81				
360	79				
370	77				
380	75				

Calculations based on copper resistance of 12.5 ohms per CM-ft at 50 C (122 F).
Reactance and impedance losses calculated for each wire.
Conductors closely grouped in metallic conduit.
For 208-v, four-wire Y feeders, multiply given length by 0.9.
For 230-v, single-phase feeders, multiply given length by 0.85.
For 460-v, three- or four-wire feeders, multiply given lengths by 2.
For aluminum wire, multiply given lengths by 0.7 or use length of copper wire which is two sizes smaller than the aluminum size under consideration.

TABLE 17 Average Circuit Lengths (Feet) for 1 Percent Voltage Drop (Continued)

Table C. Balanced Lighting Loads
Three- and four-wire, 115 v 1 percent drop from supply cabinet to first
outlet supplying permanently connected appliance or fixture*

Maximum overcurrent circuit protection†	Intermittent loads			Continuous loads		
	100% F 2–3 C	80% F 4–6 C	70% F 7–9 C	100% F 2–3 C	80% F 4–6 C	70% F 7–9 C
15 A	15 A / 1725 W	12 A / 1380 W	10.5 A / 1207 W	12 A / 1380 W	9.6 A / 1104 W	8.4 amp / 966 watts
20 A	20 A / 2300 W	16 A / 1840 W	14 A / 1610 W	16 A / 1840 W	12.8 A / 1472 W	11.2 A / 1288 W

Loads and lengths in ft for 1% drop on three- and four-wire 115-v circuits

Amp load	No. 10 wire	No. 12 wire	No. 14 wire
1	946	596	374
2	474	298	188
3	316	198	124
4	236	148	94
5	190	120	76
6	158	100	62
7	136	86	54
8	118	74	46
9	106	66	42
10	94	60	38

11	86	54	34
12	78	50	32
13	72	46	28
14	68	42	26
15	64	40	24
16	60	38	
17	56	36	
18	52	34	
19	50	32	
20	48	30	
21	46		
22	44		
23	42		
24	40		
25	38		
26	36		
27	36		
28	34		
29	32		
30	32		

*Calculations based on copper resistance of 13 ohms per CM-ft at 60 C (140 F).
For two-phase, three-wire circuits tapped off a three-phase, four-wire Y service, multiply given lengths by 0.67.
†A – amp; W – watts; C – conduit conductor; F – fills.

2. Compute the plant power demand

Table 19 lists the combined power and lighting load densities for various industries. This table shows that the load density for electronic equipment manufacture is 10 va/sq ft. Since the lighting demand was computed in Step 1 as 3.0 va/sq ft, the power demand will be the difference between 3.0 va and the tabulated total of 10.0 va/sq ft for power and light, or $10.0 - 3.0 = 7.0$ va/sq ft. With a total area of 20,000 sq ft, the power demand is $7.0(20,000) = 140,000$ va.

TABLE 18   Power Requirements for Lighting*

Fixture height, ft	Lighting-demand factor†	
	Incandescent	Fluorescent
Less than 14...........	0.12	0.060
14–35.................	0.13	0.065
35–50.................	0.15	0.070

*Based on median-characteristic fixtures and a 75-ft-wide room, 200 or more ft long.

†Va = (lighting-demand factor)(intensity of illumination, ft = c).

SOURCE: Beeman—*Industrial Power Systems Handbook*, McGraw-Hill.

3. Compute the total load demand

The total load demand is the sum of the loads computed in Steps 1 and 2, or the product of the floor area of the plant and the appropriate value from Table 19. Or $60,000 + 140,000 = 10(20,000) = 200,000$ va.

TABLE 19   Typical Power Load Densities

Factory Type	Va Demand for Power and Light, va/sq ft
Beet-sugar factory and refinery	19
Paper mills..................................	14
Textile mills, engine builders.................	12
Cigarette manufacturing	11
General manufacturing, chemicals electronic equipment	10
Small appliance manufacturing, machine-repair shops......................	7.5
Lamp manufacturing.........................	5
Small-device fea manufacturing..............	3.5

SOURCE: Beeman—*Industrial Power Systems Handbook*, McGraw-Hill.

Table 20 shows the typical va demands for a variety of plants. Based on actual plants, this table shows both average values and the ranges encountered in the plants surveyed.

4. Compute the demand amperage

The product of volts and amperes, va, is termed *apparent power* in an alternating-current circuit. In a single-phase unity-power-factor circuit, $va = EI$. In a balanced three-wire three-phase delta or wye circuit, $va = 1.732EI$. Thus, with a power supply at 13,800 v, the demand amperage is $200,000 = 13,800I$; $I = 14.5$ amp.

Related Calculations: Use this general method for preliminary estimates of the electrical load of any type of industrial plant. For exact determination of the electrical demand, use the methods given in later Calculation Procedures in this section.

TABLE 20   Recorded-Load Data for a Group of Plants (Summary)

Type of plant	Demand, va/sq ft		Annual kwhr per sq ft		Annual kwhr per va demand	
	Average	Range	Average	Range	Average	Range
Chemical	10.0	6–13	33.7	14–54	3.3	2–4
Electronics.................	10.3	3–20	25.8	11–67	2.4	1–4
Foundry	9.9	...	32.4	...	3.3	
Lamps:						
General..................	4.8	2–12	19.5	5–53	4.2	2–6
Wire works..............	8.7	6–13	30.0	13–64	2.8	2–5
Base works..............	4.0	...	25.0	...	6.0	
Glass works	4.5	2–7	24.2	13–37	5.4	4–6
Porcelain	2.0	...	10.9	...	5.4	
Printing	3.0	...	8.0	...	3.0	
Small appliance	7.4	2–13	20.5	4–46	2.9	1–7
Small device	3.6	2–7	9.5	3–27	2.4	1–5
General:						
Large – over 5,000 kva	10.0	5–17	70.0	8–195	7.0	1–19
Small – under 5,000 kva	9.8	3–18	31.0	5–50	5.2	1–27

SOURCE: Beeman – *Industrial Power Systems Handbook*, McGraw-Hill.

PLANT POWER-DISTRIBUTION-SYSTEM PLANNING

Select the power-distribution system for an industrial plant having motors ranging from $\frac{1}{2}$ to 1,000 hp. The total demand of the plant is estimated to be 15,000 kva. Choose the type of distribution system for use within this plant.

Calculation Procedure:

1. Select the system voltage rating

Table 21 lists the typical voltage ratings (classes) for industrial power equipment. Popular main generation and distribution voltages are 4,160 and 13,800 v. Utilization voltage, i.e., the voltage at which motors and similar large equipment is operated, is usually 2,400 and/or 480 v. As a general guide, the higher the voltage in a given voltage class, the lower the overall system cost.

TABLE 21   Voltage Ratings for Power Equipment

Nominal system voltage class	Generator or transformer no-load rated voltage	Utilization equipment rated voltage
240	240	220
480*	480*	440*
600	600	550
2,400*	2,400*	2,300*
4,160*	4,160*	4,000*
6,900	6,900	5,600
13,800*	13,800*	13,200*

*The voltages marked by an asterisk are the preferred ratings for most applications because of the availability of equipment and overall sound-system engineering.

Figure 11 shows typical voltages used for various loads. Based on the installed cost of all the components comprising a distribution system, a 4,160-v system has a lower first cost than a 13,800-v system for plants with a demand up to about 10,000 kva. For plants

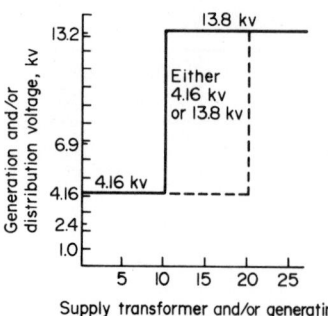

Fig. 11 Relationship of distribution voltage to load.

in the range of 10,000- to 20,000-kva capacity, costs of the two systems are comparable. Above 20,000-kva capacity, a 13,800-v system is usually preferred.

Since this plant has a 15,000-kva demand a 13,800-v distribution voltage will be chosen. This choice will permit easier future expansion without a higher initial investment. As a further verification of this voltage choice, Fig. 12 shows the approximate limitations on the amount of power that can be fed from a single source at different voltages. These curves indicate that a 13,800-v distribution voltage can adequately supply the demand. The costs shown are relative; do not use the absolute values shown. Transformer costs are not shown because the cost of a transformer is about the same for secondary voltages of 2,400, 4,160, and 13,800 v.

2. Choose the motor operating voltage

Figure 13 shows the recommended utilization voltage for motors rated at 10 to 10,000 hp. This chart shows that with a 13,800-v source a 2,300-v motor utilization voltage is recommended for motors rated 150 to 3,500 hp. Hence, this is an acceptable operating voltage for the larger motors in this plant. For motors less than 150 hp,

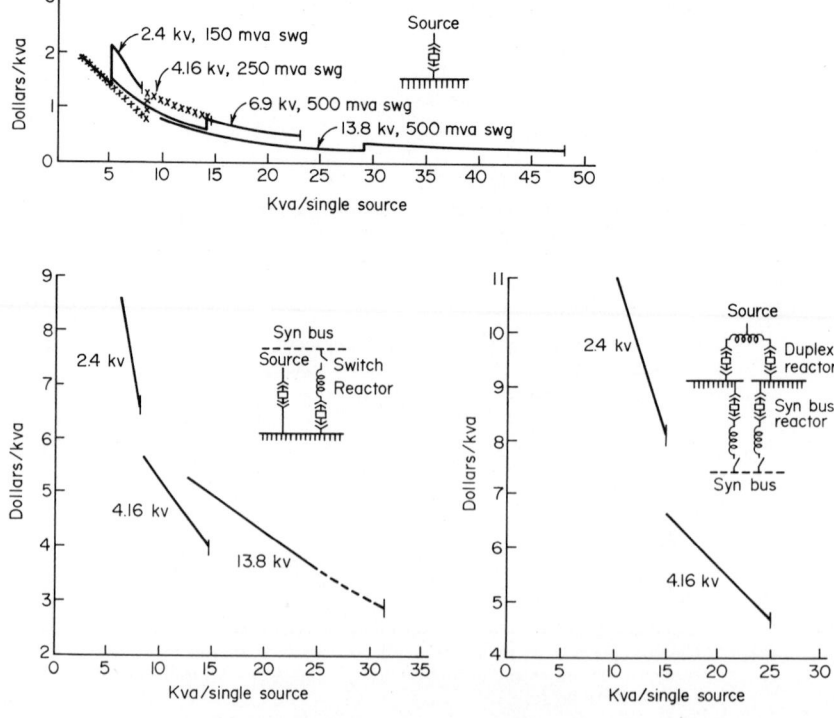

Fig. 12 Typical loads that can be fed from a single source at different voltages.

440 v would be used, as Fig. 13 indicates. This voltage could be obtained by using step-down transformers.

In the layout of a new plant it is usually best to consider first the selection of the system voltage from the standpoint of power distribution only. Too often there is a tendency to compromise between distribution voltage and motor voltage. The result is that the final voltage will be too low from the standpoint of good overall system design.

If a motor application occurs near the ends of the range for each voltage class in Fig. 13, review the motor application carefully. Other factors such as starting equipment,

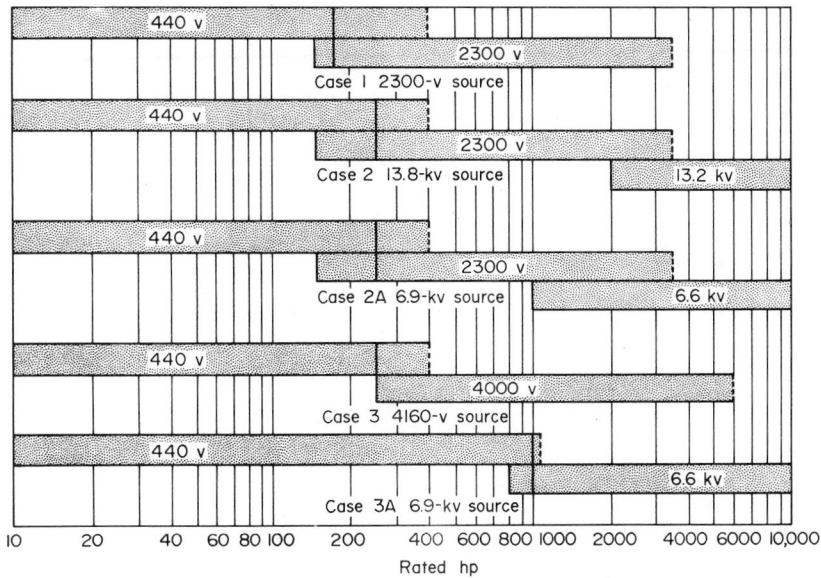

Fig. 13   Motor ratings recommended for various utilization voltages.

speed, type of motor, etc., may affect the voltage selection sufficiently to make a different voltage more desirable.

3. *Select the power-distribution system*

The *load-center* power-distribution system, Fig. 14, is probably the most economical system for this plant. For new or rearranged plants, the load-center system of power distribution is the most flexible way to supply existing loads or to meet changing load demands with the lowest investment. A unit substation can supply power to load centers at 13,800, 4,160, 2,400 v, or other voltages.

The fundamental approach is to use a relatively high voltage for transmitting power to the load centers where it is stepped down to the utilization voltage. The primary voltage, as shown in Step 1, varies as a function of the total plant load and the secondary voltage as a function of the motor size, Step 2.

Related Calculations: The general method given here is valid for all types of industrial plants—chemical, petroleum, food, textile, metalworking, etc. Note that the problem of selecting the proper voltage for a plant's main power-distribution system is similar whether power is purchased from a utility or generated within the plant. Power may be generated in industrial plants at 13,800 v or below, as desired. Power purchased from a utility may be supplied at 13,800 v, or at some higher or lower voltage, depending upon the utility system and the industrial load requirements. The recommendations given here were developed by W. B. Wilson, General Electric Company, and reported in *Chemical Engineering*.

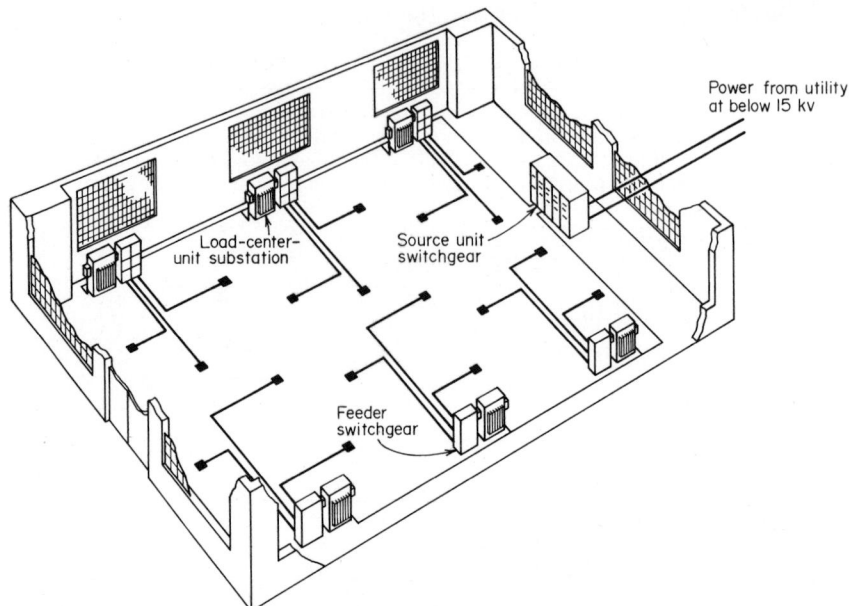

Fig. 14 Power-distribution layout with source unit, load-center substation, feeder switchgear, and cables.

INDUSTRIAL-BATTERY SELECTION AND SIZING

Choose the type of battery to use to start a diesel-engine standby generating unit for an industrial plant. The starting motor is rated at 10 hp at 24 v. It is desired to keep the battery investment as low as possible. How long could the chosen battery supply a 24-v emergency-lighting load of 4,800 watts?

Calculation Procedure:

1. Compute the starting-motor amperage

The starting motor is rated at 10-hp output, or (10 hp)(746 watts/hp) = 7,460 watts = 7.46 kw. To compute the amperage, use the dc relation $P = IE$, where P = power, watts; I = current flow, amp; E = rated circuit voltage. Solving, $I = P/E = 7,460/24 = 310.8$; say 311 amp.

2. Choose the type of battery to use

Table 22 lists the major characteristics, advantages, and disadvantages of recharge-able industrial batteries. Study of this tabulation shows that either a lead-acid or nickel-cadmium battery would be suitable for this application. Both types of batteries have a suitable voltage range.

3. Compute the number of cells required

When a battery is discharging, the voltage at its terminals decreases. To determine the number of cells required, divide the required voltage by the voltage during discharge.

Table 22 shows that the discharge voltage of a lead-acid battery is 2.1 to 1.46 v and that of a nickel-cadmium battery is 1.3 to 0.75 v. Using the average discharge voltage for each battery, the number of lead-acid cells required = 24 v/1.78 = 13.46; say 14 cells. The number of cells required for a nickel-cadmium battery = 24 v/1.03 = 23.3; say 24 cells.

4. Check the relative costs of the batteries

Table 22 shows that the lead-acid battery has the lowest first cost of all the batteries listed. A nickel-cadmium battery, according to Table 22, costs two to three times as much as a lead-acid battery of the same amp-hr rating. Since both batteries must have the same rating, the nickel-cadmium battery will cost twice as much as the lead-acid battery. To keep the battery investment as low as possible, a lead-acid battery would therefore be chosen.

5. Compute the battery operating time

The battery must supply 311 amp during the starting cycle. A 300-amp-hr lead-acid battery (i.e., a battery capable of supplying 300 amp for 1 hr) would be suitable for the engine starting load because it would have sufficient amperage capacity to supply the 311 amp for the short starting cycle—usually less than 1 min.

The emergency-lighting load of 480 watts requires $I = P/E = 4,800/24 = 200$ amp. With a 300-amp-hr battery, the operating time, hr, with this emergency load would be $T = $ amp-hr/amp $ = 300/200 = 1.5$ hr.

Related Calculations: Use this general method to choose rechargeable batteries for starting engines, powering industrial trucks, supplying emergency lighting, powering portable tools, furnishing inverter power supply, etc. Where an economic analysis of a battery installation must be made, use the methods of engineering economy to compare the various alternatives.

For industrial, automotive, and similar heavy-demand services, the output of a battery is often stated thus: 150 amp for 4.1 min; 66 amp for 20 min; 53 amp-hr for 20 hr. The actual rating of this battery is 53 amp-hr for 20 hr. Most manufacturers list three discharge rates for each battery. Figure 15 shows how the output current of three different nickel-cadmium plate designs compares with a typical lead-acid cell of the same amp-hr capacity. Two types of emergency-power battery systems are shown in Fig. 16.

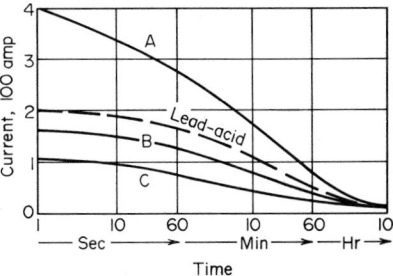

Fig. 15 Discharge rates for three nickel-cadmium plate designs (A, B, C) compared with lead-acid cells at the same amp-hr capacity.

Figure 17 shows how the output voltage of four different types (alkaline-manganese zinc, mercury, silver-zinc, and carbon-zinc) of small batteries varies with time. Small cells can be designed for optimum characteristics. These batteries find wide use in a variety of portable devices.

ELECTRIC-MOTOR SELECTION FOR A KNOWN LOAD

Select an electric motor to drive a 3,600-rpm centrifugal pump rated at 3,000 gpm of water at a 100-ft total head.

Calculation Procedure:

1. Compute the required hp input

Use a hp-input relation that is applicable to the driven unit. For a centrifugal pump the input hp is $hp_i = (gpm)(h)(s)/3,960$ (eff), where $gpm =$ quantity of fluid handled, gpm; $h =$ pump total head, ft of liquid handled; $s =$ specific gravity of liquid handled; eff $=$ efficiency of pump, expressed as a decimal.

Assume an efficiency for the driven device if it is not known or given. Thus, assume that this pump has an efficiency of 85 percent, a typical value for well-constructed, centrifugal pumps. Then, $hp_i = (3,000)(100)(1.0)/3,960(0.85) = 89.1$ hp.

TABLE 22   Battery Characteristics

Type of battery	Typical plant jobs	Typical battery voltage	Typical Amp-hr (6-hr rating)	Typical life (good mtce)	Volts per cell
Lead-acid	Truck starting and small truck motive power	12	160–300	1–4 (yrs.)	2.14 (initial); 2.1 to 1.46 under discharge
	Industrial truck motive power	12–72	200–1000	5–10	
	Packaged emergency lighting	6	5–50	4–6	
	Diesel standby generator starting	12–32	120–300	1–3	
	Switchgear control source	32–250	50–200+	6–12	
	Inverter continuous power source	24–130	50–500+	6–14	
	Portable tools or instruments	6–12 (sealed)	5–12	1–4	
Nickel-cadmium	Diesel standby generator starting	12–32	40–190	15+	1.34 (initial); 1.3 to 0.75 under discharge
	Packaged emergency lighting	6	5–50	15+	
	Switchgear control source	32–250	10–200+	15+	
	Inverter continuous power source	24–130	50–500+	15+	
	Portable tools or instruments	6–12 (sealed)	to 15+	15+	
Nickel-iron	Industrial truck motive power	24–36	340–1000	8–20	Same as Ni-Cd
Silver-zinc	Portable tools or instruments	6–12	3–8+	1–2	1.86, 1.55 to 1.1
Silver-cadmium	Portable tools or instruments	6–12	3–8+	2–3	1.34, 1.3 to 0.8

Composition	Typical Advantages	Typical Disadvantages
Lead and lead oxide electrodes and H_2SO_4 electrolyte	Lowest of all in first cost. Long term price picture relatively stable and competitive. Knowledge of maintenance requirements is widespread. Supplies high output under emergency overload conditions (at some sacrifice in overall life). Emergency replacements readily available everywhere. Easy to repair or rebuild damaged cells. Significant scrap value.	Deteriorates quickly under neglect or incompetent maintenance. Fumes during charging are corrosive, explosive, annoying—must be vented. Needs frequent addition of water. Heavy and bulky to handle, (acid hazard). Freezes readily when discharged. Cells will not be uniform unless very close quality control is maintained in manufacture. Vibration shortens life—shakes loose electrode ingredients.
Nickel oxide and cadmium electrodes and KOH electrolyte	Withstands neglect or incomptent maintenance Needs little water per year. Will not freeze. Substantially smaller and lighter than lead acid. Holds charge without attention for months or years. Very high short-time current capability (often enables use of smaller battery). Withstands vibration. Fumes do not corrode enclosure.	Up to two or three times as expensive as lead-acid for same ampere-hour rating. Requires over $\frac{1}{3}$ more cells (and extra connections) for same load as lead-acid. Small number of market sources for industrial types (virtually all plates are imported—U.S. assembled). Multi-cell sealed batteries sometimes need individual cell charging. Charger characteristics are critical on sealed cells.
Nickel oxide, iron, KOH	Extremely long life. Needs little maintenance (has most advantages of Ni-Cd).	Requires most space of all. Nearly as heavy as lead-acid.
Silver oxide, zinc, KOH	Very high energy per unit volume.	May cost about 10 times that of lead-acid. Relatively short cycle life.
Silver oxide, cadmium, KOH	High energy per unit volume. Withstands frequent cycling.	Very expensive. Limited operating life.

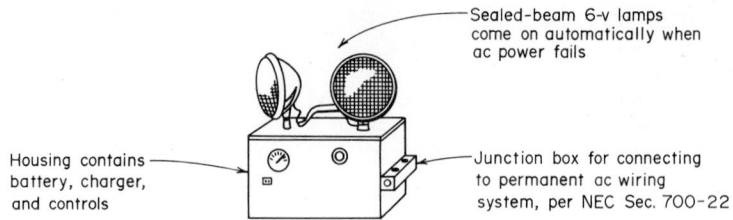

Sealed-beam 6-v lamps
come on automatically when
ac power fails

Housing contains
battery, charger,
and controls

Junction box for connecting
to permanent ac wiring
system, per NEC Sec. 700-22

Typical Unit Emergency Light
for Use on Branch Circuits

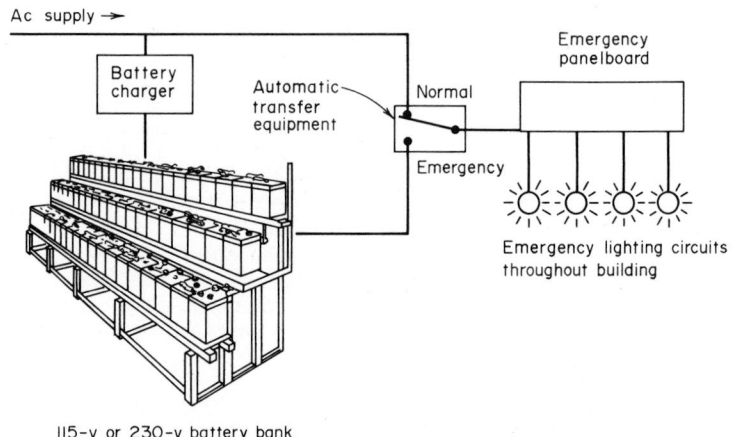

115-v or 230-v battery bank

Basic Layout of Full-system
DC Emergency Lighting

Fig. 16   Two types of emergency-power systems served by batteries.

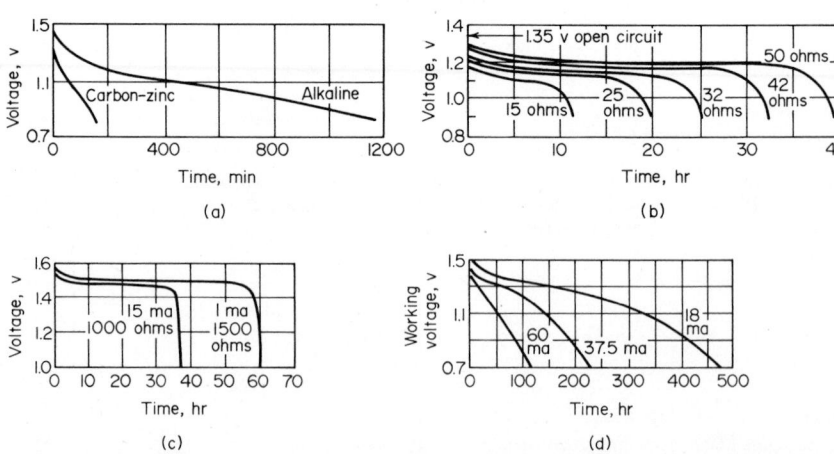

Fig. 17   Output voltage of small cells. (*a*) Alkaline-manganese-zinc; (*b*) mercury; (*c*) silver-zinc; and (*d*) carbon-zinc.

2. Select the motor hp

Checking a tabulation of standard motor hp ratings, such as Table 23, shows that the next highest standard hp rating is 100 hp. Hence, this rating will be tentatively chosen to drive the pump. (As a general guide, choose the next *larger* motor rating for usual industrial applications).

TABLE 23   Standard Motor Hp Ratings

$\frac{1}{8}$	2	30	200	700	2,250
$\frac{1}{6}$	3	40	250	800	2,500
$\frac{1}{4}$	5	50	300	900	3,000
$\frac{1}{3}$	$7\frac{1}{2}$	60	350	1,000	3,500
$\frac{1}{2}$	10	75	400	1,250	4,000
$\frac{3}{4}$	15	100	450	1,500	4,500
1	20	125	500	1,750	5,000
$1\frac{1}{2}$	25	150	600	2,000	

3. Select the motor electrical supply

In almost every situation, the electrical supply will be determined by the existing supply in the area or facility, except for small portable battery-powered motors. Thus, the availability of an alternating-current supply will make the choice of an ac motor more likely. The same is true where the local electrical supply is direct current.

Assume that the electrical supply is 60-Hz alternating current. This immediately indicates that an ac motor of some type will be used to drive the pump unless there are special design requirements that dictate use of a dc motor. Since no such design requirements are stated, an ac motor will probably be acceptable.

4. Determine the driven-machine torque requirements

The motor hp rating alone is generally insufficient for selecting large motors. Besides the required output hp, two other factors must be known. These are: (a) the speed-torque characteristic of the motor and load, and (b) the moment of inertia or WK^2 of the motor and load, where W = weight accelerated, lb; K = radius of gyration of the load, in.

Figure 18 shows that the starting torque of a typical centrifugal pump with its dis-

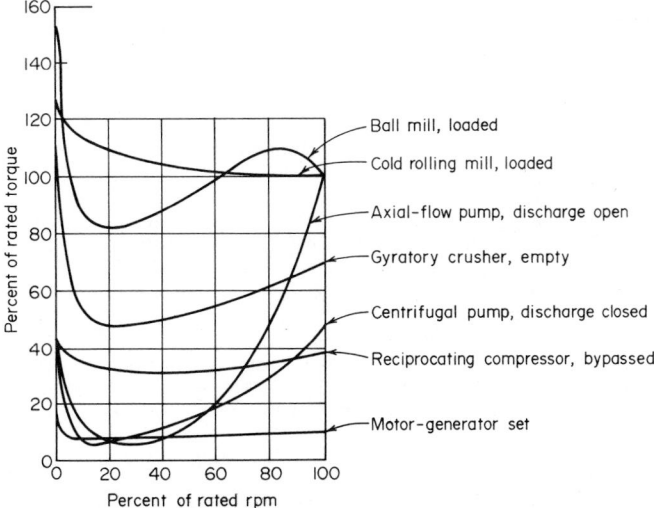

Fig. 18   Speed-torque characteristics for typical motor-driven loads. Flywheel effect and the time required for load acceleration are neglected here.

charge valve closed is about 40 percent of the rated torque. Hence, the type of motor chosen should develop at least 40 percent of the rated torque on starting.

5. *Choose the motor type for the driven load*

Figure 19 shows that induction motors are generally chosen for speeds above 500 rpm and power ratings from 0 to 1,000 hp, or more. The three-phase induction motor is also called a *squirrel-cage motor* because of the appearance of its rotor.

The starting torque of an induction motor depends on its NEMA classification. Usual starting torques range from about 60 to 150 percent of the full-load torque. Typical starting-torque ranges of ac motors are plotted in the *Standard Handbook for Electrical Engineers.* These plots show that any NEMA induction motor would be suitable for this pump drive because they all develop a starting torque greater than the driven

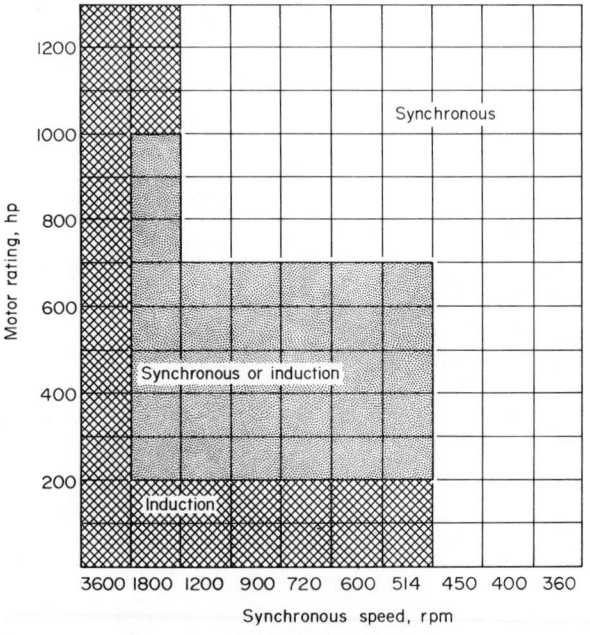

Fig. 19   Typical speed- and power-range applications for ac motors.

machine's starting torque. Also, the WK^2 value of a centrifugal pump is characteristically low, so this factor can be neglected here. The WK^2 value of the load affects the length of the starting period, which in turn determines the heating of the motor. Thus, the motor selected to drive this pump will be a general-purpose 100-hp induction type without special controls.

Related Calculations: Use this general method to select ac and dc motors for all types of industrial and commercial drives—fans, compressors, crushers, motor-generator sets, blowers, etc. Figure 20 shows the speed-torque characteristics of the three popular types of dc motors—series, shunt, and compound—that might be used for any of these loads.

As a further guide to motor selection, Table 24 summarizes ac and dc motor characteristics and applications.

Where there is a choice between synchronous and induction motors, use the following rule of thumb as a guide to relative cost: *Synchronous motors are less expensive than are induction motors if the motor-rated hp exceeds 1 hp/rpm.* This rule, however, overlooks the higher power factor and efficiency of the synchronous motor. These two factors are important in low-speed-motor choice.

As a general guide, the WK^2 of a driven unit is usually relatively high when the starting torque of the driven unit is high. Thus, the WK^2 and starting torque of crushers, ball mills, rolling mills, and similar equipment are usually high. Figure 18 shows typical starting torques for these units.

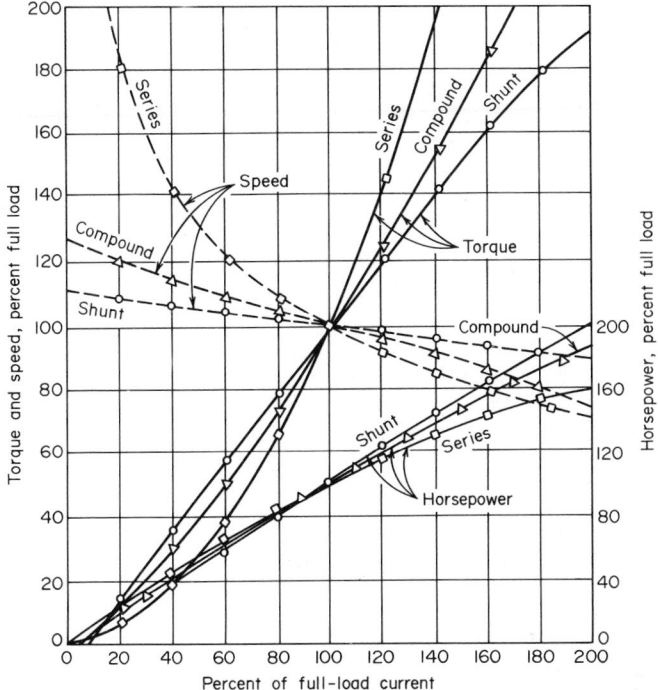

Fig. 20   Speed-torque characteristics of three types of direct-current motors.

To pick a new NEMA *rerate motor* of greater hp that will fit the mounting pad of an old motor, use the following guides:

1. The first two digits of the frame number are four times the D dimension, Fig. 21, in in. If $D = 9$, the first two digits are $4 \times 9 = 36$.

2. The third digit of the frame number depends on the value of the F dimension in in. Select this digit from the 0 to 9 headings over the F dimensions in Table 25. For example, if $D = 9$ and $F = 7$, the third digit is 6, from Table 25. The motor frame number for the rerate motor is therefore 366.

3. The letter A is added after some frame numbers to designate an industrial dc motor or generator. The A denotes that certain detailed dimensions may differ from those of an ac motor or generator having the same frame number. Usually this is not a critical factor, and dimension variables are available in the manufacturer's engineering data.

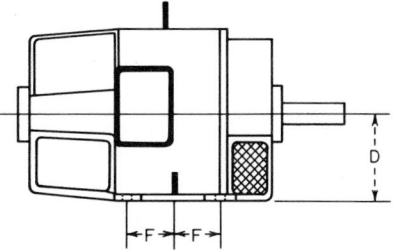

Fig. 21   Important dimensions of NEMA rerate motors.

TABLE 24 Summary of Motor Characteristics and Applications

	Polyphase motors			
Speed regulation	Speed control	Starting torque	Breakdown torque	Applications
General-purpose squirrel-cage (design B):				
Drops about 3% for large to 5% for small sizes	None, except multispeed types, designed for two to four fixed speeds	100% for large, 275% for 1-hp four-pole unit	200% of full load	Constant-speed service where starting torque is not excessive. Fans, blowers, rotary compressors, and centrifugal pumps
High-torque squirrel-cage (design C):				
Drops about 3% for large to 6% for small sizes	None, except multispeed types, designed for two to four fixed speeds	250% of full load for high-speed to 200% for low-speed designs	200% of full load	Constant-speed where fairly high starting torque is required infrequently with starting current about 550% of full load. Reciprocating pumps and compressors, crushers, etc.
High-slip squirrel-cage (design D):				
Drops about 10 to 15% from no load to full load	None, except multispeed types, designed for two to four fixed speeds	225 to 300% of full load, depending on spee pending on speed with rotor resistance	200%. Will usually not stall until loaded to maximum torque, which occurs at standstill	Constant-speed and high starting torque, if starting is not too frequent, and for high-peak loads with or without flywheels. Punch presses, shears, elevators, etc.
Low-torque squirrel-cage (design F):				
Drops about 3% for large to 5% for small sizes	None, except multispeed types, designed for two to four fixed speeds	50% of full load for high-speed to 90% for low-speed designs	135 to 170% of full load	Constant-speed service where starting duty is light. Fans, blowers, centrifugal pumps and similar loads

TABLE 24   Summary of Motor Characteristics and Applications (Continued)

Polyphase motors				
Speed regulation	Speed control	Starting torque	Breakdown torque	Applications
Wound-rotor: With rotor rings short-circuited drops about 3% for large to 5% for small sizes	Speed can be reduced to 50% by rotor resistance. Speed varies inversely as load	Up to 300% depending on external resistance in rotor circuit and how distributed	300% when rotor slip rings are short-circuited	Where high starting torque with low starting current or where limited speed control is required. Fans, centrifugal and plunger pumps, compressors, conveyors, hoists, cranes etc.
Synchronous: Constant	None, except special motors designed for two fixed speeds	40% for slow- to 160% for medium-speed 80% pf. Specials develop higher	Unity-pf motors, 170%, 80%-pf motors, 225% Specials up to 300%	For constant-speed service, direct connection to slow-speed machines and where power-factor correction is required
Dc and single-phase motors				
Series: Varies inversely as load. Races on light loads and full voltage	Zero to maximum depending on control and load.	High. Varies as square of voltage Limited by commutation, heating, capacity	High. Limited by commutation, heating and line capacity	Where high starting torque is required and speed can be regulated. Traction, bridges, hoists, gates, car dumpers, car retarders
Shunt: Drops 3 to 5% from no load to full load	Any desired range depending on design type and type of system	Good. With constant field, varies directly as voltage applied to armature	High. Limited by commutation, heating, and line capacity	Where constant or adjustable speed is required and starting conditions are not severe. Fans, blowers, centrifugal pumps, conveyors, wood and metal-working elevators

TABLE 24   Summary of Motor Characteristics and Applications (Continued)

Polyphase motors				
Speed regulation	Speed control	Starting torque	Breakdown torque	Applications
Compound:				
Drops 7 to 20% from no load to full load depending on amount of compounding	Any desired range, depending on design and type of control.	Higher than for shunt, depending on amount of compounding	High. Limited by commutation, heating, and line capacity	Where high starting torque and fairly constant speed is required. Plunger pumps, punch presses, shears, bending rolls, geared elevators, conveyors, hoists
Split-phase:				
Drops about 10% from no load to full load	None	75% for large to 175% for small sizes	150% for large to 200% for small sizes	Constant-speed service where starting is easy. Small fans, centrifugal pumps and light-running machines, where polyphase is not available
Capacitor:				
Drops about 5% for large to 10% for small sizes	None	150 to 350% of full load depending on design and size	150% for large to 200% for small sizes	Constant-speed service for any starting duty and quiet operation, where polyphase current cannot be used
Commutator:				
Drops about 5% for large to 10% for small sizes	Repulsion induction, none. Brush-shifting types, 4:1 at full full load	250% for large to 350% for small sizes	150% for large to 250% for small sizes	Constant-speed service for any starting duty where speed control is required and polyphase current cannot be used

When a new motor of given frame size, such as 366, is to be installed, the user may need the D and F dimensions to check the foundation requirements. Working with the frame number, say xyz, or 366, determine these values from $D = xy/4 = 36/4 = 9$.

From the tabulated data, Table 25, if z (the third digit) is 6, then $F = 7$ in. With this information, any shimming or relocating of the mounting bolt holes necessary for installing a new motor can be easily determined.

TABLE 25   NEMA Motor Frame Numbers

D	F Dimension Third digit in frame number										Frame series
	0	1	2	3	4	5	6	7	8	9	
$4\frac{1}{2}$	$1\frac{3}{4}$	2	$2\frac{1}{4}$	$2\frac{1}{2}$	$2\frac{3}{4}$	$3\frac{1}{8}$	$3\frac{1}{2}$	4	$4\frac{1}{2}$	5	180
5	2	$2\frac{1}{4}$	$2\frac{1}{2}$	$2\frac{3}{4}$	$3\frac{1}{4}$	$3\frac{1}{2}$	4	$4\frac{1}{2}$	5	$5\frac{1}{2}$	200
$5\frac{1}{4}$	2	$2\frac{1}{4}$	$2\frac{1}{2}$	$2\frac{3}{4}$	$3\frac{1}{8}$	$3\frac{1}{2}$	4	$4\frac{1}{2}$	5	$5\frac{1}{2}$	210
$5\frac{1}{2}$	$2\frac{1}{4}$	$2\frac{1}{2}$	$2\frac{3}{4}$	$3\frac{1}{8}$	$3\frac{3}{8}$	$3\frac{3}{4}$	$4\frac{1}{2}$	5	$5\frac{1}{2}$	$6\frac{1}{4}$	220
$6\frac{1}{4}$	$2\frac{1}{2}$	$2\frac{3}{4}$	$3\frac{1}{8}$	$3\frac{1}{2}$	$4\frac{1}{8}$	$4\frac{1}{2}$	5	$5\frac{1}{2}$	$6\frac{1}{4}$	7	250
7	$2\frac{3}{4}$	$3\frac{1}{8}$	$3\frac{1}{2}$	4	$4\frac{3}{4}$	5	$5\frac{1}{2}$	$6\frac{1}{4}$	7	8	280
8	$3\frac{1}{4}$	$3\frac{1}{2}$	4	$4\frac{1}{2}$	$5\frac{1}{4}$	$5\frac{1}{2}$	6	7	8	9	320
9	$3\frac{1}{2}$	4	$4\frac{1}{2}$	5	$5\frac{5}{8}$	$6\frac{1}{8}$	7	8	9	10	360
10	4	$4\frac{1}{2}$	5	$5\frac{1}{2}$	$6\frac{1}{8}$	$6\frac{7}{8}$	8	9	10	11	400
11	$4\frac{1}{2}$	5	$5\frac{1}{2}$	$6\frac{1}{4}$	$7\frac{1}{4}$	$8\frac{1}{4}$	9	10	11	$12\frac{1}{2}$	440
$12\frac{1}{2}$	5	$5\frac{1}{2}$	$6\frac{1}{4}$	7	8	9	10	11	$12\frac{1}{2}$	14	500
$14\frac{1}{2}$	$5\frac{1}{2}$	$6\frac{1}{4}$	7	8	9	10	11	$12\frac{1}{2}$	14	16	580
17	7	8	9	10	11	$12\frac{1}{2}$	14	16	18	20	680

STARTING TIME AND CURRENT FOR AC ELECTRIC MOTORS

A 300-hp 1,176-rpm (1,200-rpm synchronous speed) motor drives a directly connected reciprocating pump. The motor and pump have a WK^2 or inertia of 28,000 lb-ft². How long will it take to bring this load up to full speed if the starting torque is negligible? What is the energy lost in overcoming the rotor and load inertia during starting? How much starting current will this motor draw if it is a 220-v three-phase NEMA type B motor?

Calculation Procedure:

1. Compute the full-load torque

Use the relation $T = 5{,}252\ hp/rpm$, where T = full-load torque developed by the motor, lb-ft; hp = rated horsepower of the motor; rpm = motor operating speed, rpm. Thus, $T = 5{,}252(300)/1176 = 1{,}342$ lb-ft. This relation assumes that full-load torque is developed during acceleration.

2. Compute the starting time

Use the relation $S = WK^2(rpm)/308\ T$, where S = starting time, sec; other symbols as before. Thus, $S = 28{,}000(1176)/308(1{,}342) = 79.5$ sec. Hence, this motor will take approximately $1\frac{1}{3}$ min to bring this load up to the operating speed.

3. Compute the energy required to overcome inertia

Use the relation $E = 2.31\ WK^2(rpm)^2 10^{-7}$, where E = energy required to overcome the starting inertia, kw-sec; other symbols as before. Thus, $E = 2.31(28{,}000)(1176)^2 \times (10^{-7}) = 7{,}600$ kw-sec.

4. Compute the motor starting current

Use the relation $A = (KVA$ per hp$)($motor hp$)/$constant, where A = motor starting current, amp; KVA per hp is the locked-rotor current per hp from Table 26; constant = value from Table 27 for the motor operating voltage. Since this is known to be a NEMA type B motor, $A = (3.15)(300)/0.381 = 2{,}480$ amp; using the appropriate data from Tables 26 and 27. Note that the lower value of KVA per hp given in Table 26 was used in this relation. Either value can be used, depending on the anticipated current requirements — small or large.

Related Calculations: Use this general method to determine the starting time and current of any ac induction or synchronous motor. Note that the motor is assumed to exert full-load torque during the acceleration. As a general guide, the starting current of a motor can be up to about 5.5 times the full-load operating current. The NEMA code-letter designation of a motor can be determined from the catalog description of the motor or from the motor nameplate.

TABLE 26   Locked-Rotor Kva per Hp for AC Motors

NEMA Code Letter designation of motor	Locked-Rotor kva hp
A	0–3.15
B	3.15–3.55
C	3.55–4.0
D	4.0–4.5
E	4.5–5.0
F	5.0–5.6
G	5.6–6.3
H	6.3–7.1
J	7.1–8.0
K	8.0–9.0
L	9.0–10.0
M	10.0–11.2
N	11.2–12.5
P	12.5–14.0
R	14.0–16.0
S	16.0–18.0
T	18.0–20.0
U	20.0–22.4
V	22.4 and up

TABLE 27   Constants for Motor-Starting-Current Determination

Power Supply	Constant
Three-phase:	
208-v	0.360
220-v	0.381
440-v	0.762
550-v	0.952
2,300-v	3.99
Two-phase:	
Three-wire, 220-v	0.311
Three-wire, 440-v	0.622

INTERIOR-LIGHTING-SYSTEM SELECTION AND SIZING

Choose the type of lighting and number of fixtures required for a machine shop that is 80 ft long and 48 ft wide with a 14-ft-high ceiling. The work done in the shop is classified as rough-bench and machine operations. The reflection factor of the ceiling is 80 percent, the walls 50 percent. Maintenance of the fixtures and surfaces will be good.

Calculation Procedure:

1. *Determine the required illumination*

Table 28 shows the typical lamp manufacturer's and IES recommended illumination levels for various work areas. Study of this tabulation shows that 100 ft-c is the

TABLE 28   Typical In-Service Illumination Levels in Footcandles

Automobile manufacturing:	Hot-strip finishing...... 15	Plating.................. 10
Assembly line...........100	Box annealing.......... 5	
Frame assembly 30	Hot-strip process....... 5	Polishing and burnishing... 20
Body manufacturing:	Plate finishing.......... 10	
Parts................ 30	Slab-furnace building... 10	Power plants, engine room,
Assembly............. 30	Continuous pickler..... 15	Boilers:
Finishing and inspect-	Cold mill.............. 15	Boiler room (operating
ing.................200	Roll shop.............. 15	floor) 10
	Cold-mill finishing 15	Chemical laboratory 20
Chemical works:	Shear building......... 10	Coal bunker (conveyor
Hand furnaces, boiling	Mill run out building ... 15	floor) 10
tanks, stationary driers,	Mesta pickle........... 10	Coal-crusher house...... 10
stationary and gravity	Temper mill........... 15	Condenser pit (turbine
crystallizers.......... 5	Galvanize pickle 10	room)................. 10
Mechanical furnaces,	Annealing 10	Control rooms (on vertical
generators and stills,	Pipe mill.............. 10	plane) 30
mechanical driers, eva-	Chipping.............. 50	Heater gallery........... 10
porators, filtration,		Machine shop 50
mechanical crystal-	Loading platforms........ 5	Oil-pump house (turbine
lizers, bleaching 10		room)................. 10
Tanks for cooking, extrac-		Pump bay (turbine
tors, percolators, nitrators,	Machine shops:	room)................. 10
electrolytic cells...... 20	Rough bench and machine	Switchgear area 15
	work 20	Turbine room (operating
Elevators — freight and	Medium bench and ma-	floor) 30
passenger.............. 10	chine work, ordinary	Water pumps............ 10
	automatic machines,	Water-treating area 10
Forge shops (and weld-	rough grinding, med-	
ing).................... 10	ium buffing, polishing 50	Sheet metal works
	Fine bench and machine	Miscellaneous machines,
Foundries:	work, fine automatic	ordinary bench work ... 20
Annealing (furnaces) 10	machines, medium	Punches, presses, shears,
Cleaning.............. 20	grinding, fine buffing	stamps, spinning, medium
Core making:	ing and polishing.... 100	bench work 20–25
Fine 50	Extra-fine bench and	Tin-plate inspection...... 10
Medium............. 25	machine work, grinding,	
Grinding and chipping... 30	fine work 200	Stairways and passageways 10
Inspection:		
Fine 100	Outdoor storage......... 1	Storage and stock rooms
Medium fine.......... 50		Rough bulky material 5
Medium............. 30	Paper manufacturing:	Medium 10
Moulding:	Beaters, grinding, and	Fine material requiring
Medium............. 50	calendering.......... 20	care 20
Large 30	Finishing, cutting, and	
Pouring............... 10	trimming 50	Warehouse............... 5
Cupola................ 10		
Shakeout.............. 10	Paper machine 25	Welding and general illumina-
		tion 30
Iron and steel manufacturing:		
Hot mill............... 10	Parking areas............. 2	Yard lighting 0.2–0.8

TABLE 29 Light-Source-Selection Guide*

Light source	Minimum mounting ht, ft	Advantages	Disadvantages
Fluorescent: Preheat 40-watt and 90-watt; Rapid-start 40-watt; Slimline: 4 ft, 6 ft, 8 ft T-12	12 for 20-watt	General: High lamp efficiency, long life, low brightness, good color quality; slimlines and rapid-start lamps start instantly (rapid-start within one second) and require no starters	General: High initial installation cost; large number of lamps required, with accompanying maintenance problems; low system efficiency in high narrow areas
Incandescent: Standard-bulb lamps	Varies with wattage	General: Low initial installation cost, good color quality, start instantly	General: Low lamp efficiency, short lamp life in comparison with other types
Reflector lamps: 550-watt narrow-beam R57.	20	Good system efficiency in high narrow areas, relatively long life (2000 hr)	High lamp brightness may cause direct and reflected glare; relatively low vertical-surface illumination
800-watt narrow-beam R57.	24		
550-watt wide-beam R52...	15	Good vertical-surface illumination, relatively long life (2,000 hr)	High lamp brightness may cause direct and reflected glare; low system efficiency in high narrow areas
800-watt wide-beam R-52 ..	18		
550-watt medium-beam R-57	16	Relatively long life (2,000 hr), designed for use on 230-v distribution systems	High-voltage lamps are slightly less efficient than standard-voltage lamps
800-watt medium-beam R-57	20		
Mercury:	...	General: High lamp efficiency, long life, easy to maintain, good system efficiency (except for A-H9 in high narrow areas)	General: Color deficiency characteristic of mercury arc; lamps do not start at full brightness or restart instantly
400-watt A-H1..........	18	Exceptionally long life, relatively low arc brightness	Not economically competitive with other mercury lamps listed
400-watt E-H1..........	18	Exceptionally long life	
700-watt A-H18.........	26	Low transformer cost and losses	High socket voltage (460 v)
1,000-watt A-H12........	35	High lumen output	
1,000-watt A-H15........	35	Low transformer cost and losses	High socket voltage (460 v)
3,000-watt A-H9.........	40	Extremely high lumen output, minimum maintenance	Less efficient than other high-wattage mercury lamps, low system efficiency in high narrow areas

Lamp	Factor	Characteristics	Remarks
Reflector lamps:	..	Relatively low initial installation cost	High lamp brightness may cause glare
400-watt K-H1..............	20	High vertical surface illumination	Low system efficiency in high narrow areas
400-watt L-H1..............	24	High system efficiency in high narrow areas	Low vertical surface illumination
Color-improved mercury:	..	General: White light, high lamp efficiency, long life, easy to maintain, good system efficiency when used with proper reflector	General: Lamps do not start at full brightness or restart instantly; slightly less efficient than conventional mercury lamps
400-watt J-H1..............	16	Exceptionally long life, relatively low brightness	
700-watt B-H18.............	22	Relatively low brightness, low transformer cost and losses	High socket voltage (460 v)
1,000-watt C-H12..........	28	Minimum number of luminaires required	
1,000-watt B-H15..........	28	Low transformer cost and losses; minimum number of luminaires required	High socket voltage (460 v)

Power magazine.

recommended illumination level for machine shops in which rough-bench and machine work is performed.

2. Select the light sources and luminaries

Table 29 lists the various types of light sources used today. Study of this table indicates that 90-watt fluorescent lights will probably be suitable. Since the room ceiling height is 14 ft, there will be enough vertical clearance to install the lights because only a 12-ft clearance is required.

To choose the luminaire type, consider the general practice prevalent in industry today. Three types of luminaires are currently popular for industrial lighting—diffusing, semidirect, and direct. The choice of a particular type of luminaire depends on the degree of diffusion wanted and on the ability of the ceiling and walls to reflect light. Choose direct luminaires for this lighting task because they will provide the high illumination level desired.

3. Compute the room ratio and room index

Use the relation room ratio = $WL/(H[W+L])$, where W = room width, ft; L = room length, ft; H = room height, ft. For this room, the room ratio = $(48)(80)/(14[48+80])$ = 2.145.

Table 30 shows that a room ratio of 1.75:2.25 indicates a room index of E, with a center point of 2.00.

TABLE 30   Room-Ratio Ranges

Room index	Room ratio	
	Range	Center point
J	Less than 0.7	0.6
I	0.7–0.9	0.8
H	0.9–1.12	1.0
G	1.12–1.38	1.25
F	1.38–1.75	1.50
E	1.75–2.25	2.00
D	2.25–2.75	2.50
C	2.75–3.50	3.00
B	3.50–4.50	4.00
A	More than 4.50	5.00

4. Determine the coefficient of utilization

The coefficient of utilization CU, is the ratio of the lumens reaching a working plane (normally assumed to be a horizontal plane 30 in. above the floor) to the total lumens given out by the lamps. It takes into account the lighting efficiency of the luminaire, the mounting height, the room proportions, and the reflection factors of the ceiling and walls.

Coefficient-of-utilization tables are prepared for specific luminaires by the IES and manufacturers. Table 31 lists CU factors for three common general luminaire types. Enter this table with the luminaire type, Step 2, room index, Step 3, and the ceiling- and wall-reflection factors. Using these data, CU = 0.65.

5. Estimate the lamp-maintenance factor

Illumination decreases in service because dirt accumulates on the lamps and luminaires, reflection factors are reduced by the aging of paint, etc. With a clean atmosphere and good cleaning of the lamps, the maintenance factor MF = 0.70. Use this factor for this installation. Typical values of MF for direct luminaires are: good, 0.70; medium, 0.60; poor, 0.55.

TABLE 31   Coefficients of Utilization for General Types of Luminaires*

Typical luminaire distribution	Room index	Reflection factors											
		Ceiling											
		80%			70%			50%			30%		
		Walls											
		50%	30%	10%	50%	30%	10%	50%	30%	10%	50%	30%	10%
Diffusing	J	0.26	0.21	0.18	0.25	0.21	0.17	0.23	0.19	0.16	0.20	0.17	0.15
	I	0.32	0.27	0.23	0.31	0.26	0.22	0.28	0.24	0.21	0.25	0.22	0.19
	H	0.38	0.33	0.29	0.36	0.32	0.28	0.33	0.29	0.26	0.29	0.26	0.23
	G	0.43	0.38	0.34	0.41	0.36	0.33	0.37	0.33	0.30	0.33	0.30	0.27
	F	0.47	0.42	0.38	0.45	0.40	0.36	0.40	0.36	0.33	0.35	0.32	0.30
	E	0.53	0.48	0.44	0.50	0.46	0.42	0.44	0.41	0.38	0.39	0.36	0.34
	D	0.56	0.52	0.48	0.53	0.49	0.46	0.47	0.44	0.41	0.41	0.39	0.37
	C	0.59	0.55	0.51	0.55	0.52	0.49	0.49	0.46	0.44	0.43	0.41	0.39
	B	0.62	0.59	0.56	0.58	0.55	0.53	0.52	0.49	0.47	0.45	0.43	0.42
	A	0.64	0.61	0.59	0.61	0.58	0.55	0.54	0.51	0.49	0.46	0.45	0.44
Semidirect	J	0.34	0.28	0.24	0.33	0.28	0.24	0.31	0.26	0.24	0.30	0.25	0.22
	I	0.42	0.36	0.32	0.40	0.35	0.31	0.38	0.33	0.30	0.36	0.32	0.29
	H	0.48	0.42	0.38	0.47	0.41	0.37	0.44	0.39	0.36	0.41	0.37	0.34
	G	0.54	0.48	0.44	0.52	0.47	0.43	0.49	0.45	0.41	0.46	0.42	0.39
	F	0.58	0.53	0.48	0.56	0.51	0.47	0.53	0.49	0.45	0.49	0.46	0.43
	E	0.64	0.59	0.55	0.62	0.57	0.54	0.58	0.54	0.51	0.54	0.51	0.48
	D	0.67	0.63	0.59	0.65	0.61	0.58	0.60	0.57	0.54	0.56	0.54	0.52
	C	0.70	0.66	0.62	0.68	0.64	0.61	0.63	0.60	0.57	0.58	0.56	0.54
	B	0.73	0.70	0.67	0.70	0.67	0.65	0.66	0.63	0.61	0.61	0.59	0.57
	A	0.75	0.72	0.70	0.72	0.70	0.68	0.68	0.65	0.63	0.62	0.61	0.60
Direct	J	0.34	0.28	0.24	0.34	0.28	0.23	0.33	0.27	0.24	0.32	0.27	0.23
	I	0.43	0.36	0.31	0.42	0.36	0.31	0.41	0.35	0.31	0.40	0.35	0.31
	H	0.49	0.42	0.38	0.48	0.42	0.38	0.47	0.42	0.37	0.46	0.41	0.37
	G	0.55	0.49	0.44	0.55	0.48	0.44	0.53	0.48	0.44	0.52	0.47	0.44
	F	0.60	0.54	0.49	0.59	0.53	0.49	0.57	0.52	0.48	0.56	0.52	0.48
	E	0.65	0.60	0.56	0.64	0.60	0.55	0.63	0.59	0.55	0.61	0.58	0.55
	D	0.69	0.64	0.60	0.68	0.64	0.60	0.66	0.63	0.59	0.65	0.62	0.59
	C	0.72	0.67	0.64	0.71	0.67	0.63	0.69	0.66	0.63	0.67	0.65	0.62
	B	0.76	0.72	0.69	0.75	0.71	0.69	0.73	0.70	0.68	0.71	0.69	0.67
	A	0.78	0.75	0.72	0.77	0.74	0.72	0.75	0.73	0.71	0.74	0.72	0.70

*A floor reflection factor of 10% is assumed for values in table. For floors having higher reflection factors, refer to IES tables. Numbers accompanying sketches at left indicate percent of lamp lumens directed upward and downward. Sum of percentages equals luminaire light efficiency.

6. *Compute the number of lamps and luminaires required*

Use the relation number of lamps = (ft-c required) (floor area, sq ft)/(lumes per lamp) (CU)(MF). Determine the lumens per lamp from Fig. 22. Thus, a 90-watt T-17 preheat lamp is rated at 5,150 lumens per lamp. Entering the appropriate data, number of lamps = (100)(80 × 48)/(5,150)(0.65)(0.70) = 163.8; say 164 lamps.

Compute the number of luminaires from number of luminaires = number of lamps/

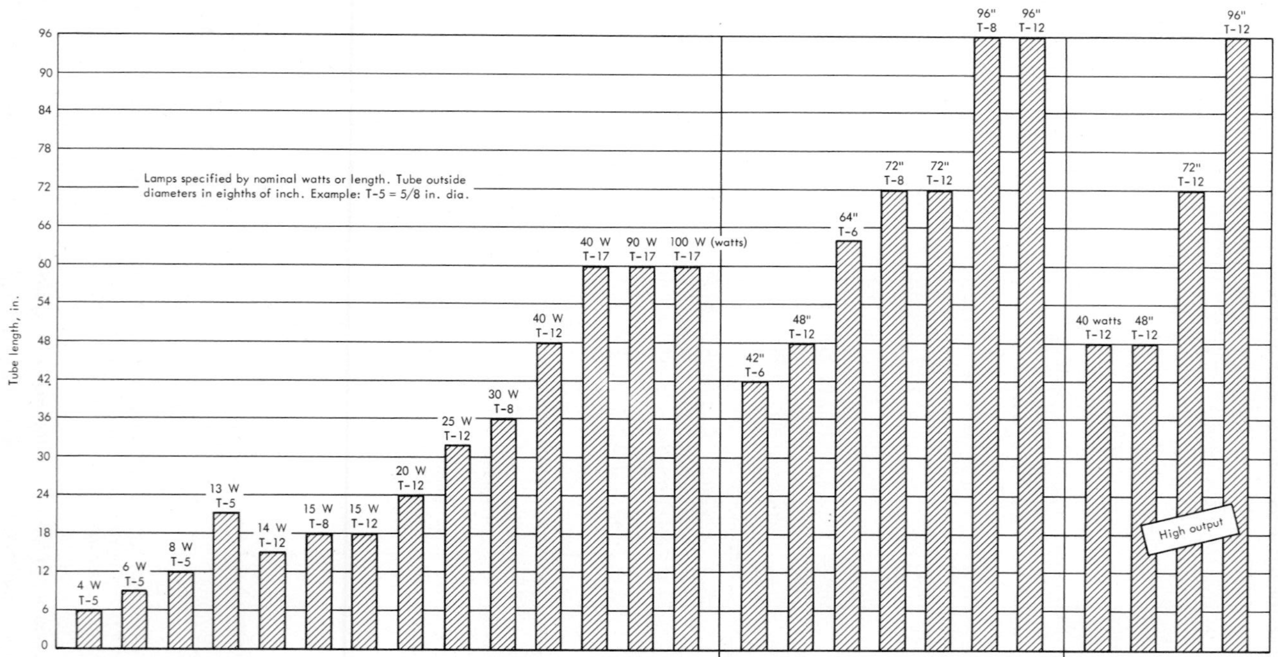

Lamps specified by nominal watts or length. Tube outside diameters in eighths of inch. Example: T-5 = 5/8 in. dia.

High output

Tube length, in.

100 W (watts)

	4 W T-5	6 W T-5	8 W T-5	13 W T-5	14 W T-12	15 W T-8	15 W T-12	20 W T-12	25 W T-12	30 W T-8	40 W T-12	40 W T-17	90 W T-17	100 W T-17
Average lamp watts	4	6.1	7.9	13	14	15	14.1	19.7	26	30	39	41	82	99
Lumens cool white	100	210	330	700	540	730	620	1000	1600	1890	2500	2500	5150	4850
Lamp amperes	.125	.145	.160	.160	.390	.300	.330	.380	.490	.355	.430	.425	1.550	1.520
Lamp volts	35	47	58	99	37.5	55	45.5	56	60	98	99	104	62	68

Preheat line

	42" T-6	48" T-12	64" T-6	72" T-8	72" T-12	96" T-8	96" T-12
Average lamp watts	25	38	37	36.5	55	49	74
Lumens cool white	1480	2300	2450	2550	3600	3550	5050
Lamp amperes	.200	.425	.200	.200	.425	.200	.425
Lamp volts	145	97	225	210	145	285	192

Slimline

	40 watts T-12	48" T-12	72" T-12	96" T-12
Average lamp watts	39	60	85	105
Lumens cool white	2500	3100	4800	7250
Lamp amperes	.420	.800	.800	.800
Lamp volts	99	80	115	148

Rapid start

Fig. 22 Performance factors for typical fluorescent lamps.

lamps per luminaire. With four lamps per luminaire, number of luminaires = 164/4 = 41.

7. *Choose the location of the luminaires*

As a general rule, equal spacing of the luminaires is used to provide uniform illumination. Such a spacing arrangement would be suitable for this installation.

Related Calculations: The procedure presented above is termed the *lumen method.* It is widely used for the design of lighting systems for general-area indoor illumination in all types of installations — industrial, commercial, residential, marine, aircraft, mobile, etc. Used with the data presented, or with additional manufacturer's engineering data, this method permits a quick, economical choice of a suitable indoor-lighting system.

A refinement of the lumen method, termed the *IES zonal-cavity method*, was recently introduced. With this refinement of the lumen method, the room must be divided into three cavities — ceiling, room, and floor — as indicated in Fig. 23. Then, by using the specific reflectances for the surfaces in each of these three cavities, more accurate calculations can be obtained than is possible by the original lumen method of calculation. In addition, with the zonal-cavity method, the effects of room proportions, luminaire suspension length, and work-plane height upon the coefficient of utilization (CU) are more accurately accounted for.

The zonal cavities are defined as follows:

Ceiling cavity is the space bounded by the ceiling, upper walls, and an imaginary plane through the luminaires.

Room cavity is the space bounded by the plane through the luminaires, the work plane, and the portion of the walls between these planes.

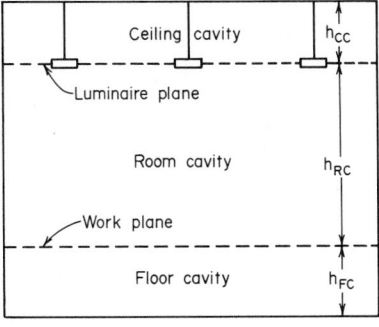

Fig. 23 Three room cavities are used in the IES zonal-cavity method of calculation.

Floor cavity is the space bounded by the work plane, the lower walls, and the floor.

In applying the zonal-cavity method of calculation, the proportions of each cavity may be represented by a *cavity ratio*. These *cavity ratios* may be obtained from a table (see Fig. 9-2, Cavity Ratios, *IES Lighting Handbook*), or from a formula, as follows

$$\text{Ceiling-cavity ratio CCR} = \frac{5h_{\text{CC}}(L+W)}{LW}$$

$$\text{Room-cavity ratio RCR} = \frac{5h_{\text{CC}}(L+W)}{LW}$$

$$\text{Floor-cavity ratio FCR} = \frac{5h_{\text{FC}}(L+W)}{LW}$$

where L = room length, ft; W = room width, ft; h = cavity height, ft.

Room-cavity ratios are always required to obtain coefficients of utilization, just as room ratios or room indices were required by the original lumen method of calculation. The *ceiling-* and *floor-cavity* ratios are required only to obtain *effective cavity reflectances*, which are not used in coefficient-of-utilization tables for specific luminaires to establish specific effective cavity-reflectance conditions. These effective ceiling- or floor-cavity reflectance percentages for various reflectance combinations are also provided in table form, by manufacturers of lighting equipment or in Fig. 9-3 in the *IES Handbook*.

Manufacturers of lighting equipment provide coefficient-of-utilization tables for each luminaire. These CU tables are for room-cavity ratios (RCR) varying from 1 to 10, based on 20 percent effective floor-cavity reflectance, and for various ceiling-cavity and wall-cavity reflectances.

When the effective floor-cavity reflectance varies from the 20 percent value normally used in manufacturers' CU tables, correction factors for 10 percent and 30 percent effective floor-cavity reflectances are provided in table form, either by lighting equipment manufacturers or in Fig. 9–5 of the *IES Lighting Handbook.*

A simple procedure may now be followed to obtain the CU value for the specific luminaire. The procedure for a suspended luminaire is typical:

1. Obtain the *room-cavity ratio* and *ceiling-cavity ratio* by formulas (above), or from table.

2. Obtain the *effective ceiling-cavity reflectance*, from manufacturers' literature or from Fig. 9–3 in the IES *Lighting Handbook* (note that expected *maintained* ceiling and wall reflectances should be used in selecting the proper column).

3. Obtain CU for expected *maintained* wall reflectance and 20 percent effective floor-cavity reflectance from the CU table for the luminaire. (Interpolate as required for the exact RCR and ceiling-cavity reflectance.)

For supplementary lighting—i.e., the lighting of special machines or small areas— the *point-by-point* calculation method is sometimes used. To apply this method, do the following. (1) Select the type of light source that will be used. (2) Find the angle in degrees between the vertical and a line to the point illuminated, Fig. 24. (3) Deter-

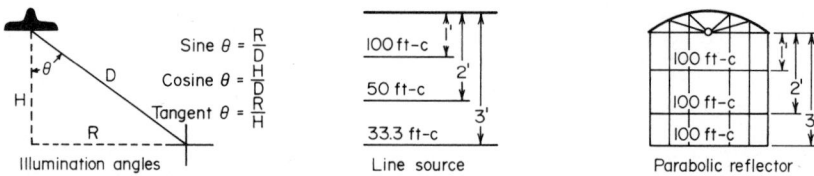

Fig. 24   Variation of illumination with the distance of the lighted area from the light source.

mine the candlepower in the direction of the point, using the angle θ, Fig. 24, and the candlepower-distribution curve of the luminaire, available from the luminaire manufacturer. (4) Compute the horizontal ft-c = (candlepower of luminaire)(cos θ)/(distance factor, and vertical ft-c) = (candlepower of luminaire)(cos θ)/distance factor. (The *distance factor* depends on the type of distribution, Step 1; for a point source it is the distance squared. Other sources are discussed in the next paragraph.) (5) Multiply each of the results in Step 4 by the maintenance factor MF. Use MF similar to those for the lumen method. However, the point-by-point method does not take the reflection of the walls and ceilings into consideration.

For line light sources, such as a continuous row of fluorescent luminaires, ft-c values near the lamp vary nearly inversely with the distance from it. As the distance increases, the ft-c variation approaches the inverse square, Fig. 24. With parallel beams, such as in spotlights and concentrators, the inverse-square law applies as the distance from the light becomes great.

OUTDOOR-LIGHTING SELECTION AND SIZING

A parking lot in an industrial plant is 400 ft long and 200 ft wide. Choose the type and location of the lighting for this lot. Lighting must be provided from sundown to sunrise.

Calculation Procedure:

1. *Choose the type of lamp to use*

Three types of lamps are popular for outdoor lighting—mercury-vapor, fluorescent, and filament. The filament-type lamp is usually more economical for installations used less than 1,000 lighting hr per year—i.e., under 3 to 4 hr per night. Since this installation requires lighting from sundown to sunrise, filament-type lamps would not be economical because they would be used more than 4 hr per night.

Thus, the lighting should be provided by either a mercury-vapor or fluorescent lamp. Select mercury-vapor floodlight-type lamps because they provide economical lighting with little maintenance.

2. Determine the recommended lighting level

Table 32 shows that the recommended lighting level for parking areas is 0.5 to 2.0 ft-c. Use a level of 2.0 ft-c for this area.

TABLE 32 Recommended Lighting Levels*

Application	Usual Recommended Level, ft-c
Building exteriors:	
Terra cotta, light marble, plaster	15; 10; 5
Buff limestone, buff brick, concrete	20; 15; 10
Brownstone, wood shingles, dark finish	50; 35; 20
Poster and bulletin boards:	
Bright surroundings, light surfaces	50
Bright surroundings, dark surfaces	100
Dark surroundings, light surfaces	20
Dark surroundings, dark surfaces	50
Industrial roadways:	
Between or adjacent to buildings	1.0
Not bordered by buildings	0.5
Loading platforms, freight docks	20
Industrial parking lots	0.5–2.0
Smoke stacks and water tanks with advertising signs	Same as poster and bulletin boards
Storage yards:	
Active	20
Inactive	1
Television surveillance:	
To indicate movement	5
For clear picture	20

*Adapted from Illuminating Engineering Society and manufacturer's recommendations.

To determine the recommended watts/sq ft, enter Fig. 25 at the illumination level in ft-c on the left and project horizontally to the appropriate lamp curve. Read at the bottom of the chart the watts/sq ft. Thus, with an illumination level of 2.0 ft-c, the recommended watts/sq ft is 0.10.

3. Select the lighting-fixture location

Figure 26 shows five different arrangements of outdoor-lighting fixtures. For large areas with lighting levels of up to 5 ft-c, center poles with multiluminaire mountings are economical. Hence, these will be chosen for this parking lot.

4. Compute the number of lights required

Use the relation:number of floodlights required = (area lighted, sq ft)(recommended ft-c)/(floodlight lumen rating)(MF), where MF = maintenance factor = 0.165 for open floodlights and 0.75 for enclosed floodlights.

Typical mercury-vapor floodlight lumen ratings range from 4,200 lumens for a 500-watt lamp to 18,000 lumens for a 1,500-watt lamp. Assume that 1,000-watt, 9,500-lumen, enclosed, clear, general-service floodlights are used. Then, number of flood-

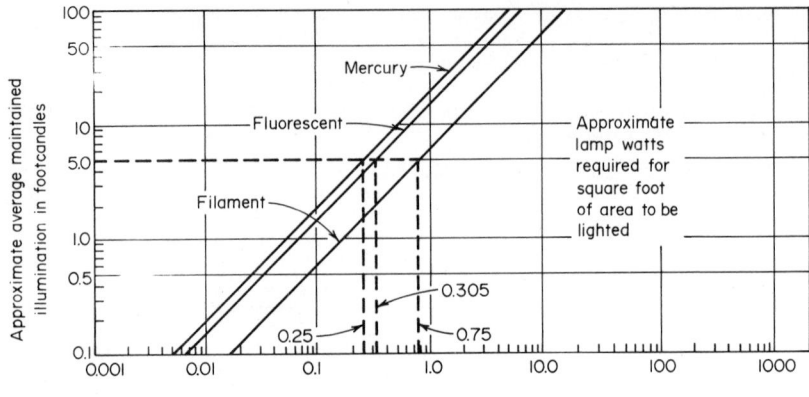

Fig. 25   Typical watts/sq ft for different illumination levels.

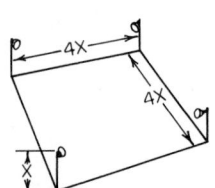

For lighting small areas don't put poles more than four times the mounting height apart. This applies regardless of number floods per pole or ft-c level.

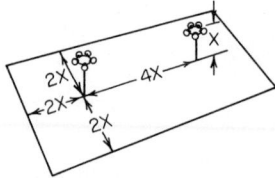

For large areas at up to 5 ft-c, center poles with multiluminaire mountings can save. Put outer poles within twice their height from area perimeter.

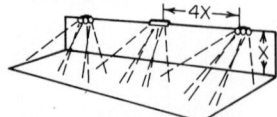

For corridor-type lighting, whether filament or fluorescent sources are used, lamps shouldn't be further apart than four times distance above surface lighted.

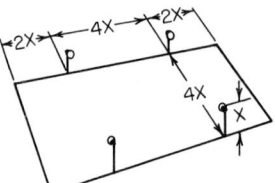

For perimeter poles set in from corners, none should be further from corner than twice mounting nor more than four times mounting height from next pole.

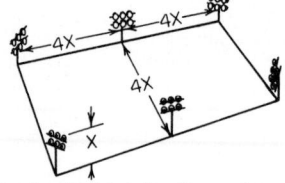

For levels above five ft-c, poles shouldn't be more than four times mounting height apart. This applies both across area lighted and between poles.

Fig. 26   Pole placements recommended for effective outdoor lighting.

lights required $= (400 \times 200)(2.0)/(9{,}500)(0.75) = 22.5$; say 24 lamps to obtain an even number of lamps to be divided between 2 supporting poles.

5. Determine the lamp arrangement

With 24 lamps a two-pole arrangement, Fig. 26, might be suitable. This would lend itself to the mounting of 12 lamps on each pole.

Refer to a lamp manufacturer's engineering data. These data show that a 40° beam floodlight mounted 55 ft above the ground will illuminate an area 104 ft long and 65 ft wide, or an elliptical area of 2,650 sq ft. Hence, 24 lamps will illuminate an area of $24(2{,}650) = 63{,}600$ sq ft. The area of the parking lot is $400 (200) = 80{,}000$ sq ft. Thus, the lamps will not illuminate the entire parking lot.

Raising the lamps to 70 ft above the ground provides an illuminated area of 4,700 sq ft per lamp, or a total area of $24(4{,}700) = 112{,}800$ sq ft. Also, the lamps will be within twice their height of $2(70) = 140$ ft from the area perimeter because they are mounted 100 ft from the perimeter, Fig. 27. Since the area covered by the floodlights exceeds the ground area, the beams will overlap. This is a desirable condition because it provides more uniform illumination.

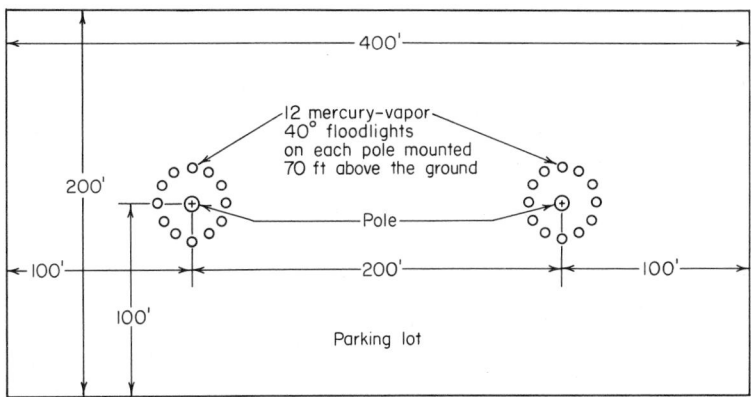

Fig. 27   Parking-lot lighting arrangement.

Related Calculations: Use the method given here for any outdoor-lighting application—building exteriors, catwalks, drill fields, gasoline service stations, piers, prison yards, quarries, railroad yards, shipyards, storage yards, baseball fields, boxing rings, skating rinks, etc. Consult the manufacturer's engineering data for the specific characteristics of the lights that will be used.

Typical output of outdoor lamp in lumens per watt are: filament, 20; mercury-vapor, 50; fluorescent, 70. Useful lamp life is usually: filament, 6 months; fluorescent, 36 months; mercury-vapor, 48 months. These lives are based on all-night every-night operation of the lamps. Give careful consideration to the lamp life. Longer lamp life means fewer maintenance tasks at elevated positions.

When choosing lamps, use the largest wattage fixtures and the fewest locations that will deliver acceptable uniform lighting. The higher the wattage per fixture, the lower the total cost on a unit-cost-of-light basis. Generally, the spacing between poles should not exceed four times the mounting height of the fixtures. Also, use the widest beam spread fixture available that is consistent with good utilization of light.†

To determine the spacing of highway-lighting poles, use the relation:spacing between luminaires, ft = (lumens required)(coefficient of utilization)(maintenance factor)/[(ft-c maintained by the lamp)(width of roadway, curb to curb, ft)].

---

†John C. Boyter and Robert E. Faucett, General Electric Company, and *Factory* magazine.

FEEDER SIZING FOR A COMBINATION ELECTRICAL LOAD

Using the *National Electrical Code*, size the feeders for a load consisting of four three-phase 220-v squirrel-cage induction motors designed for a 40-C temperature rise, marked *Code* letter H, started across the line, and made up of one 10-hp motor, one 7.5-hp motor, and two 1.5-hp motors. The feeders will also serve a 20-kw single-phase 115-v lighting load.

Calculation Procedure:

1. *Determine the motor average full-load current*

Using the motor-current table in the *Code*, the average full-load current of the motors is: 10 hp, 27 amp; 7.5 hp, 22 amp; 1.5 hp, 5 amp.

2. *Select the main feeder for the motors*

Size the feeder for a current flow of 125 percent of the average full-load current of each motor. Thus, total current flow = $1.25(27) + 1.25(22) + 1.25(5)(2) = 73.7$ amp. Referring to the *Code* motor-feeder table shows that three No. 6 RHW feeders in a 1-in. conduit are needed.

3. *Size the individual branch circuits for each motor*

Using the 125 percent current flow computed for each motor in Step 2 and *Code* wire data: 10-hp motor, 1.25 (27) = 33.75 amp; choose three No. 8 TW or RHW feeders in $\frac{3}{4}$-in. conduit, as recommended in the *Code*. For the 7.5-hp motor, 1.25(22) = 27.5 amp; use three No. 10 TW or RHW feeders in $\frac{3}{4}$-in. conduit. For each 1.5-hp motor, 1.25 (5) = 6.25 amp; use three No. 14 TW or RHW feeders in $\frac{1}{2}$-in. conduit.

4. *Select the overcurrent protection rating*

Since the same 125 percent factor applies for the overcurrent or running protection, the maximum rated thermal elements for each motor will be: 10 hp, 33.75 amp; 7.5 hp; 27.5 amp; 1.5 hp, 6.25 amp.

5. *Select the motor fuses*

The *Code* specifies a maximum of 300 percent for the fuse rating for these *Code* letter H motors, based on the average full-load current of each motor. Or, for the 10-hp motor, 3(27 amp) = 81 amp—use a 100-amp block with three 90-amp fuses. For the 7.5-hp motor, 3(22 amp) = 66 amp—use a 100-amp block with three 70-amp fuses. For each 1.5-hp motor, 3(5 amp) = 15 amp—use a 30-amp block with three 15-amp fuses.

6. *Select the main motor-feeder protection*

The maximum rating or setting for the motor-feeder protection device must not be greater than the largest rating or setting of a branch-circuit protective device for one of the motors of the group plus the sum of the full-load current of the other motors. Using the largest motor rating plus the others, $3(27) + 22 + 2(5) = 113$ amp. Use a 200-amp switch with three 125-amp fuses.

7. *Compute the lighting-load current*

Assume the lighting circuit is 115/230-v single-phase. Then, full-load current = load, watts/maximum voltage = 20(1,000)/230 = 87 amp. Use a 100-amp switch with two 90-amp fuses. If the 87-amp load is continuous, a fused switch must be selected such that the load is not over 80 percent of the fuse rating. This means that the minimum fuse size under these conditions would be (87 amp)(1.25) = 108.75, or a 110-amp fuse (the next *largest* standard rating) for each phase, in a 200-amp switch. The factor 1.25 is obtained from an assumed maximum current flow 125 percent of the full-load current. To handle the 87-amp flow, use three No. 3 RHW in $1\frac{1}{4}$-in. conduit, or three No. 2 RHW for the 108.75-amp load.

8. *Select the type of service to use*

If these two loads will be fed from separate services, then the above calculations are complete. But if a combination service is to be used, a four-wire, 240-v delta service would be suitable. The lighting load would then be fed from two of the three phases.

9. *Choose the service feeders and switches*

The motor feeder must handle a starting demand of 113 amp, as computed in Step 6. The lighting two-phase full-load current is 87 amp, Step 7. Summing these two current flows, $113 + 87 = 200$ amp.

The main switch would be a 200-amp size with two 200-amp fuses in two phases and one 125-amp fuse in the third phase. Two of the main service lines would be 2/0 RHW; one would be No. 6 RHW, and the neutral would be No. 3 RHW. These are the minimum sizes based on the *Code* rules for safe application. Local codes might require larger feeders. Figure 28 shows typical *Code* sizing of feeders and overcurrent devices for motors.

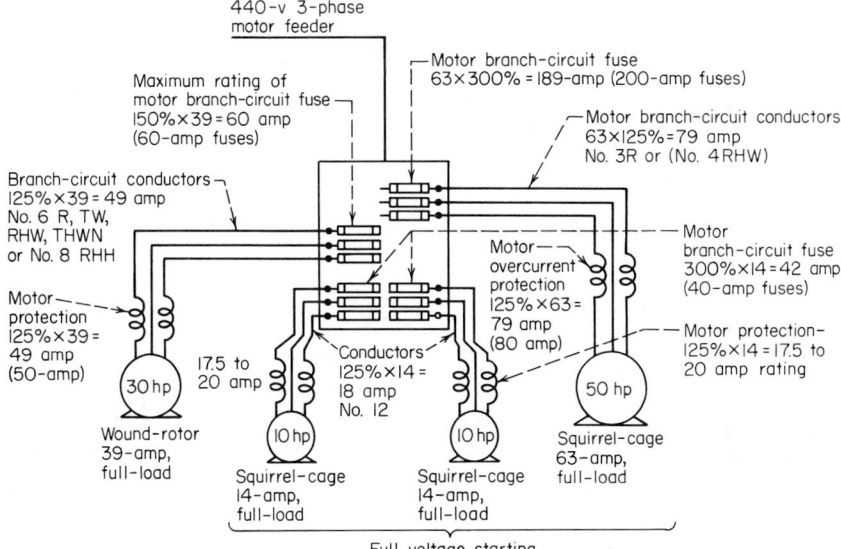

Notes: 1. Full-load current for each motor is taken from NE Code Table 430-150.
2. Running overload protection is sized on basis that nameplate values of motor full-load currents are same as values from NE Code Table 430-150. If nameplate and table values are not the same, overload protection is sized according to nameplate.

Fig. 28   National Electrical Code sizing of feeders and overcurrent devices for motors.

Related Calculations: Typical recommended limits for the voltage in various types of circuits are shown in Fig. 29. Application of load demand and diversity factors is shown in Fig. 30; and typical demand factors are listed in Table 33. Fuses of various classes are compared in Table 34. Grounding methods for interior ac wiring systems and equipment enclosures are shown in Fig. 31. The procedure given above is valid for single or combination electrical loads in a variety of installations. For specific numerical values of conductor and conduit sizes, refer to the *National Electrical Code* and the local governing code, if any.

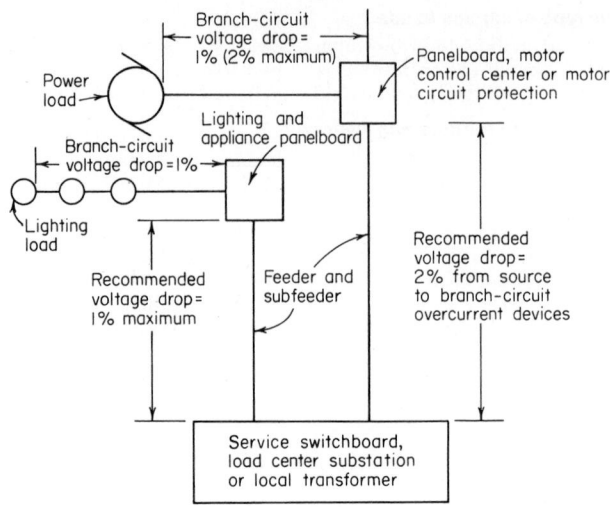

Fig. 29   Recommended limits of voltage drop in various circuits.

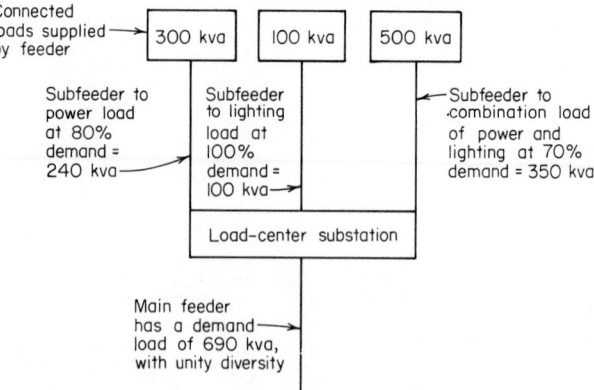

Fig. 30   Typical application of demand and diversity factors.

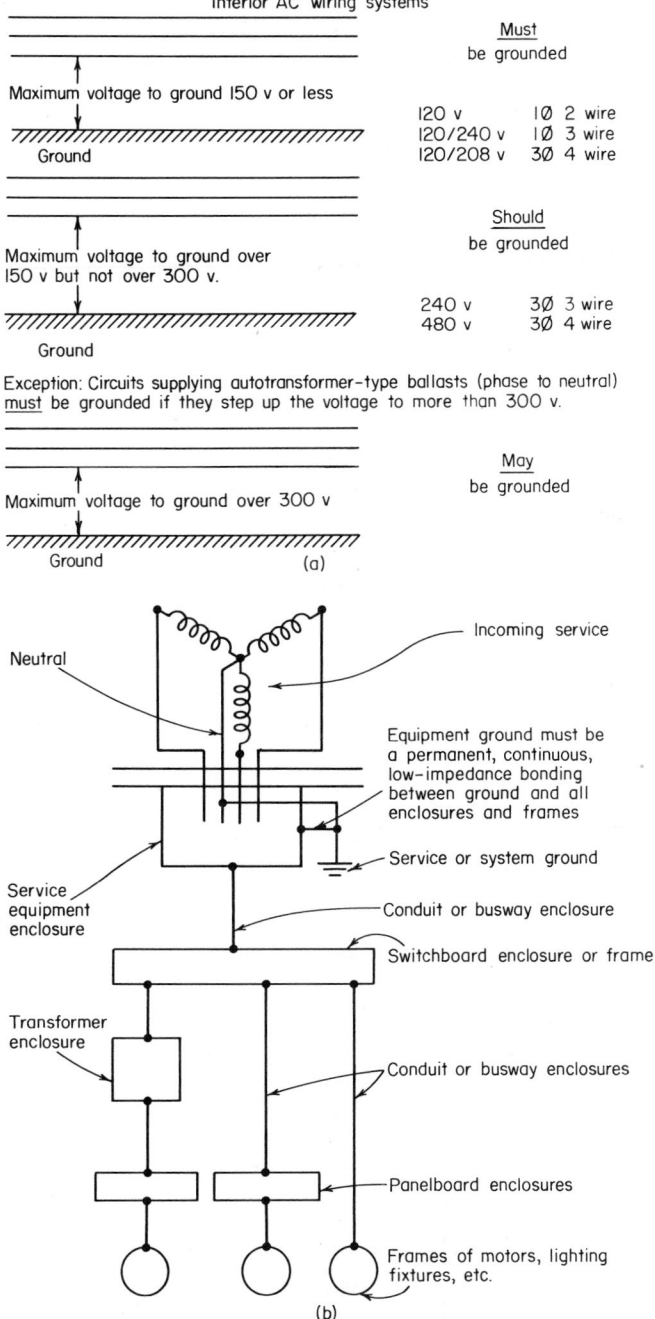

Interior AC wiring systems

<u>Must</u> be grounded

Maximum voltage to ground 150 v or less

Ground

120 v	1∅	2 wire
120/240 v	1∅	3 wire
120/208 v	3∅	4 wire

<u>Should</u> be grounded

Maximum voltage to ground over 150 v but not over 300 v.

Ground

| 240 v | 3∅ | 3 wire |
| 480 v | 3∅ | 4 wire |

Exception: Circuits supplying autotransformer-type ballasts (phase to neutral) <u>must</u> be grounded if they step up the voltage to more than 300 v.

<u>May</u> be grounded

Maximum voltage to ground over 300 v

Ground

(a)

Incoming service

Neutral

Equipment ground must be a permanent, continuous, low-impedance bonding between ground and all enclosures and frames

Service or system ground

Service equipment enclosure

Conduit or busway enclosure

Switchboard enclosure or frame

Transformer enclosure

Conduit or busway enclosures

Panelboard enclosures

Frames of motors, lighting fixtures, etc.

(b)

Fig. 31 (a) **When to ground a system conductor;** (b) **grounding methods for equipment enclosures.**

TABLE 33 Common Demand Factors for Sizing Service and Main Feeders*

Power Load Devices	Range of Common Demand Factors, %
Motors for pumps, compressors, elevators, machine tools, blowers, etc.	20–60
Motors for semicontinuous operations in various mills and process plants	50–80
Motors for continuous operations—as in textile mills	70–100
Arc furnaces	80–100
Induction furnaces	80–100
Arc welders	30–60
Resistance welders	10–40
Resistance heaters, ovens, and furnaces	80–100

*National Electrical Code.

TABLE 34 Comparison of Fuse Classes† (General-purpose cartridge-type)*

UL class	Range, amp	Interrupt capacity, amp	Maximum let-through	Time delay†	Dimensions
H	0–600	10K	None	No	Old *NEC*
J	0–600	100K or 200K	Yes‡	No	Special§
K-1	0–600	10K, 25K, 50K, 100K or 200K	Yes‡	No	Old *NEC*
K-5	0–600	10K, 25K, 50K, 100K or 200K	Yes‡ Yes	Yes	Old *NEC*
K-9	0–600	10K, 25K, 50K, or 100K	Yes‡	Yes	Old *NEC*
L	601–6,000	100K or 200K	Yes‡	No	Present NEMA sizes

*National Electrical Code.

†NEMA standards call for a minimum of 10-sec delay at 500 percent of fuse rating. No UL standards adopted.

‡UL standards state maximum peak let-through in amps and energy let-through (I^2T) for each size and type of fuse. In 600A sizes, lowest let-through is Class J, increasing slightly through Classes K-1, K-5 and K-9.

§Smaller than the noninterchangeable with *NEC*-size fuses.

SIZING RESIDENTIAL-SERVICE DEMAND LOAD

What size service is required for a 1,500 sq ft house with all-electric utilization having the loads listed in Step 1 and sized according to the *National Electrical Code*? Use the *optional method* of calculating the service demand load. The service voltage is 120/230 v, three-wire.

Calculation Procedure:

1. Compute the maximum possible demand

This house has the following loads:

1,500 watts for each of two (minimum of two required) kitchen appliance circuits......	2(1,500)	= 3,000 watts
1,500 sq ft of floor area at 3 watts/sq ft for general lighting and receptacles	3(1,500)	= 4,500
14 kw of electric space heating from more than four separately controlled units	14,000	
12-kw electric range......................	12,000	
3-kw water heater	3,000	
5-kw clothes dryer........................	5,000	
3-kw load of unit air conditioners (Because this load is less than the space-heating load and will *not* be operated simultaneously with it, no load need be added.)		

Hence, maximum possible demand .... 41,500 watts

2. Compute the probable demand load

Table 35, from the *National Electrical Code*, shows that under the optional method, the first 10 kw of all other loads, as defined in Table 35, should be taken at 100 percent, or 10,000 watts. The remainder of the load is taken at 40 percent, or $0.40(41,500 - 10,000) = 12,600$ watts. Hence, the total probable demand = $10,000 + 12,600 = 22,600$ watts.

TABLE 35   Optional Calculation for One-Family Residence*

Load, kw or kva	Percent of load
Air conditioning and cooling including heat pump compressors	100
Central electrical space heating...	100
Less than four separately controlled electrical space-heating units..............	100
First 10 kw of all other load...	100
Remainder of other load...	40

All other load shall include 1,500 watts for each 20-amp appliance-outlet circuit; lighting and portable appliances at 3 watts/sq ft; all fixed appliances (including electric space heating when there are four or more separately controlled units, ranges, wall-mounted ovens, and counter-mounted cooking units) at nameplate rated load (kva for motors and other low power-factor loads).

3. Compute the size of the service

Use the relation, size of service, amp = demand load, watts/maximum service voltage, or $22,600/230 = 98$ amp. Use the next larger standard service, or 100 amp.

Under certain load conditions, this calculation may indicate a required service capacity substantially less than 100 amp. In such cases, however, 100 amp is the minimum service that can be used, according to the *Code*. When using the alternative calculation method of *Code* Sec. 220-4, note that a calculated demand load of 10 kw or more requires that the minimum size service be 100-amp, three-wire. Further, the optional method may be used instead of the *standard method* under the following conditions: (*a*) that it is for a one-family residence only, (*b*) that it is served by a 115/230-v three-wire 100-amp or larger service, and (*c*) that the total load is supplied by one set of service conductors.

Related Calculations: To use the *standard method* of computing the service entrance for a residence, apply the following steps. (1) Multiply the floor area of the house in sq ft by 4 watts/sq ft to obtain the general-lighting and general-purpose-outlet electrical load. (2) Add the total circuit capacity, in watts, allowed for the appliance load in the kitchen, dining room, pantry, laundry, and utility area that will be served by 120-v appliance circuits. Find the total circuit capacity by multiplying the number of such circuits laid out in branch-circuit design by 2,000 watts per circuit. Or assume a load of 4,000 watts (i.e., two appliance circuits) when the exact number of such circuits is not known. Table 36 lists the typical loads and characteristics of modern circuits for appliances. (3) Take 3,000 watts of the sum of Steps 1 and 2 at 100 percent demand. (4) Add to the load in Step 3, 35 percent demand of the remaining load above 3,000 watts computed in the first three Steps. (5) Take the sum of the values computed in

TABLE 36   Modern Circuits for Appliance Loads

Load devices	Typical Load, watts	Volts	Wires	Circuit breaker or fuse	Number of outlets	Notes
For laundry areas						
Ironer	1,650	120	Two No. 12	20 amp	One	Grounding type receptacle required
Washing machine	1,200	120	Two No. 12	20 amp	One	Grounding type receptacle required
Dryer	5,000	120/240	Three No. 10	30 amp	One	Appliance may be direct-connected— must be grounded
For other loads						
Hand iron	1,000	120	Two No. 12	20 amp	Two or more	
Water heater	3,000	...	...	...	...	Consult utility code for load requirements
Workshop	1,500	120	Two No. 12	20 amp	Two or more	Separate circuit recommended
Portable heater	1,300	120	Two No. 12	20 amp	One	Should not be connected to circuit serving other heavy duty loads
Television	300	120	Two No. 12	20 amp	Two or more	Should not be connected to circuit serving appliances

TABLE 36   Modern Circuits for Appliance Loads (Continued)

Load devices	Typical Load, watts	Volts	Wires	Circuit breaker or fuse	Number of outlets	Notes
Range	12,000	120/240	Three No. 6	50–60 amp	One	Use of more than one outlet is permitted, but not recommended
Oven (built-in)	4,500	120/240	Three No. 10	30 amp	One	Appliance may be direct-connected
Range top	6,000	120/240	Three No. 10	30 amp	One	Appliance may be direct-connected
Range top	3,300	120/240	Three No. 12	20 amp	One or more	
Dishwasher	1,200	120	Two No. 12	20 amp	One	These appliances may be direct-connected on a single circuit; grounded receptacles required otherwise
Waste disposer	300	120	Two No. 12	20 amp	One	
Broiler	1,500	120	Two No. 12			Heavy duty appliances regularly used at one location should have a separate circit, only one such unit should be attached to a single circuit at a time
Fryer	1,300	120	At least two kitchen-appliance circuits	20 amp	Two or more	
Coffeemaker	1,000	120				
Refrigerator	300	120	Two No. 12	20 amp	Two	Separate circuit serving only refrigerator and freezer is recommended
Freezer	350	120	Two No. 12	20 amp	Two	

Individual circuits for unit air conditioners

Size of air conditioner	Average wattage	Circuits required	Size of circuit	Number of outlets	Remarks
$\frac{3}{4}$ hp	1,200	Separate circuit	two No. 12, 120-v	One	Use of three-wire, 120/240-v circuits to unit conditioners offers circuit flexibility for 120 or 240 v
$1\frac{1}{2}$ hp	2,400	Separate circuit	three No. 12, 120/240-v	One	

Steps 4 and 5. This is the capacity that must be provided in the service entrance conductors to supply the general-lighting and general-purpose receptacle loads. (6) Add 8,000 watts for an electric range if this is rated 12 kw or less. Consult the *Code* if the electric cooking appliances consist of a built-in oven and range top. (7) Add the rated wattage of all fixed appliances to be served by individual circuits not previously included in the calculation. If both electric heating and air conditioning will be used in the house, include only the wattage of the larger unit because the two units will not be used simultaneously. (8) Take the sum of Steps 5, 6, and 7. (9) Divide the sum found in Step 8 by 240 v (for a 120-240-v three-wire single-phase service) to obtain the required ampere rating of the service conductors.

This procedure can also be used to compute the general-lighting, general-purpose receptacle and appliance load for apartments in multiple-dwelling buildings.

ELECTRIC-COMFORT HEATING-LOAD DETERMINATION

The residence in Fig. 32 is to be heated electrically. What is the heating load for an inside design temperature of 70 F, outside design temperature of 0 F, a ground-water temperature of 50 F, and a ceiling height of 8 ft in the structure if (1) the single-floor residence is built over a ventilated crawl space, (2) the same structure is built over a heated basement, and (3) the same structure is mounted on a concrete slab? Use the data given in Tables 37 to 48.

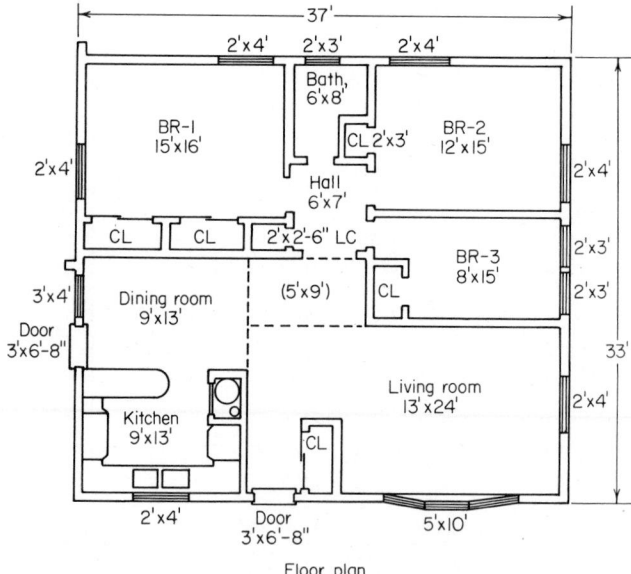

Fig. 32   Floor plan of structure whose heat loss is to be determined.

Calculation Procedure:

1.   *Compute the structure heat loss, watts*

For walls, ceilings, doors, windows, and floors over an unheated crawl space or basement, use the relation H = heat loss, watts = $WA\Delta T$, where W = heat-loss factor, watts per sq ft per degree temperature difference from Tables 41 through 46; A = surface area—wall, floor, ceiling, etc.—sq ft; ΔT = temperature difference between the outside and inside air. F.

Where the structure has a concrete floor laid at or near the grade level, use the relation $H = WL$, where W = heat-loss factor, watts per ft of exposed edge, from Table 45; L = total length of slab edge exposed to the outdoors at the foundation, ft.

When the structure has a concrete basement floor below the grade level, use the relation $H = 0.0293A_f\Delta T$, where A_f = floor area, sq ft.

To determine the infiltration loss I watts, use the relation $I = WV$, where W = heat-loss factor, watts/(cu ft)(F), from Table 44 — usual practice is to assume $\frac{3}{4}$ air change per hr; V = volume of space to be heated, cu ft.

Assemble and tabulate the heat-loss factors as shown in Table 37. Use the data presented in Tables 40 to 45.

TABLE 37 Heat-Loss Factors

Building section	Appli-cable table	Factor watts per sq ft per deg TD	Heat loss at 70 deg TD, watts per sq ft
Walls: Wood siding and sheathing: gypsum board inside; R-11 insulation; 8-ft ceiling height..................................	2	0.025	1.75
Ceiling: Gypsum board; ventilated attic above; R-19 insulation	1	0.018	1.26
Floor: Hardwood floor on subfloor; ventilated crawl space below; R-13 insulation..............	3	0.019	1.33
Windows: Tightly fitted storm sash................	5	0.132	9.24
Doors: 1½ in. solid wood with storm door...........	6	0.094	6.58
Infiltration: (¾ air change per hr)	9	0.00396*	0.28†

*Watts per cu ft per deg TD
†Watts/cu ft.

Next, compute the wall areas of the structure, as shown in Table 38. Then compute the floor and ceiling areas as shown in Table 39.

With these data available, the heat loss for each room can be computed using the equations given above. Table 40 shows the heat-loss computation for each room in this structure when the building is mounted above a ventilated crawl space. The lower portion of Table 40 shows the heat-loss computations for the same structure over a heated basement and mounted on a concrete floor slab.

TABLE 38 Wall Areas*
(Sq ft)

Room	Calculations	Glass wall area	Window area	Door area	Net wall area
BR-1	$(15 + 16) \times 8$	248	16	0	232
BR-2	$(12 + 15) \times 8$	216	16	0	200
BR-3	8×8	64	12	0	52
LR	$(13 + 24) \times 8$	296	78	20	198
DR	9×8	72	12	20	40
K	$(9 + 13) \times 8$	176	8	0	168
Bath	6×8	48	6	0	42

*Length of outside wall is multiplied by ceiling height to get gross wall area; net wall area is obtained by subtracting the window and door areas from the gross wall area.

TABLE 39   Floor and Ceiling Areas: Volume*

Room	Calculations	Area sq ft	Volume, cu ft
BR-1	240 − 5 (linen closet) + 17(hall and linen closet)...............................	252	2,016
BR-2	180 + 6 (closet) + 15 (hall and linen closet).................................	201	1,608
BR-3	120 + 15 (hall and linen closet).................	135	1,080
LR	312 + 25 (dotted area)..........................	337	2,696
DR	117 + 20 (dotted area)..........................	137	1,096
K		117	936
Bath	48 − 6 (closet)..............................	42	336
	Total......................................	1,221	9,768

*Hall and linen closet areas were divided among bedrooms. Dotted area (5 × 9 ft) was divided between living room and dining room. Room volume is obtained by multiplying the floor area by the ceiling height.

2. *Summarize the heat losses in the structure*

Table 40 shows the summary heat loss totaled horizontally and vertically for the listed areas and rooms. This summary shows that the heating capacity required for this structure is 9,161 kw when mounted over a ventilated crawl space.

When the structure is mounted over a heated basement the floor heat loss is deleted and the basement heat loss is substituted in its place. Thus, the heating capacity needed for a basement-mounted structure is 10,928 kw. With this structure mounted on a concrete slab, the total heat loss is 8,279 kw, using the data in Table 40, after deducting the floor loss computed for the crawl-space structure. The heating capacity must at least equal this heat loss.

TABLE 40   Room-by-Room Calculations

Room	Area × heat-loss factor = heat loss Sq ft × watts/sq ft = watts
Living room:	
Walls....................	$198 \times 1.75 = 347$
Ceiling..................	$337 \times 1.26 = 425$
Floor....................	$337 \times 1.33 = 448$
Windows	$78 \times 9.24 = 721$
Door	$20 \times 6.58 = 132$
Infiltration..............	$2,696^* \times 0.28\dagger = 755$
Total heat loss.............	2,828
Suggested heater rating......	3.0 kw
Dining room:	
Walls....................	$40 \times 1.75 = 70$
Ceiling..................	$137 \times 1.26 = 173$
Floor....................	$137 \times 1.33 = 182$
Windows	$12 \times 9.24 = 111$
Door	$20 \times 6.58 = 132$
Infiltration..............	$1,096^* \times 0.28\dagger = 307$
Total heat loss.............	975
Suggested heater rating......	1.0 kw

TABLE 40   Room-by-Room Calculations (Continued)

Room	Area × heat-loss factor = heat loss Sq ft × watts/sq ft = watts
Kitchen:	
Walls....................	168 × 1.75 = 294
Ceiling..................	117 × 1.26 = 147
Floor....................	117 × 1.33 = 156
Windows	8 × 9.24 = 74
Infiltration..............	936* × 0.28† = 262
Total heat loss..............	933
Suggested heater rating......	1.0 kw
Bath	
Walls....................	42 × 1.75 = 74
Ceiling..................	42 × 1.26 = 53
Floor....................	42 × 1.33 = 56
Windows	6 × 9.24 = 55
Infiltration..............	336* × 0.28† = 94
Total heat loss..............	332
Suggested heater rating......	A 500-watt heater will satisfy heating requirements; but, since quick pickup is often desirable, a 1-kw heater is recommended.

*Cu ft.
†Watts/cu ft per deg TD.

Room	Area × heat-loss factor = heat loss Sq ft × watts/sq ft = watts
Bedroom 1:	
Walls....................	232 × 1.75 = 406
Ceiling..................	252 × 1.26 = 318
Floor....................	252 × 1.33 = 335
Windows	16 × 9.24 = 148
Infiltration..............	2,016* × 0.28† = 564
Total heat loss..............	1,771
Suggested heater rating.....	1.75 kw
Bedroom 2:	
Walls....................	200 × 1.75 = 350
Ceiling..................	201 × 1.26 = 253
Floor....................	201 × 1.33 = 267
Windows	16 × 9.24 = 148
Infiltration..............	1,608* × 0.28† = 450
Total heat loss..............	1,468
Suggested heater rating......	1.5 kw
Bedroom 3:	
Walls....................	52 × 1.75 = 91
Ceiling..................	135 × 1.26 = 170
Floor....................	135 × 1.33 = 180
Windows	12 × 9.24 = 111
Infiltration..............	1,080* × 0.28† = 302
Total heat loss..............	854
Suggested heater rating.....	1.0 kw

TABLE 40   Room-by-Room Calculations (Continued)

Heat-loss summary, kw

Room	Walls	Ceiling	Floor	Wind	Doors	Infiltration	Total
Living room.......	347	425	448	721	132	755	2,828
Dining room......	70	173	182	111	132	307	975
Kitchen..........	294	147	156	74	...	262	933
Bath.............	74	53	56	55	...	94	332
Bedroom 1........	406	318	335	148	...	564	1,771
Bedroom 2........	350	253	267	148	...	450	1,468
Bedroom 3........	91	170	180	111	...	302	854
Total	1,632	1,539	1,624	1,368	264	2,734	9,161

Basement heat loss with fully heated basement instead of crawl space

Assume ground water temperature = 50 F
Basement walls are 1 ft above grade, 6 ft below
Walls are furred in and insulated to R5
There are three 2-sq-ft, wood-frame, single-glass windows in above-grade walls.
Infiltration:
   Volume = $1{,}221 \times 7 = 8{,}547$ cu ft
   At $\frac{1}{4}$ air change, heat loss = $8{,}547 \times 0.00132 \times 70 = 790$ watts
Windows:
   Area = $2 \times 3 = 6$ sq ft
   Temperature difference = $70 - 0 = 70°$
   Heat loss = $0.331 \times 6 \times 70 = 139$ watts

Walls (above grade):
   Length of outside wall = $2(33 + 37) = 140$ ft
   Area = $1 \times 40 -$ windows = $140 - 6 = 134$ sq ft
   Temperature difference = $70 - 0 = 70°$
   Heat loss = $0.037 \times 134 \times 70 = 347$ watts
Walls (below grade):
   Area = $6 \times 140 = 840$ sq ft
   Temperature difference = $70 - \frac{1}{2}(0 + 50) = 45°$
   Heat loss = $0.037 \times 840 \times 45 = 1{,}399$ watts
Floor:
   Area = $37 \times 33 = 1221$ sq ft
   Temperature difference = $70 - 50 = 20°$
   Heat loss = $0.0293 \times 20 \times 1{,}221 = 716$ watts
Total basement heat loss = 3391 watts

Floor loss with house on slab at grade level instead of over crawl space

Assumed edge of slab insulated to R6

Room	ft	× watts/ft	= watts
Living room....... $(13 + 24) = 37$		5.3	196
Dining room ...... $(9 + 0) = 9$		5.3	48
Kitchen.......... $(9 + 13) = 22$		5.3	117
Bath............. $(6 + 0) = 6$		5.3	32

TABLE 40   Room-by-Room Calculations (Continued)

Length of exposed slab × heat-loss factor = heat loss

Room	ft	× watts/ft	=	watts
Bedroom 1 . . . .(16 + 15) = 31		5.3		164
Bedroom 2 . . . . (15 + 12) = 27		5.3		143
Bedroom 3 . . . . (8 + 0) = 8		5.3		42
Total slab loss 742				

Related Calculations: Use this general procedure for any type of structure – industrial, commercial, civic, etc. Where the data given are insufficient, refer to the ASHRAE *Guide and Data Book.*

In newer buildings the heat given off by lighting fixtures is often used to reduce the load on the structure's heating system. Heated air is drawn from lighting troffers and directed to the return-air system. Figure 33 shows the heat given off by a typical modern lighting system. When the lighting level exceeds 100 ft-c, the heat from the lights is often sufficient to heat the entire building. Currently, integrated systems of lighting, heating, and air conditioning are being planned so that the heating and cooling requirements of a structure are reduced.

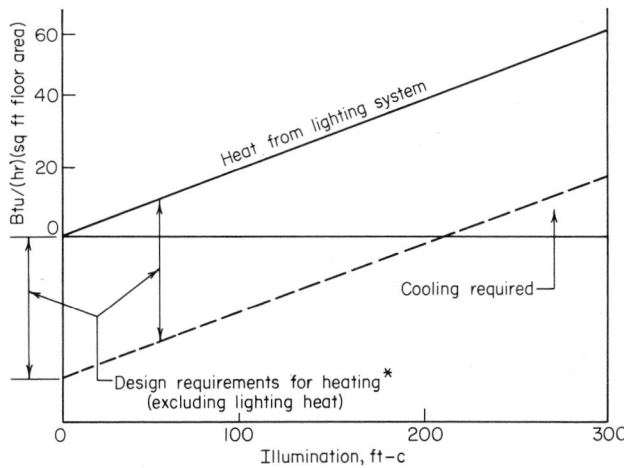

*Prototype example; varies from building to building based on existing climatic and other variable conditions.

Fig. 33   Heat recoverable from lighting systems.

Definitions: In electric heating the following definitions are in common use. R value: The amount of thermal resistance contributed by insulation when the insulation is installed in a ceiling, wall, or floor. Outdoor design temperature: The lowest outdoor temperature at which the heating system is expected to maintain the indoor design temperature. This outdoor design temperature is usually considerably higher than the lowest temperature on record for the area; the outdoor design temperature can be obtained from the local electric utility. Indoor design temperature: The temperature which is to be maintained in the heated space – usually 70 F. Design temperature difference (TD): The difference between the outdoor and indoor design temperatures.

The data and tables presented in this Calculation Procedure are based on information presented in *Electrical Construction and Maintenance* magazine.

TABLE 41   Heat Loss through Residential Ceilings

(Assuming space above insulation is ventilated and wood framing covers 15 percent of ceiling area)

Installed resistance of insulation R	Heat-loss factor, watts per sq ft per deg TD	
	Ceilings using plaster or gypsum board products	Ceilings using acoustical tile or insulating board products
15	0.022	0.020
16	0.021	0.019
17	0.020	0.018
18	0.019	0.017
19–20	0.018	0.016
21–22	0.016	0.015
23–24	0.015	0.014
25	0.014	0.013
30	0.011	0.010
35	0.009	0.009
40	0.008	0.008
45	0.007	0.007
50	0.006	0.006

TABLE 42   Heat Loss through Frame Walls

(Assuming wood framing covers 20 percent of wall area)

Installed resistance of insulation R	Heat-loss factor, watts per sq ft per deg TD		
	Masonry walls of low-density concrete 80 lb/cu ft or less	Frame walls using insulating board and insulating lath products	Frame walls using wood or metal lath and wood sheathing or masonry walls of stone, concrete block, or high-density concrete
5	0.021	0.030	0.037
6	0.020	0.028	0.034
7	0.019	0.027	0.031
8	0.018	0.025	0.029
9	0.017	0.024	0.027
10	0.016	0.022	0.026
11–12	0.015	0.021	0.025
13–14	0.014	0.019	0.022
15*	0.013	0.017	0.019
20–21*	0.012	0.014	0.016
22–23*	0.011	0.013	0.015

*Using 2 × 6-in. studs.

TABLE 43 Heat Loss through Wood Floors

(Assuming wood framing covers 15 percent of floor area. Floor consists of wood subfloor on joists and hardwood floor or tile or linoleum on suitable base.)

Installed resistance of insulation R	Heat-loss factor, watts per sq ft per deg TD
5	0.035
6	0.031
7	0.028
8	0.026
9	0.024
10	0.022
11	0.021
12	0.020
13	0.019
14	0.018
15	0.017
20	0.014
25	0.012
30	0.010
35	0.009
40	0.008

TABLE 44 Heat Loss Due to Infiltration

(Watts per cu ft per deg TD)

Number of air changes per hr	Heat-loss factor
$\frac{1}{10}$	0.00053
$\frac{1}{8}$	0.00066
$\frac{1}{4}$	0.00132
$\frac{1}{2}$	0.00264
$\frac{3}{4}$	0.00396
1	0.00527
$1\frac{1}{2}$	0.00791

TABLE 45   Heat Loss through Concrete Slab on Grade*
(Watts per ft of exposed edge)

Outdoor design temperature, F	Unheated slab				Heated slab			
	$R = 6$ to 7	$R = 5.0$	$R = 3.33$	$R = 2.50$	$R = 6$ to 7	$R = 5.0$	$R = 3.33$	$R = 2.50$
− 30 and colder	7.5	10.0	14.9	19.6	10.1	13.5	20.2	27.0
− 25 to − 29	7.0	9.4	14.0	18.8	9.7	12.9	19.3	25.8
− 20 to − 24	6.6	8.8	13.1	17.6	8.9	12.0	17.8	24.0
− 15 to − 19	6.3	8.2	12.6	16.7	8.1	11.4	17.3	22.8
− 10 to − 14	5.9	7.9	11.7	15.8	7.6	10.8	16.1	21.7
− 5 to − 9	5.6	7.3	11.1	14.9	7.0	10.2	15.2	20.5
0 to − 4	5.3	7.0	10.5	14.1	7.0	9.4	14.0	18.8
+ 5 to + 1	4.9	6.5	9.7	12.9	6.6	8.8	13.1	17.6
+ 10 to + 6†	4.6	6.2	9.1	12.3	5.6	7.3	11.1	14.6
+ 15 to + 11†	4.6	6.2	9.1	12.3	5.6	7.3	11.1	14.6
+ 20 to + 16‡	4.6	6.2	9.1	12.3	5.6	7.3	11.1	14.6

*If no edge insulation is used, calculate heat loss at 0.237 watt/ft of exposed edge per deg TD.
†Factors assume only 12 in. of edge insulation.
‡Factors assume edge insulation extends down only to bottom of slab.

TABLE 46   Heat Loss through Windows
(Watts per sq ft per deg TD)

Number of glass panes	Description	No. of air spaces	Width of each space, in.	Heat loss factor
1	Single glass	None	....	0.331
2	Window with usual storm sash		$1\frac{1}{2}$	0.220
	Sealed unit	1	$\frac{1}{4}$	0.185
	Sealed unit		$\frac{1}{2}$	0.167
	Very tightly fitted storm sash, no vents		$1\frac{1}{2}$	0.132
3	Sealed unit	5	$\frac{1}{4}$	0.126
	Sealed unit		$\frac{1}{2}$	0.111

TABLE 47   Heat Loss through Solid Wood Doors
(Watts per sq ft per deg TD)

Nominal thickness, in.	Actual thickness, in.	Heat-loss factor	
		Exposed door	With storm door*
1	$\frac{25}{32}$	0.188	0.108
$1\frac{1}{4}$	$1\frac{1}{16}$	0.161	0.100
$1\frac{1}{2}$	$1\frac{5}{16}$	0.144	0.094
$1\frac{3}{4}$	$1\frac{3}{8}$	0.141	0.091
2	$1\frac{5}{8}$	0.126	0.082
$2\frac{1}{2}$	$2\frac{1}{4}$	0.106	0.076
3	$2\frac{5}{8}$	0.091	0.067

*50 percent glass and thin wood panels.

TABLE 48   Heat Loss through Exposed-beam Ceilings with Built-up Roofing
(Watts per sq ft per deg TD)

Preformed roof insulation above deck, in.	Type of roof deck					
	Flat metal	Wood			Preformed slab; wood fiber & cement binder	
		1 in.	2 in.	3 in.	2 in.	3 in.
0	0.264	0.141	0.094	0.067	0.061	0.044
$\frac{1}{2}$	0.117	0.085	0.064	0.050		
1	0.076	0.061	0.050	0.041		
$1\frac{1}{2}$	0.056	0.047	0.041	0.035	Insulation not used	
2	0.047	0.041	0.035	0.029		
$2\frac{1}{2}$	0.038	0.032	0.029	0.026		
3	0.032	0.029	0.026	0.023		

Section **5**

Electronics Engineering

FREDERICK S. BARTON
Director, Hewlett-Packard Limited

JOSEPH MITTLEMAN
Senior Associate Editor, Design Theory, Electronics *Magazine*

ZVI PRIHAR
Scientific Advisor, Operations Research, Inc.

REFERENCES: Motorola Semiconductors—*The Semiconductor Data Book*; Kiver—*Transistors*, McGraw-Hill; Martin-Marietta Corporation—*Reliability Data Book*; Texas Instruments, Incorporated—*Transistor Circuit Design*, McGraw-Hill; Radio Corporation of America—*RCA Transistor Manual*; U.S. Government Printing Office—*Tabulation of Data on Receiving Tubes*; Martin-Marietta Corporation—*Maintainability Engineering*; Gray and Graham—*Radio Transmitters*, McGraw-Hill; Susskind—*The Encyclopedia of Electronics*, Reinhold; Blake—*Antennas*, Wiley; Landee, Davis, and Albrecht—*Electronic Designers' Handbook*, McGraw-Hill; Myers, Wong, and Gordy—*Reliability Engineering for Electronic Systems*, Wiley; Hunter—*Handbook of Semiconductor Electronics*, McGraw-Hill; Radio Corporation of America—*RCA Linear Integrated Circuits*; Shea—*Transistor Applications*, Wiley; Kraus—*Cooling Electronic Equipment*, Prentice-Hall; Terman—*Electronic and Radio Engineering*, McGraw-Hill; General Electric Company—*Transistor Manual*; Ryder—*Engineering Electronics*, McGraw-Hill; Grossner—*Transformers for Electronic Circuits*, McGraw-Hill; Mason and Zimmerman—*Electronic Circuits, Signals and Systems*, Wiley; Lee—*Electronic Transformers and Circuits*, Wiley; Angelo—*Electronic Circuits*, McGraw-Hill; Hayt and Kemmerly—*De France—Electron Tubes and Semiconductors*, Prentice-Hall; Marting—*Electronic Circuits*, Prentice-Hall; Smullin and Haus—*Noise in Electron Devices*, Wiley; Chang—*Principles of Quantum Electronics*, Addison-Wesley; Thornton—*Handbook of Basic Transistor Circuits*, Wiley; Hetterscheid—*Transistor Bandpass Amplifiers*, Philips Technical Library; Mandl—*Directory of Electronic Circuits*, Prentice-Hall; Slemon—*Magnetoelectronic Devices*, Wiley; Hetterscheid—*Designing Transistor IF Amplifiers*, Philips Technical Library; O'Donnell—*Applied Microelectronics*, American Aviation Publications; Head—*Mathematical Techniques in Electronics and Engineering Analysis*, Van Nostrand; Doetsch—*Guide to Application of Laplace Transforms*, Van Nostrand; Massey—*Threshold Decoding*, M.I.T. Press; Haviland—*Engineering Reliability and Long Life Design*, Van Nostrand; Chang—*Parametric and Tunnel Diodes*, Prentice-Hall.

SOLID-STATE-DEVICE EVALUATION

Select solid-state devices suitable for converting ac to dc (i.e., rectification) and for power switching service. The voltage drop during rectification must not exceed 10 percent with a 12-v power supply and a 0.5-amp current rating. For power switching, a control voltage of 40 v is available.

Calculation Procedure:

1. Compute the actual allowable voltage drop

The allowable voltage drop = (supply voltage)(allowable drop, percent) = (12)(0.10) = 1.2 v.

2. Choose the type of rectifier to use

Junction (i.e., solid-state) rectifier diodes are superior to vacuum rectifier diodes in many ways and are rapidly supplanting them. Hence, only the solid-state diodes will be considered here.

Three types of solid-state rectifier diodes used today are: selenium, silicon, and germanium. The voltage drop when conducting is approximately 1.5 v for selenium, 0.8 v for silicon, and 0.5 v for germanium. Of the three, the silicon diode is the most commonly used. Hence, a silicon diode will be the first choice.

3. Analyze the diode voltage drop

Step 2 shows that the usual voltage drop for a silicon rectifier diode is 0.8 v. The allowable voltage drop, Step 1, is 1.2 v for this application. Hence, a silicon rectifier diode will be acceptable from a voltage-drop standpoint.

4. Check the rectifier current capacity

Examine several specification sheets for silicon rectifier diodes. These specifications will show that a 0.5-amp rating is available from many manufacturers. Also, the ratio of reverse to forward resistance is 100 : 1 or more. With this high a ratio, the reverse leakage current is negligible.

To provide higher peak-inverse voltage ratings, several solid-state diodes can be stacked in series. When installed in matched pairs, full-wave rectification can be obtained. In the usual solid-state rectifier, the current rating depends on the temperature rise caused by the voltage drop and the ability of the diode to drain away the resulting heat. A typical upper operating temperature for a silicon diode is 135 C.

5. Select the diode case

Usual cases for solid-state diodes are glass, top-hat, and stud, Fig. 1. Select the case giving the best shock resistance for the service intended. As a general guide, the glass case is the least shock resistant and the stud case has the greatest shock resistance.

6. Choose a suitable power switching device

Table 1 summarizes the switching characteristics of a variety of solid-state devices. Use this tabulation of characteristics as a general guide to preliminary switch selection. Figure 2 shows the general arrangement of the solid-state devices listed in Table 1.

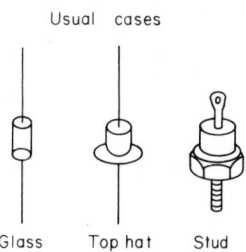

Usual cases

Glass   Top hat   Stud

Fig. 1   Typical cases used for solid-state diodes.

TABLE 1   Solid-State Switch Characteristics

Switch Type	Switching Characteristics
SCR (silicon-controlled rectifier)	Remote-controlled dc switch; once switched on it stays on until power to anode is interrupted or polarity is reversed. Switch is activated by a pulse current applied to the gate. Power applied too rapidly may trigger conduction.
DIAC (two-terminal ac diode)	Will not conduct until potential reaches the breakdown voltage of about 35 v. At this voltage, the device offers a low-resistance path useful in generating high-speed pulses for triggering SCRs or TRIACS.
TRIAC (three-terminal device for switching ac power)	May be triggered into conduction in either direction by a gate current of either polarity. Conducts a large current; when the gate current is turned off, conduction stops.
SUS (silicon unilateral switch)	Voltage-controlled dc switch for triggering SCRs. It is a small integrated circuit. When a positive potential of 8 v is applied between the anode and cathode, the SCR is triggered and the device conducts current.
SBS (silicon-bilateral switch)	Voltage-sensitive switch for ac; works with applied voltages of either polarity. It is the ac equivalent of the SUS and is useful in triggering TRIACS.

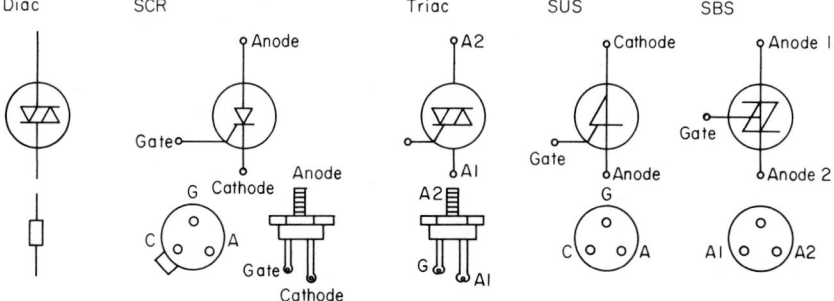

Fig. 2   General arrangement of solid-state devices.

With a control voltage of 40 v available, a DIAC is suitable for triggering an SCR. With this arrangement, the DIAC handles the small control current and the SCR handles the large power current.

Related Calculations: Zener diodes, another type of solid-state device, conduct on reverse voltage when the voltage reaches a set level. The reverse voltage is constant, regardless of the current. This characteristic makes the zener diode a good voltage regulator for levels from about 3 to 200 v or more.

The thyrector consists of two selenium rectifiers connected in opposition. It is useful for dissipating relatively high ac power for a very short time, such as when inductive circuits are energized or opened. Breakdown voltages are not as sharply defined as for zener diodes, but thyrectors are cheaper.

A recently developed process for manufacturing selenium rectifiers offers substantial improvements over conventional manufacturing methods. Selenium rectifiers produced by this new method also have advantages over silicon rectifiers for popular control-circuit voltages in cost, size, and overload capacity. Electrical performance is comparable, and the new rectifiers are not subject to damage from transient voltages. Table 2 summarizes the characteristics of these rectifiers.

TABLE 2   Selenium Rectifier Characteristics*
(Size, performance, and cost for 28-v dc bridge rectifiers for 1 and 10 amp)

	Conventional selenium	New selenium	Silicon
Potential drop, v......	2.2	2.2	2.4
Ageing life, hr.........	20,000	100,000	100,000
Needs transient protection	No	No	Yes
Overload capacity— Rating × sec.........	2	3	1
Maximum ambient temperature, C......	35	45	55
Cost, $:			
For 1-amp rating	2.00	0.80	1.00
For 10-amp rating ...	6.00	4.00	7.50
Volume, cu in.:			
Lamp..............	8.0	0.4	0.35
10 amp	100	16	32

Product Engineering.

TRANSISTOR SELECTION AND CIRCUIT ARRANGEMENT

Select a transistor suitable for producing a power gain from 10 to 300 db without a phase shift of the signal voltage and with a high output resistance.

Calculation Procedure:

1. Compute the power gain of the transistor

Use the basic power-gain relation G_o = power gain of the transistor = output power, db/input power, db = 300/10 = 30.

2. Select the type of transistor to use

Table 3 shows the principal characteristics of silicon and germanium transistors—the two types of transistors in common use today. Study of this table shows that grown silicon transistors have power gains in the range of 35. Since the desired power gain is in the range of 30, and germanium transistors normally provide a higher power gain than required here, as shown by Table 3, a grown-silicon transistor will be the initial choice for this device.

TABLE 3 Characteristics of Silicon and Germanium Transistors*

| | Transistor type | | |
| | Silicon grown | Germanium | |
		Grown	Alloy
Collector:			
Voltage, (maximum), v.........	40	40	25
Dissipation (maximum), mw ...	150	50	50
Cutoff current, μa.............	0.02	2	10
Capacitance, $\mu\mu$f.............	7	14	40
Conductance, parallel, μmhos....	0.3	0.2	1.0
Emitter:			
Current (minimum usable) ma .	1	0.01	0.1
Reverse voltage (maximum) v ..	2	10	5
Bias voltage, mv	500	160	160
Resistance, ohms	100	25	25
Base, resistance, ohms...........	500	150	300
Gain:			
Power, db...................	35	47	40
Current.....................	26	35	40

°*Electronic Design*, and Kiver—*Transistors*, McGraw-Hill.

3. *Investigate the transistor characteristics*

Table 4 shows the general characteristics of three amplifier configurations. Study of this tabulation shows that a common-base amplifier has no signal phase shift between output and input. Since a phase shift is not acceptable in this transistor and common-emitter and common-collector amplifier configurations do have a phase shift, a common-base transistor will be used.

TABLE 4 General Characteristics of Transistor* Amplifiers

| Characteristic | Amplifier configuration | | |
	Common-emitter	Common-base	Common-collector
Current gain	Large	1, approx	Large
Voltage gain..........	Large	Large	1, approx
Power gain..........	Largest	Large	Lowest
Input resistance	Low	Lowest	Highest
Output resistance.....	High	Highest	Lowest
Signal phase shift between output and input	180°	None	180°

°Kiver—*Transistors*, McGraw-Hill.

4. *Check the transistor current characteristics*

Refer to a manufacturer's transistor specification or data sheet, Fig. 3. Typical specification or data sheets list: (*a*) transistor-type number, (*b*) absolute maximum voltage
and current ratings of the transistor, (*c*) power rating of the transistor, (*d*) electrical
characteristics of the transistor including the small-signal, high-frequency, dc, cutoff,
and switching characteristics of the transistor.

The usual transistor specification or data sheet also may include characteristic curves,
Fig. 4. Use these curves to determine the current in the transistor. Thus, in Fig. 4*a*,
with a collector-base voltage V_{CB} of 20 v and a collector current of $I_C = 3$ ma, the emitter
current $I_E = -3.1$ ma.

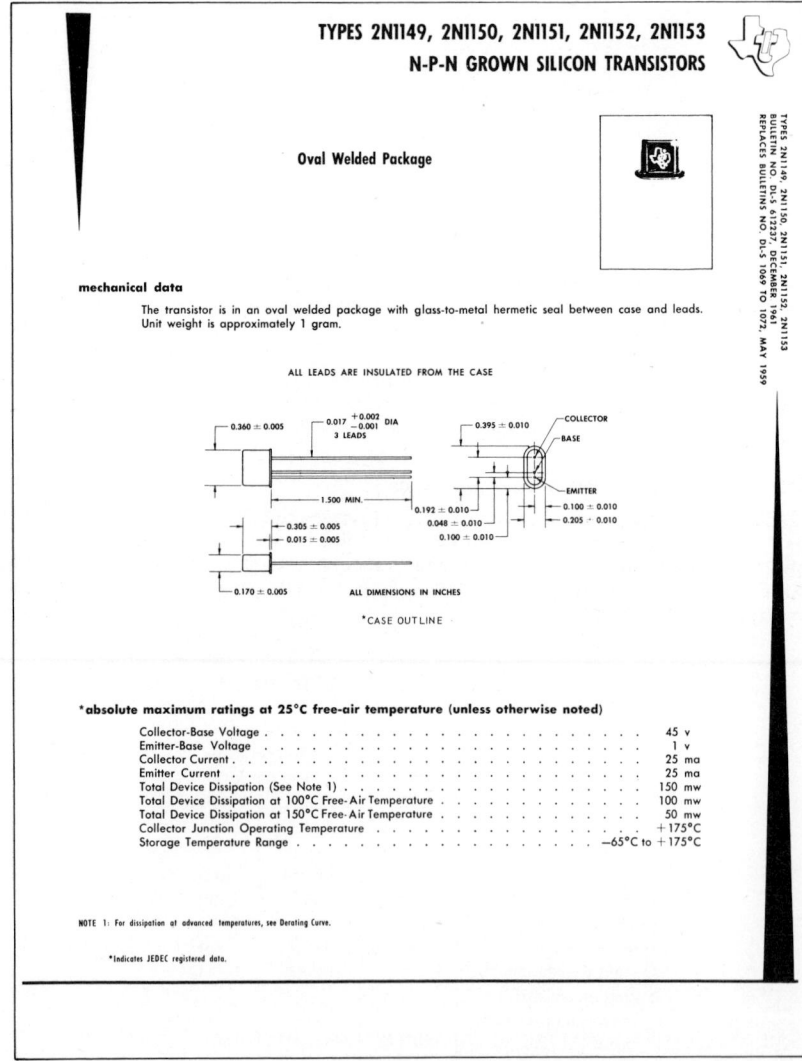

Fig. 3 Typical transistor data sheet.

Related Calculations: Use this general method to make a preliminary selection of a transistor for power, switching, detection, mixing, rectification, oscillation, amplification, and similar applications.

VACUUM-TUBE PLATE RESISTANCE, TRANSCONDUCTANCE, AMPLIFICATION FACTOR, AND LOAD LINE

A 6J5 triode vacuum tube is operated with a grid bias of -6 v. What is the plate resistance of this tube if the plate voltage is raised from 160 to 220 v? What is the tube transconductance when the grid bias is increased from -6 to -8 v while the plate

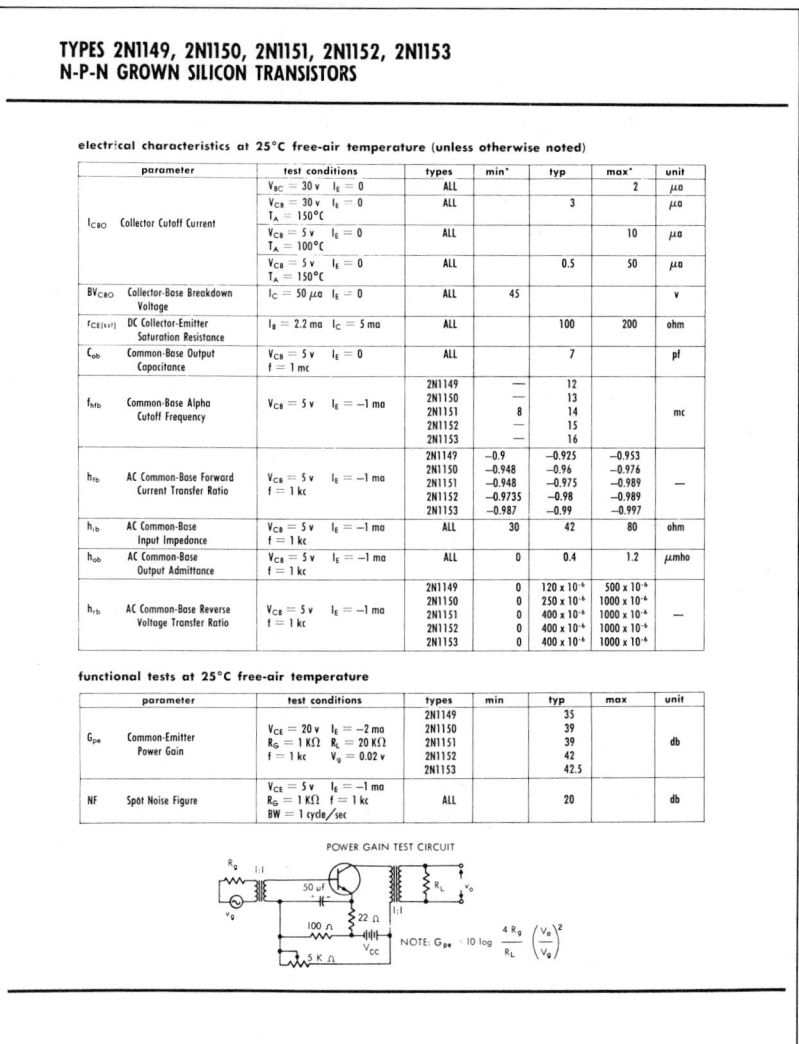

TYPES 2N1149, 2N1150, 2N1151, 2N1152, 2N1153
N-P-N GROWN SILICON TRANSISTORS

electrical characteristics at 25°C free-air temperature (unless otherwise noted)

parameter		test conditions	types	min*	typ	max*	unit
I_{CBO} Collector Cutoff Current		$V_{BC} = 30$ v $I_E = 0$	ALL			2	μa
		$V_{CB} = 30$ v $I_E = 0$ $T_A = 150°C$	ALL		3		μa
		$V_{CB} = 5$ v $I_E = 0$ $T_A = 100°C$	ALL			10	μa
		$V_{CB} = 5$ v $I_E = 0$ $T_A = 150°C$	ALL		0.5	50	μa
BV_{CBO}	Collector-Base Breakdown Voltage	$I_C = 50\ \mu a$ $I_E = 0$	ALL	45			v
$r_{CE(sat)}$	DC Collector-Emitter Saturation Resistance	$I_B = 2.2$ ma $I_C = 5$ ma	ALL		100	200	ohm
C_{ob}	Common-Base Output Capacitance	$V_{CB} = 5$ v $I_E = 0$ $f = 1$ mc	ALL		7		pf
f_{hfb}	Common-Base Alpha Cutoff Frequency	$V_{CB} = 5$ v $I_E = -1$ ma	2N1149 2N1150 2N1151 2N1152 2N1153	— — 8 — —	12 13 14 15 16		mc
h_{fb}	AC Common-Base Forward Current Transfer Ratio	$V_{CB} = 5$ v $I_E = -1$ ma $f = 1$ kc	2N1149 2N1150 2N1151 2N1152 2N1153	−0.9 −0.948 −0.948 −0.9735 −0.987	−0.925 −0.96 −0.975 −0.98 −0.99	−0.953 −0.976 −0.989 −0.989 −0.997	—
h_{ib}	AC Common-Base Input Impedance	$V_{CB} = 5$ v $I_E = -1$ ma $f = 1$ kc	ALL	30	42	80	ohm
h_{ob}	AC Common-Base Output Admittance	$V_{CB} = 5$ v $I_E = -1$ ma $f = 1$ kc	ALL	0	0.4	1.2	μmho
h_{rb}	AC Common-Base Reverse Voltage Transfer Ratio	$V_{CB} = 5$ v $I_E = -1$ ma $f = 1$ kc	2N1149 2N1150 2N1151 2N1152 2N1153	0 0 0 0 0	120×10^{-6} 250×10^{-6} 400×10^{-6} 400×10^{-6} 400×10^{-6}	500×10^{-6} 1000×10^{-6} 1000×10^{-6} 1000×10^{-6} 1000×10^{-6}	—

functional tests at 25°C free-air temperature

parameter		test conditions	types	min	typ	max	unit
G_{pe}	Common-Emitter Power Gain	$V_{CE} = 20$ v $I_E = -2$ ma $R_G = 1$ KΩ $R_L = 20$ KΩ $f = 1$ kc $V_g = 0.02$ v	2N1149 2N1150 2N1151 2N1152 2N1153		35 39 39 42 42.5		db
NF	Spot Noise Figure	$V_{CE} = 5$ v $I_E = -1$ ma $R_G = 1$ KΩ $f = 1$ kc BW = 1 cycle/sec	ALL		20		db

POWER GAIN TEST CIRCUIT

$$\text{NOTE: } G_{pe} = 10 \log \frac{4 R_g}{R_L} \left(\frac{V_o}{V_g} \right)^2$$

(Texas Instruments Incorporated.)

voltage is held at 220 v? What is the tube amplification factor if the grid bias must be changed from -6 to -10 v to maintain a constant plate current when the plate voltage is changed from 160 to 240 v? Plot the tube load line and operating point for a load of 28,000 ohms and a plate-to-cathode voltage of 280 v.

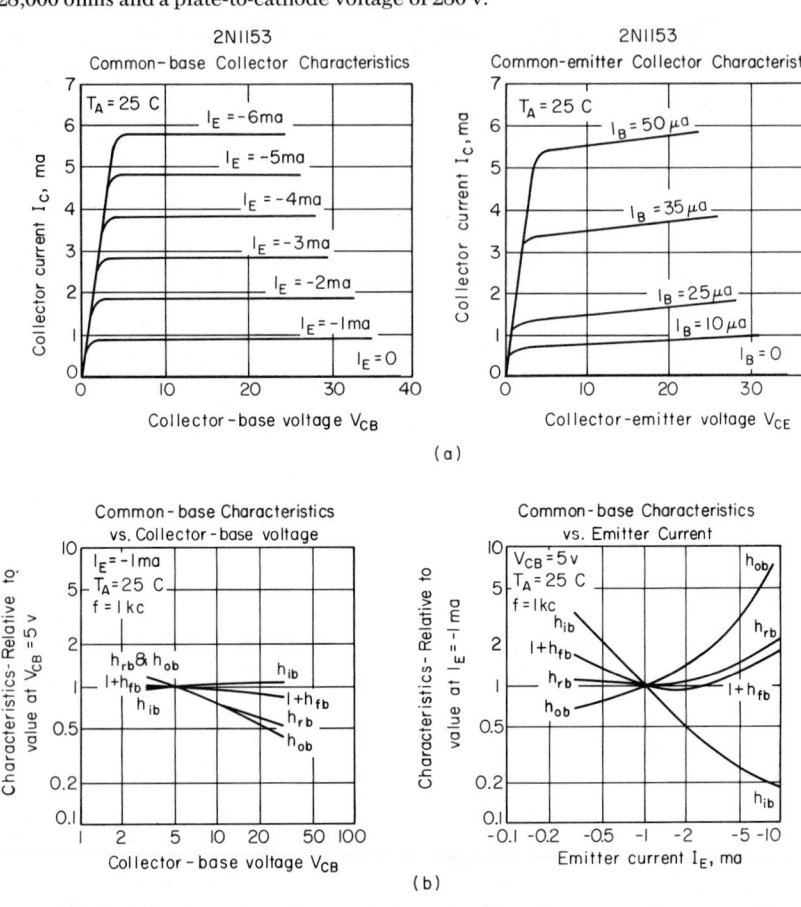

Fig. 4   Typical transistor characteristic curves. *(Texas Instruments Incorporated.)*

Calculation Procedure:

1. Determine the tube voltage and current changes

Obtain a plate characteristic curve for the tube in question. Figure 5 shows a typical set of plate characteristic curves. To determine the tube voltage and current changes, the tube characteristic curve, Fig. 5, must be used. It is known that the plate voltage E_b changes from 160 to 220 v and the grid bias is -6 v. At a plate voltage of 160 v and a grid bias of $E_c = -6$ v, Fig. 5 shows that the plate current $I_b = 3.8$ ma. Find this value of I_b by projecting vertically upward from $E_b = 160$ v to the curve $E_c = -6$ v, Fig. 5. From the intersection project horizontally to the left to read $I_b = 3.8$ ma. Using the same procedure, $I_b = 11.4$ ma at $E_b = 220$ v.

2. Compute the tube plate resistance

Use the relation $R_p = \Delta E_b/\Delta I_b$, where R_p = plate resistance, ohms; ΔE_b = plate voltage change, v; ΔI_b = plate current with E_b and I_b at a constant grid bias. For this tube at the stated conditions, $R_p = (220 - 160)/(0.0114 - 0.0038) = 7{,}890$ ohms. In the denominator of this expression, the plate currents determined in Step 1 were changed from ma to amps; that is, 1 ma $= 0.001$ amp.

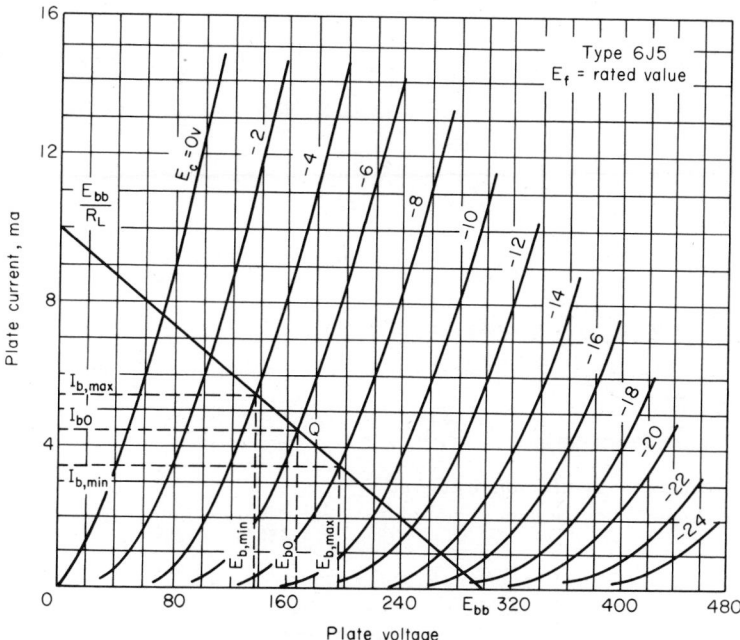

Fig. 5 Vacuum-tube plate characteristics and load line. *(General Electric Company.)*

3. Determine the plate current change

Using the plate characteristic curve, Fig. 5, find the plate current at $E_b = 220$ v and $E_c = -6$ v as $I_b = 11.4$ ma, using the procedure described in Step 2. Also, at $E_b = 220$ v and $E_c = -8$ v, $I_b = 6.0$ ma.

4. Compute the tube transconductance

Use the relation $g_m = \Delta I_b/\Delta E_c$, where $g_m =$ tube transconductance, mhos; $\Delta I_b =$ change in plate current, amps; $\Delta E_c =$ change in grid voltage, v. Thus, changing ma to amps, $g_m = (0.0114 - 0.006)/[-6 - (-8)] = 0.0027$ mho.

5. Compute the tube amplification factor

Use the relation $\mu = \Delta E_b/\Delta E_c$, where $\mu =$ tube amplification factor; $\Delta E_b =$ change in plate voltage, v; $\Delta E_c =$ change in grid voltage, v, while the plate current I_b remains constant. Thus, $\mu = (240 - 160)/[-6 - (-10)] = 20$.

6. Plot the tube load line

The tube load line is plotted on the plate characteristic curve as a straight line between a point corresponding to $I_b = 0$, $E_b = E_{bb}$, where $E_{bb} =$ plate-to-cathode voltage. Since $E_{bb} = 280$ v, the load line intersects the horizontal axis of the plate characteristic curve at $I_b = 0$, $E_{bb} = 280$ v. Plot this point, Fig. 5.

The load line intersects the vertical axis at $E_b = 0$ v and $I_b = E_{bb}/R_L$. Or $I_b = 280/28{,}000 = 0.01$ amp, or 10 ma. Plot this intersection of the load line on the vertical axis of the plate characteristic curve at $E_b = 0$, $I_b = 10$ ma. Draw a straight line between this and the previously plotted point on the horizontal axis. The resulting line, Fig. 5, is termed the *tube load line*.

The intersection of the load line with any grid-voltage curve is the tube *operating point*. Thus, with a grid voltage of $E_c = -4$ v, $I_b = 5.2$ ma, $E_b = 118$ v.

Related Calculations: Use the general procedure given here to analyze the operating characteristics of diode, triode, tetrode, and pentode vacuum tubes. When determining the operating point of a pentode, the usual procedure is to assume that the screen current is a fixed percentage of the plate current and proceed in the same manner as for a triode. A safe assumption for usual pentodes is that the total cathode current $= 1.3 I_b$ at the tube operating point.

Brophy[1] states that in many cases it is simpler to resort to the following cut-and-try procedure for determining the operating point for a pentode. Choose a point on the load line corresponding to an arbitrary grid-bias voltage and plate current. Determine the screen current from the tube-screen characteristic curve for these values of E_c and E_b. Then compare the product $(I_b + I_s)R_k$ with the chosen value of E_c, where $I_s =$ screen current, ma; $R_k =$ cathode-bias resistor resistance, ohms. If the two values being compared are equal, the original operating-point choice is satisfactory. If the values are not equal, repeat the process until the desired accuracy is obtained.

VACUUM-TUBE-AMPLIFIER SELECTION

Choose a vacuum-tube amplifier to amplify a 0.5-v input signal to a 450-v output signal. Select the class of amplifier to use.

Calculation Procedure:

1. Compute the amplification required

The amplification, or gain, required is amplification = output or plate voltage/input or grid voltage. For this amplifier, amplification = 450/0.5 = 900 times.

2. Determine the number of amplification stages required

The usual upper limit on the amplification available from one vacuum tube is 200 times. With this rule as a guide, a 0.5-v input signal could be amplified to (0.5)(200) = 100 v by one tube. Since a 450-v output signal is required, one tube or *stage* of amplification would be unsatisfactory.

If the 100-v output signal were fed to the grid of a second tube or stage that amplified 200 times, the output voltage would be (100)(200) = 20,000 v. Since this is much greater than the 900-v output desired, a two-stage amplifier will be satisfactory.

3. Choose the amplification per stage

Where possible, in multistage vacuum-tube amplifiers, an equal amplification per tube or per stage is provided. Thus, an amplification of 30 times per tube would provide $30 \times 30 = 900$ times in this two-stage amplifier. As a check, the 0.5-v input signal would be amplified to $0.5 \times 30 = 15$ v in the first stage. Feeding this output into the second stage gives a voltage of $15 \times 30 = 450$ v. This is the desired output voltage.

4. Select the class of amplifier to use

Table 5 summarizes vacuum-tube-amplifier characteristics. Study of this table shows that Class A operation is common in voltage-amplification applications. Hence, a Class A amplifier will be chosen for this application.

5. Select the type of coupling to use

RC, that is, resistance-capacitance coupling of two or more amplifier stages, provides a flat frequency response in the mid-band region. Transformer coupling provides a flat frequency response for only about one-half the mid-band region. Hence, if a flat-frequency-response characteristic throughout the mid-band region of the output is required, use an RC-coupled amplifier. If such a requirement does not exist, use a transformer-coupled amplifier.

[1]Brophy — *Basic Electronics for Scientists*, McGraw-Hill.

TABLE 5   Vacuum-Tube Amplifier Characteristics

Amplifier class	Typical applications	Important characteristics
A	Voltage amplification with small current capacity; power amplification for low-power applications	Small input signal; output current variation exactly duplicates input current variation— i.e., linear operation and little or no distortion; small amplification of input
B	Power amplification with large current capacity	Larger input signal than Class A; positive half of input signal is amplified and appears in output; amplification much greater than in class A
C	Radio-frequency power amplification with large current capacity	Largest input signal; top of output curve is flat; largest amplification of the three types

Related Calculations: When a transformer-coupled amplifier is used, the amplification obtainable = tube amplification factor × transformer step-up turns ratio. Thus, the transformer provides added amplification of the vacuum-tube input signal.

Where desired, Class A and B amplifiers can be combined to create Class AB_1 and AB_2 amplifiers. The resulting output combines the features of both Class A and Class B amplifiers. Figure 6 shows the output current waveforms of Class A, B, and C amplifiers.

Two amplifiers can be connected so that their input voltages are 180° out of phase. This is termed *push-pull*, and the plate circuit of each amplifier is connected to opposite ends of an output transformer. Push-pull vacuum-tube amplifiers are often used for power amplification. To obtain maximum power amplification, the output of a push-pull amplifier must be matched to the amplifier load. This is termed *impedance matching*. To match impedances, use the relation $Z_L = (N_s/N_p)^2 Z_p$, where Z_L = load impedance, ohms; N_s = number of turns in the output-transformer secondary winding; N_p = number of turns in the output-transformer primary winding; Z_p = impedance of output-transformer primary winding, ohms. The primary winding of the output transformer is connected in series in the amplifier-tube plate circuit, and the secondary winding of the output transformer is connected in series in the load circuit. Usually, the required secondary or load impedance Z_L is fixed by the device served by the amplifier. Knowing the required load impedance, a transformer can be selected and the primary impedance computed using the above relation.

Vacuum tubes used for voltage and power amplifiers include the triode and pentode types. The tetrode is seldom used because it has a secondary-emission disadvantage. A pentode provides more amplification than a triode vacuum tube.

TRANSISTOR-AMPLIFIER SELECTION AND ANALYSIS

Select a transistorized amplifier to amplify a 0.10-v input signal to a 29-v output signal with a 9,000-ohm load. The output signal must not undergo a phase reversal. What is the voltage gain with this load and a 30-ohm input resistance? What is the amplifier power gain? Compute the current, voltage, and power gains for a grounded-emitter amplifier configuration with a 9,000-ohm load and 3,000-ohm input resistance.

Calculation Procedure:

1. Compute the amplifier voltage gain

Use the relation gain = output voltage/input voltage = 29/0.10 = 290.

2. Select the amplifier circuit arrangement

Table 6 shows typical current, voltage, and power gains for various arrangements of transistor amplifiers. Study of this table shows that either a common-base or a common-emitter circuit configuration will give the desired voltage amplification. Further study of the table shows that a common-emmitter amplifier has a 180° phase reversal or shift. Since a phase reversal cannot be tolerated, a common-base amplifier should be used.

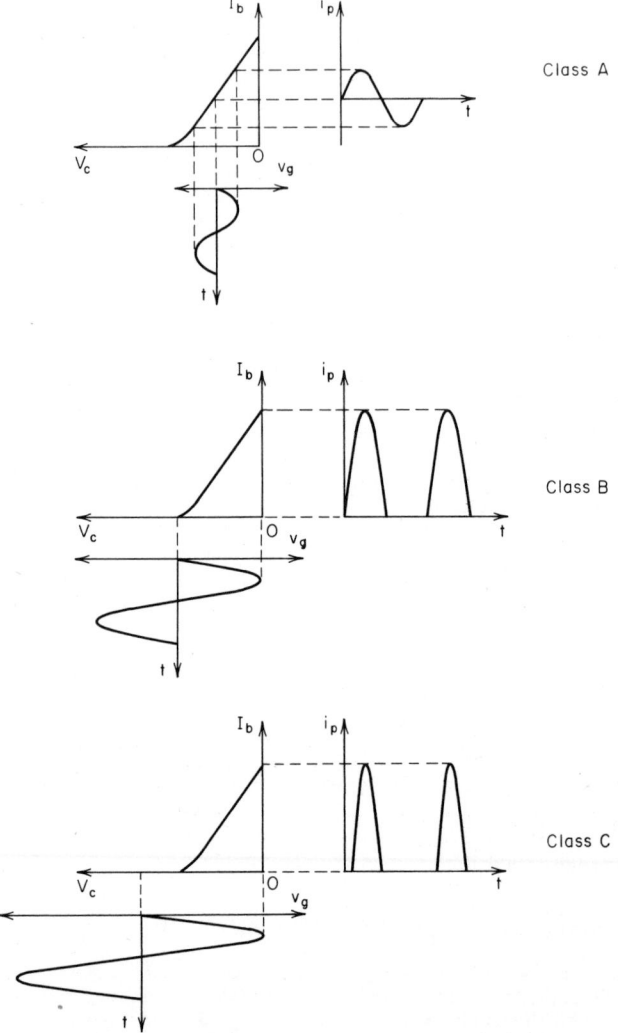

Fig. 6   Output waveforms of Class A, B, and C amplifiers.

3.   *Compute the actual voltage gain*

For a common-base transistor amplifier, voltage gain = output current, amp × load resistance, ohms/input current, amp × input resistance, ohms.

However, the ratio output current/input current = α = current gains. Table 6 shows that the current gain for a typical common-base transistor amplifier = 0.98. Hence, voltage gain = 0.98 × load resistance/input resistance = 0.98 × 9,000/30 = 294. Since a voltage gain of 290 is desired, the amplifier is suitable.

TABLE 6   Typical Transistor Amplifier Characteristics

Circuit configuration	Current gain	Voltage gain	Power gain	Phase reversal	Input resistance	Output resistance
Common-base.....	Low; 1.0 (0.98)	Large (200)*	Large (200)	No	Lowest (15 ohms)	Highest (1 meg- ohm)
Common-collector.	Large (75)	Low 1.0 (0.98)	Lowest	Yes, 180°	Highest (50,000 ohms)	Lowest (100 ohms)
Common-emitter ..	Large (75)	Large (250)	Largest	Yes, 180°	Medium (600 ohms)	High (50,000 ohms)

*Typical gain value is shown in parentheses.

4.   Compute the amplifier power gain

For an amplifier, power gain = power output, watts/power input, watts, or (output current, amp)^2(load resistance, ohms)/[input current, amp)^2(input resistance, ohms)]. As in Step 3 above, the current gain for a typical common-base transistor amplifier = 0.98. Thus, since current gain = output current/input current, the amplifier power gain using the second relation above = $(0.98)^2(9,000/30) = 288$.

5.   Compute the grounded- or common-emitter current gain

In a common emitter amplifier configuration the input signal is connected to the base. Hence, the base current = input current. Then, the current gain = β = output current, amp/input current, amp. With a typical emitter current of 6 ma, the output or collector current will be emitter current, ma−base current, ma. Assuming a base current of 0.15 ma, output or collector current = $6.0-0.15 = 5.85$ ma. Hence, power gain = $\beta = 5.85/0.15 = 39$.

6.   Compute the common-emitter voltage gain

In a common-emitter amplifier, voltage gain = current gain × resistance gain, or $\beta \times$ load resistance, ohms/input resistance, ohms. With a 9,000-ohm load and a 3,000-ohm input resistance, voltage gain = $(39)(9,000/3,000) = 117$.

7.   Compute the common-emitter power gain

As in Step 4, power gain = (current gain)^2 load resistance, ohms/input resistance, ohms = $(39)^2(9,000/3,000) = 4,560$.

Related Calculations: The current gain of a grounded- or common-collector amplifier = $\beta + 1$, whereas the voltage gain is less than unity − usually about 0.98. The power gain is computed in the same way as described in Steps 4 and 7 above.

Since the common-emitter amplifier has the largest power gain and large current and voltage gains, this configuration is widely used in transistorized electronic equipment. Even though the current and gain notations are presented in word form in this Calculation Procedure for each transistor considered, the hybrid symbols are used for hybrid characteristics or parameters in transistor manufacturer's data sheets. Thus, $\alpha = h_{FB}$, $\beta = h_{FE}$ in hybrid notation. The current values given in Step 5 are typical for common-emitter transistors.

OSCILLATOR SELECTION AND APPLICATION

Select an oscillator to operate in the 200- to 400-Hz range. At full load the oscillator output must not vary by more than 2 Hz. Ease of frequency change is required for the application. (NOTE: Hz = hertz = 1 cps).

Calculation Procedure:

1. Compute the required frequency stability

The oscillator operating range is 200 to 400 Hz, or $400 - 200 = 200$ Hz. With an allowable frequency variation of 2 Hz, the frequency stability must be allowable frequency variation, Hz/frequency operating range, Hz, or $2/200 = 0.01$, or 1 percent. This is usually considered to be a high frequency stability. Hence, an amplifier having a high frequency stability is required.

2. Select the oscillator type

Table 7 summarizes typical oscillator characteristics. Study of this table shows that a Wein-bridge oscillator has the desired characteristics — excellent frequency stability, easy frequency changing, and a frequency range of 200 to 400 Hz. Hence, this type of oscillator will be chosen for the application.

TABLE 7   Typical Oscillator Characteristics

Oscillator Type	Typical Characteristics
RC (general)............	Good frequency stability below 100 kHz; wide tuning range with constant power output.
LC (Hartley and Colpitts)	Widely used from 100 kHz to 500 MHz. At low power, frequencies to 4,000 MHz can be obtained. In Hartley, amplifier portion operates Class C; in Colpitts, Class A, although Class C is possible in the latter, if desired. Colpitts is adaptable to low-impedance loads; frequency stability is moderate.
Phase-shift (uses RC feedback)	Useful at medium and low frequencies to 1 Hz where frequency stability is not critical; frequency changing is cumbersome.
Wein-bridge............	Excellent frequency stability; easy frequency changing; typical frequency range is 10:1 or 5 Hz to 1 MHz.
Tickler (Armstrong).....	Class-C operation used to develop large power output at high frequency. Frequency stability is less than in other oscillator types.
Crystal................	Popular for use in fixed-frequency applications; frequency is constant, regardless of load.
Relaxation.............	Use nonlinear active elements; poor frequency stability.
Blocking	Used to generate pulse waveforms in digital-computer circuits.
Electron-coupled	Excellent frequency stability; can use Armstrong, Colpitts, or Hartley circuit.
Magnetron	Used for frequencies above 1,000 MHz.
Klystron...............	Used for frequencies above 1,000 MHz.

Related Calculations: As a general guide, the typical characteristics of oscillators listed in Table 7 apply whether the oscillator uses one or more vacuum tubes or transistors. Hence, the data listed can be used in preliminary selection of oscillators using either type of components. However, transistor oscillators are smaller in physical size than are vacuum-tube oscillators.

INTEGRATED-CIRCUIT SELECTION AND APPLICATION

Choose the type of integrated circuit to use for a wideband amplifier to supply a 10-v 15-ma 100-kHz output. Specify the type of integrated circuit to use.

Calculation Procedure:

1. Compute the circuit power output

Use the relation $P = IE$, where P = integrated-circuit power output, watts; I = output current, amp; E = output voltage. For this circuit, $P = (0.015)(10) = 0.15$ watt.

2. Select the type of integrated circuit to use

Use Fig. 7 to select the type of integrated circuit. Enter at the power output, watts, and project horizontally until the vertical line at 100 kHz, the output frequency, is intersected. The point of intersection is in the monolithic silicon integrated-circuit area. Hence, this type of circuit should be used.

An integrated circuit[1] consists of two or more electronic components associated on, or within, a substrate to form an electrical network. A monolithic integrated circuit has a single semiconductor body, such as a silicon crystal, as its substrate. Monolithic integrated circuits are popular for many applications.

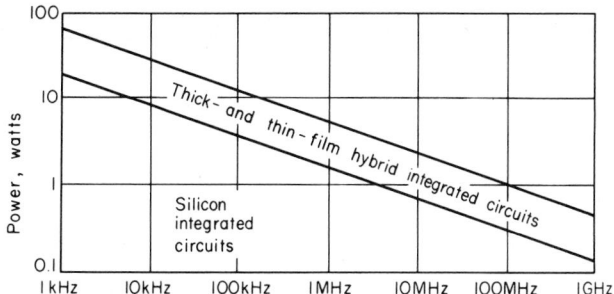

Fig. 7 Approximate performance envelopes for monolithic integrated circuits and thick-film and thin-film hybrid circuits. *(Electronics.)*

3. Check the circuit selection

Use Table 8 as a guide to the suitability of the type of circuit selected. Study of Table 8 shows that the metal-oxide silicon (MOS) monolithic integrated circuit has a limited frequency capability. Thus, if a wide frequency capability is desired, either a bipolar silicon or other type of circuit would have to be used. Since a wide frequency capability is not stated as a requirement for this circuit, the monolithic MOS circuit will probably be acceptable.

4. Choose the commercial integrated circuit

Table 9 shows a condensed listing of typical commercially available integrated circuits. Enter Table 9 at the circuit function and project horizontally to read the circuit input and output. Thus, a μ7712C wideband amplifier has a differential input (i.e., the *difference* between two input signals) and a single-ended output.

Match the desired circuit function with the desired output. When the needed function and required output are obtained from a commercially available integrated circuit, list that circuit as a tentative choice. Once several such tentative choices have been made, the final selection can be made after analysis of the available circuits.

Related Calculations: Use this general method to select integrated circuits for computers, communications equipment, amplifiers (audio, video, radio-frequency, operational, wideband, etc.), oscillators, limiters, mixers, servos, etc. For best results, keep several manufacturer's engineering data tabulations available so that comparisons of the suitable circuits can be quickly made.

TRANSISTOR WORST-CASE LEAKAGE CURRENT

What is the worst-case transistor leakage current in a transistor circuit that is part of a complex equipment if the transistor specification sheet gives the following data for an ambient temperature of 25 C: $I_{CBO,\,max} = 4\mu a\ (V_{CB} = 2\text{-v dc})$; $I_{CBO,\,max} = 5\ \mu a\ (V_{CB} = 6\text{-v dc})$. The maximum allowable junction temperature = 85 C; maximum allowable

---

[1]Donald Christiansen, "Integrated Circuits in Action: Trends and Trade Offs," *Electronics*, Nov. 14, 1966.

TABLE 8 Guide for Integrated-Circuit Selection*

(Selection of integrated-circuit types is guided by process capability.)

Integrated circuits	Monolithic		Thin film		Thick film
	Bipolar silicon	Metal oxide silicon	Evaporated thin film	Sputtered tantalum thin film	Screen printed thick film
Advantages . . .	Reliability and long life; low cost; high density; low weight; standardization; amenable to high production	Low cost; simple design; high density; resistivities to 20 kilohm/□; amenable to redundant design; direct coupling avoids capacitors; symmetrical switching capabilities	Good line definition; high reliability; high precision; wide choice of materials; permit 1 : 1 translation from discrete component design; amenable to production of interconnections	Good line definition; high reliability; high precision by anodizing	Low cost; simple production techniques; amenable to production of interconnections; resistivities of 1–20 kilohm/□; capacitance of 500–500,000 pf/sq in.; high power capability
Limitations . . .	Power-frequency product; high development cost; loose tolerances; diffused resistivities are 100–200 ohms/□	Limited frequency capability	Active devices must be added; exhibit unique failure modes; low sheet resistivity; production procedures can be complex	Active devices must be added; specialized production processes needed	Active devices must be added; trimming required for close tolerances; high voltage coefficient; nonlinear temperature characteristic

Electronics.

TABLE 9   Typical Commercial Integrated Circuits*

Circuit function	Circuit input	Circuit output	Manufacturer's circuit number
Intermediate-frequency amplifier	Differential	Single-ended	μA7703C
Operational amplifier.....................	Differential	Single-ended	μA7709C
Wideband amplifier	Differential	Single-ended	μA7712C
Radio-frequency amplifier	Differential	Single-ended	CA3004
Audio-frequency amplifier................	Single-ended or differential	PP emitter follower	CA3007
Wideband intermediate-frequency amplifier	Differential	Single-ended	CA3011
Half-adder..............................	Digital binary	....	HEP553
Bias driver	Storage element	....	HEP558
Flip-flop................................	Voltage regulator	....	HEP554

Electronics.

power dissipation at 25 C = 300 mw; $h_{\text{FE.min}} = 5$; $V_{\text{CB}} = 10$-v dc; $V_{\text{CE}} = 12$-v dc; $V_{\text{BE}} = 2$-v dc; $I_{\text{C}} = 6$ ma. The power sources for the circuit have 1 percent total tolerance ratings, whereas the resistive components have 5 percent total tolerance ratings. The equipment performance specification indicates a maximum ambient temperature requirement of 50 C. The rise in temperature caused by power dissipation within the equipment is 10 C.

Calculation Procedure:

1. Compute the transistor power dissipation

Use the relation $P_d = I_{\text{C}}[V_{\text{BE}} + (V_{\text{BE}}/h_{\text{FE}})]$, where P_d = power dissipation, mw; I_{C} = dc collector current, ma; V_{CE} = dc collector-to-emitter voltage; $h_{\text{FE}} = I_{\text{C}}/I_{\text{B}}$, where I_{B} = dc base current, ma. Substituting, $P_d = 6(12 + 2/5) = 74.4$ mv.

2. Compute the thermal resistance factor

The thermal resistance factor K is computed from $K = (T_{\text{J.max}} - T_{\text{spec}})/P_{\text{dst}}$, where $T_{\text{J.max}}$ maximum allowable junction temperature, C; T_{spec} = specified temperature, C, at which the maximum allowable power dissipation, P_{dst} mw, takes place. Substituting, $K = (85 - 25)/300 = 0.2$ mw.

3. Compute the junction-temperature increase

Use the relation $\Delta T_{pd} = KP_d$, where ΔT_{pd} = junction temperature increase, C. Or $\Delta T_{pd} = 0.2(74.4) \cong 15$ C.

4. Compute the junction operating temperature

The junction operating temperature $T_{\text{J}} = T_{\text{A}} + T_{\text{E}} + T_{pd}$, where T_{A} = maximum ambient temperature given in the performance specification, C; T_{E} = maximum expected rise in temperature due to the power dissipation within the equipment, C. Substituting, $T_{\text{J}} = 50 + 10 + 15 = 75$ C.

5. Compute the transistor thermal constant

Use the relation, $M_{\text{T}} = (T_{\text{J}} - 25 \text{ C})/10$, where M_{T} = transistor thermal constant; other symbols as before. Thus, $M_{\text{T}} = (75 - 25)/10 = 5$. Note in this relation that 25 C = the transistor-specification-sheet ambient temperature.

6. Compute the segregated leakage components

At the specified transistor ambient temperature of 25 C, $I_{\text{TO(25 C)}} + I'_{\text{SO(25 C)}} = 4$ μa, and $I_{\text{TO(25 C)}} + 3I'_{\text{SO(25 C)}} = 5$ μa, where I_{TO} = the segregated thermal component derived from the transistor-specification sheet for 25 C; $I'_{SO(25\,C)}$ = the zero-life, voltage-dependent, surface-leakage component.

Solve these simultaneous equations by subtracting the first equation from the second to obtain $I'_{SO(25\,C)} = 1$; $I'_{SO(25\,C)} = 0.5\,\mu a$. At $V_{CB} = 2$-v dc, $I_{TO} = 3.5\,\mu a$, obtained by substituting the value of $I'_{SO(25\,C)} = 0.5$ in the first equation and solving for I_{TO}.

7.   Compute the zero-life thermal-leakage component

A general rule of thumb used in transistor analysis is that the thermal-leakage component doubles for every 10-C increase in temperature above the ambient temperature given in the transistor-specification sheet. Compute the thermal-leakage component from $I_{TO} = I_{TO(25\,C)} \times 2^{M(T)}$, where I_{TO} = the zero-life thermal-leakage component; $I_{TO(25\,C)}$ = the segregated thermal component from the transistor-specification sheet; $2^{M(T)}$ = multiplying factor for the surface leakage current $I'_{SO(25\,C)}$. This multiplying factor is based on the rule of thumb given above.

Solve for I_{TO} by using the value of $I_{TO(25\,C)}$ from Step 6 and applying the rule of thumb to $I'_{SO(25\,C)}$. Thus, with a junction temperature of 75 C, the multiplying factor $2^{M(T)}$ becomes $(75 - 25)/10 = 5$; for example, $I'_{SO(25\,C)} = 0.5$ is doubled five times, or $0.5 \times 2 = 1$; $1 \times 2 = 2$; $2 \times 2 = 4$; $4 \times 2 = 8$; $8 \times 2 = 16$. Then multiplying by 2 and substituting, $I_{TO} = 3.5 \times 32 = 112\,\mu a$.

8.   Compute the zero-life, voltage-dependent, surface-leakage component

Use the relation $I_{SO} = I'_{SO} \times 2^{M(T)}/2$, where I_{SO} = zero-life surface leakage current, μa; other symbols as before. Thus, $I_{SO} = 2.5 \times 16 = 40\,\mu a$.

9.   Compute the zero-life leakage current

Use the relation $I_{LO} = I_{TO} + I_{SO}$, where I_{LO} = zero-life leakage current, μa; other symbols as before. Thus, $I_{LO} = 112 + 40 = 152\,\mu a$.

10.   Compute the effect of transistor aging

Where the increase in the leakage current caused by aging is not available from physical measurements of the transistor, assume an increase of at least 40 percent. Thus, the total leakage current $I_L\,\mu a = 1.4I_{LO}$. Or, $I_L = 1.4 \times 152 = 213\,\mu a$.

11.   Compute the total worst-case transistor leakage

With resistive components having 5 percent tolerance ratings and power sources having a 1 percent tolerance rating, the worst-case transistor leakage is $I_L(1.0 + 0.05 + 0.01) = 213 \times 1.06 = 226\,\mu a$.

Related Calculations: Use this method for any equipment having a large number of transistors as well as single isolated transistor circuits. This method and calculation procedure was developed by E. D. Peterson, Military Electronics Division, Motorola, Inc., and published in *Electronics*.

POWER-SUPPLY ANALYSIS AND SELECTION

Choose the power supply for a vacuum-tube electronic equipment to be used on a 120-v power system. High reliability is desired with a full-wave output of 400 ma. The filter ripple must not exceed 0.5 percent. What is the power-supply ripple voltage?

Calculation Procedure:

1.   Choose the type of power supply to use

Four types of power supplies are available for vacuum-tube electronic equipment: (a) transformer rectifier-filter voltage-divider assembly; (b) transformerless supply using only a rectifier and filter; (c) vibrator—synchronous or nonsynchronous; (d) dynamotor.

Of these four, the first is the most popular for high reliability service. Transformerless power supplies are used for small equipments designed to operate on ac and dc. Vibrators find some use for B+ voltage supply for radio receivers. Dynamotors deliver

1,000 v or more dc output for aircraft electronic equipment. Hence, assume that a transformer-type power supply is tentatively chosen for this application.

2. Select the transformer for the power supply

When a full-wave output is desired, a center-tap transformer is generally used. Transformers with a center-tapped high-voltage secondary winding are rated by the voltage at each end with respect to the center tap. This rating is usually 350–0–350 v at 100 ma. Low-voltage secondary windings that deliver power to tube heaters are rated by the voltage and current they can deliver—for example 6.3 v at 2 amp.

Since a full-wave power output is desired, assume that a center-tap transformer is used.

3. Select the type of rectifier to use

Table 10 summarizes the characteristics of several types of rectifiers. This table shows that a full-wave output is obtainable from several different types of rectifiers—diodes, gas-filled tubes, selenium, silicon, and bridge rectifiers. However, the vacuum-tube diode is restricted to an output of 300 ma or less. Therefore, one of the other types of rectifiers must be used.

TABLE 10   Rectifier Characteristics

Rectifier type	Output waveform	Current output	Voltage drop, v
Vacuum-tube diode (one tube)...............	Half-wave	300 ma or less	15–60
Vacuum-tube diode (two diodes or tubes).....	Full wave	300 ma or less	15–50
Gas-filled tube	Half-wave or full wave	300 or more ma	15
Selenium (dry metal)	Half-wave or full wave	1 amp or more	Low
Silicon diode..............	Half-wave or full wave	0.5 amp or more	Low
Bridge (vacuum-tube or silicon diodes)........	Full wave	0.1 amp or more	Low

For high-reliability service either a selenium or silicon diode is a good choice. Assume that a silicon-diode rectifier is used because it can be built for lower current ratings—in the range of 0.5 amp. Also, the voltage drop of the rectifier is small.

4. Compute the power-supply ripple voltage

With a transformer output of 350-v ac, and a silicon-diode rectifier, the dc output voltage will be about 325 v. Compute the dc ripple voltage from $E_r = E_{dc}E_r/100$, where E_r = dc ripple voltage; E_{dc} = rectifier output, dc v; $E_{r\%}$ = percent ripple. Substituting, $E_r = 325 (0.05)/100 = 0.1625$ v.

In general, the maximum allowable ripple in a power supply is less than 1 percent. This power supply meets that requirement.

Related Calculations: Standard, preassembled, off-the-shelf power supplies are available from a number of manufacturers. Use the method given here to check the suitability of such a power supply for a given load. Either semiconductors or vacuum tubes may be used as the diode in these power supplies. For large power outputs, the vacuum tube continues to be popular, but semiconductors are finding ever wider use for this service.

Note that Table 10 is also a useful guide for selecting rectifiers for any electronic-circuit application within the tabulated ranges.

RELIABILITY ANALYSIS OF ELECTRONIC CIRCUITS AND EQUIPMENT

Determine the reliability of the transistor control circuit shown in Fig. 8 when used in an aircraft electronic device having a mission time of 200 hr. The circuit contains 3 paper Mylar capacitors, 1 silicon diode, 2 silicon amplifier transistors, a solenoid, a toggle switch, 6 carbon-deposited resistors, a potentiometer, and 22 printed-circuit solder joints.

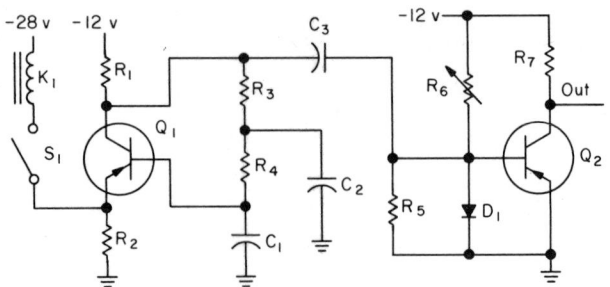

Fig. 8   Transistor control circuit. *(Electronics.)*

Calculation Procedure:

1. List the circuit components that could cause failure

With respect to reliability, a failure occurs when a circuit no longer performs within the design limits. Any of the components in this circuit could cause a failure. Hence, all the components should be listed as shown in Table 11, along with the number of each type of component in the circuit.

TABLE 11   Reliability Calculation

Component	No. $\times F_r$
Capacitors	$3 \times 0.01 = 0.03$
Diode	$1 \times 0.20 = 0.20$
Potentiometers	$1 \times 0.25 = 0.25$
Resistors	$6 \times 0.25 = 1.50$
Solenoid	$1 \times 0.05 = 0.05$
Switch	$1 \times \text{(negligible)}$
Transistors	$2 \times 0.50 = 1.00$
Printed-circuit solder joints	$22 \times 0.008 = 0.18$
	Total $= 3.21$

$$F_t = \Sigma F_r K = (3.21)(150) = 481.5$$

2. Determine the component failure rates

Table 12 lists typical failure rates per million hr for a variety of components. Use the mean failure rate listed in the center column for general-reliability analysis. Enter the respective failure rates in Table 11 opposite the name of the component.

3. Compute the total or overall failure rate

Multiply the number of each component in the circuit by the failure rate for that component. Find the sum of these products, 3.21, Table 11. This is the total or overall failure rate of the circuit per million hr of operation.

4. Apply the environmental weighting factor

Table 13 lists typical environmental weighting factors for various applications. Inspection of this table shows that for aircraft application $K = 150$. To apply this factor, take the product of it and the total or overall failure rate, or $3.21 \times 150 = 481.5$ per million hr, Table 11. This is called the total or final failure rate.

TABLE 12 Typical Electronic-Component Failure Rates

Component	Failures per 10^6 hr		
	Upper	Mean	Lower
Capacitor, paper Mylar	0.014	0.010	0.006
Diode, silicon.............	0.250	0.200	0.150
Joint, solder, printed-circuit.................	0.080	0.008	0.004
Potentiometer, carbon-deposited..............	0.750	0.250	0.100
Resistor, carbon-deposited..............	0.570	0.250	0.110
Solenoid..................	0.910	0.050	0.036
Switch toggle*	0.123	0.060	0.015
Transistor, silicon, amplifier...............	0.840	0.500	0.310

*Failures per 10^6 cycles.

TABLE 13 Typical Environmental Weighting Factors

Environment	Factor
Laboratory.......	1
Ground..........	10
Shipboard	20
Trailer	30
Rail	40
Bench...........	60
Aircraft..........	150
Missile	1,000

5. Compute the mean time between failures

Use the relation mean time between failures = MTBF = $10^6/F_t K$, where F_t = total or overall failure rate of the circuit, per million hr; K = environmental weighting factor. Thus, MTBF = $10^6/481.5 = 2,077$ hr.

6. Compute the circuit reliability

Use the relation, $R = e^{-t/m}$, where R = circuit reliability, i.e., the probability that a unit or a part will perform its intended function under design conditions for a specified period of time; t = mission time, hr; m = MTBF. Substituting, $R = e^{-200/2.077} = 0.908$, or 90.8 percent.

Related Calculations: Use this general method for computing the reliability of single components or complete circuits. Note that Table 12 lists upper, lower, and mean failure rates. For usual circuits, the mean failure rate is a safe value to use. Where a larger number of failures are anticipated or must be provided for, use the upper failure rate. For fewer failures, use the lower failure rate.

For quick reliability analyses, an initial reliability estimate for electronic equipment can be made by: (1) counting the number and type of active circuit elements in the

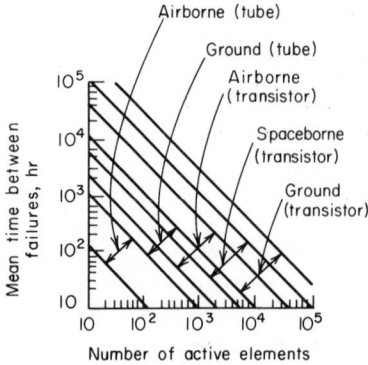

Fig. 9 Mean time between failures of circuit active elements. *(Electronics.)*

electronic package and specifying the environment in which it will operate; and (2) entering Fig. 9 on the horizontal scale at the number of active elements and projecting vertically upward to the appropriate environment curve and reading the MTBF on the left-hand scale. Compute the reliability using the equation in Step 6.

The curves in Fig. 9 apply to any heterogeneous electronic equipment using 10 or more active elements. Active elements include transistors, electron tubes, relays, capacitors, and diodes.

As a further guide, Fig. 10 shows how the total reliability R_T can be determined for series, parallel, and time-sequenced redundancy circuit arrangements. Also, Table 14 lists values of R for various values of the exponent $-t/m$.

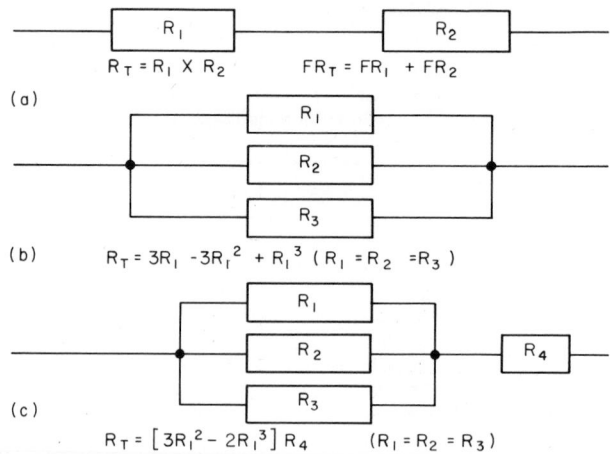

Fig. 10   Failure rates of components in series, parallel, and parallel-series. *(Electronics.)*

MAINTAINABILITY ANALYSIS OF ELECTRONIC EQUIPMENT

An electronic equipment consists of five subsystems as shown in Table 15. Determine the mean time to repair (MTTR) of this equipment if the number of subsystems and failure rates are as shown in Table 15. What is the system MTTR goal if the equipment inherent availability is 99.54 percent and the mean time between failures (MTBF) is 100 hr?

Calculation Procedure:

1. Prepare a maintainability-analysis tabulation

Set up a table listing each subsystem in the equipment. Designate each subsystem by an identifying name, number, or letter, Table 15, column 1. Next, list the number N of subsystems of each type, column 2. Then list the failure rate F_r of each subsystem. Compute this failure rate as described in the previous Calculation Procedure, or use a value based on past experience with the same or similar subsystems. Note that the F_r

TABLE 14   Exponential Values for Reliability Calculations

x	e^{-x}
0.01	0.99005
0.02	0.98020
0.05	0.95123
0.07	0.93239
0.09	0.91393
0.10	0.90484

x	$\ln x$
0.70	$-10+9.643$
0.80	$-10+9.777$
0.90	$-10+9.895$
0.95	$-10+9.949$
0.99	$-10+9.990$

value used here is the percentage of failures per 1,000 hr of operation of each subsystem.

2.   Compute the contribution to total failures

For any subsystem, the contribution that it makes to the overall failure rate of the electronic equipment $C_f = NF_r$. To obtain the C_f for each subsystem, find the product $C_f = NF_r$ and enter the value in column 4, Table 15. Take the sum of the individual C_f values and enter it in the last line of Table 15, column 4, as C_{ft}.

3.   Compute the subsystem failure-rate ratio

Set up, for each subsystem, the ratio $R = C_f/C_{fT}$, that is, the ratio of each subsystem's failure-rate contribution to the overall failure rate of the system. Enter the ratio for each system in column 5, Table 15. Find the sum of these ratios and enter it in the last line of Table 15.

4.   Apply the mean corrective-action time for each subsystem

The mean corrective-action time M_{ct} is the number of hr required to repair a defect in a subsystem. To estimate the mean corrective-action time for a subsystem when there is no previous experience to serve as a guide, use the subsystem failure rate as a source of information. Thus, subsystems contributing the highest percentage to the total failures will, ideally, have a low M_{ct}; subsystems with low contributions will have a higher M_{ct}. Using the failure rates as a guide, enter the M_{ct} value for each subsystem in column 6 of Table 15. (Typical values are shown to permit illustration of the method.)

Compute the contribution of each subsystem to the total corrective action time C_m by taking the product $C_m = C_f M_{ct}$. Enter the result in column 7, Table 15, for each subsystem. Lastly, find the sum of the individual C_m values and enter this sum in the last line of column 7.

5.   Compute the mean time to repair

Use the relation mean time to repair = MTTR = $\Sigma C_f M_{ct}/\Sigma C_f = 0.0046/0.0100 = 0.46$ hr.

6.   Compare the actual and desired MTTR values

Compute the desired MTTR from the relation MTTR = $\text{MTBF}(1 - A_i)/A_i$, where A_i = equipment inherent reliability expressed as a decimal; other symbols as before. Thus, MTTR = $100(1 - 0.9954)/0.9954 = 0.462$ hr.

The actual MTTR computed in Step 5 is 0.46 hr. This compares favorably with the MTTR goal of 0.462 hr. Hence, the M_{ct} times assumed in Table 15 are suitable.

Table 15 Computation of Mean Time to Repair*

(1) Subsystem	(2) Quantity Q	(3) Failure rate F_r	(4) Contribution to total failures C_f $C_f = QF_r$	(5) Percentage of contribution to total failures, R $R = C_f/C_{ft}$	(6) Average corrective-action time M_{ct}	(7) Contribution to total corrective-action time $C_m = C_f M_{ct}$
A	1	0.0030	0.0030	0.30	0.35	0.00105
B	1	0.0010	0.0010	0.10	0.75	0.00075
C	1	0.0035	0.0035	0.35	0.40	0.00140
D	2	0.0005	0.0010	0.10	0.50	0.00050
E	1	0.0015	0.0015	0.15	0.60	0.00090
			$C_{ft} = 0.0100$	1.00		0.00460

Martin Marietta Corporation.

Related Calculations: Should the computed MTTR for the equipment exceed the equipment goal MTTR, three steps can be taken to alter the computed MTTR. These steps are: (1) Decrease the subsystem failure rates; (2) decrease the mean corrective-maintenance times for the subsystems; (3) decrease either or both items (1) and (2) on a trade-off basis. Should it be decided to change the MTTR by decreasing the failure rate, be certain to select those subsystems where the failure rates can be reduced.

As a first choice, the subsystem having the highest failure rate appears to be the logical choice for failure-rate reduction. However, such a selection could increase the MTTR rather than reduce it. Thus, reducing the failure rate of subsystem C to 0.0028, that is, by 20 percent in the above procedure, would increase the computed MTTR.

Study of the typical reliability problem shows that the computed MTTR can be reduced by decreasing the failure rate in the subsystem having the largest M_{ct}. Another course of action is to reduce the M_{ct} of one or more of the subsystems by addition of other desirable maintainability features in the system design. Thus, reducing the M_{ct} of subsystem B by 20 percent in the above equipment will decrease the system computed MTTR.

Where a large reduction must be made in the computed MTTR, a combination of reducing F_r and M_{ct}, working by trade-off, offers good possibilities of MTTR reduction.

In a complete reliability analysis, apply the MTTR procedure described here to each subsystem. List the subsystem parts in the first column of the calculation table. Use the subsystem's estimated M_{ct} as the subsystems MTTR goal. (Adapted from the Martin Marietta Corporation publication *Maintainability Engineering*.)

SELECTION OF ELECTRONIC-EQUIPMENT COOLING SYSTEMS

Choose the type of cooling system for an electronic package $4 \times 5 \times 8$ in. that requires dissipation of 50 watts during operation.

Calculation Procedure:

1. Compute the package external surface area

The package is a six-sided box. Thus, external surface area, sq in. $= 2(L \times w) + 2(L \times H) + 2(H \times W) = 2(4 \times 5) + 2(4 \times 8) + 2(5 \times 8) = 184$ sq in. In this relation, L, W, and H represent the length, width, and height, respectively, in in., of the package.

2. Compute the package volume

Use the relation volume $= L \times H \times W = 4 \times 5 \times 8 = 160$ in.^{3}.

3. Compute the surface heat dissipation

The unit dissipation Q_U is an approximate measure of the quantity of heat w/sq in. that can be effectively dissipated from the surface of the enclosure. Compute Q_U from $Q_U = w/A$, where $w =$ heat that must be dissipated, watts; $A =$ external surface area of the package, sq in. For this package, $Q_U = 50/184 = 0.272$ w/sq in.

4. Choose the type of cooling system to use

Enter Fig. 11a at $Q_U = 0.272$ and project upward to the highest bar covering the required heat dissipation. Thus, Fig. 11a shows that natural cooling of this package will be satisfactory if the components (vacuum tubes, transistors, resistors, etc.) can withstand fairly high temperatures. If this is not the case, then either forced-air or direct-liquid cooling must be used.

5. Compute the package heat concentration

The package heat concentration $Q_c w$/in.^{3}, is an approximate measure of how many electronic components can be safely packaged together inside the enclosure. This measure is based on how effectively each of the components is mounted and what modes of heat transfer there are to conduct the heat out of the component and into the walls of the enclosure. In the method presented here, the temperature of the surroundings is assumed to be 40 C(104 F), and the average temperature rise of the components inside the enclosure is not more than 40 C(72 F). These are typical temperature ranges for a variety of electronic applications.

Compute the heat concentration in the electronic equipment from $Q_c = w/V$, where V = package volume in.^{3} Thus, $Q_c = 50/160 = 0.313$ $w/$in.^{3}.

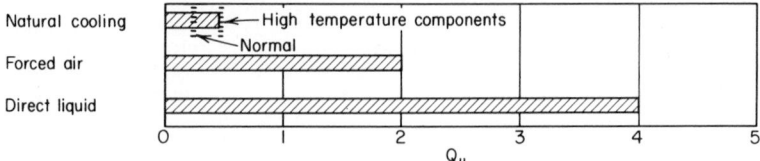

(a) Surface cooling of overall package (unit dissipation Q_U, watts/sq in.)

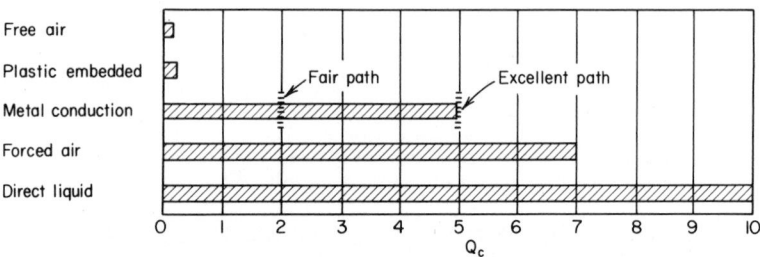

(b) Heat dissipation from crowded components (heat concentration Q_E, watts/cu in.)

Fig. 11   (a) Surface cooling of overall electronic package; (b) heat dissipation from crowded components. *(Product Engineering.)*

6.   Select the type of mounting for the electronic components

Refer to Fig. 11*b* to determine the heat dissipation from crowded electronic components. Study of Fig. 11*b* shows that plastic embedment of the components is unsuitable because the heat concentration Q_c exceeds the probable heat dissipation. Were plastic embedment used, the package would probably overheat.

Metal conduction, i.e., mounting of the components so that a direct metal conduction path is provided from the components through the mounting surfaces and into the wall of the enclosure, appears better suited for this package. Hence, this type of mounting should be specified for this package.

Related Calculations: This procedure and the charts are based on a hypothetical electronic black-box enclosure, Fig. 12, about the size of a shoe box containing heat-producing miniature components such as resistors, electron (vacuum) tubes, and power transistors. If the overall package size differs markedly from this, make more rigorous estimates of the package heat flow.

In the charts, *natural cooling* refers to an enclosure in an ordinary room, with the usual convection, conduction, and radiation to the surrounding cooler surfaces. *Free air* is assumed for natural cooling and refers to air inside the enclosure, moved by natural draft.

Forced air has two meanings in this procedure. In *unit dissipation*, it means air blown against the outside of the enclosure. In *heat concentration*, it means air blown inside the enclosure against the components. One mode of cooling affects the other, and the distinction is sometimes difficult to make.

Direct-liquid cooling means that the enclosure is partly immersed in liquid, or that the components inside the enclosure are immersed in liquid.

Plastic-embedded means that each component is encased in thermally conductive plastic. Although this plastic is twice as conductive as free air, it is not particularly efficient as a heat dissipator.

Any improvements on these assumptions will increase the heat-dissipating ability of the overall package-components plus the enclosure. For instance, if the package surface area is inadequate, it can be increased by corrugating the surface or by adding fins.

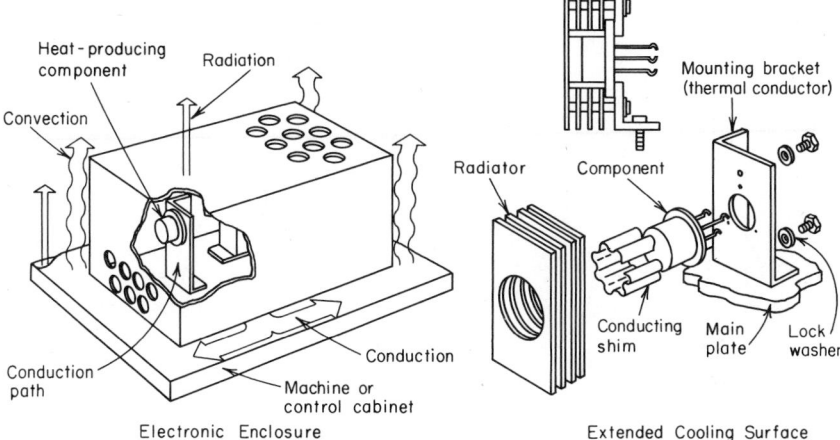

Fig. 12 Typical mounting and heat-dissipation paths in electronic equipment. *(Product Engineering.)*

This calculation procedure was developed by B. Mastisoff, electromechanical engineer, and reported in *Product Engineering.*

DETERMINATION OF TUNED-CIRCUIT Q VALUES

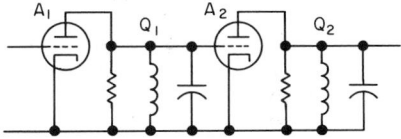

What is the Q of a two-stage triode amplifier, Fig. 13, with a frequency response having 3-db points at frequencies of 21.0 and 21.5 MHz? What Q value is necessary if one stage has a minimum Q of 35?

Fig. 13 Tuned two-stage vacuum-tube amplifier. *(Electronics.)*

Calculation Procedure:

1. Determine the amplifier $Q\Delta f/f_0$ ratio

Use Fig. 14, entering at the 3-db point for a single-tuned circuit on the vertical scale and projecting to the dashed expanded-scale curve. Project vertically downward from the intersection to read the $Q\Delta f/f_0$ ratio as 0.5 closely. In this ratio, Q = ratio of the energy stored to the energy dissipated in the circuit each half-cycle, or tuned-coil reactance, ohms/tuned-circuit coil resistance, ohms; Δf = amplifier half-bandwidth, MHz; f_0 = resonant frequency of the circuit.

2. Determine the amplifier resonant frequency

If the 3-db points are chosen at 21.0 and 21.5 MHz, then the resonant frequency $f_0 = (21.5 + 21.0)/2 = 21.25$ MHz. Also, $\Delta f = 21.25 - 21.0 = 0.25$ MHz.

3. Compute the first-stage Q

If all the selectivity were provided by the first stage $Q\Delta f/f_0 = 0.5$, then $Q = 0.5 f_0/\Delta f = 0.5(21.25)/0.25 = 42.5$.

With all the selectivity provided by the first stage, the Q of the second stage would be zero because the second stage would not provide any selectivity. This is an impractical case, because the total amplifier gain is zero.

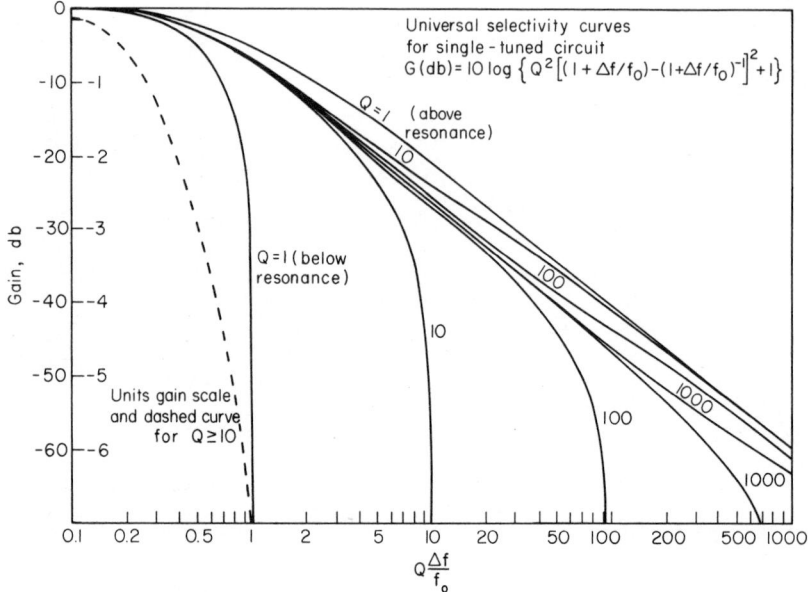

Fig. 14   Universal selectivity curves for single-tuned circuit *(Electronics.)*

4.   Compute the Q of the other stage to meet the bandpass requirements when the Q of one stage is restricted for some reason (for example, from maintaining a specific operating point under conditions of large stray capacitance)

With one stage restricted to a minimum Q or 35 with the same f_0 and Δf used in Step 2, $Q\Delta f/f_0 = 35(0.25)/21.25 = 0.412$. Entering Fig. 14 with this value of the ratio and projecting vertically upward to the dashed curve shows that this stage will contribute about 2.2 db of attenuation to the total amplifier.

With 3-db points, the other stage must then roll off $3.0 - 2.2 = 0.8$ db at the band edges. At this attenuation, the second stage requires a $Q\Delta f/f_0 = 0.225$, found from Fig. 14 by entering at 0.8 db. Then, $Q = 0.225 f_0/\Delta f$ for the second state, or 0.225 (21.25)/0.25 = 19.1.

Related Calculations: Figure 14 is equally useful for transistor amplifiers. Use a procedure similar to that given above for transistor amplifiers such as that shown in

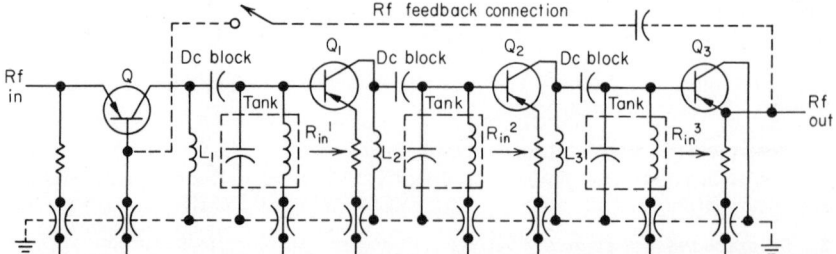

Fig. 15   Three-stage tuned transistor amplifier. Dashed line indicates the feedback connection necessary for a linear radio-frequency amplifier. *(Electronics.)*

Fig. 15. Note that the ratio $\Delta f/f_0$ is unity for the second harmonic, two for the third harmonic, three for the fourth harmonic, etc. In a multistage transistor amplifier the individual stage Q's must be equal for maximum out-of-band rejection. This procedure was developed by John D. Dunean, Montana State University, and published in *Electronics*.

ANTENNA SELECTION AND ANALYSIS

Choose a suitable antenna for a communications system operating at 100 MHz. What dimensions should the antenna have? Determine the system wavelength.

Calculation Procedure:

1. Select the type of antenna to use

Table 16 summarizes the names, types, uses, bandwidths, and wavelengths of a variety of antennas in common use. Study of this table shows that the half-wave dipole

TABLE 16 Antenna Design Characteristics*

Antenna name	Antenna type	Typical uses	Bandwidth	Typical wavelengths
Dielectric rod.......	Surface-wave	Radar feeds and arrays	10%	1–6 GHz
Yagi.........	Surface-wave	TV/FM reception	10%	1–5 GHz
Half-wave dipole	Resonant	Communications, navigation, radar, etc.	5%–40%	10 MHz–5GHz
Half-wave slot........	Resonant	Aircraft and missiles	. . .	100 MHz–35 GHz
Rhombic.....	Traveling-wave	Short-wave transmitting and receiving for long ranges	2–1	2–30 MHz
Axial mode helix	Traveling-wave	Tracking, telemetry, aerospace, ground stations	1.7–1	100 MHz–3 GHz
Log periodic	Frequency-independent	Ecm and direction finding	10–1	10 MHz–12GHz
Equiangular spiral......	Frequency-independent	Ecm, telemetry, aircraft, missiles, arrays,	10–1	100 MHz–35 GHz
Paraboloidal reflector ...	Aperture	Radar, communications, radio astronomy; other high-gain uses	Determined by feed	300 MHz–70 GHz
Conical horn .	Aperture	Radar, communications	1.6–1	300 MHz–70 GHz
Pyramidal horn	Aperture	Radar, communications	1.6–1	300 MHz–70 GHz
Dielectric lens	Aperture	Radar, communications, radio, astronomy; other high-gain uses	Determined by feed	300 MHz–70 GHz

*For additional data see R. S. Gordon and K. W. Duncan, "Ready-Reference Data Simplifies Antenna Design," *Electronics*, Dec. 21, 1962.

is a suitable antenna for communications in the 10-MHz to 5-GHz wavelength range. Hence, this type of antenna will be chosen for this application.

2. Compute the antenna dimensions

Use the relation dipole length, ft = 467.4/frequency, MHz. Or, dipole length = 467.4/100 = 4.67 ft for half-wave operation.

3. Compute the system wavelength

Use the relation $\lambda = 300 \times 10^6/f$, where λ = system wavelength, m; f = system frequency, Hz. Thus, $\lambda = 300 \times 10^6/100 \times 10^6 = 3$ m.

Related Calculations: The length of a quarter-wave antenna is $\lambda/4.2$ ft. A grounded quarter-wave antenna is termed a basic Marconi antenna.

To compute the power input to an antenna, use the relation $P = I^2R$, where P = power input to the antenna, watts; I = antenna current, amp; R = antenna resistance, ohms. Note that a properly tuned antenna is considered to be a pure resistance. Hence, the power, voltage, current, and resistance of a tuned antenna can be computed using Ohm's law and power equations.

MISMATCH EFFICIENCY AND POWER LOSS

An electronic signal source has an output impedance of 100 ohms. In attempting an impedance match with the load, the closest match that can be achieved is a load impedance of 200 ohms. What effect will this mismatch have on the power output and efficiency of the system?

Calculation Procedure:

1. Compute the mismatch ratio

The mismatch ratio for any electronic source-load combination is $R_m = Z_L/Z_s$, where R_m = mismatch ratio; R_L = load impedance, ohms; R_S = source impedance, ohms. For this system, $R_m = 200/100 = 2$.

2. Compute system power loss

Use the relation $P_L = 10[0.602 + \log R_m - 2 \log(1 + R_m)]$, where P_L = power loss, db; log = logarithm to the base 10; other symbols as before. For this system, $P_L = 10[0.602 + \log 2 - 2 \log 3] = -0.51$ db. Hence, the mismatch ratio of $R_m = 2$ causes a 0.51-db power loss.

3. Compute the system efficiency

Use the relation $E = 100[R_m/(1 + R_m)]$, where E = system efficiency, percent; other symbols as before. Thus, $E = 100[2/(1 + 2)] = 66.6$ percent.

Related Calculations: Use this procedure for any electronic equipment where an impedance match, or mismatch, is desired between a signal source and a load. If the match is between a source resistance and a load resistance, use the same method but substitute resistance for impedance in the above steps.

Where the mismatch ratio is unity — i.e., the source and load impedances match exactly — the power output and efficiency are the maximum. As the mismatch increases on either side of unity — i.e., less than or greater than unity — the power loss increases and the efficiency decreases.

PUBLIC-ADDRESS-SYSTEM DESIGN AND LAYOUT

Choose the number, type, and location of the speakers, and the amplifier rating for a factory sound system. The factory floor served by the system is 80 ft long and 100 ft wide. It is desired to use the public-address system for both voice and music amplification.

Calculation Procedure:

1. *Compute the number of speakers required*

The factory floor area is, in sq ft, $L \times W = 80 \times 100 = 8,000$ sq ft. Using this area, enter Table 17 at the line marked factories. This line shows that an 8,000 sq ft factory requires four speakers.

2. *Select the type of speaker to use*

Table 17 shows that the reentrant-type horn speaker is recommended for factories having floor areas of 8,000 or more sq ft. With only four speakers, each can operate at high output in a high-level speaker system. With a large number of speakers, eight or more, they can be the low-output type. This is termed a low-level speaker system. For this factory having four speakers, a high-level speaker system would probably be best.

3. *Select the public-address-system amplifier rating*

Table 17 shows that a 50-watt amplifier would be suitable for this factory. The output required for a public-address-system amplifier depends upon the size and type of area served by the sound system. Where music is to be played over the sound system, a

TABLE 17   Selection Guide for Public-Address Systems*

Application	Sq ft area	Amplifier rating, watts	Number of speakers	Types of speakers†
Auditoriums...	2,000	15	2	12-in. cone in wall baffles‡
	5,000	30	2	12-in. cone in wall baffles or
	15,000	50	4	12-in projector horns
Ballrooms.....	2,000	15	4	
	4,000	30	4	12-in. cone in wall baffles
	10,000	50	6	
Churches	1,000	10	2	10-in. cone in wall baffles
	4,000	15	2	12-in. cone in wall baffles
	15,000	30	4	
Classrooms,	500	10	1	8-in. cone in wall baffles
offices and	2,000	15	2	10-in. cone in wall baffles
stores	8,000	30	4	
Factories......	1,000	15	2	12-in. projector horns
	4,000	30	4	
	8,000	50	4	Re-entrant horns
	40,000	100	10	
Funeral	1,000	10	1	
parlors	4,000	15	4	12-in. cone in wall baffles
	10,000	30	8	
Restaurants	1,000	15	2	
and	5,000	30	6	12-in. projector horns
nightclubs	10,000	50	12	
Stadiums and	3,000	15	2	12-in. cone in wall baffles
gymansiums	10,000	30	4	Reentrant horns
	50,000	100	8	

*Values given in table are averages — not minimums or maximums.
†Number of speakers and amplifier power rating should be increased where background noise is higher than normal for the type of area. Acoustically "live" areas generally require lower speaker sound levels. Number of speakers will vary with shape of the plan view of the area.
‡Although wall baffles are indicated for cone speakers, ceiling-recessed or suspended baffles are frequently advantageous.

record player can be built into the amplifier housing. Where clarity of the amplified sound is critical, antifeedback controls are available on most standard amplifiers.

4. Select the amplifier power source

Standard sound amplifiers can be operated from 110–125-v 60-Hz or 115-v 25-Hz alternating current, or 115-v, 4.6-v, or 12-v direct current.

5. Choose the amplifier input devices

There are five major types of input devices. (*a*) Microphones — crystal, dynamic, or velocity, omnidirectional, bidirectional, or unidirectional (cardioid). (*b*) Record player: automatic or manual. (*c*) Tape player: single- or multiple-track. (*d*) Radio tuner: AM or FM. (*e*) Tone generator: produces a tone signal for factory work shifts, lunch periods, etc.; also used as an electronic siren for alarm applications; bell sounds for church belfry.

6. Choose the output-tap impedance values

The output-tap impedance value depends on the type of output devices selected in Step 5 because there must be an impedance match. In general three types of output taps are used. (*a*) Direction connection—4-, 8-, and 16-ohm taps. (*b*) Constant-voltage line-transformer connection with 70-, 100-, or 140-v taps. (*c*) Constant-impedance line-transformer connection with 250- and/or 500-ohm taps. Select one or more taps to give the desired impedance match with the output device or devices.

7. Choose the type of speaker connection

There are three main types of speaker connections. (*a*) Direct connection to the amplifier output taps corresponding in impedance value (ohms) to the impedance value (ohms) of a single speaker or a number of speakers in series, parallel, or series-parallel. (*b*) Connection to the amplifier constant-voltage output taps at 70, 100, 140 v, etc., through constant-voltage line-matching transformers. (*c*) Connection to amplifier high-impedance output taps of 250- or 500-ohms impedance through constant-impedance line-matching transformers.

8. Select the speaker locations

Locate the speakers on a plan of the area served. In churches, theaters, and auditoriums, place the speakers well forward of the microphones to prevent feedback (squealing).

Related Calculations: Use this general method to choose sound systems for auditoriums, ballrooms, churches, classrooms, offices, stores, factories, funeral parlors, restaurants, nightclubs, stadiums, gymnasiums, etc. For impedance values of the equipment selected, consult the manufacturer's engineering data. The method presented here is one recommended by *Electrical Construction and Maintenance* magazine.

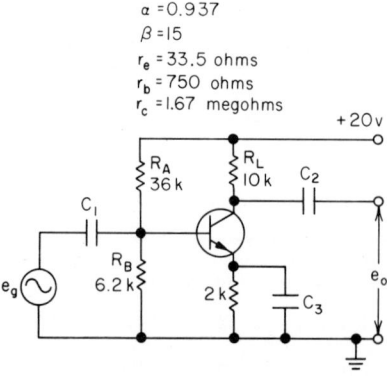

$\alpha = 0.937$
$\beta = 15$
$r_e = 33.5$ ohms
$r_b = 750$ ohms
$r_c = 1.67$ megohms

Fig. 16   Transistor voltage amplifier.

TRANSISTOR HYBRID-PARAMETER CONVERSIONS

Show how to convert the known parameter values for a common-base connected transistor to the unknown parameter values of a common-emitter transistor. How are the T-equivalent parameters related to the hybrid parameters? List the T-equivalent circuit equations for the three types of transistor amplifiers most commonly used. Compare the transistor amplifier parameters found by means of the T-equivalent circuit equations with the test-determined parameters shown in Fig. 16 for a transistor voltage amplifier.

Calculation Procedure:

1. List the transistor hybrid-parameter conversions

Table 18 lists the usual symbols employed in transistor calculations and analysis. In actual transistor circuit design a T-equivalent circuit is often used in circuit analysis in about the same way that the vacuum-tube equivalent circuit is used. The T-equivalent circuit is particularly appropriate for circuit analysis because its parameters are directly related to the basic physical structure of the transistor.

Although the T-equivalent circuit is a satisfactory representation of transistor operation, difficulty is met in determining the various resistance parameters by direct measurements on actual transistors. For this reason an equivalent circuit using *hybrid parameters* is often employed in circuit analysis.[1] Table 19 lists the usual subscript notation for hybrid, or *h*, parameters.

[1]Brophy—*Basic Electronics for Scientists*, McGraw-Hill.

TABLE 18   Usual Transistor Parameter Symbols*

h_{FE}, B	Common emitter dc current gain	f_t	Gain bandwidth product where h_{fe} equals 1
h_{fe}, β	Common emitter small-signal current gain with output ac short-circuited	I_{CBO}, I_{co}	Dc collector current where emitter is open-circuited (leakage current)
h_{fb}, α	Common-base small-signal current gain with output ac short-circuited	I_{CEO}	Dc collector current where base is open-circuited
h_{fc}	Common-collector small-signal current gain with output ac short-circuited	I_{CER}	Dc collector current where a resistor is connected from base to emitter
h_{ib}	Common-base small-signal input impedance with output ac short-circuited	I_{CES}	Dc collector current where the base is shorted to the emitter
h_{ie}	Common emitter small-signal input impedance with output ac short-circuited	BV_{CBO}	Dc breakdown voltage collector to base with emitter open-circuited
h_{ic}	Common collector small-signal input impedance with output ac short-circuited	BV_{CEO}	Dc breakdown voltage collector to emitter with base open-circuited
h_{ob}	Common base small-signal output admittance with input ac open-circuited	BV_{CER}	Dc breakdown voltage collector to emitter with a resistor connected from base to emitter
h_{oe}	Common emitter small-signal output admittance with input ac open-circuited	BV_{CES}	Dc breakdown voltage collector to emitter with base shorted to emitter
h_{oc}	Common collector small-signal output admittance with input ac open-circuited	V_{BE}	Base-to-emitter voltage
		$V_{CE(sat)}$	Collector-to-emitter saturation voltage
		r_e	Small-signal emitter resistance
h_{rb}	Common base small-signal reverse voltage transfer ratio with input ac open-circuited	r_b	Small-signal base resistance
		r_c	Small-signal collector resistance
		r_m	αr_c
h_{re}	Common emitter small-signal reverse voltage transfer ratio with input ac open-circuited	r_d	$(1-\alpha)r_c$
		r_{sat}	Collector-to-emitter saturation resistance
h_{rc}	Common collector small-signal reverse voltage transfer ratio with input ac open-circuited	C_{ob}	Output capacitance
		C_c	Collector-to-base capacitance
		θ_{JC}	Thermal resistance junction to case (C/W)
f_α, f_{hfb}	Common base small-signal current gain cutoff frequency	t_s	Storage time
		t_{on}	Turn-on time
f_β, f_{hfe}	Common emitter small-signal current gain cutoff frequency	t_{off}	Turn-off time
		t_r	Rise time
f_{max}	Maximum frequency of oscillation	t_f	Fall time

*Long—*Modern Electronic Circuit Design*, McGraw-Hill.

TABLE 19   Subscript Notation for h Parameters

Subscript	Usual Meaning
i	Input parameter
r	Reverse parameter
f	Forward parameter
o	Output parameter
e	Common emitter
b	Common base
c	Common collector

TABLE 20   Transistor Parameter Conversions*

	Common base	Common emitter	Common collector	Approximate equivalent
α	$|h_{fb}|$	$\dfrac{h_{fe}}{1+h_{fe}}$	$\dfrac{|h_{fc}|-1}{|h_{fc}|}$	$\dfrac{\beta}{\beta+1}$
r_e	$h_{ib}-\dfrac{h_{rb}(1-|h_{fb}|)}{h_{ob}}$	$\dfrac{h_{re}}{h_{oe}}$	$\dfrac{1-h_{rc}}{h_{oc}}$	
r_b	$\dfrac{h_{rb}}{h_{ob}}$	$h_{ie}-\dfrac{h_{re}(1+h_{fe})}{h_{oe}}$	$h_{ic}-\dfrac{|h_{fc}|(1-h_{rc})}{h_{oc}}$	
r_c	$\dfrac{1-h_{rb}}{h_{ob}}$	$\dfrac{1+h_{fe}}{h_{oe}}$	$\dfrac{|h_{fc}|}{h_{oc}}$	
h_{fe}	$\dfrac{|h_{fb}|}{1-|h_{fb}|}$	....	$|h_{fc}|-1$	β
h_{ie}	$\dfrac{h_{ib}}{1-|h_{fb}|}$	....	h_{ic}	$r_b+(\beta+1)r_e$
h_{re}	$\dfrac{h_{ib}h_{ob}}{1-|h_{fb}|}-h_{rb}$	....	$1-h_{rc}$	$\dfrac{r_e(\beta+1)}{r_c}$
h_{oe}	$\dfrac{h_{ob}}{1-|h_{fb}|}$	....	h_{oc}	$\dfrac{1}{r_d}$
$|h_{fb}|$	....	$\dfrac{h_{fe}}{1+h_{fe}}$	$\dfrac{|h_{fc}|-1}{|h_{fc}|}$	α
h_{ib}	....	$\dfrac{h_{ie}}{1+h_e}$	$\dfrac{h_{ic}}{|h_{fc}|}$	$r_e+\dfrac{r_b}{\beta+1}$
h_{rb}	....	$\dfrac{h_{ie}h_{oe}}{1+h_{fe}}-h_{re}$	$(h_{rc}-1)+\dfrac{h_{ic}h_{oc}}{|h_{fc}|}$	$\dfrac{r_b}{r_c}$
h_{ob}	....	$\dfrac{h_{oe}}{1+h_{fe}}$	$\dfrac{h_{oc}}{|h_{fc}|}$	$\dfrac{1}{r_c}$

*Long—*Modern Electronic Circuit Design*, McGraw-Hill.

Transistor specification or data sheets may specify hybrid parameters for only one transistor configuration—for example, common-base or common-emitter. When this occurs, the designer may have to convert from one set of parameters to another. Table 20 lists a number of hybrid-parameter conversions, including the T-equivalent parameters for three configurations—common-base, common-emitter, and common-collector.

2. List the transistor amplifier T-equivalent equations

Table 20 also lists the T-equivalent circuit equations for common-base, common-emitter, and common-collector transistor amplifiers. The type of transistor chosen for an amplifier may depend upon the types of transistors used in other parts of the equipment.† The reason for this is that most designers try to restrict the number and types of transistors used in electronic equipment because this reduces the initial and maintenance costs of the equipment.

3. Compute the T-equivalent amplifier parameters

Use the equations given in Table 21 to determine the input resistance, voltage gain, and output resistance for the common-emitter amplifier shown in Fig. 16. Thus, $R_{in} \cong r_b + \beta r_e$, where R_{in} = input resistance, ohms; other symbols as listed in Table 18. Substituting, $R_{in} \cong 750 + (15)(33.5) = 1,250$ ohms.

TABLE 21   Transistor Amplifier Parameters*

	Common base	Common emitter	Common collector
A_v	$\dfrac{\beta R_L}{r_b + \beta r_e}$	$\dfrac{-\beta R_L}{r_b + \beta r_e}$	1
A_i	α	$-\beta$	$\beta + 1$
R_o	r_c	r_d	$\dfrac{r_b + R_g}{\beta}$
R_{in}	$r_e + \dfrac{r_b}{\beta}$	$r_b + \beta r_e$	$r_b + \beta R_L$

*Long—*Modern Electronic Circuit Design*, McGraw-Hill.

The actual amplifier input resistance R'_{in} ohms equals the parallel combination of R_{in} ohms and the two base-biasing resistors of 6.2 kilohms and 36 kilohms, Fig. 16. Or, $R' = R_{in} = R_L/(1/R_{in} + 1/R_A + 1/R_B) = 10,000/(1/1.25 + 1/6.2 + 1/36) = 1.0$ kilohm. The measured value of $R'_{in} \times 1.2$ kilohms.

The voltage gain $A_V \cong -\beta R_L/(r_b + \beta r_e) = -15(10,000)/1,250 = -120$, compared with a measured voltage gain of -112. Note in Table 21 that A_i = the amplifier current gain.

The output resistance R_o ohms is approximately equal to r_d in parallel with the 10-kilohm load resistance, where $r_d = (1 - \alpha) r_c$. Using the given value of $r_c = 1.67$ megohms, $r_d = (1 - 0.937)1,670,000 = 105,000$ ohms. Then, $R_o = r_d R_L/(r_d + R_L) = (105)(10\text{ K})/115 = 9.15$ kilohms. The measured value of $R_o = 9.1$ kilohms.

Under ordinary design conditions one would not expect better than a 5 percent correlation between calculated and measured or experimental results. Resistor tolerances are usually ±5 percent, depending upon the resistor specifications. Further, transistor parameters are not known exactly. The results of this calculation show that the computed values for a particular amplifier may vary appreciably from measured or experimental results unless a method of gain stabilization is employed. Where more complete voltage-gain equations are used instead of the approximations in Table 21, a greater degree of accuracy is obtained. However, the approximate equations yield acceptable results where gain stabilization is employed.

Related Calculations: Other transistor parameters used in circuit analysis are the hybrid-pi parameters. For a complete discussion of hybrid-pi parameters, see Hunter— *Handbook of Semiconductor Electronics*, McGraw-Hill.

The "black-box" amplifier, Fig. 17, is useful in deriving the h parameters. Thus, in Fig. 17, $v_1 = i_1 h_{11} + v_2 h_{12}$, and $i_2 = i_1 h_{21} + v_2 h_{22}$. These equations can be solved for the short-circuited case for input impedance, $h_{11} = v_1/i_1$, forward current gain, $h_{21} = i_2/i_1$. With the input ac open-circuited, that is, $i_1 = 0$, $h_{12} = v_1 v_2$, and $h_{22} = i_2/v_2$.

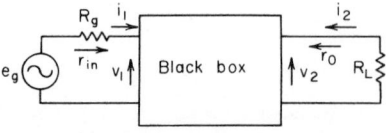

Fig. 17 Transistor "black-box" amplifier.

The general subscripts associated with h_{11}, h_{12}, h_{21}, and h_{22} are usually changed to indicate the transistor configuration. Then, h_{ie} indicates the common-emitter input impedance, h_{fe} the common-emitter current gain, h_{re} the common-emitter reverse-voltage transfer ratio, h_{oe} the common-emitter output admittance. In the case of the common-base configuration, the parameters are h_{ib}, h_{fb}, h_{rb}, and h_{ob}. Similarly, for the common-collector configuration, the parameters are h_{ic}, h_{fc}, h_{rc}, and h_{oc}. All these parameters are summarized in Table 20.

Transistor parameters are particularly useful in transistor circuit analysis. For a lucid discussion of transistor parameters, see Long—*Modern Electronic Circuit Design*, McGraw-Hill.

LARGE-SCALE INTEGRATION ANALYSIS

How many input-output pins are required for a large-scale integration package having 27 circuits if the average fan-in of the circuits in the package is 2.5? Compare the computed number of pins with actual design practice. Show how the number of pins can be reduced during circuit design.

Calculation Procedure:

1. Compute the number of input-output pins required

Use the relation $P = kN^{2/3}$, where P = number of input-output pins required; k = average fan-in of circuits plus $1 = 2.5 + 1 = 3.5$ for this package; N = number of circuits in package. Substituting, $P = 3.5(27)^{2/3} = 31.5$ input-output pins.

2. Compare the computed pins with design practice

Table 22 compares the computed number of input-output pins as found in Step 1 with the number of pins used in actual designs. Thus, with 27 circuits in partitions, actual

TABLE 22   Actual versus Computed Partitions*

Number of circuits in partitions	Computed number of pins	Number of pins in actual designs
1	3.5	
8	14.0	
27	31.0	12–24
64	55.0	24–41
125	87.0	39–52
216	125.0	
343	170.0	
512	220.0	45–96
729	285.0	45–92
1,000	350.0	62–120

Electronics

designs use 12 to 24 pins. Figure 18 shows how actual designs for large and small systems deviate from the computed number of pins.

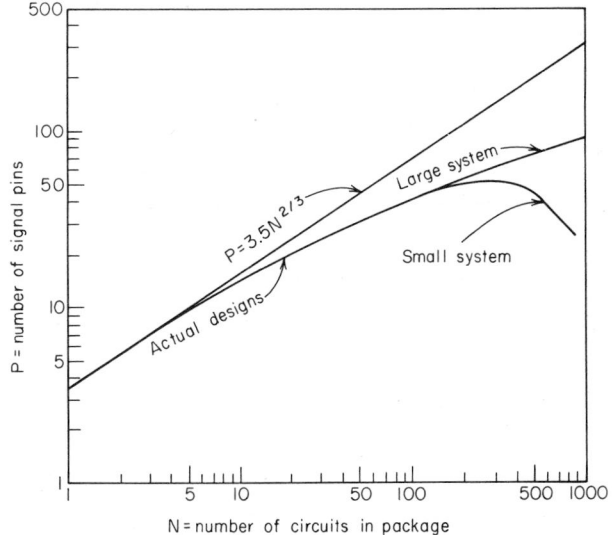

Fig. 18 Comparison of computed and actual design signal pins in large-scale integration. (*Electronics.*)

Related Calculations: If the number of chips—i.e., the first-level package count—can be reduced without increasing the total number of input-output pins, then the wiring on the second-level package will be reduced. The trade-off here presents a choice between either increasing the unit package cost at the first level or cutting the cost of the second-level package.

To achieve a minimum first-level package count while at the same time reducing the second-level package wiring, a new set of rules has been devised to augment those in general practice, which are applied at low levels of integration to partition logic into small units with few interconnections. These four rules are as follows.[1]

1. In general, to reduce the number of first-level packages and at the same time achieve high circuit-to-pin ratios, begin looking at larger collections of logic circuits as candidates for the partitions, or organize the logic into units with high circuit-to-pin ratios in mind.

2. Where possible, add supernumerary circuits to encode all nonindependent signals leaving the chip and to decode those signals at the destination chips.

3. Partition at the outputs of bit-storage cells both in data-flow sections and in sequential-circuit control sections, because here the information is usually completely coded. Try to avoid cutting within the sections of signal-combining circuitry that lie between sets of storage cells; the intercircuit connection density here is often quite high.

4. Choose as outputs from a chip those signals having large fan-out, and keep all the destination points together—on one chip if possible. The signals with large fan-out can be replicated on the destination chip as required.

FREQUENCY-CHANGER SELECTION AND APPLICATION

Choose a frequency changer suitable for a 60-Hz input and a 400-hZ 1-kva output with a voltage regulation of 0.3 percent and an output voltage of 220 v. What harmonic distor-

---

[1]Special Report, "Large Scale Integration," *Electronics*, Feb. 20, 1967.

tion, frequency regulation, and response time can be expected in the changer selected? Input and output are both single-phase. What is the weight and volume of the changer?

Calculation Procedure:

1. Make a preliminary choice of the type of changer

Figure 19 summarizes the frequency and power output ranges of three popular types of frequency changers.[1] Study of Fig. 19 shows that either a motor-generator set or a solid-state frequency changer would be suitable from the frequency and power standpoints. Hence, the final choice must be based on additional performance requirements.

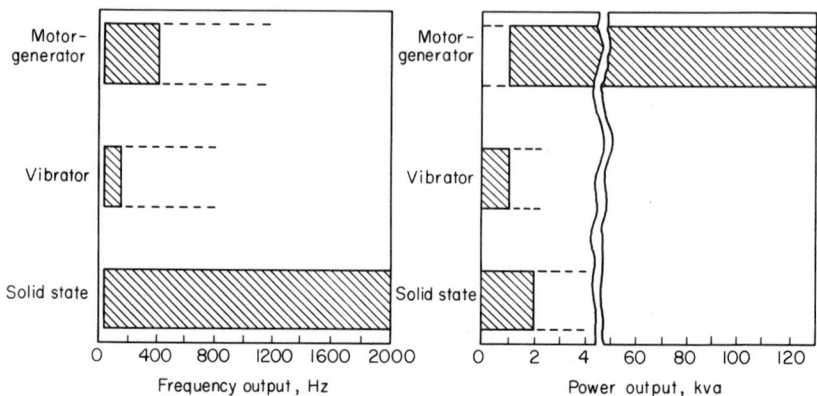

Fig. 19   Preliminary selection chart for frequency changers. (*Product Engineering.*)

2. Analyze the changer electrical performance

Table 23 summarizes the electrical and mechanical aspects of the three popular types of frequency changers — motor-generator sets, vibrators, and solid-state changers. Study of Table 23 shows that the voltage regulation of a motor-generator set varies between 0.5 and 3 percent and the voltage regulation of a solid-state frequency changer is 0.5 percent. Since the desired regulation is 0.3 percent, the solid-state changer is more suitable because it has a smaller voltage-regulation range.

Further, Table 23 shows that an output voltage of 120 to 220 v is obtainable from solid-state frequency changers at power outputs up to 2 kva. A harmonic distortion of 5 percent, frequency regulation of 0.01 to 1 percent, and a response time of 0.1 sec can be expected from the solid-state frequency changer.

3. Compute the changer volume and weight

The weight of this solid-state frequency changer will be about $0.2(1,000 \text{ va}) = 200 \text{ lb}$, using the mechanical data from Table 23. Further, the volume of the usual solid-state frequency changer is about 30 in.^{3}/lb. Hence, the volume of this changer would be $(30)(200) = 6,000$ in.^{3}.

Related Calculations: Note that for higher frequency outputs or higher power outputs (above 500 Hz and 2 kva, respectively), Fig. 19 shows that the choice of changer type is restricted. Thus, Fig. 19 indicates that a solid-state changer is needed for high frequencies and a motor-generator set is needed for high power outputs.

Vibrator frequency changers operate by rectifying alternating current to direct current, then chopping (switching) direct current into square-wave alternating current followed possibly by wave-shaping for a sinusoidal output. The output power can be used directly at low frequencies; but at frequencies above 150 Hz, the output must be amplified.

[1]Ante Lujic and Michael S. Inoue, "Frequency Changers," *Product Engineering*, Dec. 24, 1962.

TABLE 23 Frequency-Changer Characteristics

Type	Motor-Generator set	Vibrator	Solid-State
Electrical:			
Input:			
Power, kva	Up to 120	Up to 2	Up to 2
Voltage......................	120–550	10–200	12–220
Power factor................	0.8		
Phase	one, two, three	one, two, three	one, two, three
Waveform	Sine	Square	Sine or square
Frequency, cps	50/60	Dc or ac	Dc or ac
Output:			
Power, kva	Up to 120	Up to 2	Up to 2
Voltage......................	120–550	10–200	120–220
Variable voltage	Possible	Possible	Possible
Voltage regulation, %........	0.5–3	5	0.5
Efficiency, %	85	50 to 80	80
Current, amp	Up to 450	Up to 10	Up to 10
Overload, %	125	...	Very low
Phase	one, two, three	one	one, two, three
Waveform	Sine	Square	Sine or square
Frequency, cps	Up to 420	Up to 150	Up to 2,000
Harmonic distortion, %.......	2–3	5	5
Variable frequency..........	Possible	Yes	Yes
Variable regulation, %.......	1	1–5	0.01–1
Response time, sec	0.25	...	0.1
Mechanical:			
Weight, lb....................	Up to 3,000	Up to 100	0.2 lb/va output
Temperature, C..............	−40–120	−40–60	−40–60
MIL specification construction ..	Available	...	Available

Vibrator operation at frequencies much above 150 Hz is generally unreliable. Special choppers have been developed for frequencies up to 2,000 Hz but the power range is in mw.

Solid-state changers use silicon diodes in the rectifier and either power transistors or silicon-controlled rectifiers (SCRs) as switching components in the inverter. The selection and application procedure given here was developed by the writers cited in the reference below.

USES OF THE SMITH CHART IN ELECTRONICS

A lossless transmission line was tested and the voltage–standing-wave ratio measured as 3.7. With the load connected, a voltage minimum appeared at 62.7 mm; with the load shorted, the minimum moved to 79.6 mm. With the load still-shorted, the next minimum towards the load was found at 159.6 mm. Use the Smith chart to analyze this transmission line. Next, consider the case where the cable between the slotted line and the load has an attentuation of 1 db per 100 ft and the cable length is 50 ft. Determine from the Smith chart the new conditions for this transmission line.

Calculation Procedure:

1. *Draw the reflection coefficient circle*

The Smith chart, Fig. 20, has a special impedance coordinate system that portrays the impedance at any point along a transmission line in relation to the impedance at any other point.[1] The scales around the outside of the chart are calibrated in fractions of a wavelength, and the full circle (360°) corresponds to one-half wavelength. Resistances and reactances are plotted as fractions of the impedance of the transmission line. The resistance lines are circles that are tangent to the bottom of the chart, and the reactance lines are portions of circles that are tangent to the vertical line through the chart, which is called the *axis of reals.*

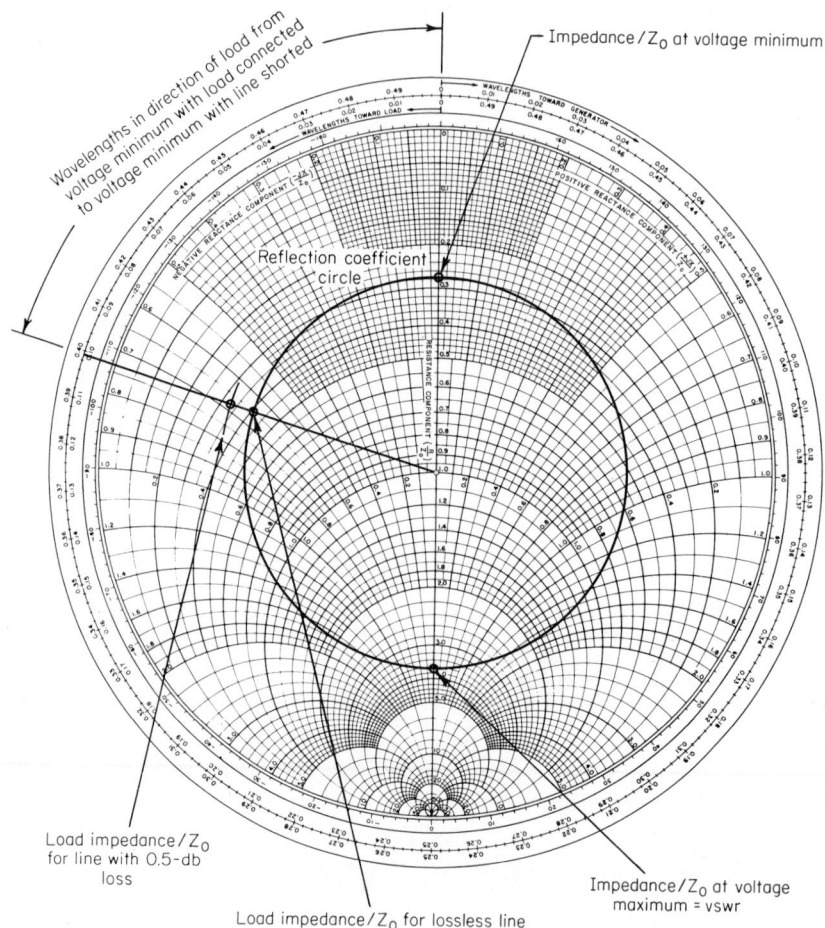

Fig. 20   Smith-chart solution of a transmission-line problem.

Draw the reflection-coefficient circle with its center at $R = 1$, $X = 0$, Fig. 20, with a radius of 3.7, that is, the circumference of the circle cuts the axis of reals at 3.7.

2.   Compute the full wavelength of the signal generator

The half-wavelength of this line is, from the measured values, $159.6 - 79.6 = 80.0$ mm. Hence, the full wavelength = $2(80.0) = 160$ mm.

[1]Gray and Graham—*Radio Transmitters*, McGraw-Hill.

3. Compute the wavelength fraction

Use the relation wavelength fraction from the "load" minimum to the "short" minimum = (load-shorted minimum, mm − voltage minimum, mm)/wavelength, mm = (79.6 − 62.7)/160.0 = 0.1057.

4. Determine the load impedance

Draw a line from the center of the reflection-coefficient circle through a value of wavelengths *toward* the load of 0.1057 on the chart circumference, Fig. 20. At the intersection of this line with the circumference of the reflection-coefficient circle, read the load impedance as $0.41 − j0.65$ ohms.

If this slotted transmission line had an impedance of 50 ohms, then the actual load impedance would be $50(0.41 − j0.65) = 20.5 − j32.5$ ohms.

5. Compute the transmission-line attenuation

Use the relation total attenuation, db = transmission-line length, ft/length, ft, for 1-db attenuation = $50/100 = 0.5$ db.

6. Determine the new load impedance

The voltage–standing-wave ratio for the lossless line was 3.7. With a transmission-line attenuation of 0.5 db towards the load, the voltage–standing-wave ratio will increase to about 4.5, as determined from a standard engineering source, such as *Reference Data for Radio Engineers*, ITT.

Plot the new reflection-coefficient circle, Fig. 20. The intersection of the previously drawn line with the new circle shows that the impedance is $0.33 − j0.68$ ohms.

Related Calculations: The Smith chart has numerous applications, including general impedance transformations, stub-tuner design, finding of the input impedance, etc. For a full discussion of the chart and its uses, see Smith − *Electronic Applications of the Smith Chart*, McGraw-Hill.

ULTRASONIC-GENERATOR SELECTION, CRYSTAL THICKNESS, AND OUTPUT

What type of ultrasonic generator should be used for an industrial-cleaning operation requiring a frequency of 40,000 Hz? Determine the power output required if the cleaning is done in a 30-gal bath. What crystal thickness is required for the generator if the crystal is an *x*-cut or a *y*-cut? Sketch a typical vacuum-tube ultrasonic-generator circuit. What wave amplitude should be used for this cleaning operation?

Calculation Procedure:

1. Select the type of ultrasonic generator to use

Magnetostriction is used for frequencies of the order of 30,000 Hz; crystals are used for most frequencies above 30,000 Hz up to about 15 MHz. Higher frequencies can be obtained by vibrating the crystal at one of its harmonics. Motor-generator sets are used for relatively low frequency applications. Since this cleaning operation requires a frequency of 40,000 Hz, a crystal generator will be used.

Figure 21 shows the circuit of a typical vacuum-tube crystal generator. Ultrasonic generators are essentially high-power oscillators such as are commonly used in radio[1] communications.

2. Determine the wave amplitude required

Table 24 shows that a high-amplitude wave is required for cleaning. The energy may be continuous, pulsed, or modulated, in various ways.

3. Compute the cleaning-bath power level required

The usual power level used for ultrasonic cleaning is 50 watts/gal of cleaning solution. With a 30-gal cleaning solution, the power level required is (50 watts/gal)(30 gal) = 1,500 watts.

[1]Carlin − *Ultrasonics*, McGraw-Hill.

4. *Compute the required crystal thickness*

For an x-cut crystal use the relation $t = 0.1126/f$, where t = crystal thickness, in.; f = ultrasonic frequency, MHz. Substituting, $t = 0.1126/0.04 = 2.815$ in. For a y-cut crystal use the relation $t = 0.0771/f = 0.0771/0.04 = 1.927$ in. Thus, a y-cut crystal would be thinner than an x-cut crystal. Table 24 lists the required crystal thickness for various common frequencies; Table 25 lists the usual frequencies for various ultrasonic applications.

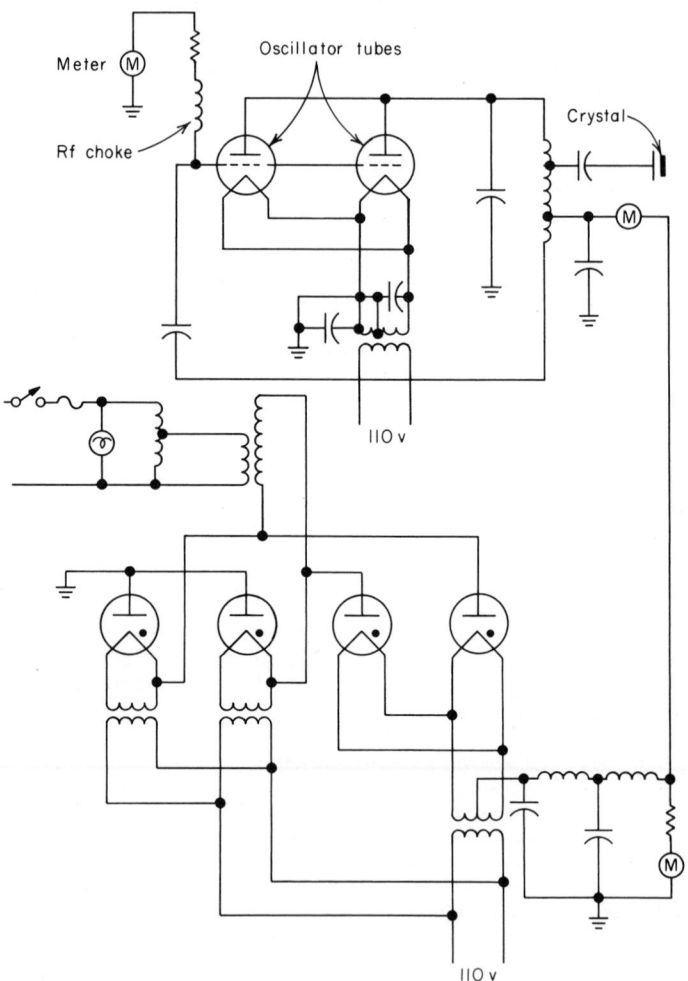

Fig. 21   Vacuum-tube ultrasonic generator.

Related Calculations: Use this general method to choose the frequency, crystal thickness, and power output for any of a large number of other ultrasonic applications — drilling, soldering, blind-guidance devices, resonance testing, emulsion production, liquid agitation, high-polymer reactions; biological work, medical therapy, etc. Tables 24 and 25 are useful guides to the amplitude, crystal thickness, and frequencies used in various ultrasonic applications.

TABLE 24   Ultrasonic Wave Amplitude and Crystal Thickness

Operation	Wave amplitude	Frequency, MHz	Thickness, in.	
			x-cut	y-cut
Cleaning	High	0.1	1.126	0.771
Welding.	High	0.5	0.224	0.154
Drilling	High	1.0	0.113	0.0788
Emulsification	High	2.0	0.0576	0.0394
Soldering.	High	3.0	0.0377	0.026
Medical therapy	High	4.0	0.0282	0.0197
Sonar	High	5.0	0.0226	0.016
Chemical.	High	6.0	0.0188	0.013
Biological	High	7.0	0.0160	0.0112
Materials testing	Low	8.0	0.0140	0.00964
Burglar alarms	Low	9.0	0.0124	0.00856
Delay lines	Low	10.0	0.0113	0.00771
Medical diagnoses . . .	Low	Varies	Varies	Varies

TABLE 25   Frequencies for Ultrasonic Applications*

Applications	Usual Frequency, kHz
Sonic altimeter.	1.0
Drilling, soldering, cleaning	16–20
Agitation of liquids, aerosol reactions	16–20
Burglar alarms.	19.2
Control apparatus and door opening	25
Cleaning (most common types)	40
Blind-guidance devices (upper limit for	
magnetostriction)	60
Galton whistle (upper limit).	100
Gas whistles in air (upper limit)	120
Resonance testing; emulsion formation.	300
Emulsion; agitation	400
Pulsed material testing (lower limit)	500
High-polymer reactions	600
Experimental biological work	750
Material tests; medical therapy; mixing;	
cleaning.	1,000
Testing of fine, homogeneous material	5,000–25,000

*After Carlin — *Ultrasonics*, McGraw-Hill.

APPLICATIONS OF SONAR EQUATIONS

What is the detection range of an active sonar gear having a source level of 100 db and a receiving directivity index of 20 db if it uses processing that produces a detection threshold of $+2$ db and is used against a target having a strength of 12 db when the noise level is -8 db? The gear produces a spherical spreading signal and there is no absorption of the signal. Give the sonar equations for active (monostatic) and passive sonar gears. What is the typical target strength for a bow-stern aspect of a fleet submarine?

Calculation Procedure:

1. *Compute the sonar-gear transmission loss*

Use the relation $TL = 0.5(SL + TS - NL + DI - DT)$, where TL = transmission loss, db; SL = source level, db; TS = target strength, db; NL = noise level, db; DI =

receiving directivity index, db; DT = detection threshold, db. Substituting, TL = 0.5(100 + 12 − (−8) + 20 − 2) = 69 db.

2. Compute the detection range

When the gear emits a signal that spreads spherically the relation between the range r, in yd, and the transmission loss is TL = 20 log r. Substituting, 69 = 20 log r; r = 2,820 yd.

Figure 22a shows the range of a sonar gear obtained under ideal conditions—i.e., a constant water temperature at all depths in which the gear is used. However, the temperature of sea water is seldom constant and refraction of the signal usually occurs. The range might then be reduced to the area shown in Fig. 22b, regardless of the power source. Refraction of the signal may reduce the range at which a submarine can be detected to almost zero.

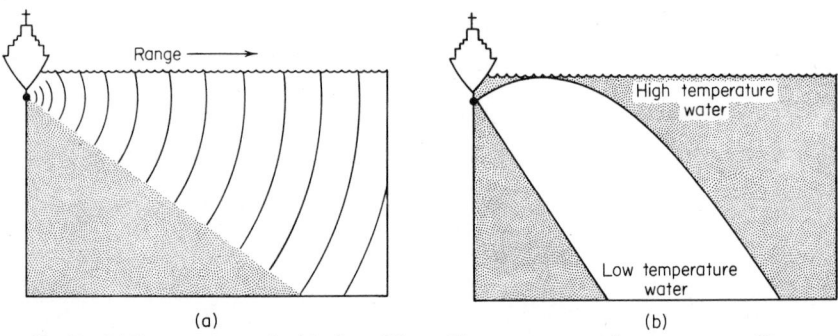

Fig. 22   (a) Sonar range under ideal conditions; (b) sonar range under average conditions.

3. List the sonar equations

The basic equations used in system design and analysis are, for *active* sonars, noise background − SL − 2TL + TS = NL + DI + DT; reverberation background − SL − 2TL + TS = RL + DT; passive sonars − SL − TL = NL − DI + DT.

4. Determine the vessel target strength

Table 26 lists nominal values of target strength for various vessels, torpedoes, and marine life. Study of Table 26 shows that a bow-stern aspect of a submarine will produce a target strength of + 10 db.

Related Calculations: Sonar calculations are less certain than many other types of engineering computations because the influencing external factors are so unpredictable. For a complete analysis of sonar calculations, see Urick—*Principles of Underwater Sound for Engineers*, McGraw-Hill.

TABLE 26   Nominal Values of Target Strength*

Target	Aspect	Target strength, db
Submarines.............	Beam	+ 25
	Bow-stern	+ 10
	Intermediate	+ 15
Surface ships	Beam	+ 25 (highly uncertain)
	Off-beam	+ 15 (highly uncertain)
Mines.................	Beam	+ 10
	Off-beam	+ 10 to − 25
Torpedoes..............	Bow	− 20
Fish of length L, ft.......	Dorsal view	− 31 + log L

*Urick—*Principles of Underwater Sound of Engineers*, McGraw-Hill.

Section **6**

Chemical Engineering

ROBERT L. DAVIDSON
Chemical Engineering

REFERENCES: Perry—*Chemical Engineers' Handbook*, McGraw-Hill; Kraus—*Pneumatic Conveying of Bulk Material*, Ronald; Box and Draper—*Evolutionary Operation*, Wiley; McCabe and Smith—*Unit Operations of Chemical Engineering*, McGraw-Hill; Warn—*Concise Chemical Thermodynamics in SI Units*, Van Nostrand, Reinhold; Weast and Selby—*Handbook of Chemistry and Physics*, Chemical Rubber; Erskine—*Chemical Conversion Factors and Yields*, Chemical Information Services; Anderson and Wenzel—*Introduction to Chemical Engineering*, McGraw-Hill; Levenspiel—*Chemical Reaction Engineering*, Wiley; Henley and Bieber—*Chemical Engineering Calculations*, McGraw-Hill; Brown—*Unit Operations*, Wiley; Peters—*Plant Design and Economics for Chemical Engineers*, McGraw-Hill; Vilbrandt and Dryden—*Chemical Engineering Plant Design*, McGraw-Hill; Foust et al.—*Principles of Unit Operations*, Wiley; Williams—*Systems Engineering for the Process Industries*, McGraw-Hill.

For additional Calculation Procedures useful in Chemical Engineering, please refer to the following sections of this Handbook: Section 3, Mechanical Engineering; Section 4, Electrical Engineering; Section 7, Control Engineering; Section 11, Sanitary Engineering; Section 12, Engineering Economics. Each of these sections contains a number of Calculation Procedures pertinent to the content of Section 6, Chemical Engineering, but size limitations prevent their repetition in Section 6.

ANALYSIS OF A SATURATED SOLUTION

If 1,000 gal of water is saturated with potassium chlorate ($KClO_3$) at 80 C, determine (*a*) the weight, lb, of $KClO_3$ that will precipitate if the solution is cooled to 30 C; and (*b*) the weight of $KClO_3$ that will precipitate if one-half the 1,000 gal of water is evaporated at 100 C.

Calculation Procedure:

1. Compute the precipitate when the solution is cooled

When a solid is dissolved in water (or any other solvent liquid), the resulting solution is termed *saturated* when at a given temperature the solvent cannot dissolve any more of the solid. Most solvents dissolve (hold) more solids at higher temperatures than at lower temperatures. Thus, when the solution temperature is lowered or a portion of the solvent is evaporated, the solution becomes *supersaturated* and solid material may precipitate. This is the basis of *crystallization*, a chemical engineering operation frequently used to produce a purer or more crystalline product.

Referring to Fig. 1, obtain these solubilities: at 80 C, $KClO_3$ solubility = 38.5 g per 100 g H_2O; at 30 C, $KClO_3$ solubility = 10.5 g per 100 g of H_2O.

The weight of the water at 80 C = (1,000 gal H_2O)(0.97183 g H_2O per cu cm H_2O) × (1 lb per 454 g) = 8,103 lb. Now, the weight of $KClO_3$ that any solvent can dissolve at a

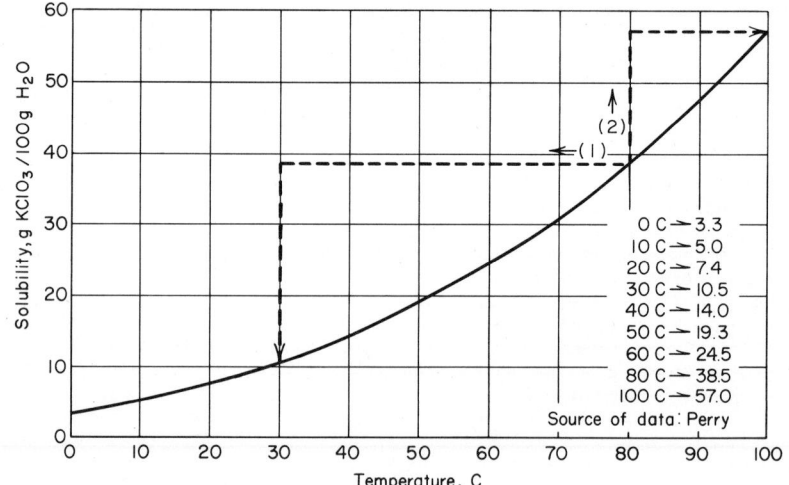

Fig. 1   Solubility of $KClO_3$.

given temperature = weight of solvent at the given temperature, lb (solubility of $KClO_3$ at the given temperature, g per 100 g of the solvent). Or, at 80 C, weight of $KClO_3$ dissolved by the water = (8,103 lb of water)(38.5 g $KClO_3$ per 100 g of H_2O) = 3,119 lb of $KClO_3$. And at 30 C with the same quantity of water but the reduced solubility, the weight of $KClO_3$ that can be dissolved = (8,103)(10.5 g per 100 g) = 851 lb of $KClO_3$.

When the temperature of the water (solvent) is reduced from 80 to 30 C, the weight of $KClO_3$ precipitated = weight of $KClO_3$ dissolved at 80 C − weight of $KClO_3$ dissolved at 30 C, or 3,119 − 851 = 2,271 lb of $KClO_3$ precipitated.

Note that the same Procedure can be followed for any similar solution — i.e., any similar solvent and solid. Neither the solvent nor the solid need be the ones considered here.

2. Compute the precipitate when a portion of the solvent is evaporated

Since half the solvent (water in this case) is evaporated, the weight of water remaining = 8,103/2 = 4,051.5 lb. Using the solubility of $KClO_3$ as before, except that the solvent temperature is 100 C, the weight of $KClO_3$ dissolved = 4,051.5(57 g $KClO_3$ per 100 g H_2O) = 2,309 lb of $KClO_3$. Then, the weight of $KClO_3$ precipitated by the evaporation = weight of $KClO_3$ dissolved in 1,000 gal of water at 80 C − weight of $KClO_3$ dissolved in 500 gal of water at 100 C = 3,119 − 2,309 = 810 lb of $KClO_3$ precipitated.

TERNARY LIQUID SYSTEM ANALYSIS

For a liquid mixture of 20 weight percent water, 30 weight percent acetic acid, and 50 weight percent isopropyl ether, determine (1) the composition of the two phases, e.g., the ether layer and the water layer; and (2) the amount of acetic acid that must be added to the system to form a one-phase (single-layer) solution.

Calculation Procedure:

1. Compute the composition of the two layers

When two pure liquids are mixed, they will dissolve in each other to some degree. If they are completely soluble in each other, such as water and acetic acid, they are *miscible*.

If their mutual solubilities are zero, they are called immiscible. Between these extremes, liquids are partially miscible.

Addition of a third liquid component often affects the mutual solubilities of the two original liquids. The third liquid may be more soluble in one liquid than in another. This difference in solubilities is the basis of the chemical engineering operation termed *liquid-liquid extraction*.

The third liquid may cause immiscible liquids to become completely miscible, or the third liquid may produce miscibility only in certain concentration ranges. Such interrelationships can be shown graphically, as with the two parts in Fig. 2.

The *phase envelope*, Fig. 2, separates the two-phase region from the one-phase region. Note, Fig. 2, that the acetic acid and water are completely miscible, as indicated by the phase envelope not touching the horizontal axis at any point. Likewise, the isopropyl ether and acetic acid are completely miscible. But water and isopropyl ether are virtually immiscible, as indicated by little of the vertical axis being free of the two-phase region of the phase envelope, Fig. 2.

The composition of the two phases, for the mixture in the two-phase region, is found on the phase envelope line itself, Fig. 2. Toward the lower part of the phase envelope, Fig. 2, is the water-rich layer, and toward the top of the phase envelope line is the ether-rich layer.

Plot on the upper portion of Fig. 2 the given values for acetic acid and isopropyl ether, that is, 30 weight percent, 50 weight percent. Through this point, draw a tie line to intersect the phase envelope at two points, line 1, Fig. 2. Read the values: *lower point* − acetic acid in water layer = 20 weight percent; *upper point* − acetic acid in isopropyl ether layer = 31.5 weight percent.

Transferring the lower intersection point to the bottom diagram for tie line 1 shows that equilibrium exists between a layer that is 20 weight percent acetic acid in water and a layer that is 9 weight percent (not 31.5 weight percent) acetic acid in isopropyl ether.

Draw a second tie line, 2, Fig. 2, as shown. Line 2 gives a check between the upper and lower diagrams, giving: *water layer* − $x_a = 0.415$; $x_c = 0.065$; $x_w = 0.520$; *ether layer* − $y_a = 0.270$; $y_c = 0.650$; $x_w = 0.080$.

2. Compute the amount of acetic acid that must be added to form a one-phase system

The water/ether ratio remains unchanged at: water/ether = 0.20/0.50 = 0.40. Then, the total system is: water + ether + acid = 1.000; ∴ ether (weight percent) = [1.000 − acid (weight percent)]/1.40.

Assume that the acid = 0.350. Then, ether = (1.000 − 0.350)/1.40 = 0.464. Checking against the upper diagram in Fig. 2, this point ($x_a = 0.350$, $x_c = 0.464$) falls inside the two-phase region. Hence, the assumption was incorrect.

As a second trial, assume acid = 0.380. Then, ether = $(1.000 - 0.380)/1.40 = 0.443$. Checking $x_a = 0.380$, $x_c = 0.443$ in the upper diagram of Fig. 2 shows that the point falls exactly on the phase envelope line. Hence, it is at the minimum one-phase region.

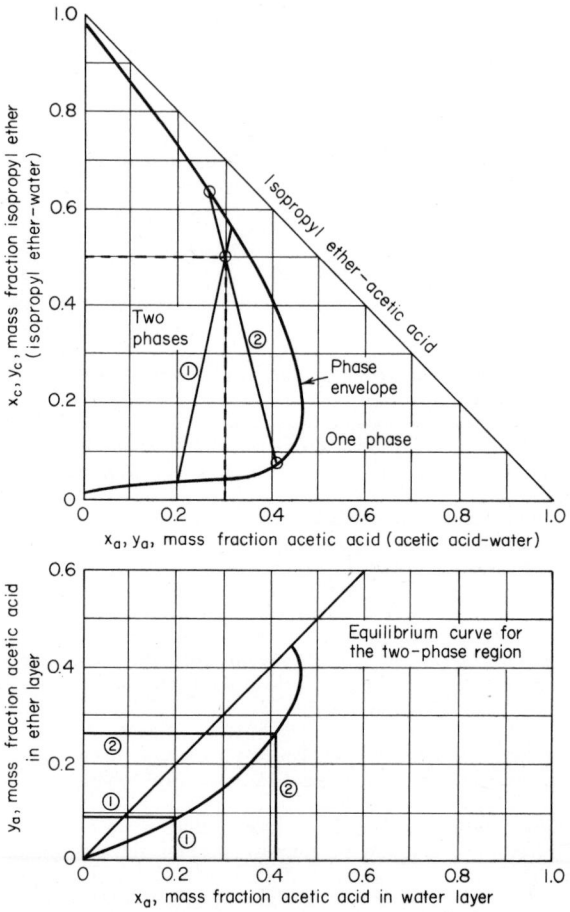

Fig. 2   Liquid-system phase-envelope plot. (After Anderson and Wenzel — *Introduction to Chemical Engineering*, McGraw-Hill.)

DETERMINING THE HEAT OF MIXING OF CHEMICALS

How many Btu of heat are released (generated) when 1,000 lb of water at 80 C is mixed with (1) 500 lb of aluminum bromide, $AlBr_3$, (2) 750 lb of barium nitrate, $Ba(NO_3)_2$, and (3) 1,000 lb of dextrin, $C_{12}H_{20}O_{10}$?

Calculation Procedure:

1.   Compute the heat released when dissolving $AlBr_3$ in water

When two or more substances are mixed, heat is usually generated or absorbed. The heat released or absorbed may be small when mixing two similar organic liquids or very large when mixing strong acids in water. The heat evolved (or absorbed) during the mixing of liquids is often called the *heat of dilution,* whereas the heat from the mixing of solids is often termed the *heat of solution.* Data for heats of solution for both organic and inorganic liquids and solids are given in Perry, Lange, and similar reference works.

Thus, at 80 C, the solubility of $AlBr_3$ in water is 126 g per 100 g of water. The weight of $AlBr_3$ that will dissolve in 1,000 lb of water $= (126/100)1,000 = 1,260$ lb. Standard references show that the heat of solution for $AlBr_3$ is 85.3 kg-cal per g mole $AlBr_3$. Since the total Btu per lb mole $= 1.8$(g-cal per g), the $AlBr_3$ in this solution can release $(85.3$ kg-cal per g mole$)(1,000$ cal per kg-cal$)1.8 = 153,540$ Btu/lb mole.

The weight of 1 lb mole of $AlBr_3 = (27 + 79.9 \times 3) = 266.7$ lb. Hence, the heat evolved when dissolving 500 lb of $AlBr_3$ in water is $[500$ lb $AlBr_3/(266.7$ lb/lb mole$)](153,540$ Btu/lb mole$) = 287,800$ Btu.

2. Compute the heat released when dissolving $Ba(NO_3)_2$ in water

At 80 C, the solubility of $Ba(NO_3)_2$ in water is 27.0 lb per lb of water. The weight of $Ba(NO_3)_2$ that will dissolve in 1,000 lb of water $= (27/100)1,000 = 270$ lb. Since 750 lb of $Ba(NO_3)_2$ is available for dissolving, the weight that will not dissolve is $750 - 270 = 480$ lb.

The heat of solution of $Ba(NO_3)_2$ is -10.2 kg-cal per g mole of $Ba(NO_3)_2$. As in Step 1, the weight of 1 lb mole of $Ba(NO_3)_2 = [137.34 + 2(14.0 + 3 \times 16)] = 261.4$. Then, as in Step 1, the heat released $= (480$ lb $Ba(NO_3)_2/261.4$ lb per lb mole$)(-10.2$ kg-cal per g mole $\times 1,000$ g-cal per kg-cal $\times 1.8$ Btu per lb mole/g-cal per g mole$) = -33,300$ Btu.

The negative heat release means that 33,300 Btu of heat must be added to the system to maintain the solution temperature at 80 C, because a fall in temperature would reduce the solubility of the $Ba(NO_3)_2$ in water and thus change the resulting solution.

3. Compute the heat released when dissolving $C_{12}H_{20}O_{10}$ in water

Perry indicates that there is no solubility limit for $C_{12}H_{20}O_{10}$ in water. Following the same Procedure as in Step 1, the heat released $= (1,000$ lb $C_{12}H_{20}O_{10}/324.2$ lb/lb mole$) \times (268$ g-cal/g mole $\times 1.8$ Btu per lb mole/g-cal per g mole$) = 1,488$ Btu released.

Related Calculations: Use the general Procedure to determine the heat of mixing of any material dissolved in another.

CHEMICAL EQUATION MATERIAL BALANCE

Ethylene oxide is produced by the catalytic reaction of ethylene and oxygen: $C_2H_4 + \frac{1}{2}O_2 \rightarrow (CH_2)_2O$. For each 100 lb of ethylene: (1) how much ethylene oxide is produced; (2) how much oxygen is required; and (3) what are the quantities of ethylene oxide and ethylene in the product if there is a 20 percent deficiency of oxygen?

Calculation Procedure:

1. Compute the quantity of ethylene oxide produced

The two most frequently met calculations in day-to-day chemical engineering are the *material balance* (discussed here) and the *energy balance,* discussed later. In a chemical process, a balance is the same as any other type of balance—i.e., an equating of input, output, and accumulation or loss: Input − output $= \pm$ accumulation.

Such a balance may be written around a single item of chemical process equipment, a portion of a process, or an entire chemical plant. A balance may be used to check experimental data or to determine an unknown quantity of some process stream.

For purposes of balance calculations, chemical processes are classified as: *steady-state,* i.e., input $=$ output, no accumulation; *unsteady-state,* i.e., input $\neq$ output, a $\pm$ accumulation; *batch process,* i.e., system is loaded, no further $\pm$ accumulation; continuous process, i.e., continuous input and output. Chemical processes may be further classed as physical, in which there is no chemical reaction, or chemical, in which a chemical change occurs. To analyze chemical reactions, the principles of chemical equation balances must be understood.

A *stoichiometrically balanced reaction* is one in which the reactants are exactly proportioned to give a product free of excess reactants, as in $C_2H_4 + \frac{1}{2}O_2 \rightarrow (CH_2)_2O$. An *excess reactant* is one present in excess of the stoichiometric quantity, such as if there were more than one-half mole of oxygen in the above equation.

The *degree of completion* is the percentage of the limiting reactant that reacts. The *limiting reactant* is the one present in less than stoichiometric proportion, so that the other reactant is in excess.

To determine how much ethylene oxide is produced, find the molecular weight of $C_2H_4 = 2(12) + 4(1) = 28$. The moles of $C_2H_4 = 100/28 = 3.571$.

Referring to the reaction equation shows that for each mole of C_2H_4, 1 mole of $(CH_2)_2O$ is produced, having a molecular weight of $2(12) + 4(1) + 16 = 44$. Then, the weight of $(CH_2)_2O = 44(3.571) = 157.14$ lb.

2. Compute the amount of oxygen required

The molecular weight of $O_2 = 16(2) = 32$. Referring to the reaction equation shows that $\frac{1}{2}$ mole of oxygen is needed for each mole of ethylene, C_2H_4. Hence, the weight of oxygen needed $= \frac{1}{2}(32)(3.571) = 57.14$ lb.

3. Compute the product mix for a reactant deficiency

Referring to Step 2, a 20 percent oxygen deficiency means that there was $0.80\ (\frac{1}{2}) = 0.40$ mole of oxygen available. Rewriting the equation, $0.2C_2H_4 + 0.8C_2H_4 + 0.40_2 \rightarrow 0.8(CH_2)O + 0.2C_2H_4$. Hence, the ethylene oxide $(CH_2)_2O$ in the product $= 0.8(157.14) = 125.71$ lb. And the ethylene, C_2H_4 in the product $= 0.2(100) = 20$ lb.

Related Calculations: Use this general Procedure for any chemical equation balance similar to that analyzed here.

BATCH PHYSICAL PROCESS BALANCE

A load of clay containing 35 percent moisture on a wet basis weighs 2,000 lb. If the clay is dried to a 15 percent moisture content (on a wet basis), how much water is evaporated in the drying process?

Calculation Procedure:

1. Compute the initial moisture content

The 2,000 lb of wet clay contains 35 percent moisture, or 2,000 lb $(0.35) = 700$ lb of water. Thus, the dry clay weighs $2,000 - 700 = 1,300$ lb.

2. Compute the weight after drying

Set up the relation y lb of wet clay $+ x$ lb of water $= 1,300$ lb of dry clay. But the final batch contains 15 percent moisture. Hence, the water $= 0.15y$. Therefore, the dry clay $= (1.00 - 0.15)y = 0.85y = 1,300$. Solving, $y = 1,529$ lb of wet (15 percent moisture) clay. And since $y + x = 2,000$ lb, $x = 2,000 - y = 2,000 - 1,529 = 471$ lb of water evaporated.

Related Calculations: Use this general Procedure for any batch physical process balance involving evaporation or drying of a solid.

Where the rate of feed is given, a steady-state physical process balance can be analyzed. Thus, if the 2,000 lb of clay in the above process were fed to the dryer in 1 hr, the rate of evaporation would be 471 lb/hr of water.

STEADY-STATE CONTINUOUS PHYSICAL BALANCE WITH RECYCLE AND BYPASS

Feed to a distillation tower is 1,000 lb mole/hr of a solution of 35 mole percent ethylene dichloride (EDC) in xylene. There is not any accumulation in the tower. The overhead distillate stream contains 90 mole percent ethylene dichloride, and the bottoms stream contains 15 mole percent ethylene dichloride. Cooling water to the overhead condenser is adjusted to give a reflux ratio of 10:1 (10 moles re-enter the column for each mole of overhead product). Heat to the reboiler, Fig. 3, is adjusted so that the recycle ratio is 5:1 (5 moles re-enter the column for each mole of bottom product), with a 2:15 bypass (2 moles bypass the reboiler for each 15 moles that pass through the reboiler). Determine the flow rate of the overhead product, bottoms product, overhead reflux re-entering the column, bottoms recycle re-entering the column, bottoms bypassing the reboiler, and the total bottoms.

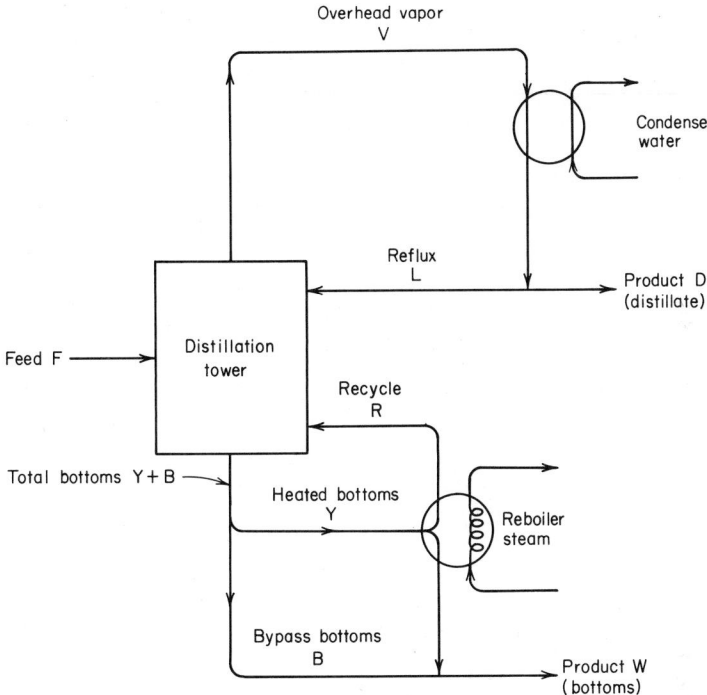

Fig. 3   Distillation tower flow.

Calculation Procedure:

1.   Compute the bottoms product

Since this is a physical system with no change in chemical composition and no accumulation, any component may be followed through the system. Having been given the important values of the ethylene dichloride (X_F, X_D, X_W), use them as the basis of the calculation, with X representing the moles of ethylene dichloride in each stream.

Set up a total material balance thus: $F = D + W = 1,000$ lb moles/hr, Eq. (1), where F = feed; D = product (i.e., distillate); W = product (i.e., bottoms), all expressed in lb moles/hr, as shown in Fig. 3.

An ethylene dichloride balance is $FX_F = DX_D + WX_W$. Substituting given values, $1,000\,(0.35) = 0.90D + 0.15W = 350$, Eq. (2). Solving Eqs. (1) and (2) simultaneously, $D = 1,000 - W$; $W = (350 - 0.90D)/0.15$; $D = 1,000 - (350 - 0.90D)/0.15$; $D = 266.67$ lb moles/hr distillate product; $W = 733.33$ lb moles/hr bottoms product.

2.   Compute the reflux flow rate

Taking the tower overhead as a separate system, Fig. 3, $X_L/X_D = 10 = L/D$. Hence, $L = 10D = 10(266.67) = 2,666.7$ lb moles/hr reflux.

3.   *Analyze the condenser*

A total material balance around the condenser is input = output; or $V = D + L$, Fig. 3. Hence, $V = 266.67 + 2,666.7 = 2,933.37$ lb moles/hr overhead vapor.

4.   *Analyze the tower reboiler*

Taking the tower reboiler as a separate system, Fig. 3, $X_R/X_W = 5 = R/W$; hence, $R = 5W = 5(733.33) = 3,666.7$ lb moles/hr bottoms recycle.

Also, $X_B/X_Y = 2/15 = B/Y$; $2Y = 15B$, Eq. (3). And a total material balance around the reboiler is input = output, or $Y = R + (W - B)$, Eq. (4). Solving Eqs. (3) and (4) simultaneously, $B = 2/15(Y)$; $Y = R + [W - 2/15(Y)]$; $Y = [15(3,666.7) + 15(733.3)]/17 = 3,882.35$ lb moles/hr reboiled bottoms. Then, $Y + B = 4,399.97$ lb moles/hr total bottoms.

Related Calculations: Use this general Procedure to analyze distillation towers handling liquids similar to those considered here.

STEADY-STATE CONTINUOUS PHYSICAL PROCESS BALANCE

The distillation tower of the previous Calculation Procedure has the temperature and thermal conditions shown in Fig. 4. The reboiler is heated by steam that condenses at 280 F. Cooling water enters the overhead condenser at 70 F and leaves at 120 F. In the condenser, the overhead vapor condenses at 184 F before being cooled to 175 F, the temperature of the liquid reflux and distillate product. The heat of condensation ΔH of the overhead vapor is 14,210 Btu/lb mole, as given in a standard reference work, and 14,820 Btu/lb mole for the tower bottoms. The heat capacity of all liquid streams in this installation is 40 Btu/(lb mole)(F). Determine the steam and cooling-water flow rates required.

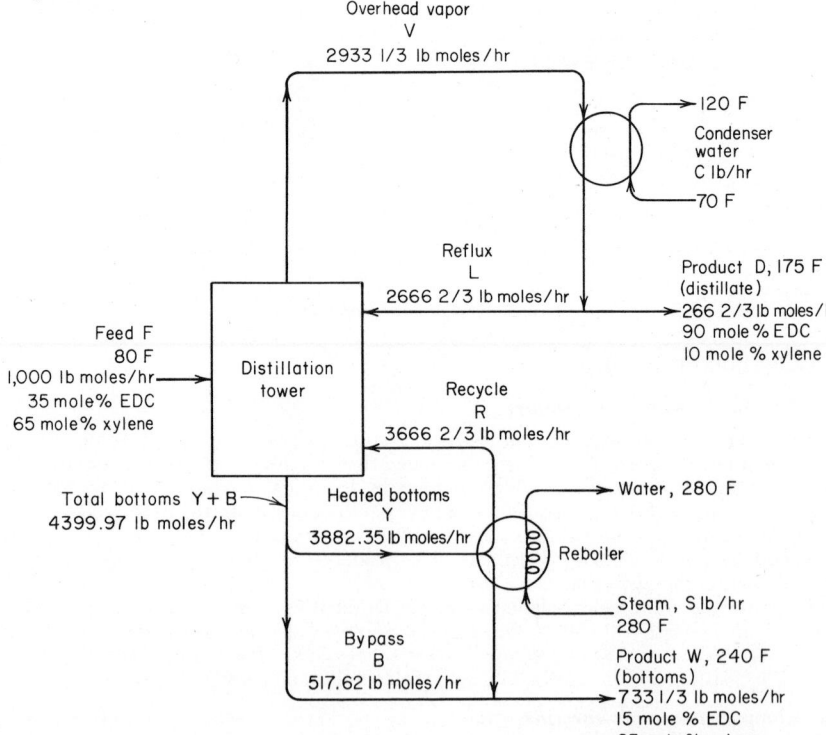

Fig. 4   Distillation-tower flow quantities and flow rates.

Calculation Procedure:

1. Set up a heat balance for the column

Thus, heat in = heat in feed + heat in steam. Let the temperature basis for the calculation = 80 F = t_b. Then, the heat in the feed = ΔH_F = (feed rate, lb/hr)[heat capacity of the feed, Btu/(lb mole)(F)](feed temperature, F – temperature basis for the calculation, F) = 1,000(40)(80 – 80) = 0.

The enthalpy of vaporization of the steam is, from the steam tables, 924.74 Btu/lb when there is complete condensation of the steam and the condensate leaves the reboiler at 280 F. Then, the heat given up by the condensation of S lb of steam is ΔH_S = 924.74S Btu/hr.

The heat out = heat in distillate + heat in bottoms + heat in water, all expressed in Btu/hr. Using the same Procedure as for the heat in the feed, the heat in the distillate = ΔH_D / $DC\Delta t_d$, where C = distillate heat capacity, Btu/(lb mole)(F); Δt_d = temperature change of the distillate, F. Or, ΔH_D = 266.7(40)(175 – 80) = 1,010,000 Btu/hr. Likewise, for the bottoms, ΔH_W = 733.3(40)(240 – 80) = 4,690,000 Btu/hr.

For the water, ΔH_w = heat to condense overhead vapor + heat absorbed when cooling the condensed vapor from 184 to 175 F, all expressed in Btu/hr. With a flow of 2,933.3 lb/hr of vapor, ΔH_w = 2,933.3[14,210 + 40(184 – 175)] = 42,700,000 Btu/hr.

2. Analyze the heat balance

The heat balance is heat in = heat out, or $\Delta H_S = \Delta H_D + \Delta H_W + \Delta H_w$. Thus, 924.74$S$ = 1,010,000 + 4,690,000 + 42,700,000; S = 52,300 lb/hr of steam.

For the water, $\Delta H_w = Cc\Delta t_c$, where C = water flow rate, lb/hr; c = specific heat of water = 1.0 Btu/(lb)(F). Substituting, ΔH_w = C(1)(120 – 70) = 42,700,000; C = 854,000 lb/hr of water.

DETERMINING THE CHARACTERISTICS OF AN IMMISCIBLE SOLUTION

For steam distillation of 2-bromoethylbenzene, the vapor temperature is 222.4 F (105.8 C). Analysis shows 0.16 lb of 2-bromoethylbenzene (BB) per lb of vapor. Saturated steam is used in the distillation process. Determine (1) the pressure in the still, and (2) how far from ideal the actual conditions are.

Calculation Procedure:

1. Compute the pressure in the still

Each component of an immiscible mixture of liquids exerts a vapor pressure that is independent of its concentration and equal to the vapor pressure of the pure substance – but only if stratification is avoided by vigorous mixing or boiling. The major industrial application of immiscible systems is in steam distillation of high-molecular weight heat-sensitive organic materials. The mixture of water (steam) and an organic substance will boil when the total solution pressure equals atmospheric pressure. Since the organic material must exert some vapor pressure, it vaporizes with the steam, at a greatly reduced temperature.

The relationship for immiscible components A and B is $w_A/w_B = y_A M_A/y_B M_B = P_{VA} M_A/P_{VB} M_B$, where $w_{A,B}$ = weight of component A, B in vapor; $M_{A,B}$ = molecular weight of component A, B; $y_{A,B}$ = vapor phase mole fraction of component A, B; $P_{VA,VB}$ = vapor pressure of component A, B.

The vapor pressure of BB at 222.4 F is, from Perry – Chemical Engineers' Handbook, 20 mm Hg, and the vapor pressure of water at 222.4 F is 938 mm Hg. Hence, the total pressure (ideal) in the still is 938 + 20 = 958 mm Hg.

2. Compare the ideal to the actual conditions

If conditions in the still were ideal (i.e., exactly according to theory), the weight of the BB in the vapor would be, according to Step 1, $w_{BB} = (P_{V,BB}/P_{V,H_2O})(M_{BB}/M_{H_2O})$ = (20/938)(185/18) = 0.219 lb versus 0.16 lb actual, as given.

Or, computing the ideal BB vapor pressure for 0.16 lb of BB per lb of vapor, using the relation in Step 1, $(P_{V,BB}/P_{H_2O})(185/18) = (P_{V,BB}/938)(185/18) = 0.16$ lb/lb; solving, $P_{V,BB} = (0.16)(938)(18/185) = 14.6$ mm Hg versus 20 mm Hg actual.

The divergence between the actual and ideal most likely means that the time of contact between the steam and the BB is insufficient to reach equilibrium. Also, the total pressure should be $938 + 14.6 = 952.6$ mm Hg, not the 958 mm Hg of the ideal case.

Related Calculations. This Procedure is valid for immiscible solutions of all types resembling the one considered here.

PUMP SELECTION FOR CHEMICAL PLANTS

Choose a pump to handle 26,000 gpm of water at 60 F in a chemical plant when the total dynamic head is 37 ft of water. What is the required hp input to the pump if the pump efficiency is 85 percent? What type of pump should be used if the rotational speed is limited to 880 rpm?

Calculation Procedure:

1. Determine the required power input to the pump

A quick way to determine the power input to a pump handling water at normal atmospheric temperatures is to use Fig. 5. Enter on the left at the total dynamic head, 37 ft, and project to the right to the required pump capacity, 26,000 gpm. At the intersection with the hp stem, read the required power input as 285 hp.

2. Select the type of pump to use

From the rotational speed, 880 rpm on the bottom stem, draw a straight line at right angles to the first construction line, as shown. At the intersection with the top stem, read the type of pump as a propeller pump having a specific speed of 9,500 rpm.

Related Calculations: Note that this pump application chart applies to rotating-type centrifugal pumps. Where a reciprocating pump is desired, use the methods given in Sec. 3 of this Handbook. The chart in Fig. 5 was developed by H. W. Hamm, and was first presented in *Power* magazine.

CRUSHER POWER INPUT DETERMINATION

A chemical process requires the crushing of 240 tons/hr of quartz. The quartz feed used is such that 80 percent passes a 3-in. screen and 80 percent of the product must pass a ¼-in. screen. Determine the power input to the crusher.

Calculation Procedure:

1. Compute the crusher capacity in tons/min

Use the relation $t_m = t_h/60$, where $t_m =$ crusher capacity, tons/min; $t_h =$ crusher capacity, tons/hr. Substituting, $t_m = 240/60 = 4$ tons/min.

2. Determine the material work index

The work index for any material that will be crushed is the total energy, kw-hr/ton, needed to reduce the feed to a size so that 80 percent of the product will pass through a 100-μ screen. Standard references such as Perry—*Chemical Engineers' Handbook* list work indexes for various materials. For quartz having a specific gravity of 2.65, Perry gives the work index $W_i = 13.57$ kw-hr/ton.

3. Compute the raw-material and product mesh sizes

Use the relation $d_r = s/12$, where $d_r =$ mesh size, ft, for feed; $s =$ mesh opening measure used, in. For the product, $d_p = s/12$, where the symbols are the same as before except that the mesh opening is that used for the product. Substituting, $d_r = 3/12 = 0.25$; $d_p = 0.25/12 = 0.0208$.

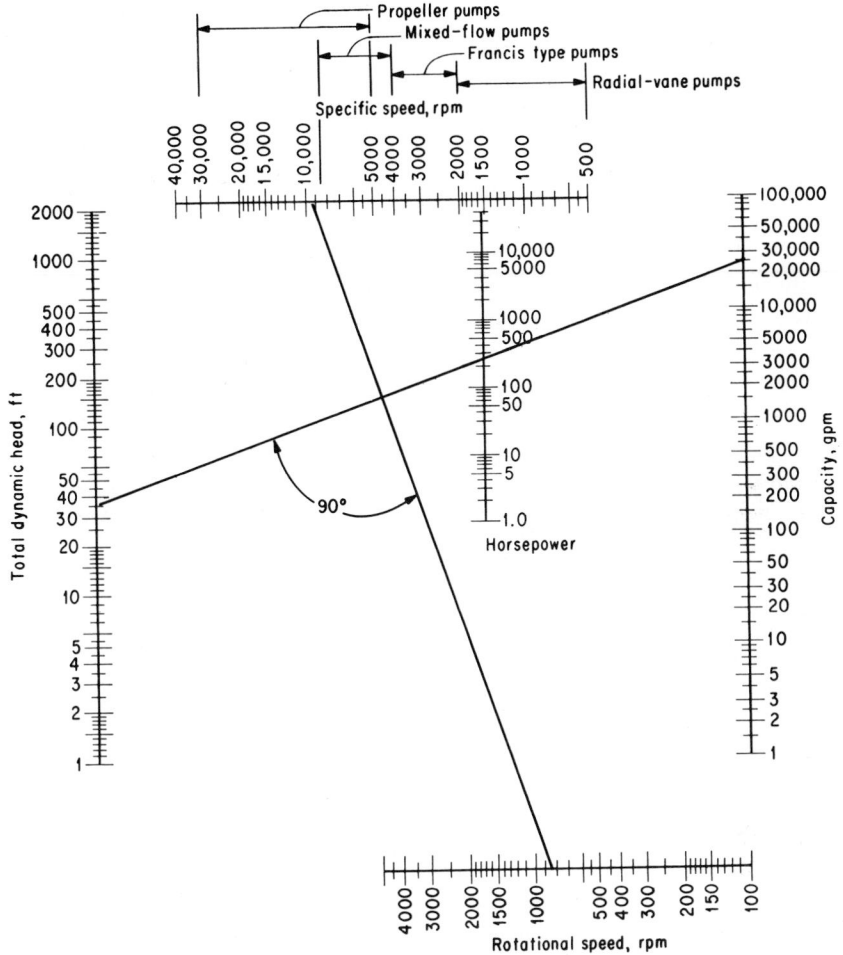

Fig. 5   Pump hp and type selection chart. (*Power.*)

4.   Compute the required power input to the crusher

Use the relation $hp = 1.46 t_m W_i (1/d_p^{0.5} - 1/d_r^{0.5})$, where the symbols are as given earlier.   Substituting, $hp = 1.46(4)(13.57)(1/0.0208^{0.5} - 1/0.25^{0.5}) = 391$ hp.   A   400-hp motor would be used to drive this crusher.

Related Calculations: Use this general Procedure, known as the bond crushing law and work index, to determine the power input required for commercially available grinders and crushers of all types. The result obtained is valid for all usual preliminary calculations.

COOLING-WATER FLOW RATE FOR CHEMICAL-PLANT MIXERS

A kneader used in a chemical plant requires 300-hp input per 1,000 gal of material kneaded. If this kneader handles 3,000 lb of a chemical having a density of 65 lb/cu ft, determine the quantity of cooling water required in gpm and gph if the maximum allowable temperature rise of the water during passage through the kneader is 25 F.

Calculation Procedure:

1. Convert the kneader load to gallons

Since the power-input requirements of chemical mixers are normally stated in hp/gal, the kneader load must be converted to gal. Use the relation, load, gal, = load weight, lb (7.48 gal/cu ft water)/load density, lb/cu ft. For this kneader, load, gal, = 3,000(7.48)/65 = 345 gal.

2. Compute the required power input

Use the relation, power input hp = hp input per 1,000 gal (load, gal)/1,000. For this kneader, power input hp = 300(345)/1,000 = 103.5 hp.

3. Compute the heat that must be removed

Since 1 hp = 2,545 Btu/hr, the heat that must be removed = 103.5 hp (2,545) = 263,407.5 Btu/hr.

4. Compute the cooling-water flow rate

With an allowable temperature rise of 25 F and a specific heat of 1 Btu/(hr)(F), the cooling-water flow rate required = (263,407.5 Btu/hr)/[(25 F)(1.0)(8.33 lb per gal of water)(60 min/hr)] = 21.1 gpm, or 21.2(60 min/hr) = 1,265 gph.

Related Calculations: Use the general Procedure given here for any of the usual chemical mixers, such as paddles, turbines, propellers, disks, cones, change cans, dispersers, tumbling mixers, mixing rolls, masticators, pug mills, and mixer-extruders. Consult Perry—*Chemical Engineers' Handbook* for suitable power-input data for mixers of various types.

LIQUID–LIQUID SEPARATION ANALYSIS

Size a liquid–liquid separator or decanter using gravitational force for continuous separation of two liquids, the first of which has a density of 47 lb/cu ft, and the second liquid a density of 81 lb/cu ft. Both liquids flow into the separator at a rate of 50 gpm. The time required for settling is 35 min. What size separator is required to handle this flow? How far above the separator bottom should overflow of the heavier liquid be located?

Calculation Procedure:

1. Compute the liquid holdup volume

Since there are two liquids, a light one and a heavy one, entering the separator, the holdup volume = (number of liquids entering)(liquid flow rate into the separator, gpm)(holdup time, min). Or for this separator, total holdup volume = 2(50)(35) = 3,500 gal.

2. Determine the separator tank volume

Usual design practice is to make the separator tank volume 10 to 25 percent greater than the required holdup volume. Using a volume 20 percent greater than the required holdup volume gives a required tank volume of 1.20(3,500 gal) = 4,200 gal.

3. Size the separator tank

Most decanter-type separator tanks are sized so that the tank diameter and height are approximately equal. Selecting a 10-ft-diameter and 10-ft-high tank gives a total tank volume of (head area, sq ft)(height, ft) = $(d^2\pi/4)h$, where d and h are the diameter and height of the tank, ft, respectively. Or, volume = $(10^2\pi/4)(10)$ = 785.4 cu ft. Since a gal of liquid occupies 0.13 cu ft, the capacity of this tank = 785.4/0.13 = 5,850 gal. This is sufficient to store the holdup liquid but somewhat oversize.

Try a 9-ft-diameter and high tank. Using the same method, the tank capacity is 4,250 gal. This is closer to the required holdup capacity. Hence a 9-ft tank will be used.

4. Compute the liquid depth in the tank

Use the relation D_1 ft $= 4$(holdup volume, gal)$/7.48\pi d^2$, where $D_1 =$ liquid depth, ft. Solving, $D_1 = 4(3,500)/7.48\pi 9^2 = 7.34$ ft.

5. Determine the height of the heavy-liquid overflow

Assume that the two liquids interface midway between the vessel bottom and the liquid surface. Then, the height of the heavy liquid $= 7.34/2 = 3.67$ ft.

To find the height of the heavy-liquid overflow, solve $H_h = H_1 + (D_1 - H_1)$(density of lighter liquid, lb/cu ft)/(density of heavier liquid, lb/cu ft), where $H_h =$ height of heavy-liquid overflow above tank bottom, ft; $H_1 =$ height of heavy liquid in tank, ft; other symbols as before. Solving, $H_h = 3.67 + (7.34 - 3.67)(47/81) = 5.80$ ft. This is the distance measured to the inside lower surface of the overflow pipe from the tank bottom.

The *continuous decanter* is a popular type of static separator for immiscible liquids of many types. This type of separator is fed from the top and vented to the open air through both the light- and heavy-liquid overflow lines.

Section 7

Control Engineering

GEORGE M. MUSCHAMP
Consulting Engineer, Honeywell, Inc.

TYLER G. HICKS, P.E.

REFERENCES: Considine—*Process Instruments and Controls Handbook*, McGraw-Hill; Merritt —*Hydraulic Control Systems*, Wiley; Considine and Ross—*Handbook of Applied Instrumentation*, McGraw-Hill; Eckman—*Automatic Process Control*, Wiley; Kallen—*Handbook of Instrumentation and Controls*, McGraw-Hill; Farrington—*Fundamentals of Automatic Control*, Wiley; ASME—*Fluid Meters*; Shinskey—*Process-Control Systems*, McGraw-Hill; Graham-McRuer—*Analysis of Nonlinear Control Systems*, Wiley; Harriott—*Process Control*, McGraw-Hill; Mesarovic—*The Control of Multivariable Systems*, Wiley; Savas—*Computer Control of Industrial Processes*, McGraw-Hill; Newton-Gould-Kaiser—*Analytical Design of Linear Feedback Controls*, Wiley; Burr-Brown Research Corp.—*Handbook of Operational Amplifier Active RC Networks*; Bower-Schultheiss—*Introduction to Design of Servomechanisms*, Wiley; Gibson-Tuteur—*Control System Components*, McGraw-Hill; Bode—*Network Analysis and Feedback Amplifier Design*, D. Van Nostrand; Doss—*Information Processing Equipment*, Reinhold, ASME—*Flowmeter Computation Handbook*; ASME—*Flow Measurement*; Dommasch and Laudeman—*Principles Underlying Systems Engineering*, Pitman.

SELECTION OF A PROCESS-CONTROL SYSTEM

A continuous industrial process contains four process centers, each of which has two variables which must be controlled. If a fast process-reaction rate is required with only small-to-moderate dead time, select a suitable mode of control. The system contains more than two resistance-capacity pairs. What type of transmission system would be suitable for this process?

Calculation Procedure:

1. Compute the number of process capacities

The number of process capacities = (number of process centers)(number of variables per center) or, for this system, $4 \times 2 = 8$ process capacities. This is defined as a *multiple* number of process capacities because the number controlled is greater than unity.

2. Analyze the process-time lags

A small-to-moderate dead time is allowed in this process-control system. With such a dead-time allowance, and with two or more resistance-capacity pairs in the system, a mode of control that provides for any number of process-time lags is desirable.

3. Select a suitable mode of control

Table 1 summarizes the forms of control suited to processes having various characteristics. This table is a *general guide*—it provides, at best, an *approximate* aid in selecting control modes. Hence, it is suitable for tentative selection of the mode of control. Final selection must be based on actual experience with similar systems.

Inspection of Table 1 shows that for a multiple number of processes with small-to-moderate dead time and any number of resistance-capacity pairs, a proportional plus reset mode of control is probably suitable. Further, this mode of control provides for any (i.e., fast or slow) reaction rate. Since a fast reaction rate is desired, the proportional plus reset method of control is suitable because it can handle any process-reaction rate.

4. Select the type of transmission system to use

Four types of transmission systems are used for process control today—pneumatic, electric, electronic, and hydraulic. The first three types are by far the most common.

Pneumatic transmission systems use air at 3 to 20 psig to convey the control signal through small-bore metal tubing at distances ranging to several thousand feet. The air used in pneumatic systems must be clean and dry. To prevent a process from getting out of control, a constant supply of air is required. Pneumatic controllers, receivers, and valve positioners usually have small air-space volumes—5 to 10 in.^{3}. Air motors of the diaphragm or piston type have relatively large volumes—100 to 5,000 in.^{3}.

Pneumatic control systems are generally considered to be spark-free. Hence, they find wide use in hazardous process areas. Also, control air is readily available and it can be "dumped" to the atmosphere safely. The response time of pneumatic control systems may be slower than electric or hydraulic systems.

Electric and electronic control systems are fast-response with the signal conveyed by a wire from the sensing point to the controller. In hazardous atmospheres the wire must be protected against abrasion and breakage.

Hydraulic control systems are also rapid-response. These systems are capable of high power actuation. Slower-acting hydraulic systems use fluid pressures in the 50- to 100-psi range; fast-acting systems use fluid pressures to 5,000 psi.

Dirt and fluid flammability are two factors that may be disadvantages in certain hydraulic-control-system applications. However, new manufacturing techniques and nonflammable fluids are overcoming these disadvantages.

Since a fast response is desired in this process-control system, either electric, electronic, or hydraulic transmission of the signals would be considered first. With long distances between the sensing points (say 1,000 ft or more), an electric or electronic system would probably be best.

TABLE 1   Process Characteristics versus Mode of Control*

Number of process capacities	Process reaction rate	Process time lags		Load changes		Suitable mode of control
		Resistance capacity (RC)	Dead time (transportation)	Size	Speed	
Single	Slow	Moderate to large	Small	Any	Any	Two-position; two-position with differential gap
				Moderate	Slow	Multiposition; proportional input
Single (self-regulating)	Fast	Small	Small	Any	Slow	Floating modes: Single speed Multispeed
					Moderate	Proportional-speed floating
Multiple	Slow to moderate	Moderate	Small	Small	Moderate	Proportional position
Multiple	Moderate	Any	Small	Small	Any	Proportional plus rate
Multiple	Any	Any	Small to moderate	Large	Slow to moderate	Proportional plus reset
Multiple	Any	Any	Small	Large	Fast	Proportional plus reset plus rate
Any...........	Faster than that of the control system	Small or nearly zero	Small to moderate	Any	Any	Wideband proportional plus fast reset

*Considine—*Process Instruments and Controls Handbook*, McGraw-Hill.

Next, determine if the systems being considered can provide the mode of control (Step 2) required. If a system cannot provide the necessary mode of control, eliminate the system from consideration.

Before a final choice of a system is made, other factors must be considered. Thus, relative cost of each type of system must be determined. Should an electric system prove too costly, the slightly slower response time of the pneumatic system might be accepted to reduce the initial investment.

Other factors influencing the choice of the type of a control system include: (*a*) type of controls, if any, currently used in the installation, (*b*) skill and experience of the operating and maintenance personnel, (*c*) type of atmosphere in which, and type of process for which, the controls will be used. Any of these factors may alter the initial choice.

Related Calculations: Use this general method to make a preliminary choice of controls for continuous processes, intermittent processes, air-conditioning systems, combustion-control systems, etc. Before making a final choice of any control system, be certain to weigh the cost, safety, operating, and maintenance factors listed above. Lastly, the system chosen *must* be able to provide the mode of control required.

PROCESS-TEMPERATURE-CONTROL ANALYSIS

A water storage tank, Fig. 1, contains 500 lb of water at 150 F when full. Water is supplied to the tank at 50 F and is withdrawn at the rate of 25 lb/min. Determine the process-time constant and the zero-frequency process gain if the thermal sensing pipe contains 15 lb of water between the tank and thermal bulb, and the maximum steam

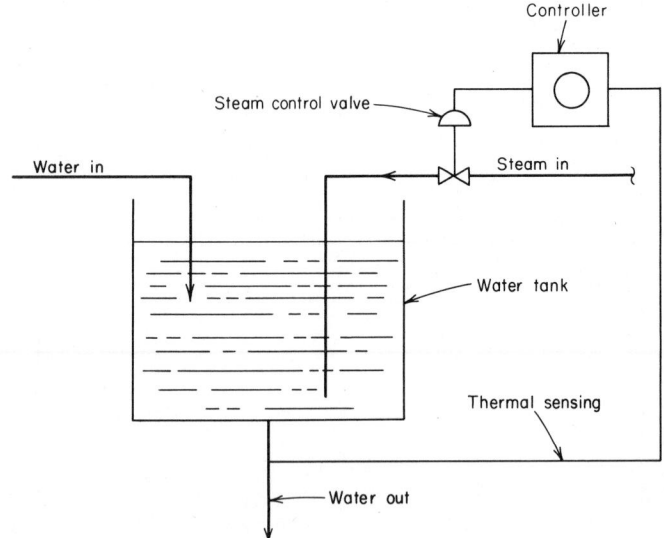

Fig. 1   Temperature control of a simple process.

flow to the tank is 8 lb/min. The steam flow to the tank is controlled by a standard linear regulating valve whose flow range is 0 to 10 lb/min when the valve operator pressure changes from 5 to 30 psi.

Calculation Procedure:

1.   *Compute the distance-velocity lag*

The time in min needed for the thermal element to detect a change in temperature in the storage tank is the *distance-velocity lag*, which is also called the *transportation lag* or *dead time*. For this process, the distance-velocity lag d is the ratio of the quantity of water in the pipe between the tank and the thermal bulb—that is, 15 gal—and the rate of flow of water out of the tank—that is, 25 lb/min—or $d = 15/25 = 0.667$ min.

2. Compute the energy input to the tank

This is a *transient-control process*—i.e., the conditions in the process are undergoing constant change instead of remaining fixed, as in *steady-state conditions*. For transient-process conditions the heat balance is $H_{in} = H_{out} + H_{stor}$, where H_{in} = heat input, Btu/min; H_{out} = heat output, Btu/min; H_{stor} = heat stored, Btu/min.

The heat input to this process is the enthalpy of vaporization h_{fg} Btu/(lb)(min) of the steam supplied to the process. Since the regulating valve is linear, its sensitivity s is the (flow-rate change, lb/min)/pressure change, psi. Or, using the known valve characteristics, $s = (10-0)/(30-5) = 0.4$ lb/min psi.

With a change in steam pressure of p psi in the valve operator, the change in the rate of energy supply to the process is $H_{in} = 0.4$ lb/min psi $\times p \times h_{fg}$. Taking h_{fg} as 938 Btu/lb, $H_{in} = 375p$ Btu/min.

3. Compute the energy output from the system

The energy output $H_{out} = $ lb per min of liquid outflow $\times$ liquid specific heat, Btu/(lb)(°F) $\times (T_a - 150$ F), where T_a = tank temperature, F, at any time. When the system is in a state of equilibrium, the temperature of the liquid in the tank is the same as that leaving the tank or, in this instance, 150 F. But when steam is supplied to the tank under equilibrium conditions, the liquid temperature will rise to $150 + T_r$, where T_r = temperature rise, F, produced by introducing steam into the water. Thus, the above equation becomes $H_{out} = 25$ lb/min $\times 1.0$ Btu/(lb)(°F) $\times T_r = 25T_r$ Btu/min.

4. Compute the energy stored in the system

With rapid mixing of the steam and water, H_{stor} = liquid storage, lb $\times$ liquid specific heat, Btu/(lb)(°F) $\times T_rq = 500 \times 1.0 \times T_rq$, where q = derivative of the tank outlet temperature with respect to time.

5. Determine the time constant and process gain

Write the process heat balance, substituting the computed values in $H_{in} = H_{out} + H_{stor}$, or $375p = 25T_r + 500T_rq$. Solving, $T_r/p = 375/(25+500q) = 15/(1+20q)$.

The denominator of this linear first-order differential equation gives the process-system time constant of 20 min in the expression $1 + 20q$. Likewise, the numerator gives the zero-frequency process gain of 15 F/psi.

Related Calculations: This general procedure is valid for any liquid using any gaseous heating medium for temperature control with a single linear lag. Likewise, this general procedure is also valid for temperature control with a double linear lag and pressure control with a single linear lag.

COMPUTER SELECTION FOR INDUSTRIAL PROCESS-CONTROL SYSTEMS

Select the type of computer and its speed of operation for use in an industrial control application. The computer will be used to monitor and control two continuous-flow process operations. Budget limitations restrict the investment in the computer to about $100,000 with a typical execution time of 10 msec or better.

Calculation Procedure:

1. Analyze the computers available; select a suitable computer

Four types of computers are available for consideration in any control problem: (a) analog; (b) digital; (c) hybrid—consisting of analog and digital; and (d) special-purpose—analog or digital computers for industrial control.

Digital computers, Fig. 2, find wide use for controlling continuous-flow processes. When used to control continuous-flow processes, the digital computer is connected *on-line*, i.e., information reflecting the activity in the process being controlled is introduced into the data-processing system as soon as it occurs, and action is immediately initiated by the system to make any needed adjustments. Since digital computers are proven machines for process control, this type will be tentatively chosen for this process and its suitability investigated further.

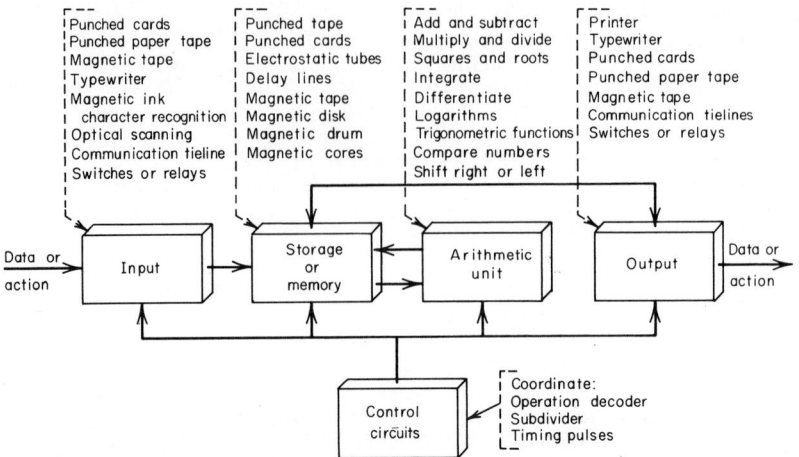

Fig. 2   Digital-computer elements used in process control.

The usual digital computer used in process control is a general-purpose digital computer that can receive and transmit analog signals. A magnetic-drum-type stored memory is often used in control applications.

2. Determine the computer operating time

The speed of a computer depends on the actions needed to perform a calculation. Thus, a drum or disk memory may have a 150-μsec add time with a memory access time of 16,000 μsec. In such a computer the memory-access time is the controlling speed factor.

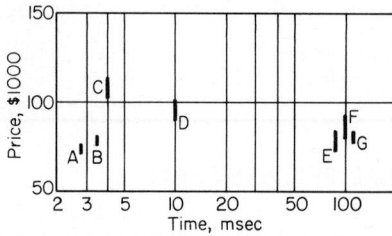

Fig. 3   Execution time for an arithmetic bench-mark problem shows the solution time required by different makes of computers.

Figure 3 shows the execution times for a variety of digital computers of different makes when solving the same *bench-mark* or *test* problem. A typical bench-mark problem consists of a pair of simultaneous equations having two unknowns.

Study of Fig. 3 shows that three machines —A, B, and D—meet the general requirements for this process with respect to speed and cost. Since each of these computers is produced by a different manufacturer, a fairly wide choice of units is available. When a chart such as Fig. 3 is not available, prepare one after computing the costs of machines available from various manufacturers.

3. Check the computer performance

Computer performance is rated according to three factors—(a) speed margin, i.e., how the computer copes with the worst-case time combination of events in the process; (b) memory-storage margin, that is, provisions for worst-case data storage, retrieval, and working space; and (c) reliability or consistency of performance.

Word length affects computer speed. Although a computer handling 24-bit words may seem to have an operation rate identical to a 12-bit-word computer, there may be as much as a 2:1 variation in the speed to the same degree of precision. In order to compare two computers, use their relative *problem-solving speeds* as a guide.

Manufacturers publish the relative problem-solving speeds of each model of computer. List these speeds for each of the suitable makes—A, B, and D, Step 2. Select that computer giving the fastest speed for the smallest investment.

4. Investigate the computer logic function

In some control applications, the logical-problem-solving ability of the computer may be more important than the arithmetic ability. In the logic function, the computer uses information transfers, manipulations, and comparisons, all of which take time. Using manufacturer's data, plot the logic speed against cost for each computer for a specific bench-mark or test problem. Usually the execution time for a logical bench-mark problem shows no clear relation to price, Fig. 4. However, the plot does indicate the price range for various operating speeds.

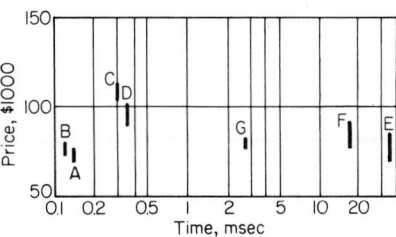

Fig. 4 Execution time for bench-mark problem shows no clear relationship between solution time and computer cost.

5. Evaluate the computer-memory size

An on-line-control computer cannot deal with a real-time problem unless all instructions defining the action are stored and available when needed. Distribution of the storage between high-speed random-access memory sections and lower-cost cyclic-access sections—such as drums and disks—influences computer speed. Good practice stores an image of the high-speed working memory in one of the slower-speed backup memories.

In machines with 12- to 15-bit words, more than one memory word may be needed to store an instruction or data item. Compare bench-mark problems to determine the real-time need of memory addresses between different computers. In general, more process-control applications are better served by machines having expandable memories. The tendency of many control-system designers is to underestimate the memory capacity needed. So choose a machine having the largest memory possible within the prevailing financial constraints.

6. Evaluate the computer reliability

The mean time between failure (MTBF) is a good measure of computer reliability. For a typical control computer, the MTBF should be in the one-thousand-hour-plus category.

Compare the MTBF for each of the computers in Step 2. Choose the machine having the highest MTBF at the desired speed. Further, the computer should be capable of operating in the 50- to 120-F ambient temperature range.

Related Calculations: Use this general method to choose a control computer for any process-type application. Where high-speed operations are involved—such as in missile-guidance applications—computers operating at nsec speeds are generally required. The selection procedure for such machines is different from that for process-control computers where msec speeds are usually satisfactory. Note that the computer prices cited here are relative; for actual prices consult the manufacturers concerned.

The program or software prepared for automatic process control can sometimes cost as much as the computer hardware. The first process applications of computers in an industry often show that the programs needed cost more than anticipated. Thus, computer programming to perform startup and shutdown sequence monitoring in a process system may require 4,000 instructions. For automatic sequence control, as many as 20,000 instructions may be needed. Further, each plant is unique, and little of the programming work done for one process can be applied to another.

Programming still uses the most time of any phase of total computer operation for problem-solving. So even if the computer operating time were cut to nearly zero, the saving in machine time would not be significant when compared with the programming time. On the other hand, saving in computer machine time is important in process-control applications because processes can be held more closely to optimum levels with reduced machine time.

Direct digital control (DDC) is replacing analog control in some process applications. The major advantage of DDC is that it removes the need for digital-to-analog converters and gives the computer more direct control over plant equipment while removing a source of spurious signals.

In process control, certain aspects of programming warrant special consideration. These aspects include real-time operation, memory capability, and operator misuse. Because a process computer functions in a real-time environment, a certain amount of "free" time must be available to allow for emergency reactions and special operator requests. A good rule of thumb is to allow 40 percent of free time within the computer; any less time may cause the computer program to fall out of step with the process situation.

Most process-control computers use one of the variants of IBMs original Fortran. This language finds its principal applications in relatively infrequent procedures, such as plant startup and shutdown. Minute-by-minute scanning is usually handled by a standard "scan, monitor, and alarm" program prepared manually for the particular scanning sequence dictated by operating requirements.

A self-checking program is an important feature of a process-control computer system. Unlike a scientific program in which the results obtained are printed out for perusal by a scientist or engineer, a closed-loop control system utilizes the computer's calculations to act directly on the process plant itself. Should either the input to the computer or the data handling within it be in error, the resulting calculated control points and output signals will be incorrect and could result in hazardous operation. Double and triple checks may be necessary to ensure that the operating data are valid. As more and more input signals are utilized, the programming necessary to provide such validity checks becomes increasingly more complicated. Continuous calculation of a process heat balance can, for example, be useful in determining the validity of the input information, since an extensive range of input data figures in the calculation.

CONTROL-VALVE SELECTION FOR PROCESS CONTROL

Select a steam control valve for a heat exchanger requiring a flow of 1,500 lb/hr of saturated steam at 80 psig at full load and 300 lb/hr at 40 psig at minimum load. Steam at 100 psig is available for heating.

Calculation Procedure:

1. Compute the valve flow coefficient

The valve flow coefficient C_v is a function of the maximum steam flow rate through the valve and the pressure drop that occurs at this flow rate. When choosing a control valve for a process-control system, the usual procedure is to assume a maximum flow rate for the valve based on a considered judgment of the overload the system may carry. Usual overloads do not exceed 25 percent of the maximum rated capacity of the system. Using this overload range as a guide, assume that the valve must handle a 20 percent overload, or 0.20 (1,500) = 300 lb/hr. Hence, the rated capacity of this valve should be 1,500 + 300 = 1,800 lb/hr.

The pressure drop across a steam control valve is a function of the valve design, size, and flow rate. The most accurate pressure-drop estimate that is usually available is that given in the valve manufacturer's engineering data for a specific valve size, type, and steam-flow rate. Without such data, assume a pressure drop of 5 to 15 percent across the valve as a first approximation. This means that the pressure loss across this valve, assuming a 10 percent drop at the maximum steam-flow rate, would be 0.10 × 80 = 8 psig.

With these data available, compute the valve flow coefficient from $C_v = WK/3(\Delta p P_2)^{0.5}$, where W = steam flow rate, lb/hr; $K = 1 + (0.0007 \times {}^\circ F$ superheat of the steam); $p =$

pressure drop across the valve at the maximum steam flow rate, psi; $P_2 =$ control-valve outlet pressure at maximum steam flow rate, psia. Since the steam is saturated it is not superheated and $K = 1$. Then, $C_v = 1,500/3(8 \times 94.7)^{0.5} = 18.1$.

2. Compute the low-load steam flow rate

Use the relation $W = 3(C_v \Delta p P_2)^{0.5}/K$, where all the symbols are as before. Thus, with a 40-psig low-load heater inlet pressure, the valve pressure drop is $80 - 40 = 40$ psig. The flow rate through the valve is then $W = 3(18.1 \times 40 \times 54.7)^{0.5}/1 = 598$ lb/hr.

Since the heater requires 300 lb/hr of steam at the minimum load, the valve is suitable. Had the flow rate of the valve been insufficient for the minimum flow rate, a different pressure drop, i.e., a larger valve, would have to be assumed and the calculation repeated until a flow rate of at least 300 lb/hr was obtained.

Related Calculations: The flow coefficient C_v of the usual 1-in.-diameter double-seated control valve is 10. For any other size valve, the approximate C_v valve can be found from the product $10 \times d^2$, where $d =$ nominal body diameter of the control valve, in. Thus, for a 2-in.-diameter valve, $C_v = 10 \times 2^2 = 40$. Using this relation and solving for d, the nominal diameter of the valve analyzed in Steps 1 and 2 is $d = (C_v/10)^{0.5} = (18.1/10)^{0.5} = 1.35$ in.; use a 1.5-in. valve because the next smaller standard control valve size, 1.25 in., is too small. Standard double-seated control-valve sizes are: $\frac{3}{4}$, 1, $1\frac{1}{4}$, $1\frac{1}{2}$, 2, $2\frac{1}{2}$, 3, 4, 6, 8, 10, and 12 in. Figure 5 shows typical flow-lift characteristics of popular types of control valves.

To size control valves for liquids, use a similar procedure and the relation $C_v = V(G/\Delta p)$, where $V =$ flow rate through the valve, gpm; $\Delta p =$ pressure drop across the valve at maximum flow rate, psi; $G =$ specific gravity of the liquid. When a liquid has a specific gravity of 100 SSU or less, the effect of viscosity on the control action is negligible.

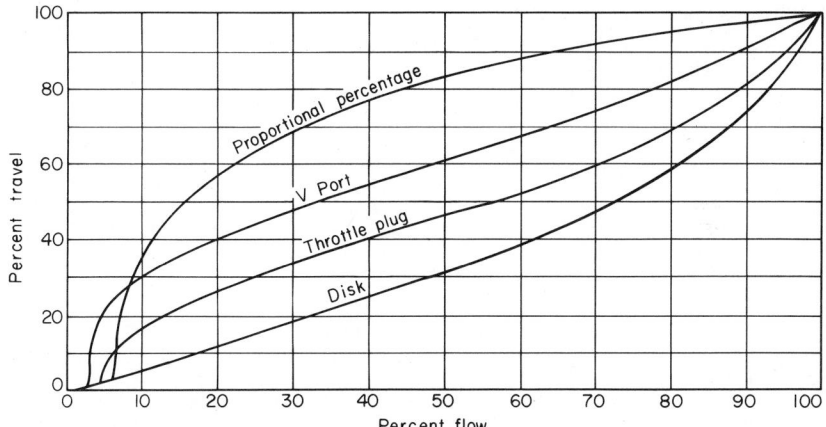

Fig. 5 Flow-lift characteristics of control valves. (*Taylor Instrument Process Control Division of Sybron Corporation.*)

To size control valves for gases, use the relation $C_v = Q(GT_a)^{0.5}/1,360(\Delta p P_2)^{0.5}$, where $Q =$ gas flow rate, cu ft/hr at 14.7 psia and 60 F; $T_a =$ temperature of the flowing gas, F abs $= 460 + F$; other symbols as before. When the valve outlet pressure P_2 is less than $0.5 P_1$, where $P_1 =$ valve inlet pressure, psia, use the value of $P_1/2$ in place $(\Delta p P_2)^{0.5}$ in the denominator of the above relation.

To size control valves for vapors other than steam, use the relation $C_v = W(v_2/\Delta p)^{0.5}/63.4$, where $W =$ vapor flow rate, lb/hr; $v_2 =$ specific volume of the vapor at the outlet pressure P_2, cu ft/lb; other symbols as before. When P_2 is less than $0.5 P_1$, use the value of $P_1/2$ in place of Δp and use the corresponding value of v_2 at $P_1/2$.

When the control valve handles a flashing mixture of water and steam, compute C_v using the relation for liquids given above after determining which pressure drop to

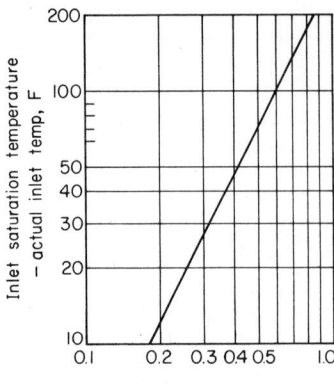

Pressure-drop factor R

Fig. 6 Pressure-drop correction factor for water in the liquid state. *(International Engineering Associates.)*

use in the equation. Use the *actual* pressure drop or the *allowable* pressure drop, whichever is smaller. Find the allowable pressure drop by taking the product of the supply pressure, psia, and the correction factor R, where R is obtained from Fig. 6. For a further discussion of control-valve sizing, see Considine — *Process Instruments and Controls Handbook*, McGraw-Hill, and G. F. Brockett and C. F. King: "Sizing Control Valves Handling Flashing Liquids," Texas A & M Symposium.

CONTROLLED-VOLUME-PUMP SELECTION FOR A CONTROL SYSTEM

Select a controlled-volume pump to deliver 80 gph of 100-F distilled water to a chemical-feed system operating at 2,000 psia. What is the net positive suction head (NPSH) at the beginning of the pump suction stroke if the supply tank produces a 2-psig suction head at the pump centerline? Compute the minimum allowable NPSH for this pump if the length of the 1.5-in. pipe between the pump and suction tank is 30 ft.

Calculation Procedure:

1. Choose the general type of pump to use

Controlled-volume pumps serve two functions when used as the final control elements in a control loop. These functions are: (*a*) deliver liquid at the required pressure and (*b*) deliver liquid in the required quantities. In its second role, the pump also serves as a meter.

Two types of controlled-volume pumps are popular: (*a*) plunger and (*b*) diaphragm. The plunger pump is of somewhat simpler construction than is the disaphragm-type pump and is often used where contact of the plunger and liquid handled is not objectionable. Since distilled water is a relatively bland liquid, a plunger pump will be the tentative first choice for this control application.

2. Determine the pump dimensions and speed

The capacity, dimensions, speed, and efficiency of plunger-type controlled-volume pumps are given by $Q = D^2 LNE/K$, where Q = pump capacity, gph; D = plunger diameter, in.; L = plunger stroke length, in.; N = number of strokes per min, i.e., the pump speed; E = volumetric efficiency of the pump; K = dimensional constant = 4.92 for Q in gph, 295 for pump capacity in gpm, and 0.0013 for pump capacity in ml/hr.

Assume a pump speed of 50 strokes/min. This is a typical speed for plunger-type controlled-volume pumps. The usual efficiency of such a pump is 90 percent. With a 3-in. stroke the plunger diameter is $D = (QK/LNE)^{0.5} = [(80 \times 4.92/(3 \times 50 \times 0.9)]^{0.5} = 1.71$ in.; say 1.75 in. This is a standard pump-plunger diameter.

3. Compute the pump NPSH

Use the relation NPSH $= P_a \pm P_h - P_v$, where NPSH = pump net positive suction head, psia; P_a = atmospheric pressure at pump location = 14.7 psia at sea level; P_h = pressure head of liquid column above (+) or below (−) the centerline of the pump suction, psig; P_v = vapor pressure of the liquid at the pumping temperature, psia.

From the steam tables, $P_v = 0.949$ psia for water at 100 F. With an atmospheric pressure of 14.7 psia, NPSH = 14.7 + 2.0 − 0.949 = 15.751 psia.

4. Compute the pump minimum NPSH

Use the relation $NPSH_{min} = sL_p LN^2 D^2/120,000 D_p^2$, where $NPSH_{min}$ = minimum net positive suction head with which the pump can operate, psia; s = specific gravity of liquid handled; L_p = length of suction pipe, ft; D_p = suction pipe inside diameter, in.; other symbols as given before. Assuming a specific gravity of 1.0 $\text{NPSH}_{min} = (1.0 \times 30 \times 3 \times 50 \times 50 \times 1.5 \times 1.5)/(120,000 \times 1.5 \times 1.5) = 1.87$ psia.

Since the available NPSH (15.751 psia, Step 3) is greater than the minimum NPSH required (1.87 psia, Step 4), the pump will operate satisfactorily and without cavitation.

Related Calculations: Use this general procedure to choose controlled-volume metering pumps for control systems requiring flows ranging from 1 ml/hr to 20 gpm, or more, at pressures ranging to 50,000 psi. Typical applications for which this procedure is valid include chemical feed, ratioing, proportioning, and control of process variables.

STEAM-BOILER-CONTROL SELECTION AND APPLICATION

Choose a suitable feedwater regulator and combustion control for an industrial boiler serving the following loads: heating, 18,000 lb/hr; process 100,000 lb/hr; miscellaneous uses, 12,000 lb/hr. The boiler will have a maximum overload of 20 percent and wide load fluctuations are expected at frequent intervals during operation. Pulverized-coal fuel will be used to fire the boiler.

Calculation Procedure:

1. Determine the required boiler rating

Find the sum of the individual loads on the boiler, or $18,000 + 100,000 + 12,000 = 130,000$ lb/hr. With a 20 percent overload, the boiler rating must be 1.2 (130,000) = 156,000 lb/hr. With a 10 percent additional reserve capacity to provide for unusual loads, the rated boiler capacity should be 1.1(156,000) = 171,500 lb/hr; say 175,000 lb/hr for selection purposes.

2. Choose the type of feedwater regulator to use

Table 2 summarizes typical feedwater regulators used for boilers of various capacities. Study of Table 2 shows that a boiler in the 75,000–200,000 lb/hr capacity range can use a relay-operated regulator with one or two elements when the load fluctuations are reasonable. With wide load swings the relay-operated three-element regulator is

TABLE 2 Boiler-Feedwater-Regulator Selector Chart*·†

Boiler capacity	Type of feedwater regulator		
	Self-operated single-element	Relay-operated single- or two-element	Relay-operated three-element
Below 75,000 lb/hr	For steady loads (building heating or continuous processes)	For irregular loads (batch processes, hoists, rolling mills, etc.)	
75,000–200,000 lb/hr	Use only in special cases	For all steady and fluctuating loads	For extreme load and water conditions and boilers with steaming economizers
Above 200,000 lb/hr	. . .	Use only on steady loads	For all types of loads

*From Kallen—*Handbook of Instrumentation and Controls*, McGraw-Hill.
†Excess pressure ahead of feedwater regulator should be at least 50 psi. and should be controlled by regulation of feed pump. Use excess-pressure valves only when excess pressure varies more than plus or minus 30 percent. Where drum level is unsteady owing to high solids concentration or boiler feed or other causes, use next higher-class feed regulator.

TABLE 3 Classification of Combustion-control Systems *

Name	*A,* series-fuel	*B,* series-air	*C,* parallel	*D,* calorimeter or steam flow–air flow
Action.............	Temperature- or pressure-actuated master adjusts fuel rate; fuel meter adjusts air flow	Temperature- or pressure-actuated master adjusts air flow; air-flow meter adjusts fuel flow	Temperature- or pressure-actuated master adjusts fuel flow and air flow simultaneously	Pressure-actuated master adjusts fuel flow; steam flow adjusts air flow
Relative speed of control	Master adjusted for fast response because fuel-rate fluctuations caused by fluctuating pressure or temperature on master do not have correspondingly fast effect on that controlled variable	Master adjusted for slow response, because air-flow fluctuations following fast fluctuating master signal have a rapid effect on controlled variable and may cause hunting action if air-flow response is too fast	Master adjusted for slow response for same reason as in series-air	Master adjusted for fast response for same reason as in series-fuel. Steam-flow–air-flow control can be relatively rapid since steam-flow fluctuations are not so rapid as pressure variations
Used on fuels	Easily metered fuels such as oil and gas	All fuels. Oil, gas, and coal, either solid or burned in suspension	Primarily on solid fuels (grate firing)	Fuels hard to meter or fuels burned simultaneously. Commonly used on pulverized-coal-fired boilers
Advantages	When fuel may be in short supply, eliminates possibility of carrying high excess air for long period	Eliminates possibility of explosive mixture in combustion space when air fails. Eliminates need of fuel cutback for this purpose	Relatively inexpensive control system. No metering necessary	Ensures proper air-fuel ratio, even though fuel cannot be accurately metered or is of varying heat content Ensures this condition even when burning a mixture of different fuels at the same time

*From Kallen – *Handbook of Instrumentation and Controls,* McGraw-Hill.

a better choice. Since this boiler will encounter wide load swings, a three-element regulator is a wise and safe choice.

3. Choose the type of combustion-control system

Table 3 summarizes the important selection features of four types of combustion-control systems. Study of Table 3 shows that a stream-flow–air-flow type combustion-control system would probably be best for the fuel and load conditions in this plant. Hence, this type of control system will be chosen.

Related Calculations: Any control system selected for a boiler using this procedure should be checked out by studying the engineering data available from the control-system manufacturer. The procedure given here is valid for heating, industrial, power, marine, and similar boilers.

CONTROL-VALVE CHARACTERISTICS AND RANGEABILITY

A flow control valve will be installed in a process system in which the flow may vary from 100 to 20 percent while the pressure drop in the system rises from 5 to 80 percent. What is the required rangeability of the control valve? What type of control-valve characteristic should be used? Show how the effective characteristic is related to the pressure drop the valve should handle.

Calculation Procedure:

1. Compute the required valve rangeability

Use the relation $R = (Q_1/Q_2)(\Delta P_2/\Delta P_1)^{0.5}$, where R = valve rangeability; Q_1 = valve initial flow, percent of total flow; Q_2 = valve final flow, percent of total flow; P_1 = initial pressure drop across the valve, percent of total pressure drop; P_2 = percent final pressure drop across the valve.

Substituting, $R = (100/20)(80/5)^{0.5} = 20$.

2. Select the type of valve characteristic to use

Table 4 lists the typical characteristics of various control valves. Study of Table 4 shows that an equal-percentage valve must be used if a rangeability of 20 is required. Such a valve has equal stem movements for equal-percentage changes in flow at a constant pressure drop based on the flow occurring just before the change is made.[1] The equal-percentage valve finds use where large rangeability is desired and where equal-percentage characteristics are necessary to match the process characteristics.

TABLE 4   Control-Valve Characteristics

Valve type	Typical flow rangeability	Stem movement
Linear..............	12–1	Equal stem movement for equal flow change
Equal-percentage ...	30–1 to 50–1	Equal stem movement for equal-percentage flow change*
On-off..............	Linear for first 25 percent of travel; on-off thereafter	Same as linear up to on-off range

*At constant pressure drop.

3. Show how the valve effective characteristic is related to pressure drop

Figure 7 shows the inherent and effective characteristics of typical linear, equal-percentage, and on-off control valves. The inherent characteristics is the theoretical performance of the valve.[1] If a valve is to operate at a constant load without changes

[1] E. Ross Forman, "Fundamentals of Process Control," *Chemical Engineering*, June 21, 1965.

in the flow rate, the characteristic of the valve is not important, since only one operating point of the valve is used.

Figure 7b and c give definite criteria for the amount of pressure drop the control valve should handle in the system. This pressure drop is not an arbitrary value such as 5 psi but rather a percent of the total dynamic drop. The control valve should take at least 33 percent of the total dynamic system pressure drop[1] if an equal-percentage valve is used and is to retain its inherent characteristics. A linear valve should not take less than a 50 percent pressure drop if its linear properties are desired.

There is an economic compromise in the selection of every control valve. Where possible, the valve pressure drop should be as high as needed to give good control. If experience or an economic study dictates that the requirement of additional horsepower to provide the needed pressure is not worth the investment in additional pumping or compressor capacity, the valve should take less pressure drop with the resulting poorer control.

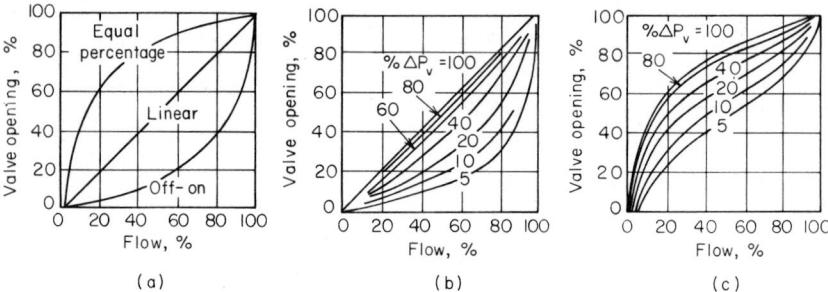

Fig. 7 (a) Inherent flow characteristics of valves at constant pressure drop; (b) effective characteristics of a linear valve; (c) effective characteristics of a 50:1 equal-percentage valve.

FLUID-AMPLIFIER SELECTION AND APPLICATION

Select a fluid amplifier to amplify the output of a fluidic sensor in a control system having a sensor output of 1 psi and requiring a control-valve operating pressure for modulating purposes of at least 10 psi.

Calculation Procedure:

1. Compute the required amplification ratio

The amplification ratio μ = the gain of the amplifier = amplifier output, psi/amplifier input, psi. For this amplifier, $\mu = 10/1 = 10$. Hence, this fluid amplifier must increase a 1-psi input signal from the sensor to at least a 10-psi signal at the amplifier output.

2. Select a suitable fluid amplifier

Refer to a manufacturer's engineering data for the characteristics of a suitable fluid amplifier. Figure 8 shows the characteristics of a typical commercially available fluid amplifier. This amplifier is available for two gains—9:1 and 18:1. Since the first gain is lower than required, the second, or 18:1, gain would be used if this amplifier were chosen. With this gain, a 1-psi input signal would be amplified in the device to an output of an $18(1) = 18$-psi signal. Since a 10-psi or larger signal is required to operate the control valve, this amplifier is acceptable.

Related Calculations: Many fluid amplifiers of the proportional type are packaged into fully operational modules—just as transistors and linear integrated circuits are. When so packaged, the device is termed an *operational amplifier*. An operational amplifier[2] is a module that contains within itself all the elements of a high-gain, accurate, and

[1]*Ibid.*
[2]Frank Yeaple, "Analog Fluidic Amplifiers Are Waiting in the Wings," *Product Engineering,*
Oct. 23, 1967.

repeatable analog amplifier. Called *op-amps* for short, these devices are useful because they can amplify low-level signals—such as the outputs of fluidic sensors—proportionally to levels high enough to modulate control valves.

A fluid amplifier can produce a gain of pressure as, in this instance, power, or flow. Compute the gain for any of these outputs by setting up the ratio of output to input, as in Step 1. Be certain to use consistent units for the output and input variables.

Where an amplifier is *not* linear, Fig. 9, the gain is different for each output. Hence, the operating points—i.e., input and output variables—must be stated for the gain being considered. Table 5 lists the selection characteristics for fluid amplifiers, control valves, and integrated circuits. Data presented in this tabulation are useful in choosing the most suitable control for a given application.

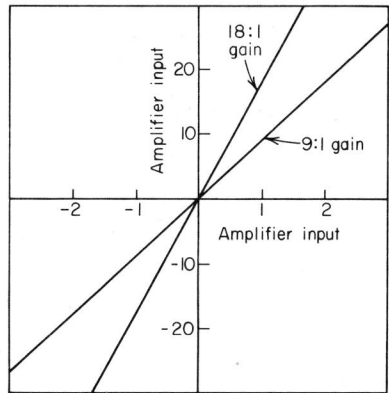

Fig. 8 Operational amplifier performance for two gains.

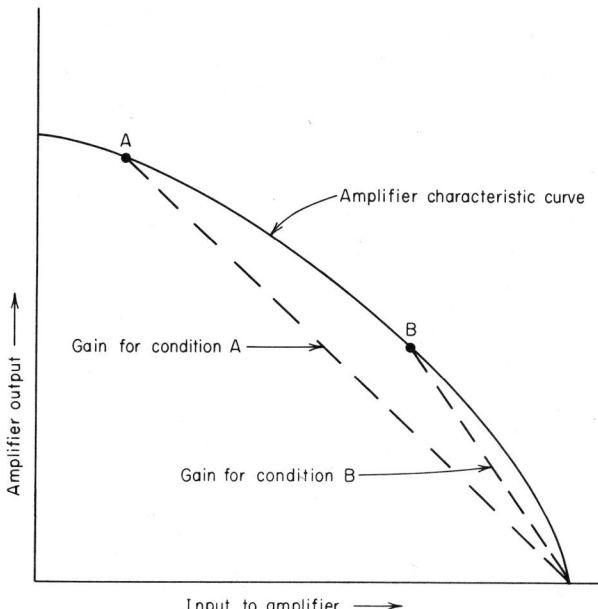

Fig. 9 Fluid-amplifier operating characteristic curve and gain for two operating conditions.

CAVITATION, SUBCRITICAL, AND CRITICAL-FLOW CONSIDERATIONS IN CONTROLLER SELECTION

Using the sizing formulas of the Fluid Controls Institute, Inc., size control valves for the cavitation, subcritical-, and critical-flow situations described below. Show how accurate the FCI formulas are.

TABLE 5   Control Selection Guide*

(Checks show which control to use)

Characteristics of device needed	Fluid amplifiers	Spools, poppets	Solid-state integrated circuits
Is environment-proof:			
Can be designed to operate at extremely high			
temperature	√		
Can be made tolerant of any atmosphere.	√	√	√
Is unhurt by nuclear radiation	√	√	
Is unhurt by heavy vibration or shock	√	. . .	√
Has no moving parts:			
No stiction, hysteresis, dead zone, or jamming; also, no			
mechanical blocking, so no fluid hammer	√	. . .	√
Is tiny, stackable, monolithic:			
Whole circuits can be made in one integrated block,			
all permanently sealed, and with no moving parts	√	. . .	√
Can be supplied from any fluid source:			
Air, gas, water, oil, or process fluids will work; even the			
water rushing by the hull of a boat, or the air slipping			
past an airfoil, can be exploited	√		
Needs no electricity:			
Is not affected in performance by radio or electrical			
interference	√	√	
Is responsive to extremely small inputs:			
Breaths of air; proximity of anything, such as tiny			
threads, specks, liquid surfaces, and air bubbles;			
motion of housing (fluid inertia effect); shock waves;			
fluid disturbances; spark discharges; sound waves;			
controlled vibration; localized heat	√	. . .	√
Is fast (millisec or better).	√	. . .	√

Characteristics of device sought

	Fluid amplifiers	Spools, poppets	Solid-state integrated circuits
Has high energy output:			
Easily transduced to mechanical movement.	Some	√	
Is widely available from many sources:			
For proportional control	Some	Some	√
For on-off control.	√	√	√
Is ultrafast (μsec)	. . .	. . .	√
Can be shut off individually when not in use:			
Power requirement drops drastically when device is			
switched off.	. . .	√	√

Product Engineering.

   Cavitation: Select a control valve for a situation where cavitation may occur. The fluid is steam condensate; inlet pressure P_1 is 167 psia; $\Delta P = 105$ psi; inlet temperature T_1 is 180 F; vapor pressure P_v is 7.5 psia.

   Subcritical gas flow: Determine the valve capacity required at these conditions: fluid is air; flow Q_g is 160,000 std cu ft/hr; inlet pressure P_1 is 275 psia; $\Delta P = 90$ psi; gas temperature T_1 is 60 F.

   Critical vapor flow: A heavy-duty angle valve is suggested for a steam pressure-reducing application. Determine the capacity required and compare an alternate valve type. The fluid is saturated steam; flow W is 78,000 lb/hr; inlet pressure P_1 is 1,260 psia; outlet pressure P_2 is 300 psia.

Calculation Procedure:

1. Choose the valve type and determine its critical flow factor for the cavitation situation

If otherwise suitable (i.e., with respect to size, materials, and space considerations), a butterfly control valve is acceptable on a steam-condensate application. Find, from Table 6, the value of the critical flow factor $C_f = 0.68$ for a butterfly valve with 60° operation.

2. Compute the maximum allowable pressure differential for the valve

Use the relation $\Delta P_m = C_f^2(P_1 - P_v)$, where ΔP_m = maximum allowable pressure differential, psi; P_1 = inlet pressure, psia; P_v = vapor pressure, psia. Substituting, $\Delta P_m = (0.68)^2(167 - 7.5) = 74$ psi. Since the actual pressure drop, 105 psi, exceeds the allowable drop, 74 psi, cavitation *will* occur.

3. Select another valve and repeat the cavitation calculation

For a single-port top-guided valve with flow to open plug, find $C_f = 0.90$ from Table 6. Then, $\Delta P_m = (0.90)^2(167 - 7.5) = 129$ psi.

In the case of the single-port top-guided valve, the allowable pressure drop, 129 psi, exceeds the actual pressure drop, 105 psi, by a comfortable margin. This valve is a better selection because cavitation will be avoided. A double-port valve might also be used, but the single-port valve offers lower seat leakage. However, the double-port valve offers the possibility of a more economical actuator, especially in larger valve sizes. This concludes the steps for choosing the valve where cavitation conditions apply.

4. Apply the FCI formula for subcritical flow

The FCI formula for subcritical gas flow is $C_v = Q_g/1,360(\Delta P/GT)^{0.5}[(P_1 + P_2)/2]^{0.5}$, where C_v = valve flow coefficient; Q_g = gas flow, std cu ft/hr; ΔP = pressure differential, psi; G = specific gravity of gas at 14.7 psia and 60 F; T = absolute temperature of the gas, R; other symbols as given earlier. Substituting, $C_v = 160,000/1,360$ $(90/520)^{0.5}$ $[(275 + 185)/2]^{0.5} = 18.6$.

5. Compute C_V using the unified gas-sizing formula

For greater accuracy, many engineers use the unified gas-sizing formula. Assuming a single-port top-guided valve installed open to flow, Table 6 shows $C_f = 0.90$. Then, $Y = (1.63/C_f)(\Delta P/P_1)^{0.5}$, where Y is defined by the equation and the other symbols are as given earlier. Substituting, $Y = (1.63/0.90)(90/275)^{0.5} = 1.04$. Figure 10 shows the flow correlation established from actual test data for many valve configurations at maximum valve opening, and relates Y and the fraction of the critical flow rate.

Find from Fig. 11 the value of $Y - 0.148Y^3 = 0.87$. Compute $C_v = Q_g(GT)^{0.5}/834C_f(Y - 0.148Y^3)$, where all the symbols are as given earlier. Or, $C_v = 160,000(520)^{0.5}/[834$ $(0.90)(275)(0.87)] = 20.4$. This value represents an error of approximately 10 percent in the use of the FCI formula.

6. Determine C_f for critical vapor flow

Assuming reduced valve trim for a heavy-duty angle valve, $C_f = 0.55$ from Table 6.

7. Compute the critical pressure drop in the valve

Use $\Delta P_c = 0.5(C_f)^2P_1$, where P_c = critical pressure drop, psi; other symbols as given earlier. Substituting, $\Delta P_c = 0.5(0.55)^2(1,260) = 191$ psi.

8. Determine the value of C_V

Use the relation $C_v = W/1.83C_fP_1$, where the symbols are as given earlier. Substituting, $C_v = 78,000/[1.83(0.55)(1,260)] = 61.5$. A lower C_v could be attained by using the valve flow to open, but a more economical choice is a single-port top-guided valve installed open to flow.

TABLE 6 Critical Flow Factors for Control Valves at 100 percent Lift*

Split body

A	Flow to close plug	0.80
	Flow to open plug	0.75
	Parabolic plug only	
B	Flow to close plug	0.50
	Flow to open plug	0.90
	Parabolic plug only	

Single-port, globe body

A	Flow to close plug	0.85
	Flow to open plug	0.90
	Parabolic plug only	
B	Flow to close plug	0.50
	Flow to open plug	0.90
	Parabolic plug only	

Angle body

A	Flow to close plug	0.40
	Flow to open plug	0.90
	Parabolic plug only	
B	Flow to close plug	0.55
	Flow to open plug	0.95
	Parabolic plug only	

Double-port, globe body

A	Parabolic plug	0.90
	V-port plug	1.00
B	Parabolic plug	0.62
	V-port plug	0.95

Butterfly

$D/d = 1$	$\alpha = 60°$	0.68
	$\alpha = 90°$	0.58
$D/d = 2$	$\alpha = 60°$	0.62
	$\alpha = 90°$	0.50

(A) Full-capacity trim, orifice diameter ~ 0.8 valve diameter

(B) Reduced capacity trim, 50% of (A) and less.

NOTE: The listed values apply for equal port-area valves only and do not include corrections for pipe friction.

*Henry W. Boger and *Chemical Engineering.*

For a single-port top-guided valve flow to open, $C_f = 0.90$ from Table 6. Hence, $C_v = 78,000/[1.83(0.90)(1,260)] = 37.6$.

A lower capacity is required at critical flow for a valve with less pressure recovery. Although this may not lead to a smaller body size because of velocity and stability

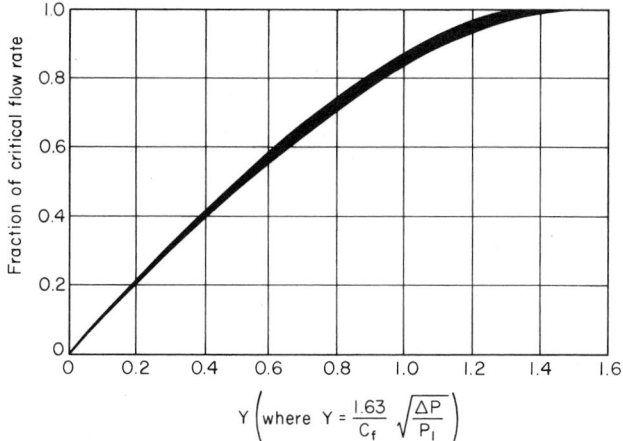

Fig. 10 Flow correction established from actual data for many valve configurations at maximum valve opening.

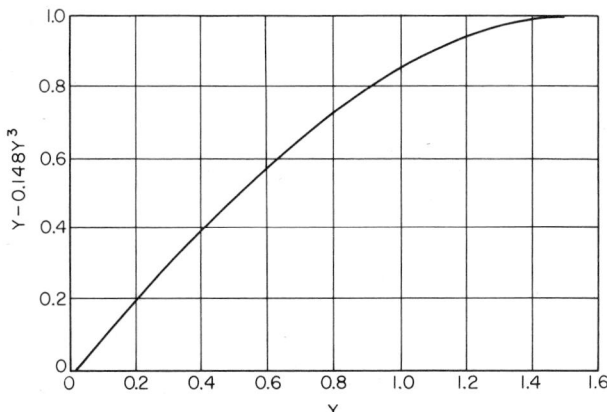

Fig. 11 Correction-factor values.

considerations, the choice of a more economical body type and a smaller actuator requirement is attractive. The heavy-duty angle valve finds its application generally on flashing-hydrocarbon liquid service with a coking tendency.

This Calculation Procedure is the work of Henry W. Boger, Engineering Technical Group Manager, Worthington Controls Co.

Section **8**

Aeronautical and Astronautical Engineering

HAROLD BECHER
President, Strato Missiles, Inc.

JOHN P. ROEDEL
Research Engineer, The Boeing Company

TYLER G. HICKS, P.E.

REFERENCES: Carroll—*The Aerodynamics of Powered Flight*, Wiley; Shields—*Air Pilot Training*, McGraw-Hill; Elsley and Devereux—*Hovercraft Design and Construction*, David & Charles (London); Koelle—*Handbook of Astronautical Engineering*, McGraw-Hill; Corliss—*Propulsion Systems for Space Flight*, McGraw-Hill; Millikan—*Aerodynamics of the Airplane*, Wiley; *Space Handbook: Astronautics and its Applications*, GPO; Allen—*Astrophysical Quantities*, Athlone Press (London); Carter—*Realities of Space Travel*, McGraw-Hill; Clarke—*Interplanetary Flight*, Temple Press (London); Puckett and Ramo—*Guided Missile Engineering*, McGraw-Hill; Sutton—*Rocket Propulsion Elements*, Wiley; von Braun—*The Mars Project*, University of Illinois Press;

Herrick—*Astrodynamics*, Van Nostrand; Moulton—*Celestial Mechanics*, Macmillan; Van Allen—*Scientific Uses of Earth Satellites*, University of Michigan Press; Stephenson—*Introduction to Nuclear Engineering*, McGraw-Hill; Cowling—*Magnetohydrodynamics*, Interscience; Blasingame—*Astronautics*, McGraw-Hill; Berkner and Odishaw—*Science in Space*, McGraw-Hill; Nokilayev—*Thermodynamic Assessment of Rocket Engines*, Pergamon; Boehm—*ROCKET: Rand's Omnibus Calculator of the Kinematics of Earth Trajectories*, Prentice-Hall; Bell—*Cryogenic Engineering*, Prentice-Hall; Pogorelov—*Fundamentals of Orbital Mechanics*, Holden-Day; Casamassa and Bent—*Jet Aircraft Power Systems*, McGraw-Hill; McClintock—*Cryogenics*, Reinhold; Henshaw—*Supersonic Engineering*, Wiley; Wolverton—*Flight Performance Handbook for Orbital Operations*, Wiley.

AIRCRAFT STALL OR LANDING SPEED

What is the sea-level stall or landing speed of a 9,300-lb airplane having a wing area of 361 sq ft if a Clark wing with a fixed slot is used? Determine these speeds for a Clark wing with a fixed slot and slotted flap. What is the stall angle of attack for each wing?

Calculation Procedure:

1. Determine the wing lift coefficient

The maximum value of the wing lift coefficient C_{Lm} is used when computing the stall or landing speed of an aircraft. Values of C_{Lm} are published in various handbooks and other references. Table 1 lists a few typical wings and the value of C_{Lm} for each wing. Thus, for the basic Clark wing with a fixed slot, $C_{Lm} = 1.77$. With a fixed slot and slotted flap, $C_{Lm} = 2.26$.

2. Compute the wing stall or landing speed

Use the relation $v_s = (2W/\rho A C_{Lm})^{0.5}$, where $v_s =$ stall or landing speed, fps; $W =$ aircraft weight, lb; $\rho =$ density of air at sea level $= 0.075$ lb/cu ft; $A =$ aircraft wing area, sq ft. Substituting for the fixed-slot Clark wing, $v_s = [2 \times 9,300/(0.075 \times 361 \times 1.77)]^{0.5} = 197$ fps, or 134.5 mph. For the fixed-slot, slotted-flap wing, $v_s = [2 \times 9,330/(0.075 \times 361 \times 2.26)] = 174$ fps, or 118.8 mph. Thus, the greater the number of high-lift devices — slots, flaps, etc.— used on a wing, the lower the landing speed. A low landing speed is desirable because it reduces the required runway length, makes for easier control of the aircraft, and lowers the impact forces between the aircraft and the runway. The most complex Clark wing has a C_{Lm} of 3.37, giving a stall or landing speed of 97.5 mph.

3. Determine the stall or landing speed angle of attack

Table 1 shows that the angle of attack at stall for the fixed-slot wing is 24°, and for the fixed-slot, slotted-flap wing 19°. For the most complex Clark wing, the stall angle of attack is 16°. Thus, high-lift devices also reduce the stall or landing angle of attack. This is a desirable feature in a wing.

Related Calculations: Use this general method to compute the minimum horizontal speed of any winged aircraft. Wing characteristics are tabulated in aeronautical engineering handbooks and wind-tunnel reports.

TABLE 1   Typical Aircraft Wing Characteristics

Wing type	C_{Lm}	$\alpha°$ at C_{Lm}
Clark, fixed-slot	1.77	24
Clark, fixed-slot, slotted-flap	2.26	19
Clark, most complex ..	3.37	16
NACA 4415	1.6	20

AIRCRAFT TRUE AIRSPEED, DRIFT, AND GROUNDSPEED

An aircraft is flying at 8,000 ft altitude at an indicated airspeed of 300 mph. What is its true airspeed? If the aircraft is on a true heading of 300° and there is a 200° wind blowing at 100 mph, find the track of the aircraft and its groundspeed.

Calculation Procedure:

1. Compute the aircraft true airspeed

Use the relation $t_a = i_a + (0.02i_a h/1,000)$, where t_a = aircraft true airspeed, mph; i_a = aircraft indicated airspeed, mph; h = flight altitude, ft. Substituting, $t_a = 300 + (0.02 \times 300 \times 8,000/1,000) = 348$ mph. The constant of 0.02 represents the average altitude error of the usual airspeed indicator of 2 percent per 1,000 ft of altitude. If the indicator being used had a different error percentage, the actual error would be substituted in the relation above.

2. Compute the aircraft groundspeed

The aircraft groundspeed—i.e., its speed on the ground—will be greater or less than the true airspeed, depending upon whether the wind, if any, at the flight altitude aids or retards the aircraft movement.

To determine the groundspeed t_g mph, draw a vector to scale representing the true heading and true airspeed of the aircraft, Fig. 1. Label this vector TH for true heading and indicate on it the angular measure of the true heading, 300°, and the true airspeed, 348 mph.

From the end of the true-heading vector lay off the wind vector at 200° with length equal to the wind velocity, or 100 mph. Connect the origin of the true-heading vector with the terminal end of the wind vector. The resulting vector is the track TR and ground-speed t_g vector of the aircraft.

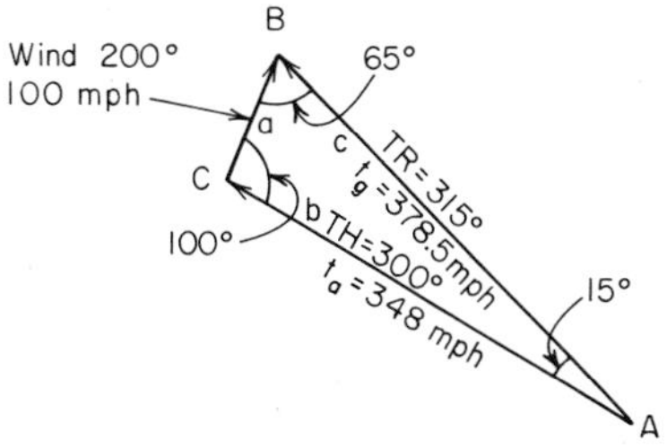

Determine the groundspeed graphically by scaling the t_g vector, or solve for this value by computation. To solve for t_g by computation, determine the included angle between the TH and the wind vector by geometry. Then, using the law of cosines, $t_g^2 = c^2 = a^2 + b^2 - 2ab \cos C$, where the letters a, b, and c represent the sides of the triangle, Fig. 1. Or $c^2 = 100^2 + 348^2 - 2(100)(348)(\cos 100°)$; $c = t_g = 378.5$ mph. Thus, the wind increases the speed of the aircraft over the ground.

Fig. 1 Velocity diagram for determining the track and groundspeed of an aircraft.

3. Compute the aircraft track

Solve for angle A, Fig. 1, using the relation $\tan A/2 = r/s - a$, where $r = [(s-a) \times (s-b)(s-c)/s]^{0.5}$, and $s = (a+b+c)/2$. Solving, $A = 15°$. Hence, the aircraft track = $300 + 15° = 315°$. This is the course made good over the ground.

Related Calculations: The flight condition met in this Calculation Procedure is termed *right drift*, i.e., the aircraft drifts to the right of its true heading because the wind is from the left. A general rule applicable to right drift is: *When the wind is from the left or the drift is to the right, the track vector will always be to the right of the heading vector and will have a larger directional (i.e., angular) value. Left drift* results from a right wind, and the track vector is always to the left of the heading vector and has a lower directional value.

Use the same general procedure to find the wind velocity, wind direction, true airspeed, or true heading. This same general procedure can also be used for determining the drift of a ship or boat resulting from the water current or surface wind, or both. The only difference is that the speed quantities involved are smaller than for aircraft. However, the difference in numerical quantities does not influence the graphical or computation procedure in any way.

AIRCRAFT RATE OF CLIMB, CLIMBING TIME, SPEED, AND RANGE

Figure 2 shows a simplified performance diagram for a 5,000-lb airplane. Using this chart, determine the rate of climb at sea level and at an 8,000-ft altitude. What is the aircraft speed at sea level and at 8,000 ft? How long will it take the aircraft to reach an altitude of 8,000 ft after takeoff? What is the absolute ceiling of the aircraft? Determine the aircraft range if it consumes 50gph of fuel at full throttle at 8,000 ft and its total fuel capacity is 500 gal.

Calculation Procedure:

1. Compute the sea-level rate of climb of the aircraft

Determine the largest difference between the power available at sea level and the power required at sea level, Fig. 2. Do this by studying the performance diagram and by using a scale to find the largest power difference at sea level on the performance diagram plot of the power available.

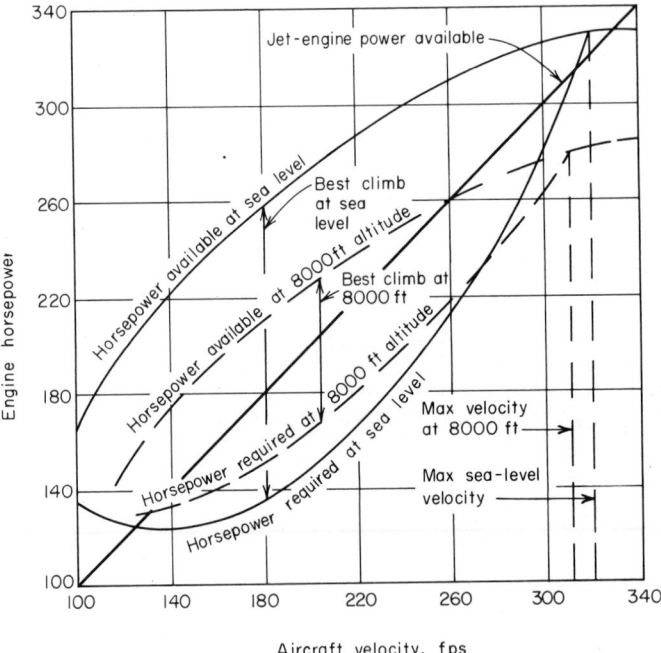

Aircraft velocity, fps

Fig. 2 Performance chart for a typical propeller aircraft. The power curve for a typical jet engine is also shown. The jet-power curve is used in the same way as the propeller-power curve.

For any airplane, the power available for climbing $hp_{ac} = hp_a - hp_r$, where hp_a = maximum available engine-horsepower output at the altitude in question; hp_r = horsepower required to overcome the aircraft drag at the altitude in question. For this airplane the maximum value of hp_{ac} at sea level is 120 hp. This value is found by inspection and scaling of Fig. 2.

Knowing hp_{ac} and the weight of the airplane, compute the maximum rate of climb at sea level from $r_c = 33,000\ hp_{ac}W$, where r_c = rate of climb, fpm, and W = aircraft weight, lb. Substituting, $r_c = 33,000(120)/5,000 = 794$ fpm.

2. Determine the maximum speed of the aircraft at sea level

Find the intersection of the power-available and power-required curves at sea level, Fig. 1. Read the velocity at this point as 320 fps, or $v = $ (320 fps)(3,600 sec/hr)/5,280 ft/mile = 218.2 mph.

3. Determine the altitude rate of climb of the aircraft

Use the same procedure as in Step 1 to find the maximum difference between hp_a and hp_r, Fig. 2, at the altitude in question, 8,000 ft. Note that there are separate hp_a and hp_r curves for the 8,000-ft altitude. These performance curves are different and distinct from the corresponding sea-level curves. Figure 2 shows that the maximum difference between hp_a and hp_r at 8,000 ft altitude is 60 hp. Hence, the maximum rate of climb at 8,000 ft is $r_c = 33,000(60)/5,000 = 396$ fpm, or about half the maximum rate of climb at sea level. The *service ceiling* of an airplane is the altitude at which the aircraft rate of climb is 100 fpm.

4. Determine the maximum aircraft speed at altitude

Use the same procedure as in Step 2 and read the maximum speed at 8,000 ft altitude as 310 fps, or $(310)(3,600)/5,280 = 211.5$ mph, closely.

5. Compute the absolute ceiling of the aircraft

At the *absolute ceiling* of an aircraft the rate of climb is zero. To compute the absolute ceiling use the relation $H = r_s h/(r_s - r)$, where $H = $ absolute ceiling of aircraft, ft; $r_s = $ rate of climb at sea level, fpm; $h = $ altitude for which another rate of climb is known, ft; $r = $ rate of climb, fpm, at altitude h. Using the sea-level and 8,000-ft values computed above, $H = 794(8,000)/(794 - 396) = 15,950$ ft.

6. Compute the time for the aircraft to climb to altitude

Use the relation $t = 2.303(H/r_s) \log H/(H-h)$, where $t = $ time to climb to altitude h, min; other symbols as before. Substituting, $t = 2.303(15,950/794) \log 15,950/(15,950 - 8,000) = 14.0$ min.

7. Compute the aircraft range

Allow 20 percent of the fuel capacity for the takeoff, climbing, and landing maneuvers, including holding patterns before landing. Hence, fuel available for cruise = 500 − 0.20(500) = 400 gal.

At 8,000 ft the speed of the aircraft is, from Step 4, 211.5 mph with a fuel consumption of 50 gph. Using the relation $R = vC/f$, where $R = $ aircraft range, miles; $v = $ aircraft speed at the altitude in question, mph; $C = $ fuel available for cruising, gal; $f = $ fuel consumption at cruise altitude, gph. Substituting, $R = 211.5(400/50) = 1,692$ miles.

The range computed here is the distance the aircraft could travel in still air—i.e., without head, tail, or off-course winds. In actual operation of the aircraft the pilot would make allowance for winds in computing his true airspeed and range.

Related Calculations: The performance curve shown in Fig. 2 is typical for a variety of aircraft. For each altitude at which the aircraft will fly, a power-available and power-required curve can be plotted. With these curves available, the maximum rate of climb and airspeed can be determined as described in Steps 1 and 2 for each of the altitudes plotted.

AIRCRAFT ENGINE THRUST DETERMINATION

An aircraft jet engine has an air inlet velocity v_i of 1,000 fps and a jet exit velocity v_e of 3,000 fps. What thrust will this engine develop when the airflow rate through it is 300 lb/sec? What is the equivalent hp of a reciprocating engine driving a propeller having an efficiency of 70 percent at a true airspeed of 400 mph? What is the rate of climb with the reciprocating power plant if the airplane weighs 100,000 lb and the power required at sea level is 27,000 hp?

Calculation Procedure:

1. Compute the jet-engine thrust

Use the relation $T = W\Delta v/g$, where T = jet-engine thrust, lb; W = rate of airflow through the engine, lb/sec; Δv = change in velocity of the air during passage through the engine, fps; $g = 32.2$ ft/sec². Substituting, $T = 300(3,000 - 1,000)/32.2 = 18,620$ lb.

2. Compute the equivalent reciprocating engine hp

Use the relation $hp_e = Tv/375e$, where hp_e = equivalent hp of the reciprocating engine; v = true airspeed at which the comparison is made; e = propeller efficiency. Substituting, $hp_e = 18,620(400)/[375(0.70)] = 28,600$ hp. This is much greater than the takeoff hp rating of any standard reciprocating aircraft engine. However, several engines whose hp sum equals or exceeds the required rating could be used if the weight conditions were acceptable.

3. Compute the rate of climb of the aircraft

Use the relation $r_c = 33,000\ hp_{ac}/W$, where r_c = rate of climb of the airplane, fpm; hp_{ac} = available power for climbing = maximum rated hp − hp required, both at sea level; W = weight of loaded aircraft, lb. Substituting, $r_c = 33,000(28,600 - 27,000)/100,000 = 528$ fpm.

Related Calculations: Use this general procedure for preliminary analysis of any jet engine whose overall characteristics are known. The equivalent hp relation, Step 2, is useful when comparing piston-engine and jet-engine aircraft.

AIR-CUSHION VEHICLE PROPORTIONS AND POWER REQUIREMENTS

Select the required power, fan airflow, and natural frequency of a plenum-type air-cushion vehicle having a total weight, including the payload, of 5,000 lb. The vehicle must have an emergency load capacity for extra passengers or dynamic loads of 1,000 lb. The available area for the plenum is 8×16 ft, and the floating height is to be 2 in. above the surface.

Calculation Procedure:

1. Compute the vehicle force increment

The vehicle force increment F is defined as the ratio SF/F_e, where SF = emergency load capacity for extra passengers or dynamic loads, lb; F_e = vehicle total weight including payload, lb. Thus, $F = SF/F_e = 1,000/5,000 = 0.2$.

2. Compute the cushion gage pressure

Use the relation $p_c = F_e/A_p$, where p_c = cushion gage pressure, psig; A_p = plenum area, sq ft; other symbols as before. Thus, $p_c = 5,000/128 = 39$ psig.

3. Compute the sealing perimeter

The sealing perimeter L is the distance around the plenum exterior, or $L = 8 + 8 + 16 + 16 = 48$ ft.

4. Compute the plenum sealing area

Use the relation $A_s = Lh_e$, where A_s perimeter sealing area, sq ft; h_e = floating height, ft; other symbols as before. Thus, $A_s = 48(2/12) = 8$ sq ft.

5. Compute the power required to drive the vehicle

Enter Fig. 3 at the force increment value $F = 0.2$ from Step 1 and project vertically upward to the power curve. From the intersection with the power curve, project horizontally to the right to read the value $P/[A_s(p_c)^{1.5}] = 0.042$, where P = required power input, hp. Solving, $P = 0.042[8(39)^{1.5}] = 81.2$ hp.

6. Compute the required airflow rate

Use the relation $W = 1.4A_s p_c^{0.5}$, where W = required airflow rate to support the vehicle, lb/sec. Substituting $W = 1.4(8)(39)^{0.5} = 70$ lb/sec.

7. Determine the vehicle natural frequency

Enter Fig. 3 at $f = 0.2$ and project vertically upward to the top scale. From here project through $h_e = 2$ in. to the natural frequency $f = 1.3$ cps.

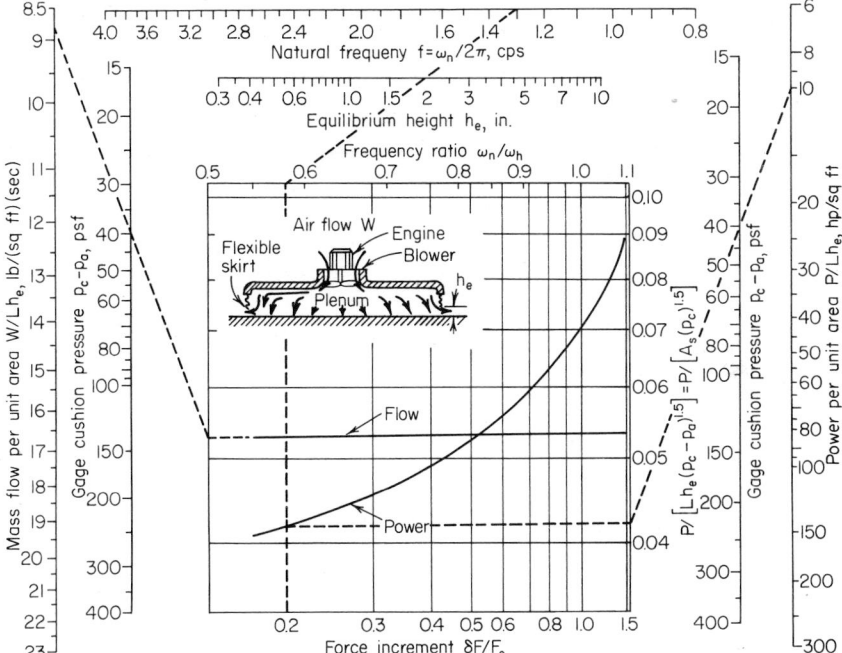

Fig. 3 Plenum-type air-cushion vehicle design chart. (*Professor H. H. Richardson, MIT; Product Engineering.*)

Related Calculations: Use Fig. 4 to analyze peripheral-jet-type air-cushion vehicles. The power equation for the peripheral-jet-type vehicle is $P = $ chart factor $(LH[\,p_c]^{1.5})$ and the airflow-rate equation is $W = PW/Lh$, where the value of W/Lh is obtained from the extreme left-hand scale of Fig. 4. The peripheral-jet air-cushion vehicle is more efficient than the simpler and more reliable plenum-type.

For a peripheral-jet air-cushion vehicle with $F_e = 40{,}000$ lb, $SF = 18{,}000$ lb, $h = 2$ in., length $= 50$ ft., width $= 10$ ft, and 80 percent of the base area available for suspension $SF/F_e = 18{,}000/40{,}000 = 0.45$. The available suspension area $A_a = 50(10)(0.80) = 400$ sq ft, and the gage cushion pressure $= F_e/A_a = 40{,}000/400 = 100$ lb/sq ft.

Assume that two suspension pads, each 10×20 ft, are used, giving a total suspension area of $2(10 \times 20) = 400$ sq ft $= A_a$. The perimeter length of each pad is $10 + 10 + 20 + 20 = 60$ ft; for two pads, the perimeter length $= 2(60) = 120$ ft. The sealing area $= A_s = Lh = 120(\frac{2}{12}) = 20$ sq ft.

To determine the power required per sq ft of sealing area and the total power required for the vehicle, enter the bottom of Fig. 4 at $SF/F_e = 0.45$ and project vertically to the power curve. From the intersection, project horizontally to the right-hand chart border, Fig. 4; then draw a straight line through $p_c - p_a = 100$ lb/sq ft and extend this line to $P/Lh = 28$ hp per sq ft of sealing area. Since the total sealing area $A_s = 20$ sq ft, $P = 28(20) = 560$ hp—the total power required for the vehicle. Note that the chart factor in the power equation given earlier is plotted on the right-hand border of the chart, Fig. 4.

To determine the air-mass flow required for the blower, enter Fig. 4 as before and project vertically to the flow curve. From the intersection project horizontally to the left-hand chart border, then through $p_c - p_a = 100$ lb/sq ft to $W/Lh = 8.05$ lb/(sq ft)(sec). Since $Lh = 20$ sq ft, $W = 8.05(20) = 161$ lb/sec.

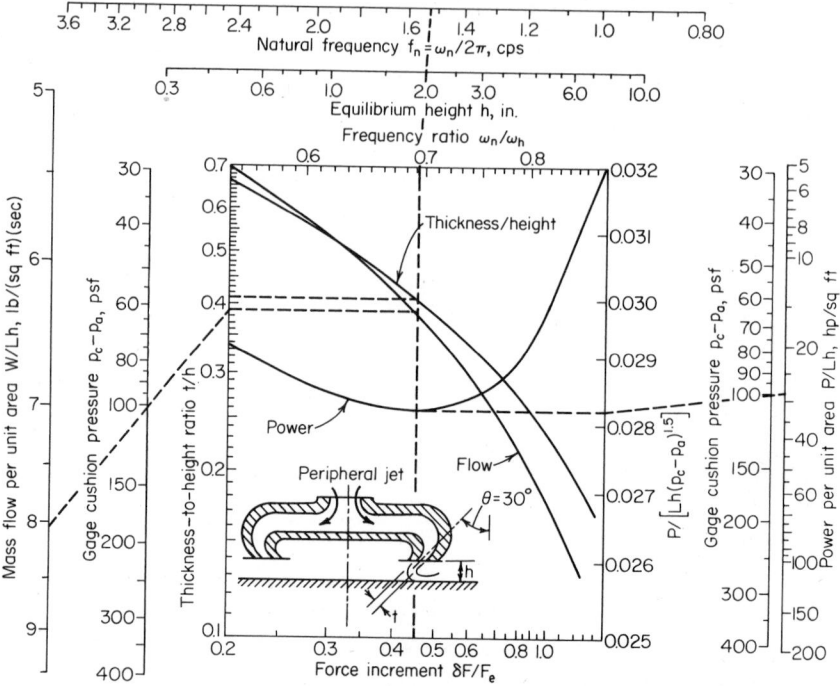

Fig. 4 Peripheral-jet-type air-cushion vehicle design chart. *(Professor H. H. Richardson, MIT; Product Engineering.)*

To determine the jet thickness/height ratio $= t/h$, enter Fig. 4 as before and project to the thickness/height curve. From the intersection, project to the left-hand chart border to read $t/h = 0.41$. Since $h = 2$ in., $t = 2(0.41) = 0.82$ in. $=$ jet thickness. To determine the vehicle natural frequency, project from the $SF/F_e = 0.45$ starting point to the frequency ratio, Fig. 4, then through $h = 2.0$ in. to the natural frequency $f_n = 1.52$ cps. Where desired, Fig. 3 can be used in a similar manner instead of the equations given earlier.

The methods given here are valid for a variety of loadings on the types of vehicles considered.

POWER REQUIRED FOR VERTICAL-TAKEOFF AIRCRAFT

Which requires a larger power output to produce a vertical takeoff—a 30,000-lb helicopter whose blades develop an air downwash velocity of 75 fps or a jet-propelled aircraft of the same weight whose jet engine produces an 800-fps downwash velocity?

Calculation Procedure:

1. Compute the helicopter takeoff power

Use the relation $hp = Wv_d/550$, where $hp =$ takeoff hp of the aircraft; $W =$ weight of aircraft fully loaded and fully fueled, lb; $v_d =$ air downwash velocity, fps.

Substituting, $hp = (30,000)(75)/550 = 4,090$ hp. The installed hp would probably be 4,200 or 4,500 hp to provide for reduced efficiency of the rotor blades, gear mechanism, or engine.

2. Compute the jet-engine power requirement

Use the same relation as in Step 1, or $hp = 30,000(800)/550 = 43,600$ hp. Thus, the jet engine requires about 10 times the power that the helicopter requires.

Related Calculations: This method is also suitable for analyzing the landing hp required by either type of aircraft. Thus, the power that must be developed during landing initially is the same as for takeoff but is gradually reduced as the aircraft approaches the ground.

COMMERCIAL AIRCRAFT OPERATING-COST ANALYSIS

Compare the direct operating costs in $ per airplane mile for ranges up to 6,000 statute miles of three proposed commercial passenger aircraft: a jumbo jet, a tri-jet, and a stretched medium jet. Determine the tri-jet direct operating costs in cents per seat mile for three cabin seating configurations: 256, 295, and 330 passengers. Plot the range-payload tradeoffs for the tri-jet. A normal flight range of 3,000 statute miles is required for the contemplated service of the aircraft.

Calculation Procedure:

1. Summarize the direct operating costs for each aircraft

The *direct operating costs* of a commercial aircraft consist of the out-of-pocket costs required to operate the aircraft between two points that are a known distance apart. These costs are: fuel, lubricants, crew wages, flight taxes and airport fees, direct maintenance, and aircraft depreciation. Figure 5a summarizes the typical direct operating costs for commercial aircraft of the jumbo-jet, tri-jet, and a stretched-jet types, as supplied by the manufacturers of these aircraft. Alternatively, the direct operating cost per mile can be computed by summing the direct costs listed above for each flight range — 300, 1,000, 2,000, 3,000, etc., statute miles.

2. Compute the seat-mile operating costs

With a 3,000-mile operating range, the aircraft choice is between the jumbo jet and the tri-jet, because the maximum range of the stretched jet is only 2,000 miles, Fig. 5a. To compute the seat-mile direct operating cost, s_m, use the relation $s_m = c_m/S$, where c_m = direct operating cost, $/mile; S = number of seats in the aircraft. For the 256-seat configuration in the tri-jet at a 3,000-mile range, $s_m = 2.1/256 = 0.82$ cent per seat mile, using the direct operating cost per mile plotted in Fig. 5a.

Compute the seat-mile cost for the tri-jet for flight ranges varying from the maximum of 3,000 miles to the minimum of 250 miles. Plot the results as shown in Fig. 5b. The jumbo jet can be ignored because its per-mile operating cost is considerably higher than the tri-jet, Fig. 5a. However, if the jumbo jet had a seating capacity approximately twice that of the tri-jet, a seat-mile cost analysis would be worth making to determine the comparable cost. But the jumbo jet being considered here has the same seating capacity as the tri-jet. Hence, the cost *differences* shown in Fig. 5a on a per-mile basis would not change significantly when converted to a seat-mile basis.

3. Plot the range-payload tradeoffs for the tri-jet

A range-payload tradeoff correlates the distance a commercial aircraft can fly carrying a given payload and the distance flown. As the range of an aircraft is increased, its payload decreases because more fuel must be carried to provide the greater flight range.

Plot, or obtain from the aircraft manufacturer, the range-payload curve. Figure 5c shows the range-payload curve for the tri-jet aircraft being considered here. Note in Fig. 5c that this aircraft has a maximum design payload of 87,381 lb when its range is 2,000 nautical miles. The upper knee of this range-payload curve, A, Fig. 5c, provides optimum operating costs for a carrier having a maximum route flight range of approxi-

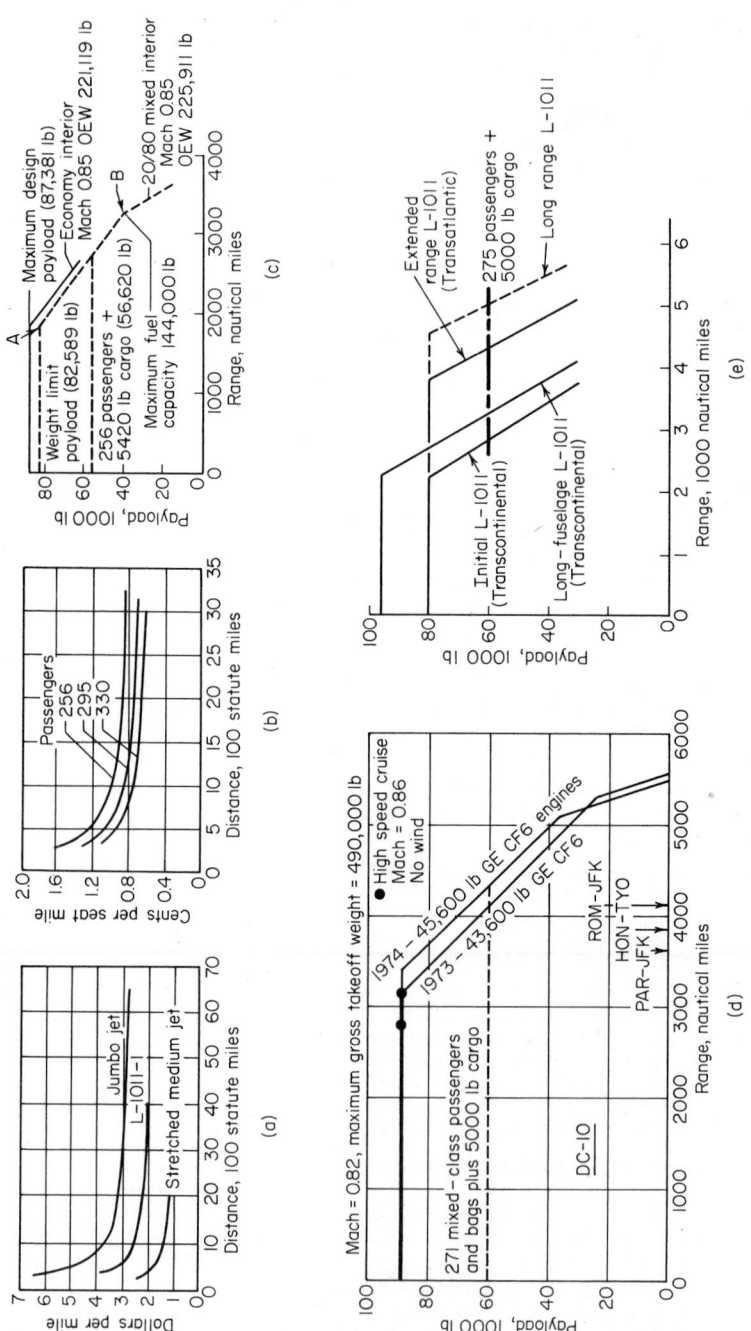

Fig. 5 (a) Aircraft direct operating cost, $/mile. (b) Aircraft direct operating cost, cents per seat mile. (c) Range-payload tradeoffs for tri-jet aircraft. (d) and (e) Range-payload charts for modern jet aircraft. (*Aviation Week.*)

mately 2,000 nautical miles. The lower knee, point B, Fig. 5c, provides optimum operating costs for a carrier having a maximum route flight range of approximately 3,500 nautical miles. Between these two ranges, a variety of payloads and flight ranges are obtainable, as Fig. 5c shows.

The tri-jet analyzed in Fig. 5c would be suitable for a United States domestic carrier with routes east of the Mississippi River if operated at point A with respect to payload and range. When operated at point B, Fig. 5c, this tri-jet would be suitable for intercontinental service, such as between the United States and Europe. The aircraft was actually chosen for these two services.

Related Calculations: The three curves shown above are typical for a variety of aircraft—large and small, pure jet, propjet, and propeller. Hence, the procedure used here can be applied to corporate, private, air-taxi, commercial, and similar aircraft carrying passengers for hire. Of course, many other factors enter the final choice of an aircraft. These factors include loading facilities, cabin arrangement, engine arrramgement, type of tail, gross takeoff weight, takeoff distance, landing distance, noise level, etc. However, the economic factors analyzed here almost always lead the list of factors considered. Figure 5d and e show two other versions of range-payload charts for modern jet aircraft.

LIFTING POWER AND MAXIMUM ALTITUDE OF BALLOONS

What is the gross and net lifting power of a 100-ft-diameter balloon containing pure helium if the balloon and its equipment weigh 1,500 lb? How high will this balloon rise if the temperature of the air and lifting gas average 32 F?

Calculation Procedure:

1. Compute the unit lifting power of the balloon

Use the relation $u = w_a - w_g$, where u = unit lifting power of balloon, lb/cu ft; w_a = weight of air, lb/cu ft; w_g = weight of gas used, lb/cu ft. For this balloon, $u = 0.07658 - 0.01058 = 0.066$ lb/cu ft. The weights of various gases used in balloons are listed in Table 2.

TABLE 2 Balloon Gas Weights

Gas	Weight, lb/cu ft	Factor a
Helium, pure......	0.01058	0.0695
Hydrogen.........	0.00530	0.0713
Coal gas..........	0.1166	0.0400
Atmospheric air ...	0.07658	

2. Compute the gross lifting power of the balloon

Use the relation $G = uV$, where G = gross lifting power of balloon, lb; V = volume of balloon, cu ft = $\pi d^3/6$, where d = balloon diameter, ft. Substituting, $G = 0.07658(\pi \times 100^3/6) = 40,100$ lb.

3. Compute the net lifting power of the balloon

Use the relation $N = G - W$, where N = net lifting power of balloon, lb; W = weight of the balloon and its equipment. Substituting, $N = 40,100 - 1,500 = 38,600$ lb.

4. Compute the height to which the balloon will rise

Use the relation $h = 60,350 \log (W/aV)$, where h = height, ft above sea level to which the balloon will rise; a = ratio from Table 2. Substituting, $h = 60,350 \log [1,500/(0.0695 \times 523,600)] = 37,200$ ft.

Related Calculations: Use this general procedure to compute the weight a balloon can lift when used to transport logs, boats, and other objects. If the balloon gondola contains ballast, the balloon will rise approximately 260 ft for each percent of the balloon weight W that is discarded. A rise of 1 F in the air temperature increases the balloon altitude by about 55 ft and the lifting power by 0.2 percent. A decrease in the air temperature of 1 F causes the reverse effect—i.e., a decrease in altitude of about 55 ft and in lifting power of 0.2 percent.

ROCKET FLIGHT VELOCITY

What is the final velocity of a rocket in which 75 percent of the rocket mass is discharged as propellant at a velocity of 10,000 fps? How does this final velocity compare with that of a rocket in which the propellant velocity is 170,000 fps? How much of a velocity increase could be produced by increasing the discharged mass to 80 percent?

Calculation Procedure:

1. Compute the rocket mass ratio

Use the relation $m_r = m_i/m_f$, where m_r = rocket mass ratio; m_i = initial mass of the rocket, lb; m_f = final mass of the rocket, lb. Where the mass ratios are given instead of the actual mass, use the appropriate ratio in the above relation. Thus, the initial mass of the rocket is 1.00, and the final mass = $m_f = m_i$ − propellant ejected, or $1.00 - 0.75 = 0.25$. Hence, $m_r = 1.00/0.25 = 4.0$.

2. Compute the final velocity of the rocket

Use the relation $v_f = v_e \ln m_r$, where v_f = final velocity of the rocket, fps; v_e = ejection velocity of the propellant, fps; ln = logarithm to the base e, or 2.71828; other symbols as before. Substituting for the 10,000-fps propellant velocity, $v_f = 10,000 \ln 4.0 = 13,863$ fps. For the 170,000-fps propellant velocity, $v_f = 170,000 \ln 4.0 = 235,700$ fps.

3. Compute the velocity increase produced by the larger mass

Using the same procedure as in Step 1, $m_r = 1.00/0.20 = 5.0$. Hence, $v_f = 10,000 \ln 5.0 = 16,094$ fps, or a velocity increase of $16,094 - 13,863 = 2,231$ fps for a 10 percent increase in the discharged mass. For the second propellant, $v_f = 170,000 \ln 5.0 = 273,600$, or a velocity increase of $273,600 - 235,700 = 37,900$ fps for a 10 percent increase in the discharged mass.

Related Calculations: In designing a rocket, a greater final velocity of the vehicle can be obtained by using a propellant having a higher discharge velocity or by ejecting a larger percentage of the initial mass of the rocket. In general, rocket designers prefer to use propellants having a higher exit velocity instead of discharging a larger portion of the original rocket mass.

MISSILE MAXIMUM RANGE AND LAUNCH ANGLE

What is the minimum-energy maximum range of a ballistic missile having a burnout velocity of 12,000 fps? What range angle must be used to obtain this range? What elevation angle should be used for the minimum-energy trajectory?

Calculation Procedure:

1. Compute the missile half-range angle

Use the relation $\cos \theta_{max} = (u_e^2 - r_{bo}u_e v_b^2)^{0.5}/(u_e - r_{bo}v^2/2)$, where θ_{max} = maximum half-range angle, deg; $u_e = 14.05 \times 10^{15}$ ft^3/sec$^2 = g_c r_e$, where g_c = acceleration of gravity at sea level = 32.17 ft/sec^2, and r_e radius of the earth = 3,963 statute miles = 20.90×10^6 ft; r_{bo} = missile burnout radius, ft; v_b = missile burnout velocity, fps. Substituting, $\cos \theta_{max} = (14.05^2 \times 10^{30} - 20.9 \times 10^6 \times 14.05 \times 10^{15} \times 12^2 \times 10^6)^{0.5}/(14.05 \times 10^{15}) - (20.9 \times 10^6 \times 12^2 \times 10^6)/2 = 0.895; \theta = 26.5° = 0.4625$ rad.

2. Compute the missile maximum range

Use the relation $R_m = r_e \theta_r$, where R_m = maximum range of the missile, nautical miles; r_e radius of earth, nautical miles; θ_r = half-range angle expressed in radians. Since the radius of the earth $r_e = 3,440$ nautical miles, $R_m = 3,440(0.4625) = 1,590$ nautical miles.

3. **Compute the required missile elevation angle**

Use the relation $\cos\phi = [u_e/2u_e - r_{bo}v_b{}^2]^{0.5}$, where ϕ = missile elevation angle; other symbols as before. Substituting, $\cos\phi = [14.05 \times 10^{15}/(2 \times 14.05 \times 10^{15}) - (20.9 \times 10^6 \times 12^2 \times 10^6)]^{0.5} = 0.749$; $\phi = 41.5°$.

Related Calculations: Reducing the missile elevation angle below that computed in Step 3 shortens the range of the missile below that computed in Step 2. Increasing the elevation angle above that computed in Step 3 also reduces the missile range below that computed in Step 2.

SATELLITE FLIGHT VELOCITY, ESCAPE VELOCITY, AND PERIOD

An unmanned satellite is placed in a circular orbit 800 miles above the earth. What is the velocity of the satellite? What is the escape velocity of this satellite at this altitude? How long will it take this satellite to make one revolution of the earth?

Calculation Procedure:

1. **Compute the satellite velocity**

Use the relation $v_s = (u_e/r_e + h)^{0.5}$, where v_s = satellite velocity, fps; $u_e = 14.05 \times 10^{15}$ ft³/sec², as defined in the previous Calculation Procedure; r_e = radius of the earth = 20.90×10^6 ft; h = satellite altitude, ft. Substituting, $v_s = [14.05 \times 10^{15}/(20.90 \times 10^6 + 800 \times 5,280)]^{0.5} = 23,650$ fps.

2. **Compute the satellite escape velocity**

Use the relation $v_e = v_s\sqrt{2}$, where v_e = satellite escape velocity, fps; other symbols as before. Substituting, $v_e = 23,650\sqrt{2} = 33,450$ fps.

3. **Compute the time required for one revolution**

The satellite revolves around the center of the earth. Since the earth has a radius of 3,963 statute miles, the radius of rotation of this satellite, which is in orbit 800 miles above the earth, is $r_s = 3,963 + 800 = 4,763$ miles. The circumference of this orbit is the distance d_s traveled by the satellite during one revolution about the earth, or $d_s = 2\pi r = 2\pi(4,763) = 29,990$ miles. Since $t = 5,280 d_s/v_s$, where t = the time required for one revolution, or the *period* of the satellite, $t = 5,280(29,990)/23,650 = 6,690$ sec, or 111.3 min.

Related Calculations: When the satellite escape velocity is known, compute the satellite velocity from $v_s = v_e/\sqrt{2}$. To compute the period of a satellite having an elliptical orbit with a major axis of length a, use the relation $t = 2\pi a^{3/2}/u_e^{1/2}$. Be certain to use consistent units—ft or miles—in both the numerator and denominator of this expression.

INTERPLANETARY FLIGHT LAUNCH VELOCITY AND FLIGHT TIME

What minimum launch velocity is required for an interplanetary satellite launched from Earth and traveling to Saturn? How long will the flight take? Compare the computed velocity and flight time with published values. Figure 6 shows the relationship of the various planets in the solar system.

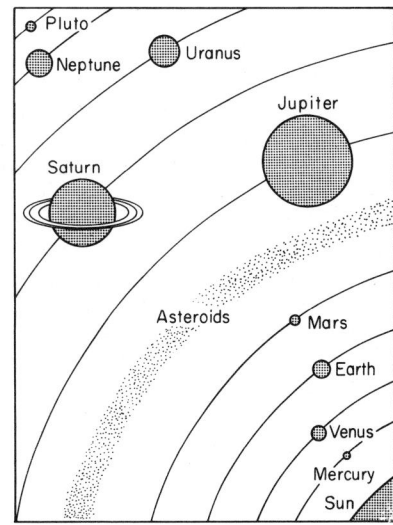

Fig. 6　The solar system. (*RAND Corporation.*)

Calculation Procedure:

1. Compute the major axis of the flight path

The flight path from Earth to another planet can be an ellipse. Assuming that the flight path of this satellite is elliptical, Fig. 7, use the relation $2m_a = d_l + d_t$, where

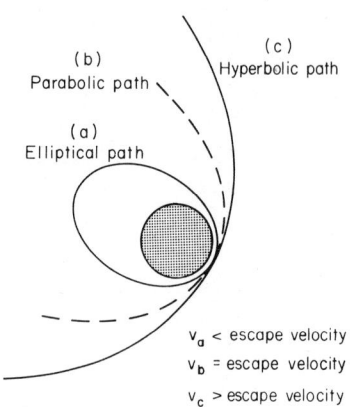

m_a = length of major axis of flight-path ellipse, miles; d_l = distance of launch body from Sun, miles; d_t = distance of target body from Sun, miles. Since Earth is the launch body and is 93×10^6 miles from the Sun and Saturn, the target body, is 886×10^6 miles from the Sun, $2m_a = 93 \times 10^6 + 886 \times 10^6 = 979 \times 10^6$ miles; $m_a = 489.5 \times 10^6$ miles.

2. Compute the specific mechanical energy of the system

Use the relation $m_a = -u_s/2E$, where $u_s = g_s r_s^2$, in which g_s = acceleration of gravity at the surface of the Sun = 900.0 ft/sec²; r_s = radius of sun, feet = 2.285×10^9 ft; thus $u_s = (900)(2.285 \times 10^9) = 4.69 \times 10^{21}$ ft³/sec²; E = specific mechanical energy of the system, per unit mass. Substituting and solving, $E = -u_s/2m_a = -4.69 \times 10^{21}/2[(489.5 \times 10^6 \times 5,280)] = -0.906 \times 10^9$.

Fig. 7 Types of space-vehicle paths.
(RAND Corporation.)

3. Compute the satellite orbit perigee velocity

Use the relation $v_p = [2E + 2u_s/r_p]^{0.5}$, where v_p = velocity at orbit perigee, fps; r_p = radius at perigee = earth radius from Sun, ft; other symbols as before. Substituting, $v_p = [2(-0.906 \times 10^9) + 4.69 \times 10^{21}/(93 \times 10^6 \times 5,280)]^{0.5} = 131,800$ fps.

4. Compute the required velocity increment

Far from the Earth, the velocity increment required from a satellite is reduced by the orbital velocity of the Earth if the satellite is orbited to take advantage of the Earth velocity. Since the orbital velocity of the Earth is 97,800 fps, the required velocity increment is $v_i = 131,800 - 97,800 = 34,000$ fps.

5. Compute the velocity required at the Earth's surface

Use the relation $E = \frac{1}{2}v_i^2 - u_e/\infty = \frac{1}{2}v_e^2 - u_e/r_e$, where $u_e = g_e r_e^2$, as defined above for the Sun; ∞ = infinity; v_e = velocity required at the Earth's surface. Substituting, $E = \frac{1}{2}(34,000)^2 - u_e/\infty$; $E = 11.56 \times 10^8 = v_e^2 - 1.344 \times 10^9$; $v_e^2 = 24.96 \times 10^8$; $v_e = 49,900$ fps. This compares favorably with the actual escape velocity shown in Table 3.

6. Compute the period of the flight

Use the relation $t = 2\pi m_a^{1.5}/u_s^{0.5}$, where t = flight time for a round trip, sec. Substituting, $t = 2\pi(489.5 \times 10^6 \times 5,280)^{1.5}/(4.69 \times 10^{21})^{0.5} = 3.81 \times 10^8$ sec = 4,405 days = 12.05 years. This is the time for a round-trip flight. A one-way flight would take $12.05/2 = 6.025$ years.

Table 3 lists minimum launching velocities and transit times for all planets. From Table 3 it can be seen that the computed one-way flight time agrees well with the published flight time for the planet Saturn.

Related Calculations: Use this general method to compute the velocities and flight time for any interplanetary probe or satellite flight. Table 3 lists the surface escape velocities for the planets. This is the velocity an interplanetary spaceship would have to develop to return to the earth from a given planet.

SPACE VEHICLE BURNOUT VELOCITY AND FUEL SELECTION

What is the burnout velocity v_b of a single-stage rocket having a mass ratio m_r of 9.25 when using a fuel having a specific impulse s_i of 250 sec? Would this rocket be suitable for a Saturn-probe launching? If the rocket payload were reduced by 200 lb, would the rocket be suitable for this probe? The rocket weighs 50,000 lb before payload reduction. What specific impulse is required for a Saturn probe when the rocket mass ratio is 9.25? Choose a suitable fuel for the rocket.

TABLE 3   Minimum Launch Velocities and Transit Times to Reach the Planets from the Earth*

Planet name	Planet surface escape velocity, fps	Minimum Earth launching velocity, fps	Transit time oneway
Mercury	13,600	44,000	110 days
Venus	33,600	38,000	150 days
Mars	16,700	38,000	260 days
Jupiter	197,000	46,000	2.7 years
Saturn	119,500	49,000	6 years
Uranus	72,400	51,000	16 years
Neptune	82,100	52,000	31 years
Pluto	31,200†	53,000	46 years
Earth	36,700		

*NASA data.
†Approximate.

Calculation Procedure:

1. Compute the rocket burnout velocity

Use the relation $v_b = s_i g \ln m_r$, where v_b = burnout velocity, fps; s_i = fuel specific impulse, sec; g = acceleration due to gravity = 32.2 ft/sec²; ln = log to the base e; m_r = rocket mass ratio. Substituting, $v_b = 250(32.2) \ln 9.25 = 17.880$ fps.

The minimum launching velocity required for a Saturn probe is 49,000 fps. Hence, a velocity of 17,880 fps is completely unsatisfactory for such a probe because it is less than half the required velocity. To overcome this velocity deficiency, a multistage

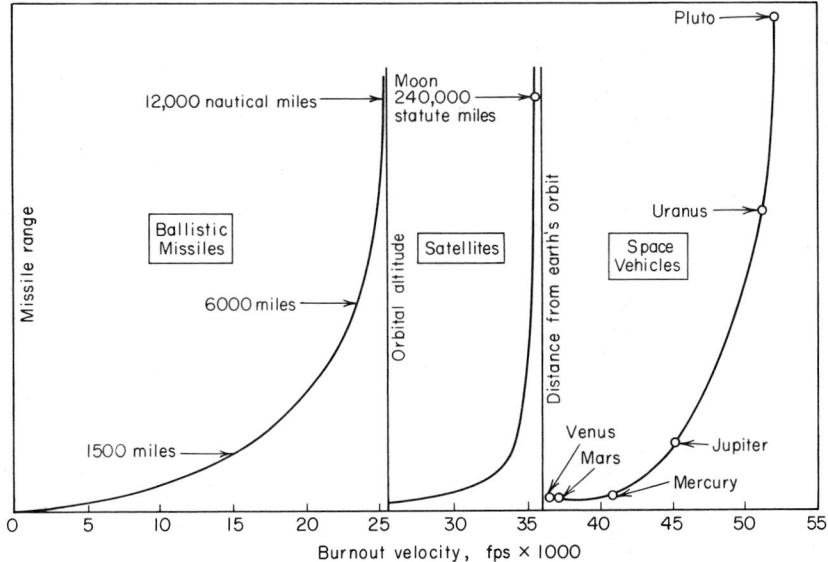

Fig. 8   Velocity requirements for ballistic-missile and space flights. (*RAND Corporation.*)

rocket is needed. Figure 8 shows the burnout velocity required for ballistic missiles, satellites, and space vehicles.

2. Compute the effect of a reduced payload

The new mass ratio $\Delta m_r = m_{rf} - m_{ri}$, where Δm_r = change in rocket mass ratio; m_{rf} = final mass ratio after payload change; m_{ri} = initial mass ratio of the rocket. However, $m_r = m_i/m_f$, where m_i = initial mass of rocket, lb; m_f = final mass of rocket, lb. Substituting for the initial condition, $m_f = m_i/m_r = 50,000/9.25 = 5,400$ lb. With a 200-lb reduction in payload, $m_i = 50,000 - 200 = 49,800$ lb; $m_f = 5,400 - 200 = 5,200$ lb; $m_r = m_i/m_f = 49,800/5,200 = 9.57$. Then, $\Delta m_r = m_{rf} - m_{ri} = 9.57 - 9.25 = 0.32$.

Compute the change in burnout velocity from $\Delta v_b = \Delta m_r s_i g/m_{ri} = 0.32(250)(32.2)/9.25 = 279$ fps. Hence the new burnout velocity is $17,880 + 279 = 18,159$ fps. This is still far below the required burnout velocity. Figure 9 shows the relationship between the propellant fraction, mass ratio, and rocket velocity for single-stage vehicles. Study of this chart shows that the high propellant fractions associated with high rocket velocities can only be achieved by severely reducing to a minimum all components of the rocket that contribute to the weight at propellant exhaustion, including the payload. Once again, a multistage rocket is needed to overcome the velocity deficiency.

Use the equation in Step 1 and solve for $s_i = v_b/g \ln m_r$. Since $v_b = 49,000$ fps for a Saturn prober, as shown in the previous Calculation Procedure, Table 3, $s_i = 49,000/[32.2(\ln 9.5)] = 676$ sec.

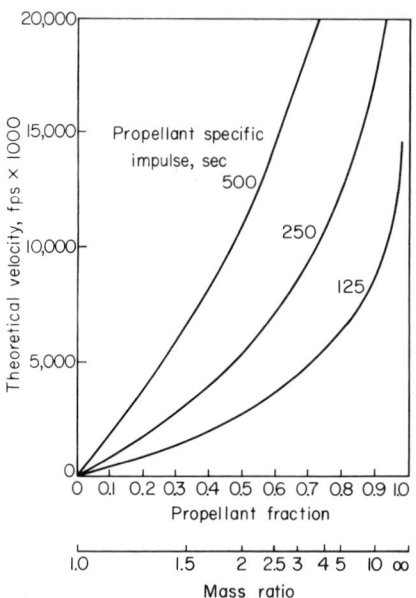

Fig. 9 Velocity characteristics of single-stage rocket vehicles. *(RAND Corporation.)*

4. Select a suitable fuel for the rocket

Table 4 lists the characteristics of typical common propulsion systems. Study of this tabulation shows that liquid fuels are unsuitable for this single-stage rocket because

TABLE 4 Typical Characteristics of Propulsion Systems

System	Specific impulse, sec	Ratio of thrust to engine weight
Liquid propellants	200–300	50 to 80
High-energy liquid propellants............	340–440	Less than 50 to 80 but of the same order of magnitude
Nuclear energy........	400–900	Less than 50 to 80 but of the same order of magnitude
Free radicals.........	400–1800	Less than 50 to 80 but of the same order of magnitude
Solar heat transfer	400–500	0.05
Ion..................	5,000–20,000	0.0005 to 0.00005
Thiokol perchlorate (solid)...............	200–215	
Rubber nitrate (solid)	180–195	

their specific impulse is too low. Only a nuclear or free-radical propellant gives the desired specific impulse for a single-stage vehicle. However, two or more stages of liquid fuel would give the desired velocity to this rocket because the specific impulse of the second stage adds to the specific impulse of the first stage. For these reasons, the type of fuel chosen depends upon the rocket mission, the propulsion-system availability, economics, and similar factors.

Related Calculations: Use the procedure given here to determine the burnout velocity and fuel for ballistic missiles, interplanetary launch vehicles, and similar applications.

To achieve the needed escape velocity required for interplanetary flights with liquid propellants, two or more rocket stages are required. Thus, if the burnout velocity of the first stage of a rocket is 17,880 fps, the second stage begins accelerating from this initial velocity. In an efficient two-stage rocket, the mass ratios of each stage are equal, or nearly so.

Figure 10 shows how the second stage of a rocket can produce a "kick in the apogee." If the second stage is fired with the rocket correctly oriented, the final orbit that can be achieved is not limited by the projection altitude of the basic booster or first-stage rocket.

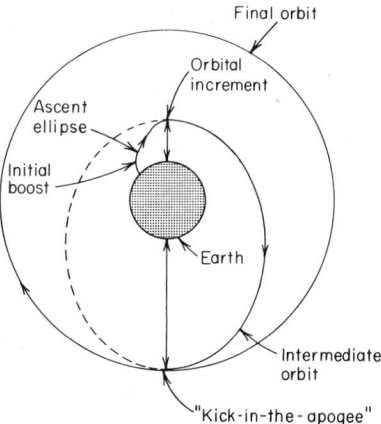

Fig. 10 "**Kick-in-the-apogee" technique of satellite launching.** *(RAND Corporation.)*

OBSERVATION-SATELLITE DETAIL DETECTION

An observation satellite carries a precision camera having a focal length of 6 in. What is the scale on pictures taken by this camera from an altitude of 150 miles? What focal length is needed for a camera at an altitude of 1,000 miles to produce a ground resolution of 20 ft? Determine the ground resolution obtainable with the 6-in.-focal-length camera if the film resolution is 100 lines per min.

1. Compute the scale number of the photograph

Factors that enter an estimate of the degree of detail that can be detected or identified by a camera include the distance between the camera and the object photographed and the focal length of the viewing lens. Using these two factors, a *scale number S* is computed from S = camera altitude, ft/lens focal length, ft. For this camera at an altitude of 150 miles, S = [(150 miles)(5,280 ft/mile)]/0.5 = 1,584,000.

A scale number of 1,584,000 means that 1 in. on the photograph taken by this camera corresponds to 1,584,000 in. on the ground area photographed. Since scale numbers are usually in miles, 1 in. on the photograph = 1,584,000 in./[(5,280 ft/mile)(12 in./ft)] = 25.05 miles.

In general, the larger the scale number, the more difficult it is to detect fine details in the photograph.

2. Determine the required focal length of the camera

Assume a film resolution of 100 lines per min. This is a typical resolution used for high-altitude photography.

Enter Fig. 11 at the satellite altitude, 1,000 miles, on the horizontal axis and project vertically upward to the desired ground resolution, 20 ft. From the intersection, project horizontally to the left to read the required focal length of 9.5 ft. With the high resolution required in cameras carried aboard observation satellites, a speed of at least $f/8$, and preferably faster, is desirable.

3. *Compute the ground resolution for the given scale number*

Use the relation $G_r = S/300R$, where $G_r =$ ground resolution for a given scale number, ft; $S =$ photograph scale number, in. on the ground per in. on the photograph; $R =$ lines per mm produced by the film-lens combination. Substituting in the relation the scale number from Step 1, $G_r = 1,584,000/[300(100)] = 56.1$ ft.

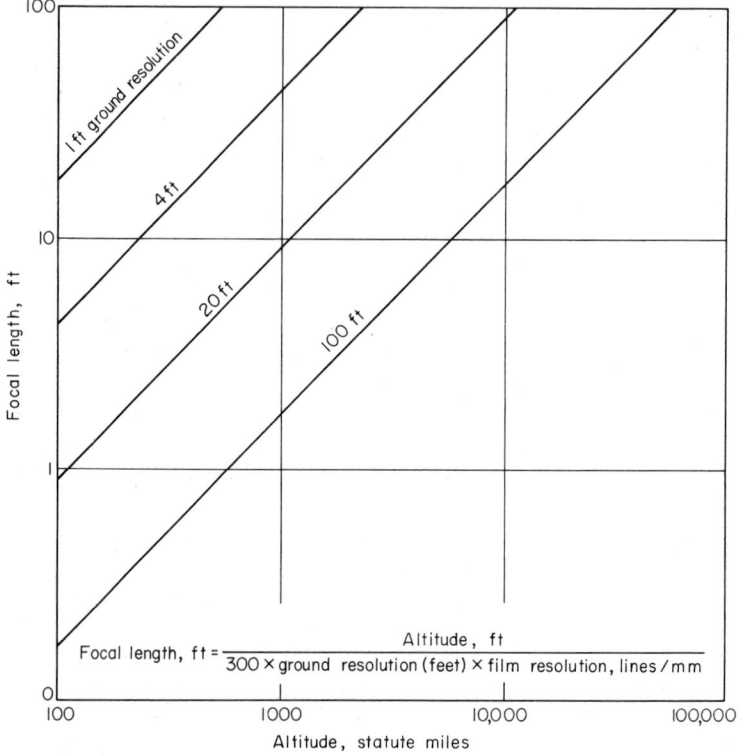

Fig. 11   Required focal-length variation with altitude for various ground resolutions. *(RAND Corporation.)*

Related Calculations: Four levels of photographic detail are generally used to define the ground resolution obtainable with observation satellites.[1] In terms of ground resolution these levels are: $A - 50$ to 200 ft; $B - 10$ to 40 ft; $C - 2$ to 8 ft; $D - 0.5$ to 2 ft. The range over a factor of 4 within each level arises from a practical inability to measure and interpret ground resolution as a fixed number and from additional detailed factors, such as the graininess of photographic emulsions.

[1] Staff Report of the Select Committee on Astronautics and Exploration — *Space Handbook: Astronautics and Its Applications,* House Document 86, GPO.

Section 9

Marine Engineering

BERNARD TICHAZ
Director, George G. Sharp Inc.

TYLER G. HICKS
International Engineering Associates

REFERENCES: Labberton—*Marine Engineering*, McGraw-Hill; Munro-Smith—*Applied Naval Architecture*, Longmans; Bell—*Petroleum Transportation Handbook*, McGraw-Hill; Osbourne— *Modern Marine Engineer's Manual*, Cornell Maritime Press; Quinn—*Design and Construction of Ports and Marine Structures*, McGraw-Hill; Baker—*Introduction to Steel Shipbuilding*, McGraw-Hill; Seward—*Marine Engineering*, Society of Naval Architects and Marine Engineers; *Rules for Building and Classing Steel Vessels*, American Bureau of Shipping; U.S. Coast Guard Regulations for Commercial Vessels; Latham—*Introduction to Marine Engineering*, U.S. Naval Institute; Van Lammeren—*Resistance, Propulsion, and Steering of Ships*, Technical Publishing Co., Haarlem, Holland; Crouch—*Nuclear Ship Propulsion*, Cornell Maritime Press; Comstock—*Principles of Naval Architecture*, Society of Naval Architects and Marine Engineers; Baxter—*Naval Architecture*, The English Universities Press, Ltd.

FORM COEFFICIENTS FOR VARIOUS VESSEL TYPES

Determine the form coefficients for a proposed 15.4-knot oceangoing freighter having a waterline length of 441 ft, a molded beam of 56.9 ft, a molded depth of 37.3 ft, a load draft of 26.0 ft, and a displacement of 13,570 tons. The area of the midship section below the water plane is 1,450 sq ft. Do the form coefficients of this vessel agree with generally accepted design practice?

Calculation Procedure:

1. Compute the vessel speed-length ratio

Use the relation $S = V/(L)^{0.5}$, where S = vessel speed-length ratio, V = vessel normal speed, knots; L = vessel waterline length, ft. For this vessel, $S = 15.4/(441)^{0.5} = 0.733$.

2. Compute the vessel block coefficient of fineness

Use the relation $b = 35D/LBH$, where b = block coefficient of fineness; D = displacement, tons of seawater; B = beam, ft; H = draft, ft. For this vessel, $b = 35(13,570)/(441 \times 56.9 \times 26.0) = 0.727$.

3. Compute the vessel midship section coefficient

Use the relation $m = M/BH$, where m = midship section coefficient; M = area of midship section below water plane, sq ft; other symbols as before. For this vessel, $m = 1,450/(56.9 \times 26) = 0.98$.

4. Compute the longitudinal prismatic coefficient

Use the relation $l = 35D/ML$, where l = longitudinal prismatic or mean-length coefficient; other symbols as before. Thus, $l = 35 \times 13,570/(1,450 \times 441) = 0.741$.

5. Compute the water-plane coefficient

Use the relation $\alpha = 0.667b + 0.333$, where α = water-plane coefficient; other symbol as before. Thus, $\alpha = 0.667(0.727) + 0.333 = 0.818$. Alternatively, $\alpha = A/BL$, where A = area of water plane at the surface of the water, sq ft; other symbols as before.

6. Compute the displacement length coefficient

Use the relation $c_d = D/(L/100)^3$, where c_d = displacement length coefficient; other symbols as before. Thus, $c_d = 13,570/[(441/100)^3] = 158$.

7. Compare the form coefficients with typical values

Table 1 lists form coefficients for a variety of typical vessels. Study of this table shows that the form coefficients computed for the freighter under consideration agree with those listed for moderate-speed freighters in all instances except one—the displacement length coefficient. However, the difference here, 158 versus 165, is insignificant. The displacement length coefficient is the displacement in tons of a mechanically similar ship that is 100 ft in length.

Related Calculations: Form coefficients are valuable to ship designers because the coefficients are dimensionless. Hence, the coefficients apply to ship forms of all sizes. A designer can thus easily compare two ships of widely different sizes. To compare two ships, simply examine their speed-length ratios and their prismatic coefficients. Using this procedure, a 300-ft vessel with a 1.04 speed-length ratio and a prismatic coefficient of 0.68 is similar to a 900-ft vessel having the same speed-length ratio and prismatic coefficient. This means that the sectional area curves of the two vessels are also similar.

The form coefficients are also useful in determining the dimensions of a certain class of ship. To use the form coefficients in this manner, select a coefficient from within the range given in Table 1. Assume one or more dimensions of the vessel and solve for the unknown dimension. Using this procedure, a vessel of desired dimensions can be developed for a given service.

TABLE 1   Typical Coefficients of Form for Various Types of Vessels

Vessel type	$V/L^{0.5}$	b	m	l	α	$D/(L+100)^3$
Great Lakes ore ships........	0.39–0.43	0.85–0.87	0.99–0.995	0.86–0.88	0.89–0.92	70–95
Slow ocean freighters........	0.45–0.50	0.77–0.82	0.99–0.995	0.78–0.83	0.85–0.88	180–200
Moderage-speed freighters...	0.55–0.75	0.67–0.76	0.98–0.99	0.68–0.78	0.78–0.84	165–195
Fast passenger liners.........	0.70–1.05	0.56–0.65	0.94–0.985	0.59–0.67	0.71–0.76	75–105
Fast cruisers	1.30–1.70	0.45–0.53	0.80–0.90	0.55–0.60	0.60–0.65	
Destroyers.................	1.8–2.5	0.44–0.53	0.72–0.83	0.62–0.71	0.67–0.73	40–65
Tugs......................	0.9–1.2	0.45–0.53	0.71–0.83	0.61–0.66	0.71–0.77	200–420

VESSEL WETTED AREA AND SHALLOW-WATER SPEED

What is the wetted area of and the frictional hp required for a 600-ft-long 16-knot vessel displacing 17,350 tons? At what water depth will there be no increase in the resistance of the vessel if its loaded draft is 32 ft? What percentage increase in resistance to vessel movement is there in 40 ft of water?

Calculation Procedure:

1. Compute the wetted area of the vessel

Use the Taylor relation $A = k(DL)^{0.5}$, where A = vessel wetted hull area, sq ft; k = the Taylor constant = 15.0 to 16.3, with the lower values of l corresponding to finely shaped vessels (destroyers, cruisers, and some yachts) and higher values of k corresponding to less finely shaped vessels (barges, bulk-cargo freighters, and small tankers); D = vessel displacement, tons; L = length of vessel on the waterline, ft. Thus, using $K = 15.6$, the approximate mid-point value, $A = 15.6(17,350 \times 600)^{0.5} = 50,400$ sq ft.

2. Compute the frictional resistance of the vessel

The frictional hp required to propel a vessel is the shp needed to overcome frictional resistance to movement of the hull through the water. Use the relation $F_r = fAV^{1.825}$, where F_r = vessel frictional resistance, lb; f = coefficient of friction for the vessel in saltwater; A = wetted hull area, sq ft; V = vessel speed, knots. Values of f for vessels of various lengths operating in saltwater are given in Table 2. Using $f = 0.008726$ for a 600-ft vessel, $F_r = 0.008726(50,400)(16)^{1.825} = 69,000$ lb.

TABLE 2   Coefficient of Friction for Vessels in Salt Water

Vessel Length, ft	Coefficient of Friction, f
100	0.009207
200	0.008992
300	0.008902
400	0.008832
500	0.008776
600	0.008726
700	0.008680
800	0.008639
900	0.008608
1,000	0.008574

3. Compute the frictional hp required

Use the relation $H_f = 101.3F_rV/33,000$, where H_f = frictional hp required; other symbols as before. The constant 101.3 converts knots to fpm; the constant 33,000 converts ft-lb/min to hp. Substituting, $H_f = 101.3(69,000)(16)/33,000 = 3,390$ hp.

As an alternate solution for frictional hp, use the appropriate Schoenherr curve from Fig. 1. This curve shows that a 600-ft 16-knot vessel requires 69 hp per 1,000 sq ft of wetted hull area to overcome the frictional resistance. Thus, $H_f = 69(50,400/1,000) = 3,475$ hp. This value agrees closely with the earlier computed value of 3,390 hp. If desired, the Schoenherr curves in Fig. 1 can be interpolated or extrapolated for vessel lengths that are not plotted.

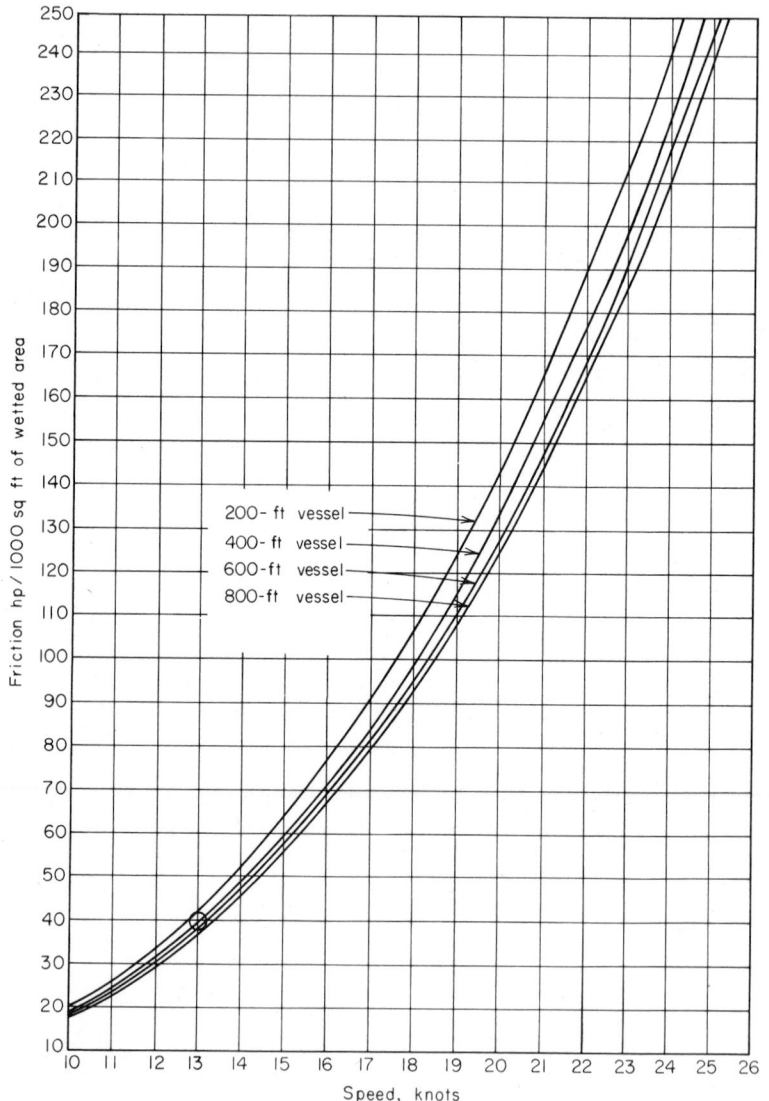

Fig. 1 Frictional hp required per 1,000 sq ft of wetted hull surface of vessels of various lengths.

4. *Compute the minimum depth of water for zero resistance increase*

The resistance to the movement of a vessel through the water increases markedly when $V = 2(H)^{0.5}$, where V = vessel speed, knots; H = water depth, ft. Solving, $H =$

$V^2/4$. Or, the minimum depth of water for zero resistance increase for a 16-knot vessel is $H = 16^2/4 = 64$ ft.

The maximum vessel speed v_{md} knots for a nonplaning hull is $V_{md} = 3.36(H)^{0.5}$ in deep water. In shallow water the maximum speed in knots is $V_{ms} = 2.5(H)^{0.5}$. For trial runs, Admiral Taylor recommends a least-water depth, ft $= 10$(vessel draft, ft) $V/(L)^{0.5}$, where $L =$ load waterline length of vessel, ft.

5. Compute the percentage increase in vessel resistance

In shallow water the percentage increase in resistance is $p = 50H/d$, where $H =$ vessel draft, ft; $d =$ water depth, ft. Thus, at a depth of 40 ft, $p = 50(32)/40 = 40$ percent. An increase in resistance of this magnitude will appreciably reduce the speed of the vessel.

Related Calculations: The total resistance of a vessel is the sum of the frictional resistance and the residual resistance. This latter resistance is comprised of the sum of eddy resistance and wave resistance. Residual resistance can be determined by test of a model of the vessel in a towing tank or from published systematic model test data such as Taylors Standard Series or Ayre's method.[1] In model tests, residual resistance of model/residual resistance of ship = (length of model, ft)³/(length of ship, ft)³.

One modern way to reduce the overall water resistance of large vessels is the use of a bulbous bow. Such a bow creates a secondary wave that partially cancels the primary wave system caused by movement of the vessel. A properly designed bulbous bow can (1) increase vessel speed 4 to 6 percent, (2) increase cargo-carrying capacity 4.5 percent without increasing shaft power or decreasing speed, and (3) reduce power and fuel consumption 10 to 15 percent without sacrificing speed or cargo capacity.

POWER REQUIRED TO PROPEL A VESSEL

What is the hp required to drive a 600-ft tanker at a cruising speed of 17.4 knots if the displacement of the vessel is 34,650 tons and the propulsion machinery is a geared turbine?

Calculation Procedure:

1. Determine the admiralty coefficient for the vessel

When a vessel is being designed, its admiralty coefficient must be approximated from known values for vessels of about the same overall length. Table 3 lists typical admiralty coefficients for various modern vessels propelled by geared turbines or diesel engines. Study of Table 3 shows that the admiralty coefficient $K = 400$ for a vessel having a length on the load waterline (LWL) of 600 ft.

TABLE 3 Admiralty Coefficients for Modern Vessels

Vessel LWL, ft.....	30	50	75	100	150	200	300	400	500	600	800
Admiralty coefficient	70	95	120	150	180	200	250	345	360	400	500

2. Compute the required power to drive the vessel

Use the relation $P = D^{2/3}V^3/K$, where $P =$ required shp to drive the vessel; $D =$ vessel displacement, long tons; $V =$ vessel cruising speed, knots; $K =$ admiralty coefficient. Using the given data, $P = 34,650^{2/3} \times 17.4^3/400 = 14,000$ hp, closely.

3. Check the power using Froude's law of comparison

Froude's law of comparison states: For ships of similar geometrical form and proportion, the shaft horsepowers required at corresponding speeds will be in the ratio of the products of the displacements by the speeds, or the power ratio will equal the

[1]Van Lammeren—*Resistance, Propulsion, and Steering of Ships*, Technical Publishing Co., Haarlem, Holland.

product of the displacements ratio by the speed ratio, or power is proportional to $L^{7/2}$ or to $D^{7/6}$. Also, corresponding speeds are in the ratio of the square roots of the vessel lengths.

To apply Froude's law of comparison, obtain the speed, LWL, and P of a similar vessel. Thus, a 445-ft tanker of modern design has a speed of 13.7 knots when developing 4,100 shp. Using Froude's speed law, the corresponding speed of the vessel being analyzed here is $13.7(600/445)^{1/2} = 15.9$ knots.

Next, applying Froude's power law, P at 15.9 knots $= 4,100(600/445)^{7/2} = 11,000$ hp. Then the shp at the design speed of 17.4 knots is, if the power varies as the cube of the speed, $11,500(17.4/15.9)^3 = 14,350$ hp. This agrees closely with the admiralty coefficient computed shp of 14,000 hp, computed in Step 2.

Related Calculations: Use this method for any size vessel, from 30 ft to 1,000 ft or more LWL, of any type—cargo, passenger, tanker, yacht, tug, naval vessel, etc. Be certain to use admiralty coefficients developed for modern vessels. Typical up-to-date values are published in *Marine Engineering/Log* (New York), *Transactions of the Institute of Marine Engineers* (London), and *Journal of the American Society of Naval Engineers* (Washington, D.C.).

For yachts, powerboats, and motorboats, the power required can be computed from $P = (VB/C)^3/L$, where $P = $ bhp to propel the boat or yacht at V statute miles/hr; $B = $ maximum waterline beam, ft; $L = $ load waterline length, ft; $C = $ a constant from Table 4.

TABLE 4   Coefficient for Powerboat Power Relation*

Powerboat type	C	Powerboat type	C
Heavy cruiser.....	8–9	Heavy runabout	11.7–14.3
Medium cruiser...	8.4–10	Average runabout	14–16.3
Light cruiser......	9.2–11.3	Racing runabout	15.8–18.2
Express cruiser ...	9.8–12.6	Hydroplane	18–20

*White—*Yachting.*

MARINE PROPELLER-SHAFT DIAMETER AND PROPELLER SLIP

What diameter line and tail or propeller shafts should be used for a 15,000-shp single-screw 17-knot ocean vessel if the propeller turns at 110 rpm at the service speed of the ship? The tail shaft will be fitted with a continuous bronze liner and the propeller is 22.5 ft diameter. What is the twisting moment on the shaft if the ship is driven by a geared steam turbine? What are the apparent and true slip of the propeller at normal speed if the propeller pitch is 15.75 ft?

Calculation Procedure:

1. Compute the line-shaft diameter

Use the American Bureau of Shipping relation $d_1 = c (KP/N)^{1/3}$, where $d_1 = $ line-shaft diameter, in.; $c = 1.0$ for line shafts, 1.05 for thrust shafts transmitting torque; $K = 64$ for ocean and coastwise service, 58 for river and harbor service; $P = $ ship shp; $N = $ propeller rpm. For this ship, $d_1 = 1.0(64 \times 15,000/110)^{1/3} = 20.6$ in.; say 20.75 in.

2. Compute the tail- or propeller-shaft diameter

Use the relation $d_p = d_1 + d/b$, where $d_p = $ propeller-or tail-shaft diameter, in.; $d_1 = $ line-shaft diameter, in.; $d = $ propeller diameter, ft; $b = 12$ for a continuous bronze liner, 8.3 if liners are fitted only at the bearings. For this vessel, $d_p = 20.75 + 22.5/12 = 22.62$ in.; say 22.75 in. Thus, the tail shaft is 2 in. larger in diameter than the line shaft.

3. Compute the shaft twisting moment

Use the relation $M_t = 63.024(P/N)$, where $M_t = $ shaft twisting moment, in.-lb; other symbols as before. For this vessel, $M_t = 63.024(15.000/110) = 8,600$ in.-lb.

4. *Compute apparent slip of the propeller*

Use the relation $s_a = (pN - 101.3V)/pN$, where s_a = apparent slip = slip ratio; p = propeller pitch, ft; N = propeller rpm; V = ship speed, knots. Solving, $s_a = (15.75 \times 110 - 101.3 \times 17)/(15.75 \times 110) = 0.006$, or 0.6 percent.

5. *Compute the true slip of the propeller*

Use the relation $s_t = (pN - 101.3V_a)/pN$, where s_t = propeller true slip; V_a = speed of advance of the propeller, knots = $V(1 - w)$, where w = wake fraction for the vessel; other symbols as before. Using $w = 0.35$, $V_a = 17(1 - 0.35) = 11.05$ knots. Then, $s_t = (15.75 \times 110 - 11.05 \times 101.3)/(15.75 \times 110) = 0.352$, or 35.2 percent.

Related Calculations: The apparent slip of a propeller can be negative, particularly if the vessel is a single-screw ship at light draft. Apparent slip varies from 5 to 30 percent, with 10 percent being a typical value.

Real or true slip is based on the pitch speed of the propeller through the moving water or wake. The real or true slip is greater than the apparent slip.

Table 5 lists typical values of the wake fraction w for various types of vessels.

TABLE 5   Typical Wake-Fraction Values*

Vessel type	Single-screw	Twin- or quadruple-screw
Tankers and slow cargo vessels	0.35	0.20
Passenger and fast cargo vessels	0.30	0.15
Yachts and bay steamers	0.25	0.10
Scout cruisers and tugs	0.20	0.05
Destroyers and motorboats. . .	0.15	0

*Baker — *Introduction to Steel Shipbuilding*, McGraw-Hill.

PROPELLER SELECTION FOR VESSELS

Choose a three-bladed propeller for a 15,000-shp single-screw 17-knot ocean vessel if the maximum allowable diameter of the screw is limited to 22.5 ft. What is the thrust on the propeller blades? Will this propeller cavitate?

Calculation Procedure:

1. *Compute the propeller speed of advance through the wake*

Use the relation $V_a = V(1 - w)$ where V_a = speed of advance of the propeller through the wake, knots; V = vessel speed, knots; w = wake fraction = wake velocity, knots/vessel velocity, knots. Table 5 lists typical values of the wake fraction for single- and multiple-screw ships of various types. Thus, for a single-screw tanker, $w = 0.35$, and $V_a = 17(1 - 0.35) = 11.05$ knots.

2. *Compute the propeller diameter factor*

The diameter factor is used in determining the best combinations of propeller diameter and pitch for a given pitch ratio. Use the relation $D = dV_a^{3/2}/P^{1/2}$, where D = propeller diameter factor; d = propeller diameter, ft; P = propeller shp. For this ship, $D = 22.5(11.05)^{3/2}/15,000^{1/2} = 6.79$; say 6.8.

3. *Determine the propeller rpm factor*

Refer to Table 6 and read the rpm factor B as 33.0 opposite the diameter factor $D = 6.8$.

4. *Compute the propeller rpm*

Use the relation, $rpm = BV_a^{5/2}/P^{1/2} = 33(11.05)^{5/2}/15,000^{1/2} = 109.9$ rpm; say 110 rpm.

5. Compute the propeller pitch

Read from Table 6 the pitch-diameter ratio $a = p/d = 0.7$, where p = propeller pitch, ft; d = propeller diameter, ft. Solving, $p = da = 22.5(0.7) = 15.75$ ft, closely.

TABLE 6*   Best Combinations of Propeller Diameter and Rpm†

Combination‡	Pitch-diameter ratio, $a = p/d$	Diameter factor	Rpm factor	Efficiency e
1	1.4	19.3	4.5	0.78
2	1.3	17.4	5.5	0.765
3	1.2	16.0	6.5	0.75
4	1.1	14.5	8.0	0.735
5	1.0	13.0	10.0	0.72
6	0.9	11.4	13.0	0.695
7	0.8	9.0	20.0	0.65
8	0.7	6.8	33.0	0.59
9	0.6	4.9	60.0	0.51

*Baker—*Introduction to Steel Shipbuilding*, McGraw-Hill.
†For even pitch ratio p/d
‡Since these are the best combinations, use them where circumstances permit. For other combinations, use the Taylor or Baker curves.

6. Determine the propeller efficiency

Read the propeller efficiency in Table 6 as 59 percent. This is an acceptable efficiency for a propeller of this type and size.

7. Compute the thrust on the propeller blades in lb

Use the relation $T = 33,000eP/101.3V_a$, where T = thrust on propeller blades, lb; e = propeller efficiency; other symbols as before. For this vessel, $T = 33,000(0.59)(15,000)/[101.3(11.05)] = 260,500$ lb.

8. Compute the thrust on the propeller blades in psi

Propeller thrust is also expressed in pressure terms—i.e, psi of actual blade area. When the developed area ratio a_r is known (0.40 is a typical value for slow- and moderate-speed vessels), the thrust in psi is $T_p = T/36\pi a_r d^2$. Using $a_r = 0.40$, $T_p = 260,500/[36\pi(0.40)(22.5)^2] = 11.36$ psi.

9. Determine if the propeller will cavitate

Compute the propeller tip speed from $t = \pi dN$, where t = propeller tip speed, fpm; N = propeller rpm; other symbols as before. Substituting, $t = \pi(22.5)(110) = 7,800$ fpm.

Table 7 lists tip speeds and pressure thrusts of cavitating propellers. To prevent cavitation, the actual pressure thrust should be about 10 percent less than the tabulated pressure thrust. Since the actual pressure thrust, 11.36 psi, is 10 percent less than the cavitating pressure thrust at a tip speed of 8,000 fpm, the nearest tabulated tip speed, this propeller will *not* cavitate. Hence, it is an acceptable propulsion unit for this

TABLE 7   Critical Thrust of Propellers at Various Tip Speeds
(Developed by Commander Irish)

Propeller tip speed, fpm ...	2,000	4,000	6,000	8,000	10,000	12,000	14,000
Trust, psi.................‡	1.2	5.6	12.0	18.2	23.6	28.5	33.0

vessel. Also, the assumed developed area ratio, 0.40, is safe, so far as cavitation is concerned.

Related Calculations: Use this general method for choosing a propeller to run at a limited rpm, or limited rpm and diameter. If desired, Taylor's diagram (Taylor – *The Speed and Power of Ships*, Ransdell, Inc., Washington, D.C.) for three-bladed propellers can be used in place of Table 6. Where desired, the wake factor or fraction can be approximated from the Taylor relation: $w = 0.5b - 0.05$ for single-screw ships; and $w = 0.55b - 0.2$ for twin-screw ships, where b = vessel block coefficient of fineness.

To determine the *approximate* optimum diameter of a three-bladed propeller use the relation $d = 50P^{0.2}/\text{rpm}^{0.6}$. Using this relation, the optimum diameter for the above propeller is 20.25 ft. A four-bladed propeller of $0.97 \times$ the diameter of a three-bladed propeller, the same pitch ratio, and $1.33 \times$ the area will absorb the same shp at the same rpm as the three-bladed propeller. Likewise, a two-bladed propeller of 5 percent greater diameter is about equivalent to a three-bladed propeller.

The *developed area ratio* of a propeller = developed area of blades, sq in./area of circle of same diameter as propeller, sq in. Even though a value of 0.40 was assumed for a_r, this value can range up to 1.0 for high-speed, high-power vessels.

The propulsive coefficient of a marine propeller = PC = ehp/shp, where ehp = the effective or towrope hp = $0.00307VR$, where R = total resistance of the vessel, lb. Typical values of PC are 0.60 for a single-screw vessel and 0.57 for a twin-screw vessel; these values can increase to 0.80 and 0.70, respectively, for moderate-power (i.e., merchant) ships with well-designed propellers and sterns.

Use the same procedure as that given above to select propellers for small craft – tugs, powerboats, auxiliaries, etc.

PROPELLER REVOLUTIONS AND VESSEL SPEED

At what rpm should a propeller having a 25-ft pitch be turned to drive a vessel at 22 knots if the apparent slip of the propeller is 12 percent? How many revolutions must the propeller turn to drive the vessel 50 nautical miles in calm weather? What is the speed of this vessel when the propeller is turning at 70 rpm?

Calculation Procedure:

1. Compute the required rpm of the propeller

Use the relation $N = 101.3V/p(1 - s_a)$, where N = required propeller rpm; V = vessel speed, knots; p = propeller pitch, ft; s_a = propeller apparent slip, expressed as a decimal. Substituting, $N = 101.3(22)/[25(1 - 0.12)] = 101.3$; say 102 rpm.

2. Compute the required number of propeller revolutions

Use the relation $n = 6,080d/p(1 - s_a)$, where n = total number of turns of the propeller to drive the vessel d nautical miles; other symbols as before. Thus, $n = 6,080(50)/[25(1 - 0.12)] = 13,820$ revolutions.

To check this result, divide the distance traveled by the vessel speed to determine the running time in hr. Or, T = distance, nautical miles/speed, knots = 50/22 = 2.27 hr. Since there are 60 min/hr and the required propeller rpm is 101.3, $n = 2.27 (60)(101.3) = 13,820$ revolutions.

3. Compute the vessel speed at the different rpm

Use the relation $N_1/N_2 = V_1/V_2$, where the subscripts 1 and 2 refer to different propeller and vessel speeds, respectively. Substituting, $101.3/70 = 22/V_2$, $V_2 = 15.2$ knots.

Related Calculations: Operating engineers aboard seagoing vessels generally use the apparent slip when computing the performance of their vessel at sea. Hence, the procedures given here are useful in propeller design and selection and vessel operation. Using the method of Step 3, the vessel speed at various rpm, or the rpm for various speeds, is easily determined. When using this method, the apparent slip is assumed to be constant. This is a valid assumption for practical calculation purposes.

VESSEL IMMERSION AND FLOODING EFFECTS

A 600-ft tanker has an 82.5-ft beam and a water-plane coefficient of fineness of 0.76 at a loaded draft of 31.9 ft. How many tons of cargo must be placed aboard this vessel to make her sink 1 in.? How many tons of water will enter this vessel if a 2×2 ft hole is stove in the hull 16 ft below the waterline? What effect will this hole have on the vessel?

Calculation Procedure:

1. Compute the vessel water-plane area

Use the relation $A_w = LB\alpha$, where A_w = waterplane area, sq ft; L = vessel length on waterline, ft; B = vessel beam at waterline, ft; α = water-plane coefficient of fineness. For this vessel, $A_w = 600(82.5)(0.76) = 37,600$ sq ft.

2. Compute the weight required to increase immersion

The weight required to increase the immersion of the vessel 1 in. is $W = A_w/420$ in saltwater, where W = weight, tons. Thus, for this vessel, $W = 37,600/420 = 89.6$ tons.

3. Compute the quantity of water entering the vessel

Use the relation $Q = 13.7(H)^{0.5}A$, where Q = weight of water entering the vessel, tons/min; H = distance of centerline of hole below the waterline, ft; A = area of hole, sq ft. For this vessel, $Q = 13.7(16)^{0.5}(4) = 219.5$ tons/min.

4. Compute the rate of vessel settling

Step 2 shows that the vessel will settle 1 in. for every 89.6 tons taken aboard. With water entering at the rate of 219.5 tons/min, the vessel will sink $219.5/89.6 = 2.44$ in./min. Assuming a constant water-plane shape (which is not quite true), the vessel will sink, in 1 hr, $60(2.44)/12 = 12.20$ ft. With a freeboard of 12 ft to the main deck, this deck would be awash about 1 hr after the hole was stove in the hull.

Related Calculations: Use this general procedure to determine the effect of loading cargo aboard any vessel whose dimensions are known. When the vessel operates in fresh water, as in the Great Lakes, compute the weight required to increase the draft 1 in. from $W = A_w/409$.

MARINE PUMP SELECTION FOR COMMERCIAL VESSELS

Choose the capacity, head, and types of pumps for a 10,000-ton cargo vessel for the following shipboard services: lube-oil, sanitary or flushing, fire-protection, ballast, bilge, fuel-oil-transfer, and refrigeration-condenser, fresh-water, ice-water-, main-condenser- and auxiliary-condenser-circulation. The shp of this steam-turbine-driven vessel is 6,000 hp with a steam flow of 37,500 lb/hr. What is the power input to the main-condenser circulating pump?

Calculation Procedure:

1. Compute the lube-oil pump size

Table 8 lists the usual type, total pressure, and capacity for various pumps used in marine service. Thus the lube-oil pump capacity $C = 36 + (7.5\,shp + 1,300)^{0.5}$, where C = pump capacity, gpm; shp = installed shp in the vessel. Substituting, $C = 36 + (7.5 \times 6,000 + 1,300)^{0.5} = 251.5$ gpm. A 260-gpm rotary pump (gear, screw, lobe, etc.) rated at a total head of 50 psi would be chosen for this ship.

2. Compute the sanitary pump size

Table 8 shows that sanitary or flushing pumps should have a capacity of 1.6 gpm per 1,000 tons displacement of the vessel. Since this is a 10,000-ton vessel, the required capacity of the sanitary pump is $C = 1.6(10,000/1,000) = 16$ gpm. A 20-gpm centrifugal pump developing a total pressure of 100 psi would probably be chosen for this vessel.

3. Select the fire-pump size

The American Bureau of Shipping and other maritime agencies publish recommendations or requirements for marine fire protection. Whereas the recommendations or requirements vary with the type of vessel and its intended service, a fire-protection pump capacity of at least 800 gpm in two pumps—usually 400 gpm each—is generally needed. Centrifugal pumps developing a total head of 125 psi or more are almost universally used for marine fire protection.

TABLE 8 Typical Pumps Used for Marine Service

Service	Usual pump type	Typical total head, psi	Capacity, gpm
Lube-oil	Rotary	50	$36 + (7.5\,shp + 1{,}300)^{0.5}$
Sanitary or flushing	Centrifugal	100	1.6 gpm/1,000 td*
Fire-protection. . . .	Centrifugal	125	At least 800 gpm in two pumps
Ballast.	Centrifugal	50	35 gpm/1,000 td
Bilge	Centrifugal	50	35 gpm/1,000 td
Fuel-oil-transfer. . .	Rotary	50	At least one 225 gpm pump
Fuel-oil	Rotary	400	Depends on steam capacity of boilers
Refrigeration-condenser	Centrifugal	25	5 gpm per ton of refrigeration
Fresh-water	Centrifugal	75	3 gpm/1,000 td
Ice-water	Centrifugal	25	0.5 gpm/1,000 td
Condenser-circulating	Centrifugal	50	1 gpm(5 lb/hr of steam condensed)

*td = tons displacement of the vessel.

4. Compute the ballast and bilge pump sizes

Table 8 shows that ballast and bilge pump capacity are based on the same relation. Or $C = 35(10{,}000/1{,}000) = 350$ gpm. Likewise, the total head developed is usually 50 psi or more. Modern vessels use centrifugal pumps for ballast and bilge service.

5. Select the fuel-transfer and fuel-service pumps

Many steam vessels use one or more 225-gpm rotary pumps for fuel-oil transfer. The total pressure developed by the pump is 50 psi or more.

To determine the capacity of the fuel-service pump, use the relation $C = W/486$, where W = weight of fuel oil pumped per hr, lb. If the pump handles 6,000 lb/hr of fuel, $C = 6{,}000/486 = 12.34$ gpm. Allowing a 30 percent reserve capacity for overloads, $C = 1.3(12.34) = 16$ gpm, closely.

Install two fuel-service pumps, each rated at 16 gpm capacity and 400 psi total pressure. Two full-capacity fuel-service pumps are needed so that one pump is available while the other is being overhauled. On some vessels a third pump rated at one-half capacity may also be installed. In modern ships, fuel-service pumps are almost always the rotary type.

6. Compute the size of the refrigeration-condenser pump

Table 8 shows that the refrigeration-condenser pump must supply 5 gpm/ton of installed refrigeration capacity. With a 500-ton refrigerating plant, $C = 5(500) = 2{,}500$ gpm. Use a centrifugal pump developing a total head of 25 psi or more.

7. Compute the size of the fresh- and ice-water pumps

Table 8 shows that the fresh-water pump should deliver 3 gpm per 1,000 tons displacement of the vessel. Or, $C = 3(10,000/1,000) = 30$ gpm. Use a centrifugal pump rated for a total head of 75 psi or more.

Likewise, using data from Table 8 for the ice-water pump, $C = 0.5(10,000/1,000) = 5$ gpm. Use a centrifugal pump rated for a total head of 25 psi or more.

8. Compute the size of the steam-condenser circulating pumps

Table 8 shows that the main- and auxiliary-condenser circulating pumps must supply 1 gpm of seawater per 5 lb of steam condensed. This flow rate is based on a 10-F rise in the seawater temperature during passage through the condenser, a specific heat of the seawater of 0.94 Btu/(lb)(F), a seawater weight of 8.58 lb/gal under average conditions, and a reserve capacity of 20 percent of the required capacity.

With a steam flow of 37,500 lb/hr to the main condenser, the circulating pump should have a capacity of $C = 37,500/5 = 7,500$ gpm. To provide for overloads and reduced heat transfer, use a 20 percent larger capacity, or $1.2(7,500) = 9,000$ gpm. Choose a centrifugal pump developing 50 psi total head. Install two full-capacity pumps per main condenser.

Compute the required capacity of the auxiliary-condenser circulating pump in the same way, using the auxiliary steam flow as the basis of the required seawater flow rate and pump capacity. Choose a pump having a 20 percent larger capacity, as was done above.

9. Compute the power input to the circulating pump

As a guide to computation of the power input to any marine pump, the main-condenser circulating pump will be analyzed.

The rated capacity of this pump is 9,000 gpm, including a 20 percent reserve. Refer to the characteristic curve of the pump, Fig. 2, to determine the efficiency of the pump at this rated flow rate. Figure 2 shows that the efficiency is 85 percent at a rated capacity of 9,000 gpm.

Compute the required power input to the pump from hp = gpm (head, psi)/1,715 (efficiency), or hp = 9,000 (50)/1715 (0.85) = 309 hp. Use a 350-hp motor to drive the pump, because this is the next largest standard-size motor.

Next, compute the percent operating capacity of the pump from actual gpm required/ pump rating, gpm. Since the pump has a 20 percent reserve capacity, actual gpm required = 9,000/1.20 = 7,500 gpm. Hence, operating capacity = 7,500/9,000 = 0.833, or 83.3 percent of the pump rated capacity.

Enter Fig. 2 at 83.3 percent of the pump rated capacity and project to the percent of rated capacity and efficiency curve. At the left read percent of rated efficiency of the pump = (percent of rated efficiency at actual capacity)(percent efficiency at pump rated capacity) = $(0.81)(0.85) = 0.688$, or 68.8 percent.

Compute the pump operating hp from hp = (actual gpm)(head, psi)/1,715 (pump actual operating efficiency), or hp = (7,500)(50)/[1,715(68.8)] = 318 hp. With a 350-hp motor the operating load is 318/350 = 0.907, or 90.7 percent of the rated motor capacity.

To convert the mechanical power input to the pump to the electrical power input to the motor—i.e., the electrical load on the ship's generator caused by the main-condenser circulating-pump drive motor: (*a*) Determine the motor efficiency at full load from the motor characteristic curve; (*b*) determine the motor efficiency at the actual operating rating, again using the motor characteristic curve; (*c*) take the product of these two efficiencies; and (*d*) convert the motor operating hp to kw and divide by the product *c*.

Thus, with $a = 0.96$ and $b = 0.85$, the kw input to this motor = 318 hp (0.746 kw/hp)/ $[(0.96)(0.85)] = 291$ kw; say 300 kw. The power input to any of the other pumps is determined in the same way.

Related Calculations: Use this general method for preliminary sizing of pumps for steam-propelled vessels of any type—tankers, dry-cargo, passenger, etc. For best results, assemble a set of typical pump and motor characteristic curves, such as Fig. 2.

These will shorten the time spent on pump sizing. Note that the discharge pressures listed in Table 8 may vary if a vessel is used for special services.

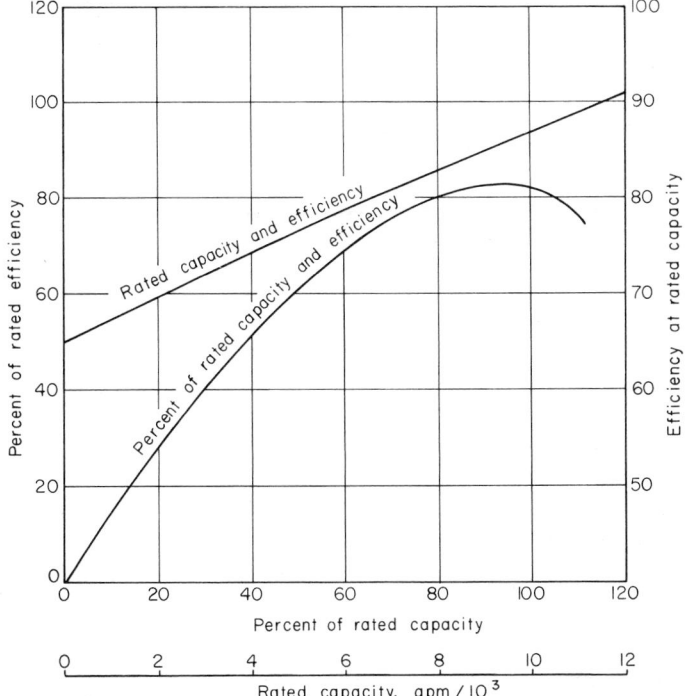

Fig. 2   Hydraulic characteristics of a typical centrifugal pump for marine service.

MARINE POWER-PLANT SELECTION

Choose a power plant suitable for a 700-ft 30,000-ton oceangoing tanker that must have a service speed of 20 knots. Indicate the factors entering the decision. The owners require a low fuel rate, small investment, and minimum maintenance cost. List the typical installation and capacity factors for the type of power plant selected.

Calculation Procedure:

1. Compute the power required to propel the vessel

Use the admiralty method given earlier in this Section. Or, $P = D^{2/3}V^3/K$, where P = shp required to drive the vessel; D = vessel displacement, long tons; V = vessel cruising or service speed, knots; K = admiralty coefficient from Table 3. With K = 450, by interpolation in Table 3, $P = (30,000)^{2/3}(20)^3/450 = 17,150$ hp. This power requirement is a typical value for vessels of this size.

2. Select the type of power plant to use

In the power range needed for this vessel—17,000 hp—the geared steam turbine, Tables 9 and 10, will provide the lowest fuel rate and lowest investment if the reciprocating steam engine is ignored. Few large vessels being built today use reciprocating steam engines.

Diesel propulsion, although suitable for this vessel, would be more expensive because: (*a*) Fuel cost is higher if diesel oil is used; (*b*) lube-oil cost is higher; (*c*) the initial investment for the engines is higher; (*d*) the weight of the engines is about 40 percent greater than that of a geared steam turbine of equal power output; (*e*) diesel-engine maintenance costs are substantially higher than are steam-turbine maintenance costs.

TABLE 9　Typical Marine Power-Plant Choices

Power Plant Type	*Typical Application*
Geared steam turbine....	Ocean and Great Lakes vessels of 6,000 shp or more
Turboelectric drive......	Ocean and Great Lakes vessels of 2,000 shp or more in service requiring much maneuvering
Reciprocating steam engines	Rarely used in new vessels today except where simple machinery is required in power ranges up to 5,000 bhp per shaft
Direct diesel drive	Ocean and Great Lakes vessels with power requirements up to 30,000 bhp per shaft, although usual installations range up to about 15,000 bhp per shaft; also popular for harbor craft—tugs, sewage vessels, etc.
Diesel-electric drive.....	Popular for harbor craft—tugs, ferries, etc.—where maneuvering is a primary consideration; also fitted to ocean and Great Lakes vessels
Geared diesel drive......	Popular for harbor craft but is also finding use in ocean and Great Lakes vessels
Nuclear power	Ocean surface and undersea vessels
Gasoline engine.........	Limited to relatively small vessels up to 60 ft plying the coastal and inland waters
Gas turbine	Limited use on large and small vessels in ocean and coastal service

TABLE 10　Comparative Costs of Marine Power Plants
(Output of 2,000 shp, or more)

Type of plant	Typical fuel rate, lb per shp-hr	Lube-oil cost	Maintenance cost	Relative cost
Geared steam turbine	0.50–0.56	Nominal	Nominal	190
Turboelectric drive	0.60	Nominal	Moderate	210
Reciprocating steam engine	0.75–0.90	Nominal	Nominal	150
Direct diesel drive	0.38	High	High	220
Diesel-electric drive	0.40	High	High	230
Geared diesel drive	0.40	High	High	225
Nuclear power	...	Nominal	High	300
Gasoline engine	0.5	High	High	
Gas turbine	0.39–0.60	High	High	210

3.　*List typical installation and capacity factors*

Watertube three-pass boilers weighing 24 to 30 lb per sq ft of heating surface are used in modern oceangoing merchant vessels. When an air heater is installed, as is often done today, it weighs about 6.5 lb/sq ft. D-type watertube boilers weigh 20 to

30 lb per sq ft of heating surface. Where an economizer is used, add its weight to that of the boiler. The economizer feedwater temperature is usually in the 400-F range. Currently, 600-psi 875-F boilers are popular for standard merchant vessels.

Usual geared steam turbine installations use two turbines—a high-pressure and a low-pressure turbine driving one propeller shaft. Large ships, and high-speed vessels, may use three turbines per propeller shaft. With this arrangement the turbine receiving steam from the boiler is called the *cruising turbine*. Monel and low-carbon steel blades are often used in modern marine turbines.

Reduction gears generally provide an 8:1 speed reduction between the turbine and the propeller shaft. Double-helical gears are used to neutralize unbalanced end thrust in the gears. The tooth opening resulting from torsional and bending deflection is limited to 0.001 in. Typical weights of complete reduction gears, lb = 1,300(shp/propeller rpm), according J. F. Nace. The usual efficiency of large reduction gears ranges from 97 to 98.5 percent.

Typical marine surface condensers have 0.9 to 1.6 sq ft of heat-transfer surface per shp. The temperature rise of the seawater during passage through a condenser serving a geared steam turbine is 6 to 10 F when the inlet temperature is 75 F.

Size the condensate-pump inlet pipe for a flow velocity of less than 2 fps. Choose the main boiler feed pump such that it has a capacity that will permit the rated cruising speed when the pump is delivering 70 percent of its rated capacity.

When a vessel is equipped with all electrically driven auxiliaries and an electric galley, the generator load kw = 0.75 (number of persons aboard the vessel) + 0.025 (shp of the vessel). With a turbine-driven main feed pump, kw = 0.75 (number of persons aboard the vessel) + 0.017 (shp of the vessel).

In air ejectors, the quantity of dry air removed, lb/hr = 7.5 + 0.00025 (condensate flow, lb/hr). With a 28.5-in. vacuum in the condenser, the weight of air and vapor removed by the air ejector is 2 to 2.5 times that given above. Usual air ejectors use 5 lb of steam per lb of air and vapor mixture removed.

Steam soot blowers use, in lb of steam, 0.008 (total heating surface of boiler, sq ft) (number of blows per day), if each soot blower blows for 45 sec.

Related Calculations: Use this general method to choose the type of power plant for any vessel—large or small. When two or more types of power plants will provide equal service, the final choice must be based on an economic comparison of the alternative plants.

Some of the newest steam-propelled vessels—such as the 25-knot roll-on roll-off trailership *Ponce de Leon*—use only one steam boiler. This boiler is usually of the reheat type. Other vessels using only one boiler are the tankers *Esso Houston, Esso New Orleans*, and the 206,000-deadweight-ton *Idemitsu Maru*.

Fuel consumption when using reheat is 0.43 lb of oil per shp-hr. The initial cost of a reheat boiler plant is 15 to 20 percent higher than a non-reheat plant. The reheat section of the boiler is used not when the vessel is maneuvering, only when it is at sea.

NUCLEAR PROPULSION FOR OCEANGOING VESSELS

A transportation firm is considering replacing its oil-fueled vessels with nuclear-powered vessels. Compare the performance of the vessels listed in Table 11 on the basis of the cargo-fuel ratios for the routes listed. What magnitude of tank-top loading can be expected in nuclear-powered vessels? List the types of reactors suitable for marine applications of nuclear power.

Calculation Procedure:

1. *Compute the vessel cargo-fuel ratios*

The cargo-fuel ratio CF of any vessel is the ratio of cargo capacity, tons/(fuel capacity, tons + steam-generating apparatus weight, tons). Table 11 shows the fuel and machinery weights for each of three ships propelled by oil and nuclear power.

Using the data in Table 11 for the tanker or liquid carrier, CF = 25,000 tons/(0.5 tons/nautical mile × 5,000 nautical miles + 250 tons machinery weight) = 9.1. For the nuclear-powered tanker, CF = 12.5. Compute the CF values for the other vessels and list the results in Table 11.

TABLE 11   Ocean-Going Vessel Performance

Propulsion	Vessel type					
	Liquid carrier		Bulk cargo		General cargo	
	Oil	Nuclear	Oil	Nuclear	Oil	Nuclear
Machinery weight, tons	250	2,000	250	2,000	250	2,000
Fuel used, tons/ nautical mile ...	0.5	...	0.5	...	0.5	
Cargo weight, tons	25,000	25,000	25,000	25,000	25,000	25,000
Voyage distance, miles	5,000	5,000	10,000	10,000	15,000	15,000
CF ratio..........	9.1	12.5	4.77	12.5	3.23	12.5

2. Evaluate the cargo-fuel ratios computed

For any ocean trade route, the higher the CF ratio, the more economical the vessel. Table 11 indicates that the nuclear-propelled vessel has a higher CF ratio for each type of cargo carried and for each voyage length. This tabulation also shows that the nuclear-propelled vessel has a greater advantage from the CF standpoint for the longer voyages. The reason for this is that the oil-fueled vessel CF decreases as the voyage length increases, whereas the nuclear-propelled vessel CF remains constant. Hence, the nuclear-propelled vessel is more economical for all the routes considered.

3. Compute the nuclear-propelled vessel tank loading

A modern pressurized-water marine reactor plant developing 20,000 shp weighs about 2,000 tons, including its fuel, shield, and related apparatus. Such a reactor will require some 2,000 sq ft of tank-top area. Thus, the load on the tank top in lb/sq ft is (2,000 tons × 2,000 lb/ton)/(2,000 sq ft) = 2,000 lb/sq ft.

Usual modern merchant vessels have a tank-top loading of 1,500 to 1,800 lb/sq ft. Hence, the hull of a nuclear-propelled vessel must be strengthened to carry the extra load of the reactor. This strengthening will, of course, increase the cost of the hull.

4. List the types of reactors suitable for the vessels

Table 12 shows the types of nuclear reactors suitable for marine propulsion systems. The major advantages and disadvantages of each type of reactor are also listed.

Related Calculations: Nuclear propulsion is still in the development stage for commercial vessels. However, the data presented here are appropriate for preliminary selection of the type of reactor suitable for a given vessel.

Nuclear vessels constructed thus far are able to operate at higher speeds over longer trade routes than are oil-fueled vessels. The increased speed produces a substantial increase in a vessel's cargo-carrying capacity over a given period of time. In a conventional merchant vessel, the hp required for higher speeds is increased by the third power of the speed ratio. To produce the higher speed, the vessel must have larger engines and a greater fuel-storage capacity. The nuclear vessel does not require a larger fuel storage capacity and can thus operate continually at the maximum level of its hull form without penalty in its cargo-carrying capacity.

TANKER CAPACITY AND CARGO-HANDLING CHARACTERISTICS

What is the T-2 equivalent of a 110,000-DWT 18-knot tanker having a capacity of 835,000 bbl of oil? How long will it take to unload such a tanker if it is fitted with the

TABLE 12   Types of Marine Reactors

Reactor type	Steam conditions	Advantages	Disadvantages
Pressurized-water (used on the N.S. *Savannah*)	460-psi saturated	Simple; easy to control	Requires large heat exchangers; boiling takes place outside the reactor; large pumps needed
Boiling-water........	460-psi saturated	Does not need heat exchangers; requires little pumping power; boiling occurs in reactor	Precise pressure and water-level control needed
Moderated	460-psi superheated	Simplest reactor design; produces superheated steam at low reactor pressures	Coolant may decompose, requiring makeup
Liquid-sodium.......	650-psi, 850 F	Reactor operates at 14.7 psia	Uses more complex coolant piping; needs more pumps; requires moderator
Gas-cooled	High pressures and temperatures	Simple; safer than other types; higher steam pressures and tempera-ture possible	Needs gas blowers; uses large pumping power

usual cargo pumps? What is the energy that must be absorbed by a dock if this ship moves at a velocity of 30 fpm during normal docking and strikes the dock at this velocity?

Calculation Procedure:

1. Compute the T-2 equivalent of the vessel

The T-2 equivalent of a tanker is a quick means of comparing a new vessel with the well-known capacity of the T-2-type tanker. To compute the T-2 equivalent of any tanker, use the relation T-2 equivalent = (new tanker DWT) (new tanker speed, knots)/ $16,000 \times 14.6$), where DWT = tanker deadweight tons = displacement of the tanker fully loaded, ready for sea, less the weight of the ship itself, or displacement, light, expressed in tons of 2,240 lb. The constants in the denominator of this expression are the DWT and speed, respectively, of the standard T-2 tanker. Substituting, T-2 equivalent = $(110,000)(18)/[16,000(14.6)] = 8.46$.

This equivalent means that the new tanker has a carrying capacity of nearly $8\frac{1}{2}$ times that of a standard T-2 tanker when the larger capacity and higher speed of the new vessel are taken into consideration. The equivalent provides an easily understood comparison for all tanker personnel afloat and ashore.

2. Determine the vessel unloading time

Modern tankers are fitted with cargo pumps having sufficient capacity to unload the vessel in 12 to 24 hr. An average unloading time for a new tanker discharging at a modern terminal is 16 hr.

Table 13 shows the pumping rates for a variety of tankers in current use. A study of this list shows that a 110,000-ton tanker can be unloaded in about 21 hr.

In some terminals the tankage or piping capacity may be small. This condition can increase the unloading time. Or if the tanker is transporting a "dirty" cargo—i.e., asphalt, heavy residual oil, fuel oil, or certain crude oils—unloading may take longer because the cargo is too viscous for the ship's pumps to handle at normal temperature.

When this situation occurs, the cargo is heated to a temperature high enough to produce a viscosity suitable for pumping. The required temperature usually ranges between 125 and 150 F. Steam coils located in the cargo tanks are used to heat the cargo.

TABLE 13   Typical Tanker Cargo-Discharge Times

Tanker type or name	DWT	Cargo capacity of 42-gal bbl	Pump capacity, bbl/hr	Discharge time, hr
T-1	16,800	141,000	8,600	16.5
T-5	26,500	204,000	17,600	11.6
T-5-5	25,000	190,000	23,400	8.1
Pensylvania	28,170	241,500	23,400	10.5
Alton Jones	38,000	336,000	30,600	11.0
World Glory......	45,500	396,000	20,000	19.8
Barracuda	60,000	430,000	34,300	12.5
Victory tankers ...	100,000	825,000	40,000	20.6
Niarchos tankers..	106,000	821,000	40,000	20.5
Super tanker	110,000	835,000	40,000	21
Universe Ireland..	312,000	2,400,000	125,000	20
Ultratanker	470,000	3,620,000	200,000	19

3. Compute the energy absorbed by the tanker dock

Use the relation $E = Wv^2/4g$, where E = energy absorbed by the dock, ft-lb, when the vessel strikes it at an assumed angle of 10° between the face of the dock and the vessel hull; W = displacement of the loaded vessel, lb; v = velocity of the vessel at impact, fps; $g = 32.2$ ft/sec². Substituting, $E = 110,000(2,240$ lb/long ton$)(0.5)^2/4(32.2) = 478,000$ ft-lb.

Related Calculations: A loaded tanker approaching a dock moves at a velocity of 0.15 to 1.0 fps. When tugs are used to dock the tanker, the velocity of approach is usually less than 0.5 fps. In exposed locations where the tanker docks without the aid of tugs, the velocity of approach may range between 0.5 and 1.0 fps. Note that the energy absorbed by the dock is a function of the vessel displacement and the velocity of approach. Hence, the relation presented in Step 3 can be used for any type of vessel.

MARINE REFRIGERATION AND AIR-CONDITIONING SYSTEMS

A marine refrigeration system has a load of 30 tons and uses a Genetron refrigerant that evaporates at 5 F and condenses at 105 F. The temperature of the refrigerant leaving the evaporator is 10 F, resulting from a slight superheating in the evaporator. Wiredrawing in the compresser suction valves causes a 5-psi pressure loss; in the discharge valves, wiredrawing causes a 10-psi pressure loss. There is a 10 percent increase in isentropic work during compression resulting from turbulence in the compressor cylinders. The volumetric efficiency of the compressor is 70 percent; its mechanical efficiency is 80 percent. Determine the quantity of refrigerant that must be circulated, the work done on the refrigerant in the compressor, the rating of the compressor driving motor, and the required capacity of the condenser circulating pump.

Calculation Procedure:

1. Compute the quantity of refrigeration circulated

Using a tabulation or a plot of Genetron refrigerant properties, list the following values for this cycle, as shown in Fig. 3: Condensing pressure = p_3, Fig. 3, = 141 psia; cylinder discharge pressure = $p_{2d} = 141 + 10 = 151$ psia; compressor suction pressure = $p_5 = 26.5$ psia; cylinder suction pressure = $p_{1s} = 26.5 - 5 = 21.5$ psia; suction vapor enthalpy = $h_1 = h_{1s} = 81.1$ Btu/lb; suction vapor specific volume = $V_1 = 1.576$ cu ft/lb; suction entropy in cylinder = $S_2 = S_{2d} = 0.17568$; discharge enthalpy, isentropic =

$h_{2d} = 95.18$ Btu/lb; evaporator liquid temperature $= t_4 = 100$ F; liquid enthalpy $= h_4 = h_5 = 31.16$ Btu/lb; evaporator exit enthalpy $= h_6 = 80.99$ Btu/lb.

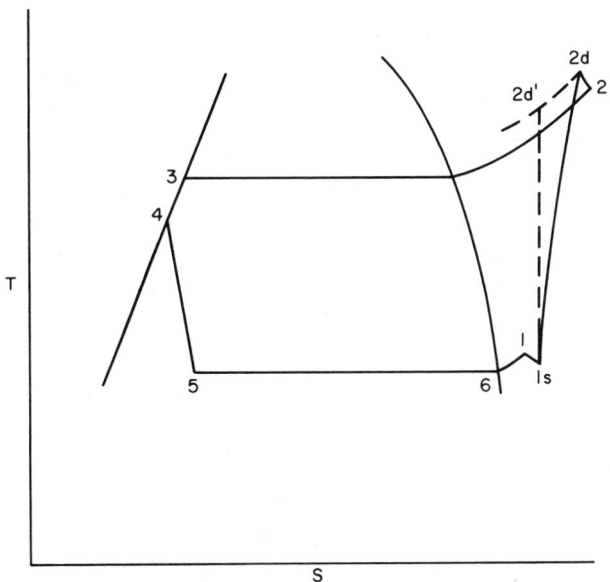

Fig. 3   Temperature-entropy diagram for a refrigeration cycle.

The refrigeration effect per lb of Genetron circulated $= h_6 - h_5 = 80.99 - 31.16 = 49.83$ Btu/lb. With a load of 30 tons, the refrigerant flow rate required in lb/min $= F = (30$ tons load$)[200$ Btu/(min)(ton)$]/[49.83] = 120.5$ lb/min.

2.   Compute the work done on the refrigerant

With a 10 percent work-input increase resulting from turbulence, the work input, Btu/lb $= W = 1.1$ $(h_{2d} - h_1) = 1.1(95.18 - 81.1) = 15.5$ Btu per lb of refrigerant circulated.

Compute the compressor indicated hp from: $ihp = WF/42.4$, where the constant in the denominator $=$ Btu/hp-min. Substituting, $ihp = 15.5(120.5)/42.4 = 44.1$ ihp.

3.   Select the size of the compressor drive motor

The mechanical efficiency of the compressor is 80 percent. Hence the power input to the compressor $= ihp/$mechanical efficiency $= 44.1/0.80 = 55.1$ hp. Use the next larger size standard motor, or 60 hp.

4.   Compute the compressor displacement required

Use the relation $D = 1,728$ FV_1/e_v, where $D =$ compressor displacement, in.^{3}; $e_v =$ compressor volumetric efficiency; other symbols as before. Substituting, $D = 1,728 \times (120.5)(1.576)/0.70 = 469,000$ in.^{3}/min.

5.   Compute the condenser heat load

The heat that must be removed in the condenser is $h = h_2 - h_4$, where $h =$ Btu per lb of refrigerant circulated; other symbols as before. Substituting, $h = 96.6 - 31.16 = 65.44$ Btu/lb. With a refrigerant flow rate of 120.5 lb/min, the total heat removed in the condenser $H = 65.44(120.5) = 7,880$ Btu/min.

6.   Compute the condensing water flow rate

With a rise in the seawater temperature of Δt during passage through the condenser, the quantity of heat absorbed per gal of water is $H_a = \Delta t s w$, where $H_a =$ heat absorbed,

Btu/gal; s = specific heat of seawater, Btu/(lb)(°F) = 0.94; w = weight of seawater, lb/gal = 8.58. Assuming a 10-F temperature rise in the water during passage through the condenser, H_a = 10(0.94)(8.58) = 80.6 Btu/gal. With a total heat load of H, the quantity of water that must be circulated is gpm = H/H_a = 7,880/80.6 = 97.8 gpm.

7. Select the condenser circulating water pump

Use a pump with a 50 percent larger capacity to guard against overloads and fouling. Thus, pump capacity = 1.5(97.8) = 146.8 gpm; say 150 gpm.

Compute the total head loss through the piping and condenser. This seldom exceeds 25 psi. For usual marine air-conditioning and refrigeration service, a centrifugal circulating pump is best because it is less subject to fouling by dirty harbor water or marine growths.

8. Compute the power input to the pump

Use the relation hp = gpm (discharge pressure, psi)/1,715 (pump efficiency). With a pump efficiency of 60 percent, hp = 150(25)/1,715(0.60) = 3.64 hp; use a 5-hp motor.

Related Calculations: Use this method to analyze refrigeration systems for cargo holds, ice-water service, air conditioning of passenger and crew quarters, and other marine applications.

Note that the general design procedures for marine refrigeration and air-conditioning systems resemble those for land service. The outdoor design conditions are, of course, determined by the areas of the world in which the vessel will sail. Carrier Air Conditioning Company recommends summer outdoor design conditions of 95-F db and 82-F wb, unless the vessel will be sailing in predominantly tropical areas. If the latter is the case, then the average warm port of call defines the summer outdoor design condition. The coldest port of call determines the winter outdoor design condition. Besides the usual sun load on the glass area of a ship, there is the added load of the radiant energy of the sun diffusely reflected from the water surface.

Ventilation needs on shipboard require a minimum of 12.5 cfm per person or 2.5 air changes per hr, whichever is greater. Deluxe passenger vessels require more ventilation for maximum comfort. Air-conditioning systems for passenger staterooms and crew quarters are limited to air-water induction, all-air reheat, and dual-duct systems. The public spaces are air-conditioned by field- or factory-assembled conventional all-air central-station fan-coil systems.

Section **10**

Nuclear Engineering

B. G. A. SKROTZKI, P.E.
Power Magazine

SAMUEL C. LIND
Consultant, United States Atomic Energy Commission

REFERENCES: El-Wakil—*Nuclear Power Engineering*, McGraw-Hill; Sachs—*Nuclear Theory*, Addison-Wesley; Hoegerton and Grass—*Reactor Handbook*, U.S. Atomic Energy Commission; Schwenk and Shannon—*Nuclear Power Engineering*, McGraw-Hill; Murphy—*Elements of Nuclear Engineering*, Wiley; Rockwell—*Reactor Shielding Design Manual*, Van Nostrand; Price—*Radiation Shielding*, Pergamon; Hollaender—*Radiation Biology*, McGraw-Hill; Glasstone and Edlund—*The Elements of Nuclear Reactor Theory*, Van Nostrand; Murray—*Nuclear Reactor Physics*, Prentice-Hall; Etherington—*Nuclear Engineering Handbook*, McGraw-Hill; Glasstone—*Principles of Nuclear Reactor Engineering*, Van Nostrand; Bonilla—*Nuclear Engineering*, McGraw-Hill; Glasstone and Lovberg—*Controlled Thermonuclear Reactions*, Van Nostrand; Schultz —*Control of Nuclear Reactors and Power Plants*, McGraw-Hill; International Atomic Energy Agency—*Directory of Nuclear Reactors*; Henley—*Advances in Nuclear Science and Technology*, Academic; Gol'denblat—*Calculation of Thermal Stresses in Nuclear Reactors*, Consultants Bureau; Greenspan—*Computing Methods in Reactor Physics*, Gordon and Breach; International Atomic Energy Agency—*Programming and Utilization of Research Reactors*; Marchuk—*Theory and Methods of Nuclear Reactor Calculations*, Consultants Bureau.

NUCLEAR POWER REACTOR SELECTION

Select a nuclear power reactor to generate 60,000 kw at a thermal efficiency of 35 percent or more. If the selected unit is a 10-ft-diameter reactor that uses a fluidized bed containing 20×10^6 fuel pellets each 0.375 in. in diameter with a density of 700 lb mass/cu ft and the reactor fluid is pressurized water at 600 F, determine the bed pressure drop when fluidized. Also, compute the reactor fuel volume, the collapsed fuel bed height, and the density of the pressurized water.

Calculation Procedure:

1. Select the type of reactor to use

Table 1 summarizes the operating characteristics of six types of power reactors. Study of this tabulation shows that a pressurized-water reactor will provide the desired thermal efficiency. Further, this type of reactor is successfully used for large-scale power generation. Hence, a pressurized-water reactor will be the first tentative choice for this plant.

TABLE 1 Nuclear Power Reactor Characteristics

Reactor type	Typical thermal efficiency, %	Typical power density, thermal, kw/cu ft	Typical reactor pressure, psig	Average heat flux Btu/(hr)(sq ft)	Typical fuel enrichment, %	Reactor coolant
Pressurized-water	36	1,600	1,500	300,000	1.5–3.0	Light water
Boiling-water.	22–30	800	1,000	100,000	1.5	Light water
Gas-cooled ...	30	200	600–1,000	...	0.70–2.5	Carbon dioxide
Liquid-metal .	33	300	100	...	...	Sodium, bismuth, lead, etc.
Fast-breeder..	32	20,000	100	650,000	...	Sodium
Fluid-fueled..	30	400	1,000–2,000.	Varies	Varies	Reactor fuel solution

2. Compute the reactor fuel volume

Use the relation $v_f = nv_p$, where v_f = fuel volume, cu ft; n = number of fuel pellets in the reactor; v_p = volume of each pellet, cu ft. Substituting, $v_f = 20 \times 10^6 \pi (0.375)^3/[6(1,728)] = 320$ cu ft.

3. Compute the fuel volume in the collapsed form

With the fuel bed not fluidized, the porosity P with packed spheres is about 0.40. Then, collapsed volume $v_c = v_f/(1-P) = 320/0.60 = 534$ cu ft.

4. Compute the collapsed fuel bed height

Use the relation $h = v_c/A_r$, where h = collapsed height of fuel bed, ft; A_r = reactor fuel bed area, sq ft. Substituting, $h = 534/[\pi 10^2/4] = 6.78$ ft.

5. Determine the density of the pressurized water

Using the steam tables, $d_w = 42.45$ lb/cu ft at 600 F for saturated liquid.

6. Compute the pressure loss through the fluidized bed

Use the relation $p = 2.9h[(1-P)d_f + Pd_w]$, where p = pressure loss through fluidized fuel bed, psf; d_f = fuel density, lb mass/cu ft; other symbols as before. Substituting, $p = 2.9[(1-0.4)700 + 0.4 \times 42.45] = 1,268$ psf, or 8.79 psi.

Related Calculations: This general procedure is valid for preliminary selection of the type of nuclear reactor to use for a given power application. Since reactors are expensive devices, a complete economic analysis must be made of the alternatives available before the final choice of the reactor is made.

NUCLEAR POWER PLANT CYCLE ANALYSIS

A nuclear power plant using two coolants, Na and NaK, is arranged as shown in Fig. 1. Sodium, the first coolant, enters the reactor at 600 F and leaves at 1,000 F; NaK, the second coolant, enters the intermediate heat exchanger at 550 F and leaves at 950 F. Neglecting heat and pressure losses in the piping, plot the enthalpy-temperature diagram for the plant if steam leaves the boiler at 1,200 psi. What are the Na and NaK flow rates with the cycle arrangement shown in Fig. 1, a reactor capacity of 400,000 kw of heat energy, and a 155,000-kw turbine output? Determine the plant thermal efficiency if the auxiliary-power needs = 12,000 kw.

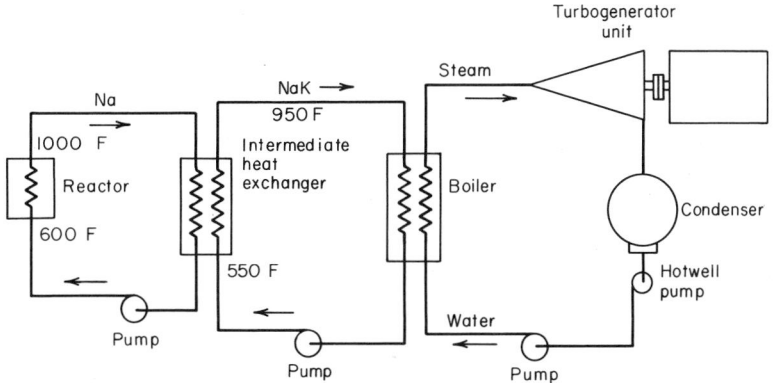

Fig. 1   Reactor plant with two-coolant system uses Na in the reactor circuit and transfers heat to the intermediate NaK circuit that acts as a buffer against making the steam circuit radioactive.

Calculation Procedure:

1. Determine the steam outlet and saturation temperature

Figure 1 shows that NaK enters the boiler at 950 F. Draw a horizontal line on the enthalpy-temperature (h-t) diagram, Fig. 2, indicating the 950-F NaK temperature entering the boiler. Also draw a horizontal line on the h-t diagram, Fig. 2, at 1,000 F, indicating the Na temperature leaving the reactor.

The steam outlet temperature from the boiler will be less than 950 F because transfer of heat between the NaK and the water and steam in the boiler provides the energy required to convert the water to steam. A temperature difference between the NaK and the steam is needed to produce the desired heat transfer.

Assume a 50-F temperature difference between the boiler outlet steam and the NaK, which is a typical temperature difference for this type of cycle. With such a temperature difference the outlet steam temperature = 950 − 50 = 900 F. From the steam tables find the saturation temperature of steam at 1,200 psia as 567.2 F. Hence the steam will be superheated when it leaves the boiler.

2. Compute the boiler evaporator coolant outlet temperature

Incoming feedwater enters the boiler evaporator section where it is heated by the NaK before entering the boiler steam section. To provide heat transfer between the NaK leaving the evaporator section of the boiler and the incoming boiler feedwater, a temperature difference between the two fluids is necessary. Assume that the NaK coolant leaves the boiler evaporator section at a temperature 40 F higher than the incoming feedwater. With the incoming feedwater at the saturation temperature, or

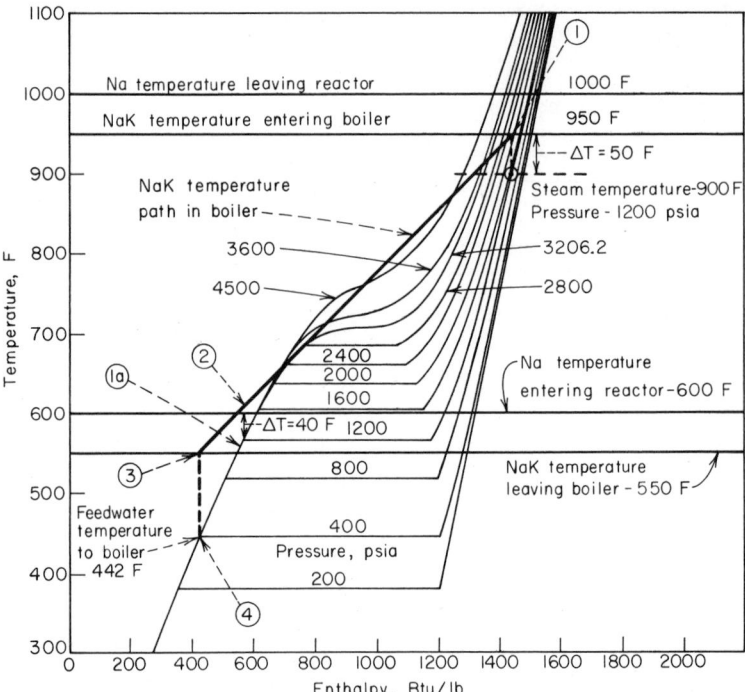

Fig. 2 Steam-water enthalpy-temperature diagram shows the relation between NaK circuit and steam circuit. Keeping the steam temperatures high raises the thermal efficiency of the plant.

567.2 F, the NaK coolant outlet temperature from the boiler evaporator $= 567.2 + 40 = 607.2$ F; say 607 F.

3. Plot the boiler coolant temperature path

Locate the boiler outlet steam state on the h-t diagram, Fig. 2, on the 1,200-psia-pressure curve and the 900-F-temperature horizontal. From this point, project vertically upward to the 950-F NaK temperature horizontal to locate point 1, the temperature of the NaK entering the boiler, Fig. 2.

Next, locate the point $1a$ where the liquid enthalpy line of the h-t diagram, Fig. 2, intersects the 1,200-psia-evaporation enthalpy line. From point $1a$, project vertically upward to 607 F, point 2, the temperature of the NaK coolant leaving the boiler evaporator section.

Points 1 and 2 are the NaK *temperature path* in the boiler evaporator and steam-generating sections. Assuming that the NaK has a constant specific heat while flowing through the boiler evaporator and steam-generating sections (a completely valid assumption), draw a straight line between points 1 and 2 and extend it to intersect the 550-F temperature line at point 3. Note that point 3 represents the temperature of the NaK entering the intermediate heat exchanger.

4. Determine the boiler feedwater inlet temperature

Feedwater enters the boiler at a yet unknown temperature. During passage between the boiler inlet and the evaporator section inlet, the feedwater absorbs heat from the NaK coolant leaving the evaporator at 607 F.

Draw a line vertically downward from point 3 until the liquid enthalpy curve is intersected, point 4. Point 4 represents the boiler feedwater inlet temperature, or 442

F, based on the valid assumption that the feedwater leaving the condenser hot well is in the saturated state.

5. Compute the reactor coolant flow rate

Sodium enters the reactor at 600 F and leaves at 1,000 F, Fig. 1. Thus, the temperature rise of the Na during passage through the reactor is $1,000 - 600 = 400$ F. Also, the average specific heat of Na is 0.306 Btu/(lb)(F), found from a tabulation of Na properties in an engineering handbook.

Compute the Na flow from $f = 3,413 \, kw/\Delta tc$, where f = Na flow rate, lb/hr; kw = reactor heat rating, kw; Δt = Na temperature rise during passage through the reactor, F; c = specific heat of the Na coolant, Btu/(lb)(F). Substituting, $f = 3,413(400,000)/[400(0.306)] = 11,130,000$ lb/hr.

6. Compute the boiler heating liquid flow rate

Use the same relation as in Step 5, substituting the temperature change and specific heat of NaK. Since the NaK enters the boiler at 950 F and leaves at 550 F, its temperature change is $950 - 550 = 400$ F. Also, the specific heat of NaK is 0.251 Btu/(lb) (F), as found from NaK properties tabulated in an engineering handbook. Substituting, $f = 3,413(400,000)/[400(0.251)] = 13,600,000$ lb/hr.

7. Compute the plant thermal efficiency

The net station output kw = gross output of turbine, kw, minus the total plant auxiliary demand, $kw = 155,000 - 12,000 = 143,000$ kw. Then, overall plant thermal efficiency = net station output, kw/reactor heat output, $kw = 143,000/400,000 = 0.357$, or 35.7 percent.

Related Calculations: This analysis is valid for a cycle in which the reactor coolant does not do work in the turbine. In general, designers prefer to avoid using the reactor coolant in the turbine. Although the thermodynamic aspects of a nuclear cycle are important, the cost of the plant must also be considered before making a final choice of a cycle. The method presented is the work of Henry C. Schwenk and Robert H. Shannon, as reported in *Power*.

REACTOR FUEL CONSUMPTION, ATOM BURNUP, AND NEUTRON FLUX

Determine the amount of fissionable material used in a 500-mw reactor having 3×10^{10} fissions/watt-sec. The reactor core has a volume of 1,360 cu ft and the fuel (99.3 percent U-238 plus 0.7 percent U-235), occupies 6 percent of the reactor volume. How much fissionable material is consumed if the plant operates 8,760 hr/year at an 80 percent load factor and the capture cross section/fission cross section ratio = 1.2? What is the maximum allowable atom burnup, the average fuel-cycle time, and the reactor neutron flux?

Calculation Procedure:

1. Compute the reactor fission rate

Use the relation $F_r = P_T C$, where F_r = reactor fission rate, fissions/watt-sec; P_T = total reactor power, watts; C = fissions/watt-sec. Substituting, $F_r = 500 \times 10^6(3 \times 10^{10}) = 1.5 \times 10^{19}$ fissions/sec.

2. Compute the total volume of the fuel

Since the fuel occupies 6 percent of the reactor volume, the fuel volume $V_f = 0.06 \times (1,360) = 81.6$ cu ft. As reactor fuel quantities are often expressed in cu cm, convert the fuel volume in cu ft by multiplying by the conversion factor 2.832×10^4, or $V_{fc} = 2.832 \times 10^4(81.6) = 2.31 \times 10^6$ cu cm.

3. Compute the U-235 nuclei in the reactor

First determine the uranium nuclei per cu cm N_U using the relation $N_U = $ [(uranium density, g/cu cm)/uranium atomic weight)](Avogadro's constant) = $(18.68/238.07)(6.023$

$\times 10^{23}) = 0.0472 \times 10^{24}$ nuclei/cu cm. In this relation the following constants are used: uranium density $= 18.68$ g/cu cm; uranium atomic weight $= 238.07$; Avogadro's constant $= N_m = 6.023 \times 10^{23}$ atoms/g-atom.

With the uranium nuclei per cu cm known, compute the U-235 nuclei in the reactor from $N_{\text{u-235}} = 0.007 N_U V_{fc} = 0.007(0.0472 \times 10^{24})(2.31 \times 10^6) = 7.64 \times 10^{26}$ U-235 nuclei in the reactor.

4. Compute the U-235 fissionable material consumed

Use the relation $F_{\text{u-235}} = F_r G_m / N_m$, where $F_{\text{u-235}} =$ fissionable U-235 material consumed or burned up for power only, g/sec; $G_m =$ g/mole of the fissionable material; other symbols as before. Substituting, $F_{\text{u-235}} = (1.5 \times 10^{19})(235)/6.023 = 5.85 \times 10^{-3}$ g/sec.

5. Compute the annual consumption of fissionable material

Use the relation $A_c = F_{\text{u-235}} YL/1,000$, where $A_c =$ annual consumption of fissionable material, kg; $Y =$ sec/year; $L =$ load factor; other symbols as before. Substituting, $A_c = 5.85 \times 10^{-3}(3,600 \times 8,760)(0.8)/1,000 = 147.4$ kg/year.

6. Compute the U-235 annual consumption

The U-235 is consumed by fissioning for power and is also lost by absorption. The proportion of these two forms of consumption is expressed by $\alpha =$ U-235 total capture cross section/U-235 fission cross section. With $\alpha = 1.2$ for a typical reactor, the total annual U-235 consumption $= 1.2(147.4) = 177$ kg/year.

7. Compute the maximum allowable atom burnup

Both U-235 and U-238 are regarded as reactor fuel. The allowable percent burnup depends on the total integrated radiation dosage and radiation energy level, and the effect on fuel material dimensional stability, thermal conductivity, and reduction in effective multiplication factor. Assuming a maximum allowable burnup of 20 percent, which is a typical value, compute $B_{ma} =$ (percent burnup)(fuel atoms per cu cm)(total cu cm of fuel), where $B_{ma} =$ maximum allowable atom burnup, atoms. Substituting, $B_{ma} = (0.002)(0.0472 \times 10^{24})(2.31 \times 10^6) = 2.18 \times 10^{26}$ atoms.

8. Compute the average fuel-cycle time

Use the relation $A_f = B_{ma}/F_r$, where $A_f =$ average fuel-cycle time, sec. Substituting, $A_f = 2.18 \times 10^{26}/[1.5 \times 10^{19}] = 1.45 \times 10^7$ sec $= 4,040$ hr $= 30$ weeks, approximately.

9. Compute the reactor neutron flux

Use the reaction $N_f = P_T C/\Sigma f V_f$, where $N_f =$ reactor neutron flux; $\Sigma f = N_{\text{U-235}} \times \sigma_{f\text{-235}}$, where $\sigma_{f\text{-235}} =$ total microscopic absorption cross section for U-235; other symbols as before. Substituting, $N_f = 500 \times 10^6 (3 \times 10^{10})/(0.00033 \times 10^{24})(549 \times 10^{-24})(2.31 \times 10^6) = 3.57 \times 10^{13}$. Note that values of $\sigma_{f\text{-235}}$ are obtained from nuclear data sources.

Related Calculations: Use this general method for any reactor designed to generate power. The method presented is the work of Henry C. Schwenk and Robert H. Shannon, as reported in *Power*.

VALUE OF FISSIONABLE MATERIAL FOR POWER GENERATION

How many tons of coal are required to produce the heat equivalent of 1 lb of fissionable U-235? If heat is worth 40 cents/million Btu, what is 1 g of fissionable U-235 worth? One ton of coal contains 24×10^6 Btu.

Calculation Procedure:

1. Compute the heat produced by 1 lb of fissionable material

When all the nuclei in the atoms of 1 lb of fissionable U-235 fission, about 0.001 lb of material converts to heat energy. Since by Einstein's mass-energy equation, 1 lb mass $= 11.3 \times 10^9$ kw-hr of energy, 1 lb of fissioning U-235 produces $0.001(3,413)(11.3 \times 10^9) = 39.5 \times 10^9$ Btu/lb. In this relation, the constant 3,413 converts kw to Btu.

2. Compute the heat equivalent of the fissionable material

Use the relation equivalent tons of coal per lb of U-235 = heat released per lb of U-235, Btu/heat released by 1 ton of coal, Btu, $= 39.5 \times 10^9/(24 \times 10^6) = 1,645$ tons of coal per lb of U-235. Thus, it takes 1,645 tons of coal to equal the potential heat produced by 1 lb of U-235 in a nuclear reactor.

3. Compute the monetary worth of the nuclear material

Since heat is worth 40 cents/million Btu in this plant, the value of 1 lb of U-235 is $(39.5 \times 10^9)(0.4/10^6) = \$15,800$, or about \$34.80 per g of U-235.

Related Calculations: Use this general procedure for other fissionable materials used for fuels in nuclear plants. The method presented is the work of Henry C. Schwenk and Robert H. Shannon, as reported in *Power*.

EFFECT OF NUCLEAR RADIATION ON HUMAN BEINGS

What is the total radiation dose in rems for a worker exposed to 0.3 rad of 1.0 Mev beta particles and 0.05 rad of 1.0 Mev neutrons each day? Is the total dose dangerous to this worker? Use National Bureau of Standards data (Tables 2 to 4) in the analysis.

Calculation Procedure:

1. Compute the total radiation dose

Use the relation, total dose, rem = Σ(dose, rad) (RBE), where rem = roentgen equivalent per man; rad = radiation absorbed dose; RBE = relative biological effectiveness. Table 2 lists the RBE values for various types of radiation. Substituting the appropriate values from Table 2, total dose = $(0.3)(1.0) + 0.05(10.5) = 0.825$ rem.

TABLE 2   Conversion: Rad to Rem*

Based on most detrimental chronic biological effects for continuous low dose exposures

---

Radiation Effects on Man, Definitions

One r (roentgen) is the quantity of gamma or x-radiation that produces an energy absorption of 83 ergs/g of dry air.

One rep (roentgen equivalent physical) is the quantity of radiation that produces an energy absorption of 93 ergs/g of aqueous tissue.

One rad (radiation absorbed dose) is required to deposit 100 ergs/g in any material by any kind of radiation.

One rem (roentgen equivalent man) is the unit of particulate radiation that produces tissue damage in man.

The conversion factor from rad to rem is the RBE (relative biological effectiveness), i.e., dose in rem = dose in rad $\times$ RBE.

Type of Radiation	RBE*	Type of Radiation	RBE*
X-rays	1	Neutrons, 0.5 Mev	10.2
Gamma rays	1	Neutrons, 1.0 Mev	10.5
Beta particles, 1.0 Mev	1	Neutrons, 10 Mev	6.4
Beta particles, 0.1 Mev	1.08	Protons, 100 Mev	1–2
Neutrons, thermal	2.8	Protons, 1 Mev	8.5
Neutrons, 0.0001 Mev	2.2	Protons, 0.1 Mev	10
Neutrons, 0.005 Mev	2.4	Alpha particles, 5 Mev	15
Neutrons, 0.02 Mev	5	Alpha particles, 1 Mev	20

*Example for total dose: For a given exposure time, a dose of 0.2 rad of γ radiation plus 0.04 rad of thermal neutrons gives a total dose of $(0.2 \times 1\text{RBE}) + (0.04 \times 2.8 \text{ RBE}) = 0.312$ rem.

2. Determine if the dose is dangerous

Table 3 lists the exposure tolerance of the human body. This listing shows that a dose of 1 rem/day is believed to cause debilitation within 3 to 6 months; death within 3 to 6 years. Since the daily dose to which this worker is exposed—0.825 rem—is close to the 1.0-rem danger level, the dose is excessive and is dangerous.

Table 4 lists the recommended weekly maximum dosage for various types of radiation on different parts of the body. Study of this list also indicates that the radiation to which this worker is exposed is dangerous.

Related Calculations: The effects of radiation can be fatal to all living organisms. Hence, extreme care must be used in computing the dose received by anyone exposed to radiation. Since the allowable dose and the effects of various doses are under constant study, be certain to refer to the latest available data from the United States Atomic Energy Commission before permitting exposure of any worker to radiation of any kind.

TABLE 3   Exposure Tolerance Values for Man
(Whole-body radiation doses)

0.001 rem/day	Natural-background radiation
0.01 rem/day	Permissible dose range, 1957
0.1 rem/day	Permissible dose range, 1930 to 1950
1 rem/day	Debilitation 3 to 6 months; death 3 to 6 years (projected from animal data)
10 rem/day	Debilitation 3 to 6 weeks; death 3 to 6 years (projected from animal data)
100 rem—1 day 150 rem—1 week 300 rem—1 month	Survivable emergency exposure dose but permitting no further exposure for life
25 rem	Single emergency exposure
100 rem	Twenty-year career allowance
500 rem	Maximum permissible 20-year-career allowance

TABLE 4   Maximum Weekly Dosage
(Rems/week)

Radiation	Skin		Lens of eye	Gonads	Blood-forming organs	Intermediate tissue (0.07–5.0 cm depth)
	Total body	Appendages				
X-rays or γ-rays < 3 Mev	0.45	1.5	0.45	0.3	0.4	0.4–0.45
Electrons or β	0.6	1.5	0.3	0.3	0.3	0.3–0.6
Protons	0.6	1.5	0.3	0.3	0.3	0.3–0.6
Fast neutrons	0.3–0.6	0.75–1.5	0.3	0.3	0.3	0.3–0.6
Thermal neutrons	0.5	1.2	0.3	0.1	0.17	0.17–0.5
Alpha particles	1.5	1.5	0.3	0.3	0.3	0.3–1.5
Heavy nuclei (O, N, C, locally generated)	1.5	1.5	0.3	0.3	...	0.3–1.6

ANALYSIS OF NUCLEAR POWER AND DESALTING PLANTS

Analyze the feasibility of building and operating nuclear-powered combined electric-generating and water-desalting plants. Sketch the different types of cycles that might be used. Determine the cycle to use for a water production of 100×10^6 gal/day, electric power net output of 500 mw and a desalting heat performance of 100.

Calculation Procedure:

1. *Draw the cycle diagrams*

Three cycles will be considered. These are the back-pressure, extraction, and multishaft cycles.

Figure 3 shows the back-pressure cycle in which the entire exhaust steam flow from the turbine is used to heat brine in the water-desalting system. For a given amount of water produced, this cycle generates large quantities of electric power.

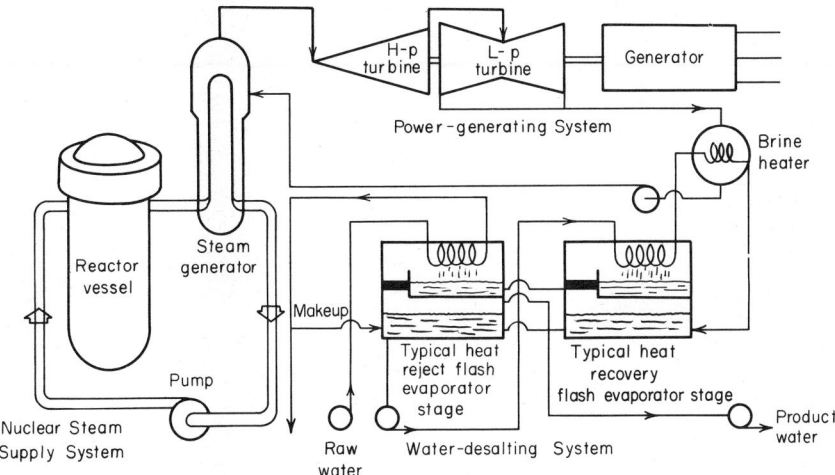

Fig. 3   Back-pressure cycle in which the entire exhaust steam from the turbines is used to heat brine in the water-desalting system.

In the extraction cycle, Fig. 4, the steam for brine heating is removed from the turbines at some midpoint during expansion. The exhaust steam goes to a standard condenser. This cycle can have a high product ratio (PR), that is, the ratio of the electric power to desalted water. If desired, large amounts of water can be produced when needed.

The multishaft cycle, Fig. 5, is fundamentally the same as Fig. 3, but it uses parallel condensing and noncondensing turbines. The electrical output can vary over a wide

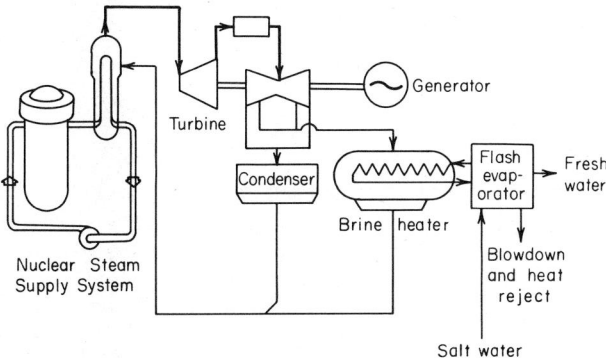

Fig. 4   Extraction cycle in which the steam for brine heating is removed from the turbines at some midpoint during expansion.

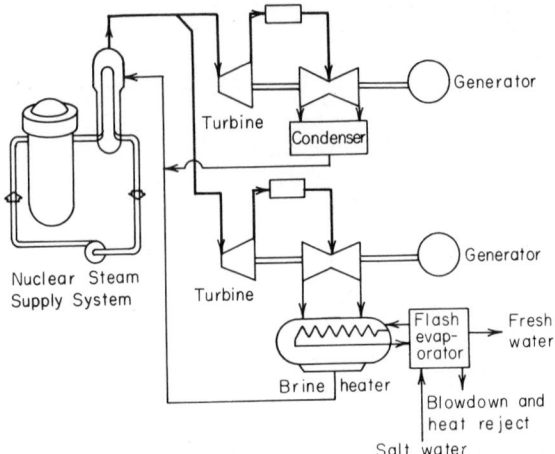

Fig. 5   Multishaft cycle is the same as the back-pressure cycle but it uses parallel condensing and noncondensing turbines.

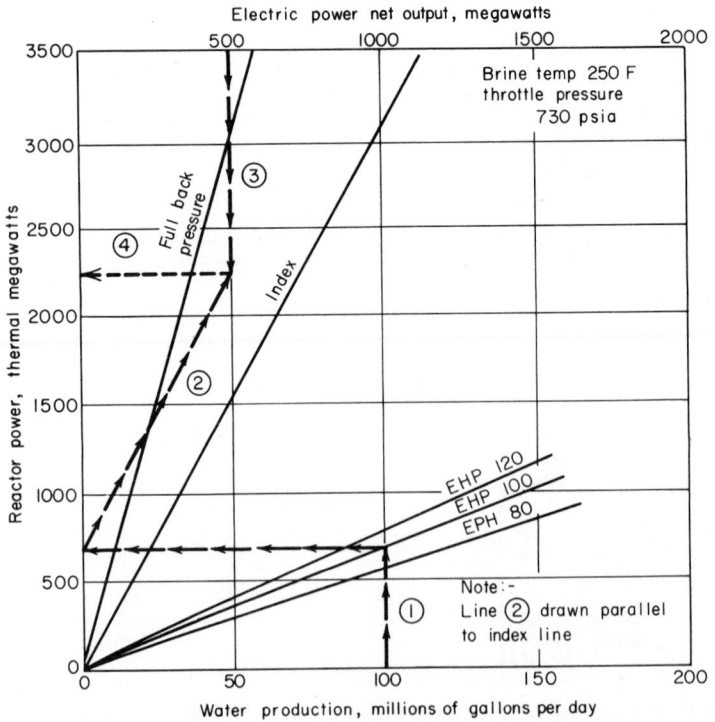

Fig. 6   Nomogram for plant ratings relates the four variables important in desalting when combined with power generation.

range without changing the water-desalting production. Although many other cycles are possible, all are variations of the three basic arrangements described above.

2. Choose the type of cycle and reactor size to use

Figure 6 allows quick *estimates* of the type of cycle and reactor size. Any of the four plotted quantities can be determined from Fig. 6 when the other three are known.

Enter Fig. 6 at the bottom at the water production rate of 100×10^6 gal/day and project vertically upward (1) to the desalting heat performance of 100. From the intersection with the appropriate curve, project horizontally to the left-hand scale of Fig. 6. Next project upward (2) parallel to the index scale. Then project vertically downward (3) from the net electric power output, 500 mw, on the top scale. From the intersection between lines 2 and 3, draw line 4 horizontally to the left-hand scale. At the intersection, read the reactor power as 2,250 thermal mw.

The *type of cycle* is determined by the location of the point of intersection between lines 2 and 3. If lines 2 and 3 intersect to the *right* of the full back-pressure line (FBP), the cycle used is the extraction or multishaft type. When the intersection falls directly on the FBP line, a back-pressure cycle is indicated. An intersection to the left of the FBP line indicates that some of the steam to the brine heater is bypassed around the turbine regardless of the cycle used. Since the intersection in Fig. 6 occurs to the right of the FBP line, either an extraction or multishaft type of cycle could be used. The final choice of a cycle would depend on the water output required.

Related Calculations: The data presented here were developed by W. H. Comtois, Westinghouse Electric Corp., and were reported in *Mechanical Engineering*. Studies made at Westinghouse show that:

1. The fixed-annual-charge rate exerts the greatest single influence on water cost, increasing the cost by about two-thirds for a factor of 2 increase in the rate. This effect is moderated somewhat for large plant sizes.

2. The plant load factor gives the expected result of decreasing product costs with increasing load factor. The effect is a 1 to 2 percent decrease (increase) for every percent increase (decrease) in load factor in the range from 75 to 95 percent.

3. Plant design life is of little consequence in the range normally considered (30 to 40 years).

4. The range of maximum brine temperatures studied was 200 to 250 F. Without exception, the computed optimum brine temperature was 250 F.

5. The single-shaft cycles (backpressure or extraction) enjoy a small (5 to 10 percent) water cost advantage over the multishaft cycle.

Section **11**

Sanitary Engineering

EDMUND B. BESSELIEVRE, P.E.
Consultant, Forrest & Cotton, Inc.

TYLER G. HICKS, P.E.
Consulting Engineer

MAX KURTZ, P.E.
Consulting Engineer

REFERENCES: Fair—*Sewage Treatment*, Wiley; Steel—*Water Supply and Sewerage*, McGraw-Hill; Gurnham—*Principles of Industrial Waste Treatment*, Wiley; Babbitt, Dolan, and Cleasby—*Water Supply Engineering*, McGraw-Hill; Wright—*Rural Water Supply and Sanitation*, Wiley; Ehlers and Steel—*Municipal and Rural Sanitation*, McGraw-Hill; Babbitt and Baumann—*Sewerage and Sewage Treatment*, Wiley; Chow—*Handbook of Applied Hydrology*, McGraw-Hill; American Society of Civil Engineers—*Design and Construction of Sanitary and Storm Sewers*; King and Brater—*Handbook of Hydraulics*, McGraw-Hill; Woods—*Highway Engineering Handbook*, McGraw-Hill; Hicks—*Pump Application Engineering*, McGraw-Hill; Fair, Geyer, and Okun—*Water Supply and Wastewater Engineering*, Wiley; Besselievre—*Industrial Waste Treatment*, McGraw-Hill; Federation of Sewage and Industrial Wastes Association—*Chlorination of Sewage and Industrial Wastes*; American Society of Civil Engineers—*Sewage Treatment Plant*

Design; Imhoff and Fair—*Sewage Treatment*, Wiley; American Society of Civil Engineers—*Filtering Materials for Sewage Treatment Plants*; Mahlie—*Manual for Sewage Plant Operators*, Texas Water & Sewage Works Association; Escritt—*Sewerage and Sewage Disposal*, Contractors Record Ltd. (London); Escritt—*Pumping Station Equipment and Design*, C. R. Books, Ltd.

WATER-SUPPLY SYSTEM FLOW-RATE AND PRESSURE-LOSS ANALYSIS

A water-supply system will serve a city of 100,000 population. Two water mains arranged in a parallel configuration, Fig. 1*a*, will supply this city. Determine the flow rate, size, and head loss of each pipe in this system. If the configuration in Fig. 1*a* were replaced by the single pipe shown in Fig. 1*b*, what would the total head loss be if $C = 100$ and the flow rate were reduced to 2,000 gpm? Explain how the Hardy Cross method is applied to the water-supply piping system in Fig. 3.

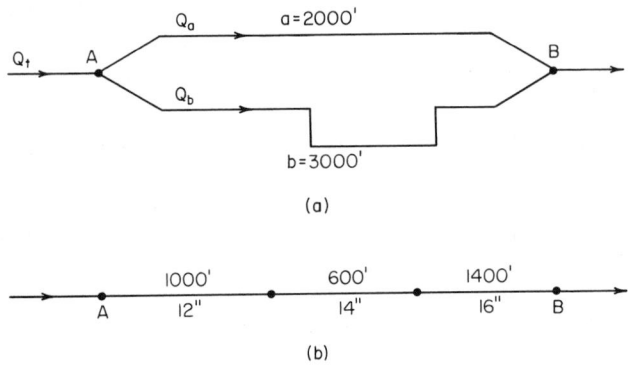

Fig. 1 (*a*) Parallel water distribution system; (*b*) Single-pipe distribution system.

Calculation Procedure:

1. Compute the domestic water flow rate in the system

Use an average annual domestic water consumption of 150 gal per capita per day (gcd). Hence, domestic water consumption = (150 gcd)(100,000 persons) = 15,000,000 gal/day. To this domestic flow, the flow required for fire protection must be added to determine the total flow required.

2. Compute the required flow rate for fire protection

Use the relation $Q_f = 1,020 \, (P)^{0.5}[1 - 0.01(P)^{0.5}]$, where Q_f = fire flow, gpm; P = population in thousands. Substituting, $Q_f = 1,020(100)^{0.5}[1 - 0.01(100)^{0.5}] = 9,180$ gpm; say 9,200 gpm.

3. Apply a load factor to the domestic consumption

To provide for unusual water demands, many design engineers apply a 200 to 250 percent load factor to the average hourly consumption that is determined from the average annual consumption. Thus, the average daily total consumption determined in Step 1 is based on an average annual daily demand. Convert the average daily total consumption in Step 1 to an average hourly consumption by dividing by 24 hr, or 15,000,000/24 = 625,000 gph. Next, apply a 200 percent load factor. Or design hourly demand = 2.00 (625,000) = 1,250,000 gph, or 1,250,000/60 min/hr = 20,850 gpm; say 20,900 gpm.

4. Compute the total water flow required

The total water flow required = domestic flow, gpm + fire flow, gpm = 20,900 + 9,200 = 30,100 gpm. If this system were required to supply water to one or more industrial plants in addition to the domestic and fire flows, the quantity needed by the industrial plants would be added to the total flow computed above.

5. Select the flow rate for each pipe

The flow rate is not known for either pipe, Fig. 1a. Assume that the shorter pipe a has a flow rate Q_a of 12,100 gpm, and the longer pipe b a flow rate Q_b of 18,000 gpm. Thus, $Q_a + Q_b = Q_t = 12,100 + 18,000 = 30,100$, where Q = flow, gpm, in the pipe identified by the subscript a or b; Q_t = total flow in the system, gpm.

6. Select the sizes of the pipes in the system

Since neither pipe size is yet known, some assumptions must be made about the system. First, assume that a friction-head loss of 10 ft of water per 1,000 ft of pipe is suitable for this system. This is a typical allowable friction-head loss for water-supply systems.

Second, assume that the pipe is sized using the Hazen-Williams equation with the coefficient $C = 100$. Most water-supply systems are designed using this equation and this value of C.

Enter Fig. 2 with the assumed friction-head loss of 10 ft per 1,000 ft of pipe on the right-hand scale and project through the assumed Hazen-Williams coefficient $C = 100$. Extend this straight line until it intersects the pivot axis. Next, enter Fig. 2 on the left-hand scale at the flow rate in pipe a, 12,100 gpm, and project to the previously found intersection on the pivot axis. At the intersection with the pipe-diameter scale, read the required pipe size as 27 in. diameter. Note that if the required pipe size falls between two plotted sizes, the next *larger* size is used.

Now in any parallel piping system, the friction-head loss through any branch connecting two common points equals the friction-head loss in any other branch connecting the same two points. Using Fig. 2 for a 27-in. pipe, find the actual friction-head loss as 8 ft per 1,000 ft of pipe. Hence, the total friction-head loss in pipe a is (2,000 ft long) (8 ft/1,000 ft) = 16 ft of water. This is also the friction-head loss in pipe b.

Since pipe b is 3,000 ft long, the friction-head loss per 1,000 ft is total head loss, ft/length of pipe, thousands of ft = 16/3 = 5.33 ft/1,000 ft. Enter Fig. 2 at this friction-head loss and $C = 100$. Project in the same manner as described for pipe a and find the required size of pipe b as 33 in.

If the district being supplied by either pipe required a specific flow rate, this flow would be used instead of assuming a flow rate. The pipe would then be sized in the same manner as described above.

7. Compute the single-pipe equivalent length

When dealing with several different sizes of pipe having the same flow rate, it is often convenient to convert each pipe to an *equivalent length* of a common-size pipe. Many design engineers use 8-in. pipe as the common size. Table 1 shows the equivalent length of 8-in. pipe for various other sizes of pipe with $C = 90$, 100, and 110 in the Hazen-Williams equation.

From Table 1, for 12-in. pipe, the equivalent length of 8-in. pipe is 0.14 ft/ft when $C = 100$. Thus, total equivalent length of 8-in. pipe = (1,000 ft of 12-in. pipe)(0.14 ft/ft) = 140 ft of 8-in. pipe. For the 14-in. pipe, total equivalent length = (600)(0.066) = 39.6 ft, using similar data from Table 1. For the 16-in. pipe, total equivalent length = (1,400)(0.034) = 47.6 ft. Hence, total equivalent length of 8-in. pipe = 140 + 39.6 + 47.6 = 227.2 ft.

8. Determine the friction-head loss in the pipe

Enter Fig. 2 at the flow rate of 2,000 gpm and project through 8 in. diameter to the pivot axis. From this intersection, project through $C = 100$ to read the friction-head loss as 100 ft/1,000 ft, due to the friction of the water in the pipe. Since the equivalent length of the pipe is 227.2 ft, the friction-head loss in the compound pipe is (227.2/1,000)(110) = 25 ft of water.

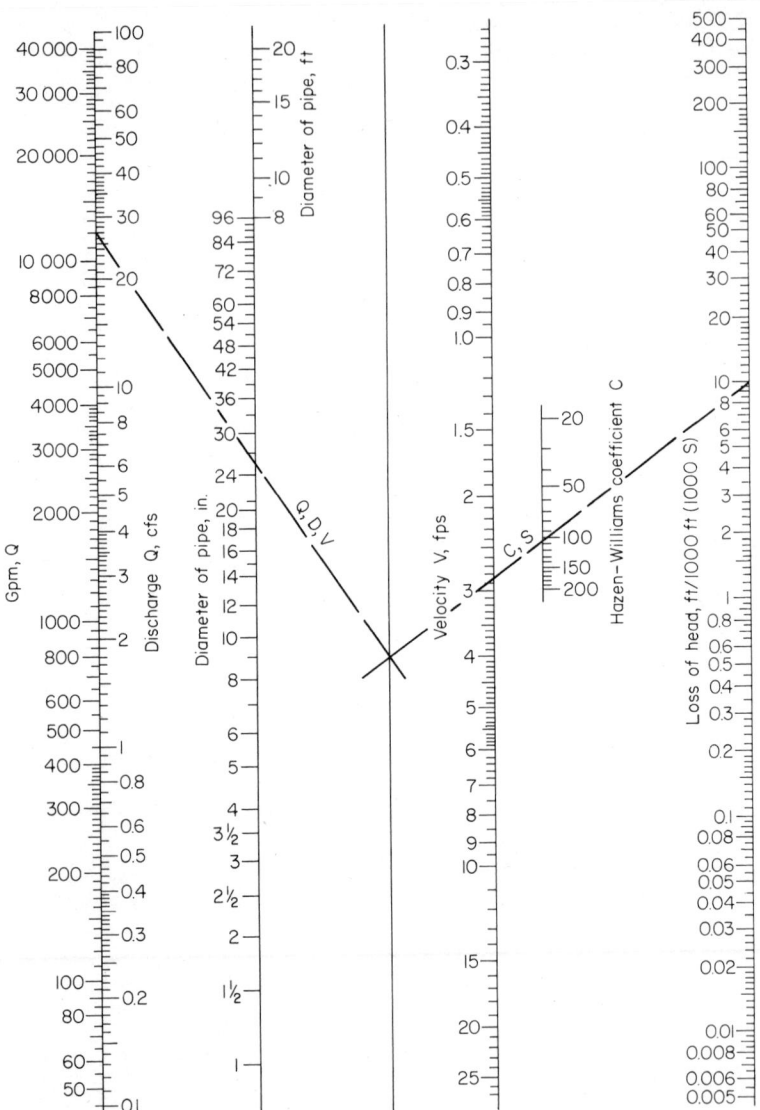

Fig. 2   Nomogram for solution of the Hazen-Williams equation for pipes flowing full.

Related Calculations: Two pipes, two piping systems, or a single pipe and a system of pipes are said to be *equivalent* when the losses of head due to friction for equal rates of flow in the pipes are equal.

To determine the flow rates and friction-head losses in complex waterworks distribution systems, the Hardy Cross method of network analysis is often used. This method[1] uses trial and error to obtain successively more accurate approximations of the flow rate through a piping system. To apply the Hardy Cross method: (1) Sketch the piping

[1]O'Rourke—*General Engineering Handbook*, McGraw-Hill.

system layout as in Fig. 3. (2) Assume a flow quantity, in terms of percent of total flow, for each part of the piping system. When assuming a flow quantity note that (*a*) the loss of head due to friction between any two points of a closed circuit must be the same by any path by which the water may flow, and (*b*) the rate of inflow into any section of the piping system must equal the outflow. (3) Compute the loss of head due to friction between two points in each part of the system, based on the assumed flow in (*a*) the

TABLE 1   Equivalent Length of 8-in. Pipe for C = 100

Pipe diameter, in.	C = 90	C = 100	C = 110
2	1,012	851	712
4	34	29	24.3
6	4.8	4.06	3.4
8	1.19	1.00	0.84
10	0.40	0.34	0.285
12	0.17	0.14	0.117
14	0.078	0.066	0.055
16	0.040	0.034	0.029
18	0.023	0.019	0.016
20	0.0137	0.0115	0.0096
24	0.0056	0.0047	0.0039
30	0.0019	0.0016	0.0013
36	0.00078	0.00066	0.00055

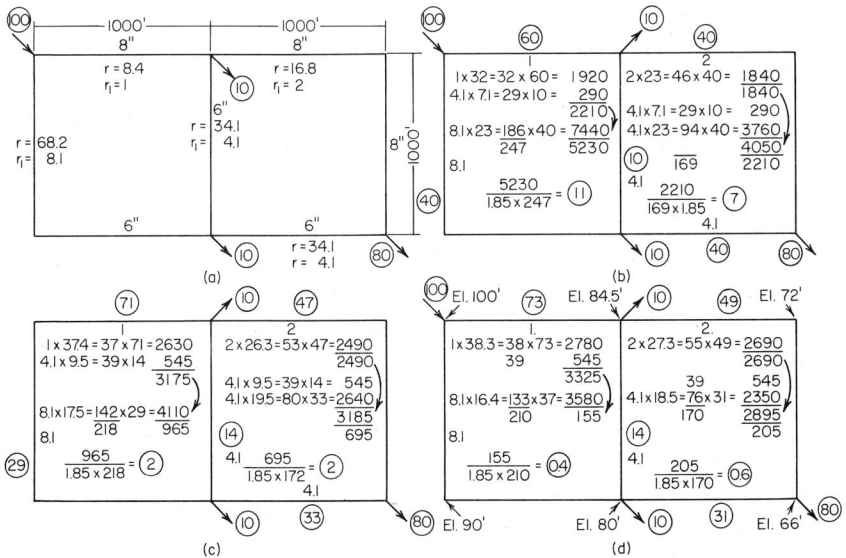

Fig. 3   Application of the Hardy Cross method to a water distribution system.

clockwise direction and (b) the counterclockwise direction. A difference in the calculated friction-head losses in the two directions indicates an error in the assumed direction of flow. (4) Compute a counterflow correction by dividing the difference in head, Δh ft, by $n(Q)^{n-1}$, where $n = 1.85$ and $Q =$ flow, gpm. Indicate the direction of this counterflow in the pipe by an arrow starting at the right side of the smaller value of h and curving toward the larger value, Fig. 3. (5) Add or subtract the counterflow to or from the assumed flow, depending on whether its direction is the same or opposite. (6) Repeat this process on each circuit in the system until a satisfactory balance of flow is obtained.

To compute the loss of head due to friction, Step 3 of the Hardy Cross method, use any standard formula, such as the Hazen-Williams, that can be reduced to the form $h = rQ^nL$, where $h =$ head loss due to friction, ft of water; $r =$ a coefficient depending on the diameter and roughness of the pipe; $Q =$ flow rate, gpm; $n = 1.85$; $L =$ length of pipe, ft. Table 2 gives values of r for 1,000-ft lengths of various sizes of pipe, and for different values of the Hazen-Williams' coefficient C. When the percentage of total flow is used for computing Σh, Fig. 3, the loss of head due to friction in ft between any two points for any flow in gpm is computed from $h = [\Sigma h$ (by percent flow)/100,000] $(gpm/100)^{0.85}$. Figure 3 shows the details of the solution using the Hardy Cross method. The circled numbers represent the flow quantities. Table 3 lists values of numbers between 0 and 100 to the 0.85 power.

TABLE 2   Values of r for 1,000 ft of Pipe Based on the Hazen–Williams Formula

(Head loss in ft $= r \times 10^{-5} \times Q^{1.85}$ per 1,000 ft, Q representing gpm)

d, in.	C = 90	C = 100	C = 110	C = 120	C = 130	C = 140
4	340	246	206	176	151	135
6	47.1	34.1	28.6	24.3	21.0	18.7
8	11.1	8.4	7.0	6.0	5.2	4.6
10	3.7	2.8	2.3	2.0	1.7	1.5
12	1.6	1.2	1.0	0.85	0.74	0.65
14	0.72	0.55	0.46	0.39	0.34	0.30
16	0.38	0.29	0.24	0.21	0.18	0.15
18	0.21	0.16	0.13	0.11	0.10	0.09
20	0.13	0.10	0.08	0.07	0.06	0.05
24	0.052	0.04	0.03	0.03	0.02	0.02
30	0.017	0.013	0.011	0.009	0.008	0.007

EXAMPLE: r for 12-in. pipe 4,000 ft long, with C = 100, is 1.2 × 4.0 = 4.8.

TABLE 3   Value of the 0.85 Power of Numbers

N	0	1	2	3	4	5	6	7	8	9
0	0	1.0	1.8	2.5	3.2	3.9	4.6	5.2	5.9	6.5
10	7.1	7.7	8.3	8.9	9.5	10.0	10.6	11.1	11.6	12.2
20	12.8	13.3	13.8	14.4	14.9	15.4	15.9	16.4	16.9	17.5
30	18.0	18.5	19.0	19.5	20.0	20.5	21.0	21.5	22.0	22.5
40	23.0	23.4	23.9	24.3	24.8	25.3	25.8	26.3	26.8	27.3
50	27.8	28.2	28.7	29.1	29.6	30.0	30.5	31.0	31.4	31.9
60	32.4	32.9	33.3	33.8	34.2	34.7	35.1	35.6	36.0	36.5
70	37.0	37.4	37.9	38.3	38.7	39.1	39.6	40.0	40.5	41.0
80	41.5	42.0	42.4	42.8	43.3	43.7	44.1	44.5	45.0	45.4
90	45.8	46.3	46.7	47.1	47.6	48.0	48.4	48.8	49.2	49.6

WATER-SUPPLY SYSTEM SELECTION

Choose the type of water-supply system for a city having a population of 100,000 persons. Indicate which type of system would be suitable for such a city today and 20 years hence. The city is located in an area of numerous lakes.

Calculation Procedure:

1. Compute the domestic water flow rate in the system

Use an average annual domestic water consumption of 150 gal per capita day (gcd). Hence, domestic water consumption = (150 gcd)(100,000 persons) = 15,000,000 gal/day. To this domestic flow, the flow required for fire protection must be added to determine the total flow required.

2. Compute the required flow rate for fire protection

Use the relation $Q_f = 1,020(P)^{0.5}[1 - 0.01(P)^{0.5}]$, where Q_f = fire flow, gpm; P = population in thousands. Substituting, $Q_f = 1,020(100)^{0.5}[1 - 0.01 \times (100)^{0.5}] = 9,180$ gpm; say 9,200 gpm.

3. Apply a load factor to the domestic consumption

To provide for unusual water demands, many design engineers apply a 200 to 250 percent load factor to the average hourly consumption that is determined from the average annual consumption. Thus, the average daily total consumption determined in Step 1 is based on an average annual daily demand. Convert the average daily total consumption in Step 1 to an average hourly consumption by dividing by 24 hr, or 15,000,000/24 = 625,000 gph. Next, apply a 200 percent load factor. Or design hourly demand = 2.00(625,000) = 1,250,000 gph, or 1,250,000/60 min/hr = 20,850 gpm; say 20,900 gpm.

4. Compute the total water flow required

The total water flow required = domestic flow, gpm + fire flow, gpm = 20,900 + 9,200 = 30,100 gpm. If this system were required to supply water to one or more industrial plants in addition to the domestic and fire flows, the quantity needed by the industrial plants would be added to the total flow computed above.

5. Study the water supplies available

Table 4 lists the principal sources of domestic water supplies. Wells that are fed by groundwater are popular in areas having sandy or porous soils. To determine if a well is suitable for supplying water in sufficient quantity, its specific capacity — i.e., the yield in gpm per ft of drawdown — must be determined.

Wells for municipal water sources may be dug, driven, or drilled. Dug wells seldom exceed 60 ft deep. Each such well should be protected from surface-water leakage by being lined with impervious concrete to a depth of 15 ft.

Driven wells seldom are more than 40 ft deep or more than 2 in. in diameter when

TABLE 4 Typical Municipal Water Sources

Source	Collection method	Remarks
Groundwater	Wells (artesian, ordinary, galleries)	30 to 40 percent of an area's rainfall becomes groundwater
Surface freshwater (lakes, rivers, streams, impounding reservoirs)	Pumping or gravity flow from submerged intakes, tower intakes, or surface intakes	Surface supplies are important in many areas
Surface saltwater.	Desalting	Wide-scale application under study at present

used for small water supplies. Bigger driven wells are constructed by driving large-diameter casings into the ground.

Drilled wells can be several thousand ft deep, if required. The yield of a driven well is usually greater than any other type of well because the well can be sunk to a depth where sufficient ground water is available. Almost all wells require a pump of some kind to lift the water from its subsurface location and discharge it to the water-supply system.

Surface freshwater can be collected from lakes, rivers, streams, or reservoirs by submerged-, tower-, or crib-type intakes. The intake leads to one or more pumps that discharge the water to the distribution system or intermediate pumping stations. Locate intakes as far below the water surface as possible. Where an intake is placed less that 20 ft below the surface of the water, it may become clogged by sand, mud, or ice.

Choose the source of water for this system after studying the local area to determine the most economical source today and 20 years hence. With a rapidly expanding population, the future water demand may dictate the type of water source chosen. Since this city is in an area of many lakes, a surface supply would probably be most economical, if the water table is not falling rapidly.

6. Select the type of pipe to use

Four types of pipes are popular for municipal water-supply systems: cast-iron, asbestos-cement, steel, and concrete. Wood-stave pipe was once popular but it is now obsolete. Some communities also use copper or lead pipes. However, the use of both types is extremely small when compared with the other types. The same is true of plastic pipe, although this type is slowly gaining some acceptance.

In general, cast-iron pipe proves dependable and long-lasting in water-supply systems that are not subject to galvanic or acidic soil conditions.

Steel pipe is generally used for long, large-diameter lines. Thus, the typical steel pipe used in water-supply systems is 36 or 48 in. in diameter. Use steel pipe for river crossings, on bridges, and for similar installations where light weight and high strength are required. Steel pipe may last 50 years or more under favorable soil conditions. Where unfavorable soil conditions exist, the life of steel pipe may be about 20 years.

Concrete-pipe use is generally confined to large, long lines, such as aqueducts. Concrete pipe is suitable for conveying relatively pure water through neutral soil. However, corrosion may occur when the soil contains an alkali or an acid.

Asbestos-cement pipe has a number of important advantages over other types of pipes. However, it does not flex readily, it can be easily punctured, and it may corrode in acidic soils.

Select the pipe to use after a study of the local soil conditions, length of runs required, and the quantity of water that must be conveyed. Usual water velocities in municipal water systems are in the 5-fps range. However, the velocities in aqueducts range from 10 to 20 fps. Earthen canals have much lower velocities—1 to 3 fps. Rock- and concrete-lined canals have velocities of 8 to 15 fps.

In cold northern areas, keep in mind the occasional need to thaw frozen pipes during the winter. Nonmetallic pipes—concrete, plastic, etc., as well as nonconducting metals—cannot be thawed by electrical means. Since electrical thawing is probably the most practical method of thawing available today, pipes that prevent its use may put the water system at a disadvantage if subfreezing temperatures are common in the area served.

7. Select the method for pressurizing the water system

Water-supply systems can be pressurized in three different ways: (*a*) by gravity or natural elevation head, (*b*) by pumps that produce a pressure head, and (*c*) by a combination of the first two ways.

Gravity systems are suitable where the water storage reservoir or receiver is high enough above the distribution system to produce the needed pressure at the farthest outlet. The operating cost of a gravity system is lower than a pumped system, but the first cost of the former is usually higher. However, the reliability of the gravity system is usually higher because there are fewer parts that may fail.

Pumping systems generally use centrifugal pumps that discharge either directly to the water main or to an elevated tank, a reservoir, or a standpipe. The water then flows from the storage chamber to the distribution system. In general, most sanitary engineers prefer to use a reservoir or storage tank between the pumps and distribution mains because this arrangement provides greater reliability and fewer pressure surges.

Surface reservoirs should store at least a 1-day water supply. Most surface reservoirs are designed to store a supply for 30 days or longer. Elevated tanks should have a capacity of at least 25 gal of water per person served, *plus* a reserve for fire protection. The capacity of typical elevated tanks ranges from a low of 40,000 gal for a 20-ft-diameter tank to a high of 2,000,000 gal for an 80-ft-diameter tank.

Choose the type of distribution system after studying the typography, water demand, and area served. In general, a pumped system is preferred today. To ensure continuity of service, duplicate pumps are generally used.

8. Choose the system operating pressure

In domestic water supply the minimum pressure required at the highest fixture in a building is usually assumed to be 15 psi. The maximum pressure allowed at a fixture in a domestic water system is usually 65 psi. High-rise buildings — i.e., those above six stories — are generally required to furnish the pressure increase needed to supply water to the upper stories. A pump and overhead storage tank are usually installed in such buildings to provide the needed pressure.

Commercial and industrial buildings require a minimum water pressure of 75 psi at the street level for fire-hydrant service. This hydrant should deliver at least 250 gpm of water for fire-fighting purposes.

Most water-supply systems served by centrifugal pumps in a central pumping station operate in the 100-psi-pressure range. In areas of one- and two-story structures, a lower pressure — say 65 psi — is permissible. Where the pressure in a system falls to too low a level, auxiliary or booster pumps may be used. These pumps increase the pressure in the main to the desired level.

Choose the system pressure based on the terrain served, quantity of water required, allowable pressure loss, and size of pipe used in the system. Usual pressures required will be in the ranges cited above, although small systems serving one-story residences may operate at pressures as low as 30 psi. Pressures over 100 psi are seldom used because heavier piping is required. As a rule, distribution pressures of 50 to 75 psi are acceptable.

9. Determine the number of hydrants for fire protection

Table 5 shows the required fire flow, number of standard hose streams of 250 gpm discharged through a $1\frac{1}{8}$-in.-diameter smooth nozzle, and the average area served by a hydrant in a high-value district. A standard hydrant may have two or three outlets.

Table 5 indicates that a city of 100,000 persons requires 36 standard hose streams. This means that 36 single-outlet or 18 dual-outlet hydrants are required. More, of course, could be used if better protection were desired in the area. Note that the required fire flow listed in Table 5 agrees closely with that computed in Step 2 above.

Related Calculations: Use this general method for any water-supply system — municipal or industrial. Note, however, that the required fire-protection quantities vary from one type of municipal area to another and among different industrial exposures. Refer to *NFPA Handbook of Fire Protection*, available from NFPA, 60 Batterymarch Street, Boston, Massachusetts, 02110, for specific fire-protection requirements for a variety of industries. In choosing a water-supply system, the wise designer looks ahead for at least 10 years when the water demand will usually exceed the present demand. Hence, the system may be designed so it is oversized for the present population but just adequate for the future population. Table 6 lists the usual water requirements for a variety of industries.

To determine the storage capacity required at the present time, proceed as follows: (1) Compute the flow needed to meet 50 percent of the present domestic daily (that is, 24-hr) demand. (2) Compute the 4-hr fire demand. (3) Find the sum of (1) and (2).

For this city, procedure (1) = (20,900 gpm)(60 min/hr)(24 hr/day)(0.5) = 15,048,000 gal, using the data computed in Step 3, above. Also procedure (2) = (4 hr)(60 min/hr) (9,200 gpm) = 2,208,000 gal, using the data computed in Step 2, above. Then, total storage capacity required = 15,048,000 + 2,208,000 = 17,256,000 gal. Where one or more reliable wells will produce a significant flow for 4 hr or longer, the storage capacity can be reduced by the 4-hr productive capacity of the wells.

TABLE 5   Required Fire Flow and Hydrant Spacing*

Population	Required fire flow, gpm	Number of standard hose streams	Average area served per hydrant, sq ft†	
			Direct streams	Engine streams
22,000	4,500	18	55,000	90,000
28,000	5,000	20	40,000	85,000
40,000	6,000	24	40,000	80,000
60,000	7,000	28	40,000	70,000
80,000	8,000	32	40,000	60,000
100,000	9,000	36	40,000	55,000
125,000	10,000	40	40,000	48,000
150,000	11,000	44	40,000	43,000
200,000	12,000	48	40,000	40,000

*National Board of Fire Underwriters.
†High-value districts.

TABLE 6   Industrial Water and Steam Requirements*

	Water	Steam
Acetic acid from carbide	....	7,300 lb per ton HAc
Acetic acid from pyroligneous acid	100,000 gal per ton HAc	15,700 lb per ton HAc
Acetic acid from pyroligneous liquor	240 M gal per ton HAc	64,000 to 74,000 lb per ton HAc
Acetic acid, direct (Othmer process)	....	54,200 lb per ton HAc
Alcohol, industrial..............	120 gal per gal 100-proof alcohol	50 lb per gal 190-proof alcohol
	52 gal per gal 190-proof alcohol	
	100 gal per gal alcohol	
	20,000 gal per ton grain	
	600,000 gal per 1,000 bu grain mashed	
Alumina (Bayer process)	6,300 gal per ton $Al_2O_3 \cdot 3H_2O$	15,000 lb per ton $Al_2O_3 \cdot 3H_2O$
Ammonia, synthetic..............	31,000 gal per ton liquid NH_3	
Ammoniated superphosphate.....	27 to 30 gal per ton ammoniated super phosphate	
Ammonium sulfate	200,000 gal per ton salt	
Buna S........................	173,000,000 gal/day for 100,000 tons Buna S per year	
Butadiene	320,000 gal per ton butadiene	
Calcium metaphosphate.........	4,000 gal per ton $Ca(PO_3)_2$	
Carbon dioxide.................	23,000 gal per ton CO_2	
	20,000 gal per ton solid CO_2 from 18% flue gas	20,000 lb per ton solid CO_2 from 18% flue gas
Casein (grain-curd process)	....	2,400 lb per ton casein
Caustic soda (lime-soda process)..	18,000 lb per ton NaOH in 11% solution	2,700 lb per ton NaOH in 11% solution
	21,000 gal per ton NaOH in 11% solution	

*Courtesy of American Society for Testing and Materials.

TABLE 6 Industrial Water and Steam Requirements (Continued)

	Water	Steam
Caustic soda (electrolytic)........	....	20,000 lb per ton 76% NaOH
Cellulose nitrate	50 gal per lb cellulose nitrate	
	10,000 gal per ton cellulose nitrate	
Charcoal and wood chemicals	65,000 gal per ton crude $CaAc_2$	64,000 lb per ton crude $CaAc_2$
Cottonseed oil..................	20 gal per gal oil	15 lb per gal oil
	0.6 gal per gal hardened oil	0.5 lb per gal hardened oil
Coumarin (synthetic)	....	3,000 lb per ton coumarin or 0.75 ton salicylaldehyde
Cuprammonium rayon...........	90,000 to 160,000 gal per ton 11% moisture rayon	
Fatty acid refining, continuous....	....	1,390 lb per ton stock charged
Gelatin......................		400 lb per ton gelatin
Glycerine.....................	1,100 gal per ton glycerine	8,000 lb per ton glycerine
Gunpowder...................	200,000 gal per ton gunpowder or explosives	
Hydrochloric acid (slat process)...	2,900 gal per ton 20 Bé HCl	
Hydrochloric acid (synthetic process......................	500 to 1,000 gal per ton 20 Bé HCl	
Hydrogen.....................	660,000 gal per ton H_2	
Lactose (milk sugar)	200,000 to 220,000 gal per ton lactose	80,000 lb per ton lactose
Magnesium carbonate, basic......	4,320 gal per ton basic $MgCO_3$	18,000 lb per ton basic $MgCO_3$
	39,000 gal per ton $MgCO_3$	
Magnesium hydroxide from sea water and dolomite	Seawater 58,000 gal and freshwater 500 gal per ton $Mg(OH)_2$	800 lb per ton $Mg(OH)_2$
Oxygen, liquid	2,000 gal per 1000 cu ft O_2	
Phenol, synthetic...............	....	4,000 lb per ton phenol
Phosphoric acid (blast furnace)	75,000 gal per ton 100% H_3PO_4	
Phosphoric acid (Dorr strong-acid process)	7,500 gal per ton 35% P_2O_5 acid	780 lb per ton 35% P_2O_5 acid
Potassium chloride from Sylvinite.....................	40,000 to 50,000 gal per ton KCl	2,500 lb per ton KCl
Soap, laundry	230 gal per ton soap	4,000 lb per ton soap
	500 gal per ton soap	
Soda ash (ammonia-soda process).........................	15,000 to 18,000 gal per ton 58% soda ash	
Sodium bichromate.............	....	6,000 lb per ton sodium bichromate
Sodium chlorate................	60,000 gal per ton sodium chlorate	11,000 lb per ton sodium chlorate
Sodium silicate.................	160 gal per ton 40 Bé water glass	1040 lb per ton 40 Bé water glass
Sodium sulfate, natural	....	3,650 lb per ton anhydrous Na_2SO_4 (95 + percent)
Stearic acid and red oil..........	....	
Sulfur dioxide, liquid	18,000 gal per ton liquid SO_2	18,000 lb per ton stearic acid
Sulfuric acid (chamber process)...	2,500 gal per ton 100% H_2SO_4	6,800 lb per ton liquid SO_2
Sulfuric acid (contact process)	4,000 gal per ton 100% H_2SO_4	
	5,000 gal per ton H_2SO_4	
Trisodium orthophosphate	....	150 lb per ton $Na_3PO_4 \cdot 12H_2O$
Vanillin (synthetic)	....	30,800 lb per ton vanillin
Viscose rayon	180,000 to 200,000 gal per ton viscose yarn	140,000 lb per ton viscose yarn

Food Industry

	Water	Steam
Bread	500 to 1,000 gal per ton bread	600 to 1,000 lb per ton bread
Brewing:		
Beer.........................	470 gal per bbl beer	
Whiskey.....................	80 gal per gal whiskey	

TABLE 6   Industrial Water and Steam Requirements (Continued)

	Water	Steam
Canning:		
Apricots	8,000 gal per 100 cases No. 2 cans	
Asparagus	7,000 gal per 100 cases No. 2 cans	
Beans:		
Green	3,500 gal per 100 cases No. 2 cans	
Lima	25,000 gal per 100 cases No. 2 cans	
Pork and beans	3,500 gal per 100 cases No. 2 cans	
Beets.	2,500 gal per 100 cases No. 2 cans	
Corn	2,500 gal per 100 cases No. 2 cans	
Cream or whole	4,000 gal per 100 cases No. 2 cans	
Peas.	3,000 gal per 100 cases No. 2 cans	
Sauerkraut.	300 gal per 100 cases No. 2 cans	
Spinach	16,000 gal per 100 cases No. 2 cans	
Succotash.	12,500 gal per 100 cases No. 2 cans	
Tomatoes:		
Products.	7,000 gal per 100 cases No. 2 cans	
Whole	750 gal per 100 cases No. 2 cans	
Corn refining	333 gal per ton corn	
Edible gelatin.	13,200 to 20,000 gal per ton gelatin	
Edible oil.	22 gal per gal oil	
Meat packing:		
Packing house.	55,000 gal per 100 hog units	
Poultry	4,400 gal per ton live weight	
Slaughter house	16,000 gal per 100 hog units	
Stockyards.	160 gal per acre	
Milk and milk products:		
Butter	5,000 gal per ton butter	
Cheese	4,000 gal per ton cheese	
Dairies	3 gal per qt milk	
Receiving and bottling	450 gal per 100 gal milk	
Creamery.	220 gal per ton raw	
Restaurants	0.5 to 4.0 gal per meal	
Sugar:		
Beet.	2,160 gal per ton refined sugar	
	20,000 to 25,000 gal per ton sugar	
	2,600 to 3,200 gal per ton beets	
Refined cane	1,000 gal per ton sugar	
	Condensing 4,800 to 8,400 gal per ton	3,500 lb per ton refined sugar
	Pure water 1,400 gal per ton refined	
	sugar	
Vegetable dehydration:		
Beets.	37,400 gal per ton product	
Cabbage.	15,000 gal per ton product	
Carrots	31,600 gal per ton product	
Potatoes	11,200 to 25,000 gal per ton product	
Rutabagas	30,400 gal per ton product	
Sweet potatoes	18,000 gal per ton product	

	Textile Industry	
Cotton:		
Bleaching.	25 to 38 gal per yd	
Dyeing	1,000 to 2,000 gal per 100 lb goods	
Finishing.	10 to 15 gal per yd	
Processing.	3,800 gal per 100 lb goods	
Knit goods, bleaching.	16,000 gal per ton goods	
Linen	200,000 gal per ton goods	
Rayon:		
Cuprammonium yarn	160,000 gal per ton yarn	
Dissolving pump	190,000 gal per ton pulp	
Viscose yarn	200,000 gal per ton yarn	

TABLE 6   Industrial Water and Steam Requirements (Continued)

	Water	Steam
Silk, hosiery dyeing..............	6,000 to 8,000 gal per ton goods	
Wool:		
Scouring.....................	2,000 to 15,000 gal per 100 lb raw wool	
Scouring and bleaching........	40,000 gal per ton goods	

Miscellaneous Industries

	Water	Steam
Air conditioning................	6,000 to 15,000 gal per person per season	
Aluminum.....................	1,920,000 gal per ton aluminum	
Buildings, office................	27 to 45 gal per day per capita	
Cement, portland...............	750 gal per ton cement	
Cement rock, beneficiation.......	720 gal per ton raw rock	
Coal:		
By-product coke..............	1,430 to 2,860 gal per ton coke	570 to 860 lb per ton coke
Carbonizing	3,500 gal per ton coal carbonized	
Washing.....................	125 gal per ton coal	
Electricity.....................	80 gal per kw electricity	
	120,000 gal per ton coal burned	
Hospitals	135 to 350 gal per day per bed	
Hotels.........................	300 to 525 gal per day per guest room	
Laundries:		
Commercial	8,600 to 11,400 gal per ton "work"	
Institutional	6,000 gal per ton "work"	
Leather tannery................	375 gal per ton vegetable tan	
	600 gal per ton chrome tan	
	6,000 to 16,000 gal per ton leather	
	16,000 gal per ton hides	
Petroleum:		
Airplane engine (to test)........	125,000 gal per airplane engine	
Gasoline.....................	7 to 10 gal per gal gasoline	
Gasoline, aviation	25 gal per gal aviation gasoline	
Gasoline, natural	20 gal per gal gasoline and 2,000 cu ft stripped gas at 150-lb pressure	6 lb per gal gasoline and 2,000 cu ft stripped gas at 150-lb pressure
Gasoline, polymerization	34 gal per gal polymer gasoline	2.7 lb per gal polymer gasoline
Oil, Fischer-Tropsch synthesis		
Oil fields.....................	18,000 gal per 100 bbl crude oil	
Oil refinery	77,000 gal per 100 bbl crude oil	
Pulp and paper mills:...........	50,000 to 150,000 gal per ton pulp	
De-inking paper...............	38,000 gal per ton paper	
Paperboard...................	14,000 gal per ton paper board	
Soda pulp....................	....	13,000 lb per ton dried soda pulp
Strawboard...................	26,000 gal per ton strawboard	
Sulfate pulp (Kraft)	....	10,000 lb per ton dried sulfate pulp
Sulfate pulp bleaching........	60,224 gal to bleach 1 ton dry pulp of 80 to 85 G.E. brightness	3120 lb to bleach 1 ton dry pulp of 80 to 85 G.E. brightness
Sulfate pulp...................	....	5,000 to 7,000 lb per ton dried pulp
Rock wool	4,000 to 5,000 gal per ton rock wool	3,000 lb per ton rock wool
Rubber (auto tire)	....	120 lb per auto tire
Steel plant....................	20,000 to 35,000 gal per ton steel	
Fabricated steel...............	42,000 gal per ton steel	
Ingot steel	18,000 gal per ton steel	
Pig iron......................	4,000 gal per ton pig iron	
Sulfur mining..................	3,000 gal per ton sulfur	

SELECTION OF TREATMENT METHOD FOR A WATER-SUPPLY SYSTEM

Choose a treatment method for a water-supply system for a city having a population of 100,000 persons. The water must be filtered, disinfected, and softened to make it suitable for domestic use.

Calculation Procedure:

1. Compute the domestic water flow rate in the system

When water is treated for domestic consumption, only the drinking water passes through the filtration plant. Fire-protection water is seldom treated unless it is so turbid that it will clog fire pumps or hoses. Assuming that the fire-protection water is acceptable for use without treatment, only the drinking water will be considered here.

Use the same method as in Steps 1 and 3 of the previous Calculation Procedure to determine the required domestic water flow of 20,900 gpm for this city.

2. Select the type of water-treatment system to use

Water supplies are treated by a number of methods including sedimentation, coagulation, filtration, softening, and disinfection. Other treatments include disinfection, taste and odor control, and miscellaneous methods.

Since the water must be filtered, disinfected, and softened, each of these steps must be considered separately.

3. Choose the type of filtration to use

Slow sand filters operate at an average rate of 3 million gal/(acre)(day). This type of filter removes about 99 percent of the bacterial content of the water and most tastes and odors.

Rapid sand filters operate at an average rate of 150 million gal/(acre)(day). But the raw water must be treated before entering the rapid sand filter. This preliminary treatment often includes chemical coagulation and sedimentation. A high percentage of bacterial content—up to 99.98 percent—is removed by the preliminary treatment and the filtration. But color and turbidity removed is not as dependable as with slow sand filters. Table 7 lists the typical limits for certain impurities in water supplies.

TABLE 7   Typical Limits for Impurities in Water Supplies

Impurity	Limit, ppm	Impurity	Limit, ppm
Turbidity	10	Iron plus manganese	0.3
Color	20	Magnesium	125
Lead	0.1	Total solids	500
Fluoride	1.0	Total hardness	100
Copper	3.0	Ca + Mg salts	

The daily water flow rate for this city is, from Step 1, (20,900 gpm)(24 hr/day)(60 min/hr) = 30,096,000 gal/day. If a slow sand filter were used, the required area would be (30.096 million gal/day/3 million gal/(acre)(day) = 10 + acres.

A rapid sand filter would require 30.096/150 = 0.2 acre. Hence, if space were scarce in this city—and it usually is—a rapid sand filter would be used. With this choice of filtration, chemical coagulation and sedimentation are almost a necessity. Hence, these two additional steps would be included in the treatment process.

Table 8 gives pertinent data on both slow and rapid sand filters. These data are useful in filter selection.

4. Select the softening process to use

The principal water-softening processes use: (*a*) lime and sodium carbonate followed by sedimentation or filtration, or both, to remove the precipitates; (*b*) zeolites of the sodium type in a pressure filter. Zeolite softening is popular and is widely used in municipal water-supply systems today. Based on its proven usefulness and economy, zeolite softening will be chosen for this installation.

TABLE 8   Typical Sand-Filter Characteristics

Slow sand filters
Usual filtration rate............... 2.5 to 6.0×10^6 gal/(acre)(day)
Sand depth .... 30 to 36 in.
Sand size........................ 35 mm
Sand uniformity coefficient........ 1.75
Water depth .... 3 to 5 ft
Water velocity in underdrains..... 2 fps
Cleaning frequency required ..... 2 to 11 times per year
Units required................... At least two to permit alternate cleaning

Fast sand filters
Usual filtration rate............... 100 to 200×10^6 gal/(acre)(day)
Sand depth .... 30 in.
Gravel depth .... 18 in.
Sand size........................ 0.4 to 0.5 mm
Sand uniformity coefficient........ 1.7, or less
Units required................... At least three to permit cleaning one unit while the other two are operating

5.   Select the disinfection method to use

Chlorination by the addition of chlorine to the water is the principal method of disinfection used today. To reduce the unpleasant effects that may result from using chlorine alone, a mixture of chlorine and ammonia, known as chloramine, may be used. The ammonia dosage is generally 0.25 ppm or less. Assume that the chloramine method is chosen for this installation.

6.   Select the method of taste and odor control

The methods used for taste and odor control are: (*a*) aeration, (*b*) activated carbon, (*c*) prechlorination, and (*d*) chloramine. Aeration is popular for groundwaters containing hydrogen sulfide and odors caused by microscopic organisms.

Activated carbon absorbs impurities that cause tastes, odors, or color. Generally, 10 to 20 lb of activated carbon per million gal of water are used, but larger quantities — from 50 to 60 lb—may be specified. In recent years, some 2,000 municipal water systems have installed activated carbon devices for taste and odor control.

Prechlorination and chloramine are also used in some installations for taste and odor control. Of the two methods, chloramine appears more popular at present.

Based on the data given for this water-supply system, method *b*, *c*, or *d* would probably be suitable. Because method *b* has proven highly effective, it will be chosen tentatively, pending later investigation of the economic factors.

Related Calculations: Use this general procedure to choose the treatment method for all types of water-supply systems where the water will be used for human consumption. Thus, the procedure is suitable for municipal, commercial, and industrial systems.

STORM-WATER RUNOFF RATE AND RAINFALL INTENSITY

What is the storm-water runoff rate from a 40-acre industrial site having an imperviousness of 50 percent if the time of concentration is 15 min? What would be the effect of planting a lawn over 75 percent of the site?

Calculation Procedure:

1. Compute the hourly rate of rainfall

Two common relations, called the Talbot formulas, used to compute the hourly rate of rainfall R in./hr are $R = 360/(t + 30)$ for the heaviest storms, and $R = 105/(t + 15)$ for ordinary storms, where t = time of concentration, min. Using the equation for the heaviest storms because this relation gives a larger flow rate and produces a more conservative design, $R = 360/(15 + 30) = 8$ in./hr.

2. Compute the storm-water runoff rate

Apply the *rational method* to compute the runoff rate. This method uses the relation $Q = AIR$, where Q = storm-water runoff rate, cu ft/sec; A = area served by sewer, acres; I = coefficient of runoff or percent of imperviousness of the area; other symbols as before. Substituting, $Q = (40)(0.50)(8) = 160$ cfs.

3. Compute the effect of changed imperviousness

Planting a lawn on a large part of the site will increase the imperviousness of the soil. This means that less rainwater will reach the sewer because the coefficient of imperviousness of a lawn is lower. Table 9 lists typical coefficients of imperviousness for various surfaces. This tabulation shows that the coefficient for lawns varies from 0.05 to 0.25. Using a value of $I = 0.10$ for the $40(0.75) = 30$ acres of lawn, $Q = (30)(0.10)(8) = 24$ cfs.

TABLE 9   Coefficient of Runoff for Various Surfaces

Surface	Coefficient
Parks, gardens, lawns, meadows	0.05–0.25
Gravel roads and walks..................	0.15–0.30
Macadamized roadways	0.25–0.60
Inferior block pavements with uncemented	0.40–0.50
joints	0.40–0.50
Stone, brick, and wood-block pavements	
with tightly cemented joints	0.75–0.85
Same with uncemented joints.............	0.50–0.70
Asphaltic pavements in good condition	0.85–0.90
Watertight roof surfaces	0.70–0.95

The runoff for the remaining 10 acres is, as in Step 2, $Q = (10)(0.5)(8) = 40$ cfs. Hence, the total runoff is $24 + 40 = 64$ cfs. This is $160 - 64 = 96$ cfs less than when the lawn was not used.

Related Calculations: The time of concentration for any area being drained by a sewer is the time required for the maximum runoff rate to develop. It is also defined as the time for a drop of water to drain from the farthest point of the watershed to the sewer.

When rainfall continues for an extended period, T min, the coefficient of imperviousness changes. For impervious surfaces such as watertight roofs, $I = T/(8 + T)$. For improved pervious surfaces, $I = 0.3T/(20 + T)$. These relations can be used to compute the coefficient in areas of heavy rainfall.

Equations for R for various areas of the United States are available in Steel—*Water Supply and Sewerage*, McGraw-Hill. The Talbot formulas, however, are widely used and have proven reliable.

The time of concentration for a given area can be approximated from $t = I(L/Si^2)^{1/3}$, where L = distance of overland flow of the rainfall from the most remote part of the site, ft; S = slope of the land, ft/ft; i = rainfall intensity, in./hr; other symbols as before. For portions of the flow carried in ditches, the time of flow to the inlet can be computed using the Manning formula.

Table 10 lists the coefficient of runoff for specific types of built-up and industrial areas. Use these coefficients in the same way as shown above. Tables 9 and 10 present data developed by Kuichling and ASCE.

TABLE 10   Coefficient of Runoff for Various Areas

Area	Coefficient
Business:	
Downtown	0.70–0.95
Neighborhood...........	0.50–0.70
Residential:	
Single-family	0.30–0.50
Multiunits, detached	0.40–0.60
Multiunits, attached......	0.60–0.75
Residential (suburban)	0.25–0.40
Apartment dwelling........	0.50–0.70
Industrial:	
Light industry	0.50–0.80
Heavy industry	0.60–0.90
Playgrounds..............	0.20–0.35
Railroad yards	0.20–0.40
Unimproved..............	0.10–0.30

SIZING SEWER PIPES FOR VARIOUS FLOW RATES

Determine the size, flow rate, and depth of flow from a 1,000-ft-long sewer which slopes 5 ft between inlet and outlet and which must carry a flow of 5 million gal/day. The sewer will flow about half full. Will this sewer provide the desired flow rate?

Calculation Procedure:

1. Compute the flow rate in the half-full sewer

A flow of 1 million gal/day = 1.55 cfs. Hence, a flow of 5 million gal/day = 5(1.55) = 7.75 cfs in a *half-full* sewer.

2. Compute the full-sewer flow rate

In a *full sewer*, the flow rate is twice that in a half-full sewer, or 2(7.75) = 15.50 cfs for this sewer. This is equivalent to 15.50/1.55 = 10 million gal/day. Full-sewer flow rates are used because pipes are sized on the basis of being full of liquid.

3. Compute the sewer-pipe slope

The pipe slope S ft/ft $= (E_i - E_o)/L$, where E_i = inlet elevation, ft above the site datum; E_o = outlet elevation, ft above site datum; L = pipe length between inlet and outlet, ft. Substituting, $S = 5/1,000 = 0.005$ ft/ft.

4. Determine the pipe size to use

The Manning formula $v = (1.486/n)R^{2/3}S^{1/2}$ is often used for sizing sewer pipes. In this formula, v = flow velocity, fps; n = a factor that is a function of the pipe roughness; R = pipe hydraulic radius = 0.25 pipe diameter, ft; S = pipe slope, ft/ft. Table 11 lists values of n for various types of sewer pipe. In sewer design, the value $n = 0.013$ for pipes flowing full.

Since the Manning formula is complex, numerous charts have been designed to simplify its solution. Figure 4 is one such typical chart designed specifically for sewers.

Enter Fig. 4 at 15.5 cfs on the left and project through the slope ratio of 0.005. On the central scale between the flow rate and slope scales read the *next larger* standard sewer-pipe diameter as 24 in. When using this chart, always read the next larger pipe size.

5. Determine the fluid flow velocity

Continue the solution line of Step 4 to read the fluid flow velocity as 5 fps on the extreme right-hand scale of Fig. 4. This is for a sewer flowing *full*.

TABLE 11   Values of *n* for the Manning Formula

Type of Surface of Pipe	n
Ditches and rivers, rough bottoms with much vegetation	0.040
Ditches and rivers in good condition with some stones and weeds	0.030
Smooth earth or firm gravel	0.020
Rough brick; tuberculated iron pipe	0.017
Vitrified tile and concrete pipe poorly jointed and unevenly settled; average brickwork	0.015
Good concrete; riveted steel pipe; well-laid vitrified tile or brickwork	0.013*
Cast-iron pipe of ordinary roughness; unplaned timber	0.012
Smoothest pipes; neat cement	0.010
Well-planed timber evenly laid	0.009

*Probably the most frequently used value.

6. Compute the half-full flow depth

Determine the full-flow capacity of this 24-in. sewer by entering Fig. 4 at the slope ratio, 0.005, and projecting through the pipe diameter, 24 in. At the left read the full-flow capacity as 16 cfs.

The required half-flow capacity is 7.75 cfs, from Step 1. Determine the ratio of the required half-flow capacity to the full-flow capacity, both expressed in cfs. Or 7.75/16.0 = 0.484.

Enter Fig. 5 on the bottom at 0.484 and project vertically upward to the discharge curve. From the intersection, project horizontally to the left to read the depth-of-flow ratio as 0.49. This means that the depth of liquid in the sewer at a flow of 7.75 cfs is 0.49(24 in.) = 11.75 in. Hence, the sewer will be just slightly less than half full when handling the designed flow quantity.

7. Compute the half-full flow velocity

Project horizontally to the right along the previously found 0.49 depth-of-flow ratio until the velocity curve is intersected. From this intersection, project vertically downward to the bottom scale to read the ratio of hydraulic elements as 0.99. Hence, the fluid velocity when flowing half-full is 0.99(5.0 fps) = 4.95 fps.

Related Calculations: The minimum flow velocity required in sanitary sewers is 2 fps. At 2 fps, solids will not settle out of the fluid. Since the velocity in this sewer is 4.95 fps, as computed in Step 7, the sewer meets, and exceeds, the minimum required flow velocity.

Certain localities have minimum slope requirements for sanitary sewers. The required slope produces a minimum flow velocity of 2 fps with an *n* value of 0.013.

Storm sewers handling rainwater and other surface drainage require a higher flow velocity than sanitary sewers because sand and grit often enter a storm sewer. The usual minimum allowable velocity for a storm sewer is 2.5 fps; where possible, the sewer should be designed for 3.0 fps. If the sewer designed above were used for storm service, it would be acceptable because the fluid velocity is 4.95 fps. To prevent excessive wear of the sewer, the fluid velocity should not exceed 8 fps.

Note that Figs. 4 and 5 can be used whenever two variables are known. When a sewer flows at 0.8, or more, full, the partial-flow diagram, Fig. 5, may not give accurate results, especially at high flow velocities.

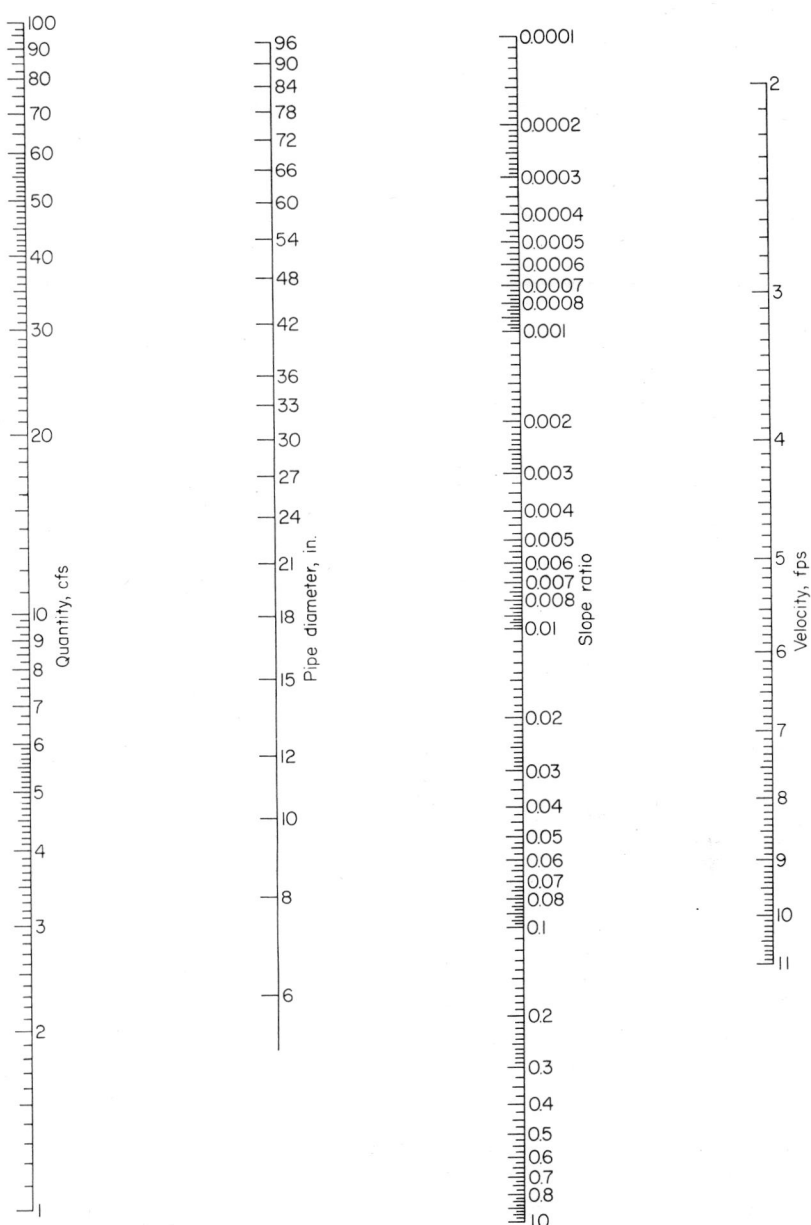

Fig. 4   Nomogram for solving the Manning formula for circular pipes flowing full, and *n* = 0.013.

SEWER-PIPE EARTH LOAD AND BEDDING REQUIREMENTS

A 36-in.-diameter clay sewer pipe is placed in a 15-ft-deep trench in damp sand. What is the earth load on this sewer pipe? What bedding should be used for the pipe? If a 5-ft-wide drainage trench weighing 2,000 lb per ft of length crosses the sewer pipe at right angles to the pipe, what load is transmitted to the pipe? The bottom of the flume is 11 ft above the top of the sewer pipe.

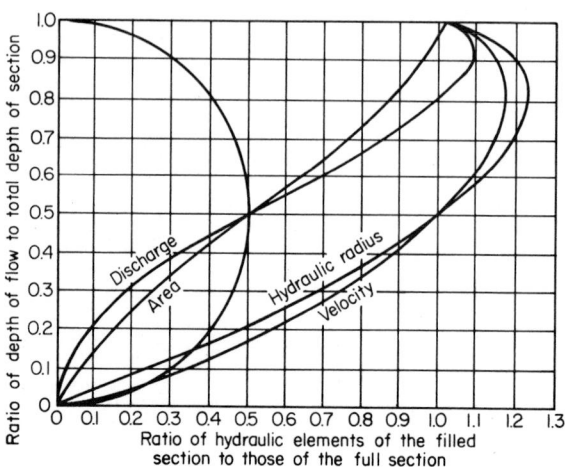

Fig. 5   Hydraulic elements of a circular pipe.

Calculation Procedure:

1. Compute the width of the pipe trench

Compute the trench width from $w = 1.5d + 12$, where $w =$ trench width, in.; $d =$ sewer-pipe diameter, in. Substituting, $w = 1.5(36) + 12 = 66$ in., or 5 ft 6 in.

2. Compute the trench depth-to-width ratio

To determine this ratio, subtract the pipe diameter from the depth in ft and divide the result by the trench width in ft. Or, $(15 - 3)/5.5 = 2.18$.

3. Compute the load on the pipe

Use the relation $L = kWw^2$, where $L =$ pipe load, lb per lin ft of trench; $k =$ a constant from Table 12; $W =$ weight of the fill material used in the trench, lb/cu ft; other symbol as before.

Enter Table 12 at the depth-to-width ratio of 2.18. Since this particular value is not tabulated, use the next higher value, 2.5. Opposite this, read $k = 1.70$ for a sand filling.

Enter Table 13 at damp sand and read the weight as 115 lb/cu ft. With these data the pipe load relation can be solved.

Substituting in $L = kWw^2$, $L = 1.70(115)(5.5)^2 = 5,920$ lb/ft. Study of the properties of clay pipe, Table 14, shows that 36-in. extra-strength clay pipe has a minimum average crushing strength of 6,000 lb by the three-edge-bearing method.

4. Apply the loading safety factor

ASTM recommends a factor of safety of 1.5 for clay sewers. To apply this factor of safety, divide it into the tabulated three-edge-bearing strength found in Step 3. Or, $6,000/1.5 = 4,000$ lb.

5. Compute the pipe load-to-strength ratio

Use the strength found in Step 4. Or pipe load-to-strength ratio, also called the *load factor*, $= 5,920/4,000 = 1.48$.

TABLE 12   Values of k for Use in the Pipe Load Equation*

Ratio of trench depth to width	Sand and damp topsoil	Saturated topsoil	Damp clay	Saturated clay
0.5	0.46	0.46	0.47	0.47
1.0	0.85	0.86	0.88	0.90
1.5	1.18	1.21	1.24	1.28
2.0	1.46	1.50	1.56	1.62
2.5	1.70	1.76	1.84	1.92
3.0	1.90	1.98	2.08	2.20
3.5	2.08	2.17	2.30	2.44
4.0	2.22	2.33	2.49	2.66
4.5	2.34	2.47	2.65	2.87
5.0	2.45	2.59	2.80	3.03
5.5	2.54	2.69	2.93	3.19
6.0	2.61	2.78	3.04	3.33
6.5	2.68	2.86	3.14	3.46
7.0	2.73	2.93	3.22	3.57
7.5	2.78	2.98	3.30	3.67

*Iowa State Univ. Eng. Exp. Sta. Bull. 47.

TABLE 13   Weight of Pipe-Trench Fill

Fill	lb/cu ft
Dry sand	100
Damp sand	115
Wet sand	120
Damp clay.............	120
Saturated clay..........	130
Saturated topsoil	115
Sand and damp topsoil...	100

TABLE 14   Clay Pipe Strength

Pipe size, in.	Minimum average strength, lb/lin ft	
	Three-edge-bearing	Sand-bearing
4	1,000	1,500
6	1,100	1,650
8	1,300	1,950
10	1,400	2,100
12	1,500	2,250
15	1,750	2,625
18	2,000	3,000
21	2,200	3,300
24	2,400	3,600
27	2,750	4,125
30	3,200	4,800
33	3,500	5,250
36	3,900	5,850

6. Select the bedding method for the pipe

Figure 6 shows methods for bedding sewer pipe and the strength developed. Thus, earth embedment, type-2 bedding, develops a load factor of 1.5. Since the computed load factor, Step 5, is 1.48, this type of bedding is acceptable. (When choosing a type of bedding be certain that the load factor of the actual pipe is less than, or equals, the developed load factor for the three-edge-bearing strength.)

The type-2 earth embedment, Fig. 6, is a highly satisfactory method, except that the shaping of the lower part of the trench to fit the pipe may be expensive. Type-3 granular embedment, may be less expensive, particularly if the crushed stone, gravel, or shell is placed by machine.

7. Compute the direct load transmitted to the sewer pipe

The weight of the drainage flume is carried by the soil over the sewer pipes. Hence, a portion of this weight may reach the sewer pipe. To determine how much of the flume weight reaches the pipe, find the weight of the flume per ft of width, or 2,000 lb/5 ft = 400 lb per ft of width.

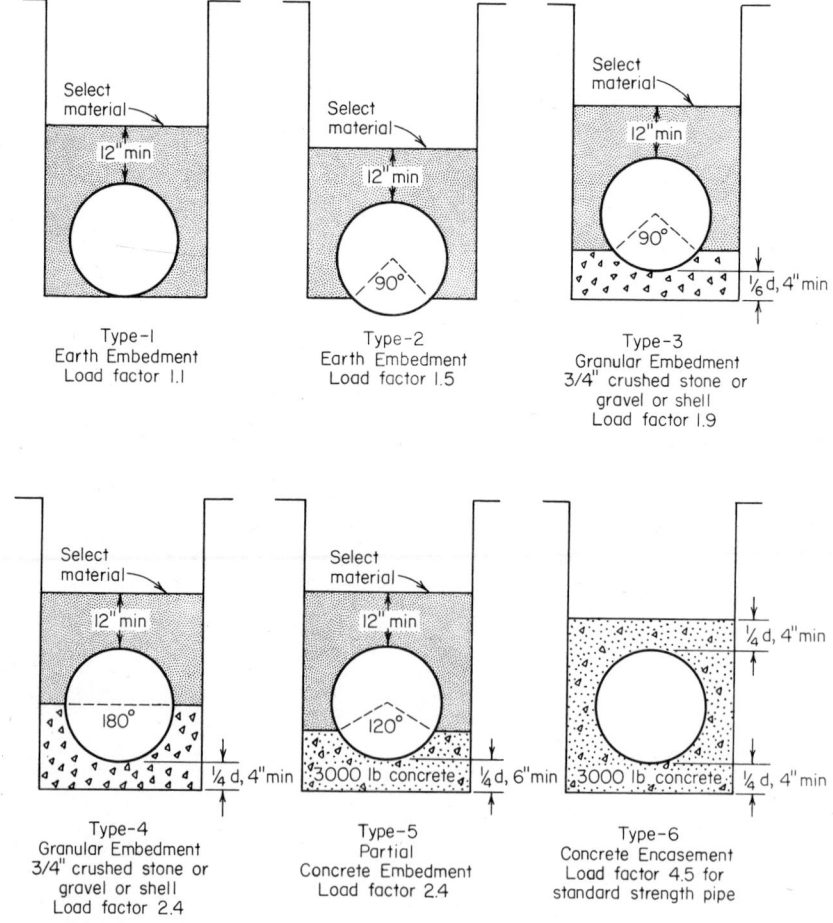

Fig. 6 Strengths developed for various methods of bedding sewer pipes. (*W. S. Dickey Clay Manufacturing Co.*)

Since the pipe trench is 5.5 ft wide, Step 1, the 1-ft-wide section of the flume imposes a total load of 5.5(400) = 2,200 lb on the soil beneath it.

To determine what portion of the flume load reaches the sewer pipe, compute the ratio of the depth of the flume bottom to the width of the sewer-pipe trench, or 11/5.5 = 2.0.

Enter Table 15 at a value of 2.0 and read the load proportion for sand and damp topsoil as 0.35. Hence, the load of the flume reaching each ft of sewer pipe in 0.35(2,200) = 770 lb.

TABLE 15   Proportion of Short Loads Reaching Pipe in Trenches

Depth-to-width ratio	Sand and damp topsoil	Saturated topsoil	Damp clay	Saturated clay
0.0	1.00	1.00	1.00	1.00
0.5	0.77	0.78	0.79	0.81
1.0	0.59	0.61	0.63	0.66
1.5	0.46	0.48	0.51	0.54
2.0	0.35	0.38	0.40	0.44
2.5	0.27	0.29	0.32	0.35
3.0	0.21	0.23	0.25	0.29
4.0	0.12	0.14	0.16	0.19
5.0	0.07	0.09	0.10	0.13
6.0	0.04	0.05	0.06	0.08
8.0	0.02	0.02	0.03	0.04
10.0	0.01	0.01	0.01	0.02

Related Calculations: A load such as that in Step 7 is termed a *short load*—i.e., it is shorter than the pipe-trench width. Typical short loads result from auto and truck traffic, road rollers, building foundations, etc. *Long loads* are imposed by weights that are longer than the trench is wide. Typical long loads are stacks of lumber, steel, and poles, and piles of sand, coal, gravel, etc. Table 16 shows the proportion of long loads transmitted to buried pipes. Use the same procedure as in Step 7 to compute the load reaching the buried pipe.

When a sewer pipe is placed on undisturbed ground and covered with fill, compute the load on the pipe from $L = kWd^2$, where $d =$ pipe diameter, ft; other symbols as in Step 3. Tables 15 and 16 are the work of Prof. Anson Marston, Iowa State University.

TABLE 16   Proportion of Long Loads Reaching Pipe in Trenches

Depth-to-width ratio	Sand and damp topsoil	Saturated topsoil	Damp yellow clay	Saturated yellow clay
0.0	1.00	1.00	1.00	1.00
0.5	0.85	0.86	0.88	0.89
1.0	0.72	0.75	0.77	0.80
1.5	0.61	0.64	0.67	0.72
2.0	0.52	0.55	0.59	0.64
2.5	0.44	0.48	0.52	0.57
3.0	0.37	0.41	0.45	0.51
4.0	0.27	0.31	0.35	0.41
5.0	0.19	0.23	0.27	0.33
6.0	0.14	0.17	0.20	0.26
8.0	0.07	0.09	0.12	0.17
10.00	0.04	0.05	0.07	0.11

To find the total load on trenched or surface-level buried pipes subjected to both fill and long or short loads, add the proportion of the long or short load reaching the pipe to the load produced by the fill.

Note that sewers may have several cross-sectional shapes—circular, egg, rectangular, square, etc. The circular sewer is the most common shape because it has a number of advantages, including economy. Egg-shaped sewers are not as popular as circular and are less often used today because of their higher costs.

Rectangular and square sewers are often used for storm service. However, their hydraulic characteristics are not as desirable as circular sewers.

SANITARY SEWER SYSTEM DESIGN

What size main sanitary sewer is required for a midwestern city 30-acre residential area containing six-story apartment houses if the hydraulic gradient is 0.0035 and the pipe roughness factor $n = 0.013$? One-third of the area is served by a branch sewer. What should the size of this sewer be? If the branch and main sewers must also handle groundwater infiltration, determine the required sewer size. The sewer is below the normal groundwater level.

Calculation Procedure:

1. Compute the sanitary sewage flow rate

Table 17 shows the typical population per acre for various residential areas and the flow rate used in sewer design. Using the typical population of 500 persons per acre given in Table 17, the total population of the area served is (30 acres)(500 persons per acre) = 15,000 persons.

TABLE 17   Sanitary Sewer Design Factors

Population data	
Type of Area	Typical Population, persons per acre
Light residential	15
Closely built residential..................	55
Single-family residential.................	100
Six-story apartment district...............	500

Sewage-flow data	
City	Sewer Design Basis, gcd°
Berkeley, California......................	92
Cranston, Rhode island	167
Des Moines, Iowa	200
Las Vegas, Nevada.......................	250
Little Rock, Arkansas	100
Shreveport, Louisiana	150

Typical sewer design practice	
Sewer Type	Design Flow, gcd
Laterals and sub-mains	400
Main, trunk, and outfall	250
New sewers......................	Never < 100

°Gallons per capita per day

Since this is a midwestern city, the sewer design basis used for Des Moines, Iowa, 200 gcd, Table 17, appears to be an appropriate value. Checking with the minimum flow recommended in Table 17, 100 gcd, the value of 200 gcd seems to be well justified. Hence, the sanitary sewage flow rate that the main sewer must handle is (15,000 persons)(200 gcd) = 3,000,000 gal/day.

2. Convert the flow rate to cfs

Use the relation $cfs = 1.55 \, (gpd/10^6)$, where cfs = flow rate, cfs; gpd = flow rate, gal/24 hr. Substituting, $cfs = 1.55 \, (3,000,000/1,000,000) = 4.65$ cfs.

3. Compute the required size of the main sewer

Size the main sewer on the basis of its flowing full. This is the usual design procedure followed by experienced sanitary engineers.

Two methods can be used to size the sewer pipe. (a) Use the chart in Fig. 4 for the Manning formula, entering with the flow rate of 4.65 cfs and projecting to the slope ratio or hydraulic gradient of 0.0035. Read the required pipe diameter as 18 in.

(b) Use the Manning formula and the appropriate *conveyance factor* from Table 18. When the conveyance factor C_f is used, the Manning formula becomes $Q = C_f S^{1/2}$, where Q = flow rate through the pipe, cfs; C_f = conveyance factor corresponding to a specific n value listed in Table 18; S = pipe slope or hydraulic gradient, ft/ft. Since Q and S are known, substitute and solve for C_f, or $C_f = Q/S^{1/2} = 4.65/(0.0035)^{1/2} = 78.5$. Enter Table 18 at $n = 0.013$ and $C_f = 78.5$ and project to the exact or next higher value of C_f. Table 18 shows that $C_f = 64.70$ for 15-in. pipe and 105.1 for 18-in. pipe. Since the actual value of C_f is 78.5, a 15-in. pipe would be too small. Hence, an 18-in. pipe would be used. This size agrees with that found in procedure a.

TABLE 18   Manning Formula Conveyance Factor

Pipe diameter, in.	Pipe cross-sectional area, sq ft	0.011	n 0.013	0.015	0.017
6	0.196	6.62	5.60	4.85	4.28
8	0.349	14.32	12.12	10.50	9.27
10	0.545	25.80	21.83	18.92	16.70
12	0.785	42.15	35.66	30.91	27.27
15	1.227	76.46	64.70	56.07	49.48
18	1.767	124.2	105.1	91.04	80.33
21	2.405	187.1	158.3	137.2	121.1
24	3.142	267.4	226.2	196.1	173.0
27	3.976	365.8	309.6	268.3	236.7
30	4.909	484.7	410.1	355.5	313.6
36	7.069	788	667	578	510
42	9.621	1,189	1,006	872	770
48	12.566	1,698	1,436	1,245	1,098
54	15.904	2,325	1,967	1,705	1,504
60	19.635	3,077	2,604	2,256	1,991

4. Compute the size of the lateral sewer

The lateral sewer serves one-third of the total area. Since the total sanitary flow from the entire area is 4.65 cfs, the flow from one-third of the area, assuming an even distribution of population and the same pipe slope, is 4.65/3 = 1.55 cfs. Using either procedure in Step 3, the required pipe size = 12 in. Hence, three 12-in. laterals will discharge into the main sewer, assuming that each lateral serves an equal area and has the same slope.

5. Check the suitability of the main sewer size

Compute the value of $d^{2.5}$ for each of the lateral sewer pipes discharging into the main sewer pipe. Thus, for one 12-in. lateral line, where d = smaller pipe diameter, in., $d^{2.5} = 12^{2.5} = 496$. For three pipes of equal diameter, $3d^{2.5} = 1{,}488 = D^{2.5}$, where D = larger pipe diameter, $3d^{2.5} = 1{,}488 = D^{2.5}$, where D = larger pipe diameter, in. Solving, $D^{2.5} = 1{,}488$ and $D = 17.5$ in. Hence, the 18-in. sewer main has sufficient capacity to handle the discharge of three 12 in. sewers. Note that Fig. 4 shows that the flow velocity in both the lateral and main sewers exceeds the minimum required velocity of 2 fps.

6. Compute the sewer size with infiltration

Infiltration is the groundwater that enters a sewer. The quantity and rate of infiltration depends upon the character of the soil in which the sewer is laid, the relative position of the groundwater level and the sewer, the diameter and length of the sewer, and the material and care with which the sewer is constructed. With tile and other jointed sewers, infiltration depends largely upon the type of joint used in the pipes. In large concrete or brick sewers, the infiltration depends on the type of waterproofing applied.

Infiltration is usually expressed in gal per day per mile of sewer. With very careful construction, infiltration can be kept down to 5,000 gal per day per mile of pipe even when the ground water level is above the pipe. With poor construction, porous soil, and high ground-water level, infiltration may amount to 100,000 gal(day)(mile) or more. Sewers laid in dense soil where the groundwater level is below the sewer do not experience infiltration except during and immediately after a rainfall. Even then, the infiltration will be in small amounts.

Assuming an infiltration rate of 20,000 gal per day per mile of sewer and a sewer length of 1.2 miles for this city, the daily infiltration is $1.2\,(20{,}000) = 24{,}000$ gal.

Checking the pipe size by either method in Step 3 shows that both the 12-in. laterals and the 18-in. main are of sufficient size to handle both the sanitary and infiltration flow.

Related Calculations: Where a sewer must also handle the runoff from fire-fighting apparatus, compute the quantity of fire-fighting water for cities of less than 200,000 population from $Q = 1{,}020\,(P)^{0.5}[1 - 0.01(P)^{0.5}]$, where Q = fire demand, gpm; P = city population in thousands. Add the fire demand to the sanitary sewage and infiltration flows to determine the maximum quantity of liquid the sewer must handle. For cities having a population of more than 200,000 persons, consult the fire department headquarters to determine the water flow quantities anticipated.

Some sanitary engineers apply a demand factor to the average daily water requirements per capita before computing the flow rate into the sewer. Thus, the maximum monthly water consumption is generally about 125 percent of the average annual demand but may range up to 200 percent of the average annual demand. Maximum daily demands of 150 percent of the average annual demand and maximum hourly demands of 200 to 250 percent of the annual average demand are commonly used for design by some sanitary engineers. To apply a demand factor, simply multiply the flow rate computed in Step 2 by the appropriate factor. Current practice in the use of demand factors varies; sewers designed without demand factors are generally adequate. Applying a demand factor simply provides a margin of safety in the design and the sewer is likely to give service for a longer period before becoming overloaded.

Most local laws and many sewer authorities recommend that no sewer be less than 8 in. in diameter. The sewer should be sloped sufficiently to give a flow velocity of 2 fps or more when flowing full. This velocity prevents the deposit of solids in the pipe. Manholes serving sewers should not be more than 400 ft apart.

Where industrial sewage is discharged into a sanitary sewer, the industrial flow quantity must be added to the domestic sewage flow quantity before choosing the pipe size. Swimming pools may also be drained into sanitary sewers and may cause temporary overflowing because the sewer capacity is inadequate. The sanitary sewage flow rate from an industrial area may be less than from a residential area of the same size because the industrial population is smaller.

Many localities and cities restrict the quantity of commercial and industrial sewage that may be discharged into public sewers. Thus, one city restricts commercial sewage from stores, garages, beauty salons, etc., to 135 gcd. Another city restricts industrial sewage from factories and plants to 50,000 gal(day)(acre). In other cities each proposed installation must be studied separately. Still other cities prohibit any discharge of commercial or industrial sewage into sanitary sewers. For these reasons, the local authorities and sanitary codes, if any, must be consulted before beginning the design of any sewer.

Before starting a sewer design, do the following. (*a*) Prepare a profile diagram of the area that will be served by the sewer. Indicate on the diagram the elevation above grade of each profile. (*b*) Compile data on the soil, groundwater level, type of paving, number and type of foundations, underground services (gas, electric, sewage, water-supply, etc.), and other characteristics of the area that will be served by the sewer. (*c*) Sketch the main sewer and lateral sewers on the profile diagram. Indicate the proposed direction of sewage flow by arrows. With these steps finished, start the sewer design.

To design the sewers, proceed as follows. (*a*) Size the sewers using the procedure given in Steps 1 through 6 above. (*b*) Check the sewage flow rate to see that it is 2 fps, or more. (*c*) Check the plot to see that the required slope for the pipes can be obtained without expensive blasting or rock removal.

Where the outlet of a building plumbing system is below the level of the sewer serving the building, a pump must be used to deliver the sewage to the sewer. Compute the pump capacity using the discharge from the various plumbing fixtures in the building as the source of the liquid flow to the pump. The head on the pump is the difference between the level of the sewage in the pump intake and the centerline of the sewer into which the pump discharges, plus any friction losses in the piping.

STORM-SEWER INLET SIZE AND FLOW RATE

What size storm-sewer inlet is required to handle a flow of 2 cfs if the gutter is sloped $\frac{1}{4}$ in./ft across the inlet and 0.05 in./ft along the length of the inlet? The maximum depth of flow in the gutter is estimated to be 0.2 ft, and the gutter is depressed 4 in. below the normal street level.

Calculation Procedure:

1. Compute the reciprocal of the gutter transverse slope

The *transverse slope* of the gutter across the inlet is $\frac{1}{4}$ in./ft. Expressing the reciprocal of this slope in ft as r, compute the value for this gutter as $r = 4 \times 12/1 = 48$.

2. Determine the inlet capacity per ft of length

Enter Table 19 at the flow depth of 0.2 ft and project to the depth of depression of the gutter of 4 in. Opposite this depth, read the inlet capacity per ft of length as 0.50 cfs.

3. Compute the required gutter inlet length

The gutter must handle a maximum flow of 2 cfs. Since the inlet has a capacity of 0.50 cfs per ft of length, the required length, ft = maximum required capacity, cfs/capacity per ft, cfs = 2.0/0.50 = 4.0 ft. A length of 4.0 ft will be satisfactory. Were a length of 4.2 or 4.4 ft required, a 4.5-ft-long inlet would be chosen. The reasoning behind the choice of a longer length is that the extra initial investment for the longer length is small compared with the extra capacity obtained.

4. Determine how far the water will extend from the curb

Use the relation, $l = rd$, where l = distance water will extend from the curb, ft; d = depth of water in the gutter at the curb line, ft; other symbol as before. Substituting, $l = 48(0.2) = 9.6$ ft. This distance is acceptable because the water would extend out this far only during the heaviest of storms.

Related Calculations: To compute the flow rate in a gutter, use the relation $F = 0.56$ $(r/n)s^{0.5}d^{8/3}$, where F = flow rate in gutter, cfs; n = roughness coefficient, usually taken as 0.015; s = gutter slope, in./ft; other symbols as before. Where the computed inlet length is 5 ft or more, some engineers assume that a portion of the water will pass the first inlet and enter the next one along the street.

TABLE 19 Storm-Sewer Inlet Capacity per Ft of Length

Flow depth in gutter, ft	Depression depth, in.	Capacity per ft length, cfs
0.1	0	0.022
	1	0.086
	2	0.180
	3	0.290
	4	0.420
0.2	0	0.062
	1	0.141
	2	0.245
	3	0.358
	4	0.500
0.3	0	0.115
	1	0.205
	2	0.320
	3	0.450
	4	0.590
0.4	0	0.175
	1	0.277
	2	0.400
	3	0.540
	4	0.682

STORM-SEWER DESIGN

Design a storm-sewer system for a 30-acre residential area in which the storm-water runoff rate is computed to be 24 cfs. The total area is divided into 10 plots of equal area having similar soil and runoff conditions.

Calculation Procedure:

1. Sketch a plan of the sewer system

Sketch the area and the 10 plots as in Fig. 7. A scale of 1 in. = 100 ft is generally suitable. Indicate the terrain elevations by drawing the profile curves on the plot plan. Since the profiles, Fig. 7, show that the terrain slopes from north to south, the main sewer can probably be best run from north to south. The sewer would also slope downward from north to south, following the general slope of the terrain.

Indicate a storm-water inlet for each of the areas served by the sewer. With the terrain sloping from north to south, each inlet will probably give best service if it is located on the southern border of the plot.

Since the plots are equal in area, the main sewer can be run down the center of the plot with each inlet feeding into it. Use arrows to indicate the flow direction in the laterals and main sewer.

2. Compute the lateral sewer size

Each lateral sewer handles 24 cfs/10 plots = 2.4 cfs of storm water. Size each lateral using the Manning formula with $n = 0.013$ and full flow in the pipe. Assume a slope ratio of 0.05 for each inlet pipe between the inlet and the main sewer. This means that the inlet pipe will slope 1 ft in 20 ft of length. In an installation such as this, a slope ratio of 0.05 is adequate.

Using Fig. 4 for a flow of 2.4 cfs and a slope of 0.05, an 8-in. pipe is required for each lateral. The fluid velocity is, from Fig. 4, 7.45 fps. This is a high enough velocity to prevent solids from settling out of the water. (The flow velocity should not be less than 2 fps.)

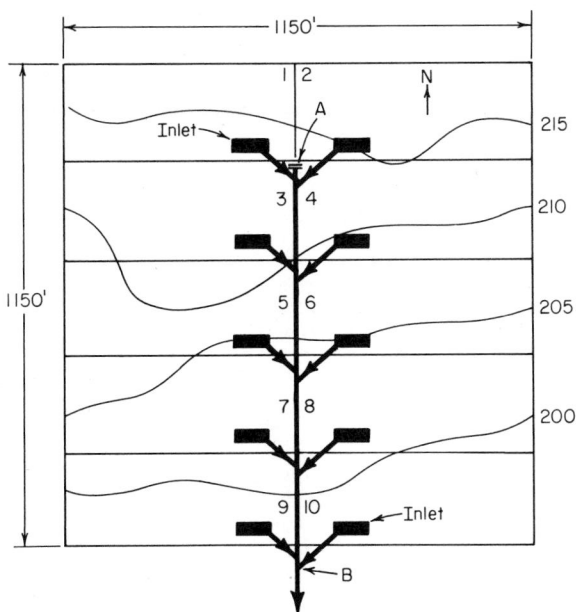

Fig. 7   Typical storm-sewer plot plan and layout diagram.

3. *Compute the size of the main sewer*

There are four sections of the main sewer, Fig. 7. The first section, section 3-4, Fig. 7, serves the two northernmost plots. Since the flow from each plot is 2.4 cfs, the quantity of storm water that this portion of the main sewer must handle is $2(2.4) = 4.8$ cfs.

The main sewer begins at point A, which has an elevation of about 213 ft, as shown by the profile, Fig. 7. At point B the terrain elevation is about 190 ft. Hence, the slope between points A and B is about $213 - 190 = 23$ ft, and the distance between the two points is about 920 ft.

Assume a slope of 1 ft per 100 ft of length, or $1/100 = 0.01$ for the main sewer. This is a typical slope used for main sewers and it is within the range permitted by a pipe run along the surface of this terrain. Table 20 shows the minimum slope required to produce a flow velocity of 2 fps.

Using Fig. 4 for a flow of 4.8 cfs and a slope of 0.01, the required size for section 3-4 of the main sewer is 15 in. The flow velocity in the pipe is 4.88 fps. The size of this sewer is in keeping with general design practice, which seldom uses a storm sewer less than 12 in. in diameter.

Section 5-6 conveys 9.6 cfs. Using Fig. 4 again, the required pipe size is 18 in. and the flow velocity is 5.75 fps. Likewise, section 7-8 must handle 14.4 cfs. The required pipe size is 21 in. and the flow velocity in the pipe is 6.35 fps. Section 9-10 of the main sewer handles 19.2 cfs and must be 24 in. in diameter. The velocity in this section of the sewer pipe will be 6.9 fps. The last section of the main sewer handles the total flow, or 24 cfs. Its size must be 27 in., Fig. 4, although a 24-in. pipe would suffice if the slope at point B could be increased to 0.012.

Related Calculations: Most new sewers built today are the *separate* type—i.e., one sewer for sanitary service and another sewer for storm service. Sanitary sewers are usually installed first because they are generally smaller than storm sewers and cost

TABLE 20   Minimum Slope of Sewers*

Sewer Diameter, in.	Minimum Slope, ft per 100 ft of length
4	1.20
6	0.60
8	0.40
10	0.29
12	0.22
15	0.15
18	0.12
20	0.10
24	0.08

*Based on the Manning formula with $n = 0.13$ and the sewer flowing either full or half full.

less. *Combined sewers* handle both sanitary and storm flows and are used where expensive excavation for underground sewers is necessary. Many older cities have combined sewers.

To size a combined sewer, compute the sum of the maximum sanitary and storm water flow for each section of the sewer. Then use the method given in this Procedure after having assumed a value for n in the Manning formula and for the slope of the sewer main.

Where a continuous slope cannot be provided for a sewer main, a pumping station to lift the sewage must be installed. Most cities require one or more pumping stations because the terrain does not permit an unrestricted slope for the sewer mains. Motor-driven centrifugal pumps are generally used to handle sewage. For unscreened sewage, the suction inlet of the pump should not be less than 3 in. in diameter.

SELECTION OF SEWAGE-TREATMENT METHOD

A city of 100,000 population is considering installing a new sewage-treatment plant. Select a suitable treatment method. Local ordinances require that suspended matter in the sewage be reduced 80 percent, that bacteria be reduced 60 percent, and that the biochemical oxygen demand be reduced 90 percent. The plant will handle only domestic sanitary sewage. What is the daily oxygen demand and the daily suspended-solids content of the sewage? If an industrial plant discharges into this system sewage requiring 4,500 lb of oxygen per day, determine the population equivalent of the industrial sewage.

Calculation Procedure:

1. Compute the daily sewage flow

Using an average flow of 200 gcd, this sewage-treatment plant must handle (200 gcd)(100,000 population) = 20,000,000 gal/day.

2. Compute the sewage oxygen demand

Usual domestic sewage shows a 5-day oxygen demand of 0.12 to 0.17 lb per day per person. Using an average of 0.15 lb per person per day, the daily oxygen demand of the sewage is (0.15)(100,000) = 15,000 lb/day.

3. *Compute the suspended-solids content of the sewage*

Usual domestic sewage contains about 0.25 lb of suspended solids per person per day. Using this average, the total quantity of suspended solids that must be handled is $(0.25)(100,000) = 25,000$ lb/day.

4. *Select the sewage-treatment method*

Table 21 shows the efficiency of various sewage-treatment methods. Since the desired reduction in suspended matter, BOD, and bacteria is known, this will serve as a guide to the initial choice of the equipment.

Study of Table 21 shows that a number of treatments are available which will reduce the suspended matter by 80 percent. Hence, any one of these methods might be used. The same is true for the desired reduction in bacteria and BOD. Thus, the system choice resolves to selection of the most economical group of treatment units.

TABLE 21   Typical Efficiencies of Sewage-Treatment Methods*

Treatment	Percent reduction		
	Suspended matter	BOD	Bacteria
Fine screens..................	5–20	· · ·	10–20
Plain sedimentation	35–65	25–40	50–60
Chemical precipitation	75–90	60–85	70–90
Low-rate trickling filter with pre- and final sedimentation...............	70–90+	75–90	90+
High-rate trickling filter with pre- and final sedimentation...............	70–90	65–95	70–95
Conventional activated sludge with pre- and final sedimentation...............	80–95	80–95	90–95+
High-rate activated sludge with pre- and final sedimentation...............	70–90	70–95	80–95
Contact aeration with pre- and final sedimentation	80–95	80–95	90–95+
Intermittent sand filtration with presedimentation........	90–95	85–95	95+
Chlorination:			
Settled sewage..............	· · ·	†	90–95
Biologically treated sewage ...	· · ·	†	98–99

*Steel—*Water Supply and Sewerage,* McGraw-Hill.
†Reduction is dependent upon dosage.

For a city of this size, four steps of sewage treatment would be advisable. The first step, *preliminary* treatment, could include (*a*) screening to remove large suspended solids, (*b*) grit removal, and (*c*) grease removal. The next step, *primary treatment,* could include sedimentation or chemical precipitation. *Secondary treatment,* the next step, might be of a biological type such as the activated-sludge process or the trickling filter. In the final step, the sewage might be treated by chlorination. Treated sewage can then be disposed of in fields, streams, or other suitable areas.

Choose the following units for this sewage-treatment plant, using the data in Table 21 as a guide: (*a*) rocks or screens to remove large suspended solids, (*b*) grit chambers to

TABLE 22 Sludge and Other Products of Sewage-Treatment Processes per Million Gal of Sewage Treated*

Data	Treatment process							
	Racks	Fine screens	Grit chambers	Plain sedimentation	Septic tanks	Imhoff or separate tanks	Activated sludge	Trickling filter humus tanks
Character of product.........	Screenings	Screenings	Grit	Raw sludge	Digested sludge	Digested sludge	Raw sludge	Raw sludge
Average amount per million gal...................	4 to 8 cu ft	10 to 30 cu ft	2.5 cu ft	2,500 gal	900 gal	500 gal	13,500 gal	500 gal
Average moisture content, percent.................	80	80	15	95	90	85	99	92.5
Specific gravity.............	....	....	....	1,020	1,040	1,040	1,005	1,025
Usual disposal methods.......	Burying, burning, or shredding and digestion with sludge	Burying, burning, or digesting with sludge	Filling land	Processing, digestion, or drying	Drying	Drying	Processing, digestion, or lagooning	Digestion and drying

*O'Rourke – *General Engineering Handbook*, McGraw-Hill.

remove grit, (c) skimming tanks for grease removal, (d) plain sedimentation, (e) activated-sludge process, and (f) chlorination.

Reference to Table 21 shows that screens and plain sedimentation will reduce the suspended solids by the desired amount. Likewise, the activated-sludge process reduces the BOD by up to 95 percent and the bacteria up to 95+ percent. Hence, the chosen system satisfies the design requirements.

5. Compute the population equivalent of the industrial sewage

Use the relation $P_e = R/D$, where P_e = population equivalent of the industrial sewage, persons; R = required oxygen of the sewage, lb/day; D = daily oxygen demand, lb per person per day. Substituting, $P_e = 4,500/0.15 = 30,000$ persons.

Related Calculations: Where sewage is combined—i.e., sanitary and storm sewage mixed—the 5-day per-capita oxygen demand is about 0.25 lb/day. Where large quantities of industrial waste are part of combined sewage, the per-capita oxygen demand is usually about 0.5 lb per person per day. To convert the strength of an industrial waste to the same base used for sanitary waste, apply the population equivalent relation in Step 5. Some cities use the population equivalent as a means of evaluating the load placed on the sewage-treatment works by industrial plants.

Table 22 shows the products resulting from various sewage-treatment processes per million gal of sewage treated. The tabulated data are useful for computing the volume of product each process produces.

Section **12**

Engineering Economics

MAX KURTZ, P. E.
Consulting Engineer

REFERENCES: Kurtz—*Engineering Economics for Professional Engineers' Examinations*, McGraw-Hill; Whittaker—*Economic Analysis*, Wiley; Grant—*Principles of Engineering Economy*, Ronald; Barish—*Economic Analysis*, McGraw-Hill; Thuesen-Fabrycky—*Engineering Economy*, Prentice-Hall; Riggs—*Economic Decision Models*, McGraw-Hill; Starr—*Product Design and Decision Theory*, Prentice-Hall; Thrall, et al.—*Decision Processes*, Wiley; Eilon—*Elements of Production Planning and Control*, Crowell Collier & Macmillan; Moore—*Manufacturing Management*, Irwin; Schweyer—*Analytic Models for Managerial and Engineering Economics*, Reinhold; Chung—*Linear Programming*, Merrill; Ferguson and Sargent—*Linear Programming*, McGraw-Hill; Sasieni and Yaspan—*Operations Research*, Wiley; Shuchman—*Scientific Decision-making in Business*, Holt; Miller—*Schedule, Cost and Profit Control with PERT*, McGraw-Hill; Ford and Fulkerson—*Flow in Networks*, Princeton; Muth and Thompson—*Industrial Scheduling*, Prentice-Hall; Moder and Phillips—*Project Management with CPM and PERT*, Reinhold; Taylor—*Managerial and Engineering Economy*, Van Nostrand.

Calculation of Interest, Principal, and Payments

Symbols and Abbreviations

General: i = interest rate per period, percent; n = number of interest periods.

Simple and compound interest – single payment: P = value of payment at beginning of first interest period, also termed *present worth* of payment; S = value of payment at end of nth interest period, also termed *future value* of payment.

Compound interest – uniform-payment series: R = sum paid at end of each interest period for n periods; P = value of payments at beginning of first interest period, also termed *present worth* of payments; S = value of payments at end of nth interest period, also termed *future value* of payments.

Compound-interest factors:

Single payment – S/P = SPCA = single-payment compound-amount factor; P/S = SPPW = single-payment present-worth factor. *Uniform-payment series* – S/R = USCA = uniform-series compound-amount factor; R/S = SFP = sinking-fund-payment factor; P/R = USPW = uniform-series present-worth factor; R/P = CR = capital-recovery factor.

Basic Equations

Simple interest, single payment

$$S = P(1+ni) \tag{1}$$

Compound interest

$$\text{SPCA} = (1+i)^n \tag{2}$$

$$\text{SPPW} = \frac{1}{(1+i)^n} \tag{3}$$

$$\text{USCA} = \frac{(1+i)^n - 1}{i} \tag{4}$$

$$\text{SFP} = \frac{i}{(1+i)^n - 1} \tag{5}$$

$$\text{USPW} = \frac{(1+i)^n - 1}{i(1+i)^n} \tag{6}$$

$$\text{CR} = \frac{i(1+i)^n}{(1+i)^n - 1} \tag{7}$$

A uniform-payment series that continues indefinitely is termed a *perpetuity*. For this case,

$$\text{USPW} = \frac{1}{i} \tag{6a}$$

$$\text{CR} = i \tag{7a}$$

DETERMINATION OF SIMPLE INTEREST

A company borrows $4,000 at 6 percent per annum simple interest. What payment must be made to retire the debt at the end of 5 years?

Calculation Procedure

1. **Apply the equation for simple interest**

   This equation is $S = P(1 + ni) = \$4,000(1 + 5 \times 0.06) = \$5,200$.

   NOTE: See the introduction to this Section for the symbols used throughout the Section.

COMPOUND INTEREST; FUTURE VALUE OF SINGLE PAYMENT

The sum of \$2,600 was deposited in a fund that earned interest at 8 percent per annum compounded quarterly. What was the principal in the fund at the end of 3 years?

Calculation Procedure:

1. **Compute the true interest rate and number of interest periods**

   Since there are four interest periods per year, the interest rate i per period is $i = 8$ percent/4 = 2 percent per period. With a 3-yr-deposit period, the number n of interest periods is $n = 3 \times 4 = 12$.

2. **Apply the SPCA value given in a compound-interest table**

   Look up the SPCA value for the interest rate, 2 percent, and the number of interest periods, 12. Then substitute in $S = P(\text{SPCA}) = \$2,600(1.268) = \$3,296.80$.

PRESENT WORTH OF SINGLE PAYMENT

On January 1 of a certain year, a deposit was made in a fund that earns interest at 6 percent per annum. On December 31, 7 years later, the principal resulting from this deposit was \$1,082. What sum was deposited?

Calculation Procedure:

1. **Apply the SPPW relation**

   Obtain the SPPW factor for $i = 6$ percent, $n = 7$ years from the interest table. Substitute in the relation $P = S(\text{SPPW}) = \$1,082(0.6651) = \$719.64$.

PRINCIPAL IN SINKING FUND

To accumulate capital for an expansion program, a corporation made a deposit of \$200,000 at the end of each year for 5 years in a fund earning interest at 4 percent per annum. What was the principal in the fund immediately after the fifth deposit was made?

Calculation Procedure:

1. **Apply the USCA factor**

   Obtain the USCA factor for $i = 4$ percent, $n = 5$ from the interest table. Substitute in the relation $S = R(\text{USCA}) = \$200,000(5.416) = \$1,083,200$.

DETERMINATION OF SINKING-FUND DEPOSIT

The XYZ Corporation borrows \$65,000, which it is required to repay at the end of 5 years at 8 percent interest. To accumulate this sum, XYZ will make five equal annual deposits in a fund that earns interest at 3 percent, the first deposit being made 1 year after negotiation of the loan. What is the amount of the annual deposit required?

Calculation Procedure:

1. **Compute the sum to be paid at the expiration of the loan**

   Obtain the SPCA factor from the interest table for $i = 8$ percent, $n = 5$. Then substitute in the relation $S = P(\text{SPCA}) = \$65,000(1.469) = \$95,485$.

2. Compute the annual deposit corresponding to this future value

Obtain the SFP factor from the interest table for $i = 3$ percent, $n = 5$ and substitute in the relation $R = S(\text{SFP}) = \$95,485(0.18835) = \$17,985$.

PRESENT WORTH OF A UNIFORM SERIES

An inventor is negotiating with two firms for assignment of his rights to a patent. The ABC Corp. offers him an annuity of 12 annual payments of $15,000 each, the first payment to be made 1 year after sale of the patent. The DEF Corp. proposes to buy the patent by making an immediate lump-sum payment of $120,000. If the inventor considers that he can invest his capital at 10 percent, which offer should be accept?

Calculation Procedure

1. Compute the present worth of the annuity using an interest rate of 10 percent

Obtain the USPW factor from an interest table for $i = 10$ percent, $n = 12$ and substitute in the relation $P = R(\text{USPW}) = \$15,000(6.814) = \$102,210$. Since the DEF Corp. offered an immediate payment of $120,000, its offer is more attractive than the offer made by ABC Corp.

CAPITAL-RECOVERY DETERMINATION

On January 1 of a certain year a company had a bank balance of $58,000. The company decided to allot this money to an improvement program by making a series of equal payments 4 times a year for 5 years, beginning on April 1 of the same year. If the account earned interest at 4 percent compounded quarterly, what was the amount of the periodic payment?

Calculation Procedure:

1. Compute the true interest rate and number of interest periods

Since the annual rate = 4 percent and there are four interest periods per year, the rate per period is $i = 4$ percent$/4 = 1$ percent. And with a 5-year pay period, the number of interest periods = 5 years (4 periods per year) = 20 periods.

2. Compute the uniform payment, i.e., capital recovery

The present worth of the sum is $58,000. Obtain the CR factor from an interest table for $i = 1$ percent, $n = 20$ and substitute in the relation $R = P(\text{CR}) = \$58,000(0.05542) = \$3,214.36$.

EFFECTIVE INTEREST RATE

An account earns interest at the rate of 6 percent per annum, compounded quarterly. Compute the effective interest rate to four significant figures.

Calculation Procedure:

1. Compute the interest earned by $1/year

With four interest periods per year, the interest rate per period $= i = 6$ percent$/4 = 1.5$ percent. In 1 year there are four interest periods for this account.

Find the compounded value of $1 at the end of 1 year from $S = (1+i)^n = (1+0.015)^4 = \$1.06136$. Thus, the interest earned by $1 in 1 year $= \$1.06136 - 1.00000 = \$0.06136$. Hence, the effective interest rate = 6.136 percent.

PERPETUITY DETERMINATION

What sum must be deposited to provide annual payments of $10,000 that are to continue indefinitely if the endowment fund earns interest of 4 percent compounded semi-annually?

Calculation Procedure:

1. Compute the effective interest rate

Using the same procedure as in the previous Calculation for $1, the effective interest rate $i_e = (1.02)^2 - 1 = 0.04040$, or 4.04 percent.

2. Apply the USPW relation

The endowment or principal required $= P =$ payment/i_e, or $P = \$10,000/0.0404 = \$247,525$.

DETERMINATION OF EQUIVALENT SUMS

Jones Corp. borrowed $900 from Brown Corp. on January 1, 1968, and $1,200 on January 1, 1970. Jones Corp. made a partial payment of $700 on January 1, 1971. It was agreed that the balance of the loan would be amortized by two payments, one on January 1, 1972, and the other on January 1, 1973, the second payment being 50 percent larger than the first. If the interest rate is 6 percent, what is the amount of each payment?

Calculation Procedure:

1. Construct a line diagram indicating the loan data

Figure 1 shows the line diagram for these loans and is typical of the diagrams that can be prepared for any similar set of loans.

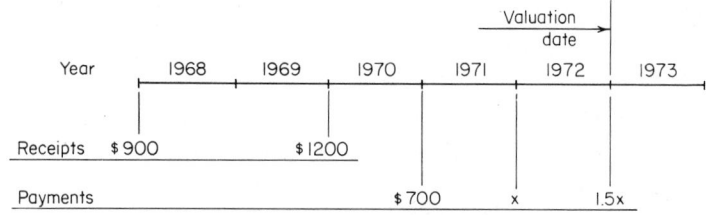

Fig. 1 Time, receipt, and payment diagram.

2. Select a convenient date for evaluating all the sums

For this situation, select January 1, 1973. Mark the valuation date on Fig. 1, as shown.

3. Evaluate each sum at the date selected

Use the applicable interest rate, 6 percent, and the equivalence equation, value of money borrowed = value of money paid. Substituting the applicable SPCA factor from the interest table for each of the interest periods involved, or $n = 5$, $n = 3$, $n = 2$, and $n = 1$, respectively, $\$900(\text{SPCA}) + \$1,200(\text{SPCA}) = \$700(\text{SPCA}) + x(\text{SPCA}) + 1.5x$, where $x =$ payment made on January 1, 1972, and $1.5x =$ payment made on January 1, 1973. Substituting, $\$900(1.338) + \$1,200(1.191) = \$700(1.124) + 1.06x + 1.5x$; $x = \$721.30$. Hence, $1.5x = \$1,081.95$.

Related Calculations: Note that this procedure can be used for more than two loans and for payments of any type that retire a debt.

ANALYSIS OF A NONUNIFORM SERIES

On January 1 of a certain year, ABC Corp. borrowed $1,450,000 for 12 years at 6 percent interest. The terms of the loan obliged the firm to establish a sinking fund in which the following deposits were to be made: $200,000 at the end of the second to the sixth year; $250,000 at the end of the seventh to the eleventh year; and one for the balance of the loan at the end of the twelfth year. The interest rate earned by the sinking fund was 3 percent. Adverse financial conditions prevented the firm from making the deposit of $200,000 at the end of the fifth year. What was the amount of the final deposit?

Calculation Procedure:

1. *Prepare a money-time diagram*

Figure 2 shows the money-time diagram for this situation. Record on Fig. 2 the actual series of deposits, except the final one. Designate this as series A, Fig. 2.

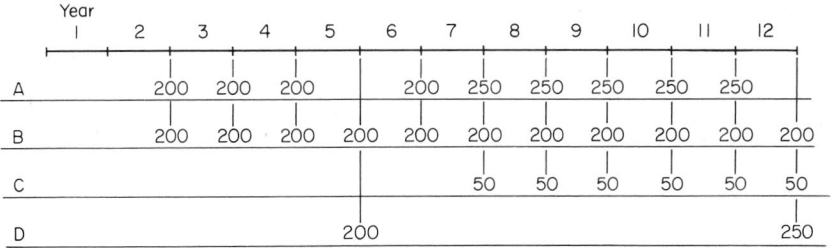

All sums in units of $1000

Fig. 2 Money-time diagram.

2. *Construct two series of uniform deposits approximating, in combination, the actual deposits*

Construct each series so it ends at the close of the twelfth year, Fig. 2. Designate these series as B and C, Fig. 2.

3. *Construct a third series of payments*

This, or series D, Fig. 2, should be constructed such that series D = series B + series C − series A.

4. *Compute the principal of the loan at the end of the twelfth year*

Use the relation $S = P(\text{SPCA})$ for $i = 6$ percent, $n = 12$. Obtain the SPCA value from an interest table and substitute in the above relation, or $S = \$1,450,000(2.012) = \$2,917,400$.

5. *Compute the principal in the sinking fund at the end of the twelfth year*

The principal in the sinking fund at the end of the twelfth year is the value of series A. And, from Step 3, series A = series B + series C − series D. Hence, series B, C, and D must be computed to determine series A.

The value of series $B = R(\text{USCA})$ for $i = 3$ percent, $n = 11$. Or, substituting the USCA factor, series $B = \$200,000(12.808) = \$2,561,600$. Likewise, series $C = \$50,000(6.468) = \$323,400$, for $i = 3$ percent, $n = 6$. Since two payments are made in series D, Fig. 2, series $D = \$200,000(1.230) + \$250,000 = \$496,000$, for $i = 3$ percent, $n = 7$. Then, principal in the sinking fund at the end of the twelfth year = series $A = \$2,561,600 + \$323,400 − \$496,000 = \$2,389,000$.

6. *Compute the final deposit*

To determine the final deposit, subtract the principal in the fund from the principal of the loan. Or, deposit at the end of the twelfth year = $\$2,917,400 − \$2,389,000 = \$528,400$.

Alternate Method: A nonuniform series can also be analyzed as follows. (1) To find the principal in the sinking fund at the end of its life, evaluate all amounts contained in series A, Fig. 2, individually. Use the SPCA relation and the series A payments to accomplish this. (2) Compute the final deposit by deducting the principal in the fund from principal of the loan.

UNIFORM SERIES WITH PAYMENT PERIOD DIFFERENT FROM INTEREST PERIOD

Deposits of $2,000 each were made in a fund earning interest at 4 percent per annum compounded quarterly. The interval between deposits was 18 months. What was the balance in the account immediately after the fifth deposit was made?

Calculation Procedure:

1. Compute the actual interest rate

Replace the interest rate i_3 for the quarterly period with an equivalent rate i_{18} for the 18-month period. Or, $i_{18} = (1 + i_3)^n - 1 = (1.01)^6 - 1 = 6.15$ percent.

2. Determine the appropriate USCA value

Refer to the compound-interest table for values of USCA and interpolate linearly. Thus, if $i = 6$ percent, USCA = 5.637; if $i = 6.5$ percent, USCA = 5.694. Interpolating for $i = 6.15$ percent, USCA = 5.654.

3. Calculate the principal in the fund

Use the relation $S = R(\text{USCA}) = \$2,000(5.654) = \$11,308$.

CONTINUOUS COMPOUNDING

If $1,000 is invested at 6 percent per annum compounded continuously, what will it amount to in 5 years?

Calculation Procedure:

1. Apply the continuous compounding equation

Use the relation $\text{SPCA} = e^{jn}$, where e = base of the natural logarithm system = $2.71828\ldots$; j = nominal interest rate; n = number of years. Substituting, $\text{SPCA} = (2.718)^{0.30} = 1.350$. Then, $S = P(\text{SPCA}) = \$1,000(1.350) = \$1,350$.

UNIFORM-GRADIENT SERIES; CONVERSION TO UNIFORM SERIES

A loan was to be amortized by a group of six end-of-year payments forming an ascending arithmetic progression. The initial payment was to be $5,000, and the difference between successive payments was to be $400, as shown in Fig. 3. But the loan was renegotiated to provide for the payment of equal rather than uniformly varying sums. If the interest rate of the loan was 8 percent, what was the annual payment?

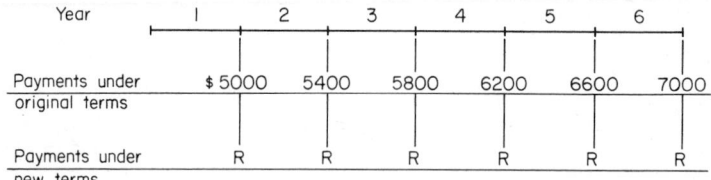

Fig. 3   Diagram showing changed payment plan.

Calculation Procedure:

1. Apply the equivalent-uniform-series equation

Let P = initial payment in a uniform-gradient series; g = difference between successive payments; n = number of payments; R = periodic payment in an equivalent uniform series. Then, $R = P + (g/i)(1 - n\text{SFP})$. Substituting with $P = \$5,000$, $g = 400$, $n = 6$, $i = 8$ percent, $R = \$5,000 + (400/0.08)(1 - 6 \times 0.13632) = \$5,911$.

2. As an alternative, use the uniform-gradient conversion factor UGC

With $n = 6$, $i = 8$ percent, UGC $= 2.28$. Then, $R = \$5,000 + 400(2.28) = \$5,912$.

PRESENT WORTH OF UNIFORM-GRADIENT SERIES

Under the terms of a contract, Brown Corp. was to receive a payment at the end of each year from 1970 to 1976, the payments varying uniformly from \$8,000 in 1970 to \$5,000 in 1976. At the beginning of 1970, Brown Corp. assigned its annuity to Edwards Corp. at a price that yielded Edwards a 6 percent investment rate. What did Edwards pay for the annuity?

Calculation Procedure:

1. Apply the relation of Step 1 of the previous Calculation Procedure

The relation referred to above converts a uniform-gradient series to an equivalent uniform series. Thus, with $g = -500$, $n = 7$, $i = 6$ percent, $R = \$8,000 + (-500/0.06) \times (1 - 7 \times 0.11914) = \$6,617$.

2. Compute the present worth of the equivalent annuity

Use the relation $P = R(\text{USPW}) = \$6,617(5.582) = \$36,936$.

Depreciation and Depletion

Notational System

$D_U =$ depreciation charge for Uth year; $D =$ annual depreciation; $\Sigma D_U =$ cumulative depreciation at end of Uth year $= D_1 + D_2 + D_3 + \cdots + D_U$, where the subscript numbers refer to the year numbers; $P_0 =$ original cost of asset; $P_U =$ book value of asset at end of Uth year $= P_0 - \Sigma D_U$; IRS = Internal Revenue Service; $L =$ salvage value; $W =$ wearing value, or total depreciation $= P_0 - L$; $N =$ longevity or life of asset, years.

STRAIGHT-LINE DEPRECIATION

The initial cost of a machine, including its installation, is \$15,000. The IRS life of this machine is 10 years. The estimated salvage value of the machine is \$1,000 and the cost of dismantling the machine is estimated to be \$200. Using straight-line depreciation, what is the annual depreciation charge? What is the book value of the machine at the end of the seventh year?

Calculation Procedure:

1. Compute the annual depreciation charge

When straight-line depreciation is used, the annual depreciation charge is constant, $D = W/N$. Since $P_0 = \$15,000$, $L = \$1,000 - \$200 = \$800$, $W = \$15,000 - \$800 = \$14,200$, $N = 10$. Then, $D = \$14,200/10 = \$1,420$.

2. Compute the book value of the machine at the end of the seventh year

$\Sigma D_7 = 7D = 7(\$1,420) = \$9,940$. Then, $P_7 = \$15,000 - \$9,940 = \$5,060$.

STRAIGHT-LINE DEPRECIATION WITH TWO RATES

An asset having an initial cost of \$30,000 has a life expectancy of 15 years and an estimated salvage value of \$5,000. What are the depreciation charges under a modified straight-line method in which 60 percent of the total depreciation is considered to occur during the first 5 years of the life of the asset?

Calculation Procedure:

1. Proportion the total wearing value of the asset

Divide the asset's life-span into the two specified intervals and proportion the total wearing value between them. Thus, $W = \$30,000 - \$5,000 = \$25,000$; $N = 15$; $W_1 =$ first-period wearing value $= 0.60(\$25,000) = \$15,000$; $W_2 =$ second-period wearing value $= 0.40(\$25,000) = \$10,000$.

2. Compute the annual depreciation charge

For the first 5 years, $D = \$15,000/5 = \$3,000$. For the next 10 years, $D = \$10,000/10 = \$1,000$.

SINKING-FUND METHOD; ASSET BOOK VALUE

A factory constructed at a cost of \$9,000,000 has an anticipated salvage value of \$400,000 at the end of 30 years. What is the book value of this factory at the end of the tenth year if depreciation is charged by the sinking-fund method with an interest rate of 5 percent?

Calculation Procedure:

1. Compute the cumulative depreciation

This method of depreciation accounting assumes that when the asset is retired, it is replaced by an exact duplicate and that replacement capital is accumulated by making uniform end-of-year deposits in a reserve fund. The cumulative depreciation ΣD_U is therefore equated to the principal in the fund at the end of the Uth year. Or, $\Sigma D_U = W(\text{SFP})(\text{USCA})$. Substituting $W = \$9,000,000 - \$400,000 = \$8,600,000$, $\text{SFP} = 0.01505$ for 30 years, $U = 10$ years, and $i = 5$ percent, $\Sigma D_{10} = \$8,600,000(0.01505)(12.578) = \$1,628,000$.

2. Compute the book value

At the end of 10 years, the book value $P_{10} = P_0 - \Sigma D_{10} = \$9,000,000 - \$1,628,000 = \$7,372,000$.

SINKING-FUND METHOD DEPRECIATION CHARGES

An asset costing \$20,000 is expected to remain serviceable for 5 years and to have a salvage value of \$3,000. Compute the depreciation charges using the sinking-fund method and an interest rate of 4 percent.

Calculation Procedure:

1. Compute the annual sinking-fund payment

Use the relation $R = W(\text{SFP})$. With $W = \$20,000 - \$3,000 = \$17,000$, $N = 5$ years, $i = 4$ percent, $\text{SFP} = 0.18463$. Then, $R = \$17,000(0.18463) = \$3,139$.

2. Compute the annual depreciation charges

Use the relation $D_U = R(\text{SPCA})$, or $D_1 = \$3,139(1.000) = \$3,139$; $D_2 = \$3,139(1.040) = \$3,265$; $D_3 = \$3,139(1.082) = \$3,396$; $D_4 = \$3,139(1.125) = \$3,531$; $D_5 = \$3,139(1.170) = \$3,673$. Then, $\Sigma D_5 = \$17,004$.

FIXED-PERCENTAGE (DECLINING-BALANCE) METHOD

An asset cost \$5,000 and has a life expectancy of 6 years and an estimated salvage value of \$800. Construct a depreciation schedule for this asset using the fixed-percentage method.

Calculation Procedure:

1. Compute the rate of depreciation

Use the relation $\log(1-h) = (\log L - \log P_0)/N$, where $h =$ rate of depreciation. Substituting, $\log(1-h) = (\log 800 - \log 5,000)/6$; $h = 0.2632$, or 26.32 percent.

2. *Compute the end-of-year book value*

Use the relation $D_1 = hP_0 = 0.2632(\$5,000) = \$1,316$. Then, $P_1 = P_0 - D_1 = \$5,000 - \$1,316 = \$3,684$. Likewise, $D_2 = 0.2632(\$3,684) = \$969.63$; $P_2 = \$3,684 - \$969.63 = \$2,714.37$. In a similar manner, $D_3 = \$714.42$, $P_3 = \$1,999.95$, $D_4 = \$526.39$, $P_4 = \$1,473.56$, $D_5 = \$387.84$, $P_5 = \$1,085.72$, $D_6 = \$285.76$, $P_6 = \$799.96$.

COMBINATION OF FIXED-PERCENTAGE AND STRAIGHT-LINE METHODS

An asset cost $5,000 and has a life of 8 years and a salvage value of $1,000. The IRS permits use of the double-declining-balance method to charge depreciation. Compute the depreciation charges.

Calculation Procedure:

1. *Compute the rate of depreciation*

Under the double-declining-balance method, depreciation is initially charged on a fixed-percentage basis, taking $0.02N$ as the rate of depreciation. Thus, rate of depreciation $= 0.02(8) = 0.16$, or 16 percent.

2. *Compute the depreciation charge for each year by the fixed-percentage method*

For the first year, depreciation charge $= 0.16(\$5,000) = \$800$. Then the book value at the end of the first year $= \$5,000 - \$800 = \$4,200$. Following this procedure, construct Table 1.

3. *Compute the depreciation for the transfer study*

Assume that the transfer in depreciation accounting from the fixed-percentage to the straight-line method is made at the end of a particular year. Calculate the annual depreciation charge D' that applies for the remaining life of the asset.

For example, at the end of the third year the book value is $2,964, and the depreciation that remains to be charged during the last 5 years is $1,964. Then, $D' = \$1,964/5 = \$393$. Tabulate the values found in this manner in Table 1.

4. *Determine the transfer date*

To establish the transfer date, compare each value of D' with the depreciation charge that will occur the following year if the fixed-percentage method is used. This comparison shows that the method should be revised at the end of the fifth year because after that time the fixed-percentage method results in a smaller depreciation charge. The depreciation charges, Table 1, are therefore: $D_1 = \$800$; $D_2 = \$672$; $D_3 = \$564$; $D_4 = \$475$; $D_5 = \$398$; $D_6 = D_7 = D_8 = \$364$.

TABLE 1 Depreciation by the Double-Declining-Balance Method

Year	Depreciation charge, $	Book value at year end, $	D', $
0	...	5,000	
1	800	4,200	457
2	672	3,528	421
3	564	2,964	393
4	475	2,489	372
5	398	2,091	364
6	334	1,757	379
7	282	1,475	475
8	236	1,239	

CONSTANT-UNIT-USE METHOD OF DEPRECIATION

A machine cost \$38,000 and has a life of 5 years and a salvage value of \$800. The production output of this machine in units per year is: first year, 2,000; second year, 2,500; third year, 2,250; fourth year, 1,750; fifth year, 1,500 units. If the depreciation is ascribable to use rather than the effects of time, and the units produced are of uniform quality, what are the annual depreciation charges?

Calculation Procedure:

1. Determine the depreciation charge per production unit

Proportion the wearing value on the basis of annual production. Since $W = \$38,000 - \$800 = \$37,200$, and 10,000 units are produced in 5 years, the depreciation charge per production unit = \$37,200/10,000 = \$3.72.

2. Compute the annual depreciation charge

Since the annual depreciation charge is a function of the production rate, take the product of the depreciation charge per production unit and the annual production. Or, $D_1 = \$3.72(2,000) = \$7,440$; $D_2 = \$9,300$; $D_3 = \$8,370$; $D_4 = \$6,510$; $D_5 = \$5,580$.

DECLINING-UNIT-USE METHOD OF DEPRECIATION

Using the same data as in the previous Calculation Procedure, assume that depreciation will be charged by weighting the units produced according to their relative quality. This method reflects the quality loss resulting from increased use of the machine. The quality weights assigned this machine are: first 4,000 units produced, 2.0; next 3,000 units, 1.5; remainder, 1.0. Compute the depreciation charges for this machine.

Calculation Procedure:

1. Compute the number of depreciation units

The depreciation units are related to the annual production by applying the assigned quality rates. Thus:

Year	Depreciation Units
1	$2,000 \times 2 = 4,000$
2	$\begin{cases} 2,000 \times 2 = 4,000 \\ 500 \times 1.5 = 750 \end{cases}$
3	$2,250 \times 1.5 = 3,375$
4	$\begin{cases} 250 \times 1.5 = 375 \\ 1,500 \times 1 = 1,500 \end{cases}$
5	$1,500 \times 1 = 1,500$
Total.	15,500

2. Proportion the wearing value

Consider the number of depreciation units as the criterion. Or, depreciation charge per depreciation unit = \$37,200/15,500 = \$2.40.

3. Compute the annual depreciation

Take the product of the depreciation charge per depreciation unit and the annual depreciation units. Or $D_1 = \$2.40(4,000) = \$9,600$; likewise, $D_2 = \$11,400$; $D_3 = \$8,100$; $D_4 = \$4,500$; $D_5 = \$3,600$. Taking the sum of these charges, the total depreciation = \$37,200.

SUM-OF-THE-DIGITS METHOD OF DEPRECIATION

A machine costing $15,000 is expected to remain serviceable for 7 years. The machine will have a salvage value of $1,000. What are the annual depreciation charges based on the sum-of-the-digits method?

Calculation Procedure:

1. Compute the machine wearing value

The wearing value W or total depreciation $= \$15,000 - \$1,000 = \$14,000$.

2. Compute the annual depreciation

Use the relation $D_U = W(N-U+1)/0.5[N(N+1)]$. Thus, for $N = 1$, $D_1 = \$3,500$. Likewise, for $N = 2$, $D_2 = \$3,000$; for $N = 3$, $D_3 = \$2,500$; for $N = 4$, $D_4 = \$2,000$; for $N = 5$, $D_5 = \$1,500$; for $N = 6$, $D_6 = \$1,000$; for $N = 7$, $D_7 = \$500$.

The sum-of-the-digits method is widely used and is approved by the IRS for tax purposes.

COMBINATION OF TIME- AND USE-DEPRECIATION METHODS

A machine cost $38,000 and has a life of 5 years and a salvage value of $800. Studies show that one-third of the total depreciation stems from the effects of time, and two-thirds stems from use. Compute the annual depreciation charges if time depreciation is based on sum of the digits and use depreciation on a production basis with all units of equal quality. Use the same production as in the third previous Procedure.

Calculation Procedure:

1. Divide the wearing value into its two elements

Knowing the respective depreciation proportions, let the subscripts t and u refer to time and use, respectively. Also, $W = \$38,000 - \$800 = \$37,200$, and $W_t = \frac{1}{3}(\$37,200) = \$12,400$; $W_u = \frac{2}{3}(\$37,200) = \$24,800$.

2. Compute the annual depreciation charge

For the first year, $D_{t1} = W_t N/[N(N+1)/2] = \$12,400(5)/[5(6/2)] = \$4,133$. Also, $D_{u1} = (\$24,800/10,000 \text{ units})(2,000 \text{ units the first year}) = \$4,960$. Thus, the total depreciation for the first year is $D_1 = \$4,133 + \$4,960 = \$9,093$.

EFFECTS OF DEPRECIATION ACCOUNTING ON TAXES AND EARNINGS

The QRS Corp. purchased capital equipment for use in a 5-year venture. The equipment cost $240,000 and had zero salvage value. If the income tax rate was 52 percent and the annual income from the investment was $83,000 before taxes and depreciation, what was the average rate of earnings if the profits after taxes were invested in tax-free bonds yielding 3 percent? Compare the results obtained if depreciation is computed by the straight-line and sum-of-the-digits methods.

Calculation Procedure:

1. Compute the taxable income

With straight-line depreciation, the depreciation charge is $\$240,000/5 = \$48,000/$ year. Then, the taxable income $= \$83,000 - \$48,000 = \$35,000$, because depreciation is fully deductible from gross income.

2. Compute the annual tax payment

With a tax rate of 52 percent, the annual tax payment, excluding other deductions, $= 0.52(\$35,000) = \$18,200$.

3. Compute the net income

The net cash income $=$ gross income $-$ tax payment, if there are no other expenses. Or, net income $= \$83,000 - \$18,200 = \$64,800$.

4. Determine the capital accumulated by investing the net income in bonds

Use the USCA factor for $i = 3$ percent, $n = 5$ years. Or, $S = R(\text{USCA}) = \$64,800$ $(5.309) = \$344,000$.

5. Compute the average earnings rate on the venture

Use the relation $\text{SPCA} = (1 + i)^n$, where $\text{SPCA} = \$344,000/\$240,000 = (1 + i)^5$; $i = 7.47$ percent.

6. Compute the sum-of-the-digits annual depreciation

Using the previously developed procedure for sum-of-the-digits depreciation charges, $D_1 = \$80,000$; $D_2 = \$64,000$; $D_3 = \$48,000$; $D_4 = \$32,000$; $D_5 = \$16,000$.

7. Compute the annual tax and net income

Using the same method as in Steps 2 and 3, the annual net income R is $R_1 = \$81,440$; $R_2 = \$73,120$; $R_3 = \$64,800$; $R_4 = \$56,480$; $R_5 = \$48,160$.

8. Determine the capital accumulated

Use the respective SPCA values for $i = 3$ percent and the years 1 through 5 for the income earned in each year. Or, $S = \$81,440(1.126) + \$73,120(1.093) + \$64,800(1.061) + \$56,480(1.030) + \$48,160 = \$346,700$.

9. Compute the average earnings rate on the venture

Using the method of Step 5, $(\$346,700/\$240,000) = (1 + i)^5$; $i = 7.63$ percent.

The computed interest rates apply to a composite investment—the purchase and operation of the capital equipment and the purchase of bonds. The total income accruing from the first element is \$324,000, regardless of the depreciation method used. However, the *timing* as well as the amount of this income is important.

The straight-line method produces a uniform annual depreciation charge, tax payment, and net income. Under the sum-of-the-digits method, these amounts are nonuniform; the net income is highest in the first year and then gradually declines. Therefore, the interest earned through the purchase of bonds is higher if the firm adopts the sum-of-the-digits method.

If the interest rate associated with the second element of this composite investment had been higher—say 4 or 5 percent—the disparity between the two average returns would have been correspondingly higher.

DEPLETION ACCOUNTING BY THE SINKING-FUND METHOD

An oil field is anticipated to yield an annual income, before depletion allowances, of \$120,000. The field will be dry after 5 years, when the land will then have a residual value of \$60,000. If a firm desires a return of 10 percent on its investment, what is the maximum amount it should invest in this oil field? Use a 4 percent interest rate for the sinking fund.

Calculation Procedure:

1. Determine the replacement cost of the asset

In this method of depletion accounting, the firm deposits a portion of the annual income in a reserve fund to accumulate the capital needed to replace the asset. Let C denote the investment required. Then the replacement cost $r = C - \$60,000$ for this venture.

2. Compute the annual deposit required

Let $d = $ annual deposit required. Then, $d = r(\text{SFP})$ for this venture, or any similar situation. With $i = 4$ percent, $n = 5$, $d = (C - \$60,000)(0.18463) = 0.18463C - 11,077.80$.

3. Compute the investment required

Set the residual income equal to 10 percent of the investment and solve for C. Or, $\$120,000 - (0.18463C - 11,077.80) = 0.10C$; $C = \$460,520$.

Related Calculations: Note that this method can be applied to any situation where there is a gradual depletion of a valuable, profit-generating asset. Further, the method given here is homologous to the sinking-fund method of depreciation accounting.

INCOME FROM A DEPLETING ASSET

An oil field purchased for $800,000 is expected to be dry at the end of 4 years. If the resale value of the land is $20,000, what annual income is required to yield an investment rate of 8 percent? Use a sinking-fund rate of 3 percent.

Calculation Procedure:

1. Compute the annual deposit required to accumulate the replacement capital

The replacement cost = $800,000 − $20,000 = $780,000 = r. Use the relation annual deposit $d = r(\text{SFP})$. With $i = 3$ percent, $n = 4$, $d = \$780,000(0.23903) = \$186,440$.

2. Compute the annual income required

Combine the annual return on the invested capital with the reserve-fund deposit to obtain the required annual income from the asset. Or, annual return on investment = 0.08($800,000) = $64,000. Then, the required annual income = $64,000 + $186,440 = $250,440.

Cost Comparisons of Alternative Proposals

Annual Cost

For analytical purposes, it is desirable to convert the estimated costs associated with a proposed scheme to an equivalent series of uniform annual payments. The annual payment thus obtained is termed the *annual cost* of the scheme. The interest rate applied in making this conversion is the minimum investment rate that is considered acceptable by the organization making the investment or incurring the costs.

Where alternative schemes are being evaluated on the basis of their annual cost, the usual procedure is to exclude those expenses which are identical for all schemes, since they do not affect the comparison.

Notational System

P = initial cost of asset acquired in the proposed scheme; L = salvage value of the asset; N = life of asset, years; i_1 = interest rate; c = sum of annual costs of operation, maintenance, etc., that are assumed to remain constant for the asset life; A = annual cost = $(P-L)(\text{CR}) + Li_1 + c$, where CR is the capital-recovery factor from the compound-interest tables for $i = i_1$, $n = N$; other symbols as defined earlier. Also, $A = (P-L)$ $(\text{SFP}) + Pi_1 + c$, for the same interest and life as the above annual-cost relation.

DETERMINATION OF ANNUAL COST OF AN ASSET

A firm contemplates building a new warehouse. A choice is to be made between a brick and a galvanized-iron structure. The cost data associated with each structure are as follows:

	Brick	Galvanized iron
First cost, $.	80,000	36,000
Salvage value, $	15,000	4,000
Life, years	40	15
Annual maintenance cost, $	1,000	2,300
Annual taxes, $/$100	1.30	1.30
Annual insurance, $/$1,000	2	5

If this firm earns 6 percent on its invested capital, which type of structure is the more economical one?

Calculation Procedure:

1. Compute the operating and maintenance costs

For the brick building, the annual operating and maintenance cost = c = maintenance cost per year, \$ + annual taxes, \$ + annual insurance cost, \$, or $1,000 + 0.013 ($80,000) + 0.002($80,000) = $2,200.

For the galvanized-iron building, $c = $2,300 + 0.013($36,000) + 0.005($36,000) = $2,948.

2. Compute the annual cost of each building

Use the capital-recovery equation. Thus, for the brick building, $A = ($80,000 - $15,000)(0.06646) + $15,000(0.06) + $2,200 = $7,420.

For the galvanized-iron building, $A = ($36,000 - $4,000)(0.10296) + $4,000(0.06) + $2,948 = $6,483.

Since the galvanized-iron building has a lower annual cost, it is the more economical structure.

Related Calculations: This general method of computing annual costs can be used for any number of industrial or commercial assets regardless of whether they are stationary, moving, or water- or air-borne. The key fact to keep in mind is that accurate costs are required if the annual cost comparison is to have validity.

MINIMUM ASSET LIFE TO JUSTIFY A HIGHER INVESTMENT

The timber floor of a bridge is to be replaced, and consideration is being given to treating the timber to prolong its life and reduce maintenance costs. An untreated timber floor costs $5,000 and has an annual maintenance cost of $500 and a life of 10 years. Treated timber costs $8,500 and has an annual maintenance cost of $300 and a life of 20 years. The salvage value of both floors is zero. Which is the more economical floor? Use a 5 percent interest rate.

Calculation Procedure:

1. Compute the annual costs of the alternatives

Using the capital-recovery factor, the annual cost of the untreated timber floor is $5,000(0.12950) + $500 = $1,147.50. Likewise, the annual cost of the treated timber floor = $8,500(CR) + $300. Or, $A = $8,500(0.08024) + $300 = $982.04.

2. Equate the annual costs to determine the minimum life for equal annual costs

If the life of one of the alternatives were unknown, as is sometimes the case, the procedure to follow is to equate the annual costs and compute the CR value for the unknown life. Thus, if the treated timber life were unknown, $1,147.50 = $8,500(CR) + $300; CR = 0.09971. Interpolating in the compound-interest table for 5 percent, $N = 14.3$ years.

If the life of the second alternative exceeds 14.3 years, it is the more economical investment. Since this is true here, the treated timber would be used.

Related Calculations: Use this general method to compare two or more alternatives. The method given in Step 2 is particularly useful where the life of an alternative is not known.

COMPARISON OF EQUIPMENT COST AND INCOME GENERATED

A firm is considering purchasing equipment that will reduce annual labor costs by $4,000. The equipment costs $30,000 and has a salvage value of $5,000 and a life of 7 years. The annual maintenance cost is $600. While not in use by the firm, the equipment can be rented to others to generate an income of $1,000/year. If money can be invested for an 8 percent return, is the firm justified in buying the equipment?

Calculation Procedure

1. Compute the annual cost of using the equipment

Using the capital-recovery-factor annual cost, $A = (\$30,000 - \$5,000)(0.19207) + \$5,000(0.08) + \$600 = \$5,802$.

2. Compute the annual cost of not purchasing the equipment

If the equipment is not purchased, the firm will incur an extra labor cost of $4,000 over that with the equipment. Also, the rental income that would be obtained from the equipment will be lost. Hence, the total annual cost without the equipment would be $A = \$4,000 + \$1,000 = \$5,000$.

Since the annual cost with the equipment would be \$5,802, the firm should not purchase the equipment because without it the annual cost is only \$5,000.

SELECTION OF RELEVANT DATA IN ANNUAL-COST STUDIES

An existing factory must be enlarged or replaced to accommodate new production machinery. The structure was built at a cost of \$130,000. Its present book value, based on straight-line depreciation, is \$35,000, but it has been appraised at \$40,000. If the structure is altered, the cost will be \$80,000 and its service life will be extended 8 years, with a salvage value of \$30,000. A new factory could be purchased for \$250,000. It would have a life of 20 years and a salvage value of \$35,000. Annual maintenance costs of the new building would be \$8,000, compared with \$5,000 in the enlarged structure. However, the improved layout in the new building would reduce annual production costs by \$12,000. All other expenses for the two structures are estimated as being equal. Using an investment rate of 8 percent, determine which is the more attractive investment for this firm.

Calculation Procedure:

1. Segregate the relevant data for the existing structure

Relevant data—present resale value. Irrelevant data—cost of construction and present book value.

2. Record the pertinent cost data for each scheme

Classify the income that would accrue from one scheme as a "cost" of its alternative. Thus:

	Enlarged building	New building
Initial cost or payment, $...........	80,000	250,000
Resale value existing building, $......	40,000	
Total first cost, $	120,000	250,000
Salvage value, $.....................	30,000	35,000
Life, years..........................	8	20
Operating cost, $....................	5,000	8,000
Production "cost," $.................	12,000	

3. Compute the annual cost of the enlarged building

Using the capital-recovery factor for $i_1 = 8$ percent, $n = 8$ years, $A = (\$120,000 - \$30,000)(0.17401) + \$30,000(0.08) + \$5,000 + \$12,000 = \$35,061$.

4. Compute the annual cost of the new building

Using the capital-recovery factor for $i = 8$ percent, $n = 20$ years, $A = (\$250,000 - \$35,000)(0.10185) + \$35,000(0.08) + \$8,000 = \$32,698$.

Since the new building has an annual cost almost \$2,400 less than the enlarged existing structure, the new building is the more economical choice.

Related Calculations: This general procedure can be used to compare any two or more alternatives having characteristics similar to those described above.

DETERMINATION OF MANUFACTURING BREAK-EVEN POINT

A manufacturing firm has a choice between two machines to produce a product. The relevant data are as follows:

	Machine A	Machine B
First cost, $...............	20,000	28,000
Salvage value, $...........	2,000	
Life, years................	10	6
Annual operating cost, $ ...	3,000 + 5.00 per unit	2,500 + 1.50 per unit

If money is worth 7 percent, what annual production is required to justify purchase of machine B?

Calculation Procedure:

1. Compute the annual cost of the first machine

Let x denote the number of units produced annually. Then, using the capital-recovery factor, $A = (\$20,000 - \$2,000)(0.14238) + \$2,000(0.07) + \$3,000 + 5x = \$5,703 + 5x$ for machine A.

2. Compute the annual cost of the second machine

Using the same procedure for machine B, $A = (\$28,000)(0.20980) + \$2,500 + 1.5x = \$8,374 + 1.5x$.

3. Equate the annual costs and solve for the unknown

Substituting the annual costs from Steps 1 and 2, $\$5,703 + 5x = \$8,374 + 1.5x$; $x = 763$ units.

This is the break-even point at which the costs of each machine are equal. If production is expected to exceed this volume, machine B is the economical choice.

EFFECT OF NONUNIFORM OPERATING COSTS

An existing asset that cost $30,000 has a remaining life of 6 years. If extensive repairs are made now, the annual operating cost will be reduced by $2,000 during the next 4 years and by $1,400 during the last 2 years. The life of the asset will not be affected, but the salvage value will be increased by $1,300. What amount may be spent on repairs if the invested capital is to return 10 percent?

Calculation Procedure:

1. Construct a money-time diagram

The nonuniform savings in operating cost may be replaced with an equivalent uniform annual saving s. Construct the money-time diagram, Fig. 4, and record the actual savings in series A.

2. Find the total present worth of the savings

Resolve series A, Fig. 4, into series B and C, Fig. 4, of equal annual amounts. Find the total present worth of these series using an interest rate of 10 percent and a life of 6 years for series B and a life of 4 years for series C. Thus, $1,400(USPW) + $600 (USPW) = $7,999.

3. *Compute the annual savings of an equivalent uniform series*

Use the capital-recovery factor for $i = 10$ percent, $n = 6$ years, or $s = \$7,999(0.22961)$ = \$1,837.

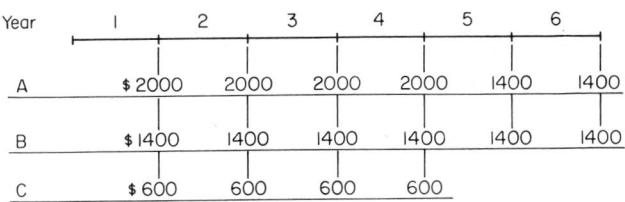

Fig. 4   Money-time diagram.

4. *Record the cost data associated with the alternative schemes*

Let C denote the cost of repairs. Record the cost data associated with the alternative schemes. Note that the true salvage values and operating costs are immaterial; only their differences are significant. Thus:

	Without repairs	With repairs
First cost, $..........	...	C
Salvage value, $......	...	1,300
Life, years...........	6	6
Operating cost, $.....	1,837	

5. *Compute the annual cost of each scheme; equate the results to find C*

The annual cost without repairs = \$1,837. Using the capital-recovery factors with $i = 10$ percent, $n = 6$, the annual cost with repairs $A = (C - \$1,300)(0.22961) + \$1,300$ $(0.10) = 0.22961C - 168$. Solving, $C = \$8,732$. This is the amount that may be spent on repairs.

Related Calculations: As an alternative solution of this type of problem, determine the limiting value of C by calculating the present worth of the savings that accrue if the repairs are undertaken. From the preceding calculations, present worth of savings in operating cost = \$7,999. Present worth of increased salvage value = \$1,300(0.5645) = \$734. Then, $C = \$7,999 + \$734 = \$8,733$, as before.

This Procedure illustrates the point that the annual-cost method of solution is not the simplest in all instances.

ECONOMICS OF EQUIPMENT REPLACEMENT

A machine having an installed cost of \$10,000 was used for 5 years. During that time its trade-in value and operating costs changed as follows:

End of year	Salvage value, $	Operating cost, $/year
1	6,000	2,300
2	4,000	2,500
3	3,200	3,300
4	2,500	4,800
5	2,000	6,800

If the cost of a new machine remained constant during this time, at what date would it be most economical to replace the machine with a duplicate? Use a 7 percent interest rate.

Calculation Procedure:

1. Compute the present worth of all payments on the asset

Let P' denote the present worth (i.e., the value at the date of purchase) of all expenditures ascribable to an asset. In the capital-recovery annual-cost equation, substitute P' for P and set $c = 0$ to obtain the following alternative equation: $A = (P' - L)(CR) + Li_1$. Using $i_1 = 7$ percent, compute the present worth of the operating costs. Or:

Year	
1	PW = ($2,300)(0.9346) = $2,150
2	PW = ($2,500)(0.8734) = $2,184
3	PW = ($3,300)(0.8163) = $2,694
4	PW = ($4,800)(0.7629) = $3,662
5	PW = ($6,800)(0.7130) = $4,848

2. Determine the present worth for each life-span

Take the sum of the installed cost, $10,000, and the present worth of the operating cost found in Step 1. Or:

Life, years	
0	$P' = \$10,000 + \$0 = \$10,000$
1	$P' = \$10,000 + \$2,150 = \$12,150$
2	$P' = \$12,150 + \$2,184 = \$14,334$
3	$P' = \$14,334 + \$2,694 = \$17,028$
4	$P' = \$17,028 + \$3,662 = \$20,690$
5	$P' = \$20,690 + \$4,848 = \$25,538$

3. Apply the annual-cost equation developed in Step 1

Life, years	
1	$A = \$6,150(1.07000) + \$6,000(0.07) = \$7,001$
2	$A = \$10,334(0.55309) + \$4,000(0.07) = \$5,996$
3	$A = \$13,828(0.38105) + \$3,200(0.07) = \$5,493$
4	$A = \$18,190(0.29523) + \$2,500(0.07) = \$5,545$
5	$A = \$23,538(0.24389) + \$2,000(0.07) = \$5,881$

Inspect these annual costs to determine when the minimum annual cost occurs. Since the annual cost is a minimum when $N = 3$, the asset should be retired at the end of the third year.

ANNUAL COST BY THE AMORTIZATION (SINKING-FUND-DEPRECIATION) METHOD

A machine costs $30,000 and will be retired at the end of 8 years with a salvage value of $5,000. The annual operating cost is $3,200. Determine the annual cost by the amortization method if the interest rate on the loan is 6 percent and that of the sinking fund is 3 percent.

Calculation Procedure:

1. Compute the annual cost of the asset

The amortization method is based on the following assumptions: the asset is purchased with borrowed funds; interest on the loan is paid annually; the loan principal is paid as a lump sum at the retirement of the asset; the funds required to retire the debt are accumulated by uniform annual deposits in a reserve fund. This assumed method of financing is unrealistic; the amortization method is therefore approximate.

Let i_1 = interest rate on loan; i_2 = interest rate on sinking fund. Then $A = (P-L)$ (SFP) $+ Pi_1 + c$. In this equation the SFP factor is based on i_2. Apply this equation using $P = \$30,000$, $L = \$5,000$, $N = 8$, $c = \$3,200$, $i_1 = 6$ percent, $i_2 = 3$ percent. Thus, $A = \$7,812$.

ANNUAL COST BY THE STRAIGHT-LINE-DEPRECIATION METHOD

The director of a corporation recommends that a firm buy a computer instead of renting at the rate of \$50/hr. A new computer costs \$120,000; annual operating, maintenance, and insurance costs total \$8,500. The computer will be traded at the end of 10 years for \$30,000. The director forecasts computer usage for 480 hr/year. He bases his calculation of annual cost on the straight-line-depreciation method with an interest rate of 6 percent. Is his recommendation sound?

Calculation Procedure:

1. Compute the annual cost of owning the asset

The straight-line method is an approximate one which assumes that the asset is purchased with borrowed funds. However, the method disregards the timing of payments and considers only their arithmetical average. Thus, the annual cost $A = [(P-L)/N] + (P-L)i_1(N+1)/2N + Li_1 + c$. Substituting, $A = (\$90,000/10) + \$90,000(0.06)(11/20) + \$30,000(0.06) + \$8,500 = \$22,270$.

2. Compute the annual cost of renting the asset

The annual cost of renting the asset = (hourly rate, \$)(annual hr of use) = (\$50)(480) = \$24,000.

Since the annual cost of owning the asset is less than the annual cost of renting it, the firm would save money by owning the asset. Note that this is an approximate method.

Present Worth of Future Costs

A cost analysis of alternative schemes may be performed by computing the present worth of all expenses incurred in each scheme during a stipulated period of time called the *analysis period*. This period should encompass an integral number of lives of each asset required under the alternative schemes.

PRESENT WORTH OF FUTURE COSTS OF AN INSTALLATION

A city contemplates increasing the capacity of existing water-transmission lines. Two plans are under consideration: Plan A requires construction of a parallel pipeline, flow being maintained by gravity. The initial cost is \$800,000, and the life is 60 years with an annual operating cost of \$1,000. Plan B requires construction of a booster pumping station costing \$210,000 with a life of 30 years. The pumping equipment costs an additional \$50,000; it has a life of 15 years and a salvage value of \$10,000. The annual operating cost is \$35,000. Which is the more economical plan if the interest rate is 6 percent?

Calculation Procedure:

1. Construct a money-time diagram of the situation

Figure 5 shows the money-time diagram. Note that this diagram uses 60 years as the analysis period. Record on the money-time diagram the capital expenditures during this 60-year period.

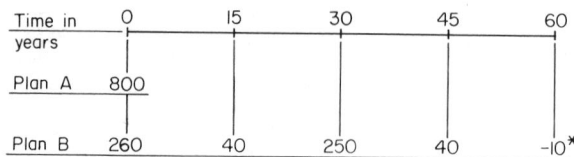

*Income from disposal of equipment.
All sums in units of $1000

Fig. 5   Money-time diagram.

2. Compute the total present worth of the payments

For plan A, using the USPW factor for $n = 60$ years, PW = \$800,000 + \$1,000(16.161) = \$816,160. For plan B, using the SPPW factor for the payments shown in Fig. 5, and the uniform series present-worth factor for the operating cost, PW = \$260,000 + \$40,000(SPPW) + \$250,000(SPPW) + \$40,000(SPPW) + \$35,000(USPW) − \$10,000 (SPPW) = \$260,000 + \$40,000(0.4173) + \$250,000(0.1741) + \$40,000(0.0727) + \$35,000 (16.161) − \$10,000(0.0303) = \$888,460.

Since the present worth of plan A is less than that of plan B, the scheme for plan A should be adopted because it is the more economical.

Capitalized Cost

When computing the present worth of the costs associated with a proposed scheme, it is often advantageous to select an analysis period of infinite duration. The present worth of the future costs is then referred to as the *capitalized cost* of the scheme.

Since each expenditure recurs indefinitely during the analysis period, the various costs constitute a group of perpetuities. Thus, the capitalized cost C_c, is $C_c = [(P-L)/i_1]$ $(CR) + L + c/i_1$, or $C_c = [(P-L)/i_1](SFP) + P + c/i_1$.

If an asset is considered to have an infinite life-span, these equations reduce to $C_c = P + c/i_1$. In these equations, $i = i_1$, $n = N$.

DETERMINATION OF CAPITALIZED COST

Two methods of conveying water for an industrial plant are being analyzed. Method A uses a tunnel and method B a ditch and flume. The costs are as follows:

	Method A	Method B	
	Tunnel	Ditch	Flume
First cost, $.............	180,000	50,000	40,000
Salvage value, $.........	...	...	5,000
Life, years..............	Infinite	50	15
Operating cost, $/year...	2,300	2,000	3,600

Evaluate these two alternatives on the basis of capitalized cost using a 5 percent interest rate.

Calculation Procedure:

1. *Compute the capitalized cost of the first alternative*

Since the tunnel has an infinite life, $C_c = P + c/i_1 = \$180,000 + \$2,300/0.05 = \$226,000$.

2. *Compute the capitalized cost of the second alternative*

Using the capital-recovery factor for $n = 50$ years, $i = 5$ percent, for the ditch, $C_c = [\$50,000/0.05](0.05478) + \$2,000/0.05 = \$94,780$.

Using a similar procedure for the flume, which has a 15-year life, $C_c = (\$35,000/0.05)(0.09634) + \$5,000 + \$3,600/0.05 = \$144,440$. The total capitalized cost for method B = the sum of the flume and ditch costs, or $\$239,220$. Since method A costs less, it is the more economical choice.

CAPITALIZED COST OF ASSET WITH UNIFORM INTERMITTENT PAYMENTS

What is the capitalized cost of a bridge costing $85,000 and having a 25-year life, a $10,000 salvage value, $400 annual maintenance cost, and repairs at 5-year intervals of $2,000, if the interest rate is 5 percent?

Calculation Procedure

1. *Convert the assumed repair costs to an equivalent series of uniform annual payments*

Assume that the repairs are made at the end of every 5-year interval, including the replacement date. Using the SFP factor, the equivalent series of uniform annual payments $R_1 = \$2,000(\text{SFP})$ for $i = 5$ percent, $n = 5$ years, or $R_1 = \$2,000(0.18097) = \$362$.

2. *Convert the true repair costs to an equivalent series of uniform annual payments*

Repairs are omitted when the bridge is scrapped at the end of 25 years, thereby saving $2,000 in the final 5-year period. Convert this amount to an equivalent series of uniform annual payments—i.e., savings—and subtract from the result in Step 1. Or, $R_2 = \$2,000(\text{SFP})$, for $i = 5$ percent, $n = 25$ years. Or $R_2 = \$2,000(0.02095) = \$42$. Thus the annual cost of the repairs = $\$362 - \$42 = \$320$.

3. *Compute the capitalized cost*

Using the capital-recovery factor for $i = 5$ percent, $n = 25$ years, $C_c = (\$75,000/0.05)(0.07095) + \$10,000 + \$400/0.05 + \$320/0.05 = \$130,830$.

Related Calculations: An alternative solution could be worked as follows: Since the $2,000 saving at the end of every 25-year interval coincides in timing with the income of $10,000 from the sale of the old bridge as scrap, this saving can be combined with the salvage value to obtain an effective value of $12,000 for salvage. The annual cost of repairs is therefore taken as $362, the value of R_1, Step 1. Applying the capital-recovery factor, $C_c = [(\$85,000 - \$12,000)/0.05](0.07095) + \$12,000 + \$400/0.05 + \$362/0.05 = \$130,830$. This agrees with the previously determined value.

CAPITALIZED COST OF AN ASSET WITH NONUNIFORM INTERMITTENT PAYMENTS

A bridge has the same cost data as in the previous Calculation Procedure except for the repairs which are as follows:

End of Year	Repair Cost, $
10	2,000
15	3,500
20	1,500

What is the capitalized cost of the bridge if the interest rate is 5 percent?

Calculation Procedure:

1. Compute the present worth of the repairs for one life-span

Use the single-payment present-worth factor for each of the repair periods. Or, $PW = \$2,000(0.6139) + \$3,500(0.4810) + \$1,500(0.3769) = \$3,477$.

2. Convert the result of Step 1 to an equivalent series of uniform annual payments

Using the capital-recovery factor, annual cost of repairs $= \$3,477(CR)$, where $i = 5$ percent, $n = 25$ years. Or, $c_a = \$3,477(0.07095) = \$247$.

3. Compute the capitalized cost

Using the same method as in Step 3 of the previous Calculation Procedure, $C_c = \$106,430 + \$10,000 + \$8,900 + \$247/0.05 = \$129,370$.

Related Calculations: An alternative way of solving this problem is to combine the present worth of the payments for repairs ($3,477) with the initial cost ($85,000) to obtain an equivalent initial cost P'. Then, $P' = \$88,477$, and $P' - L = \$88,477 - \$10,000 = \$78,477$. Applying the capital-recovery factor, $C_c = \$129,370$, as before.

STEPPED-PROGRAM CAPITALIZED COST

A firm plans to build a new warehouse with provision for anticipated growth. Two alternative plans are available.

	Plan A	Plan B
First cost, $.................	100,000	80,000
Salvage value, $.................	10,000	15,000
Life, years.......................	25	30
Annual maintenance, $	1,400	1,200 first 10 years, 1,800 thereafter
Cost of enlarging structure 10 years hence, $	...	40,000

If money is worth 10 percent, which is the more economical plan?

Calculation Procedure:

1. Compute the total present worth of the second plan costs

Let P' represent the total present worth of the costs associated with plan B for one life-span. Using the SPPW for $i = 10$ percent, $n = 10$ years, for the cost of enlarging the structure, and the USPW for the annual maintenance *after* expansion, and the *difference* between the annual maintenance costs of this structure and the original structure, $P' = \$80,000 + \$40,000(0.3855) + \$1,800(9.427) - \$600(6.144) = \$108,700$.

2. Compute the capitalized cost of each alternative

Using the capital-recovery factor for plan A with $i = 10$ percent, $n = 25$ years, $C_c = [(\$100,000 - \$10,000)/0.10](0.11017) + \$10,000 + \$1,400/0.10 = \$123,150$.

For plan B, using the present worth from Step 1, $C_c = [(\$108,700 - \$15,000)/0.10](0.10608) + \$15,000 = \$114,400$. Note that the capital-recovery factor for plan B is for 30 years.

Since plan B has the lower capitalized cost, it is the more economical.

Evaluation of Investments

PREMIUM WORTH METHOD OF INVESTMENT EVALUATION

A firm contemplates investing in a depleting asset and has a choice between two enterprises. Project A requires the investment of $57,500; project B requires the investment of $63,000. The forecast end-of-year dividends are as follows:

Year	Project A, $	Project B, $
1	10,000	15,000
2	15,000	25,000
3	25,000	30,000
4	20,000	20,000
5	10,000	

After weighing the risks involved, the firm decides that the minimum acceptable rate of return on project A is 10 percent; on project B, 12 percent. Evaluate these investments by the premium-worth method. If both investments are satisfactory, determine which is more satisfactory.

Calculation Procedure:

1. Compute the present worth of the dividends from both investments

The *present* generally refers to the date on which the investment is made. Where the present worth is greater than the sum invested, the excess is termed the *premium worth*. Such a result signifies that the true investment rate exceeds the minimum acceptable rate.

For any year, PW = (dividend, $)(PW factor for 10 percent and the number of years involved). Thus, for project A:

Year	PW
1	$10,000(0.9091) = $9,091
2	15,000(0.8264) = 12,396
3	25,000(0.7513) = 18,783
4	20,000(0.6830) = 13,660
5	10,000(0.6209) = 6,209
Total	.... $60,139

Then, the premium worth = $60,139 − $57,500 = $2,639.

Using a similar procedure, the present worth of project B at 12 percent is as follows:

Year	PW
1	$15,000(0.8929) = $13,394
2	25,000(0.7972) = 19,930
3	30,000(0.7118) = 21,354
4	20,000(0.6355) = 12,710
Total	.... $67,388

Then, the premium worth = $67,388 − $63,000 = $4,388.

2. Determine the relative values of the investments

Since both investments satisfy the minimum requirements, determine their relative values by computing the premium-worth percentage — i.e., the ratio of the premium worth to the capital invested. Thus, for project A, the premium-worth percentage is $2,639(100)/$57,500 = 4.6 percent. For project B the premium-worth percentage is $4,388(100)/$63,000 = 7.0 percent. Thus, project B is the more attractive because it has a higher premium-worth percentage.

VALUATION OF CORPORATE BONDS

A $10,000, 4 percent corporation bond paying semiannual dividends is redeemable at 102 at the end of 15 years. What is the maximum price an investor should pay for this bond if he desires a return of 6 percent compounded semiannually?

Calculation Procedure:

1. Determine the semiannual dividend and redemption payment

The dividend = (principal, $$)$i/2$ = $10,000(0.04/2)$ = $200. Also, the redemption payment = (redemption price/100)(principal) = (102/100)($10,000) = $10,200.

2. Compute the purchase price

Using an interest rate of $6/2 = 3$ percent per semiannual period, compute the present worth of the dividends and the redemption payment. Equate the present worth to the purchase price of the bond. Or, purchase price = (dividend, $$)(USPW) + (redemption payment, $$)(SPPW), for $i = 3$ percent, $n = 30$. Hence, purchase price = ($200)(19.60) + ($10,200)(0.4120) = $8,122.

RATE OF RETURN ON BOND INVESTMENT

A $10,000, 6 percent, 20-year bond paid dividends semiannually and was redeemed at par. An investor bought the bond for $11,500 at its date of issue and held it to maturity. What interest rate did the holder earn?

Calculation Procedure:

1. Record the payment and receipts associated with the investment

The *payment* was $11,500 at the date of issue. The *receipt* for each semiannual interest period was (6 percent/2)($10,000) = $300 for 40 periods. Also, $10,000 was received at the end of the 40 periods. The correct interest rate is that which will make the payment equal the receipts.

2. Select a trial interest rate and compute the results

Selecting an interest rate of 2.5 percent as a trial, compute the value of the receipts at the date of issue using the USPW factor for the dividends and the SPPW factor for the principal repayment. Or, ($300)(25.103) + $10,000(0.3724) = $11,255. Since the purchase price exceeded this value, the true interest rate was less than 2.5 percent.

3. Select another trial interest rate

Repeat the previous calculation using a 2 percent rate. Or, $300(27.355) + $10,000 (0.4529) = $12,736.

4. Interpolate linearly between the trial values

Interest Rate, %	Purchase Price, $
2.5	11,255
i	11,500
2	12,736

This interpolation gives $i = 2.42$ percent per semiannual period, or 4.84 percent per annum compounded semiannually.

INVESTMENT-RATE CALCULATION AS ALTERNATIVE TO ANNUAL-COST CALCULATION

In the Comparison of Equipment Cost and Income Generated Procedure in this section it was concluded that the proposed investment in labor-saving equipment could not be justified because it failed to yield the minimum acceptable rate of 8 percent. Determine the actual rate of return for this investment.

Calculation Procedure:

1. Compute the net annual dividend

Labor saving........	$4,000
Rental income.......	1,000
Total.............	$5,000
Less maintenance ...	600
Net dividend........	$4,400

2. Select a trial interest rate

Using an interest rate of 5 percent, determine the present worth of the dividends and the equipment salvage value. Thus, (net dividend, $)(USPW)+(salvage value, $)(SPPW), for $i = 5$ percent, $n = 7$ year. Or, $4,400(5.786) + $5,000(0.7107) = $29,012. Since the investment was $30,000, the actual interest rate is smaller.

3. Test another trial interest rate

Using a 4 percent interest rate and repeating the calculation in Step 2, $4,400(6.002)+ $5,000(0.7599) = $30,208.

4. Interpolate linearly to obtain the actual interest rate

Linear interpolation yields a rate of $i = 4.2$ percent. This verifies that the earlier results were valid.

ALLOCATION OF INVESTMENT CAPITAL

In divising a program for investment of $8,000 in surplus funds, a firm has a choice between two plans. Each plan pays an annual dividend and repayment of the invested capital when the venture terminates. Under plan A the dividend varies with the sum invested in the manner shown below. Under plan B the dividend rate is 10 percent, irrespective of the sum invested. In what manner should this firm divide its investment capital to secure the maximum return?

Calculation Procedure:

1. List the annual dividends obtainable

The table below shows the dividend that can be expected under plan A.

Investment, $	Annual dividend, $	Dividend rate, %
1,000	300	30.0
2,000	540	27.0
3,000	720	24.0
4,000	900	22.5
5,000	950	19.0
6,000	1,020	17.0
7,000	1,220	17.4
8,000	1,300	16.3

2. Construct a dividend-investment diagram

Figure 6 shows the dividend-investment diagram for this situation. Points A to H represent the sets of values under plan A. The slope of a line connecting any two points represents the rate of return on the incremental investment. For example, the slope of line EG = $270/$2,000 = 0.135, or 13.5 percent represents the rate obtained on the $2,000 investment added in going from E to G, Fig. 6.

3. Determine the investments to make

Draw line OJ, Fig. 6, having a slope of 10 percent, the dividend rate under plan B. Next, determine which of the points, A to H, is most distant from line OJ. Do this by scaling the vertical offsets or by drawing lines through these points parallel to OJ.

Point G, which has a vertical offset of $520, is the most distant one. Therefore, $7,000 is the appropriate sum to invest in plan A because of the following (a) When the investment is extended from some lower level, such as $5,000, to the stipulated level, the rate of return on this incremental investment, which is represented by the slope of line EG, exceeds 10 percent. (b) If the investment is carried beyond G, the rate of return on this incremental investment, represented by the slope of line GH, is less than 10 percent. Hence, this firm should allocate $7,000 to plan A and $1,000 to plan B.

Related Calculations: As an alternative, construct the following tabulation to determine the total annual dividend corresponding to every possible division of the capital. Study of this table shows that the maximum dividend of $1,320 accrues when $7,000 is allocated to plan A and $1,000 to plan B.

Investment, $		Dividend, $		Total dividend, $
Plan A	Plan B	Plan A	Plan B	
	8,000	· · ·	800	800
1,000	7,000	300	700	1,000
2,000	6,000	540	600	1,140
3,000	5,000	720	500	1,220
4,000	4,000	900	400	1,300
5,000	3,000	950	300	1,250
6,000	2,000	1,020	200	1,220
7,000	1,000	1,220	100	1,320
8,000	· · ·	1,300	· · ·	1,300

ALLOCATION OF CAPITAL TO TWO INVESTMENTS WITH VARIABLE RATES OF RETURN

Suppose that the dividend under plan B in the previous Procedure is 15 percent of the first $3,000 invested and 10 percent of the excess. Determine the optimal division of the $8,000 investment between the two plans.

Calculation Procedure:

1. Construct a dividend-investment diagram

Use Fig. 6 and draw line OK having a slope of 10 percent and line KL having a slope of 15 percent, where K has an abscissa of $5,000. The ordinate of each point on the line OKL represents the prospective plan B dividend that is forfeited by allocating part of the investment capital to plan A. The optimal division of the investment capital is that for which the excess of plan-A dividends over forfeited plan-B dividends is the maximum.

2. Determine the investment allocation

Find which of the points from A to H is most distant from the line OKL. This is point

D, which has a vertical offset of $500. Therefore, the firm should allocate $4,000 to plan *A* and $4,000 to plan *B*.

Related Calculations: Alternatively, calculate the total dividend corresponding to every possible division of the available capital. For example, if $1,000 is allocated to plan *A* and $7,000 to plan *B*: Dividend under plan *A* = $300; dividend under plan *B* = $3,000(0.15) + $4,000(0.10) = $850; total dividend = $1,150. Using this technique, the maximum total dividend is found to be $1,450; this occurs when the capital is divided equally between the two plans. Thus, the previous findings are verified.

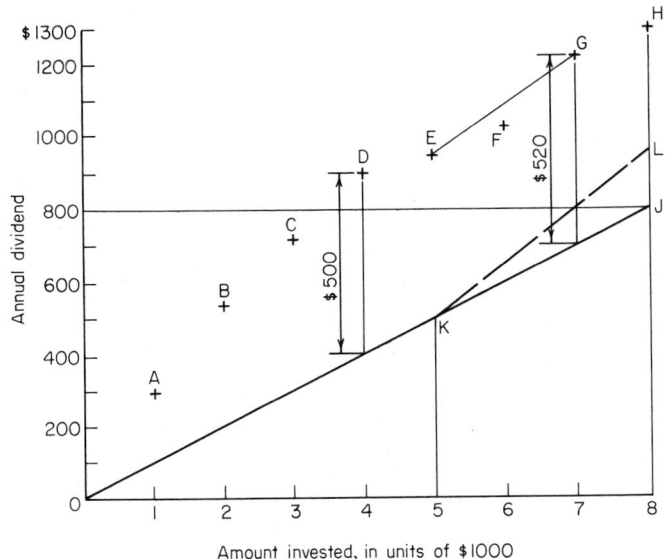

Fig. 6   Dividend-investment diagram.

This Procedure and the previous one show two methods of establishing the optimal division of available capital—i.e., computing the rate of return on an incremental investment and computing the total dividend. The latter represents a more straightforward approach, particularly where both alternative investments yield a variable rate of return.

ECONOMIC LEVEL OF INVESTMENT

A firm planned to purchase and improve property in the expectation that land values in the area would appreciate in the near future. The question arose as to how large an investment should be made. The following data were compiled for five alternative plans, each representing a different level of investment.

	Plan				
	A	B	C	D	E
Investment, $	200,000	270,000	340,000	410,000	460,000
Rate of return, %	14.1	13.8	12.5	11.6	12.3

If 12 percent is considered the minimum acceptable rate of return, determine the most attractive plan.

Calculation Procedure:

1. Establish a basis of comparison for the investments

To establish a basis of comparison, assume that each investment pays an annual dividend and that the invested capital remains intact until the venture terminates. (Although these assumptions are not realistic, they are entirely valid for comparative purposes.)

2. Calculate the annual dividend under each plan

Thus, for plan A, the annual dividend = $200,000(0.141) = $28,200. Compute the dividends for the other plans in the same manner.

3. Construct a dividend-investment diagram

Use the same procedure as for Fig. 6. Plot the points representing the sets of values associated with the five plans, A through E.

Draw a line through the origin of the dividend-investment diagram with a slope of 12 percent. Determine which point is most distant from this line. This point corresponds to the most profitable rate of return. Study of the plot shows that plan B is the one that should be adopted.

Related Calculations: To compare these five plans algebraically, assume that the firm has a total available capital of $460,000 (the investment required under plan E) and that the amount remaining after investment in one of the five plans will be allocated to another investment yielding an annual dividend of 12 percent.

Next, calculate the total annual dividend corresponding to each plan. Or:

Plan	Dividend, $
A	200,000(0.141) + 260,000(0.12) = 59,400
B	270,000(0.138) + 190,000(0.12) = 60,060
C	340,000(0.125) + 120,000(0.12) = 56,900
D	410,000(0.116) + 50,000(0.12) = 53,560
E	460,000(0.123) = 56,580

Since plan B yields the highest total dividend, it is the best choice.

APPARENT RATES OF RETURN ON A CONTINUING INVESTMENT

A firm leasing construction equipment purchased an asset for $24,000, charging depreciation on a straight-line basis. The life used was 4 years; salvage value zero. The asset was used for 6 years and scrapped for $800. Net revenues obtained from this asset are listed in Table 2. The firm's normal income was taxed at 50 percent, but the

TABLE 2   Determination of Apparent Rates of Return

Year	Net revenue, $	Depreciation charge, $	Net profit before tax, $	Net profit after tax, $	Book value beginning of year, $	Apparent rate of return, %
1	10,000	6,000	4,000	2,000	24,000	8.3
2	9,600	6,000	3,600	1,800	18,000	10.0
3	8,000	6,000	2,000	1,000	12,000	8.3
4	6,400	6,000	400	200	6,000	3.3
5	4,400	· · ·	4,400	2,200	· · ·	Infinite
6	2,400	· · ·	2,400	1,200	· · ·	Infinite
	800*	· · ·	800	600		

*Income from sale of asset.

proceeds from the salvage sale were taxed at 25 percent. What were the apparent rates of return on this asset investment, after taxes, computed during the life of the asset?

Calculation Procedure:

1. Compute the annual depreciation charge

Using the straight-line method and a 4-year life, annual depreciation = $24,000/4 = $6,000, assuming zero salvage. Record the depreciation charge in the third column of Table 2.

2. Compute the net profit before taxes

Deduct from the annual net revenue the annual depreciation charge, Table 2, and enter the result in column 4. Thus, for the first year with a revenue of $10,000, the net income before taxes = $10,000 − $6,000 = $4,000.

3. Compute the after-tax profit

With a tax of 50 percent of the profit before taxes, multiply the value in Table 2, column 4, by 0.50 to determine the after-tax profit. Thus, for year 1, the after-tax profit = $4,000(0.50) = $2,000.

4. Record the asset book value at the beginning of the year

In this type of calculation, the book value = the unrecovered capital investment for that year. Or, for year 1, the book value = $24,000. For year 2, the book value = $24,000 − $6,000 = $18,000. In this relation, the $6,000 is the depreciation during year 1.

5. Compute the apparent rate of return

Divide the after-tax profit for any year by the book value of the asset at the beginning of the year to determine the apparent rate of return. Or, for year 2, apparent rate of return = $1,800/$18,000 = 0.10, or 10.0 percent.

TRUE RATE OF RETURN ON A COMPLETED INVESTMENT

Referring to the previous Calculation Procedure, what was the actual after-tax rate of return yielded by this investment, computed at the conclusion of the venture?

Calculation Procedure:

1. Determine the after-tax income

In this situation, it is only the actual disbursements and receipts, as well as their timing, that are pertinent. The depreciation charges, which arise from bookkeeping entries, are irrelevant.

To determine the after-tax income, deduct the tax payment from the net revenue listed in Table 2 to obtain the after-tax income. List this income in Table 3.

TABLE 3 Determination of True Rate of Return

Year	Net revenue, $	Tax payment, $	After-tax income, $
1	10,000	2,000	8,000
2	9,600	1,800	7,800
3	8,000	1,000	7,000
4	6,400	200	6,200
5	4,400	2,200	2,200
6	3,200	1,400	1,800

2. Determine the present worth of the annual receipts

Compute the present worth of each year's after-tax income for years 1 through 6 and take the sum. To perform this computation, assume an interest rate that is believed to approximate the actual rate of return on the $24,000 investment.

At 12 percent, present worth of the after-tax income = $24,444. At 15 percent, present worth of the after-tax income = $22,874. By linear interpolation for a present worth of $24,000, i = rate of return = 12.8 percent.

AVERAGE RATE OF RETURN ON COMPOSITE INVESTMENT

Suppose that the income in the previous Calculation Procedure were reinvested at 8 percent, until the end of the fourth year. Thereafter, the income received was reinvested at 10 percent. What was the average rate of return on the $24,000 capital during the 6-year period?

Calculation Procedure:

1. Compute the value of the original capital at the end of the sixth year

Thus, using the SPCA factor for the after-tax income listed for each year in Table 3 for i = 8 percent or 10 percent, the value of the original capital at the end of the sixth year = $8,000(1.469) + $7,800(1.360) + $7,000(1.260) + $6,200(1.166) + $2,200(1.100) + $1,800 = $42,629.

2. Compute the average investment rate

Let i' = average investment rate. Equate the original investment to $42,629 at a date 6 years in the future, and solve for i'. Thus, $24,000(SPCA for i') = $42,629; i' = 10.1 percent.

RATE OF RETURN ON A SPECULATIVE INVESTMENT

A firm purchased a parcel of land for $25,000 and spent $600 during the first year to improve the property. (This investment for improvements should be considered a lump-sum end-of-year payment.) The expenses for real estate tax, insurance, and maintenance totaled $1,200/year. At the end of 5 years, the firm sold the property at a price that yielded $48,700 after payment of legal fees and commissions. In computing the Federal income tax, the firm deducted the ordinary expenses of holding this property from the income derived from other sources. This income was subject to a 53 percent tax rate. The profit on the sale of the land was taxed at the 25 percent capital-gain rate. What was the rate of return on the investment?

Calculation Procedure:

1. Determine the effective annual payment

The expenses related to possession of the land served to reduce the income tax payments. Therefore, the *effective* cost of holding the property (or any similar asset) was less than the actual expenses. To obtain the effective annual payment, deduct the annual income tax saving from the annual payment related to the asset. Thus, effective annual payment = $1,200(1.00 − 0.53) = $564.

2. Compute the net proceeds from the sale of the asset

Deduct the capital-gain tax from the selling price of the asset to obtain the net proceeds. This is often called the *effective selling price*. Thus, capital gain = $48,700 − ($25,000 + $600) = $23,100. The capital-gain tax = $23,100(0.25) = $5,775. Hence, net proceeds = $48,700 − $5,775 = $42,925.

3. Set up an equation for the rate of return

Selecting the date at which the asset was sold as the reference date, express the value of every sum of money, and equate the total effective payments to the income. Thus, $25,000 (SPCA for n = 5 years, i = ?) + $600 (SPCA for n = 4 years, i = ?) + $564 (USCA for n = 5 years, i = ?) = $42,925.

4. Solve the rate-of-return equation using trial values

As a trial, set i = 10 percent, and evaluate the left-hand side of the relation in Step 3. Thus, $25,000(1.611) + $600(1.464) + $564(6.105) = $44,597.

Since the actual income, $42,925, was less than $44,597, the assumed rate of return is too high. Try 8 percent. Then, $25,000(1.469) + $600(1.360) + $564(5.867) = $40,850. This is less than the actual income. Interpolating linearly between the two trial values, $i = 9.1$ percent.

Related Calculations: As a general guide for selecting trial rate-of-return values, choose a higher value and a lower value around the estimated true rate of return. Check the result by computing the $ return. Interpolate linearly when higher and lower $ returns are obtained.

INVESTMENT AT AN INTERMEDIATE DATE (AMBIGUOUS CASE)

A firm purchased an oil-producing property under terms which did not require an immediate payment to the seller but which did require payment of royalties on income from sale of the oil. By the end of the third year the primary reserves were nearly exhausted and the firm spent $2,830,000 on a water-injection program to extend the oil yield. Operations were continued until the end of the sixth year. The income from the venture is listed in Table 4. Compute the rate of return on this investment. Is more than one solution obtained? How may the ambiguity inherent in this type of investment be resolved?

TABLE 4 Income from an Asset

Year	Net Income, $
1	600,000
2	300,000
3	100,000
4	1,220,000
5	500,000
6	200,000

Calculation Procedure:

1. Set up an equation for the rate of return

Selecting the end of the third year as the reference date, express the value of every sum of money. Consider receipts to be positive and expenditures to be negative. Then, $600,000(SPCA for $n = 2$, $i = ?$) + $300,000(SPCA for $n = 1$, $i = ?$) + $100,000 + $1,220,000(SPPW for $n = 1$, $i = ?$) + $500,000 (SPPW for $n = 2$, $i = ?$) + $200,000(SPPW for $n = 3$, $i = ?$) − $2,830,000 = 0.

2. Solve the rate-of-return equation using trial values

Assign a series of trial values for i in the equation in Step 1. Record the results in Table 5. Then, by linear interpolation, $i = 9.8$ percent, or $i = 30.1$ percent.

TABLE 5 Trial Calculations for Rate-of-Return Equation

Interest Rate, %	Value of Polynomial, $
8	10,600
10	−1,400
15	−21,300
20	−26,400
25	−19,400
30	−700
40	65,400

3. Evaluate the rates of return obtained

The polynomial in Step 1 resembles a quadratic polynomial since it contains either two real roots or none. That there are two values of i which satisfy this equation is explained as follows.

First, consider that $i = 9.8$ percent, causing the polynomial to assume the value of zero. Then replace 9.8 percent with the higher rate of 30.1 percent. Second, when this substitution is made, the value of the income received prior to the end of the third year is increased by a certain amount. The value of the income received after that date is decreased by the same amount. Hence, the value of the polynomial remains zero.

4. Make a realistic appraisal of the investment

A realistic appraisal of an investment of this type requires consideration of the reinvestment rate earned either by the entire income or that part of the income received prior to the expenditure. In the present instance, assume that the income received up to the end of the third year was reinvested at 8 percent. Its value at the date of the expenditure for water injection is, using the equation from Step 1, $600,000(1.166) + $300,000(1.080) + $100,000 = $1,123,600$.

Then, the effective investment $= $2,830,000 - $1,123,600 = $1,706,400$. To determine the rate of return, set the effective investment $1,706,400 = $1,220,000(SPPW for $n = 1$) + $500,000(SPPW for $n = 2$) + $200,000(SPPW for $n = 3$) and solve for i. The result of this solution is $i = 8.5$ percent.

Related Calculations: This Procedure illustrates the fact that in financial analyses it is not possible to place exclusive reliance on mathematical results. However rigorous the mathematical solution may appear to be, the results must be interpreted in a rational manner. Note that this procedure may be used for any type of asset.

Analysis of Business Operations

LINEAR REGRESSION APPLIED TO SALES FORECASTING

In 1970 a firm decided to expand its production facilities in anticipation of continued growth. A forecast of future sales was needed to formulate the expansion program. The sales during preceeding years were as follows:

Year	Sales, $000
1965	348
1966	377
1967	418
1968	475
1969	500

Apply linear regression to discern the sales trend. What is the projected sales volume for 1974?

Calculation Procedure:

1. Plot a scatter diagram for the given data

Regression analysis is applied where a causal relationship exists between two variables, although the relationship is obscured by the influence of random factors. The problem is to establish this relationship on the basis of observed data. In the present instance, it will be assumed that the sales volume is a linear function of time.

Plot the given sales data as shown in Fig. 7. The aggregate of points is termed a *scatter diagram*. Replace the scatter diagram by a straight line that most closely approaches the plotted points. This straight line is called the *regression line*, or *line of best fit*.

2. *Draw the regression line*

Draw the line shown in Fig. 7, using judgement of the best fit. Consider the vertical deviation e of a point in the scatter diagram, Fig. 7, from the arbitrary line drawn. The regression line is that line for which the sum of the squares of the deviations is minimum.

Let Y denote the ordinate of a point in the scatter diagram and Y_R the corresponding ordinate on the regression line. By definition, $\Sigma e^2 = \Sigma(Y - Y_R)^2 = $ minimum.

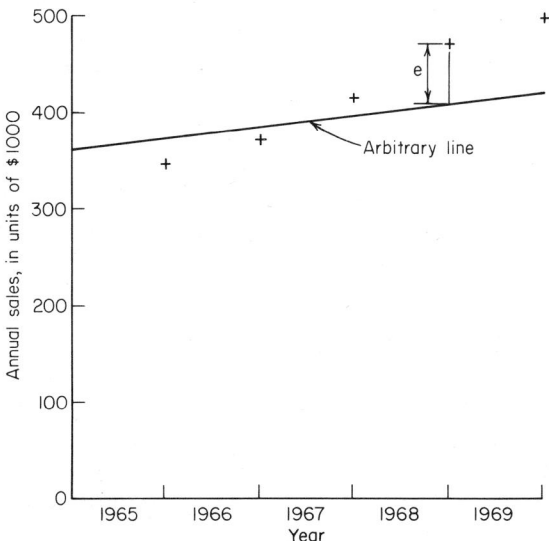

Fig. 7 Regression line, or line of best fit.

3. *Write the equation of the regression line*

Let n denote the number of points in the scatter diagram, and let $Y_R = a + bx$ be the equation of the regression line. To find the regression line, the parameters a and b must be evaluated.

Since Σe^2 is to have a minimum value, express the partial derivatives of Σe^2 with respect to a and b, and set these both equal to zero. Then derive the following simultaneous equations containing the unknown quantities a and b.

$$\Sigma Y = an + b\Sigma X$$

$$\Sigma XY = a\Sigma X + b\Sigma X^2$$

4. *Simplify the calculation*

To simplify the calculations in the present instance, select the median date (the end of 1967) as datum, and consider the annual sales income to be a lump sum received at the end of the given year. This step causes the term ΣX to vanish.

5. *Determine the values of ΣX^2 and ΣXY*

Prepare a tabulation such as Table 6. Use the data for each year in question.

6. *Solve for the parameters a and b*

Substitute in the equations in Step 3, and solve for a and b. Thus: $2,118 = 5a$; $402 = 10b$; $a = 423.6$; $b = 40.2$.

7. Write the regression equation; extrapolate for the year in question

$Y_R = 423.6 + 40.2X$. For 1974: $X = 7$; $Y_R = 423.6 + 40.2(7) = 705$. Hence, the forecast sales for 1974 = \$705,000. For comparative purposes, the past sales volumes as determined by the regression line are listed in Table 6.

TABLE 6   Locating a Regression Line

Year	X	Y	X^2	XY	Y_R
1965	−2	348	4	−696	343.2
1966	−1	377	1	−377	383.4
1967	0	418	0	0	423.6
1968	1	475	1	475	463.8
1969	2	500	4	1,000	504.0
Total	0	2,118	10	402	

STANDARD DEVIATION FROM REGRESSION LINE

Using the data in the previous Calculation Procedure, appraise the reliability of the regression line in forecasting future sales by computing the standard deviation of the points in the scatter diagram, using the regression line as the datum from which the deviation is measured.

Calculation Procedure:

1. Calculate the deviation of each point; square the result

The standard deviation serves as an index of the dispersion of the points in the scatter diagram. The standard deviation $\sigma = \sqrt{\Sigma e^2 / n}$.

Calculate the value of e for each point, Fig. 7. Enter the results for each year in a tabulation such as Table 7. Then, $\sigma = \sqrt{236.8/5} = 6.9$.

TABLE 7   Determining the Standard Deviation

Year	$Y - Y_R = e$	e^2
1965	$348 - 343.2 = 4.8$	23.0
1966	$377 - 383.4 = -6.4$	41.0
1967	$418 - 423.6 = -5.6$	31.4
1968	$475 - 463.8 = 11.2$	125.4
1969	$500 - 504.0 = -4.0$	16.0
Total...	....	236.8

2. Determine the monetary value represented by the standard deviation

Since the given monetary values are expressed in thousands of \$, the value of the standard deviation = 6.9(\$1,000) = \$6,900.

LINEAR PROGRAMMING TO MAXIMIZE INCOME FROM JOINT PRODUCTS

A firm manufactures two articles—A and B. The unit cost of production, exclusive of fixed costs, is \$10 for A and \$7 for B. The unit selling price is \$16 for A and \$13.50 for B. The estimated maximum monthly sales potential of A is 9,000 units; of B, 7,000 units. It is the policy of the firm to produce only as many units as can readily be sold. If production is restricted to one article, the factory can turn out 13,000 units of A or 8,500 units of B per month. The capital allotted to monthly production after payment of fixed costs is $100,000. What monthly production of each article will yield the maximum profit?

Calculation Procedure:

1. Express the production constraints imposed by sales and capital

Let N_A and N_B denote the number of articles A and B, respectively, produced monthly. Then, potential sales: $N_A \leq 9,000$, Eq. (a). $N_B \leq 7,000$, Eq. (b). Available capital: $10N_A + 7N_B \leq \$100,000$, Eq. (c).

2. Determine the production constraint imposed by the plant capacity

The number of months required to produce N_A units of A is $N_A/13,000$. Likewise, to produce N_B units of B would be $N_B/8,500$. Then, $N_A/13,000 + N_B/8,500 \leq 1$, or $8.5N_A + 13N_B \leq 110,500$, Eq. (d).

3. Express the monthly profit in equation form

Before deducting fixed costs, the profit $P = (16-10)N_A + (13.5-7)N_B$, or $P = 6N_A + 6.5N_B$, Eq. (e).

4. Construct a monthly production chart

Considering the expressions (a) to (d) above to be equalities, plot the straight lines representing them, Fig. 8.

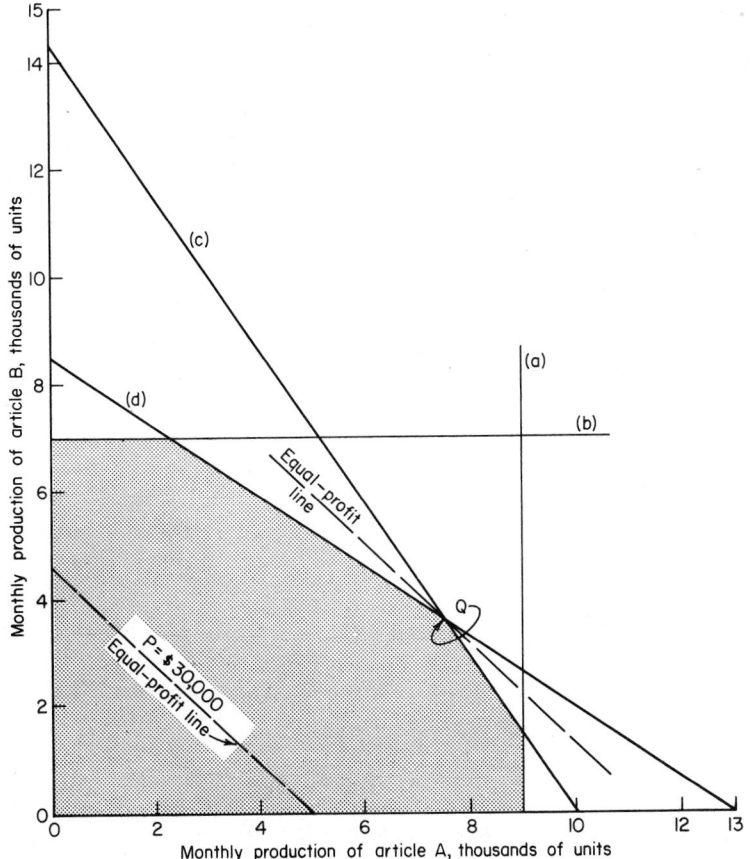

Fig. 8   Linear programming solution.

Since these expressions actually establish upper limits to the values of N_A and N_B, it follows that the point representing the joint production of articles A and B must lie either within the shaded area, which is termed the *feasible region*, or on one of its boundary lines.

5. Plot an equal-profit line

Assign the arbitrary value of $30,000 to P, and plot the straight line corresponding to Eq. (*e*) above. Every point on this line, Fig. 8, represents a set of values for N_A and N_B for which the profit is $30,000. This line is therefore termed an *equal-profit line*.

Next, consider that P assumes successively greater values. As P does so, the equal-profit line moves away from the origin while remaining parallel to its initial position.

6. Maximize the profit potential

To maximize the profit, locate the point in Fig. 8 at which the equal-profit line, in its outward displacement, is on the verge of leaving the feasible region. This is point Q, which lies at the intersection of the lines representing the equalities (*c*) and (*d*).

7. Determine the number of units for maximum profit

Establish the coordinates of the maximum-profit point Q either by reading them from the chart, Fig. 8, or by solving the equalities (*c*) and (*d*) simultaneously. The results are: $N_A = 7{,}468$ units; $N_B = 3{,}617$ units.

Related Calculations: Note that this method can be used for any type of product manufactured by any means.

OPTIMAL INVENTORY LEVEL

A firm is under contract to supply 41,600 parts per year and plans to produce them in equal lots spaced at equal intervals. The production capacity is 800 parts per day. Setup and teardown cost for the production machines is $550 for each run. The cost of storage, insurance, and interest on the investment is $1.40 per part for each year the part is carried in inventory. The regular production cost, exclusive of setup and teardown, is $5 per part. A reserve stock of parts is not needed. Determine the most economical lot size and the corresponding cost of production.

Calculation Procedure:

1. Compute the parts delivery rate

Assume that the parts are delivered to the buyer at a uniform rate and compute the daily delivery rate. Since there are approximately 260 working days per year, the rate of delivery = 41,600 parts/260 days = 160 parts per day.

2. Construct an inventory-time diagram

Figure 9 shows such a diagram, starting with zero inventory.

3. Compute the peak inventory for a lot size of N

The time OA required to produce 1 lot, Fig. 9, = lot size/maximum production rate = $N/800$, days. The slope of OB, Fig. 9, = rate of production − rate of delivery = 800 − 160 = 640 parts per day. Then, AB, Fig. 9, = $(N/800)(640) = 0.8N$ parts.

4. Compute the total annual cost in terms of N

The number of runs per year = 41,600/N. Also, the annual cost of setup and teardown = $550(41,600/N) = 22{,}880{,}000/N$. Further, the average inventory, Fig. 9, = $0.5AB = 0.4N$ parts. Taking the product of the carrying cost and the average inventory, the annual cost of carrying the inventory = $1.40(0.4N) = 0.56N$. With a $5 per unit

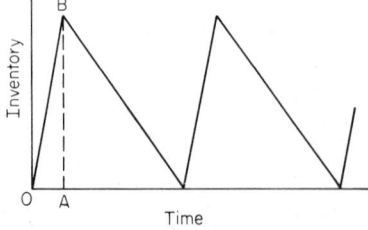

Fig. 9   Variation in inventory level.

regular cost, the annual regular cost = $5(41,600 units) = $208,000. Then, the annual total cost $C = (22,880,000/N) + 0.56N + 208,000$.

5. Find the economical lot size

To minimize C, set the derivative of C with respect to N equal to zero; solve for N to find the economical lot size. Thus, $dC/dN = -(22,880,000/N^2) + 0.56 = 0$; $N = 6,392$ parts per lot.

6. Compute the total annual cost

Substitute the number of parts from Step 5 in the total annual-cost equation in Step 4. Or, $C = 3,580 + 3,580 + 208,000 = $215,160$.

EFFECT OF QUANTITY DISCOUNT ON OPTIMAL INVENTORY LEVEL

Using the data from the previous Calculation Procedure, the firm finds that it can obtain quantity discounts if the parts are produced in lots of 7,500 or more. These discounts reduce the regular production cost from $5 to $4.80 per part; this saving reduces the interest cost on inventory by $0.02 per part. Determine the most economical lot size under these conditions.

Calculation Procedure:

1. Determine the number of parts for minimum cost

Assume that $N \geq 7,500$. Proceeding as before, express C in terms of N, and set $dC/dN = 0$.

The annual cost of carrying the inventory = $1.38(0.4N) = 0.552N$. Also, the annual regular cost = $41,600($4.80) = $199,680$. Also, $C = (22,880,000/N) + 0.552N + 199,680$. And, $dC/dN = -(22,880,000/N^2) + 0.552 = 0$. For minimum cost, $N = 6,438$ parts. However, since discounts are not obtained until N reaches 7,500, the last calculation lacks significance for this situation.

2. Compute the economical lot size

Set $N = 7,500$ and substitute in the cost equation above. Then, $C = 3,051 + 4,140 + 199,680 = $206,871$. Since this result is less than the value of $215,160 computed in the previous Calculation Procedure corresponding to 6,392 parts, the economical lot size is 7,500 parts.

SIMULATION OF COMMERCIAL ACTIVITY BY THE MONTE CARLO TECHNIQUE

A firm sells and delivers a standard commodity. The terms of sale require that the firm deliver the product within 1 day after an order is placed. In the past, the volume of orders received averaged 3,315 units per week, with the variation in volume shown in Table 8.

TABLE 8 Frequency Distribution

Orders received per week			Weekly shipping capacity		
Number of units	Relative frequency	Median value	Number of units	Relative frequency	Median value
3,000–3,099	0.05	3,050	3,300–3,349	0.15	3,325
3,100–3,199	0.10	3,150	3,350–3,399	0.30	3,375
3,200–3,299	0.35	3,250	3,400–3,449	0.35	3,425
3,300–3,399	0.25	3,350	3,450–3,499	0.20	3,475
3,400–3,499	0.15	3,450	Total.....	1.00	
3,500–3,599	0.10	3,550			
Total.....	1.00				

The firm currently employs a trucking company. But the firm contemplates purchasing its own fleet of trucks to make deliveries. It is therefore necessary to decide how many trucks are to be purchased. Several plans are under consideration. The shipping facilities under plan A have an estimated average capacity of 3,405 units per week. Experience indicates that this capacity may be expected to vary in the manner shown in Table 8.

When the volume of daily orders exceeds the shipping capacity, sales will be lost; when the reverse condition occurs, trucks will be idle. Lost sales are valued at $2.40 per unit, which includes an allowance for partial loss of goodwill. Unused shipping capacity is valued at $1.10 per unit. Applying the Monte Carlo technique, estimate the amount of these losses if plan A is adopted.

Calculation Procedure:

1. Determine the average weekly losses

In Table 8, record the *median* value for each range, as shown in the third and sixth columns. For convenience, apply only these median values in the calculations. This procedure is equivalent to assuming, for example, that the volume of prospective sales varies discretely from 3,050 to 3,550 units with an interval of 100 units between consecutive values.

Analysis of Table 8 reveals that the excess of weekly shipping capacity over delivery requirements may range between 425 units, $(3,475-3,050)$; and -225 units, $(3,325-3,550)$, and that it may assume any of the following values:

425	375	325	275	225
175	125	75	25	-25
-75	-125	-175	-225	

To evaluate the average weekly losses, it is necessary to evaluate the frequency with which these values are likely to exist. The Monte Carlo technique is a probabilistic device that circumvents the mathematical complexity inherent in a rigorous solution by resorting to a set of numbers generated in a purely random manner. Tables of random numbers are published in books listed in the references for this section.

2. Compute the cumulative frequency of the prospective sales

The cumulative frequency of each value of prospective sales is the relative frequency with which orders of the designated magnitude, or less, are received. The results of this calculation appear in Table 9.

TABLE 9 Cumulative Frequency of Prospective Sales

Number of Units Ordered	Cumulative Frequency
3,050	0.05
3,150	0.15
3,250	0.50
3,350	0.75
3,450	0.90
3,550	1.00

3. Prepare a histogram of the frequency distributions

Plot the cumulative-frequency values in Fig. 10. Draw horizontal and vertical lines as shown. The relative frequency of a given value of the prospective sales is represented in this drawing by the length of the vertical line directly above the value.

4. Select random numbers for the solution

Refer to a table of random numbers. Select the first 10 numbers found in the table. Enter these numbers in the second column of Table 10. (In actual practice, a larger quantity of random numbers would be selected.)

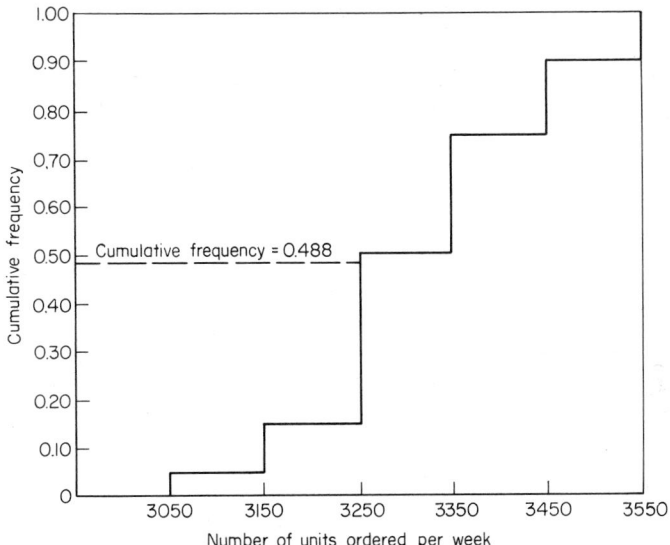

Fig. 10   Cumulative-frequency histogram of prospective sales.

TABLE 10   Simulated Values of Prospective Sales and Shipping Capacity

Week	Random number	Number of units ordered	Random number	Shipping capacity
1	0.488	3,250	0.339	3,375
2	0.322	3,250	0.697	3,425
3	0.274	3,250	0.031	3,325
4	0.557	3,350	0.052	3,325
5	0.931	3,550	0.506	3,425
6	0.986	3,550	0.865	3,475
7	0.682	3,350	0.948	3,475
8	0.179	3,250	0.308	3,375
9	0.881	3,450	0.218	3,375
10	0.834	3,450	0.367	3,375

5. Use the random numbers in the solution

Consider each random number as a cumulative frequency. Refer to the histogram, Fig. 10, to find the volume of orders corresponding to this value of the random number. Thus, draw a horizontal through the random-number value of 0.488 on the vertical axis of Fig. 10. This line intersects the vertical that lies above the value of 3,250 on the horizontal axis. Therefore, enter in Table 10 the value of 3,250 opposite the random number 0.488.

6. Repeat Steps 3 to 5 for the shipping capacity

Enter the results in Table 10 in the same manner as for the units ordered, Step 5.

7. Evaluate the loss on sales and unused capacity

Compare the simulated prospective sales with the simulated capacity. For example, during week 1, the loss on unused capacity = 1.10(3,375 − 3,250) = \$137.50, using the data from Table 10. Likewise, during week 4, loss on lost sales = 2.40(3,350 − 3,325) = \$60.00.

8. Determine the average weekly losses

Total the computed losses obtained in Step 7, and divide by 10 to obtain the following average weekly values:

Loss on unused capacity ...	\$90.00
Loss on forfeited sales	68.75
Total..................	\$158.75

If more trucking facilities are procured, the forfeited sales will be reduced. But the unused capacity will be increased. The optimal number of trucks to be purchased is that for which the total loss is a minimum.

PROJECT PLANNING USING CRITICAL PATH METHOD (CPM)

Table 11 lists the activities performed in preparing a building site and installing the utilities. Assuming that the estimated durations are precise, determine the minimum time needed to complete the project. Identify the critical path. Upon completion of the project, it was found that each activity was undertaken at the earliest possible date and that its duration coincided with the estimate, except for the following: activity D was started 3 days late; activity H required 9 days instead of 6; activity I required 5 days instead of 4. Determine the duration of the project.

TABLE 11   Project Activities

Mark	Activity	Estimated duration, days
A	Clear site	4
B	Survey and lay out site	3
C	Rough grade	3
D	Excavate for sewer	8
E	Excavate for electrical manholes	1
F	Install sewer and backfill	4
G	Install electrical manholes	6
H	Install overhead pole line	6
I	Install electrical duct bank	4
J	Pull in power feeder	5
K	Construct foundations for water tank	3
L	Erect water tank	8
M	Install piping and valves for water tank	12
N	Drill well	14
O	Install well pump	2
P	Install underground water piping	9
Q	Connect all piping	2

Calculation Procedure:

1. *Identify the predecessor(s) of each activity; tabulate results*

The critical path method offers a systematic means of scheduling activities in a project and analyzing the consequences of departures from the schedule. The procedure consists of devising a logical concatenation of activities after ascertaining the relationships that exist among them. For example, activities D and E, Table 11, are independent of one another and may therefore be performed concurrently. But activities D and F are sequentially related—F cannot commence until D is finished. Thus, D is the immediate predecessor of F. Using these principles, list the related activities as shown in Table 12.

TABLE 12   Related CPM Activities

Activity	Predecessor	Activity	Predecessor
B	A	J	H and I
C	B	K	C
D	C	L	K
E	C	M	L
F	D	N	C
G	E	O	N
H	C	P	O
I	F and G	Q	M and P

2. *Construct the network for the project, Fig. 11*

The network is a delineation of the sequence in which the activities are to be performed. Each activity is represented by a horizontal arrow, which may or may not be to scale. The arrow representing a given activity is placed to the right of its immediate

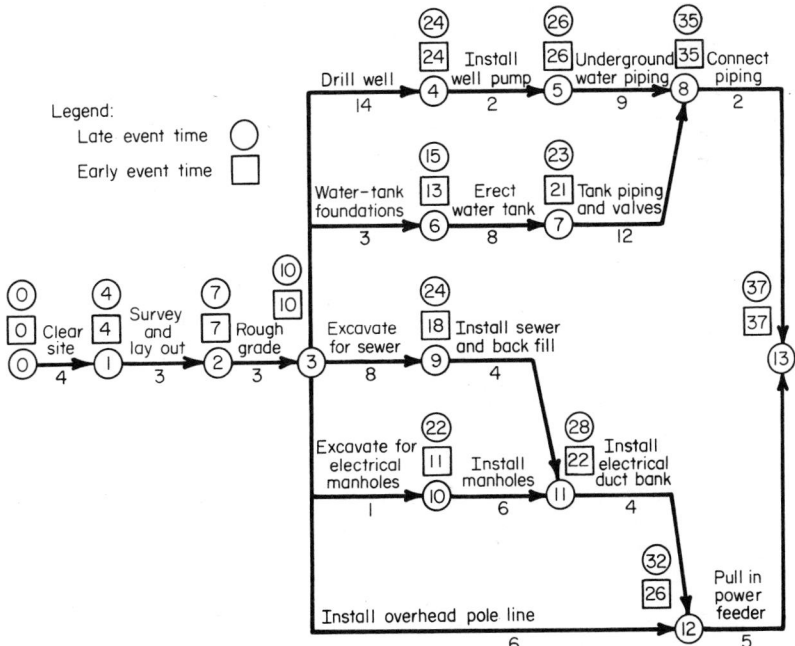

Fig. 11   CPM network for site-preparation project.

predecessor activity. Where there are multiple predecessors or successors, broken arrows are used to transfer from one activity to another. The duration of each activity is recorded under its corresponding arrow.

Completion of an activity and the start of its successor constitute an *event*. Commencement of a given activity is termed its *i event*; completion of an activity is termed its *j event*. A number is assigned to each event and is recorded in the network in a circle between consecutive arrows. Each activity is identified by the events it separates. For instance, 6-7 designates *erect water tank*. A chain of activities extending from inception to completion of the project is termed a *path*. In this project there are the following five paths.

```
0-1-2-3-4-5-8-13
0-1-2-3-6-7-8-13
0-1-2-3-9-11-12-13
0-1-2-3-10-11-12-13
0-1-2-3-12-13
```

3. Compute the early event time T_E of each event

The *early event time* is the earliest possible date at which the event may occur. Compute the early event time from $T_{E(n)} = T_{E(n-1)} + D$, where $T_{E(n)}$ = early event time of a given event; $T_{E(n-1)}$ = early event time of preceeding event; D = duration of intervening activity. Where an event has multiple immediate predecessors, this equation yields multiple values of T_E; the correct value is the maximum value.

Table 13 shows the calculations for the early event times. In the calculations, the starting date of the project was used as the datum. Record the early event times on the network by entering the time (usually in days) in a square above each event. From Table 13, the minimum duration of this project is 37 days.

TABLE 13   Calculation of Early Event Times

Event	T_E, days
0	0
1	4
2	$4+3 = 7$
3	$7+3 = 10$
4	$10+14 = 24$
5	$24+2 = 26$
6	$10+3 = 13$
7	$13+8 = 21$
8	$26+9 = 35$
	or $21 + 12 = 33$ (disregard)
9	$10+8 = 18$
10	$10+1 = 11$
11	$18+4 = 22$
	or $11+6 = 17$ (disregard)
12	$22+4 = 26$
	or $10+6 = 16$ (disregard)
13	$35+2 = 37$
	or $26+5 = 31$ (disregard)

4. Compute the late event time T_L of each event

The late event time of each event is the latest date at which the event may occur without extending the duration of the project beyond the minimum time. Use the relation $T_{L(n)} = T_{L(n+1)} - D$, where $T_{L(n)}$ = late event time of a given event; $T_{L(n+1)}$ = late event time of succeeding event; D = duration of intervening activity. Both early and late event times are usually measured in days, but on unusually long projects they may be measured in months. Where an event has multiple immediate successors, the equation above yields multiple values of T_L; the correct result is the minimum value.

Table 14 shows the calculations of the late event times. Enter the late event time on the network in a circle above each event.

TABLE 14 Calculation of Late Event Times

Event	T_L, days
13	37
12	$37-5 = 32$
11	$32-4 = 28$
10	$28-6 = 22$
9	$28-4 = 24$
8	$37-2 = 35$
7	$35-12 = 23$
6	$23-8 = 15$
5	$35-9 = 26$
4	$26-2 = 24$
3	$24-14 = 10$
	or $15-3 = 12$ (disregard)
	or $24-8 = 16$ (disregard)
	or $22-1 = 21$ (disregard)
	or $32-6 = 26$ (disregard)
2	$10-3 = 7$
1	$7-3 = 4$
0	$4-4 = 0$

5. Compute the float of each activity

Float is the time, usually in days, that the completion of each activity may be delayed without extending the duration of the project, with the understanding that no other delays will occur. Compute the float from $F = T_{L(j)} - [T_{E(i)}+D]$, where F = float, usually in days; $T_{L(j)}$ = late event time of completion in the same time units as F; $T_{E(i)}$ = early event starting time, in the same time units as F; D = activity duration, in the same time units as F. The expression in parentheses represents the earliest possible date at which the activity may be completed. Table 15 shows the float calculations.

TABLE 15 Calculation of Project Float*

Activity	$T_{L(j)}$	$T_{E(i)}$	D	F
0-1	4	0	4	0
1-2	7	4	3	0
2-3	10	7	3	0
3-4	24	10	14	0
4-5	26	24	2	0
5-8	35	26	9	0
3-6	15	10	3	2
6-7	23	13	8	2
7-8	35	21	12	2
8-13	37	35	2	0
3-9	24	10	8	6
9-11	28	18	4	6
3-10	22	10	1	11
10-11	28	11	6	11
11-12	32	22	4	6
3-12	32	10	6	16
12-13	37	26	5	6

*Measured in days.

6. *Identify the critical path*

An activity is *critical* if any delay in its completion will extend the duration of the project. The path on which the critical activities are located is termed the *critical path*. (There may be several critical paths associated with a project.) In the terminology of CPM, a critical activity is one having zero float. The critical path for this project is therefore 0-1-2-3-4-5-8-13.

7. *Verify the results of Step 6*

Plot the project activities on a time scale, Fig. 12. This diagram was constructed assuming that each activity commences at the earliest possible date. However, a diagram based on another assumed date would show the same results.

Note in Fig. 12 that the float of a given activity equals the total gap in the chain extending from the completion of that activity to the completion of the project. For instance, 6-7, Fig. 12, has a float of 2 days, and 10-11 has a float of $5 + 6 = 11$ days.

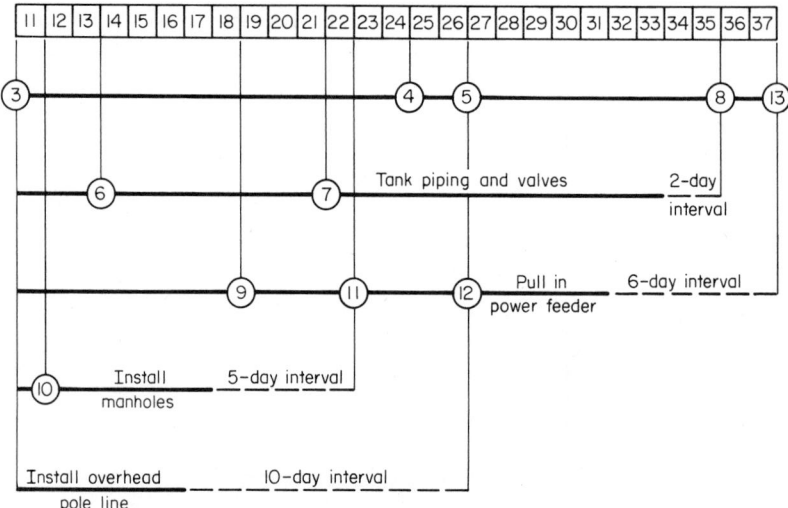

Fig. 12 Activity-time diagram.

8. *Indicate where the actual schedule departed from the forecast*

List the data as follows:

Old mark	New mark	Delay in completion, days
D	3-9	3
H	3-12	3
I	11-12	1

9. *Determine the true duration of the project*

Treating each departure individually, deduce the effects of that departure by referring to Fig. 12. Combine these effects where they are cumulative. On the basis of these results, establish the true duration of the project. Thus: activity 3-9—events 11 and 12 would be delayed 3 days; activity 3-12—event 12 would not be delayed, since there is a latitude of 10 days along this path; activity 11-12–event 12 would be delayed 1 day. *Summary*: Event 12 is delayed 4 days; event 13 is not delayed, since there is a latitude of 6 days along this path. Therefore, the true duration of the project = 37 days, as forecast.

PROJECT PLANNING BASED ON AVAILABLE MANPOWER

A manufacturing firm has a contract to build a pilot model of a newly invented machine. To plan the work, the firm constructed the CPM network in Fig. 13 and compiled the data in Table 16. The following activities must be performed as a unit, without loss of continuity: 0-1-2-4; 2-3-4; 5-7-8. Activities 4-8 and 6-8 may be performed piecemeal, if this proves convenient. The manpower available for assignment to this project is: 15 men for the first 8 days; 25 men for the remaining time. Each employee is capable of performing all 11 activities. But the constraints imposed by the available facilities limit the number of men for each activity to that shown in Table 16. Overtime is not permissible. Devise a schedule that will allow completion of this project at the earliest possible date.

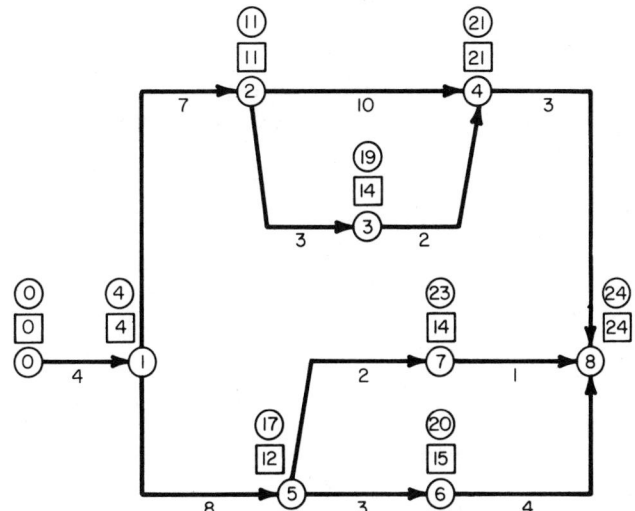

Fig. 13   CPM network for construction of pilot model.

TABLE 16   Time and Manpower Requirements for a Technical Project

Activity, Fig. 13	Duration, days	Manpower required
0-1	4	3
1-2	7	5
2-4	10	9
2-3	3	9
3-4	2	12
4-8	3	4
1-5	8	6
5-6	3	10
6-8	4	8
5-7	2	6
7-8	1	7

Calculation Procedure:

1. Compute the early and late event times

The results of these calculations, made in accordance with the previous Calculation Procedure, are shown in Fig. 13.

2. Identify the critical path

Since the critical path is the longest path through the network, Fig. 13 shows that this is 0-1-2-4-8. Note that the critical path sets a lower limit of 24 days on the duration of the project. But the manpower limitations may lengthen the project beyond 24 days. Thus, the objective will be to devise a schedule that fits noncritical activities into the 24-day period while satisfying the manpower availability.

3. Schedule the project assuming unlimited manpower

As a first trial, assume that unlimited manpower is available and schedule each activity to start at the *earliest* possible date. Construct the manpower-time diagram for this condition, as shown in Fig. 14a. Study of the diagram shows that this schedule is

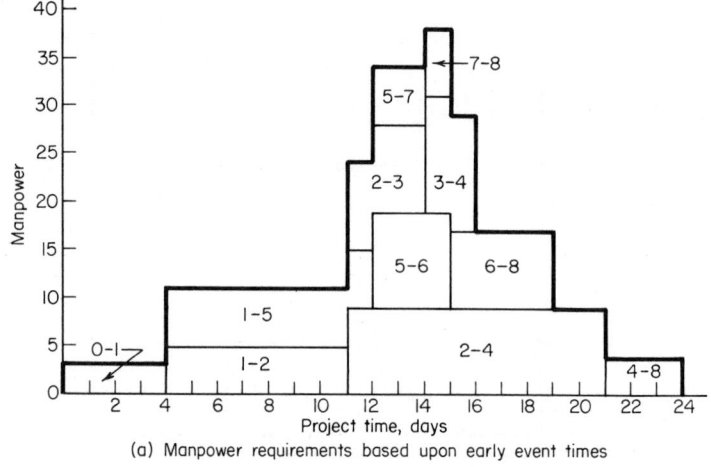

(a) Manpower requirements based upon early event times

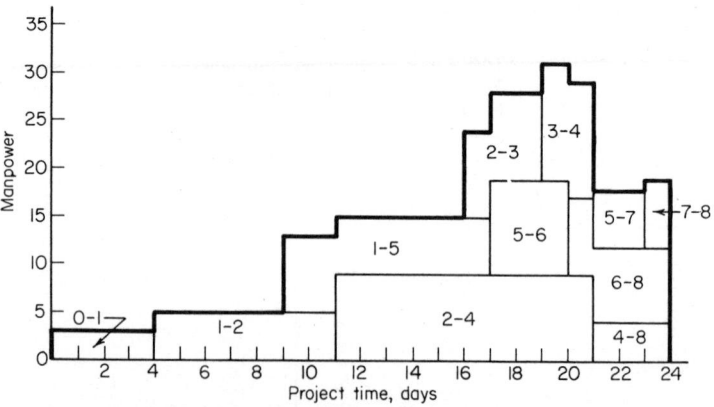

(b) Manpower requirements based upon late event times

Fig. 14 **(a) Manpower requirements based upon early event times; (b) manpower requirements based upon late event times.**

unsatisfactory because the manpower requirements exceed the available manpower on days 12 to 15, inclusive.

4. Schedule the project with the latest possible start

Assume unlimited manpower again, and schedule each activity to start at the *latest* possible date. Construct the corresponding manpower-time diagram, Fig. 14*b*. Study of the diagram shows that this schedule is also unsatisfactory, but it is an improvement over the schedule in Step 3.

Although both these schedules are unsatisfactory, they are useful because their manpower-time diagrams reveal the boundaries of each activity or chain of activities based on a project duration of 24 days. For example, the chain 5-7-8 may start at any time between days 12 and 21, Fig. 14. This information is not explicitly supplied by the network, since the late event time for event 5 is determined by the chain 5-6-8.

5. Shift noncritical activities to obtain a suitable schedule

Using the allowable manpower limits, shift noncritical activities such that the project can be completed in 24 days using the available manpower.

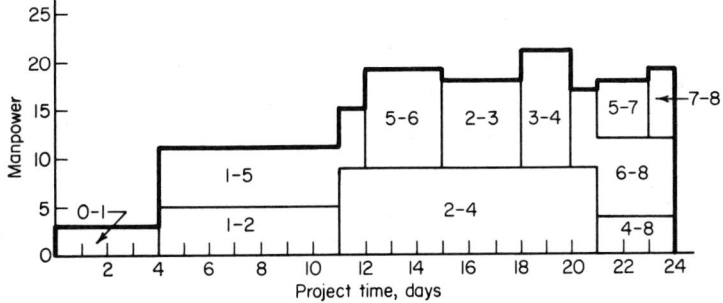

Fig. 15   Manpower requirements based upon project schedule.

Construct the schedule shown in Fig. 15. Note that the manpower requirements are less than the available personnel. Further, the schedule preserves the integrity of the three chains of activities mentioned earlier. Although fragmentation of activity 6-8 is permissible, it proved unnecessary. With the schedule shown, the project will be finished in 24 days.

Related Calculations: Note that this procedure can be used for any type of project requiring allocation of manpower or other resources.

Index

Index